Wheel Alignment Specifications

2017 Edition

IMPORTANT NOTICE

ALL OF THE DATA IN THIS MANUAL HAS BEEN COMPILED WITH THE UTMOST CARE AND ATTENTION. HOWEVER, DUE TO THE LARGE VOLUME AND NUMERICAL NATURE OF THE DATA, THE POSSIBILITY OF ERRORS, OMISSIONS, OR SUBSEQUENT CHANGES EXISTS. THEREFORE, NEITHER THE PUBLISHER NOR THE DISTRIBUTOR OF THIS MANUAL CAN ASSUME ANY LIABILITY FOR DAMAGES, WHETHER DIRECT OR INDIRECT, INCIDENTAL OR CONSEQUENTIAL, THAT RESULT FROM YOUR USE OF THE INFORMATION CONTAINED WITHIN THIS MANUAL.

This manual is intended for use by professional automotive service technicians. Your experience and common sense are crucial to a successful alignment, and to your customer's satisfaction. Always follow the procedures recommended by the manufacturer of your alignment equipment.

SHOP SAFETY FIRST.......ALWAYS!

Contents

PLEASE ADDRESS YOUR QUESTIONS OR COMMENTS TO:

Naas Publishing Company
4166 Cleveland Avenue
St. Louis, Missouri 63110
www.AlignmentSpecs.com
ph. (314)865-3219
e-mail editor@AlignmentSpecs.com

A guide to

the ALIGNMENT SPECIFICATIONS page.

All specification data in this manual was obtained from manufacturer sources. For vehicles that have different tolerances for *checking* and *setting*, the tighter (setting) tolerances are listed. All specifications are accurate for use with both computerized and mechanical alignment equipment.

Models are listed alphabetically across the top half of the page, then continuing across the bottom half, from oldest to newest. The first three lines show basic model information. The fourth line shows the model years covered.

Chevrolet

MODEL INFORMATION	Corvette
YEARS	1983
INSTRUCTIONS	p.221
LUGNUT TORQUE	100 ft.lbs.
FRONT	
CASTER (Degrees)	3.5 ±0.75 [1] 3½ ± ¾ [1]
CROSS | TOL.	0.0 (0) | 0.75 (¾)
CAMBER (Degrees)	-0.5 ±0.75 -½ ± ¾
CROSS | TOL.	0.0 (0) | 0.75 (¾)
FRONT TOE (Degrees)	0.15° ±0.25° 5/32° ± ¼°
FRONT TOE (Inches)	0.08" ±0.13" 1/16" ± ⅛"
SAI	9.5 ± 0.5 (9½ ± ½)
TURNING ANGLE	20.0 / 19.7 ± 0.5
ILLUSTRATION #	23 ▲
REAR	
CAMBER (Degrees)	-0.5 ±0.5 -½ ± ½
REAR TOE (Degrees)	0.13° ±0.25° ⅛° ± ¼°
REAR TOE (Inches)	0.06" ±0.13" 1/16" ± ⅛"
THRUST ANGLE	0.15 (5/32)
ILLUSTRATION #	40 39
NOTES	[1] w/MS 2.25 (2¼)

INSTRUCTIONS refers to a page where additional information about certain vehicles is detailed. You should read that section before performing an alignment.

FRONT CAMBER and tolerance in decimal degrees (°) with the fractional equivalent below.

FRONT TOTAL TOE and tolerance in decimal degrees(°) with the fractional equivalent below. Negative numbers (indicating toe-out) are in **bold**.

FRONT TOTAL TOE and tolerance in decimal inches(") with the fractional equivalent below. Negative numbers (indicating toe-out) are in **bold**.

TURNING ANGLE (inside / outside) of wheel and tolerance in decimal degrees.

REAR CAMBER and tolerance in decimal degrees (°) with the fractional equivalent below.

THRUST ANGLE tolerance with fractional equivalent in parenthesis. (Spec is 0°.)

Additional NOTES, exceptions, or other information is listed here.

LUGNUT TORQUE in foot pounds.

FRONT CASTER and tolerance in decimal degrees (°) with the fractional equivalent below.

CROSS Caster and CROSS Camber spec and tolerance, followed by the fractional equivalent in parenthesis. Also known as "split," cross caster and cross camber are calculated as left side minus right side.

SAI and tolerance in decimal degrees (°) followed by the fractional equivalent in parenthesis.

REAR TOTAL TOE and tolerance in decimal degrees(°) with the fractional equivalent below. Negative numbers (indicating toe-out) are in **bold**.

REAR TOTAL TOE and tolerance in decimal inches(") with the fractional equivalent below. Negative numbers (indicating toe-out) are in **bold**.

Adjustment ILLUSTRATION #'s, front and rear, if applicable. Refer to the appropriate illustration # in the adjustment methods section.
The symbol "▲" indicates that engine cradle or suspension carrier adjustments may be possible.

Abbreviations

exc/ = "except"	PS = power steering	4X4 = four wheel drive	DOHC = dual overhead cam
w/ = "with"	MS = manual steering	4X2 = two wheel drive	TCS = traction control system
w/o = "without"	A/T = automatic transmission	RWD = rear wheel drive	ECS = elect. controlled susp.
ABS = anti-lock brakes	M/T = manual transmission	L = liter	IRS = independent rear susp.
AWD = all wheel drive	2WD = two wheel drive	2WS = two wheel steering	IFS = independent front susp.
FWD = front wheel drive	4WD = four wheel drive	4WS = four wheel steering	ARC = automatic ride control
n/a = not adjustable	n/s = not specified	TC = turbocharger	HD, LD = heavy duty, light duty
SRW = single rear wheels	DRW = dual rear wheels		

Acura

MODEL	CL	CL	CL	CSX (Canada)	EL (Canada)	EL (Canada)
INFORMATION	2.2, 2.3	3.0				
YEARS	1997-99	1997-99	2001-2003	2006-2010	1997-2000	2001-2002
INSTRUCTIONS						
LUGNUT TORQUE	80 ft.lbs.	80 ft.lbs.	80 ft.lbs.	80 ft.lbs.	80 ft.lbs.	80 ft.lbs.
FRONT						
CASTER (Degrees)	3.35 ± 1.0 3⅜ ± 1	3.3 ± 1.0 3 5/16 ± 1	2.8 ± 1.0 2 13/16 ± 1	7.0 ± 1.0 7 ± 1	1.67 ± 1.0 1 11/16 ± 1	1.55 ± 1.0 1 9/16 ± 1
CROSS | TOL.	0.0 (0) | n/s	0.0 (0) | n/s	0.0 (0) | n/s	0.0 (0) | n/s	0.0 (0) | n/s	0.0 (0) | n/s
CAMBER (Degrees)	0.12 ± 1.0 ⅛ ± 1	0.0 ± 1.0 0 ± 1	0.0 ± 1.0 0 ± 1	0.0 ± 0.5 0 ± ½	0.0 ± 1.0 0 ± 1	0.0 ± 0.75 0 ± ¾
CROSS | TOL.	0.0 (0) | n/s	0.0 (0) | n/s	0.0 (0) | n/s	0.0 (0) | 0.58 (9/16)	0.0 (0) | n/s	0.0 (0) | n/s
FRONT TOE (Degrees)	0.0° ± 0.16° 0° ± 5/32°	0.0° ± 0.16° 0° ± 5/32°	0.0° ± 0.16° 0° ± 5/32°	0.0° ± 0.16° 0° ± 5/32°	0.08° ± 0.16° 3/32° ± 5/32°	0.0° ± 0.24° 0° ± ¼°
FRONT TOE (Inches)	0.0" ± 0.08" 0" ± 3/32"	0.0" ± 0.08" 0" ± 3/32"	0.0" ± 0.08" 0" ± 3/32"	0.0" ± 0.08" 0" ± 3/32"	0.04" ± 0.08" 1/32" ± 3/32"	0.0" ± 0.12" 0" ± ⅛"
SAI						
TURNING ANGLE	34.5 / 27.5 ±2.0	34.5 / 27.5 ±2.0	38.5 / 31.1 ±2.0 [1]	38.8 / 31.2 ±2.0	35.8 / 30.8 ±2.0	40.0 / 31.0 ±2.0
ILLUSTRATION #	03	03	03	26		26
REAR						
CAMBER (Degrees)	-1.2 ± 0.5 -1 3/16 ± ½	-0.5 ± 0.5 -½ ± ½	-0.5 ± 1.0 -½ ± 1	-1.5 +1.08 -0.75 [1] -1½ +1 1/16 -¾ [1]	-1.0 ± 1.0 -1 ± 1	-0.75 ± 0.75 -¾ ± ¾
REAR TOE (Degrees)	0.16° ± 0.16° 5/32° ± 5/32°	0.16° ± 0.16° 5/32° ± 5/32°	0.16° ± 0.16° 5/32° ± 5/32°	0.16° +0.16° -0.08° 5/32° +5/32° -3/32°	0.16° +0.16° -0.08° 5/32° +5/32° -3/32°	0.16° +0.16° -0.08° 5/32° +5/32° -3/32°
REAR TOE (Inches)	0.08" ± 0.08" 3/32" ± 3/32"	0.08" ± 0.08" 3/32" ± 3/32"	0.08" ± 0.08" 3/32" ± 3/32"	0.08" +0.08" -0.04" 3/32" +3/32" -1/32"	0.08" +0.08" -0.04" 3/32" +3/32" -1/32"	0.08" +0.08" -0.04" 3/32" +3/32" -1/32"
THRUST ANGLE						
ILLUSTRATION #	05	05	06	140	02	140
NOTES			[1] Type S = 35.3°/29.2°	[1] '08-'09 mods. with "C" on rear arm = -0.75° (-¾°)		
	01001	01002	01003	01004	01005	01006

Acura

MODEL	EL (Canada)	ILX	Integra	Integra	Integra	Legend Coupe
INFORMATION		incl/Hybrid		exc/Type R	Type R	
YEARS	2003-2005	2013-2017	1990-93	1994-2001	1998-2001	1990
INSTRUCTIONS						
LUGNUT TORQUE	80 ft.lbs.	80 ft.lbs.	80 ft.lbs.	80 ft.lbs.	80 ft.lbs.	80 ft.lbs.
FRONT						
CASTER (Degrees)	2.23 ± 1.0 2¼ ± 1	5.25 ± 0.5 5¼ ± ½	1.5 ± 1.0 1½ ± 1	1.17 ± 1.0 1 3/16 ± 1	1.17 ± 1.0 1 3/16 ± 1	1.67 ± 1.0 1 11/16 ± 1
CROSS | TOL.	0.0 (0) | n/s	0.0 (0) | n/s	0.0 (0) | n/s	0.0 (0) | n/s	0.0 (0) | n/s	0.0 (0) | n/s
CAMBER (Degrees)	0.0 ± 0.75 0 ± ¾	0.0 ± 0.5 0 ± ½	0.0 ± 1.0 0 ± 1	-0.17 ± 1.0 -3/16 ± 1	-0.5 ± 1.0 -½ ± 1	0.0 ± 1.0 0 ± 1
CROSS | TOL.	0.0 (0) | n/s	0.0 (0) | 0.5 (½)	0.0 (0) | n/s	0.0 (0) | n/s	0.0 (0) | n/s	0.0 (0) | n/s
FRONT TOE (Degrees)	0.0° ± 0.24° 0° ± ¼°	0.0° ± 0.16° 0° ± 5/32°	0.0° ± 0.16° 0° ± 5/32°	0.0° ± 0.16° 0° ± 5/32°	0.0° ± 0.16° 0° ± 5/32°	0.0° ± 0.24° 0° ± ¼°
FRONT TOE (Inches)	0.0" ± 0.12" 0" ± ⅛"	0.0" ± 0.08" 0" ± 3/32"	0.0" ± 0.08" 0" ± 3/32"	0.0" ± 0.08" 0" ± 3/32"	0.0" ± 0.08" 0" ± 3/32"	0.0" ± 0.12" 0" ± ⅛"
SAI						
TURNING ANGLE	40.0 / 31.0 ±2.0	37.7 / 30.4 ±1.0 [1]	40.5 / 32.0 ±2.0	36.0 / 30.5 ±2.0	36.0 / 30.5 ±2.0	
ILLUSTRATION #	26	26				03
REAR						
CAMBER (Degrees)	-0.75 ± 0.75 -¾ ± ¾	-0.75 ± 0.75 -¾ ± ¾	-0.67 ± 1.0 -11/16 ± 1	-0.75 +0.75 -1.25 -¾ +¾ -1¼	-0.75 +0.75 -1.25 -¾ +¾ -1¼	0.0 ± 1.0 0 ± 1
REAR TOE (Degrees)	0.16° +0.16° -0.08° 5/32° +5/32° -3/32°	0.16° +0.16° -0.08° 5/32° +5/32° -3/32°	0.16° +0.08° -0.16° 5/32° +3/32° -5/32°	0.16° +0.16° -0.08° 5/32° +5/32° -3/32°	0.16° +0.16° -0.08° 5/32° +5/32° -3/32°	0.16° ± 0.16° 5/32° ± 5/32°
REAR TOE (Inches)	0.08" +0.08" -0.04" 3/32" +3/32" -1/32"	0.08" +0.08" -0.04" 3/32" +3/32" -1/32"	0.08" +0.04" -0.08" 3/32" +1/32" -3/32"	0.08" +0.08" -0.04" 3/32" +3/32" -1/32"	0.08" +0.08" -0.04" 3/32" +3/32" -1/32"	0.08" ± 0.08" 3/32" ± 3/32"
THRUST ANGLE						
ILLUSTRATION #	140	140	02	02	02	05
NOTES		[1] 2016-17 = 35.9°/29.5°				
	01007	01008	01010	01011	01012	01013

Acura

Legend Coupe	Legend Sedan	Legend Sedan	MODEL INFORMATION	MDX	MDX	MDX
1991-95	1990	1991-95	YEARS	2001-2006	2007-2013	2014-2016
			INSTRUCTIONS			
80 ft.lbs.	80 ft.lbs.	80 ft.lbs.	LUGNUT TORQUE	80 ft.lbs.	80 ft.lbs.	80 ft.lbs.
			FRONT			
3.75 ± 1.0 3¾ ± 1	1.67 ± 1.0 1 11/16 ± 1	3.75 ± 1.0 3¾ ± 1	CASTER (Degrees)	1.88 ± 1.0 1⅞ ± 1	4.2 ± 0.58 4 3/16 ± 9/16	3.73 ± 0.6 3¾ ± ⅝
0.0 (0) | n/s	0.0 (0) | n/s	0.0 (0) | n/s	CROSS | TOL.	0.0 (0) | n/s	0.0 (0) | n/s	0.0 (0) | n/s
0.0 ± 1.0 0 ± 1	0.0 ± 1.0 0 ± 1	0.0 ± 1.0 0 ± 1	CAMBER (Degrees)	-0.5 ± 1.0 -½ ± 1	-0.5 ± 1.0 -½ ± 1	-0.5 ± 0.75 -½ ± ¾
0.0 (0) | n/s	0.0 (0) | n/s	0.0 (0) | n/s	CROSS | TOL.	0.0 (0) | n/s	0.0 (0) | n/s	0.0 (0) | n/s
-0.08° ± 0.16° **-3/32°** ± 5/32°	0.0° ± 0.24° 0° ± ¼°	**-0.08°** ± 0.16° **-3/32°** ± 5/32°	FRONT TOE (Degrees)	0.0° ± 0.16° 0° ± 5/32°	0.0° ± 0.16° 0° ± 5/32°	0.0° ± 0.16° 0° ± 5/32°
-0.04" ± 0.08" **-1/32"** ± 3/32"	0.0" ± 0.12" 0" ± ⅛"	**-0.04"** ± 0.08" **-1/32"** ± 3/32"	FRONT TOE (Inches)	0.0" ± 0.08" 0" ± 3/32"	0.0" ± 0.08" 0" ± 3/32"	0.0" ± 0.08" 0" ± 3/32"
			SAI			
44.0 / 35.0 ±2.0 [1]		44.0 / 35.0 ±2.0	TURNING ANGLE	38.1 / 30.3 ±2.0	38.2 / 31.7 ±2.0	38.5 / 32.4 ±2.0
△	03	△	ILLUSTRATION #	102	26	
			REAR			
-0.33 ± 1.0 -5/16 ± 1	0.0 ± 1.0 0 ± 1	-0.33 ± 1.0 -5/16 ± 1	CAMBER (Degrees)	-0.5 ± 0.75 -½ ± ¾	-0.5 ± 0.75 -½ ± ¾	-0.5 ± 0.75 -½ ± ¾
0.16° ± 0.16° 5/32° ± 5/32°	0.16° ± 0.16° 5/32° ± 5/32°	0.16° ± 0.16° 5/32° ± 5/32°	REAR TOE (Degrees)	0.0° ± 0.16° 0° ± 5/32°	0.16° ± 0.16° 5/32° ± 5/32°	0.16° ± 0.16° 5/32° ± 5/32°
0.08" ± 0.08" 3/32" ± 3/32"	0.08" ± 0.08" 3/32" ± 3/32"	0.08" ± 0.08" 3/32" ± 3/32"	REAR TOE (Inches)	0.0" ± 0.08" 0" ± 3/32"	0.08" ± 0.08" 3/32" ± 3/32"	0.08" ± 0.08" 3/32" ± 3/32"
			THRUST ANGLE			
06	05	06	ILLUSTRATION #	06	06	174
[1] 1993-95 w/TCS = 40°/33°			NOTES			
01014	01016	01017		01018	01019	01020

Acura

MDX	NSX, NSX-T exc/Type S	NSX, NSX-T Type S	MODEL INFORMATION	RDX	RDX	RL
2017	1991-2005	1999-2001	YEARS	2007-2012	2013-2017	1996-2004
			INSTRUCTIONS			
94 ft.lbs.	80 ft.lbs.	80 ft.lbs.	LUGNUT TORQUE	80 ft.lbs.	80 ft.lbs.	80 ft.lbs.
			FRONT			
3.73 ± 0.6 3¾ ± ⅝	8.0 ± 0.25 8 ± ¼	8.23 ± 0.25 8¼ ± ¼	CASTER (Degrees)	2.95 ± 1.0 2 15/16 ± 1	2.85 ± 1.0 2⅞ ± 1	2.83 ± 1.0 2 13/16 ± 1
0.0 (0) | n/s	0.0 (0) | n/s	0.0 (0) | n/s	CROSS | TOL.	0.0 (0) | n/s	0.0 (0) | n/s	0.0 (0) | n/s
-0.5 ± 0.75 -½ ± ¾	-0.33 ± 0.17 -5/16 ± 3/16	-0.5 ± 0.17 -½ ± 3/16	CAMBER (Degrees)	0.17 ± 0.75 3/16 ± ¾	0.0 ± 0.5 0 ± ½	0.0 ± 1.0 0 ± 1
0.0 (0) | n/s	0.0 (0) | n/s	0.0 (0) | n/s	CROSS | TOL.	0.0 (0) | 0.5 (½)	0.0 (0) | 0.5 (½)	0.0 (0) | n/s
0.0° ± 0.16° [1] 0° ± 5/32° [1]	**-0.28°** ± 0.08° **-9/32°** ± 3/32°	**-0.28°** ± 0.08° **-9/32°** ± 3/32°	FRONT TOE (Degrees)	0.0° ± 0.16° 0° ± 5/32°	0.0° ± 0.16° 0° ± 5/32°	0.0° ± 0.16° 0° ± 5/32°
0.0" ± 0.08" [1] 0" ± 3/32" [1]	**-0.14"** ± 0.04" **-⅛"** ± 1/32"	**-0.14"** ± 0.04" **-⅛"** ± 1/32"	FRONT TOE (Inches)	0.0" ± 0.08" 0" ± 3/32"	0.0" ± 0.08" 0" ± 3/32"	0.0" ± 0.08" 0" ± 3/32"
			SAI			
38.5 / 32.4 ±2.0	33.1 / 26.6 ±2.0	33.0 / 26.5 ±2.0	TURNING ANGLE	35.4 / 30.5 ±2.0	35.5 / 30.3 ±2.0	41.5 / 33.9 ±2.0
	07	07	ILLUSTRATION #	26	26	
			REAR			
-0.5 ± 0.75 -½ ± ¾	-1.5 ± 0.17 -1½ ± 3/16	-2.0 ± 0.17 -2 ± 3/16	CAMBER (Degrees)	-1.0 ± 0.75 -1 ± ¾	-1.0 ± 0.75 -1 ± ¾	-0.5 ± 1.0 -½ ± 1
0.16° ± 0.16° 5/32° ± 5/32°	0.32° ± 0.08° 5/16° ± 3/32°	0.32° ± 0.08° 5/16° ± 3/32°	REAR TOE (Degrees)	0.2° ± 0.16° 3/16° ± 5/32°	0.16° ± 0.16° 5/32° ± 5/32°	0.16° ± 0.16° 5/32° ± 5/32°
0.08" ± 0.08" 3/32" ± 3/32"	0.16" ± 0.04" 5/32" ± 1/32"	0.16" ± 0.04" 5/32" ± 1/32"	REAR TOE (Inches)	0.1" ± 0.08" 3/32" ± 3/32"	0.08" ± 0.08" 3/32" ± 3/32"	0.08" ± 0.08" 3/32" ± 3/32"
			THRUST ANGLE			
174	08	08	ILLUSTRATION #	06	06	06
[1] 18" wheels = -0.1°(-0.05")			NOTES			
01021	01022	01023		01024	01025	01026

Acura

RL	RL	RLX	MODEL INFORMATION	RLX Hybrid	RSX	RSX
2005-2008	2009-2012	2014-2017	YEARS	2014-2017	2002-2004	2005-2006
			INSTRUCTIONS			
94 ft.lbs.	94 ft.lbs.	94 ft.lbs.	LUGNUT TORQUE	94 ft.lbs.	80 ft.lbs.	80 ft.lbs.
			FRONT			
2.17 ±0.5 2 3/16 ± ½	2.17 ±0.5 2 3/16 ± ½	3.7 ±0.5 3 11/16 ± ½	CASTER (Degrees)	3.65 ±0.5 3⅝ ± ½	1.5 ±1.0 1½ ±1	2.88 ±1.0 2⅞ ±1
0.0 (0) | 0.75 (¾)	0.0 (0) | 0.75 (¾)	0.0 (0) | n/s	CROSS | TOL.	0.0 (0) | n/s	0.0 (0) | n/s	0.0 (0) | n/s
-0.13 ±0.5 -⅛ ± ½	-0.13 ±0.5 -⅛ ± ½	-0.5 ±0.67 -½ ± 11/16	CAMBER (Degrees)	-0.47 ±0.67 -½ ± 11/16	0.0 ±0.75 0 ± ¾	-0.5 ±0.75 -½ ± ¾
0.0 (0) | 0.5 (½)	0.0 (0) | 0.5 (½)	0.0 (0) | 0.67 (11/16)	CROSS | TOL.	0.0 (0) | 0.67 (11/16)	0.0 (0) | n/s	0.0 (0) | n/s
0.0° ±0.16° 0° ± 5/32°	0.0° ±0.16° 0° ± 5/32°	-0.08° ±0.12° -3/32° ± ⅛°	FRONT TOE (Degrees)	-0.08° ±0.12° -3/32° ± ⅛°	0.0° ±0.24° 0° ± ¼°	0.0° ±0.24° 0° ± ¼°
0.0" ±0.08" 0" ± 3/32"	0.0" ±0.08" 0" ± 3/32"	-0.04" ±0.06" -1/32" ± 1/16"	FRONT TOE (Inches)	-0.04" ±0.06" -1/32" ± 1/16"	0.0" ±0.12" 0" ± ⅛"	0.0" ±0.12" 0" ± ⅛"
			SAI			
37.2 / 30.0 ±2.0	37.3 / 30.7 ±2.0	36.4 / 30.1 ±2.0	TURNING ANGLE	36.4 / 30.2 ±2.0	35.0 / 28.0 ±2.0	35.0 / 28.0 ±2.0
			ILLUSTRATION #		26	26
			REAR			
-1.25 ±0.5 -1¼ ± ½	-1.25 ±0.5 -1¼ ± ½	-1.0 ±0.67 -1 ± 11/16	CAMBER (Degrees)	-0.88 ±0.67 -⅞ ± 11/16	-0.75 ±0.75 -¾ ± ¾	-1.0 ±0.75 -1 ± ¾
0.16° ±0.16° 5/32° ± 5/32°	0.16° ±0.16° 5/32° ± 5/32°	0.16° ±0.09° 5/32° ± 3/32°	REAR TOE (Degrees)	0.16° ±0.09° 5/32° ± 3/32°	0.16° +0.16° -0.08° 5/32° +5/32° -3/32°	0.16° +0.16° -0.08° 5/32° +5/32° -3/32°
0.08" ±0.08" 3/32" ± 3/32"	0.08" ±0.08" 3/32" ± 3/32"	0.08" ±0.05" 3/32" ± 1/16"	REAR TOE (Inches)	0.08" ±0.05" 3/32" ± 1/16"	0.08" +0.08" -0.04" 3/32" +3/32" -1/32"	0.08" +0.08" -0.04" 3/32" +3/32" -1/32"
			THRUST ANGLE			
06	06	174	ILLUSTRATION #	174	140	140
			NOTES			
01027	01028	01029		01030	01031	01032

Acura

SLX	TL 2.5	TL 3.2	MODEL INFORMATION	TL	TL	TL
1996-99	1995-98	1996-98	YEARS	1999-2003	2004-2008	2009-2014
			INSTRUCTIONS			
87 ft.lbs.	80 ft.lbs.	80 ft.lbs.	LUGNUT TORQUE	80 ft.lbs.	80 ft.lbs.	94 ft.lbs.
			FRONT			
2.17 ±0.75 2 3/16 ± ¾	2.12 ±1.0 2⅛ ±1	3.67 ±1.0 3 11/16 ±1	CASTER (Degrees)	2.8 ±1.0 2 13/16 ±1	3.28 ±0.75 3 5/16 ± ¾	3.8 ±0.75 3 13/16 ± ¾
0.0 (0) | 0.5 (½)	0.0 (0) | n/s	0.0 (0) | n/s	CROSS | TOL.	0.0 (0) | n/s	0.0 (0) | 0.75 (¾)	0.0 (0) | 0.5 (½)
0.0 ±0.5 0 ± ½	0.0 ±1.0 0 ±1	0.0 ±1.0 0 ±1	CAMBER (Degrees)	0.0 ±1.0 0 ±1	-0.5 ±0.5 -½ ± ½	-0.5 +0.5 -0.75 -½ + ½ -¾
0.0 (0) | 0.5 (½)	0.0 (0) | n/s	0.0 (0) | n/s	CROSS | TOL.	0.0 (0) | n/s	0.0 (0) | 0.75 (¾)	0.0 (0) | 0.5 (½)
0.0° ±0.16° 0° ± 5/32°	-0.08° ±0.16° [1] -3/32° ± 5/32° [1]	-0.08° ±0.16° -3/32° ± 5/32°	FRONT TOE (Degrees)	0.0° ±0.16° 0° ± 5/32°	0.0° ±0.16° 0° ± 5/32°	0.0° ±0.16° 0° ± 5/32°
0.0" ±0.08" 0" ± 3/32"	-0.04" ±0.08" [1] -1/32" ± 3/32" [1]	-0.04" ±0.08" -1/32" ± 3/32"	FRONT TOE (Inches)	0.0" ±0.08" 0" ± 3/32"	0.0" ±0.08" 0" ± 3/32"	0.0" ±0.08" 0" ± 3/32"
12.5 ±0.5 (12½ ± ½)			SAI			
33.0 / 32.0 ±1.0	40.0 / 33.3 ±2.0	47.5 / 34.5 ±2.0	TURNING ANGLE	38.5 / 31.1 ±2.0 [1]	35.1 / 30.2 ±2.0	38.1 / 30.4 ±2.0
23	03		ILLUSTRATION #	03		
			REAR			
	-0.5 ±1.0 -½ ±1	-0.5 ±1.0 -½ ±1	CAMBER (Degrees)	-0.5 ±0.5 -½ ± ½	-1.0 ±0.5 -1 ± ½	-1.0 +0.5 -0.75 -1 + ½ -¾
	0.16° ±0.16° 5/32° ± 5/32°	0.24° ±0.16° ¼° ± 5/32°	REAR TOE (Degrees)	0.16° ±0.16° 5/32° ± 5/32°	0.0° ±0.16° 0° ± 5/32°	0.16° ±0.16° 5/32° ± 5/32°
	0.08" ±0.08" 3/32" ± 3/32"	0.12" ±0.08" ⅛" ± 3/32"	REAR TOE (Inches)	0.08" ±0.08" 3/32" ± 3/32"	0.0" ±0.08" 0" ± 3/32"	0.08" ±0.08" 3/32" ± 3/32"
			THRUST ANGLE			
	05	05	ILLUSTRATION #	178	178	174
	[1] 1995-96 = 0° (0")		NOTES	[1] Type S = 35.3°/29.2°		
01033	01034	01035		01036	01037	01038

Acura

TLX	TLX	TSX	MODEL	TSX	TSX	Vigor
FWD	AWD		INFORMATION	L4	V6	
2015-2017	2015-2017	2004-2008	YEARS	2009-2014	2010-2014	1992-94
			INSTRUCTIONS			
80 ft.lbs.	80 ft.lbs.	80 ft.lbs.	LUGNUT TORQUE	80 ft.lbs.	80 ft.lbs.	80 ft.lbs.
			FRONT			
4.18 ±0.6 4 3/16 ± 5/8	4.18 ±0.6 4 3/16 ± 5/8	3.22 ±0.75 3 3/16 ± 3/4	CASTER (Degrees)	3.78 +0.17 -0.83 3 13/16 + 3/16 - 13/16	3.87 +0.17 -0.83 3 7/8 + 3/16 - 13/16	1.63 ±1.0 1 5/8 ±1
0.0 (0) | 0.75 (3/4)	0.0 (0) | 0.75 (3/4)	0.0 (0) | 0.75 (3/4)	CROSS | TOL.	0.0 (0) | 0.5 (1/2)	0.0 (0) | 0.5 (1/2)	0.0 (0) | n/s
-0.5 ±0.75 -1/2 ± 3/4	-0.5 ±0.75 -1/2 ± 3/4	0.0 ±0.75 0 ± 3/4	CAMBER (Degrees)	0.0 ±0.5 0 ± 1/2	-0.05 ±0.5 -1/16 ± 1/2	0.0 ±1.0 0 ±1
0.0 (0) | 0.5 (1/2)	0.0 (0) | 0.5 (1/2)	0.0 (0) | 0.75 (3/4)	CROSS | TOL.	0.0 (0) | 0.5 (1/2)	0.0 (0) | 0.5 (1/2)	0.0 (0) | n/s
0.0° ±0.16° 0° ± 5/32°	0.0° ±0.16° 0° ± 5/32°	0.0° ±0.16° 0° ± 5/32°	FRONT TOE (Degrees)	0.0° ±0.16° 0° ± 5/32°	0.0° ±0.16° 0° ± 5/32°	0.0° ±0.16° 0° ± 5/32°
0.0" ±0.08" 0" ± 3/32"	0.0" ±0.08" 0" ± 3/32"	0.0" ±0.08" 0" ± 3/32"	FRONT TOE (Inches)	0.0" ±0.08" 0" ± 3/32"	0.0" ±0.08" 0" ± 3/32"	0.0" ±0.08" 0" ± 3/32"
			SAI			
38.0 / 30.8 ±2.0	37.3 / 29.5 ±2.0	35.3 / 29.6 ±2.0	TURNING ANGLE	39.0 / 31.5 ±2.0	37.0 / 30.4 ±2.0	39.4 / 33.6 ±2.0
197	197		ILLUSTRATION #			03 ▲
			REAR			
-1.0 ±0.5 -1 ± 1/2	-1.0 ±0.5 -1 ± 1/2	-1.0 ±0.5 -1 ± 1/2	CAMBER (Degrees)	-1.0 ±0.5 -1 ± 1/2	-1.2 ±0.5 -1 3/16 ± 1/2	-0.5 ±0.5 -1/2 ± 1/2
0.16° ±0.16° 5/32° ± 5/32°	0.16° ±0.16° 5/32° ± 5/32°	0.16° ±0.16° 5/32° ± 5/32°	REAR TOE (Degrees)	0.16° ±0.16° 5/32° ± 5/32°	0.16° ±0.16° 5/32° ± 5/32°	0.24° ±0.16° 1/4° ± 5/32°
0.08" ±0.08" 3/32" ± 3/32"	0.08" ±0.08" 3/32" ± 3/32"	0.08" ±0.08" 3/32" ± 3/32"	REAR TOE (Inches)	0.08" ±0.08" 3/32" ± 3/32"	0.08" ±0.08" 3/32" ± 3/32"	0.12" ±0.08" 1/8" ± 3/32"
			THRUST ANGLE			
174	174	178	ILLUSTRATION #	174	174	06
			NOTES			
01039	01040	01041		01042	01043	01044

Acura

ZDX	Notes	Notes	MODEL	Notes	Notes	Notes
			INFORMATION			
2010-2013			YEARS			
			INSTRUCTIONS			
94 ft.lbs.			LUGNUT TORQUE			
			FRONT			
4.63 ±0.6 4 5/8 ± 5/8			CASTER (Degrees)			
0.0 (0) | 0.75 (3/4)			CROSS | TOL.			
-0.5 ±0.75 -1/2 ± 3/4			CAMBER (Degrees)			
0.0 (0) | 1.0 (1)			CROSS | TOL.			
0.0° ±0.16° 0° ± 5/32°			FRONT TOE (Degrees)			
0.0" ±0.08" 0" ± 3/32"			FRONT TOE (Inches)			
			SAI			
38.2 / 31.7 ±2.0			TURNING ANGLE			
26			ILLUSTRATION #			
			REAR			
-0.5 ±0.75 -1/2 ± 3/4			CAMBER (Degrees)			
0.16° ±0.16° 5/32° ± 5/32°			REAR TOE (Degrees)			
0.08" ±0.08" 3/32" ± 3/32"			REAR TOE (Inches)			
			THRUST ANGLE			
06			ILLUSTRATION #			
			NOTES			
01045	01046	01047		01048	01049	01050

Alfa Romeo

164	4C	Giulia RWD exc/Quadrifoglio	MODEL INFORMATION	Giulia RWD Quadrifoglio	Giulia AWD	Spider
1991-94	2016-2017	2017	YEARS	2017	2017	1990-94
			INSTRUCTIONS			p. 304
74 ft.lbs.			LUGNUT TORQUE			
			FRONT			
2.3 ±0.33 2 5/16 ± 5/16	2.18 ±0.5 2 3/16 ± 1/2	6.13 ±0.3 6 1/8 ± 5/16	CASTER (Degrees)	6.3 ±0.3 6 5/16 ± 5/16	5.62 ±0.3 5 5/8 ± 5/16	1.5 ±0.5 1 1/2 ± 1/2
0.0 (0) | n/s	0.0 (0) | n/s	0.0 (0) | n/s	CROSS | TOL.	0.0 (0) | n/s	0.0 (0) | n/s	0.0 (0) | 0.33 (5/16)
-1.6 ±0.33 -1 5/8 ± 5/16	-0.75 ±0.17 -3/4 ± 3/16	-0.5 ±0.33 -1/2 ± 5/16	CAMBER (Degrees)	-0.58 ±0.33 -9/16 ± 5/16	-0.5 ±0.33 -1/2 ± 5/16	0.33 ±0.5 5/16 ± 1/2
0.0 (0) | n/s	0.0 (0) | 0.25 (1/4)	0.0 (0) | n/s	CROSS | TOL.	0.0 (0) | n/s	0.0 (0) | n/s	0.0 (0) | 0.67 (11/16)
-0.16° ±0.16° -5/32° ± 5/32°	-0.16° ±0.4° -5/32° ± 13/32°	0.0° ±0.14° 0° ± 1/8°	FRONT TOE (Degrees)	0.0° ±0.14° 0° ± 1/8°	0.0° ±0.14° 0° ± 1/8°	0.47° ±0.16° 15/32° ± 5/32°
-0.08" ±0.08" -1/16" ± 1/16"	-0.08" ±0.2" -3/32" ± 3/16"	0.0" ±0.07" 0" ± 1/16"	FRONT TOE (Inches)	0.0" ±0.07" 0" ± 1/16"	0.0" ±0.07" 0" ± 1/16"	0.23" ±0.08" 7/32" ± 1/16"
			SAI			
			TURNING ANGLE			
			ILLUSTRATION #			09
			REAR			
-0.25 ±0.33 -1/4 ± 5/16	-1.75 ±0.25 -1 3/4 ± 1/4	-1.5 ±0.33 -1 1/2 ± 5/16	CAMBER (Degrees)	-1.5 ±0.33 -1 1/2 ± 5/16	-1.5 ±0.33 -1 1/2 ± 5/16	
0.75° ±0.15° 3/4° ± 5/32°	0.67° ±0.13° 21/32° ± 1/8°	0.44° ±0.26° 7/16° ± 1/4°	REAR TOE (Degrees)	0.44° ±0.26° 7/16° ± 1/4°	0.44° ±0.26° 7/16° ± 1/4°	
0.38" ±0.08" 3/8" ± 1/16"	0.34" ±0.07" 11/32" ± 1/16"	0.22" ±0.13" 7/32" ± 1/8"	REAR TOE (Inches)	0.22" ±0.13" 7/32" ± 1/8"	0.22" ±0.13" 7/32" ± 1/8"	
			THRUST ANGLE			
10	49		ILLUSTRATION #			
			NOTES			
02001	02002	02004		02005	02006	02010

Audi

80/90	90	80/90 Quattro (except 20V)	MODEL INFORMATION	90 Quattro 20V	80 Quattro	90 Quattro
1990-92	1993-95	1990-91	YEARS	1990-91	1992	1993-95
			INSTRUCTIONS			
81 ft.lbs.	81 ft.lbs.	81 ft.lbs.	LUGNUT TORQUE	81 ft.lbs.	81 ft.lbs.	81 ft.lbs.
			FRONT			
1.25 ±0.5 1 1/4 ± 1/2	2.17 ±0.5 2 3/16 ± 1/2	1.25 ±0.5 1 1/4 ± 1/2	CASTER (Degrees)	1.42 ±0.5 1 7/16 ± 1/2	1.25 ±0.5 1 1/4 ± 1/2	2.17 ±0.5 2 3/16 ± 1/2
0.0 (0) | 0.5 (1/2)	0.0 (0) | 0.5 (1/2)	0.0 (0) | 0.5 (1/2)	CROSS | TOL.	0.0 (0) | 0.5 (1/2)	0.0 (0) | 0.5 (1/2)	0.0 (0) | 0.5 (1/2)
-0.75 ±0.5 -3/4 ± 1/2	-0.75 ±0.5 -3/4 ± 1/2	-0.83 ±0.5 -13/16 ± 1/2	CAMBER (Degrees)	-0.83 ±0.5 -13/16 ± 1/2	-0.83 ±0.5 -13/16 ± 1/2	-0.75 ±0.5 -3/4 ± 1/2
0.0 (0) | 0.5 (1/2)	0.0 (0) | 0.5 (1/2)	0.0 (0) | 0.5 (1/2)	CROSS | TOL.	0.0 (0) | 0.5 (1/2)	0.0 (0) | 0.5 (1/2)	0.0 (0) | 0.5 (1/2)
0.17° ±0.17° 5/32° ± 5/32°	0.17° ±0.17° 5/32° ± 5/32°	0.17° ±0.17° 5/32° ± 5/32°	FRONT TOE (Degrees)	0.17° ±0.17° 5/32° ± 5/32°	0.17° ±0.17° 5/32° ± 5/32°	0.17° ±0.17° 5/32° ± 5/32°
0.08" ±0.08" 3/32" ± 3/32"	0.08" ±0.08" 3/32" ± 3/32"	0.08" ±0.08" 3/32" ± 3/32"	FRONT TOE (Inches)	0.08" ±0.08" 3/32" ± 3/32"	0.08" ±0.08" 3/32" ± 3/32"	0.08" ±0.08" 3/32" ± 3/32"
			SAI			
20.0 / 19.1 ±0.5	20.0 / 19.0 ±0.5	20.0 / 19.0 ±0.5	TURNING ANGLE	20.0 / 18.9 ±0.5	20.0 / 19.0 ±0.5	20.0 / 19.0 ±0.5
15	15	15	ILLUSTRATION #	15	15	15
			REAR			
-1.0 ±0.33 -1 ± 5/16	-1.5 ±0.33 -1 1/2 ± 5/16	-0.75 ±0.5 -3/4 ± 1/2	CAMBER (Degrees)	-0.75 ±0.5 -3/4 ± 1/2	-0.75 ±0.5 -3/4 ± 1/2	-0.67 ±0.5 -11/16 ± 1/2
0.33° ±0.33° 11/32° ± 11/32°	0.5° ±0.17° 1/2° ± 5/32°	0.33° ±0.17° 11/32° ± 5/32°	REAR TOE (Degrees)	0.33° ±0.17° 11/32° ± 5/32°	0.33° ±0.17° 11/32° ± 5/32°	0.17° ±0.17° 5/32° ± 5/32°
0.17" ±0.17" 5/32" ± 5/32"	0.25" ±0.08" 1/4" ± 3/32"	0.17" ±0.08" 5/32" ± 3/32"	REAR TOE (Inches)	0.17" ±0.08" 5/32" ± 3/32"	0.17" ±0.08" 5/32" ± 3/32"	0.08" ±0.08" 3/32" ± 3/32"
0.23 (7/32)	0.23 (7/32)		THRUST ANGLE		0.23 (7/32)	0.23 (7/32)
		15 16	ILLUSTRATION #	15 16	83 06	83 06
			NOTES			
04001	04002	04003		04004	04005	04006

Audi

100/200	100/200	100/200 Quattro	MODEL INFORMATION	100/200 Quattro S4 Quattro	A3 w/Std. Susp. (UA0/UA4)	A3 w/Sport Susp. (UA1/UA5/UA9/UA6)
1990-91	1992-94	1990-91	YEARS	1992-94	2006-2013	2006-2013
			INSTRUCTIONS			
81 ft.lbs.	81 ft.lbs.	81 ft.lbs.	LUGNUT TORQUE	81 ft.lbs.	89 ft.lbs.	89 ft.lbs.
			FRONT			
1.0 ±0.67 1 ± 11/16	1.17 ±0.67 1 3/16 ± 11/16	1.0 ±0.67 1 ± 11/16	CASTER (Degrees)	0.92 ±0.67 15/16 ± 11/16	7.57 ±0.5 7 9/16 ± ½	7.78 ±0.5 7¾ ± ½
0.0 (0) | 1.0 (1)	0.0 (0) | 1.0 (1)	0.0 (0) | 1.0 (1)	CROSS | TOL.	0.0 (0) | 1.0 (1)	0.0 (0) | 0.5 (½)	0.0 (0) | 0.5 (½)
-0.5 ±0.5 -½ ± ½	-0.67 +0.58 -0.42 [1] -11/16 + 9/16 - 7/16 [1]	-0.5 ±0.5 -½ ± ½	CAMBER (Degrees)	-0.67 +0.58 -0.42 -11/16 + 9/16 - 7/16	-0.5 ±0.5 -½ ± ½	-0.68 ±0.5 -11/16 ± ½
0.0 (0) | 0.5 (½)	0.0 (0) | 0.5 (½)	0.0 (0) | 0.5 (½)	CROSS | TOL.	0.0 (0) | 0.5 (½)	0.0 (0) | 0.5 (½)	0.0 (0) | 0.5 (½)
0.0° +0.08° -0.17° 0° +3/32° -5/32°	0.25° +0.08° -0.17° ¼° +3/32° -5/32°	0.08° +0.08° -0.17° 3/32° +3/32° -5/32°	FRONT TOE (Degrees)	0.25° +0.08° -0.17° ¼° +3/32° -5/32°	0.17° ±0.17° 5/32° ± 5/32°	0.17° ±0.17° 5/32° ± 5/32°
0.0" +0.04" -0.08" 0" +1/32" -3/32"	0.13" +0.04" -0.08" ⅛" +1/32" -3/32"	0.04" +0.04" -0.08" 1/32" +1/32" -3/32"	FRONT TOE (Inches)	0.13" +0.04" -0.08" ⅛" +1/32" -3/32"	0.08" ±0.08" 3/32" ± 3/32"	0.08" ±0.08" 3/32" ± 3/32"
			SAI			
20.0 / 18.3 ±0.5	20.0 / 18.3 ±0.5	20.0 / 18.3 ±0.5	TURNING ANGLE	20.0 / 18.8 ±0.5	20.0 / 18.4 ±0.33	20.0 / 18.3 ±0.33
17	17	17	ILLUSTRATION #	17	△	△
			REAR			
-0.83 ±0.5 -13/16 ± ½	-0.83 ±0.5 -13/16 ± ½	-0.25 ±0.25 -¼ ± ¼	CAMBER (Degrees)	-0.25 ±0.25 -¼ ± ¼	-1.33 ±0.5 -1 5/16 ± ½	-1.33 ±0.5 -1 5/16 ± ½
0.33° ±0.17° 11/32° ± 5/32°	0.33° ±0.17° 11/32° ± 5/32°	0.17° ±0.17° 5/32° ± 5/32°	REAR TOE (Degrees)	0.17° ±0.17° 5/32° ± 5/32°	0.42° ±0.17° 13/32° ± 5/32°	0.42° ±0.17° 13/32° ± 5/32°
0.17" ±0.08" 5/32" ± 3/32"	0.17" ±0.08" 5/32" ± 3/32"	0.08" ±0.08" 3/32" ± 3/32"	REAR TOE (Inches)	0.08" ±0.08" 3/32" ± 3/32"	0.21" ±0.08" 7/32" ± 3/32"	0.21" ±0.08" 7/32" ± 3/32"
0.23 (7/32)	0.23 (7/32)		THRUST ANGLE		0.33 (11/32)	0.33 (11/32)
		18 19	ILLUSTRATION #	18 19	127	127
	[1] with M/T -0.83° (-13/16°)		NOTES			
04009	04010	04011		04012	04025	04026

Audi

A3 Quattro w/Std. Susp. (UA0/UA4)	A3 Quattro w/Sport Susp. (UA1/UA5/UA9)	A3 Quattro GMBH w/Sport Susp. (UA3/UB3)	MODEL INFORMATION	A3/A3 Quattro Std. Susp. (2UA) w/IRS	A3 (incl e-tron) Std. Susp. (2UA) w/Beam Axle	A3/A3 Quattro Sport Susp. (2UC) w/IRS
2006-2013	2006-2013	2006-2013	YEARS	2015-2017	2015-2017	2015-2017
			INSTRUCTIONS			
89 ft.lbs.	89 ft.lbs.	89 ft.lbs.	LUGNUT TORQUE	89 ft.lbs.	89 ft.lbs.	89 ft.lbs.
			FRONT			
7.57 ±0.5 7 9/16 ± ½	7.78 ±0.5 7¾ ± ½	8.12 ±0.5 8⅛ ± ½	CASTER (Degrees)	7.42 ±0.5 7 7/16 ± ½	7.42 ±0.5 7 7/16 ± ½	7.67 ±0.5 7 11/16 ± ½
0.0 (0) | 0.5 (½)	0.0 (0) | 0.5 (½)	0.0 (0) | 0.5 (½)	CROSS | TOL.	0.0 (0) | 0.5 (½)	0.0 (0) | 0.5 (½)	0.0 (0) | 0.5 (½)
-0.5 ±0.5 -½ ± ½	-0.68 ±0.5 -11/16 ± ½	-0.77 ±0.5 -¾ ± ½	CAMBER (Degrees)	-0.58 ±0.5 -9/16 ± ½	-0.58 ±0.5 -9/16 ± ½	-0.77 ±0.5 -¾ ± ½
0.0 (0) | 0.5 (½)	0.0 (0) | 0.5 (½)	0.0 (0) | 0.5 (½)	CROSS | TOL.	0.0 (0) | 0.5 (½)	0.0 (0) | 0.5 (½)	0.0 (0) | 0.5 (½)
0.17° ±0.17° 5/32° ± 5/32°	0.17° ±0.17° 5/32° ± 5/32°	0.17° ±0.17° 5/32° ± 5/32°	FRONT TOE (Degrees)	0.17° ±0.17° 5/32° ± 5/32°	0.17° ±0.17° 5/32° ± 5/32°	0.17° ±0.17° 5/32° ± 5/32°
0.08" ±0.08" 3/32" ± 3/32"	0.08" ±0.08" 3/32" ± 3/32"	0.08" ±0.08" 3/32" ± 3/32"	FRONT TOE (Inches)	0.08" ±0.08" 3/32" ± 3/32"	0.08" ±0.08" 3/32" ± 3/32"	0.08" ±0.08" 3/32" ± 3/32"
			SAI			
20.0 / 18.4 ±0.33	20.0 / 18.3 ±0.33	20.0 / 18.0 ±0.33	TURNING ANGLE	20.0 / 18.7 ±0.33	20.0 / 18.7 ±0.33	20.0 / 18.5 ±0.33
△	△	△	ILLUSTRATION #			
			REAR			
-1.33 ±0.5 -1 5/16 ± ½	-1.33 ±0.5 -1 5/16 ± ½	-1.33 ±0.5 -1 5/16 ± ½	CAMBER (Degrees)	-1.33 ±0.5 -1 5/16 ± ½	-1.0 ±0.17 -1 ± 3/16	-1.33 ±0.5 -1 5/16 ± ½
0.42° ±0.17° 13/32° ± 5/32°	0.42° ±0.17° 13/32° ± 5/32°	0.42° ±0.17° 13/32° ± 5/32°	REAR TOE (Degrees)	0.42° ±0.17° 13/32° ± 5/32°	0.33° ±0.17° 13/32° ± 5/32°	0.42° ±0.17° 13/32° ± 5/32°
0.21" ±0.08" 7/32" ± 3/32"	0.21" ±0.08" 7/32" ± 3/32"	0.21" ±0.08" 7/32" ± 3/32"	REAR TOE (Inches)	0.21" ±0.08" 7/32" ± 3/32"	0.17" ±0.08" 5/32" ± 3/32"	0.21" ±0.08" 7/32" ± 3/32"
0.33 (11/32)	0.33 (11/32)	0.33 (11/32)	THRUST ANGLE	0.2 (3/16)	0.33 (11/32)	0.2 (3/16)
127	127	127	ILLUSTRATION #	127	127	127
			NOTES			
04027	04028	04029		04030	04031	04032

Audi

A3 (incl e-tron)	A3/A3 Quattro	A3	MODEL	A4	A4	A4 Quattro
Sport Susp. (2UC)	S-line Sport Susp. (2UG)	S-line Sport Susp. (2UG)	INFORMATION	exc/Sport Susp.	Sport Susp.	exc/Sport Susp.
w/Beam Axle	w/IRS	w/Beam Axle				
2015-2017	2015-2017	2015-2017	YEARS	1996-2001	1996-2001	1996-2001
			INSTRUCTIONS			
89 ft.lbs.	89 ft.lbs.	89 ft.lbs.	LUGNUT TORQUE	89 ft.lbs.	89 ft.lbs.	89 ft.lbs.
			FRONT			
7.67 ±0.5 7 11/16 ± 1/2	7.85 ±0.5 7 7/8 ± 1/2	7.85 ±0.5 7 7/8 ± 1/2	CASTER (Degrees)	Caster is not specified by the manufacturer.	Caster is not specified by the manufacturer.	Caster is not specified by the manufacturer.
0.0 (0) | 0.5 (1/2)	0.0 (0) | 0.5 (1/2)	0.0 (0) | 0.5 (1/2)	CROSS | TOL.			
-0.77 ±0.5 -3/4 ± 1/2	-0.87 ±0.5 -7/8 ± 1/2	-0.87 ±0.5 -7/8 ± 1/2	CAMBER (Degrees)	-0.58 ±0.42 [1] -9/16 ± 7/16 [1]	-0.83 ±0.42 [1] -13/16 ± 7/16 [1]	-0.58 ±0.42 [1] -9/16 ± 7/16 [1]
0.0 (0) | 0.5 (1/2)	0.0 (0) | 0.5 (1/2)	0.0 (0) | 0.5 (1/2)	CROSS | TOL.	0.0 (0) | 0.5 (1/2)	0.0 (0) | 0.5 (1/2)	0.0 (0) | 0.5 (1/2)
0.17° ±0.17° 5/32° ± 5/32°	0.17° ±0.17° 5/32° ± 5/32°	0.17° ±0.17° 5/32° ± 5/32°	FRONT TOE (Degrees)	0.33° ±0.07° 11/32° ± 1/16°	0.33° ±0.07° 11/32° ± 1/16°	0.33° ±0.07° 11/32° ± 1/16°
0.08" ±0.08" 3/32" ± 3/32"	0.08" ±0.08" 3/32" ± 3/32"	0.08" ±0.08" 3/32" ± 3/32"	FRONT TOE (Inches)	0.17" ±0.03" 5/32" ± 1/32"	0.17" ±0.03" 5/32" ± 1/32"	0.17" ±0.03" 5/32" ± 1/32"
			SAI			
20.0 / 18.5 ±0.33	20.0 / 18.3 ±0.33	20.0 / 18.3 ±0.33	TURNING ANGLE	20.0 / 18.7 ±0.5	20.0 / 18.7 ±0.5	20.0 / 18.7 ±0.5
			ILLUSTRATION #	▲	▲	▲
			REAR			
-1.0 ±0.17 -1 ± 3/16	-1.33 ±0.5 -1 5/16 ± 1/2	-1.0 ±0.17 -1 ± 3/16	CAMBER (Degrees)	-1.5 ±0.33 -1 1/2 ± 5/16	-1.5 ±0.33 -1 1/2 ± 5/16	-0.67 ±0.5 -11/16 ± 1/2
0.4° ±0.17° 13/32° ± 5/32°	0.42° ±0.17° 13/32° ± 5/32°	0.45° ±0.17° 7/16° ± 5/32°	REAR TOE (Degrees)	0.33° +0.25° -0.17° 11/32° + 1/4° - 5/32°	0.47° +0.25° -0.17° 15/32° + 1/4° - 5/32°	0.27° ±0.17° 9/32° ± 5/32°
0.2" ±0.08" 3/16" ± 3/32"	0.21" ±0.08" 7/32" ± 3/32"	0.23" ±0.08" 7/32" ± 3/32"	REAR TOE (Inches)	0.17" +0.13" -0.08" 5/32" + 1/8" - 3/32"	0.23" +0.13" -0.08" 7/32" + 1/8" - 3/32"	0.13" ±0.08" 1/8" ± 3/32"
0.33 (11/32)	0.2 (3/16)	0.33 (11/32)	THRUST ANGLE	0.25 (1/4)	0.25 (1/4)	0.17 (5/32)
	127		ILLUSTRATION #	114	114	83 06
			NOTES	[1] 1996-97 = -0.42° (-7/16°)	[1] 1996-97 = -0.67° (-11/16°)	[1] 1996-97 = -0.42° (-7/16°)
04033	04034	04035		04036	04037	04038

Audi

A4 Quattro	A4 (B6/B7)	A4 (B6/B7)	MODEL	A4 (B6/B7)	A4 Quattro (B6/B7)	A4 Quattro (B6/B7)
Sport Susp.	Sedan & Wagon	Sedan & Wagon	INFORMATION	Convertible	Sedan & Wagon	Sedan & Wagon
	exc/Sport Susp. (1BA)	Sport Susp. (1BE, 1BV)		(1BA/1BE)	exc/Sport Susp.	Sport Susp.
1996-2001	2002-2008	2002-2008	YEARS	2003-2009	2002-2008	2002-2008
			INSTRUCTIONS			
89 ft.lbs.	89 ft.lbs.	89 ft.lbs.	LUGNUT TORQUE	89 ft.lbs.	89 ft.lbs.	89 ft.lbs.
			FRONT			
Caster is not specified by the manufacturer.	Caster is not specified by the manufacturer.	Caster is not specified by the manufacturer.	CASTER (Degrees)	Caster is not specified by the manufacturer.	Caster is not specified by the manufacturer.	Caster is not specified by the manufacturer.
			CROSS | TOL.			
-0.83 ±0.42 [1] -13/16 ± 7/16 [1]	-0.55 ±0.42 -9/16 ± 7/16	-0.78 ±0.42 -13/16 ± 7/16	CAMBER (Degrees)	-0.78 ±0.42 -13/16 ± 7/16	-0.55 ±0.42 -9/16 ± 7/16	-0.78 ±0.42 -13/16 ± 7/16
0.0 (0) | 0.5 (1/2)	0.0 (0) | 0.5 (1/2)	0.0 (0) | 0.5 (1/2)	CROSS | TOL.	0.0 (0) | 0.5 (1/2)	0.0 (0) | 0.5 (1/2)	0.0 (0) | 0.5 (1/2)
0.33° ±0.07° 11/32° ± 1/16°	0.33° ±0.07° 11/32° ± 1/16°	0.33° ±0.07° 11/32° ± 1/16°	FRONT TOE (Degrees)	0.33° ±0.07° 11/32° ± 1/16°	0.33° ±0.07° 11/32° ± 1/16°	0.33° ±0.07° 11/32° ± 1/16°
0.17" ±0.03" 5/32" ± 1/32"	0.17" ±0.03" 5/32" ± 1/32"	0.17" ±0.03" 5/32" ± 1/32"	FRONT TOE (Inches)	0.17" ±0.03" 5/32" ± 1/32"	0.17" ±0.03" 5/32" ± 1/32"	0.17" ±0.03" 5/32" ± 1/32"
			SAI			
20.0 / 18.7 ±0.5	20.0 / 18.9 ±0.5	20.0 / 18.9 ±0.5	TURNING ANGLE	20.0 / 18.9 ±0.5	20.0 / 18.8 ±0.5	20.0 / 18.8 ±0.5
▲	▲	▲	ILLUSTRATION #	▲	▲	▲
			REAR			
-0.67 ±0.5 -11/16 ± 1/2	-1.33 ±0.5 -1 5/16 ± 1/2	-1.33 ±0.5 -1 5/16 ± 1/2	CAMBER (Degrees)	-1.33 ±0.5 -1 5/16 ± 1/2	-1.33 ±0.5 -1 5/16 ± 1/2	-1.33 ±0.5 -1 5/16 ± 1/2
0.27° ±0.17° 9/32° ± 5/32°	0.33° +0.17° -0.08° 11/32° + 5/32° - 3/32°	0.33° +0.17° -0.08° 11/32° + 5/32° - 3/32°	REAR TOE (Degrees)	0.33° +0.17° -0.08° 11/32° + 5/32° - 3/32°	0.33° +0.17° -0.08° 11/32° + 5/32° - 3/32°	0.33° +0.17° -0.08° 11/32° + 5/32° - 3/32°
0.13" ±0.08" 1/8" ± 3/32"	0.17" +0.08" -0.04" 5/32" + 3/32" - 1/32"	0.17" +0.08" -0.04" 5/32" + 3/32" - 1/32"	REAR TOE (Inches)	0.17" +0.08" -0.04" 5/32" + 3/32" - 1/32"	0.17" +0.08" -0.04" 5/32" + 3/32" - 1/32"	0.17" +0.08" -0.04" 5/32" + 3/32" - 1/32"
0.17 (5/32)	0.12 (1/8)	0.12 (1/8)	THRUST ANGLE	0.17 (5/32)	0.12 (1/8)	0.12 (1/8)
83 06	142	142	ILLUSTRATION #	142	142	142
[1] 1996-97 = -0.67° (-11/16°)			NOTES			
04039	04040	04041		04042	04043	04044

Audi

A4 Quattro (B6/B7) Convertible (1BA/1BE)	A4/A4 Quattro (B8) Std. Susp. (1BA)	A4/A4 Quattro (B8) Sport/Air Susp. (1BD/1BE/1BL)	MODEL INFORMATION	A4/A4 Quattro (B8) Sport Susp. (1BV) S-Line	A4 allroad (B8) Std. Susp. (1BP) Steel Springs	A4/A4 Quattro (B9) 1BA/1BB/1BM Susp.
2002-2009	2009-2016	2009-2016	YEARS	2009-2016	2013-2016	2017
			INSTRUCTIONS			
89 ft.lbs.	89 ft.lbs.	89 ft.lbs.	LUGNUT TORQUE	89 ft.lbs.	89 ft.lbs.	89 ft.lbs.
			FRONT			
Caster is not specified by the manufacturer.	Caster is not specified by the manufacturer.	Caster is not specified by the manufacturer.	CASTER (Degrees)	Caster is not specified by the manufacturer.	Caster is not specified by the manufacturer.	Caster is not specified by the manufacturer.
			CROSS | TOL.			
-0.78 ±0.42 -13/16 ± 7/16	-0.72 ±0.38 -11/16 ± 3/8	-1.08 ±0.38 -1 1/16 ± 3/8	CAMBER (Degrees)	-1.22 ±0.38 -1 3/16 ± 3/8	-0.33 ±0.38 -5/16 ± 3/8	-0.72 ±0.38 -11/16 ± 3/8
0.0 (0) | 0.5 (½)	0.0 (0) | 0.5 (½)	0.0 (0) | 0.5 (½)	CROSS | TOL.	0.0 (0) | 0.5 (½)	0.0 (0) | 0.5 (½)	0.0 (0) | 0.5 (½)
0.33° ±0.07° 11/32° ± 1/16°	0.33° ±0.17° 11/32° ± 5/32°	0.33° ±0.17° 11/32° ± 5/32°	FRONT TOE (Degrees)	0.33° ±0.17° 11/32° ± 5/32°	0.33° ±0.17° 11/32° ± 5/32°	0.33° ±0.17° 11/32° ± 5/32°
0.17" ±0.03" 5/32" ± 1/32"	0.17" ±0.08" 5/32" ± 3/32"	0.17" ±0.08" 5/32" ± 3/32"	FRONT TOE (Inches)	0.17" ±0.08" 5/32" ± 3/32"	0.17" ±0.08" 5/32" ± 3/32"	0.17" ±0.08" 5/32" ± 3/32"
			SAI			
20.0 / 18.8 ±0.5	20.0 / 18.2 ±0.5	20.0 / 18.2 ±0.5	TURNING ANGLE	20.0 / 18.2 ±0.5	20.0 / 18.2 ±0.5	20.0 / 18.1 ±0.5
▲			ILLUSTRATION #			
			REAR			
-1.33 ±0.5 -1 5/16 ± 1/2	-1.33 ±0.42 -1 5/16 ± 7/16	-1.33 ±0.42 -1 5/16 ± 7/16	CAMBER (Degrees)	-1.33 ±0.42 -1 5/16 ± 7/16	-1.33 ±0.42 -1 5/16 ± 7/16	-0.83 ±0.42 -13/16 ± 7/16
0.33° +0.17° -0.08° 11/32° +5/32° -3/32°	0.33° ±0.17° 11/32° ± 5/32°	0.33° ±0.17° 11/32° ± 5/32°	REAR TOE (Degrees)	0.33° ±0.17° 11/32° ± 5/32°	0.33° ±0.17° 11/32° ± 5/32°	0.33° ±0.17° 11/32° ± 5/32°
0.17" +0.08" -0.04" 5/32" +3/32" -1/32"	0.17" ±0.08" 5/32" ± 3/32"	0.17" ±0.08" 5/32" ± 3/32"	REAR TOE (Inches)	0.17" ±0.08" 5/32" ± 3/32"	0.17" ±0.08" 5/32" ± 3/32"	0.17" ±0.08" 5/32" ± 3/32"
0.17 (5/32)	0.17 (5/32)	0.17 (5/32)	THRUST ANGLE	0.17 (5/32)	0.17 (5/32)	0.17 (5/32)
142	115 174	115 174	ILLUSTRATION #	115 174	115 174	177
			NOTES			
04045	04046	04047		04048	04049	04050

Audi

A4/A4 Quattro (B9) 1BD/1BE Susp.	A4/A4 Quattro (B9) 1BR Susp.	A4 allroad (B9) 1BP/2MN Susp.	MODEL INFORMATION	A5 Coupe & Conv. exc/S-Line	A5 Coupe & Conv. S-Line (1BV)	A6 A6 Wagon
2017	2017	2017	YEARS	2008-2017	2008-2017	1995-98
			INSTRUCTIONS			
89 ft.lbs.	89 ft.lbs.	89 ft.lbs.	LUGNUT TORQUE	89 ft.lbs.	89 ft.lbs.	81 ft.lbs.
			FRONT			
Caster is not specified by the manufacturer.	Caster is not specified by the manufacturer.	Caster is not specified by the manufacturer.	CASTER (Degrees)	Caster is not specified by the manufacturer.	Caster is not specified by the manufacturer.	1.17 ±0.67 1 3/16 ± 11/16
			CROSS | TOL.			0.0 (0) | 1.0 (1)
-1.05 ±0.38 -1 1/16 ± 3/8	-0.57 ±0.38 -9/16 ± 3/8	-0.47 ±0.38 -7/16 ± 3/8	CAMBER (Degrees)	-1.08 ±0.38 -1 1/16 ± 3/8	-1.22 ±0.38 -1 1/4 ± 3/8	-0.67 +0.58 -0.42 [1] -11/16 +9/16 -7/16 [1]
0.0 (0) | 0.5 (½)	0.0 (0) | 0.5 (½)	0.0 (0) | 0.5 (½)	CROSS | TOL.	0.0 (0) | 0.5 (½)	0.0 (0) | 0.5 (½)	0.0 (0) | 0.5 (½)
0.33° ±0.17° 11/32° ± 5/32°	0.33° ±0.17° 11/32° ± 5/32°	0.33° ±0.17° 11/32° ± 5/32°	FRONT TOE (Degrees)	0.33° ±0.17° 11/32° ± 5/32°	0.33° ±0.17° 11/32° ± 5/32°	0.25° +0.08° -0.17° 1/4° +3/32° -5/32°
0.17" ±0.08" 5/32" ± 3/32"	0.17" ±0.08" 5/32" ± 3/32"	0.17" ±0.08" 5/32" ± 3/32"	FRONT TOE (Inches)	0.17" ±0.08" 5/32" ± 3/32"	0.17" ±0.08" 5/32" ± 3/32"	0.13" +0.04" -0.08" 1/8" +1/32" -3/32"
			SAI			
20.0 / 18.1 ±0.5	20.0 / 18.1 ±0.5	20.0 / 18.2 ±0.5	TURNING ANGLE	20.0 / 18.2 ±0.5	20.0 / 18.2 ±0.5	20.0 / 18.3 ±0.67
			ILLUSTRATION #			17
			REAR			
-0.83 ±0.42 -13/16 ± 7/16	-0.83 ±0.42 -13/16 ± 7/16	-0.83 ±0.42 -13/16 ± 7/16	CAMBER (Degrees)	-1.33 ±0.42 -1 5/16 ± 7/16	-1.33 ±0.42 -1 5/16 ± 7/16	-0.83 ±0.5 -13/16 ± 1/2
0.33° ±0.17° 11/32° ± 5/32°	0.33° ±0.17° 11/32° ± 5/32°	0.33° ±0.17° 11/32° ± 5/32°	REAR TOE (Degrees)	0.33° ±0.17° 11/32° ± 5/32°	0.33° ±0.17° 11/32° ± 5/32°	0.33° ±0.17° 11/32° ± 5/32°
0.17" ±0.08" 5/32" ± 3/32"	0.17" ±0.08" 5/32" ± 3/32"	0.17" ±0.08" 5/32" ± 3/32"	REAR TOE (Inches)	0.17" ±0.08" 5/32" ± 3/32"	0.17" ±0.08" 5/32" ± 3/32"	0.17" ±0.08" 5/32" ± 3/32"
0.17 (5/32)	0.17 (5/32)	0.17 (5/32)	THRUST ANGLE	0.17 (5/32)	0.17 (5/32)	0.23 (7/32)
177	177	177	ILLUSTRATION #	115 174	115 174	
			NOTES			[1] with M/T -0.83° (-13/16°)
04051	04052	04053		04054	04055	04056

Audi

A6 exc/Sport Susp.	A6 Sport Susp.	A6 Std. Susp. (1BA, 1BB)	MODEL INFORMATION	A6 Sport Susp. (1BE)	A6 Air Susp. (1BK)	A6 Quattro A6 Quattro Wgn.
1998-2004	2002-2004	2006-2011	YEARS	2006-2011	2006-2011	1995-98
			INSTRUCTIONS			
81 ft.lbs.	81 ft.lbs.	81 ft.lbs.	LUGNUT TORQUE	81 ft.lbs.	81 ft.lbs.	81 ft.lbs.
			FRONT			
Caster is not specified by the manufacturer.	Caster is not specified by the manufacturer.	Caster is not specified by the manufacturer.	CASTER (Degrees)	Caster is not specified by the manufacturer.	Caster is not specified by the manufacturer.	0.92 ± 0.67 15/16 ± 11/16
			CROSS | TOL.			0.0 (0) | 1.0 (1)
-0.83 ± 0.42 -13/16 ± 7/16	-1.08 ± 0.42 -1 1/16 ± 7/16	-0.87 ± 0.42 -7/8 ± 7/16	CAMBER (Degrees)	-1.12 ± 0.42 -1 1/8 ± 7/16	-1.03 ± 0.42 -1 ± 7/16	-0.67 +0.58 -0.42 -11/16 + 9/16 - 7/16
0.0 (0) | 0.5 (½)	0.0 (0) | 0.5 (½)	0.0 (0) | 0.5 (½)	CROSS | TOL.	0.0 (0) | 0.5 (½)	0.0 (0) | 0.5 (½)	0.0 (0) | 0.5 (½)
0.33° ± 0.07° 11/32° ± 1/16°	0.33° ± 0.07° 11/32° ± 1/16°	0.3° ± 0.13° 5/16° ± 1/8°	FRONT TOE (Degrees)	0.33° ± 0.13° 11/32° ± 1/8°	0.37° ± 0.13° 3/8° ± 1/8°	0.25° +0.08° -0.17° 1/4° + 3/32° - 5/32°
0.17" ± 0.03" 5/32" ± 1/32"	0.17" ± 0.03" 5/32" ± 1/32"	0.15" ± 0.07" 5/32" ± 1/16"	FRONT TOE (Inches)	0.17" ± 0.07" 5/32" ± 1/16"	0.18" ± 0.07" 3/16" ± 1/16"	0.13" +0.04" -0.08" 1/8" + 1/32" - 3/32"
			SAI			
20.0 / 18.5 ±0.5	20.0 / 18.5 ±0.5	20.0 / 18.8 ±0.5	TURNING ANGLE	20.0 / 18.7 ±0.5	20.0 / 19.0 ±0.5	20.0 / 18.8 ±0.5
			ILLUSTRATION #			17
			REAR			
-1.5 ± 0.33 -1½ ± 5/16	-1.5 ± 0.33 -1½ ± 5/16	-0.83 ± 0.33 -13/16 ± 5/16	CAMBER (Degrees)	-0.83 ± 0.33 -13/16 ± 5/16	-1.25 ± 0.33 -1¼ ± 5/16	-0.25 ± 0.25 -¼ ± ¼
0.33° +0.25° -0.17° 11/32° + 1/4° - 5/32°	0.47° +0.25° -0.17° 15/32° + 1/4° - 5/32°	0.3° ± 0.17° 5/16° ± 5/32°	REAR TOE (Degrees)	0.3° ± 0.17° 5/16° ± 5/32°	0.3° ± 0.17° 5/16° ± 5/32°	0.17° ± 0.17° 5/32° ± 5/32°
0.17" +0.13" -0.08" 5/32" + 1/8" - 3/32"	0.23" +0.13" -0.08" 7/32" + 1/8" - 3/32"	0.15" ± 0.08" 5/32" ± 3/32"	REAR TOE (Inches)	0.15" ± 0.08" 5/32" ± 3/32"	0.15" ± 0.08" 5/32" ± 3/32"	0.08" ± 0.08" 3/32" ± 3/32"
0.25 (¼)	0.25 (¼)	0.08 (3/32)	THRUST ANGLE	0.08 (3/32)	0.08 (3/32)	
		106 170	ILLUSTRATION #	106 170	106 170	18 19
			NOTES			
04057	04058	04059		04060	04061	04062

Audi

A6 Quattro exc/V8 Engine	A6 Quattro V8 Engine	A6 Quattro Avant	MODEL INFORMATION	A6 allroad quattro	A6 Quattro Sport Susp. (1BE)	A6 Quattro Air Susp. (1BY)
1998-2004	2000-2004	1999-2004	YEARS	2001-2005	2005-2011	2005-2011
			INSTRUCTIONS			
81 ft.lbs.	81 ft.lbs.	81 ft.lbs.	LUGNUT TORQUE	81 ft.lbs.	81 ft.lbs.	81 ft.lbs.
			FRONT			
Caster is not specified by the manufacturer.	Caster is not specified by the manufacturer.	Caster is not specified by the manufacturer.	CASTER (Degrees)	Caster is not specified by the manufacturer.	Caster is not specified by the manufacturer.	Caster is not specified by the manufacturer.
			CROSS | TOL.			
-0.83 ± 0.42 -13/16 ± 7/16	-1.0 ± 0.42 [1] -1 ± 7/16 [1]	-0.83 ± 0.42 -13/16 ± 7/16	CAMBER (Degrees)	-1.0 ± 0.42 -1 ± 7/16	-1.12 ± 0.42 -1 1/8 ± 7/16	-1.03 ± 0.42 -1 ± 7/16
0.0 (0) | 0.5 (½)	0.0 (0) | 0.5 (½)	0.0 (0) | 0.5 (½)	CROSS | TOL.	0.0 (0) | 0.5 (½)	0.0 (0) | 0.5 (½)	0.0 (0) | 0.5 (½)
0.33° ± 0.07° 11/32° ± 1/16°	0.33° ± 0.07° 11/32° ± 1/16°	0.33° ± 0.07° 11/32° ± 1/16°	FRONT TOE (Degrees)	0.33° ± 0.07° 11/32° ± 1/16°	0.33° ± 0.13° 11/32° ± 1/8°	0.37° ± 0.13° 3/8° ± 1/8°
0.17" ± 0.03" 5/32" ± 1/32"	0.17" ± 0.03" 5/32" ± 1/32"	0.17" ± 0.03" 5/32" ± 1/32"	FRONT TOE (Inches)	0.17" ± 0.03" 5/32" ± 1/32"	0.17" ± 0.07" 5/32" ± 1/16"	0.18" ± 0.07" 3/16" ± 1/16"
			SAI			
20.0 / 18.5 ±0.5	20.0 / 18.5 ±0.5	20.0 / 18.5 ±0.5	TURNING ANGLE	20.0 / 18.6 ±0.5	20.0 / 18.5 ±0.5	20.0 / 19.0 ±0.5
			ILLUSTRATION #			
			REAR			
-0.67 ± 0.5 -11/16 ± ½	-0.67 ± 0.5 [2] -11/16 ± ½ [2]	-0.67 ± 0.5 -11/16 ± ½	CAMBER (Degrees)	-1.0 ± 0.5 -1 ± ½	-0.83 ± 0.33 -13/16 ± 5/16	-1.25 ± 0.33 -1¼ ± 5/16
0.27° ± 0.17° 9/32° ± 5/32°	0.27° ± 0.17° 9/32° ± 5/32°	0.27° ± 0.17° 9/32° ± 5/32°	REAR TOE (Degrees)	0.3° ± 0.13° 5/16° ± 1/8°	0.3° ± 0.17° 5/16° ± 5/32°	0.3° ± 0.17° 5/16° ± 5/32°
0.13" ± 0.08" 1/8" ± 3/32"	0.13" ± 0.08" 1/8" ± 3/32"	0.13" ± 0.08" 1/8" ± 3/32"	REAR TOE (Inches)	0.15" ± 0.07" 5/32" ± 1/16"	0.15" ± 0.08" 5/32" ± 3/32"	0.15" ± 0.08" 5/32" ± 3/32"
0.17 (5/32)	0.17 (5/32)	0.17 (5/32)	THRUST ANGLE	0.17 (5/32)	0.08 (3/32)	0.08 (3/32)
83 06	83 06	83 06	ILLUSTRATION #	83 06	106 170	106 170
	[1] Sport susp. = -1.08° (-1 1/16°) [1] Sport susp. = -1.0° (-1°)		NOTES			
04063	04064	04065		04066	04067	04068

Audi

A6/A6 Quattro	A6/A6 Quattro	A6/A6 Quattro	MODEL	A7/A7 Quattro	A7/A7 Quattro	A7/A7 Quattro
Std. Susp.	Sport Susp.	Air Susp.	INFORMATION	Std. Susp.	Sport Susp.	Air Susp.
(1BA)	(1BE, 1BV)	(1BK, 2MB)		(1BA)	(1BE, 1BV)	(1BK, 2MB)
2012-2017	2012-2017	2012-2017	YEARS	2012-2017	2012-2017	2012-2017
			INSTRUCTIONS			
81 ft.lbs.	81 ft.lbs.	81 ft.lbs.	LUGNUT TORQUE	81 ft.lbs.	81 ft.lbs.	81 ft.lbs.
			FRONT			
Caster is not specified by the manufacturer.	Caster is not specified by the manufacturer.	Caster is not specified by the manufacturer.	CASTER (Degrees)	Caster is not specified by the manufacturer.	Caster is not specified by the manufacturer.	Caster is not specified by the manufacturer.
			CROSS | TOL.			
-0.72 ±0.38 -11/16 ± 3/8	-1.08 ±0.38 [1] -1 1/16 ± 3/8 [1]	-1.08 ±0.38 -1 1/16 ± 3/8	CAMBER (Degrees)	-0.88 ±0.38 -7/8 ± 3/8	-1.08 ±0.38 -1 1/16 ± 3/8	-1.08 ±0.38 -1 1/16 ± 3/8
0.0 (0) | 0.5 (1/2)	0.0 (0) | 0.5 (1/2)	0.0 (0) | 0.5 (1/2)	CROSS | TOL.	0.0 (0) | 0.5 (1/2)	0.0 (0) | 0.5 (1/2)	0.0 (0) | 0.5 (1/2)
0.33° ±0.17° 11/32° ± 5/32°	0.33° ±0.17° 11/32° ± 5/32°	0.33° ±0.17° 11/32° ± 5/32°	FRONT TOE (Degrees)	0.33° ±0.17° 11/32° ± 5/32°	0.33° ±0.17° 11/32° ± 5/32°	0.33° ±0.17° 11/32° ± 5/32°
0.17" ±0.08" 5/32" ± 3/32"	0.17" ±0.08" 5/32" ± 3/32"	0.17" ±0.08" 5/32" ± 3/32"	FRONT TOE (Inches)	0.17" ±0.08" 5/32" ± 3/32"	0.17" ±0.08" 5/32" ± 3/32"	0.17" ±0.08" 5/32" ± 3/32"
			SAI			
20.0 / 18.2 ±0.5	20.0 / 18.2 ±0.5	20.0 / 18.2 ±0.5	TURNING ANGLE	20.0 / 18.2 ±0.5	20.0 / 18.2 ±0.5	20.0 / 18.2 ±0.5
			ILLUSTRATION #			
			REAR			
-1.33 ±0.42 -1 5/16 ± 7/16	-1.33 ±0.42 -1 5/16 ± 7/16	-1.33 ±0.42 -1 5/16 ± 7/16	CAMBER (Degrees)	-1.33 ±0.42 -1 5/16 ± 7/16	-1.33 ±0.42 -1 5/16 ± 7/16	-1.33 ±0.42 -1 5/16 ± 7/16
0.33° ±0.17° 11/32° ± 5/32°	0.33° ±0.17° 11/32° ± 5/32°	0.33° ±0.17° 11/32° ± 5/32°	REAR TOE (Degrees)	0.33° ±0.17° 11/32° ± 5/32°	0.33° ±0.17° 11/32° ± 5/32°	0.33° ±0.17° 11/32° ± 5/32°
0.17" ±0.08" 5/32" ± 3/32"	0.17" ±0.08" 5/32" ± 3/32"	0.17" ±0.08" 5/32" ± 3/32"	REAR TOE (Inches)	0.17" ±0.08" 5/32" ± 3/32"	0.17" ±0.08" 5/32" ± 3/32"	0.17" ±0.08" 5/32" ± 3/32"
0.17 (5/32)	0.17 (5/32)	0.17 (5/32)	THRUST ANGLE	0.17 (5/32)	0.17 (5/32)	0.17 (5/32)
106 174	106 174	106 174	ILLUSTRATION #	106 174	106 174	106 174
	[1] 1BV Susp. = -1.3° (-1 5/16°)		NOTES			
04069	04070	04071		04072	04073	04074

Audi

A8/A8 Quattro	A8 Quattro	A8/A8L Quattro	MODEL	A8/A8L Quattro	A8/A8L Quattro	A8L W12 Quattro
		Sport Susp.	INFORMATION	Std. Susp.	Sport Susp.	Std. Susp.
		(2MA)		(1BK)	(2MA)	(1BK)
1997-99	2000-2003	2004-2010	YEARS	2011-2017	2011-2017	2012-2016
			INSTRUCTIONS			
89 ft.lbs.	89 ft.lbs.	89 ft.lbs.	LUGNUT TORQUE	89 ft.lbs.	89 ft.lbs.	89 ft.lbs.
			FRONT			
Caster is not specified by the manufacturer.	Caster is not specified by the manufacturer.	Caster is not specified by the manufacturer.	CASTER (Degrees)	Caster is not specified by the manufacturer.	Caster is not specified by the manufacturer.	Caster is not specified by the manufacturer.
			CROSS | TOL.			
-0.5 ±0.5 -1/2 ± 1/2	-0.83 ±0.5 -13/16 ± 1/2	-1.25 ±0.42 -1 1/4 ± 7/16	CAMBER (Degrees)	-0.93 ±0.38 -15/16 ± 3/8	-1.12 ±0.38 -1 1/8 ± 3/8	-0.93 ±0.38 -15/16 ± 3/8
0.0 (0) | 0.5 (1/2)	0.0 (0) | 0.5 (1/2)	0.0 (0) | 0.5 (1/2)	CROSS | TOL.	0.0 (0) | 0.5 (1/2)	0.0 (0) | 0.5 (1/2)	0.0 (0) | 0.5 (1/2)
0.33° ±0.07° 11/32° ± 1/16°	0.33° ±0.07° 11/32° ± 1/16°	0.27° ±0.13° 9/32° ± 1/8°	FRONT TOE (Degrees)	0.33° ±0.17° 11/32° ± 5/32°	0.33° ±0.17° 11/32° ± 5/32°	0.4° ±0.17° 13/32° ± 5/32°
0.17" ±0.03" 5/32" ± 1/32"	0.17" ±0.03" 5/32" ± 1/32"	0.13" ±0.07" 1/8" ± 1/16"	FRONT TOE (Inches)	0.17" ±0.08" 5/32" ± 3/32"	0.17" ±0.08" 5/32" ± 3/32"	0.2" ±0.08" 3/16" ± 3/32"
			SAI			
20.0 / 18.6 ±0.5	20.0 / 18.6 ±0.5	20.0 / 18.4 ±0.5	TURNING ANGLE	20.0 / 18.2 ±0.5	20.0 / 18.2 ±0.5	20.0 / 18.2 ±0.5
△	△	△	ILLUSTRATION #			
			REAR			
-0.67 ±0.5 -11/16 ± 1/2	-0.67 ±0.5 -11/16 ± 1/2	-1.25 ±0.33 -1 1/4 ± 5/16	CAMBER (Degrees)	-1.25 ±0.42 -1 1/4 ± 7/16	-1.33 ±0.42 -1 5/16 ± 7/16	-1.33 ±0.42 -1 5/16 ± 7/16
0.2° ±0.17° 3/16° ± 5/32°	0.2° ±0.17° 3/16° ± 5/32°	0.4° ±0.17° 13/32° ± 5/32°	REAR TOE (Degrees)	0.4° ±0.17° 13/32° ± 5/32°	0.4° ±0.17° 13/32° ± 5/32°	0.4° ±0.17° 13/32° ± 5/32°
0.1" ±0.08" 3/32" ± 3/32"	0.1" ±0.08" 3/32" ± 3/32"	0.2" ±0.08" 3/16" ± 3/32"	REAR TOE (Inches)	0.2" ±0.08" 3/16" ± 3/32"	0.2" ±0.08" 3/16" ± 3/32"	0.2" ±0.08" 3/16" ± 3/32"
0.17 (5/32)	0.17 (5/32)	0.17 (5/32)	THRUST ANGLE	0.17 (5/32)	0.17 (5/32)	0.17 (5/32)
106 143	106 143	142	ILLUSTRATION #	115 174	115 174	115 174
			NOTES			
04075	04076	04078		04079	04080	04081

Audi

A8L W12 Quattro	Cabriolet	Coupe Quattro	MODEL	Q3/Q3 Quattro	Q3/Q3 Quattro	Q5 Quattro
Sport Susp.			INFORMATION	Std. Susp.	Sport Susp.	All versions
(2MA, 2MB)						
2012-2016	1994-98	1990-91	YEARS	2015-2017	2015-2017	2009-2017
			INSTRUCTIONS			
89 ft.lbs.	81 ft.lbs.	81 ft.lbs.	LUGNUT TORQUE	89 ft.lbs.	89 ft.lbs.	103 ft.lbs.
			FRONT			
Caster is not specified by the manufacturer.	2.17 ±0.5 2 3/16 ± 1/2	2.33 ±0.5 2 5/16 ± 1/2	CASTER (Degrees)	7.73 ±0.5 7 3/4 ± 1/2	8.12 ±0.5 8 1/8 ± 1/2	Caster is not specified by the manufacturer.
	0.0 (0) | 0.5 (1/2)	0.0 (0) | 0.5 (1/2)	CROSS | TOL.	0.0 (0) | 0.5 (1/2)	0.0 (0) | 0.5 (1/2)	
-1.12 ±0.38 -1 1/8 ± 3/8	-0.75 ±0.5 -3/4 ± 1/2	-0.83 ±0.5 -13/16 ± 1/2	CAMBER (Degrees)	-0.6 ±0.5 -5/8 ± 1/2	-0.87 ±0.5 -7/8 ± 1/2	-0.4 ±0.33 [1] -3/8 ± 5/16 [1]
0.0 (0) | 0.5 (1/2)	0.0 (0) | 0.5 (1/2)	0.0 (0) | 0.5 (1/2)	CROSS | TOL.	0.0 (0) | 0.5 (1/2)	0.0 (0) | 0.5 (1/2)	0.0 (0) | 0.5 (1/2)
0.4° ±0.17° 13/32° ± 5/32°	0.17° ±0.17° 5/32° ± 5/32°	0.17° ±0.17° 5/32° ± 5/32°	FRONT TOE (Degrees)	0.17° ±0.17° 5/32° ± 5/32°	0.17° ±0.17° 5/32° ± 5/32°	0.33° ±0.17° 11/32° ± 5/32°
0.2" ±0.08" 3/16" ± 3/32"	0.08" ±0.08" 3/32" ± 3/32"	0.08" ±0.08" 3/32" ± 3/32"	FRONT TOE (Inches)	0.08" ±0.08" 3/32" ± 3/32"	0.08" ±0.08" 3/32" ± 3/32"	0.17" ±0.08" 5/32" ± 3/32"
			SAI			
20.0 / 18.2 ±0.5	20.0 / 19.0 ±0.5	20.0 / 18.9 ±0.5	TURNING ANGLE	20.0 / 18.4 ±0.33	20.0 / 18.5 ±0.33	20.0 / 18.5 ±0.5
	15	15	ILLUSTRATION #			
			REAR			
-1.33 ±0.42 -1 5/16 ± 7/16	-1.0 ±0.33 -1 ± 5/16	-0.75 ±0.5 -3/4 ± 1/2	CAMBER (Degrees)	-1.33 ±0.5 -1 5/16 ± 1/2	-1.33 ±0.5 -1 5/16 ± 1/2	-1.33 ±0.42 -1 5/16 ± 7/16
0.4° ±0.17° 13/32° ± 5/32°	0.33° ±0.33° 11/32° ± 11/32°	0.33° ±0.17° 11/32° ± 5/32°	REAR TOE (Degrees)	0.17° ±0.17° 5/32° ± 5/32°	0.17° ±0.17° 5/32° ± 5/32°	0.33° ±0.17° 11/32° ± 5/32°
0.2" ±0.08" 3/16" ± 3/32"	0.17" ±0.17" 5/32" ± 5/32"	0.17" ±0.08" 5/32" ± 3/32"	REAR TOE (Inches)	0.08" ±0.08" 3/32" ± 3/32"	0.08" ±0.08" 3/32" ± 3/32"	0.17" ±0.08" 5/32" ± 3/32"
0.17 (5/32)	0.17 (5/32)		THRUST ANGLE	0.17 (5/32)	0.17 (5/32)	0.17 (5/32)
115 174		15 16	ILLUSTRATION #	127	127	115 174
			NOTES			[1] 2009-12 = -0.42° (-7/16°)
04082	04083	04086		04089	04090	04092

Audi

Q7 Quattro (4L)	Q7 Quattro (4L)	Q7 Quattro (4M)	MODEL	Q7 Quattro (4M)	RS4	RS4
Std./Comfort/Sport	Air Susp.	Std. Susp. (1BA)	INFORMATION	Air Susp.	Std. Susp.	Sport Susp.
(1BA, 1BW, 1BV)	(1BK, 2MB)			(1BK, 2MA, 2MB)	(2MC)	(2ME)
2007-2016	2007-2016	2016-2017	YEARS	2016-2017	2007-2008	2007-2008
			INSTRUCTIONS			
118 ft.lbs.	118 ft.lbs.	118 ft.lbs.	LUGNUT TORQUE	118 ft.lbs.	89 ft.lbs.	89 ft.lbs.
			FRONT			
8.17 +0.17 -0.5 8 3/16 +3/16 -1/2	8.58 +0.17 -0.5 8 9/16 +3/16 -1/2	Caster is not specified by the manufacturer.	CASTER (Degrees)	Caster is not specified by the manufacturer.	Caster is not specified by the manufacturer.	Caster is not specified by the manufacturer.
0.0 (0) | 0.5 (1/2)	0.0 (0) | 0.5 (1/2)		CROSS | TOL.			
-0.17 ±0.25 -3/16 ± 1/4	-0.17 ±0.25 -3/16 ± 1/4	-0.42 ±0.38 -7/16 ± 3/8	CAMBER (Degrees)	-0.72 ±0.38 [1] -11/16 ± 3/8 [1]	-0.83 ±0.42 -13/16 ± 7/16	-1.27 ±0.42 -1 1/4 ± 7/16
0.0 (0) | 0.17 (3/16)	0.0 (0) | 0.17 (3/16)	0.0 (0) | 0.5 (1/2)	CROSS | TOL.	0.0 (0) | 0.5 (1/2)	0.0 (0) | 0.5 (1/2)	0.0 (0) | 0.5 (1/2)
0.23° ±0.17° 7/32° ± 5/32°	0.23° ±0.17° 7/32° ± 5/32°	0.33° ±0.17° 11/32° ± 5/32°	FRONT TOE (Degrees)	0.33° ±0.17° 11/32° ± 5/32°	0.33° ±0.07° 11/32° ± 1/16°	0.33° ±0.07° 11/32° ± 1/16°
0.12" ±0.08" 1/8" ± 3/32"	0.12" ±0.08" 1/8" ± 3/32"	0.17" ±0.08" 5/32" ± 3/32"	FRONT TOE (Inches)	0.17" ±0.08" 5/32" ± 3/32"	0.17" ±0.03" 5/32" ± 1/32"	0.17" ±0.03" 5/32" ± 1/32"
			SAI			
20.0 / 18.7 ±0.5	20.0 / 18.5 ±0.5	20.0 / 18.5 ±0.5	TURNING ANGLE	20.0 / 18.5 ±0.5	20.0 / 18.9 ±0.5	20.0 / 18.9 ±0.5
52	52		ILLUSTRATION #			
			REAR			
-1.33 ±0.33 -1 5/16 ± 5/16	-1.33 ±0.33 -1 5/16 ± 5/16	-0.83 ±0.42 -13/16 ± 7/16	CAMBER (Degrees)	-0.83 ±0.42 -13/16 ± 7/16	-1.33 ±0.5 -1 5/16 ± 1/2	-1.33 ±0.5 -1 5/16 ± 1/2
0.4° ±0.17° 13/32° ± 5/32°	0.4° ±0.17° 13/32° ± 5/32°	0.33° ±0.17° 11/32° ± 5/32°	REAR TOE (Degrees)	0.33° ±0.17° 11/32° ± 5/32°	0.33° +0.17° -0.08° 11/32° +5/32° -3/32°	0.33° +0.17° -0.08° 11/32° +5/32° -3/32°
0.2" ±0.08" 3/16" ± 3/32"	0.2" ±0.08" 3/16" ± 3/32"	0.17" ±0.08" 5/32" ± 3/32"	REAR TOE (Inches)	0.17" ±0.08" 5/32" ± 3/32"	0.17" +0.08" -0.04" 5/32" +3/32" -1/32"	0.17" +0.08" -0.04" 5/32" +3/32" -1/32"
0.07 (1/16)	0.07 (1/16)	0.17 (5/32)	THRUST ANGLE	0.17 (5/32)	0.12 (1/8)	0.12 (1/8)
162	162	162	ILLUSTRATION #	162	142	142
			NOTES	2MA, 2MB Susp. = -0.92° (-15/16°)		
04093	04094	04095		04096	04097	04098

Audi

Model	RS5	RS6	RS6	RS7	R8 Quattro	S3 Quattro
Information	Sport Susp. (1BU)		Sport Susp. (2MS)	Air Susp. (2MC)	All versions	
Years	2013-2015	2003	2009	2014-2017	2008-2015	2015-2017
Instructions						
LugNut Torque	89 ft.lbs.	89 ft.lbs.	81 ft.lbs.	81 ft.lbs.	89 ft.lbs.	89 ft.lbs.
FRONT						
Caster (Degrees)	Caster is not specified by the manufacturer.	Caster is not specified by the manufacturer.	Caster is not specified by the manufacturer.	Caster is not specified by the manufacturer.	Caster is not specified by the manufacturer.	7.85 ± 0.5 7 7/8 ± 1/2
Cross | Tol.						0.0 (0) | 0.5 (1/2)
Camber (Degrees)	-1.52 ± 0.38 -1 1/2 ± 3/8	-1.33 ± 0.42 -1 5/16 ± 7/16	-0.63 ± 0.42 -5/8 ± 7/16	-1.03 ± 0.38 -1 ± 3/8	-1.0 ± 0.17 -1 ± 3/16	-0.87 ± 0.5 -7/8 ± 1/2
Cross | Tol.	0.0 (0) | 0.5 (1/2)	0.0 (0) | 0.5 (1/2)	0.0 (0) | 0.5 (1/2)	0.0 (0) | 0.5 (1/2)	0.0 (0) | 0.33 (5/16)	0.0 (0) | 0.5 (1/2)
Front Toe (Degrees)	0.33° ± 0.17° 11/32° ± 5/32°	0.33° ± 0.07° 11/32° ± 1/16°	0.33° ± 0.13° 11/32° ± 1/8°	0.33° ± 0.17° 11/32° ± 5/32°	0.33° ± 0.07° 11/32° ± 1/16°	0.17° ± 0.17° 5/32° ± 5/32°
Front Toe (Inches)	0.17" ± 0.08" 5/32" ± 3/32"	0.17" ± 0.03" 5/32" ± 1/32"	0.17" ± 0.07" 5/32" ± 1/16"	0.17" ± 0.08" 5/32" ± 3/32"	0.17" ± 0.03" 5/32" ± 1/32"	0.08" ± 0.08" 3/32" ± 3/32"
SAI						
Turning Angle	20.0 / 18.2 ± 0.5	20.0 / 18.5 ± 0.5	20.0 / 18.5 ± 0.5	20.0 / 18.2 ± 0.5	20.0 / 19.1 ± 0.17	20.0 / 18.3 ± 0.33
Illustration #		▲			186	
REAR						
Camber (Degrees)	-1.33 ± 0.42 -1 5/16 ± 7/16	-1.5 ± 0.33 -1 1/2 ± 5/16	-0.83 ± 0.33 -13/16 ± 5/16	-1.33 ± 0.42 -1 5/16 ± 7/16	-1.0 ± 0.25 -1 ± 1/4	-1.33 ± 0.5 -1 5/16 ± 1/2
Rear Toe (Degrees)	0.33° ± 0.17° 11/32° ± 5/32°	0.27° +0.25° -0.17° 9/32° +1/4° -5/32°	0.3° ± 0.17° 5/16° ± 5/32°	0.33° ± 0.17° 11/32° ± 5/32°	0.33° ± 0.1° 11/32° ± 3/32°	0.42° ± 0.17° 13/32° ± 5/32°
Rear Toe (Inches)	0.17" ± 0.08" 5/32" ± 3/32"	0.13" +0.13" -0.08" 1/8" +1/8" -3/32"	0.15" ± 0.08" 5/32" ± 3/32"	0.17" ± 0.08" 5/32" ± 3/32"	0.17" ± 0.05" 5/32" ± 1/16"	0.21" ± 0.08" 7/32" ± 3/32"
Thrust Angle	0.17 (5/32)	0.25 (1/4)	0.17 (5/32)	0.17 (5/32)	0.17 (5/32)	0.2 (3/16)
Illustration #	115 174	83 06	106 170	106 174	185	127
Notes						
	04099	04100	04101	04102	04103	04105

Audi

Model	S4 Quattro	S4 Quattro (B6, B7)	S4 Quattro (B8)	S5 Quattro	S6 Quattro	S6 Quattro
Information			1BD Susp.	1BD Susp.		1BD Susp.
Years	2000-2002	2004-2009	2010-2016	2008-2017	1995	2007-2011
Instructions						
LugNut Torque	89 ft.lbs.	89 ft.lbs.	89 ft.lbs.	89 ft.lbs.	81 ft.lbs.	89 ft.lbs.
FRONT						
Caster (Degrees)	Caster is not specified by the manufacturer.	Caster is not specified by the manufacturer.	Caster is not specified by the manufacturer.	Caster is not specified by the manufacturer.	0.92 ± 0.67 15/16 ± 11/16	Caster is not specified by the manufacturer.
Cross | Tol.					0.0 (0) | 1.0 (1)	
Camber (Degrees)	-0.83 ± 0.42 -13/16 ± 7/16	-0.83 ± 0.42 -13/16 ± 7/16	-1.08 ± 0.38 -1 1/16 ± 3/8	-1.08 ± 0.38 -1 1/16 ± 3/8	-0.83 +0.58 -0.42 -13/16 +9/16 -7/16	-1.12 ± 0.42 -1 1/8 ± 7/16
Cross | Tol.	0.0 (0) | 0.5 (1/2)	0.0 (0) | 0.5 (1/2)	0.0 (0) | 0.5 (1/2)	0.0 (0) | 0.5 (1/2)	0.0 (0) | 0.5 (1/2)	0.0 (0) | 0.5 (1/2)
Front Toe (Degrees)	0.33° ± 0.07° 11/32° ± 1/16°	0.33° ± 0.07° 11/32° ± 1/16°	0.33° ± 0.17° 11/32° ± 5/32°	0.33° ± 0.17° 11/32° ± 5/32°	0.25° +0.08° -0.17° 1/4° +3/32° -5/32°	0.33° ± 0.13° 11/32° ± 1/8°
Front Toe (Inches)	0.17" ± 0.03" 5/32" ± 1/32"	0.17" ± 0.03" 5/32" ± 1/32"	0.17" ± 0.08" 5/32" ± 3/32"	0.17" ± 0.08" 5/32" ± 3/32"	0.13" +0.04" -0.08" 1/8" +1/32" -3/32"	0.17" ± 0.07" 5/32" ± 1/16"
SAI						
Turning Angle	20.0 / 18.5 ± 0.5	20.0 / 18.9 ± 0.5	20.0 / 18.2 ± 0.5	20.0 / 18.2 ± 0.5	20.0 / 18.8 ± 0.5	20.0 / 18.5 ± 0.5
Illustration #	▲	▲			17	▲
REAR						
Camber (Degrees)	-0.92 ± 0.25 -15/16 ± 1/4	-1.33 ± 0.5 -1 5/16 ± 1/2	-1.33 ± 0.42 -1 5/16 ± 7/16	-1.33 ± 0.42 -1 5/16 ± 7/16	-0.25 ± 0.25 -1/4 ± 1/4	-0.83 ± 0.33 -13/16 ± 5/16
Rear Toe (Degrees)	0.27° ± 0.17° 9/32° ± 5/32°	0.33° +0.17° -0.08° 11/32° +5/32° -3/32°	0.33° ± 0.17° 11/32° ± 5/32°	0.33° ± 0.17° 11/32° ± 5/32°	0.17° ± 0.17° 5/32° ± 5/32°	0.3° ± 0.17° 5/16° ± 5/32°
Rear Toe (Inches)	0.13" ± 0.08" 1/8" ± 3/32"	0.17" +0.08" -0.04" 5/32" +3/32" -1/32"	0.17" ± 0.08" 5/32" ± 3/32"	0.17" ± 0.08" 5/32" ± 3/32"	0.08" ± 0.08" 3/32" ± 3/32"	0.15" ± 0.08" 5/32" ± 3/32"
Thrust Angle	0.17 (5/32)	0.17 (5/32)	0.17 (5/32)	0.17 (5/32)		0.08 (3/32)
Illustration #	83 06	142	115 174	115 174	18 19	106 170
Notes						
	04106	04107	04108	04109	04110	04111

Audi

	S6 Quattro	S7 Quattro	S8 Quattro	S8 Quattro	S8 Quattro	SQ5 Quattro
MODEL INFORMATION	Air Susp. (2MB)	Air Susp. (2MB)			Sport Susp. (2MB)	Sport Susp. (1BD)
YEARS	2013-2017	2013-2017	2001-2003	2007-2010	2013-2017	2014-2017
INSTRUCTIONS						
LUGNUT TORQUE	81 ft.lbs.	81 ft.lbs.	89 ft.lbs.	89 ft.lbs.	89 ft.lbs.	103 ft.lbs.
FRONT						
CASTER (Degrees)	Caster is not specified by the manufacturer.	Caster is not specified by the manufacturer.	Caster is not specified by the manufacturer.	Caster is not specified by the manufacturer.	Caster is not specified by the manufacturer.	Caster is not specified by the manufacturer.
CROSS | TOL.						
CAMBER (Degrees)	-1.08 ± 0.38 -1 1/16 ± 3/8	-1.08 ± 0.38 -1 1/16 ± 3/8	-1.08 ± 0.5 -1 1/16 ± 1/2	-1.25 ± 0.42 -1 1/4 ± 7/16	-1.12 ± 0.38 -1 1/8 ± 3/8	-0.92 ± 0.38 -15/16 ± 3/8
CROSS | TOL.	0.0 (0) | 0.5 (1/2)	0.0 (0) | 0.5 (1/2)	0.0 (0) | 0.5 (1/2)	0.0 (0) | 0.5 (1/2)	0.0 (0) | 0.5 (1/2)	0.0 (0) | 0.5 (1/2)
FRONT TOE (Degrees)	0.33° ± 0.17° 11/32° ± 5/32°	0.33° ± 0.17° 11/32° ± 5/32°	0.33° ± 0.07° 11/32° ± 1/16°	0.27° ± 0.13° 9/32° ± 1/8°	0.33° ± 0.17° 11/32° ± 5/32°	0.33° ± 0.17° 11/32° ± 5/32°
FRONT TOE (Inches)	0.17" ± 0.08" 5/32" ± 3/32"	0.17" ± 0.08" 5/32" ± 3/32"	0.17" ± 0.03" 5/32" ± 1/32"	0.13" ± 0.07" 1/8" ± 1/16"	0.17" ± 0.08" 5/32" ± 3/32"	0.17" ± 0.08" 5/32" ± 3/32"
SAI						
TURNING ANGLE	20.0 / 18.2 ±0.5	20.0 / 18.2 ±0.5	20.0 / 18.6 ±0.5	20.0 / 18.4 ±0.5	20.0 / 18.2 ±0.5	20.0 / 18.2 ±0.5
ILLUSTRATION #			△	△		
REAR						
CAMBER (Degrees)	-1.33 ± 0.42 -1 5/16 ± 7/16	-1.33 ± 0.42 -1 5/16 ± 7/16	-0.67 ± 0.5 -11/16 ± 1/2	-1.25 ± 0.33 -1 1/4 ± 5/16	-1.33 ± 0.42 -1 5/16 ± 7/16	-1.33 ± 0.42 -1 5/16 ± 7/16
REAR TOE (Degrees)	0.33° ± 0.17° 11/32° ± 5/32°	0.33° ± 0.17° 11/32° ± 5/32°	0.2° ± 0.17° 3/16° ± 5/32°	0.4° ± 0.17° 13/32° ± 5/32°	0.4° ± 0.17° 13/32° ± 5/32°	0.33° ± 0.17° 11/32° ± 5/32°
REAR TOE (Inches)	0.17" ± 0.08" 5/32" ± 3/32"	0.17" ± 0.08" 5/32" ± 3/32"	0.1" ± 0.08" 3/32" ± 3/32"	0.2" ± 0.08" 3/16" ± 3/32"	0.2" ± 0.08" 3/16" ± 3/32"	0.17" ± 0.08" 5/32" ± 3/32"
THRUST ANGLE	0.17 (5/32)	0.17 (5/32)	0.17 (5/32)	0.17 (5/32)	0.17 (5/32)	0.17 (5/32)
ILLUSTRATION #	106 174	106 174	106 143	142	195	115 174
NOTES						
	04112	04113	04114	04115	04116	04117

Audi

	SQ5 Quattro	TT	TT Quattro	TT Quattro	TT	V8 Quattro
MODEL INFORMATION	Sport Susp. (2MT)			All Versions		
YEARS	2014-2017	2000-2006	2000-2006	2008-2015	2016-2017	1990-94
INSTRUCTIONS						
LUGNUT TORQUE	103 ft.lbs.	89 ft.lbs.	89 ft.lbs.	89 ft.lbs.	89 ft.lbs.	81 ft.lbs.
FRONT						
CASTER (Degrees)	Caster is not specified by the manufacturer.	7.97 ± 0.5 7 15/16 ± 1/2	7.97 ± 0.5 7 15/16 ± 1/2	Caster is not specified by the manufacturer.	7.38 ± 0.5 7 3/8 ± 1/2	1.25 ± 0.67 1 1/4 ± 11/16
CROSS | TOL.		0.0 (0) | 0.5 (1/2)	0.0 (0) | 0.5 (1/2)		0.0 (0) | 0.5 (1/2)	0.0 (0) | 1.0 (1)
CAMBER (Degrees)	-0.4 ± 0.38 -3/8 ± 3/8	-0.75 ± 0.5 -3/4 ± 1/2	-0.75 ± 0.5 -3/4 ± 1/2	-0.68 ± 0.5 -11/16 ± 1/2	-0.73 ± 0.5 [1] -3/4 ± 1/2 [1]	-0.5 ± 0.5 -1/2 ± 1/2
CROSS | TOL.	0.0 (0) | 0.5 (1/2)	0.0 (0) | 0.5 (1/2)	0.0 (0) | 0.5 (1/2)	0.0 (0) | 0.5 (1/2)	0.0 (0) | 0.5 (1/2)	0.0 (0) | 0.5 (1/2)
FRONT TOE (Degrees)	0.33° ± 0.17° 11/32° ± 5/32°	0.13° ± 0.12° 1/8° ± 1/8°	0.13° ± 0.12° 1/8° ± 1/8°	0.17° ± 0.17° 5/32° ± 5/32°	0.17° ± 0.17° 5/32° ± 5/32°	0.17° +0.17° -0.08° 5/32° +5/32° -3/32°
FRONT TOE (Inches)	0.17" ± 0.08" 5/32" ± 3/32"	0.07" ± 0.06" 1/16" ± 1/16"	0.07" ± 0.06" 1/16" ± 1/16"	0.08" ± 0.08" 3/32" ± 3/32"	0.08" ± 0.08" 3/32" ± 3/32"	0.08" +0.08" -0.04" 3/32" +3/32" -1/32"
SAI						
TURNING ANGLE	20.0 / 18.2 ±0.5	20.0 / 18.5 ±0.33	20.0 / 18.5 ±0.33	20.0 / 18.7 ±0.33	20.0 / 18.7 ±0.33	20.0 / 18.3 ±0.5
ILLUSTRATION #		△	△	15	15	17
REAR						
CAMBER (Degrees)	-1.33 ± 0.42 -1 5/16 ± 7/16	-2.0 ± 0.33 -2 ± 5/16	-1.5 ± 0.33 -1 1/2 ± 5/16	-1.33 ± 0.5 -1 5/16 ± 1/2	-1.33 ± 0.5 -1 5/16 ± 1/2	-0.25 ± 0.25 -1/4 ± 1/4
REAR TOE (Degrees)	0.33° ± 0.17° 11/32° ± 5/32°	0.47° ± 0.17° 15/32° ± 5/32°	0.25° ± 0.17° 1/4° ± 5/32°	0.42° ± 0.17° 13/32° ± 5/32°	0.43° ± 0.17° 7/16° ± 5/32°	0.17° ± 0.17° 5/32° ± 5/32°
REAR TOE (Inches)	0.17" ± 0.08" 5/32" ± 3/32"	0.23" ± 0.08" 7/32" ± 3/32"	0.13" ± 0.08" 1/8" ± 3/32"	0.21" ± 0.08" 7/32" ± 3/32"	0.22" ± 0.08" 7/32" ± 3/32"	0.08" ± 0.08" 3/32" ± 3/32"
THRUST ANGLE	0.17 (5/32)	0.25 (1/4)	0.33 (11/32)	0.17 (5/32)	0.17 (5/32)	
ILLUSTRATION #	115 174	114		127	127	18 19
NOTES					[1] Sport Susp. = -0.88° (-7/8°)	
	04118	04120	04121	04122	04123	04124

BMW

1 Series (E82, E88)	1 Series (E82, E88)	1 Series (E82, E88)	MODEL	1 Series (E82, E88)	1 Series (E82, E88)	1 Series (E82)
128i Cpe./Conv.	128i Cpe./Conv.	135i Cpe./Conv.	INFORMATION	135i Cpe./Conv.	135iS Cpe./Conv.	M Coupe
Reg. Suspension	Sport Suspension	Reg. Suspension		Sport Suspension	All Models	
2008-2014	2008-2014	2008-2014	YEARS	2008-2014	2013-2014	2011-2012
p. 304	p. 304	p. 304	INSTRUCTIONS	p. 304	p. 304	p. 304
89 ft.lbs.	89 ft.lbs.	89 ft.lbs.	LUGNUT TORQUE	89 ft.lbs.	89 ft.lbs.	89 ft.lbs.
			FRONT			
Caster is not specified by the manufacturer.	Caster is not specified by the manufacturer.	Caster is not specified by the manufacturer.	CASTER (Degrees)	Caster is not specified by the manufacturer.	Caster is not specified by the manufacturer.	Caster is not specified by the manufacturer.
0.0 (0) | 0.5 (1/2)	0.0 (0) | 0.5 (1/2)	0.0 (0) | 0.5 (1/2)	CROSS | TOL.	0.0 (0) | 0.5 (1/2)	0.0 (0) | 0.5 (1/2)	0.0 (0) | 0.5 (1/2)
-0.2 ±0.42 -3/16 ± 7/16	-0.47 ±0.42 -7/16 ± 7/16	-0.2 ±0.42 -3/16 ± 7/16	CAMBER (Degrees)	-0.47 ±0.42 -7/16 ± 7/16	-0.47 ±0.42 -7/16 ± 7/16	-1.33 ±0.33 -1 5/16 ± 5/16
0.0 (0) | 0.5 (1/2)	0.0 (0) | 0.5 (1/2)	0.0 (0) | 0.5 (1/2)	CROSS | TOL.	0.0 (0) | 0.5 (1/2)	0.0 (0) | 0.5 (1/2)	0.0 (0) | 0.5 (1/2)
0.27° ±0.07° 9/32° ± 1/16°	0.27° ±0.07° 9/32° ± 1/16°	0.27° ±0.07° 9/32° ± 1/16°	FRONT TOE (Degrees)	0.27° ±0.07° 9/32° ± 1/16°	0.27° ±0.07° 9/32° ± 1/16°	0.17° ±0.07° 5/32° ± 1/16°
0.13" ±0.03" 1/8" ± 1/32"	0.13" ±0.03" 1/8" ± 1/32"	0.13" ±0.03" 1/8" ± 1/32"	FRONT TOE (Inches)	0.13" ±0.03" 1/8" ± 1/32"	0.13" ±0.03" 1/8" ± 1/32"	0.08" ±0.03" 3/32" ± 1/32"
			SAI			
20.0 / 18.4 ±0.5	20.0 / 18.4 ±0.5	20.0 / 18.4 ±0.5	TURNING ANGLE	20.0 / 18.4 ±0.5	20.0 / 18.4 ±0.5	20.0 / 17.8 ±0.5
119	119	119	ILLUSTRATION #	119	119	119
			REAR			
-1.5 ±0.08 -1 1/2 ± 1/16	-1.83 ±0.08 -1 13/16 ± 1/16	-1.5 ±0.08 -1 1/2 ± 1/16	CAMBER (Degrees)	-1.83 ±0.08 -1 13/16 ± 1/16	-1.83 ±0.08 -1 13/16 ± 1/16	-1.75 ±0.25 -1 3/4 ± 1/4
0.3° ±0.07° 5/16° ± 1/16°	0.3° ±0.07° 5/16° ± 1/16°	0.3° ±0.07° 5/16° ± 1/16°	REAR TOE (Degrees)	0.3° ±0.07° 5/16° ± 1/16°	0.3° ±0.07° 5/16° ± 1/16°	0.17° ±0.07° 5/32° ± 1/16°
0.15" ±0.03" 5/32" ± 1/32"	0.15" ±0.03" 5/32" ± 1/32"	0.15" ±0.03" 5/32" ± 1/32"	REAR TOE (Inches)	0.15" ±0.03" 5/32" ± 1/32"	0.15" ±0.03" 5/32" ± 1/32"	0.08" ±0.03" 3/32" ± 1/32"
0.2 (3/16)	0.2 (3/16)	0.2 (3/16)	THRUST ANGLE	0.2 (3/16)	0.2 (3/16)	0.07 (1/16)
179 178	179 178	179 178	ILLUSTRATION #	179 178	179 178	179 178
			NOTES			
05005	05006	05007		05008	05009	05010

BMW

2 Series (F22, F23)	2 Series (F22, F23)	2 Series (F22, F23)	MODEL	2 Series (F22, F23)	2 Series (F22, F23)	2 Series (F22, F23)
228i, 230i Cpe./Conv.	228i, 230i Cpe./Conv.	228i, 230i xDrive Cpe./Conv.	INFORMATION	228i, 230i xDrive Cpe./Conv.	M235i, M240i.	M235i, M240i xDrive
Reg. Suspension	Sport Suspension	Reg. Suspension		Sport Suspension	Cpe./Conv.	Cpe./Conv.
2015-2017	2015-2017	2015-2017	YEARS	2015-2017	2015-2017	2015-2017
p. 304	p. 304	p. 304	INSTRUCTIONS	p. 304	p. 304	p. 304
103 ft.lbs.	103 ft.lbs.	103 ft.lbs.	LUGNUT TORQUE	103 ft.lbs.	103 ft.lbs.	103 ft.lbs.
			FRONT			
Caster is not specified by the manufacturer.	Caster is not specified by the manufacturer.	Caster is not specified by the manufacturer.	CASTER (Degrees)	Caster is not specified by the manufacturer.	Caster is not specified by the manufacturer.	Caster is not specified by the manufacturer.
0.0 (0) | 0.5 (1/2)	0.0 (0) | 0.5 (1/2)	0.0 (0) | 0.5 (1/2)	CROSS | TOL.	0.0 (0) | 0.5 (1/2)	0.0 (0) | 0.5 (1/2)	0.0 (0) | 0.5 (1/2)
-0.33 ±0.42 -5/16 ± 7/16	-0.48 ±0.42 -1/2 ± 7/16	-0.23 ±0.42 -1/4 ± 7/16	CAMBER (Degrees)	-0.33 ±0.42 -5/16 ± 7/16	-0.88 ±0.42 -7/8 ± 7/16	-0.83 ±0.42 -13/16 ± 7/16
0.0 (0) | 0.5 (1/2)	0.0 (0) | 0.5 (1/2)	0.0 (0) | 0.5 (1/2)	CROSS | TOL.	0.0 (0) | 0.5 (1/2)	0.0 (0) | 0.5 (1/2)	0.0 (0) | 0.5 (1/2)
0.23° ±0.07° 7/32° ± 1/16°	0.23° ±0.07° 7/32° ± 1/16°	0.23° ±0.07° 7/32° ± 1/16°	FRONT TOE (Degrees)	0.23° ±0.07° 7/32° ± 1/16°	0.23° ±0.07° 7/32° ± 1/16°	0.23° ±0.07° 7/32° ± 1/16°
0.12" ±0.03" 1/8" ± 1/32"	0.12" ±0.03" 1/8" ± 1/32"	0.12" ±0.03" 1/8" ± 1/32"	FRONT TOE (Inches)	0.12" ±0.03" 1/8" ± 1/32"	0.12" ±0.03" 1/8" ± 1/32"	0.12" ±0.03" 1/8" ± 1/32"
			SAI			
20.0 / 18.5 ±0.5	20.0 / 18.5 ±0.5	20.0 / 18.3 ±0.5	TURNING ANGLE	20.0 / 18.3 ±0.5	20.0 / 18.5 ±0.5	20.0 / 18.3 ±0.5
			ILLUSTRATION #			
			REAR			
-1.5 ±0.08 -1 1/2 ± 1/16	-1.5 ±0.08 -1 1/2 ± 1/16	-1.5 ±0.08 -1 1/2 ± 1/16	CAMBER (Degrees)	-1.75 ±0.08 -1 3/4 ± 1/16	-1.75 ±0.08 -1 3/4 ± 1/16	-1.75 ±0.08 -1 3/4 ± 1/16
0.3° ±0.07° 5/16° ± 1/16°	0.3° ±0.07° 5/16° ± 1/16°	0.3° ±0.07° 5/16° ± 1/16°	REAR TOE (Degrees)	0.3° ±0.07° 5/16° ± 1/16°	0.3° ±0.07° 5/16° ± 1/16°	0.3° ±0.07° 5/16° ± 1/16°
0.15" ±0.03" 5/32" ± 1/32"	0.15" ±0.03" 5/32" ± 1/32"	0.15" ±0.03" 5/32" ± 1/32"	REAR TOE (Inches)	0.15" ±0.03" 5/32" ± 1/32"	0.15" ±0.03" 5/32" ± 1/32"	0.15" ±0.03" 5/32" ± 1/32"
0.2 (3/16)	0.2 (3/16)	0.2 (3/16)	THRUST ANGLE	0.2 (3/16)	0.2 (3/16)	0.2 (3/16)
179 178	179 178	179 178	ILLUSTRATION #	179 178	179 178	179 178
			NOTES			
05011	05012	05013		05014	05015	05016

BMW

2 Series (F87)	3 Series (E30)	3 Series (E30)	MODEL	3 Series (E30)	3 Series (E30)	3 Series (E30)
M2	325, 325e/es 325i/is, 318i/is	318i/325i Conv. exc/M-Technic	INFORMATION	325i Convertible M-Technic	325iX	M3
2016-2017	1990-91	1990-93	YEARS	1992	1990-91	1990-91
p. 304	p. 304	p. 304	INSTRUCTIONS	p. 304	p. 304	p. 304
103 ft.lbs.	89 ft.lbs.	89 ft.lbs.	LUGNUT TORQUE	89 ft.lbs.	89 ft.lbs.	89 ft.lbs.
			FRONT			
Caster is not specified by the manufacturer.	8.5 ±0.5 8½ ± ½	8.5 ±0.5 8½ ± ½	CASTER (Degrees)	8.77 ±0.5 8¾ ± ½	1.55 ±0.5 1 9/16 ± ½	9.13 ±0.5 9⅛ ± ½
0.0 (0) | 0.5 (½)	0.0 (0) | 0.5 (½)	0.0 (0) | 0.5 (½)	CROSS | TOL.	0.0 (0) | 0.5 (½)	0.0 (0) | 0.5 (½)	0.0 (0) | 0.5 (½)
-1.48 ±0.42 -1½ ± 7/16	-0.67 ±0.5 -11/16 ± ½	-0.67 ±0.5 -11/16 ± ½	CAMBER (Degrees)	-1.17 ±0.5 -1 3/16 ± ½	-1.0 ±0.5 -1 ± ½	-0.7 ±0.5 -11/16 ± ½
0.0 (0) | 0.5 (½)	0.0 (0) | 0.5 (½)	0.0 (0) | 0.5 (½)	CROSS | TOL.	0.0 (0) | 0.5 (½)	0.0 (0) | 0.5 (½)	0.0 (0) | 0.5 (½)
0.17° ±0.07° 5/32° ± 1/16°	0.3° ±0.08° 5/16° ± 3/32°	0.3° ±0.08° 5/16° ± 3/32°	FRONT TOE (Degrees)	0.3° ±0.08° 5/16° ± 3/32°	0.0° ±0.08° 0° ± 3/32°	0.28° ±0.08° 9/32° ± 3/32°
0.08" ±0.03" 3/32" ± 1/32"	0.15" ±0.04" 5/32" ± 1/32"	0.15" ±0.04" 5/32" ± 1/32"	FRONT TOE (Inches)	0.15" ±0.04" 5/32" ± 1/32"	0.0" ±0.04" 0" ± 1/32"	0.14" ±0.04" 5/32" ± 1/32"
			SAI			
40.0 / 32.0 ±0.5	20.0 / 18.3 ±0.5	20.0 / 18.3 ±0.5	TURNING ANGLE	20.0 / 18.2 ±0.5	20.0 / 18.8 ±0.5	20.0 / 18.2 ±0.5
			ILLUSTRATION #			
			REAR			
-1.67 ±0.08 -1 11/16 ± 1/16	-2.0 ±0.5 -2 ± ½	-1.83 ±0.5 -1 13/16 ± ½	CAMBER (Degrees)	-2.5 ±0.5 -2½ ± ½	-2.0 ±0.5 -2 ± ½	-2.33 ±0.5 -2 5/16 ± ½
0.27° ±0.07° 9/32° ± 1/16°	0.37° ±0.12° 3/8° ± 1/8°	0.38° ±0.12° 3/8° ± 1/8°	REAR TOE (Degrees)	0.52° ±0.12° 17/32° ± 1/8°	0.42° ±0.12° 13/32° ± 1/8°	0.47° ±0.12° 15/32° ± 1/8°
0.13" ±0.03" 1/8" ± 1/32"	0.18" ±0.06" 3/16" ± 1/16"	0.19" ±0.06" 3/16" ± 1/16"	REAR TOE (Inches)	0.26" ±0.06" 1/4" ± 1/16"	0.21" ±0.06" 7/32" ± 1/16"	0.23" ±0.06" 7/32" ± 1/16"
0.2 (3/16)	0.25 (¼)	0.25 (¼)	THRUST ANGLE	0.25 (¼)	0.25 (¼)	0.25 (¼)
			ILLUSTRATION #			
			NOTES			
05017	05020	05021		05022	05023	05024

BMW

3 Series (E36)	3 Series (E36)	3 Series (E36)	MODEL	3 Series (E36)	3 Series (E36)	3 Series (E36)
318i/is, 325i/is (Conv. 1994 only)	318i/is, 323is, 325i/is 318i/323i/325i Conv.	328i, 328is 328i Convertible	INFORMATION	Sport Suspension (exc/Z3, Ti)	M3	M3
1992-94	1995-99	1996-99	YEARS	1992-99	1995	1996-99
p. 304	p. 304	p. 304	INSTRUCTIONS	p. 304	p. 304	p. 304
89 ft.lbs.	89 ft.lbs.	89 ft.lbs.	LUGNUT TORQUE	89 ft.lbs.	89 ft.lbs.	89 ft.lbs.
			FRONT			
3.73 ±0.5 3¾ ± ½	3.73 ±0.5 3¾ ± ½	3.73 ±0.5 3¾ ± ½	CASTER (Degrees)	3.83 ±0.5 3 13/16 ± ½	6.63 ±0.5 6⅝ ± ½	7.58 ±0.5 7 9/16 ± ½
0.0 (0) | 0.5 (½)	0.0 (0) | 0.5 (½)	0.0 (0) | 0.5 (½)	CROSS | TOL.	0.0 (0) | 0.5 (½)	0.0 (0) | 0.5 (½)	0.0 (0) | 0.5 (½)
-0.67 ±0.5 -11/16 ± ½	-0.5 ±0.5 -½ ± ½	-0.5 ±0.5 -½ ± ½	CAMBER (Degrees)	-0.85 ±0.5 [1] -⅞ ± ½ [1]	-0.92 ±0.5 -15/16 ± ½	-0.77 ±0.5 -¾ ± ½
0.0 (0) | 0.5 (½)	0.0 (0) | 0.5 (½)	0.0 (0) | 0.5 (½)	CROSS | TOL.	0.0 (0) | 0.5 (½)	0.0 (0) | 0.5 (½)	0.0 (0) | 0.5 (½)
0.3° ±0.13° 5/16° ± 1/8°	0.3° ±0.13° 5/16° ± 1/8°	0.3° ±0.13° 5/16° ± 1/8°	FRONT TOE (Degrees)	0.3° ±0.13° 5/16° ± 1/8°	0.17° ±0.08° 5/32° ± 3/32°	0.17° ±0.08° 5/32° ± 3/32°
0.15" ±0.07" 5/32" ± 1/16"	0.15" ±0.07" 5/32" ± 1/16"	0.15" ±0.07" 5/32" ± 1/16"	FRONT TOE (Inches)	0.15" ±0.07" 5/32" ± 1/16"	0.08" ±0.04" 3/32" ± 1/32"	0.08" ±0.04" 3/32" ± 1/32"
			SAI			
20.0 / 18.5 ±0.5	20.0 / 18.5 ±0.5	20.0 / 18.5 ±0.5	TURNING ANGLE	20.0 / 18.5 ±0.5	20.0 / 18.4 ±0.5	20.0 / 18.4 ±0.5
			ILLUSTRATION #			
			REAR			
-1.67 ±0.25 -1 11/16 ± ¼	-1.67 ±0.25 -1 11/16 ± ¼	-1.67 ±0.25 -1 11/16 ± ¼	CAMBER (Degrees)	-2.0 ±0.25 -2 ± ¼	-1.75 ±0.17 -1¾ ± 3/16	-1.75 ±0.17 -1¾ ± 3/16
0.4° ±0.1° 13/32° ± 3/32°	0.4° ±0.1° 13/32° ± 3/32°	0.4° ±0.1° 13/32° ± 3/32°	REAR TOE (Degrees)	0.4° ±0.1° 13/32° ± 3/32°	0.5° ±0.08° ½° ± 3/32°	0.33° ±0.08° 11/32° ± 3/32°
0.2" ±0.05" 3/16" ± 1/16"	0.2" ±0.05" 3/16" ± 1/16"	0.2" ±0.05" 3/16" ± 1/16"	REAR TOE (Inches)	0.2" ±0.05" 3/16" ± 1/16"	0.25" ±0.04" ¼" ± 1/32"	0.17" ±0.04" 5/32" ± 1/32"
0.05 (1/16)	0.05 (1/16)	0.05 (1/16)	THRUST ANGLE	0.05 (1/16)	0.05 (1/16)	0.05 (1/16)
29 28	29 28	29 28	ILLUSTRATION #	29 28	29 28	29 28
			NOTES	[1] 1992-94 = -0.97°(-15/16°)		
05025	05026	05027		05028	05029	05030

BMW

MODEL INFORMATION	3 Series (E36)	3 Series (E36)	3 Series (E36)	3 Series (E36)	3 Series (E46)	3 Series (E46)
	M Coupe	M Roadster	318Ti	318Ti	323i, 323Ci, 328i, 328Ci	323i, 323Ci, 328i, 328Ci
			Reg. Suspension	Sport Suspension	Reg. Suspension	Sport Suspension
YEARS	1998-2002	1999-2002	1995-99	1995-99	1999-2000	1999-2000
INSTRUCTIONS	p. 304	p. 304	p. 304	p. 304	p. 304	p. 304
LUGNUT TORQUE	89 ft.lbs.	89 ft.lbs.	89 ft.lbs.	89 ft.lbs.	89 ft.lbs.	89 ft.lbs.
FRONT						
CASTER (Degrees)	7.58 ±0.5 7 9/16 ± 1/2	7.58 ±0.5 7 9/16 ± 1/2	3.73 ±0.5 3 3/4 ± 1/2	3.83 ±0.5 3 13/16 ± 1/2	5.43 ±0.5 5 7/16 ± 1/2	5.6 ±0.5 5 5/8 ± 1/2
CROSS | TOL.	0.0 (0) | 0.5 (1/2)	0.0 (0) | 0.5 (1/2)	0.0 (0) | 0.5 (1/2)	0.0 (0) | 0.5 (1/2)	0.0 (0) | 0.5 (1/2)	0.0 (0) | 0.5 (1/2)
CAMBER (Degrees)	-0.77 ±0.5 -3/4 ± 1/2	-0.77 ±0.5 -3/4 ± 1/2	-0.5 ±0.5 -1/2 ± 1/2	-0.85 ±0.5 -7/8 ± 1/2	-0.33 ±0.33 -5/16 ± 5/16	-0.72 ±0.33 -11/16 ± 5/16
CROSS | TOL.	0.0 (0) | 0.5 (1/2)	0.0 (0) | 0.5 (1/2)	0.0 (0) | 0.5 (1/2)	0.0 (0) | 0.5 (1/2)	0.0 (0) | 0.5 (1/2)	0.0 (0) | 0.5 (1/2)
FRONT TOE (Degrees)	0.33° ±0.08° 11/32° ± 3/32°	0.33° ±0.08° 11/32° ± 3/32°	0.3° ±0.13° 5/16° ± 1/8°	0.3° ±0.13° 5/16° ± 1/8°	0.23° ±0.13° 7/32° ± 1/8°	0.23° ±0.13° 7/32° ± 1/8°
FRONT TOE (Inches)	0.17" ±0.04" 5/32" ± 1/32"	0.17" ±0.04" 5/32" ± 1/32"	0.15" ±0.07" 5/32" ± 1/16"	0.15" ±0.07" 5/32" ± 1/16"	0.12" ±0.07" 1/8" ± 1/16"	0.12" ±0.07" 1/8" ± 1/16"
SAI						
TURNING ANGLE	20.0 / 18.4 ±0.5	20.0 / 18.4 ±0.5	20.0 / 18.5 ±0.5	20.0 / 18.5 ±0.5	20.0 / 18.5 ±0.5	20.0 / 18.5 ±0.5
ILLUSTRATION #					119	119
REAR						
CAMBER (Degrees)	-1.75 ±0.17 -1 3/4 ± 3/16	-1.75 ±0.17 -1 3/4 ± 3/16	-2.0 ±0.5 -2 ± 1/2	-2.5 ±0.5 -2 1/2 ± 1/2	-1.5 ±0.25 -1 1/2 ± 1/4	-2.07 ±0.25 -2 1/16 ± 1/4
REAR TOE (Degrees)	0.5° ±0.22° 1/2° ± 7/32°	0.5° ±0.22° 1/2° ± 7/32°	0.5° ±0.22° 1/2° ± 7/32°	0.6° ±0.22° 19/32° ± 7/32°	0.27° ±0.1° 9/32° ± 3/32°	0.27° ±0.1° 9/32° ± 3/32°
REAR TOE (Inches)	0.25" ±0.11" 1/4" ± 3/32"	0.25" ±0.11" 1/4" ± 3/32"	0.25" ±0.11" 1/4" ± 3/32"	0.3" ±0.11" 5/16" ± 3/32"	0.13" ±0.05" 1/8" ± 1/16"	0.13" ±0.05" 1/8" ± 1/16"
THRUST ANGLE	0.25 (1/4)	0.25 (1/4)	0.25 (1/4)	0.25 (1/4)	0.1 (3/32)	0.1 (3/32)
ILLUSTRATION #					29 125	29 125
NOTES						
	05031	05032	05033	05034	05035	05036

BMW

MODEL INFORMATION	3 Series (E46)	3 Series (E46)	3 Series (E46)	3 Series (E46)	3 Series (E46)	3 Series (E46)
	325i, 330i Sedan	325i, 330i Sedan	325Ci, 330Ci Cpe./Conv.	325Ci, 330Ci Cpe./Conv.	325xi, 330xi	325xi, 330xi
	Reg. Suspension	Sport Suspension	Reg. Suspension	Sport Suspension	to March, 2002	after March, 2002
YEARS	2001-2005	2001-2005	2001-2006	2001-2006	2001-2002	2002-2005
INSTRUCTIONS	p. 304	p. 304	p. 304	p. 304	p. 304	p. 304
LUGNUT TORQUE	89 ft.lbs.	89 ft.lbs.	89 ft.lbs.	89 ft.lbs.	89 ft.lbs.	89 ft.lbs.
FRONT						
CASTER (Degrees)	5.43 ±0.5 5 7/16 ± 1/2	5.6 ±0.5 5 5/8 ± 1/2	5.43 ±0.5 5 7/16 ± 1/2	5.6 ±0.5 5 5/8 ± 1/2	5.45 ±0.5 5 7/16 ± 1/2	5.45 ±0.5 5 7/16 ± 1/2
CROSS | TOL.	0.0 (0) | 0.5 (1/2)	0.0 (0) | 0.5 (1/2)	0.0 (0) | 0.5 (1/2)	0.0 (0) | 0.5 (1/2)	0.0 (0) | 0.5 (1/2)	0.0 (0) | 0.5 (1/2)
CAMBER (Degrees)	-0.33 ±0.33 -5/16 ± 5/16	-0.72 ±0.33 -11/16 ± 5/16	-0.33 ±0.33 -5/16 ± 5/16	-0.72 ±0.33 -11/16 ± 5/16	-0.33 ±0.33 -5/16 ± 5/16	-0.83 ±0.33 -13/16 ± 5/16
CROSS | TOL.	0.0 (0) | 0.5 (1/2)	0.0 (0) | 0.5 (1/2)	0.0 (0) | 0.5 (1/2)	0.0 (0) | 0.5 (1/2)	0.0 (0) | 0.5 (1/2)	0.0 (0) | 0.5 (1/2)
FRONT TOE (Degrees)	0.23° ±0.13° 7/32° ± 1/8°	0.23° ±0.13° 7/32° ± 1/8°	0.23° ±0.13° 7/32° ± 1/8°	0.23° ±0.13° 7/32° ± 1/8°	0.23° ±0.13° 7/32° ± 1/8°	0.1° ±0.23° 3/32° ± 7/32°
FRONT TOE (Inches)	0.12" ±0.07" 1/8" ± 1/16"	0.12" ±0.07" 1/8" ± 1/16"	0.12" ±0.07" 1/8" ± 1/16"	0.12" ±0.07" 1/8" ± 1/16"	0.12" ±0.07" 1/8" ± 1/16"	0.05" ±0.12" 1/16" ± 1/8"
SAI						
TURNING ANGLE	20.0 / 18.5 ±0.5	20.0 / 18.5 ±0.5	20.0 / 18.5 ±0.5	20.0 / 18.5 ±0.5	20.0 / 19.1 ±0.5	20.0 / 19.1 ±0.5
ILLUSTRATION #	119	119	119	119	119	119
REAR						
CAMBER (Degrees)	-1.5 ±0.25 -1 1/2 ± 1/4	-2.07 ±0.25 -2 1/16 ± 1/4	-1.5 ±0.25 -1 1/2 ± 1/4	-2.07 ±0.25 -2 1/16 ± 1/4	-1.25 ±0.25 -1 1/4 ± 1/4	-1.25 ±0.25 -1 1/4 ± 1/4
REAR TOE (Degrees)	0.27° ±0.1° 9/32° ± 3/32°	0.27° ±0.1° 9/32° ± 3/32°	0.27° ±0.1° 9/32° ± 3/32°	0.27° ±0.1° 9/32° ± 3/32°	0.27° ±0.1° 9/32° ± 3/32°	0.27° ±0.1° 9/32° ± 3/32°
REAR TOE (Inches)	0.13" ±0.05" 1/8" ± 1/16"	0.13" ±0.05" 1/8" ± 1/16"	0.13" ±0.05" 1/8" ± 1/16"	0.13" ±0.05" 1/8" ± 1/16"	0.13" ±0.05" 1/8" ± 1/16"	0.13" ±0.05" 1/8" ± 1/16"
THRUST ANGLE	0.1 (3/32)	0.1 (3/32)	0.1 (3/32)	0.1 (3/32)	0.1 (3/32)	0.1 (3/32)
ILLUSTRATION #	29 125	29 125	29 125	29 125	29 125	29 125
NOTES						
	05037	05038	05039	05040	05041	05042

BMW

MODEL INFORMATION	3 Series (E46) M3 Coupe & Conv.	3 Series (E90) 325i, 330i Sedan Reg. Suspension	3 Series (E90) 325i, 330i Sedan Sport Suspension	3 Series (E90) 328i, 335i, 335d Sedan Reg. Suspension	3 Series (E90) 328i, 335i, 335d Sedan Sport Suspension	3 Series (E90) 325xi, 330xi, 328xi, 335xi Sedan
YEARS	2001-2006	2006	2006	2007-2011	2007-2011	2006-2008
INSTRUCTIONS	p. 304	p. 304	p. 304	p. 304	p. 304	p. 304
LUGNUT TORQUE	89 ft.lbs.	89 ft.lbs.	89 ft.lbs.	89 ft.lbs.	89 ft.lbs.	89 ft.lbs.
FRONT						
CASTER (Degrees)	7.42 ±0.5 7 7/16 ± 1/2	Caster is not specified by the manufacturer.	Caster is not specified by the manufacturer.	Caster is not specified by the manufacturer.	Caster is not specified by the manufacturer.	Caster is not specified by the manufacturer.
CROSS | TOL.	0.0 (0) | 0.5 (1/2)	0.0 (0) | 0.5 (1/2)	0.0 (0) | 0.5 (1/2)	0.0 (0) | 0.5 (1/2)	0.0 (0) | 0.5 (1/2)	0.0 (0) | 0.5 (1/2)
CAMBER (Degrees)	-1.0 ±0.33 -1 ± 5/16	-0.3 ±0.42 - 5/16 ± 7/16	-0.5 ±0.42 - 1/2 ± 7/16	-0.3 ±0.42 - 5/16 ± 7/16	-0.5 ±0.42 - 1/2 ± 7/16	-0.45 ±0.42 - 7/16 ± 7/16
CROSS | TOL.	0.0 (0) | 0.5 (1/2)	0.0 (0) | 0.5 (1/2)	0.0 (0) | 0.5 (1/2)	0.0 (0) | 0.5 (1/2)	0.0 (0) | 0.5 (1/2)	0.0 (0) | 0.5 (1/2)
FRONT TOE (Degrees)	0.27° ±0.12° 9/32° ± 1/8°	0.23° ±0.07° 7/32° ± 1/16°	0.23° ±0.07° 7/32° ± 1/16°	0.23° ±0.07° 7/32° ± 1/16°	0.23° ±0.07° 7/32° ± 1/16°	0.07° ±0.07° 1/16° ± 1/16°
FRONT TOE (Inches)	0.13" ±0.06" 1/8" ± 1/16"	0.12" ±0.03" 1/8" ± 1/32"	0.12" ±0.03" 1/8" ± 1/32"	0.12" ±0.03" 1/8" ± 1/32"	0.12" ±0.03" 1/8" ± 1/32"	0.03" ±0.03" 1/32" ± 1/32"
SAI						
TURNING ANGLE	20.0 / 18.5 ±0.5	20.0 / 18.3 ±0.5	20.0 / 18.3 ±0.5	20.0 / 18.3 ±0.5	20.0 / 18.3 ±0.5	20.0 / 18.1 ±0.5
ILLUSTRATION #	119	119	119	119	119	119
REAR						
CAMBER (Degrees)	-1.75 ±0.25 -1 3/4 ± 1/4	-1.5 ±0.08 -1 1/2 ± 1/16	-1.83 ±0.08 -1 13/16 ± 1/16	-1.5 ±0.08 -1 1/2 ± 1/16	-1.83 ±0.08 -1 13/16 ± 1/16	-1.5 ±0.08 -1 1/2 ± 1/16
REAR TOE (Degrees)	0.37° ±0.1° 3/8° ± 3/32°	0.3° ±0.07° 5/16° ± 1/16°	0.3° ±0.07° 5/16° ± 1/16°	0.3° ±0.07° 5/16° ± 1/16°	0.3° ±0.07° 5/16° ± 1/16°	0.17° ±0.07° 5/32° ± 1/16°
REAR TOE (Inches)	0.18" ±0.05" 3/16" ± 1/16"	0.15" ±0.03" 5/32" ± 1/32"	0.15" ±0.03" 5/32" ± 1/32"	0.15" ±0.03" 5/32" ± 1/32"	0.15" ±0.03" 5/32" ± 1/32"	0.08" ±0.03" 3/32" ± 1/32"
THRUST ANGLE	0.07 (1/16)	0.2 (3/16)	0.2 (3/16)	0.2 (3/16)	0.2 (3/16)	0.2 (3/16)
ILLUSTRATION #	29 125	179 178	179 178	179 178	179 178	179 178
NOTES						
	05043	05044	05045	05046	05047	05048

BMW

MODEL INFORMATION	3 Series (E90) 328i xDrive Sedan, 335i xDrive Sedan	3 Series (E90) M3 Sedan	3 Series (E91) 328i Wagon Reg. Suspension	3 Series (E91) 328i Wagon Sport Suspension	3 Series (E91) 325xi, 328xi, 328i xDrive Wagon	3 Series (E92) 328i, 335i Coupe Reg. Suspension
YEARS	2009-2011	2008-2011	2007-2012	2007-2012	2006-2012	2007-2012
INSTRUCTIONS	p. 304	p. 304	p. 304	p. 304	p. 304	p. 304
LUGNUT TORQUE	89 ft.lbs.	89 ft.lbs.	89 ft.lbs.	89 ft.lbs.	89 ft.lbs.	89 ft.lbs.
FRONT						
CASTER (Degrees)	Caster is not specified by the manufacturer.	Caster is not specified by the manufacturer.	Caster is not specified by the manufacturer.	Caster is not specified by the manufacturer.	Caster is not specified by the manufacturer.	Caster is not specified by the manufacturer.
CROSS | TOL.	0.0 (0) | 0.5 (1/2)	0.0 (0) | 0.5 (1/2)	0.0 (0) | 0.5 (1/2)	0.0 (0) | 0.5 (1/2)	0.0 (0) | 0.5 (1/2)	0.0 (0) | 0.5 (1/2)
CAMBER (Degrees)	-0.45 ±0.42 - 7/16 ± 7/16	-1.17 ±0.42 -1 3/16 ± 7/16	-0.3 ±0.42 - 5/16 ± 7/16	-0.5 ±0.42 - 1/2 ± 7/16	-0.45 ±0.42 - 7/16 ± 7/16	-0.3 ±0.42 - 5/16 ± 7/16
CROSS | TOL.	0.0 (0) | 0.5 (1/2)	0.0 (0) | 0.5 (1/2)	0.0 (0) | 0.5 (1/2)	0.0 (0) | 0.5 (1/2)	0.0 (0) | 0.5 (1/2)	0.0 (0) | 0.5 (1/2)
FRONT TOE (Degrees)	0.07° ±0.07° 1/16° ± 1/16°	0.17° ±0.13° 5/32° ± 1/8°	0.23° ±0.07° 7/32° ± 1/16°	0.23° ±0.07° 7/32° ± 1/16°	0.07° ±0.07° 1/16° ± 1/16°	0.23° ±0.07° 7/32° ± 1/16°
FRONT TOE (Inches)	0.03" ±0.03" 1/32" ± 1/32"	0.08" ±0.07" 3/32" ± 1/16"	0.12" ±0.03" 1/8" ± 1/32"	0.12" ±0.03" 1/8" ± 1/32"	0.03" ±0.03" 1/32" ± 1/32"	0.12" ±0.03" 1/8" ± 1/32"
SAI						
TURNING ANGLE	20.0 / 18.1 ±0.5	20.0 / 17.8 ±0.5	20.0 / 18.3 ±0.5	20.0 / 18.3 ±0.5	20.0 / 18.1 ±0.5	20.0 / 18.3 ±0.5
ILLUSTRATION #	119	119	119	119	119	119
REAR						
CAMBER (Degrees)	-1.5 ±0.08 -1 1/2 ± 1/16	-1.75 ±0.2 -1 3/4 ± 3/16	-1.5 ±0.08 -1 1/2 ± 1/16	-1.83 ±0.08 -1 13/16 ± 1/16	-1.5 ±0.08 -1 1/2 ± 1/16	-1.5 ±0.08 -1 1/2 ± 1/16
REAR TOE (Degrees)	0.17° ±0.07° 5/32° ± 1/16°	0.17° ±0.08° 5/32° ± 3/32°	0.3° ±0.07° 5/16° ± 1/16°	0.3° ±0.07° 5/16° ± 1/16°	0.17° ±0.07° 5/32° ± 1/16°	0.3° ±0.07° 5/16° ± 1/16°
REAR TOE (Inches)	0.08" ±0.03" 3/32" ± 1/32"	0.08" ±0.04" 3/32" ± 1/32"	0.15" ±0.03" 5/32" ± 1/32"	0.15" ±0.03" 5/32" ± 1/32"	0.08" ±0.03" 3/32" ± 1/32"	0.15" ±0.03" 5/32" ± 1/32"
THRUST ANGLE	0.2 (3/16)	0.07 (1/16)	0.2 (3/16)	0.2 (3/16)	0.2 (3/16)	0.2 (3/16)
ILLUSTRATION #	179 178	179 178	179 178	179 178	179 178	179 178
NOTES						
	05049	05050	05051	05052	05053	05054

BMW

3 Series (E92) 328i, 335i Coupe Sport Suspension	3 Series (E92) 335is Coupe	3 Series (E92) 328xi, 335xi Coupe 328i xDrive, 335i xDrive	MODEL INFORMATION	3 Series (E92) 328i xDrive, 335i xDrive	3 Series (E92) M3 Coupe	3 Series (E93) 328i, 335i Conv. Reg. Suspension
2007-2012	2011-2012	2007-2012	YEARS	2007-2012	2008-2012	2007-2012
p. 304	p. 304	p. 304	INSTRUCTIONS	p. 304	p. 304	p. 304
89 ft.lbs.	89 ft.lbs.	89 ft.lbs.	LUGNUT TORQUE	89 ft.lbs.	89 ft.lbs.	89 ft.lbs.
			FRONT			
Caster is not specified by the manufacturer.	Caster is not specified by the manufacturer.	Caster is not specified by the manufacturer.	CASTER (Degrees)	Caster is not specified by the manufacturer.	Caster is not specified by the manufacturer.	Caster is not specified by the manufacturer.
0.0 (0) | 0.5 (½)	0.0 (0) | 0.5 (½)	0.0 (0) | 0.5 (½)	CROSS | TOL.	0.0 (0) | 0.5 (½)	0.0 (0) | 0.5 (½)	0.0 (0) | 0.5 (½)
-0.5 ±0.42 -½ ± 7/16	-0.5 ±0.42 -½ ± 7/16	-0.45 ±0.42 -7/16 ± 7/16	CAMBER (Degrees)	-0.45 ±0.42 -7/16 ± 7/16	-1.17 ±0.42 -1 3/16 ± 7/16	-0.3 ±0.42 -5/16 ± 7/16
0.0 (0) | 0.5 (½)	0.0 (0) | 0.5 (½)	0.0 (0) | 0.5 (½)	CROSS | TOL.	0.0 (0) | 0.5 (½)	0.0 (0) | 0.5 (½)	0.0 (0) | 0.5 (½)
0.23° ±0.07° 7/32° ± 1/16°	0.23° ±0.07° 7/32° ± 1/16°	0.07° ±0.07° 1/16° ± 1/16°	FRONT TOE (Degrees)	0.07° ±0.07° 1/16° ± 1/16°	0.17° ±0.07° 5/32° ± 1/16°	0.23° ±0.07° 7/32° ± 1/16°
0.12" ±0.03" ⅛" ± 1/32"	0.12" ±0.03" ⅛" ± 1/32"	0.03" ±0.03" 1/32" ± 1/32"	FRONT TOE (Inches)	0.03" ±0.03" 1/32" ± 1/32"	0.08" ±0.03" 3/32" ± 1/32"	0.12" ±0.03" ⅛" ± 1/32"
			SAI			
20.0 / 18.3 ±0.5	20.0 / 18.3 ±0.5	20.0 / 18.1 ±0.5	TURNING ANGLE	20.0 / 18.1 ±0.5	20.0 / 17.8 ±0.5	20.0 / 18.3 ±0.5
119	119	119	ILLUSTRATION #	119	119	119
			REAR			
-1.83 ±0.08 -1 13/16 ± 1/16	-1.83 ±0.08 -1 13/16 ± 1/16	-1.5 ±0.08 -1½ ± 1/16	CAMBER (Degrees)	-1.5 ±0.08 -1½ ± 1/16	-1.75 ±0.2 -1¾ ± 3/16	-1.5 ±0.08 -1½ ± 1/16
0.3° ±0.07° 5/16° ± 1/16°	0.3° ±0.07° 5/16° ± 1/16°	0.17° ±0.07° 5/32° ± 1/16°	REAR TOE (Degrees)	0.17° ±0.07° 5/32° ± 1/16°	0.17° ±0.07° 5/32° ± 1/16°	0.3° ±0.07° 5/16° ± 1/16°
0.15" ±0.03" 5/32" ± 1/32"	0.15" ±0.03" 5/32" ± 1/32"	0.08" ±0.03" 3/32" ± 1/32"	REAR TOE (Inches)	0.08" ±0.03" 3/32" ± 1/32"	0.08" ±0.03" 3/32" ± 1/32"	0.15" ±0.03" 5/32" ± 1/32"
0.2 (3/16)	0.2 (3/16)	0.2 (3/16)	THRUST ANGLE	0.2 (3/16)	0.07 (1/16)	0.2 (3/16)
179 178	179 178	179 178	ILLUSTRATION #	179 178	179 178	179 178
			NOTES			
05055	05056	05057		05058	05059	05060

BMW

3 Series (E93) 328i, 335i Conv. Sport Suspension	3 Series (E93) 335is Convertible	3 Series (E93) M3 Convertible	MODEL INFORMATION	3 Series (F30) All Sedans Reg. Suspension	3 Series (F30) All Sedans Sport Suspension	3 Series (F30) All xDrive Sedans
2007-2012	2011-2012	2008-2012	YEARS	2012-2017	2012-2017	2012-2017
p. 304	p. 304	p. 304	INSTRUCTIONS	p. 304	p. 304	p. 304
89 ft.lbs.	89 ft.lbs.	89 ft.lbs.	LUGNUT TORQUE	89 ft.lbs.	89 ft.lbs.	89 ft.lbs.
			FRONT			
Caster is not specified by the manufacturer.	Caster is not specified by the manufacturer.	Caster is not specified by the manufacturer.	CASTER (Degrees)	Caster is not specified by the manufacturer.	Caster is not specified by the manufacturer.	Caster is not specified by the manufacturer.
0.0 (0) | 0.5 (½)	0.0 (0) | 0.5 (½)	0.0 (0) | 0.5 (½)	CROSS | TOL.	0.0 (0) | 0.5 (½)	0.0 (0) | 0.5 (½)	0.0 (0) | 0.5 (½)
-0.5 ±0.42 -½ ± 7/16	-0.5 ±0.42 -½ ± 7/16	-1.17 ±0.33 -1 3/16 ± 5/16	CAMBER (Degrees)	-0.33 ±0.42 -5/16 ± 7/16	-0.48 ±0.42 -½ ± 7/16	-0.33 ±0.42 -5/16 ± 7/16
0.0 (0) | 0.5 (½)	0.0 (0) | 0.5 (½)	0.0 (0) | 0.5 (½)	CROSS | TOL.	0.0 (0) | 0.5 (½)	0.0 (0) | 0.5 (½)	0.0 (0) | 0.5 (½)
0.23° ±0.07° 7/32° ± 1/16°	0.23° ±0.07° 7/32° ± 1/16°	0.17° ±0.07° 5/32° ± 1/16°	FRONT TOE (Degrees)	0.23° ±0.07° 7/32° ± 1/16°	0.23° ±0.07° 7/32° ± 1/16°	0.23° ±0.07° 7/32° ± 1/16°
0.12" ±0.03" ⅛" ± 1/32"	0.12" ±0.03" ⅛" ± 1/32"	0.08" ±0.03" 3/32" ± 1/32"	FRONT TOE (Inches)	0.12" ±0.03" ⅛" ± 1/32"	0.12" ±0.03" ⅛" ± 1/32"	0.12" ±0.03"
			SAI			
20.0 / 18.3 ±0.5	20.0 / 18.3 ±0.5	20.0 / 17.8 ±0.5	TURNING ANGLE	20.0 / 18.6 ±0.5	20.0 / 18.6 ±0.5	20.0 / 18.6 ±0.5
119	119	119	ILLUSTRATION #	119	119	119
			REAR			
-1.83 ±0.08 -1 13/16 ± 1/16	-1.83 ±0.08 -1 13/16 ± 1/16	-1.75 ±0.2 -1¾ ± 3/16	CAMBER (Degrees)	-1.5 ±0.08 -1½ ± 1/16	-1.75 ±0.08 -1¾ ± 1/16	-1.38 ±0.08 -1⅜ ± 1/16
0.3° ±0.07° 5/16° ± 1/16°	0.3° ±0.07° 5/16° ± 1/16°	0.33° ±0.07° 11/32° ± 1/16°	REAR TOE (Degrees)	0.3° ±0.07° 5/16° ± 1/16°	0.3° ±0.07° 5/16° ± 1/16°	0.3° ±0.07° 5/16° ± 1/16°
0.15" ±0.03" 5/32" ± 1/32"	0.15" ±0.03" 5/32" ± 1/32"	0.17" ±0.03" 5/32" ± 1/32"	REAR TOE (Inches)	0.15" ±0.03" 5/32" ± 1/32"	0.15" ±0.03" 5/32" ± 1/32"	0.15" ±0.03" 5/32" ± 1/32"
0.2 (3/16)	0.2 (3/16)	0.07 (1/16)	THRUST ANGLE	0.2 (3/16)	0.2 (3/16)	0.2 (3/16)
179 178	179 178	179 178	ILLUSTRATION #	179 178	179 178	179 178
			NOTES			
05061	05062	05063		05064	05065	05066

BMW

3 Series (F30)	3 Series (F30)	3 Series (F30)	MODEL	3 Series (F31)	3 Series (F34)	3 Series (F80)
ActiveHybrid 3	ActiveHybrid 3	330e iPerformance	INFORMATION	All xDrive Wagons	All xDrive GT	M3 Sedan
Reg. Suspension	Sport Suspension					
2013-2015	2013-2015	2017	YEARS	2014-2017	2014-2017	2015-2017
p. 304	p. 304		INSTRUCTIONS	p. 304	p. 304	p. 304
89 ft.lbs.	89 ft.lbs.	103 ft.lbs.	LUGNUT TORQUE	89 ft.lbs.	89 ft.lbs.	103 ft.lbs.
			FRONT			
Caster is not specified by the manufacturer.	Caster is not specified by the manufacturer.	Caster is not specified by the manufacturer.	CASTER (Degrees)	Caster is not specified by the manufacturer.	Caster is not specified by the manufacturer.	Caster is not specified by the manufacturer.
0.0 (0) | 0.5 (½)	0.0 (0) | 0.5 (½)	0.0 (0) | 0.5 (½)	CROSS | TOL.	0.0 (0) | 0.5 (½)	0.0 (0) | 0.5 (½)	0.0 (0) | 0.5 (½)
-0.33 ±0.42 -5/16 ± 7/16	-0.48 ±0.42 -½ ± 7/16	-0.33 ±0.42 -5/16 ± 7/16	CAMBER (Degrees)	-0.33 ±0.42 -5/16 ± 7/16	-0.33 ±0.42 -5/16 ± 7/16	-1.48 ±0.42 -1½ ± 7/16
0.0 (0) | 0.5 (½)	0.0 (0) | 0.5 (½)	0.0 (0) | 0.5 (½)	CROSS | TOL.	0.0 (0) | 0.5 (½)	0.0 (0) | 0.5 (½)	0.0 (0) | 0.5 (½)
0.23° ±0.07° 7/32° ± 1/16°	0.23° ±0.07° 7/32° ± 1/16°	0.23° ±0.07° 7/32° ± 1/16°	FRONT TOE (Degrees)	0.23° ±0.07° 7/32° ± 1/16°	0.23° ±0.07° 7/32° ± 1/16°	0.17° ±0.07° 5/32° ± 1/16°
0.12" ±0.03" ⅛" ± 1/32"	0.12" ±0.03" ⅛" ± 1/32"	0.12" ±0.03" ⅛" ± 1/32"	FRONT TOE (Inches)	0.12" ±0.03"	0.12" ±0.03" ⅛" ± 1/32"	0.08" ±0.03" 3/32" ± 1/32"
			SAI			
20.0 / 18.6 ±0.5	20.0 / 18.6 ±0.5	20.0 / 18.6 ±0.5	TURNING ANGLE	20.0 / 18.6 ±0.5	20.0 / 18.6 ±0.5	20.0 / 18.0 ±0.5
119	119		ILLUSTRATION #	119	119	
			REAR			
-1.5 ±0.08 -1½ ± 1/16	-1.75 ±0.08 -1¾ ± 1/16	-1.38 ±0.08 -1⅜ ± 1/16	CAMBER (Degrees)	-1.38 ±0.08 -1⅜ ± 1/16	-1.38 ±0.08 -1⅜ ± 1/16	-1.83 ±0.2 -1 13/16 ± 3/16
0.3° ±0.07° 5/16° ± 1/16°	0.3° ±0.07° 5/16° ± 1/16°	0.3° ±0.07° 5/16° ± 1/16°	REAR TOE (Degrees)	0.3° ±0.07° 5/16° ± 1/16°	0.3° ±0.07° 5/16° ± 1/16°	0.27° ±0.07° 9/32° ± 1/16°
0.15" ±0.03" 5/32" ± 1/32"	0.15" ±0.03" 5/32" ± 1/32"	0.15" ±0.03" 5/32" ± 1/32"	REAR TOE (Inches)	0.15" ±0.03" 5/32" ± 1/32"	0.15" ±0.03" 5/32" ± 1/32"	0.13" ±0.03" ⅛" ± 1/32"
0.2 (3/16)	0.2 (3/16)	0.2 (3/16)	THRUST ANGLE	0.2 (3/16)	0.2 (3/16)	0.2 (3/16)
179 178	179 178	179 178	ILLUSTRATION #	179 178	179 178	179 178
			NOTES			
05067	05068	05069		05070	05071	05072

BMW

4 Series (F32, F33, F36)	4 Series (F32, F33, F36)	4 Series (F82, F83)	MODEL	5 Series (E34)	5 Series (E34)	5 Series (E34)
Coupe, Conv., GT	All xDrive incl. GT	M4 Coupe & Conv.	INFORMATION	525i, 530i, 535i,	525i/530i Touring	M5
All Suspensions	All Suspensions	All Suspensions		540i		
2014-2017	2014-2017	2015-2017	YEARS	1990-95	1992-95	1991-93
p. 304	p. 304	p. 304	INSTRUCTIONS	p. 304	p. 304	p. 304
89 ft.lbs.	89 ft.lbs.	89 ft.lbs.	LUGNUT TORQUE	89 ft.lbs.	89 ft.lbs.	89 ft.lbs.
			FRONT			
Caster is not specified by the manufacturer.	Caster is not specified by the manufacturer.	Caster is not specified by the manufacturer.	CASTER (Degrees)	7.88 ±0.5 7⅞ ± ½	7.88 ±0.5 7⅞ ± ½	8.17 ±0.5 8 3/16 ± ½
0.0 (0) | 0.5 (½)	0.0 (0) | 0.5 (½)	0.0 (0) | 0.5 (½)	CROSS | TOL.	0.0 (0) | 0.5 (½)	0.0 (0) | 0.5 (½)	0.0 (0) | 0.5 (½)
-0.48 ±0.42 -½ ± 7/16	-0.33 ±0.42 -5/16 ± 7/16	-1.48 ±0.42 -1½ ± 7/16	CAMBER (Degrees)	-0.22 ±0.5 -3/16 ± ½	-0.22 ±0.5 -3/16 ± ½	-0.5 ±0.5 -½ ± ½
0.0 (0) | 0.5 (½)	0.0 (0) | 0.5 (½)	0.0 (0) | 0.5 (½)	CROSS | TOL.	0.0 (0) | 0.5 (½)	0.0 (0) | 0.5 (½)	0.0 (0) | 0.5 (½)
0.23° ±0.07° 7/32° ± 1/16°	0.23° ±0.07° 7/32° ± 1/16°	0.17° ±0.07° 5/32° ± 1/16°	FRONT TOE (Degrees)	0.3° ±0.08° 5/16° ± 3/32°	0.3° ±0.08° 5/16° ± 3/32°	0.33° ±0.08° 11/32° ± 3/32°
0.12" ±0.03" ⅛" ± 1/32"	0.12" ±0.03" ⅛" ± 1/32"	0.08" ±0.03" 3/32" ± 1/32"	FRONT TOE (Inches)	0.15" ±0.04" 5/32" ± 1/32"	0.15" ±0.04" 5/32" ± 1/32"	0.17" ±0.04" 5/32" ± 1/32"
			SAI			
20.0 / 18.2 ±0.5	20.0 / 18.6 ±0.5	20.0 / 18.0 ±0.5	TURNING ANGLE	20.0 / 18.2 ±0.5	20.0 / 18.2 ±0.5	20.0 / 18.0 ±0.5
	119	119	ILLUSTRATION #			
			REAR			
-1.75 ±0.08 -1¾ ± 1/16	-1.38 ±0.08 -1⅜ ± 1/16	-1.83 ±0.08 -1 13/16 ± 1/16	CAMBER (Degrees)	-2.33 ±0.5 -2 5/16 ± ½	-2.33 ±0.5 -2 5/16 ± ½	-2.25 ±0.5 -2¼ ± ½
0.3° ±0.07° 5/16° ± 1/16°	0.3° ±0.07° 5/16° ± 1/16°	0.27° ±0.07° 9/32° ± 1/16°	REAR TOE (Degrees)	0.42° ±0.12° 13/32° ± ⅛°	0.42° ±0.12° 13/32° ± ⅛°	0.27° ±0.06° 9/32° ± 1/16°
0.15" ±0.03" 5/32" ± 1/32"	0.15" ±0.03" 5/32" ± 1/32"	0.13" ±0.03" ⅛" ± 1/32"	REAR TOE (Inches)	0.21" ±0.06" 7/32" ± 1/16"	0.21" ±0.06" 7/32" ± 1/16"	0.13" ±0.03" ⅛" ± 1/32"
0.2 (3/16)	0.2 (3/16)	0.2 (3/16)	THRUST ANGLE	0.25 (¼)	0.25 (¼)	0.25 (¼)
179 178	179 178	179 178	ILLUSTRATION #			
			NOTES			
05073	05074	05075		05079	05080	05081

BMW

5 Series (E39)	5 Series (E39)	5 Series (E39)	MODEL	5 Series (E39)	5 Series (E39)	5 Series (E60)
525i, 528i, 530i	525i, 528i, 530i	540i	INFORMATION	540i	M5	525i, 528i, 530i, 535i,
Reg. Suspension	Sport Suspension	Reg. Suspension		Sport Suspension		545i, 550i Reg. Susp.
1997-2003	1997-2003	1997-2003	YEARS	1997-2003	2000-2003	2004-2010
p. 304	p. 304	p. 304	INSTRUCTIONS	p. 304	p. 304	p. 304
89 ft.lbs.	89 ft.lbs.	89 ft.lbs.	LUGNUT TORQUE	89 ft.lbs.	89 ft.lbs.	89 ft.lbs.
			FRONT			
6.47 ±0.5	6.68 ±0.5	6.3 ±0.5	CASTER	6.57 ±0.5	6.45 ±0.5	Caster is not specified
6 7/16 ± 1/2	6 11/16 ± 1/2	6 5/16 ± 1/2	(Degrees)	6 9/16 ± 1/2	6 7/16 ± 1/2	by the manufacturer.
0.0 (0) | 0.5 (1/2)	0.0 (0) | 0.5 (1/2)	0.0 (0) | 0.5 (1/2)	CROSS | TOL.	0.0 (0) | 0.5 (1/2)	0.0 (0) | 0.5 (1/2)	0.0 (0) | 0.5 (1/2)
-0.22 ±0.5	-0.6 ±0.5	-0.22 ±0.5	CAMBER	-0.62 ±0.5	-0.5 ±0.5	-0.2 ±0.33
-3/16 ± 1/2	-5/8 ± 1/2	-3/16 ± 1/2	(Degrees)	-5/8 ± 1/2	-1/2 ± 1/2	-3/16 ± 5/16
0.0 (0) | 0.67 (11/16)	0.0 (0) | 0.67 (11/16)	0.0 (0) | 0.67 (11/16)	CROSS | TOL.	0.0 (0) | 0.67 (11/16)	0.0 (0) | 0.5 (1/2)	0.0 (0) | 0.5 (1/2)
0.08° ±0.17°	0.08° ±0.17°	0.23° ±0.17°	FRONT TOE	0.23° ±0.17°	0.17° ±0.17°	0.13° ±0.17°
3/32° ± 5/32°	3/32° ± 5/32°	7/32° ± 5/32°	(Degrees)	7/32° ± 5/32°	5/32° ± 5/32°	1/8° ± 5/32°
0.04" ±0.08"	0.04" ±0.08"	0.12" ±0.08"	FRONT TOE	0.12" ±0.08"	0.08" ±0.08"	0.07" ±0.08"
1/32" ± 3/32"	1/32" ± 3/32"	1/8" ± 3/32"	(Inches)	1/8" ± 3/32"	3/32" ± 3/32"	1/16" ± 3/32"
			SAI			
20.0 / 18.1 ±0.5	20.0 / 18.1 ±0.5	20.0 / 18.4 ±0.5	TURNING ANGLE	20.0 / 18.3 ±0.5	20.0 / 18.3 ±0.5	20.0 / 18.3 ±0.5
119	119	119	ILLUSTRATION #	119	119	119
			REAR			
-2.07 ±0.08 [1]	-2.07 ±0.08 [1]	-2.07 ±0.08 [1]	CAMBER	-2.07 ±0.08 [1]	-1.82 ±0.25	-2.0 ±0.33
-2 1/16 ± 1/16 [1]	-2 1/16 ± 1/16 [1]	-2 1/16 ± 1/16 [1]	(Degrees)	-2 1/16 ± 1/16 [1]	-1 13/16 ± 1/4	-2 ± 5/16
0.37° ±0.07° [2]	0.37° ±0.07° [2]	0.37° ±0.07° [2]	REAR TOE	0.37° ±0.07° [2]	0.17° ±0.13°	0.3° ±0.17°
3/8° ± 1/16° [2]	3/8° ± 1/16° [2]	3/8° ± 1/16° [2]	(Degrees)	3/8° ± 1/16° [2]	5/32° ± 1/8°	5/16° ± 5/32°
0.18" ±0.03" [2]	0.18" ±0.03" [2]	0.18" ±0.03" [2]	REAR TOE	0.18" ±0.03" [2]	0.08" ±0.07"	0.15" ±0.08"
3/16" ± 1/32" [2]	3/16" ± 1/32" [2]	3/16" ± 1/32" [2]	(Inches)	3/16" ± 1/32" [2]	3/32" ± 1/16"	5/32" ± 3/32"
0.2 (3/16)	0.2 (3/16)	0.2 (3/16)	THRUST ANGLE	0.2 (3/16)	0.2 (3/16)	0.2 (3/16)
144	144	144	ILLUSTRATION #	144	144	144
[1] Sport Wagon = -1.83° (-1 13/16°)	[1] Sport Wagon = -1.83° (-1 13/16°)	[1] Sport Wagon = -1.83° (-1 13/16°)	NOTES	[1] Sport Wagon = -1.83° (-1 13/16°)		
[2] Set sedan to 0.37° (0.18")	[2] Set sedan to 0.37° (0.18")	[2] Set sedan to 0.37° (0.18")		[2] Set sedan to 0.37° (0.18")		
05082	05083	05084		05085	05086	05087

BMW

5 Series (E60)	5 Series (E60)	5 Series (E60)	MODEL	5 Series (E61)	5 Series (F07)	5 Series (F07)
525i, 528i, 530i, 535i,	525xi, 528xi, 530xi,	M5	INFORMATION	525xi, 530xi, 535xi	535i GT	535i xDrive GT
545i, 550i Sport Susp.	535xi (xDrive) Sedan			(xDrive) Wagon	550i GT	550i xDrive GT
2004-2010	2006-2010	2006-2010	YEARS	2006-2010	2010-2017	2011-2017
p. 304	p. 304	p. 304	INSTRUCTIONS	p. 304	p. 304	p. 304
89 ft.lbs.	89 ft.lbs.	89 ft.lbs.	LUGNUT TORQUE	89 ft.lbs.	103 ft.lbs.	103 ft.lbs.
			FRONT			
Caster is not specified	Caster is not specified	Caster is not specified	CASTER	Caster is not specified	Caster is not specified	Caster is not specified
by the manufacturer.	by the manufacturer.	by the manufacturer.	(Degrees)	by the manufacturer.	by the manufacturer.	by the manufacturer.
0.0 (0) | 0.5 (1/2)	0.0 (0) | 0.5 (1/2)	0.0 (0) | 0.5 (1/2)	CROSS | TOL.	0.0 (0) | 0.5 (1/2)	0.0 (0) | 0.5 (1/2)	0.0 (0) | 0.5 (1/2)
-0.5 ±0.33	-0.33 +0.5 -0.33	-1.0 ±0.33	CAMBER	-0.33 +0.5 -0.33	-0.2 ±0.42 [1]	-0.2 ±0.42 [1]
-1/2 ± 5/16	-5/16 +1/2 -5/16	-1 ± 5/16	(Degrees)	-5/16 +1/2 -5/16	-3/16 ± 7/16 [1]	-3/16 ± 7/16 [1]
0.0 (0) | 0.5 (1/2)	0.0 (0) | 0.5 (1/2)	0.0 (0) | 0.5 (1/2)	CROSS | TOL.	0.0 (0) | 0.5 (1/2)	0.0 (0) | 0.5 (1/2)	0.0 (0) | 0.5 (1/2)
0.13° ±0.17°	0.2° ±0.17°	0.07° ±0.17°	FRONT TOE	0.2° ±0.17°	0.23° ±0.07°	0.23° ±0.07°
1/8° ± 5/32°	3/16° ± 5/32°	1/16° ± 5/32°	(Degrees)	3/16° ± 5/32°	7/32° ± 1/16°	7/32° ± 1/16°
0.07" ±0.08"	0.1" ±0.08"	0.03" ±0.08"	FRONT TOE	0.1" ±0.08"	0.12" ±0.03"	0.12" ±0.03"
1/16" ± 3/32"	3/32" ± 3/32"	1/32" ± 3/32"	(Inches)	3/32" ± 3/32"	1/8" ± 1/32"	1/8" ± 1/32"
			SAI			
20.0 / 18.2 ±0.5	20.0 / 18.1 ±0.5	20.0 / 18.2 ±0.5	TURNING ANGLE	20.0 / 18.1 ±0.5	20.0 / 18.1 ±0.5	20.0 / 18.1 ±0.5
119	119	119	ILLUSTRATION #	119		
			REAR			
-2.0 ±0.33	-2.0 ±0.33	-1.67 ±0.33	CAMBER	-2.0 ±0.33	-1.83 ±0.08	-1.83 ±0.08
-2 ± 5/16	-2 ± 5/16	-1 11/16 ± 5/16	(Degrees)	-2 ± 5/16	-1 13/16 ± 1/16	-1 13/16 ± 1/16
0.3° ±0.17°	0.3° ±0.17°	0.23° ±0.17°	REAR TOE	0.3° ±0.17°	0.23° ±0.07°	0.23° ±0.07°
5/16° ± 5/32°	5/16° ± 5/32°	7/32° ± 5/32°	(Degrees)	5/16° ± 5/32°	7/32° ± 1/16°	7/32° ± 1/16°
0.15" ±0.08"	0.15" ±0.08"	0.12" ±0.08"	REAR TOE	0.15" ±0.08"	0.12" ±0.03"	0.12" ±0.03"
5/32" ± 3/32"	5/32" ± 3/32"	1/8" ± 3/32"	(Inches)	5/32" ± 3/32"	1/8" ± 1/32"	1/8" ± 1/32"
0.2 (3/16)	0.2 (3/16)	0.2 (3/16)	THRUST ANGLE	0.2 (3/16)	0.2 (3/16)	0.2 (3/16)
144	144	144	ILLUSTRATION #	144	194	194
			NOTES		[1] Sport Susp. = -0.48° (1/2°)	[1] Sport Susp. = -0.48° (1/2°)
05088	05089	05090		05091	05092	05093

BMW

5 Series (F10)	5 Series (F10)	5 Series (F10)	MODEL	5 Series (F10)	5 Series (G30)	5 Series (G30)
528i, 535i, 550i	528i xDrive, 535i xDrive, 550i xDrive	M5	INFORMATION	ActiveHybrid 5	530i, 540i Reg. Suspension	530i, 540i Sport Suspension
2011-2016	2011-2016	2012-2016	YEARS	2014-2016	2017	2017
p. 304	p. 304	p. 304	INSTRUCTIONS	p. 304		
103 ft.lbs.	103 ft.lbs.	103 ft.lbs.	LUGNUT TORQUE	103 ft.lbs.	103 ft.lbs.	103 ft.lbs.
			FRONT			
Caster is not specified by the manufacturer.	Caster is not specified by the manufacturer.	Caster is not specified by the manufacturer.	CASTER (Degrees)	Caster is not specified by the manufacturer.	Caster is not specified by the manufacturer.	Caster is not specified by the manufacturer.
0.0 (0) | 0.5 (1/2)	0.0 (0) | 0.5 (1/2)	0.0 (0) | 0.5 (1/2)	CROSS | TOL.	0.0 (0) | 0.5 (1/2)	0.0 (0) | 0.5 (1/2)	0.0 (0) | 0.5 (1/2)
-0.2 ±0.42 [1] -3/16 ± 7/16 [1]	-0.2 ±0.42 -3/16 ± 7/16	-0.67 ±0.42 -11/16 ± 7/16	CAMBER (Degrees)	-0.2 ±0.42 [1] -3/16 ± 7/16 [1]	-0.15 ±0.42 -1/8 ± 7/16	-0.32 ±0.42 -5/16 ± 7/16
0.0 (0) | 0.5 (1/2)	0.0 (0) | 0.5 (1/2)	0.0 (0) | 0.5 (1/2)	CROSS | TOL.	0.0 (0) | 0.5 (1/2)	0.0 (0) | 0.5 (1/2)	0.0 (0) | 0.5 (1/2)
0.17° ±0.07° 5/32° ± 1/16°	0.17° ±0.07° 5/32° ± 1/16°	0.17° ±0.07° 5/32° ± 1/16°	FRONT TOE (Degrees)	0.17° ±0.07° 5/32° ± 1/16°	0.17° ±0.07° 5/32° ± 1/16°	0.17° ±0.07° 5/32° ± 1/16°
0.08" ±0.03" 3/32" ± 1/32"	0.08" ±0.03" 3/32" ± 1/32"	0.08" ±0.03" 3/32" ± 1/32"	FRONT TOE (Inches)	0.08" ±0.03" 3/32" ± 1/32"	0.08" ±0.03" 3/32" ± 1/32"	0.08" ±0.03" 3/32" ± 1/32"
			SAI		8.9 ±0.5 (8 7/8 ± 1/2)	9.1 ±0.5 (9 1/10 ± 1/2)
20.0 / 18.9 ±0.5	20.0 / 18.9 ±0.5	20.0 / 17.3 ±0.5	TURNING ANGLE	20.0 / 18.9 ±0.5	40.8 / 33.0 ±0.5	40.8 / 33.0 ±0.5
			ILLUSTRATION #			
			REAR			
-1.83 ±0.08 -1 13/16 ± 1/16	-1.83 ±0.08 -1 13/16 ± 1/16	-1.5 ±0.08 -1 1/2 ± 1/16	CAMBER (Degrees)	-1.83 ±0.08 -1 13/16 ± 1/16	-1.67 ±0.08 -1 11/16 ± 1/16	-1.85 ±0.08 -1 7/8 ± 1/16
0.3° ±0.07° 5/16° ± 1/16°	0.3° ±0.07° 5/16° ± 1/16°	0.2° ±0.07° 3/16° ± 1/16°	REAR TOE (Degrees)	0.3° ±0.07° 5/16° ± 1/16°	0.33° ±0.07° 5/16° ± 1/16°	0.33° ±0.07° 5/16° ± 1/16°
0.15" ±0.03" 5/32" ± 1/32"	0.15" ±0.03" 5/32" ± 1/32"	0.1" ±0.03" 3/32" ± 1/32"	REAR TOE (Inches)	0.15" ±0.03" 5/32" ± 1/32"	0.17" ±0.03" 5/32" ± 1/32"	0.17" ±0.03" 5/32" ± 1/32"
0.2 (3/16)	0.2 (3/16)	0.2 (3/16)	THRUST ANGLE	0.2 (3/16)	0.2 (3/16)	0.2 (3/16)
194	194	194	ILLUSTRATION #	194		
[1] Sport Susp. = -0.48° (1/2°)			NOTES	[1] Sport Susp. = -0.48°(-1/2°)		
05094	05095	05096		05097	05098	05099

BMW

5 Series (G30)	5 Series (G30)	6 Series (E63, E64)	MODEL	6 Series (E63, E64)	6 Series (F12, F13)	6 Series (F06)
530i xDrive, 540i xDrive Reg. Suspension	530i xDrive, 540i xDrive Sport Suspension	645Ci Cpe./Conv. 650i Cpe./Conv.	INFORMATION	M6 Coupe M6 Convertible	640i, 650i, M6, Cpe./Conv. incl. xDrive models	640i, 650i, Alpina Gran Coupe incl. xDrive models
2017	2017	2004-2011	YEARS	2006-2011	2012-2017	2013-2017
		p. 304	INSTRUCTIONS	p. 304	p. 304	p. 304
103 ft.lbs.	103 ft.lbs.	89 ft.lbs.	LUGNUT TORQUE	89 ft.lbs.	89 ft.lbs.	89 ft.lbs.
			FRONT			
Caster is not specified by the manufacturer.	Caster is not specified by the manufacturer.	Caster is not specified by the manufacturer.	CASTER (Degrees)	Caster is not specified by the manufacturer.	Caster is not specified by the manufacturer.	Caster is not specified by the manufacturer.
0.0 (0) | 0.5 (1/2)	0.0 (0) | 0.5 (1/2)	0.0 (0) | 0.5 (1/2)	CROSS | TOL.	0.0 (0) | 0.5 (1/2)	0.0 (0) | 0.5 (1/2)	0.0 (0) | 0.5 (1/2)
-0.2 ±0.42 -3/16 ± 7/16	-0.37 ±0.42 -3/8 ± 7/16	-0.2 ±0.42 -3/16 ± 7/16	CAMBER (Degrees)	-1.0 ±0.42 -1 ± 7/16	-0.48 ±0.42 -1/2 ± 7/16	-0.48 ±0.42 -1/2 ± 7/16
0.0 (0) | 0.5 (1/2)	0.0 (0) | 0.5 (1/2)	0.0 (0) | 0.5 (1/2)	CROSS | TOL.	0.0 (0) | 0.5 (1/2)	0.0 (0) | 0.5 (1/2)	0.0 (0) | 0.5 (1/2)
0.17° ±0.07° 5/32° ± 1/16°	0.17° ±0.07° 5/32° ± 1/16°	0.17° ±0.07° 5/32° ± 1/16°	FRONT TOE (Degrees)	0.07° ±0.07° 1/16° ± 1/16°	0.17° ±0.07° 5/32° ± 1/16°	0.17° ±0.07° 5/32° ± 5/32°
0.08" ±0.03" 3/32" ± 1/32"	0.08" ±0.03" 3/32" ± 1/32"	0.08" ±0.03" 3/32" ± 1/32"	FRONT TOE (Inches)	0.03" ±0.03" 1/32" ± 1/32"	0.08" ±0.03" 3/32" ± 1/32"	0.08" ±0.03" 3/32" ± 1/32"
9.0 ±0.5 (9 ± 1/2)	9.2 ±0.5 (3/16 ± 1/2)		SAI			
39.8 / 32.8 ±0.5	39.8 / 32.8 ±0.5	20.0 / 18.1 ±0.5	TURNING ANGLE	20.0 / 18.1 ±0.5	20.0 / 18.2 ±0.5 [1]	20.0 / 18.2 ±0.5 [1]
		119	ILLUSTRATION #	119		
			REAR			
-1.67 ±0.08 -1 11/16 ± 1/16	-1.85 ±0.08 -1 7/8 ± 1/16	-2.0 ±0.33 -2 ± 5/16	CAMBER (Degrees)	-1.67 ±0.5 -1 11/16 ± 1/2	-1.83 ±0.08 -1 13/16 ± 1/16	-1.83 ±0.08 -1 13/16 ± 1/16
0.33° ±0.07° 5/16° ± 1/16°	0.33° ±0.07° 5/16° ± 1/16°	0.3° ±0.07° 5/16° ± 1/16°	REAR TOE (Degrees)	0.23° ±0.07° 7/32° ± 1/16°	0.3° ±0.07° 5/16° ± 1/16°	0.3° ±0.07° 5/16° ± 1/16°
0.17" ±0.03" 5/32" ± 1/32"	0.17" ±0.03" 5/32" ± 1/32"	0.15" ±0.03" 5/32" ± 1/32"	REAR TOE (Inches)	0.12" ±0.03" 1/8" ± 1/32"	0.15" ±0.03" 5/32" ± 1/32"	0.15" ±0.03" 5/32" ± 1/32"
0.2 (3/16)	0.2 (3/16)	0.2 (3/16)	THRUST ANGLE	0.2 (3/16)	0.2 (3/16)	0.2 (3/16)
		144	ILLUSTRATION #	144	194	194
			NOTES		[1] xDrive = 20°/18.25°	[1] xDrive = 20°/18.25°
05100	05101	05105		05106	05107	05108

Alignment Specifications

BMW

MODEL INFORMATION	7 Series (E32) 735i, 740i 735iL, 740iL, 750iL	7 Series (E38) 740i, 740iL, 750iL Reg. Suspension	7 Series (E38) 740i Sport Suspension	7 Series (E65) 745i, 760i Reg. & Air Susp.	7 Series (E65) 745i Sport Suspension	7 Series (E65) 750i, 760i Reg. & Air Susp.
YEARS	1990-94	1995-2001	1999-2001	2002-2005	2002-2005	2006-2008
INSTRUCTIONS	p. 304	p. 304	p. 304	p. 304	p. 304	p. 304
LUGNUT TORQUE	89 ft.lbs.	89 ft.lbs.	89 ft.lbs.	103 ft.lbs.	103 ft.lbs.	103 ft.lbs.
FRONT						
CASTER (Degrees)	7.98 ±0.5 8 ± ½	Caster is not specified by the manufacturer.	Caster is not specified by the manufacturer.	Caster is not specified by the manufacturer.	Caster is not specified by the manufacturer.	Caster is not specified by the manufacturer.
CROSS | TOL.	0.0 (0) | 0.5 (½)	0.0 (0) | 0.5 (½)	0.0 (0) | 0.5 (½)	0.0 (0) | 0.5 (½)	0.0 (0) | 0.5 (½)	0.0 (0) | 0.5 (½)
CAMBER (Degrees)	-0.18 ±0.5 -3/16 ± ½	-0.22 ±0.5 -3/16 ± ½	-0.6 ±0.5 -5/8 ± ½	-0.1 ±0.42 -1/8 ± 7/16	-0.5 ±0.42 -½ ± 7/16	-0.1 ±0.42 -1/8 ± 7/16
CROSS | TOL.	0.0 (0) | 0.5 (½)	0.0 (0) | 0.5 (½)	0.0 (0) | 0.5 (½)	0.0 (0) | 0.5 (½)	0.0 (0) | 0.5 (½)	0.0 (0) | 0.5 (½)
FRONT TOE (Degrees)	0.3° ±0.08° 5/16° ± 3/32°	0.23° ±0.17° 7/32° ± 5/32°	0.23° ±0.17° 7/32° ± 5/32°	0.17° ±0.07° 5/32° ± 1/16°	0.17° ±0.07° 5/32° ± 1/16°	0.2° ±0.07° 3/16° ± 1/16°
FRONT TOE (Inches)	0.15" ±0.04" 5/32" ± 1/32"	0.12" ±0.08" 1/8" ± 3/32"	0.12" ±0.08" 1/8" ± 3/32"	0.08" ±0.03" 3/32" ± 1/32"	0.08" ±0.03" 3/32" ± 1/32"	0.1" ±0.03" 3/32" ± 1/32"
SAI	12.1 ±0.5 (12 1/16 ± ½)					
TURNING ANGLE	20.0 / 18.3 ±0.5	20.0 / 18.4 ±0.5	20.0 / 18.3 ±0.5	20.0 / 18.6 ±0.5	20.0 / 18.6 ±0.5	20.0 / 18.6 ±0.5
ILLUSTRATION #				119	119	119
REAR						
CAMBER (Degrees)	-2.33 ±0.5 -2 5/16 ± ½	-1.45 ±0.33 -1 7/16 ± 5/16	-1.45 ±0.33 -1 7/16 ± 5/16	-1.5 ±0.08 -1 ½ ± 1/16	-1.5 ±0.08 -1 ½ ± 1/16	-1.83 ±0.08 -1 13/16 ± 1/16
REAR TOE (Degrees)	0.3° ±0.12° 5/16° ± 1/8°	0.3° ±0.17° 5/16° ± 5/32°	0.3° ±0.17° 5/16° ± 5/32°	0.17° ±0.07° 5/32° ± 1/16°	0.17° ±0.07° 5/32° ± 1/16°	0.3° ±0.07° 5/16° ± 1/16°
REAR TOE (Inches)	0.15" ±0.06" 5/32" ± 1/16"	0.15" ±0.08" 5/32" ± 3/32"	0.15" ±0.08" 5/32" ± 3/32"	0.08" ±0.03" 3/32" ± 1/32"	0.08" ±0.03" 3/32" ± 1/32"	0.15" ±0.03" 5/32" ± 1/32"
THRUST ANGLE	0.25 (¼)	0.2 (3/16)	0.2 (3/16)	0.2 (3/16)	0.2 (3/16)	0.2 (3/16)
ILLUSTRATION #		129 143	129 143	143	144	144
NOTES						
	05110	05111	05112	05113	05114	05115

BMW

MODEL INFORMATION	7 Series (E65) 750i Sport Suspension	7 Series (E65) Alpina B7	7 Series (E66) 745Li, 760Li Reg. & Air Susp.	7 Series (E66) 750Li, 760Li Reg. & Air Susp.	7 Series (E66) 750Li Sport Susp.	7 Series (F01, F02) 740i, 740Li, 750i, 750Li, 760Li
YEARS	2006-2008	2006-2008	2002-2005	2006-2008	2006-2008	2009-2015
INSTRUCTIONS	p. 304	p. 304	p. 304	p. 304	p. 304	p. 304
LUGNUT TORQUE	103 ft.lbs.	103 ft.lbs.	103 ft.lbs.	103 ft.lbs.	103 ft.lbs.	103 ft.lbs.
FRONT						
CASTER (Degrees)	Caster is not specified by the manufacturer.	Caster is not specified by the manufacturer.	Caster is not specified by the manufacturer.	Caster is not specified by the manufacturer.	Caster is not specified by the manufacturer.	Caster is not specified by the manufacturer.
CROSS | TOL.	0.0 (0) | 0.5 (½)	0.0 (0) | 0.5 (½)	0.0 (0) | 0.5 (½)	0.0 (0) | 0.5 (½)	0.0 (0) | 0.5 (½)	0.0 (0) | 0.5 (½)
CAMBER (Degrees)	-0.5 ±0.42 -½ ± 7/16	-0.1 ±0.33 -1/8 ± 5/16	-0.1 ±0.42 -1/8 ± 7/16	-0.1 ±0.42 -1/8 ± 7/16	-0.5 ±0.42 -½ ± 7/16	-0.18 ±0.42 -3/16 ± 7/16
CROSS | TOL.	0.0 (0) | 0.5 (½)	0.0 (0) | 0.5 (½)	0.0 (0) | 0.5 (½)	0.0 (0) | 0.5 (½)	0.0 (0) | 0.5 (½)	0.0 (0) | 0.5 (½)
FRONT TOE (Degrees)	0.2° ±0.07° 3/16° ± 1/16°	0.17° ±0.13° 5/32° ± 1/8°	0.17° ±0.07° 5/32° ± 1/16°	0.2° ±0.07° 3/16° ± 1/16°	0.2° ±0.07° 3/16° ± 1/16°	0.2° ±0.07° 3/16° ± 1/16°
FRONT TOE (Inches)	0.1" ±0.03" 3/32" ± 1/32"	0.08" ±0.07" 3/32" ± 1/16"	0.08" ±0.03" 3/32" ± 1/32"	0.1" ±0.03" 3/32" ± 1/32"	0.1" ±0.03" 3/32" ± 1/32"	0.1" ±0.03" 3/32" ± 1/32"
SAI						
TURNING ANGLE	20.0 / 18.6 ±0.5	20.0 / 18.6 ±0.5	20.0 / 18.6 ±0.5	20.0 / 18.6 ±0.5	20.0 / 18.6 ±0.5	20.0 / 18.0 ±0.5
ILLUSTRATION #	119	119	119	119	119	
REAR						
CAMBER (Degrees)	-1.83 ±0.08 -1 13/16 ± 1/16	-1.5 ±0.33 -1 ½ ± 5/16	-1.5 ±0.08 -1 ½ ± 1/16	-1.83 ±0.08 -1 13/16 ± 1/16	-1.83 ±0.08 -1 13/16 ± 1/16	-1.67 ±0.08 -1 11/16 ± 1/16
REAR TOE (Degrees)	0.3° ±0.07° 5/16° ± 1/16°	0.17° ±0.33° 5/32° ± 11/32°	0.17° ±0.07° 5/32° ± 1/16°	0.3° ±0.07° 5/16° ± 1/16°	0.3° ±0.07° 5/16° ± 1/16°	0.2° ±0.07° 3/16° ± 1/16°
REAR TOE (Inches)	0.15" ±0.03" 5/32" ± 1/32"	0.08" ±0.17" 3/32" ± 5/32"	0.08" ±0.03" 3/32" ± 1/32"	0.15" ±0.03" 5/32" ± 1/32"	0.15" ±0.03" 5/32" ± 1/32"	0.1" ±0.03" 3/32" ± 1/32"
THRUST ANGLE	0.2 (3/16)	0.2 (3/16)	0.2 (3/16)	0.2 (3/16)	0.2 (3/16)	0.2 (3/16)
ILLUSTRATION #	144	144	144	144	144	194
NOTES						
	05116	05117	05118	05119	05120	05121

BMW

7 Series (F01, F02)	7 Series (F01, F02)	7 Series (F01, F02)	MODEL	7 Series (F04)	7 Series (G12)	7 Series (G12)
740Li xDrive, 750i/Li xDrive	Alpina B7	Alpina B7 xDrive	INFORMATION	ActiveHybrid 750i/Li	All RWD Sedans	All xDrive Sedans incl. iPerformance
2009-2015	2011-2015	2011-2015	YEARS	2011-2015	2016-2017	2016-2017
p. 304	p. 304	p. 304	INSTRUCTIONS	p. 304	p. 304	p. 304
103 ft.lbs.	103 ft.lbs.	103 ft.lbs.	LUGNUT TORQUE	103 ft.lbs.	103 ft.lbs.	103 ft.lbs.
			FRONT			
Caster is not specified by the manufacturer.	Caster is not specified by the manufacturer.	Caster is not specified by the manufacturer.	CASTER (Degrees)	Caster is not specified by the manufacturer.	Caster is not specified by the manufacturer.	Caster is not specified by the manufacturer.
0.0 (0) | 0.5 (½)	0.0 (0) | 0.5 (½)	0.0 (0) | 0.5 (½)	CROSS | TOL.	0.0 (0) | 0.5 (½)	0.0 (0) | 0.5 (½)	0.0 (0) | 0.5 (½)
-0.17 ±0.42 -3/16 ± 7/16	-0.18 ±0.42 -3/16 ± 7/16	-0.17 ±0.42 -3/16 ± 7/16	CAMBER (Degrees)	-0.2 ±0.42 -3/16 ± 7/16	-0.72 ±0.42 -3/4 ± 7/16	-0.78 ±0.42 -13/16 ± 7/16
0.0 (0) | 0.5 (½)	0.0 (0) | 0.5 (½)	0.0 (0) | 0.5 (½)	CROSS | TOL.	0.0 (0) | 0.5 (½)	0.0 (0) | 0.5 (½)	0.0 (0) | 0.5 (½)
0.27° ±0.07° 9/32° ± 1/16°	0.2° ±0.07° 3/16° ± 1/16°	0.27° ±0.07° 9/32° ± 1/16°	FRONT TOE (Degrees)	0.2° ±0.07° 3/16° ± 1/16°	0.33° ±0.07° 11/32° ± 1/16°	0.33° ±0.07° 11/32° ± 1/16°
0.13" ±0.03" 1/8" ± 1/32"	0.1" ±0.03" 3/32" ± 1/32"	0.13" ±0.03" 1/8" ± 1/32"	FRONT TOE (Inches)	0.1" ±0.03" 3/32" ± 1/32"	0.17" ±0.03" 5/32" ± 1/32"	0.17" ±0.03" 5/32" ± 1/32"
			SAI		8.9 ±0.5 (8 15/16 ± ½)	9.0 ±0.5 (9 ± ½)
20.0 / 18.1 ±0.5	20.0 / 18.0 ±0.5		TURNING ANGLE	20.0 / 18.0 ±0.5	41.1 / 33.2 ±0.5	40.1 / 33.1 ±0.5
			ILLUSTRATION #			
			REAR			
-1.67 ±0.08 -1 11/16 ± 1/16	-1.67 ±0.08 -1 11/16 ± 1/16	-1.67 ±0.08 -1 11/16 ± 1/16	CAMBER (Degrees)	-1.67 ±0.08 -1 11/16 ± 1/16	-1.5 ±0.08 -1 ½ ± 1/16	-1.5 ±0.08 -1 ½ ± 1/16
0.2° ±0.07° 3/16° ± 1/16°	0.2° ±0.07° 3/16° ± 1/16°	0.2° ±0.07° 3/16° ± 1/16°	REAR TOE (Degrees)	0.2° ±0.07° 3/16° ± 1/16°	0.27° ±0.07° 3/16° ± 1/16°	0.27° ±0.07° 3/16° ± 1/16°
0.1" ±0.03" 3/32" ± 1/32"	0.1" ±0.03" 3/32" ± 1/32"	0.1" ±0.03" 3/32" ± 1/32"	REAR TOE (Inches)	0.1" ±0.03" 3/32" ± 1/32"	0.1" ±0.03" 3/32" ± 1/32"	0.1" ±0.03" 3/32" ± 1/32"
0.2 (3/16)	0.2 (3/16)	0.2 (3/16)	THRUST ANGLE	0.2 (3/16)	0.2 (3/16)	0.2 (3/16)
194	194	194	ILLUSTRATION #	194	194	194
			NOTES			
05122	05123	05124		05125	05126	05127

BMW

8 Series (E31)	8 Series (E31)	BMW i	MODEL	BMW i	X1 (E84)	X1 (F48)
840Ci, 850i, 850Ci	850CSi	i3	INFORMATION	i8		
1991-97	1994-95	2014-2017	YEARS	2015-2017	2012-2015	2016-2017
p. 304	p. 304		INSTRUCTIONS		p. 304	p. 304
89 ft.lbs.	89 ft.lbs.		LUGNUT TORQUE		103 ft.lbs.	103 ft.lbs.
			FRONT			
7.98 ±0.5 8 ± ½	8.13 ±0.5 8 1/8 ± ½	Caster is not specified by the manufacturer.	CASTER (Degrees)	Caster is not specified by the manufacturer.	Caster is not specified by the manufacturer.	Caster is not specified by the manufacturer.
0.0 (0) | 0.5 (½)	0.0 (0) | 0.5 (½)	0.0 (0) | 0.5 (½)	CROSS | TOL.	0.0 (0) | 0.5 (½)	0.0 (0) | 0.5 (½)	0.0 (0) | 0.5 (½)
-0.18 ±0.5 -3/16 ± ½	-0.4 ±0.5 -3/8 ± ½	-0.33 ±0.42 -5/16 ± 7/16	CAMBER (Degrees)	-0.15 ±0.42 -1/8 ± 7/16	-0.45 ±0.42 [1] -7/16 ± 7/16 [1]	-0.67 ±0.42 -11/16 ± 7/16
0.0 (0) | 0.5 (½)	0.0 (0) | 0.5 (½)	0.0 (0) | 0.5 (½)	CROSS | TOL.	0.0 (0) | 0.5 (½)	0.0 (0) | 0.5 (½)	0.0 (0) | 0.5 (½)
0.3° ±0.08° 5/16° ± 3/32°	0.43° ±0.08° 7/16° ± 3/32°	0.23° ±0.07° 7/32° ± 1/16°	FRONT TOE (Degrees)	0.23° ±0.07° 7/32° ± 1/16°	0.23° ±0.07° 7/32° ± 1/16°	0.2° ±0.07° 3/16° ± 1/16°
0.15" ±0.04" 5/32" ± 1/32"	0.22" ±0.04" 7/32" ± 1/32"	0.12" ±0.03" 1/8" ± 1/32"	FRONT TOE (Inches)	0.12" ±0.03" 1/8" ± 1/32"	0.12" ±0.03" 1/8" ± 1/32"	0.1" ±0.03" 3/32" ± 1/32"
			SAI	9.6 ±0.5 (9 9/16 ± ½)		11.6 ±0.5 (11 5/8 ± ½)
20.0 / 18.3 ±0.5	20.0 / 18.3 ±0.5	20.0 / 17.7 ±0.5	TURNING ANGLE	20.0 / 18.1 ±0.5	20.0 / 18.4 ±0.5	20.0 / 18.3 ±0.5
			ILLUSTRATION #		119	119
			REAR			
-1.17 ±0.25 -1 3/16 ± ¼	-1.38 ±0.25 -1 3/8 ± ¼	-1.67 ±0.08 -1 11/16 ± 1/16	CAMBER (Degrees)	-2.0 ±0.08 -2 ± 1/16	-1.5 ±0.42 -1 ½ ± 7/16	-1.5 ±0.42 -1 ½ ± 7/16
0.3° ±0.1° 5/16° ± 3/32°	0.33° ±0.08° 11/32° ± 3/32°	0.3° ±0.07° 5/16° ± 1/16°	REAR TOE (Degrees)	0.33° ±0.07° 11/32° ± 1/16°	0.3° ±0.07° 5/16° ± 1/16°	0.3° ±0.07° 5/16° ± 1/16°
0.15" ±0.05" 5/32" ± 1/16"	0.17" ±0.04" 5/32" ± 1/32"	0.15" ±0.03" 5/32" ± 1/32"	REAR TOE (Inches)	0.17" ±0.03" 5/32" ± 1/32"	0.15" ±0.03" 5/32" ± 1/32"	0.15" ±0.03" 5/32" ± 1/32"
0.05 (1/16)	0.05 (1/16)	0.2 (3/16)	THRUST ANGLE	0.2 (3/16)	0.2 (3/16)	0.2 (3/16)
		179 178	ILLUSTRATION #	194	179 178	129 143
			NOTES		[1] 2WD = -0.3° (-5/16°)	
05128	05129	05130		05131	05132	05133

BMW

X3 (E83)	X3 (F25)	X4 (F26) exc/2017 M40i	MODEL INFORMATION	X4 (F26) M40i	X5 (E53) w/Steel Springs	X5 (E53) w/Air Springs (Rear axle only)
2004-2010	2011-2017	2015-2017	YEARS	2017	2000-2006	2000-2006
p. 304	p. 304	p. 304	INSTRUCTIONS	p. 304	p. 304	p. 304
103 ft.lbs.	103 ft.lbs.	103 ft.lbs.	LUGNUT TORQUE	103 ft.lbs.	103 ft.lbs.	103 ft.lbs.
			FRONT			
Caster is not specified by the manufacturer.	Caster is not specified by the manufacturer.	Caster is not specified by the manufacturer.	CASTER (Degrees)	Caster is not specified by the manufacturer.	Caster is not specified by the manufacturer.	Caster is not specified by the manufacturer.
0.0 (0) | 0.5 (½)	0.0 (0) | 0.5 (½)	0.0 (0) | 0.5 (½)	CROSS | TOL.	0.0 (0) | 0.5 (½)	0.0 (0) | 0.5 (½)	0.0 (0) | 0.5 (½)
-0.33 ±0.42 -5/16 ± 7/16	-0.33 ±0.42 -5/16 ± 7/16	-0.33 ±0.42 -5/16 ± 7/16	CAMBER (Degrees)	-0.83 ±0.42 -13/16 ± 7/16	-0.2 ±0.42 -3/16 ± 7/16	-0.2 ±0.42 -3/16 ± 7/16
0.0 (0) | 0.5 (½)	0.0 (0) | 0.5 (½)	0.0 (0) | 0.5 (½)	CROSS | TOL.	0.0 (0) | 0.5 (½)	0.0 (0) | 0.5 (½)	0.0 (0) | 0.5 (½)
0.1° ±0.07° 3/32° ± 1/16°	0.1° ±0.07° 3/32° ± 1/16°	0.1° ±0.07° 3/32° ± 1/16°	FRONT TOE (Degrees)	0.17° ±0.07° 5/32° ± 1/16°	0.3° ±0.07° 5/16° ± 1/16°	0.3° ±0.07° 5/16° ± 1/16°
0.05" ±0.03" 1/16" ± 1/32"	0.05" ±0.03" 1/16" ± 1/32"	0.05" ±0.03" 1/16" ± 1/32"	FRONT TOE (Inches)	0.08" ±0.03" 3/32" ± 1/32"	0.15" ±0.03" 5/32" ± 1/32"	0.15" ±0.03" 5/32" ± 1/32"
			SAI			
20.0 / 17.7 ±0.5	20.0 / 18.5 ±0.5	20.0 / 18.5 ±0.5	TURNING ANGLE	20.0 / 18.5 ±0.5	20.0 / 18.4 ±0.5	20.0 / 18.4 ±0.5
119	119	119	ILLUSTRATION #	119	119	119
			REAR			
-2.0 ±0.08 -2 ± 1/16	-1.5 ±0.08 -1 ½ ± 1/16	-1.5 ±0.08 -1 ½ ± 1/16	CAMBER (Degrees)	-1.5 ±0.08 -1 ½ ± 1/16	-1.83 ±0.08 -1 13/16 ± 1/16	-1.83 ±0.08 -1 13/16 ± 1/16
0.27° ±0.07° 9/32° ± 1/16°	0.2° ±0.07° 3/16° ± 1/16°	0.2° ±0.07° 3/16° ± 1/16°	REAR TOE (Degrees)	0.3° ±0.07° 5/16° ± 1/16°	0.3° ±0.07° 5/16° ± 1/16°	0.3° ±0.07° 5/16° ± 1/16°
0.13" ±0.03" 1/8" ± 1/32"	0.1" ±0.03" 3/32" ± 1/32"	0.1" ±0.03" 3/32" ± 1/32"	REAR TOE (Inches)	0.15" ±0.03" 5/32" ± 1/32"	0.15" ±0.03" 5/32" ± 1/32"	0.15" ±0.03" 5/32" ± 1/32"
0.2 (3/16)	0.2 (3/16)	0.2 (3/16)	THRUST ANGLE	0.2 (3/16)	0.2 (3/16)	0.2 (3/16)
29 125	179 178	179 178	ILLUSTRATION #	179 178	144	144
			NOTES			
05134	05135	05136		05137	05138	05139

BMW

X5 (E53) w/Air Springs (F&R) Reg. Susp	X5 (E53) w/Air Springs (F&R) Sport Susp.	X5 (E70) All X5 exc/X5 M	MODEL INFORMATION	X5 (E70) X5 M	X5 (F15) All X5 exc/X5 M	X5 (F85) X5 M
2000-2006	2000-2006	2007-2013	YEARS	2010-2013	2014-2017	2015-2017
p. 304	p. 304	p. 304	INSTRUCTIONS	p. 304	p. 304	p. 304
103 ft.lbs.	103 ft.lbs.	103 ft.lbs.	LUGNUT TORQUE	103 ft.lbs.	103 ft.lbs.	103 ft.lbs.
			FRONT			
Caster is not specified by the manufacturer.	Caster is not specified by the manufacturer.	Caster is not specified by the manufacturer.	CASTER (Degrees)	Caster is not specified by the manufacturer.	Caster is not specified by the manufacturer.	Caster is not specified by the manufacturer.
0.0 (0) | 0.5 (½)	0.0 (0) | 0.5 (½)	0.0 (0) | 0.5 (½)	CROSS | TOL.	0.0 (0) | 0.5 (½)	0.0 (0) | 0.5 (½)	0.0 (0) | 0.5 (½)
-0.2 ±0.42 -3/16 ± 7/16	-0.38 ±0.42 -3/8 ± 7/16	-0.33 ±0.42 -5/16 ± 7/16	CAMBER (Degrees)	-0.52 ±0.42 -½ ± 7/16	-0.33 ±0.42 -5/16 ± 7/16	-1.0 ±0.42 -1 ± 7/16
0.0 (0) | 0.5 (½)	0.0 (0) | 0.5 (½)	0.0 (0) | 0.5 (½)	CROSS | TOL.	0.0 (0) | 0.5 (½)	0.0 (0) | 0.5 (½)	0.0 (0) | 0.5 (½)
0.3° ±0.07° 5/16° ± 1/16°	0.3° ±0.07° 5/16° ± 1/16°	0.17° ±0.07° 5/32° ± 1/16°	FRONT TOE (Degrees)	0.23° ±0.07° 7/32° ± 1/16°	0.23° ±0.07° 7/32° ± 1/16°	0.23° ±0.07° 7/32° ± 1/16°
0.15" ±0.03" 5/32" ± 1/32"	0.15" ±0.03" 5/32" ± 1/32"	0.08" ±0.03" 3/32" ± 1/32"	FRONT TOE (Inches)	0.12" ±0.03" 1/8" ± 1/32"	0.12" ±0.03" 1/8" ± 1/32"	0.12" ±0.03" 1/8" ± 1/32"
			SAI			
20.0 / 18.4 ±0.5	20.0 / 18.4 ±0.5	20.0 / 17.9 ±0.5	TURNING ANGLE	37.3 / 31.0 ±0.5	37.4 / 31.0 ±0.5	38.0 / 31.0 ±0.5
119	119		ILLUSTRATION #			
			REAR			
-1.83 ±0.08 -1 13/16 ± 1/16	-1.83 ±0.08 -1 13/16 ± 1/16	-1.5 ±0.08 -1 ½ ± 1/16	CAMBER (Degrees)	-1.5 ±0.08 -1 ½ ± 1/16	-1.5 ±0.08 -1 ½ ± 1/16	-1.5 ±0.08 -1 ½ ± 1/16
0.3° ±0.07° 5/16° ± 1/16°	0.3° ±0.07° 5/16° ± 1/16°	0.17° ±0.07° 5/32° ± 1/16°	REAR TOE (Degrees)	0.17° ±0.07° 5/32° ± 1/16°	0.17° ±0.07° 5/32° ± 1/16°	0.17° ±0.07° 5/32° ± 1/16°
0.15" ±0.03" 5/32" ± 1/32"	0.15" ±0.03" 5/32" ± 1/32"	0.08" ±0.03" 3/32" ± 1/32"	REAR TOE (Inches)	0.08" ±0.03" 3/32" ± 1/32"	0.08" ±0.03" 3/32" ± 1/32"	0.08" ±0.03" 3/32" ± 1/32"
0.2 (3/16)	0.2 (3/16)	0.07 (1/16)	THRUST ANGLE	0.07 (1/16)	0.2 (3/16)	0.2 (3/16)
144	144	144	ILLUSTRATION #	144	144	
			NOTES			
05140	05141	05142		05143	05144	05145

BMW

X6 (E71)	X6 (E71)	X6 (E71)	MODEL	X6 (E72)	X6 (F16)	X6 (F86)
xDrive35i/50i	xDrive35i/50i	X6 M	INFORMATION	ActiveHybrid X6	All X6 exc/X6 M	X6 M
w/o Sport Pkg.	w/Sport Pkg.					
2008-2014	2008-2014	2010-2014	YEARS	2010-2011	2015-2017	2015-2017
p. 304	p. 304	p. 304	INSTRUCTIONS	p. 304	p. 304	p. 304
103 ft.lbs.	103 ft.lbs.	103 ft.lbs.	LUGNUT TORQUE	103 ft.lbs.	103 ft.lbs.	103 ft.lbs.
			FRONT			
Caster is not specified by the manufacturer.	Caster is not specified by the manufacturer.	Caster is not specified by the manufacturer.	CASTER (Degrees)	Caster is not specified by the manufacturer.	Caster is not specified by the manufacturer.	Caster is not specified by the manufacturer.
0.0 (0) | 0.5 (½)	0.0 (0) | 0.5 (½)	0.0 (0) | 0.5 (½)	CROSS | TOL.	0.0 (0) | 0.5 (½)	0.0 (0) | 0.5 (½)	0.0 (0) | 0.5 (½)
-0.52 ±0.42 -½ ± 7/16	-0.52 ±0.42 -½ ± 7/16	-0.52 ±0.42 -½ ± 7/16	CAMBER (Degrees)	-0.33 ±0.42 -5/16 ± 7/16	-0.33 ±0.42 -5/16 ± 7/16	-1.0 ±0.42 -1 ± 7/16
0.0 (0) | 0.5 (½)	0.0 (0) | 0.5 (½)	0.0 (0) | 0.5 (½)	CROSS | TOL.	0.0 (0) | 0.5 (½)	0.0 (0) | 0.5 (½)	0.0 (0) | 0.5 (½)
0.23° ±0.07° 7/32° ± 1/16°	0.23° ±0.07° 7/32° ± 1/16°	0.23° ±0.07° 7/32° ± 1/16°	FRONT TOE (Degrees)	0.23° ±0.07° 7/32° ± 1/16°	0.23° ±0.07° 7/32° ± 1/16°	0.23° ±0.07° 7/32° ± 1/16°
0.12" ±0.03" 1/8" ± 1/32"	0.12" ±0.03" 1/8" ± 1/32"	0.12" ±0.03" 1/8" ± 1/32"	FRONT TOE (Inches)	0.12" ±0.03" 1/8" ± 1/32"	0.12" ±0.03" 1/8" ± 1/32"	0.12" ±0.03" 1/8" ± 1/32"
			SAI			
20.0 / 17.7 ±0.5	20.0 / 17.9 ±0.5	37.3 / 31.0 ±0.5	TURNING ANGLE	20.0 / 17.9 ±0.5	37.5 / 31.0 ±0.5	38.0 / 31.0 ±0.5
			ILLUSTRATION #			
			REAR			
-1.5 ±0.08 -1½ ± 1/16	-1.5 ±0.08 -1½ ± 1/16	-1.5 ±0.08 -1½ ± 1/16	CAMBER (Degrees)	-1.5 ±0.08 -1½ ± 1/16	-1.5 ±0.08 -1½ ± 1/16	-1.5 ±0.08 -1½ ± 1/16
0.17° ±0.1° 5/32° ± 3/32°	0.17° ±0.1° 5/32° ± 3/32°	0.17° ±0.1° 5/32° ± 3/32°	REAR TOE (Degrees)	0.17° ±0.1° 5/32° ± 3/32°	0.17° ±0.07° 5/32° ± 1/16°	0.17° ±0.07° 5/32° ± 1/16°
0.08" ±0.05" 3/32" ± 1/16"	0.08" ±0.05" 3/32" ± 1/16"	0.08" ±0.05" 3/32" ± 1/16"	REAR TOE (Inches)	0.08" ±0.05" 3/32" ± 1/16"	0.08" ±0.03" 3/32" ± 1/32"	0.08" ±0.03" 3/32" ± 1/32"
0.07 (1/16)	0.07 (1/16)	0.2 (3/16)	THRUST ANGLE	0.07 (1/16)	0.2 (3/16)	0.2 (3/16)
144	144	144	ILLUSTRATION #	144	144	
			NOTES			
05146	05147	05148		05149	05150	05151

BMW

Z3 (E36)	Z3 (E36)	Z3 (E36)	MODEL	Z4 (E86)	Z4 (E86)	Z4 (E86)
Coupe	Roadster	Roadster	INFORMATION	Coupe	Coupe	M Coupe
	Reg. Suspension	Sport Suspension		Reg. Suspension	Sport Suspension	
1999-2002	1996-2002	1996-2002	YEARS	2006-2009	2006-2009	2007-2009
p. 304	p. 304	p. 304	INSTRUCTIONS	p. 304	p. 304	p. 304
89 ft.lbs.	89 ft.lbs.	89 ft.lbs.	LUGNUT TORQUE	89 ft.lbs.	89 ft.lbs.	89 ft.lbs.
			FRONT			
3.83 ±0.5 3 13/16 ± ½	3.68 ±0.5 3 11/16 ± ½	3.8 ±0.5 3 13/16 ± ½	CASTER (Degrees)	Caster is not specified by the manufacturer.	Caster is not specified by the manufacturer.	Caster is not specified by the manufacturer.
0.0 (0) | 0.5 (½)	0.0 (0) | 0.5 (½)	0.0 (0) | 0.5 (½)	CROSS | TOL.	0.0 (0) | 0.5 (½)	0.0 (0) | 0.5 (½)	0.0 (0) | 0.5 (½)
-0.85 ±0.5 -7/8 ± ½	-0.73 ±0.5 -¾ ± ½	-1.07 ±0.5 -1 1/16 ± ½	CAMBER (Degrees)	-0.58 ±0.42 -9/16 ± 7/16	-0.95 ±0.42 -15/16 ± 7/16	-1.0 ±0.42 -1 ± 7/16
0.0 (0) | 0.5 (½)	0.0 (0) | 0.5 (½)	0.0 (0) | 0.5 (½)	CROSS | TOL.	0.0 (0) | 0.5 (½)	0.0 (0) | 0.5 (½)	0.0 (0) | 0.5 (½)
0.3° ±0.13° 5/16° ± 1/8°	0.3° ±0.13° 5/16° ± 1/8°	0.3° ±0.13° 5/16° ± 1/8°	FRONT TOE (Degrees)	0.23° ±0.07° 7/32° ± 1/16°	0.23° ±0.07° 7/32° ± 1/16°	0.17° ±0.07° 5/32° ± 1/16°
0.15" ±0.07" 5/32" ± 1/16"	0.15" ±0.07" 5/32" ± 1/16"	0.15" ±0.07" 5/32" ± 1/16"	FRONT TOE (Inches)	0.12" ±0.03" 1/8" ± 1/32"	0.12" ±0.03" 1/8" ± 1/32"	0.08" ±0.03" 3/32" ± 1/32"
			SAI			
20.0 / 18.4 ±0.5	20.0 / 18.4 ±0.5	20.0 / 18.4 ±0.5	TURNING ANGLE	20.0 / 18.4 ±0.5	20.0 / 18.4 ±0.5	20.0 / 18.4 ±0.5
			ILLUSTRATION #	119	119	119
			REAR			
-2.52 ±0.5 -2½ ± ½	-2.33 ±0.5 -2 5/16 ± ½	-2.83 ±0.5 -2 13/16 ± ½	CAMBER (Degrees)	-2.25 ±0.08 -2¼ ± 1/16	-2.83 ±0.08 -2 13/16 ± 1/16	-1.83 ±0.08 -1 13/16 ± 1/16
0.63° ±0.22° 5/8° ± 7/32°	0.33° ±0.22° [1] 11/32° ± 7/32° [1]	0.5° ±0.22° [1] ½° ± 7/32° [1]	REAR TOE (Degrees)	0.27° ±0.07° 9/32° ± 1/16°	0.27° ±0.07° 9/32° ± 1/16°	0.47° ±0.07° 15/32° ± 1/16°
0.32" ±0.11" 5/16" ± 3/32"	0.17" ±0.11" [1] 5/32" ± 3/32" [1]	0.25" ±0.11" [1] ¼" ± 3/32" [1]	REAR TOE (Inches)	0.13" ±0.03" 1/8" ± 1/32"	0.13" ±0.03" 1/8" ± 1/32"	0.23" ±0.03" 7/32" ± 1/32"
0.25 (¼)	0.25 (¼)	0.25 (¼)	THRUST ANGLE	0.2 (3/16)	0.2 (3/16)	0.2 (3/16)
			ILLUSTRATION #	29 125	29 125	29 125
	[1] 1.8L/1.9L= 0.57° (0.29")	[1] 1.8L/1.9L= 0.73° (0.37")	NOTES			
05152	05153	05154		05155	05156	05157

ALIGNMENT SPECIFICATIONS

BMW

MODEL INFORMATION	Z4 (E85) Roadster Reg. Suspension	Z4 (E85) Roadster Sport Suspension	Z4 (E85) M Roadster	Z4 (E89) sDrive28i/30i, sDrive35i	Z4 (E89) sDrive35is	Z8 (E52) Roadster
YEARS	2003-2009	2003-2009	2007-2009	2009-2016	2011-2016	2000-2003
INSTRUCTIONS	p. 304	p. 304	p. 304	p. 304	p. 304	p. 304
LUGNUT TORQUE	89 ft.lbs.	89 ft.lbs.	89 ft.lbs.	89 ft.lbs.	89 ft.lbs.	89 ft.lbs.
FRONT						
CASTER (Degrees)	Caster is not specified by the manufacturer.	Caster is not specified by the manufacturer.	Caster is not specified by the manufacturer.	Caster is not specified by the manufacturer.	Caster is not specified by the manufacturer.	6.42 ±0.5 6 7/16 ± 1/2
CROSS | TOL.	0.0 (0) | 0.5 (1/2)	0.0 (0) | 0.5 (1/2)	0.0 (0) | 0.5 (1/2)	0.0 (0) | 0.5 (1/2)	0.0 (0) | 0.5 (1/2)	0.0 (0) | 0.5 (1/2)
CAMBER (Degrees)	-0.58 ±0.42 - 9/16 ± 7/16	-0.95 ±0.42 - 15/16 ± 7/16	-1.0 ±0.42 -1 ± 7/16	-0.38 ±0.42 [1] - 3/8 ± 7/16 [1]	-0.55 ±0.42 - 9/16 ± 7/16	-0.5 ±0.5 - 1/2 ± 1/2
CROSS | TOL.	0.0 (0) | 0.5 (1/2)	0.0 (0) | 0.5 (1/2)	0.0 (0) | 0.5 (1/2)	0.0 (0) | 0.5 (1/2)	0.0 (0) | 0.5 (1/2)	0.0 (0) | 0.67 (11/16)
FRONT TOE (Degrees)	0.23° ±0.07° 7/32° ± 1/16°	0.23° ±0.07° 7/32° ± 1/16°	0.07° ±0.07° 1/16° ± 1/16°	0.23° ±0.07° 7/32° ± 1/16°	0.23° ±0.07° 7/32° ± 1/16°	0.13° ±0.1° 1/8° ± 3/32°
FRONT TOE (Inches)	0.12" ±0.03" 1/8" ± 1/32"	0.12" ±0.03" 1/8" ± 1/32"	0.03" ±0.03" 1/32" ± 1/32"	0.12" ±0.03" 1/8" ± 1/32"	0.12" ±0.03" 1/8" ± 1/32"	0.07" ±0.05" 1/16" ± 1/16"
SAI						
TURNING ANGLE	20.0 / 18.4 ±0.5	20.0 / 18.4 ±0.5	20.0 / 18.4 ±0.5	20.0 / 17.9 ±0.5	20.0 / 17.9 ±0.5	20.0 / 18.1 ±0.5
ILLUSTRATION #	119	119	119	119	119	
REAR						
CAMBER (Degrees)	-2.25 ±0.08 -2 1/4 ± 1/16	-2.83 ±0.08 -2 13/16 ± 1/16	-1.83 ±0.08 -1 13/16 ± 1/16	-2.33 ±0.08 -2 5/16 ± 1/16	-2.33 ±0.08 -2 5/16 ± 1/16	-1.45 ±0.08 -1 7/16 ± 1/16
REAR TOE (Degrees)	0.37° ±0.07° 3/8° ± 1/16°	0.37° ±0.07° 3/8° ± 1/16°	0.37° ±0.07° 3/8° ± 1/16°	0.3° ±0.07° 5/16° ± 1/16°	0.3° ±0.07° 5/16° ± 1/16°	0.4° ±0.17° 13/32° ± 5/32°
REAR TOE (Inches)	0.18" ±0.03" 3/16" ± 1/32"	0.18" ±0.03" 3/16" ± 1/32"	0.18" ±0.03" 3/16" ± 1/32"	0.15" ±0.03" 5/32" ± 1/32"	0.15" ±0.03" 5/32" ± 1/32"	0.2" ±0.08" 3/16" ± 3/32"
THRUST ANGLE	0.2 (3/16)	0.2 (3/16)	0.2 (3/16)	0.2 (3/16)	0.2 (3/16)	0.2 (3/16)
ILLUSTRATION #	29 125	29 125	29 125	29 125	29 125	129 143
NOTES				[1] M-Susp. = -0.55° (- 9/16°)		
	05158	05159	05160	05161	05162	05163

Buick

MODEL INFORMATION	Cascada	Century	Century	Century	Century	Century
YEARS	2016-2017	1990-92	1993-96	1997-2000	2001-2003	2004
INSTRUCTIONS						
LUGNUT TORQUE	103 ft.lbs.	100 ft.lbs.	100 ft.lbs.	100 ft.lbs.	100 ft.lbs.	100 ft.lbs.
FRONT						
CASTER (Degrees)	6.6 ±0.75 6 5/8 ± 3/4	1.7 ±0.5 1 11/16 ± 1/2	3.0 ±0.75 3 ± 3/4	3.0 ±0.5 3 ± 1/2	3.0 ±0.5 3 ± 1/2	2.99 ±0.5 3 ± 1/2
CROSS | TOL.	0.0 (0) | 0.75 (3/4)	0.0 (0) | 1.0 (1)	0.0 (0) | 0.75 (3/4)	0.0 (0) | 0.75 (3/4)	0.0 (0) | 0.75 (3/4)	0.0 (0) | 0.75 (3/4)
CAMBER (Degrees)	-0.45 ±0.75 - 7/16 ± 3/4	0.0 ±0.5 0 ± 1/2	0.0 ±0.5 0 ± 1/2	-0.9 ±0.5 - 7/8 ± 1/2	-0.9 ±0.5 - 7/8 ± 1/2	-0.82 ±0.5 - 13/16 ± 1/2
CROSS | TOL.	0.0 (0) | 0.75 (3/4)	0.0 (0) | 0.5 (1/2)	0.0 (0) | 0.75 (3/4)	0.0 (0) | 0.75 (3/4)	0.0 (0) | 0.75 (3/4)	0.0 (0) | 0.75 (3/4)
FRONT TOE (Degrees)	0.2° ±0.2° 3/16° ± 3/16°	0.0° ±0.2° 0° ± 3/16°	0.0° ±0.2° 0° ± 3/16°	0.1° ±0.2° 3/32° ± 3/16°	0.1° ±0.2° 3/32° ± 3/16°	0.1° ±0.2° 3/32° ± 3/16°
FRONT TOE (Inches)	0.1" ±0.1" 3/32" ± 3/32"	0.0" ±0.1" 0" ± 3/32"	0.0" ±0.1" 0" ± 3/32"	0.05" ±0.1" 1/16" ± 3/32"	0.05" ±0.1" 1/16" ± 3/32"	0.05" ±0.1" 1/16" ± 3/32"
SAI						
TURNING ANGLE	20.0 / 17.9 ±0.75					
ILLUSTRATION #	192	102 ▲	102 ▲	25 ▲	25 ▲	25 ▲
REAR						
CAMBER (Degrees)	-1.4 ±0.5 -1 3/8 ± 1/2	0.0 ±0.25 0 ± 1/4	0.0 ±0.3 0 ± 5/16	-0.9 ±0.5 - 7/8 ± 1/2	-0.6 ±0.5 - 5/8 ± 1/2	-0.46 ±0.5 - 7/16 ± 1/2
REAR TOE (Degrees)	0.2° ±0.4° 3/16° ± 13/32°	0.0° ±0.2° 0° ± 3/16°	0.0° ±0.3° 0° ± 5/16°	0.1° ±0.2° 3/32° ± 3/16°	0.1° ±0.2° 3/32° ± 3/16°	0.1° ±0.2° 3/32° ± 3/16°
REAR TOE (Inches)	0.1" ±0.2" 3/32" ± 3/16"	0.0" ±0.1" 0" ± 3/32"	0.0" ±0.15" 0" ± 5/32"	0.05" ±0.1" 1/16" ± 3/32"	0.05" ±0.1" 1/16" ± 3/32"	0.05" ±0.1" 1/16" ± 3/32"
THRUST ANGLE	0.3 (5/16)	0.15 (5/32)	0.15 (5/32)	0.15 (5/32)	0.15 (5/32)	0.15 (5/32)
ILLUSTRATION #				108 10	108 10	108 10
NOTES						
	06002	06008	06009	06011	06012	06013

Buick

MODEL INFORMATION	Century	Enclave All Models	Encore	Envision exc/Hiper-Strut	Envision w/Hiper-Strut	Estate Wagon (LeSabre & Electra)
YEARS	2005	2008-2017	2013-2017	2017	2017	1990
INSTRUCTIONS						
LUGNUT TORQUE	100 ft.lbs.	140 ft.lbs.	103 ft.lbs.	103 ft.lbs.	103 ft.lbs.	100 ft.lbs.
FRONT						
CASTER (Degrees)	3.0 ± 0.75 3 ± 3/4	4.7 ± 0.8 4 11/16 ± 13/16	4.5 ± 0.75 4 1/2 ± 3/4	4.5 ± 0.8 4 1/2 ± 13/16	6.6 ± 0.8 6 5/8 ± 13/16	2.8 ± 0.5 2 13/16 ± 1/2
CROSS │ TOL.	0.0 (0) │ 0.75 (3/4)	0.0 (0) │ 0.8 (13/16)	0.0 (0) │ 0.75 (3/4)	0.0 (0) │ 0.9 (7/8)	0.0 (0) │ 0.9 (7/8)	0.0 (0) │ 0.5 (1/2)
CAMBER (Degrees)	-0.9 ± 0.75 -7/8 ± 3/4	-0.9 ± 0.8 -7/8 ± 13/16	-0.6 ± 0.75 -5/8 ± 3/4	-0.5 ± 0.7 -1/2 ± 11/16	-0.2 ± 0.7 -3/16 ± 11/16	0.8 ± 0.5 13/16 ± 1/2
CROSS │ TOL.	0.0 (0) │ 0.75 (3/4)	0.0 (0) │ 0.8 (13/16)	0.0 (0) │ 0.75 (3/4)	0.0 (0) │ 0.8 (13/16)	0.0 (0) │ 0.8 (13/16)	0.0 (0) │ 1.0 (1)
FRONT TOE (Degrees)	0.1° ± 0.2° 3/32° ± 3/16°	0.0° ± 0.2° 0° ± 3/16°	0.16° ± 0.2° 5/32° ± 3/16°	0.2° ± 0.2° 3/16° ± 3/16°	0.2° ± 0.2° 3/16° ± 3/16°	0.1° ± 0.2° 3/32° ± 3/16°
FRONT TOE (Inches)	0.05" ± 0.1" 1/16" ± 3/32"	0.0" ± 0.1" 0" ± 3/32"	0.08" ± 0.1" 3/32" ± 3/32"	0.1" ± 0.1" 3/32" ± 3/32"	0.1" ± 0.1" 3/32" ± 3/32"	0.05" ± 0.1" 1/16" ± 3/32"
SAI						
TURNING ANGLE						
ILLUSTRATION #	25 ▲	25		26	192	23
REAR						
CAMBER (Degrees)	-0.8 ± 0.5 -13/16 ± 1/2	-0.9 ± 0.8 -7/8 ± 13/16	-1.0 ± 0.5 -1 ± 1/2	-0.5 ± 0.8 -1/2 ± 13/16	-0.5 ± 0.8 -1/2 ± 13/16	
REAR TOE (Degrees)	0.1° ± 0.2° 3/32° ± 3/16°	0.1° ± 0.2° 3/32° ± 3/16°	0.2° ± 0.4° 3/16° ± 13/32°	0.1° ± 0.2° 3/32° ± 3/16°	0.1° ± 0.2° 3/32° ± 3/16°	
REAR TOE (Inches)	0.05" ± 0.1" 1/16" ± 3/32"	0.05" ± 0.1" 1/16" ± 3/32"	0.1" ± 0.2" 3/32" ± 3/16"	0.05" ± 0.1" 1/16" ± 3/32"	0.05" ± 0.1" 1/16" ± 3/32"	
THRUST ANGLE	0.15 (5/32)	0.3 (5/16)	0.3 (5/16)			
ILLUSTRATION #	108 10	107 143		177	177	
NOTES						
	06014	06020	06021	06022	06023	06026

Buick

MODEL INFORMATION	LaCrosse (USA) Allure (Canada)	LaCrosse (USA) Allure (Canada)	LaCrosse FWD exc/Hiper Strut	LaCrosse FWD w/Hiper-Strut	LaCrosse AWD	LaCrosse exc/Hiper-Strut exc/RPO QQI
YEARS	2005-2007	2008-2009	2010-2016	2010-2016	2010-2016	2017
INSTRUCTIONS						
LUGNUT TORQUE	100 ft.lbs.	111 ft.lbs.	111 ft.lbs.	111 ft.lbs.	111 ft.lbs.	103 ft.lbs.
FRONT						
CASTER (Degrees)	3.4 ± 0.75 3 3/8 ± 3/4	3.0 ± 0.75 3 ± 3/4	3.9 ± 0.75 3 7/8 ± 3/4	6.0 ± 0.75 6 ± 3/4	3.9 ± 0.75 3 7/8 ± 3/4	6.3 ± 0.8 6 5/16 ± 13/16
CROSS │ TOL.	0.0 (0) │ 0.75 (3/4)	0.0 (0) │ 0.75 (3/4)	0.0 (0) │ 0.75 (3/4)	0.0 (0) │ 0.75 (3/4)	0.0 (0) │ 0.75 (3/4)	0.0 (0) │ 0.9 (7/8)
CAMBER (Degrees)	-0.8 ± 0.75 -13/16 ± 3/4	-0.8 ± 0.75 [1] -13/16 ± 3/4 [1]	-0.4 ± 0.75 -3/8 ± 3/4	-0.4 ± 0.75 -3/8 ± 3/4	-0.4 ± 0.75 -3/8 ± 3/4	-0.3 ± 0.8 -5/16 ± 13/16
CROSS │ TOL.	0.0 (0) │ 0.75 (3/4)	0.0 (0) │ 0.75 (3/4)	0.0 (0) │ 0.75 (3/4)	0.0 (0) │ 0.75 (3/4)	0.0 (0) │ 0.75 (3/4)	0.0 (0) │ 0.8 (13/16)
FRONT TOE (Degrees)	0.1° ± 0.2° 3/32° ± 3/16°	0.1° ± 0.2° 3/32° ± 3/16°	0.2° ± 0.2° 3/16° ± 3/16°	0.2° ± 0.2° 3/16° ± 3/16°	0.2° ± 0.2° 3/16° ± 3/16°	0.2° ± 0.2° 3/16° ± 3/16°
FRONT TOE (Inches)	0.05" ± 0.1" 1/16" ± 3/32"	0.05" ± 0.1" 1/16" ± 3/32"	0.1" ± 0.1" 3/32" ± 3/32"	0.1" ± 0.1" 3/32" ± 3/32"	0.1" ± 0.1" 3/32" ± 3/32"	0.1" ± 0.1" 3/32" ± 3/32"
SAI						
TURNING ANGLE						
ILLUSTRATION #	25 ▲	25 ▲	25	192	25	202
REAR						
CAMBER (Degrees)	-0.8 ± 0.5 -13/16 ± 1/2	-0.8 ± 0.75 [2] -13/16 ± 3/4 [2]	-0.8 ± 0.75 -13/16 ± 3/4	-0.8 ± 0.75 -13/16 ± 3/4	-0.8 ± 0.75 -13/16 ± 3/4	-0.8 ± 0.8 -13/16 ± 13/16
REAR TOE (Degrees)	0.1° ± 0.2° 3/32° ± 3/16°	0.1° ± 0.2° 3/32° ± 3/16°	0.1° ± 0.2° 3/32° ± 3/16°	0.1° ± 0.2° 3/32° ± 3/16°	0.1° ± 0.2° 3/32° ± 3/16°	0.1° ± 0.2° 3/32° ± 3/16°
REAR TOE (Inches)	0.05" ± 0.1" 1/16" ± 3/32"	0.05" ± 0.1" 1/16" ± 3/32"	0.05" ± 0.1" 1/16" ± 3/32"	0.05" ± 0.1" 1/16" ± 3/32"	0.05" ± 0.1" 1/16" ± 3/32"	0.05" ± 0.1" 1/16" ± 3/32"
THRUST ANGLE	0.2 (3/16)	0.2 (3/16)	0.3 (5/16)	0.3 (5/16)	0.3 (5/16)	0.3 (5/16)
ILLUSTRATION #	108 10	108 10	177	177	188	177
NOTES		[1] Super = -1.0° (-1°) [2] Super = -1.3° (-1 5/16°)				
	06028	06029	06030	06031	06032	06033

Buick

LaCrosse exc/Hiper-Strut w/RPO QQI	LaCrosse w/Hiper-Strut	LeSabre	MODEL INFORMATION	LeSabre	Lucerne	Park Avenue Ultra Electra
2017	2017	1990-99	YEARS	2000-2005	2006-2011	1990-96
			INSTRUCTIONS			
103 ft.lbs.	103 ft.lbs.	100 ft.lbs.	LUGNUT TORQUE	100 ft.lbs.	100 ft.lbs.	100 ft.lbs.
			FRONT			
6.3 ± 0.8 6 5/16 ± 13/16	6.6 ± 0.8 6 5/8 ± 13/16	3.0 ± 0.5 3 ± 1/2	CASTER (Degrees)	5.0 ± 0.75[1] 5 ± 3/4[1]	5.4 ± 0.75 5 3/8 ± 3/4	3.0 ± 0.5 3 ± 1/2
0.0 (0) | 0.9 (7/8)	0.0 (0) | 0.9 (7/8)	0.0 (0) | 0.75 (3/4)	CROSS | TOL.	0.0 (0) | 0.75 (3/4)	0.0 (0) | 0.75 (3/4)	0.0 (0) | 0.75 (3/4)
-0.2 ± 0.8 -3/16 ± 13/16	-0.2 ± 0.8 -3/16 ± 13/16	0.2 ± 0.5 3/16 ± 1/2	CAMBER (Degrees)	-0.2 ± 0.75 -3/16 ± 3/4	0.0 ± 0.75 0 ± 3/4	0.2 ± 0.75 3/16 ± 3/4
0.0 (0) | 0.8 (13/16)	0.0 (0) | 0.8 (13/16)	0.0 (0) | 0.75 (3/4)	CROSS | TOL.	0.0 (0) | 0.75 (3/4)	0.0 (0) | 0.75 (3/4)	0.0 (0) | 0.75 (3/4)
0.2° ± 0.2° 3/16° ± 3/16°	0.2° ± 0.2° 3/16° ± 3/16°	0.0° ± 0.2° 0° ± 3/16°	FRONT TOE (Degrees)	0.2° ± 0.2° 3/16° ± 3/16°	0.2° ± 0.2° 3/16° ± 3/16°	0.0° ± 0.2° 0° ± 3/16°
0.1" ± 0.1" 3/32" ± 3/32"	0.1" ± 0.1" 3/32" ± 3/32"	0.0" ± 0.1" 0" ± 3/32"	FRONT TOE (Inches)	0.1" ± 0.1" 3/32" ± 3/32"	0.1" ± 0.1" 3/32" ± 3/32"	0.0" ± 0.1" 0" ± 3/32"
			SAI			
			TURNING ANGLE			
202	192	31 102 △	ILLUSTRATION #	31 102	31 102	31 102 △
			REAR			
-0.8 ± 0.8 -13/16 ± 13/16	-0.8 ± 0.8 -13/16 ± 13/16	-0.3 ± 0.5 -5/16 ± 1/2	CAMBER (Degrees)	-0.3 ± 0.5 -5/16 ± 1/2	0.05 ± 0.75 1/16 ± 3/4	-0.3 ± 0.75 -5/16 ± 3/4
0.1° ± 0.2° 3/32° ± 3/16°	0.1° ± 0.2° 3/32° ± 3/16°	0.1° ± 0.2° 3/32° ± 3/16°	REAR TOE (Degrees)	0.2° ± 0.2° 3/16° ± 3/16°	0.1° ± 0.2° 3/32° ± 3/16°	0.1° ± 0.2° 3/32° ± 3/16°
0.05" ± 0.1" 1/16" ± 3/32"	0.05" ± 0.1" 1/16" ± 3/32"	0.05" ± 0.1" 1/16" ± 3/32"	REAR TOE (Inches)	0.1" ± 0.1" 3/32" ± 3/32"	0.05" ± 0.1" 1/16" ± 3/32"	0.05" ± 0.1" 1/16" ± 3/32"
0.3 (5/16)	0.3 (5/16)	0.1 (3/32)	THRUST ANGLE	0.15 (5/32)	0.2 (3/16)	0.15 (5/32)
177	177	102 32	ILLUSTRATION #	06	06	102 32
			NOTES	[1] 2000 = 6.0° (6°)		
06034	06035	06042		06043	06044	06046

Buick

Park Avenue Ultra	Park Avenue Ultra	Rainier	MODEL INFORMATION	Rainier	Rainier	Reatta
1997-2000	2001-2005	2004	YEARS	2005	2006-2007	1990-91
			INSTRUCTIONS			
		103 ft.lbs.	LUGNUT TORQUE	103 ft.lbs.	103 ft.lbs.	
			FRONT			
6.0 ± 0.5 6 ± 1/2	5.0 ± 0.5 5 ± 1/2	L= 3.5, R= 4.0 ± 0.5 L= 3 1/2, R= 4 ± 1/2	CASTER (Degrees)	L= 4.0, R= 4.5 ± 0.6 L= 4, R= 4 1/2 ± 5/8	L= 4.15, R= 4.35 ± 0.6 L= 4 1/8, R= 4 3/8 ± 5/8	2.8 ± 1.0 2 13/16 ± 1
0.0 (0) | 0.5 (1/2)	0.0 (0) | 0.5 (1/2)	-0.5 (-1/2) | 0.3 (5/16)	CROSS | TOL.	-0.5 (-1/2) | 0.6 (5/8)	-0.2 (-3/16) | 0.6 (5/8)	0.0 (0) | 1.0 (1)
-0.2 ± 0.5 -3/16 ± 1/2	-0.2 ± 0.5 -3/16 ± 1/2	-0.5 ± 0.5 -1/2 ± 1/2	CAMBER (Degrees)	-0.5 ± 0.6 -1/2 ± 5/8	-0.5 ± 0.6 -1/2 ± 5/8	0.0 ± 0.8 0 ± 13/16
0.0 (0) | 0.5 (1/2)	0.0 (0) | 0.5 (1/2)	0.0 (0) | 0.3 (5/16)	CROSS | TOL.	0.0 (0) | 0.6 (5/8)	0.0 (0) | 0.6 (5/8)	0.0 (0) | 1.0 (1)
0.2° ± 0.2° 3/16° ± 3/16°	0.2° ± 0.2° 3/16° ± 3/16°	0.1° ± 0.2° 3/32° ± 3/16°	FRONT TOE (Degrees)	0.1° ± 0.2° 3/32° ± 3/16°	0.1° ± 0.2° 3/32° ± 3/16°	0.2° ± 0.2° 3/16° ± 3/16°
0.1" ± 0.1" 3/32" ± 3/32"	0.1" ± 0.1" 3/32" ± 3/32"	0.05" ± 0.1" 1/16" ± 3/32"	FRONT TOE (Inches)	0.05" ± 0.1" 1/16" ± 3/32"	0.05" ± 0.1" 1/16" ± 3/32"	0.1" ± 0.1" 3/32" ± 3/32"
			SAI			
			TURNING ANGLE			
31 102 △	31 102 △	133	ILLUSTRATION #	133	133	31 102 △
			REAR			
-0.3 ± 0.5 -5/16 ± 1/2	-0.3 ± 0.5 -5/16 ± 1/2		CAMBER (Degrees)			-0.1 ± 0.5 -1/8 ± 1/2
0.2° ± 0.2° 3/16° ± 3/16°	0.2° ± 0.2° 3/16° ± 3/16°		REAR TOE (Degrees)			0.2° ± 0.2° 3/16° ± 3/16°
0.1" ± 0.1" 3/32" ± 3/32"	0.1" ± 0.1" 3/32" ± 3/32"		REAR TOE (Inches)			0.1" ± 0.1" 3/32" ± 3/32"
0.15 (5/32)	0.15 (5/32)		THRUST ANGLE			0.15 (5/32)
06	06		ILLUSTRATION #			30
			NOTES			
06047	06048	06049		06050	06051	06052

Buick

Regal	Regal	Regal	MODEL INFORMATION	Regal	Regal exc/Performance	Regal Performance Hiper-Strut
1990-96	1997½-2000	2001-2003	YEARS	2004	2011-2017	2011-2017
			INSTRUCTIONS			
100 ft.lbs.	100 ft.lbs.	100 ft.lbs.	LUGNUT TORQUE	100 ft.lbs.	111 ft.lbs.	111 ft.lbs.
			FRONT			
2.0 ±0.5 2 ± 1/2	3.0 ±0.5 3 ± 1/2	3.0 ±0.5 3 ± 1/2	CASTER (Degrees)	3.07 ±0.5 3 1/16 ± 1/2	3.8 ±0.75 3 13/16 ± 3/4	6.0 ±0.75 6 ± 3/4
0.0 (0) | 0.75 (3/4)	0.0 (0) | 0.75 (3/4)	0.0 (0) | 0.75 (3/4)	CROSS | TOL.	0.0 (0) | 0.75 (3/4)	0.0 (0) | 0.75 (3/4)	0.0 (0) | 0.75 (3/4)
0.7 ±0.5 11/16 ± 1/2	-0.9 ±0.5 - 7/8 ± 1/2	-0.9 ±0.5 - 7/8 ± 1/2	CAMBER (Degrees)	-0.94 ±0.5 - 15/16 ± 1/2	-0.4 ±0.75 -3/8 ± 3/4	-0.4 ±0.75 -3/8 ± 3/4
0.0 (0) | 0.75 (3/4)	0.0 (0) | 0.75 (3/4)	0.0 (0) | 0.75 (3/4)	CROSS | TOL.	0.0 (0) | 0.75 (3/4)	0.0 (0) | 0.75 (3/4)	0.0 (0) | 0.75 (3/4)
0.0° ±0.2° 0° ± 3/16°	0.1° ±0.2° 3/32° ± 3/16°	0.1° ±0.2° 3/32° ± 3/16°	FRONT TOE (Degrees)	0.1° ±0.2° 3/32° ± 3/16°	0.2° ±0.2° 3/16° ± 3/16°	0.2° ±0.2° 3/16° ± 3/16°
0.0" ±0.1" 0" ± 3/32"	0.05" ±0.1" 1/16" ± 3/32"	0.05" ±0.1" 1/16" ± 3/32"	FRONT TOE (Inches)	0.05" ±0.1" 1/16" ± 3/32"	0.1" ±0.1" 3/32" ± 3/32"	0.1" ±0.1" 3/32" ± 3/32"
			SAI			
			TURNING ANGLE			
33 ▲	25 ▲	25 ▲	ILLUSTRATION #	25 ▲	25	192
			REAR			
0.1 ±0.5 1/8 ± 1/2	-0.9 ±0.5 - 7/8 ± 1/2	-0.6 ±0.5 -5/8 ± 1/2	CAMBER (Degrees)	-0.87 ±0.5 - 7/8 ± 1/2	-0.8 ±0.75 - 13/16 ± 3/4	-0.8 ±0.75 - 13/16 ± 3/4
-0.1° ±0.3° -3/32° ± 5/16°	0.1° ±0.2° 3/32° ± 3/16°	0.1° ±0.2° 3/32° ± 3/16°	REAR TOE (Degrees)	-0.1° ±0.2° -3/32° ± 3/16°	0.1° ±0.2° 3/32° ± 3/16°	0.1° ±0.2° 3/32° ± 3/16°
-0.05" ±0.15" -1/16" ± 5/32"	0.05" ±0.1" 1/16" ± 3/32"	0.05" ±0.1" 1/16" ± 3/32"	REAR TOE (Inches)	-0.05" ±0.1" -1/16" ± 3/32"	0.05" ±0.1" 1/16" ± 3/32"	0.05" ±0.1" 1/16" ± 3/32"
0.15 (5/32)	0.15 (5/32)	0.15 (5/32)	THRUST ANGLE	0.15 (5/32)	0.3 (5/16)	0.3 (5/16)
25 34	108 10	108 10	ILLUSTRATION #	108 10	188	188
			NOTES			
06059	06060	06061		06062	06063	06064

Buick

Rendezvous	Riviera	Riviera	MODEL INFORMATION	Riviera	Roadmaster	Roadmaster
2002-2007	1990-93	1995-96	YEARS	1997-99	1991-94	1995-96
			INSTRUCTIONS			
100 ft.lbs.	100 ft.lbs.	100 ft.lbs.	LUGNUT TORQUE	100 ft.lbs.	100 ft.lbs.	100 ft.lbs.
			FRONT			
2.4 ±0.75 2 3/8 ± 3/4	2.3 ±1.0 2 5/16 ±1	6.0 ±0.5 6 ± 1/2	CASTER (Degrees)	6.0 ±0.5 6 ± 1/2	3.5 ±1.0 3 1/2 ±1	L= 3.25, R= 3.75 ± 1.0 L= 3 1/4, R= 3 3/4 ± 1
0.0 (0) | 0.75 (3/4)	0.0 (0) | 1.0 (1)	0.0 (0) | 1.0 (1)	CROSS | TOL.	0.0 (0) | 1.0 (1)	0.0 (0) | 1.0 (1)	-0.5 (-1/2) | 1.0 (1)
-0.65 ±0.75 -5/8 ± 3/4	0.0 ±0.8 0 ± 13/16	0.2 ±1.0 3/16 ±1	CAMBER (Degrees)	-0.2 ±1.0 - 3/16 ±1	0.0 ±0.8 0 ± 13/16	0.0 ±1.0 0 ±1
0.0 (0) | 0.75 (3/4)	0.0 (0) | 1.0 (1)	0.0 (0) | 1.0 (1)	CROSS | TOL.	0.0 (0) | 1.0 (1)	0.0 (0) | 1.0 (1)	0.0 (0) | 1.0 (1)
0.0° ±0.2° 0° ± 3/16°	0.2° ±0.2° 3/16° ± 3/16°	0.0° ±0.2° 0° ± 3/16°	FRONT TOE (Degrees)	0.2° ±0.2° 3/16° ± 3/16°	0.16° ±0.2° 5/32° ± 3/16°	0.0° ±0.2° 0° ± 3/16°
0.0" ±0.1" 0" ± 3/32"	0.1" ±0.1" 3/32" ± 3/32"	0.0" ±0.1" 0" ± 3/32"	FRONT TOE (Inches)	0.1" ±0.1" 3/32" ± 3/32"	0.08" ±0.1" 3/32" ± 3/32"	0.0" ±0.1" 0" ± 3/32"
			SAI			
			TURNING ANGLE			
25	31 102 ▲	31 102	ILLUSTRATION #	31 102	23	23
			REAR			
-0.3 ±0.6 - 5/16 ± 5/8	0.0 ±0.5 0 ± 1/2	-0.3 ±0.5 - 5/16 ± 1/2	CAMBER (Degrees)	-0.3 ±0.5 - 5/16 ± 1/2		
0.0° ±0.2° 0° ± 3/16°	0.2° ±0.2° 3/16° ± 3/16°	0.2° ±0.2° 3/16° ± 3/16°	REAR TOE (Degrees)	0.2° ±0.2° 3/16° ± 3/16°		
0.0" ±0.1" 0" ± 3/32"	0.1" ±0.1" 3/32" ± 3/32"	0.1" ±0.1" 3/32" ± 3/32"	REAR TOE (Inches)	0.1" ±0.1" 3/32" ± 3/32"		
0.2 (3/16)	0.15 (5/32)	0.1 (3/32)	THRUST ANGLE	0.1 (3/32)	0.15 (5/32)	0.15 (5/32)
134 135	30	06	ILLUSTRATION #	06		
			NOTES			
06065	06073	06074		06075	06076	06077

Buick

MODEL INFORMATION	Skylark	Skylark	Skylark	Terraza	Terraza	Verano
YEARS	1990-92	1993-95	1996-98	2005-2006	2007	2012-2017
INSTRUCTIONS						
LUGNUT TORQUE	100 ft.lbs.	100 ft.lbs.	100 ft.lbs.	100 ft.lbs.	100 ft.lbs.	103 ft.lbs.
FRONT						
CASTER (Degrees)	1.7 ±1.0 1 11/16 ±1	1.45 ±1.0 1 7/16 ±1	1.45 ±1.0 1 7/16 ±1	2.7 ±0.75 2 11/16 ± 3/4	2.7 ±0.75 2 11/16 ± 3/4	4.35 ±0.75 4 3/8 ± 3/4
CROSS | TOL.	0.0 (0) | 0.7 (11/16)	0.0 (0) | 0.7 (11/16)	0.0 (0) | 0.7 (11/16)	0.0 (0) | 0.75 (3/4)	0.0 (0) | 0.75 (3/4)	0.0 (0) | 0.75 (3/4)
CAMBER (Degrees)	0.0 ±0.7 0 ± 11/16	0.0 ±0.7 0 ± 11/16	0.0 ±0.7 0 ± 11/16	-0.65 ±0.75 -5/8 ± 3/4	-0.65 ±0.75 -5/8 ± 3/4	-0.45 ±0.75 -7/16 ± 3/4
CROSS | TOL.	0.0 (0) | 1.0 (1)	0.0 (0) | 1.0 (1)	0.0 (0) | 1.0 (1)	0.0 (0) | 0.75 (3/4)	0.0 (0) | 0.75 (3/4)	0.0 (0) | 0.75 (3/4)
FRONT TOE (Degrees)	0.0° ±0.2° 0° ± 3/16°	0.0° ±0.2° 0° ± 3/16°	0.0° ±0.2° 0° ± 3/16°	0.0° ±0.2° 0° ± 3/16°	0.0° ±0.2° 0° ± 3/16°	0.1° ±0.2° 3/32° ± 3/16°
FRONT TOE (Inches)	0.0" ±0.1" 0" ± 3/32"	0.0" ±0.1" 0" ± 3/32"	0.0" ±0.1" 0" ± 3/32"	0.0" ±0.1" 0" ± 3/32"	0.0" ±0.1" 0" ± 3/32"	0.05" ±0.1" 1/16" ± 3/32"
SAI						
TURNING ANGLE						20.0 / 18.3
ILLUSTRATION #	25	25	25	25	25	25
REAR						
CAMBER (Degrees)	-0.25 ±0.5 -1/4 ± 1/2	-0.25 ±0.5 -1/4 ± 1/2	-0.4 ±0.75 -3/8 ± 3/4	-0.4 ±0.6 -3/8 ± 5/8	-1.0 ±0.5 -1 ± 1/2	-1.45 ±0.75 -1 7/16 ± 3/4
REAR TOE (Degrees)	0.25° ±0.2° 1/4° ± 3/16°	0.0° ±0.3° 0° ± 5/16°	0.2° ±0.3° 3/16° ± 5/16°	0.0° ±0.2° 0° ± 3/16°	0.0° ±0.3° 0° ± 5/16°	0.15° ±0.4° 5/32° ± 13/32°
REAR TOE (Inches)	0.13" ±0.1" 1/8" ± 3/32"	0.0" ±0.15" 0" ± 5/32"	0.1" ±0.15" 3/32" ± 5/32"	0.0" ±0.1" 0" ± 3/32"	0.0" ±0.15" 0" ± 5/32"	0.08" ±0.2" 1/16" ± 3/16"
THRUST ANGLE	0.35 (11/32)	0.25 (1/4)	0.25 (1/4)	0.3 (5/16)	0.3 (5/16)	0.3 (5/16)
ILLUSTRATION #				134 135		
NOTES						
	06092	06093	06094	06096	06097	06098

Cadillac

MODEL INFORMATION	Allante'	Allante'	ATS Sedan AWD	ATS Sedan RWD	ATS-V Sedan	Catera
YEARS	1990-92	1993	2013-2017	2013-2017	2016-2017	1997-2001
INSTRUCTIONS						
LUGNUT TORQUE	100 ft.lbs.	100 ft.lbs.	103 ft.lbs.	103 ft.lbs.	103 ft.lbs.	80 ft.lbs.
FRONT						
CASTER (Degrees)	2.3 ±1.0 2 5/16 ±1	2.5 ±1.0 2 1/2 ±1	6.6 ±0.6 6 5/8 ± 5/8	6.1 ±0.6 [1] 6 1/8 ± 5/8 [1]	6.5 ±0.6 6 1/2 ± 5/8	5.0 ±0.5 5 ± 1/2
CROSS | TOL.	0.0 (0) | 1.0 (1)	0.0 (0) | 1.0 (1)	0.0 (0) | 0.6 (5/8)	0.0 (0) | 0.6 (5/8)	0.0 (0) | 0.6 (5/8)	0.0 (0) | 0.5 (1/2)
CAMBER (Degrees)	0.0 ±0.8 0 ± 13/16	0.0 ±0.8 0 ± 13/16	-0.2 ±0.8 -3/16 ± 13/16	-0.2 ±0.8 [2] -3/16 ± 13/16 [2]	-1.3 ±0.8 -1 5/16 ± 13/16	-0.6 ±0.5 -5/8 ± 1/2
CROSS | TOL.	0.0 (0) | 1.0 (1)	0.0 (0) | 1.0 (1)	0.0 (0) | 0.6 (5/8)	0.0 (0) | 0.6 (5/8)	0.0 (0) | n/s	0.0 (0) | 0.5 (1/2)
FRONT TOE (Degrees)	0.2° ±0.2° 3/16° ± 3/16°	0.2° ±0.2° 3/16° ± 3/16°	0.2° ±0.2° 3/16° ± 3/16°	0.2° ±0.2° 3/16° ± 3/16°	0.2° ±0.2° 3/16° ± 3/16°	0.1° ±0.16° 3/32° ± 5/32°
FRONT TOE (Inches)	0.1" ±0.1" 3/32" ± 3/32"	0.1" ±0.1" 3/32" ± 3/32"	0.1" ±0.1" 3/32" ± 3/32"	0.1" ±0.1" 3/32" ± 3/32"	0.1" ±0.1" 3/32" ± 3/32"	0.05" ±0.08" 1/16" ± 3/32"
SAI						
TURNING ANGLE						
ILLUSTRATION #	31 102 △	31 102 △	192	192	192	102
REAR						
CAMBER (Degrees)	0.2 ±0.3 3/16 ± 5/16	0.0 ±0.5 0 ± 1/2	-0.9 ±0.8 -7/8 ± 13/16	-0.9 ±0.8 -7/8 ± 13/16	-1.0 ±0.5 -1 ± 1/2	-1.75 ±0.4 -1 3/4 ± 3/8
REAR TOE (Degrees)	0.2° ±0.2° 3/16° ± 3/16°	0.2° ±0.2° 3/16° ± 3/16°	0.3° ±0.2° 5/16° ± 3/16°	0.3° ±0.2° 5/16° ± 3/16°	0.2° ±0.2° 3/16° ± 3/16°	0.1° ±0.16° 3/32° ± 5/32°
REAR TOE (Inches)	0.1" ±0.1" 3/32" ± 3/32"	0.1" ±0.1" 3/32" ± 3/32"	0.15" ±0.1" 5/32" ± 3/32"	0.15" ±0.1" 5/32" ± 3/32"	0.1" ±0.1" 3/32" ± 3/32"	0.05" ±0.08" 1/16" ± 3/32"
THRUST ANGLE	0.15 (5/32)	0.15 (5/32)	0.3 (5/16)	0.3 (5/16)	0.3 (5/16)	0.15 (5/32)
ILLUSTRATION #	30	35	150	150	150	10
NOTES				[1] RPO FE3 = 6.3° (6 5/16°) [2] RPO FE3 = -0.5° (-1/2°)		
	07001	07002	07003	07004	07005	07010

Cadillac

CT6	CT6	CTS	MODEL	CTS	CTS	CTS
RWD	AWD		INFORMATION	Base Susp.	Sport Susp.	Perf. Susp.
				RPO FE1	RPO FE2	RPO FE5
2016-2017	2016-2017	2003-2005	YEARS	2006-2007	2006-2007	2006-2007
			INSTRUCTIONS			
111 ft.lbs.	111 ft.lbs.	100 ft.lbs.	LUGNUT TORQUE	100 ft.lbs.	100 ft.lbs.	100 ft.lbs.
			FRONT			
6.1 ±0.9 6⅛ ± ⅞	6.4 ±0.8 6⅜ ± 13/16	5.1 ±0.5 5⅛ ± ½	CASTER (Degrees)	5.1 ±0.6 5⅛ ± ⅝	5.1 ±0.6 5⅛ ± ⅝	5.3 ±0.6 5 5/16 ± ⅝
0.0 (0) | 1.0 (1)	0.0 (0) | 1.0 (1)	0.0 (0) | 0.5 (½)	CROSS | TOL.	0.0 (0) | 0.6 (⅝)	0.0 (0) | 0.6 (⅝)	0.0 (0) | 0.6 (⅝)
-0.5 ±0.8 -½ ± 13/16	-0.5 ±0.8 -½ ± 13/16	-0.5 ±0.5 -½ ± ½	CAMBER (Degrees)	-0.35 ±0.6 -⅜ ± ⅝	-0.35 ±0.6 -⅜ ± ⅝	-0.6 ±0.6 -⅝ ± ⅝
0.0 (0) | 1.0 (1)	0.0 (0) | 1.0 (1)	0.0 (0) | 0.5 (½)	CROSS | TOL.	0.0 (0) | 0.6 (⅝)	0.0 (0) | 0.6 (⅝)	0.0 (0) | 0.6 (⅝)
0.2° ±0.2° 3/16° ± 3/16°	0.2° ±0.2° 3/16° ± 3/16°	0.2° ±0.2° 3/16° ± 3/16°	FRONT TOE (Degrees)	0.2° ±0.2° 3/16° ± 3/16°	0.2° ±0.2° 3/16° ± 3/16°	0.2° ±0.2° 3/16° ± 3/16°
0.1" ±0.1" 3/32" ± 3/32"	0.1" ±0.1" 3/32" ± 3/32"	0.1" ±0.1" 3/32" ± 3/32"	FRONT TOE (Inches)	0.1" ±0.1" 3/32" ± 3/32"	0.1" ±0.1" 3/32" ± 3/32"	0.1" ±0.1" 3/32" ± 3/32"
			SAI			
			TURNING ANGLE			
201	201	149	ILLUSTRATION #	149	149	149
			REAR			
-1.4 ±0.8 -1⅜ ± 13/16	-1.4 ±0.8 -1⅜ ± 13/16	-1.0 ±0.4 [1] -1 ± ⅜ [1]	CAMBER (Degrees)	-1.0 ±0.5 -1 ± ½	-1.4 ±0.4 -1⅜ ± ⅜	-1.4 ±0.4 -1⅜ ± ⅜
0.1° ±0.2° 3/32° ± 3/16°	0.1° ±0.2° 3/32° ± 3/16°	0.2° ±0.2° 3/16° ± 3/16°	REAR TOE (Degrees)	0.2° ±0.2° 3/16° ± 3/16°	0.2° ±0.2° 3/16° ± 3/16°	0.2° ±0.2° 3/16° ± 3/16°
0.05" ±0.1" 1/16" ± 3/32"	0.05" ±0.1" 1/16" ± 3/32"	0.1" ±0.1" 3/32" ± 3/32"	REAR TOE (Inches)	0.1" ±0.1" 3/32" ± 3/32"	0.1" ±0.1" 3/32" ± 3/32"	0.1" ±0.1" 3/32" ± 3/32"
0.3 (5/16)	0.3 (5/16)	0.2 (3/16)	THRUST ANGLE	0.2 (3/16)	0.2 (3/16)	0.2 (3/16)
06	06	150	ILLUSTRATION #	150	150	150
		[1] FE2, FE3 (Sport, Perf.) = -1.4° (-1⅜°)	NOTES			
07013	07014	07015		07016	07017	07018

Cadillac

CTS	CTS	CTS	MODEL	CTS	CTS	CTS
RWD	AWD	All exc/2014 Sedan	INFORMATION	All exc/2014 Sedan	Sedan RWD	Sedan AWD
		RWD		AWD		
2008-2009	2008-2009	2010-2014	YEARS	2010-2014	2014-2017	2014-2017
			INSTRUCTIONS			
140 ft.lbs.	140 ft.lbs.	140 ft.lbs.	LUGNUT TORQUE	140 ft.lbs.	103 ft.lbs.	103 ft.lbs.
			FRONT			
5.6 ±0.6 5⅝ ± ⅝	5.4 ±0.6 5⅜ ± ⅝	5.6 ±0.6 5⅝ ± ⅝	CASTER (Degrees)	5.4 ±0.6 5⅜ ± ⅝	6.1 ±0.8 [1] 6⅛ ± 13/16 [1]	6.6 ±0.8 6⅝ ± 13/16
0.0 (0) | 0.6 (⅝)	0.0 (0) | 0.6 (⅝)	0.0 (0) | 0.6 (⅝)	CROSS | TOL.	0.0 (0) | 0.6 (⅝)	0.0 (0) | 0.8 (13/16)	0.0 (0) | 0.8 (13/16)
-0.5 ±0.6 -½ ± ⅝	-0.5 ±0.6 -½ ± ⅝	-0.5 ±0.6 -½ ± ⅝	CAMBER (Degrees)	-0.5 ±0.6 -½ ± ⅝	-0.2 ±0.8 [2] -3/16 ± 13/16 [2]	-0.2 ±0.8 -3/16 ± 13/16
0.0 (0) | 0.6 (⅝)	0.0 (0) | 0.6 (⅝)	0.0 (0) | 0.6 (⅝)	CROSS | TOL.	0.0 (0) | 0.6 (⅝)	0.0 (0) | 0.8 (13/16)	0.0 (0) | 0.8 (13/16)
0.2° ±0.2° 3/16° ± 3/16°	0.2° ±0.2° 3/16° ± 3/16°	0.2° ±0.2° 3/16° ± 3/16°	FRONT TOE (Degrees)	0.2° ±0.2° 3/16° ± 3/16°	0.2° ±0.2° 3/16° ± 3/16°	0.2° ±0.2° 3/16° ± 3/16°
0.1" ±0.1" 3/32" ± 3/32"	0.1" ±0.1" 3/32" ± 3/32"	0.1" ±0.1" 3/32" ± 3/32"	FRONT TOE (Inches)	0.1" ±0.1" 3/32" ± 3/32"	0.1" ±0.1" 3/32" ± 3/32"	0.1" ±0.1" 3/32" ± 3/32"
			SAI			
			TURNING ANGLE			
149	149	149	ILLUSTRATION #	149	200	200
			REAR			
-1.0 ±0.5 [1] -1 ± ½ [1]	-1.0 ±0.5 [1] -1 ± ½ [1]	-1.0 ±0.5 -1 ± ½	CAMBER (Degrees)	-1.0 ±0.5 -1 ± ½	-0.9 ±0.8 -⅞ ± 13/16	-0.9 ±0.8 -⅞ ± 13/16
0.2° ±0.2° 3/16° ± 3/16°	0.2° ±0.2° 3/16° ± 3/16°	0.2° ±0.2° 3/16° ± 3/16°	REAR TOE (Degrees)	0.2° ±0.2° 3/16° ± 3/16°	0.3° ±0.2° 5/16° ± 3/16°	0.3° ±0.2° 5/16° ± 3/16°
0.1" ±0.1" 3/32" ± 3/32"	0.1" ±0.1" 3/32" ± 3/32"	0.1" ±0.1" 3/32" ± 3/32"	REAR TOE (Inches)	0.1" ±0.1" 3/32" ± 3/32"	0.15" ±0.1" 5/32" ± 3/32"	0.15" ±0.1" 5/32" ± 3/32"
0.2 (3/16)	0.2 (3/16)	0.2 (3/16)	THRUST ANGLE	0.2 (3/16)	0.3 (5/16)	0.3 (5/16)
150	150	150	ILLUSTRATION #	150	162	162
[1] FE2, FE3 (Sport, Perf.) = -1.4° (-1⅜°)	[1] FE2, FE3 (Sport, Perf.) = -1.4° (-1⅜°)		NOTES		[1] RPO FE3 = 6.3° (6 5/16°) [2] RPO FE3 = -0.5° (-½°)	
07019	07020	07021		07022	07023	07024

Cadillac

CTS-V	CTS-V	CTS-V	MODEL INFORMATION	CTS-V	DeVille Fleetwood Sixty Special	DeVille Concours
2004-2005	2006-2007	2009-2015	YEARS	2016-2017	1990-93	1994-97
			INSTRUCTIONS			
100 ft.lbs.	100 ft.lbs.	140 ft.lbs.[1]	LUGNUT TORQUE	111 ft.lbs.	100 ft.lbs.	100 ft.lbs.
			FRONT			
5.1 ± 0.5 5⅛ ± ½	5.3 ± 0.6 5 5⁄16 ± ⅝	5.8 ± 0.6 5 13⁄16 ± ⅝	CASTER (Degrees)	6.5 ± 0.8 6½ ± 13⁄16	3.0 ± 0.5 3 ± ½	2.5 ± 1.0 2½ ± 1
0.0 (0) | 0.6 (⅝)	0.0 (0) | 0.6 (⅝)	0.0 (0) | 0.6 (⅝)	CROSS | TOL.	0.0 (0) | n/s	0.0 (0) | 0.75 (¾)	0.0 (0) | 1.0 (1)
-0.5 ± 0.5 -½ ± ½	-0.6 ± 0.6 -⅝ ± ⅝	-0.9 ± 0.6 -⅞ ± ⅝	CAMBER (Degrees)	-1.3 ± 0.8 -1 5⁄16 ± 13⁄16	L= -0.5, R= 0.5 ± 0.5 L= -½, R= ½ ± ½	0.0 ± 0.5 0 ± ½
0.0 (0) | 0.5 (½)	0.0 (0) | 0.6 (⅝)	0.0 (0) | 0.6 (⅝)	CROSS | TOL.	0.0 (0) | n/s	-1.0 (-1) | 0.75 (¾)	0.0 (0) | 0.6 (⅝)
0.2° ± 0.2° 3⁄16° ± 3⁄16°	0.2° ± 0.2° 3⁄16° ± 3⁄16°	0.2° ± 0.2° 3⁄16° ± 3⁄16°	FRONT TOE (Degrees)	0.2° ± 0.2° 3⁄16° ± 3⁄16°	0.0° ± 0.2° 0° ± 3⁄16°	0.2° ± 0.2° 3⁄16° ± 3⁄16°
0.1" ± 0.1" 3⁄32" ± 3⁄32"	0.1" ± 0.1" 3⁄32" ± 3⁄32"	0.1" ± 0.1" 3⁄32" ± 3⁄32"	FRONT TOE (Inches)	0.1" ± 0.1" 3⁄32" ± 3⁄32"	0.0" ± 0.1" 0" ± 3⁄32"	0.1" ± 0.1" 3⁄32" ± 3⁄32"
			SAI			
			TURNING ANGLE			
149	149	149	ILLUSTRATION #	200	31 102 △	31 102 △
			REAR			
-1.4 ± 0.4 -1⅜ ± ⅜	-1.4 ± 0.4 -1⅜ ± ⅜	-1.4 ± 0.5 -1⅜ ± ½	CAMBER (Degrees)	-1.0 ± 0.5 -1 ± ½	-0.3 ± 0.5 -5⁄16 ± ½	0.0 ± 0.5 0 ± ½
0.2° ± 0.2° 3⁄16° ± 3⁄16°	0.2° ± 0.2° 3⁄16° ± 3⁄16°	0.2° ± 0.2° 3⁄16° ± 3⁄16°	REAR TOE (Degrees)	0.2° ± 0.2° 3⁄16° ± 3⁄16°	0.1° ± 0.2° 3⁄32° ± 3⁄16°	0.1° ± 0.2° 3⁄32° ± 3⁄16°
0.1" ± 0.1" 3⁄32" ± 3⁄32"	0.1" ± 0.1" 3⁄32" ± 3⁄32"	0.1" ± 0.1" 3⁄32" ± 3⁄32"	REAR TOE (Inches)	0.1" ± 0.1" 3⁄32" ± 3⁄32"	0.05" ± 0.1" 1⁄16" ± 3⁄32"	0.05" ± 0.1" 1⁄16" ± 3⁄32"
0.2 (3⁄16)	0.15 (5⁄32)	0.2 (3⁄16)	THRUST ANGLE	0.3 (5⁄16)	0.1 (3⁄32)	0.15 (5⁄32)
150	150	150	ILLUSTRATION #	162	102 32	35
		[1] 2014-15 = 122 ft.lbs.	NOTES			
07025	07026	07027		07028	07034	07035

Cadillac

DeVille Concours	Deville	DTS	MODEL INFORMATION	Eldorado	Eldorado	Eldorado
1998-99	2000-2005	2006-2011	YEARS	1990-91	1992	1993-97
			INSTRUCTIONS			
100 ft.lbs.	80 ft.lbs.	103 ft.lbs.	LUGNUT TORQUE	100 ft.lbs.	100 ft.lbs.	100 ft.lbs.
			FRONT			
3.0 ± 1.0 3 ± 1	5.0 ± 0.5[1] 5 ± ½[1]	5.4 ± 0.75 5⅜ ± ¾	CASTER (Degrees)	2.3 ± 1.0 2 5⁄16 ± 1	2.3 ± 1.0 2 5⁄16 ± 1	2.5 ± 1.0 2½ ± 1
0.0 (0) | 1.0 (1)	0.0 (0) | 0.5 (½)	0.0 (0) | 0.75 (¾)	CROSS | TOL.	0.0 (0) | 1.0 (1)	0.0 (0) | 1.0 (1)	0.0 (0) | 1.0 (1)
0.0 ± 0.5 0 ± ½	-0.2 ± 0.5 -3⁄16 ± ½	0.0 ± 0.75 0 ± ¾	CAMBER (Degrees)	0.0 ± 0.8 0 ± 13⁄16	L= -0.5, R= 0.5 ± 0.5 L= -½, R= ½ ± ½	0.0 ± 0.5 0 ± ½
0.0 (0) | 0.6 (⅝)	0.0 (0) | 0.5 (½)	0.0 (0) | 0.75 (¾)	CROSS | TOL.	0.0 (0) | 1.0 (1)	-1.0 (-1) | 0.75 (¾)	0.0 (0) | 0.6 (⅝)
0.0° ± 0.2° 0° ± 3⁄16°	0.2° ± 0.2° 3⁄16° ± 3⁄16°	0.2° ± 0.2° 3⁄16° ± 3⁄16°	FRONT TOE (Degrees)	0.2° ± 0.2° 3⁄16° ± 3⁄16°	0.2° ± 0.2° 3⁄16° ± 3⁄16°	0.2° ± 0.2° 3⁄16° ± 3⁄16°
0.0" ± 0.1" 0" ± 3⁄32"	0.1" ± 0.1" 3⁄32" ± 3⁄32"	0.1" ± 0.1" 3⁄32" ± 3⁄32"	FRONT TOE (Inches)	0.1" ± 0.1" 3⁄32" ± 3⁄32"	0.1" ± 0.1" 3⁄32" ± 3⁄32"	0.1" ± 0.1" 3⁄32" ± 3⁄32"
			SAI			
			TURNING ANGLE			
31 102 △	31 102	31 102	ILLUSTRATION #	31 102 △	31 102 △	31 102 △
			REAR			
0.0 ± 0.5 0 ± ½	-0.3 ± 0.5 -5⁄16 ± ½	0.05 ± 0.75 1⁄16 ± ¾	CAMBER (Degrees)	0.0 ± 0.5 0 ± ½	0.0 ± 0.5 0 ± ½	0.0 ± 0.5 0 ± ½
0.2° ± 0.2° 3⁄16° ± 3⁄16°	0.2° ± 0.2° 3⁄16° ± 3⁄16°	0.1° ± 0.2° 3⁄32° ± 3⁄16°	REAR TOE (Degrees)	0.2° ± 0.2° 3⁄16° ± 3⁄16°	0.2° ± 0.2° 3⁄16° ± 3⁄16°	0.2° ± 0.2° 3⁄16° ± 3⁄16°
0.1" ± 0.1" 3⁄32" ± 3⁄32"	0.1" ± 0.1" 3⁄32" ± 3⁄32"	0.05" ± 0.1" 1⁄16" ± 3⁄32"	REAR TOE (Inches)	0.1" ± 0.1" 3⁄32" ± 3⁄32"	0.1" ± 0.1" 3⁄32" ± 3⁄32"	0.1" ± 0.1" 3⁄32" ± 3⁄32"
0.15 (5⁄32)	0.15 (5⁄32)	0.2 (3⁄16)	THRUST ANGLE	0.1 (3⁄32)	0.1 (3⁄32)	0.15 (5⁄32)
35	06	06	ILLUSTRATION #	30	30	35
	[1] 2000 = 6.0 (6°)		NOTES			
07036	07037	07038		07047	07048	07049

Cadillac

MODEL INFORMATION	Eldorado	ELR	Escalade	Escalade	Escalade/ESV/EXT 2WD	Escalade/ESV/EXT 2WD
YEARS	1998-2002	2014-2016	1999	2000	2002-2003	2004-2005
INSTRUCTIONS			p. 305	p. 305	p. 305	p. 305
LUGNUT TORQUE	100 ft.lbs.	100 ft.lbs.	125 ft.lbs.	125 ft.lbs.	140 ft.lbs.	140 ft.lbs.
FRONT						
CASTER (Degrees)	2.5 ± 1.0 2½ ± 1	6.3 ± 0.8 6 5/16 ± 13/16	3.0 ± 1.0 3 ± 1	L= 2.6, R= 3.5 ± 0.5 L= 2⅝, R= 3½ ± ½	L= 4.1, R= 4.7 ± 1.0 L= 4⅛, R= 4 11/16 ± 1	L= 3.9, R= 4.7 ± 1.0 L= 3⅞, R= 4 11/16 ± 1
CROSS | TOL.	0.0 (0) | 1.0 (1)	0.0 (0) | 0.8 (13/16)	0.0 (0) | 0.5 (½)	-0.9 (-⅞) | 0.5 (½)	-0.6 (-⅝) | 0.5 (½)	-0.8 (-13/16) | 0.75 (¾)
CAMBER (Degrees)	0.0 ± 0.5 0 ± ½	-0.9 ± 0.8 -⅞ ± 13/16	0.65 ± 0.5 ⅝ ± ½	L= 0.3, R= -0.2 ± 0.5 L= 5/16, R= -3/16 ± ½	0.5 ± 0.5 ½ ± ½	0.5 ± 0.6 ½ ± ⅝
CROSS | TOL.	0.0 (0) | 0.6 (⅝)	0.0 (0) | 0.8 (13/16)	0.0 (0) | 0.5 (½)	0.5 (½) | 0.5 (½)	0.0 (0) | 0.5 (½)	0.0 (0) | 0.6 (⅝)
FRONT TOE (Degrees)	0.2° ± 0.2° 3/16° ± 3/16°	0.2° ± 0.2° 3/16° ± 3/16°	0.24° ± 0.2° ¼° ± 3/16°	0.24° ± 0.2° ¼° ± 3/16°	0.1° ± 0.2° 3/32° ± 3/16°	0.1° ± 0.2° 3/32° ± 3/16°
FRONT TOE (Inches)	0.1" ± 0.1" 3/32" ± 3/32"	0.1" ± 0.1" 3/32" ± 3/32"	0.12" ± 0.1" ⅛" ± 3/32"	0.12" ± 0.1" ⅛" ± 3/32"	0.05" ± 0.1" 1/16" ± 3/32"	0.05" ± 0.1" 1/16" ± 3/32"
SAI						
TURNING ANGLE						
ILLUSTRATION #	31 102 △	192	43	43	126	126
REAR						
CAMBER (Degrees)	-0.3 ± 0.5 [1] -5/16 ± ½ [1]	-1.4 ± 0.8 -1⅜ ± 13/16				
REAR TOE (Degrees)	0.2° ± 0.2° 3/16° ± 3/16°	0.15° ± 0.4° 5/32° ± 13/32°				
REAR TOE (Inches)	0.1" ± 0.1" 3/32" ± 3/32"	0.08" ± 0.2" 1/16" ± 3/16"				
THRUST ANGLE	0.15 (5/32)	0.4 (13/32)				
ILLUSTRATION #	35					
NOTES	[1] 1998-99 = 0.0° (0°)					
	07050	07051	07052	07053	07054	07055

Cadillac

MODEL INFORMATION	Escalade/ESV/EXT AWD with 17" tires	Escalade/ESV/EXT AWD with 20" tires	Escalade/ESV/EXT 2WD	Escalade/ESV/EXT AWD with 17" tires	Escalade/ESV/EXT AWD with 20" tires	Escalade/ESV/EXT 2WD & AWD with 18" tires
YEARS	2002-2005	2004-2005	2006	2006	2006	2007
INSTRUCTIONS	p. 305	p. 305				
LUGNUT TORQUE	140 ft.lbs.	140 ft.lbs.	140 ft.lbs.	140 ft.lbs.	140 ft.lbs.	140 ft.lbs.
FRONT						
CASTER (Degrees)	L= 3.8, R= 4.5 ± 1.0 L= 3 13/16, R= 4½ ± 1	L= 4.1, R= 5.1 ± 1.0 L= 4⅛, R= 5⅛ ± 1	L= 2.9, R= 3.7 ± 1.0 [1] L= 2⅞, R= 3 11/16 ± 1 [1]	L= 2.8, R= 3.5 ± 1.0 [1] L= 2 13/16, R= 3½ ± 1 [1]	L= 3.45, R= 3.75 ± 1.0 [1] L= 3 7/16, R= 3¾ ± 1 [1]	L= 3.45, R= 3.35 ± 1.0 [1] L= 3 7/16, R= 3⅜ ± 1 [1]
CROSS | TOL.	-0.7 (-11/16) | 0.75 (¾)	-1.0 (-1) | 0.75 (¾)	-0.8 (-13/16) | 0.75 (¾)	-0.7 (-11/16) | 0.75 (¾)	-0.3 (-5/16) | 0.75 (¾)	0.1 (⅛) | 0.75 (¾)
CAMBER (Degrees)	0.25 ± 0.6 ¼ ± ⅝	-0.1 ± 0.6 -⅛ ± ⅝	0.5 ± 0.6 ½ ± ⅝	0.25 ± 0.6 ¼ ± ⅝	-0.1 ± 0.6 -⅛ ± ⅝	-0.1 ± 0.5 -⅛ ± ½
CROSS | TOL.	0.0 (0) | 0.6 (⅝)	0.0 (0) | 0.6 (⅝)	0.0 (0) | 0.6 (⅝)	0.0 (0) | 0.6 (⅝)	0.0 (0) | 0.6 (⅝)	0.0 (0) | 0.5 (½)
FRONT TOE (Degrees)	0.1° ± 0.2° 3/32° ± 3/16°	0.1° ± 0.2° 3/32° ± 3/16°	0.1° ± 0.2° 3/32° ± 3/16°	0.1° ± 0.2° 3/32° ± 3/16°	0.1° ± 0.2° 3/32° ± 3/16°	0.1° ± 0.2° 3/32° ± 3/16°
FRONT TOE (Inches)	0.05" ± 0.1" 1/16" ± 3/32"	0.05" ± 0.1" 1/16" ± 3/32"	0.05" ± 0.1" 1/16" ± 3/32"	0.05" ± 0.1" 1/16" ± 3/32"	0.05" ± 0.1" 1/16" ± 3/32"	0.05" ± 0.1" 1/16" ± 3/32"
SAI						
TURNING ANGLE						
ILLUSTRATION #	126	126	126	126	126	126
REAR						
CAMBER (Degrees)						
REAR TOE (Degrees)						
REAR TOE (Inches)						
THRUST ANGLE						
ILLUSTRATION #						
NOTES			[1] Caster relative to ground	[1] Caster relative to ground	[1] Caster relative to ground	[1] Caster relative to ground
	07056	07057	07058	07059	07060	07061

ALIGNMENT SPECIFICATIONS

ALIGNMENT SPECIFICATIONS

Cadillac

MODEL INFORMATION	Escalade/ESV/EXT 2WD & AWD with 22" tires	Escalade/ESV/EXT 2WD & AWD with 18" tires	Escalade/ESV/EXT 2WD & AWD with 22" tires	Escalade/ESV 2WD & 4WD w/20" B'stones (QSS)	Escalade/ESV 2WD & 4WD w/20" Cont. (QSS)	Escalade/ESV 2WD & 4WD w/22" tires (QST)
YEARS	2007	2008-2014	2008-2014	2015-2017	2015-2017	2015-2017
INSTRUCTIONS						
LUGNUT TORQUE	140 ft.lbs.	140 ft.lbs.	140 ft.lbs.	140 ft.lbs.	140 ft.lbs.	140 ft.lbs.
FRONT						
CASTER (Degrees)	L= 3.3, R= 3.5 ± 1.0 [1] L= 3 5/16, R= 3 1/2 ± 1 [1]	L= 3.0, R= 2.9 ± 1.0 [1] L= 3, R= 2 7/8 ± 1 [1]	L= 2.9, R= 3.1 ± 1.0 [1] L= 2 7/8, R= 3 1/8 ± 1 [1]	3.1 ± 1.0 3 1/8 ± 1	L= 3.2, R= 2.9 ± 1.0 L= 3 3/16, R= 2 7/8 ± 1	3.1 ± 1.0 3 1/8 ± 1
CROSS | TOL.	-0.2 (-3/16) | 0.75 (3/4)	0.1 (1/8) | 0.75 (3/4)	-0.2 (-3/16) | 0.75 (3/4)	0.0 (0) | 0.75 (3/4)	0.3 (5/16) | 0.75 (3/4)	0.0 (0) | 0.75 (3/4)
CAMBER (Degrees)	-0.1 ± 0.5 -1/8 ± 1/2	-0.1 ± 0.5 -1/8 ± 1/2	-0.1 ± 0.5 -1/8 ± 1/2	-0.1 ± 0.8 -1/8 ± 13/16	-0.1 ± 0.8 -1/8 ± 13/16	-0.1 ± 0.8 -1/8 ± 13/16
CROSS | TOL.	0.0 (0) | 0.5 (1/2)	0.0 (0) | 0.5 (1/2)	0.0 (0) | 0.5 (1/2)	0.0 (0) | 0.8 (13/16)	0.0 (0) | 0.8 (13/16)	0.0 (0) | 0.8 (13/16)
FRONT TOE (Degrees)	0.1° ± 0.2° 3/32° ± 3/16°	0.1° ± 0.2° [2] 3/32° ± 3/16° [2]	0.1° ± 0.2° [2] 3/32° ± 3/16° [2]	0.1° ± 0.2° 3/32° ± 3/16°	0.1° ± 0.2° [1] 3/32° ± 3/16° [1]	0.1° ± 0.2° 3/32° ± 3/16°
FRONT TOE (Inches)	0.05" ± 0.1" 1/16" ± 3/32"	0.05" ± 0.1" [2] 1/16" ± 3/32" [2]	0.05" ± 0.1" [2] 1/16" ± 3/32" [2]	0.05" ± 0.1" 1/16" ± 3/32"	0.05" ± 0.1" [1] 1/16" ± 3/32" [1]	0.05" ± 0.1" 1/16" ± 3/32"
SAI						
TURNING ANGLE						
ILLUSTRATION #	126	126	126	126	126	126
REAR						
CAMBER (Degrees)						
REAR TOE (Degrees)						
REAR TOE (Inches)						
THRUST ANGLE		0.4 (13/32)	0.4 (13/32)			
ILLUSTRATION #						
NOTES	[1] Caster relative to ground	[1] Caster relative to ground [2] Hybrids = 0.2° (0.1")	[1] Caster relative to ground [2] Hybrids = 0.2° (0.1")		[1] 2015 = 0.2° (0.1")	
	07062	07063	07064	07065	07066	07067

Cadillac

MODEL INFORMATION	Fleetwood (FWD)	Fleetwood (RWD) Brougham	Fleetwood (RWD) Brougham	Seville	Seville	Seville Seville STS
YEARS	1990-93	1990-92	1993-96	1990-91	1992	1993-97
INSTRUCTIONS						
LUGNUT TORQUE	100 ft.lbs.	100 ft.lbs.	100 ft.lbs.	100 ft.lbs.	100 ft.lbs.	100 ft.lbs.
FRONT						
CASTER (Degrees)	3.0 ± 0.5 3 ± 1/2	3.0 ± 1.0 3 ± 1	3.5 ± 1.0 3 1/2 ± 1	2.3 ± 1.0 2 5/16 ± 1	2.3 ± 1.0 2 5/16 ± 1	2.5 ± 1.0 2 1/2 ± 1
CROSS | TOL.	0.0 (0) | 0.75 (3/4)	0.0 (0) | 1.0 (1)	0.0 (0) | 1.0 (1)	0.0 (0) | 1.0 (1)	0.0 (0) | 1.0 (1)	0.0 (0) | 1.0 (1)
CAMBER (Degrees)	L= -0.5, R= 0.5 ± 0.5 L= -1/2, R= 1/2 ± 1/2	0.0 ± 0.5 0 ± 1/2	0.0 ± 1.0 0 ± 1	0.0 ± 0.8 0 ± 13/16	L= -0.5, R= 0.5 ± 0.5 L= -1/2, R= 1/2 ± 1/2	0.0 ± 0.5 0 ± 1/2
CROSS | TOL.	-1.0 (-1) | 0.75 (3/4)	0.0 (0) | 1.0 (1)	0.0 (0) | 1.0 (1)	0.0 (0) | 1.0 (1)	-1.0 (-1) | 0.75 (3/4)	0.0 (0) | 0.6 (5/8)
FRONT TOE (Degrees)	0.0° ± 0.2° 0° ± 3/16°	0.0° ± 0.2° 0° ± 3/16°	0.16° ± 0.2° 5/32° ± 3/16°	0.2° ± 0.2° 3/16° ± 3/16°	0.2° ± 0.2° 3/16° ± 3/16°	0.2° ± 0.2° 3/16° ± 3/16°
FRONT TOE (Inches)	0.0" ± 0.1" 0" ± 3/32"	0.0" ± 0.1" 0" ± 3/32"	0.08" ± 0.1" 3/32" ± 3/32"	0.1" ± 0.1" 3/32" ± 3/32"	0.1" ± 0.1" 3/32" ± 3/32"	0.1" ± 0.1" 3/32" ± 3/32"
SAI						
TURNING ANGLE						
ILLUSTRATION #	31 102 ▲	23	23	31 102 ▲	31 102 ▲	31 102 ▲
REAR						
CAMBER (Degrees)	-0.3 ± 0.5 -5/16 ± 1/2			0.0 ± 0.5 0 ± 1/2	0.0 ± 0.5 0 ± 1/2	0.0 ± 0.5 0 ± 1/2
REAR TOE (Degrees)	0.1° ± 0.2° 3/32° ± 3/16°			0.2° ± 0.2° 3/16° ± 3/16°	0.2° ± 0.2° 3/16° ± 3/16°	0.2° ± 0.2° 3/16° ± 3/16°
REAR TOE (Inches)	0.05" ± 0.1" 1/16" ± 3/32"			0.1" ± 0.1" 3/32" ± 3/32"	0.1" ± 0.1" 3/32" ± 3/32"	0.1" ± 0.1" 3/32" ± 3/32"
THRUST ANGLE	0.1 (3/32)			0.1 (3/32)	0.1 (3/32)	0.15 (5/32)
ILLUSTRATION #	102 32			30	30	35
NOTES						
	07069	07074	07075	07079	07080	07081

Cadillac

Seville Seville STS	Seville SLS Seville STS	SRX	MODEL INFORMATION	SRX	STS RWD	STS AWD
1998-2000	2001-2004	2004-2009	YEARS	2010-2016	2005-2011	2005-2011
			INSTRUCTIONS			
		100 ft.lbs.	LUGNUT TORQUE	111 ft.lbs.[1]	100 ft.lbs.	100 ft.lbs.
			FRONT			
6.0 ±0.5 6 ± ½	5.0 ±0.5 5 ± ½	4.1 ±0.6 4⅛ ± ⅝	CASTER (Degrees)	4.5 ±0.75 4½ ± ¾	5.85 ±0.6 5⅞ ± ⅝	5.5 ±0.6 5½ ± ⅝
0.0 (0) | 0.5 (½)	0.0 (0) | 0.5 (½)	0.0 (0) | 0.6 (⅝)	CROSS | TOL.	0.0 (0) | 0.75 (¾)	0.0 (0) | 0.6 (⅝)	0.0 (0) | 0.6 (⅝)
-0.2 ±0.5 -3⁄16 ± ½	-0.2 ±0.5 -3⁄16 ± ½	-0.5 ±0.6 -½ ± ⅝	CAMBER (Degrees)	-0.6 ±0.75 -⅝ ± ¾	-0.5 ±0.6 -½ ± ⅝	-0.5 ±0.6 -½ ± ⅝
0.0 (0) | 0.5 (½)	0.0 (0) | 0.5 (½)	0.0 (0) | 0.6 (⅝)	CROSS | TOL.	0.0 (0) | 0.75 (¾)	0.0 (0) | 0.6 (⅝)	0.0 (0) | 0.6 (⅝)
0.2° ±0.2° 3⁄16° ± 3⁄16°	0.2° ±0.2° 3⁄16° ± 3⁄16°	0.2° ±0.2° 3⁄16° ± 3⁄16°	FRONT TOE (Degrees)	0.2° ±0.2° 3⁄16° ± 3⁄16°	0.2° ±0.2° 3⁄16° ± 3⁄16°	0.2° ±0.2° 3⁄16° ± 3⁄16°
0.1" ±0.1" 3⁄32" ± 3⁄32"	0.1" ±0.1" 3⁄32" ± 3⁄32"	0.1" ±0.1" 3⁄32" ± 3⁄32"	FRONT TOE (Inches)	0.1" ±0.1" 3⁄32" ± 3⁄32"	0.1" ±0.1" 3⁄32" ± 3⁄32"	0.1" ±0.1" 3⁄32" ± 3⁄32"
			SAI			
			TURNING ANGLE			
31 102	31 102	149	ILLUSTRATION #		149	149
			REAR			
-0.3 ±0.5 -5⁄16 ± ½	-0.3 ±0.5 -5⁄16 ± ½	-1.0 ±0.5 -1 ± ½	CAMBER (Degrees)	-1.0 ±0.75 -1 ± ¾	-0.9 ±0.5 -⅞ ± ½	-0.9 ±0.5 -⅞ ± ½
0.2° ±0.2° 3⁄16° ± 3⁄16°	0.2° ±0.2° 3⁄16° ± 3⁄16°	0.2° ±0.2° 3⁄16° ± 3⁄16°	REAR TOE (Degrees)	0.2° ±0.2° 3⁄16° ± 3⁄16°	0.2° ±0.2° 3⁄16° ± 3⁄16°	0.2° ±0.2° 3⁄16° ± 3⁄16°
0.1" ±0.1" 3⁄32" ± 3⁄32"	0.1" ±0.1" 3⁄32" ± 3⁄32"	0.1" ±0.1" 3⁄32" ± 3⁄32"	REAR TOE (Inches)	0.1" ±0.1" 3⁄32" ± 3⁄32"	0.1" ±0.1" 3⁄32" ± 3⁄32"	0.1" ±0.1" 3⁄32" ± 3⁄32"
0.15 (5⁄32)	0.15 (5⁄32)	0.2 (3⁄16)	THRUST ANGLE	0.3 (5⁄16)	0.2 (3⁄16)	0.2 (3⁄16)
06	06	150	ILLUSTRATION #	130 135	150	150
			NOTES	[1] Silver finish lugs = 140 ft.lbs.		
07082	07083	07085		07086	07087	07088

Cadillac — Chevrolet

XLR	XT5	XTS	MODEL INFORMATION	Aveo	Beretta	Beretta
2004-2009	2017	2013-2017	YEARS	2004-2011	1990-92	1993-96
			INSTRUCTIONS			
100 ft.lbs.	140 ft.lbs.	103 ft.lbs.	LUGNUT TORQUE	88 ft.lbs.	100 ft.lbs.	100 ft.lbs.
			FRONT			
7.4 ±0.6 [1] 7⅜ ± ⅝ [1]	4.3 ±0.8 4 5⁄16 ± 13⁄16	6.0 ±0.8 [1] 6 ± 13⁄16 [1]	CASTER (Degrees)	2.5 ±0.75 2½ ± ¾	1.15 ±0.75 1⅛ ± ¾	1.2 ±1.0 1 3⁄16 ± 1
0.0 (0) | 0.6 (⅝)	0.0 (0) | 0.8 (13⁄16)	0.0 (0) | 0.8 (13⁄16)	CROSS | TOL.	0.0 (0) | 1.0 (1)	0.0 (0) | 0.7 (11⁄16)	0.0 (0) | 0.7 (11⁄16)
-0.5 ±0.6 -½ ± ⅝	-0.6 ±0.8 -⅝ ± 13⁄16	-0.4 ±0.8 -⅜ ± 13⁄16	CAMBER (Degrees)	-0.4 ±0.75 -⅜ ± ¾	0.6 ±0.6 ⅝ ± ⅝	0.0 ±0.7 0 ± 11⁄16
0.0 (0) | 0.6 (⅝)	0.0 (0) | 0.8 (13⁄16)	0.0 (0) | 0.75 (13⁄16)	CROSS | TOL.	0.0 (0) | 0.5 (½)	0.0 (0) | 1.0 (1)	0.0 (0) | 0.7 (11⁄16)
0.1° ±0.2° 3⁄32° ± 3⁄16°	0.2° ±0.2° 3⁄16° ± 3⁄16°	0.2° ±0.2° 3⁄16° ± 3⁄16°	FRONT TOE (Degrees)	0.07° ±0.17° 1⁄16° ± 5⁄32°	0.0° ±0.2° 0° ± 3⁄16°	0.0° ±0.2° 0° ± 3⁄16°
0.05" ±0.1" 1⁄16" ± 3⁄32"	0.1" ±0.1" 3⁄32" ± 3⁄32"	0.1" ±0.1" 3⁄32" ± 3⁄32"	FRONT TOE (Inches)	0.04" ±0.08" 1⁄32" ± 3⁄32"	0.0" ±0.1" 0" ± 3⁄32"	0.0" ±0.1" 0" ± 3⁄32"
			SAI			
			TURNING ANGLE			
12	202	192	ILLUSTRATION #		25	25
			REAR			
-0.45 ±0.5 -7⁄16 ± ½	-0.6 ±0.8 -⅝ ± 13⁄16	-0.9 ±0.8 -⅞ ± 13⁄16	CAMBER (Degrees)	-1.5 ±0.5 [1] -1½ ± ½ [1]	-0.25 ±0.6 -¼ ± ⅝	-0.25 ±0.6 [1] -¼ ± ⅝ [1]
-0.05° ±0.2° -1⁄16° ± 3⁄16°	0.2° ±0.2° 3⁄16° ± 3⁄16°	0.1° ±0.2° 3⁄32° ± 3⁄16°	REAR TOE (Degrees)	0.5° ±0.66° ½° ± 21⁄32°	0.25° ±0.25° ¼° ± ¼°	0.0° ±0.3° [2] 0° ± 5⁄16° [2]
-0.03" ±0.1" -1⁄32" ± 3⁄32"	0.1" ±0.1" 3⁄32" ± 3⁄32"	0.05" ±0.1" 1⁄16" ± 3⁄32"	REAR TOE (Inches)	0.25" ±0.33" ¼" ± 11⁄32"	0.13" ±0.13" ⅛" ± ⅛"	0.0" ±0.15" [2] 0" ± 5⁄32" [2]
0.1 (3⁄32)	0.2 (3⁄16)	0.3 (5⁄16)	THRUST ANGLE	0.25 (¼)	0.35 (11⁄32)	0.35 (11⁄32)
134 135	177	150	ILLUSTRATION #			
[1] 2004-06 = 7.7° (7 11⁄16°)		[1] non-Hiperstrut mods. = 3.9° (3⅞°)	NOTES	[1] 2004-07 = 0.25° (¼°)		[1] 1996 = -0.4° (-⅜°) [2] 1996 = 0.2° (0.1")
07089	07090	07091		08012	08013	08014

ALIGNMENT SPECIFICATIONS

Chevrolet

Beretta GTZ	Beretta GTZ Beretta Z26	Bolt EV	MODEL INFORMATION	Camaro	Camaro	Camaro with Goodyear Eagle RSA Tires
1990-92	1993-96	2017	YEARS	1990-92	1993-2002	1995-96
			INSTRUCTIONS			
100 ft.lbs.	100 ft.lbs.	103 ft.lbs.	LUGNUT TORQUE	100 ft.lbs.	100 ft.lbs.	100 ft.lbs.
			FRONT			
1.15 ± 0.75 1 1/8 ± 3/4	1.2 ± 1.0 1 3/16 ± 1	5.5 ± 0.75 5 1/2 ± 3/4	CASTER (Degrees)	4.8 ± 0.5 4 13/16 ± 1/2	4.8 ± 0.5 4 13/16 ± 1/2	L= 4.0, R= 4.4 ± 0.5 L= 4, R= 4 3/8 ± 1/2
0.0 (0) | 0.7 (11/16)	0.0 (0) | 0.7 (11/16)	0.0 (0) | 0.75 (3/4)	CROSS | TOL.	0.0 (0) | 0.5 (1/2)	0.0 (0) | 0.7 (11/16)	-0.4 (-3/8) | 0.7 (11/16)
-0.2 ± 0.6 -3/16 ± 5/8	-0.15 ± 0.7 -1/8 ± 11/16	-0.55 ± 0.75 -9/16 ± 3/4	CAMBER (Degrees)	0.3 ± 0.5 5/16 ± 1/2	0.4 ± 0.5 3/8 ± 1/2	L= 0.2, R= -0.2 ± 0.5 L= 3/16, R= -3/16 ± 1/2
0.0 (0) | 1.0 (1)	0.0 (0) | 1.0 (1)	0.0 (0) | 0.75 (3/4)	CROSS | TOL.	0.0 (0) | 0.7 (11/16)	0.0 (0) | 0.7 (11/16)	0.4 (3/8) | 0.7 (11/16)
0.0° ± 0.2° 0° ± 3/16°	0.0° ± 0.2° 0° ± 3/16°	0.2° ± 0.2° 3/16° ± 3/16°	FRONT TOE (Degrees)	0.0° ± 0.2° 0° ± 3/16°	0.0° ± 0.2° 0° ± 3/16°	0.0° ± 0.2° 0° ± 3/16°
0.0" ± 0.1" 0" ± 3/32"	0.0" ± 0.1" 0" ± 3/32"	0.1" ± 0.1" 3/32" ± 3/32"	FRONT TOE (Inches)	0.0" ± 0.1" 0" ± 3/32"	0.0" ± 0.1" 0" ± 3/32"	0.0" ± 0.1" 0" ± 3/32"
			SAI			
			TURNING ANGLE			
25	25		ILLUSTRATION #	36	37	37
			REAR			
-0.25 ± 0.6 -1/4 ± 5/8	-0.25 ± 0.6 [1] -1/4 ± 5/8 [1]	-1.4 ± 0.5 -1 3/8 ± 1/2	CAMBER (Degrees)			
0.25° ± 0.25° 1/4° ± 1/4°	0.0° ± 0.3° [2] 0° ± 5/16° [2]	0.2° ± 0.35° 3/16° ± 11/32°	REAR TOE (Degrees)			
0.13" ± 0.13" 1/8" ± 1/8"	0.0" ± 0.15" [2] 0" ± 5/32" [2]	0.1" ± 0.18" 3/32" ± 3/16"	REAR TOE (Inches)			
0.35 (11/32)	0.35 (11/32)	0.3 (5/16)	THRUST ANGLE	0.25 (1/4)	0.25 (1/4)	0.25 (1/4)
			ILLUSTRATION #			
	[1] 1996 = -0.4° (-3/8°) [2] 1996 = 0.2° (0.1")		NOTES			
08015	08016	08017		08026	08027	08028

Chevrolet

Camaro LS/LT	Camaro SS	Camaro ZL1	MODEL INFORMATION	Camaro Z/28	Camaro RPO FE2	Camaro RPO FE3
2010-2015	2010-2015	2012-2015	YEARS	2014-2015	2016-2017	2016-2017
			INSTRUCTIONS			
			LUGNUT TORQUE			
			FRONT			
5.9 ± 0.75 5 7/8 ± 3/4	5.9 ± 0.75 5 7/8 ± 3/4	5.6 ± 0.75 5 5/8 ± 3/4	CASTER (Degrees)	5.6 ± 0.75 5 5/8 ± 3/4	7.2 ± 0.8 7 3/16 ± 13/16	7.2 ± 0.8 7 3/16 ± 13/16
0.0 (0) | 0.75 (3/4)	0.0 (0) | 0.75 (3/4)	0.0 (0) | 0.75 (3/4)	CROSS | TOL.	0.0 (0) | 0.75 (3/4)	0.0 (0) | 0.8 (13/16)	0.0 (0) | 0.8 (13/16)
-0.2 ± 0.75 -3/16 ± 3/4	-0.8 ± 0.75 -13/16 ± 3/4	-1.0 ± 0.75 -1 ± 3/4	CAMBER (Degrees)	-1.0 ± 0.75 -1 ± 3/4	-0.5 ± 0.8 -1/2 ± 13/16	-0.75 ± 0.8 -3/4 ± 13/16
0.0 (0) | 0.75 (3/4)	0.0 (0) | 0.75 (3/4)	0.0 (0) | 0.75 (3/4)	CROSS | TOL.	0.0 (0) | 0.75 (3/4)	0.0 (0) | 0.8 (13/16)	0.0 (0) | 0.8 (13/16)
0.2° ± 0.2° 3/16° ± 3/16°	0.2° ± 0.2° 3/16° ± 3/16°	0.2° ± 0.2° 3/16° ± 3/16°	FRONT TOE (Degrees)	0.1° ± 0.2° 3/32° ± 3/16°	0.2° ± 0.2° 3/16° ± 3/16°	0.2° ± 0.2° 3/16° ± 3/16°
0.1" ± 0.1" 3/32" ± 3/32"	0.1" ± 0.1" 3/32" ± 3/32"	0.1" ± 0.1" 3/32" ± 3/32"	FRONT TOE (Inches)	0.05" ± 0.1" 1/16" ± 3/32"	0.1" ± 0.1" 3/32" ± 3/32"	0.1" ± 0.1" 3/32" ± 3/32"
			SAI			
			TURNING ANGLE			
25	25	25	ILLUSTRATION #	25	102	102
			REAR			
-0.75 ± 0.75 [1] -3/4 ± 3/4 [1]	-0.4 ± 0.75 -3/8 ± 3/4	-0.75 ± 0.75 [1] -3/4 ± 3/4 [1]	CAMBER (Degrees)	-1.0 ± 0.75 -1 ± 3/4	-1.0 ± 0.8 -1 ± 13/16	-0.9 ± 0.8 -7/8 ± 13/16
0.2° ± 0.2° 3/16° ± 3/16°	0.2° ± 0.2° 3/16° ± 3/16°	0.2° ± 0.2° 3/16° ± 3/16°	REAR TOE (Degrees)	0.1° ± 0.2° 3/32° ± 3/16°	0.1° ± 0.2° 3/32° ± 3/16°	0.2° ± 0.2° 3/16° ± 3/16°
0.1" ± 0.1" 3/32" ± 3/32"	0.1" ± 0.1" 3/32" ± 3/32"	0.1" ± 0.1" 3/32" ± 3/32"	REAR TOE (Inches)	0.05" ± 0.1" 1/16" ± 3/32"	0.05" ± 0.1" 1/16" ± 3/32"	0.1" ± 0.1" 3/32" ± 3/32"
0.3 (5/16)	0.3 (5/16)	0.3 (5/16)	THRUST ANGLE	0.3 (5/16)	0.3 (5/16)	0.3 (5/16)
177	177	177	ILLUSTRATION #	177	194	194
[1] w/RS Pkg. = -0.4° (-3/8°)		[1] 2012 = -0.5° (-1/2°)	NOTES			
08029	08030	08031		08032	08033	08034

Chevrolet

Caprice	Caprice	Caprice	MODEL	Caprice PPV	Cavalier	Cavalier
Impala	Impala SS	Impala SS	INFORMATION	All Police Models		
1990	1991-94	1995-96	YEARS	2011-2017	1990-92	1993-94
			INSTRUCTIONS			
80 ft.lbs.		100 ft.lbs.	LUGNUT TORQUE	103 ft.lbs.	100 ft.lbs.	100 ft.lbs.
			FRONT			
2.8 ±0.5 2 13⁄16 ± ½	3.5 ±1.0 3½ ±1	L= 3.25, R= 3.75 ± 1.0 L= 3¼, R= 3¾ ±1	CASTER (Degrees)	6.0 ±1.25 6 ±1¼	1.7 ±1.0 1 11⁄16 ±1	1.3 ±1.0 1 5⁄16 ±1
0.0 (0) | 0.5 (½)	0.0 (0) | 1.0 (1)	-0.5 (-½) | 1.0 (1)	CROSS | TOL.	0.0 (0) | 1.0 (1)	0.0 (0) | 0.7 (11⁄16)	0.0 (0) | 0.7 (11⁄16)
0.8 ±0.5 13⁄16 ± ½	0.0 ±0.8 0 ± 13⁄16	0.0 ±1.0 0 ±1	CAMBER (Degrees)	-0.1 ±0.33 -⅛ ± 5⁄16	0.0 ±0.7 0 ± 11⁄16	-0.15 ±0.7 -⅛ ± 11⁄16
0.0 (0) | 0.5 (½)	0.0 (0) | 1.0 (1)	0.0 (0) | 1.0 (1)	CROSS | TOL.	0.0 (0) | 0.33 (5⁄16)	0.0 (0) | 1.0 (1)	0.0 (0) | 0.7 (11⁄16)
0.1° ±0.2° 3⁄32° ± 3⁄16°	0.16° ±0.2° 5⁄32° ± 3⁄16°	0.16° ±0.2° 5⁄32° ± 3⁄16°	FRONT TOE (Degrees)	0.17° ±0.17° 5⁄32° ± 5⁄32°	0.0° ±0.2° 0° ± 3⁄16°	0.0° ±0.2° 0° ± 3⁄16°
0.05" ±0.1" 1⁄16" ± 3⁄32"	0.08" ±0.1" 3⁄32" ± 3⁄32"	0.08" ±0.1" 3⁄32" ± 3⁄32"	FRONT TOE (Inches)	0.09" ±0.09" 3⁄32" ± 3⁄32"	0.0" ±0.1" 0" ± 3⁄32"	0.0" ±0.1" 0" ± 3⁄32"
			SAI			
			TURNING ANGLE			
23	23	23	ILLUSTRATION #	102	25	25
			REAR			
			CAMBER (Degrees)	-0.5 ±0.45 -½ ± 7⁄16	-0.25 ±0.5 -¼ ± ½	-0.25 ±0.5 -¼ ± ½
			REAR TOE (Degrees)	0.2° ±0.2° 3⁄16° ± 3⁄16°	0.25° ±0.16° ¼° ± 5⁄32°	0.25° ±0.35° ¼° ± 11⁄32°
			REAR TOE (Inches)	0.1" ±0.1" 3⁄32" ± 3⁄32"	0.13" ±0.08" ⅛" ± 3⁄32"	0.13" ±0.18" ⅛" ± 3⁄16"
	0.15 (5⁄32)		THRUST ANGLE	0.2 (3⁄16)	0.31 (5⁄16)	
			ILLUSTRATION #	177		
			NOTES			
08037	08038	08039		08040	08043	08044

Chevrolet

Cavalier	Cavalier	Cavalier	MODEL	Celebrity	Cobalt exc/SS	Cobalt SS
	USA Assembly	Mexico Assembly	INFORMATION	Celebrity Wagon		
		VIN Code S (11th)				
1995-99	2000-2005	2000-2005	YEARS	1990	2005-2010	2005-2010
			INSTRUCTIONS			
100 ft.lbs.	100 ft.lbs.	100 ft.lbs.	LUGNUT TORQUE	100 ft.lbs.	100 ft.lbs.	100 ft.lbs.
			FRONT			
4.3 ±1.0 4 5⁄16 ±1	4.1 ±0.75 4⅛ ± ¾	4.1 ±0.75 4⅛ ± ¾	CASTER (Degrees)	1.7 ±1.0 1 11⁄16 ±1	3.0 ±0.75 [1] 3 ± ¾ [1]	3.65 ±0.75 3⅝ ± ¾
0.0 (0) | 0.7 (11⁄16)	0.0 (0) | 0.75 (¾)	0.0 (0) | 0.75 (¾)	CROSS | TOL.	0.0 (0) | 1.0 (1)	0.0 (0) | 0.75 (¾)	0.0 (0) | 0.75 (¾)
-0.2 ±1.0 -3⁄16 ±1	-0.2 ±0.75 -3⁄16 ± ¾	-0.35 ±0.75 -⅜ ± ¾	CAMBER (Degrees)	0.0 ±0.5 0 ± ½	-1.0 ±0.75 -1 ± ¾	-1.25 ±0.75 [1] -1¼ ± ¾ [1]
0.0 (0) | 0.7 (11⁄16)	0.0 (0) | 0.75 (¾)	0.0 (0) | 0.75 (¾)	CROSS | TOL.	0.0 (0) | 0.5 (½)	0.0 (0) | 0.75 (¾)	0.0 (0) | 0.75 (¾)
0.1° ±0.2° [1] 3⁄32° ± 3⁄16° [1]	0.1° ±0.2° 3⁄32° ± 3⁄16°	0.0° ±0.2° 0° ± 3⁄16°	FRONT TOE (Degrees)	0.0° ±0.2° 0° ± 3⁄16°	0.2° ±0.2° 3⁄16° ± 3⁄16°	0.2° ±0.2° 3⁄16° ± 3⁄16°
0.05" ±0.1" [1] 1⁄16" ± 3⁄32" [1]	0.05" ±0.1" 1⁄16" ± 3⁄32"	0.0" ±0.1" 0" ± 3⁄32"	FRONT TOE (Inches)	0.0" ±0.1" 0" ± 3⁄32"	0.1" ±0.1" 3⁄32" ± 3⁄32"	0.1" ±0.1" 3⁄32" ± 3⁄32"
			SAI			
			TURNING ANGLE			
25	25	25	ILLUSTRATION #	25 ▲	25	25
			REAR			
-0.25 ±0.75 -¼ ± ¾	-0.25 ±0.5 -¼ ± ½	-0.25 ±0.5 -¼ ± ½	CAMBER (Degrees)	0.0 ±0.25 0 ± ¼	-0.8 ±0.75 -13⁄16 ± ¾	-0.8 ±0.75 -13⁄16 ± ¾
0.2° ±0.3° 3⁄16° ± 5⁄16°	0.2° ±0.2° 3⁄16° ± 3⁄16°	0.2° ±0.2° 3⁄16° ± 3⁄16°	REAR TOE (Degrees)	0.0° ±0.2° 0° ± 3⁄16°	0.25° ±0.3° ¼° ± 5⁄16°	0.25° ±0.3° ¼° ± 5⁄16°
0.1" ±0.15" 3⁄32" ± 5⁄32"	0.1" ±0.1" 3⁄32" ± 3⁄32"	0.1" ±0.1" 3⁄32" ± 3⁄32"	REAR TOE (Inches)	0.0" ±0.1" 0" ± 3⁄32"	0.13" ±0.15" ⅛" ± 5⁄32"	0.13" ±0.15" ⅛" ± 5⁄32"
0.25 (¼)	0.25 (¼)	0.25 (¼)	THRUST ANGLE		0.3 (5⁄16)	0.3 (5⁄16)
			ILLUSTRATION #			
[1] 1995 = 0.0° (0")			NOTES		[1] FE3 Susp. = 3.65° (3⅝°)	[1] 2005-08 = -1.05° (-1 1⁄16°)
08046	08047	08048		08050	08062	08063

ALIGNMENT SPECIFICATIONS

Chevrolet

Corsica	Corsica	Corvette	MODEL INFORMATION	Corvette	Corvette exc/Z06	Corvette Z06
1990-92	1993-96	1990-96	YEARS	1997-2000	2001-2004	2001-2004
			INSTRUCTIONS			
100 ft.lbs.	100 ft.lbs.	100 ft.lbs.	LUGNUT TORQUE	100 ft.lbs.	100 ft.lbs.	100 ft.lbs.
			FRONT			
1.15±0.75 1⅛ ± ¾	1.2±1.0 1 3/16 ± 1	6.0±0.5 6 ± ½	CASTER (Degrees)	6.5±0.5 6½ ± ½	7.4±0.5 7⅜ ± ½	6.9±0.5 6⅞ ± ½
0.0 (0) | 0.7 (11/16)	0.0 (0) | 0.7 (11/16)	0.0 (0) | 0.7 (11/16)	CROSS | TOL.	0.0 (0) | 0.5 (½)	0.0 (0) | 0.5 (½)	0.0 (0) | 0.5 (½)
0.6±0.6 ⅝ ± ⅝	0.0±0.7 0 ± 11/16	0.5±0.5 ½ ± ½	CAMBER (Degrees)	-0.25±0.5 -¼ ± ½	-0.2±0.5 -3/16 ± ½	-0.7±0.5 -11/16 ± ½
0.0 (0) | 1.0 (1)	0.0 (0) | 0.7 (11/16)	0.0 (0) | 0.5 (½)	CROSS | TOL.	0.0 (0) | 0.5 (½)	0.0 (0) | 0.5 (½)	0.0 (0) | 0.5 (½)
0.0° ±0.2° 0° ± 3/16°	0.0° ±0.2° 0° ± 3/16°	0.0° ±0.2° 0° ± 3/16°	FRONT TOE (Degrees)	0.1° ±0.2° 3/32° ± 3/16°	0.08° ±0.2° 3/32° ± 3/16°	0.08° ±0.2° 3/32° ± 3/16°
0.0" ±0.1" 0" ± 3/32"	0.0" ±0.1" 0" ± 3/32"	0.0" ±0.1" 0" ± 3/32"	FRONT TOE (Inches)	0.05" ±0.1" 1/16" ± 3/32"	0.04" ±0.1" 1/32" ± 3/32"	0.04" ±0.1" 1/32" ± 3/32"
			SAI			
			TURNING ANGLE			
25	25	41	ILLUSTRATION #	12	12	12
			REAR			
-0.25±0.6 -¼ ± ⅝	-0.25±0.6 [1] -¼ ± ⅝ [1]	0.0±0.5 0 ± ½	CAMBER (Degrees)	-0.25±0.5 -¼ ± ½	-0.18±0.5 -3/16 ± ½	-0.68±0.5 -11/16 ± ½
0.25° ±0.25° ¼° ± ¼°	0.0° ±0.3° [2] 0° ± 5/16° [2]	0.0° ±0.2° 0° ± 3/16°	REAR TOE (Degrees)	0.0° ±0.2° 0° ± 3/16°	-0.02° ±0.2° -1/32° ± 3/16°	-0.02° ±0.2° -1/32° ± 3/16°
0.13" ±0.13" ⅛" ± ⅛"	0.0" ±0.15" [2] 0" ± 5/32" [2]	0.0" ±0.1" 0" ± 3/32"	REAR TOE (Inches)	0.0" ±0.1" 0" ± 3/32"	-0.01" ±0.1" -1/64" ± 3/32"	-0.01" ±0.1" -1/64" ± 3/32"
0.35 (11/32)	0.35 (11/32)	0.1 (3/32)	THRUST ANGLE	0.1 (3/32)	0.1 (3/32)	0.1 (3/32)
		40 10	ILLUSTRATION #	111 32	111 32	111 32
	[1] 1996 = -0.4° (-⅜°) [2] 1996 = 0.2° (0.1")		NOTES			
08064	08065	08082		08083	08084	08085

Chevrolet

Corvette	Corvette except Z06	Corvette Z06	MODEL INFORMATION	Corvette except Z06, ZR1	Corvette Z06 exc/PS CUP Tires	Corvette ZR1 exc/PS CUP Tires
2005	2006-2007	2006-2007	YEARS	2008-2013	2008-2013	2009-2013
			INSTRUCTIONS			
100 ft.lbs.	100 ft.lbs.	100 ft.lbs.	LUGNUT TORQUE	100 ft.lbs.	100 ft.lbs.	100 ft.lbs.
			FRONT			
7.9±0.6 7⅞ ± ⅝	7.5±0.6 7½ ± ⅝	7.6±0.6 7⅝ ± ⅝	CASTER (Degrees)	7.5±0.6 7½ ± ⅝	7.6±0.6 7⅝ ± ⅝	7.7±0.6 7 11/16 ± ⅝
0.0 (0) | 0.6 (⅝)	0.0 (0) | 0.6 (⅝)	0.0 (0) | 0.6 (⅝)	CROSS | TOL.	0.0 (0) | 0.6 (⅝)	0.0 (0) | 0.6 (⅝)	0.0 (0) | 0.6 (⅝)
-0.45±0.6 -7/16 ± ⅝	-0.45±0.6 -7/16 ± ⅝	-1.0±0.6 -1 ± ⅝	CAMBER (Degrees)	-0.45±0.6 -7/16 ± ⅝	-0.95±0.6 -15/16 ± ⅝	-1.0±0.6 -1 ± ⅝
0.0 (0) | 0.6 (⅝)	0.0 (0) | 0.6 (⅝)	0.0 (0) | 0.6 (⅝)	CROSS | TOL.	0.0 (0) | 0.6 (⅝)	0.0 (0) | 0.6 (⅝)	0.0 (0) | 0.6 (⅝)
0.1° ±0.2° 3/32° ± 3/16°	0.1° ±0.2° 3/32° ± 3/16°	0.1° ±0.2° 3/32° ± 3/16°	FRONT TOE (Degrees)	0.1° ±0.2° 3/32° ± 3/16°	0.1° ±0.2° 3/32° ± 3/16°	0.1° ±0.2° 3/32° ± 3/16°
0.05" ±0.1" 1/16" ± 3/32"	0.05" ±0.1" 1/16" ± 3/32"	0.05" ±0.1" 1/16" ± 3/32"	FRONT TOE (Inches)	0.05" ±0.1" 1/16" ± 3/32"	0.05" ±0.1" 1/16" ± 3/32"	0.05" ±0.1" 1/16" ± 3/32"
			SAI			
			TURNING ANGLE			
12	12	12	ILLUSTRATION #	12	12	12
			REAR			
-0.45±0.5 -7/16 ± ½	-0.45±0.5 -7/16 ± ½	-1.0±0.5 -1 ± ½	CAMBER (Degrees)	-0.45±0.6 -7/16 ± ⅝	-1.2±0.6 -1 3/16 ± ⅝	-1.25±0.6 -1¼ ± ⅝
0.0° ±0.2° 0° ± 3/16°	0.0° ±0.2° 0° ± 3/16°	0.0° ±0.2° 0° ± 3/16°	REAR TOE (Degrees)	0.0° ±0.2° 0° ± 3/16°	0.0° ±0.2° 0° ± 3/16°	0.1° ±0.2° 3/32° ± 3/16°
0.0" ±0.1" 0" ± 3/32"	0.0" ±0.1" 0" ± 3/32"	0.0" ±0.1" 0" ± 3/32"	REAR TOE (Inches)	0.0" ±0.1" 0" ± 3/32"	0.0" ±0.1" 0" ± 3/32"	0.05" ±0.1" 1/16" ± 3/32"
0.2 (3/16)	0.2 (9/16)	0.2 (3/16)	THRUST ANGLE	0.2 (3/10)	0.2 (3/16)	0.2 (3/16)
134 135	134 135	134 135	ILLUSTRATION #	134 135	134 135	134 135
			NOTES			
08086	08087	08088		08089	08090	08091

Chevrolet

Corvette Z06 & ZR1 w/Michelin PS CUP Tires	Corvette	Corvette exc/Z06 & Z07 Performance Pkg.	MODEL INFORMATION	Corvette w/Z06 & Z07 Performance Pkg.	Cruze/Cruze Limited exc/Sport Susp.	Cruze/Cruze Limited w/Sport Susp. (RPO XJ2)
2012-2013	2014	2015-2017	YEARS	2015-2017	2011-2016	2011-2016
			INSTRUCTIONS			
100 ft.lbs.	100 ft.lbs.	100 ft.lbs.	LUGNUT TORQUE	100 ft.lbs.	103 ft.lbs.	103 ft.lbs.
			FRONT			
7.2 ±0.6 7 3/16 ± 5/8	7.4 ±0.6 7 3/8 ± 5/8	7.4 ±0.6 7 3/8 ± 5/8	CASTER (Degrees)	7.4 ±0.6 7 3/8 ± 5/8	4.5 ±0.75 4 1/2 ± 3/4	4.5 ±0.75 4 1/2 ± 3/4
0.0 (0) | 0.6 (5/8)	0.0 (0) | 0.6 (5/8)	0.0 (0) | 0.6 (5/8)	CROSS | TOL.	0.0 (0) | 0.6 (5/8)	0.0 (0) | 0.75 (3/4)	0.0 (0) | 0.75 (3/4)
-1.3 ±0.6 -1 5/16 ± 5/8	-0.5 ±0.6 - 1/2 ± 5/8	L= -0.2, R= -0.5 ±0.6 L= -3/16, R= - 1/2 ± 5/8	CAMBER (Degrees)	L= -0.6, R= -0.9 ±0.6 L= -5/8, R= - 7/8 ± 5/8	-0.5 ±0.75 - 1/2 ± 3/4	-0.55 ±0.75 - 9/16 ± 3/4
0.0 (0) | 0.6 (5/8)	0.0 (0) | 0.6 (5/8)	0.3 (5/16) | 0.6 (5/8)	CROSS | TOL.	0.3 (5/16) | 0.6 (5/8)	0.0 (0) | 0.75 (3/4)	0.0 (0) | 0.75 (3/4)
0.1° ±0.2° 3/32° ± 3/16°	0.1° ±0.2° 3/32° ± 3/16°	0.1° ±0.2° 3/32° ± 3/16°	FRONT TOE (Degrees)	0.1° ±0.2° 3/32° ± 3/16°	0.2° ±0.2° 3/16° ± 3/16°	0.08° ±0.2° 3/32° ± 3/16°
0.05" ±0.1" 1/16" ± 3/32"	0.05" ±0.1" 1/16" ± 3/32"	0.05" ±0.1" 1/16" ± 3/32"	FRONT TOE (Inches)	0.05" ±0.1" 1/16" ± 3/32"	0.1" ±0.1" 3/32" ± 3/32"	0.04" ±0.1" 1/32" ± 3/32"
			SAI			
			TURNING ANGLE			
12	12	12	ILLUSTRATION #	12	25	25
			REAR			
-1.3 ±0.6 -1 5/16 ± 5/8	-0.5 ±0.6 - 1/2 ± 5/8	-0.5 ±0.6 - 1/2 ± 5/8	CAMBER (Degrees)	-1.1 ±0.6 -1 1/8 ± 5/8	-1.45 ±0.75 -1 7/16 ± 3/4	-1.45 ±0.75 -1 7/16 ± 3/4
0.1° ±0.2° 3/32° ± 3/16°	0.0° ±0.2° 0° ± 3/16°	0.0° ±0.2° 0° ± 3/16°	REAR TOE (Degrees)	0.0° ±0.2° 0° ± 3/16°	0.15° ±0.4° 5/32° ± 13/32°	0.15° ±0.4° 5/32° ± 13/32°
0.05" ±0.1" 1/16" ± 3/32"	0.0" ±0.1" 0" ± 3/32"	0.0" ±0.1" 0" ± 3/32"	REAR TOE (Inches)	0.0" ±0.1" 0" ± 3/32"	0.08" ±0.2" 1/16" ± 3/16"	0.08" ±0.2" 1/16" ± 3/16"
0.2 (3/16)	0.2 (3/16)	0.2 (3/16)	THRUST ANGLE	0.2 (3/16)	0.3 (5/16)	0.3 (5/16)
134 135	134 06	134 06	ILLUSTRATION #	134 06		
			NOTES			
08092	08093	08094		08095	08096	08097

Chevrolet

Cruze (VIN B) USA & Canada	HHR exc/SS	HHR SS	MODEL INFORMATION	HHR SS	Impala	Impala
2016-2017	2006-2011	2008	YEARS	2009-2011	2000	2001-2004
			INSTRUCTIONS			
103 ft.lbs.	100 ft.lbs.	100 ft.lbs.	LUGNUT TORQUE	100 ft.lbs.	100 ft.lbs.	100 ft.lbs.
			FRONT			
4.65 ±0.75 4 5/8 ± 3/4	3.7 ±0.75 3 11/16 ± 3/4	3.7 ±0.75 3 11/16 ± 3/4	CASTER (Degrees)	3.65 ±0.75 3 5/8 ± 3/4	3.21 ±0.5 3 3/16 ± 1/2	3.21 ±0.5 3 3/16 ± 1/2
0.0 (0) | 0.75 (3/4)	0.0 (0) | 0.75 (3/4)	0.0 (0) | 0.75 (3/4)	CROSS | TOL.	0.0 (0) | 0.75 (3/4)	0.0 (0) | 0.75 (3/4)	0.0 (0) | 0.75 (3/4)
-0.3 ±0.75 - 5/16 ± 3/4	-0.95 ±0.75 - 15/16 ± 3/4	-1.05 ±0.75 -1 1/16 ± 3/4	CAMBER (Degrees)	-1.25 ±0.75 -1 1/4 ± 3/4	-0.85 ±0.5 - 7/8 ± 1/2	-0.78 ±0.5 - 3/4 ± 1/2
0.0 (0) | 0.8 (13/16)	0.0 (0) | 0.75 (3/4)	0.0 (0) | 0.75 (3/4)	CROSS | TOL.	0.0 (0) | 0.75 (3/4)	0.0 (0) | 0.75 (3/4)	0.0 (0) | 0.75 (3/4)
0.3° ±0.2° 5/16° ± 3/16°	0.2° ±0.2° 3/16° ± 3/16°	0.2° ±0.2° 3/16° ± 3/16°	FRONT TOE (Degrees)	0.2° ±0.2° 3/16° ± 3/16°	0.1° ±0.2° 3/32° ± 3/16°	0.1° ±0.2° 3/32° ± 3/16°
0.15" ±0.1" 5/32" ± 3/32"	0.1" ±0.1" 3/32" ± 3/32"	0.1" ±0.1" 3/32" ± 3/32"	FRONT TOE (Inches)	0.1" ±0.1" 3/32" ± 3/32"	0.05" ±0.1" 1/16" ± 3/32"	0.05" ±0.1" 1/16" ± 3/32"
			SAI			
			TURNING ANGLE			
	25	25	ILLUSTRATION #	25	33	33
			REAR			
-1.3 ±0.5 -1 5/16 ± 1/2	-0.8 ±0.75 - 13/16 ± 3/4	-0.8 ±0.75 - 13/16 ± 3/4	CAMBER (Degrees)	-0.8 ±0.75 - 13/16 ± 3/4	0.0 ±0.5 0 ± 1/2	-0.5 ±0.5 - 1/2 ± 1/2
0.1° ±0.4° 3/32° ± 13/32°	0.25° ±0.3° 1/4° ± 5/16°	0.25° ±0.3° 1/4° ± 5/16°	REAR TOE (Degrees)	0.25° ±0.3° 1/4° ± 5/16°	0.1° ±0.2° 3/32° ± 3/16°	0.1° ±0.2° 3/32° ± 3/16°
0.05" ±0.2" 1/16" ± 3/16"	0.13" ±0.15" 1/8" ± 5/32"	0.13" ±0.15" 1/8" ± 5/32"	REAR TOE (Inches)	0.13" ±0.15" 1/8" ± 5/32"	0.05" ±0.1" 1/16" ± 3/32"	0.05" ±0.1" 1/16" ± 3/32"
0.3 (5/16)	0.3 (5/16)	0.3 (5/16)	THRUST ANGLE	0.3 (5/16)	0.15 (5/32)	0.15 (5/32)
			ILLUSTRATION #		108 10	108 10
			NOTES			
08098	08105	08106		08107	08109	08110

ALIGNMENT SPECIFICATIONS

Chevrolet

MODEL INFORMATION	Impala exc/SS, Police	Impala SS	Impala Police Models	Impala (VIN W) exc/SS, LTZ, and Police	Impala (VIN W) SS, LTZ	Impala (VIN 1)
YEARS	2005	2005	2005	2006-2016	2006-2016	2014-2017
INSTRUCTIONS						
LUGNUT TORQUE	100 ft.lbs.	100 ft.lbs.	100 ft.lbs.	103 ft.lbs.	103 ft.lbs.	122 ft.lbs.
FRONT						
CASTER (Degrees)	3.0 ±0.75 3 ± ¾	3.5 ±0.75 3½ ± ¾	3.5 ±0.75 3½ ± ¾	2.9 ±0.75 2⅞ ± ¾	3.15 ±0.75 3⅛ ± ¾	3.8 ±0.8 3 13/16 ± 13/16
CROSS | TOL.	0.0 (0) | 0.75 (¾)	0.0 (0) | 0.75 (¾)	0.0 (0) | 0.75 (¾)	0.0 (0) | 0.75 (¾)	0.0 (0) | 0.75 (¾)	0.0 (0) | 0.8 (13/16)
CAMBER (Degrees)	-0.75 ±0.75 -¾ ± ¾	-0.7 ±0.75 -11/16 ± ¾	-0.6 ±0.75 -⅝ ± ¾	-0.8 ±0.75 -13/16 ± ¾	-0.7 ±0.75 -11/16 ± ¾	-0.4 ±0.8 -⅜ ± 13/16
CROSS | TOL.	0.0 (0) | 0.75 (¾)	0.0 (0) | 0.75 (¾)	0.0 (0) | 0.75 (¾)	0.0 (0) | 0.75 (¾)	0.0 (0) | 0.75 (¾)	0.0 (0) | 0.8 (13/16)
FRONT TOE (Degrees)	0.1° ±0.2° 3/32° ± 3/16°	0.1° ±0.2° 3/32° ± 3/16°	0.1° ±0.2° 3/32° ± 3/16°	0.1° ±0.2° 3/32° ± 3/16°	0.1° ±0.2° 3/32° ± 3/16°	0.2° ±0.2° 3/16° ± 3/16°
FRONT TOE (Inches)	0.05" ±0.1" 1/16" ± 3/32"	0.05" ±0.1" 1/16" ± 3/32"	0.05" ±0.1" 1/16" ± 3/32"	0.05" ±0.1" 1/16" ± 3/32"	0.05" ±0.1" 1/16" ± 3/32"	0.1" ±0.1" 3/32" ± 3/32"
SAI						
TURNING ANGLE						
ILLUSTRATION #	33	33	33	33	33	25
REAR						
CAMBER (Degrees)	-0.65 ±0.5 -⅝ ± ½	-0.9 ±0.5 -⅞ ± ½	-0.7 ±0.5 -11/16 ± ½	-0.65 ±0.5 -⅝ ± ½	-0.9 ±0.5 -⅞ ± ½	-0.8 ±0.8 -13/16 ± 13/16
REAR TOE (Degrees)	-0.1° ±0.2° -3/32° ± 3/16°	0.1° ±0.2° 3/32° ± 3/16°	0.1° ±0.2° 3/32° ± 3/16°	0.1° ±0.2° 3/32° ± 3/16°	0.1° ±0.2° 3/32° ± 3/16°	0.1° ±0.2° 3/32° ± 3/16°
REAR TOE (Inches)	-0.05" ±0.1" -1/16" ± 3/32"	0.05" ±0.1" 1/16" ± 3/32"	0.05" ±0.1" 1/16" ± 3/32"	0.05" ±0.1" 1/16" ± 3/32"	0.05" ±0.1" 1/16" ± 3/32"	0.05" ±0.1" 1/16" ± 3/32"
THRUST ANGLE	0.2 (3/16)	0.2 (3/16)	0.2 (3/16)	0.2 (3/16)	0.2 (3/16)	0.3 (5/16)
ILLUSTRATION #	108 10	108 10	108 10	108 10	108 10	179 174
NOTES						
	08111	08112	08113	08114	08115	08116

Chevrolet

MODEL INFORMATION	Impala Police Models	Impala Police Models	Impala Police Models	Lumina APV	Lumina APV	Lumina
YEARS	2006-2007	2008-2012	2013-2016	1990-92	1993-96	1990
INSTRUCTIONS						
LUGNUT TORQUE	103 ft.lbs.	103 ft.lbs.	103 ft.lbs.	100 ft.lbs.	100 ft.lbs.	103 ft.lbs.
FRONT						
CASTER (Degrees)	3.2 ±0.75 [1] 3 3/16 ± ¾ [1]	3.15 ±0.75 3⅛ ± ¾	3.0 ±0.75 3 ± ¾	1.7 ±0.7 1 11/16 ± 11/16	3.0 ±0.7 3 ± 11/16	2.0 ±0.5 2 ± ½
CROSS | TOL.	0.0 (0) | 0.75 (¾)	0.0 (0) | 0.75 (¾)	0.0 (0) | 0.75 (¾)	0.0 (0) | 0.7 (11/16)	0.0 (0) | 0.7 (11/16)	0.0 (0) | 0.75 (¾)
CAMBER (Degrees)	-0.5 ±0.75 [2] -½ ± ¾ [2]	-0.7 ±0.75 -11/16 ± ¾	-0.6 ±0.75 -⅝ ± ¾	0.0 ±0.5 0 ± ½	0.0 ±0.5 0 ± ½	0.7 ±0.5 11/16 ± ½
CROSS | TOL.	0.0 (0) | 0.75 (¾)	0.0 (0) | 0.75 (¾)	0.0 (0) | 0.75 (¾)	0.0 (0) | 0.7 (11/16)	0.0 (0) | 0.7 (11/16)	0.0 (0) | 0.75 (¾)
FRONT TOE (Degrees)	0.1° ±0.2° 3/32° ± 3/16°	0.1° ±0.2° 3/32° ± 3/16°	0.1° ±0.2° 3/32° ± 3/16°	0.0° ±0.2° 0° ± 3/16°	0.0° ±0.2° 0° ± 3/16°	0.0° ±0.2° 0° ± 3/16°
FRONT TOE (Inches)	0.05" ±0.1" 1/16" ± 3/32"	0.05" ±0.1" 1/16" ± 3/32"	0.05" ±0.1" 1/16" ± 3/32"	0.0" ±0.1" 0" ± 3/32"	0.0" ±0.1" 0" ± 3/32"	0.0" ±0.1" 0" ± 3/32"
SAI						
TURNING ANGLE						
ILLUSTRATION #	33	33	33	102 ▲	102 ▲	33 ▲
REAR						
CAMBER (Degrees)	-0.7 ±0.5 -11/16 ± ½	-0.8 ±0.5 -13/16 ± ½	-0.7 ±0.5 -11/16 ± ½	-0.1 ±0.3 -⅛ ± 5/16	-0.1 ±0.3 -⅛ ± 5/16	-0.1 ±0.5 [1] -⅛ ± ½ [1]
REAR TOE (Degrees)	0.1° ±0.2° 3/32° ± 3/16°	0.1° ±0.2° 3/32° ± 3/16°	0.1° ±0.2° 3/32° ± 3/16°	0.0° ±0.3° 0° ± 5/16°	0.0° ±0.3° 0° ± 5/16°	0.0° ±0.3° 0° ± 5/16°
REAR TOE (Inches)	0.05" ±0.1" 1/16" ± 3/32"	0.05" ±0.1" 1/16" ± 3/32"	0.05" ±0.1" 1/16" ± 3/32"	0.0" ±0.15" 0" ± 5/32"	0.0" ±0.15" 0" ± 5/32"	0.0" ±0.15" 0" ± 5/32"
THRUST ANGLE	0.2 (3/16)	0.2 (3/16)	0.2 (3/16)	0.15 (5/32)	0.15 (5/32)	0.15 (5/32)
ILLUSTRATION #	108 10	108 10	108 10			25 34
NOTES	[1] RPO 9C3 = 3.05° (3 1/16°) [1] RPO 9C3 = -0.65° (-⅝°)					[1] w/16" tires -0.15° (-⅛°)
	08117	08118	08119	08120	08121	08122

Chevrolet

Lumina	Lumina	Lumina	MODEL INFORMATION	Lumina	Malibu	Malibu Classic
1991	1992-94	1995-96	YEARS	1997-2000	1999-2003	2004-2005
			INSTRUCTIONS			
103 ft.lbs.	103 ft.lbs.	100 ft.lbs.	LUGNUT TORQUE	100 ft.lbs.	100 ft.lbs.	100 ft.lbs.
			FRONT			
1.8 ±0.5 1 13/16 ± 1/2	2.0 ±0.5 2 ± 1/2	1.8 ±0.5 1 13/16 ± 1/2	CASTER (Degrees)	1.8 ±0.5 1 13/16 ± 1/2	4.1 ±0.75 4 1/8 ± 3/4	4.1 ±0.75 4 1/8 ± 3/4
0.0 (0) | 0.75 (3/4)	0.0 (0) | 0.75 (3/4)	0.0 (0) | 0.75 (3/4)	CROSS | TOL.	0.0 (0) | 0.75 (3/4)	0.0 (0) | 0.75 (3/4)	0.0 (0) | 0.75 (3/4)
0.7 ±0.5 11/16 ± 1/2	0.7 ±0.5 11/16 ± 1/2	0.7 ±0.5 11/16 ± 1/2	CAMBER (Degrees)	0.7 ±0.5 11/16 ± 1/2	-0.2 ±0.7 - 3/16 ± 11/16	-0.2 ±0.7 - 3/16 ± 11/16
0.0 (0) | 0.75 (3/4)	0.0 (0) | 0.75 (3/4)	0.0 (0) | 0.75 (3/4)	CROSS | TOL.	0.0 (0) | 0.75 (3/4)	0.0 (0) | 0.7 (11/16)	0.0 (0) | 0.7 (11/16)
0.0° ±0.2° 0° ± 3/16°	0.0° ±0.2° 0° ± 3/16°	0.0° ±0.2° 0° ± 3/16°	FRONT TOE (Degrees)	0.0° ±0.2° 0° ± 3/16°	0.15° ±0.2° 5/32° ± 3/16°	0.15° ±0.2° 5/32° ± 3/16°
0.0" ±0.1" 0" ± 3/32"	0.0" ±0.1" 0" ± 3/32"	0.0" ±0.1" 0" ± 3/32"	FRONT TOE (Inches)	0.0" ±0.1" 0" ± 3/32"	0.08" ±0.1" 1/16" ± 3/32"	0.08" ±0.1" 1/16" ± 3/32"
			SAI			
			TURNING ANGLE			
33 ▲	33 ▲	33 ▲	ILLUSTRATION #	33 ▲	102	102
			REAR			
-0.1 ±0.5 -1/8 ± 1/2	0.1 ±0.5 [1] 1/8 ± 1/2 [1]	-0.35 ±0.5 [1] -3/8 ± 1/2 [1]	CAMBER (Degrees)	-0.15 ±0.5 -1/8 ± 1/2	-0.2 ±0.5 - 3/16 ± 1/2	-0.2 ±0.5 - 3/16 ± 1/2
-0.1° ±0.3° -3/32° ± 5/16°	-0.1° ±0.3° -3/32° ± 5/16°	0.0° ±0.3° 0° ± 5/16°	REAR TOE (Degrees)	-0.1° ±0.3° -3/32° ± 5/16°	0.1° ±0.2° 3/32° ± 3/16°	0.1° ±0.2° 3/32° ± 3/16°
-0.05" ±0.15" -1/16" ± 5/32"	-0.05" ±0.15" -1/16" ± 5/32"	0.0" ±0.15" 0" ± 5/32"	REAR TOE (Inches)	-0.05" ±0.15" -1/16" ± 5/32"	0.05" ±0.1" 1/16" ± 3/32"	0.05" ±0.1" 1/16" ± 3/32"
0.15 (5/32)	0.15 (5/32)	0.15 (5/32)	THRUST ANGLE	0.15 (5/32)	0.2 (3/16)	0.2 (3/16)
25 34	25 34	108 34	ILLUSTRATION #	108 34	06	06
	[1] 1992 w/14" tires= 0.32° (5/16°)	[1] w/16" tires -0.45°(- 7/16°)	NOTES			
08123	08124	08125		08126	08131	08132

Chevrolet

Malibu Classic	Malibu	Malibu	MODEL INFORMATION	Malibu Malibu Hybrid	Malibu (Incl. Limited) Malibu Hybrid	Malibu (VIN Z) except Hybrid
2008	2004	2005-2007	YEARS	2008-2012	2013-2016	2016-2017
			INSTRUCTIONS			
100 ft.lbs.	100 ft.lbs.	100 ft.lbs.	LUGNUT TORQUE	100 ft.lbs.	111 ft.lbs.	111 ft.lbs.
			FRONT			
2.8 ±0.75 2 13/16 ± 3/4	3.25 ±0.5 3 1/4 ± 1/2	3.0 ±0.75 3 ± 3/4	CASTER (Degrees)	2.9 ±0.75 2 7/8 ± 3/4	3.9 ±0.75 3 7/8 ± 3/4	6.3 ±0.8 6 5/16 ± 13/16
0.0 (0) | 0.75 (3/4)	0.0 (0) | 0.5 (1/2)	0.0 (0) | 0.75 (3/4)	CROSS | TOL.	0.0 (0) | 0.75 (3/4)	0.0 (0) | 0.75 (3/4)	0.0 (0) | 0.8 (13/16)
L= -0.7, R= -0.4 ±0.75 L= -11/16, R= -3/8 ± 3/4	-1.0 ±0.5 -1 ± 1/2	L= -0.8, R= -0.4 ±0.75 L= -13/16, R= -3/8 ± 3/4	CAMBER (Degrees)	L= -1.0, R= -0.7 ±0.75 L= -1, R= -11/16 ± 3/4	-0.4 ±0.75 -3/8 ± 3/4	-0.3 ±0.8 - 5/16 ± 13/16
-0.3 (- 5/16) | 0.75 (3/4)	0.0 (0) | 0.5 (1/2)	-0.4 (-3/8) | 0.5 (1/2)	CROSS | TOL.	-0.3 (- 5/16) | 0.75 (3/4)	0.0 (0) | 0.75 (3/4)	0.0 (0) | 0.8 (13/16)
0.2° ±0.2° 3/16° ± 3/16°	0.2° ±0.2° 3/16° ± 3/16°	0.2° ±0.2° 3/16° ± 3/16°	FRONT TOE (Degrees)	0.2° ±0.2° 3/16° ± 3/16°	0.2° ±0.2° 3/16° ± 3/16°	0.2° ±0.2° 3/16° ± 3/16°
0.1" ±0.1" 3/32" ± 3/32"	0.1" ±0.1" 3/32" ± 3/32"	0.1" ±0.1" 3/32" ± 3/32"	FRONT TOE (Inches)	0.1" ±0.1" 3/32" ± 3/32"	0.1" ±0.1" 3/32" ± 3/32"	0.1" ±0.1" 3/32" ± 3/32"
			SAI			
			TURNING ANGLE			
25	25	25	ILLUSTRATION #	25	25	200
			REAR			
-0.8 ±0.6 - 13/16 ± 5/8	-0.8 ±0.5 - 13/16 ± 1/2	-0.8 ±0.5 - 13/16 ± 1/2	CAMBER (Degrees)	-0.8 ±0.6 - 13/16 ± 5/8	-0.8 ±0.75 - 13/16 ± 3/4	-0.8 ±0.8 - 13/16 ± 13/16
0.2° ±0.2° 3/16° ± 3/16°	0.2° ±0.2° 3/16° ± 3/16°	0.2° ±0.2° 3/16° ± 3/16°	REAR TOE (Degrees)	0.2° ±0.2° 3/16° ± 3/16°	0.1° ±0.2° 3/32° ± 3/16°	0.1° ±0.2° 3/32° ± 3/16°
0.1" ±0.1" 3/32" ± 3/32"	0.1" ±0.1" 3/32" ± 3/32"	0.1" ±0.1" 3/32" ± 3/32"	REAR TOE (Inches)	0.1" ±0.1" 3/32" ± 3/32"	0.05" ±0.1" 1/16" ± 3/32"	0.05" ±0.1" 1/16" ± 3/32"
0.3 (5/16)	0.15 (5/32)	0.3 (5/16)	THRUST ANGLE	0.3 (5/16)	0.3 (5/16)	0.3 (5/16)
177	177	177	ILLUSTRATION #	177	177	179 178
			NOTES			
08133	08134	08135		08136	08137	08138

ALIGNMENT SPECIFICATIONS

Chevrolet

MODEL INFORMATION	Malibu (VIN Z) Hybrid	Metro	Monte Carlo	Monte Carlo	Monte Carlo	Monte Carlo
YEARS	2016-2017	1998-2001	1995-96	1997-99	2000-2001	2002-2004
INSTRUCTIONS						
LUGNUT TORQUE	111 ft.lbs.	44 ft.lbs.	100 ft.lbs.	100 ft.lbs.	100 ft.lbs.	100 ft.lbs.
FRONT						
CASTER (Degrees)	6.3 ± 0.8 6 5/16 ± 13/16	3.0 ± 2.0 3 ± 2	1.8 ± 0.5 1 13/16 ± 1/2	1.8 ± 0.5 1 13/16 ± 1/2	3.21 ± 0.5 3 3/16 ± 1/2	3.21 ± 0.5 3 3/16 ± 1/2
CROSS | TOL.	0.0 (0) | 0.8 (13/16)	0.0 (0) | 1.0 (1)	0.0 (0) | 0.75 (3/4)	0.0 (0) | 0.75 (3/4)	0.0 (0) | 0.75 (3/4)	0.0 (0) | 0.75 (3/4)
CAMBER (Degrees)	-0.4 ± 0.8 -3/8 ± 13/16	0.0 ± 1.0 0 ± 1	0.7 ± 0.5 11/16 ± 1/2	0.7 ± 0.5 11/16 ± 1/2	-0.7 ± 0.25 -11/16 ± 1/4	-0.85 ± 0.5 -7/8 ± 1/2
CROSS | TOL.	0.0 (0) | 0.8 (13/16)	0.0 (0) | 0.5 (1/2)	0.0 (0) | 0.75 (3/4)	0.0 (0) | 0.75 (3/4)	0.0 (0) | 0.75 (3/4)	0.0 (0) | 0.75 (3/4)
FRONT TOE (Degrees)	0.2° ± 0.2° 3/16° ± 3/16°	0.0° ± 0.16° 0° ± 5/32°	0.0° ± 0.2° 0° ± 3/16°	0.0° ± 0.2° 0° ± 3/16°	0.1° ± 0.2° 3/32° ± 3/16°	0.1° ± 0.2° 3/32° ± 3/16°
FRONT TOE (Inches)	0.1" ± 0.1" 3/32" ± 3/32"	0.0" ± 0.08" 0" ± 3/32"	0.0" ± 0.1" 0" ± 3/32"	0.0" ± 0.1" 0" ± 3/32"	0.05" ± 0.1" 1/16" ± 3/32"	0.05" ± 0.1" 1/16" ± 3/32"
SAI						
TURNING ANGLE		38.0 / 32.0				
ILLUSTRATION #	200		33 △	33 △	33	33
REAR						
CAMBER (Degrees)	-0.9 ± 0.8 -7/8 ± 13/16	0.0 ± 1.0 0 ± 1	-0.35 ± 0.5 [1] -3/8 ± 1/2 [1]	-0.15 ± 0.5 -1/8 ± 1/2	0.0 ± 0.7 0 ± 11/16	-0.7 ± 0.7 -11/16 ± 11/16
REAR TOE (Degrees)	0.1° ± 0.2° 3/32° ± 3/16°	0.16° ± 0.16° 5/32° ± 5/32°	0.0° ± 0.3° 0° ± 5/16°	-0.1° ± 0.3° -3/32° ± 5/16°	0.1° ± 0.2° 3/32° ± 3/16°	0.1° ± 0.2° 3/32° ± 3/16°
REAR TOE (Inches)	0.05" ± 0.1" 1/16" ± 3/32"	0.08" ± 0.08" 3/32" ± 3/32"	0.0" ± 0.15" 0" ± 5/32"	-0.05" ± 0.15" -1/16" ± 5/32"	0.05" ± 0.1" 1/16" ± 3/32"	0.05" ± 0.1" 1/16" ± 3/32"
THRUST ANGLE	0.3 (5/16)	0.15 (5/32)	0.15 (5/32)	0.15 (5/32)	0.15 (5/32)	0.15 (5/32)
ILLUSTRATION #	179 178	06	108 34	108 34	108 10	108 10
NOTES			[1] w/16" tires -0.45°(-7/16°)			
	08139	08140	08147	08148	08149	08150

Chevrolet

MODEL INFORMATION	Monte Carlo	Monte Carlo exc/SS	Monte Carlo SS	Optra	Orlando (Canada)	Prizm
YEARS	2005	2006-2007	2006-2007	2004-2009	2012-2014	1998-2002
INSTRUCTIONS						
LUGNUT TORQUE	100 ft.lbs.	100 ft.lbs.	100 ft.lbs.	74 ft.lbs.	103 ft.lbs.	76 ft.lbs.
FRONT						
CASTER (Degrees)	3.4 ± 0.75 3 3/8 ± 3/4	3.0 ± 0.5 3 ± 1/2	3.15 ± 0.5 3 1/8 ± 1/2	4.0 ± 0.75 4 ± 3/4	4.45 ± 0.75 4 7/16 ± 3/4	1.32 ± 0.75 1 5/16 ± 3/4
CROSS | TOL.	0.0 (0) | 0.75 (3/4)	0.0 (0) | 0.5 (1/2)	0.0 (0) | 0.5 (1/2)	0.0 (0) | n/s	0.0 (0) | 0.75 (3/4)	0.0 (0) | 0.75 (3/4)
CAMBER (Degrees)	-0.85 ± 0.75 -7/8 ± 3/4	-0.85 ± 0.5 -7/8 ± 1/2	-0.8 ± 0.5 -13/16 ± 1/2	-0.33 ± 0.75 -5/16 ± 3/4	-0.3 ± 0.75 -5/16 ± 3/4	-0.18 ± 0.75 -3/16 ± 3/4
CROSS | TOL.	0.0 (0) | 0.75 (3/4)	0.0 (0) | 0.5 (1/2)	0.0 (0) | 0.5 (1/2)	0.0 (0) | n/s	0.0 (0) | 0.75 (3/4)	0.0 (0) | 0.75 (3/4)
FRONT TOE (Degrees)	0.1° ± 0.2° 3/32° ± 3/16°	0.1° ± 0.2° 3/32° ± 3/16°	0.1° ± 0.2° 3/32° ± 3/16°	0.0° ± 0.17° 0° ± 5/32°	0.15° ± 0.2° 5/32° ± 3/16°	0.1° ± 0.2° 3/32° ± 3/16°
FRONT TOE (Inches)	0.05" ± 0.1" 1/16" ± 3/32"	0.05" ± 0.1" 1/16" ± 3/32"	0.05" ± 0.1" 1/16" ± 3/32"	0.0" ± 0.08" 0" ± 3/32"	0.08" ± 0.1" 1/16" ± 3/32"	0.05" ± 0.1" 1/16" ± 3/32"
SAI						
TURNING ANGLE						39.2 / 33.4 ± 2.0
ILLUSTRATION #	33	33	33		25	
REAR						
CAMBER (Degrees)	-0.85 ± 0.5 -7/8 ± 1/2	-0.85 ± 0.5 -7/8 ± 1/2	-0.9 ± 0.5 -7/8 ± 1/2	-1.0 ± 0.75 -1 ± 3/4	-1.35 ± 0.75 -1 3/8 ± 3/4	-0.92 ± 0.75 -15/16 ± 3/4
REAR TOE (Degrees)	-0.1° ± 0.2° -3/32° ± 3/16°	0.1° ± 0.2° 3/32° ± 3/16°	0.1° ± 0.2° 3/32° ± 3/16°	0.2° ± 0.17° 3/16° ± 5/32°	0.15° ± 0.4° 5/32° ± 13/32°	0.4° ± 0.2° 13/32° ± 3/16°
REAR TOE (Inches)	-0.05" ± 0.1" -1/16" ± 3/32"	0.05" ± 0.1" 1/16" ± 3/32"	0.05" ± 0.1" 1/16" ± 3/32"	0.1" ± 0.08" 3/32" ± 3/32"	0.08" ± 0.2" 1/16" ± 3/16"	0.2" ± 0.1" 3/16" ± 3/32"
THRUST ANGLE	0.2 (3/16)	0.15 (5/32)	0.15 (5/32)		0.3 (5/16)	0.15 (5/32)
ILLUSTRATION #	108 10	108 10	108 10	06		10
NOTES						
	08151	08152	08153	08159	08160	08161

Chevrolet

Sonic	Sonic (RPO FE3, ECO)	Sonic (RPO FE1, FE4)	MODEL INFORMATION	Spark	Spark EV	Spark
2012	2013-2017	2013-2017	YEARS	2013-2015	2014-2016	2016-2017
			INSTRUCTIONS			
103 ft.lbs.	103 ft.lbs.	103 ft.lbs.	LUGNUT TORQUE	103 ft.lbs.	89 ft.lbs.	103 ft.lbs.
			FRONT			
4.4 ± 0.75 4⅜ ± ¾	4.55 ± 0.75 4 9/16 ± ¾	4.4 ± 0.75 4⅜ ± ¾	CASTER (Degrees)	3.55 ± 0.75 3 9/16 ± ¾	3.65 ± 0.75 3⅝ ± ¾	5.17 ± 0.75 5 3/16 ± ¾
0.0 (0) | 0.75 (¾)	0.0 (0) | 0.75 (¾)	0.0 (0) | 0.75 (¾)	CROSS | TOL.	0.0 (0) | 0.75 (¾)	0.0 (0) | 0.75 (¾)	0.0 (0) | 0.75 (¾)
-0.8 ± 0.75 -13/16 ± ¾	-0.95 ± 0.75 -15/16 ± ¾	-0.75 ± 0.75 -¾ ± ¾	CAMBER (Degrees)	-0.2 ± 0.75 -3/16 ± ¾	-0.42 ± 0.75 -7/16 ± ¾	-0.2 ± 0.75 -3/16 ± ¾
0.0 (0) | 0.75 (¾)	0.0 (0) | 0.75 (¾)	0.0 (0) | 0.75 (¾)	CROSS | TOL.	0.0 (0) | 0.75 (¾)	0.0 (0) | 0.75 (¾)	0.0 (0) | 0.75 (¾)
0.1° ± 0.2° 3/32° ± 3/16°	0.04° ± 0.2° 1/16° ± 3/16°	0.12° ± 0.2° ⅛° ± 3/16°	FRONT TOE (Degrees)	0.1° ± 0.2° 3/32° ± 3/16°	0.04° ± 0.2° 1/16° ± 3/16°	0.1° ± 0.2° 3/32° ± 3/16°
0.05" ± 0.1" 1/16" ± 3/32"	0.02" ± 0.1" 1/32" ± 3/32"	0.06" ± 0.1" 1/16" ± 3/32"	FRONT TOE (Inches)	0.05" ± 0.1" 1/16" ± 3/32"	0.02" ± 0.1" 1/32" ± 3/32"	0.05" ± 0.1" 1/16" ± 3/32"
			SAI			
			TURNING ANGLE			
26	26	26	ILLUSTRATION #			
			REAR			
-1.0 ± 0.5 -1 ± ½	-1.0 ± 0.5 -1 ± ½	-1.0 ± 0.5 -1 ± ½	CAMBER (Degrees)	-1.45 ± 0.5 -1 7/16 ± ½	-1.0 ± 0.5 -1 ± ½	-1.2 ± 0.5 -1 3/16 ± ½
0.4° ± 0.2° 13/32° ± 3/16°	0.4° ± 0.5° 13/32° ± ½°	0.4° ± 0.5° 13/32° ± ½°	REAR TOE (Degrees)	0.24° ± 0.35° ¼° ± 11/32°	0.05° ± 0.35° 1/16° ± 11/32°	0.3° ± 0.3° 5/16° ± 5/16°
0.2" ± 0.1" 3/16" ± 3/32"	0.2" ± 0.25" 3/16" ± ¼"	0.2" ± 0.25" 3/16" ± ¼"	REAR TOE (Inches)	0.12" ± 0.18" ⅛" ± 3/16"	0.03" ± 0.18" 1/32" ± 3/16"	0.15" ± 0.15" 5/32" ± 5/32"
0.4 (13/32)	0.4 (13/32)	0.4 (13/32)	THRUST ANGLE	0.3 (5/16)	0.3 (5/16)	0.3 (5/16)
			ILLUSTRATION #			
			NOTES			
08162	08163	08164		08165	08166	08167

Chevrolet

Spark ACTIV	SS	Uplander FWD	MODEL INFORMATION	Uplander AWD	Venture FWD	Venture AWD
2017	2014-2017	2005-2009	YEARS	2005-2006	1997-2005	2002-2004
			INSTRUCTIONS			
103 ft.lbs.	140 ft.lbs.	100 ft.lbs.	LUGNUT TORQUE	100 ft.lbs.	100 ft.lbs.	100 ft.lbs.
			FRONT			
5.09 ± 0.75 5 1/16 ± ¾	6.0 ± 1.25 6 ± 1¼	2.7 ± 0.75 2 11/16 ± ¾	CASTER (Degrees)	2.7 ± 0.75 2 11/16 ± ¾	2.9 ± 0.5 [1] 2⅞ ± ½ [1]	2.9 ± 0.5 [1] 2⅞ ± ½ [1]
0.0 (0) | 0.75 (¾)	0.0 (0) | 1.0 (1)	0.0 (0) | 0.75 (¾)	CROSS | TOL.	0.0 (0) | 0.75 (¾)	0.0 (0) | 0.75 (¾)	0.0 (0) | 0.75 (¾)
-0.2 ± 0.75 -3/16 ± ¾	-0.3 ± 0.33 -5/16 ± 5/16	-0.65 ± 0.75 -⅝ ± ¾	CAMBER (Degrees)	-0.65 ± 0.75 -⅝ ± ¾	-0.7 ± 0.5 -11/16 ± ½	-0.7 ± 0.5 -11/16 ± ½
0.0 (0) | 0.75 (¾)	0.0 (0) | 0.33 (5/16)	0.0 (0) | 0.75 (¾)	CROSS | TOL.	0.0 (0) | 0.75 (¾)	0.0 (0) | 0.75 (¾)	0.0 (0) | 0.75 (¾)
0.26° ± 0.2° ¼° ± 3/16°	0.17° ± 0.17° 5/32° ± 5/32°	0.0° ± 0.2° 0° ± 3/16°	FRONT TOE (Degrees)	0.0° ± 0.2° 0° ± 3/16°	0.0° ± 0.2° 0° ± 3/16°	0.0° ± 0.2° 0° ± 3/16°
0.13" ± 0.1" ⅛" ± 3/32"	0.09" ± 0.09" 3/32" ± 3/32"	0.0" ± 0.1" 0" ± 3/32"	FRONT TOE (Inches)	0.0" ± 0.1" 0" ± 3/32"	0.0" ± 0.1" 0" ± 3/32"	0.0" ± 0.1" 0" ± 3/32"
			SAI			
			TURNING ANGLE			
	102	25 △	ILLUSTRATION #	25 △	25 △	25 △
			REAR			
-1.21 ± 0.5 -1 3/16 ± ½	-0.7 ± 0.45 -11/16 ± 7/16	-1.0 ± 0.5 -1 ± ½	CAMBER (Degrees)	-0.4 ± 0.6 -⅜ ± ⅝	-1.0 ± 0.25 -1 ± ¼	-0.25 ± 0.6 -¼ ± ⅝
0.27° ± 0.3° 9/32° ± 5/16°	0.2° ± 0.2° 3/16° ± 3/16°	0.0° ± 0.3° 0° ± 5/16°	REAR TOE (Degrees)	0.0° ± 0.2° 0° ± 3/16°	0.0° ± 0.3° 0° ± 5/16°	0.0° ± 0.2° 0° ± 3/16°
0.14" ± 0.15" ⅛" ± 5/32"	0.1" ± 0.1" 3/32" ± 3/32"	0.0" ± 0.15" 0" ± 5/32"	REAR TOE (Inches)	0.0" ± 0.1" 0" ± 3/32"	0.0" ± 0.15" 0" ± 5/32"	0.0" ± 0.1" 0" ± 3/32"
0.3 (5/16)	0.2 (3/16)	0.3 (5/16)	THRUST ANGLE	0.3 (5/16)	0.15 (5/32)	0.15 (5/32)
	177		ILLUSTRATION #	134 135		134 135
			NOTES		[1] 2004-05 = 3.1° (3⅛°)	[1] 2004 = 3.1° (3⅛°)
08168	08172	08173		08174	08177	08178

ALIGNMENT SPECIFICATIONS

Chevrolet — Chevrolet/GMC Truck & SUV

Volt	Volt	Notes	MODEL	C-Pickup 4X2	C-Pickup 4X2	R-Pickup 4X2
			INFORMATION	1500/2500/3500	3500 HD, C3500	10
				exc/3500 HD	w/ I-Beam F. Axle	1500
2011-2015	2016-2017		YEARS	1990-2000	1991-2006	1990-91
			INSTRUCTIONS	p. 304, 305	p. 304, 305	p. 304-306
100 ft.lbs.	100 ft.lbs.		LUGNUT TORQUE	140 ft.lbs.	177 ft.lbs.	
			FRONT			
4.4 ± 0.8 4⅜ ± 13/16	4.2 ± 0.8 4 3/16 ± 13/16		CASTER (Degrees)	3.75 ± 1.0 3¾ ± 1	5.0 ± 1.0 5 ± 1	see table p. 306 ±0.5 (½)
0.0 (0) | 0.8 (13/16)	0.0 (0) | 0.9 (⅞)		CROSS | TOL.	0.0 (0) | 0.5 (½)	0.0 (0) | n/s	0.0 (0) | 0.5 (½)
-0.4 ± 0.8 -⅜ ± 13/16	-0.35 ± 0.8 -⅜ ± 13/16		CAMBER (Degrees)	0.5 ± 0.5 ½ ± ½	1.25 ± 0.5 1¼ ± ½	0.7 ± 0.5 11/16 ± ½
0.0 (0) | 0.8 (13/16)	0.0 (0) | 0.8 (13/16)		CROSS | TOL.	0.0 (0) | 0.5 (½)	0.0 (0) | n/s	0.0 (0) | 0.5 (½)
0.2° ± 0.2° 3/16° ± 3/16°	0.2° ± 0.2° 3/16° ± 3/16°		FRONT TOE (Degrees)	0.24° ± 0.2° ¼° ± 3/16°	0.12° ± 0.12° ⅛° ± ⅛°	0.15° ± 0.1° 5/32° ± 3/32°
0.1" ± 0.1" 3/32" ± 3/32"	0.1" ± 0.1" 3/32" ± 3/32"		FRONT TOE (Inches)	0.12" ± 0.1" ⅛" ± 3/32"	0.06" ± 0.06" 1/16" ± 1/16"	0.08" ± 0.05" 1/16" ± 1/16"
			SAI			
			TURNING ANGLE			
25	25		ILLUSTRATION #	43		41
			REAR			
-1.4 ± 0.8 -1⅜ ± 13/16			CAMBER (Degrees)			
0.15° ± 0.4° 5/32° ± 13/32°	0.2° ± 0.4° 3/16° ± 13/32°		REAR TOE (Degrees)			
0.08" ± 0.2" 1/16" ± 3/16"	0.1" ± 0.2" 3/32" ± 3/16"		REAR TOE (Inches)			
0.4 (13/32)	0.4 (13/32)		THRUST ANGLE			
			ILLUSTRATION #			
			NOTES			
08179	08180	08197		09012	09013	09016

Chevrolet/GMC Truck & SUV

R-Pickup 4X2	K-Pickup 4X4	K-Pickup 4X4	MODEL	K-Pickup 4X4	V-Pickup 4X4	V-Pickup 4X4
20/30	1500/2500	3500	INFORMATION	3500	10/20	30
2500/3500	Chevy & GMC	Chevy & GMC		Chevy & GMC	1500/2500	3500
1990-91	1990-99	1990-92	YEARS	1993-2000	1990-91	1990-91
p. 304-306	p. 304, 305	p. 304, 305	INSTRUCTIONS	p. 304, 305	p. 304, 305	p. 304, 305
	140 ft.lbs.	140 ft.lbs.	LUGNUT TORQUE	140 ft.lbs.		
			FRONT			
see table p. 306 ±0.5 (½)	3.0 ± 1.0 3 ± 1	3.0 ± 1.0 3 ± 1	CASTER (Degrees)	3.0 ± 1.0 3 ± 1	8.0 ± 1.0 8 ± 1	8.0 ± 1.0 8 ± 1
0.0 (0) | 0.5 (½)	0.0 (0) | 0.5 (½)	0.0 (0) | 0.5 (½)	CROSS | TOL.	0.0 (0) | 0.5 (½)	0.0 (0) | 0.5 (½)	0.0 (0) | 0.5 (½)
0.25 ± 0.5 ¼ ± ½	0.65 ± 0.5 ⅝ ± ½	0.5 ± 0.5 ½ ± ½	CAMBER (Degrees)	0.65 ± 0.6 ⅝ ± ⅝	1.0 ± 0.75 1 ± ¾	0.0 ± 0.75 0 ± ¾
0.0 (0) | 0.5 (½)	0.0 (0) | 0.5 (½)	0.0 (0) | 0.5 (½)	CROSS | TOL.	0.0 (0) | 0.5 (½)	0.0 (0) | 0.5 (½)	0.0 (0) | 0.5 (½)
0.15° ± 0.1° 5/32° ± 3/32°	0.24° ± 0.2° ¼° ± 3/16°	0.24° ± 0.2° ¼° ± 3/16°	FRONT TOE (Degrees)	0.24° ± 0.2° ¼° ± 3/16°	0.15° ± 0.1° 5/32° ± 3/32°	0.15° ± 0.1° 5/32° ± 3/32°
0.08" ± 0.05" 1/16" ± 1/16"	0.12" ± 0.1" ⅛" ± 3/32"	0.12" ± 0.1" ⅛" ± 3/32"	FRONT TOE (Inches)	0.12" ± 0.1" ⅛" ± 3/32"	0.08" ± 0.05" 1/16" ± 1/16"	0.08" ± 0.05" 1/16" ± 1/16"
			SAI			
			TURNING ANGLE			
41	43	43	ILLUSTRATION #	43		
			REAR			
			CAMBER (Degrees)			
			REAR TOE (Degrees)			
			REAR TOE (Inches)			
			THRUST ANGLE			
			ILLUSTRATION #			
			NOTES			
09017	09028	09029		09030	09032	09033

Chevrolet/GMC Truck & SUV

<table>
<tr><th>MODEL INFORMATION</th><th>Acadia/Limited
All Models</th><th>Acadia
All Models</th><th>Astro Van RWD
Safari Van RWD</th><th>Astro Van RWD
Safari Van RWD</th><th>Astro Van AWD
Safari Van AWD</th><th>Astro Van AWD
Safari Van AWD</th></tr>
<tr><td>YEARS</td><td>2008-2017</td><td>2017</td><td>1990-98</td><td>1999-2005</td><td>1990-91</td><td>1992-98</td></tr>
<tr><td>INSTRUCTIONS</td><td></td><td></td><td></td><td></td><td></td><td></td></tr>
<tr><td>LUGNUT TORQUE</td><td>140 ft.lbs.</td><td>140 ft.lbs.</td><td>90-100 ft.lbs.</td><td>90-100 ft.lbs.</td><td>90-100 ft.lbs.</td><td>90-100 ft.lbs.</td></tr>
<tr><td>FRONT</td><td></td><td></td><td></td><td></td><td></td><td></td></tr>
<tr><td>CASTER (Degrees)</td><td>4.7 ±0.8
4 11/16 ± 13/16</td><td>4.3 ±0.8
4 5/16 ± 13/16</td><td>2.7 ±0.5
2 11/16 ± ½</td><td>L= 3.0, R= 3.5 ± 1.0
L= 3, R= 3½ ±1</td><td>3.5 ±0.5
3½ ± ½</td><td>2.0 ±0.5
2 ± ½</td></tr>
<tr><td>CROSS | TOL.</td><td>0.0 (0) | 0.8 (13/16)</td><td>0.0 (0) | 0.8 (13/16)</td><td>0.0 (0) | 0.5 (½)</td><td>-0.5 (-½) | 0.5 (½)</td><td>0.0 (0) | 0.5 (½)</td><td>0.0 (0) | 0.5 (½)</td></tr>
<tr><td>CAMBER (Degrees)</td><td>-0.9 ±0.8
-⅞ ± 13/16</td><td>-0.6 ±0.8
-⅝ ± 13/16</td><td>0.8 ±0.5
13/16 ± ½</td><td>0.6 ±0.5
⅝ ± ½</td><td>0.9 ±0.5
⅞ ± ½</td><td>0.9 ±0.5
⅞ ± ½</td></tr>
<tr><td>CROSS | TOL.</td><td>0.0 (0) | 0.8 (13/16)</td><td>0.0 (0) | 0.8 (13/16)</td><td>0.0 (0) | 0.5 (½)</td><td>0.0 (0) | 0.5 (½)</td><td>0.0 (0) | 0.5 (½)</td><td>0.0 (0) | 0.5 (½)</td></tr>
<tr><td>FRONT TOE (Degrees)</td><td>0.0° ±0.2°
0° ± 3/16°</td><td>0.2° ±0.2°
3/16° ± 3/16°</td><td>0.2° ±0.1°
3/16° ± 3/32°</td><td>0.0° ±0.2° [1]
0° ± 3/16° [1]</td><td>-0.1° ±0.1°
-3/32° ± 3/32°</td><td>-0.1° ±0.1°
-3/32° ± 3/32°</td></tr>
<tr><td>FRONT TOE (Inches)</td><td>0.0" ±0.1"
0" ± 3/32"</td><td>0.1" ±0.1"
3/32" ± 3/32"</td><td>0.1" ±0.05"
3/32" ± 1/16"</td><td>0.0" ±0.1" [1]
0" ± 3/32" [1]</td><td>-0.05" ±0.05"
-1/16" ± 1/16"</td><td>-0.05" ±0.05"
-1/16" ± 1/16"</td></tr>
<tr><td>SAI</td><td></td><td></td><td></td><td></td><td></td><td></td></tr>
<tr><td>TURNING ANGLE</td><td></td><td></td><td></td><td></td><td></td><td></td></tr>
<tr><td>ILLUSTRATION #</td><td>25</td><td>202</td><td>23</td><td>23</td><td>24</td><td>24</td></tr>
<tr><td>REAR</td><td></td><td></td><td></td><td></td><td></td><td></td></tr>
<tr><td>CAMBER (Degrees)</td><td>-0.9 ±0.8
-⅞ ± 13/16</td><td>-0.6 ±0.8
-⅝ ± 13/16</td><td></td><td></td><td></td><td></td></tr>
<tr><td>REAR TOE (Degrees)</td><td>0.1° ±0.2°
3/32° ± 3/16°</td><td>0.2° ±0.2°
3/16° ± 3/16°</td><td></td><td></td><td></td><td></td></tr>
<tr><td>REAR TOE (Inches)</td><td>0.05" ±0.1"
1/16" ± 3/32"</td><td>0.1" ±0.1"
3/32" ± 3/32"</td><td></td><td></td><td></td><td></td></tr>
<tr><td>THRUST ANGLE</td><td>0.3 (5/16)</td><td>0.2 (3/16)</td><td></td><td></td><td></td><td></td></tr>
<tr><td>ILLUSTRATION #</td><td>107 143</td><td>177</td><td></td><td></td><td></td><td></td></tr>
<tr><td>NOTES</td><td></td><td></td><td></td><td>[1] 2005 = 0.1° (0.05")</td><td></td><td></td></tr>
<tr><td></td><td>09034</td><td>09035</td><td>09037</td><td>09038</td><td>09039</td><td>09040</td></tr>
</table>

Chevrolet/GMC Truck & SUV

<table>
<tr><th>MODEL INFORMATION</th><th>Astro Van AWD
Safari Van AWD</th><th>Avalanche 1500
4X2</th><th>Avalanche 1500
4X4</th><th>Avalanche 1500
4X2</th><th>Avalanche 1500
4X4</th><th>Avalanche 2500
4X2</th></tr>
<tr><td>YEARS</td><td>1999-2005</td><td>2002-2005</td><td>2002-2005</td><td>2006</td><td>2006</td><td>2002-2005</td></tr>
<tr><td>INSTRUCTIONS</td><td></td><td>p. 304, 305</td><td>p. 304, 305</td><td></td><td></td><td>p. 304, 305</td></tr>
<tr><td>LUGNUT TORQUE</td><td>90-100 ft.lbs.</td><td></td><td></td><td></td><td></td><td></td></tr>
<tr><td>FRONT</td><td></td><td></td><td></td><td></td><td></td><td></td></tr>
<tr><td>CASTER (Degrees)</td><td>L= 3.5, R= 4.5 ± 1.0
L= 3½, R= 4½ ±1</td><td>L= 3.9, R= 4.7 ± 1.0 [1]
L= 3⅞, R= 4 11/16 ±1 [1]</td><td>L= 3.6, R= 4.4 ± 1.0 [1]
L= 3⅝, R= 4⅜ ±1 [1]</td><td>L= 2.9, R= 3.7 ± 1.0 [1 2]
L= 2⅞, R= 3 11/16 ±1 [1 2]</td><td>L= 2.6, R= 3.4 ± 1.0 [1 2]
L= 2⅝, R= 3⅜ ±1 [1 2]</td><td>L= 4.5, R= 4.75 ± 1.0
L= 4½, R= 4¾ ±1</td></tr>
<tr><td>CROSS | TOL.</td><td>-1.0 (-1) | 0.5 (½)</td><td>-0.8 (-13/16) | 0.75 (¾)</td><td>-0.8 (-13/16) | 0.75 (¾)</td><td>-0.8 (-13/16) | 0.75 (¾)</td><td>-0.8 (-13/16) | 0.75 (¾)</td><td>-0.25 (-¼) | 0.75 (¾)</td></tr>
<tr><td>CAMBER (Degrees)</td><td>0.0 ±0.5
0 ± ½</td><td>0.25 ±0.5
¼ ± ½</td><td>0.25 ±0.5
¼ ± ½</td><td>0.25 ±0.6
¼ ± ⅝</td><td>0.25 ±0.6
¼ ± ⅝</td><td>0.25 ±0.5
¼ ± ½</td></tr>
<tr><td>CROSS | TOL.</td><td>0.0 (0) | 0.5 (½)</td><td>0.0 (0) | 0.5 (½)</td><td>0.0 (0) | 0.5 (½)</td><td>0.0 (0) | 0.6 (⅝)</td><td>0.0 (0) | 0.6 (⅝)</td><td>0.0 (0) | 0.5 (½)</td></tr>
<tr><td>FRONT TOE (Degrees)</td><td>0.0° ±0.2°
0° ± 3/16°</td><td>0.1° ±0.2°
3/32° ± 3/16°</td><td>0.1° ±0.2°
3/32° ± 3/16°</td><td>0.1° ±0.2°
3/32° ± 3/16°</td><td>0.1° ±0.2°
3/32° ± 3/16°</td><td>0.1° ±0.2°
3/32° ± 3/16°</td></tr>
<tr><td>FRONT TOE (Inches)</td><td>0.0" ±0.1"
0" ± 3/32"</td><td>0.05" ±0.1"
1/16" ± 3/32"</td><td>0.05" ±0.1"
1/16" ± 3/32"</td><td>0.05" ±0.1"
1/16" ± 3/32"</td><td>0.05" ±0.1"
1/16" ± 3/32"</td><td>0.05" ±0.1"
1/16" ± 3/32"</td></tr>
<tr><td>SAI</td><td></td><td></td><td></td><td></td><td></td><td></td></tr>
<tr><td>TURNING ANGLE</td><td></td><td></td><td></td><td></td><td></td><td></td></tr>
<tr><td>ILLUSTRATION #</td><td>24</td><td>126</td><td>126</td><td>126</td><td>126</td><td>126</td></tr>
<tr><td>REAR</td><td></td><td></td><td></td><td></td><td></td><td></td></tr>
<tr><td>CAMBER (Degrees)</td><td></td><td></td><td></td><td></td><td></td><td></td></tr>
<tr><td>REAR TOE (Degrees)</td><td></td><td></td><td></td><td></td><td></td><td></td></tr>
<tr><td>REAR TOE (Inches)</td><td></td><td></td><td></td><td></td><td></td><td></td></tr>
<tr><td>THRUST ANGLE</td><td></td><td></td><td></td><td></td><td></td><td></td></tr>
<tr><td>ILLUSTRATION #</td><td></td><td></td><td></td><td></td><td></td><td></td></tr>
<tr><td>NOTES</td><td></td><td>[1] w/17" Tires L=4.1° R=4.7°</td><td>[1] w/17" Tires L=3.8° R=4.5°</td><td>[1] w/17" Tires L=3.1° R=3.7°
[2] Relative to ground</td><td>[1] w/17" Tires L=2.8° R=3.5°
[2] Relative to ground</td><td></td></tr>
<tr><td></td><td>09041</td><td>09042</td><td>09043</td><td>09044</td><td>09045</td><td>09046</td></tr>
</table>

ALIGNMENT SPECIFICATIONS

Chevrolet/GMC Truck & SUV

Avalanche 2500	Avalanche 2500	Avalanche 2500	MODEL	Avalanche (All)	Avalanche (All)	Avalanche (All)
4X4	4X2	4X4	INFORMATION	4X2 & 4X4	4X2 & 4X4	4X2 & 4X4
				17" B'stone Tires	17" G'year Tires	18" Tires
2002-2005	2006	2006	YEARS	2007	2007	2007
p. 304, 305			INSTRUCTIONS			
	140 ft.lbs.	140 ft.lbs.	LUGNUT TORQUE	140 ft.lbs.	140 ft.lbs.	140 ft.lbs.
			FRONT			
L= 4.5, R= 4.75 ± 1.0 L= 4½, R= 4¾ ± 1	L= 3.5, R= 3.75 ± 1.0 [1] L= 3½, R= 3¾ ± 1 [1]	L= 3.5, R= 3.75 ± 1.0 [1] L= 3½, R= 3¾ ± 1 [1]	CASTER (Degrees)	L= 3.35, R= 3.45 ± 1.0 [1] L= 3⅜, R= 3 7⁄16 ± 1 [1]	L= 3.2, R= 3.6 ± 1.0 [1] L= 3 3⁄16, R= 3⅝ ± 1 [1]	L= 3.3, R= 3.5 ± 1.0 [1] L= 3 5⁄16, R= 3½ ± 1 [1]
-0.25 (-¼) | 0.75 (¾)	-0.25 (-¼) | 0.75 (¾)	-0.25 (-¼) | 0.75 (¾)	CROSS | TOL.	-0.1 (-⅛) | 0.5 (½)	-0.4 (-⅜) | 0.5 (½)	-0.2 (-3⁄16) | 0.5 (½)
0.25 ± 0.5 ¼ ± ½	0.25 ± 0.6 ¼ ± ⅝	0.25 ± 0.6 ¼ ± ⅝	CAMBER (Degrees)	-0.1 ± 0.5 -⅛ ± ½	-0.1 ± 0.5 -⅛ ± ½	-0.1 ± 0.5 -⅛ ± ½
0.0 (0) | 0.5 (½)	0.0 (0) | 0.6 (⅝)	0.0 (0) | 0.6 (⅝)	CROSS | TOL.	0.0 (0) | 0.5 (½)	0.0 (0) | 0.5 (½)	0.0 (0) | 0.5 (½)
0.1° ± 0.2° 3⁄32° ± 3⁄16°	0.1° ± 0.2° 3⁄32° ± 3⁄16°	0.1° ± 0.2° 3⁄32° ± 3⁄16°	FRONT TOE (Degrees)	0.1° ± 0.2° 3⁄32° ± 3⁄16°	0.1° ± 0.2° 3⁄32° ± 3⁄16°	0.1° ± 0.2° 3⁄32° ± 3⁄16°
0.05" ± 0.1" 1⁄16" ± 3⁄32"	0.05" ± 0.1" 1⁄16" ± 3⁄32"	0.05" ± 0.1" 1⁄16" ± 3⁄32"	FRONT TOE (Inches)	0.05" ± 0.1" 1⁄16" ± 3⁄32"	0.05" ± 0.1" 1⁄16" ± 3⁄32"	0.05" ± 0.1" 1⁄16" ± 3⁄32"
			SAI			
			TURNING ANGLE			
126	126	126	ILLUSTRATION #	126	126	126
			REAR			
			CAMBER (Degrees)			
			REAR TOE (Degrees)			
			REAR TOE (Inches)			
			THRUST ANGLE			
			ILLUSTRATION #			
	[1] Relative to ground	[1] Relative to ground	NOTES	[1] Relative to ground	[1] Relative to ground	[1] Relative to ground
09047	09048	09049		09050	09051	09052

Chevrolet/GMC Truck & SUV

Avalanche (All)	Avalanche	Avalanche	MODEL	Avalanche	Avalanche	Avalanche
4X2 & 4X4	4X2 & 4X4	4X2 & 4X4	INFORMATION	4X2 & 4X4	4X2 & 4X4	4X2 & 4X4
20" Tires	17" B'stone Tires	17" G'year Tires		18" Tires	20" B'stone Tires	20" G'year Tires
2007	2008-2010	2008-2010	YEARS	2008-2010	2008-2010	2008-2010
			INSTRUCTIONS			
140 ft.lbs.	140 ft.lbs.	140 ft.lbs.	LUGNUT TORQUE	140 ft.lbs.	140 ft.lbs.	140 ft.lbs.
			FRONT			
L= 3.55, R= 3.25 ± 1.0 [1] L= 3 9⁄16, R= 3¼ ± 1 [1]	L= 2.95, R= 3.05 ± 1.0 [1] L= 2 15⁄16, R= 3 1⁄16 ± 1 [1]	L= 2.8, R= 3.2 ± 1.0 [1] L= 2 13⁄16, R= 3 3⁄16 ± 1 [1]	CASTER (Degrees)	L= 3.05, R= 2.95 ± 1.0 [1] L= 3 1⁄16, R= 2 15⁄16 ± 1 [1]	L= 3.15, R= 2.85 ± 1.0 [1] L= 3⅛, R= 2⅞ ± 1 [1]	3.1 ± 1.0 [1] 3⅛ ± 1 [1]
0.3 (5⁄16) | 0.5 (½)	-0.1 (-⅛) | 0.75 (¾)	-0.4 (-⅜) | 0.75 (¾)	CROSS | TOL.	0.1 (⅛) | 0.75 (¾)	0.3 (5⁄16) | 0.75 (¾)	0.0 (0) | 0.75 (¾)
-0.1 ± 0.5 -⅛ ± ½	-0.1 ± 0.6 -⅛ ± ⅝	-0.1 ± 0.6 -⅛ ± ⅝	CAMBER (Degrees)	-0.1 ± 0.6 -⅛ ± ⅝	-0.1 ± 0.6 -⅛ ± ⅝	-0.1 ± 0.6 -⅛ ± ⅝
0.0 (0) | 0.5 (½)	0.0 (0) | 0.6 (⅝)	0.0 (0) | 0.6 (⅝)	CROSS | TOL.	0.0 (0) | 0.6 (⅝)	0.0 (0) | 0.6 (⅝)	0.0 (0) | 0.6 (⅝)
0.1° ± 0.2° 3⁄32° ± 3⁄16°	0.1° ± 0.2° 3⁄32° ± 3⁄16°	0.1° ± 0.2° 3⁄32° ± 3⁄16°	FRONT TOE (Degrees)	0.1° ± 0.2° 3⁄32° ± 3⁄16°	0.1° ± 0.2° 3⁄32° ± 3⁄16°	0.1° ± 0.2° 3⁄32° ± 3⁄16°
0.05" ± 0.1" 1⁄16" ± 3⁄32"	0.05" ± 0.1" 1⁄16" ± 3⁄32"	0.05" ± 0.1" 1⁄16" ± 3⁄32"	FRONT TOE (Inches)	0.05" ± 0.1" 1⁄16" ± 3⁄32"	0.05" ± 0.1" 1⁄16" ± 3⁄32"	0.05" ± 0.1" 1⁄16" ± 3⁄32"
			SAI			
			TURNING ANGLE			
126	126	126	ILLUSTRATION #	126	126	126
			REAR			
			CAMBER (Degrees)			
			REAR TOE (Degrees)			
			REAR TOE (Inches)			
			THRUST ANGLE			
			ILLUSTRATION #			
[1] Relative to ground	[1] Relative to ground	[1] Relative to ground	NOTES	[1] Relative to ground	[1] Relative to ground	[1] Relative to ground
09053	09054	09055		09056	09057	09058

Chevrolet/GMC Truck & SUV

MODEL INFORMATION	Avalanche 4X2 & 4X4 w/17" B'stone Tires	Avalanche 4X2 & 4X4 w/17" G'year Tires	Avalanche 4X2 & 4X4 with 18" Tires	Avalanche 4X2 & 4X4 with 20" Tires	Blazer 4X4 Jimmy 4X4 (Full-size)	Blazer 4X4 (GMC see Yukon) (Full-size)
YEARS	2011-2013	2011-2013	2011-2013	2011-2013	1990-91	1992-94
INSTRUCTIONS					p. 304, 305	p. 304, 305
LUGNUT TORQUE	140 ft.lbs.	140 ft.lbs.	140 ft.lbs.	140 ft.lbs.		
FRONT						
CASTER (Degrees)	L= 2.9, R= 3.0 ± 1.0 L= 2 ⅞, R= 3 ± 1	L= 2.8, R= 3.2 ± 1.0 L= 2 13/16, R= 3 3/16 ± 1	L= 3.0, R= 2.9 ± 1.0 L= 3, R= 2 ⅞ ± 1	3.1 ± 1.0 3 ⅛ ± 1	8.0 ± 1.0 8 ± 1	3.0 ± 1.0 3 ± 1
CROSS | TOL.	-0.1 (-⅛) | 0.75 (¾)	-0.4 (-⅜) | 0.75 (¾)	0.1 (⅛) | 0.75 (¾)	0.0 (0) | 0.75 (¾)	0.0 (0) | 0.5 (½)	0.0 (0) | 0.5 (½)
CAMBER (Degrees)	-0.1 ± 0.8 -⅛ ± 13/16	-0.1 ± 0.8 -⅛ ± 13/16	-0.1 ± 0.8 -⅛ ± 13/16	-0.1 ± 0.8 -⅛ ± 13/16	1.0 ± 0.75 1 ± ¾	0.65 ± 0.5 ⅝ ± ½
CROSS | TOL.	0.0 (0) | 0.8 (13/16)	0.0 (0) | 0.8 (13/16)	0.0 (0) | 0.8 (13/16)	0.0 (0) | 0.8 (13/16)	0.0 (0) | 0.5 (½)	0.0 (0) | 0.5 (½)
FRONT TOE (Degrees)	0.1° ± 0.2° 3/32° ± 3/16°	0.1° ± 0.2° 3/32° ± 3/16°	0.1° ± 0.2° 3/32° ± 3/16°	0.1° ± 0.2° 3/32° ± 3/16°	0.15° ± 0.1° 5/32° ± 3/32°	0.24° ± 0.2° ¼° ± 3/16°
FRONT TOE (Inches)	0.05" ± 0.1" 1/16" ± 3/32"	0.05" ± 0.1" 1/16" ± 3/32"	0.05" ± 0.1" 1/16" ± 3/32"	0.05" ± 0.1" 1/16" ± 3/32"	0.08" ± 0.05" 1/16" ± 1/16"	0.12" ± 0.1" ⅛" ± 3/32"
SAI						
TURNING ANGLE						
ILLUSTRATION #	126	126	126	126		43
REAR						
CAMBER (Degrees)						
REAR TOE (Degrees)						
REAR TOE (Inches)						
THRUST ANGLE	0.4 (13/32)	0.4 (13/32)	0.4 (13/32)	0.4 (13/32)		
ILLUSTRATION #						
NOTES						
	09059	09060	09061	09062	09074	09075

Chevrolet/GMC Truck & SUV

MODEL INFORMATION	Captiva Sport 2WD	Captiva Sport AWD	City Express Van	Colorado 4X2 Canyon 4X2 All exc/Sport, Z71	Colorado 4X2 Canyon 4X2 ZQ8 Sport	Colorado 4X2 Canyon 4X2 Z71 (All)
YEARS	2012-2016	2012-2016	2015-2017	2004-2008	2004-2008	2004-2008
INSTRUCTIONS						
LUGNUT TORQUE	100 ft.lbs.	100 ft.lbs.	83 ft.lbs.	103 ft.lbs.	103 ft.lbs.	103 ft.lbs.
FRONT						
CASTER (Degrees)	2.4 ± 0.8 2 ⅜ ± 13/16	2.7 ± 0.8 2 11/16 ± 13/16	4.17 ± 0.75 4 3/16 ± ¾	L= 4.3, R= 4.5 ± 0.75 L= 4 5/16, R= 4 ½ ± ¾	L= 4.2, R= 4.7 ± 0.75 L= 4 3/16, R= 4 11/16 ± ¾	L= 3.6, R= 4.0 ± 0.75 L= 3 ⅝, R= 4 ± ¾
CROSS | TOL.	0.0 (0) | 0.8 (13/16)	0.0 (0) | 0.8 (13/16)	0.0 (0) | 0.75 (¾)	-0.2 (-3/16) | 0.75 (¾)	-0.5 (-½) | 0.75 (¾)	-0.4 (-⅜) | 0.75 (¾)
CAMBER (Degrees)	-0.8 ± 0.8 [1] -13/16 ± 13/16 [1]	-0.6 ± 0.8 -⅝ ± 13/16	-0.33 ± 0.75 -5/16 ± ¾	0.0 ± 0.75 0 ± ¾	0.0 ± 0.75 0 ± ¾	0.0 ± 0.75 0 ± ¾
CROSS | TOL.	0.0 (0) | 0.8 (13/16)	0.0 (0) | 0.8 (13/16)	0.0 (0) | 0.75 (¾)	0.0 (0) | 0.75 (¾)	0.0 (0) | 0.75 (¾)	0.0 (0) | 0.75 (¾)
FRONT TOE (Degrees)	0.2° ± 0.2° 3/16° ± 3/16°	0.2° ± 0.2° 3/16° ± 3/16°	0.1° ± 0.08° 3/32° ± 3/32°	0.0° ± 0.2° 0° ± 3/16°	0.0° ± 0.2° 0° ± 3/16°	0.0° ± 0.2° 0° ± 3/16°
FRONT TOE (Inches)	0.1" ± 0.1" 3/32" ± 3/32"	0.1" ± 0.1" 3/32" ± 3/32"	0.05" ± 0.04" 1/16" ± 1/32"	0.0" ± 0.1" 0" ± 3/32"	0.0" ± 0.1" 0" ± 3/32"	0.0" ± 0.1" 0" ± 3/32"
SAI			11.9 (11 15/16)			
TURNING ANGLE						
ILLUSTRATION #	25	25		156	156	126
REAR						
CAMBER (Degrees)	-0.4 ± 0.8 -⅜ ± 13/16	-0.4 ± 0.8 -⅜ ± 13/16	-1.5 ± 0.5 1 ½ ± ½			
REAR TOE (Degrees)	0.2° ± 0.2° 3/16° ± 3/16°	0.2° ± 0.2° 3/16° ± 3/16°	0.5° ± 25.0° ½° ± ¼°			
REAR TOE (Inches)	0.1" ± 0.1" 3/32" ± 3/32"	0.1" ± 0.1" 3/32" ± 3/32"	0.25" ± 0.13" ¼" ± ⅛"			
THRUST ANGLE	0.3 (5/16)	0.3 (5/16)				
ILLUSTRATION #	107 143	107 143				
NOTES	[1] 6 Cyl. = -0.7° (-11/16°)					
	09076	09077	09078	09079	09080	09081

ALIGNMENT SPECIFICATIONS

Chevrolet/GMC Truck & SUV

MODEL INFORMATION	Colorado 4X2 Canyon 4X2 V-8 w/235-75R16 Off-Rd	Colorado 4X2 Canyon 4X2 16" tires exc/Sport, Z71	Colorado 4X2 Canyon 4X2 18" tires exc/Sport, Z71	Colorado 4X2 Canyon 4X2 ZQ8 Sport	Colorado 4X2 Canyon 4X2 Z71 exc/235-75R16	Colorado 4X2 Canyon 4X2 Z71 w/235-75R16
YEARS	2009-2012	2009-2012	2009-2012	2009-2012	2009-2012	2009-2012
INSTRUCTIONS						
LUGNUT TORQUE	103 ft.lbs.	103 ft.lbs.	103 ft.lbs.	103 ft.lbs.	103 ft.lbs.	103 ft.lbs.
FRONT						
CASTER (Degrees)	2.9 ± 0.75 2⅞ ± ¾	L= 2.8, R= 3.3 ± 0.75 L= 2 13/16, R= 3 5/16 ± ¾	L= 2.9, R= 3.1 ± 0.75 L= 2⅞, R= 3⅛ ± ¾	L= 3.2, R= 3.5 ± 0.75 L= 3 3/16, R= 3½ ± ¾	2.8 ± 0.75 2 13/16 ± ¾	L= 2.6, R= 3.0 ± 0.75 L= 2⅝, R= 3 ± ¾
CROSS | TOL.	0.0 (0) | 0.75 (¾)	-0.5 (-½) | 0.75 (¾)	-0.2 (-3/16) | 0.75 (¾)	-0.3 (-5/16) | 0.75 (¾)	0.0 (0) | 0.75 (¾)	-0.4 (-⅜) | 0.75 (¾)
CAMBER (Degrees)	0.0 ± 0.75 0 ± ¾	0.0 ± 0.75 0 ± ¾	0.0 ± 0.75 0 ± ¾	0.0 ± 0.75 0 ± ¾	0.0 ± 0.75 0 ± ¾	0.0 ± 0.75 0 ± ¾
CROSS | TOL.	0.0 (0) | 0.75 (¾)	0.0 (0) | 0.75 (¾)	0.0 (0) | 0.75 (¾)	0.0 (0) | 0.75 (¾)	0.0 (0) | 0.75 (¾)	0.0 (0) | 0.75 (¾)
FRONT TOE (Degrees)	0.0° ± 0.2° 0° ± 3/16°	0.0° ± 0.2° 0° ± 3/16°	0.0° ± 0.2° 0° ± 3/16°	0.0° ± 0.2° 0° ± 3/16°	0.0° ± 0.2° 0° ± 3/16°	0.0° ± 0.2° 0° ± 3/16°
FRONT TOE (Inches)	0.0" ± 0.1" 0" ± 3/32"	0.0" ± 0.1" 0" ± 3/32"	0.0" ± 0.1" 0" ± 3/32"	0.0" ± 0.1" 0" ± 3/32"	0.0" ± 0.1" 0" ± 3/32"	0.0" ± 0.1" 0" ± 3/32"
SAI						
TURNING ANGLE						
ILLUSTRATION #	156	156	156	156	126	126
REAR						
CAMBER (Degrees)						
REAR TOE (Degrees)						
REAR TOE (Inches)						
THRUST ANGLE						
ILLUSTRATION #						
NOTES						
	09082	09083	09084	09085	09086	09087

Chevrolet/GMC Truck & SUV

MODEL INFORMATION	Colorado 4X4 Canyon 4X4 All exc/Z71	Colorado 4X4 Canyon 4X4 Z71 (All)	Colorado 4X4 Canyon 4X4 All exc/Z71	Colorado 4X4 Canyon 4X4 Z71 exc/235-75R16	Colorado 4X4 Canyon 4X4 Z71 w/235-75R16	Colorado 4X2 Canyon 4X2
YEARS	2004-2008	2004-2008	2009-2012	2009-2012	2009-2012	2015-2017
INSTRUCTIONS						
LUGNUT TORQUE	103 ft.lbs.	103 ft.lbs.	103 ft.lbs.	103 ft.lbs.	103 ft.lbs.	140 ft.lbs.
FRONT						
CASTER (Degrees)	L= 3.8, R= 4.0 ± 0.75 L= 3 13/16, R= 4 ± ¾	L= 3.6, R= 4.0 ± 0.75 L= 3⅝, R= 4 ± ¾	2.9 ± 0.75 2⅞ ± ¾	2.8 ± 0.75 32 13/16 ± ¾	L= 2.6, R= 3.0 ± 0.75 L= 2⅝, R= 3 ± ¾	2.8 ± 1.0 2 13/16 ± 1
CROSS | TOL.	-0.2 (-3/16) | 0.75 (¾)	-0.4 (-⅜) | 0.75 (¾)	0.0 (0) | 0.75 (¾)	0.0 (0) | 0.75 (¾)	-0.4 (-⅜) | 0.75 (¾)	0.0 (0) | 1.0 (1)
CAMBER (Degrees)	0.0 ± 0.75 0 ± ¾	0.0 ± 0.75 0 ± ¾	0.0 ± 0.75 0 ± ¾	0.0 ± 0.75 0 ± ¾	0.0 ± 0.75 0 ± ¾	-0.1 ± 0.8 -⅛ ± 13/16
CROSS | TOL.	0.0 (0) | 0.75 (¾)	0.0 (0) | 0.75 (¾)	0.0 (0) | 0.75 (¾)	0.0 (0) | 0.75 (¾)	0.0 (0) | 0.75 (¾)	0.0 (0) | 0.8 (13/16)
FRONT TOE (Degrees)	0.0° ± 0.2° 0° ± 3/16°	0.0° ± 0.2° 0° ± 3/16°	0.0° ± 0.2° 0° ± 3/16°	0.0° ± 0.2° 0° ± 3/16°	0.0° ± 0.2° 0° ± 3/16°	0.1° ± 0.2° 3/32° ± 3/16°
FRONT TOE (Inches)	0.0" ± 0.1" 0" ± 3/32"	0.0" ± 0.1" 0" ± 3/32"	0.0" ± 0.1" 0" ± 3/32"	0.0" ± 0.1" 0" ± 3/32"	0.0" ± 0.1" 0" ± 3/32"	0.05" ± 0.1" 1/16" ± 3/32"
SAI						
TURNING ANGLE						
ILLUSTRATION #	126	126	126	126	126	126
REAR						
CAMBER (Degrees)						
REAR TOE (Degrees)						
REAR TOE (Inches)						
THRUST ANGLE						0.4 (13/32)
ILLUSTRATION #						
NOTES						
	09088	09089	09090	09091	09092	09093

Chevrolet/GMC Truck & SUV

MODEL INFORMATION	Colorado 4X4 Canyon 4X4	Envoy	Envoy	Envoy All models	Envoy All models with coil spring r. susp.	Envoy All models with air spring r. susp.
YEARS	2015-2017	1999	2000-2001	2002-2004	2005	2005
INSTRUCTIONS						
LUGNUT TORQUE	140 ft.lbs.	100 ft.lbs.	100 ft.lbs.	103 ft.lbs.	103 ft.lbs.	103 ft.lbs.
FRONT						
CASTER (Degrees)	2.8 ± 1.0 2 13/16 ± 1	3.0 ± 0.5 3 ± ½	L= 2.8, R= 3.3 ± 1.0 L= 2 13/16, R= 3 5/16 ± 1	L= 3.5, R= 4.0 ± 0.5 L= 3½, R= 4 ± ½	L= 3.75, R= 4.25 ± 0.6 L= 3¾, R= 4¼ ± 5/8	L= 4.0, R= 4.5 ± 0.6 L= 4, R= 4½ ± 5/8
CROSS | TOL.	0.0 (0) | 1.0 (1)	0.0 (0) | 0.5 (½)	-0.5 (-½) | 0.5 (½)	-0.5 (-½) | 0.3 (5/16)	-0.5 (-½) | 0.6 (5/8)	-0.5 (-½) | 0.6 (5/8)
CAMBER (Degrees)	-0.1 ± 0.8 -1/8 ± 13/16	0.0 ± 0.5 0 ± ½	0.0 ± 1.0 0 ± 1	-0.5 ± 0.5 -½ ± ½	-0.5 ± 0.6 -½ ± 5/8	-0.5 ± 0.6 -½ ± 5/8
CROSS | TOL.	0.0 (0) | 0.8 (13/16)	0.0 (0) | 0.5 (½)	0.0 (0) | 0.5 (½)	0.0 (0) | 0.3 (5/16)	0.0 (0) | 0.6 (5/8)	0.0 (0) | 0.6 (5/8)
FRONT TOE (Degrees)	0.1° ± 0.2° 3/32° ± 3/16°	0.1° ± 0.1° 3/32° ± 3/32°	0.1° ± 0.2° 3/32° ± 3/16°	0.1° ± 0.2° 3/32° ± 3/16°	0.1° ± 0.2° 3/32° ± 3/16°	0.1° ± 0.2° 3/32° ± 3/16°
FRONT TOE (Inches)	0.05" ± 0.1" 1/16" ± 3/32"	0.05" ± 0.05" 1/16" ± 1/16"	0.05" ± 0.1" 1/16" ± 3/32"	0.05" ± 0.1" 1/16" ± 3/32"	0.05" ± 0.1" 1/16" ± 3/32"	0.05" ± 0.1" 1/16" ± 3/32"
SAI						
TURNING ANGLE						
ILLUSTRATION #	126	24	24	133	133	133
REAR						
CAMBER (Degrees)						
REAR TOE (Degrees)						
REAR TOE (Inches)						
THRUST ANGLE	0.4 (13/32)					
ILLUSTRATION #						
NOTES						
	09094	09098	09099	09100	09101	09102

Chevrolet/GMC Truck & SUV

MODEL INFORMATION	Envoy All models with coil spring r. susp.	Envoy All models with air spring r. susp.	Envoy XL All models with coil spring r. susp.	Envoy XL All models with air spring r. susp.	Equinox FWD & AWD	Equinox exc/Sport FWD & AWD
YEARS	2006-2009	2006-2009	2006	2006	2005-2007	2008-2009
INSTRUCTIONS						
LUGNUT TORQUE	103 ft.lbs.	103 ft.lbs.	103 ft.lbs.	103 ft.lbs.	100 ft.lbs.	100 ft.lbs.
FRONT						
CASTER (Degrees)	L= 3.9, R= 4.1 ± 0.6 L= 3 7/8, R= 4 1/8 ± 5/8	L= 4.15, R= 4.35 ± 0.6 L= 4 1/8, R= 4 3/8 ± 5/8	L= 4.1, R= 3.9 ± 0.6 L= 4 1/8, R= 3 7/8 ± 5/8	L= 4.35, R= 4.15 ± 0.6 L= 4 3/8, R= 4 1/8 ± 5/8	3.0 ± 0.75 3 ± ¾	3.0 ± 0.75 3 ± ¾
CROSS | TOL.	-0.2 (-3/16) | 0.6 (5/8)	-0.2 (-3/16) | 0.6 (5/8)	0.2 (3/16) | 0.6 (5/8)	0.2 (3/16) | 0.6 (5/8)	0.0 (0) | 0.5 (½)	0.0 (0) | 0.5 (½)
CAMBER (Degrees)	-0.5 ± 0.6 -½ ± 5/8	-0.5 ± 0.6 -½ ± 5/8	-0.5 ± 0.6 -½ ± 5/8	-0.5 ± 0.6 -½ ± 5/8	-0.6 ± 0.75 -5/8 ± ¾	L= -0.4, R= -0.6 ± 0.75 L= -3/8, R= -5/8 ± ¾
CROSS | TOL.	0.0 (0) | 0.6 (5/8)	0.0 (0) | 0.6 (5/8)	0.0 (0) | 0.6 (5/8)	0.0 (0) | 0.6 (5/8)	0.0 (0) | 0.75 (¾)	0.2 (3/16) | 0.75 (¾)
FRONT TOE (Degrees)	0.1° ± 0.2° 3/32° ± 3/16°	0.1° ± 0.2° 3/32° ± 3/16°	0.1° ± 0.2° 3/32° ± 3/16°	0.1° ± 0.2° 3/32° ± 3/16°	0.15° ± 0.2° 5/32° ± 3/16°	0.15° ± 0.2° 5/32° ± 3/16°
FRONT TOE (Inches)	0.05" ± 0.1" 1/16" ± 3/32"	0.05" ± 0.1" 1/16" ± 3/32"	0.05" ± 0.1" 1/16" ± 3/32"	0.05" ± 0.1" 1/16" ± 3/32"	0.08" ± 0.1" 1/16" ± 3/32"	0.08" ± 0.1" 1/16" ± 3/32"
SAI						
TURNING ANGLE						
ILLUSTRATION #	133	133	133	133	25	25
REAR						
CAMBER (Degrees)					-0.5 ± 0.75 -½ ± ¾	-0.5 ± 0.75 -½ ± ¾
REAR TOE (Degrees)					0.2° ± 0.2° 3/16° ± 3/16°	0.2° ± 0.2° 3/16° ± 3/16°
REAR TOE (Inches)					0.1" ± 0.1" 3/32" ± 3/32"	0.1" ± 0.1" 3/32" ± 3/32"
THRUST ANGLE					0.2 (3/16)	0.2 (3/16)
ILLUSTRATION #					107 143	107 143
NOTES						
	09103	09104	09105	09106	09107	09108

ALIGNMENT SPECIFICATIONS

Chevrolet/GMC Truck & SUV

Equinox	Equinox	G-Van	MODEL	G-Van	G-Van RWD	G-Van RWD
Sport		10/20	INFORMATION	30	Express/Savana	Express/Savana
FWD & AWD		1500/2500		3500	1500/2500/3500	1500, 2500 LD
2008-2009	2010-2017	1990-96	YEARS	1990-96	1996-2002	2003-2004
		p. 304-306	INSTRUCTIONS	p. 304-306	p. 304, 305	p. 304, 305
100 ft.lbs.	140 ft.lbs.	100 ft.lbs.	LUGNUT TORQUE		140 ft.lbs.	140 ft.lbs.
			FRONT			
3.2 ±0.75 3 3/16 ± ¾	2.7 ±0.6 2 11/16 ± ⅝	see table p. 306 ±0.5 (½)	CASTER (Degrees)	see table p. 306 ±0.5 (½)	3.75 ±1.0 3 ¾ ±1	L= 4.2, R= 4.5 ± 1.0 L= 4 3/16, R= 4 ½ ±1
0.0 (0) | 0.5 (½)	0.0 (0) | 0.9 (⅞)	0.0 (0) | 0.5 (½)	CROSS | TOL.	0.0 (0) | 0.5 (½)	0.0 (0) | 0.5 (½)	-0.3 (-5/16) | 0.75 (¾)
L= -0.6, R= -0.8 ±0.75 L= -⅝, R= -13/16 ± ¾	-0.5 ±0.8 -½ ± 13/16	0.5 ±0.5 ½ ± ½	CAMBER (Degrees)	0.25 ±0.5 ¼ ± ½	0.5 ±0.5 ½ ± ½	0.15 ±0.5 ⅛ ± ½
0.2 (3/16) | 0.75 (¾)	0.0 (0) | 0.8 (13/16)	0.0 (0) | 0.5 (½)	CROSS | TOL.	0.0 (0) | 0.5 (½)	0.0 (0) | 0.5 (½)	0.0 (0) | 0.5 (½)
0.15° ±0.2° 5/32° ± 3/16°	0.2° ±0.2° 3/16° ± 3/16°	0.0° ±0.1° 0° ± 3/32°	FRONT TOE (Degrees)	0.0° ±0.1° 0° ± 3/32°	0.24° ±0.2° ¼° ± 3/16°	0.1° ±0.2° 3/32° ± 3/16°
0.08" ±0.1" 1/16" ± 3/32"	0.1" ±0.1" 3/32" ± 3/32"	0.0" ±0.05" 0" ± 1/16"	FRONT TOE (Inches)	0.0" ±0.05" 0" ± 1/16"	0.12" ±0.1" ⅛" ± 3/32"	0.05" ±0.1" 1/16" ± 3/32"
			SAI			
			TURNING ANGLE			
25	25	41	ILLUSTRATION #	41	43	126
			REAR			
-0.5 ±0.75 -½ ± ¾	-0.4 ±0.8 -⅜ ± 13/16		CAMBER (Degrees)			
0.2° ±0.2° 3/16° ± 3/16°	0.1° ±0.2° 3/32° ± 3/16°		REAR TOE (Degrees)			
0.1" ±0.1" 3/32" ± 3/32"	0.05" ±0.1" 1/16" ± 3/32"		REAR TOE (Inches)			
0.2 (3/16)	0.3 (5/16)		THRUST ANGLE			
107 143	107 143		ILLUSTRATION #			
			NOTES			
09109	09110	09121		09122	09123	09124

Chevrolet/GMC Truck & SUV

G-Van RWD	G-Van RWD	G-Van RWD	MODEL	G-Van RWD	G-Van RWD	G-Van RWD
Express/Savana	Express/Savana	Express/Savana	INFORMATION	Express/Savana	Express/Savana	Express/Savana
2500 HD	3500 <10000 GVW	3500 10000> GVW		1500, 2500 LD	2500 HD 3500	1500, 2500 LD
2003-2004	2003-2004	2003-2004	YEARS	2005-2007	2005-2007	2008-2014
p. 304, 305	p. 304, 305	p. 304, 305	INSTRUCTIONS	p. 304, 305	p. 304, 305	p. 304, 305
140 ft.lbs.	140 ft.lbs.[1]	140 ft.lbs.[1]	LUGNUT TORQUE	140 ft.lbs.	140 ft.lbs.	140 ft.lbs.
			FRONT			
L= 4.6, R= 5.0 ± 1.0 L= 4⅝, R= 5 ±1	L= 4.6, R= 5.0 ± 1.0 L= 4⅝, R= 5 ±1	L= 4.6, R= 4.9 ± 1.0 L= 4⅝, R= 4⅞ ±1	CASTER (Degrees)	L= 4.2, R= 4.5 ± 1.0 L= 4 3/16, R= 4 ½ ±1	L= 4.6, R= 5.0 ± 1.0 L= 4⅝, R= 5 ±1	L= 4.0, R= 4.3 ± 1.0 L= 4, R= 4 5/16 ±1
-0.4 (-⅜) | 0.75 (¾)	-0.4 (-⅜) | 0.75 (¾)	-0.3 (-5/16) | 0.75 (¾)	CROSS | TOL.	-0.3 (-5/16) | 0.75 (¾)	-0.4 (-⅜) | 0.75 (¾)	-0.3 (-5/16) | 0.75 (¾)
0.25 ±0.5 ¼ ± ½	0.25 ±0.5 ¼ ± ½	0.25 ±0.5 ¼ ± ½	CAMBER (Degrees)	0.15 ±0.5 ⅛ ± ½	0.25 ±0.5 ¼ ± ½	0.15 ±0.6 ⅛ ± ⅝
0.0 (0) | 0.5 (½)	0.0 (0) | 0.5 (½)	0.0 (0) | 0.5 (½)	CROSS | TOL.	0.0 (0) | 0.5 (½)	0.0 (0) | 0.5 (½)	0.0 (0) | 0.6 (⅝)
0.1° ±0.2° 3/32° ± 3/16°	0.1° ±0.2° 3/32° ± 3/16°	0.1° ±0.2° 3/32° ± 3/16°	FRONT TOE (Degrees)	0.1° ±0.2° 3/32° ± 3/16°	0.1° ±0.2° 3/32° ± 3/16°	0.1° ±0.2° 3/32° ± 3/16°
0.05" ±0.1" 1/16" ± 3/32"	0.05" ±0.1" 1/16" ± 3/32"	0.05" ±0.1" 1/16" ± 3/32"	FRONT TOE (Inches)	0.05" ±0.1" 1/16" ± 3/32"	0.05" ±0.1" 1/16" ± 3/32"	0.05" ±0.1" 1/16" ± 3/32"
			SAI			
			TURNING ANGLE			
126	126	126	ILLUSTRATION #	126	126	126
			REAR			
			CAMBER (Degrees)			
			REAR TOE (Degrees)			
			REAR TOE (Inches)			
			THRUST ANGLE			
			ILLUSTRATION #			
	[1] Duals = 175 ft.lbs	[1] Duals = 175 ft.lbs	NOTES			
09125	09126	09127		09128	09129	09130

Chevrolet/GMC Truck & SUV

G-Van RWD Express/Savana 2500 HD, 3500, 4500	Express/Savana Van RWD Cargo & Pass. Mods	Express/Savana Van RWD Cutaway Models	MODEL INFORMATION	H-Van AWD Express/Savana All exc/RV Conv.	H-Van AWD Express/Savana All RV Conv.	H-Van AWD Express/Savana All models
2008-2015	2016-2017	2016-2017	YEARS	2003	2003	2004-2007
p. 304, 305			INSTRUCTIONS	p. 304, 305	p. 304, 305	p. 304, 305
140 ft.lbs.	140 ft.lbs.	140 ft.lbs.	LUGNUT TORQUE	140 ft.lbs.[1]	140 ft.lbs.[1]	140 ft.lbs.[1]
			FRONT			
L= 4.3, R= 4.7 ± 1.0 L= 4 5/16, R= 4 11/16 ± 1	L= 4.6, R= 5.0 ± 0.75 L= 4⅝, R= 5 ± ¾	L= 4.6, R= 4.9 ± 0.75 L= 4⅝, R= 4⅞ ± ¾	CASTER (Degrees)	L= 3.6, R= 4.0 ± 1.0 L= 3⅝, R= 4 ± 1	L= 3.6, R= 4.0 ± 1.5 L= 3⅝, R= 4 ± 1½	L= 4.1, R= 4.5 ± 1.0 L= 4⅛, R= 4½ ± 1
-0.4 (-⅜)	-0.4 (-⅜)	-0.3 (-5/16)	CROSS	-0.4 (-⅜)	-0.4 (-⅜)	-0.4 (-⅜)
0.75 (¾)	0.75 (¾)	0.75 (¾)	TOL.	0.5 (½)	0.5 (½)	0.75 (¾)
0.25 ± 0.6 ¼ ± ⅝	0.25 ± 0.75 ¼ ± ¾	0.25 ± 0.75 ¼ ± ¾	CAMBER (Degrees)	0.1 ± 0.75 ⅛ ± ¾	0.1 ± 0.5 ⅛ ± ½	0.15 ± 0.5 ⅛ ± ½
0.0 (0)	0.0 (0)	0.0 (0)	CROSS	0.0 (0)	0.0 (0)	0.0 (0)
0.6 (⅝)	0.75 (¾)	0.75 (¾)	TOL.	0.5 (½)	0.5 (½)	0.5 (½)
0.1° ± 0.2° 3/32° ± 3/16°	0.1° ± 0.2° 3/32° ± 3/16°	0.1° ± 0.2° 3/32° ± 3/16°	FRONT TOE (Degrees)	0.2° ± 0.2° 3/16° ± 3/16°	0.3° ± 0.2° 5/16° ± 3/16°	0.1° ± 0.2° 3/32° ± 3/16°
0.05" ± 0.1" 1/16" ± 3/32"	0.05" ± 0.1" 1/16" ± 3/32"	0.05" ± 0.1" 1/16" ± 3/32"	FRONT TOE (Inches)	0.1" ± 0.1" 3/32" ± 3/32"	0.15" ± 0.1" 5/32" ± 3/32"	0.05" ± 0.1" 1/16" ± 3/32"
			SAI			
			TURNING ANGLE			
126	126	126	ILLUSTRATION #	126	126	126
			REAR			
			CAMBER (Degrees)			
			REAR TOE (Degrees)			
			REAR TOE (Inches)			
			THRUST ANGLE			
			ILLUSTRATION #			
			NOTES	[1] Duals = 175 ft.lbs	[1] Duals = 175 ft.lbs	[1] Duals = 175 ft.lbs
09131	09132	09133		09134	09135	09136

Chevrolet/GMC Truck & SUV

H-Van AWD Express/Savana All models	S10, S15 4X2	S10, S15 4X4	MODEL INFORMATION	S10, Sonoma 4X2	S10, Sonoma 4X4	S10, Sonoma 4X2
2008-2014	1990	1990	YEARS	1991-96	1991-96	1997-99
p. 304, 305			INSTRUCTIONS			
140 ft.lbs.[1]	95 ft.lbs.	95 ft.lbs.	LUGNUT TORQUE	95 ft.lbs.	95 ft.lbs.	100 ft.lbs.
			FRONT			
L= 3.9, R= 4.3 ± 1.0 L= 3⅞, R= 4 5/16 ± 1	2.0 ± 0.5 2 ± ½	2.0 ± 0.5 2 ± ½	CASTER (Degrees)	3.0 ± 0.5[1] 3 ± ½[1]	3.0 ± 0.5[1] 3 ± ½[1]	3.0 ± 0.5[1] 3 ± ½[1]
-0.4 (-⅜)	0.0 (0)	0.0 (0)	CROSS	0.0 (0)	0.0 (0)	0.0 (0)
0.75 (¾)	0.5 (½)	0.5 (½)	TOL.	0.5 (½)	0.5 (½)	0.5 (½)
0.15 ± 0.6 ⅛ ± ⅝	0.8 ± 0.5 13/16 ± ½	0.8 ± 0.5 13/16 ± ½	CAMBER (Degrees)	0.0 ± 0.5 0 ± ½	0.0 ± 0.5 0 ± ½	0.0 ± 0.5 0 ± ½
0.0 (0)	0.0 (0)	0.0 (0)	CROSS	0.0 (0)	0.0 (0)	0.0 (0)
0.6 (⅝)	0.5 (½)	0.5 (½)	TOL.	0.5 (½)	0.5 (½)	0.5 (½)
0.1° ± 0.2° 3/32° ± 3/16°	0.3° ± 0.2° 5/16° ± 3/16°	0.3° ± 0.2° 5/16° ± 3/16°	FRONT TOE (Degrees)	0.2° ± 0.1° 3/16° ± 3/32°	0.2° ± 0.1° 3/16° ± 3/32°	0.1° ± 0.1° 3/32° ± 3/32°
0.05" ± 0.1" 1/16" ± 3/32"	0.15" ± 0.1" 5/32" ± 3/32"	0.15" ± 0.1" 5/32" ± 3/32"	FRONT TOE (Inches)	0.1" ± 0.05" 3/32" ± 1/16"	0.1" ± 0.05" 3/32" ± 1/16"	0.05" ± 0.05" 1/16" ± 1/16"
			SAI			
			TURNING ANGLE			
126	23	24	ILLUSTRATION #	23	24	23
			REAR			
			CAMBER (Degrees)			
			REAR TOE (Degrees)			
			REAR TOE (Inches)			
			THRUST ANGLE			
			ILLUSTRATION #			
[1] Duals = 175 ft.lbs			NOTES	[1] ZR2 Sport Pkg., Z85, ZM6 firm-ride pkg.= 2°	[1] ZR2 Sport Pkg., Z85, ZM6 firm-ride pkg.= 2°	[1] ZR2 Sport Pkg., Z85, ZM6 firm-ride pkg.= 2°
09137	09142	09143		09144	09145	09146

Chevrolet/GMC Truck & SUV

S10, Sonoma 4X4	S10, Sonoma 4X2	S10, Sonoma 4X4	MODEL INFORMATION	S10 Blazer 4X2 S15 Jimmy 4X2	S10 Blazer 4X4 S15 Jimmy 4X4	S10 Blazer 4X2 S15 Jimmy 4X2
1997-99	2000-2003	2000-2004	YEARS	1990-94	1990-94	1995-96
			INSTRUCTIONS			
100 ft.lbs.	100 ft.lbs.	100 ft.lbs.	LUGNUT TORQUE	95 ft.lbs.	95 ft.lbs.	95 ft.lbs.
			FRONT			
3.0 ±0.5 [1] 3 ± ½ [1]	L= 2.8, R= 3.3 ± 1.0 [1] L= 2 13/16, R= 3 5/16 ± 1 [1]	L= 2.8, R= 3.3 ± 1.0 L= 2 13/16, R= 3 5/16 ± 1	CASTER (Degrees)	2.0 ±0.5 2 ± ½	2.0 ±0.5 2 ± ½	3.0 ±0.5 3 ± ½
0.0 (0) | 0.5 (½)	-0.5 (-½) | 0.5 (½)	-0.5 (-½) | 0.5 (½)	CROSS | TOL.	0.0 (0) | 0.5 (½)	0.0 (0) | 0.5 (½)	0.0 (0) | 0.5 (½)
0.0 ±0.5 0 ± ½	0.0 ±1.0 0 ±1	0.0 ±1.0 0 ±1	CAMBER (Degrees)	0.8 ±0.5 13/16 ± ½	0.8 ±0.5 13/16 ± ½	0.0 ±0.5 0 ± ½
0.0 (0) | 0.5 (½)	0.0 (0) | 0.5 (½)	0.0 (0) | 0.5 (½)	CROSS | TOL.	0.0 (0) | 0.5 (½)	0.0 (0) | 0.5 (½)	0.0 (0) | 0.5 (½)
0.1° ±0.1° 3/32° ± 3/32°	0.1° ±0.2° 3/32° ± 3/16°	0.1° ±0.2° 3/32° ± 3/16°	FRONT TOE (Degrees)	0.3° ±0.2° 5/16° ± 3/16°	0.3° ±0.2° 5/16° ± 3/16°	0.2° ±0.1° 3/16° ± 3/32°
0.05" ±0.05" 1/16" ± 1/16"	0.05" ±0.1" 1/16" ± 3/32"	0.05" ±0.1" 1/16" ± 3/32"	FRONT TOE (Inches)	0.15" ±0.1" 5/32" ± 3/32"	0.15" ±0.1" 5/32" ± 3/32"	0.1" ±0.05" 3/32" ± 1/16"
			SAI			
			TURNING ANGLE			
24	23	24	ILLUSTRATION #	23	24	23
			REAR			
			CAMBER (Degrees)			
			REAR TOE (Degrees)			
			REAR TOE (Inches)			
			THRUST ANGLE			
			ILLUSTRATION #			
[1] ZR2 Sport Pkg., Z85, ZM6 firm-ride pkg.= 2°	[1] Xtreme L=4.7°, R=5.2°		NOTES			
09147	09148	09149		09150	09151	09152

Chevrolet/GMC Truck & SUV

S10 Blazer 4X4 S15 Jimmy 4X4	S10 Blazer 4X2 S15 Jimmy 4X2	S10 Blazer 4X4 S15 Jimmy 4X4	MODEL INFORMATION	S10 Blazer 4X2 S15 Jimmy 4X2	S10 Blazer 4X4 S15 Jimmy 4X4	(S)Syclone
1995-96	1997-99	1997-99	YEARS	2000-2005	2000-2005	1991-92
			INSTRUCTIONS			
95 ft.lbs.	100 ft.lbs.	100 ft.lbs.	LUGNUT TORQUE	100 ft.lbs.	100 ft.lbs.	
			FRONT			
3.0 ±0.5 3 ± ½	3.0 ±0.5 3 ± ½	3.0 ±0.5 3 ± ½	CASTER (Degrees)	L= 2.8, R= 3.3 ± 1.0 [1] L= 2 13/16, R= 3 5/16 ± 1 [1]	L= 2.8, R= 3.3 ± 1.0 L= 2 13/16, R= 3 5/16 ± 1	3.5 ±1.0 3½ ±1
0.0 (0) | 0.5 (½)	0.0 (0) | 0.5 (½)	0.0 (0) | 0.5 (½)	CROSS | TOL.	-0.5 (-½) | 0.5 (½)	-0.5 (-½) | 0.5 (½)	0.0 (0) | 0.5 (½)
0.0 ±0.5 0 ± ½	0.0 ±0.5 0 ± ½	0.0 ±0.5 0 ± ½	CAMBER (Degrees)	0.0 ±1.0 [2] 0 ±1 [2]	0.0 ±1.0 [1] 0 ±1 [1]	0.0 ±0.5 0 ± ½
0.0 (0) | 0.5 (½)	0.0 (0) | 0.5 (½)	0.0 (0) | 0.5 (½)	CROSS | TOL.	0.0 (0) | 0.5 (½)	0.0 (0) | 0.5 (½)	0.0 (0) | 0.5 (½)
0.2° ±0.1° 3/16° ± 3/32°	0.1° ±0.1° 3/32° ± 3/32°	0.1° ±0.1° 3/32° ± 3/32°	FRONT TOE (Degrees)	0.1° ±0.2° 3/32° ± 3/16°	0.1° ±0.2° 3/32° ± 3/16°	0.2° ±0.2° 3/16° ± 3/16°
0.1" ±0.05" 3/32" ± 1/16"	0.05" ±0.05" 1/16" ± 1/16"	0.05" ±0.05" 1/16" ± 1/16"	FRONT TOE (Inches)	0.05" ±0.1" 1/16" ± 3/32"	0.05" ±0.1" 1/16" ± 3/32"	0.1" ±0.1" 3/32" ± 3/32"
			SAI			
			TURNING ANGLE			
24	23	24	ILLUSTRATION #	23	24	24
			REAR			
			CAMBER (Degrees)			
			REAR TOE (Degrees)			
			REAR TOE (Inches)			
			THRUST ANGLE			
			ILLUSTRATION #			
			NOTES	[1] Xtreme L=4.7°, R=5.2° [2] ZR2 = 1.25° (1¼°)	[1] ZR2 = 1.25° (1¼°)	
09153	09154	09155		09156	09157	09158

Chevrolet/GMC Truck & SUV

(S)Typhoon	Sierra Denali K1500 (4X4) 17" Tires	Sierra Denali K1500 (4X4) 20" Tires	MODEL INFORMATION	Silverado SS C1500 (4X2)	Silverado SS[1] C1500 (4X2)	Silverado SS K1500 (4X4)
1992-93	2005	2005	YEARS	2005	2006-2007	2003-2005
	p. 304, 305	p. 304, 305	INSTRUCTIONS	p. 304, 305		p. 304, 305
	140 ft.lbs.	140 ft.lbs.	LUGNUT TORQUE	140 ft.lbs.	140 ft.lbs.	140 ft.lbs.
			FRONT			
3.5 ± 1.0 3½ ± 1	L= 3.6, R= 4.3 ± 1.0 L= 3⅝, R= 4 5⁄16 ± 1	L= 3.5, R= 4.5 ± 1.0 L= 3½, R= 4½ ± 1	CASTER (Degrees)	L= 4.1, R= 5.1 ± 1.0 L= 4⅛, R= 5⅛ ± 1	L= 3.55, R= 3.85 ± 1.0 [2] L= 3 9⁄16, R= 3⅞ ± 1 [2]	L= 4.1, R= 5.1 ± 1.0 L= 4⅛, R= 5⅛ ± 1
0.0 (0) | 0.5 (½)	-0.7 (-11⁄16) | 0.75 (¾)	-1.0 (-1) | 0.75 (¾)	CROSS | TOL.	-1.0 (-1) | 0.75 (¾)	-0.3 (-5⁄16) | 0.75 (¾)	-1.0 (-1) | 0.5 (½)
0.0 ± 0.5 0 ± ½	0.25 ± 0.6 ¼ ± ⅝	-0.25 ± 0.6 -¼ ± ⅝	CAMBER (Degrees)	-0.1 ± 0.6 -⅛ ± ⅝	-0.1 ± 0.6 -⅛ ± ⅝	-0.1 ± 0.5 [1] -⅛ ± ½ [1]
0.0 (0) | 0.5 (½)	0.0 (0) | 0.6 (⅝)	0.0 (0) | 0.6 (⅝)	CROSS | TOL.	0.0 (0) | 0.6 (⅝)	0.0 (0) | 0.6 (⅝)	0.0 (0) | 1.0 (1)
0.2° ± 0.2° 3⁄16° ± 3⁄16°	0.1° ± 0.2° 3⁄32° ± 3⁄16°	0.1° ± 0.2° 3⁄32° ± 3⁄16°	FRONT TOE (Degrees)	0.1° ± 0.2° 3⁄32° ± 3⁄16°	0.1° ± 0.2° 3⁄32° ± 3⁄16°	0.1° ± 0.2° 3⁄32° ± 3⁄16°
0.1" ± 0.1" 3⁄32" ± 3⁄32"	0.05" ± 0.1" 1⁄16" ± 3⁄32"	0.05" ± 0.1" 1⁄16" ± 3⁄32"	FRONT TOE (Inches)	0.05" ± 0.1" 1⁄16" ± 3⁄32"	0.05" ± 0.1" 1⁄16" ± 3⁄32"	0.05" ± 0.1" 1⁄16" ± 3⁄32"
			SAI			
			TURNING ANGLE			
24	126	126	ILLUSTRATION #	126	126	126
			REAR			
			CAMBER (Degrees)			
			REAR TOE (Degrees)			
			REAR TOE (Inches)			
			THRUST ANGLE			
			ILLUSTRATION #			
			NOTES		[1] Incl. 2007 Classic [2] Relative to ground	[1] 2003 = +0.25° (¼°)
09159	09160	09161		09162	09163	09164

Chevrolet/GMC Truck & SUV

Silverado/Sierra C1500 (4X2)	Silverado/Sierra K1500 (4X4)	Silverado/Sierra C2500 (4X2)	MODEL INFORMATION	Silverado/Sierra K2500 (4X4)	Silverado/Sierra C1500 (4X2) exc/Crew Cab	Silverado/Sierra C1500 (4X2) Crew Cab
1999-2000	1999-2000	1999-2000	YEARS	1999-2000	2001-2003	2001-2003
p. 304, 305	p. 304, 305	p. 304, 305	INSTRUCTIONS	p. 304, 305	p. 304, 305	p. 304, 305
140 ft.lbs.	140 ft.lbs.	140 ft.lbs.	LUGNUT TORQUE	140 ft.lbs.	140 ft.lbs.	140 ft.lbs.
			FRONT			
L= 3.75, R= 4.0 ± 1.0 L= 3¾, R= 4 ± 1	L= 4.25, R= 4.75 ± 1.0 L= 4¼, R= 4¾ ± 1	L= 4.5, R= 4.75 ± 1.0 L= 4½, R= 4¾ ± 1	CASTER (Degrees)	4.5 ± 1.0 4½ ± 1	L= 3.5, R= 4.0 ± 1.0 [1] L= 3½, R= 4 ± 1 [1]	L= 4.5, R= 4.75 ± 1.0 L= 4½, R= 4¾ ± 1
-0.25 (-¼) | 0.5 (½)	-0.5 (-½) | 0.5 (½)	-0.25 (-¼) | 0.5 (½)	CROSS | TOL.	0.0 (0) | 1.0 (1)	-0.5 (-½) | 0.5 (½)	-0.25 (-¼) | 0.5 (½)
0.25 ± 0.5 ¼ ± ½	0.25 ± 0.5 ¼ ± ½	0.25 ± 0.5 ¼ ± ½	CAMBER (Degrees)	0.25 ± 0.5 ¼ ± ½	0.25 ± 0.5 ¼ ± ½	0.25 ± 0.5 ¼ ± ½
0.0 (0) | 1.0 (1)	0.0 (0) | 1.0 (1)	0.0 (0) | 1.0 (1)	CROSS | TOL.	0.0 (0) | 1.0 (1)	0.0 (0) | 0.5 (½)	0.0 (0) | 0.5 (½)
0.1° ± 0.2° 3⁄32° ± 3⁄16°	0.1° ± 0.2° 3⁄32° ± 3⁄16°	0.1° ± 0.2° 3⁄32° ± 3⁄16°	FRONT TOE (Degrees)	0.1° ± 0.2° 3⁄32° ± 3⁄16°	0.1° ± 0.2° 3⁄32° ± 3⁄16°	0.1° ± 0.2° 3⁄32° ± 3⁄16°
0.05" ± 0.1" 1⁄16" ± 3⁄32"	0.05" ± 0.1" 1⁄16" ± 3⁄32"	0.05" ± 0.1" 1⁄16" ± 3⁄32"	FRONT TOE (Inches)	0.05" ± 0.1" 1⁄16" ± 3⁄32"	0.05" ± 0.1" 1⁄16" ± 3⁄32"	0.05" ± 0.1" 1⁄16" ± 3⁄32"
			SAI			
			TURNING ANGLE			
126	126	126	ILLUSTRATION #	126	126	126
			REAR			
			CAMBER (Degrees)		-0.5 ± 0.5 [2] -½ ± ½ [2]	-0.5 ± 0.5 [1] -½ ± ½ [1]
			REAR TOE (Degrees)		0.0° ± 0.2° [2] 0° ± 3⁄16° [2]	0.0° ± 0.2° [1] 0° ± 3⁄16° [1]
			REAR TOE (Inches)		0.0" ± 0.1" [2] 0" ± 3⁄32" [2]	0.0" ± 0.1" [1] 0" ± 3⁄32" [1]
			THRUST ANGLE			
			ILLUSTRATION #			
			NOTES		[1] 2001 L = 3.75° (3¾°) [2] Quadrasteer Only	[1] Quadrasteer Only
09165	09166	09167		09168	09169	09170

Chevrolet/GMC Truck & SUV

Model Information	Silverado/Sierra K1500 (4X4) exc/Crew, SS VIN 1GC...[1]	Silverado/Sierra K1500 (4X4) exc/Crew, SS VIN 2GC...[1]	Silverado/Sierra K1500 (4X4) exc/Crew, SS VIN 1GC...[1]	Silverado/Sierra K1500 (4X4) exc/Crew, SS VIN 2GC...[1]	Silverado/Sierra K1500 (4X4) Crew Cab	Silverado/Sierra C2500 (4X2) Light Duty
Years	2001-2003	2001-2003	2003	2003	2001-2003	2001-2003
Instructions	p. 304, 305	p. 304, 305	p. 304, 305	p. 304, 305	p. 304, 305	p. 304, 305
LugNut Torque	140 ft.lbs.	140 ft.lbs.	140 ft.lbs.	140 ft.lbs.	140 ft.lbs.	140 ft.lbs.
FRONT						
Caster (Degrees)	L= 3.5, R= 4.5 ± 1.0 L= 3½, R= 4½ ± 1	L= 3.5, R= 4.5 ± 1.0 L= 3½, R= 4½ ± 1	L= 3.8, R= 4.1 ± 1.0 L= 3 13/16, R= 4⅛ ± 1	L= 3.8, R= 4.1 ± 1.0 L= 3 13/16, R= 4⅛ ± 1	L= 4.25, R= 4.75 ± 1.0 L= 4¼, R= 4¾ ± 1	L= 4.5, R= 4.75 ± 1.0 L= 4½, R= 4¾ ± 1
Cross	-1.0 (-1)	-1.0 (-1)	-0.3 (-5/16)	-0.3 (-5/16)	-0.5 (-½)	-0.25 (-¼)
Tol.	0.5 (½)	0.5 (½)	0.5 (½)	0.5 (½)	0.5 (½)	0.5 (½)
Camber (Degrees)	0.25 ± 0.5 ¼ ± ½	0.25 ± 0.5 ¼ ± ½	0.25 ± 0.5 ¼ ± ½	0.25 ± 0.5 ¼ ± ½	0.25 ± 0.5 ¼ ± ½	0.25 ± 0.5 ¼ ± ½
Cross	0.0 (0)	0.0 (0)	0.0 (0)	0.0 (0)	0.0 (0)	0.0 (0)
Tol.	1.0 (1)	1.0 (1)	1.0 (1)	1.0 (1)	0.5 (½)	0.5 (½)
Front Toe (Degrees)	0.1° ± 0.2° 3/32° ± 3/16°	0.1° ± 0.2° 3/32° ± 3/16°	0.1° ± 0.2° 3/32° ± 3/16°	0.1° ± 0.2° 3/32° ± 3/16°	0.1° ± 0.2° 3/32° ± 3/16°	0.1° ± 0.2° 3/32° ± 3/16°
Front Toe (Inches)	0.05" ± 0.1" 1/16" ± 3/32"	0.05" ± 0.1" 1/16" ± 3/32"	0.05" ± 0.1" 1/16" ± 3/32"	0.05" ± 0.1" 1/16" ± 3/32"	0.05" ± 0.1" 1/16" ± 3/32"	0.05" ± 0.1" 1/16" ± 3/32"
SAI						
Turning Angle						
Illustration #	126	126	126	126	126	126
REAR						
Camber (Degrees)	-0.5 ± 0.5 [2] -½ ± ½ [2]	-0.5 ± 0.5 [2] -½ ± ½ [2]	-0.5 ± 0.5 [2] -½ ± ½ [2]	-0.5 ± 0.5 [2] -½ ± ½ [2]	-0.5 ± 0.5 [1] -½ ± ½ [1]	
Rear Toe (Degrees)	0.0° ± 0.2° [2] 0° ± 3/16° [2]	0.0° ± 0.2° [2] 0° ± 3/16° [2]	0.0° ± 0.2° [2] 0° ± 3/16° [2]	0.0° ± 0.2° [2] 0° ± 3/16° [2]	0.0° ± 0.2° [1] 0° ± 3/16° [1]	
Rear Toe (Inches)	0.0" ± 0.1" [2] 0" ± 3/32" [2]	0.0" ± 0.1" [2] 0" ± 3/32" [2]	0.0" ± 0.1" [2] 0" ± 3/32" [2]	0.0" ± 0.1" [2] 0" ± 3/32" [2]	0.0" ± 0.1" [1] 0" ± 3/32" [1]	
Thrust Angle						
Illustration #						
Notes	[1] Build date to 11/13/02 (US) [2] Quadrasteer Only	[1] Build date to 11/11/02 (CAN) [2] Quadrasteer Only	[1] Build date 11/14/02 up (US) [2] Quadrasteer Only	[1] Build date 11/12/02 up (CAN) [2] Quadrasteer Only	[1] Quadrasteer Only	
	09171	09172	09173	09174	09175	09176

Chevrolet/GMC Truck & SUV

Model Information	Silverado/Sierra K2500 (4X4) Light Duty	Silverado/Sierra C2500 (4X2) Heavy Duty	Silverado/Sierra K2500 (4X4) Heavy Duty	Silverado/Sierra C3500 (4X2)	Silverado/Sierra K3500 (4X4)	Silverado/Sierra C1500 (4X2) 16" Tires, exc/Crew
Years	2001-2003	2001-2003	2001-2003	2001-2003	2001-2003	2004-2005
Instructions	p. 304, 305	p. 304, 305	p. 304, 305	p. 304, 305	p. 304, 305	p. 304, 305
LugNut Torque	140 ft.lbs.	140 ft.lbs.	140 ft.lbs.	140 ft.lbs.	140 ft.lbs.	140 ft.lbs.
FRONT						
Caster (Degrees)	L= 4.25, R= 4.75 ± 1.0 L= 4¼, R= 4¾ ± 1	L= 4.25, R= 4.75 ± 1.0 L= 4¼, R= 4¾ ± 1	L= 4.0, R= 4.75 ± 1.0 L= 4, R= 4¾ ± 1	L= 4.25, R= 4.75 ± 1.0 L= 4¼, R= 4¾ ± 1	L= 4.0, R= 4.75 ± 1.0 L= 4, R= 4¾ ± 1	L= 3.75, R= 4.0 ± 1.0 L= 3¾, R= 4 ± 1
Cross	-0.5 (-½)	-0.5 (-½)	-0.75 (-¾)	-0.5 (-½)	-0.75 (-¾)	-0.25 (-¼)
Tol.	0.5 (½)	0.5 (½)	0.5 (½)	0.5 (½)	0.5 (½)	0.5 (½)
Camber (Degrees)	0.25 ± 0.5 ¼ ± ½	0.25 ± 0.5 ¼ ± ½	0.25 ± 0.5 ¼ ± ½	0.25 ± 0.5 ¼ ± ½	0.25 ± 0.5 ¼ ± ½	0.25 ± 0.5 ¼ ± ½
Cross	0.0 (0)	0.0 (0)	0.0 (0)	0.0 (0)	0.0 (0)	0.0 (0)
Tol.	0.5 (½)	0.5 (½)	0.5 (½)	0.5 (½)	0.5 (½)	0.5 (½)
Front Toe (Degrees)	0.1° ± 0.2° 3/32° ± 3/16°	0.1° ± 0.2° 3/32° ± 3/16°	0.1° ± 0.2° 3/32° ± 3/16°	0.1° ± 0.2° 3/32° ± 3/16°	0.1° ± 0.2° 3/32° ± 3/16°	0.1° ± 0.2° 3/32° ± 3/16°
Front Toe (Inches)	0.05" ± 0.1" 1/16" ± 3/32"	0.05" ± 0.1" 1/16" ± 3/32"	0.05" ± 0.1" 1/16" ± 3/32"	0.05" ± 0.1" 1/16" ± 3/32"	0.05" ± 0.1" 1/16" ± 3/32"	0.05" ± 0.1" 1/16" ± 3/32"
SAI						
Turning Angle						
Illustration #	126	126	126	126	126	126
REAR						
Camber (Degrees)						-0.5 ± 0.5 [1] -½ ± ½ [1]
Rear Toe (Degrees)						0.0° ± 0.2° [1] 0° ± 3/16° [1]
Rear Toe (Inches)						0.0" ± 0.1" [1] 0" ± 3/32" [1]
Thrust Angle						
Illustration #						
Notes						[1] Quadrasteer Only
	09177	09178	09179	09180	09181	09182

Chevrolet/GMC Truck & SUV

Silverado/Sierra C1500 (4X2) 17" Tires, exc/Crew	Silverado/Sierra C1500 (4X2) 20" Tires, exc/Crew	Silverado/Sierra K1500 (4X4) 16" Tires, exc/Crew	MODEL INFORMATION	Silverado/Sierra K1500 (4X4) 17" exc/Crew, Denali	Silverado/Sierra K1500 (4X4) 20" exc/Crew, Denali	Silverado/Sierra C1500 (4X2) Crew Cab
2005	2004-2005	2004-2005	YEARS	2004-2005	2004-2005	2004-2005
p. 304, 305	p. 304, 305	p. 304, 305	INSTRUCTIONS	p. 304, 305	p. 304, 305	p. 304, 305
140 ft.lbs.	140 ft.lbs.	140 ft.lbs.	LUGNUT TORQUE	140 ft.lbs.	140 ft.lbs.	140 ft.lbs.
			FRONT			
L= 3.85, R= 4.0 ± 1.0 L= 3⅞, R= 4 ± 1	L= 4.1, R= 5.1 ± 1.0 L= 4⅛, R= 5⅛ ± 1	L= 3.8, R= 4.1 ± 1.0 L= 3 13/16, R= 4⅛ ± 1	CASTER (Degrees)	L= 3.6, R= 4.3 ± 1.0 L= 3⅝, R= 4 5/16 ± 1	L= 4.1, R= 5.1 ± 1.0 L= 4⅛, R= 5⅛ ± 1	L= 4.5, R= 4.75 ± 1.0 L= 4½, R= 4¾ ± 1
-0.15 (-⅛) | 0.5 (½)	-1.0 (-1) | 0.5 (½)	-0.3 (-5/16) | 0.5 (½)	CROSS | TOL.	-0.7 (-11/16) | 0.5 (½)	-1.0 (-1) | 0.75 (¾)	-0.25 (-¼) | 0.5 (½)
0.25 ± 0.5 ¼ ± ½	-0.1 ± 0.5 -⅛ ± ½	0.25 ± 0.5 ¼ ± ½	CAMBER (Degrees)	0.25 ± 0.5 ¼ ± ½	-0.1 ± 0.6 -⅛ ± ⅝	0.25 ± 0.5 ¼ ± ½
0.0 (0) | 0.5 (½)	0.0 (0) | 0.5 (½)	0.0 (0) | 1.0 (1)	CROSS | TOL.	0.0 (0) | 1.0 (1)	0.0 (0) | 0.6 (⅝)	0.0 (0) | 0.5 (½)
0.1° ± 0.2° 3/32° ± 3/16°	0.1° ± 0.2° 3/32° ± 3/16°	0.1° ± 0.2° 3/32° ± 3/16°	FRONT TOE (Degrees)	0.1° ± 0.2° 3/32° ± 3/16°	0.1° ± 0.2° 3/32° ± 3/16°	0.1° ± 0.2° 3/32° ± 3/16°
0.05" ± 0.1" 1/16" ± 3/32"	0.05" ± 0.1" 1/16" ± 3/32"	0.05" ± 0.1" 1/16" ± 3/32"	FRONT TOE (Inches)	0.05" ± 0.1" 1/16" ± 3/32"	0.05" ± 0.1" 1/16" ± 3/32"	0.05" ± 0.1" 1/16" ± 3/32"
			SAI			
			TURNING ANGLE			
126	126	126	ILLUSTRATION #	126	126	126
			REAR			
-0.5 ± 0.5 [1] -½ ± ½ [1]	-0.5 ± 0.5 [1] -½ ± ½ [1]	-0.5 ± 0.5 [1] -½ ± ½ [1]	CAMBER (Degrees)	-0.5 ± 0.5 [1] -½ ± ½ [1]	-0.5 ± 0.5 [1] -½ ± ½ [1]	-0.5 ± 0.5 [1] -½ ± ½ [1]
0.0° ± 0.2° [1] 0° ± 3/16° [1]	0.0° ± 0.2° [1] 0° ± 3/16° [1]	0.0° ± 0.2° [1] 0° ± 3/16° [1]	REAR TOE (Degrees)	0.0° ± 0.2° [1] 0° ± 3/16° [1]	0.0° ± 0.2° [1] 0° ± 3/16° [1]	0.0° ± 0.2° [1] 0° ± 3/16° [1]
0.0" ± 0.1" [1] 0" ± 3/32" [1]	0.0" ± 0.1" [1] 0" ± 3/32" [1]	0.0" ± 0.1" [1] 0" ± 3/32" [1]	REAR TOE (Inches)	0.0" ± 0.1" [1] 0" ± 3/32" [1]	0.0" ± 0.1" [1] 0" ± 3/32" [1]	0.0" ± 0.1" [1] 0" ± 3/32" [1]
			THRUST ANGLE			
			ILLUSTRATION #			
[1] Quadrasteer Only	[1] Quadrasteer Only	[1] Quadrasteer Only	NOTES	[1] Quadrasteer Only	[1] Quadrasteer Only	[1] Quadrasteer Only
09183	09184	09185		09186	09187	09188

Chevrolet/GMC Truck & SUV

Silverado/Sierra K1500 (4X4) Crew Cab	Silverado/Sierra C2500 (4X2) Light Duty	Silverado/Sierra K2500 (4X4) Light Duty	MODEL INFORMATION	Silverado/Sierra C2500 (4X2) Heavy Duty	Silverado/Sierra K2500 (4X4) Heavy Duty	Silverado/Sierra C3500 (4X2)
2004-2005	2004-2005	2004-2005	YEARS	2004-2005	2004-2005	2004-2005
p. 304, 305	p. 304, 305	p. 304, 305	INSTRUCTIONS	p. 304, 305	p. 304, 305	p. 304, 305
140 ft.lbs.	140 ft.lbs.	140 ft.lbs.	LUGNUT TORQUE	140 ft.lbs.	140 ft.lbs.	140 ft.lbs.
			FRONT			
L= 4.25, R= 4.5 ± 1.0 L= 4¼, R= 4½ ± 1	L= 4.5, R= 4.75 ± 1.0 L= 4½, R= 4¾ ± 1	L= 4.25, R= 4.5 ± 1.0 L= 4¼, R= 4½ ± 1	CASTER (Degrees)	L= 4.25, R= 4.5 ± 1.0 L= 4¼, R= 4½ ± 1	L= 4.25, R= 4.5 ± 1.0 L= 4¼, R= 4½ ± 1	L= 4.25, R= 4.5 ± 1.0 L= 4¼, R= 4½ ± 1
-0.25 (-¼) | 0.5 (½)	-0.25 (-¼) | 0.5 (½)	-0.25 (-¼) | 0.5 (½)	CROSS | TOL.	-0.25 (-¼) | 0.5 (½)	-0.25 (-¼) | 0.5 (½)	-0.25 (-¼) | 0.5 (½)
0.25 ± 0.5 ¼ ± ½	0.25 ± 0.5 ¼ ± ½	0.25 ± 0.5 ¼ ± ½	CAMBER (Degrees)	0.25 ± 0.5 ¼ ± ½	0.25 ± 0.5 ¼ ± ½	0.25 ± 0.5 ¼ ± ½
0.0 (0) | 0.5 (½)	0.0 (0) | 0.5 (½)	0.0 (0) | 0.5 (½)	CROSS | TOL.	0.0 (0) | 0.5 (½)	0.0 (0) | 0.5 (½)	0.0 (0) | 0.5 (½)
0.1° ± 0.2° 3/32° ± 3/16°	0.1° ± 0.2° 3/32° ± 3/16°	0.1° ± 0.2° 3/32° ± 3/16°	FRONT TOE (Degrees)	0.1° ± 0.2° 3/32° ± 3/16°	0.1° ± 0.2° 3/32° ± 3/16°	0.1° ± 0.2° 3/32° ± 3/16°
0.05" ± 0.1" 1/16" ± 3/32"	0.05" ± 0.1" 1/16" ± 3/32"	0.05" ± 0.1" 1/16" ± 3/32"	FRONT TOE (Inches)	0.05" ± 0.1" 1/16" ± 3/32"	0.05" ± 0.1" 1/16" ± 3/32"	0.05" ± 0.1" 1/16" ± 3/32"
			SAI			
			TURNING ANGLE			
126	126	126	ILLUSTRATION #	126	126	126
			REAR			
-0.5 ± 0.5 [1] -½ ± ½ [1]	-0.5 ± 0.5 [1] -½ ± ½ [1]	-0.5 ± 0.5 [1] -½ ± ½ [1]	CAMBER (Degrees)			
0.0° ± 0.2° [1] 0° ± 3/16° [1]	0.0° ± 0.2° [1] 0° ± 3/16° [1]	0.0° ± 0.2° [1] 0° ± 3/16° [1]	REAR TOE (Degrees)			
0.0" ± 0.1" [1] 0" ± 3/32" [1]	0.0" ± 0.1" [1] 0" ± 3/32" [1]	0.0" ± 0.1" [1] 0" ± 3/32" [1]	REAR TOE (Inches)			
			THRUST ANGLE			
			ILLUSTRATION #			
[1] Quadrasteer Only	[1] Quadrasteer Only	[1] Quadrasteer Only	NOTES			
09189	09190	09191		09192	09193	09194

ALIGNMENT SPECIFICATIONS

Chevrolet/GMC Truck & SUV

Silverado/Sierra	Silverado/Sierra	Silverado/Sierra[1]	MODEL	Silverado/Sierra[1]	Silverado/Sierra[1]	Silverado/Sierra[1]
K3500 (4X4)	K3500 (4X4)	C1500 4X2	INFORMATION	C1500 4X2	C1500 4X2	C1500 4X2
except Dually	Dually	16" Tires		17" Tires	20", Crew w/Torsion Bars	20", Ext. w/Torsion Bars
2004-2005	2004-2005	2006-2007	YEARS	2006-2007	2006-2007	2006-2007
p. 304, 305	p. 304, 305		INSTRUCTIONS			
140 ft.lbs.	140 ft.lbs.	140 ft.lbs.	LUGNUT TORQUE	140 ft.lbs.	140 ft.lbs.	140 ft.lbs.
			FRONT			
L= 4.5, R= 5.0 ± 1.0 L= 4½, R= 5 ± 1	L= 4.25, R= 4.5 ± 1.0 L= 4¼, R= 4½ ± 1	L= 2.35, R= 2.6 ± 1.0[2] L= 2⅜, R= 2⅝ ± 1[2]	CASTER (Degrees)	L= 2.1, R= 2.35 ± 1.0[2] L= 2⅛, R= 2⅜ ± 1[2]	L= 2.45, R= 2.75 ± 1.0[2] L= 2 7/16, R= 2¾ ± 1[2]	L= 3.55, R= 3.85 ± 1.0[2] L= 3 9/16, R= 3⅞ ± 1[2]
-0.5 (-½) | 0.5 (½)	-0.25 (-¼) | 0.5 (½)	-0.25 (-¼) | 0.75 (¾)	CROSS | TOL.	-0.25 (-¼) | 0.75 (¾)	-0.3 (-5/16) | 0.75 (¾)	-0.3 (-5/16) | 0.75 (¾)
0.25 ± 0.5 ¼ ± ½	0.25 ± 0.5 ¼ ± ½	0.25 ± 0.6 ¼ ± ⅝	CAMBER (Degrees)	0.25 ± 0.6 ¼ ± ⅝	-0.1 ± 0.6 -⅛ ± ⅝	-0.1 ± 0.6 -⅛ ± ⅝
0.0 (0) | 0.5 (½)	0.0 (0) | 0.5 (½)	0.0 (0) | 0.6 (⅝)	CROSS | TOL.	0.0 (0) | 0.6 (⅝)	0.0 (0) | 0.6 (⅝)	0.0 (0) | 0.6 (⅝)
0.1° ± 0.2° 3/32° ± 3/16°	0.1° ± 0.2° 3/32° ± 3/16°	0.1° ± 0.2° 3/32° ± 3/16°	FRONT TOE (Degrees)	0.1° ± 0.2° 3/32° ± 3/16°	0.1° ± 0.2° 3/32° ± 3/16°	0.1° ± 0.2° 3/32° ± 3/16°
0.05" ± 0.1" 1/16" ± 3/32"	0.05" ± 0.1" 1/16" ± 3/32"	0.05" ± 0.1" 1/16" ± 3/32"	FRONT TOE (Inches)	0.05" ± 0.1" 1/16" ± 3/32"	0.05" ± 0.1" 1/16" ± 3/32"	0.05" ± 0.1" 1/16" ± 3/32"
			SAI			
			TURNING ANGLE			
126	126	126	ILLUSTRATION #	126	126	126
			REAR			
			CAMBER (Degrees)			
			REAR TOE (Degrees)			
			REAR TOE (Inches)			
			THRUST ANGLE			
			ILLUSTRATION #			
		[1] Incl. 2007 Classic [2] Relative to ground	NOTES	[1] Incl. 2007 Classic [2] Relative to ground	[1] Incl. 2007 Classic [2] Relative to ground	[1] Incl. 2007 Classic [2] Relative to ground
09195	09196	09197		09198	09199	09200

Chevrolet/GMC Truck & SUV

Silverado/Sierra[1]	Silverado/Sierra[1]	Silverado/Sierra[1]	MODEL	Silverado/Sierra[1]	Silverado/Sierra[1]	Silverado/Sierra[1]
K1500 4X4	K1500 4X4	K1500 4X4	INFORMATION	C1500HD 4X2	K1500HD 4X4	C2500 4X2
16" Tires	17" Tires	20" Tires		Crew Cab	Crew Cab	Light Duty
2006-2007	2006-2007	2006-2007	YEARS	2006-2007	2006-2007	2006-2007
			INSTRUCTIONS			
140 ft.lbs.	140 ft.lbs.	140 ft.lbs.	LUGNUT TORQUE	140 ft.lbs.	140 ft.lbs.	140 ft.lbs.
			FRONT			
L= 2.4, R= 2.7 ± 1.0[2] L= 2⅜, R= 2 11/16 ± 1[2]	L= 2.1, R= 2.35 ± 1.0[2] L= 2⅛, R= 2⅜ ± 1[2]	L= 2.45, R= 2.75 ± 1.0[2] L= 2 7/16, R= 2¾ ± 1[2]	CASTER (Degrees)	L= 3.2, R= 3.45 ± 1.0[2] L= 3 3/16, R= 3 7/16 ± 1[2]	L= 2.95, R= 3.2 ± 1.0[2] L= 2 15/16, R= 3 3/16 ± 1[2]	L= 3.2, R= 3.45 ± 1.0[2] L= 3 3/16, R= 3 7/16 ± 1[2]
-0.3 (-5/16) | 0.75 (¾)	-0.25 (-¼) | 0.75 (¾)	-0.3 (-5/16) | 0.75 (¾)	CROSS | TOL.	-0.25 (-¼) | 0.75 (¾)	-0.25 (-¼) | 0.75 (¾)	-0.25 (-¼) | 0.75 (¾)
0.25 ± 0.6 ¼ ± ⅝	0.25 ± 0.6 ¼ ± ⅝	-0.25 ± 0.6 -¼ ± ⅝	CAMBER (Degrees)	0.25 ± 0.6 ¼ ± ⅝	0.25 ± 0.6 ¼ ± ⅝	0.25 ± 0.6 ¼ ± ⅝
0.0 (0) | 0.6 (⅝)	0.0 (0) | 0.6 (⅝)	0.0 (0) | 0.6 (⅝)	CROSS | TOL.	0.0 (0) | 0.6 (⅝)	0.0 (0) | 0.6 (⅝)	0.0 (0) | 0.6 (⅝)
0.1° ± 0.2° 3/32° ± 3/16°	0.1° ± 0.2° 3/32° ± 3/16°	0.1° ± 0.2° 3/32° ± 3/16°	FRONT TOE (Degrees)	0.1° ± 0.2° 3/32° ± 3/16°	0.1° ± 0.2° 3/32° ± 3/16°	0.1° ± 0.2° 3/32° ± 3/16°
0.05" ± 0.1" 1/16" ± 3/32"	0.05" ± 0.1" 1/16" ± 3/32"	0.05" ± 0.1" 1/16" ± 3/32"	FRONT TOE (Inches)	0.05" ± 0.1" 1/16" ± 3/32"	0.05" ± 0.1" 1/16" ± 3/32"	0.05" ± 0.1" 1/16" ± 3/32"
			SAI			
			TURNING ANGLE			
126	126	126	ILLUSTRATION #	126	126	126
			REAR			
			CAMBER (Degrees)			
			REAR TOE (Degrees)			
			REAR TOE (Inches)			
			THRUST ANGLE			
			ILLUSTRATION #			
[1] Incl. 2007 Classic [2] Relative to ground	[1] Incl. 2007 Classic [2] Relative to ground	[1] Incl. 2007 Classic [2] Relative to ground	NOTES	[1] Incl. 2007 Classic [2] Relative to ground	[1] Incl. 2007 Classic [2] Relative to ground	[1] Incl. 2007 Classic [2] Relative to ground
09201	09202	09203		09204	09205	09206

Chevrolet/GMC Truck & SUV

Silverado/Sierra[1] C2500 4X2 Heavy Duty	Silverado/Sierra[1] K2500 4X4 Light Duty	Silverado/Sierra[1] K2500 4X4 Heavy Duty	MODEL INFORMATION	Silverado/Sierra[1] C3500 4X2	Silverado/Sierra[1] K3500 4X4 except Dually	Silverado/Sierra[1] K3500 4X4 Dually
2006-2007	2006-2007	2006-2007	YEARS	2006-2007	2006-2007	2006-2007
			INSTRUCTIONS			
140 ft.lbs.	140 ft.lbs.	140 ft.lbs.	LUGNUT TORQUE	140 ft.lbs.	140 ft.lbs.	
			FRONT			
L= 2.95, R= 3.2 ± 1.0[2] L= 2 15/16, R= 3 3/16 ±1[2]	L= 2.95, R= 3.2 ± 1.0[2] L= 2 15/16, R= 3 3/16 ±1[2]	L= 2.95, R= 3.2 ± 1.0[2] L= 2 15/16, R= 3 3/16 ±1[2]	CASTER (Degrees)	L= 2.5, R= 2.85 ± 1.0[2] L= 2½, R= 2⅞ ±1[2]	L= 2.85, R= 3.35 ± 1.0[2] L= 2⅞, R= 3⅜ ±1[2]	L= 2.5, R= 2.85 ± 1.0[2] L= 2½, R= 2⅞ ±1[2]
-0.25 (-¼) | 0.75 (¾)	-0.25 (-¼) | 0.75 (¾)	-0.25 (-¼) | 0.75 (¾)	CROSS | TOL.	-0.35 (-⅜) | 0.75 (¾)	-0.5 (-½) | 0.75 (¾)	-0.35 (-⅜) | 0.75 (¾)
0.25 ±0.6 ¼ ± ⅝	0.25 ±0.5 ¼ ± ½	0.25 ±0.5 ¼ ± ½	CAMBER (Degrees)	0.25 ±0.6 ¼ ± ⅝	0.25 ±0.6 ¼ ± ⅝	0.25 ±0.6 ¼ ± ⅝
0.0 (0) | 0.6 (⅝)	0.0 (0) | 0.5 (½)	0.0 (0) | 0.5 (½)	CROSS | TOL.	0.0 (0) | 0.6 (⅝)	0.0 (0) | 0.6 (⅝)	0.0 (0) | 0.6 (⅝)
0.1° ±0.2° 3/32° ± 3/16°	0.1° ±0.2° 3/32° ± 3/16°	0.1° ±0.2° 3/32° ± 3/16°	FRONT TOE (Degrees)	0.1° ±0.2° 3/32° ± 3/16°	0.1° ±0.2° 3/32° ± 3/16°	0.1° ±0.2° 3/32° ± 3/16°
0.05" ±0.1" 1/16" ± 3/32"	0.05" ±0.1" 1/16" ± 3/32"	0.05" ±0.1" 1/16" ± 3/32"	FRONT TOE (Inches)	0.05" ±0.1" 1/16" ± 3/32"	0.05" ±0.1" 1/16" ± 3/32"	0.05" ±0.1" 1/16" ± 3/32"
			SAI			
			TURNING ANGLE			
126	126	126	ILLUSTRATION #	126	126	126
			REAR			
			CAMBER (Degrees)			
			REAR TOE (Degrees)			
			REAR TOE (Inches)			
			THRUST ANGLE			
			ILLUSTRATION #			
1 Incl. 2007 Classic 2 Relative to ground	1 Incl. 2007 Classic 2 Relative to ground	1 Incl. 2007 Classic 2 Relative to ground	NOTES	1 Incl. 2007 Classic 2 Relative to ground	1 Incl. 2007 Classic 2 Relative to ground	1 Incl. 2007 Classic 2 Relative to ground
09207	09208	09209		09210	09211	09212

Chevrolet/GMC Truck & SUV

Silverado/Sierra C/K 1500 Reg. Cab, Std. Bed RPO QXR, QXN, QXQ	Silverado/Sierra C/K 1500 Reg. Cab, Std. Bed RPO QVL, QVM, QSS	Silverado/Sierra C/K 1500 Reg. Cab, Std. Bed RPO QXK, QJP, QJM	MODEL INFORMATION	Silverado/Sierra C/K 1500 Reg. Cab, Std. Bed RPO QNM, QPR, QBL	Silverado/Sierra C/K 1500 Reg. Cab, Long Bed RPO QXR (LT45/70R17)	Silverado/Sierra C/K 1500 Reg. Cab, Long Bed RPO QXQ, QVM, QSS
2007-2008	2007-2008	2007-2008	YEARS	2007-2008	2007-2008	2007-2008
p. 305	p. 305	p. 305	INSTRUCTIONS	p. 305	p. 305	p. 305
140 ft.lbs.	140 ft.lbs.	140 ft.lbs.	LUGNUT TORQUE	140 ft.lbs.	140 ft.lbs.	140 ft.lbs.
			FRONT			
L= 2.75, R= 2.65 ± 1.0 L= 2¾, R= 2⅝ ±1	L= 2.85, R= 2.6 ± 1.0 L= 2⅞, R= 2⅝ ±1	L= 2.85, R= 2.6 ± 1.0 L= 2⅞, R= 2⅝ ±1	CASTER (Degrees)	L= 2.95, R= 2.5 ± 1.0 L= 2 15/16, R= 2½ ±1	L= 3.45, R= 3.55 ± 1.0 L= 3 7/16, R= 3 9/16 ±1	L= 3.55, R= 3.45 ± 1.0 L= 3 9/16, R= 3 7/16 ±1
0.1 (⅛) | 0.75 (¾)	0.25 (¼) | 0.75 (¾)	0.25 (¼) | 0.75 (¾)	CROSS | TOL.	0.45 (7/16) | 0.75 (¾)	-0.1 (-⅛) | 0.75 (¾)	0.1 (⅛) | 0.75 (¾)
-0.1 ±0.6 -⅛ ± ⅝	-0.1 ±0.6 -⅛ ± ⅝	-0.1 ±0.6 -⅛ ± ⅝	CAMBER (Degrees)	-0.1 ±0.6 -⅛ ± ⅝	-0.1 ±0.6 -⅛ ± ⅝	-0.1 ±0.6 -⅛ ± ⅝
0.0 (0) | 0.6 (⅝)	0.0 (0) | 0.6 (⅝)	0.0 (0) | 0.6 (⅝)	CROSS | TOL.	0.0 (0) | 0.6 (⅝)	0.0 (0) | 0.6 (⅝)	0.0 (0) | 0.6 (⅝)
0.1° ±0.2° 3/32° ± 3/16°	0.1° ±0.2° 3/32° ± 3/16°	0.1° ±0.2° 3/32° ± 3/16°	FRONT TOE (Degrees)	0.1° ±0.2° 3/32° ± 3/16°	0.1° ±0.2° 3/32° ± 3/16°	0.1° ±0.2° 3/32° ± 3/16°
0.05" ±0.1" 1/16" ± 3/32"	0.05" ±0.1" 1/16" ± 3/32"	0.05" ±0.1" 1/16" ± 3/32"	FRONT TOE (Inches)	0.05" ±0.1" 1/16" ± 3/32"	0.05" ±0.1" 1/16" ± 3/32"	0.05" ±0.1" 1/16" ± 3/32"
			SAI			
			TURNING ANGLE			
126	126	126	ILLUSTRATION #	126	126	126
			REAR			
			CAMBER (Degrees)			
			REAR TOE (Degrees)			
			REAR TOE (Inches)			
			THRUST ANGLE			
			ILLUSTRATION #			
			NOTES			
09213	09214	09215		09216	09217	09218

ALIGNMENT SPECIFICATIONS

Chevrolet/GMC Truck & SUV

MODEL INFORMATION	Silverado/Sierra C/K 1500 Reg. Cab, Long Bed RPO QXN, QVL	Silverado/Sierra C/K 1500 Reg. Cab, Long Bed RPO QNM, QXK, QJP	Silverado/Sierra C/K 1500 Reg. Cab, Long Bed RPO QJM, QPR, QBL	Silverado/Sierra C/K 1500 Ext. Cab, Short Bed RPO QXR (LT45/70R17)	Silverado/Sierra C/K 1500 Ext. Cab, Short Bed RPO QXQ, QSS, QVM	Silverado/Sierra C/K 1500 Ext. Cab, Short Bed RPO QXN, QVL
YEARS	2007-2008	2007-2008	2007-2008	2007-2008	2007-2008	2007-2008
INSTRUCTIONS	p. 305	p. 305	p. 305	p. 305	p. 305	p. 305
LUGNUT TORQUE	140 ft.lbs.	140 ft.lbs.	140 ft.lbs.	140 ft.lbs.	140 ft.lbs.	140 ft.lbs.
FRONT						
CASTER (Degrees)	L= 3.55, R= 3.45 ± 1.0 L= 3 9/16, R= 3 7/16 ± 1	L= 3.65, R= 3.4 ± 1.0 L= 3 5/8, R= 3 3/8 ± 1	L= 3.65, R= 3.4 ± 1.0 L= 3 5/8, R= 3 3/8 ± 1	L= 3.45, R= 3.55 ± 1.0 L= 3 7/16, R= 3 9/16 ± 1	L= 3.55, R= 3.45 ± 1.0 L= 3 9/16, R= 3 7/16 ± 1	L= 3.55, R= 3.45 ± 1.0 L= 3 9/16, R= 3 7/16 ± 1
CROSS | TOL.	0.1 (1/8) | 0.75 (3/4)	0.25 (1/4) | 0.75 (3/4)	0.25 (1/4) | 0.75 (3/4)	-0.1 (-1/8) | 0.75 (3/4)	0.1 (1/8) | 0.75 (3/4)	0.1 (1/8) | 0.75 (3/4)
CAMBER (Degrees)	-0.1 ± 0.6 -1/8 ± 5/8	-0.1 ± 0.6 -1/8 ± 5/8	-0.1 ± 0.6 -1/8 ± 5/8	-0.1 ± 0.6 -1/8 ± 5/8	-0.1 ± 0.6 -1/8 ± 5/8	-0.1 ± 0.6 -1/8 ± 5/8
CROSS | TOL.	0.0 (0) | 0.6 (5/8)	0.0 (0) | 0.6 (5/8)	0.0 (0) | 0.6 (5/8)	0.0 (0) | 0.6 (5/8)	0.0 (0) | 0.6 (5/8)	0.0 (0) | 0.6 (5/8)
FRONT TOE (Degrees)	0.1° ± 0.2° 3/32° ± 3/16°	0.1° ± 0.2° 3/32° ± 3/16°	0.1° ± 0.2° 3/32° ± 3/16°	0.1° ± 0.2° 3/32° ± 3/16°	0.1° ± 0.2° 3/32° ± 3/16°	0.1° ± 0.2° 3/32° ± 3/16°
FRONT TOE (Inches)	0.05" ± 0.1" 1/16" ± 3/32"	0.05" ± 0.1" 1/16" ± 3/32"	0.05" ± 0.1" 1/16" ± 3/32"	0.05" ± 0.1" 1/16" ± 3/32"	0.05" ± 0.1" 1/16" ± 3/32"	0.05" ± 0.1" 1/16" ± 3/32"
SAI						
TURNING ANGLE						
ILLUSTRATION #	126	126	126	126	126	126
REAR						
CAMBER (Degrees)						
REAR TOE (Degrees)						
REAR TOE (Inches)						
THRUST ANGLE						
ILLUSTRATION #						
NOTES						
Ref.	09219	09220	09221	09222	09223	09224

Chevrolet/GMC Truck & SUV

MODEL INFORMATION	Silverado/Sierra C/K 1500 Ext. Cab, Short Bed RPO QNM, QXK	Silverado/Sierra C/K 1500 Ext. Cab, Short Bed RPO QJP, QJM, QBL	Silverado/Sierra C/K 1500 Ext. Cab, Std. Bed RPO QXR (LT45/70R17)	Silverado/Sierra C/K 1500 Ext. Cab, Std. Bed RPO QVL, QVM, QSS	Silverado/Sierra C/K 1500 Ext. Cab, Std. Bed RPO QXQ, QXN	Silverado/Sierra C/K 1500 Ext. Cab, Std. Bed RPO QNM, QPR
YEARS	2007-2008	2007-2008	2007-2008	2007-2008	2007-2008	2007-2008
INSTRUCTIONS	p. 305	p. 305	p. 305	p. 305	p. 305	p. 305
LUGNUT TORQUE	140 ft.lbs.	140 ft.lbs.	140 ft.lbs.	140 ft.lbs.	140 ft.lbs.	140 ft.lbs.
FRONT						
CASTER (Degrees)	L= 3.65, R= 3.4 ± 1.0 L= 3 5/8, R= 3 3/8 ± 1	L= 3.65, R= 3.4 ± 1.0 L= 3 5/8, R= 3 3/8 ± 1	L= 3.4, R= 3.55 ± 1.0 L= 3 3/8, R= 3 9/16 ± 1	L= 3.55, R= 3.45 ± 1.0 L= 3 9/16, R= 3 7/16 ± 1	L= 3.55, R= 3.45 ± 1.0 L= 3 9/16, R= 3 7/16 ± 1	L= 3.65, R= 3.4 ± 1.0 L= 3 5/8, R= 3 3/8 ± 1
CROSS | TOL.	0.25 (1/4) | 0.75 (3/4)	0.25 (1/4) | 0.75 (3/4)	-0.15 (-1/8) | 0.75 (3/4)	0.1 (1/8) | 0.75 (3/4)	0.1 (1/8) | 0.75 (3/4)	0.25 (1/4) | 0.75 (3/4)
CAMBER (Degrees)	-0.1 ± 0.6 -1/8 ± 5/8	-0.1 ± 0.6 -1/8 ± 5/8	-0.1 ± 0.6 -1/8 ± 5/8	-0.1 ± 0.6 -1/8 ± 5/8	-0.1 ± 0.6 -1/8 ± 5/8	-0.1 ± 0.6 -1/8 ± 5/8
CROSS | TOL.	0.0 (0) | 0.6 (5/8)	0.0 (0) | 0.6 (5/8)	0.0 (0) | 0.6 (5/8)	0.0 (0) | 0.6 (5/8)	0.0 (0) | 0.6 (5/8)	0.0 (0) | 0.6 (5/8)
FRONT TOE (Degrees)	0.1° ± 0.2° 3/32° ± 3/16°	0.1° ± 0.2° 3/32° ± 3/16°	0.1° ± 0.2° 3/32° ± 3/16°	0.1° ± 0.2° 3/32° ± 3/16°	0.1° ± 0.2° 3/32° ± 3/16°	0.1° ± 0.2° 3/32° ± 3/16°
FRONT TOE (Inches)	0.05" ± 0.1" 1/16" ± 3/32"	0.05" ± 0.1" 1/16" ± 3/32"	0.05" ± 0.1" 1/16" ± 3/32"	0.05" ± 0.1" 1/16" ± 3/32"	0.05" ± 0.1" 1/16" ± 3/32"	0.05" ± 0.1" 1/16" ± 3/32"
SAI						
TURNING ANGLE						
ILLUSTRATION #	126	126	126	126	126	126
REAR						
CAMBER (Degrees)						
REAR TOE (Degrees)						
REAR TOE (Inches)						
THRUST ANGLE						
ILLUSTRATION #						
NOTES						
Ref.	09225	09226	09227	09228	09229	09230

Chevrolet/GMC Truck & SUV

MODEL INFORMATION	Silverado/Sierra	Silverado/Sierra	Silverado/Sierra	Silverado/Sierra	Silverado/Sierra	Silverado/Sierra
	C/K 1500 Ext. Cab, Std. Bed	C/K 1500 Ext. Cab, Std. Bed	C/K 1500 Ext. Cab, Long Bed	C/K 1500 Ext. Cab, Long Bed	C/K 1500 Ext. Cab, Long Bed	C/K 1500 Ext. Cab, Long Bed
	RPO QPO, QBL	RPO QJP, QJM, QXK	RPO QXN, QVL	RPO QVM, QXR	RPO QJP, QJM, QBL	RPO QXK, QXQ, QSS
YEARS	2007-2008	2007-2008	2007-2008	2007-2008	2007-2008	2007-2008
INSTRUCTIONS	p. 305	p. 305	p. 305	p. 305	p. 305	p. 305
LUGNUT TORQUE	140 ft.lbs.	140 ft.lbs.	140 ft.lbs.	140 ft.lbs.	140 ft.lbs.	140 ft.lbs.
FRONT						
CASTER (Degrees)	L= 3.65, R= 3.4 ± 1.0 L= 3⅝, R= 3⅜ ± 1	L= 3.65, R= 3.4 ± 1.0 L= 3⅝, R= 3⅜ ± 1	L= 3.45, R= 3.55 ± 1.0 L= 3 7/16, R= 3 9/16 ± 1	L= 3.45, R= 3.55 ± 1.0 L= 3 7/16, R= 3 9/16 ± 1	L= 3.55, R= 3.45 ± 1.0 L= 3 9/16, R= 3 7/16 ± 1	L= 3.55, R= 3.45 ± 1.0 L= 3 9/16, R= 3 7/16 ± 1
CROSS | TOL.	0.25 (¼) | 0.75 (¾)	0.25 (¼) | 0.75 (¾)	-0.1 (-⅛) | 0.75 (¾)	-0.1 (-⅛) | 0.75 (¾)	0.1 (⅛) | 0.75 (¾)	0.1 (⅛) | 0.75 (¾)
CAMBER (Degrees)	-0.1 ±0.6 -⅛ ± ⅝	-0.1 ±0.6 -⅛ ± ⅝	-0.1 ±0.6 -⅛ ± ⅝	-0.1 ±0.6 -⅛ ± ⅝	-0.1 ±0.6 -⅛ ± ⅝	-0.1 ±0.6 -⅛ ± ⅝
CROSS | TOL.	0.0 (0) | 0.6 (⅝)	0.0 (0) | 0.6 (⅝)	0.0 (0) | 0.6 (⅝)	0.0 (0) | 0.6 (⅝)	0.0 (0) | 0.6 (⅝)	0.0 (0) | 0.6 (⅝)
FRONT TOE (Degrees)	0.1° ±0.2° 3/32° ± 3/16°	0.1° ±0.2° 3/32° ± 3/16°	0.1° ±0.2° 3/32° ± 3/16°	0.1° ±0.2° 3/32° ± 3/16°	0.1° ±0.2° 3/32° ± 3/16°	0.1° ±0.2° 3/32° ± 3/16°
FRONT TOE (Inches)	0.05" ±0.1" 1/16" ± 3/32"	0.05" ±0.1" 1/16" ± 3/32"	0.05" ±0.1" 1/16" ± 3/32"	0.05" ±0.1" 1/16" ± 3/32"	0.05" ±0.1" 1/16" ± 3/32"	0.05" ±0.1" 1/16" ± 3/32"
SAI						
TURNING ANGLE						
ILLUSTRATION #	126	126	126	126	126	126
REAR						
CAMBER (Degrees)						
REAR TOE (Degrees)						
REAR TOE (Inches)						
THRUST ANGLE						
ILLUSTRATION #						
NOTES						
	09231	09232	09233	09234	09235	09236

Chevrolet/GMC Truck & SUV

MODEL INFORMATION	Silverado/Sierra	Silverado/Sierra	Silverado/Sierra	Silverado/Sierra	Silverado/Sierra	Silverado/Sierra
	C/K 1500 Ext. Cab, Long Bed	C/K 1500 Crew Cab	C/K 1500 Crew Cab	C/K 1500 Crew Cab	C/K 1500 Crew Cab	C/K 1500 Crew Cab
	RPO QPR, QNM, QPO	RPO QXR (LT45/70R17)	RPO QVL, QVM, QSS	RPO QXQ, QXN	RPO QNM, QPR	RPO QPO, QBL
YEARS	2007-2008	2007-2008	2007-2008	2007-2008	2007-2008	2007-2008
INSTRUCTIONS	p. 305	p. 305	p. 305	p. 305	p. 305	p. 305
LUGNUT TORQUE	140 ft.lbs.	140 ft.lbs.	140 ft.lbs.	140 ft.lbs.	140 ft.lbs.	140 ft.lbs.
FRONT						
CASTER (Degrees)	L= 3.65, R= 3.35 ± 1.0 L= 3⅝, R= 3⅜ ± 1	L= 3.4, R= 3.55 ± 1.0 L= 3⅜, R= 3 9/16 ± 1	L= 3.55, R= 3.45 ± 1.0 L= 3 9/16, R= 3 7/16 ± 1	L= 3.55, R= 3.45 ± 1.0 L= 3 9/16, R= 3 7/16 ± 1	L= 3.65, R= 3.4 ± 1.0 L= 3⅝, R= 3⅜ ± 1	L= 3.65, R= 3.4 ± 1.0 L= 3⅝, R= 3⅜ ± 1
CROSS | TOL.	0.3 (5/16) | 0.75 (¾)	-0.15 (-⅛) | 0.75 (¾)	0.1 (⅛) | 0.75 (¾)	0.1 (⅛) | 0.75 (¾)	0.25 (¼) | 0.75 (¾)	0.25 (¼) | 0.75 (¾)
CAMBER (Degrees)	-0.1 ±0.6 -⅛ ± ⅝	-0.1 ±0.6 -⅛ ± ⅝	-0.1 ±0.6 -⅛ ± ⅝	-0.1 ±0.6 -⅛ ± ⅝	-0.1 ±0.6 -⅛ ± ⅝	-0.1 ±0.6 -⅛ ± ⅝
CROSS | TOL.	0.0 (0) | 0.6 (⅝)	0.0 (0) | 0.6 (⅝)	0.0 (0) | 0.6 (⅝)	0.0 (0) | 0.6 (⅝)	0.0 (0) | 0.6 (⅝)	0.0 (0) | 0.6 (⅝)
FRONT TOE (Degrees)	0.1° ±0.2° 3/32° ± 3/16°	0.1° ±0.2° 3/32° ± 3/16°	0.1° ±0.2° 3/32° ± 3/16°	0.1° ±0.2° 3/32° ± 3/16°	0.1° ±0.2° 3/32° ± 3/16°	0.1° ±0.2° 3/32° ± 3/16°
FRONT TOE (Inches)	0.05" ±0.1" 1/16" ± 3/32"	0.05" ±0.1" 1/16" ± 3/32"	0.05" ±0.1" 1/16" ± 3/32"	0.05" ±0.1" 1/16" ± 3/32"	0.05" ±0.1" 1/16" ± 3/32"	0.05" ±0.1" 1/16" ± 3/32"
SAI						
TURNING ANGLE						
ILLUSTRATION #	126	126	126	126	126	126
REAR						
CAMBER (Degrees)						
REAR TOE (Degrees)						
REAR TOE (Inches)						
THRUST ANGLE						
ILLUSTRATION #						
NOTES						
	09237	09238	09239	09240	09241	09242

Chevrolet/GMC Truck & SUV

Silverado/Sierra	Silverado/Sierra	Silverado/Sierra	MODEL	Silverado/Sierra	Silverado/Sierra	Silverado/Sierra
C/K 1500 Crew Cab	C/K 1500 Reg. Cab, Std. Bed	C/K 1500 Reg. Cab, Std. Bed	INFORMATION	C/K 1500 Reg. Cab, Std. Bed	C/K 1500 Reg. Cab, Std. Bed	C/K 1500 Reg. Cab, Std. Bed
RPO QJP, QJM, QXK	RPO QXR, QXN	RPO QXQ, QXQ, RBX		RPO QVL, QVM, QSS	RPO QXK, QJP, QJM	RPO QMG, QMH
2007-2008	2009-2013	2009-2013	YEARS	2009-2013	2009-2013	2009-2013
p. 305	p. 305	p. 305	INSTRUCTIONS	p. 305	p. 305	p. 305
140 ft.lbs.	140 ft.lbs.	140 ft.lbs.	LUGNUT TORQUE	140 ft.lbs.	140 ft.lbs.	140 ft.lbs.
			FRONT			
L= 3.65, R= 3.4 ± 1.0 L= 3⅝, R= 3⅜ ± 1	L= 2.3, R= 2.2 ± 1.0 L= 2 5/16, R= 2 3/16 ± 1	L= 2.3, R= 2.2 ± 1.0 L= 2 5/16, R= 2 3/16 ± 1	CASTER (Degrees)	L= 2.4, R= 2.2 ± 1.0 L= 2⅜, R= 2 3/16 ± 1	L= 2.4, R= 2.2 ± 1.0 L= 2⅜, R= 2 3/16 ± 1	L= 2.4, R= 2.2 ± 1.0 L= 2⅜, R= 2 3/16 ± 1
0.25 (¼) | 0.75 (¾)	0.1 (⅛) | 0.75 (¾)	0.1 (⅛) | 0.75 (¾)	CROSS | TOL.	0.2 (3/16) | 0.75 (¾)	0.2 (3/16) | 0.75 (¾)	0.2 (3/16) | 0.75 (¾)
-0.1 ±0.75 -⅛ ± ¾	-0.1 ±0.75 -⅛ ± ¾	-0.1 ±0.75 -⅛ ± ¾	CAMBER (Degrees)	-0.1 ±0.75 -⅛ ± ¾	-0.1 ±0.75 -⅛ ± ¾	-0.1 ±0.75 -⅛ ± ¾
0.0 (0) | 0.75 (¾)	0.0 (0) | 0.75 (¾)	0.0 (0) | 0.75 (¾)	CROSS | TOL.	0.0 (0) | 0.75 (¾)	0.0 (0) | 0.75 (¾)	0.0 (0) | 0.75 (¾)
0.1° ±0.2° 3/32° ± 3/16°	0.1° ±0.2° 3/32° ± 3/16°	0.1° ±0.2° 3/32° ± 3/16°	FRONT TOE (Degrees)	0.1° ±0.2° 3/32° ± 3/16°	0.1° ±0.2° 3/32° ± 3/16°	0.1° ±0.2° 3/32° ± 3/16°
0.05" ±0.1" 1/16" ± 3/32"	0.05" ±0.1" 1/16" ± 3/32"	0.05" ±0.1" 1/16" ± 3/32"	FRONT TOE (Inches)	0.05" ±0.1" 1/16" ± 3/32"	0.05" ±0.1" 1/16" ± 3/32"	0.05" ±0.1" 1/16" ± 3/32"
			SAI			
			TURNING ANGLE			
126	126	126	ILLUSTRATION #	126	126	126
			REAR			
			CAMBER (Degrees)			
			REAR TOE (Degrees)			
			REAR TOE (Inches)			
			THRUST ANGLE			
			ILLUSTRATION #			
			NOTES			
09243	09244	09245		09246	09247	09248

Chevrolet/GMC Truck & SUV

Silverado/Sierra	Silverado/Sierra	Silverado/Sierra	MODEL	Silverado/Sierra	Silverado/Sierra	Silverado/Sierra
C/K 1500 Reg. Cab, Std. Bed	C/K 1500 Reg. Cab, Std. Bed	C/K 1500 Reg. Cab, Long Bed	INFORMATION	C/K 1500 Reg. Cab, Long Bed	C/K 1500 Reg. Cab, Long Bed	C/K 1500 Reg. Cab, Long Bed
RPO QNM, QPR	RPO QBL, QU1	RPO QXR (LT45/70R17)		RPO QXQ, QVM, QSS	RPO QXN, QVL, RBX	RPO QNM, QXK, QJP
2009-2013	2009-2013	2009-2013	YEARS	2009-2013	2009-2013	2009-2013
p. 305	p. 305	p. 305	INSTRUCTIONS	p. 305	p. 305	p. 305
140 ft.lbs.	140 ft.lbs.	140 ft.lbs.	LUGNUT TORQUE	140 ft.lbs.	140 ft.lbs.	140 ft.lbs.
			FRONT			
L= 2.5, R= 2.1 ± 1.0 L= 2½, R= 2⅛ ± 1	L= 2.5, R= 2.1 ± 1.0 L= 2½, R= 2⅛ ± 1	L= 3.0, R= 3.1 ± 1.0 L= 3, R= 3⅛ ± 1	CASTER (Degrees)	L= 3.1, R= 3.0 ± 1.0 L= 3⅛, R= 3 ± 1	L= 3.1, R= 3.0 ± 1.0 L= 3⅛, R= 3 ± 1	L= 3.3, R= 3.0 ± 1.0 L= 3 5/16, R= 3 ± 1
0.4 (⅜) | 0.75 (¾)	0.4 (⅜) | 0.75 (¾)	-0.1 (-⅛) | 0.75 (¾)	CROSS | TOL.	0.1 (⅛) | 0.75 (¾)	0.1 (⅛) | 0.75 (¾)	0.3 (5/16) | 0.75 (¾)
-0.1 ±0.75 -⅛ ± ¾	-0.1 ±0.75 -⅛ ± ¾	-0.1 ±0.75 -⅛ ± ¾	CAMBER (Degrees)	-0.1 ±0.75 -⅛ ± ¾	-0.1 ±0.75 -⅛ ± ¾	-0.1 ±0.75 -⅛ ± ¾
0.0 (0) | 0.75 (¾)	0.0 (0) | 0.75 (¾)	0.0 (0) | 0.75 (¾)	CROSS | TOL.	0.0 (0) | 0.75 (¾)	0.0 (0) | 0.75 (¾)	0.0 (0) | 0.75 (¾)
0.1° ±0.2° 3/32° ± 3/16°	0.1° ±0.2° 3/32° ± 3/16°	0.1° ±0.2° 3/32° ± 3/16°	FRONT TOE (Degrees)	0.1° ±0.2° 3/32° ± 3/16°	0.1° ±0.2° 3/32° ± 3/16°	0.1° ±0.2° 3/32° ± 3/16°
0.05" ±0.1" 1/16" ± 3/32"	0.05" ±0.1" 1/16" ± 3/32"	0.05" ±0.1" 1/16" ± 3/32"	FRONT TOE (Inches)	0.05" ±0.1" 1/16" ± 3/32"	0.05" ±0.1" 1/16" ± 3/32"	0.05" ±0.1" 1/16" ± 3/32"
			SAI			
			TURNING ANGLE			
126	126	126	ILLUSTRATION #	126	126	126
			REAR			
			CAMBER (Degrees)			
			REAR TOE (Degrees)			
			REAR TOE (Inches)			
			THRUST ANGLE			
			ILLUSTRATION #			
			NOTES			
09249	09250	09251		09252	09253	09254

Chevrolet/GMC Truck & SUV

MODEL INFORMATION	Silverado/Sierra C/K 1500 Reg. Cab, Long Bed RPO QJM, QPR	Silverado/Sierra C/K 1500 Reg. Cab, Long Bed RPO QBL, QU1	Silverado/Sierra C/K 1500 Ext. Cab, Short Bed RPO QXR, QGI	Silverado/Sierra C/K 1500 Ext. Cab, Short Bed RPO QBL, QXK	Silverado/Sierra C/K 1500 Ext. Cab, Short Bed RPO QJP, QJM	Silverado/Sierra C/K 1500 Ext. Cab, Short Bed RPO QPR, QNM
YEARS	2009-2013	2009-2013	2009-2013	2009-2013	2009-2013	2009-2013
INSTRUCTIONS	p. 305	p. 305	p. 305	p. 305	p. 305	p. 305
LUGNUT TORQUE	140 ft.lbs.	140 ft.lbs.	140 ft.lbs.	140 ft.lbs.	140 ft.lbs.	140 ft.lbs.
FRONT						
CASTER (Degrees)	L= 3.3, R= 3.0 ± 1.0 L= 3 5/16, R= 3 ±1	L= 3.3, R= 3.0 ± 1.0 L= 3 5/16, R= 3 ±1	L= 3.0, R= 3.1 ± 1.0 L= 3, R= 3 1/8 ±1	L= 3.3, R= 3.0 ± 1.0 L= 3 5/16, R= 3 ±1	L= 3.3, R= 3.0 ± 1.0 L= 3 5/16, R= 3 ±1	L= 3.3, R= 3.0 ± 1.0 L= 3 5/16, R= 3 ±1
CROSS	0.3 (5/16)	0.3 (5/16)	-0.1 (-1/8)	0.3 (5/16)	0.3 (5/16)	0.3 (5/16)
TOL.	0.75 (3/4)	0.75 (3/4)	0.75 (3/4)	0.75 (3/4)	0.75 (3/4)	0.75 (3/4)
CAMBER (Degrees)	-0.1 ±0.75 -1/8 ± 3/4	-0.1 ±0.75 -1/8 ± 3/4	-0.1 ±0.75 -1/8 ± 3/4	-0.1 ±0.75 -1/8 ± 3/4	-0.1 ±0.75 -1/8 ± 3/4	-0.1 ±0.75 -1/8 ± 3/4
CROSS	0.0 (0)	0.0 (0)	0.0 (0)	0.0 (0)	0.0 (0)	0.0 (0)
TOL.	0.75 (3/4)	0.75 (3/4)	0.75 (3/4)	0.75 (3/4)	0.75 (3/4)	0.75 (3/4)
FRONT TOE (Degrees)	0.1° ±0.2° 3/32° ± 3/16°	0.1° ±0.2° 3/32° ± 3/16°	0.1° ±0.2° 3/32° ± 3/16°	0.1° ±0.2° 3/32° ± 3/16°	0.1° ±0.2° 3/32° ± 3/16°	0.1° ±0.2° 3/32° ± 3/16°
FRONT TOE (Inches)	0.05" ±0.1" 1/16" ± 3/32"	0.05" ±0.1" 1/16" ± 3/32"	0.05" ±0.1" 1/16" ± 3/32"	0.05" ±0.1" 1/16" ± 3/32"	0.05" ±0.1" 1/16" ± 3/32"	0.05" ±0.1" 1/16" ± 3/32"
SAI						
TURNING ANGLE						
ILLUSTRATION #	126	126	126	126	126	126
REAR						
CAMBER (Degrees)						
REAR TOE (Degrees)						
REAR TOE (Inches)						
THRUST ANGLE						
ILLUSTRATION #						
NOTES						
	09255	09256	09257	09258	09259	09260

Chevrolet/GMC Truck & SUV

MODEL INFORMATION	Silverado/Sierra C/K 1500 Ext. Cab, Short Bed RPO QU1	Silverado/Sierra C/K 1500 Ext. Cab, Short Bed RPO QMG, QMH	Silverado/Sierra C/K 1500 Ext. Cab, Std. Bed RPO QXR (LT45/70R17)	Silverado/Sierra C/K 1500 Ext. Cab, Std. Bed RPO QVL, QGI, QVM	Silverado/Sierra C/K 1500 Ext. Cab, Std. Bed RPO QSS, QXQ, QXN	Silverado/Sierra C/K 1500 Ext. Cab, Std. Bed RPO QNM, QPR
YEARS	2009-2013	2009-2013	2009-2013	2009-2013	2009-2013	2009-2013
INSTRUCTIONS	p. 305	p. 305	p. 305	p. 305	p. 305	p. 305
LUGNUT TORQUE	140 ft.lbs.	140 ft.lbs.	140 ft.lbs.	140 ft.lbs.	140 ft.lbs.	140 ft.lbs.
FRONT						
CASTER (Degrees)	L= 3.3, R= 3.0 ± 1.0 L= 3 5/16, R= 3 ±1	L= 3.3, R= 3.0 ± 1.0 L= 3 5/16, R= 3 ±1	L= 3.0, R= 3.2 ± 1.0 L= 3, R= 3 3/16 ±1	L= 3.1, R= 3.0 ± 1.0 L= 3 1/8, R= 3 ±1	L= 3.1, R= 3.0 ± 1.0 L= 3 1/8, R= 3 ±1	L= 3.1, R= 3.0 ± 1.0 L= 3 1/8, R= 3 ±1
CROSS	0.3 (5/16)	0.3 (5/16)	-0.2 (-3/16)	0.1 (1/8)	0.1 (1/8)	0.1 (1/8)
TOL.	0.75 (3/4)	0.75 (3/4)	0.75 (3/4)	0.75 (3/4)	0.75 (3/4)	0.75 (3/4)
CAMBER (Degrees)	-0.1 ±0.75 -1/8 ± 3/4	-0.1 ±0.75 -1/8 ± 3/4	-0.1 ±0.75 -1/8 ± 3/4	-0.1 ±0.75 -1/8 ± 3/4	-0.1 ±0.75 -1/8 ± 3/4	-0.1 ±0.75 -1/8 ± 3/4
CROSS	0.0 (0)	0.0 (0)	0.0 (0)	0.0 (0)	0.0 (0)	0.0 (0)
TOL.	0.75 (3/4)	0.75 (3/4)	0.75 (3/4)	0.75 (3/4)	0.75 (3/4)	0.75 (3/4)
FRONT TOE (Degrees)	0.1° ±0.2° 3/32° ± 3/16°	0.1° ±0.2° 3/32° ± 3/16°	0.1° ±0.2° 3/32° ± 3/16°	0.1° ±0.2° 3/32° ± 3/16°	0.1° ±0.2° 3/32° ± 3/16°	0.1° ±0.2° 3/32° ± 3/16°
FRONT TOE (Inches)	0.05" ±0.1" 1/16" ± 3/32"	0.05" ±0.1" 1/16" ± 3/32"	0.05" ±0.1" 1/16" ± 3/32"	0.05" ±0.1" 1/16" ± 3/32"	0.05" ±0.1" 1/16" ± 3/32"	0.05" ±0.1" 1/16" ± 3/32"
SAI						
TURNING ANGLE						
ILLUSTRATION #	126	126	126	126	126	126
REAR						
CAMBER (Degrees)						
REAR TOE (Degrees)						
REAR TOE (Inches)						
THRUST ANGLE						
ILLUSTRATION #						
NOTES						
	09261	09262	09263	09264	09265	09266

Chevrolet/GMC Truck & SUV

Silverado/Sierra	Silverado/Sierra	Silverado/Sierra	MODEL	Silverado/Sierra	Silverado/Sierra	Silverado/Sierra
C/K 1500 Ext. Cab, Std. Bed	C/K 1500 Ext. Cab, Std. Bed	C/K 1500 Ext. Cab, Std. Bed	INFORMATION	C/K 1500 Ext. Cab, Std. Bed	C/K 1500 Ext. Cab, Std. Bed	C/K 1500 Ext. Cab, Std. Bed
RPO QST, RBX	RPO QMG, QMH	RPO QNM, QPR, QBL		RPO QJP, QJM, QXK	RPO QU1, QPO	RPO QXT
2009-2013	2009-2013	2009-2013	YEARS	2009-2013	2009-2013	2009-2013
p. 305	p. 305	p. 305	INSTRUCTIONS	p. 305	p. 305	p. 305
140 ft.lbs.	140 ft.lbs.	140 ft.lbs.	LUGNUT TORQUE	140 ft.lbs.	140 ft.lbs.	140 ft.lbs.
			FRONT			
L= 3.1, R= 3.0 ± 1.0 L= 3⅛, R= 3 ± 1	L= 3.2, R= 3.0 ± 1.0 L= 3 3/16, R= 3 ± 1	L= 3.3, R= 3.0 ± 1.0 L= 3 5/16, R= 3 ± 1	CASTER (Degrees)	L= 3.3, R= 3.0 ± 1.0 L= 3 5/16, R= 3 ± 1	L= 3.3, R= 3.0 ± 1.0 L= 3 5/16, R= 3 ± 1	L= 3.6, R= 2.6 ± 1.0 L= 3⅝, R= 2⅝ ± 1
0.1 (⅛) | 0.75 (¾)	0.2 (3/16) | 0.75 (¾)	0.3 (5/16) | 0.75 (¾)	CROSS | TOL.	0.3 (5/16) | 0.75 (¾)	0.3 (5/16) | 0.75 (¾)	1.0 (1) | 0.75 (¾)
-0.1 ± 0.75 -⅛ ± ¾	-0.1 ± 0.75 -⅛ ± ¾	-0.1 ± 0.75 -⅛ ± ¾	CAMBER (Degrees)	-0.1 ± 0.75 -⅛ ± ¾	-0.1 ± 0.75 -⅛ ± ¾	-0.1 ± 0.75 -⅛ ± ¾
0.0 (0) | 0.75 (¾)	0.0 (0) | 0.75 (¾)	0.0 (0) | 0.75 (¾)	CROSS | TOL.	0.0 (0) | 0.75 (¾)	0.0 (0) | 0.75 (¾)	0.0 (0) | 0.75 (¾)
0.1° ± 0.2° 3/32° ± 3/16°	0.1° ± 0.2° 3/32° ± 3/16°	0.1° ± 0.2° 3/32° ± 3/16°	FRONT TOE (Degrees)	0.1° ± 0.2° 3/32° ± 3/16°	0.1° ± 0.2° 3/32° ± 3/16°	0.1° ± 0.2° 3/32° ± 3/16°
0.05" ± 0.1" 1/16" ± 3/32"	0.05" ± 0.1" 1/16" ± 3/32"	0.05" ± 0.1" 1/16" ± 3/32"	FRONT TOE (Inches)	0.05" ± 0.1" 1/16" ± 3/32"	0.05" ± 0.1" 1/16" ± 3/32"	0.05" ± 0.1" 1/16" ± 3/32"
			SAI			
			TURNING ANGLE			
126	126	126	ILLUSTRATION #	126	126	126
			REAR			
			CAMBER (Degrees)			
			REAR TOE (Degrees)			
			REAR TOE (Inches)			
			THRUST ANGLE			
			ILLUSTRATION #			
			NOTES			
09267	09268	09269		09270	09271	09272

Chevrolet/GMC Truck & SUV

Silverado/Sierra	Silverado/Sierra	Silverado/Sierra	MODEL	Silverado/Sierra	Silverado/Sierra	Silverado/Sierra
C/K 1500 Ext. Cab, Long Bed	C/K 1500 Ext. Cab, Long Bed	C/K 1500 Ext. Cab, Long Bed	INFORMATION	C/K 1500 Ext. Cab, Long Bed	C/K 1500 Ext. Cab, Long Bed	C/K 1500 Ext. Cab, Long Bed
RPO QXR, QGI, QXN	RPO QVL, QVM	RPO QXQ, QSS, QJP		RPO QJM, QBL, QXK	RPO QST, QMG, QMH	RPO QPR, QNM
2009-2013	2009-2013	2009-2013	YEARS	2009-2013	2009-2013	2009-2013
p. 305	p. 305	p. 305	INSTRUCTIONS	p. 305	p. 305	p. 305
140 ft.lbs.	140 ft.lbs.	140 ft.lbs.	LUGNUT TORQUE	140 ft.lbs.	140 ft.lbs.	140 ft.lbs.
			FRONT			
L= 3.0, R= 3.1 ± 1.0 L= 3, R= 3⅛ ± 1	L= 3.0, R= 3.1 ± 1.0 L= 3, R= 3⅛ ± 1	L= 3.1, R= 3.0 ± 1.0 L= 3⅛, R= 3 ± 1	CASTER (Degrees)	L= 3.1, R= 3.0 ± 1.0 L= 3⅛, R= 3 ± 1	L= 3.1, R= 3.0 ± 1.0 L= 33⅛, R= 3 ± 1	L= 3.2, R= 2.9 ± 1.0 L= 3 3/16, R= 2⅞ ± 1
-0.1 (-⅛) | 0.75 (¾)	-0.1 (-⅛) | 0.75 (¾)	0.1 (⅛) | 0.75 (¾)	CROSS | TOL.	0.1 (⅛) | 0.75 (¾)	0.1 (⅛) | 0.75 (¾)	0.3 (5/16) | 0.75 (¾)
-0.1 ± 0.75 -⅛ ± ¾	-0.1 ± 0.75 -⅛ ± ¾	-0.1 ± 0.75 -⅛ ± ¾	CAMBER (Degrees)	-0.1 ± 0.75 -⅛ ± ¾	-0.1 ± 0.75 -⅛ ± ¾	-0.1 ± 0.75 -⅛ ± ¾
0.0 (0) | 0.75 (¾)	0.0 (0) | 0.75 (¾)	0.0 (0) | 0.75 (¾)	CROSS | TOL.	0.0 (0) | 0.75 (¾)	0.0 (0) | 0.75 (¾)	0.0 (0) | 0.75 (¾)
0.1° ± 0.2° 3/32° ± 3/16°	0.1° ± 0.2° 3/32° ± 3/16°	0.1° ± 0.2° 3/32° ± 3/16°	FRONT TOE (Degrees)	0.1° ± 0.2° 3/32° ± 3/16°	0.1° ± 0.2° 3/32° ± 3/16°	0.1° ± 0.2° 3/32° ± 3/16°
0.05" ± 0.1" 1/16" ± 3/32"	0.05" ± 0.1" 1/16" ± 3/32"	0.05" ± 0.1" 1/16" ± 3/32"	FRONT TOE (Inches)	0.05" ± 0.1" 1/16" ± 3/32"	0.05" ± 0.1" 1/16" ± 3/32"	0.05" ± 0.1" 1/16" ± 3/32"
			SAI			
			TURNING ANGLE			
126	126	126	ILLUSTRATION #	126	126	126
			REAR			
			CAMBER (Degrees)			
			REAR TOE (Degrees)			
			REAR TOE (Inches)			
			THRUST ANGLE			
			ILLUSTRATION #			
			NOTES			
09273	09274	09275		09276	09277	09278

Chevrolet/GMC Truck & SUV

Silverado/Sierra C/K 1500 Ext. Cab, Long Bed RPO QU1, QPO	Silverado/Sierra C/K 1500 Crew Cab RPO QXR (LT45/70R17)	Silverado/Sierra C/K 1500 Crew Cab RPO QVL, QVM, QSS	MODEL INFORMATION	Silverado/Sierra C/K 1500 Crew Cab RPO QXQ, QXN	Silverado/Sierra C/K 1500 Crew Cab RPO QST, QGI	Silverado/Sierra C/K 1500 Crew Cab RPO QNM, QU1, QPO
2009-2013	2009-2013	2009-2013	YEARS	2009-2013	2009-2013	2009-2013
p. 305	p. 305	p. 305	INSTRUCTIONS	p. 305	p. 305	p. 305
140 ft.lbs.	140 ft.lbs.	140 ft.lbs.	LUGNUT TORQUE	140 ft.lbs.	140 ft.lbs.	140 ft.lbs.
			FRONT			
L= 3.2, R= 2.9 ± 1.0 L= 3 3/16, R= 2 7/8 ±1	L= 3.0, R= 3.2 ± 1.0 L= 3, R= 3 3/16 ±1	L= 3.1, R= 3.0 ± 1.0 L= 3 1/8, R= 3 ±1	CASTER (Degrees)	L= 3.1, R= 3.0 ± 1.0 L= 3 1/8, R= 3 ±1	L= 3.1, R= 3.0 ± 1.0 L= 3 1/8, R= 3 ±1	L= 3.3, R= 3.0 ± 1.0 L= 3 5/16, R= 3 ±1
0.3 (5/16) | 0.75 (3/4)	-0.2 (-3/16) | 0.75 (3/4)	0.1 (1/8) | 0.75 (3/4)	CROSS | TOL.	0.1 (1/8) | 0.75 (3/4)	0.1 (1/8) | 0.75 (3/4)	0.3 (5/16) | 0.75 (3/4)
-0.1 ±0.75 -1/8 ± 3/4	-0.1 ±0.75 -1/8 ± 3/4	-0.1 ±0.75 -1/8 ± 3/4	CAMBER (Degrees)	-0.1 ±0.75 -1/8 ± 3/4	-0.1 ±0.75 -1/8 ± 3/4	-0.1 ±0.75 -1/8 ± 3/4
0.0 (0) | 0.75 (3/4)	0.0 (0) | 0.75 (3/4)	0.0 (0) | 0.75 (3/4)	CROSS | TOL.	0.0 (0) | 0.75 (3/4)	0.0 (0) | 0.75 (3/4)	0.0 (0) | 0.75 (3/4)
0.1° ±0.2° 3/32° ± 3/16°	0.1° ±0.2° 3/32° ± 3/16°	0.1° ±0.2° 3/32° ± 3/16°	FRONT TOE (Degrees)	0.1° ±0.2° 3/32° ± 3/16°	0.1° ±0.2° 3/32° ± 3/16°	0.1° ±0.2° 3/32° ± 3/16°
0.05" ±0.1" 1/16" ± 3/32"	0.05" ±0.1" 1/16" ± 3/32"	0.05" ±0.1" 1/16" ± 3/32"	FRONT TOE (Inches)	0.05" ±0.1" 1/16" ± 3/32"	0.05" ±0.1" 1/16" ± 3/32"	0.05" ±0.1" 1/16" ± 3/32"
			SAI			
			TURNING ANGLE			
126	126	126	ILLUSTRATION #	126	126	126
			REAR			
			CAMBER (Degrees)			
			REAR TOE (Degrees)			
			REAR TOE (Inches)			
			THRUST ANGLE			
			ILLUSTRATION #			
			NOTES			
09279	09280	09281		09282	09283	09284

Chevrolet/GMC Truck & SUV

Silverado/Sierra C/K 1500 Crew Cab RPO QPR, QBL, QJP	Silverado/Sierra C/K 1500 Crew Cab RPO QJM, QXK	Silverado/Sierra C/K 1500 Crew Cab RPO QMG, QMH	MODEL INFORMATION	Silverado/Sierra C/K 1500 Crew Cab RPO QXT	Silverado/Sierra C/K 1500 Reg. Cab, Std. Bed RPO RBW, RBX, RHH, QBL, RC7	Silverado/Sierra C/K 1500 Reg. Cab, Std. Bed RPO RBZ, QNM, RC3, RC5, RKX, QXN
2009-2013	2009-2013	2009-2013	YEARS	2009-2013	2014-2017	2014-2017
p. 305	p. 305	p. 305	INSTRUCTIONS	p. 305	p. 305	p. 305
140 ft.lbs.	140 ft.lbs.	140 ft.lbs.	LUGNUT TORQUE	140 ft.lbs.	140 ft.lbs.	140 ft.lbs.
			FRONT			
L= 3.3, R= 3.0 ± 1.0 L= 3 5/16, R= 3 ±1	L= 3.3, R= 3.0 ± 1.0 L= 3 5/16, R= 3 ±1	L= 3.2, R= 3.0 ± 1.0 L= 3 3/16, R= 3 ±1	CASTER (Degrees)	L= 3.6, R= 2.6 ± 1.0 L= 3 5/8, R= 2 5/8 ±1	2.4 ±1.0 2 3/8 ±1	L= 2.5, R= 2.3 ± 1.0 L= 2 1/2, R= 2 5/16 ±1
0.3 (5/16) | 0.75 (3/4)	0.3 (5/16) | 0.75 (3/4)	0.2 (3/16) | 0.75 (3/4)	CROSS | TOL.	1.0 (1) | 0.75 (3/4)	0.0 (0) | 1.0 (1)	0.2 (3/16) | 1.0 (1)
-0.1 ±0.75 -1/8 ± 3/4	-0.1 ±0.75 -1/8 ± 3/4	-0.1 ±0.75 -1/8 ± 3/4	CAMBER (Degrees)	-0.1 ±0.75 -1/8 ± 3/4	-0.1 ±0.8 -1/8 ± 13/16	-0.1 ±0.8 -1/8 ± 13/16
0.0 (0) | 0.75 (3/4)	0.0 (0) | 0.75 (3/4)	0.0 (0) | 0.75 (3/4)	CROSS | TOL.	0.0 (0) | 0.75 (3/4)	0.0 (0) | 0.8 (13/16)	0.0 (0) | 0.8 (13/16)
0.1° ±0.2° 3/32° ± 3/16°	0.1° ±0.2° 3/32° ± 3/16°	0.1° ±0.2° 3/32° ± 3/16°	FRONT TOE (Degrees)	0.1° ±0.2° 3/32° ± 3/16°	0.1° ±0.2° 3/32° ± 3/16°	0.1° ±0.2° 3/32° ± 3/16°
0.05" ±0.1" 1/16" ± 3/32"	0.05" ±0.1" 1/16" ± 3/32"	0.05" ±0.1" 1/16" ± 3/32"	FRONT TOE (Inches)	0.05" ±0.1" 1/16" ± 3/32"	0.05" ±0.1" 1/16" ± 3/32"	0.05" ±0.1" 1/16" ± 3/32"
			SAI			
			TURNING ANGLE			
126	126	126	ILLUSTRATION #	126	126	126
			REAR			
			CAMBER (Degrees)			
			REAR TOE (Degrees)			
			REAR TOE (Inches)			
			THRUST ANGLE		0.4 (13/32)	0.4 (13/32)
			ILLUSTRATION #			
			NOTES			
09285	09286	09287		09288	09289	09290

ALIGNMENT SPECIFICATIONS

Chevrolet/GMC Truck & SUV

Silverado/Sierra C/K 1500 Reg. Cab, Std. Bed RPO QXT (LT265/70R17E)	Silverado/Sierra C/K 1500 Reg. Cab, Std. Bed RPO QHX (P285/50R20 111H)	Silverado/Sierra C/K 1500 Reg. Cab, Std. Bed RPO QSS (P275/55R20 111S)	MODEL INFORMATION	Silverado/Sierra C/K 1500 Reg. Cab, Long Bed RPO RBW, RBX, RHH	Silverado/Sierra C/K 1500 Reg. Cab, Long Bed RPO QBL, RC7	Silverado/Sierra C/K 1500 Reg. Cab, Long Bed RPO RBZ, QNM, RC3
2014-2017	2014-2017	2014-2017	YEARS	2014-2017	2014-2017	2014-2017
p. 305	p. 305	p. 305	INSTRUCTIONS	p. 305	p. 305	p. 305
140 ft.lbs.	140 ft.lbs.	140 ft.lbs.	LUGNUT TORQUE	140 ft.lbs.	140 ft.lbs.	140 ft.lbs.
			FRONT			
L= 2.9, R= 1.9 ± 1.0 L= 2⅞, R= 1⅞ ± 1	L= 2.2, R= 2.5 ± 1.0 L= 2 3/16, R= 2½ ± 1	L= 2.6, R= 2.2 ± 1.0 [1] L= 2⅝, R= 2 3/16 ± 1 [1]	CASTER (Degrees)	2.6 ± 1.0 2⅝ ± 1	2.6 ± 1.0 2⅝ ± 1	L= 2.7, R= 2.5 ± 1.0 L= 2 11/16, R= 2½ ± 1
1.0 (1) | 1.0 (1)	-0.3 (- 5/16) | 1.0 (1)	0.4 (⅜) | 1.0 (1)	CROSS | TOL.	0.0 (0) | 1.0 (1)	0.0 (0) | 1.0 (1)	0.2 (3/16) | 1.0 (1)
-0.1 ± 0.8 -⅛ ± 13/16	-0.1 ± 0.8 -⅛ ± 13/16	-0.1 ± 0.8 -⅛ ± 13/16	CAMBER (Degrees)	-0.1 ± 0.8 -⅛ ± 13/16	-0.1 ± 0.8 -⅛ ± 13/16	-0.1 ± 0.8 -⅛ ± 13/16
0.0 (0) | 0.8 (13/16)	0.0 (0) | 0.8 (13/16)	0.0 (0) | 0.8 (13/16)	CROSS | TOL.	0.0 (0) | 0.8 (13/16)	0.0 (0) | 0.8 (13/16)	0.0 (0) | 0.8 (13/16)
0.1° ± 0.2° 3/32° ± 3/16°	0.1° ± 0.2° 3/32° ± 3/16°	0.1° ± 0.2° 3/32° ± 3/16°	FRONT TOE (Degrees)	0.1° ± 0.2° 3/32° ± 3/16°	0.1° ± 0.2° 3/32° ± 3/16°	0.1° ± 0.2° 3/32° ± 3/16°
0.05" ± 0.1" 1/16" ± 3/32"	0.05" ± 0.1" 1/16" ± 3/32"	0.05" ± 0.1" 1/16" ± 3/32"	FRONT TOE (Inches)	0.05" ± 0.1" 1/16" ± 3/32"	0.05" ± 0.1" 1/16" ± 3/32"	0.05" ± 0.1" 1/16" ± 3/32"
			SAI			
			TURNING ANGLE			
126	126	126	ILLUSTRATION #	126	126	126
			REAR			
			CAMBER (Degrees)			
			REAR TOE (Degrees)			
			REAR TOE (Inches)			
0.4 (13/32)	0.4 (13/32)	0.4 (13/32)	THRUST ANGLE	0.4 (13/32)	0.4 (13/32)	0.4 (13/32)
			ILLUSTRATION #			
		[1] 2016 L=2.6° R=2.4°	NOTES			
09291	09292	09293		09294	09295	09296

Chevrolet/GMC Truck & SUV

Silverado/Sierra C/K 1500 Reg. Cab, Long Bed RPO RC5, RKX, QXN	Silverado/Sierra C/K 1500 Reg. Cab, Long Bed RPO QXT (LT265/70R17E)	Silverado/Sierra C/K 1500 Reg. Cab, Long Bed RPO QHX (P285/50R20 111H)	MODEL INFORMATION	Silverado/Sierra C/K 1500 Reg. Cab, Long Bed RPO QSS (P275/55R20 111S)	Silverado/Sierra C/K 1500 Double Cab RPO RBW, RBX, RHH	Silverado/Sierra C/K 1500 Double Cab RPO QBL, RC7, RBZ
2014-2017	2014-2017	2014-2017	YEARS	2014-2017	2014-2017	2014-2017
p. 305	p. 305	p. 305	INSTRUCTIONS	p. 305	p. 305	p. 305
140 ft.lbs.	140 ft.lbs.	140 ft.lbs.	LUGNUT TORQUE	140 ft.lbs.	140 ft.lbs.	140 ft.lbs.
			FRONT			
L= 2.7, R= 2.5 ± 1.0 L= 2 11/16, R= 2½ ± 1	L= 3.1, R= 2.1 ± 1.0 L= 3⅛, R= 2⅛ ± 1	L= 2.4, R= 2.7 ± 1.0 L= 2⅜, R= 2 11/16 ± 1	CASTER (Degrees)	L= 2.8, R= 2.4 ± 1.0 L= 2 13/16, R= 2⅜ ± 1	2.7 ± 1.0 2 11/16 ± 1	2.7 ± 1.0 2 11/16 ± 1
0.2 (3/16) | 1.0 (1)	1.0 (1) | 1.0 (1)	-0.3 (- 5/16) | 1.0 (1)	CROSS | TOL.	0.4 (⅜) | 1.0 (1)	0.0 (0) | 1.0 (1)	0.0 (0) | 1.0 (1)
-0.1 ± 0.8 -⅛ ± 13/16	-0.1 ± 0.8 -⅛ ± 13/16	-0.1 ± 0.8 -⅛ ± 13/16	CAMBER (Degrees)	-0.1 ± 0.8 -⅛ ± 13/16	-0.1 ± 0.8 -⅛ ± 13/16	-0.1 ± 0.8 -⅛ ± 13/16
0.0 (0) | 0.8 (13/16)	0.0 (0) | 0.8 (13/16)	0.0 (0) | 0.8 (13/16)	CROSS | TOL.	0.0 (0) | 0.8 (13/16)	0.0 (0) | 0.8 (13/16)	0.0 (0) | 0.8 (13/16)
0.1° ± 0.2° 3/32° ± 3/16°	0.1° ± 0.2° 3/32° ± 3/16°	0.1° ± 0.2° 3/32° ± 3/16°	FRONT TOE (Degrees)	0.1° ± 0.2° 3/32° ± 3/16°	0.1° ± 0.2° 3/32° ± 3/16°	0.1° ± 0.2° 3/32° ± 3/16°
0.05" ± 0.1" 1/16" ± 3/32"	0.05" ± 0.1" 1/16" ± 3/32"	0.05" ± 0.1" 1/16" ± 3/32"	FRONT TOE (Inches)	0.05" ± 0.1" 1/16" ± 3/32"	0.05" ± 0.1" 1/16" ± 3/32"	0.05" ± 0.1" 1/16" ± 3/32"
			SAI			
			TURNING ANGLE			
126	126	126	ILLUSTRATION #	126	126	126
			REAR			
			CAMBER (Degrees)			
			REAR TOE (Degrees)			
			REAR TOE (Inches)			
0.4 (13/32)	0.4 (13/32)	0.4 (13/32)	THRUST ANGLE	0.4 (13/32)	0.4 (13/32)	0.4 (13/32)
			ILLUSTRATION #			
			NOTES			
09297	09298	09299		09300	09301	09302

Chevrolet/GMC Truck & SUV

Silverado/Sierra	Silverado/Sierra	Silverado/Sierra	MODEL	Silverado/Sierra	Silverado/Sierra	Silverado/Sierra
C/K 1500 Double Cab	C/K 1500 Double Cab	C/K 1500 Double Cab	INFORMATION	C/K 1500 Double Cab	C/K 1500 Double Cab	C/K 1500 Crew Cab, Short Bed
RPO QNM, RC3	RPO RC5, RKX, QXN	RPO QXT (LT265/70R17E)		RPO QHX (P285/50R20 111H)	RPO QSS (P275/55R20 111S)	RPO RBW, RBX, RHH
2014-2017	2014-2017	2014-2017	YEARS	2014-2017	2014-2017	2014-2017
p. 305	p. 305	p. 305	INSTRUCTIONS	p. 305	p. 305	p. 305
140 ft.lbs.	140 ft.lbs.	140 ft.lbs.	LUGNUT TORQUE	140 ft.lbs.	140 ft.lbs.	140 ft.lbs.
			FRONT			
L= 2.8, R= 2.6 ± 1.0 L= 2 13/16, R= 2 5/8 ±1	L= 2.8, R= 2.6 ± 1.0 L= 2 13/16, R= 2 5/8 ±1	L= 3.2, R= 2.2 ± 1.0 L= 3 3/16, R= 2 3/16 ±1	CASTER (Degrees)	L= 2.5, R= 2.8 ± 1.0 L= 2 1/2, R= 2 13/16 ±1	L= 2.9, R= 2.5 ± 1.0 L= 2 7/8, R= 2 1/2 ±1	2.9 ±1.0 2 7/8 ±1
0.2 (3/16) | 1.0 (1)	0.2 (3/16) | 1.0 (1)	1.0 (1) | 1.0 (1)	CROSS | TOL.	-0.3 (-5/16) | 1.0 (1)	0.4 (3/8) | 1.0 (1)	0.0 (0) | 1.0 (1)
-0.1 ±0.8 -1/8 ± 13/16	-0.1 ±0.8 -1/8 ± 13/16	-0.1 ±0.8 -1/8 ± 13/16	CAMBER (Degrees)	-0.1 ±0.8 -1/8 ± 13/16	-0.1 ±0.8 -1/8 ± 13/16	-0.1 ±0.8 -1/8 ± 13/16
0.0 (0) | 0.8 (13/16)	0.0 (0) | 0.8 (13/16)	0.0 (0) | 0.8 (13/16)	CROSS | TOL.	0.0 (0) | 0.8 (13/16)	0.0 (0) | 0.8 (13/16)	0.0 (0) | 0.8 (13/16)
0.1° ±0.2° 3/32° ± 3/16°	0.1° ±0.2° 3/32° ± 3/16°	0.1° ±0.2° 3/32° ± 3/16°	FRONT TOE (Degrees)	0.1° ±0.2° 3/32° ± 3/16°	0.1° ±0.2° 3/32° ± 3/16°	0.1° ±0.2° 3/32° ± 3/16°
0.05" ±0.1" 1/16" ± 3/32"	0.05" ±0.1" 1/16" ± 3/32"	0.05" ±0.1" 1/16" ± 3/32"	FRONT TOE (Inches)	0.05" ±0.1" 1/16" ± 3/32"	0.05" ±0.1" 1/16" ± 3/32"	0.05" ±0.1" 1/16" ± 3/32"
			SAI			
			TURNING ANGLE			
126	126	126	ILLUSTRATION #	126	126	126
			REAR			
			CAMBER (Degrees)			
			REAR TOE (Degrees)			
			REAR TOE (Inches)			
0.4 (13/32)	0.4 (13/32)	0.4 (13/32)	THRUST ANGLE	0.4 (13/32)	0.4 (13/32)	0.4 (13/32)
			ILLUSTRATION #			
			NOTES			
09303	09304	09305		09306	09307	09308

Chevrolet/GMC Truck & SUV

Silverado/Sierra	Silverado/Sierra	Silverado/Sierra	MODEL	Silverado/Sierra	Silverado/Sierra	Silverado/Sierra
C/K 1500 Crew Cab, Short Bed	C/K 1500 Crew Cab, Short Bed	C/K 1500 Crew Cab, Short Bed	INFORMATION	C/K 1500 Crew Cab, Short Bed	C/K 1500 Crew Cab, Short Bed	C/K 1500 Crew Cab, Short Bed
RPO QBL, RC7, RBZ	RPO QNM, RC3	RPO RC5, RKX, QXN		RPO QXT (LT265/70R17E)	RPO QHX (P285/50R20 111H)	RPO QSS (P275/55R20 111S)
2014-2017	2014-2017	2014-2017	YEARS	2014-2017	2014-2017	2014-2017
p. 305	p. 305	p. 305	INSTRUCTIONS	p. 305	p. 305	p. 305
140 ft.lbs.	140 ft.lbs.	140 ft.lbs.	LUGNUT TORQUE	140 ft.lbs.	140 ft.lbs.	140 ft.lbs.
			FRONT			
2.9 ±1.0 2 7/8 ±1	L= 3.0, R= 2.8 ± 1.0 L= 3, R= 2 13/16 ±1	L= 3.0, R= 2.8 ± 1.0 L= 3, R= 2 13/16 ±1	CASTER (Degrees)	L= 3.4, R= 2.4 ± 1.0 L= 3 3/8, R= 2 3/8 ±1	L= 2.7, R= 3.0 ± 1.0 L= 2 11/16, R= 3 ±1	L= 3.1, R= 2.7 ± 1.0 L= 3 1/8, R= 2 11/16 ±1
0.0 (0) | 1.0 (1)	0.2 (3/16) | 1.0 (1)	0.2 (3/16) | 1.0 (1)	CROSS | TOL.	1.0 (1) | 1.0 (1)	-0.3 (-5/16) | 1.0 (1)	0.4 (3/8) | 1.0 (1)
-0.1 ±0.8 -1/8 ± 13/16	-0.1 ±0.8 -1/8 ± 13/16	-0.1 ±0.8 -1/8 ± 13/16	CAMBER (Degrees)	-0.1 ±0.8 -1/8 ± 13/16	-0.1 ±0.8 -1/8 ± 13/16	-0.1 ±0.8 -1/8 ± 13/16
0.0 (0) | 0.8 (13/16)	0.0 (0) | 0.8 (13/16)	0.0 (0) | 0.8 (13/16)	CROSS | TOL.	0.0 (0) | 0.8 (13/16)	0.0 (0) | 0.8 (13/16)	0.0 (0) | 0.8 (13/16)
0.1° ±0.2° 3/32° ± 3/16°	0.1° ±0.2° 3/32° ± 3/16°	0.1° ±0.2° 3/32° ± 3/16°	FRONT TOE (Degrees)	0.1° ±0.2° 3/32° ± 3/16°	0.1° ±0.2° 3/32° ± 3/16°	0.1° ±0.2° 3/32° ± 3/16°
0.05" ±0.1" 1/16" ± 3/32"	0.05" ±0.1" 1/16" ± 3/32"	0.05" ±0.1" 1/16" ± 3/32"	FRONT TOE (Inches)	0.05" ±0.1" 1/16" ± 3/32"	0.05" ±0.1" 1/16" ± 3/32"	0.05" ±0.1" 1/16" ± 3/32"
			SAI			
			TURNING ANGLE			
126	126	126	ILLUSTRATION #	126	126	126
			REAR			
			CAMBER (Degrees)			
			REAR TOE (Degrees)			
			REAR TOE (Inches)			
0.4 (13/32)	0.4 (13/32)	0.4 (13/32)	THRUST ANGLE	0.4 (13/32)	0.4 (13/32)	0.4 (13/32)
			ILLUSTRATION #			
			NOTES			
09309	09310	09311		09312	09313	09314

Chevrolet/GMC Truck & SUV

Silverado/Sierra	Silverado/Sierra	Silverado/Sierra	MODEL	Silverado/Sierra	Silverado/Sierra	Silverado/Sierra
C/K 1500 Crew Cab, Std. Bed	C/K 1500 Crew Cab, Std. Bed	C/K 1500 Crew Cab, Std. Bed	INFORMATION	C/K 1500 Crew Cab, Std. Bed	C/K 1500 Crew Cab, Std. Bed	C/K 1500 Crew Cab, Std. Bed
RPO RBW, RBX, RHH	RPO QBL, RC7, RBZ	RPO QNM, RC3		RPO RC5, RKX, QXN	RPO QXT (LT265/70R17E)	RPO QHX (P285/50R20 111H)
2014-2017	2014-2017	2014-2017	YEARS	2014-2017	2014-2017	2014-2017
p. 305	p. 305	p. 305	INSTRUCTIONS	p. 305	p. 305	p. 305
140 ft.lbs.	140 ft.lbs.	140 ft.lbs.	LUGNUT TORQUE	140 ft.lbs.	140 ft.lbs.	140 ft.lbs.
			FRONT			
3.0 ± 1.0 3 ± 1	3.0 ± 1.0 3 ± 1	L= 3.1, R= 2.9 ± 1.0 L= 3 1/8, R= 2 7/8 ± 1	CASTER (Degrees)	L= 3.1, R= 2.9 ± 1.0 L= 3 1/8, R= 2 7/8 ± 1	L= 3.5, R= 2.5 ± 1.0 L= 3 1/2, R= 2 1/2 ± 1	L= 2.8, R= 3.1 ± 1.0 L= 2 13/16, R= 3 1/8 ± 1
0.0 (0) | 1.0 (1)	0.0 (0) | 1.0 (1)	0.2 (3/16) | 1.0 (1)	CROSS | TOL.	0.2 (3/16) | 1.0 (1)	1.0 (1) | 1.0 (1)	-0.3 (-5/16) | 1.0 (1)
-0.1 ± 0.8 -1/8 ± 13/16	-0.1 ± 0.8 -1/8 ± 13/16	-0.1 ± 0.8 -1/8 ± 13/16	CAMBER (Degrees)	-0.1 ± 0.8 -1/8 ± 13/16	-0.1 ± 0.8 -1/8 ± 13/16	-0.1 ± 0.8 -1/8 ± 13/16
0.0 (0) | 0.8 (13/16)	0.0 (0) | 0.8 (13/16)	0.0 (0) | 0.8 (13/16)	CROSS | TOL.	0.0 (0) | 0.8 (13/16)	0.0 (0) | 0.8 (13/16)	0.0 (0) | 0.8 (13/16)
0.1° ± 0.2° 3/32° ± 3/16°	0.1° ± 0.2° 3/32° ± 3/16°	0.1° ± 0.2° 3/32° ± 3/16°	FRONT TOE (Degrees)	0.1° ± 0.2° 3/32° ± 3/16°	0.1° ± 0.2° 3/32° ± 3/16°	0.1° ± 0.2° 3/32° ± 3/16°
0.05" ± 0.1" 1/16" ± 3/32"	0.05" ± 0.1" 1/16" ± 3/32"	0.05" ± 0.1" 1/16" ± 3/32"	FRONT TOE (Inches)	0.05" ± 0.1" 1/16" ± 3/32"	0.05" ± 0.1" 1/16" ± 3/32"	0.05" ± 0.1" 1/16" ± 3/32"
			SAI			
			TURNING ANGLE			
126	126	126	ILLUSTRATION #	126	126	126
			REAR			
			CAMBER (Degrees)			
			REAR TOE (Degrees)			
			REAR TOE (Inches)			
0.4 (13/32)	0.4 (13/32)	0.4 (13/32)	THRUST ANGLE	0.4 (13/32)	0.4 (13/32)	0.4 (13/32)
			ILLUSTRATION #			
			NOTES			
09315	09316	09317		09318	09319	09320

Chevrolet/GMC Truck & SUV

Silverado/Sierra	Silverado/Sierra 2500HD	Silverado/Sierra 2500HD	MODEL	Silverado/Sierra 2500HD	Silverado/Sierra 2500HD	Silverado/Sierra 2500HD
C/K 1500 Crew Cab, Std. Bed	All Models	All Models	INFORMATION	Reg. Cab,	Reg. Cab,	Ext. Cab,
RPO QSS (P275/55R20 111S)				exc/LT265/70R17E tires	w/LT265/70R17E tires	exc/LT265/70R17E tires
2014-2017	2007	2008	YEARS	2009-2010	2009-2010	2009-2010
p. 305			INSTRUCTIONS			
140 ft.lbs.	140 ft.lbs.	140 ft.lbs.	LUGNUT TORQUE	140 ft.lbs.	140 ft.lbs.	140 ft.lbs.
			FRONT			
L= 3.2, R= 2.8 ± 1.0 L= 3 3/16, R= 2 13/16 ± 1	L= 3.3, R= 3.6 ± 1.0 [1] L= 3 5/16, R= 3 5/8 ± 1 [1]	L= 3.25, R= 3.5 ± 1.0 [1] L= 3 1/4, R= 3 1/2 ± 1 [1]	CASTER (Degrees)	L= 3.0, R= 2.7 ± 1.0 L= 3, R= 2 11/16 ± 1	L= 2.5, R= 2.2 ± 1.0 L= 2 1/2, R= 2 3/16 ± 1	L= 3.2, R= 3.4 ± 1.0 L= 3 3/16, R= 3 3/8 ± 1
0.4 (3/8) | 1.0 (1)	-0.3 (-5/16) | 0.75 (3/4)	-0.25 (-1/4) | 0.75 (3/4)	CROSS | TOL.	0.3 (5/16) | 0.75 (3/4)	0.3 (5/16) | 0.75 (3/4)	-0.2 (-3/16) | 0.75 (3/4)
-0.1 ± 0.8 -1/8 ± 13/16	0.25 ± 0.6 1/4 ± 5/8	0.25 ± 0.6 1/4 ± 5/8	CAMBER (Degrees)	0.25 ± 0.75 1/4 ± 3/4	0.25 ± 0.75 1/4 ± 3/4	0.25 ± 0.75 1/4 ± 3/4
0.0 (0) | 0.8 (13/16)	0.0 (0) | 0.6 (5/8)	0.0 (0) | 0.6 (5/8)	CROSS | TOL.	0.0 (0) | 0.75 (3/4)	0.0 (0) | 0.75 (3/4)	0.0 (0) | 0.75 (3/4)
0.1° ± 0.2° 3/32° ± 3/16°	0.1° ± 0.2° 3/32° ± 3/16°	0.1° ± 0.2° 3/32° ± 3/16°	FRONT TOE (Degrees)	0.2° ± 0.2° 3/16° ± 3/16°	0.2° ± 0.2° 3/16° ± 3/16°	0.2° ± 0.2° 3/16° ± 3/16°
0.05" ± 0.1" 1/16" ± 3/32"	0.05" ± 0.1" 1/16" ± 3/32"	0.05" ± 0.1" 1/16" ± 3/32"	FRONT TOE (Inches)	0.1" ± 0.1" 3/32" ± 3/32"	0.1" ± 0.1" 3/32" ± 3/32"	0.1" ± 0.1" 3/32" ± 3/32"
			SAI			
			TURNING ANGLE			
126	126	126	ILLUSTRATION #	126	126	126
			REAR			
			CAMBER (Degrees)			
			REAR TOE (Degrees)			
			REAR TOE (Inches)			
0.4 (13/32)			THRUST ANGLE			
			ILLUSTRATION #			
	[1] Relative to ground	[1] Relative to ground	NOTES			
09321	09322	09323		09324	09325	09326

Chevrolet/GMC Truck & SUV

Silverado/Sierra 2500HD Ext. Cab, w/LT265/70R17E tires	Silverado/Sierra 2500HD Crew Cab, Std. Bed, exc/LT265/70R17E tires	Silverado/Sierra 2500HD Crew Cab, Std. Bed, w/LT265/70R17E tires	MODEL INFORMATION	Silverado/Sierra 2500HD Crew Cab, Long Bed, exc/LT265/70R17E tires	Silverado/Sierra 2500HD Crew Cab, Long Bed, w/LT265/70R17E tires	Silverado/Sierra 2500HD Reg. Cab, Gas, exc/LT265/70R17E tires
2009-2010	2009-2010	2009-2010	YEARS	2009-2010	2009-2010	2011-2017
			INSTRUCTIONS			
140 ft.lbs.	140 ft.lbs.	140 ft.lbs.	LUGNUT TORQUE	140 ft.lbs.	140 ft.lbs.	140 ft.lbs.
			FRONT			
L= 3.1, R= 2.8 ± 1.0 L= 3 1/8, R= 2 13/16 ± 1	L= 3.1, R= 3.4 ± 1.0 L= 3 1/8, R= 3 3/8 ± 1	L= 3.3, R= 3.0 ± 1.0 L= 3 5/16, R= 3 ± 1	CASTER (Degrees)	L= 3.3, R= 3.6 ± 1.0 L= 3 5/16, R= 3 5/8 ± 1	L= 3.4, R= 3.1 ± 1.0 L= 3 3/8, R= 3 1/8 ± 1	L= 2.4, R= 2.3 ± 1.0 [1] L= 2 3/8, R= 2 5/16 ± 1 [1]
0.3 (5/16) | 0.75 (3/4)	-0.3 (-5/16) | 0.75 (3/4)	0.3 (5/16) | 0.75 (3/4)	CROSS | TOL.	-0.3 (-5/16) | 0.75 (3/4)	0.3 (5/16) | 0.75 (3/4)	0.1 (1/8) | 0.75 (3/4)
0.25 ± 0.75 1/4 ± 3/4	0.25 ± 0.75 1/4 ± 3/4	0.25 ± 0.75 1/4 ± 3/4	CAMBER (Degrees)	0.25 ± 0.75 1/4 ± 3/4	0.25 ± 0.75 1/4 ± 3/4	0.5 ± 0.75 1/2 ± 3/4
0.0 (0) | 0.75 (3/4)	0.0 (0) | 0.75 (3/4)	0.0 (0) | 0.75 (3/4)	CROSS | TOL.	0.0 (0) | 0.75 (3/4)	0.0 (0) | 0.75 (3/4)	0.0 (0) | 0.75 (3/4)
0.2° ± 0.2° 3/16° ± 3/16°	0.2° ± 0.2° 3/16° ± 3/16°	0.2° ± 0.2° 3/16° ± 3/16°	FRONT TOE (Degrees)	0.2° ± 0.2° 3/16° ± 3/16°	0.2° ± 0.2° 3/16° ± 3/16°	0.1° ± 0.2° 3/32° ± 3/16°
0.1" ± 0.1" 3/32" ± 3/32"	0.1" ± 0.1" 3/32" ± 3/32"	0.1" ± 0.1" 3/32" ± 3/32"	FRONT TOE (Inches)	0.1" ± 0.1" 3/32" ± 3/32"	0.1" ± 0.1" 3/32" ± 3/32"	0.05" ± 0.1" 1/16" ± 3/32"
			SAI			
			TURNING ANGLE			
126	126	126	ILLUSTRATION #	126	126	126
			REAR			
			CAMBER (Degrees)			
			REAR TOE (Degrees)			
			REAR TOE (Inches)			
			THRUST ANGLE			
			ILLUSTRATION #			
			NOTES			[1] BoxOff L=2.0° R=1.9°
09327	09328	09329		09330	09331	09332

Chevrolet/GMC Truck & SUV

Silverado/Sierra 2500HD Reg. Cab, Gas, w/LT265/70R17E tires	Silverado/Sierra 2500HD Reg. Cab, Diesel, exc/LT265/70R17E tires	Silverado/Sierra 2500HD Reg. Cab, Diesel, w/LT265/70R17E tires	MODEL INFORMATION	Silverado/Sierra 2500HD Ext. Cab, Std. Box, exc/LT265/70R17E tires	Silverado/Sierra 2500HD Ext. Cab, Std. Box, w/LT265/70R17E tires	Silverado/Sierra 2500HD Ext. Cab, Long Box, Diesel, exc/LT265/70R17E tires
2011-2017	2011-2017	2011-2017	YEARS	2011-2013	2011-2013	2011-2013
			INSTRUCTIONS			
140 ft.lbs.	140 ft.lbs.	140 ft.lbs.	LUGNUT TORQUE	140 ft.lbs.	140 ft.lbs.	140 ft.lbs.
			FRONT			
L= 2.5, R= 2.2 ± 1.0 [1] L= 2 1/2, R= 2 3/16 ± 1 [1]	L= 2.4, R= 2.3 ± 1.0 [1] L= 2 3/8, R= 2 5/16 ± 1 [1]	L= 2.5, R= 2.2 ± 1.0 [1] L= 2 1/2, R= 2 3/16 ± 1 [1]	CASTER (Degrees)	L= 2.5, R= 2.4 ± 1.0 [1] L= 2 1/2, R= 2 3/8 ± 1 [1]	L= 2.6, R= 2.3 ± 1.0 [1] L= 2 5/8, R= 2 5/16 ± 1 [1]	L= 2.4, R= 2.3 ± 1.0 [1] L= 2 3/8, R= 2 5/16 ± 1 [1]
0.3 (5/16) | 0.75 (3/4)	0.1 (1/8) | 0.75 (3/4)	0.3 (5/16) | 0.75 (3/4)	CROSS | TOL.	0.1 (1/8) | 0.75 (3/4)	0.3 (5/16) | 0.75 (3/4)	0.1 (1/8) | 0.75 (3/4)
0.5 ± 0.75 1/2 ± 3/4	0.5 ± 0.75 1/2 ± 3/4	0.5 ± 0.75 1/2 ± 3/4	CAMBER (Degrees)	0.5 ± 0.75 1/2 ± 3/4	0.5 ± 0.75 1/2 ± 3/4	0.5 ± 0.75 1/2 ± 3/4
0.0 (0) | 0.75 (3/4)	0.0 (0) | 0.75 (3/4)	0.0 (0) | 0.75 (3/4)	CROSS | TOL.	0.0 (0) | 0.75 (3/4)	0.0 (0) | 0.75 (3/4)	0.0 (0) | 0.75 (3/4)
0.2° ± 0.2° [2] 3/16° ± 3/16° [2]	0.1° ± 0.2° 3/32° ± 3/16°	0.1° ± 0.2° 3/32° ± 3/16°	FRONT TOE (Degrees)	0.2° ± 0.2° 3/16° ± 3/16°	0.1° ± 0.2° 3/32° ± 3/16°	0.1° ± 0.2° 3/32° ± 3/16°
0.1" ± 0.1" [2] 3/32" ± 3/32" [2]	0.05" ± 0.1" 1/16" ± 3/32"	0.05" ± 0.1" 1/16" ± 3/32"	FRONT TOE (Inches)	0.1" ± 0.1" 3/32" ± 3/32"	0.05" ± 0.1" 1/16" ± 3/32"	0.05" ± 0.1" 1/16" ± 3/32"
			SAI			
			TURNING ANGLE			
126	126	126	ILLUSTRATION #	126	126	126
			REAR			
			CAMBER (Degrees)			
			REAR TOE (Degrees)			
			REAR TOE (Inches)			
			THRUST ANGLE			
			ILLUSTRATION #			
[1] BoxOff L=2.1° R=1.8° [2] BoxOff = 0.1° (3/32°)	[1] BoxOff L=2.0° R=1.9°	[1] BoxOff L=2.1° R=1.8°	NOTES	[1] BoxOff L=2.5° R=2.4°	[1] BoxOff L=2.5° R=2.2°	[1] BoxOff L=2.2° R=2.1°
09333	09334	09335		09336	09337	09338

ALIGNMENT SPECIFICATIONS

Chevrolet/GMC Truck & SUV

Silverado/Sierra 2500HD	Silverado/Sierra 2500HD	Silverado/Sierra 2500HD	MODEL	Silverado/Sierra 2500HD	Silverado/Sierra 2500HD	Silverado/Sierra 2500HD
Ext. Cab, Long Box,	Ext. Cab, Long Box,	Ext. Cab, Long Box,	INFORMATION	Double Cab, Std. Box,	Double Cab, Std. Box,	Double Cab, Long Box,
Diesel, w/LT265/70R17E tires	Gas, exc/LT265/70R17E tires	Gas, w/LT265/70R17E tires		except LT265/70R17E tires (QXT)	with LT265/70R17E tires (QXT)	exc/LT265/70R17E tires (QXT), Diesel
2011-2013	2011-2013	2011-2013	YEARS	2015-2017	2015-2017	2015-2017
			INSTRUCTIONS			
140 ft.lbs.	140 ft.lbs.	140 ft.lbs.	LUGNUT TORQUE	140 ft.lbs.	140 ft.lbs.	140 ft.lbs.
			FRONT			
L= 2.9, R= 2.6 ± 1.0 [1] L= 2⅞, R= 2⅝ ± 1 [1]	L= 2.7, R= 2.6 ± 1.0 [1] L= 2 11/16, R= 2⅝ ± 1 [1]	L= 2.6, R= 2.3 ± 1.0 [1] L= 2⅝, R= 2 5/16 ± 1 [1]	CASTER (Degrees)	L= 2.5, R= 2.4 ± 1.0 [1] L= 2½, R= 2⅜ ± 1 [1]	L= 2.6, R= 2.3 ± 1.0 [1] L= 2⅝, R= 2 5/16 ± 1 [1]	L= 2.4, R= 2.3 ± 1.0 [1] L= 2⅜, R= 2 5/16 ± 1 [1]
0.3 (5/16) | 0.75 (¾)	0.1 (⅛) | 0.75 (¾)	0.3 (5/16) | 0.75 (¾)	CROSS | TOL.	0.1 (⅛) | 0.75 (¾)	0.3 (5/16) | 0.75 (¾)	0.1 (⅛) | 0.75 (¾)
0.5 ± 0.75 ½ ± ¾	0.5 ± 0.75 ½ ± ¾	0.5 ± 0.75 ½ ± ¾	CAMBER (Degrees)	0.5 ± 0.75 ½ ± ¾	0.5 ± 0.75 ½ ± ¾	0.5 ± 0.75 ½ ± ¾
0.0 (0) | 0.75 (¾)	0.0 (0) | 0.75 (¾)	0.0 (0) | 0.75 (¾)	CROSS | TOL.	0.0 (0) | 0.75 (¾)	0.0 (0) | 0.75 (¾)	0.0 (0) | 0.75 (¾)
0.1° ± 0.2° 3/32° ± 3/16°	0.1° ± 0.2° 3/32° ± 3/16°	0.1° ± 0.2° 3/32° ± 3/16°	FRONT TOE (Degrees)	0.2° ± 0.2° 3/16° ± 3/16°	0.1° ± 0.2° 3/32° ± 3/16°	0.1° ± 0.2° 3/32° ± 3/16°
0.05" ± 0.1" 1/16" ± 3/32"	0.05" ± 0.1" 1/16" ± 3/32"	0.05" ± 0.1" 1/16" ± 3/32"	FRONT TOE (Inches)	0.1" ± 0.1" 3/32" ± 3/32"	0.05" ± 0.1" 1/16" ± 3/32"	0.05" ± 0.1" 1/16" ± 3/32"
			SAI			
			TURNING ANGLE			
126	126	126	ILLUSTRATION #	126	126	126
			REAR			
			CAMBER (Degrees)			
			REAR TOE (Degrees)			
			REAR TOE (Inches)			
			THRUST ANGLE			
			ILLUSTRATION #			
[1] BoxOff L=2.6° R=2.3°	[1] BoxOff L=2.5° R=2.4°	[1] BoxOff L=2.5° R=2.2°	NOTES	[1] BoxOff L=2.5° R=2.4°	[1] BoxOff L=2.5° R=2.2°	[1] BoxOff L=2.2° R=2.1°
09339	09340	09341		09342	09343	09344

Chevrolet/GMC Truck & SUV

Silverado/Sierra 2500HD	Silverado/Sierra 2500HD	Silverado/Sierra 2500HD	MODEL	Silverado/Sierra 2500HD	Silverado/Sierra 2500HD	Silverado/Sierra 2500HD
Double Cab, Long Box,	Double Cab, Long Box,	Crew Cab, Std. Box,	INFORMATION	Crew Cab, Std. Box,	Crew Cab, Std. Box,	Crew Cab, Long Box,
exc/LT265/70R17E tires (QXT), Gas	All w/LT265/70R17E tires (QXT)	Diesel, exc/LT265/70R17E tires		Gas, exc/LT265/70R17E tires	All w/LT265/70R17E tires (QXT)	exc/LT265/70R17E tires (QXT), Diesel
2015-2017	2015-2017	2011-2017	YEARS	2011-2017	2011-2017	2011-2017
			INSTRUCTIONS			
140 ft.lbs.	140 ft.lbs.	140 ft.lbs.	LUGNUT TORQUE	140 ft.lbs.	140 ft.lbs.	140 ft.lbs.
			FRONT			
L= 2.7, R= 2.6 ± 1.0 [1] L= 2 11/16, R= 2⅝ ± 1 [1]	L= 2.6, R= 2.3 ± 1.0 [1] L= 2⅝, R= 2 5/16 ± 1 [1]	2.4 ± 1.0 2⅜ ± 1	CASTER (Degrees)	2.7 ± 1.0 2 11/16 ± 1	L= 2.8, R= 2.5 ± 1.0 L= 2 13/16, R= 2½ ± 1	2.5 ± 1.0 [1] 2½ ± 1 [1]
0.1 (⅛) | 0.75 (¾)	0.3 (5/16) | 0.75 (¾)	0.0 (0) | 0.75 (¾)	CROSS | TOL.	0.0 (0) | 0.75 (¾)	0.3 (5/16) | 0.75 (¾)	0.0 (0) | 0.75 (¾)
0.5 ± 0.75 ½ ± ¾	0.5 ± 0.75 ½ ± ¾	0.5 ± 0.75 ½ ± ¾	CAMBER (Degrees)	0.5 ± 0.75 ½ ± ¾	0.5 ± 0.75 ½ ± ¾	0.5 ± 0.75 ½ ± ¾
0.0 (0) | 0.75 (¾)	0.0 (0) | 0.75 (¾)	0.0 (0) | 0.75 (¾)	CROSS | TOL.	0.0 (0) | 0.75 (¾)	0.0 (0) | 0.75 (¾)	0.0 (0) | 0.75 (¾)
0.1° ± 0.2° 3/32° ± 3/16°	0.1° ± 0.2° 3/32° ± 3/16°	0.1° ± 0.2° 3/32° ± 3/16°	FRONT TOE (Degrees)	0.1° ± 0.2° 3/32° ± 3/16°	0.1° ± 0.2° 3/32° ± 3/16°	0.1° ± 0.2° 3/32° ± 3/16°
0.05" ± 0.1" 1/16" ± 3/32"	0.05" ± 0.1" 1/16" ± 3/32"	0.05" ± 0.1" 1/16" ± 3/32"	FRONT TOE (Inches)	0.05" ± 0.1" 1/16" ± 3/32"	0.05" ± 0.1" 1/16" ± 3/32"	0.05" ± 0.1" 1/16" ± 3/32"
			SAI			
			TURNING ANGLE			
126	126	126	ILLUSTRATION #	126	126	126
			REAR			
			CAMBER (Degrees)			
			REAR TOE (Degrees)			
			REAR TOE (Inches)			
			THRUST ANGLE			
			ILLUSTRATION #			
[1] BoxOff L=2.5° R=2.4°	[1] BoxOff L=2.6° R=2.3°		NOTES			[1] BoxOff = 2.3° (2 5/16°)
09345	09346	09347		09348	09349	09350

Chevrolet/GMC Truck & SUV

Silverado/Sierra 2500HD	Silverado/Sierra 2500HD	Silverado/Sierra 3500HD	MODEL	Silverado/Sierra 3500HD	Silverado/Sierra 3500HD	Silverado/Sierra 3500HD
Crew Cab, Long Box, exc/LT265/70R17E tires (QXT), Gas	Crew Cab, Long Box, All w/LT265/70R17E tires (QXT)	All Models	INFORMATION	All C (4X2) Models	All K (4X4) SRW	All K (4X4) DRW
2011-2017	2011-2017	2007	YEARS	2008	2008	2008
			INSTRUCTIONS			
140 ft.lbs.	140 ft.lbs.	140 ft.lbs.	LUGNUT TORQUE	140 ft.lbs.	140 ft.lbs.	140 ft.lbs.
			FRONT			
2.8 ± 1.0[1] 2 13/16 ± 1[1]	L= 2.9, R= 2.6 ± 1.0[1] L= 2⅞, R= 2⅝ ± 1[1]	L= 3.3, R= 3.6 ± 1.0 L= 3 5/16, R= 3⅝ ± 1	CASTER (Degrees)	L= 2.9, R= 3.15 ± 1.0 L= 2⅞, R= 3⅛ ± 1	L= 3.15, R= 3.65 ± 1.0 L= 3⅛, R= 3⅝ ± 1	L= 2.9, R= 3.15 ± 1.0 L= 2⅞, R= 3⅛ ± 1
0.0 (0)	0.3 (5/16)	-0.3 (-5/16)	CROSS	-0.25 (-¼)	-0.5 (-½)	-0.25 (-¼)
0.75 (¾)	0.75 (¾)	0.75 (¾)	TOL.	0.75 (¾)	0.75 (¾)	0.75 (¾)
0.5 ± 0.75 ½ ± ¾	0.5 ± 0.75 ½ ± ¾	0.25 ± 0.6 ¼ ± ⅝	CAMBER (Degrees)	0.25 ± 0.6 ¼ ± ⅝	0.25 ± 0.6 ¼ ± ⅝	0.25 ± 0.6 ¼ ± ⅝
0.0 (0)	0.0 (0)	0.0 (0)	CROSS	0.0 (0)	0.0 (0)	0.0 (0)
0.75 (¾)	0.75 (¾)	0.6 (⅝)	TOL.	0.6 (⅝)	0.6 (⅝)	0.6 (⅝)
0.1° ± 0.2° 3/32° ± 3/16°	0.1° ± 0.2° 3/32° ± 3/16°	0.1° ± 0.2° 3/32° ± 3/16°	FRONT TOE (Degrees)	0.1° ± 0.2° 3/32° ± 3/16°	0.1° ± 0.2° 3/32° ± 3/16°	0.1° ± 0.2° 3/32° ± 3/16°
0.05" ± 0.1" 1/16" ± 3/32"	0.05" ± 0.1" 1/16" ± 3/32"	0.05" ± 0.1" 1/16" ± 3/32"	FRONT TOE (Inches)	0.05" ± 0.1" 1/16" ± 3/32"	0.05" ± 0.1" 1/16" ± 3/32"	0.05" ± 0.1" 1/16" ± 3/32"
			SAI			
			TURNING ANGLE			
126	126	126	ILLUSTRATION #	126	126	126
			REAR			
			CAMBER (Degrees)			
			REAR TOE (Degrees)			
			REAR TOE (Inches)			
			THRUST ANGLE			
			ILLUSTRATION #			
[1] BoxOff = 2.6° (2⅝°)	[1] BoxOff L=2.7° R=2.4°		NOTES			
09351	09352	09353		09354	09355	09356

Chevrolet/GMC Truck & SUV

Silverado/Sierra 3500HD	Silverado/Sierra 3500HD	Silverado/Sierra 3500HD	MODEL	Silverado/Sierra 3500HD	Silverado/Sierra 3500HD	Silverado/Sierra 3500HD
All C (4X2) Models	All K (4X4) SRW	All K (4X4) DRW	INFORMATION	Reg. Cab, SRW	Reg. Cab, DRW	Ext. Cab, SRW, Gas
2009-2010	2009-2010	2009-2010	YEARS	2011-2017	2011-2017	2011-2013
			INSTRUCTIONS			
140 ft.lbs.	140 ft.lbs.	140 ft.lbs.	LUGNUT TORQUE	140 ft.lbs.	140 ft.lbs.	140 ft.lbs.
			FRONT			
L= 3.0, R= 3.1 ± 1.0 L= 3, R= 3⅛ ± 1	L= 3.15, R= 3.25 ± 1.0 L= 3⅛, R= 3¼ ± 1	L= 3.0, R= 3.1 ± 1.0 L= 3, R= 3⅛ ± 1	CASTER (Degrees)	L= 2.6, R= 2.5 ± 1.0[1] L= 2⅝, R= 2½ ± 1[1]	2.8 ± 1.0[1] 2 13/16 ± 1[1]	L= 2.9, R= 2.8 ± 1.0[1] L= 2⅞, R= 2 13/16 ± 1[1]
-0.1 (-⅛)	-0.1 (-⅛)	-0.1 (-⅛)	CROSS	0.1 (⅛)	0.0 (0)	0.1 (⅛)
0.75 (¾)	0.75 (¾)	0.75 (¾)	TOL.	0.75 (¾)	0.75 (¾)	0.75 (¾)
0.25 ± 0.6 ¼ ± ⅝	0.25 ± 0.6 ¼ ± ⅝	0.25 ± 0.6 ¼ ± ⅝	CAMBER (Degrees)	0.5 ± 0.75 ½ ± ¾	0.5 ± 0.75 ½ ± ¾	0.5 ± 0.75 ½ ± ¾
0.0 (0)	0.0 (0)	0.0 (0)	CROSS	0.0 (0)	0.0 (0)	0.0 (0)
0.6 (⅝)	0.6 (⅝)	0.6 (⅝)	TOL.	0.75 (¾)	0.75 (¾)	0.75 (¾)
0.2° ± 0.2° 3/16° ± 3/16°	0.2° ± 0.2° 3/16° ± 3/16°	0.2° ± 0.2° 3/16° ± 3/16°	FRONT TOE (Degrees)	0.1° ± 0.2° 3/32° ± 3/16°	0.1° ± 0.2° 3/32° ± 3/16°	0.1° ± 0.2° 3/32° ± 3/16°
0.1" ± 0.1" 3/32" ± 3/32"	0.1" ± 0.1" 3/32" ± 3/32"	0.1" ± 0.1" 3/32" ± 3/32"	FRONT TOE (Inches)	0.05" ± 0.1" 1/16" ± 3/32"	0.05" ± 0.1" 1/16" ± 3/32"	0.05" ± 0.1" 1/16" ± 3/32"
			SAI			
			TURNING ANGLE			
126	126	126	ILLUSTRATION #	126	126	126
			REAR			
			CAMBER (Degrees)			
			REAR TOE (Degrees)			
			REAR TOE (Inches)			
			THRUST ANGLE			
			ILLUSTRATION #			
			NOTES	[1] BoxOff L=2.3° R=2.1°	[1] BoxOff = 2.6° (2⅝°)	[1] BoxOff L=2.6° R=2.5°
09357	09358	09359		09360	09361	09362

Chevrolet/GMC Truck & SUV

Silverado/Sierra 3500HD Ext. Cab, SRW, Diesel	Silverado/Sierra 3500HD Ext. Cab, DRW, Gas	Silverado/Sierra 3500HD Ext. Cab, DRW, Diesel	MODEL INFORMATION	Silverado/Sierra 3500HD Double Cab, SRW, Gas, Long Bed	Silverado/Sierra 3500HD Double Cab, SRW, Diesel, Long Bed	Silverado/Sierra 3500HD Double Cab, DRW, Gas, Long Bed
2011-2013	2011-2013	2011-2013	YEARS	2015-2017	2015-2017	2015-2017
			INSTRUCTIONS			
140 ft.lbs.	140 ft.lbs.	140 ft.lbs.	LUGNUT TORQUE	140 ft.lbs.	140 ft.lbs.	140 ft.lbs.
			FRONT			
L= 2.6, R= 2.5 ± 1.0 [1] L= 2⅝, R= 2½ ± 1 [1]	L= 2.8, R= 2.7 ± 1.0 [1] L= 2 13/16, R= 2 11/16 ± 1 [1]	L= 2.5, R= 2.4 ± 1.0 [1] L= 2½, R= 2⅜ ± 1 [1]	CASTER (Degrees)	L= 2.9, R= 2.8 ± 1.0 [1] L= 2⅞, R= 2 13/16 ± 1 [1]	L= 2.6, R= 2.5 ± 1.0 [1] L= 2⅝, R= 2½ ± 1 [1]	L= 2.8, R= 2.7 ± 1.0 [1] L= 2 13/16, R= 2 11/16 ± 1 [1]
0.1 (⅛) | 0.75 (¾)	0.1 (⅛) | 0.75 (¾)	0.1 (⅛) | 0.75 (¾)	CROSS | TOL.	0.1 (⅛) | 0.75 (¾)	0.1 (⅛) | 0.75 (¾)	0.1 (⅛) | 0.75 (¾)
0.5 ± 0.75 ½ ± ¾	0.5 ± 0.75 ½ ± ¾	0.5 ± 0.75 ½ ± ¾	CAMBER (Degrees)	0.5 ± 0.75 ½ ± ¾	0.5 ± 0.75 ½ ± ¾	0.5 ± 0.75 ½ ± ¾
0.0 (0) | 0.75 (¾)	0.0 (0) | 0.75 (¾)	0.0 (0) | 0.75 (¾)	CROSS | TOL.	0.0 (0) | 0.75 (¾)	0.0 (0) | 0.75 (¾)	0.0 (0) | 0.75 (¾)
0.1° ± 0.2° 3/32° ± 3/16°	0.1° ± 0.2° 3/32° ± 3/16°	0.1° ± 0.2° 3/32° ± 3/16°	FRONT TOE (Degrees)	0.1° ± 0.2° 3/32° ± 3/16°	0.1° ± 0.2° 3/32° ± 3/16°	0.1° ± 0.2° 3/32° ± 3/16°
0.05" ± 0.1" 1/16" ± 3/32"	0.05" ± 0.1" 1/16" ± 3/32"	0.05" ± 0.1" 1/16" ± 3/32"	FRONT TOE (Inches)	0.05" ± 0.1" 1/16" ± 3/32"	0.05" ± 0.1" 1/16" ± 3/32"	0.05" ± 0.1" 1/16" ± 3/32"
			SAI			
			TURNING ANGLE			
126	126	126	ILLUSTRATION #	126	126	126
			REAR			
			CAMBER (Degrees)			
			REAR TOE (Degrees)			
			REAR TOE (Inches)			
			THRUST ANGLE			
			ILLUSTRATION #			
[1] BoxOff L=2.3° R=2.1°	[1] BoxOff L=2.7° R=2.6°	[1] BoxOff L=2.4° R=2.3°	NOTES	[1] BoxOff L=2.6° R=2.5°	[1] BoxOff L=2.3° R=2.1°	[1] BoxOff L=2.7° R=2.6°
09363	09364	09365		09366	09367	09368

Chevrolet/GMC Truck & SUV

Silverado/Sierra 3500HD Double Cab, DRW, Diesel, Long Bed	Silverado/Sierra 3500HD Crew Cab, 154" WB, Gas	Silverado/Sierra 3500HD Crew Cab, 154" WB, Diesel	MODEL INFORMATION	Silverado/Sierra 3500HD Crew Cab, 168" WB, SRW, Gas	Silverado/Sierra 3500HD Crew Cab, 168" WB, SRW, Diesel	Silverado/Sierra 3500HD Crew Cab, 168" WB, DRW, Gas
2015-2017	2011-2017	2011-2017	YEARS	2011-2017	2011-2017	2011-2017
			INSTRUCTIONS			
140 ft.lbs.	140 ft.lbs.	140 ft.lbs.	LUGNUT TORQUE	140 ft.lbs.	140 ft.lbs.	140 ft.lbs.
			FRONT			
L= 2.5, R= 2.4 ± 1.0 [1] L= 2½, R= 2⅜ ± 1 [1]	2.9 ± 1.0 2⅞ ± 1	2.6 ± 1.0 2⅝ ± 1	CASTER (Degrees)	3.0 ± 1.0 [1] 3 ± 1 [1]	2.7 ± 1.0 [1] 2 11/16 ± 1 [1]	L= 2.9, R= 3.0 ± 1.0 L= 2⅞, R= 3 ± 1
0.1 (⅛) | 0.75 (¾)	0.0 (0) | 0.75 (¾)	0.0 (0) | 0.75 (¾)	CROSS | TOL.	0.0 (0) | 0.75 (¾)	0.0 (0) | 0.75 (¾)	-0.1 (-⅛) | 0.75 (¾)
0.5 ± 0.75 ½ ± ¾	0.5 ± 0.75 ½ ± ¾	0.5 ± 0.75 ½ ± ¾	CAMBER (Degrees)	0.5 ± 0.75 ½ ± ¾	0.5 ± 0.75 ½ ± ¾	0.5 ± 0.75 ½ ± ¾
0.0 (0) | 0.75 (¾)	0.0 (0) | 0.75 (¾)	0.0 (0) | 0.75 (¾)	CROSS | TOL.	0.0 (0) | 0.75 (¾)	0.0 (0) | 0.75 (¾)	0.0 (0) | 0.75 (¾)
0.1° ± 0.2° 3/32° ± 3/16°	0.1° ± 0.2° 3/32° ± 3/16°	0.1° ± 0.2° 3/32° ± 3/16°	FRONT TOE (Degrees)	0.1° ± 0.2° 3/32° ± 3/16°	0.1° ± 0.2° 3/32° ± 3/16°	0.1° ± 0.2° 3/32° ± 3/16°
0.05" ± 0.1" 1/16" ± 3/32"	0.05" ± 0.1" 1/16" ± 3/32"	0.05" ± 0.1" 1/16" ± 3/32"	FRONT TOE (Inches)	0.05" ± 0.1" 1/16" ± 3/32"	0.05" ± 0.1" 1/16" ± 3/32"	0.05" ± 0.1" 1/16" ± 3/32"
			SAI			
			TURNING ANGLE			
126	126	126	ILLUSTRATION #	126	126	126
			REAR			
			CAMBER (Degrees)			
			REAR TOE (Degrees)			
			REAR TOE (Inches)			
			THRUST ANGLE			
			ILLUSTRATION #			
[1] BoxOff L=2.4° R=2.3°			NOTES	[1] BoxOff = 2.7° (2 11/16°)	[1] BoxOff = 2.4° (2⅜°)	
09369	09370	09371		09372	09373	09374

Chevrolet/GMC Truck & SUV

Silverado/Sierra 3500HD Crew Cab, 168" WB, DRW, Diesel	Silverado/Sierra 3500HD Chassis Cab, Reg. Cab, 138" WB	Silverado/Sierra 3500HD Chassis Cab, Reg. Cab, 162" WB, Gas	MODEL INFORMATION	Silverado/Sierra 3500HD Chassis Cab, Reg. Cab, 162" WB, Diesel	Silverado/Sierra 3500HD Chassis Cab, Crew Cab, 172" WB, Gas	Silverado/Sierra 3500HD Chassis Cab, Crew Cab, 172" WB, Diesel
2011-2017	2011-2017	2011-2017	YEARS	2011-2017	2011-2017	2011-2017
			INSTRUCTIONS			
140 ft.lbs.	140 ft.lbs.	140 ft.lbs.	LUGNUT TORQUE	140 ft.lbs.	140 ft.lbs.	140 ft.lbs.
			FRONT			
L= 2.6, R= 2.7 ± 1.0 [1] L= 2⅝, R= 2 11/16 ± 1 [1]	2.4 ± 1.0 2⅜ ± 1	L= 2.6, R= 2.7 ± 1.0 L= 2⅝, R= 2 11/16 ± 1	CASTER (Degrees)	L= 2.3, R= 2.4 ± 1.0 L= 2 5/16, R= 2⅜ ± 1	L= 2.7, R= 2.8 ± 1.0 L= 2 11/16, R= 2 13/16 ± 1	L= 2.4, R= 2.5 ± 1.0 L= 2⅜, R= 2½ ± 1
-0.1 (-⅛) / 0.75 (¾)	0.0 (0) / 0.75 (¾)	-0.1 (-⅛) / 0.75 (¾)	CROSS / TOL.	-0.1 (-⅛) / 0.75 (¾)	-0.1 (-⅛) / 0.75 (¾)	-0.1 (-⅛) / 0.75 (¾)
0.5 ± 0.75 ½ ± ¾	0.5 ± 0.75 ½ ± ¾	0.5 ± 0.75 ½ ± ¾	CAMBER (Degrees)	0.5 ± 0.75 ½ ± ¾	0.5 ± 0.75 ½ ± ¾	0.5 ± 0.75 ½ ± ¾
0.0 (0) / 0.75 (¾)	0.0 (0) / 0.75 (¾)	0.0 (0) / 0.75 (¾)	CROSS / TOL.	0.0 (0) / 0.75 (¾)	0.0 (0) / 0.75 (¾)	0.0 (0) / 0.75 (¾)
0.1° ± 0.2° 3/32° ± 3/16°	0.1° ± 0.2° 3/32° ± 3/16°	0.1° ± 0.2° 3/32° ± 3/16°	FRONT TOE (Degrees)	0.1° ± 0.2° 3/32° ± 3/16°	0.1° ± 0.2° 3/32° ± 3/16°	0.1° ± 0.2° 3/32° ± 3/16°
0.05" ± 0.1" 1/16" ± 3/32"	0.05" ± 0.1" 1/16" ± 3/32"	0.05" ± 0.1" 1/16" ± 3/32"	FRONT TOE (Inches)	0.05" ± 0.1" 1/16" ± 3/32"	0.05" ± 0.1" 1/16" ± 3/32"	0.05" ± 0.1" 1/16" ± 3/32"
			SAI			
			TURNING ANGLE			
126	126	126	ILLUSTRATION #	126	126	126
			REAR			
			CAMBER (Degrees)			
			REAR TOE (Degrees)			
			REAR TOE (Inches)			
			THRUST ANGLE			
			ILLUSTRATION #			
[1] BoxOff L=2.4° R=2.5°			NOTES			
09375	09376	09377		09378	09379	09380

Chevrolet/GMC Truck & SUV

Suburban 1500 4X2 Chevy & GMC	Suburban 1500 4X2 Chevy & GMC	Suburban 1500 4X2	MODEL INFORMATION	Suburban 1500 4X2	Suburban 1500 4X2	Suburban 2500 4X2 Chevy & GMC
1990-91	1992-99	2000	YEARS	2001-2005	2006	1990-91
p. 304-306	p. 304, 305	p. 304, 305	INSTRUCTIONS	p. 304, 305		p. 304-306
			LUGNUT TORQUE			
			FRONT			
see table p. 306 ±0.5 (½)	3.75 ± 1.0 3¾ ± 1	L= 4.25, R= 4.75 ± 1.0 L= 4¼, R= 4¾ ± 1	CASTER (Degrees)	L= 3.9, R= 4.7 ± 1.0 L= 3⅞, R= 4 11/16 ± 1	L= 2.9, R= 3.7 ± 1.0 [1] L= 2⅞, R= 3 11/16 ± 1 [1]	see table p. 306 ±0.5 (½)
0.0 (0) / 0.5 (½)	0.0 (0) / 0.5 (½)	-0.5 (-½) / 0.5 (½)	CROSS / TOL.	-0.8 (-13/16) / 0.75 (¾)	-0.8 (-13/16) / 0.75 (¾)	0.0 (0) / 0.5 (½)
0.7 ± 0.5 11/16 ± ½	0.5 ± 0.5 ½ ± ½	0.25 ± 0.5 ¼ ± ½	CAMBER (Degrees)	0.25 ± 0.5 ¼ ± ½	0.25 ± 0.6 ¼ ± ⅝	0.25 ± 0.5 ¼ ± ½
0.0 (0) / 0.5 (½)	0.0 (0) / 0.5 (½)	0.0 (0) / 0.5 (½)	CROSS / TOL.	0.0 (0) / 0.5 (½)	0.0 (0) / 0.6 (⅝)	0.0 (0) / 0.5 (½)
0.15° ± 0.1° 5/32° ± 3/32°	0.24° ± 0.2° ¼° ± 3/16°	0.1° ± 0.2° 3/32° ± 3/16°	FRONT TOE (Degrees)	0.1° ± 0.2° 3/32° ± 3/16°	0.1° ± 0.2° 3/32° ± 3/16°	0.15° ± 0.1° 5/32° ± 3/32°
0.08" ± 0.05" 1/16" ± 1/16"	0.12" ± 0.1" ⅛" ± 3/32"	0.05" ± 0.1" 1/16" ± 3/32"	FRONT TOE (Inches)	0.05" ± 0.1" 1/16" ± 3/32"	0.05" ± 0.1" 1/16" ± 3/32"	0.08" ± 0.05" 1/16" ± 1/16"
			SAI			
			TURNING ANGLE			
41	43	126	ILLUSTRATION #	126	126	41
			REAR			
			CAMBER (Degrees)			
			REAR TOE (Degrees)			
			REAR TOE (Inches)			
			THRUST ANGLE			
			ILLUSTRATION #			
			NOTES		[1] Relative to ground	
09388	09389	09390		09391	09392	09394

Chevrolet/GMC Truck & SUV

Suburban 2500	Suburban 1500	Suburban 1500	MODEL	Suburban 1500	Suburban 1500	Suburban 1500
4X2	4X4	4X4	INFORMATION	4X4	4X4	4X4
Chevy & GMC	Chevy & GMC	Chevy & GMC				
1992-99	1990-91	1992-99	YEARS	2000	2001-2003	2004-2005
p. 304, 305	p. 304, 305	p. 304, 305	INSTRUCTIONS	p. 304, 305	p. 304, 305	p. 304, 305
			LUGNUT TORQUE			
			FRONT			
3.75 ± 1.0 3¾ ± 1	8.0 ± 1.0 8 ± 1	3.0 ± 1.0 3 ± 1	CASTER (Degrees)	L= 4.25, R= 4.75 ± 1.0 L= 4¼, R= 4¾ ± 1	L= 3.5, R= 4.5 ± 1.0 L= 3½, R= 4½ ± 1	L= 3.6, R= 4.4 ± 1.0 L= 3⅝, R= 4⅜ ± 1
0.0 (0) | 0.5 (½)	0.0 (0) | 0.5 (½)	0.0 (0) | 0.5 (½)	CROSS | TOL.	-0.5 (-½) | 0.5 (½)	-1.0 (-1) | 0.5 (½)	-0.8 (-13/16) | 0.75 (¾)
0.5 ± 0.5 ½ ± ½	1.0 ± 0.75 1 ± ¾	0.65 ± 0.5 ⅝ ± ½	CAMBER (Degrees)	0.25 ± 0.5 ¼ ± ½	0.25 ± 0.5 ¼ ± ½	0.25 ± 0.5 ¼ ± ½
0.0 (0) | 0.5 (½)	0.0 (0) | 0.5 (½)	0.0 (0) | 0.5 (½)	CROSS | TOL.	0.0 (0) | 0.5 (½)	0.0 (0) | 0.5 (½)	0.0 (0) | 0.5 (½)
0.24° ± 0.2° ¼° ± 3/16°	0.15° ± 0.1° 5/32° ± 3/32°	0.24° ± 0.2° ¼° ± 3/16°	FRONT TOE (Degrees)	0.1° ± 0.2° 3/32° ± 3/16°	0.1° ± 0.2° 3/32° ± 3/16°	0.1° ± 0.2° 3/32° ± 3/16°
0.12" ± 0.1" ⅛" ± 3/32"	0.08" ± 0.05" 1/16" ± 1/16"	0.12" ± 0.1" ⅛" ± 3/32"	FRONT TOE (Inches)	0.05" ± 0.1" 1/16" ± 3/32"	0.05" ± 0.1" 1/16" ± 3/32"	0.05" ± 0.1" 1/16" ± 3/32"
			SAI			
			TURNING ANGLE			
43		43	ILLUSTRATION #	126	126	126
			REAR			
			CAMBER (Degrees)			
			REAR TOE (Degrees)			
			REAR TOE (Inches)			
			THRUST ANGLE			
			ILLUSTRATION #			
			NOTES			
09395	09405	09406		09407	09408	09409

Chevrolet/GMC Truck & SUV

Suburban 1500	Suburban 1500	Suburban 1500	MODEL	Suburban 1500	Suburban 1500	Suburban 1500
4X4	4X4	4X2 & 4X4	INFORMATION	4X2 & 4X4	4X2 & 4X4	4X2 & 4X4
exc/20" Tires	w/20" Tires	17" Bridgestones		17" Goodyears	18" Tires	20" Tires
2006	2006	2007	YEARS	2007	2007	2007
			INSTRUCTIONS			
		140 ft.lbs.	LUGNUT TORQUE	140 ft.lbs.	140 ft.lbs.	140 ft.lbs.
			FRONT			
L= 2.6, R= 3.4 ± 1.0 [1,2] L= 2⅝, R= 3⅜ ± 1 [1,2]	L= 2.85, R= 3.15 ± 1.0 [1] L= 2⅞, R= 3⅛ ± 1 [1]	L= 3.35, R= 3.45 ± 1.0 [1] L= 3⅜, R= 3 7/16 ± 1 [1]	CASTER (Degrees)	L= 3.2, R= 3.6 ± 1.0 [1] L= 3 3/16, R= 3⅝ ± 1 [1]	L= 3.3, R= 3.5 ± 1.0 [1] L= 3 5/16, R= 3½ ± 1 [1]	L= 3.55, R= 3.25 ± 1.0 [1] L= 3 9/16, R= 3¼ ± 1 [1]
-0.8 (-13/16) | 0.75 (¾)	-0.3 (-5/16) | 0.75 (¾)	-0.1 (-⅛) | 0.75 (¾)	CROSS | TOL.	-0.4 (-⅜) | 0.75 (¾)	-0.2 (-3/16) | 0.75 (¾)	0.3 (5/16) | 0.75 (¾)
0.25 ± 0.6 ¼ ± ⅝	0.25 ± 0.6 ¼ ± ⅝	-0.1 ± 0.6 -⅛ ± ⅝	CAMBER (Degrees)	-0.1 ± 0.6 -⅛ ± ⅝	-0.1 ± 0.6 -⅛ ± ⅝	-0.1 ± 0.6 -⅛ ± ⅝
0.0 (0) | 0.6 (⅝)	0.0 (0) | 0.6 (⅝)	0.0 (0) | 0.6 (⅝)	CROSS | TOL.	0.0 (0) | 0.6 (⅝)	0.0 (0) | 0.6 (⅝)	0.0 (0) | 0.6 (⅝)
0.1° ± 0.2° 3/32° ± 3/16°	0.1° ± 0.2° 3/32° ± 3/16°	0.1° ± 0.2° 3/32° ± 3/16°	FRONT TOE (Degrees)	0.1° ± 0.2° 3/32° ± 3/16°	0.1° ± 0.2° 3/32° ± 3/16°	0.1° ± 0.2° 3/32° ± 3/16°
0.05" ± 0.1" 1/16" ± 3/32"	0.05" ± 0.1" 1/16" ± 3/32"	0.05" ± 0.1" 1/16" ± 3/32"	FRONT TOE (Inches)	0.05" ± 0.1" 1/16" ± 3/32"	0.05" ± 0.1" 1/16" ± 3/32"	0.05" ± 0.1" 1/16" ± 3/32"
			SAI			
			TURNING ANGLE			
126	126	126	ILLUSTRATION #	126	126	126
			REAR			
			CAMBER (Degrees)			
			REAR TOE (Degrees)			
			REAR TOE (Inches)			
			THRUST ANGLE			
			ILLUSTRATION #			
[1] w/17" Tires L=2.8° R=3.5° [2] Relative to ground	[1] Relative to ground	[1] Relative to ground	NOTES	[1] Relative to ground	[1] Relative to ground	[1] Relative to ground
09410	09411	09412		09413	09414	09415

Chevrolet/GMC Truck & SUV

Suburban 1500 4X2 & 4X4 17" Bridgestones	Suburban 1500 4X2 & 4X4 17" Goodyears	Suburban 1500 4X2 & 4X4 18" Tires	MODEL INFORMATION	Suburban 1500 4X2 & 4X4 20" exc/G'year QSS	Suburban 1500 4X2 & 4X4 20" G'year QSS	Suburban 1500 4X2 & 4X4 20" Tires (All)
2008-2014	2008-2014	2008-2014	YEARS	2008-2010	2008-2010	2011-2014
			INSTRUCTIONS			
140 ft.lbs.	140 ft.lbs.	140 ft.lbs.	LUGNUT TORQUE	140 ft.lbs.	140 ft.lbs.	140 ft.lbs.
			FRONT			
L= 2.9, R= 3.0 ± 1.0 [1] L= 2⅞, R= 3 ± 1 [1]	L= 2.8, R= 3.2 ± 1.0 [1] L= 2 13/16, R= 3 3/16 ± 1 [1]	L= 3.0, R= 2.9 ± 1.0 [1] L= 3, R= 2⅞ ± 1 [1]	CASTER (Degrees)	L= 3.15, R= 2.85 ± 1.0 [1] L= 3⅛, R= 2⅞ ± 1 [1]	3.1 ± 1.0 [1] 3⅛ ± 1 [1]	3.1 ± 1.0 [1] 3⅛ ± 1 [1]
-0.1 (-⅛) | 0.75 (¾)	-0.4 (-⅜) | 0.75 (¾)	0.1 (⅛) | 0.75 (¾)	CROSS | TOL.	0.3 (5/16) | 0.75 (¾)	0.0 (0) | 0.75 (¾)	0.0 (0) | 0.75 (¾)
-0.1 ± 0.6 -⅛ ± ⅝	-0.1 ± 0.6 -⅛ ± ⅝	-0.1 ± 0.6 -⅛ ± ⅝	CAMBER (Degrees)	-0.1 ± 0.6 -⅛ ± ⅝	-0.1 ± 0.6 -⅛ ± ⅝	-0.1 ± 0.6 -⅛ ± ⅝
0.0 (0) | 0.6 (⅝)	0.0 (0) | 0.6 (⅝)	0.0 (0) | 0.6 (⅝)	CROSS | TOL.	0.0 (0) | 0.6 (⅝)	0.0 (0) | 0.6 (⅝)	0.0 (0) | 0.6 (⅝)
0.1° ± 0.2° 3/32° ± 3/16°	0.1° ± 0.2° 3/32° ± 3/16°	0.1° ± 0.2° 3/32° ± 3/16°	FRONT TOE (Degrees)	0.1° ± 0.2° 3/32° ± 3/16°	0.1° ± 0.2° 3/32° ± 3/16°	0.1° ± 0.2° 3/32° ± 3/16°
0.05" ± 0.1" 1/16" ± 3/32"	0.05" ± 0.1" 1/16" ± 3/32"	0.05" ± 0.1" 1/16" ± 3/32"	FRONT TOE (Inches)	0.05" ± 0.1" 1/16" ± 3/32"	0.05" ± 0.1" 1/16" ± 3/32"	0.05" ± 0.1" 1/16" ± 3/32"
			SAI			
			TURNING ANGLE			
126	126	126	ILLUSTRATION #	126	126	126
			REAR			
			CAMBER (Degrees)			
			REAR TOE (Degrees)			
			REAR TOE (Inches)			
			THRUST ANGLE			
			ILLUSTRATION #			
[1] Relative to ground	[1] Relative to ground	[1] Relative to ground	NOTES	[1] Relative to ground	[1] Relative to ground	[1] Relative to ground
09416	09417	09418		09419	09420	09421

Chevrolet/GMC Truck & SUV

Suburban 1500 4X2 & 4X4 w/17" tires (RC3)	Suburban 1500 4X2 & 4X4 w/18" tires (RKX)	Suburban 1500 4X2 & 4X4 w/20" B'stones (QSS)	MODEL INFORMATION	Suburban 1500 4X2 & 4X4 w/20" Cont. (QSS)	Suburban 1500 4X2 & 4X4 w/22" tires (QST)	Suburban 2500 4X4 Chevy & GMC
2015-2017	2015-2017	2015-2017	YEARS	2015-2017	2015-2017	1990-91
			INSTRUCTIONS			p. 304, 305
140 ft.lbs.	140 ft.lbs.	140 ft.lbs.	LUGNUT TORQUE	140 ft.lbs.	140 ft.lbs.	
			FRONT			
3.1 ± 1.0 3⅛ ± 1	L= 3.2, R= 3.0 ± 1.0 L= 3 3/16, R= 3 ± 1	3.1 ± 1.0 3⅛ ± 1	CASTER (Degrees)	L= 3.2, R= 2.9 ± 1.0 L= 3 3/16, R= 2⅞ ± 1	3.1 ± 1.0 3⅛ ± 1	8.0 ± 1.0 8 ± 1
0.0 (0) | 0.75 (¾)	0.2 (3/16) | 0.75 (¾)	0.0 (0) | 0.75 (¾)	CROSS | TOL.	0.3 (5/16) | 0.75 (¾)	0.0 (0) | 0.75 (¾)	0.0 (0) | 0.5 (½)
-0.1 ± 0.8 -⅛ ± 13/16	-0.1 ± 0.8 -⅛ ± 13/16	-0.1 ± 0.8 -⅛ ± 13/16	CAMBER (Degrees)	-0.1 ± 0.8 -⅛ ± 13/16	-0.1 ± 0.8 -⅛ ± 13/16	1.0 ± 0.75 1 ± ¾
0.0 (0) | 0.8 (13/16)	0.0 (0) | 0.8 (13/16)	0.0 (0) | 0.8 (13/16)	CROSS | TOL.	0.0 (0) | 0.8 (13/16)	0.0 (0) | 0.8 (13/16)	0.0 (0) | 0.5 (½)
0.1° ± 0.2° 3/32° ± 3/16°	0.1° ± 0.2° 3/32° ± 3/16°	0.1° ± 0.2° 3/32° ± 3/16°	FRONT TOE (Degrees)	0.1° ± 0.2° [1] 3/32° ± 3/16° [1]	0.1° ± 0.2° 3/32° ± 3/16°	0.15° ± 0.1° 5/32° ± 3/32°
0.05" ± 0.1" 1/16" ± 3/32"	0.05" ± 0.1" 1/16" ± 3/32"	0.05" ± 0.1" 1/16" ± 3/32"	FRONT TOE (Inches)	0.05" ± 0.1" [1] 1/16" ± 3/32" [1]	0.05" ± 0.1" 1/16" ± 3/32"	0.08" ± 0.05" 1/16" ± 1/16"
			SAI			
			TURNING ANGLE			
126	126	126	ILLUSTRATION #	126	126	
			REAR			
			CAMBER (Degrees)			
			REAR TOE (Degrees)			
			REAR TOE (Inches)			
			THRUST ANGLE			
			ILLUSTRATION #			
			NOTES	[1] 2015 = 0.2° (0.1")		
09422	09423	09424		09425	09426	09428

ALIGNMENT SPECIFICATIONS

Chevrolet/GMC Truck & SUV

Suburban 2500	Suburban 2500	Suburban 2500	MODEL	Suburban 2500	Suburban 3500HD	SSR
4X4	4X2 & 4X4	4X2 & 4X4	INFORMATION	4X2 & 4X4	4X4	
Chevy & GMC					w/17" tires (RPO QXT)	
1992-99	2000-2005	2006-2007	YEARS	2008-2013	2016-2017	2003-2006
p. 304, 305	p. 304, 305		INSTRUCTIONS			
	140 ft.lbs.	140 ft.lbs.	LUGNUT TORQUE	140 ft.lbs.	140 ft.lbs.	100 ft.lbs.
			FRONT			
3.0 ±1.0 3 ±1	L= 4.5, R= 4.75 ± 1.0 L= 4½, R= 4¾ ±1	L= 3.5, R= 3.75 ± 1.0 [1] L= 3½, R= 3¾ ±1 [1]	CASTER (Degrees)	L= 3.1, R= 3.3 ± 1.0 [1] L= 3⅛, R= 3 5/16 ±1 [1]	L= 4.7, R= 4.4 ± 1.0 L= 4 11/16, R= 4⅜ ±1	4.25 ±0.5 4¼ ± ½
0.0 (0) | 0.5 (½)	-0.25 (-¼) | 0.75 (¾)	-0.25 (-¼) | 0.5 (½)	CROSS | TOL.	-0.2 (-3/16) | 0.75 (¾)	0.3 (5/16) | 0.75 (¾)	0.0 (0) | 0.3 (5/16)
0.65 ±0.5 ⅝ ± ½	0.25 ±0.5 ¼ ± ½	0.25 ±0.5 ¼ ± ½	CAMBER (Degrees)	0.25 ±0.6 ¼ ± ⅝	-0.2 ±0.8 -3/16 ± 13/16	-0.5 ±0.5 -½ ± ½
0.0 (0) | 0.5 (½)	0.0 (0) | 0.5 (½)	0.0 (0) | 0.6 (⅝)	CROSS | TOL.	0.0 (0) | 0.6 (⅝)	0.0 (0) | 0.8 (13/16)	0.0 (0) | 0.3 (5/16)
0.24° ±0.2° ¼° ± 3/16°	0.1° ±0.2° 3/32° ± 3/16°	0.1° ±0.2° 3/32° ± 3/16°	FRONT TOE (Degrees)	0.1° ±0.2° 3/32° ± 3/16°	0.1° ±0.2° 3/32° ± 3/16°	0.1° ±0.2° 3/32° ± 3/16°
0.12" ±0.1" ⅛" ± 3/32"	0.05" ±0.1" 1/16" ± 3/32"	0.05" ±0.1" 1/16" ± 3/32"	FRONT TOE (Inches)	0.05" ±0.1" 1/16" ± 3/32"	0.05" ±0.1" 1/16" ± 3/32"	0.05" ±0.1" 1/16" ± 3/32"
			SAI			
			TURNING ANGLE			
43	126	126	ILLUSTRATION #	126	126	133
			REAR			
	-0.5 ±0.5 [1] -½ ± ½ [1]		CAMBER (Degrees)			
	0.0° ±0.2° [1] 0° ± 3/16° [1]		REAR TOE (Degrees)			
	0.0" ±0.1" [1] 0" ± 3/32" [1]		REAR TOE (Inches)			
			THRUST ANGLE			0.4 (13/32)
			ILLUSTRATION #			
	[1] Quadrasteer Only	[1] Relative to ground	NOTES	[1] Relative to ground		
09429	09430	09431		09432	09433	09434

Chevrolet/GMC Truck & SUV

Tahoe	Tahoe	Tahoe	MODEL	Tahoe	Tahoe	Tahoe
4X2	4X4	4X2	INFORMATION	4X2	4X4	4X4
		(GTM800)		Limited (GMT400)	(GTM800)	Z71 (GMT400)
1995-99	1995-99	2000	YEARS	2000	2000	2000
p. 304, 305	p. 304, 305	p. 304, 305	INSTRUCTIONS	p. 304, 305	p. 304, 305	p. 304, 305
			LUGNUT TORQUE			
			FRONT			
3.75 ±1.0 3¾ ±1	3.0 ±1.0 3 ±1	L= 4.25, R= 4.75 ± 1.0 L= 4¼, R= 4¾ ±1	CASTER (Degrees)	3.75 ±1.0 3¾ ±1	L= 4.25, R= 4.75 ± 1.0 L= 4¼, R= 4¾ ±1	3.0 ±1.0 3 ±1
0.0 (0) | 0.5 (½)	0.0 (0) | 0.5 (½)	-0.5 (-½) | 0.5 (½)	CROSS | TOL.	0.0 (0) | 0.5 (½)	-0.5 (-½) | 0.5 (½)	0.0 (0) | 0.5 (½)
0.5 ±0.5 ½ ± ½	0.65 ±0.5 ⅝ ± ½	0.25 ±0.5 ¼ ± ½	CAMBER (Degrees)	0.5 ±0.5 ½ ± ½	0.25 ±0.5 ¼ ± ½	0.65 ±0.5 ⅝ ± ½
0.0 (0) | 0.5 (½)	0.0 (0) | 0.5 (½)	0.0 (0) | 0.5 (½)	CROSS | TOL.	0.0 (0) | 0.5 (½)	0.0 (0) | 0.5 (½)	0.0 (0) | 0.5 (½)
0.24° ±0.2° ¼° ± 3/16°	0.24° ±0.2° ¼° ± 3/16°	0.1° ±0.2° 3/32° ± 3/16°	FRONT TOE (Degrees)	0.24° ±0.2° ¼° ± 3/16°	0.1° ±0.2° 3/32° ± 3/16°	0.24° ±0.2° ¼° ± 3/16°
0.12" ±0.1" ⅛" ± 3/32"	0.12" ±0.1" ⅛" ± 3/32"	0.05" ±0.1" 1/16" ± 3/32"	FRONT TOE (Inches)	0.12" ±0.1" ⅛" ± 3/32"	0.05" ±0.1" 1/16" ± 3/32"	0.12" ±0.1" ⅛" ± 3/32"
			SAI			
			TURNING ANGLE			
43	43	126	ILLUSTRATION #	43	126	43
			REAR			
			CAMBER (Degrees)			
			REAR TOE (Degrees)			
			REAR TOE (Inches)			
			THRUST ANGLE			
			ILLUSTRATION #			
			NOTES			
09435	09436	09437		09438	09439	09440

Chevrolet/GMC Truck & SUV

Tahoe 4X2 exc/Police	Tahoe 4X2 Police Pursuit	Tahoe 4X4	MODEL INFORMATION	Tahoe 4X4	Tahoe 4X2 exc/Police	Tahoe 4X2 Police Pursuit
2001-2005	2004-2005	2001-2003	YEARS	2004-2005	2006	2006
p. 304, 305	p. 304, 305	p. 304, 305	INSTRUCTIONS	p. 304, 305		
			LUGNUT TORQUE		140 ft.lbs.	140 ft.lbs.
			FRONT			
L= 3.9, R= 4.7 ± 1.0 L= 3⅞, R= 4 11/16 ± 1	L= 4.35, R= 4.85 ± 1.0 L= 4⅜, R= 4⅞ ± 1	L= 3.5, R= 4.5 ± 1.0 [1] L= 3½, R= 3½ ± 1 [1]	CASTER (Degrees)	L= 3.6, R= 4.4 ± 1.0 [1] L= 3⅝, R= 4⅜ ± 1 [1]	L= 2.9, R= 3.7 ± 1.0 [1] L= 2⅞, R= 3 11/16 ± 1 [1]	L= 3.35, R= 3.85 ± 1.0 [1] L= 3⅜, R= 3⅞ ± 1 [1]
-0.8 (-13/16) | 0.75 (¾)	-0.5 (-½) | 0.75 (¾)	-1.0 (-1) | 0.5 (½)	CROSS | TOL.	-0.8 (-13/16) | 0.5 (½)	-0.8 (-13/16) | 0.75 (¾)	-0.5 (-½) | 0.75 (¾)
0.25 ± 0.5 ¼ ± ½	-0.1 ± 0.6 -⅛ ± ⅝	0.25 ± 0.5 ¼ ± ½	CAMBER (Degrees)	0.25 ± 0.5 ¼ ± ½	0.25 ± 0.6 ¼ ± ⅝	-0.1 ± 0.6 -⅛ ± ⅝
0.0 (0) | 0.5 (½)	0.0 (0) | 0.6 (⅝)	0.0 (0) | 0.5 (½)	CROSS | TOL.	0.0 (0) | 0.5 (½)	0.0 (0) | 0.6 (⅝)	0.0 (0) | 0.6 (⅝)
0.1° ± 0.2° 3/32° ± 3/16°	0.1° ± 0.2° 3/32° ± 3/16°	0.1° ± 0.2° 3/32° ± 3/16°	FRONT TOE (Degrees)	0.1° ± 0.2° 3/32° ± 3/16°	0.1° ± 0.2° 3/32° ± 3/16°	0.1° ± 0.2° 3/32° ± 3/16°
0.05" ± 0.1" 1/16" ± 3/32"	0.05" ± 0.1" 1/16" ± 3/32"	0.05" ± 0.1" 1/16" ± 3/32"	FRONT TOE (Inches)	0.05" ± 0.1" 1/16" ± 3/32"	0.05" ± 0.1" 1/16" ± 3/32"	0.05" ± 0.1" 1/16" ± 3/32"
			SAI			
			TURNING ANGLE			
126	126	126	ILLUSTRATION #	126	126	126
			REAR			
			CAMBER (Degrees)			
			REAR TOE (Degrees)			
			REAR TOE (Inches)			
			THRUST ANGLE			
			ILLUSTRATION #			
		[1] w/17" Tires L=3.8° R=4.5°	NOTES	[1] w/17" Tires L=3.8° R=4.5°	[1] Relative to ground	[1] Relative to ground
09441	09442	09443		09444	09445	09446

Chevrolet/GMC Truck & SUV

Tahoe 4X4	Tahoe 4X2 & 4X4 w/17" B'stones	Tahoe 4X2 & 4X4 w/17" G'years (incl. Police)	MODEL INFORMATION	Tahoe 4X2 & 4X4 w/18" Tires	Tahoe 4X2 & 4X4 w/20" Tires	Tahoe 4X2 & 4X4 17" B'stones
2006	2007	2007	YEARS	2007	2007	2008-2010
			INSTRUCTIONS			
140 ft.lbs.	140 ft.lbs.	140 ft.lbs.	LUGNUT TORQUE	140 ft.lbs.	140 ft.lbs.	140 ft.lbs.
			FRONT			
L= 2.6, R= 3.4 ± 1.0 [1,2] L= 2⅝, R= 3⅜ ± 1 [1,2]	L= 3.35, R= 3.45 ± 1.0 [1] L= 3⅜, R= 3 7/16 ± 1 [1]	L= 3.2, R= 3.6 ± 1.0 [1] L= 3 3/16, R= 3⅝ ± 1 [1]	CASTER (Degrees)	L= 3.3, R= 3.5 ± 1.0 [1] L= 3 5/16, R= 3½ ± 1 [1]	L= 3.55, R= 3.25 ± 1.0 [1] L= 3 9/16, R= 3¼ ± 1 [1]	L= 2.95, R= 3.05 ± 1.0 [1] L= 2 15/16, R= 3 1/16 ± 1 [1]
-0.8 (-13/16) | 0.75 (¾)	-0.1 (-⅛) | 0.75 (¾)	-0.4 (-⅜) | 0.75 (¾)	CROSS | TOL.	-0.2 (-3/16) | 0.75 (¾)	0.3 (5/16) | 0.75 (¾)	-0.1 (-⅛) | 0.75 (¾)
0.25 ± 0.6 ¼ ± ⅝	-0.1 ± 0.6 -⅛ ± ⅝	-0.1 ± 0.6 -⅛ ± ⅝	CAMBER (Degrees)	-0.1 ± 0.6 -⅛ ± ⅝	-0.1 ± 0.6 -⅛ ± ⅝	-0.1 ± 0.6 -⅛ ± ⅝
0.0 (0) | 0.6 (⅝)	0.0 (0) | 0.6 (⅝)	0.0 (0) | 0.6 (⅝)	CROSS | TOL.	0.0 (0) | 0.6 (⅝)	0.0 (0) | 0.6 (⅝)	0.0 (0) | 0.6 (⅝)
0.1° ± 0.2° 3/32° ± 3/16°	0.1° ± 0.2° 3/32° ± 3/16°	0.1° ± 0.2° 3/32° ± 3/16°	FRONT TOE (Degrees)	0.1° ± 0.2° 3/32° ± 3/16°	0.1° ± 0.2° 3/32° ± 3/16°	0.1° ± 0.2° 3/32° ± 3/16°
0.05" ± 0.1" 1/16" ± 3/32"	0.05" ± 0.1" 1/16" ± 3/32"	0.05" ± 0.1" 1/16" ± 3/32"	FRONT TOE (Inches)	0.05" ± 0.1" 1/16" ± 3/32"	0.05" ± 0.1" 1/16" ± 3/32"	0.05" ± 0.1" 1/16" ± 3/32"
			SAI			
			TURNING ANGLE			
126	126	126	ILLUSTRATION #	126	126	126
			REAR			
			CAMBER (Degrees)			
			REAR TOE (Degrees)			
			REAR TOE (Inches)			
			THRUST ANGLE			
			ILLUSTRATION #			
[1] w/17" Tires L=2.8° R=3.5° [2] Relative to ground	[1] Relative to ground	[1] Relative to ground	NOTES	[1] Relative to ground	[1] Relative to ground	[1] Relative to ground
09447	09448	09449		09450	09451	09452

Chevrolet/GMC Truck & SUV

Tahoe 4X2 & 4X4 17" G'years	Tahoe 4X2 & 4X4 17" Police Models	Tahoe 4X2 & 4X4 18" Tires	MODEL INFORMATION	Tahoe 4X2 & 4X4 20" exc/G'year QSS	Tahoe 4X2 & 4X4 20" G'year QSS	Tahoe 4X2 & 4X4 with 17" Tires
2008-2010	2008-2010	2008-2010	YEARS	2008-2010	2008-2010	2011-2014
			INSTRUCTIONS			
140 ft.lbs.	140 ft.lbs.	140 ft.lbs.	LUGNUT TORQUE	140 ft.lbs.	140 ft.lbs.	140 ft.lbs.
			FRONT			
L= 2.8, R= 3.2 ± 1.0 [1] L= 2 13/16, R= 3 3/16 ± 1 [1]	L= 2.95, R= 3.05 ± 1.0 [1] L= 2 15/16, R= 3 1/16 ± 1 [1]	L= 3.05, R= 2.95 ± 1.0 [1] L= 3 1/16, R= 2 15/16 ± 1 [1]	CASTER (Degrees)	L= 3.15, R= 2.85 ± 1.0 [1] L= 3⅛, R= 2⅞ ± 1 [1]	3.1 ± 1.0 [1] 3⅛ ± 1 [1]	L= 3.3, R= 3.5 ± 1.0 L= 3 5/16, R= 3½ ± 1
-0.4 (-⅜) | 0.75 (¾)	-0.1 (-⅛) | 0.75 (¾)	0.1 (⅛) | 0.75 (¾)	CROSS | TOL.	0.3 (5/16) | 0.75 (¾)	0.0 (0) | 0.75 (¾)	-0.2 (-3/16) | 0.75 (¾)
-0.1 ± 0.6 -⅛ ± ⅝	-0.1 ± 0.6 -⅛ ± ⅝	-0.1 ± 0.6 -⅛ ± ⅝	CAMBER (Degrees)	-0.1 ± 0.6 -⅛ ± ⅝	-0.1 ± 0.6 -⅛ ± ⅝	-0.1 ± 0.8 -⅛ ± 13/16
0.0 (0) | 0.6 (⅝)	0.0 (0) | 0.6 (⅝)	0.0 (0) | 0.6 (⅝)	CROSS | TOL.	0.0 (0) | 0.6 (⅝)	0.0 (0) | 0.6 (⅝)	0.0 (0) | 0.8 (13/16)
0.1° ± 0.2° 3/32° ± 3/16°	0.1° ± 0.2° 3/32° ± 3/16°	0.1° ± 0.2° 3/32° ± 3/16°	FRONT TOE (Degrees)	0.1° ± 0.2° 3/32° ± 3/16°	0.1° ± 0.2° 3/32° ± 3/16°	0.1° ± 0.2° 3/32° ± 3/16°
0.05" ± 0.1" 1/16" ± 3/32"	0.05" ± 0.1" 1/16" ± 3/32"	0.05" ± 0.1" 1/16" ± 3/32"	FRONT TOE (Inches)	0.05" ± 0.1" 1/16" ± 3/32"	0.05" ± 0.1" 1/16" ± 3/32"	0.05" ± 0.1" 1/16" ± 3/32"
			SAI			
			TURNING ANGLE			
126	126	126	ILLUSTRATION #	126	126	126
			REAR			
			CAMBER (Degrees)			
			REAR TOE (Degrees)			
			REAR TOE (Inches)			
			THRUST ANGLE			
			ILLUSTRATION #			
[1] Relative to ground	[1] Relative to ground	[1] Relative to ground	NOTES	[1] Relative to ground	[1] Relative to ground	
09453	09454	09455		09456	09457	09458

Chevrolet/GMC Truck & SUV

Tahoe 4X2 & 4X4 with 17" Police	Tahoe 4X2 & 4X4 with 18" Tires	Tahoe 4X2 & 4X4 with 20" Tires	MODEL INFORMATION	Tahoe 4X2 & 4X4 w/17" tires (incl. Police)	Tahoe 4X2 & 4X4 w/18" tires (RKX)	Tahoe 4X2 & 4X4 w/20" B'stones (QSS)
2011-2014	2011-2014	2011-2014	YEARS	2015-2017	2015-2017	2015-2017
			INSTRUCTIONS			
140 ft.lbs.	140 ft.lbs.	140 ft.lbs.	LUGNUT TORQUE	140 ft.lbs.	140 ft.lbs.	140 ft.lbs.
			FRONT			
L= 2.9, R= 3.1 ± 1.0 L= 2⅞, R= 3⅛ ± 1	L= 3.0, R= 2.9 ± 1.0 L= 3, R= 2⅞ ± 1	3.1 ± 1.0 [1] 3⅛ ± 1 [1]	CASTER (Degrees)	3.1 ± 1.0 3⅛ ± 1	L= 3.2, R= 3.0 ± 1.0 L= 3 3/16, R= 3 ± 1	3.1 ± 1.0 3⅛ ± 1
-0.2 (-3/16) | 0.75 (¾)	0.1 (⅛) | 0.75 (¾)	0.0 (0) | 0.75 (¾)	CROSS | TOL.	0.0 (0) | 0.75 (¾)	0.2 (3/16) | 0.75 (¾)	0.0 (0) | 0.75 (¾)
-0.1 ± 0.8 -⅛ ± 13/16	-0.1 ± 0.8 -⅛ ± 13/16	-0.1 ± 0.6 -⅛ ± ⅝	CAMBER (Degrees)	-0.1 ± 0.8 -⅛ ± 13/16	-0.1 ± 0.8 -⅛ ± 13/16	-0.1 ± 0.8 -⅛ ± 13/16
0.0 (0) | 0.8 (13/16)	0.0 (0) | 0.8 (13/16)	0.0 (0) | 0.6 (⅝)	CROSS | TOL.	0.0 (0) | 0.8 (13/16)	0.0 (0) | 0.8 (13/16)	0.0 (0) | 0.8 (13/16)
0.1° ± 0.2° 3/32° ± 3/16°	0.1° ± 0.2° [1] 3/32° ± 3/16° [1]	0.1° ± 0.2° 3/32° ± 3/16°	FRONT TOE (Degrees)	0.1° ± 0.2° 3/32° ± 3/16°	0.1° ± 0.2° 3/32° ± 3/16°	0.1° ± 0.2° 3/32° ± 3/16°
0.05" ± 0.1" 1/16" ± 3/32"	0.05" ± 0.1" [1] 1/16" ± 3/32" [1]	0.05" ± 0.1" 1/16" ± 3/32"	FRONT TOE (Inches)	0.05" ± 0.1" 1/16" ± 3/32"	0.05" ± 0.1" 1/16" ± 3/32"	0.05" ± 0.1" 1/16" ± 3/32"
			SAI			
			TURNING ANGLE			
126	126	126	ILLUSTRATION #	126 126	126 126	126 126
			REAR			
			CAMBER (Degrees)			
			REAR TOE (Degrees)			
			REAR TOE (Inches)			
			THRUST ANGLE			
			ILLUSTRATION #			
	[1] Hybrids = 0.2° (0.1")	[1] Relative to ground	NOTES			
09459	09460	09461		09462	09463	09464

Chevrolet/GMC Truck & SUV

Tahoe	Tahoe	Terrain	MODEL	Tracker	Tracker	Trailblazer
4X2 & 4X4	4X2 & 4X4		INFORMATION			
w/20" Cont. (QSS)	w/22" tires (QST)					
2015-2017	2015-2017	2010-2017	YEARS	1998	1999-2004	2002-2004
			INSTRUCTIONS			
140 ft.lbs.	140 ft.lbs.	140 ft.lbs.	LUGNUT TORQUE	70 ft.lbs.	70 ft.lbs.	103 ft.lbs.
			FRONT			
L= 3.2, R= 2.9 ± 1.0 L= 3 3/16, R= 2 7/8 ± 1	3.1 ± 1.0 3⅛ ± 1	2.7 ± 0.6 2 11/16 ± ⅝	CASTER (Degrees)	1.5 ± 1.0 1½ ± 1	2.65 ± 1.0 2⅝ ± 1	L= 3.5, R= 4.0 ± 0.5 L= 3½, R= 4 ± ½
0.3 (5/16) | 0.75 (¾)	0.0 (0) | 0.75 (¾)	0.0 (0) | 0.9 (⅞)	CROSS | TOL.	0.0 (0) | 1.0 (1)	0.0 (0) | 1.0 (1)	-0.5 (-½) | 0.3 (5/16)
-0.1 ± 0.8 -⅛ ± 13/16	-0.1 ± 0.8 -⅛ ± 13/16	-0.5 ± 0.8 -½ ± 13/16	CAMBER (Degrees)	0.5 ± 1.0 ½ ± 1	0.0 ± 1.0 [1] 0 ± 1 [1]	-0.5 ± 0.5 -½ ± ½
0.0 (0) | 0.8 (13/16)	0.0 (0) | 0.8 (13/16)	0.0 (0) | 0.8 (13/16)	CROSS | TOL.	0.0 (0) | 1.0 (1)	0.0 (0) | 1.0 (1)	0.0 (0) | 0.3 (5/16)
0.1° ± 0.2° [1] 3/32° ± 3/16° [1]	0.1° ± 0.2° 3/32° ± 3/16°	0.2° ± 0.2° 3/16° ± 3/16°	FRONT TOE (Degrees)	0.31° ± 0.16° 5/16° ± 5/32°	0.0° ± 0.16° 0° ± 5/32°	0.1° ± 0.2° 3/32° ± 3/16°
0.05" ± 0.1" [1] 1/16" ± 3/32" [1]	0.05" ± 0.1" 1/16" ± 3/32"	0.1" ± 0.1" 3/32" ± 3/32"	FRONT TOE (Inches)	0.16" ± 0.08" 5/32" ± 1/16"	0.0" ± 0.08" 0" ± 1/16"	0.05" ± 0.1" 1/16" ± 3/32"
			SAI			
			TURNING ANGLE	32.5 / 30.5	34.5 / 32.5	
126 126	126 126	25	ILLUSTRATION #			133
			REAR			
		-0.4 ± 0.8 -⅜ ± 13/16	CAMBER (Degrees)			
		0.1° ± 0.2° 3/32° ± 3/16°	REAR TOE (Degrees)			
		0.05" ± 0.1" 1/16" ± 3/32"	REAR TOE (Inches)			
		0.3 (5/16)	THRUST ANGLE			
		107 143	ILLUSTRATION #			
[1] 2015 = 0.2° (0.1")			NOTES		[1] 1999 = 0.5° (½°)	
09465	09466	09467		09468	09469	09470

Chevrolet/GMC Truck & SUV

Trailblazer	Trailblazer EXT	Trailblazer	MODEL	Trailblazer SS	Traverse	Trax
			INFORMATION		All Models	
2005	2006	2006-2009	YEARS	2006-2009	2009-2017	2013-2017
			INSTRUCTIONS			
103 ft.lbs.	103 ft.lbs.	103 ft.lbs.	LUGNUT TORQUE	103 ft.lbs.	140 ft.lbs.	103 ft.lbs.
			FRONT			
L= 3.75, R= 4.25 ± 0.6 L= 3¾, R= 4¼ ± ⅝	L= 4.1, R= 3.9 ± 0.6 L= 4⅛, R= 3⅞ ± ⅝	L= 3.9, R= 4.1 ± 0.6 L= 3⅞, R= 4⅛ ± ⅝	CASTER (Degrees)	4.25 ± 0.6 4¼ ± ⅝	4.7 ± 0.8 4 11/16 ± 13/16	4.5 ± 0.75 4½ ± ¾
-0.5 (-½) | 0.6 (⅝)	0.2 (3/16) | 0.6 (⅝)	-0.2 (-3/16) | 0.6 (⅝)	CROSS | TOL.	0.0 (0) | 0.6 (⅝)	0.0 (0) | 0.8 (13/16)	0.0 (0) | 0.75 (¾)
-0.5 ± 0.6 -½ ± ⅝	-0.5 ± 0.6 -½ ± ⅝	-0.5 ± 0.6 -½ ± ⅝	CAMBER (Degrees)	-0.5 ± 0.6 -½ ± ⅝	-0.9 ± 0.8 -⅞ ± 13/16	-0.6 ± 0.75 -⅝ ± ¾
0.0 (0) | 0.6 (⅝)	0.0 (0) | 0.6 (⅝)	0.0 (0) | 0.6 (⅝)	CROSS | TOL.	0.0 (0) | 0.6 (⅝)	0.0 (0) | 0.8 (13/16)	0.0 (0) | 0.75 (¾)
0.1° ± 0.2° 3/32° ± 3/16°	0.1° ± 0.2° 3/32° ± 3/16°	0.1° ± 0.2° 3/32° ± 3/16°	FRONT TOE (Degrees)	0.2° ± 0.2° 3/16° ± 3/16°	0.0° ± 0.2° 0° ± 3/16°	0.16° ± 0.2° 5/32° ± 3/16°
0.05" ± 0.1" 1/16" ± 3/32"	0.05" ± 0.1" 1/16" ± 3/32"	0.05" ± 0.1" 1/16" ± 3/32"	FRONT TOE (Inches)	0.1" ± 0.1" 3/32" ± 3/32"	0.0" ± 0.1" 0" ± 3/32"	0.08" ± 0.1" 3/32" ± 3/32"
			SAI			
			TURNING ANGLE			
133	133	133	ILLUSTRATION #	133	25	
			REAR			
			CAMBER (Degrees)		-0.9 ± 0.8 -⅞ ± 13/16	-1.0 ± 0.5 -1 ± ½
			REAR TOE (Degrees)		0.1° ± 0.2° [1] 3/32° ± 3/16° [1]	0.2° ± 0.4° 3/16° ± 13/32°
			REAR TOE (Inches)		0.05" ± 0.1" [1] 1/16" ± 3/32" [1]	0.1" ± 0.2" 3/32" ± 3/16"
	0.4 (13/32)	0.4 (13/32)	THRUST ANGLE	0.4 (13/32)	0.3 (5/16)	0.3 (5/16)
			ILLUSTRATION #		107 143	
			NOTES		[1] w/RPO QPR= 0.0° (0") (245/70R17BW OE)	
09471	09472	09473		09474	09475	09476

Chevrolet/GMC Truck & SUV

Yukon 4X2	Yukon 4X4 exc/Denali	Yukon 4X2	MODEL INFORMATION	Yukon 4X4 exc/Denali	Yukon 4X2	Yukon 4X4 exc/Denali
1994-99	1992-99	2000	YEARS	2000	2001-2005	2001-2003
p. 304, 305	p. 304, 305	p. 304, 305	INSTRUCTIONS	p. 304, 305	p. 304, 305	p. 304, 305
125 ft.lbs.	125 ft.lbs.	140 ft.lbs.	LUGNUT TORQUE	140 ft.lbs.	140 ft.lbs.	140 ft.lbs.
			FRONT			
3.75 ± 1.0 3¾ ± 1	3.0 ± 1.0 3 ± 1	L= 4.25, R= 4.75 ± 1.0 L= 4¼, R= 4¾ ± 1	CASTER (Degrees)	L= 4.25, R= 4.75 ± 1.0 L= 4¼, R= 4¾ ± 1	L= 3.9, R= 4.7 ± 1.0 L= 3⅞, R= 4 11/16 ± 1	L= 3.5, R= 4.5 ± 1.0 [1] L= 3½, R= 3½ ± 1 [1]
0.0 (0) | 0.5 (½)	0.0 (0) | 0.5 (½)	-0.5 (-½) | 0.5 (½)	CROSS | TOL.	-0.5 (-½) | 0.5 (½)	-0.8 (-13/16) | 0.75 (¾)	-1.0 (-1) | 0.5 (½)
0.5 ± 0.5 ½ ± ½	0.65 ± 0.5 ⅝ ± ½	0.25 ± 0.5 ¼ ± ½	CAMBER (Degrees)	0.25 ± 0.5 ¼ ± ½	0.25 ± 0.5 ¼ ± ½	0.25 ± 0.5 ¼ ± ½
0.0 (0) | 0.5 (½)	0.0 (0) | 0.5 (½)	0.0 (0) | 0.5 (½)	CROSS | TOL.	0.0 (0) | 0.5 (½)	0.0 (0) | 0.5 (½)	0.0 (0) | 0.5 (½)
0.24° ± 0.2° ¼° ± 3/16°	0.24° ± 0.2° ¼° ± 3/16°	0.1° ± 0.2° 3/32° ± 3/16°	FRONT TOE (Degrees)	0.1° ± 0.2° 3/32° ± 3/16°	0.1° ± 0.2° 3/32° ± 3/16°	0.1° ± 0.2° 3/32° ± 3/16°
0.12" ± 0.1" ⅛" ± 3/32"	0.12" ± 0.1" ⅛" ± 3/32"	0.05" ± 0.1" 1/16" ± 3/32"	FRONT TOE (Inches)	0.05" ± 0.1" 1/16" ± 3/32"	0.05" ± 0.1" 1/16" ± 3/32"	0.05" ± 0.1" 1/16" ± 3/32"
			SAI			
			TURNING ANGLE			
43	43	126	ILLUSTRATION #	126	126	126
			REAR			
			CAMBER (Degrees)			
			REAR TOE (Degrees)			
			REAR TOE (Inches)			
			THRUST ANGLE			
			ILLUSTRATION #			
			NOTES			[1] w/17" Tires L=3.8° R=4.5°
09477	09478	09479		09480	09481	09482

Chevrolet/GMC Truck & SUV

Yukon 4X4 exc/Denali	Yukon 4X2	Yukon 4X4	MODEL INFORMATION	Yukon 4X2 & 4X4 w/17" B'stones	Yukon 4X2 & 4X4 w/17" G'years	Yukon 4X2 & 4X4 w/18" Tires
2004-2005	2006	2006	YEARS	2007	2007	2007
p. 304, 305			INSTRUCTIONS			
140 ft.lbs.	140 ft.lbs.	140 ft.lbs.	LUGNUT TORQUE	140 ft.lbs.	140 ft.lbs.	140 ft.lbs.
			FRONT			
L= 3.6, R= 4.4 ± 1.0 [1] L= 3⅝, R= 4⅜ ± 1 [1]	L= 2.9, R= 3.7 ± 1.0 [1] L= 2⅞, R= 3 11/16 ± 1 [1]	L= 2.6, R= 3.4 ± 1.0 [1,2] L= 2⅝, R= 3⅜ ± 1 [1,2]	CASTER (Degrees)	L= 3.35, R= 3.45 ± 1.0 [1] L= 3⅜, R= 3 7/16 ± 1 [1]	L= 3.2, R= 3.6 ± 1.0 [1] L= 3 3/16, R= 3⅝ ± 1 [1]	L= 3.3, R= 3.5 ± 1.0 [1] L= 3 5/16, R= 3½ ± 1 [1]
-0.8 (-13/16) | 0.75 (¾)	-0.8 (-13/16) | 0.75 (¾)	-0.8 (-13/16) | 0.75 (¾)	CROSS | TOL.	-0.1 (-⅛) | 0.5 (½)	-0.4 (-⅜) | 0.5 (½)	-0.2 (-3/16) | 0.5 (½)
0.25 ± 0.5 ¼ ± ½	0.25 ± 0.6 ¼ ± ⅝	0.25 ± 0.6 ¼ ± ⅝	CAMBER (Degrees)	-0.1 ± 0.5 -⅛ ± ½	-0.1 ± 0.5 -⅛ ± ½	-0.1 ± 0.5 -⅛ ± ½
0.0 (0) | 0.5 (½)	0.0 (0) | 0.6 (⅝)	0.0 (0) | 0.6 (⅝)	CROSS | TOL.	0.0 (0) | 0.5 (½)	0.0 (0) | 0.5 (½)	0.0 (0) | 0.5 (½)
0.1° ± 0.2° 3/32° ± 3/16°	0.1° ± 0.2° 3/32° ± 3/16°	0.1° ± 0.2° 3/32° ± 3/16°	FRONT TOE (Degrees)	0.1° ± 0.2° 3/32° ± 3/16°	0.1° ± 0.2° 3/32° ± 3/16°	0.1° ± 0.2° 3/32° ± 3/16°
0.05" ± 0.1" 1/16" ± 3/32"	0.05" ± 0.1" 1/16" ± 3/32"	0.05" ± 0.1" 1/16" ± 3/32"	FRONT TOE (Inches)	0.05" ± 0.1" 1/16" ± 3/32"	0.05" ± 0.1" 1/16" ± 3/32"	0.05" ± 0.1" 1/16" ± 3/32"
			SAI			
			TURNING ANGLE			
126	126	126	ILLUSTRATION #	126	126	126
			REAR			
			CAMBER (Degrees)			
			REAR TOE (Degrees)			
			REAR TOE (Inches)			
			THRUST ANGLE			
			ILLUSTRATION #			
[1] w/17" Tires L=3.8° R=4.5°	[1] Relative to ground	[1] w/17" Tires L=2.8° R=3.5° [2] Relative to ground	NOTES	[1] Relative to ground	[1] Relative to ground	[1] Relative to ground
09483	09484	09485		09486	09487	09488

Chevrolet/GMC Truck & SUV

Yukon 4X2 & 4X4 w/20" Tires	Yukon 4X2 & 4X4 17" B'stones	Yukon 4X2 & 4X4 17" G'years	MODEL INFORMATION	Yukon 4X2 & 4X4 18" Tires (incl. Hybrids)	Yukon 4X2 & 4X4 20" exc/G'year QSS	Yukon 4X2 & 4X4 20" G'year QSS
2007	2008-2010	2008-2010	YEARS	2008-2010	2008-2010	2008-2010
			INSTRUCTIONS			
140 ft.lbs.	140 ft.lbs.	140 ft.lbs.	LUGNUT TORQUE	140 ft.lbs.	140 ft.lbs.	140 ft.lbs.
			FRONT			
L= 3.55, R= 3.25 ± 1.0 [1] L= 3 9⁄16, R= 3 1⁄4 ± 1 [1]	L= 2.95, R= 3.05 ± 1.0 [1] L= 2 15⁄16, R= 3 1⁄16 ± 1 [1]	L= 2.8, R= 3.2 ± 1.0 [1] L= 2 13⁄16, R= 3 3⁄16 ± 1 [1]	CASTER (Degrees)	L= 3.05, R= 2.95 ± 1.0 [1] L= 3 1⁄16, R= 2 15⁄16 ± 1 [1]	L= 3.15, R= 2.85 ± 1.0 [1] L= 3 1⁄8, R= 2 7⁄8 ± 1 [1]	3.1 ± 1.0 [1] 3 1⁄8 ± 1 [1]
0.3 (5⁄16) | 0.5 (1⁄2)	-0.1 (-1⁄8) | 0.75 (3⁄4)	-0.4 (-3⁄8) | 0.75 (3⁄4)	CROSS | TOL.	0.1 (1⁄8) | 0.75 (3⁄4)	0.3 (5⁄16) | 0.75 (3⁄4)	0.0 (0) | 0.75 (3⁄4)
-0.1 ± 0.5 -1⁄8 ± 1⁄2	-0.1 ± 0.6 -1⁄8 ± 5⁄8	-0.1 ± 0.6 -1⁄8 ± 5⁄8	CAMBER (Degrees)	-0.1 ± 0.6 -1⁄8 ± 5⁄8	-0.1 ± 0.6 -1⁄8 ± 5⁄8	-0.1 ± 0.6 -1⁄8 ± 5⁄8
0.0 (0) | 0.5 (1⁄2)	0.0 (0) | 0.6 (5⁄8)	0.0 (0) | 0.6 (5⁄8)	CROSS | TOL.	0.0 (0) | 0.6 (5⁄8)	0.0 (0) | 0.6 (5⁄8)	0.0 (0) | 0.6 (5⁄8)
0.1° ± 0.2° 3⁄32° ± 3⁄16°	0.1° ± 0.2° 3⁄32° ± 3⁄16°	0.1° ± 0.2° 3⁄32° ± 3⁄16°	FRONT TOE (Degrees)	0.1° ± 0.2° 3⁄32° ± 3⁄16°	0.1° ± 0.2° 3⁄32° ± 3⁄16°	0.1° ± 0.2° 3⁄32° ± 3⁄16°
0.05" ± 0.1" 1⁄16" ± 3⁄32"	0.05" ± 0.1" 1⁄16" ± 3⁄32"	0.05" ± 0.1" 1⁄16" ± 3⁄32"	FRONT TOE (Inches)	0.05" ± 0.1" 1⁄16" ± 3⁄32"	0.05" ± 0.1" 1⁄16" ± 3⁄32"	0.05" ± 0.1" 1⁄16" ± 3⁄32"
			SAI			
			TURNING ANGLE			
126	126	126	ILLUSTRATION #	126	126	126
			REAR			
			CAMBER (Degrees)			
			REAR TOE (Degrees)			
			REAR TOE (Inches)			
			THRUST ANGLE			
			ILLUSTRATION #			
[1] Relative to ground	[1] Relative to ground	[1] Relative to ground	NOTES	[1] Relative to ground	[1] Relative to ground	[1] Relative to ground
09489	09490	09491		09492	09493	09494

Chevrolet/GMC Truck & SUV

Yukon 4X2 & 4X4 with 17" Tires	Yukon 4X2 & 4X4 with 18" Tires	Yukon 4X2 & 4X4 with 20" Tires	MODEL INFORMATION	Yukon 4X2 & 4X4 w/18" tires (RKX)	Yukon 4X2 & 4X4 w/20" B'stones (QSS)	Yukon 4X2 & 4X4 w/20" Cont. (QSS)
2011-2014	2011-2014	2011-2014	YEARS	2015-2017	2015-2017	2015-2017
			INSTRUCTIONS			
140 ft.lbs.	140 ft.lbs.	140 ft.lbs.	LUGNUT TORQUE	140 ft.lbs.	140 ft.lbs.	140 ft.lbs.
			FRONT			
L= 3.3, R= 3.5 ± 1.0 L= 3 5⁄16, R= 3 1⁄2 ± 1	L= 3.0, R= 2.9 ± 1.0 L= 3, R= 2 7⁄8 ± 1	3.1 ± 1.0 3 1⁄8 ± 1	CASTER (Degrees)	L= 3.2, R= 3.0 ± 1.0 L= 3 3⁄16, R= 3 ± 1	3.1 ± 1.0 3 1⁄8 ± 1	L= 3.2, R= 2.9 ± 1.0 L= 3 3⁄16, R= 2 7⁄8 ± 1
-0.2 (-3⁄16) | 0.75 (3⁄4)	0.1 (1⁄8) | 0.75 (3⁄4)	0.0 (0) | 0.75 (3⁄4)	CROSS | TOL.	0.2 (3⁄16) | 0.75 (3⁄4)	0.0 (0) | 0.75 (3⁄4)	0.3 (5⁄16) | 0.75 (3⁄4)
-0.1 ± 0.8 -1⁄8 ± 13⁄16	-0.1 ± 0.8 -1⁄8 ± 13⁄16	-0.1 ± 0.6 -1⁄8 ± 5⁄8	CAMBER (Degrees)	-0.1 ± 0.8 -1⁄8 ± 13⁄16	-0.1 ± 0.8 -1⁄8 ± 13⁄16	-0.1 ± 0.8 -1⁄8 ± 13⁄16
0.0 (0) | 0.8 (13⁄16)	0.0 (0) | 0.8 (13⁄16)	0.0 (0) | 0.6 (5⁄8)	CROSS | TOL.	0.0 (0) | 0.8 (13⁄16)	0.0 (0) | 0.8 (13⁄16)	0.0 (0) | 0.8 (13⁄16)
0.1° ± 0.2° 3⁄32° ± 3⁄16°	0.1° ± 0.2° [1] 3⁄32° ± 3⁄16° [1]	0.1° ± 0.2° 3⁄32° ± 3⁄16°	FRONT TOE (Degrees)	0.1° ± 0.2° 3⁄32° ± 3⁄16°	0.1° ± 0.2° 3⁄32° ± 3⁄16°	0.1° ± 0.2° [1] 3⁄32° ± 3⁄16° [1]
0.05" ± 0.1" 1⁄16" ± 3⁄32"	0.05" ± 0.1" [1] 1⁄16" ± 3⁄32" [1]	0.05" ± 0.1" 1⁄16" ± 3⁄32"	FRONT TOE (Inches)	0.05" ± 0.1" 1⁄16" ± 3⁄32"	0.05" ± 0.1" 1⁄16" ± 3⁄32"	0.05" ± 0.1" [1] 1⁄16" ± 3⁄32" [1]
			SAI			
			TURNING ANGLE			
126	126	126	ILLUSTRATION #	126	126	126
			REAR			
			CAMBER (Degrees)			
			REAR TOE (Degrees)			
			REAR TOE (Inches)			
			THRUST ANGLE			
			ILLUSTRATION #			
	[1] Hybrids = 0.2° (0.1")		NOTES			[1] 2015 = 0.2° (0.1")
09495	09496	09497		09498	09499	09500

ALIGNMENT SPECIFICATIONS

Chevrolet/GMC Truck & SUV

MODEL INFORMATION	Yukon 4X2 & 4X4 w/22" Tires (QST)	Yukon XL 1500 4X2	Yukon XL 1500 4X4	Yukon XL 1500 4X2	Yukon XL 1500 4X4 exc/Denali	Yukon XL 1500 4X2
YEARS	2015-2017	2000	2000	2001-2005	2001-2005	2006
INSTRUCTIONS		p. 304, 305	p. 304, 305	p. 304, 305	p. 304, 305	
LUGNUT TORQUE	140 ft.lbs.	140 ft.lbs.	140 ft.lbs.	140 ft.lbs.	140 ft.lbs.	140 ft.lbs.
FRONT						
CASTER (Degrees)	3.1 ± 1.0 3⅛ ± 1	L= 4.25, R= 4.75 ± 1.0 L= 4¼, R= 4¾ ± 1	L= 4.25, R= 4.75 ± 1.0 L= 4¼, R= 4¾ ± 1	L= 3.9, R= 4.7 ± 1.0 L= 3⅞, R= 4 11/16 ± 1	L= 3.6, R= 4.4 ± 1.0 [1] L= 3⅝, R= 4⅜ ± 1 [1]	L= 2.9, R= 3.7 ± 1.0 [1] L= 2⅞, R= 3 11/16 ± 1 [1]
CROSS | TOL.	0.0 (0) | 0.75 (¾)	-0.5 (-½) | 0.5 (½)	-0.5 (-½) | 0.5 (½)	-0.8 (-13/16) | 0.75 (¾)	-0.8 (-13/16) | 0.75 (¾)	-0.8 (-13/16) | 0.75 (¾)
CAMBER (Degrees)	-0.1 ± 0.8 -⅛ ± 13/16	0.25 ± 0.5 ¼ ± ½	0.25 ± 0.5 ¼ ± ½	0.25 ± 0.5 ¼ ± ½	0.25 ± 0.5 ¼ ± ½	0.25 ± 0.6 ¼ ± ⅝
CROSS | TOL.	0.0 (0) | 0.8 (13/16)	0.0 (0) | 0.5 (½)	0.0 (0) | 0.5 (½)	0.0 (0) | 0.5 (½)	0.0 (0) | 0.5 (½)	0.0 (0) | 0.6 (⅝)
FRONT TOE (Degrees)	0.1° ± 0.2° 3/32° ± 3/16°	0.1° ± 0.2° 3/32° ± 3/16°	0.1° ± 0.2° 3/32° ± 3/16°	0.1° ± 0.2° 3/32° ± 3/16°	0.1° ± 0.2° 3/32° ± 3/16°	0.1° ± 0.2° 3/32° ± 3/16°
FRONT TOE (Inches)	0.05" ± 0.1" 1/16" ± 3/32"	0.05" ± 0.1" 1/16" ± 3/32"	0.05" ± 0.1" 1/16" ± 3/32"	0.05" ± 0.1" 1/16" ± 3/32"	0.05" ± 0.1" 1/16" ± 3/32"	0.05" ± 0.1" 1/16" ± 3/32"
SAI						
TURNING ANGLE						
ILLUSTRATION #	126	126	126	126	126	126
REAR						
CAMBER (Degrees)						
REAR TOE (Degrees)						
REAR TOE (Inches)						
THRUST ANGLE						
ILLUSTRATION #						
NOTES					[1] 2001-03 L=3.5°, R-4.5°	[1] Relative to ground
	09501	09502	09503	09504	09505	09506

Chevrolet/GMC Truck & SUV

MODEL INFORMATION	Yukon XL 1500 4X4 exc/Denali	Yukon XL 1500 4X2 & 4X4 w/17" B'stones	Yukon XL 1500 4X2 & 4X4 w/17" G'years	Yukon XL 1500 4X2 & 4X4 w/18" Tires	Yukon XL 1500 4X2 & 4X4 w/20" Tires	Yukon XL 1500 4X2 & 4X4 17" B'stones
YEARS	2006	2007	2007	2007	2007	2008-2014
INSTRUCTIONS						
LUGNUT TORQUE	140 ft.lbs.	140 ft.lbs.	140 ft.lbs.	140 ft.lbs.	140 ft.lbs.	140 ft.lbs.
FRONT						
CASTER (Degrees)	L= 2.6, R= 3.4 ± 1.0 [1,2] L= 2⅝, R= 3⅜ ± 1 [1,2]	L= 3.35, R= 3.45 ± 1.0 [1] L= 3⅜, R= 3 7/16 ± 1 [1]	L= 3.2, R= 3.6 ± 1.0 [1] L= 3 3/16, R= 3⅝ ± 1 [1]	L= 3.3, R= 3.5 ± 1.0 [1] L= 3 5/16, R= 3½ ± 1 [1]	L= 3.55, R= 3.25 ± 1.0 [1] L= 3 9/16, R= 3¼ ± 1 [1]	L= 2.9, R= 3.0 ± 1.0 [1] L= 2⅞, R= 3 ± 1 [1]
CROSS | TOL.	-0.8 (-13/16) | 0.75 (¾)	-0.1 (-⅛) | 0.5 (½)	-0.4 (-⅜) | 0.5 (½)	-0.2 (-3/16) | 0.5 (½)	0.3 (5/16) | 0.5 (½)	-0.1 (-⅛) | 0.75 (¾)
CAMBER (Degrees)	0.25 ± 0.6 ¼ ± ⅝	-0.1 ± 0.5 -⅛ ± ½	-0.1 ± 0.5 -⅛ ± ½	-0.1 ± 0.5 -⅛ ± ½	-0.1 ± 0.5 -⅛ ± ½	-0.1 ± 0.6 -⅛ ± ⅝
CROSS | TOL.	0.0 (0) | 0.6 (⅝)	0.0 (0) | 0.5 (½)	0.0 (0) | 0.5 (½)	0.0 (0) | 0.5 (½)	0.0 (0) | 0.5 (½)	0.0 (0) | 0.6 (⅝)
FRONT TOE (Degrees)	0.1° ± 0.2° 3/32° ± 3/16°	0.1° ± 0.2° 3/32° ± 3/16°	0.1° ± 0.2° 3/32° ± 3/16°	0.1° ± 0.2° 3/32° ± 3/16°	0.1° ± 0.2° 3/32° ± 3/16°	0.1° ± 0.2° 3/32° ± 3/16°
FRONT TOE (Inches)	0.05" ± 0.1" 1/16" ± 3/32"	0.05" ± 0.1" 1/16" ± 3/32"	0.05" ± 0.1" 1/16" ± 3/32"	0.05" ± 0.1" 1/16" ± 3/32"	0.05" ± 0.1" 1/16" ± 3/32"	0.05" ± 0.1" 1/16" ± 3/32"
SAI						
TURNING ANGLE						
ILLUSTRATION #	126	126	126	126	126	126
REAR						
CAMBER (Degrees)						
REAR TOE (Degrees)						
REAR TOE (Inches)						
THRUST ANGLE						
ILLUSTRATION #						
NOTES	[1] w/17" Tires L=2.8° R=3.5° [2] Relative to ground	[1] Relative to ground	[1] Relative to ground	[1] Relative to ground	[1] Relative to ground	[1] Relative to ground
	09507	09508	09509	09510	09511	09512

ALIGNMENT SPECIFICATIONS

Chevrolet/GMC Truck & SUV

Yukon XL 1500 4X2 & 4X4 17" G'years	Yukon XL 1500 4X2 & 4X4 18" Tires	Yukon XL 1500 4X2 & 4X4 20" exc/G'year	MODEL INFORMATION	Yukon XL 1500 4X2 & 4X4 20" G'year	Yukon XL 1500 4X2 & 4X4 20" Tires (all)	Yukon XL 1500 4X2 & 4X4 w/18" tires (RKX)
2008-2014	2008-2010	2008-2010	YEARS	2008-2010	2011-2014	2015-2017
			INSTRUCTIONS			
140 ft.lbs.	140 ft.lbs.	140 ft.lbs.	LUGNUT TORQUE	140 ft.lbs.	140 ft.lbs.	140 ft.lbs.
			FRONT			
L= 2.8, R= 3.2 ± 1.0 [1] L= 2 13/16, R= 3 3/16 ± 1 [1]	L= 3.05, R= 2.95 ± 1.0 [1] L= 3 1/16, R= 2 15/16 ± 1 [1]	L= 3.15, R= 2.85 ± 1.0 [1] L= 3⅛, R= 2⅞ ± 1 [1]	CASTER (Degrees)	3.1 ± 1.0 [1] 3⅛ ± 1 [1]	3.1 ± 1.0 [1] 3⅛ ± 1 [1]	L= 3.2, R= 3.0 ± 1.0 L= 3 3/16, R= 3 ± 1
-0.4 (-⅜) | 0.75 (¾)	0.1 (⅛) | 0.75 (¾)	0.3 (5/16) | 0.75 (¾)	CROSS | TOL.	0.0 (0) | 0.75 (¾)	0.0 (0) | 0.75 (¾)	0.2 (3/16) | 0.75 (¾)
-0.1 ± 0.6 -⅛ ± ⅝	-0.1 ± 0.6 -⅛ ± ⅝	-0.1 ± 0.6 -⅛ ± ⅝	CAMBER (Degrees)	-0.1 ± 0.6 -⅛ ± ⅝	-0.1 ± 0.6 -⅛ ± ⅝	-0.1 ± 0.8 -⅛ ± 13/16
0.0 (0) | 0.6 (⅝)	0.0 (0) | 0.6 (⅝)	0.0 (0) | 0.6 (⅝)	CROSS | TOL.	0.0 (0) | 0.6 (⅝)	0.0 (0) | 0.6 (⅝)	0.0 (0) | 0.8 (13/16)
0.1° ± 0.2° 3/32° ± 3/16°	0.1° ± 0.2° 3/32° ± 3/16°	0.1° ± 0.2° 3/32° ± 3/16°	FRONT TOE (Degrees)	0.1° ± 0.2° 3/32° ± 3/16°	0.1° ± 0.2° 3/32° ± 3/16°	0.1° ± 0.2° 3/32° ± 3/16°
0.05" ± 0.1" 1/16" ± 3/32"	0.05" ± 0.1" 1/16" ± 3/32"	0.05" ± 0.1" 1/16" ± 3/32"	FRONT TOE (Inches)	0.05" ± 0.1" 1/16" ± 3/32"	0.05" ± 0.1" 1/16" ± 3/32"	0.05" ± 0.1" 1/16" ± 3/32"
			SAI			
			TURNING ANGLE			
126	126	126	ILLUSTRATION #	126	126	126
			REAR			
			CAMBER (Degrees)			
			REAR TOE (Degrees)			
			REAR TOE (Inches)			
			THRUST ANGLE			
			ILLUSTRATION #			
[1] Relative to ground	[1] Relative to ground	[1] Relative to ground	NOTES	[1] Relative to ground	[1] Relative to ground	
09513	09514	09515		09516	09517	09518

Chevrolet/GMC Truck & SUV

Yukon XL 1500 4X2 & 4X4 w/20" B'stones (QSS)	Yukon XL 1500 4X2 & 4X4 w/20" Cont. (QSS)	Yukon XL 1500 4X2 & 4X4 w/22" tires (QST)	MODEL INFORMATION	Yukon XL 2500 4X2 & 4X4	Yukon XL 2500 4X2 & 4X4	Yukon XL 2500 4X2 & 4X4
2015-2017	2015-2017	2015-2017	YEARS	2000-2005	2006-2007	2008-2013
			INSTRUCTIONS	p. 304, 305		
140 ft.lbs.	140 ft.lbs.	140 ft.lbs.	LUGNUT TORQUE	140 ft.lbs.	140 ft.lbs.	140 ft.lbs.
			FRONT			
3.1 ± 1.0 3⅛ ± 1	L= 3.2, R= 2.9 ± 1.0 L= 3 3/16, R= 2⅞ ± 1	3.1 ± 1.0 3⅛ ± 1	CASTER (Degrees)	L= 4.5, R= 4.75 ± 1.0 L= 4½, R= 4¾ ± 1	L= 3.5, R= 3.75 ± 1.0 [1] L= 3½, R= 3¾ ± 1 [1]	L= 3.1, R= 3.3 ± 1.0 [1] L= 3⅛, R= 3 5/16 ± 1 [1]
0.0 (0) | 0.75 (¾)	0.3 (5/16) | 0.75 (¾)	0.0 (0) | 0.75 (¾)	CROSS | TOL.	-0.25 (-¼) | 0.5 (½)	-0.25 (-¼) | 0.5 (½)	-0.2 (-3/16) | 0.75 (¾)
-0.1 ± 0.8 -⅛ ± 13/16	-0.1 ± 0.8 -⅛ ± 13/16	-0.1 ± 0.8 -⅛ ± 13/16	CAMBER (Degrees)	0.25 ± 0.5 ¼ ± ½	0.25 ± 0.5 ¼ ± ½	0.25 ± 0.6 ¼ ± ⅝
0.0 (0) | 0.8 (13/16)	0.0 (0) | 0.8 (13/16)	0.0 (0) | 0.8 (13/16)	CROSS | TOL.	0.0 (0) | 0.5 (½)	0.0 (0) | 0.6 (⅝)	0.0 (0) | 0.6 (⅝)
0.1° ± 0.2° 3/32° ± 3/16°	0.1° ± 0.2° [1] 3/32° ± 3/16° [1]	0.1° ± 0.2° 3/32° ± 3/16°	FRONT TOE (Degrees)	0.1° ± 0.2° 3/32° ± 3/16°	0.1° ± 0.2° 3/32° ± 3/16°	0.1° ± 0.2° 3/32° ± 3/16°
0.05" ± 0.1" 1/16" ± 3/32"	0.05" ± 0.1" [1] 1/16" ± 3/32" [1]	0.05" ± 0.1" 1/16" ± 3/32"	FRONT TOE (Inches)	0.05" ± 0.1" 1/16" ± 3/32"	0.05" ± 0.1" 1/16" ± 3/32"	0.05" ± 0.1" 1/16" ± 3/32"
			SAI			
			TURNING ANGLE			
126	126	126	ILLUSTRATION #	126	126	126
			REAR			
			CAMBER (Degrees)	-0.5 ± 0.5 [1] -½ ± ½ [1]		
			REAR TOE (Degrees)	0.0° ± 0.2° [1] 0° ± 3/16° [1]		
			REAR TOE (Inches)	0.0" ± 0.1" [1] 0" ± 3/32" [1]		
			THRUST ANGLE			
			ILLUSTRATION #			
	[1] 2015 = 0.2° (0.1")		NOTES	[1] Quadrasteer Only	[1] Relative to ground	[1] Relative to ground
09519	09520	09521		09522	09523	09524

ALIGNMENT SPECIFICATIONS

Chevrolet/GMC Truck & SUV

Yukon Denali	Yukon Denali	Yukon Denali	MODEL INFORMATION	Yukon Denali	Yukon Denali	Yukon Denali
		Yukon XL Denali		Yukon XL Denali	Yukon XL Denali	Yukon XL Denali
				exc/20" tires	w/20" tires	exc/20" tires
1999	2000	2001	YEARS	2002-2005	2005	2006
p. 304, 305	p. 304, 305	p. 304, 305	INSTRUCTIONS	p. 304, 305	p. 304, 305	
			LUGNUT TORQUE			
			FRONT			
3.0 ± 1.0 3 ± 1	L= 2.6, R= 3.5 ± 0.5 L= 2⅝, R= 3½ ± ½	L= 3.5, R= 4.5 ± 1.0 L= 3½, R= 4½ ± 1	CASTER (Degrees)	L= 3.8, R= 4.5 ± 1.0 L= 3 13/16, R= 4½ ± 1	L= 4.1, R= 5.1 ± 1.0 L= 4⅛, R= 5⅛ ± 1	L= 2.8, R= 3.5 ± 1.0 [1] L= 2 13/16, R= 3½ ± 1 [1]
0.0 (0) | 0.5 (½)	-0.9 (-⅞) | 0.5 (½)	-1.0 (-1) | 0.5 (½)	CROSS | TOL.	-0.7 (-11/16) | 0.75 (¾)	-1.0 (-1) | 0.75 (¾)	-0.7 (-11/16) | 0.75 (¾)
0.65 ± 0.5 ⅝ ± ½	L= 0.3, R= -0.2 ± 0.5 L= 5/16, R= -3/16 ± ½	0.25 ± 0.5 ¼ ± ½	CAMBER (Degrees)	0.25 ± 0.5 ¼ ± ½	-0.1 ± 0.6 -⅛ ± ⅝	0.25 ± 0.6 ¼ ± ⅝
0.0 (0) | 0.5 (½)	0.5 (½) | 0.5 (½)	0.0 (0) | 0.5 (½)	CROSS | TOL.	0.0 (0) | 0.5 (½)	0.0 (0) | 0.6 (⅝)	0.0 (0) | 0.6 (⅝)
0.24° ± 0.2° ¼° ± 3/16°	0.24° ± 0.2° ¼° ± 3/16°	0.1° ± 0.2° 3/32° ± 3/16°	FRONT TOE (Degrees)	0.1° ± 0.2° 3/32° ± 3/16°	0.1° ± 0.2° 3/32° ± 3/16°	0.1° ± 0.2° 3/32° ± 3/16°
0.12" ± 0.1" ⅛" ± 3/32"	0.12" ± 0.1" ⅛" ± 3/32"	0.05" ± 0.1" 1/16" ± 3/32"	FRONT TOE (Inches)	0.05" ± 0.1" 1/16" ± 3/32"	0.05" ± 0.1" 1/16" ± 3/32"	0.05" ± 0.1" 1/16" ± 3/32"
			SAI			
			TURNING ANGLE			
43	43	126	ILLUSTRATION #	126	126	126
			REAR			
			CAMBER (Degrees)			
			REAR TOE (Degrees)			
			REAR TOE (Inches)			
			THRUST ANGLE			
			ILLUSTRATION #			
			NOTES			[1] Relative to ground
09525	09526	09527		09528	09529	09530

Chevrolet/GMC Truck & SUV

Yukon Denali	Yukon Denali	Yukon Denali	MODEL INFORMATION	Yukon Denali	Yukon Denali	Yukon Denali
Yukon XL Denali	Yukon XL Denali	Yukon XL Denali		Yukon XL Denali	Yukon XL Denali	Yukon XL Denali
w/20" tires	except 20" tires	with 20" tires		w/18" tires	w/20" exc/G'year QSS	w/20" G'year QSS
2006	2007	2007	YEARS	2008-2010	2008-2010	2008-2010
			INSTRUCTIONS			
	140 ft.lbs.	140 ft.lbs.	LUGNUT TORQUE	140 ft.lbs.	140 ft.lbs.	140 ft.lbs.
			FRONT			
L= 3.45, R= 3.75 ± 1.0 [1] L= 3 7/16, R= 3¾ ± 1 [1]	L= 3.45, R= 3.35 ± 1.0 [1] L= 3 7/16, R= 3⅜ ± 1 [1]	L= 3.55, R= 3.25 ± 1.0 [1] L= 3 9/16, R= 3¼ ± 1 [1]	CASTER (Degrees)	L= 3.05, R= 2.95 ± 1.0 [1] L= 3 1/16, R= 2 15/16 ± 1 [1]	L= 3.15, R= 2.85 ± 1.0 [1] L= 3⅛, R= 2⅞ ± 1 [1]	3.1 ± 1.0 [1] 3⅛ ± 1 [1]
-0.3 (-5/16) | 0.75 (¾)	0.1 (⅛) | 0.5 (½)	0.3 (5/16) | 0.5 (½)	CROSS | TOL.	0.1 (⅛) | 0.5 (½)	0.3 (5/16) | 0.5 (½)	0.0 (0) | 0.5 (½)
-0.1 ± 0.6 -⅛ ± ⅝	-0.1 ± 0.5 -⅛ ± ½	-0.1 ± 0.5 -⅛ ± ½	CAMBER (Degrees)	-0.1 ± 0.5 -⅛ ± ½	-0.1 ± 0.5 -⅛ ± ½	-0.1 ± 0.5 -⅛ ± ½
0.0 (0) | 0.6 (⅝)	0.0 (0) | 0.5 (½)	0.0 (0) | 0.5 (½)	CROSS | TOL.	0.0 (0) | 0.5 (½)	0.0 (0) | 0.5 (½)	0.0 (0) | 0.5 (½)
0.1° ± 0.2° 3/32° ± 3/16°	0.1° ± 0.2° 3/32° ± 3/16°	0.1° ± 0.2° 3/32° ± 3/16°	FRONT TOE (Degrees)	0.1° ± 0.2° 3/32° ± 3/16°	0.1° ± 0.2° 3/32° ± 3/16°	0.1° ± 0.2° 3/32° ± 3/16°
0.05" ± 0.1" 1/16" ± 3/32"	0.05" ± 0.1" 1/16" ± 3/32"	0.05" ± 0.1" 1/16" ± 3/32"	FRONT TOE (Inches)	0.05" ± 0.1" 1/16" ± 3/32"	0.05" ± 0.1" 1/16" ± 3/32"	0.05" ± 0.1" 1/16" ± 3/32"
			SAI			
			TURNING ANGLE			
126	126	126	ILLUSTRATION #	126	126	126
			REAR			
			CAMBER (Degrees)			
			REAR TOE (Degrees)			
			REAR TOE (Inches)			
			THRUST ANGLE			
			ILLUSTRATION #			
[1] Relative to ground	[1] Relative to ground	[1] Relative to ground	NOTES	[1] Relative to ground	[1] Relative to ground	[1] Relative to ground
09531	09532	09533		09534	09535	09536

Chevrolet/GMC Truck & SUV

Yukon Denali Yukon XL Denali w/20" tires (All)	Yukon Denali Yukon XL Denali w/22" tires	Yukon Denali Yukon XL Denali w/20" B'stones (QSS)	MODEL INFORMATION	Yukon Denali Yukon XL Denali w/20" Cont.(QSS)	Yukon Denali Yukon XL Denali w/22" tires (QST)	LLV (U.S. Postal Vehicle)
2011-2014	2011-2014	2015-2017	YEARS	2015-2017	2015-2017	1990-95
			INSTRUCTIONS			
140 ft.lbs.	140 ft.lbs.	140 ft.lbs.	LUGNUT TORQUE	140 ft.lbs.	140 ft.lbs.	
			FRONT			
3.1 ± 1.0 [1] 3⅛ ± 1 [1]	L= 2.9, R= 3.1 ± 1.0 [1] L= 2⅞, R= 3⅛ ± 1 [1]	3.1 ± 1.0 3⅛ ± 1	CASTER (Degrees)	L= 3.2, R= 2.9 ± 1.0 L= 3 3/16, R= 2⅞ ± 1	3.1 ± 1.0 3⅛ ± 1	2.0 ± 0.5 2 ± ½
0.0 (0) | 0.5 (½)	-0.2 (-3/16) | 0.5 (½)	0.0 (0) | 0.75 (¾)	CROSS | TOL.	0.3 (5/16) | 0.75 (¾)	0.0 (0) | 0.75 (¾)	0.0 (0) | 0.5 (½)
-0.1 ± 0.5 -⅛ ± ½	-0.1 ± 0.5 -⅛ ± ½	-0.1 ± 0.8 -⅛ ± 13/16	CAMBER (Degrees)	-0.1 ± 0.8 -⅛ ± 13/16	-0.1 ± 0.8 -⅛ ± 13/16	-0.2 ± 0.5 -3/16 ± ½
0.0 (0) | 0.5 (½)	0.0 (0) | 0.5 (½)	0.0 (0) | 0.8 (13/16)	CROSS | TOL.	0.0 (0) | 0.8 (13/16)	0.0 (0) | 0.8 (13/16)	0.0 (0) | 0.5 (½)
0.1° ± 0.2° 3/32° ± 3/16°	0.1° ± 0.2° 3/32° ± 3/16°	0.1° ± 0.2° 3/32° ± 3/16°	FRONT TOE (Degrees)	0.1° ± 0.2° [1] 3/32° ± 3/16° [1]	0.1° ± 0.2° 3/32° ± 3/16°	-0.2° ± 0.1° -3/16° ± 3/32°
0.05" ± 0.1" 1/16" ± 3/32"	0.05" ± 0.1" 1/16" ± 3/32"	0.05" ± 0.1" 1/16" ± 3/32"	FRONT TOE (Inches)	0.05" ± 0.1" [1] 1/16" ± 3/32" [1]	0.05" ± 0.1" 1/16" ± 3/32"	-0.1" ± 0.05" -3/32" ± 1/16"
			SAI			
			TURNING ANGLE			
126	126	126	ILLUSTRATION #	126	126	23
			REAR			
			CAMBER (Degrees)			
			REAR TOE (Degrees)			
			REAR TOE (Inches)			
			THRUST ANGLE			
			ILLUSTRATION #			
[1] Relative to ground	[1] Relative to ground		NOTES	[1] 2015 = 0.2° (0.1")		
09537	09538	09539		09540	09541	09551

Chrysler

200	200	300M	MODEL INFORMATION	300/300C RWD except SRT8	300/300C RWD except SRT8	300/300C RWD except SRT8
2011-2014	2015-2017	1999-2004	YEARS	2005-2008	2009-2010	2011-2017
			INSTRUCTIONS			
100 ft.lbs.	100 ft.lbs.	100 ft.lbs.	LUGNUT TORQUE	110 ft.lbs.	110 ft.lbs.	110 ft.lbs.
			FRONT			
3.0 ± 1.0 3 ± 1	3.83 ± 0.4 3 13/16 ± ⅜	3.0 ± 1.0 3 ± 1	CASTER (Degrees)	L= 9.4, R=10.4 ± 1.0 L= 9⅜, R= 10⅜ ± 1	L= 8.0, R= 9.0 ± 1.0 L= 8, R= 9 ± 1	L= 8.0, R= 8.7 ± 1.0 L= 8, R= 8 11/16 ± 1
0.0 (0) | 1.0 (1)	0.0 (0) | 0.5 (½)	0.0 (0) | 1.0 (1)	CROSS | TOL.	-1.0 (-1) | 0.6 (⅝)	-1.0 (-1) | 0.6 (⅝)	-0.7 (-11/16) | 0.6 (⅝)
-0.62 ± 0.55 -⅝ ± 9/16	-0.96 ± 0.4 -15/16 ± ⅜	0.0 ± 0.6 0 ± ⅝	CAMBER (Degrees)	L= 0.15, R= -0.25 ± 0.65 L= ⅛, R= -¼ ± ⅝	L= 0.15, R= -0.25 ± 0.65 L= ⅛, R= -¼ ± ⅝	L= -0.85, R= -1.25 ± 0.55 L= -⅞, R= -1¼ ± 9/16
0.0 (0) | 0.8 (13/16)	0.0 (0) | 0.5 (½)	0.0 (0) | 0.7 (11/16)	CROSS | TOL.	0.4 (⅜) | 0.5 (½)	0.4 (⅜) | 0.5 (½)	0.4 (⅜) | 0.55 (9/16)
0.2° ± 0.3° 3/16° ± 5/16°	0.08° ± 0.15° 3/32° ± 5/32°	0.0° ± 0.2° [1] 0° ± 3/16° [1]	FRONT TOE (Degrees)	0.2° ± 0.1° 3/16° ± 3/32°	0.2° ± 0.1° 3/16° ± 3/32°	0.2° ± 0.2° 3/16° ± 3/16°
0.1" ± 0.15" 3/32" ± 5/32"	0.04" ± 0.08" 1/32" ± 1/16"	0.0" ± 0.1" [1] 0" ± 3/32" [1]	FRONT TOE (Inches)	0.1" ± 0.05" 3/32" ± 1/16"	0.1" ± 0.05" 3/32" ± 1/16"	0.1" ± 0.1" 3/32" ± 3/32"
		13.7 (13 11/16)	SAI			
			TURNING ANGLE			
161		26	ILLUSTRATION #	164 165 ▲	164 165 ▲	165 ▲
			REAR			
-0.78 ± 0.65 -¾ ± ⅝	-1.13 ± 0.4 -1⅛ ± ⅜	-0.2 ± 0.5 -3/16 ± ½	CAMBER (Degrees)	-0.55 ± 0.65 -9/16 ± ⅝	-0.55 ± 0.65 -9/16 ± ⅝	-1.75 ± 0.55 -1¾ ± 9/16
0.2° ± 0.3° 3/16° ± 5/16°	0.08° ± 0.15° 3/32° ± 5/32°	0.1° ± 0.3° 3/32° ± 5/16°	REAR TOE (Degrees)	0.2° ± 0.15° [1] 3/16° ± 5/32° [1]	0.2° ± 0.15° [1] 3/16° ± 5/32° [1]	0.2° ± 0.15° 3/16° ± 5/32°
0.1" ± 0.15" 3/32" ± 5/32"	0.04" ± 0.08" 1/32" ± 1/16"	0.05" ± 0.15" 1/16" ± 5/32"	REAR TOE (Inches)	0.1" ± 0.08" [1] 3/32" ± 1/16" [1]	0.1" ± 0.08" [1] 3/32" ± 1/16" [1]	0.1" ± 0.08" 3/32" ± 1/16"
0.15 (5/32)	0.1 (3/32)	0.15 (5/32)	THRUST ANGLE	0.15 (5/32)	0.15 (5/32)	0.15 (5/32)
174	174	10	ILLUSTRATION #	171	171	171
		[1] 1999 = 0.1° (0.05")	NOTES	[1] Set Left to 0.17° (0.09") [1] Set Right to 0.03° (0.015")	[1] Set Left to 0.17° (0.09") [1] Set Right to 0.03° (0.015")	
10001	10002	10005		10006	10007	10008

Chrysler

300/300C	300/300C	300/300C	MODEL	300/300C	300/300C	300/300C
RWD	RWD	RWD	INFORMATION	AWD	AWD	AWD
SRT8	SRT8	SRT8				
2005-2008	2009-2010	2011-2017	YEARS	2005-2008	2009-2010	2011-2017
			INSTRUCTIONS			
110 ft.lbs.	110 ft.lbs.	110 ft.lbs.	LUGNUT TORQUE	110 ft.lbs.	110 ft.lbs.	110 ft.lbs.
			FRONT			
L= 9.7, R=10.3 ± 1.0 L= 9 11/16, R= 10 5/16 ± 1	L= 8.3, R= 9.3 ± 1.0 L= 8 5/16, R= 9 5/16 ± 1	L= 8.3, R= 9.0 ± 1.0 L= 8 5/16, R= 9 ± 1	CASTER (Degrees)	L= 4.5, R= 5.1 ± 1.0 L= 4 1/2, R= 5 1/8 ± 1	L= 4.5, R= 5.1 ± 1.0 L= 4 1/2, R= 5 1/8 ± 1	5.4 ± 1.0 5 3/8 ± 1
-0.6 (-5/8) | 0.6 (5/8)	-1.0 (-1) | 0.6 (5/8)	-0.7 (-11/16) | 0.6 (5/8)	CROSS | TOL.	-0.6 (-5/8) | 0.6 (5/8)	-0.6 (-5/8) | 0.6 (5/8)	0.0 (0) | 0.6 (5/8)
L= -0.05, R= -0.35 ± 0.5 L= -1/16, R= -3/8 ± 1/2	L= -0.05, R= $-0.35^{+0.45}_{-0.65}$ L= -1/16, R= $-3/8^{+7/16}_{-5/8}$	L= -1.05, R= -1.35 ± 0.55 L= -1 1/16, R= -1 3/8 ± 9/16	CAMBER (Degrees)	L= 0.15, R= -0.25 ± 0.65 L= 1/8, R= -1/4 ± 5/8	L= 0.15, R= -0.25 ± 0.65 L= 1/8, R= -1/4 ± 5/8	L= 0.0, R= -0.4 ± 0.55 L= 0, R= -3/8 ± 9/16
0.3 (5/16) | 0.5 (1/2)	0.3 (5/16) | 0.5 (1/2)	0.3 (5/16) | 0.55 (9/16)	CROSS | TOL.	0.4 (3/8) | 0.5 (1/2)	0.4 (3/8) | 0.5 (1/2)	0.4 (3/8) | 0.55 (9/16)
0.2° ± 0.2° 3/16° ± 3/16°	0.2° ± 0.1° 3/16° ± 3/32°	0.2° ± 0.2° 3/16° ± 3/16°	FRONT TOE (Degrees)	0.1° ± 0.05° 3/32° ± 1/16°	0.2° ± 0.1° 3/16° ± 3/32°	0.1° ± 0.2° 3/32° ± 3/16°
0.1" ± 0.1" 3/32" ± 3/32"	0.1" ± 0.05" 3/32" ± 1/16"	0.1" ± 0.1" 3/32" ± 3/32"	FRONT TOE (Inches)	0.05" ± 0.03" 1/16" ± 1/32"	0.1" ± 0.05" 3/32" ± 1/16"	0.05" ± 0.1" 1/16" ± 3/32"
			SAI			
			TURNING ANGLE			
164 165 △	164 165 △	165 △	ILLUSTRATION #	165 △	165 △	165 △
			REAR			
-0.75 ± 0.5 -3/4 ± 1/2	$-0.75^{+0.6}_{-0.5}$ $-3/4^{+5/8}_{-1/2}$	-0.75 ± 0.55 -3/4 ± 9/16	CAMBER (Degrees)	-0.55 ± 0.65 -9/16 ± 5/8	-0.55 ± 0.65 -9/16 ± 5/8	-0.75 ± 0.55 -3/4 ± 9/16
0.2° ± 0.15° [1] 3/16° ± 5/32° [1]	0.2° ± 0.15° [1] 3/16° ± 5/32° [1]	0.2° ± 0.15° 3/16° ± 5/32°	REAR TOE (Degrees)	0.2° ± 0.15° [1] 3/16° ± 5/32° [1]	0.2° ± 0.15° [1] 3/16° ± 5/32° [1]	0.2° ± 0.15° 3/16° ± 5/32°
0.1" ± 0.08" [1] 3/32" ± 1/16" [1]	0.1" ± 0.08" [1] 3/32" ± 1/16" [1]	0.1" ± 0.08" 3/32" ± 1/16"	REAR TOE (Inches)	0.1" ± 0.08" [1] 3/32" ± 1/16" [1]	0.1" ± 0.08" [1] 3/32" ± 1/16" [1]	0.1" ± 0.08" 3/32" ± 1/16"
0.15 (5/32)	0.15 (5/32)	0.15 (5/32)	THRUST ANGLE	0.15 (5/32)	0.15 (5/32)	0.15 (5/32)
171	171	171	ILLUSTRATION #	171	171	171
[1] Set Left to 0.17° (0.09") [1] Set Right to 0.03° (0.015")	[1] Set Left to 0.17° (0.09") [1] Set Right to 0.03° (0.015")		NOTES	[1] Set Left to 0.17° (0.09") [1] Set Right to 0.03° (0.015")	[1] Set Left to 0.17° (0.09") [1] Set Right to 0.03° (0.015")	
10009	10010	10011		10012	10013	10014

Chrysler

Aspen	Aspen	Cirrus	MODEL	Concorde	Concorde	Crossfire
4X2	4X4		INFORMATION			
2007-2009	2007-2009	1995-2000	YEARS	1993-98	1999-2004	2004-2008
			INSTRUCTIONS			p. 309
135 ft.lbs.	135 ft.lbs.	95-100 ft.lbs.	LUGNUT TORQUE	100 ft.lbs.	100 ft.lbs.	81 ft.lbs.
			FRONT			
L= 3.2, R= 3.5 ± 0.5 L= 3 3/16, R= 3 1/2 ± 1/2	3.5 ± 0.5 3 1/2 ± 1/2	3.3 ± 1.0 3 5/16 ± 1	CASTER (Degrees)	3.0 ± 1.0 3 ± 1	3.0 ± 1.0 3 ± 1	5.2 ± 0.42 5 3/16 ± 7/16
-0.3 (-5/16) | 0.5 (1/2)	0.0 (0) | 0.5 (1/2)	0.0 (0) | 1.0 (1)	CROSS | TOL.	0.0 (0) | 1.0 (1)	0.0 (0) | 1.0 (1)	0.0 (0) | n/s
0.0 ± 0.5 0 ± 1/2	L= 0.15, R= -0.15 ± 0.5 L= 1/8, R= -1/8 ± 1/2	0.0 ± 0.6 [1] 0 ± 5/8 [1]	CAMBER (Degrees)	0.0 ± 0.6 0 ± 5/8	0.0 ± 0.6 0 ± 5/8	-1.22 ± 0.42 -1 1/4 ± 7/16
0.0 (0) | 0.5 (1/2)	0.3 (5/16) | 0.5 (1/2)	0.0 (0) | 0.7 (11/16)	CROSS | TOL.	0.0 (0) | 0.7 (11/16)	0.0 (0) | 0.7 (11/16)	0.0 (0) | 0.33 (5/16)
0.1° ± 0.05° 3/32° ± 1/16°	0.1° ± 0.05° 3/32° ± 1/16°	0.1° ± 0.2° 3/32° ± 3/16°	FRONT TOE (Degrees)	0.2° ± 0.2° [1] 3/16° ± 3/16° [1]	0.0° ± 0.2° [1] 0° ± 3/16° [1]	0.33° ± 0.17° 11/32° ± 5/32°
0.05" ± 0.03" 1/16" ± 1/32"	0.05" ± 0.03" 1/16" ± 1/32"	0.05" ± 0.1" 1/16" ± 3/32"	FRONT TOE (Inches)	0.1" ± 0.1" [1] 3/32" ± 3/32" [1]	0.0" ± 0.1" [1] 0" ± 3/32" [1]	0.17" ± 0.08" 5/32" ± 3/32"
		7.0 (7)	SAI	14.0 (14)	13.7 (13 11/16)	
			TURNING ANGLE			20.0 / 18.6 ± 0.5
158	158	113	ILLUSTRATION #		26	88
			REAR			
		-0.2 ± 0.4 -3/16 ± 3/8	CAMBER (Degrees)	-0.1 ± 0.5 -1/8 ± 1/2	-0.2 ± 0.5 -3/16 ± 1/2	-1.13 ± 0.5 -1 1/8 ± 1/2
		0.1° ± 0.2° 3/32° ± 3/16°	REAR TOE (Degrees)	0.1° ± 0.3° 3/32° ± 5/16°	0.1° ± 0.3° 3/32° ± 5/16°	0.55° ± 0.12° 9/16° ± 1/8°
		0.05" ± 0.1" 1/16" ± 3/32"	REAR TOE (Inches)	0.05" ± 0.15" 1/16" ± 5/32"	0.05" ± 0.15" 1/16" ± 5/32"	0.28" ± 0.06" 9/32" ± 1/16"
0.4 (13/32)	0.4 (13/32)	0.15 (5/32)	THRUST ANGLE	0.15 (5/32)	0.15 (5/32)	
		01	ILLUSTRATION #	10	10	86
		[1] 1995-96 = 0.3° (5/16°)	NOTES	[1] 1998 = 0.1° (0.05")	[1] 1999 = 0.1° (0.05")	
10015	10016	10018		10020	10022	10026

Chrysler

Dynasty (Canada)	Fifth Avenue	Fifth Avenue	MODEL INFORMATION	Imperial	Imperial	Intrepid (Canada)
1991-93	1990	1991-93	YEARS	1990	1991-93	1999-2004
			INSTRUCTIONS			
95 ft.lbs.		95 ft.lbs.	LUGNUT TORQUE	95 ft.lbs.	95 ft.lbs.	100 ft.lbs.
			FRONT			
2.7 ± 0.75 2 11/16 ± ¾	1.2 ± 0.75 [1] 1 3/16 ± ¾ [1]	2.7 ± 0.75 [1] 2 11/16 ± ¾ [1]	CASTER (Degrees)	1.2 ± 0.75 [1] 1 3/16 ± ¾ [1]	2.7 ± 0.75 [1] 2 11/16 ± ¾ [1]	3.0 ± 1.0 3 ± 1
0.0 (0) | 1.5 (1½)	0.0 (0) | 1.5 (1½)	0.0 (0) | 1.5 (1½)	CROSS | TOL.	0.0 (0) | 1.5 (1½)	0.0 (0) | 1.5 (1½)	0.0 (0) | 1.0 (1)
0.3 ± 0.5 5/16 ± ½	0.3 ± 0.5 [2] 5/16 ± ½ [2]	0.3 ± 0.5 [2] 5/16 ± ½ [2]	CAMBER (Degrees)	0.3 ± 0.5 [2] 5/16 ± ½ [2]	0.3 ± 0.5 [2] 5/16 ± ½ [2]	0.0 ± 0.6 0 ± ⅝
0.0 (0) | n/s	0.0 (0) | n/s	0.0 (0) | n/s	CROSS | TOL.	0.0 (0) | n/s	0.0 (0) | n/s	0.0 (0) | 0.7 (11/16)
0.1° ± 0.1° 3/32° ± 3/32°	0.1° ± 0.1° 3/32° ± 3/32°	0.1° ± 0.1° 3/32° ± 3/32°	FRONT TOE (Degrees)	0.1° ± 0.1° 3/32° ± 3/32°	0.1° ± 0.1° 3/32° ± 3/32°	0.0° ± 0.2° [1] 0° ± 3/16° [1]
0.05" ± 0.05" 1/16" ± 1/16"	0.05" ± 0.05" 1/16" ± 1/16"	0.05" ± 0.05" 1/16" ± 1/16"	FRONT TOE (Inches)	0.05" ± 0.05" 1/16" ± 1/16"	0.05" ± 0.05" 1/16" ± 1/16"	0.0" ± 0.1" [1] 0" ± 3/32" [1]
			SAI			13.7 (13 11/16)
			TURNING ANGLE			
[22]	[22]	[22]	ILLUSTRATION #	[22]	[22]	[26]
			REAR			
-0.5 ± 0.5 -½ ± ½	-0.5 ± 0.5 -½ ± ½	-0.5 ± 0.5 -½ ± ½	CAMBER (Degrees)	-0.5 ± 0.5 -½ ± ½	-0.5 ± 0.5 -½ ± ½	-0.2 ± 0.5 -3/16 ± ½
0.0° ± 0.25° 0° ± ¼°	0.0° ± 0.25° 0° ± ¼°	0.0° ± 0.25° 0° ± ¼°	REAR TOE (Degrees)	0.0° ± 0.25° 0° ± ¼°	0.0° ± 0.25° 0° ± ¼°	0.1° ± 0.3° 3/32° ± 5/16°
0.0" ± 0.13" 0" ± ⅛"	0.0" ± 0.13" 0" ± ⅛"	0.0" ± 0.13" 0" ± ⅛"	REAR TOE (Inches)	0.0" ± 0.13" 0" ± ⅛"	0.0" ± 0.13" 0" ± ⅛"	0.05" ± 0.15" 1/16" ± 5/32"
0.4 (13/32)	0.4 (13/32)	0.4 (13/32)	THRUST ANGLE	0.4 (13/32)	0.4 (13/32)	0.15 (5/32)
[76]	[76]	[76]	ILLUSTRATION #	[76]	[76]	[10]
	[1] w/Air Susp. 1.3°(1 5/16°) [2] w/Air Susp. -0.3° to +0.5°	[1] w/Air Susp. 3.0° (3°) [2] w/Air Susp. -0.3° to +0.5°	NOTES	[1] w/Air Susp. 1.3°(1 5/16°) [2] w/Air Susp. -0.3° to +0.5°	[1] w/Air Susp. 3.0° (3°) [2] w/Air Susp. -0.3° to +0.5°	[1] 1999 = 0.1° (0.05")
10027	10030	10031		10037	10038	10039

Chrysler

LeBaron	LeBaron	Neon (Canada)	MODEL INFORMATION	New Yorker	New Yorker	New Yorker
1990	1991-95	2000-2002	YEARS	1990	1991-93	1994-96
			INSTRUCTIONS			
95 ft.lbs.	95 ft.lbs.	85-115 ft.lbs.	LUGNUT TORQUE	95 ft.lbs.	95 ft.lbs.	95 ft.lbs.
			FRONT			
1.2 ± 0.75 [1] 1 3/16 ± ¾ [1]	2.8 ± 1.0 2 13/16 ± 1	2.6 ± 1.0 2⅝ ± 1	CASTER (Degrees)	1.2 ± 0.75 1 3/16 ± ¾	2.7 ± 0.75 2 11/16 ± ¾	3.0 ± 1.0 3 ± 1
0.0 (0) | 1.5 (1½)	0.0 (0) | 1.5 (1½)	0.0 (0) | 1.0 (1)	CROSS | TOL.	0.0 (0) | 1.5 (1½)	0.0 (0) | 1.5 (1½)	0.0 (0) | 1.0 (1)
0.3 ± 0.5 5/16 ± ½	0.3 ± 0.5 5/16 ± ½	0.0 ± 0.4 0 ± ⅜	CAMBER (Degrees)	0.3 ± 0.5 5/16 ± ½	0.3 ± 0.5 5/16 ± ½	0.0 ± 0.6 0 ± ⅝
0.0 (0) | n/s	0.0 (0) | n/s	0.0 (0) | 0.5 (½)	CROSS | TOL.	0.0 (0) | n/s	0.0 (0) | n/s	0.0 (0) | 0.7 (11/16)
0.1° ± 0.1° 3/32° ± 3/32°	0.1° ± 0.1° 3/32° ± 3/32°	0.2° ± 0.2° [1] 3/16° ± 3/16° [1]	FRONT TOE (Degrees)	0.1° ± 0.1° 3/32° ± 3/32°	0.1° ± 0.1° 3/32° ± 3/32°	0.2° ± 0.2° 3/16° ± 3/16°
0.05" ± 0.05" 1/16" ± 1/16"	0.05" ± 0.05" 1/16" ± 1/16"	0.1" ± 0.1" [1] 3/32" ± 3/32" [1]	FRONT TOE (Inches)	0.05" ± 0.05" 1/16" ± 1/16"	0.05" ± 0.05" 1/16" ± 1/16"	0.1" ± 0.1" 3/32" ± 3/32"
			SAI			
			TURNING ANGLE			
[22]	[22]	[26]	ILLUSTRATION #	[22]	[22]	
			REAR			
-0.5 ± 0.5 -½ ± ½	-0.5 ± 0.5 -½ ± ½	-0.25 ± 0.4 -¼ ± ⅜	CAMBER (Degrees)	-0.5 ± 0.5 -½ ± ½	-0.5 ± 0.5 -½ ± ½	-0.1 ± 0.5 -⅛ ± ½
0.0° ± 0.25° 0° ± ¼°	0.0° ± 0.25° 0° ± ¼°	0.3° ± 0.2° 5/16° ± 3/16°	REAR TOE (Degrees)	0.0° ± 0.25° 0° ± ¼°	0.0° ± 0.25° 0° ± ¼°	0.1° ± 0.3° 3/32° ± 5/16°
0.0" ± 0.13" 0" ± ⅛"	0.0" ± 0.13" 0" ± ⅛"	0.15" ± 0.1" 5/32" ± 3/32"	REAR TOE (Inches)	0.0" ± 0.13" 0" ± ⅛"	0.0" ± 0.13" 0" ± ⅛"	0.05" ± 0.15" 1/16" ± 5/32"
0.4 (13/32)	0.4 (13/32)	0.1 (3/32)	THRUST ANGLE	0.4 (13/32)	0.4 (13/32)	0.15 (5/32)
[76]	[76]	[26] [06]	ILLUSTRATION #	[76]	[76]	[10]
[1] Wagon = 0.9° (7/8°)		[1] 2000 = 0.1° (0.05")	NOTES			
10043	10046	10048		10055	10056	10059

Chrysler

LHS	LHS	Pacifica	MODEL INFORMATION	Pacifica incl. PHEV	PT Cruiser	PT Cruiser
1994-97	1999-2001	2004-2008	YEARS	2017	2001-2005	2006-2010
			INSTRUCTIONS			
95 ft.lbs.	100 ft.lbs.	100 ft.lbs.	LUGNUT TORQUE	100 ft.lbs.	100 ft.lbs.	100 ft.lbs.
			FRONT			
3.0 ± 1.0 3 ± 1	3.0 ± 1.0 3 ± 1	4.5 ± 0.5 4 1/2 ± 1/2	CASTER (Degrees)	6.9 ± 0.4 6 7/8 ± 3/8	2.45 ± 1.0 2 7/16 ± 1	2.45 ± 1.0 2 7/16 ± 1
0.0 (0) | 1.0 (1)	0.0 (0) | 1.0 (1)	0.0 (0) | 1.0 (1)	CROSS | TOL.	0.0 (0) | 0.5 (1/2)	0.0 (0) | 1.0 (1)	0.0 (0) | 1.0 (1)
0.0 ± 0.6 0 ± 5/8	0.0 ± 0.6 0 ± 5/8	-0.2 ± 0.4 -3/16 ± 3/8	CAMBER (Degrees)	-0.6 ± 0.4 -5/8 ± 3/8	0.0 ± 0.4 0 ± 3/8	0.0 ± 0.4 0 ± 3/8
0.0 (0) | 0.7 (11/16)	0.0 (0) | 0.7 (11/16)	0.0 (0) | 0.5 (1/2)	CROSS | TOL.	0.3 (5/16) | 0.55 (9/16)	0.0 (0) | 0.5 (1/2)	0.0 (0) | 0.5 (1/2)
0.2° ± 0.2° 3/16° ± 3/16°	0.0° ± 0.2° [1] 0° ± 3/16° [1]	0.1° ± 0.2° 3/32° ± 3/16°	FRONT TOE (Degrees)	0.1° ± 0.1° 3/32° ± 3/32°	0.1° ± 0.1° 3/32° ± 3/32°	0.1° ± 0.1° 3/32° ± 3/32°
0.1" ± 0.1" 3/32" ± 3/32"	0.0" ± 0.1" [1] 0" ± 3/32" [1]	0.05" ± 0.1" 1/16" ± 3/32"	FRONT TOE (Inches)	0.05" ± 0.05" 1/16" ± 1/16"	0.05" ± 0.05" 1/16" ± 1/16"	0.05" ± 0.05" 1/16" ± 1/16"
14.0 (14)	13.7 (13 11/16)		SAI			
			TURNING ANGLE			
	26	161	ILLUSTRATION #		26	26
			REAR			
-0.1 ± 0.5 -1/8 ± 1/2	-0.2 ± 0.5 -3/16 ± 1/2	-0.4 ± 0.5 -3/8 ± 1/2	CAMBER (Degrees)	-1.0 ± 0.45 -1 ± 7/16	0.0 ± 0.25 0 ± 1/4	0.0 ± 0.25 0 ± 1/4
0.1° ± 0.3° 3/32° ± 5/16°	0.1° ± 0.3° 3/32° ± 5/16°	0.1° ± 0.12° 3/32° ± 1/8°	REAR TOE (Degrees)	0.1° ± 0.2° 3/32° ± 3/16°	0.2° ± 0.2° 3/16° ± 3/16°	0.3° ± 0.2° 5/16° ± 3/16°
0.05" ± 0.15" 1/16" ± 5/32"	0.05" ± 0.15" 1/16" ± 5/32"	0.05" ± 0.06" 1/16" ± 1/16"	REAR TOE (Inches)	0.05" ± 0.1" 1/16" ± 3/32"	0.1" ± 0.1" 3/32" ± 3/32"	0.15" ± 0.1" 5/32" ± 3/32"
0.15 (5/32)	0.15 (5/32)	0.2 (3/16)	THRUST ANGLE	0.3 (5/16)	0.3 (5/16)	0.3 (5/16)
10	10	06	ILLUSTRATION #	177		
	[1] 1999 = 0.1° (0.05")		NOTES			
10060	10061	10068		10069	10071	10072

Chrysler

Sebring	Sebring Sedan & Conv.	Sebring Sedan & Conv. FWD & AWD	MODEL INFORMATION	Sebring Conv.	Sebring Conv.	Sebring Coupe
1995-2000	2001-2006	2007-2010	YEARS	1996	1997-2000	2001
			INSTRUCTIONS			
65-80 ft.lbs.[1]	100 ft.lbs.	100 ft.lbs.	LUGNUT TORQUE	65-80 ft.lbs.	65-80 ft.lbs.	73 ft.lbs.
			FRONT			
4.33 ± 1.5 [2] 4 5/16 ± 1 1/2 [2]	3.3 ± 1.0 3 5/16 ± 1	3.0 ± 1.0 3 ± 1	CASTER (Degrees)	3.3 ± 1.0 3 5/16 ± 1	3.1 ± 1.0 3 1/8 ± 1	3.0 ± 0.5 3 ± 1/2
0.0 (0) | n/s	0.0 (0) | 1.0 (1)	0.0 (0) | 1.0 (1)	CROSS | TOL.	0.0 (0) | 1.0 (1)	0.0 (0) | 1.0 (1)	0.0 (0) | 0.5 (1/2)
0.0 ± 0.5 0 ± 1/2	-0.3 ± 0.6 -5/16 ± 5/8	0.0 ± 0.55 0 ± 9/16	CAMBER (Degrees)	0.3 ± 0.6 5/16 ± 5/8	0.1 ± 0.6 1/8 ± 5/8	0.0 ± 0.5 0 ± 1/2
0.0 (0) | n/s	0.0 (0) | 0.7 (11/16)	0.0 (0) | 0.8 (13/16)	CROSS | TOL.	0.0 (0) | 0.7 (11/16)	0.0 (0) | 0.7 (11/16)	0.0 (0) | 0.5 (1/2)
0.0° ± 0.24° 0° ± 1/4°	0.24° ± 0.2° 1/4° ± 3/16°	0.1° ± 0.2° 3/32° ± 3/16°	FRONT TOE (Degrees)	0.1° ± 0.2° 3/32° ± 3/16°	0.1° ± 0.2° 3/32° ± 3/16°	0.0° ± 0.24° 0° ± 1/4°
0.0" ± 0.12" 0" ± 1/8"	0.12" ± 0.1" 1/8" ± 3/32"	0.05" ± 0.1" 1/16" ± 3/32"	FRONT TOE (Inches)	0.05" ± 0.1" 1/16" ± 3/32"	0.05" ± 0.1" 1/16" ± 3/32"	0.0" ± 0.12" 0" ± 1/8"
7.3 (7 5/16)			SAI	7.3 ± 0.5 (7 5/16 ± 1/2)	7.3 ± 0.5 (7 5/16 ± 1/2)	
32.5 / 27.0 ± 2.0			TURNING ANGLE			36.6 / 30.4 ± 2.0 [1]
	113	161 △	ILLUSTRATION #	113	113	26
			REAR			
-1.33 ± 0.5 -1 5/16 ± 1/2	-0.5 ± 0.5 -1/2 ± 1/2	-0.6 ± 0.65 -5/8 ± 5/8	CAMBER (Degrees)	-0.2 ± 0.4 -3/16 ± 3/8	-0.1 ± 0.4 -1/8 ± 3/8	-1.33 ± 0.5 -1 5/16 ± 1/2
0.24° ± 0.24° 1/4° ± 1/4°	0.1° ± 0.2° 3/32° ± 3/16°	0.2° ± 0.2° 3/16° ± 3/16°	REAR TOE (Degrees)	0.1° ± 0.2° 3/32° ± 3/16°	0.1° ± 0.2° 3/32° ± 3/16°	0.24° ± 0.24° 1/4° ± 1/4°
0.12" ± 0.12" 1/8" ± 1/8"	0.05" ± 0.1" 1/16" ± 3/32"	0.1" ± 0.1" 3/32" ± 3/32"	REAR TOE (Inches)	0.05" ± 0.1" 1/16" ± 3/32"	0.05" ± 0.1" 1/16" ± 3/32"	0.12" ± 0.12" 1/8" ± 1/8"
	0.15 (5/32)	0.15 (5/32)	THRUST ANGLE		0.15 (5/32)	
06	01	174	ILLUSTRATION #	01	01	06
[1] 1996-98 = 87-101 ft.lbs. [2] 1997 w/17" Wheels 4.67°			NOTES			[1] 3.0L 33.1°/28.5°
10076	10077	10078		10079	10080	10081

Chrysler

Sebring Coupe	TC (by Maserati)	Town & Country FWD Minivan	MODEL INFORMATION	Town & Country FWD Minivan	Town & Country AWD Minivan	Town & Country Minivan
2002-2005	1990-91	1990	YEARS	1991-95	1992-95	1996-2000
			INSTRUCTIONS			
73 ft.lbs.	95 ft.lbs.	95 ft.lbs.	LUGNUT TORQUE	95 ft.lbs.	95 ft.lbs.	100 ft.lbs.
			FRONT			
3.0 ± 0.5 3 ± ½	1.2 ± 0.75 1 3/16 ± ¾	0.7 ± 0.5 11/16 ± ½	CASTER (Degrees)	1.3 ± 0.5 [1] 1 5/16 ± ½ [1]	1.3 ± 0.5 1 5/16 ± ½	1.4 ± 1.0 1⅜ ± 1
0.0 (0) | 0.5 (½)	0.0 (0) | 1.5 (1½)	0.0 (0) | 1.5 (1½)	CROSS | TOL.	0.0 (0) | 1.5 (1½)	0.0 (0) | 1.5 (1½)	0.0 (0) | 1.0 (1)
0.0 ± 0.5 0 ± ½	0.3 ± 0.5 5/16 ± ½	0.3 ± 0.5 5/16 ± ½	CAMBER (Degrees)	0.3 ± 0.5 5/16 ± ½	0.3 ± 0.5 5/16 ± ½	0.15 ± 0.4 [1] ⅛ ± ⅜ [1]
0.0 (0) | 0.5 (½)	0.0 (0) | n/s	0.0 (0) | n/s	CROSS | TOL.	0.0 (0) | n/s	0.0 (0) | n/s	0.0 (0) | 0.5 (½)
0.0° ± 0.24° 0° ± ¼°	0.1° ± 0.1° 3/32° ± 3/32°	0.1° ± 0.1° 3/32° ± 3/32°	FRONT TOE (Degrees)	0.1° ± 0.1° 3/32° ± 3/32°	0.1° ± 0.1° 3/32° ± 3/32°	0.1° ± 0.2° 3/32° ± 3/16°
0.0" ± 0.12" 0" ± ⅛"	0.05" ± 0.05" 1/16" ± 1/16"	0.05" ± 0.05" 1/16" ± 1/16"	FRONT TOE (Inches)	0.05" ± 0.05" 1/16" ± 1/16"	0.05" ± 0.05" 1/16" ± 1/16"	0.05" ± 0.1" 1/16" ± 3/32"
			SAI			12.6 (12⅝)
36.2 / 30.4 ± 2.0 [1]			TURNING ANGLE			
26	22	22	ILLUSTRATION #	22	22	22
			REAR			
-1.33 ± 0.5 -1 5/16 ± ½	-0.5 ± 0.5 -½ ± ½	-0.2 ± 0.5 -3/16 ± ½	CAMBER (Degrees)	-0.2 ± 0.6 -3/16 ± ⅝	-0.2 ± 0.6 -3/16 ± ⅝	0.0 ± 0.25 [2] 0 ± ¼ [2]
0.24° ± 0.24° ¼° ± ¼°	0.0° ± 0.25° 0° ± ¼°	0.0° ± 0.6° 0° ± 19/32°	REAR TOE (Degrees)	0.0° ± 0.6° 0° ± 19/32°	0.0° ± 0.6° 0° ± 19/32°	0.0° ± 0.4° 0° ± 13/32°
0.12" ± 0.12" ⅛" ± ⅛"	0.0" ± 0.13" 0" ± ⅛"	0.0" ± 0.3" 0" ± 5/16"	REAR TOE (Inches)	0.0" ± 0.3" 0" ± 5/16"	0.0" ± 0.3" 0" ± 5/16"	0.0" ± 0.2" 0" ± 3/16"
	0.4 (13/32)	0.4 (13/32)	THRUST ANGLE	0.4 (13/32)	0.4 (13/32)	0.3 (5/16)
06	76	76	ILLUSTRATION #	76		
[1] 3.0L 31°/27°			NOTES	[1] 1991 = 1.7° (1 11/16°)		[1] 205/15" or 215/16"=.05°(1/16°) [2] CNG or Electric = -0.1°(-1/8°)
10082	10083	10089		10090	10091	10092

Chrysler

Town & Country Minivan	Town & Country Minivan	Voyager Minivan	MODEL INFORMATION	Voyager Minivan	Notes	Notes
2001-2007	2008-2016	2000	YEARS	2001-2003		
			INSTRUCTIONS			
100 ft.lbs.	100 ft.lbs.	100 ft.lbs.	LUGNUT TORQUE	100 ft.lbs.		
			FRONT			
2.31 ± 1.0 2 5/16 ± 1	2.5 ± 1.0 2½ ± 1	1.4 ± 1.0 1⅜ ± 1	CASTER (Degrees)	2.31 ± 1.0 2 5/16 ± 1		
0.0 (0) | 1.0 (1)	0.0 (0) | 1.0 (1)	0.0 (0) | 1.0 (1)	CROSS | TOL.	0.0 (0) | 1.0 (1)		
0.1 ± 0.4 [1] ⅛ ± ⅜ [1]	L= -0.1, R= -0.4 ± 0.4 L= -⅛, R= -⅜ ± ⅜	0.15 ± 0.4 [1] ⅛ ± ⅜ [1]	CAMBER (Degrees)	0.1 ± 0.4 [1] ⅛ ± ⅜ [1]		
0.0 (0) | 0.5 (½)	0.3 (5/16) | 0.5 (½)	0.0 (0) | 0.5 (½)	CROSS | TOL.	0.0 (0) | 0.5 (½)		
0.1° ± 0.2° 3/32° ± 3/16°	0.26° ± 0.2° ¼° ± 3/16°	0.1° ± 0.2° 3/32° ± 3/16°	FRONT TOE (Degrees)	0.1° ± 0.2° 3/32° ± 3/16°		
0.05" ± 0.1" 1/16" ± 3/32"	0.13" ± 0.1" ⅛" ± 3/32"	0.05" ± 0.1" 1/16" ± 3/32"	FRONT TOE (Inches)	0.05" ± 0.1" 1/16" ± 3/32"		
		12.6 (12⅝)	SAI			
			TURNING ANGLE			
26	161 △	22	ILLUSTRATION #	26		
			REAR			
0.0 ± 0.25 0 ± ¼	-0.06 ± 0.4 -1/16 ± ⅜	0.0 ± 0.25 [2] 0 ± ¼ [2]	CAMBER (Degrees)	0.0 ± 0.25 0 ± ¼		
0.0° ± 0.2° 0° ± 3/16°	0.15° ± 0.25° [1] 5/32° ± ¼° [1]	0.0° ± 0.4° 0° ± 13/32°	REAR TOE (Degrees)	0.0° ± 0.2° 0° ± 3/16°		
0.0" ± 0.1" 0" ± 3/32"	0.08" ± 0.13" [1] 1/16" ± ⅛" [1]	0.0" ± 0.2" 0" ± 3/16"	REAR TOE (Inches)	0.0" ± 0.1" 0" ± 3/32"		
0.3 (5/16)	0.3 (5/16)	0.3 (5/16)	THRUST ANGLE	0.3 (5/16)		
			ILLUSTRATION #			
[1] Cam kit required	[1] 2008-11 = 0.1° (0.05")	[1] 205/15" or 215/16"=.05°(1/16°) [2] CNG or Electric = -0.1°(-1/8°)	NOTES	[1] Cam kit required		
10093	10094	10095		10096	10099	10100

ALIGNMENT SPECIFICATIONS

Daewoo / Daihatsu

Lanos	Leganza	Nubira	MODEL INFORMATION	Charade	Charade GTX	Rocky
1999-2002	1999-2002	1999-2002	YEARS	1990-92	1993-95	1990-92
			INSTRUCTIONS			
	74 ft.lbs.	66 ft.lbs.	LUGNUT TORQUE			
			FRONT			
2.75 ± 1.0 [1] 2¾ ± 1 [1]	3.0 ± 1.0 3 ± 1	3.0 ± 0.75 3 ± ¾	CASTER (Degrees)	2.92 ± 0.5 2 15/16 ± ½	3.0 ± 0.5 3 ± ½	2.0 ± 0.5 2 ± ½
0.0 (0) | 1.0 (1)	0.0 (0) | 1.0 (1)	0.0 (0) | 1.0 (1)	CROSS | TOL.	0.0 (0) | 1.0 (1)	0.0 (0) | 1.0 (1)	0.0 (0) | 1.0 (1)
-0.42 ± 0.75 -7/16 ± ¾	-0.2 ± 1.0 -3/16 ± 1	-0.4 ± 0.75 -⅜ ± ¾	CAMBER (Degrees)	0.33 +0.67 -0.33 5/16 +11/16 -5/16	0.25 ± 0.67 ¼ ± 11/16	1.0 +0.0 -1.0 1 +0 -1
0.0 (0) | 1.0 (1)	0.0 (0) | 1.0 (1)	0.0 (0) | 1.0 (1)	CROSS | TOL.	0.0 (0) | 0.5 (½)	0.0 (0) | 0.5 (½)	0.0 (0) | 0.5 (½)
0.0° ± 0.17° [2] 0° ± 5/32° [2]	-0.2° ± 0.16° [1] -3/16° ± 5/32° [1]	0.0° ± 0.17° [1] 0° ± 5/32° [1]	FRONT TOE (Degrees)	0.12° ± 0.12° ⅛° ± ⅛°	0.2° ± 0.12° 3/16° ± ⅛°	0.2° ± 0.12° 3/16° ± ⅛°
0.0" ± 0.08" [2] 0" ± 3/32" [2]	-0.1" ± 0.08" [1] -3/32" ± 3/32" [1]	0.0" ± 0.08" [1] 0" ± 3/32" [1]	FRONT TOE (Inches)	0.06" ± 0.06" 1/16" ± 1/16"	0.1" ± 0.06" 3/32" ± 1/16"	0.1" ± 0.06" 3/32" ± 1/16"
			SAI	12.0 ± 0.5 (12 ± ½)	12.5 ± 0.5 (12½ ± ½)	9.5 ± 0.5 (9½ ± ½)
			TURNING ANGLE	39.9 / 35.0 ± 2.0	39.0 / 36.0 ± 2.0	27.1 / 23.9 ± 2.0
			ILLUSTRATION #			23
			REAR			
-1.67 ± 0.5 -1 11/16 ± ½	-0.8 ± 1.0 -13/16 ± 1	-0.83 ± 0.75 -13/16 ± ¾	CAMBER (Degrees)	-0.67 ± 0.58 -11/16 ± 9/16	-1.5 ± 0.5 -1½ ± ½	
0.25° ± 0.42° [2] ¼° ± 13/32° [2]	0.32° ± 0.09° [1] 5/16° ± 3/32° [1]	0.05° ± 0.75° [1] 1/16° ± ¾° [1]	REAR TOE (Degrees)	0.47° ± 0.15° 15/32° ± 5/32°	0.3° ± 0.15° 5/16° ± 5/32°	
0.13" ± 0.21" [2] ⅛" ± 7/32" [2]	0.16" ± 0.04" [1] 5/32" ± 1/32" [1]	0.03" ± 0.38" [1] 1/32" ± ⅜" [1]	REAR TOE (Inches)	0.24" ± 0.08" ¼" ± 1/16"	0.15" ± 0.08" 5/32" ± 1/16"	
			THRUST ANGLE			
	10	06	ILLUSTRATION #	06	06	
[1] w/MS 1.5° (1½°) [2] w/2 person load	[1] w/2 person load	[1] w/2 person load	NOTES			
54001	54002	54003		11001	11002	11003

Dodge

Avenger	Avenger	Avenger	MODEL INFORMATION	Caliber w/15" Wheels	Caliber w/17" Wheels	Caliber w/18" Wheels
1995-2000	2008-2010	2011-2014	YEARS	2007-2011	2007-2011	2007-2011
			INSTRUCTIONS			
65-80 ft.lbs. [1]	100 ft.lbs.	100 ft.lbs.	LUGNUT TORQUE	100 ft.lbs.	100 ft.lbs.	100 ft.lbs.
			FRONT			
4.33 ± 1.5 4 5/16 ± 1½	3.0 ± 1.0 3 ± 1	3.0 ± 1.0 3 ± 1	CASTER (Degrees)	L= 2.9, R= 2.6 ± 1.0 L= 2⅞, R= 2⅝ ± 1	L= 2.9, R= 2.6 ± 1.0 L= 2⅞, R= 2⅝ ± 1	L= 3.0, R= 2.7 ± 1.0 L= 3, R= 2 11/16 ± 1
0.0 (0) | n/s	0.0 (0) | 1.0 (1)	0.0 (0) | 1.0 (1)	CROSS | TOL.	0.3 (5/16) | 1.0 (1)	0.3 (5/16) | 1.0 (1)	0.3 (5/16) | 1.0 (1)
0.0 ± 0.5 0 ± ½	0.0 ± 0.55 0 ± 9/16	-0.62 ± 0.55 -⅝ ± 9/16	CAMBER (Degrees)	-0.5 ± 0.4 -½ ± ⅜	-0.7 ± 0.4 -11/16 ± ⅜	-0.7 ± 0.4 -11/16 ± ⅜
0.0 (0) | n/s	0.0 (0) | 0.8 (13/16)	0.0 (0) | 0.8 (13/16)	CROSS | TOL.	0.0 (0) | 0.5 (½)	0.0 (0) | 0.5 (½)	0.0 (0) | 0.5 (½)
0.0° ± 0.24° 0° ± ¼°	0.1° ± 0.2° 3/32° ± 3/16°	0.2° ± 0.3° 3/16° ± 5/16°	FRONT TOE (Degrees)	0.2° ± 0.2° 3/16° ± 3/16°	0.2° ± 0.2° 3/16° ± 3/16°	0.2° ± 0.2° 3/16° ± 3/16°
0.0" ± 0.12" 0" ± ⅛"	0.05" ± 0.1" 1/16" ± 3/32"	0.1" ± 0.15" 3/32" ± 5/32"	FRONT TOE (Inches)	0.1" ± 0.1" 3/32" ± 3/32"	0.1" ± 0.1" 3/32" ± 3/32"	0.1" ± 0.1" 3/32" ± 3/32"
			SAI			
32.5 / 27.0 ± 2.0			TURNING ANGLE			
	161	161	ILLUSTRATION #	26 △	26 △	26 △
			REAR			
-1.33 ± 0.5 -1 5/16 ± ½	-0.6 ± 0.65 -⅝ ± ⅝	-0.78 ± 0.65 -¾ ± ⅝	CAMBER (Degrees)	-0.4 ± 0.4 -⅜ ± ⅜	-0.6 ± 0.4 -⅝ ± ⅜	-0.7 ± 0.4 -11/16 ± ⅜
0.24° ± 0.24° ¼° ± ¼°	0.2° ± 0.2° 3/16° ± 3/16°	0.2° ± 0.3° 3/16° ± 5/16°	REAR TOE (Degrees)	0.2° ± 0.2° 3/16° ± 3/16°	0.2° ± 0.2° 3/16° ± 3/16°	0.2° ± 0.2° 3/16° ± 3/16°
0.12" ± 0.12" ⅛" ± ⅛"	0.1" ± 0.1" 3/32" ± 3/32"	0.1" ± 0.15" 3/32" ± 5/32"	REAR TOE (Inches)	0.1" ± 0.1" 3/32" ± 3/32"	0.1" ± 0.1" 3/32" ± 3/32"	0.1" ± 0.1" 3/32" ± 3/32"
	0.15 (5/32)	0.15 (5/32)	THRUST ANGLE	0.1 (3/32)	0.1 (3/32)	0.1 (3/32)
06	174	174	ILLUSTRATION #	174	174	174
[1] 1996-98 = 87-101 ft.lbs.			NOTES			
12010	12011	12012		12013	12014	12015

ALIGNMENT SPECIFICATIONS

Dodge

Caliber SRT4	Caliber	Caravan Minivan	MODEL INFORMATION	Caravan Minivan FWD	Caravan Minivan AWD	Caravan Minivan All Models
2008-2009	2012	1990	YEARS	1991-95	1991-95	1996-2000
			INSTRUCTIONS			
100 ft.lbs.	100 ft.lbs.	95 ft.lbs.	LUGNUT TORQUE			100 ft.lbs.
			FRONT			
L= 2.44, R= 2.1 ± 1.0 L= 2 7/16, R= 2⅛ ± 1	L= 3.0, R= 2.7 ± 1.0 L= 3, R= 2 11/16 ± 1	0.7 ± 0.5 11/16 ± ½	CASTER (Degrees)	1.3 ± 0.5 [1] 1 5/16 ± ½ [1]	1.3 ± 0.5 [1] 1 5/16 ± ½ [1]	1.4 ± 1.0 1⅜ ± 1
0.34 (5/16) | 1.0 (1)	0.3 (5/16) | 1.0 (1)	0.0 (0) | 1.5 (1½)	CROSS | TOL.	0.0 (0) | 1.5 (1½)	0.0 (0) | 1.5 (1½)	0.0 (0) | 1.0 (1)
-0.25 ± 0.4 -¼ ± ⅜	-0.64 ± 0.55 -⅝ ± 9/16	0.3 ± 0.5 5/16 ± ½	CAMBER (Degrees)	0.3 ± 0.5 5/16 ± ½	0.3 ± 0.5 5/16 ± ½	0.15 ± 0.4 [1] ⅛ ± ⅜ [1]
0.0 (0) | 0.5 (½)	0.0 (0) | 0.6 (⅝)	0.0 (0) | n/s	CROSS | TOL.	0.0 (0) | n/s	0.0 (0) | n/s	0.0 (0) | 0.5 (½)
0.1° ± 0.2° 3/32° ± 3/16°	0.2° ± 0.2° 3/16° ± 3/16°	0.1° ± 0.1° 3/32° ± 3/32°	FRONT TOE (Degrees)	0.1° ± 0.1° 3/32° ± 3/32°	0.1° ± 0.1° 3/32° ± 3/32°	0.1° ± 0.2° 3/32° ± 3/16°
0.05" ± 0.1" 1/16" ± 3/32"	0.1" ± 0.1" 3/32" ± 3/32"	0.05" ± 0.05" 1/16" ± 1/16"	FRONT TOE (Inches)	0.05" ± 0.05" 1/16" ± 1/16"	0.05" ± 0.05" 1/16" ± 1/16"	0.05" ± 0.1" 1/16" ± 3/32"
			SAI			12.6 (12⅝)
			TURNING ANGLE			
26	26	22	ILLUSTRATION #	22	22	22
			REAR			
-0.85 ± 0.4 -⅞ ± ⅜	-0.61 ± 0.6 -⅝ ± ⅝	-0.2 ± 0.6 -3/16 ± ⅝	CAMBER (Degrees)	-0.2 ± 0.6 -3/16 ± ⅝	-0.2 ± 0.6 -3/16 ± ⅝	0.0 ± 0.25 [2] 0 ± ¼ [2]
0.2° ± 0.2° 3/16° ± 3/16°	0.2° ± 0.25° 3/16° ± ¼°	0.0° ± 0.6° 0° ± 19/32°	REAR TOE (Degrees)	0.0° ± 0.6° 0° ± 19/32°	0.0° ± 0.6° 0° ± 19/32°	0.0° ± 0.4° 0° ± 13/32°
0.1" ± 0.1" 3/32" ± 3/32"	0.1" ± 0.13" 3/32" ± ⅛"	0.0" ± 0.3" 0" ± 5/16"	REAR TOE (Inches)	0.0" ± 0.3" 0" ± 5/16"	0.0" ± 0.3" 0" ± 5/16"	0.0" ± 0.2" 0" ± 3/16"
0.1 (3/32)	0.15 (5/32)	0.4 (13/32)	THRUST ANGLE	0.4 (13/32)	0.4 (13/32)	0.3 (5/16)
174	174	76	ILLUSTRATION #	76		
			NOTES	[1] 1991 = 1.7° (1 11/16°)	[1] 1991 = 1.7° (1 11/16°)	[1] 205/15", 215/16", 17" = .05° (1/16°) [2] CNG or Electric = -0.1° (-⅛°)
12016	12017	12019		12020	12021	12022

Dodge

Caravan Minivan All Models	Grand Caravan	Challenger SRT8	MODEL INFORMATION	Challenger exc/SRT8	Challenger SRT8	Challenger exc/SRT8
2001-2007	2008-2017	2008	YEARS	2009-2010	2009-2010	2011-2014
			INSTRUCTIONS			
100 ft.lbs.	100 ft.lbs.	110 ft.lbs.	LUGNUT TORQUE	110 ft.lbs.	110 ft.lbs.	130 ft.lbs.
			FRONT			
2.31 ± 1.0 2 5/16 ± 1	2.5 ± 1.0 2½ ± 1	L= 9.7, R= 10.3 ± 1.0 L= 9 11/16, R= 10 5/16 ± 1	CASTER (Degrees)	L= 8.0, R= 9.0 ± 1.0 L= 8, R= 9 ± 1	L= 8.3, R= 9.3 ± 1.0 L= 8 5/16, R= 9 5/16 ± 1	L= 8.0, R= 8.7 ± 1.0 L= 8, R= 8 11/16 ± 1
0.0 (0) | 1.0 (1)	0.0 (0) | 1.0 (1)	-0.6 (-⅝) | 0.6 (⅝)	CROSS | TOL.	-1.0 (-1) | 0.6 (⅝)	-1.0 (-1) | 0.6 (⅝)	-0.7 (-11/16) | 0.6 (⅝)
0.1 ± 0.4 [1] ⅛ ± ⅜ [1]	L= -0.1, R= -0.4 ± 0.4 L= -⅛, R= -⅜ ± ⅜	L= -0.05, R= -0.35 ± 0.5 L= -1/16, R= -⅜ ± ½	CAMBER (Degrees)	L= 0.15, R= -0.25 ± 0.65 L= ⅛, R= -¼ ± ⅝	L= -0.05, R= -0.35 ± 0.65 L= -1/16, R= -⅜ ± ⅝	L= -0.85, R= -1.25 ± 0.65 L= -⅞, R= -1¼ ± ⅝
0.0 (0) | 0.5 (½)	0.3 (5/16) | 0.5 (½)	0.3 (5/16) | 0.5 (½)	CROSS | TOL.	0.4 (⅜) | 0.55 (9/16)	0.3 (5/16) | 0.55 (9/16)	0.4 (⅜) | 0.55 (9/16)
0.1° ± 0.2° 3/32° ± 3/16°	0.26° ± 0.2° ¼° ± 3/16°	0.2° ± 0.2° 3/16° ± 3/16°	FRONT TOE (Degrees)	0.2° ± 0.1° 3/16° ± 3/32°	0.2° ± 0.1° 3/16° ± 3/32°	0.2° ± 0.1° 3/16° ± 3/32°
0.05" ± 0.1" 1/16" ± 3/32"	0.13" ± 0.1" ⅛" ± 3/32"	0.1" ± 0.1" 3/32" ± 3/32"	FRONT TOE (Inches)	0.1" ± 0.05" 3/32" ± 1/16"	0.1" ± 0.05" 3/32" ± 1/16"	0.1" ± 0.05" 3/32" ± 1/16"
			SAI			
			TURNING ANGLE			
161	161 △	164 165 △	ILLUSTRATION #	164 165 △	164 165 △	165 △
			REAR			
0.0 ± 0.25 0 ± ¼	-0.06 ± 0.4 -1/16 ± ⅜	-0.75 ± 0.5 -¾ ± ½	CAMBER (Degrees)	-1.75 ± 0.55 -1¾ ± 9/16	-0.75 ± 0.55 -¾ ± 9/16	-1.75 ± 0.55 -1¾ ± 9/16
0.0° ± 0.2° 0° ± 3/16°	0.15° ± 0.25° [1] 5/32° ± ¼° [1]	0.2° ± 0.15° 3/16° ± 5/32°	REAR TOE (Degrees)	0.2° ± 0.15° 3/16° ± 5/32°	0.2° ± 0.15° 3/16° ± 5/32°	0.2° ± 0.15° 3/16° ± 5/32°
0.0" ± 0.1" 0" ± 3/32"	0.08" ± 0.13" [1] 1/16" ± ⅛" [1]	0.1" ± 0.08" 3/32" ± 1/16"	REAR TOE (Inches)	0.1" ± 0.08" 3/32" ± 1/16"	0.1" ± 0.08" 3/32" ± 1/16"	0.1" ± 0.08" 3/32" ± 1/16"
0.3 (5/16)	0.3 (5/16)	0.15 (5/32)	THRUST ANGLE	0.15 (5/32)	0.15 (5/32)	0.15 (5/32)
		171	ILLUSTRATION #	171	171	171
[1] Cam kit required	[1] 2008-11 = 0.1° (0.05")		NOTES			
12023	12024	12028		12029	12030	12031

Dodge

Challenger	Challenger	Challenger	MODEL	Challenger	Challenger	Challenger
SRT8	exc/SRT8	SRT8	INFORMATION	RWD	RWD	RWD
				exc/SRT8, Police	SRT8	Police
2011-2014	2015-2016	2015-2016	YEARS	2017	2017	2017
			INSTRUCTIONS			
111 ft.lbs.	130 ft.lbs.	111 ft.lbs.	LUGNUT TORQUE	130 ft.lbs.	111 ft.lbs.	130 ft.lbs.
			FRONT			
L= 8.3, R= 9.0 ± 1.0 L= 8 5/16, R= 9 ± 1	L= 8.0, R= 8.7 ± 1.0 L= 8, R= 8 11/16 ± 1	L= 8.3, R= 9.0 ± 1.0 L= 8 5/16, R= 9 ± 1	CASTER (Degrees)	L= 8.0, R= 8.7 ± 1.0 L= 8, R= 8 11/16 ± 1	L= 8.3, R= 9.0 ± 1.0 L= 8 5/16, R= 9 ± 1	L= 7.7, R= 8.4 ± 1.0 L= 7 11/16, R= 8 3/8 ± 1
-0.7 (-11/16) | 0.6 (5/8)	-0.7 (-11/16) | 0.6 (5/8)	-0.7 (-11/16) | 0.6 (5/8)	CROSS | TOL.	-0.7 (-11/16) | 0.6 (5/8)	-0.7 (-11/16) | 0.6 (5/8)	-0.7 (-11/16) | 0.6 (5/8)
L= -1.05, R= -1.35 +0.45 -0.65 L= -1 1/16, R= -1 3/8 +7/16 -5/8	L= -0.85, R= -1.25 ± 0.55 L= -7/8, R= -1 1/4 ± 9/16	L= -1.05, R= -1.35 ± 0.55 L= -1 1/16, R= -1 3/8 ± 9/16	CAMBER (Degrees)	L= -0.75, R= -1.15 ± 0.55 L= -3/4, R= -1 1/8 ± 9/16	L= -0.95, R= -1.35 ± 0.55 L= -1 1/16, R= -1 3/8 ± 9/16	L= -0.65, R= -1.05 ± 0.55 L= -5/8, R= -1 1/16 ± 9/16
0.3 (5/16) | 0.55 (9/16)	0.4 (3/8) | 0.55 (9/16)	0.3 (5/16) | 0.55 (9/16)	CROSS | TOL.	0.4 (3/8) | 0.55 (9/16)	0.4 (3/8) | 0.55 (9/16)	0.4 (3/8) | 0.55 (9/16)
0.2° ±0.1° 3/16° ± 3/32°	0.2° ±0.1° 3/16° ± 3/32°	0.2° ±0.2° 3/16° ± 3/16°	FRONT TOE (Degrees)	0.2° ±0.1° 3/16° ± 3/32°	0.2° ±0.2° 3/16° ± 3/16°	0.2° ±0.2° 3/16° ± 3/16°
0.1" ±0.05" 3/32" ± 1/16"	0.1" ±0.05" 3/32" ± 1/16"	0.1" ±0.1" 3/32" ± 3/32"	FRONT TOE (Inches)	0.1" ±0.05" 3/32" ± 1/16"	0.1" ±0.1" 3/32" ± 3/32"	0.1" ±0.1" 3/32" ± 3/32"
			SAI			
			TURNING ANGLE			
165 ▲	165 ▲	165 ▲	ILLUSTRATION #	165	165	165
			REAR			
-0.75 ±0.65 -3/4 ± 5/8	-1.75 ±0.55 -1 3/4 ± 9/16	-0.75 ±0.55 -3/4 ± 9/16	CAMBER (Degrees)	-1.75 ±0.55 -1 3/4 ± 9/16	-0.75 ±0.55 -3/4 ± 9/16	-0.75 ±0.55 -3/4 ± 9/16
0.2° ±0.15° 3/16° ± 5/32°	0.2° ±0.15° 3/16° ± 5/32°	0.2° ±0.15° 3/16° ± 5/32°	REAR TOE (Degrees)	0.2° ±0.15° 3/16° ± 5/32°	0.2° ±0.15° 3/16° ± 5/32°	0.2° ±0.15° 3/16° ± 5/32°
0.1" ±0.08" 3/32" ± 1/16"	0.1" ±0.08" 3/32" ± 1/16"	0.1" ±0.08" 3/32" ± 1/16"	REAR TOE (Inches)	0.1" ±0.08" 3/32" ± 1/16"	0.1" ±0.08" 3/32" ± 1/16"	0.1" ±0.08" 3/32" ± 1/16"
0.15 (5/32)	0.15 (5/32)	0.15 (5/32)	THRUST ANGLE	0.15 (5/32)	0.15 (5/32)	0.15 (5/32)
171	171	171	ILLUSTRATION #	171	171	171
			NOTES			
12032	12033	12034		12035	12036	12037

Dodge

Challenger	Challenger	Charger	MODEL	Charger	Charger	Charger
AWD	AWD	RWD	INFORMATION	RWD	RWD	RWD
exc/Police	Police	exc/SRT8		SRT8	exc/SRT8, Police	SRT8
2017	2017	2006-2008	YEARS	2006-2008	2009-2010	2009-2010
			INSTRUCTIONS			
130 ft.lbs.	130 ft.lbs.	110 ft.lbs.	LUGNUT TORQUE	110 ft.lbs.	110 ft.lbs.	110 ft.lbs.
			FRONT			
5.4 ±1.0 5 3/8 ±1	5.4 ±1.0 5 3/8 ±1	L= 9.4, R=10.4 ± 1.0 L= 9 3/8, R= 10 3/8 ± 1	CASTER (Degrees)	L= 9.7, R=10.3 ± 1.0 L= 9 11/16, R= 10 5/16 ± 1	L= 8.0, R= 9.0 ± 1.0 L= 8, R= 9 ± 1	L= 8.3, R= 9.3 ± 1.0 L= 8 5/16, R= 9 5/16 ± 1
0.0 (0) | 0.6 (5/8)	0.0 (0) | 0.6 (5/8)	-1.0 (-1) | 0.6 (5/8)	CROSS | TOL.	-0.6 (-5/8) | 0.6 (5/8)	-1.0 (-1) | 0.6 (5/8)	-1.0 (-1) | 0.6 (5/8)
L= 0.0, R= -0.4 ± 0.55 L= 0, R= -3/8 ± 9/16	L= -0.2, R= -0.6 ± 0.55 L= -3/16, R= -5/8 ± 9/16	L= 0.15, R= -0.25 ± 0.65 L= 1/8, R= -1/4 ± 5/8	CAMBER (Degrees)	L= -0.05, R= -0.35 ± 0.5 L= -1/16, R= -3/8 ± 1/2	L= 0.15, R= -0.25 ± 0.65 L= 1/8, R= -1/4 ± 5/8	L= -0.05, R= -0.35 ± 0.5 L= -1/16, R= -3/8 ± 1/2
0.4 (3/8) | 0.55 (9/16)	0.4 (3/8) | 0.55 (9/16)	0.4 (3/8) | 0.5 (1/2)	CROSS | TOL.	0.3 (5/16) | 0.5 (1/2)	0.4 (3/8) | 0.5 (1/2)	0.3 (5/16) | 0.5 (1/2)
0.3° ±0.2° 5/16° ± 3/16°	0.3° ±0.2° 5/16° ± 3/16°	0.2° ±0.1° 3/16° ± 3/32°	FRONT TOE (Degrees)	0.2° ±0.2° 3/16° ± 3/16°	0.2° ±0.1° 3/16° ± 3/32°	0.2° ±0.1° 3/16° ± 3/32°
0.15" ±0.1" 5/32" ± 3/32"	0.15" ±0.1" 5/32" ± 3/32"	0.1" ±0.05" 3/32" ± 1/16"	FRONT TOE (Inches)	0.1" ±0.1" 3/32" ± 3/32"	0.1" ±0.05" 3/32" ± 1/16"	0.1" ±0.05" 3/32" ± 1/16"
			SAI			
			TURNING ANGLE			
165	165	164 165 ▲	ILLUSTRATION #	164 165 ▲	164 165 ▲	164 165 ▲
			REAR			
-1.75 ±0.55 -1 3/4 ± 9/16	-1.75 ±0.55 -1 3/4 ± 9/16	-0.55 ±0.65 -9/16 ± 5/8	CAMBER (Degrees)	-0.75 ±0.5 -3/4 ± 1/2	-0.55 ±0.65 -9/16 ± 5/8	-0.75 ±0.5 -3/4 ± 1/2
0.2° ±0.15° 3/16° ± 5/32°	0.2° ±0.15° 3/16° ± 5/32°	0.2° ±0.15° [1] 3/16° ± 5/32° [1]	REAR TOE (Degrees)	0.2° ±0.15° [1] 3/16° ± 5/32° [1]	0.2° ±0.15° [1] 3/16° ± 5/32° [1]	0.2° ±0.15° [1] 3/16° ± 5/32° [1]
0.1" ±0.08" 3/32" ± 1/16"	0.1" ±0.08" 3/32" ± 1/16"	0.1" ±0.08" [1] 3/32" ± 1/16" [1]	REAR TOE (Inches)	0.1" ±0.08" [1] 3/32" ± 1/16" [1]	0.1" ±0.08" [1] 3/32" ± 1/16" [1]	0.1" ±0.08" [1] 3/32" ± 1/16" [1]
0.15 (5/32)	0.15 (5/32)	0.5 (1/2)	THRUST ANGLE	0.15 (5/32)	0.5 (1/2)	0.15 (5/32)
171	171	171	ILLUSTRATION #	171	171	171
		[1] Set Left to 0.17° (0.09") [1] Set Right to 0.03° (0.015")	NOTES	[1] Set Left to 0.17° (0.09") [1] Set Right to 0.03° (0.015")	[1] Set Left to 0.17° (0.09") [1] Set Right to 0.03° (0.015")	[1] Set Left to 0.17° (0.09") [1] Set Right to 0.03° (0.015")
12038	12039	12044		12045	12046	12047

Dodge

Charger RWD Police Package	Charger RWD exc/SRT8, Police	Charger RWD Police Package	MODEL INFORMATION	Charger RWD SRT8	Charger AWD	Charger AWD
2009-2010	2011-2017	2011-2017	YEARS	2011-2017	2007-2008	2009-2010
			INSTRUCTIONS			
110 ft.lbs.	130 ft.lbs.	130 ft.lbs.	LUGNUT TORQUE	111 ft.lbs.	110 ft.lbs.	130 ft.lbs.
			FRONT			
L= 7.7, R= 8.7 ± 1.0 L= 7 11/16, R= 8 11/16 ± 1	L= 8.0, R= 8.7 ± 1.0 L= 8, R= 8 11/16 ± 1	L= 7.7, R= 8.4 ± 1.0 L= 7 11/16, R= 8 3/8 ± 1	CASTER (Degrees)	L= 8.3, R= 9.0 ± 1.0 L= 8 5/16, R= 9 ± 1	L= 4.5, R= 5.1 ± 1.0 L= 4 1/2, R= 5 1/8 ± 1	L= 4.5, R= 5.1 ± 1.0 L= 4 1/2, R= 5 1/8 ± 1
-1.0 (-1) | 0.6 (5/8)	-0.7 (-11/16) | 0.6 (5/8)	-0.7 (-11/16) | 0.6 (5/8)	CROSS | TOL.	-0.7 (-11/16) | 0.6 (5/8)	-0.6 (-5/8) | 0.6 (5/8)	-0.6 (-5/8) | 0.6 (5/8)
L= 0.35, R= 0.05 ± 0.5 L= 3/8, R= 1/16 ± 1/2	L= -0.85, R= -1.25 ± 0.55 L= -7/8, R= -1 1/4 ± 9/16	L= -0.85, R= -1.25 ± 0.55 L= -7/8, R= -1 1/4 ± 9/16	CAMBER (Degrees)	L= -1.05, R= -1.35 ± 0.55 L= -1 1/16, R= -1 3/8 ± 9/16	L= 0.15, R= -0.25 ± 0.65 L= 1/8, R= -1/4 ± 5/8	L= 0.15, R= -0.25 ± 0.65 L= 1/8, R= -1/4 ± 5/8
0.3 (5/16) | 0.5 (1/2)	0.4 (3/8) | 0.5 (1/2)	0.4 (3/8) | 0.5 (1/2)	CROSS | TOL.	0.3 (5/16) | 0.5 (1/2)	0.4 (3/8) | 0.5 (1/2)	0.4 (3/8) | 0.5 (1/2)
0.2° ±0.1° 3/16° ± 3/32°	0.2° ±0.2° 3/16° ± 3/16°	0.2° ±0.2° 3/16° ± 3/16°	FRONT TOE (Degrees)	0.2° ±0.2° 3/16° ± 3/16°	0.1° ±0.05° 3/32° ± 1/16°	0.2° ±0.1° 3/16° ± 3/32°
0.1" ±0.05" 3/32" ± 1/16"	0.1" ±0.1" 3/32" ± 3/32"	0.1" ±0.1" 3/32" ± 3/32"	FRONT TOE (Inches)	0.1" ±0.1" 3/32" ± 3/32"	0.05" ±0.03" 1/16" ± 1/32"	0.1" ±0.05" 3/32" ± 1/16"
			SAI			
			TURNING ANGLE			
164 165 △	166 △	166 △	ILLUSTRATION #	166 △	165	165
			REAR			
-0.25 ±0.65 -1/4 ± 5/8	-1.75 ±0.55 -1 3/4 ± 9/16	-1.65 ±0.55 -1 5/8 ± 9/16	CAMBER (Degrees)	-0.75 ±0.55 -3/4 ± 9/16	-0.55 ±0.65 -9/16 ± 5/8	-0.55 ±0.65 -9/16 ± 5/8
0.2° ±0.15° [1] 3/16° ± 5/32° [1]	0.2° ±0.15° 3/16° ± 5/32°	0.2° ±0.15° 3/16° ± 5/32°	REAR TOE (Degrees)	0.2° ±0.15° 3/16° ± 5/32°	0.2° ±0.15° [1] 3/16° ± 5/32° [1]	0.2° ±0.15° [1] 3/16° ± 5/32° [1]
0.1" ±0.08" [1] 3/32" ± 1/16" [1]	0.1" ±0.08" 3/32" ± 1/16"	0.1" ±0.08" 3/32" ± 1/16"	REAR TOE (Inches)	0.1" ±0.08" 3/32" ± 1/16"	0.1" ±0.08" [1] 3/32" ± 1/16" [1]	0.1" ±0.08" [1] 3/32" ± 1/16" [1]
0.15 (5/32)	0.15 (5/32)	0.15 (5/32)	THRUST ANGLE	0.15 (5/32)	0.15 (5/32)	0.15 (5/32)
171	171	171	ILLUSTRATION #	171	171	171
[1] Set Left to 0.17° (0.09") [1] Set Right to 0.03° (0.015")			NOTES		[1] Set Left to 0.17° (0.09") [1] Set Right to 0.03° (0.015")	[1] Set Left to 0.17° (0.09") [1] Set Right to 0.03° (0.015")
12048	12049	12050		12051	12052	12053

Dodge

Charger AWD incl. Police Equip.	Charger AWD Police w/o Equip.	Colt exc/Wagon	MODEL INFORMATION	Colt Wagon 4X2	Colt Wagon 4X4	Colt (1995 Canada)
2011-2017	2015-2017	1990-92	YEARS	1990	1990	1993-95
			INSTRUCTIONS			
130 ft.lbs.	130 ft.lbs.		LUGNUT TORQUE		65-80 ft.lbs.	65-80 ft.lbs.
			FRONT			
5.4 ±1.0 5 3/8 ± 1	5.35 ±1.0 5 3/8 ± 1	2.33 ±0.5 2 5/16 ± 1/2	CASTER (Degrees)	1.33 ±0.5 [1] 1 5/16 ± 1/2 [1]	1.0 ±0.5 1 ± 1/2	2.25 ±0.5 2 1/4 ± 1/2
0.0 (0) | 0.6 (5/8)	0.0 (0) | 0.6 (5/8)	0.0 (0) | n/s	CROSS | TOL.	0.0 (0) | n/s	0.0 (0) | n/s	0.0 (0) | n/s
L= 0.0, R= -0.4 ± 0.65 L= 0, R= -3/8 ± 5/8	L= 0.1, R= -0.3 ± 0.65 L= 1/8, R= -5/16 ± 5/8	0.0 ±0.5 0 ± 1/2	CAMBER (Degrees)	0.0 ±0.5 0 ± 1/2	0.0 ±0.5 0 ± 1/2	0.0 ±0.5 0 ± 1/2
0.4 (3/8) | 0.5 (1/2)	0.4 (3/8) | 0.5 (1/2)	0.0 (0) | n/s	CROSS | TOL.	0.0 (0) | n/s	0.0 (0) | n/s	0.0 (0) | n/s
0.1° ±0.2° 3/32° ± 3/16°	0.1° ±0.2° 3/32° ± 3/16°	0.0° ±0.24° 0° ± 1/4°	FRONT TOE (Degrees)	0.0° ±0.24° 0° ± 1/4°	0.0° ±0.24° 0° ± 1/4°	0.0° ±0.24° 0° ± 1/4°
0.05" ±0.1" 1/16" ± 3/32"	0.05" ±0.1" 1/16" ± 3/32"	0.0" ±0.12" 0" ± 1/8"	FRONT TOE (Inches)	0.0" ±0.12" 0" ± 1/8"	0.0" ±0.12" 0" ± 1/8"	0.0" ±0.12" 0" ± 1/8"
			SAI			
		37.0 / 32.5 ±1.5 [1]	TURNING ANGLE	37.0 / 32.0 ±1.0	37.0 / 32.0 ±1.0	37.3 / 31.0 ±1.0
166	166		ILLUSTRATION #			
			REAR			
-0.75 ±0.65 -3/4 ± 5/8	-0.65 ±0.65 -5/8 ± 5/8	-0.67 ±0.58 -11/16 ± 9/16	CAMBER (Degrees)	0.0 ±0.58 0 ± 9/16		-0.67 ±0.5 -11/16 ± 1/2
0.2° ±0.15° 3/16° ± 5/32°	0.2° ±0.15° 3/16° ± 5/32°	0.0° ±0.36° 0° ± 3/8°	REAR TOE (Degrees)	0.0° ±0.36° 0° ± 3/8°		0.24° ±0.16° 1/4° ± 5/32°
0.1" ±0.08" 3/32" ± 1/16"	0.1" ±0.08" 3/32" ± 1/16"	0.0" ±0.18" 0" ± 3/16"	REAR TOE (Inches)	0.0" ±0.18" 0" ± 3/16"		0.12" ±0.08" 1/8" ± 3/32"
0.15 (5/32)	0.15 (5/32)		THRUST ANGLE			
171	171		ILLUSTRATION #			40 05
		[1] w/MS 41°/35°	NOTES	[1] w/MS 0.67° (11/16°)		
12054	12055	12068		12069	12070	12071

ALIGNMENT SPECIFICATIONS

Dodge

Colt Vista 4X2	Colt Vista 4X4	Colt 200 (Canada)	MODEL INFORMATION	Colt Wagon FWD (Canada)	Colt Wagon AWD (Canada)	Dart exc/Limited, GT
1990-91	1990-91	1990-92	YEARS	1992-96	1992-96	2013-2016
			INSTRUCTIONS			
50-57 ft.lbs.[1]	50-57 ft.lbs.[1]		LUGNUT TORQUE	65-80 ft.lbs.	65-80 ft.lbs.	83 ft.lbs.
			FRONT			
0.8 ±0.5 13/16 ± ½	0.83 ±0.5 13/16 ± ½	2.33 ±0.5 2 5/16 ± ½	CASTER (Degrees)	2.17 ±0.67 2 3/16 ± 11/16	2.08 ±0.67 2 1/16 ± 11/16	3.9 ±0.55 3 ⅞ ± 9/16
0.0 (0) | n/s	0.0 (0) | n/s	0.0 (0) | n/s	CROSS | TOL.	0.0 (0) | n/s	0.0 (0) | n/s	0.0 (0) | 0.55 (9/16)
0.42 ±0.5 7/16 ± ½	0.83 ±0.5 13/16 ± ½	0.0 ±0.5 0 ± ½	CAMBER (Degrees)	0.33 ±0.5 5/16 ± ½	0.67 ±0.5 11/16 ± ½	-0.66 ±0.55 -11/16 ± 9/16
0.0 (0) | n/s	0.0 (0) | n/s	0.0 (0) | n/s	CROSS | TOL.	0.0 (0) | n/s	0.0 (0) | n/s	0.0 (0) | 0.55 (9/16)
0.0° ±0.24° 0° ± ¼°	0.0° ±0.24° 0° ± ¼°	0.0° ±0.24° 0° ± ¼°	FRONT TOE (Degrees)	0.0° ±0.24° 0° ± ¼°	0.0° ±0.24° 0° ± ¼°	0.1° ±0.15° 3/32° ± 5/32°
0.0" ±0.12" 0" ± ⅛"	0.0" ±0.12" 0" ± ⅛"	0.0" ±0.12" 0" ± ⅛"	FRONT TOE (Inches)	0.0" ±0.12" 0" ± ⅛"	0.0" ±0.12" 0" ± ⅛"	0.05" ±0.08" 1/16" ± 1/16"
			SAI			
37.8 / 30.7 ±1.5	37.8 / 30.7 ±1.5	37.0 / 32.5 ±1.5[1]	TURNING ANGLE	36.5 / 30.5 ±2.0	36.5 / 30.5 ±2.0	
03	03		ILLUSTRATION #			
			REAR			
-0.58 ±0.58 -9/16 ± 9/16	0.0 ±0.5 0 ± ½	-0.67 ±0.58 -11/16 ± 9/16	CAMBER (Degrees)	-0.5 ±0.5 -½ ± ½	-0.5 ±0.5 -½ ± ½	-1.0 ±0.45 -1 ± 7/16
0.0° ±0.24° 0° ± ¼°	0.0° ±0.24° 0° ± ¼°	0.0° ±0.36° 0° ± ⅜°	REAR TOE (Degrees)	0.16° +0.24° -0.16° 5/32° +¼° -5/32°	0.16° +0.24° -0.16° 5/32° +¼° -5/32°	0.36° ±0.2° ⅜° ± 3/16°
0.0" ±0.12" 0" ± ⅛"	0.0" ±0.12" 0" ± ⅛"	0.0" ±0.18" 0" ± 3/16"	REAR TOE (Inches)	0.08" +0.12" -0.08" 3/32" +⅛" -3/32"	0.08" +0.12" -0.08" 3/32" +⅛" -3/32"	0.18" ±0.1" 3/16" ± 3/32"
			THRUST ANGLE			0.15 (5/32)
	77		ILLUSTRATION #	27	27	06
[1] Alum. Whls. 65-80 ft.lbs.	[1] Alum. Whls. 65-80 ft.lbs.	[1] w/MS 41°/35°	NOTES			
12072	12073	12074		12075	12076	12085

Dodge

Dart Limited	Dart GT	Daytona	MODEL INFORMATION	Daytona	Dynasty	Dynasty
2013-2016	2013-2016	1990	YEARS	1991-93	1990	1991-93
			INSTRUCTIONS			
83 ft.lbs.	83 ft.lbs.	95 ft.lbs.	LUGNUT TORQUE	95 ft.lbs.	95 ft.lbs.	95 ft.lbs.
			FRONT			
3.8 ±0.55 3 13/16 ± 9/16	3.9 ±0.55 3 ⅞ ± 9/16	1.2 ±0.75 1 3/16 ± ¾	CASTER (Degrees)	2.8 ±0.75 2 13/16 ± ¾	1.2 ±0.75 1 3/16 ± ¾	2.8 ±0.75 2 13/16 ± ¾
0.0 (0) | 0.55 (9/16)	0.0 (0) | 0.55 (9/16)	0.0 (0) | 1.5 (1½)	CROSS | TOL.	0.0 (0) | 1.5 (1½)	0.0 (0) | 1.5 (1½)	0.0 (0) | 1.5 (1½)
-0.54 ±0.55 -9/16 ± 9/16	-0.66 ±0.55 -11/16 ± 9/16	0.3 ±0.5 5/16 ± ½	CAMBER (Degrees)	0.3 ±0.5 5/16 ± ½	0.3 ±0.5 5/16 ± ½	0.3 ±0.5 5/16 ± ½
0.0 (0) | 0.55 (9/16)	0.0 (0) | 0.55 (9/16)	0.0 (0) | n/s	CROSS | TOL.	0.0 (0) | n/s	0.0 (0) | n/s	0.0 (0) | n/s
0.1° ±0.15° 3/32° ± 5/32°	0.1° ±0.15° 3/32° ± 5/32°	0.1° ±0.1° 3/32° ± 3/32°	FRONT TOE (Degrees)	0.1° ±0.1° 3/32° ± 3/32°	0.1° ±0.1° 3/32° ± 3/32°	0.1° ±0.1° 3/32° ± 3/32°
0.05" ±0.08" 1/16" ± 1/16"	0.05" ±0.08" 1/16" ± 1/16"	0.05" ±0.05" 1/16" ± 1/16"	FRONT TOE (Inches)	0.05" ±0.05" 1/16" ± 1/16"	0.05" ±0.05" 1/16" ± 1/16"	0.05" ±0.05" 1/16" ± 1/16"
			SAI			
			TURNING ANGLE			
		22	ILLUSTRATION #	22	22	22
			REAR			
-1.0 ±0.45 -1 ± 7/16	-1.6 ±0.45 -1 ⅝ ± 7/16	-0.5 ±0.5 -½ ± ½	CAMBER (Degrees)	-0.5 ±0.5 -½ ± ½	-0.5 ±0.5 -½ ± ½	-0.5 ±0.5 -½ ± ½
0.36° ±0.2° ⅜° ± 3/16°	0.36° ±0.2° ⅜° ± 3/16°	0.0° ±0.25° 0° ± ¼°	REAR TOE (Degrees)	0.0° ±0.25° 0° ± ¼°	0.0° ±0.25° 0° ± ¼°	0.0° ±0.25° 0° ± ¼°
0.18" ±0.1" 3/16" ± 3/32"	0.18" ±0.1" 3/16" ± 3/32"	0.0" ±0.13" 0" ± ⅛"	REAR TOE (Inches)	0.0" ±0.13" 0" ± ⅛"	0.0" ±0.13" 0" ± ⅛"	0.0" ±0.13" 0" ± ⅛"
0.15 (5/32)	0.15 (5/32)	0.4 (13/32)	THRUST ANGLE	0.4 (13/32)	0.4 (13/32)	0.4 (13/32)
06	06	76	ILLUSTRATION #	76	76	76
			NOTES			
12086	12087	12088		12089	12093	12094

ALIGNMENT SPECIFICATIONS

Dodge

<table>
<tr><td>2000GTX FWD
(Canada)</td><td>2000GTX AWD
(Canada)</td><td>Intrepid</td><td>MODEL
INFORMATION</td><td>Intrepid</td><td>Intrepid</td><td>Magnum
RWD
exc/SRT8</td></tr>
<tr><td>1990-93</td><td>1990-93</td><td>1993-97</td><td>YEARS</td><td>1998</td><td>1999-2004</td><td>2005-2008</td></tr>
<tr><td></td><td></td><td></td><td>INSTRUCTIONS</td><td></td><td></td><td></td></tr>
<tr><td>65-80 ft.lbs.</td><td>65-80 ft.lbs.</td><td>95 ft.lbs.</td><td>LUGNUT TORQUE</td><td>100 ft.lbs.</td><td>100 ft.lbs.</td><td>110 ft.lbs.</td></tr>
<tr><td></td><td></td><td></td><td>FRONT</td><td></td><td></td><td></td></tr>
<tr><td>2.0 ±0.5
2 ± ½</td><td>1.93 ±0.5
1 15/16 ± ½</td><td>3.0 ±1.0
3 ±1</td><td>CASTER
(Degrees)</td><td>3.0 ±1.0
3 ±1</td><td>3.0 ±1.0
3 ±1</td><td>L= 9.4, R=10.4 ± 1.0
L= 9⅜, R= 10⅜ ± 1</td></tr>
<tr><td>0.0 (0) | n/s</td><td>0.0 (0) | n/s</td><td>0.0 (0) | 1.0 (1)</td><td>CROSS | TOL.</td><td>0.0 (0) | 1.0 (1)</td><td>0.0 (0) | 1.0 (1)</td><td>-1.0 (-1) | 0.6 (⅝)</td></tr>
<tr><td>0.37 ±0.5
⅜ ± ½</td><td>0.52 ±0.5
½ ± ½</td><td>0.0 ±0.6
0 ± ⅝</td><td>CAMBER
(Degrees)</td><td>0.0 ±0.6
0 ± ⅝</td><td>0.0 ±0.6
0 ± ⅝</td><td>L= 0.15, R= -0.25 ± 0.65
L= ⅛, R= -¼ ± ⅝</td></tr>
<tr><td>0.0 (0) | n/s</td><td>0.0 (0) | n/s</td><td>0.0 (0) | 0.7 (11/16)</td><td>CROSS | TOL.</td><td>0.0 (0) | 0.7 (11/16)</td><td>0.0 (0) | 0.7 (11/16)</td><td>0.4 (⅜) | 0.5 (½)</td></tr>
<tr><td>0.0° ±0.24°
0° ± ¼°</td><td>0.0° ±0.24°
0° ± ¼°</td><td>0.2° ±0.2°
3/16° ± 3/16°</td><td>FRONT TOE
(Degrees)</td><td>0.1° ±0.1°
3/32° ± 3/32°</td><td>0.0° ±0.1° [1]
0° ± 3/32° [1]</td><td>0.2° ±0.1°
3/16° ± 3/32°</td></tr>
<tr><td>0.0" ±0.12"
0" ± ⅛"</td><td>0.0" ±0.12"
0" ± ⅛"</td><td>0.1" ±0.1"
3/32" ± 3/32"</td><td>FRONT TOE
(Inches)</td><td>0.05" ±0.05"
1/16" ± 1/16"</td><td>0.0" ±0.05" [1]
0" ± 1/16" [1]</td><td>0.1" ±0.05"
3/32" ± 1/16"</td></tr>
<tr><td></td><td></td><td></td><td>SAI</td><td></td><td></td><td></td></tr>
<tr><td>37.5 / 30.8 ±2.0</td><td>37.5 / 30.8 ±2.0 [1]</td><td></td><td>TURNING ANGLE</td><td></td><td></td><td></td></tr>
<tr><td></td><td></td><td></td><td>ILLUSTRATION #</td><td></td><td></td><td>164 165 ▲</td></tr>
<tr><td></td><td></td><td></td><td>REAR</td><td></td><td></td><td></td></tr>
<tr><td>-0.75 ±0.5
-¾ ± ½</td><td>-1.0 ±0.5
-1 ± ½</td><td>-0.1 ±0.5
-⅛ ± ½</td><td>CAMBER
(Degrees)</td><td>-0.1 ±0.5
-⅛ ± ½</td><td>-0.2 ±0.5
-3/16 ± ½</td><td>-0.55 ±0.65
-9/16 ± ⅝</td></tr>
<tr><td>0.0° ±0.24°
0° ± ¼°</td><td>0.24° ±0.24°
¼° ± ¼°</td><td>0.1° ±0.3°
3/32° ± 5/16°</td><td>REAR TOE
(Degrees)</td><td>0.1° ±0.3°
3/32° ± 5/16°</td><td>0.1° ±0.3°
3/32° ± 5/16°</td><td>0.2° ±0.15° [1]
3/16° ± 5/32° [1]</td></tr>
<tr><td>0.0" ±0.12"
0" ± ⅛"</td><td>0.12" ±0.12"
⅛" ± ⅛"</td><td>0.05" ±0.15"
1/16" ± 5/32"</td><td>REAR TOE
(Inches)</td><td>0.05" ±0.15"
1/16" ± 5/32"</td><td>0.05" ±0.15"
1/16" ± 5/32"</td><td>0.1" ±0.08" [1]
3/32" ± 1/16" [1]</td></tr>
<tr><td></td><td></td><td>0.15 (5/32)</td><td>THRUST ANGLE</td><td>0.15 (5/32)</td><td>0.15 (5/32)</td><td>0.15 (5/32)</td></tr>
<tr><td></td><td>44</td><td>10</td><td>ILLUSTRATION #</td><td>10</td><td>10</td><td>171</td></tr>
<tr><td></td><td>[1] w/AT 35°/29.3°</td><td></td><td>NOTES</td><td></td><td>[1] 1999 = 0.1° (0.05")</td><td>[1] Set Left to 0.17° (0.09")
[1] Set Right to 0.03° (0.015")</td></tr>
<tr><td>12095</td><td>12096</td><td>12097</td><td></td><td>12098</td><td>12099</td><td>12103</td></tr>
</table>

Dodge

<table>
<tr><td>Magnum
RWD
SRT8</td><td>Magnum
AWD</td><td>Monaco</td><td>MODEL
INFORMATION</td><td>Neon</td><td>Neon</td><td>Neon w/ACR [1 2]
Competition Pkg.</td></tr>
<tr><td>2005-2008</td><td>2005-2008</td><td>1990-92</td><td>YEARS</td><td>1995-99</td><td>2000-2005</td><td>2001-2004</td></tr>
<tr><td></td><td></td><td></td><td>INSTRUCTIONS</td><td></td><td></td><td></td></tr>
<tr><td>110 ft.lbs.</td><td>110 ft.lbs.</td><td>85 ft.lbs.[1]</td><td>LUGNUT TORQUE</td><td>85-115 ft.lbs.</td><td>100 ft.lbs.</td><td></td></tr>
<tr><td></td><td></td><td></td><td>FRONT</td><td></td><td></td><td></td></tr>
<tr><td>L= 9.7, R=10.3 ± 1.0
L= 9 11/16, R= 10 5/16 ± 1</td><td>L= 4.5, R= 5.1 ± 1.0
L= 4½, R= 5⅛ ± 1</td><td>2.0 ±0.5
2 ± ½</td><td>CASTER
(Degrees)</td><td>2.8 ±1.0
2 13/16 ± 1</td><td>2.6 ±1.0
2⅝ ± 1</td><td>2.6 ±1.0
2⅝ ± 1</td></tr>
<tr><td>-0.6 (-⅝) | 0.6 (⅝)</td><td>-0.6 (-⅝) | 0.6 (⅝)</td><td>0.0 (0) | n/s</td><td>CROSS | TOL.</td><td>0.0 (0) | 1.0 (1)</td><td>0.0 (0) | 1.0 (1)</td><td>0.0 (0) | 1.0 (1)</td></tr>
<tr><td>L= -0.05, R= -0.35 ± 0.5
L= -1/16, R= -⅜ ± ½</td><td>L= 0.15, R= -0.25 ± 0.65
L= ⅛, R= -¼ ± ⅝</td><td>0.33 ±0.25
5/16 ± ¼</td><td>CAMBER
(Degrees)</td><td>0.0 ±0.4
0 ± ⅜</td><td>0.0 ±0.4
0 ± ⅜</td><td>0.0 +0.4 -2.4
0 +⅜ -2⅜</td></tr>
<tr><td>0.3 (5/16) | 0.5 (½)</td><td>0.4 (⅜) | 0.5 (½)</td><td>0.0 (0) | n/s</td><td>CROSS | TOL.</td><td>0.0 (0) | n/s</td><td>0.0 (0) | 0.5 (½)</td><td>0.0 (0) | 0.5 (½)</td></tr>
<tr><td>0.2° ±0.2°
3/16° ± 3/16°</td><td>0.1° ±0.05°
3/32° ± 1/16°</td><td>0.0° ±0.3°
0° ± 5/16°</td><td>FRONT TOE
(Degrees)</td><td>0.1° ±0.2°
3/32° ± 3/16°</td><td>0.2° ±0.2° [1]
3/16° ± 3/16° [1]</td><td>0.2° +0.1° -0.9° [1]
3/16° +3/32° -29/32° [1]</td></tr>
<tr><td>0.1" ±0.1"
3/32" ± 3/32"</td><td>0.05" ±0.03"
1/16" ± 1/32"</td><td>0.0" ±0.15"
0" ± 5/32"</td><td>FRONT TOE
(Inches)</td><td>0.05" ±0.1"
1/16" ± 3/32"</td><td>0.1" ±0.1" [1]
3/32" ± 3/32" [1]</td><td>0.1" +0.05" -0.45" [1]
3/32" +1/16" -7/16" [1]</td></tr>
<tr><td></td><td></td><td></td><td>SAI</td><td>13.2 (13 3/16)</td><td></td><td></td></tr>
<tr><td></td><td></td><td></td><td>TURNING ANGLE</td><td></td><td></td><td></td></tr>
<tr><td>164 165 ▲</td><td>166</td><td></td><td>ILLUSTRATION #</td><td>26</td><td>26</td><td>26</td></tr>
<tr><td></td><td></td><td></td><td>REAR</td><td></td><td></td><td></td></tr>
<tr><td>-0.75 ±0.5
-¾ ± ½</td><td>-0.55 ±0.65
-9/16 ± ⅝</td><td>-0.5 ±0.5
-½ ± ½</td><td>CAMBER
(Degrees)</td><td>-0.25 ±0.5
-¼ ± ½</td><td>-0.25 ±0.4
-¼ ± ⅜</td><td>-0.25 +0.45 -0.75
-¼ +7/16 -¾</td></tr>
<tr><td>0.2° ±0.15° [1]
3/16° ± 5/32° [1]</td><td>0.2° ±0.15° [1]
3/16° ± 5/32° [1]</td><td>0.0° ±0.25°
0° ± ¼°</td><td>REAR TOE
(Degrees)</td><td>0.1° ±0.2°
3/32° ± 3/16°</td><td>0.3° ±0.2°
5/16° ± 3/16°</td><td>0.3° +0.0° -1.0°
5/16° +0° -1°</td></tr>
<tr><td>0.1" ±0.08" [1]
3/32" ± 1/16" [1]</td><td>0.1" ±0.08" [1]
3/32" ± 1/16" [1]</td><td>0.0" ±0.13"
0" ± ⅛"</td><td>REAR TOE
(Inches)</td><td>0.05" ±0.1"
1/16" ± 3/32"</td><td>0.15" ±0.1"
5/32" ± 3/32"</td><td>0.15" +0.0" -0.5"
5/32" +0" -½"</td></tr>
<tr><td>0.15 (5/32)</td><td>0.15 (5/32)</td><td></td><td>THRUST ANGLE</td><td>0.1 (3/32)</td><td>0.1 (3/32)</td><td>0.1 (3/32)</td></tr>
<tr><td>171</td><td>171</td><td></td><td>ILLUSTRATION #</td><td>26 06</td><td>26 06</td><td>26 06</td></tr>
<tr><td>[1] Set Left to 0.17° (0.09")
[1] Set Right to 0.03° (0.05")</td><td>[1] Set Left to 0.17° (0.09")
[1] Set Right to 0.03° (0.015")</td><td>[1] Alum. Whls. 90 ft.lbs.</td><td>NOTES</td><td></td><td>[1] 2000 = 0.1° (0.05")</td><td>[1] Competition specs and tolerances.
[2] See std. Neon for street use.</td></tr>
<tr><td>12104</td><td>12105</td><td>12112</td><td></td><td>12113</td><td>12115</td><td>12116</td></tr>
</table>

Dodge

Model Information	Omni	Shadow	Shadow	Spirit	SRT-4	Stealth FWD
Years	1990	1990	1991-94	1990-95	2003-2005	1991-96
Instructions						
LugNut Torque	95 ft.lbs.	95 ft.lbs.	95 ft.lbs.	95 ft.lbs.	100 ft.lbs.	87-101 ft.lbs.
FRONT						
Caster (Degrees)	1.4±0.75 1⅜±¾	1.2±0.75 1 3/16±¾	2.8±0.75 2 13/16±¾	2.8±1.0 [1] 2 13/16±1 [1]	2.6±1.0 2⅝±1	3.92±0.5 3 15/16±½
Cross | Tol.	0.0 (0) | 1.5 (1½)	0.0 (0) | 1.5 (1½)	0.0 (0) | 1.5 (1½)	0.0 (0) | 1.5 (1½)	0.0 (0) | 1.0 (1)	0.0 (0) | n/s
Camber (Degrees)	0.3±0.5 5/16±½	0.3±0.5 5/16±½	0.3±0.5 5/16±½	0.3±0.5 5/16±½	0.0±0.4 0±⅜	0.0±0.5 0±½
Cross | Tol.	0.0 (0) | n/s	0.0 (0) | n/s	0.0 (0) | n/s	0.0 (0) | n/s	0.0 (0) | 0.5 (½)	0.0 (0) | 0.5 (½)
Front Toe (Degrees)	0.1°±0.1° 3/32°±3/32°	0.1°±0.1° 3/32°±3/32°	0.1°±0.1° 3/32°±3/32°	0.1°±0.1° 3/32°±3/32°	0.2°±0.2° 3/16°±3/16°	0.0°±0.24° 0°±¼°
Front Toe (Inches)	0.05"±0.05" 1/16"±1/16"	0.05"±0.05" 1/16"±1/16"	0.05"±0.05" 1/16"±1/16"	0.05"±0.05" 1/16"±1/16"	0.1"±0.1" 3/32"±3/32"	0.0"±0.12" 0"±⅛"
SAI						
Turning Angle						33.8 / 28.4
Illustration #	22	22	22	22	26	22
REAR						
Camber (Degrees)	-0.75±0.5 -¾±½	-0.5±0.5 -½±½	-0.5±0.5 -½±½	-0.5±0.5 -½±½	-1.0±0.4 [1] -1±⅜ [1]	0.0±0.5 0±½
Rear Toe (Degrees)	0.2°±0.25° 3/16°±¼°	0.0°±0.25° 0°±¼°	0.0°±0.25° 0°±¼°	0.0°±0.25° 0°±¼°	0.3°±0.2° 5/16°±3/16°	0.04°±0.2° 1/32°±3/16°
Rear Toe (Inches)	0.1"±0.13" 3/32"±⅛"	0.0"±0.13" 0"±⅛"	0.0"±0.13" 0"±⅛"	0.0"±0.13" 0"±⅛"	0.15"±0.1" 5/32"±3/32"	0.02"±0.1" 1/64"±3/32"
Thrust Angle	0.4 (13/32)	0.4 (13/32)	0.4 (13/32)	0.4 (13/32)	0.1 (3/32)	
Illustration #	76	76	76	76	26 06	40 05
Notes				[1] 1990 = 1.2° (1 3/16°)	[1] 2003-04 = -0.25° (-¼°)	
	12118	12122	12123	12125	12126	12129

Dodge

Model Information	Stealth AWD	Stratus	Stratus	Stratus Coupe	Stratus Coupe	Stratus Sedan exc/2006 R/T
Years	1991-96	1995-96	1997-2000	2001	2002-2005	2001-2006
Instructions						
LugNut Torque	87-101 ft.lbs.	95-100 ft.lbs.	95-100 ft.lbs.	73 ft.lbs.	73 ft.lbs.	100 ft.lbs.
FRONT						
Caster (Degrees)	3.92±0.5 3 15/16±½	3.3±1.0 3 5/16±1	3.3±1.0 3 5/16±1	3.0±0.5 3±½	3.0±0.5 3±½	3.3±1.0 3 5/16±1
Cross | Tol.	0.0 (0) | n/s	0.0 (0) | 1.0 (1)	0.0 (0) | 1.0 (1)	0.0 (0) | 0.5 (½)	0.0 (0) | 0.5 (½)	0.0 (0) | 1.0 (1)
Camber (Degrees)	0.0±0.5 0±½	0.3±0.6 5/16±⅝	0.0±0.6 0±⅝	0.0±0.5 0±½	0.0±0.5 0±½	-0.3±0.6 -5/16±⅝
Cross | Tol.	0.0 (0) | 0.5 (½)	0.0 (0) | 0.7 (11/16)	0.0 (0) | 0.7 (11/16)	0.0 (0) | 0.5 (½)	0.0 (0) | 0.5 (½)	0.0 (0) | 0.7 (11/16)
Front Toe (Degrees)	0.0°±0.24° 0°±¼°	0.1°±0.2° 3/32°±3/16°	0.1°±0.2° 3/32°±3/16°	0.0°±0.24° 0°±¼°	0.0°±0.24° 0°±¼°	0.24°±0.2° ¼°±3/16°
Front Toe (Inches)	0.0"±0.12" 0"±⅛"	0.05"±0.1" 1/16"±3/32"	0.05"±0.1" 1/16"±3/32"	0.0"±0.12" 0"±⅛"	0.0"±0.12" 0"±⅛"	0.12"±0.1" ⅛"±3/32"
SAI		7.0 (7)	7.0 (7)			
Turning Angle	33.8 / 28.4			36.6 / 30.4 ±2.0 [1]	36.2 / 30.4 ±2.0 [1]	
Illustration #	22	113	113	26	26	113
REAR						
Camber (Degrees)	-0.17±0.5 -3/16±½	-0.2±0.4 -3/16±⅜	-0.2±0.4 -3/16±⅜	-1.33±0.5 -1 5/16±½	-1.33±0.5 -1 5/16±½	-0.5±0.5 -½±½
Rear Toe (Degrees)	0.04°±0.2° 1/32°±3/16°	0.1°±0.2° 3/32°±3/16°	0.1°±0.2° 3/32°±3/16°	0.24°±0.24° ¼°±¼°	0.24°±0.24° ¼°±¼°	0.1°±0.2° 3/32°±3/16°
Rear Toe (Inches)	0.02"±0.1" 1/64"±3/32"	0.05"±0.1" 1/16"±3/32"	0.05"±0.1" 1/16"±3/32"	0.12"±0.12" ⅛"±⅛"	0.12"±0.12" ⅛"±⅛"	0.05"±0.1" 1/16"±3/32"
Thrust Angle		0.15 (5/32)	0.15 (5/32)			0.15 (5/32)
Illustration #	44	01	01	06	06	01
Notes				[1] 3.0L 33.1°/28.5°	[1] 3.0L 31°/27°	
	12130	12131	12132	12133	12134	12135

Dodge — Dodge (Ram) Truck & SUV

Stratus Sedan	SX 2.0	Notes	MODEL	Ram Pickup 1500	Ram Pickup 1500	Ram Pickup 1500
R/T	(Canada)		INFORMATION	(4X2)	(4X2)	(4X2)
						exc/SRT-10
2006	2003-2005		YEARS	1994-97	1998-2001	2002-2005
			INSTRUCTIONS			
100 ft.lbs.	100 ft.lbs.		LUGNUT TORQUE	95 ft.lbs.	95 ft.lbs.	135 ft.lbs.
			FRONT			
3.5±1.0 3½±1	2.6±1.0 2⅝±1		CASTER (Degrees)	3.75±1.0 3¾±1	3.65±1.0 [1] 3⅝±1 [1]	4.2±0.75 [1,2] 4 3⁄16 ± ¾ [1,2]
0.0 (0) | 1.0 (1)	0.0 (0) | 1.0 (1)		CROSS | TOL.	0.0 (0) | 0.5 (½)	0.0 (0) | 0.5 (½)	0.0 (0) | 0.4 (⅜)
-0.5±0.6 -½ ± ⅝	0.0±0.4 0 ± ⅜		CAMBER (Degrees)	0.5±0.5 ½ ± ½	0.5±0.5 ½ ± ½	0.0±0.5 0 ± ½
0.0 (0) | 0.7 (11⁄16)	0.0 (0) | 0.5 (½)		CROSS | TOL.	0.0 (0) | 0.5 (½)	0.0 (0) | 0.5 (½)	0.0 (0) | 0.5 (½)
0.24° ±0.2° ¼° ± 3⁄16°	0.2° ±0.2° 3⁄16° ± 3⁄16°		FRONT TOE (Degrees)	0.1° ±0.1° 3⁄32° ± 3⁄32°	0.1° ±0.1° 3⁄32° ± 3⁄32°	0.1° ±0.1° 3⁄32° ± 3⁄32°
0.12" ±0.1" ⅛" ± 3⁄32"	0.1" ±0.1" 3⁄32" ± 3⁄32"		FRONT TOE (Inches)	0.05" ±0.05" 1⁄16" ± 1⁄16"	0.05" ±0.05" 1⁄16" ± 1⁄16"	0.05" ±0.05" 1⁄16" ± 1⁄16"
			SAI			
			TURNING ANGLE			
113	26		ILLUSTRATION #	42	42	139
			REAR			
-0.5±0.5 -½ ± ½	-0.25±0.4 -¼ ± ⅜		CAMBER (Degrees)			
0.1° ±0.2° 3⁄32° ± 3⁄16°	0.3° ±0.2° 5⁄16° ± 3⁄16°		REAR TOE (Degrees)			
0.05" ±0.1" 1⁄16" ± 3⁄32"	0.15" ±0.1" 5⁄32" ± 3⁄32"		REAR TOE (Inches)			
0.15 (5⁄32)	0.1 (3⁄32)		THRUST ANGLE	0.4 (13⁄32)	0.4 (13⁄32)	0.4 (13⁄32)
01	26 06		ILLUSTRATION #			
			NOTES		[1] Club Cab/Quad Cab = 4.0° (4°)	[1] 120.5" WB = 4.0° (4°) [2] 160.5" WB = 4.4° (4⅜°)
12136	12137	12141		13001	13002	13003

Dodge (Ram) Truck & SUV

Ram Pickup 1500	Ram Pickup 1500	Ram Pickup 1500	MODEL	Ram Pickup 1500	Ram Pickup 1500	Ram Pickup 1500
(4X2)	(4X2)	(4X2)	INFORMATION	(4X2)	(4X2)	(4X4)
SRT-10	exc/SRT-10, Megacab	SRT-10		exc/Megacab	Megacab	
2004-2005	2006	2006	YEARS	2007-2008	2006-2008	1994-97
			INSTRUCTIONS			
135 ft.lbs.	130 ft.lbs.	130 ft.lbs.	LUGNUT TORQUE	130 ft.lbs.	130 ft.lbs.	95 ft.lbs.
			FRONT			
5.75±0.75 5¾ ± ¾	L= 3.35, R= 3.75 ± 0.75 L= 3⅜, R= 3¾ ± ¾	4.5±0.5 4½ ± ½	CASTER (Degrees)	L= 3.5, R= 3.9 ± 0.5 L= 3½, R= 3⅞ ± ½	4.0±0.75 4 ± ¾	3.4±1.0 [1] 3⅜±1 [1]
0.0 (0) | 0.4 (⅜)	-0.4 (-⅜) | 0.5 (½)	0.0 (0) | 0.3 (5⁄16)	CROSS | TOL.	-0.4 (-⅜) | 0.5 (½)	0.0 (0) | 0.5 (½)	0.0 (0) | 0.5 (½)
0.0±0.5 0 ± ½	0.1±0.5 ⅛ ± ½	$-0.15^{+0.15}_{-0.25}$ $-\frac{1}{8}^{+\frac{1}{8}}_{-\frac{1}{4}}$	CAMBER (Degrees)	L= 0.1, R= -0.1±0.5 L= ⅛, R= -⅛ ± ½	0.0±0.5 0 ± ½	0.25±0.5 ¼ ± ½
0.0 (0) | 0.5 (½)	0.0 (0) | 0.5 (½)	0.0 (0) | 0.3 (5⁄16)	CROSS | TOL.	0.2 (3⁄16) | 0.5 (½)	0.0 (0) | 0.3 (5⁄16)	0.0 (0) | 0.5 (½)
0.24° ±0.16° ¼° ± 5⁄32°	0.1° ±0.1° 3⁄32° ± 3⁄32°	0.24° ±0.16° ¼° ± 5⁄32°	FRONT TOE (Degrees)	0.24° ±0.1° [1] ¼° ± 3⁄32° [1]	0.2° ±0.1° 3⁄16° ± 3⁄32°	0.1° ±0.1° 3⁄32° ± 3⁄32°
0.12" ±0.08" ⅛" ± 3⁄32"	0.05" ±0.05" 1⁄16" ± 1⁄16"	0.12" ±0.08" ⅛" ± 3⁄32"	FRONT TOE (Inches)	0.12" ±0.05" [1] ⅛" ± 1⁄16" [1]	0.1" ±0.05" 3⁄32" ± 1⁄16"	0.05" ±0.05" 1⁄16" ± 1⁄16"
			SAI			
			TURNING ANGLE			
139	12	12	ILLUSTRATION #	12	12	78
			REAR			
			CAMBER (Degrees)			
			REAR TOE (Degrees)			
			REAR TOE (Inches)			
0.17 (5⁄32)	0.4 (13⁄32)	0.17 (5⁄32)	THRUST ANGLE	0.4 (13⁄32)	0.2 (3⁄16)	0.4 (13⁄32)
			ILLUSTRATION #			
			NOTES	[1] 2007 = 0.36° (0.18")		[1] Club Cab/Quad Cab = 3.48° (3½°)
13004	13005	13006		13007	13008	13009

ALIGNMENT SPECIFICATIONS

Dodge (Ram) Truck & SUV

Ram Pickup 1500	Ram Pickup 1500	Ram Pickup 1500	MODEL	Ram Pickup 1500	Ram Pickup 1500	Ram Pickup 1500
(4X4)	(4X4)	(4X4)	INFORMATION	(4X4)	(4X4)	(4X4)
				exc/Megacab	exc/Megacab	Megacab
1998-99	2000-2001	2002-2005	YEARS	2006	2007-2008	2006-2008
			INSTRUCTIONS			
95 ft.lbs.	95 ft.lbs.	135 ft.lbs.	LUGNUT TORQUE	130 ft.lbs.	130 ft.lbs.	130 ft.lbs.
			FRONT			
3.2 ± 1.0 [1] 3 3/16 ± 1 [1]	2.95 ± 1.0 [1] 2 15/16 ± 1 [1]	4.4 ± 0.75 [1 2] 4⅜ ± ¾ [1 2]	CASTER (Degrees)	L= 3.35, R= 3.75 ± 0.75 L= 3⅜, R= 3¾ ± ¾	L= 3.5, R= 3.9 ± 0.5 L= 3½, R= 3⅞ ± ½	4.5 ± 0.5 4½ ± ½
0.0 (0) | 0.5 (½)	0.0 (0) | 0.5 (½)	0.0 (0) | 0.4 (⅜)	CROSS | TOL.	-0.4 (-⅜) | 0.5 (½)	-0.4 (-⅜) | 0.5 (½)	0.0 (0) | 0.5 (½)
0.25 ± 0.5 ¼ ± ½	Camber is not specified by the manufacturer.	0.0 ± 0.5 0 ± ½	CAMBER (Degrees)	0.1 ± 0.5 ⅛ ± ½	L= 0.1, R= -0.1 ± 0.5 L= ⅛, R= -⅛ ± ½	0.25 ± 0.5 ¼ ± ½
0.0 (0) | 0.5 (½)	0.0 (0) | 0.5 (½)	0.0 (0) | 0.5 (½)	CROSS | TOL.	0.0 (0) | 0.5 (½)	0.2 (3/16) | 0.5 (½)	0.0 (0) | 0.5 (½)
0.1° ± 0.1° 3/32° ± 3/32°	0.1° ± 0.1° 3/32° ± 3/32°	0.1° ± 0.1° 3/32° ± 3/32°	FRONT TOE (Degrees)	0.1° ± 0.1° 3/32° ± 3/32°	0.24° ± 0.1° [1] ¼° ± 3/32° [1]	0.2° ± 0.1° 3/16° ± 3/32°
0.05" ± 0.05" 1/16" ± 1/16"	0.05" ± 0.05" 1/16" ± 1/16"	0.05" ± 0.05" 1/16" ± 1/16"	FRONT TOE (Inches)	0.05" ± 0.05" 1/16" ± 1/16"	0.12" ± 0.05" [1] ⅛" ± 1/16" [1]	0.1" ± 0.05" 3/32" ± 1/16"
			SAI			
			TURNING ANGLE			
78	78	139	ILLUSTRATION #	12	12	78
			REAR			
			CAMBER (Degrees)			
			REAR TOE (Degrees)			
			REAR TOE (Inches)			
0.4 (13/32)	0.4 (13/32)	0.4 (13/32)	THRUST ANGLE	0.4 (13/32)	0.4 (13/32)	0.2 (3/16)
			ILLUSTRATION #			
[1] Club Cab/Quad Cab = 3.48° (3½°)	[1] Club Cab/Quad Cab = 3.28° (3¼°)	[1] 120.5" WB = 4.2° (4 3/16°) [2] 160.5" WB = 4.6° (4⅝°)	NOTES		[1] 2007 = 0.36° (0.18")	
13010	13011	13012		13013	13014	13015

Dodge (Ram) Truck & SUV

Ram Pickup 1500	Ram Pickup 1500	Ram Pickup 1500	MODEL	Ram Pickup 1500	Ram Pickup 1500	Ram Pickup 1500
(4X2 & 4X4) exc/AIR Susp.	(4X2 & 4X4) exc/AIR Susp.	(4X2 & 4X4) exc/AIR Susp.	INFORMATION	(4X2 & 4X4) exc/AIR Susp.	(4X2 & 4X4) w/AIR Susp.	(4X2 & 4X4) w/AIR Susp.
w/17" tires (OE)	w/20" & 22" tires (OE)	exc/Rebel		Rebel	exc/Rebel	Rebel
2009-2010	2009-2010	2011-2017	YEARS	2011-2017	2013-2017	2013-2017
			INSTRUCTIONS			
130 ft.lbs.	130 ft.lbs.	130 ft.lbs.	LUGNUT TORQUE	130 ft.lbs.	130 ft.lbs.	130 ft.lbs.
			FRONT			
L= 3.5, R= 3.75 ± 0.5 L= 3½, R= 3¾ ± ½	L= 3.5, R= 3.9 ± 0.5 L= 3½, R= 3⅞ ± ½	L= 3.5, R= 3.75 ± 0.5 L= 3½, R= 3¾ ± ½	CASTER (Degrees)	L= 3.0, R= 3.25 ± 0.5 L= 3, R= 3¼ ± ½	L= 3.67, R= 3.92 ± 0.5 [1] L= 3 11/16, R= 3 15/16 ± ½ [1]	L= 3.0, R= 3.25 ± 0.5 [1] L= 3, R= 3¼ ± ½ [1]
-0.25 (-¼) | 0.5 (½)	-0.4 (-⅜) | 0.5 (½)	-0.25 (-¼) | 0.5 (½)	CROSS | TOL.	-0.25 (-¼) | 0.5 (½)	-0.25 (-¼) | 0.5 (½)	-0.25 (-¼) | 0.5 (½)
L= 0.1, R= -0.1 ± 0.5 L= ⅛, R= -⅛ ± ½	L= 0.1, R= -0.1 ± 0.5 L= ⅛, R= -⅛ ± ½	L= 0.1, R= -0.1 ± 0.5 L= ⅛, R= -⅛ ± ½	CAMBER (Degrees)	L= 0.1, R= -0.1 ± 0.5 L= ⅛, R= -⅛ ± ½	L=-0.24, R=-0.44 ± 0.5 [1] L= -¼, R= -7/16 ± ½ [1]	L= 0.1, R= -0.1 ± 0.5 [1] L= ⅛, R= -⅛ ± ½ [1]
0.2 (3/16) | 0.5 (½)	0.2 (3/16) | 0.5 (½)	0.2 (3/16) | 0.5 (½)	CROSS | TOL.	0.2 (3/16) | 0.5 (½)	0.2 (3/16) | 0.5 (½)	0.2 (3/16) | 0.5 (½)
0.1° ± 0.1° 3/32° ± 3/32°	0.1° ± 0.1° 3/32° ± 3/32°	0.1° ± 0.1° 3/32° ± 3/32°	FRONT TOE (Degrees)	0.3° ± 0.1° 5/16° ± 3/32°	0.22° ± 0.1° [1] 7/32° ± 3/32° [1]	0.15° ± 0.05° [1] 5/32° ± 1/16° [1]
0.05" ± 0.05" 1/16" ± 1/16"	0.05" ± 0.05" 1/16" ± 1/16"	0.05" ± 0.05" 1/16" ± 1/16"	FRONT TOE (Inches)	0.15" ± 0.05" 5/32" ± 1/16"	0.11" ± 0.05" [1] ⅛" ± 1/16" [1]	0.08" ± 0.03" [1] 1/16" ± 1/32" [1]
			SAI			
			TURNING ANGLE			
12	12	12	ILLUSTRATION #	12	12	12
			REAR			
			CAMBER (Degrees)			
			REAR TOE (Degrees)			
			REAR TOE (Inches)			
0.4 (13/32)	0.4 (13/32)	0.4 (13/32)	THRUST ANGLE	0.4 (13/32)	0.4 (13/32)	0.4 (13/32)
			ILLUSTRATION #			
			NOTES		[1] NRH = Normal Ride Height	[1] NRH = Normal Ride Height
13016	13017	13018		13019	13020	13021

Dodge (Ram) Truck & SUV

Ram Pickup 2500 (4X2)	Ram Pickup 2500 (4X2)	Ram Pickup 2500 (4X2)	MODEL INFORMATION	Ram Pickup 2500 (4X2)	Ram Pickup 2500 (4X2) exc/AIR, Box Off	Ram Pickup 2500 (4X2) w/AIR susp.
1994-97	1998-2002	2003-2005	YEARS	2006-2008	2009-2017	2014-2017
			INSTRUCTIONS			
135 ft.lbs.	135 ft.lbs.	135 ft.lbs.	LUGNUT TORQUE	140 ft.lbs.	140 ft.lbs.	140 ft.lbs.
			FRONT			
3.55 ± 1.0 [1] 3 9⁄16 ± 1 [1]	3.5 ± 1.0 [1] 3 ½ ± 1 [1]	4.0 ± 0.75 [1] 4 ± ¾ [1]	CASTER (Degrees)	4.0 ± 0.75 4 ± ¾	L= 4.15, R= 4.0 ± 0.75 L= 4 ⅛, R= 4 ± ¾	L= 5.0, R= 5.3 ± 0.5 L= 5, R= 5 5⁄16 ± ½
0.0 (0) | 0.5 (½)	0.0 (0) | 0.5 (½)	0.0 (0) | 0.4 (⅜)	CROSS | TOL.	0.0 (0) | 0.5 (½)	0.15 (⅛) | 0.5 (½)	-0,3 (- 5⁄16) | 0.5 (½)
0.5 ± 0.5 ½ ± ½	0.5 ± 0.5 ½ ± ½	0.0 ± 0.5 0 ± ½	CAMBER (Degrees)	0.0 ± 0.5 0 ± ½	0.0 ± 0.5 0 ± ½	0.25 ± 0.5 ¼ ± ½
0.0 (0) | 0.5 (½)	0.0 (0) | 0.5 (½)	0.0 (0) | 0.6 (⅝)	CROSS | TOL.	0.0 (0) | 0.5 (½)	0.0 (0) | 0.5 (½)	0.0 (0) | 0.5 (½)
0.1° ± 0.1° 3⁄32° ± 3⁄32°	0.1° ± 0.1° 3⁄32° ± 3⁄32°	0.2° ± 0.1° 3⁄16° ± 3⁄32°	FRONT TOE (Degrees)	0.2° ± 0.1° 3⁄16° ± 3⁄32°	0.2° ± 0.1° 3⁄16° ± 3⁄32°	0.1° ± 0.1° 3⁄32° ± 3⁄32°
0.05" ± 0.05" 1⁄16" ± 1⁄16"	0.05" ± 0.05" 1⁄16" ± 1⁄16"	0.1" ± 0.05" 3⁄32" ± 1⁄16"	FRONT TOE (Inches)	0.1" ± 0.05" 3⁄32" ± 1⁄16"	0.1" ± 0.05" 3⁄32" ± 1⁄16"	0.05" ± 0.05" 1⁄16" ± 1⁄16"
			SAI			
			TURNING ANGLE			
42	42	139	ILLUSTRATION #	12	12	12
			REAR			
			CAMBER (Degrees)			
			REAR TOE (Degrees)			
			REAR TOE (Inches)			
0.4 (13⁄32)	0.4 (13⁄32)	0.4 (13⁄32)	THRUST ANGLE	0.2 (3⁄16)	0.4 (13⁄32)	0.4 (13⁄32)
			ILLUSTRATION #			
[1] Club Cab/Quad Cab = 3.65° (3 ⅝°)	[1] Club Cab/Quad Cab = 3.6° (3 ⅝°)	[1] 160" WB = 4.3° (4 5⁄16°)	NOTES			
13022	13023	13024		13025	13026	13027

Dodge (Ram) Truck & SUV

Ram Pickup 2500 (4X2) Box Off	Ram Pickup 2500 (4X4)	Ram Pickup 2500 (4X4)	MODEL INFORMATION	Ram Pickup 2500 (4X4)	Ram Pickup 2500 (4X4)	Ram Pickup 2500 (4X4) exc/Box Off
2009-2017	1994-99	2000-2002	YEARS	2003-2005	2006-2008	2009-2010
			INSTRUCTIONS			
140 ft.lbs.	135 ft.lbs.	135 ft.lbs.	LUGNUT TORQUE	135 ft.lbs.	140 ft.lbs.	140 ft.lbs.
			FRONT			
3.75 ± 0.5 3 ¾ ± ½	3.05 ± 1.0 [1] 3 1⁄16 ± 1 [1]	2.68 ± 1.0 [1] 2 11⁄16 ± 1 [1]	CASTER (Degrees)	4.5 ± 0.75 [1] 4 ½ ± ¾ [1]	4.5 ± 0.5 4 ½ ± ½	4.5 ± 0.5 4 ½ ± ½
0.0 (0) | 0.5 (½)	0.0 (0) | 0.5 (½)	0.0 (0) | 0.5 (½)	CROSS | TOL.	0.0 (0) | 0.5 (½)	0.0 (0) | 0.5 (½)	0.0 (0) | 0.5 (½)
0.0 ± 0.5 0 ± ½	0.25 ± 0.5 ¼ ± ½	Camber is not specified by the manufacturer.	CAMBER (Degrees)	0.25 ± 0.5 ¼ ± ½	0.25 ± 0.5 ¼ ± ½	0.25 ± 0.5 ¼ ± ½
0.0 (0) | 0.5 (½)	0.0 (0) | 0.5 (½)	0.0 (0) | 0.5 (½)	CROSS | TOL.	0.0 (0) | 0.5 (½)	0.0 (0) | 0.5 (½)	0.0 (0) | 0.5 (½)
0.2° ± 0.1° 3⁄16° ± 3⁄32°	0.1° ± 0.1° 3⁄32° ± 3⁄32°	0.1° ± 0.1° 3⁄32° ± 3⁄32°	FRONT TOE (Degrees)	0.2° ± 0.1° 3⁄16° ± 3⁄32°	0.2° ± 0.1° 3⁄16° ± 3⁄32°	0.2° ± 0.1° 3⁄16° ± 3⁄32°
0.1" ± 0.05" 3⁄32" ± 1⁄16"	0.05" ± 0.05" 1⁄16" ± 1⁄16"	0.05" ± 0.05" 1⁄16" ± 1⁄16"	FRONT TOE (Inches)	0.1" ± 0.05" 3⁄32" ± 1⁄16"	0.1" ± 0.05" 3⁄32" ± 1⁄16"	0.1" ± 0.05" 3⁄32" ± 1⁄16"
			SAI			
			TURNING ANGLE			
12	78	78	ILLUSTRATION #	78	78	78
			REAR			
			CAMBER (Degrees)			
			REAR TOE (Degrees)			
			REAR TOE (Inches)			
0.4 (13⁄32)	0.4 (13⁄32)	0.4 (13⁄32)	THRUST ANGLE	0.4 (13⁄32)	0.2 (3⁄16)	0.4 (13⁄32)
			ILLUSTRATION #			
	[1] Club Cab/Quad Cab = 3.2° (3 3⁄16°)	[1] Club Cab/Quad Cab = 2.8° (2 13⁄16°)	NOTES	[1] 160" WB = 4.7° (4 11⁄16°)		
13028	13029	13030		13031	13032	13033

ALIGNMENT SPECIFICATIONS

ALIGNMENT SPECIFICATIONS

Dodge (Ram) Truck & SUV

Ram Pickup 2500 (4X4) exc/AIR, Box Off	Ram Pickup 2500 (4X4) w/AIR susp.	Ram Pickup 2500 (4X4) Box Off	MODEL INFORMATION	Ram Pickup 3500 (4X2)	Ram Pickup 3500 (4X2)	Ram Pickup 3500 (4X2)
2011-2017	2014-2017	2009-2017	YEARS	1994-2002	2003-2005	2006-2008
			INSTRUCTIONS			
140 ft.lbs.	140 ft.lbs.	140 ft.lbs.	LUGNUT TORQUE	145 ft.lbs.	145 ft.lbs.	145 ft.lbs.
			FRONT			
L= 4.5, R= 4.8 ± 0.5 L= 4½, R= 4 13/16 ± ½	L= 5.0, R= 5.3 ± 0.5 L= 5, R= 5 5/16 ± ½	3.75±0.5 3¾ ± ½	CASTER (Degrees)	3.3±1.0 [1] 3 5/16 ±1 [1]	4.0±0.75 [1] 4 ± ¾ [1]	4.0±0.75 4 ± ¾
-0.3 (-5/16) | 0.5 (½)	-0.3 (-5/16) | 0.5 (½)	0.0 (0) | 0.5 (½)	CROSS | TOL.	0.0 (0) | 0.5 (½)	0.0 (0) | 0.4 (⅜)	0.0 (0) | 0.5 (½)
0.25±0.5 ¼ ± ½	0.25±0.5 ¼ ± ½	0.25±0.5 ¼ ± ½	CAMBER (Degrees)	0.5±0.5 ½ ± ½	0.0±0.5 0 ± ½	0.0±0.5 0 ± ½
0.0 (0) | 0.5 (½)	0.0 (0) | 0.5 (½)	0.0 (0) | 0.5 (½)	CROSS | TOL.	0.0 (0) | 0.5 (½)	0.0 (0) | 0.6 (⅝)	0.0 (0) | 0.5 (½)
0.2° ±0.1° 3/16° ± 3/32°	0.1° ±0.1° 3/32° ± 3/32°	0.2° ±0.1° 3/16° ± 3/32°	FRONT TOE (Degrees)	0.1° ±0.1° 3/32° ± 3/32°	0.2° ±0.1° 3/16° ± 3/32°	0.2° ±0.1° 3/16° ± 3/32°
0.1" ±0.05" 3/32" ± 1/16"	0.05" ±0.05" 1/16" ± 1/16"	0.1" ±0.05" 3/32" ± 1/16"	FRONT TOE (Inches)	0.05" ±0.05" 1/16" ± 1/16"	0.1" ±0.05" 3/32" ± 1/16"	0.1" ±0.05" 3/32" ± 1/16"
			SAI			
			TURNING ANGLE			
78		78	ILLUSTRATION #	42	139	12
			REAR			
			CAMBER (Degrees)			
			REAR TOE (Degrees)			
			REAR TOE (Inches)			
0.4 (13/32)	0.4 (13/32)	0.4 (13/32)	THRUST ANGLE	0.4 (13/32)	0.4 (13/32)	0.2 (3/16)
			ILLUSTRATION #			
			NOTES	[1] Club Cab/Quad Cab = 3.5° (3½°)	[1] 160" WB = 4.3° (4 5/16°)	
13034	13035	13036		13037	13038	13039

Dodge (Ram) Truck & SUV

Ram Pickup 3500 (4X2) exc/Box Off	Ram Pickup 3500 (4X2) Box Off	Ram Pickup 3500 (4X4)	MODEL INFORMATION	Ram Pickup 3500 (4X4)	Ram Pickup 3500 (4X4)	Ram Pickup 3500 (4X4)
2009-2017	2009-2017	1994-99	YEARS	2000-2002	2003-2005	2006-2008
			INSTRUCTIONS			
145 ft.lbs.	140 ft.lbs.	145 ft.lbs.	LUGNUT TORQUE	145 ft.lbs.	145 ft.lbs.	145 ft.lbs.
			FRONT			
L= 4.15, R= 4.0 ± 0.5 L= 4⅛, R= 4 ± ½	3.75±0.5 3¾ ± ½	2.9±1.0 [1] 2⅞±1 [1]	CASTER (Degrees)	2.48±1.0 [1] 2½±1 [1]	4.5±0.75 [1] 4½ ± ¾ [1]	4.5±0.5 4½ ± ½
0.15 (⅛) | 0.5 (½)	0.0 (0) | 0.5 (½)	0.0 (0) | 0.5 (½)	CROSS | TOL.	0.0 (0) | 0.5 (½)	0.0 (0) | 0.5 (½)	0.0 (0) | 0.5 (½)
0.0±0.5 0 ± ½	0.0±0.5 0 ± ½	0.25±0.5 ¼ ± ½	CAMBER (Degrees)	Camber is not specified by the manufacturer.	0.25±0.5 ¼ ± ½	0.25±0.5 ¼ ± ½
0.0 (0) | 0.5 (½)	0.0 (0) | 0.5 (½)	0.0 (0) | 0.5 (½)	CROSS | TOL.	0.0 (0) | 0.5 (½)	0.0 (0) | 0.5 (½)	0.0 (0) | 0.5 (½)
0.2° ±0.1° 3/16° ± 3/32°	0.2° ±0.1° 3/16° ± 3/32°	0.1° ±0.1° 3/32° ± 3/32°	FRONT TOE (Degrees)	0.1° ±0.1° 3/32° ± 3/32°	0.2° ±0.1° 3/16° ± 3/32°	0.2° ±0.1° 3/16° ± 3/32°
0.1" ±0.05" 3/32" ± 1/16"	0.1" ±0.05" 3/32" ± 1/16"	0.05" ±0.05" 1/16" ± 1/16"	FRONT TOE (Inches)	0.05" ±0.05" 1/16" ± 1/16"	0.1" ±0.05" 3/32" ± 1/16"	0.1" ±0.05" 3/32" ± 1/16"
			SAI			
			TURNING ANGLE			
12	12	78	ILLUSTRATION #	78	78	78
			REAR			
			CAMBER (Degrees)			
			REAR TOE (Degrees)			
			REAR TOE (Inches)			
0.4 (13/32)	0.4 (13/32)	0.4 (13/32)	THRUST ANGLE	0.4 (13/32)	0.4 (13/32)	0.2 (3/16)
			ILLUSTRATION #			
		[1] Club Cab/Quad Cab = 3.0° (3°)	NOTES	[1] Club Cab/Quad Cab = 2.6° (2 5/8°)	[1] 160" WB = 4.7° (4 11/16°)	
13040	13041	13042		13043	13044	13045

Dodge (Ram) Truck & SUV

Ram Pickup 3500 (4X4) exc/Box Off	Ram Pickup 3500 (4X4) exc/Box Off	Ram Pickup 3500 (4X4) Box Off	MODEL INFORMATION	Ram HD Chassis-Cab All Models	Ram 3500 HD Chassis-Cab All Models	Ram 3500 HD Chassis-Cab All Models
2009-2010	2011-2017	2009-2017	YEARS	1994-2002	2007-2008	2009-2017
			INSTRUCTIONS			
145 ft.lbs.	145 ft.lbs.	140 ft.lbs.	LUGNUT TORQUE		145 ft.lbs.	145 ft.lbs.
			FRONT			
4.5 ±0.5 4½ ± ½	L= 4.5, R= 4.8 ± 0.5 L= 4½, R= 4 13/16 ± ½	3.75 ±0.5 3¾ ± ½	CASTER (Degrees)	MIN.= 3.8 MAX.= 6.8 MIN.= 3¾ MAX.= 6¾	3.5 ±0.5[1] 3½ ± ½[1]	3.75 ±0.5[1] 3¾ ± ½[1]
0.0 (0) | 0.5 (½)	-0.3 (-5/16) | 0.5 (½)	0.0 (0) | 0.5 (½)	CROSS | TOL.	0.0 (0) | 0.5 (½)	0.0 (0) | 0.5 (½)	0.0 (0) | 0.5 (½)
0.25 ±0.5 ¼ ± ½	0.25 ±0.5 ¼ ± ½	0.25 ±0.5 ¼ ± ½	CAMBER (Degrees)	Camber is not specified by the manufacturer.	0.25 ±0.5 ¼ ± ½	0.25 ±0.5 ¼ ± ½
0.0 (0) | 0.5 (½)	0.0 (0) | 0.5 (½)	0.0 (0) | 0.5 (½)	CROSS | TOL.	0.0 (0) | 0.5 (½)	0.0 (0) | 0.5 (½)	0.0 (0) | 0.5 (½)
0.2° ±0.1° 3/16° ± 3/32°	0.2° ±0.1° 3/16° ± 3/32°	0.2° ±0.1° 3/16° ± 3/32°	FRONT TOE (Degrees)	0.1° ±0.1° 3/32° ± 3/32°	0.2° ±0.1° 3/16° ± 3/32°	0.2° ±0.1° 3/16° ± 3/32°
0.1" ±0.05" 3/32" ± 1/16"	0.1" ±0.05" 3/32" ± 1/10"	0.1" ±0.05" 3/32" ± 1/16"	FRONT TOE (Inches)	0.05" ±0.05" 1/16" ± 1/16"	0.1" ±0.05" 3/32" ± 1/16"	0.1" ±0.05" 3/32" ± 1/16"
			SAI			
			TURNING ANGLE			
78	78	78	ILLUSTRATION #	42	78	78
			REAR			
			CAMBER (Degrees)			
			REAR TOE (Degrees)			
			REAR TOE (Inches)			
0.4 (13/32)	0.4 (13/32)	0.4 (13/32)	THRUST ANGLE	0.4 (13/32)		0.4 (13/32)
			ILLUSTRATION #			
			NOTES		[1] 2009 = 3.75° (3¾°)	[1] 2009 = 3.75° (3¾°)
13046	13047	13048		13049	13050	13051

Dodge (Ram) Truck & SUV

Ram 4500 HD Chassis-Cab All Models	Ram 5500 HD Chassis-Cab All Models	Ram 4500/5500 HD Chassis-Cab 144", 164", 168" WB	MODEL INFORMATION	Ram 4500/5500 HD Chassis-Cab 188", 192" WB	Ram 4500/5500 HD Chassis-Cab 204" WB	Ram 4500/5500 HD Chassis-Cab 144", 168", 173" WB
2008	2008	2009-2010	YEARS	2009-2010	2009-2010	2011-2012
			INSTRUCTIONS			
145 ft.lbs.	145 ft.lbs.	145 ft.lbs.	LUGNUT TORQUE	145 ft.lbs.	145 ft.lbs.	145 ft.lbs.
			FRONT			
3.0 ±1.5 3 ±1½	3.5 ±1.5 3½ ±1½	L= 3.0, R= 3.5 ± 1.0 L= 3, R= 3½ ±1	CASTER (Degrees)	L= 3.5, R= 4.0 ± 1.0 L= 3½, R= 4 ±1	L= 3.8, R= 4.0 ± 1.0 L= 3 13/16, R= 4 ±1	L= 5.5, R= 6.0 ± 1.0 L= 5½, R= 6 ±1
0.0 (0) | 0.5 (½)	0.0 (0) | 0.5 (½)	-0.5 (-½) | 0.5 (½)	CROSS | TOL.	-0.5 (-½) | 0.5 (½)	-0.2 (-3/16) | 0.5 (½)	-0.5 (-½) | 0.5 (½)
0.0 ±0.5 0 ± ½	0.0 ±0.5 0 ± ½	0.1 ±0.5 ⅛ ± ½	CAMBER (Degrees)	0.1 ±0.5 ⅛ ± ½	0.1 ±0.5 ⅛ ± ½	0.1 ±0.5 ⅛ ± ½
0.0 (0) | 0.5 (½)	0.0 (0) | 0.5 (½)	0.0 (0) | 0.5 (½)	CROSS | TOL.	0.0 (0) | 0.5 (½)	0.0 (0) | 0.5 (½)	0.0 (0) | 0.5 (½)
0.2° ±0.1° 3/16° ± 3/32°	0.2° ±0.1° 3/16° ± 3/32°	0.2° ±0.1° 3/16° ± 3/32°	FRONT TOE (Degrees)	0.2° ±0.1° 3/16° ± 3/32°	0.2° ±0.1° 3/16° ± 3/32°	0.2° ±0.1° 3/16° ± 3/32°
0.1" ±0.05" 3/32" ± 1/16"	0.1" ±0.05" 3/32" ± 1/16"	0.1" ±0.05" 3/32" ± 1/16"	FRONT TOE (Inches)	0.1" ±0.05" 3/32" ± 1/16"	0.1" ±0.05" 3/32" ± 1/16"	0.1" ±0.05" 3/32" ± 1/16"
12.8 ±0.8 (12 13/16 ± ¾)	12.8 ±0.8 (12 13/16 ± ¾)	12.8 ±0.8 (12 13/16 ± ¾)	SAI	12.8 ±0.8 (12 13/16 ± ¾)	12.8 ±0.8 (12 13/16 ± ¾)	12.8 ±0.8 (12 13/16 ± ¾)
			TURNING ANGLE			
			ILLUSTRATION #			
			REAR			
			CAMBER (Degrees)			
			REAR TOE (Degrees)			
			REAR TOE (Inches)			
0.2 (3/16)	0.2 (3/16)	0.4 (13/32)	THRUST ANGLE	0.4 (13/32)	0.4 (13/32)	0.4 (13/32)
			ILLUSTRATION #			
			NOTES			
13052	13053	13054		13055	13056	13057

ALIGNMENT SPECIFICATIONS

Dodge (Ram) Truck & SUV

Ram 4500/5500	Ram 4500/5500	Ram 4500/5500	MODEL	Ram 4500/5500	Ram 4500/5500	Ram Promaster
HD Chassis-Cab	HD Chassis-Cab	HD Chassis-Cab	INFORMATION	HD Chassis-Cab	HD Chassis-Cab	1500 Cargo 118" WB,
192", 197" WB	204" WB	144", 164", 168", 173" WB		192" & 197" WB	204" WB	8550 GVWR, Diesel
2011-2012	2011-2012	2013-2017	YEARS	2013-2017	2013-2017	2014-2017
			INSTRUCTIONS			
145 ft.lbs.	145 ft.lbs.	145 ft.lbs.	LUGNUT TORQUE	145 ft.lbs.	145 ft.lbs.	146 ft.lbs.
			FRONT			
L=6.3, R=6.8 ± 1.0 L=6 5/16, R=6 13/16 ±1	L=6.5, R=7.0 ± 1.0 L=6 1/2, R=7 ±1	L=7.8, R=8.1 ± 1.0 L=7 13/16, R=8 1/8 ±1	CASTER (Degrees)	L=8.3, R=8.6 ± 1.0 L=8 5/16, R=8 5/8 ±1	L=8.6, R=8.9 ± 1.0 L=8 5/8, R=8 7/8 ±1	0.28 ±0.5 [1] 1/4 ± 1/2 [1]
-0.5 (-1/2) | 0.5 (1/2)	-0.5 (-1/2) | 0.5 (1/2)	-0.3 (-5/16) | 0.5 (1/2)	CROSS | TOL.	-0.3 (-5/16) | 0.5 (1/2)	-0.3 (-5/16) | 0.5 (1/2)	0.0 (0) | 0.5 (1/2)
0.1 ±0.5 1/8 ± 1/2	0.1 ±0.5 1/8 ± 1/2	0.1 ±0.5 1/8 ± 1/2	CAMBER (Degrees)	0.1 ±0.5 1/8 ± 1/2	0.1 ±0.5 1/8 ± 1/2	0.0 ±0.5 0 ± 1/2
0.0 (0) | 0.5 (1/2)	0.0 (0) | 0.5 (1/2)	0.0 (0) | 0.5 (1/2)	CROSS | TOL.	0.0 (0) | 0.5 (1/2)	0.0 (0) | 0.5 (1/2)	0.0 (0) | 0.45 (7/16)
0.2° ±0.1° 3/16° ± 3/32°	0.2° ±0.1° 3/16° ± 3/32°	0.2° ±0.1° 3/16° ± 3/32°	FRONT TOE (Degrees)	0.2° ±0.1° 3/16° ± 3/32°	0.2° ±0.1° 3/16° ± 3/32°	0.15° ±0.15° 5/32° ± 5/32°
0.1" ±0.05" 3/32" ± 1/16"	0.1" ±0.05" 3/32" ± 1/16"	0.1" ±0.05" 3/32" ± 1/16"	FRONT TOE (Inches)	0.1" ±0.05" 3/32" ± 1/16"	0.1" ±0.05" 3/32" ± 1/16"	0.08" ±0.08" 1/16" ± 1/16"
12.8 ±0.8 (12 13/16 ± 3/4)	12.8 ±0.8 (12 13/16 ± 3/4)	12.8 ±0.8 (12 13/16 ± 3/4)	SAI	12.8 ±0.8 (12 13/16 ± 3/4)	12.8 ±0.8 (12 13/16 ± 3/4)	
			TURNING ANGLE			
			ILLUSTRATION #			26
			REAR			
			CAMBER (Degrees)			
			REAR TOE (Degrees)			
			REAR TOE (Inches)			
0.4 (13/32)	0.4 (13/32)	0.4 (13/32)	THRUST ANGLE	0.4 (13/32)	0.4 (13/32)	0.3 (5/16)
			ILLUSTRATION #			
			NOTES			[1] Gas mods. = 0.44° (7/16°)
13058	13059	13060		13061	13062	13063

Dodge (Ram) Truck & SUV

Ram Promaster	Ram Promaster	Ram Promaster	MODEL	Ram Promaster	Ram Promaster	Ram Promaster
1500 Cargo 136" WB,	2500 Cargo 136" WB,	2500 Cargo 159" WB,	INFORMATION	3500 Cargo L 159" WB,	3500 Cargo XL 159" WB,	2500 Window Van
8550 GVWR, Diesel	8900 GVWR, Diesel	8900 GVWR, Diesel		9350 GVWR, Diesel	9350 GVWR, Diesel	8900 GVWR, Diesel
2014-2017	2014-2017	2014-2017	YEARS	2014-2017	2014-2017	2014-2017
			INSTRUCTIONS			
146 ft.lbs.	146 ft.lbs.	146 ft.lbs.	LUGNUT TORQUE	146 ft.lbs.	146 ft.lbs.	146 ft.lbs.
			FRONT			
0.28 ±0.5 [1] 1/4 ± 1/2 [1]	0.4 ±0.5 [1] 3/8 ± 1/2 [1]	1.14 ±0.5 [1] 1 1/8 ± 1/2 [1]	CASTER (Degrees)	1.14 ±0.5 [1] 1 1/8 ± 1/2 [1]	1.18 ±0.5 [1] 1 3/16 ± 1/2 [1]	1.14 ±0.5 [1] 1 1/8 ± 1/2 [1]
0.0 (0) | 0.5 (1/2)	0.0 (0) | 0.5 (1/2)	0.0 (0) | 0.5 (1/2)	CROSS | TOL.	0.0 (0) | 0.5 (1/2)	0.0 (0) | 0.5 (1/2)	0.0 (0) | 0.5 (1/2)
0.0 ±0.5 0 ± 1/2	0.0 ±0.5 0 ± 1/2	0.0 ±0.5 0 ± 1/2	CAMBER (Degrees)	0.0 ±0.5 0 ± 1/2	0.0 ±0.5 0 ± 1/2	0.0 ±0.5 0 ± 1/2
0.0 (0) | 0.45 (7/16)	0.0 (0) | 0.45 (7/16)	0.0 (0) | 0.45 (7/16)	CROSS | TOL.	0.0 (0) | 0.45 (7/16)	0.0 (0) | 0.45 (7/16)	0.0 (0) | 0.45 (7/16)
0.15° ±0.15° 5/32° ± 5/32°	0.15° ±0.15° 5/32° ± 5/32°	0.15° ±0.15° 5/32° ± 5/32°	FRONT TOE (Degrees)	0.15° ±0.15° 5/32° ± 5/32°	0.15° ±0.15° 5/32° ± 5/32°	0.15° ±0.15° 5/32° ± 5/32°
0.08" ±0.08" 1/16" ± 1/16"	0.08" ±0.08" 1/16" ± 1/16"	0.08" ±0.08" 1/16" ± 1/16"	FRONT TOE (Inches)	0.08" ±0.08" 1/16" ± 1/16"	0.08" ±0.08" 1/16" ± 1/16"	0.08" ±0.08" 1/16" ± 1/16"
			SAI			
			TURNING ANGLE			
26	26	26	ILLUSTRATION #	26	26	26
			REAR			
			CAMBER (Degrees)			
			REAR TOE (Degrees)			
			REAR TOE (Inches)			
0.3 (5/16)	0.3 (5/16)	0.3 (5/16)	THRUST ANGLE	0.3 (5/16)	0.3 (5/16)	0.3 (5/16)
			ILLUSTRATION #			
[1] Gas mods. = 0.44° (7/16°)	[1] Gas mods. = 0.48° (1/2°)	[1] Gas mods. = 1.23° (1 1/4°)	NOTES	[1] Gas mods. = 1.18° (1 3/16°)	[1] Gas mods. = 1.24° (1 1/4°)	[1] Gas mods. = 1.18° (1 3/16°)
13064	13065	13066		13067	13068	13069

Dodge (Ram) Truck & SUV

Ram Promaster	Ram Promaster	Ram Promaster City	MODEL	Ram Promaster City	Ram Tradesman	B Series Vans
3500 Window Van	Chassis Cab & Cutaway	Cargo Van	INFORMATION	Window Van		
9350 GVWR, Diesel	All Models					
2014-2017	2014-2017	2015-2017	YEARS	2015-2017	2013-2016	1990-97
			INSTRUCTIONS			
146 ft.lbs.	146 ft.lbs.	64 ft.lbs.	LUGNUT TORQUE	89 ft.lbs.	100 ft.lbs.	80-110 ft.lbs.[1]
			FRONT			
1.11 ±0.5 [1] 1⅛ ± ½ [1]	1.5 ±0.5 1½ ± ½	2.8 ±0.5 2 13/16 ± ½	CASTER (Degrees)	2.9 ±0.5 2⅞ ± ½	2.5 ±1.0 2½ ±1	2.5 ±1.25 2½ ±1¼
0.0 (0) | 0.5 (½)	0.0 (0) | 0.5 (½)	0.0 (0) | 0.5 (½)	CROSS | TOL.	0.0 (0) | 0.5 (½)	0.0 (0) | 1.0 (1)	0.0 (0) | 1.25 (1¼)
0.0 ±0.5 0 ± ½	0.0 ±0.5 0 ± ½	-0.62 ±0.33 -⅝ ± 5/16	CAMBER (Degrees)	-0.62 ±0.33 -⅝ ± 5/16	L= -0.1, R= -0.4 ±0.4 L= -⅛, R= -⅜ ± ⅜	0.0 ±0.6 0 ± ⅝
0.0 (0) | 0.45 (7/16)	0.0 (0) | 0.45 (7/16)	0.0 (0) | 0.5 (½)	CROSS | TOL.	0.0 (0) | 0.5 (½)	0.3 (5/16) | 0.5 (½)	0.0 (0) | n/s
0.15° ±0.15° 5/32° ± 5/32°	0.15° ±0.15° 5/32° ± 5/32°	0.22° ±0.14° 7/32° ± ⅛°	FRONT TOE (Degrees)	0.2° ±0.14° 3/16° ± ⅛°	0.26° ±0.2° ¼° ± 3/16°	0.0° ±0.25° 0° ± ¼°
0.08" ±0.08" 1/16" ± 1/16"	0.08" ±0.08" 1/16" ± 1/16"	0.11" ±0.07" ⅛" ± 1/16"	FRONT TOE (Inches)	0.1" ±0.07" 3/32" ± 1/16"	0.13" ±0.1" ⅛" ± 3/32"	0.0" ±0.13" 0" ± ⅛"
			SAI			
			TURNING ANGLE			
26	26		ILLUSTRATION #		161	42
			REAR			
		0.2 ±0.52 3/16 ± ½	CAMBER (Degrees)	0.2 ±0.52 3/16 ± ½	-0.06 ±0.3 -1/16 ± 5/16	
		0.5° ±0.36° ½° ± ⅜°	REAR TOE (Degrees)	0.36° ±0.36° ⅜° ± ⅜°	0.15° ±0.25° 5/32° ± ¼°	
		0.25" ±0.18" ¼" ± 3/16"	REAR TOE (Inches)	0.18" ±0.18" 3/16" ± 3/16"	0.08" ±0.13" 1/16" ± ⅛"	
0.3 (5/16)	0.3 (5/16)	0.18 (3/16)	THRUST ANGLE	0.18 (3/16)	0.3 (5/16)	
			ILLUSTRATION #			
[1] Gas mods. = 1.24° (1¼°)			NOTES			[1] ½ x 20 lug nuts only
13070	13071	13072		13073	13074	13079

Dodge (Ram) Truck & SUV

B Series Vans	B Series Vans	D150-350 4X2	MODEL	D150-350 4X4	Dakota 4X2	Dakota 4X2
			INFORMATION	(W-Series)		
1998-2001	2002-2003	1990-93	YEARS	1990-93	1990-96	1997
			INSTRUCTIONS			
80-110 ft.lbs.[1]	80-110 ft.lbs.[1]	85-115 ft.lbs.[1]	LUGNUT TORQUE	105 ft.lbs.[1]	85-115 ft.lbs.	85-115 ft.lbs.
			FRONT			
2.75 ±1.5 2¾ ±1½	3.5 ±1.0 3½ ±1	0.5 ±1.5 ½ ±1½	CASTER (Degrees)	2.0 ±1.5 2 ±1½	1.5 ±1.0 1½ ±1	3.9 ±0.7 3⅞ ± 11/16
0.0 (0) | 0.5 (½)	0.0 (0) | 0.5 (½)	0.0 (0) | n/s	CROSS | TOL.	0.0 (0) | n/s	0.0 (0) | 1.25 (1¼)	0.0 (0) | 0.5 (½)
0.0 ±0.3 0 ± 5/16	0.0 ±0.3 0 ± 5/16	0.5 ±0.5 ½ ± ½	CAMBER (Degrees)	0.0 ±0.5 0 ± ½	0.5 ±0.5 ½ ± ½	0.38 ±0.5 ⅜ ± ½
0.0 (0) | n/s	0.0 (0) | n/s	0.0 (0) | n/s	CROSS | TOL.	0.0 (0) | n/s	0.0 (0) | n/s	0.0 (0) | 0.5 (½)
0.25° ±0.25° ¼° ± ¼°	0.25° ±0.25° ¼° ± ¼°	0.25° ±0.25° ¼° ± ¼°	FRONT TOE (Degrees)	0.2° ±0.25° 3/16° ± ¼°	0.25° ±0.1° ¼° ± 3/32°	0.15° ±0.1° 5/32° ± 3/32°
0.13" ±0.13" ⅛" ± ⅛"	0.13" ±0.13" ⅛" ± ⅛"	0.13" ±0.13" ⅛" ± ⅛"	FRONT TOE (Inches)	0.1" ±0.13" 3/32" ± ⅛"	0.13" ±0.05" ⅛" ± 1/16"	0.08" ±0.05" 1/16" ± 1/16"
			SAI	8.5 (8½)		
			TURNING ANGLE			
42	42	24	ILLUSTRATION #		42	42
			REAR			
			CAMBER (Degrees)			
			REAR TOE (Degrees)			
			REAR TOE (Inches)			
0.4 (13/32)	0.4 (13/32)		THRUST ANGLE			
			ILLUSTRATION #			
[1] ½ x 20 lug nuts only	[1] 8 stud wheel = 120-150	[1] ½ x 20 lug nuts only	NOTES	[1] ½ x 20 lug nuts only		
13080	13081	13088		13095	13097	13098

Dodge (Ram) Truck & SUV

Dakota 4X2	Dakota 4X2 (exc R/T)	Dakota 4X2 (exc R/T)	MODEL INFORMATION	Dakota 4X2 R/T	Dakota 4X2 R/T	Dakota 4X2
1998	1999-2001	2002-2004	YEARS	1999-2001	2002-2003	2005-2006
			INSTRUCTIONS			
85-115 ft.lbs.	85-115 ft.lbs.	85-115 ft.lbs.	LUGNUT TORQUE	85-115 ft.lbs.	85-115 ft.lbs.	135 ft.lbs.
			FRONT			
3.05 ± 0.5 [1] 3 1/16 ± 1/2 [1]	3.0 ± 0.5 [1] 3 ± 1/2 [1]	3.0 ± 0.5 [1] 3 ± 1/2 [1]	CASTER (Degrees)	3.67 ± 0.5 [1] 3 11/16 ± 1/2 [1]	3.67 ± 0.5 [1] 3 11/16 ± 1/2 [1]	L= 3.5, R= 3.7 ± 0.5 L= 3 1/2, R= 3 11/16 ± 1/2
0.0 (0) | 0.5 (1/2)	0.0 (0) | 0.5 (1/2)	0.0 (0) | 0.5 (1/2)	CROSS | TOL.	0.0 (0) | 0.5 (1/2)	0.0 (0) | 0.5 (1/2)	-0.2 (-3/16) | 0.5 (1/2)
0.38 ± 0.5 3/8 ± 1/2	-0.25 ± 0.5 -1/4 ± 1/2	0.0 ± 0.5 0 ± 1/2	CAMBER (Degrees)	-0.34 ± 0.5 -5/16 ± 1/2	-0.09 ± 0.5 -1/16 ± 1/2	-0.25 ± 0.5 -1/4 ± 1/2
0.0 (0) | 0.5 (1/2)	0.0 (0) | 0.5 (1/2)	0.0 (0) | 0.5 (1/2)	CROSS | TOL.	0.0 (0) | 0.5 (1/2)	0.0 (0) | 0.5 (1/2)	0.0 (0) | 0.5 (1/2)
0.15° ± 0.1° 5/32° ± 3/32°	0.1° ± 0.06° 3/32° ± 1/16°	0.1° ± 0.06° 3/32° ± 1/16°	FRONT TOE (Degrees)	0.1° ± 0.06° 3/32° ± 1/16°	0.1° ± 0.06° 3/32° ± 1/16°	0.2° ± 0.1° 3/16° ± 3/32°
0.08" ± 0.05" 1/16" ± 1/16"	0.05" ± 0.03" 1/16" ± 1/32"	0.05" ± 0.03" 1/16" ± 1/32"	FRONT TOE (Inches)	0.05" ± 0.03" 1/16" ± 1/32"	0.05" ± 0.03" 1/16" ± 1/32"	0.1" ± 0.05" 3/32" ± 1/16"
			SAI			
			TURNING ANGLE			
42	42	42	ILLUSTRATION #	42	42	158
			REAR			
			CAMBER (Degrees)			
			REAR TOE (Degrees)			
			REAR TOE (Inches)			
			THRUST ANGLE			0.4 (13/32)
			ILLUSTRATION #			
[1] Club Cab = 3.15° (3 1/8°)	[1] Club Cab = 3.13° (3 1/8°)	[1] Club Cab = 3.13° (3 1/8°)	NOTES	[1] Club Cab = 3.81° (3 13/16°)	[1] Club Cab = 3.81° (3 13/16°)	
13099	13100	13101		13102	13103	13104

Dodge (Ram) Truck & SUV

Dakota 4X2	Dakota 4X4	Dakota 4X4	MODEL INFORMATION	Dakota 4X4	Dakota 4X4	Dakota 4X4
2007-2011	1990-96	1997	YEARS	1998-99	2000-2001	2002-2004
			INSTRUCTIONS			
135 ft.lbs.	85-115 ft.lbs.	85-115 ft.lbs.	LUGNUT TORQUE	85-115 ft.lbs.	85-115 ft.lbs.	85-115 ft.lbs.
			FRONT			
L= 3.5, R= 3.8 ± 0.5 L= 3 1/2, R= 3 13/16 ± 1/2	1.5 ± 1.0 1 1/2 ± 1	3.7 ± 0.6 3 11/16 ± 5/8	CASTER (Degrees)	3.2 ± 0.5 [1] 3 3/16 ± 1/2 [1]	3.16 ± 0.5 [1] 3 3/16 ± 1/2 [1]	3.16 ± 0.5 [1] 3 3/16 ± 1/2 [1]
-0.3 (-5/16) | 0.5 (1/2)	0.0 (0) | 1.25 (1 1/4)	0.0 (0) | 0.5 (1/2)	CROSS | TOL.	0.0 (0) | 0.5 (1/2)	0.0 (0) | 0.5 (1/2)	0.0 (0) | 0.5 (1/2)
L= 0.1, R= -0.1 ± 0.5 L= 1/8, R= -1/8 ± 1/2	0.5 ± 0.5 1/2 ± 1/2	0.38 ± 0.5 3/8 ± 1/2	CAMBER (Degrees)	0.38 ± 0.5 3/8 ± 1/2	-0.25 ± 0.5 -1/4 ± 1/2	0.0 ± 0.5 0 ± 1/2
0.2 (3/16) | 0.5 (1/2)	0.0 (0) | n/s	0.0 (0) | 0.5 (1/2)	CROSS | TOL.	0.0 (0) | 0.5 (1/2)	0.0 (0) | 0.5 (1/2)	0.0 (0) | 0.5 (1/2)
0.2° ± 0.1° 3/16° ± 3/32°	0.25° ± 0.1° 1/4° ± 3/32°	0.15° ± 0.1° 5/32° ± 3/32°	FRONT TOE (Degrees)	0.13° ± 0.1° 1/8° ± 3/32°	0.1° ± 0.06° 3/32° ± 1/16°	0.1° ± 0.06° 3/32° ± 1/16°
0.1" ± 0.05" 3/32" ± 1/16"	0.13" ± 0.05" 1/8" ± 1/16"	0.08" ± 0.05" 1/16" ± 1/16"	FRONT TOE (Inches)	0.06" ± 0.05" 1/16" ± 1/16"	0.05" ± 0.03" 1/16" ± 1/32"	0.05" ± 0.03" 1/16" ± 1/32"
			SAI			
			TURNING ANGLE			
158	42	42	ILLUSTRATION #	42	42	42
			REAR			
			CAMBER (Degrees)			
			REAR TOE (Degrees)			
			REAR TOE (Inches)			
0.4 (13/32)			THRUST ANGLE		0.4 (13/32)	0.4 (13/32)
			ILLUSTRATION #			
			NOTES	[1] Club Cab = 3.27° (3 1/4°)	[1] Club Cab = 3.27° (3 1/4°)	[1] Club Cab = 3.27° (3 1/4°)
13105	13106	13107		13108	13109	13110

Dodge (Ram) Truck & SUV

Dakota 4X4	Dakota 4X4	Durango 4X2	MODEL INFORMATION	Durango 4X2	Durango 4X2	Durango 4X2
2005-2006	2007-2011	1999	YEARS	2000-2003	2004-2005	2006-2009
			INSTRUCTIONS			
135 ft.lbs.	135 ft.lbs.	85-115 ft.lbs.	LUGNUT TORQUE	85-115 ft.lbs.	135 ft.lbs.	135 ft.lbs.
			FRONT			
L= 3.5, R= 3.7 ± 0.5 L= 3 1/2, R= 3 11/16 ± 1/2	L= 3.5, R= 3.8 ± 0.5 L= 3 1/2, R= 3 13/16 ± 1/2	3.31 ± 1.0 3 5/16 ± 1	CASTER (Degrees)	3.1 ± 0.5 3 1/8 ± 1/2	3.5 ± 0.5 3 1/2 ± 1/2	L= 3.2, R= 3.5 ± 0.5 L= 3 3/16, R= 3 1/2 ± 1/2
-0.2 (-3/16) | 0.5 (1/2)	-0.3 (-5/16) | 0.5 (1/2)	0.0 (0) | 0.5 (1/2)	CROSS | TOL.	0.0 (0) | 0.5 (1/2)	0.0 (0) | 0.5 (1/2)	-0.3 (-5/16) | 0.5 (1/2)
-0.25 ± 0.5 -1/4 ± 1/2	L= 0.1, R= -0.1 ± 0.5 L= 1/8, R= -1/8 ± 1/2	-0.25 ± 0.5 -1/4 ± 1/2	CAMBER (Degrees)	-0.25 ± 0.5 -1/4 ± 1/2	0.0 ± 0.5 0 ± 1/2	0.0 ± 0.5 0 ± 1/2
0.0 (0) | 0.5 (1/2)	0.2 (3/16) | 0.5 (1/2)	0.0 (0) | 0.5 (1/2)	CROSS | TOL.	0.0 (0) | 0.5 (1/2)	0.0 (0) | 0.6 (5/8)	0.0 (0) | 0.5 (1/2)
0.2° ± 0.1° 3/16° ± 3/32°	0.2° ± 0.1° 3/16° ± 3/32°	0.1° ± 0.1° 3/32° ± 3/32°	FRONT TOE (Degrees)	0.1° ± 0.06° 3/32° ± 1/16°	0.1° ± 0.1° 3/32° ± 3/32°	0.1° ± 0.05° 3/32° ± 1/16°
0.1" ± 0.05" 3/32" ± 1/16"	0.1" ± 0.05" 3/32" ± 1/16"	0.05" ± 0.05" 1/16" ± 1/16"	FRONT TOE (Inches)	0.05" ± 0.03" 1/16" ± 1/32"	0.05" ± 0.05" 1/16" ± 1/16"	0.05" ± 0.03" 1/16" ± 1/32"
			SAI			
			TURNING ANGLE			
158	158	42	ILLUSTRATION #	42	158	158
			REAR			
			CAMBER (Degrees)			
			REAR TOE (Degrees)			
			REAR TOE (Inches)			
0.4 (13/32)	0.4 (13/32)	0.4 (13/32)	THRUST ANGLE	0.4 (13/32)	0.4 (13/32)	0.4 (13/32)
			ILLUSTRATION #			
			NOTES			
13111	13112	13113		13114	13115	13116

Dodge (Ram) Truck & SUV

Durango 4X4	Durango 4X4	Durango 4X4	MODEL INFORMATION	Durango 4X4	Durango 4X4	Durango STD. Susp.
1998-99	2000-2001	2002-2003	YEARS	2004-2005	2006-2009	2011-2013
			INSTRUCTIONS			
85-115 ft.lbs.	85-115 ft.lbs.	85-115 ft.lbs.	LUGNUT TORQUE	135 ft.lbs.	135 ft.lbs.	100 ft.lbs.
			FRONT			
3.31 ± 0.5 3 5/16 ± 1/2	3.3 ± 0.5 3 5/16 ± 1/2	3.3 ± 0.5 3 5/16 ± 1/2	CASTER (Degrees)	3.5 ± 0.5 3 1/2 ± 1/2	3.5 ± 0.5 3 1/2 ± 1/2	5.0 ± 0.6 5 ± 5/8
0.0 (0) | 0.5 (1/2)	0.0 (0) | 0.5 (1/2)	0.0 (0) | 0.5 (1/2)	CROSS | TOL.	0.0 (0) | 0.5 (1/2)	0.0 (0) | 0.5 (1/2)	0.0 (0) | 1.0 (1)
0.38 ± 0.5 3/8 ± 1/2	-0.25 ± 0.5 -1/4 ± 1/2	0.0 ± 0.5 [1] 0 ± 1/2 [1]	CAMBER (Degrees)	0.0 ± 0.5 0 ± 1/2	L= 0.15, R= -0.15 ± 0.5 L= 1/8, R= -1/8 ± 1/2	L= -0.12, R= -0.62 ± 0.6 L= -1/8, R= -5/8 ± 5/8
0.0 (0) | 0.5 (1/2)	0.0 (0) | 0.5 (1/2)	0.0 (0) | 0.5 (1/2)	CROSS | TOL.	0.0 (0) | 0.6 (5/8)	0.3 (5/16) | 0.5 (1/2)	0.5 (1/2) | 0.5 (1/2)
0.13° ± 0.1° 1/8° ± 3/32°	0.1° ± 0.06° 3/32° ± 1/16°	0.1° ± 0.06° 3/32° ± 1/16°	FRONT TOE (Degrees)	0.1° ± 0.1° 3/32° ± 3/32°	0.1° ± 0.05° 3/32° ± 1/16°	0.2° ± 0.13° 3/16° ± 1/8°
0.06" ± 0.05" 1/16" ± 1/16"	0.05" ± 0.03" 1/16" ± 1/32"	0.05" ± 0.03" 1/16" ± 1/32"	FRONT TOE (Inches)	0.05" ± 0.05" 1/16" ± 1/16"	0.05" ± 0.03" 1/16" ± 1/32"	0.1" ± 0.06" 3/32" ± 1/16"
			SAI			
			TURNING ANGLE			
42	42	42	ILLUSTRATION #	158	158	193
			REAR			
			CAMBER (Degrees)			-0.54 ± 0.55 -9/16 ± 9/16
			REAR TOE (Degrees)			0.2° ± 0.3° 3/16° ± 5/16°
			REAR TOE (Inches)			0.1" ± 0.15" 3/32" ± 5/32"
0.4 (13/32)	0.4 (13/32)	0.4 (13/32)	THRUST ANGLE	0.4 (13/32)	0.4 (13/32)	0.15 (5/32)
			ILLUSTRATION #			130 06
		[1] R/T Cross Camber = 0.2° (3/16°)	NOTES			
13117	13118	13119		13120	13121	13122

Dodge (Ram) Truck & SUV

Durango Sport Susp.	Durango STD. Susp.	Durango Aero & Sport Susp.	MODEL INFORMATION	Durango STD. Susp.	Durango Aero & Sport Susp.	Journey
2011-2013	2014-2015	2014-2015	YEARS	2016-2017	2016-2017	2009-2010
			INSTRUCTIONS			
100 ft.lbs.	100 ft.lbs.	100 ft.lbs.	LUGNUT TORQUE	100 ft.lbs.	100 ft.lbs.	100 ft.lbs.
			FRONT			
5.45 ±0.6 5 7/16 ± 5/8	4.84 ±0.6 4 13/16 ± 5/8	5.37 ±0.6 5 3/8 ± 5/8	CASTER (Degrees)	6.95 ±0.5 6 15/16 ± ½	7.46 ±0.5 7 7/16 ± ½	3.0 ±1.0 3 ±1
0.0 (0) | 1.0 (1)	0.0 (0) | 1.0 (1)	0.0 (0) | 1.0 (1)	CROSS | TOL.	0.0 (0) | 0.5 (½)	0.0 (0) | 0.5 (½)	0.0 (0) | 0.5 (½)
L= -0.6, R= -1.1 ±0.6 L= -5/8, R= -1 1/8 ± 5/8	L= -0.17, R= -0.57 ±0.6 L= -1/8, R= -5/8 ± 5/8	L= -0.61, R= -1.01 ±0.6 L= -5/8, R= -1 ± 5/8	CAMBER (Degrees)	L= -0.38, R= -0.78 ±0.5 L= -3/8, R= -3/4 ± ½	L= -0.72, R= -1.12 ±0.5 L= -3/4, R= -1 1/8 ± ½	L= 0.15, R= -0.25 ±0.55 L= 1/8, R= -1/4 ± 9/16
0.5 (½) | 0.5 (½)	0.4 (3/8) | 0.6 (5/8)	0.4 (3/8) | 0.6 (5/8)	CROSS | TOL.	0.4 (3/8) | 0.5 (½)	0.4 (3/8) | 0.5 (½)	0.4 (3/8) | 0.55 (9/16)
0.2° ±0.13° 3/16° ± 1/8°	0.2° ±0.13° 3/16° ± 1/8°	0.2° ±0.13° 3/16° ± 1/8°	FRONT TOE (Degrees)	0.2° ±0.13° 3/16° ± 1/8°	0.2° ±0.13° 3/16° ± 1/8°	0.1° ±0.35° 3/32° ± 11/32°
0.1" ±0.06" 3/32" ± 1/16"	0.1" ±0.06" 3/32" ± 1/16"	0.1" ±0.06" 3/32" ± 1/16"	FRONT TOE (Inches)	0.1" ±0.06" 3/32" ± 1/16"	0.1" ±0.06" 3/32" ± 1/16"	0.05" ±0.18" 1/16" ± 3/16"
			SAI			
			TURNING ANGLE			
193	193	193	ILLUSTRATION #	193	193	161
			REAR			
-1.02 ±0.55 -1 ± 9/16	-0.54 ±0.55 - 9/16 ± 9/16	-0.96 ±0.55 - 15/16 ± 9/16	CAMBER (Degrees)	-0.54 ±0.55 - 9/16 ± 9/16	-0.96 ±0.55 - 15/16 ± 9/16	-0.6 ±0.65 -5/8 ± 5/8
0.2° ±0.3° 3/16° ± 5/16°	0.2° ±0.3° 3/16° ± 5/16°	0.2° ±0.3° 3/16° ± 5/16°	REAR TOE (Degrees)	0.2° ±0.3° 3/16° ± 5/16°	0.2° ±0.3° 3/16° ± 5/16°	0.0° ±0.3° 0° ± 5/16°
0.1" ±0.15" 3/32" ± 5/32"	0.1" ±0.15" 3/32" ± 5/32"	0.1" ±0.15" 3/32" ± 5/32"	REAR TOE (Inches)	0.1" ±0.15" 3/32" ± 5/32"	0.1" ±0.15" 3/32" ± 5/32"	0.0" ±0.15" 0" ± 5/32"
0.15 (5/32)	0.15 (5/32)	0.15 (5/32)	THRUST ANGLE	0.15 (5/32)	0.15 (5/32)	0.15 (5/32)
130 06	130 06	130 06	ILLUSTRATION #	130 06	130 06	174
			NOTES			
13123	13124	13125		13126	13127	13128

Dodge (Ram) Truck & SUV

Journey FWD	Journey AWD	Mini Ram Van Cargo Van	MODEL INFORMATION	Nitro	Ram 50 (D50) 4X2	Ram 50 (D50) 4X4
2011-2017	2011-2017	1990	YEARS	2007-2011	1990-93	1990-93
			INSTRUCTIONS			
100 ft.lbs.	100 ft.lbs.	95 ft.lbs.	LUGNUT TORQUE	85-115 ft.lbs.	87-101 ft.lbs.	87-101 ft.lbs.
			FRONT			
3.0 ±0.6 3 ± 5/8	3.0 ±0.6 3 ± 5/8	0.4 ±0.5 3/8 ± ½	CASTER (Degrees)	3.3 ±0.5 3 5/16 ± ½	2.5 ±1.0 2½ ±1	2.0 ±1.0 2 ±1
0.0 (0) | 0.6 (5/8)	0.0 (0) | 0.6 (5/8)	0.0 (0) | 1.5 (1½)	CROSS | TOL.	0.0 (0) | 0.5 (½)	0.0 (0) | 0.5 (½)	0.0 (0) | 0.5 (½)
L= 0.15, R= -0.25 ±0.5 L= 1/8, R= -1/4 ± ½	L= 0.15, R= -0.35 ±0.5 L= 1/8, R= -3/8 ± ½	0.3 ±0.5 5/16 ± ½	CAMBER (Degrees)	-0.38 ±0.38 -3/8 ± 3/8	0.67 ±0.5 11/16 ± ½	1.0 ±0.5 1 ± ½
0.4 (3/8) | 0.5 (½)	0.5 (½) | 0.5 (½)	0.0 (0) | n/s	CROSS | TOL.	0.0 (0) | 0.7 (11/16)	0.0 (0) | 0.5 (½)	0.0 (0) | 0.5 (½)
0.1° ±0.25° 3/32° ± 1/4°	0.1° ±0.25° 3/32° ± 1/4°	0.1° ±0.1° 3/32° ± 3/32°	FRONT TOE (Degrees)	0.2° ±0.13° 3/16° ± 1/8°	0.43° ±0.27° 7/16° ± 9/32°	0.43° ±0.27° 7/16° ± 9/32°
0.05" ±0.13" 1/16" ± 1/8"	0.05" ±0.13" 1/16" ± 1/8"	0.05" ±0.05" 1/16" ± 1/16"	FRONT TOE (Inches)	0.1" ±0.06" 3/32" ± 1/16"	0.22" ±0.14" 7/32" ± 1/8"	0.22" ±0.14" 7/32" ± 1/8"
			SAI			
			TURNING ANGLE		37.0 / 30.4 ±1.5	30.0 / 27.0 ±1.5
161	161	22	ILLUSTRATION #	12	23	23
			REAR			
-0.5 ±0.55 -½ ± 9/16	-0.36 ±0.55 -3/8 ± 9/16	-0.2 ±0.5 - 3/16 ± ½	CAMBER (Degrees)			
0.2° ±0.3° 3/16° ± 5/16°	0.2° ±0.3° 3/16° ± 5/16°	0.0° ±0.6° 0° ± 19/32°	REAR TOE (Degrees)			
0.1" ±0.15" 3/32" ± 5/32"	0.1" ±0.15" 3/32" ± 5/32"	0.0" ±0.3" 0" ± 5/16"	REAR TOE (Inches)			
0.15 (5/32)	0.15 (5/32)	0.4 (13/32)	THRUST ANGLE	0.25 (¼)		
174	174	76	ILLUSTRATION #			
			NOTES			
13129	13130	13132		13135	13139	13141

Dodge (Ram) Truck & SUV

Ramcharger 4X2	Ramcharger 4X4	Sprinter 2500 118" Wheelbase	MODEL INFORMATION	Sprinter 2500 140" & 158" W'base	Sprinter 3500	Sprinter 2500, 3500 w/o reinforced f. axle
1990-93	1990-93	2003-2006	YEARS	2003-2006	2003-2006	2007-2009
			INSTRUCTIONS			
85-115 ft.lbs.			LUGNUT TORQUE			177 ft.lbs.[1]
			FRONT			
0.5 ± 1.5 ½ ± 1½	2.0 ± 1.5 2 ± 1½	1.0 ± 0.75 1 ± ¾	CASTER (Degrees)	3.0 ± 0.75 3 ± ¾	0.5 ± 0.75 ½ ± ¾	3.6 ± 1.5 3⅝ ± 1½
0.0 (0) | n/s	0.0 (0) | n/s	0.0 (0) | n/s	CROSS | TOL.	0.0 (0) | n/s	0.0 (0) | n/s	0.0 (0) | n/s
0.5 ± 0.5 ½ ± ½	0.0 ± 0.5 0 ± ½	0.5 ± 0.75 ½ ± ¾	CAMBER (Degrees)	0.5 ± 0.75 ½ ± ¾	1.0 ± 0.75 1 ± ¾	1.0 ± 0.3 1 ± 5/16
0.0 (0) | n/s	0.0 (0) | n/s	0.0 (0) | n/s	CROSS | TOL.	0.0 (0) | n/s	0.0 (0) | n/s	0.0 (0) | n/s
0.25° ± 0.25° ¼° ± ¼°	0.2° ± 0.25° 3/16° ± ¼°	0.25° ± 0.1° ¼° ± 3/32°	FRONT TOE (Degrees)	0.25° ± 0.1° ¼° ± 3/32°	0.3° ± 0.1° 5/16° ± 3/32°	0.2° ± 0.1° 3/16° ± 3/32°
0.13" ± 0.13" ⅛" ± ⅛"	0.1" ± 0.13" 3/32" ± ⅛"	0.13" ± 0.05" ⅛" ± 1/16"	FRONT TOE (Inches)	0.13" ± 0.05" ⅛" ± 1/16"	0.15" ± 0.05" 5/32" ± 1/16"	0.1" ± 0.05" 3/32" ± 1/16"
	8.5 (8½)	0.0 ± 0.5 (0 ± ½)	SAI	0.0 ± 0.5 (0 ± ½)	0.0 ± 0.5 (0 ± ½)	
			TURNING ANGLE			
24			ILLUSTRATION #			102
			REAR			
			CAMBER (Degrees)			
			REAR TOE (Degrees)			
			REAR TOE (Inches)			
			THRUST ANGLE			
			ILLUSTRATION #			
			NOTES			[1] Alum. Whls.= 130ft.lbs
13144	13148	13151		13152	13153	13154

Dodge (Ram) Truck & SUV — Eagle

Sprinter 3500 w/reinforced f. axle	Notes	Notes	MODEL INFORMATION	2000GTX FWD (Canada)	2000GTX AWD (Canada)	Premier
2007-2009			YEARS	1990-93	1990-93	1990-92
			INSTRUCTIONS			
177 ft.lbs.[1]			LUGNUT TORQUE	65-80 ft.lbs.	65-80 ft.lbs.	63 ft.lbs.[1]
			FRONT			
3.75 ± 1.5 3¾ ± 1½			CASTER (Degrees)	2.0 ± 0.5 2 ± ½	1.93 ± 0.5 1 15/16 ± ½	2.0 ± 0.5 2 ± ½
0.0 (0) | n/s			CROSS | TOL.	0.0 (0) | n/s	0.0 (0) | n/s	0.0 (0) | n/s
1.3 ± 0.3 1 5/16 ± 5/16			CAMBER (Degrees)	0.37 ± 0.5 ⅜ ± ½	0.52 ± 0.5 ½ ± ½	0.33 ± 0.25 5/16 ± ¼
0.0 (0) | n/s			CROSS | TOL.	0.0 (0) | n/s	0.0 (0) | n/s	0.0 (0) | n/s
0.25° ± 0.1° ¼° ± 3/32°			FRONT TOE (Degrees)	0.0° ± 0.24° 0° ± ¼°	0.0° ± 0.24° 0° ± ¼°	0.0° ± 0.3° 0° ± 5/16°
0.13" ± 0.05" ⅛" ± 1/16"			FRONT TOE (Inches)	0.0" ± 0.12" 0" ± ⅛"	0.0" ± 0.12" 0" ± ⅛"	0.0" ± 0.15" 0" ± 5/32"
			SAI			
			TURNING ANGLE	37.5 / 30.8 ±2.0	37.5 / 30.8 ±2.0 [1]	
102			ILLUSTRATION #	△	△	
			REAR			
			CAMBER (Degrees)	-0.75 ± 0.5 -¾ ± ½	-1.0 ± 0.5 -1 ± ½	-0.5 ± 0.5 -½ ± ½
			REAR TOE (Degrees)	0.0° ± 0.24° 0° ± ¼°	0.24° ± 0.24° ¼° ± ¼°	0.0° ± 0.25° 0° ± ¼°
			REAR TOE (Inches)	0.0" ± 0.12" 0" ± ⅛"	0.12" ± 0.12" ⅛" ± ⅛"	0.0" ± 0.13" 0" ± ⅛"
			THRUST ANGLE			
			ILLUSTRATION #		44	
[1] Alum. Whls.= 130ft.lbs			NOTES		[1] w/AT 35°/29.3°	[1] Alum. Whls. 90 ft.lbs.
13155	13159	13160		31001	31002	31005

Eagle

Summit	Summit	Summit Wagon FWD	MODEL INFORMATION	Summit Wagon AWD	Talon FWD 2.0L	Talon FWD 1.8L
1990-92	1993-96	1992-96	YEARS	1992-96	1990-94	1993-94
			INSTRUCTIONS			
65-80 ft.lbs.	65-80 ft.lbs.	65-80 ft.lbs.	LUGNUT TORQUE	65-80 ft.lbs.	87-101 ft.lbs.	87-101 ft.lbs.
			FRONT			
2.33 ±0.5 2 5⁄16 ± ½	2.25 ±0.5 2¼ ± ½	2.17 ±0.67 2 3⁄16 ± 11⁄16	CASTER (Degrees)	2.08 ±0.67 2 1⁄16 ± 11⁄16	2.4 ±0.5 2⅜ ± ½	2.33 ±0.5 2 5⁄16 ± ½
0.0 (0) | n/s	0.0 (0) | n/s	0.0 (0) | n/s	CROSS | TOL.	0.0 (0) | n/s	0.0 (0) | n/s	0.0 (0) | n/s
0.0 ±0.5 0 ± ½	0.0 ±0.5 0 ± ½	0.33 ±0.5 5⁄16 ± ½	CAMBER (Degrees)	0.67 ±0.5 11⁄16 ± ½	0.08 ±0.5 1⁄16 ± ½	0.23 ±0.5 ¼ ± ½
0.0 (0) | n/s	0.0 (0) | n/s	0.0 (0) | n/s	CROSS | TOL.	0.0 (0) | n/s	0.0 (0) | n/s	0.0 (0) | n/s
0.0° ±0.24° 0° ± ¼°	0.0° ±0.24° 0° ± ¼°	0.0° ±0.24° 0° ± ¼°	FRONT TOE (Degrees)	0.0° ±0.24° 0° ± ¼°	0.0° ±0.24° 0° ± ¼°	0.0° ±0.24° 0° ± ¼°
0.0" ±0.12" 0" ± ⅛"	0.0" ±0.12" 0" ± ⅛"	0.0" ±0.12" 0" ± ⅛"	FRONT TOE (Inches)	0.0" ±0.12" 0" ± ⅛"	0.0" ±0.12" 0" ± ⅛"	0.0" ±0.12" 0" ± ⅛"
			SAI			
37.0 / 32.5 ±1.5 [1]	37.3 / 31.0 ±1.0	36.5 / 30.5 ±2.0	TURNING ANGLE	36.5 / 30.5 ±2.0	35.0 / 29.3 ±2.0 [1,2]	35.0 / 29.3 ±2.0 [1]
			ILLUSTRATION #			
			REAR			
-0.67 ±0.58 -1 1⁄16 ± 9⁄16	-0.67 ±0.5 -1 1⁄16 ± ½	-0.5 ±0.5 -½ ± ½	CAMBER (Degrees)	-0.5 ±0.5 -½ ± ½	-0.75 ±0.5 -¾ ± ½	-0.75 ±0.5 -¾ ± ½
0.0° ±0.36° 0° ± ⅜°	0.24° ±0.16° ¼° ± 5⁄32°	0.16° +0.24° -0.16° 5⁄32° + ¼° - 5⁄32°	REAR TOE (Degrees)	0.16° +0.24° -0.16° 5⁄32° + ¼° - 5⁄32°	0.0° ±0.24° 0° ± ¼°	0.0° ±0.24° 0° ± ¼°
0.0" ±0.18" 0" ± 3⁄16"	0.12" ±0.08" ⅛" ± 3⁄32"	0.08" +0.12" -0.08" 3⁄32" + ⅛" - 3⁄32"	REAR TOE (Inches)	0.08" +0.12" -0.08" 3⁄32" + ⅛" - 3⁄32"	0.0" ±0.12" 0" ± ⅛"	0.0" ±0.12" 0" ± ⅛"
			THRUST ANGLE			
	40 05	27	ILLUSTRATION #	27		
[1] w/MS 41º/35º			NOTES		[1] w/MS 37º/30º [2] w/Turbo A/T 31.5º/27º	[1] w/MS 37º/30º
31006	31007	31008		31009	31010	31011

Eagle

Talon AWD	Talon FWD	Talon AWD	MODEL INFORMATION	Talon FWD & AWD	Talon FWD	Talon AWD
1990-94	1995	1995	YEARS	1996	1997-98	1997-98
			INSTRUCTIONS			
87-101 ft.lbs.	65-80 ft.lbs.	87-101 ft.lbs.	LUGNUT TORQUE	87-101 ft.lbs.	87-101 ft.lbs.	87-101 ft.lbs.
			FRONT			
2.3 ±0.5 2 5⁄16 ± ½	4.67 ±1.5 4 11⁄16 ±1½	4.67 ±1.5 4 11⁄16 ±1½	CASTER (Degrees)	4.67 ±1.5 [1] 4 11⁄16 ±1½ [1]	4.67 ±1.5 4 11⁄16 ±1½	4.67 ±1.5 4 11⁄16 ±1½
0.0 (0) | n/s	0.0 (0) | n/s	0.0 (0) | n/s	CROSS | TOL.	0.0 (0) | n/s	0.0 (0) | n/s	0.0 (0) | n/s
0.17 ±0.5 3⁄16 ± ½	-0.08 ±0.5 -1⁄16 ± ½	-0.08 ±0.5 -1⁄16 ± ½	CAMBER (Degrees)	-0.08 ±0.5 [2] -1⁄16 ± ½ [2]	-0.08 ±0.5 -1⁄16 ± ½	-0.08 ±0.5 -1⁄16 ± ½
0.0 (0) | n/s	0.0 (0) | n/s	0.0 (0) | n/s	CROSS | TOL.	0.0 (0) | n/s	0.0 (0) | n/s	0.0 (0) | n/s
0.0° ±0.24° 0° ± ¼°	0.0° ±0.24° 0° ± ¼°	0.0° ±0.24° 0° ± ¼°	FRONT TOE (Degrees)	0.0° ±0.24° 0° ± ¼°	0.0° ±0.24° 0° ± ¼°	0.0° ±0.24° 0° ± ¼°
0.0" ±0.12" 0" ± ⅛"	0.0" ±0.12" 0" ± ⅛"	0.0" ±0.12" 0" ± ⅛"	FRONT TOE (Inches)	0.0" ±0.12" 0" ± ⅛"	0.0" ±0.12" 0" ± ⅛"	0.0" ±0.12" 0" ± ⅛"
		7.3 (7 5⁄16)	SAI	7.3 (7 5⁄16)	7.3 (7 5⁄16)	7.3 (7 5⁄16)
31.5 / 27.0 ±2.0 [1,2]	32.5 / 27.0 ±2.0	32.5 / 27.0 ±2.0	TURNING ANGLE	32.5 / 27.0 ±2.0	32.5 / 27.0 ±2.0	31.5 / 26.5 ±2.0
			ILLUSTRATION #			
			REAR			
-1.55 ±0.5 -1 9⁄16 ± ½	-1.33 ±0.5 [1] -1 5⁄16 ± ½ [1]	-1.33 ±0.5 [1] -1 5⁄16 ± ½ [1]	CAMBER (Degrees)	-1.33 ±0.5 -1 5⁄16 ± ½	-1.33 ±0.5 [1] -1 5⁄16 ± ½ [1]	-1.33 ±0.5 -1 5⁄16 ± ½
0.28° ±0.24° 9⁄32° ± ¼°	0.24° ±0.24° ¼° ± ¼°	0.24° ±0.24° ¼° ± ¼°	REAR TOE (Degrees)	0.24° ±0.24° ¼° ± ¼°	0.24° ±0.24° ¼° ± ¼°	0.24° ±0.24° ¼° ± ¼°
0.14" ±0.12" ⅛" ± ⅛"	0.12" ±0.12" ⅛" ± ⅛"	0.12" ±0.12" ⅛" ± ⅛"	REAR TOE (Inches)	0.12" ±0.12" ⅛" ± ⅛"	0.12" ±0.12" ⅛" ± ⅛"	0.12" ±0.12" ⅛" ± ⅛"
			THRUST ANGLE			
44	06	06	ILLUSTRATION #	06	06	06
[1] w/MS 37º/30º [2] w/Man Trans. 35º/29.3º	[1] w/16" tires -1.7º (-1 11⁄16º)	[1] w/MT -1.7º (-1 11⁄16º)	NOTES	[1] w/14" tires 4.3º (4 5⁄16º) [2] w/14" tires 0.0º (0º)	[1] w/16" tires -1.7º (-1 11⁄16º)	
31012	31013	31014		31015	31016	31017

Eagle — Fiat

Vision	Vista Wagon 4X2 (Canada)	Vista Wagon 4X4 (Canada)	MODEL INFORMATION	124 Spider	500 exc/Abarth	500 Abarth
1993-97	1990-91	1990-91	YEARS	2017	2012-2017	2012-2017
			INSTRUCTIONS			
95 ft.lbs.	50-57 ft.lbs.[1]	50-57 ft.lbs.[1]	LUGNUT TORQUE	94 ft.lbs.	74 ft.lbs.[1]	74 ft.lbs.[1]
			FRONT			
3.0 ± 1.0 3 ± 1	0.8 ± 0.5 13/16 ± ½	0.83 ± 0.5 13/16 ± ½	CASTER (Degrees)	7.85 ± 0.25 7⅞ ± ¼	2.5 ± 0.5 2½ ± ½	2.65 ± 0.5 2⅝ ± ½
0.0 (0) | 1.0 (1)	0.0 (0) | n/s	0.0 (0) | n/s	CROSS | TOL.	0.0 (0) | n/s	0.0 (0) | 0.5 (½)	0.0 (0) | 0.5 (½)
0.0 ± 0.6 0 ± ⅝	0.42 ± 0.5 7/16 ± ½	0.83 ± 0.5 13/16 ± ½	CAMBER (Degrees)	-0.12 ± 0.17 -⅛ ± 3/16	-0.6 ± 0.5 -⅝ ± ½	-1.5 +0.5 -1.0 -1½ +½ -1
0.0 (0) | 0.7 (11/16)	0.0 (0) | n/s	0.0 (0) | n/s	CROSS | TOL.	0.0 (0) | n/s	0.0 (0) | 0.55 (9/16)	0.0 (0) | 0.55 (9/16)
0.2° ± 0.2° 3/16° ± 3/16°	0.0° ± 0.24° 0° ± ¼°	0.0° ± 0.24° 0° ± ¼°	FRONT TOE (Degrees)	0.15° ± 0.33° 5/32° ± 11/32°	0.0° ± 0.25° 0° ± ¼°	0.0° ± 0.25° 0° ± ¼°
0.1" ± 0.1" 3/32" ± 3/32"	0.0" ± 0.12" 0" ± ⅛"	0.0" ± 0.12" 0" ± ⅛"	FRONT TOE (Inches)	0.08" ± 0.17" 1/16" ± 5/32"	0.0" ± 0.13" 0" ± ⅛"	0.0" ± 0.13" 0" ± ⅛"
14.0 (14)			SAI			
	37.8 / 30.7 ± 1.5	37.8 / 30.7 ± 1.5	TURNING ANGLE			
	03	03	ILLUSTRATION #	173		
			REAR			
-0.1 ± 0.5 -⅛ ± ½	-0.58 ± 0.58 -9/16 ± 9/16	0.0 ± 0.5 0 ± ½	CAMBER (Degrees)	-0.95 ± 0.2 -15/16 ± 3/16	-0.9 ± 0.5 -⅞ ± ½	-0.93 ± 0.5 -15/16 ± ½
0.1° ± 0.3° 3/32° ± 5/16°	0.0° ± 0.24° 0° ± ¼°	0.0° ± 0.24° 0° ± ¼°	REAR TOE (Degrees)	0.28° ± 0.33° 9/32° ± 11/32°	0.5° ± 0.25° ½° ± ¼°	0.56° ± 0.25° 9/16° ± ¼°
0.05" ± 0.15" 1/16" ± 5/32"	0.0" ± 0.12" 0" ± ⅛"	0.0" ± 0.12" 0" ± ⅛"	REAR TOE (Inches)	0.14" ± 0.17" ⅛" ± 5/32"	0.25" ± 0.13" ¼" ± ⅛"	0.28" ± 0.13" 9/32" ± ⅛"
0.15 (5/32)			THRUST ANGLE		0.18 (3/16)	0.18 (3/16)
10		77	ILLUSTRATION #	160		
	[1] Alum. Whls. 65-80 ft.lbs.	[1] Alum. Whls. 65-80 ft.lbs.	NOTES		[1] Steel Wheels = 66 ft.lbs	[1] Steel Wheels = 66 ft.lbs
31018	31019	31020		15006	15001	15002

Fiat — Ford

500L	500X 2WD	500X 4WD	MODEL INFORMATION	Aspire	C-MAX All models	Contour GS/LS
2014-2017	2016-2017	2016-2017	YEARS	1994-97	2013-2017	1995-96
			INSTRUCTIONS			
74 ft.lbs.[1]	89 ft.lbs.	89 ft.lbs.	LUGNUT TORQUE	65-87 ft.lbs.	100 ft.lbs.	85-105 ft.lbs.
			FRONT			
5.07 ± 0.33 5 1/16 ± 5/16	5.12 ± 0.5 5⅛ ± ½	4.93 ± 0.5 4 15/16 ± ½	CASTER (Degrees)	1.67 ± 0.75 1 11/16 ± ¾	4.42 ± 0.75 4 7/16 ± ¾	2.3 ± 1.0 2 5/16 ± 1
0.0 (0) | 0.5 (½)	0.0 (0) | n/s	0.0 (0) | n/s	CROSS | TOL.	0.0 (0) | n/s	0.0 (0) | 0.75 (¾)	0.0 (0) | 0.85 (⅞)
-0.5 ± 0.33 -½ ± 5/16	-0.5 ± 0.5 -½ ± ½	-0.75 ± 0.5 -¾ ± ½	CAMBER (Degrees)	0.83 ± 0.75 13/16 ± ¾	L= -0.57, R= -0.42 ± 0.75 L= -9/16, R= -7/16 ± ¾	-0.4 ± 1.0 -⅜ ± 1
0.0 (0) | 0.55 (9/16)	0.0 (0) | n/s	0.0 (0) | n/s	CROSS | TOL.	0.0 (0) | n/s	-0.15 (-⅛) | 0.75 (¾)	0.0 (0) | 0.7 (11/16)
-0.13° ± 0.13° -⅛° ± ⅛°	0.06° ± 0.14° 1/16° ± ⅛°	-0.04° ± 0.14° -1/32° ± ⅛°	FRONT TOE (Degrees)	0.28° ± 0.24° 9/32° ± ¼°	0.2° ± 0.2° 3/16° ± 3/16°	-0.34° ± 0.17° -11/32° ± 5/32°
-0.07" ± 0.07" -1/16" ± 1/16"	0.03" ± 0.07" 1/32" ± 1/16"	-0.02" ± 0.07" -1/64" ± 1/16"	FRONT TOE (Inches)	0.14" ± 0.12" ⅛" ± ⅛"	0.1" ± 0.1" 3/32" ± 3/32"	-0.17" ± 0.09" -5/32" ± 3/32"
			SAI	14.0 ± 0.4 (14 ± ⅜)		
			TURNING ANGLE			20.0 / 18.3
			ILLUSTRATION #	47	190	▲
			REAR			
-1.17 ± 0.33 -1 3/16 ± 5/16	-0.92 ± 0.5 -15/16 ± ½	-1.08 ± 0.5 -1 1/16 ± ½	CAMBER (Degrees)	-0.25 ± 0.75 -¼ ± ¾	-1.61 ± 0.75 -1⅝ ± ¾	-0.4 ± 1.0 -⅜ ± 1
0.27° ± 0.27° 9/32° ± 9/32°	0.2° ± 0.14° 3/16° ± ⅛°	0.2° ± 0.14° 3/16° ± ⅛°	REAR TOE (Degrees)	0.24° ± 0.24° ¼° ± ¼°	0.38° ± 0.2° ⅜° ± 3/16°	0.3° ± 0.2° 5/16° ± 3/16°
0.14" ± 0.14" ⅛" ± ⅛"	0.1" ± 0.07" 3/32" ± 1/16"	0.1" ± 0.07" 3/32" ± 1/16"	REAR TOE (Inches)	0.12" ± 0.12" ⅛" ± ⅛"	0.19" ± 0.1" 3/16" ± 3/32"	0.15" ± 0.1" 5/32" ± 3/32"
0.18 (3/16)			THRUST ANGLE		0.5 (½)	
	06	06	ILLUSTRATION #		174	06
[1] Steel Wheels = 63 ft.lbs			NOTES			
15003	15004	15005		16001	16002	16003

ALIGNMENT SPECIFICATIONS

Ford

MODEL INFORMATION	Contour GS/LS	Contour SE	Contour SE	Crown Victoria, LTD Sedan and Wagon	Crown Victoria, LTD Sedan and Wagon	Crown Victoria, LTD
YEARS	1997-99	1995-96	1997-2000	1990	1991-94	1995-97
INSTRUCTIONS						
LUGNUT TORQUE	85-105 ft.lbs.	85-105 ft.lbs.	85-105 ft.lbs.	85-105 ft.lbs.	85-105 ft.lbs.	85-105 ft.lbs.
FRONT						
CASTER (Degrees)	2.3 ± 1.0 2 5⁄16 ± 1	2.4 ± 1.0 2⅜ ± 1	2.4 ± 1.0 2⅜ ± 1	3.5 ± 1.0 3½ ± 1	5.5 ± 0.75 [1,2] 5½ ± ¾ [1,2]	5.5 ± 0.75 5½ ± ¾
CROSS | TOL.	0.0 (0) | 1.0 (1)	0.0 (0) | 0.85 (⅞)	0.0 (0) | 1.0 (1)	0.0 (0) | 0.75 (¾)	0.0 (0) | 0.75 (¾)	0.0 (0) | 0.75 (¾)
CAMBER (Degrees)	-0.46 ± 1.25 - 7⁄16 ± 1¼	-0.53 ± 1.0 - ½ ± 1	-0.55 ± 1.25 - 9⁄16 ± 1¼	-0.5 ± 0.75 - ½ ± ¾	-0.5 ± 0.75 [2] - ½ ± ¾ [2]	-0.5 ± 0.75 - ½ ± ¾
CROSS | TOL.	0.0 (0) | 1.0 (1)	0.0 (0) | 0.7 (11⁄16)	0.0 (0) | 1.0 (1)	0.0 (0) | 0.75 (¾)	0.0 (0) | 0.75 (¾)	0.0 (0) | 0.75 (¾)
FRONT TOE (Degrees)	-0.17° ± 0.17° - 5⁄32° ± 5⁄32°	-0.34° ± 0.17° - 11⁄32° ± 5⁄32°	-0.17° ± 0.17° - 5⁄32° ± 5⁄32°	0.13° ± 0.25° ⅛° ± ¼°	-0.13° ± 0.25° - ⅛° ± ¼°	-0.13° ± 0.25° - ⅛° ± ¼°
FRONT TOE (Inches)	-0.09" ± 0.09" - 3⁄32" ± 3⁄32"	-0.17" ± 0.09" - 5⁄32" ± 3⁄32"	-0.09" ± 0.09" - 3⁄32" ± 3⁄32"	0.06" ± 0.13" 1⁄16" ± ⅛"	-0.06" ± 0.13" - 1⁄16" ± ⅛"	-0.06" ± 0.13" - 1⁄16" ± ⅛"
SAI						
TURNING ANGLE	20.0 / 18.3	20.0 / 18.3	20.0 / 18.3	20.0 / 18.5	20.0 / 18.5	20.0 / 18.5
ILLUSTRATION #	▲	▲	▲	42	70	70
REAR						
CAMBER (Degrees)	-0.4 ± 1.0 - ⅜ ± 1	-0.53 ± 1.0 - ½ ± 1	-0.53 ± 1.0 - ½ ± 1			
REAR TOE (Degrees)	0.28° ± 0.2° 9⁄32° ± 3⁄16°	0.3° ± 0.2° 5⁄16° ± 3⁄16°	0.28° ± 0.2° 9⁄32° ± 3⁄16°			
REAR TOE (Inches)	0.14" ± 0.1" ⅛" ± 3⁄32"	0.15" ± 0.1" 5⁄32" ± 3⁄32"	0.14" ± 0.1" ⅛" ± 3⁄32"			
THRUST ANGLE						
ILLUSTRATION #	06	06	06			
NOTES					[1] 1991 = 3.5° (3½°) [2] 1991 see img. #42	
	16004	16005	16006	16011	16014	16015

Ford

MODEL INFORMATION	Crown Victoria, LTD	Crown Victoria Base Suspension	Crown Victoria Air Level Susp.	Crown Victoria Air Level Susp.	Crown Victoria Police Models	Crown Victoria Base Suspension
YEARS	1998-2002	2003-2005	2003-2004	2005	2005	2006-2007
INSTRUCTIONS						
LUGNUT TORQUE	85-105 ft.lbs.	95 ft.lbs.	95 ft.lbs.	95 ft.lbs.	95 ft.lbs.	100 ft.lbs.
FRONT						
CASTER (Degrees)	5.5 ± 0.75 5½ ± ¾	6.2 ± 0.75 6 3⁄16 ± ¾	6.5 ± 0.75 6½ ± ¾	6.5 ± 0.75 6½ ± ¾	6.0 ± 0.75 6 ± ¾	L= 6.2, R= 6.5 ± 0.75 L= 6 3⁄16, R= 6½ ± ¾
CROSS | TOL.	0.0 (0) | 0.75 (¾)	0.0 (0) | 0.75 (¾)	0.0 (0) | 0.75 (¾)	0.0 (0) | 0.75 (¾)	0.0 (0) | 0.75 (¾)	-0.3 (- 5⁄16) | 0.75 (¾)
CAMBER (Degrees)	-0.5 ± 0.75 - ½ ± ¾	-0.4 ± 0.75 - ⅜ ± ¾	-0.7 ± 0.75 - 11⁄16 ± ¾	-0.4 ± 0.75 - ⅜ ± ¾	-0.4 ± 0.75 - ⅜ ± ¾	-0.4 ± 0.75 - ⅜ ± ¾
CROSS | TOL.	0.0 (0) | 0.75 (¾)	0.0 (0) | 0.75 (¾)	0.0 (0) | 0.75 (¾)	0.0 (0) | 0.75 (¾)	0.0 (0) | 0.75 (¾)	0.0 (0) | 0.75 (¾)
FRONT TOE (Degrees)	-0.13° ± 0.25° - ⅛° ± ¼°	-0.15° ± 0.2° - 5⁄32° ± 3⁄16°	-0.15° ± 0.2° - 5⁄32° ± 3⁄16°	-0.15° ± 0.2° - 5⁄32° ± 3⁄16°	-0.15° ± 0.2° - 5⁄32° ± 3⁄16°	-0.15° ± 0.2° - 5⁄32° ± 3⁄16°
FRONT TOE (Inches)	-0.06" ± 0.13" - 1⁄16" ± ⅛"	-0.08" ± 0.1" - 1⁄16" ± 3⁄32"	-0.08" ± 0.1" - 1⁄16" ± 3⁄32"	-0.08" ± 0.1" - 1⁄16" ± 3⁄32"	-0.08" ± 0.1" - 1⁄16" ± 3⁄32"	-0.08" ± 0.1" - 1⁄16" ± 3⁄32"
SAI						
TURNING ANGLE	20.0 / 18.5					
ILLUSTRATION #	70	148	148	148	148	148
REAR						
CAMBER (Degrees)						
REAR TOE (Degrees)						
REAR TOE (Inches)						
THRUST ANGLE						
ILLUSTRATION #						
NOTES						
	16016	16017	16018	16019	16020	16021

Ford

Crown Victoria Air Level Susp.	Crown Victoria Police/Taxi/Fleet	Crown Victoria Base Suspension	MODEL INFORMATION	Crown Victoria Air Level Susp.	Crown Victoria Police/Taxi/Fleet	Escort
2006-2007	2006-2007	2008-2011	YEARS	2008-2011	2008-2011	1990
			INSTRUCTIONS			
100 ft.lbs.	100 ft.lbs.	100 ft.lbs.	LUGNUT TORQUE	100 ft.lbs.	100 ft.lbs.	80-105 ft.lbs.
			FRONT			
L= 6.5, R= 6.8 ± 0.75 L= 6½, R= 6 13⁄16 ± ¾	L= 6.0, R= 6.3 ± 0.75 L= 6, R= 6 5⁄16 ± ¾	L= 5.9, R= 6.1 ± 0.75 [1] L= 5⅞, R= 6⅛ ± ¾ [1]	CASTER (Degrees)	L= 6.2, R= 6.4 ± 0.75 [1] L= 6 3⁄16, R= 6⅜ ± ¾ [1]	L= 5.7, R= 5.9 ± 0.75 [1] L= 5 11⁄16, R= 5⅞ ± ¾ [1]	2.3 ±0.75 2 5⁄16 ± ¾
-0.3 (-5⁄16) | 0.75 (¾)	-0.3 (-5⁄16) | 0.75 (¾)	-0.2 (-3⁄16) | 0.75 (¾)	CROSS | TOL.	-0.2 (-3⁄16) | 0.75 (¾)	-0.2 (-3⁄16) | 0.75 (¾)	0.0 (0) | 0.75 (¾)
-0.4 ±0.75 -⅜ ± ¾	-0.4 ±0.75 -⅜ ± ¾	-0.4 ±0.75 -⅜ ± ¾	CAMBER (Degrees)	-0.4 ±0.75 -⅜ ± ¾	-0.4 ±0.75 -⅜ ± ¾	L= 1.35, R= 0.95 ±0.75 [1] L= 1⅜, R= 15⁄16 ± ¾ [1]
0.0 (0) | 0.75 (¾)	0.0 (0) | 0.75 (¾)	0.0 (0) | 0.75 (¾)	CROSS | TOL.	0.0 (0) | 0.75 (¾)	0.0 (0) | 0.75 (¾)	0.4 (⅜) | 0.75 (¾)
-0.15° ±0.2° -5⁄32° ± 3⁄16°	-0.15° ±0.2° -5⁄32° ± 3⁄16°	-0.15° ±0.2° -5⁄32° ± 3⁄16°	FRONT TOE (Degrees)	-0.15° ±0.2° -5⁄32° ± 3⁄16°	-0.15° ±0.2° -5⁄32° ± 3⁄16°	-0.2° ±0.25° -3⁄16° ± ¼°
-0.08" ±0.1" -1⁄16" ± 3⁄32"	-0.08" ±0.1" -1⁄16" ± 3⁄32"	-0.08" ±0.1" -1⁄16" ± 3⁄32"	FRONT TOE (Inches)	-0.08" ±0.1" -1⁄16" ± 3⁄32"	-0.08" ±0.1" -1⁄16" ± 3⁄32"	-0.1" ±0.13" -3⁄32" ± ⅛"
			SAI			
			TURNING ANGLE			20.0 / 18.2
148	148	148	ILLUSTRATION #	148	148	
			REAR			
			CAMBER (Degrees)			-0.35 ±0.85 -⅜ ± ⅞
			REAR TOE (Degrees)			0.36° ±0.36° ⅜° ± ⅜°
			REAR TOE (Inches)			0.18" ±0.18" 3⁄16" ± 3⁄16"
			THRUST ANGLE			
			ILLUSTRATION #			54
		[1] 2008 L=5.9°, R=6.2°	NOTES	[1] 2008 L=6.2°, R=6.5°	[1] 2008 L=5.7°, R=6.0°	[1] GT L=1.0°, R=0.56°
16022	16023	16024		16025	16026	16037

Ford

Escort	Escort exc/Coupe	Escort ZX2 Coupe	MODEL INFORMATION	Festiva	Fiesta except ST	Fiesta ST
1991-96	1997-2002	1998-2003	YEARS	1990-93	2011-2017	2014-2017
			INSTRUCTIONS			
65-87 ft.lbs.	65-87 ft.lbs. [1]	74-100 ft.lbs. [1]	LUGNUT TORQUE	65-87 ft.lbs.	100 ft.lbs.	100 ft.lbs.
			FRONT			
1.25 ±0.75 1¼ ± ¾	2.2 ±1.0 2 3⁄16 ±1	2.2 ±1.0 2 3⁄16 ±1	CASTER (Degrees)	1.58 +0.25 -1.25 1 9⁄16 +¼ -1¼	3.37 ±0.75 3⅜ ± ¾	3.78 ±0.75 3¾ ± ¾
0.0 (0) | 0.75 (¾)	0.0 (0) | 1.0 (1)	0.0 (0) | 1.0 (1)	CROSS | TOL.	0.0 (0) | 0.75 (¾)	0.0 (0) | 0.75 (¾)	0.0 (0) | 0.75 (¾)
0.2 ±0.75 3⁄16 ± ¾	-0.4 ±1.0 -⅜ ±1	-0.4 ±1.0 -⅜ ±1	CAMBER (Degrees)	0.67 ±0.92 11⁄16 ± 15⁄16	-0.7 ±0.75 -11⁄16 ± ¾	-1.18 ±0.75 -1 3⁄16 ± ¾
0.0 (0) | 0.75 (¾)	0.0 (0) | 1.0 (1)	0.0 (0) | 1.0 (1)	CROSS | TOL.	0.0 (0) | 0.92 (15⁄16)	0.0 (0) | 0.75 (¾)	0.0 (0) | 0.75 (¾)
0.2° ±0.3° 3⁄16° ± 5⁄16°	0.2° ±0.25° 3⁄16° ± ¼°	0.2° ±0.25° 3⁄16° ± ¼°	FRONT TOE (Degrees)	0.35° ±0.3° 11⁄32° ± 5⁄16°	0.15° ±0.2° 5⁄32° ± 3⁄16°	0.2° ±0.2° 3⁄16° ± 3⁄16°
0.1" ±0.15" 3⁄32" ± 5⁄32"	0.1" ±0.13" 3⁄32" ± ⅛"	0.1" ±0.13" 3⁄32" ± ⅛"	FRONT TOE (Inches)	0.18" ±0.15" 3⁄16" ± 5⁄32"	0.08" ±0.1" 1⁄16" ± 3⁄32"	0.1" ±0.1" 3⁄32" ± 3⁄32"
	12.9 (12⅞)	12.9 (12⅞)	SAI	14.2 (14 3⁄16)		
40.0 / 33.0 ±2.0			TURNING ANGLE			
46	46	46	ILLUSTRATION #	47		
			REAR			
-0.5 ±1.0 -½ ±1	-1.18 ±1.0 -1 3⁄16 ±1	-1.18 ±1.0 -1 3⁄16 ±1	CAMBER (Degrees)	-0.25 ±0.75 -¼ ± ¾	-1.52 ±0.75 -1½ ± ¾	-0.64 ±0.75 -⅝ ± ¾
0.2° ±0.3° 3⁄16° ± 5⁄16°	0.2° ±0.3° 3⁄16° ± 5⁄16°	0.2° ±0.3° 3⁄16° ± 5⁄16°	REAR TOE (Degrees)	0.3° ±0.3° 5⁄16° ± 5⁄16°	0.31° ±0.3° 5⁄16° ± 5⁄16°	0.31° ±0.3° 5⁄16° ± 5⁄16°
0.1" ±0.15" 3⁄32" ± 5⁄32"	0.1" ±0.15" 3⁄32" ± 5⁄32"	0.1" ±0.15" 3⁄32" ± 5⁄32"	REAR TOE (Inches)	0.15" ±0.15" 5⁄32" ± 5⁄32"	0.16" ±0.15" 5⁄32" ± 5⁄32"	0.16" ±0.15" 5⁄32" ± 5⁄32"
0.4 (13⁄32)			THRUST ANGLE		0.5 (½)	0.5 (½)
10	06	06	ILLUSTRATION #			
	[1] Alloy wheels = 74-100 ft.lbs.	[1] Steel wheels = 65-87 ft.lbs.	NOTES			
16039	16040	16041		16056	16058	16059

ALIGNMENT SPECIFICATIONS

Ford

Five-Hundred FWD	Five-Hundred AWD	Five-Hundred AWD	MODEL INFORMATION	Focus exc/Wagon, SVT	Focus exc/Wagon	Focus SVT
2005-2007	2005	2006-2007	YEARS	2000-2004	2005	2002-2004
			INSTRUCTIONS			
95 ft.lbs.	95 ft.lbs.	95 ft.lbs.	LUGNUT TORQUE	94 ft.lbs.	94 ft.lbs.	94 ft.lbs.
			FRONT			
3.3 ± 0.75 3 5/16 ± ¾	3.5 ± 0.75 3½ ± ¾	3.3 ± 0.75 3 5/16 ± ¾	CASTER (Degrees)	3.0 ± 1.0 3 ± 1	2.75 ± 1.0 2¾ ± 1	3.0 ± 1.0 3 ± 1
0.0 (0) | 0.75 (¾)	0.0 (0) | 0.75 (¾)	0.0 (0) | 0.75 (¾)	CROSS | TOL.	0.0 (0) | 1.0 (1)	0.0 (0) | 1.0 (1)	0.0 (0) | 1.0 (1)
-0.6 ± 0.75 -⅝ ± ¾	-0.4 ± 0.75 -⅜ ± ¾	-0.4 ± 0.75 -⅜ ± ¾	CAMBER (Degrees)	-0.56 ± 1.25 -9/16 ± 1¼	-0.5 ± 1.25 -½ ± 1¼	-0.56 ± 1.25 -9/16 ± 1¼
0.0 (0) | 0.75 (¾)	0.0 (0) | 0.75 (¾)	0.0 (0) | 0.75 (¾)	CROSS | TOL.	0.0 (0) | 1.25 (1¼)	0.0 (0) | 1.25 (1¼)	0.0 (0) | 1.25 (1¼)
0.2° ± 0.2° 3/16° ± 3/16°	0.2° ± 0.2° 3/16° ± 3/16°	0.2° ± 0.2° 3/16° ± 3/16°	FRONT TOE (Degrees)	0.0° ± 0.2° 0° ± 3/16°	0.0° ± 0.2° 0° ± 3/16°	-0.15° ± 0.15° -5/32° ± 5/32°
0.1" ± 0.1" 3/32" ± 3/32"	0.1" ± 0.1" 3/32" ± 3/32"	0.1" ± 0.1" 3/32" ± 3/32"	FRONT TOE (Inches)	0.0" ± 0.1" 0" ± 3/32"	0.0" ± 0.1" 0" ± 3/32"	-0.08" ± 0.08" -1/16" ± 1/16"
			SAI			
			TURNING ANGLE			
167	167	167	ILLUSTRATION #			
			REAR			
-0.35 ± 0.75 -⅜ ± ¾	-0.7 ± 0.75 -11/16 ± ¾	-0.45 ± 0.75 -7/16 ± ¾	CAMBER (Degrees)	-1.0 ± 1.25 -1 ± 1¼	-1.02 ± 1.25 -1 ± 1¼	-1.0 ± 1.25 -1 ± 1¼
0.1° ± 0.2° 3/32° ± 3/16°	0.1° ± 0.2° 3/32° ± 3/16°	0.1° ± 0.2° 3/32° ± 3/16°	REAR TOE (Degrees)	0.5° ± 0.2° ½° ± 3/16°	0.36° ± 0.18° ⅜° ± 3/16°	0.3° ± 0.15° 5/16° ± 5/32°
0.05" ± 0.1" 1/16" ± 3/32"	0.05" ± 0.1" 1/16" ± 3/32"	0.05" ± 0.1" 1/16" ± 3/32"	REAR TOE (Inches)	0.25" ± 0.1" ¼" ± 3/32"	0.18" ± 0.09" 3/16" ± 3/32"	0.15" ± 0.08" 5/32" ± 1/16"
0.3 (5/16)	0.3 (5/16)	0.3 (5/16)	THRUST ANGLE			
06	06	06	ILLUSTRATION #	06	06	06
			NOTES			
16060	16061	16062		16063	16064	16065

Ford

Focus Wagon	Focus Wagon	Focus All models	MODEL INFORMATION	Focus All models	Focus Sedan	Focus Hatchback
2000-2004	2005	2006-2007	YEARS	2008-2011	2012-2017	2013-2017
			INSTRUCTIONS			
94 ft.lbs.	94 ft.lbs.	94 ft.lbs.	LUGNUT TORQUE	94 ft.lbs.	100 ft.lbs.	100 ft.lbs.
			FRONT			
2.88 ± 1.0 2⅞ ± 1	2.75 ± 1.0 2¾ ± 1	2.75 ± 0.75 2¾ ± ¾	CASTER (Degrees)	2.5 ± 0.75 2½ ± ¾	4.28 ± 0.75 4¼ ± ¾	4.23 ± 0.75 4¼ ± ¾
0.0 (0) | 1.0 (1)	0.0 (0) | 1.0 (1)	0.0 (0) | 0.75 (¾)	CROSS | TOL.	0.0 (0) | 0.75 (¾)	0.0 (0) | 0.75 (¾)	0.0 (0) | 0.75 (¾)
-0.63 ± 1.25 -⅝ ± 1¼	-0.5 ± 1.25 -½ ± 1¼	-0.5 ± 0.75 -½ ± ¾	CAMBER (Degrees)	-0.5 ± 0.75 -½ ± ¾	L= -0.52, R= -0.62 ± 0.75 L= -½, R= -⅝ ± ¾	L= -0.83, R= -0.93 ± 0.75 L= -13/16, R= -15/16 ± ¾
0.0 (0) | 1.25 (1¼)	0.0 (0) | 1.25 (1¼)	0.0 (0) | 0.75 (¾)	CROSS | TOL.	0.0 (0) | 0.75 (¾)	0.1 (⅛) | 0.75 (¾)	0.1 (⅛) | 0.75 (¾)
0.0° ± 0.2° 0° ± 3/16°	0.0° ± 0.2° 0° ± 3/16°	0.0° ± 0.2° 0° ± 3/16°	FRONT TOE (Degrees)	0.0° ± 0.2° 0° ± 3/16°	0.2° ± 0.2° 3/16° ± 3/16°	0.1° ± 0.2° 3/32° ± 3/16°
0.0" ± 0.1" 0" ± 3/32"	0.0" ± 0.1" 0" ± 3/32"	0.0" ± 0.1" 0" ± 3/32"	FRONT TOE (Inches)	0.0" ± 0.1" 0" ± 3/32"	0.1" ± 0.1" 3/32" ± 3/32"	0.05" ± 0.1" 1/16" ± 3/32"
			SAI			
			TURNING ANGLE			
			ILLUSTRATION #	190 ▲	190	190
			REAR			
-0.94 ± 1.25 -15/16 ± 1¼	-0.94 ± 1.25 -15/16 ± 1¼	-1.0 ± 0.75 -1 ± ¾	CAMBER (Degrees)	-1.0 ± 1.0 -1 ± 1	-1.41 ± 0.75 -1 7/16 ± ¾	-1.35 ± 0.75 -1⅜ ± ¾
0.5° ± 0.2° ½° ± 3/16°	0.36° ± 0.18° ⅜° ± 3/16°	0.36° ± 0.2° ⅜° ± 3/16°	REAR TOE (Degrees)	0.36° ± 0.2° ⅜° ± 3/16°	0.38° ± 0.2° ⅜° ± 3/16°	0.38° ± 0.2° ⅜° ± 3/16°
0.25" ± 0.1" ¼" ± 3/32"	0.18" ± 0.09" 3/16" ± 3/32"	0.18" ± 0.1" 3/16" ± 3/32"	REAR TOE (Inches)	0.18" ± 0.1" 3/16" ± 3/32"	0.19" ± 0.1" 3/16" ± 3/32"	0.19" ± 0.1" 3/16" ± 3/32"
			THRUST ANGLE		0.5 (½)	0.5 (½)
06	06	06	ILLUSTRATION #	06	06	06
			NOTES			
16066	16067	16068		16069	16070	16071

Ford

Focus RS	Focus EV	Freestar	MODEL INFORMATION	Freestar	Freestyle FWD	Freestyle FWD
2017	2013-2017	2004-2005	YEARS	2006-2007	2005	2006-2007
			INSTRUCTIONS			
100 ft.lbs.	100 ft.lbs.	100 ft.lbs.	LUGNUT TORQUE	100 ft.lbs.	95 ft.lbs.	95 ft.lbs.
			FRONT			
4.79 ±1.0 4 13/16 ±1	4.42 ±0.75 4 7/16 ± ¾	3.6 ±0.75 3⅝ ± ¾	CASTER (Degrees)	3.5 ±0.5 3½ ± ½	3.1 ±0.75 3⅛ ± ¾	3.0 ±0.75 3 ± ¾
0.0 (0) | 1.25 (1¼)	0.0 (0) | 0.75 (¾)	0.0 (0) | 0.5 (½)	CROSS | TOL.	0.0 (0) | 0.75 (¾)	0.0 (0) | 0.75 (¾)	0.0 (0) | 0.75 (¾)
-1.51 ±0.75 -1½ ± ¾	L= -0.46, R= -0.56 ±0.75 L= -7/16, R= -9/16 ± ¾	-0.4 ±0.75 -⅜ ± ¾	CAMBER (Degrees)	-0.44 ±0.5 -7/16 ± ½	-0.28 ±0.75 -¼ ± ¾	-0.28 ±0.75 -¼ ± ¾
0.1 (⅛) | 1.25 (1¼)	0.1 (⅛) | 0.75 (¾)	0.0 (0) | 0.75 (¾)	CROSS | TOL.	0.0 (0) | 0.75 (¾)	0.0 (0) | 0.75 (¾)	0.0 (0) | 0.75 (¾)
0.2° ±0.2° 3/16° ± 3/16°	0.2° ±0.2° 3/16° ± 3/16°	**-0.15°** ±0.2° **-5/32°** ± 3/16°	FRONT TOE (Degrees)	**-0.15°** ±0.2° **-5/32°** ± 3/16°	0.2° ±0.2° 3/16° ± 3/16°	0.2° ±0.2° 3/16° ± 3/16°
0.1" ±0.1" 3/32" ± 3/32"	0.1" ±0.1" 3/32" ± 3/32"	**-0.08"** ±0.1" **-1/16"** ± 3/32"	FRONT TOE (Inches)	**-0.08"** ±0.1" **-1/16"** ± 3/32"	0.1" ±0.1" 3/32" ± 3/32"	0.1" ±0.1" 3/32" ± 3/32"
			SAI			
			TURNING ANGLE			
190	190	67	ILLUSTRATION #	67	167	167
			REAR			
-0.97 ±1.25 -1 ±1¼	-1.61 ±0.75 -1⅝ ± ¾	-0.3 ±0.75 -5/16 ± ¾	CAMBER (Degrees)	-0.25 ±0.58 -¼ ± 9/16	-0.57 ±0.75 -9/16 ± ¾	-0.27 ±0.75 -¼ ± ¾
0.3° ±0.2° 5/16° ± 3/16°	0.38° ±0.2° ⅜° ± 3/16°	0.14° ±0.25° ⅛° ± ¼°	REAR TOE (Degrees)	0.14° ±0.25° ⅛° ± ¼°	0.1° ±0.2° 3/32° ± 3/16°	0.1° ±0.2° 3/32° ± 3/16°
0.15" ±0.1" 5/32" ± 3/32"	0.19" ±0.1" 3/16" ± 3/32"	0.07" ±0.13" 1/16" ± ⅛"	REAR TOE (Inches)	0.07" ±0.13" 1/16" ± ⅛"	0.05" ±0.1" 1/16" ± 3/32"	0.05" ±0.1" 1/16" ± 3/32"
0.5 (½)	0.5 (½)		THRUST ANGLE		0.3 (5/16)	0.3 (5/16)
174	174		ILLUSTRATION #		06	06
			NOTES			
16072	16073	16074		16075	16076	16077

Ford

Freestyle AWD	Fusion	Fusion Fusion Hybrid	MODEL INFORMATION	Fusion Fusion Hybrid	GT	Mustang exc/1993 Cobra
2005-2007	2006-2009	2010-2012	YEARS	2013-2017	2005-2006	1990-93
			INSTRUCTIONS			
95 ft.lbs.	98 ft.lbs.	100 ft.lbs.	LUGNUT TORQUE	100 ft.lbs.[1]	111 ft.lbs.	85-105 ft.lbs.
			FRONT			
3.0 ±0.75 3 ± ¾	4.0 ±1.0 4 ±1	L= 3.75, R= 4.25 ± 1.0 L= 3¾, R= 4¼ ±1	CASTER (Degrees)	3.35 ±0.75 3⅜ ± ¾	5.0 ±0.3 5 ± 5/16	1.9 ±0.75 1⅞ ± ¾
0.0 (0) | 0.75 (¾)	0.0 (0) | 1.0 (1)	-0.5 (-½) | 1.0 (1)	CROSS | TOL.	0.0 (0) | 0.75 (¾)	0.0 (0) | 0.4 (⅜)	0.0 (0) | 0.75 (¾)
-0.32 ±0.75 -5/16 ± ¾	-0.03 ±0.75 -1/32 ± ¾	-0.03 ±0.75 -1/32 ± ¾	CAMBER (Degrees)	-0.8 ±0.75 -13/16 ± ¾	L= -0.4, R= -0.35 ±0.3 L= -⅜, R= -⅜ ± 5/16	-0.5 ±0.75 [1] -½ ± ¾ [1]
0.0 (0) | 0.75 (¾)	0.0 (0) | 1.0 (1)	0.0 (0) | 1.0 (1)	CROSS | TOL.	0.0 (0) | 0.75 (¾)	-0.05 (-1/16) | 0.4 (⅜)	0.0 (0) | 0.75 (¾)
0.2° ±0.2° 3/16° ± 3/16°	0.2° ±0.2° 3/16° ± 3/16°	0.2° ±0.2° 3/16° ± 3/16°	FRONT TOE (Degrees)	0.2° ±0.2° 3/16° ± 3/16°	**-0.08°** ±0.1° **-3/32°** ± 3/32°	**-0.25°** ±0.25° **-¼°** ± ¼°
0.1" ±0.1" 3/32" ± 3/32"	0.1" ±0.1" 3/32" ± 3/32"	0.1" ±0.1" 3/32" ± 3/32"	FRONT TOE (Inches)	0.1" ±0.1" 3/32" ± 3/32"	**-0.04"** ±0.05" **-1/32"** ± 1/16"	**-0.13"** ±0.13" **-⅛"** ± ⅛"
			SAI			
			TURNING ANGLE			20.0 / 19.8
167			ILLUSTRATION #	26	168	72
			REAR			
0.0 ±0.75 0 ± ¾	-1.25 ±0.75 -1¼ ± ¾	-1.25 ±0.75 -1¼ ± ¾	CAMBER (Degrees)	-1.2 ±0.75 -1 3/16 ± ¾	-1.44 ±0.3 -1 7/16 ± 5/16	
0.1° ±0.2° 3/32° ± 3/16°	0.24° ±0.2° ¼° ± 3/16°	0.24° ±0.2° ¼° ± 3/16°	REAR TOE (Degrees)	0.3° ±0.2° 5/16° ± 3/16°	0.22° ±0.1° 7/32° ± 3/32°	
0.05" ±0.1" 1/16" ± 3/32"	0.12" ±0.1" ⅛" ± 3/32"	0.12" ±0.1" ⅛" ± 3/32"	REAR TOE (Inches)	0.15" ±0.1" 5/32" ± 3/32"	0.11" ±0.05" ⅛" ± 1/16"	
0.3 (5/16)	0.1 (3/32)	0.5 (½)	THRUST ANGLE	0.5 (½)		
06	175	175	ILLUSTRATION #	174	169 10	
			NOTES	[1] '15-'17 w/V6 = 150 ft.lbs		[1] 5.0L = -0.6° (-⅝°)
16078	16079	16080		16081	16086	16118

Ford

MODEL INFORMATION	Mustang Cobra	Mustang exc/1994-95 Cobra	Mustang Cobra	Mustang exc/Cobra	Mustang Cobra	Mustang
YEARS	1993	1994-98	1994-95	1999-2001	1999-2001	2002
INSTRUCTIONS						
LUGNUT TORQUE	85-105 ft.lbs.	85-105 ft.lbs.	85-105 ft.lbs.	95 ft.lbs.	95 ft.lbs.	95 ft.lbs.
FRONT						
CASTER (Degrees)	1.9 ±0.75 1⅞ ± ¾	3.2 ±0.75 [1] 3 3/16 ± ¾ [1]	3.4 ±0.75 3⅜ ± ¾	3.4 ±0.75 3⅜ ± ¾	3.4 ±0.75 3⅜ ± ¾	3.2 ±0.75 3 3/16 ± ¾
CROSS | TOL.	0.0 (0) | 0.75 (¾)	0.0 (0) | 0.75 (¾)	0.0 (0) | 0.75 (¾)	0.0 (0) | 0.75 (¾)	0.0 (0) | 0.75 (¾)	0.0 (0) | 0.75 (¾)
CAMBER (Degrees)	-0.6 ±0.75 -⅝ ± ¾	-0.6 ±0.75 -⅝ ± ¾	-0.6 ±0.75 -⅝ ± ¾	-0.6 ±0.75 [1] -⅝ ± ¾ [1]	-0.6 ±0.75 -⅝ ± ¾	-0.6 ±0.4 -⅝ ± ⅜
CROSS | TOL.	0.0 (0) | 0.75 (¾)	0.0 (0) | 0.75 (¾)	0.0 (0) | 0.75 (¾)	0.0 (0) | 0.75 (¾)	0.0 (0) | 0.75 (¾)	0.0 (0) | 0.4 (⅜)
FRONT TOE (Degrees)	0.0° ±0.25° 0° ± ¼°	0.25° ±0.25° ¼° ± ¼°	0.0° ±0.25° 0° ± ¼°	0.25° ±0.25° ¼° ± ¼°	0.25° ±0.25° ¼° ± ¼°	0.25° ±0.15° ¼° ± 5/32°
FRONT TOE (Inches)	0.0" ±0.13" 0" ± ⅛"	0.13" ±0.13" ⅛" ± ⅛"	0.0" ±0.13" 0" ± ⅛"	0.13" ±0.13" ⅛" ± ⅛"	0.13" ±0.13" ⅛" ± ⅛"	0.13" ±0.08" ⅛" ± 1/16"
SAI						
TURNING ANGLE	20.0 / 19.8	20.0 / 19.8	20.0 / 19.8			
ILLUSTRATION #	72	72	72	116	116	116
REAR						
CAMBER (Degrees)					-0.7 ±0.2 -1 1/16 ± 3/16	
REAR TOE (Degrees)					0.25° ±0.15° ¼° ± 5/32°	
REAR TOE (Inches)					0.13" ±0.08" ⅛" ± 1/16"	
THRUST ANGLE						
ILLUSTRATION #					115 10	
NOTES		[1] 1994-96=3.4° (3⅜°)		[1] 2001 Bullitt = -0.9° (-⅞°)		
	16119	16120	16121	16122	16123	16124

Ford

MODEL INFORMATION	Mustang except Cobra	Mustang Cobra	Mustang Cobra	Mustang	Mustang exc/Shelby & Boss	Mustang Shelby Models
YEARS	2003-2004	2003	2004	2005-2009	2010-2014	2010
INSTRUCTIONS						
LUGNUT TORQUE	95 ft.lbs.	95 ft.lbs.	95 ft.lbs.	98 ft.lbs.	98 ft.lbs.	98 ft.lbs.
FRONT						
CASTER (Degrees)	3.2 ±0.75 3 3/16 ± ¾	3.2 ±0.75 3 3/16 ± ¾	L= 3.1, R= 3.3 ± 0.2 L= 3⅛, R= 3 5/16 ± 3/16	7.1 ±0.75 7⅛ ± ¾	7.1 ±0.75 7⅛ ± ¾	7.1 ±0.75 7⅛ ± ¾
CROSS | TOL.	0.0 (0) | 0.75 (¾)	0.0 (0) | 0.75 (¾)	-0.2 (-3/16) | 0.2 (3/16)	0.0 (0) | 0.75 (¾)	0.0 (0) | 0.75 (¾)	0.0 (0) | 0.75 (¾)
CAMBER (Degrees)	-0.5 ±0.75 -½ ± ¾	-0.9 ±0.4 -⅞ ± ⅜	L= -0.8, R= -1.0 ± 0.2 L= -13/16, R= -1 ± 3/16	-0.75 ±0.75 -¾ ± ¾	-0.75 ±0.75 -¾ ± ¾	-0.75 ±0.75 -¾ ± ¾
CROSS | TOL.	0.0 (0) | 0.75 (¾)	0.0 (0) | 0.4 (⅜)	0.2 (3/16) | 0.2 (3/16)	0.0 (0) | 0.75 (¾)	0.0 (0) | 0.75 (¾)	0.0 (0) | 0.75 (¾)
FRONT TOE (Degrees)	0.25° ±0.25° ¼° ± ¼°	0.25° ±0.25° ¼° ± ¼°	0.25° ±0.08° ¼° ± 3/32°	0.1° ±0.2° 3/32° ± 3/16°	0.2° ±0.2° 3/16° ± 3/16°	0.0° ±0.2° 0° ± 3/16°
FRONT TOE (Inches)	0.13" ±0.13" ⅛" ± ⅛"	0.13" ±0.13" ⅛" ± ⅛"	0.13" ±0.04" ⅛" ± 1/32"	0.05" ±0.1" 1/16" ± 3/32"	0.1" ±0.1" 3/32" ± 3/32"	0.0" ±0.1" 0" ± 3/32"
SAI						
TURNING ANGLE						
ILLUSTRATION #	116	116	116	161	161	161
REAR						
CAMBER (Degrees)		-0.8 ±0.2 -13/16 ± 3/16	-0.8 ±0.2 -13/16 ± 3/16			
REAR TOE (Degrees)		0.2° ±0.15° 3/16° ± 5/32°	0.2° ±0.08° 3/16° ± 3/32°			
REAR TOE (Inches)		0.1" ±0.08" 3/32" ± 1/16"	0.1" ±0.04" 3/32" ± 1/32"			
THRUST ANGLE					0.5 (½)	
ILLUSTRATION #		115 10	115 10			
NOTES						
	16125	16126	16127	16128	16129	16130

Ford

Mustang Shelby Models	Mustang Boss exc/Laguna Seca	Mustang Boss Laguna Seca	MODEL INFORMATION	Mustang exc Perf., Track	Mustang w/Perf. or Track Pkg.	Mustang GT350
2011	2012-2013	2012-2013	YEARS	2015-2017	2015-2017	2017
			INSTRUCTIONS			
98 ft.lbs.	98 ft.lbs.	100 ft.lbs.	LUGNUT TORQUE	148 ft.lbs.	148 ft.lbs.	148 ft.lbs.
			FRONT			
7.2 ±0.75 7 3/16 ± ¾	7.0 ±0.75 7 ± ¾	7.0 ±0.75 7 ± ¾	CASTER (Degrees)	7.12 ±0.75 7 1/8 ± ¾	6.91 ±0.75 6 15/16 ± ¾	6.81 ±0.75 6 13/16 ± ¾
0.0 (0) | 0.75 (¾)	0.0 (0) | 0.75 (¾)	0.0 (0) | 0.75 (¾)	CROSS | TOL.	0.0 (0) | 0.75 (¾)	0.0 (0) | 0.75 (¾)	0.0 (0) | 0.75 (¾)
-1.0 ±0.75 -1 ± ¾	-0.95 ±0.75 -15/16 ± ¾	-1.05 ±0.75 -1 1/16 ± ¾	CAMBER (Degrees)	-0.72 ±0.75 -¾ ± ¾	-1.03 ±0.75 -1 ± ¾	-1.05 ±0.75 -1 1/16 ± ¾
0.0 (0) | 0.75 (¾)	0.0 (0) | 0.75 (¾)	0.0 (0) | 0.75 (¾)	CROSS | TOL.	0.0 (0) | 0.75 (¾)	0.0 (0) | 0.75 (¾)	0.0 (0) | 0.75 (¾)
0.0° ±0.2° 0° ± 3/16°	0.2° ±0.2° 3/16° ± 3/16°	0.2° ±0.2° 3/16° ± 3/16°	FRONT TOE (Degrees)	0.0° ±0.2° 0° ± 3/16°	0.0° ±0.2° 0° ± 3/16°	-0.1° ±0.2° -3/32° ± 3/16°
0.0" ±0.1" 0" ± 3/32"	0.1" ±0.1" 3/32" ± 3/32"	0.1" ±0.1" 3/32" ± 3/32"	FRONT TOE (Inches)	0.0" ±0.1" 0" ± 3/32"	0.0" ±0.1" 0" ± 3/32"	-0.05" ±0.1" -1/16" ± 3/32"
			SAI			
			TURNING ANGLE			
161	161	161	ILLUSTRATION #	108	108	108
			REAR			
			CAMBER (Degrees)	-1.5 ±0.75 -1 ½ ± ¾	-1.5 ±0.75 -1 ½ ± ¾	-0.75 ±0.75 -¾ ± ¾
			REAR TOE (Degrees)	0.23° ±0.2° 7/32° ± 3/16°	0.23° ±0.2° 7/32° ± 3/16°	0.3° ±0.2° 5/16° ± 3/16°
			REAR TOE (Inches)	0.12" ±0.1" 1/8" ± 3/32"	0.12" ±0.1" 1/8" ± 3/32"	0.15" ±0.1" 5/32" ± 3/32"
	0.5 (½)	0.5 (½)	THRUST ANGLE			
			ILLUSTRATION #	183	183	183
			NOTES			
16131	16132	16133		16134	16135	16136

Ford

Mustang GT350R	Probe up to VIN# 1ZVPT22L2L5170310	Probe from VIN# 1ZVPT22L2L5170311	MODEL INFORMATION	Probe 2.0L	Probe 2.5L	Probe 2.0L
2017	1990	1990-92	YEARS	1993-95	1993-95	1996-97
			INSTRUCTIONS			
148 ft.lbs.	65-87 ft.lbs.	65-87 ft.lbs.	LUGNUT TORQUE	65-87 ft.lbs.	65-87 ft.lbs.	65-87 ft.lbs.
			FRONT			
6.92 ±0.75 6 15/16 ± ¾	1.22 ±0.75 1 3/16 ± ¾	1.68 ±0.75 1 11/16 ± ¾	CASTER (Degrees)	3.02 ±1.0 3 ±1	3.02 ±1.0 3 ±1	2.88 ±1.0 2 7/8 ±1
0.0 (0) | 0.75 (¾)	0.0 (0) | 0.75 (¾)	0.0 (0) | 0.75 (¾)	CROSS | TOL.	0.0 (0) | 1.5 (1½)	0.0 (0) | 1.5 (1½)	0.0 (0) | 1.5 (1½)
-1.12 ±0.75 -1 1/8 ± ¾	-0.27 ±0.75 -¼ ± ¾	-0.27 ±0.75 -¼ ± ¾	CAMBER (Degrees)	-0.7 ±0.5 -11/16 ± ½	-0.9 ±0.5 -7/8 ± ½	-0.7 ±0.75 -11/16 ± ¾
0.0 (0) | 0.75 (¾)	0.0 (0) | 0.5 (½)	0.0 (0) | 0.5 (½)	CROSS | TOL.	0.0 (0) | 0.5 (½)	0.0 (0) | 0.5 (½)	0.0 (0) | 0.5 (½)
0.06° ±0.2° 1/16° ± 3/16°	0.3° ±0.3° 5/16° ± 5/16°	0.0° ±0.3° 0° ± 5/16°	FRONT TOE (Degrees)	0.28° ±0.1° 9/32° ± 3/32°	0.28° ±0.1° 9/32° ± 3/32°	0.28° ±0.1° 9/32° ± 3/32°
0.03" ±0.1" 1/32" ± 3/32"	0.15" ±0.15" 5/32" ± 5/32"	0.0" ±0.15" 0" ± 5/32"	FRONT TOE (Inches)	0.14" ±0.05" 5/32" ± 1/16"	0.14" ±0.05" 5/32" ± 1/16"	0.14" ±0.05" 5/32" ± 1/16"
	12.8 (12 13/16)	12.8 (12 13/16)	SAI	15.4 (15 7/16)	15.7 (15¾)	15.4 (15 7/16)
	36.0 / 31.0 ±2.0	36.0 / 31.0 ±2.0	TURNING ANGLE	37.0 / 32.0 ±2.0	37.0 / 32.0 ±2.0	37.0 / 32.0 ±2.0
108	51	51	ILLUSTRATION #	51	51	51
			REAR			
-0.7 ±0.75 -11/16 ± ¾	-0.43 ±0.75 -7/16 ± ¾	-0.43 ±0.75 -7/16 ± ¾	CAMBER (Degrees)	-0.33 ±1.0 -5/16 ±1	-0.33 ±1.0 -5/16 ±1	-0.7 ±1.0 -11/16 ±1
0.3° ±0.2° 5/16° ± 3/16°	0.3° ±0.3° 5/16° ± 5/16°	0.3° ±0.3° 5/16° ± 5/16°	REAR TOE (Degrees)	0.28° ±0.1° 9/32° ± 3/32°	0.28° ±0.1° 9/32° ± 3/32°	0.28° ±0.1° 9/32° ± 3/32°
0.15" ±0.1" 5/32" ± 3/32"	0.15" ±0.15" 5/32" ± 5/32"	0.15" ±0.15" 5/32" ± 5/32"	REAR TOE (Inches)	0.14" ±0.05" 5/32" ± 1/16"	0.14" ±0.05" 5/32" ± 1/16"	0.14" ±0.05" 5/32" ± 1/16"
			THRUST ANGLE	0.1 (3/32)	0.1 (3/32)	0.1 (3/32)
183	10	10	ILLUSTRATION #	06	06	06
			NOTES			
16137	16146	16147		16148	16149	16150

Ford

MODEL INFORMATION	Probe 2.5L	Taurus Sedan	Taurus Wagon	Taurus Sedan	Taurus Wagon	Taurus Sedan
YEARS	1996-97	1990-95	1990-95	1996-99	1996-99	2000-2007
INSTRUCTIONS						
LUGNUT TORQUE	65-87 ft.lbs.	85-105 ft.lbs.	85-105 ft.lbs.	85-105 ft.lbs.	85-105 ft.lbs.	
FRONT						
CASTER (Degrees)	2.9 ± 1.0 2⅞ ± 1	3.8 ± 1.0 3 13/16 ± 1	3.7 ± 1.0 3 11/16 ± 1	3.8 ± 1.0 3 13/16 ± 1	3.8 ± 1.0 3 13/16 ± 1	3.8 ± 1.0 3 13/16 ± 1
CROSS | TOL.	0.0 (0) | 1.5 (1½)	0.0 (0) | 0.85 (⅞)	0.0 (0) | 0.85 (⅞)	0.0 (0) | 0.85 (⅞)	0.0 (0) | 0.85 (⅞)	0.0 (0) | 0.85 (⅞)
CAMBER (Degrees)	-0.9 ± 0.75 -⅞ ± ¾	-0.5 ± 0.6 -½ ± ⅝	-0.4 ± 0.6 -⅜ ± ⅝	-0.5 ± 0.6 -½ ± ⅝	-0.5 ± 0.6 -½ ± ⅝	-0.5 ± 0.6 -½ ± ⅝
CROSS | TOL.	0.0 (0) | 0.5 (½)	0.0 (0) | 0.7 (11/16)	0.0 (0) | 0.7 (11/16)	0.0 (0) | 0.7 (11/16)	0.0 (0) | 0.7 (11/16)	0.0 (0) | 0.7 (11/16)
FRONT TOE (Degrees)	0.28° ± 0.1° 9/32° ± 3/32°	**-0.2°** +0.5° -0.25° -3/16° +½° -¼°	**-0.2°** +0.5° -0.25° -3/16° +½° -¼°	**-0.2°** +0.5° -0.25° -3/16° +½° -¼°	**-0.2°** +0.5° -0.25° -3/16° +½° -¼°	**-0.2°** ± 0.25° -3/16° ± ¼°
FRONT TOE (Inches)	0.14" ± 0.05" 5/32" ± 1/16"	**-0.1"** +0.25" -0.13" -3/32" +¼" -⅛"	**-0.1"** +0.25" -0.13" -3/32" +¼" -⅛"	**-0.1"** +0.25" -0.13" -3/32" +¼" -⅛"	**-0.1"** +0.25" -0.13" -3/32" +¼" -⅛"	**-0.1"** ± 0.13" -3/32" ± ⅛"
SAI	15.7 (15¾)					
TURNING ANGLE	37.0 / 32.0 ± 2.0	20.0 / 18.3	20.0 / 18.3	20.0 / 18.3	20.0 / 18.3	
ILLUSTRATION #	51	68 ▲	68 ▲	68	68	67
REAR						
CAMBER (Degrees)	-0.9 ± 1.0 -⅞ ± 1	-0.9 ± 0.7 -⅞ ± 11/16	-0.9 ± 1.0 [1] -⅞ ± 1 [1]	-0.6 ± 0.7 -⅝ ± 11/16	-0.6 ± 0.7 -⅝ ± 11/16	-1.0 ± 0.7 -1 ± 11/16
REAR TOE (Degrees)	0.28° ± 0.1° 9/32° ± 3/32°	0.12° ± 0.25° ⅛° ± ¼°	0.12° ± 0.25° ⅛° ± ¼°	0.12° ± 0.25° ⅛° ± ¼°	0.12° ± 0.25° ⅛° ± ¼°	0.36° ± 0.25° ⅜° ± ¼°
REAR TOE (Inches)	0.14" ± 0.05" 5/32" ± 1/16"	0.06" ± 0.13" 1/16" ± ⅛"	0.06" ± 0.13" 1/16" ± ⅛"	0.06" ± 0.13" 1/16" ± ⅛"	0,06" ± 0.13" 1/16" ± ⅛"	0.18" ± 0.13" 3/16" ± ⅛"
THRUST ANGLE	0.1 (3/32)					
ILLUSTRATION #	06	69	64	69	64	05
NOTES			[1] Cam kit available.			
	16151	16154	16155	16156	16157	16158

Ford

MODEL INFORMATION	Taurus Wagon	Taurus FWD	Taurus AWD	Taurus	Taurus exc/SE w/17" wheels	Taurus SE w/17" wheels
YEARS	2000-2005	2008-2009	2008-2009	2010	2011-2012	2011-2012
INSTRUCTIONS						
LUGNUT TORQUE		95 ft.lbs.	95 ft.lbs.	100 ft.lbs.	100 ft.lbs.	100 ft.lbs.
FRONT						
CASTER (Degrees)	3.8 ± 1.0 3 13/16 ± 1	L= 3.1, R= 3.3 ± 0.75 L= 3⅛, R= 3 5/16 ± ¾	L= 3.1, R= 3.3 ± 0.75 L= 3⅛, R= 3 5/16 ± ¾	L= 3.1, R= 3.3 ± 0.75 L= 3⅛, R= 3 5/16 ± ¾	L= 3.2, R= 3.4 ± 0.75 L= 3 3/16, R= 3⅜ ± ¾	L= 3.1, R= 3.3 ± 0.75 L= 3⅛, R= 3 5/16 ± ¾
CROSS | TOL.	0.0 (0) | 0.85 (⅞)	-0.2 (-3/16) | 0.75 (¾)	-0.2 (-3/16) | 0.75 (¾)	-0.2 (-3/16) | 0.75 (¾)	-0.2 (-3/16) | 0.75 (¾)	-0.2 (-3/16) | 0.75 (¾)
CAMBER (Degrees)	-0.5 ± 0.6 -½ ± ⅝	-0.6 ± 0.75 -⅝ ± ¾	-0.4 ± 0.75 -⅜ ± ¾	L= -0.4, R= -0.6 ± 0.75 L= -⅜, R= -⅝ ± ¾	L= -0.4, R= -0.6 ± 0.75 L= -⅜, R= -⅝ ± ¾	L= -0.5, R= -0.7 ± 0.75 L= -½, R= -11/16 ± ¾
CROSS | TOL.	0.0 (0) | 0.7 (11/16)	0.0 (0) | 0.75 (¾)	0.0 (0) | 0.75 (¾)	0.2 (3/16) | 0.75 (¾)	0.2 (3/16) | 0.75 (¾)	0.2 (3/16) | 0.75 (¾)
FRONT TOE (Degrees)	**-0.2°** ± 0.25° -3/16° ± ¼°	0.2° ± 0.2° 3/16° ± 3/16°	0.2° ± 0.2° 3/16° ± 3/16°	0.2° ± 0.2° 3/16° ± 3/16°	0.2° ± 0.2° 3/16° ± 3/16°	0.2° ± 0.2° 3/16° ± 3/16°
FRONT TOE (Inches)	**-0.1"** ± 0.13" -3/32" ± ⅛"	0.1" ± 0.1" 3/32" ± 3/32"	0.1" ± 0.1" 3/32" ± 3/32"	0.1" ± 0.1" 3/32" ± 3/32"	0.1" ± 0.1" 3/32" ± 3/32"	0.1" ± 0.1" 3/32" ± 3/32"
SAI						
TURNING ANGLE						
ILLUSTRATION #	67	167	167	167	167	167
REAR						
CAMBER (Degrees)	-0.6 ± 0.7 -⅝ ± 11/16	-0.35 ± 0.75 -⅜ ± ¾	-0.45 ± 0.75 -7/16 ± ¾	-1.0 ± 0.75 -1 ± ¾	-1.0 ± 0.75 -1 ± ¾	-1.0 ± 0.75 -1 ± ¾
REAR TOE (Degrees)	0.36° ± 0.25° ⅜° ± ¼°	0.1° ± 0.2° 3/32° ± 3/16°	0.1° ± 0.2° 3/32° ± 3/16°	0.24° ± 0.2° ¼° ± 3/16°	0.24° ± 0.2° ¼° ± 3/16°	0.24° ± 0.2° ¼° ± 3/16°
REAR TOE (Inches)	0.18" ± 0.13" 3/16" ± ⅛"	0.05" ± 0.1" 1/16" ± 3/32"	0.05" ± 0.1" 1/16" ± 3/32"	0.12" ± 0.1" ⅛" ± 3/32"	0.12" ± 0.1" ⅛" ± 3/32"	0.12" ± 0.1" ⅛" ± 3/32"
THRUST ANGLE		0.3 (5/16)	0.3 (5/16)	0.3 (5/16)	0.3 (5/16)	0.3 (5/16)
ILLUSTRATION #	64	06	06	187	187	187
NOTES						
	16159	16160	16161	16162	16163	16164

Ford

MODEL INFORMATION	Taurus except Turbo, Police	Taurus Turbo	Taurus Police	Taurus X FWD	Taurus X AWD	Tempo FWD
YEARS	2013-2017	2013-2017	2013-2017	2008-2009	2008-2009	1990-91
INSTRUCTIONS						
LUGNUT TORQUE	100 ft.lbs.	100 ft.lbs.	110 ft.lbs.	95 ft.lbs.	95 ft.lbs.	85-105 ft.lbs.
FRONT						
CASTER (Degrees)	L= 3.1, R= 3.3 ± 0.75 L= 3 1/8, R= 3 5/16 ± 3/4	L= 3.2, R= 3.4 ± 0.75 L= 3 3/16, R= 3 3/8 ± 3/4	L= 2.9, R= 3.1 ± 0.75 L= 2 7/8, R= 3 1/8 ± 3/4	L= 2.8, R= 3.0 ± 0.75 L= 2 13/16, R= 3 ± 3/4	L= 2.8, R= 3.0 ± 0.75 L= 2 13/16, R= 3 ± 3/4	2.45 ± 0.75 2 7/16 ± 3/4
CROSS | TOL.	-0.2 (-3/16) | 0.75 (3/4)	-0.2 (-3/16) | 0.75 (3/4)	-0.2 (-3/16) | 0.75 (3/4)	-0.2 (-3/16) | 0.75 (3/4)	-0.2 (-3/16) | 0.75 (3/4)	0.0 (0) | 0.75 (3/4)
CAMBER (Degrees)	L= -0.55, R= -0.75 ± 0.75 L= -9/16, R= -3/4 ± 3/4	L= -0.65, R= -0.85 ± 0.75 L= -5/8, R= -7/8 ± 3/4	L= -0.55, R= -0.75 ± 0.75 L= -9/16, R= -3/4 ± 3/4	-0.28 ± 0.75 -1/4 ± 3/4	-0.32 ± 0.75 -5/16 ± 3/4	L= 1.4, R= 0.95 ± 0.75 L= 1 3/8, R= 15/16 ± 3/4
CROSS | TOL.	0.2 (3/16) | 0.75 (3/4)	0.2 (3/16) | 0.75 (3/4)	0.2 (3/16) | 0.75 (3/4)	0.0 (0) | 0.75 (3/4)	0.0 (0) | 0.75 (3/4)	0.45 (7/16) | 0.75 (3/4)
FRONT TOE (Degrees)	0.2° ± 0.2° 3/16° ± 3/16°	0.2° ± 0.2° 3/16° ± 3/16°	0.2° ± 0.2° 3/16° ± 3/16°	0.2° ± 0.2° 3/16° ± 3/16°	0.2° ± 0.2° 3/16° ± 3/16°	-0.2° ± 0.25° -3/16° ± 1/4°
FRONT TOE (Inches)	0.1" ± 0.1" 3/32" ± 3/32"	0.1" ± 0.1" 3/32" ± 3/32"	0.1" ± 0.1" 3/32" ± 3/32"	0.1" ± 0.1" 3/32" ± 3/32"	0.1" ± 0.1" 3/32" ± 3/32"	-0.1" ± 0.13" -3/32" ± 1/8"
SAI						
TURNING ANGLE						20.0 / 18.3
ILLUSTRATION #	167	167	167	167	167	
REAR						
CAMBER (Degrees)	-1.0 ± 0.75 -1 ± 3/4	-1.3 ± 0.75 -1 5/16 ± 3/4	-0.8 ± 0.75 -13/16 ± 3/4	-0.27 ± 0.75 -1/4 ± 3/4	0.0 ± 0.75 0 ± 3/4	-0.15 ± 0.75 -1/8 ± 3/4
REAR TOE (Degrees)	0.24° ± 0.2° 1/4° ± 3/16°	0.24° ± 0.2° 1/4° ± 3/16°	0.24° ± 0.2° 1/4° ± 3/16°	0.1° ± 0.2° 3/32° ± 3/16°	0.1° ± 0.2° 3/32° ± 3/16°	0.0° ± 0.33° 0° ± 11/32°
REAR TOE (Inches)	0.12" ± 0.1" 1/8" ± 3/32"	0.12" ± 0.1" 1/8" ± 3/32"	0.12" ± 0.1" 1/8" ± 3/32"	0.05" ± 0.1" 1/16" ± 3/32"	0.05" ± 0.1" 1/16" ± 3/32"	0.0" ± 0.17" 0" ± 5/32"
THRUST ANGLE	0.3 (5/16)	0.3 (5/16)	0.3 (5/16)	0.3 (5/16)	0.3 (5/16)	
ILLUSTRATION #	187	187	187	06	06	06
NOTES						
	16165	16166	16167	16168	16169	16171

Ford

MODEL INFORMATION	Tempo AWD	Tempo	Thunderbird	Thunderbird	Windstar	Windstar
YEARS	1990-91	1992-94	1990-97	2002-2005	1995-98	1999-2003
INSTRUCTIONS						
LUGNUT TORQUE	85-105 ft.lbs.	85-105 ft.lbs.	85-105 ft.lbs.	100 ft.lbs.	85-105 ft.lbs.	85-105 ft.lbs.
FRONT						
CASTER (Degrees)	2.45 ± 0.75 2 7/16 ± 3/4	2.45 ± 0.75 2 7/16 ± 3/4	5.5 ± 0.75 5 1/2 ± 3/4	8.1 ± 0.5 8 1/8 ± 1/2	L= 3.3, R= 3.8 ± 0.5 L= 3 5/16, R= 3 13/16 ± 1/2	L= 3.3, R= 3.8 ± 0.75 L= 3 5/16, R= 3 13/16 ± 3/4
CROSS | TOL.	0.0 (0) | 0.75 (3/4)	0.0 (0) | 0.75 (3/4)	0.0 (0) | 0.75 (3/4)	0.0 (0) | 0.75 (3/4)	-0.5 (-1/2) | 0.5 (1/2)	-0.5 (-1/2) | 0.5 (1/2)
CAMBER (Degrees)	L= 1.4, R= 0.95 ± 0.75 L= 1 3/8, R= 15/16 ± 3/4	L= 1.4, R= 0.95 ± 0.75 L= 1 3/8, R= 15/16 ± 3/4	-0.5 ± 0.75 -1/2 ± 3/4	-0.15 ± 0.5 -1/8 ± 1/2	-0.25 ± 0.5 -1/4 ± 1/2	-0.44 ± 0.5 -7/16 ± 1/2
CROSS | TOL.	0.45 (7/16) | 0.75 (3/4)	0.45 (7/16) | 0.75 (3/4)	0.0 (0) | 0.75 (3/4)	0.0 (0) | 0.75 (3/4)	0.0 (0) | 0.5 (1/2)	0.0 (0) | 0.5 (1/2)
FRONT TOE (Degrees)	-0.2° ± 0.25° -3/16° ± 1/4°	-0.2° ± 0.25° -3/16° ± 1/4°	0.15° ± 0.25° [1] 5/32° ± 1/4° [1]	0.16° ± 0.25° 5/32° ± 1/4°	-0.1° ± 0.3° -3/32° ± 5/16°	-0.15° ± 0.25° -5/32° ± 1/4°
FRONT TOE (Inches)	-0.1" ± 0.13" -3/32" ± 1/8"	-0.1" ± 0.13" -3/32" ± 1/8"	0.08" ± 0.13" [1] 1/16" ± 1/8" [1]	0.08" ± 0.13" 3/32" ± 1/8"	-0.05" ± 0.15" -1/16" ± 5/32"	-0.08" ± 0.13" -1/16" ± 1/8"
SAI						
TURNING ANGLE	20.0 / 18.3	20.0 / 18.3	20.0 / 19.7		20.0 / 18.3	
ILLUSTRATION #			03 14 △	123 124	67 △	67 △
REAR						
CAMBER (Degrees)	0.35 ± 0.75 3/8 ± 3/4	-0.55 ± 0.75 -9/16 ± 3/4	-0.5 ± 0.5 -1/2 ± 1/2	-1.0 ± 0.75 -1 ± 3/4	0.0 ± 0.4 0 ± 3/8	-0.1 ± 0.5 -1/8 ± 1/2
REAR TOE (Degrees)	0.0° ± 0.33° 0° ± 11/32°	0.0° ± 0.33° 0° ± 11/32°	0.13° ± 0.25° 1/8° ± 1/4°	0.24° ± 0.25° 1/4° ± 1/4°	0.06° ± 0.4° [1] 1/16° ± 13/32° [1]	0.06° ± 0.34° 1/16° ± 11/32°
REAR TOE (Inches)	0.0" ± 0.17" 0" ± 5/32"	0.0" ± 0.17" 0" ± 5/32"	0.06" ± 0.13" 1/16" ± 1/8"	0.12" ± 0.13" 1/8" ± 1/8"	0.03" ± 0.2" [1] 1/32" ± 3/16" [1]	0.03" ± 0.17" 1/32" ± 5/32"
THRUST ANGLE					0.3 (5/16)	0.3 (5/16)
ILLUSTRATION #	06	06	71	135		
NOTES			[1] 1990-91 = 0.25° (0.13")		[1] 1997-98 = 0° (0")	
	16172	16173	16189	16190	16195	16196

Ford Truck & SUV

MODEL	Aerostar 4X2	Aerostar 4X4	Bronco	Bronco	Bronco II 4X2	Bronco II 4X4
INFORMATION						Dana 28 axle Axle tag #E87A
YEARS	1990-97	1990-97	1990-92	1993-96	1990	1990
INSTRUCTIONS			p. 306, 307	p. 306, 307	p. 306, 307	p. 306, 307
LUGNUT TORQUE	100 ft.lbs.	100 ft.lbs.	100 ft.lbs.	100 ft.lbs.	100 ft.lbs.	100 ft.lbs.
FRONT						
CASTER (Degrees)	4.0 ±2.0 4 ±2	4.0 ±2.0 4 ±2	MIN.= 2.0 MAX.= 6.0 MIN.= 2 MAX.= 6	MIN.= 2.0 MAX.= 6.0 MIN.= 2 MAX.= 6	MIN.= 4.0 MAX.= 6.0 MIN.= 4 MAX.= 6	MIN.= 2.5 MAX.= 4.5 MIN.= 2½ MAX.= 4½
CROSS | TOL.	0.0 (0) | 0.5 (½)	0.0 (0) | 0.5 (½)	0.0 (0) | 0.75 (¾)	-0.5 (-½) | 0.5 (½)	-0.5 (-½) | 0.5 (½)	-0.5 (-½) | 0.5 (½)
CAMBER (Degrees)	0.25 ±0.5 ¼ ± ½	0.25 ±0.5 [1] ¼ ± ½ [1]	0.25 ±0.5 ¼ ± ½	0.25 ±0.5 ¼ ± ½	0.25 ±1.0 ¼ ±1	0.5 ±0.75 ½ ± ¾
CROSS | TOL.	0.0 (0) | 0.5 (½)	0.0 (0) | 0.5 (½)	0.0 (0) | 0.5 (½)	0.0 (0) | 0.5 (½)	0.0 (0) | 1.0 (1)	0.0 (0) | 1.0 (1)
FRONT TOE (Degrees)	0.06° ±0.25° 1/16° ± ¼°	-0.06° ±0.25° -1/16° ± ¼°	0.06° ±0.25° 1/16° ± ¼°	0.06° ±0.25° 1/16° ± ¼°	0.0° ±0.25° 0° ± ¼°	0.0° ±0.25° 0° ± ¼°
FRONT TOE (Inches)	0.03" ±0.13" 1/32" ± ⅛"	-0.03" ±0.13" -1/32" ± ⅛"	0.03" ±0.13" 1/32" ± ⅛"	0.03" ±0.13" 1/32" ± ⅛"	0.0" ±0.13" 0" ± ⅛"	0.0" ±0.13" 0" ± ⅛"
SAI						
TURNING ANGLE						
ILLUSTRATION #	41	41	75	75	56	56
REAR						
CAMBER (Degrees)						
REAR TOE (Degrees)						
REAR TOE (Inches)						
THRUST ANGLE						
ILLUSTRATION #						
NOTES		[1] 1990-92 = 0.0° (0°)				
	17001	17002	17009	17010	17012	17014

Ford Truck & SUV

MODEL	Bronco II 4X4	E150	E150	E150	E150	E250/E350
INFORMATION	Dana 35 axle Axle tag #F07A					
YEARS	1990	1990-91	1992-97	1998-2006	2007	1990-91
INSTRUCTIONS	p. 306, 307	p. 306, 307	p. 306, 307	p. 306, 307	p. 306, 307	p. 306, 307
LUGNUT TORQUE	100 ft.lbs.	100 ft.lbs.	74-133 ft.lbs.	74-133 ft.lbs.	74-133 ft.lbs.	140 ft.lbs.
FRONT						
CASTER (Degrees)	MIN.= 2.5 MAX.= 6.0 MIN.= 2½ MAX.= 6	MIN.= 2.0 MAX.= 7.5 MIN.= 2 MAX.= 7½	MIN.= 2.0 MAX.= 7.5 MIN.= 2 MAX.= 7½	L= 4.0, R= 4.5 ± 2.75 L= 4, R= 4½ ±2¾	L= 4.1, R= 4.5 ± 2.75 L= 4⅛, R= 4½ ±2¾	MIN.= 2.0 MAX.= 7.5 MIN.= 2 MAX.= 7½
CROSS | TOL.	-0.5 (-½) | 0.5 (½)	-0.75 (-¾) | 0.5 (½)	-1.0 (-1) | 0.5 (½)	-0.5 (-½) | 1.0 (1)	-0.4 (-⅜) | 1.0 (1)	-0.75 (-¾) | 0.5 (½)
CAMBER (Degrees)	0.25 ±1.0 ¼ ±1	0.25 ±0.5 ¼ ± ½	0.25 ±0.5 ¼ ± ½	0.25 ±0.75 ¼ ± ¾	0.5 ±1.0 ½ ±1	0.5 ±0.5 ½ ± ½
CROSS | TOL.	0.0 (0) | 1.0 (1)	0.0 (0) | 0.5 (½)	0.0 (0) | 0.5 (½)	0.0 (0) | 0.75 (¾)	0.0 (0) | 1.0 (1)	0.0 (0) | 0.5 (½)
FRONT TOE (Degrees)	0.0° ±0.25° 0° ± ¼°	0.06° ±0.25° 1/16° ± ¼°	0.06° ±0.25° 1/16° ± ¼°	0.06° ±0.25° 1/16° ± ¼°	0.06° ±0.25° 1/16° ± ¼°	0.06° ±0.25° 1/16° ± ¼°
FRONT TOE (Inches)	0.0" ±0.13" 0" ± ⅛"	0.03" ±0.13" 1/32" ± ⅛"	0.03" ±0.13" 1/32" ± ⅛"	0.03" ±0.13" 1/32" ± ⅛"	0.03" ±0.13" 1/32" ± ⅛"	0.03" ±0.13" 1/32" ± ⅛"
SAI						
TURNING ANGLE						
ILLUSTRATION #	56		56	56	56	
REAR						
CAMBER (Degrees)						
REAR TOE (Degrees)						
REAR TOE (Inches)						
THRUST ANGLE						
ILLUSTRATION #						
NOTES						
	17015	17029	17030	17031	17032	17039

Ford Truck & SUV

MODEL INFORMATION	E250/E350 includes E-Super Duty	E250/E350 includes E-Super Duty	E250/E350/E450	E-Series E-150/E-250/E-350 exc/Cutaway/Stripped Chas.	E-Series E-250 Cutaway	E-Series E-350/E-450 Cutaway/Stripped Chas.
YEARS	1992-97	1998-2005	2006-2007	2008-2014	2008-2015	2008-2015
INSTRUCTIONS	p. 306, 307	p. 306, 307	p. 306, 307			
LUGNUT TORQUE	74-133 ft.lbs.[1]	74-133 ft.lbs.[1]	103 ft.lbs.[1]	148 ft.lbs.[1]	148 ft.lbs.[1]	148 ft.lbs.[1]
FRONT						
CASTER (Degrees)	MIN.= 2.0 MAX.= 7.5 MIN.= 2 MAX.= 7½	L= 4.0, R= 4.5 ± 2.75 L= 4, R= 4½ ±2¾	L= 4.1, R= 4.5 ± 2.75 L= 4⅛, R= 4½ ±2¾	L= 3.5, R= 3.9 ± 1.3 L= 3½, R= 3⅞ ± 1 5/16	L= 4.5, R= 4.9 ± 2.4 L= 4½, R= 4⅞ ±2⅜	L= 3.9, R= 4.3 ± 2.9 L= 3⅞, R= 4 5/16 ±2⅞
CROSS | TOL.	-1.0 (-1) | 0.5 (½)	-0.5 (-½) | 1.0 (1)	-0.4 (-⅜) | 1.0 (1)	-0.4 (-⅜) | 1.0 (1)	-0.4 (-⅜) | 1.0 (1)	-0.4 (-⅜) | 1.0 (1)
CAMBER (Degrees)	0.5 ± 0.5 ½ ± ½	0.5 ± 1.0 ½ ± 1	0.5 ± 1.0 ½ ± 1	0.5 ± 0.75 ½ ± ¾	0.5 ± 0.75 ½ ± ¾	0.5 ± 0.75 ½ ± ¾
CROSS | TOL.	0.0 (0) | 0.5 (½)	0.0 (0) | 0.5 (½)	0.0 (0) | 0.5 (½)	0.0 (0) | 0.75 (¾)	0.0 (0) | 0.75 (¾)	0.0 (0) | 0.75 (¾)
FRONT TOE (Degrees)	0.06° ±0.25° 1/16° ± ¼°	0.06° ±0.25° 1/16° ± ¼°	0.06° ±0.25° 1/16° ± ¼°	0.06° ±0.25° 1/16° ± ¼°	0.06° ±0.25° 1/16° ± ¼°	0.06° ±0.25° 1/16° ± ¼°
FRONT TOE (Inches)	0.03" ±0.13" 1/32" ± ⅛"	0.03" ±0.13" 1/32" ± ⅛"	0.03" ±0.13" 1/32" ± ⅛"	0.03" ±0.13" 1/32" ± ⅛"	0.03" ±0.13" 1/32" ± ⅛"	0.03" ±0.13" 1/32" ± ⅛"
SAI						
TURNING ANGLE						
ILLUSTRATION #	56	56	56	56	56	56
REAR						
CAMBER (Degrees)						
REAR TOE (Degrees)						
REAR TOE (Inches)						
THRUST ANGLE				0.5 (½)	0.5 (½)	0.5 (½)
ILLUSTRATION #						
NOTES	[1] 8 Lug = 126-170 ft.lbs	[1] 8 Lug = 126-170 ft.lbs	[1] 8 Lug = 148 ft.lbs	[1] DRW = 140 ft.lbs	[1] DRW = 140 ft.lbs	[1] DRW = 140 ft.lbs
	17040	17041	17042	17043	17044	17045

Ford Truck & SUV

MODEL INFORMATION	Edge All models	Edge All models	Edge exc/EcoBoost	Edge EcoBoost	Edge exc/21" wheels	Edge w/21" wheels
YEARS	2007	2008-2010	2011-2014	2012-2014	2015-2017	2015-2017
INSTRUCTIONS						
LUGNUT TORQUE	100 ft.lbs.	100 ft.lbs.	100 ft.lbs.	100 ft.lbs.	162 ft.lbs.	162 ft.lbs.
FRONT						
CASTER (Degrees)	4.4 ±0.75 4⅜ ± ¾	L= 4.3, R= 4.5 ± 0.75 L= 4 5/16, R= 4½ ± ¾	L= 4.3, R= 4.5 ± 0.75 L= 4 5/16, R= 4½ ± ¾	L= 4.4, R= 4.6 ± 0.75 L= 4⅜, R= 4⅝ ± ¾	2.9 ±0.75 2⅞ ± ¾	2.9 ±0.75 2⅞ ± ¾
CROSS | TOL.	0.0 (0) | 0.75 (¾)	-0.2 (-3/16) | 0.75 (¾)	-0.2 (-3/16) | 0.75 (¾)	-0.2 (-3/16) | 0.75 (¾)	0.0 (0) | 0.75 (¾)	0.0 (0) | 0.75 (¾)
CAMBER (Degrees)	-0.3 ±0.75 - 5/16 ± ¾	L= -0.1, R= -0.6 ±0.75 L= -⅛, R= -⅝ ± ¾	L= -0.05, R= -0.55 ±0.75 L= -1/16, R= - 9/16 ± ¾	L= -0.15, R= -0.65 ±0.75 L= -⅛, R= -⅝ ± ¾	-0.71 ±0.75 - 11/16 ± ¾	-0.71 ±0.75 - 11/16 ± ¾
CROSS | TOL.	0.0 (0) | 0.75 (¾)	0.5 (½) | 0.75 (¾)	0.5 (½) | 0.75 (¾)	0.5 (½) | 0.75 (¾)	0.0 (0) | 0.75 (¾)	0.0 (0) | 0.75 (¾)
FRONT TOE (Degrees)	0.1° ±0.2° 3/32° ± 3/16°	0.1° ±0.2° 3/32° ± 3/16°	0.1° ±0.2° 3/32° ± 3/16°	0.1° ±0.2° 3/32° ± 3/16°	0.2° ±0.14° 3/16° ± ⅛°	0.2° ±0.14° 3/16° ± ⅛°
FRONT TOE (Inches)	0.05" ±0.1" 1/16" ± 3/32"	0.05" ±0.1" 1/16" ± 3/32"	0.05" ±0.1" 1/16" ± 3/32"	0.05" ±0.1" 1/16" ± 3/32"	0.1" ±0.07" 3/32" ± 1/16"	0.1" ±0.07" 3/32" ± 1/16"
SAI						
TURNING ANGLE						
ILLUSTRATION #	167	167	167	167	26	26
REAR						
CAMBER (Degrees)	-0.4 ±0.75 -⅜ ± ¾	-0.45 ±0.75 - 7/16 ± ¾	-0.45 ±0.75 - 7/16 ± ¾	-0.45 ±0.75 - 7/16 ± ¾	-1.2 ±0.75 -1 3/16 ± ¾	-1.0 ±0.75 -1 ± ¾
REAR TOE (Degrees)	0.1° ±0.2° 3/32° ± 3/16°	0.1° ±0.2° 3/32° ± 3/16°	0.1° ±0.2° 3/32° ± 3/16°	0.1° ±0.2° 3/32° ± 3/16°	0.26° ±0.14° ¼° ± ⅛°	0.26° ±0.14° ¼° ± ⅛°
REAR TOE (Inches)	0.05" ±0.1" 1/16" ± 3/32"	0.05" ±0.1" 1/16" ± 3/32"	0.05" ±0.1" 1/16" ± 3/32"	0.05" ±0.1" 1/16" ± 3/32"	0.13" ±0.07" ⅛" ± 1/16"	0.13" ±0.07" ⅛" ± 1/16"
THRUST ANGLE	0.3 (5/16)	0.3 (5/16)	0.5 (½)	0.5 (½)		
ILLUSTRATION #	177	177	177	177	174	174
NOTES						
	17046	17047	17048	17049	17050	17051

ALIGNMENT SPECIFICATIONS

Ford Truck & SUV

Escape 2.0L	Escape 3.0L	Escape	MODEL	Escape	Escape	Escape
		Early-prior to 8/02/04 at KC (letter K 11th digit of VIN)	INFORMATION	Early-prior to 8/16/04 at Ohio (letter D 11th digit of VIN)	All build dates after those listed previously	Escape Hybrid
2001-2004	2001-2004	2005	YEARS	2005	2005	2006-2007
			INSTRUCTIONS			
98 ft.lbs.	98 ft.lbs.	98 ft.lbs.	LUGNUT TORQUE	98 ft.lbs.	98 ft.lbs.	98 ft.lbs.
			FRONT			
1.79 ± 1.0 1 13⁄16 ± 1	1.79 ± 1.0 1 13⁄16 ± 1	1.6 ± 1.0 1⅝ ± 1	CASTER (Degrees)	1.6 ± 1.0 1⅝ ± 1	1.6 ± 1.0 1⅝ ± 1	L= 1.8, R= 1.5 ± 1.0 L= 1 13⁄16, R= 1½ ± 1
0.0 (0) | 1.0 (1)	0.0 (0) | 1.0 (1)	0.0 (0) | 0.75 (¾)	CROSS | TOL.	0.0 (0) | 0.75 (¾)	0.0 (0) | 0.75 (¾)	0.3 (5⁄16) | 0.75 (¾)
-0.48 ± 1.0 -½ ± 1	-0.84 ± 1.0 -13⁄16 ± 1	-1.0 ± 0.75 -1 ± ¾	CAMBER (Degrees)	-1.0 ± 0.75 -1 ± ¾	-1.0 ± 0.75 -1 ± ¾	-1.0 ± 0.75 -1 ± ¾
0.0 (0) | 1.0 (1)	0.0 (0) | 1.0 (1)	0.0 (0) | 0.75 (¾)	CROSS | TOL.	0.0 (0) | 0.75 (¾)	0.0 (0) | 0.75 (¾)	0.0 (0) | 0.75 (¾)
-0.08° ± 0.32° -3⁄32° ± 5⁄16°	0.23° ± 0.32° 7⁄32° ± 5⁄16°	0.23° ± 0.23° 7⁄32° ± 7⁄32°	FRONT TOE (Degrees)	0.23° ± 0.23° 7⁄32° ± 7⁄32°	0.23° ± 0.23° 7⁄32° ± 7⁄32°	0.23° ± 0.23° 7⁄32° ± 7⁄32°
-0.04" ± 0.16" -1⁄32" ± 5⁄32"	0.12" ± 0.16" ⅛" ± 5⁄32"	0.12" ± 0.12" ⅛" ± ⅛"	FRONT TOE (Inches)	0.12" ± 0.12" ⅛" ± ⅛"	0.12" ± 0.12" ⅛" ± ⅛"	0.12" ± 0.12" ⅛" ± ⅛"
			SAI			
			TURNING ANGLE			
51	51	51	ILLUSTRATION #	51	51	51
			REAR			
0.04 ± 1.0 1⁄16 ± 1	0.04 ± 1.0 1⁄16 ± 1	0.2 ± 0.75 3⁄16 ± ¾	CAMBER (Degrees)	0.2 ± 0.75 3⁄16 ± ¾	0.0 ± 0.75 0 ± ¾	0.0 ± 0.75 0 ± ¾
0.08° ± 0.34° 3⁄32° ± 11⁄32°	0.2° ± 0.34° 3⁄16° ± 11⁄32°	0.2° ± 0.2° 3⁄16° ± 3⁄16°	REAR TOE (Degrees)	0.2° ± 0.2° 3⁄16° ± 3⁄16°	-0.06° ± 0.2° -1⁄16° ± 3⁄16°	-0.14° ± 0.2° -⅛° ± 3⁄16°
0.04" ± 0.17" 1⁄32" ± 5⁄32"	0.1" ± 0.17" 3⁄32" ± 5⁄32"	0.1" ± 0.1" 3⁄32" ± 3⁄32"	REAR TOE (Inches)	0.1" ± 0.1" 3⁄32" ± 3⁄32"	-0.03" ± 0.1" -1⁄32" ± 3⁄32"	-0.07" ± 0.1" -1⁄16" ± 3⁄32"
		0.3 (5⁄16)	THRUST ANGLE	0.3 (5⁄16)	0.3 (5⁄16)	0.3 (5⁄16)
27	27	27	ILLUSTRATION #	27	27	27
			NOTES			
17052	17053	17054		17055	17056	17057

Ford Truck & SUV

Escape	Escape Hybrid	Escape	MODEL	Escape	Excursion	Excursion
		FWD	INFORMATION	AWD	4X2	4X4
2008-2012	2008-2012	2013-2017	YEARS	2013-2017	2000-2004	2000-2004
			INSTRUCTIONS			
100 ft.lbs.	100 ft.lbs.	100 ft.lbs.	LUGNUT TORQUE	100 ft.lbs.	155-165 ft.lbs.	155-165 ft.lbs.
			FRONT			
1.6 ± 0.5 1⅝ ± ½	1.6 ± 0.5 1⅝ ± ½	5.0 ± 0.75 5 ± ¾	CASTER (Degrees)	4.95 ± 0.75 4 15⁄16 ± ¾	4.0 ± 2.0 4 ± 2	3.5 ± 2.0 3½ ± 2
0.0 (0) | 0.7 (11⁄16)	0.0 (0) | 0.7 (11⁄16)	0.0 (0) | 0.75 (¾)	CROSS | TOL.	0.0 (0) | 0.75 (¾)	0.0 (0) | 1.0 (1)	0.0 (0) | 1.0 (1)
-0.84 ± 0.5 -13⁄16 ± ½	-0.84 ± 0.5 -13⁄16 ± ½	-0.73 ± 0.75 -¾ ± ¾	CAMBER (Degrees)	-0.65 ± 0.75 -⅝ ± ¾	0.62 ± 1.0 ⅝ ± 1	0.25 ± 1.0 ¼ ± 1
0.0 (0) | 0.7 (11⁄16)	0.0 (0) | 0.7 (11⁄16)	0.0 (0) | 0.75 (¾)	CROSS | TOL.	0.0 (0) | 0.75 (¾)	0.0 (0) | 1.0 (1)	0.0 (0) | 1.0 (1)
0.23° ± 0.23° 7⁄32° ± 7⁄32°	0.23° ± 0.23° 7⁄32° ± 7⁄32°	0.2° ± 0.2° 3⁄16° ± 3⁄16°	FRONT TOE (Degrees)	0.2° ± 0.2° 3⁄16° ± 3⁄16°	0.03° ± 0.25° 1⁄32° ± ¼°	0.03° ± 0.25° 1⁄32° ± ¼°
0.12" ± 0.12" ⅛" ± ⅛"	0.12" ± 0.12" ⅛" ± ⅛"	0.1" ± 0.1" 3⁄32" ± 3⁄32"	FRONT TOE (Inches)	0.1" ± 0.1" 3⁄32" ± 3⁄32"	0.02" ± 0.13" 1⁄64" ± ⅛"	0.02" ± 0.13" 1⁄64" ± ⅛"
			SAI			
			TURNING ANGLE			
51	51	196	ILLUSTRATION #	196	56	56
			REAR			
0.0 ± 0.65 0 ± ⅝	-0.1 ± 0.65 -⅛ ± ⅝	-1.38 ± 0.75 -1⅜ ± ¾	CAMBER (Degrees)	-1.34 ± 0.75 -1 5⁄16 ± ¾		
-0.18° ± 0.2° -3⁄16° ± 3⁄16°	-0.12° ± 0.2° -⅛° ± 3⁄16°	0.38° ± 0.2° ⅜° ± 3⁄16°	REAR TOE (Degrees)	0.38° ± 0.2° ⅜° ± 3⁄16°		
-0.09" ± 0.1" -3⁄32" ± 3⁄32"	-0.06" ± 0.1" -1⁄16" ± 3⁄32"	0.19" ± 0.1" 3⁄16" ± 3⁄32"	REAR TOE (Inches)	0.19" ± 0.1" 3⁄16" ± 3⁄32"		
		0.5 (½)	THRUST ANGLE	0.5 (½)		
27	27	06	ILLUSTRATION #	06		
			NOTES			
17058	17059	17060		17061	17062	17063

Ford Truck & SUV

Excursion 4X2	Excursion 4X4	Expedition 4X2 All Models	MODEL INFORMATION	Expedition 4X4 exc/Air Susp.	Expedition 4X4 w/Air Susp.	Expedition exc/Air Susp.
2005	2005	1997-2002	YEARS	1997-2002	1997-2002	2003-2006
			INSTRUCTIONS			
155-165 ft.lbs.	155-165 ft.lbs.	83-112 ft.lbs.	LUGNUT TORQUE	83-112 ft.lbs.	83-112 ft.lbs.	150 ft.lbs.
			FRONT			
L= 4.4, R= 4.6 ± 1.4 L= 4⅜, R= 4⅝ ± 1⅜	L= 3.63, R= 4.0 ± 2.0 L= 3⅝, R= 4 ± 2	L= 6.3, R= 6.6 ± 1.0 [1] L= 6 5/16, R= 6⅝ ± 1 [1]	CASTER (Degrees)	L= 5.1, R= 5.8 ± 1.0 [1] L= 5⅛, R= 5 13/16 ± 1 [1]	L= 5.0, R= 5.7 ± 1.0 [1] L= 5, R= 5 11/16 ± 1 [1]	5.3 ± 1.0 [1] 5 5/16 ± 1 [1]
-0.2 (-3/16) | 0.75 (¾)	-0.37 (-⅜) | 0.5 (½)	-0.3 (-5/16) | 0.7 (11/16)	CROSS | TOL.	-0.7 (-11/16) | 0.7 (11/16)	-0.7 (-11/16) | 0.7 (11/16)	0.0 (0) | 1.0 (1)
0.62 ± 0.75 ⅝ ± ¾	0.25 ± 0.75 ¼ ± ¾	-0.3 ± 0.7 [1] - 5/16 ± 11/16 [1]	CAMBER (Degrees)	-0.2 ± 0.7 [1] - 3/16 ± 11/16 [1]	-0.4 ± 0.7 [1] - ⅜ ± 11/16 [1]	-0.3 ± 0.75 [1] - 5/16 ± ¾ [1]
0.0 (0) | 1.0 (1)	0.0 (0) | 1.0 (1)	0.0 (0) | 0.7 (11/16)	CROSS | TOL.	0.0 (0) | 0.7 (11/16)	0.0 (0) | 0.7 (11/16)	0.0 (0) | 0.75 (¾)
0.1° ± 0.25° 3/32° ± ¼°	0.01° ± 0.25° 1/64° ± ¼°	0.06° ± 0.25° 1/16° ± ¼°	FRONT TOE (Degrees)	0.3° ± 0.25° 5/16° ± ¼°	0.2° ± 0.25° 3/16° ± ¼°	0.14° ± 0.2° ⅛° ± 3/16°
0.05" ± 0.13" 1/16" ± ⅛"	0.01" ± 0.13" 0" ± ⅛"	0.03" ± 0.13" 1/32" ± ⅛"	FRONT TOE (Inches)	0.15" ± 0.13" 5/32" ± ⅛"	0.1" ± 0.13" 3/32" ± ⅛"	0.07" ± 0.1" 1/16" ± 3/32"
			SAI			
			TURNING ANGLE			
56	56	128	ILLUSTRATION #	128	128	12
			REAR			
			CAMBER (Degrees)			-0.9 ± 0.75 [1] - ⅞ ± ¾ [1]
			REAR TOE (Degrees)			0.2° ± 0.2° 3/16° ± 3/16°
			REAR TOE (Inches)			0.1" ± 0.1" 3/32" ± 3/32"
			THRUST ANGLE			
			ILLUSTRATION #			147
		[1] Service cam kit reqd.	NOTES	[1] Service cam kit reqd.	[1] Service cam kit reqd.	[1] Service cam kit reqd.
17064	17065	17066		17067	17068	17069

Ford Truck & SUV

Expedition w/Air Susp.	Expedition exc/Air Susp.	Expedition w/Air Susp. RWD Models	MODEL INFORMATION	Expedition w/Air Susp. 4WD Models	Expedition All models	Explorer 4X2
2003-2006	2007-2014	2007-2014	YEARS	2007-2014	2015-2017	1991-94
			INSTRUCTIONS			p. 306, 307
150 ft.lbs.	150 ft.lbs.	150 ft.lbs.	LUGNUT TORQUE	150 ft.lbs.	150 ft.lbs.	100 ft.lbs.
			FRONT			
5.5 ± 1.0 [1] 5½ ± 1 [1]	L= 3.9, R= 4.5 ± 1.0 [1] L= 3⅞, R= 4½ ± 1 [1]	L= 4.1, R= 4.7 ± 1.0 [1] L= 4⅛, R= 4 11/16 ± 1 [1]	CASTER (Degrees)	L= 4.0, R= 4.6 ± 1.0 [1] L= 4, R= 4⅝ ± 1 [1]	L= 3.9, R= 4.5 ± 1.0 L= 3⅞, R= 4½ ± 1	MIN.= 2.0 MAX.= 7.0 MIN.= 2 MAX.= 7
0.0 (0) | 1.0 (1)	-0.6 (-⅝) | 0.75 (¾)	-0.6 (-⅝) | 0.75 (¾)	CROSS | TOL.	-0.6 (-⅝) | 0.75 (¾)	-0.6 (-⅝) | 0.75 (¾)	-0.75 (-¾) | 0.5 (½)
-0.3 ± 0.75 [1] - 5/16 ± ¾ [1]	-0.3 ± 0.75 - 5/16 ± ¾	-0.3 ± 0.75 - 5/16 ± ¾	CAMBER (Degrees)	-0.3 ± 0.75 - 5/16 ± ¾	-0.3 ± 0.75 - 5/16 ± ¾	0.25 ± 0.5 ¼ ± ½
0.0 (0) | 0.75 (¾)	0.0 (0) | 0.75 (¾)	0.0 (0) | 0.75 (¾)	CROSS | TOL.	0.0 (0) | 0.75 (¾)	0.0 (0) | 0.75 (¾)	0.0 (0) | 0.5 (½)
0.14° ± 0.2° ⅛° ± 3/16°	0.14° ± 0.2° ⅛° ± 3/16°	0.14° ± 0.2° ⅛° ± 3/16°	FRONT TOE (Degrees)	0.14° ± 0.2° ⅛° ± 3/16°	0.14° ± 0.2° ⅛° ± 3/16°	0.06° ± 0.25° 1/16° ± ¼°
0.07" ± 0.1" 1/16" ± 3/32"	0.07" ± 0.1" 1/16" ± 3/32"	0.07" ± 0.1" 1/16" ± 3/32"	FRONT TOE (Inches)	0.07" ± 0.1" 1/16" ± 3/32"	0.07" ± 0.1" 1/16" ± 3/32"	0.03" ± 0.13" 1/32" ± ⅛"
			SAI			
			TURNING ANGLE			
12	12	12	ILLUSTRATION #	12	12	56
			REAR			
-1.0 ± 0.75 [1] -1 ± ¾ [1]	-0.5 ± 0.75 - ½ ± ¾	-0.5 ± 0.75 - ½ ± ¾	CAMBER (Degrees)	-0.5 ± 0.75 - ½ ± ¾	-0.5 ± 0.75 - ½ ± ¾	
0.2° ± 0.2° 3/16° ± 3/16°	0.2° ± 0.2° 3/16° ± 3/16°	0.2° ± 0.2° 3/16° ± 3/16°	REAR TOE (Degrees)	0.2° ± 0.2° 3/16° ± 3/16°	0.2° ± 0.2° 3/16° ± 3/16°	
0.1" ± 0.1" 3/32" ± 3/32"	0.1" ± 0.1" 3/32" ± 3/32"	0.1" ± 0.1" 3/32" ± 3/32"	REAR TOE (Inches)	0.1" ± 0.1" 3/32" ± 3/32"	0.1" ± 0.1" 3/32" ± 3/32"	
			THRUST ANGLE		0.5 (½)	
147	183	183	ILLUSTRATION #	183	183	
[1] Service cam kit reqd.	[1] EL L=4.0°, R=4.6°	[1] EL L=4.2°, R=4.8°	NOTES	[1] EL L=4.1°, R=4.7°		
17070	17071	17072		17073	17074	17075

ALIGNMENT SPECIFICATIONS

Ford Truck & SUV

Explorer 4X4	Explorer 2WD & AWD except ARC	Explorer AWD with ARC	MODEL INFORMATION	Explorer 2WD & AWD except ARC	Explorer AWD with ARC	Explorer
1991-94	1995-97	1995-97	YEARS	1998-2001	1998-2001	2002-2005
p. 306, 307	p. 306, 307	p. 306, 307	INSTRUCTIONS	p. 306, 307	p. 306, 307	
100 ft.lbs.	100 ft.lbs.	100 ft.lbs.	LUGNUT TORQUE	100 ft.lbs.	100 ft.lbs.	100 ft.lbs.
			FRONT			
MIN.=2.0 MAX.=7.0 MIN.=2 MAX.=7	MIN.=3.2 MAX.=5.2 [1] MIN.=3 3/16 MAX.=5 3/16 [1]	MIN.=3.5 MAX.=5.5 [1] MIN.=3½ MAX.=5½ [1]	CASTER (Degrees)	MIN.=3.2 MAX.=5.2 [1] MIN.=3 3/16 MAX.=5 3/16 [1]	MIN.=3.5 MAX.=5.5 [1] MIN.=3½ MAX.=5½ [1]	L=5.1, R=5.3±1.0 L=5⅛, R=5 5/16±1
-0.75 (-¾) | 0.5 (½)	-0.5 (-½) | 0.75 (¾)	-0.5 (-½) | 0.75 (¾)	CROSS | TOL.	-0.75 (-¾) | 0.75 (¾)	-0.75 (-¾) | 0.75 (¾)	-0.2 (-3/16) | 0.4 (⅜)
0.25±0.5 ¼±½	-0.5±0.75 [1] -½±¾ [1]	-0.75±0.75 [1] -¾±¾ [1]	CAMBER (Degrees)	-0.5±0.75 [1] -½±¾ [1]	-0.75±0.75 [1] -¾±¾ [1]	-0.5±0.8 -½±13/16
0.0 (0) | 0.5 (½)	0.0 (0) | 0.75 (¾)	0.0 (0) | 0.75 (¾)	CROSS | TOL.	0.0 (0) | 0.75 (¾)	0.0 (0) | 0.75 (¾)	0.0 (0) | 0.8 (13/16)
0.06°±0.25° 1/16°±¼°	0.3°±0.25° 5/16°±¼°	0.3°±0.25° 5/16°±¼°	FRONT TOE (Degrees)	0.13°±0.25° ⅛°±¼°	0.13°±0.25° ⅛°±¼°	0.2°±0.25° 3/16°±¼°
0.03"±0.13" 1/32"±⅛"	0.15"±0.13" 5/32"±⅛"	0.15"±0.13" 5/32"±⅛"	FRONT TOE (Inches)	0.06"±0.13" 1/16"±⅛"	0.06"±0.13" 1/16"±⅛"	0.1"±0.13" 3/32"±⅛"
			SAI			
			TURNING ANGLE			
56	73	73	ILLUSTRATION #	73	73	137
			REAR			
			CAMBER (Degrees)			-0.9±0.8 -⅞±13/16
			REAR TOE (Degrees)			0.2°±0.25° 3/16°±¼°
			REAR TOE (Inches)			0.1"±0.13" 3/32"±⅛"
	0.3 (5/16)	0.3 (5/16)	THRUST ANGLE	0.3 (5/16)	0.3 (5/16)	
			ILLUSTRATION #			136
	[1] Service cam kit reqd.	[1] Service cam kit reqd.	NOTES	[1] Service cam kit reqd.	[1] Service cam kit reqd.	
17076	17077	17078		17079	17080	17081

Ford Truck & SUV

Explorer	Explorer	Explorer FWD	MODEL INFORMATION	Explorer FWD	Explorer AWD	Explorer Police Vehicle
2006-2007	2008-2010	2011-2012	YEARS	2013-2015	2011-2015	2013-2015
			INSTRUCTIONS			
100 ft.lbs.	100 ft.lbs.	100 ft.lbs.	LUGNUT TORQUE	100 ft.lbs.	100 ft.lbs.	108 ft.lbs.
			FRONT			
4.6±0.75 4⅝±¾	L=4.4, R=4.6±0.75 L=4⅜, R=4⅝±¾	L=3.4, R=3.6±0.75 L=3⅜, R=3⅝±¾	CASTER (Degrees)	L=3.4, R=3.6±0.75 L=3⅜, R=3⅝±¾	L=3.3, R=3.5±0.75 L=3 5/16, R=3½±¾	L=3.1, R=3.3±0.75 L=3⅛, R=3 5/16±¾
0.0 (0) | 0.75 (¾)	-0.2 (-3/16) | 0.75 (¾)	-0.2 (-3/16) | 0.75 (¾)	CROSS | TOL.	-0.2 (-3/16) | 0.75 (¾)	-0.2 (-3/16) | 0.75 (¾)	-0.2 (-3/16) | 0.75 (¾)
-0.5±0.75 -½±¾	-0.5±0.75 -½±¾	L=-0.4, R=-0.6±0.75 L=-⅜, R=-⅝±¾	CAMBER (Degrees)	L=-0.5, R=-0.7±0.75 L=-½, R=-11/16±¾	L=-0.4, R=-0.6±0.75 L=-⅜, R=-⅝±¾	L=-0.45, R=-0.65±0.75 L=-7/16, R=-⅝±¾
0.0 (0) | 0.75 (¾)	0.0 (0) | 0.75 (¾)	0.2 (3/16) | 0.75 (¾)	CROSS | TOL.	0.2 (3/16) | 0.75 (¾)	0.2 (3/16) | 0.75 (¾)	0.2 (3/16) | 0.75 (¾)
0.2°±0.2° 3/16°±3/16°	0.2°±0.2° 3/16°±3/16°	0.2°±0.2° 3/16°±3/16°	FRONT TOE (Degrees)	0.2°±0.2° 3/16°±3/16°	0.2°±0.2° 3/16°±3/16°	0.2°±0.2° 3/16°±3/16°
0.1"±0.1" 3/32"±3/32"	0.1"±0.1" 3/32"±3/32"	0.1"±0.1" 3/32"±3/32"	FRONT TOE (Inches)	0.1"±0.1" 3/32"±3/32"	0.1"±0.1" 3/32"±3/32"	0.1"±0.1" 3/32"±3/32"
			SAI			
			TURNING ANGLE			
139	139	167	ILLUSTRATION #	167	167	167
			REAR			
-0.5±0.75 -½±¾	-0.5±0.75 -½±¾	-0.85±0.75 -⅞±¾	CAMBER (Degrees)	-0.85±0.75 -⅞±¾	-0.7±0.75 -11/16±¾	-0.5±0.75 -½±¾
0.1°±0.2° 3/32°±3/16°	0.1°±0.2° 3/32°±3/16°	0.24°±0.2° ¼°±3/16°	REAR TOE (Degrees)	0.24°±0.2° ¼°±3/16°	0.24°±0.2° ¼°±3/16°	0.24°±0.2° ¼°±3/16°
0.05"±0.1" 1/16"±3/32"	0.05"±0.1" 1/16"±3/32"	0.13"±0.1" ⅛"±3/32"	REAR TOE (Inches)	0.13"±0.1" ⅛"±3/32"	0.13"±0.1" ⅛"±3/32"	0.13"±0.1" ⅛"±3/32"
			THRUST ANGLE	0.3 (5/16)	0.3 (5/16)	0.3 (5/16)
176	176	187	ILLUSTRATION #	187	187	187
			NOTES			
17082	17083	17084		17085	17086	17087

Ford Truck & SUV

Explorer FWD	Explorer AWD	Explorer Police Vehicle	MODEL INFORMATION	Explorer Sport	Explorer Sport Trac	Explorer Sport Trac
2016-2017	2016-2017	2016-2017	YEARS	2001-2003	2001	2002-2005
			INSTRUCTIONS	p. 306, 307	p. 306, 307	p. 306, 307
100 ft.lbs.	100 ft.lbs.	110 ft.lbs.	LUGNUT TORQUE	100 ft.lbs.	100 ft.lbs.	100 ft.lbs.
			FRONT			
L= 3.2, R= 3.4 ± 0.75 L= 3 3/16, R= 3 3/8 ± 3/4	L= 3.1, R= 3.3 ± 0.75 L= 3 1/8, R= 3 5/16 ± 3/4	L= 2.9, R= 3.1 ± 0.75 L= 2 7/8, R= 3 1/8 ± 3/4	CASTER (Degrees)	L= 3.95, R= 4.45 ± 1.0 [1] L= 3 15/16, R= 4 7/16 ± 1 [1]	L= 3.15, R= 3.65 ± 1.0 [1] L= 3 1/8, R= 3 5/8 ± 1 [1]	L= 3.15, R= 4.05 ± 1.0 [1] L= 3 1/8, R= 4 1/16 ± 1 [1]
-0.2 (-3/16) | 0.75 (3/4)	-0.2 (-3/16) | 0.75 (3/4)	-0.2 (-3/16) | 0.75 (3/4)	CROSS | TOL.	-0.5 (-1/2) | 0.75 (3/4)	-0.5 (-1/2) | 0.75 (3/4)	-0.9 (-7/8) | 0.75 (3/4)
L= -0.5, R= -0.7 ±0.75 L= -1/2, R= -11/16 ± 3/4	L= -0.4, R= -0.6 ±0.75 L= -3/8, R= -5/8 ± 3/4	L= -0.45, R= -0.65 ±0.75 L= -7/16, R= -5/8 ± 3/4	CAMBER (Degrees)	-0.5 ±0.75 [1] -1/2 ± 3/4 [1]	-0.5 ±0.75 [1] -1/2 ± 3/4 [1]	-0.5 ±0.75 [1] -1/2 ± 3/4 [1]
0.2 (3/16) | 0.75 (3/4)	0.2 (3/16) | 0.75 (3/4)	0.2 (3/16) | 0.75 (3/4)	CROSS | TOL.	0.0 (0) | 0.75 (3/4)	0.0 (0) | 0.75 (3/4)	0.0 (0) | 0.75 (3/4)
0.2° ±0.2° 3/16° ± 3/16°	0.2° ±0.2° 3/16° ± 3/16°	0.2° ±0.2° 3/16° ± 3/16°	FRONT TOE (Degrees)	0.13° ±0.25° 1/8° ± 1/4°	0.13° ±0.25° 1/8° ± 1/4°	0.13° ±0.25° 1/8° ± 1/4°
0.1" ±0.1" 3/32" ± 3/32"	0.1" ±0.1" 3/32" ± 3/32"	0.1" ±0.1" 3/32" ± 3/32"	FRONT TOE (Inches)	0.06" ±0.13" 1/16" ± 1/8"	0.06" ±0.13" 1/16" ± 1/8"	0.06" ±0.13" 1/16" ± 1/8"
			SAI			
			TURNING ANGLE			
167	167	167	ILLUSTRATION #	73	73	73
			REAR			
-0.85 ±0.75 -7/8 ± 3/4	-0.7 ±0.75 -11/16 ± 3/4	-0.5 ±0.75 -1/2 ± 3/4	CAMBER (Degrees)			
0.3° ±0.2° 5/16° ± 3/16°	0.3° ±0.2° 5/16° ± 3/16°	0.3° ±0.2° 5/16° ± 3/16°	REAR TOE (Degrees)			
0.15" ±0.1" 5/32" ± 3/32"	0.15" ±0.1" 5/32" ± 3/32"	0.15" ±0.1" 5/32" ± 3/32"	REAR TOE (Inches)			
0.3 (5/16)	0.3 (5/16)	0.3 (5/16)	THRUST ANGLE	0.3 (5/16)	0.3 (5/16)	0.3 (5/16)
187 187	187 187	187 187	ILLUSTRATION #			
			NOTES	[1] Service cam kit reqd.	[1] Service cam kit reqd.	[1] Service cam kit reqd.
17088	17089	17090		17091	17092	17093

Ford Truck & SUV

Explorer Sport Trac	Explorer Sport Trac	Explorer Sport Trac	MODEL INFORMATION	F150 4X2 exc/Lightning	F150 4X2 Lightning	F150 4X2
2007	2008	2009-2010	YEARS	1990-96	1994-96	1997-98
			INSTRUCTIONS	p. 306, 307	p. 306, 307	
100 ft.lbs.	100 ft.lbs.	100 ft.lbs.	LUGNUT TORQUE	100 ft.lbs.	100 ft.lbs.	83-112 ft.lbs.
			FRONT			
4.6 ±0.75 4 5/8 ± 3/4	L= 4.5, R= 4.7 ± 0.75 L= 4 1/2, R= 4 11/16 ± 3/4	L= 4.3, R= 4.7 ± 0.75 L= 4 5/16, R= 4 11/16 ± 3/4	CASTER (Degrees)	MIN.= 2.0 MAX.= 6.0 MIN.= 2 MAX.= 6	MIN.= 2.0 MAX.= 6.0 MIN.= 2 MAX.= 6	L= 6.25, R= 6.75 ± 1.0 L= 6 1/4, R= 6 3/4 ± 1
0.0 (0) | 0.75 (3/4)	-0.2 (-3/16) | 0.75 (3/4)	-0.4 (-3/8) | 0.75 (3/4)	CROSS | TOL.	-0.5 (-1/2) | 0.5 (1/2)	-0.4 (-3/8) | 0.5 (1/2)	-0.5 (-1/2) | 0.75 (3/4)
-0.5 ±0.75 -1/2 ± 3/4	-0.5 ±0.75 -1/2 ± 3/4	-0.5 ±0.75 -1/2 ± 3/4	CAMBER (Degrees)	0.25 ±0.5 1/4 ± 1/2	0.0 ±0.5 0 ± 1/2	-0.3 ±0.7 -5/16 ± 11/16
0.0 (0) | 0.75 (3/4)	0.0 (0) | 0.75 (3/4)	0.0 (0) | 0.75 (3/4)	CROSS | TOL.	0.0 (0) | 0.5 (1/2)	0.0 (0) | 0.5 (1/2)	0.0 (0) | 0.7 (11/16)
0.2° ±0.2° 3/16° ± 3/16°	0.2° ±0.2° 3/16° ± 3/16°	0.2° ±0.2° 3/16° ± 3/16°	FRONT TOE (Degrees)	0.06° ±0.25° 1/16° ± 1/4°	0.06° ±0.25° 1/16° ± 1/4°	0.06° ±0.25° 1/16° ± 1/4°
0.1" ±0.1" 3/32" ± 3/32"	0.1" ±0.1" 3/32" ± 3/32"	0.1" ±0.1" 3/32" ± 3/32"	FRONT TOE (Inches)	0.03" ±0.13" 1/32" ± 1/8"	0.03" ±0.13" 1/32" ± 1/8"	0.03" ±0.13" 1/32" ± 1/8"
			SAI			
			TURNING ANGLE			
139	139	139	ILLUSTRATION #	56	56	128
			REAR			
-0.5 ±0.75 -1/2 ± 3/4	-0.5 ±0.75 -1/2 ± 3/4	-0.5 ±0.75 -1/2 ± 3/4	CAMBER (Degrees)			
0.1° ±0.2° 3/32° ± 3/16°	0.1° ±0.2° 3/32° ± 3/16°	0.1° ±0.2° 3/32° ± 3/16°	REAR TOE (Degrees)			
0.05" ±0.1" 1/16" ± 3/32"	0.05" ±0.1" 1/16" ± 3/32"	0.05" ±0.1" 1/16" ± 3/32"	REAR TOE (Inches)			
			THRUST ANGLE			
176	176	176	ILLUSTRATION #			
			NOTES			
17094	17095	17096		17109	17110	17111

Ford Truck & SUV

F150 4X2	F150 4X2	F150 Heritage 4X2	MODEL	F150 Heritage 4X2	F150 4X2	F150 4X2
exc/Lightning	Lightning	exc/Lightning	INFORMATION	Lightning	Regular Cab	Super/Crew Cab
1999-2003	1999-2003	2004	YEARS	2004	2004-2005	2004-2005
			INSTRUCTIONS			
100 ft.lbs.	100 ft.lbs.	100 ft.lbs.	LUGNUT TORQUE	100 ft.lbs.	150 ft.lbs.	150 ft.lbs.
			FRONT			
L= 6.25, R= 6.75 ± 1.0 L= 6¼, R= 6¾ ± 1	L= 6.7, R= 7.2 ± 1.0 L= 6 11/16, R= 7 3/16 ± 1	L= 6.25, R= 6.75 ± 1.0 L= 6¼, R= 6¾ ± 1	CASTER (Degrees)	L= 6.7, R= 7.2 ± 1.0 L= 6 11/16, R= 7 3/16 ± 1	4.4 ± 1.0 [1] 4 3/8 ± 1 [1]	4.6 ± 1.0 [1] 4 5/8 ± 1 [1]
-0.5 (-½) | 0.75 (¾)	-0.5 (-½) | 0.75 (¾)	-0.5 (-½) | 0.75 (¾)	CROSS | TOL.	-0.5 (-½) | 0.75 (¾)	0.0 (0) | 0.75 (¾)	0.0 (0) | 0.75 (¾)
-0.3 ± 0.7 -5/16 ± 11/16	-0.5 ± 0.7 -½ ± 11/16	-0.3 ± 0.7 -5/16 ± 11/16	CAMBER (Degrees)	-0.5 ± 0.7 -½ ± 11/16	-0.2 ± 0.75 [1] -3/16 ± ¾ [1]	-0.2 ± 0.75 [1] -3/16 ± ¾ [1]
0.0 (0) | 0.7 (11/16)	0.0 (0) | 0.7 (11/16)	0.0 (0) | 0.7 (11/16)	CROSS | TOL.	0.0 (0) | 0.7 (11/16)	0.0 (0) | 0.75 (¾)	0.0 (0) | 0.75 (¾)
0.06° ± 0.25° 1/16° ± ¼°	-0.1° ± 0.25° -3/32° ± ¼°	0.06° ± 0.25° 1/16° ± ¼°	FRONT TOE (Degrees)	-0.1° ± 0.25° -3/32° ± ¼°	0.15° ± 0.25° 5/32° ± ¼°	0.15° ± 0.25° 5/32° ± ¼°
0.03" ± 0.13" 1/32" ± 1/8"	-0.05" ± 0.13" -1/16" ± 1/8"	0.03" ± 0.13" 1/32" ± 1/8"	FRONT TOE (Inches)	-0.05" ± 0.13" -1/16" ± 1/8"	0.08" ± 0.13" 1/16" ± 1/8"	0.08" ± 0.13" 1/16" ± 1/8"
			SAI			
			TURNING ANGLE			
128	128	128	ILLUSTRATION #	128	12	12
			REAR			
			CAMBER (Degrees)			
			REAR TOE (Degrees)			
			REAR TOE (Inches)			
			THRUST ANGLE		0.3 (5/16)	0.3 (5/16)
			ILLUSTRATION #			
			NOTES		[1] Service cam kit reqd.	[1] Service cam kit reqd.
17112	17113	17114		17115	17116	17117

Ford Truck & SUV

F150 4X2	F150 4X2	F150 4X2	MODEL	F150 4X2	F150 4X2	F150 4X2
Regular Cab	Super Cab	Super Cab	INFORMATION	Crew Cab	Crew Cab	
	exc/20" or 22" tires	w/20" or 22" tires		exc/20" or 22" tires	w/20" or 22" tires	
2006-2008	2006-2008	2006-2008	YEARS	2006-2008	2006-2008	2009-2014
			INSTRUCTIONS			
150 ft.lbs.	150 ft.lbs.	150 ft.lbs.	LUGNUT TORQUE	150 ft.lbs.	150 ft.lbs.	150 ft.lbs.
			FRONT			
L= 4.1, R= 4.5 ± 1.0 [1] L= 4 1/8, R= 4½ ± 1 [1]	L= 4.3, R= 4.7 ± 1.0 [1] L= 4 5/16, R= 4 11/16 ± 1 [1]	L= 3.9, R= 4.3 ± 1.0 [1] L= 3 7/8, R= 4 5/16 ± 1 [1]	CASTER (Degrees)	L= 4.4, R= 4.8 ± 1.0 [1] L= 4 3/8, R= 4 13/16 ± 1 [1]	L= 4.0, R= 4.4 ± 1.0 [1] L= 4, R= 4 3/8 ± 1 [1]	L= 3.9, R= 4.3 ± 1.0 [1] L= 3 7/8, R= 4 5/16 ± 1 [1]
-0.4 (-3/8) | 0.75 (¾)	-0.4 (-3/8) | 0.75 (¾)	-0.4 (-3/8) | 0.75 (¾)	CROSS | TOL.	-0.4 (-3/8) | 0.75 (¾)	-0.4 (-3/8) | 0.75 (¾)	-0.4 (-3/8) | 0.75 (¾)
-0.2 ± 0.75 [1] -3/16 ± ¾ [1]	-0.2 ± 0.75 [1] -3/16 ± ¾ [1]	-0.2 ± 0.75 [1] -3/16 ± ¾ [1]	CAMBER (Degrees)	-0.2 ± 0.75 [1] -3/16 ± ¾ [1]	-0.2 ± 0.75 [1] -3/16 ± ¾ [1]	L= -0.2, R= -0.4 ± 0.75 [1] L= -3/16, R= -3/8 ± ¾ [1]
0.0 (0) | 0.75 (¾)	0.0 (0) | 0.75 (¾)	0.0 (0) | 0.75 (¾)	CROSS | TOL.	0.0 (0) | 0.75 (¾)	0.0 (0) | 0.75 (¾)	0.2 (3/16) | 0.75 (¾)
0.2° ± 0.2° [2] 3/16° ± 3/16° [2]	0.2° ± 0.2° [2] 3/16° ± 3/16° [2]	0.2° ± 0.2° [2] 3/16° ± 3/16° [2]	FRONT TOE (Degrees)	0.2° ± 0.2° [2] 3/16° ± 3/16° [2]	0.2° ± 0.2° [2] 3/16° ± 3/16° [2]	0.2° ± 0.2° 3/16° ± 3/16°
0.1" ± 0.1" [2] 3/32" ± 3/32" [2]	0.1" ± 0.1" [2] 3/32" ± 3/32" [2]	0.1" ± 0.1" [2] 3/32" ± 3/32" [2]	FRONT TOE (Inches)	0.1" ± 0.1" [2] 3/32" ± 3/32" [2]	0.1" ± 0.1" [2] 3/32" ± 3/32" [2]	0.1" ± 0.1" 3/32" ± 3/32"
			SAI			
			TURNING ANGLE			
12	12	12	ILLUSTRATION #	12	12	12
			REAR			
			CAMBER (Degrees)			
			REAR TOE (Degrees)			
			REAR TOE (Inches)			
0.3 (5/16)	0.3 (5/16)	0.3 (5/16)	THRUST ANGLE	0.3 (5/16)	0.3 (5/16)	0.5 (½)
			ILLUSTRATION #			
[1] Service cam kit reqd. [2] 2006 = 0.15° (0.075")	[1] Service cam kit reqd. [2] 2006 = 0.15° (0.075")	[1] Service cam kit reqd. [2] 2006 = 0.15° (0.075")	NOTES	[1] Service cam kit reqd. [2] 2006 = 0.15° (0.075")	[1] Service cam kit reqd. [2] 2006 = 0.15° (0.075")	[1] Service cam kit reqd.
17118	17119	17120		17121	17122	17123

Ford Truck & SUV

MODEL INFORMATION	F150 4X2 Reg. Cab	F150 4X2 Super/Crew Cab	F150 4X4	F150 4X4	F150 Heritage 4X4	F150 4X4 Regular Cab
YEARS	2015-2017	2015-2017	1990-96	1997-2003	2004	2004-2005
INSTRUCTIONS			p. 306, 307			
LUGNUT TORQUE	150 ft.lbs.	150 ft.lbs.	100 ft.lbs.	83-112 ft.lbs.	83-112 ft.lbs.	150 ft.lbs.
FRONT						
CASTER (Degrees)	L= 3.5, R= 3.9 ± 0.75 L= 3½, R= 3⅞ ± ¾	L= 3.8, R= 4.2 ± 0.75 L= 3 13/16, R= 4 3/16 ± ¾	MIN.= 2.0 MAX.= 6.0 MIN.= 2 MAX.= 6	L= 4.6, R= 5.3 ± 1.0 L= 4⅝, R= 5 5/16 ± 1	L= 4.6, R= 5.3 ± 1.0 L= 4⅝, R= 5 5/16 ± 1	4.3 ± 1.0 [1] 4 5/16 ± 1 [1]
CROSS | TOL.	-0.4 (-⅜) | 0.75 (¾)	-0.4 (-⅜) | 0.75 (¾)	-0.5 (-½) | 0.5 (½)	-0.7 (-11/16) | 0.7 (11/16)	-0.7 (-11/16) | 0.7 (11/16)	0.0 (0) | 0.75 (¾)
CAMBER (Degrees)	-0.1 ± 0.75 -⅛ ± ¾	-0.1 ± 0.75 -⅛ ± ¾	0.25 ± 0.5 ¼ ± ½	-0.1 ± 0.7 -⅛ ± 11/16	-0.1 ± 0.7 -⅛ ± 11/16	-0.2 ± 0.75 [1] -3/16 ± ¾ [1]
CROSS | TOL.	0.2 (3/16) | 0.75 (¾)	0.2 (3/16) | 0.75 (¾)	0.0 (0) | 0.5 (½)	0.0 (0) | 0.7 (11/16)	0.0 (0) | 0.7 (11/16)	0.0 (0) | 0.7 (11/16)
FRONT TOE (Degrees)	0.2° ± 0.2° 3/16° ± 3/16°	0.2° ± 0.2° 3/16° ± 3/16°	0.06° ± 0.25° 1/16° ± ¼°	0.2° ± 0.25° 3/16° ± ¼°	0.2° ± 0.25° 3/16° ± ¼°	0.15° ± 0.25° 5/32° ± ¼°
FRONT TOE (Inches)	0.1" ± 0.1" 3/32" ± 3/32"	0.1" ± 0.1" 3/32" ± 3/32"	0.03" ± 0.13" 1/32" ± ⅛"	0.1" ± 0.13" 3/32" ± ⅛"	0.1" ± 0.13" 3/32" ± ⅛"	0.08" ± 0.13" 1/16" ± ⅛"
SAI						
TURNING ANGLE						
ILLUSTRATION #	12	12	75	128	128	12
REAR						
CAMBER (Degrees)						
REAR TOE (Degrees)						
REAR TOE (Inches)						
THRUST ANGLE	0.5 (½)	0.5 (½)				0.3 (5/16)
ILLUSTRATION #						
NOTES						[1] Service cam kit reqd.
	17124	17125	17136	17137	17138	17139

Ford Truck & SUV

MODEL INFORMATION	F150 4X4 Super/Crew Cab	F150 4X4 Regular Cab	F150 4X4 Super Cab	F150 4X4 Crew Cab	F150 4X4 exc/SVT Raptor	F150 4X4 SVT Raptor
YEARS	2004-2005	2006-2008	2006-2008	2006-2008	2009-2014	2011-2014
INSTRUCTIONS						
LUGNUT TORQUE	150 ft.lbs.	150 ft.lbs.	150 ft.lbs.	150 ft.lbs.	150 ft.lbs.	150 ft.lbs.
FRONT						
CASTER (Degrees)	4.6 ± 1.0 [1] 4⅝ ± 1 [1]	L= 4.1, R= 4.5 ± 1.0 [1] L= 4⅛, R= 4½ ± 1 [1]	L= 4.4, R= 4.8 ± 1.0 [1] L= 4⅜, R= 4 13/16 ± 1 [1]	L= 4.5, R= 4.9 ± 1.0 [1] L= 4½, R= 4⅞ ± 1 [1]	L= 3.8, R= 4.2 ± 1.0 [1] L= 3 13/16, R= 4 3/16 ± 1 [1]	L= 3.7, R= 4.1 ± 1.0 L= 3 11/16, R= 4⅛ ± 1
CROSS | TOL.	0.0 (0) | 0.75 (¾)	-0.4 (-⅜) | 0.75 (¾)	-0.4 (-⅜) | 0.75 (¾)	-0.4 (-⅜) | 0.75 (¾)	-0.4 (-⅜) | 0.75 (¾)	-0.4 (-⅜) | 0.75 (¾)
CAMBER (Degrees)	-0.2 ± 0.75 [1] -3/16 ± ¾ [1]	-0.2 ± 0.75 [1] -3/16 ± ¾ [1]	-0.2 ± 0.75 [1] -3/16 ± ¾ [1]	-0.2 ± 0.75 [1] -3/16 ± ¾ [1]	L= -0.2, R= -0.4 ± 0.75 [1] L= -3/16, R= -⅜ ± ¾ [1]	-0.45 ± 0.75 -7/16 ± ¾
CROSS | TOL.	0.0 (0) | 0.7 (11/16)	0.0 (0) | 0.75 (¾)	0.0 (0) | 0.75 (¾)	0.0 (0) | 0.75 (¾)	0.2 (3/16) | 0.75 (¾)	0.0 (0) | 0.75 (¾)
FRONT TOE (Degrees)	0.15° ± 0.25° 5/32° ± ¼°	0.2° ± 0.2° [2] 3/16° ± 3/16° [2]	0.2° ± 0.2° [2] 3/16° ± 3/16° [2]	0.2° ± 0.2° [2] 3/16° ± 3/16° [2]	0.2° ± 0.2° 3/16° ± 3/16°	0.2° ± 0.2° 3/16° ± 3/16°
FRONT TOE (Inches)	0.08" ± 0.13" 1/16" ± ⅛"	0.1" ± 0.1" [2] 3/32" ± 3/32" [2]	0.1" ± 0.1" [2] 3/32" ± 3/32" [2]	0.1" ± 0.1" [2] 3/32" ± 3/32" [2]	0.1" ± 0.1" 3/32" ± 3/32"	0.1" ± 0.1" 3/32" ± 3/32"
SAI						
TURNING ANGLE						
ILLUSTRATION #	12	12	12	12	12	12
REAR						
CAMBER (Degrees)						
REAR TOE (Degrees)						
REAR TOE (Inches)						
THRUST ANGLE	0.3 (5/16)	0.3 (5/16)	0.3 (5/16)	0.3 (5/16)	0.5 (½)	0.5 (½)
ILLUSTRATION #						
NOTES	[1] Service cam kit reqd.	[1] Service cam kit reqd. [2] 2006 = 0.15° (0.075")	[1] Service cam kit reqd. [2] 2006 = 0.15° (0.075")	[1] Service cam kit reqd. [2] 2006 = 0.15° (0.075")	[1] Service cam kit reqd.	
	17140	17141	17142	17143	17144	17145

ALIGNMENT SPECIFICATIONS

Ford Truck & SUV

MODEL	F150 4X4	F150 4X4	F150 4X4	F250 4X2	F250 4X2	F250 4X2
INFORMATION	Regular Cab	Super/Crew Cab exc/2017 Raptor	Raptor	(incl.1997 F250 HD)	(exc 1997 HD)	Super Duty Pickup
YEARS	2015-2017	2015-2017	2017	1990-97	1997-99	1999-2004
INSTRUCTIONS				p. 306, 307	p. 306, 307	
LUGNUT TORQUE	150 ft.lbs.	150 ft.lbs.	150 ft.lbs.	140 ft.lbs.		
FRONT						
CASTER (Degrees)	L= 3.1, R= 3.5 ± 0.75 L= 3⅛, R= 3½ ± ¾	L= 3.4, R= 3.8 ± 0.75 L= 3⅜, R= 3 13/16 ± ¾	3.5 ± 0.75 3½ ± ¾	MIN.= 2.0 MAX.= 6.0 MIN.= 2 MAX.= 6	L= 5.9, R= 6.4 ± 1.0 [1,2] L= 5⅞, R= 6⅜ ± 1 [1,2]	4.0 ± 2.0 4 ± 2
CROSS | TOL.	-0.4 (-⅜) | 0.75 (¾)	-0.4 (-⅜) | 0.75 (¾)	0.0 (0) | 0.75 (¾)	-0.5 (-½) | 0.5 (½)	-0.5 (-½) | 0.7 (11/16)	0.0 (0) | 1.0 (1)
CAMBER (Degrees)	-0.1 ± 0.75 -⅛ ± ¾	-0.1 ± 0.75 -⅛ ± ¾	-0.5 ± 0.75 -½ ± ¾	0.25 ± 0.5 ¼ ± ½	-0.3 ± 0.7 [1] -5/16 ± 11/16 [1]	0.62 ± 1.0 ⅝ ± 1
CROSS | TOL.	0.2 (3/16) | 0.75 (¾)	0.2 (3/16) | 0.75 (¾)	0.2 (3/16) | 0.75 (¾)	0.0 (0) | 0.5 (½)	0.0 (0) | 0.7 (11/16)	0.0 (0) | 1.0 (1)
FRONT TOE (Degrees)	0.2° ± 0.2° 3/16° ± 3/16°	0.2° ± 0.2° 3/16° ± 3/16°	0.2° ± 0.2° 3/16° ± 3/16°	0.06° ± 0.25° 1/16° ± ¼°	0.06° ± 0.25° 1/16° ± ¼°	0.03° ± 0.25° 1/32° ± ¼°
FRONT TOE (Inches)	0.1" ± 0.1" 3/32" ± 3/32"	0.1" ± 0.1" 3/32" ± 3/32"	0.1" ± 0.1" 3/32" ± 3/32"	0.03" ± 0.13" 1/32" ± ⅛"	0.03" ± 0.13" 1/32" ± ⅛"	0.02" ± 0.13" 1/64" ± ⅛"
SAI						
TURNING ANGLE						
ILLUSTRATION #	12	12	12	75	128	75
REAR						
CAMBER (Degrees)						
REAR TOE (Degrees)						
REAR TOE (Inches)						
THRUST ANGLE	0.5 (½)	0.5 (½)	0.5 (½)			
ILLUSTRATION #						
NOTES					[1] Service cam kit reqd. [2] w/Load Level L=6.9°, R=7.4°	
	17146	17147	17148	17166	17167	17168

Ford Truck & SUV

MODEL	F250 4X2	F250 4X2	F250 4X2	F250 4X4	F250 4X4	F250 4X4
INFORMATION	Super Duty Pickup	Super Duty Pickup	Super Duty Pickup	(incl.1997 F250 Super Duty)	(exc Super Duty)	Super Duty Pickup
YEARS	2005-2007	2008-2016	2017	1990-97	1997-99	1999-2004
INSTRUCTIONS				p. 306, 307		
LUGNUT TORQUE		165 ft.lbs.	165 ft.lbs.	140 ft.lbs.		
FRONT						
CASTER (Degrees)	3.8 ± 1.2 3 13/16 ± 1 3/16	4.0 ± 1.2 [1] 4 ± 1 3/16 [1]	4.0 ± 1.2 [1] 4 ± 1 3/16 [1]	MIN.= 2.0 MAX.= 6.0 MIN.= 2 MAX.= 6	L= 4.3, R= 5.0 ± 1.0 [1,2] L= 4 5/16, R= 5 ± 1 [1,2]	3.5 ± 2.0 3½ ± 2
CROSS | TOL.	0.0 (0) | 0.75 (¾)	0.0 (0) | 0.75 (¾)	0.0 (0) | 0.75 (¾)	-0.5 (-½) | 0.5 (½)	-0.7 (-11/16) | 0.7 (11/16)	0.0 (0) | 1.0 (1)
CAMBER (Degrees)	0.62 ± 0.75 ⅝ ± ¾	0.62 ± 0.75 ⅝ ± ¾	0.62 ± 0.75 ⅝ ± ¾	0.25 ± 0.5 [1] ¼ ± ½ [1]	-0.1 ± 0.7 [1] -⅛ ± 11/16 [1]	0.25 ± 1.0 ¼ ± 1
CROSS | TOL.	0.0 (0) | 0.75 (¾)	0.0 (0) | 0.75 (¾)	0.0 (0) | 0.75 (¾)	0.0 (0) | 0.5 (½)	0.0 (0) | 0.7 (11/16)	0.0 (0) | 1.0 (1)
FRONT TOE (Degrees)	0.1° ± 0.25° 3/32° ± ¼°	0.1° ± 0.25° 3/32° ± ¼°	0.1° ± 0.25° 3/32° ± ¼°	0.06° ± 0.25° 1/16° ± ¼°	0.2° ± 0.25° 3/16° ± ¼°	0.03° ± 0.25° 1/32° ± ¼°
FRONT TOE (Inches)	0.05" ± 0.13" 1/16" ± ⅛"	0.05" ± 0.13" 1/16" ± ⅛"	0.05" ± 0.13" 1/16" ± ⅛"	0.03" ± 0.13" 1/32" ± ⅛"	0.1" ± 0.13" 3/32" ± ⅛"	0.02" ± 0.13" 1/64" ± ⅛"
SAI						
TURNING ANGLE						
ILLUSTRATION #	75	75	75	75	128	75
REAR						
CAMBER (Degrees)						
REAR TOE (Degrees)						
REAR TOE (Inches)						
THRUST ANGLE		0.5 (½)	0.5 (½)			
ILLUSTRATION #						
NOTES		[1] HD Susp. = 3.6° (3⅝°)	[1] HD Susp. = 3.6° (3⅝°)	[1] 1993-94 = 0.5° (½°)	[1] Service cam kit reqd. [2] w/Load Level L=5.3°, R=6.0°	
	17169	17170	17171	17182	17183	17184

Ford Truck & SUV

F250 4X4 Super Duty Pickup	F250 4X4 Super Duty Pickup	F250 4X4 Super Duty Pickup	MODEL INFORMATION	F350 4X2 Single Rear Wheels	F350 4X2 Dual Rear Wheels	F350 4X2 Super Duty Pickup
2005-2007	2008-2016	2017	YEARS	1990-97	1990-97	1999-2004
			INSTRUCTIONS	p. 306, 307	p. 306, 307	
	165 ft.lbs.	165 ft.lbs.	LUGNUT TORQUE	140 ft.lbs.	140 ft.lbs.	
			FRONT			
3.3 ± 1.25[1] 3 5/16 ± 1 ¼[1]	3.5 ± 1.2[1] 3 ½ ± 1 3/16[1]	L= 3.5, R= 3.8 ± 1.2[1] L= 3 ½, R= 3 13/16 ± 1 3/16[1]	CASTER (Degrees)	MIN.= 2.0 MAX.= 6.0 MIN.= 2 MAX.= 6	MIN.= 2.0 MAX.= 4.5 MIN.= 2 MAX.= 4 ½	4.0 ± 2.0 4 ± 2
0.0 (0) | 0.5 (½)	0.0 (0) | 0.55 (9/16)	-0.3 (-5/16) | 0.55 (9/16)	CROSS | TOL.	-0.5 (-½) | 0.5 (½)	-0.5 (-½) | 0.5 (½)	0.0 (0) | 1.0 (1)
0.15 ± 0.75 ⅛ ± ¾	0.15 ± 0.75 ⅛ ± ¾	0.15 ± 0.75 ⅛ ± ¾	CAMBER (Degrees)	0.5 ± 0.5 ½ ± ½	0.5 ± 0.5 ½ ± ½	0.62 ± 1.0 ⅝ ± 1
0.0 (0) | 0.75 (¾)	0.0 (0) | 0.75 (¾)	0.0 (0) | 0.75 (¾)	CROSS | TOL.	0.0 (0) | 0.5 (½)	0.0 (0) | 0.5 (½)	0.0 (0) | 1.0 (1)
0.1° ± 0.25° 3/32° ± ¼°	0.1° ± 0.25° 3/32° ± ¼°	0.1° ± 0.25° 3/32° ± ¼°	FRONT TOE (Degrees)	0.06° ± 0.25° 1/16° ± ¼°	0.06° ± 0.25° 1/16° ± ¼°	0.03° ± 0.25° 1/32° ± ¼°
0.05" ± 0.13" 1/16" ± ⅛"	0.05" ± 0.13" 1/16" ± ⅛"	0.05" ± 0.13" 1/16" ± ⅛"	FRONT TOE (Inches)	0.03" ± 0.13" 1/32" ± ⅛"	0.03" ± 0.13" 1/32" ± ⅛"	0.02" ± 0.13" 1/64" ± ⅛"
			SAI			
			TURNING ANGLE			
75	75	75	ILLUSTRATION #	75	75	75
			REAR			
			CAMBER (Degrees)			
			REAR TOE (Degrees)			
			REAR TOE (Inches)			
	0.5 (½)	0.5 (½)	THRUST ANGLE			
			ILLUSTRATION #			
[1] HD Susp. = 2.6° (2 5/8°)	[1] HD Susp. = 2.8° (2 13/16°)	[1] HD Susp. = L=2.8°, R=3.1°	NOTES			
17185	17186	17187		17200	17201	17202

Ford Truck & SUV

F350 4X2 Super Duty Pickup	F350 4X2 Super Duty Pickup	F350 4X2 Super Duty Pickup	MODEL INFORMATION	F350 4X2 Chassis Cab w/STD Susp.	F350 4X2 Chassis Cab w/HD Susp.	F350 4X2 Chassis Cab w/STD Susp.
2005-2007	2008-2016	2017	YEARS	2008-2016	2008-2016	2017
			INSTRUCTIONS			
	165 ft.lbs.	165 ft.lbs.	LUGNUT TORQUE	165 ft.lbs.	165 ft.lbs.	165 ft.lbs.
			FRONT			
3.8 ± 1.2 3 13/16 ± 1 3/16	4.0 ± 1.2[1] 4 ± 1 3/16[1]	4.0 ± 1.2[1] 4 ± 1 3/16[1]	CASTER (Degrees)	3.5 ± 1.2[1] 3 ½ ± 1 3/16[1]	3.2 ± 1.3[1] 3 3/16 ± 1 5/16[1]	3.5 ± 1.2[1] 3 ½ ± 1 3/16[1]
0.0 (0) | 0.75 (¾)	0.0 (0) | 0.75 (¾)	0.0 (0) | 0.75 (¾)	CROSS | TOL.	0.0 (0) | 0.75 (¾)	0.0 (0) | 0.75 (¾)	0.0 (0) | 0.75 (¾)
0.62 ± 0.75 ⅝ ± ¾	0.62 ± 0.75 ⅝ ± ¾	0.62 ± 0.75 ⅝ ± ¾	CAMBER (Degrees)	0.62 ± 0.75 ⅝ ± ¾	0.62 ± 0.75 ⅝ ± ¾	0.62 ± 0.75 ⅝ ± ¾
0.0 (0) | 0.75 (¾)	0.0 (0) | 0.75 (¾)	0.0 (0) | 0.75 (¾)	CROSS | TOL.	0.0 (0) | 0.75 (¾)	0.0 (0) | 0.75 (¾)	0.0 (0) | 0.75 (¾)
0.1° ± 0.25° 3/32° ± ¼°	0.1° ± 0.25° 3/32° ± ¼°	0.1° ± 0.25° 3/32° ± ¼°	FRONT TOE (Degrees)	0.1° ± 0.25° 3/32° ± ¼°	0.1° ± 0.25° 3/32° ± ¼°	0.1° ± 0.25° 3/32° ± ¼°
0.05" ± 0.13" 1/16" ± ⅛"	0.05" ± 0.13" 1/16" ± ⅛"	0.05" ± 0.13" 1/16" ± ⅛"	FRONT TOE (Inches)	0.05" ± 0.13" 1/16" ± ⅛"	0.05" ± 0.13" 1/16" ± ⅛"	0.05" ± 0.13" 1/16" ± ⅛"
			SAI			
			TURNING ANGLE			
75	75	75	ILLUSTRATION #	75	75	75
			REAR			
			CAMBER (Degrees)			
			REAR TOE (Degrees)			
			REAR TOE (Inches)			
	0.5 (½)	0.5 (½)	THRUST ANGLE	0.5 (½)	0.5 (½)	0.5 (½)
			ILLUSTRATION #			
	[1] HD Susp. = 3.6° (3 5/8°)	[1] HD Susp. = 3.6° (3 5/8°)	NOTES	[1] w/Duals = 2.9° (2 7/8°)	[1] w/Duals = 2.5° (2 ½°)	[1] w/Duals = 2.9° (2 7/8°)
17203	17204	17205		17206	17207	17208

Ford Truck & SUV

MODEL	F350 4X2	F350 4X4	F350 4X4	F350 4X4	F350 4X4	F350 4X4
INFORMATION	Chassis Cab w/HD Susp.		Super Duty Pickup	Super Duty Pickup w/STD Susp.	Super Duty Pickup w/HD Susp.	Super Duty Pickup w/STD Susp.
YEARS	2017	1990-97	1999-2004	2005-2007	2005-2007	2008-2016
INSTRUCTIONS		p. 306, 307				
LUGNUT TORQUE	165 ft.lbs.	140 ft.lbs.				165 ft.lbs.
FRONT						
CASTER (Degrees)	3.2 ± 1.3 [1] 3 3/16 ± 1 5/16 [1]	MIN.= 2.0 MAX.= 4.8 MIN.= 2 MAX.= 4 3/4	3.5 ± 2.0 3 1/2 ± 2	2.5 ± 1.25 [1] 2 1/2 ± 1 1/4 [1]	1.8 ± 1.25 [1] 1 13/16 ± 1 1/4 [1]	2.7 ± 1.2 [1] 2 11/16 ± 1 3/16 [1]
CROSS	0.0 (0)	-0.5 (-1/2)	0.0 (0)	0.0 (0)	0.0 (0)	0.0 (0)
TOL.	0.75 (3/4)	0.5 (1/2)	1.0 (1)	0.5 (1/2)	0.5 (1/2)	0.55 (9/16)
CAMBER (Degrees)	0.62 ± 0.75 5/8 ± 3/4	0.0 ± 0.5 0 ± 1/2	0.25 ± 1.0 1/4 ± 1	0.15 ± 0.75 1/8 ± 3/4	0.15 ± 0.75 1/8 ± 3/4	0.15 ± 0.75 1/8 ± 3/4
CROSS	0.0 (0)	0.0 (0)	0.0 (0)	0.0 (0)	0.0 (0)	0.0 (0)
TOL.	0.75 (3/4)	0.5 (1/2)	1.0 (1)	0.75 (3/4)	0.75 (3/4)	0.75 (3/4)
FRONT TOE (Degrees)	0.1° ± 0.25° 3/32° ± 1/4°	0.06° ± 0.25° 1/16° ± 1/4°	0.03° ± 0.25° 1/32° ± 1/4°	0.1° ± 0.25° 3/32° ± 1/4°	0.1° ± 0.25° 3/32° ± 1/4°	0.1° ± 0.25° 3/32° ± 1/4°
FRONT TOE (Inches)	0.05" ± 0.13" 1/16" ± 1/8"	0.03" ± 0.13" 1/32" ± 1/8"	0.02" ± 0.13" 1/64" ± 1/8"	0.05" ± 0.13" 1/16" ± 1/8"	0.05" ± 0.13" 1/16" ± 1/8"	0.05" ± 0.13" 1/16" ± 1/8"
SAI						
TURNING ANGLE						
ILLUSTRATION #	75		75	75	75	75
REAR						
CAMBER (Degrees)						
REAR TOE (Degrees)						
REAR TOE (Inches)						
THRUST ANGLE	0.5 (1/2)					0.5 (1/2)
ILLUSTRATION #						
NOTES	[1] w/Duals = 2.5° (2 1/2°)			[1] w/Duals = 3.0° (3°)	[1] w/Duals = 2.4° (2 3/8°)	[1] w/Duals = L=3.3°, R=3.6°
	17209	17214	17215	17216	17217	17218

Ford Truck & SUV

MODEL	F350 4X4	F350 4X4	F350 4X4	F350 4X4	F350 4X4	F350 4X4
INFORMATION	Super Duty Pickup w/HD Susp.	Super Duty Pickup w/STD Susp.	Super Duty Pickup w/HD Susp.	Chassis Cab w/STD Susp.	Chassis Cab w/HD Susp.	Chassis Cab w/STD Susp.
YEARS	2008-2016	2017	2017	2005-2007	2005-2007	2008-2016
INSTRUCTIONS						
LUGNUT TORQUE	165 ft.lbs.	165 ft.lbs.	165 ft.lbs.			165 ft.lbs.
FRONT						
CASTER (Degrees)	2.1 ± 1.3 [1] 2 1/8 ± 1 5/16 [1]	L= 3.5, R= 3.8 ± 1.2 L= 3 1/2, R= 3 13/16 ± 1 3/16	L= 2.8, R= 3.1 ± 1.3 L= 2 13/16, R= 3 1/8 ± 1 5/16	2.8 ± 1.25 [1] 2 13/16 ± 1 1/4 [1]	2.3 ± 1.25 [1] 2 5/16 ± 1 1/4 [1]	2.9 ± 1.2 [1] 2 7/8 ± 1 3/16 [1]
CROSS	0.0 (0)	-0.3 (-5/16)	-0.3 (-5/16)	0.0 (0)	0.0 (0)	0.0 (0)
TOL.	0.55 (9/16)	0.55 (9/16)	0.55 (9/16)	0.5 (1/2)	0.5 (1/2)	0.55 (9/16)
CAMBER (Degrees)	0.15 ± 0.75 1/8 ± 3/4	0.15 ± 0.75 1/8 ± 3/4	0.15 ± 0.75 1/8 ± 3/4	0.15 ± 0.75 1/8 ± 3/4	0.15 ± 0.75 1/8 ± 3/4	0.15 ± 0.75 1/8 ± 3/4
CROSS	0.0 (0)	0.0 (0)	0.0 (0)	0.0 (0)	0.0 (0)	0.0 (0)
TOL.	0.75 (3/4)	0.75 (3/4)	0.75 (3/4)	0.75 (3/4)	0.75 (3/4)	0.75 (3/4)
FRONT TOE (Degrees)	0.1° ± 0.25° 3/32° ± 1/4°	0.1° ± 0.25° 3/32° ± 1/4°	0.1° ± 0.25° 3/32° ± 1/4°	0.1° ± 0.25° 3/32° ± 1/4°	0.1° ± 0.25° 3/32° ± 1/4°	0.1° ± 0.25° 3/32° ± 1/4°
FRONT TOE (Inches)	0.05" ± 0.13" 1/16" ± 1/8"	0.05" ± 0.13" 1/16" ± 1/8"	0.05" ± 0.13" 1/16" ± 1/8"	0.05" ± 0.13" 1/16" ± 1/8"	0.05" ± 0.13" 1/16" ± 1/8"	0.05" ± 0.13" 1/16" ± 1/8"
SAI						
TURNING ANGLE						
ILLUSTRATION #	75	75	75	75	75	75
REAR						
CAMBER (Degrees)						
REAR TOE (Degrees)						
REAR TOE (Inches)						
THRUST ANGLE	0.5 (1/2)	0.5 (1/2)	0.5 (1/2)			0.5 (1/2)
ILLUSTRATION #						
NOTES	[1] w/Duals = L=2.6°, R=2.9°			[1] w/Duals = 2.5° (2 1/2°)	[1] w/Duals = 1.9° (1 7/8°)	[1] w/Duals = 2.5° (2 1/2°)
	17219	17220	17221	17222	17223	17224

Ford Truck & SUV

F350 4X4 Chassis Cab w/HD Susp.	F350 4X4 Chassis Cab w/STD Susp.	F350 4X4 Chassis Cab w/HD Susp.	MODEL INFORMATION	F450 4X4 Super Duty Pickup	F450 4X4 Super Duty Pickup	F450/F550 4X2 Chassis Cab
2008-2016	2017	2017	YEARS	2008-2016	2017	1999-2004
			INSTRUCTIONS			
165 ft.lbs.	165 ft.lbs.	165 ft.lbs.	LUGNUT TORQUE	165 ft.lbs.	165 ft.lbs.	
			FRONT			
2.4 ± 1.3 [1] 2⅜ ± 1 5/16 [1]	L= 3.5, R= 3.8 ± 1.2 L= 3½, R= 2 13/16 ± 1 3/16	L= 2.8, R= 3.1 ± 1.3 L= 2 13/16, R= 3⅛ ± 1 5/16	CASTER (Degrees)	L= 3.1, R= 3.5 ± 1.2 [1] L= 3⅛, R= 3½ ± 1 3/16 [1]	L= 3.1, R= 3.4 ± 1.2 [1] L= 3⅛, R= 3⅜ ± 1 3/16 [1]	3.5 ± 2.0 3½ ± 2
0.0 (0) | 0.55 (9/16)	-0.3 (-5/16) | 0.55 (9/16)	-0.3 (-5/16) | 0.55 (9/16)	CROSS | TOL.	-0.4 (-⅜) | 0.55 (9/16)	-0.3 (-5/16) | 0.55 (9/16)	0.0 (0) | 1.0 (1)
0.15 ± 0.75 ⅛ ± ¾	0.15 ± 0.75 ⅛ ± ¾	0.15 ± 0.75 ⅛ ± ¾	CAMBER (Degrees)	0.15 ± 0.75 ⅛ ± ¾	0.15 ± 0.75 ⅛ ± ¾	0.25 ± 1.0 ¼ ± 1
0.0 (0) | 0.75 (¾)	0.0 (0) | 0.75 (¾)	0.0 (0) | 0.75 (¾)	CROSS | TOL.	0.0 (0) | 0.75 (¾)	0.0 (0) | 0.75 (¾)	0.0 (0) | 1.0 (1)
0.1° ± 0.25° 3/32° ± ¼°	0.1° ± 0.25° 3/32° ± ¼°	0.1° ± 0.25° 3/32° ± ¼°	FRONT TOE (Degrees)	0.1° ± 0.25° 3/32° ± ¼°	0.1° ± 0.25° 3/32° ± ¼°	0.03° ± 0.25° 1/32° ± ¼°
0.05" ± 0.13" 1/16" ± ⅛"	0.05" ± 0.13" 1/16" ± ⅛"	0.05" ± 0.13" 1/16" ± ⅛"	FRONT TOE (Inches)	0.05" ± 0.13" 1/16" ± ⅛"	0.05" ± 0.13" 1/16" ± ⅛"	0.02" ± 0.13" 1/64" ± ⅛"
			SAI			
			TURNING ANGLE			
75	75	75	ILLUSTRATION #	75	75	75
			REAR			
			CAMBER (Degrees)			
			REAR TOE (Degrees)			
			REAR TOE (Inches)			
0.5 (½)	0.5 (½)	0.5 (½)	THRUST ANGLE	0.5 (½)	0.5 (½)	
			ILLUSTRATION #			
[1] w/Duals = 2.1° (2⅛°)			NOTES	[1] HD Susp. = L=2.5°, R=2.9°	[1] HD Susp. = L=2.4°, R=2.7°	
17225	17226	17227		17228	17229	17230

Ford Truck & SUV

F450/F550 4X2 Chassis Cab	F450/F550 4X2 Chassis Cab	F450/F550 4X2 Chassis Cab	MODEL INFORMATION	F450/F550 4X2 Chassis Cab	F450/F550 4X4 Chassis Cab	F450/F550 4X4 Chassis Cab
2005-2007	2008-2010	2011-2016	YEARS	2017	1999-2004	2005-2007
			INSTRUCTIONS			
	165 ft.lbs.	165 ft.lbs.	LUGNUT TORQUE	165 ft.lbs.		
			FRONT			
2.2 ± 1.25 [1,2] 2 3/16 ± 1¼ [1,2]	L= 2.6, R= 3.0 ± 1.2 [1] L= 2⅝, R= 3 ± 1 3/16 [1]	L= 2.6, R= 2.9 ± 1.2 L= 2⅝, R= 2⅞ ± 1 3/16	CASTER (Degrees)	L= 2.5, R= 2.8 ± 1.2 [1] L= 2½, R= 2 13/16 ± 1 3/16 [1]	3.5 ± 2.0 3½ ± 2	2.2 ± 1.25 [1] 2 3/16 ± 1¼ [1]
0.0 (0) | 0.5 (½)	-0.4 (-⅜) | 0.55 (9/16)	-0.3 (-5/16) | 0.55 (9/16)	CROSS | TOL.	-0.3 (-5/16) | 0.55 (9/16)	0.0 (0) | 1.0 (1)	0.0 (0) | 0.5 (½)
0.15 ± 0.75 ⅛ ± ¾	0.15 ± 0.75 ⅛ ± ¾	0.15 ± 0.75 ⅛ ± ¾	CAMBER (Degrees)	0.15 ± 0.75 ⅛ ± ¾	0.25 ± 1.0 ¼ ± 1	0.15 ± 0.75 ⅛ ± ¾
0.0 (0) | 0.75 (¾)	0.0 (0) | 0.75 (¾)	0.0 (0) | 0.75 (¾)	CROSS | TOL.	0.0 (0) | 0.75 (¾)	0.0 (0) | 1.0 (1)	0.0 (0) | 0.75 (¾)
0.1° ± 0.25° 3/32° ± ¼°	0.1° ± 0.25° 3/32° ± ¼°	0.1° ± 0.25° 3/32° ± ¼°	FRONT TOE (Degrees)	0.1° ± 0.25° 3/32° ± ¼°	0.03° ± 0.25° 1/32° ± ¼°	0.1° ± 0.25° 3/32° ± ¼°
0.05" ± 0.13" 1/16" ± ⅛"	0.05" ± 0.13" 1/16" ± ⅛"	0.05" ± 0.13" 1/16" ± ⅛"	FRONT TOE (Inches)	0.05" ± 0.13" 1/16" ± ⅛"	0.02" ± 0.13" 1/64" ± ⅛"	0.05" ± 0.13" 1/16" ± ⅛"
			SAI			
			TURNING ANGLE			
75	75	75	ILLUSTRATION #	75	75	75
			REAR			
			CAMBER (Degrees)			
			REAR TOE (Degrees)			
			REAR TOE (Inches)			
	0.5 (½)	0.5 (½)	THRUST ANGLE	0.5 (½)		
			ILLUSTRATION #			
[1] HD = 1.9° (1⅞°) [2] HD Incl. XHD, Camper, Ambul.	[1] HD Susp. = L=2.1°, R=2.5°		NOTES	[1] HD Susp. = L=2.0°, R=2.3°		[1] HD = 1.9° (1⅞°) [2] HD incl. XHD, Camper, Ambul.
17231	17232	17233		17234	17235	17236

ALIGNMENT SPECIFICATIONS

ALIGNMENT SPECIFICATIONS

Ford Truck & SUV

F450/F550 4X4 Chassis Cab	F450/F550 4X4 Chassis Cab	F450/F550 4X4 Chassis Cab	MODEL INFORMATION	F-Super Duty Chassis Cab	F-Super Duty Chassis Cab	F-Super Duty Commercial Strip
2008-2010	2011-2016	2017	YEARS	1990-93	1994-97	1991-97
			INSTRUCTIONS	p. 306, 307	p. 306, 307	p. 306, 307
165 ft.lbs.	165 ft.lbs.	165 ft.lbs.	LUGNUT TORQUE	140 ft.lbs.	140 ft.lbs.	
			FRONT			
L=2.6, R=3.0±1.2 [1] L=2⅝, R=3±1 3/16 [1]	L=2.6, R=2.9±1.2 L=2⅝, R=2⅞±1 3/16	L=2.5, R=2.8±1.2 [1] L=2½, R=2 13/16±1 3/16 [1]	CASTER (Degrees)	MIN.=2.0 MAX.=5.0 MIN.=2 MAX.=5	MIN.=2.0 MAX.=5.0 MIN.=2 MAX.=5	MIN.=2.0 MAX.=5.5 MIN.=2 MAX.=5½
-0.4 (-⅜) | 0.55 (9/16)	-0.3 (-5/16) | 0.55 (9/16)	-0.3 (-5/16) | 0.55 (9/16)	CROSS | TOL.	-0.5 (-½) | 0.5 (½)	0.0 (0) | 0.5 (½)	0.0 (0) | 0.5 (½)
0.15±0.75 ⅛±¾	0.15±0.75 ⅛±¾	0.15±0.75 ⅛±¾	CAMBER (Degrees)	0.0±0.5 0±½	0.0±0.3 0±5/16	0.0±0.3 0±5/16
0.0 (0) | 0.75 (¾)	0.0 (0) | 0.75 (¾)	0.0 (0) | 0.75 (¾)	CROSS | TOL.	0.0 (0) | 0.75 (¾)	0.0 (0) | 0.5 (½)	0.0 (0) | 0.5 (½)
0.1°±0.25° 3/32°±¼°	0.1°±0.25° 3/32°±¼°	0.1°±0.25° 3/32°±¼°	FRONT TOE (Degrees)	0.06°±0.25° 1/16°±¼°	0.06°±0.25° 1/16°±¼°	0.06°±0.25° 1/16°±¼°
0.05"±0.13" 1/16"±⅛"	0.05"±0.13" 1/16"±⅛"	0.05"±0.13" 1/16"±⅛"	FRONT TOE (Inches)	0.03"±0.13" 1/32"±⅛"	0.03"±0.13" 1/32"±⅛"	0.03"±0.13" 1/32"±⅛"
			SAI			
			TURNING ANGLE			
75	75	75	ILLUSTRATION #			
			REAR			
			CAMBER (Degrees)			
			REAR TOE (Degrees)			
			REAR TOE (Inches)			
0.5 (½)	0.5 (½)	0.5 (½)	THRUST ANGLE			
			ILLUSTRATION #			
[1] HD Susp. = L=2.1°, R=2.5°		[1] HD Susp. = L=2.0°, R=2.3°	NOTES			
17237	17238	17239		17242	17243	17244

Ford Truck & SUV

Flex	Flex exc/Turbo	Flex Turbo	MODEL INFORMATION	Flex exc/Turbo	Flex Turbo	Ranger 4X2
2009	2010-2012	2010-2012	YEARS	2013-2017	2013-2017	1990-92
			INSTRUCTIONS			p. 306, 307
98 ft.lbs.	100 ft.lbs.	100 ft.lbs.	LUGNUT TORQUE	100 ft.lbs.	100 ft.lbs.	100 ft.lbs.
			FRONT			
3.1±0.5 3⅛±½	L=3.3, R=3.5±0.75 L=3 5/16, R=3½±¾	L=3.3, R=3.5±0.75 L=3 5/16, R=3½±¾	CASTER (Degrees)	L=3.25, R=3.45±0.75 L=3¼, R=3 7/16±¾	L=3.25, R=3.45±0.75 L=3¼, R=3 7/16±¾	MIN.=3.5 MAX.=6.0 MIN.=3½ MAX.=6
0.0 (0) | 0.7 (11/16)	-0.2 (-3/16) | 0.75 (¾)	-0.2 (-3/16) | 0.75 (¾)	CROSS | TOL.	-0.2 (-3/16) | 0.75 (¾)	-0.2 (-3/16) | 0.75 (¾)	-0.5 (-½) | 0.5 (½)
-0.5±0.5 -½±½	L=-0.4, R=-0.5±0.75 L=-⅜, R=-½±¾	L=-0.5, R=-0.6±0.75 L=-½, R=-⅝±¾	CAMBER (Degrees)	L=-0.5, R=-0.6±0.75 L=-½, R=-⅝±¾	L=-0.6, R=-0.7±0.75 L=-⅝, R=-11/16±¾	0.25±1.0 ¼±1
0.0 (0) | 0.7 (11/16)	0.1 (⅛) | 0.75 (¾)	0.1 (⅛) | 0.75 (¾)	CROSS | TOL.	0.1 (⅛) | 0.75 (¾)	0.1 (⅛) | 0.75 (¾)	0.0 (0) | 1.0 (1)
0.1°±0.2° 3/32°±3/16°	0.1°±0.2° 3/32°±3/16°	0.1°±0.2° 3/32°±3/16°	FRONT TOE (Degrees)	0.1°±0.2° 3/32°±3/16°	0.1°±0.2° 3/32°±3/16°	0.0°±0.25° 0°±¼°
0.05"±0.1" 1/16"±3/32"	0.05"±0.1" 1/16"±3/32"	0.05"±0.1" 1/16"±3/32"	FRONT TOE (Inches)	0.05"±0.1" 1/16"±3/32"	0.05"±0.1" 1/16"±3/32"	0.0"±0.13" 0"±⅛"
			SAI			
			TURNING ANGLE			
167	167	167	ILLUSTRATION #	167	167	56
			REAR			
-0.25±0.65 -¼±⅝	-0.8±0.75 -13/16±¾	-1.0±0.75 -1±¾	CAMBER (Degrees)	-0.8±0.75 -13/16±¾	-1.0±0.75 -1±¾	
0.36°±0.2° ⅜°±3/16°	0.24°±0.2° ¼°±3/16°	0.24°±0.2° ¼°±3/16°	REAR TOE (Degrees)	0.24°±0.2° ¼°±3/16°	0.24°±0.2° ¼°±3/16°	
0.18"±0.1" 3/16"±3/32"	0.12"±0.1" ⅛"±3/32"	0.12"±0.1" ⅛"±3/32"	REAR TOE (Inches)	0.12"±0.1" ⅛"±3/32"	0.12"±0.1" ⅛"±3/32"	
0.3 (5/16)	0.3 (5/16)	0.3 (5/16)	THRUST ANGLE	0.3 (5/16)	0.3 (5/16)	
187	187	187	ILLUSTRATION #	187	187	
			NOTES			
17245	17246	17247		17248	17249	17258

Ford Truck & SUV

Ranger 4X2	Ranger 4X2	Ranger 4X2	MODEL	Ranger 4X2	Ranger 4X2	Ranger 4X2
	exc/Splash	Splash	INFORMATION	Coil spring suspension	Coil spring suspension	Coil spring suspension
1993-97	1998	1998	YEARS	1999-2005	2006-2007	2008-2009
p. 306, 307	p. 306, 307	p. 306, 307	INSTRUCTIONS			
100 ft.lbs.	100 ft.lbs.	100 ft.lbs.	LUGNUT TORQUE	100 ft.lbs.	100 ft.lbs.	100 ft.lbs.
			FRONT			
MIN.= 2.0 MAX.= 6.0 MIN.= 2 MAX.= 6	L= 4.1, R= 4.5 ± 1.0 L= 4⅛, R= 4½ ± 1	L= 4.6, R= 5.0 ± 1.0 L= 4⅝, R= 5 ± 1	CASTER (Degrees)	L= 3.0, R= 3.4 ± 1.0 L= 3, R= 3⅜ ± 1	L= 3.5, R= 3.9 ± 1.0 L= 3½, R= 3⅞ ± 1	L= 3.0, R= 3.4 ± 1.0 L= 3, R= 3⅜ ± 1
-0.75 (-¾) | 0.5 (½)	-0.4 (-⅜) | 0.7 (11/16)	-0.4 (-⅜) | 0.7 (11/16)	CROSS | TOL.	-0.4 (-⅜) | 0.7 (11/16)	-0.4 (-⅜) | 0.7 (11/16)	-0.4 (-⅜) | 0.75 (¾)
0.25 ± 0.5 ¼ ± ½	-0.5 ± 0.7 -½ ± 11/16	-0.7 ± 0.7 -11/16 ± 11/16	CAMBER (Degrees)	-0.5 ± 0.7 -½ ± 11/16	-0.5 ± 0.7 -½ ± 11/16	-0.5 ± 0.7 -½ ± 11/16
0.0 (0) | 0.5 (½)	0.0 (0) | 0.7 (11/16)	0.0 (0) | 0.7 (11/16)	CROSS | TOL.	0.0 (0) | 0.7 (11/16)	0.0 (0) | 0.7 (11/16)	0.0 (0) | 0.75 (¾)
0.06° ± 0.25° 1/16° ± ¼°	0.06° ± 0.25° 1/16° ± ¼°	0.06° ± 0.25° 1/16° ± ¼°	FRONT TOE (Degrees)	0.06° ± 0.25° 1/16° ± ¼°	0.06° ± 0.25° 1/16° ± ¼°	0.06° ± 0.25° 1/16° ± ¼°
0.03" ± 0.13" 1/32" ± ⅛"	0.03" ± 0.13" 1/32" ± ⅛"	0.03" ± 0.13" 1/32" ± ⅛"	FRONT TOE (Inches)	0.03" ± 0.13" 1/32" ± ⅛"	0.03" ± 0.13" 1/32" ± ⅛"	0.03" ± 0.13" 1/32" ± ⅛"
			SAI			
			TURNING ANGLE			
56	24	24	ILLUSTRATION #	24	24	24
			REAR			
			CAMBER (Degrees)			
			REAR TOE (Degrees)			
			REAR TOE (Inches)			
			THRUST ANGLE			
			ILLUSTRATION #			
			NOTES			
17259	17260	17261		17262	17263	17264

Ford Truck & SUV

Ranger 4X2	Ranger 4X2	Ranger 4X2	MODEL	Ranger 4X2	Ranger 4X4	Ranger 4X4
Coil spring suspension	Torsion bar suspension	Torsion bar suspension	INFORMATION	Torsion bar suspension	Dana 28 axle Axle tag #E87A	Dana 35 axle Axle tag #F07A
2010-2011	1999-2005	2006-2007	YEARS	2008-2011	1990	1990-92
			INSTRUCTIONS		p. 306, 307	p. 306, 307
100 ft.lbs.	100 ft.lbs.	100 ft.lbs.	LUGNUT TORQUE	100 ft.lbs.	100 ft.lbs.	100 ft.lbs.
			FRONT			
L= 3.0, R= 3.2 ± 1.0 L= 3, R= 3 3/16 ± 1	L= 2.9, R= 3.6 ± 1.0 L= 2⅞, R= 3⅝ ± 1	L= 2.8, R= 3.5 ± 1.0 L= 2 13/16, R= 3½ ± 1	CASTER (Degrees)	L= 2.8, R= 3.3 ± 1.0 L= 2 13/16, R= 3 5/16 ± 1	MIN.= 2.5 MAX.= 4.5 MIN.= 2½ MAX.= 4½	MIN.= 2.5 MAX.= 6.0 MIN.= 2½ MAX.= 6
-0.2 (-3/16) | 0.75 (¾)	-0.7 (-11/16) | 0.7 (11/16)	-0.7 (-11/16) | 0.7 (11/16)	CROSS | TOL.	-0.5 (-½) | 0.75 (¾)	-0.5 (-½) | 0.5 (½)	-0.5 (-½) | 0.5 (½)
-0.5 ± 0.7 -½ ± 11/16	-0.5 ± 0.7 -½ ± 11/16	-0.5 ± 0.7 -½ ± 11/16	CAMBER (Degrees)	0.0 ± 0.7 0 ± 11/16	0.5 ± 0.75 [1] ½ ± ¾ [1]	0.25 ± 1.0 [1] ¼ ± 1 [1]
0.0 (0) | 0.75 (¾)	0.0 (0) | 0.7 (11/16)	0.0 (0) | 0.7 (11/16)	CROSS | TOL.	0.0 (0) | 0.75 (¾)	0.0 (0) | 1.0 (1)	0.0 (0) | 1.0 (1)
0.06° ± 0.25° 1/16° ± ¼°	0.12° ± 0.25° ⅛° ± ¼°	0.12° ± 0.25° ⅛° ± ¼°	FRONT TOE (Degrees)	0.12° ± 0.25° ⅛° ± ¼°	0.0° ± 0.25° 0° ± ¼°	0.0° ± 0.25° 0° ± ¼°
0.03" ± 0.13" 1/32" ± ⅛"	0.06" ± 0.13" 1/16" ± ⅛"	0.06" ± 0.13" 1/16" ± ⅛"	FRONT TOE (Inches)	0.06" ± 0.13" 1/16" ± ⅛"	0.0" ± 0.13" 0" ± ⅛"	0.0" ± 0.13" 0" ± ⅛"
			SAI			
			TURNING ANGLE			
24	73	73	ILLUSTRATION #	73	74	56
			REAR			
			CAMBER (Degrees)			
			REAR TOE (Degrees)			
			REAR TOE (Inches)			
			THRUST ANGLE			
			ILLUSTRATION #			
			NOTES		[1] STX = 0.25° (¼°)	[1] STX = ±0.75° (± ¾°)
17265	17266	17267		17268	17270	17271

ALIGNMENT SPECIFICATIONS

Ford Truck & SUV

Ranger 4X4	Ranger 4X4	Ranger 4X4	MODEL	Ranger 4X4	Ranger 4X4	Transit (Cargo)
			INFORMATION			Gas, 130" WB
						> 8600 GVW
1993-97	1998	1999-2005	YEARS	2006-2007	2008-2011	2015-2017
p. 306, 307	p. 306, 307		INSTRUCTIONS			
100 ft.lbs.	100 ft.lbs.	100 ft.lbs.	LUGNUT TORQUE	100 ft.lbs.	100 ft.lbs.	148 ft.lbs.
			FRONT			
MIN.= 2.0 MAX.= 6.0	L= 4.0, R= 4.5 ± 1.0	L= 2.9, R= 3.6 ± 1.0	CASTER	L= 2.8, R= 3.5 ± 1.0	L= 2.8, R= 3.3 ± 1.0	2.27 ± 0.6
MIN.= 2 MAX.= 6	L= 4, R= 4½ ± 1	L= 2⅞, R= 3⅝ ± 1	(Degrees)	L= 2 13/16, R= 3½ ± 1	L= 2 13/16, R= 3 5/16 ± 1	2¼ ± ⅝
-0.75 (-¾) | 0.5 (½)	-0.5 (-½) | 0.7 (11/16)	-0.7 (-11/16) | 0.7 (11/16)	CROSS | TOL.	-0.7 (-11/16) | 0.7 (11/16)	-0.5 (-½) | 0.75 (¾)	0.0 (0) | 0.6 (⅝)
0.25 ± 0.5	-0.5 ± 0.7	-0.5 ± 0.7	CAMBER	-0.5 ± 0.7	0.0 ± 0.7	-0.13 ± 0.5
¼ ± ½	-½ ± 11/16	-½ ± 11/16	(Degrees)	-½ ± 11/16	0 ± 11/16	-⅛ ± ½
0.0 (0) | 0.5 (½)	0.0 (0) | 0.7 (11/16)	0.0 (0) | 0.7 (11/16)	CROSS | TOL.	0.0 (0) | 0.7 (11/16)	0.0 (0) | 0.75 (¾)	0.0 (0) | 0.5 (½)
0.06° ± 0.25°	0.12° ± 0.25°	0.12° ± 0.25°	FRONT TOE	0.12° ± 0.25°	0.12° ± 0.25°	0.27° ± 0.25°
1/16° ± ¼°	⅛° ± ¼°	⅛° ± ¼°	(Degrees)	⅛° ± ¼°	⅛° ± ¼°	¼° ± ¼°
0.03" ± 0.13"	0.06" ± 0.13"	0.06" ± 0.13"	FRONT TOE	0.06" ± 0.13"	0.06" ± 0.13"	0.14" ± 0.13"
1/32" ± ⅛"	1/16" ± ⅛"	1/16" ± ⅛"	(Inches)	1/16" ± ⅛"	1/16" ± ⅛"	⅛" ± ⅛"
			SAI			
			TURNING ANGLE			
56	73	73	ILLUSTRATION #	73	73	
			REAR			
			CAMBER (Degrees)			
			REAR TOE (Degrees)			
			REAR TOE (Inches)			
			THRUST ANGLE			0.45 (7/16)
			ILLUSTRATION #			
			NOTES			
17272	17273	17274		17275	17276	17277

Ford Truck & SUV

Transit (Cargo)	Transit (Cargo)	Transit (Cargo)	MODEL	Transit (Cargo)	Transit (Cargo)	Transit (Cargo)
Gas, 130" WB	Gas, 148" WB	Gas, 148" WB	INFORMATION	Gas, 148" WB	Gas, 148" WB	Diesel, 130" WB
9000 GVW >	> 8600 GVW	9000 GVW >		Ext. Body, SRW	Ext. Body, DRW	> 8600 GVW
2015-2017	2015-2017	2015-2017	YEARS	2015-2017	2015-2017	2015-2017
			INSTRUCTIONS			
148 ft.lbs.	148 ft.lbs.	148 ft.lbs.	LUGNUT TORQUE	148 ft.lbs.	148 ft.lbs.	148 ft.lbs.
			FRONT			
1.87 ± 0.6	2.35 ± 0.6	1.97 ± 0.6	CASTER	2.2 ± 0.6	2.11 ± 0.6	1.88 ± 0.6
1⅞ ± ⅝	2⅜ ± ⅝	2 ± ⅝	(Degrees)	2 3/16 ± ⅝	2⅛ ± ⅝	1⅞ ± ⅝
0.0 (0) | 0.6 (⅝)	0.0 (0) | 0.6 (⅝)	0.0 (0) | 0.6 (⅝)	CROSS | TOL.	0.0 (0) | 0.6 (⅝)	0.0 (0) | 0.6 (⅝)	0.0 (0) | 0.6 (⅝)
0.51 ± 0.5	0.36 ± 0.5	0.4 ± 0.5	CAMBER	0.51 ± 0.5	0.71 ± 0.5	0.29 ± 0.5
½ ± ½	⅜ ± ½	⅜ ± ½	(Degrees)	½ ± ½	11/16 ± ½	5/16 ± ½
0.0 (0) | 0.5 (½)	0.0 (0) | 0.5 (½)	0.0 (0) | 0.5 (½)	CROSS | TOL.	0.0 (0) | 0.5 (½)	0.0 (0) | 0.5 (½)	0.0 (0) | 0.5 (½)
0.27° ± 0.25°	0.27° ± 0.25°	0.27° ± 0.25°	FRONT TOE	0.27° ± 0.25°	0.27° ± 0.25°	0.27° ± 0.25°
¼° ± ¼°	¼° ± ¼°	¼° ± ¼°	(Degrees)	¼° ± ¼°	¼° ± ¼°	¼° ± ¼°
0.14" ± 0.13"	0.14" ± 0.13"	0.14" ± 0.13"	FRONT TOE	0.14" ± 0.13"	0.14" ± 0.13"	0.14" ± 0.13"
⅛" ± ⅛"	⅛" ± ⅛"	⅛" ± ⅛"	(Inches)	⅛" ± ⅛"	⅛" ± ⅛"	⅛" ± ⅛"
			SAI			
			TURNING ANGLE			
			ILLUSTRATION #			
			REAR			
			CAMBER (Degrees)			
			REAR TOE (Degrees)			
			REAR TOE (Inches)			
0.45 (7/16)	0.45 (7/16)	0.45 (7/16)	THRUST ANGLE	0.45 (7/16)	0.45 (7/16)	0.45 (7/16)
			ILLUSTRATION #			
			NOTES			
17278	17279	17280		17281	17282	17283

Ford Truck & SUV

Transit (Cargo)	Transit (Cargo)	Transit (Cargo)	MODEL	Transit (Cargo)	Transit (Cargo)	Transit (Passenger)
Diesel, 130" WB	Diesel, 148" WB	Diesel, 148" WB	INFORMATION	Diesel, 148" WB	Diesel, 148" WB	Gas, 130" WB
9000 GVW >	> 8600 GVW	9000 GVW >		Ext. Body, SRW	Ext. Body, DRW	
2015-2017	2015-2017	2015-2017	YEARS	2015-2017	2015-2017	2015-2017
			INSTRUCTIONS			
148 ft.lbs.	148 ft.lbs.	148 ft.lbs.	LUGNUT TORQUE	148 ft.lbs.	148 ft.lbs.	148 ft.lbs.
			FRONT			
1.89 ±0.6 1 7/8 ± 5/8	2.3 ±0.6 2 5/16 ± 5/8	1.87 ±0.6 1 7/8 ± 5/8	CASTER (Degrees)	2.2 ±0.6 2 3/16 ± 5/8	2.11 ±0.6 2 1/8 ± 5/8	2.57 ±0.6 2 9/16 ± 5/8
0.0 (0) | 0.6 (5/8)	0.0 (0) | 0.6 (5/8)	0.0 (0) | 0.6 (5/8)	CROSS | TOL.	0.0 (0) | 0.6 (5/8)	0.0 (0) | 0.6 (5/8)	0.0 (0) | 0.6 (5/8)
0.26 ±0.5 1/4 ± 1/2	0.22 ±0.5 1/4 ± 1/2	0.2 ±0.5 3/16 ± 1/2	CAMBER (Degrees)	0.29 ±0.5 5/16 ± 1/2	0.48 ±0.5 1/2 ± 1/2	-0.23 ±0.5 -1/4 ± 1/2
0.0 (0) | 0.5 (1/2)	0.0 (0) | 0.5 (1/2)	0.0 (0) | 0.5 (1/2)	CROSS | TOL.	0.0 (0) | 0.5 (1/2)	0.0 (0) | 0.5 (1/2)	0.0 (0) | 0.5 (1/2)
0.27° ±0.25° 1/4° ± 1/4°	0.27° ±0.25° 1/4° ± 1/4°	0.27° ±0.25° 1/4° ± 1/4°	FRONT TOE (Degrees)	0.27° ±0.25° 1/4° ± 1/4°	0.27° ±0.25° 1/4° ± 1/4°	0.27° ±0.25° 1/4° ± 1/4°
0.14" ±0.13" 1/8" ± 1/8"	0.14" ±0.13" 1/8" ± 1/8"	0.14" ±0.13" 1/8" ± 1/8"	FRONT TOE (Inches)	0.14" ±0.13" 1/8" ± 1/8"	0.14" ±0.13" 1/8" ± 1/8"	0.14" ±0.13" 1/8" ± 1/8"
			SAI			
			TURNING ANGLE			
			ILLUSTRATION #			
			REAR			
			CAMBER (Degrees)			
			REAR TOE (Degrees)			
			REAR TOE (Inches)			
0.45 (7/16)	0.45 (7/16)	0.45 (7/16)	THRUST ANGLE	0.45 (7/16)	0.45 (7/16)	0.45 (7/16)
			ILLUSTRATION #			
			NOTES			
17284	17285	17286		17287	17288	17289

Ford Truck & SUV

Transit (Passenger)	Transit (Passenger)	Transit (Passenger)	MODEL	Transit (Passenger)	Transit (Passenger)	Transit (Passenger)
Gas, 148" WB	Gas, 148" WB	Gas, 148" WB	INFORMATION	Diesel, 148" WB	Diesel, 148" WB	Diesel, 148" WB
Low Roof	Medium Roof	Ext. Body, DRW		Low Roof	Medium Roof	Ext. Body, DRW
2015-2017	2015-2017	2015-2017	YEARS	2015-2017	2015-2017	2015-2017
			INSTRUCTIONS			
148 ft.lbs.	148 ft.lbs.	148 ft.lbs.	LUGNUT TORQUE	148 ft.lbs.	148 ft.lbs.	148 ft.lbs.
			FRONT			
2.41 ±0.6 2 7/16 ± 5/8	2.69 ±0.6 2 11/16 ± 5/8	2.34 ±0.6 2 5/16 ± 5/8	CASTER (Degrees)	2.32 ±0.6 2 5/16 ± 5/8	2.72 ±0.6 2 3/4 ± 5/8	2.3 ±0.6 2 5/16 ± 5/8
0.0 (0) | 0.6 (5/8)	0.0 (0) | 0.6 (5/8)	0.0 (0) | 0.6 (5/8)	CROSS | TOL.	0.0 (0) | 0.6 (5/8)	0.0 (0) | 0.6 (5/8)	0.0 (0) | 0.6 (5/8)
0.31 ±0.5 5/16 ± 1/2	-0.36 ±0.5 -3/8 ± 1/2	0.15 ±0.5 1/8 ± 1/2	CAMBER (Degrees)	0.09 ±0.5 1/16 ± 1/2	-0.56 ±0.5 -9/16 ± 1/2	-0.17 ±0.5 -3/16 ± 1/2
0.0 (0) | 0.5 (1/2)	0.0 (0) | 0.5 (1/2)	0.0 (0) | 0.5 (1/2)	CROSS | TOL.	0.0 (0) | 0.5 (1/2)	0.0 (0) | 0.5 (1/2)	0.0 (0) | 0.5 (1/2)
0.27° ±0.25° 1/4° ± 1/4°	0.27° ±0.25° 1/4° ± 1/4°	0.27° ±0.25° 1/4° ± 1/4°	FRONT TOE (Degrees)	0.27° ±0.25° 1/4° ± 1/4°	0.27° ±0.25° 1/4° ± 1/4°	0.27° ±0.25° 1/4° ± 1/4°
0.14" ±0.13" 1/8" ± 1/8"	0.14" ±0.13" 1/8" ± 1/8"	0.14" ±0.13" 1/8" ± 1/8"	FRONT TOE (Inches)	0.14" ±0.13" 1/8" ± 1/8"	0.14" ±0.13" 1/8" ± 1/8"	0.14" ±0.13" 1/8" ± 1/8"
			SAI			
			TURNING ANGLE			
			ILLUSTRATION #			
			REAR			
			CAMBER (Degrees)			
			REAR TOE (Degrees)			
			REAR TOE (Inches)			
0.45 (7/16)	0.45 (7/16)	0.45 (7/16)	THRUST ANGLE	0.45 (7/16)	0.45 (7/16)	0.45 (7/16)
			ILLUSTRATION #			
			NOTES			
17290	17291	17292		17293	17294	17295

Ford Truck & SUV

Transit Connect	Transit Connect Wagon 105" WB	Transit Connect Wagon 121" WB	MODEL INFORMATION	Transit Connect Cargo Van 105" WB	Transit Connect Cargo Van 121" WB	Notes
2010-2013	2014-2017	2014-2017	YEARS	2014-2017	2014-2017	
			INSTRUCTIONS			
66 ft.lbs.	100 ft.lbs.	100 ft.lbs.	LUGNUT TORQUE	100 ft.lbs.	100 ft.lbs.	
			FRONT			
1.8 ±1.0 1 13/16 ±1	3.8 ±1.0 3 13/16 ±1	4.17 ±0.75 4 3/16 ± 3/4	CASTER (Degrees)	3.3 ±0.75 3 5/16 ± 3/4	3.63 ±1.0 3 5/8 ±1	
0.0 (0) | 1.0 (1)	0.0 (0) | 1.0 (1)	0.0 (0) | 1.0 (1)	CROSS | TOL.	0.0 (0) | 1.0 (1)	0.0 (0) | 1.0 (1)	
-0.61 ±1.0 -5/8 ±1	-0.55 ±1.25 -9/16 ±1 1/4	-0.53 ±0.75 -1/2 ± 3/4	CAMBER (Degrees)	-0.5 ±0.75 -1/2 ± 3/4	-0.57 ±1.25 -9/16 ±1 1/4	
0.0 (0) | 0.5 (1/2)	0.0 (0) | 0.5 (1/2)	0.0 (0) | 0.5 (1/2)	CROSS | TOL.	0.0 (0) | 0.5 (1/2)	0.0 (0) | 0.5 (1/2)	
0.43° ±0.15° 7/16° ± 5/32°	0.2° ±0.2° 3/16° ± 3/16°	0.2° ±0.2° 3/16° ± 3/16°	FRONT TOE (Degrees)	0.2° ±0.2° 3/16° ± 3/16°	0.2° ±0.2° 3/16° ± 3/16°	
0.22" ±0.08" 7/32" ± 1/16"	0.1" ±0.1" 3/32" ± 3/32"	0.1" ±0.1" 3/32" ± 3/32"	FRONT TOE (Inches)	0.1" ±0.1" 3/32" ± 3/32"	0.1" ±0.1" 3/32" ± 3/32"	
			SAI			
			TURNING ANGLE			
190 △	190 △	190 △	ILLUSTRATION #	190 △	190 △	
			REAR			
-1.0 ±1.0 -1 ±1	-1.42 ±1.25 -1 7/16 ±1 1/4	-1.37 ±0.75 -1 3/8 ± 3/4	CAMBER (Degrees)	-1.32 ±0.75 -1 5/16 ± 3/4	-1.35 ±1.25 -1 3/8 ±1 1/4	
0.3° ±0.3° 5/16° ± 5/16°	0.37° ±0.3° 3/8° ± 5/16°	0.45° ±0.3° 7/16° ± 5/16°	REAR TOE (Degrees)	0.35° ±0.3° 11/32° ± 5/16°	0.25° ±0.3° 1/4° ± 5/16°	
0.15" ±0.15" 5/32" ± 5/32"	0.18" ±0.15" 3/16" ± 5/32"	0.22" ±0.15" 7/32" ± 5/32"	REAR TOE (Inches)	0.18" ±0.15" 3/16" ± 5/32"	0.13" ±0.15" 1/8" ± 5/32"	
	0.5 (1/2)	0.5 (1/2)	THRUST ANGLE	0.5 (1/2)	0.5 (1/2)	
			ILLUSTRATION #			
			NOTES			
17296	17297	17298		17299	17300	17305

Genesis / Geo

Genesis G80 RWD	Genesis G80 AWD	Genesis G90	MODEL INFORMATION	Metro	Metro	Metro
2017	2017	2017	YEARS	1990-94	1995-96	1997
			INSTRUCTIONS			
65-80 ft.lbs.	65-80 ft.lbs.	80-94 ft.lbs.	LUGNUT TORQUE	44 ft.lbs.	44 ft.lbs.	44 ft.lbs.
			FRONT			
8.35 ±0.75 8 3/8 ± 3/4	5.65 ±0.75 5 5/8 ± 3/4	5.87 ±0.75 5 7/8 ± 3/4	CASTER (Degrees)	3.0 ±2.0 3 ±2	3.0 ±2.0 3 ±2	3.0 ±2.0 3 ±2
0.0 (0) | n/s	0.0 (0) | n/s	0.0 (0) | n/s	CROSS | TOL.	0.0 (0) | 1.0 (1)	0.0 (0) | 1.0 (1)	0.0 (0) | 1.0 (1)
-0.5 ±0.5 -1/2 ± 1/2	-0.5 ±0.5 -1/2 ± 1/2	-0.67 ±0.5 -11/16 ± 1/2	CAMBER (Degrees)	0.0 ±1.0 0 ±1	0.5 ±1.0 1/2 ±1	0.0 ±1.0 0 ±1
0.0 (0) | n/s	0.0 (0) | n/s	0.0 (0) | n/s	CROSS | TOL.	0.0 (0) | 0.5 (1/2)	0.0 (0) | 0.5 (1/2)	0.0 (0) | 0.5 (1/2)
0.2° ±0.2° 3/16° ± 3/16°	0.2° ±0.2° 3/16° ± 3/16°	0.2° ±0.2° 3/16° ± 3/16°	FRONT TOE (Degrees)	0.0° ±0.2° 0° ± 3/16°	0.16° ±0.16° 5/32° ± 5/32°	0.0° ±0.16° 0° ± 5/32°
0.1" ±0.1" 3/32" ± 3/32"	0.1" ±0.1" 3/32" ± 3/32"	0.1" ±0.1" 3/32" ± 3/32"	FRONT TOE (Inches)	0.0" ±0.1" 0" ± 3/32"	0.08" ±0.08" 3/32" ± 3/32"	0.0" ±0.08" 0" ± 3/32"
7.2 (7 3/16)	6.5 (6 7/16)	6.7 (6 11/16)	SAI			
43.8 / 34.5 ±1.3	43.8 / 34.5 ±1.3	43.8 / 34.5 ±1.5	TURNING ANGLE	38.0 / 32.0	38.0 / 32.0	38.0 / 32.0
			ILLUSTRATION #			
			REAR			
-1.5 ±0.5 -1 1/2 ± 1/2	-1.5 ±0.5 -1 1/2 ± 1/2	-1.4 ±0.6 -1 3/8 ± 5/8	CAMBER (Degrees)	0.0 ±1.0 0 ±1	0.0 ±1.0 0 ±1	0.0 ±1.0 0 ±1
0.4° ±0.2° 13/32° ± 3/16°	0.4° ±0.2° 13/32° ± 3/16°	0.46° ±0.2° 15/32° ± 3/16°	REAR TOE (Degrees)	0.2° ±0.2° 3/16° ± 3/16°	0.45° ±0.15° 7/16° ± 5/32°	0.16° ±0.16° 5/32° ± 5/32°
0.2" ±0.1" 3/16" ± 3/32"	0.2" ±0.1" 3/16" ± 3/32"	0.23" ±0.1" 7/32" ± 3/32"	REAR TOE (Inches)	0.1" ±0.1" 3/32" ± 3/32"	0.23" ±0.08" 7/32" ± 1/16"	0.08" ±0.08" 3/32" ± 3/32"
			THRUST ANGLE	0.15 (5/32)	0.15 (5/32)	0.15 (5/32)
177	177	179 174	ILLUSTRATION #	06	06	06
			NOTES			
62001	62002	62003		18001	18002	18003

Geo

Prizm w/Auto. Trans.	Prizm w/Man. Trans.	Prizm GSi	MODEL INFORMATION	Prizm	Storm	Tracker
1990-92	1990-92	1990-92	YEARS	1993-97	1990-93	1990-97
			INSTRUCTIONS			
76 ft.lbs.	76 ft.lbs.	76 ft.lbs.	LUGNUT TORQUE	76 ft.lbs.	76-87 ft.lbs.	70 ft.lbs.[1]
			FRONT			
1.4 ±0.75 1 3/8 ± 3/4	1.3 ±0.75 1 5/16 ± 3/4	1.45 ±0.75 [1] 1 7/16 ± 3/4 [1]	CASTER (Degrees)	1.33 ±0.75 1 5/16 ± 3/4	4.5 ±1.0 [1] 4 1/2 ±1 [1]	1.5 ±1.0 1 1/2 ±1
0.0 (0) | 0.5 (1/2)	0.0 (0) | 0.5 (1/2)	0.0 (0) | 0.5 (1/2)	CROSS | TOL.	0.0 (0) | 0.5 (1/2)	0.0 (0) | 1.0 (1)	0.0 (0) | 1.0 (1)
-0.17 ±0.75 -3/16 ± 3/4	-0.25 ±0.75 -1/4 ± 3/4	-0.25 ±0.75 -1/4 ± 3/4	CAMBER (Degrees)	-0.17 ±0.75 -3/16 ± 3/4	-0.5 ±0.75 -1/2 ± 3/4	0.5 ±1.0 1/2 ±1
0.0 (0) | 0.5 (1/2)	0.0 (0) | 0.5 (1/2)	0.0 (0) | 0.5 (1/2)	CROSS | TOL.	0.0 (0) | 0.5 (1/2)	0.0 (0) | n/s	0.0 (0) | 1.0 (1)
0.08° ±0.08° 1/16° ± 1/16°	0.08° ±0.08° 1/16° ± 1/16°	0.08° ±0.08° 1/16° ± 1/16°	FRONT TOE (Degrees)	0.1° ±0.2° 3/32° ± 3/16°	0.0° ±0.08° 0° ± 1/16°	0.31° ±0.16° 5/16° ± 5/32°
0.04" ±0.04" 1/32" ± 1/32"	0.04" ±0.04" 1/32" ± 1/32"	0.04" ±0.04" 1/32" ± 1/32"	FRONT TOE (Inches)	0.05" ±0.1" 1/16" ± 3/32"	0.0" ±0.04" 0" ± 1/32"	0.16" ±0.08" 5/32" ± 1/16"
			SAI	12.6 ±0.8 (12 5/8 ± 3/4)		
39.0 / 32.5 ±1.5	39.0 / 32.5 ±1.5	39.0 / 32.5 ±1.5	TURNING ANGLE	39.0 / 33.0 ±2.0		32.5 / 30.5
			ILLUSTRATION #		102	
			REAR			
-0.6 ±0.75 -5/8 ± 3/4	-0.6 ±0.75 -5/8 ± 3/4	-0.7 ±0.75 -11/16 ± 3/4	CAMBER (Degrees)	-0.92 ±0.75 -15/16 ± 3/4	-0.5 ±0.75 -1/2 ± 3/4	
0.31° ±0.16° 5/16° ± 5/32°	0.31° ±0.16° 5/16° ± 5/32°	0.31° ±0.16° 5/16° ± 5/32°	REAR TOE (Degrees)	0.4° ±0.2° 13/32° ± 3/16°	0.16° ±0.08° 5/32° ± 1/16°	
0.16" ±0.08" 5/32" ± 1/16"	0.16" ±0.08" 5/32" ± 1/16"	0.16" ±0.08" 5/32" ± 1/16"	REAR TOE (Inches)	0.2" ±0.1" 3/16" ± 3/32"	0.08" ±0.04" 3/32" ± 1/32"	
0.15 (5/32)	0.15 (5/32)		THRUST ANGLE	0.15 (5/32)	0.15 (5/32)	
06	06	06	ILLUSTRATION #	10	10	
		[1] with M/T 1.3° (1 5/16°)	NOTES		[1] 1990-91 = 3.0° (3°)	[1] 1990-91 = 60 ft.lbs.
18004	18005	18006		18007	18010	18011

Honda

Accord	Accord	Accord	MODEL INFORMATION	Accord exc/DX, VP, Hybrid	Accord DX, VP	Accord exc/VP, V6, Hybrid
1990-93	1994-97	1998-2002	YEARS	2003-2005	2003-2005	2006-2007
			INSTRUCTIONS			
80 ft.lbs.	80 ft.lbs.	80 ft.lbs.	LUGNUT TORQUE	80 ft.lbs.	80 ft.lbs.	80 ft.lbs.
			FRONT			
3.0 ±1.0 3 ±1	3.0 ±1.0 3 ±1	2.8 ±1.0 2 13/16 ±1	CASTER (Degrees)	3.25 ±0.75 3 1/4 ± 3/4	3.17 ±0.75 3 3/16 ± 3/4	3.25 ±0.75 3 1/4 ± 3/4
0.0 (0) | n/s	0.0 (0) | n/s	0.0 (0) | n/s	CROSS | TOL.	0.0 (0) | n/s	0.0 (0) | n/s	0.0 (0) | n/s
0.0 ±1.0 0 ±1	0.0 ±1.0 0 ±1	0.0 ±1.0 0 ±1	CAMBER (Degrees)	0.0 ±0.75 0 ± 3/4	0.05 ±0.75 1/16 ± 3/4	0.0 ±0.75 0 ± 3/4
0.0 (0) | n/s	0.0 (0) | n/s	0.0 (0) | n/s	CROSS | TOL.	0.0 (0) | n/s	0.0 (0) | n/s	0.0 (0) | n/s
0.0° ±0.16° 0° ± 5/32°	0.0° ±0.24° 0° ± 1/4°	0.0° ±0.16° 0° ± 5/32°	FRONT TOE (Degrees)	0.0° ±0.16° 0° ± 5/32°	0.0° ±0.16° 0° ± 5/32°	0.0° ±0.16° 0° ± 5/32°
0.0" ±0.08" 0" ± 3/32"	0.0" ±0.12" 0" ± 1/8"	0.0" ±0.08" 0" ± 3/32"	FRONT TOE (Inches)	0.0" ±0.08" 0" ± 3/32"	0.0" ±0.08" 0" ± 3/32"	0.0" ±0.08" 0" ± 3/32"
			SAI			
39.1 / 29.5 ±2.0	39.0 / 30.0 ±2.0	38.5 / 31.1 ±2.0	TURNING ANGLE	38.8 / 31.7 ±2.0	39.0 / 31.8 ±2.0	38.8 / 31.7 ±2.0
03	03	03	ILLUSTRATION #			
			REAR			
-0.5 ±0.5 -1/2 ± 1/2	-0.42 ±0.5 -7/16 ± 1/2	-0.5 ±0.5 -1/2 ± 1/2	CAMBER (Degrees)	-1.0 ±0.5 -1 ± 1/2	-0.83 ±0.5 -13/16 ± 1/2	-1.0 ±0.5 -1 ± 1/2
0.16° ±0.16° [1] 5/32° ± 5/32° [1]	0.16° ±0.16° 5/32° ± 5/32°	0.16° ±0.16° 5/32° ± 5/32°	REAR TOE (Degrees)	0.16° ±0.16° 5/32° ± 5/32°	0.16° ±0.16° 5/32° ± 5/32°	0.16° ±0.16° 5/32° ± 5/32°
0.08" ±0.08" [1] 3/32" ± 3/32" [1]	0.08" ±0.08" 3/32" ± 3/32"	0.08" ±0.08" 3/32" ± 3/32"	REAR TOE (Inches)	0.08" ±0.08" 3/32" ± 3/32"	0.08" ±0.08" 3/32" ± 3/32"	0.08" ±0.08" 3/32" ± 3/32"
			THRUST ANGLE			
05	05	06	ILLUSTRATION #	06	06	178
[1] Wagon = 0.32° (0.16")			NOTES			
19008	19009	19010		19011	19012	19013

Honda

Accord VP (Value Pkg.)	Accord V6	Accord L4 Engine	MODEL INFORMATION	Accord V6 Engine w/17" wheels	Accord V6 Engine w/18" wheels	Accord exc/18", 19" tires
2006-2007	2006-2007	2008-2012	YEARS	2008-2012	2008-2012	2013-2017
			INSTRUCTIONS			
80 ft.lbs.	80 ft.lbs.	80 ft.lbs.	LUGNUT TORQUE	80 ft.lbs.	80 ft.lbs.	80 ft.lbs.
			FRONT			
3.17 ± 0.75 3 3/16 ± 3/4	3.25 ± 0.75 3 1/4 ± 3/4	3.8 +0.42 -1.08 3 13/16 + 7/16 -1 1/16	CASTER (Degrees)	3.8 +0.42 -1.08 3 13/16 + 7/16 -1 1/16	3.8 +0.42 -1.08 3 13/16 + 7/16 -1 1/16	3.88 ± 0.5 3 7/8 ± 1/2
0.0 (0) | n/s	0.0 (0) | n/s	0.0 (0) | 0.75 (3/4)	CROSS | TOL.	0.0 (0) | 0.75 (3/4)	0.0 (0) | 0.75 (3/4)	0.0 (0) | 0.5 (1/2)
0.05 ± 0.75 1/16 ± 3/4	0.0 ± 0.75 0 ± 3/4	0.0 +0.5 -0.75 0 + 1/2 - 3/4	CAMBER (Degrees)	0.0 +0.5 -0.75 0 + 1/2 - 3/4	-0.08 +0.5 -0.75 - 1/16 + 1/2 - 3/4	-0.33 ± 0.5 - 5/16 ± 1/2
0.0 (0) | n/s	0.0 (0) | n/s	0.0 (0) | 0.5 (1/2)	CROSS | TOL.	0.0 (0) | 0.5 (1/2)	0.0 (0) | 0.5 (1/2)	0.0 (0) | 0.5 (1/2)
0.0° ± 0.16° 0° ± 5/32°	0.0° ± 0.16° 0° ± 5/32°	0.0° ± 0.16° 0° ± 5/32°	FRONT TOE (Degrees)	0.0° ± 0.16° 0° ± 5/32°	0.0° ± 0.16° 0° ± 5/32°	0.0° ± 0.16° 0° ± 5/32°
0.0" ± 0.08" 0" ± 3/32"	0.0" ± 0.08" 0" ± 3/32"	0.0" ± 0.08" 0" ± 3/32"	FRONT TOE (Inches)	0.0" ± 0.08" 0" ± 3/32"	0.0" ± 0.08" 0" ± 3/32"	0.0" ± 0.08" 0" ± 3/32"
			SAI			
39.0 / 31.8 ±2.0	35.3 / 29.7 ±2.0	39.0 / 31.8 ±2.0	TURNING ANGLE	39.0 / 31.8 ±2.0	37.0 / 30.3 ±2.0	38.2 / 30.9 ±2.0
			ILLUSTRATION #			197
			REAR			
-0.83 ± 0.5 - 13/16 ± 1/2	-1.0 ± 0.5 -1 ± 1/2	-1.0 +0.5 -0.75 -1 + 1/2 - 3/4	CAMBER (Degrees)	-1.0 +0.5 -0.75 -1 + 1/2 - 3/4	-1.32 +0.5 -0.75 -1 5/16 + 1/2 - 3/4	-1.25 ± 0.5 -1 1/4 ± 1/2
0.16° ± 0.16° 5/32° ± 5/32°	0.16° ± 0.16° 5/32° ± 5/32°	0.16° ± 0.16° 5/32° ± 5/32°	REAR TOE (Degrees)	0.16° ± 0.16° 5/32° ± 5/32°	0.16° ± 0.16° 5/32° ± 5/32°	0.16° ± 0.16° 5/32° ± 5/32°
0.08" ± 0.08" 3/32" ± 3/32"	0.08" ± 0.08" 3/32" ± 3/32"	0.08" ± 0.08" 3/32" ± 3/32"	REAR TOE (Inches)	0.08" ± 0.08" 3/32" ± 3/32"	0.08" ± 0.08" 3/32" ± 3/32"	0.08" ± 0.08" 3/32" ± 3/32"
			THRUST ANGLE			
06	178	174	ILLUSTRATION #	174	174	174
			NOTES			
19014	19015	19016		19017	19018	19019

Honda

Accord w/18", 19" tires	Accord Hybrid	Accord Hybrid incl. Plug-in	MODEL INFORMATION	Accord Hybrid	Civic exc/Wagon	Civic Wagon 4X2
2013-2017	2005-2007	2014-2015	YEARS	2017	1990-91	1990-92
			INSTRUCTIONS			
80 ft.lbs.	80 ft.lbs.	80 ft.lbs.	LUGNUT TORQUE	80 ft.lbs.	80 ft.lbs.	80 ft.lbs.
			FRONT			
3.95 ± 0.5 3 15/16 ± 1/2	3.25 ± 0.75 3 1/4 ± 3/4	3.93 ± 0.5 3 15/16 ± 1/2	CASTER (Degrees)	3.93 ± 0.5 3 15/16 ± 1/2	3.0 ± 1.0 3 ± 1	3.0 ± 1.0 3 ± 1
0.0 (0) | 0.5 (1/2)	0.0 (0) | n/s	0.0 (0) | 0.5 (1/2)	CROSS | TOL.	0.0 (0) | 0.5 (1/2)	0.0 (0) | n/s	0.0 (0) | n/s
-0.38 ± 0.5 -3/8 ± 1/2	0.0 ± 0.75 0 ± 3/4	-0.38 ± 0.5 -3/8 ± 1/2	CAMBER (Degrees)	-0.38 ± 0.5 -3/8 ± 1/2	0.0 ± 1.0 0 ± 1	0.32 ± 1.0 5/16 ± 1
0.0 (0) | 0.5 (1/2)	0.0 (0) | n/s	0.0 (0) | 0.5 (1/2)	CROSS | TOL.	0.0 (0) | 0.5 (1/2)	0.0 (0) | n/s	0.0 (0) | n/s
0.0° ± 0.16° 0° ± 5/32°	0.0° ± 0.16° 0° ± 5/32°	0.0° ± 0.16° 0° ± 5/32°	FRONT TOE (Degrees)	0.0° ± 0.16° 0° ± 5/32°	0.16° +0.08° -0.16° 5/32° + 3/32° - 5/32°	0.16° +0.08° -0.16° 5/32° + 3/32° - 5/32°
0.0" ± 0.08" 0" ± 3/32"	0.0" ± 0.08" 0" ± 3/32"	0.0" ± 0.08" 0" ± 3/32"	FRONT TOE (Inches)	0.0" ± 0.08" 0" ± 3/32"	0.08" +0.04" -0.08" 3/32" + 1/32" - 3/32"	0.08" +0.04" -0.08" 3/32" + 1/32" - 3/32"
			SAI		7.6 ± 1.0 (7 9/16 ± 1)	7.2 ± 1.0 (7 1/4 ± 1)
36.2 / 29.8 ±2.0	38.8 / 31.7 ±2.0	38.0 / 30.8 ±2.0	TURNING ANGLE	38.0 / 30.8 ±2.0	41.5 / 33.5 ±2.0	41.7 / 33.7 ±2.0
197		197	ILLUSTRATION #	197		
			REAR			
-1.57 ± 0.5 -1 9/16 ± 1/2	-1.0 ± 0.5 -1 ± 1/2	-1.45 ± 0.5 -1 7/16 ± 1/2	CAMBER (Degrees)	-1.45 ± 0.5 -1 7/16 ± 1/2	-0.43 ± 1.0 - 7/16 ± 1	-0.38 ± 1.0 -3/8 ± 1
0.16° ± 0.16° 5/32° ± 5/32°	0.16° ± 0.16° 5/32° ± 5/32°	0.16° ± 0.16° 5/32° ± 5/32°	REAR TOE (Degrees)	0.16° ± 0.16° 5/32° ± 5/32°	0.16° +0.16° -0.08° 5/32° + 5/32° - 3/32°	0.16° +0.16° -0.08° 5/32° + 5/32° - 3/32°
0.08" ± 0.08" 3/32" ± 3/32"	0.08" ± 0.08" 3/32" ± 3/32"	0.08" ± 0.08" 3/32" ± 3/32"	REAR TOE (Inches)	0.08" ± 0.08" 3/32" ± 3/32"	0.08" +0.08" -0.04" 3/32" + 3/32" - 1/32"	0.08" +0.08" -0.04" 3/32" + 3/32" - 1/32"
			THRUST ANGLE			
174	06	174	ILLUSTRATION #	174	02	02
			NOTES			
19020	19021	19022		19023	19039	19040

Honda

Civic Wagon 4X4	Civic exc/Wagon	Civic	MODEL INFORMATION	Civic exc/Si & Hybrid	Civic exc/Si & Hybrid	Civic DX, GX
1990-91	1992-95	1996-2000	YEARS	2001-2002	2003-2005	2006-2011
			INSTRUCTIONS			
80 ft.lbs.	80 ft.lbs.	80 ft.lbs.	LUGNUT TORQUE	80 ft.lbs.	80 ft.lbs.	80 ft.lbs.
			FRONT			
2.93 ± 1.0 2 15/16 ± 1	1.17 ± 1.0 1 3/16 ± 1	1.67 ± 1.0 1 11/16 ± 1	CASTER (Degrees)	1.55 ± 1.0 1 9/16 ± 1	2.23 ± 1.0 2 ¼ ± 1	7.0 ± 1.0 7 ± 1
0.0 (0) | n/s	0.0 (0) | n/s	0.0 (0) | n/s	CROSS | TOL.	0.0 (0) | n/s	0.0 (0) | n/s	0.0 (0) | n/s
0.58 ± 1.0 9/16 ± 1	0.0 ± 1.0 0 ± 1	0.0 ± 1.0 0 ± 1	CAMBER (Degrees)	0.0 ± 0.75 0 ± ¾	0.0 ± 0.75 0 ± ¾	0.0 ± 0.5 0 ± ½
0.0 (0) | n/s	0.0 (0) | n/s	0.0 (0) | n/s	CROSS | TOL.	0.0 (0) | n/s	0.0 (0) | n/s	0.0 (0) | 0.58 (9/16)
0.16° +0.08° -0.16° 5/32° +3/32° -5/32°	0.0° ± 0.16° [1] 0° ± 5/32° [1]	0.08° ± 0.16° 3/32° ± 5/32°	FRONT TOE (Degrees)	0.0° ± 0.24° 0° ± ¼°	0.0° ± 0.24° 0° ± ¼°	0.0° ± 0.16° 0° ± 5/32°
0.08" +0.04" -0.08" 3/32" +1/32" -3/32"	0.0" ± 0.08" [1] 0" ± 3/32" [1]	0.04" ± 0.08" 1/32" ± 3/32"	FRONT TOE (Inches)	0.0" ± 0.12" 0" ± ⅛"	0.0" ± 0.12" 0" ± ⅛"	0.0" ± 0.08" 0" ± 3/32"
			SAI			
40.5 / 33.0 ±2.0 [1]	40.5 / 33.0 ±2.0	39.8 / 33.2 ±2.0	TURNING ANGLE	40.0 / 31.0 ±2.0	40.0 / 31.0 ±2.0 [1]	40.1 / 31.9 ±2.0
			ILLUSTRATION #	26	26	26
			REAR			
0.0 ± 1.0 0 ± 1	-0.33 ± 1.0 -5/16 ± 1	-1.0 ± 1.0 -1 ± 1	CAMBER (Degrees)	-0.75 ± 0.75 -¾ ± ¾	-0.75 ± 0.75 -¾ ± ¾	-1.5 +1.08 -0.75 [1] -1 ½ +1 1/16 -¾ [1]
0.16° +0.16° -0.08° 5/32° +5/32° -3/32°	0.16° +0.16° -0.08° 5/32° +5/32° -3/32°	0.16° +0.16° -0.08° 5/32° +5/32° -3/32°	REAR TOE (Degrees)	0.16° +0.16° -0.08° 5/32° +5/32° -3/32°	0.16° +0.16° -0.08° 5/32° +5/32° -3/32°	0.16° +0.16° -0.08° 5/32° +5/32° -3/32°
0.08" +0.08" -0.04" 3/32" +3/32" -1/32"	0.08" +0.08" -0.04" 3/32" +3/32" -1/32"	0.08" +0.08" -0.04" 3/32" +3/32" -1/32"	REAR TOE (Inches)	0.08" +0.08" -0.04" 3/32" +3/32" -1/32"	0.08" +0.08" -0.04" 3/32" +3/32" -1/32"	0.08" +0.08" -0.04" 3/32" +3/32" -1/32"
			THRUST ANGLE			
02	02	02	ILLUSTRATION #	140	140	140
[1] w/MT 42°/33.8°	[1] set to POS. limit if inside edge wear.		NOTES		[1] P195/60R15 tires = 38°/30°	[1] models with "C" stamp on rear upper arm = -0.75° (-¾°)
19041	19042	19043		19044	19045	19046

Honda

Civic EX, LX, Si	Civic except Si	Civic Si	MODEL INFORMATION	Civic except Si	Civic Si exc/18" wheels	Civic Si w/18" wheels
2006-2011	2012-2013	2012-2013	YEARS	2014-2015	2014-2015	2014-2015
			INSTRUCTIONS			
80 ft.lbs.	80 ft.lbs.	80 ft.lbs.	LUGNUT TORQUE	80 ft.lbs.	80 ft.lbs.	80 ft.lbs.
			FRONT			
7.0 ± 1.0 7 ± 1	5.32 ± 0.5 5 5/16 ± ½	5.37 ± 0.5 5 ⅜ ± ½	CASTER (Degrees)	5.32 ± 0.5 5 5/16 ± ½	5.37 ± 0.5 5 ⅜ ± ½	5.25 ± 0.5 5 ¼ ± ½
0.0 (0) | n/s	0.0 (0) | n/s	0.0 (0) | n/s	CROSS | TOL.	0.0 (0) | n/s	0.0 (0) | n/s	0.0 (0) | n/s
0.0 ± 0.5 0 ± ½	0.0 ± 0.5 0 ± ½	-0.07 ± 0.5 -1/16 ± ½	CAMBER (Degrees)	0.0 ± 0.5 0 ± ½	-0.07 ± 0.5 -1/16 ± ½	-0.3 ± 0.5 -5/16 ± ½
0.0 (0) | 0.58 (9/16)	0.0 (0) | 0.5 (½)	0.0 (0) | 0.5 (½)	CROSS | TOL.	0.0 (0) | 0.5 (½)	0.0 (0) | 0.5 (½)	0.0 (0) | 0.5 (½)
0.0° ± 0.16° 0° ± 5/32°	0.0° ± 0.16° 0° ± 5/32°	0.0° ± 0.16° 0° ± 5/32°	FRONT TOE (Degrees)	0.0° ± 0.16° 0° ± 5/32°	0.0° ± 0.16° 0° ± 5/32°	0.0° ± 0.16° 0° ± 5/32°
0.0" ± 0.08" 0" ± 3/32"	0.0" ± 0.08" 0" ± 3/32"	0.0" ± 0.08" 0" ± 3/32"	FRONT TOE (Inches)	0.0" ± 0.08" 0" ± 3/32"	0.0" ± 0.08" 0" ± 3/32"	0.0" ± 0.08" 0" ± 3/32"
			SAI			
38.8 / 31.2 ±2.0	38.7 / 30.9 ±2.0	38.6 / 30.5 ±2.0	TURNING ANGLE	38.5 / 30.8 ±2.0	38.4 / 30.7 ±2.0	38.6 / 29.9 ±2.0
26	26	26	ILLUSTRATION #	26	26	26
			REAR			
-1.5 +1.08 -0.75 [1] -1 ½ +1 1/16 -¾ [1]	-0.75 ± 0.5 -¾ ± ½	-0.87 ± 0.5 -⅞ ± ½	CAMBER (Degrees)	-0.75 ± 0.5 -¾ ± ½	-0.87 ± 0.5 -⅞ ± ½	-0.75 ± 0.5 -¾ ± ½
0.16° +0.16° -0.08° 5/32° +5/32° -3/32°	0.16° +0.16° -0.08° 5/32° +5/32° -3/32°	0.16° +0.16° -0.08° 5/32° +5/32° -3/32°	REAR TOE (Degrees)	0.16° +0.16° -0.08° 5/32° +5/32° -3/32°	0.16° +0.16° -0.08° 5/32° +5/32° -3/32°	0.16° +0.16° -0.08° 5/32° +5/32° -3/32°
0.08" +0.08" -0.04" 3/32" +3/32" -1/32"	0.08" +0.08" -0.04" 3/32" +3/32" -1/32"	0.08" +0.08" -0.04" 3/32" +3/32" -1/32"	REAR TOE (Inches)	0.08" +0.08" -0.04" 3/32" +3/32" -1/32"	0.08" +0.08" -0.04" 3/32" +3/32" -1/32"	0.08" +0.08" -0.04" 3/32" +3/32" -1/32"
			THRUST ANGLE			
140	140	140	ILLUSTRATION #	140	140	140
[1] models with "C" stamp on rear upper arm = -0.75° (-¾°)			NOTES			
19047	19048	19049		19050	19051	19052

Honda

MODEL INFORMATION	Civic Sedan & Coupe exc/Si	Civic Hatchback exc/Type-R	Civic SI	Civic Type-R	Civic Si Civic SiR	Civic Si Civic SiR
YEARS	2016-2017	2017	2017	2017	2002-2003	2004-2005
INSTRUCTIONS						
LUGNUT TORQUE	80 ft.lbs.	80 ft.lbs.	80 ft.lbs.	80 ft.lbs.	80 ft.lbs.	80 ft.lbs.
FRONT						
CASTER (Degrees)	5.28 ±0.6 5 5/16 ± 5/8	5.3 ±0.5 5 5/16 ± 1/2	5.4 ±0.5 5 3/8 ± 1/2	8.0 ±0.5 8 ± 1/2	1.05 ±1.0 1 1/16 ±1	1.6 ±1.0 1 5/8 ±1
CROSS | TOL.	0.0 (0) | n/s	0.0 (0) | n/s	0.0 (0) | n/s	0.0 (0) | n/s	0.0 (0) | n/s	0.0 (0) | n/s
CAMBER (Degrees)	-0.3 ±0.5 - 5/16 ± 1/2	-0.3 ±0.5 - 5/16 ± 1/2	-0.33 ±0.5 - 5/16 ± 1/2	-1.33 ±0.5 -1 5/16 ± 1/2	0.0 ±0.75 0 ± 3/4	-0.15 ±0.75 - 1/8 ± 3/4
CROSS | TOL.	0.0 (0) | 0.75 (3/4)	0.0 (0) | 0.75 (3/4)	0.0 (0) | 0.75 (3/4)	0.0 (0) | 0.75 (3/4)	0.0 (0) | n/s	0.0 (0) | n/s
FRONT TOE (Degrees)	0.0° ±0.16° 0° ± 5/32°	0.0° ±0.16° 0° ± 5/32°	0.0° ±0.16° 0° ± 5/32°	0.0° ±0.16° 0° ± 5/32°	0.0° ±0.24° 0° ± 1/4°	0.0° ±0.24° 0° ± 1/4°
FRONT TOE (Inches)	0.0" ±0.08" 0" ± 3/32"	0.0" ±0.08" 0" ± 3/32"	0.0" ±0.08" 0" ± 3/32"	0.0" ±0.08" 0" ± 3/32"	0.0" ±0.12" 0" ± 1/8"	0.0" ±0.12" 0" ± 1/8"
SAI						
TURNING ANGLE	40.1 / 33.0 ±2.0	40.1 / 33.0 ±2.0 [1]	36.9 / 31.1 ±2.0	35.0 / 29.9 ±2.0	34.0 / 28.0 ±2.0	34.7 / 28.2 ±2.0
ILLUSTRATION #	197	197	197	197	26	26
REAR						
CAMBER (Degrees)	-1.2 ±0.75 -1 3/16 ± 3/4	-1.2 ±0.75 -1 3/16 ± 3/4	-1.33 ±0.75 -1 5/16 ± 3/4	-1.5 ±0.75 -1 1/2 ± 3/4	-0.75 ±0.75 - 3/4 ± 3/4	-0.85 ±0.75 - 7/8 ± 3/4
REAR TOE (Degrees)	0.16° +0.16° -0.08° 5/32° + 5/32° - 3/32°	0.16° +0.16° -0.08° 5/32° + 5/32° - 3/32°	0.16° +0.16° -0.08° 5/32° + 5/32° - 3/32°	0.16° +0.16° -0.08° 5/32° + 5/32° - 3/32°	0.16° +0.16° -0.08° 5/32° + 5/32° - 3/32°	0.16° +0.16° -0.08° 5/32° + 5/32° - 3/32°
REAR TOE (Inches)	0.08" +0.08" -0.04" 3/32" + 3/32" - 1/32"	0.08" +0.08" -0.04" 3/32" + 3/32" - 1/32"	0.08" +0.08" -0.04" 3/32" + 3/32" - 1/32"	0.08" +0.08" -0.04" 3/32" + 3/32" - 1/32"	0.08" +0.08" -0.04" 3/32" + 3/32" - 1/32"	0.08" +0.08" -0.04" 3/32" + 3/32" - 1/32"
THRUST ANGLE						
ILLUSTRATION #	06	06	06	06	140	140
NOTES		[1] 18" wheels = 37.0°/31.3°				
	19053	19054	19055	19056	19059	19060

Honda

MODEL INFORMATION	Civic Hybrid	Civic Hybrid	Civic Hybrid	Civic del Sol except VTEC Engine	Civic del Sol with VTEC Engine	Clarity
YEARS	2003-2005	2006-2011	2012-2015	1993-97	1994-97	2017
INSTRUCTIONS						
LUGNUT TORQUE	80 ft.lbs.	80 ft.lbs.	80 ft.lbs.	80 ft.lbs.	80 ft.lbs.	80 ft.lbs.
FRONT						
CASTER (Degrees)	2.33 ±1.0 [1] 2 5/16 ±1 [1]	7.1 ±1.0 7 1/8 ±1	5.38 ±0.5 5 3/8 ± 1/2	1.17 ±1.0 1 3/16 ±1	1.17 ±1.0 1 3/16 ±1	5.38 ±0.5 5 3/8 ± 1/2
CROSS | TOL.	0.0 (0) | n/s	0.0 (0) | n/s	0.0 (0) | n/s	0.0 (0) | n/s	0.0 (0) | n/s	0.0 (0) | n/s
CAMBER (Degrees)	-0.2 ±0.75 - 3/16 ± 3/4	-0.05 ±0.5 - 1/16 ± 1/2	-0.07 ±0.5 - 1/16 ± 1/2	-0.25 ±1.0 - 1/4 ±1	-0.33 ±1.0 - 5/16 ±1	-0.3 ±0.5 - 5/16 ± 1/2
CROSS | TOL.	0.0 (0) | n/s	0.0 (0) | 0.58 (9/16)	0.0 (0) | 0.5 (1/2)	0.0 (0) | n/s	0.0 (0) | n/s	0.0 (0) | 0.5 (1/2)
FRONT TOE (Degrees)	0.0° ±0.24° 0° ± 1/4°	0.0° ±0.16° 0° ± 5/32°	0.0° ±0.16° 0° ± 5/32°	0.0° ±0.16° [1] 0° ± 5/32° [1]	0.0° ±0.16° [1] 0° ± 5/32° [1]	0.0° ±0.16° 0° ± 5/32°
FRONT TOE (Inches)	0.0" ±0.12" 0" ± 1/8"	0.0" ±0.08" 0" ± 3/32"	0.0" ±0.08" 0" ± 3/32"	0.0" ±0.08" [1] 0" ± 3/32" [1]	0.0" ±0.08" [1] 0" ± 3/32" [1]	0.0" ±0.08" 0" ± 3/32"
SAI						
TURNING ANGLE	39.0 / 31.0 ±2.0	39.9 / 31.8 ±2.0	38.3 / 30.7 ±2.0 [1]	40.0 / 33.0 ±2.0	36.0 / 30.5 ±2.0	37.5 / 30.8 ±2.0
ILLUSTRATION #	26	26	26			197
REAR						
CAMBER (Degrees)	-1.0 ±0.75 -1 ± 3/4	-1.65 +1.08 -0.75 [1] -1 5/8 +1 1/16 - 3/4 [1]	-0.9 ±0.5 - 7/8 ± 1/2	-0.5 ±1.0 - 1/2 ±1	-0.5 ±1.0 - 1/2 ±1	-1.25 ±0.5 -1 1/4 ± 1/2
REAR TOE (Degrees)	0.16° +0.16° -0.08° 5/32° + 5/32° - 3/32°	0.16° +0.16° -0.08° 5/32° + 5/32° - 3/32°	0.16° +0.16° -0.08° 5/32° + 5/32° - 3/32°	0.16° +0.16° -0.08° 5/32° + 5/32° - 3/32°	0.16° +0.16° -0.08° 5/32° + 5/32° - 3/32°	0.16° +0.16° -0.08° 5/32° + 5/32° - 3/32°
REAR TOE (Inches)	0.08" +0.08" -0.04" 3/32" + 3/32" - 1/32"	0.08" +0.08" -0.04" 3/32" + 3/32" - 1/32"	0.08" +0.08" -0.04" 3/32" + 3/32" - 1/32"	0.08" +0.08" -0.04" 3/32" + 3/32" - 1/32"	0.08" +0.08" -0.04" 3/32" + 3/32" - 1/32"	0.08" +0.08" -0.04" 3/32" + 3/32" - 1/32"
THRUST ANGLE						
ILLUSTRATION #	140	140	140	02	02	06
NOTES	[1] Calif. Models = 2.17° (2 3/16°)	[1] '09 mods. with "C" stamp on rear upper arm = -0.9° (- 7/8°)	[1] 2012 = 38.6°/30.5°	[1] set to POS. limit if inside edge wear.	[1] set to POS. limit if inside edge wear.	
	19061	19062	19063	19064	19065	19066

Honda

MODEL	CR-V	CR-V	CR-V	CR-V	CR-V	CR-V
INFORMATION					FWD	FWD
					except 16" wheels	with 16" wheels
YEARS	1997-2001	2002-2006	2007-2009	2010-2011	2012-2014	2012-2014
INSTRUCTIONS						
LUGNUT TORQUE	80 ft.lbs.	80 ft.lbs.	80 ft.lbs.	80 ft.lbs.	80 ft.lbs.	80 ft.lbs.
FRONT						
CASTER (Degrees)	2.17 ± 1.0 2 3/16 ± 1	1.75 ± 1.0 1¾ ± 1	3.03 ± 1.0 3 ± 1	2.73 ± 1.0 2¾ ± 1	2.83 ± 0.5 2 13/16 ± ½	2.73 ± 0.5 2¾ ± ½
CROSS | TOL.	0.0 (0) | n/s	0.0 (0) | n/s	0.0 (0) | n/s	0.0 (0) | n/s	0.0 (0) | n/s	0.0 (0) | n/s
CAMBER (Degrees)	0.0 ± 1.0 0 ± 1	0.0 ± 0.75 0 ± ¾	0.0 ± 0.5 0 ± ½	0.0 +0.75 -0.5 0 +¾ -½	-0.17 ± 1.0 -3/16 ± 1	0.0 ± 1.0 0 ± 1
CROSS | TOL.	0.0 (0) | n/s	0.0 (0) | n/s	0.0 (0) | n/s	0.0 (0) | 0.58 (9/16)	0.0 (0) | 1.0 (1)	0.0 (0) | 1.0 (1)
FRONT TOE (Degrees)	0.0° ± 0.24° 0° ± ¼°	0.0° ± 0.16° 0° ± 5/32°	0.0° ± 0.16° 0° ± 5/32°	0.0° ± 0.16° 0° ± 5/32°	0.0° ± 0.16° 0° ± 5/32°	0.0° ± 0.16° 0° ± 5/32°
FRONT TOE (Inches)	0.0" ± 0.12" 0" ± ⅛"	0.0" ± 0.08" 0" ± 3/32"	0.0" ± 0.08" 0" ± 3/32"	0.0" ± 0.08" 0" ± 3/32"	0.0" ± 0.08" 0" ± 3/32"	0.0" ± 0.08" 0" ± 3/32"
SAI						
TURNING ANGLE	37.0 / 31.5 ±2.0	39.0 / 32.0 ±2.0[1]	36.5 / 31.2 ±2.0	36.5 / 31.2 ±2.0	36.2 / 31.0 ±2.0	36.5 / 31.2 ±2.0
ILLUSTRATION #		26	26	26	26	26
REAR						
CAMBER (Degrees)	-1.0 ± 1.0 -1 ± 1	-1.0 ± 0.75 -1 ± ¾	-1.0 ± 0.75 -1 ± ¾	-1.0 ± 0.75 -1 ± ¾	-1.2 ± 0.75 -1 3/16 ± ¾	-1.1 ± 0.75 -1⅛ ± ¾
REAR TOE (Degrees)	0.16° +0.16° -0.08° 5/32° +5/32° -3/32°	0.16° +0.16° -0.08° 5/32° +5/32° -3/32°	0.16° +0.16° -0.08° 5/32° +5/32° -3/32°	0.16° +0.16° -0.08° 5/32° +5/32° -3/32°	0.16° +0.16° -0.08° 5/32° +5/32° -3/32°	0.16° +0.16° -0.08° 5/32° +5/32° -3/32°
REAR TOE (Inches)	0.08" +0.08" -0.04" 3/32" +3/32" -1/32"	0.08" +0.08" -0.04" 3/32" +3/32" -1/32"	0.08" +0.08" -0.04" 3/32" +3/32" -1/32"	0.08" +0.08" -0.04" 3/32" +3/32" -1/32"	0.08" +0.08" -0.04" 3/32" +3/32" -1/32"	0.08" +0.08" -0.04" 3/32" +3/32" -1/32"
THRUST ANGLE						
ILLUSTRATION #	02	140	140	140	140	140
NOTES		[1] 2002-04 = 39.8°/32.5°				
	19067	19068	19069	19070	19071	19072

Honda

MODEL	CR-V	CR-V	CR-V	CR-V	CR-V	CR-V
INFORMATION	AWD	AWD	FWD	FWD	AWD	AWD
	except 16" wheels	with 16" wheels	except 16" wheels	with 16" wheels	except 16" wheels	with 16" wheels
YEARS	2012-2014	2012-2014	2015-2016	2015-2016	2015-2016	2015-2016
INSTRUCTIONS						
LUGNUT TORQUE	80 ft.lbs.	80 ft.lbs.	80 ft.lbs.	80 ft.lbs.	80 ft.lbs.	80 ft.lbs.
FRONT						
CASTER (Degrees)	2.73 ± 0.5 2¾ ± ½	2.63 ± 0.5 2⅝ ± ½	2.83 ± 0.5 2 13/16 ± ½	2.73 ± 0.5 2¾ ± ½	2.73 ± 0.5 2¾ ± ½	2.65 ± 0.5 2⅝ ± ½
CROSS | TOL.	0.0 (0) | n/s	0.0 (0) | n/s	0.0 (0) | 0.5 (½)	0.0 (0) | 0.5 (½)	0.0 (0) | 0.5 (½)	0.0 (0) | 0.5 (½)
CAMBER (Degrees)	0.0 ± 1.0 0 ± 1	-0.13 ± 1.0 -⅛ ± 1	-0.67 ± 0.75 -11/16 ± ¾	-0.17 ± 0.75 -3/16 ± ¾	-0.5 ± 0.75 -½ ± ¾	-0.67 ± 0.75 -11/16 ± ¾
CROSS | TOL.	0.0 (0) | 1.0 (1)	0.0 (0) | 1.0 (1)	0.0 (0) | 0.75 (¾)	0.0 (0) | 0.75 (¾)	0.0 (0) | 0.75 (¾)	0.0 (0) | 0.75 (¾)
FRONT TOE (Degrees)	0.0° ± 0.16° 0° ± 5/32°	0.0° ± 0.16° 0° ± 5/32°	0.0° ± 0.16° 0° ± 5/32°	0.0° ± 0.16° 0° ± 5/32°	0.0° ± 0.16° 0° ± 5/32°	0.0° ± 0.16° 0° ± 5/32°
FRONT TOE (Inches)	0.0" ± 0.08" 0" ± 3/32"	0.0" ± 0.08" 0" ± 3/32"	0.0" ± 0.08" 0" ± 3/32"	0.0" ± 0.08" 0" ± 3/32"	0.0" ± 0.08" 0" ± 3/32"	0.0" ± 0.08" 0" ± 3/32"
SAI						
TURNING ANGLE	36.5 / 31.2 ±2.0	36.8 / 31.3 ±2.0	36.5 / 31.1 ±2.0	36.5 / 31.4 ±2.0	36.7 / 31.3 ±2.0	36.6 / 31.5 ±2.0
ILLUSTRATION #	26	26	26	26	26	26
REAR						
CAMBER (Degrees)	-1.0 ± 0.75 -1 ± ¾	-0.83 ± 0.75 -13/16 ± ¾	-1.22 ± 0.75 -1 ¼ ± ¾	-1.07 ± 0.75 -1 1/16 ± ¾	-1.0 ± 0.75 -1 ± ¾	-1.23 ± 0.75 -1 ¼ ± ¾
REAR TOE (Degrees)	0.16° +0.16° -0.08° 5/32° +5/32° -3/32°	0.16° +0.16° -0.08° 5/32° +5/32° -3/32°	0.16° ± 0.16° 5/32° ± 5/32°	0.16° ± 0.16° 5/32° ± 5/32°	0.16° ± 0.16° 5/32° ± 5/32°	0.16° ± 0.16° 5/32° ± 5/32°
REAR TOE (Inches)	0.08" +0.08" -0.04" 3/32" +3/32" -1/32"	0.08" +0.08" -0.04" 3/32" +3/32" -1/32"	0.08" ± 0.08" 3/32" ± 3/32"	0.08" ± 0.08" 3/32" ± 3/32"	0.08" ± 0.08" 3/32" ± 3/32"	0.08" ± 0.08" 3/32" ± 3/32"
THRUST ANGLE						
ILLUSTRATION #	140	140	140	140	140	140
NOTES						
	19073	19074	19075	19076	19077	19078

Honda

CR-V	CR-V	CR-Z	MODEL	Crosstour	CRX	Element
FWD	AWD		INFORMATION	(Accord Crosstour)		exc/SC
2017	2017	2011-2016	YEARS	2010-2015	1990-91	2003-2011
			INSTRUCTIONS			
80 ft.lbs.	80 ft.lbs.	80 ft.lbs.	LUGNUT TORQUE	80 ft.lbs.	80 ft.lbs.	80 ft.lbs.
			FRONT			
5.17 ± 0.5 5 3/16 ± 1/2	5.07 ± 0.5 5 1/16 ± 1/2	3.37 ± 1.0 3 3/8 ± 1	CASTER (Degrees)	3.5 ± 0.5 3 1/2 ± 1/2	3.0 ± 1.0 3 ± 1	1.83 ± 1.0 1 13/16 ± 1
0.0 (0) | n/s	0.0 (0) | n/s	0.0 (0) | n/s	CROSS | TOL.	0.0 (0) | 0.5 (1/2)	0.0 (0) | n/s	0.0 (0) | n/s
-0.3 ± 0.5 -5/16 ± 1/2	-0.2 ± 0.5 -3/16 ± 1/2	0.0 ± 1.0 0 ± 1	CAMBER (Degrees)	0.0 ± 0.5 0 ± 1/2	0.0 ± 1.0 0 ± 1	-0.22 ± 0.75 -3/16 ± 3/4
0.0 (0) | 0.75 (3/4)	0.0 (0) | 0.75 (3/4)	0.0 (0) | 0.75 (3/4)	CROSS | TOL.	0.0 (0) | 0.5 (1/2)	0.0 (0) | n/s	0.0 (0) | 1.0 (1)
0.0° ± 0.16° 0° ± 5/32°	0.0° ± 0.16° 0° ± 5/32°	0.0° ± 0.24° 0° ± 1/4°	FRONT TOE (Degrees)	0.0° ± 0.16° 0° ± 5/32°	0.0° ± 0.16° [1] 0° ± 5/32° [1]	0.0° ± 0.24° 0° ± 1/4°
0.0" ± 0.08" 0" ± 3/32"	0.0" ± 0.08" 0" ± 3/32"	0.0" ± 0.12" 0" ± 1/8"	FRONT TOE (Inches)	0.0" ± 0.08" 0" ± 3/32"	0.0" ± 0.08" [1] 0" ± 3/32" [1]	0.0" ± 0.12" 0" ± 1/8"
			SAI		7.6 (7 9/16)	
37.3 / 30.9 ±2.0	37.7 / 31.0 ±2.0	38.1 / 31.8 ±2.0	TURNING ANGLE	36.6 / 28.8 ±2.0	41.5 / 33.5 ±2.0	40.1 / 32.3 ±2.0
197	197	26	ILLUSTRATION #			26
			REAR			
-1.5 ± 0.75 -1 1/2 ± 3/4	-1.2 ± 0.75 -1 3/16 ± 3/4	-1.5 ± 1.0 -1 1/2 ± 1	CAMBER (Degrees)	-1.0 ± 0.5 -1 ± 1/2	-0.43 ± 1.0 -7/16 ± 1	-1.0 ± 0.75 -1 ± 3/4
0.28° +0.08° -0.16° 9/32° +3/32° -5/32°	0.28° +0.08° -0.16° 9/32° +3/32° -5/32°	0.24° +0.16° -0.24° 1/4° +5/32° -1/4°	REAR TOE (Degrees)	0.16° ± 0.16° 5/32° ± 5/32°	0.16° +0.16° -0.08° 5/32° +5/32° -3/32°	0.16° +0.16° -0.08° 5/32° +5/32° -3/32°
0.14" +0.04" -0.08" 1/8" +1/32" -3/32"	0.14" +0.04" -0.08" 1/8" +1/32" -3/32"	0.12" +0.08" -0.12" 1/8" +3/32" -1/8"	REAR TOE (Inches)	0.08" ± 0.08" 3/32" ± 3/32"	0.08" +0.08" -0.04" 3/32" +3/32" -1/32"	0.08" +0.08" -0.04" 3/32" +3/32" -1/32"
			THRUST ANGLE			
06	06		ILLUSTRATION #	174	02	140
			NOTES		[1] set to POS. limit if inside edge wear.	
19079	19080	19081		19082	19085	19086

Honda

Element	Fit	Fit	MODEL	Fit	Fit EV	HR-V
SC			INFORMATION			
2007-2011	2007-2008	2009-2013	YEARS	2015-2017	2013-2014	2016-2017
			INSTRUCTIONS			
80 ft.lbs.	80 ft.lbs.	80 ft.lbs.	LUGNUT TORQUE	80 ft.lbs.	80 ft.lbs.	80 ft.lbs.
			FRONT			
1.83 ± 1.0 1 13/16 ± 1	3.75 ± 1.0 3 3/4 ± 1	3.33 ± 1.0 3 5/16 ± 1	CASTER (Degrees)	4.97 ± 1.0 5 ± 1	3.1 ± 0.5 3 1/8 ± 1/2	5.0 ± 1.0 5 ± 1
0.0 (0) | n/s	0.0 (0) | n/s	0.0 (0) | n/s	CROSS | TOL.	0.0 (0) | n/s	0.0 (0) | 0.75 (3/4)	0.0 (0) | n/s
-0.4 ± 0.75 -3/8 ± 3/4	0.0 ± 1.0 0 ± 1	0.0 ± 1.0 0 ± 1	CAMBER (Degrees)	0.0 ± 1.0 0 ± 1	0.0 ± 0.5 0 ± 1/2	0.0 ± 1.0 0 ± 1
0.0 (0) | 1.0 (1)	0.0 (0) | 1.0 (1)	0.0 (0) | 0.75 (3/4)	CROSS | TOL.	0.0 (0) | 1.0 (1)	0.0 (0) | 0.75 (3/4)	0.0 (0) | 1.0 (1)
0.0° ± 0.24° 0° ± 1/4°	0.0° ± 0.24° 0° ± 1/4°	0.0° ± 0.24° 0° ± 1/4°	FRONT TOE (Degrees)	0.0° ± 0.24° 0° ± 1/4°	0.0° ± 0.16° 0° ± 5/32°	0.0° ± 0.24° 0° ± 1/4°
0.0" ± 0.12" 0" ± 1/8"	0.0" ± 0.12" 0" ± 1/8"	0.0" ± 0.12" 0" ± 1/8"	FRONT TOE (Inches)	0.0" ± 0.12" 0" ± 1/8"	0.0" ± 0.08" 0" ± 3/32"	0.0" ± 0.12" 0" ± 1/8"
			SAI			
38.4 / 31.4 ±2.0	38.0 / 32.0 ±2.0	38.1 / 32.0 ±2.0	TURNING ANGLE	37.0 / 32.2 ±2.0	37.9 / 32.1 ±2.0	38.2 / 31.9 ±2.0
26	26	26	ILLUSTRATION #	26	26	26
			REAR			
-1.1 ± 0.75 -1 1/8 ± 3/4	-1.5 ± 1.0 -1 1/2 ± 1	-1.0 ± 1.0 -1 ± 1	CAMBER (Degrees)	-1.5 ± 1.0 -1 1/2 ± 1	-0.75 ± 1.0 -3/4 ± 1	-1.5 ± 1.0 -1 1/2 ± 1
0.16° +0.16° -0.08° 5/32° +5/32° -3/32°	0.2° ± 0.2° 3/16° ± 3/16°	0.2° ± 0.2° 3/16° ± 3/16°	REAR TOE (Degrees)	0.2° +0.28° -0.2° 3/16° +9/32° -3/16°	0.16° +0.16° -0.08° 5/32° +5/32° -3/32°	0.24° ± 0.24° 1/4° ± 1/4°
0.08" +0.08" -0.04" 3/32" +3/32" -1/32"	0.1" ± 0.1" 3/32" ± 3/32"	0.1" ± 0.1" 3/32" ± 3/32"	REAR TOE (Inches)	0.1" +0.14" -0.1" 3/32" +1/8" -3/32"	0.08" +0.08" -0.04" 3/32" +3/32" -1/32"	0.12" ± 0.12" 1/8" ± 1/8"
			THRUST ANGLE			
140			ILLUSTRATION #		140	
			NOTES			
19087	19088	19089		19090	19091	19092

Honda

Insight	Insight	Insight	MODEL	Odyssey	Odyssey	Odyssey
			INFORMATION			
2000-2006	2010-2011	2012-2014	YEARS	1995-98	1999-2004	2005-2010
			INSTRUCTIONS			
80 ft.lbs.	80 ft.lbs.	80 ft.lbs.	LUGNUT TORQUE	80 ft.lbs.	80 ft.lbs.	94 ft.lbs.
			FRONT			
2.0 ± 1.0 2 ± 1	3.3 ± 1.0 3 5⁄16 ± 1	3.3 ± 0.5 3 5⁄16 ± ½	CASTER (Degrees)	3.0 ± 1.0 3 ± 1	2.12 ± 1.0 2⅛ ± 1	2.53 ± 1.0 2 9⁄16 ± 1
0.0 (0) | n/s	0.0 (0) | n/s	0.0 (0) | n/s	CROSS | TOL.	0.0 (0) | n/s	0.0 (0) | n/s	0.0 (0) | n/s
0.0 ± 1.0 0 ± 1	0.0 ± 1.0 0 ± 1	0.0 ± 1.0 0 ± 1	CAMBER (Degrees)	0.0 ± 1.0 0 ± 1	0.0 ± 1.0 0 ± 1	0.0 ± 0.5 0 ± ½
0.0 (0) | n/s	0.0 (0) | 0.75 (¾)	0.0 (0) | 0.75 (¾)	CROSS | TOL.	0.0 (0) | n/s	0.0 (0) | n/s	0.0 (0) | 0.5 (½)
0.0° ± 0.16° 0° ± 5⁄32°	0.0° ± 0.24° 0° ± ¼°	0.0° ± 0.24° 0° ± ¼°	FRONT TOE (Degrees)	0.0° ± 0.16° 0° ± 5⁄32°	0.0° ± 0.16° 0° ± 5⁄32°	0.0° ± 0.16° 0° ± 5⁄32°
0.0" ± 0.08" 0" ± 3⁄32"	0.0" ± 0.12" 0" ± ⅛"	0.0" ± 0.12" 0" ± ⅛"	FRONT TOE (Inches)	0.0" ± 0.08" 0" ± 3⁄32"	0.0" ± 0.08" 0" ± 3⁄32"	0.0" ± 0.08" 0" ± 3⁄32"
			SAI			
40.0 / 33.0 ± 2.0	38.5 / 32.4 ± 2.0	38.3 / 32.4 ± 2.0	TURNING ANGLE	39.0 / 30.0 ± 2.0	41.8 / 33.4 ± 2.0	41.9 / 34.5 ± 2.0
26	26	26	ILLUSTRATION #	03	26	26
			REAR			
-1.0 ± 1.0 -1 ± 1	-1.5 ± 1.0 [1] -1½ ± 1 [1]	-1.5 ± 1.0 -1½ ± 1	CAMBER (Degrees)	-0.5 ± 0.5 -½ ± ½	-0.5 ± 0.75 -½ ± ¾	-0.5 ± 0.75 -½ ± ¾
0.24° ± 0.24° ¼° ± ¼°	0.2° ± 0.2° 3⁄16° ± 3⁄16°	0.2° ± 0.2° 3⁄16° ± 3⁄16°	REAR TOE (Degrees)	0.0° ± 0.16° 0° ± 5⁄32°	0.0° ± 0.16° 0° ± 5⁄32°	0.16° ± 0.16° 5⁄32° ± 5⁄32°
0.12" ± 0.12" ⅛" ± ⅛"	0.1" ± 0.1" 3⁄32" ± 3⁄32"	0.1" ± 0.1" 3⁄32" ± 3⁄32"	REAR TOE (Inches)	0.0" ± 0.08" 0" ± 3⁄32"	0.0" ± 0.08" 0" ± 3⁄32"	0.08" ± 0.08" 3⁄32" ± 3⁄32"
			THRUST ANGLE			
			ILLUSTRATION #	06	06	06
	[1] 2010 = -1.0° (-1°)		NOTES			
19093	19094	19095		19096	19097	19098

Honda

Odyssey	Passport	Passport	MODEL	Pilot	Pilot	Pilot
			INFORMATION	FWD	AWD	
2011-2017	1994-95	1996-2002	YEARS	2006-2008	2003-2008	2009-2015
			INSTRUCTIONS			
94 ft.lbs.	66 ft.lbs. [1]	87 ft.lbs. [1]	LUGNUT TORQUE	80 ft.lbs.	80 ft.lbs.	94 ft.lbs.
			FRONT			
2.6 ± 0.6 2⅝ ± ⅝	2.33 ± 0.75 2 5⁄16 ± ¾	2.5 ± 0.75 [2] 2½ ± ¾ [2]	CASTER (Degrees)	1.98 ± 1.0 2 ± 1	1.88 ± 1.0 1⅞ ± 1	4.28 ± 0.58 [1] 4 5⁄16 ± 9⁄16 [1]
0.0 (0) | n/s	0.0 (0) | 0.58 (9⁄16)	0.0 (0) | 0.5 (½)	CROSS | TOL.	0.0 (0) | n/s	0.0 (0) | n/s	0.0 (0) | n/s
0.0 ± 0.75 0 ± ¾	0.5 ± 1.0 ½ ± 1	0.0 ± 0.5 0 ± ½	CAMBER (Degrees)	-0.67 ± 1.0 -11⁄16 ± 1	-0.5 ± 1.0 -½ ± 1	-0.5 ± 0.75 -½ ± ¾
0.0 (0) | 0.75 (¾)	0.0 (0) | 0.75 (¾)	0.0 (0) | 0.5 (½)	CROSS | TOL.	0.0 (0) | n/s	0.0 (0) | n/s	0.0 (0) | n/s
0.0° ± 0.16° 0° ± 5⁄32°	0.16° ± 0.16° 5⁄32° ± 5⁄32°	0.0° ± 0.16° 0° ± 5⁄32°	FRONT TOE (Degrees)	0.0° ± 0.16° 0° ± 5⁄32°	0.0° ± 0.16° 0° ± 5⁄32°	0.0° ± 0.16° 0° ± 5⁄32°
0.0" ± 0.08" 0" ± 3⁄32"	0.08" ± 0.08" 3⁄32" ± 3⁄32"	0.0" ± 0.08" 0" ± 3⁄32"	FRONT TOE (Inches)	0.0" ± 0.08" 0" ± 3⁄32"	0.0" ± 0.08" 0" ± 3⁄32"	0.0" ± 0.08" 0" ± 3⁄32"
	10.0 ± 1.0 (10 ± 1)	12.5 ± 0.5 (12½ ± ½)	SAI	11.9 (11⅞)	11.9 (11⅞)	11.9 (11⅞)
42.1 / 34.7 ± 2.0	35.0 / 33.0 ± 1.0	34.0 / 32.0 ± 1.0	TURNING ANGLE	37.9 / 30.2 ± 2.0	38.1 / 30.3 ± 2.0	38.2 / 31.7 ± 2.0
26	23	23	ILLUSTRATION #	102	102	26
			REAR			
-0.5 ± 0.75 -½ ± ¾			CAMBER (Degrees)	-0.67 ± 0.75 -11⁄16 ± ¾	-0.5 ± 0.75 -½ ± ¾	-0.5 ± 0.75 -½ ± ¾
0.16° ± 0.16° 5⁄32° ± 5⁄32°			REAR TOE (Degrees)	0.16° ± 0.16° 5⁄32° ± 5⁄32°	0.0° ± 0.16° 0° ± 5⁄32°	0.16° ± 0.16° 5⁄32° ± 5⁄32°
0.08" ± 0.08" 3⁄32" ± 3⁄32"			REAR TOE (Inches)	0.08" ± 0.08" 3⁄32" ± 3⁄32"	0.0" ± 0.08" 0" ± 3⁄32"	0.08" ± 0.08" 3⁄32" ± 3⁄32"
			THRUST ANGLE			
06			ILLUSTRATION #	06	06	06
	[1] w/Alum. whls 87 ft lbs.	[1] w/Alum. whls 87 ft lbs. [2] 1996-97 = 2.75° (2¾°)	NOTES			[1] 2009-11 = 4.18° (4 3⁄16°)
19099	19101	19102		19103	19104	19105

Honda

MODEL INFORMATION	Pilot	Prelude 2WS	Prelude 4WS	Prelude 2WS	Prelude 4WS	Prelude exc/Type SH, SVR-R
YEARS	2016-2017	1990-91	1990-91	1992-96	1992-95	1997-2001
INSTRUCTIONS						
LUGNUT TORQUE	94 ft.lbs.	80 ft.lbs.	80 ft.lbs.	80 ft.lbs.	80 ft.lbs.	80 ft.lbs.
FRONT						
CASTER (Degrees)	3.73 ± 0.6 3¾ + 9⁄16 5⁄8 - 5⁄8	2.33 ± 0.5 2 5⁄16 ± ½	2.33 ± 0.5 2 5⁄16 ± ½	2.67 ± 1.0 2 11⁄16 ± 1	2.67 ± 1.0 2 11⁄16 ± 1	2.67 ± 1.0 2 11⁄16 ± 1
CROSS | TOL.	0.0 (0) | n/s	0.0 (0) | n/s	0.0 (0) | n/s	0.0 (0) | n/s	0.0 (0) | n/s	0.0 (0) | n/s
CAMBER (Degrees)	-0.5 ± 0.75 -½ ± ¾	0.0 ± 1.0 0 ± 1	0.0 ± 1.0 0 ± 1	0.0 ± 1.0 0 ± 1	0.0 ± 1.0 0 ± 1	0.0 ± 1.0 0 ± 1
CROSS | TOL.	0.0 (0) | 0.75 (¾)	0.0 (0) | n/s	0.0 (0) | n/s	0.0 (0) | n/s	0.0 (0) | n/s	0.0 (0) | n/s
FRONT TOE (Degrees)	0.1° ± 0.16° 3⁄32° ± 5⁄32°	0.0° ± 0.16° 0° ± 5⁄32°	0.0° ± 0.16° 0° ± 5⁄32°	0.0° ± 0.16° 0° ± 5⁄32°	0.0° ± 0.16° 0° ± 5⁄32°	0.0° ± 0.16° 0° ± 5⁄32°
FRONT TOE (Inches)	0.05" ± 0.08" 1⁄16" ± 3⁄32"	0.0" ± 0.08" 0" ± 3⁄32"	0.0" ± 0.08" 0" ± 3⁄32"	0.0" ± 0.08" 0" ± 3⁄32"	0.0" ± 0.08" 0" ± 3⁄32"	0.0" ± 0.08" 0" ± 3⁄32"
SAI						
TURNING ANGLE	38.2 / 31.7 ±2.0	37.3 / 30.3 ±2.0	37.3 / 30.3 ±2.0	36.3 / 29.7 ±2.0	36.3 / 29.7 ±2.0	36.3 / 29.7 ±2.0
ILLUSTRATION #		03 58	03 58	03	03	03
REAR						
CAMBER (Degrees)	-0.5 ± 0.75 -½ ± ¾	-0.33 ± 1.0 -5⁄16 ± 1	-0.33 ± 0.5 -5⁄16 ± ½	-0.75 ± 1.0 -¾ ± 1	-0.75 ± 0.5 -¾ ± ½	-0.75 ± 1.0 -¾ ± 1
REAR TOE (Degrees)	0.16° ± 0.16° 5⁄32° ± 5⁄32°	0.16° ± 0.16° 5⁄32° ± 5⁄32°	0.24° ± 0.08° ¼° ± 3⁄32°	0.16° ± 0.16° 5⁄32° ± 5⁄32°	0.16° ± 0.16° 5⁄32° ± 5⁄32°	0.16° ± 0.16° 5⁄32° ± 5⁄32°
REAR TOE (Inches)	0.08" ± 0.08" 3⁄32" ± 3⁄32"	0.08" ± 0.08" 3⁄32" ± 3⁄32"	0.12" ± 0.04" ⅛" ± 1⁄32"	0.08" ± 0.08" 3⁄32" ± 3⁄32"	0.08" ± 0.08" 3⁄32" ± 3⁄32"	0.08" ± 0.08" 3⁄32" ± 3⁄32"
THRUST ANGLE						
ILLUSTRATION #	174	58 06	58 49	06	49	06
NOTES						
Code	19106	19109	19110	19111	19112	19113

Honda

MODEL INFORMATION	Prelude Type SH, SVR-R	Ridgeline	Ridgeline 2WD	Ridgeline AWD	S2000	S2000
YEARS	1997-2001	2006-2014	2017	2017	2000-2003	2004-2009
INSTRUCTIONS						
LUGNUT TORQUE	80 ft.lbs.	94 ft.lbs.	94 ft.lbs.	94 ft.lbs.	80 ft.lbs.	80 ft.lbs.
FRONT						
CASTER (Degrees)	4.33 ± 0.75 4 5⁄16 ± ¾	1.88 ± 1.0 1 ⅞ ± 1	3.37 ± 0.6 3⅜ ± ⅝	3.23 ± 0.6 3¼ ± ⅝	6.0 ± 0.25 6 ± ¼	6.0 ± 0.25 6 ± ¼
CROSS | TOL.	0.0 (0) | n/s	0.0 (0) | n/s	0.0 (0) | n/s	0.0 (0) | n/s	0.0 (0) | 0.25 (¼)	0.0 (0) | 0.25 (¼)
CAMBER (Degrees)	0.0 ± 1.0 0 ± 1	-0.5 ± 1.0 -½ ± 1	-0.5 ± 0.75 -½ ± ¾	-0.32 ± 0.75 -5⁄16 ± ¾	-0.5 ± 0.17 -½ ± 3⁄16	-0.5 ± 0.17 -½ ± 3⁄16
CROSS | TOL.	0.0 (0) | n/s	0.0 (0) | 1.0 (1)	0.0 (0) | 0.75 (¾)	0.0 (0) | 0.75 (¾)	0.0 (0) | 0.17 (3⁄16)	0.0 (0) | 0.17 (3⁄16)
FRONT TOE (Degrees)	0.0° ± 0.16° 0° ± 5⁄32°	0.0° ± 0.16° 0° ± 5⁄32°	0.0° ± 0.16° 0° ± 5⁄32°	0.0° ± 0.16° 0° ± 5⁄32°	0.0° ± 0.16° 0° ± 5⁄32°	0.0° ± 0.16° 0° ± 5⁄32°
FRONT TOE (Inches)	0.0" ± 0.08" 0" ± 3⁄32"	0.0" ± 0.08" 0" ± 3⁄32"	0.0" ± 0.08" 0" ± 3⁄32"	0.0" ± 0.08" 0" ± 3⁄32"	0.0" ± 0.08" 0" ± 3⁄32"	0.0" ± 0.08" 0" ± 3⁄32"
SAI						
TURNING ANGLE	34.8 / 28.8 ±2.0	37.9 / 30.3 ±2.0	36.4 / 31.8 ±2.0	36.9 / 32.2 ±2.0	34.0 / 29.0 ±2.0	34.0 / 29.0 ±2.0
ILLUSTRATION #		102			121	121
REAR						
CAMBER (Degrees)	-0.75 ± 1.0 -¾ ± 1	-0.5 ± 0.75 -½ ± ¾	-0.5 ± 0.75 -½ ± ¾	-0.02 ± 0.75 -0 ± ¾	-1.5 ± 0.17 -1 ½ ± 3⁄16	-1.5 ± 0.17 -1 ½ ± 3⁄16
REAR TOE (Degrees)	0.16° ± 0.16° 5⁄32° ± 5⁄32°	0.0° ± 0.16° 0° ± 5⁄32°	0.16° ± 0.16° 5⁄32° ± 5⁄32°	0.32° ± 0.16° 5⁄16° ± 5⁄32°	0.5° ± 0.16° ½° ± 5⁄32°	0.28° ± 0.16° [1] 9⁄32° ± 5⁄32° [1]
REAR TOE (Inches)	0.08" ± 0.08" 3⁄32" ± 3⁄32"	0.0" ± 0.08" 0" ± 3⁄32"	0.08" ± 0.08" 3⁄32" ± 3⁄32"	0.16" ± 0.08" 5⁄32" ± 3⁄32"	0.25" ± 0.08" ¼" ± 3⁄32"	0.14" ± 0.08" [1] ⅛" ± 3⁄32" [1]
THRUST ANGLE						
ILLUSTRATION #	06	06	174	174	122	122
NOTES						[1] CR = 0.44° (0.22")
Code	19114	19115	19116	19117	19118	19119

Hummer

H1	H1	H2	MODEL	H2	H2	H3
(normal load)	(normal load)		INFORMATION			
1992-97	1998-2006	2003-2005	YEARS	2006-2007	2008-2010	2006-2010
		p. 305	INSTRUCTIONS			
		125 ft.lbs.	LUGNUT TORQUE	140 ft.lbs.	140 ft.lbs.	103 ft.lbs.
			FRONT			
3.0 ±0.5 3 ± ½	3.0 ±0.5 3 ± ½	L= 4.25, R= 4.75 ± 1.0 L= 4¼, R= 4¾ ± 1	CASTER (Degrees)	L= 3.25, R= 3.75 ± 1.0 [1] L= 3¼, R= 3¾ ± 1 [1]	L= 3.35, R= 3.65 ± 1.0 [1] L= 3⅜, R= 3⅝ ± 1 [1]	L= 2.8, R= 3.1 ± 1.0 [1] L= 2 13/16, R= 3⅛ ± 1 [1]
0.0 (0) | 0.5 (½)	0.0 (0) | 0.5 (½)	-0.5 (-½) | 0.5 (½)	CROSS | TOL.	-0.5 (-½) | 0.75 (¾)	-0.3 (-5/16) | 0.75 (¾)	-0.3 (-5/16) | 0.5 (½)
0.25 ±0.25 ¼ ± ¼	0.25 ±0.25 ¼ ± ¼	0.25 ±0.5 ¼ ± ½	CAMBER (Degrees)	0.25 ±0.6 ¼ ± ⅝	0.25 ±0.6 ¼ ± ⅝	0.0 ±0.6 0 ± ⅝
0.0 (0) | 0.5 (½)	0.0 (0) | 0.5 (½)	0.0 (0) | 0.5 (½)	CROSS | TOL.	0.0 (0) | 0.6 (⅝)	0.0 (0) | 0.6 (⅝)	0.0 (0) | 0.5 (½)
0.13° ±0.13° ⅛° ± ⅛°	0.1° ±0.1° 3/32° ± 3/32°	0.1° ±0.2° 3/32° ± 3/16°	FRONT TOE (Degrees)	0.1° ±0.2° 3/32° ± 3/16°	0.1° ±0.2° 3/32° ± 3/16°	0.1° ±0.2° 3/32° ± 3/16°
0.06" ±0.06" 1/16" ± 1/16"	0.05" ±0.05" 1/16" ± 1/16"	0.05" ±0.1" 1/16" ± 3/32"	FRONT TOE (Inches)	0.05" ±0.1" 1/16" ± 3/32"	0.05" ±0.1" 1/16" ± 3/32"	0.05" ±0.1" 1/16" ± 3/32"
			SAI			
			TURNING ANGLE			
41	24	126	ILLUSTRATION #	126	126	126
			REAR			
0.5 ±0.25 ½ ± ¼	0.5 ±0.25 ½ ± ¼		CAMBER (Degrees)			
-0.13° ±0.13° -⅛° ± ⅛°	-0.1° ±0.1° -3/32° ± 3/32°		REAR TOE (Degrees)			
-0.06" ±0.06" -1/16" ± 1/16"	-0.05" ±0.05" -1/16" ± 1/16"		REAR TOE (Inches)			
			THRUST ANGLE			
10	155 10		ILLUSTRATION #			
			NOTES	[1] Relative to ground	[1] Relative to ground	[1] RHD L=3.1°, R=2.8°
58001	58002	58003		58004	58005	58006

Hyundai

Accent	Accent	Accent	MODEL	Accent	Accent	Accent
			INFORMATION			
1995-97	1998	1999	YEARS	2000-2005	2006-2011	2012-2017
			INSTRUCTIONS			
65-80 ft.lbs.	65-80 ft.lbs.	65-80 ft.lbs.	LUGNUT TORQUE	65-80 ft.lbs.	65-80 ft.lbs.	65-80 ft.lbs.
			FRONT			
2.0 ±0.5 2 ± ½	2.0 ±0.5 2 ± ½	1.8 ±0.5 1 13/16 ± ½	CASTER (Degrees)	1.8 ±0.5 1 13/16 ± ½	4.0 ±0.5 4 ± ½	4.1 ±0.5 4⅛ ± ½
0.0 (0) | n/s	0.0 (0) | n/s	0.0 (0) | n/s	CROSS | TOL.	0.0 (0) | n/s	0.0 (0) | n/s	0.0 (0) | n/s
0.0 ±0.5 0 ± ½	0.0 ±0.5 0 ± ½	0.0 ±0.5 0 ± ½	CAMBER (Degrees)	0.0 ±0.5 0 ± ½	0.0 ±0.5 0 ± ½	-0.5 ±0.5 -½ ± ½
0.0 (0) | n/s	0.0 (0) | n/s	0.0 (0) | n/s	CROSS | TOL.	0.0 (0) | n/s	0.0 (0) | n/s	0.0 (0) | n/s
0.0° ±0.24° 0° ± ¼°	0.0° ±0.24° 0° ± ¼°	0.0° ±0.24° 0° ± ¼°	FRONT TOE (Degrees)	0.0° ±0.24° 0° ± ¼°	0.0° ±0.2° 0° ± 3/16°	0.1° ±0.2° [1] 3/32° ± 3/16° [1]
0.0" ±0.12" 0" ± ⅛"	0.0" ±0.12" 0" ± ⅛"	0.0" ±0.12" 0" ± ⅛"	FRONT TOE (Inches)	0.0" ±0.12" 0" ± ⅛"	0.0" ±0.1" 0" ± 3/32"	0.05" ±0.1" [1] 1/16" ± 3/32" [1]
12.1 ±0.5 (12⅛ ± ½)	12.1 ±0.5 (12⅛ ± ½)	12.1 ±0.5 (12⅛ ± ½)	SAI	11.1 ±0.5 (11 1/16 ± ½)	13.0 ±0.5 (13 ± ½)	13.8 ±0.5 (13 13/16 ± ½)
39.0 / 33.3 ±1.5	39.0 / 33.3 ±1.5	39.0 / 33.3 ±1.5	TURNING ANGLE	37.6 / 31.9 ±1.5	40.0 / 32.9 ±1.5	39.0 / 31.0 ±1.5
			ILLUSTRATION #			
			REAR			
-0.68 ±0.5 -11/16 ± ½	-0.68 ±0.5 -11/16 ± ½	-0.68 ±0.5 -11/16 ± ½	CAMBER (Degrees)	-0.68 ±0.5 -11/16 ± ½	-1.0 ±0.5 -1 ± ½	-1.5 ±0.5 -1½ ± ½
0.48° ±0.16° 15/32° ± 5/32°	0.24° ±0.16° ¼° ± 5/32°	0.24° ±0.16° ¼° ± 5/32°	REAR TOE (Degrees)	0.24° ±0.16° ¼° ± 5/32°	0.4° ±0.2° 13/32° ± 3/16°	0.4° +0.4° -0.3° [2] 13/32° + 13/32° - 5/16° [2]
0.24" ±0.08" ¼" ± 3/32"	0.12" ±0.08" ⅛" ± 3/32"	0.12" ±0.08" ⅛" ± 3/32"	REAR TOE (Inches)	0.12" ±0.08" ⅛" ± 3/32"	0.2" ±0.1" 3/16" ± 3/32"	0.2" +0.2" -0.15" [2] 3/16" + 3/16" - 5/32" [2]
			THRUST ANGLE			
30	30	30	ILLUSTRATION #	30		
			NOTES			[1] 2012-14 = 0.15° (0.08") [2] 2012-14 = 0.5° (0.25")
20001	20002	20003		20004	20005	20007

ALIGNMENT SPECIFICATIONS

Hyundai

Azera	Azera	Elantra	MODEL INFORMATION	Elantra	Elantra	Elantra
2006-2011	2012-2017	1992-95	YEARS	1996-2000	2001-2003	2004-2006
			INSTRUCTIONS			
65-80 ft.lbs.	65-80 ft.lbs.	65-80 ft.lbs.	LUGNUT TORQUE	65-80 ft.lbs.	65-80 ft.lbs.	65-80 ft.lbs.
			FRONT			
4.83 ±0.75 4 13/16 ± ¾	4.38 ±0.5 4⅜ ± ½	2.57 ±0.5 2 9/16 ± ½	CASTER (Degrees)	2.35 ±0.5 2⅜ ± ½	2.82 ±0.5 2 13/16 ± ½	2.82 ±0.5 2 13/16 ± ½
0.0 (0) | 0.5 (½)	0.0 (0) | 0.5 (½)	0.0 (0) | n/s	CROSS | TOL.	0.0 (0) | n/s	0.0 (0) | 0.5 (½)	0.0 (0) | 0.5 (½)
0.0 ±0.5 0 ± ½	L= -0.5, R= -0.8 ±0.5 L= -½, R= -13/16 ± ½	0.0 ±0.5 0 ± ½	CAMBER (Degrees)	0.0 ±0.5 0 ± ½	0.0 ±0.5 0 ± ½	0.0 ±0.5 0 ± ½
0.0 (0) | 0.5 (½)	0.3 (5/16) | 0.5 (½)	0.0 (0) | n/s	CROSS | TOL.	0.0 (0) | n/s	0.0 (0) | 0.5 (½)	0.0 (0) | 0.5 (½)
0.0° ±0.16° 0° ± 5/32°	0.1° ±0.2° 3/32° ± 3/16°	0.0° ±0.24° 0° ± ¼°	FRONT TOE (Degrees)	0.0° ±0.24° 0° ± ¼°	0.0° ±0.16° 0° ± 5/32°	0.0° ±0.16° 0° ± 5/32°
0.0" ±0.08" 0" ± 3/32"	0.05" ±0.1" 1/16" ± 3/32"	0.0" ±0.12" 0" ± ⅛"	FRONT TOE (Inches)	0.0" ±0.12" 0" ± ⅛"	0.0" ±0.08" 0" ± 3/32"	0.0" ±0.08" 0" ± 3/32"
9.5 ±0.5 (9 7/16 ± ½)	12.7 ±0.5 (12¾ ± ½)	12.5 ±0.5 (12½ ± ½)	SAI	12.4 ±0.5 (12⅜ ± ½)	12.2 ±0.5 (12 3/16 ± ½)	12.2 ±0.5 (12 3/16 ± ½)
37.6 / 30.7 ±2.0	40.5 / 33.0 ±2.0	39.4 / 34.0 ±1.5	TURNING ANGLE	39.3 / 33.3 ±1.5	40.0 / 33.8 ±2.0	40.0 / 33.8 ±2.0
			ILLUSTRATION #			
			REAR			
-0.5 ±0.5 -½ ± ½	-1.0 ±0.5 -1 ± ½	-0.67 ±0.25 -11/16 ± ¼	CAMBER (Degrees)	-0.7 ±0.5 -11/16 ± ½	-0.92 ±0.5 -15/16 ± ½	-0.92 ±0.5 -15/16 ± ½
0.16° ±0.16° 5/32° ± 5/32°	0.17° ±0.2° 5/32° ± 3/16°	0.0° ±0.24° 0° ± ¼°	REAR TOE (Degrees)	0.32° +0.24° -0.08° 5/16° +¼° -3/32°	0.32° +0.24° -0.08° 5/16° +¼° -3/32°	0.4° ±0.16° 13/32° ± 5/32°
0.08" ±0.08" 3/32" ± 3/32"	0.09" ±0.1" 3/32" ± 3/32"	0.0" ±0.12" 0" ± ⅛"	REAR TOE (Inches)	0.16" +0.12" -0.04" 5/32" +⅛" -1/32"	0.16" +0.12" -0.04" 5/32" +⅛" -1/32"	0.2" ±0.08" 3/16" ± 3/32"
			THRUST ANGLE			0.25 (¼)
177	177		ILLUSTRATION #	06	06	10
			NOTES			
20008	20009	20010		20011	20012	20013

Hyundai

Elantra	Elantra Sedan (MD)	Elantra Sedan (AD)	MODEL INFORMATION	Elantra Coupe	Elantra GT	Elantra Touring
2007-2010	2011-2016	2017	YEARS	2013-2015	2013-2017	2009-2012
			INSTRUCTIONS			
65-80 ft.lbs.	65-80 ft.lbs.	65-80 ft.lbs.	LUGNUT TORQUE	65-80 ft.lbs.	65-80 ft.lbs.	65-80 ft.lbs.
			FRONT			
4.32 ±0.5 4 5/16 ± ½	4.03 ±0.5 4 ± ½	4.5 ±0.5 4½ ± ½	CASTER (Degrees)	4.24 ±0.5 4¼ ± ½	4.24 ±0.5 4¼ ± ½	4.43 ±0.5 4 7/16 ± ½
0.0 (0) | n/s	0.0 (0) | 0.5 (½)	0.0 (0) | 0.5 (½)	CROSS | TOL.	0.0 (0) | 0.5 (½)	0.0 (0) | 0.5 (½)	0.0 (0) | n/s
-0.6 ±0.5 -⅝ ± ½	-0.5 ±0.5 -½ ± ½	-0.55 ±0.5 -9/16 ± ½	CAMBER (Degrees)	-0.5 ±0.5 -½ ± ½	-0.5 ±0.5 -½ ± ½	-0.6 ±0.5 -⅝ ± ½
0.0 (0) | n/s	0.0 (0) | n/s	0.0 (0) | n/s	CROSS | TOL.	0.0 (0) | n/s	0.0 (0) | n/s	0.0 (0) | n/s
0.0° ±0.16° 0° ± 5/32°	0.1° ±0.2° 3/32° ± 3/16°	0.1° ±0.2° 3/32° ± 3/16°	FRONT TOE (Degrees)	0.0° ±0.2° 0° ± 3/16°	0.12° ±0.2° ⅛° ± 3/16°	0.0° ±0.2° 0° ± 3/16°
0.0" ±0.08" 0" ± 3/32"	0.05" ±0.1" 1/16" ± 3/32"	0.05" ±0.1" 1/16" ± 3/32"	FRONT TOE (Inches)	0.0" ±0.1" 0" ± 3/32"	0.06" ±0.1" 1/16" ± 3/32"	0.0" ±0.1" 0" ± 3/32"
13.4 ±0.5 (13⅜ ± ½)	13.8 ±0.5 (13 13/16 ± ½)	14.0 ±0.5 (14 ± ½)	SAI	13.8 ±0.5 (13 13/16 ± ½)	13.8 ±0.5 (13 13/16 ± ½)	13.7 ±0.5 (13⅝ ± ½)
	39.9 / 32.8 ±1.5	39.4 / 32.6 ±1.5	TURNING ANGLE	39.8 / 32.8 ±1.5	39.6 / 32.4 ±1.5	
			ILLUSTRATION #			
			REAR			
-1.0 ±0.5 -1 ± ½	-1.5 ±0.5 -1½ ± ½	-1.2 ±0.5 -1 3/16 ± ½	CAMBER (Degrees)	-1.5 ±0.5 -1½ ± ½	-1.5 ±0.5 -1½ ± ½	-1.15 ±0.5 -1⅛ ± ½
0.17° ±0.16° 3/16° ± 5/32°	0.4° +0.5° -0.4° 13/32° +½° -13/32°	0.34° ±0.3° 11/32° ± 5/16°	REAR TOE (Degrees)	0.5° +0.4° -0.3° ½° +13/32° -5/16°	0.5° +0.5° -0.4° ½° +½° -13/32°	0.1° ±0.2° 3/32° ± 3/16°
0.09" ±0.08" 3/32" ± 3/32"	0.2" +0.25" -0.2" 3/16" +¼" -3/16"	0.17" ±0.15" 5/32" ± 5/32"	REAR TOE (Inches)	0.25" +0.2" -0.15" ¼" +3/16" -5/32"	0.25" +0.25" -0.2" ¼" +¼" -3/16"	0.05" ±0.1" 1/16" ± 3/32"
			THRUST ANGLE			
174			ILLUSTRATION #			174
			NOTES			
20014	20015	20016		20017	20018	20019

Hyundai

Entourage	Equus	Excel	MODEL INFORMATION	Genesis Sedan	Genesis Sedan RWD	Genesis Sedan AWD
2007-2009	2011-2016	1990-94	YEARS	2009-2014	2015-2016	2015-2016
			INSTRUCTIONS			
65-80 ft.lbs.	65-80 ft.lbs.	65-80 ft.lbs.	LUGNUT TORQUE	65-80 ft.lbs.	65-80 ft.lbs.	65-80 ft.lbs.
			FRONT			
4.08 ±0.5 4 1/16 ± ½	7.8 ±0.75 7 13/16 ± ¾	1.67 ±0.5 [1] 1 11/16 ± ½ [1]	CASTER (Degrees)	7.78 ±0.75 7¾ ± ¾	8.35 ±0.75 8⅜ ± ¾	5.65 ±0.75 5⅝ ± ¾
0.0 (0) | 0.5 (½)	0.0 (0) | 0.5 (½)	0.0 (0) | n/s	CROSS | TOL.	0.0 (0) | 0.5 (½)	0.0 (0) | 0.5 (½)	0.0 (0) | 0.5 (½)
0.0 ±0.5 0 ± ½	-0.55 ±0.5 - 9/16 ± ½	0.0 ±0.5 0 ± ½	CAMBER (Degrees)	-0.53 ±0.5 -½ ± ½	-0.5 ±0.5 -½ ± ½	-0.5 ±0.5 -½ ± ½
0.0 (0) | 0.5 (½)	0.0 (0) | n/s	0.0 (0) | 0.5 (½)	CROSS | TOL.	0.0 (0) | 0.5 (½)	0.0 (0) | 0.5 (½)	0.0 (0) | 0.5 (½)
0.0° ±0.16° 0° ± 5/32°	0.0° ±0.2° 0° ± 3/16°	0.08° ±0.24° 3/32° ± ¼°	FRONT TOE (Degrees)	0.1° ±0.2° 3/32° ± 3/16°	0.2° ±0.2° 3/16° ± 3/16°	0.2° ±0.2° 3/16° ± 3/16°
0.0" ±0.08" 0" ± 3/32"	0.0" ±0.1" 0" + 3/32"	0.04" ±0.12" 1/32" ± ⅛"	FRONT TOE (Inches)	0.05" ±0.1" 1/16" ± 3/32"	0.1" ±0.1" 3/32" ± 3/32"	0.1" ±0.1" 3/32" ± 3/32"
12.5 ±0.5 (12 7/16 ± ½)	7.0 (7)	13.0 (13)	SAI	6.9 ±0.5 (6 ⅞ ± ½)	7.2 ±0.5 (7 3/16 ± ½)	6.5 ±0.5 (6 7/16 ± ½)
38.6 / 31.8 ±1.5	42.0 / 33.5 ±1.5	37.4 / 31.5 ±1.5	TURNING ANGLE	41.6 / 34.1 ±1.5 [1]	43.8 / 34.5 ±1.3	43.8 / 34.5 ±1.3
			ILLUSTRATION #			
			REAR			
-0.33 ±0.5 - 5/16 ± ½	-1.45 ±0.5 -1 7/16 ± ½	-0.67 +0.5 -0.33 - 11/16 + ½ - 5/16	CAMBER (Degrees)	-1.43 ±0.5 -1 7/16 ± ½	-1.5 ±0.5 -1½ ± ½	-1.5 ±0.5 -1½ ± ½
0.28° ±0.16° 9/32° ± 5/32°	0.4° ±0.2° 13/32° ± 3/16°	0.0° +0.48° -0.24° 0° + 15/32° - ¼°	REAR TOE (Degrees)	0.4° ±0.2° 13/32° ± 3/16°	0.4° ±0.2° 13/32° ± 3/16°	0.4° ±0.2° 13/32° ± 3/16°
0.14" ±0.08" ⅛" ± 3/32"	0.2" ±0.1" 3/16" ± 3/32"	0.0" +0.24" -0.12" 0" + ¼" - ⅛"	REAR TOE (Inches)	0.2" ±0.1" 3/16" ± 3/32"	0.2" ±0.1" 3/16" ± 3/32"	0.2" ±0.1" 3/16" ± 3/32"
			THRUST ANGLE			
174	179 174		ILLUSTRATION #	177	177	177
		[1] w/MS 1.03° (1 1/16°)	NOTES	[1] w/18" tires = 42.7°/34°		
20020	20021	20023		20024	20025	20026

Hyundai

Genesis Coupe	Santa Fe up to 11/08/04 production	Santa Fe from 11/09/04 production	MODEL INFORMATION	Santa Fe	Santa Fe	Santa Fe Sport
2010-2016	2001-2005	2005-2006	YEARS	2007-2012	2013-2017	2013-2017
			INSTRUCTIONS			
65-80 ft.lbs.	66-81 ft.lbs.	66-81 ft.lbs.	LUGNUT TORQUE	65-80 ft.lbs.	65-80 ft.lbs.	65-80 ft.lbs.
			FRONT			
7.45 ±0.5 7 7/16 ± ½	2.5 ±0.5 2½ ± ½	2.5 ±0.5 2½ ± ½	CASTER (Degrees)	4.5 ±0.5 4½ ± ½	4.14 ±0.5 4⅛ ± ½	4.14 ±0.5 4⅛ ± ½
0.0 (0) | 0.5 (½)	0.0 (0) | 0.5 (½)	0.0 (0) | 0.5 (½)	CROSS | TOL.	0.0 (0) | n/s	0.0 (0) | n/s	0.0 (0) | n/s
-0.5 ±0.5 [1] -½ ± ½ [1]	0.0 ±0.5 0 ± ½	-0.25 ±0.5 -¼ ± ½	CAMBER (Degrees)	-0.5 ±0.5 -½ ± ½	-0.5 ±0.5 -½ ± ½	-0.5 ±0.5 -½ ± ½
0.0 (0) | 0.5 (½)	0.0 (0) | 0.5 (½)	0.0 (0) | 0.5 (½)	CROSS | TOL.	0.0 (0) | n/s	0.0 (0) | n/s	0.0 (0) | n/s
0.28° ±0.2° 9/32° ± 3/16°	-0.16° ±0.16° -5/32° ± 5/32°	-0.16° ±0.16° -5/32° ± 5/32°	FRONT TOE (Degrees)	0.0° ±0.2° 0° ± 3/16°	0.1° ±0.2° 3/32° ± 3/16°	0.1° ±0.2° 3/32° ± 3/16°
0.14" ±0.1" ⅛" ± 3/32"	-0.08" ±0.08" -3/32" ± 3/32"	-0.08" ±0.08" -3/32" ± 3/32"	FRONT TOE (Inches)	0.0" ±0.1" 0" ± 3/32"	0.05" ±0.1" 1/16" ± 3/32"	0.05" ±0.1" 1/16" ± 3/32"
13.7 ±0.5 (13 11/16 ± ½)	13.0 (13)	13.0 (13)	SAI	13.0 ±0.5 (13 ± ½)	13.6 ±0.5 (13⅝ ± ½)	13.6 ±0.5 (13⅝ ± ½)
38.0 / 31.6 ±1.5	37.4 / 32.7 ±1.5	37.4 / 32.7 ±1.5	TURNING ANGLE	39.2 / 31.6 ±1.5	38.8 / 31.9 ±1.5	38.8 / 31.9 ±1.5
			ILLUSTRATION #			
			REAR			
-1.5 ±0.5 -1½ ± ½	0.0 ±0.5 0 ± ½	0.0 ±0.5 0 ± ½	CAMBER (Degrees)	-1.0 ±0.5 -1 ± ½	-1.0 ±0.5 -1 ± ½	-1.0 ±0.5 -1 ± ½
0.16° ±0.2° 5/32° ± 3/16°	0.0° ±0.16° 0° ± 5/32°	0.0° ±0.16° 0° ± 5/32°	REAR TOE (Degrees)	0.2° ±0.2° 3/16° ± 3/16°	0.2° ±0.2° 3/16° ± 3/16°	0.2° ±0.2° 3/16° ± 3/16°
0.08" ±0.1" 3/32" ± 3/32"	0.0" ±0.08" 0" ± 3/32"	0.0" ±0.08" 0" ± 3/32"	REAR TOE (Inches)	0.1" ±0.1" 3/32" ± 3/32"	0.1" ±0.1" 3/32" ± 3/32"	0.1" ±0.1" 3/32" ± 3/32"
			THRUST ANGLE			
177	127	127	ILLUSTRATION #	177	174	174
[1] 2013 = -0.7° (-11/16°)			NOTES			
20027	20029	20030		20031	20032	20033

ALIGNMENT SPECIFICATIONS

Hyundai

MODEL INFORMATION	Scoupe	Sonata	Sonata	Sonata	Sonata	Sonata
YEARS	1991-96	1990-94	1995-98	1999-2005	2006-2010	2011-2014
INSTRUCTIONS						
LUGNUT TORQUE	65-80 ft.lbs.	65-80 ft.lbs.	65-80 ft.lbs.	66-81 ft.lbs.	65-80 ft.lbs.	65-80 ft.lbs.
FRONT						
CASTER (Degrees)	1.73 ±0.5 [1] 1¾ ± ½ [1]	2.0 ±0.5 2 ± ½	2.67 ±0.5 2 11/16 ± ½	3.25 ±1.0 3¼ ±1	4.83 ±0.75 4 13/16 ± ¾	4.44 ±0.5 4 7/16 ± ½
CROSS | TOL.	0.0 (0) | n/s	0.0 (0) | n/s	0.0 (0) | n/s	0.0 (0) | 0.5 (½)	0.0 (0) | 0.5 (½)	0.0 (0) | 0.5 (½)
CAMBER (Degrees)	-0.17 ±0.5 -3/16 ± ½	0.5 ±0.5 ½ ± ½	0.0 ±0.5 0 ± ½	0.0 ±0.5 0 ± ½	0.0 ±0.5 0 ± ½	-0.5 ±0.5 -½ ± ½
CROSS | TOL.	0.0 (0) | 0.5 (½)	0.0 (0) | 0.5 (½)	0.0 (0) | 0.5 (½)	0.0 (0) | 0.5 (½)	0.0 (0) | 0.5 (½)	0.0 (0) | 0.5 (½)
FRONT TOE (Degrees)	0.0° +0.32° -0.16° 0° + 5/16° - 5/32°	0.0° ±0.24° 0° ± ¼°	0.0° ±0.24° 0° ± ¼°	0.0° ±0.16° 0° ± 5/32°	0.0° ±0.16° 0° ± 5/32°	0.16° ±0.2° 5/32° ± 3/16°
FRONT TOE (Inches)	0.0" +0.16" -0.08" 0" + 5/32" - 3/32"	0.0" ±0.12" 0" ± ⅛"	0.0" ±0.12" 0" ± ⅛"	0.0" ±0.08" 0" ± 3/32"	0.0" ±0.08" 0" ± 3/32"	0.08" ±0.1" 3/32" ± 3/32"
SAI	13.2 (13¼)	13.4 (13 7/16)	13.5 (13½)	8.5 (8½)	9.5 (9 7/16)	12.7 ±0.5 (12¾ ± ½)
TURNING ANGLE	37.4 / 31.5 ±1.5	39.9 / 32.0 ±2.0	39.7 / 32.2 ±2.0	40.7 / 32.4 ±2.0	39.2 / 31.6 ±2.0	39.8 / 33.0 ±1.5
ILLUSTRATION #						
REAR						
CAMBER (Degrees)	-0.67 +0.5 -0.33 -11/16 + ½ - 5/16	0.0 ±0.5 0 ± ½	-0.5 ±0.5 -½ ± ½	-0.5 ±0.5 -½ ± ½	-0.5 ±0.5 -½ ± ½	-1.0 ±0.5 -1 ± ½
REAR TOE (Degrees)	0.0° +0.48° -0.24° 0° + 15/32° - ¼°	0.0° ±0.24° 0° ± ¼°	0.05° ±0.16° 1/16° ± 5/32°	0.16° ±0.16° 5/32° ± 5/32°	0.16° ±0.16° 5/32° ± 5/32°	0.17° ±0.2° 5/32° ± 3/16°
REAR TOE (Inches)	0.0" +0.24" -0.12" 0" + ¼" - ⅛"	0.0" ±0.12" 0" ± ⅛"	0.03" ±0.08" 1/32" ± 3/32"	0.08" ±0.08" 3/32" ± 3/32"	0.08" ±0.08" 3/32" ± 3/32"	0.09" ±0.1" 3/32" ± 3/32"
THRUST ANGLE						
ILLUSTRATION #			40 05	06	177	177
NOTES	[1] w/MS 1.08° (1 1/16°)					
	20034	20035	20036	20037	20038	20039

Hyundai

MODEL INFORMATION	Sonata	Sonata HEV	Sonata HEV	Sonata Hybrids	Tiburon	Tiburon
YEARS	2015-2017	2011-2012	2013-2015	2016-2017	1997-98	1999-2001
INSTRUCTIONS						
LUGNUT TORQUE	65-80 ft.lbs.	65-80 ft.lbs.	65-80 ft.lbs.	65-80 ft.lbs.	65-80 ft.lbs.	65-80 ft.lbs.
FRONT						
CASTER (Degrees)	4.68 ±0.5 4 11/16 ± ½	4.34 ±0.5 4 5/16 ± ½	4.34 ±0.5 4 5/16 ± ½	4.68 ±0.5 4 11/16 ± ½	2.35 ±0.5 2⅜ ± ½	2.45 ±0.5 2 7/16 ± ½
CROSS | TOL.	0.0 (0) | 0.5 (½)	0.0 (0) | 0.5 (½)	0.0 (0) | 0.5 (½)	0.0 (0) | 0.5 (½)	0.0 (0) | n/s	0.0 (0) | n/s
CAMBER (Degrees)	-0.5 ±0.5 -½ ± ½	-0.5 ±0.5 -½ ± ½	L= -0.8, R= -0.5 ±0.5 [1] L= -13/16, R= -½ ± ½ [1]	-0.5 ±0.5 -½ ± ½	0.0 ±0.5 0 ± ½	-0.17 ±0.5 -3/16 ± ½
CROSS | TOL.	0.0 (0) | 0.5 (½)	0.0 (0) | 0.5 (½)	-0.3 (-5/16) | 0.5 (½)	0.0 (0) | 0.5 (½)	0.0 (0) | n/s	0.0 (0) | n/s
FRONT TOE (Degrees)	0.0° ±0.2° 5/320° ± 3/16°	0.16° ±0.2° 5/32° ± 3/16°	0.16° ±0.2° 5/32° ± 3/16°	0.0° ±0.2° 0° ± 3/16°	0.0° ±0.24° 0° ± ¼°	0.0° ±0.24° 0° ± ¼°
FRONT TOE (Inches)	0.0" ±0.1" 0" ± 3/32"	0.08" ±0.1" 3/32" ± 3/32"	0.08" ±0.1" 3/32" ± 3/32"	0.0" ±0.1" 0" ± 3/32"	0.0" ±0.12" 0" ± ⅛"	0.0" ±0.12" 0" ± ⅛"
SAI	13.9 ±0.5 (13⅞ ± ½)	12.7 ±0.5 (12¾ ± ½)	12.8 ±0.5 (12 13/16 ± ½)	13.9 ±0.5 (13⅞ ± ½)	12.4 ±0.5 (12⅜ ± ½)	12.7 ±0.5 (12¾ ± ½)
TURNING ANGLE	40.5 / 33.5 ±2.0	39.8 / 33.0 ±1.5	39.8 / 33.0 ±1.5	40.5 / 33.5 ±2.0	39.3 / 33.3 ±1.5	35.6 / 30.5 ±1.5
ILLUSTRATION #						
REAR						
CAMBER (Degrees)	-1.0 ±0.5 -1 ± ½	-1.0 ±0.5 -1 ± ½	-1.0 ±0.5 -1 ± ½	-1.0 ±0.5 -1 ± ½	-0.7 ±0.5 -11/16 ± ½	-0.9 ±0.5 -⅞ ± ½
REAR TOE (Degrees)	0.17° ±0.2° 5/32° ± 3/16°	0.17° ±0.2° 5/32° ± 3/16°	0.17° ±0.2° 5/32° ± 3/16°	0.17° ±0.2° 5/32° ± 3/16°	0.32° ±0.16° 5/16° ± 5/32°	0.32° +0.24° -0.08° 5/16° + ¼° - 3/32°
REAR TOE (Inches)	0.09" ±0.1" 3/32" ± 3/32"	0.09" ±0.1" 3/32" ± 3/32"	0.09" ±0.1" 3/32" ± 3/32"	0.09" ±0.1" 3/32" ± 3/32"	0.16" ±0.08" 5/32" ± 3/32"	0.16" +0.12" -0.04" 5/32" + ⅛" - 1/32"
THRUST ANGLE						
ILLUSTRATION #	177	177	177	177	06	06
NOTES			[1] 16" tires = -0.5° (-½°) (L & R)			
	20040	20041	20042	20043	20046	20047

Hyundai

Tiburon	Tiburon	Tucson	MODEL INFORMATION	Tucson	Tucson	Tucson
2003-2005	2006-2008	2005-2009	YEARS	2010-2013	2014-2015	2016-2017
			INSTRUCTIONS			
65-80 ft.lbs.	67-82 ft.lbs.	66-81 ft.lbs.	LUGNUT TORQUE	65-80 ft.lbs.	65-80 ft.lbs.	65-80 ft.lbs.
			FRONT			
2.38 ±0.5 [1] 2 3/8 ± 1/2 [1]	3.38 ±0.5 3 3/8 ± 1/2	3.53 ±0.5 3 1/2 ± 1/2	CASTER (Degrees)	4.02 ±0.5 4 ± 1/2	4.05 ±0.5 4 1/16 ± 1/2	4.71 ±0.5 4 11/16 ± 1/2
0.0 (0) | 0.5 (1/2)	0.0 (0) | 0.5 (1/2)	0.0 (0) | 0.5 (1/2)	CROSS | TOL.	0.0 (0) | 0.5 (1/2)	0.0 (0) | 0.5 (1/2)	0.0 (0) | 0.5 (1/2)
0.22 ±0.5 3/16 ± 1/2	-0.5 ±0.5 -1/2 ± 1/2	0.0 ±0.5 0 ± 1/2	CAMBER (Degrees)	-0.5 ±0.5 -1/2 ± 1/2	-0.5 ±0.5 -1/2 ± 1/2	-0.5 ±0.5 -1/2 ± 1/2
0.0 (0) | 0.5 (1/2)	0.0 (0) | n/s	0.0 (0) | 0.5 (1/2)	CROSS | TOL.	0.0 (0) | 0.5 (1/2)	0.0 (0) | 0.5 (1/2)	0.0 (0) | 0.5 (1/2)
0.0° ±0.16° 0° ± 5/32°	0.0° ±0.16° 0° ± 5/32°	0.0° ±0.16° 0° ± 5/32°	FRONT TOE (Degrees)	0.0° ±0.2° 0° ± 3/16°	0.1° ±0.2° 3/32° ± 3/16°	0.1° ±0.2° 3/32° ± 3/16°
0.0" ±0.08" 0" ± 3/32"	0.0" ±0.08" 0" ± 3/32"	0.0" ±0.08" 0" ± 3/32"	FRONT TOE (Inches)	0.0" ±0.1" 0" ± 3/32"	0.05" ±0.1" 1/16" ± 3/32"	0.05" ±0.1" 1/16" ± 3/32"
12.7 ±0.5 (12 5/8 ± 1/2)	12.4 ±0.5 (12 7/16 ± 1/2)	12.8 ±0.5 (12 3/4 ± 1/2)	SAI	12.9 ±0.5 (12 15/16 ± 1/2)	13.4 ±0.5 (13 7/16 ± 1/2)	14.0 ±0.5 (13 15/16 ± 1/2)
35.0 / 30.3 ±2.0	35.0 / 30.3 ±2.0	36.8 / 31.8 ±1.5	TURNING ANGLE	39.5 / 31.9 ±1.5	39.4 / 31.9 ±1.5	39.5 / 32.3 ±1.0
			ILLUSTRATION #			
			REAR			
-1.18 ±0.5 -1 3/16 ± 1/2	-1.18 ±0.5 -1 3/16 ± 1/2	-0.92 ±0.5 -15/16 ± 1/2	CAMBER (Degrees)	-1.0 ±0.5 -1 ± 1/2	-1.0 ±0.5 -1 ± 1/2	-1.0 ±0.5 -1 ± 1/2
0.32° +0.24° -0.08° 5/16° + 1/4° - 3/32°	0.32° +0.24° -0.16° 5/16° + 1/4° - 5/32°	0.36° +0.24° -0.08° 3/8° + 1/4° - 3/32°	REAR TOE (Degrees)	0.2° ±0.2° 3/16° ± 3/16°	0.2° ±0.2° 3/16° ± 3/16°	0.2° ±0.2° 3/16° ± 3/16°
0.16" +0.12" -0.04" 5/32" + 1/8" - 1/32"	0.16" +0.12" -0.08" 5/32" + 1/8" - 3/32"	0.18" +0.12" -0.04" 3/16" + 1/8" - 1/32"	REAR TOE (Inches)	0.1" ±0.1" 3/32" ± 3/32"	0.1" ±0.1" 3/32" ± 3/32"	0.1" ±0.1" 3/32" ± 3/32"
			THRUST ANGLE			
10	10	10	ILLUSTRATION #	174	174	174
[1] 2003-04 = 3.0° (3°)			NOTES			
20048	20049	20050		20051	20052	20053

Hyundai

Veloster	Veloster	Veracruz	MODEL INFORMATION	XG300, XG350	Notes	Notes
2012	2013-2017	2007-2013	YEARS	2001-2005		
			INSTRUCTIONS			
65-80 ft.lbs.	65-80 ft.lbs.	65-80 ft.lbs.	LUGNUT TORQUE	66-81 ft.lbs.		
			FRONT			
4.14 ±0.5 4 1/8 ± 1/2	4.1 ±0.5 4 1/8 ± 1/2	4.33 ±0.5 4 5/16 ± 1/2	CASTER (Degrees)	3.25 ±1.0 3 1/4 ±1		
0.0 (0) | n/s	0.0 (0) | n/s	0.0 (0) | n/s	CROSS | TOL.	0.0 (0) | 0.5 (1/2)		
-0.5 ±0.5 -1/2 ± 1/2	-0.5 ±0.5 -1/2 ± 1/2	-0.5 ±0.5 -1/2 ± 1/2	CAMBER (Degrees)	0.0 ±0.5 0 ± 1/2		
0.0 (0) | n/s	0.0 (0) | n/s	0.0 (0) | n/s	CROSS | TOL.	0.0 (0) | 0.5 (1/2)		
0.0° ±0.2° 0° ± 3/16°	0.1° ±0.2° 3/32° ± 3/16°	0.0° ±0.2° 0° ± 3/16°	FRONT TOE (Degrees)	0.0° ±0.16° 0° ± 5/32°		
0.0" ±0.1" 0" ± 3/32"	0.05" ±0.1" 1/16" ± 3/32"	0.0" ±0.1" 0" ± 3/32"	FRONT TOE (Inches)	0.0" ±0.08" 0" ± 3/32"		
13.7 ±0.5 (13 11/16 ± 1/2)	14.0 ±0.5 (13 15/16 ± 1/2)	12.9 ±0.5 (12 15/16 ± 1/2)	SAI	8.5 (8 1/2)		
39.9 / 32.8 ±1.5	39.0 / 32.8 ±1.5 [1]	39.2 / 31.6 ±2.0	TURNING ANGLE	40.0 / 32.0 ±2.0		
			ILLUSTRATION #			
			REAR			
-1.5 ±0.5 -1 1/2 ± 1/2	-1.5 ±0.5 -1 1/2 ± 1/2	-1.0 ±0.5 -1 ± 1/2	CAMBER (Degrees)	-0.5 ±0.5 -1/2 ± 1/2		
0.5° +0.5° -0.4° 1/2° + 1/2° - 13/32°	0.4° +0.5° -0.4° 13/32° + 1/2° - 13/32°	0.2° ±0.2° 3/16° ± 3/16°	REAR TOE (Degrees)	0.16° ±0.16° 5/32° ± 5/32°		
0.25" +0.25" -0.2" 1/4" + 1/4" - 3/16"	0.2" +0.25" -0.2" 3/16" + 1/4" - 3/16"	0.1" ±0.1" 3/32" ± 3/32"	REAR TOE (Inches)	0.08" ±0.08" 3/32" ± 3/32"		
			THRUST ANGLE			
		177	ILLUSTRATION #	06		
	[1] 2017 = 39.4°/32.3°		NOTES			
20054	20055	20056		20057	20058	20059

ALIGNMENT SPECIFICATIONS

Infiniti

	21001	21002	21003	21004	21005	21006
MODEL INFORMATION	EX35 RWD	EX35 AWD	FX35 (RWD & AWD)	FX35	FX37	FX45
YEARS	2008-2013	2008-2013	2003-2008	2009-2012	2013	2003-2008
INSTRUCTIONS						
LUGNUT TORQUE	80 ft.lbs.	80 ft.lbs.	80 ft.lbs.	80 ft.lbs.	80 ft.lbs.	80 ft.lbs.
FRONT						
CASTER (Degrees)	4.25 ±0.75 4¼ ± ¾	4.17 ±0.75 4 3/16 ± ¾	3.78 ±0.75 4¾ ± ¾	3.67 ±0.75 3 11/16 ± ¾	3.67 ±0.75 3 11/16 ± ¾	3.78 ±0.75 4¾ ± ¾
CROSS	0.0 (0)	0.0 (0)	0.0 (0)	0.0 (0)	0.0 (0)	0.0 (0)
TOL.	0.75 (¾)	0.75 (¾)	0.75 (¾)	0.75 (¾)	0.75 (¾)	0.75 (¾)
CAMBER (Degrees)	0.08 ±0.75 1/16 ± ¾	-0.33 ±0.75 - 5/16 ± ¾	-0.73 ±0.75 -¾ ± ¾	-0.33 ±0.75 - 5/16 ± ¾	-0.33 ±0.75 - 5/16 ± ¾	-0.73 ±0.75 -¾ ± ¾
CROSS	0.0 (0)	0.0 (0)	0.0 (0)	0.0 (0)	0.0 (0)	0.0 (0)
TOL.	0.75 (¾)	0.75 (¾)	0.75 (¾)	0.75 (¾)	0.75 (¾)	0.75 (¾)
FRONT TOE (Degrees)	0.08° ±0.15° 3/32° ± 5/32°	0.08° ±0.15° 3/32° ± 5/32°	0.13° ±0.08° ⅛° ± 3/32°	0.15° ±0.15° 5/32° ± 5/32°	0.15° ±0.15° 5/32° ± 5/32°	0.13° ±0.08° ⅛° ± 3/32°
FRONT TOE (Inches)	0.04" ±0.08" 1/32" ± 1/16"	0.04" ±0.08" 1/32" ± 1/16"	0.06" ±0.04" 1/16" ± 1/32"	0.08" ±0.08" 1/16" ± 1/16"	0.08" ±0.08" 1/16" ± 1/16"	0.06" ±0.04" 1/16" ± 1/32"
SAI	6.8 ±0.8 (6 13/16 ± ¾)	7.3 ±0.8 (7 5/16 ± ¾)	13.1 ±0.8 (13 1/16 ± ¾)	8.7 ±0.8 (8 11/16 ± ¾)	8.7 ±0.8 (8 11/16 ± ¾)	13.1 ±0.8 (13 1/16 ± ¾)
TURNING ANGLE	39.4 / 33.6 ±1.5	39.3 / 31.9 ±1.5		41.0 / 32.3 ±1.0	41.0 / 32.3 ±1.0	
ILLUSTRATION #						
REAR						
CAMBER (Degrees)	-0.58 ±0.5 - 9/16 ± ½	-0.58 ±0.5 - 9/16 ± ½	-0.8 ±0.5 - 13/16 ± ½	-1.17 ±0.5 -1 3/16 ± ½	-1.17 ±0.5 -1 3/16 ± ½	-0.8 ±0.5 - 13/16 ± ½
REAR TOE (Degrees)	0.23° ±0.23° 7/32° ± 7/32°	0.23° ±0.23° 7/32° ± 7/32°	0.37° ±0.2° ⅜° ± 3/16°	0.23° ±0.23° 7/32° ± 7/32°	0.23° ±0.23° 7/32° ± 7/32°	0.37° ±0.2° ⅜° ± 3/16°
REAR TOE (Inches)	0.12" ±0.12" ⅛" ± ⅛"	0.12" ±0.12" ⅛" ± ⅛"	0.19" ±0.1" 3/16" ± 3/32"	0.12" ±0.12" ⅛" ± ⅛"	0.12" ±0.12" ⅛" ± ⅛"	0.19" ±0.1" 3/16" ± 3/32"
THRUST ANGLE						
ILLUSTRATION #	151	151	151	151	151	151
NOTES						

Infiniti

	21007	21008	21009	21010	21011	21012
MODEL INFORMATION	FX50	G20	G20	G20	G20	G25 Sedan RWD
YEARS	2009-2013	1991-93	1994-95	1996	1999-2002	2011-2012
INSTRUCTIONS						
LUGNUT TORQUE	80 ft.lbs.				73-86 ft.lbs.	80 ft.lbs.
FRONT						
CASTER (Degrees)	3.67 ±0.75 3 11/16 ± ¾	1.83 ±0.75 1 13/16 ± ¾	1.83 ±0.75 1 13/16 ± ¾	1.83 ±0.75 1 13/16 ± ¾	1.92 ±0.75 1 15/16 ± ¾	4.58 ±0.75 4 9/16 ± ¾
CROSS	0.0 (0)	0.0 (0)	0.0 (0)	0.0 (0)	0.0 (0)	0.0 (0)
TOL.	0.75 (¾)	0.75 (¾)	0.75 (¾)	0.75 (¾)	0.75 (¾)	0.65 (⅝)
CAMBER (Degrees)	-0.33 ±0.75 - 5/16 ± ¾	0.0 ±0.75 0 ± ¾	0.0 ±0.75 0 ± ¾	0.0 ±0.75 0 ± ¾	0.0 ±0.75 0 ± ¾	-0.33 ±0.75 - 5/16 ± ¾
CROSS	0.0 (0)	0.0 (0)	0.0 (0)	0.0 (0)	0.0 (0)	0.0 (0)
TOL.	0.75 (¾)	0.75 (¾)	0.75 (¾)	0.75 (¾)	0.75 (¾)	0.55 (9/16)
FRONT TOE (Degrees)	0.15° ±0.15° 5/32° ± 5/32°	0.1° ±0.1° 3/32° ± 3/32°	0.1° ±0.1° 3/32° ± 3/32°	0.1° ±0.1° 3/32° ± 3/32°	0.1° ±0.1° 3/32° ± 3/32°	0.08° ±0.08° 3/32° ± 3/32°
FRONT TOE (Inches)	0.08" ±0.08" 1/16" ± 1/16"	0.05" ±0.05" 1/16" ± 1/16"	0.05" ±0.05" 1/16" ± 1/16"	0.05" ±0.05" 1/16" ± 1/16"	0.05" ±0.05" 1/16" ± 1/16"	0.04" ±0.04" 1/32" ± 1/32"
SAI	8.7 ±0.8 (8 11/16 ± ¾)	14.5 ±0.8 (14½ ± ¾)	14.5 ±0.8 (14½ ± ¾)	14.5 ±0.8 (14½ ± ¾)	14.5 ±0.8 (14½ ± ¾)	7.3 ±0.8 (7 5/16 ± ¾)
TURNING ANGLE	41.0 / 32.3 ±1.0	35.0 / 30.0 ±2.0	35.0 / 30.0 ±2.0	35.0 / 30.0 ±2.0	34.0 / 29.0 ±1.0	39.8 / 33.5 ±1.0
ILLUSTRATION #						
REAR						
CAMBER (Degrees)	-1.17 ±0.5 -1 3/16 ± ½	-1.0 ±0.75 -1 ± ¾	-1.25 ±0.58 -1¼ ± 9/16	-1.08 ±0.75 -1 1/16 ± ¾	-1.3 ±0.75 -1 5/16 ± ¾	-0.83 ±0.5 - 13/16 ± ½
REAR TOE (Degrees)	0.23° ±0.23° 7/32° ± 7/32°	0.0° ±0.2° 0° ± 3/16°	0.1° ±0.2° 3/32° ± 3/16°	0.08° ±0.17° 3/32° ± 5/32°	0.4° ±0.4° 13/32° ± 13/32°	0.23° ±0.23° 7/32° ± 7/32°
REAR TOE (Inches)	0.12" ±0.12" ⅛" ± ⅛"	0.0" ±0.1" 0" ± 3/32"	0.05" ±0.1" 1/16" ± 3/32"	0.04" ±0.08" 1/32" ± 3/32"	0.2" ±0.2" 3/16" ± 3/16"	0.12" ±0.12" ⅛" ± ⅛"
THRUST ANGLE						
ILLUSTRATION #	151	06	06	06		151
NOTES						

Infiniti

G25x	G25x	G35 Sedan	MODEL	G35x Sedan	G35 Sedan	G35 Sedan
Sedan AWD	Sedan AWD	RWD	INFORMATION	AWD	RWD, 17" tires	RWD, 18" tires
with 17" tires (OE)	with 18" tires (OE)					
2011-2012	2011-2012	2003-2006	YEARS	2004-2006	2007-2008	2007-2008
			INSTRUCTIONS			
80 ft.lbs.	80 ft.lbs.	72-87 ft.lbs.	LUGNUT TORQUE	72-87 ft.lbs.	80 ft.lbs.	80 ft.lbs.
			FRONT			
4.08 ±0.75 4 1/16 ± ¾	4.17 ±0.75 4 3/16 ± ¾	7.75 ±0.75 7 ¾ ± ¾	CASTER (Degrees)	6.67 ±0.75 6 11/16 ± ¾	4.58 ±0.75 4 9/16 ± ¾	4.67 ±0.75 4 11/16 ± ¾
0.0 (0) | 0.65 (5/8)	0.0 (0) | 0.65 (5/8)	0.0 (0) | 0.75 (¾)	CROSS | TOL.	0.0 (0) | 0.75 (¾)	0.0 (0) | 0.75 (¾)	0.0 (0) | 0.75 (¾)
-0.42 ±0.75 - 7/16 ± ¾	-0.42 ±0.75 - 7/16 ± ¾	-0.08 ±0.75 - 1/16 ± ¾	CAMBER (Degrees)	-0.25 ±0.75 - ¼ ± ¾	-0.33 ±0.75 - 5/16 ± ¾	-0.33 ±0.75 - 5/16 ± ¾
0.0 (0) | 0.55 (9/16)	0.0 (0) | 0.55 (9/16)	0.0 (0) | 0.75 (¾)	CROSS | TOL.	0.0 (0) | 0.75 (¾)	0.0 (0) | 0.75 (¾)	0.0 (0) | 0.75 (¾)
0.08° ±0.08° 3/32° ± 3/32°	0.08° ±0.08° 3/32° ± 3/32°	0.08° ±0.08° 3/32° ± 3/32°	FRONT TOE (Degrees)	0.08° ±0.08° 3/32° ± 3/32°	0.08° ±0.08° 3/32° ± 3/32°	0.08° ±0.08° 3/32° ± 3/32°
0.04" ±0.04" 1/32" ± 1/32"	0.04" ±0.04" 1/32" ± 1/32"	0.04" ±0.04" 1/32" ± 1/32"	FRONT TOE (Inches)	0.04" ±0.04" 1/32" ± 1/32"	0.04" ±0.04" 1/32" ± 1/32"	0.04" ±0.04" 1/32" ± 1/32"
7.4 ±0.8 (7 7/16 ± ¾)	7.4 ±0.8 (7 7/16 ± ¾)	4.5 ±0.8 (4 ½ ± ¾)	SAI	6.0 ±0.8 (6 ± ¾)	7.3 ±0.8 (7 5/16 ± ¾)	7.3 ±0.8 (7 5/16 ± ¾)
40.8 / 31.5 ±1.0	40.1 / 31.2 ±1.0	40.5 / 32.5 ±1.0 [1]	TURNING ANGLE	41.0 / 31.7 ±1.0	39.8 / 33.6 ±1.0	39.8 / 33.6 ±1.0
			ILLUSTRATION #			
			REAR			
-0.92 ±0.5 - 15/16 ± ½	-0.92 ±0.5 - 15/16 ± ½	-0.58 ±0.5 [2] - 9/16 ± ½ [2]	CAMBER (Degrees)	-0.58 ±0.5 - 9/16 ± ½	-0.83 ±0.5 [1] - 13/16 ± ½ [1]	-0.83 ±0.5 [1] - 13/16 ± ½ [1]
0.23° ±0.23° 7/32° ± 7/32°	0.23° ±0.23° 7/32° ± 7/32°	0.22° ±0.22° 7/32° ± 7/32°	REAR TOE (Degrees)	0.22° ±0.22° 7/32° ± 7/32°	0.22° ±0.22° 7/32° ± 7/32°	0.22° ±0.22° 7/32° ± 7/32°
0.12" ±0.12" ⅛" ± ⅛"	0.12" ±0.12" ⅛" ± ⅛"	0.11" ±0.11" ⅛" ± ⅛"	REAR TOE (Inches)	0.11" ±0.11" ⅛" ± ⅛"	0.11" ±0.11" ⅛" ± ⅛"	0.11" ±0.11" ⅛" ± ⅛"
			THRUST ANGLE			
151	151	151	ILLUSTRATION #	151	151	151
		[1] w/18" tires = 38.6°/31° [2] w/18" tires = -0.67° (-11/16°)	NOTES		[1] 4WAS = -0.92° (-15/16°) (4 wheel active steering)	[1] 4WAS = -0.92° (-15/16°) (4 wheel active steering)
21013	21014	21015		21016	21017	21018

Infiniti

G35x Sedan	G37 Sedan	G37 Sedan	MODEL	G37x Sedan	G37x Sedan	G35 Coupe
AWD	RWD	RWD	INFORMATION	AWD	AWD	
	exc/Sports Models	Sports Models		with 17" tires (OE)	with 18" tires (OE)	
2007-2008	2009-2014	2009-2014	YEARS	2009-2014	2009-2014	2003
			INSTRUCTIONS			
80 ft.lbs.	80 ft.lbs.	80 ft.lbs.	LUGNUT TORQUE	80 ft.lbs.	80 ft.lbs.	72-87 ft.lbs.
			FRONT			
4.0 ±0.75 4 ± ¾	4.58 ±0.75 4 9/16 ± ¾	4.67 ±0.75 4 11/16 ± ¾	CASTER (Degrees)	4.08 ±0.75 4 1/16 ± ¾	4.17 ±0.75 4 3/16 ± ¾	8.17 ±0.75 [1] 8 3/16 ± ¾ [1]
0.0 (0) | 0.75 (¾)	0.0 (0) | 0.75 (¾)	0.0 (0) | 0.75 (¾)	CROSS | TOL.	0.0 (0) | 0.65 (5/8)	0.0 (0) | 0.65 (5/8)	0.0 (0) | 0.75 (¾)
-0.33 ±0.75 - 5/16 ± ¾	-0.33 ±0.75 - 5/16 ± ¾	-0.33 ±0.75 - 5/16 ± ¾	CAMBER (Degrees)	-0.42 ±0.75 - 7/16 ± ¾	-0.42 ±0.75 - 7/16 ± ¾	-0.5 ±0.75 - ½ ± ¾
0.0 (0) | 0.75 (¾)	0.0 (0) | 0.75 (¾)	0.0 (0) | 0.75 (¾)	CROSS | TOL.	0.0 (0) | 0.55 (9/16)	0.0 (0) | 0.55 (9/16)	0.0 (0) | 0.75 (¾)
0.08° ±0.08° 3/32° ± 3/32°	0.08° ±0.17° 3/32° ± 5/32°	0.08° ±0.17° 3/32° ± 5/32°	FRONT TOE (Degrees)	0.08° ±0.17° 3/32° ± 5/32°	0.08° ±0.17° 3/32° ± 5/32°	0.08° ±0.08° 3/32° ± 3/32°
0.04" ±0.04" 1/32" ± 1/32"	0.04" ±0.08" 1/32" ± 3/32"	0.04" ±0.08" 1/32" ± 3/32"	FRONT TOE (Inches)	0.04" ±0.08" 1/32" ± 3/32"	0.04" ±0.08" 1/32" ± 3/32"	0.04" ±0.04" 1/32" ± 1/32"
7.3 ±0.8 (7 5/16 ± ¾)	7.3 ±0.8 (7 5/16 ± ¾)	7.3 ±0.8 (7 5/16 ± ¾)	SAI	7.4 ±0.8 (7 7/16 ± ¾)	7.4 ±0.8 (7 7/16 ± ¾)	4.9 ±0.8 (4 15/16 ± ¾)
40.9 / 31.5 ±1.0	39.8 / 33.5 ±1.0	39.8 / 33.5 ±1.0	TURNING ANGLE	40.8 / 31.5 ±1.0	40.1 / 31.2 ±1.0	
			ILLUSTRATION #			
			REAR			
-0.25 ±0.5 - ¼ ± ½	-0.83 ±0.5 - 13/16 ± ½	-0.92 ±0.5 - 15/16 ± ½	CAMBER (Degrees)	-0.92 ±0.5 - 15/16 ± ½	-0.92 ±0.5 - 15/16 ± ½	-1.5 ±0.5 -1 ½ ± ½
0.22° ±0.22° 7/32° ± 7/32°	0.23° ±0.23° 7/32° ± 7/32°	0.23° ±0.23° 7/32° ± 7/32°	REAR TOE (Degrees)	0.23° ±0.23° 7/32° ± 7/32°	0.23° ±0.23° 7/32° ± 7/32°	0.22° ±0.22° 7/32° ± 7/32°
0.11" ±0.11" ⅛" ± ⅛"	0.12" ±0.12" ⅛" ± ⅛"	0.12" ±0.12" ⅛" ± ⅛"	REAR TOE (Inches)	0.12" ±0.12" ⅛" ± ⅛"	0.12" ±0.12" ⅛" ± ⅛"	0.11" ±0.11" ⅛" ± ⅛"
			THRUST ANGLE			
151	151	151	ILLUSTRATION #	151	151	151
			NOTES			[1] 18" tires = 8.0° (8°)
21019	21020	21021		21022	21023	21024

Infiniti

G35 Coupe with 17" tires	G35 Coupe with 18" tires	G35 Coupe with 19" tires	MODEL INFORMATION	G37 Coupe RWD	G37x Coupe AWD	G37 Conv. w/225/50R18 tires
2004-2007	2004-2007	2004-2007	YEARS	2008-2013	2009-2013	2009-2013
			INSTRUCTIONS			
72-87 ft.lbs.	72-87 ft.lbs.	72-87 ft.lbs.	LUGNUT TORQUE	80 ft.lbs.	80 ft.lbs.	80 ft.lbs.
			FRONT			
8.17±0.75 8 3/16 ± ¾	8.0±0.75 8 ± ¾	8.08±0.75 8 1/16 ± ¾	CASTER (Degrees)	4.83±0.75 4 13/16 ± ¾	4.25±0.75 4 ¼ ± ¾	4.83±0.75 4 13/16 ± ¾
0.0 (0) | 0.75 (¾)	0.0 (0) | 0.75 (¾)	0.0 (0) | 0.75 (¾)	CROSS | TOL.	0.0 (0) | 0.75 (¾)	0.0 (0) | 0.75 (¾)	0.0 (0) | 0.65 (⅝)
-0.5±0.75 -½ ± ¾	-0.5±0.75 -½ ± ¾	-0.42±0.75 - 7/16 ± ¾	CAMBER (Degrees)	-0.42±0.75 - 7/16 ± ¾	-0.42±0.75 - 7/16 ± ¾	-0.42±0.75 - 7/16 ± ¾
0.0 (0) | 0.75 (¾)	0.0 (0) | 0.75 (¾)	0.0 (0) | 0.75 (¾)	CROSS | TOL.	0.0 (0) | 0.75 (¾)	0.0 (0) | 0.75 (¾)	0.0 (0) | 0.55 (9/16)
0.08°±0.08° 3/32° ± 3/32°	0.08°±0.08° 3/32° ± 3/32°	0.08°±0.08° 3/32° ± 3/32°	FRONT TOE (Degrees)	0.08°±0.17° 3/32° ± 5/32°	0.08°±0.17° 3/32° ± 5/32°	0.08°±0.17° 3/32° ± 5/32°
0.04"±0.04" 1/32" ± 1/32"	0.04"±0.04" 1/32" ± 1/32"	0.04"±0.04" 1/32" ± 1/32"	FRONT TOE (Inches)	0.04"±0.08" 1/32" ± 3/32"	0.04"±0.08" 1/32" ± 3/32"	0.04"±0.08" 1/32" ± 3/32"
4.9±0.8 (4 15/16 ± ¾)	4.9±0.8 (4 15/16 ± ¾)	4.9±0.8 (4 15/16 ± ¾)	SAI	7.4±0.8 (7 7/16 ± ¾)	7.4±0.8 (7 7/16 ± ¾)	7.4±0.8 (7 7/16 ± ¾)
			TURNING ANGLE	39.8 / 33.5 ±1.0	40.6 / 31.5 ±1.0	39.8 / 33.5 ±1.0
			ILLUSTRATION #			
			REAR			
-1.25±0.5 [1] -1 ¼ ± ½ [1]	-1.25±0.5 [1] -1 ¼ ± ½ [1]	-1.25±0.5 [1] -1 ¼ ± ½ [1]	CAMBER (Degrees)	-1.25±0.5 -1 ¼ ± ½	-1.25±0.5 -1 ¼ ± ½	-1.25±0.5 -1 ¼ ± ½
0.22°±0.22° 7/32° ± 7/32°	0.22°±0.22° 7/32° ± 7/32°	0.22°±0.22° 7/32° ± 7/32°	REAR TOE (Degrees)	0.23°±0.23° 7/32° ± 7/32°	0.23°±0.23° 7/32° ± 7/32°	0.23°±0.23° 7/32° ± 7/32°
0.11"±0.11" ⅛" ± ⅛"	0.11"±0.11" ⅛" ± ⅛"	0.11"±0.11" ⅛" ± ⅛"	REAR TOE (Inches)	0.12"±0.12" ⅛" ± ⅛"	0.12"±0.12" ⅛" ± ⅛"	0.12"±0.12" ⅛" ± ⅛"
			THRUST ANGLE			
151	151	151	ILLUSTRATION #	151	151	151
[1] 2006 = -1.42° (-1 7/16°)	[1] 2006 = -1.42° (-1 7/16°)	[1] 2004-06 = -1.42° (-1 7/16°)	NOTES			
21025	21026	21027		21028	21029	21030

Infiniti

G37 Conv. w/225/45R19 tires	I30	I35	MODEL INFORMATION	J30	JX35	M30
2009-2013	1996-2001	2002-2004	YEARS	1993-97	2013	1990-92
			INSTRUCTIONS			
80 ft.lbs.		72-87 ft.lbs.	LUGNUT TORQUE		83 ft.lbs.	
			FRONT			
4.92±0.75 4 15/16 ± ¾	2.75±0.75 2¾ ± ¾	2.75±0.75 2¾ ± ¾	CASTER (Degrees)	6.58±0.75 6 9/16 ± ¾	4.67±0.75 4 11/16 ± ¾	4.67±0.75 4 11/16 ± ¾
0.0 (0) | 0.65 (⅝)	0.0 (0) | 0.75 (¾)	0.0 (0) | 0.75 (¾)	CROSS | TOL.	0.0 (0) | 0.75 (¾)	0.0 (0) | 0.5 (½)	0.0 (0) | 0.75 (¾)
-0.42±0.75 - 7/16 ± ¾	-0.25±0.75 -¼ ± ¾	-0.25±0.75 -¼ ± ¾	CAMBER (Degrees)	-0.75±0.75 -¾ ± ¾	L= -0.25, R= -0.5±0.75 L= -¼, R= -½ ± ¾	0.17±0.75 3/16 ± ¾
0.0 (0) | 0.55 (9/16)	0.0 (0) | 0.75 (¾)	0.0 (0) | 0.75 (¾)	CROSS | TOL.	0.0 (0) | 0.75 (¾)	0.25 (¼) | 0.55 (9/16)	0.0 (0) | 0.75 (¾)
0.08°±0.17° 3/32° ± 5/32°	0.1°±0.1° [1] 3/32° ± 3/32° [1]	0.1°±0.1° 3/32° ± 3/32°	FRONT TOE (Degrees)	0.09°±0.09° 3/32° ± 3/32°	0.1°±0.15° 3/32° ± 5/32°	0.0°±0.1° 0° ± 3/32°
0.04"±0.08" 1/32" ± 3/32"	0.05"±0.05" [1] 1/16" ± 1/16" [1]	0.05"±0.05" 1/16" ± 1/16"	FRONT TOE (Inches)	0.05"±0.05" 1/32" ± 1/32"	0.05"±0.08" 1/16" ± 1/16"	0.0"±0.05" 0" ± 1/16"
7.4±0.8 (7 7/16 ± ¾)	14.3±0.8 (14 ¼ ± ¾)	14.3±0.8 (14 ¼ ± ¾)	SAI	13.4±0.8 (13 7/16 ± ¾)	12.8±0.8 (12 13/16 ± ¾)	12.7±0.8 (12 11/16 ± ¾)
39.8 / 33.5 ±1.0	39.5 / 32.0 ±2.0 [2]	39.5 / 32.0 ±2.0	TURNING ANGLE	37.3 / 31.8 ±2.0	38.0 / 33.0 ±1.0	42.5 / 33.5 ±2.0
			ILLUSTRATION #			
			REAR			
-1.25±0.5 -1 ¼ ± ½	-1.0±0.75 -1 ± ¾	-1.0±0.75 -1 ± ¾	CAMBER (Degrees)	-1.0±0.5 -1 ± ½	-0.58±0.5 - 9/16 ± ½	-0.33±0.75 - 5/16 ± ¾
0.23°±0.23° 7/32° ± 7/32°	0.09°±0.34° 3/32° ± 11/32°	0.09°±0.34° 3/32° ± 11/32°	REAR TOE (Degrees)	0.18°±0.18° 3/16° ± 3/16°	0.16°±0.2° 5/32° ± 3/16°	-0.18°±0.18° -3/16° ± 3/16°
0.12"±0.12" ⅛" ± ⅛"	0.05"±0.17" 1/32" ± 5/32"	0.05"±0.17" 1/32" ± 5/32"	REAR TOE (Inches)	0.09"±0.09" 3/32" ± 3/32"	0.08"±0.1" 3/32" ± 3/32"	-0.09"±0.09" -3/32" ± 3/32"
			THRUST ANGLE			
151			ILLUSTRATION #	65	151	27
	[1] 1996-99 = 0.18° (.09") [2] w/17" wheels 33°/28.5°		NOTES			
21031	21032	21033		21034	21035	21036

Infiniti

M45	M35/M45	M35/M45	MODEL	M37/M56	M37/M56	M37x/M56x
	RWD	AWD	INFORMATION	RWD	RWD	AWD
				exc/Sport Pkg.	w/Sport Pkg.	
2003-2004	2006-2010	2006-2010	YEARS	2011-2013	2011-2013	2011-2013
			INSTRUCTIONS			
72-87 ft.lbs.	83 ft.lbs.	83 ft.lbs.	LUGNUT TORQUE	80 ft.lbs.	80 ft.lbs.	80 ft.lbs.
			FRONT			
6.58 ±0.75 6 9/16 ± ¾	4.5 ±0.75 [1] 4 ½ ± ¾ [1]	3.83 ±0.75 3 13/16 ± ¾	CASTER (Degrees)	4.5 ±1.33 4 ½ ±1 5/16	4.5 ±1.33 4 ½ ±1 5/16	4.0 ±1.33 4 ±1 5/16
0.0 (0) | 0.75 (¾)	0.0 (0) | 0.65 (5/8)	0.0 (0) | 0.65 (5/8)	CROSS | TOL.	0.0 (0) | 0.65 (5/8)	0.0 (0) | 0.65 (5/8)	0.0 (0) | 0.65 (5/8)
-0.67 ±0.75 -11/16 ± ¾	-0.25 ±0.75 -¼ ± ¾	-0.25 ±0.75 -¼ ± ¾	CAMBER (Degrees)	-0.17 ±0.75 -3/16 ± ¾	-0.25 ±0.75 -¼ ± ¾	-0.08 ±0.75 -1/16 ± ¾
0.0 (0) | 0.75 (¾)	0.0 (0) | 0.55 (9/16)	0.0 (0) | 0.55 (9/16)	CROSS | TOL.	0.0 (0) | 0.55 (9/16)	0.0 (0) | 0.55 (9/16)	0.0 (0) | 0.55 (9/16)
0.16° ±0.08° 5/32° ± 3/32°	0.1° ±0.1° 3/32° ± 3/32°	0.1° ±0.1° 3/32° ± 3/32°	FRONT TOE (Degrees)	0.08° ±0.17° 3/32° ± 5/32°	0.08° ±0.17° 3/32° ± 5/32°	0.08° ±0.17° 3/32° ± 5/32°
0.08" ±0.04" 3/32" ± 1/32"	0.05" ±0.05" 1/10" ± 1/16"	0.05" ±0.05" 1/16" ± 1/16"	FRONT TOE (Inches)	0.04" ±0.08" 1/32" ± 3/32"	0.04" ±0.08" 1/32" ± 3/32"	0.04" ±0.08" 1/32" ± 3/32"
13.4 ±0.8 (13 7/16 ± ¾)	7.3 ±0.8 (7 ¼ ± ¾)	7.3 ±0.8 (7 ¼ ± ¾)	SAI	7.2 ±0.8 (7 3/16 ± ¾)	7.3 ±0.8 (7 ¼ ± ¾)	7.1 ±0.8 (7 1/16 ± ¾)
33.9 / 28.8 ±1.0	39.3 / 33.4 ±1.0	42.8 / 32.5 ±1.0	TURNING ANGLE	39.8 / 33.7 ±1.0	39.8 / 33.7 ±1.0	40.3 / 33.6 ±1.0
			ILLUSTRATION #			
			REAR			
-0.67 ±0.5 -11/16 ± ½	-0.67 ±0.5 [2] -11/16 ± ½ [2]	-0.17 ±0.5 -3/16 ± ½	CAMBER (Degrees)	-1.0 ±0.5 -1 ± ½	-1.0 ±0.5 -1 ± ½	-0.5 ±0.5 -½ ± ½
0.2° ±0.2° 3/16° ± 3/16°	0.23° ±0.23° 7/32° ± 7/32°	0.23° ±0.23° 7/32° ± 7/32°	REAR TOE (Degrees)	0.23° ±0.23° 7/32° ± 7/32°	0.23° ±0.23° 7/32° ± 7/32°	0.23° ±0.23° 7/32° ± 7/32°
0.1" ±0.1" 3/32" ± 3/32"	0.12" ±0.12" ⅛" ± ⅛"	0.12" ±0.12" ⅛" ± ⅛"	REAR TOE (Inches)	0.12" ±0.12" ⅛" ± ⅛"	0.12" ±0.12" ⅛" ± ⅛"	0.12" ±0.12" ⅛" ± ⅛"
			THRUST ANGLE			
151	151	151	ILLUSTRATION #	151	191	151
	[1] w/19" tires = 4.58° (4 9/16°) [2] w/19" tires = -0.83° (-13/16°)		NOTES			
21037	21038	21039		21040	21041	21042

Infiniti

M Hybrid	Q40	Q40	MODEL	Q45	Q45	Q45
	RWD	AWD	INFORMATION		STD. Susp.	Active Susp.
2012-2013	2015	2015	YEARS	1990	1991-96	1991-96
			INSTRUCTIONS			
80 ft.lbs.	80 ft.lbs.	80 ft.lbs.	LUGNUT TORQUE			
			FRONT			
4.5 ±1.33 4 ½ ±1 5/16	4.58 ±0.75 4 9/16 ± ¾	4.08 ±0.75 4 1/16 ± ¾	CASTER (Degrees)	6.5 ±0.75 6 ½ ± ¾	6.5 ±0.75 6 ½ ± ¾	6.92 ±0.75 [1] 6 15/16 ± ¾ [1]
0.0 (0) | 0.65 (5/8)	0.0 (0) | 0.65 (5/8)	0.0 (0) | 0.65 (5/8)	CROSS | TOL.	0.0 (0) | 0.75 (¾)	0.0 (0) | 0.75 (¾)	0.0 (0) | 0.75 (¾)
-0.17 ±0.75 -3/16 ± ¾	-0.33 ±0.75 -5/16 ± ¾	-0.42 ±0.75 -7/16 ± ¾	CAMBER (Degrees)	-0.83 ±0.75 -13/16 ± ¾	-0.83 ±0.75 -13/16 ± ¾	-0.92 ±0.75 [1] -15/16 ± ¾ [1]
0.0 (0) | 0.55 (9/16)	0.0 (0) | 0.55 (9/16)	0.0 (0) | 0.55 (9/16)	CROSS | TOL.	0.0 (0) | 0.75 (¾)	0.0 (0) | 0.75 (¾)	0.0 (0) | 0.75 (¾)
0.08° ±0.17° 3/32° ± 5/32°	0.08° ±0.17° 3/32° ± 5/32°	0.08° ±0.17° 3/32° ± 5/32°	FRONT TOE (Degrees)	0.08° ±0.08° 3/32° ± 3/32°	0.08° ±0.08° 3/32° ± 3/32°	0.0° ±0.08° [1] 0° ± 3/32° [1]
0.04" ±0.08" 1/32" ± 3/32"	0.04" ±0.08" 1/32" ± 3/32"	0.04" ±0.08" 1/32" ± 3/32"	FRONT TOE (Inches)*	0.04" ±0.04" 1/32" ± 1/32"	0.04" ±0.04" 1/32" ± 1/32"	0.0" ±0.04" [1] 0" ± 1/32" [1]
7.2 ±0.8 (7 3/16 ± ¾)	7.3 ±0.8 (7 5/16 ± ¾)	7.4 ±0.8 (7 7/16 ± ¾)	SAI	12.8 ±0.8 (12 ¾ ± ¾)	12.8 ±0.8 (12 ¾ ± ¾)	12.9 ±0.8 (12 15/16 ± ¾)
39.8 / 33.7 ±1.0	39.8 / 33.5 ±1.0	40.8 / 31.5 ±1.0	TURNING ANGLE	37.5 / 32.0 ±2.0	37.5 / 32.0 ±2.0	37.0 / 32.0 ±2.0
			ILLUSTRATION #			
			REAR			
-1.0 ±0.5 -1 ± ½	-0.83 ±0.5 -13/16 ± ½	-0.92 ±0.5 -15/16 ± ½	CAMBER (Degrees)	-1.08 ±0.5 -1 1/16 ± ½	-1.08 ±0.5 -1 1/16 ± ½	-1.5 ±0.5 [1] -1 ½ ± ½ [1]
0.23° ±0.23° 7/32° ± 7/32°	0.23° ±0.23° 7/32° ± 7/32°	0.23° ±0.23° 7/32° ± 7/32°	REAR TOE (Degrees)	0.25° ±0.25° ¼° ± ¼°	0.18° ±0.18° 3/16° ± 3/16°	0.18° ±0.18° [1] 3/16° ± 3/16° [1]
0.12" ±0.12" ⅛" ± ⅛"	0.12" ±0.12" ⅛" ± ⅛"	0.12" ±0.12" ⅛" ± ⅛"	REAR TOE (Inches)	0.13" ±0.13" ⅛" ± ⅛"	0.09" ±0.09" 3/32" ± 3/32"	0.09" ±0.09" [1] 3/32" ± 3/32" [1]
			THRUST ANGLE			
151	151	151	ILLUSTRATION #	65	65	65 49
			NOTES			[1] All specs are for running engine.
21043	21044	21045		21046	21047	21048

ALIGNMENT SPECIFICATIONS

Infiniti

Q45	Q45	Q45	MODEL	Q50	Q50	Q50
			INFORMATION	Sedan RWD	Sedan RWD	Sedan AWD
				exc/Sports Models	Sports Models	with 17" tires (OE)
1997-2001	2002	2003-2006	YEARS	2014-2015	2014-2015	2014-2015
			INSTRUCTIONS			
	72-87 ft.lbs.	72-87 ft.lbs.	LUGNUT TORQUE	80 ft.lbs.	80 ft.lbs.	80 ft.lbs.
			FRONT			
6.42 ±0.75 6 7⁄16 ± ¾	6.17 ±0.75 6 3⁄16 ± ¾	6.17 ±0.75 6 3⁄16 ± ¾	CASTER (Degrees)	4.67 ±0.75 4 11⁄16 ± ¾	4.67 ±0.75 4 11⁄16 ± ¾	4.25 ±0.75 4 ¼ ± ¾
0.0 (0) | 1.0 (1)	0.0 (0) | 0.75 (¾)	0.0 (0) | 0.75 (¾)	CROSS | TOL.	0.0 (0) | 0.5 (½)	0.0 (0) | 0.5 (½)	0.0 (0) | 0.5 (½)
-0.67 ±0.75 -1 1⁄16 ± ¾	-0.75 ±0.75 -¾ ± ¾	-0.75 ±0.75 -¾ ± ¾	CAMBER (Degrees)	-0.42 ±0.75 -7⁄16 ± ¾	-0.33 ±0.75 -5⁄16 ± ¾	-0.58 ±0.75 -9⁄16 ± ¾
0.0 (0) | 1.0 (1)	0.0 (0) | 0.75 (¾)	0.0 (0) | 0.75 (¾)	CROSS | TOL.	0.0 (0) | 0.5 (½)	0.0 (0) | 0.5 (½)	0.0 (0) | 0.5 (½)
0.17° ±0.1° 5⁄32° ± 3⁄32°	0.08° ±0.08° 3⁄32° ± 3⁄32°	0.08° ±0.08° 3⁄32° ± 3⁄32°	FRONT TOE (Degrees)	0.08° ±0.17° 3⁄32° ± 5⁄32°	0.08° ±0.17° 3⁄32° ± 5⁄32°	0.08° ±0.17° 3⁄32° ± 5⁄32°
0.08" ±0.05" 3⁄32" ± 1⁄16"	0.04" ±0.04" 1⁄32" ± 1⁄32"	0.04" ±0.04" 1⁄32" ± 1⁄32"	FRONT TOE (Inches)	0.04" ±0.08" 1⁄32" ± 3⁄32"	0.04" ±0.08" 1⁄32" ± 3⁄32"	0.04" ±0.08" 1⁄32" ± 3⁄32"
13.2 ±0.8 (13 3⁄16 ± ¾)	14.0 ±0.8 (14 ± ¾)	14.0 ±0.8 (14 ± ¾)	SAI	7.4 ±0.8 (7 7⁄16 ± ¾)	7.3 ±0.8 (7 5⁄16 ± ¾)	7.5 ±0.8 (7 ½ ± ¾)
39.8 / 32.4 ±1.0	42.8 / 33.8 ±1.0	42.8 / 33.8 ±1.0	TURNING ANGLE	39.8 / 33.5 ±1.0	39.8 / 33.5 ±1.0	40.8 / 31.5 ±1.0
			ILLUSTRATION #			
			REAR			
-0.75 ±0.5 -¾ ± ½	-0.5 ±0.5 [1] -½ ± ½ [1]	-0.42 ±0.5 [1] -7⁄16 ± ½ [1]	CAMBER (Degrees)	-1.17 ±0.5 -1 3⁄16 ± ½	-1.17 ±0.5 -1 3⁄16 ± ½	-1.17 ±0.5 -1 3⁄16 ± ½
0.23° ±0.23° 7⁄32° ± 7⁄32°	0.06° ±0.22° 1⁄16° ± 7⁄32°	0.06° ±0.22° [2] 1⁄16° ± 7⁄32° [2]	REAR TOE (Degrees)	0.24° ±0.23° ¼° ± 7⁄32°	0.24° ±0.23° ¼° ± 7⁄32°	0.24° ±0.23° ¼° ± 7⁄32°
0.12" ±0.12" ⅛" ± ⅛"	0.03" ±0.11" 1⁄32" ± ⅛"	0.03" ±0.11" [2] 1⁄32" ± ⅛" [2]	REAR TOE (Inches)	0.12" ±0.12" ⅛" ± ⅛"	0.12" ±0.12" ⅛" ± ⅛"	0.12" ±0.12" ⅛" ± ⅛"
			THRUST ANGLE			
107 06	63	63	ILLUSTRATION #	177	177	177
	[1] 18" tires = -0.58° (- 9⁄16°)	[1] 18" tires = -0.58° (- 9⁄16°) [2] 18" tires = 0.09° (0.045")	NOTES			
21049	21050	21051		21052	21053	21054

Infiniti

Q50	Q50 HEV	Q50 HEV	MODEL	Q50	Q50	Q50
Sedan AWD	Sedan RWD	Sedan AWD	INFORMATION	Sedan RWD	Sedan RWD	Sedan AWD
with 18" tires (OE)				except Turbo	Turbo	except Turbo
2014-2015	2014-2017	2014-2017	YEARS	2016-2017	2016-2017	2016-2017
			INSTRUCTIONS			
80 ft.lbs.	80 ft.lbs.	80 ft.lbs.	LUGNUT TORQUE	80 ft.lbs.	80 ft.lbs.	80 ft.lbs.
			FRONT			
4.25 ±0.75 4 ¼ ± ¾	4.67 ±1.33 4 11⁄16 ±1 5⁄16	4.3 ±1.33 4 5⁄16 ±1 5⁄16	CASTER (Degrees)	4.33 ±1.33 4 5⁄16 ±1 5⁄16	4.75 ±1.33 4 ¾ ±1 5⁄16	4.33 ±1.33 4 5⁄16 ±1 5⁄16
0.0 (0) | 0.5 (½)	0.0 (0) | 0.65 (⅝)	0.0 (0) | 0.65 (⅝)	CROSS | TOL.	0.0 (0) | 0.5 (½)	0.0 (0) | 0.5 (½)	0.0 (0) | 0.5 (½)
-0.58 ±0.75 -9⁄16 ± ¾	-0.33 ±0.75 -5⁄16 ± ¾	-0.5 ±0.75 -½ ± ¾	CAMBER (Degrees)	-0.58 ±0.75 -9⁄16 ± ¾	-0.33 ±0.75 -5⁄16 ± ¾	-0.58 ±0.75 -9⁄16 ± ¾
0.0 (0) | 0.5 (½)	0.0 (0) | 0.55 (9⁄16)	0.0 (0) | 0.55 (9⁄16)	CROSS | TOL.	0.0 (0) | 0.5 (½)	0.0 (0) | 0.5 (½)	0.0 (0) | 0.5 (½)
0.08° ±0.17° 3⁄32° ± 5⁄32°	0.17° ±0.17° 5⁄32° ± 5⁄32°	0.17° ±0.17° 5⁄32° ± 5⁄32°	FRONT TOE (Degrees)	0.17° ±0.17° 5⁄32° ± 5⁄32°	0.17° ±0.17° 5⁄32° ± 5⁄32°	0.17° ±0.17° 5⁄32° ± 5⁄32°
0.04" ±0.08" 1⁄32" ± 3⁄32"	0.08" ±0.08" 3⁄32" ± 3⁄32"	0.08" ±0.08" 3⁄32" ± 3⁄32"	FRONT TOE (Inches)	0.08" ±0.08" 3⁄32" ± 3⁄32"	0.08" ±0.08" 3⁄32" ± 3⁄32"	0.08" ±0.08" 3⁄32" ± 3⁄32"
7.5 ±0.8 (7 ½ ± ¾)	7.3 ±0.8 (7 5⁄16 ± ¾)	7.5 ±0.8 (7 ½ ± ¾)	SAI	7.5 ±0.8 (7 ½ ± ¾)	7.3 ±0.8 (7 5⁄16 ± ¾)	7.5 ±0.8 (7 ½ ± ¾)
40.1 / 31.2 ±1.0	38.3 / 32.6 ±1.0	39.0 / 30.7 ±1.0	TURNING ANGLE	38.3 / 32.6 ±1.0	38.3 / 32.6 ±1.0	39.0 / 30.7 ±1.0
			ILLUSTRATION #			
			REAR			
-1.17 ±0.5 -1 3⁄16 ± ½	-1.17 ±0.5 -1 3⁄16 ± ½	-1.17 ±0.5 -1 3⁄16 ± ½	CAMBER (Degrees)	-0.92 ±0.5 -15⁄16 ± ½	-1.17 ±0.5 -1 3⁄16 ± ½	-1.17 ±0.5 -1 3⁄16 ± ½
0.24° ±0.23° ¼° ± 7⁄32°	0.23° ±0.23° 7⁄32° ± 7⁄32°	0.23° ±0.23° 7⁄32° ± 7⁄32°	REAR TOE (Degrees)	0.23° ±0.23° 7⁄32° ± 7⁄32°	0.23° ±0.23° 7⁄32° ± 7⁄32°	0.23° ±0.23° 7⁄32° ± 7⁄32°
0.12" ±0.12" ⅛" ± ⅛"	0.12" ±0.12" ⅛" ± ⅛"	0.12" ±0.12" ⅛" ± ⅛"	REAR TOE (Inches)	0.12" ±0.12" ⅛" ± ⅛"	0.12" ±0.12" ⅛" ± ⅛"	0.12" ±0.12" ⅛" ± ⅛"
			THRUST ANGLE			
177	177	177	ILLUSTRATION #	177	177	177
			NOTES			
21055	21056	21057		21058	21059	21060

Infiniti

Q50	Q60 Conv.	Q60 Conv.	MODEL	Q60 Coupe	Q60 Coupe	Q60
Sedan AWD	with 18" tires (OE)	with 19" tires (OE)	INFORMATION	RWD	AWD	RWD 2.0L
Turbo						
2016-2017	2014-2015	2014-2015	YEARS	2014-2015	2014-2015	2017
			INSTRUCTIONS			
80 ft.lbs.	80 ft.lbs.	80 ft.lbs.	LUGNUT TORQUE	80 ft.lbs.	80 ft.lbs.	80 ft.lbs.
			FRONT			
4.25 ± 1.33 4 1/4 ± 1 5/16	4.83 ± 0.75 4 13/16 ± 3/4	4.92 ± 0.75 4 15/16 ± 3/4	CASTER (Degrees)	4.83 ± 0.75 4 13/16 ± 3/4	4.25 ± 0.75 4 1/4 ± 3/4	4.5 ± 1.33 4 1/2 ± 1 5/16
0.0 (0) | 0.5 (1/2)	0.0 (0) | 0.65 (5/8)	0.0 (0) | 0.65 (5/8)	CROSS | TOL.	0.0 (0) | 0.65 (5/8)	0.0 (0) | 0.65 (5/8)	0.0 (0) | 0.5 (1/2)
-0.58 ± 0.75 - 9/16 ± 3/4	-0.42 ± 0.75 - 7/16 ± 3/4	-0.42 ± 0.75 - 7/16 ± 3/4	CAMBER (Degrees)	-0.42 ± 0.75 - 7/16 ± 3/4	-0.42 ± 0.75 - 7/16 ± 3/4	-0.58 ± 0.75 - 9/16 ± 3/4
0.0 (0) | 0.5 (1/2)	0.0 (0) | 0.55 (9/16)	0.0 (0) | 0.55 (9/16)	CROSS | TOL.	0.0 (0) | 0.55 (9/16)	0.0 (0) | 0.55 (9/16)	0.0 (0) | 0.5 (1/2)
0.17° ± 0.17° 5/32° ± 5/32°	0.08° ± 0.17° 3/32° ± 5/32°	0.08° ± 0.17° 3/32° ± 5/32°	FRONT TOE (Degrees)	0.08° ± 0.17° 3/32° ± 5/32°	0.08° ± 0.17° 3/32° ± 5/32°	0.08° ± 0.17° 3/32° ± 5/32°
0.08" ± 0.08" 3/32" ± 3/32"	0.04" ± 0.08" 1/32" ± 3/32"	0.04" ± 0.08" 1/32" ± 3/32"	FRONT TOE (Inches)	0.04" ± 0.08" 1/32" ± 3/32"	0.04" ± 0.08" 1/32" ± 3/32"	0.04" ± 0.08" 1/32" ± 3/32"
7.5 ± 0.8 (7 1/2 ± 3/4)	7.4 ± 0.8 (7 7/16 ± 3/4)	7.4 ± 0.8 (7 7/16 ± 3/4)	SAI	7.4 ± 0.8 (7 7/16 ± 3/4)	7.4 ± 0.8 (7 7/16 ± 3/4)	7.5 ± 0.8 (7 1/2 ± 3/4)
39.0 / 30.7 ±1.0	39.8 / 33.5 ±1.0	39.8 / 33.5 ±1.0	TURNING ANGLE	39.8 / 33.5 ±1.0	40.6 / 31.5 ±1.0	38.3 / 32.6 ±1.0
			ILLUSTRATION #			
			REAR			
-1.17 ± 0.5 -1 3/16 ± 1/2	-1.33 ± 0.5 -1 5/16 ± 1/2	-1.33 ± 0.5 -1 5/16 ± 1/2	CAMBER (Degrees)	-1.25 ± 0.5 -1 1/4 ± 1/2	-1.25 ± 0.5 -1 1/4 ± 1/2	-1.08 ± 0.5 -1 1/16 ± 1/2
0.23° ± 0.23° 7/32° ± 7/32°	0.23° ± 0.23° 7/32° ± 7/32°	0.23° ± 0.23° 7/32° ± 7/32°	REAR TOE (Degrees)	0.23° ± 0.23° 7/32° ± 7/32°	0.23° ± 0.23° 7/32° ± 7/32°	0.23° ± 0.23° 7/32° ± 7/32°
0.12" ± 0.12" 1/8" ± 1/8"	0.12" ± 0.12" 1/8" ± 1/8"	0.12" ± 0.12" 1/8" ± 1/8"	REAR TOE (Inches)	0.12" ± 0.12" 1/8" ± 1/8"	0.12" ± 0.12" 1/8" ± 1/8"	0.12" ± 0.12" 1/8" ± 1/8"
			THRUST ANGLE			
177	151	151	ILLUSTRATION #	151	151	177
			NOTES			
21061	21062	21063		21064	21065	21066

Infiniti

Q60	Q60	Q60	MODEL	Q70	Q70	Q70
RWD 3.0L	AWD 2.0L	AWD 3.0L	INFORMATION	RWD	RWD	AWD
				exc/Sport	Sport	
2017	2017	2017	YEARS	2014-2017	2014-2017	2014-2016
			INSTRUCTIONS			
80 ft.lbs.	80 ft.lbs.	80 ft.lbs.	LUGNUT TORQUE	80 ft.lbs.	80 ft.lbs.	80 ft.lbs.
			FRONT			
4.83 ± 1.33 4 13/16 ± 1 5/16	4.5 ± 1.33 [1] 4 1/2 ± 1 5/16 [1]	4.42 ± 1.33 4 7/16 ± 1 5/16	CASTER (Degrees)	4.5 ± 1.33 4 1/2 ± 1 5/16	4.5 ± 1.33 4 1/2 ± 1 5/16	4.0 ± 1.33 4 ± 1 5/16
0.0 (0) | 0.5 (1/2)	0.0 (0) | 0.5 (1/2)	0.0 (0) | 0.5 (1/2)	CROSS | TOL.	0.0 (0) | 0.65 (5/8)	0.0 (0) | 0.65 (5/8)	0.0 (0) | 0.65 (5/8)
-0.42 ± 0.75 - 7/16 ± 3/4	-0.58 ± 0.75 - 9/16 ± 3/4	-0.58 ± 0.75 - 9/16 ± 3/4	CAMBER (Degrees)	-0.17 ± 0.75 - 3/16 ± 3/4	-0.25 ± 0.75 - 1/4 ± 3/4	-0.08 ± 0.75 - 1/16 ± 3/4
0.0 (0) | 0.5 (1/2)	0.0 (0) | 0.5 (1/2)	0.0 (0) | 0.5 (1/2)	CROSS | TOL.	0.0 (0) | 0.55 (9/16)	0.0 (0) | 0.55 (9/16)	0.0 (0) | 0.55 (9/16)
0.08° ± 0.17° 3/32° ± 5/32°	0.08° ± 0.17° 3/32° ± 5/32°	0.08° ± 0.17° 3/32° ± 5/32°	FRONT TOE (Degrees)	0.08° ± 0.17° 3/32° ± 5/32°	0.08° ± 0.17° 3/32° ± 5/32°	0.08° ± 0.17° 3/32° ± 5/32°
0.04" ± 0.08" 1/32" ± 3/32"	0.04" ± 0.08" 1/32" ± 3/32"	0.04" ± 0.08" 1/32" ± 3/32"	FRONT TOE (Inches)	0.04" ± 0.08" 1/32" ± 3/32"	0.04" ± 0.08" 1/32" ± 3/32"	0.04" ± 0.08" 1/32" ± 3/32"
7.4 ± 0.8 (7 7/16 ± 3/4)	7.5 ± 0.8 (7 1/2 ± 3/4)	7.5 ± 0.8 (7 1/2 ± 3/4)	SAI	7.2 ± 0.8 (7 3/16 ± 3/4)	7.3 ± 0.8 (7 1/4 ± 3/4)	7.1 ± 0.8 (7 1/16 ± 3/4)
38.3 / 32.6 ±1.0	39.0 / 30.6 ±1.0	39.0 / 30.6 ±1.0	TURNING ANGLE	39.8 / 33.7 ±1.0	39.8 / 33.7 ±1.0	40.3 / 33.6 ±1.0
			ILLUSTRATION #			
			REAR			
-1.42 ± 0.5 [1] -1 7/16 ± 1/2 [1]	-1.08 ± 0.5 -1 1/16 ± 1/2	-1.42 ± 0.5 [1] -1 7/16 ± 1/2 [1]	CAMBER (Degrees)	-1.0 ± 0.5 -1 ± 1/2	-1.0 ± 0.5 -1 ± 1/2	-0.5 ± 0.5 -1/2 ± 1/2
0.23° ± 0.23° 7/32° ± 7/32°	0.23° ± 0.23° 7/32° ± 7/32°	0.23° ± 0.23° 7/32° ± 7/32°	REAR TOE (Degrees)	0.23° ± 0.23° 7/32° ± 7/32°	0.23° ± 0.23° 7/32° ± 7/32°	0.23° ± 0.23° 7/32° ± 7/32°
0.12" ± 0.12" 1/8" ± 1/8"	0.12" ± 0.12" 1/8" ± 1/8"	0.12" ± 0.12" 1/8" ± 1/8"	REAR TOE (Inches)	0.12" ± 0.12" 1/8" ± 1/8"	0.12" ± 0.12" 1/8" ± 1/8"	0.12" ± 0.12" 1/8" ± 1/8"
			THRUST ANGLE			
177	177	177	ILLUSTRATION #	177	177	177
[1] Canada = -1.33°	[1] Canada = 4.42°	[1] Canada = -1.33°	NOTES	[1] 2017 = 0.13°(0.07")	[1] 2017 = 0.13°(0.07")	
21067	21068	21069		21070	21071	21072

ALIGNMENT SPECIFICATIONS

Infiniti

Q70	Q70	Q70 HEV	MODEL	QX4	QX30	QX30
AWD w/18" wheels	AWD w/20" wheels		INFORMATION		RWD exc/Sport	RWD Sport
2017	2017	2014-2017	YEARS	1997-2003	2017	2017
			INSTRUCTIONS			
80 ft.lbs.	80 ft.lbs.	80 ft.lbs.	LUGNUT TORQUE	87-108 ft.lbs.	96 ft.lbs.	96 ft.lbs.
			FRONT			
4.0 ± 1.33 4 ± 1 5/16	4.0 ± 1.33 4 ± 1 5/16	4.5 ± 1.33 4 1/2 ± 1 5/16	CASTER (Degrees)	3.0 ± 0.75 3 ± 3/4	6.25 ± 0.75 6 1/4 ± 3/4	6.58 ± 0.75 6 9/16 ± 3/4
0.0 (0) | 0.65 (5/8)	0.0 (0) | 0.65 (5/8)	0.0 (0) | 0.65 (5/8)	CROSS | TOL.	0.0 (0) | 0.75 (3/4)	0.0 (0) | 0.5 (1/2)	0.0 (0) | 0.5 (1/2)
-0.08 ± 0.75 -1/16 ± 3/4	-0.08 ± 0.75 -1/16 ± 3/4	-0.17 ± 0.75 -3/16 ± 3/4	CAMBER (Degrees)	0.17 ± 0.75 3/16 ± 3/4	-0.83 ± 0.75 -13/16 ± 3/4	-1.0 ± 0.75 -1 ± 3/4
0.0 (0) | 0.55 (9/16)	0.0 (0) | 0.55 (9/16)	0.0 (0) | 0.55 (9/16)	CROSS | TOL.	0.0 (0) | 0.75 (3/4)	0.0 (0) | 0.5 (1/2)	0.0 (0) | 0.5 (1/2)
0.13° ± 0.17° 1/8° ± 5/32°	0.13° ± 0.17° 1/8° ± 5/32°	0.08° ± 0.17° [1] 3/32° ± 5/32° [1]	FRONT TOE (Degrees)	0.17° ± 0.08° 5/32° ± 3/32°	0.23° ± 0.08° 7/32° ± 3/32°	0.22° ± 0.08° 7/32° ± 5/32°
0.07" ± 0.08" 1/32" ± 3/32"	0.07" ± 0.08" 1/32" ± 3/32"	0.04" ± 0.08" [1] 1/32" ± 3/32" [1]	FRONT TOE (Inches)	0.08" ± 0.04" 3/32" ± 1/32"	0.12" ± 0.04" 1/8" ± 1/32"	0.11" ± 0.04" 1/8" ± 1/32"
7.1 ± 0.8 (7 1/16 ± 3/4)	7.1 ± 0.8 (7 1/16 ± 3/4)	7.2 ± 0.8 (7 3/16 ± 3/4)	SAI	14.3 ± 0.8 (14 5/16 ± 3/4)	10.5 ± 0.8 (10 1/2 ± 3/4)	10.7 ± 0.8 (10 11/16 ± 3/4)
40.3 / 33.6 ± 1.0	40.3 / 33.6 ± 1.0	39.8 / 33.7 ± 1.0	TURNING ANGLE	33.0 / 31.0 ± 1.0	36.8 / 31.8 ± 1.0	36.5 / 31.5 ± 1.0
			ILLUSTRATION #			
			REAR			
-0.5 ± 0.5 -1/2 ± 1/2	-0.5 ± 0.5 -1/2 ± 1/2	-1.0 ± 0.5 -1 ± 1/2	CAMBER (Degrees)		-0.75 ± 0.5 -3/4 ± 1/2	-1.25 ± 0.5 -1 1/4 ± 1/2
0.23° ± 0.23° 7/32° ± 7/32°	0.39° +0.08° -0.23° 3/8° +3/32° -7/32°	0.23° ± 0.23° 7/32° ± 7/32°	REAR TOE (Degrees)		0.03° ± 0.07° 1/32° ± 1/16°	0.1° ± 0.07° 3/32° ± 1/16°
0.12" ± 0.12" 1/8" ± 1/8"	0.2" +0.04" -0.12" 3/16" +1/32" -1/8"	0.12" ± 0.12" 1/8" ± 1/8"	REAR TOE (Inches)		0.02" ± 0.04" 1/64" ± 1/32"	0.05" ± 0.04" 1/16" ± 1/32"
			THRUST ANGLE			
177	177	177	ILLUSTRATION #		174	174
		[1] 2017 = 0.13°(0.07")	NOTES			
21073	21074	21075		21076	21077	21078

Infiniti

QX30	QX50	QX50	MODEL	QX50	QX50	QX56
AWD	RWD	AWD	INFORMATION	RWD	AWD	2WD
2017	2014-2015	2014-2015	YEARS	2016-2017	2016-2017	2004-2010
			INSTRUCTIONS			
96 ft.lbs.	80 ft.lbs.	80 ft.lbs.	LUGNUT TORQUE	80 ft.lbs.	80 ft.lbs.	98 ft.lbs.
			FRONT			
5.92 ± 0.75 5 15/16 ± 3/4	4.25 ± 0.75 4 1/4 ± 3/4	4.17 ± 0.75 4 3/16 ± 3/4	CASTER (Degrees)	3.83 ± 0.75 3 13/16 ± 3/4	3.67 ± 0.75 3 11/16 ± 3/4	4.0 ± 0.75 4 ± 3/4
0.0 (0) | 0.5 (1/2)	0.0 (0) | 0.75 (3/4)	0.0 (0) | 0.75 (3/4)	CROSS | TOL.	0.0 (0) | 0.75 (3/4)	0.0 (0) | 0.75 (3/4)	0.0 (0) | 0.75 (3/4)
-0.83 ± 0.75 -13/16 ± 3/4	0.08 ± 0.75 1/16 ± 3/4	-0.33 ± 0.75 -5/16 ± 3/4	CAMBER (Degrees)	0.58 ± 0.75 9/16 ± 3/4	0.08 ± 0.75 1/16 ± 3/4	-0.1 ± 0.75 -1/8 ± 3/4
0.0 (0) | 0.5 (1/2)	0.0 (0) | 0.75 (3/4)	0.0 (0) | 0.75 (3/4)	CROSS | TOL.	0.0 (0) | 0.75 (3/4)	0.0 (0) | 0.75 (3/4)	0.0 (0) | 0.75 (3/4)
0.23° ± 0.08° 7/32° ± 5/32°	0.08° ± 0.15° 3/32° ± 5/32°	0.08° ± 0.15° 3/32° ± 5/32°	FRONT TOE (Degrees)	0.08° ± 0.15° 3/32° ± 5/32°	0.08° ± 0.15° 3/32° ± 5/32°	0.17° ± 0.1° 5/32° ± 3/32°
0.12" ± 0.04" 1/8" ± 1/32"	0.04" ± 0.08" 1/32" ± 1/16"	0.04" ± 0.08" 1/32" ± 1/16"	FRONT TOE (Inches)	0.04" ± 0.08" 1/32" ± 1/16"	0.04" ± 0.08" 1/32" ± 1/16"	0.08" ± 0.05" 3/32" ± 1/16"
10.2 ± 0.8 (10 3/16 ± 3/4)	6.8 ± 0.8 (6 13/16 ± 3/4)	7.3 ± 0.8 (7 5/16 ± 3/4)	SAI	6.4 ± 0.8 (6 7/16 ± 3/4)	6.9 ± 0.8 (6 15/16 ± 3/4)	13.5 ± 0.8 (13 1/2 ± 3/4)
37.3 / 32.3 ± 1.0	38.6 / 33.1 ± 1.0	38.0 / 30.3 ± 1.0	TURNING ANGLE	38.8 / 33.2 ± 1.0	38.2 / 30.4 ± 1.0	37.5 / 34.0 ± 1.0
			ILLUSTRATION #			12
			REAR			
-0.83 ± 0.5 -13/16 ± 1/2	-0.58 ± 0.5 -9/16 ± 1/2	-0.58 ± 0.5 -9/16 ± 1/2	CAMBER (Degrees)	-0.58 ± 0.5 -9/16 ± 1/2	-0.58 ± 0.5 -9/16 ± 1/2	-0.5 ± 0.5 -1/2 ± 1/2
0.04° ± 0.07° 1/32° ± 1/16°	0.23° ± 0.23° 7/32° ± 7/32°	0.23° ± 0.23° 7/32° ± 7/32°	REAR TOE (Degrees)	0.23° ± 0.23° 7/32° ± 7/32°	0.23° ± 0.23° 7/32° ± 7/32°	0.23° ± 0.23° 7/32° ± 7/32°
0.02" ± 0.04" 1/64" ± 1/32"	0.12" ± 0.12" 1/8" ± 1/8"	0.12" ± 0.12" 1/8" ± 1/8"	REAR TOE (Inches)	0.12" ± 0.12" 1/8" ± 1/8"	0.12" ± 0.12" 1/8" ± 1/8"	0.12" ± 0.12" 1/8" ± 1/8"
			THRUST ANGLE			
174	151	151	ILLUSTRATION #	151	151	154
			NOTES			
21079	21080	21081		21082	21083	21084

ALIGNMENT SPECIFICATIONS

Infiniti

Model	QX56	QX56	QX60	QX60 HEV	QX70	QX80
Information	4WD					
Years	2004-2010	2011-2013	2014-2017	2014-2015	2014-2017	2014-2017
Instructions						
LugNut Torque	98 ft.lbs.	98 ft.lbs.	83 ft.lbs.	83 ft.lbs.	80 ft.lbs.	98 ft.lbs.
FRONT						
Caster (Degrees)	3.5 ±0.75 3½ ± ¾	L= 3.08, R= 3.42 ± 0.75 L= 3 1/16, R= 3 7/16 ± ¾	4.67 ±0.75 4 11/16 ± ¾	4.75 ±0.75 4¾ ± ¾	3.67 ±0.75 3 11/16 ± ¾	L= 3.08, R= 3.42 ± 0.75 L= 3 1/16, R= 3 7/16 ± ¾
Cross | Tol.	0.0 (0) | 0.75 (¾)	-0.33 (-5/16) | 0.75 (¾)	0.0 (0) | 0.5 (½)	0.0 (0) | 0.5 (½)	0.0 (0) | 0.75 (¾)	-0.33 (-5/16) | 0.75 (¾)
Camber (Degrees)	0.2 ±0.75 3/16 ± ¾	L= 0.0, R= -0.17 ±0.75 L= 0, R= -3/16 ± ¾	L= -0.25, R= -0.5 ±0.75 L= -¼, R= -½ ± ¾	L= -0.25, R= -0.5 ±0.75 L= -¼, R= -½ ± ¾	-0.33 ±0.75 -5/16 ± ¾	L= 0.0, R= -0.17 ±0.75 L= 0, R= -3/16 ± ¾
Cross | Tol.	0.0 (0) | 0.75 (¾)	0.17 (3/16) | 0.75 (¾)	0.25 (¼) | 0.55 (9/16)	0.25 (¼) | 0.55 (9/16)	0.0 (0) | 0.75 (¾)	0.17 (3/16) | 0.75 (¾)
Front Toe (Degrees)	0.17° ±0.1° 5/32° ± 3/32°	0.17° ±0.14° 5/32° ± 1/8°	0.1° ±0.15° 3/32° ± 5/32°	0.1° ±0.15° 3/32° ± 5/32°	0.15° ±0.15° 5/32° ± 5/32°	0.22° ±0.18° ¼° ± 3/16°
Front Toe (Inches)	0.08" ±0.05" 3/32" ± 1/16"	0.09" ±0.07" 3/32" ± 1/16"	0.05" ±0.08" 1/16" ± 1/16"	0.05" ±0.08" 1/16" ± 1/16"	0.08" ±0.08" 1/16" ± 1/16"	0.11" ±0.09" 1/8" ± 3/32"
SAI	13.2 ±0.8 (13 3/16 ± ¾)	13.4 ±0.8 (13 3/8 ± ¾)	12.8 ±0.8 (12 13/16 ± ¾)	12.7 ±0.8 (12 11/16 ± ¾)	8.7 ±0.8 (8 11/16 ± ¾)	13.4 ±0.8 (13 3/8 ± ¾)
Turning Angle	37.7 / 33.5 ±1.0	36.6 / 32.5 ±1.0	38.0 / 33.0 ±1.0	38.0 / 33.0 ±1.0	41.0 / 32.3 ±1.0	37.0 / 32.5 ±1.0
Illustration #	12	12				12
REAR						
Camber (Degrees)	-0.5 ±0.5 -½ ± ½	-0.5 ±0.5 -½ ± ½	-0.58 ±0.5 -9/16 ± ½	-0.58 ±0.5 -9/16 ± ½	-1.17 ±0.5 -1 3/16 ± ½	-0.5 ±0.5 -½ ± ½
Rear Toe (Degrees)	0.23° ±0.23° 7/32° ± 7/32°	0.23° ±0.23° 7/32° ± 7/32°	0.16° ±0.2° 5/32° ± 3/16°	0.16° ±0.2° 5/32° ± 3/16°	0.23° ±0.23° 7/32° ± 7/32°	0.23° ±0.23° 7/32° ± 7/32°
Rear Toe (Inches)	0.12" ±0.12" 1/8" ± 1/8"	0.12" ±0.12" 1/8" ± 1/8"	0.08" ±0.1" 3/32" ± 3/32"	0.08" ±0.1" 3/32" ± 3/32"	0.12" ±0.12" 1/8" ± 1/8"	0.12" ±0.12" 1/8" ± 1/8"
Thrust Angle						
Illustration #	154	154	151	151	151	154
Notes						
	21085	21086	21087	21088	21089	21090

Isuzu

Model	Amigo	Amigo	Ascender	Ascender	Axiom	Hombre
Information	All Models					4X2
Years	1990-94	1998-2000	2003-2006	2007-2008	2002-2004	1996-2000
Instructions						
LugNut Torque	65 ft.lbs.[1]	87 ft.lbs.	103 ft.lbs.	103 ft.lbs.	87 ft.lbs.	95 ft.lbs.
FRONT						
Caster (Degrees)	2.5 ±0.75 2½ ± ¾	2.5 ±0.75 2½ ± ¾	L= 3.75, R= 4.25 ± 0.5 [1 2] L= 3¾, R= 4¼ ± ½ [1 2]	L= 3.9, R= 4.1 ± 0.6 L= 3 7/8, R= 1/8 ± 5/8	2.5 ±0.75 2½ ± ¾	2.0 ±0.5 [1] 2 ± ½ [1]
Cross | Tol.	0.0 (0) | 0.58 (9/16)	0.0 (0) | 0.5 (½)	-0.5 (-½) | 0.5 (½)	-0.2 (-3/16) | 0.6 (5/8)	0.0 (0) | 0.5 (½)	0.0 (0) | n/s
Camber (Degrees)	0.5 ±1.0 ½ ± 1	0.0 ±0.5 0 ± ½	-0.5 ±0.5 -½ ± ½	-0.5 ±0.6 -½ ± 5/8	0.0 ±0.5 0 ± ½	0.0 ±0.5 0 ± ½
Cross | Tol.	0.0 (0) | 0.75 (¾)	0.0 (0) | 0.5 (½)	0.0 (0) | 0.3 (5/16)	0.0 (0) | 0.6 (5/8)	0.0 (0) | 0.5 (½)	0.0 (0) | n/s
Front Toe (Degrees)	0.16° ±0.16° 5/32° ± 5/32°	0.0° ±0.16° 0° ± 5/32°	0.1° ±0.2° 3/32° ± 3/16°	0.1° ±0.2° 3/32° ± 3/16°	0.08° ±0.08° 3/32° ± 3/32°	0.1° ±0.1° [2] 3/32° ± 3/32° [2]
Front Toe (Inches)	0.08" ±0.08" 3/32" ± 3/32"	0.0" ±0.08" 0" ± 3/32"	0.05" ±0.1" 1/16" ± 3/32"	0.05" ±0.1" 1/16" ± 3/32"	0.04" ±0.04" 1/32" ± 1/32"	0.05" ±0.05" [2] 1/16" ± 1/16" [2]
SAI	10.0 ±1.0 (10 ± 1)	12.5 ±0.5 (12½ ± ½)			12.5 ±0.5 (12½ ± ½)	
Turning Angle	35.0 / 33.0 ±1.0				32.6 / 31.8 ±1.5	
Illustration #	23	23	133	133	23	23
REAR						
Camber (Degrees)						
Rear Toe (Degrees)						
Rear Toe (Inches)						
Thrust Angle						
Illustration #						
Notes	[1] w/Alum. whls. 87 ft.lbs.		[1] 2003 L&R = 3.75° (3¾°) [2] 2004 L = 3.5°, R = 4.0°			[1] 1996 = 3.0° (3°) [2] 1996-97 = 0.2° (0.1")
	22001	22002	22003	22004	22005	22006

Isuzu

MODEL INFORMATION	Hombre 4X4	i-Series Pickup 4X2	i-Series Pickup 4X4	Impulse	Oasis	Pickup 4X2 Short Wheelbase
YEARS	1998-2000	2006-2008	2006-2008	1990-93	1996-99	1990-95
INSTRUCTIONS						
LUGNUT TORQUE	95 ft.lbs.	103 ft.lbs.	103 ft.lbs.	87 ft.lbs.	80 ft.lbs.	58-87 ft.lbs.
FRONT						
CASTER (Degrees)	2.0 ± 0.5 2 ± ½	L= 4.3, R= 4.5 ± 1.0 L= 4 5/16, R= 4 ½ ± 1	L= 3.8, R= 4.0 ± 1.0 L= 3 13/16, R= 4 ± 1	3.0 ± 1.0 3 ± 1	3.0 ± 1.0 3 ± 1	1.58 ± 0.75 1 9/16 ± ¾
CROSS | TOL.	0.0 (0) | n/s	-0.2 (-3/16) | 0.5 (½)	-0.2 (-3/16) | 0.5 (½)	0.0 (0) | n/s	0.0 (0) | n/s	0.0 (0) | 0.58 (9/16)
CAMBER (Degrees)	0.0 ± 0.5 0 ± ½	0.0 ± 0.5 0 ± ½	0.0 ± 0.5 0 ± ½	-0.5 ± 0.75 -½ ± ¾	0.0 ± 1.0 0 ± 1	0.5 ± 1.0 ½ ± 1
CROSS | TOL.	0.0 (0) | n/s	0.0 (0) | 0.5 (½)	0.0 (0) | 0.5 (½)	0.0 (0) | n/s	0.0 (0) | n/s	0.0 (0) | 0.75 (¾)
FRONT TOE (Degrees)	0.1° ± 0.1° 3/32° ± 3/32°	0.0° ± 0.1° 0° ± 3/32°	0.0° ± 0.1° 0° ± 3/32°	0.04° ± 0.12° 1/32° ± 1/8°	0.0° ± 0.16° 0° ± 5/32°	0.16° ± 0.16° 5/32° ± 5/32°
FRONT TOE (Inches)	0.05" ± 0.05" 1/16" ± 1/16"	0.0" ± 0.05" 0" ± 1/16"	0.0" ± 0.05" 0" ± 1/16"	0.02" ± 0.06" 1/32" ± 1/16"	0.0" ± 0.08" 0" ± 3/32"	0.08" ± 0.08" 3/32" ± 3/32"
SAI						10.0 ± 1.0 (10 ± 1)
TURNING ANGLE					39.0 / 30.0 ± 2.0	37.0 / 33.0 ± 1.0
ILLUSTRATION #	24	156	126		03	03 23
REAR						
CAMBER (Degrees)				-0.5 ± 0.75 -½ ± ¾	-0.5 ± 0.5 -½ ± ½	
REAR TOE (Degrees)				0.32° ± 0.16° 5/16° ± 5/32°	0.0° ± 0.16° 0° ± 5/32°	
REAR TOE (Inches)				0.16" ± 0.08" 5/32" ± 3/32"	0.0" ± 0.08" 0" ± 3/32"	
THRUST ANGLE						
ILLUSTRATION #				10	06	
NOTES						
	22007	22011	22012	22015	22016	22019

Isuzu

MODEL INFORMATION	Pickup 4X2 Long Wheelbase	Pickup 4X4 Short Wheelbase	Pickup 4X4 Long Wheelbase	Rodeo	Rodeo	Rodeo
YEARS	1990-95	1990-95	1990-95	1991-95	1996-97	1998-2004
INSTRUCTIONS						
LUGNUT TORQUE	58-87 ft.lbs.	58-87 ft.lbs.[1]	58-87 ft.lbs.[1]	66 ft.lbs.[1]	66 ft.lbs.	87 ft.lbs.
FRONT						
CASTER (Degrees)	1.83 ± 0.75 1 13/16 ± ¾	1.92 ± 0.75 1 15/16 ± ¾	2.17 ± 0.75 2 3/16 ± ¾	2.33 ± 0.75 2 5/16 ± ¾	2.75 ± 0.75 2 ¾ ± ¾	2.5 ± 0.75 2 ½ ± ¾
CROSS | TOL.	0.0 (0) | 0.58 (9/16)	0.0 (0) | 0.58 (9/16)	0.0 (0) | 0.58 (9/16)	0.0 (0) | 0.58 (9/16)	0.0 (0) | 0.5 (½)	0.0 (0) | 0.5 (½)
CAMBER (Degrees)	0.5 ± 1.0 ½ ± 1	0.5 ± 0.5 ½ ± ½	0.5 ± 0.5 ½ ± ½	0.5 ± 1.0 ½ ± 1	0.0 ± 0.5 0 ± ½	0.0 ± 0.5 0 ± ½
CROSS | TOL.	0.0 (0) | 0.75 (¾)	0.0 (0) | 0.75 (¾)	0.0 (0) | 0.75 (¾)	0.0 (0) | 0.75 (¾)	0.0 (0) | 0.5 (½)	0.0 (0) | 0.5 (½)
FRONT TOE (Degrees)	0.16° ± 0.16° 5/32° ± 5/32°	0.16° ± 0.16° 5/32° ± 5/32°	0.16° ± 0.16° 5/32° ± 5/32°	0.16° ± 0.16° 5/32° ± 5/32°	0.0° ± 0.16° 0° ± 5/32°	0.08° ± 0.08° 3/32° ± 3/32°
FRONT TOE (Inches)	0.08" ± 0.08" 3/32" ± 3/32"	0.08" ± 0.08" 3/32" ± 3/32"	0.08" ± 0.08" 3/32" ± 3/32"	0.08" ± 0.08" 3/32" ± 3/32"	0.0" ± 0.08" 0" ± 3/32"	0.04" ± 0.04" 1/32" ± 1/32"
SAI	10.0 ± 1.0 (10 ± 1)	10.0 ± 1.0 (10 ± 1)	10.0 ± 1.0 (10 ± 1)	10.0 ± 1.0 (10 ± 1)	12.5 ± 0.5 (12 ½ ± ½)	12.5 ± 0.5 (12 ½ ± ½)
TURNING ANGLE	37.0 / 33.0 ± 1.0	35.0 / 33.0 ± 1.0[2]	35.0 / 33.0 ± 1.0[2]	35.0 / 33.0 ± 1.0	34.0 / 32.0 ± 1.0	32.6 / 31.8 ± 1.5
ILLUSTRATION #	03 23	23	23	23	23	23
REAR						
CAMBER (Degrees)						
REAR TOE (Degrees)						
REAR TOE (Inches)						
THRUST ANGLE						
ILLUSTRATION #						
NOTES		[1] w/Alum. whls. 80-94 ft.lbs. [2] w/10.5R15 tires 32°/29°	[1] w/Alum. whls. 80-94 ft.lbs. [2] w/10.5R15 tires 32°/29°	[1] w/Alum. whls. 87 ft.lbs.		
	22020	22021	22022	22023	22024	22025

Isuzu

Rodeo Sport	Stylus	Trooper	MODEL INFORMATION	Trooper (All Models)	VehiCROSS	Notes
1998-2003	1990-93	1990-91	YEARS	1992-2002	1999-2001	
			INSTRUCTIONS			
87 ft.lbs.	87 ft.lbs.	58-87 ft.lbs.[1]	LUGNUT TORQUE	87 ft.lbs.	87 ft.lbs.	
			FRONT			
2.5 ± 0.75 2½ ± ¾	3.0 ± 1.0 3 ± 1	2.5 ± 0.5 2½ ± ½	CASTER (Degrees)	2.17 ± 0.75 2 3/16 ± ¾	2.08 ± 0.75 2 1/16 ± ¾	
0.0 (0) | 0.5 (½)	0.0 (0) | n/s	0.0 (0) | n/s	CROSS | TOL.	0.0 (0) | 0.5 (½)	0.0 (0) | 0.5 (½)	
0.0 ± 0.5 0 ± ½	-0.5 ± 0.75 -½ ± ¾	0.5 ± 0.5 ½ ± ½	CAMBER (Degrees)	0.0 ± 0.5 0 ± ½	0.0 ± 0.5 0 ± ½	
0.0 (0) | 0.5 (½)	0.0 (0) | n/s	0.0 (0) | n/s	CROSS | TOL.	0.0 (0) | 0.5 (½)	0.0 (0) | 0.5 (½)	
0.08° ± 0.08° 3/32° ± 3/32°	0.0° ± 0.16° 0° ± 5/32°	0.16° ± 0.16° 5/32° ± 5/32°	FRONT TOE (Degrees)	0.0° ± 0.16° 0° ± 5/32°	0.0° ± 0.16° 0° ± 5/32°	
0.04" ± 0.04" 1/32" ± 1/32"	0.0" ± 0.08" 0" ± 3/32"	0.08" ± 0.08" 3/32" ± 3/32"	FRONT TOE (Inches)	0.0" ± 0.08" 0" ± 3/32"	0.0" ± 0.08" 0" ± 3/32"	
12.5 ± 0.5 (12½ ± ½)		10.0 ± 1.0 (10 ± 1)	SAI	12.5 ± 0.5 (12½ ± ½)	12.5 ± 0.5 (12½ ± ½)	
32.6 / 31.8 ± 1.5			TURNING ANGLE	34.0 / 32.0 ± 1.0	33.0 / 31.0	
23		23	ILLUSTRATION #	23	23	
			REAR			
	-0.5 ± 0.75 -½ ± ¾		CAMBER (Degrees)			
	0.32° +0.16° -0.08° 5/16° +5/32° -3/32°		REAR TOE (Degrees)			
	0.16" +0.08" -0.04" 5/32" +3/32" -1/32"		REAR TOE (Inches)			
			THRUST ANGLE			
	10		ILLUSTRATION #			
		[1] w/Alum. whls. 80-94 ft.lbs.	NOTES			
22026	22027	22029		22030	22031	22034

Jaguar

F-Pace	F-Type	F-Type All Models	MODEL INFORMATION	S-Type	S-Type	X-Type
2017	2014-2015	2016-2017	YEARS	2000-2002	2003-2008	2002-2009
p. 307, 308	p. 307, 308	p. 307, 308	INSTRUCTIONS	p. 307, 308	p. 307, 308	p. 307, 308
92 ft.lbs.	92 ft.lbs.	92 ft.lbs.	LUGNUT TORQUE	85-100 ft.lbs.	85-100 ft.lbs.	65-87 ft.lbs.[1]
			FRONT			
7.52 ± 0.75 7½ ± ¾	6.48 ± 0.75 6½ ± ¾	6.9 ± 0.75 6⅞ ± ¾	CASTER (Degrees)	7.8 ± 0.8 7 13/16 ± 13/16	L= 6.6, R= 6.75 ± 0.75 L= 6⅝, R= 6¾ ± ¾	2.48 ± 0.75 2½ ± ¾
0.0 (0) | 0.9 (⅞)	0.0 (0) | 0.75 (¾)	0.0 (0) | 0.75 (¾)	CROSS | TOL.	0.0 (0) | 0.7 (11/16)	-0.15 (-⅛) | 0.75 (¾)	0.0 (0) | 0.75 (¾)
-0.38 ± 0.75 -⅜ ± ¾	L= -0.7, R= -1.2 ± 0.75 L= -11/16, R= -1 3/16 ± ¾	-0.87 ± 0.75 -⅞ ± ¾	CAMBER (Degrees)	L= 0.0, R= -0.6 ± 0.5 L= 0, R= -⅝ ± ½	L= -0.2, R= -0.55 ± 0.75 L= -3/16, R= -9/16 ± ¾	-0.54 ± 0.75 -9/16 ± ¾
0.0 (0) | 0.75 (¾)	0.5 (½) | 0.75 (¾)		CROSS | TOL.	0.6 (⅝) | 0.7 (11/16)	0.35 (⅜) | 0.75 (¾)	0.0 (0) | 0.75 (¾)
0.26° ± 0.2° ¼° ± 3/16°	0.27° ± 0.2° 9/32° ± 3/16°	0.27° ± 0.2° [1] 9/32° ± 3/16° [1]	FRONT TOE (Degrees)	0.17° ± 0.17° 5/32° ± 5/32°	-0.2° ± 0.17° -3/16° ± 5/32°	-0.1° ± 0.2° -3/32° ± 3/16°
0.13" ± 0.1" ⅛" ± 3/32"	0.14" ± 0.1" ⅛" ± 3/32"	0.14" ± 0.1" [1] ⅛" ± 3/32" [1]	FRONT TOE (Inches)	0.08" ± 0.08" 3/32" ± 3/32"	-0.1" ± 0.08" -3/32" ± 3/32"	-0.05" ± 0.1" -1/16" ± 3/32"
			SAI			
			TURNING ANGLE			
181	181	181	ILLUSTRATION #	123 124	123 124	
			REAR			
-0.86 ± 0.75 -⅞ ± ¾	-1.5 ± 0.75 -1½ ± ¾	-1.42 ± 0.75 -1 7/16 ± ¾	CAMBER (Degrees)	-1.0 ± 0.75 -1 ± ¾	-0.54 ± 0.75 -9/16 ± ¾	-0.7 ± 0.75 -11/16 ± ¾
0.2° ± 0.13° 3/16° ± ⅛°	0.28° ± 0.2° 9/32° ± 3/16°	0.3° ± 0.13° 5/16° ± ⅛°	REAR TOE (Degrees)	0.25° ± 0.17° ¼° ± 5/32°	0.17° ± 0.2° 5/32° ± 3/16°	0.25° ± 0.2° ¼° ± 3/16°
0.1" ± 0.07" 3/32" ± 1/16"	0.14" ± 0.1" ⅛" ± 3/32"	0.15" ± 0.07" 5/32" ± 1/16"	REAR TOE (Inches)	0.13" ± 0.08" ⅛" ± 3/32"	0.08" ± 0.1" 3/32" ± 3/32"	0.13" ± 0.1" ⅛" ± 3/32"
0.14 (⅛)	0.13 (⅛)	0.13 (⅛)	THRUST ANGLE	0.1 (3/32)	0.1 (3/32)	0.14 (⅛)
174	10	10	ILLUSTRATION #	135	135	06
		[1] SVR = 0.25° (0.13")	NOTES			[1] w/Steel wheels 50-68 ft.lbs
23005	23006	23007		23008	23009	23010

Jaguar

MODEL INFORMATION	XE	XF Sedan	XF Sedan RWD (Start VIN# R47154)	XF Sedan AWD Models	XF Sedan	XFR Sedan
YEARS	2017	2009-2010	2010-2015	2013-2015	2016-2017	2010-2015
INSTRUCTIONS	p. 307, 308	p. 307, 308	p. 307, 308	p. 307, 308	p. 307, 308	p. 307, 308
LUGNUT TORQUE	92 ft.lbs.	85-100 ft.lbs.	85-100 ft.lbs.	85-100 ft.lbs.	92 ft.lbs.	85-100 ft.lbs.
FRONT						
CASTER (Degrees)	7.19±0.75 7 3/16 ± ¾	L= 6.61, R= 6.74 ± 0.75 L= 6⅝, R= 6¾ ± ¾	L= 6.61, R= 6.74 ± 0.75 L= 6⅝, R= 6¾ ± ¾	L= 6.04, R= 6.17 ± 0.75 L= 6 1/16, R= 6 3/16 ± ¾	7.19±0.75 7 3/16 ± ¾	L= 6.61, R= 6.74 ± 0.75 L= 6⅝, R= 6¾ ± ¾
CROSS | TOL.	0.0 (0) | 0.9 (⅞)	-0.13 (-⅛) | 0.75 (¾)	-0.13 (-⅛) | 0.75 (¾)	-0.13 (-⅛) | 0.75 (¾)	0.0 (0) | 0.9 (⅞)	-0.13 (-⅛) | 0.75 (¾)
CAMBER (Degrees)	-0.32±0.5 - 5/16 ± ½	L= -0.2, R= -0.55±0.75 L= - 3/16, R= - 9/16 ± ¾	L= -0.2, R= -0.55±0.75 L= - 3/16, R= - 9/16 ± ¾	L= -0.19, R= -0.54±0.75 L= - 3/16, R= - 9/16 ± ¾	-0.32±0.75 - 5/16 ± ¾	L= -0.2, R= -0.55±0.75 L= - 3/16, R= - 9/16 ± ¾
CROSS | TOL.		0.35 (⅜) | 0.75 (¾)	0.35 (⅜) | 0.75 (¾)	0.35 (⅜) | 0.75 (¾)	0.0 (0) | 0.75 (¾)	0.35 (⅜) | 0.75 (¾)
FRONT TOE (Degrees)	0.33° ±0.06° 11/32° ± 1/16°	0.22° ±0.2° 7/32° ± 3/16°	0.22° ±0.2° 7/32° ± 3/16°	0.27° ±0.2° 9/32° ± 3/16°	0.31° ±0.06° 5/16° ± 1/16°	0.22° ±0.2° 7/32° ± 3/16°
FRONT TOE (Inches)	0.17" ±0.03" 5/32" ± 1/32"	0.11" ±0.1" 3/32" ± 3/32"	0.11" ±0.1" 3/32" ± 3/32"	0.13" ±0.1" ⅛" ± 3/32"	0.16" ±0.03" 5/32" ± 1/32"	0.11" ±0.1" 3/32" ± 3/32"
SAI						
TURNING ANGLE						
ILLUSTRATION #		181	181	181		181
REAR						
CAMBER (Degrees)	-0.81±0.65 - 13/16 ± ⅝	-0.69±0.75 - 11/16 ± ¾	-0.78±0.75 -¾ ± ¾	-0.98±0.75 -1 ± ¾	-0.81±0.75 - 13/16 ± ¾	-0.98±0.75 -1 ± ¾
REAR TOE (Degrees)	0.12° ±0.06° 5/16° ± 1/16°	0.17° ±0.2° 5/32° ± 3/16°	0.17° ±0.2° 5/32° ± 3/16°	0.15° ±0.2° 5/32° ± 3/16°	0.12° ±0.06° ⅛° ± 1/16°	0.17° ±0.2° 5/32° ± 3/16°
REAR TOE (Inches)	0.06" ±0.03" 5/32" ± 1/32"	0.08" ±0.1" 3/32" ± 3/32"	0.08" ±0.1" 3/32" ± 3/32"	0.08" ±0.1" 1/16" ± 3/32"	0.06" ±0.03" 1/16" ± 1/32"	0.08" ±0.1" 3/32" ± 3/32"
THRUST ANGLE	0.05 (1/16)	0.14 (⅛)	0.14 (⅛)	0.14 (⅛)	0.05 (1/16)	0.14 (⅛)
ILLUSTRATION #	174	10	10	10	174	10
NOTES						
	23011	23013	23014	23015	23016	23018

Jaguar

MODEL INFORMATION	XJ Sedans up to VIN# ..696458	XJ Sedans from VIN# ..696459 (exc/XJR)	XJR Sedan	XJ Sedan	XJR Sedan	XJ Sedan
YEARS	1990-94	1995-97	1995-97	1998-2003	1998-2003	2004-2010
INSTRUCTIONS	p. 307, 308	p. 307, 308	p. 307, 308	p. 307, 308	p. 307, 308	p. 307, 308
LUGNUT TORQUE				65-75 ft.lbs.[1]	65-75 ft.lbs.[1]	85-100 ft.lbs.
FRONT						
CASTER (Degrees)	4.5±2.5 4½±2½	4.5±2.5 4½±2½	4.5±2.5 4½±2½	6.0±2.0 6±2	6.0±2.0 6±2	7.48±0.75 7½ ± ¾
CROSS | TOL.	0.0 (0) | 0.5 (½)	0.0 (0) | 1.0 (1)	0.0 (0) | 1.0 (1)	0.0 (0) | 1.2 (1 3/16)	0.0 (0) | 1.2 (1 3/16)	0.0 (0) | 0.75 (¾)
CAMBER (Degrees)	-0.3±0.7 - 5/16 ± 11/16	-0.3±0.7 - 5/16 ± 11/16	-0.3±0.7 - 5/16 ± 11/16	L= -0.03, R= -0.53±0.5 L= - 1/32, R= -½ ± ½	L= -0.03, R= -0.53±0.5 L= - 1/32, R= -½ ± ½	L= -0.25, R= -0.75±0.75 L= -¼, R= -¾ ± ¾
CROSS | TOL.	0.0 (0) | n/s	0.0 (0) | n/s	0.0 (0) | n/s	0.5 (½) | 1.0 (1)	0.5 (½) | 1.0 (1)	0.5 (½) | 0.75 (¾)
FRONT TOE (Degrees)	0.06° ±0.06° 1/16° ± 1/16°	0.09° ±0.17° 3/32° ± 5/32°	0.09° ±0.17° 3/32° ± 5/32°	0.25° ±0.17° ¼° ± 5/32°	0.25° ±0.17° ¼° ± 5/32°	0.1° ±0.2° 3/32° ± 3/16°
FRONT TOE (Inches)	0.03" ±0.03" 1/32" ± 1/32"	0.04" ±0.08" 1/32" ± 3/32"	0.04" ±0.08" 1/32" ± 3/32"	0.13" ±0.08" ⅛" ± 3/32"	0.13" ±0.08" ⅛" ± 3/32"	0,05" ±0.1" 1/16" ± 3/32"
SAI						
TURNING ANGLE						
ILLUSTRATION #	84	84	84	118	118	118
REAR						
CAMBER (Degrees)	-0.75±0.25 -¾ ± ¼	-0.75±0.4 -¾ ± ⅜	-1.6±0.4 -1⅝ ± ⅜	-0.65±0.55 -⅝ ± 9/16	-0.65±0.55 -⅝ ± 9/16	-0.89±0.75 -⅞ ± ¾
REAR TOE (Degrees)	0.0° ±0.08° 0° ± 3/32°	0.25° ±0.33° ¼° ± 11/32°	0.25° ±0.33° ¼° ± 11/32°	0.33° ±0.17° 11/32° ± 5/32°	0.33° ±0.17° 11/32° ± 5/32°	0.17° ±0.2° 5/32° ± 3/16°
REAR TOE (Inches)	0.0" ±0.04" 0" ± 1/32"	0.13" ±0.17" ⅛" ± 5/32"	0.13" ±0.17" ⅛" ± 5/32"	0.17" ±0.08" 5/32" ± 3/32"	0.17" ±0.08" 5/32" ± 3/32"	0.08" ±0.1" 3/32" ± 3/32"
THRUST ANGLE				0.08 (3/32)	0.08 (3/32)	0.1 (3/32)
ILLUSTRATION #	85	85 64	85 64	85 64	85 64	85 64
NOTES				[1] w/Steel wheels 50-60 ft.lbs	[1] w/Steel wheels 50-60 ft.lbs	
	23022	23023	23024	23025	23026	23027

Jaguar

XJR Sedan	XJ Sedan RWD	XJ Sedan AWD	MODEL INFORMATION	XJS Coupe XJS Convertible	XJS Coupe XJS Convertible	XK8 XKR
2004-2009	2011-2017	2013-2017	YEARS	1990-93	1994-96	1997-2006
p. 307, 308	p. 307, 308	p. 307, 308	INSTRUCTIONS	p. 307, 308	p. 307, 308	p. 307, 308
85-100 ft.lbs.	92 ft.lbs.	92 ft.lbs.	LUGNUT TORQUE			65-75 ft.lbs.[1]
			FRONT			
7.48 ±0.75 7½ ± ¾	6.89 ±0.75 6⅞ ± ¾	5.29 ±0.75 5 5⁄16 ± ¾	CASTER (Degrees)	3.5 ±0.25 3½ ± ¼	4.0 ±2.0 4 ±2	L= 6.57, R= 7.33 ± 0.7 L= 6 9⁄16, R= 7 5⁄16 ± 11⁄16
0.0 (0) | 0.75 (¾)	0.0 (0) | 0.75 (¾)	0.0 (0) | 0.75 (¾)	CROSS | TOL.	0.0 (0) | 1.0 (1)	0.0 (0) | 1.0 (1)	-0.76 (-¾) | 0.7 (11⁄16)
L= -0.25, R= -0.75 ±0.75 L= -¼, R= -¾ ± ¾	L= -0.05, R= -0.55 ±0.75 L= -1⁄16, R= -9⁄16 ± ¾	L= -0.05, R= -0.55 ±0.75 L= -1⁄16, R= -9⁄16 ± ¾	CAMBER (Degrees)	-0.5 ±0.25 -½ ± ¼	-0.5 ±0.4 -½ ± ⅜	L= -0.67, R= -0.8 ±0.5 L= -11⁄16, R= -13⁄16 ± ½
0.5 (½) | 0.75 (¾)	0.5 (½) | 0.75 (¾)	0.5 (½) | 0.75 (¾)	CROSS | TOL.	0.0 (0) | n/s	0.0 (0) | n/s	0.13 (⅛) | 0.7 (11⁄16)
0.1° ±0.2° 3⁄32° ± 3⁄16°	0.17° ±0.2° 5⁄32° ± 3⁄16°	0.17° ±0.2° 5⁄32° ± 3⁄16°	FRONT TOE (Degrees)	0.13° ±0.13° ⅛° ± ⅛°	0.33° ±0.17° 11⁄32° ± 5⁄32°	0.25° ±0.17° ¼° ± 5⁄32°
0.05" ±0.1" 1⁄16" ± 3⁄32"	0.08" ±0.1" 3⁄32" ± 3⁄32"	0.08" ±0.1" 3⁄32" ± 3⁄32"	FRONT TOE (Inches)	0.06" ±0.06" 1⁄16" ± 1⁄16"	0.17" ±0.08" 5⁄32" ± 3⁄32"	0.13" ±0.08" ⅛" ± 3⁄32"
			SAI			
			TURNING ANGLE			
118	181	181	ILLUSTRATION #	84	84	112
			REAR			
-0.89 ±0.75 -⅞ ± ¾	-0.67 ±0.75 -11⁄16 ± ¾	-0.67 ±0.75 -11⁄16 ± ¾	CAMBER (Degrees)	-0.75 ±0.25 -¾ ± ¼	-0.75 ±0.25 -¾ ± ¼	-0.5 ±0.4 -½ ± ⅜
0.17° ±0.2° 5⁄32° ± 3⁄16°	0.2° ±0.2° 3⁄16° ± 3⁄16°	0.2° ±0.2° 3⁄16° ± 3⁄16°	REAR TOE (Degrees)	0.13° ±0.13° ⅛° ± ⅛°	0.0° ±0.33° 0° ± 11⁄32°	0.33° ±0.17° 11⁄32° ± 5⁄32°
0.08" ±0.1" 3⁄32" ± 3⁄32"	0.1" ±0.1" 3⁄32" ± 3⁄32"	0.1" ±0.1" 3⁄32" ± 3⁄32"	REAR TOE (Inches)	0.06" ±0.06" 1⁄16" ± 1⁄16"	0.0" ±0.17" 0" ± 5⁄32"	0.17" ±0.08" 5⁄32" ± 3⁄32"
0.1 (3⁄32)	0.14 (⅛)	0.14 (⅛)	THRUST ANGLE			0.08 (3⁄32)
85 64	49	49	ILLUSTRATION #	85	85	85 64
			NOTES			[1] w/Steel wheels 50-60 ft.lbs
23028	23029	23030		23035	23036	23037

Jaguar

XK	XKR	XKR exc/Dynamics Pack	MODEL INFORMATION	XKR w/Dynamics Pack	XKR-S	XKR-S GT
2007-2015	2007-2009	2010-2015	YEARS	2010-2015	2010-2015	2014
p. 307, 308	p. 307, 308	p. 307, 308	INSTRUCTIONS	p. 307, 308	p. 307, 308	p. 307, 308
92 ft.lbs.	92 ft.lbs.	92 ft.lbs.	LUGNUT TORQUE	92 ft.lbs.	92 ft.lbs.	85-100 ft.lbs.
			FRONT			
6.78 ±0.75 6¾ ± ¾	7.0 ±0.75 7 ± ¾	6.78 ±0.75 6¾ ± ¾	CASTER (Degrees)	6.93 ±0.75 6 15⁄16 ± ¾	6.78 ±0.75 6¾ ± ¾	6.62 ±0.75 6⅝ ± ¾
0.0 (0) | 0.75 (¾)	0.0 (0) | 0.75 (¾)	0.0 (0) | 0.75 (¾)	CROSS | TOL.	0.0 (0) | 0.75 (¾)	0.0 (0) | 0.75 (¾)	0.0 (0) | 0.75 (¾)
L= -0.25, R= -0.75 ±0.75 L= -¼, R= -¾ ± ¾	L= -0.25, R= -0.75 ±0.75 L= -¼, R= -¾ ± ¾	L= -0.25, R= -0.75 ±0.75 L= -¼, R= -¾ ± ¾	CAMBER (Degrees)	L= -0.55, R= -1.05 ±0.75 L= -9⁄16, R= -1 1⁄16 ± ¾	L= -0.25, R= -0.75 ±0.75 L= -¼, R= -¾ ± ¾	-1.22 ±0.75 -1¼ ± ¾
0.5 (½) | 0.75 (¾)	0.5 (½) | 0.75 (¾)	0.5 (½) | 0.75 (¾)	CROSS | TOL.	0.5 (½) | 0.75 (¾)	0.5 (½) | 0.75 (¾)	0.0 (0) | 0.75 (¾)
0.17° ±0.2° 5⁄32° ± 3⁄16°	0.17° ±0.2° 5⁄32° ± 3⁄16°	0.17° ±0.2° 5⁄32° ± 3⁄16°	FRONT TOE (Degrees)	0.25° ±0.2° ¼° ± 3⁄16°	0.17° ±0.2° 5⁄32° ± 3⁄16°	0.13° ±0.2° ⅛° ± 3⁄16°
0.08" ±0.1" 3⁄32" ± 3⁄32"	0.08" ±0.1" 3⁄32" ± 3⁄32"	0.08" ±0.1" 3⁄32" ± 3⁄32"	FRONT TOE (Inches)	0.13" ±0.1" ⅛" ± 3⁄32"	0.08" ±0.1" 3⁄32" ± 3⁄32"	0.07" ±0.1" 1⁄16" ± 3⁄32"
			SAI			
			TURNING ANGLE			
181	181	181	ILLUSTRATION #	181	181	181
			REAR			
-0.6 ±0.75[1] -⅝ ± ¾[1]	-0.63 ±0.75 -⅝ ± ¾	-0.72 ±0.75 -¾ ± ¾	CAMBER (Degrees)	-1.0 ±0.75 -1 ± ¾	-0.72 ±0.75 -¾ ± ¾	-1.76 ±0.75 -1¾ ± ¾
0.25° ±0.2° ¼° ± 3⁄16°	0.25° ±0.2° ¼° ± 3⁄16°	0.25° ±0.2° ¼° ± 3⁄16°	REAR TOE (Degrees)	0.25° ±0.2° ¼° ± 3⁄16°	0.25° ±0.2° ¼° ± 3⁄16°	0.13° ±0.2° ⅛° ± 3⁄16°
0.13" ±0.1" ⅛" ± 3⁄32"	0.13" ±0.1" ⅛" ± 3⁄32"	0.13" ±0.1" ⅛" ± 3⁄32"	REAR TOE (Inches)	0.13" ±0.1" ⅛" ± 3⁄32"	0.13" ±0.1" ⅛" ± 3⁄32"	0.07" ±0.1" 1⁄16" ± 3⁄32"
0.08 (3⁄32)	0.08 (3⁄32)	0.08 (3⁄32)	THRUST ANGLE	0.08 (3⁄32)	0.08 (3⁄32)	0.13 (⅛)
10	10	10	ILLUSTRATION #	10	10	10
[1] 2007-09 = -0.53° (-½°)			NOTES			
23038	23039	23040		23041	23042	23043

ALIGNMENT SPECIFICATIONS

Jeep

Notes	Cherokee (XJ) Wagoneer (XJ)	Cherokee (XJ)	MODEL INFORMATION	Cherokee (XJ)	Cherokee (XJ)	Cherokee (KL) FWD
	1990-94	1995-96	YEARS	1997-98	1999-2001	2014-2017
			INSTRUCTIONS			
	80-110 ft.lbs.[1]	80-110 ft.lbs.	LUGNUT TORQUE	80-110 ft.lbs.	85-115 ft.lbs.	100 ft.lbs.
			FRONT			
	6.0 +3.0 -1.0 6 +3 -1	7.0 +1.0 -1.75 7 ± 1¾	CASTER (Degrees)	7.0 +1.5 -1.75 7 ± 1¾	7.0 +1.5 -1.75 7 ± 1¾	4.0 ± 0.45 [1] 4 ± 7/16 [1]
	0.0 (0) | n/s	0.0 (0) | n/s	CROSS | TOL.	0.0 (0) | 1.25 (1¼)	0.0 (0) | 1.25 (1¼)	0.0 (0) | 0.5 (½)
	0.0 +0.5 -0.75 0 +½ -¾	-0.25 +0.75 -0.5 -¼ +¾ -½	CAMBER (Degrees)	-0.25 +0.75 -0.5 -¼ +¾ -½	-0.25 +0.75 -0.5 -¼ +¾ -½	-0.54 ± 0.4 - 9/16 ± ⅜
	0.0 (0) | n/s	0.0 (0) | n/s	CROSS | TOL.	0.0 (0) | 1.0 (1)	0.0 (0) | 1.0 (1)	0.0 (0) | 0.5 (½)
	0.0° ± 0.13° 0° ± ⅛°	0.0° ± 0.13° 0° ± ⅛°	FRONT TOE (Degrees)	0.0° +0.5° -0.2° 0° +½° -3/16°	0.25° +0.2° -0.25° ¼° +3/16° -¼°	0.1° ± 0.1° 3/32° ± 3/32°
	0.0" ± 0.06" 0" ± 1/16"	0.0" ± 0.06" 0" ± 1/16"	FRONT TOE (Inches)	0.0" +0.25" -0.1" 0" +¼" -3/32"	0.13" +0.1" -0.13" ⅛" +3/32" -⅛"	0.05" ± 0.05" 1/16" ± 1/16"
			SAI			
			TURNING ANGLE			
	78	78	ILLUSTRATION #	78	78	
			REAR			
			CAMBER (Degrees)			-1.0 ± 0.45 -1 ± 7/16
			REAR TOE (Degrees)			0.0° ± 0.15° 0° ± 5/32°
			REAR TOE (Inches)			0.0" ± 0.08" 0" ± 1/16"
			THRUST ANGLE		0.15 (5/32)	0.1 (3/32)
			ILLUSTRATION #			179 174
	[1] 1990-91 = 65-90 ft.lbs.		NOTES			[1] to 09/15/13 = 3.5° (3½°)
24001	24002	24003		24004	24005	24006

Jeep

Cherokee (KL) AWD	Cherokee (KL) Lifted	Comanche (MJ)	MODEL INFORMATION	Commander (XK)	Compass (MK) All w/17" Tires & AWL Off-Rd	Compass (MK) All w/17" Tires w/o AWL
2014-2017	2014-2017	1990-92	YEARS	2006-2010	2007-2010	2007-2010
			INSTRUCTIONS			
100 ft.lbs.	100 ft.lbs.	80-110 ft.lbs.[1]	LUGNUT TORQUE	100 ft.lbs.	100 ft.lbs.	100 ft.lbs.
			FRONT			
4.26 ± 0.45 [1] 4¼ ± 7/16 [1]	4.28 ± 0.45 [1] 4¼ ± 7/16 [1]	6.0 +3.0 -1.0 6 +3 -1	CASTER (Degrees)	4.0 ± 0.45 4 ± 7/16	L= 2.9, R= 2.6 ± 1.0 L= 2⅞, R= 2⅝ ± 1	L= 3.0, R= 2.7 ± 1.0 L= 3, R= 2 11/16 ± 1
0.0 (0) | 0.5 (½)	0.0 (0) | 0.5 (½)	0.0 (0) | n/s	CROSS | TOL.	0.0 (0) | 0.5 (½)	0.3 (5/16) | 1.0 (1)	0.3 (5/16) | 1.0 (1)
-0.54 ± 0.4 - 9/16 ± ⅜	-0.45 ± 0.4 - 7/16 ± ⅜	0.0 +0.5 -0.75 0 +½ -¾	CAMBER (Degrees)	-0.25 ± 0.45 -¼ ± 7/16	-0.7 ± 0.4 - 11/16 ± ⅜	-0.9 ± 0.4 - ⅞ ± ⅜
0.0 (0) | 0.5 (½)	0.0 (0) | 0.5 (½)	0.0 (0) | n/s	CROSS | TOL.	0.0 (0) | 0.5 (½)	0.0 (0) | 0.5 (½)	0.0 (0) | 0.5 (½)
0.1° ± 0.1° 3/32° ± 3/32°	0.1° ± 0.1° 3/32° ± 3/32°	0.0° ± 0.13° 0° ± ⅛°	FRONT TOE (Degrees)	0.25° ± 0.25° ¼° ± ¼°	0.2° ± 0.2° 3/16° ± 3/16°	0.2° ± 0.2° 3/16° ± 3/16°
0.05" ± 0.05" 1/16" ± 1/16"	0.05" ± 0.05" 1/16" ± 1/16"	0.0" ± 0.06" 0" ± 1/16"	FRONT TOE (Inches)	0.13" ± 0.13" ⅛" ± ⅛"	0.1" ± 0.1" 3/32" ± 3/32"	0.1" ± 0.1" 3/32" ± 3/32"
			SAI			
			TURNING ANGLE			
		78	ILLUSTRATION #	172	26 ▲	26 ▲
			REAR			
-1.0 ± 0.45 -1 ± 7/16	-1.05 ± 0.45 -1 1/16 ± 7/16		CAMBER (Degrees)		-0.6 ± 0.4 -⅝ ± ⅜	-0.8 ± 0.4 - 13/16 ± ⅜
0.0° ± 0.15° 0° ± 5/32°	0.0° ± 0.15° 0° ± 5/32°		REAR TOE (Degrees)		0.2° ± 0.2° 3/16° ± 3/16°	0.2° ± 0.2° 3/16° ± 3/16°
0.0" ± 0.08" 0" ± 1/16"	0.0" ± 0.08" 0" ± 1/16"		REAR TOE (Inches)		0.1" ± 0.1" 3/32" ± 3/32"	0.1" ± 0.1" 3/32" ± 3/32"
0.1 (3/32)	0.1 (3/32)		THRUST ANGLE	0.25 (¼)	0.1 (3/32)	0.1 (3/32)
179 174	179 174		ILLUSTRATION #		174	174
[1] to 09/15/13 = 3.8° (3 13/16°)	[1] to 09/15/13 = 4.1° (4⅛°)	[1] 1990-91 = 65-90 ft.lbs.	NOTES			
24007	24008	24010		24011	24013	24014

Jeep

Compass (MK)	Compass (MK)	Compass (MK)	MODEL	Grand Cherokee (ZJ)	Grand Cherokee (ZJ)	Grand Cherokee (WJ)
All w/18" Tires	4X2, 4X4	4X4	INFORMATION			
	w/o AWL Off-Road	w/AWL Off-Road				
2007-2010	2011-2017	2011-2017	YEARS	1993-96	1997-98	1999-2004
			INSTRUCTIONS			
100 ft.lbs.	100 ft.lbs.	100 ft.lbs.	LUGNUT TORQUE	80-110 ft.lbs.	80-110 ft.lbs.	85-115 ft.lbs.
			FRONT			
L= 3.0, R= 2.7 ± 1.0 L= 3, R= 2 11/16 ± 1	L= 3.0, R= 2.7 ± 1.0 L= 3, R= 2 11/16 ± 1	L= 3.0, R= 2.7 ± 1.0 L= 3, R= 2 11/16 ± 1	CASTER (Degrees)	7.0 ± 0.5 7 ± 1/2	7.0 ± 0.5 7 ± 1/2	6.75 ± 0.75 [1] 6 3/4 ± 3/4 [1]
0.3 (5/16) | 1.0 (1)	0.3 (5/16) | 1.0 (1)	0.3 (5/16) | 1.0 (1)	CROSS | TOL.	0.0 (0) | n/s	0.0 (0) | n/s	0.0 (0) | 0.5 (1/2)
-0.7 ± 0.4 -11/16 ± 3/8	-0.64 ± 0.55 -5/8 ± 9/16	-0.71 ± 0.55 -11/16 ± 9/16	CAMBER (Degrees)	-0.25 +0.75 -0.5 -1/4 +3/4 -1/2	-0.5 ± 0.63 -1/2 ± 5/8	-0.37 ± 0.38 -3/8 ± 3/8
0.0 (0) | 0.5 (1/2)	0.0 (0) | 0.6 (5/8)	0.0 (0) | 0.6 (5/8)	CROSS | TOL.	0.0 (0) | n/s	0.0 (0) | n/s	0.0 (0) | 0.5 (1/2)
0.2° ± 0.2° 3/16° ± 3/16°	0.2° ± 0.2° 3/16° ± 3/16°	0.2° ± 0.2° 3/16° ± 3/16°	FRONT TOE (Degrees)	0.24° +0.2° -0.24° 1/4° +3/16° -1/4°	0.24° +0.2° -0.24° 1/4° +3/16° -1/4°	0.2° ± 0.06° 3/16° ± 1/16°
0.1" ± 0.1" 3/32" ± 3/32"	0.1" ± 0.1" 3/32" ± 3/32"	0.1" ± 0.1" 3/32" ± 3/32"	FRONT TOE (Inches)	0.12" +0.1" -0.12" 1/8" +3/32" -1/8"	0.12" +0.1" -0.12" 1/8" +3/32" -1/8"	0.1" ± 0.03" 3/32" ± 1/32"
			SAI			
			TURNING ANGLE			
26 ▲	26	26	ILLUSTRATION #	78	78	
			REAR			
-0.7 ± 0.4 -11/16 ± 3/8	-0.61 ± 0.65 -5/8 ± 5/8	-0.8 ± 0.65 -13/16 ± 5/8	CAMBER (Degrees)			
0.2° ± 0.2° 3/16° ± 3/16°	0.2° ± 0.25° 3/16° ± 1/4°	0.2° ± 0.25° 3/16° ± 1/4°	REAR TOE (Degrees)			
0.1" ± 0.1" 3/32" ± 3/32"	0.1" ± 0.13" 3/32" ± 1/8"	0.1" ± 0.13" 3/32" ± 1/8"	REAR TOE (Inches)			
0.1 (3/32)	0.1 (3/32)	0.1 (3/32)	THRUST ANGLE			0.25 (1/4)
174	174	174	ILLUSTRATION #			
			NOTES			[1] Up Country = 6.5° (6 1/2°)
24015	24016	24017		24019	24020	24021

Jeep

Grand Cherokee (WK)	Grand Cherokee (WK)	Grand Cherokee (WK2)	MODEL	Grand Cherokee (WK2)	Grand Cherokee (WK2)	Grand Cherokee (WK2)
exc/SRT8	SRT8	STD. Susp.	INFORMATION	Sport Susp.	Air Susp.	STD. Susp.
					(in Auto mode)	
2005-2010	2006-2010	2011-2013	YEARS	2011-2013	2011-2013	2014-2015
			INSTRUCTIONS			
100 ft.lbs.	100 ft.lbs.	100 ft.lbs.	LUGNUT TORQUE	100 ft.lbs.	100 ft.lbs.	100 ft.lbs.
			FRONT			
4.0 ± 0.45 4 ± 7/16	4.1 ± 0.45 4 1/8 ± 7/16	5.0 ± 0.6 5 ± 5/8	CASTER (Degrees)	5.45 ± 0.6 5 7/16 ± 5/8	5.5 ± 0.6 5 1/2 ± 5/8	4.84 ± 0.6 4 13/16 ± 5/8
0.0 (0) | 0.5 (1/2)	0.0 (0) | 0.5 (1/2)	0.0 (0) | 1.0 (1)	CROSS | TOL.	0.0 (0) | 1.0 (1)	0.0 (0) | 1.0 (1)	0.0 (0) | 1.0 (1)
-0.25 ± 0.45 -1/4 ± 7/16	-0.5 ± 0.45 -1/2 ± 7/16	L= -0.12, R= -0.62 ± 0.6 L= -1/8, R= -5/8 ± 5/8	CAMBER (Degrees)	L= -0.6, R= -1.1 ± 0.6 L= -5/8, R= -1 1/8 ± 5/8	L= -0.64, R= -1.14 ± 0.6 L= -5/8, R= -1 1/8 ± 5/8	L= -0.17, R= -0.57 ± 0.6 L= -3/16, R= -9/16 ± 5/8
0.0 (0) | 0.5 (1/2)	0.0 (0) | 0.5 (1/2)	0.5 (1/2) | 0.6 (5/8)	CROSS | TOL.	0.5 (1/2) | 0.6 (5/8)	0.5 (1/2) | 0.6 (5/8)	0.4 (3/8) | 0.6 (5/8)
0.25° ± 0.25° 1/4° ± 1/4°	0.25° ± 0.25° 1/4° ± 1/4°	0.2° ± 0.13° 3/16° ± 1/8°	FRONT TOE (Degrees)	0.2° ± 0.13° 3/16° ± 1/8°	0.2° ± 0.13° 3/16° ± 1/8°	0.2° ± 0.13° 3/16° ± 1/8°
0.13" ± 0.13" 1/8" ± 1/8"	0.13" ± 0.13" 1/8" ± 1/8"	0.1" ± 0.06" 3/32" ± 1/16"	FRONT TOE (Inches)	0.1" ± 0.06" 3/32" ± 1/16"	0.1" ± 0.06" 3/32" ± 1/16"	0.1" ± 0.06" 3/32" ± 1/16"
			SAI			
			TURNING ANGLE			
172	172	193	ILLUSTRATION #	193	193	193
			REAR			
		-0.54 ± 0.55 -9/16 ± 9/16	CAMBER (Degrees)	-1.02 ± 0.55 -1 ± 9/16	-1.14 ± 0.55 -1 1/8 ± 9/16	-0.54 ± 0.55 -9/16 ± 9/16
		0.2° ± 0.25° 3/16° ± 1/4°	REAR TOE (Degrees)	0.2° ± 0.25° 3/16° ± 1/4°	0.4° ± 0.3° [1] 13/32° ± 5/16° [1]	0.2° ± 0.35° 3/16° ± 11/32°
		0.1" ± 0.13" 3/32" ± 1/8"	REAR TOE (Inches)	0.1" ± 0.13" 3/32" ± 1/8"	0.2" ± 0.15" [1] 3/16" ± 5/32" [1]	0.1" ± 0.18" 3/32" ± 3/16"
0.25 (1/4)	0.25 (1/4)	0.15 (5/32)	THRUST ANGLE	0.15 (5/32)	0.15 (5/32)	0.15 (5/32)
		130 06	ILLUSTRATION #	130 06	130 06	130 06
			NOTES		[1] 2011 = 0.2° (0.1")	
24022	24023	24024		24025	24026	24027

ALIGNMENT SPECIFICATIONS

Jeep

Grand Cherokee (WK2)	Grand Cherokee (WK2)	Grand Cherokee (WK2)	MODEL	Grand Cherokee (WK2)	Grand Cherokee (WK2)	Grand Cherokee (WK2)
Air Susp. (in Auto mode)	SRT(8) w/4 season tires	STD. Suspension	INFORMATION	Aero Suspension	Air Suspension (Aero mode)	Sport (Aero mode)
2014-2015	2014-2015	2016-2017	YEARS	2016-2017	2016-2017	2016-2017
			INSTRUCTIONS			
100 ft.lbs.	100 ft.lbs.	100 ft.lbs.	LUGNUT TORQUE	100 ft.lbs.	100 ft.lbs.	100 ft.lbs.
			FRONT			
5.27 ±0.6 5¼ ± ⅝	5.2 ±0.55 5 3/16 ± 9/16	6.95 ±0.5 6 15/16 ± ½	CASTER (Degrees)	7.32 ±0.5 7 5/16 ± ½	7.55 ±0.5 7 9/16 ± ½	7.46 ±0.5 7 7/16 ± ½
0.0 (0) | 1.0 (1)	0.0 (0) | 0.5 (½)	0.0 (0) | 0.5 (½)	CROSS | TOL.	0.0 (0) | 0.5 (½)	0.0 (0) | 0.5 (½)	0.0 (0) | 0.5 (½)
L= -0.64, R= -1.04 ±0.6 L= -⅝, R= -1 1/16 ± ⅝	L= -1.3, R= -1.6 ±0.55 [1] L= -1⅝, R= -1⅝ ± 9/16 [1]	L= -0.38, R= -0.78 ±0.6 L= -⅜, R= -¾ ± ⅝	CAMBER (Degrees)	L= -0.63, R= -1.03 ±0.6 L= -⅝, R= -1 ± ⅝	L= -0.75, R= -1.15 ±0.6 L= -¾, R= -1⅛ ± ⅝	L= -0.72, R= -1.12 ±0.6 L= -¾, R= -1⅛ ± ⅝
0.4 (⅜) | 0.6 (⅝)	0.3 (5/16) | 0.5 (½)	0.4 (⅜) | 0.6 (⅝)	CROSS | TOL.	0.4 (⅜) | 0.6 (⅝)	0.4 (⅜) | 0.6 (⅝)	0.4 (⅜) | 0.6 (⅝)
0.2° ±0.13° 3/16° ± ⅛°	0.2° ±0.13° 3/16° ± ⅛°	0.2° ±0.13° 3/16° ± ⅛°	FRONT TOE (Degrees)	0.2° ±0.13° 3/16° ± ⅛°	0.2° ±0.13° 3/16° ± ⅛°	0.2° ±0.13° 3/16° ± ⅛°
0.1" ±0.06" 3/32" ± 1/16"	0.1" ±0.06" 3/32" ± 1/16"	0.1" ±0.06" 3/32" ± 1/16"	FRONT TOE (Inches)	0.1" ±0.06" 3/32" ± 1/16"	0.1" ±0.06" 3/32" ± 1/16"	0.1" ±0.06" 3/32" ± 1/16"
			SAI			
			TURNING ANGLE			
193	193	193	ILLUSTRATION #	193	193	193
			REAR			
-1.09 ±0.55 -1 1/16 ± 9/16	-1.3 ±0.5 -1 5/16 ± ½	-0.54 ±0.55 -9/16 ± 9/16	CAMBER (Degrees)	-0.88 ±0.55 -⅞ ± 9/16	-1.09 ±0.55 -1 1/16 ± 9/16	-1.09 ±0.55 -1 1/16 ± 9/16
0.4° ±0.3° 13/32° ± 5/16°	0.25° ±0.2° ¼° ± 3/16°	0.2° ±0.35° 3/16° ± 11/32°	REAR TOE (Degrees)	0.2° ±0.35° 3/16° ± 11/32°	0.4° ±0.3° 13/32° ± 5/16°	0.4° ±0.3° 13/32° ± 5/16°
0.2" ±0.15" 3/32" ± 5/32"	0.13" ±0.1" ⅛" ± 3/32"	0.1" ±0.18" 3/32" ± 3/16"	REAR TOE (Inches)	0.1" ±0.18" 3/32" ± 3/16"	0.2" ±0.15" 3/32" ± 5/32"	0.2" ±0.15" 3/32" ± 5/32"
0.15 (5/32)	0.1 (3/32)	0.15 (5/32)	THRUST ANGLE	0.15 (5/32)	0.15 (5/32)	0.15 (5/32)
130 06	130 06	130 06	ILLUSTRATION #	130 06	130 06	130 06
	[1] w/3 season tires L&R = -1.6° (-1⅝°)		NOTES			
24028	24029	24030		24031	24032	24033

Jeep

Grand Cherokee (WK2)	Grand Wagoneer (SJ)	Liberty	MODEL	Liberty	Patriot (MK)	Patriot (MK)
SRT(8)			INFORMATION		All w/16" Tires	All w/17" Tires & AWL Off-Rd
2016-2017	1990-91	2002	YEARS	2003-2012	2007-2010	2007-2010
			INSTRUCTIONS			
100 ft.lbs.	65-90 ft.lbs.	85-115 ft.lbs.	LUGNUT TORQUE	85-115 ft.lbs.	100 ft.lbs.	100 ft.lbs.
			FRONT			
6.67 ±0.55 6 11/16 ± 9/16	4.0 ±0.5 4 ± ½	3.5 ±0.6 3½ ± ⅝	CASTER (Degrees)	3.3 ±0.5 [1] 3 5/16 ± ½ [1]	L= 3.0, R= 2.7 ± 1.0 L= 3, R= 2 11/16 ± 1	L= 2.9, R= 2.6 ± 1.0 L= 2⅞, R= 2⅝ ± 1
0.0 (0) | 0.55 (9/16)	0.0 (0) | n/s	0.0 (0) | 0.5 (½)	CROSS | TOL.	0.0 (0) | 0.5 (½)	0.3 (5/16) | 1.0 (1)	0.3 (5/16) | 1.0 (1)
L= -1.68, R= -1.98 ±0.55 L= -1 11/16, R= -2 ± 9/16	0.0 ±0.5 0 ± ½	0.0 ±0.38 0 ± ⅜	CAMBER (Degrees)	-0.38 ±0.38 -⅜ ± ⅜	-0.8 ±0.4 -13/16 ± ⅜	-0.7 ±0.4 -11/16 ± ⅜
0.3 (5/16) | 0.5 (½)	0.0 (0) | n/s	0.0 (0) | 0.7 (11/16)	CROSS | TOL.	0.0 (0) | 0.7 (11/16)	0.0 (0) | 0.5 (½)	0.0 (0) | 0.5 (½)
0.2° ±0.13° 3/16° ± ⅛°	0.31° ±0.31° 5/16° ± 5/16°	0.2° ±0.13° 3/16° ± ⅛°	FRONT TOE (Degrees)	0.2° ±0.13° 3/16° ± ⅛°	0.2° ±0.2° 3/16° ± 3/16°	0.2° ±0.2° 3/16° ± 3/16°
0.1" ±0.06" 3/32" ± 1/16"	0.16" ±0.16" 5/32" ± 5/32"	0.1" ±0.06" 3/32" ± 1/16"	FRONT TOE (Inches)	0.1" ±0.06" 3/32" ± 1/16"	0.1" ±0.1" 3/32" ± 3/32"	0.1" ±0.1" 3/32" ± 3/32"
			SAI			
			TURNING ANGLE			
193		12	ILLUSTRATION #		26	26
			REAR			
-1.3 ±0.5 -1 5/16 ± ½			CAMBER (Degrees)		-0.7 ±0.4 -11/16 ± ⅜	-0.6 ±0.4 -⅝ ± ⅜
0.25° ±0.2° ¼° ± 3/16°			REAR TOE (Degrees)		0.2° ±0.2° 3/16° ± 3/16°	0.2° ±0.2° 3/16° ± 3/16°
0.13" ±0.1" ⅛" ± 3/32"			REAR TOE (Inches)		0.1" ±0.1" 3/32" ± 3/32"	0.1" ±0.1" 3/32" ± 3/32"
0.1 (3/32)		0.25 (¼)	THRUST ANGLE	0.25 (¼)	0.1 (3/32)	0.1 (3/32)
130 06			ILLUSTRATION #		174	174
			NOTES	[1] 2003-07 = 3.9° (3⅞°)		
24034	24035	24043		24044	24045	24046

Jeep

Patriot (MK) All w/17" Tires w/o AWL Off-Rd	Patriot (MK) 4X2, 4X4 w/o AWL Off-Road	Patriot (MK) 4X4 w/AWL Off-Road	MODEL INFORMATION	Renegade (BU) 4X2	Renegade (BU) 4X4 exc/TrailHawk	Renegade (BU) 4X4 TrailHawk
2007-2010	2011-2017	2011-2017	YEARS	2015-2017	2015-2017	2015-2017
			INSTRUCTIONS			
100 ft.lbs.	100 ft.lbs.	100 ft.lbs.	LUGNUT TORQUE	89 ft.lbs.	89 ft.lbs.	89 ft.lbs.
			FRONT			
L= 3.0, R= 2.7 ± 1.0 L= 3, R= 2 11/16 ± 1	L= 3.0, R= 2.7 ± 1.0 L= 3, R= 2 11/16 ± 1	L= 3.0, R= 2.7 ± 1.0 L= 3, R= 2 11/16 ± 1	CASTER (Degrees)	5.27 ± 0.5 5¼ ± ½	4.98 ± 0.5 5 ± ½	4.93 ± 0.5 4 15/16 ± ½
0.3 (5/16) | 1.0 (1)	0.3 (5/16) | 1.0 (1)	0.3 (5/16) | 1.0 (1)	CROSS | TOL.	0.0 (0) | n/s	0.0 (0) | n/s	0.0 (0) | n/s
-0.9 ± 0.4 - ⅞ ± ⅜	-0.64 ± 0.55 - ⅝ ± 9/16	-0.71 ± 0.55 - 11/16 ± 9/16	CAMBER (Degrees)	-0.63 ± 0.5 - ⅝ ± ½	-0.67 ± 0.5 - 11/16 ± ½	-0.58 ± 0.5 - 9/16 ± ½
0.0 (0) | 0.5 (½)	0.0 (0) | 0.6 (⅝)	0.0 (0) | 0.6 (⅝)	CROSS | TOL.	0.0 (0) | 0.5 (½)	0.0 (0) | 0.5 (½)	0.0 (0) | 0.5 (½)
0.2° ± 0.2° 3/16° ± 3/16°	0.2° ± 0.2° 3/16° ± 3/16°	0.2° ± 0.2° 3/16° ± 3/16°	FRONT TOE (Degrees)	**-0.16°** ± 0.14° **- 5/16- 5/32°** ± ⅛°	**-0.14°** ± 0.14° **- ⅛°** ± ⅛°	**-0.06°** ± 0.14° **- 1/16°** ± ⅛°
0.1" ± 0.1" 3/32" ± 3/32"	0.1" ± 0.1" 3/32" ± 3/32"	0.1" ± 0.1" 3/32" ± 3/32"	FRONT TOE (Inches)	**-0.08"** ± 0.07" **- 3/32"** ± 1/16"	**-0.07"** ± 0.07" **- 1/16"** ± 1/16"	**-0.03"** ± 0.07" **- 1/32"** ± 1/16"
			SAI			
			TURNING ANGLE			
26	26	26	ILLUSTRATION #			
			REAR			
-0.8 ± 0.4 - 13/16 ± ⅜	-0.61 ± 0.65 - ⅝ ± ⅝	-0.8 ± 0.65 - 13/16 ± ⅝	CAMBER (Degrees)	-1.17 ± 0.5 -1 3/16 ± ½	-1.13 ± 0.5 -1 ⅛ ± ½	-1.05 ± 0.5 -1 1/16 ± ½
0.2° ± 0.2° 3/16° ± 3/16°	0.2° ± 0.25° 3/16° ± ¼°	0.2° ± 0.25° 3/16° ± ¼°	REAR TOE (Degrees)	0.34° ± 0.14° 11/32° ± ⅛°	0.2° ± 0.14° 3/16° ± ⅛°	0.16° ± 0.14° 5/32° ± ⅛°
0.1" ± 0.1" 3/32" ± 3/32"	0.1" ± 0.13" 3/32" ± ⅛"	0.1" ± 0.13" 3/32" ± ⅛"	REAR TOE (Inches)	0.17" ± 0.07" 5/32" ± 1/16"	0.1" ± 0.07" 3/32" ± 1/16"	0.08" ± 0.07" 3/32" ± 1/16"
0.1 (3/32)	0.1 (3/32)	0.1 (3/32)	THRUST ANGLE	0.25 (¼)	0.25 (¼)	0.25 (¼)
174	174	174	ILLUSTRATION #	06	06	06
			NOTES			
24047	24049	24050		24051	24052	24053

Jeep

Wrangler (YJ)	Wrangler (TJ)	Wrangler (JK)	MODEL INFORMATION	Wrangler (JK)	Wrangler (JK)	Notes
1990-95	1997-2006	2007-2008	YEARS	2009-2014	2015-2017	
			INSTRUCTIONS			
80-110 ft.lbs.	85-115 ft.lbs.	92-132 ft.lbs.	LUGNUT TORQUE	92-132 ft.lbs.[1]	122 ft.lbs.	
			FRONT			
6.5 +0.75 -1.25 [1] 6½ +¾ -1¼ [1]	7.0 ± 1.0 7 ± 1	4.2 ± 0.5 4 3/16 ± ½	CASTER (Degrees)	L= 4.2, R= 4.4 ± 1.0 L= 4 3/16, R= 4⅜ ± 1	L= 4.2, R= 4.4 ± 1.0 L= 4 3/16, R= 4⅜ ± 1	
0.0 (0) | n/s	0.0 (0) | 0.65 (⅝)	0.0 (0) | 0.65 (⅝)	CROSS | TOL.	-0.2 (-3/16) | 0.5 (½)	-0.2 (-3/16) | 0.5 (½)	
0.0 ± 0.5 0 ± ½	-0.25 ± 0.63 - ¼ ± ⅝	-0.25 ± 0.37 - ¼ ± ⅜	CAMBER (Degrees)	-0.25 ± 0.37 - ¼ ± ⅜	-0.25 ± 0.37 - ¼ ± ⅜	
0.0 (0) | n/s	0.0 (0) | 1.0 (1)	0.0 (0) | 0.5 (½)	CROSS | TOL.	0.0 (0) | 0.5 (½)	0.0 (0) | 0.5 (½)	
0.0° ± 0.13° 0° ± ⅛°	0.3° ± 0.13° 5/16° ± ⅛°	0.2° ± 0.03° 3/16° ± 1/32°	FRONT TOE (Degrees)	0.2° ± 0.03° 3/16° ± 1/32°	0.2° ± 0.15° 3/16° ± 5/32°	
0.0" ± 0.06" 0" ± 1/16"	0.15" ± 0.06" 5/32" ± 1/16"	0.1" ± 0.02" 3/32" ± 1/64"	FRONT TOE (Inches)	0.1" ± 0.02" 3/32" ± 1/64"	0.1" ± 0.08" 3/32" ± 1/16"	
			SAI			
			TURNING ANGLE			
79	78	78	ILLUSTRATION #	78	78	
			REAR			
			CAMBER (Degrees)			
			REAR TOE (Degrees)			
			REAR TOE (Inches)			
	0.25 (¼)	0.25 (¼)	THRUST ANGLE	0.25 (¼)	0.25 (¼)	
			ILLUSTRATION #			
[1] w/Auto Trans. 8.0° (8°)			NOTES	[1] 2013-14 = 122 ft.lbs.		
24057	24058	24059		24060	24061	24062

ALIGNMENT SPECIFICATIONS

Kia

Amanti	Amanti	Borrego	MODEL INFORMATION	Cadenza (VG)	Cadenza (YG)	Forte (TD)
						Sedan, Hatch
						except SX
2004-2006	2007-2009	2009	YEARS	2014-2016	2017	2010-2013
			INSTRUCTIONS			
65-87 ft.lbs.	65-80 ft.lbs.	65-80 ft.lbs.	LUGNUT TORQUE	65-80 ft.lbs.	65-80 ft.lbs.	65-80 ft.lbs.
			FRONT			
3.25 ±1.0 3¼ ±1	4.53 ±0.75 4 9/16 ± ¾	3.8 ±0.5 3 13/16 ± ½	CASTER (Degrees)	3.88 ±0.5 3⅞ ± ½	4.75 ±0.5 4¾ ± ½	4.38 ±0.5 4⅜ ± ½
0.0 (0) | 0.5 (½)	0.0 (0) | 0.5 (½)	0.0 (0) | 0.5 (½)	CROSS | TOL.	0.0 (0) | n/s	0.0 (0) | n/s	0.0 (0) | n/s
0.0 ±0.5 0 ± ½	-0.5 ±0.5 -½ ± ½	-0.5 ±0.5 -½ ± ½	CAMBER (Degrees)	L= -0.5, R= -0.8 ±0.5 L= -½, R= -13/16 ± ½	-0.54 ±0.5 -9/16 ± ½	-0.64 ±0.5 -⅝ ± ½
0.0 (0) | 0.5 (½)	0.0 (0) | 0.5 (½)	0.0 (0) | 0.5 (½)	CROSS | TOL.	0.3 (⅜) | n/s	0.0 (0) | n/s	0.0 (0) | n/s
0.0° ±0.16° 0° ± 5/32°	0.0° ±0.16° [1] 0° ± 5/32° [1]	0.0° ±0.16° 0° ± 5/32°	FRONT TOE (Degrees)	0.0° ±0.2° 0° ± 3/16°	0.0° ±0.2° 0° ± 3/16°	0.12° ±0.18° ⅛° ± 3/16°
0.0" ±0.08" 0" ± 3/32"	0.0" ±0.08" [1] 0" ± 3/32" [1]	0.0" ±0.08" 0" ± 3/32"	FRONT TOE (Inches)	0.0" ±0.1" 0" ± 3/32"	0.0" ±0.1" 0" ± 3/32"	0.06" ±0.09" 1/16" ± 3/32"
9.2 (9⅛)	9.7 ±0.5 (¾ ± ½)	13.9 (13 15/16)	SAI	12.7 ±0.5 (12¾ ± ½)	13.9 ±0.5 (13⅞ ± ½)	13.6 ±0.5 (13⅝ ± ½)
38.2 / 30.8 ±2.0	38.1 / 30.8 ±2.0	39.8 / 34.3 ±1.5	TURNING ANGLE	39.9 / 32.8 ±1.5	38.8 / 32.5 ±2.5	41.0 / 32.3 ±1.5
		181	ILLUSTRATION #			
			REAR			
-0.5 ±0.5 -½ ± ½	-1.0 ±0.5 -1 ± ½	-1.0 ±0.5 -1 ± ½	CAMBER (Degrees)	-1.0 ±0.5 -1 ± ½	-1.0 ±0.5 -1 ± ½	-1.5 ±0.5 -1½ ± ½
0.16° ±0.16° 5/32° ± 5/32°	0.16° ±0.16° 5/32° ± 5/32°	0.16° ±0.16° 5/32° ± 5/32°	REAR TOE (Degrees)	0.17° ±0.2° 5/32° ± 3/16°	0.17° ±0.2° 5/32° ± 3/16°	0.4° ±0.2° 13/32° ± 3/16°
0.08" ±0.08" 3/32" ± 3/32"	0.08" ±0.08" 3/32" ± 3/32"	0.08" ±0.08" 3/32" ± 3/32"	REAR TOE (Inches)	0.09" ±0.1" 3/32" ± 3/32"	0.09" ±0.1" 3/32" ± 3/32"	0.2" ±0.1" 3/16" ± 3/32"
			THRUST ANGLE			
06	177	179 174	ILLUSTRATION #	174	174	
	[1] 2009 = 0.16° (0.08")		NOTES			
25001	25002	25003		25004	25005	25006

Kia

Forte (TD)	Forte Koup (TD)	Forte Koup (TD)	MODEL INFORMATION	Forte (YD)	K900	Niro HEV
Sedan, Hatch	2 Door	2 Door		Forte Koup (YD)		
SX	except SX	SX		Forte5 (YD)		
2010-2013	2010-2013	2010-2013	YEARS	2014-2017	2015-2017	2017
			INSTRUCTIONS			
65-80 ft.lbs.	65-80 ft.lbs.	65-80 ft.lbs.	LUGNUT TORQUE	65-80 ft.lbs.	65-80 ft.lbs.	65-80 ft.lbs.
			FRONT			
4.38 ±0.5 4⅜ ± ½	4.94 ±0.5 4 15/16 ± ½	4.94 ±0.5 4 15/16 ± ½	CASTER (Degrees)	4.13 ±0.5 4⅛ ± ½	7.95 ±0.75 7 15/16 ± ¾	4.4 ±0.5 4⅜ ± ½
0.0 (0) | n/s	0.0 (0) | n/s	0.0 (0) | n/s	CROSS | TOL.	0.0 (0) | n/s	0.0 (0) | n/s	0.0 (0) | n/s
-0.64 ±0.5 -⅝ ± ½	-0.81 ±0.5 -13/16 ± ½	-0.81 ±0.5 -13/16 ± ½	CAMBER (Degrees)	-0.5 ±0.5 -½ ± ½	-0.68 ±0.5 -11/16 ± ½	-0.4 ±0.5 -⅜ ± ½
0.0 (0) | n/s	0.0 (0) | n/s	0.0 (0) | n/s	CROSS | TOL.	0.0 (0) | n/s	0.0 (0) | n/s	0.0 (0) | n/s
0.12° ±0.18° ⅛° ± 3/16°	0.4° ±0.2° 13/32° ± 3/16°	0.4° ±0.2° 13/32° ± 3/16°	FRONT TOE (Degrees)	0.1° ±0.2° 3/32° ± 3/16°	0.0° ±0.2° 0° ± 3/16°	0.1° ±0.2° 3/32° ± 3/16°
0.06" ±0.09" 1/16" ± 3/32"	0.2" ±0.1" 3/16" ± 3/32"	0.2" ±0.1" 3/16" ± 3/32"	FRONT TOE (Inches)	0.05" ±0.1" 1/16" ± 3/32"	0.0" ±0.1" 0" ± 3/32"	0.05" ±0.1" 1/16" ± 3/32"
13.6 ±0.5 (13⅝ ± ½)	14.0 ±0.5 (14 ± ½)	14.0 ±0.5 (14 ± ½)	SAI	13.9 ±0.5 (13⅞ ± ½)	7.1 ±0.5 (7 1/16 ± ½)	13.7 ±0.5 (13¾ ± ½)
38.4 / 30.9 ±1.5	41.0 / 32.3 ±1.5	38.4 / 30.9 ±1.5	TURNING ANGLE	39.6 / 32.5 ±1.0	42.7 / 33.3 ±1.5	39.4 / 32.6 ±1.5
			ILLUSTRATION #			
			REAR			
-1.5 ±0.5 -1½ ± ½	-1.5 ±0.5 -1½ ± ½	-1.5 ±0.5 -1½ ± ½	CAMBER (Degrees)	-1.5 ±0.5 -1½ ± ½	-1.45 ±0.5 -1 7/16 ± ½	-1.2 ±0.5 -1 3/16 ± ½
0.4° ±0.2° 13/32° ± 3/16°	0.4° ±0.2° 13/32° ± 3/16°	0.4° ±0.2° 13/32° ± 3/16°	REAR TOE (Degrees)	0.4° +0.5° -0.4° 13/32° +½° -13/32°	0.4° ±0.2° 13/32° ± 3/16°	0.14° ±0.2° ⅛° ± 3/16°
0.2" ±0.1" 3/16" ± 3/32"	0.2" ±0.1" 3/16" ± 3/32"	0.2" ±0.1" 3/16" ± 3/32"	REAR TOE (Inches)	0.2" +0.25" -0.2" 3/16" +¼" -3/16"	0.2" ±0.1" 3/16" ± 3/32"	0.07" ±0.1" 1/16" ± 3/32"
			THRUST ANGLE			
			ILLUSTRATION #		129 174	174
			NOTES			
25007	25008	25009		25010	25011	25012

Kia

Optima Magentis	Optima Magentis	Optima up to 11/29/2011	MODEL INFORMATION	Optima from 11/30/2011	Optima (JFA)	Optima Hybrid (TF)
2001-2006	2007-2010	2011-2012	YEARS	2012-2015	2016-2017	2011-2016
			INSTRUCTIONS			
65-80 ft.lbs.	65-80 ft.lbs.	65-80 ft.lbs.	LUGNUT TORQUE	65-80 ft.lbs.	65-80 ft.lbs.	65-80 ft.lbs.
			FRONT			
3.25 ±1.0 3¼ ±1	4.75 ±0.5 4¾ ± ½	4.44 +0.5 -0.8 4 7⁄16 + ½ - 13⁄16	CASTER (Degrees)	4.0 ±0.5 4 ± ½	4.68 ±0.5 4 11⁄16 ± ½	3.84 ±0.5 [1] 3 13⁄16 ± ½ [1]
0.0 (0) | 0.5 (½)	0.0 (0) | 0.5 (½)	0.0 (0) | 0.5 (½)	CROSS | TOL.	0.0 (0) | 0.5 (½)	0.0 (0) | 0.5 (½)	0.0 (0) | 0.5 (½)
0.0 ±0.5 0 ± ½	-0.5 ±0.5 - ½ ± ½	-0.5 ±0.5 - ½ ± ½	CAMBER (Degrees)	-0.5 ±0.5 - ½ ± ½	-0.5 ±0.5 - ½ ± ½	-0.5 ±0.5 - ½ ± ½
0.0 (0) | 0.5 (½)	0.0 (0) | 0.5 (½)	0.0 (0) | 0.5 (½)	CROSS | TOL.	0.0 (0) | 0.5 (½)	0.0 (0) | 0.5 (½)	0.0 (0) | 0.5 (½)
0.0° ±0.18° 0° ± 3⁄16°	0.0° ±0.2° 0° ± 3⁄16°	0.16° ±0.2° 5⁄32° ± 3⁄16°	FRONT TOE (Degrees)	0.16° ±0.2° 5⁄32° ± 3⁄16°	0.0° ±0.2° 0° ± 3⁄16°	0.16° ±0.2° 5⁄32° ± 3⁄16°
0.0" ±0.09" 0" ± 3⁄32"	0.0" ±0.1" 0" ± 3⁄32"	0.08" ±0.1" 3⁄32" ± 3⁄32"	FRONT TOE (Inches)	0.08" ±0.1" 3⁄32" ± 3⁄32"	0.0" ±0.1" 0" ± 3⁄32"	0.08" ±0.1" 3⁄32" ± 3⁄32"
8.5 (8½)	13.1 ±0.5 (13⅛ ± ½)	12.7 ±0.5 (12¾ ± ½)	SAI	12.7 ±0.5 (12¾ ± ½)	13.9 ±0.5 (13 ⅞ ± ½)	12.8 ±0.5 (12 13⁄16 ± ½)
40.7 / 32.4 ±2.0	39.5 / 32.2	40.0 / 33.0 ±1.5	TURNING ANGLE	40.0 / 33.0 ±1.5	40.5 / 33.5 ±2.0	40.0 / 33.0 ±1.5
			ILLUSTRATION #			
			REAR			
-0.5 ±0.5 - ½ ± ½	-1.0 ±0.5 -1 ± ½	-1.0 ±0.5 -1 ± ½	CAMBER (Degrees)	-1.0 ±0.5 -1 ± ½	-1.0 ±0.5 -1 ± ½	-1.0 ±0.5 -1 ± ½
0.18° ±0.18° 3⁄16° ± 3⁄16°	0.2° ±0.2° 3⁄16° ± 3⁄16°	0.17° ±0.2° 5⁄32° ± 3⁄16°	REAR TOE (Degrees)	0.17° ±0.2° 5⁄32° ± 3⁄16°	0.17° ±0.2° 5⁄32° ± 3⁄16°	0.17° ±0.2° 5⁄32° ± 3⁄16°
0.09" ±0.09" 3⁄32" ± 3⁄32"	0.1" ±0.1" 3⁄32" ± 3⁄32"	0.09" ±0.1" 3⁄32" ± 3⁄32"	REAR TOE (Inches)	0.09" ±0.1" 3⁄32" ± 3⁄32"	0.09" ±0.1" 3⁄32" ± 3⁄32"	0.09" ±0.1" 3⁄32" ± 3⁄32"
			THRUST ANGLE			
06	174	174	ILLUSTRATION #	174	174	174
			NOTES			[1] 2011-12 = 4.34° (4 5⁄16°)
25013	25014	25015		25016	25017	25018

Kia

Optima Hybrid (JF) Optima PHEV (JF)	Rio	Rio (JB)	MODEL INFORMATION	Rio (UB) Sedan & Hatch	Rondo	Rondo (Canada)
2017	2001-2005	2006-2012	YEARS	2012-2017	2007-2012	2014-2016
			INSTRUCTIONS			
65-80 ft.lbs.	65-87 ft.lbs.	65-79 ft.lbs.	LUGNUT TORQUE	65-80 ft.lbs.	65-80 ft.lbs.	65-80 ft.lbs.
			FRONT			
4.68 ±0.5 4 11⁄16 ± ½	1.92 ±0.75 1 15⁄16 ± ¾	4.0 ±0.5 4 ± ½	CASTER (Degrees)	4.1 ±0.5 4⅛ ± ½	4.74 ±0.5 4¾ ± ½	4.24 ±0.5 4¼ ± ½
0.0 (0) | 0.5 (½)	0.0 (0) | n/s	0.0 (0) | n/s	CROSS | TOL.	0.0 (0) | 0.5 (½)	0.0 (0) | 0.5 (½)	0.0 (0) | 0.5 (½)
-0.5 ±0.5 - ½ ± ½	0.53 ±0.75 9⁄16 ± ¾	0.0 ±0.5 0 ± ½	CAMBER (Degrees)	-0.5 ±0.5 - ½ ± ½	-0.5 ±0.5 - ½ ± ½	-0.5 ±0.5 - ½ ± ½
0.0 (0) | 0.5 (½)	0.0 (0) | n/s	0.0 (0) | n/s	CROSS | TOL.	0.0 (0) | 0.5 (½)	0.0 (0) | 0.5 (½)	0.0 (0) | 0.5 (½)
0.0° ±0.2° 0° ± 3⁄16°	0.4° ±0.3° 13⁄32° ± 5⁄16°	0.0° ±0.2° 0° ± 3⁄16°	FRONT TOE (Degrees)	0.0° ±0.2° 0° ± 3⁄16°	0.0° ±0.2° 0° ± 3⁄16°	0.0° ±0.2° 0° ± 3⁄16°
0.0" ±0.1" 0" ± 3⁄32"	0.2" ±0.15" 3⁄16" ± 5⁄32"	0.0" ±0.1" 0" ± 3⁄32"	FRONT TOE (Inches)	0.0" ±0.1" 0" ± 3⁄32"	0.0" ±0.1" 0" ± 3⁄32"	0.0" ±0.1" 0" ± 3⁄32"
13.9 ±0.5 (13 ⅞ ± ½)	13.1 ±0.8 (13 1⁄16 ± ¾)	13.0 ±0.5 (13 ± ½)	SAI	13.8 ±0.5 (13 13⁄16 ± ½)	13.1 ±0.5 (13⅛ ± ½)	14.2 ±0.5 (14¼ ± ½)
40.5 / 33.5 ±2.0		40.0 / 32.9 ±1.5 [1]	TURNING ANGLE	38.4 / 30.8 ±2.0	40.3 / 31.5 ±1.5 [1]	39.2 / 32.1 ±1.5
			ILLUSTRATION #			
			REAR			
-1.0 ±0.5 -1 ± ½	-0.88 ±0.3 - ⅞ ± 5⁄16	-1.0 ±0.5 -1 ± ½	CAMBER (Degrees)	-1.5 ±0.5 -1½ ± ½	-1.0 ±0.5 -1 ± ½	-1.5 ±0.5 -1½ ± ½
0.17° ±0.2° 5⁄32° ± 3⁄16°	0.5° ±0.6° ½° ± 19⁄32°	0.4° ±0.2° 13⁄32° ± 3⁄16°	REAR TOE (Degrees)	0.5° +0.4° -0.5° ½° + 13⁄32° - ½°	0.2° ±0.2° 3⁄16° ± 3⁄16°	0.2° +0.4° -0.5° [1] 3⁄16° + 13⁄32° - ½° [1]
0.09" ±0.1" 3⁄32" ± 3⁄32"	0.25" ±0.3" ¼" ± 5⁄16"	0.2" ±0.1" 3⁄16" ± 3⁄32"	REAR TOE (Inches)	0.25" +0.2" -0.25" ¼" + 3⁄16" - ¼"	0.1" ±0.1" 3⁄32" ± 3⁄32"	0.1" +0.2" -0.25" [1] 3⁄32" + 3⁄16" - ¼" [1]
			THRUST ANGLE			
174 174			ILLUSTRATION #		174	
		[1] w/MS = 39°/32.6°	NOTES		[1] w/17" tires 38.4°/30.5°	[1] w/18" wheels ±0.3°
25019	25020	25021		25022	25023	25024

ALIGNMENT SPECIFICATIONS

Kia

Sedona	Sedona	Sedona	MODEL INFORMATION	Sephia	Sephia	Sorento
2002-2005	2006-2014	2015-2017	YEARS	1993-97	1998-2001	2003-2006
			INSTRUCTIONS			
80-101 ft.lbs.	65-80 ft.lbs.	65-80 ft.lbs.	LUGNUT TORQUE	65-87 ft.lbs.	65-87 ft.lbs.	65-86 ft.lbs.
			FRONT			
1.88 ±0.5 1⅞ ± ½	3.77 ±0.5 3¾ ± ½	4.29 ±0.5 4 5⁄16 ± ½	CASTER (Degrees)	2.0 ±0.75 2 ± ¾	2.45 ±0.75 2 7⁄16 ± ¾	3.3 ±0.5 3 5⁄16 ± ½
0.0 (0) | n/s	0.0 (0) | 0.5 (½)	0.0 (0) | 0.5 (½)	CROSS | TOL.	0.0 (0) | 0.75 (¾)	0.0 (0) | 0.75 (¾)	0.0 (0) | 0.5 (½)
0.51 ±0.5 ½ ± ½	0.0 ±0.5 0 ± ½	-0.5 ±0.5 -½ ± ½	CAMBER (Degrees)	0.0 ±0.5 0 ± ½	0.0 ±0.5 0 ± ½	0.39 ±0.5 ⅜ ± ½
0.0 (0) | n/s	0.0 (0) | 0.5 (½)	0.0 (0) | 0.5 (½)	CROSS | TOL.	0.0 (0) | 0.5 (½)	0.0 (0) | 0.5 (½)	0.0 (0) | 0.5 (½)
-0.07° ±0.2° -1⁄16° ± 3⁄16°	0.0° ±0.2° 0° ± 3⁄16°	0.1° ±0.1° 3⁄32° ± 3⁄32°	FRONT TOE (Degrees)	-0.28° ±0.3° -9⁄32° ± 5⁄16°	0.27° ±0.33° 9⁄32° ± 11⁄32°	0.2° ±0.2° 3⁄16° ± 3⁄16°
-0.04" ±0.1" -1⁄32" ± 3⁄32"	0.0" ±0.1" 0" ± 3⁄32"	0.05" ±0.05" 1⁄16" ± 1⁄16"	FRONT TOE (Inches)	-0.14" ±0.15" -⅛" ± 5⁄32"	0.13" ±0.17" ⅛" ± 5⁄32"	0.1" ±0.1" 3⁄32" ± 3⁄32"
	12.5 ±0.5 (12 7⁄16 ± ½)	13.2 ±0.5 (13 3⁄16 ± ½)	SAI	11.0 ±1.0 (11 ± 15⁄16)	12.6 ±0.8 (12 9⁄16 ± ¾)	12.1 (12 1⁄16)
34.2 / 29.5 ±2.0	38.6 / 31.8 ±1.5	41.9 / 32.8 ±1.5	TURNING ANGLE	39.5 / 32.5 ±2.0	39.5 / 32.5 ±2.0	37.2 / 32.1 ±2.0 [1]
03 50			ILLUSTRATION #	46	46	12
			REAR			
	0.1 ±0.5 ⅛ ± ½	-1.0 ±0.5 -1 ± ½	CAMBER (Degrees)	0.5 ±0.5 ½ ± ½	-0.53 ±0.5 -½ ± ½	
	0.26° ±0.2° ¼° ± 3⁄16°	0.2° ±0.2° 3⁄16° ± 3⁄16°	REAR TOE (Degrees)	-0.04° ±0.25° -1⁄32° ± ¼°	0.17° ±0.33° 5⁄32° ± 11⁄32°	
	0.13" ±0.1" ⅛" ± 3⁄32"	0.1" ±0.1" 3⁄32" ± 3⁄32"	REAR TOE (Inches)	-0.02" ±0.13" -1⁄32" ± ⅛"	0.08" ±0.17" 3⁄32" ± 5⁄32"	
			THRUST ANGLE			
	174	174	ILLUSTRATION #	10	10	
			NOTES			[1] 2WD = 36.9°/32°
25025	25026	25027		25028	25029	25030

Kia

Sorento	Sorento	Sorento	MODEL INFORMATION	Sorento	Soul	Soul Soul EV
2007-2009	2011-2013	2014-2015	YEARS	2016-2017	2010-2013	2014-2017
			INSTRUCTIONS			
65-86 ft.lbs.	65-80 ft.lbs.	65-80 ft.lbs.	LUGNUT TORQUE	65-80 ft.lbs.	65-79 ft.lbs.	65-80 ft.lbs.
			FRONT			
3.89 ±0.5 3⅞ ± ½	4.82 ±0.5 4 13⁄16 ± ½	4.16 ±0.5 4 3⁄16 ± ½	CASTER (Degrees)	4.38 ±0.5 4⅜ ± ½	3.4 ±0.5 3⅜ ± ½	5.1 ±0.5 5⅛ ± ½
0.0 (0) | 0.5 (½)	0.0 (0) | n/s	0.0 (0) | n/s	CROSS | TOL.	0.0 (0) | n/s	0.0 (0) | n/s	0.0 (0) | n/s
0.0 ±0.5 0 ± ½	-0.63 ±0.5 -⅝ ± ½	-0.5 ±0.5 -½ ± ½	CAMBER (Degrees)	-0.5 ±0.5 -½ ± ½	-0.5 ±0.5 -½ ± ½	-0.5 ±0.5 -½ ± ½
0.0 (0) | 0.5 (½)	0.0 (0) | n/s	0.0 (0) | n/s	CROSS | TOL.	0.0 (0) | n/s	0.0 (0) | n/s	0.0 (0) | n/s
0.0° ±0.16° 0° ± 5⁄32°	0.0° ±0.2° 0° ± 3⁄16°	0.1° ±0.2° 3⁄32° ± 3⁄16°	FRONT TOE (Degrees)	0.1° ±0.2° 3⁄32° ± 3⁄16°	0.0° ±0.2° 0° ± 3⁄16°	0.1° ±0.2° 3⁄32° ± 3⁄16°
0.0" ±0.08" 0" ± 3⁄32"	0.0" ±0.1" 0" ± 3⁄32"	0.05" ±0.1" 1⁄16" ± 3⁄32"	FRONT TOE (Inches)	0.05" ±0.1" 1⁄16" ± 3⁄32"	0.0" ±0.1" 0" ± 3⁄32"	0.05" ±0.1" 1⁄16" ± 3⁄32"
12.1 ±0.5 (12⅛ ± ½)	13.0 ±0.5 (13 ± ½)	13.4 ±0.5 (13 7⁄16 ± ½)	SAI	13.3 ±0.5 (13 5⁄16 ± ½)	14.5 ±0.5 (14 9⁄16 ± ½)	13.4 ±0.5 (13⅜ ± ½)
36.9 / 32.0 ±2.0 [1]	39.3 / 31.6 ±1.5	39.1 / 31.9 ±1.5	TURNING ANGLE	37.4 / 32.0 ±1.5	38.4 / 30.8 ±1.5	37.5 / 31.4 ±1.5
12			ILLUSTRATION #			
			REAR			
	-1.0 ±0.5 -1 ± ½	-1.0 ±0.5 -1 ± ½	CAMBER (Degrees)	-1.0 ±0.5 -1 ± ½	-1.3 ±0.5 [1] -1 5⁄16 ± ½ [1]	-1.5 ±0.5 -1½ ± ½
	0.2° ±0.2° 3⁄16° ± 3⁄16°	0.2° ±0.2° 3⁄16° ± 3⁄16°	REAR TOE (Degrees)	0.2° ±0.2° 3⁄16° ± 3⁄16°	0.4° ±0.2° 13⁄32° ± 3⁄16°	0.3° +0.5° -0.3° 5⁄16° +½° - 5⁄16°
	0.1" ±0.1" 3⁄32" ± 3⁄32"	0.1" ±0.1" 3⁄32" ± 3⁄32"	REAR TOE (Inches)	0.1" ±0.1" 3⁄32" ± 3⁄32"	0.2" ±0.1" 3⁄16" ± 3⁄32"	0.15" +0.25" -0.15" 5⁄32" +¼" - 3⁄32"
			THRUST ANGLE			
	177	177	ILLUSTRATION #	174		
[1] 4WD = 37.2°/32°			NOTES		[1] 2010-11 = -1.5° (-1½°)	
25031	25032	25033		25034	25035	25036

Kia

Spectra 1.8L	Spectra 2.0L Spectra5	Sportage	MODEL INFORMATION	Sportage	Sportage	Sportage
2000-2004	2004.5-2009	1995-2002	YEARS	2005-2010	2011-2016	2017
			INSTRUCTIONS			
65-87 ft.lbs.	65-80 ft.lbs.	74 ft.lbs.	LUGNUT TORQUE	66-81 ft.lbs.	65-80 ft.lbs.	80-94 ft.lbs.
			FRONT			
2.45 ± 0.75 2 7/16 ± 3/4	2.6 ± 0.5 2 5/8 ± 1/2	3.58 ± 0.75 3 9/16 ± 3/4	CASTER (Degrees)	3.6 ± 0.5 3 5/8 ± 1/2	4.02 ± 0.5 4 ± 1/2	4.6 ± 0.5 4 5/8 ± 1/2
0.0 (0) | 0.75 (3/4)	0.0 (0) | 0.5 (1/2)	0.0 (0) | 0.5 (1/2)	CROSS | TOL.	0.0 (0) | 0.5 (1/2)	0.0 (0) | 0.5 (1/2)	0.0 (0) | 0.5 (1/2)
0.0 ± 0.5 0 ± 1/2	0.0 ± 0.5 0 ± 1/2	0.44 ± 0.75 7/16 ± 3/4	CAMBER (Degrees)	0.0 ± 0.5 0 ± 1/2	-0.5 ± 0.5 - 1/2 ± 1/2	-0.5 ± 0.5 - 1/2 ± 1/2
0.0 (0) | 0.5 (1/2)	0.0 (0) | 0.5 (1/2)	0.0 (0) | 0.5 (1/2)	CROSS | TOL.	0.0 (0) | 0.5 (1/2)	0.0 (0) | 0.5 (1/2)	0.0 (0) | 0.5 (1/2)
-0.08° ± 0.25° - 3/32° ± 1/4°	0.0° ± 0.16° 0° ± 5/32°	0.25° ± 0.25° 1/4° ± 1/4°	FRONT TOE (Degrees)	0.0° ± 0.16° 0° ± 5/32°	0.0° ± 0.2° 0° ± 3/16°	0.1° ± 0.1° 3/32° ± 3/32°
-0.04" ± 0.13" - 1/32" ± 1/8"	0.0" ± 0.08" 0" ± 3/32"	0.13" ± 0.13" 1/8" ± 1/8"	FRONT TOE (Inches)	0.0" ± 0.08" 0" ± 3/32"	0.0" ± 0.1" 0" ± 3/32"	0.05" ± 0.05" 1/16" ± 1/16"
12.6 ± 0.8 (12 9/16 ± 3/4)	12.2 ± 0.5 (12 3/16 ± 1/2)		SAI	12.8 ± 0.5 (12 3/4 ± 1/2)	12.9 ± 0.5 (12 15/16 ± 1/2)	14.0 ± 0.5 (14 ± 1/2)
39.5 / 32.5 ± 2.0	37.0 / 31.8 ± 0.5		TURNING ANGLE	36.8 / 31.8 ± 1.5	39.5 / 31.9 ± 1.5	39.5 / 32.3 ± 1.5
46		24	ILLUSTRATION #			
			REAR			
-0.53 ± 0.5 - 1/2 ± 1/2	-0.92 ± 0.5 - 15/16 ± 1/2		CAMBER (Degrees)	-0.92 ± 0.5 - 15/16 ± 1/2	-1.0 ± 0.5 -1 ± 1/2	-1.0 ± 0.5 -1 ± 1/2
0.25° ± 0.25° 1/4° ± 1/4°	0.32° ± 0.16° 5/16° ± 5/32°		REAR TOE (Degrees)	0.36° +0.24° -0.08° 3/8° + 1/4° - 3/32°	0.2° ± 0.2° 3/16° ± 3/16°	0.2° ± 0.2° 3/16° ± 3/16°
0.13" ± 0.13" 1/8" ± 1/8"	0.16" ± 0.08" 5/32" ± 3/32"		REAR TOE (Inches)	0.18" +0.12" -0.04" 3/16" + 1/8" - 1/32"	0.1" ± 0.1" 3/32" ± 3/32"	0.1" ± 0.1" 3/32" ± 3/32"
			THRUST ANGLE			
10	10		ILLUSTRATION #	10	179 174	174
			NOTES			
25037	25038	25039		25040	25041	25042

Land Rover

Defender	Discovery I	Discovery II to VIN..754807	MODEL INFORMATION	Discovery II from VIN..754808	Discovery Sport	Discovery Coil Susp.
1993-97	1994-99	1999-2002	YEARS	2002-2004	2015-2016	2017
			INSTRUCTIONS			
95 ft.lbs.	96 ft.lbs.	103 ft.lbs.	LUGNUT TORQUE	103 ft.lbs.	98 ft.lbs.	103 ft.lbs.
			FRONT			
3.25 ± 0.75 3 1/4 ± 3/4	3.25 ± 0.75 3 1/4 ± 3/4	3.75 ± 0.75 3 3/4 ± 3/4	CASTER (Degrees)	3.75 ± 0.75 3 3/4 ± 3/4	3.2 ± 0.75 3 3/16 ± 3/4	4.02 ± 0.75 4 ± 3/4
0.0 (0) | 1.0 (1)	0.0 (0) | 1.0 (1)	0.0 (0) | 1.0 (1)	CROSS | TOL.	0.0 (0) | 1.0 (1)	0.0 (0) | 0.75 (3/4)	0.0 (0) | 0.75 (3/4)
-0.17 ± 0.75 - 3/16 ± 3/4	-0.17 ± 0.75 - 3/16 ± 3/4	-0.17 ± 0.5 - 3/16 ± 1/2	CAMBER (Degrees)	0.33 ± 0.5 5/16 ± 1/2	-0.55 ± 0.75 - 9/16 ± 3/4	-0.41 ± 0.75 - 7/16 ± 3/4
0.0 (0) | 1.0 (1)	0.0 (0) | 1.0 (1)	0.0 (0) | 0.75 (3/4)	CROSS | TOL.	0.0 (0) | 0.75 (3/4)	0.0 (0) | 0.75 (3/4)	0.0 (0) | 0.75 (3/4)
-0.17° ± 0.17° - 5/32° ± 5/32°	-0.14° ± 0.14° - 1/8° ± 1/8°	-0.17° ± 0.17° - 5/32° ± 5/32°	FRONT TOE (Degrees)	-0.17° ± 0.17° - 5/32° ± 5/32°	0.24° ± 0.2° 1/4° ± 3/16°	0.31° ± 0.2° 5/16° ± 3/16°
-0.08" ± 0.08" - 3/32" ± 3/32"	-0.07" ± 0.07" - 1/16" ± 1/16"	-0.08" ± 0.08" - 3/32" ± 3/32"	FRONT TOE (Inches)	-0.08" ± 0.08" - 3/32" ± 3/32"	0.12" ± 0.1" 1/8" ± 3/32"	0.16" ± 0.1" 5/32" ± 3/32"
7.0 (7)	7.0 (7)	13.2 (13 3/16)	SAI	13.2 (13 3/16)		
			TURNING ANGLE			
			ILLUSTRATION #			
			REAR			
-0.17 ± 0.75 - 3/16 ± 3/4	-0.17 ± 0.75 - 3/16 ± 3/4	-0.17 ± 0.5 - 3/16 ± 1/2	CAMBER (Degrees)	-0.17 ± 0.5 - 3/16 ± 1/2	-1.56 ± 0.75 -1 9/16 ± 3/4	-1.24 ± 0.75 -1 1/4 ± 3/4
0.08° ± 0.25° 3/32° ± 1/4°	0.08° ± 0.25° 3/32° ± 1/4°	0.08° ± 0.25° 3/32° ± 1/4°	REAR TOE (Degrees)	0.08° ± 0.25° 3/32° ± 1/4°	0.22° ± 0.14° 7/32° ± 1/8°	0.28° ± 0.14° 9/32° ± 1/8°
0.04" ± 0.13" 1/32" ± 1/8"	0.04" ± 0.13" 1/32" ± 1/8"	0.04" ± 0.13" 1/32" ± 1/8"	REAR TOE (Inches)	0.04" ± 0.13" 1/32" ± 1/8"	0.11" ± 0.07" 1/8" ± 1/16"	0.14" ± 0.07" 1/8" ± 1/16"
0.3 (5/16)	0.3 (5/16)	0.33 (11/32)	THRUST ANGLE	0.33 (11/32)	0.13 (1/8)	0.13 (1/8)
			ILLUSTRATION #		174	174
			NOTES			
48001	48002	48003		48004	48005	48006

Land Rover

MODEL	Discovery	Evoque	Freelander	Range Rover Classic	LR2	LR3
INFORMATION	Air Susp.					
YEARS	2017	2012-2017	2002-2005	1990-95	2008-2015	2005-2009
INSTRUCTIONS						
LUGNUT TORQUE	103 ft.lbs.	98 ft.lbs.	85 ft.lbs.	96 ft.lbs.	98 ft.lbs.	96-110 ft.lbs.
FRONT						
CASTER (Degrees)	4.43 ±0.75 4 7/16 ± ¾	3.17 ±0.75 3 3/16 ± ¾	3.5 ±1.0 3½ ±1	3.25 ±0.75 3¼ ± ¾	3.25 ±0.75 3¼ ± ¾	L= 3.86, R= 4.17 ± 0.75 L= 3⅞, R= 4 3/16 ± ¾
CROSS | TOL.	0.0 (0) | 0.75 (¾)	0.0 (0) | 0.75 (¾)	0.0 (0) | 1.0 (1)	0.0 (0) | 1.0 (1)	0.0 (0) | 0.75 (¾)	-0.31 (-5/16) | 0.75 (¾)
CAMBER (Degrees)	-0.58 ±0.75 -9/16 ± ¾	-0.47 ±0.75 -½ ± ¾	-0.25 ±0.75 -¼ ± ¾	-0.17 ±0.75 -3/16 ± ¾	-0.5 ±0.75 -½ ± ¾	L= -0.15, R= -0.5 ±0.75 L= -⅛, R= -½ ± ¾
CROSS | TOL.	0.0 (0) | 0.75 (¾)	0.0 (0) | 0.75 (¾)	0.0 (0) | 1.0 (1)	0.0 (0) | 1.0 (1)	0.0 (0) | 0.75 (¾)	0.35 (⅜) | 0.75 (¾)
FRONT TOE (Degrees)	0.28° ±0.2° 9/32° ± 3/16°	0.14° ±0.2° [1] ⅛° ± 3/16° [1]	-0.23° ±0.25° -7/32° ± ¼°	-0.14° ±0.14° -⅛° ± ⅛°	0.2° ±0.2° 3/16° ± 3/16°	0.17° ±0.17° 5/32° ± 5/32°
FRONT TOE (Inches)	0.14" ±0.1" ⅛" ± 3/32"	0.07" ±0.1" [1] 1/16" ± 3/32" [1]	-0.12" ±0.13" -⅛" ± ⅛"	-0.07" ±0.07" -1/16" ± 1/16"	0.1" ±0.1" 3/32" ± 3/32"	0.08" ±0.08" 3/32" ± 3/32"
SAI			12.3 (12 5/16)	7.0 (7)	13.9 (13⅞)	13.9 (13⅞)
TURNING ANGLE						
ILLUSTRATION #						52
REAR						
CAMBER (Degrees)	-1.4 ±0.75 -1⅜ ± ¾	-1.18 ±0.75 -1 3/16 ± ¾	-0.5 ±0.75 -½ ± ¾	-0.17 ±0.75 -3/16 ± ¾	-1.95 ±0.75 -1 15/16 ± ¾	-0.75 ±0.75 [1] -¾ ± ¾ [1]
REAR TOE (Degrees)	0.3° ±0.14° 5/16° ± ⅛°	0.16° ±0.2° 5/32° ± 3/16°	0.33° ±0.25° 11/32° ± ¼°	0.08° ±0.25° 3/32° ± ¼°	0.33° ±0.2° 11/32° ± 3/16°	0.23° ±0.13° 7/32° ± ⅛°
REAR TOE (Inches)	0.15" ±0.07" 5/32" ± 1/16"	0.08" ±0.1" 3/32" ± 3/32"	0.17" ±0.13" 5/32" ± ⅛"	0.04" ±0.13" 1/32" ± ⅛"	0.17" ±0.1" 5/32" ± 3/32"	0.12" ±0.06" ⅛" ± 1/16"
THRUST ANGLE	0.13 (⅛)	0.06 (1/16)	0.1 (3/32)	0.3 (5/16)	0.13 (⅛)	0.13 (⅛)
ILLUSTRATION #	174	06	10		06	147
NOTES		[1] Conv. = 0.21° (0.1")				[1] Coil Susp. = -0.5° (-½°)
	48007	48008	48009	48010	48011	48012

Land Rover

MODEL	LR4	Range Rover	Range Rover	Range Rover	Range Rover	Range Rover
INFORMATION			exc/Supercharged	Supercharged	exc/Supercharged	Supercharged
YEARS	2010-2016	1995-2002	2003-2009	2006-2009	2010-2012	2010-2012
INSTRUCTIONS						
LUGNUT TORQUE	103 ft.lbs.	80 ft.lbs.	96-110 ft.lbs.	96-110 ft.lbs.	96-110 ft.lbs.	96-110 ft.lbs.
FRONT						
CASTER (Degrees)	L= 3.86, R= 4.17 ± 0.75 L= 3⅞, R= 4 3/16 ± ¾	4.0 ±1.0 4 ±1	6.08 ±0.75 6 1/16 ± ¾	6.08 ±0.75 6 1/16 ± ¾	6.08 ±0.75 6 1/16 ± ¾	6.08 ±0.75 6 1/16 ± ¾
CROSS | TOL.	-0.31 (-5/16) | 0.75 (¾)	0.0 (0) | 1.0 (1)	0.0 (0) | 0.75 (¾)	0.0 (0) | 0.75 (¾)	0.0 (0) | 0.75 (¾)	0.0 (0) | 0.75 (¾)
CAMBER (Degrees)	L= -0.15, R= -0.5 ±0.75 L= -⅛, R= -½ ± ¾	-0.17 ±0.75 -3/16 ± ¾	-0.33 ±0.75 -5/16 ± ¾	-0.33 ±0.75 -5/16 ± ¾	-0.33 ±0.75 -5/16 ± ¾	-0.33 ±0.75 -5/16 ± ¾
CROSS | TOL.	0.35 (⅜) | 0.75 (¾)	0.0 (0) | 1.0 (1)	0.0 (0) | 0.75 (¾)	0.0 (0) | 0.75 (¾)	0.0 (0) | 0.75 (¾)	0.0 (0) | 0.75 (¾)
FRONT TOE (Degrees)	0.17° ±0.17° 5/32° ± 5/32°	-0.17° ±0.17° -5/32° ± 5/32°	0.0° ±0.2° 0° ± 3/16°	-0.15° ±0.2° -5/32° ± 3/16°	0.0° ±0.2° 0° ± 3/16°	0.15° ±0.2° 5/32° ± 3/16°
FRONT TOE (Inches)	0.08" ±0.08" 3/32" ± 3/32"	-0.08" ±0.08" -3/32" ± 3/32"	0.0" ±0.1" 0" ± 3/32"	-0.08" ±0.1" -1/16" ± 3/32"	0.0" ±0.1" 0" ± 3/32"	0.08" ±0.1" 1/16" ± 3/32"
SAI	13.9 (13⅞)	8.0 (8)	11.8 (11¾)	11.8 (11¾)	11.8 (11¾)	11.8 (11¾)
TURNING ANGLE						
ILLUSTRATION #	52		152	152	152	152
REAR						
CAMBER (Degrees)	-0.75 ±0.75 [1] -¾ ± ¾ [1]	-0.17 ±0.75 -3/16 ± ¾	-1.0 ±0.75 -1 ± ¾	-1.0 ±0.75 -1 ± ¾	-1.0 ±0.75 -1 ± ¾	-1.0 ±0.75 -1 ± ¾
REAR TOE (Degrees)	0.23° ±0.13° 7/32° ± ⅛°	0.08° ±0.25° 3/32° ± ¼°	0.33° ±0.2° 11/32° ± 3/16°	0.2° ±0.2° 3/16° ± 3/16°	0.3° ±0.2° 5/16° ± 3/16°	0.2° ±0.2° 3/16° ± 3/16°
REAR TOE (Inches)	0.12" ±0.06" ⅛" ± 1/16"	0.04" ±0.13" 1/32" ± ⅛"	0.17" ±0.1" 5/32" ± 3/32"	0.1" ±0.1" 3/32" ± 3/32"	0.15" ±0.1" 5/32" ± 3/32"	0.1" ±0.1" 3/32" ± 3/32"
THRUST ANGLE	0.13 (⅛)	0.3 (5/16)	0.13 (⅛)	0.13 (⅛)	0.13 (⅛)	0.13 (⅛)
ILLUSTRATION #	147		115 143	115 143	115 143	115 143
NOTES	[1] Coil Susp. = -0.5° (-½°)					
	48013	48014	48015	48016	48017	48018

Land Rover

MODEL INFORMATION	Range Rover	Range Rover Sport	Range Rover Sport exc/Supercharged	Range Rover Sport Supercharged	Range Rover Sport	Notes
YEARS	2013-2017	2006-2009	2010-2013	2010-2013	2014-2017	
INSTRUCTIONS						
LUGNUT TORQUE	103 ft.lbs.	96-110 ft.lbs.	96-110 ft.lbs.	96-110 ft.lbs.	103 ft.lbs.	
FRONT						
CASTER (Degrees)	4.12 ±0.75 [1] 4⅛ ± ¾ [1]	L= 3.96, R= 4.07 ± 0.75 L= 3 15/16, R= 4 1/16 ± ¾	L= 3.96, R= 4.07 ± 0.75 L= 3 15/16, R= 4 1/16 ± ¾	L= 3.96, R= 4.07 ± 0.75 L= 3 15/16, R= 4 1/16 ± ¾	4.12 ±0.75 4⅛ ± ¾	
CROSS / TOL.	0.0 (0) / 0.75 (¾)	-0.11 (-⅛) / 0.75 (¾)	-0.11 (-⅛) / 0.75 (¾)	-0.11 (-⅛) / 0.75 (¾)	0.0 (0) / 0.75 (¾)	
CAMBER (Degrees)	-0.68 ±0.75 -11/16 ± ¾	L= -0.38, R= -0.5 ±0.5 L= -⅜, R= -½ ± ½	L= -0.38, R= -0.5 ±0.5 L= -⅜, R= -½ ± ½	L= -0.63, R= -0.75 ±0.5 L= -⅝, R= -¾ ± ½	-0.68 ±0.75 -11/16 ± ¾	
CROSS / TOL.	0.0 (0) / 0.75 (¾)	0.12 (⅛) / 0.75 (¾)	0.12 (⅛) / 0.75 (¾)	0.12 (⅛) / 0.75 (¾)	0.0 (0) / 0.75 (¾)	
FRONT TOE (Degrees)	0.14° ±0.2° ⅛° ± 3/16°	-0.17° ±0.2° -5/32° ± 3/16°	0.1° ±0.2° 3/32° ± 3/16°	0.15° ±0.2° 5/32° ± 3/16°	0.14° ±0.2° ⅛° ± 3/16°	
FRONT TOE (Inches)	0.07" ±0.1" 1/16" ± 3/32"	-0.08" ±0.1" -3/32" ± 3/32"	0.05" ±0.1" 1/16" ± 3/32"	0.08" ±0.1" 1/16" ± 3/32"	0.07" ±0.1" 1/16" ± 3/32"	
SAI		13.9 (13 7/8)	13.9 (13 7/8)	13.9 (13 7/8)		
TURNING ANGLE						
ILLUSTRATION #		173	173	173		
REAR						
CAMBER (Degrees)	-1.42 ±0.75 -1 7/16 ± ¾	-1.0 ±0.75 -1 ± ¾	-1.0 ±0.75 -1 ± ¾	-1.0 ±0.75 -1 ± ¾	-1.42 ±0.75 -1 7/16 ± ¾	
REAR TOE (Degrees)	0.3° ±0.2° 5/16° ± 3/16°	0.23° ±0.13° 7/32° ± ⅛°	0.18° ±0.13° 3/16° ± ⅛°	0.18° ±0.13° 3/16° ± ⅛°	0.3° ±0.2° 5/16° ± 3/16°	
REAR TOE (Inches)	0.15" ±0.1" 5/32" ± 3/32"	0.12" ±0.07" ⅛" ± 1/16"	0.09" ±0.07" 3/32" ± 1/16"	0.09" ±0.07" 3/32" ± 1/16"	0.15" ±0.1" 5/32" ± 3/32"	
THRUST ANGLE	0.13 (⅛)	0.1 (3/32)	0.1 (3/32)	0.1 (3/32)	0.13 (⅛)	
ILLUSTRATION #	174	147	147	147	174	
NOTES	[1] 2013 = 3.87° (3 7/8°)					
	48019	48020	48021	48022	48023	48024

Lexus

MODEL INFORMATION	CT200H exc/Sport	CT200H Sport	ES 250	ES 250	ES 300	ES 300
YEARS	2011-2017	2011-2017	1990	1991	1992-96	1997-2001
INSTRUCTIONS						
LUGNUT TORQUE	76 ft.lbs.	76 ft.lbs.				76 ft.lbs.
FRONT						
CASTER (Degrees)	5.95 ±0.75 5 15/16 ± ¾	5.85 ±0.75 5 7/8 ± ¾	1.67 ±0.5 1 11/16 ± ½	1.58 ±0.5 1 9/16 ± ½	1.25 ±0.75 1¼ ± ¾	2.3 ±0.75 2 5/16 ± ¾
CROSS / TOL.	0.0 (0) / 0.75 (¾)	0.0 (0) / 0.75 (¾)	0.0 (0) / 0.5 (½)	0.0 (0) / 0.5 (½)	0.0 (0) / 0.75 (¾)	0.0 (0) / 0.75 (¾)
CAMBER (Degrees)	-0.22 ±0.75 -¼ ± ¾	-0.23 ±0.75 -¼ ± ¾	0.5 ±0.75 ½ ± ¾	0.42 ±0.75 7/16 ± ¾	-0.67 ±0.75 -11/16 ± ¾	-0.62 ±0.75 -⅝ ± ¾
CROSS / TOL.	0.0 (0) / 0.75 (¾)	0.0 (0) / 0.75 (¾)	0.0 (0) / 0.5 (½)	0.0 (0) / 0.5 (½)	0.0 (0) / 0.75 (¾)	0.0 (0) / 0.75 (¾)
FRONT TOE (Degrees)	0.2° ±0.2° 3/16° ± 3/16°	0.2° ±0.2° 3/16° ± 3/16°	0.1° ±0.2° 3/32° ± 3/16°	0.1° ±0.2° 3/32° ± 3/16°	0.0° ±0.2° 0° ± 3/16°	0.0° ±0.2° 0° ± 3/16°
FRONT TOE (Inches)	0.1" ±0.1" 3/32" ± 3/32"	0.1" ±0.1" 3/32" ± 3/32"	0.05" ±0.1" 1/16" ± 3/32"	0.05" ±0.1" 1/16" ± 3/32"	0.0" ±0.1" 0" ± 3/32"	0.0" ±0.1" 0" ± 3/32"
SAI	12.2 ±0.8 (12 3/16 ± ¾)	12.3 ±0.8 (12¼ ± ¾)	12.9 ±0.8 (12 15/16 ± ¾)	12.9 ±0.8 (12 15/16 ± ¾)	13.1 ±0.8 (13 1/16 ± ¾)	13.1 ±0.8 (13 1/16 ± ¾)
TURNING ANGLE	39.4 / 32.9 ±2.0	39.2 / 32.9 ±2.0	35.8 / 29.8 ±1.0	35.8 / 29.8 ±1.0	35.9 / 31.3 ±2.0	35.8 / 31.4 ±2.0
ILLUSTRATION #	26	26	62	62	26	26
REAR						
CAMBER (Degrees)	-1.17 ±0.75 -1 3/16 ± ¾	-1.17 ±0.75 -1 3/16 ± ¾	-0.67 ±0.75 -11/16 ± ¾	-0.67 ±0.75 -11/16 ± ¾	-0.5 ±0.75 -½ ± ¾	-0.8 ±0.75 -13/16 ± ¾
REAR TOE (Degrees)	0.2° ±0.2° 3/16° ± 3/16°	0.2° ±0.2° 3/16° ± 3/16°	0.4° ±0.2° 13/32° ± 3/16°	0.4° ±0.2° 13/32° ± 3/16°	0.4° ±0.2° 13/32° ± 3/16°	0.4° ±0.2° 13/32° ± 3/16°
REAR TOE (Inches)	0.1" ±0.1" 3/32" ± 3/32"	0.1" ±0.1" 3/32" ± 3/32"	0.2" ±0.1" 3/16" ± 3/32"	0.2" ±0.1" 3/16" ± 3/32"	0.2" ±0.1" 3/16" ± 3/32"	0.2" ±0.1" 3/16" ± 3/32"
THRUST ANGLE						
ILLUSTRATION #	05	05	06	06	10	10
NOTES						
	26001	26002	26003	26004	26005	26006

Lexus

ES 300	ES 330	ES 350	MODEL INFORMATION	ES 350	ES 300h	GS 300
2002-2003	2004-2006	2007-2012	YEARS	2013-2017	2013-2017	1993-97
			INSTRUCTIONS			
76 ft.lbs.	76 ft.lbs.	76 ft.lbs.	LUGNUT TORQUE	76 ft.lbs.	76 ft.lbs.	76 ft.lbs.
			FRONT			
2.77 ±0.75 2¾ ± ¾	2.77 ±0.75 2¾ ± ¾	2.95 ±0.75 2 15/16 ± ¾	CASTER (Degrees)	2.92 ±0.75 2 15/16 ± ¾	2.92 ±0.75 2 15/16 ± ¾	7.08 ±0.75 7 1/16 ± ¾
0.0 (0) | 0.75 (¾)	0.0 (0) | 0.75 (¾)	0.0 (0) | 0.75 (¾)	CROSS | TOL.	0.0 (0) | 0.75 (¾)	0.0 (0) | 0.75 (¾)	0.0 (0) | 0.5 (½)
-0.72 ±0.75 -¾ ± ¾	-0.72 ±0.75 -¾ ± ¾	-0.72 ±0.75 -¾ ± ¾	CAMBER (Degrees)	-0.67 ±0.75 -11/16 ± ¾	-0.58 ±0.75 - 9/16 ± ¾	-0.25 ±0.75 -¼ ± ¾
0.0 (0) | 0.75 (¾)	0.0 (0) | 0.75 (¾)	0.0 (0) | 0.75 (¾)	CROSS | TOL.	0.0 (0) | 0.75 (¾)	0.0 (0) | 0.75 (¾)	0.0 (0) | 0.5 (½)
0.0° ±0.2° 0° ± 3/16°	0.0° ±0.2° 0° ± 3/16°	0.0° ±0.2° 0° ± 3/16°	FRONT TOE (Degrees)	0.0° ±0.17° 0° ± 5/32°	0.0° ±0.17° 0° ± 5/32°	0.1° ±0.2° 3/32° ± 3/16°
0.0" ±0.1" 0" ± 3/32"	0.0" ±0.1" 0" ± 3/32"	0.0" ±0.1" 0" ± 3/32"	FRONT TOE (Inches)	0.0" ±0.08" 0" ± 3/32"	0.0" ±0.08" 0" ± 3/32"	0.05" ±0.1" 1/16" ± 3/32"
11.5 ±0.8 (11 7/16 ± ¾)	11.5 ±0.8 (11 7/16 ± ¾)	12.3 ±0.8 (12 5/16 ± ¾)	SAI	12.3 ±0.8 (12 ¼ ± ¾)	12.2 ±0.8 (12 3/16 ± ¾)	8.8 ±0.8 (8 13/16 ± ¾)
36.7 / 32.2 ±2.0	36.7 / 32.2 ±2.0	38.5 / 33.7 ±2.0	TURNING ANGLE	37.3 / 32.9 ±2.0	37.3 / 32.9 ±2.0	38.8 / 32.0 ±1.5
26	26	26	ILLUSTRATION #	26	26	121
			REAR			
-1.38 ±0.75 -1⅜ ± ¾	-1.38 ±0.75 -1⅜ ± ¾	-1.35 ±0.75 -1⅜ ± ¾	CAMBER (Degrees)	-1.17 ±0.75 -1 3/16 ± ¾	-1.17 ±0.75 -1 3/16 ± ¾	-1.0 ±0.75 -1 ± ¾
0.4° ±0.2° 13/32° ± 3/16°	0.4° ±0.2° 13/32° ± 3/16°	0.32° ±0.16° 5/16° ± 5/32°	REAR TOE (Degrees)	0.22° ±0.17° 7/32° ± 5/32°	0.22° ±0.17° 7/32° ± 5/32°	0.4° ±0.2° 13/32° ± 3/16°
0.2" ±0.1" 3/16" ± 3/32"	0.2" ±0.1" 3/16" ± 3/32"	0.16" ±0.08" 5/32" ± 3/32"	REAR TOE (Inches)	0.11" ±0.08" 3/32" ± 3/32"	0.11" ±0.08" 3/32" ± 3/32"	0.2" ±0.1" 3/16" ± 3/32"
			THRUST ANGLE			
10	10	10	ILLUSTRATION #	10	10	63
			NOTES			
26007	26008	26009		26010	26011	26012

Lexus

GS 300/GS 400 GS 430	GS 300/GS 350 RWD	GS 300/GS 350 AWD	MODEL INFORMATION	GS 430/GS 460	GS 450h	GS 350 RWD
1998-2005	2006-2011	2006-2011	YEARS	2006-2011	2007-2011	2013-2015
			INSTRUCTIONS			
76 ft.lbs.	76 ft.lbs.	76 ft.lbs.	LUGNUT TORQUE	76 ft.lbs.	76 ft.lbs.	76 ft.lbs.
			FRONT			
7.55 ±0.5 7 9/16 ± ½	7.38 ±0.75 7⅜ ± ¾	4.88 ±0.75 4 ⅞ ± ¾	CASTER (Degrees)	7.43 ±0.75 7 7/16 ± ¾	7.43 ±0.75 7 7/16 ± ¾	7.87 ±0.75 7 ⅞ ± ¾
0.0 (0) | 0.5 (½)	0.0 (0) | 0.5 (½)	0.0 (0) | 0.5 (½)	CROSS | TOL.	0.0 (0) | 0.5 (½)	0.0 (0) | 0.5 (½)	0.0 (0) | 0.75 (¾)
-0.27 ±0.5 -¼ ± ½	-0.38 ±0.75 -⅜ ± ¾	-0.42 ±0.75 - 7/16 ± ¾	CAMBER (Degrees)	-0.43 ±0.75 - 7/16 ± ¾	-0.43 ±0.75 - 7/16 ± ¾	-0.22 ±0.75 -¼ ± ¾
0.0 (0) | 0.5 (½)	0.0 (0) | 0.5 (½)	0.0 (0) | 0.5 (½)	CROSS | TOL.	0.0 (0) | 0.5 (½)	0.0 (0) | 0.5 (½)	0.0 (0) | 0.75 (¾)
0.15° ±0.2° 5/32° ± 3/16°	0.08° ±0.16° 3/32° ± 5/32°	0.08° ±0.16° 3/32° ± 5/32°	FRONT TOE (Degrees)	0.08° ±0.16° 3/32° ± 5/32°	0.08° ±0.16° 3/32° ± 5/32°	0.09° ±0.17° 3/32° ± 5/32°
0.08" ±0.1" 1/16" ± 3/32"	0.04" ±0.08" 1/32" ± 3/32"	0.04" ±0.08" 1/32" ± 3/32"	FRONT TOE (Inches)	0.04" ±0.08" 1/32" ± 3/32"	0.04" ±0.08" 1/32" ± 3/32"	0.04" ±0.08" 1/32" ± 3/32"
8.9 ±0.5 (8 ⅞ ± ½)	9.4 ±0.8 (9⅜ ± ¾)	11.2 ±0.8 (11 3/16 ± ¾)	SAI	9.4 ±0.8 (9 7/16 ± ¾)	9.4 ±0.8 (9 7/16 ± ¾)	10.5 ±0.8 (10½ ± ¾)
38.8 / 32.1 ±1.5	41.8 / 36.2 ±1.5	39.4 / 36.0 ±1.5	TURNING ANGLE	41.8 / 36.1 ±1.5	41.8 / 36.1 ±1.5	41.6 / 36.0 ±2.0
14		14	ILLUSTRATION #			
			REAR			
-0.78 ±0.5 - 13/16 ± ½	-1.17 ±0.75 -1 3/16 ± ¾	-1.05 ±0.75 -1 1/16 ± ¾	CAMBER (Degrees)	-1.32 ±0.75 -1 5/16 ± ¾	-1.32 ±0.75 -1 5/16 ± ¾	-1.48 ±0.75 -1½ ± ¾
0.13° ±0.2° ⅛° ± 3/16°	0.24° ±0.16° ¼° ± 5/32°	0.24° ±0.16° ¼° ± 5/32°	REAR TOE (Degrees)	0.24° ±0.16° ¼° ± 5/32°	0.24° ±0.16° ¼° ± 5/32°	0.24° ±0.16° ¼° ± 5/32°
0.07" ±0.1" 1/16" ± 3/32"	0.12" ±0.08" ⅛" ± 3/32"	0.12" ±0.08" ⅛" ± 3/32"	REAR TOE (Inches)	0.12" ±0.08" ⅛" ± 3/32"	0.12" ±0.08" ⅛" ± 3/32"	0.12" ±0.08" ⅛" ± 3/32"
			THRUST ANGLE			
117	174	174	ILLUSTRATION #	174	174	195
			NOTES			
26013	26014	26015		26016	26017	26018

Lexus

GS 350/GS 200t RWD	GS 350 AWD	GS 450h	MODEL INFORMATION	GS F	GX 460	GX 460 (start 08/2013)
2016-2017	2013-2017	2013-2017	YEARS	2016-2017	2010-2013	2014-2017
			INSTRUCTIONS			
76 ft.lbs.	76 ft.lbs.	76 ft.lbs.	LUGNUT TORQUE	76 ft.lbs.	82 ft.lbs.	82 ft.lbs.
			FRONT			
7.83 ±0.75 7⅞ ± ¾	4.68 ±0.75 4 11/16 ± ¾	7.9 ±0.75 7 15/16 ± ¾	CASTER (Degrees)	7.63 ±0.75 7⅝ ± ¾	3.45 ±0.5 3 7/16 ± ½	3.48 ±0.5 3½ ± ½
0.0 (0) | 0.75 (¾)	0.0 (0) | 0.75 (¾)	0.0 (0) | 0.75 (¾)	CROSS | TOL.	0.0 (0) | 0.75 (¾)	0.0 (0) | 0.5 (½)	0.0 (0) | 0.5 (½)
-0.18 ±0.75 -3/16 ± ¾	-0.47 ±0.75 -½ ± ¾	-0.2 ±0.75 -3/16 ± ¾	CAMBER (Degrees)	-0.47 ±0.75 -½ ± ¾	0.03 ±0.5 1/16 ± ½	0.03 ±0.5 0 ± ½
0.0 (0) | 0.75 (¾)	0.0 (0) | 0.75 (¾)	0.0 (0) | 0.75 (¾)	CROSS | TOL.	0.0 (0) | 0.75 (¾)	0.0 (0) | 0.5 (½)	0.0 (0) | 0.5 (½)
0.09° ±0.17° 3/32° ± 5/32°	0.09° ±0.17° 3/32° ± 5/32°	0.08° ±0.16° 3/32° ± 5/32°	FRONT TOE (Degrees)	0.0° ±0.17° 0° ± 5/32°	0.2° ±0.17° [1] 3/16° ± 5/32° [1]	0.24° ±0.15° ¼° ± 5/32°
0.04" ±0.08" 1/32" ± 3/32"	0.04" ±0.08" 1/32" ± 3/32"	0.04" ±0.08" 1/32" ± 3/32"	FRONT TOE (Inches)	0.0" ±0.08" 0" ± 3/32"	0.1" ±0.08" [1] 3/32" ± 3/32" [1]	0.12" ±0.08" ⅛" ± 1/16"
10.4 ±0.8 (10 7/16 ± ¾)	10.8 ±0.8 (10 13/16 ± ¾)	10.5 ±0.8 (10½ ± ¾)	SAI	9.8 ±0.8 (9.8 ± ¾)	12.1 ±0.8 (12⅛ ± ¾)	12.1 ±0.5 (12⅛ ± ½)
41.6 / 36.0 ±2.0	39.4 / 36.0 ±2.0	41.6 / 36.0 ±2.0	TURNING ANGLE	39.2 / 34.0 ±2.0	35.7 / 32.5 ±1.5	35.5 / 32.5 ±1.5
	14		ILLUSTRATION #		12	12
			REAR			
-1.43 ±0.75 -1 7/16 ± ¾	-1.25 ±0.75 -1¼ ± ¾	-1.5 ±0.75 -1½ ± ¾	CAMBER (Degrees)	-1.27 ±0.75 -1¼ ± ¾		
0.24° ±0.16° ¼° ± 5/32°	0.24° ±0.16° ¼° ± 5/32°	0.24° ±0.16° ¼° ± 5/32°	REAR TOE (Degrees)	0.25° ±0.16° ¼° ± 5/32°		
0.12" ±0.08" ⅛" ± 3/32"	0.12" ±0.08" ⅛" ± 3/32"	0.12" ±0.08" ⅛" ± 3/32"	REAR TOE (Inches)	0.13" ±0.08" ⅛" ± 3/32"		
			THRUST ANGLE			
195	195	195	ILLUSTRATION #	195		
			NOTES		[1] 2013 = 0.06°(0.03")	
26019	26020	26021		26022	26023	26024

Lexus

GX 470	HS 250h w/17" Tires	HS 250h w/18" Tires	MODEL INFORMATION	IS 300 Sedan USA	IS 300 Sedan Canada	IS 300 Sedan Canada
2003-2009	2010-2012	2010-2012	YEARS	2001-2005	2001	2002-2005
			INSTRUCTIONS			
	76 ft.lbs.	76 ft.lbs.	LUGNUT TORQUE	76 ft.lbs.	76 ft.lbs.	76 ft.lbs.
			FRONT			
3.28 ±0.75 3¼ ± ¾	6.07 ±0.75 6 1/16 ± ¾	5.87 ±0.75 5⅞ ± ¾	CASTER (Degrees)	6.12 ±0.5 6⅛ ± ½	5.55 ±0.5 5 9/16 ± ½	5.77 ±0.5 5¾ ± ½
0.0 (0) | 0.75 (¾)	0.0 (0) | 0.75 (¾)	0.0 (0) | 0.75 (¾)	CROSS | TOL.	0.0 (0) | 0.5 (½)	0.0 (0) | 0.5 (½)	0.0 (0) | 0.5 (½)
0.0 ±0.75 0 ± ¾	-0.22 ±0.75 -¼ ± ¾	-0.22 ±0.75 -¼ ± ¾	CAMBER (Degrees)	-0.5 ±0.5 -½ ± ½	-0.1 ±0.5 -⅛ ± ½	-0.35 ±0.5 -⅜ ± ½
0.0 (0) | 0.75 (¾)	0.0 (0) | 0.75 (¾)	0.0 (0) | 0.75 (¾)	CROSS | TOL.	0.0 (0) | 0.5 (½)	0.0 (0) | 0.5 (½)	0.0 (0) | 0.5 (½)
0.08° ±0.17° 3/32° ± 5/32°	0.18° ±0.18° 3/16° ± 3/16°	0.18° ±0.18° 3/16° ± 3/16°	FRONT TOE (Degrees)	0.1° ±0.2° 3/32° ± 3/16°	0.1° ±0.2° 3/32° ± 3/16°	0.1° ±0.2° 3/32° ± 3/16°
0.04" ±0.08" 1/32" ± 3/32"	0.09" ±0.09" 3/32" ± 3/32"	0.09" ±0.09" 3/32" ± 3/32"	FRONT TOE (Inches)	0.05" ±0.1" 1/16" ± 3/32"	0.05" ±0.1" 1/16" ± 3/32"	0.05" ±0.1" 1/16" ± 3/32"
12.5 ±0.8 (12½ ± ¾)	12.0 ±0.8 (12 ± ¾)	12.0 ±0.8 (12 ± ¾)	SAI	9.4 ±0.5 (9 7/16 ± ½)	9.0 ±0.5 (9 ± ½)	9.3 ±0.5 (9¼ ± ½)
36.6 / 33.0 ±1.5	36.9 / 31.7 ±2.0	36.9 / 31.7 ±2.0	TURNING ANGLE	41.0 / 33.4 ±1.5	40.5 / 33.7 ±1.5	41.0 / 33.5 ±1.5
12	26	26	ILLUSTRATION #	14	14	14
			REAR			
	-1.02 ±0.75 -1 ± ¾	-0.85 ±0.75 -⅞ ± ¾	CAMBER (Degrees)	-0.92 ±0.5 -15/16 ± ½	-0.07 ±0.5 -1/16 ± ½	-0.38 ±0.5 -⅜ ± ½
	0.18° ±0.18° 3/16° ± 3/16°	0.18° ±0.18° 3/16° ± 3/16°	REAR TOE (Degrees)	0.2° ±0.2° 3/16° ± 3/16°	0.2° ±0.2° 3/16° ± 3/16°	0.2° ±0.2° 3/16° ± 3/16°
	0.09" ±0.09" 3/32" ± 3/32"	0.09" ±0.09" 3/32" ± 3/32"	REAR TOE (Inches)	0.1" ±0.1" 3/32" ± 3/32"	0.1" ±0.1" 3/32" ± 3/32"	0.1" ±0.1" 3/32" ± 3/32"
			THRUST ANGLE			
	10	10	ILLUSTRATION #	117	117	117
			NOTES			
26025	26026	26027		26028	26029	26030

ALIGNMENT SPECIFICATIONS

Lexus

IS 300 Wagon USA	IS 300 Wagon Canada	IS 250/IS 350 RWD Normal Susp.	MODEL INFORMATION	IS 250/IS 350 RWD Normal Susp.	IS 250/IS 350 RWD Sport Susp.	IS 250/IS 350 AWD
2002-2005	2002-2005	2006-2008	YEARS	2009-2013	2006-2013	2006-2013
			INSTRUCTIONS			
76 ft.lbs.	76 ft.lbs.	76 ft.lbs.	LUGNUT TORQUE	76 ft.lbs.	76 ft.lbs.	76 ft.lbs.
			FRONT			
5.77 ±0.5 5¾ ± ½	5.52 ±0.5 5½ ± ½	8.2 ±0.75 [1] 8 3/16 ± ¾ [1]	CASTER (Degrees)	8.2 ±0.75 [1] 8 3/16 ± ¾ [1]	8.07 ±0.75 8 1/16 ± ¾	4.63 ±0.75 4⅝ ± ¾
0.0 (0) | 0.5 (½)	0.0 (0) | 0.5 (½)	0.0 (0) | 0.5 (½)	CROSS | TOL.	0.0 (0) | 0.5 (½)	0.0 (0) | 0.5 (½)	0.0 (0) | 0.5 (½)
-0.35 ±0.5 -⅜ ± ½	-0.08 ±0.5 -1/16 ± ½	-0.38 ±0.75 -⅜ ± ¾	CAMBER (Degrees)	-0.38 ±0.75 -⅜ ± ¾	-0.6 ±0.75 -⅝ ± ¾	-0.42 ±0.75 -7/16 ± ¾
0.0 (0) | 0.5 (½)	0.0 (0) | 0.5 (½)	0.0 (0) | 0.5 (½)	CROSS | TOL.	0.0 (0) | 0.5 (½)	0.0 (0) | 0.5 (½)	0.0 (0) | 0.5 (½)
0.1° ±0.2° 3/32° ± 3/16°	0.1° ±0.2° 3/32° ± 3/16°	0.08° ±0.16° 3/32° ± 5/32°	FRONT TOE (Degrees)	0.08° ±0.16° 3/32° ± 5/32°	0.08° ±0.16° 3/32° ± 5/32°	0.08° ±0.16° 3/32° ± 5/32°
0.05" ±0.1" 1/16" ± 3/32"	0.05" ±0.1" 1/16" ± 3/32"	0.04" ±0.08" 1/32" ± 3/32"	FRONT TOE (Inches)	0.04" ±0.08" 1/32" ± 3/32"	0.04" ±0.08" 1/32" ± 3/32"	0.04" ±0.08" 1/32" ± 3/32"
9.3 ±0.5 (9¼ ± ½)	9.0 ±0.5 (9 ± ½)	10.7 ±0.8 (10 11/16 ± ¾)	SAI	10.7 ±0.8 (10 11/16 ± ¾)	10.9 ±0.8 (10 15/16 ± ¾)	11.3 ±0.8 (11¼ ± ¾)
41.0 / 33.5 ±1.5	41.0 / 33.6 ±1.5	41.6 / 36.2 ±1.5	TURNING ANGLE	41.6 / 36.2 ±1.5	41.4 / 35.9 ±1.5	39.3 / 35.9 ±1.5
14	14		ILLUSTRATION #			14
			REAR			
-0.38 ±0.5 -⅜ ± ½	-0.07 ±0.5 -1/16 ± ½	-1.23 ±0.75 [2] -1¼ ± ¾ [2]	CAMBER (Degrees)	-1.17 ±0.75 [2] -1 3/16 ± ¾ [2]	-1.23 ±0.75 [1] -1¼ ± ¾ [1]	-0.83 ±0.75 [1] -13/16 ± ¾ [1]
0.2° ±0.2° 3/16° ± 3/16°	0.2° ±0.2° 3/16° ± 3/16°	0.24° ±0.16° ¼° ± 5/32°	REAR TOE (Degrees)	0.24° ±0.16° ¼° ± 5/32°	0.24° ±0.16° ¼° ± 5/32°	0.24° ±0.16° ¼° ± 5/32°
0.1" ±0.1" 3/32" ± 3/32"	0.1" ±0.1" 3/32" ± 3/32"	0.12" ±0.08" ⅛" ± 3/32"	REAR TOE (Inches)	0.12" ±0.08" ⅛" ± 3/32"	0.12" ±0.08" ⅛" ± 3/32"	0.12" ±0.08" ⅛" ± 3/32"
			THRUST ANGLE			
117	117	174	ILLUSTRATION #	174	174	174
		[1] 16" or 18" wheels = 8.1° (8⅛°) [2] 16" wheels = -1.0° (-1°)	NOTES	[1] 16" or 18" wheels = 8.1° (8⅛°) [2] 16" wheels = -0.95° (-15/16°)	[1] 2009-10 = -1.17° (-1 3/16°)	[1] 2009-10 = -0.75° (-¾°)
26031	26032	26033		26034	26035	26036

Lexus

IS 250/IS 350 RWD with 16" wheels	IS 250/IS 350 RWD with 17" or 18" wheels	IS 250/IS 350 AWD with 16" wheels	MODEL INFORMATION	IS 250/IS 350 AWD with 17" or 18" wheels	IS 200t/IS 350 RWD	IS 300/IS 350 AWD
2014-2015	2014-2015	2014-2015	YEARS	2014-2015	2016-2017	2016-2017
			INSTRUCTIONS			
76 ft.lbs.	76 ft.lbs.	76 ft.lbs.	LUGNUT TORQUE	76 ft.lbs.	76 ft.lbs.	76 ft.lbs.
			FRONT			
8.25 ±0.75 8¼ ± ¾	8.3 ±0.75 8 5/16 ± ¾	4.87 ±0.75 4⅞ ± ¾	CASTER (Degrees)	4.93 ±0.75 4 15/16 ± ¾	8.22 ±0.75 8¼ ± ¾	4.93 ±0.75 4 15/16 ± ¾
0.0 (0) | 0.5 (½)	0.0 (0) | 0.5 (½)	0.0 (0) | 0.5 (½)	CROSS | TOL.	0.0 (0) | 0.5 (½)	0.0 (0) | 0.5 (½)	0.0 (0) | 0.5 (½)
-0.18 ±0.75 -3/16 ± ¾	-0.28 ±0.75 -¼ ± ¾	-0.38 ±0.75 -⅜ ± ¾	CAMBER (Degrees)	-0.48 ±0.75 -½ ± ¾	-0.22 ±0.75 -¼ ± ¾	-0.48 ±0.75 -½ ± ¾
0.0 (0) | 0.5 (½)	0.0 (0) | 0.5 (½)	0.0 (0) | 0.5 (½)	CROSS | TOL.	0.0 (0) | 0.5 (½)	0.0 (0) | 0.5 (½)	0.0 (0) | 0.5 (½)
0.05° ±0.2° 1/16° ± 3/16°	0.05° ±0.2° 1/16° ± 3/16°	0.05° ±0.2° 1/16° ± 3/16°	FRONT TOE (Degrees)	0.05° ±0.2° 1/16° ± 3/16°	0.0° ±0.18° 0° ± 3/16°	0.05° ±0.18° 1/16° ± 3/16°
0.03" ±0.1" 1/32" ± 3/32"	0.03" ±0.1" 1/32" ± 3/32"	0.03" ±0.1" 1/32" ± 3/32"	FRONT TOE (Inches)	0.03" ±0.1" 1/32" ± 3/32"	0.0" ±0.09" 0" ± 3/32"	0.03" ±0.09" 1/32" ± 3/32"
10.1 ±0.8 (10⅛ ± ¾)	10.2 ±0.8 (10 3/16 ± ¾)	11.2 ±0.8 (11 3/16 ± ¾)	SAI	11.3 ±0.8 (11 5/16 ± ¾)	10.4 ±0.8 (10 7/16 ± ¾)	11.3 ±0.8 (11 5/16 ± ¾)
41.5 / 36.2 ±1.5	41.8 / 36.6 ±1.5	39.2 / 35.8 ±1.5	TURNING ANGLE	39.5 / 35.2 ±1.5	41.8 / 36.4 ±1.5	39.5 / 36.2 ±1.5
		14	ILLUSTRATION #	14		14
			REAR			
-1.43 ±0.75 -1 7/16 ± ¾	-1.58 ±0.75 -1 9/16 ± ¾	-1.48 ±0.75 -1½ ± ¾	CAMBER (Degrees)	-1.63 ±0.75 -1⅝ ± ¾	-1.58 ±0.75 -1 9/16 ± ¾	-1.63 ±0.75 -1⅝ ± ¾
0.18° ±0.2° 3/16° ± 3/16°	0.18° ±0.2° 3/16° ± 3/16°	0.18° ±0.2° 3/16° ± 3/16°	REAR TOE (Degrees)	0.18° ±0.2° 3/16° ± 3/16°	0.18° ±0.18° 3/16° ± 3/16°	0.18° ±0.18° 3/16° ± 3/16°
0.09" ±0.1" 3/32" ± 3/32"	0.09" ±0.1" 3/32" ± 3/32"	0.09" ±0.1" 3/32" ± 3/32"	REAR TOE (Inches)	0.09" ±0.1" 3/32" ± 3/32"	0.09" ±0.09" 3/32" ± 3/32"	0.09" ±0.09" 3/32" ± 3/32"
			THRUST ANGLE			
174	174	174	ILLUSTRATION #	174	174	174
			NOTES			
26037	26038	26039		26040	26041	26042

Lexus

			MODEL INFORMATION			
IS 250 C/IS 350 C	IS F	IS F	MODEL INFORMATION	LS 400 without Air Suspension	LS 400 without Air Suspension	LS 400 without Air Suspension
2010-2014	2008-2010	2011-2014	YEARS	1990	1991-92	1993-94
			INSTRUCTIONS			
76 ft.lbs.	76 ft.lbs.	76 ft.lbs.	LUGNUT TORQUE			
			FRONT			
7.92 ± 0.75 [1] 7 15/16 ± ¾ [1]	8.7 ± 0.75 8 11/16 ± ¾	8.7 ± 0.75 8 11/16 ± ¾	CASTER (Degrees)	9.25 ± 0.75 9 ¼ ± ¾	9.17 ± 0.75 9 3/16 ± ¾	9.33 ± 0.75 9 5/16 ± ¾
0.0 (0) | 0.5 (½)	0.0 (0) | 0.5 (½)	0.0 (0) | 0.5 (½)	CROSS | TOL.	0.0 (0) | 0.5 (½)	0.0 (0) | 0.5 (½)	0.0 (0) | 0.5 (½)
-0.18 ± 0.75 -3/16 ± ¾	-0.87 ± 0.75 -7/8 ± ¾	-0.87 ± 0.75 -7/8 ± ¾	CAMBER (Degrees)	0.08 ± 0.5 1/16 ± ½	0.0 ± 0.75 0 ± ¾	0.08 ± 0.75 1/16 ± ¾
0.0 (0) | 0.5 (½)	0.0 (0) | 0.5 (½)	0.0 (0) | 0.5 (½)	CROSS | TOL.	0.0 (0) | 0.5 (½)	0.0 (0) | 0.5 (½)	0.0 (0) | 0.5 (½)
0.1° ± 0.18° 3/32° ± 3/16°	0.0° ± 0.16° 0° ± 5/32°	0.02° ± 0.16° 1/32° ± 5/32°	FRONT TOE (Degrees)	0.2° ± 0.2° 3/16° ± 3/16°	0.2° ± 0.2° 3/16° ± 3/16°	0.3° ± 0.2° 5/16° ± 3/16°
0.05" ± 0.09" 1/16" ± 3/32"	0.0" ± 0.08" 0" ± 3/32"	0.01" ± 0.08" 1/64" ± 3/32"	FRONT TOE (Inches)	0.1" ± 0.1" 3/32" ± 3/32"	0.1" ± 0.1" 3/32" ± 3/32"	0.15" ± 0.1" 5/32" ± 3/32"
10.5 ± 0.8 (10 ½ ± ¾)	11.2 ± 0.8 (11 3/16 ± ¾)	11.2 ± 0.8 (11 3/16 ± ¾)	SAI	8.8 ± 0.8 (8 13/16 ± ¾)	8.8 ± 0.8 (8 ¾ ± ¾)	8.8 ± 0.8 (8 ¾ ± ¾)
41.8 / 36.3 ± 2.0	41.8 / 35.8 ± 2.0	41.8 / 35.8 ± 2.0	TURNING ANGLE	39.3 / 32.5 ± 1.0	39.3 / 32.7 ± 1.0	39.3 / 32.7 ± 1.0
			ILLUSTRATION #	03 14	03 14	03 14
			REAR			
-0.97 ± 0.75 -1 ± ¾	-0.78 ± 0.75 -¾ ± ¾	-1.5 ± 0.75 -1 ½ ± ¾	CAMBER (Degrees)	0.0 ± 0.5 0 ± ½	0.0 ± 0.75 0 ± ¾	-0.17 ± 0.75 -3/16 ± ¾
0.28° ± 0.18° 9/32° ± 3/16°	0.24° ± 0.16° ¼° ± 5/32°	0.24° ± 0.16° ¼° ± 5/32°	REAR TOE (Degrees)	0.2° ± 0.2° 3/16° ± 3/16°	0.2° ± 0.2° 3/16° ± 3/16°	0.2° ± 0.2° 3/16° ± 3/16°
0.14" ± 0.09" 5/32" ± 3/32"	0.12" ± 0.08" 1/8" ± 3/32"	0.12" ± 0.08" 1/8" ± 3/32"	REAR TOE (Inches)	0.1" ± 0.1" 3/32" ± 3/32"	0.1" ± 0.1" 3/32" ± 3/32"	0.1" ± 0.1" 3/32" ± 3/32"
			THRUST ANGLE			
174	174	174	ILLUSTRATION #	63	63	63
[1] w/18" tires = 7.8° (7 13/16°)			NOTES			
26043	26044	26045		26047	26048	26049

Lexus

			MODEL INFORMATION			
LS 400 with Air Suspension	LS 400 without Air Suspension	LS 400 with Air Suspension	MODEL INFORMATION	LS 430 without Air Suspension	LS 430 with Air Suspension	LS 430 without Air Suspension
1990-94	1995-2000	1995-2000	YEARS	2001-2003	2001-2003	2004-2006
			INSTRUCTIONS			
			LUGNUT TORQUE	76 ft.lbs.	76 ft.lbs.	76 ft.lbs.
			FRONT			
9.83 ± 0.75 9 13/16 ± ¾	7.0 ± 0.75 7 ± ¾	7.42 ± 0.75 7 7/16 ± ¾	CASTER (Degrees)	6.75 ± 0.75 [1] 6 ¾ ± ¾ [1]	7.25 ± 0.75 7 ¼ ± ¾	6.75 ± 0.75 6 ¾ ± ¾
0.0 (0) | 0.5 (½)	0.0 (0) | 0.5 (½)	0.0 (0) | 0.5 (½)	CROSS | TOL.	0.0 (0) | 0.5 (½)	0.0 (0) | 0.5 (½)	0.0 (0) | 0.5 (½)
-0.08 ± 0.75 -1/16 ± ¾	0.33 ± 0.75 5/16 ± ¾	0.08 ± 0.75 1/16 ± ¾	CAMBER (Degrees)	-0.08 ± 0.75 -1/16 ± ¾	-0.25 ± 0.75 -¼ ± ¾	-0.08 ± 0.75 -1/16 ± ¾
0.0 (0) | 0.5 (½)	0.0 (0) | 0.5 (½)	0.0 (0) | 0.5 (½)	CROSS | TOL.	0.0 (0) | 0.5 (½)	0.0 (0) | 0.5 (½)	0.0 (0) | 0.5 (½)
0.1° ± 0.2° 3/32° ± 3/16°	0.3° ± 0.2° 5/16° ± 3/16°	0.1° ± 0.2° 3/32° ± 3/16°	FRONT TOE (Degrees)	0.1° ± 0.2° 3/32° ± 3/16°	0.1° ± 0.2° 3/32° ± 3/16°	0.1° ± 0.2° 3/32° ± 3/16°
0.05" ± 0.1" 1/16" ± 3/32"	0.15" ± 0.1" 5/32" ± 3/32"	0.05" ± 0.1" 1/16" ± 3/32"	FRONT TOE (Inches)	0.05" ± 0.1" 1/16" ± 3/32"	0.05" ± 0.1" 1/16" ± 3/32"	0.05" ± 0.1" 1/16" ± 3/32"
8.8 ± 0.8 (8 13/16 ± ¾)	8.4 ± 0.8 (8 7/16 ± ¾)	8.7 ± 0.8 (8 11/16 ± ¾)	SAI	9.0 ± 0.8 (9 ± ¾)	9.3 ± 0.8 (9 ¼ ± ¾)	9.0 ± 0.8 (9 ± ¾)
38.8 / 32.5 ± 1.5	42.0 / 34.3 ± 1.5	42.0 / 34.0 ± 1.5	TURNING ANGLE	43.5 / 37.3 ± 1.5	43.5 / 37.3 ± 1.5	42.2 / 36.3 ± 1.5
03 14	03 14	03 14	ILLUSTRATION #	14	14	14
			REAR			
-0.75 ± 0.5 -¾ ⊥ ½	-0.83 ± 0.75 13/10 ± ¾	-1.42 ± 0.75 -1 7/16 ± ¾	CAMBER (Degrees)	-1.0 ± 0.75 -1 ± ¾	-1.55 ± 0.75 -1 9/16 ± ¾	-1.25 ± 0.75 -1 ¼ ± ¾
0.3° ± 0.2° 5/16° ± 3/16°	0.2° ± 0.2° 3/16° ± 3/16°	0.3° ± 0.2° 5/16° ± 3/16°	REAR TOE (Degrees)	0.3° ± 0.2° 5/16° ± 3/16°	0.3° ± 0.2° 5/16° ± 3/16°	0.3° ± 0.2° 5/16° ± 3/16°
0.15" ± 0.1" 5/32" ± 3/32"	0.1" ± 0.1" 3/32" ± 3/32"	0.15" ± 0.1" 5/32" ± 3/32"	REAR TOE (Inches)	0.15" ± 0.1" 5/32" ± 3/32"	0.15" ± 0.1" 5/32" ± 3/32"	0.15" ± 0.1" 5/32" ± 3/32"
			THRUST ANGLE			
63	63	63	ILLUSTRATION #	174	174	174
			NOTES	[1] with Sport Susp. 6.67° (6 11/16°)		
26050	26051	26052		26053	26054	26055

Lexus

LS 430 with Air Suspension	LS 460/460L RWD Coil Suspension	LS 460/460L RWD Air Suspension	MODEL INFORMATION	LS 460/460L AWD Coil Suspension	LS 460/460L AWD Air Suspension	LS 460/460L RWD Coil Suspension
2004-2006	2007-2012	2007-2012	YEARS	2009-2012	2009-2012	2013-2017
			INSTRUCTIONS			
76 ft.lbs.	103 ft.lbs.	103 ft.lbs.	LUGNUT TORQUE	103 ft.lbs.	103 ft.lbs.	103 ft.lbs.
			FRONT			
7.25 ±0.75 7¼ ± ¾	6.62 ±0.75 6⅝ ± ¾	7.02 ±0.75 7 ± ¾	CASTER (Degrees)	4.6 ±0.75 4⅝ ± ¾	4.97 ±0.75 5 ± ¾	6.47 ±0.75 6½ ± ¾
0.0 (0) | 0.5 (½)	0.0 (0) | 0.75 (¾)	0.0 (0) | 0.75 (¾)	CROSS | TOL.	0.0 (0) | 0.75 (¾)	0.0 (0) | 0.75 (¾)	0.0 (0) | 0.75 (¾)
-0.25 ±0.75 -¼ ± ¾	-0.28 ±0.75 -¼ ± ¾	-0.43 ±0.75 -7/16 ± ¾	CAMBER (Degrees)	-0.42 ±0.75 -7/16 ± ¾	-0.67 ±0.75 -11/16 ± ¾	-0.27 ±0.75 -¼ ± ¾
0.0 (0) | 0.5 (½)	0.0 (0) | 0.5 (½)	0.0 (0) | 0.5 (½)	CROSS | TOL.	0.0 (0) | 0.5 (½)	0.0 (0) | 0.5 (½)	0.0 (0) | 0.5 (½)
0.1° ±0.2° 3/32° ± 3/16°	0.08° ±0.16° 3/32° ± 5/32°	0.0° ±0.16° 0° ± 5/32°	FRONT TOE (Degrees)	0.0° ±0.16° 0° ± 5/32°	0.0° ±0.16° 0° ± 5/32°	0.0° ±0.17° 0° ± 5/32°
0.05" ±0.1" 1/16" ± 3/32"	0.04" ±0.08" 1/32" ± 3/32"	0.0" ±0.08" 0" ± 3/32"	FRONT TOE (Inches)	0.0" ±0.08" 0" ± 3/32"	0.0" ±0.08" 0" ± 3/32"	0.0" ±0.08" 0" ± 3/32"
9.3 ±0.8 (9¼ ± ¾)	9.1 ±0.8 (9⅛ ± ¾)	9.3 ±0.8 (9 5/16 ± ¾)	SAI	9.0 ±0.8 (9 ± ¾)	9.7 ±0.8 (9 11/16 ± ¾)	9.2 ±0.8 (9¼ ± ¾)
42.2 / 36.3 ±1.5	41.8 / 35.3 ±1.5 [1]	41.6 / 35.2 ±1.5 [1]	TURNING ANGLE	38.1 / 33.9 ±2.0	38.0 / 33.7 ±2.0	41.3 / 35.2 ±2.0 [1]
14	▲	▲	ILLUSTRATION #	173	173	
			REAR			
-1.58 ±0.75 -1 9/16 ± ¾	-1.63 ±0.75 -1⅝ ± ¾	-1.83 ±0.75 -1 13/16 ± ¾	CAMBER (Degrees)	-1.53 ±0.75 -1½ ± ¾	-1.83 ±0.75 -1 13/16 ± ¾	-1.63 ±0.75 -1⅝ ± ¾
0.3° ±0.2° 5/16° ± 3/16°	0.24° ±0.16° ¼° ± 5/32°	0.24° ±0.16° ¼° ± 5/32°	REAR TOE (Degrees)	0.24° ±0.16° ¼° ± 5/32°	0.24° ±0.16° ¼° ± 5/32°	0.25° ±0.17° ¼° ± 5/32°
0.15" ±0.1" 5/32" ± 3/32"	0.12" ±0.08" ⅛" ± 3/32"	0.12" ±0.08" ⅛" ± 3/32"	REAR TOE (Inches)	0.12" ±0.08" ⅛" ± 3/32"	0.12" ±0.08" ⅛" ± 3/32"	0.13" ±0.08" ⅛" ± 3/32"
			THRUST ANGLE			
174	178	178	ILLUSTRATION #	178	178	178
	[1] 19" wheels = 40.8°/34.7°	[1] 19" wheels = 40.6°/34.6°	NOTES			[1] 19" wheels = 40.4°/34.6°
26056	26057	26058		26059	26060	26061

Lexus

LS 460/460L RWD AIR Suspension	LS 460/460L RWD F-Sport Pkg.	LS 460/460L AWD Coil Suspension	MODEL INFORMATION	LS 460/460L AWD AIR Suspension	LS 460/460L AWD F-Sport Pkg.	LS 600h L
2013-2017	2013-2017	2013-2017	YEARS	2013-2017	2013-2017	2008-2012
			INSTRUCTIONS			
103 ft.lbs.	103 ft.lbs.	103 ft.lbs.	LUGNUT TORQUE	103 ft.lbs.	103 ft.lbs.	103 ft.lbs.
			FRONT			
6.95 ±0.75 6 15/16 ± ¾	7.35 ±0.75 7⅜ ± ¾	4.5 ±0.75 4½ ± ¾	CASTER (Degrees)	4.97 ±0.75 5 ± ¾	5.33 ±0.75 5 5/16 ± ¾	4.97 ±0.75 5 ± ¾
0.0 (0) | 0.75 (¾)	0.0 (0) | 0.75 (¾)	0.0 (0) | 0.75 (¾)	CROSS | TOL.	0.0 (0) | 0.75 (¾)	0.0 (0) | 0.75 (¾)	0.0 (0) | 0.75 (¾)
-0.42 ±0.75 -7/16 ± ¾	-0.65 ±0.75 -⅝ ± ¾	-0.42 ±0.75 -7/16 ± ¾	CAMBER (Degrees)	-0.67 ±0.75 -11/16 ± ¾	-0.92 ±0.75 -15/16 ± ¾	-0.67 ±0.5 -11/16 ± ½
0.0 (0) | 0.5 (½)	0.0 (0) | 0.5 (½)	0.0 (0) | 0.5 (½)	CROSS | TOL.	0.0 (0) | 0.5 (½)	0.0 (0) | 0.5 (½)	0.0 (0) | 0.5 (½)
0.0° ±0.17° 0° ± 5/32°	0.0° ±0.17° 0° ± 5/32°	0.0° ±0.17° 0° ± 5/32°	FRONT TOE (Degrees)	0.0° ±0.17° 0° ± 5/32°	0.0° ±0.17° 0° ± 5/32°	0.0° ±0.17° 0° ± 5/32°
0.0" ±0.08" 0" ± 3/32"	0.0" ±0.08" 0" ± 3/32"	0.0" ±0.08" 0" ± 3/32"	FRONT TOE (Inches)	0.0" ±0.08" 0" ± 3/32"	0.0" ±0.08" 0" ± 3/32"	0.0" ±0.08" 0" ± 3/32"
9.4 ±0.8 (9 7/16 ± ¾)	9.7 ±0.8 (9¾ ± ¾)	8.6 ±0.8 (9⅝ ± ¾)	SAI	8.9 ±0.8 (9⅞ ± ¾)	9.2 ±0.8 (9¼ ± ¾)	9.7 ±0.8 (9 11/16 ± ¾)
41.2 / 35.1 ±2.0 [1]	40.0 / 34.2 ±2.0	38.1 / 33.9 ±2.0	TURNING ANGLE	38.0 / 33.7 ±2.0	37.3 / 32.9 ±2.0	38.0 / 33.7 ±2.0
		173	ILLUSTRATION #	173	173	173 ▲
			REAR			
-1.83 ±0.75 -1 13/16 ± ¾	-1.98 ±0.75 -2 ± ¾	-1.53 ±0.75 -1½ ± ¾	CAMBER (Degrees)	-1.68 ±0.75 -1 11/16 ± ¾	-1.83 ±0.75 -1 13/16 ± ¾	-1.75 ±0.75 -1¾ ± ¾
0.25° ±0.17° ¼° ± 5/32°	0.25° ±0.17° ¼° ± 5/32°	0.25° ±0.17° ¼° ± 5/32°	REAR TOE (Degrees)	0.25° ±0.17° ¼° ± 5/32°	0.25° ±0.17° ¼° ± 5/32°	0.25° ±0.17° ¼° ± 5/32°
0.13" ±0.08" ⅛" ± 3/32"	0.13" ±0.08" ⅛" ± 3/32"	0.13" ±0.08" ⅛" ± 3/32"	REAR TOE (Inches)	0.13" ±0.08" ⅛" ± 3/32"	0.13" ±0.08" ⅛" ± 3/32"	0.13" ±0.08" ⅛" ± 3/32"
			THRUST ANGLE			
178	178	178	ILLUSTRATION #	178	178	178
[1] 19" wheels = 40.2°/34.4°			NOTES			
26062	26063	26064		26065	26066	26067

Lexus

MODEL INFORMATION	LS 600h L	LX 450	LX 470	LX 570	LX 570	NX200T with 17" tires (OE)
YEARS	2013-2017	1996-97	1998-2007	2008-2011	2013-2017	2015-2017
INSTRUCTIONS						
LUGNUT TORQUE	103 ft.lbs.	109 ft.lbs.[1]	97 ft.lbs.	97 ft.lbs.	97 ft.lbs.	76 ft.lbs.
FRONT						
CASTER (Degrees)	4.97 ±0.75 5 ± ¾	3.0 ±1.0 3 ±1	3.08 ±0.75 3 1/16 ± ¾	3.38 ±0.75 3⅜ ± ¾	3.37 ±0.75 3⅜ ± ¾	6.0 ±0.75 6 ± ¾
CROSS	0.0 (0)	0.0 (0)	0.0 (0)	0.0 (0)	0.0 (0)	0.0 (0)
TOL.	0.75 (¾)	0.75 (¾)	0.5 (½)	0.75 (¾)	0.75 (¾)	0.5 (½)
CAMBER (Degrees)	-0.67 ±0.75 -11/16 ± ¾	1.0 ±0.75 1 ± ¾	0.0 ±0.75 0 ± ¾	0.0 ±0.75 0 ± ¾	0.0 ±0.75 0 ± ¾	-0.17 ±0.75 -3/16 ± ¾
CROSS	0.0 (0)	0.0 (0)	0.0 (0)	0.0 (0)	0.0 (0)	0.0 (0)
TOL.	0.5 (½)	0.75 (¾)	0.5 (½)	0.75 (¾)	0.75 (¾)	0.5 (½)
FRONT TOE (Degrees)	0.0° ±0.17° 0° ± 5/32°	0.2° ±0.2° 3/16° ± 3/16°	0.0° ±0.2° 0° ± 3/16°	0.0° ±0.17° 0° ± 5/32°	0.0° ±0.17° 0° ± 5/32°	0.17° ±0.17° 5/32° ± 5/32°
FRONT TOE (Inches)	0.0" ±0.08" 0" ± 3/32"	0.1" ±0.1" 3/32" ± 3/32"	0.0" ±0.1" 0" ± 3/32"	0.0" ±0.08" 0" ± 3/32"	0.0" ±0.08" 0" ± 3/32"	0.08" ±0.08" 3/32" ± 3/32"
SAI	8.9 ±0.8 (9⅞ ± ¾)	13.0 ±0.8 (13 ± ¾)	12.3 ±0.8 (12¼ ± ¾)	13.0 ±0.8 (13 ± ¾)	13.0 ±0.8 (13 ± ¾)	11.5 ±0.8 (11½ ± ¾)
TURNING ANGLE	38.0 / 33.7 ±2.0	33.5 / 31.0 ±1.5	35.7 / 32.6 ±1.0	35.0 / 32.0 ±1.5	35.0 / 32.0 ±1.5	35.8 / 30.5 ±2.0
ILLUSTRATION #	173		24	12	12	26
REAR						
CAMBER (Degrees)	-1.68 ±0.75 -1 11/16 ± ¾					-1.25 ±0.75 -1¼ ± ¾
REAR TOE (Degrees)	0.25° ±0.17° ¼° ± 5/32°					0.17° ±0.17° 5/32° ± 5/32°
REAR TOE (Inches)	0.13" ±0.08" ⅛" ± 3/32"					0.08" ±0.08" 3/32" ± 3/32"
THRUST ANGLE						
ILLUSTRATION #	178					05
NOTES		[1] Alum. Wheels = 76 ft.lbs.				
	26068	26069	26070	26071	26072	26073

Lexus

MODEL INFORMATION	NX200T with 225/60R18" tires (OE)	NX200T with 235/55R18" tires (OE)	NX300h	RC 200t RWD	RC 300 AWD	RC 350 RWD
YEARS	2015-2017	2015-2017	2015-2017	2016-2017	2016-2017	2015-2017
INSTRUCTIONS						
LUGNUT TORQUE	76 ft.lbs.	76 ft.lbs.	76 ft.lbs.	76 ft.lbs.	76 ft.lbs.	76 ft.lbs.
FRONT						
CASTER (Degrees)	6.08 ±0.75 6 1/16 ± ¾	6.08 ±0.75 6 1/16 ± ¾	6.0 ±0.75 6 ± ¾	8.4 ±0.75 8⅜ ± ¾	5.17 ±0.75 5 3/16 ± ¾	8.4 ±0.75 8⅜ ± ¾
CROSS	0.0 (0)	0.0 (0)	0.0 (0)	0.0 (0)	0.0 (0)	0.0 (0)
TOL.	0.5 (½)	0.5 (½)	0.5 (½)	0.5 (½)	0.5 (½)	0.5 (½)
CAMBER (Degrees)	-0.17 ±0.75 -3/16 ± ¾	-0.17 ±0.75 -3/16 ± ¾	-0.25 ±0.75 -¼ ± ¾	-0.33 ±0.75 -5/16 ± ¾	-0.43 ±0.75 -7/16 ± ¾	-0.33 ±0.75 -5/16 ± ¾
CROSS	0.0 (0)	0.0 (0)	0.0 (0)	0.0 (0)	0.0 (0)	0.0 (0)
TOL.	0.5 (½)	0.5 (½)	0.5 (½)	0.75 (¾)	0.75 (¾)	0.75 (¾)
FRONT TOE (Degrees)	0.17° ±0.17° 5/32° ± 5/32°	0.17° ±0.17° 5/32° ± 5/32°	0.17° ±0.17° 5/32° ± 5/32°	0.08° ±0.18° 3/32° ± 3/16°	0.0° ±0.18° 0° ± 3/16°	0.08° ±0.18° 3/32° ± 3/16°
FRONT TOE (Inches)	0.08" ±0.08" 3/32" ± 3/32"	0.08" ±0.08" 3/32" ± 3/32"	0.08" ±0.08" 3/32" ± 3/32"	0.04" ±0.09" 1/32" ± 3/32"	0.0" ±0.09" 0" ± 3/32"	0.04" ±0.09" 1/32" ± 3/32"
SAI	11.6 ±0.8 (11 9/16 ± ¾)	11.5 ±0.8 (11½ ± ¾)	11.5 ±0.8 (11½ ± ¾)	10.6 ±0.8 (10 9/16 ± ¾)	11.1 ±0.8 (11 1/16 ± ¾)	10.6 ±0.8 (10 9/16 ± ¾)
TURNING ANGLE	35.8 / 30.5 ±2.0	35.8 / 30.5 ±2.0	35.6 / 30.2 ±2.0	41.5 / 35.9 ±1.5	39.2 / 35.7 ±1.5	41.5 / 35.9 ±1.5
ILLUSTRATION #	26	26	26		14	
REAR						
CAMBER (Degrees)	-1.33 ±0.75 -1 5/16 ± ¾	-1.25 ±0.75 -1¼ ± ¾	-1.25 ±0.75 [1] -1¼ ± ¾ [1]	-1.88 ±0.75 -1⅞ ± ¾	-1.88 ±0.75 -1⅞ ± ¾	-1.88 ±0.75 -1⅞ ± ¾
REAR TOE (Degrees)	0.17° ±0.17° 5/32° ± 5/32°	0.17° ±0.17° 5/32° ± 5/32°	0.17° ±0.17° 5/32° ± 5/32°	0.18° ±0.18° 3/16° ± 3/16°	0.18° ±0.18° 3/16° ± 3/16°	0.18° ±0.18° 3/16° ± 3/16°
REAR TOE (Inches)	0.08" ±0.08" 3/32" ± 3/32"	0.08" ±0.08" 3/32" ± 3/32"	0.08" ±0.08" 3/32" ± 3/32"	0.09" ±0.09" 3/32" ± 3/32"	0.09" ±0.09" 3/32" ± 3/32"	0.09" ±0.09" 3/32" ± 3/32"
THRUST ANGLE						
ILLUSTRATION #	05	05	05	174	174	174
NOTES			[1] 18" tires = -1.33° (-1 5/16°)			
	26074	26075	26076	26077	26078	26079

Lexus

RC 350 AWD	RC F	RX 300 FWD	MODEL INFORMATION	RX 300 AWD	RX 330 FWD RX 350 FWD Coil Suspension	RX 330 FWD RX 350 FWD Air Suspension
2015-2017	2015-2017	1999-2003	YEARS	1999-2003	2004-2009	2004-2009
			INSTRUCTIONS			
76 ft.lbs.	76 ft.lbs.	76 ft.lbs.	LUGNUT TORQUE	76 ft.lbs.	76 ft.lbs.	76 ft.lbs.
			FRONT			
5.17 ±0.75 5 3⁄16 ± 3⁄4	7.63 ±0.75 7 5⁄8 ± 3⁄4	2.08 ±0.75 2 1⁄16 ± 3⁄4	CASTER (Degrees)	2.08 ±0.75 2 1⁄16 ± 3⁄4	2.75 ±0.75 [1] 2 3⁄4 ± 3⁄4 [1]	2.85 ±0.75 2 7⁄8 ± 3⁄4
0.0 (0) | 0.5 (1⁄2)	0.0 (0) | 0.5 (1⁄2)	0.0 (0) | 0.75 (3⁄4)	CROSS | TOL.	0.0 (0) | 0.75 (3⁄4)	0.0 (0) | 0.75 (3⁄4)	0.0 (0) | 0.75 (3⁄4)
-0.43 ±0.75 - 7⁄16 ± 3⁄4	-0.47 ±0.75 - 1⁄2 ± 3⁄4	-0.33 ±0.75 - 5⁄16 ± 3⁄4	CAMBER (Degrees)	-0.33 ±0.75 - 5⁄16 ± 3⁄4	-0.67 ±0.75 - 11⁄16 ± 3⁄4	-0.67 ±0.75 - 11⁄16 ± 3⁄4
0.0 (0) | 0.75 (3⁄4)	0.0 (0) | 0.75 (3⁄4)	0.0 (0) | 0.5 (1⁄2)	CROSS | TOL.	0.0 (0) | 0.5 (1⁄2)	0.0 (0) | 0.5 (1⁄2)	0.0 (0) | 0.5 (1⁄2)
0.0° ±0.18° 0° ± 3⁄16°	0.0° ±0.17° 0° ± 5⁄32°	0.1° ±0.2° 3⁄32° ± 3⁄16°	FRONT TOE (Degrees)	0.1° ±0.2° 3⁄32° ± 3⁄16°	0.0° ±0.2° 0° ± 3⁄16°	0.0° ±0.2° 0° ± 3⁄16°
0.0" ±0.09" 0" ± 3⁄32"	0.0" ±0.08" 0" ± 3⁄32"	0.05" ±0.1" 1⁄16" ± 3⁄32"	FRONT TOE (Inches)	0.05" ±0.1" 1⁄16" ± 3⁄32"	0.0" ±0.1" 0" ± 3⁄32"	0.0" ±0.1" 0" ± 3⁄32"
11.1 ±0.8 (11 1⁄16 ± 3⁄4)	9.8 ±0.8 (9 13⁄16 ± 3⁄4)	12.2 ±0.8 (12 3⁄16 ± 3⁄4)	SAI	12.2 ±0.8 (12 3⁄16 ± 3⁄4)	10.8 ±0.8 (10 3⁄4 ± 3⁄4)	10.7 ±0.8 (10 11⁄16 ± 3⁄4)
39.2 / 35.7 ±1.5	39.2 / 34.0 ±1.5	31.1 / 27.8 ±1.5	TURNING ANGLE	31.1 / 27.8 ±1.5	35.5 / 31.2 ±2.0	35.7 / 31.3 ±2.0
14		26	ILLUSTRATION #	26	26	26
			REAR			
-1.88 ±0.75 -1 7⁄8 ± 3⁄4	-1.53 ±0.75 -1 1⁄2 ± 3⁄4	-0.5 ±0.75 - 1⁄2 ± 3⁄4	CAMBER (Degrees)	-0.58 ±0.75 - 9⁄16 ± 3⁄4	-1.33 ±0.75 -1 5⁄16 ± 3⁄4	-1.35 ±0.75 -1 3⁄8 ± 3⁄4
0.18° ±0.18° 3⁄16° ± 3⁄16°	0.25° ±0.17° 1⁄4° ± 5⁄32°	0.2° ±0.2° 3⁄16° ± 3⁄16°	REAR TOE (Degrees)	0.3° ±0.2° 5⁄16° ± 3⁄16°	0.23° ±0.15° 7⁄32° ± 5⁄32°	0.3° ±0.2° 5⁄16° ± 3⁄16°
0.09" ±0.09" 3⁄32" ± 3⁄32"	0.13" ±0.08" 1⁄8" ± 3⁄32"	0.1" ±0.1" 3⁄32" ± 3⁄32"	REAR TOE (Inches)	0.15" ±0.1" 5⁄32" ± 3⁄32"	0.12" ±0.08" 1⁄8" ± 1⁄16"	0.15" ±0.1" 5⁄32" ± 3⁄32"
			THRUST ANGLE			
174	174	06	ILLUSTRATION #	06	10	10
			NOTES		[1] 2004-06 = 2.85° (2 7⁄8°)	
26080	26081	26082		26083	26084	26085

Lexus

RX 330 AWD RX 350 AWD Coil Suspension	RX 330 AWD RX 350 AWD Air Suspension	RX 400h FWD	MODEL INFORMATION	RX 400h AWD	RX 350 FWD RX 450h FWD	RX 350 FWD RX 450h FWD
2004-2009	2004-2009	2006-2008	YEARS	2006-2008	2010-2015	2016
			INSTRUCTIONS			
76 ft.lbs.	76 ft.lbs.	76 ft.lbs.	LUGNUT TORQUE	76 ft.lbs.	76 ft.lbs.	76 ft.lbs.
			FRONT			
2.75 ±0.75 [1] 2 3⁄4 ± 3⁄4 [1]	2.83 ±0.75 2 13⁄16 ± 3⁄4	2.75 ±0.75 2 3⁄4 ± 3⁄4	CASTER (Degrees)	2.5 ±0.75 2 1⁄2 ± 3⁄4	2.83 ±0.75 2 13⁄16 ± 3⁄4	2.53 ±0.75 2 1⁄2 ± 3⁄4
0.0 (0) | 0.75 (3⁄4)	0.0 (0) | 0.75 (3⁄4)	0.0 (0) | 0.75 (3⁄4)	CROSS | TOL.	0.0 (0) | 0.75 (3⁄4)	0.0 (0) | 0.75 (3⁄4)	0.0 (0) | 0.75 (3⁄4)
-0.58 ±0.75 - 9⁄16 ± 3⁄4	-0.62 ±0.75 - 5⁄8 ± 3⁄4	-0.58 ±0.75 - 9⁄16 ± 3⁄4	CAMBER (Degrees)	-0.58 ±0.75 - 9⁄16 ± 3⁄4	-0.67 ±0.75 - 11⁄16 ± 3⁄4	-0.58 ±0.75 - 9⁄16 ± 3⁄4
0.0 (0) | 0.5 (1⁄2)	0.0 (0) | 0.5 (1⁄2)	0.0 (0) | 0.5 (1⁄2)	CROSS | TOL.	0.0 (0) | 0.5 (1⁄2)	0.0 (0) | 0.75 (3⁄4)	0.0 (0) | 0.75 (3⁄4)
0.0° ±0.2° 0° ± 3⁄16°	0.0° ±0.2° 0° ± 3⁄16°	0.0° ±0.2° 0° ± 3⁄16°	FRONT TOE (Degrees)	0.0° ±0.2° 0° ± 3⁄16°	0.08° ±0.17° 3⁄32° ± 5⁄32°	0.08° ±0.17° 3⁄32° ± 5⁄32°
0.0" ±0.1" 0" ± 3⁄32"	0.0" ±0.1" 0" ± 3⁄32"	0.0" ±0.1" 0" ± 3⁄32"	FRONT TOE (Inches)	0.0" ±0.1" 0" ± 3⁄32"	0.04" ±0.08" 1⁄32" ± 3⁄32"	0.04" ±0.08" 1⁄32" ± 3⁄32"
10.8 ±0.8 (10 3⁄4 ± 3⁄4)	10.6 ±0.8 (10 9⁄16 ± 3⁄4)	10.6 ±0.8 (10 9⁄16 ± 3⁄4)	SAI	10.6 ±0.8 (10 9⁄16 ± 3⁄4)	10.9 ±0.8 (10 15⁄16 ± 3⁄4)	10.8 ±0.8 (10 3⁄4 ± 3⁄4)
35.7 / 31.4 ±2.0	35.8 / 31.3 ±2.0	35.7 / 31.4 ±2.0	TURNING ANGLE	35.7 / 31.4 ±2.0	36.1 / 32.0 ±1.5	36.1 / 32.0 ±1.5 [1]
26	26	26	ILLUSTRATION #	26	26	26
			REAR			
-0.83 ±0.75 - 13⁄16 ± 3⁄4	-0.92 ±0.75 - 15⁄16 ± 3⁄4	-1.17 ±0.75 -1 3⁄16 ± 3⁄4	CAMBER (Degrees)	-0.67 ±0.75 - 11⁄16 ± 3⁄4	-0.67 ±0.75 - 11⁄16 ± 3⁄4	-0.08 ±0.75 [2] - 1⁄16 ± 3⁄4 [2]
0.23° ±0.15° 7⁄32° ± 5⁄32°	0.3° ±0.2° 5⁄16° ± 3⁄16°	0.23° ±0.15° 7⁄32° ± 5⁄32°	REAR TOE (Degrees)	0.23° ±0.15° 7⁄32° ± 5⁄32°	0.08° ±0.17° 3⁄32° ± 5⁄32°	**-0.23°** ±0.15° **- 7⁄32°** ± 5⁄32°
0.12" ±0.08" 1⁄8" ± 1⁄16"	0.15" ±0.1" 5⁄32" ± 3⁄32"	0.12" ±0.08" 1⁄8" ± 1⁄16"	REAR TOE (Inches)	0.12" ±0.08" 1⁄8" ± 1⁄16"	0.04" ±0.08" 1⁄32" ± 3⁄32"	**-0.12"** ±0.08" **- 1⁄8"** ± 1⁄16"
			THRUST ANGLE			
06	06	10	ILLUSTRATION #	06	06	06
[1] 2004-06 = 2.83° (2 13⁄16°)			NOTES			[1] RX450h = 35.9°/31.3° [2] RX450h = -0.17° (- 3⁄16°)
26086	26087	26088		26089	26090	26091

Lexus

RX 350 FWD exc/F-Sport	RX 350 FWD F-Sport	RX 350 AWD RX 450h AWD Air Suspension	MODEL INFORMATION	RX 350 AWD RX 450h AWD Coil Suspension	RX 350 AWD RX 450h AWD exc/F-Sport	RX 350 AWD F-Sport
2017	2017	2010	YEARS	2011-2015	2016-2017	2016-2017
			INSTRUCTIONS			
76 ft.lbs.	76 ft.lbs.	76 ft.lbs.	LUGNUT TORQUE	76 ft.lbs.	76 ft.lbs.	76 ft.lbs.
			FRONT			
2.53 ±0.75 2½ ± ¾	2.73 ±0.75 2¾ ± ¾	2.88 ±0.75 2⅞ ± ¾	CASTER (Degrees)	2.92 ±0.75 2 15/16 ± ¾	2.55 ±0.75 2 9/16 ± ¾	2.75 ±0.75 2¾ ± ¾
0.0 (0) | 0.75 (¾)	0.0 (0) | 0.75 (¾)	0.0 (0) | 0.75 (¾)	CROSS | TOL.	0.0 (0) | 0.75 (¾)	0.0 (0) | 0.75 (¾)	0.0 (0) | 0.75 (¾)
-0.58 ±0.75 - 9/16 ± ¾	-0.68 ±0.75 - 11/16 ± ¾	-0.63 ±0.75 - ⅝ ± ¾	CAMBER (Degrees)	-0.67 ±0.75 - 11/16 ± ¾	-0.58 ±0.75 - 9/16 ± ¾	-0.68 ±0.75 - 11/16 ± ¾
0.0 (0) | 0.75 (¾)	0.0 (0) | 0.75 (¾)	0.0 (0) | 0.75 (¾)	CROSS | TOL.	0.0 (0) | 0.75 (¾)	0.0 (0) | 0.75 (¾)	0.0 (0) | 0.75 (¾)
0.08° ±0.17° 3/32° ± 5/32°	0.08° ±0.17° 3/32° ± 5/32°	0.08° ±0.17° 3/32° ± 5/32°	FRONT TOE (Degrees)	0.08° ±0.17° 3/32° ± 5/32°	0.08° ±0.17° 3/32° ± 5/32°	0.08° ±0.17° 3/32° ± 5/32°
0.04" ±0.08" 1/32" ± 3/32"	0.04" ±0.08" 1/32" ± 3/32"	0.04" ±0.08" 1/32" ± 3/32"	FRONT TOE (Inches)	0.04" ±0.08" 1/32" ± 3/32"	0.04" ±0.08" 1/32" ± 3/32"	0.04" ±0.08" 1/32" ± 3/32"
10.8 ±0.8 (10¾ ± ¾)	11.0 ±0.8 (11 ± ¾)	10.9 ±0.8 (10⅞ ± ¾)	SAI	11.0 ±0.8 (11 ± ¾)	10.8 ±0.8 (10¾ ± ¾)	11.0 ±0.8 (11 ± ¾)
36.1 / 32.0 ±1.5	35.5 / 31.2 ±1.5	36.2 / 32.1 ±1.5	TURNING ANGLE	36.1 / 32.0 ±1.5	36.1 / 32.0 ±1.5	36.1 / 32.0 ±1.5
26	26	26	ILLUSTRATION #	26	26	6
			REAR			
-0.08 ±0.75 [1] - 1/16 ± ¾ [1]	-0.58 ±0.75 - 9/16 ± ¾	-0.82 ±0.75 - 13/16 ± ¾	CAMBER (Degrees)	-0.75 ±0.75 - ¾ ± ¾	-0.17 ±0.75 - 3/16 ± ¾	-0.58 ±0.75 - 9/16 ± ¾
-0.23° ±0.15° **- 7/32°** ± 5/32°	0.0° ±0.15° 0° ± 5/32°	0.22° ±0.17° 7/32° ± 5/32°	REAR TOE (Degrees)	0.08° ±0.17° 3/32° ± 5/32°	**-0.23°** ±0.15° [1] **- 7/32°** ± 5/32° [1]	0.0° ±0.17° 0° ± 5/32°
-0.12" ±0.08" **- ⅛"** ± 1/16"	0.0" ±0.08" 0" ± 1/16"	0.11" ±0.08" 3/32" ± 3/32"	REAR TOE (Inches)	0.04" ±0.08" 1/32" ± 3/32"	**-0.12"** ±0.08" [1] **- ⅛"** ± 1/16" [1]	0.0" ±0.08" 0" ± 3/32"
			THRUST ANGLE			
06	06	06	ILLUSTRATION #	06	06	06
[1] TMC made = -0.17° (-3/16°)			NOTES		[1] RX450h = -0.15° (-.08")	
26092	26093	26094		26095	26096	26097

Lexus

RX 450h AWD F-Sport	SC 300 SC 400	SC 300 SC 400	MODEL INFORMATION	SC 430	Notes	Notes
2016-2017	1992-96	1997-2000	YEARS	2002-2010		
			INSTRUCTIONS			
76 ft.lbs.			LUGNUT TORQUE	76 ft.lbs.		
			FRONT			
2.78 ±0.75 2¾ ± ¾	2.75 ±0.75 [1] 2¾ ± ¾ [1]	3.02 ±0.75 3 ± ¾	CASTER (Degrees)	7.92 ±0.75 7 15/16 ± ¾		
0.0 (0) | 0.75 (¾)	0.0 (0) | 0.5 (½)	0.0 (0) | 0.5 (½)	CROSS | TOL.	0.0 (0) | 0.5 (½)		
-0.68 ±0.75 - 11/16 ± ¾	0.02 ±0.75 1/32 ± ¾	0.03 ±0.75 1/16 ± ¾	CAMBER (Degrees)	-0.58 ±0.75 - 9/16 ± ¾		
0.0 (0) | 0.75 (¾)	0.0 (0) | 0.5 (½)	0.0 (0) | 0.5 (½)	CROSS | TOL.	0.0 (0) | 0.5 (½)		
0.08° ±0.17° 3/32° ± 5/32°	0.1° ±0.2° 3/32° ± 3/16°	0.1° ±0.2° 3/32° ± 3/16°	FRONT TOE (Degrees)	0.15° ±0.2° 5/32° ± 3/16°		
0.04" ±0.08" 1/32" ± 3/32"	0.05" ±0.1" 1/16" ± 3/32"	0.05" ±0.1" 1/16" ± 3/32"	FRONT TOE (Inches)	0.08" ±0.1" 1/16" ± 3/32"		
11.1 ±0.8 (11 1/16 ± ¾)	9.0 ±0.8 (8 15/16 ± ¾)	9.0 ±0.8 (9 ± ¾)	SAI	9.2 ±0.8 (9 3/16 ± ¾)		
35.5 / 31.2 ±1.5	37.1 / 32.4 ±1.0	37.1 / 32.4 ±1.0	TURNING ANGLE	36.7 / 30.8 ±1.5		
26	52	52	ILLUSTRATION #	14		
			REAR			
-0.67 ±0.75 - 11/16 ± ¾	-0.88 ±0.75 - ⅞ ± ¾	-1.08 ±0.75 -1 1/16 ± ¾	CAMBER (Degrees)	-1.17 ±0.5 -1 3/16 ± ½		
0.08° ±0.15° 3/32° ± 5/32°	0.4° +0.1° -0.2° 13/32° + 3/32° - 3/16°	0.5° +0.1° -0.3° ½° + 3/32° - 5/16°	REAR TOE (Degrees)	0.13° ±0.2° ⅛° ± 3/16°		
0.04" ±0.08" 1/32" ± 1/16"	0.2" +0.05" -0.1" 3/16" + 1/16" - 3/32"	0.25" +0.05" -0.15" ¼" + 1/16" - 5/32"	REAR TOE (Inches)	0.07" ±0.1" 1/16" ± 3/32"		
			THRUST ANGLE			
06	63	63	ILLUSTRATION #	117		
	[1] 1996 = 2.9° (2 15/16°)		NOTES			
26098	26099	26100		26101	26102	26103

ALIGNMENT SPECIFICATIONS

Lincoln

Aviator	Blackwood	Continental	MODEL	Continental	Continental	Continental
			INFORMATION			
2003-2005	2002-2003	1990-94	YEARS	1995-97	1998-2002	2017
			INSTRUCTIONS			
100 ft.lbs.	148 ft.lbs.	85-105 ft.lbs.	LUGNUT TORQUE	85-104 ft.lbs.	85-104 ft.lbs.	162 ft.lbs.
			FRONT			
5.3 ± 1.0 5 5/16 ± 1	L= 6.1, R= 6.6 ± 1.0 [1] L= 6⅛, R= 6⅝ ± 1 [1]	4.5 ± 1.0 4½ ± 1	CASTER (Degrees)	4.5 ± 1.0 4½ ± 1	4.4 ± 1.0 4⅜ ± 1	3.35 ± 0.75 3⅜ ± ¾
0.0 (0) | 0.5 (½)	-0.5 (-½) | 0.7 (11/16)	0.0 (0) | 0.75 (¾)	CROSS | TOL.	0.0 (0) | 0.75 (¾)	0.0 (0) | 0.75 (¾)	0.0 (0) | 0.75 (¾)
-0.5 ± 0.8 -½ ± 13/16	-0.3 ± 0.7 [1] -5/16 ± 11/16 [1]	-1.1 ± 0.6 -1⅛ ± ⅝	CAMBER (Degrees)	-0.7 ± 0.6 -11/16 ± ⅝	-0.6 ± 0.6 -⅝ ± ⅝	-0.8 ± 0.75 -13/16 ± ¾
0.0 (0) | 0.8 (13/16)	0.0 (0) | 0.7 (11/16)	0.0 (0) | 0.7 (11/16)	CROSS | TOL.	0.0 (0) | 0.7 (11/16)	0.0 (0) | 0.7 (11/16)	0.0 (0) | 0.75 (¾)
0.24° ± 0.25° ¼° ± ¼°	0.06° ± 0.25° 1/16° ± ¼°	-0.2° ± 0.25° -3/16° ± ¼°	FRONT TOE (Degrees)	-0.2° ± 0.25° -3/16° ± ¼°	-0.2° ± 0.25° -3/16° ± ¼°	0.2° ± 0.2° 3/16° ± 3/16°
0.12" ± 0.13" ⅛" ± ⅛"	0.03" ± 0.13" 1/32" ± ⅛"	-0.1" ± 0.13" -3/32" ± ⅛"	FRONT TOE (Inches)	-0.1" ± 0.13" -3/32" ± ⅛"	-0.1" ± 0.13" -3/32" ± ⅛"	0.1" ± 0.1" 3/32" ± 3/32"
			SAI			
		20.0 / 18.2	TURNING ANGLE	20.0 / 18.2	20.0 / 18.2	
137	128	68 ▲	ILLUSTRATION #	68	68	26
			REAR			
-0.95 ± 0.8 -15/16 ± 13/16		-1.3 ± 0.7 -1 5/16 ± 11/16	CAMBER (Degrees)	-0.7 ± 0.5 -11/16 ± ½	-0.7 ± 0.5 -11/16 ± ½	-1.2 ± 0.75 -1 3/16 ± ¾
0.16° ± 0.25° 5/32° ± ¼°		0.2° ± 0.25° 3/16° ± ¼°	REAR TOE (Degrees)	0.2° ± 0.25° 3/16° ± ¼°	0.2° ± 0.25° 3/16° ± ¼°	0.3° ± 0.2° 5/16° ± 3/16°
0.08" ± 0.13" 3/32" ± ⅛"		0.1" ± 0.13" 3/32" ± ⅛"	REAR TOE (Inches)	0.1" ± 0.13" 3/32" ± ⅛"	0.1" ± 0.13" 3/32" ± ⅛"	0.15" ± 0.1" 5/32" ± 3/32"
			THRUST ANGLE			0.5 (½)
136		06	ILLUSTRATION #	40 05	40 05	130 174
	[1] Service cam kit available.		NOTES			
27001	27002	27015		27016	27017	27018

Lincoln

LS	Mark VII	Mark VIII	MODEL	Mark LT	Mark LT	Mark LT
			INFORMATION	4X2 w/18" Tires	4X2 w/20" Tires	4X4
2000-2006	1990-92	1993-98	YEARS	2006-2008	2006-2008	2006-2008
			INSTRUCTIONS			
100 ft.lbs.	85-105 ft.lbs.	85-105 ft.lbs.	LUGNUT TORQUE	150 ft.lbs.	150 ft.lbs.	150 ft.lbs.
			FRONT			
8.1 ± 0.5 8⅛ ± ½	1.5 ± 1.0 1½ ± 1	5.5 ± 0.75 5½ ± ¾	CASTER (Degrees)	L= 4.4, R= 4.8 ± 1.0 [1] L= 4⅜, R= 4 13/16 ± 1 [1]	L= 4.0, R= 4.4 ± 1.0 [1] L= 4, R= 4⅜ ± 1 [1]	L= 4.5, R= 4.9 ± 1.0 [1] L= 4½, R= 4⅞ ± 1 [1]
0.0 (0) | 0.75 (¾)	0.0 (0) | 0.88 (⅞)	0.0 (0) | 0.75 (¾)	CROSS | TOL.	-0.4 (-⅜) | 0.75 (¾)	-0.4 (-⅜) | 0.75 (¾)	-0.4 (-⅜) | 0.75 (¾)
-0.15 ± 0.5 -⅛ ± ½	0.0 ± 0.75 0 ± ¾	-0.5 ± 0.75 -½ ± ¾	CAMBER (Degrees)	-0.2 ± 0.75 [1] -3/16 ± ¾ [1]	-0.2 ± 0.75 [1] -3/16 ± ¾ [1]	-0.2 ± 0.75 [1] -3/16 ± ¾ [1]
0.0 (0) | 0.75 (¾)	0.0 (0) | 0.9 (⅞)	0.0 (0) | 0.75 (¾)	CROSS | TOL.	0.0 (0) | 0.75 (¾)	0.0 (0) | 0.75 (¾)	0.0 (0) | 0.75 (¾)
0.16° ± 0.25° 5/32° ± ¼°	0.25° ± 0.25° ¼° ± ¼°	0.25° ± 0.25° ¼° ± ¼°	FRONT TOE (Degrees)	0.2° ± 0.2° [2] 3/16° ± 3/16° [2]	0.2° ± 0.2° [2] 3/16° ± 3/16° [2]	0.2° ± 0.2° [2] 3/16° ± 3/16° [2]
0.08" ± 0.13" 3/32" ± ⅛"	0.13" ± 0.13" ⅛" ± ⅛"	0.13" ± 0.13" ⅛" ± ⅛"	FRONT TOE (Inches)	0.1" ± 0.1" [2] 3/32" ± 3/32" [2]	0.1" ± 0.1" [2] 3/32" ± 3/32" [2]	0.1" ± 0.1" [2] 3/32" ± 3/32" [2]
			SAI			
	20.0 / 17.1	20.0 / 19.7	TURNING ANGLE			
123 124	72	03 14 ▲	ILLUSTRATION #	12	12	12
			REAR			
-1.0 ± 0.75 -1 ± ¾		-0.5 ± 0.5 -½ ± ½	CAMBER (Degrees)			
0.24° ± 0.25° ¼° ± ¼°		0.12° ± 0.25° ⅛° ± ¼°	REAR TOE (Degrees)			
0.12" ± 0.13" ⅛" ± ⅛"		0.06" ± 0.13" 1/16" ± ⅛"	REAR TOE (Inches)			
			THRUST ANGLE	0.3 (5/16)	0.3 (5/16)	0.3 (5/16)
135		71	ILLUSTRATION #			
			NOTES	[1] Service cam kit available. [2] 2006 = 0.15º (0.075")	[1] Service cam kit available. [2] 2006 = 0.15º (0.075")	[1] Service cam kit available. [2] 2006 = 0.15º (0.075")
27019	27027	27028		27029	27030	27031

Lincoln

MKC	MKC	MKS	MODEL	MKS	MKT	MKT
FWD	AWD		INFORMATION		exc/Turbo	Turbo
2015-2017	2015-2017	2009-2012	YEARS	2013-2016	2010-2012	2010-2012
			INSTRUCTIONS			
100 ft.lbs.	100 ft.lbs.	100 ft.lbs.	LUGNUT TORQUE	100 ft.lbs.	100 ft.lbs.	100 ft.lbs.
			FRONT			
4.75 ± 0.75 4¾ ± ¾	4.77 ± 0.75 4¾ ± ¾	L= 3.1, R= 3.3 ± 0.75 [1] L= 3⅛, R= 3 5/16 ± ¾ [1]	CASTER (Degrees)	L= 3.1, R= 3.3 ± 0.75 L= 3⅛, R= 3 5/16 ± ¾	L= 3.3, R= 3.5 ± 0.75 L= 3 5/16, R= 3½ ± ¾	L= 3.3, R= 3.5 ± 0.75 L= 3 5/16, R= 3½ ± ¾
0.0 (0) │ 0.75 (¾)	0.0 (0) │ 0.75 (¾)	-0.2 (-3/16) │ 0.75 (¾)	CROSS │ TOL.	-0.2 (-3/16) │ 0.75 (¾)	-0.2 (-3/16) │ 0.75 (¾)	-0.2 (-3/16) │ 0.75 (¾)
-0.74 ± 0.75 -¾ ± ¾	-0.67 ± 0.75 -11/16 ± ¾	L= -0.4, R= -0.6 ± 0.75 L= -⅜, R= -⅝ ± ¾	CAMBER (Degrees)	L= -0.55, R= -0.75 ± 0.75 L= -9/16, R= -¾ ± ¾	L= -0.4, R= -0.5 ± 0.75 L= -⅜, R= -½ ± ¾	L= -0.5, R= -0.6 ± 0.75 L= -½, R= -⅝ ± ¾
0.0 (0) │ 0.75 (¾)	0.0 (0) │ 0.75 (¾)	0.2 (3/16) │ 0.75 (¾)	CROSS │ TOL.	0.2 (3/16) │ 0.75 (¾)	0.1 (⅛) │ 0.75 (¾)	0.1 (⅛) │ 0.75 (¾)
0.2° ± 0.2° 3/16° ± 3/16°	0.2° ± 0.2° 3/16° ± 3/16°	0.2° ± 0.2° 3/16° ± 3/16°	FRONT TOE (Degrees)	0.2° ± 0.2° 3/16° ± 3/16°	0.1° ± 0.2° 3/32° ± 3/16°	0.1° ± 0.2° 3/32° ± 3/16°
0.1" ± 0.1" 3/32" ± 3/32"	0.1" ± 0.1" 3/32" ± 3/32"	0.1" ± 0.1" 3/32" ± 3/32"	FRONT TOE (Inches)	0.1" ± 0.1" 3/32" ± 3/32"	0.05" ± 0.1" 1/16" ± 3/32"	0.05" ± 0.1" 1/16" ± 3/32"
			SAI			
			TURNING ANGLE			
190	190	167	ILLUSTRATION #	167	167	167
			REAR			
-1.44 ± 0.75 -1 7/16 ± ¾	-1.45 ± 0.75 -1 7/16 ± ¾	-1.0 ± 0.75 -1 ± ¾	CAMBER (Degrees)	-1.0 ± 0.75 -1 ± ¾	-0.8 ± 0.75 -13/16 ± ¾	-1.0 ± 0.75 -1 ± ¾
0.38° ± 0.2° ⅜° ± 3/16°	0.38° ± 0.2° ⅜° ± 3/16°	0.24° ± 0.2° ¼° ± 3/16°	REAR TOE (Degrees)	0.24° ± 0.2° ¼° ± 3/16°	0.24° ± 0.2° ¼° ± 3/16°	0.24° ± 0.2° ¼° ± 3/16°
0.19" ± 0.1" 3/16" ± 3/32"	0.19" ± 0.1" 3/16" ± 3/32"	0.12" ± 0.1" ⅛" ± 3/32"	REAR TOE (Inches)	0.12" ± 0.1" ⅛" ± 3/32"	0.12" ± 0.1" ⅛" ± 3/32"	0.12" ± 0.1" ⅛" ± 3/32"
0.5 (½)	0.5 (½)	0.3 (5/16)	THRUST ANGLE	0.3 (5/16)	0.3 (5/16)	0.3 (5/16)
06	06	187	ILLUSTRATION #	187	187	187
		[1] 2011 L=3.2°, R=3.4°	NOTES			
27032	27033	27034		27035	27036	27037

Lincoln

MKT	MKT	MKT	MODEL	MKT	MKT	MKX
exc/Turbo, Limo	Turbo	Limo	INFORMATION	Limo	Limo	All models
		44 in.		80 in.	120 in.	
2013-2017	2013-2017	2013-2017	YEARS	2013-2017	2013-2017	2007
			INSTRUCTIONS			
100 ft.lbs.	100 ft.lbs.	100 ft.lbs.	LUGNUT TORQUE	100 ft.lbs.	100 ft.lbs.	
			FRONT			
L= 3.25, R= 3.45 ± 0.75 L= 3¼, R= 3 7/16 ± ¾	L= 3.25, R= 3.45 ± 0.75 L= 3¼, R= 3 7/16 ± ¾	L= 3.2, R= 3.4 ± 0.75 L= 3 3/16, R= 3⅜ ± ¾	CASTER (Degrees)	L= 3.3, R= 3.5 ± 0.75 L= 3 5/16, R= 3½ ± ¾	L= 3.4, R= 3.6 ± 0.75 L= 3⅜, R= 3⅝ ± ¾	4.4 ± 0.75 4⅜ ± ¾
-0.2 (-3/16) │ 0.75 (¾)	-0.2 (-3/16) │ 0.75 (¾)	-0.2 (-3/16) │ 0.75 (¾)	CROSS │ TOL.	-0.2 (-3/16) │ 0.75 (¾)	-0.2 (-3/16) │ 0.75 (¾)	0.0 (0) │ 0.75 (¾)
L= -0.5, R= -0.6 ± 0.75 L= -½, R= -⅝ ± ¾	L= -0.6, R= -0.7 ± 0.75 L= -⅝, R= -11/16 ± ¾	L= -0.3, R= -0.4 ± 0.75 L= -5/16, R= -⅜ ± ¾	CAMBER (Degrees)	L= -0.3, R= -0.4 ± 0.75 L= -5/16, R= -⅜ ± ¾	L= -0.3, R= -0.4 ± 0.75 L= -5/16, R= -⅜ ± ¾	-0.3 ± 0.75 -5/16 ± ¾
0.1 (⅛) │ 0.75 (¾)	0.1 (⅛) │ 0.75 (¾)	0.1 (⅛) │ 0.75 (¾)	CROSS │ TOL.	0.1 (⅛) │ 0.75 (¾)	0.1 (⅛) │ 0.75 (¾)	0.0 (0) │ 0.75 (¾)
0.1° ± 0.2° 3/32° ± 3/16°	0.1° ± 0.2° 3/32° ± 3/16°	0.1° ± 0.2° 3/32° ± 3/16°	FRONT TOE (Degrees)	0.1° ± 0.2° 3/32° ± 3/16°	0.1° ± 0.2° 3/32° ± 3/16°	0.1° ± 0.2° 3/32° ± 3/16°
0.05" ± 0.1" 1/16" ± 3/32"	0.05" ± 0.1" 1/16" ± 3/32"	0.05" ± 0.1" 1/16" ± 3/32"	FRONT TOE (Inches)	0.05" ± 0.1" 1/16" ± 3/32"	0.05" ± 0.1" 1/16" ± 3/32"	0.05" ± 0.1" 1/16" ± 3/32"
			SAI			
			TURNING ANGLE			
167	167	167	ILLUSTRATION #	167	167	167
			REAR			
-0.8 ± 0.75 -13/16 ± ¾	-1.0 ± 0.75 -1 ± ¾	-0.4 ± 0.75 -⅜ ± ¾	CAMBER (Degrees)	-0.5 ± 0.75 -½ ± ¾	-0.6 ± 0.75 -⅝ ± ¾	-0.4 ± 0.75 -⅜ ± ¾
0.24° ± 0.2° ¼° ± 3/16°	0.24° ± 0.2° ¼° ± 3/16°	0.24° ± 0.2° ¼° ± 3/16°	REAR TOE (Degrees)	0.24° ± 0.2° ¼° ± 3/16°	0.24° ± 0.2° ¼° ± 3/16°	0.1° ± 0.2° 3/32° ± 3/16°
0.12" ± 0.1" ⅛" ± 3/32"	0.12" ± 0.1" ⅛" ± 3/32"	0.12" ± 0.1" ⅛" ± 3/32"	REAR TOE (Inches)	0.12" ± 0.1" ⅛" ± 3/32"	0.12" ± 0.1" ⅛" ± 3/32"	0.05" ± 0.1" 1/16" ± 3/32"
0.3 (5/16)	0.3 (5/16)	0.3 (5/16)	THRUST ANGLE	0.3 (5/16)	0.3 (5/16)	0.3 (5/16)
187	187	187	ILLUSTRATION #	187	187	177
			NOTES			
27038	27039	27040		27041	27042	27043

ALIGNMENT SPECIFICATIONS

Lincoln

MKX All models	MKX exc/2015 EcoBoost	MKX EcoBoost	MODEL INFORMATION	MKX	MKZ All models	MKZ All models
2008-2010	2011-2015	2015	YEARS	2016-2017	2007-2009	2010-2012
			INSTRUCTIONS			
100 ft.lbs.	100 ft.lbs.	100 ft.lbs.	LUGNUT TORQUE	162 ft.lbs.	98 ft.lbs.	100 ft.lbs.
			FRONT			
L= 4.3, R= 4.5 ± 0.75 L= 4 5/16, R= 4 1/2 ± 3/4	L= 4.3, R= 4.5 ± 0.75 L= 4 5/16, R= 4 1/2 ± 3/4	L= 4.4, R= 4.6 ± 0.75 L= 4 3/8, R= 4 5/8 ± 3/4	CASTER (Degrees)	3.1 ± 0.75 3 1/8 ± 3/4	4.0 ± 1.0 4 ± 1	L= 3.75, R= 4.25 ± 1.0 L= 3 3/4, R= 4 1/4 ± 1
-0.2 (-3/16) | 0.75 (3/4)	-0.2 (-3/16) | 0.75 (3/4)	-0.2 (-3/16) | 0.75 (3/4)	CROSS | TOL.	0.0 (0) | 0.75 (3/4)	0.0 (0) | 1.0 (1)	-0.5 (-1/2) | 1.0 (1)
L= -0.1, R= -0.6 ± 0.75 L= -1/8, R= -5/8 ± 3/4	L= -0.05, R= -0.55 ± 0.75 L= -1/16, R= -9/16 ± 3/4	L= -0.15, R= -0.65 ± 0.75 L= -1/8, R= -5/8 ± 3/4	CAMBER (Degrees)	-0.71 ± 0.75 -11/16 ± 3/4	-0.03 ± 0.75 -1/32 ± 3/4	-0.03 ± 0.75 -1/32 ± 3/4
0.5 (1/2) | 0.75 (3/4)	0.5 (1/2) | 0.75 (3/4)	0.5 (1/2) | 0.75 (3/4)	CROSS | TOL.	0.0 (0) | 0.75 (3/4)	0.0 (0) | 1.0 (1)	0.0 (0) | 1.0 (1)
0.1° ± 0.2° 3/32° ± 3/16°	0.1° ± 0.2° 3/32° ± 3/16°	0.1° ± 0.2° 3/32° ± 3/16°	FRONT TOE (Degrees)	0.2° ± 0.14° 3/16° ± 1/8°	0.2° ± 0.2° 3/16° ± 3/16°	0.2° ± 0.2° 3/16° ± 3/16°
0.05" ± 0.1" 1/16" ± 3/32"	0.05" ± 0.1" 1/16" ± 3/32"	0.05" ± 0.1" 1/16" ± 3/32"	FRONT TOE (Inches)	0.1" ± 0.07" 3/32" ± 1/16"	0.1" ± 0.1" 3/32" ± 3/32"	0.1" ± 0.1" 3/32" ± 3/32"
			SAI			
			TURNING ANGLE			
167	167	167	ILLUSTRATION #	26		
			REAR			
-0.45 ± 0.75 -7/16 ± 3/4	-0.45 ± 0.75 -7/16 ± 3/4	-0.45 ± 0.75 -7/16 ± 3/4	CAMBER (Degrees)	-1.2 ± 0.75 [1] -1 3/16 ± 3/4 [1]	-1.25 ± 0.75 -1 1/4 ± 3/4	-1.25 ± 0.75 -1 1/4 ± 3/4
0.1° ± 0.2° 3/32° ± 3/16°	0.1° ± 0.2° 3/32° ± 3/16°	0.1° ± 0.2° 3/32° ± 3/16°	REAR TOE (Degrees)	0.26° ± 0.14° 1/4° ± 1/8°	0.24° ± 0.2° 1/4° ± 3/16°	0.24° ± 0.2° 1/4° ± 3/16°
0.05" ± 0.1" 1/16" ± 3/32"	0.05" ± 0.1" 1/16" ± 3/32"	0.05" ± 0.1" 1/16" ± 3/32"	REAR TOE (Inches)	0.13" ± 0.07" 1/8" ± 1/16"	0.12" ± 0.1" 1/8" ± 3/32"	0.12" ± 0.1" 1/8" ± 3/32"
0.3 (5/16)	0.3 (5/16)	0.5 (1/2)	THRUST ANGLE	0.5 (1/2)	0.1 (3/32)	0.5 (1/2)
177	177	177	ILLUSTRATION #		175	175
			NOTES	[1] 21" wheels = -1.0°		
27044	27045	27046		27047	27048	27049

Lincoln

MKZ MKZ Hybrid	Navigator 4X2	Navigator 4X4	MODEL INFORMATION	Navigator All Models	Navigator exc/Air Susp.	Navigator w/Air Susp. RWD
2013-2017	1998-2002	1998-2002	YEARS	2003-2006	2007-2014	2007-2014
			INSTRUCTIONS			
100 ft.lbs. [1]	83-112 ft.lbs.	83-112 ft.lbs.	LUGNUT TORQUE	150 ft.lbs.	150 ft.lbs.	150 ft.lbs.
			FRONT			
3.35 ± 0.75 3 3/8 ± 3/4	L= 6.3, R= 6.6 ± 1.0 [1] L= 6 5/16, R= 6 5/8 ± 1 [1]	L= 5.0, R= 5.7 ± 1.0 [1] L= 5, R= 5 11/16 ± 1 [1]	CASTER (Degrees)	5.5 ± 1.0 [1] 5 1/2 ± 1 [1]	L= 3.9, R= 4.5 ± 1.0 [1] L= 3 7/8, R= 4 1/2 ± 1 [1]	L= 4.1, R= 4.7 ± 1.0 [1] L= 4 1/8, R= 4 11/16 ± 1 [1]
0.0 (0) | 0.75 (3/4)	-0.3 (-5/16) | 0.7 (11/16)	-0.7 (-11/16) | 0.7 (11/16)	CROSS | TOL.	0.0 (0) | 1.0 (1)	-0.6 (-5/8) | 0.75 (3/4)	-0.6 (-5/8) | 0.75 (3/4)
-0.8 ± 0.75 -13/16 ± 3/4	-0.3 ± 0.7 [1] -5/16 ± 11/16 [1]	-0.4 ± 0.7 [1] -3/8 ± 11/16 [1]	CAMBER (Degrees)	-0.6 ± 0.75 [1] -5/8 ± 3/4 [1]	-0.3 ± 0.75 -5/16 ± 3/4	-0.3 ± 0.75 -5/16 ± 3/4
0.0 (0) | 0.75 (3/4)	0.0 (0) | 0.7 (11/16)	0.0 (0) | 0.7 (11/16)	CROSS | TOL.	0.0 (0) | 0.75 (3/4)	0.0 (0) | 0.75 (3/4)	0.0 (0) | 0.75 (3/4)
0.2° ± 0.2° 3/16° ± 3/16°	0.06° ± 0.25° 1/16° ± 1/4°	0.2° ± 0.25° 3/16° ± 1/4°	FRONT TOE (Degrees)	0.14° ± 0.2° 1/8° ± 3/16°	0.14° ± 0.2° 1/8° ± 3/16°	0.14° ± 0.2° 1/8° ± 3/16°
0.1" ± 0.1" 3/32" ± 3/32"	0.03" ± 0.13" 1/32" ± 1/8"	0.1" ± 0.13" 3/32" ± 1/8"	FRONT TOE (Inches)	0.07" ± 0.1" 1/16" ± 3/32"	0.07" ± 0.1" 1/16" ± 3/32"	0.07" ± 0.1" 1/16" ± 3/32"
			SAI			
			TURNING ANGLE			
26	128	128	ILLUSTRATION #	12	12	12
			REAR			
-1.2 ± 0.75 -1 3/16 ± 3/4			CAMBER (Degrees)	-1.0 ± 0.75 [1] -1 ± 3/4 [1]	-0.5 ± 0.75 -1/2 ± 3/4	-0.5 ± 0.75 -1/2 ± 3/4
0.3° ± 0.2° 5/16° ± 3/16°			REAR TOE (Degrees)	0.2° ± 0.2° 3/16° ± 3/16°	0.2° ± 0.2° 3/16° ± 3/16°	0.2° ± 0.2° 3/16° ± 3/16°
0.15" ± 0.1" 5/32" ± 3/32"			REAR TOE (Inches)	0.1" ± 0.1" 3/32" ± 3/32"	0.1" ± 0.1" 3/32" ± 3/32"	0.1" ± 0.1" 3/32" ± 3/32"
0.5 (1/2)			THRUST ANGLE			
174			ILLUSTRATION #	147	183	183
[1] '15-'17 w/V6 = 150 ft.lbs	[1] Service cam kit available.	[1] Service cam kit available.	NOTES	[1] Service cam kit available.	[1] L Mods. L=4.1°, R=4.7°	[1] L Mods. L=4.2°, R=4.8°
27050	27051	27052		27053	27054	27055

Lincoln

Navigator	Navigator	Town Car	MODEL	Town Car	Town Car	Town Car
w/Air Susp.	All models		INFORMATION			
4WD						
2007-2014	2015-2017	1990	YEARS	1991	1992-2002	2003-2004
			INSTRUCTIONS			
150 ft.lbs.	150 ft.lbs.	85-105 ft.lbs.	LUGNUT TORQUE	85-105 ft.lbs.	85-105 ft.lbs.	85-105 ft.lbs.
			FRONT			
L= 4.0, R= 4.6 ± 1.0 [1] L= 4, R= 4⅝ ± 1 [1]	L= 3.9, R= 4.5 ± 1.0 L= 3⅞, R= 4½ ± 1	3.5 ± 1.0 3½ ± 1	CASTER (Degrees)	5.5 ± 0.75 5½ ± ¾	6.0 ± 0.75 6 ± ¾	6.2 ± 0.75 6 3/16 ± ¾
-0.6 (-⅝) | 0.75 (¾)	-0.6 (-⅝) | 0.75 (¾)	0.0 (0) | 0.75 (¾)	CROSS | TOL.	0.0 (0) | 0.75 (¾)	0.0 (0) | 0.75 (¾)	0.0 (0) | 0.75 (¾)
-0.3 ± 0.75 -5/16 ± ¾	-0.3 ± 0.75 -5/16 ± ¾	-0.5 ± 0.75 -½ ± ¾	CAMBER (Degrees)	-0.5 ± 0.75 -½ ± ¾	-0.5 ± 0.75 -½ ± ¾	-0.6 ± 0.75 -⅝ ± ¾
0.0 (0) | 0.75 (¾)	0.0 (0) | 0.75 (¾)	0.0 (0) | 0.75 (¾)	CROSS | TOL.	0.0 (0) | 0.75 (¾)	0.0 (0) | 0.75 (¾)	0.0 (0) | 0.75 (¾)
0.14° ± 0.2° ⅛° ± 3/16°	0.14° ± 0.2° ⅛° ± 3/16°	0.13° ± 0.25° ⅛° ± ¼°	FRONT TOE (Degrees)	-0.13° ± 0.25° -⅛° ± ¼°	-0.13° ± 0.25° -⅛° ± ¼°	-0.13° ± 0.25° -⅛° ± ¼°
0.07" ± 0.1" 1/16" ± 3/32"	0.07" ± 0.1" 1/16" ± 3/32"	0.06" ± 0.13" 1/16" ± ⅛"	FRONT TOE (Inches)	-0.06" ± 0.13" -1/16" ± ⅛"	-0.06" ± 0.13" -1/16" ± ⅛"	-0.06" ± 0.13" -1/16" ± ⅛"
			SAI			
		20.0 / 18.5	TURNING ANGLE	20.0 / 18.5	20.0 / 18.5	
12	12	42	ILLUSTRATION #	70	70	148
			REAR			
-0.5 ± 0.75 -½ ± ¾	-0.5 ± 0.75 -½ ± ¾		CAMBER (Degrees)			
0.2° ± 0.2° 3/16° ± 3/16°	0.2° ± 0.2° 3/16° ± 3/16°		REAR TOE (Degrees)			
0.1" ± 0.1" 3/32" ± 3/32"	0.1" ± 0.1" 3/32" ± 3/32"		REAR TOE (Inches)			
	0.5 (½)		THRUST ANGLE			
183	183		ILLUSTRATION #			
[1] L Mods. L=4.1°, R=4.7°			NOTES			
27056	27057	27063		27064	27065	27066

Lincoln / Mazda

Town Car	Town Car	Zephyr	MODEL	323/Protegé	323/Protegé	626
			INFORMATION	4X2	4X4	
2005-2007	2008-2011	2006	YEARS	1990-94	1990-91	1990
			INSTRUCTIONS			
100 ft.lbs.	100 ft.lbs.	98 ft.lbs.	LUGNUT TORQUE	66-87 ft.lbs.	66-87 ft.lbs.	
			FRONT			
6.5 ± 0.75 6½ ± ¾	L= 6.2, R= 6.4 ± 0.75 L= 6 3/16, R= 6⅜ ± ¾	4.0 ± 1.0 4 ± 1	CASTER (Degrees)	1.92 ± 0.92 [1] 1 15/16 ± 15/16 [1]	2.58 ± 0.92 2 9/16 ± 15/16	1.22 ± 0.75 1 3/16 ± ¾
0.0 (0) | 0.75 (¾)	-0.2 (-3/16) | 0.75 (¾)	0.0 (0) | 1.0 (1)	CROSS | TOL.	0.0 (0) | 1.0 (1)	0.0 (0) | n/s	0.0 (0) | 0.75 (¾)
-0.4 ± 0.75 -⅜ ± ¾	-0.4 ± 0.75 [1] -⅜ ± ¾ [1]	-0.03 ± 0.75 -1/32 ± ¾	CAMBER (Degrees)	-0.08 ± 0.75 -1/16 ± ¾	-0.75 ± 0.75 -¾ ± ¾	0.28 ± 0.75 5/16 ± ¾
0.0 (0) | 0.75 (¾)	0.0 (0) | 0.75 (¾)	0.0 (0) | 1.0 (1)	CROSS | TOL.	0.0 (0) | 1.0 (1)	0.0 (0) | 1.0 (1)	0.0 (0) | 0.5 (½)
-0.13° ± 0.2° -⅛° ± 3/16°	-0.15° ± 0.15° -5/32° ± 5/32°	0.2° ± 0.2° 3/16° ± 3/16°	FRONT TOE (Degrees)	0.2° ± 0.3° 3/16° ± 5/16°	0.2° ± 0.3° 3/16° ± 5/16°	0.0° ± 0.3° 0° ± 5/16°
-0.06" ± 0.1" -1/16" ± 3/32"	-0.08" ± 0.08" -1/16" ± 1/16"	0.1" ± 0.1" 3/32" ± 3/32"	FRONT TOE (Inches)	0.1" ± 0.15" 3/32" ± 5/32"	0.1" ± 0.15" 3/32" ± 5/32"	0.0" ± 0.15" 0" ± 5/32"
			SAI	12.4 (12 7/16)	12.2 (12 3/16)	12.8 (12 13/16)
			TURNING ANGLE	40.0 / 33.0 ± 2.0	38.0 / 30.0 ± 2.0	36.0 / 31.0 ± 2.0 [1]
148	148		ILLUSTRATION #	46	46	51
			REAR			
		-1.25 ± 0.75 -1¼ ± ¾	CAMBER (Degrees)	-0.33 ± 0.75 -5/16 ± ¾	-0.33 ± 0.75 -5/16 ± ¾	-0.5 ± 0.75 [2] -½ ± ¾ [2]
		0.24° ± 0.2° ¼° ± 3/16°	REAR TOE (Degrees)	0.2° ± 0.3° 3/16° ± 5/16°	0.2° ± 0.3° 3/16° ± 5/16°	0.3° ± 0.3° 5/16° ± 5/16°
		0.12" ± 0.1" ⅛" ± 3/32"	REAR TOE (Inches)	0.1" ± 0.15" 3/32" ± 5/32"	0.1" ± 0.15" 3/32" ± 5/32"	0.15" ± 0.15" 5/32" ± 5/32"
		0.1 (3/32)	THRUST ANGLE			
		175	ILLUSTRATION #	10	10	10
	[1] Limo = 0.5° (½°)		NOTES	[1] 1990 = 2.08° (2 1/16°)		[1] w/ Turbo A/T 34°/29° [2] w/4WS 0.0° (0°)
27067	27068	27070		28002	28004	28009

ALIGNMENT SPECIFICATIONS

Mazda

MODEL INFORMATION	626	626	626 (Fuel tank full)	626 (Fuel tank full)	929	929
YEARS	1991-92	1993-94	1995-97	1998-2002	1990-91	1992-94
INSTRUCTIONS						
LUGNUT TORQUE					66-87 ft.lbs.	66-87 ft.lbs.
FRONT						
CASTER (Degrees)	1.68 ± 0.75 1 11/16 ± 3/4	2.62 ± 0.75 2 5/8 ± 3/4	2.62 ± 1.0 2 5/8 ± 1	2.15 ± 1.0 2 1/8 ± 1	4.5 ± 0.75 4 1/2 ± 3/4	5.25 ± 0.75 5 1/4 ± 3/4
CROSS | TOL.	0.0 (0) | 0.75 (3/4)	0.0 (0) | 1.5 (1 1/2)	0.0 (0) | 1.5 (1 1/2)	0.0 (0) | 1.5 (1 1/2)	0.0 (0) | 0.7 (11/16)	0.0 (0) | 1.0 (1)
CAMBER (Degrees)	0.28 ± 0.75 5/16 ± 3/4	-0.6 ± 0.75 -5/8 ± 3/4	-0.6 ± 1.0 -5/8 ± 1	-0.72 ± 1.0 -11/16 ± 1	1.0 ± 0.75 1 ± 3/4	0.17 ± 0.75 3/16 ± 3/4
CROSS | TOL.	0.0 (0) | 0.5 (1/2)	0.0 (0) | 1.5 (1 1/2)	0.0 (0) | 1.5 (1 1/2)	0.0 (0) | 1.5 (1 1/2)	0.0 (0) | 0.5 (1/2)	0.0 (0) | 1.0 (1)
FRONT TOE (Degrees)	0.0° ± 0.3° 0° ± 5/16°	0.28° ± 0.28° 9/32° ± 9/32°	0.28° ± 0.38° 9/32° ± 3/8°	0.28° ± 0.38° 9/32° ± 3/8°	0.4° ± 0.3° 13/32° ± 5/16°	0.16° ± 0.24° 5/32° ± 1/4°
FRONT TOE (Inches)	0.0" ± 0.15" 0" ± 5/32"	0.14" ± 0.14" 5/32" ± 5/32"	0.14" ± 0.19" 5/32" ± 3/16"	0.14" ± 0.19" 5/32" ± 3/16"	0.2" ± 0.15" 3/16" ± 5/32"	0.08" ± 0.12" 3/32" ± 1/8"
SAI	12.8 (12 13/16)	15.1 (15 1/16)	15.1 (15 1/16)	15.0 (15 1/16)	12.7 (12 11/16)	10.6 (10 9/16)
TURNING ANGLE	36.0 / 31.0 ± 2.0 [1]	37.0 / 32.0 ± 2.0	37.0 / 32.0 ± 2.0	37.0 / 32.0 ± 3.0	39.0 / 32.0 ± 2.0	39.0 / 32.0 ± 2.0
ILLUSTRATION #	51	51	51	51	50	48
REAR						
CAMBER (Degrees)	-0.5 ± 0.75 -1/2 ± 3/4	-0.15 ± 0.75 -1/8 ± 3/4	-0.15 ± 1.0 -1/8 ± 1	-0.1 ± 1.0 -1/8 ± 1	-0.25 ± 0.5 -1/4 ± 1/2	-0.42 ± 0.75 -7/16 ± 3/4
REAR TOE (Degrees)	0.3° ± 0.3° 5/16° ± 5/16°	0.28° ± 0.28° 9/32° ± 9/32°	0.28° ± 0.38° 9/32° ± 3/8°	0.28° ± 0.38° 9/32° ± 3/8°	0.22° ± 0.2° 7/32° ± 3/16°	0.24° ± 0.24° 1/4° ± 1/4°
REAR TOE (Inches)	0.15" ± 0.15" 5/32" ± 5/32"	0.14" ± 0.14" 5/32" ± 5/32"	0.14" ± 0.19" 5/32" ± 3/16"	0.14" ± 0.19" 5/32" ± 3/16"	0.11" ± 0.1" 3/32" ± 3/32"	0.12" ± 0.12" 1/8" ± 1/8"
THRUST ANGLE						0.33 (11/32)
ILLUSTRATION #	10	06	06	06	06	48 49
NOTES	[1] w/ Turbo A/T 34°/29°					
	28010	28011	28012	28013	28019	28020

Mazda

MODEL INFORMATION	929	B Series Pickup 4X2	B Series Pickup 4X2	B Series Pickup 4X2 Coil spring susp.	B Series Pickup 4X2 Coil spring susp.	B Series Pickup 4X2 Torsion bar susp.
YEARS	1995	1990-93	1994-97	1998	1999-2003	1999-2003
INSTRUCTIONS		p. 306, 307	p. 306, 307	p. 306, 307		
LUGNUT TORQUE	66-87 ft.lbs.		100 ft.lbs.	100 ft.lbs.	100 ft.lbs.	100 ft.lbs.
FRONT						
CASTER (Degrees)	5.35 ± 1.0 5 3/8 ± 1	1.83 ± 0.75 [1] 1 13/16 ± 3/4 [1]	MIN.= 2.0 MAX.= 6.0 MIN.= 2 MAX.= 6	L= 4.0, R= 4.5 ± 1.0 L= 4, R= 4 1/2 ± 1	L= 4.0, R= 4.5 ± 1.0 L= 4, R= 4 1/2 ± 1	L= 4.0, R= 4.5 ± 1.0 L= 4, R= 4 1/2 ± 1
CROSS | TOL.	0.0 (0) | 1.5 (1 1/2)	0.0 (0) | 0.75 (3/4)	-1.0 (-1) | 0.4 (3/8)	-0.5 (-1/2) | 0.7 (11/16)	-0.5 (-1/2) | 0.7 (11/16)	-0.5 (-1/2) | 0.7 (11/16)
CAMBER (Degrees)	0.27 ± 1.0 1/4 ± 1	0.75 +0.5 -0.33 3/4 +1/2 -5/16	0.25 ± 0.5 1/4 ± 1/2	-0.5 ± 0.7 -1/2 ± 11/16	-0.5 ± 0.7 -1/2 ± 11/16	-0.5 ± 0.7 -1/2 ± 11/16
CROSS | TOL.	0.0 (0) | 1.5 (1 1/2)	0.0 (0) | 0.5 (1/2)	0.0 (0) | 0.4 (3/8)	0.0 (0) | 0.7 (11/16)	0.0 (0) | 0.7 (11/16)	0.0 (0) | 0.7 (11/16)
FRONT TOE (Degrees)	0.32° ± 0.32° 5/16° ± 5/16°	0.3° ± 0.3° 5/16° ± 5/16°	0.06° ± 0.25° 1/16° ± 1/4°	0.06° ± 0.25° 1/16° ± 1/4°	0.06° ± 0.25° 1/16° ± 1/4°	0.12° ± 0.25° 1/8° ± 1/4°
FRONT TOE (Inches)	0.16" ± 0.16" 5/32" ± 5/32"	0.15" ± 0.15" 5/32" ± 5/32"	0.03" ± 0.13" 1/32" ± 1/8"	0.03" ± 0.13" 1/32" ± 1/8"	0.03" ± 0.13" 1/32" ± 1/8"	0.06" ± 0.13" 1/16" ± 1/8"
SAI	10.5 (10 7/16)	8.3 (8 1/4)				
TURNING ANGLE	39.0 / 32.0 ± 2.0	35.0 / 33.0 ± 2.0				
ILLUSTRATION #	48	23	56	24	24	73
REAR						
CAMBER (Degrees)	-0.42 ± 1.0 -7/16 ± 1					
REAR TOE (Degrees)	0.2° ± 0.32° 3/16° ± 5/16°					
REAR TOE (Inches)	0.1" ± 0.16" 3/32" ± 5/32"					
THRUST ANGLE						
ILLUSTRATION #	48 49					
NOTES		[1] w/MS 0.83° (13/16°)				
	28021	28024	28025	28026	28027	28028

Mazda

MODEL INFORMATION	B Series Pickup 4X2 Coil spring susp.	B Series Pickup 4X2 Torsion bar susp.	B Series Pickup 4X2 Coil spring susp.	B Series Pickup 4X2 Torsion bar susp.	B Series Pickup 4X2 Coil spring susp.	B Series Pickup 4X2 Torsion bar susp.
YEARS	2004-2006	2004-2006	2007	2007	2008-2009	2008-2009
INSTRUCTIONS						
LUGNUT TORQUE	100 ft.lbs.	100 ft.lbs.	100 ft.lbs.	100 ft.lbs.	100 ft.lbs.	100 ft.lbs.
FRONT						
CASTER (Degrees)	L= 2.9, R= 3.6 ± 1.0 L= 2 7/8, R= 3 5/8 ± 1	L= 3.9, R= 4.5 ± 1.0 L= 3 7/8, R= 4 1/2 ± 1	L= 3.5, R= 3.9 ± 1.0 L= 3 1/2, R= 3 7/8 ± 1	L= 2.8, R= 3.5 ± 1.0 L= 2 13/16, R= 3 1/2 ± 1	L= 3.0, R= 3.4 ± 1.0 L= 3, R= 3 3/8 ± 1	L= 2.8, R= 3.5 ± 1.0 L= 2 13/16, R= 3 1/2 ± 1
CROSS / TOL.	-0.7 (-11/16) / 0.7 (11/16)	-0.6 (-5/8) / 0.7 (11/16)	-0.4 (-3/8) / 0.7 (11/16)	-0.7 (-11/16) / 0.7 (11/16)	-0.4 (-3/8) / 0.75 (3/4)	-0.7 (-11/16) / 0.75 (3/4)
CAMBER (Degrees)	-0.5 ±0.7 -1/2 ± 11/16	-0.5 ±0.7 -1/2 ± 11/16	-0.5 ±0.7 -1/2 ± 11/16	-0.5 ±0.7 -1/2 ± 11/16	-0.5 ±0.75 -1/2 ± 3/4	0.0 ±0.7 0 ± 11/16
CROSS / TOL.	0.0 (0) / 0.7 (11/16)	0.0 (0) / 0.7 (11/16)	0.0 (0) / 0.7 (11/16)	0.0 (0) / 0.7 (11/16)	0.0 (0) / 0.75 (3/4)	0.0 (0) / 0.75 (3/4)
FRONT TOE (Degrees)	0.06° ±0.25° 1/16° ± 1/4°	0.12° ±0.25° 1/8° ± 1/4°	0.06° ±0.25° 1/16° ± 1/4°	0.12° ±0.25° 1/8° ± 1/4°	0.06° ±0.25° 1/16° ± 1/4°	0.12° ±0.25° 1/8° ± 1/4°
FRONT TOE (Inches)	0.03" ±0.13" 1/32" ± 1/8"	0.06" ±0.13" 1/16" ± 1/8"	0.03" ±0.13" 1/32" ± 1/8"	0.06" ±0.13" 1/16" ± 1/8"	0.03" ±0.13" 1/32" ± 1/8"	0.06" ±0.13" 1/16" ± 1/8"
SAI						
TURNING ANGLE						
ILLUSTRATION #	24	73	24	73	24	73
REAR						
CAMBER (Degrees)						
REAR TOE (Degrees)						
REAR TOE (Inches)						
THRUST ANGLE						
ILLUSTRATION #						
NOTES						
	28029	28030	28031	28032	28033	28034

Mazda

MODEL INFORMATION	B Series Pickup 4X4	B Series Pickup 4X4	B Series Pickup 4X4	B Series Pickup 4X4	B Series Pickup 4X4	B Series Pickup 4X4
YEARS	1990-93	1994-97	1998	1999-2003	2004-2006	2007
INSTRUCTIONS	p. 306, 307	p. 306, 307	p. 306, 307			
LUGNUT TORQUE		100 ft.lbs.	100 ft.lbs.	100 ft.lbs.	100 ft.lbs.	100 ft.lbs.
FRONT						
CASTER (Degrees)	2.0 ±0.75 2 ± 3/4	MIN.= 2.0 MAX.= 6.0 MIN.= 2 MAX.= 6	L= 4.0, R= 4.5 ± 1.0 L= 4, R= 4 1/2 ± 1	L= 4.0, R= 4.5 ± 1.0 L= 4, R= 4 1/2 ± 1	L= 3.9, R= 4.5 ± 1.0 L= 3 7/8, R= 4 1/2 ± 1	L= 2.8, R= 3.5 ± 1.0 L= 2 13/16, R= 3 1/2 ± 1
CROSS / TOL.	0.0 (0) / 0.75 (3/4)	-1.0 (-1) / 0.4 (3/8)	-0.5 (-1/2) / 0.7 (11/16)	-0.5 (-1/2) / 0.7 (11/16)	-0.6 (-5/8) / 0.7 (11/16)	-0.7 (-11/16) / 0.7 (11/16)
CAMBER (Degrees)	1.0 +0.5 -0.33 1 +1/2 -5/16	0.25 ±0.5 1/4 ± 1/2	-0.5 ±0.7 -1/2 ± 11/16	-0.5 ±0.7 -1/2 ± 11/16	-0.5 ±0.7 -1/2 ± 11/16	-0.5 ±0.7 -1/2 ± 11/16
CROSS / TOL.	0.0 (0) / 0.5 (1/2)	0.0 (0) / 0.4 (3/8)	0.0 (0) / 0.7 (11/16)	0.0 (0) / 0.7 (11/16)	0.0 (0) / 0.7 (11/16)	0.0 (0) / 0.7 (11/16)
FRONT TOE (Degrees)	0.3° ±0.3° 5/16° ± 5/16°	0.06° ±0.25° 1/16° ± 1/4°	0.12° ±0.25° 1/8° ± 1/4°	0.12° ±0.25° 1/8° ± 1/4°	0.12° ±0.25° 1/8° ± 1/4°	0.12° ±0.25° 1/8° ± 1/4°
FRONT TOE (Inches)	0.15" ±0.15" 5/32" ± 5/32"	0.03" ±0.13" 1/32" ± 1/8"	0.06" ±0.13" 1/16" ± 1/8"	0.06" ±0.13" 1/16" ± 1/8"	0.06" ±0.13" 1/16" ± 1/8"	0.06" ±0.13" 1/16" ± 1/8"
SAI	10.3 (10 5/16)					
TURNING ANGLE	33.5 / 30.0 ±2.0					
ILLUSTRATION #	23	56	73	73	73	73
REAR						
CAMBER (Degrees)						
REAR TOE (Degrees)						
REAR TOE (Inches)						
THRUST ANGLE						
ILLUSTRATION #						
NOTES						
	28035	28036	28037	28038	28039	28040

Mazda

B Series Pickup 4X4 2008-2009	CX-3 2WD (Fuel tank full) 2016-2017	CX-3 AWD (Fuel tank full) 2016-2017	MODEL INFORMATION YEARS	CX-5 All models (Fuel tank full) 2013-2016	CX-5 All models (Fuel tank full) 2017	CX-7 All models (Fuel tank full) 2007-2009
			INSTRUCTIONS			
100 ft.lbs.	80-108 ft.lbs.	80-108 ft.lbs.	LUGNUT TORQUE	80-108 ft.lbs.	80-108 ft.lbs.	65-87 ft.lbs.
			FRONT			
L= 2.8, R= 3.5 ± 1.0 L= 2 13/16, R= 3 1/2 ± 1	5.05 ± 1.0 5 1/16 ± 1	5.05 ± 1.0 5 1/16 ± 1	CASTER (Degrees)	6.48 ± 1.0 6 1/2 ± 1	5.95 ± 1.0 5 15/16 ± 1	2.95 ± 1.0 2 15/16 ± 1
-0.7 (-11/16) | 0.75 (3/4)	0.0 (0) | 1.5 (1 1/2)	0.0 (0) | 1.5 (1 1/2)	CROSS | TOL.	0.0 (0) | 1.5 (1 1/2)	0.0 (0) | 1.5 (1 1/2)	0.0 (0) | 1.5 (1 1/2)
0.0 ± 0.7 0 ± 11/16	-0.27 ± 1.0 -1/4 ± 1	-0.27 ± 1.0 -1/4 ± 1	CAMBER (Degrees)	-0.37 ± 1.0 -3/8 ± 1	-0.43 ± 1.0 -7/16 ± 1	-0.33 ± 1.0 -5/16 ± 1
0.0 (0) | 0.75 (3/4)	0.0 (0) | 1.5 (1 1/2)	0.0 (0) | 1.5 (1 1/2)	CROSS | TOL.	0.0 (0) | 1.5 (1 1/2)	0.0 (0) | 1.5 (1 1/2)	0.0 (0) | 1.5 (1 1/2)
0.12° ± 0.25° 1/8° ± 1/4°	0.08° ± 0.17° 3/32° ± 5/32°	0.08° ± 0.17° 3/32° ± 5/32°	FRONT TOE (Degrees)	0.17° ± 0.33° 5/32° ± 11/32°	0.17° ± 0.17° 5/32° ± 5/32°	0.0° ± 0.15° 0° ± 5/32°
0.06" ± 0.13" 1/16" ± 1/8"	0.04" ± 0.08" 1/32" ± 3/32"	0.04" ± 0.08" 1/32" ± 3/32"	FRONT TOE (Inches)	0.08" ± 0.17" 3/32" ± 5/32"	0.08" ± 0.08" 3/32" ± 3/32"	0.0" ± 0.08" 0" ± 1/16"
	13.0 (13)	13.0 (13)	SAI	12.0 (12)	11.6 (11 9/16)	11.7 (11 3/4)
	37.0 / 30.0 ±3.0	37.0 / 30.0 ±3.0	TURNING ANGLE	37.8 / 30.9 ±3.0	37.9 / 30.9 ±3.0	36.1 / 30.9 ±3.0
73			ILLUSTRATION #			51
			REAR			
	-1.58 ± 1.0 -1 9/16 ± 1	-1.43 ± 1.0 -1 9/16 ± 1	CAMBER (Degrees)	-1.02 ± 1.0 -1 ± 1	-0.73 ± 1.0 -3/4 ± 1	-1.05 ± 1.0 -1 1/16 ± 1
	0.38° ± 0.37° 3/8° ± 3/8°	0.33° ± 0.37° 11/32° ± 3/8°	REAR TOE (Degrees)	0.17° ± 0.33° 5/32° ± 11/32°	0.17° ± 0.17° 5/32° ± 5/32°	0.17° ± 0.33° 5/32° ± 11/32°
	0.19" ± 0.18" 3/16" ± 3/16"	0.17" ± 0.18" 5/32" ± 3/16"	REAR TOE (Inches)	0.08" ± 0.17" 3/32" ± 5/32"	0.08" ± 0.08" 3/32" ± 3/32"	0.08" ± 0.17" 3/32" ± 5/32"
	0.8 (13/16)	0.8 (13/16)	THRUST ANGLE	0.8 (13/16)	0.8 (13/16)	0.8 (13/16)
			ILLUSTRATION #	06	06	06
			NOTES			
28041	28043	28044		28045	28046	28047

Mazda

CX-7 All models (Fuel tank full) 2010-2012	CX-9 All models (Fuel tank full) 2007-2015	CX-9 FWD (Fuel tank full) 2016-2017	MODEL INFORMATION YEARS	CX-9 AWD (Fuel tank full) 2016-2017	Mazda 2 (Fuel tank full) 2011-2014	Mazda 3 (Fuel tank full) 2004-2006
			INSTRUCTIONS			
65-87 ft.lbs.	80-108 ft.lbs.	80-108 ft.lbs.	LUGNUT TORQUE	80-108 ft.lbs.	65-87 ft.lbs.	65-87 ft.lbs.
			FRONT			
2.95 ± 1.0 2 15/16 ± 1	3.08 ± 1.0 3 1/16 ± 1	5.95 ± 1.0 5 15/16 ± 1	CASTER (Degrees)	5.95 ± 1.0 5 15/16 ± 1	3.72 ± 1.0 3 11/16 ± 1	3.03 ± 1.0 3 1/16 ± 1
0.0 (0) | 1.5 (1 1/2)	0.0 (0) | 1.5 (1 1/2)	0.0 (0) | 1.5 (1 1/2)	CROSS | TOL.	0.0 (0) | 1.5 (1 1/2)	0.0 (0) | 1.5 (1 1/2)	0.0 (0) | 1.5 (1 1/2)
-0.33 ± 1.0 -5/16 ± 1	-0.35 ± 1.0 -3/8 ± 1	-0.3 ± 1.0 -5/16 ± 1	CAMBER (Degrees)	-0.3 ± 1.0 -5/16 ± 1	-0.75 ± 1.0 -3/4 ± 1	-0.68 ± 1.0 -11/16 ± 1
0.0 (0) | 1.5 (1 1/2)	0.0 (0) | 1.5 (1 1/2)	0.0 (0) | 1.5 (1 1/2)	CROSS | TOL.	0.0 (0) | 1.5 (1 1/2)	0.0 (0) | 1.5 (1 1/2)	0.0 (0) | 1.5 (1 1/2)
0.0° ± 0.3° 0° ± 5/16°	0.15° ± 0.3° 5/32° ± 5/16°	0.15° ± 0.3° 5/32° ± 5/16°	FRONT TOE (Degrees)	0.15° ± 0.3° 5/32° ± 5/16°	0.32° ± 0.4° 5/16° ± 13/32°	0.18° ± 0.37° 3/16° ± 3/8°
0.0" ± 0.15" 0" ± 5/32"	0.08" ± 0.15" 1/16" ± 5/32"	0.08" ± 0.15" 1/16" ± 5/32"	FRONT TOE (Inches)	0.08" ± 0.15" 1/16" ± 5/32"	0.16" ± 0.2" 5/32" ± 3/16"	0.09" ± 0.18" 3/32" ± 3/16"
11.7 (11 3/4)	11.6 (11 9/16)	12.9 (12 7/8)	SAI	12.9 (12 7/8)	13.5 (13 1/2)	14.1 (14 1/16)
36.1 / 30.9 ±3.0	38.0 / 32.0 ±3.0	37.6 / 31.6 ±3.0	TURNING ANGLE	37.6 / 31.6 ±3.0	40.2 / 33.2 ±3.0	39.8 / 32.8 ±3.0
51			ILLUSTRATION #			
			REAR			
-1.05 ± 1.0 -1 1/16 ± 1	-0.55 ± 1.0 -9/16 ± 1	-0.9 ± 1.0 -7/8 ± 1	CAMBER (Degrees)	-0.9 ± 1.0 -7/8 ± 1	-1.5 ± 1.0 -1 1/2 ± 1	-1.3 ± 1.0 -1 5/16 ± 1
0.15° ± 0.3° 5/32° ± 5/16°	0.15° ± 0.3° 5/32° ± 5/16°	0.15° ± 0.3° 5/32° ± 5/16°	REAR TOE (Degrees)	0.15° ± 0.3° 5/32° ± 5/16°	0.37° ± 0.4° 3/8° ± 13/32°	0.18° ± 0.37° 3/16° ± 3/8°
0.08" ± 0.15" 1/16" ± 5/32"	0.08" ± 0.15" 1/16" ± 5/32"	0.08" ± 0.15" 1/16" ± 5/32"	REAR TOE (Inches)	0.08" ± 0.15" 1/16" ± 5/32"	0.18" ± 0.2" 3/16" ± 3/16"	0.09" ± 0.18" 3/32" ± 3/16"
0.8 (13/16)	0.8 (13/16)	0.8 (13/16)	THRUST ANGLE	0.8 (13/16)	0.8 (13/16)	0.8 (13/16)
06	177	06	ILLUSTRATION #	06		06
			NOTES			
28048	28049	28050		28051	28056	28057

Mazda

Mazda 3	Mazda 3	Mazda 3	MODEL INFORMATION	MAZDASPEED3	MAZDASPEED3	Mazda 5
(Fuel tank full)	(Fuel tank full)	(Fuel tank full)		(Fuel tank full)	(Fuel tank full)	(Fuel tank full)
2007-2009	2010-2013	2014-2017	YEARS	2007-2009	2010-2013	2006-2007
			INSTRUCTIONS			
65-87 ft.lbs.	65-87 ft.lbs.	80-108 ft.lbs.	LUGNUT TORQUE	65-87 ft.lbs.	65-87 ft.lbs.	65-87 ft.lbs.
			FRONT			
3.03 ±1.0 3 1/16 ±1	3.12 ±1.0 3 1/8 ±1	6.48 ±1.0 6 1/2 ±1	CASTER (Degrees)	3.13 ±1.0 3 1/8 ±1	3.2 ±1.0 3 3/16 ±1	3.02 ±1.0 3 1/32 ±1
0.0 (0) | 1.5 (1 1/2)	0.0 (0) | 1.5 (1 1/2)	0.0 (0) | 1.5 (1 1/2)	CROSS | TOL.	0.0 (0) | 1.5 (1 1/2)	0.0 (0) | 1.5 (1 1/2)	0.0 (0) | 1.5 (1 1/2)
-0.68 ±1.0 -11/16 ±1	-0.63 ±1.0 -5/8 ±1	-0.2 ±1.0 -3/16 ±1	CAMBER (Degrees)	-0.95 ±1.0 -15/16 ±1	-0.9 ±1.0 -7/8 ±1	-0.7 ±1.0 -11/16 ±1
0.0 (0) | 1.5 (1 1/2)	0.0 (0) | 1.5 (1 1/2)	0.0 (0) | 1.5 (1 1/2)	CROSS | TOL.	0.0 (0) | 1.5 (1 1/2)	0.0 (0) | 1.5 (1 1/2)	0.0 (0) | 1.5 (1 1/2)
0.18° ±0.37° 3/16° ± 3/8°	0.18° ±0.37° 3/16° ± 3/8°	0.18° ±0.37° 3/16° ± 3/8°	FRONT TOE (Degrees)	0.18° ±0.37° 3/16° ± 3/8°	0.18° ±0.37° 3/16° ± 3/8°	0.18° ±0.37° 3/16° ± 3/8°
0.09" ±0.18" 3/32" ± 3/16"	0.09" ±0.18" 3/32" ± 3/16"	0.09" ±0.18" 3/32" ± 3/16"	FRONT TOE (Inches)	0.09" ±0.18" 3/32" ± 3/16"	0.09" ±0.18" 3/32" ± 3/16"	0.09" ±0.18" 3/32" ± 3/16"
13.9 (13 15/16)	13.9 (13 15/16)	14.0 (14)	SAI	14.4 (14 3/8)	14.3 (14 5/16)	14.0 (14)
39.8 / 32.8 ±3.0	41.0 / 34.0 ±3.0	39.4 / 32.1 ±3.0	TURNING ANGLE	36.0 / 30.2 ±3.0	36.0 / 30.0 ±3.0	40.5 / 33.8 ±3.0
			ILLUSTRATION #			
			REAR			
-1.45 ±1.0 -1 7/16 ±1	-1.5 ±1.0 -1 1/2 ±1	-1.0 ±1.0 -11 ±1	CAMBER (Degrees)	-1.7 ±1.0 -1 11/16 ±1	-1.68 ±1.0 -1 11/16 ±1	-1.3 ±1.0 -1 5/16 ±1
0.18° ±0.37° 3/16° ± 3/8°	0.27° ±0.35° 9/32° ± 11/32°	0.17° ±0.33° 5/32° ± 11/32°	REAR TOE (Degrees)	0.18° ±0.37° 3/16° ± 3/8°	0.27° ±0.35° 9/32° ± 11/32°	0.18° ±0.37° 3/16° ± 3/8°
0.09" ±0.18" 3/32" ± 3/16"	0.13" ±0.18" 1/8" ± 3/16"	0.08" ±0.17" 3/32" ± 5/32"	REAR TOE (Inches)	0.09" ±0.18" 3/32" ± 3/16"	0.13" ±0.18" 1/8" ± 3/16"	0.09" ±0.18" 3/32" ± 3/16"
0.8 (13/16)	0.8 (13/16)	0.8 (13/16)	THRUST ANGLE	0.8 (13/16)	0.8 (13/16)	0.8 (13/16)
06	06	06	ILLUSTRATION #	06	06	06
			NOTES			
28058	28059	28060		28061	28062	28063

Mazda

Mazda 5	Mazda 5	Mazda 6 Sedan & Hatch	MODEL INFORMATION	Mazda 6 Wagon	Mazda 6 Sedan & Hatch	Mazda 6 2.5L 4 Cyl.
(Fuel tank full)	(Fuel tank full)	(Fuel tank full)		(Fuel tank full)	(Fuel tank full)	(Fuel tank full)
2008-2010	2012-2017	2003-2007	YEARS	2004-2007	2008	2009-2013
			INSTRUCTIONS			
65-87 ft.lbs.	65-87 ft.lbs.	65-87 ft.lbs.	LUGNUT TORQUE	65-87 ft.lbs.	65-87 ft.lbs.	65-87 ft.lbs.
			FRONT			
3.02 ±1.0 3 1/32 ±1	3.05 ±1.0 3 1/16 ±1	3.78 ±1.0 3 3/4 ±1	CASTER (Degrees)	3.65 ±1.0 3 5/8 ±1	4.3 ±1.0 4 5/16 ±1	3.7 ±1.0 3 11/16 ±1
0.0 (0) | 1.5 (1 1/2)	0.0 (0) | 1.5 (1 1/2)	0.0 (0) | 1.5 (1 1/2)	CROSS | TOL.	0.0 (0) | 1.5 (1 1/2)	0.0 (0) | 1.5 (1 1/2)	0.0 (0) | 1.5 (1 1/2)
-0.7 ±1.0 -11/16 ±1	-0.68 ±1.0 -11/16 ±1	-0.28 ±1.0 [1] -5/16 ±1 [1]	CAMBER (Degrees)	-0.28 ±1.0 -5/16 ±1	-0.17 ±1.0 [1] -3/16 ±1 [1]	-0.25 ±1.0 -1/4 ±1
0.0 (0) | 1.5 (1 1/2)	0.0 (0) | 1.5 (1 1/2)	0.0 (0) | 1.5 (1 1/2)	CROSS | TOL.	0.0 (0) | 1.5 (1 1/2)	0.0 (0) | 1.5 (1 1/2)	0.0 (0) | 1.5 (1 1/2)
0.18° ±0.37° 3/16° ± 3/8°	0.18° ±0.37° 3/16° ± 3/8°	0.18° ±0.37° 3/16° ± 3/8°	FRONT TOE (Degrees)	0.18° ±0.37° 3/16° ± 3/8°	0.18° ±0.37° 3/16° ± 3/8°	0.17° ±0.33° 5/32° ± 11/32°
0.09" ±0.18" 3/32" ± 3/16"	0.09" ±0.18" 3/32" ± 3/16"	0.09" ±0.18" 3/32" ± 3/16"	FRONT TOE (Inches)	0.09" ±0.18" 3/32" ± 3/16"	0.09" ±0.18" 3/32" ± 3/16"	0.08" ±0.17" 3/32" ± 5/32"
14.0 (14)	14.0 (14)	5.5 (5 1/2)	SAI	5.5 (5 7/16)	5.3 (5 5/16)	6.6 ±1.0 (6 5/8 ±1)
40.5 / 33.8 ±3.0	41.0 / 34.0 ±3.0	34.0 / 29.0 ±3.0	TURNING ANGLE	34.0 / 29.0 ±3.0	34.0 / 29.0 ±3.0	38.9 / 32.6 ±3.0
			ILLUSTRATION #			
			REAR			
-1.3 ±1.0 -1 5/16 ±1	-1.23 ±1.0 -1 1/4 ±1	-1.22 ±1.0 -1 3/16 ±1	CAMBER (Degrees)	-1.1 ±1.0 -1 1/8 ±1	-1.35 ±1.0 -1 3/8 ±1	-0.95 ±1.0 -15/16 ±1
0.27° +0.18° -0.27° 9/32° +3/16° -9/32°	0.27° ±0.37° 9/32° ± 3/8°	0.18° ±0.37° 3/16° ± 3/8°	REAR TOE (Degrees)	0.18° ±0.37° 3/16° ± 3/8°	0.18° ±0.37° 3/16° ± 3/8°	0.17° ±0.33° 5/32° ± 11/32°
0.13" +0.09" -0.13" 1/8" +3/32" -1/8"	0.13" ±0.18" 1/8" ± 3/16"	0.09" ±0.18" 3/32" ± 3/16"	REAR TOE (Inches)	0.09" ±0.18" 3/32" ± 3/16"	0.09" ±0.18" 3/32" ± 3/16"	0.08" ±0.17" 3/32" ± 5/32"
0.8 (13/16)	0.8 (13/16)	0.8 (13/16)	THRUST ANGLE	0.8 (13/16)	0.8 (13/16)	0.8 (13/16)
06	06	175	ILLUSTRATION #	175	175	189
		[1] 6 Cyl. = -0.3° (-5/16°)	NOTES		[1] 6 Cyl. = -0.18° (-3/16°)	
28064	28065	28066		28067	28068	28069

ALIGNMENT SPECIFICATIONS

Mazda

Mazda 6 3.7L 6 Cyl. (Fuel tank full)	Mazda 6 (Fuel tank full)	MAZDASPEED6 (Fuel tank full)	MODEL INFORMATION	Millenia	Millenia (Fuel tank full)	MPV 4X2
2009-2013	2014-2017	2006-2007	YEARS	1995-2000	2001-2002	1990-93
			INSTRUCTIONS			
65-87 ft.lbs.	80-108 ft.lbs.	65-87 ft.lbs.	LUGNUT TORQUE	65-94 ft.lbs.	65-94 ft.lbs.	65-87 ft.lbs.
			FRONT			
3.67 ± 1.0 3 11/16 ± 1	6.28 ± 1.0 6 5/16 ± 1	3.88 ± 1.0 3 7/8 ± 1	CASTER (Degrees)	2.58 ± 1.0 [1,2] 2 9/16 ± 1 [1,2]	2.95 ± 1.0 2 15/16 ± 1	5.45 ± 0.75 [1] 5 7/16 ± 3/4 [1]
0.0 (0) | 1.5 (1 1/2)	0.0 (0) | 1.5 (1 1/2)	0.0 (0) | 1.5 (1 1/2)	CROSS | TOL.	0.0 (0) | 1.5 (1 1/2)	0.0 (0) | 1.5 (1 1/2)	0.0 (0) | n/s
-0.27 ± 1.0 - 1/4 ± 1	-0.32 ± 1.0 - 5/16 ± 1	-0.28 ± 1.0 - 5/16 ± 1	CAMBER (Degrees)	-0.18 ± 1.0 - 3/16 ± 1	-0.18 ± 1.0 - 3/16 ± 1	0.37 ± 0.5 3/8 ± 1/2
0.0 (0) | 1.5 (1 1/2)	0.0 (0) | 1.5 (1 1/2)	0.0 (0) | 1.5 (1 1/2)	CROSS | TOL.	0.0 (0) | 1.5 (1 1/2)	0.0 (0) | 1.5 (1 1/2)	0.0 (0) | n/s
0.17° ± 0.33° 5/32° ± 11/32°	0.17° ± 0.33° 5/32° ± 11/32°	0.18° ± 0.37° 3/16° ± 3/8°	FRONT TOE (Degrees)	0.28° ± 0.4° 9/32° ± 13/32°	0.28° ± 0.4° 9/32° ± 13/32°	0.4° ± 0.3° 13/32° ± 5/16°
0.08" ± 0.17" 3/32" ± 5/32"	0.08" ± 0.17" 3/32" ± 5/32"	0.09" ± 0.18" 3/32" ± 3/16"	FRONT TOE (Inches)	0.14" ± 0.2" 5/32" ± 3/16"	0.14" ± 0.2" 5/32" ± 3/16"	0.2" ± 0.15" 3/16" ± 5/32"
6.6 ± 1.0 (6 5/8 ± 1)	15.2 ± 1.0 (15 3/16 ± 1)	5.5 (5 1/2)	SAI	9.1 ± 0.8 (9 1/8 ± 3/4)	9.1 ± 0.8 (9 1/8 ± 3/4)	12.9 (12 15/16)
38.9 / 32.6 ± 3.0	38.6 / 31.9 ± 3.0	36.0 / 30.0 ± 3.0	TURNING ANGLE	36.0 / 30.0 ± 3.0	36.0 / 30.0 ± 3.0	40.8 / 31.8 ± 2.0
			ILLUSTRATION #	48	48	51
			REAR			
-0.95 ± 1.0 - 15/16 ± 1	-1.05 ± 1.0 -1 1/16 ± 1	-1.62 ± 1.0 -1 5/8 ± 1	CAMBER (Degrees)	-0.32 ± 1.0 - 5/16 ± 1	-0.28 ± 1.0 - 5/16 ± 1	
0.17° ± 0.33° 5/32° ± 11/32°	0.17° ± 0.33° 5/32° ± 11/32°	0.18° ± 0.37° 3/16° ± 3/8°	REAR TOE (Degrees)	0.28° ± 0.4° 9/32° ± 13/32°	0.28° ± 0.4° 9/32° ± 13/32°	
0.08" ± 0.17" 3/32" ± 5/32"	0.08" ± 0.17" 3/32" ± 5/32"	0.09" ± 0.18" 3/32" ± 3/16"	REAR TOE (Inches)	0.14" ± 0.2" 5/32" ± 3/16"	0.14" ± 0.2" 5/32" ± 3/16"	
0.8 (13/16)	0.8 (13/16)	0.8 (13/16)	THRUST ANGLE	0.8 (13/16)	0.1 (3/32)	
189	174	175	ILLUSTRATION #	48 49	48 49	
			NOTES	[1] 1999-00 = 2.23° (2 1/4°) [2] Fuel tank FULL		[1] 1993 = 4.8° (4 13/16°)
28070	28071	28072		28073	28074	28077

Mazda

MPV 4X2	MPV 4X2 (Fuel tank full)	MPV 4X4	MODEL INFORMATION	MPV 4X4 (Fuel tank full)	MPV	MPV
1994-95	1996-98	1990-93	YEARS	1994-98	2000-2001	2002-2006
			INSTRUCTIONS			
		65-87 ft.lbs.	LUGNUT TORQUE		80-108 ft.lbs.	80-108 ft.lbs.
			FRONT			
L= 5.45, R= 5.95 ± 0.75 L= 5 7/16, R= 5 15/16 ± 3/4	L= 4.95, R= 5.45 ± 1.0 L= 4 15/16, R= 5 7/16 ± 1	5.52 ± 0.75 [1] 5 1/2 ± 3/4 [1]	CASTER (Degrees)	L= 5.02, R= 5.52 ± 1.0 L= 5, R= 5 1/2 ± 1	1.75 ± 1.0 1 3/4 ± 1	2.03 ± 1.0 2 1/16 ± 1
-0.5 (-1/2) | 1.5 (1 1/2)	-0.5 (-1/2) | 1.5 (1 1/2)	0.0 (0) | n/s	CROSS | TOL.	-0.5 (-1/2) | 1.5 (1 1/2)	0.0 (0) | 1.5 (1 1/2)	0.0 (0) | 1.5 (1 1/2)
0.37 ± 0.75 3/8 ± 3/4	0.37 ± 1.0 3/8 ± 1	0.17 ± 0.5 3/16 ± 1/2	CAMBER (Degrees)	0.17 ± 1.0 3/16 ± 1	-0.9 ± 1.0 - 7/8 ± 1	-0.23 ± 1.0 - 1/4 ± 1
0.0 (0) | 1.5 (1 1/2)	0.0 (0) | 1.5 (1 1/2)	0.0 (0) | n/s	CROSS | TOL.	0.0 (0) | 1.5 (1 1/2)	0.0 (0) | 1.5 (1 1/2)	0.0 (0) | 1.5 (1 1/2)
0.4° ± 0.3° 13/32° ± 5/16°	0.4° ± 0.4° 13/32° ± 13/32°	0.4° ± 0.3° 13/32° ± 5/16°	FRONT TOE (Degrees)	0.4° ± 0.4° 13/32° ± 13/32°	0.18° ± 0.37° 3/16° ± 3/8°	0.18° ± 0.37° 3/16° ± 3/8°
0.2" ± 0.15" 3/16" ± 5/32"	0.2" ± 0.2" 3/16" ± 3/16"	0.2" ± 0.15" 3/16" ± 5/32"	FRONT TOE (Inches)	0.2" ± 0.2" 3/16" ± 3/16"	0.09" ± 0.18" 3/32" ± 3/16"	0.09" ± 0.18" 3/32" ± 3/16"
12.9 (12 15/16)	12.9 (12 15/16)	11.8 (11 13/16)	SAI	11.8 (11 13/16)	11.2 (11 3/16)	11.2 (11 3/16)
40.8 / 31.8 ± 2.0	40.8 / 31.8 ± 2.0	33.8 / 31.1 ± 2.0	TURNING ANGLE	33.8 / 31.1 ± 2.0	37.1 / 32.0 ± 3.0	37.1 / 32.0 ± 3.0
51	51	51	ILLUSTRATION #	51	51 26	51 26
			REAR			
			CAMBER (Degrees)		-1.0 ± 1.0 -1 ± 1	-1.0 ± 1.0 -1 ± 1
			REAR TOE (Degrees)		0.27° ± 0.37° 9/32° ± 3/8°	0.28° ± 0.37° 9/32° ± 3/8°
			REAR TOE (Inches)		0.13" ± 0.18" 1/8" ± 3/16"	0.14" ± 0.18" 5/32" ± 3/16"
			THRUST ANGLE			
			ILLUSTRATION #			
		[1] 1993 = 5.1° (5 1/8°)	NOTES			
28078	28079	28080		28082	28083	28084

Mazda

MX-3	MX-3	MX-5 Miata	MODEL	MX-5 Miata	MX-5 Miata	MX-5 Miata
			INFORMATION	exc/MAZDASPEED	MAZDASPEED	16" tires
		(Fuel tank full)		(Fuel tank full)	(Fuel tank full)	(Fuel tank full)
1992-94	1995-96	1990-97	YEARS	1999-2005	2004-2005	2006-2008
			INSTRUCTIONS			
			LUGNUT TORQUE	66-86 ft.lbs.	66-86 ft.lbs.	
			FRONT			
2.63 ± 0.75 2⅝ ± ¾	2.77 ± 1.0 2¾ ± 1	4.43 ± 0.75 4 7/16 ± ¾	CASTER (Degrees)	5.8 ± 1.0 5 13/16 ± 1	6.3 ± 1.0 6 5/16 ± 1	5.98 ± 1.0 6 ± 1
0.0 (0) | 0.67 (11/16)	0.0 (0) | 1.5 (1½)	0.0 (0) | 1.5 (1½)	CROSS | TOL.	0.0 (0) | 1.5 (1½)	0.0 (0) | 1.5 (1½)	0.0 (0) | 1.0 (1)
-0.8 ± 0.75 -13/16 ± ¾	-0.97 ± 1.0 -15/16 ± 1	0.4 ± 0.75 ⅜ ± ¾	CAMBER (Degrees)	0.1 ± 1.0 ⅛ ± 1	-0.17 ± 1.0 -3/16 ± 1	-0.1 ± 1.0 -⅛ ± 1
0.0 (0) | 0.5 (½)	0.0 (0) | 1.5 (1½)	0.0 (0) | 1.0 (1)	CROSS | TOL.	0.0 (0) | 1.0 (1)	0.0 (0) | 1.0 (1)	0.0 (0) | 1.0 (1)
0.3° ± 0.3° 5/16° ± 5/16°	0.3° ± 0.4° 5/16° ± 13/32°	0.3° ± 0.3° 5/16° ± 5/16°	FRONT TOE (Degrees)	0.3° ± 0.4° 5/16° ± 13/32°	0.3° ± 0.4° 5/16° ± 13/32°	0.18° ± 0.37° 3/16° ± ⅜°
0.15" ± 0.15" 5/32" ± 5/32"	0.15" ± 0.2" 5/32" ± 3/16"	0.15" ± 0.15" 5/32" ± 5/32"	FRONT TOE (Inches)	0.15" ± 0.2" 5/32" ± 3/16"	0.15" ± 0.2" 5/32" ± 3/16"	0.09" ± 0.18" 3/32" ± 3/16"
13.8 (13¾)	14.3 (14¼)	11.3 (11 5/16)	SAI	11.6 (11⅝)	11.7 (11¾)	10.7 (10⅝)
39.0 / 33.0 ±2.0	39.5 / 32.9 ±2.0	37.4 / 32.5 ±2.0	TURNING ANGLE	38.0 / 33.0 ±3.0	32.0 / 29.0 ±3.0	38.7 / 32.9 ±3.0
46	46	52	ILLUSTRATION #	52	52	121
			REAR			
-0.92 ± 0.75 -15/16 ± ¾	-0.95 ± 1.0 -15/16 ± 1	-0.72 ± 0.5 -11/16 ± ½	CAMBER (Degrees)	-0.78 ± 1.0 -13/16 ± 1	-1.0 ± 1.0 -1 ± 1	-1.07 ± 1.0 -1 1/16 ± 1
0.2° ± 0.3° 3/16° ± 5/16°	0.2° ± 0.4° 3/16° ± 13/32°	0.3° ± 0.3° 5/16° ± 5/16°	REAR TOE (Degrees)	0.3° ± 0.4° 5/16° ± 13/32°	0.3° ± 0.4° 5/16° ± 13/32°	0.28° ± 0.37° 9/32° ± ⅜°
0.1" ± 0.15" 3/32" ± 5/32"	0.1" ± 0.2" 3/32" ± 3/16"	0.15" ± 0.15" 5/32" ± 5/32"	REAR TOE (Inches)	0.15" ± 0.2" 5/32" ± 3/16"	0.15" ± 0.2" 5/32" ± 3/16"	0.14" ± 0.18" 5/32" ± 3/16"
0.1 (3/32)	0.8 (13/16)		THRUST ANGLE			
10	10	53	ILLUSTRATION #	53	53	160
			NOTES			
28085	28086	28087		28088	28089	28090

Mazda

MX-5 Miata	MX-5 Miata	MX-5 Miata	MODEL	MX-5 Miata	MX-6	MX-6
17" tires	16" tires	17" tires	INFORMATION	(Fuel tank full)		
(Fuel tank full)	(Fuel tank full)	(Fuel tank full)				
2006-2008	2009-2015	2009-2015	YEARS	2016-2017	1990	1991-92
			INSTRUCTIONS			
			LUGNUT TORQUE	80-108 ft.lbs.		
			FRONT			
6.1 ± 1.0 6⅛ ± 1	5.93 ± 1.0 5 15/16 ± 1	6.07 ± 1.0 6 1/16 ± 1	CASTER (Degrees)	7.85 ± 1.0 7⅞ ± 1	1.22 ± 0.75 1 3/16 ± ¾	1.68 ± 0.75 1 11/16 ± ¾
0.0 (0) | 1.0 (1)	0.0 (0) | 1.0 (1)	0.0 (0) | 1.0 (1)	CROSS | TOL.	0.0 (0) | 1.0 (1)	0.0 (0) | 0.75 (¾)	0.0 (0) | 0.75 (¾)
-0.23 ± 1.0 -¼ ± 1	-0.12 ± 1.0 -⅛ ± 1	-0.22 ± 1.0 -3/16 ± 1	CAMBER (Degrees)	-0.12 ± 1.0 -⅛ ± 1	0.28 ± 0.75 5/16 ± ¾	0.28 ± 0.75 5/16 ± ¾
0.0 (0) | 1.0 (1)	0.0 (0) | 1.0 (1)	0.0 (0) | 1.0 (1)	CROSS | TOL.	0.0 (0) | 1.0 (1)	0.0 (0) | 0.5 (½)	0.0 (0) | 0.5 (½)
0.18° ± 0.37° 3/16° ± ⅜°	0.15° ± 0.37° 5/32° ± ⅜°	0.15° ± 0.37° 5/32° ± ⅜°	FRONT TOE (Degrees)	0.15° ± 0.33° 5/32° ± 11/32°	0.0° ± 0.3° 0° ± 5/16°	0.0° ± 0.3° 0° ± 5/16°
0.09" ± 0.18" 3/32" ± 3/16"	0.08" ± 0.18" 1/16" ± 3/16"	0.08" ± 0.18" 1/16" ± 3/16"	FRONT TOE (Inches)	0.08" ± 0.17" 1/16" ± 5/32"	0.0" ± 0.15" 0" ± 5/32"	0.0" ± 0.15" 0" ± 5/32"
10.8 (10¾)	11.0 (10 15/16)	11.1 (11 1/16)	SAI	13.5 (13½)	12.8 (12 13/16)	12.8 (12 13/16)
38.7 / 32.9 ±3.0	38.7 / 32.9 ±3.0	38.7 / 32.9 ±3.0	TURNING ANGLE	37.7 / 32.1 ±3.0	36.0 / 31.0 ±2.0 [1]	36.0 / 31.0 ±2.0 [1]
121	121	121	ILLUSTRATION #	173	51	51
			REAR			
-1.18 ± 1.0 -1 3/16 ± 1	-1.07 ± 1.0 -1 1/16 ± 1	-1.18 ± 1.0 -1 3/16 ± 1	CAMBER (Degrees)	-0.95 ± 1.0 -15/16 ± 1	-0.5 ± 0.75 [2] -½ ± ¾ [2]	-0.5 ± 0.75 -½ ± ¾
0.28° ± 0.37° 9/32° ± ⅜°	0.28° ± 0.37° 9/32° ± ⅜°	0.28° ± 0.37° 9/32° ± ⅜°	REAR TOE (Degrees)	0.28° ± 0.33° 9/32° ± 11/32°	0.3° ± 0.3° 5/16° ± 5/16°	0.3° ± 0.3° 5/16° ± 5/16°
0.14" ± 0.18" 5/32" ± 3/16"	0.14" ± 0.18" 5/32" ± 3/16"	0.14" ± 0.18" 5/32" ± 3/16"	REAR TOE (Inches)	0.14" ± 0.17" 5/32" ± 5/32"	0.15" ± 0.15" 5/32" ± 5/32"	0.15" ± 0.15" 5/32" ± 5/32"
			THRUST ANGLE			
160	160	160	ILLUSTRATION #	160	10	10
			NOTES		[1] w/ Turbo A/T 34°/29° [2] w/4WS 0.0° (0°)	[1] w/ Turbo A/T 34°/29°
28091	28092	28093		28094	28095	28096

Mazda

<table>
<tr><td>MX-6</td><td>MX-6</td><td>Navajo 4X2</td><td>MODEL</td><td>Navajo 4X4</td><td>Protegé</td><td>Protegé</td></tr>
<tr><td></td><td></td><td></td><td>INFORMATION</td><td></td><td></td><td></td></tr>
<tr><td></td><td>(Fuel tank full)</td><td></td><td></td><td></td><td>(Fuel tank full)</td><td>(Fuel tank full)</td></tr>
<tr><td>1993-95</td><td>1996-97</td><td>1992-94</td><td>YEARS</td><td>1991-94</td><td>1995-98</td><td>1999-2000</td></tr>
<tr><td></td><td></td><td>p. 306, 307</td><td>INSTRUCTIONS</td><td>p. 306, 307</td><td></td><td></td></tr>
<tr><td></td><td></td><td></td><td>LUGNUT TORQUE</td><td></td><td>66-94 ft.lbs.</td><td></td></tr>
<tr><td></td><td></td><td></td><td>FRONT</td><td></td><td></td><td></td></tr>
<tr><td>3.02 ±0.75 [1]
3 ± ¾ [1]</td><td>3.02 ±0.75
3 ± ¾</td><td>MIN.=2.5 MAX.=6.0
MIN.=2½ MAX.=6</td><td>CASTER
(Degrees)</td><td>MIN.=2.5 MAX.=6.0
MIN.=2½ MAX.=6</td><td>1.92 ±1.0
1 15/16 ±1</td><td>1.88 ±1.0
1 7/8 ±1</td></tr>
<tr><td>0.0 (0) | 1.5 (1½)</td><td>0.0 (0) | 1.5 (1½)</td><td>-0.5 (-½) | 0.5 (½)</td><td>CROSS | TOL.</td><td>-0.5 (-½) | 0.5 (½)</td><td>0.0 (0) | 1.5 (1½)</td><td>0.0 (0) | 1.5 (1½)</td></tr>
<tr><td>-0.72 ±0.75 [1]
-11/16 ± ¾ [1]</td><td>-0.72 ±1.0
-11/16 ±1</td><td>0.25 ±0.5
¼ ± ½</td><td>CAMBER
(Degrees)</td><td>0.25 ±0.5
¼ ± ½</td><td>-0.73 ±1.0
-¾ ±1</td><td>-0.78 ±1.0
-13/16 ±1</td></tr>
<tr><td>0.0 (0) | 1.5 (1½)</td><td>0.0 (0) | 1.5 (1½)</td><td>0.0 (0) | 0.5 (½)</td><td>CROSS | TOL.</td><td>0.0 (0) | 0.5 (½)</td><td>0.0 (0) | 1.5 (1½)</td><td>0.0 (0) | 1.5 (1½)</td></tr>
<tr><td>0.28° ±0.28°
9/32° ± 9/32°</td><td>0.28° ±0.38°
9/32° ± 3/8°</td><td>0.06° ±0.25°
1/16° ± ¼°</td><td>FRONT TOE
(Degrees)</td><td>0.0° ±0.25°
0° ± ¼°</td><td>0.2° ±0.3°
3/16° ± 5/16°</td><td>0.2° ±0.4°
3/16° ± 13/32°</td></tr>
<tr><td>0.14" ±0.14"
5/32" ± 5/32"</td><td>0.14" ±0.19"
5/32" ± 3/16"</td><td>0.03" ±0.13"
1/32" ± 1/8"</td><td>FRONT TOE
(Inches)</td><td>0.0" ±0.13"
0" ± 1/8"</td><td>0.1" ±0.15"
3/32" ± 5/32"</td><td>0.1" ±0.2"
3/32" ± 3/16"</td></tr>
<tr><td>15.4 (15 7/16)</td><td>15.4 (15 7/16)</td><td></td><td>SAI</td><td></td><td>13.4 (13 3/8)</td><td>12.6 (12 9/16)</td></tr>
<tr><td>37.0 / 32.0 ±2.0</td><td>37.0 / 32.0 ±2.0</td><td></td><td>TURNING ANGLE</td><td></td><td>40.0 / 33.0 ±3.0</td><td>37.0 / 33.0 ±2.0</td></tr>
<tr><td>51</td><td>51</td><td>56</td><td>ILLUSTRATION #</td><td>56</td><td>26</td><td>46</td></tr>
<tr><td></td><td></td><td></td><td>REAR</td><td></td><td></td><td></td></tr>
<tr><td>-0.37 ±0.75 [1]
-3/8 ± ¾ [1]</td><td>-0.9 ±1.0
-7/8 ±1</td><td></td><td>CAMBER
(Degrees)</td><td></td><td>-0.82 ±1.0
-13/16 ±1</td><td>-0.52 ±1.0
-½ ±1</td></tr>
<tr><td>0.28° ±0.28°
9/32° ± 9/32°</td><td>0.28° ±0.38°
9/32° ± 3/8°</td><td></td><td>REAR TOE
(Degrees)</td><td></td><td>0.2° ±0.3°
3/16° ± 5/16°</td><td>0.2° ±0.4°
3/16° ± 13/32°</td></tr>
<tr><td>0.14" ±0.14"
5/32" ± 5/32"</td><td>0.14" ±0.19"
5/32" ± 3/16"</td><td></td><td>REAR TOE
(Inches)</td><td></td><td>0.1" ±0.15"
3/32" ± 5/32"</td><td>0.1" ±0.2"
3/32" ± 3/16"</td></tr>
<tr><td></td><td></td><td></td><td>THRUST ANGLE</td><td></td><td>0.8 (13/16)</td><td>0.1 (3/32)</td></tr>
<tr><td>06</td><td>06</td><td></td><td>ILLUSTRATION #</td><td></td><td>26 06</td><td>06</td></tr>
<tr><td>[1] 1995 = ±1.0° (1°)</td><td></td><td></td><td>NOTES</td><td></td><td></td><td></td></tr>
<tr><td>28097</td><td>28098</td><td>28099</td><td></td><td>28100</td><td>28101</td><td>28102</td></tr>
</table>

Mazda

<table>
<tr><td>Protegé</td><td>Protegé</td><td>Protegé 5</td><td>MODEL</td><td>RX-7</td><td>RX-7</td><td>RX-7</td></tr>
<tr><td></td><td>MAZDASPEED</td><td></td><td>INFORMATION</td><td></td><td></td><td></td></tr>
<tr><td>(Fuel tank full)</td><td>(Fuel tank full)</td><td>(Fuel tank full)</td><td></td><td></td><td></td><td></td></tr>
<tr><td>2001-2003</td><td>2003</td><td>2002-2003</td><td>YEARS</td><td>1990-91</td><td>1993-94</td><td>1995</td></tr>
<tr><td></td><td></td><td></td><td>INSTRUCTIONS</td><td></td><td></td><td></td></tr>
<tr><td></td><td></td><td></td><td>LUGNUT TORQUE</td><td></td><td></td><td></td></tr>
<tr><td></td><td></td><td></td><td>FRONT</td><td></td><td></td><td></td></tr>
<tr><td>1.93 ±1.0
1 15/16 ±1</td><td>2.02 ±1.0
2 ±1</td><td>1.93 ±1.0
1 15/16 ±1</td><td>CASTER
(Degrees)</td><td>4.67 ±0.75
4 11/16 ± ¾</td><td>6.08 ±0.75
6 1/16 ± ¾</td><td>6.67 ±1.0
6 11/16 ±1</td></tr>
<tr><td>0.0 (0) | 1.5 (1½)</td><td>0.0 (0) | 1.5 (1½)</td><td>0.0 (0) | 1.5 (1½)</td><td>CROSS | TOL.</td><td>0.0 (0) | 0.75 (¾)</td><td>0.0 (0) | 1.0 (1)</td><td>0.0 (0) | 1.5 (1½)</td></tr>
<tr><td>-0.82 ±1.0
-13/16 ±1</td><td>-0.97 ±1.0
-15/16 ±1</td><td>-0.82 ±1.0
-13/16 ±1</td><td>CAMBER
(Degrees)</td><td>0.33 ±0.5
5/16 ± ½</td><td>0.08 ±0.75
1/16 ± ¾</td><td>0.1 ±1.0
1/8 ±1</td></tr>
<tr><td>0.0 (0) | 1.5 (1½)</td><td>0.0 (0) | 1.5 (1½)</td><td>0.0 (0) | 1.5 (1½)</td><td>CROSS | TOL.</td><td>0.0 (0) | 0.5 (½)</td><td>0.0 (0) | 1.0 (1)</td><td>0.0 (0) | 1.5 (1½)</td></tr>
<tr><td>0.2° ±0.4°
3/16° ± 13/32°</td><td>-0.1° ±0.1°
-3/32° ± 3/32°</td><td>0.2° ±0.4°
3/16° ± 13/32°</td><td>FRONT TOE
(Degrees)</td><td>0.3° ±0.3°
5/16° ± 5/16°</td><td>0.2° ±0.2° [1]
3/16° ± 3/16° [1]</td><td>0.08° ±0.24°
3/32° ± ¼°</td></tr>
<tr><td>0.1" ±0.2"
3/32" ± 3/16"</td><td>-0.05" ±0.05"
-1/16" ± 1/16"</td><td>0.1" ±0.2"
3/32" ± 3/16"</td><td>FRONT TOE
(Inches)</td><td>0.15" ±0.15"
5/32" ± 5/32"</td><td>0.1" ±0.1" [1]
3/32" ± 3/32" [1]</td><td>0.04" ±0.12"
1/32" ± 1/8"</td></tr>
<tr><td>12.6 (12 5/8)</td><td>13.0 (13)</td><td>12.6 (12 5/8)</td><td>SAI</td><td>13.8 (13 ¾)</td><td>13.9 (13 15/16)</td><td>14.1 (14 1/16)</td></tr>
<tr><td>37.0 / 33.0 ±2.0</td><td>37.0 / 33.0 ±2.0</td><td>37.0 / 33.0 ±2.0</td><td>TURNING ANGLE</td><td>36.0 / 32.0 ±2.0</td><td>36.0 / 32.0 ±2.0</td><td>36.0 / 32.0 ±2.0</td></tr>
<tr><td>46</td><td>46</td><td>46</td><td>ILLUSTRATION #</td><td>51</td><td>52</td><td>52</td></tr>
<tr><td></td><td></td><td></td><td>REAR</td><td></td><td></td><td></td></tr>
<tr><td>-0.53 ±1.0
-9/16 ±1</td><td>-0.8 ±1.0
-13/16 ±1</td><td>-0.53 ±1.0
-9/16 ±1</td><td>CAMBER
(Degrees)</td><td>-0.73 ±0.5
-¾ ± ½</td><td>-1.22 ±0.75
-1 ¼ ± ¾</td><td>-1.22 ±1.0
-1 ¼ ±1</td></tr>
<tr><td>0.2° ±0.4°
3/16° ± 13/32°</td><td>-0.1° ±0.1°
-3/32° ± 3/32°</td><td>0.2° ±0.4°
3/16° ± 13/32°</td><td>REAR TOE
(Degrees)</td><td>0.3° ±0.3°
5/16° ± 5/16°</td><td>0.2° ±0.3°
3/16° ± 5/16°</td><td>0.16° ±0.24°
5/32° ± ¼°</td></tr>
<tr><td>0.1" ±0.2"
3/32" ± 3/16"</td><td>-0.05" ±0.05"
-1/16" ± 1/16"</td><td>0.1" ±0.2"
3/32" ± 3/16"</td><td>REAR TOE
(Inches)</td><td>0.15" ±0.15"
5/32" ± 5/32"</td><td>0.1" ±0.15"
3/32" ± 5/32"</td><td>0.08" ±0.12"
3/32" ± 1/8"</td></tr>
<tr><td>0.8 (13/16)</td><td>0.8 (13/16)</td><td>0.8 (13/16)</td><td>THRUST ANGLE</td><td></td><td>0.1 (3/32)</td><td>0.8 (13/16)</td></tr>
<tr><td>06</td><td>06</td><td>06</td><td>ILLUSTRATION #</td><td>54</td><td>55</td><td>55</td></tr>
<tr><td></td><td></td><td></td><td>NOTES</td><td></td><td>[1] 1993 = 0.1° (.05")</td><td></td></tr>
<tr><td>28103</td><td>28104</td><td>28105</td><td></td><td>28115</td><td>28116</td><td>28117</td></tr>
</table>

Mazda

MODEL INFORMATION	RX-8 Std. Susp. (Fuel tank full)	RX-8 Sport Susp. (Fuel tank full)	RX-8 Std. Susp. (Fuel tank full)	RX-8 Sport Susp. (Fuel tank full)	Tribute	Tribute 4 Cyl.
YEARS	2004-2008	2004-2008	2009-2011	2009-2011	2001	2002-2004
INSTRUCTIONS						
LUGNUT TORQUE	65-87 ft.lbs.	65-87 ft.lbs.	65-87 ft.lbs.	65-87 ft.lbs.		
FRONT						
CASTER (Degrees)	6.1 ± 1.0 6 1/8 ± 1	6.27 ± 1.0 6 1/4 ± 1	6.33 ± 1.0 6 5/16 ± 1	6.37 ± 1.0 6 3/8 ± 1	1.93 ± 0.75 1 15/16 ± 3/4	1.79 ± 0.75 1 13/16 ± 3/4
CROSS | TOL.	0.0 (0) | 1.0 (1)	0.0 (0) | 1.0 (1)	0.0 (0) | 1.0 (1)	0.0 (0) | 1.0 (1)	0.0 (0) | 1.0 (1)	0.0 (0) | 1.0 (1)
CAMBER (Degrees)	0.07 ± 1.0 1/16 ± 1	-0.1 ± 1.0 - 1/8 ± 1	-0.05 ± 1.0 - 1/16 ± 1	-0.25 ± 1.0 - 1/4 ± 1	-0.84 ± 0.75 - 13/16 ± 3/4	-0.48 ± 0.75 - 1/2 ± 3/4
CROSS | TOL.	0.0 (0) | 1.0 (1)	0.0 (0) | 1.0 (1)	0.0 (0) | 1.0 (1)	0.0 (0) | 1.0 (1)	0.0 (0) | 1.0 (1)	0.0 (0) | 1.0 (1)
FRONT TOE (Degrees)	0.18° ±0.35° 3/16° ± 11/32°	0.18° ±0.35° 3/16° ± 11/32°	0.18° ±0.35° 3/16° ± 11/32°	0.18° ±0.35° 3/16° ± 11/32°	0.25° ±0.25° 1/4° ± 1/4°	0.08° ±0.25° 3/32° ± 1/4°
FRONT TOE (Inches)	0.09" ±0.18" 3/32" ± 3/16"	0.09" ±0.18" 3/32" ± 3/16"	0.09" ±0.18" 3/32" ± 3/16"	0.09" ±0.18" 3/32" ± 3/16"	0.13" ±0.13" 1/8" ± 1/8"	0.04" ±0.13" 1/32" ± 1/8"
SAI	10.9 (10 7/8)	11.0 (11)	11.0 (11)	11.2 (11 3/16)	11.4 (11 3/8)	
TURNING ANGLE	38.7 / 33.3 ±3.0	38.6 / 33.1 ±3.0	38.6 / 33.3 ±3.0	38.6 / 33.3 ±3.0		
ILLUSTRATION #	121	121	121	121	51	51
REAR						
CAMBER (Degrees)	-0.93 ± 1.0 - 15/16 ± 1	-1.12 ± 1.0 -1 1/8 ± 1	-0.75 ± 1.0 - 3/4 ± 1	-0.8 ± 1.0 - 13/16 ± 1	0.04 ± 0.75 1/16 ± 3/4	0.04 ± 0.75 1/16 ± 3/4
REAR TOE (Degrees)	0.27° ±0.33° 9/32° ± 11/32°	0.27° ±0.33° 9/32° ± 11/32°	0.32° ±0.33° 5/16° ± 11/32°	0.32° ±0.33° 5/16° ± 11/32°	0.19° ±0.25° 3/16° ± 1/4°	0.08° ±0.34° 3/32° ± 11/32°
REAR TOE (Inches)	0.13" ±0.17" 1/8" ± 5/32"	0.13" ±0.17" 1/8" ± 5/32"	0.16" ±0.17" 5/32" ± 5/32"	0.16" ±0.17" 5/32" ± 5/32"	0.09" ±0.13" 3/32" ± 1/8"	0.04" ±0.17" 1/32" ± 5/32"
THRUST ANGLE						
ILLUSTRATION #	160	160	160	160	27	27
NOTES						
	28118	28119	28120	28121	28125	28126

Mazda

MODEL INFORMATION	Tribute 6 Cyl.	Tribute Early build models prior to 8/02/04	Tribute Late build models from 8/02/04	Tribute (incl. Hybrid)	Tribute	Tribute Hybrid
YEARS	2002-2004	2005	2005-2006	2008	2009-2011	2009-2011
INSTRUCTIONS						
LUGNUT TORQUE		98 ft.lbs.	98 ft.lbs.	98 ft.lbs.	98 ft.lbs.	98 ft.lbs.
FRONT						
CASTER (Degrees)	1.79 ± 0.75 1 13/16 ± 3/4	1.7 ± 1.0 1 11/16 ± 1	1.7 ± 1.0 1 11/16 ± 1	L= 1.8, R= 1.5 ± 0.5 L= 1 13/16, R= 1 1/2 ± 1/2	1.6 ± 0.5 1 5/8 ± 1/2	1.6 ± 0.5 1 5/8 ± 1/2
CROSS | TOL.	0.0 (0) | 1.0 (1)	0.0 (0) | 0.75 (3/4)	0.0 (0) | 0.75 (3/4)	0.3 (5/16) | 0.7 (11/16)	0.0 (0) | 0.7 (11/16)	0.0 (0) | 0.7 (11/16)
CAMBER (Degrees)	-0.84 ± 0.75 - 13/16 ± 3/4	-1.0 ± 0.75 -1 ± 3/4	-1.0 ± 0.75 -1 ± 3/4	-1.0 ± 0.5 -1 ± 1/2	-0.84 ± 0.5 - 13/16 ± 1/2	-0.84 ± 0.5 - 13/16 ± 1/2
CROSS | TOL.	0.0 (0) | 1.0 (1)	0.0 (0) | 0.75 (3/4)	0.0 (0) | 0.75 (3/4)	0.0 (0) | 0.7 (11/16)	0.0 (0) | 0.7 (11/16)	0.0 (0) | 0.7 (11/16)
FRONT TOE (Degrees)	0.25° ±0.25° 1/4° ± 1/4°	0.23° ±0.23° 7/32° ± 7/32°	0.23° ±0.23° 7/32° ± 7/32°	0.23° ±0.23° 7/32° ± 7/32°	0.23° ±0.23° 7/32° ± 7/32°	0.23° ±0.23° 7/32° ± 7/32°
FRONT TOE (Inches)	0.13" ±0.13" 1/8" ± 1/8"	0.12" ±0.12" 1/8" ± 1/8"	0.12" ±0.12" 1/8" ± 1/8"	0.12" ±0.12" 1/8" ± 1/8"	0.12" ±0.12" 1/8" ± 1/8"	0.12" ±0.12" 1/8" ± 1/8"
SAI						
TURNING ANGLE						
ILLUSTRATION #	51	51	51	51	51	51
REAR						
CAMBER (Degrees)	0.04 ± 0.75 1/16 ± 3/4	0.1 ± 0.75 1/8 ± 3/4	0.0 ± 0.75 0 ± 3/4	0.0 ± 0.65 0 ± 5/8	0.0 ± 0.65 0 ± 5/8	0.1 ± 0.65 1/8 ± 5/8
REAR TOE (Degrees)	0.2° ±0.34° 3/16° ± 11/32°	0.2° ±0.2° 3/16° ± 3/16°	-0.06° ±0.2° -1/16° ± 3/16°	-0.18° ±0.2° -3/16° ± 3/16°	-0.18° ±0.2° -3/16° ± 3/16°	-0.18° ±0.2° -3/16° ± 3/16°
REAR TOE (Inches)	0.1" ±0.17" 3/32" ± 5/32"	0.1" ±0.1" 3/32" ± 3/32"	-0.03" ±0.1" -1/32" ± 3/32"	-0.09" ±0.1" -3/32" ± 3/32"	-0.09" ±0.1" -3/32" ± 3/32"	-0.09" ±0.1" -3/32" ± 3/32"
THRUST ANGLE						
ILLUSTRATION #	27	27	27	27	27	27
NOTES						
	28127	28128	28129	28130	28131	28132

ALIGNMENT SPECIFICATIONS

Mercedes-Benz

190E, 190D	B Class (245.*) STD. Suspension	B Class (245.*) Sport Suspension	MODEL INFORMATION	B Class (246.*) STD. Suspension	B Class (246.*) Sport Suspension	B Class (242.*) Electric Vehicle
1990-93	2005-2011	2005-2011	YEARS	2012-2017	2012-2017	2014-2017
p. 308, 309	p. 308, 309	p. 308, 309	INSTRUCTIONS	p. 308, 309	p. 308, 309	p. 308, 309
81 ft.lbs.	96 ft.lbs.	96 ft.lbs.	LUGNUT TORQUE	96 ft.lbs.	96 ft.lbs.	96 ft.lbs.
			FRONT			
10.12 ±0.5 [1] 10 1/8 ± 1/2 [1]	3.61 ±0.5 3 5/8 ± 1/2	3.72 ±0.5 3 11/16 ± 1/2	CASTER (Degrees)	6.69 ±0.5 6 11/16 ± 1/2	6.98 ±0.5 7 ± 1/2	6.18 ±0.5 6 3/16 ± 1/2
0.0 (0) | 0.5 (1/2)	0.0 (0) | 0.5 (1/2)	0.0 (0) | 0.5 (1/2)	CROSS | TOL.	0.0 (0) | 0.5 (1/2)	0.0 (0) | 0.5 (1/2)	0.0 (0) | 0.5 (1/2)
-0.42 +0.17 -0.33 [2] - 7/16 + 3/16 - 5/16 [2]	-0.67 ±0.33 - 11/16 ± 5/16	-0.74 ±0.33 - 3/4 ± 5/16	CAMBER (Degrees)	-0.91 ±0.33 - 15/16 ± 5/16	-1.03 ±0.33 -1 ± 5/16	-0.5 ±0.32 - 1/2 ± 5/16
0.0 (0) | 0.33 (5/16)	0.0 (0) | 0.33 (5/16)	0.0 (0) | 0.33 (5/16)	CROSS | TOL.	0.0 (0) | 0.33 (5/16)	0.0 (0) | 0.33 (5/16)	0.0 (0) | 0.32 (5/16)
0.33° ±0.17° 11/32° ± 5/32°	0.2° ±0.17° 3/16° ± 5/32°	0.2° ±0.17° 3/16° ± 5/32°	FRONT TOE (Degrees)	0.2° ±0.17° 3/16° ± 5/32°	0.2° ±0.17° 3/16° ± 5/32°	0.22° ±0.17° 7/32° ± 5/32°
0.17" ±0.08" 5/32" ± 3/32"	0.1" ±0.08" 3/32" ± 3/32"	0.1" ±0.08" 3/32" ± 3/32"	FRONT TOE (Inches)	0.1" ±0.08" 3/32" ± 3/32"	0.1" ±0.08" 3/32" ± 3/32"	0.11" ±0.08" 3/32" ± 3/32"
			SAI			
43.0 / 34.0 ±2.0			TURNING ANGLE			20.0 / 19.1 ±0.5
88	102	102	ILLUSTRATION #	102	102	102
			REAR			
	-1.5 ±0.5 -1 1/2 ± 1/2	-1.5 ±0.5 -1 1/2 ± 1/2	CAMBER (Degrees)	-1.19 ±0.5 -1 3/16 ± 1/2	-1.63 ±0.5 -1 5/8 ± 1/2	-1.0 ±0.5 -1 ± 1/2
0.42° +0.17° -0.08° 13/32° + 5/32° - 3/32°	0.1° ±0.17° 3/32° ± 5/32°	0.18° ±0.17° 3/16° ± 5/32°	REAR TOE (Degrees)	0.1° ±0.17° 3/32° ± 5/32°	0.18° ±0.17° 3/16° ± 5/32°	0.07° ±0.12° 1/16° ± 1/8°
0.21" +0.08" -0.04" 7/32" + 3/32" - 1/32"	0.05" ±0.08" 1/16" ± 3/32"	0.09" ±0.08" 3/32" ± 3/32"	REAR TOE (Inches)	0.05" ±0.08" 1/16" ± 3/32"	0.09" ±0.08" 3/32" ± 3/32"	0.03" ±0.06" 1/32" ± 1/16"
			THRUST ANGLE			
86	174	174	ILLUSTRATION #	174	174	174
[1] Sport Susp.=10.3°(10 5/16°) [2] Sport Susp.= -0.84°(-13/16°)			NOTES			
29009	29011	29012		29014	29015	29016

Mercedes-Benz

C-Class (202.*) Classic/Elegance exc/Level Control	C-Class (202.*) Classic/Elegance w/Level Control	C-Class (202.*) Sport/Esprit	MODEL INFORMATION	C-Class (203.*) Sport Coupe	C-Class (203.*) Luxury Sedan STD. Susp. (485)	C-Class (203.*) Sport Sdn. (incl. AMG) Sport Susp. (486)
1994-2000	1994-2000	1994-2000	YEARS	2002-2006	2001-2007	2001-2007
p. 308, 309	p. 308, 309	p. 308, 309	INSTRUCTIONS	p. 308, 309	p. 308, 309	p. 308, 309
81 ft.lbs.	81 ft.lbs.	81 ft.lbs.	LUGNUT TORQUE	81 ft.lbs.	81 ft.lbs.	81 ft.lbs.
			FRONT			
4.57 ±0.5 4 9/16 ± 1/2	4.72 ±0.5 4 3/4 ± 1/2	4.46 ±0.5 4 7/16 ± 1/2	CASTER (Degrees)	10.3 ±0.5 10 5/16 ± 1/2	10.04 ±0.5 10 1/16 ± 1/2	10.53 ±0.5 10 9/16 ± 1/2
0.0 (0) | 0.5 (1/2)	0.0 (0) | 0.5 (1/2)	0.0 (0) | 0.5 (1/2)	CROSS | TOL.	0.0 (0) | 0.5 (1/2)	0.0 (0) | 0.5 (1/2)	0.0 (0) | 0.5 (1/2)
-0.53 ±0.33 - 1/2 ± 5/16	-0.53 ±0.33 - 1/2 ± 5/16	-0.49 ±0.33 - 1/2 ± 5/16	CAMBER (Degrees)	-0.41 +0.26 -0.46 - 7/16 + 1/4 - 7/16	-0.27 +0.26 -0.46 - 1/4 + 1/4 - 7/16	-0.56 +0.26 -0.46 - 9/16 + 1/4 - 7/16
0.0 (0) | 0.33 (5/16)	0.0 (0) | 0.33 (5/16)	0.0 (0) | 0.33 (5/16)	CROSS | TOL.	0.0 (0) | 0.4 (3/8)	0.0 (0) | 0.4 (3/8)	0.0 (0) | 0.4 (3/8)
0.4° ±0.17° 13/32° ± 5/32°	0.4° ±0.17° 13/32° ± 5/32°	0.4° ±0.17° 13/32° ± 5/32°	FRONT TOE (Degrees)	0.2° ±0.17° 3/16° ± 5/32°	0.2° ±0.17° 3/16° ± 5/32°	0.2° ±0.17° 3/16° ± 5/32°
0.2" ±0.08" 3/16" ± 3/32"	0.2" ±0.08" 3/16" ± 3/32"	0.2" ±0.08" 3/16" ± 3/32"	FRONT TOE (Inches)	0.1" ±0.08" 3/32" ± 3/32"	0.1" ±0.08" 3/32" ± 3/32"	0.1" ±0.08" 3/32" ± 3/32"
			SAI			
20.0 / 19.0 ±0.5	20.0 / 19.0 ±0.5	20.0 / 19.0 ±0.5	TURNING ANGLE	20.0 / 18.4 ±0.5	20.0 / 18.5 ±0.5	20.0 / 18.3 ±0.5
88	88	88	ILLUSTRATION #	164 165	164 165	164 165
			REAR			
-0.94 ±0.5 - 15/16 ± 1/2	-1.12 ±0.5 -1 1/8 ± 1/2	-0.83 ±0.5 -1 13/16 ± 1/2	CAMBER (Degrees)	-1.1 ±0.5 -1 1/8 ± 1/2	-0.93 ±0.5 - 15/16 ± 1/2	-1.27 ±0.5 -1 1/4 ± 1/2
0.55° ±0.25° 9/16° ± 1/4°	0.55° ±0.25° 9/16° ± 1/4°	0.55° ±0.25° 9/16° ± 1/4°	REAR TOE (Degrees)	0.55° ±0.25° 9/16° ± 1/4°	0.54° ±0.25° 17/32° ± 1/4°	0.55° ±0.25° 9/16° ± 1/4°
0.28" ±0.13" 9/32" ± 1/8"	0.28" ±0.13" 9/32" ± 1/8"	0.28" ±0.13" 9/32" ± 1/8"	REAR TOE (Inches)	0.28" ±0.13" 9/32" ± 1/8"	0.27" ±0.13" 9/32" ± 1/8"	0.28" ±0.13" 9/32" + 1/8"
0.25 (1/4)	0.25 (1/4)	0.25 (1/4)	THRUST ANGLE	0.25 (1/4)	0.25 (1/4)	0.25 (1/4)
86	86	86	ILLUSTRATION #	86	86	86
			NOTES			
29018	29019	29020		29022	29023	29024

ALIGNMENT SPECIFICATIONS

Mercedes-Benz

C-Class (203.*) Luxury Wagon STD. Susp. (485)	C-Class (203.*) Sport Wgn. (incl. AMG) Sport Susp. (486)	C-Class 4MATIC (203.*) All Models	MODEL INFORMATION	C-Class (204.*) Luxury Models Cpe. & Sed.(to 2014)	C-Class (204.*) Luxury Models 4MATIC Sedan	C-Class (204.*) Sport Models Cpe. & Sed.(to 2014)
2001-2006	2001-2006	2003-2007	YEARS	2008-2016	2008-2014	2008-2016
p. 308, 309	p. 308, 309	p. 308, 309	INSTRUCTIONS	p. 308, 309	p. 308, 309	p. 308, 309
81 ft.lbs.	81 ft.lbs.	81 ft.lbs.	LUGNUT TORQUE	96 ft.lbs.	96 ft.lbs.	96 ft.lbs.
			FRONT			
10.08 ±0.5 10 1/16 ± 1/2	10.42 ±0.5 10 7/16 ± 1/2	11.23 ±0.5 11 1/4 ± 1/2	CASTER (Degrees)	10.31 ±0.5 10 5/16 ± 1/2	10.06 ±0.5 10 1/16 ± 1/2	10.75 ±0.5 10 3/4 ± 1/2
0.0 (0) | 0.5 (1/2)	0.0 (0) | 0.5 (1/2)	0.0 (0) | 0.5 (1/2)	CROSS | TOL.	0.0 (0) | 0.5 (1/2)	0.0 (0) | 0.5 (1/2)	0.0 (0) | 0.5 (1/2)
-0.27 +0.26 -0.46 -1/4 +1/4 -7/16	-0.62 +0.26 -0.46 -5/8 +1/4 -7/16	-0.2 +0.26 -0.46 -3/16 +1/4 -7/16	CAMBER (Degrees)	-0.34 +0.36 -0.46 -5/16 +3/8 -7/16	-0.42 +0.36 -0.46 -7/16 +3/8 -7/16	-0.51 +0.36 -0.46 -1/2 +3/8 -7/16
0.0 (0) | 0.4 (3/8)	0.0 (0) | 0.4 (3/8)	0.0 (0) | 0.4 (3/8)	CROSS | TOL.	0.0 (0) | 0.4 (3/8)	0.0 (0) | 0.4 (3/8)	0.0 (0) | 0.4 (3/8)
0.2° ±0.17° 3/16° ± 5/32°	0.2° ±0.17° 3/16° ± 5/32°	0.0° ±0.17° 0° ± 5/32°	FRONT TOE (Degrees)	0.17° ±0.17° 5/32° ± 5/32°	0.17° ±0.17° 5/32° ± 5/32°	0.17° ±0.17° 5/32° ± 5/32°
0.1" ±0.08" 3/32" ± 3/32"	0.1" ±0.08" 3/32" ± 3/32"	0.0" ±0.08" 0" ± 3/32"	FRONT TOE (Inches)	0.08" ±0.08" 3/32" ± 3/32"	0.08" ±0.08" 3/32" ± 3/32"	0.08" ±0.08" 3/32" ± 3/32"
			SAI			
20.0 / 18.5 ±0.5	20.0 / 18.3 ±0.5	20.0 / 18.8 ±0.5	TURNING ANGLE	20.0 / 18.3 ±0.5	20.0 / 18.3 ±0.5	20.0 / 18.3 ±0.5
164 165	164 165	164 165	ILLUSTRATION #	198 199	198 199	198 199
			REAR			
-0.96 ±0.5 -15/16 ± 1/2	-0.96 ±0.5 -15/16 ± 1/2	-1.06 ±0.5 -1 1/16 ± 1/2	CAMBER (Degrees)	-1.07 ±0.5 -1 1/16 ± 1/2	-1.02 ±0.5 -1 ± 1/2	-1.48 ±0.5 -1 1/2 ± 1/2
0.55° ±0.25° 9/16° ± 1/4°	0.55° ±0.25° 9/16° ± 1/4°	0.54° ±0.25° 17/32° ± 1/4°	REAR TOE (Degrees)	0.4° ±0.25° 13/32° ± 1/4°	0.39° ±0.13° 13/32° ± 1/8°	0.42° ±0.25° 13/32° ± 1/4°
0.28" ±0.13" 9/32" ± 1/8"	0.28" ±0.13" 9/32" ± 1/8"	0.27" ±0.13" 9/32" ± 1/8"	REAR TOE (Inches)	0.2" ±0.13" 3/16" ± 1/8"	0.2" ±0.06" 3/16" ± 1/16"	0.21" ±0.13" 7/32" ± 1/8"
0.25 (1/4)	0.25 (1/4)	0.25 (1/4)	THRUST ANGLE	0.25 (1/4)	0.25 (1/4)	0.25 (1/4)
86	86	86	ILLUSTRATION #	174	174	174
			NOTES	[1] 2015 Coupe only		[1] 2015 Coupe only
29025	29026	29027		29029	29030	29031

Mercedes-Benz

C-Class (204.*) Sport Models 4MATIC Sedan	C-Class (204.*) C63 AMG Sedan to 02/2011	C-Class (204.*) C63 AMG Sedan 03/2011 up	MODEL INFORMATION	C-Class (204.*) C63 AMG Coupe	GLK350 (x204.9**) All Models	C-Class (205.*) RWD Sedan exc/AMG STD Suspension
2008-2014	2008-2011	2011-2014	YEARS	2012-2016	2010-2015	2015-2016
p. 308, 309	p. 308, 309	p. 308, 309	INSTRUCTIONS	p. 308, 309	p. 308, 309	p. 308, 309
96 ft.lbs.	96 ft.lbs.	96 ft.lbs.	LUGNUT TORQUE	96 ft.lbs.	111 ft.lbs.	
			FRONT			
10.47 ±0.5 10 1/2 ± 1/2	10.73 ±0.5 10 3/4 ± 1/2	10.73 ±0.5 10 3/4 ± 1/2	CASTER (Degrees)	11.01 ±0.5 11 ± 1/2	9.18 ±0.5 9 3/16 ± 1/2	8.8 ±0.5 8 13/16 ± 1/2
0.0 (0) | 0.5 (1/2)	0.0 (0) | 0.5 (1/2)	0.0 (0) | 0.5 (1/2)	CROSS | TOL.	0.0 (0) | 0.5 (1/2)	0.0 (0) | 0.5 (1/2)	0.0 (0) | 0.5 (1/2)
-0.55 +0.36 -0.46 -9/16 +3/8 -7/16	-1.4 +0.36 -0.46 -1 3/8 +3/8 -7/16	-1.75 +0.36 -0.46 -1 3/4 +3/8 -7/16	CAMBER (Degrees)	-1.93 +0.36 -0.46 -1 15/16 +3/8 -7/16	-0.23 ±0.35 -1/4 ± 7/16	-0.23 ±0.35 -1/4 ± 3/8
0.0 (0) | 0.4 (3/8)	0.0 (0) | 0.4 (3/8)	0.0 (0) | 0.4 (3/8)	CROSS | TOL.	0.0 (0) | 0.4 (3/8)	0.0 (0) | 0.4 (3/8)	0.0 (0) | 0.4 (3/8)
0.17° ±0.17° 5/32° ± 5/32°	0.17° ±0.17° 5/32° ± 5/32°	0.37° ±0.17° 3/8° ± 5/32°	FRONT TOE (Degrees)	0.26° ±0.17° 1/4° ± 5/32°	0.17° ±0.17° 5/32° ± 5/32°	0.48° ±0.17° 1/2° ± 5/32°
0.08" ±0.08" 3/32" ± 3/32"	0.08" ±0.08" 3/32" ± 3/32"	0.18" ±0.08" 3/16" ± 3/32"	FRONT TOE (Inches)	0.13" ±0.08" 1/8" ± 3/32"	0.08" ±0.08" 3/32" ± 3/32"	0.24" ±0.08" 1/4" ± 3/32"
			SAI			
20.0 / 18.3 ±0.5	20.0 / 18.5 ±0.5	20.0 / 18.5 ±0.5	TURNING ANGLE	20.0 / 18.5 ±0.5	20.0 / 18.5 ±0.5	20.0 / 18.5 ±0.5
198 199	198 199	198 199	ILLUSTRATION #	198 199	198 199	
			REAR			
-1.43 ±0.5 -1 7/16 ± 1/2	-1.42 ±0.5 -1 7/16 ± 1/2	-1.38 ±0.5 -1 3/8 ± 1/2	CAMBER (Degrees)	-2.18 ±0.5 -2 3/16 ± 1/2	-1.13 ±0.5 -1 1/8 ± 1/2	-1.3 ±0.5 -1 5/16 ± 1/2
0.41° ±0.13° 13/32° ± 1/8°	0.8° ±0.13° 13/16° ± 1/8°	0.26° ±0.13° 1/4° ± 1/8°	REAR TOE (Degrees)	0.27° ±0.13° 9/32° ± 1/8°	0.38° ±0.13° 3/8° ± 1/8°	0.28° ±0.12° 9/32° ± 1/8°
0.21" ±0.06" 7/32" ± 1/16"	0.4" ±0.06" 13/32" ± 1/16"	0.13" ±0.06" 1/8" ± 1/16"	REAR TOE (Inches)	0.13" ±0.06" 1/8" ± 1/16"	0.19" ±0.06" 3/16" ± 1/16"	0.14" ±0.06" 5/32" ± 1/16"
0.25 (1/4)	0.25 (1/4)	0.25 (1/4)	THRUST ANGLE	0.25 (1/4)	0.25 (1/4)	0.25 (1/4)
174	174	174	ILLUSTRATION #	174	174	174
			NOTES			
29032	29033	29034		29035	29036	29038

Mercedes-Benz

C-Class (205.*) RWD Sedan exc/AMG Sport Suspension	C-Class (205.*) RWD Sedan exc/AMG AIR Suspension	C-Class (205.*) RWD Sedan AMG	MODEL INFORMATION	C-Class (205.*) 4MATIC Sedan STD Suspension	C-Class (205.*) 4MATIC Sedan Sport Suspension	C-Class (205.*) 4MATIC Sedan AIR Suspension
2015-2016	2015-2016	2015-2016	YEARS	2015-2016	2015-2016	2015-2016
p. 308, 309	p. 308, 309	p. 308, 309	INSTRUCTIONS	p. 308, 309	p. 308, 309	p. 308, 309
			LUGNUT TORQUE			
			FRONT			
9.37 ±0.5 9 3/8 ± 1/2	9.45 ±0.5 9 7/16 ± 1/2	9.38 ±0.5 9 3/8 ± 1/2	CASTER (Degrees)	8.9 ±0.5 8 7/8 ± 1/2	9.38 ±0.5 9 3/8 ± 1/2	9.68 ±0.5 9 11/16 ± 1/2
0.0 (0) | 0.5 (1/2)	0.0 (0) | 0.5 (1/2)	0.0 (0) | 0.5 (1/2)	CROSS | TOL.	0.0 (0) | 0.5 (1/2)	0.0 (0) | 0.5 (1/2)	0.0 (0) | 0.5 (1/2)
-0.53 ±0.35 - 9/16 ± 3/8	-0.5 ±0.35 - 1/2 ± 3/8	-0.52 ±0.35 - 1/2 ± 3/8	CAMBER (Degrees)	-0.23 ±0.35 - 1/4 ± 3/8	-0.53 ±0.35 - 9/16 ± 3/8	-0.62 ±0.35 - 5/8 ± 3/8
0.0 (0) | 0.4 (3/8)	0.0 (0) | 0.4 (3/8)	0.0 (0) | 0.4 (3/8)	CROSS | TOL.	0.0 (0) | 0.4 (3/8)	0.0 (0) | 0.4 (3/8)	0.0 (0) | 0.4 (3/8)
0.47° ±0.17° 15/32° ± 5/32°	0.33° ±0.17° 11/32° ± 5/32°	0.32° ±0.17° 5/16° ± 5/32°	FRONT TOE (Degrees)	0.37° ±0.17° 3/8° ± 5/32°	0.35° ±0.17° 11/32° ± 5/32°	0.33° ±0.17° 11/32° ± 5/32°
0.23" ±0.08" 7/32" ± 3/32"	0.17" ±0.08" 5/32" ± 3/32"	0.16" ±0.08" 5/32" ± 3/32"	FRONT TOE (Inches)	0.18" ±0.08" 3/16" ± 3/32"	0.18" ±0.08" 3/16" ± 3/32"	0.17" ±0.08" 5/32" ± 3/32"
			SAI			
20.0 / 18.4 ±0.5	20.0 / 18.4 ±0.5	20.0 / 18.4 ±0.5	TURNING ANGLE	20.0 / 18.4 ±0.5	20.0 / 18.4 ±0.5	20.0 / 18.4 ±0.5
			ILLUSTRATION #			
			REAR			
-1.65 ±0.5 -1 5/8 ± 1/2	-1.8 ±0.5 -1 13/16 ± 1/2	-1.97 ±0.5 -1 15/16 ± 1/2	CAMBER (Degrees)	-1.17 ±0.5 -1 3/16 ± 1/2	-1.5 ±0.5 -1 1/2 ± 1/2	-1.8 ±0.5 -1 13/16 ± 1/2
0.37° ±0.12° 3/8° ± 1/8°	0.4° ±0.12° 13/32° ± 1/8°	0.2° ±0.12° 3/16° ± 1/8°	REAR TOE (Degrees)	0.25° ±0.12° 1/4° ± 1/8°	0.33° ±0.12° 11/32° ± 1/8°	0.45° ±0.12° 7/16° ± 1/8°
0.18" ±0.06" 3/16" ± 1/16"	0.2" ±0.06" 3/16" ± 1/16"	0.1" ±0.06" 3/32" ± 1/16"	REAR TOE (Inches)	0.13" ±0.06" 1/8" ± 1/16"	0.17" ±0.06" 5/32" ± 1/16"	0.23" ±0.06" 7/32" ± 1/16"
0.25 (1/4)	0.25 (1/4)	0.25 (1/4)	THRUST ANGLE	0.25 (1/4)	0.25 (1/4)	0.25 (1/4)
174	174	174	ILLUSTRATION #	174	174	174
			NOTES			
29039	29040	29041		29042	29043	29044

Mercedes-Benz

C-Class (205.*) 4MATIC Sedan C450 AMG	CLA-Class (117.*) Std. Suspension	CLA-Class (117.*) Sport Suspension	MODEL INFORMATION	CLA-Class (117.*) AMG Models	GLA-Class (156.*) All exc/AMG	GLA-Class (156.*) AMG Models
2015-2016	2014-2017	2014-2017	YEARS	2014-2017	2015-2017	2015-2017
p. 308, 309	p. 308, 309	p. 308, 309	INSTRUCTIONS	p. 308, 309	p. 308, 309	p. 308, 309
	96 ft.lbs.	96 ft.lbs.	LUGNUT TORQUE	96 ft.lbs.		
			FRONT			
9.38 ±0.5 9 3/8 ± 1/2	6.78 ±0.5 6 3/4 ± 1/2	7.03 ±0.5 7 ± 1/2	CASTER (Degrees)	7.26 ±0.5 7 1/4 ± 1/2	6.12 ±0.5 6 3/4 ± 1/2	7.0 ±0.5 7 ± 1/2
0.0 (0) | 0.5 (1/2)	0.0 (0) | 0.5 (1/2)	0.0 (0) | 0.5 (1/2)	CROSS | TOL.	0.0 (0) | 0.5 (1/2)	0.0 (0) | 0.5 (1/2)	0.0 (0) | 0.5 (1/2)
-0.53 ±0.35 - 9/16 ± 3/8	-0.93 ±0.32 - 15/16 ± 5/16	-1.03 ±0.32 -1 ± 5/16	CAMBER (Degrees)	-1.8 ±0.32 -1 13/16 ± 5/16	-0.43 ±0.32 - 7/16 ± 5/16	-1.75 ±0.32 -1 3/4 ± 5/16
0.0 (0) | 0.4 (3/8)	0.0 (0) | 0.32 (5/16)	0.0 (0) | 0.32 (5/16)	CROSS | TOL.	0.0 (0) | 0.32 (5/16)	0.0 (0) | 0.32 (5/16)	0.0 (0) | 0.32 (5/16)
0.35° ±0.17° 11/32° ± 5/32°	0.2° ±0.17° 3/16° ± 5/32°	0.2° ±0.17° 3/16° ± 5/32°	FRONT TOE (Degrees)	0.35° ±0.17° 11/32° ± 5/32°	0.22° ±0.17° 7/32° ± 5/32°	0.35° ±0.17° 11/32° ± 5/32°
0.18" ±0.08" 3/16" ± 3/32"	0.1" ±0.08" 3/32" ± 3/32"	0.1" ±0.08" 3/32" ± 3/32"	FRONT TOE (Inches)	0.18" ±0.08" 3/16" ± 3/32"	0.11" ±0.08" 3/32" ± 3/32"	0.18" ±0.08" 3/16" ± 3/32"
			SAI			
20.0 / 18.4 ±0.5	20.0 / 19.1 ±0.5	20.0 / 19.1 ±0.5	TURNING ANGLE	20.0 / 19.1 ±0.5	20.0 / 19.1 ±0.5	20.0 / 19.1 ±0.5
	102	102	ILLUSTRATION #	102	102	102
			REAR			
-1.5 ±0.5 -1 1/2 ± 1/2	-1.32 ±0.5 -1 5/16 ± 1/2	-1.7 ±0.5 -1 11/16 ± 1/2	CAMBER (Degrees)	-2.05 ±0.5 -2 1/16 ± 1/2	-0.88 ±0.5 - 7/8 ± 1/2	-1.58 ±0.5 -1 9/16 ± 1/2
0.33° ±0.12° 11/32° ± 1/8°	0.12° ±0.12° 1/8° ± 1/8°	0.18° ±0.12° 3/16° ± 1/8°	REAR TOE (Degrees)	0.42° ±0.23° 13/32° ± 7/32°	0.05° ±0.12° 1/16° ± 1/8°	0.28° ±0.23° 9/32° ± 7/32°
0.17" ±0.06" 5/32" ± 1/16"	0.06" ±0.06" 1/16" ± 1/16"	0.09" ±0.06" 3/32" ± 1/16"	REAR TOE (Inches)	0.14" ±0.12" 5/32" ± 1/8"	0.03" ±0.06" 1/32" ± 1/16"	0.14" ±0.12" 5/32" ± 1/8"
0.25 (1/4)	0.25 (1/4)	0.25 (1/4)	THRUST ANGLE	0.25 (1/4)	0.25 (1/4)	0.25 (1/4)
174	174	174	ILLUSTRATION #	174	174	174
			NOTES			
29045	29047	29048		29049	29051	29052

Mercedes-Benz

GLC-Class SUV (253.*)	GLC-Class SUV (253.*)	CLK Coupe (208.*)	MODEL	CLK Conv. (208.*)	CLK55 AMG (208.*)	CLK-Class (209.*)
w/Air Body Control	except ABC		INFORMATION		(Coupe & Cabriolet)	Coupe/Conv.
						Std. Susp.
2016-2017	2016-2017	1998-2002	YEARS	1999-2003	2001-2002	2003-2009
p. 308, 309	p. 308, 309	p. 308, 309	INSTRUCTIONS	p. 308, 309	p. 308, 309	p. 308, 309
		81 ft.lbs.	LUGNUT TORQUE	81 ft.lbs.	81 ft.lbs.	81 ft.lbs.
			FRONT			
8.4 ±0.5 8 3/8 ± 1/2	8.07 ±0.5 8 1/16 ± 1/2	4.95 ±0.5 4 15/16 ± 1/2	CASTER (Degrees)	4.9 ±0.5 4 7/8 ± 1/2	5.14 ±0.5 5 1/8 ± 1/2	10.31 ±0.5 10 5/16 ± 1/2
0.0 (0) | 0.5 (1/2)	0.0 (0) | 0.5 (1/2)	0.0 (0) | 0.5 (1/2)	CROSS | TOL.	0.0 (0) | 0.5 (1/2)	0.0 (0) | 0.5 (1/2)	0.0 (0) | 0.5 (1/2)
-0.5 ±0.35 -1/2 ± 3/8	-0.4 ±0.35 -3/8 ± 3/8	-0.95 ±0.33 -15/16 ± 5/16	CAMBER (Degrees)	-0.95 ±0.33 -15/16 ± 5/16	-1.14 ±0.33 -1 1/8 ± 5/16	-0.39 +0.26 -0.46 -3/8 +1/4 -7/16
0.0 (0) | 0.4 (3/8)	0.0 (0) | 0.4 (3/8)	0.0 (0) | 0.33 (5/16)	CROSS | TOL.	0.0 (0) | 0.33 (5/16)	0.0 (0) | 0.33 (5/16)	0.0 (0) | 0.4 (3/8)
0.17° ±0.17° 5/32° ± 5/32°	0.2° ±0.17° 3/16° ± 5/32°	0.4° ±0.17° 13/32° ± 5/32°	FRONT TOE (Degrees)	0.4° ±0.17° 13/32° ± 5/32°	0.4° ±0.17° 13/32° ± 5/32°	0.2° ±0.17° 3/16° ± 5/32°
0.08" ±0.08" 3/32" ± 3/32"	0.1" ±0.08" 3/32" ± 3/32"	0.2" ±0.08" 3/16" ± 3/32"	FRONT TOE (Inches)	0.2" ±0.08" 3/16" ± 3/32"	0.2" ±0.08" 3/16" ± 3/32"	0.1" ±0.08" 3/32" ± 3/32"
			SAI			
		20.0 / 19.3 ±0.5	TURNING ANGLE	20.0 / 19.3 ±0.5	20.0 / 19.3 ±0.5	20.0 / 18.4 ±0.5
		88	ILLUSTRATION #	88	88	164 165
			REAR			
-1.55 ±0.5 -1 9/16 ± 1/2	-1.13 ±0.5 -1 1/8 ± 1/2	-1.43 ±0.5 -1 7/16 ± 1/2	CAMBER (Degrees)	-1.23 ±0.5 -1 1/4 ± 1/2	-1.63 ±0.5 -1 5/8 ± 1/2	-1.16 ±0.5 -1 3/16 ± 1/2
0.4° ±0.12° 13/32° ± 1/8°	0.38° ±0.12° 3/8° ± 1/8°	0.55° ±0.25° 9/16° ± 1/4°	REAR TOE (Degrees)	0.55° ±0.25° 9/16° ± 1/4°	0.55° ±0.25° 9/16° ± 1/4°	0.55° ±0.25° 9/16° ± 1/4°
0.2" ±0.06" 3/16" ± 1/16"	0.19" ±0.06" 3/16" ± 1/16"	0.28" ±0.13" 9/32" ± 1/8"	REAR TOE (Inches)	0.28" ±0.13" 9/32" ± 1/8"	0.28" ±0.13" 9/32" ± 1/8"	0.28" ±0.13" 9/32" ± 1/8"
0.25 (1/4)	0.25 (1/4)		THRUST ANGLE			0.25 (1/4)
174	174	86	ILLUSTRATION #	86	86	86
			NOTES			
29054	29055	29057		29058	29059	29061

Mercedes-Benz

CLK-Class (209.*)	300E 4MATIC	E300 Sedan	MODEL	E320 Coupe	E320 Sedan	E320 Convertible
Coupe/Conv.	300TE 4MATIC	300D	INFORMATION	300CE	260E, 300E	300CE Conv.
Sport Susp. (incl AMG)						
2003-2009	1990-93	1990-95	YEARS	1990-95	1990-95	1993-95
p. 308, 309	p. 308, 309	p. 308, 309	INSTRUCTIONS	p. 308, 309	p. 308, 309	p. 308, 309
81 ft.lbs.	81 ft.lbs.	81 ft.lbs.	LUGNUT TORQUE	81 ft.lbs.	81 ft.lbs.	81 ft.lbs.
			FRONT			
10.59 ±0.5 10 9/16 ± 1/2	10.05 ±0.5 10 1/16 ± 1/2	10.17 ±0.5 [1] 10 3/16 ± 1/2 [1]	CASTER (Degrees)	10.12 ±0.5 [1] 10 1/8 ± 1/2 [1]	10.12 ±0.5 [1] 10 1/8 ± 1/2 [1]	10.22 ±0.5 [1] 10 3/16 ± 1/2 [1]
0.0 (0) | 0.5 (1/2)	0.0 (0) | 0.5 (1/2)	0.0 (0) | 0.5 (1/2)	CROSS | TOL.	0.0 (0) | 0.5 (1/2)	0.0 (0) | 0.5 (1/2)	0.0 (0) | 0.5 (1/2)
-0.59 +0.26 -0.46 -9/16 +1/4 -7/16	-0.58 +0.17 -0.33 -9/16 +3/16 -5/16	-0.42 +0.17 -0.33 [2] -7/16 +3/16 -5/16 [2]	CAMBER (Degrees)	-0.42 +0.17 -0.33 [2] -7/16 +3/16 -5/16 [2]	-0.42 +0.17 -0.33 [2] -7/16 +3/16 -5/16 [2]	-0.63 +0.17 -0.33 [2] -5/8 +3/16 -5/16 [2]
0.0 (0) | 0.4 (3/8)	0.0 (0) | 0.33 (5/16)	0.0 (0) | 0.33 (5/16)	CROSS | TOL.	0.0 (0) | 0.33 (5/16)	0.0 (0) | 0.33 (5/16)	0.0 (0) | 0.33 (5/16)
0.33° ±0.17° 11/32° ± 5/32°	0.33° ±0.17° 11/32° ± 5/32°	0.33° ±0.17° 11/32° ± 5/32°	FRONT TOE (Degrees)	0.33° ±0.17° 11/32° ± 5/32°	0.33° ±0.17° 11/32° ± 5/32°	0.33° ±0.17° 11/32° ± 5/32°
0.17" ±0.08" 5/32" ± 3/32"	0.17" ±0.08" 5/32" ± 3/32"	0.17" ±0.08" 5/32" ± 3/32"	FRONT TOE (Inches)	0.17" ±0.08" 5/32" ± 3/32"	0.17" ±0.08" 5/32" ± 3/32"	0.17" ±0.08" 5/32" ± 3/32"
			SAI			
20.0 / 18.4 ±0.5	37.2 / 30.8 ±2.0	43.0 / 33.6 ±2.0	TURNING ANGLE	43.0 / 33.6 ±2.0	43.0 / 33.6 ±2.0	43.0 / 33.6 ±2.0
164 165	88	88	ILLUSTRATION #	88	88	88
			REAR			
-1.29 ±0.5 -1 5/16 ± 1/2			CAMBER (Degrees)			
0.55° ±0.25° 9/16° ± 1/4°	0.42° +0.17° -0.08° 13/32° +5/32° -3/32°	0.42° +0.17° -0.08° 13/32° +5/32° -3/32°	REAR TOE (Degrees)	0.42° +0.17° -0.08° 13/32° +5/32° -3/32°	0.42° +0.17° -0.08° 13/32° +5/32° -3/32°	0.42° +0.17° -0.08° 13/32° +5/32° -3/32°
0.28" ±0.13" 9/32" ± 1/8"	0.21" +0.08" -0.04" 7/32" +3/32" -1/32"	0.21" +0.08" -0.04" 7/32" +3/32" -1/32"	REAR TOE (Inches)	0.21" +0.08" -0.04" 7/32" +3/32" -1/32"	0.21" +0.08" -0.04" 7/32" +3/32" -1/32"	0.21" +0.08" -0.04" 7/32" +3/32" -1/32"
0.25 (1/4)			THRUST ANGLE			
86	86	86	ILLUSTRATION #	86	86	86
		[1] Sport Susp.=10.3°(10 5/16°) [2] Sport Susp.= -0.78°(-3/4°)	NOTES	[1] Sport Susp.=10.3°(10 5/16°) [2] Sport Susp.= -0.78°(-3/4°)	[1] Sport Susp.=10.3°(10 5/16°) [2] Sport Susp.= -0.78°(-3/4°)	[1] Sport Susp.=10.4°(10 3/8°) [2] Sport Susp.= -0.95°(-15/16°)
29062	29069	29070		29071	29072	29073

ALIGNMENT SPECIFICATIONS

Mercedes-Benz

E320 Wagon 300TE	E420 400E	E500 500E	MODEL INFORMATION	E-Class 4MATIC (All) E320 4MATIC Wagon (2003)	E300, E320, E420, E430 E320 Wagon (2003)	E320, E430 Sport Susp.
1990-95	1992-95	1992-95	YEARS	1998-2003	1996-2003	2001-2002
p. 308, 309	p. 308, 309	p. 308, 309	INSTRUCTIONS	p. 308, 309	p. 308, 309	p. 308, 309
81 ft.lbs.	81 ft.lbs.	81 ft.lbs.	LUGNUT TORQUE	81 ft.lbs.	81 ft.lbs.	81 ft.lbs.
			FRONT			
9.92 ±0.5 [1] 9 15/16 ± ½ [1]	10.22 ±0.5 [1] 10 3/16 ± ½ [1]	10.47 ±0.5 [1] 10 7/16 ± ½ [1]	CASTER (Degrees)	5.11 ±0.5 5⅛ ± ½	5.3 ±0.5 [1] 5 5/16 ± ½ [1]	5.75 ±0.5 [1] 5¾ ± ½ [1]
0.0 (0) | 0.5 (½)	0.0 (0) | 0.5 (½)	0.0 (0) | 0.5 (½)	CROSS | TOL.	0.0 (0) | 0.5 (½)	0.0 (0) | 0.5 (½)	0.0 (0) | 0.5 (½)
-0.08 +0.17 -0.33 [2] -1/16 +3/16 -5/16 [2]	-0.63 +0.17 -0.33 [2] -⅝ +3/16 -5/16 [2]	-1.0 +0.17 -0.33 [2] -1 +3/16 -5/16 [2]	CAMBER (Degrees)	-1.29 ±0.33 -1 5/16 ± 5/16	-0.66 ±0.33 -11/16 ± 5/16	-1.25 ±0.33 -1 ¼ ± 5/16
0.0 (0) | 0.33 (5/16)	0.0 (0) | 0.33 (5/16)	0.0 (0) | 0.33 (5/16)	CROSS | TOL.	0.0 (0) | 0.33 (5/16)	0.0 (0) | 0.33 (5/16)	0.0 (0) | 0.33 (5/16)
0.33° ±0.17° 11/32° ± 5/32°	0.33° ±0.17° 11/32° ± 5/32°	0.33° ±0.17° 11/32° ± 5/32°	FRONT TOE (Degrees)	0.43° ±0.17° 7/16° ± 5/32°	0.17° ±0.17° 5/32° ± 5/32°	0.17° ±0.17° 5/32° ± 5/32°
0.17" ±0.08" 5/32" ± 3/32"	0.17" ±0.08" 5/32" ± 3/32"	0.17" ±0.08" 5/32" ± 3/32"	FRONT TOE (Inches)	0.22" ±0.08" 7/32" ± 3/32"	0.08" ±0.08" 3/32" ± 3/32"	0.08" ±0.08" 3/32" ± 3/32"
			SAI			
43.0 / 33.2 ±2.0	43.0 / 33.2 ±2.0	40.0 / 32.2 ±2.0	TURNING ANGLE	20.0 / 18.5 ±0.5	20.0 / 18.5 ±0.5	20.0 / 18.5 ±0.5
88	88	88	ILLUSTRATION #	88	88	88
			REAR			
			CAMBER (Degrees)	-1.06 ±0.5 -1 1/16 ± ½	-1.18 ±0.5 [2] -1 3/16 ± ½ [2]	-1.62 ±0.5 [2] -1⅝ ± ½ [2]
0.42° +0.17° -0.08° 13/32° +5/32° -3/32°	0.42° +0.17° -0.08° 13/32° +5/32° -3/32°	0.42° +0.17° -0.08° 13/32° +5/32° -3/32°	REAR TOE (Degrees)	0.55° ±0.25° 9/16° ± ¼°	0.55° ±0.25° 9/16° ± ¼°	0.55° ±0.25° 9/16° ± ¼°
0.21" +0.08" -0.04" 7/32" +3/32" -1/32"	0.21" +0.08" -0.04" 7/32" +3/32" -1/32"	0.21" +0.08" -0.04" 7/32" +3/32" -1/32"	REAR TOE (Inches)	0.28" ±0.13" 9/32" ± ⅛"	0.28" ±0.13" 9/32" ± ⅛"	0.28" ±0.13" 9/32" ± ⅛"
			THRUST ANGLE			
86	86	86	ILLUSTRATION #	86	86	86
[1] Sport Susp.=10.3º(10 5/16º) [2] Sport Susp.= -0.78º(-¾º)	[1] Sport Susp.=10.4º(10⅜º) [2] Sport Susp.= -0.95º(-15/16º)	[1] Sport Susp.=10.3º(10 5/16º) [2] Sport Susp.= -0.78º(-¾º)	NOTES		[1] w/Level Control 5.4º (5⅜º) [2] w/Level Control -1.38º (-1⅜º)	[1] w/Level Control 5.9º (5⅞º) [2] w/Level Control -1.75º (-1¾º)
29074	29075	29076		29078	29079	29080

Mercedes-Benz

E55 AMG	E-Class (211.*) Sdn. w/Coil Susp. (F&R) STD.	E-Class (211.*) Sdn. w/Coil Susp. (F&R) Sport	MODEL INFORMATION	E-Class (211.*) Sdn. Full Air Susp. except AMG	E-Class (211.*) Wgn. Coil Front/Rear Air STD.	E-Class (211.*) Wgn. Coil Front/Rear Air Sport
1999-2002	2003-2009	2003-2009	YEARS	2003-2009	2003-2009	2003-2009
p. 308, 309	p. 308, 309	p. 308, 309	INSTRUCTIONS	p. 308, 309	p. 308, 309	p. 308, 309
81 ft.lbs.	96 ft.lbs.	96 ft.lbs.	LUGNUT TORQUE	96 ft.lbs.	96 ft.lbs.	96 ft.lbs.
			FRONT			
5.75 ±0.5 5¾ ± ½	10.77 ±0.5 10¾ ± ½	11.24 ±0.5 11 ¼ ± ½	CASTER (Degrees)	11.3 ±0.5 11 5/16 ± ½	10.72 ±0.5 10¾ ± ½	11.25 ±0.5 11 ¼ ± ½
0.0 (0) | 0.5 (½)	0.0 (0) | 0.5 (½)	0.0 (0) | 0.5 (½)	CROSS | TOL.	0.0 (0) | 0.5 (½)	0.0 (0) | 0.5 (½)	0.0 (0) | 0.5 (½)
-1.25 ±0.33 -1 ¼ ± 5/16	-0.61 +0.17 -0.5 -⅝ +3/16 -½	-0.95 +0.17 -0.5 -15/16 +3/16 -½	CAMBER (Degrees)	-0.99 +0.17 -0.5 -1 +3/16 -½	-0.57 +0.17 -0.5 -9/16 +3/16 -½	-0.95 +0.17 -0.5 -15/16 +3/16 -½
0.0 (0) | 0.33 (5/16)	0.0 (0) | 0.33 (5/16)	0.0 (0) | 0.33 (5/16)	CROSS | TOL.	0.0 (0) | 0.33 (5/16)	0.0 (0) | 0.33 (5/16)	0.0 (0) | 0.33 (5/16)
0.17° ±0.17° 5/32° ± 5/32°	0.26° ±0.17° [1] ¼° ± 5/32° [1]	0.26° ±0.17° [1] ¼° ± 5/32° [1]	FRONT TOE (Degrees)	0.26° ±0.17° [1] ¼° ± 5/32° [1]	0.26° ±0.17° [1] ¼° ± 5/32° [1]	0.26° ±0.17° [1] ¼° ± 5/32° [1]
0.08" ±0.08" 3/32" ± 3/32"	0.13" ±0.08" [1] ⅛" ± 3/32" [1]	0.13" ±0.08" [1] ⅛" ± 3/32" [1]	FRONT TOE (Inches)	0.13" ±0.08" [1] ⅛" ± 3/32" [1]	0.13" ±0.08" [1] ⅛" ± 3/32" [1]	0.13" ±0.08" [1] ⅛" ± 3/32" [1]
			SAI			
20.0 / 18.5 ±0.5	20.0 / 18.2 ±0.5	20.0 / 19.0 ±0.5	TURNING ANGLE	20.0 / 18.2 ±0.5	20.0 / 18.2 ±0.5	20.0 / 19.0 ±0.5
88	164 165	164 165	ILLUSTRATION #	164 165	164 165	164 165
			REAR			
-1.62 ±0.5 -1⅝ ± ½	-1.03 ±0.5 -1 ± ½	-1.39 ±0.5 -1⅜ ± ½	CAMBER (Degrees)	-1.5 ±0.5 -1 ½ ± ½	-1.43 ±0.5 -1 7/16 ± ½	-1.73 ±0.5 -1¾ ± ½
0.55° ±0.25° 9/16° ± ¼°	0.33° ±0.13° 11/32° ± ⅛°	0.33° ±0.13° 11/32° ± ⅛°	REAR TOE (Degrees)	0.33° ±0.13° 11/32° ± ⅛°	0.33° ±0.13° 11/32° ± ⅛°	0.33° ±0.13° 11/32° ± ⅛°
0.28" ±0.13" 9/32" ± ⅛"	0.17" ±0.06" 5/32" ± 1/16"	0.17" ±0.06" 5/32" ± 1/16"	REAR TOE (Inches)	0.17" ±0.06" 5/32" ± 1/16"	0.17" ±0.06" 5/32" ± 1/16"	0.17" ±0.06" 5/32" ± 1/16"
	0.25 (¼)	0.25 (¼)	THRUST ANGLE	0.25 (¼)	0.25 (¼)	0.25 (¼)
86	06	06	ILLUSTRATION #	06	06	06
	[1] w/16" wheels = 0.33º (0.17")	[1] w/16" wheels = 0.33º (0.17")	NOTES	[1] w/16" wheels = 0.33º (0.17")	[1] w/16" wheels = 0.33º (0.17")	[1] w/16" wheels = 0.33º (0.17")
29081	29083	29084		29085	29086	29087

Mercedes-Benz

E-Class (211.*)	E-Class (211.*)	E-Class 4Matic (211.*)	MODEL	E-Class 4Matic (211.*)	E-Class 4Matic (211.*)	E-Class (212.*)
Wgn. Full Air Susp. except AMG	All AMG Models Sedan & Wagon	Coil Susp. (F & R)	INFORMATION	Coil Front/Rear Air	Full Air Suspension	RWD Sedan AMG
2003-2009	2003-2009	2003-2009	YEARS	2003-2009	2003-2009	2010-2016
p. 308, 309	p. 308, 309	p. 308, 309	INSTRUCTIONS	p. 308, 309	p. 308, 309	p. 308, 309
96 ft.lbs.	96 ft.lbs.	96 ft.lbs.	LUGNUT TORQUE	96 ft.lbs.	96 ft.lbs.	
			FRONT			
11.3 ± 0.5 11 5/16 ± 1/2	11.6 ± 0.5 11 5/8 ± 1/2	7.09 ± 0.5 7 1/16 ± 1/2	CASTER (Degrees)	7.43 ± 0.5 7 7/16 ± 1/2	7.66 ± 0.5 7 11/16 ± 1/2	8.38 ± 0.5 8 3/8 ± 1/2
0.0 (0) | 0.5 (1/2)	0.0 (0) | 0.5 (1/2)	0.0 (0) | 0.5 (1/2)	CROSS | TOL.	0.0 (0) | 0.5 (1/2)	0.0 (0) | 0.5 (1/2)	0.0 (0) | 0.5 (1/2)
-0.99 +0.17 -0.5 -1 +3/16 -1/2	-1.22 +0.17 -0.5 -1 1/4 +3/16 -1/2	-0.71 ± 0.33 -11/16 ± 5/16	CAMBER (Degrees)	-0.71 ± 0.33 -11/16 ± 5/16	-1.07 ± 0.33 -1 1/16 ± 5/16	-1.54 ± 0.35 -1 9/16 ± 3/8
0.0 (0) | 0.33 (5/16)	0.0 (0) | 0.33 (5/16)	0.0 (0) | 0.33 (5/16)	CROSS | TOL.	0.0 (0) | 0.33 (5/16)	0.0 (0) | 0.33 (5/16)	0.0 (0) | 0.4 (3/8)
0.26° ± 0.17° [1] 1/4° ± 5/32° [1]	0.26° ± 0.17° 1/4° ± 5/32°	0.25° ± 0.17° 1/4° ± 5/32°	FRONT TOE (Degrees)	0.25° ± 0.17° 1/4° ± 5/32°	0.25° ± 0.17° 1/4° ± 5/32°	0.5° ± 0.12° 1/2° ± 1/8°
0.13" ± 0.08" [1] 1/8" ± 3/32" [1]	0.13" ± 0.08" 1/8" ± 3/32"	0.13" ± 0.08" 1/8" ± 3/32"	FRONT TOE (Inches)	0.13" ± 0.08" 1/8" ± 3/32"	0.13" ± 0.08" 1/8" ± 3/32"	0.25" ± 0.06" 1/4" ± 1/16"
			SAI			
20.0 / 18.2 ± 0.5	20.0 / 18.2 ± 0.5	20.0 / 19.0 ± 0.5	TURNING ANGLE	20.0 / 19.0 ± 0.5	20.0 / 19.0 ± 0.5	20.0 / 18.4 ± 0.5
164 165	164 165	164 165	ILLUSTRATION #	164 165	164 165	198 199
			REAR			
-1.5 ± 0.5 -1 1/2 ± 1/2	-1.73 ± 0.5 -1 3/4 ± 1/2	-0.87 ± 0.5 -7/8 ± 1/2	CAMBER (Degrees)	-1.22 ± 0.5 -1 1/4 ± 1/2	-1.37 ± 0.5 -1 3/8 ± 1/2	-2.07 ± 0.5 -2 1/16 ± 1/2
0.33° ± 0.13° 11/32° ± 1/8°	0.33° ± 0.13° 11/32° ± 1/8°	0.4° ± 0.13° 13/32° ± 1/8°	REAR TOE (Degrees)	0.4° ± 0.13° 13/32° ± 1/8°	0.4° ± 0.13° 13/32° ± 1/8°	0.24° ± 0.12° 1/4° ± 1/8°
0.17" ± 0.06" 5/32" ± 1/16"	0.17" ± 0.06" 5/32" ± 1/16"	0.2" ± 0.06" 3/16" ± 1/16"	REAR TOE (Inches)	0.2" ± 0.06" 3/16" ± 1/16"	0.2" ± 0.06" 3/16" ± 1/16"	0.12" ± 0.06" 1/8" ± 1/16"
0.25 (1/4)	0.25 (1/4)	0.25 (1/4)	THRUST ANGLE	0.25 (1/4)	0.25 (1/4)	0.25 (1/4)
06	06	06	ILLUSTRATION #	06	06	174
[1] w/16" wheels = 0.33° (0.17")			NOTES			
29088	29089	29090		29091	29092	29094

Mercedes-Benz

E-Class (212.*)	E-Class (212.*)	E-Class (212.*)	MODEL	E-Class (212.*)	E-Class (212.*)	E-Class (212.*)
RWD Sedan exc/AMG Steel Spring Susp.	RWD Sedan exc/AMG Steel Spring Sport Susp.	RWD Sedan exc/AMG 4-Corner Air Susp.	INFORMATION	RWD Sedan exc/AMG Steel Spring Rough Road	RWD Wagon exc/AMG Luxury Version	RWD Wagon exc/AMG Sport Version
2010-2016	2010-2016	2010-2016	YEARS	2010-2016	2010-2017	2010-2017
p. 308, 309	p. 308, 309	p. 308, 309	INSTRUCTIONS	p. 308, 309	p. 308, 309	p. 308, 309
			LUGNUT TORQUE			
			FRONT			
6.75 ± 0.5 6 3/4 ± 1/2	6.96 ± 0.5 6 15/16 ± 1/2	L= 7.0, R= 7.01 ± 0.5 L= 7, R= 7 ± 1/2	CASTER (Degrees)	6.59 ± 0.5 6 9/16 ± 1/2	7.06 ± 0.5 7 1/16 ± 1/2	7.2 ± 0.5 7 3/16 ± 1/2
0.0 (0) | 0.5 (1/2)	0.0 (0) | 0.5 (1/2)	0.0 (0) | 0.5 (1/2)	CROSS | TOL.	0.0 (0) | 0.5 (1/2)	0.0 (0) | 0.5 (1/2)	0.0 (0) | 0.5 (1/2)
-0.39 ± 0.35 -3/8 ± 3/8	-0.56 ± 0.35 -9/16 ± 3/8	-0.55 ± 0.35 -9/16 ± 3/8	CAMBER (Degrees)	-0.18 ± 0.35 -3/16 ± 3/8	-0.39 ± 0.35 -3/8 ± 3/8	-0.56 ± 0.35 -9/16 ± 3/8
0.0 (0) | 0.4 (3/8)	0.0 (0) | 0.4 (3/8)	0.0 (0) | 0.4 (3/8)	CROSS | TOL.	0.0 (0) | 0.4 (3/8)	0.0 (0) | 0.4 (3/8)	0.0 (0) | 0.4 (3/8)
0.15° ± 0.17° 5/32° ± 5/32°	0.15° ± 0.17° 5/32° ± 5/32°	0.15° ± 0.17° 5/32° ± 5/32°	FRONT TOE (Degrees)	0.15° ± 0.17° 5/32° ± 5/32°	0.15° ± 0.17° 5/32° ± 5/32°	0.15° ± 0.17° 5/32° ± 5/32°
0.08" ± 0.08" 1/16" ± 3/32"	0.08" ± 0.08" 1/16" ± 3/32"	0.08" ± 0.08" 1/16" ± 3/32"	FRONT TOE (Inches)	0.08" ± 0.08" 1/16" ± 3/32"	0.08" ± 0.08" 1/16" ± 3/32"	0.08" ± 0.08" 1/16" ± 3/32"
			SAI			
20.0 / 18.1 ± 0.5	20.0 / 18.1 ± 0.5	20.0 / 18.1 ± 0.5	TURNING ANGLE	20.0 / 18.1 ± 0.5	20.0 / 18.1 ± 0.5	20.0 / 18.1 ± 0.5
198 199	198 199	198 199	ILLUSTRATION #	198 199	198 199	198 199
			REAR			
-1.05 ± 0.5 -1 1/16 ± 1/2	-1.45 ± 0.5 -1 7/16 ± 1/2	-1.5 ± 0.5 -1 1/2 ± 1/2	CAMBER (Degrees)	-0.71 ± 0.5 -11/16 ± 1/2	-1.43 ± 0.5 -1 7/16 ± 1/2	-1.75 ± 0.5 -1 3/4 ± 1/2
0.36° ± 0.12° 3/8° ± 1/8°	0.41° ± 0.12° 13/32° ± 1/8°	0.42° ± 0.12° 13/32° ± 1/8°	REAR TOE (Degrees)	0.32° ± 0.12° 5/16° ± 1/8°	0.41° ± 0.12° 13/32° ± 1/8°	0.44° ± 0.12° 7/16° ± 1/8°
0.18" ± 0.06" 3/16" ± 1/16"	0.21" ± 0.06" 7/32" ± 1/16"	0.21" ± 0.06" 7/32" ± 1/16"	REAR TOE (Inches)	0.16" ± 0.06" 5/32" ± 1/16"	0.21" ± 0.06" 7/32" ± 1/16"	0.22" ± 0.06" 7/32" ± 1/16"
0.25 (1/4)	0.25 (1/4)	0.25 (1/4)	THRUST ANGLE	0.25 (1/4)	0.25 (1/4)	0.25 (1/4)
174	174	174	ILLUSTRATION #	174	174	174
			NOTES			
29095	29096	29097		29098	29099	29100

Mercedes-Benz

	29101	29102	29103	29104	29105	29106
MODEL	E-Class (212.*)	E-Class (212.*)	E-Class (212.*)	E-Class (212.*)	E-Class (212.*)	E-Class (212.*)
INFORMATION	4MATIC Sed. exc/AMG STD Susp.	4MATIC Sed. exc/AMG Sport Susp.	4MATIC Sed. exc/AMG Rear Axle Air STD. Susp.	4MATIC Sed. exc/AMG Rear Axle Air Sport Susp.	4MATIC Sed. exc/AMG 4-Corner Air Susp.	4MATIC Sed. AMG E63
YEARS	2010-2016	2010-2016	2010-2016	2010-2016	2010-2016	2010-2016
INSTRUCTIONS	p. 308, 309	p. 308, 309	p. 308, 309	p. 308, 309	p. 308, 309	p. 308, 309
LUGNUT TORQUE						
FRONT						
CASTER (Degrees)	6.66 ±0.5 6 11/16 ± ½	6.95 ±0.5 6 15/16 ± ½	7.04 ±0.5 7 1/16 ± ½	7.19 ±0.5 7 3/16 ± ½	6.99 ±0.5 7 ± ½	8.35 ±0.5 8⅜ ± ½
CROSS | TOL.	0.0 (0) | 0.5 (½)	0.0 (0) | 0.5 (½)	0.0 (0) | 0.5 (½)	0.0 (0) | 0.5 (½)	0.0 (0) | 0.5 (½)	0.0 (0) | 0.5 (½)
CAMBER (Degrees)	-0.38 ±0.35 -⅜ ± ⅜	-0.51 ±0.35 -½ ± ⅜	-0.38 ±0.35 -⅜ ± ⅜	-0.51 ±0.35 -½ ± ⅜	-0.51 ±0.35 -½ ± ⅜	-1.79 ±0.35 -1 13/16 ± ⅜
CROSS | TOL.	0.0 (0) | 0.4 (⅜)	0.0 (0) | 0.4 (⅜)	0.0 (0) | 0.4 (⅜)	0.0 (0) | 0.4 (⅜)	0.0 (0) | 0.4 (⅜)	0.0 (0) | 0.4 (⅜)
FRONT TOE (Degrees)	0.15° ±0.17° 5/32° ± 5/32°	0.15° ±0.17° 5/32° ± 5/32°	0.15° ±0.17° 5/32° ± 5/32°	0.15° ±0.12° 5/32° ± ⅛°	0.15° ±0.17° 5/32° ± 5/32°	0.3° ±0.17° 5/16° ± 5/32°
FRONT TOE (Inches)	0.08" ±0.08" 1/16" ± 3/32"	0.08" ±0.08" 1/16" ± 3/32"	0.08" ±0.08" 1/16" ± 3/32"	0.08" ±0.06" 1/16" ± 1/16"	0.08" ±0.08" 1/16" ± 3/32"	0.15" ±0.08" 5/32" ± 3/32"
SAI						
TURNING ANGLE	20.0 / 18.1 ±0.5	20.0 / 18.1 ±0.5	20.0 / 18.1 ±0.5	20.0 / 18.1 ±0.5	20.0 / 18.1 ±0.5	20.0 / 18.1 ±0.5
ILLUSTRATION #	198 199	198 199	198 199	198 199	198 199	198 199
REAR						
CAMBER (Degrees)	-1.05 ±0.5 -1 1/16 ± ½	-1.45 ±0.5 -1 7/16 ± ½	-1.5 ±0.5 -1½ ± ½	-1.75 ±0.5 -1¾ ± ½	-1.5 ±0.5 -1½ ± ½	-1.75 ±0.5 -1¾ ± ½
REAR TOE (Degrees)	0.37° ±0.12° ⅜° ± ⅛°	0.41° ±0.12° 13/32° ± ⅛°	0.42° ±0.12° 13/32° ± ⅛°	0.44° ±0.12° 7/16° ± ⅛°	0.42° ±0.12° 13/32° ± ⅛°	0.44° ±0.12° 7/16° ± ⅛°
REAR TOE (Inches)	0.19" ±0.06" 3/16" ± 1/16"	0.21" ±0.06" 7/32" ± 1/16"	0.21" ±0.06" 7/32" ± 1/16"	0.22" ±0.06" 7/32" ± 1/16"	0.21" ±0.06" 7/32" ± 1/16"	0.22" ±0.06" 7/32" ± 1/16"
THRUST ANGLE	0.05 (1/16)	0.05 (1/16)	0.05 (1/16)	0.05 (1/16)	0.05 (1/16)	0.05 (1/16)
ILLUSTRATION #	174	174	174	174	174	174
NOTES						

Mercedes-Benz

	29107	29109	29110	29111	29116	29117
MODEL	E-Class (212.*)	E-Class Coupe (207.*)	E-Class Coupe (207.*)	E-Class Cabriolet (207.*)	300SDL	300SE, 300SEL
INFORMATION	4MATIC Wagon AMG	exc/4MATIC	4MATIC	All Models	350SD, 350SDL	420SEL 560SEL, 560SEC
YEARS	2010-2017	2010-2017	2012-2017	2012-2017	1990-91	1990-91
INSTRUCTIONS	p. 308, 309	p. 308, 309	p. 308, 309	p. 308, 309	p. 308, 309	p. 308, 309
LUGNUT TORQUE		96 ft.lbs.	96 ft.lbs.	96 ft.lbs.	81 ft.lbs.	81 ft.lbs.
FRONT						
CASTER (Degrees)	8.34 ±0.5 8⅜ ± ½	10.82 ±0.5 10 13/16 ± ½	10.47 ±0.5 10½ ± ½	10.82 ±0.5 10 13/16 ± ½	10.0 ±0.5 10 ± ½	10.0 ±0.5 10 ± ½
CROSS | TOL.	0.0 (0) | 0.5 (½)	0.0 (0) | 0.5 (½)	0.0 (0) | 0.5 (½)	0.0 (0) | 0.5 (½)	0.0 (0) | 0.5 (½)	0.0 (0) | 0.5 (½)
CAMBER (Degrees)	-1.79 ±0.35 -1 13/16 ± ⅜	-0.52 +0.05 -0.65 -½ +1/16 -⅝	-0.58 +0.25 -0.45 -9/16 +¼ -7/16	-0.52 +0.05 -0.65 -½ +1/16 -⅝	0.0 +0.17 -0.33 0 +3/16 -5/16	0.0 +0.17 -0.33 0 +3/16 -5/16
CROSS | TOL.	0.0 (0) | 0.4 (⅜)	0.0 (0) | 0.4 (⅜)	0.0 (0) | 0.4 (⅜)	0.0 (0) | 0.4 (⅜)	0.0 (0) | 0.33 (5/16)	0.0 (0) | 0.33 (5/16)
FRONT TOE (Degrees)	0.3° ±0.12° 5/16° ± ⅛°	0.17° ±0.17° 5/32° ± 5/32°	0.17° ±0.17° 5/32° ± 5/32°	0.17° ±0.17° 5/32° ± 5/32°	0.42° ±0.17° 13/32° ± 5/32°	0.42° ±0.17° 13/32° ± 5/32°
FRONT TOE (Inches)	0.15" ±0.06" 5/32" ± 1/16"	0.08" ±0.08" 3/32" ± 3/32"	0.08" ±0.08" 3/32" ± 3/32"	0.08" ±0.08" 3/32" ± 3/32"	0.21" ±0.08" 7/32" ± 3/32"	0.21" ±0.08" 7/32" ± 3/32"
SAI						
TURNING ANGLE	20.0 / 18.4 ±0.5	20.0 / 18.3 ±0.5	20.0 / 18.3 ±0.5	20.0 / 18.3 ±0.5	43.0 / 32.6 ±2.0	43.0 / 32.6 ±2.0
ILLUSTRATION #	198 199	198 199	198 199	198 199	87	87
REAR						
CAMBER (Degrees)	-2.38 ±0.5 -2⅜ ± ½	-1.55 ±0.5 -1 9/16 ± ½	-1.42 ±0.5 -1 7/16 ± ½	-1.55 ±0.5 -1 9/16 ± ½		
REAR TOE (Degrees)	0.23° ±0.12° ¼° ± ⅛°	0.42° ±0.12° 13/32° ± ⅛°	0.41° ±0.12° 13/32° ± ⅛°	0.42° ±0.12° 13/32° ± ⅛°		
REAR TOE (Inches)	0.12" ±0.06" ⅛" ± 1/16"	0.21" ±0.06" 7/32" ± 1/16"	0.21" ±0.06" 7/32" ± 1/16"	0.21" ±0.06" 7/32" ± 1/16"		
THRUST ANGLE	0.05 (1/16)		0.12 (⅛)			
ILLUSTRATION #	174	174	174	174		
NOTES						

Mercedes-Benz

MODEL INFORMATION	420SEL, 560SEL w/Hydropneumatic Suspension	300SE, 300SD 400SE, 400SEL	500SEL, 500SEC 600SEL, 600SEC	S320, S420 S500, S600 S350 Turbodiesel	CL500, CL600	S-Class Sedan (220.*) exc/V-12 & AMG S55
YEARS	1990-91	1992-93	1992-93	1994-99	1998-99	2000-2006
INSTRUCTIONS	p. 308, 309	p. 308, 309	p. 308, 309	p. 308, 309	p. 308, 309	p. 308, 309
LUGNUT TORQUE	81 ft.lbs.	111 ft.lbs.	111 ft.lbs.	111 ft.lbs.	111 ft.lbs.	111 ft.lbs.
FRONT						
CASTER (Degrees)	10.25 ±0.5 10¼ ± ½	10.17 ±0.5 10 3/16 ± ½	10.17 ±0.5 10 3/16 ± ½	10.33 ±0.5 10 5/16 ± ½	10.33 ±0.5 10 5/16 ± ½	8.93 ±0.5 8 15/16 ± ½
CROSS | TOL.	0.0 (0) | 0.5 (½)	0.0 (0) | 0.5 (½)	0.0 (0) | 0.5 (½)	0.0 (0) | 0.5 (½)	0.0 (0) | 0.5 (½)	0.0 (0) | 0.5 (½)
CAMBER (Degrees)	-0.33 +0.17 -0.33 -5/16 +3/16 -5/16	-0.17 +0.17 -0.33 -3/16 +3/16 -5/16	-0.17 +0.17 -0.33 -3/16 +3/16 -5/16	-0.33 +0.17 -0.33 -5/16 +3/16 -5/16	-0.33 +0.17 -0.33 -5/16 +3/16 -5/16	-0.77 ±0.33 -¾ ± 5/16
CROSS | TOL.	0.0 (0) | 0.33 (5/16)	0.0 (0) | 0.33 (5/16)	0.0 (0) | 0.33 (5/16)	0.0 (0) | 0.33 (5/16)	0.0 (0) | 0.33 (5/16)	0.0 (0) | 0.33 (5/16)
FRONT TOE (Degrees)	0.33° ±0.17° 11/32° ± 5/32°	0.58° ±0.17° 19/32° ± 5/32°	0.58° ±0.17° 19/32° ± 5/32°	0.58° ±0.17° 19/32° ± 5/32°	0.58° ±0.17° 19/32° ± 5/32°	0.24° ±0.17° ¼° ± 5/32°
FRONT TOE (Inches)	0.17" ±0.08" 5/32" ± 3/32"	0.29" ±0.08" 9/32" ± 3/32"	0.29" ±0.08" 9/32" ± 3/32"	0.29" ±0.08" 9/32" ± 3/32"	0.29" ±0.08" 9/32" ± 3/32"	0.12" ±0.08" 1/8" ± 3/32"
SAI						
TURNING ANGLE	43.0 / 32.6 ±2.0	43.0 / 33.1 ±2.0	43.0 / 33.1 ±2.0	43.0 / 33.1 ±2.0	43.0 / 33.1 ±2.0	20.0 / 18.4 ±0.5
ILLUSTRATION #	87	88	88	88	88	164 165
REAR						
CAMBER (Degrees)						-1.05 ±0.5 -1 1/16 ± ½
REAR TOE (Degrees)		0.58° +0.17° -0.08° 19/32° +5/32° -3/32°	0.58° +0.17° -0.08° 19/32° +5/32° -3/32°	0.58° +0.17° -0.08° 19/32° +5/32° -3/32°	0.58° +0.17° -0.08° 19/32° +5/32° -3/32°	0.72° ±0.25° 23/32° ± ¼°
REAR TOE (Inches)		0.29" +0.08" -0.04" 9/32" +3/32" -1/32"	0.29" +0.08" -0.04" 9/32" +3/32" -1/32"	0.29" +0.08" -0.04" 9/32" +3/32" -1/32"	0.29" +0.08" -0.04" 9/32" +3/32" -1/32"	0.36" ±0.13" 3/8" ± 1/8"
THRUST ANGLE						0.25 (¼)
ILLUSTRATION #		86	86	86	86	174
NOTES						
	29118	29120	29121	29122	29123	29125

Mercedes-Benz

MODEL INFORMATION	S-Class Sedan (220.*) V-12 Turbo	S-Class Sedan (220.*) AMG S55	S-Class Sedan (220.*) 4MATIC AWD Models	S-Class Sedan (221.*) RWD exc/V-12 incl. AMG S63	S-Class Sedan (221.*) RWD V-12 incl. AMG S65	S-Class Sedan (221.*) All 4MATIC
YEARS	2000-2006	2000-2006	2000-2006	2007-2013	2007-2013	2007-2013
INSTRUCTIONS	p. 308, 309	p. 308, 309	p. 308, 309	p. 308, 309	p. 308, 309	p. 308, 309
LUGNUT TORQUE	111 ft.lbs.	111 ft.lbs.	111 ft.lbs.	111 ft.lbs.	111 ft.lbs.	111 ft.lbs.
FRONT						
CASTER (Degrees)	8.91 ±0.5 8 7/8 ± ½	9.18 ±0.5 9 3/16 ± ½	5.59 ±0.5 5 9/16 ± ½	9.4 ±0.5 9 3/8 ± ½	9.28 ±0.5 9 5/16 ± ½	8.18 ±0.5 8 3/16 ± ½
CROSS | TOL.	0.0 (0) | 0.5 (½)	0.0 (0) | 0.5 (½)	0.0 (0) | 0.5 (½)	0.0 (0) | 0.5 (½)	0.0 (0) | 0.5 (½)	0.0 (0) | 0.5 (½)
CAMBER (Degrees)	-0.77 ±0.33 -¾ ± 5/16	-1.0 ±0.33 -1 ± 5/16	-0.99 ±0.33 -1 ± 5/16	-0.69 +0.26 -0.4 -11/16 +¼ -3/8	-0.69 +0.26 -0.4 -11/16 +¼ -3/8	-0.66 +0.26 -0.4 -11/16 +¼ -3/8
CROSS | TOL.	0.0 (0) | 0.33 (5/16)	0.0 (0) | 0.33 (5/16)	0.0 (0) | 0.33 (5/16)	0.0 (0) | 0.33 (5/16)	0.0 (0) | 0.33 (5/16)	0.0 (0) | 0.33 (5/16)
FRONT TOE (Degrees)	0.24° ±0.17° ¼° ± 5/32°	0.24° ±0.17° ¼° ± 5/32°	0.24° ±0.17° ¼° ± 5/32°	0.24° ±0.17° ¼° ± 5/32°	0.24° ±0.17° ¼° ± 5/32°	0.24° ±0.17° ¼° ± 5/32°
FRONT TOE (Inches)	0.12" ±0.08" 1/8" ± 3/32"	0.12" ±0.08" 1/8" ± 3/32"	0.12" ±0.08" 1/8" ± 3/32"	0.12" ±0.08" 1/8" ± 3/32"	0.12" ±0.08" 1/8" ± 3/32"	0.12" ±0.08" 1/8" ± 3/32"
SAI						
TURNING ANGLE	20.0 / 18.4 ±0.5	20.0 / 18.4 ±0.5	20.0 / 18.5 ±0.5	20.0 / 18.3 ±0.5	20.0 / 18.3 ±0.5	20.0 / 18.3 ±0.5
ILLUSTRATION #	164 165	164 165	164 165	198 199	198 199	198 199
REAR						
CAMBER (Degrees)	-1.24 ±0.5 -1¼ ± ½	-1.43 ±0.5 -1 7/16 ± ½	-1.24 ±0.5 -1¼ ± ½	-1.87 ±0.5 -1 7/8 ± ½	-1.87 ±0.5 -1 7/8 ± ½	-1.21 ±0.5 -1 3/16 ± ½
REAR TOE (Degrees)	0.46° ±0.25° 15/32° ± ¼°	0.78° ±0.25° 25/32° ± ¼°	0.72° ±0.25° 23/32° ± ¼°	0.51° ±0.13° ½° ± 1/8°	0.51° ±0.13° ½° ± 1/8°	0.41° ±0.13° 13/32° ± 1/8°
REAR TOE (Inches)	0.23" ±0.13" 7/32" ± 1/8"	0.39" ±0.13" 3/8" ± 1/8"	0.36" ±0.13" 3/8" ± 1/8"	0.25" ±0.06" ¼" ± 1/16"	0.25" ±0.06" ¼" ± 1/16"	0.21" ±0.06" 7/32" ± 1/16"
THRUST ANGLE	0.25 (¼)	0.25 (¼)	0.25 (¼)	0.25 (¼)	0.25 (¼)	0.25 (¼)
ILLUSTRATION #	174	174	174	174	174	174
NOTES						
	29126	29127	29128	29130	29131	29132

ALIGNMENT SPECIFICATIONS

Mercedes-Benz

S-Class Sedan (222.*)	S-Class Sedan (222.*)	S-Class Sedan (222.*)	MODEL	S-Class Sedan (222.*)	S-Class Sedan (222.*)	CL-Class Coupe (215.*)
S 550	S 550	S 550 4MATIC	INFORMATION	S 65 AMG	S 63 AMG 4MATIC	exc/V-12 & AMG CL55
exc/ABC	w/ABC					
2014-2016	2014-2016	2014-2016	YEARS	2014-2016	2014-2016	2000-2006
p. 308, 309	p. 308, 309	p. 308, 309	INSTRUCTIONS	p. 308, 309	p. 308, 309	p. 308, 309
111 ft.lbs.	111 ft.lbs.	111 ft.lbs.	LUGNUT TORQUE	111 ft.lbs.	111 ft.lbs.	111 ft.lbs.
			FRONT			
9.45 ± 0.5 9 7/16 ± ½	9.45 ± 0.5 9 7/16 ± ½	9.1 ± 0.5 9⅛ ± ½	CASTER (Degrees)	9.45 ± 0.5 9 7/16 ± ½	8.65 ± 0.5 8⅝ ± ½	7.96 ± 0.5 7 15/16 ± ½
0.0 (0) | 0.5 (½)	0.0 (0) | 0.5 (½)	0.0 (0) | 0.5 (½)	CROSS | TOL.	0.0 (0) | 0.5 (½)	0.0 (0) | 0.5 (½)	0.0 (0) | 0.5 (½)
-0.4 +0.25 -0.45 -⅜ +¼ -7/16	-0.4 +0.25 -0.45 -⅜ +¼ -7/16	-0.4 +0.25 -0.45 -⅜ +¼ -7/16	CAMBER (Degrees)	-0.4 +0.25 -0.45 -⅜ +¼ -7/16	-1.0 +0.25 -0.45 -1 +¼ -7/16	-1.09 ± 0.33 -1 1/16 ± 5/16
0.0 (0) | 0.4 (⅜)	0.0 (0) | 0.4 (⅜)	0.0 (0) | 0.4 (⅜)	CROSS | TOL.	0.0 (0) | 0.4 (⅜)	0.0 (0) | 0.4 (⅜)	0.0 (0) | 0.33 (5/16)
0.17° ± 0.17° 5/32° ± 5/32°	0.17° ± 0.17° 5/32° ± 5/32°	0.17° ± 0.17° 5/32° ± 5/32°	FRONT TOE (Degrees)	0.17° ± 0.17° 5/32° ± 5/32°	0.15° ± 0.17° 5/32° ± 5/32°	0.24° ± 0.1° ¼° ± 3/32°
0.08" ± 0.08" 3/32" ± 3/32"	0.08" ± 0.08" 3/32" ± 3/32"	0.08" ± 0.08" 3/32" ± 3/32"	FRONT TOE (Inches)	0.08" ± 0.08" 3/32" ± 3/32"	0.08" ± 0.08" 1/16" ± 3/32"	0.12" ± 0.05" ⅛" ± 1/16"
			SAI			
20.0 / 18.5 ± 0.5	20.0 / 18.5 ± 0.5	20.0 / 18.5 ± 0.5	TURNING ANGLE	20.0 / 18.5 ± 0.5	20.0 / 18.8 ± 0.5	20.0 / 18.4 ± 0.5
			ILLUSTRATION #			164 165
			REAR			
-1.85 ± 0.5 -1 ⅞ ± ½	-1.85 ± 0.5 -1 ⅞ ± ½	-1.85 ± 0.5 -1 ⅞ ± ½	CAMBER (Degrees)	-1.85 ± 0.5 -1 ⅞ ± ½	-1.85 ± 0.5 -1 ⅞ ± ½	-1.32 ± 0.5 -1 5/16 ± ½
0.4° ± 0.12° 13/32° ± ⅛°	0.4° ± 0.12° 13/32° ± ⅛°	0.4° ± 0.12° 13/32° ± ⅛°	REAR TOE (Degrees)	0.4° ± 0.12° 13/32° ± ⅛°	0.4° ± 0.12° 13/32° ± ⅛°	0.8° ± 0.25° 13/16° ± ¼°
0.2" ± 0.06" 7/32" ± 1/16"	0.2" ± 0.06" 7/32" ± 1/16"	0.2" ± 0.06" 7/32" ± 1/16"	REAR TOE (Inches)	0.2" ± 0.06" 7/32" ± 1/16"	0.2" ± 0.06" 7/32" ± 1/16"	0.4" ± 0.13" 13/32" ± ⅛"
0.25 (¼)	0.25 (¼)	0.25 (¼)	THRUST ANGLE	0.25 (¼)	0.25 (¼)	0.25 (¼)
174	174	174	ILLUSTRATION #	174	174	174
			NOTES			
29134	29135	29136		29137	29138	29140

Mercedes-Benz

CL-Class Coupe (215.*)	CL-Class Coupe (215.*)	CL-Class Coupe (216.*)	MODEL	CL-Class Coupe (216.*)	CL-Class Coupe (216.*)	S-Class Coupe (217.*)
V-12	AMG CL55	V-8	INFORMATION	V-12	All 4MATIC	S 550 4MATIC
(incl. Bi-Turbo AMG)		(incl. CL63 AMG)		(incl. CL65 AMG)		
2000-2006	2000-2006	2007-2014	YEARS	2007-2014	2008-2014	2015-2017
p. 308, 309	p. 308, 309	p. 308, 309	INSTRUCTIONS	p. 308, 309	p. 308, 309	p. 308, 309
111 ft.lbs.	111 ft.lbs.	111 ft.lbs.	LUGNUT TORQUE	111 ft.lbs.	111 ft.lbs.	111 ft.lbs.
			FRONT			
8.09 ± 0.5 8 1/16 ± ½	8.21 ± 0.5 8 3/16 ± ½	9.5 ± 0.5 9½ ± ½	CASTER (Degrees)	9.37 ± 0.5 9⅜ ± ½	9.26 ± 0.5 9¼ ± ½	9.27 ± 0.5 9¼ ± ½
0.0 (0) | 0.5 (½)	0.0 (0) | 0.5 (½)	0.0 (0) | 0.5 (½)	CROSS | TOL.	0.0 (0) | 0.5 (½)	0.0 (0) | 0.5 (½)	0.0 (0) | 0.5 (½)
-1.09 ± 0.33 -1 1/16 ± 5/16	-1.32 ± 0.33 -1 5/16 ± 5/16	-0.78 +0.26 -0.4 -¾ +¼ -⅜	CAMBER (Degrees)	-0.78 +0.26 -0.4 -¾ +¼ -⅜	-0.82 +0.26 -0.4 -13/16 +¼ -⅜	-0.5 +0.25 -0.45 -½ +¼ -7/16
0.0 (0) | 0.33 (5/16)	0.0 (0) | 0.33 (5/16)	0.0 (0) | 0.33 (5/16)	CROSS | TOL.	0.0 (0) | 0.33 (5/16)	0.0 (0) | 0.33 (5/16)	0.0 (0) | 0.4 (⅜)
0.24° ± 0.1° ¼° ± 3/32°	0.24° ± 0.1° ¼° ± 3/32°	0.24° ± 0.17° ¼° ± 5/32°	FRONT TOE (Degrees)	0.24° ± 0.17° ¼° ± 5/32°	0.24° ± 0.17° ¼° ± 5/32°	0.24° ± 0.17° ¼° ± 5/32°
0.12" ± 0.05" ⅛" ± 1/16"	0.12" ± 0.05" ⅛" ± 1/16"	0.12" ± 0.08" ⅛" ± 3/32"	FRONT TOE (Inches)	0.12" ± 0.08" ⅛" ± 3/32"	0.12" ± 0.08" ⅛" ± 3/32"	0.13" ± 0.08" ⅛" ± 3/32"
			SAI			
20.0 / 18.4 ± 0.5	20.0 / 18.3 ± 0.5	20.0 / 18.3 ± 0.5	TURNING ANGLE	20.0 / 18.3 ± 0.5		20.0 / 18.5 ± 0.5
164 165	164 165	198 199	ILLUSTRATION #	198 199	198 199	
			REAR			
-1.48 ± 0.5 -1½ ± ½	-1.73 ± 0.5 -1¾ ± ½	-1.99 ± 0.5 -2 ± ½	CAMBER (Degrees)	-1.99 ± 0.5 -2 ± ½	-1.87 ± 0.5 -1 ⅞ ± ½	-1.99 ± 0.5 -2 ± ½
0.44° ± 0.25° 7/16° ± ¼°	0.85° ± 0.25° 25/32° ± ¼°	0.52° ± 0.13° 17/32° ± ⅛°	REAR TOE (Degrees)	0.52° ± 0.13° 17/32° ± ⅛°	0.51° ± 0.13° ½° ± ⅛°	0.43° ± 0.12° 13/32° ± ⅛°
0.22" ± 0.13" 7/32" ± ⅛"	0.43" ± 0.13" 7/16" ± ⅛"	0.26" ± 0.06" ¼" ± 1/16"	REAR TOE (Inches)	0.26" ± 0.06" ¼" ± 1/16"	0.25" ± 0.06" ¼" ± 1/16"	0.22" ± 0.06" 7/32" ± 1/16"
0.25 (¼)	0.25 (¼)	0.25 (¼)	THRUST ANGLE	0.25 (¼)	0.25 (¼)	0.25 (¼)
174	174	174	ILLUSTRATION #	174	174	174
			NOTES			
29141	29142	29144		29145	29146	29148

Mercedes-Benz

MODEL	S-Class Coupe (217.*)	S-Class Coupe (217.*)	S-Class Coupe (217.*)	SL-Class Roadster (129.*)	SL-Class Roadster (129.*)	SL-Class Roadster (230.*)
INFORMATION	S 63 AMG 4MATIC exc/California Models	S 63 AMG 4MATIC California Models	S 65 AMG	300SL, 500SL, 600SL	SL320, SL500, SL600	ABC
YEARS	2015-2017	2015-2017	2015-2017	1990-93	1994-2002	2003-2012
INSTRUCTIONS	p. 308, 309	p. 308, 309	p. 308, 309	p. 308, 309	p. 308, 309	p. 308, 309
LUGNUT TORQUE	111 ft.lbs.	111 ft.lbs.	111 ft.lbs.	81 ft.lbs.	81 ft.lbs.	96 ft.lbs.
FRONT						
CASTER (Degrees)	9.13 ±0.5 9⅛ ± ½	8.82 ±0.5 8 13/16 ± ½	9.62 ±0.5 9⅝ ± ½	9.93 ±0.5 [1] 9 15/16 ± ½ [1]	9.93 ±0.5 [1] 9 15/16 ± ½ [1]	11.85 ±0.5 11 ⅞ ± ½
CROSS | TOL.	0.0 (0) | 0.5 (½)	0.0 (0) | 0.5 (½)	0.0 (0) | 0.5 (½)	0.0 (0) | 0.5 (½)	0.0 (0) | 0.5 (½)	0.0 (0) | 0.5 (½)
CAMBER (Degrees)	-1.82 +0.25 -0.45 -1 13/16 +¼ -7/16	-1.6 +0.25 -0.45 -1 ⅝ +¼ -7/16	-0.5 +0.25 -0.45 -½ +¼ -7/16	-0.83 +0.17 -0.33 [2] -13/16 +3/16 -5/16 [2]	-0.83 +0.17 -0.33 [2] -13/16 +3/16 -5/16 [2]	-1.34 +0.24 -0.42 -1 5/16 +¼ -7/16
CROSS | TOL.	0.0 (0) | 0.4 (⅜)	0.0 (0) | 0.4 (⅜)	0.0 (0) | 0.4 (⅜)	0.0 (0) | 0.33 (5/16)	0.0 (0) | 0.33 (5/16)	0.0 (0) | 0.33 (5/16)
FRONT TOE (Degrees)	0.03° ±0.17° 1/32° ± 5/32°	0.24° ±0.17° ¼° ± 5/32°	0.24° ±0.17° ¼° ± 5/32°	0.33° ±0.17° 11/32° ± 5/32°	0.33° ±0.17° 11/32° ± 5/32°	0.2° ±0.17° 3/16° ± 5/32°
FRONT TOE (Inches)	0.02" ±0.08" 1/64" ± 3/32"	0.13" ±0.08" ⅛" ± 3/32"	0.13" ±0.08" ⅛" ± 3/32"	0.17" ±0.08" 5/32" ± 3/32"	0.17" ±0.08" 5/32" ± 3/32"	0.1" ±0.08" 3/32" ± 3/32"
SAI						
TURNING ANGLE	20.0 / 18.9 ±0.5	20.0 / 18.9 ±0.5	20.0 / 18.5 ±0.5	40.0 / 32.5 ±2.0	40.0 / 32.5 ±2.0	20.0 / 18.1 ±0.5
ILLUSTRATION #				88	88	164 165
REAR						
CAMBER (Degrees)	-2.27 ±0.5 -2 ¼ ± ½	-1.98 ±0.5 -2 ± ½	-1.99 ±0.5 -2 ± ½			-1.66 ±0.5 -1 11/16 ± ½
REAR TOE (Degrees)	0.42° ±0.12° 13/32° ± ⅛°	0.42° ±0.12° 13/32° ± ⅛°	0.43° ±0.12° 13/32° ± ⅛°	0.42° +0.17° -0.08° 13/32° +5/32° -3/32°	0.42° +0.17° -0.08° 13/32° +5/32° -3/32°	0.2° ±0.25° 3/16° ± ¼°
REAR TOE (Inches)	0.21" ±0.06" 7/32" ± 1/16"	0.21" ±0.06" 7/32" ± 1/16"	0.22" ±0.06" 7/32" ± 1/16"	0.21" +0.08" -0.04" 7/32" +3/32" -1/32"	0.21" +0.08" -0.04" 7/32" +3/32" -1/32"	0.1" ±0.13" 3/32" ± ⅛"
THRUST ANGLE	0.25 (¼)	0.25 (¼)	0.25 (¼)			0.25 (¼)
ILLUSTRATION #	174	174	174	86	86	06
NOTES				[1] w/Level Control 10.0º(10º) [2] w/Level Control -1.0º (-1º)	[1] w/Level Control 10.0º(10º) [2] w/Level Control -1.0º (-1º)	
	29149	29150	29151	29162	29163	29165

Mercedes-Benz

MODEL	SL-Class Roadster (230.*)	SL-Class Roadster (230.*)	SL-Class Roadster (230.*)	SL-Class Roadster (231.*)	SL-Class Roadster (231.*)	SL-Class Roadster (231.*)
INFORMATION	ABC Level 1	ABC Level 2	Black Series	Std. Susp.	Sport Susp.	All w/ABC
YEARS	2003-2012	2003-2012	2009	2013-2017	2013-2017	2013-2017
INSTRUCTIONS	p. 308, 309	p. 308, 309	p. 308, 309	p. 308, 309	p. 308, 309	p. 308, 309
LUGNUT TORQUE	96 ft.lbs.	96 ft.lbs.	96 ft.lbs.	96 ft.lbs.	96 ft.lbs.	96 ft.lbs.
FRONT						
CASTER (Degrees)	11.35 ±0.5 11⅜ ± ½	11.02 ±0.5 11 ± ½	11.67 ±0.5 11 11/16 ± ½	11.44 ±0.5 11 7/16 ± ½	11.82 ±0.5 11 13/16 ± ½	11.83 ±0.5 11 13/16 ± ½
CROSS | TOL.	0.0 (0) | 0.5 (½)	0.0 (0) | 0.5 (½)	0.0 (0) | 0.5 (½)	0.0 (0) | 0.5 (½)	0.0 (0) | 0.5 (½)	0.0 (0) | 0.5 (½)
CAMBER (Degrees)	-0.96 +0.24 -0.42 -15/16 +¼ -7/16	-0.71 +0.24 -0.42 -11/16 +¼ -7/16	L=-1.34, R=-1.69 +0.24 -0.42 L=-1 5/16, R=-1 11/16 +¼ -7/16	-0.95 +0.24 -0.42 -15/16 +¼ -7/16	-1.23 +0.24 -0.42 -1 ¼ +¼ -7/16	-1.2 +0.24 -0.42 -1 3/16 +¼ -7/16
CROSS | TOL.	0.0 (0) | 0.33 (5/16)	0.0 (0) | 0.33 (5/16)	0.35 (⅜) | 0.33 (5/16)	0.0 (0) | 0.33 (5/16)	0.0 (0) | 0.33 (5/16)	0.0 (0) | 0.33 (5/16)
FRONT TOE (Degrees)	0.2° ±0.17° 3/16° ± 5/32°	0.2° ±0.17° 3/16° ± 5/32°	0.25° ±0.17° ¼° ± 5/32°	0.25° ±0.17° ¼° ± 5/32°	0.25° ±0.17° ¼° ± 5/32°	0.2° ±0.17° 3/16° ± 5/32°
FRONT TOE (Inches)	0.1" ±0.08" 3/32" ± 3/32"	0.1" ±0.08" 3/32" ± 3/32"	0.12" ±0.08" ⅛" ± 3/32"	0.13" ±0.08" ⅛" ± 3/32"	0.12" ±0.08" ⅛" ± 3/32"	0.1" ±0.08" 3/32" ± 3/32"
SAI						
TURNING ANGLE	20.0 / 18.1 ±0.5	20.0 / 18.1 ±0.5	20.0 / 18.1 ±0.5	20.0 / 18.1 ±0.5	20.0 / 18.1 ±0.5	20.0 / 18.1 ±0.5
ILLUSTRATION #	164 165	164 165	139	164	164	164
REAR						
CAMBER (Degrees)	-1.32 ±0.5 -1 5/16 ± ½	-1.11 ±0.5 -1 ⅛ ± ½	-1.44 ±0.5 -1 7/16 ± ½	-1.79 ±0.5 -1 13/16 ± ½	-2.05 ±0.5 -2 1/16 ± ½	-2.07 ±0.5 -2 1/16 ± ½
REAR TOE (Degrees)	0.2° ±0.13° 3/16° ± ⅛°	0.2° ±0.13° 3/16° ± ⅛°	0.3° ±0.13° 5/16° ± ⅛°	0.32° ±0.13° 5/16° ± ⅛°	0.39° ±0.13° 13/32° ± ⅛°	0.4° ±0.13° 13/32° ± ⅛°
REAR TOE (Inches)	0.1" ±0.06" 3/32" ± 1/16"	0.1" ±0.06" 3/32" ± 1/16"	0.15" ±0.06" 5/32" ± 1/16"	0.16" ±0.06" 5/32" ± 1/16"	0.2" ±0.06" 3/16" ± 1/16"	0.2" ±0.06" 3/16" ± 1/16"
THRUST ANGLE	0.25 (¼)	0.25 (¼)	0.25 (¼)	0.25 (¼)	0.25 (¼)	0.25 (¼)
ILLUSTRATION #	06	06	135	174	174	174
NOTES						
	29166	29167	29168	29170	29171	29172

Mercedes-Benz

MODEL INFORMATION	SL-Class Roadster (231.*) AMG Models	SLK230 SLK230 Kompressor (up to 02/2000)	SLK230, SLK320 SLK32 AMG (03/2000 up)	SLK280, SLK300, SLK350 Std. Susp.	SLK280, SLK300, SLK350 Sport Susp. SLK55 AMG	SLK Class (172.*) w/Std. Susp.
YEARS	2013-2017	1998-2000	2000-2004	2005-2011	2005-2011	2012-2016
INSTRUCTIONS	p. 308, 309	p. 308, 309	p. 308, 309	p. 308, 309	p. 308, 309	p. 308, 309
LUGNUT TORQUE	96 ft.lbs.	81 ft.lbs.	81 ft.lbs.	81 ft.lbs.	81 ft.lbs.	
FRONT						
CASTER (Degrees)	11.8 ± 0.5 11 13/16 ± ½	5.23 ± 0.5 5¼ ± ½	5.27 ± 0.5 5¼ ± ½	10.69 ± 0.5 10 11/16 ± ½	11.0 ± 0.5 11 ± ½	10.8 ± 0.5 10 13/16 ± ½
CROSS | TOL.	0.0 (0) | 0.5 (½)	0.0 (0) | 0.5 (½)	0.0 (0) | 0.5 (½)	0.0 (0) | 0.5 (½)	0.0 (0) | 0.5 (½)	0.0 (0) | 0.5 (½)
CAMBER (Degrees)	-1.89 +0.24 -0.42 -1 ⅞ +¼ -7/16	-0.97 ± 0.33 -1 ± 5/16	-1.15 ± 0.33 -1⅛ ± 5/16	-1.48 +0.32 -0.4 -1½ +5/16 -⅜	-1.5 +0.32 -0.4 -1½ +5/16 -⅜	-1.44 ± 0.4 -1 7/16 ± ⅜
CROSS | TOL.	0.0 (0) | 0.33 (5/16)	0.0 (0) | 0.33 (5/16)	0.0 (0) | 0.33 (5/16)	0.0 (0) | 0.33 (5/16)	0.0 (0) | 0.33 (5/16)	0.0 (0) | 0.4 (⅜)
FRONT TOE (Degrees)	0.17° ± 0.17° 5/32° ± 5/32°	0.33° ± 0.17° 11/32° ± 5/32°	0.33° ± 0.17° 11/32° ± 5/32°	0.2° ± 0.17° 3/16° ± 5/32°	0.2° ± 0.17° 3/16° ± 5/32°	0.17° ± 0.17° 5/32° ± 5/32°
FRONT TOE (Inches)	0.09" ± 0.08" 3/32" ± 3/32"	0.17" ± 0.08" 5/32" ± 3/32"	0.17" ± 0.08" 5/32" ± 3/32"	0.1" ± 0.08" 3/32" ± 3/32"	0.1" ± 0.08" 3/32" ± 3/32"	0.08" ± 0.08" 3/32" ± 3/32"
SAI						
TURNING ANGLE	20.0 / 18.3 ±0.5	20.0 / 18.6 ±0.5	20.0 / 18.6 ±0.5	20.0 / 18.1 ±0.5	20.0 / 18.1 ±0.5	20.0 / 18.1 ±0.5
ILLUSTRATION #	164			164 165	164 165	198 199
REAR						
CAMBER (Degrees)	-2.2 ± 0.5 -2 3/16 ± ½	-1.44 ± 0.5 -1 7/16 ± ½	-1.49 ± 0.5 -1½ ± ½	-1.12 ± 0.5 -1⅛ ± ½	-1.34 ± 0.5 -1 5/16 ± ½	-1.22 ± 0.5 -1¼ ± ½
REAR TOE (Degrees)	0.24° ± 0.13° ¼° ± ⅛°	0.55° ± 0.25° 9/16° ± ¼°	0.55° ± 0.25° 9/16° ± ¼°	0.55° ± 0.25° 9/16° ± ¼°	0.55° ± 0.25° 9/16° ± ¼°	0.57° ± 0.12° 9/16° ± ⅛°
REAR TOE (Inches)	0.12" ± 0.06" ⅛" ± 1/16"	0.28" ± 0.13" 9/32" ± ⅛"	0.28" ± 0.13" 9/32" ± ⅛"	0.28" ± 0.13" 9/32" ± ⅛"	0.28" ± 0.13" 9/32" ± ⅛"	0.28" ± 0.06" 9/32" ± 1/16"
THRUST ANGLE	0.05 (1/16)	0.25 (¼)	0.25 (¼)	0.25 (¼)	0.25 (¼)	0.25 (¼)
ILLUSTRATION #	174	86	86	86	86	174
NOTES						
	29173	29175	29176	29178	29179	29181

Mercedes-Benz

MODEL INFORMATION	SLK Class (172.*) w/Sport Susp.	SLK Class (172.*) AMG	SLC Class Roadster (172.4*) w/Std. Susp.	SLC Class Roadster (172.4*) w/Sport Susp.	SLC Class Roadster (172.4*) AMG Models (172.466)	SLC Class Roadster (172.4*) AMG Models (172.475)
YEARS	2012-2016	2012-2016	2017	2017	2017	2017
INSTRUCTIONS	p. 308, 309	p. 308, 309				
LUGNUT TORQUE						
FRONT						
CASTER (Degrees)	11.11 ± 0.5 11⅛ ± ½	11.09 ± 0.5 11⅛ ± ½	10.8 ± 0.5 10 13/16 ± ½	11.1 ± 0.5 11⅛ ± ½	11.08 ± 0.5 11 1/16 ± ½	11.08 ± 0.5 11 1/16 ± ½
CROSS | TOL.	0.0 (0) | 0.5 (½)	0.0 (0) | 0.5 (½)	0.0 (0) | 0.5 (½)	0.0 (0) | 0.5 (½)	0.0 (0) | 0.5 (½)	0.0 (0) | 0.5 (½)
CAMBER (Degrees)	-1.48 ± 0.4 -1½ ± ⅜	-1.46 ± 0.4 -1 7/16 ± ⅜	-1.44 +0.25 -0.4 -1 7/16 +¼ -⅜	-1.48 +0.25 -0.4 -1½ +¼ -⅜	-1.93 ± 0.4 -1 15/16 ± ⅜	-1.45 ± 0.4 -1 7/16 ± ⅜
CROSS | TOL.	0.0 (0) | 0.4 (⅜)	0.0 (0) | 0.4 (⅜)	0.0 (0) | 0.4 (⅜)	0.0 (0) | 0.4 (⅜)	0.0 (0) | 0.4 (⅜)	0.0 (0) | 0.4 (⅜)
FRONT TOE (Degrees)	0.17° ± 0.17° 5/32° ± 5/32°	0.22° ± 0.17° 7/32° ± 5/32°	0.17° ± 0.17° 5/32° ± 5/32°	0.17° ± 0.17° 5/32° ± 5/32°	0.28° ± 0.17° 9/32° ± 5/32°	0.22° ± 0.17° 7/32° ± 5/32°
FRONT TOE (Inches)	0.08" ± 0.08" 3/32" ± 3/32"	0.11" ± 0.08" ⅛" ± 3/32"	0.08" ± 0.08" 3/32" ± 3/32"	0.08" ± 0.08" 3/32" ± 3/32"	0.14" ± 0.08" 5/32" ± 3/32"	0.11" ± 0.08" 3/32" ± 3/32"
SAI						
TURNING ANGLE	20.0 / 18.1 ±0.5	20.0 / 18.3 ±0.5	20.0 / 18.1 ±0.5	20.0 / 18.1 ±0.5	20.0 / 18.7 ±0.5	20.0 / 18.3 ±0.5
ILLUSTRATION #	198 199	198 199	198 199	198 199	198 199	198 199
REAR						
CAMBER (Degrees)	-1.4 ± 0.5 -1⅜ ± ½	-1.72 ± 0.5 -1¾ ± ½	-1.22 ± 0.5 -1 3/16 ± ½	-1.4 ± 0.5 -1⅜ ± ½	-2.0 ± 0.5 -2 ± ½	-1.72 ± 0.5 -1¾ ± ½
REAR TOE (Degrees)	0.56° ± 0.12° 9/16° ± ⅛°	0.55° ± 0.12° 9/16° ± ⅛°	0.57° ± 0.12° 9/16° ± ⅛°	0.56° ± 0.12° 9/16° ± ⅛°	0.55° ± 0.12° 9/16° ± ⅛°	0.55° ± 0.12° 9/16° ± ⅛°
REAR TOE (Inches)	0.28" ± 0.06" 9/32" ± 1/16"	0.28" ± 0.06" 9/32" ± 1/16"	0.28" ± 0.06" 9/32" ± 1/16"	0.28" ± 0.06" 9/32" ± 1/16"	0.28" ± 0.06" 9/32" ± 1/16"	0.28" ± 0.06" 9/32" ± 1/16"
THRUST ANGLE	0.25 (¼)	0.25 (¼)	0.25 (¼)	0.25 (¼)	0.25 (¼)	0.25 (¼)
ILLUSTRATION #	174	174	174	174	174	174
NOTES						
	29182	29183	29185	29186	29187	29188

Mercedes-Benz

M-Class SUV (163.*)	M-Class SUV (163.*)	ML-Class SUV (164.*)	MODEL	ML-Class SUV (164.*)	ML-Class SUV (164.*)	ML-Class SUV (166.*)
All exc/ML55	ML55 (incl. AMG)	RWD Models	INFORMATION	GL-Class SUV (x164.*)	GL-Class SUV (x164.*)	Steel Suspension
				except Air Susp.	with Air Susp.	
1998-2005	2000-2003	2009-2011	YEARS	2006-2011	2006-2011	2012-2015
p. 308, 309	p. 308, 309	p. 308, 309	INSTRUCTIONS	p. 308, 309	p. 308, 309	p. 308, 309
111 ft.lbs.	111 ft.lbs.	111 ft.lbs.	LUGNUT TORQUE	111 ft.lbs.	111 ft.lbs.	110 ft.lbs.
			FRONT			
6.1 ±0.5 6 1/8 ± 1/2	6.55 ±0.5 6 9/16 ± 1/2	6.06 ±0.5 6 1/16 ± 1/2	CASTER (Degrees)	5.73 ±0.5 5 3/4 ± 1/2	6.56 ±0.5 6 9/16 ± 1/2	5.47 ±0.5 5 7/16 ± 1/2
0.0 (0) | 0.5 (1/2)	0.0 (0) | 0.5 (1/2)	0.0 (0) | 0.5 (1/2)	CROSS | TOL.	0.0 (0) | 0.5 (1/2)	0.0 (0) | 0.5 (1/2)	0.0 (0) | 0.5 (1/2)
-0.63 ±0.33 -5/8 ± 5/16	-0.46 ±0.33 -7/16 ± 5/16	-0.92 ±0.5 -15/16 ± 1/2	CAMBER (Degrees)	-0.52 ±0.5 -1/2 ± 1/2	-1.38 ±0.5 -1 3/8 ± 1/2	-0.72 ±0.5 -11/16 ± 1/2
0.0 (0) | 0.33 (5/16)	0.0 (0) | 0.33 (5/16)	0.0 (0) | 0.33 (5/16)	CROSS | TOL.	0.0 (0) | 0.33 (5/16)	0.0 (0) | 0.33 (5/16)	0.0 (0) | 0.32 (5/16)
0.32° ±0.16° 5/16° ± 5/32°	0.37° ±0.16° 3/8° ± 5/32°	0.38° ±0.13° 3/8° ± 1/8°	FRONT TOE (Degrees)	0.42° ±0.13° 13/32° ± 1/8°	0.24° ±0.13° 1/4° ± 1/8°	0.42° ±0.12° 7/16° ± 1/8°
0.16" ±0.08" 5/32" ± 3/32"	0.19" ±0.08" 3/16" ± 3/32"	0.19" ±0.06" 3/16" ± 1/16"	FRONT TOE (Inches)	0.21" ±0.06" 7/32" ± 1/16"	0.12" ±0.06" 1/8" ± 1/16"	0.21" ±0.06" 7/32" ± 1/16"
			SAI			
20.0 / 17.8 ±0.15	20.0 / 17.8 ±0.15	20.0 / 18.4 ±0.5	TURNING ANGLE	20.0 / 18.4 ±0.5	20.0 / 18.1 ±0.5	20.0 / 18.4 ±0.5
			ILLUSTRATION #			
			REAR			
-0.61 ±0.33 -5/8 ± 5/16	-0.53 ±0.33 -1/2 ± 5/16	-1.18 ±0.5 -1 3/16 ± 1/2	CAMBER (Degrees)	-1.18 ±0.5 -1 3/16 ± 1/2	-2.07 ±0.5 -2 1/16 ± 1/2	-1.32 ±0.5 -1 5/16 ± 1/2
0.35° ±0.25° 11/32° ± 1/4°	0.35° ±0.25° 11/32° ± 1/4°	0.4° ±0.13° 13/32° ± 1/8°	REAR TOE (Degrees)	0.4° ±0.13° 13/32° ± 1/8°	0.52° ±0.13° 17/32° ± 1/8°	0.42° ±0.12° 13/32° ± 1/8°
0.17" ±0.13" 3/16" ± 1/8"	0.18" ±0.13" 3/16" ± 1/8"	0.2" ±0.06" 3/16" ± 1/16"	REAR TOE (Inches)	0.2" ±0.06" 3/16" ± 1/16"	0.26" ±0.06" 1/4" ± 1/16"	0.21" ±0.06" 7/32" ± 1/16"
0.25 (1/4)	0.25 (1/4)	0.25 (1/4)	THRUST ANGLE	0.25 (1/4)	0.25 (1/4)	0.25 (1/4)
		151	ILLUSTRATION #	151	151	151
			NOTES			
29190	29191	29193		29194	29195	29197

Mercedes-Benz

ML-Class SUV (166.*)	GL-Class SUV (x166.8)	GLE-Class SUV (x166.*)	MODEL	GLE-Class SUV (x166.*)	GLE-Class SUV (x166.*)	GLS-Class SUV (x166.8*)
Air Suspension	All Models	Steel Susp.	INFORMATION	AIR Susp.	AMG Models	exc/AMG Model
incl. AMG						
2012-2015	2013-2016	2016-2017	YEARS	2016-2017	2016-2017	2016-2017
p. 308, 309	p. 308, 309	p. 308, 309	INSTRUCTIONS	p. 308, 309	p. 308, 309	
110 ft.lbs.	110 ft.lbs.	110 ft.lbs.	LUGNUT TORQUE	110 ft.lbs.	110 ft.lbs.	110 ft.lbs.
			FRONT			
6.1 ±0.5 6 1/8 ± 1/2	6.1 ±0.5 6 1/8 ± 1/2	5.47 ±0.5 5 7/16 ± 1/2	CASTER (Degrees)	6.1 ±0.5 6 1/8 ± 1/2	6.1 ±0.5 6 1/8 ± 1/2	6.1 ±0.5 6 1/8 ± 1/2
0.0 (0) | 0.5 (1/2)	0.0 (0) | 0.5 (1/2)	0.0 (0) | 0.5 (1/2)	CROSS | TOL.	0.0 (0) | 0.5 (1/2)	0.0 (0) | 0.5 (1/2)	0.0 (0) | 0.5 (1/2)
-1.4 ±0.5 -1 3/8 ± 1/2	-1.4 ±0.5 -1 3/8 ± 1/2	-0.72 ±0.5 -11/16 ± 1/2	CAMBER (Degrees)	-1.4 ±0.5 -1 3/8 ± 1/2	-1.4 ±0.5 -1 3/8 ± 1/2	-1.4 ±0.5 -1 3/8 ± 1/2
0.0 (0) | 0.32 (5/16)	0.0 (0) | 0.32 (5/16)	0.0 (0) | 0.32 (5/16)	CROSS | TOL.	0.0 (0) | 0.32 (5/16)	0.0 (0) | 0.32 (5/16)	0.0 (0) | 0.32 (5/16)
0.23° ±0.12° 1/4° ± 1/8°	0.23° ±0.12° 1/4° ± 1/8°	0.42° ±0.12° 7/16° ± 1/8°	FRONT TOE (Degrees)	0.23° ±0.12° 1/4° ± 1/8°	0.23° ±0.12° 1/4° ± 1/8°	0.23° ±0.12° 1/4° ± 1/8°
0.12" ±0.06" 1/8" ± 1/16"	0.12" ±0.06" 1/8" ± 1/16"	0.21" ±0.06" 7/32" ± 1/16"	FRONT TOE (Inches)	0.12" ±0.06" 1/8" ± 1/16"	0.12" ±0.06" 1/8" ± 1/16"	0.12" ±0.06" 1/8" ± 1/16"
			SAI			
20.0 / 18.1 ±0.5	20.0 / 18.1 ±0.5	20.0 / 18.4 ±0.5	TURNING ANGLE	20.0 / 18.1 ±0.5	20.0 / 18.1 ±0.5	20.0 / 18.1 ±0.5
			ILLUSTRATION #			
			REAR			
-2.13 ±0.5 -2 1/8 ± 1/2	-2.13 ±0.5 -2 1/8 ± 1/2	-1.32 ±0.5 -1 5/16 ± 1/2	CAMBER (Degrees)	-2.13 ±0.5 -2 1/8 ± 1/2	-2.13 ±0.5 -2 1/8 ± 1/2	-2.13 ±0.5 -2 1/8 ± 1/2
0.53° ±0.12° 17/32° ± 1/8°	0.53° ±0.12° 17/32° ± 1/8°	0.42° ±0.12° 13/32° ± 1/8°	REAR TOE (Degrees)	0.53° ±0.12° 17/32° ± 1/8°	0.53° ±0.12° 17/32° ± 1/8°	0.53° ±0.12° 17/32° ± 1/8°
0.27" ±0.06" 9/32" ± 1/16"	0.27" ±0.06" 9/32" ± 1/16"	0.21" ±0.06" 7/32" ± 1/16"	REAR TOE (Inches)	0.27" ±0.06" 9/32" ± 1/16"	0.27" ±0.06" 9/32" ± 1/16"	0.27" ±0.06" 9/32" ± 1/16"
0.25 (1/4)	0.25 (1/4)	0.25 (1/4)	THRUST ANGLE	0.25 (1/4)	0.25 (1/4)	0.25 (1/4)
151	151	151	ILLUSTRATION #	151	151	151 151
			NOTES			
29198	29199	29200		29201	29202	29203

Mercedes-Benz

GLS-Class SUV (x166.8*)	GLE-Class Coupe (292.*)	GLE-Class Coupe (292.*)	MODEL	R-Class SUV (251.*)	R-Class SUV (251.*)	R-Class SUV (251.*)
AMG Models	Steel Susp.	AIR Susp.	INFORMATION	Coil Springs F&R	Coil Front/Air Rear	Air Susp. F&R
						exc/AMG
2016-2017	2016-2017	2016-2017	YEARS	2006-2012	2006-2012	2006-2012
	p. 308, 309	p. 308, 309	INSTRUCTIONS	p. 308, 309	p. 308, 309	p. 308, 309
110 ft.lbs.	110 ft.lbs.	110 ft.lbs.	LUGNUT TORQUE	111 ft.lbs.	111 ft.lbs.	111 ft.lbs.
			FRONT			
6.1 ±0.5 6⅛ ± ½	5.82 ±0.5 5 13/16 ± ½	6.1 ±0.5 6⅛ ± ½	CASTER (Degrees)	5.36 ±0.5 5⅜ ± ½	5.64 ±0.5 5⅝ ± ½	5.9 ±0.5 5⅞ ± ½
0.0 (0) | 0.5 (½)	0.0 (0) | 0.5 (½)	0.0 (0) | 0.5 (½)	CROSS | TOL.	0.0 (0) | 0.5 (½)	0.0 (0) | 0.5 (½)	0.0 (0) | 0.5 (½)
-1.4 ±0.5 -1⅜ ± ½	-1.08 ±0.5 -1 1/16 ± ½	-1.4 ±0.5 -1⅜ ± ½	CAMBER (Degrees)	-0.67 ±0.5 -11/16 ± ½	-0.46 ±0.5 -7/16 ± ½	-1.0 ±0.5 -1 ± ½
0.0 (0) | 0.32 (5/16)	0.0 (0) | 0.32 (5/16)	0.0 (0) | 0.32 (5/16)	CROSS | TOL.	0.0 (0) | 0.33 (5/16)	0.0 (0) | 0.33 (5/16)	0.0 (0) | 0.33 (5/16)
0.23° ±0.12° ¼° ± ⅛°	0.42° ±0.12° 7/16° ± ⅛°	0.23° ±0.12° ¼° ± ⅛°	FRONT TOE (Degrees)	0.47° ±0.13° 15/32° ± ⅛°	0.66° ±0.13° 21/32° ± ⅛°	0.3° ±0.13° 5/16° ± ⅛°
0.12" ±0.06" ⅛" ± 1/16"	0.21" ±0.06" 7/32" ± 1/16"	0.12" ±0.06" ⅛" ± 1/16"	FRONT TOE (Inches)	0.24" ±0.06" ¼" ± 1/16"	0.33" ±0.06" 5/16" ± 1/16"	0.15" ±0.06" 5/32" ± 1/16"
			SAI			
20.0 / 18.1 ±0.5	20.0 / 18.4 ±0.5	20.0 / 18.1 ±0.5	TURNING ANGLE	20.0 / 18.6 ±0.5	20.0 / 18.7 ±0.5	20.0 / 18.4 ±0.5
			ILLUSTRATION #			
			REAR			
-2.13 ±0.5 -2⅛ ± ½	-2.13 ±0.5 -2⅛ ± ½	-1.75 ±0.5 -1¾ ± ½	CAMBER (Degrees)	-0.02 ±0.5 0 ± ½	-0.66 ±0.5 -11/16 ± ½	-0.79 ±0.5 -13/16 ± ½
0.53° ±0.12° 17/32° ± ⅛°	0.48° ±0.12° ½° ± ⅛°	0.53° ±0.12° 17/32° ± ⅛°	REAR TOE (Degrees)	0.31° ±0.13° 5/16° ± ⅛°	0.38° ±0.13° ⅜° ± ⅛°	0.4° ±0.13° 13/32° ± ⅛°
0.27" ±0.06" 9/32" ± 1/16"	0.24" ±0.06" ¼" ± 1/16"	0.27" ±0.06" 9/32" ± 1/16"	REAR TOE (Inches)	0.16" ±0.06" 5/32" ± 1/16"	0.19" ±0.06" 3/16" ± 1/16"	0.2" ±0.06" 3/16" ± 1/16"
0.25 (¼)	0.25 (¼)	0.25 (¼)	THRUST ANGLE	0.25 (¼)	0.25 (¼)	0.25 (¼)
151 151	151	151	ILLUSTRATION #			
			NOTES			
29204	29206	29207		29209	29210	29211

Mercedes-Benz

R-Class SUV (251.*)	CLS-Class Coupe (219.*)	CLS-Class Coupe (219.*)	MODEL	CLS-Class Coupe (219.*)	CLS-Class Coupe (218.*)	CLS-Class Coupe (218.*)
Air Susp. F&R	Coil Suspension	Air Susp. exc/AMG	INFORMATION	AMG models	RWD Models exc/AMG	RWD Models exc/AMG
AMG					with Air Susp.	with Steel Susp.
2006-2012	2006-2011	2006-2011	YEARS	2006-2011	2012-2017	2012-2017
p. 308, 309	p. 308, 309	p. 308, 309	INSTRUCTIONS	p. 308, 309	p. 308, 309	p. 308, 309
111 ft.lbs.	96 ft.lbs.	96 ft.lbs.	LUGNUT TORQUE	96 ft.lbs.	96 ft.lbs.	96 ft.lbs.
			FRONT			
5.97 ±0.5 6 ± ½	11.53 ±0.5 11 9/16 ± ½	11.3 ±0.5 11 5/16 ± ½	CASTER (Degrees)	11.6 ±0.5 11⅝ ± ½	7.31 ±0.5 7 5/16 ± ½	7.29 ±0.5 7 5/16 ± ½
0.0 (0) | 0.5 (½)	0.0 (0) | 0.5 (½)	0.0 (0) | 0.5 (½)	CROSS | TOL.	0.0 (0) | 0.5 (½)	0.0 (0) | 0.5 (½)	0.0 (0) | 0.5 (½)
-1.06 ±0.5 -1 1/16 ± ½	-1.17 +0.17 -0.5 -1 3/16 +3/16 -½	-0.99 +0.17 -0.5 -1 +3/16 -½	CAMBER (Degrees)	-1.22 +0.17 -0.5 -1¼ +3/16 -½	-0.8 ±0.4 -13/16 ± ⅜	-0.8 ±0.4 -13/16 ± ⅜
0.0 (0) | 0.33 (5/16)	0.0 (0) | 0.33 (5/16)	0.0 (0) | 0.33 (5/16)	CROSS | TOL.	0.0 (0) | 0.33 (5/16)	0.0 (0) | 0.4 (⅜)	0.0 (0) | 0.4 (⅜)
0.28° ±0.13° 9/32° ± ⅛°	0.26° ±0.17° ¼° ± 5/32°	0.26° ±0.17° ¼° ± 5/32°	FRONT TOE (Degrees)	0.26° ±0.17° ¼° ± 5/32°	0.17° ±0.17° 5/32° ± 5/32°	0.17° ±0.17° 5/32° ± 5/32°
0.14" ±0.06" ⅛" ± 1/16"	0.13" ±0.08" ⅛" ± 3/32"	0.13" ±0.08" ⅛" ± 3/32"	FRONT TOE (Inches)	0.13" ±0.08" ⅛" ± 3/32"	0.08" ±0.08" 3/32" ± 3/32"	0.08" ±0.08" 3/32" ± 3/32"
			SAI			
20.0 / 18.4 ±0.5	20.0 / 18.2 ±0.5	20.0 / 18.2 ±0.5	TURNING ANGLE	20.0 / 18.2 ±0.5	20.0 / 18.1 ±0.5	20.0 / 18.1 ±0.5
	164 165	164 165	ILLUSTRATION #	164 165	198 199	198 199
			REAR			
-0.9 ±0.5 -⅞ ± ½	-1.39 ±0.5 -1⅜ ± ½	-1.5 ±0.5 -1½ ± ½	CAMBER (Degrees)	-1.73 ±0.5 -1¾ ± ½	-1.5 ±0.5 -1½ ± ½	-1.48 ±0.5 -1½ ± ½
0.41° ±0.13° 13/32° ± ⅛°	0.33° ±0.13° 11/32° ± ⅛°	0.33° ±0.13° 11/32° ± ⅛°	REAR TOE (Degrees)	0.33° ±0.13° 11/32° ± ⅛°	0.42° ±0.12° 13/32° ± ⅛°	0.41° ±0.12° 13/32° ± ⅛°
0.21" ±0.06" 7/32" ± 1/16"	0.17" ±0.06" 5/32" ± 1/16"	0.17" ±0.06" 5/32" ± 1/16"	REAR TOE (Inches)	0.17" ±0.06" 5/32" ± 1/16"	0.21" ±0.06" 7/32" ± 1/16"	0.21" ±0.06" 7/32" ± 1/16"
0.25 (¼)	0.25 (¼)	0.25 (¼)	THRUST ANGLE	0.25 (¼)		
	06	06	ILLUSTRATION #	06	174	174
			NOTES			
29212	29214	29215		29216	29218	29219

ALIGNMENT SPECIFICATIONS

Mercedes-Benz

CLS-Class Coupe (218.*) RWD AMG Models	CLS-Class Coupe (218.*) 4MATIC Models exc/AMG with Air Susp.	CLS-Class Coupe (218.*) 4MATIC Models exc/AMG with Steel Susp.	MODEL INFORMATION	CLS-Class Coupe (218.*) 4MATIC AMG Models	Notes	Notes
2012-2017	2012-2017	2012-2017	YEARS	2012-2017		
p. 308, 309	p. 308, 309	p. 308, 309	INSTRUCTIONS	p. 308, 309		
96 ft.lbs.	96 ft.lbs.	96 ft.lbs.	LUGNUT TORQUE	96 ft.lbs.		
			FRONT			
8.4 ±0.5 8⅜ ± ½	7.33 ±0.5 7 5/16 ± ½	7.36 ±0.5 7⅜ ± ½	CASTER (Degrees)	8.35 ±0.5 8⅜ ± ½		
0.0 (0) | 0.5 (½)	0.0 (0) | 0.5 (½)	0.0 (0) | 0.5 (½)	CROSS | TOL.	0.0 (0) | 0.5 (½)		
-1.54 ±0.4 -1 9/16 ± ⅜	-0.8 ±0.4 - 13/16 ± ⅜	-0.8 ±0.4 - 13/16 ± ⅜	CAMBER (Degrees)	-1.79 ±0.4 -1 13/16 ± ⅜		
0.0 (0) | 0.4 (⅜)	0.0 (0) | 0.4 (⅜)	0.0 (0) | 0.4 (⅜)	CROSS | TOL.	0.0 (0) | 0.4 (⅜)		
0.33° ±0.17° 11/32° ± 5/32°	0.17° ±0.17° 5/32° ± 5/32°	0.17° ±0.17° 5/32° ± 5/32°	FRONT TOE (Degrees)	0.3° ±0.17° 5/16° ± 5/32°		
0.17" ±0.08" 5/32" ± 3/32"	0.08" ±0.08" 3/32" ± 3/32"	0.08" ±0.08" 3/32" ± 3/32"	FRONT TOE (Inches)	0.15" ±0.08" 5/32" ± 3/32"		
			SAI			
20.0 / 18.0 ±0.5	20.0 / 18.1 ±0.5	20.0 / 18.1 ±0.5	TURNING ANGLE	20.0 / 18.0 ±0.5		
198 199	198 199	198 199	ILLUSTRATION #	198 199		
			REAR			
-2.26 ±0.5 -2¼ ± ½	-1.5 ±0.5 -1½ ± ½	-1.55 ±0.5 -1 9/16 ± ½	CAMBER (Degrees)	-2.38 ±0.5 -2⅜ ± ½		
0.24° ±0.12° ¼° ± ⅛°	0.42° ±0.12° 13/32° ± ⅛°	0.42° ±0.12° 13/32° ± ⅛°	REAR TOE (Degrees)	0.23° ±0.12° 7/32° ± ⅛°		
0.12" ±0.06" ⅛" ± 1/16"	0.21" ±0.06" 7/32" ± 1/16"	0.21" ±0.06" 7/32" ± 1/16"	REAR TOE (Inches)	0.12" ±0.06" ⅛" ± 1/16"		
			THRUST ANGLE			
174	174	174	ILLUSTRATION #	174		
			NOTES			
29220	29221	29222		29223	29224	29225

Mercury

Capri	Cougar	Cougar	MODEL INFORMATION	Cougar	Grand Marquis Colony Park Wgn.	Grand Marquis Colony Park Wgn.
1991-94	1990-91	1992-97	YEARS	1999-2002	1990	1991-94
			INSTRUCTIONS			
67-88 ft.lbs.	85-105 ft.lbs.	85-105 ft.lbs.	LUGNUT TORQUE	94 ft.lbs.	85-105 ft.lbs.	85-105 ft.lbs.
			FRONT			
1.58 ±0.75 1 9/16 ± ¾	5.5 ±0.75 5½ ± ¾	5.5 ±0.75 5½ ± ¾	CASTER (Degrees)	2.5 ±1.0 [1] 2½ ±1 [1]	3.5 ±1.0 3½ ±1	5.5 ±0.75 [1,2] 5½ ± ¾ [1,2]
0.0 (0) | 0.67 (11/16)	0.0 (0) | 0.75 (¾)	0.0 (0) | 0.75 (¾)	CROSS | TOL.	0.0 (0) | 1.0 (1)	0.0 (0) | 0.75 (¾)	0.0 (0) | 0.75 (¾)
0.8 ±0.75 13/16 ± ¾	-0.5 ±0.75 -½ ± ¾	-0.5 ±0.75 -½ ± ¾	CAMBER (Degrees)	-0.61 ±1.25 -⅝ ±1¼	-0.5 ±0.75 -½ ± ¾	-0.5 ±0.75 [2] -½ ± ¾ [2]
0.0 (0) | 0.5 (½)	0.0 (0) | 0.75 (¾)	0.0 (0) | 0.75 (¾)	CROSS | TOL.	0.0 (0) | 1.25 (1¼)	0.0 (0) | 0.75 (¾)	0.0 (0) | 0.75 (¾)
0.3° ±0.3° 5/16° ± 5/16°	0.25° ±0.25° ¼° ± ¼°	0.15° ±0.25° 5/32° ± ¼°	FRONT TOE (Degrees)	0.0° ±0.2° 0° ± 3/16°	0.13° ±0.25° ⅛° ± ¼°	-0.13° ±0.25° -⅛° ± ¼°
0.15" ±0.15" 5/32" ± 5/32"	0.13" ±0.13" ⅛" ± ⅛"	0.08" ±0.13" 1/16" ± ⅛"	FRONT TOE (Inches)	0.0" ±0.1" 0" ± 3/32"	0.06" ±0.13" 1/16" ± ⅛"	-0.06" ±0.13" -1/16" ± ⅛"
12.4 (12⅜)			SAI			
	20.0 / 19.7	20.0 / 19.7	TURNING ANGLE		20.0 / 18.5	20.0 / 18.5
46	03 14 △	03 14 △	ILLUSTRATION #		42	70
			REAR			
0.0 ±0.75 0 ± ¾	-0.5 ±0.5 -½ ± ½	-0.5 ±0.5 -½ ± ½	CAMBER (Degrees)	-0.7 ±1.25 -11/16 ±1¼		
0.3° ±0.3° 5/16° ± 5/16°	0.13° ±0.25° ⅛° ± ¼°	0.13° ±0.25° ⅛° ± ¼°	REAR TOE (Degrees)	0.32° ±0.2° 5/16° ± 3/16°		
0.15" ±0.15" 5/32" ± 5/32"	0.06" ±0.13" 1/16" ± ⅛"	0.06" ±0.13" 1/16" ± ⅛"	REAR TOE (Inches)	0.16" ±0.1" 5/32" ± 3/32"		
			THRUST ANGLE			
10	71	71	ILLUSTRATION #	06		
			NOTES	[1] 1999-00 = 2.25º (2¼º)		[1] 1991 = 3.5º (3½º) [2] 1991 see img. #42
30008	30028	30029		30030	30034	30035

ALIGNMENT SPECIFICATIONS

Mercury

Grand Marquis	Grand Marquis	Grand Marquis Base Suspension	MODEL INFORMATION	Grand Marquis Air Level Susp.	Grand Marquis Air Level Susp.	Grand Marquis Police Models
1995-97	1998-2002	2003-2005	YEARS	2003-2004	2005	2003-2005
			INSTRUCTIONS			
85-105 ft.lbs.	85-105 ft.lbs.	95 ft.lbs.	LUGNUT TORQUE	95 ft.lbs.	95 ft.lbs.	95 ft.lbs.
			FRONT			
5.5 ±0.75 5½ ± ¾	5.5 ±0.75 5½ ± ¾	6.2 ±0.75 6 3/16 ± ¾	CASTER (Degrees)	6.5 ±0.75 6½ ± ¾	6.5 ±0.75 6½ ± ¾	6.0 ±0.75 6 ± ¾
0.0 (0) | 0.75 (¾)	0.0 (0) | 0.75 (¾)	0.0 (0) | 0.75 (¾)	CROSS | TOL.	0.0 (0) | 0.75 (¾)	0.0 (0) | 0.75 (¾)	0.0 (0) | 0.75 (¾)
-0.5 ±0.75 -½ ± ¾	-0.5 ±0.75 -½ ± ¾	-0.4 ±0.75 -⅜ ± ¾	CAMBER (Degrees)	-0.7 ±0.75 -11/16 ± ¾	-0.4 ±0.75 -⅜ ± ¾	-0.4 ±0.75 -⅜ ± ¾
0.0 (0) | 0.75 (¾)	0.0 (0) | 0.75 (¾)	0.0 (0) | 0.75 (¾)	CROSS | TOL.	0.0 (0) | 0.75 (¾)	0.0 (0) | 0.75 (¾)	0.0 (0) | 0.75 (¾)
-0.13° ±0.25° -⅛° ± ¼°	**-0.13° ±0.25°** -⅛° ± ¼°	**-0.15° ±0.2°** -5/32° ± 3/16°	FRONT TOE (Degrees)	**-0.15° ±0.2°** -5/32° ± 3/16°	**-0.15° ±0.2°** -5/32° ± 3/16°	**-0.15° ±0.2°** -5/32° ± 3/16°
-0.06" ±0.13" -1/16" ± ⅛"	**-0.06" ±0.13"** -1/16" ± ⅛"	**-0.08" ±0.1"** -1/16" ± 3/32"	FRONT TOE (Inches)	**-0.08" ±0.1"** -1/16" ± 3/32"	**-0.08" ±0.1"** -1/16" ± 3/32"	**-0.08" ±0.1"** -1/16" ± 3/32"
			SAI	10.7 (10 11/16)		10.7 (10 11/16)
20.0 / 18.5	20.0 / 18.5		TURNING ANGLE			
70	70	148	ILLUSTRATION #	148	148	148
			REAR			
			CAMBER (Degrees)			
			REAR TOE (Degrees)			
			REAR TOE (Inches)			
			THRUST ANGLE			
			ILLUSTRATION #			
			NOTES			
30036	30037	30038		30039	30040	30041

Mercury

Grand Marquis Base Suspension	Grand Marquis Air Level Susp.	Grand Marquis Police/Taxi/Fleet	MODEL INFORMATION	Grand Marquis Base Suspension	Grand Marquis Air Level Susp.	Grand Marquis Police/Taxi/Fleet
2006-2007	2006-2007	2006-2007	YEARS	2008-2011	2008-2011	2008-2011
			INSTRUCTIONS			
100 ft.lbs.	100 ft.lbs.	100 ft.lbs.	LUGNUT TORQUE	100 ft.lbs.	100 ft.lbs.	100 ft.lbs.
			FRONT			
L= 6.2, R= 6.5 ± 0.75 L= 6 3/16, R= 6½ ± ¾	L= 6.5, R= 6.8 ± 0.75 L= 6½, R= 6 13/16 ± ¾	L= 6.0, R= 6.3 ± 0.75 L= 6, R= 6 5/16 ± ¾	CASTER (Degrees)	L= 5.9, R= 6.1 ± 0.75 [1] L= 5 ⅞, R= 6⅛ ± ¾ [1]	L= 6.2, R= 6.4 ± 0.75 [1] L= 6 3/16, R= 6⅜ ± ¾ [1]	L= 5.7, R= 5.9 ± 0.75 [1] L= 5 11/16, R= 5 ⅞ ± ¾ [1]
-0.3 (-5/16) | 0.75 (¾)	-0.3 (-5/16) | 0.75 (¾)	-0.3 (-5/16) | 0.75 (¾)	CROSS | TOL.	-0.2 (-3/16) | 0.75 (¾)	-0.2 (-3/16) | 0.75 (¾)	-0.2 (-3/16) | 0.75 (¾)
-0.4 ±0.75 -⅜ ± ¾	-0.4 ±0.75 -⅜ ± ¾	-0.4 ±0.75 -⅜ ± ¾	CAMBER (Degrees)	-0.4 ±0.75 -⅜ ± ¾	-0.4 ±0.75 -⅜ ± ¾	-0.4 ±0.75 -⅜ ± ¾
0.0 (0) | 0.75 (¾)	0.0 (0) | 0.75 (¾)	0.0 (0) | 0.75 (¾)	CROSS | TOL.	0.0 (0) | 0.75 (¾)	0.0 (0) | 0.75 (¾)	0.0 (0) | 0.75 (¾)
-0.15° ±0.2° -5/32° ± 3/16°	**-0.15° ±0.2°** -5/32° ± 3/16°	**-0.15° ±0.2°** -5/32° ± 3/16°	FRONT TOE (Degrees)	**-0.15° ±0.2°** -5/32° ± 3/16°	**-0.15° ±0.2°** -5/32° ± 3/16°	**-0.15° ±0.2°** -5/32° ± 3/16°
-0.08" ±0.1" -1/16" ± 3/32"	**-0.08" ±0.1"** -1/16" ± 3/32"	**-0.08" ±0.1"** -1/16" ± 3/32"	FRONT TOE (Inches)	**-0.08" ±0.1"** -1/16" ± 3/32"	**-0.08" ±0.1"** -1/16" ± 3/32"	**-0.08" ±0.1"** -1/16" ± 3/32"
			SAI			
			TURNING ANGLE			
148	148	148	ILLUSTRATION #	148	148	148
			REAR			
			CAMBER (Degrees)			
			REAR TOE (Degrees)			
			REAR TOE (Inches)			
			THRUST ANGLE			
			ILLUSTRATION #			
			NOTES	[1] 2008 L=5.9°, R=6.2°	[1] 2008 L=6.2°, R=6.5°	[1] 2008 L=5.7°, R=6.0°
30042	30043	30044		30045	30046	30047

Mercury

MODEL INFORMATION	Marauder	Mariner	Mariner Mariner Hybrid	Mariner	Mariner Hybrid	Milan
YEARS	2003-2004	2005	2006-2007	2008-2011	2008-2011	2006-2009
INSTRUCTIONS						
LUGNUT TORQUE	95 ft.lbs.	98 ft.lbs.	98 ft.lbs.	100 ft.lbs.	100 ft.lbs.	98 ft.lbs.
FRONT						
CASTER (Degrees)	5.5±0.75 5½±¾	1.6±1.0 1⅝±1	L= 1.8, R= 1.5± 1.0 L= 1 13/16, R= 1½±1	1.6±0.5 1⅝±½	1.6±0.5 1⅝±½	4.0±1.0 4±1
CROSS | TOL.	0.0 (0) | 0.75 (¾)	0.0 (0) | 0.75 (¾)	0.3 (5/16) | 0.75 (¾)	0.0 (0) | 0.7 (11/16)	0.0 (0) | 0.7 (11/16)	0.0 (0) | 1.0 (1)
CAMBER (Degrees)	-0.6±0.75 -⅝±¾	-1.0±0.75 -1±¾	-1.0±0.75 -1±¾	-0.84±0.5 -13/16±½	-0.84±0.5 -13/16±½	-0.03±0.75 -1/32±¾
CROSS | TOL.	0.0 (0) | 0.75 (¾)	0.0 (0) | 0.75 (¾)	0.0 (0) | 0.75 (¾)	0.0 (0) | 0.7 (11/16)	0.0 (0) | 0.7 (11/16)	0.0 (0) | 1.0 (1)
FRONT TOE (Degrees)	-0.13° ±0.25° -⅛° ± ¼°	0.23° ±0.23° 7/32° ± 7/32°	0.23° ±0.23° 7/32° ± 7/32°	0.23° ±0.23° 7/32° ± 7/32°	0.23° ±0.23° 7/32° ± 7/32°	0.2° ±0.2° 3/16° ± 3/16°
FRONT TOE (Inches)	-0.06" ±0.13" -1/16" ± ⅛"	0.12" ±0.12" ⅛" ± ⅛"	0.12" ±0.12" ⅛" ± ⅛"	0.12" ±0.12" ⅛" ± ⅛"	0.12" ±0.12" ⅛" ± ⅛"	0.1" ±0.1" 3/32" ± 3/32"
SAI	11.0 (11)					
TURNING ANGLE						
ILLUSTRATION #	148	51	51	51	51	
REAR						
CAMBER (Degrees)		0.0±0.75 0±¾	0.0±0.75 0±¾	0.0±0.65 0±⅝	-0.1±0.65 -⅛±⅝	-1.25±0.75 -1¼±¾
REAR TOE (Degrees)		-0.06° ±0.2° -1/16° ± 3/16°	-0.14° ±0.2° -⅛° ± 3/16°	-0.18° ±0.2° -3/16° ± 3/16°	-0.12° ±0.2° -⅛° ± 3/16°	0.24° ±0.2° ¼° ± 3/16°
REAR TOE (Inches)		-0.03" ±0.1" -1/32" ± 3/32"	-0.07" ±0.1" -1/16" ± 3/32"	-0.09" ±0.1" -3/32" ± 3/32"	-0.06" ±0.1" -1/16" ± 3/32"	0.12" ±0.1" ⅛" ± 3/32"
THRUST ANGLE		0.3 (5/16)	0.3 (5/16)			0.1 (3/32)
ILLUSTRATION #		27	27	27	27	175
NOTES						
	30052	30053	30054	30055	30056	30076

Mercury

MODEL INFORMATION	Milan Milan Hybrid	Montego FWD	Montego AWD	Montego AWD	Monterey	Monterey
YEARS	2010-2011	2005-2007	2005	2006-2007	2004-2005	2006-2007
INSTRUCTIONS						
LUGNUT TORQUE	100 ft.lbs.	95 ft.lbs.	95 ft.lbs.	95 ft.lbs.	100 ft.lbs.	100 ft.lbs.
FRONT						
CASTER (Degrees)	L= 3.75, R= 4.25± 1.0 L= 3¾, R= 4¼±1	3.3±0.75 3 5/16±¾	3.5±0.75 3½±¾	3.3±0.75 3 5/16±¾	3.6±0.75 3⅝±¾	3.5±0.5 3½±½
CROSS | TOL.	-0.5 (-½) | 1.0 (1)	0.0 (0) | 0.75 (¾)	0.0 (0) | 0.75 (¾)	0.0 (0) | 0.75 (¾)	0.0 (0) | 0.5 (½)	0.0 (0) | 0.75 (¾)
CAMBER (Degrees)	-0.03±0.75 -1/32±¾	-0.6±0.75 -⅝±¾	-0.4±0.75 -⅜±¾	-0.4±0.75 -⅜±¾	-0.4±0.75 -⅜±¾	-0.44±0.5 -7/16±½
CROSS | TOL.	0.0 (0) | 1.0 (1)	0.0 (0) | 0.75 (¾)	0.0 (0) | 0.75 (¾)	0.0 (0) | 0.75 (¾)	0.0 (0) | 0.75 (¾)	0.0 (0) | 0.75 (¾)
FRONT TOE (Degrees)	0.2° ±0.2° 3/16° ± 3/16°	0.2° ±0.2° 3/16° ± 3/16°	0.2° ±0.2° 3/16° ± 3/16°	0.2° ±0.2° 3/16° ± 3/16°	-0.15° ±0.2° -5/32° ± 3/16°	-0.15° ±0.2° -5/32° ± 3/16°
FRONT TOE (Inches)	0.1" ±0.1" 3/32" ± 3/32"	0.1" ±0.1" 3/32" ± 3/32"	0.1" ±0.1" 3/32" ± 3/32"	0.1" ±0.1" 3/32" ± 3/32"	-0.08" ±0.1" -1/16" ± 3/32"	-0.08" ±0.1" -1/16" ± 3/32"
SAI						
TURNING ANGLE						
ILLUSTRATION #		167	167	167	67	67
REAR						
CAMBER (Degrees)	-1.25±0.75 -1¼±¾	-0.35±0.75 -⅜±¾	-0.7±0.75 -11/16±¾	-0.45±0.75 -7/16±¾	-0.3±0.75 -5/16±¾	-0.25±0.58 -¼±9/16
REAR TOE (Degrees)	0.24° ±0.2° ¼° ± 3/16°	0.1° ±0.2° 3/32° ± 3/16°	0.1° ±0.2° 3/32° ± 3/16°	0.1° ±0.2° 3/32° ± 3/16°	0.14° ±0.25° ⅛° ± ¼°	0.14° ±0.25° ⅛° ± ¼°
REAR TOE (Inches)	0.12" ±0.1" ⅛" ± 3/32"	0.05" ±0.1" 1/16" ± 3/32"	0.05" ±0.1" 1/16" ± 3/32"	0.05" ±0.1" 1/16" ± 3/32"	0.07" ±0.13" 1/16" ± ⅛"	0.07" ±0.13" 1/16" ± ⅛"
THRUST ANGLE	0.5 (½)	0.3 (5/16)	0.3 (5/16)	0.3 (5/16)		
ILLUSTRATION #	175	06	06	06		
NOTES						
	30077	30083	30084	30085	30086	30087

ALIGNMENT SPECIFICATIONS

Mercury

	30088	30089	30090	30091	30092	30093
MODEL INFORMATION	Mountaineer 2WD, AWD except ARC	Mountaineer AWD with ARC	Mountaineer 2WD, AWD except ARC	Mountaineer AWD with ARC	Mountaineer	Mountaineer
YEARS	1997	1997	1998-2001	1998-2001	2002-2005	2006-2007
INSTRUCTIONS	p. 306, 307	p. 306, 307	p. 306, 307	p. 306, 307		
LUGNUT TORQUE	100 ft.lbs.	100 ft.lbs.	100 ft.lbs.	100 ft.lbs.	100 ft.lbs.	100 ft.lbs.
FRONT						
CASTER (Degrees)	MIN.= 3.2 MAX.= 5.2 [1] MIN.= 3 3/16 MAX.= 5 3/16 [1]	MIN.= 3.5 MAX.= 5.5 [1] MIN.= 3½ MAX.= 5½ [1]	MIN.= 3.2 MAX.= 5.2 [1] MIN.= 3 3/16 MAX.= 5 3/16 [1]	MIN.= 3.5 MAX.= 5.5 [1] MIN.= 3½ MAX.= 5½ [1]	L= 5.1, R= 5.3 ± 1.0 L= 5⅛, R= 5 5/16 ± 1	4.6 ± 0.75 4⅝ ± ¾
CROSS | TOL.	-0.5 (-½) | 0.75 (¾)	-0.5 (-½) | 0.75 (¾)	-0.75 (-¾) | 0.75 (¾)	-0.75 (-¾) | 0.75 (¾)	-0.2 (-3/16) | 0.4 (⅜)	0.0 (0) | 0.75 (¾)
CAMBER (Degrees)	-0.5 ± 0.75 [1] -½ ± ¾ [1]	-0.75 ± 0.75 [1] -¾ ± ¾ [1]	-0.5 ± 0.75 [1] -½ ± ¾ [1]	-0.75 ± 0.75 [1] -¾ ± ¾ [1]	-0.5 ± 0.8 -½ ± 13/16	-0.5 ± 0.75 -½ ± ¾
CROSS | TOL.	0.0 (0) | 0.75 (¾)	0.0 (0) | 0.75 (¾)	0.0 (0) | 0.75 (¾)	0.0 (0) | 0.75 (¾)	0.0 (0) | 0.8 (13/16)	0.0 (0) | 0.75 (¾)
FRONT TOE (Degrees)	0.3° ± 0.25° 5/16° ± ¼°	0.3° ± 0.25° 5/16° ± ¼°	0.13° ± 0.25° ⅛° ± ¼°	0.13° ± 0.25° ⅛° ± ¼°	0.2° ± 0.25° 3/16° ± ¼°	0.2° ± 0.2° 3/16° ± 3/16°
FRONT TOE (Inches)	0.15" ± 0.13" 5/32" ± ⅛"	0.15" ± 0.13" 5/32" ± ⅛"	0.06" ± 0.13" 1/16" ± ⅛"	0.06" ± 0.13" 1/16" ± ⅛"	0.1" ± 0.13" 3/32" ± ⅛"	0.1" ± 0.1" 3/32" ± 3/32"
SAI						
TURNING ANGLE						
ILLUSTRATION #	73	73	73	73	137	139
REAR						
CAMBER (Degrees)					-0.9 ± 0.8 -⅞ ± 13/16	-0.5 ± 0.75 -½ ± ¾
REAR TOE (Degrees)					0.2° ± 0.25° 3/16° ± ¼°	0.1° ± 0.2° 3/32° ± 3/16°
REAR TOE (Inches)					0.1" ± 0.13" 3/32" ± ⅛"	0.05" ± 0.1" 1/16" ± 3/32"
THRUST ANGLE	0.3 (5/16)	0.3 (5/16)	0.3 (5/16)	0.3 (5/16)		
ILLUSTRATION #					136	176
NOTES	[1] Service cam kit reqd.	[1] Service cam kit reqd.	[1] Service cam kit reqd.	[1] Service cam kit reqd.		

Mercury

	30094	30095	30096	30099	30100	30101
MODEL INFORMATION	Mountaineer	Mystique	Mystique	Sable Sedan	Sable Wagon	Sable Sedan
YEARS	2008-2010	1995-96	1997-2000	1990-95	1990-95	1996-99
INSTRUCTIONS						
LUGNUT TORQUE	100 ft.lbs.	85-105 ft.lbs.	85-105 ft.lbs.	85-105 ft.lbs.	85-105 ft.lbs.	85-105 ft.lbs.
FRONT						
CASTER (Degrees)	L= 4.4, R= 4.6 ± 0.75 [1] L= 4⅜, R= 4⅝ ± ¾ [1]	2.3 ± 1.0 2 5/16 ± 1	2.3 ± 1.0 2 5/16 ± 1	3.8 ± 1.0 3 13/16 ± 1	3.7 ± 1.0 3 11/16 ± 1	3.8 ± 1.0 3 13/16 ± 1
CROSS | TOL.	-0.2 (-3/16) | 0.75 (¾)	0.0 (0) | 0.85 (⅞)	0.0 (0) | 0.85 (⅞)	0.0 (0) | 0.85 (⅞)	0.0 (0) | 0.85 (⅞)	0.0 (0) | 0.85 (⅞)
CAMBER (Degrees)	-0.5 ± 0.75 -½ ± ¾	-0.4 ± 1.0 -⅜ ± 1	-0.46 ± 1.25 -7/16 ± 1¼	-0.5 ± 0.6 -½ ± ⅝	-0.4 ± 0.6 -⅜ ± ⅝	-0.5 ± 0.6 -½ ± ⅝
CROSS | TOL.	0.0 (0) | 0.75 (¾)	0.0 (0) | 0.7 (11/16)	0.0 (0) | 0.7 (11/16)	0.0 (0) | 0.7 (11/16)	0.0 (0) | 0.7 (11/16)	0.0 (0) | 0.7 (11/16)
FRONT TOE (Degrees)	0.2° ± 0.2° 3/16° ± 3/16°	-0.34° ± 0.17° -11/32° ± 5/32°	-0.17° ± 0.17° -5/32° ± 5/32°	-0.2° +0.5° -0.25° -3/16° +½° -¼°	-0.2° +0.5° -0.25° -3/16° +½° -¼°	-0.2° +0.5° -0.25° -3/16° +½° -¼°
FRONT TOE (Inches)	0.1" ± 0.1" 3/32" ± 3/32"	-0.17" ± 0.09" -5/32" ± 3/32"	-0.09" ± 0.09" -3/32" ± 3/32"	-0.1" +0.25" -0.13" -3/32" +¼" -⅛"	-0.1" +0.25" -0.13" -3/32" +¼" -⅛"	-0.1" +0.25" -0.13" -3/32" +¼" -⅛"
SAI						
TURNING ANGLE		20.0 / 18.3	20.0 / 18.3	20.0 / 18.3	20.0 / 18.3	20.0 / 18.3
ILLUSTRATION #	139	▲	▲	68 ▲	68 ▲	68
REAR						
CAMBER (Degrees)	-0.5 ± 0.75 -½ ± ¾	-0.4 ± 1.0 -⅜ ± 1	-0.4 ± 1.0 -⅜ ± 1	-0.9 ± 0.7 -⅞ ± 11/16	-0.9 ± 1.0 [1] -⅞ ± 1 [1]	-0.6 ± 0.7 -⅝ ± 11/16
REAR TOE (Degrees)	0.1° ± 0.2° 3/32° ± 3/16°	0.3° ± 0.2° 5/16° ± 3/16°	0.28° ± 0.28° 9/32° ± 9/32°	0.12° ± 0.25° ⅛° ± ¼°	0.12° ± 0.25° ⅛° ± ¼°	0.12° ± 0.25° ⅛° ± ¼°
REAR TOE (Inches)	0.05" ± 0.1" 1/16" ± 3/32"	0.15" ± 0.1" 5/32" ± 3/32"	0.14" ± 0.14" ⅛" ± ⅛"	0.06" ± 0.13" 1/16" ± ⅛"	0.06" ± 0.13" 1/16" ± ⅛"	0.06" ± 0.13" 1/16" ± ⅛"
THRUST ANGLE						
ILLUSTRATION #	176	06	06	69	64	69
NOTES	[1] 2008 L=4.4°, R=4.6°				[1] Cam kit available.	

Mercury

Sable Wagon	Sable Sedan	Sable Wagon	MODEL INFORMATION	Sable FWD	Sable AWD	Topaz
1996-99	2000-2005	2000-2005	YEARS	2008-2009	2008-2009	1990-91
			INSTRUCTIONS			
85-105 ft.lbs.	95 ft.lbs.	95 ft.lbs.	LUGNUT TORQUE	95 ft.lbs.	95 ft.lbs.	85-105 ft.lbs.
			FRONT			
3.8 ± 1.0 3 13/16 ± 1	3.8 ± 1.0 3 13/16 ± 1	3.8 ± 1.0 3 13/16 ± 1	CASTER (Degrees)	L= 3.1, R= 3.3 ± 0.75 L= 3⅛, R= 3 5/16 ± ¾	L= 3.1, R= 3.3 ± 0.75 L= 3⅛, R= 3 5/16 ± ¾	2.45 ± 0.75 2 7/16 ± ¾
0.0 (0) | 0.85 (⅞)	0.0 (0) | 0.85 (⅞)	0.0 (0) | 0.85 (⅞)	CROSS | TOL.	-0.2 (-3/16) | 0.75 (¾)	-0.2 (-3/16) | 0.75 (¾)	0.0 (0) | 0.75 (¾)
-0.5 ± 0.6 -½ ± ⅝	-0.5 ± 0.6 -½ ± ⅝	-0.5 ± 0.6 -½ ± ⅝	CAMBER (Degrees)	-0.6 ± 0.75 -⅝ ± ¾	-0.4 ± 0.75 -⅜ ± ¾	L= 1.4, R= 0.95 ± 0.75 L= 1⅜, R= 15/16 ± ¾
0.0 (0) | 0.7 (11/16)	0.0 (0) | 0.7 (11/16)	0.0 (0) | 0.7 (11/16)	CROSS | TOL.	0.0 (0) | 0.75 (¾)	0.0 (0) | 0.75 (¾)	0.45 (7/16) | 0.75 (¾)
-0.2° +0.5° -0.25° **-3/16°** +½° -¼°	**-0.2°** ± 0.25° **-3/16°** ± ¼°	**-0.2°** ± 0.25° **-3/16°** ± ¼°	FRONT TOE (Degrees)	0.2° ± 0.2° 3/16° ± 3/16°	0.2° ± 0.2° 3/16° ± 3/16°	**-0.2°** ± 0.25° **-3/16°** ± ¼°
-0.1" +0.25" -0.13" **-3/32"** +¼" -⅛"	**-0.1"** ± 0.13" **-3/32"** ± ⅛"	**-0.1"** ± 0.13" **-3/32"** ± ⅛"	FRONT TOE (Inches)	0.1" ± 0.1" 3/32" ± 3/32"	0.1" ± 0.1" 3/32" ± 3/32"	**-0.1"** ± 0.13" **-3/32"** ± ⅛"
			SAI			
20.0 / 18.3			TURNING ANGLE			20.0 / 18.3
68	67	67	ILLUSTRATION #	167	167	

Sable Wagon	Sable Sedan	Sable Wagon	MODEL INFORMATION	Sable FWD	Sable AWD	Topaz
			REAR			
-0.6 ± 0.7 -⅝ ± 11/16	-1.0 ± 0.7 -1 ± 11/16	-0.6 ± 0.7 -⅝ ± 11/16	CAMBER (Degrees)	-0.35 ± 0.75 -⅜ ± ¾	-0.45 ± 0.75 -7/16 ± ¾	-0.15 ± 0.75 -⅛ ± ¾
0.12° ± 0.25° ⅛° ± ¼°	0.36° ± 0.25° ⅜° ± ¼°	0.36° ± 0.25° ⅜° ± ¼°	REAR TOE (Degrees)	0.1° ± 0.2° 3/32° ± 3/16°	0.1° ± 0.2° 3/32° ± 3/16°	0.0° ± 0.33° 0° ± 11/32°
0.06" ± 0.13" 1/16" ± ⅛"	0.18" ± 0.13" 3/16" ± ⅛"	0.18" ± 0.13" 3/16" ± ⅛"	REAR TOE (Inches)	0.05" ± 0.1" 1/16" ± 3/32"	0.05" ± 0.1" 1/16" ± 3/32"	0.0" ± 0.17" 0" ± 5/32"
			THRUST ANGLE	0.3 (5/16)	0.3 (5/16)	
64	05	64	ILLUSTRATION #	06	06	06
			NOTES			
30102	30103	30104		30105	30106	30108

Mercury

Topaz AWD	Topaz	Tracer	MODEL INFORMATION	Tracer	Villager	Notes
1990-91	1992-94	1991-96	YEARS	1997-99	1993-2002	
			INSTRUCTIONS			
85-105 ft.lbs.	85-105 ft.lbs.	65-87 ft.lbs.	LUGNUT TORQUE	65-87 ft.lbs.	72-87 ft.lbs.	
			FRONT			
2.45 ± 0.75 2 7/16 ± ¾	2.45 ± 0.75 2 7/16 ± ¾	1.25 ± 0.75 1¼ ± ¾	CASTER (Degrees)	2.2 ± 0.75 2 3/16 ± ¾	0.8 ± 0.75 13/16 ± ¾	
0.0 (0) | 0.75 (¾)	0.0 (0) | 0.75 (¾)	0.0 (0) | 0.75 (¾)	CROSS | TOL.	0.0 (0) | 1.0 (1)	0.0 (0) | 0.75 (¾)	
L= 1.4, R= 0.95 ± 0.75 L= 1⅜, R= 15/16 ± ¾	L= 1.4, R= 0.95 ± 0.75 L= 1⅜, R= 15/16 ± ¾	0.2 ± 0.75 3/16 ± ¾	CAMBER (Degrees)	-0.4 ± 0.75 -⅜ ± ¾	0.3 ± 0.75 5/16 ± ¾	
0.45 (7/16) | 0.75 (¾)	0.45 (7/16) | 0.75 (¾)	0.0 (0) | 0.75 (¾)	CROSS | TOL.	0.0 (0) | 0.75 (¾)	0.0 (0) | 0.75 (¾)	
-0.2° ± 0.25° **-3/16°** ± ¼°	**-0.2°** ± 0.25° **-3/16°** ± ¼°	0.2° ± 0.3° 3/16° ± 5/16°	FRONT TOE (Degrees)	0.2° ± 0.25° 3/16° ± ¼°	0.28° ± 0.09° 9/32° ± 3/32°	
-0.1" ± 0.13" **-3/32"** ± ⅛"	**-0.1"** ± 0.13" **-3/32"** ± ⅛"	0.1" ± 0.15" 3/32" ± 5/32"	FRONT TOE (Inches)	0.1" ± 0.13" 3/32" ± ⅛"	0.14" ± 0.05" ⅛" ± 1/32"	
			SAI		13.6 ± 0.8 (13 9/16 ± ¾)	
20.0 / 18.3	20.0 / 18.3	40.0 / 33.0 ± 2.0	TURNING ANGLE		38.0 / 30.0 ± 2.0	
		46	ILLUSTRATION #			

Topaz AWD	Topaz	Tracer	MODEL INFORMATION	Tracer	Villager	Notes
			REAR			
0.35 ± 0.75 ⅜ ± ¾	-0.55 ± 0.75 -9/16 ± ¾	-0.5 ± 1.0 -½ ± 1	CAMBER (Degrees)	-0.7 ± 0.75 -11/16 ± ¾	0.0 ± 0.25 0 ± ¼	
0.0° ± 0.33° 0° ± 11/32°	0.0° ± 0.33° 0° ± 11/32°	0.2° ± 0.3° 3/16° ± 5/16°	REAR TOE (Degrees)	0.2° ± 0.3° 3/16° ± 5/16°	0.0° ± 0.36° 0° ± ⅜°	
0.0" ± 0.17" 0" ± 5/32"	0.0" ± 0.17" 0" ± 5/32"	0.1" ± 0.15" 3/32" ± 5/32"	REAR TOE (Inches)	0.1" ± 0.15" 3/32" ± 5/32"	0.0" ± 0.18" 0" ± 3/16"	
		0.4 (13/32)	THRUST ANGLE			
06	06	10	ILLUSTRATION #			
			NOTES			
30109	30110	30112		30114	30115	30154

MINI

(R50) MINI ONE	(R50) MINI ONE	(R50) MINI ONE	MODEL	(R50) MINI ONE	(R50) Cooper	(R50) Cooper
Std. Susp. -> 11/2004	Std. Susp. 12/2004 ->	Sport Susp. -> 11/2004	INFORMATION	Sport Susp. 12/2004 ->	-> 11/2004	12/2004 ->
(non-adj. r. camber)	(adj. rear camber)	(non-adj. r. camber)		(adj. rear camber)	(non-adj. r. camber)	(adj. rear camber)
2002-2005	2005-2006	2002-2005	YEARS	2005-2006	2002-2005	2005-2006
p. 304	p. 304	p. 304	INSTRUCTIONS	p. 304	p. 304	p. 304
			LUGNUT TORQUE			
			FRONT			
Caster is not specified by the manufacturer.	Caster is not specified by the manufacturer.	Caster is not specified by the manufacturer.	CASTER (Degrees)	Caster is not specified by the manufacturer.	Caster is not specified by the manufacturer.	Caster is not specified by the manufacturer.
0.0 (0) | 0.5 (½)	0.0 (0) | 0.5 (½)	0.0 (0) | 0.5 (½)	CROSS | TOL.	0.0 (0) | 0.5 (½)	0.0 (0) | 0.5 (½)	0.0 (0) | 0.5 (½)
-0.5 ±0.42 -½ ± 7/16	-0.5 ±0.42 -½ ± 7/16	-0.5 ±0.42 -½ ± 7/16	CAMBER (Degrees)	-0.5 ±0.42 -½ ± 7/16	-0.5 ±0.42 -½ ± 7/16	-0.5 ±0.42 -½ ± 7/16
0.0 (0) | 0.5 (½)	0.0 (0) | 0.5 (½)	0.0 (0) | 0.5 (½)	CROSS | TOL.	0.0 (0) | 0.5 (½)	0.0 (0) | 0.5 (½)	0.0 (0) | 0.5 (½)
0.3° ±0.08° 5/16° ± 3/32°	0.3° ±0.08° 5/16° ± 3/32°	0.3° ±0.08° 5/16° ± 3/32°	FRONT TOE (Degrees)	0.3° ±0.08° 5/16° ± 3/32°	0.3° ±0.08° 5/16° ± 3/32°	0.3° ±0.08° 5/16° ± 3/32°
0.15" ±0.04" 5/32" ± 1/32"	0.15" ±0.04" 5/32" ± 1/32"	0.15" ±0.04" 5/32" ± 1/32"	FRONT TOE (Inches)	0.15" ±0.04" 5/32" ± 1/32"	0.15" ±0.04" 5/32" ± 1/32"	0.15" ±0.04" 5/32" ± 1/32"
			SAI			
20.0 / 18.7 ±0.5	20.0 / 18.7 ±0.5	20.0 / 18.7 ±0.5	TURNING ANGLE	20.0 / 18.7 ±0.5	20.0 / 18.7 ±0.5	20.0 / 18.7 ±0.5
			ILLUSTRATION #			
			REAR			
-1.3 ±0.5 -1 5/16 ± ½	-1.52 ±0.33 -1½ ± 5/16	-1.53 ±0.5 -1 9/16 ± ½	CAMBER (Degrees)	-1.75 ±0.33 -1¾ ± 5/16	-1.53 ±0.5 -1 9/16 ± ½	-1.75 ±0.33 -1¾ ± 5/16
0.4° ±0.13° 13/32° ± 1/8°	0.4° ±0.13° 13/32° ± 1/8°	0.4° ±0.13° 13/32° ± 1/8°	REAR TOE (Degrees)	0.4° ±0.13° 13/32° ± 1/8°	0.4° ±0.13° 13/32° ± 1/8°	0.4° ±0.13° 13/32° ± 1/8°
0.2" ±0.07" 3/16" ± 1/16"	0.2" ±0.07" 3/16" ± 1/16"	0.2" ±0.07" 3/16" ± 1/16"	REAR TOE (Inches)	0.2" ±0.07" 3/16" ± 1/16"	0.2" ±0.07" 3/16" ± 1/16"	0.2" ±0.07" 3/16" ± 1/16"
0.2 (3/16)	0.2 (3/16)	0.17 (5/32)	THRUST ANGLE	0.17 (5/32)	0.17 (5/32)	0.17 (5/32)
28	29 28	28	ILLUSTRATION #	29 28	28	29 28
			NOTES			
57001	57002	57003		57004	57005	57006

MINI

(R50) JCW	(R50) JCW	(R52) Cooper, ONE Conv.	MODEL	(R52) Cooper, ONE Conv.	(R52) Cooper Convertible	(R52) Cooper Convertible
-> 11/2004	12/2004 ->	Std. Susp. -> 11/2004	INFORMATION	Std. Susp. 12/2004 ->	Sport Susp. -> 11/2004	Sport Susp. 12/2004 ->
(non-adj. r. camber)	(adj. rear camber)	(non-adj. r. camber)		(adj. rear camber)	(non-adj. r. camber)	(adj. rear camber)
2002-2005	2005-2006	2004-2005	YEARS	2005-2008	2004-2005	2005-2008
p. 304	p. 304	p. 304	INSTRUCTIONS	p. 304	p. 304	p. 304
			LUGNUT TORQUE			
			FRONT			
Caster is not specified by the manufacturer.	Caster is not specified by the manufacturer.	Caster is not specified by the manufacturer.	CASTER (Degrees)	Caster is not specified by the manufacturer.	Caster is not specified by the manufacturer.	Caster is not specified by the manufacturer.
0.0 (0) | 0.5 (½)	0.0 (0) | 0.5 (½)	0.0 (0) | 0.5 (½)	CROSS | TOL.	0.0 (0) | 0.5 (½)	0.0 (0) | 0.5 (½)	0.0 (0) | 0.5 (½)
-0.58 ±0.42 - 9/16 ± 7/16	-0.58 ±0.42 - 9/16 ± 7/16	-0.43 ±0.42 - 7/16 ± 7/16	CAMBER (Degrees)	-0.43 ±0.42 - 7/16 ± 7/16	-0.5 ±0.42 -½ ± 7/16	-0.5 ±0.42 -½ ± 7/16
0.0 (0) | 0.5 (½)	0.0 (0) | 0.5 (½)	0.0 (0) | 0.5 (½)	CROSS | TOL.	0.0 (0) | 0.5 (½)	0.0 (0) | 0.5 (½)	0.0 (0) | 0.5 (½)
0.3° ±0.08° 5/16° ± 3/32°	0.3° ±0.08° 5/16° ± 3/32°	0.3° ±0.08° 5/16° ± 3/32°	FRONT TOE (Degrees)	0.3° ±0.08° 5/16° ± 3/32°	0.3° ±0.08° 5/16° ± 3/32°	0.3° ±0.08° 5/16° ± 3/32°
0.15" ±0.04" 5/32" ± 1/32"	0.15" ±0.04" 5/32" ± 1/32"	0.15" ±0.04" 5/32" ± 1/32"	FRONT TOE (Inches)	0.15" ±0.04" 5/32" ± 1/32"	0.15" ±0.04" 5/32" ± 1/32"	0.15" ±0.04" 5/32" ± 1/32"
			SAI			
20.0 / 18.7 ±0.5	20.0 / 18.7 ±0.5	20.0 / 18.7 ±0.5	TURNING ANGLE	20.0 / 18.7 ±0.5	20.0 / 18.7 ±0.5	20.0 / 18.7 ±0.5
			ILLUSTRATION #			
			REAR			
-1.87 ±0.5 -1⅞ ± ½	-2.1 ±0.5 -2⅛ ± ½	-1.3 ±0.5 -1 5/16 ± ½	CAMBER (Degrees)	-1.52 ±0.33 -1½ ± 5/16	-1.53 ±0.5 -1 9/16 ± ½	-1.75 ±0.33 -1¾ ± 5/16
0.4° ±0.13° 13/32° ± 1/8°	0.4° ±0.13° 13/32° ± 1/8°	0.4° ±0.13° 13/32° ± 1/8°	REAR TOE (Degrees)	0.4° ±0.13° 13/32° ± 1/8°	0.4° ±0.13° 13/32° ± 1/8°	0.4° ±0.13° 13/32° ± 1/8°
0.2" ±0.07" 3/16" ± 1/16"	0.2" ±0.07" 3/16" ± 1/16"	0.2" ±0.07" 3/16" ± 1/16"	REAR TOE (Inches)	0.2" ±0.07" 3/16" ± 1/16"	0.2" ±0.07" 3/16" ± 1/16"	0.2" ±0.07" 3/16" ± 1/16"
0.17 (5/32)	0.17 (5/32)	0.2 (3/16)	THRUST ANGLE	0.2 (3/16)	0.2 (3/16)	0.17 (5/32)
29 28	29 28	28	ILLUSTRATION #	29 28	28	29 28
			NOTES			
57007	57008	57009		57010	57011	57012

MINI

(R52) Cooper S Convertible -> 11/2004 (non-adj. r. camber)	(R52) Cooper S Convertible 12/2004 -> (adj. rear camber)	(R53) Cooper S Hardtop -> 11/2004 (non-adj. r. camber)	MODEL INFORMATION	(R53) Cooper S Hardtop 12/2004 -> (adj. rear camber)	(R53) JCW Hardtop -> 11/2004 (non-adj. r. camber)	(R53) JCW Hardtop 12/2004 -> (adj. rear camber)
2004-2005	2005-2008	2002-2005	YEARS	2005-2006	2002-2005	2005-2006
p. 304	p. 304	p. 304	INSTRUCTIONS	p. 304	p. 304	p. 304
			LUGNUT TORQUE			
			FRONT			
Caster is not specified by the manufacturer.	Caster is not specified by the manufacturer.	Caster is not specified by the manufacturer.	CASTER (Degrees)	Caster is not specified by the manufacturer.	Caster is not specified by the manufacturer.	Caster is not specified by the manufacturer.
0.0 (0) | 0.5 (½)	0.0 (0) | 0.5 (½)	0.0 (0) | 0.5 (½)	CROSS | TOL.	0.0 (0) | 0.5 (½)	0.0 (0) | 0.5 (½)	0.0 (0) | 0.5 (½)
-0.5 ±0.42 -½ ± 7/16	-0.5 ±0.42 -½ ± 7/16	-0.5 ±0.42 -½ ± 7/16	CAMBER (Degrees)	-0.5 ±0.42 -½ ± 7/16	-0.58 ±0.42 - 9/16 ± 7/16	-0.58 ±0.42 - 9/16 ± 7/16
0.0 (0) | 0.5 (½)	0.0 (0) | 0.5 (½)	0.0 (0) | 0.5 (½)	CROSS | TOL.	0.0 (0) | 0.5 (½)	0.0 (0) | 0.5 (½)	0.0 (0) | 0.5 (½)
0.3° ±0.08° 5/16° ± 3/32°	0.3° ±0.08° 5/16° ± 3/32°	0.3° ±0.08° 5/16° ± 3/32°	FRONT TOE (Degrees)	0.3° ±0.08° 5/16° ± 3/32°	0.3° ±0.08° 5/16° ± 3/32°	0.3° ±0.08° 5/16° ± 3/32°
0.15" ±0.04" 5/32" ± 1/32"	0.15" ±0.04" 5/32" ± 1/32"	0.15" ±0.04" 5/32" ± 1/32"	FRONT TOE (Inches)	0.15" ±0.04" 5/32" ± 1/32"	0.15" ±0.04" 5/32" ± 1/32"	0.15" ±0.04" 5/32" ± 1/32"
			SAI			
20.0 / 18.7 ±0.5	20.0 / 18.7 ±0.5	20.0 / 18.7 ±0.5	TURNING ANGLE	20.0 / 18.7 ±0.5	20.0 / 18.7 ±0.5	20.0 / 18.7 ±0.5
			ILLUSTRATION #			
			REAR			
-1.53 ±0.5 -1 9/16 ± ½	-1.75 ±0.33 -1 ¾ ± 5/16	-1.53 ±0.5 -1 9/16 ± ½	CAMBER (Degrees)	-1.75 ±0.33 -1 ¾ ± 5/16	-1.87 ±0.5 -1 7/8 ± ½	-2.1 ±0.5 -2 1/8 ± ½
0.4° ±0.13° 13/32° ± 1/8°	0.4° ±0.13° 13/32° ± 1/8°	0.4° ±0.13° 13/32° ± 1/8°	REAR TOE (Degrees)	0.4° ±0.13° 13/32° ± 1/8°	0.4° ±0.13° 13/32° ± 1/8°	0.4° ±0.13° 13/32° ± 1/8°
0.2" ±0.07" 3/16" ± 1/16"	0.2" ±0.07" 3/16" ± 1/16"	0.2" ±0.07" 3/16" ± 1/16"	REAR TOE (Inches)	0.2" ±0.07" 3/16" ± 1/16"	0.2" ±0.07" 3/16" ± 1/16"	0.2" ±0.07" 3/16" ± 1/16"
0.17 (5/32)	0.17 (5/32)	0.17 (5/32)	THRUST ANGLE	0.17 (5/32)	0.17 (5/32)	0.17 (5/32)
28	29 28	28	ILLUSTRATION #	29 28	28	29 28
			NOTES			
57013	57014	57015		57016	57017	57018

MINI

(R55) Cooper/S Clubman exc/JCW models	(R55) JCW Clubman Std. Susp.	(R55) JCW Clubman Sport Susp.	MODEL INFORMATION	(R56) Classic Hardtop (Canada)	(R56) MINI ONE Hardtop	(R56) Cooper Hardtop
2008-2015	2008-2015	2008-2015	YEARS	2007-2013	2007-2013	2007-2013
p. 304	p. 304	p. 304	INSTRUCTIONS	p. 304	p. 304	p. 304
			LUGNUT TORQUE			
			FRONT			
Caster is not specified by the manufacturer.	Caster is not specified by the manufacturer.	Caster is not specified by the manufacturer.	CASTER (Degrees)	Caster is not specified by the manufacturer.	Caster is not specified by the manufacturer.	Caster is not specified by the manufacturer.
0.0 (0) | 0.5 (½)	0.0 (0) | 0.5 (½)	0.0 (0) | 0.5 (½)	CROSS | TOL.	0.0 (0) | 0.5 (½)	0.0 (0) | 0.5 (½)	0.0 (0) | 0.5 (½)
-0.5 ±0.08 -½ ± 1/16	-0.5 ±0.08 -½ ± 1/16	-0.58 ±0.08 - 9/16 ± 1/16	CAMBER (Degrees)	-0.5 ±0.42 -½ ± 7/16	-0.5 ±0.42 -½ ± 7/16	-0.5 ±0.42 -½ ± 7/16
0.0 (0) | 0.5 (½)	0.0 (0) | 0.5 (½)	0.0 (0) | 0.5 (½)	CROSS | TOL.	0.0 (0) | 0.5 (½)	0.0 (0) | 0.5 (½)	0.0 (0) | 0.5 (½)
0.2° ±0.07° 3/16° ± 1/16°	0.2° ±0.07° 3/16° ± 1/16°	0.2° ±0.07° 3/16° ± 1/16°	FRONT TOE (Degrees)	0.2° ±0.07° 3/16° ± 1/16°	0.2° ±0.07° 3/16° ± 1/16°	0.2° ±0.07° 3/16° ± 1/16°
0.1" ±0.03" 3/32" ± 1/32"	0.1" ±0.03" 3/32" ± 1/32"	0.1" ±0.03" 3/32" ± 1/32"	FRONT TOE (Inches)	0.1" ±0.03" 3/32" ± 1/32"	0.1" ±0.03" 3/32" ± 1/32"	0.1" ±0.03" 3/32" ± 1/32"
			SAI			
20.0 / 18.2 ±0.5	20.0 / 18.2 ±0.5	20.0 / 18.2 ±0.5	TURNING ANGLE	20.0 / 18.2 ±0.5	20.0 / 18.2 ±0.5	20.0 / 18.2 ±0.5
119	119	119	ILLUSTRATION #	119	119	119
			REAR			
-1.75 ±0.08 -1 ¾ ± 1/16	-1.75 ±0.08 -1 ¾ ± 1/16	-2.0 ±0.08 -2 ± 1/16	CAMBER (Degrees)	-1.75 ±0.08 -1 ¾ ± 1/16	-1.75 ±0.08 -1 ¾ ± 1/16	-1.75 ±0.08 -1 ¾ ± 1/16
0.4° ±0.2° 13/32° ± 3/16°	0.4° ±0.2° 13/32° ± 3/16°	0.4° ±0.2° 13/32° ± 3/16°	REAR TOE (Degrees)	0.4° ±0.07° 13/32° ± 1/16°	0.4° ±0.07° 13/32° ± 1/16°	0.4° ±0.07° 13/32° ± 1/16°
0.2" ±0.1" 3/16" ± 3/32"	0.2" ±0.1" 3/16" ± 3/32"	0.2" ±0.1" 3/16" ± 3/32"	REAR TOE (Inches)	0.2" ±0.03" 3/16" ± 1/32"	0.2" ±0.03" 3/16" ± 1/32"	0.2" ±0.03" 3/16" ± 1/32"
0.2 (3/16)	0.2 (3/16)	0.2 (3/16)	THRUST ANGLE	0.2 (3/16)	0.2 (3/16)	0.2 (3/16)
29 28	29 28	29 28	ILLUSTRATION #	29 28	29 28	29 28
			NOTES			
57019	57020	57021		57022	57023	57024

MINI

(R56) Cooper S Hardtop	(R56) JCW Hardtop Std. Susp.	(R56) JCW Hardtop Sport Susp.	MODEL INFORMATION	(R57) Cooper Convertible (R57) Cooper S Convertible	(R57) JCW Convertible	(R58) Cooper Coupe (R58) Cooper S Coupe All Suspensions
2007-2013	2007-2013	2007-2013	YEARS	2009-2015	2009-2015	2012-2015
p. 304	p. 304	p. 304	INSTRUCTIONS	p. 304	p. 304	p. 304
			LUGNUT TORQUE			
			FRONT			
Caster is not specified by the manufacturer.	Caster is not specified by the manufacturer.	Caster is not specified by the manufacturer.	CASTER (Degrees)	Caster is not specified by the manufacturer.	Caster is not specified by the manufacturer.	Caster is not specified by the manufacturer.
0.0 (0) | 0.5 (½)	0.0 (0) | 0.5 (½)	0.0 (0) | 0.5 (½)	CROSS | TOL.	0.0 (0) | 0.5 (½)	0.0 (0) | 0.5 (½)	0.0 (0) | 0.5 (½)
-0.5 ±0.42 -½ ± 7/16	-0.5 ±0.42 -½ ± 7/16	-0.58 ±0.42 - 9/16 ± 7/16	CAMBER (Degrees)	-0.5 ±0.42 -½ ± 7/16	-0.5 ±0.42 -½ ± 7/16	-0.5 ±0.42 -½ ± 7/16
0.0 (0) | 0.5 (½)	0.0 (0) | 0.5 (½)	0.0 (0) | 0.5 (½)	CROSS | TOL.	0.0 (0) | 0.5 (½)	0.0 (0) | 0.5 (½)	0.0 (0) | 0.5 (½)
0.2° ±0.07° 3/16° ± 1/16°	0.2° ±0.07° 3/16° ± 1/16°	0.2° ±0.07° 3/16° ± 1/16°	FRONT TOE (Degrees)	0.2° ±0.07° 3/16° ± 1/16°	0.2° ±0.07° 3/16° ± 1/16°	0.2° ±0.07° 3/16° ± 1/16°
0.1" ±0.03" 3/32" ± 1/32"	0.1" ±0.03" 3/32" ± 1/32"	0.1" ±0.03" 3/32" ± 1/32"	FRONT TOE (Inches)	0.1" ±0.03" 3/32" ± 1/32"	0.1" ±0.03" 3/32" ± 1/32"	0.1" ±0.03" 3/32" ± 1/32"
			SAI			
20.0 / 18.2 ±0.5	20.0 / 18.2 ±0.5	20.0 / 18.2 ±0.5	TURNING ANGLE	20.0 / 18.2 ±0.5	20.0 / 18.2 ±0.5	20.0 / 18.2 ±0.5
119	119	119	ILLUSTRATION #	119	119	119
			REAR			
-1.75 ±0.08 -1¾ ± 1/16	-1.75 ±0.08 -1¾ ± 1/16	-2.0 ±0.08 -2 ± 1/16	CAMBER (Degrees)	-1.75 ±0.08 -1¾ ± 1/16	-1.75 ±0.08 -1¾ ± 1/16	-1.75 ±0.08 -1¾ ± 1/16
0.4° ±0.07° 13/32° ± 1/16°	0.4° ±0.07° 13/32° ± 1/16°	0.4° ±0.07° 13/32° ± 1/16°	REAR TOE (Degrees)	0.4° ±0.07° 13/32° ± 1/16°	0.4° ±0.07° 13/32° ± 1/16°	0.4° ±0.07° 13/32° ± 1/16°
0.2" ±0.03" 3/16" ± 1/32"	0.2" ±0.03" 3/16" ± 1/32"	0.2" ±0.03" 3/16" ± 1/32"	REAR TOE (Inches)	0.2" ±0.03" 3/16" ± 1/32"	0.2" ±0.03" 3/16" ± 1/32"	0.2" ±0.03" 3/16" ± 1/32"
0.2 (3/16)	0.2 (3/16)	0.2 (3/16)	THRUST ANGLE	0.2 (3/16)	0.2 (3/16)	0.2 (3/16)
29 28	29 28	29 28	ILLUSTRATION #	29 28	29 28	29 28
			NOTES			
57025	57026	57027		57028	57029	57030

MINI

(R58) JCW Coupe	(R59) Cooper Roadster (R59) Cooper S Roadster All Suspensions	(R59) JCW Roadster	MODEL INFORMATION	(R60) Cooper Countryman (R60) Cooper S Countryman Std. Susp.	(R60) Cooper Countryman (R60) Cooper S Countryman Sport Susp.	(R60) JCW Countryman
2012-2015	2012-2015	2012-2015	YEARS	2011-2016	2011-2016	2013-2016
p. 304	p. 304	p. 304	INSTRUCTIONS	p. 304	p. 304	p. 304
			LUGNUT TORQUE			
			FRONT			
Caster is not specified by the manufacturer.	Caster is not specified by the manufacturer.	Caster is not specified by the manufacturer.	CASTER (Degrees)	Caster is not specified by the manufacturer.	Caster is not specified by the manufacturer.	Caster is not specified by the manufacturer.
0.0 (0) | 0.5 (½)	0.0 (0) | 0.5 (½)	0.0 (0) | 0.5 (½)	CROSS | TOL.	0.0 (0) | 0.5 (½)	0.0 (0) | 0.5 (½)	0.0 (0) | 0.5 (½)
-0.58 ±0.42 - 9/16 ± 7/16	-0.5 ±0.42 -½ ± 7/16	-0.58 ±0.42 - 9/16 ± 7/16	CAMBER (Degrees)	-0.5 ±0.42 -½ ± 7/16	-0.58 ±0.42 - 9/16 ± 7/16	-0.58 ±0.42 - 9/16 ± 7/16
0.0 (0) | 0.5 (½)	0.0 (0) | 0.5 (½)	0.0 (0) | 0.5 (½)	CROSS | TOL.	0.0 (0) | 0.5 (½)	0.0 (0) | 0.5 (½)	0.0 (0) | 0.5 (½)
0.2° ±0.07° 3/16° ± 1/16°	0.2° ±0.07° 3/16° ± 1/16°	0.2° ±0.07° 3/16° ± 1/16°	FRONT TOE (Degrees)	0.3° ±0.07° 5/16° ± 1/16°	0.3° ±0.07° 5/16° ± 1/16°	0.3° ±0.07° 5/16° ± 1/16°
0.1" ±0.03" 3/32" ± 1/32"	0.1" ±0.03" 3/32" ± 1/32"	0.1" ±0.03" 3/32" ± 1/32"	FRONT TOE (Inches)	0.15" ±0.03" 5/32" ± 1/32"	0.15" ±0.03" 5/32" ± 1/32"	0.15" ±0.03" 5/32" ± 1/32"
			SAI			
20.0 / 18.2 ±0.5	20.0 / 18.2 ±0.5	20.0 / 18.2 ±0.5	TURNING ANGLE	20.0 / 18.3 ±0.5	20.0 / 18.3 ±0.5	20.0 / 18.3 ±0.5
119	119	119	ILLUSTRATION #	119	119	119
			REAR			
-2.0 ±0.08 -2 ± 1/16	-1.75 ±0.08 -1¾ ± 1/16	-2.0 ±0.08 -2 ± 1/16	CAMBER (Degrees)	-1.75 ±0.08 -1¾ ± 1/16	-2.1 ±0.08 -2⅛ ± 1/16	-2.1 ±0.08 -2⅛ ± 1/16
0.4° ±0.07° 13/32° ± 1/16°	0.4° ±0.07° 13/32° ± 1/16°	0.4° ±0.07° 13/32° ± 1/16°	REAR TOE (Degrees)	0.33° ±0.07° 11/32° ± 1/16°	0.33° ±0.07° 11/32° ± 1/16°	0.33° ±0.07° 11/32° ± 1/16°
0.2" ±0.03" 3/16" ± 1/32"	0.2" ±0.03" 3/16" ± 1/32"	0.2" ±0.03" 3/16" ± 1/32"	REAR TOE (Inches)	0.17" ±0.03" 5/32" ± 1/32"	0.17" ±0.03" 5/32" ± 1/32"	0.17" ±0.03" 5/32" ± 1/32"
0.2 (3/16)	0.2 (3/16)	0.2 (3/16)	THRUST ANGLE	0.2 (3/16)	0.2 (3/16)	0.2 (3/16)
29 28	29 28	29 28	ILLUSTRATION #	29 28	29 28	29 28
			NOTES			
57031	57032	57033		57034	57035	57036

MINI

(R61) Cooper Paceman	(R61) Cooper Paceman	(R61) Cooper S Paceman	MODEL	(R61) Cooper S Paceman	(R61) JCW Paceman	(F54) Cooper Clubman
Std. Susp.	Sport Susp.	Std. Susp.	INFORMATION	Sport Susp.		(F54) Cooper S Clubman
		(incl. All4)		(incl. All4)		All Suspensions
2014-2016	2014-2016	2014-2016	YEARS	2014-2016	2014-2016	2016-2017
p. 304	p. 304	p. 304	INSTRUCTIONS	p. 304	p. 304	p. 304
			LUGNUT TORQUE			
			FRONT			
Caster is not specified by the manufacturer.	Caster is not specified by the manufacturer.	Caster is not specified by the manufacturer.	CASTER (Degrees)	Caster is not specified by the manufacturer.	Caster is not specified by the manufacturer.	Caster is not specified by the manufacturer.
0.0 (0) | 0.5 (½)	0.0 (0) | 0.5 (½)	0.0 (0) | 0.5 (½)	CROSS | TOL.	0.0 (0) | 0.5 (½)	0.0 (0) | 0.5 (½)	0.0 (0) | 0.5 (½)
-0.5 ±0.42 -½ ± 7/16	-0.6 ±0.42 -5/8 ± 7/16	-0.5 ±0.42 -½ ± 7/16	CAMBER (Degrees)	-0.6 ±0.42 -5/8 ± 7/16	-0.6 ±0.42 -5/8 ± 7/16	-0.78 ±0.42 -13/16 ± 7/16
0.0 (0) | 0.5 (½)	0.0 (0) | 0.5 (½)	0.0 (0) | 0.5 (½)	CROSS | TOL.	0.0 (0) | 0.5 (½)	0.0 (0) | 0.5 (½)	0.0 (0) | 0.5 (½)
0.3° ±0.07° 5/16° ± 1/16°	0.3° ±0.07° 5/16° ± 1/16°	0.3° ±0.07° 5/16° ± 1/16°	FRONT TOE (Degrees)	0.3° ±0.07° 5/16° ± 1/16°	0.3° ±0.07° 5/16° ± 1/16°	0.2° ±0.07° 3/16° ± 1/16°
0.15" ±0.03" 5/32" ± 1/32"	0.15" ±0.03" 5/32" ± 1/32"	0.15" ±0.03" 5/32" ± 1/32"	FRONT TOE (Inches)	0.15" ±0.03" 5/32" ± 1/32"	0.15" ±0.03" 5/32" ± 1/32"	0.1" ±0.03" 3/32" ± 1/32"
			SAI			11.9 ±0.5 (11 7/8 ± ½)
20.0 / 18.3 ±0.5	20.0 / 18.3 ±0.5	20.0 / 18.3 ±0.5	TURNING ANGLE	20.0 / 18.3 ±0.5	20.0 / 18.3 ±0.5	20.0 / 18.3 ±0.5
119	119	119	ILLUSTRATION #	119	119	119
			REAR			
-1.75 ±0.08 -1 3/4 ± 1/16	-2.1 ±0.08 -2 1/8 ± 1/16	-1.75 ±0.08 -1 3/4 ± 1/16	CAMBER (Degrees)	-2.1 ±0.08 -2 1/8 ± 1/16	-2.1 ±0.08 -2 1/8 ± 1/16	-1.75 ±0.08 -1 3/4 ± 1/16
0.33° ±0.07° 11/32° ± 1/16°	0.33° ±0.07° 11/32° ± 1/16°	0.33° ±0.07° 11/32° ± 1/16°	REAR TOE (Degrees)	0.33° ±0.07° 11/32° ± 1/16°	0.33° ±0.07° 11/32° ± 1/16°	0.3° ±0.07° 5/16° ± 1/16°
0.17" ±0.03" 5/32" ± 1/32"	0.17" ±0.03" 5/32" ± 1/32"	0.17" ±0.03" 5/32" ± 1/32"	REAR TOE (Inches)	0.17" ±0.03" 5/32" ± 1/32"	0.17" ±0.03" 5/32" ± 1/32"	0.15" ±0.03" 5/32" ± 1/32"
0.2 (3/16)	0.2 (3/16)	0.2 (3/16)	THRUST ANGLE	0.2 (3/16)	0.2 (3/16)	0.2 (3/16)
29 28	29 28	29 28	ILLUSTRATION #	29 28	29 28	29 28
			NOTES			
57037	57038	57039		57040	57041	57042

MINI

(F55) Cooper 4-Dr. Hardtop	(F56) Cooper 2-Dr. Hardtop	(F57) Cooper Convertible	MODEL	(F60) Cooper Countryman	(F60) Cooper Countryman	Notes
(F55) Cooper S 4-Dr. Hardtop	(F56) Cooper S 2-Dr. Hardtop	(F57) Cooper S Convertible	INFORMATION	(F60) Cooper S Countryman	(F60) Cooper S Countryman	
All Suspensions	All Suspensions	All Suspensions		Std. Susp. (incl. All4)	Sport Susp. (incl. All4)	
2015-2017	2015-2017	2016-2017	YEARS	2017	2017	
p. 304	p. 304	p. 304	INSTRUCTIONS	p. 304	p. 304	
			LUGNUT TORQUE			
			FRONT			
Caster is not specified by the manufacturer.	Caster is not specified by the manufacturer.	Caster is not specified by the manufacturer.	CASTER (Degrees)	Caster is not specified by the manufacturer.	Caster is not specified by the manufacturer.	
0.0 (0) | 0.5 (½)	0.0 (0) | 0.5 (½)	0.0 (0) | 0.5 (½)	CROSS | TOL.	0.0 (0) | 0.5 (½)	0.0 (0) | 0.5 (½)	
-0.5 ±0.42 -½ ± 7/16	-0.5 ±0.42 -½ ± 7/16	-0.5 ±0.42 -½ ± 7/16	CAMBER (Degrees)	-0.5 ±0.42 -½ ± 7/16	-0.6 ±0.42 -5/8 ± 7/16	
0.0 (0) | 0.5 (½)	0.0 (0) | 0.5 (½)	0.0 (0) | 0.5 (½)	CROSS | TOL.	0.0 (0) | 0.5 (½)	0.0 (0) | 0.5 (½)	
0.2° ±0.07° 3/16° ± 1/16°	0.2° ±0.07° 3/16° ± 1/16°	0.2° ±0.07° 3/16° ± 1/16°	FRONT TOE (Degrees)	0.3° ±0.07° 5/16° ± 1/16°	0.3° ±0.07° 5/16° ± 1/16°	
0.1" ±0.03" 3/32" ± 1/32"	0.1" ±0.03" 3/32" ± 1/32"	0.1" ±0.03" 3/32" ± 1/32"	FRONT TOE (Inches)	0.15" ±0.03" 5/32" ± 1/32"	0.15" ±0.03" 5/32" ± 1/32"	
11.1 ±0.5 (11 1/8 ± ½)	11.1 ±0.5 (11 1/8 ± ½)	11.1 ±0.5 (11 1/8 ± ½)	SAI	12.3 ±0.5 (12 1/4 ± ½)	12.5 ±0.5 (12 ½ ± ½)	
20.0 / 18.8 ±0.5	20.0 / 18.8 ±0.5	20.0 / 18.7 ±0.5	TURNING ANGLE	36.7 / 30.7 ±0.5	36.7 / 30.7 ±0.5	
119	119	119	ILLUSTRATION #	119	119	
			REAR			
-1.75 ±0.08 -1 3/4 ± 1/16	-1.75 ±0.08 -1 3/4 ± 1/16	-1.75 ±0.08 -1 3/4 ± 1/16	CAMBER (Degrees)	-1.75 ±0.08 -1 3/4 ± 1/16	-2.1 ±0.08 -2 1/8 ± 1/16	
0.3° ±0.07° 5/16° ± 1/16°	0.3° ±0.07° 5/16° ± 1/16°	0.3° ±0.07° 5/16° ± 1/16°	REAR TOE (Degrees)	0.3° ±0.07° 5/16° ± 1/16°	0.3° ±0.07° 5/16° ± 1/16°	
0.15" ±0.03" 5/32" ± 1/32"	0.15" ±0.03" 5/32" ± 1/32"	0.15" ±0.03" 5/32" ± 1/32"	REAR TOE (Inches)	0.15" ±0.03" 5/32" ± 1/32"	0.15" ±0.03" 5/32" ± 1/32"	
0.2 (3/16)	0.2 (3/16)	0.2 (3/16)	THRUST ANGLE	0.2 (3/16)	0.2 (3/16)	
29 28	29 28	29 28	ILLUSTRATION #	29 28	29 28	
			NOTES			
57043	57044	57045		57042	57042	57046

Mitsubishi

3000GT FWD	3000GT AWD (incl. VR-4)	Diamante Sedan	MODEL INFORMATION	Diamante Wagon	Diamante	Eclipse FWD 1.8L
1991-99	1991-99	1992-96	YEARS	1993-96	1997-2004	1990-94
			INSTRUCTIONS			
87-101 ft.lbs.	87-101 ft.lbs.	65-80 ft.lbs.	LUGNUT TORQUE	65-80 ft.lbs.	66-80 ft.lbs.	87-101 ft.lbs.
			FRONT			
3.92 ±0.5 3 15/16 ± ½	3.92 ±0.5 3 15/16 ± ½	2.75 ±0.5 [1] 2¾ ± ½ [1]	CASTER (Degrees)	2.75 ±0.5 2¾ ± ½	3.0 ±0.5 3 ± ½	2.33 ±0.5 2 5/16 ± ½
0.0 (0) / n/s	0.0 (0) / n/s	0.0 (0) / n/s	CROSS / TOL.	0.0 (0) / n/s	0.0 (0) / 0.5 (½)	0.0 (0) / n/s
0.0 ±0.5 0 ± ½	0.0 ±0.5 0 ± ½	0.0 ±0.5 [2] 0 ± ½ [2]	CAMBER (Degrees)	0.0 ±0.5 0 ± ½	0.0 ±0.5 0 ± ½	0.23 ±0.5 ¼ ± ½
0.0 (0) / 0.5 (½)	0.0 (0) / 0.5 (½)	0.0 (0) / n/s	CROSS / TOL.	0.0 (0) / n/s	0.0 (0) / 0.5 (½)	0.0 (0) / n/s
0.0° ±0.24° 0° ± ¼°	0.0° ±0.24° 0° ± ¼°	0.0° ±0.24° 0° ± ¼°	FRONT TOE (Degrees)	0.0° ±0.24° 0° ± ¼°	0.0° ±0.24° 0° ± ¼°	0.0° ±0.24° 0° ± ¼°
0.0" ±0.12" 0" ± ⅛"	0.0" ±0.12" 0" ± ⅛"	0.0" ±0.12" 0" ± ⅛"	FRONT TOE (Inches)	0.0" ±0.12" 0" ± ⅛"	0.0" ±0.12" 0" ± ⅛"	0.0" ±0.12" 0" ± ⅛"
			SAI			
33.8 / 28.4	33.8 / 28.4		TURNING ANGLE		39.0 / 30.5 ±2.0	35.0 / 29.3 ±2.0 [1]
22	22		ILLUSTRATION #			
			REAR			
0.0 ±0.5 0 ± ½	-0.17 ±0.5 -3/16 ± ½	0.0 ±0.5 0 ± ½	CAMBER (Degrees)	0.0 ±0.25 0 ± ¼	-0.67 ±0.5 [1] -11/16 ± ½ [1]	-0.75 ±0.5 -¾ ± ½
0.04° ±0.2° 1/32° ± 3/16°	0.04° ±0.2° 1/32° ± 3/16°	0.0° ±0.24° 0° ± ¼°	REAR TOE (Degrees)	0.0° ±0.24° 0° ± ¼°	0.24° ±0.24° ¼° ± ¼°	0.0° ±0.24° 0° ± ¼°
0.02" ±0.1" 1/64" ± 3/32"	0.02" ±0.1" 1/64" ± 3/32"	0.0" ±0.12" 0" ± ⅛"	REAR TOE (Inches)	0.0" ±0.12" 0" ± ⅛"	0.12" ±0.12" ⅛" ± ⅛"	0.0" ±0.12" 0" ± ⅛"
			THRUST ANGLE			
40 05	44	40 05	ILLUSTRATION #		06	
		[1] 1992-93 w/ECS = 2.0º (2º) [2] 1992-93 w/ECS=0.33º (5/16º)	NOTES		[1] w/16" tires= -0.8º (-13/16º)	[1] w/MS 37º/30º
31001	31002	31004		31005	31006	31007

Mitsubishi

Eclipse FWD 2.0L	Eclipse AWD	Eclipse FWD with 14" tires	MODEL INFORMATION	Eclipse FWD with 16" tires	Eclipse AWD AWD	Eclipse
1990-94	1990-94	1995-99	YEARS	1995-99	1995-99	2000-2005
			INSTRUCTIONS			
87-101 ft.lbs.	87-101 ft.lbs.	87-101 ft.lbs.	LUGNUT TORQUE	87-101 ft.lbs.	87-101 ft.lbs.	66-80 ft.lbs.
			FRONT			
2.4 ±0.5 2⅜ ± ½	2.3 ±0.5 2 5/16 ± ½	4.67 ±1.5 4 11/16 ± 1½	CASTER (Degrees)	4.67 ±1.5 4 11/16 ± 1½	4.67 ±1.5 4 11/16 ± 1½	3.0 ±0.5 3 ± ½
0.0 (0) / n/s	0.0 (0) / n/s	0.0 (0) / n/s	CROSS / TOL.	0.0 (0) / n/s	0.0 (0) / n/s	0.0 (0) / 0.5 (½)
0.08 ±0.5 1/16 ± ½	0.17 ±0.5 3/16 ± ½	-0.08 ±0.5 -1/16 ± ½	CAMBER (Degrees)	-0.33 ±0.5 -5/16 ± ½	-0.08 ±0.5 -1/16 ± ½	0.0 ±0.5 0 ± ½
0.0 (0) / n/s	0.0 (0) / n/s	0.0 (0) / n/s	CROSS / TOL.	0.0 (0) / n/s	0.0 (0) / n/s	0.0 (0) / 0.5 (½)
0.0° ±0.24° 0° ± ¼°	0.0° ±0.24° 0° ± ¼°	0.0° ±0.24° 0° ± ¼°	FRONT TOE (Degrees)	0.0° ±0.24° 0° ± ¼°	0.0° ±0.24° 0° ± ¼°	0.0° ±0.24° 0° ± ¼°
0.0" ±0.12" 0" ± ⅛"	0.0" ±0.12" 0" ± ⅛"	0.0" ±0.12" 0" ± ⅛"	FRONT TOE (Inches)	0.0" ±0.12" 0" ± ⅛"	0.0" ±0.12" 0" ± ⅛"	0.0" ±0.12" 0" ± ⅛"
			SAI			
35.0 / 29.3 ±2.0 [1,2]	31.5 / 27.0 ±2.0 [1,2]	32.5 / 27.0 ±2.0	TURNING ANGLE	32.5 / 27.0 ±2.0	32.5 / 27.0 ±2.0	36.2 / 30.4 ±2.0 [1]
			ILLUSTRATION #			26
			REAR			
-0.75 ±0.5 -¾ ± ½	-1.55 ±0.5 -1 9/16 ± ½	-1.33 ±0.5 -1 5/16 ± ½	CAMBER (Degrees)	-1.67 ±0.5 -1 11/16 ± ½	-1.33 ±0.5 -1 5/16 ± ½	-1.33 ±0.5 -1 5/16 ± ½
0.0° ±0.24° 0° ± ¼°	0.28° ±0.24° 9/32° ± ¼°	0.24° ±0.24° ¼° ± ¼°	REAR TOE (Degrees)	0.24° ±0.24° ¼° ± ¼°	0.24° ±0.24° ¼° ± ¼°	0.24° ±0.24° ¼° ± ¼°
0.0" ±0.12" 0" ± ⅛"	0.14" ±0.12" ⅛" ± ⅛"	0.12" ±0.12" ⅛" ± ⅛"	REAR TOE (Inches)	0.12" ±0.12" ⅛" ± ⅛"	0.12" ±0.12" ⅛" ± ⅛"	0.12" ±0.12" ⅛" ± ⅛"
		0.15 (5/32)	THRUST ANGLE	0.15 (5/32)	0.15 (5/32)	
	44	06	ILLUSTRATION #	06	06	06
[1] w/MS 37º/30º [2] w/Turbo A/T 31.5º/27º	[1] w/MS 37º/30º [2] w/Man Trans. 35º/29.3º		NOTES			[1] 3.0L=31º/27º
31008	31009	31010		31011	31012	31013

ALIGNMENT SPECIFICATIONS

Mitsubishi

Eclipse Spyder	Eclipse Eclipse Spyder	Eclipse Eclipse Spyder	MODEL INFORMATION	Endeavor	Endeavor	Expo/Expo LRV FWD
2001-2005	2006-2010	2011-2012	YEARS	2004-2008	2010-2011	1992-96
			INSTRUCTIONS			
66-80 ft.lbs.	66-80 ft.lbs.	66-80 ft.lbs.	LUGNUT TORQUE	66-80 ft.lbs.	66-80 ft.lbs.	65-80 ft.lbs.
			FRONT			
3.0 ±0.5 3 ± ½	3.0 ±0.5 3 ± ½	3.17 ±0.5 3 3/16 ± ½	CASTER (Degrees)	3.0 ±0.5 3 ± ½	3.0 ±0.5 3 ± ½	2.17 ±0.67 2 3/16 ± 11/16
0.0 (0) | 0.5 (½)	0.0 (0) | 0.5 (½)	0.0 (0) | 0.5 (½)	CROSS | TOL.	0.0 (0) | 0.5 (½)	0.0 (0) | 0.5 (½)	0.0 (0) | n/s
0.0 ±0.5 0 ± ½	0.0 ±0.5 0 ± ½	-0.17 ±0.5 - 3/16 ± ½	CAMBER (Degrees)	0.0 ±0.5 0 ± ½	0.0 ±0.5 0 ± ½	0.33 ±0.5 5/16 ± ½
0.0 (0) | 0.5 (½)	0.0 (0) | 0.5 (½)	0.0 (0) | 0.5 (½)	CROSS | TOL.	0.0 (0) | 0.5 (½)	0.0 (0) | 0.5 (½)	0.0 (0) | n/s
0.0° ±0.24° 0° ± ¼°	0.0° ±0.24° 0° ± ¼°	0.0° ±0.24° 0° ± ¼°	FRONT TOE (Degrees)	0.0° ±0.24° 0° ± ¼°	0.0° ±0.24° 0° ± ¼°	0.0° ±0.24° 0° ± ¼°
0.0" ±0.12" 0" ± ⅛"	0.0" ±0.12" 0" ± ⅛"	0.0" ±0.12" 0" ± ⅛"	FRONT TOE (Inches)	0.0" ±0.12" 0" ± ⅛"	0.0" ±0.12" 0" ± ⅛"	0.0" ±0.12" 0" ± ⅛"
	12.9 ±1.5 (12 ⅞ ± 1 ½)	13.3 ±1.5 (13 ¼ ± 1 ½)	SAI	12.8 ±1.5 (12 13/16 + 1 ½)	12.8 ±1.5 (12 13/16 ± 1 ½)	
36.2 / 30.4 ±2.0 [1]	31.8 / 27.1 ±2.0	31.8 / 27.1 ±2.0	TURNING ANGLE	35.6 / 30.0 ±1.5	35.6 / 30.0 ±2.0 [1]	36.5 / 30.5 ±2.0
26	26	26	ILLUSTRATION #	26	26	
			REAR			
-1.17 ±0.5 -1 3/16 ± ½	-0.83 ±0.5 - 13/16 ± ½	-1.0 ±0.5 -1 ± ½	CAMBER (Degrees)	-0.83 ±0.5 - 13/16 ± ½	-0.83 ±0.5 - 13/16 ± ½	-0.5 ±0.5 - ½ ± ½
0.24° ±0.24° ¼° ± ¼°	0.24° ±0.24° ¼° ± ¼°	0.24° ±0.24° ¼° ± ¼°	REAR TOE (Degrees)	0.24° ±0.24° ¼° ± ¼°	0.24° ±0.24° ¼° ± ¼°	0.16° +0.24° -0.16° 5/32° + ¼° - 5/32°
0.12" ±0.12" ⅛" ± ⅛"	0.12" ±0.12" ⅛" ± ⅛"	0.12" ±0.12" ⅛" ± ⅛"	REAR TOE (Inches)	0.12" ±0.12" ⅛" ± ⅛"	0.12" ±0.12" ⅛" ± ⅛"	0.08" +0.12" -0.08" 3/32" + ⅛" - 3/32"
			THRUST ANGLE			
06	151	151	ILLUSTRATION #	151	151	27
[1] 3.0L=33.8º/28.5º			NOTES		[1] w/18" wheels = 33º/28.2º	
31014	31015	31016		31017	31018	31019

Mitsubishi

Expo/Expo LRV AWD	Galant FWD	Galant AWD	MODEL INFORMATION	Galant	Galant	Galant
1992-96	1990-93	1990-93	YEARS	1994-98	1999-2003	2004-2010
			INSTRUCTIONS			
65-80 ft.lbs.	65-80 ft.lbs.	65-80 ft.lbs.	LUGNUT TORQUE	65-80 ft.lbs.	66-81 ft.lbs.	66-80 ft.lbs.
			FRONT			
2.08 ±0.67 2 1/16 ± 11/16	2.0 ±0.5 2 ± ½	1.93 ±0.5 1 15/16 ± ½	CASTER (Degrees)	4.33 ±1.5 4 5/16 ± 1 ½	3.0 ±0.5 3 ± ½	3.0 ±0.5 3 ± ½
0.0 (0) | n/s	0.0 (0) | n/s	0.0 (0) | n/s	CROSS | TOL.	0.0 (0) | 0.5 (½)	0.0 (0) | 0.5 (½)	0.0 (0) | 0.5 (½)
0.67 ±0.5 11/16 ± ½	0.37 ±0.5 ⅜ ± ½	0.52 ±0.5 ½ ± ½	CAMBER (Degrees)	0.0 ±0.5 0 ± ½	0.0 ±0.5 0 ± ½	0.0 ±0.5 0 ± ½
0.0 (0) | n/s	0.0 (0) | n/s	0.0 (0) | n/s	CROSS | TOL.	0.0 (0) | 0.5 (½)	0.0 (0) | 0.5 (½)	0.0 (0) | 0.5 (½)
0.0° ±0.24° 0° ± ¼°	0.0° ±0.24° 0° ± ¼°	0.0° ±0.24° 0° ± ¼°	FRONT TOE (Degrees)	0.0° ±0.24° 0° ± ¼°	0.0° ±0.24° 0° ± ¼°	0.0° ±0.24° 0° ± ¼°
0.0" ±0.12" 0" ± ⅛"	0.0" ±0.12" 0" ± ⅛"	0.0" ±0.12" 0" ± ⅛"	FRONT TOE (Inches)	0.0" ±0.12" 0" ± ⅛"	0.0" ±0.12" 0" ± ⅛"	0.0" ±0.12" 0" ± ⅛"
			SAI			12.9 ±1.5 (12 ⅞ ± 1 ½)
36.5 / 30.5 ±2.0	37.5 / 30.8 ±2.0	37.5 / 30.8 ±2.0 [1]	TURNING ANGLE	39.0 / 30.5 ±2.0	36.6 / 30.7 ±2.0 [1]	37.2 / 30.3 ±2.0 [1]
			ILLUSTRATION #			26
			REAR			
-0.5 ±0.5 - ½ ± ½	-0.75 ±0.5 - ¾ ± ½	-1.0 ±0.5 -1 ± ½	CAMBER (Degrees)	-1.33 ±0.5 -1 5/16 ± ½	-1.33 ±0.5 [2] -1 5/16 ± ½ [2]	-0.83 ±0.5 - 13/16 ± ½
0.16° +0.24° -0.16° 5/32° + ¼° - 5/32°	0.0° ±0.24° 0° ± ¼°	0.24° ±0.24° ¼° ± ¼°	REAR TOE (Degrees)	0.24° ±0.24° ¼° ± ¼°	0.24° ±0.24° ¼° ± ¼°	0.24° ±0.24° ¼° ± ¼°
0.08" +0.12" -0.08" 3/32" + ⅛" - 3/32"	0.0" ±0.12" 0" ± ⅛"	0.12" ±0.12" ⅛" ± ⅛"	REAR TOE (Inches)	0.12" ±0.12" ⅛" ± ⅛"	0.12" ±0.12" ⅛" ± ⅛"	0.12" ±0.12" ⅛" ± ⅛"
			THRUST ANGLE			
27		44	ILLUSTRATION #	06	06	151
		[1] w/AT 35º/29.3º	NOTES		[1] 3.0L=33.1º/28.5º [2] 1999 = -1.0º (-1º)	[1] w/18" wheels = 32.9º/27.8º
31020	31022	31023		31024	31025	31026

Mitsubishi

Galant	I-MiEV	Lancer	MODEL INFORMATION	Lancer exc/Evolution w/14" wheels	Lancer exc/Evolution w/15" wheels	Lancer exc/Evolution w/16" wheels
2011-2012	2012-2017	2002-2003	YEARS	2004-2006	2004-2006	2004-2006
			INSTRUCTIONS			
66-80 ft.lbs.	66-80 ft.lbs.	66-80 ft.lbs.	LUGNUT TORQUE	66-80 ft.lbs.	66-80 ft.lbs.	66-80 ft.lbs.
			FRONT			
3.08 ±0.5 3 1/16 ± 1/2	3.83 ±0.75 3 13/16 ± 3/4	2.83 ±0.5 2 13/16 ± 1/2	CASTER (Degrees)	2.92 ±0.5 2 15/16 ± 1/2	2.75 ±0.5 [1] 2 3/4 ± 1/2 [1]	2.92 ±0.5 [1] 2 15/16 ± 1/2 [1]
0.0 (0) | 0.5 (1/2)	0.0 (0) | 0.5 (1/2)	0.0 (0) | 0.5 (1/2)	CROSS | TOL.	0.0 (0) | 0.5 (1/2)	0.0 (0) | 0.5 (1/2)	0.0 (0) | 0.5 (1/2)
-0.17 ±0.5 -3/16 ± 1/2	0.0 ±0.75 0 ± 3/4	0.0 ±0.5 0 ± 1/2	CAMBER (Degrees)	0.0 ±0.5 0 ± 1/2	0.08 ±0.5 1/16 ± 1/2	-0.08 ±0.5 -1/16 ± 1/2
0.0 (0) | 0.5 (1/2)	0.0 (0) | 0.5 (1/2)	0.0 (0) | 0.5 (1/2)	CROSS | TOL.	0.0 (0) | 0.5 (1/2)	0.0 (0) | 0.5 (1/2)	0.0 (0) | 0.5 (1/2)
0.0° ±0.24° 0° ± 1/4°	0.2° ±0.2° 3/16° ± 3/16°	0.08° ±0.16° 3/32° ± 5/32°	FRONT TOE (Degrees)	0.08° ±0.16° 3/32° ± 5/32°	0.08° ±0.16° 3/32° ± 5/32°	0.08° ±0.16° 3/32° ± 5/32°
0.0" ±0.12" 0" ± 1/8"	0.1" ±0.1" 3/32" ± 3/32"	0.04" ±0.08" 1/32" ± 3/32"	FRONT TOE (Inches)	0.04" ±0.08" 1/32" ± 3/32"	0.04" ±0.08" 1/32" ± 3/32"	0.04" ±0.08" 1/32" ± 3/32"
12.9 ±1.5 (12 7/8 ±1 1/2)	15.3 ±1.5 (15 1/4 ±1 1/2)	12.6 ±1.5 (12 9/16 ±1 1/2)	SAI	12.7 ±1.5 (12 11/16 ±1 1/2)	12.5 ±1.5 (12 1/2 ±1 1/2)	12.7 ±1.5 (12 11/16 ±1 1/2)
37.2 / 30.3 ±2.0 [1]	43.0 / 35.0 ±2.0	40.7 / 33.3 ±1.5 [1]	TURNING ANGLE	40.7 / 33.3 ±1.5	39.5 / 32.5 ±1.5	33.7 / 28.8 ±1.5
26			ILLUSTRATION #			
			REAR			
-1.17 ±0.5 -1 3/16 ± 1/2	0.0 ±0.5 0 ± 1/2	-0.67 ±0.5 -11/16 ± 1/2	CAMBER (Degrees)	-0.67 ±0.5 -11/16 ± 1/2	-0.67 ±0.5 -11/16 ± 1/2	-0.67 ±0.5 -11/16 ± 1/2
0.24° ±0.24° 1/4° ± 1/4°	0.0° ±0.2° 0° ± 3/16°	0.24° ±0.16° 1/4° ± 5/32°	REAR TOE (Degrees)	0.24° ±0.16° 1/4° ± 5/32°	0.24° ±0.16° 1/4° ± 5/32°	0.24° ±0.16° 1/4° ± 5/32°
0.12" ±0.12" 1/8" ± 1/8"	0.0" ±0.1" 0" ± 3/32"	0.12" ±0.08" 1/8" ± 3/32"	REAR TOE (Inches)	0.12" ±0.08" 1/8" ± 3/32"	0.12" ±0.08" 1/8" ± 3/32"	0.12" ±0.08" 1/8" ± 3/32"
		0.15 (5/32)	THRUST ANGLE	0.15 (5/32)	0.15 (5/32)	0.15 (5/32)
151		141	ILLUSTRATION #	141	141	141
[1] w/18" wheels = 32.9º/27.8º		[1] w/15" tires 39.5º/32.5º	NOTES		[1] H'back = 2.58º (2 9/16º)	[1] H'back = 2.75º (2 3/4º)
31027	31028	31029		31030	31031	31032

Mitsubishi

Lancer exc/Evolution	Lancer Evolution	Lancer Evolution	MODEL INFORMATION	Mirage	Mirage	Mirage
2008-2017	2003-2006	2008-2015	YEARS	1990-92	1993-96	1997-2002
			INSTRUCTIONS			
66-80 ft.lbs.	66-80 ft.lbs.	66-80 ft.lbs.	LUGNUT TORQUE	65-80 ft.lbs.	65-80 ft.lbs.	65-80 ft.lbs.
			FRONT			
2.67 ±0.5 2 11/16 ± 1/2	3.92 ±0.5 3 15/16 ± 1/2	4.42 ±0.5 4 7/16 ± 1/2	CASTER (Degrees)	2.33 ±0.5 2 5/16 ± 1/2	2.25 ±0.5 2 1/4 ± 1/2	2.85 ±0.5 2 7/8 ± 1/2
0.0 (0) | 0.5 (1/2)	0.0 (0) | 0.5 (1/2)	0.0 (0) | 0.5 (1/2)	CROSS | TOL.	0.0 (0) | n/s	0.0 (0) | 0.5 (1/2)	0.0 (0) | 0.5 (1/2)
-0.08 ±0.5 -1/16 ± 1/2	-1.0 ±0.5 [1] -1 ± 1/2 [1]	-1.0 ±0.5 -1 ± 1/2	CAMBER (Degrees)	0.0 ±0.5 0 ± 1/2	0.0 ±0.5 0 ± 1/2	0.0 ±0.5 0 ± 1/2
0.0 (0) | 0.5 (1/2)	0.0 (0) | 0.5 (1/2)	0.0 (0) | 0.5 (1/2)	CROSS | TOL.	0.0 (0) | n/s	0.0 (0) | n/s	0.0 (0) | 0.5 (1/2)
0.08° ±0.16° 3/32° ± 5/32°	0.0° ±0.16° 0° ± 5/32°	0.0° ±0.16° 0° ± 5/32°	FRONT TOE (Degrees)	0.0° ±0.24° 0° ± 1/4°	0.0° ±0.24° 0° ± 1/4°	0.0° ±0.24° 0° ± 1/4°
0.04" ±0.08" 1/32" ± 3/32"	0.0" ±0.08" 0" ± 3/32"	0.0" ±0.08" 0" ± 3/32"	FRONT TOE (Inches)	0.0" ±0.12" 0" ± 1/8"	0.0" ±0.12" 0" ± 1/8"	0.0" ±0.12" 0" ± 1/8"
13.5 ±1.5 (13 1/2 ±1 1/2)	13.8 ±1.5 (13 3/4 ±1 1/2)	13.8 ±1.5 (13 13/16 ±1 1/2)	SAI			
40.8 / 33.8 ±1.5	31.8 / 27.3 ±1.5	33.2 / 28.2 ±1.5	TURNING ANGLE	41.0 / 35.0 ±1.5 [1]	37.3 / 31.0 ±1.0	37.5 / 31.7 ±2.0
	157		ILLUSTRATION #			
			REAR			
-0.92 ±0.5 -15/16 ± 1/2	-1.0 ±0.5 -1 ± 1/2	-1.0 ±0.5 -1 ± 1/2	CAMBER (Degrees)	-0.67 ±0.58 -11/16 ± 9/16	-0.67 ±0.5 -11/16 ± 1/2	-0.67 ±0.5 -11/16 ± 1/2
0.24° ±0.16° 1/4° ± 5/32°	0.24° ±0.16° 1/4° ± 5/32°	0.24° ±0.16° 1/4° ± 5/32°	REAR TOE (Degrees)	0.0° ±0.36° 0° ± 3/8°	0.24° ±0.16° 1/4° ± 5/32°	0.24° ±0.16° 1/4° ± 5/32°
0.12" ±0.08" 1/8" ± 3/32"	0.12" ±0.08" 1/8" ± 3/32"	0.12" ±0.08" 1/8" ± 3/32"	REAR TOE (Inches)	0.0" ±0.18" 0" ± 3/16"	0.12" ±0.08" 1/8" ± 3/32"	0.12" ±0.08" 1/8" ± 3/32"
	0.15 (5/32)		THRUST ANGLE			0.15 (5/32)
109	141	177	ILLUSTRATION #		40 05	40 109
	[1] optional -2.0º (-2º)		NOTES	[1] w/PS&175/7013 37º/32.5º		
31033	31034	31035		31039	31040	31041

Mitsubishi

Mirage	Montero	Montero	MODEL INFORMATION	Montero	Montero Sport	Outlander
2014-2017	1990-91	1992-2000	YEARS	2001-2006	1997-2004	2003-2006
72 ft.lbs.	72-87 ft.lbs.	72-87 ft.lbs.	INSTRUCTIONS LUGNUT TORQUE	80 ft.lbs.	72-87 ft.lbs.	66-80 ft.lbs.
			FRONT			
4.0 ± 0.75 4 ± 3/4	2.95 ± 0.5 2 15/16 ± 1/2	3.0 ± 1.0 3 ± 1	CASTER (Degrees)	3.83 ± 0.5 3 13/16 ± 1/2	2.67 ± 1.0 2 11/16 ± 1	3.17 ± 0.5 3 3/16 ± 1/2
0.0 (0) | 0.5 (1/2)	0.0 (0) | 0.5 (1/2)	0.0 (0) | 0.5 (1/2)	CROSS | TOL.	0.0 (0) | 0.5 (1/2)	0.0 (0) | 0.5 (1/2)	0.0 (0) | 0.5 (1/2)
0.08 ± 0.75 1/16 ± 3/4	1.0 ± 0.5 1 ± 1/2	0.67 ± 0.5 11/16 ± 1/2	CAMBER (Degrees)	0.0 ± 0.5 0 ± 1/2	L= 0.92, R= 0.42 ± 0.5 [1] L= 15/16, R= 7/16 ± 1/2 [1]	0.0 ± 0.5 0 ± 1/2
0.0 (0) | 0.5 (1/2)	0.0 (0) | 0.5 (1/2)	0.0 (0) | 0.5 (1/2)	CROSS | TOL.	0.0 (0) | 0.5 (1/2)	0.5 (1/2) | 0.5 (1/2)	0.0 (0) | 0.5 (1/2)
0.0° ± 0.24° 0° ± 1/4°	0.43° ± 0.27° 7/16° ± 9/32°	0.28° ± 0.28° 9/32° ± 9/32°	FRONT TOE (Degrees)	0.2° ± 0.2° 3/16° ± 3/16°	0.28° ± 0.28° 9/32° ± 9/32°	0.08° ± 0.16° 3/32° ± 5/32°
0.0" ± 0.12" 0" ± 1/8"	0.22" ± 0.14" 7/32" ± 1/8"	0.14" ± 0.14" 5/32" ± 5/32"	FRONT TOE (Inches)	0.1" ± 0.1" 3/32" ± 3/32"	0.14" ± 0.14" 5/32" ± 5/32"	0.04" ± 0.08" 1/32" ± 3/32"
11.6 ± 1.5 (11 9/16 ± 1 1/2)		14.9 (14 7/8)	SAI			12.1 ± 1.5 (12 1/16 ± 1 1/2)
41.3 / 35.2 ± 1.5	32.5 / 29.0 ± 2.0	21.9 / 20.0	TURNING ANGLE	36.5 / 31.7 ± 1.5	31.0 / 29.5 ± 1.5	34.8 / 29.3 ± 1.5
	45 23	23	ILLUSTRATION #	12	23	
			REAR			
-1.0 ± 0.75 -1 ± 3/4			CAMBER (Degrees)	0.0 ± 0.5 0 ± 1/2		-0.67 ± 0.5 -11/16 ± 1/2
0.2° ± 0.24° 3/16° ± 1/4°			REAR TOE (Degrees)	0.24° ± 0.24° 1/4° ± 1/4°		0.24° ± 0.16° 1/4° ± 5/32°
0.1" ± 0.12" 3/32" ± 1/8"			REAR TOE (Inches)	0.12" ± 0.12" 1/8" ± 1/8"		0.12" ± 0.08" 1/8" ± 3/32"
			THRUST ANGLE			
			ILLUSTRATION #	141		109
			NOTES		[1] 1997-98 = 0.67° (11/16°)	
31042	31043	31044		31045	31046	31047

Mitsubishi

Outlander	Outlander	Outlander Sport (USA) RVR (Canada)	MODEL INFORMATION	Outlander Sport (USA) RVR (Canada)	Pickup 4X2	Pickup 4X4
2007-2013	2014-2017	2016-2017	YEARS	2011-2015	1990-96	1990-95
66-80 ft.lbs.	72 ft.lbs.	66-80 ft.lbs.	INSTRUCTIONS LUGNUT TORQUE	66-80 ft.lbs.	87-101 ft.lbs.	87-101 ft.lbs.
			FRONT			
2.58 ± 0.5 2 9/16 ± 1/2	2.75 ± 0.5 [1] 2 3/4 ± 1/2 [1]	2.75 ± 0.5 2 3/4 ± 1/2	CASTER (Degrees)	2.75 ± 0.5 2 3/4 ± 1/2	2.5 ± 1.0 2 1/2 ± 1	2.0 ± 1.0 2 ± 1
0.0 (0) | 0.5 (1/2)	0.0 (0) | 0.5 (1/2)	0.0 (0) | 0.5 (1/2)	CROSS | TOL.	0.0 (0) | 0.5 (1/2)	0.0 (0) | 0.5 (1/2)	0.0 (0) | 0.5 (1/2)
0.33 ± 0.5 5/16 ± 1/2	0.33 ± 0.5 5/16 ± 1/2	0.33 ± 0.5 5/16 ± 1/2	CAMBER (Degrees)	0.33 ± 0.5 5/16 ± 1/2	0.67 ± 0.5 11/16 ± 1/2	1.0 ± 0.5 1 ± 1/2
0.0 (0) | 0.5 (1/2)	0.0 (0) | 0.5 (1/2)	0.0 (0) | 0.5 (1/2)	CROSS | TOL.	0.0 (0) | 0.5 (1/2)	0.0 (0) | 0.5 (1/2)	0.0 (0) | 0.5 (1/2)
0.08° ± 0.16° 3/32° ± 5/32°	0.07° ± 0.13° 1/16° ± 1/8°	0.08° ± 0.16° 3/32° ± 5/32°	FRONT TOE (Degrees)	0.08° ± 0.16° 3/32° ± 5/32°	0.43° ± 0.27° 7/16° ± 9/32°	0.43° ± 0.27° 7/16° ± 9/32°
0.04" ± 0.08" 1/32" ± 3/32"	0.03" ± 0.07" 1/32" ± 1/16"	0.04" ± 0.08" 1/32" ± 3/32"	FRONT TOE (Inches)	0.04" ± 0.08" 1/32" ± 3/32"	0.22" ± 0.14" 7/32" ± 1/8"	0.22" ± 0.14" 7/32" ± 1/8"
12.7 ± 1.5 (12 3/4 ± 1 1/2)	12.8 ± 1.5 (12 3/4 ± 1 1/2)	12.8 ± 1.5 (12 3/4 ± 1 1/2)	SAI	12.8 ± 1.5 (12 3/4 ± 1 1/2)		
38.7 / 32.3 ± 1.5	38.8 / 32.3 ± 1.5	38.8 / 32.3 ± 1.5	TURNING ANGLE	38.8 / 32.3 ± 1.5	37.0 / 30.5 ± 2.0	30.0 / 27.0 ± 2.0
			ILLUSTRATION #		23	23
			REAR			
-0.42 ± 0.5 -7/16 ± 1/2	-0.5 ± 0.75 -1/2 ± 3/4	-0.5 ± 0.5 -1/2 ± 1/2	CAMBER (Degrees)	-0.42 ± 0.5 -7/16 ± 1/2		
0.24° ± 0.16° 1/4° ± 5/32°	0.23° ± 0.17° 7/32° ± 5/32°	0.24° ± 0.16° 1/4° ± 5/32°	REAR TOE (Degrees)	0.24° ± 0.16° 1/4° ± 5/32°		
0.12" ± 0.08" 1/8" ± 3/32"	0.12" ± 0.08" 1/8" ± 3/32"	0.12" ± 0.08" 1/8" ± 3/32"	REAR TOE (Inches)	0.12" ± 0.08" 1/8" ± 3/32"		
			THRUST ANGLE			
109	109	109	ILLUSTRATION #	109		
	[1] 2016 tol. - 0.75° (3/4°)		NOTES			
31048	31049	31050		31050	31053	31054

ALIGNMENT SPECIFICATIONS

Mitsubishi

Precis	Raider 4X2 & 4X4	Sigma	MODEL INFORMATION	Van/Wagon	Notes	Notes
1990-93	2006-2009	1990	YEARS	1990		
			INSTRUCTIONS			
65-80 ft.lbs.	135 ft.lbs.	65-80 ft.lbs.	LUGNUT TORQUE	87-101 ft.lbs.		
			FRONT			
1.67 ±0.5 [1] 1 11/16 ± 1/2 [1]	L= 3.5, R= 3.7 ± 0.5 L= 3 1/2, R= 3 11/16 ± 1/2	0.67 ±0.5 11/16 ± 1/2	CASTER (Degrees)	3.0 ±1.0 3 ±1		
0.0 (0) | n/s	-0.2 (-3/16) | 0.5 (1/2)	0.0 (0) | n/s	CROSS | TOL.	0.0 (0) | 0.5 (1/2)		
0.0 ±0.5 0 ± 1/2	-0.25 ±0.5 -1/4 ± 1/2	0.5 ±0.5 1/2 ± 1/2	CAMBER (Degrees)	0.5 ±0.75 1/2 ± 3/4		
0.0 (0) | n/s	0.0 (0) | 0.5 (1/2)	0.0 (0) | n/s	CROSS | TOL.	0.0 (0) | 0.5 (1/2)		
0.08° ±0.24° 3/32° ± 1/4°	0.2° ±0.1° 3/16° ± 3/32°	0.0° ±0.24° 0° ± 1/4°	FRONT TOE (Degrees)	0.08° ±0.24° 1/16° ± 1/4°		
0.04" ±0.12" 1/32" ± 1/8"	0.1" ±0.05" 3/32" ± 1/16"	0.0" ±0.12" 0" ± 1/8"	FRONT TOE (Inches)	0.04" ±0.12" 1/32" ± 1/8"		
13.0 (13)			SAI			
37.4 / 31.5 ±1.5			TURNING ANGLE			
03	158		ILLUSTRATION #	14		
			REAR			
-0.67 +0.5 -0.33 -11/16 +1/2 -5/16		-0.75 ±0.25 -3/4 ± 1/4	CAMBER (Degrees)			
0.0° +0.47° -0.23° 0° +15/32° -7/32°		0.0° ±0.24° 0° ± 1/4°	REAR TOE (Degrees)			
0.0" +0.23" -0.12" 0" +7/32" -1/8"		0.0" ±0.12" 0" ± 1/8"	REAR TOE (Inches)			
			THRUST ANGLE			
			ILLUSTRATION #			
[1] w/MS 1.03° (1°)			NOTES			
31056	31058	31059		31064	31065	31066

Nissan

200SX	240SX	240SX	MODEL INFORMATION	300ZX	350Z	350Z
1995-98	1990-94	1995-98	YEARS	1990-96	2003-2005	2006-2009
			INSTRUCTIONS			
72-87 ft.lbs.		72-87 ft.lbs.	LUGNUT TORQUE		73-93 ft.lbs.[1]	80 ft.lbs.
			FRONT			
1.42 ±0.75 1 7/16 ± 3/4	6.75 ±0.75 6 3/4 ± 3/4	6.75 ±0.75 6 3/4 ± 3/4	CASTER (Degrees)	9.67 ±0.75 9 11/16 ± 3/4	8.17 ±0.75 8 3/16 ± 3/4	8.17 ±0.75 8 3/16 ± 3/4
0.0 (0) | 1.0 (1)	0.0 (0) | n/s	0.0 (0) | 0.75 (3/4)	CROSS | TOL.	0.0 (0) | 0.75 (3/4)	0.0 (0) | 0.75 (3/4)	0.0 (0) | 0.75 (3/4)
-0.58 ±0.75 -9/16 ± 3/4	-0.75 ±0.75 -3/4 ± 3/4	-0.75 ±0.75 -3/4 ± 3/4	CAMBER (Degrees)	-0.83 ±0.75 -13/16 ± 3/4	-0.58 ±0.75 -9/16 ± 3/4	-0.58 ±0.75 -9/16 ± 3/4
0.0 (0) | 1.0 (1)	0.0 (0) | n/s	0.0 (0) | 0.75 (3/4)	CROSS | TOL.	0.0 (0) | 0.75 (3/4)	0.0 (0) | 0.75 (3/4)	0.0 (0) | 0.75 (3/4)
0.2° ±0.2° 3/16° ± 3/16°	0.13° ±0.09° 1/8° ± 3/32°	0.23° ±0.1° 7/32° ± 3/32°	FRONT TOE (Degrees)	0.08° ±0.08° [1] 3/32° ± 3/32° [1]	0.1° ±0.1° 3/32° ± 3/32°	0.08° ±0.08° 3/32° ± 3/32°
0.1" ±0.1" 3/32" ± 3/32"	0.06" ±0.05" 1/16" ± 1/32"	0.12" ±0.05" 1/8" ± 1/16"	FRONT TOE (Inches)	0.04" ±0.04" [1] 1/32" ± 1/32" [1]	0.05" ±0.05" 1/16" ± 1/16"	0.04" ±0.04" 1/32" ± 1/32"
14.8 ±0.8 (14 3/4 ± 3/4)	13.3 ±0.8 (13 1/4 ± 3/4)	13.7 ±0.8 (13 11/16 ± 3/4)	SAI	12.9 ±0.8 (12 15/16 ± 3/4)	5.1 ±0.8 (5 1/16 ± 3/4)	5.1 ±0.8 (5 1/16 ± 3/4)
37.0 / 31.0 ±1.0 [1]	41.0 / 33.0 ±2.0	42.0 / 33.0 ±1.0	TURNING ANGLE	34.5 / 28.5 ±2.0	38.9 / 30.7 ±1.0	38.9 / 30.7 ±1.0
			ILLUSTRATION #			
			REAR			
-1.0 ±0.75 -1 ± 3/4	-1.1 ±0.5 -1 1/8 ± 1/2	-1.17 ±0.5 -1 3/16 ± 1/2	CAMBER (Degrees)	-1.1 ±0.5 -1 1/8 ± 1/2	-1.58 ±0.5 -1 9/16 ± 1/2	-1.58 ±0.5 -1 9/16 ± 1/2
0.1° ±0.4° 3/32° ± 13/32°	0.23° ±0.18° 7/32° ± 3/16°	0.23° ±0.23° 7/32° ± 7/32°	REAR TOE (Degrees)	0.22° ±0.18° 7/32° ± 3/16°	0.1° ±0.08° [2] 3/32° ± 3/32° [2]	0.15° ±0.06° 5/32° ± 1/16°
0.05" ±0.2" 1/16" ± 3/16"	0.12" ±0.09" 1/8" ± 3/32"	0.12" ±0.12" 1/8" ± 1/8"	REAR TOE (Inches)	0.11" ±0.09" 3/32" ± 3/32"	0.05" ±0.04" [2] 1/16" ± 1/32" [2]	0.08" ±0.03" 1/16" ± 1/32"
			THRUST ANGLE			
	65	65	ILLUSTRATION #	65	151	151
[1] w/MS 41°/34°			NOTES	[1] 2+2 = 0.18° (.09")	[1] 2005 = 80 ft.lbs [2] 18" tires = 0.2° (0.1")	
32004	32007	32008		32017	32019	32020

Nissan

	32021	32022	32040	32041	32042	32043
MODEL	370Z	370Z	Altima	Altima	Altima	Altima
INFORMATION	Regular Grade	Nismo Models				16" Tires (OE)
YEARS	2009-2017	2009-2017	1993-98	1999-2001	2002	2003-2006
INSTRUCTIONS						
LUGNUT TORQUE	80 ft.lbs.	80 ft.lbs.	72-86 ft.lbs.	72-86 ft.lbs.	73-86 ft.lbs.	73-86 ft.lbs.
FRONT						
CASTER (Degrees)	5.17 ±0.75 5 3/16 ± 3/4	5.25 ±0.75 5 1/4 ± 3/4	2.67 ±0.75 2 11/16 ± 3/4	2.67 ±0.75 2 11/16 ± 3/4	2.83 ±0.75 2 13/16 ± 3/4	2.83 ±0.75 2 13/16 ± 3/4
CROSS | TOL.	0.0 (0) | 0.65 (5/8)	0.0 (0) | 0.65 (5/8)	0.0 (0) | 0.75 (3/4)	0.0 (0) | 0.75 (3/4)	0.0 (0) | 0.75 (3/4)	0.0 (0) | 0.75 (3/4)
CAMBER (Degrees)	-0.67 ±0.75 -11/16 ± 3/4	-0.75 ±0.75 -3/4 ± 3/4	-0.1 ±0.75 -1/8 ± 3/4	-0.1 ±0.75 -1/8 ± 3/4	-0.25 ±0.75 -1/4 ± 3/4	-0.25 ±0.75 -1/4 ± 3/4
CROSS | TOL.	0.0 (0) | 0.55 (9/16)	0.0 (0) | 0.55 (9/16)	0.0 (0) | 0.75 (3/4)	0.0 (0) | 0.75 (3/4)	0.0 (0) | 0.75 (3/4)	0.0 (0) | 0.75 (3/4)
FRONT TOE (Degrees)	0.17° ±0.07° 5/32° ± 1/16°	0.17° ±0.07° 5/32° ± 1/16°	0.1° ±0.1° 3/32° ± 3/32°	0.1° ±0.1° 3/32° ± 3/32°	0.04° ±0.08° 1/32° ± 3/32°	0.04° ±0.08° 1/32° ± 3/32°
FRONT TOE (Inches)	0.08" ±0.03" 3/32" ± 1/32"	0.08" ±0.03" 3/32" ± 1/32"	0.05" ±0.05" 1/16" ± 1/16"	0.05" ±0.05" 1/16" ± 1/16"	0.02" ±0.04" 1/64" ± 1/32"	0.02" ±0.04" 1/64" ± 1/32"
SAI	7.7 ±0.8 (7 11/16 ± 3/4)	7.8 ±0.8 (7 3/4 ± 3/4)	14.1 ±0.8 (14 1/16 ± 3/4)	14.1 ±0.8 (14 1/16 ± 3/4)	14.6 ±0.8 (14 9/16 ± 3/4)	14.6 ±0.8 (14 9/10 ± 3/4)
TURNING ANGLE	39.7 / 30.3 ±1.0 [1]	37.2 / 31.8 ±1.0	34.5 / 28.6 ±1.5	35.1 / 29.3 ±1.5	38.0 / 30.5 ±1.5 [1]	38.0 / 30.5 ±1.5
ILLUSTRATION #						
REAR						
CAMBER (Degrees)	-1.67 ±0.5 -1 11/16 ± 1/2	-1.67 ±0.5 -1 11/16 ± 1/2	-1.25 ±0.75 -1 1/4 ± 3/4	-1.25 ±0.75 -1 1/4 ± 3/4	-0.25 ±0.5 -1/4 ± 1/2	-0.57 ±0.5 -9/16 ± 1/2
REAR TOE (Degrees)	0.3° ±0.13° 5/16° ± 1/8°	0.3° ±0.13° 5/16° ± 1/8°	0.2° ±0.1° 3/16° ± 3/32°	0.2° ±0.1° 3/16° ± 3/32°	0.28° ±0.12° 9/32° ± 1/8°	0.3° ±0.12° 5/16° ± 1/8°
REAR TOE (Inches)	0.15" ±0.07" 5/32" ± 1/16"	0.15" ±0.07" 5/32" ± 1/16"	0.1" ±0.05" 3/32" ± 1/16"	0.1" ±0.05" 3/32" ± 1/16"	0.14" ±0.06" 1/8" ± 1/16"	0.15" ±0.06" 5/32" ± 1/16"
THRUST ANGLE						
ILLUSTRATION #	151	151	10	10	151	151
NOTES	[1] 19" tires = 37.2°/31.8°				[1] w/215-55R17 tires 35.5°/29°	

Nissan

	32044	32045	32046	32047	32048	32049
MODEL	Altima	Altima	Altima	Altima	Altima	Altima
INFORMATION	17" Tires (OE)	18" Tires (OE)	Sedan w/16" tires (OE)	Sedan w/17" tires (OE)	Coupe w/17" tires (OE)	Coupe w/18" tires (OE)
YEARS	2003-2006	2005-2006	2007-2012	2007-2012	2008-2013	2008-2013
INSTRUCTIONS						
LUGNUT TORQUE	73-86 ft.lbs.	73-86 ft.lbs.	83 ft.lbs.	83 ft.lbs.	83 ft.lbs.	83 ft.lbs.
FRONT						
CASTER (Degrees)	2.83 ±0.75 2 13/16 ± 3/4	2.83 ±0.75 2 13/16 ± 3/4	4.9 ±0.75 [1] 4 7/8 ± 3/4 [1]	4.9 ±0.75 [1] 4 7/8 ± 3/4 [1]	5.0 ±0.75 5 ± 3/4	5.0 ±0.75 5 ± 3/4
CROSS | TOL.	0.0 (0) | 0.75 (3/4)	0.0 (0) | 0.75 (3/4)	0.0 (0) | 0.55 (9/16)	0.0 (0) | 0.55 (9/16)	0.0 (0) | 0.55 (9/16)	0.0 (0) | 0.55 (9/16)
CAMBER (Degrees)	-0.25 ±0.75 -1/4 ± 3/4	-0.25 ±0.75 -1/4 ± 3/4	L= -0.5, R= -0.75 ±0.75 L= -1/2, R= -3/4 ± 3/4	L= -0.5, R= -0.75 ±0.75 L= -1/2, R= -3/4 ± 3/4	L= -0.5, R= -0.75 ±0.75 L= -1/2, R= -3/4 ± 3/4	L= -0.5, R= -0.75 ±0.75 L= -1/2, R= -3/4 ± 3/4
CROSS | TOL.	0.0 (0) | 0.75 (3/4)	0.0 (0) | 0.75 (3/4)	0.25 (1/4) | 0.55 (9/16)	0.25 (1/4) | 0.55 (9/16)	0.25 (1/4) | 0.55 (9/16)	0.25 (1/4) | 0.55 (9/16)
FRONT TOE (Degrees)	0.04° ±0.08° 1/32° ± 3/32°	0.04° ±0.08° 1/32° ± 3/32°	0.1° ±0.1° 3/32° ± 3/32°	0.1° ±0.1° 3/32° ± 3/32°	0.1° ±0.1° 3/32° ± 3/32°	0.1° ±0.1° 3/32° ± 3/32°
FRONT TOE (Inches)	0.02" ±0.04" 1/64" ± 1/32"	0.02" ±0.04" 1/64" ± 1/32"	0.05" ±0.05" 1/16" ± 1/16"	0.05" ±0.05" 1/16" ± 1/16"	0.05" ±0.05" 1/16" ± 1/16"	0.05" ±0.05" 1/16" ± 1/16"
SAI	14.6 ±0.8 (14 9/16 ± 3/4)	14.6 ±0.8 (14 9/16 ± 3/4)	12.8 ±0.8 (12 3/4 ± 3/4)	12.8 ±0.8 (12 3/4 ± 3/4)	12.8 ±0.8 (12 3/4 ± 3/4)	12.8 ±0.8 (12 3/4 ± 3/4)
TURNING ANGLE	35.5 / 29.0 ±1.5	35.5 / 29.0 ±1.5	38.5 / 31.5 ±1.0	36.5 / 30.5 ±1.0	38.5 / 31.5 ±1.0	35.0 / 29.5 ±1.0
ILLUSTRATION #						
REAR						
CAMBER (Degrees)	-0.53 ±0.5 -9/16 ± 1/2	-0.72 ±0.5 -11/16 ± 1/2	-0.62 ±0.5 -5/8 ± 1/2	-0.62 ±0.5 -5/8 ± 1/2	-0.08 ±0.5 -1/16 ± 1/2	-0.08 ±0.5 -1/16 ± 1/2
REAR TOE (Degrees)	0.3° ±0.12° 5/16° ± 1/8°	0.3° ±0.12° 5/16° ± 1/8°	0.22° ±0.08° 7/32° ± 3/32°	0.22° ±0.08° 7/32° ± 3/32°	0.22° ±0.08° 7/32° ± 3/32°	0.22° ±0.08° 7/32° ± 3/32°
REAR TOE (Inches)	0.15" ±0.06" 5/32" ± 1/16"	0.15" ±0.06" 5/32" ± 1/16"	0.11" ±0.04" 1/8" ± 1/32"	0.11" ±0.04" 1/8" ± 1/32"	0.11" ±0.04" 1/8" ± 1/32"	0.11" ±0.04" 1/8" ± 1/32"
THRUST ANGLE						
ILLUSTRATION #	151	151	177	177	177	177
NOTES			[1] 2007 = 5.0° (5°)	[1] 2007 = 5.0° (5°)		

Alignment Specifications

Nissan

Altima HEV	Altima USA	Altima Canada	MODEL INFORMATION	Altima	Altima	Armada 2WD except Air Susp.
2007-2011	2013	2013	YEARS	2014-2015	2016-2017	2004-2015
			INSTRUCTIONS			
83 ft.lbs.	83 ft.lbs.	83 ft.lbs.	LUGNUT TORQUE	83 ft.lbs.	83 ft.lbs.	98 ft.lbs.
			FRONT			
5.0 ±0.75 5 ± ¾	4.92 ±0.75 [1] 4 15/16 ± ¾ [1]	4.92 ±0.75 [1] 4 15/16 ± ¾ [1]	CASTER (Degrees)	4.92 ±0.75 4 15/16 ± ¾	4.92 ±0.75 4 15/16 ± ¾	3.4 ±0.75 3⅜ ± ¾
0.0 (0) | 0.55 (9/16)	0.0 (0) | 0.75 (¾)	0.0 (0) | 0.75 (¾)	CROSS | TOL.	0.0 (0) | 0.75 (¾)	0.0 (0) | 0.75 (¾)	0.0 (0) | 0.75 (¾)
L= -0.4, R= -0.65 ±0.75 L= -⅜, R= -⅝ ± ¾	L= -0.33, R= -0.58 ±0.75 L= - 5/16, R= - 9/16 ± ¾	L= -0.33, R= -0.58 ±0.75 L= - 5/16, R= - 9/16 ± ¾	CAMBER (Degrees)	L= -0.33, R= -0.58 ±0.75 L= - 5/16, R= - 9/16 ± ¾	-0.33 ±0.75 - 5/16 ± ¾	-0.1 ±0.75 - ⅛ ± ¾
0.25 (¼) | 0.55 (9/16)	0.25 (¼) | 0.75 (¾)	0.25 (¼) | 0.75 (¾)	CROSS | TOL.	0.25 (¼) | 0.75 (¾)	0.0 (0) | 0.75 (¾)	0.0 (0) | 0.75 (¾)
0.1° ±0.07° 3/32° ± 1/16°	0.06° ±0.15° 1/16° ± 5/32°	0.06° ±0.15° 1/16° ± 5/32°	FRONT TOE (Degrees)	0.06° ±0.15° 1/16° ± 5/32°	0.06° ±0.15° 1/16° ± 5/32°	0.17° ±0.07° 5/32° ± 1/16°
0.05" ±0.03" 1/16" ± 1/32"	0.03" ±0.08" 1/32" ± 1/16"	0.03" ±0.08" 1/32" ± 1/16"	FRONT TOE (Inches)	0.03" ±0.08" 1/32" ± 1/16"	0.03" ±0.08" 1/32" ± 1/16"	0.08" ±0.03" 3/32" ± 1/32"
12.6 ±0.8 (12 9/16 ± ¾)	14.5 ±0.8 (14 7/16 ± ¾)	14.5 ±0.8 (14 7/16 ± ¾)	SAI	14.4 ±0.8 (14⅜ ± ¾)	14.4 ±0.8 (14⅜ ± ¾)	13.5 ±0.8 (13½ ± ¾)
38.5 / 31.5 ±1.0	39.5 / 32.5 ±1.0 [2]	39.5 / 32.5 ±1.0	TURNING ANGLE	37.5 / 31.5 ±1.0 [1]	37.5 / 31.5 ±1.0 [1]	37.5 / 34.0 ±1.0
			ILLUSTRATION #			12
			REAR			
-0.25 ±0.5 - ¼ ± ½	-0.67 ±0.5 - 11/16 ± ½	-0.58 ±0.5 - 9/16 ± ½	CAMBER (Degrees)	-0.67 ±0.5 - 11/16 ± ½	-0.67 ±0.5 - 11/16 ± ½	0.08 ±0.75 1/16 ± ¾
0.2° ±0.08° 3/16° ± 3/32°	0.07° ±0.2° 1/16° ± 3/16°	0.07° ±0.2° 1/16° ± 3/16°	REAR TOE (Degrees)	0.12° ±0.2° ⅛° ± 3/16°	0.12° ±0.2° ⅛° ± 3/16°	0.06° ±0.24° 1/16° ± ¼°
0.1" ±0.04" 3/32" ± 1/32"	0.03" ±0.1" 1/32" ± 3/32"	0.03" ±0.1" 1/32" ± 3/32"	REAR TOE (Inches)	0.06" ±0.1" 1/16" ± 3/32"	0.06" ±0.1" 1/16" ± 3/32"	0.03" ±0.12" 1/32" ± ⅛"
			THRUST ANGLE			
177	177	177	ILLUSTRATION #	177	177	154
	[1] 18" tires = 5.0° (5°) [2] 17" & 18" tires = 37.5°/31.5°	[1] 17" & 18" tires = 37.5°/31.5°	NOTES	[1] w/16" tires = 39.5°/32.5°	[1] w/16" tires = 39.5°/32.5°	
32050	32051	32052		32053	32054	32055

Nissan

Armada 2WD with Air Susp.	Armada 4WD except Air Susp.	Armada 4WD with Air Susp.	MODEL INFORMATION	Armada 2WD	Armada 4WD	Axxess 4X2
2004-2015	2004-2015	2004-2015	YEARS	2017	2017	1990
			INSTRUCTIONS			
98 ft.lbs.	98 ft.lbs.	98 ft.lbs.	LUGNUT TORQUE	98 ft.lbs.	98 ft.lbs.	58-72 ft.lbs.
			FRONT			
4.0 ±0.75 4 ± ¾	3.0 ±0.75 3 ± ¾	3.5 ±0.75 3½ ± ¾	CASTER (Degrees)	L= 3.08, R= 3.42 ± 0.75 L= 3 1/16, R= 3 7/16 ± ¾	L= 3.08, R= 3.42 ± 0.75 L= 3 1/16, R= 3 7/16 ± ¾	0.5 ±0.75 ½ ± ¾
0.0 (0) | 0.75 (¾)	0.0 (0) | 0.75 (¾)	0.0 (0) | 0.75 (¾)	CROSS | TOL.	0.0 (0) | 0.75 (¾)	0.0 (0) | 0.75 (¾)	0.0 (0) | n/s
-0.1 ±0.75 - ⅛ ± ¾	0.2 ±0.75 3/16 ± ¾	0.2 ±0.75 3/16 ± ¾	CAMBER (Degrees)	L= 0.0, R= -0.17 ±0.75 L= 0, R= - 3/16 ± ¾	L= 0.0, R= -0.17 ±0.75 L= 0, R= - 3/16 ± ¾	0.33 ±0.75 5/16 ± ¾
0.0 (0) | 0.75 (¾)	0.0 (0) | 0.75 (¾)	0.0 (0) | 0.75 (¾)	CROSS | TOL.	0.17 (3/16) | 0.55 (9/16)	0.17 (3/16) | 0.55 (9/16)	0.0 (0) | n/s
0.17° ±0.07° 5/32° ± 1/16°	0.17° ±0.07° 5/32° ± 1/16°	0.17° ±0.07° 5/32° ± 1/16°	FRONT TOE (Degrees)	0.22° ±0.18° 7/32° ± 3/16°	0.22° ±0.18° 7/32° ± 3/16°	0.2° ±0.2° 3/16° ± 3/16°
0.08" ±0.03" 3/32" ± 1/32"	0.08" ±0.03" 3/32" ± 1/32"	0.08" ±0.03" 3/32" ± 1/32"	FRONT TOE (Inches)	0.11" ±0.09" 3/32" ± 3/32"	0.11" ±0.09" 3/32" ± 3/32"	0.1" ±0.1" 3/32" ± 3/32"
13.5 ±0.8 (13½ ± ¾)	13.2 ±0.8 (13 3/16 ± ¾)	13.2 ±0.8 (13 3/16 ± ¾)	SAI	13.4 ±0.8 (13⅜ ± ¾)	13.4 ±0.8 (13⅜ ± ¾)	14.3 ±0.8 (14 5/16 ± ¾)
37.5 / 34.0 ±1.0	37.7 / 33.5 ±1.0	37.7 / 33.5 ±1.0	TURNING ANGLE	36.6 / 32.5 ±1.5	36.6 / 32.5 ±1.5	39.0 / 30.0 ±2.0
12	12	12	ILLUSTRATION #	12 12	12 12	
			REAR			
-0.5 ±0.75 - ½ ± ¾	0.08 ±0.75 1/16 ± ¾	-0.5 ±0.75 - ½ ± ¾	CAMBER (Degrees)	-0.5 ±0.5 - ½ ± ½	-0.5 ±0.5 - ½ ± ½	0.0 ±0.75 0 ± ¾
0.22° ±0.24° 7/32° ± ¼°	0.06° ±0.24° 1/16° ± ¼°	0.22° ±0.24° 7/32° ± ¼°	REAR TOE (Degrees)	0.23° ±0.23° 7/32° ± 7/32°	0.23° ±0.23° 7/32° ± 7/32°	-0.4° ±0.2° - 13/32° ± 3/16°
0.11" ±0.12" ⅛" ± ⅛"	0.03" ±0.12" 1/32" ± ⅛"	0.11" ±0.12" ⅛" ± ⅛"	REAR TOE (Inches)	0.12" ±0.12" ⅛" ± ⅛"	0.12" ±0.12" ⅛" ± ⅛"	-0.2" ±0.1" - 3/16" ± 3/32"
			THRUST ANGLE			
154	154	154	ILLUSTRATION #	154 154	154 154	10
			NOTES			
32056	32057	32058		32059	32060	32061

Nissan

Axxess 4X4	Cube	Frontier 2WD w/2.4L (KA)	MODEL INFORMATION	Frontier 2WD w/3.3L (VG)	Frontier 2WD w/3.3L (VG)	Frontier 4WD w/o Supercharger
1990	2009-2014	1998-2004	YEARS	2000	2001-2004	1998-2004
			INSTRUCTIONS			
58-72 ft.lbs.	80 ft.lbs.	87-108 ft.lbs.	LUGNUT TORQUE	87-108 ft.lbs.	87-108 ft.lbs.	87-108 ft.lbs.
			FRONT			
0.67 ±0.75 11/16 ± 3/4	4.67 ±0.75 4 11/16 ± 3/4	0.6 ±0.5 5/8 ± 1/2	CASTER (Degrees)	2.17 ±0.5 2 3/16 ± 1/2	2.57 ±0.5 2 9/16 ± 1/2	2.17 ±0.5 2 3/16 ± 1/2
0.0 (0) | n/s	0.0 (0) | 0.75 (3/4)	0.0 (0) | 0.75 (3/4)	CROSS | TOL.	0.0 (0) | 0.75 (3/4)	0.0 (0) | 0.75 (3/4)	0.0 (0) | 0.75 (3/4)
0.17 ±0.75 3/16 ± 3/4	-0.17 ±0.75 -3/16 ± 3/4	0.42 ±0.5 7/16 ± 1/2	CAMBER (Degrees)	0.6 ±0.5 5/8 ± 1/2	0.55 ±0.5 9/16 ± 1/2	0.6 ±0.5 5/8 ± 1/2
0.0 (0) | n/s	0.0 (0) | 0.75 (3/4)	0.0 (0) | 0.75 (3/4)	CROSS | TOL.	0.0 (0) | 0.75 (3/4)	0.0 (0) | 0.75 (3/4)	0.0 (0) | 0.75 (3/4)
0.2° ±0.2° 3/16° ± 3/16°	0.1° ±0.2° 3/32° ± 3/16°	0.27° ±0.08° 9/32° ± 3/32°	FRONT TOE (Degrees)	0.33° ±0.08° 11/32° ± 3/32°	0.33° ±0.08° 11/32° ± 3/32°	0.33° ±0.08° 11/32° ± 3/32°
0.1" ±0.1" 3/32" ± 3/32"	0.05" ±0.1" 1/16" ± 3/32"	0.13" ±0.04" 1/8" ± 1/32"	FRONT TOE (Inches)	0.17" ±0.04" 5/32" ± 1/32"	0.17" ±0.04" 5/32" ± 1/32"	0.17" ±0.04" 5/32" ± 1/32"
14.0 ±0.8 (14 ± 3/4)	9.9 ±0.8 (9 15/16 ± 3/4)	9.1 ±0.5 (9 1/16 ± 1/2)	SAI	10.8 ±0.5 (10 13/16 ± 1/2)	10.8 ±0.5 (10 13/16 ± 1/2)	10.8 ±0.5 (10 13/16 ± 1/2)
38.0 / 30.0 ±2.0	38.0 / 33.0 ±1.0	37.0 / 34.0 ±1.0 [1]	TURNING ANGLE	35.1 / 33.2 [1]	33.0 / 31.0 ±1.0 [1]	33.0 / 31.0 ±1.0 [1]
		24	ILLUSTRATION #	24	24	24
			REAR			
-0.17 ±0.75 -3/16 ± 3/4	-1.51 ±0.5 -1 1/2 ± 1/2		CAMBER (Degrees)			
-0.2° ±0.2° -3/16° ± 3/16°	0.28° ±0.38° 9/32° ± 3/8°		REAR TOE (Degrees)			
-0.1" ±0.1" -3/32" ± 3/32"	0.14" ±0.19" 1/8" ± 3/16"		REAR TOE (Inches)			
			THRUST ANGLE			
10			ILLUSTRATION #			
		[1] 2003-04 = 34°/31°	NOTES	[1] w255-65R16 tires 33°/31°	[1] w/Supercharger 32.8°/30.7°	[1] 1998-00 = 34°/32° ±2°
32062	32064	32066		32067	32068	32069

Nissan

Frontier 4WD with Supercharger	Frontier 2WD	Frontier 4WD	MODEL INFORMATION	Frontier 2WD	Frontier 4WD	Juke 2WD USA
2001-2004	2005-2013	2005-2013	YEARS	2014-2017	2014-2017	2011-2012
			INSTRUCTIONS			
87-108 ft.lbs.	98 ft.lbs.	98 ft.lbs.	LUGNUT TORQUE	98 ft.lbs.	98 ft.lbs.	80 ft.lbs.
			FRONT			
2.57 ±0.5 2 9/16 ± 1/2	3.0 ±0.5 3 ± 1/2	2.75 ±0.5 2 3/4 ± 1/2	CASTER (Degrees)	2.75 ±0.5 2 3/4 ± 1/2	2.5 ±0.5 2 1/2 ± 1/2	4.5 ±0.75 4 1/2 ± 3/4
0.0 (0) | 0.75 (3/4)	0.0 (0) | 0.75 (3/4)	0.0 (0) | 0.75 (3/4)	CROSS | TOL.	0.0 (0) | 0.75 (3/4)	0.0 (0) | 0.75 (3/4)	0.0 (0) | 0.75 (3/4)
0.55 ±0.5 9/16 ± 1/2	0.25 ±0.5 1/4 ± 1/2	0.5 ±0.5 1/2 ± 1/2	CAMBER (Degrees)	0.25 ±0.5 1/4 ± 1/2	0.5 ±0.5 1/2 ± 1/2	-0.42 ±0.75 -7/16 ± 3/4
0.0 (0) | 0.75 (3/4)	0.0 (0) | 0.75 (3/4)	0.0 (0) | 0.75 (3/4)	CROSS | TOL.	0.0 (0) | 0.75 (3/4)	0.0 (0) | 0.75 (3/4)	0.0 (0) | 0.75 (3/4)
0.33° ±0.08° 11/32° ± 3/32°	0.24° ±0.08° 1/4° ± 3/32°	0.24° ±0.08° 1/4° ± 3/32°	FRONT TOE (Degrees)	0.3° ±0.06° 5/16° ± 1/16°	0.3° ±0.06° 5/16° ± 1/16°	0.17° ±0.1° 5/32° ± 3/32°
0.17" ±0.04" 5/32" ± 1/32"	0.12" ±0.04" 1/8" ± 1/32"	0.12" ±0.04" 1/8" ± 1/32"	FRONT TOE (Inches)	0.15" ±0.03" 5/32" ± 1/32"	0.15" ±0.03" 5/32" ± 1/32"	0.08" ±0.05" 3/32" ± 1/16"
10.8 ±0.5 (10 13/16 ± 1/2)	13.0 (13)	12.8 (12 3/4)	SAI	13.0 (13)	12.8 (12 3/4)	11.4 ±0.8 (11 7/16 ± 3/4)
32.8 / 30.7 ±1.0	35.4 / 31.4 ±1.0	35.6 / 31.7 ±1.0	TURNING ANGLE	35.4 / 31.4 ±1.0	35.6 / 31.8 ±1.0	36.0 / 31.0 ±1.0
24	12	12	ILLUSTRATION #	12	12	
			REAR			
			CAMBER (Degrees)			-1.52 ±0.5 -1 1/2 ± 1/2
			REAR TOE (Degrees)			0.35° ±0.33° 11/32° ± 11/32°
			REAR TOE (Inches)			0.18" ±0.17" 3/16" ± 5/32"
			THRUST ANGLE			
			ILLUSTRATION #			
			NOTES			
32070	32071	32072		32073	32074	32075

Nissan

MODEL INFORMATION	Juke 2WD Canada	Juke AWD USA	Juke AWD Canada	Juke 2WD USA	Juke AWD USA	Juke 2WD Canada
YEARS	2011-2012	2011-2012	2011-2012	2013-2017	2013-2017	2013-2017
INSTRUCTIONS						
LUGNUT TORQUE	80 ft.lbs.	80 ft.lbs.	80 ft.lbs.	80 ft.lbs.	80 ft.lbs.	80 ft.lbs.
FRONT						
CASTER (Degrees)	4.5 ±0.75 4½ ± ¾	4.58 ±0.75 4 9/16 ± ¾	4.58 ±0.75 4 9/16 ± ¾	4.5 ±0.75 4½ ± ¾	4.58 ±0.75 4 9/16 ± ¾	4.5 ±0.75 4½ ± ¾
CROSS | TOL.	0.0 (0) | 0.75 (¾)	0.0 (0) | 0.75 (¾)	0.0 (0) | 0.75 (¾)	0.0 (0) | 0.75 (¾)	0.0 (0) | 0.75 (¾)	0.0 (0) | 0.75 (¾)
CAMBER (Degrees)	-0.42 ±0.75 - 7/16 ± ¾	-0.42 ±0.75 - 7/16 ± ¾	-0.42 ±0.75 - 7/16 ± ¾	-0.42 ±0.75 - 7/16 ± ¾	-0.42 ±0.75 - 7/16 ± ¾	-0.42 ±0.75 - 7/16 ± ¾
CROSS | TOL.	0.0 (0) | 0.75 (¾)	0.0 (0) | 0.75 (¾)	0.0 (0) | 0.75 (¾)	0.0 (0) | 0.75 (¾)	0.0 (0) | 0.75 (¾)	0.0 (0) | 0.75 (¾)
FRONT TOE (Degrees)	0.17° ±0.1° 5/32° ± 3/32°	0.17° ±0.1° 5/32° ± 3/32°	0.17° ±0.1° 5/32° ± 3/32°	0.17° ±0.1° 5/32° ± 3/32°	0.17° ±0.1° 5/32° ± 3/32°	0.17° ±0.1° 5/32° ± 3/32°
FRONT TOE (Inches)	0.08" ±0.05" 3/32" ± 1/16"	0.08" ±0.05" 3/32" ± 1/16"	0.08" ±0.05" 3/32" ± 1/16"	0.08" ±0.05" 3/32" ± 1/16"	0.08" ±0.05" 3/32" ± 1/16"	0.08" ±0.05" 3/32" ± 1/16"
SAI	11.4 ±0.8 (11 7/16 ± ¾)	11.4 ±0.8 (11 7/16 ± ¾)	11.4 ±0.8 (11 7/16 ± ¾)	11.4 ±0.8 (11 7/16 ± ¾)	11.4 ±0.8 (11 7/16 ± ¾)	11.4 ±0.8 (11 7/16 ± ¾)
TURNING ANGLE	36.0 / 31.0 ±1.0	36.0 / 31.0 ±1.0	36.0 / 31.0 ±1.0	36.0 / 31.0 ±1.0	36.0 / 31.0 ±1.0	36.0 / 31.0 ±1.0
ILLUSTRATION #						
REAR						
CAMBER (Degrees)	-1.52 ±0.5 -1½ ± ½	0.0 ±0.5 0 ± ½	0.03 ±0.5 1/16 ± ½	-1.52 ±0.5 -1½ ± ½	0.0 ±0.5 [1] 0 ± ½ [1]	-1.52 ±0.5 -1½ ± ½
REAR TOE (Degrees)	0.35° ±0.33° 11/32° ± 11/32°	0.27° ±0.16° 9/32° ± 5/32°	0.23° ±0.16° 7/32° ± 5/32°	0.25° ±0.33° ¼° ± 11/32°	0.27° ±0.18° [2] 9/32° ± 3/16° [2]	0.24° ±0.33° ¼° ± 11/32°
REAR TOE (Inches)	0.18" ±0.17" 3/16" ± 5/32"	0.13" ±0.08" 1/8" ± 3/32"	0.12" ±0.08" 1/8" ± 3/32"	0.13" ±0.17" 1/8" ± 5/32"	0.14" ±0.09" [2] 1/8" ± 3/32" [2]	0.12" ±0.17" 1/8" ± 5/32"
THRUST ANGLE						
ILLUSTRATION #						
NOTES					[1] Nismo= 0.07° (1/16°) [2] Nismo 0.35° (0.18")	
	32076	32077	32078	32079	32080	32081

Nissan

MODEL INFORMATION	Juke AWD Canada	LEAF	LEAF	Maxima	Maxima	Maxima w/15" Tires
YEARS	2013-2017	2011-2012	2013-2017	1990-94	1995-99	2000-2001
INSTRUCTIONS						
LUGNUT TORQUE	80 ft.lbs.	80 ft.lbs.	80 ft.lbs.	72-87 ft.lbs.	72-87 ft.lbs.	
FRONT						
CASTER (Degrees)	4.58 ±0.75 4 9/16 ± ¾	4.83 ±0.75 4 13/16 ± ¾	4.83 ±0.75 4 13/16 ± ¾	1.25 ±0.75 1¼ ± ¾	2.75 ±0.75 2¾ ± ¾	2.75 ±0.75 2¾ ± ¾
CROSS | TOL.	0.0 (0) | 0.75 (¾)	0.0 (0) | 0.75 (¾)	0.0 (0) | 0.75 (¾)	0.0 (0) | n/s	0.0 (0) | 0.75 (¾)	0.0 (0) | 0.75 (¾)
CAMBER (Degrees)	-0.42 ±0.75 - 7/16 ± ¾	-0.42 ±0.75 - 7/16 ± ¾	-0.42 ±0.75 - 7/16 ± ¾	-0.25 ±0.75 - ¼ ± ¾	-0.25 ±0.75 - ¼ ± ¾	-0.33 ±0.75 - 5/16 ± ¾
CROSS | TOL.	0.0 (0) | 0.75 (¾)	0.0 (0) | 0.75 (¾)	0.0 (0) | 0.75 (¾)	0.0 (0) | n/s	0.0 (0) | 0.75 (¾)	0.0 (0) | 0.75 (¾)
FRONT TOE (Degrees)	0.17° ±0.1° 5/32° ± 3/32°	0.16° ±0.08° 5/32° ± 3/32°	0.16° ±0.16° 5/32° ± 5/32°	0.18° ±0.08° 3/16° ± 3/32°	0.18° ±0.08° 3/16° ± 3/32°	0.1° ±0.1° 3/32° ± 3/32°
FRONT TOE (Inches)	0.08" ±0.05" 3/32" ± 1/16"	0.08" ±0.04" 3/32" ± 1/32"	0.08" ±0.08" 3/32" ± 3/32"	0.09" ±0.04" 3/32" ± 1/32"	0.09" ±0.04" 3/32" ± 1/32"	0.05" ±0.05" 1/16" ± 1/16"
SAI	11.4 ±0.8 (11 7/16 ± ¾)	11.9 ±0.8 (11 15/16 ± ¾)	11.9 ±0.8 (11 15/16 ± ¾)	14.3 ±0.8 (14 5/16 ± ¾)	14.3 ±0.8 (14¼ ± ¾)	14.3 ±0.8 (14 5/16 ± ¾)
TURNING ANGLE	36.0 / 31.0 ±1.0	40.0 / 33.0 ±1.0	40.0 / 33.0 ±1.0 [1]	37.0 / 30.0 ±2.0	39.5 / 32.0 ±1.0	39.5 / 32.0 ±1.0
ILLUSTRATION #						
REAR						
CAMBER (Degrees)	0.03 ±0.5 1/16 ± ½	-1.48 ±0.5 -1½ ± ½	-1.48 ±0.5 -1½ ± ½	-0.58 ±0.75 - 9/16 ± ¾	-1.0 ±0.75 -1 ± ¾	-1.0 ±0.75 -1 ± ¾
REAR TOE (Degrees)	0.25° ±0.18° [1] ¼° ± 3/16° [1]	0.44° ±0.24° 7/16° ± ¼°	0.26° ±0.4° ¼° ± 13/32°	-0.17° ±0.08° [1] -5/32° ± 3/32° [1]	0.1° ±0.4° 3/32° ± 13/32°	0.1° ±0.4° 3/32° ± 13/32°
REAR TOE (Inches)	0.13" ±0.09" [1] 1/8" ± 3/32" [1]	0.22" ±0.12" 7/32" ± 1/8"	0.13" ±0.2" 1/8" ± 3/16"	-0.08" ±0.04" [1] -3/32" ± 1/32" [1]	0.05" ±0.2" 1/16" ± 3/16"	0.05" ±0.2" 1/16" ± 3/16"
THRUST ANGLE						
ILLUSTRATION #				10		
NOTES	[1] Nismo 0.28° (0.14")		[1] w/17" tires = 37°/32°	[1] 1990-91 = -0.19° (-0.1")		
	32082	32083	32084	32088	32089	32090

Nissan

Maxima exc/15" Tires	Maxima	Maxima USA w/18" Tires (OE)	MODEL INFORMATION	Maxima USA w/19" Tires (OE)	Maxima Canada w/18" Tires (OE)	Maxima Canada w/19" Tires (OE)
2000-2003	2004-2008	2009-2014	YEARS	2009-2014	2009-2014	2009-2014
			INSTRUCTIONS			
72-87 ft.lbs.	73-86 ft.lbs.	83 ft.lbs.	LUGNUT TORQUE	83 ft.lbs.	83 ft.lbs.	83 ft.lbs.
			FRONT			
2.75 ±0.75 2¾ ± ¾	2.83 ±0.75 2 13/16 ± ¾	4.92 ±0.75 4 15/16 ± ¾	CASTER (Degrees)	5.0 ±0.75 5 ± ¾	4.92 ±0.75 4 15/16 ± ¾	5.0 ±0.75 5 ± ¾
0.0 (0) | 0.75 (¾)	0.0 (0) | 0.75 (¾)	0.0 (0) | 0.55 (9/16)	CROSS | TOL.	0.0 (0) | 0.55 (9/16)	0.0 (0) | 0.55 (9/16)	0.0 (0) | 0.55 (9/16)
-0.25 ±0.75 -¼ ± ¾	-0.25 ±0.75 -¼ ± ¾	L= -0.33, R= -0.58 ±0.75 L= -5/16, R= -9/16 ± ¾	CAMBER (Degrees)	L= -0.42, R= -0.67 ±0.75 L= -7/16, R= -11/16 ± ¾	L= -0.33, R= -0.58 ±0.75 L= -5/16, R= -9/16 ± ¾	L= -0.42, R= -0.67 ±0.75 L= -7/16, R= -11/16 ± ¾
0.0 (0) | 0.75 (¾)	0.0 (0) | 0.75 (¾)	0.25 (¼) | 0.55 (9/16)	CROSS | TOL.	0.25 (¼) | 0.55 (9/16)	0.25 (¼) | 0.55 (9/16)	0.25 (¼) | 0.55 (9/16)
0.1° ±0.1° 3/32° ± 3/32°	0.04° ±0.08° 1/32° ± 3/32°	0.08° ±0.15° 3/32° ± 5/32°	FRONT TOE (Degrees)	0.08° ±0.15° 3/32° ± 5/32°	0.08° ±0.15° 3/32° ± 5/32°	0.08° ±0.15° 3/32° ± 5/32°
0.05" ±0.05" 1/16" ± 1/10"	0.02" ±0.04" 1/32" ± 1/32"	0.04" ±0.08" 1/32" ± 1/16"	FRONT TOE (Inches)	0.04" ±0.08" 1/32" ± 1/16"	0.04" ±0.08" 1/32" ± 1/16"	0.04" ±0.08" 1/32" ± 1/16"
14.3 ±0.8 (14¼ ± ¾)	14.6 ±0.8 (14 9/16 ± ¾)	14.4 ±0.8 (14 7/16 ± ¾)	SAI	14.4 ±0.8 (14 7/16 ± ¾)	14.4 ±0.8 (14 7/16 ± ¾)	14.4 ±0.8 (14 7/16 ± ¾)
39.5 / 32.0 ±1.0 [1]	35.5 / 29.0 ±1.5 [1]	36.5 / 31.0 ±1.5	TURNING ANGLE	36.5 / 31.0 ±1.5	36.5 / 31.0 ±1.5	36.5 / 31.0 ±1.5
			ILLUSTRATION #			
			REAR			
-1.0 ±0.75 -1 ± ¾	-0.63 ±0.5 -⅝ ± ½	-0.42 ±0.5 -7/16 ± ½	CAMBER (Degrees)	-0.53 ±0.5 -9/16 ± ½	-0.37 ±0.5 -⅜ ± ½	-0.53 ±0.5 -9/16 ± ½
0.1° ±0.4° 3/32° ± 13/32°	0.3° ±0.12° 5/16° ± ⅛°	0.13° ±0.13° ⅛° ± ⅛°	REAR TOE (Degrees)	0.17° ±0.13° 5/32° ± ⅛°	0.13° ±0.13° ⅛° ± ⅛°	0.17° ±0.13° 5/32° ± ⅛°
0.05" ±0.2" 1/16" ± 3/16"	0.15" ±0.06" 5/32" ± 1/16"	0.07" ±0.07" 1/16" ± 1/16"	REAR TOE (Inches)	0.08" ±0.07" 3/32" ± 1/16"	0.07" ±0.07" 1/16" ± 1/16"	0.08" ±0.07" 3/32" ± 1/16"
			THRUST ANGLE			
	151	151	ILLUSTRATION #	151	151	151
[1] w/225-50R17 tires 33°/28.5°	[1] w/18" tires = 34.5°/28.5°		NOTES			
32091	32092	32093		32094	32095	32096

Nissan

Maxima USA & Canada with 18" Tires (OE)	Maxima USA & Canada with 19" Tires (OE)	Micra with 185/60R15 tires (OE)	MODEL INFORMATION	Murano exc SE	Murano SE	Murano
2016-2017	2016-2017	2015-2017	YEARS	2003-2005	2003-2005	2006-2007
			INSTRUCTIONS			
83 ft.lbs.	83 ft.lbs.	83 ft.lbs.	LUGNUT TORQUE	73-93 ft.lbs.	73-93 ft.lbs.	80 ft.lbs.
			FRONT			
4.83 ±0.75 4 13/16 ± ¾	4.91 ±0.75 4 15/16 ± ¾	3.83 ±0.75 [1] 3 13/16 ± ¾ [1]	CASTER (Degrees)	2.58 ±0.75 2 9/16 ± ¾	2.58 ±0.75 2 9/16 ± ¾	2.58 ±0.75 [1] 2 9/16 ± ¾ [1]
0.0 (0) | 0.5 (½)	0.0 (0) | 0.5 (½)	0.0 (0) | 0.75 (¾)	CROSS | TOL.	0.0 (0) | 0.75 (¾)	0.0 (0) | 0.75 (¾)	0.0 (0) | 0.75 (¾)
-0.39 ±0.75 -⅜ ± ¾	-0.39 ±0.75 -⅜ ± ¾	-0.08 ±0.75 -1/16 ± ¾	CAMBER (Degrees)	-0.33 ±0.75 -5/16 ± ¾	-0.33 ±0.75 -5/16 ± ¾	-0.33 ±0.75 -5/16 ± ¾
0.0 (0) | 0.55 (9/16)	0.0 (0) | 0.55 (9/16)	0.0 (0) | 0.75 (¾)	CROSS | TOL.	0.0 (0) | 0.75 (¾)	0.0 (0) | 0.75 (¾)	0.0 (0) | 0.75 (¾)
0.07° ±0.18° 1/16° ± 3/16°	0.07° ±0.18° 1/16° ± 3/16°	0.1° ±0.1° 3/32° ± 3/32°	FRONT TOE (Degrees)	0.04° ±0.08° 1/32° ± 3/32°	0.04° ±0.08° 1/32° ± 3/32°	0.04° ±0.08° 1/32° ± 3/32°
0.04" ±0.09" 1/32" ± 3/32"	0.04" ±0.09" 1/32" ± 3/32"	0.05" ±0.05" 1/16" ± 1/16"	FRONT TOE (Inches)	0.02" ±0.04" 1/32" ± 1/32"	0.02" ±0.04" 1/32" ± 1/32"	0.02" ±0.04" 1/32" ± 1/32"
14.3 ±0.8 (14¼ ± ¾)	14.3 ±0.8 (14¼ ± ¾)	11.7 ±0.8 (11 11/16 ± ¾)	SAI	14.3 ±0.8 (14 5/16 ± ¾)	14.3 ±0.8 (14 5/16 ± ¾)	14.3 ±0.8 (14 5/16 ± ¾)
37.0 / 31.0 ±1.5	37.0 / 31.0 ±1.5	43.5 / 35.5 ±1.0 [2]	TURNING ANGLE	38.0 / 31.5 ±1.0	38.0 / 31.5 ±1.0	38.0 / 31.5 ±1.0
			ILLUSTRATION #			
			REAR			
-0.38 ±0.75 -⅜ ± ¾	-0.52 ±0.75 -½ ± ¾	-1.42 ±0.5 -1 7/16 ± ½	CAMBER (Degrees)	-0.77 ±0.5 -¾ ± ½	-0.8 ±0.5 -13/16 ± ½	-0.77 ±0.5 -¾ ± ½
0.22° ±0.2° 7/32° ± 3/16°	0.26° ±0.2° ¼° ± 3/16°	0.17° ±0.42° 5/32° ± 7/16°	REAR TOE (Degrees)	0.25° ±0.14° ¼° ± ⅛°	0.28° ±0.14° 9/32° ± ⅛°	0.25° ±0.14° ¼° ± ⅛°
0.11" ±0.1" ⅛" ± 3/32"	0.13" ±0.1" ⅛" ± 3/32"	0.08" ±0.21" 3/32" ± 7/32"	REAR TOE (Inches)	0.13" ±0.07" ⅛" ± 1/16"	0.14" ±0.07" ⅛" ± 1/16"	0.13" ±0.07" ⅛" ± 1/16"
			THRUST ANGLE			
151	151		ILLUSTRATION #	63	63	63
		[1] 16" tires = 3.92° (3 15/16°) [2] 2015 = 43.5°/35.5°	NOTES			[1] 2WD = 2.5° (2½°)
32097	32098	32099		32100	32101	32102

Nissan

Murano	Murano	Murano	MODEL INFORMATION	Murano CrossCabriolet	NV 2500/NV 3500	NV200 Compact Cargo & Taxi
2009-2014	2015-2016	2017	YEARS	2012-2014	2012-2017	2013-2017
			INSTRUCTIONS			
80 ft.lbs.	83 ft.lbs.	83 ft.lbs.	LUGNUT TORQUE	80 ft.lbs.	138 ft.lbs.	83 ft.lbs.
			FRONT			
L= 4.67, R= 5.0 ± 0.75 L= 4 11/16, R= 5 ± ¾	4.75 ±0.75 4 ¾ ± ¾	4.75 ±0.75 4 ¾ ± ¾	CASTER (Degrees)	L= 4.75, R= 4.92 ± 0.75 L= 4 ¾, R= 4 15/16 ± ¾	6.17 ±0.5 [1] 6 3/16 ± ½ [1]	4.17 ±0.75 4 3/16 ± ¾
-0.33 (- 5/16) | 0.55 (9/16)	0.0 (0) | 0.5 (½)	0.0 (0) | 0.5 (½)	CROSS | TOL.	-0.17 (- 5/16) | 0.55 (9/16)	0.0 (0) | 0.75 (¾)	0.0 (0) | 0.75 (¾)
L= -0.25, R= -0.5 ±0.75 L= -¼, R= -½ ± ¾	L= -0.25, R= -0.5 ±0.75 L= -¼, R= -½ ± ¾	L= -0.25, R= -0.5 ±0.75 L= -¼, R= -½ ± ¾	CAMBER (Degrees)	L= -0.25, R= -0.5 ±0.75 L= -¼, R= -½ ± ¾	0.0 ±0.5 0 ± ½	-0.33 ±0.75 - 5/16 ± ¾
0.25 (¼) | 0.55 (9/16)	0.25 (¼) | 0.55 (9/16)	0.25 (¼) | 0.55 (9/16)	CROSS | TOL.	0.25 (¼) | 0.55 (9/16)	0.0 (0) | 0.75 (¾)	0.0 (0) | 0.75 (¾)
0.1° ±0.13° 3/32° ± 1/8°	0.12° ±0.16° 3/32° ± 5/32°	0.12° ±0.08° 3/32° ± 3/32°	FRONT TOE (Degrees)	0.1° ±0.13° 3/32° ± 1/8°	0.44° ±0.16° 7/16° ± 5/32°	0.1° ±0.08° 3/32° ± 3/32°
0.05" ±0.07" 1/16" ± 1/16"	0.06" ±0.08" 1/16" ± 3/32"	0.06" ±0.04" 1/16" ± 1/32"	FRONT TOE (Inches)	0.05" ±0.07" 1/16" ± 1/16"	0.22" ±0.08" 7/32" ± 3/32"	0.05" ±0.04" 1/16" ± 1/32"
12.7 ±0.8 (12 11/16 ± ¾)	12.9 ±0.8 (12 7/8 ± ¾)	12.9 ±0.8 (12 7/8 ± ¾)	SAI	12.8 ±0.8 (12 ¾ ± ¾)	8.9 (8 15/16)	11.9 ±0.8 (11 15/16 ± ¾)
36.5 / 31.5 ±1.0 [1]	37.8 / 32.8 ±1.0	37.5 / 32.5 ±1.0	TURNING ANGLE	35.0 / 30.5 ±1.0	38.5 / 35.5 ±1.0	39.9 / 33.6 ±1.0
			ILLUSTRATION #		12	
			REAR			
-0.7 ±0.5 - 11/16 ± ½	-0.75 ±0.5 - ¾ ± ½	-0.75 ±0.5 - ¾ ± ½	CAMBER (Degrees)	-0.8 ±0.5 - 13/16 ± ½		-1.5 ±0.5 -1 ½ ± ½
0.2° ±0.14° 3/16° ± 1/8°	0.26° ±0.22° ¼° ± 7/32°	0.26° ±0.14° ¼° ± 1/8°	REAR TOE (Degrees)	0.23° ±0.13° 7/32° ± 1/8°		0.28° ±0.4° 9/32° ± 13/32°
0.1" ±0.07" 3/32" ± 1/16"	0.13" ±0.11" 1/8" ± 1/8"	0.13" ±0.07" 1/8" ± 1/16"	REAR TOE (Inches)	0.12" ±0.07" 1/8" ± 1/16"		0.14" ±0.2" 1/8" ± 3/16"
			THRUST ANGLE			
151	151	151	ILLUSTRATION #	151		
[1] 20" tires = 35°/31.5°			NOTES		[1] 3500 = 5.92° (5 15/16°)	
32103	32104	32105		32106	32107	32108

Nissan

NX Coupe	Pathfinder 4X2	Pathfinder 4X4 (1990-92 with Bias Tires)	MODEL INFORMATION	Pathfinder 4X4	Pathfinder All Models	Pathfinder 2WD
1991-93	1990-95	1990-91	YEARS	1992-95	1996-2004	2005-2012
			INSTRUCTIONS			
72-87 ft.lbs.	87-108 ft.lbs.	87-108 ft.lbs.	LUGNUT TORQUE	87-108 ft.lbs.	87-108 ft.lbs.	98 ft.lbs.
			FRONT			
1.83 ±0.75 1 13/16 ± ¾	0.37 ±0.5 3/8 ± ½	1.3 ±0.5 1 5/16 ± ½	CASTER (Degrees)	1.3 ±0.5 1 5/16 ± ½	3.0 ±0.75 3 ± ¾	3.0 ±0.5 3 ± ½
0.0 (0) | n/s	0.0 (0) | 0.75 (¾)	0.0 (0) | 0.75 (¾)	CROSS | TOL.	0.0 (0) | 0.75 (¾)	0.0 (0) | 0.75 (¾)	0.0 (0) | 0.75 (¾)
-0.25 ±0.75 - ¼ ± ¾	0.42 ±0.5 7/16 ± ½	0.67 ±0.5 11/16 ± ½	CAMBER (Degrees)	0.67 ±0.5 11/16 ± ½	0.17 ±0.75 3/16 ± ¾	0.25 ±0.5 ¼ ± ½
0.0 (0) | n/s	0.0 (0) | 0.75 (¾)	0.0 (0) | 0.75 (¾)	CROSS | TOL.	0.0 (0) | 0.75 (¾)	0.0 (0) | 0.75 (¾)	0.0 (0) | 0.75 (¾)
0.2° ±0.1° 3/16° ± 3/32°	0.25° ±0.08° [1] ¼° ± 3/32° [1]	0.37° ±0.08° [1] 3/8° ± 3/32° [1]	FRONT TOE (Degrees)	0.32° ±0.08° 5/16° ± 3/32°	0.17° ±0.08° 5/32° ± 3/32°	0.24° ±0.08° ¼° ± 3/32°
0.1" ±0.05" 3/32" ± 1/16"	0.13" ±0.04" [1] 1/8" ± 1/32" [1]	0.18" ±0.04" [1] 3/16" ± 1/32" [1]	FRONT TOE (Inches)	0.16" ±0.04" 5/32" ± 1/32"	0.08" ±0.04" 3/32" ± 1/32"	0.12" ±0.04" 1/8" ± 1/32"
14.0 ±0.8 (14 ± ¾)	9.1 ±0.5 (9 1/16 ± ½)	8.1 ±0.5 (8 1/8 ± ½)	SAI	8.1 ±0.5 (8 1/8 ± ½)	14.3 ±0.8 (14 5/16 ± ¾)	13.0 (13)
35.0 / 33.0 ±2.0 [1]	37.0 / 34.0 ±1.0	34.0 / 32.0 ±1.0 [2]	TURNING ANGLE	34.0 / 32.0 ±1.0 [1]	33.0 / 31.0 ±1.0 [1]	35.4 / 31.4 ±1.0
	41	41	ILLUSTRATION #	41		12
			REAR			
-1.17 ±0.75 -1 3/16 ± ¾			CAMBER (Degrees)			-0.03 ±0.5 0 ± ½
0.1° ±0.2° 3/32° ± 3/16°			REAR TOE (Degrees)			0.13° ±0.23° 1/8° ± 7/32°
0.05" ±0.1" 1/16" ± 3/32"			REAR TOE (Inches)			0.07" ±0.12" 1/16" ± 1/8"
			THRUST ANGLE			
64			ILLUSTRATION #			154
[1] 2.0L = 35°/30°	[1] w/bias tires 0.42° (0.21")	[1] w/bias tires 0.42° (0.21") [2] w/31x10.5R15 tires 28°/26°	NOTES	[1] w/31x10.5R15 tires 28°/26°	[1] w/235-70R15 tires 35°/33°	
32109	32110	32112		32113	32114	32115

Nissan

MODEL INFORMATION	Pathfinder 4WD	Pathfinder USA	Pathfinder Canada	Pathfinder USA	Pathfinder Canada	Pathfinder HEV USA
YEARS	2005-2012	2013-2015	2013-2015	2016-2017	2016-2017	2014-2015
INSTRUCTIONS						
LUGNUT TORQUE	98 ft.lbs.	83 ft.lbs.	83 ft.lbs.	83 ft.lbs.	83 ft.lbs.	83 ft.lbs.
FRONT						
CASTER (Degrees)	2.75 ±0.5 2¾ ± ½	4.67 ±0.75 4 11/16 ± ¾	4.75 ±0.75 4¾ ± ¾	4.67 ±0.75 4 11/16 ± ¾	4.75 ±0.75 4¾ ± ¾	4.75 ±0.75 4¾ ± ¾
CROSS | TOL.	0.0 (0) | 0.75 (¾)	0.0 (0) | 0.5 (½)	0.0 (0) | 0.5 (½)	0.0 (0) | 0.5 (½)	0.0 (0) | 0.5 (½)	0.0 (0) | 0.5 (½)
CAMBER (Degrees)	0.5 ±0.5 ½ ± ½	L= -0.25, R= -0.5 ±0.75 L= -¼, R= -½ ± ¾	L= -0.25, R= -0.5 ±0.75 L= -¼, R= -½ ± ¾	L= -0.25, R= -0.5 ±0.75 L= -¼, R= -½ ± ¾	L= -0.25, R= -0.5 ±0.75 L= -¼, R= -½ ± ¾	L= -0.25, R= -0.5 ±0.75 L= -¼, R= -½ ± ¾
CROSS | TOL.	0.0 (0) | 0.75 (¾)	0.25 (¼) | 0.55 (9/16)	0.25 (¼) | 0.55 (9/16)	0.25 (¼) | 0.55 (9/16)	0.25 (¼) | 0.55 (9/16)	0.25 (¼) | 0.55 (9/16)
FRONT TOE (Degrees)	0.24° ±0.08° ¼° ± 3/32°	0.1° ±0.16° 3/32° ± 5/32°	0.1° ±0.16° 3/32° ± 5/32°	0.08° ±0.12° 3/32° ± 1/8°	0.08° ±0.12° 3/32° ± 1/8°	0.1° ±0.16° 3/32° ± 5/32°
FRONT TOE (Inches)	0.12" ±0.04" 1/8" ± 1/32"	0.05" ±0.08" 1/16" ± 3/32"	0.05" ±0.08" 1/16" ± 3/32"	0.04" ±0.06" 1/16" ± 1/16"	0.04" ±0.06" 1/16" ± 1/16"	0.05" ±0.08" 1/16" ± 3/32"
SAI	12.8 (12¾)	12.8 (12 13/16)	12.8 (12¾)	12.8 (12 13/16)	12.8 (12 13/16)	12.8 (12¾)
TURNING ANGLE	35.6 / 31.6 ±1.0	38.0 / 33.0 ±1.0	38.0 / 33.0 ±1.0	38.0 / 33.0 ±1.0	38.0 / 33.0 ±1.0	38.0 / 33.0 ±1.0
ILLUSTRATION #	12					
REAR						
CAMBER (Degrees)	-0.03 ±0.5 0 ± ½	-0.58 ±0.5 - 9/16 ± ½	-0.58 ±0.5 - 9/16 ± ½	-0.58 ±0.5 - 9/16 ± ½	-0.58 ±0.5 - 9/16 ± ½	-0.58 ±0.5 - 9/16 ± ½
REAR TOE (Degrees)	0.13° ±0.23° 1/8° ± 7/32°	0.17° ±0.24° 3/16° ± ¼°	0.17° ±0.24° 3/16° ± ¼°	0.17° ±0.16° 3/16° ± 5/32°	0.17° ±0.16° 3/16° ± 5/32°	0.17° ±0.24° 3/16° ± ¼°
REAR TOE (Inches)	0.07" ±0.12" 1/16" ± 1/8"	0.09" ±0.12" 3/32" ± 1/8"	0.09" ±0.12" 3/32" ± 1/8"	0.09" ±0.08" 3/32" ± 3/32"	0.09" ±0.08" 3/32" ± 3/32"	0.09" ±0.12" 3/32" ± 1/8"
THRUST ANGLE						
ILLUSTRATION #	154	151	151	151	151	151
NOTES						
	32116	32117	32118	32119	32120	32121

Nissan

MODEL INFORMATION	Pathfinder HEV Canada	Pickup 2WD	Pickup 4WD	Pickup 4WD	Pulsar/NX	Quest
YEARS	2014-2015	1990-97	1990-91	1992-97	1990	1993-2002
INSTRUCTIONS						
LUGNUT TORQUE	83 ft.lbs.	87-108 ft.lbs.	87-108 ft.lbs.	87-108 ft.lbs.	72-87 ft.lbs.	72-87 ft.lbs.
FRONT						
CASTER (Degrees)	4.67 ±0.75 4 11/16 ± ¾	0.37 ±0.5 3/8 ± ½	1.3 ±0.5 1 5/16 ± ½	1.3 ±0.5 1 5/16 ± ½	1.92 ±0.75 1 15/16 ± ¾	0.8 ±0.75 13/16 ± ¾
CROSS | TOL.	0.0 (0) | 0.5 (½)	0.0 (0) | 0.75 (¾)	0.0 (0) | 0.75 (¾)	0.0 (0) | 0.75 (¾)	0.0 (0) | n/s	0.0 (0) | 0.75 (¾)
CAMBER (Degrees)	L= -0.25, R= -0.5 ±0.75 L= -¼, R= -½ ± ¾	0.42 ±0.5 7/16 ± ½	0.67 ±0.5 11/16 ± ½	0.67 ±0.5 11/16 ± ½	-0.5 ±0.75 -½ ± ¾	0.3 ±0.75 5/16 ± ¾
CROSS | TOL.	0.25 (¼) | 0.55 (9/16)	0.0 (0) | 0.75 (¾)	0.0 (0) | 0.75 (¾)	0.0 (0) | 0.75 (¾)	0.0 (0) | n/s	0.0 (0) | 0.75 (¾)
FRONT TOE (Degrees)	0.1° ±0.16° 3/32° ± 5/32°	0.25° ±0.08° [1] ¼° ± 3/32° [1]	0.37° ±0.08° [1] 3/8° ± 3/32° [1]	0.32° ±0.08° 5/16° ± 3/32°	0.0° ±0.1° 0° ± 3/32°	0.28° ±0.09° 9/32° ± 3/32°
FRONT TOE (Inches)	0.05" ±0.08" 1/16" ± 3/32"	0.13" ±0.04" [1] 1/8" ± 1/32" [1]	0.18" ±0.04" [1] 3/16" ± 1/32" [1]	0.16" ±0.04" 5/32" ± 1/32"	0.0" ±0.05" 0" ± 1/16"	0.14" ±0.05" 1/8" ± 1/32"
SAI	12.7 (12 11/16)	9.1 ±0.5 (9 1/16 ± ½)	8.1 ±0.5 (8 1/8 ± ½)	8.1 ±0.5 (8 1/8 ± ½)	14.5 ±0.8 (14 7/16 ± ¾)	13.6 ±0.8 (13 9/16 ± ¾)
TURNING ANGLE	38.0 / 33.0 ±1.0	37.0 / 34.0 ±1.0	34.0 / 32.0 ±1.0 [2]	34.0 / 32.0 ±1.0 [1]	39.0 / 32.5 ±2.0 [1]	38.0 / 30.0 ±2.0
ILLUSTRATION #		41	41	41	22	
REAR						
CAMBER (Degrees)	-0.58 ±0.5 - 9/16 ± ½				-1.25 ±0.75 -1¼ ± ¾	0.0 ±0.25 0 ± ¼
REAR TOE (Degrees)	0.17° ±0.24° 3/16° ± ¼°				0.05° ±0.2° 1/16° ± 3/16°	0.0° ±0.36° 0° ± 3/8°
REAR TOE (Inches)	0.07" ±0.12" 3/32" ± 1/8"				0.03" ±0.1" 1/32" ± 3/32"	0.0" ±0.18" 0" ± 3/16"
THRUST ANGLE						
ILLUSTRATION #	151				06	
NOTES		[1] w/bias tires 0.42° (0.21")	[1] w/bias tires 0.42° (0.21") [2] w/31x10.5R15 tires 28°/26°	[1] w/31x10.5R15 tires 28°/26°	[1] SE model = 37.5°/30°	
	32122	32128	32131	32132	32135	32136

Nissan

Quest	Quest USA	Quest Canada	MODEL INFORMATION	Quest	Rogue Select	Rogue AWD
2004-2009	2011-2016	2011-2016	YEARS	2017	2008-2015	2014-2017
			INSTRUCTIONS			
83 ft.lbs.	80 ft.lbs.	80 ft.lbs.	LUGNUT TORQUE	80 ft.lbs.	80 ft.lbs.	83 ft.lbs.
			FRONT			
2.7 ±0.75 2 11/16 ± 3/4	L=4.67, R=4.83 ± 0.75 L=4 11/16, R=4 13/16 ± 3/4	L=4.6, R=4.8 ± 0.75 L=4 5/8, R=4 13/16 ± 3/4	CASTER (Degrees)	L=4.67, R=4.83 ± 0.75 L=4 11/16, R=4 13/16 ± 3/4	4.75 ±0.75 [1] 4 3/4 ± 3/4 [1]	5.6 ±0.75 5 5/8 ± 3/4
0.0 (0) | 0.75 (3/4)	-0.17 (-3/16) | 0.5 (1/2)	-0.2 (-3/16) | 0.5 (1/2)	CROSS | TOL.	-0.17 (-3/16) | 0.5 (1/2)	0.0 (0) | 0.6 (5/8)	0.0 (0) | 0.6 (5/8)
-0.5 ±0.75 -1/2 ± 3/4	L=-0.25, R=-0.5 ±0.75 L=-1/4, R=-1/2 ± 3/4	L=-0.25, R=-0.5 ±0.75 L=-1/4, R=-1/2 ± 3/4	CAMBER (Degrees)	L=-0.25, R=-0.5 ±0.75 L=-1/4, R=-1/2 ± 3/4	L=-0.25, R=-0.5 ±0.75 L=-1/4, R=-1/2 ± 3/4	-0.48 ±0.75 -1/2 ± 3/4
0.0 (0) | 0.75 (3/4)	0.25 (1/4) | 0.5 (1/2)	0.25 (1/4) | 0.5 (1/2)	CROSS | TOL.	0.25 (1/4) | 0.5 (1/2)	0.25 (1/4) | 0.55 (9/16)	0.0 (0) | 0.6 (5/8)
0.02° ±0.08° 1/32° ± 3/32°	0.05° ±0.08° 1/16° ± 3/32°	0.05° ±0.08° 1/16° ± 3/32°	FRONT TOE (Degrees)	0.05° ±0.16° 1/16° ± 5/32°	0.17° ±0.08° 5/32° ± 3/32°	0.12° ±0.2° 1/8° ± 3/16°
0.01" ±0.04" 1/64" ± 1/32"	0.02" ±0.04" 1/32" ± 1/32"	0.02" ±0.04" 1/32" ± 1/32"	FRONT TOE (Inches)	0.02" ±0.08" 1/32" ± 3/32"	0.08" ±0.04" 3/32" ± 1/32"	0.06" ±0.1" 1/16" ± 3/32"
14.4 ±0.8 (14 3/8 ± 3/4)	12.8 ±0.8 (12 3/4 ± 3/4)	12.8 ±0.8 (12 3/4 ± 3/4)	SAI	12.8 ±0.8 (12 3/4 ± 3/4)	10.5 ±0.8 (10 1/2 ± 3/4)	11.6 ±0.8 (11 5/8 ± 3/4)
38.8 / 32.5 ±1.0	42.0 / 36.0 ±1.0	42.0 / 36.0 ±1.0	TURNING ANGLE	42.0 / 36.0 ±1.0	36.0 / 31.0 ±1.0	36.4 / 30.8 ±1.0
			ILLUSTRATION #			
			REAR			
-0.55 ±0.5 -9/16 ± 1/2	-0.6 ±0.5 -5/8 ± 1/2	-0.58 ±0.5 -9/16 ± 1/2	CAMBER (Degrees)	-0.6 ±0.5 -5/8 ± 1/2	-0.92 ±0.5 -15/16 ± 1/2	-0.58 ±0.75 -9/16 ± 3/4
0.25° ±0.13° 1/4° ± 1/8°	0.22° ±0.13° 7/32° ± 1/8°	0.19° ±0.13° 3/16° ± 1/8°	REAR TOE (Degrees)	0.22° ±0.13° 7/32° ± 1/8°	0.17° ±0.17° 5/32° ± 5/32°	0.33° ±0.33° [1] 11/32° ± 11/32° [1]
0.13" ±0.06" 1/8" ± 1/16"	0.11" ±0.06" 1/8" ± 1/16"	0.1" ±0.06" 3/32" ± 1/16"	REAR TOE (Inches)	0.11" ±0.06" 1/8" ± 1/16"	0.08" ±0.08" 3/32" ± 3/32"	0.17" ±0.17" [1] 5/32" ± 5/32" [1]
			THRUST ANGLE			
151	151	151	ILLUSTRATION #	151	184	184
			NOTES		[1] FWD = 4.7° (4 11/16°)	[1] w/19" wheels = 0.42°(0.21")
32137	32138	32139		32140	32141	32142

Nissan

Rogue FWD	Sentra Sedan, Hatch, 2WD Wagon	Sentra Coupe	MODEL INFORMATION	Sentra 4WD Wagon	Sentra	Sentra
2014-2017	1990	1990	YEARS	1990	1991-94	1995-99
			INSTRUCTIONS			
83 ft.lbs.	72-87 ft.lbs.	72-87 ft.lbs.	LUGNUT TORQUE	72-87 ft.lbs.	72-87 ft.lbs.	72-87 ft.lbs.
			FRONT			
5.7 ±0.75 5 11/16 ± 3/4	1.5 ±0.75 1 1/2 ± 3/4	1.67 ±0.75 1 11/16 ± 3/4	CASTER (Degrees)	0.92 ±0.75 15/16 ± 3/4	1.42 ±0.75 1 7/16 ± 3/4	1.42 ±0.75 1 7/16 ± 3/4
0.0 (0) | 0.6 (5/8)	0.0 (0) | n/s	0.0 (0) | n/s	CROSS | TOL.	0.0 (0) | n/s	0.0 (0) | n/s	0.0 (0) | 1.0 (1)
-0.57 ±0.75 -9/16 ± 3/4	-0.17 ±0.75 -3/16 ± 3/4	-0.33 ±0.75 -5/16 ± 3/4	CAMBER (Degrees)	-0.08 ±0.75 -1/16 ± 3/4	-0.25 ±0.75 -1/4 ± 3/4	-0.58 ±0.75 -9/16 ± 3/4
0.0 (0) | 0.6 (5/8)	0.0 (0) | n/s	0.0 (0) | n/s	CROSS | TOL.	0.0 (0) | n/s	0.0 (0) | n/s	0.0 (0) | 1.0 (1)
0.12° ±0.2° 1/8° ± 3/16°	0.05° ±0.1° 1/16° ± 3/32°	0.03° ±0.1° 1/32° ± 3/32°	FRONT TOE (Degrees)	0.05° ±0.1° 1/16° ± 3/32°	0.2° ±0.1° 3/16° ± 3/32°	0.2° ±0.2° 3/16° ± 3/16°
0.06" ±0.1" 1/16" ± 3/32"	0.03" ±0.05" 1/32" ± 1/16"	0.02" ±0.05" 1/32" ± 1/16"	FRONT TOE (Inches)	0.03" ±0.05" 1/32" ± 1/16"	0.1" ±0.05" 3/32" ± 1/16"	0.1" ±0.1" 3/32" ± 3/32"
11.9 ±0.8 (11 7/8 ± 3/4)	14.0 ±0.8 (14 ± 3/4)	14.3 ±0.8 (14 5/16 ± 3/4)	SAI	13.6 ±0.8 (13 9/16 ± 3/4)	14.0 ±0.8 (14 ± 3/4)	14.8 ±0.8 (14 3/4 ± 3/4)
36.4 / 30.8 ±1.0	41.0 / 34.0 ±1.5	41.0 / 34.0 ±1.5 [1]	TURNING ANGLE	39.0 / 31.0 ±1.5 [1]	41.0 / 34.0 ±2.0 [1]	37.0 / 31.0 ±1.0 [1]
	22	22	ILLUSTRATION #	22		
			REAR			
-0.92 ±0.75 -15/16 ± 3/4	-1.08 ±0.75 -1 1/16 ± 3/4	-1.17 ±0.75 -1 3/16 ± 3/4	CAMBER (Degrees)	-0.17 ±0.75 -3/16 ± 3/4	-0.92 ±0.75 -15/16 ± 3/4	-1.0 ±0.75 -1 ± 3/4
0.33° ±0.33° [1] 11/32° ± 11/32° [1]	-0.2° ±0.2° -3/16° ± 3/16°	-0.1° ±0.2° -3/32° ± 3/16°	REAR TOE (Degrees)	-0.2° ±0.2° -3/16° ± 3/16°	0.1° ±0.2° 3/32° ± 3/16°	0.1° ±0.4° 3/32° ± 13/32°
0.17" ±0.17" [1] 5/32" ± 5/32" [1]	-0.1" ±0.1" -3/32" ± 3/32"	-0.05" ±0.1" -1/16" ± 3/32"	REAR TOE (Inches)	-0.1" ±0.1" -3/32" ± 3/32"	0.05" ±0.1" 1/16" ± 3/32"	0.05" ±0.2" 1/16" ± 3/16"
			THRUST ANGLE			
184	06	06	ILLUSTRATION #	64	64	
[1] w/19" wheels = 0.42°(0.21")		[1] SE model = 38.5°/32.5°	NOTES	[1] w/MS 41°/34°	[1] 2.0L = 35°/30°	[1] w/MS 41°/34°
32143	32145	32146		32147	32148	32149

Nissan

Sentra exc/2.5L	Sentra 2.5L	Sentra exc/2.5L	MODEL INFORMATION	Sentra 2.5L	Sentra exc/SE-R	Sentra SE-R exc/SPEC-V
2000-2004	2002-2004	2005-2006	YEARS	2005-2006	2007-2009	2007-2009
			INSTRUCTIONS			
73-86 ft.lbs.	73-86 ft.lbs.	73-86 ft.lbs.	LUGNUT TORQUE	73-86 ft.lbs.	83 ft.lbs.	83 ft.lbs.
			FRONT			
1.6 ± 0.75 1⅝ ± ¾	1.72 ± 0.75 1¾ ± ¾	1.58 ± 0.75 1 9/16 ± ¾	CASTER (Degrees)	1.72 ± 0.75 1¾ ± ¾	L= 4.8, R= 5.03 ± 0.63 L= 4 13/16, R= 5 ± ⅝	L= 4.8, R= 5.03 ± 0.63 L= 4 13/16, R= 5 ± ⅝
0.0 (0) | 0.75 (¾)	0.0 (0) | 0.75 (¾)	0.0 (0) | 0.75 (¾)	CROSS | TOL.	0.0 (0) | 0.75 (¾)	-0.23 (-¼) | 0.75 (¾)	-0.23 (-¼) | 0.75 (¾)
-0.42 ± 0.75 - 7/16 ± ¾	-0.45 ± 0.75 - 7/16 ± ¾	-0.33 ± 0.75 - 5/16 ± ¾	CAMBER (Degrees)	-0.45 ± 0.75 - 7/16 ± ¾	L= -0.05, R= -0.28 ± 0.63 L= - 1/16, R= - ¼ ± ⅝	L= -0.05, R= -0.28 ± 0.63 L= - 1/16, R= - ¼ ± ⅝
0.0 (0) | 0.75 (¾)	0.0 (0) | 0.75 (¾)	0.0 (0) | 0.75 (¾)	CROSS | TOL.	0.0 (0) | 0.75 (¾)	0.23 (¼) | 0.75 (¾)	0.23 (¼) | 0.75 (¾)
0.2° ± 0.1° 3/16° ± 3/32°	0.2° ± 0.1° 3/16° ± 3/32°	0.18° ± 0.1° 3/16° ± 3/32°	FRONT TOE (Degrees)	0.18° ± 0.1° 3/16° ± 3/32°	0.0° ± 0.1° 0° ± 3/32°	0.0° ± 0.1° 0° ± 3/32°
0.1" ± 0.05" 3/32" ± 1/16"	0.1" ± 0.05" 3/32" ± 1/16"	0.09" ± 0.05" 3/32" ± 1/16"	FRONT TOE (Inches)	0.09" ± 0.05" 3/32" ± 1/16"	0.0" ± 0.05" 0" ± 1/16"	0.0" ± 0.05" 0" ± 1/16"
14.7 ± 0.8 (14¾ ± ¾)	14.8 ± 0.8 (14¾ ± ¾)	14.6 ± 0.8 (14⅝ ± ¾)	SAI	14.8 ± 0.8 (14¾ ± ¾)	11.1 ± 0.8 (11 1/16 ± ¾)	11.1 ± 0.8 (11 1/16 ± ¾)
37.0 / 31.0 ± 1.0	32.0 / 27.0 ± 1.0	37.0 / 31.0 ± 1.0	TURNING ANGLE	32.0 / 27.0 ± 1.0	39.0 / 33.0 ± 1.0	36.0 / 30.0 ± 1.0
			ILLUSTRATION #			
			REAR			
-1.0 ± 0.75 -1 ± ¾	-1.0 ± 0.75 -1 ± ¾	-1.0 ± 0.75 -1 ± ¾	CAMBER (Degrees)	-1.0 ± 0.75 -1 ± ¾	-1.5 ± 0.5 -1½ ± ½	-1.5 ± 0.5 -1½ ± ½
0.1° ± 0.4° 3/32° ± 13/32°	0.1° ± 0.4° 3/32° ± 13/32°	0.09° ± 0.35° 3/32° ± 11/32°	REAR TOE (Degrees)	0.09° ± 0.35° 3/32° ± 11/32°	0.1° ± 0.37° 3/32° ± ⅜°	0.2° ± 0.37° 3/16° ± ⅜°
0.05" ± 0.2" 1/16" ± 3/16"	0.05" ± 0.2" 1/16" ± 3/16"	0.05" ± 0.18" 1/32" ± 3/16"	REAR TOE (Inches)	0.05" ± 0.18" 1/32" ± 3/16"	0.05" ± 0.18" 1/16" ± 3/16"	0.1" ± 0.18" 3/32" ± 3/16"
			THRUST ANGLE			
			ILLUSTRATION #			
			NOTES			
32150	32151	32152		32153	32154	32155

Nissan

Sentra SE-R SPEC-V	Sentra exc/SE-R	Sentra SE-R	MODEL INFORMATION	Sentra exc/SE-R	Sentra SE-R exc/Spec-V	Sentra SE-R SPEC-V
2007-2009	2010	2010	YEARS	2011-2012	2011-2012	2011-2012
			INSTRUCTIONS			
83 ft.lbs.	83 ft.lbs.	83 ft.lbs.	LUGNUT TORQUE	83 ft.lbs.	83 ft.lbs.	83 ft.lbs.
			FRONT			
L= 5.05, R= 5.28 ± 0.63 L= 5 1/16, R= 5¼ ± ⅝	L= 4.72, R= 4.95 ± 0.63 L= 4¾, R= 4 15/16 ± ⅝	L= 5.05, R= 5.28 ± 0.63 L= 5 1/16, R= 5¼ ± ⅝	CASTER (Degrees)	L= 4.8, R= 5.03 ± 0.63 L= 4 13/16, R= 5 ± ⅝	L= 4.8, R= 5.03 ± 0.63 L= 4 13/16, R= 5 ± ⅝	L= 5.05, R= 5.28 ± 0.63 L= 5 1/16, R= 5¼ ± ⅝
-0.23 (-¼) | 0.75 (¾)	-0.23 (-¼) | 0.75 (¾)	-0.23 (-¼) | 0.75 (¾)	CROSS | TOL.	-0.23 (-¼) | 0.75 (¾)	-0.23 (-¼) | 0.75 (¾)	-0.23 (-¼) | 0.75 (¾)
L= -0.13, R= -0.37 ± 0.63 L= - ⅛, R= - ⅜ ± ⅝	L= 0.12, R= -0.12 ± 0.63 L= ⅛, R= - ⅛ ± ⅝	L= -0.13, R= -0.37 ± 0.63 L= - ⅛, R= - ⅜ ± ⅝	CAMBER (Degrees)	-0.17 ± 0.55 - 3/16 ± 9/16	-0.17 ± 0.55 - 3/16 ± 9/16	-0.25 ± 0.55 - ¼ ± 9/16
0.24 (¼) | 0.75 (¾)	0.23 (¼) | 0.75 (¾)	0.24 (¼) | 0.75 (¾)	CROSS | TOL.	0.0 (0) | 0.75 (¾)	0.0 (0) | 0.75 (¾)	0.0 (0) | 0.75 (¾)
0.0° ± 0.1° 0° ± 3/32°	0.0° ± 0.1° 0° ± 3/32°	0.0° ± 0.1° 0° ± 3/32°	FRONT TOE (Degrees)	0.0° ± 0.1° 0° ± 3/32°	0.0° ± 0.1° 0° ± 3/32°	0.0° ± 0.1° 0° ± 3/32°
0.0" ± 0.05" 0" ± 1/16"	0.0" ± 0.05" 0" ± 1/16"	0.0" ± 0.05" 0" ± 1/16"	FRONT TOE (Inches)	0.0" ± 0.05" 0" ± 1/16"	0.0" ± 0.05" 0" ± 1/16"	0.0" ± 0.05" 0" ± 1/16"
11.3 ± 0.8 (11 5/16 ± ¾)	10.9 ± 0.8 (10 15/16 ± ¾)	11.2 ± 0.8 (11¼ ± ¾)	SAI	10.9 ± 0.8 (10 15/16 ± ¾)	11.2 ± 0.8 (11¼ ± ¾)	11.2 ± 0.8 (11¼ ± ¾)
36.0 / 30.0 ± 1.0	39.0 / 33.0 ± 1.0	36.0 / 30.0 ± 1.0	TURNING ANGLE	39.0 / 33.0 ± 1.0	36.0 / 30.0 ± 1.0	36.0 / 30.0 ± 1.0
			ILLUSTRATION #			
			REAR			
-1.5 ± 0.5 -1½ ± ½	-1.5 ± 0.5 -1½ ± ½	-1.5 ± 0.5 -1½ ± ½	CAMBER (Degrees)	-1.5 ± 0.5 -1½ ± ½	-1.5 ± 0.5 -1½ ± ½	-1.5 ± 0.5 -1½ ± ½
0.27° ± 0.37° 9/32° ± ⅜°	0.1° ± 0.37° 3/32° ± ⅜°	0.18° ± 0.37° [1] 3/16° ± ⅜° [1]	REAR TOE (Degrees)	0.1° ± 0.37° 3/32° ± ⅜°	0.18° ± 0.37° 3/16° ± ⅜°	0.28° ± 0.37° 9/32° ± ⅜°
0.13" ± 0.18" ⅛" ± 3/16"	0.05" ± 0.18" 1/16" ± 3/16"	0.09" ± 0.18" [1] 3/32" ± 3/16" [1]	REAR TOE (Inches)	0.05" ± 0.18" 1/16" ± 3/16"	0.09" ± 0.18" 3/32" ± 3/16"	0.14" ± 0.18" ⅛" ± 3/16"
			THRUST ANGLE			
			ILLUSTRATION #			
		[1] SPEC-V = 0.28° (0.14")	NOTES			
32156	32157	32158		32159	32160	32161

ALIGNMENT SPECIFICATIONS

Nissan

MODEL INFORMATION	Sentra All models	Stanza	Titan 4X2	Titan 4X4	Titan 4X2	Titan 4X4
YEARS	2013-2017	1990-92	2004-2005	2004-2005	2006-2015	2006-2015
INSTRUCTIONS						
LUGNUT TORQUE	83 ft.lbs.	72-87 ft.lbs.	98 ft.lbs.	98 ft.lbs.	98 ft.lbs.	98 ft.lbs.
FRONT						
CASTER (Degrees)	4.83 ±0.75 4 13⁄16 ± ¾	1.33 ±0.75 1 5⁄16 ± ¾	3.27 ±0.75 3 ¼ ± ¾	2.37 ±0.75 2 ⅜ ± ¾	3.0 ±0.75 3 ± ¾	2.2 ±0.75 2 3⁄16 ± ¾
CROSS | TOL.	0.0 (0) | 0.75 (¾)	0.0 (0) | n/s	0.0 (0) | 0.75 (¾)	0.0 (0) | 0.75 (¾)	0.0 (0) | 0.75 (¾)	0.0 (0) | 0.75 (¾)
CAMBER (Degrees)	-0.42 ±0.65 - 7⁄16 ± ⅝	0.25 ±0.75 ¼ ± ¾	-0.12 ±0.75 - ⅛ ± ¾	0.43 ±0.75 7⁄16 ± ¾	-0.2 ±0.75 - 3⁄16 ± ¾	0.3 ±0.75 5⁄16 ± ¾
CROSS | TOL.	0.0 (0) | 0.58 (9⁄16)	0.0 (0) | n/s	0.0 (0) | 0.75 (¾)	0.0 (0) | 0.75 (¾)	0.0 (0) | 0.75 (¾)	0.0 (0) | 0.75 (¾)
FRONT TOE (Degrees)	0.17° ±0.16° 5⁄32° ± 5⁄32°	0.2° ±0.1° 3⁄16° ± 3⁄32°	0.22° ±0.08° 7⁄32° ± 3⁄32°	0.22° ±0.08° 7⁄32° ± 3⁄32°	0.17° ±0.17° 5⁄32° ± 5⁄32°	0.17° ±0.17° 5⁄32° ± 5⁄32°
FRONT TOE (Inches)	0.08" ±0.08" 3⁄32" ± 3⁄32"	0.1" ±0.05" 3⁄32" ± 1⁄16"	0.11" ±0.04" ⅛" ± 1⁄32"	0.11" ±0.04" ⅛" ± 1⁄32"	0.08" ±0.08" 3⁄32" ± 3⁄32"	0.08" ±0.08" 3⁄32" ± 3⁄32"
SAI	12.1 ±0.8 (12 1⁄16 ± ¾)	14.5 ±0.8 (14 ½ ± ¾)	13.6 (13 9⁄16)	13.0 (13)	13.6 (13 9⁄16)	13.0 (13)
TURNING ANGLE	39.5 / 33.5 ±1.0	36.0 / 29.0 ±2.0	37.5 / 33.5 ±1.5	37.9 / 34.0 ±1.5	37.5 / 34.0 ±1.5	37.9 / 34.0 ±1.5
ILLUSTRATION #			12	12	12	12
REAR						
CAMBER (Degrees)	-1.5 ±0.5 -1 ½ ± ½	-0.58 ±0.75 - 9⁄16 ± ¾				
REAR TOE (Degrees)	0.33° ±0.38° 11⁄32° ± ⅜°	-0.19° ±0.19° - 3⁄16° ± 3⁄16°				
REAR TOE (Inches)	0.17" ±0.19" 5⁄32" ± 3⁄16"	-0.1" ±0.1" - 3⁄32" ± 3⁄32"				
THRUST ANGLE						
ILLUSTRATION #		10				
NOTES						
	32162	32166	32169	32170	32171	32172

Nissan

MODEL INFORMATION	Titan XD 4X2 & 4X4	Titan XD 4X2 & 4X4	Titan 4X2	Titan 4X4 S, SV, & Platinum	Titan 4X4 P4X, SL	Van XE, GXE
YEARS	2016	2017	2017	2017	2017	1990
INSTRUCTIONS						
LUGNUT TORQUE	130 ft.lbs.	130 ft.lbs.	98 ft.lbs.	98 ft.lbs.	98 ft.lbs.	72-87 ft.lbs.
FRONT						
CASTER (Degrees)	5.92 ±0.5 5 15⁄16 ± ½	L= 4.9, R= 5.2 ± 0.92 L= 4 ⅞, R= 5 3⁄16 ± 15⁄16	3.08 ±0.92 3 1⁄16 ± 15⁄16	2.83 ±0.92 2 13⁄16 ± 15⁄16	2.33 ±0.92 2 5⁄16 ± 15⁄16	1.33 ±0.75 1 5⁄16 ± ¾
CROSS | TOL.	0.0 (0) | 0.75 (¾)	-0.3 (- 5⁄16) | 0.75 (¾)	0.0 (0) | 0.75 (¾)	0.0 (0) | 0.75 (¾)	0.0 (0) | 0.75 (¾)	0.0 (0) | 0.75 (¾)
CAMBER (Degrees)	0.08 ±0.5 1⁄16 ± ½	L= -0.08, R= -0.26 ±0.83 L= - 1⁄16, R= - ¼ ± 13⁄16	-0.17 ±0.83 - 3⁄16 ± 13⁄16	-0.08 ±0.83 - 1⁄16 ± 13⁄16	0.3 ±0.83 5⁄16 ± 13⁄16	0.25 ±0.75 ¼ ± ¾
CROSS | TOL.	0.0 (0) | 0.75 (¾)	0.18 (3⁄16) | 0.75 (¾)	0.0 (0) | 0.75 (¾)	0.0 (0) | 0.75 (¾)	0.0 (0) | 0.75 (¾)	0.0 (0) | n/s
FRONT TOE (Degrees)	0.5° ±0.17° ½° ± 5⁄32°	0.3° ±0.2° 5⁄16° ± 3⁄16°	0.3° ±0.2° 5⁄16° ± 3⁄16°	0.3° ±0.2° 5⁄16° ± 3⁄16°	0.3° ±0.2° 5⁄16° ± 3⁄16°	0.33° ±0.17° 11⁄32° ± 5⁄32°
FRONT TOE (Inches)	0.25" ±0.08" ¼" ± 3⁄32"	0.15" ±0.1" 5⁄32" ± 3⁄32"	0.15" ±0.1" 5⁄32" ± 3⁄32"	0.15" ±0.1" 5⁄32" ± 3⁄32"	0.15" ±0.1" 5⁄32" ± 3⁄32"	0.17" ±0.08" 5⁄32" ± 3⁄32"
SAI	8.9 (8 15⁄16)	9.0 (9)	13.6 (13 9⁄16)	13.3 (13 5⁄16)	13.0 (13)	9.5 ±0.8 (9 ½ ± ¾)
TURNING ANGLE	36.4 / 32.2 ±1.5	36.4 / 32.2 ±1.5	37.5 / 33.9 ±1.0	37.9 / 34.0 ±1.0	37.9 / 34.0 ±1.0	37.0 / 34.0 ±1.0
ILLUSTRATION #	12	12	12	12	12	41
REAR						
CAMBER (Degrees)						
REAR TOE (Degrees)						
REAR TOE (Inches)						
THRUST ANGLE						
ILLUSTRATION #						
NOTES						
	32173	32174	32175	32176	32177	32179

Nissan

MODEL INFORMATION	Versa	Versa 14" tires (OE)	Versa	Versa 15" & 16" tires (OE)	Versa Sedan	Versa Hatch
YEARS	2007	2008	2009-2011	2009-2011	2012-2017	2012
INSTRUCTIONS						
LUGNUT TORQUE	83 ft.lbs.	83 ft.lbs.	83 ft.lbs.	83 ft.lbs.	83 ft.lbs.	83 ft.lbs.
FRONT						
CASTER (Degrees)	L= 4.67, R= 4.83 ± 0.75 L= 4 11/16, R= 4 13/16 ± ¾	L= 4.67, R= 4.83 ± 0.75 L= 4 11/16, R= 4 13/16 ± ¾	L= 4.33, R= 4.5 ± 0.75 L= 4 5/16, R= 4½ ± ¾	L= 4.67, R= 4.83 ± 0.75 L= 4 11/16, R= 4 13/16 ± ¾	3.67 ±0.75 3 11/16 ± ¾	L= 4.67, R= 4.83 ± 0.75 L= 4 11/16, R= 4 13/16 ± ¾
CROSS | TOL.	-0.17 (-3/16) | 0.5 (½)	-0.17 (-3/16) | 0.5 (½)	-0.17 (-3/16) | 0.5 (½)	-0.17 (-3/16) | 0.5 (½)	0.0 (0) | 0.75 (¾)	-0.17 (-3/16) | 0.5 (½)
CAMBER (Degrees)	L= -0.17, R= -0.33 ±0.75 L= -3/16, R= -5/16 ± ¾	L= -0.17, R= -0.08 ±0.75 L= -3/16, R= -1/16 ± ¾	-0.08 ±0.75 -1/16 ± ¾	-0.17 ±0.75 -3/16 ± ¾	-0.08 ±0.75 -1/16 ± ¾	-0.17 ±0.75 -3/16 ± ¾
CROSS | TOL.	0.17 (3/16) | 0.5 (½)	-0.08 (-1/16) | 0.5 (½)	0.0 (0) | 0.5 (½)	0.0 (0) | 0.5 (½)	0.0 (0) | 0.58 (9/16)	0.0 (0) | 0.5 (½)
FRONT TOE (Degrees)	0.1° ±0.1° 3/32° ± 3/32°	0.1° ±0.1° 3/32° ± 3/32°	0.1° ±0.1° 3/32° ± 3/32°	0.1° ±0.1° 3/32° ± 3/32°	0.08° ±0.16° 3/32° ± 5/32°	0.1° ±0.1° 3/32° ± 3/32°
FRONT TOE (Inches)	0.05" ±0.05" 1/16" ± 1/16"	0.05" ±0.05" 1/16" ± 1/16"	0.05" ±0.05" 1/16" ± 1/16"	0.05" ±0.05" 1/16" ± 1/16"	0.04" ±0.08" 1/32" ± 3/32"	0.05" ±0.05" 1/16" ± 1/16"
SAI	9.9 ±0.8 (9 15/16 ± ¾)	9.9 ±0.8 (9 15/16 ± ¾)	9.8 ±0.8 (9 13/16 ± ¾)	9.9 ±0.8 (9 15/16 ± ¾)	11.8 ±0.8 (11¾ ± ¾)	9.9 ±0.8 (9 15/16 ± ¾)
TURNING ANGLE	38.0 / 33.0 ±1.0	38.0 / 33.0 ±1.0	38.0 / 33.0 ±1.0	38.0 / 33.0 ±1.0	37.5 / 32.0 ±1.0	38.0 / 33.0 ±1.0
ILLUSTRATION #						
REAR						
CAMBER (Degrees)	-1.5 ±0.5 -1½ ± ½	-1.5 ±0.5 -1½ ± ½	-1.5 ±0.5 -1½ ± ½	-1.5 ±0.5 -1½ ± ½	-1.42 ±0.5 -1 7/16 ± ½	-1.5 ±0.5 -1½ ± ½
REAR TOE (Degrees)	0.3° ±0.4° 5/16° ± 13/32°	0.3° ±0.4° 5/16° ± 13/32°	0.17° ±0.4° 5/32° ± 13/32°	0.3° ±0.4° 5/16° ± 13/32°	0.16° ±0.32° 5/32° ± 5/16°	0.24° ±0.32° ¼° ± 5/16°
REAR TOE (Inches)	0.15" ±0.2" 5/32" ± 3/16"	0.15" ±0.2" 5/32" ± 3/16"	0.09" ±0.2" 3/32" ± 3/16"	0.15" ±0.2" 5/32" ± 3/16"	0.08" ±0.16" 3/32" ± 5/32"	0.12" ±0.16" 1/8" ± 5/32"
THRUST ANGLE						
ILLUSTRATION #						
NOTES						
	32180	32181	32182	32183	32184	32185

Nissan

MODEL INFORMATION	Versa Note	Xterra 4X2	Xterra 4X4	Xterra 4X2	Xterra 4X4	Xterra 4X2
YEARS	2014-2017	2000-2004	2000-2004	2005-2008	2005-2008	2009-2015
INSTRUCTIONS						
LUGNUT TORQUE	83 ft.lbs.	87-108 ft.lbs.	87-108 ft.lbs.	98 ft.lbs.	98 ft.lbs.	98 ft.lbs.
FRONT						
CASTER (Degrees)	3.75 ±0.75 3¾ ± ¾	2.57 ±0.5 2 9/16 ± ½	2.17 ±0.5 2 3/16 ± ½	3.0 ±0.5 3 ± ½	2.75 ±0.5 2¾ ± ½	3.0 ±0.75 3 ± ¾
CROSS | TOL.	0.0 (0) | 0.75 (¾)	0.0 (0) | 0.75 (¾)	0.0 (0) | 0.75 (¾)	0.0 (0) | 0.75 (¾)	0.0 (0) | 0.75 (¾)	0.0 (0) | 0.75 (¾)
CAMBER (Degrees)	-0.08 ±0.75 -1/16 ± ¾	0.55 ±0.5 9/16 ± ½	0.6 ±0.5 5/8 ± ½	0.25 ±0.5 ¼ ± ½	0.5 ±0.5 ½ ± ½	0.25 ±0.75 ¼ ± ¾
CROSS | TOL.	0.0 (0) | 0.75 (¾)	0.0 (0) | 0.75 (¾)	0.0 (0) | 0.75 (¾)	0.0 (0) | 0.75 (¾)	0.0 (0) | 0.75 (¾)	0.0 (0) | 0.75 (¾)
FRONT TOE (Degrees)	0.08° ±0.16° 3/32° ± 5/32°	0.33° ±0.08° 11/32° ± 3/32°	0.33° ±0.08° 11/32° ± 3/32°	0.32° ±0.08° 5/16° ± 3/32°	0.32° ±0.08° 5/16° ± 3/32°	0.24° ±0.08° ¼° ± 3/32°
FRONT TOE (Inches)	0.04" ±0.08" 1/32" ± 3/32"	0.17" ±0.04" 5/32" ± 1/32"	0.17" ±0.04" 5/32" ± 1/32"	0.16" ±0.04" 5/32" ± 1/32"	0.16" ±0.04" 5/32" ± 1/32"	0.12" ±0.04" 1/8" ± 1/32"
SAI	11.8 ±0.8 (11¾ ± ¾)	10.9 ±0.5 (10 7/8 ± ½)	10.8 ±0.5 (10 13/16 ± ½)	13.0 (13)	12.8 (12¾)	13.0 (13)
TURNING ANGLE	37.5 / 32.0 ±1.0	34.8 / 33.0 ±1.0 [1]	35.1 / 33.2 ±1.0 [1]	35.4 / 31.4 ±1.0	35.7 / 32.0 ±1.0	35.4 / 31.4 ±1.0
ILLUSTRATION #		24	24	12	12	12
REAR						
CAMBER (Degrees)	-1.42 ±0.5 -1 7/16 ± ½					
REAR TOE (Degrees)	0.16° ±0.32° 5/32° ± 5/16°					
REAR TOE (Inches)	0.08" ±0.16" 3/32" ± 5/32"					
THRUST ANGLE						
ILLUSTRATION #						
NOTES		[1] w/265-70R15 tires 32.8°/30.7°	[1] w/265-70R15 tires 33°/31°			
	32186	32187	32188	32189	32190	32191

Nissan / Oldsmobile

Xterra 4X4	X-Trail (FWD & AWD)	Notes	MODEL INFORMATION	Achieva	Achieva	Alero
2009-2015	2005-2006		YEARS	1992-95	1996-98	1999-2003
			INSTRUCTIONS			
98 ft.lbs.	83 ft.lbs.		LUGNUT TORQUE	100 ft.lbs.	100 ft.lbs.	100 ft.lbs.
			FRONT			
2.75 ± 0.75 2¾ ± ¾	2.45 ± 0.75 2 7/16 ± ¾		CASTER (Degrees)	1.45 ± 1.0 1 7/16 ± 1	1.45 ± 1.0 1 7/16 ± 1	4.1 ± 1.0 4⅛ ± 1
0.0 (0) | 0.75 (¾)	0.0 (0) | 0.75 (¾)		CROSS | TOL.	0.0 (0) | 0.7 (11/16)	0.0 (0) | 0.7 (11/16)	0.0 (0) | 1.0 (1)
0.5 ± 0.75 ½ ± ¾	-0.4 +1.0 -0.5 -⅜ +1 -½		CAMBER (Degrees)	0.0 ± 0.7 0 ± 11/16	0.0 ± 0.7 0 ± 11/16	-0.2 ± 1.0 -3/16 ± 1
0.0 (0) | 0.75 (¾)	0.0 (0) | 0.75 (¾)		CROSS | TOL.	0.0 (0) | 1.0 (1)	0.0 (0) | 1.0 (1)	0.0 (0) | 1.0 (1)
0.24° ± 0.08° ¼° ± 3/32°	0.08° ± 0.08° 3/32° ± 3/32°		FRONT TOE (Degrees)	0.0° ± 0.2° 0° ± 3/16°	0.0° ± 0.2° 0° ± 3/16°	0.1° ± 0.2° 3/32° ± 3/16°
0.12" ± 0.04" ⅛" ± 1/32"	0.04" ± 0.04" 1/32" ± 1/32"		FRONT TOE (Inches)	0.0" ± 0.1" 0" ± 3/32"	0.0" ± 0.1" 0" ± 3/32"	0.05" ± 0.1" 1/16" ± 3/32"
12.8 (12¾)	13.5 (13½)		SAI			
35.7 / 32.0 ± 1.0	39.0 / 31.0 ± 1.0		TURNING ANGLE			
12			ILLUSTRATION #	25	25	102
			REAR			
	-0.67 ± 0.75 -11/16 ± ¾		CAMBER (Degrees)	-0.25 ± 0.6 -¼ ± ⅝	-0.4 ± 0.75 -⅜ ± ¾	-0.2 ± 0.5 -3/16 ± ½
	0.08° ± 0.08° 3/32° ± 3/32°		REAR TOE (Degrees)	0.0° ± 0.35° 0° ± 11/32°	0.2° ± 0.3° 3/16° ± 5/16°	0.06° ± 0.2° 1/16° ± 3/16°
	0.04" ± 0.04" 1/32" ± 1/32"		REAR TOE (Inches)	0.0" ± 0.18" 0" ± 3/16"	0.1" ± 0.15" 3/32" ± 5/32"	0.03" ± 0.1" 1/32" ± 3/32"
			THRUST ANGLE	0.35 (11/32)	0.25 (¼)	0.2 (3/16)
	06		ILLUSTRATION #			06
			NOTES			
32192	32193	32204		33003	33004	33005

Oldsmobile

Alero	Aurora	Aurora	MODEL INFORMATION	Aurora	Bravada	Bravada
2004	1995-96	1997-99	YEARS	2001-2003	1991-99	2000-2001
			INSTRUCTIONS			
100 ft.lbs.	100 ft.lbs.	100 ft.lbs.	LUGNUT TORQUE	100 ft.lbs.	100 ft.lbs.	100 ft.lbs.
			FRONT			
4.1 ± 0.75 4⅛ ± ¾	6.0 ± 1.0 6 ± 1	6.0 ± 0.5 6 ± ½	CASTER (Degrees)	5.0 ± 0.5 [1] 5 ± ½ [1]	3.0 ± 0.5 3 ± ½	L= 2.8, R= 3.3 ± 1.0 L= 2 13/16, R= 3 5/16 ± 1
0.0 (0) | 0.5 (½)	0.0 (0) | 1.0 (1)	0.0 (0) | 0.5 (½)	CROSS | TOL.	0.0 (0) | 0.5 (½)	0.0 (0) | 0.5 (½)	-0.5 (-½) | 0.5 (½)
-0.2 ± 0.7 -3/16 ± 11/16	0.2 ± 1.0 3/16 ± 1	-0.2 ± 0.5 -3/16 ± ½	CAMBER (Degrees)	-0.2 ± 0.5 -3/16 ± ½	0.0 ± 0.5 0 ± ½	0.0 ± 1.0 0 ± 1
0.0 (0) | 0.7 (11/16)	0.0 (0) | 1.0 (1)	0.0 (0) | 0.5 (½)	CROSS | TOL.	0.0 (0) | 0.5 (½)	0.0 (0) | 0.5 (½)	0.0 (0) | 0.5 (½)
0.14° ± 0.2° [1] ⅛° ± 3/16° [1]	0.0° ± 0.2° 0° ± 3/16°	0.2° ± 0.2° 3/16° ± 3/16°	FRONT TOE (Degrees)	0.2° ± 0.2° 3/16° ± 3/16°	0.1° ± 0.1° [1] 3/32° ± 3/32° [1]	0.1° ± 0.2° 3/32° ± 3/16°
0.07" ± 0.1" [1] 1/16" ± 3/32" [1]	0.0" ± 0.1" 0" ± 3/32"	0.1" ± 0.1" 3/32" ± 3/32"	FRONT TOE (Inches)	0.1" ± 0.1" 3/32" ± 3/32"	0.05" ± 0.05" [1] 1/16" ± 1/16" [1]	0.05" ± 0.1" 1/16" ± 3/32"
			SAI			
			TURNING ANGLE			
102	31 102	31 25	ILLUSTRATION #	31 102	24	24
			REAR			
-0.2 ± 0.7 -3/16 ± 11/16	-0.3 ± 0.5 -5/16 ± ½	-0.3 ± 0.5 -5/16 ± ½	CAMBER (Degrees)	-0.3 ± 0.5 -5/16 ± ½		
0.1° ± 0.2° 3/32° ± 3/16°	0.2° ± 0.2° 3/16° ± 3/16°	0.2° ± 0.2° 3/16° ± 3/16°	REAR TOE (Degrees)	0.2° ± 0.2° 3/16° ± 3/16°		
0.05" ± 0.1" 1/16" ± 3/32"	0.1" ± 0.1" 3/32" ± 3/32"	0.1" ± 0.1" 3/32" ± 3/32"	REAR TOE (Inches)	0.1" ± 0.1" 3/32" ± 3/32"		
0.2 (3/16)	0.1 (3/32)	0.15 (5/32)	THRUST ANGLE	0.15 (5/32)		
06	06	06	ILLUSTRATION #	06		
[1] 16" tires = 0.2º (0.1")			NOTES	[1] up to VIN..14103349 Caster = 6.0º (6º)	[1] 1991-96 = 0.2º (0.1")	
33006	33007	33008		33009	33011	33012

Oldsmobile

Bravada	Calais	Custom Cruiser	MODEL INFORMATION	Custom Cruiser	Cutlass Ciera Cutlass Cruiser	Cutlass Ciera Cutlass Cruiser
2002-2004	1990-91	1990	YEARS	1991-92	1990-92	1993-96
			INSTRUCTIONS			
103 ft.lbs.	100 ft.lbs.	80-100 ft.lbs.	LUGNUT TORQUE	100 ft.lbs.	100 ft.lbs.	100 ft.lbs.
			FRONT			
L= 3.5, R= 4.0 ± 0.5 L= 3½, R= 4 ± ½	1.7 ± 1.0 1 11/16 ± 1	3.0 ± 0.5 3 ± ½	CASTER (Degrees)	3.5 ± 1.0 3½ ± 1	1.7 ± 0.7 1 11/16 ± 11/16	3.0 ± 0.75 3 ± ¾
-0.5 (-½) | 0.3 (5/16)	0.0 (0) | 0.7 (11/16)	0.0 (0) | 0.5 (½)	CROSS | TOL.	0.0 (0) | 1.0 (1)	0.0 (0) | 0.7 (11/16)	0.0 (0) | 0.75 (¾)
-0.5 ± 0.5 -½ ± ½	0.0 ± 0.7 0 ± 11/16	0.8 ± 0.5 13/16 ± ½	CAMBER (Degrees)	0.0 ± 0.8 0 ± 13/16	0.0 ± 0.5 0 ± ½	0.0 ± 0.5 0 ± ½
0.0 (0) | 0.3 (5/16)	0.0 (0) | 1.0 (1)	0.0 (0) | 0.5 (½)	CROSS | TOL.	0.0 (0) | 1.0 (1)	0.0 (0) | 0.7 (11/16)	0.0 (0) | 0.75 (¾)
0.1° ± 0.2° 3/32° ± 3/16°	0.0° ± 0.2° 0° ± 3/16°	0.1° ± 0.2° 3/32° ± 3/16°	FRONT TOE (Degrees)	0.16° ± 0.2° 5/32° ± 3/16°	0.0° ± 0.2° 0° ± 3/16°	0.0° ± 0.2° 0° ± 3/16°
0.05" ± 0.1" 1/16" ± 3/32"	0.0" ± 0.1" 0" ± 3/32"	0.05" ± 0.1" 1/16" ± 3/32"	FRONT TOE (Inches)	0.08" ± 0.1" 3/32" ± 3/32"	0.0" ± 0.1" 0" ± 3/32"	0.0" ± 0.1" 0" ± 3/32"
			SAI			
			TURNING ANGLE			
133	25	23	ILLUSTRATION #	23	102 ▲	102 ▲
			REAR			
	-0.25 ± 0.5 -¼ ± ½		CAMBER (Degrees)		0.0 ± 0.3 0 ± 5/16	0.0 ± 0.3 0 ± 5/16
	0.25° ± 0.16° ¼° ± 5/32°		REAR TOE (Degrees)		0.0° ± 0.3° 0° ± 5/16°	0.0° ± 0.3° 0° ± 5/16°
	0.13" ± 0.08" ⅛" ± 3/32"		REAR TOE (Inches)		0.0" ± 0.15" 0" ± 5/32"	0.0" ± 0.15" 0" ± 5/32"
	0.31 (5/16)		THRUST ANGLE	0.15 (5/32)	0.15 (5/32)	0.15 (5/32)
			ILLUSTRATION #			
			NOTES			
33013	33016	33019		33020	33031	33032

Oldsmobile

Cutlass	Cutlass	Cutlass Supreme FWD	MODEL INFORMATION	Cutlass Supreme FWD	Eighty-Eight/LSS	Intrigue
1997	1998-99	1990-91	YEARS	1992-97	1990-99	1998-2002
			INSTRUCTIONS			
100 ft.lbs.	100 ft.lbs.	103 ft.lbs.	LUGNUT TORQUE	100 ft.lbs.	100 ft.lbs.	100 ft.lbs.
			FRONT			
4.2 ± 1.0 4 3/16 ± 1	4.1 ± 1.0 4⅛ ± 1	2.0 ± 0.5 2 ± ½	CASTER (Degrees)	2.0 ± 0.5 2 ± ½	3.0 ± 0.5 3 ± ½	3.0 ± 0.5 3 ± ½
0.0 (0) | 0.75 (¾)	0.0 (0) | 0.75 (¾)	0.0 (0) | 0.75 (¾)	CROSS | TOL.	0.0 (0) | 0.75 (¾)	0.0 (0) | 0.75 (¾)	0.0 (0) | 0.75 (¾)
-0.3 ± 1.0 -5/16 ± 1	-0.2 ± 1.0 [1] -3/16 ± 1 [1]	0.7 ± 0.5 11/16 ± ½	CAMBER (Degrees)	0.7 ± 0.5 11/16 ± ½	0.2 ± 0.5 3/16 ± ½	-0.9 ± 0.5 -⅞ ± ½
0.0 (0) | 0.75 (¾)	0.0 (0) | 0.75 (¾)	0.0 (0) | 0.75 (¾)	CROSS | TOL.	0.0 (0) | 0.75 (¾)	0.0 (0) | 0.75 (¾)	0.0 (0) | 0.75 (¾)
0.1° ± 0.2° 3/32° ± 3/16°	0.1° ± 0.2° 3/32° ± 3/16°	0.0° ± 0.2° 0° ± 3/16°	FRONT TOE (Degrees)	0.0° ± 0.2° 0° ± 3/16°	0.0° ± 0.2° 0° ± 3/16°	0.1° ± 0.2° 3/32° ± 3/16°
0.05" ± 0.1" 1/16" ± 3/32"	0.05" ± 0.1" 1/16" ± 3/32"	0.0" ± 0.1" 0" ± 3/32"	FRONT TOE (Inches)	0.0" ± 0.1" 0" ± 3/32"	0.0" ± 0.1" 0" ± 3/32"	0.05" ± 0.1" 1/16" ± 3/32"
			SAI			
			TURNING ANGLE			
25	25	33 ▲	ILLUSTRATION #	33 ▲	31 102 ▲	25 ▲
			REAR			
-0.3 ± 0.5 -5/16 ± ½	-0.2 ± 0.5 -3/16 ± ½	0.1 ± 0.5 [1] ⅛ ± ½ [1]	CAMBER (Degrees)	0.1 ± 0.5 ⅛ ± ½	-0.3 ± 0.5 -5/16 ± ½	-0.6 ± 0.5 [1] -⅝ ± ½ [1]
0.05° ± 0.2° 1/16° ± 3/16°	0.06° ± 0.2° 1/16° ± 3/16°	-0.1° ± 0.3° [2] -3/32° ± 5/16° [2]	REAR TOE (Degrees)	-0.1° ± 0.3° -3/32° ± 5/16°	0.1° ± 0.2° 3/32° ± 3/16°	0.1° ± 0.2° 3/32° ± 3/16°
0.03" ± 0.1" 1/32" ± 3/32"	0.03" ± 0.1" 1/32" ± 3/32"	-0.05" ± 0.15" [2] -1/16" ± 5/32" [2]	REAR TOE (Inches)	-0.05" ± 0.15" -1/16" ± 5/32"	0.05" ± 0.1" 1/16" ± 3/32"	0.05" ± 0.1" 1/16" ± 3/32"
0.2 (3/16)	0.2 (3/16)	0.15 (5/32)	THRUST ANGLE	0.15 (5/32)	0.1 (3/32)	0.15 (5/32)
10	06	25 34	ILLUSTRATION #	25 34	102 32	108 10
	[1] 1999 = 0.3° (5/16°)	[1] w/FE3 Susp. -0.15° (-⅛°) [2] w/FE3 Susp. 0.0° (0")	NOTES			[1] 1998-00 = -0.9° (-⅞°)
33033	33034	33036		33037	33048	33055

Oldsmobile

Model Information	Ninety-Eight	Regency	Silhouette	Silhouette w/beam axle susp.	Silhouette w/IRS	Toronado Trofeo
Years	1990-97	1997-98	1990-96	1997-2004	2002-2004	1990-92
Instructions						
LugNut Torque	100 ft.lbs.	100 ft.lbs.	100 ft.lbs.	100 ft.lbs.	100 ft.lbs.	100 ft.lbs.
FRONT						
Caster (Degrees)	3.0 ±0.5 3 ± 1/2	3.0 ±0.5 3 ± 1/2	3.0 ±0.7 [1] 3 ± 11/16 [1]	2.9 ±0.5 2 7/8 ± 1/2	2.9 ±0.5 2 7/8 ± 1/2	2.3 ±1.0 2 5/16 ± 1
Cross	0.0 (0)	0.0 (0)	0.0 (0)	0.0 (0)	0.0 (0)	0.0 (0)
Tol.	0.75 (3/4)	0.75 (3/4)	0.7 (11/16)	0.75 (3/4)	0.75 (3/4)	1.0 (1)
Camber (Degrees)	0.2 ±0.5 3/16 ± 1/2	0.2 ±0.5 3/16 ± 1/2	0.0 ±0.5 0 ± 1/2	-0.7 ±0.5 -11/16 ± 1/2	-0.7 ±0.5 -11/16 ± 1/2	L= -0.5, R= 0.5 ±0.5 L= -1/2, R= 1/2 ± 1/2
Cross	0.0 (0)	0.0 (0)	0.0 (0)	0.0 (0)	0.0 (0)	-1.0 (-1)
Tol.	0.75 (3/4)	0.75 (3/4)	0.7 (11/16)	0.75 (3/4)	0.75 (3/4)	0.75 (3/4)
Front Toe (Degrees)	0.0° ±0.2° 0° ± 3/16°	0.0° ±0.2° 0° ± 3/16°	0.0° ±0.2° 0° ± 3/16°	0.0° ±0.2° 0° ± 3/16°	0.0° ±0.2° 0° ± 3/16°	0.2° ±0.2° 3/16° ± 3/16°
Front Toe (Inches)	0.0" ±0.1" 0" ± 3/32"	0.0" ±0.1" 0" ± 3/32"	0.0" ±0.1" 0" ± 3/32"	0.0" ±0.1" 0" ± 3/32"	0.0" ±0.1" 0" ± 3/32"	0.1" ±0.1" 3/32" ± 3/32"
SAI						
Turning Angle						
Illustration #	31 102 ▲	31 102	102 ▲	25	25	31 102 ▲
REAR						
Camber (Degrees)	-0.3 ±0.5 -5/16 ± 1/2	-0.3 ±0.5 -5/16 ± 1/2	-0.1 ±0.2 -1/8 ± 3/16	-1.0 ±0.25 -1 ± 1/4	-0.25 ±0.6 -1/4 ± 5/8	-0.2 ±0.5 -3/16 ± 1/2
Rear Toe (Degrees)	0.1° ±0.2° 3/32° ± 3/16°	0.1° ±0.2° 3/32° ± 3/16°	0.0° ±0.3° 0° ± 5/16°	0.0° ±0.3° 0° ± 5/16°	0.0° ±0.2° 0° ± 3/16°	0.2° ±0.2° 3/16° ± 3/16°
Rear Toe (Inches)	0.05" ±0.1" 1/16" ± 3/32"	0.05" ±0.1" 1/16" ± 3/32"	0.0" ±0.15" 0" ± 5/32"	0.0" ±0.15" 0" ± 5/32"	0.0" ±0.1" 0" ± 3/32"	0.1" ±0.1" 3/32" ± 3/32"
Thrust Angle	0.1 (3/32)	0.1 (3/32)	0.15 (5/32)	0.15 (5/32)	0.15 (5/32)	0.1 (3/32)
Illustration #	102 32	102 32			134 135	30
Notes			[1] 1990-92 = 1.7° (1 11/16°)			
	33066	33072	33073	33074	33075	33084

Peugeot

Model Information	405 Sedan exc/Mi 16	405 Mi 16	405 DL Wagon 405 S Wagon	505 Sedan exc/V6, Turbo, ABS	505 Sedan V6	505 Sedan All w/ABS
Years	1990-91	1990-91	1990-91	1990	1990	1990
Instructions						
LugNut Torque						
FRONT						
Caster (Degrees)	2.0 ±0.5 2 ± 1/2	2.58 ±0.5 2 9/16 ± 1/2	2.0 ±0.5 2 ± 1/2	2.67 ±0.5 2 11/16 ± 1/2	3.0 ±0.5 3 ± 1/2	3.0 ±0.5 3 ± 1/2
Cross	0.0 (0)	0.0 (0)	0.0 (0)	0.0 (0)	0.0 (0)	0.0 (0)
Tol.	n/s	n/s	n/s	n/s	n/s	n/s
Camber (Degrees)	0.16 ±0.5 1/8 ± 1/2	0.08 ±0.5 1/16 ± 1/2	0.16 ±0.5 1/8 ± 1/2	-0.75 ±0.5 -3/4 ± 1/2	-1.33 +0.5 -1.0 -1 5/16 +1/2 -1	-1.33 +0.5 -1.0 -1 5/16 +1/2 -1
Cross	0.0 (0)	0.0 (0)	0.0 (0)	0.0 (0)	0.0 (0)	0.0 (0)
Tol.	n/s	n/s	n/s	n/s	n/s	n/s
Front Toe (Degrees)	0.5° ±0.16° 1/2° ± 5/32°	0.5° ±0.16° 1/2° ± 5/32°	0.5° ±0.16° 1/2° ± 5/32°	0.42° +0.17° -0.08° 13/32° +5/32° -3/32°	0.5° ±0.17° 1/2° ± 5/32°	0.42° ±0.17° 13/32° ± 5/32°
Front Toe (Inches)	0.25" ±0.08" 1/4" ± 1/16"	0.25" ±0.08" 1/4" ± 1/16"	0.25" ±0.08" 1/4" ± 1/16"	0.21" +0.08" -0.04" 7/32" +3/32" -1/32"	0.25" ±0.08" 1/4" ± 3/32"	0.21" ±0.08" 7/32" ± 3/32"
SAI						
Turning Angle						
Illustration #						
REAR						
Camber (Degrees)	-1.33 ±0.5 -1 5/16 ± 1/2	-1.33 ±0.5 -1 5/16 ± 1/2	-1.33 ±0.5 -1 5/16 ± 1/2	-1.0 ±0.5 -1 ± 1/2	-1.0 ±0.5 -1 ± 1/2	-1.0 ±0.5 -1 ± 1/2
Rear Toe (Degrees)	-0.13° ±0.22° -1/8° ± 7/32°	0.31° ±0.22° 5/16° ± 7/32°	-0.13° ±0.22° -1/8° ± 7/32°	0.42° +0.17° -0.08° 13/32° +5/32° -3/32°	0.42° ±0.17° 13/32° ± 5/32°	0.42° ±0.17° 13/32° ± 5/32°
Rear Toe (Inches)	-0.06" ±0.11" -1/16" ± 1/8"	0.16" ±0.11" 5/32" ± 1/8"	-0.06" ±0.11" -1/16" ± 1/8"	0.21" +0.08" -0.04" 7/32" +3/32" -1/32"	0.21" ±0.08" 7/32" ± 3/32"	0.21" ±0.08" 7/32" ± 3/32"
Thrust Angle						
Illustration #						
Notes						
	34003	34004	34005	34010	34011	34013

Peugeot / Plymouth

505 Wagon exc/Turbo	505 Wagon Turbo	Notes	MODEL INFORMATION	Acclaim	Breeze	Colt Colt 200 (Canada) except Wagon
1990-91	1990-91		YEARS	1990-95	1996-2000	1990-92
			INSTRUCTIONS			
			LUGNUT TORQUE	95 ft.lbs.	95-100 ft.lbs.	65-80 ft.lbs.
			FRONT			
2.67 ±0.5 2 11/16 ± 1/2	2.0 ±0.5 2 ± 1/2		CASTER (Degrees)	2.8 ±1.0 [1] 2 13/16 ±1 [1]	3.3 ±1.0 3 5/16 ±1	2.33 ±0.5 2 5/16 ± 1/2
0.0 (0) | n/s	0.0 (0) | n/s		CROSS | TOL.	0.0 (0) | 1.5 (1 1/2)	0.0 (0) | 1.0 (1)	0.0 (0) | n/s
-0.75 ±0.5 -3/4 ± 1/2	-1.0 +0.5 -1.0 -1 +1/2 -1		CAMBER (Degrees)	0.3 ±0.5 5/16 ± 1/2	0.0 ±0.6 [1] 0 ± 5/8 [1]	0.0 ±0.5 0 ± 1/2
0.0 (0) | n/s	0.0 (0) | n/s		CROSS | TOL.	0.0 (0) | n/s	0.0 (0) | 0.7 (11/16)	0.0 (0) | n/s
0.5° ±0.17° 1/2° ± 5/32°	0.58° ±0.17° 19/32° ± 5/32°		FRONT TOE (Degrees)	0.1° ±0.1° 3/32° ± 3/32°	0.1° ±0.2° 3/32° ± 3/16°	0.0° ±0.24° 0° ± 1/4°
0.25" ±0.08" 1/4" ± 3/32"	0.29" ±0.08" 9/32" ± 3/32"		FRONT TOE (Inches)	0.05" ±0.05" 1/16" ± 1/16"	0.05" ±0.1" 1/16" ± 3/32"	0.0" ±0.12" 0" ± 1/8"
			SAI		7.0 ±0.5 (7 ± 1/2)	
			TURNING ANGLE			37.0 / 32.5 ±1.5 [1]
			ILLUSTRATION #	22	113	
			REAR			
			CAMBER (Degrees)	-0.5 ±0.5 -1/2 ± 1/2	-0.2 ±0.4 -3/16 ± 3/8	-0.67 ±0.58 -11/16 ± 9/16
			REAR TOE (Degrees)	0.0° ±0.25° 0° ± 1/4°	0.1° ±0.2° 3/32° ± 3/16°	0.0° ±0.36° 0° ± 3/8°
			REAR TOE (Inches)	0.0" ±0.13" 0" ± 1/8"	0.05" ±0.1" 1/16" ± 3/32"	0.0" ±0.18" 0" ± 3/16"
			THRUST ANGLE	0.4 (13/32)	0.15 (5/32)	
			ILLUSTRATION #	76	01	
			NOTES	[1] 1990 = 1.2° (1 3/16°)	[1] 1996 = 0.3° (5/16°)	[1] w/MS 41°/35°
34018	34019	34021		35001	35007	35015

Plymouth

Colt Wagon 4X2	Colt Wagon 4X4	Colt (1995 Canada)	MODEL INFORMATION	Colt Vista 4X2	Colt Vista 4X4	Colt Vista 2WD
1990	1990	1993-95	YEARS	1990-91	1990-91	1992-94
			INSTRUCTIONS			
65-80 ft.lbs.	65-80 ft.lbs.	65-80 ft.lbs.	LUGNUT TORQUE	50-57 ft.lbs. [1]	50-57 ft.lbs. [1]	65-80 ft.lbs.
			FRONT			
1.33 ±0.5 [1] 1 5/16 ± 1/2 [1]	1.0 ±0.5 1 ± 1/2	2.25 ±0.5 2 1/4 ± 1/2	CASTER (Degrees)	0.8 ±0.5 13/16 ± 1/2	0.83 ±0.5 13/16 ± 1/2	2.17 ±0.67 2 3/16 ± 11/16
0.0 (0) | n/s	0.0 (0) | n/s	0.0 (0) | n/s	CROSS | TOL.	0.0 (0) | n/s	0.0 (0) | n/s	0.0 (0) | n/s
0.0 ±0.5 0 ± 1/2	0.0 ±0.5 0 ± 1/2	0.0 ±0.5 0 ± 1/2	CAMBER (Degrees)	0.42 ±0.5 7/16 ± 1/2	0.83 ±0.5 13/16 ± 1/2	0.33 ±0.5 5/16 ± 1/2
0.0 (0) | n/s	0.0 (0) | n/s	0.0 (0) | n/s	CROSS | TOL.	0.0 (0) | n/s	0.0 (0) | n/s	0.0 (0) | n/s
0.0° ±0.24° 0° ± 1/4°	0.0° ±0.24° 0° ± 1/4°	0.0° ±0.24° 0° ± 1/4°	FRONT TOE (Degrees)	0.0° ±0.24° 0° ± 1/4°	0.0° ±0.24° 0° ± 1/4°	0.0° ±0.24° 0° ± 1/4°
0.0" ±0.12" 0" ± 1/8"	0.0" ±0.12" 0" ± 1/8"	0.0" ±0.12" 0" ± 1/8"	FRONT TOE (Inches)	0.0" ±0.12" 0" ± 1/8"	0.0" ±0.12" 0" ± 1/8"	0.0" ±0.12" 0" ± 1/8"
			SAI			
37.0 / 32.0 ±1.0	37.0 / 32.0 ±1.0	37.3 / 31.0 ±1.0	TURNING ANGLE	37.8 / 30.7 ±1.5	37.8 / 30.7 ±1.5	36.5 / 30.5 ±2.0
			ILLUSTRATION #	03	03	
			REAR			
0.0 ±0.58 0 ± 9/16		-0.67 ±0.5 -11/16 ± 1/2	CAMBER (Degrees)	-0.58 ±0.58 -9/16 ± 9/16	0.0 ±0.5 0 ± 1/2	-0.5 ±0.5 -1/2 ± 1/2
0.0° ±0.36° 0° ± 3/8°		0.24° ±0.16° 1/4° ± 5/32°	REAR TOE (Degrees)	0.0° ±0.24° 0° ± 1/4°	0.0° ±0.24° 0° ± 1/4°	0.16° +0.24° -0.16° 5/32° +1/4° -5/32°
0.0" ±0.18" 0" ± 3/16"		0.12" ±0.08" 1/8" ± 3/32"	REAR TOE (Inches)	0.0" ±0.12" 0" ± 1/8"	0.0" ±0.12" 0" ± 1/8"	0.08" +0.12" -0.08" 3/32" +1/8" -3/32"
			THRUST ANGLE			
		40 05	ILLUSTRATION #		77	27
[1] w/MS 0.67° (11/16°)			NOTES	[1] Alum. Whls. 65-80 ft.lbs.	[1] Alum. Whls. 65-80 ft.lbs.	
35016	35017	35018		35019	35020	35021

Plymouth

Colt Vista AWD	Colt Wagon 2WD Wagon (Canada)	Colt Wagon AWD Wagon (Canada)	MODEL INFORMATION	Horizon	Laser 1.8L FWD	Laser 2.0L FWD
1992-94	1992-96	1992-96	YEARS	1990	1990-94	1990-94
			INSTRUCTIONS			
65-80 ft.lbs.	65-80 ft.lbs.	65-80 ft.lbs.	LUGNUT TORQUE	95 ft.lbs.	87-101 ft.lbs.	87-101 ft.lbs.
			FRONT			
2.08 ±0.67 2 1/16 ± 11/16	2.17 ±0.67 2 3/16 ± 11/16	2.08 ±0.67 2 1/16 ± 11/16	CASTER (Degrees)	1.4 ±0.75 1⅜ ± ¾	2.33 ±0.5 2 5/16 ± ½	2.4 ±0.5 2⅜ ± ½
0.0 (0) | n/s	0.0 (0) | n/s	0.0 (0) | n/s	CROSS | TOL.	0.0 (0) | 1.5 (1½)	0.0 (0) | n/s	0.0 (0) | n/s
0.67 ±0.5 11/16 ± ½	0.33 ±0.5 5/16 ± ½	0.67 ±0.5 11/16 ± ½	CAMBER (Degrees)	0.3 ±0.5 5/16 ± ½	0.23 ±0.5 ¼ ± ½	0.08 ±0.5 1/16 ± ½
0.0 (0) | n/s	0.0 (0) | n/s	0.0 (0) | n/s	CROSS | TOL.	0.0 (0) | n/s	0.0 (0) | n/s	0.0 (0) | n/s
0.0° ±0.24° 0° ± ¼°	0.0° ±0.24° 0° ± ¼°	0.0° ±0.24° 0° ± ¼°	FRONT TOE (Degrees)	0.1° ±0.1° 3/32° ± 3/32°	0.0° ±0.24° 0° ± ¼°	0.0° ±0.24° 0° ± ¼°
0.0" ±0.12" 0" ± ⅛"	0.0" ±0.12" 0" ± ⅛"	0.0" ±0.12" 0" ± ⅛"	FRONT TOE (Inches)	0.05" ±0.05" 1/16" ± 1/16"	0.0" ±0.12" 0" ± ⅛"	0.0" ±0.12" 0" ± ⅛"
			SAI			
36.5 / 30.5 ±2.0	36.5 / 30.5 ±2.0	36.5 / 30.5 ±2.0	TURNING ANGLE		35.0 / 29.3 ±2.0 [1]	35.0 / 29.3 ±2.0 [1 2]
			ILLUSTRATION #	22		
			REAR			
-0.5 ±0.5 -½ ± ½	-0.5 ±0.5 -½ ± ½	-0.5 ±0.5 -½ ± ½	CAMBER (Degrees)	-0.75 ±0.5 -¾ ± ½	-0.75 ±0.5 -¾ ± ½	-0.75 ±0.5 -¾ ± ½
0.16° +0.24° -0.16° 5/32° +¼° -5/32°	0.16° +0.24° -0.16° 5/32° +¼° -5/32°	0.16° +0.24° -0.16° 5/32° +¼° -5/32°	REAR TOE (Degrees)	0.2° ±0.25° 3/16° ± ¼°	0.0° ±0.24° 0° ± ¼°	0.0° ±0.24° 0° ± ¼°
0.08" +0.12" -0.08" 3/32" +⅛" -3/32"	0.08" +0.12" -0.08" 3/32" +⅛" -3/32"	0.08" +0.12" -0.08" 3/32" +⅛" -3/32"	REAR TOE (Inches)	0.1" ±0.13" 3/32" ± ⅛"	0.0" ±0.12" 0" ± ⅛"	0.0" ±0.12" 0" ± ⅛"
			THRUST ANGLE	0.4 (13/32)		
27	27	27	ILLUSTRATION #	76		
			NOTES		[1] w/MS 37º/30º	[1] w/MS 37º/30º [2] w/Turbo A/T 31.5º/27º
35022	35024	35025		35041	35042	35043

Plymouth

Laser AWD	Neon	Neon	MODEL INFORMATION	Prowler [1]	Sundance	Voyager Minivan
1992-94	1995-99	2000-2001	YEARS	1997-2002	1990-94	1990
			INSTRUCTIONS			
87-101 ft.lbs.	85-115 ft.lbs.	85-115 ft.lbs.	LUGNUT TORQUE		95 ft.lbs.	95 ft.lbs.
			FRONT			
2.3 ±0.5 2 5/16 ± ½	2.8 ±1.0 2 13/16 ±1	2.6 ±1.0 2⅝ ±1	CASTER (Degrees)	5.0 ±0.5 5 ± ½	2.8 ±0.75 [1] 2 13/16 ± ¾ [1]	0.7 ±0.5 11/16 ± ½
0.0 (0) | n/s	0.0 (0) | 1.0 (1)	0.0 (0) | 1.0 (1)	CROSS | TOL.	0.0 (0) | 0.5 (½)	0.0 (0) | 1.5 (1½)	0.0 (0) | 1.5 (1½)
0.17 ±0.5 3/16 ± ½	0.0 ±0.4 0 ± ⅜	0.0 ±0.4 0 ± ⅜	CAMBER (Degrees)	0.0 +0.05 -0.2 0 +1/16 -3/16	0.3 ±0.5 5/16 ± ½	0.3 ±0.5 5/16 ± ½
0.0 (0) | n/s	0.0 (0) | n/s	0.0 (0) | 0.5 (½)	CROSS | TOL.	0.0 (0) | 1.25 (1¼)	0.0 (0) | n/s	0.0 (0) | n/s
0.0° ±0.24° 0° ± ¼°	0.1° ±0.2° 3/32° ± 3/16°	0.2° ±0.2° [1] 3/16° ± 3/16° [1]	FRONT TOE (Degrees)	0.2° ±0.1° 3/16° ± 3/32°	0.1° ±0.1° 3/32° ± 3/32°	0.1° ±0.1° 3/32° ± 3/32°
0.0" ±0.12" 0" ± ⅛"	0.05" ±0.1" 1/16" ± 3/32"	0.1" ±0.1" [1] 3/32" ± 3/32" [1]	FRONT TOE (Inches)	0.1" ±0.05" 3/32" ± 1/16"	0.05" ±0.05" 1/16" ± 1/16"	0.05" ±0.05" 1/16" ± 1/16"
	13.2 (13 3/16)		SAI			
31.5 / 27.0 ±2.0 [1 2]			TURNING ANGLE			
	26	26	ILLUSTRATION #	24	22	22
			REAR			
-1.55 ±0.5 -1 9/16 ± ½	-0.25 ±0.5 -¼ ± ½	-0.25 ±0.4 -¼ ± ⅜	CAMBER (Degrees)	-0.23 ±0.13 -¼ ± ⅛	-0.5 ±0.5 -½ ± ½	-0.2 ±0.5 -3/16 ± ½
0.28° ±0.24° 9/32° ± ¼°	0.1° ±0.2° 3/32° ± 3/16°	0.3° ±0.2° 5/16° ± 3/16°	REAR TOE (Degrees)	0.2° ±0.1° 3/16° ± 3/32°	0.0° ±0.25° 0° ± ¼°	0.0° ±0.6° 0° ± 19/32°
0.14" ±0.12" ⅛" ± ⅛"	0.05" ±0.1" 1/16" ± 3/32"	0.15" ±0.1" 5/32" ± 3/32"	REAR TOE (Inches)	0.1" ±0.05" 3/32" ± 1/16"	0.0" ±0.13" 0" ± ⅛"	0.0" ±0.3" 0" ± 5/16"
	0.1 (3/32)	0.1 (3/32)	THRUST ANGLE	0.1 (3/32)	0.4 (13/32)	0.4 (13/32)
44	26 06	26 06	ILLUSTRATION #	110	76	76
[1] w/MS 37º/30º [2] w/Man Trans. 35º/29.3º		[1] 2000 = 0.1º (0.05")	NOTES	[1] 2002 = Chrysler	[1] 1990 = 1.2º (1 3/16º)	
35044	35045	35046		35047	35058	35069

Plymouth / Pontiac

Voyager Minivan FWD	Voyager Minivan AWD	Voyager Minivan All Models	MODEL INFORMATION	6000 exc/AWD	6000 AWD	Aztek FWD
1991-95	1991-95	1996-2000	YEARS	1990-91	1990	2001-2005
			INSTRUCTIONS			
95 ft.lbs.	95 ft.lbs.	100 ft.lbs.	LUGNUT TORQUE	100 ft.lbs.	100 ft.lbs.	100 ft.lbs.
			FRONT			
1.3 ±0.5 [1] 1 5/16 ± 1/2 [1]	1.3 ±0.5 [1] 1 5/16 ± 1/2 [1]	1.4 ±1.0 1 3/8 ±1	CASTER (Degrees)	1.7 ±1.0 1 11/16 ±1	1.7 ±1.0 1 11/16 ±1	2.4 ±0.5 2 3/8 ± 1/2
0.0 (0) | 1.5 (1 1/2)	0.0 (0) | 1.5 (1 1/2)	0.0 (0) | 1.0 (1)	CROSS | TOL.	0.0 (0) | 2.0 (2)	0.0 (0) | 2.0 (2)	0.0 (0) | 0.75 (3/4)
0.3 ±0.5 5/16 ± 1/2	0.3 ±0.5 5/16 ± 1/2	0.15 ±0.4 [1] 1/8 ± 3/8 [1]	CAMBER (Degrees)	0.0 ±0.5 0 ± 1/2	0.0 ±0.5 0 ± 1/2	-0.6 ±0.5 -5/8 ± 1/2
0.0 (0) | n/s	0.0 (0) | n/s	0.0 (0) | 0.5 (1/2)	CROSS | TOL.	0.0 (0) | 0.5 (1/2)	0.0 (0) | 0.5 (1/2)	0.0 (0) | 0.75 (3/4)
0.1° ±0.1° 3/32° ± 3/32°	0.1° ±0.1° 3/32° ± 3/32°	0.1° ±0.2° 3/32° ± 3/16°	FRONT TOE (Degrees)	0.0° ±0.2° 0° ± 3/16°	0.0° ±0.2° 0° ± 3/16°	0.0° ±0.2° 0° ± 3/16°
0.05" ±0.05" 1/16" ± 1/16"	0.05" ±0.05" 1/16" ± 1/16"	0.05" ±0.1" 1/16" ± 3/32"	FRONT TOE (Inches)	0.0" ±0.1" 0" ± 3/32"	0.0" ±0.1" 0" ± 3/32"	0.0" ±0.1" 0" ± 3/32"
			SAI			
			TURNING ANGLE			
22	22	22	ILLUSTRATION #	102 ▲	102 ▲	25
			REAR			
-0.2 ±0.6 -3/16 ± 5/8	-0.2 ±0.6 -3/16 ± 5/8	0.0 ±0.25 [2] 0 ± 1/4 [2]	CAMBER (Degrees)	0.0 ±0.3 0 ± 5/16	0.4 ±0.4 3/8 ± 3/8	-0.9 ±0.25 -7/8 ± 1/4
0.0° ±0.6° 0° ± 19/32°	0.0° ±0.6° 0° ± 19/32°	0.0° ±0.4° 0° ± 13/32°	REAR TOE (Degrees)	0.0° ±0.2° 0° ± 3/16°	0.2° ±0.2° 3/16° ± 3/16°	0.0° ±0.2° 0° ± 3/16°
0.0" ±0.3" 0" ± 5/16"	0.0" ±0.3" 0" ± 5/16"	0.0" ±0.2" 0" ± 3/16"	REAR TOE (Inches)	0.0" ±0.1" 0" ± 3/32"	0.1" ±0.1" 3/32" ± 3/32"	0.0" ±0.1" 0" ± 3/32"
0.4 (13/32)	0.4 (13/32)	0.3 (5/16)	THRUST ANGLE	0.15 (5/32)	0.15 (5/32)	0.15 (5/32)
76			ILLUSTRATION #		102 10	
[1] 1991 = 1.7° (1 11/16°)	[1] 1991 = 1.7° (1 11/16°)	[1] 205/15" or 215/16"=.05(1/16) [2] CNG or Electric = -0.1° (-1/8°)	NOTES			
35070	35071	35072		36002	36003	36005

Pontiac

Aztek AWD	Bonneville	Bonneville	MODEL INFORMATION	Firebird incl. Trans Am	Firebird incl. Trans Am	Firebird & TA with Goodyear Eagle RSA Tires
2001-2005	1990-99	2000-2005	YEARS	1990-92	1993-2002	1995-96
			INSTRUCTIONS			
100 ft.lbs.	100 ft.lbs.	100 ft.lbs.	LUGNUT TORQUE	100 ft.lbs.	100 ft.lbs.	100 ft.lbs.
			FRONT			
2.4 ±0.5 2 3/8 ± 1/2	3.0 ±0.5 3 ± 1/2	5.0 ±0.5 [1] 5 ± 1/2 [1]	CASTER (Degrees)	4.8 ±0.5 4 13/16 ± 1/2	4.8 ±0.5 4 13/16 ± 1/2	L= 4.0, R= 4.4 ± 0.5 L= 4, R= 4 3/8 ± 1/2
0.0 (0) | 0.75 (3/4)	0.0 (0) | 0.75 (3/4)	0.0 (0) | 0.5 (1/2)	CROSS | TOL.	0.0 (0) | 0.7 (11/16)	0.0 (0) | 0.7 (11/16)	-0.4 (-3/8) | 0.7 (11/16)
-0.6 ±0.5 -5/8 ± 1/2	0.2 ±0.5 3/16 ± 1/2	-0.2 ±0.5 -3/16 ± 1/2	CAMBER (Degrees)	0.3 ±0.5 5/16 ± 1/2	0.4 ±0.5 3/8 ± 1/2	L= 0.2, R= -0.2 ±0.5 L= 3/16, R= -3/16 ± 1/2
0.0 (0) | 0.75 (3/4)	0.0 (0) | 0.75 (3/4)	0.0 (0) | 0.5 (1/2)	CROSS | TOL.	0.0 (0) | 0.7 (11/16)	0.0 (0) | 0.7 (11/16)	0.4 (3/8) | 0.7 (11/16)
0.0° ±0.2° 0° ± 3/16°	0.0° ±0.2° 0° ± 3/16°	0.2° ±0.2° 3/16° ± 3/16°	FRONT TOE (Degrees)	0.0° ±0.2° 0° ± 3/16°	0.0° ±0.2° 0° ± 3/16°	0.0° ±0.2° 0° ± 3/16°
0.0" ±0.1" 0" ± 3/32"	0.0" ±0.1" 0" ± 3/32"	0.1" ±0.1" 3/32" ± 3/32"	FRONT TOE (Inches)	0.0" ±0.1" 0" ± 3/32"	0.0" ±0.1" 0" ± 3/32"	0.0" ±0.1" 0" ± 3/32"
			SAI			
			TURNING ANGLE			
25	31 102 ▲	31 102	ILLUSTRATION #	36	37	37
			REAR			
-0.25 ±0.6 -1/4 ± 5/8	-0.3 ±0.5 -5/16 ± 1/2	-0.3 ±0.5 -5/16 ± 1/2	CAMBER (Degrees)			
0.0° ±0.2° 0° ± 3/16°	0.1° ±0.2° 3/32° ± 3/16°	0.2° ±0.2° 3/16° ± 3/16°	REAR TOE (Degrees)			
0.0" ±0.1" 0" ± 3/32"	0.05" ±0.1" 1/16" ± 3/32"	0.1" ±0.1" 3/32" ± 3/32"	REAR TOE (Inches)			
0.15 (5/32)	0.1 (3/32)	0.15 (5/32)	THRUST ANGLE	0.25 (1/4)	0.25 (1/4)	0.25 (1/4)
134 135	102 32	06	ILLUSTRATION #			
		[1] 2000 = 6.0° (6°)	NOTES			
36006	36015	36016		36034	36035	36036

Pontiac

Firefly (Canada)	Firefly (Canada)	Firefly (Canada)	MODEL INFORMATION	G3 (Mexico) 2007-2009 G3 (USA)	G5 exc/GT	G5 GT
1990-94	1995-96	1997-2001	YEARS	2009	2007-2009	2007-2009
			INSTRUCTIONS			
44 ft.lbs.	44 ft.lbs.	44 ft.lbs.	LUGNUT TORQUE	88 ft.lbs.	100 ft.lbs.	100 ft.lbs.
			FRONT			
3.0 ± 2.0 3 ± 2	3.0 ± 2.0 3 ± 2	3.0 ± 1.0 3 ± 1	CASTER (Degrees)	2.5 ± 0.75 2½ ± ¾	3.0 ± 0.75 3 ± ¾	3.65 ± 0.75 3⅝ ± ¾
0.0 (0) | 1.0 (1)	0.0 (0) | 1.0 (1)	0.0 (0) | 1.0 (1)	CROSS | TOL.	0.0 (0) | 1.0 (1)	0.0 (0) | 0.75 (¾)	0.0 (0) | 0.75 (¾)
0.0 ± 1.0 0 ± 1	0.5 ± 1.0 ½ ± 1	0.0 ± 1.0 0 ± 1	CAMBER (Degrees)	-0.4 ± 0.75 -⅜ ± ¾	-1.0 ± 0.75 -1 ± ¾	-1.05 ± 0.75 -1 1/16 ± ¾
0.0 (0) | 0.5 (½)	0.0 (0) | 0.5 (½)	0.0 (0) | 0.5 (½)	CROSS | TOL.	0.0 (0) | 0.5 (½)	0.0 (0) | 0.75 (¾)	0.0 (0) | 0.75 (¾)
0.0° ± 0.2° 0° ± 3/16°	0.16° ± 0.16° 5/32° ± 5/32°	0.0° ± 0.16° 0° ± 5/32°	FRONT TOE (Degrees)	0.07° ± 0.17° 1/16° ± 5/32°	0.2° ± 0.2° 3/16° ± 3/16°	0.2° ± 0.2° 3/16° ± 3/16°
0.0" ± 0.1" 0" ± 3/32"	0.08" ± 0.08" 3/32" ± 3/32"	0.0" ± 0.08" 0" ± 3/32"	FRONT TOE (Inches)	0.04" ± 0.08" 1/32" ± 3/32"	0.1" ± 0.1" 3/32" ± 3/32"	0.1" ± 0.1" 3/32" ± 3/32"
			SAI			
38.0 / 32.0	38.0 / 32.0	38.0 / 32.0	TURNING ANGLE			
▲	▲		ILLUSTRATION #		25	25
			REAR			
0.0 ± 1.0 0 ± 1	0.0 ± 1.0 0 ± 1	0.0 ± 1.0 0 ± 1	CAMBER (Degrees)	-1.5 ± 0.5 -1½ ± ½	-0.8 ± 0.75 -13/16 ± ¾	-0.8 ± 0.75 -13/16 ± ¾
0.2° ± 0.2° 3/16° ± 3/16°	0.45° ± 0.15° 7/16° ± 5/32°	0.16° ± 0.16° 5/32° ± 5/32°	REAR TOE (Degrees)	0.25° ± 0.33° ¼° ± 11/32°	0.25° ± 0.3° ¼° ± 5/16°	0.25° ± 0.3° ¼° ± 5/16°
0.1" ± 0.1" 3/32" ± 3/32"	0.23" ± 0.08" 7/32" ± 1/16"	0.08" ± 0.08" 3/32" ± 3/32"	REAR TOE (Inches)	0.13" ± 0.17" ⅛" ± 5/32"	0.13" ± 0.15" ⅛" ± 5/32"	0.13" ± 0.15" ⅛" ± 5/32"
0.15 (5/32)	0.15 (5/32)	0.15 (5/32)	THRUST ANGLE	0.25 (¼)	0.25 (¼)	0.25 (¼)
06	06	06	ILLUSTRATION #			
			NOTES			
36037	36038	36039		36040	36041	36042

Pontiac

G6	G6	G8	MODEL INFORMATION	GTO	Grand Am	Grand Am
2005-2007	2008-2009	2008-2009	YEARS	2004-2006	1990-92	1993-95
			INSTRUCTIONS			
100 ft.lbs.	100 ft.lbs.	125 ft.lbs.	LUGNUT TORQUE	100 ft.lbs.	100 ft.lbs.	100 ft.lbs.
			FRONT			
3.1 ± 0.75 3⅛ ± ¾	2.9 ± 0.75 2⅞ ± ¾	6.0 ± 1.25 6 ± 1¼	CASTER (Degrees)	7.75 ± 1.25 7¾ ± 1¼	1.7 ± 1.0 1 11/16 ± 1	1.45 ± 1.0 1 7/16 ± 1
0.0 (0) | 0.75 (¾)	0.0 (0) | 0.75 (¾)	0.0 (0) | n/s	CROSS | TOL.	0.0 (0) | 0.6 (⅝)	0.0 (0) | 0.7 (11/16)	0.0 (0) | 0.7 (11/16)
L= -0.9, R= -0.7 ± 0.75 L= -⅞, R= -11/16 ± ¾	L= -1.0, R= -0.7 ± 0.75 L= -1, R= -11/16 ± ¾	-0.34 ± 0.33 -5/16 ± 5/16	CAMBER (Degrees)	-0.2 ± 0.5 -3/16 ± ½	0.0 ± 0.7 0 ± 11/16	0.0 ± 0.7 0 ± 11/16
-0.2 (-3/16) | 0.75 (¾)	-0.3 (-5/16) | 0.75 (¾)	0.0 (0) | 0.6 (⅝)	CROSS | TOL.	0.0 (0) | 0.8 (13/16)	0.0 (0) | 1.0 (1)	0.0 (0) | 1.0 (1)
0.2° ± 0.2° 3/16° ± 3/16°	0.2° ± 0.2° 3/16° ± 3/16°	0.17° ± 0.17° 5/32° ± 5/32°	FRONT TOE (Degrees)	0.17° ± 0.17° 5/32° ± 5/32°	0.0° ± 0.2° 0° ± 3/16°	0.0° ± 0.2° 0° ± 3/16°
0.1" ± 0.1" 3/32" ± 3/32"	0.1" ± 0.1" 3/32" ± 3/32"	0.09" ± 0.09" 3/32" ± 3/32"	FRONT TOE (Inches)	0.09" ± 0.09" 3/32" ± 3/32"	0.0" ± 0.1" 0" ± 3/32"	0.0" ± 0.1" 0" ± 3/32"
			SAI			
			TURNING ANGLE			
25	25	03 102	ILLUSTRATION #	159 102	25	25
			REAR			
-0.8 ± 0.5 -13/16 ± ½	-0.8 ± 0.6 -13/16 ± ⅝	-0.8 +0.35 -0.55 -13/16 +⅜ -9/16	CAMBER (Degrees)	-1.05 ± 0.63 -1 1/16 ± ⅝	-0.25 ± 0.5 -¼ ± ½	-0.25 ± 0.6 -¼ ± ⅝
0.2° ± 0.2° 3/16° ± 3/16°	0.2° ± 0.2° 3/16° ± 3/16°	0.33° ± 0.2° 11/32° ± 3/16°	REAR TOE (Degrees)	0.4° ± 0.34° 13/32° ± 11/32°	0.25° ± 0.2° ¼° ± 3/16°	0.0° ± 0.3° 0° ± 5/16°
0.1" ± 0.1" 3/32" ± 3/32"	0.1" ± 0.1" 3/32" ± 3/32"	0.17" ± 0.1" 5/32" ± 3/32"	REAR TOE (Inches)	0.2" ± 0.17" 3/16" ± 5/32"	0.13" ± 0.1" ⅛" ± 3/32"	0.0" ± 0.15" 0" ± 5/32"
0.3 (5/16)	0.3 (5/16)	0.2 (3/16)	THRUST ANGLE	0.17 (5/32)	0.31 (5/16)	0.25 (¼)
177	177	175	ILLUSTRATION #	10		
			NOTES			
36043	36044	36045		36048	36054	36055

Pontiac

MODEL INFORMATION	Grand Am	Grand Am	Grand Am	Grand Prix FWD	Grand Prix FWD	Grand Prix
YEARS	1996-98	1999-2003	2004-2005	1990	1991-96	1997-2001
INSTRUCTIONS						
LUGNUT TORQUE	100 ft.lbs.	100 ft.lbs.	100 ft.lbs.	100 ft.lbs.	100 ft.lbs.	100 ft.lbs.
FRONT						
CASTER (Degrees)	1.45 ± 1.0 1 7/16 ± 1	4.1 ± 1.0 4 1/8 ± 1	4.1 ± 0.75 4 1/8 ± 3/4	1.8 ± 0.5 1 13/16 ± 1/2	2.0 ± 0.5 2 ± 1/2	3.0 ± 0.5 3 ± 1/2
CROSS / TOL.	0.0 (0) / 0.7 (11/16)	0.0 (0) / 1.0 (1)	0.0 (0) / 0.5 (1/2)	0.0 (0) / 0.75 (3/4)	0.0 (0) / 0.75 (3/4)	0.0 (0) / 0.75 (3/4)
CAMBER (Degrees)	0.0 ± 1.0 0 ± 1	-0.2 ± 1.0 -3/16 ± 1	-0.2 ± 0.7 -3/16 ± 11/16	0.7 ± 0.5 11/16 ± 1/2	0.7 ± 0.5 11/16 ± 1/2	-0.9 ± 0.5 -7/8 ± 1/2
CROSS / TOL.	0.0 (0) / 1.0 (1)	0.0 (0) / 1.0 (1)	0.0 (0) / 0.7 (11/16)	0.0 (0) / 0.75 (3/4)	0.0 (0) / 0.75 (3/4)	0.0 (0) / 0.75 (3/4)
FRONT TOE (Degrees)	0.0° ± 0.2° 0° ± 3/16°	0.1° ± 0.2° 3/32° ± 3/16°	0.15° ± 0.2° [1] 5/32° ± 3/16° [1]	0.0° ± 0.2° 0° ± 3/16°	0.0° ± 0.2° 0° ± 3/16°	0.1° ± 0.2° 3/32° ± 3/16°
FRONT TOE (Inches)	0.0" ± 0.1" 0" ± 3/32"	0.05" ± 0.1" 1/16" ± 3/32"	0.08" ± 0.1" [1] 1/16" ± 3/32" [1]	0.0" ± 0.1" 0" ± 3/32"	0.0" ± 0.1" 0" ± 3/32"	0.05" ± 0.1" 1/16" ± 3/32"
SAI						
TURNING ANGLE						
ILLUSTRATION #	25	102	102	33 ▲	33 ▲	25 ▲
REAR						
CAMBER (Degrees)	-0.4 ± 0.75 -3/8 ± 3/4	-0.2 ± 0.5 -3/16 ± 1/2	-0.2 ± 0.7 -3/16 ± 11/16	0.25 ± 0.5 [1] 1/4 ± 1/2 [1]	0.1 ± 0.5 1/8 ± 1/2	-0.9 ± 0.5 -7/8 ± 1/2
REAR TOE (Degrees)	0.2° ± 0.3° 3/16° ± 5/16°	0.06° ± 0.2° 1/16° ± 3/16°	0.1° ± 0.2° 3/32° ± 3/16°	0.0° ± 0.3° 0° ± 5/16°	**-0.1°** ± 0.3° **-3/32°** ± 5/16°	0.1° ± 0.2° 3/32° ± 3/16°
REAR TOE (Inches)	0.1" ± 0.15" 3/32" ± 5/32"	0.03" ± 0.1" 1/32" ± 3/32"	0.05" ± 0.1" 1/16" ± 3/32"	0.0" ± 0.15" 0" ± 5/32"	**-0.05"** ± 0.15" **-1/16"** ± 5/32"	0.05" ± 0.1" 1/16" ± 3/32"
THRUST ANGLE	0.25 (1/4)	0.2 (3/16)	0.2 (3/16)	0.15 (5/32)	0.15 (5/32)	0.15 (5/32)
ILLUSTRATION #		06	06	25 34	25 34	108 10
NOTES			[1] 16" tires = 0.2° (0.1") (2004 only)	[1] 1990 w/16" tires 0.1° (1/8°)		
	36056	36057	36058	36069	36070	36071

Pontiac

MODEL INFORMATION	Grand Prix	Grand Prix	Grand Prix except GXP	Grand Prix GXP	LeMans	Montana/Trans Sport FWD
YEARS	2002-2003	2004	2005-2008	2005-2008	1990-93	1997-2005
INSTRUCTIONS						
LUGNUT TORQUE	100 ft.lbs.	100 ft.lbs.	100 ft.lbs.	100 ft.lbs.	66 ft.lbs.	100 ft.lbs.
FRONT						
CASTER (Degrees)	3.0 ± 0.5 3 ± 1/2	3.07 ± 0.5 3 1/16 ± 1/2	2.75 ± 0.75 2 3/4 ± 3/4	2.85 ± 0.75 2 7/8 ± 3/4	1.75 ± 1.0 1 3/4 ± 1	2.9 ± 0.5 [1] 2 7/8 ± 1/2 [1]
CROSS / TOL.	0.0 (0) / 0.75 (3/4)	0.0 (0) / 0.75 (3/4)	0.0 (0) / 0.75 (3/4)	0.0 (0) / 0.75 (3/4)	0.0 (0) / 1.0 (1)	0.0 (0) / 0.75 (3/4)
CAMBER (Degrees)	-0.9 ± 0.5 -7/8 ± 1/2	-0.94 ± 0.5 -15/16 ± 1/2	-0.8 ± 0.75 -13/16 ± 3/4	-1.0 ± 0.75 -1 ± 3/4	-0.33 ± 0.75 -5/16 ± 3/4	-0.7 ± 0.5 -11/16 ± 1/2
CROSS / TOL.	0.0 (0) / 0.75 (3/4)	0.0 (0) / 0.75 (3/4)	0.0 (0) / 0.75 (3/4)	0.0 (0) / 0.75 (3/4)	0.0 (0) / 1.5 (1 1/2)	0.0 (0) / 0.75 (3/4)
FRONT TOE (Degrees)	0.1° ± 0.2° 3/32° ± 3/16°	0.1° ± 0.2° 3/32° ± 3/16°	0.1° ± 0.2° 3/32° ± 3/16°	0.1° ± 0.2° 3/32° ± 3/16°	0.0° ± 0.33° 0° ± 11/32°	0.0° ± 0.2° 0° ± 3/16°
FRONT TOE (Inches)	0.05" ± 0.1" 1/16" ± 3/32"	0.05" ± 0.1" 1/16" ± 3/32"	0.05" ± 0.1" 1/16" ± 3/32"	0.05" ± 0.1" 1/16" ± 3/32"	0.0" ± 0.17" 0" ± 5/32"	0.0" ± 0.1" 0" ± 3/32"
SAI						
TURNING ANGLE						
ILLUSTRATION #	25 ▲	25	25	25		25 ▲
REAR						
CAMBER (Degrees)	-0.9 ± 0.5 -7/8 ± 1/2	-0.87 ± 0.5 -7/8 ± 1/2	-0.95 ± 0.5 -15/16 ± 1/2	-1.15 ± 0.7 -1 1/8 ± 11/16	-0.5 ± 0.5 -1/2 ± 1/2	-1.0 ± 0.3 -1 ± 5/16
REAR TOE (Degrees)	0.06° ± 0.2° 1/16° ± 3/16°	0.1° ± 0.2° 3/32° ± 3/16°	0.1° ± 0.2° 3/32° ± 3/16°	0.1° ± 0.2° 3/32° ± 3/16°	0.33° ± 0.33° 11/32° ± 11/32°	0.0° ± 0.3° 0° ± 5/16°
REAR TOE (Inches)	0.03" ± 0.1" 1/32" ± 3/32"	0.05" ± 0.1" 1/16" ± 3/32"	0.05" ± 0.1" 1/16" ± 3/32"	0.05" ± 0.1" 1/16" ± 3/32"	0.17" ± 0.17" 5/32" ± 5/32"	0.0" ± 0.15" 0" ± 5/32"
THRUST ANGLE	0.15 (5/32)	0.15 (5/32)	0.2 (3/16)	0.15 (5/32)	0.13 (1/8)	0.15 (5/32)
ILLUSTRATION #	108 10	108 10	108 10	108 10		
NOTES						[1] 2004-05 = 3.1° (3 1/8°)
	36072	36073	36074	36075	36087	36090

Pontiac

Montana/Trans Sport AWD (or FWD w/IRS)	Montana SV6 FWD	Montana SV6 AWD	MODEL INFORMATION	Pursuit (Canada)	Solstice	Solstice
2002-2005	2005-2009	2005-2006	YEARS	2005-2006	2006-2007	2008-2009
			INSTRUCTIONS			
100 ft.lbs.	100 ft.lbs.	100 ft.lbs.	LUGNUT TORQUE	100 ft.lbs.	100 ft.lbs.	100 ft.lbs.
			FRONT			
2.9 ±0.5 [1] 2⅞ ± ½ [1]	2.7 ±0.75 2 11/16 ± ¾	2.7 ±0.75 2 11/16 ± ¾	CASTER (Degrees)	3.0 ±0.75 [1] 3 ± ¾ [1]	8.0 ±0.6 8 ± ⅝	7.5 ±0.6 7½ ± ⅝
0.0 (0) / 0.75 (¾)	0.0 (0) / 0.75 (¾)	0.0 (0) / 0.75 (¾)	CROSS / TOL.	0.0 (0) / 0.75 (¾)	0.0 (0) / 0.6 (⅝)	0.0 (0) / 0.6 (⅝)
-0.7 ±0.5 -1 1/16 ± ½	-0.65 ±0.75 -⅝ ± ¾	-0.65 ±0.75 -⅝ ± ¾	CAMBER (Degrees)	-1.0 ±0.75 [2] -1 ± ¾ [2]	-0.5 ±0.6 -½ ± ⅝	-0.5 ±0.6 -½ ± ⅝
0.0 (0) / 0.75 (¾)	0.0 (0) / 0.75 (¾)	0.0 (0) / 0.75 (¾)	CROSS / TOL.	0.0 (0) / 0.75 (¾)	0.0 (0) / 0.6 (⅝)	0.0 (0) / 0.6 (⅝)
0.0° ±0.2° 0° ± 3/16°	0.0° ±0.2° 0° ± 3/16°	0.0° ±0.2° 0° ± 3/16°	FRONT TOE (Degrees)	0.2° ±0.2° 3/16° ± 3/16°	0.1° ±0.2° 3/32° ± 3/16°	0.1° ±0.2° 3/32° ± 3/16°
0.0" ±0.1" 0" ± 3/32"	0.0" ±0.1" 0" ± 3/32"	0.0" ±0.1" 0" ± 3/32"	FRONT TOE (Inches)	0.1" ±0.1" 3/32" ± 3/32"	0.05" ±0.1" 1/16" ± 3/32"	0.05" ±0.1" 1/16" ± 3/32"
			SAI			
			TURNING ANGLE			
25 △	25 △	25 △	ILLUSTRATION #	25	12	12
			REAR			
-0.25 ±0.6 -¼ ± ⅝	-1.0 ±0.5 -1 ± ½	-0.4 ±0.6 -⅜ ± ⅝	CAMBER (Degrees)	-0.8 ±0.75 -13/16 ± ¾	-0.5 ±0.5 -½ ± ½	-0.5 ±0.6 -½ ± ⅝
0.0° ±0.2° 0° ± 3/16°	0.0° ±0.3° 0° ± 5/16°	0.0° ±0.2° 0° ± 3/16°	REAR TOE (Degrees)	0.25° ±0.3° ¼° ± 5/16°	0.1° ±0.2° 3/32° ± 3/16°	0.1° ±0.2° 3/32° ± 3/16°
0.0" ±0.1" 0" ± 3/32"	0.0" ±0.15" 0" ± 5/32"	0.0" ±0.1" 0" ± 3/32"	REAR TOE (Inches)	0.13" ±0.15" ⅛" ± 5/32"	0.05" ±0.1" 1/16" ± 3/32"	0.05" ±0.1" 1/16" ± 3/32"
0.15 (5/32)	0.3 (5/16)	0.3 (5/16)	THRUST ANGLE	0.25 (¼)	0.2 (3/16)	0.2 (3/16)
134 135		134 135	ILLUSTRATION #		180 135	180 135
[1] 2004-05 = 3.1° (3⅛°)			NOTES	[1] GT = 3.65° (3⅝°) [2] GT = -1.05° (-1 1/16°)		
36091	36092	36093		36098	36100	36101

Pontiac

Sunbird	Sunbird	Sunfire	MODEL INFORMATION	Sunfire USA Assembly	Sunfire Mexico Assy. VIN ..S.. (11th char)	Torrent FWD & AWD
1990-92	1993-94	1995-99	YEARS	2000-2005	2000-2005	2006-2007
			INSTRUCTIONS			
100 ft.lbs.	100 ft.lbs.	100 ft.lbs.	LUGNUT TORQUE	100 ft.lbs.	100 ft.lbs.	100 ft.lbs.
			FRONT			
1.7 ±1.0 1 11/16 ± 1	1.3 ±1.0 1 5/16 ± 1	4.3 ±1.0 4 5/16 ± 1	CASTER (Degrees)	4.1 ±0.75 4⅛ ± ¾	4.1 ±0.75 4⅛ ± ¾	3.0 ±0.75 3 ± ¾
0.0 (0) / 0.7 (11/16)	0.0 (0) / 0.7 (11/16)	0.0 (0) / 0.7 (11/16)	CROSS / TOL.	0.0 (0) / 0.75 (¾)	0.0 (0) / 0.75 (¾)	0.0 (0) / 0.5 (½)
0.0 ±0.7 0 ± 11/16	-0.15 ±0.7 -⅛ ± 11/16	-0.2 ±1.0 -3/16 ± 1	CAMBER (Degrees)	-0.2 ±0.75 -3/16 ± ¾	-0.35 ±0.75 -⅜ ± ¾	-0.6 ±0.75 -⅝ ± ¾
0.0 (0) / 1.0 (1)	0.0 (0) / 0.7 (11/16)	0.0 (0) / 0.7 (11/16)	CROSS / TOL.	0.0 (0) / 0.75 (¾)	0.0 (0) / 0.75 (¾)	0.0 (0) / 0.5 (½)
0.0° ±0.2° 0° ± 3/16°	0.0° ±0.2° 0° ± 3/16°	0.1° ±0.2° [1] 3/32° ± 3/16° [1]	FRONT TOE (Degrees)	0.1° ±0.2° 3/32° ± 3/16°	0.0° ±0.2° 0° ± 3/16°	0.15° ±0.2° 5/32° ± 3/16°
0.0" ±0.1" 0" ± 3/32"	0.0" ±0.1" 0" ± 3/32"	0.05" ±0.1" [1] 1/16" ± 3/32" [1]	FRONT TOE (Inches)	0.05" ±0.1" 1/16" ± 3/32"	0.0" ±0.1" 0" ± 3/32"	0.08" ±0.1" 1/16" ± 3/32"
			SAI			
			TURNING ANGLE			
25	25	25	ILLUSTRATION #	25	25	25
			REAR			
-0.25 ±0.5 -¼ ± ½	-0.25 ±0.5 -¼ ± ½	-0.25 ±0.75 -¼ ± ¾	CAMBER (Degrees)	-0.25 ±0.5 -¼ ± ½	-0.25 ±0.5 -¼ ± ½	-0.5 ±0.75 -½ ± ¾
0.25° ±0.16° ¼° ± 5/32°	0.25° ±0.31° ¼° ± 5/16°	0.2° ±0.3° 3/16° ± 5/16°	REAR TOE (Degrees)	0.2° ±0.2° 3/16° ± 3/16°	0.2° ±0.2° 3/16° ± 3/16°	0.2° ±0.2° 3/16° ± 3/16°
0.13" ±0.08" ⅛" ± 3/32"	0.13" ±0.16" ⅛" ± 5/32"	0.1" ±0.15" 3/32" ± 5/32"	REAR TOE (Inches)	0.1" ±0.1" 3/32" ± 3/32"	0.1" ±0.1" 3/32" ± 3/32"	0.1" ±0.1" 3/32" ± 3/32"
0.31 (5/16)	0.35 (11/32)	0.25 (¼)	THRUST ANGLE	0.25 (¼)	0.25 (¼)	0.2 (3/16)
			ILLUSTRATION #			107 143
		[1] 1995 = 0.0° (0")	NOTES			
36107	36108	36109		36110	36111	36121

Pontiac

MODEL INFORMATION	Torrent exc/GXP FWD & AWD	Torrent GXP FWD & AWD	Trans Sport	Vibe FWD	Vibe AWD	Vibe FWD (non-IRS)
YEARS	2008-2009	2008-2009	1990-96	2003-2008	2003-2006	2009-2010
INSTRUCTIONS						
LUGNUT TORQUE	100 ft.lbs.	100 ft.lbs.	100 ft.lbs.	76 ft.lbs.	76 ft.lbs.	76 ft.lbs.
FRONT						
CASTER (Degrees)	3.0 ±0.75 3 ± ¾	3.2 ±0.75 3 3/16 ± ¾	3.0 ±0.7 [1] 3 ± 11/16 [1]	2.75 ±0.75 2¾ ± ¾	2.75 ±0.75 2¾ ± ¾	2.9 ±0.75 2 7/8 ± ¾
CROSS | TOL.	0.0 (0) | 0.5 (½)	0.0 (0) | 0.5 (½)	0.0 (0) | 0.7 (11/16)	0.0 (0) | 0.75 (¾)	0.0 (0) | 0.75 (¾)	0.0 (0) | 0.75 (¾)
CAMBER (Degrees)	L= -0.4, R= -0.6 ±0.75 L= -⅜, R= -⅝ ± ¾	L= -0.6, R= -0.8 ±0.75 L= -⅝, R= -13/16 ± ¾	0.0 ±0.5 0 ± ½	-0.57 ±0.75 -9/16 ± ¾	-0.48 ±0.75 -½ ± ¾	-0.55 ±0.75 -9/16 ± ¾
CROSS | TOL.	0.2 (3/16) | 0.75 (¾)	0.2 (3/16) | 0.75 (¾)	0.0 (0) | 0.7 (11/16)	0.0 (0) | 0.75 (¾)	0.0 (0) | 0.75 (¾)	0.0 (0) | 0.75 (¾)
FRONT TOE (Degrees)	0.15° ±0.2° 5/32° ± 3/16°	0.15° ±0.2° 5/32° ± 3/16°	0.0° ±0.2° 0° ± 3/16°	0.0° ±0.2° 0° ± 3/16°	0.0° ±0.2° 0° ± 3/16°	0.0° ±0.2° 0° ± 3/16°
FRONT TOE (Inches)	0.08" ±0.1" 1/16" ± 3/32"	0.08" ±0.1" 1/16" ± 3/32"	0.0" ±0.1" 0" ± 3/32"	0.0" ±0.1" 0" ± 3/32"	0.0" ±0.1" 0" ± 3/32"	0.0" ±0.1" 0" ± 3/32"
SAI						
TURNING ANGLE						
ILLUSTRATION #	25	25	102 △	26	26	26
REAR						
CAMBER (Degrees)	-0.5 ±0.75 -½ ± ¾	-0.5 ±0.75 -½ ± ¾	-0.1 ±0.3 -⅛ ± 5/16	-1.45 ±0.5 -1 7/16 ± ½	-0.73 ±0.5 -¾ ± ½	-1.45 ±0.75 -1 7/16 ± ¾
REAR TOE (Degrees)	0.2° ±0.2° 3/16° ± 3/16°	0.2° ±0.2° 3/16° ± 3/16°	0.0° ±0.3° 0° ± 5/16°	0.27° ±0.27° 9/32° ± 9/32°	0.2° ±0.2° 3/16° ± 3/16°	0.3° ±0.25° 5/16° ± ¼°
REAR TOE (Inches)	0.1" ±0.1" 3/32" ± 3/32"	0.1" ±0.1" 3/32" ± 3/32"	0.0" ±0.15" 0" ± 5/32"	0.13" ±0.13" ⅛" ± ⅛"	0.1" ±0.1" 3/32" ± 3/32"	0.15" ±0.13" 5/32" ± ⅛"
THRUST ANGLE	0.2 (3/16)	0.2 (3/16)	0.15 (5/32)			0.3 (5/16)
ILLUSTRATION #	107 143	107 143			120	
NOTES			[1] 1990-92 = 1.7° (1 11/16°)			
	36122	36123	36124	36128	36129	36130

Pontiac / Porsche

MODEL INFORMATION	Vibe AWD & GT (IRS)	Wave (Canada)	Wave (Canada)	911 Carrera 2/4	911 Carrera Speedster	911 Carrera RS America
YEARS	2009-2010	2005-2007	2008-2009	1990-94	1994	1993-94
INSTRUCTIONS						
LUGNUT TORQUE	76 ft.lbs.	88 ft.lbs.	88 ft.lbs.			
FRONT						
CASTER (Degrees)	3.0 ±0.75 3 ± ¾	2.5 ±0.75 2½ ± ¾	2.5 ±0.75 2½ ± ¾	4.42 +0.25 -0.5 4 7/16 +¼ -½	4.42 +0.25 -0.5 4 7/16 +¼ -½	4.42 +0.25 -0.5 4 7/16 +¼ -½
CROSS | TOL.	0.0 (0) | 0.75 (¾)	0.0 (0) | 1.0 (1)	0.0 (0) | 1.0 (1)	0.0 (0) | 0.25 (¼)	0.0 (0) | 0.25 (¼)	0.0 (0) | 0.25 (¼)
CAMBER (Degrees)	-0.6 ±0.75 -⅝ ± ¾	-0.4 ±0.75 -⅜ ± ¾	-0.4 ±0.75 -⅜ ± ¾	0.0 ±0.17 [1] 0 ± 3/16 [1]	-0.25 ±0.17 [1] -¼ ± 3/16 [1]	-0.25 ±0.17 [1] -¼ ± 3/16 [1]
CROSS | TOL.	0.0 (0) | 0.75 (¾)	0.0 (0) | 0.5 (½)	0.0 (0) | 0.5 (½)	0.0 (0) | 0.17 (3/16)	0.0 (0) | 0.17 (3/16)	0.0 (0) | 0.17 (3/16)
FRONT TOE (Degrees)	0.0° ±0.2° 0° ± 3/16°	0.07° ±0.17° 1/16° ± 5/32°	0.07° ±0.17° 1/16° ± 5/32°	0.42° ±0.08° 13/32° ± 3/32°	0.25° ±0.08° ¼° ± 3/32°	0.25° ±0.08° ¼° ± 3/32°
FRONT TOE (Inches)	0.0" ±0.1" 0" ± 3/32"	0.04" ±0.08" 1/32" ± 3/32"	0.04" ±0.08" 1/32" ± 3/32"	0.21" ±0.04" 7/32" ± 1/32"	0.13" ±0.04" ⅛" ± 1/32"	0.13" ±0.04" ⅛" ± 1/32"
SAI						
TURNING ANGLE				20.0 / 19.3 ±0.5	20.0 / 19.3 ±0.5	20.0 / 19.3 ±0.5
ILLUSTRATION #	26			97	97	97
REAR						
CAMBER (Degrees)	-1.05 ±0.75 -1 1/16 ± ¾	-1.5 ±0.5 -1½ ± ½	-1.5 ±0.5 -1½ ± ½	-0.67 ±0.17 -11/16 ± 3/16	-0.75 ±0.17 -¾ ± 3/16	-0.75 ±0.17 -¾ ± 3/16
REAR TOE (Degrees)	0.25° ±0.25° ¼° ± ¼°	0.25° ±0.33° ¼° ± 11/32°	0.5° ±0.66° ½° ± 21/32°	0.33° ±0.17° 11/32° ± 5/32°	0.5° ±0.17° ½° ± 5/32°	0.5° ±0.17° ½° ± 5/32°
REAR TOE (Inches)	0.13" ±0.13" ⅛" ± ⅛"	0.13" ±0.17" ⅛" ± 5/32"	0.25" ±0.33" ¼" ± 11/32"	0.17" ±0.08" 5/32" ± 3/32"	0.25" ±0.08" ¼" ± 3/32"	0.25" ±0.08" ¼" ± 3/32"
THRUST ANGLE	0.2 (3/16)	0.25 (¼)	0.25 (¼)	0.17 (5/32)	0.17 (5/32)	0.17 (5/32)
ILLUSTRATION #	120			98	98	98
NOTES				[1] Adjusting tool reqd.	[1] Adjusting tool reqd.	[1] Adjusting tool reqd.
	36131	36132	36133	37007	37008	37009

Porsche

911 America Roadster	911 Carrera Turbo (incl. Turbo-Look)	(993) 911 Carrera 911 Carrera Sport	MODEL INFORMATION	(993) 911 Carrera RS	(996) 911 Carrera	(996) 911 Carrera 4S
1992-93	1991-94	1994-98	YEARS	1995-98	1999-2005	2002-2005
		p. 310	INSTRUCTIONS			
			LUGNUT TORQUE		96 ft.lbs.	96 ft.lbs.
			FRONT			
4.42 ±0.25 4 7/16 ± 1/4	4.42 ±0.25 4 7/16 ± 1/4	5.33 +0.25 -0.5 5 5/16 +1/4 -1/2	CASTER (Degrees)	5.33 +0.25 -0.5 5 5/16 +1/4 -1/2	8.0 ±0.5 8 ± 1/2	8.0 ±0.5 8 ± 1/2
0.0 (0) | 0.25 (1/4)	0.0 (0) | 0.25 (1/4)	0.0 (0) | 0.25 (1/4)	CROSS | TOL.	0.0 (0) | 0.25 (1/4)	0.0 (0) | 0.67 (11/16)	0.0 (0) | 0.67 (11/16)
0.0 ±0.17 [1] 0 ± 3/16 [1]	0.0 ±0.17 [1] 0 ± 3/16 [1]	-0.33 ±0.17 - 5/16 ± 3/16	CAMBER (Degrees)	-1.0 ±0.17 -1 ± 3/16	-0.25 ±0.25 - 1/4 ± 1/4	-0.5 ±0.25 - 1/2 ± 1/4
0.0 (0) | 0.17 (3/16)	0.0 (0) | 0.17 (3/16)	0.0 (0) | 0.17 (3/16)	CROSS | TOL.	0.0 (0) | 0.17 (3/16)	0.0 (0) | 0.33 (5/16)	0.0 (0) | 0.33 (5/16)
0.42° ±0.08° 13/32° ± 3/32°	0.42° ±0.08° 13/32° ± 3/32°	0.08° ±0.08° 3/32° ± 3/32°	FRONT TOE (Degrees)	0.08° ±0.08° 3/32° ± 3/32°	0.08° ±0.08° 3/32° ± 3/32°	0.08° ±0.08° 3/32° ± 3/32°
0.21" ±0.04" 7/32" ± 1/32"	0.21" ±0.04" 7/32" ± 1/32"	0.04" ±0.04" 1/32" ± 1/32"	FRONT TOE (Inches)	0.04" ±0.04" 1/32" ± 1/32"	0.04" ±0.04" 1/32" ± 1/32"	0.04" ±0.04" 1/32" ± 1/32"
			SAI			
20.0 / 18.7 ±0.5	20.0 / 18.7 ±0.5	20.0 / 19.3 ±0.5	TURNING ANGLE	20.0 / 18.6 ±0.5	20.0 / 18.2 ±0.5	20.0 / 18.2 ±0.5
97	97	97	ILLUSTRATION #	97	103	103
			REAR			
-0.75 +0.5 -0.17 -3/4 +1/2 -3/16	-0.75 +0.5 -0.17 -3/4 +1/2 -3/16	-1.17 ±0.25 -1 3/16 ± 1/4	CAMBER (Degrees)	-1.33 ±0.25 -1 5/16 ± 1/4	-1.17 ±0.25 -1 3/16 ± 1/4	-1.42 ±0.25 -1 7/16 ± 1/4
0.5° ±0.17° 1/2° ± 5/32°	0.5° ±0.17° 1/2° ± 5/32°	0.33° ±0.17° 11/32° ± 5/32°	REAR TOE (Degrees)	0.5° ±0.17° 1/2° ± 5/32°	0.33° ±0.17° 11/32° ± 5/32°	0.33° ±0.17° 11/32° ± 5/32°
0.25" ±0.08" 1/4" ± 3/32"	0.25" ±0.08" 1/4" ± 3/32"	0.17" ±0.08" 5/32" ± 3/32"	REAR TOE (Inches)	0.25" ±0.08" 1/4" ± 3/32"	0.17" ±0.08" 5/32" ± 3/32"	0.17" ±0.08" 5/32" ± 3/32"
0.17 (5/32)	0.17 (5/32)	0.17 (5/32)	THRUST ANGLE	0.17 (5/32)	0.17 (5/32)	0.17 (5/32)
98	98	99	ILLUSTRATION #	99	104	104
[1] Adjusting tool reqd.	[1] Adjusting tool reqd.		NOTES			
37010	37011	37012		37013	37014	37015

Porsche

(996) 911 GT2	(996) 911 GT3	(996) 911 GT3 RS	MODEL INFORMATION	(996) 911 Turbo	(997) 911 Carrera 2 911 Carrera 2S exc/PASM	(997) 911 Carrera 2 911 Carrera 2S w/PASM
2002-2005	2004-2005	2004-2005	YEARS	2001-2005	2005-2012	2005-2012
			INSTRUCTIONS			
96 ft.lbs.	96 ft.lbs.	96 ft.lbs.	LUGNUT TORQUE	96 ft.lbs.	96 ft.lbs.	96 ft.lbs.
			FRONT			
8.0 ±0.5 8 ± 1/2	8.0 ±0.5 8 ± 1/2	8.0 ±0.5 8 ± 1/2	CASTER (Degrees)	8.0 ±0.5 8 ± 1/2	8.0 ±0.5 8 ± 1/2	8.17 ±0.5 8 3/16 ± 1/2
0.0 (0) | 0.67 (11/16)	0.0 (0) | 0.67 (11/16)	0.0 (0) | 0.67 (11/16)	CROSS | TOL.	0.0 (0) | 0.67 (11/16)	0.0 (0) | 0.67 (11/16)	0.0 (0) | 0.67 (11/16)
-1.0 ±0.08 -1 ± 1/16	-1.0 ±0.08 -1 ± 1/16	-0.83 ±0.08 - 13/16 ± 1/16	CAMBER (Degrees)	0.0 ±0.25 0 ± 1/4	-0.25 ±0.25 - 1/4 ± 1/4	-0.25 ±0.25 - 1/4 ± 1/4
0.0 (0) | 0.17 (3/16)	0.0 (0) | 0.17 (3/16)	0.0 (0) | 0.17 (3/16)	CROSS | TOL.	0.0 (0) | 0.33 (5/16)	0.0 (0) | 0.33 (5/16)	0.0 (0) | 0.33 (5/16)
0.13° ±0.03° 1/8° ± 1/32°	0.13° ±0.03° 1/8° ± 1/32°	0.03° ±0.03° 1/32° ± 1/32°	FRONT TOE (Degrees)	0.08° ±0.08° 3/32° ± 3/32°	0.08° ±0.08° 3/32° ± 3/32°	0.08° ±0.08° 3/32° ± 3/32°
0.07" ±0.02" 1/16" ± 1/32"	0.07" ±0.02" 1/16" ± 1/32"	0.02" ±0.02" 1/64" ± 1/64"	FRONT TOE (Inches)	0.04" ±0.04" 1/32" ± 1/32"	0.04" ±0.04" 1/32" ± 1/32"	0.04" ±0.04" 1/32" ± 1/32"
			SAI			
20.0 / 18.2 ±0.5	20.0 / 18.5 ±0.5	20.0 / 18.5 ±0.5	TURNING ANGLE	20.0 / 18.7 ±0.5	20.0 / 18.3 ±0.5	20.0 / 18.2 ±0.5
103	103	103	ILLUSTRATION #	103	103	103
			REAR			
-1.83 ±0.08 -1 13/16 ± 1/16	-1.83 ±0.08 -1 13/16 ± 1/16	-1.83 ±0.08 -1 13/16 ± 1/16	CAMBER (Degrees)	-1.42 ±0.25 -1 7/16 ± 1/4	-1.5 ±0.25 -1 1/2 ± 1/4	-1.5 ±0.25 -1 1/2 ± 1/4
0.43° ±0.07° 7/16° ± 1/16°	0.43° ±0.07° 7/16° ± 1/16°	0.43° ±0.07° 7/16° ± 1/16°	REAR TOE (Degrees)	0.33° ±0.17° 11/32° ± 5/32°	0.33° ±0.17° 11/32° ± 5/32°	0.33° ±0.17° 11/32° ± 5/32°
0.22" ±0.03" 7/32" ± 1/32"	0.22" ±0.03" 7/32" ± 1/32"	0.22" ±0.03" 7/32" ± 1/32"	REAR TOE (Inches)	0.17" ±0.08" 5/32" ± 3/32"	0.17" ±0.08" 5/32" ± 3/32"	0.17" ±0.08" 5/32" ± 3/32"
0.17 (5/32)	0.17 (5/32)	0.17 (5/32)	THRUST ANGLE	0.17 (5/32)	0.17 (5/32)	0.17 (5/32)
104	104	104	ILLUSTRATION #	104	104	104
			NOTES			
37016	37017	37018		37019	37020	37021

Porsche

MODEL INFORMATION	(997) 911 All C2 Performance Models	(997) 911 Carrera 4 911 Carrera 4S exc/PASM	(997) 911 Carrera 4 911 Carrera 4S w/PASM	(997) 911 All C4 Performance Models	(997) 911 GT3	(997) 911 GT3 RS
YEARS	2005-2013	2006-2012	2006-2012	2006-2013	2007-2009	2008-2009
INSTRUCTIONS						
LUGNUT TORQUE	96 ft.lbs.	96 ft.lbs.		96 ft.lbs.		
FRONT						
CASTER (Degrees)	8.33 +0.5 -0.75 8 5/16 +½ -¾	7.78 ±0.5 7¾ ± ½	7.92 ±0.5 7 15/16 ± ½	8.05 +0.5 -0.75 8 1/16 +½ -¾	8.0 ±0.5 8 ± ½	8.0 ±0.5 8 ± ½
CROSS | TOL.	0.0 (0) | 0.67 (11/16)	0.0 (0) | 0.67 (11/16)	0.0 (0) | 0.67 (11/16)	0.0 (0) | 0.67 (11/16)	0.0 (0) | 0.67 (11/16)	0.0 (0) | 0.67 (11/16)
CAMBER (Degrees)	-0.5 ±0.25 -½ ± ¼	-0.5 ±0.25 -½ ± ¼	-0.67 ±0.25 -11/16 ± ¼	-0.75 ±0.25 -¾ ± ¼	-1.5 ±0.08 -1½ ± 1/16	-1.5 ±0.08 -1½ ± 1/16
CROSS | TOL.	0.0 (0) | 0.33 (5/16)	0.0 (0) | 0.33 (5/16)	0.0 (0) | 0.33 (5/16)	0.0 (0) | 0.33 (5/16)	0.0 (0) | 0.17 (3/16)	0.0 (0) | 0.17 (3/16)
FRONT TOE (Degrees)	0.08° ±0.08° 3/32° ± 3/32°	0.08° ±0.08° 3/32° ± 3/32°	0.08° ±0.08° 3/32° ± 3/32°	0.08° ±0.08° 3/32° ± 3/32°	0.1° ±0.03° 3/32° ± 1/32°	0.1° ±0.03° 3/32° ± 1/32°
FRONT TOE (Inches)	0.04" ±0.04" 1/32" ± 1/32"	0.04" ±0.04" 1/32" ± 1/32"	0.04" ±0.04" 1/32" ± 1/32"	0.04" ±0.04" 1/32" ± 1/32"	0.05" ±0.02" 1/16" ± 1/64"	0.05" ±0.02" 1/16" ± 1/64"
SAI						
TURNING ANGLE	20.0 / 17.8 ±0.5	20.0 / 18.8 ±0.5	20.0 / 18.4 ±0.5	20.0 / 18.0 ±0.5	20.0 / 18.2 ±0.5	20.0 / 18.2 ±0.5
ILLUSTRATION #	103	103	103	103	103	103
REAR						
CAMBER (Degrees)	-1.67 ±0.25 -1 11/16 ± ¼	-1.5 ±0.25 -1½ ± ¼	-1.5 ±0.25 -1½ ± ¼	-1.67 ±0.25 -1 11/16 ± ¼	-1.5 ±0.08 -1½ ± 1/16	-1.5 ±0.08 -1½ ± 1/16
REAR TOE (Degrees)	0.33° ±0.17° 11/32° ± 5/32°	0.33° ±0.17° 11/32° ± 5/32°	0.33° ±0.17° 11/32° ± 5/32°	0.33° ±0.17° 11/32° ± 5/32°	0.43° ±0.07° 7/16° ± 1/16°	0.43° ±0.07° 7/16° ± 1/16°
REAR TOE (Inches)	0.17" ±0.08" 5/32" ± 3/32"	0.17" ±0.08" 5/32" ± 3/32"	0.17" ±0.08" 5/32" ± 3/32"	0.17" ±0.08" 5/32" ± 3/32"	0.22" ±0.04" 7/32" ± 1/32"	0.22" ±0.04" 7/32" ± 1/32"
THRUST ANGLE	0.17 (5/32)	0.17 (5/32)	0.17 (5/32)	0.17 (5/32)	0.08 (3/32)	0.08 (3/32)
ILLUSTRATION #	104	104	104	104	104	104
NOTES						
	37022	37023	37024	37025	37026	37027

Porsche

MODEL INFORMATION	(997) 911 GT2	(997) 911 Turbo	(991) 911 Carrera 2, 2S Base Models	(991) 911 Carrera 2, 2S Base Models w/Active Susp.	(991) 911 Carrera 2, 2S Sport Models w/Active Susp.	(991) 911 Carrera 4, 4S Base Models
YEARS	2008-2009	2007-2013	2012-2017	2012-2017	2012-2017	2013-2017
INSTRUCTIONS						
LUGNUT TORQUE						
FRONT						
CASTER (Degrees)	8.0 ±0.5 8 ± ½	7.92 ±0.5 7 15/16 ± ½	8.0 +0.5 -0.75 8 +½ -¾	8.17 +0.5 -0.75 8 3/16 +½ -¾	8.33 +0.5 -0.75 8 5/16 +½ -¾	8.0 +0.5 -0.75 8 +½ -¾
CROSS | TOL.	0.0 (0) | 0.67 (11/16)	0.0 (0) | 0.67 (11/16)	0.0 (0) | 0.67 (11/16)	0.0 (0) | 0.67 (11/16)	0.0 (0) | 0.67 (11/16)	0.0 (0) | 0.67 (11/16)
CAMBER (Degrees)	-1.5 ±0.08 -1½ ± 1/16	-0.67 ±0.25 -11/16 ± ¼	-0.17 ±0.25 [1] -3/16 ± ¼ [1]	-0.58 ±0.25 [1] -9/16 ± ¼ [1]	-1.0 ±0.25 [1] -1 ± ¼ [1]	-0.17 ±0.25 [1] -3/16 ± ¼ [1]
CROSS | TOL.	0.0 (0) | 0.17 (3/16)	0.0 (0) | 0.33 (5/16)	0.0 (0) | 0.33 (5/16)	0.0 (0) | 0.33 (5/16)	0.0 (0) | 0.33 (5/16)	0.0 (0) | 0.33 (5/16)
FRONT TOE (Degrees)	0.1° ±0.03° 3/32° ± 1/32°	0.08° ±0.08° 3/32° ± 3/32°	0.03° ±0.08° 1/32° ± 3/32°	0.03° ±0.08° 1/32° ± 3/32°	0.03° ±0.08° 1/32° ± 3/32°	0.03° ±0.08° 1/32° ± 3/32°
FRONT TOE (Inches)	0.05" ±0.02" 1/16" ± 1/64"	0.04" ±0.04" 1/32" ± 1/32"	0.02" ±0.04" 1/64" ± 1/32"	0.02" ±0.04" 1/64" ± 1/32"	0.02" ±0.04" 1/64" ± 1/32"	0.02" ±0.04" 1/64" ± 1/32"
SAI						
TURNING ANGLE	20.0 / 18.2 ±0.5	20.0 / 18.4 ±0.5	20.0 / 18.8 ±0.5 [2]	20.0 / 18.8 ±0.5 [2]	20.0 / 18.8 ±0.5	20.0 / 18.8 ±0.5 [2]
ILLUSTRATION #	103	103	103	103	103	103
REAR						
CAMBER (Degrees)	-1.5 ±0.08 -1½ ± 1/16	-1.67 ±0.25 -1 11/16 ± ¼	-1.5 ±0.25 -1½ ± ¼	-1.5 ±0.25 -1½ ± ¼	-1.67 ±0.25 -1 11/16 ± ¼	-1.5 ±0.25 -1½ ± ¼
REAR TOE (Degrees)	0.43° ±0.07° 7/16° ± 1/16°	0.33° ±0.17° 11/32° ± 5/32°	0.33° ±0.17° 11/32° ± 5/32°	0.33° ±0.17° 11/32° ± 5/32°	0.33° ±0.17° 11/32° ± 5/32°	0.33° ±0.17° 11/32° ± 5/32°
REAR TOE (Inches)	0.22" ±0.04" 7/32" ± 1/32"	0.17" ±0.08" 5/32" ± 3/32"	0.17" ±0.08" 5/32" ± 3/32"	0.17" ±0.08" 5/32" ± 3/32"	0.17" ±0.08" 5/32" ± 3/32"	0.17" ±0.08" 5/32" ± 3/32"
THRUST ANGLE	0.08 (3/32)	0.17 (5/32)	0.17 (5/32)	0.17 (5/32)	0.17 (5/32)	0.17 (5/32)
ILLUSTRATION #	104	104	104	104	104	104
NOTES			[1] 2017 = -0.58° (-9/16°) [2] 2017 = 20°/18.77°	[1] 2017 = -0.75° (-¾°) [2] 2017 = 20°/18.77°	[1] 2017 = 20°/18.77°	[1] 2017 = -0.58° (-9/16°) [2] 2017 = 20°/18.77°
	37028	37029	37030	37031	37032	37033

Porsche

(991) 911 Carrera 4, 4S Base Models w/Active Susp.	(991) 911 Carrera 4, 4S Sport Models w/Active Susp.	(991) 911 GT3	MODEL INFORMATION	(991) 911 Turbo/S exc/PDCC	(991) 911 Turbo/S w/PDCC	928
2013-2017	2013-2017	2014-2015	YEARS	2014-2015	2014-2015	1990
			INSTRUCTIONS			
			LUGNUT TORQUE			
			FRONT			
8.17 +0.5 -0.75 8 3/16 +1/2 -3/4	8.33 +0.5 -0.75 8 5/16 +1/2 -3/4	8.0 ±0.5 8 ± 1/2	CASTER (Degrees)	8.17 +0.5 -0.75 8 3/16 +1/2 -3/4	8.17 +0.5 -0.75 8 3/16 +1/2 -3/4	3.5 +0.5 -0.1 3 1/2 +1/2 -1/8
0.0 (0) | 0.67 (11/16)	0.0 (0) | 0.67 (11/16)	0.0 (0) | 0.67 (11/16)	CROSS | TOL.	0.0 (0) | 0.67 (11/16)	0.0 (0) | 0.67 (11/16)	0.0 (0) | 0.33 (5/16)
-0.58 ±0.25 [1] -9/16 ± 1/4 [1]	-1.0 ±0.25 [1] -1 ± 1/4 [1]	-1.5 ±0.08 -1 1/2 ± 1/16	CAMBER (Degrees)	-0.83 ±0.25 -13/16 ± 1/4	-0.58 ±0.25 -9/16 ± 1/4	-0.5 ±0.17 -1/2 ± 3/16
0.0 (0) | 0.33 (5/16)	0.0 (0) | 0.33 (5/16)	0.0 (0) | 0.17 (3/16)	CROSS | TOL.	0.0 (0) | 0.33 (5/16)	0.0 (0) | 0.33 (5/16)	0.0 (0) | 0.17 (3/16)
0.03° ±0.08° 1/32° ± 3/32°	0.03° ±0.08° 1/32° ± 3/32°	0.03° ±0.03° 1/32° ± 1/32°	FRONT TOE (Degrees)	0.08° ±0.08° 3/32° ± 3/32°	0.08° ±0.08° 3/32° ± 3/32°	0.25° ±0.08° 1/4° ± 3/32°
0.02" ±0.04" 1/64" ± 1/32"	0.02" ±0.04" 1/64" ± 1/32"	0.02" ±0.02" 1/64" ± 1/32"	FRONT TOE (Inches)	0.04" ±0.04" 1/32" ± 1/32"	0.04" ±0.04" 1/32" ± 1/32"	0.13" ±0.04" 1/8" ± 1/32"
			SAI			
20.0 / 18.8 ±0.5 [2]	20.0 / 18.8 ±0.5	20.0 / 18.2 ±0.5	TURNING ANGLE	20.0 / 18.4 ±0.5	20.0 / 18.4 ±0.5	20.0 / 19.0 ±0.33
103	103	103	ILLUSTRATION #	103	103	93
			REAR			
-1.5 ±0.25 -1 1/2 ± 1/4	-1.67 ±0.25 -1 11/16 ± 1/4	-1.5 ±0.5 -1 1/2 ± 1/2	CAMBER (Degrees)	-1.75 ±0.25 -1 3/4 ± 1/4	-1.75 ±0.25 -1 3/4 ± 1/4	-0.67 ±0.17 -11/16 ± 3/16
0.33° ±0.17° 11/32° ± 5/32°	0.33° ±0.17° 11/32° ± 5/32°	0.33° ±0.07° 11/32° ± 1/16°	REAR TOE (Degrees)	0.33° ±0.17° 11/32° ± 5/32°	0.33° ±0.17° 11/32° ± 5/32°	0.33° ±0.17° 11/32° ± 5/32°
0.17" ±0.08" 5/32" ± 3/32"	0.17" ±0.08" 5/32" ± 3/32"	0.17" ±0.04" 5/32" ± 1/32"	REAR TOE (Inches)	0.17" ±0.08" 5/32" ± 3/32"	0.17" ±0.08" 5/32" ± 3/32"	0.17" ±0.08" 5/32" ± 3/32"
0.17 (5/32)	0.17 (5/32)	0.17 (5/32)	THRUST ANGLE	0.17 (5/32)	0.17 (5/32)	0.17 (5/32)
104	104	104	ILLUSTRATION #	104	104	94
[1] 2017 = -0.75° (-3/4°) [2] 2017 = 20°/18.77°	[1] 2017 = 20°/18.77°		NOTES			
37034	37035	37036		37037	37038	37043

Porsche

928	944 S2, Turbo	968	MODEL INFORMATION	Boxster Boxster S	(987) Boxster (987) Boxster S exc/PASM	(987) Boxster (987) Boxster S w/PASM
1991-95	1990-91	1992-95	YEARS	1997-2004	2005-2012	2005-2012
			INSTRUCTIONS			
			LUGNUT TORQUE	96 ft.lbs.		
			FRONT			
4.0 +1.0 -0.1 4 +1 -1/8	2.5 +0.5 -0.25 2 1/2 +1/2 -1/4	3.25 +0.1 -0.75 3 1/4 +1/8 -3/4	CASTER (Degrees)	8.0 ±0.5 [1] 8 ± 1/2 [1]	8.0 ±0.5 8 ± 1/2	8.0 ±0.5 8 ± 1/2
0.0 (0) | 0.33 (5/16)	0.0 (0) | 0.5 (1/2)	0.0 (0) | 0.5 (1/2)	CROSS | TOL.	0.0 (0) | 0.67 (11/16)	0.0 (0) | 0.67 (11/16)	0.0 (0) | 0.67 (11/16)
-0.5 ±0.17 -1/2 ± 3/16	0.0 ±0.17 0 ± 3/16	0.0 ±0.17 0 ± 3/16	CAMBER (Degrees)	0.08 ±0.5 1/16 ± 1/2	-0.17 ±0.5 -3/16 ± 1/2	-0.25 ±0.5 -1/4 ± 1/2
0.0 (0) | 0.17 (3/16)	0.0 (0) | 0.33 (5/16)	0.0 (0) | 0.33 (5/16)	CROSS | TOL.	0.0 (0) | 0.33 (5/16)	0.0 (0) | 0.33 (5/16)	0.0 (0) | 0.33 (5/16)
0.25° ±0.08° 1/4° ± 3/32°	0.17° ±0.08° 5/32° ± 3/32°	0.17° ±0.08° 5/32° ± 3/32°	FRONT TOE (Degrees)	0.08° ±0.08° 3/32° ± 3/32°	0.08° ±0.08° 3/32° ± 3/32°	0.08° ±0.08° 3/32° ± 3/32°
0.13" ±0.04" 1/8" ± 1/32"	0.08" ±0.04" 3/32" ± 1/32"	0.08" ±0.04" 3/32" ± 1/32"	FRONT TOE (Inches)	0.04" ±0.04" 1/32" ± 1/32"	0.04" ±0.04" 1/32" ± 1/32"	0.04" ±0.04" 1/32" ± 1/32"
			SAI			
20.0 / 19.0 ±0.33	20.0 / 19.0 ±0.33	20.0 / 18.8 ±0.58	TURNING ANGLE	20.0 / 18.7 ±0.5	20.0 / 18.3 ±0.5	20.0 / 18.2 ±0.5
93	90 22	90 22	ILLUSTRATION #	103	103	103
			REAR			
-0.67 ±0.17 -11/16 ± 3/16	-1.0 ±0.33 -1 ± 5/16	-0.75 ±0.33 -3/4 ± 5/16	CAMBER (Degrees)	-1.33 ±0.5 -1 5/16 ± 1/2	-1.33 ±0.5 -1 5/16 ± 1/2	-1.5 ±0.5 -1 1/2 ± 1/2
0.33° ±0.17° 11/32° ± 5/32°	0.33° ±0.33° 11/32° ± 11/32°	0.33° ±0.33° 11/32° ± 11/32°	REAR TOE (Degrees)	0.17° ±0.17° 5/32° ± 5/32°	0.17° ±0.17° 5/32° ± 5/32°	0.17° ±0.17° 5/32° ± 5/32°
0.17" ±0.08" 5/32" ± 3/32"	0.17" ±0.17" 5/32" ± 5/32"	0.17" ±0.17" 5/32" ± 5/32"	REAR TOE (Inches)	0.08" ±0.08" 3/32" ± 3/32"	0.08" ±0.08" 3/32" ± 3/32"	0.08" ±0.08" 3/32" ± 3/32"
0.17 (5/32)	0.17 (5/32)	0.17 (5/32)	THRUST ANGLE	0.17 (5/32)	0.17 (5/32)	0.17 (5/32)
94	91 92	91 92	ILLUSTRATION #	104	104	104
			NOTES	[1] Adjusting kit avail. from Porsche		
37044	37046	37047		37048	37049	37050

Porsche

(987) Boxster Spyder	(981) Boxster	(981) Boxster	MODEL	Carrera GT (980)	Cayenne	Cayenne
	(981) Boxster S	(981) Boxster S	INFORMATION		with Steel Springs	with Air Springs [1]
	exc/PASM	w/PASM (@10mm)				
2011-2012	2013-2016	2013-2016	YEARS	2004-2006	2003-2010	2003-2010
			INSTRUCTIONS			p. 310
			LUGNUT TORQUE		118 ft.lbs.	118 ft.lbs.
			FRONT			
8.0 +0.5 -0.75 8 +½ -¾	8.0 +0.5 -0.75 8 +½ -¾	8.17 +0.5 -0.75 8 3/16 +½ -¾	CASTER (Degrees)	11.5 ±0.5 11½ ± ½	8.58 ±0.5 8 9/16 ± ½	8.58 ±0.5 8 9/16 ± ½
0.0 (0) | 0.67 (11/16)	0.0 (0) | 0.67 (11/16)	0.0 (0) | 0.67 (11/16)	CROSS | TOL.	0.0 (0) | 0.67 (11/16)	0.0 (0) | 0.5 (½)	0.0 (0) | 0.5 (½)
-0.75 ±0.25 -¾ ± ¼	-0.33 ±0.25 - 5/16 ± ¼	-0.5 ±0.25 -½ ± ¼	CAMBER (Degrees)	-1.67 ±0.17 -1 11/16 ± 3/16	-0.25 ±0.33 -¼ ± 5/16	-0.25 ±0.33 -¼ ± 5/16
0.0 (0) | 0.33 (5/16)	0.0 (0) | 0.33 (5/16)	0.0 (0) | 0.33 (5/16)	CROSS | TOL.	0.0 (0) | 0.17 (3/16)	0.0 (0) | 0.33 (5/16)	0.0 (0) | 0.33 (5/16)
0.08° ±0.08° 3/32° ± 3/32°	0.03° ±0.08° 1/32° ± 3/32°	0.03° ±0.08° 1/32° ± 3/32°	FRONT TOE (Degrees)	0.05° +0.03° -0.05° 1/16° +1/32° -1/16°	0.08° ±0.08° 3/32° ± 3/32°	0.17° ±0.08° 5/32° ± 3/32°
0.04" ±0.04" 1/32" ± 1/32"	0.02" ±0.04" 1/64" ± 1/32"	0.02" ±0.04" 1/64" ± 1/32"	FRONT TOE (Inches)	0.03" +0.02" -0.03" 1/32" ± 1/32"	0.04" ±0.04" 1/32" ± 1/32"	0.08" ±0.04" 3/32" ± 1/32"
			SAI		10.8 (10¾)	10.8 (10¾)
20.0 / 18.2 ±0.5	20.0 / 18.8 ±0.5	20.0 / 18.8 ±0.5	TURNING ANGLE	20.0 / 18.3 ±0.5	20.0 / 18.5 ±0.5	20.0 / 18.5 ±0.5
103			ILLUSTRATION #		52	52
			REAR			
-1.83 ±0.25 -1 13/16 ± ¼	-1.33 ±0.25 -1 5/16 ± ¼	-1.5 ±0.25 -1½ ± ¼	CAMBER (Degrees)	-1.5 ±0.17 -1½ ± 3/16	-1.33 ±0.33 -1 5/16 ± 5/16	-1.33 ±0.33 -1 5/16 ± 5/16
0.17° ±0.17° 5/32° ± 5/32°	0.27° ±0.17° 9/32° ± 5/32°	0.27° ±0.17° 9/32° ± 5/32°	REAR TOE (Degrees)	0.33° +0.1° -0.07° 11/32° +3/32° -1/16°	0.33° ±0.17° 11/32° ± 5/32°	0.33° ±0.17° 11/32° ± 5/32°
0.08" ±0.08" 3/32" ± 3/32"	0.13" ±0.08" 1/8" ± 3/32"	0.13" ±0.08" 1/8" ± 3/32"	REAR TOE (Inches)	0.17" +0.05" -0.03" 5/32" +1/16" -1/32"	0.17" ±0.08" 5/32" ± 3/32"	0.17" ±0.08" 5/32" ± 3/32"
0.17 (5/32)	0.17 (5/32)	0.17 (5/32)	THRUST ANGLE	0.17 (5/32)	0.17 (5/32)	0.17 (5/32)
104			ILLUSTRATION #		162	162
			NOTES			[1] Check & adjust with air susp. in normal (on) position.
37051	37052	37053		37054	37055	37056

Porsche

Cayenne	Cayenne GTS	(987) Cayman	MODEL	(987) Cayman	(987) Cayman R	(981) Cayman
		(987) Cayman S	INFORMATION	(987) Cayman S		(981) Cayman S
		exc/PASM		w/PASM		exc/PASM
2011-2016	2013	2006-2012	YEARS	2006-2012	2012	2014-2016
			INSTRUCTIONS			
			LUGNUT TORQUE			
			FRONT			
8.58 ±0.5 8 9/16 ± ½	9.08 ±0.5 9 1/16 ± ½	8.0 ±0.5 8 ± ½	CASTER (Degrees)	8.0 ±0.5 8 ± ½	8.0 +0.5 -0.75 8 +½ -¾	8.0 +0.5 -0.75 8 +½ -¾
0.0 (0) | 0.33 (5/16)	0.0 (0) | 0.33 (5/16)	0.0 (0) | 0.67 (11/16)	CROSS | TOL.	0.0 (0) | 0.67 (11/16)	0.0 (0) | 0.67 (11/16)	0.0 (0) | 0.67 (11/16)
-0.25 ±0.33 -¼ ± 5/16	-0.42 ±0.33 - 7/16 ± 5/16	-0.17 ±0.5 - 3/16 ± ½	CAMBER (Degrees)	-0.25 ±0.5 -¼ ± ½	-0.75 ±0.25 -¾ ± ¼	-0.33 ±0.25 - 5/16 ± ¼
0.0 (0) | 0.33 (5/16)	0.0 (0) | 0.33 (5/16)	0.0 (0) | 0.33 (5/16)	CROSS | TOL.	0.0 (0) | 0.33 (5/16)	0.0 (0) | 0.33 (5/16)	0.0 (0) | 0.33 (5/16)
0.17° ±0.08° 5/32° ± 3/32°	0.17° ±0.08° 5/32° ± 3/32°	0.08° ±0.08° 3/32° ± 3/32°	FRONT TOE (Degrees)	0.08° ±0.08° 3/32° ± 3/32°	0.08° ±0.08° 3/32° ± 3/32°	0.03° ±0.08° 1/32° ± 3/32°
0.08" ±0.04" 3/32" ± 1/32"	0.08" ±0.04" 3/32" ± 1/32"	0.04" ±0.04" 1/32" ± 1/32"	FRONT TOE (Inches)	0.04" ±0.04" 1/32" ± 1/32"	0.04" ±0.04" 1/32" ± 1/32"	0.02" ±0.04" 1/64" ± 1/32"
11.0 (11)	11.0 (11)		SAI			
20.0 / 18.5 ±0.5	20.0 / 18.5 ±0.5	20.0 / 18.3 ±0.5	TURNING ANGLE	20.0 / 18.2 ±0.5	20.0 / 18.2 ±0.5	20.0 / 18.8 ±0.5
52	52	103	ILLUSTRATION #	103	103	
			REAR			
-1.33 ±0.33 -1 5/16 ± 5/16	-1.33 ±0.33 -1 5/16 ± 5/16	-1.33 ±0.5 -1 5/16 ± ½	CAMBER (Degrees)	-1.5 ±0.5 -1½ ± ½	-1.83 ±0.25 -1 13/16 ± ¼	-1.33 ±0.25 -1 5/16 ± ¼
0.33° ±0.17° 11/32° ± 5/32°	0.33° ±0.17° 11/32° ± 5/32°	0.17° ±0.17° 5/32° ± 5/32°	REAR TOE (Degrees)	0.17° ±0.17° 5/32° ± 5/32°	0.17° ±0.17° 5/32° ± 5/32°	0.27° ±0.17° 9/32° ± 5/32°
0.17" ±0.08" 5/32" ± 3/32"	0.17" ±0.08" 5/32" ± 3/32"	0.08" ±0.08" 3/32" ± 3/32"	REAR TOE (Inches)	0.08" ±0.08" 3/32" ± 3/32"	0.08" ±0.08" 3/32" ± 3/32"	0.13" ±0.08" 1/8" ± 3/32"
0.12 (1/8)	0.12 (1/8)	0.17 (5/32)	THRUST ANGLE	0.17 (5/32)	0.17 (5/32)	0.17 (5/32)
162	162	104	ILLUSTRATION #	104	104	
			NOTES			
37057	37058	37059		37060	37061	37062

Porsche

(981) Cayman (981) Cayman S w/PASM (@10mm)	(95B) Macan with Steel Springs	(95B) Macan with Air Springs	MODEL INFORMATION	(970) Panamera All Models (incl. Hybrid)	(971) Panamera with Steel Springs	(971) Panamera with Air Springs
2014-2016	2014-2016	2014-2016	YEARS	2010-2016	2017	2017
			INSTRUCTIONS			
			LUGNUT TORQUE			
			FRONT			
8.17 +0.5 -0.75 8 3/16 +½ -¾	3.8 ± 0.5 3 13/16 ± ½	4.2 ± 0.5 4 3/16 ± ½	CASTER (Degrees)	7.08 ± 0.5 7 1/16 ± ½	7.5 ± 0.5 7 ½ ± ½	7.5 ± 0.5 7 ½ ± ½
0.0 (0) | 0.67 (11/16)	0.0 (0) | 0.33 (5/16)	0.0 (0) | 0.33 (5/16)	CROSS | TOL.	0.0 (0) | 0.25 (¼)	0.0 (0) | 0.25 (¼)	0.0 (0) | 0.25 (¼)
-0.5 ± 0.25 -½ ± ¼	-0.5 ± 0.33 -½ ± 5/16	-0.73 ± 0.33 -¾ ± 5/16	CAMBER (Degrees)	-0.33 ± 0.25 -5/16 ± ¼	-0.75 ± 0.25 -¾ ± ¼	-0.75 ± 0.25 -¾ ± ¼
0.0 (0) | 0.33 (5/16)	0.0 (0) | 0.33 (5/16)	0.0 (0) | 0.33 (5/16)	CROSS | TOL.	0.0 (0) | 0.33 (5/16)	0.0 (0) | 0.33 (5/16)	0.0 (0) | 0.33 (5/16)
0.03° ± 0.08° 1/32° ± 3/32°	0.17° ± 0.17° 5/32° ± 5/32°	0.17° ± 0.17° 5/32° ± 5/32°	FRONT TOE (Degrees)	0.25° ± 0.08° ¼° ± 3/32°	0.17° ± 0.08° 5/32° ± 3/32°	0.17° ± 0.08° 5/32° ± 3/32°
0.02" ± 0.04" 1/64" ± 1/32"	0.08" ± 0.08" 3/32" ± 3/32"	0.08" ± 0.08" 3/32" ± 3/32"	FRONT TOE (Inches)	0.13" ± 0.04" ⅛" ± 1/32"	0.08" ± 0.04" 3/32" ± 1/32"	0.08" ± 0.04" 3/32" ± 1/32"
			SAI	11.4 (11 7/16)		
20.0 / 18.8 ±0.5	20.0 / 18.2 ±0.5	20.0 / 18.2 ±0.5	TURNING ANGLE	20.0 / 19.1 ±0.5	20.0 / 18.9 ±0.5	20.0 / 18.9 ±0.5
			ILLUSTRATION #			
			REAR			
-1.5 ± 0.25 -1 ½ ± ¼	-1.25 ± 0.33 -1 ¼ ± 5/16	-1.58 ± 0.33 -1 9/16 ± 5/16	CAMBER (Degrees)	-1.5 ± 0.25 -1 ½ ± ¼	-1.17 ± 0.25 -1 3/16 ± ¼	-1.5 ± 0.25 -1 ½ ± ¼
0.27° ± 0.17° 9/32° ± 5/32°	0.33° ± 0.17° 11/32° ± 5/32°	0.33° ± 0.17° 11/32° ± 5/32°	REAR TOE (Degrees)	0.17° ± 0.17° 5/32° ± 5/32°	0.17° ± 0.17° 5/32° ± 5/32°	0.17° ± 0.17° 5/32° ± 5/32°
0.13" ± 0.08" ⅛" ± 3/32"	0.17" ± 0.08" 5/32" ± 3/32"	0.17" ± 0.08" 5/32" ± 3/32"	REAR TOE (Inches)	0.08" ± 0.08" 3/32" ± 3/32"	0.08" ± 0.08" 3/32" ± 3/32"	0.08" ± 0.08" 3/32" ± 3/32"
0.17 (5/32)	0.12 (⅛)	0.12 (⅛)	THRUST ANGLE	0.08 (3/32)	0.08 (3/32)	0.08 (3/32)
			ILLUSTRATION #			
			NOTES			
37063	37064	37065		37066	37067	37068

Saab

900	900	900 w/SPG Sport Chassis	MODEL INFORMATION	900 Conv.	900 exc/1994 Conv.	9000
1990-91	1992-93	1990-91	YEARS	1994	1994-98	1990-98
			INSTRUCTIONS			
77-92 ft.lbs.	77-92 ft.lbs.	77-92 ft.lbs.	LUGNUT TORQUE	77-92 ft.lbs.	78-96 ft.lbs.	76-90 ft.lbs.
			FRONT			
2.0 ± 0.5 [1] 2 ± ½ [1]	2.0 ± 0.5 [1] 2 ± ½ [1]	2.0 ± 0.25 2 ± ¼	CASTER (Degrees)	2.0 ± 0.5 [1] 2 ± ½ [1]	1.95 ± 0.7 [1] 1 15/16 ± 11/16 [1]	1.65 ± 0.5 1 ⅝ ± ½
0.0 (0) | n/s	0.0 (0) | n/s	0.0 (0) | n/s	CROSS | TOL.	0.0 (0) | n/s	0.0 (0) | 0.5 (½)	0.0 (0) | n/s
0.5 ± 0.5 ½ ± ½	-0.25 ± 0.5 -¼ ± ½	0.25 ± 0.25 ¼ ± ¼	CAMBER (Degrees)	-0.25 ± 0.5 -¼ ± ½	-0.8 ± 0.8 -13/16 ± 13/16	-0.65 ± 0.5 -⅝ ± ½
0.0 (0) | n/s	0.0 (0) | n/s	0.0 (0) | n/s	CROSS | TOL.	0.0 (0) | n/s	0.0 (0) | n/s	0.0 (0) | n/s
0.29° ± 0.14° 9/32° ± ⅛°	0.29° ± 0.14° 9/32° ± ⅛°	0.29° ± 0.14° 9/32° ± ⅛°	FRONT TOE (Degrees)	0.29° ± 0.14° 9/32° ± ⅛°	0.21° ± 0.07° 7/32° ± 1/16°	0.22° ± 0.07° 7/32° ± 1/16°
0.14" ± 0.07" 5/32" ± 1/16"	0.14" ± 0.07" 5/32" ± 1/16"	0.14" ± 0.07" 5/32" ± 1/16"	FRONT TOE (Inches)	0.14" ± 0.07" 5/32" ± 1/16"	0.11" ± 0.04" 3/32" ± 1/32"	0.11" ± 0.04" 3/32" ± 1/32"
11.5 ± 1.0 (11 ½ ± 1)	11.5 ± 1.0 (11 ½ ± 1)	11.5 ± 1.0 (11 ½ ± 1)	SAI	11.5 ± 1.0 (11 ½ ± 1)	13.3 (13 5/16)	11.3 ± 0.5 (11 5/16 ± ½)
20.8 / 20.0 ±0.5	20.8 / 20.0 ±0.5	20.8 / 20.0 ±0.5	TURNING ANGLE	20.8 / 20.0 ±0.5	20.9 / 20.0 ±0.5	21.0 / 20.0 ±0.5
60	60	60	ILLUSTRATION #	60		
			REAR			
-0.5 ± 0.25 -½ ± ¼	-0.5 ± 0.25 -½ ± ¼	-0.5 ± 0.25 -½ ± ¼	CAMBER (Degrees)	-0.5 ± 0.25 -½ ± ¼	-1.7 ± 0.3 -1 11/16 ± 5/16	-0.25 ± 0.25 -¼ ± ¼
0.57° ± 0.29° 9/16° ± 9/32°	0.57° ± 0.29° 9/16° ± 9/32°	0.57° ± 0.29° 9/16° ± 9/32°	REAR TOE (Degrees)	0.57° ± 0.29° 9/16° ± 9/32°	0.14° ± 0.28° ⅛° ± 9/32°	0.36° ± 0.22° 11/32° ± 7/32°
0.29" ± 0.14" 9/32" ± ⅛"	0.29" ± 0.14" 9/32" ± ⅛"	0.29" ± 0.14" 9/32" ± ⅛"	REAR TOE (Inches)	0.29" ± 0.14" 9/32" ± ⅛"	0.07" ± 0.14" 1/16" ± ⅛"	0.18" ± 0.11" 3/16" ± 3/32"
			THRUST ANGLE			
			ILLUSTRATION #			
[1] w/MS 1.0° (1°)	[1] w/MS 1.0° (1°)		NOTES	[1] w/MS 1.0° (1°)	[1] 1996 Between #VIN...T2000001 to #VIN...T2006524 = 1.6° (1 ⅝°)	
30004	30005	30006		30007	30008	30009

Saab

9-2X	9-2X	9-3	MODEL	9-3	9-3 FWD	9-3 FWD
Linear	Aero	Sedan & Conv.	INFORMATION	Convertible	(Conv. 2004-08 only) exc/Sport Chassis	(Conv. 2004-08 only) Sport Chassis
2005-2006	2005-2006	1999-2002	YEARS	2003	2003-2009	2003-2009
			INSTRUCTIONS			
66 ft.lbs.	66 ft.lbs.	78-96 ft.lbs.	LUGNUT TORQUE	78-96 ft.lbs.		
			FRONT			
3.42 ±0.75 3 7/16 ± 3/4	3.5 ±0.75 3 1/2 ± 3/4	1.95 ±0.7 1 15/16 ± 11/16	CASTER (Degrees)	1.95 ±0.7 1 15/16 ± 11/16	2.9 ±0.5 2 7/8 ± 1/2	2.9 ±0.5 2 7/8 ± 1/2
0.0 (0) | 0.75 (3/4)	0.0 (0) | 0.75 (3/4)	0.0 (0) | 0.5 (1/2)	CROSS | TOL.	0.0 (0) | 0.5 (1/2)	0.0 (0) | n/s	0.0 (0) | n/s
-0.17 ±0.5 -3/16 ± 1/2	-0.33 ±0.5 -5/16 ± 1/2	-0.8 ±0.8 -13/16 ± 13/16	CAMBER (Degrees)	-0.8 ±0.8 -13/16 ± 13/16	-0.8 ±0.5 -13/16 ± 1/2	-0.9 ±0.5 -7/8 ± 1/2
0.0 (0) | n/s	0.0 (0) | n/s	0.0 (0) | n/s	CROSS | TOL.	0.0 (0) | n/s	0.0 (0) | 0.5 (1/2)	0.0 (0) | 0.5 (1/2)
0.17° ±0.17° 5/32° ± 5/32°	0.17° ±0.17° 5/32° ± 5/32°	0.21° ±0.07° [1] 7/32° ± 1/16° [1]	FRONT TOE (Degrees)	0.21° ±0.07° 7/32° ± 1/16°	0.28° ±0.08° 9/32° ± 3/32°	0.28° ±0.08° 9/32° ± 3/32°
0.08" ±0.08" 3/32" ± 3/32"	0.08" ±0.08" 3/32" ± 3/32"	0.11" ±0.04" [1] 3/32" ± 1/32" [1]	FRONT TOE (Inches)	0.11" ±0.04" 3/32" ± 1/32"	0.14" ±0.04" 1/8" ± 1/32"	0.14" ±0.04" 1/8" ± 1/32"
13.5 (13 1/2)	13.8 (13 3/4)	13.3 (13 5/16)	SAI	13.3 (13 5/16)		
34.5 / 30.3 ±1.5	34.5 / 30.3 ±1.5	20.9 / 20.0 ±0.5	TURNING ANGLE	20.9 / 20.0 ±0.5	21.3 / 20.0 ±0.5	21.3 / 20.0 ±0.5
22	22		ILLUSTRATION #			
			REAR			
-1.25 ±0.5 -1 1/4 ± 1/2	-1.33 ±0.5 -1 5/16 ± 1/2	-1.7 ±0.3 -1 11/16 ± 5/16	CAMBER (Degrees)	-1.7 ±0.3 -1 11/16 ± 5/16	-0.7 ±0.25 -11/16 ± 1/4	-1.0 ±0.25 -1 ± 1/4
0.0° ±0.17° 0° ± 5/32°	0.0° ±0.17° 0° ± 5/32°	0.14° ±0.28° 1/8° ± 9/32°	REAR TOE (Degrees)	0.14° ±0.28° 1/8° ± 9/32°	0.24° ±0.08° 1/4° ± 3/32°	0.24° ±0.08° 1/4° ± 3/32°
0.0" ±0.08" 0" ± 3/32"	0.0" ±0.08" 0" ± 3/32"	0.07" ±0.14" 1/16" ± 1/8"	REAR TOE (Inches)	0.07" ±0.14" 1/16" ± 1/8"	0.12" ±0.04" 1/8" ± 1/32"	0.12" ±0.04" 1/8" ± 1/32"
0.33 (11/32)	0.33 (11/32)		THRUST ANGLE			
06	06		ILLUSTRATION #		177	177
		[1] Viggen = 0.14° (0.07")	NOTES			
30010	30011	30012		30013	30014	30015

Saab

9-3 AWD	9-3 FWD	9-3 FWD	MODEL	9-3 AWD	9-3X	9-5
	exc/Aero	Aero	INFORMATION			exc/Aero
2008-2009	2010-2011	2010-2011	YEARS	2010-2011	2010-2011	1999-2009
			INSTRUCTIONS			
81 ft.lbs.	81 ft.lbs.	81 ft.lbs.	LUGNUT TORQUE	81 ft.lbs.	81 ft.lbs.	90 ft.lbs.
			FRONT			
2.9 ±0.5 2 7/8 ± 1/2	3.2 ±0.5 3 3/16 ± 1/2	3.3 ±0.5 3 5/16 ± 1/2	CASTER (Degrees)	3.2 ±0.5 3 3/16 ± 1/2	3.1 ±0.5 3 1/8 ± 1/2	2.9 ±0.5 2 7/8 ± 1/2
0.0 (0) | 0.5 (1/2)	0.0 (0) | 0.5 (1/2)	0.0 (0) | 0.5 (1/2)	CROSS | TOL.	0.0 (0) | 0.5 (1/2)	0.0 (0) | 0.5 (1/2)	0.0 (0) | n/s
-1.0 ±0.5 -1 ± 1/2	-0.8 ±0.5 -13/16 ± 1/2	-0.9 ±0.5 -7/8 ± 1/2	CAMBER (Degrees)	-1.1 ±0.5 -1 1/8 ± 1/2	-1.05 ±0.5 -1 1/16 ± 1/2	-0.8 ±0.5 -13/16 ± 1/2
0.0 (0) | 0.5 (1/2)	0.0 (0) | 0.5 (1/2)	0.0 (0) | 0.5 (1/2)	CROSS | TOL.	0.0 (0) | 0.5 (1/2)	0.0 (0) | 0.5 (1/2)	0.0 (0) | n/s
0.24° ±0.08° 1/4° ± 3/32°	0.24° ±0.08° 1/4° ± 3/32°	0.24° ±0.08° 1/4° ± 3/32°	FRONT TOE (Degrees)	0.24° ±0.08° 1/4° ± 3/32°	0.24° ±0.08° 1/4° ± 3/32°	0.28° ±0.08° 9/32° ± 3/32°
0.12" ±0.04" 1/8" ± 1/32"	0.12" ±0.04" 1/8" ± 1/32"	0.12" ±0.04" 1/8" ± 1/32"	FRONT TOE (Inches)	0.12" ±0.04" 1/8" ± 1/32"	0.12" ±0.04" 1/8" ± 1/32"	0.14" ±0.04" 1/8" ± 1/32"
			SAI			12.6 ±0.5 (12 5/8 ± 1/2)
21.3 / 20.0 ±0.5	21.3 / 20.0 ±0.5	21.3 / 20.0 ±0.5	TURNING ANGLE	21.3 / 20.0 ±0.5	21.3 / 20.0 ±0.5	21.3 / 20.0 ±0.5
			ILLUSTRATION #			
			REAR			
-1.0 ±0.3 -1 ± 5/16	-0.7 ±0.3 -11/16 ± 5/16	-1.0 ±0.3 -1 ± 5/16	CAMBER (Degrees)	-0.7 ±0.3 -11/16 ± 5/16	-0.7 ±0.3 -11/16 ± 5/16	-0.8 ±0.25 -13/16 ± 1/4
0.2° ±0.08° 3/16° ± 3/32°	0.24° ±0.1° [1] 1/4° ± 3/32° [1]	0.24° ±0.1° [1] 1/4° ± 3/32° [1]	REAR TOE (Degrees)	0.2° ±0.1° 3/16° ± 3/32°	0.24° ±0.1° 1/4° ± 3/32°	0.42° ±0.2° 13/32° ± 3/16°
0.1" ±0.04" 3/32" ± 1/32"	0.12" ±0.05" [1] 1/8" ± 1/16" [1]	0.12" ±0.05" [1] 1/8" ± 1/16" [1]	REAR TOE (Inches)	0.1" ±0.05" 3/32" ± 1/16"	0.12" ±0.05" 1/8" ± 1/16"	0.21" ±0.1" 7/32" ± 3/32"
			THRUST ANGLE			
177	177	177	ILLUSTRATION #	177	177	182
	[1] Conv. = 0.16° (5/32")	[1] Conv. = 0.16° (5/32")	NOTES			
30016	30017	30018		30019	30020	30021

ALIGNMENT SPECIFICATIONS

Saab / Saturn

9-5	9-4X	9-7X	MODEL	Astra	Astra	Astra
Aero			INFORMATION	XE	XR	XR
				5-Door	5-Door	3-Door
2000-2009	2011	2005-2009	YEARS	2008	2008	2008
			INSTRUCTIONS			
90 ft.lbs.	111 ft.lbs.	103 ft.lbs.	LUGNUT TORQUE	81 ft.lbs.	81 ft.lbs.	81 ft.lbs.
			FRONT			
2.9 ± 0.5 2⅞ ± ½	4.5 ± 0.75 4½ ± ¾	L= 4.65, R= 4.4 ± 0.6 L= 4⅝, R= 4⅜ ± ⅝	CASTER (Degrees)	4.0 ± 0.75 4 ± ¾	4.0 ± 0.75 4 ± ¾	4.0 ± 0.75 4 ± ¾
0.0 (0) | n/s	0.0 (0) | 0.75 (¾)	0.25 (¼) | 0.6 (⅝)	CROSS | TOL.	0.0 (0) | 1.0 (1)	0.0 (0) | 1.0 (1)	0.0 (0) | 1.0 (1)
-0.9 ± 0.5 -⅞ ± ½	-0.6 ± 0.75 -⅝ ± ¾	-0.5 ± 0.6 -½ ± ⅝	CAMBER (Degrees)	-0.4 ± 0.75 -⅜ ± ¾	-0.4 +0.75 -0.85 [1] -⅜ +¾ -⅞ [1]	-0.5 ± 0.75 -½ ± ¾
0.0 (0) | n/s	0.0 (0) | 0.75 (¾)	0.0 (0) | 0.6 (⅝)	CROSS | TOL.	0.0 (0) | 1.0 (1)	0.0 (0) | 1.0 (1)	0.0 (0) | 1.0 (1)
0.28° ± 0.08° 9/32° ± 3/32°	0.2° ± 0.2° 3/16° ± 3/16°	0.2° ± 0.2° 3/16° ± 3/16°	FRONT TOE (Degrees)	0.2° ± 0.15° 3/16° ± 5/32°	0.2° ± 0.15° 3/16° ± 5/32°	0.2° ± 0.15° 3/16° ± 5/32°
0.14" ± 0.04" 1/8" ± 1/32"	0.1" ± 0.1" 3/32" ± 3/32"	0.1" ± 0.1" 3/32" ± 3/32"	FRONT TOE (Inches)	0.1" ± 0.08" 3/32" ± 1/16"	0.1" ± 0.08" 3/32" ± 1/16"	0.1" ± 0.08" 3/32" ± 1/16"
12.6 ± 0.5 (12⅝ ± ½)			SAI			
21.3 / 20.0 ± 0.5			TURNING ANGLE			
		133	ILLUSTRATION #	26	26	26
			REAR			
-1.1 ± 0.25 -1⅛ ± ¼	-1.0 ± 0.75 -1 ± ¾		CAMBER (Degrees)	-1.25 ± 0.75 -1¼ ± ¾	-1.25 ± 0.75 -1¼ ± ¾	-1.25 ± 0.75 -1¼ ± ¾
0.42° ± 0.2° 13/32° ± 3/16°	0.2° ± 0.2° 3/16° ± 3/16°		REAR TOE (Degrees)	0.2° +0.5° -0.3° 3/16° +½° -5/16°	0.15° +0.5° -0.3° 5/32° +½° -5/16°	0.15° +0.5° -0.3° 5/32° +½° -5/16°
0.21" ± 0.1" 7/32" ± 3/32"	0.1" ± 0.1" 3/32" ± 3/32"		REAR TOE (Inches)	0.1" +0.25" -0.15" 3/32" +¼" -5/32"	0.08" +0.25" -0.15" 1/16" +¼" -5/32"	0.08" +0.25" -0.15" 1/16" +¼" -5/32"
	0.3 (5/16)		THRUST ANGLE	0.3 (5/16)	0.3 (5/16)	0.3 (5/16)
182	130 135		ILLUSTRATION #			
			NOTES		[1] w/Sport Susp. -0.5° (-½°)	
30022	30023	30024		40001	40002	40003

Saturn

Aura	Aura	ION	MODEL	ION	ION	L Series
		exc/Red Line, Tuner	INFORMATION	exc/Red Line, Tuner	Red Line	
		to VIN#...5Z150137		from VIN#...5Z150138	and Tuner susp.	
2007	2008-2009	2003-2005	YEARS	2005-2007	2003-2007	2000-2002
			INSTRUCTIONS			
100 ft.lbs.	100 ft.lbs.	100 ft.lbs.	LUGNUT TORQUE	100 ft.lbs.	100 ft.lbs.	92 ft.lbs.
			FRONT			
3.1 ± 0.75 3⅛ ± ¾	2.9 ± 0.75 2⅞ ± ¾	3.25 ± 0.75 3¼ ± ¾	CASTER (Degrees)	3.25 ± 0.75 3¼ ± ¾	3.6 ± 0.75 3⅝ ± ¾	3.7 ± 0.5 3 11/16 ± ½
0.0 (0) | 0.75 (¾)	0.0 (0) | 0.75 (¾)	0.0 (0) | 0.75 (¾)	CROSS | TOL.	0.0 (0) | 0.75 (¾)	0.0 (0) | 0.75 (¾)	0.0 (0) | 1.0 (1)
L= -0.9, R= -0.7 ± 0.75 L= -⅞, R= -11/16 ± ¾	L= -1.0, R= -0.7 ± 0.75 L= -1, R= -11/16 ± ¾	-1.0 ± 0.75 -1 ± ¾	CAMBER (Degrees)	-1.0 ± 0.75 -1 ± ¾	-0.9 ± 0.75 -⅞ ± ¾	-1.0 ± 0.5 -1 ± ½
-0.2 (-3/16) | 0.75 (¾)	-0.3 (-5/16) | 0.75 (¾)	0.0 (0) | 0.75 (¾)	CROSS | TOL.	0.0 (0) | 0.75 (¾)	0.0 (0) | 0.75 (¾)	0.0 (0) | 1.0 (1)
0.2° ± 0.2° 3/16° ± 3/16°	0.2° ± 0.2° 3/16° ± 3/16°	0.2° ± 0.2° 3/16° ± 3/16°	FRONT TOE (Degrees)	0.2° ± 0.2° 3/16° ± 3/16°	0.2° ± 0.2° 3/16° ± 3/16°	0.2° ± 0.16° 3/16° ± 5/32°
0.1" ± 0.1" 3/32" ± 3/32"	0.1" ± 0.1" 3/32" ± 3/32"	0.1" ± 0.1" 3/32" ± 3/32"	FRONT TOE (Inches)	0.1" ± 0.1" 3/32" ± 3/32"	0.1" ± 0.1" 3/32" ± 3/32"	0.1" ± 0.08" 3/32" ± 3/32"
			SAI			
			TURNING ANGLE			
25	25	25	ILLUSTRATION #	25	25	25
			REAR			
-0.8 ± 0.5 -13/16 ± ½	-0.8 ± 0.6 -13/16 ± ⅝	-1.15 ± 0.75 -1⅛ ± ¾	CAMBER (Degrees)	-0.8 ± 0.75 -13/16 ± ¾	-0.8 ± 0.75 -13/16 ± ¾	-1.0 ± 0.5 -1 ± ½
0.2° ± 0.2° 3/16° ± 3/16°	0.2° ± 0.2° 3/16° ± 3/16°	0.0° ± 0.3° 0° ± 5/16°	REAR TOE (Degrees)	0.25° ± 0.3° ¼° ± 5/16°	0.25° ± 0.3° ¼° ± 5/16°	0.2° ± 0.1° 3/16° ± 3/32°
0.1" ± 0.1" 3/32" ± 3/32"	0.1" ± 0.1" 3/32" ± 3/32"	0.0" ± 0.15" 0" ± 5/32"	REAR TOE (Inches)	0.13" ± 0.15" ⅛" ± 5/32"	0.13" ± 0.15" ⅛" ± 5/32"	0.1" ± 0.05" 3/32" ± 1/16"
0.3 (5/16)	0.3 (5/16)	0.3 (5/16)	THRUST ANGLE	0.3 (5/16)	0.3 (5/16)	0.15 (5/32)
177	177		ILLUSTRATION #			153
			NOTES			
40004	40005	40007		40008	40009	40010

Saturn

<table>
<tr><td>L Series</td><td>OUTLOOK</td><td>Relay</td><td>MODEL</td><td>Relay</td><td>S Series</td><td>SKY</td></tr>
<tr><td></td><td>All Models</td><td>FWD</td><td>INFORMATION</td><td>AWD</td><td></td><td></td></tr>
<tr><td></td><td></td><td></td><td></td><td></td><td></td><td></td></tr>
<tr><td>2003-2005</td><td>2007-2010</td><td>2005-2007</td><td>YEARS</td><td>2005-2006</td><td>1991-2002</td><td>2007</td></tr>
<tr><td></td><td></td><td></td><td>INSTRUCTIONS</td><td></td><td></td><td></td></tr>
<tr><td>92 ft.lbs.</td><td>140 ft.lbs.</td><td>100 ft.lbs.</td><td>LUGNUT TORQUE</td><td>100 ft.lbs.</td><td>103 ft.lbs.</td><td>100 ft.lbs.</td></tr>
<tr><td></td><td></td><td></td><td>FRONT</td><td></td><td></td><td></td></tr>
<tr><td>3.7 ±0.5
3 11/16 ± ½</td><td>4.65 ±0.75
4⅝ ± ¾</td><td>2.7 ±0.75
2 11/16 ± ¾</td><td>CASTER
(Degrees)</td><td>2.7 ±0.75
2 11/16 ± ¾</td><td>1.7 ±0.6
1 11/16 ± ⅝</td><td>8.0 ±0.6
8 ± ⅝</td></tr>
<tr><td>0.0 (0) | 1.0 (1)</td><td>0.0 (0) | 0.75 (¾)</td><td>0.0 (0) | 0.75 (¾)</td><td>CROSS | TOL.</td><td>0.0 (0) | 0.75 (¾)</td><td>0.0 (0) | 0.4 (⅜)</td><td>0.0 (0) | 0.6 (⅝)</td></tr>
<tr><td>-1.0 ±0.5
-1 ± ½</td><td>-0.9 ±0.75
-⅞ ± ¾</td><td>-0.65 ±0.75
-⅝ ± ¾</td><td>CAMBER
(Degrees)</td><td>-0.65 ±0.75
-⅝ ± ¾</td><td>-0.5 ±0.7
-½ ± 11/16</td><td>-0.5 ±0.6
-½ ± ⅝</td></tr>
<tr><td>0.0 (0) | 1.0 (1)</td><td>0.0 (0) | 0.75 (¾)</td><td>0.0 (0) | 0.75 (¾)</td><td>CROSS | TOL.</td><td>0.0 (0) | 0.75 (¾)</td><td>0.0 (0) | 0.4 (⅜)</td><td>0.0 (0) | 0.6 (⅝)</td></tr>
<tr><td>0.2° ±0.16°
3/16° ± 5/32°</td><td>0.0° ±0.2°
0° ± 3/16°</td><td>0.0° ±0.2°
0° ± 3/16°</td><td>FRONT TOE
(Degrees)</td><td>0.0° ±0.2°
0° ± 3/16°</td><td>0.2° ±0.1°
3/16° ± 3/32°</td><td>0.1° ±0.2°
3/32° ± 3/16°</td></tr>
<tr><td>0.1" ±0.08"
3/32" ± 3/32"</td><td>0.0" ±0.1"
0" ± 3/32"</td><td>0.0" ±0.1"
0" ± 3/32"</td><td>FRONT TOE
(Inches)</td><td>0.0" ±0.1"
0" ± 3/32"</td><td>0.1" ±0.05"
3/32" ± 1/16"</td><td>0.05" ±0.1"
1/16" ± 3/32"</td></tr>
<tr><td></td><td></td><td></td><td>SAI</td><td></td><td></td><td></td></tr>
<tr><td></td><td></td><td></td><td>TURNING ANGLE</td><td></td><td></td><td></td></tr>
<tr><td>25</td><td>25</td><td>25</td><td>ILLUSTRATION #</td><td>25</td><td>31 25 ▲</td><td>12</td></tr>
<tr><td></td><td></td><td></td><td>REAR</td><td></td><td></td><td></td></tr>
<tr><td>-1.0 ±0.5
-1 ± ½</td><td>-0.85 ±0.75
-⅞ ± ¾</td><td>-1.0 ±0.5
-1 ± ½</td><td>CAMBER
(Degrees)</td><td>-0.4 ±0.6
-⅜ ± ⅝</td><td>-0.7 ±0.7
-11/16 ± 11/16</td><td>-0.5 ±0.5
-½ ± ½</td></tr>
<tr><td>0.1° ±0.14°
3/32° ± ⅛°</td><td>0.1° ±0.2°
3/32° ± 3/16°</td><td>0.0° ±0.3°
0° ± 5/16°</td><td>REAR TOE
(Degrees)</td><td>0.0° ±0.2°
0° ± 3/16°</td><td>0.2° ±0.1°
3/16° ± 3/32°</td><td>0.1° ±0.2°
3/32° ± 3/16°</td></tr>
<tr><td>0.05" ±0.07"
1/16" ± 1/16"</td><td>0.05" ±0.1"
1/16" ± 3/32"</td><td>0.0" ±0.15"
0" ± 5/32"</td><td>REAR TOE
(Inches)</td><td>0.0" ±0.1"
0" ± 3/32"</td><td>0.1" ±0.05"
3/32" ± 1/16"</td><td>0.05" ±0.1"
1/16" ± 3/32"</td></tr>
<tr><td>0.15 (5/32)</td><td>0.2 (3/16)</td><td>0.3 (5/16)</td><td>THRUST ANGLE</td><td>0.25 (¼)</td><td>0.15 (5/32)</td><td>0.2 (3/16)</td></tr>
<tr><td>153</td><td>107 143</td><td></td><td>ILLUSTRATION #</td><td>134 135</td><td>25 34</td><td>180 135</td></tr>
<tr><td></td><td></td><td></td><td>NOTES</td><td></td><td></td><td></td></tr>
<tr><td>40011</td><td>40012</td><td>40013</td><td></td><td>40014</td><td>40015</td><td>40016</td></tr>
</table>

Saturn

<table>
<tr><td>SKY</td><td>VUE</td><td>VUE</td><td>MODEL</td><td>VUE</td><td>VUE</td><td>Notes</td></tr>
<tr><td></td><td>exc/Red Line</td><td>Red Line</td><td>INFORMATION</td><td>exc/Red, Green Line</td><td>Red, Green Line</td><td></td></tr>
<tr><td></td><td>FWD & AWD</td><td>FWD & AWD</td><td></td><td>FWD & AWD</td><td>FWD & AWD</td><td></td></tr>
<tr><td>2008-2009</td><td>2002-2006</td><td>2004-2006</td><td>YEARS</td><td>2007-2009</td><td>2007-2009</td><td></td></tr>
<tr><td></td><td></td><td></td><td>INSTRUCTIONS</td><td></td><td></td><td></td></tr>
<tr><td>100 ft.lbs.</td><td>92 ft.lbs.</td><td>92 ft.lbs.</td><td>LUGNUT TORQUE</td><td>92 ft.lbs.</td><td>92 ft.lbs.</td><td></td></tr>
<tr><td></td><td></td><td></td><td>FRONT</td><td></td><td></td><td></td></tr>
<tr><td>7.5 ±0.6
7½ ± ⅝</td><td>L= 3.15, R= 2.9 ± 0.75
L= 3⅛, R= 2⅞ ± ¾</td><td>L= 3.35, R= 3.1 ± 0.75
L= 3⅜, R= 3⅛ ± ¾</td><td>CASTER
(Degrees)</td><td>3.0 ±0.75
3 ± ¾</td><td>3.15 ±0.75
3⅛ ± ¾</td><td></td></tr>
<tr><td>0.0 (0) | 0.6 (⅝)</td><td>0.25 (¼) | 0.75 (¾)</td><td>0.25 (¼) | 0.75 (¾)</td><td>CROSS | TOL.</td><td>0.0 (0) | 0.75 (¾)</td><td>0.0 (0) | 0.75 (¾)</td><td></td></tr>
<tr><td>-0.5 ±0.6
-½ ± ⅝</td><td>-0.6 ±0.75
-⅝ ± ¾</td><td>-0.9 ±0.75
-⅞ ± ¾</td><td>CAMBER
(Degrees)</td><td>-0.6 ±0.75
-⅝ ± ¾</td><td>-0.75 ±0.75
-¾ ± ¾</td><td></td></tr>
<tr><td>0.0 (0) | 0.6 (⅝)</td><td>0.0 (0) | 0.75 (¾)</td><td>0.0 (0) | 0.75 (¾)</td><td>CROSS | TOL.</td><td>0.0 (0) | 0.75 (¾)</td><td>0.0 (0) | 0.75 (¾)</td><td></td></tr>
<tr><td>0.1° ±0.2°
3/32° ± 3/16°</td><td>0.15° ±0.15°
5/32° ± 5/32°</td><td>0.15° ±0.15°
5/32° ± 5/32°</td><td>FRONT TOE
(Degrees)</td><td>0.2° ±0.2°
3/16° ± 3/16°</td><td>0.2° ±0.2°
3/16° ± 3/16°</td><td></td></tr>
<tr><td>0.05" ±0.1"
1/16" ± 3/32"</td><td>0.08" ±0.08"
1/16" ± 1/16"</td><td>0.08" ±0.08"
1/16" ± 1/16"</td><td>FRONT TOE
(Inches)</td><td>0.1" ±0.1"
3/32" ± 3/32"</td><td>0.1" ±0.1"
3/32" ± 3/32"</td><td></td></tr>
<tr><td></td><td></td><td></td><td>SAI</td><td></td><td></td><td></td></tr>
<tr><td></td><td></td><td></td><td>TURNING ANGLE</td><td></td><td></td><td></td></tr>
<tr><td>12</td><td>25</td><td>25</td><td>ILLUSTRATION #</td><td>25</td><td>25</td><td></td></tr>
<tr><td></td><td></td><td></td><td>REAR</td><td></td><td></td><td></td></tr>
<tr><td>-0.5 ±0.6
-½ ± ⅝</td><td>-0.35 ±0.75 [1]
-⅜ ± ¾ [1]</td><td>-0.35 ±0.75
-⅜ ± ¾</td><td>CAMBER
(Degrees)</td><td>-0.45 ±0.75
-7/16 ± ¾</td><td>-0.45 ±0.75
-7/16 ± ¾</td><td></td></tr>
<tr><td>0.1° ±0.2°
3/32° ± 3/16°</td><td>0.2° ±0.2°
3/16° ± 3/16°</td><td>0.2° ±0.2°
3/16° ± 3/16°</td><td>REAR TOE
(Degrees)</td><td>0.2° ±0.2°
3/16° ± 3/16°</td><td>0.2° ±0.2°
3/16° ± 3/16°</td><td></td></tr>
<tr><td>0.05" ±0.1"
1/16" ± 3/32"</td><td>0.1" ±0.1"
3/32" ± 3/32"</td><td>0.1" ±0.1"
3/32" ± 3/32"</td><td>REAR TOE
(Inches)</td><td>0.1" ±0.1"
3/32" ± 3/32"</td><td>0.1" ±0.1"
3/32" ± 3/32"</td><td></td></tr>
<tr><td>0.2 (3/16)</td><td>0.2 (3/16)</td><td>0.2 (3/16)</td><td>THRUST ANGLE</td><td>0.2 (3/16)</td><td>0.2 (3/16)</td><td></td></tr>
<tr><td>180 135</td><td>107 143</td><td>107 143</td><td>ILLUSTRATION #</td><td>107 143</td><td>107 143</td><td></td></tr>
<tr><td></td><td>[1] 2002-05 = -0.5° (-½°)</td><td></td><td>NOTES</td><td></td><td></td><td></td></tr>
<tr><td>40017</td><td>40018</td><td>40019</td><td></td><td>40020</td><td>40021</td><td>40022</td></tr>
</table>

Scion

FR-S	iA	iM	MODEL INFORMATION	iQ	iQ EV	tC
2013-2016	2016	2016	YEARS	2012-2015	2013	2005-2010
			INSTRUCTIONS			
89 ft.lbs.	94 ft.lbs.	76 ft.lbs.	LUGNUT TORQUE	76 ft.lbs.	76 ft.lbs.	76 ft.lbs.
			FRONT			
5.9 ± 0.75 5 ⅞ ± ¾	5.23 ± 0.75 5 ¼ ± ¾	5.87 ± 0.75 5 ⅞ ± ¾	CASTER (Degrees)	7.45 ± 0.75 7 7/16 ± ¾	7.38 ± 0.75 7 ⅜ ± ¾	3.03 ± 0.75 3 1/16 ± ¾
0.0 (0) | 0.5 (½)	0.0 (0) | 0.75 (¾)	0.0 (0) | 0.75 (¾)	CROSS | TOL.	0.0 (0) | 0.5 (½)	0.0 (0) | 0.5 (½)	0.0 (0) | 0.75 (¾)
0.0 ± 0.75 0 ± ¾	-0.43 ± 0.75 - 7/16 ± ¾	-0.23 ± 0.75 - ¼ ± ¾	CAMBER (Degrees)	-0.43 ± 0.75 - 7/16 ± ¾	-0.43 ± 0.75 - 7/16 ± ¾	-0.52 ± 0.75 - ½ ± ¾
0.0 (0) | 0.5 (½)	0.0 (0) | 0.75 (¾)	0.0 (0) | 0.75 (¾)	CROSS | TOL.	0.0 (0) | 0.5 (½)	0.0 (0) | 0.5 (½)	0.0 (0) | 0.75 (¾)
0.0° ± 0.18° 0° ± 3/16°	0.18° ± 0.17° 3/16° ± 3/16°	0.18° ± 0.18° 3/16° ± 3/16°	FRONT TOE (Degrees)	0.17° ± 0.3° 5/32° ± 5/16°	0.17° ± 0.3° 5/32° ± 5/16°	0.0° ± 0.2° 0° ± 3/16°
0.0" ± 0.09" 0" ± 3/32"	0.09" ± 0.08" 3/32" ± 3/32"	0.09" ± 0.09" 3/32" ± 3/32"	FRONT TOE (Inches)	0.08" ± 0.15" 3/32" ± 5/32"	0.08" ± 0.15" 3/32" ± 5/32"	0.0" ± 0.1" 0" ± 3/32"
15.5 ± 0.8 (15 ½ ± ¾)	12.8 ± 0.8 (12 13/16 ± ¾)	12.3 ± 0.8 (12 ¼ ± ¾)	SAI	10.5 ± 0.8 (10 ½ ± ¾)	10.5 ± 0.8 (10 ½ ± ¾)	12.5 ± 0.8 (12 ½ ± ¾)
36.9 / 31.2 ±2.0	41.0 / 34.0 ±3.0	37.4 / 31.8 ±2.0	TURNING ANGLE	43.5 / 35.6 ±2.0	40.4 / 33.8 ±2.0	38.2 / 32.8 ±2.0
26		26	ILLUSTRATION #	26	26	26
			REAR			
-1.2 ± 0.75 -1 3/16 ± ¾	-1.48 ± 1.0 -1 ½ ± 1	-1.0 ± 0.75 -1 ± ¾	CAMBER (Degrees)	-0.87 ± 0.5 - ⅞ ± ½	-0.87 ± 0.5 - ⅞ ± ½	-0.9 ± 0.75 - ⅞ ± ¾
0.16° ± 0.24° 5/32° ± ¼°	0.4° ± 0.37° 13/32° ± ⅜°	0.17° ± 0.18° 5/32° ± 3/16°	REAR TOE (Degrees)	0.3° ± 0.25° 5/16° ± ¼°	0.3° ± 0.25° 5/16° ± ¼°	0.3° ± 0.2° 5/16° ± 3/16°
0.08" ± 0.12" 3/32" ± ⅛"	0.2" ± 0.18" 3/16" ± 3/16"	0.08" ± 0.09" 3/32" ± 3/32"	REAR TOE (Inches)	0.15" ± 0.13" 5/32" ± ⅛"	0.15" ± 0.13" 5/32" ± ⅛"	0.15" ± 0.1" 5/32" ± 3/32"
	0.8 (13/16)		THRUST ANGLE			
174			ILLUSTRATION #			107 06
			NOTES			
59001	59002	59003		59004	59005	59006

Scion

tC	xA	xB	MODEL INFORMATION	xB	xD	xD
2011-2016	2004-2006	2004-2006	YEARS	2008-2015	2008-2010	2011-2014
			INSTRUCTIONS			
76 ft.lbs.	76 ft.lbs.	76 ft.lbs.	LUGNUT TORQUE	76 ft.lbs.	76 ft.lbs.	76 ft.lbs.
			FRONT			
6.0 ± 0.75 6 ± ¾	1.78 ± 0.75 1 13/16 ± ¾	1.75 ± 0.75 1 ¾ ± ¾	CASTER (Degrees)	5.75 ± 0.75 5 ¾ ± ¾	4.85 ± 0.75 4 ⅞ ± ¾	5.07 ± 0.75 5 1/16 ± ¾
0.0 (0) | 0.75 (¾)	0.0 (0) | 0.5 (½)	0.0 (0) | 0.5 (½)	CROSS | TOL.	0.0 (0) | 0.75 (¾)	0.0 (0) | 0.75 (¾)	0.0 (0) | 0.75 (¾)
-0.22 ± 0.75 - ¼ ± ¾	-0.65 ± 0.75 - ⅝ ± ¾	-0.57 ± 0.75 - 9/16 ± ¾	CAMBER (Degrees)	-0.17 ± 0.75 - 3/16 ± ¾	-0.18 ± 0.75 - 3/16 ± ¾	-0.18 ± 0.75 - 3/16 ± ¾
0.0 (0) | 0.75 (¾)	0.0 (0) | 0.5 (½)	0.0 (0) | 0.5 (½)	CROSS | TOL.	0.0 (0) | 0.75 (¾)	0.0 (0) | 0.75 (¾)	0.0 (0) | 0.75 (¾)
0.1° ± 0.2° 3/32° ± 3/16°	0.0° ± 0.18° 0° ± 3/16°	0.0° ± 0.18° 0° ± 3/16°	FRONT TOE (Degrees)	0.17° ± 0.17° 5/32° ± 5/32°	0.12° ± 0.16° ⅛° ± 5/32°	0.13° ± 0.2° ⅛° ± 3/16°
0.05" ± 0.1" 1/16" ± 3/32"	0.0" ± 0.09" 0" ± 3/32"	0.0" ± 0.09" 0" ± 3/32"	FRONT TOE (Inches)	0.08" ± 0.08" 3/32" ± 3/32"	0.06" ± 0.08" 1/16" ± 3/32"	0.06" ± 0.1" 1/16" ± 3/32"
12.3 ± 0.8 (12 ¼ ± ¾)	10.2 ± 0.8 (10 3/16 ± ¾)	10.0 ± 0.8 (10 ± ¾)	SAI	12.2 ± 0.8 (12 3/16 ± ¾)	11.4 ± 0.8 (11 ⅜ ± ¾)	11.4 ± 0.8 (11 ⅜ ± ¾)
36.9 / 31.8 ±2.0	33.6 / 29.9 ±2.0	33.7 / 30.1 ±2.0	TURNING ANGLE	39.3 / 33.0 ±2.0	32.6 / 29.6 ±2.0	32.6 / 29.6 ±2.0
26	26	26	ILLUSTRATION #	26	26	26
			REAR			
-1.15 ± 0.75 -1 ⅛ ± ¾	-0.95 ± 0.75 - 15/16 ± ¾	-0.95 ± 0.75 - 15/16 ± ¾	CAMBER (Degrees)	-1.42 ± 0.5 -1 7/16 ± ½	-0.95 ± 0.5 - 15/16 ± ½	-0.93 ± 0.5 - 15/16 ± ½
0.3° ± 0.2° 5/16° ± 3/16°	0.35° ± 0.33° 11/32° ± 11/32°	0.33° ± 0.33° 11/32° ± 11/32°	REAR TOE (Degrees)	0.2° ± 0.25° 3/16° ± ¼°	0.26° ± 0.22° ¼° ± 7/32°	0.24° ± 0.24° ¼° ± ¼°
0.15" ± 0.1" 5/32" ± 3/32"	0.18" ± 0.17" 3/16" ± 5/32"	0.17" ± 0.17" 5/32" ± 5/32"	REAR TOE (Inches)	0.1" ± 0.13" 3/32" ± ⅛"	0.13" ± 0.11" ⅛" ± ⅛"	0.12" ± 0.12" ⅛" ± ⅛"
			THRUST ANGLE			
174			ILLUSTRATION #			
			NOTES			
59007	59008	59009		59010	59011	59012

Smart

Fortwo	Fortwo (Model 453.1)	Fortwo (Model 453.3/4)	MODEL INFORMATION	Brabus	Electric Drive	Notes
2008-2015	2016-2017	2016-2017	YEARS	2008-2015	2012-2015	
			INSTRUCTIONS			
			LUGNUT TORQUE			
			FRONT			
7.89 ±0.5 7 7/8 ± 1/2	7.67 ±1.0 7 11/16 ±1	7.67 ±1.0 7 11/16 ±1	CASTER (Degrees)	8.18 ±0.5 8 3/16 ± 1/2	7.93 ±0.5 7 15/16 ± 1/2	
0.0 (0) | 0.33 (5/16)	0.0 (0) | n/s	0.0 (0) | n/s	CROSS | TOL.	0.0 (0) | 0.33 (5/16)	0.0 (0) | 0.33 (5/16)	
0.31 ±0.5 5/16 ± 1/2	-0.3 ±1.0 -5/16 ±1	-0.3 ±1.0 -5/16 ±1	CAMBER (Degrees)	0.31 ±0.5 5/16 ± 1/2	0.48 ±0.5 1/2 ± 1/2	
0.0 (0) | 0.33 (5/16)	0.0 (0) | n/s	0.0 (0) | n/s	CROSS | TOL.	0.0 (0) | 0.33 (5/16)	0.0 (0) | 0.33 (5/16)	
0.6° ±0.17° 19/32° ± 5/32°	-0.15° ±0.25° -5/32° ± 1/4°	-0.15° ±0.25° -5/32° ± 1/4°	FRONT TOE (Degrees)	0.6° ±0.17° 19/32° ± 5/32°	0.6° ±0.17° 19/32° ± 5/32°	
0.3" ±0.08" 5/16" ± 3/32"	-0.08" ±0.13" -1/16" ± 1/8"	-0.08" ±0.13" -1/16" ± 1/8"	FRONT TOE (Inches)	0.3" ±0.08" 5/16" ± 3/32"	0.3" ±0.08" 5/16" ± 3/32"	
			SAI			
			TURNING ANGLE			
			ILLUSTRATION #			
			REAR			
-2.0 ±0.5 -2 ± 1/2	-1.2 ±0.5 -1 3/16 ± 1/2	-1.2 ±0.5 -1 3/16 ± 1/2	CAMBER (Degrees)	-2.0 ±0.5 -2 ± 1/2	-2.0 ±0.5 -2 ± 1/2	
0.09° ±0.25° 3/32° ± 1/4°	0.17° ±0.25° 5/32° ± 1/4°	0.25° ±0.25° 1/4° ± 1/4°	REAR TOE (Degrees)	0.29° ±0.25° 9/32° ± 1/4°	-0.03° ±0.25° -1/32° ± 1/4°	
0.04" ±0.13" 1/32" ± 1/8"	0.08" ±0.13" 3/32" ± 1/8"	0.13" ±0.13" 1/8" ± 1/8"	REAR TOE (Inches)	0.15" ±0.13" 5/32" ± 1/8"	-0.02" ±0.13" -1/64" ± 1/8"	
			THRUST ANGLE			
			ILLUSTRATION #			
			NOTES			
60001	60002	60003		60004	60005	60006

Sterling / Subaru

All models	Notes	Notes	MODEL INFORMATION	Baja	Baja	BRZ
1990-91			YEARS	2003	2004-2006	2013-2017
			INSTRUCTIONS			
			LUGNUT TORQUE			89 ft.lbs.
			FRONT			
2.5 ±1.0 2 1/2 ±1			CASTER (Degrees)	2.67 ±0.75 2 11/16 ± 3/4	2.75 ±0.75 2 3/4 ± 3/4	5.9 ±0.75 5 7/8 ± 3/4
0.0 (0) | n/s			CROSS | TOL.	0.0 (0) | n/s	0.0 (0) | n/s	0.0 (0) | n/s
0.42 ±1.0 7/16 ±1			CAMBER (Degrees)	0.33 ±0.5 5/16 ± 1/2	0.67 ±0.5 11/16 ± 1/2	0.0 ±0.5 0 ± 1/2
0.0 (0) | n/s			CROSS | TOL.	0.0 (0) | n/s	0.0 (0) | n/s	0.0 (0) | 0.5 (1/2)
0.0° ±0.13° 0° ± 1/8°			FRONT TOE (Degrees)	0.0° ±0.3° 0° ± 5/16°	0.0° ±0.17° 0° ± 5/32°	0.0° ±0.25° 0° ± 1/4°
0.0" ±0.06" 0" ± 1/16"			FRONT TOE (Inches)	0.0" ±0.15" 0" ± 5/32"	0.0" ±0.08" 0" ± 3/32"	0.0" ±0.13" 0" ± 1/8"
8.0 ±1.2 (8 ±1 3/16)			SAI	13.5 (13 1/2)	13.0 ±1.0 (13 ±1)	15.5 (15 1/2)
39.6 / 31.1			TURNING ANGLE	34.5 / 30.3 ±1.5	34.5 / 30.3 ±1.5	36.9 / 31.2 ±1.5
			ILLUSTRATION #	22	22	26
			REAR			
0.3 ±1.2 5/16 ± 1 3/16			CAMBER (Degrees)	-0.17 ±0.75 -3/16 ± 3/4	0.08 ±0.5 1/16 ± 1/2	-1.17 ±0.75 -1 3/16 ± 3/4
0.33° ±0.13° 11/32° ± 1/8°			REAR TOE (Degrees)	0.0° ±0.3° 0° ± 5/16°	0.0° ±0.17° 0° ± 5/32°	0.17° ±0.25° 5/32° ± 1/4°
0.17" ±0.07" 5/32" ± 1/16"			REAR TOE (Inches)	0.0" ±0.15" 0" ± 5/32"	0.0" ±0.08" 0" ± 3/32"	0.08" ±0.13" 3/32" ± 1/8"
0.15 (5/32)			THRUST ANGLE	0.33 (11/32)	0.33 (11/32)	0.33 (11/32)
80			ILLUSTRATION #	06	06	174
			NOTES			
41001	41002	41003		42004	42005	42010

Subaru

Crosstrek	Crosstrek	Forester	MODEL	Forester	Forester	Forester
XV Crosstrek	XV Crosstrek		INFORMATION			
(incl.Hybrid)	(incl.Hybrid)					
2013-2014	2015-2017	1998-99	YEARS	2000	2001	2002
			INSTRUCTIONS			
89 ft.lbs.	89 ft.lbs.	66 ft.lbs.	LUGNUT TORQUE	66 ft.lbs.	66 ft.lbs.	66 ft.lbs.
			FRONT			
6.0 ± 0.75 6 ± ¾	6.0 ± 0.75 6 ± ¾	2.83 ± 0.75 2 13/16 ± ¾	CASTER (Degrees)	2.83 ± 0.75 2 13/16 ± ¾	2.58 ± 0.75 2 9/16 ± ¾	2.61 ± 0.75 2⅝ ± ¾
0.0 (0) | n/s	0.0 (0) | n/s	0.0 (0) | n/s	CROSS | TOL.	0.0 (0) | n/s	0.0 (0) | n/s	0.0 (0) | n/s
0.17 ± 0.5 3/16 ± ½	0.17 ± 0.5 3/16 ± ½	-0.25 ± 0.5 -¼ ± ½	CAMBER (Degrees)	-0.25 ± 0.5 -¼ ± ½	-0.25 ± 0.5 -¼ ± ½	-0.25 ± 0.5 -¼ ± ½
0.0 (0) | 0.75 (¾)	0.0 (0) | 0.75 (¾)	0.0 (0) | n/s	CROSS | TOL.	0.0 (0) | n/s	0.0 (0) | n/s	0.0 (0) | n/s
0.0° ± 0.17° 0° ± 5/32°	0.0° ± 0.17° 0° ± 5/32°	0.0° ± 0.3° 0° ± 5/16°	FRONT TOE (Degrees)	0.0° ± 0.3° 0° ± 5/16°	0.0° ± 0.3° 0° ± 5/16°	0.0° ± 0.3° 0° ± 5/16°
0.0" ± 0.08" 0" ± 3/32"	0.0" ± 0.08" 0" ± 3/32"	0.0" ± 0.15" 0" ± 5/32"	FRONT TOE (Inches)	0.0" ± 0.15" 0" ± 5/32"	0.0" ± 0.15" 0" ± 5/32"	0.0" ± 0.15" 0" ± 5/32"
13.1 ± 0.8 (13⅛ ± ¾)	13.1 ± 0.8 (13⅛ ± ¾)	13.4 ± 1.0 (13 7/16 ± 1)	SAI	13.4 ± 1.0 (13 7/16 ± 1)	13.4 ± 1.0 (13 7/16 ± 1)	13.4 ± 1.0 (13 7/16 ± 1)
38.5 / 34.0 ±1.5	38.5 / 34.0 ±1.5	34.4 / 30.2 ±1.5	TURNING ANGLE	34.4 / 30.2 ±1.5	34.4 / 30.2 ±1.5	34.4 / 30.2 ±1.5
22	22	22	ILLUSTRATION #	22	22	22
			REAR			
-0.33 ± 0.5 -5/16 ± ½	-0.42 ± 0.5 -7/16 ± ½	-0.58 ± 0.75 -9/16 ± ¾	CAMBER (Degrees)	-0.25 ± 0.75 -¼ ± ¾	-0.58 ± 0.75 -9/16 ± ¾	-0.58 ± 0.75 -9/16 ± ¾
0.17° ± 0.17° 5/32° ± 5/32°	0.0° ± 0.17° 0° ± 5/32°	0.25° ± 0.15° ¼° ± 5/32°	REAR TOE (Degrees)	0.25° ± 0.15° ¼° ± 5/32°	0.25° ± 0.15° ¼° ± 5/32°	0.2° ± 0.3° 3/16° ± 5/16°
0.08" ± 0.08" 3/32" ± 3/32"	0.0" ± 0.08" 0" ± 3/32"	0.13" ± 0.08" ⅛" ± 1/16"	REAR TOE (Inches)	0.13" ± 0.08" ⅛" ± 1/16"	0.13" ± 0.08" ⅛" ± 1/16"	0.1" ± 0.15" 3/32" ± 5/32"
0.33 (11/32)	0.33 (11/32)	0.33 (11/32)	THRUST ANGLE	0.33 (11/32)	0.33 (11/32)	0.33 (11/32)
174	174	06	ILLUSTRATION #	06	06	06
			NOTES			
42011	42012	42013		42014	42015	42016

Subaru

Forester	Forester	Forester	MODEL	Forester	Forester	Impreza FWD
exc/XT Turbo	XT Turbo		INFORMATION			
2003-2008	2004-2008	2009-2013	YEARS	2014-2016	2017	1993-96
			INSTRUCTIONS			
66 ft.lbs.[1]	66 ft.lbs.[1]	74 ft.lbs.	LUGNUT TORQUE	74 ft.lbs.	74 ft.lbs.	65 ft.lbs.
			FRONT			
3.05 ± 0.75 3 1/16 ± ¾	3.05 ± 0.75 3 1/16 ± ¾	5.42 ± 0.75 5 7/16 ± ¾	CASTER (Degrees)	5.67 ± 0.75 5 11/16 ± ¾	5.67 ± 0.75 5 11/16 ± ¾	3.0 ± 1.0 3 ± 1
0.0 (0) | n/s	0.0 (0) | n/s	0.0 (0) | n/s	CROSS | TOL.	0.0 (0) | n/s	0.0 (0) | n/s	0.0 (0) | n/s
-0.42 ± 0.75 -7/16 ± ¾	-0.42 ± 0.75 -7/16 ± ¾	0.0 ± 0.5 0 ± ½	CAMBER (Degrees)	0.0 ± 0.5 0 ± ½	0.0 ± 0.5 0 ± ½	0.0 ± 0.5 0 ± ½
0.0 (0) | 0.75 (¾)	0.0 (0) | 0.75 (¾)	0.0 (0) | 0.75 (¾)	CROSS | TOL.	0.0 (0) | 0.58 (9/16)	0.0 (0) | 0.58 (9/16)	0.0 (0) | n/s
0.0° ± 0.25° 0° ± ¼°	0.0° ± 0.25° 0° ± ¼°	0.0° ± 0.16° 0° ± 5/32°	FRONT TOE (Degrees)	0.0° ± 0.16° 0° ± 5/32°	0.0° ± 0.16° 0° ± 5/32°	0.0° ± 0.3° 0° ± 5/16°
0.0" ± 0.13" 0" ± ⅛"	0.0" ± 0.13" 0" ± ⅛"	0.0" ± 0.08" 0" ± 3/32"	FRONT TOE (Inches)	0.0" ± 0.08" 0" ± 3/32"	0.0" ± 0.08" 0" ± 3/32"	0.0" ± 0.15" 0" ± 5/32"
13.2 (13 3/16)	13.2 (13 3/16)	13.3 (13 5/16)	SAI	14.2 (14 3/16)	14.2 (14 3/16)	14.0 (14)
36.4 / 32.0 ±1.5[2]	36.4 / 32.0 ±1.5	38.3 / 33.8 ±1.5	TURNING ANGLE	38.4 / 33.8 ±1.5	38.4 / 33.8 ±1.5	39.0 / 33.5 ±1.5
22	22	22	ILLUSTRATION #	22	22	22
			REAR			
-0.83 ± 0.75 -13/16 ± ¾	-0.92 ± 0.75 -15/16 ± ¾	0.0 ± 0.75 0 ± ¾	CAMBER (Degrees)	-1.0 ± 0.75 -1 ± ¾	-1.0 ± 0.75 -1 ± ¾	-0.83 ± 0.75 -13/16 ± ¾
0.17° ± 0.25° 5/32° ± ¼°	0.17° ± 0.25° 5/32° ± ¼°	0.0° ± 0.16° 0° ± 5/32°	REAR TOE (Degrees)	0.25° ± 0.16° ¼° ± 5/32°	0.0° ± 0.16° 0° ± 5/32°	0.0° ± 0.3° 0° ± 5/16°
0.08" ± 0.13" 3/32" ± ⅛"	0.08" ± 0.13" 3/32" ± ⅛"	0.0" ± 0.08" 0" ± 3/32"	REAR TOE (Inches)	0.13" ± 0.08" ⅛" ± 3/32"	0.0" ± 0.08" 0" ± 3/32"	0.0" ± 0.15" 0" ± 5/32"
0.33 (11/32)	0.33 (11/32)	0.33 (11/32)	THRUST ANGLE	0.33 (11/32)	0.33 (11/32)	0.33 (11/32)
06	06	174	ILLUSTRATION #	174	174	06
[1] 2003-06 = 66 ft.lbs [2] 2008 = 35.0°/30.9°	[1] 2003-06 = 66 ft.lbs		NOTES			
42017	42018	42019		42020	42021	42022

Subaru

Impreza AWD exc/RS Models	Impreza RS Coupe & Sedan	Impreza Sedan exc/Turbo	MODEL INFORMATION	Impreza Sedan Turbo exc/STI	Impreza Sedan WRX STI	Impreza Wagon exc/Turbo exc/Outback Sport
1993-2001	1998-2001	2002-2007	YEARS	2002-2007	2004-2007	2002-2007
			INSTRUCTIONS			
65 ft.lbs.	65 ft.lbs.	66 ft.lbs.	LUGNUT TORQUE	66 ft.lbs.	66 ft.lbs.	66 ft.lbs.
			FRONT			
3.0 ± 1.0 3 ± 1	3.08 ± 1.0 3 1/16 ± 1	3.42 ± 0.75 3 7/16 ± 3/4	CASTER (Degrees)	3.5 ± 0.75 3 1/2 ± 3/4	4.83 ± 0.75 [1] 4 13/16 ± 3/4 [1]	3.42 ± 0.75 3 7/16 ± 3/4
0.0 (0) | n/s	0.0 (0) | n/s	0.0 (0) | n/s	CROSS | TOL.	0.0 (0) | n/s	0.0 (0) | n/s	0.0 (0) | n/s
0.0 ± 0.5 0 ± 1/2	-0.42 ± 0.5 - 7/16 ± 1/2	-0.25 ± 0.5 - 1/4 ± 1/2	CAMBER (Degrees)	-0.42 ± 0.5 - 7/16 ± 1/2	-0.5 ± 0.5 - 1/2 ± 1/2	-0.17 ± 0.5 - 3/16 ± 1/2
0.0 (0) | n/s	0.0 (0) | n/s	0.0 (0) | n/s	CROSS | TOL.	0.0 (0) | n/s	0.0 (0) | n/s	0.0 (0) | n/s
0.0° ± 0.3° 0° ± 5/16°	0.0° ± 0.3° 0° ± 5/16°	0.0° ± 0.17° 0° ± 5/32°	FRONT TOE (Degrees)	0.0° ± 0.17° 0° ± 5/32°	0.0° ± 0.17° 0° ± 5/32°	0.0° ± 0.17° 0° ± 5/32°
0.0" ± 0.15" 0" ± 5/32"	0.0" ± 0.15" 0" ± 5/32"	0.0" ± 0.08" 0" ± 3/32"	FRONT TOE (Inches)	0.0" ± 0.08" 0" ± 3/32"	0.0" ± 0.08" 0" ± 3/32"	0.0" ± 0.08" 0" ± 3/32"
14.0 (14)	14.0 (14)	14.3 ± 0.8 (14 5/16 ± 3/4)	SAI	14.6 ± 0.8 (14 9/16 ± 3/4)	14.8 ± 0.8 (14 13/16 ± 3/4)	13.5 ± 0.8 (13 1/2 ± 3/4)
37.4 / 32.5 ± 1.5	37.4 / 32.5 ± 1.5	34.5 / 30.3 ± 1.5	TURNING ANGLE	34.5 / 30.3 ± 1.5	32.9 / 28.5 ± 1.5	34.5 / 30.3 ± 1.5
22	22	22	ILLUSTRATION #	22	22	22
			REAR			
-0.92 ± 0.75 - 15/16 ± 3/4	-1.17 ± 0.75 -1 3/16 ± 3/4	-1.42 ± 0.5 -1 7/16 ± 1/2	CAMBER (Degrees)	-1.5 ± 0.5 -1 1/2 ± 1/2	-1.67 ± 0.5 -1 11/16 ± 1/2	-1.25 ± 0.5 -1 1/4 ± 1/2
0.0° ± 0.3° 0° ± 5/16°	0.0° ± 0.3° 0° ± 5/16°	0.0° ± 0.17° 0° ± 5/32°	REAR TOE (Degrees)	0.0° ± 0.17° 0° ± 5/32°	0.0° ± 0.17° 0° ± 5/32°	0.0° ± 0.17° 0° ± 5/32°
0.0" ± 0.15" 0" ± 5/32"	0.0" ± 0.15" 0" ± 5/32"	0.0" ± 0.08" 0" ± 3/32"	REAR TOE (Inches)	0.0" ± 0.08" 0" ± 3/32"	0.0" ± 0.08" 0" ± 3/32"	0.0" ± 0.08" 0" ± 3/32"
0.33 (11/32)	0.33 (11/32)	0.33 (11/32)	THRUST ANGLE	0.33 (11/32)	0.33 (11/32)	0.33 (11/32)
06	06	06	ILLUSTRATION #	06	06	06
			NOTES		[1] 2007 = 3.7° (3 11/16°)	
42023	42024	42025		42026	42027	42028

Subaru

Impreza Wagon Turbo	Impreza exc/Outback Sport	Impreza Sedan & Hatch	MODEL INFORMATION	Impreza Sedan & Hatch	Impreza WRX exc/STI	Impreza WRX STI
2002-2007	2008-2011	2012-2016	YEARS	2017	2008-2010	2008-2010
			INSTRUCTIONS			
66 ft.lbs.	74 ft.lbs.	89 ft.lbs.	LUGNUT TORQUE	89 ft.lbs.	74 ft.lbs.	74 ft.lbs.
			FRONT			
3.5 ± 0.75 3 1/2 ± 3/4	5.92 ± 0.75 5 15/16 ± 3/4	6.0 ± 0.75 6 ± 3/4	CASTER (Degrees)	5.93 ± 0.75 5 15/16 ± 3/4	5.92 ± 0.75 5 15/16 ± 3/4	6.42 ± 0.75 6 7/16 ± 3/4
0.0 (0) | n/s	0.0 (0) | n/s	0.0 (0) | n/s	CROSS | TOL.	0.0 (0) | n/s	0.0 (0) | n/s	0.0 (0) | n/s
-0.33 ± 0.5 - 5/16 ± 1/2	-0.17 ± 0.5 - 3/16 ± 1/2	-0.3 ± 0.5 - 5/16 ± 1/2	CAMBER (Degrees)	-0.5 ± 0.5 - 1/2 ± 1/2	-0.17 ± 0.5 - 3/16 ± 1/2	-0.67 ± 0.5 - 11/16 ± 1/2
0.0 (0) | n/s	0.0 (0) | 0.75 (3/4)	0.0 (0) | 0.75 (3/4)	CROSS | TOL.	0.0 (0) | 0.58 (9/16)	0.0 (0) | 0.75 (3/4)	0.0 (0) | 0.75 (3/4)
0.0° ± 0.17° 0° ± 5/32°	0.0° ± 0.16° 0° ± 5/32°	0.0° ± 0.16° 0° ± 5/32°	FRONT TOE (Degrees)	0.0° ± 0.2° 0° ± 3/16°	0.0° ± 0.16° 0° ± 5/32°	0.0° ± 0.16° 0° ± 5/32°
0.0" ± 0.08" 0" ± 3/32"	0.0" ± 0.08" 0" ± 3/32"	0.0" ± 0.08" 0" ± 3/32"	FRONT TOE (Inches)	0.0" ± 0.1" 0" ± 3/32"	0.0" ± 0.08" 0" ± 3/32"	0.0" ± 0.08" 0" ± 3/32"
13.8 ± 0.8 (13 3/4 ± 3/4)	13.6 ± 0.8 (13 9/16 ± 3/4)	13.8 ± 0.8 (13 13/16 ± 3/4)	SAI	14.2 (14 1/4)	13.6 ± 0.8 (13 9/16 ± 3/4)	15.1 ± 0.8 (15 1/16 ± 3/4)
34.5 / 30.3 ± 1.5	37.8 / 33.4 ± 1.5	38.3 / 33.7 ± 1.5	TURNING ANGLE	37.8 / 32.9 ± 1.5	37.8 / 33.4 ± 1.5	36.6 / 32.2 ± 1.5
22	22	22	ILLUSTRATION #	22	22	22
			REAR			
-1.33 ± 0.5 -1 5/16 ± 1/2	-1.08 ± 0.5 -1 1/16 ± 1/2	-1.3 ± 0.5 -1 5/16 ± 1/2	CAMBER (Degrees)	-1.33 ± 0.5 -1 5/16 ± 1/2	-1.25 ± 0.5 -1 1/4 ± 1/2	-1.58 ± 0.5 -1 9/16 ± 1/2
0.0° ± 0.17° 0° ± 5/32°	0.0° ± 0.16° 0° ± 5/32°	0.24° ± 0.16° 1/4° ± 5/32°	REAR TOE (Degrees)	0.25° ± 0.25° 1/4° ± 1/4°	0.0° ± 0.16° 0° ± 5/32°	0.0° ± 0.16° 0° ± 5/32°
0.0" ± 0.08" 0" ± 3/32"	0.0" ± 0.08" 0" ± 3/32"	0.12" ± 0.08" 1/8" ± 3/32"	REAR TOE (Inches)	0.13" ± 0.13" 1/8" ± 1/8"	0.0" ± 0.08" 0" ± 3/32"	0.0" ± 0.08" 0" ± 3/32"
0.33 (11/32)	0.33 (11/32)	0.33 (11/32)	THRUST ANGLE	0.5 (1/2)	0.33 (11/32)	0.33 (11/32)
06	174	174	ILLUSTRATION #	174	174	174
			NOTES			
42029	42030	42031		42032	42033	42034

Subaru

Impreza WRX exc/STI	Impreza WRX STI	Impreza WRX exc/STI exc/18" wheels (OE)	MODEL INFORMATION	Impreza WRX STI 18" wheels (OE)	Impreza Outback Sport	Impreza Outback Sport
2011-2014	2011-2014	2015-2017	YEARS	2015-2017	1996-2001	2002-2007
			INSTRUCTIONS			
74 ft.lbs.	74 ft.lbs.	74 ft.lbs.	LUGNUT TORQUE	74 ft.lbs.	65 ft.lbs.	66 ft.lbs.
			FRONT			
6.5±0.75 6½±¾	6.5±0.75 6½±¾	6.5±0.75 6½±¾	CASTER (Degrees)	6.55±0.75 6 9/16±¾	3.0±1.0 3±1	3.42±0.75 3 7/16±¾
0.0 (0) | n/s	0.0 (0) | n/s	0.0 (0) | n/s	CROSS | TOL.	0.0 (0) | n/s	0.0 (0) | n/s	0.0 (0) | n/s
-0.83±0.5 -13/16±½	-0.75±0.5 -¾±½	-0.75±0.5 -¾±½	CAMBER (Degrees)	-0.75±0.5 -¾±½	0.0±0.5 0±½	-0.08±0.5 -1/16±½
0.0 (0) | 0.75 (¾)	0.0 (0) | 0.75 (¾)	0.0 (0) | 0.58 (9/16)	CROSS | TOL.	0.0 (0) | 0.58 (9/16)	0.0 (0) | n/s	0.0 (0) | n/s
0.0°±0.16° 0°±5/32°	0.0°±0.16° 0°±5/32°	0.0°±0.16° 0°±5/32°	FRONT TOE (Degrees)	0.0°±0.16° 0°±5/32°	0.0°±0.3° 0°±5/16°	0.0°±0.17° 0°±5/32°
0.0"±0.08" 0"±3/32"	0.0"±0.08" 0"±3/32"	0.0"±0.08" 0"±3/32"	FRONT TOE (Inches)	0.0"±0.08" 0"±3/32"	0.0"±0.15" 0"±5/32"	0.0"±0.08" 0"±3/32"
15.3±0.8 (15¼±¾)	15.3±0.8 (15¼±¾)	15.2±0.8 (15¼±¾)	SAI	15.4±0.8 (15⅜±¾)	14.0 (14)	13.3±0.8 (13 5/16±¾)
37.8 / 33.4 ±1.5	36.6 / 32.2 ±1.5	37.5 / 32.9 ±1.5	TURNING ANGLE	36.3 / 32.0 ±1.5	37.4 / 32.5 ±1.5	34.5 / 30.3 ±1.5
22	22	22	ILLUSTRATION #	22	22	22
			REAR			
-1.58±0.5 -1 9/16±½	-1.5±0.5 -1 9/16±½	-1.5±0.75 -1½ +¾ -½¾	CAMBER (Degrees)	-1.67±0.75 -1 11/16±¾	-0.92±0.75 -15/16±¾	-1.17±0.5 -1 3/16±½
0.0°±0.16° 0°±5/32°	0.0°±0.16° 0°±5/32°	0.24°±0.16° ¼°±5/32°	REAR TOE (Degrees)	0.24°±0.16° ¼°±5/32°	0.0°±0.3° 0°±5/16°	0.0°±0.17° 0°±5/32°
0.0"±0.08" 0"±3/32"	0.0"±0.08" 0"±3/32"	0.12"±0.08" ⅛"±3/32"	REAR TOE (Inches)	0.12"±0.08" ⅛"±3/32"	0.0"±0.15" 0"±5/32"	0.0"±0.08" 0"±3/32"
0.33 (11/32)	0.33 (11/32)	0.33 (11/32)	THRUST ANGLE	0.33 (11/32)	0.33 (11/32)	0.33 (11/32)
174	174	174	ILLUSTRATION #	174	06	06
			NOTES			
42035	42036	42037		42038	42039	42040

Subaru

Impreza Outback Sport	Justy	Legacy w/Air Ride All Models	MODEL INFORMATION	Legacy Turbo 4X4	Legacy Sedan 4X2	Legacy Sedan 4X4
2008-2011	1990-94	1990-94	YEARS	1991-94	1990-94	1990-94
			INSTRUCTIONS			
	58-72 ft.lbs.	65 ft.lbs.	LUGNUT TORQUE	65 ft.lbs.	65 ft.lbs.	65 ft.lbs.
			FRONT			
5.83±0.75 5 13/16±¾	2.5±1.0 2½±1	3.0±1.0 3±1	CASTER (Degrees)	2.83±1.0 2 13/16±1	3.08±1.0 3 1/16±1	3.0±1.0 3±1
0.0 (0) | n/s	0.0 (0) | n/s	0.0 (0) | n/s	CROSS | TOL.	0.0 (0) | n/s	0.0 (0) | n/s	0.0 (0) | n/s
-0.08±0.5 -1/16±½	0.67±1.0 11/16±1	0.0±0.75 0±¾	CAMBER (Degrees)	-0.25±0.75 -¼±¾	-0.25±0.75 -¼±¾	0.0±0.75 0±¾
0.0 (0) | 0.75 (¾)	0.0 (0) | n/s	0.0 (0) | n/s	CROSS | TOL.	0.0 (0) | n/s	0.0 (0) | n/s	0.0 (0) | n/s
0.0°±0.16° 0°±5/32°	0.2°±0.1° 3/16°±3/32°	0.0°±0.3° 0°±5/16°	FRONT TOE (Degrees)	0.0°±0.3° 0°±5/16°	0.0°±0.3° 0°±5/16°	0.0°±0.3° 0°±5/16°
0.0"±0.08" 0"±3/32"	0.1"±0.05" 3/32"±1/16"	0.0"±0.15" 0"±5/32"	FRONT TOE (Inches)	0.0"±0.15" 0"±5/32"	0.0"±0.15" 0"±5/32"	0.0"±0.15" 0"±5/32"
13.5±0.8 (13½±¾)			SAI			
37.9 / 33.5 ±1.5	39.5 / 33.5	37.6 / 32.6 ±1.5 [1]	TURNING ANGLE	37.6 / 32.6 ±1.5	37.6 / 32.6 ±1.5 [1]	37.6 / 32.6 ±1.5 [1]
22		22	ILLUSTRATION #	22	22	22
			REAR			
-1.08±0.5 -1 1/16±½	0.0±1.0 0±1	-1.0±1.0 [2] -1±1 [2]	CAMBER (Degrees)	-1.0±1.0 -1±1	-1.0±1.0 [2] -1±1 [2]	-1.0±1.0 [2] -1±1 [2]
0.0°±0.16° 0°±5/32°	0.0°±0.3° 0°±5/16°	0.0°±0.3° 0°±5/16°	REAR TOE (Degrees)	0.0°±0.3° 0°±5/16°	0.0°±0.3° 0°±5/16°	0.0°±0.3° 0°±5/16°
0.0"±0.08" 0"±3/32"	0.0"±0.15" 0"±5/32"	0.0"±0.15" 0"±5/32"	REAR TOE (Inches)	0.0"±0.15" 0"±5/32"	0.0"±0.15" 0"±5/32"	0.0"±0.15" 0"±5/32"
0.33 (11/32)		0.33 (11/32)	THRUST ANGLE	0.33 (11/32)	0.33 (11/32)	0.33 (11/32)
174	54	06	ILLUSTRATION #	06	06	06
		[1] 1990-91 = 39°/33.5° [2] 1990-91 = -0.5° (-½°)	NOTES		[1] 1990-91 = 39°/33.5° [2] 1990-91 = -0.5° (-½°)	[1] 1990-91 = 39°/33.5° [2] 1990-91 = -0.5° (-½°)
42041	42042	42043		42044	42045	42046

Subaru

MODEL INFORMATION	Legacy Wagon 4X2	Legacy Wagon 4X4	Legacy Wagon 4X2	Legacy Wagon 4X4	Legacy Sedan 4X2	Legacy Sedan 4X4
YEARS	1990-91	1990-91	1992-94	1992-94	1995-96	1995-99
INSTRUCTIONS						
LUGNUT TORQUE	65 ft.lbs.	65 ft.lbs.	65 ft.lbs.	65 ft.lbs.	65 ft.lbs.	65 ft.lbs.
FRONT						
CASTER (Degrees)	2.83 ± 1.0 2 13/16 ± 1	2.75 ± 1.0 2¾ ± 1	2.83 ± 1.0 2 13/16 ± 1	2.75 ± 1.0 2¾ ± 1	3.08 ± 1.0 3 1/16 ± 1	3.08 ± 1.0 3 1/16 ± 1
CROSS / TOL.	0.0 (0) / n/s	0.0 (0) / n/s	0.0 (0) / n/s	0.0 (0) / n/s	0.0 (0) / n/s	0.0 (0) / n/s
CAMBER (Degrees)	-0.25 ± 0.5 -¼ ± ½	0.0 ± 0.5 0 ± ½	-0.25 ± 0.75 -¼ ± ¾	0.0 ± 0.75 0 ± ¾	-0.08 ± 0.5 -1/16 ± ½	-0.08 ± 0.5 -1/16 ± ½
CROSS / TOL.	0.0 (0) / n/s	0.0 (0) / n/s	0.0 (0) / n/s	0.0 (0) / n/s	0.0 (0) / n/s	0.0 (0) / n/s
FRONT TOE (Degrees)	0.0° ± 0.3° 0° ± 5/16°	0.0° ± 0.3° 0° ± 5/16°	0.0° ± 0.3° 0° ± 5/16°	0.0° ± 0.3° 0° ± 5/16°	0.0° ± 0.3° 0° ± 5/16°	0.0° ± 0.3° 0° ± 5/16°
FRONT TOE (Inches)	0.0" ± 0.15" 0" ± 5/32"	0.0" ± 0.15" 0" ± 5/32"	0.0" ± 0.15" 0" ± 5/32"	0.0" ± 0.15" 0" ± 5/32"	0.0" ± 0.15" 0" ± 5/32"	0.0" ± 0.15" 0" ± 5/32"
SAI					14.3 ± 1.0 (14¼ ± 1)	14.3 ± 1.0 (14¼ ± 1)
TURNING ANGLE	39.0 / 33.5	39.0 / 33.5	37.6 / 32.6 ± 1.5	37.6 / 32.6 ± 1.5		
ILLUSTRATION #	22	22	22	22	22	22
REAR						
CAMBER (Degrees)	-0.33 ± 1.0 -5/16 ± 1	-0.33 ± 1.0 -5/16 ± 1	-0.83 ± 1.0 -13/16 ± 1	-0.83 ± 1.0 -13/16 ± 1	-0.92 ± 0.75 -15/16 ± ¾	-1.0 ± 0.75 -1 ± ¾
REAR TOE (Degrees)	0.0° ± 0.3° 0° ± 5/16°	0.0° ± 0.3° 0° ± 5/16°	0.0° ± 0.3° 0° ± 5/16°	0.0° ± 0.3° 0° ± 5/16°	0.0° ± 0.3° 0° ± 5/16°	0.0° ± 0.3° 0° ± 5/16°
REAR TOE (Inches)	0.0" ± 0.15" 0" ± 5/32"	0.0" ± 0.15" 0" ± 5/32"	0.0" ± 0.15" 0" ± 5/32"	0.0" ± 0.15" 0" ± 5/32"	0.0" ± 0.15" 0" ± 5/32"	0.0" ± 0.15" 0" ± 5/32"
THRUST ANGLE	0.33 (11/32)	0.33 (11/32)	0.33 (11/32)	0.33 (11/32)	0.33 (11/32)	0.33 (11/32)
ILLUSTRATION #	06	06	06	06	06	06
NOTES						
	42047	42048	42049	42050	42051	42052

Subaru

MODEL INFORMATION	Legacy Wagon 4X2	Legacy Wagon 4X4 exc/Outback	Legacy Sedan exc/Outback	Legacy Wagon exc/Outback	Legacy Sedan exc/Outback	Legacy Wagon exc/Outback
YEARS	1995-96	1995-99	2000-2004	2000-2004	2005-2006	2005-2006
INSTRUCTIONS						
LUGNUT TORQUE	65 ft.lbs.	65 ft.lbs.	65 ft.lbs.	65 ft.lbs.	89 ft.lbs.	89 ft.lbs.
FRONT						
CASTER (Degrees)	3.08 ± 1.0 3 1/16 ± 1	3.08 ± 1.0 3 1/16 ± 1	3.08 ± 0.75 3 1/16 ± ¾	2.83 ± 0.75 2 13/16 ± ¾	5.92 ± 0.75 5 15/16 ± ¾	5.67 ± 0.75 5 11/16 ± ¾
CROSS / TOL.	0.0 (0) / n/s	0.0 (0) / n/s	0.0 (0) / n/s	0.0 (0) / n/s	0.0 (0) / n/s	0.0 (0) / n/s
CAMBER (Degrees)	-0.08 ± 0.5 -1/16 ± ½	-0.08 ± 0.5 -1/16 ± ½	-0.08 ± 0.5 -1/16 ± ½	-0.08 ± 0.5 -1/16 ± ½	-0.25 ± 0.5 -¼ ± ½	-0.25 ± 0.5 -¼ ± ½
CROSS / TOL.	0.0 (0) / n/s	0.0 (0) / n/s	0.0 (0) / 0.75 (¾)	0.0 (0) / 0.75 (¾)	0.0 (0) / 0.75 (¾)	0.0 (0) / 0.75 (¾)
FRONT TOE (Degrees)	0.0° ± 0.3° 0° ± 5/16°	0.0° ± 0.3° 0° ± 5/16°	0.0° ± 0.17° 0° ± 5/32°	0.0° ± 0.17° 0° ± 5/32°	0.0° ± 0.17° 0° ± 5/32°	0.0° ± 0.17° 0° ± 5/32°
FRONT TOE (Inches)	0.0" ± 0.15" 0" ± 5/32"	0.0" ± 0.15" 0" ± 5/32"	0.0" ± 0.08" 0" ± 3/32"	0.0" ± 0.08" 0" ± 3/32"	0.0" ± 0.08" 0" ± 3/32"	0.0" ± 0.08" 0" ± 3/32"
SAI	14.3 ± 1.0 (14¼ ± 1)	14.3 ± 1.0 (14¼ ± 1)	14.3 ± 1.0 (14¼ ± 1)	14.3 ± 1.0 (14¼ ± 1)	13.8 (13¾)	13.8 (13¾)
TURNING ANGLE			36.3 / 31.6 ± 1.5	36.3 / 31.6 ± 1.5	37.2 / 33.0 ± 1.5	37.2 / 33.0 ± 1.5
ILLUSTRATION #	22	22	22	22	22	22
REAR						
CAMBER (Degrees)	-0.75 ± 0.75 -¾ ± ¾	-0.92 ± 0.75 -15/16 ± ¾	-0.5 ± 0.75 -½ ± ¾	-0.33 ± 0.75 -5/16 ± ¾	-0.67 ± 0.75 -11/16 ± ¾	-0.5 ± 0.75 -½ ± ¾
REAR TOE (Degrees)	0.0° ± 0.3° 0° ± 5/16°	0.0° ± 0.3° 0° ± 5/16°	0.0° ± 0.16° 0° ± 5/32°	0.0° ± 0.16° 0° ± 5/32°	0.0° ± 0.16° 0° ± 5/32°	0.0° ± 0.16° 0° ± 5/32°
REAR TOE (Inches)	0.0" ± 0.15" 0" ± 5/32"	0.0" ± 0.15" 0" ± 5/32"	0.0" ± 0.08" 0" ± 3/32"	0.0" ± 0.08" 0" ± 3/32"	0.0" ± 0.08" 0" ± 3/32"	0.0" ± 0.08" 0" ± 3/32"
THRUST ANGLE	0.33 (11/32)	0.33 (11/32)	0.33 (11/32)	0.33 (11/32)	0.33 (11/32)	0.33 (11/32)
ILLUSTRATION #	06	06	06	06	06	06
NOTES						
	42053	42054	42055	42056	42057	42058

ALIGNMENT SPECIFICATIONS

Subaru

Legacy Sedan exc/spec.B, 3.0R -LTD, & '07 Outback	Legacy Sedan 2.5 GT spec.B 3.0R -LTD	Legacy Wagon exc/Outback	MODEL INFORMATION	Legacy w/16" or P215/50R17 tires	Legacy w/18" or P225/50R17 tires	Legacy Sedan
2007-2009	2007-2009	2007	YEARS	2010-2014	2010-2014	2015-2017
			INSTRUCTIONS			
89 ft.lbs.	89 ft.lbs.	89 ft.lbs.	LUGNUT TORQUE	89 ft.lbs.	89 ft.lbs.	89 ft.lbs.
			FRONT			
5.92 ±0.75 5 15/16 ± ¾	5.92 ±0.75 5 15/16 ± ¾	5.67 ±0.75 5 11/16 ± ¾	CASTER (Degrees)	5.92 ±0.75 5 15/16 ± ¾	6.0 ±0.75 6 ± ¾	5.83 ±0.75 5 13/16 ± ¾
0.0 (0) | n/s	0.0 (0) | n/s	0.0 (0) | n/s	CROSS | TOL.	0.0 (0) | n/s	0.0 (0) | n/s	0.0 (0) | n/s
-0.25 ±0.5 -¼ ± ½	-0.25 ±0.5 -¼ ± ½	-0.25 ±0.5 -¼ ± ½	CAMBER (Degrees)	-0.5 ±0.5 -½ ± ½	-0.25 ±0.5 -¼ ± ½	-0.25 ±0.5 -¼ ± ½
0.0 (0) | 0.75 (¾)	0.0 (0) | 0.75 (¾)	0.0 (0) | 0.75 (¾)	CROSS | TOL.	0.0 (0) | 0.75 (¾)	0.0 (0) | 0.75 (¾)	0.0 (0) | 0.75 (¾)
0.0° ±0.17° 0° ± 5/32°	0.0° ±0.17° 0° ± 5/32°	0.0° ±0.17° 0° ± 5/32°	FRONT TOE (Degrees)	0.08° ±0.16° [1] 3/32° ± 5/32° [1]	0.08° ±0.16° [1] 3/32° ± 5/32° [1]	0.0° ±0.16° 0° ± 5/32°
0.0" ±0.08" 0" ± 3/32"	0.0" ±0.08" 0" ± 3/32"	0.0" ±0.08" 0" ± 3/32"	FRONT TOE (Inches)	0.04" ±0.08" [1] 1/32" ± 3/32" [1]	0.04" ±0.08" [1] 1/32" ± 3/32" [1]	0.0" ±0.08" 0" ± 3/32"
13.8 (13¾)	13.8 (13¾)	13.8 (13¾)	SAI	14.0 (14)	14.3 (14 ¼)	13.9 (13 15/16)
37.6 / 33.0 ±1.5	33.8 / 30.2 ±1.5	37.6 / 33.0 ±1.5	TURNING ANGLE	37.6 / 33.3 ±1.5	37.4 / 33.0 ±1.5	37.6 / 32.8 ±1.5
22	22	22	ILLUSTRATION #	22	22	22
			REAR			
-0.67 ±0.75 -11/16 ± ¾	-0.67 ±0.75 -11/16 ± ¾	-0.5 ±0.75 -½ ± ¾	CAMBER (Degrees)	-1.0 ±0.75 -1 ± ¾	-1.17 ±0.75 -1 3/16 ± ¾	-1.42 ±0.75 -1 7/16 ± ¾
0.0° ±0.16° 0° ± 5/32°	0.0° ±0.16° 0° ± 5/32°	0.0° ±0.16° 0° ± 5/32°	REAR TOE (Degrees)	0.0° ±0.16° 0° ± 5/32°	0.0° ±0.16° 0° ± 5/32°	0.24° ±0.16° ¼° ± 5/32°
0.0" ±0.08" 0" ± 3/32"	0.0" ±0.08" 0" ± 3/32"	0.0" ±0.08" 0" ± 3/32"	REAR TOE (Inches)	0.0" ±0.08" 0" ± 3/32"	0.0" ±0.08" 0" ± 3/32"	0.12" ±0.08" ⅛" ± 3/32"
0.33 (11/32)	0.33 (11/32)	0.33 (11/32)	THRUST ANGLE	0.33 (11/32)	0.33 (11/32)	0.33 (11/32)
06	06	06	ILLUSTRATION #	174	174	174
			NOTES	[1] 2010-11 = 0° (0")	[1] 2010-11 = 0° (0")	
42059	42060	42061		42062	42063	42064

Subaru

Outback	Outback	Outback	MODEL INFORMATION	Outback	Outback	Outback
1995-99	2000-2004	2005-2006	YEARS	2007-2009	2010-2014	2015-2017
			INSTRUCTIONS			
	65 ft.lbs.	89 ft.lbs.	LUGNUT TORQUE	89 ft.lbs.	89 ft.lbs.	89 ft.lbs.
			FRONT			
2.83 ±1.0 2 13/16 ±1	2.67 ±0.75 2 11/16 ± ¾	4.92 ±0.75 5 15/16 ± ¾	CASTER (Degrees)	4.92 ±0.75 5 15/16 ± ¾	5.08 ±0.75 5 1/16 ± ¾	5.0 ±0.75 5 ± ¾
0.0 (0) | n/s	0.0 (0) | n/s	0.0 (0) | n/s	CROSS | TOL.	0.0 (0) | n/s	0.0 (0) | n/s	0.0 (0) | n/s
0.33 ±0.5 5/16 ± ½	0.33 ±0.5 5/16 ± ½	0.67 ±0.5 11/16 ± ½	CAMBER (Degrees)	0.67 ±0.5 11/16 ± ½	0.0 ±0.5 0 ± ½	0.0 ±0.5 0 ± ½
0.0 (0) | n/s	0.0 (0) | n/s	0.0 (0) | 0.75 (¾)	CROSS | TOL.	0.0 (0) | 0.75 (¾)	0.0 (0) | 0.75 (¾)	0.0 (0) | 0.75 (¾)
0.0° ±0.3° 0° ± 5/16°	0.0° ±0.16° 0° ± 5/32°	0.0° ±0.16° 0° ± 5/32°	FRONT TOE (Degrees)	0.0° ±0.16° 0° ± 5/32°	0.08° ±0.16° [1] 3/32° ± 5/32° [1]	0.0° ±0.16° 0° ± 5/32°
0.0" ±0.15" 0" ± 5/32"	0.0" ±0.08" 0" ± 3/32"	0.0" ±0.08" 0" ± 3/32"	FRONT TOE (Inches)	0.0" ±0.08" 0" ± 3/32"	0.04" ±0.08" [1] 1/32" ± 3/32" [1]	0.0" ±0.08" 0" ± 3/32"
13.5 ±1.0 (13 ½ ±1)	13.5 ±1.0 (13 ½ ±1)	12.1 (12 1/16)	SAI	12.1 (12 1/16)	12.3 (12 5/16)	12.3 (12 ¼)
	34.5 / 30.3 ±1.5	38.0 / 33.7 ±1.5	TURNING ANGLE	37.6 / 33.0 ±1.5	38.4 / 34.1 ±1.5	38.6 / 33.8 ±1.5
22	22	22	ILLUSTRATION #	22	22	22
			REAR			
-0.58 ±0.75 -9/16 ± ¾	-0.17 ±0.75 -3/16 ± ¾	-0.17 ±0.75 -3/16 ± ¾	CAMBER (Degrees)	-0.17 ±0.75 -3/16 ± ¾	0.0 ±0.75 0 ± ¾	-1.0 ±0.75 -1 ± ¾
0.0° ±0.3° 0° ± 5/16°	0.0° ±0.17° 0° ± 5/32°	**-0.13°** ±0.13° **-⅛°** ± ⅛°	REAR TOE (Degrees)	**-0.13°** ±0.13° **-⅛°** ± ⅛°	0.17° ±0.25° 5/32° ± ¼°	0.24° ±0.16° ¼° ± 5/32°
0.0" ±0.15" 0" ± 5/32"	0.0" ±0.08" 0" ± 3/32"	**-0.06"** ±0.06" **-1/16"** ± 1/16"	REAR TOE (Inches)	**-0.06"** ±0.06" **-1/16"** ± 1/16"	0.08" ±0.13" 3/32" ± ⅛"	0.12" ±0.08" ⅛" ± 3/32"
0.33 (11/32)	0.33 (11/32)	0.33 (11/32)	THRUST ANGLE	0.33 (11/32)	0.5 (½)	0.33 (11/32)
06	06	06	ILLUSTRATION #	06	174	174
			NOTES		[1] 2010-11 = 0° (0")	
42065	42066	42067		42068	42069	42070

Subaru

Subaru 2WD -89 Loyale 2WD 90- exc/Wagon	Subaru 4WD -89 Loyale 4WD 90- exc/Wagon	Subaru Wagon Loyale Wagon 2WD	MODEL INFORMATION	Subaru Wagon Loyale Wagon 4WD	SVX	Tribeca (B9 Tribeca)
1990-93	1990-93	1990-94	YEARS	1990-94	1992-97	2006-2009
			INSTRUCTIONS			
58-72 ft.lbs.	58-72 ft.lbs.	58-72 ft.lbs.	LUGNUT TORQUE	58-72 ft.lbs.	72-86 ft.lbs.	89 ft.lbs.[1]
			FRONT			
2.5 ± 0.75 2½ ± ¾	1.83 ± 0.75 1 13/16 ± ¾	2.08 ± 0.75 2 1/16 ± ¾	CASTER (Degrees)	1.58 ± 0.75 1 9/16 ± ¾	4.83 ± 0.75 4 13/16 ± ¾	4.07 ± 0.75 4 1/16 ± ¾
0.0 (0) | n/s	0.0 (0) | n/s	0.0 (0) | n/s	CROSS | TOL.	0.0 (0) | n/s	0.0 (0) | n/s	0.0 (0) | n/s
0.75 ± 0.75 ¾ ± ¾	1.67 ± 0.75 1 11/16 ± ¾	1.0 ± 0.75 1 ± ¾	CAMBER (Degrees)	1.75 ± 0.75 1¾ ± ¾	-0.42 ± 0.75 -7/16 ± ¾	0.0 ± 0.5 0 ± ½
0.0 (0) | n/s	0.0 (0) | n/s	0.0 (0) | n/s	CROSS | TOL.	0.0 (0) | n/s	0.0 (0) | n/s	0.0 (0) | 0.75 (¾)
0.2° ± 0.1° 3/16° ± 3/32°	-0.5° ± 0.1° -½° ± 3/32°	0.2° ± 0.1° 3/16° ± 3/32°	FRONT TOE (Degrees)	-0.5° ± 0.1° -½° ± 3/32°	0.0° ± 0.3° 0° ± 5/16°	0.0° ± 0.16° 0° ± 5/32°
0.1" ± 0.05" 3/32" ± 1/16"	-0.25" ± 0.05" -¼" ± 1/16"	0.1" ± 0.05" 3/32" ± 1/16"	FRONT TOE (Inches)	-0.25" ± 0.05" -¼" ± 1/16"	0.0" ± 0.15" 0" ± 5/32"	0.0" ± 0.08" 0" ± 3/32"
			SAI		16.0 (16)	12.2 (12 3/16)
			TURNING ANGLE		36.2 / 31.6 ±1.5	37.0 / 32.0 ±1.5
			ILLUSTRATION #		22	22
			REAR			
0.0 ± 0.5 0 ± ½	0.0 ± 0.5 0 ± ½	0.0 ± 0.5 0 ± ½	CAMBER (Degrees)	0.0 ± 0.5 0 ± ½	-0.67 ± 0.75 -11/16 ± ¾	-0.52 ± 0.75 -½ ± ¾
0.0° ± 0.2° 0° ± 3/16°	0.0° ± 0.2° 0° ± 3/16°	0.0° ± 0.2° 0° ± 3/16°	REAR TOE (Degrees)	0.0° ± 0.2° 0° ± 3/16°	0.0° ± 0.3° 0° ± 5/16°	0.16° ± 0.12° 5/32° ± ⅛°
0.0" ± 0.1" 0" ± 3/32"	0.0" ± 0.1" 0" ± 3/32"	0.0" ± 0.1" 0" ± 3/32"	REAR TOE (Inches)	0.0" ± 0.1" 0" ± 3/32"	0.0" ± 0.15" 0" ± 5/32"	0.08" ± 0.06" 3/32" ± 1/16"
0.33 (11/32)	0.33 (11/32)	0.33 (11/32)	THRUST ANGLE	0.33 (11/32)	0.33 (11/32)	0.33 (11/32)
61	61	61	ILLUSTRATION #	61	06	174
			NOTES			[1] Chrome wheel = 111 ft.lbs
42085	42086	42087		42088	42089	42090

Subaru / Suzuki

Tribeca	XT 4X2	XT 4X4	MODEL INFORMATION	Aerio FWD & AWD	Aerio FWD & AWD	Equator 2WD
2010-2014	1990-91	1990-91	YEARS	2002-2003	2004-2007	2009-2012
			INSTRUCTIONS			
89 ft.lbs.[1]			LUGNUT TORQUE	62 ft.lbs.	62 ft.lbs.	98 ft.lbs.
			FRONT			
3.67 ± 0.75 3 11/16 ± ¾	4.08 ± 0.75 4 1/16 ± ¾	3.42 ± 0.75 [1] 3 7/16 ± ¾ [1]	CASTER (Degrees)	2.18 ± 2.0 2 3/16 ± 2	2.18 ± 2.0 2 3/16 ± 2	3.0 ± 0.75 3 ± ¾
0.0 (0) | n/s	0.0 (0) | n/s	0.0 (0) | n/s	CROSS | TOL.	0.0 (0) | 1.0 (1)	0.0 (0) | 1.0 (1)	0.0 (0) | 0.75 (¾)
0.17 ± 0.5 3/16 ± ½	0.0 ± 0.75 0 ± ¾	0.67 ± 0.75 [2] 11/16 ± ¾ [2]	CAMBER (Degrees)	0.17 ± 1.0 3/16 ± 1	-0.17 ± 1.0 -3/16 ± 1	0.25 ± 0.75 ¼ ± ¾
0.0 (0) | 0.75 (¾)	0.0 (0) | n/s	0.0 (0) | n/s	CROSS | TOL.	0.0 (0) | n/s	0.0 (0) | n/s	0.0 (0) | 0.75 (¾)
0.0° ± 0.16° 0° ± 5/32°	0.0° ± 0.1° 0° ± 3/32°	-0.5° ± 0.1° -½° ± 3/32°	FRONT TOE (Degrees)	0.0° ± 0.16° 0° ± 5/32°	0.08° ± 0.16° 3/32° ± 5/32°	0.24° ± 0.08° ¼° ± 3/32°
0.0" ± 0.08" 0" ± 3/32"	0.0" ± 0.05" 0" ± 1/16"	-0.25" ± 0.05" -¼" ± 1/16"	FRONT TOE (Inches)	0.0" ± 0.08" 0" ± 3/32"	0.04" ± 0.08" 1/32" ± 3/32"	0.12" ± 0.04" ⅛" ± 1/32"
12.0 (11 15/16)			SAI	12.9 ± 3.0 (12 7/8 ± 3)	12.9 ± 3.0 (12 7/8 ± 3)	13.0 (13)
37.0 / 32.0 ±1.5			TURNING ANGLE	37.5 / 32.5 ±3.0	37.5 / 32.5 ±3.0	35.4 / 31.4 ±1.0
22			ILLUSTRATION #			12
			REAR			
-0.25 ± 0.75 -¼ ± ¾	0.0 ± 0.5 0 ± ½	-0.17 ± 0.75 -3/16 ± ¾	CAMBER (Degrees)	0.6 ± 1.0 ⅝ ± 1	-1.25 ± 1.0 -1¼ ± 1	
0.16° ± 0.12° 5/32° ± ⅛°	0.0° ± 0.2° 0° ± 3/16°	0.0° ± 0.2° 0° ± 3/16°	REAR TOE (Degrees)	0.2° ± 0.16° 3/16° ± 5/32°	0.2° ± 0.16° 3/16° ± 5/32°	
0.08" ± 0.06" 3/32" ± 1/16"	0.0" ± 0.1" 0" ± 3/32"	0.0" ± 0.1" 0" ± 3/32"	REAR TOE (Inches)	0.1" ± 0.08" 3/32" ± 3/32"	0.1" ± 0.08" 3/32" ± 3/32"	
0.33 (11/32)	0.33 (11/32)	0.33 (11/32)	THRUST ANGLE			
174	61	61	ILLUSTRATION #	06	06	
[1] Chrome wheel = 111 ft.lbs		[1] 6-cyl = 3.5° (3½°) [2] 6-cyl = 0.8° (13/16°)	NOTES			
42091	42092	42094		43001	43002	43003

Suzuki

MODEL INFORMATION	Equator 4WD	Esteem	Forenza	Grand Vitara RWD & 4WD	Grand Vitara RWD & 4WD	Grand Vitara
YEARS	2009-2012	1995-2002	2004-2008	1999-2002	2003-2005	2006-2013
INSTRUCTIONS						
LUGNUT TORQUE	98 ft.lbs.	62 ft.lbs.	88 ft.lbs.	73 ft.lbs.	73 ft.lbs.	73 ft.lbs.
FRONT						
CASTER (Degrees)	2.75±0.75 2¾±¾	2.7±2.0 2 11/16±2	4.0±0.75 4±¾	2.67±1.0 2 11/16±1	2.67±1.0 2 11/16±1	2.5±1.0 2½±1
CROSS | TOL.	0.0 (0) | 0.75 (¾)	0.0 (0) | n/s	0.0 (0) | n/s	0.0 (0) | n/s	0.0 (0) | n/s	0.0 (0) | n/s
CAMBER (Degrees)	0.5±0.75 ½±¾	0.0±1.0 0±1	-0.33±0.75 - 5/16±¾	0.0±1.0 0±1	0.0±1.0 0±1	0.0±1.0 0±1
CROSS | TOL.	0.0 (0) | 0.75 (¾)	0.0 (0) | n/s	0.0 (0) | n/s	0.0 (0) | n/s	0.0 (0) | n/s	0.0 (0) | n/s
FRONT TOE (Degrees)	0.24°±0.08° ¼°±3/32°	0.0°±0.16° 0°±5/32°	0.0°±0.17° 0°±5/32°	0.0°±0.16° 0°±5/32°	0.1°±0.16° 3/32°±5/32°	0.0°±0.16° 0°±5/32°
FRONT TOE (Inches)	0.12"±0.04" ⅛"±1/32"	0.0"±0.08" 0"±3/32"	0.0"±0.08" 0"±3/32"	0.0"±0.08" 0"±3/32"	0.05"±0.08" 1/16"±3/32"	0.0"±0.08" 0"±3/32"
SAI	12.8 (12¾)	12.6±3.0 (12 5/8±3)				
TURNING ANGLE	35.6 / 31.7 ±1.0	39.0 / 34.0 ±3.0		34.5 / 32.5 ±3.0	34.5 / 32.5 ±3.0	37.0 / 32.0 ±3.0
ILLUSTRATION #	12					
REAR						
CAMBER (Degrees)		0.5±1.0 ½±1	-1.0±0.75 -1±¾			-1.25±0.5 -1¼±½
REAR TOE (Degrees)		0.2°±0.16° 3/16°±5/32°	0.2°±0.17° 3/16°±5/32°			0.48°±0.16° 15/32°±5/32°
REAR TOE (Inches)		0.1"±0.08" 3/32"±3/32"	0.1"±0.08" 3/32"±3/32"			0.24"±0.08" ¼"±3/32"
THRUST ANGLE						
ILLUSTRATION #		06	06			177
NOTES						
	43004	43005	43006	43007	43008	43009

Suzuki

MODEL INFORMATION	Kizashi	Reno	Samurai	Sidekick All Models	Swift	Swift
YEARS	2010-2013	2005-2008	1990-95	1990-98	1990-94	1995-99
INSTRUCTIONS						
LUGNUT TORQUE	104 ft.lbs.	88 ft.lbs.		70 ft.lbs.[1]	44 ft.lbs.	44 ft.lbs.
FRONT						
CASTER (Degrees)	4.17±1.0 4 3/16±1	4.0±0.75 4±¾	3.5±1.0 3½±1	1.5±1.0 1½±1	3.0±2.0 3±2	3.0±2.0 3±2
CROSS | TOL.	0.0 (0) | n/s	0.0 (0) | n/s	0.0 (0) | n/s	0.0 (0) | n/s	0.0 (0) | 1.0 (1)	0.0 (0) | 1.0 (1)
CAMBER (Degrees)	-0.5±1.0 -½±1	-0.33±0.75 - 5/16±¾	1.0±0.75 1±¾	0.5±1.0 ½±1	0.0±1.0 0±1	0.0±1.0 0±1
CROSS | TOL.	0.0 (0) | n/s	0.0 (0) | n/s	0.0 (0) | n/s	0.0 (0) | n/s	0.0 (0) | 0.5 (½)	0.0 (0) | 0.5 (½)
FRONT TOE (Degrees)	0.0°±0.17° 0°±5/32°	0.0°±0.17° 0°±5/32°	0.3°±0.16° 5/16°±5/32°	0.3°±0.16° 5/16°±5/32°	0.0°±0.16° 0°±5/32°	0.16°±0.16° 5/32°±5/32°
FRONT TOE (Inches)	0.0"±0.08" 0"±3/32"	0.0"±0.08" 0"±3/32"	0.15"±0.08" 5/32"±3/32"	0.15"±0.08" 5/32"±3/32"	0.0"±0.08" 0"±3/32"	0.08"±0.08" 3/32"±3/32"
SAI	13.3±2.0 (13¼±2)					
TURNING ANGLE	38.8 / 30.6 ±2.0			32.5 / 30.5 ±2.0	38.0 / 32.0 ±2.0	38.0 / 32.0 ±2.0
ILLUSTRATION #						
REAR						
CAMBER (Degrees)	-1.25±0.5 -1¼±½	-1.0±0.75 -1±¾			0.0±1.0 0±1	0.0±1.0 0±1
REAR TOE (Degrees)	0.16°±0.08° 5/32°±3/32°	0.2°±0.17° 3/16°±5/32°			0.16°±0.16° 5/32°±5/32°	0.16°±0.16° 5/32°±5/32°
REAR TOE (Inches)	0.08"±0.04" 3/32"±1/32"	0.1"±0.08" 3/32"±3/32"			0.08"±0.08" 3/32"±3/32"	0.08"±0.08" 3/32"±3/32"
THRUST ANGLE					0.15 (5/32)	0.15 (5/32)
ILLUSTRATION #	177	06			06	06
NOTES				[1] 1990-91 = 60 ft.lbs.		
	43010	43011	43012	43013	43014	43015

Suzuki

Swift	Swift/Swift+	SX4 All exc/Sedan w/17" wheels	MODEL INFORMATION	SX4 Sedan w/17" wheels	Verona	Vitara RWD & 4WD
2000-2001	2004-2009	2007-2013	YEARS	2008-2013	2004-2006	1999-2004
			INSTRUCTIONS			
44-48 ft.lbs.	88 ft.lbs.	62 ft.lbs.	LUGNUT TORQUE	62 ft.lbs.	81 ft.lbs.	73 ft.lbs.
			FRONT			
3.0 ±2.0 3 ±2	2.5 ±0.75 [1] 2½ ± ¾ [1]	3.67 ±2.0 3 11/16 ±2	CASTER (Degrees)	3.77 ±2.0 3¾ ±2	3.0 ±1.0 [1] 3 ±1 [1]	2.67 ±1.0 2 11/16 ±1
0.0 (0) | 1.0 (1)	0.0 (0) | 1.0 (1)	0.0 (0) | n/s	CROSS | TOL.	0.0 (0) | n/s	0.0 (0) | 1.0 (1)	0.0 (0) | n/s
0.5 ±1.0 ½ ±1	-0.4 ±0.75 [1] -⅜ ± ¾ [1]	-0.17 ±1.0 -3/16 ±1	CAMBER (Degrees)	-0.28 ±1.0 -5/16 ±1	-0.5 ±1.0 [1] -½ ±1 [1]	0.0 ±1.0 0 ±1
0.0 (0) | n/s	0.0 (0) | 0.5 (½)	0.0 (0) | n/s	CROSS | TOL.	0.0 (0) | n/s	0.0 (0) | 1.0 (1)	0.0 (0) | n/s
0.16° ±0.16° 5/32° ± 5/32°	0.07° ±0.17° [1] 1/16° ± 5/32° [1]	0.03° ±0.07° 1/32° ± 1/16°	FRONT TOE (Degrees)	0.1° ±0.07° 3/32° ± 1/16°	0.2° ±0.33° [1] 3/16° ± 11/32° [1]	0.0° ±0.2° 0° ± 3/16°
0.08" ±0.08" 3/32" ± 3/32"	0.04" ±0.08" [1] 1/32" ± 3/32" [1]	0.02" ±0.03" 1/64" ± 1/32"	FRONT TOE (Inches)	0.05" ±0.03" 1/16" ± 1/32"	0.1" ±0.17" [1] 3/32" ± 5/32" [1]	0.0" ±0.1" 0" ± 3/32"
12.9 (12 15/16)		12.3 (12¼)	SAI	12.5 (12½)		
38.0 / 32.0 ±2.0		36.0 / 32.1 ±2.0	TURNING ANGLE	35.7 / 31.9 ±2.0		34.5 / 32.5 ±3.0
			ILLUSTRATION #			
			REAR			
0.0 ±1.0 0 ±1	-1.5 ±0.5 [1] -1½ ± ½ [1]	-1.0 ±1.0 -1 ±1	CAMBER (Degrees)	-1.0 ±1.0 -1 ±1	-1.5 ±1.0 [1] -1½ ±1 [1]	
0.24° ±0.16° ¼° ± 5/32°	0.25° ±0.33° [1] ¼° ± 11/32° [1]	0.57° ±0.57° 9/16° ± 9/16°	REAR TOE (Degrees)	0.57° ±0.57° 9/16° ± 9/16°	0.3° ±0.33° [1] 5/16° ± 11/32° [1]	
0.12" ±0.08" ⅛" ± 3/32"	0.13" ±0.17" [1] ⅛" ± 5/32" [1]	0.28" ±0.28" 9/32" ± 9/32"	REAR TOE (Inches)	0.28" ±0.28" 9/32" ± 9/32"	0.15" ±0.17" [1] 5/32" ± 5/32" [1]	
0.15 (5/32)			THRUST ANGLE		0.25 (¼)	
06			ILLUSTRATION #		05	
	[1] Load 150lbs each fr. seat.		NOTES		[1] Load 150lbs each fr. seat.	
43016	43017	43018		43019	43020	43021

Suzuki / Toyota

X90	XL7 RWD & 4WD	XL7	MODEL INFORMATION	86	Avalon	Avalon
1996-98	2001-2006	2007-2009	YEARS	2017	1995-96	1997-99
			INSTRUCTIONS			
70 ft.lbs.	73 ft.lbs.	100 ft.lbs.	LUGNUT TORQUE	89 ft.lbs.		
			FRONT			
1.5 ±1.0 1½ ±1	2.5 ±1.0 2½ ±1	3.0 ±0.75 3 ± ¾	CASTER (Degrees)	5.9 ±0.75 5⅞ ± ¾	1.17 ±0.75 1 3/16 ± ¾	2.18 ±0.75 2 3/16 ± ¾
0.0 (0) | n/s	0.0 (0) | n/s	0.0 (0) | 0.75 (¾)	CROSS | TOL.	0.0 (0) | 0.5 (½)	0.0 (0) | 0.75 (¾)	0.0 (0) | 0.75 (¾)
0.5 ±1.0 ½ ±1	0.0 ±1.0 0 ±1	-0.6 ±0.75 -⅝ ± ¾	CAMBER (Degrees)	0.0 ±0.75 0 ± ¾	-0.58 ±0.75 -9/16 ± ¾	-0.62 ±0.75 -⅝ ± ¾
0.0 (0) | n/s	0.0 (0) | n/s	0.0 (0) | 0.75 (¾)	CROSS | TOL.	0.0 (0) | 0.75 (¾)	0.0 (0) | 0.75 (¾)	0.0 (0) | 0.75 (¾)
0.3° ±0.2° 5/16° ± 3/16°	0.1° ±0.2° 3/32° ± 3/16°	0.15° ±0.2° 5/32° ± 3/16°	FRONT TOE (Degrees)	0.0° ±0.18° 0° ± 3/16°	0.0° ±0.2° 0° ± 3/16°	0.0° ±0.2° 0° ± 3/16°
0.15" ±0.1" 5/32" ± 3/32"	0.05" ±0.1" 1/16" ± 3/32"	0.08" ±0.1" 1/16" ± 3/32"	FRONT TOE (Inches)	0.0" ±0.09" 0" ± 3/32"	0.0" ±0.1" 0" ± 3/32"	0.0" ±0.1" 0" ± 3/32"
			SAI		13.1 ±0.8 (13 1/16 ± ¾)	13.1 ±0.8 (13 1/16 ± ¾)
32.5 / 30.5 ±2.0	35.0 / 33.0 ±3.0		TURNING ANGLE	36.9 / 31.2 ±2.0	36.0 / 31.3 ±2.0	35.8 / 31.4 ±2.0
		25	ILLUSTRATION #	26	26	26
			REAR			
		-0.5 ±0.75 -½ ± ¾	CAMBER (Degrees)	-1.2 ±0.75 -1 3/16 ± ¾	-0.75 ±0.75 -¾ ± ¾	-0.77 ±0.75 -¾ ± ¾
		0.2° ±0.2° 3/16° ± 3/16°	REAR TOE (Degrees)	0.17° ±0.25° 5/32° ± ¼°	0.4° ±0.2° 13/32° ± 3/16°	0.4° ±0.2° 13/32° ± 3/16°
		0.1" ±0.1" 3/32" ± 3/32"	REAR TOE (Inches)	0.08" ±0.13" 3/32" ± ⅛"	0.2" ±0.1" 3/16" ± 3/32"	0.2" ±0.1" 3/16" ± 3/32"
		0.2 (3/16)	THRUST ANGLE			
		107 143	ILLUSTRATION #	174	10	10
			NOTES			
43022	43023	43024		44001	44002	44003

ALIGNMENT SPECIFICATIONS

Toyota

Avalon	Avalon XL	Avalon XLS	MODEL INFORMATION	Avalon Touring	Avalon Limited	Avalon Avalon HV
2000-2004	2005-2010	2005-2012	YEARS	2005-2008	2005-2012	2013-2017
			INSTRUCTIONS			
			LUGNUT TORQUE			
			FRONT			
2.17 ±0.75 2 3/16 ± 3/4	2.65 ±0.75 2 5/8 ± 3/4	2.7 ±0.75 2 11/16 ± 3/4	CASTER (Degrees)	2.72 ±0.75 2 3/4 ± 3/4	2.8 ±0.75 2 13/16 ± 3/4	2.87 ±0.75 2 7/8 ± 3/4
0.0 (0) | 0.75 (3/4)	0.0 (0) | 0.75 (3/4)	0.0 (0) | 0.75 (3/4)	CROSS | TOL.	0.0 (0) | 0.75 (3/4)	0.0 (0) | 0.75 (3/4)	0.0 (0) | 0.75 (3/4)
-0.62 ±0.75 -5/8 ± 3/4	-0.67 ±0.75 -11/16 ± 3/4	-0.72 ±0.75 -3/4 ± 3/4	CAMBER (Degrees)	-0.72 ±0.75 -3/4 ± 3/4	-0.72 ±0.75 --3/4 ± 3/4	-0.72 ±0.75 -3/4 ± 3/4
0.0 (0) | 0.75 (3/4)	0.0 (0) | 0.75 (3/4)	0.0 (0) | 0.75 (3/4)	CROSS | TOL.	0.0 (0) | 0.75 (3/4)	0.0 (0) | 0.75 (3/4)	0.0 (0) | 0.75 (3/4)
0.0° ±0.2° 0° ± 3/16°	0.0° ±0.1° 0° ± 3/32°	0.0° ±0.1° 0° ± 3/32°	FRONT TOE (Degrees)	0.0° ±0.1° 0° ± 3/32°	0.0° ±0.1° 0° ± 3/32°	0.02° ±0.18° 1/32° ± 3/16°
0.0" ±0.1" 0" ± 3/32"	0.0" ±0.05" 0" ± 1/16"	0.0" ±0.05" 0" ± 1/16"	FRONT TOE (Inches)	0.0" ±0.05" 0" ± 1/16"	0.0" ±0.05" 0" ± 1/16"	0.01" ±0.09" 1/64" ± 3/32"
13.1 ±0.8 (13 1/16 ± 3/4)	12.3 ±0.8 (12 1/4 ± 3/4)	12.3 ±0.8 (12 5/16 ± 3/4)	SAI	12.4 ±0.8 (12 3/8 ± 3/4)	12.3 ±0.8 (12 5/16 ± 3/4)	12.3 ±0.8 (12 5/16 ± 3/4)
35.8 / 31.4 ±2.0	38.6 / 33.8 ±2.0	38.5 / 33.7 ±2.0	TURNING ANGLE	38.5 / 33.7 ±2.0	38.5 / 33.7 ±2.0	37.2 / 32.8 ±2.0
26	26	26	ILLUSTRATION #	26	26	26
			REAR			
-0.72 ±0.75 -3/4 ± 3/4	-1.15 ±0.75 -1 1/8 ± 3/4	-1.22 ±0.75 -1 1/4 ± 3/4	CAMBER (Degrees)	-1.22 ±0.75 -1 1/4 ± 3/4	-1.25 ±0.75 -1 1/4 ± 3/4	-1.18 ±0.75 -1 3/16 ± 3/4
0.4° ±0.2° 13/32° ± 3/16°	0.4° ±0.2° 13/32° ± 3/16°	0.4° ±0.2° 13/32° ± 3/16°	REAR TOE (Degrees)	0.4° ±0.2° 13/32° ± 3/16°	0.4° ±0.2° 13/32° ± 3/16°	0.22° ±0.18° 7/32° ± 3/16°
0.2" ±0.1" 3/16" ± 3/32"	0.2" ±0.1" 3/16" ± 3/32"	0.2" ±0.1" 3/16" ± 3/32"	REAR TOE (Inches)	0.2" ±0.1" 3/16" ± 3/32"	0.2" ±0.1" 3/16" ± 3/32"	0.11" ±0.09" 1/8" ± 3/32"
			THRUST ANGLE			
10	10	10	ILLUSTRATION #	10	10	10
			NOTES			
44004	44005	44006		44007	44008	44009

Toyota

Camry Sedan	Camry Wagon	Camry Alltrac	MODEL INFORMATION	Camry Sedan, Coupe	Camry Wagon	Camry
1990-91	1990-91	1990-91	YEARS	1992-96	1992-96	1997-2001
			INSTRUCTIONS			
			LUGNUT TORQUE			76 ft.lbs.
			FRONT			
1.67 ±0.5 1 11/16 ± 1/2	1.0 ±0.5 1 ± 1/2	1.67 ±0.5 1 11/16 ± 1/2	CASTER (Degrees)	1.17 ±0.75 1 3/16 ± 3/4	1.08 ±0.75 1 1/16 ± 3/4	2.17 ±0.75 2 3/16 ± 3/4
0.0 (0) | 0.5 (1/2)	0.0 (0) | 0.5 (1/2)	0.0 (0) | 0.5 (1/2)	CROSS | TOL.	0.0 (0) | 0.75 (3/4)	0.0 (0) | 0.75 (3/4)	0.0 (0) | 0.75 (3/4)
0.58 ±0.5 [1] 9/16 ± 1/2 [1]	0.5 ±0.75 1/2 ± 3/4	0.58 ±0.5 9/16 ± 1/2	CAMBER (Degrees)	-0.58 ±0.75 - 9/16 ± 3/4	-0.58 ±0.75 - 9/16 ± 3/4	-0.6 ±0.75 -5/8 ± 3/4
0.0 (0) | 0.5 (1/2)	0.0 (0) | 0.5 (1/2)	0.0 (0) | 0.5 (1/2)	CROSS | TOL.	0.0 (0) | 0.75 (3/4)	0.0 (0) | 0.75 (3/4)	0.0 (0) | 0.75 (3/4)
0.1° ±0.1° 3/32° ± 3/32°	0.1° ±0.1° 3/32° ± 3/32°	0.1° ±0.1° 3/32° ± 3/32°	FRONT TOE (Degrees)	0.0° ±0.2° 0° ± 3/16°	0.0° ±0.2° 0° ± 3/16°	0.0° ±0.2° 0° ± 3/16°
0.05" ±0.05" 1/16" ± 1/16"	0.05" ±0.05" 1/16" ± 1/16"	0.05" ±0.05" 1/16" ± 1/16"	FRONT TOE (Inches)	0.0" ±0.1" 0" ± 3/32"	0.0" ±0.1" 0" ± 3/32"	0.0" ±0.1" 0" ± 3/32"
12.8 ±0.5 (12 3/4 ± 1/2)	12.8 ±0.8 (12 13/16 ± 3/4)	12.8 ±0.8 (12 3/4 ± 3/4)	SAI	13.1 ±0.8 (13 1/16 ± 3/4)	13.0 ±0.8 (13 ± 3/4)	13.1 ±0.8 (13 1/16 ± 3/4)
37.5 / 30.8 ±1.0 [2]	37.5 / 30.8 ±1.0 [1]	37.5 / 30.8 ±1.0	TURNING ANGLE	37.3 / 32.2 ±2.0 [1]	37.3 / 32.3 ±2.0 [1]	35.8 / 31.4 ±2.0 [1]
62 26	62 26	26	ILLUSTRATION #	26	26	26
			REAR			
-0.58 ±0.75 - 9/16 ± 3/4	-0.08 ±0.5 -1/16 ± 1/2	-0.5 ±0.5 -1/2 ± 1/2	CAMBER (Degrees)	-0.43 ±0.75 [2] - 7/16 ± 3/4 [2]	-0.25 ±0.75 -1/4 ± 3/4	-0.7 ±0.75 -11/16 ± 3/4
0.4° ±0.1° 13/32° ± 3/32°	0.4° ±0.1° 13/32° ± 3/32°	0.3° ±0.1° 5/16° ± 3/32°	REAR TOE (Degrees)	0.4° ±0.2° 13/32° ± 3/16°	0.4° ±0.2° 13/32° ± 3/16°	0.4° ±0.2° 13/32° ± 3/16°
0.2" ±0.05" 3/16" ± 1/16"	0.2" ±0.05" 3/16" ± 1/16"	0.15" ±0.05" 5/32" ± 1/16"	REAR TOE (Inches)	0.2" ±0.1" 3/16" ± 3/32"	0.2" ±0.1" 3/16" ± 3/32"	0.2" ±0.1" 3/16" ± 3/32"
			THRUST ANGLE			
06	06	05	ILLUSTRATION #	10	10	10
[1] Eccentric adjustable up to 12/87 [2] 6 cyl = 35.8°/29.8°	[1] 6 cyl = 35.8°/29.8°		NOTES	[1] w/205/65R15 36°/31.3° [2] 1992-93 = -0.5° (-1/2°)	[1] w/205/65R15 36°/31.3°	[1] 1997-99 = 37.2°/32.4°
44012	44014	44015		44016	44017	44018

Toyota

Camry	Camry	Camry	MODEL INFORMATION	Camry	Camry HV	Camry HV
		L4		V-6		
2002-2006	2007-2011	2012-2017	YEARS	2012-2017	2007-2011	2012-2017
			INSTRUCTIONS			
76 ft.lbs.	76 ft.lbs.	76 ft.lbs.	LUGNUT TORQUE	76 ft.lbs.	76 ft.lbs.	76 ft.lbs.
			FRONT			
2.65 ±0.75 2⅝ ± ¾	2.95 ±0.75 2 15/16 ± ¾	2.92 ±0.75 2 15/16 ± ¾	CASTER (Degrees)	2.95 ±0.75 2 15/16 ± ¾	2.92 ±0.75 2 15/16 ± ¾	3.0 ±0.75 3 ± ¾
0.0 (0) | 0.75 (¾)	0.0 (0) | 0.75 (¾)	0.0 (0) | 0.75 (¾)	CROSS | TOL.	0.0 (0) | 0.75 (¾)	0.0 (0) | 0.75 (¾)	0.0 (0) | 0.75 (¾)
-0.72 ±0.75 -¾ ± ¾	-0.67 ±0.75 -11/16 ± ¾	-0.69 ±0.75 -11/16 ± ¾	CAMBER (Degrees)	-0.66 ±0.75 -11/16 ± ¾	-0.67 ±0.75 -11/16 ± ¾	-0.67 ±0.75 -11/16 ± ¾
0.0 (0) | 0.75 (¾)	0.0 (0) | 0.75 (¾)	0.0 (0) | 0.75 (¾)	CROSS | TOL.	0.0 (0) | 0.75 (¾)	0.0 (0) | 0.75 (¾)	0.0 (0) | 0.75 (¾)
0.0° ±0.2° 0° ± 3/16°	0.0° ±0.2° 0° ± 3/16°	0.0° ±0.17° 0° ± 5/32°	FRONT TOE (Degrees)	0.0° ±0.17° 0° ± 5/32°	0.0° ±0.2° 0° ± 3/16°	0.0° ±0.17° 0° ± 5/32°
0.0" ±0.1" 0" ± 3/32"	0.0" ±0.1" 0" ± 3/32"	0.0" ±0.08" 0" ± 3/32"	FRONT TOE (Inches)	0.0" ±0.08" 0" ± 3/32"	0.0" ±0.1" 0" ± 3/32"	0.0" ±0.08" 0" ± 3/32"
11.5 ±0.8 (11 7/16 ± ¾)	12.3 ±0.8 (12¼ ± ¾)	12.3 ±0.8 (12¼ ± ¾)	SAI	12.3 ±0.8 (12¼ ± ¾)	12.3 ±0.8 (12¼ ± ¾)	12.3 ±0.8 (12¼ ± ¾)
39.1 / 33.7 ±2.0 [1]	38.4 / 33.6 ±2.0 [1 2]	37.3 / 32.8 ±2.0	TURNING ANGLE	37.3 / 32.9 ±2.0	38.4 / 33.6 ±2.0	37.2 / 32.8 ±2.0
26	26	26	ILLUSTRATION #	26	26	26
			REAR			
-1.27 ±0.75 -1¼ ± ¾	-1.27 ±0.75 -1¼ ± ¾	-1.2 ±0.75 -1 3/16 ± ¾	CAMBER (Degrees)	-1.24 ±0.75 -1¼ ± ¾	-1.33 ±0.75 -1 5/16 ± ¾	-1.2 ±0.75 -1 3/16 ± ¾
0.4° ±0.2° 13/32° ± 3/16°	0.4° ±0.2° 13/32° ± 3/16°	0.22° ±0.17° 7/32° ± 5/32°	REAR TOE (Degrees)	0.22° ±0.17° 7/32° ± 5/32°	0.4° ±0.2° 13/32° ± 3/16°	0.21° ±0.17° 7/32° ± 5/32°
0.2" ±0.1" 3/16" ± 3/32"	0.2" ±0.1" 3/16" ± 3/32"	0.11" ±0.08" ⅛" ± 3/32"	REAR TOE (Inches)	0.11" ±0.08" ⅛" ± 3/32"	0.2" ±0.1" 3/16" ± 3/32"	0.11" ±0.08" 3/32" ± 3/32"
			THRUST ANGLE			
10	10	10	ILLUSTRATION #	10	10	10
[1] w/16" tires 36.7°/32.2°	[1] SE 4 Cyl. = 38.7°/33.8° [2] SE 6 Cyl. = 38.3°/33.5°		NOTES			
44019	44020	44021		44022	44023	44024

Toyota

Celica	Celica	Celica	MODEL INFORMATION	Celica	Celica ST	Celica GT, GT-S
exc/Alltrac	Alltrac	exc/Alltrac		Alltrac		
1990-91	1990-91	1992-93	YEARS	1992-93	1994-97	1994-99
			INSTRUCTIONS			
			LUGNUT TORQUE		76 ft.lbs.	76 ft.lbs.
			FRONT			
0.92 ±0.75 [1] 15/16 ± ¾ [1]	0.83 ±0.75 [1] 13/16 ± ¾ [1]	0.92 ±0.75 15/16 ± ¾	CASTER (Degrees)	0.83 ±0.75 13/16 ± ¾	2.12 ±0.75 2⅛ ± ¾	2.08 ±0.75 2 1/16 ± ¾
0.0 (0) | 0.5 (½)	0.0 (0) | 0.5 (½)	0.0 (0) | 0.5 (½)	CROSS | TOL.	0.0 (0) | 0.5 (½)	0.0 (0) | 0.75 (¾)	0.0 (0) | 0.75 (¾)
-0.17 ±0.75 -3/16 ± ¾	-0.17 ±0.75 -3/16 ± ¾	-0.17 ±0.75 -3/16 ± ¾	CAMBER (Degrees)	-0.17 ±0.75 -3/16 ± ¾	-0.85 ±0.75 -⅞ ± ¾	-0.77 ±0.75 -¾ ± ¾
0.0 (0) | 0.5 (½)	0.0 (0) | 0.5 (½)	0.0 (0) | 0.5 (½)	CROSS | TOL.	0.0 (0) | 0.5 (½)	0.0 (0) | 0.75 (¾)	0.0 (0) | 0.75 (¾)
0.0° ±0.1° [2] 0° ± 3/32° [2]	0.0° ±0.1° [2] 0° ± 3/32° [2]	0.0° ±0.2° 0° ± 3/16°	FRONT TOE (Degrees)	0.0° ±0.2° 0° ± 3/16°	0.0° ±0.1° 0° ± 3/32°	0.0° ±0.2° 0° ± 3/16°
0.0" ±0.05" [2] 0" ± 1/16" [2]	0.0" ±0.05" [2] 0" ± 1/16" [2]	0.0" ±0.1" 0" ± 3/32"	FRONT TOE (Inches)	0.0" ±0.1" 0" ± 3/32"	0.0" ±0.05" 0" ± 1/16"	0.0" ±0.1" 0" ± 3/32"
14.2 ±0.8 (14 3/16 ± ¾)	14.2 ±0.8 (14 3/16 ± ¾)	14.2 ±0.8 (14 3/16 ± ¾)	SAI	14.2 ±0.8 (14 3/16 ± ¾)	15.1 ±0.8 (15⅛ ± ¾)	15.0 ±0.8 (15 ± ¾)
33.5 / 29.5	33.5 / 29.5	33.5 / 29.5	TURNING ANGLE	33.5 / 29.5	36.8 / 30.6 ±2.0	36.9 / 30.8 ±2.0
			ILLUSTRATION #		26	26
			REAR			
-0.75 ±0.5 -¾ ± ½	-0.75 ±0.5 -¾ ± ½	-1.25 ±0.75 -1¼ ± ¾	CAMBER (Degrees)	-1.25 ±0.75 -1¼ ± ¾	-1.27 ±0.75 -1¼ ± ¾	-1.17 ±0.75 -1 3/16 ± ¾
0.5° ±0.1° ½° ± 3/32°	0.5° ±0.1° ½° ± 3/32°	0.5° ±0.2° ½° ± 3/16°	REAR TOE (Degrees)	0.5° ±0.2° ½° ± 3/16°	0.35° ±0.1° 11/32° ± 3/32°	0.35° ±0.2° 11/32° ± 3/16°
0.25" ±0.05" ¼" ± 1/16"	0.25" ±0.05" ¼" ± 1/16"	0.25" ±0.1" ¼" ± 3/32"	REAR TOE (Inches)	0.25" ±0.1" ¼" ± 3/32"	0.18" ±0.05" 3/16" ± 1/16"	0.18" ±0.1" 3/16" ± 3/32"
			THRUST ANGLE			
06	05	06	ILLUSTRATION #	05	10	10
[1] 1990 = 1.0° (1°) [2] 1990 = 0.2° (0.1")	[1] 1990 = 1.0° (1°) [2] 1990 = 0.2° (0.1")		NOTES			
44035	44036	44037		44038	44039	44040

ALIGNMENT SPECIFICATIONS

Toyota

MODEL INFORMATION	Celica GT	Celica GT-S w/15" tires	Celica GT-S w/16" tires	Corolla FWD w/4A-F engine exc Coupe	Corolla FWD Coupe exc/4A-GE engine	Corolla FWD w/4A-GE engine All models
YEARS	2000-2005	2000-2005	2000-2005	1990	1990-91	1990
INSTRUCTIONS						
LUGNUT TORQUE	76 ft.lbs.	76 ft.lbs.	76 ft.lbs.	76 ft.lbs.	76 ft.lbs.	76 ft.lbs.
FRONT						
CASTER (Degrees)	2.12 ±0.75 2⅛ ± ¾	2.02 ±0.75 2 ± ¾	2.03 ±0.75 2 1/16 ± ¾	1.33 ±0.75 1 5/16 ± ¾	1.5 ±0.75 1½ ± ¾	1.33 ±0.75 1 5/16 ± ¾
CROSS | TOL.	0.0 (0) | 0.75 (¾)	0.0 (0) | 0.75 (¾)	0.0 (0) | 0.75 (¾)	0.0 (0) | 0.5 (½)	0.0 (0) | 0.5 (½)	0.0 (0) | 0.5 (½)
CAMBER (Degrees)	-0.47 ±0.75 -7/16 ± ¾	-0.42 ±0.75 -7/16 ± ¾	-0.48 ±0.75 -½ ± ¾	-0.17 ±0.75 -3/16 ± ¾	-0.17 ±0.75 [1] -3/16 ± ¾ [1]	-0.25 ±0.75 -¼ ± ¾
CROSS | TOL.	0.0 (0) | 0.75 (¾)	0.0 (0) | 0.75 (¾)	0.0 (0) | 0.75 (¾)	0.0 (0) | 0.5 (½)	0.0 (0) | 0.5 (½)	0.0 (0) | 0.5 (½)
FRONT TOE (Degrees)	0.0° ±0.2° 0° ± 3/16°	0.0° ±0.2° 0° ± 3/16°	0.0° ±0.2° 0° ± 3/16°	0.1° ±0.1° 3/32° ± 3/32°	0.1° ±0.1° 3/32° ± 3/32°	0.1° ±0.1° 3/32° ± 3/32°
FRONT TOE (Inches)	0.0" ±0.1" 0" ± 3/32"	0.0" ±0.1" 0" ± 3/32"	0.0" ±0.1" 0" ± 3/32"	0.05" ±0.05" 1/16" ± 1/16"	0.05" ±0.05" 1/16" ± 1/16"	0.05" ±0.05" 1/16" ± 1/16"
SAI	13.2 ±0.8 (13⅛ ± ¾)	13.1 ±0.8 (13 1/16 ± ¾)	13.2 ±0.8 (13 3/16 ± ¾)	12.7 ±0.8 (12 11/16 ± ¾)	12.7 ±0.8 (12 11/16 ± ¾)	12.8 ±0.8 (12 13/16 ± ¾)
TURNING ANGLE	38.7 / 33.3 ±2.0	38.8 / 33.4 ±2.0	38.6 / 33.3 ±2.0	37.5 / 32.5 ±2.0	37.5 / 32.5 ±2.0	37.0 / 32.0 ±2.0 [1]
ILLUSTRATION #	26	26	26			
REAR						
CAMBER (Degrees)	-1.18 ±0.75 -1 3/16 ± ¾	-1.18 ±0.75 -1 3/16 ± ¾	-1.18 ±0.75 -1 3/16 ± ¾	-0.58 ±0.75 -9/16 ± ¾	-0.68 ±0.75 -11/16 ± ¾	-0.52 ±0.75 -½ ± ¾
REAR TOE (Degrees)	0.3° ±0.2° 5/16° ± 3/16°	0.3° ±0.2° 5/16° ± 3/16°	0.3° ±0.2° 5/16° ± 3/16°	0.4° ±0.1° 13/32° ± 3/32°	0.4° ±0.1° 13/32° ± 3/32°	0.4° ±0.1° 13/32° ± 3/32°
REAR TOE (Inches)	0.15" ±0.1" 5/32" ± 3/32"	0.15" ±0.1" 5/32" ± 3/32"	0.15" ±0.1" 5/32" ± 3/32"	0.2" ±0.05" 3/16" ± 1/16"	0.2" ±0.05" 3/16" ± 1/16"	0.2" ±0.05" 3/16" ± 1/16"
THRUST ANGLE						
ILLUSTRATION #	120	120	120	06	06	
NOTES					[1] 1991 = -0.25° (-¼°)	[1] w/MS 37.5°/32.5°
	44041	44042	44043	44053	44054	44055

Toyota

MODEL INFORMATION	Corolla FWD w/4A-GE engine All models	Corolla FWD w/4A-FE engine exc Coupe (91)	Corolla Alltrac	Corolla All models	Corolla exc/2005 XRS	Corolla XRS
YEARS	1991	1991-92	1990-92	1993-2002	2003-2006	2005-2006
INSTRUCTIONS						
LUGNUT TORQUE	76 ft.lbs.	76 ft.lbs.	76 ft.lbs.	76 ft.lbs.	76 ft.lbs.	76 ft.lbs.
FRONT						
CASTER (Degrees)	1.5 ±0.75 [1] 1½ ± ¾ [1]	1.42 ±0.75 1 7/16 ± ¾	1.25 ±0.5 1¼ ± ½	1.33 ±0.75 1 5/16 ± ¾	2.83 ±0.75 2 13/16 ± ¾	3.08 ±0.75 3 1/16 ± ¾
CROSS | TOL.	0.0 (0) | 0.5 (½)	0.0 (0) | 0.5 (½)	0.0 (0) | 0.5 (½)	0.0 (0) | 0.5 (½)	0.0 (0) | 0.75 (¾)	0.0 (0) | 0.75 (¾)
CAMBER (Degrees)	-0.25 ±0.75 -¼ ± ¾	-0.17 ±0.75 -3/16 ± ¾	0.17 ±0.5 3/16 ± ½	-0.18 ±0.75 -3/16 ± ¾	-0.53 ±0.75 -9/16 ± ¾	-0.63 ±0.75 -⅝ ± ¾
CROSS | TOL.	0.0 (0) | 0.5 (½)	0.0 (0) | 0.5 (½)	0.0 (0) | 0.5 (½)	0.0 (0) | 0.5 (½)	0.0 (0) | 0.75 (¾)	0.0 (0) | 0.75 (¾)
FRONT TOE (Degrees)	0.1° ±0.2° 3/32° ± 3/16°	0.1° ±0.2° 3/32° ± 3/16°	0.1° ±0.1° 3/32° ± 3/32°	0.1° ±0.2° 3/32° ± 3/16°	0.0° ±0.2° 0° ± 3/16°	0.0° ±0.2° 0° ± 3/16°
FRONT TOE (Inches)	0.05" ±0.1" 1/16" ± 3/32"	0.05" ±0.1" 1/16" ± 3/32"	0.05" ±0.05" 1/16" ± 1/16"	0.05" ±0.1" 1/16" ± 3/32"	0.0" ±0.1" 0" ± 3/32"	0.0" ±0.1" 0" ± 3/32"
SAI	12.8 ±0.8 (12 13/16 ± ¾)	12.7 ±0.8 (12 11/16 ± ¾)	12.1 ±0.5 (12 1/16 ± ½)	12.6 ±0.8 (12⅝ ± ¾)	11.4 ±0.8 (11⅜ ± ¾)	11.6 ±0.8 (11 9/16 ± ¾)
TURNING ANGLE	37.0 / 32.0 ±2.0 [2]	37.5 / 32.5 ±2.0	38.0 / 32.0 ±2.0	39.2 / 33.3 ±2.0 [1]	37.1 / 31.8 ±2.0	37.0 / 31.6 ±2.0
ILLUSTRATION #				26	26	26
REAR						
CAMBER (Degrees)	-0.67 ±0.75 -11/16 ± ¾	-0.58 ±0.75 -9/16 ± ¾		-0.92 ±0.75 -15/16 ± ¾	-1.45 ±0.5 -1 7/16 ± ½	-1.47 ±0.5 -1½ ± ½
REAR TOE (Degrees)	0.4° ±0.2° 13/32° ± 3/16°	0.4° ±0.2° 13/32° ± 3/16°		0.4° ±0.2° [2] 13/32° ± 3/16° [2]	0.25° ±0.25° ¼° ± ¼°	0.33° ±0.25° 11/32° ± ¼°
REAR TOE (Inches)	0.2" ±0.1" 3/16" ± 3/32"	0.2" ±0.1" 3/16" ± 3/32"		0.2" ±0.1" [2] 3/16" ± 3/32" [2]	0.13" ±0.13" ⅛" ± ⅛"	0.17" ±0.13" 5/32" ± ⅛"
THRUST ANGLE						
ILLUSTRATION #		06		10		
NOTES	[1] w/ M/T 1.3° (1 5/16°) [2] w/MS 37.5°/32.5°			[1] w/MS 39°/33° [2] 1993-95=0.3° (0.15")		
	44056	44057	44058	44059	44060	44061

Toyota

Corolla	Corolla 1.8L	Corolla 2.4L	MODEL INFORMATION	Corolla	Cressida	Echo
2007-2008	2009-2013	2009-2013	YEARS	2014-2017	1990-92	2000
			INSTRUCTIONS			
76 ft.lbs.	76 ft.lbs.	76 ft.lbs.	LUGNUT TORQUE	76 ft.lbs.		
			FRONT			
2.83 ±0.75 2 13⁄16 ± ¾	2.85 ±0.75 2 ⅞ ± ¾	2.92 ±0.75 2 15⁄16 ± ¾	CASTER (Degrees)	3.0 ±0.75 3 ± ¾	7.33 ±0.5 7 5⁄16 ± ½	1.58 ±0.75 [1] 1 9⁄16 ± ¾ [1]
0.0 (0) | 0.75 (¾)	0.0 (0) | 0.75 (¾)	0.0 (0) | 0.75 (¾)	CROSS | TOL.	0.0 (0) | 0.75 (¾)	0.0 (0) | 0.5 (½)	0.0 (0) | 0.75 (¾)
-0.53 ±0.75 - 9⁄16 ± ¾	-0.6 ±0.75 -⅝ ± ¾	-0.6 ±0.75 -⅝ ± ¾	CAMBER (Degrees)	-0.6 ±0.75 -⅝ ± ¾	0.5 ±0.75 ½ ± ¾	-0.58 ±0.75 - 9⁄16 ± ¾
0.0 (0) | 0.75 (¾)	0.0 (0) | 0.75 (¾)	0.0 (0) | 0.75 (¾)	CROSS | TOL.	0.0 (0) | 0.75 (¾)	0.0 (0) | 0.5 (½)	0.0 (0) | 0.75 (¾)
0.0° ±0.2° 0° ± 3⁄16°	0.0° ±0.2° 0° ± 3⁄16°	0.0° ±0.2° 0° ± 3⁄16°	FRONT TOE (Degrees)	0.0° ±0.2° 0° ± 3⁄16°	0.2° ±0.1° 3⁄16° ± 3⁄32°	0.0° ±0.2° 0° ± 3⁄16°
0.0" ±0.1" 0" ± 3⁄32"	0.0" ±0.1" 0" ± 3⁄32"	0.0" ±0.1" 0" ± 3⁄32"	FRONT TOE (Inches)	0.0" ±0.1" 0" ± 3⁄32"	0.1" ±0.05" 3⁄32" ± 1⁄16"	0.0" ±0.1" 0" ± 3⁄32"
11.4 ±0.8 (11⅜ ± ¾)	12.5 ±0.8 (12½ ± ¾)	12.6 ±0.8 (12⅝ ± ¾)	SAI	12.6 ±0.8 (12⅝ ± ¾)	13.2 ±0.8 (13 3⁄16 ± ¾)	10.1 ±0.8 (10 1⁄16 ± ¾)
37.1 / 31.8 ±2.0	38.2 / 32.8 ±2.0	37.0 / 32.0 ±2.0	TURNING ANGLE	38.1 / 32.7 ±2.0	41.5 / 34.3 ±1.0	36.9 / 32.2 ±2.0
26	26	26	ILLUSTRATION #	26	03	26
			REAR			
-1.45 ±0.5 -1 7⁄16 ± ½	-1.5 ±0.5 -1½ ± ½	-1.5 ±0.5 -1½ ± ½	CAMBER (Degrees)	-1.5 ±0.5 -1½ ± ½	0.0 ±0.5 0 ± ½	-0.93 ±0.75 - 15⁄16 ± ¾
0.25° ±0.25° [1] ¼° ± ¼° [1]	0.27° ±0.25° 9⁄32° ± ¼°	0.27° ±0.25° 9⁄32° ± ¼°	REAR TOE (Degrees)	0.3° ±0.25° 5⁄16° ± ¼°	0.2° ±0.1° 3⁄16° ± 3⁄32°	0.32° ±0.3° 5⁄16° ± 5⁄16°
0.13" ±0.13" [1] ⅛" ± ⅛" [1]	0.14" ±0.13" ⅛" ± ⅛"	0.14" ±0.13" ⅛" ± ⅛"	REAR TOE (Inches)	0.15" ±0.13" 5⁄32" ± ⅛"	0.1" ±0.05" 3⁄32" ± 1⁄16"	0.16" ±0.15" 5⁄32" ± 5⁄32"
			THRUST ANGLE			
10			ILLUSTRATION #		63	
[1] w/16" tires 0.34° (0.17")			NOTES			[1] w/MS 0.6° (⅝°)
44062	44063	44064		44065	44081	44082

Toyota

Echo	Echo w/Power Steering	Echo w/Manual Steering	MODEL INFORMATION	Echo Sedan	Echo Coupe/Hatch	Matrix FWD
2001-2002	2003	2003	YEARS	2004-2005	2004-2005	2003-2008
			INSTRUCTIONS			
			LUGNUT TORQUE			76 ft.lbs.
			FRONT			
1.77 ±0.75 [1] 1¾ ± ¾ [1]	2.0 ±0.75 [1] 2 ± ¾ [1]	1.5 ±0.75 1½ ± ¾	CASTER (Degrees)	1.93 ±0.75 [1] 1 15⁄16 ± ¾ [1]	1.85 ±0.75 [1] 1⅞ ± ¾ [1]	2.78 ±0.75 2¾ ± ¾
0.0 (0) | 0.75 (¾)	0.0 (0) | 0.75 (¾)	0.0 (0) | 0.75 (¾)	CROSS | TOL.	0.0 (0) | 0.75 (¾)	0.0 (0) | 0.75 (¾)	0.0 (0) | 0.75 (¾)
-0.62 ±0.75 -⅝ ± ¾	-0.6 ±0.75 -⅝ ± ¾	-0.6 ±0.75 -⅝ ± ¾	CAMBER (Degrees)	-0.57 ±0.75 - 9⁄16 ± ¾	-0.63 ±0.75 -⅝ ± ¾	-0.57 ±0.75 - 9⁄16 ± ¾
0.0 (0) | 0.75 (¾)	0.0 (0) | 0.75 (¾)	0.0 (0) | 0.75 (¾)	CROSS | TOL.	0.0 (0) | 0.75 (¾)	0.0 (0) | 0.75 (¾)	0.0 (0) | 0.75 (¾)
0.0° ±0.2° 0° ± 3⁄16°	0.0° ±0.2° 0° ± 3⁄16°	0.0° ±0.2° 0° ± 3⁄16°	FRONT TOE (Degrees)	0.0° ±0.2° 0° ± 3⁄16°	0.0° ±0.2° 0° ± 3⁄16°	0.0° ±0.2° 0° ± 3⁄16°
0.0" ±0.1" 0" ± 3⁄32"	0.0" ±0.1" 0" ± 3⁄32"	0.0" ±0.1" 0" ± 3⁄32"	FRONT TOE (Inches)	0.0" ±0.1" 0" ± 3⁄32"	0.0" ±0.1" 0" ± 3⁄32"	0.0" ±0.1" 0" ± 3⁄32"
10.1 ±0.8 (10⅛ ± ¾)	10.1 ±0.8 (10 1⁄16 ± ¾)	10.1 ±0.8 (10 1⁄16 ± ¾)	SAI	10.0 ±0.8 (10 ± ¾)	10.2 ±0.8 (10 3⁄16 ± ¾)	12.5 ±0.8 (12½ ± ¾)
36.9 / 32.2 ±2.0	33.5 / 29.7 ±2.0	36.7 / 31.7 ±2.0	TURNING ANGLE	33.5 / 29.7 ±2.0 [2]	33.6 / 30.0 ±2.0 [2]	36.7 / 31.9 ±2.0
26	26	26	ILLUSTRATION #	26	26	26
			REAR			
-0.95 ±0.75 - 15⁄16 ± ¾	-0.95 ±0.75 - 15⁄16 ± ¾	-0.95 ±0.75 - 15⁄16 ± ¾	CAMBER (Degrees)	-0.95 ±0.75 - 15⁄16 ± ¾	-0.95 ±0.75 - 15⁄16 ± ¾	-1.45 ±0.5 -1 7⁄16 ± ½
0.35° ±0.3° 11⁄32° ± 5⁄16°	0.35° ±0.3° 11⁄32° ± 5⁄16°	0.35° ±0.3° 11⁄32° ± 5⁄16°	REAR TOE (Degrees)	0.35° ±0.3° 11⁄32° ± 5⁄16°	0.35° ±0.3° 11⁄32° ± 5⁄16°	0.27° ±0.27° 9⁄32° ± 9⁄32°
0.18" ±0.15" 3⁄16" ± 5⁄32"	0.18" ±0.15" 3⁄16" ± 5⁄32"	0.18" ±0.15" 3⁄16" ± 5⁄32"	REAR TOE (Inches)	0.18" ±0.15" 3⁄16" ± 5⁄32"	0.18" ±0.15" 3⁄16" ± 5⁄32"	0.13" ±0.13" ⅛" ± ⅛"
			THRUST ANGLE			
			ILLUSTRATION #			
[1] w/MS 0.8° (13⁄16°)	[1] w/15" tires 2.05° (2 1⁄16°)		NOTES	[1] w/MS 1.0° (1°) [2] w/MS 37°/32.2°	[1] w/MS 0.9° (⅞°) [2] w/MS 36.9°/32.1	
44083	44084	44085		44086	44087	44088

Toyota

MODEL INFORMATION	Matrix AWD	Matrix FWD	Matrix AWD	MR2	MR2	MR2
YEARS	2003-2006	2009-2014	2009-2013	1991-92	1993	1994-95
INSTRUCTIONS						
LUGNUT TORQUE	76 ft.lbs.	76 ft.lbs.	76 ft.lbs.			
FRONT						
CASTER (Degrees)	2.77 ±0.75 2¾ ± ¾	2.9 ±0.75 2⅞ ± ¾	3.0 ±0.75 3 ± ¾	2.75 ±0.75 2¾ ± ¾	2.83 ±0.75 2 13/16 ± ¾	3.25 ±0.75 3¼ ± ¾
CROSS | TOL.	0.0 (0) | 0.75 (¾)	0.0 (0) | 0.75 (¾)	0.0 (0) | 0.75 (¾)	0.0 (0) | 0.5 (½)	0.0 (0) | 0.5 (½)	0.0 (0) | 0.5 (½)
CAMBER (Degrees)	-0.48 ±0.75 -½ ± ¾	-0.57 ±0.75 -9/16 ± ¾	-0.6 ±0.75 -⅝ ± ¾	-0.92 ±0.75 -15/16 ± ¾	-1.0 ±0.75 -1 ± ¾	-1.0 ±0.75 -1 ± ¾
CROSS | TOL.	0.0 (0) | 0.75 (¾)	0.0 (0) | 0.75 (¾)	0.0 (0) | 0.75 (¾)	0.0 (0) | 0.5 (½)	0.0 (0) | 0.5 (½)	0.0 (0) | 0.5 (½)
FRONT TOE (Degrees)	0.0° ±0.2° 0° ± 3/16°	0.0° ±0.2° 0° ± 3/16°	0.0° ±0.2° 0° ± 3/16°	0.1° ±0.1° 3/32° ± 3/32°	0.1° ±0.2° 3/32° ± 3/16°	0.1° ±0.2° 3/32° ± 3/16°
FRONT TOE (Inches)	0.0" ±0.1" 0" ± 3/32"	0.0" ±0.1" 0" ± 3/32"	0.0" ±0.1" 0" ± 3/32"	0.05" ±0.05" 1/16" ± 1/16"	0.05" ±0.1" 1/16" ± 3/32"	0.05" ±0.1" 1/16" ± 3/32"
SAI	12.2 ±0.8 (12¼ ± ¾)	12.5 ±0.8 (12½ ± ¾)	12.3 ±0.8 (12 5/16 ± ¾)	13.6 ±0.8 (13 9/16 ± ¾)	13.8 ±0.8 (13 13/16 ± ¾)	13.8 ±0.8 (13 13/16 ± ¾)
TURNING ANGLE	37.1 / 32.0 ±2.0	37.1 / 32.1 ±2.0	36.5 / 32.0 ±2.0	37.5 / 32.0 ±1.5	37.0 / 32.0 ±1.5	37.0 / 32.0 ±1.5
ILLUSTRATION #	26	26	26	03	03	03
REAR						
CAMBER (Degrees)	-0.73 ±0.5 -¾ ± ½	-1.47 ±0.75 -1½ ± ¾	-1.05 ±0.75 -1 1/16 ± ¾	-1.33 ±0.75 -1 5/16 ± ¾	-1.58 ±0.75 -1 9/16 ± ¾	-1.58 ±0.75 -1 9/16 ± ¾
REAR TOE (Degrees)	0.2° ±0.2° 3/16° ± 3/16°	0.3° ±0.25° 5/16° ± ¼°	0.25° ±0.25° ¼° ± ¼°	0.5° ±0.1° ½° ± 3/32°	0.6° ±0.1° 19/32° ± 3/32°	0.4° ±0.2° 13/32° ± 3/16°
REAR TOE (Inches)	0.1" ±0.1" 3/32" ± 3/32"	0.15" ±0.13" 5/32" ± ⅛"	0.13" ±0.13" ⅛" ± ⅛"	0.25" ±0.05" ¼" ± 1/16"	0.3" ±0.05" 5/16" ± 1/16"	0.2" ±0.1" 3/16" ± 3/32"
THRUST ANGLE		0.3 (5/16)	0.2 (3/16)			
ILLUSTRATION #	120		120	10	10	10
NOTES						
	44089	44090	44091	44094	44095	44096

Toyota

MODEL INFORMATION	MR2 Spyder	Paseo	Previa 4X2	Previa 4X4	Prius	Prius
YEARS	2000-2005	1992-97	1991-97	1991-97	2001-2003	2004-2009
INSTRUCTIONS						
LUGNUT TORQUE					76 ft.lbs.	76 ft.lbs.
FRONT						
CASTER (Degrees)	3.1 ±0.75 3⅛ ± ¾	1.5 ±0.75 1½ ± ¾	5.5 ±0.5 5½ ± ½	5.33 ±0.5 5 5/16 ± ½	1.03 ±0.75 1 1/16 ± ¾	3.17 ±0.75 3 3/16 ± ¾
CROSS | TOL.	0.0 (0) | 0.75 (¾)	0.0 (0) | 0.5 (½)	0.0 (0) | 0.5 (½)	0.0 (0) | 0.5 (½)	0.0 (0) | 0.75 (¾)	0.0 (0) | 0.75 (¾)
CAMBER (Degrees)	-0.8 ±0.75 -13/16 ± ¾	-0.42 ±0.75 -7/16 ± ¾	0.08 ±0.5 1/16 ± ½	0.25 ±0.5 ¼ ± ½	-0.43 ±0.75 -7/16 ± ¾	-0.58 ±0.75 -9/16 ± ¾
CROSS | TOL.	0.0 (0) | 0.75 (¾)	0.0 (0) | 0.5 (½)	0.0 (0) | 0.5 (½)	0.0 (0) | 0.5 (½)	0.0 (0) | 0.75 (¾)	0.0 (0) | 0.75 (¾)
FRONT TOE (Degrees)	0.15° ±0.2° 5/32° ± 3/16°	0.1° ±0.2° 3/32° ± 3/16°	0.2° ±0.2° 3/16° ± 3/16°	0.3° ±0.2° 5/16° ± 3/16°	0.1° ±0.2° 3/32° ± 3/16°	0.0° ±0.2° 0° ± 3/16°
FRONT TOE (Inches)	0.08" ±0.1" 1/16" ± 3/32"	0.05" ±0.1" 1/16" ± 3/32"	0.1" ±0.1" 3/32" ± 3/32"	0.15" ±0.1" 5/32" ± 3/32"	0.05" ±0.1" 1/16" ± 3/32"	0.0" ±0.1" 0" ± 3/32"
SAI	14.9 ±0.8 (14⅞ ± ¾)	12.3 ±0.8 (12¼ ± ¾)	10.6 ±0.8 (10 9/16 ± ¾)	10.3 ±0.8 (10 5/16 ± ¾)	9.9 ±0.8 (9⅞ ± ¾)	12.6 ±0.8 (12 9/16 ± ¾)
TURNING ANGLE	38.0 / 32.9 ±2.0	35.8 / 32.1 ±2.0	35.6 / 32.8 ±2.0 [1]	35.6 / 33.0 ±2.0 [1]	42.7 / 35.9 ±2.0	40.6 / 34.3 ±2.0
ILLUSTRATION #	26	26	26	26	26	26
REAR						
CAMBER (Degrees)	-1.08 ±0.75 -1 1/16 ± ¾	-0.5 ±0.75 -½ ± ¾			-1.5 ±0.75 -1½ ± ¾	-1.5 ±0.5 -1½ ± ½
REAR TOE (Degrees)	0.3° ±0.2° 5/16° ± 3/16°	0.3° ±0.3° 5/16° ± 5/16°			0.1° ±0.33° 3/32° ± 11/32°	0.3° ±0.25° 5/16° ± ¼°
REAR TOE (Inches)	0.15" ±0.1" 5/32" ± 3/32"	0.15" ±0.15" 5/32" ± 5/32"			0.05" ±0.17" 1/16" ± 5/32"	0.15" ±0.13" 5/32" ± ⅛"
THRUST ANGLE						
ILLUSTRATION #	06					
NOTES			[1] 1991 = 36°/33.1°	[1] 1991 = 36°/32.5°		
	44097	44098	44099	44100	44101	44102

Toyota

Prius	Prius	Prius	MODEL	Prius	Prius PHV	Prius PHV
195/65R15 tires (OE)	215/45R17 tires (OE)	195/65R15 tires (OE)	INFORMATION	215/45R17 tires (OE)		
2010-2015	2010-2015	2016-2017	YEARS	2016-2017	2010	2012-2015
			INSTRUCTIONS			
76 ft.lbs.	76 ft.lbs.	76 ft.lbs.	LUGNUT TORQUE	76 ft.lbs.	76 ft.lbs.	76 ft.lbs.
			FRONT			
5.88 ±0.75 5 7/8 ± 3/4	5.83 ±0.75 5 13/16 ± 3/4	6.85 ±0.75 6 7/8 ± 3/4	CASTER (Degrees)	6.88 ±0.75 6 7/8 ± 3/4	6.07 ±0.75 6 1/16 ± 3/4	5.88 ±0.75 5 7/8 ± 3/4
0.0 (0) | 0.75 (3/4)	0.0 (0) | 0.75 (3/4)	0.0 (0) | 0.75 (3/4)	CROSS | TOL.	0.0 (0) | 0.75 (3/4)	0.0 (0) | 0.75 (3/4)	0.0 (0) | 0.75 (3/4)
-0.22 ±0.75 -1/4 ± 3/4	-0.2 ±0.75 -3/16 ± 3/4	-0.15 ±0.75 -1/8 ± 3/4	CAMBER (Degrees)	-0.15 ±0.75 -1/8 ± 3/4	-0.22 ±0.75 -1/4 ± 3/4	-0.22 ±0.75 -1/4 ± 3/4
0.0 (0) | 0.75 (3/4)	0.0 (0) | 0.75 (3/4)	0.0 (0) | 0.75 (3/4)	CROSS | TOL.	0.0 (0) | 0.75 (3/4)	0.0 (0) | 0.75 (3/4)	0.0 (0) | 0.75 (3/4)
0.2° ±0.2° 3/16° ± 3/16°	0.2° ±0.2° 3/16° ± 3/16°	0.12° ±0.18° 1/8° ± 3/16°	FRONT TOE (Degrees)	0.12° ±0.18° 1/8° ± 3/16°	0.2° ±0.2° 3/16° ± 3/16°	0.2° ±0.2° 3/16° ± 3/16°
0.1" ±0.1" 3/32" ± 3/32"	0.1" ±0.1" 3/32" ± 3/32"	0.06" ±0.09" 1/16" ± 3/32"	FRONT TOE (Inches)	0.06" ±0.09" 1/16" ± 3/32"	0.1" ±0.1" 3/32" ± 3/32"	0.1" ±0.1" 3/32" ± 3/32"
12.3 ±0.8 (12 1/4 ± 3/4)	12.2 ±0.8 (12 3/16 ± 3/4)	13.3 ±0.8 (13 5/16 ± 3/4)	SAI	13.3 ±0.8 (13 5/16 ± 3/4)	12.3 ±0.8 (12 1/4 ± 3/4)	12.3 ±0.8 (12 1/4 ± 3/4)
40.8 / 33.8 ±2.0	37.5 / 31.9 ±2.0	41.7 / 34.7 ±2.0	TURNING ANGLE	37.7 / 32.4 ±2.0	37.4 / 31.9 ±2.0	40.8 / 33.8 ±2.0
26	26	26	ILLUSTRATION #	26	26	26
			REAR			
-1.48 ±0.5 -1 1/2 ± 1/2	-1.47 ±0.5 -1 1/2 ± 1/2	-0.97 ±0.75 -1 ± 3/4	CAMBER (Degrees)	-0.97 ±0.75 -1 ± 3/4	-1.5 ±0.5 -1 1/2 ± 1/2	-1.48 ±0.5 -1 1/2 ± 1/2
0.3° ±0.3° 5/16° ± 5/16°	0.25° ±0.3° 1/4° ± 5/16°	0.17° ±0.18° 5/32° ± 3/16°	REAR TOE (Degrees)	0.17° ±0.18° 5/32° ± 3/16°	0.33° ±0.3° 11/32° ± 5/16°	0.29° ±0.3° 9/32° ± 5/16°
0.15" ±0.15" 5/32" ± 5/32"	0.13" ±0.15" 1/8" ± 5/32"	0.08" ±0.09" 3/32" ± 3/32"	REAR TOE (Inches)	0.08" ±0.09" 3/32" ± 3/32"	0.17" ±0.15" 5/32" ± 5/32"	0.14" ±0.15" 1/8" ± 5/32"
			THRUST ANGLE			
		06	ILLUSTRATION #	06		
			NOTES			
44103	44104	44105		44106	44107	44108

Toyota

Prius C	Prius Prime	Prius V	MODEL	Prius V	Sienna	Sienna FWD
		16" tires	INFORMATION	17" tires		
2012-2017	2017	2012-2017	YEARS	2012-2017	1998-2003	2004-2006
			INSTRUCTIONS			
76 ft.lbs.	76 ft.lbs.	76 ft.lbs.	LUGNUT TORQUE	76 ft.lbs.	76 ft.lbs.	76 ft.lbs.
			FRONT			
5.0 ±0.75 5 ± 3/4	7.12 ±0.75 7 1/8 ± 3/4	5.93 ±0.75 5 15/16 ± 3/4	CASTER (Degrees)	5.93 ±0.75 5 15/16 ± 3/4	1.53 ±0.75 1 9/16 ± 3/4	2.8 ±0.75 2 13/16 ± 3/4
0.0 (0) | 0.75 (3/4)	0.0 (0) | 0.75 (3/4)	0.0 (0) | 0.75 (3/4)	CROSS | TOL.	0.0 (0) | 0.75 (3/4)	0.0 (0) | 0.75 (3/4)	0.0 (0) | 0.5 (1/2)
-0.25 ±0.75 -1/4 ± 3/4	-0.17 ±0.75 -3/16 ± 3/4	-0.22 ±0.75 -1/4 ± 3/4	CAMBER (Degrees)	-0.22 ±0.75 -1/4 ± 3/4	-0.53 ±0.75 -9/16 ± 3/4	-0.28 ±0.75 -1/4 ± 3/4
0.0 (0) | 0.75 (3/4)	0.0 (0) | 0.75 (3/4)	0.0 (0) | 0.75 (3/4)	CROSS | TOL.	0.0 (0) | 0.75 (3/4)	0.0 (0) | 0.75 (3/4)	0.0 (0) | 0.5 (1/2)
0.1° ±0.2° 3/32° ± 3/16°	0.11° ±0.18° 1/8° ± 3/16°	0.17° ±0.2° 5/32° ± 3/16°	FRONT TOE (Degrees)	0.17° ±0.2° 5/32° ± 3/16°	0.27° ±0.2° 9/32° ± 3/16°	0.0° ±0.2° 0° ± 3/16°
0.05" ±0.1" 1/16" ± 3/32"	0.06" ±0.09" 1/16" ± 3/32"	0.08" ±0.1" 3/32" ± 3/32"	FRONT TOE (Inches)	0.08" ±0.1" 3/32" ± 3/32"	0.13" ±0.1" 1/8" ± 3/32"	0.0" ±0.1" 0" ± 3/32"
11.5 ±0.8 (11 1/2 ± 3/4)	13.3 (13 5/16)	12.2 ±0.8 (12 3/16 ± 3/4)	SAI	12.2 ±0.8 (12 3/16 ± 3/4)	12.6 ±0.8 (12 9/16 ± 3/4)	10.4 ±0.8 (10 7/16 ± 3/4)
40.8 / 35.1 ±2.0 [1]	41.7 / 34.7 ±2.0	39.2 / 33.0 ±2.0	TURNING ANGLE	35.9 / 31.0 ±2.0	34.3 / 30.3 ±1.0	42.8 / 36.4 ±2.0
26	26	26	ILLUSTRATION #	26	26	26
			REAR			
-0.96 ±0.5 -1 ± 1/2	-1.12 ±0.75 -1 1/8 ± 3/4	-1.43 ±0.5 -1 7/16 ± 1/2	CAMBER (Degrees)	-1.43 ±0.5 -1 7/16 ± 1/2	-0.92 ±0.75 -15/16 ± 3/4	-1.37 ±0.5 -1 3/8 ± 1/2
0.35° ±0.25° 11/32° ± 1/4°	0.23° ±0.18° 7/32° ± 3/16°	0.23° ±0.3° 7/32° ± 5/16°	REAR TOE (Degrees)	0.23° ±0.3° 7/32° ± 5/16°	0.22° ±0.3° 7/32° ± 5/16°	0.25° ±0.3° 1/4° ± 5/16°
0.16" ±0.13" 5/32" ± 1/8"	0.12" ±0.09" 1/8" ± 3/32"	0.12" ±0.15" 1/8" ± 5/32"	REAR TOE (Inches)	0.12" ±0.15" 1/8" ± 5/32"	0.11" ±0.15" 3/32" ± 5/32"	0.13" ±0.15" 1/8" ± 5/32"
			THRUST ANGLE			
	06		ILLUSTRATION #			
[1] 16" tires = 33.5°/29.5°			NOTES			
44109	44110	44111		44112	44113	44114

Toyota

MODEL INFORMATION	Sienna AWD	Sienna FWD	Sienna AWD	Sienna FWD Base & LE (incl. Cargo Van)	Sienna FWD SE	Sienna FWD XLE, Limited
YEARS	2004-2006	2007-2010	2007-2010	2011-2017	2011-2017	2011-2017
INSTRUCTIONS						
LUGNUT TORQUE	76 ft.lbs.	76 ft.lbs.	76 ft.lbs.	76 ft.lbs.	76 ft.lbs.	76 ft.lbs.
FRONT						
CASTER (Degrees)	2.67 ±0.75 2 11/16 ± 3/4	3.15 ±0.75 3 1/8 ± 3/4	3.0 ±0.75 3 ± 3/4	2.83 ±0.75 2 13/16 ± 3/4	3.0 ±0.75 3 ± 3/4	2.96 ±0.75 2 15/16 ± 3/4
CROSS | TOL.	0.0 (0) | 0.5 (1/2)	0.0 (0) | 0.5 (1/2)	0.0 (0) | 0.5 (1/2)	0.0 (0) | 0.5 (1/2)	0.0 (0) | 0.5 (1/2)	0.0 (0) | 0.5 (1/2)
CAMBER (Degrees)	-0.28 ±0.75 -1/4 ± 3/4	-0.28 ±0.75 -1/4 ± 3/4	-0.28 ±0.75 -1/4 ± 3/4	-0.42 ±0.75 -7/16 ± 3/4	-0.52 ±0.75 -1/2 ± 3/4	-0.46 ±0.75 -7/16 ± 3/4
CROSS | TOL.	0.0 (0) | 0.5 (1/2)	0.0 (0) | 0.5 (1/2)	0.0 (0) | 0.5 (1/2)	0.0 (0) | 0.5 (1/2)	0.0 (0) | 0.5 (1/2)	0.0 (0) | 0.5 (1/2)
FRONT TOE (Degrees)	0.0° ±0.2° 0° ± 3/16°	-0.02° ±0.2° -1/32° ± 3/16°	0.04° ±0.2° 1/32° ± 3/16°	0.1° ±0.2° 3/32° ± 3/16°	0.07° ±0.2° 1/16° ± 3/16°	0.07° ±0.2° 1/16° ± 3/16°
FRONT TOE (Inches)	0.0" ±0.1" 0" ± 3/32"	-0.01" ±0.1" -1/64" ± 3/32"	0.02" ±0.1" 1/64" ± 3/32"	0.05" ±0.1" 1/16" ± 3/32"	0.04" ±0.1" 1/32" ± 3/32"	0.04" ±0.1" 1/32" ± 3/32"
SAI	10.2 ±0.8 (10 3/16 ± 3/4)	10.4 ±0.8 (10 7/16 ± 3/4)	10.2 ±0.8 (10 3/16 ± 3/4)	11.1 ±0.8 (11 1/16 ± 3/4)	11.4 ±0.8 (11 3/8 ± 3/4)	11.2 ±0.8 (11 3/16 ± 3/4)
TURNING ANGLE	42.7 / 36.3 ±2.0	42.8 / 36.4 ±2.0	42.7 / 36.3 ±2.0	42.5 / 36.6 ±2.0	42.1 / 36.3 ±2.0	42.5 / 36.3 ±2.0
ILLUSTRATION #	26	26	26	26	26	26
REAR						
CAMBER (Degrees)	-1.42 ±0.5 -1 7/16 ± 1/2	-1.37 ±0.5 -1 3/8 ± 1/2	-1.42 ±0.5 -1 7/16 ± 1/2	-1.05 ±0.75 -1 1/16 ± 3/4	-1.18 ±0.75 -1 3/16 ± 3/4	-1.15 ±0.75 -1 1/8 ± 3/4
REAR TOE (Degrees)	0.13° ±0.3° 1/8° ± 5/16°	0.27° ±0.27° 9/32° ± 9/32°	0.21° ±0.27° 7/32° ± 9/32°	0.33° ±0.27° 11/32° ± 9/32°	0.35° ±0.27° 11/32° ± 9/32°	0.35° ±0.27° 11/32° ± 9/32°
REAR TOE (Inches)	0.06" ±0.15" 1/16" ± 5/32"	0.14" ±0.14" 1/8" ± 1/8"	0.11" ±0.14" 3/32" ± 1/8"	0.17" ±0.14" 5/32" ± 1/8"	0.18" ±0.14" 3/16" ± 1/8"	0.18" ±0.14" 3/16" ± 1/8"
THRUST ANGLE						
ILLUSTRATION #						
NOTES						
	44115	44116	44117	44118	44119	44120

Toyota

MODEL INFORMATION	Sienna AWD Base & LE	Sienna AWD XLE & Limited	Solara	Solara exc/Convertible	Solara Convertible	Supra
YEARS	2011-2017	2011-2017	1999-2003	2004-2008	2004-2008	1990-92
INSTRUCTIONS						
LUGNUT TORQUE	76 ft.lbs.	76 ft.lbs.	76 ft.lbs.	76 ft.lbs.	76 ft.lbs.	
FRONT						
CASTER (Degrees)	2.82 ±0.75 2 13/16 ± 3/4	2.93 ±0.75 2 15/16 ± 3/4	2.13 ±0.75 2 1/8 ± 3/4	2.9 ±0.75 2 7/8 ± 3/4	2.85 ±0.75 2 7/8 ± 3/4	7.58 ±0.5 7 9/16 ± 1/2
CROSS | TOL.	0.0 (0) | 0.5 (1/2)	0.0 (0) | 0.5 (1/2)	0.0 (0) | 0.75 (3/4)	0.0 (0) | 0.75 (3/4)	0.0 (0) | 0.75 (3/4)	0.0 (0) | 0.5 (1/2)
CAMBER (Degrees)	-0.45 ±0.75 -7/16 ± 3/4	-0.47 ±0.75 -1/2 ± 3/4	-0.52 ±0.75 -1/2 ± 3/4	-0.75 ±0.75 -3/4 ± 3/4	-0.73 ±0.75 -3/4 ± 3/4	-0.17 ±0.5 -3/16 ± 1/2
CROSS | TOL.	0.0 (0) | 0.5 (1/2)	0.0 (0) | 0.5 (1/2)	0.0 (0) | 0.75 (3/4)	0.0 (0) | 0.75 (3/4)	0.0 (0) | 0.75 (3/4)	0.0 (0) | 0.5 (1/2)
FRONT TOE (Degrees)	0.07° ±0.2° 1/16° ± 3/16°	0.05° ±0.2° 1/16° ± 3/16°	0.0° ±0.2° 0° ± 3/16°	0.0° ±0.2° 0° ± 3/16°	0.0° ±0.2° 0° ± 3/16°	0.0° ±0.2° 0° ± 3/16°
FRONT TOE (Inches)	0.04" ±0.1" 1/32" ± 3/32"	0.03" ±0.1" 1/32" ± 3/32"	0.0" ±0.1" 0" ± 3/32"	0.0" ±0.1" 0" ± 3/32"	0.0" ±0.1" 0" ± 3/32"	0.0" ±0.1" 0" ± 3/32"
SAI	11.1 ±0.8 (11 1/16 ± 3/4)	11.1 ±0.8 (11 1/16 ± 3/4)	12.9 ±0.8 (12 7/8 ± 3/4)	11.5 ±0.8 (11 1/2 ± 3/4)	11.5 ±0.8 (11 1/2 ± 3/4)	10.9 ±0.8 (10 15/16 ± 3/4)
TURNING ANGLE	41.9 / 36.3 ±2.0	41.9 / 36.3 ±2.0	34.3 / 30.4 ±2.0	36.8 / 32.1 ±2.0	36.6 / 32.1 ±2.0	34.5 / 31.8 ±2.0
ILLUSTRATION #	26	26	26	26	26	52
REAR						
CAMBER (Degrees)	-1.38 ±0.75 -1 3/8 ± 3/4	-1.43 ±0.75 -1 7/16 ± 3/4	-0.65 ±0.75 -5/8 ± 3/4	-1.35 ±0.75 -1 3/8 ± 3/4	-1.33 ±0.75 -1 5/16 ± 3/4	-0.83 ±0.75 -13/16 ± 3/4
REAR TOE (Degrees)	0.25° ±0.27° 1/4° ± 9/32°	0.27° ±0.27° 9/32° ± 9/32°	0.4° ±0.2° 13/32° ± 3/16°	0.4° ±0.2° 13/32° ± 3/16°	0.43° ±0.2° 7/16° ± 3/16°	0.4° ±0.2° 13/32° ± 3/16°
REAR TOE (Inches)	0.13" ±0.14" 1/8" ± 1/8"	0.14" ±0.14" 1/8" ± 1/8"	0.2" ±0.1" 3/16" ± 3/32"	0.2" ±0.1" 3/16" ± 3/32"	0.22" ±0.1" 7/32" ± 3/32"	0.2" ±0.1" 3/16" ± 3/32"
THRUST ANGLE						
ILLUSTRATION #			10	10	10	63
NOTES						
	44121	44122	44123	44124	44125	44130

Toyota

Supra	Supra	Tercel Liftback	MODEL	Tercel	Tercel	Tercel
except Turbo	Turbo	Coupe	INFORMATION			
		4X2				
1993-98	1993-98	1990	YEARS	1991	1992-93	1994
			INSTRUCTIONS			
			LUGNUT TORQUE			
			FRONT			
3.33 ± 0.5 3 5/16 ± ½	3.5 ± 0.5 3½ ± ½	2.5 ± 0.75 [1] 2½ ± ¾ [1]	CASTER (Degrees)	1.25 ± 0.75 1¼ ± ¾	1.75 ± 0.75 1¾ ± ¾	1.25 ± 0.75 1¼ ± ¾
0.0 (0) | 0.5 (½)	0.0 (0) | 0.5 (½)	0.0 (0) | 0.5 (½)	CROSS | TOL.	0.0 (0) | 0.5 (½)	0.0 (0) | 0.5 (½)	0.0 (0) | 0.75 (¾)
-0.33 ± 0.5 -5/16 ± ½	-0.5 ± 0.5 -½ ± ½	0.0 ± 0.75 0 ± ¾	CAMBER (Degrees)	-0.25 ± 0.75 -¼ ± ¾	-0.42 ± 0.75 -7/16 ± ¾	-0.25 ± 0.75 -¼ ± ¾
0.0 (0) | 0.5 (½)	0.0 (0) | 0.5 (½)	0.0 (0) | 0.5 (½)	CROSS | TOL.	0.0 (0) | 0.5 (½)	0.0 (0) | 0.5 (½)	0.0 (0) | 0.75 (¾)
0.0° ± 0.1° 0° ± 3/32°	0.0° ± 0.1° 0° ± 3/32°	0.0° ± 0.1° 0° ± 3/32°	FRONT TOE (Degrees)	0.1° ± 0.2° 3/32° ± 3/16°	0.1° ± 0.2° 3/32° ± 3/16°	0.1° ± 0.2° 3/32° ± 3/16°
0.0" ± 0.05" 0" ± 1/16"	0.0" ± 0.05" 0" ± 1/16"	0.0" ± 0.05" 0" ± 1/16"	FRONT TOE (Inches)	0.05" ± 0.1" 1/16" ± 3/32"	0.05" ± 0.1" 1/16" ± 3/32"	0.05" ± 0.1" 1/16" ± 3/32"
9.6 ± 0.8 (9 9/16 ± ¾)	9.8 ± 0.8 (9¾ ± ¾)	11.5 ± 0.8 (11½ ± ¾)	SAI	12.0 ± 0.8 (12 ± ¾)	12.3 ± 0.8 (12¼ ± ¾)	12.0 ± 0.8 (12 ± ¾)
35.0 / 30.8 ± 2.0	34.9 / 30.6 ± 2.0	37.8 / 32.8 ± 2.0	TURNING ANGLE	36.2 / 32.3 ± 2.0	36.2 / 32.3 ± 2.0	36.2 / 32.3 ± 2.0
52	52		ILLUSTRATION #			
			REAR			
-1.58 ± 0.5 [1] -1 9/16 ± ½ [1]	-1.5 ± 0.5 -1½ ± ½		CAMBER (Degrees)	-0.5 ± 0.75 -½ ± ¾	-0.5 ± 0.75 -½ ± ¾	-0.5 ± 0.75 -½ ± ¾
0.3° ± 0.1° 5/16° ± 3/32°	0.3° ± 0.1° 5/16° ± 3/32°		REAR TOE (Degrees)	0.3° ± 0.3° 5/16° ± 5/16°	0.3° ± 0.3° 5/16° ± 5/16°	0.3° ± 0.3° 5/16° ± 5/16°
0.15" ± 0.05" 5/32" ± 1/16"	0.15" ± 0.05" 5/32" ± 1/16"		REAR TOE (Inches)	0.15" ± 0.15" 5/32" ± 5/32"	0.15" ± 0.15" 5/32" ± 5/32"	0.15" ± 0.15" 5/32" ± 5/32"
			THRUST ANGLE			
63	63		ILLUSTRATION #			
[1] 1998 = 1.5° (1½°)		[1] w/MS 1.0° (1°)	NOTES			
44131	44132	44141		44142	44143	44144

Toyota

Tercel	Venza	Venza	MODEL	Yaris	Yaris	Yaris
	FWD	AWD	INFORMATION	Sedan	Sedan (VIN NPC)	Sedan (VIN 3**)
	(2016 Canada)	(2016 Canada)				(Fuel tank full)
1995-98	2009-2016	2009-2016	YEARS	2007-2013	2014-2016	2017
			INSTRUCTIONS			
	76 ft.lbs.	76 ft.lbs.	LUGNUT TORQUE	76 ft.lbs.		94 ft.lbs.
			FRONT			
1.33 ± 0.75 1 5/16 ± ¾	2.65 ± 0.75 2⅝ ± ¾	2.65 ± 0.75 2⅝ ± ¾	CASTER (Degrees)	4.65 ± 0.75 4⅝ ± ¾	4.38 ± 0.75 4⅜ ± ¾	5.23 ± 1.0 5¼ ± 1
0.0 (0) | 0.75 (¾)	0.0 (0) | 0.75 (¾)	0.0 (0) | 0.75 (¾)	CROSS | TOL.	0.0 (0) | 0.75 (¾)	0.0 (0) | 0.75 (¾)	0.0 (0) | 1.5 (1½)
-0.33 ± 0.75 -5/16 ± ¾	-0.6 ± 0.75 -⅝ ± ¾	-0.6 ± 0.75 -⅝ ± ¾	CAMBER (Degrees)	-0.17 ± 0.75 -3/16 ± ¾	0.1 ± 0.75 ⅛ ± ¾	-0.43 ± 1.0 -7/16 ± 1
0.0 (0) | 0.75 (¾)	0.0 (0) | 0.75 (¾)	0.0 (0) | 0.75 (¾)	CROSS | TOL.	0.0 (0) | 0.75 (¾)	0.0 (0) | 0.75 (¾)	0.0 (0) | 1.5 (1½)
0.1° ± 0.2° 3/32° ± 3/16°	0.08° ± 0.2° 3/32° ± 3/16°	0.08° ± 0.2° 3/32° ± 3/16°	FRONT TOE (Degrees)	0.2° ± 0.2° 3/16° ± 3/16°	0.15° ± 0.2° 5/32° ± 3/16°	0.18° ± 0.17° 3/16° ± 5/32°
0.05" ± 0.1" 1/16" ± 3/32"	0.04" ± 0.1" 1/32" ± 3/32"	0.04" ± 0.1" 1/32" ± 3/32"	FRONT TOE (Inches)	0.1" ± 0.1" 3/32" ± 3/32"	0.08" ± 0.1" 1/16" ± 3/32"	0.09" ± 0.08" 3/32" ± 3/32"
12.2 ± 0.8 (12 3/16 ± ¾)	10.7 ± 0.8 (10 11/16 ± ¾)	10.7 ± 0.8 (10 11/16 ± ¾)	SAI	11.3 ± 0.8 (11¼ ± ¾)	10.8 ± 0.8 (10¾ ± ¾)	12.8 (12 13/16)
35.9 / 32.2 ± 2.0	35.2 / 31.4 ± 2.0	35.2 / 31.4 ± 2.0	TURNING ANGLE	41.0 / 35.3 ± 2.0	41.4 / 35.8 ± 2.0	41.0 / 34.0 ± 3.0
26	26	26	ILLUSTRATION #	26	26	
			REAR			
-0.5 ± 0.75 -½ ± ¾	-1.2 ± 0.75 -1 3/16 ± ¾	-0.8 ± 0.75 -13/16 ± ¾	CAMBER (Degrees)	-0.93 ± 0.75 -15/16 ± ¾	-0.9 ± 0.75 -⅞ ± ¾	-1.48 ± 1.0 -1½ ± 1
0.3° ± 0.3° 5/16° ± 5/16°	0.28° ± 0.17° 9/32° ± 5/32°	0.28° ± 0.17° 9/32° ± 5/32°	REAR TOE (Degrees)	0.28° ± 0.3° 9/32° ± 5/16°	0.22° ± 0.3° 7/32° ± 5/16°	0.4° ± 0.37° 13/32° ± ⅜°
0.15" ± 0.15" 5/32" ± 5/32"	0.14" ± 0.08" ⅛" ± 3/32"	0.14" ± 0.08" ⅛" ± 3/32"	REAR TOE (Inches)	0.14" ± 0.15" ⅛" ± 5/32"	0.11" ± 0.15" ⅛" ± 5/32"	0.2" ± 0.18" 3/16" ± 3/16"
			THRUST ANGLE			0.8 (13/16)
	10	06	ILLUSTRATION #			
			NOTES			
44145	44146	44147		44148	44149	44150

ALIGNMENT SPECIFICATIONS

ALIGNMENT SPECIFICATIONS

Toyota

Yaris	Yaris	Yaris	MODEL	Yaris iA	Yaris iM	Notes
Liftback	Hatchback	Hatchback	INFORMATION	(Fuel tank full)		
	P175/65R15" tires	P195/50R16" tires				
2006-2011	2012-2017	2012-2017	YEARS	2017	2017	
			INSTRUCTIONS			
76 ft.lbs.	76 ft.lbs.	76 ft.lbs.	LUGNUT TORQUE	94 ft.lbs.	76 ft.lbs.	
			FRONT			
4.77 ±0.75 4¾ ± ¾	5.0 ±0.75 5 ± ¾	5.0 ±0.75 5 ± ¾	CASTER (Degrees)	5.23 ±1.0 5¼ ±1	5.87 ±0.75 5¼ ⅞ ± ¾	
0.0 (0) | 0.75 (¾)	0.0 (0) | 0.75 (¾)	0.0 (0) | 0.75 (¾)	CROSS | TOL.	0.0 (0) | 1.5 (1½)	0.0 (0) | 0.75 (¾)	
-0.17 ±0.75 -3/16 ± ¾	-0.35 ±0.75 -⅜ ± ¾	-0.4 ±0.75 -⅜ ± ¾	CAMBER (Degrees)	-0.43 ±1.0 -7/16 ±1	-0.23 ±0.75 -7/16 ± ¾	
0.0 (0) | 0.75 (¾)	0.0 (0) | 0.75 (¾)	0.0 (0) | 0.75 (¾)	CROSS | TOL.	0.0 (0) | 1.5 (1½)	0.0 (0) | 0.75 (¾)	
0.15° ±0.2° [1] 5/32° ± 3/16° [1]	0.1° ±0.08° 3/32° ± 3/32°	0.08° ±0.08° 3/32° ± 3/32°	FRONT TOE (Degrees)	0.18° ±0.17° 3/16° ± 5/32°	0.18° ±0.17° 3/16° ± 5/32°	
0.08" ±0.1" [1] 1/16" ± 3/32" [1]	0.05" ±0.04" 1/16" ± 1/32"	0.04" ±0.04" 1/32" ± 1/32"	FRONT TOE (Inches)	0.09" ±0.08" 3/32" ± 3/32"	0.09" ±0.08" 3/32" ± 3/32"	
11.3 ±0.8 (11 5/16 ± ¾)	11.4 ±0.8 (11⅜ ± ¾)	11.4 ±0.8 (11⅜ ± ¾)	SAI	12.8 (12 13/16)	12.3 (12¼)	
41.0 / 35.8 ±2.0	40.7 / 35.1 ±2.0	32.4 / 29.4 ±2.0	TURNING ANGLE	41.0 / 34.0 ±3.0	37.4 / 31.8 ±2.0	
26	26	26	ILLUSTRATION #			
			REAR			
-0.95 ±0.75 -15/16 ± ¾	-0.95 ±0.75 -15/16 ± ¾	-0.95 ±0.75 -15/16 ± ¾	CAMBER (Degrees)	-1.48 ±1.0 -1½ ±1	-1.0 ±0.75 -1 ± ¾	
0.28° ±0.3° 9/32° ± 5/16°	0.3° ±0.24° 5/16° ± ¼°	0.3° ±0.24° 5/16° ± ¼°	REAR TOE (Degrees)	0.4° ±0.37° 13/32° ± ⅜°	0.18° ±0.3° 3/16° ± 5/16°	
0.14" ±0.15" ⅛" ± 5/32"	0.15" ±0.12" 5/32" ± ⅛"	0.15" ±0.12" 5/32" ± ⅛"	REAR TOE (Inches)	0.2" ±0.18" 3/16" ± 3/16"	0.09" ±0.15" 3/32" ± 5/32"	
			THRUST ANGLE	0.8 (13/16)		
			ILLUSTRATION #			
[1] 2006 = 0.2° (0.1")			NOTES			
44151	44152	44153		44154	44155	44156

Toyota Truck & SUV

4Runner	4Runner	4Runner	MODEL	4Runner	4Runner	4Runner
4X2	4X4	All models with	INFORMATION	All models with	4X2	4X4
		225/75R15 tires		265/70R16 tires		
1990-95	1990-95	1996	YEARS	1996	1997-99	1997-99
			INSTRUCTIONS			
			LUGNUT TORQUE			
			FRONT			
1.78 ±0.75 1¾ ± ¾	1.72 ±0.75 1¾ ± ¾	3.37 ±0.75 3⅜ ± ¾	CASTER (Degrees)	2.64 ±0.75 2⅝ ± ¾	3.08 ±0.75 3 1/16 ± ¾	2.42 ±0.75 2 7/16 ± ¾
0.0 (0) | 0.5 (½)	0.0 (0) | 0.5 (½)	0.0 (0) | 0.5 (½)	CROSS | TOL.	0.0 (0) | 0.5 (½)	0.0 (0) | 0.5 (½)	0.0 (0) | 0.5 (½)
0.72 ±0.75 11/16 ± ¾	0.72 ±0.75 11/16 ± ¾	-0.22 ±0.75 -¼ ± ¾	CAMBER (Degrees)	0.17 ±0.75 3/16 ± ¾	-0.25 ±0.75 -¼ ± ¾	0.17 ±0.75 3/16 ± ¾
0.0 (0) | 0.5 (½)	0.0 (0) | 0.5 (½)	0.0 (0) | 0.5 (½)	CROSS | TOL.	0.0 (0) | 0.5 (½)	0.0 (0) | 0.5 (½)	0.0 (0) | 0.5 (½)
0.19° ±0.2° 3/16° ± 3/16°	0.19° ±0.2° 3/16° ± 3/16°	0.16° ±0.2° 5/32° ± 3/16°	FRONT TOE (Degrees)	0.2° ±0.2° 3/16° ± 3/16°	0.2° ±0.2° 3/16° ± 3/16°	0.2° ±0.2° 3/16° ± 3/16°
0.1" ±0.1" 3/32" ± 3/32"	0.1" ±0.1" 3/32" ± 3/32"	0.08" ±0.1" 3/32" ± 3/32"	FRONT TOE (Inches)	0.1" ±0.1" 3/32" ± 3/32"	0.1" ±0.1" 3/32" ± 3/32"	0.1" ±0.1" 3/32" ± 3/32"
11.9 ±0.8 (11⅞ ± ¾)	11.9 ±0.8 (11⅞ ± ¾)	11.0 ±0.8 (11 ± ¾)	SAI	10.6 ±0.8 (10⅝ ± ¾)	11.0 ±0.8 (11 ± ¾)	10.6 ±0.8 (10⅝ ± ¾)
32.0 / 31.0 ±2.0	32.0 / 31.0 ±2.0	35.5 / 31.4 ±2.0	TURNING ANGLE	35.8 / 31.9 ±2.0	35.0 / 31.0 ±1.0	36.0 / 32.0 ±1.0
23	12	12	ILLUSTRATION #	12	12	12
			REAR			
			CAMBER (Degrees)			
			REAR TOE (Degrees)			
			REAR TOE (Inches)			
			THRUST ANGLE			
			ILLUSTRATION #			
			NOTES			
49002	49003	49004		49005	49006	49007

Toyota Truck & SUV

4Runner 4X2	4Runner 4X4	4Runner 4X2 w/o Air Susp.	MODEL INFORMATION	4Runner 4X2 w/Air Suspension	4Runner 4X4 w/o Air Susp.	4Runner 4X4 w/Air Suspension
2000-2002	2000-2002	2003-2009	YEARS	2003-2009	2003-2009	2003-2009
			INSTRUCTIONS			
			LUGNUT TORQUE			
			FRONT			
3.25±0.75 3¼±¾	3.08±0.75 3¹⁄₁₆±¾	3.38±0.75 3⅜±¾	CASTER (Degrees)	3.55±0.75 3 ⁹⁄₁₆±¾	3.22±0.75 3¼±¾	3.37±0.75 3⅜±¾
0.0 (0) | 0.5 (½)	0.0 (0) | 0.5 (½)	0.0 (0) | 0.75 (¾)	CROSS | TOL.	0.0 (0) | 0.75 (¾)	0.0 (0) | 0.75 (¾)	0.0 (0) | 0.75 (¾)
-0.25±0.75 -¼±¾	-0.25±0.75 -¼±¾	-0.47±0.75 -½±¾	CAMBER (Degrees)	-0.5±0.75 -½±¾	-0.15±0.75 -⅛±¾	-0.17±0.75 -³⁄₁₆±¾
0.0 (0) | 0.5 (½)	0.0 (0) | 0.5 (½)	0.0 (0) | 0.75 (¾)	CROSS | TOL.	0.0 (0) | 0.75 (¾)	0.0 (0) | 0.75 (¾)	0.0 (0) | 0.75 (¾)
0.2° ±0.2° ³⁄₁₆° ± ³⁄₁₆°	0.2° ±0.2° ³⁄₁₆° ± ³⁄₁₆°	0.08° ±0.17° ³⁄₃₂° ± ⁵⁄₃₂°	FRONT TOE (Degrees)	0.08° ±0.17° ³⁄₃₂° ± ⁵⁄₃₂°	0.08° ±0.17° ³⁄₃₂° ± ⁵⁄₃₂°	0.08° ±0.17° ³⁄₃₂° ± ⁵⁄₃₂°
0.1" ±0.1" ³⁄₃₂" ± ³⁄₃₂"	0.1" ±0.1" ³⁄₃₂" ± ³⁄₃₂"	0.04" ±0.08" ¹⁄₃₂" ± ³⁄₃₂"	FRONT TOE (Inches)	0.04" ±0.08" ¹⁄₃₂" ± ³⁄₃₂"	0.04" ±0.08" ¹⁄₃₂" ± ³⁄₃₂"	0.04" ±0.08" ¹⁄₃₂" ± ³⁄₃₂"
11.0±0.8 (11±¾)	11.0±0.8 (11±¾)	13.0±0.8 (13±¾)	SAI	13.0±0.8 (13±¾)	12.7±0.8 (12⅝+¾)	12.7±0.8 (12¹¹⁄₁₆±¾)
35.0 / 31.0 ±1.5	35.0 / 31.0 ±1.5	37.0 / 32.5 ±1.5	TURNING ANGLE	37.0 / 32.5 ±1.5	36.7 / 32.9 ±1.5	36.7 / 32.9 ±1.5
12	12	12	ILLUSTRATION #	12	12	12
			REAR			
			CAMBER (Degrees)			
			REAR TOE (Degrees)			
			REAR TOE (Inches)			
			THRUST ANGLE			
			ILLUSTRATION #			
			NOTES			
49008	49009	49010		49011	49012	49013

Toyota Truck & SUV

4Runner SR5 4X2 4 Cyl	4Runner SR5 4X2 4 Cyl.	4Runner SR5 4X2 6 Cyl	MODEL INFORMATION	4Runner SR5 4X4	4Runner Limited 4X2	4Runner Limited 4X4
2010	2011	2010-2017	YEARS	2010-2017	2010-2017	2010-2017
			INSTRUCTIONS			
83 ft.lbs.	83 ft.lbs.	83 ft.lbs.	LUGNUT TORQUE	83 ft.lbs.	83 ft.lbs.	83 ft.lbs.
			FRONT			
3.65±0.75 3⅝±¾	3.78±0.75 3¾±¾	3.67±0.75 3¹¹⁄₁₆±¾	CASTER (Degrees)	3.25±0.75 3¼±¾	3.75±0.75 3¾±¾	3.3±0.75 3 ⁵⁄₁₆±¾
0.0 (0) | 0.75 (¾)	0.0 (0) | 0.75 (¾)	0.0 (0) | 0.75 (¾)	CROSS | TOL.	0.0 (0) | 0.75 (¾)	0.0 (0) | 0.75 (¾)	0.0 (0) | 0.75 (¾)
-0.57±0.75 -⁹⁄₁₆±¾	-0.67±0.75 -¹¹⁄₁₆±¾	L= -0.6, R= -0.56±0.75 L= -⅝, R= -⅝±¾	CAMBER (Degrees)	0.0±0.75 0±¾	-0.62±0.75 -⅝±¾	-0.05±0.75 -¹⁄₁₆±¾
0.0 (0) | 0.75 (¾)	0.0 (0) | 0.75 (¾)	0.0 (0) | 0.75 (¾)	CROSS | TOL.	0.0 (0) | 0.75 (¾)	0.0 (0) | 0.75 (¾)	0.0 (0) | 0.75 (¾)
0.1° ±0.17° ³⁄₃₂° ± ⁵⁄₃₂°	0.15° ±0.17° ⁵⁄₃₂° ± ⁵⁄₃₂°	0.17° ±0.15° ⁵⁄₃₂° ± ⁵⁄₃₂°	FRONT TOE (Degrees)	0.17° ±0.15° ⁵⁄₃₂° ± ⁵⁄₃₂°	0.16° ±0.15° ⁵⁄₃₂° ± ⁵⁄₃₂°	0.15° ±0.15° ⁵⁄₃₂° ± ⁵⁄₃₂°
0.05" ±0.08" ¹⁄₁₆" ± ³⁄₃₂"	0.08" ±0.08" ¹⁄₁₆" ± ³⁄₃₂"	0.08" ±0.08" ³⁄₃₂" ± ¹⁄₁₆"	FRONT TOE (Inches)	0.08" ±0.08" ³⁄₃₂" ± ¹⁄₁₆"	0.08" ±0.08" ³⁄₃₂" ± ¹⁄₁₆"	0.08" ±0.08" ¹⁄₁₆" ± ¹⁄₁₆"
12.7±0.8 (12¹¹⁄₁₆±¾)	12.8±0.8 (12¾±¾)	12.7±0.8 (12¹¹⁄₁₆±¾)	SAI	12.2±0.8 (12³⁄₁₆±¾)	12.7±0.8 (12¹¹⁄₁₆±¾)	12.2±0.8 (12³⁄₁₆±¾)
31.3 / 29.6 ±1.5	31.3 / 29.1 ±1.5	31.3 / 29.6 ±1.5	TURNING ANGLE	31.9 / 29.5 ±1.5	31.3 / 29.5 ±1.5	31.8 / 29.5 ±1.5
12	12	12	ILLUSTRATION #	12	12	12
			REAR			
			CAMBER (Degrees)			
			REAR TOE (Degrees)			
			REAR TOE (Inches)			
			THRUST ANGLE			
			ILLUSTRATION #			
			NOTES			
49014	49015	49016		49017	49018	49019

Toyota Truck & SUV

4Runner Trail w/o KDSS (Dynamic Susp.)	4Runner Trail w/KDSS (Dynamic Susp.)	FJ Cruiser 2WD	MODEL INFORMATION	FJ Cruiser 4WD	Highlander FWD	Highlander AWD
2010-2017	2010-2017	2007-2014	YEARS	2007-2014	2001-2007	2001-2007
			INSTRUCTIONS			
83 ft.lbs.	83 ft.lbs.		LUGNUT TORQUE		76 ft.lbs.	76 ft.lbs.
			FRONT			
3.27 ±0.75 3¼ ± ¾	3.35 ±0.75 3⅜ ± ¾	3.57 ±0.5 [1] 3 9/16 ± ½ [1]	CASTER (Degrees)	2.82 ±0.5 [1] 2 13/16 ± ½ [1]	2.75 ±0.75 2¾ ± ¾	2.75 ±0.75 2¾ ± ¾
0.0 (0) | 0.75 (¾)	0.0 (0) | 0.75 (¾)	0.0 (0) | 0.5 (½)	CROSS | TOL.	0.0 (0) | 0.5 (½)	0.0 (0) | 0.75 (¾)	0.0 (0) | 0.75 (¾)
-0.05 ±0.75 -1/16 ± ¾	-0.12 ±0.75 -⅛ ± ¾	-0.57 ±0.5 -9/16 ± ½	CAMBER (Degrees)	0.15 ±0.5 ⅛ ± ½	-0.67 ±0.75 -11/16 ± ¾	-0.58 ±0.75 -9/16 ± ¾
0.0 (0) | 0.75 (¾)	0.0 (0) | 0.75 (¾)	0.0 (0) | 0.5 (½)	CROSS | TOL.	0.0 (0) | 0.5 (½)	0.0 (0) | 0.75 (¾)	0.0 (0) | 0.75 (¾)
0.16° ±0.15° 5/32° ± 5/32°	0.1° ±0.15° 3/32° ± 5/32°	0.08° ±0.17° 3/32° ± 5/32°	FRONT TOE (Degrees)	0.08° ±0.17° 3/32° ± 5/32°	0.0° ±0.17° 0° ± 5/32°	0.0° ±0.17° 0° ± 5/32°
0.08" ±0.08" 3/32" ± 1/16"	0.05" ±0.08" 1/16" ± 1/16"	0.04" ±0.08" 1/32" ± 3/32"	FRONT TOE (Inches)	0.04" ±0.08" 1/32" ± 3/32"	0.0" ±0.08" 0" ± 3/32"	0.0" ±0.08" 0" ± 3/32"
12.2 ±0.8 (12 3/16 ± ¾)	12.3 ±0.8 (12¼ ± ¾)	13.0 ±0.5 (13 ± ½)	SAI	12.4 ±0.5 (12⅜ ± ½)	10.8 ±0.8 (10¾ ± ¾)	10.6 ±0.8 (10 9/16 ± ¾)
31.9 / 29.5 ±1.5	31.8 / 29.4 ±1.5	32.8 / 28.9 ±1.5	TURNING ANGLE	33.2 / 29.6 ±1.5	35.5 / 31.2 ±1.5	35.7 / 31.4 ±1.5
12	12	12	ILLUSTRATION #	12	26	26
			REAR			
			CAMBER (Degrees)		-1.33 ±0.75 -1 5/16 ± ¾	-0.75 ±0.75 -¾ ± ¾
			REAR TOE (Degrees)		0.3° ±0.2° 5/16° ± 3/16°	0.3° ±0.2° 5/16° ± 3/16°
			REAR TOE (Inches)		0.15" ±0.1" 5/32" ± 3/32"	0.15" ±0.1" 5/32" ± 3/32"
			THRUST ANGLE			
			ILLUSTRATION #		10	06
		[1] 2011-12 = 4.22° (4¼°)	NOTES	[1] 2011-12 = 3.28° (3¼°)		
49020	49021	49031		49032	49034	49035

Toyota Truck & SUV

Highlander FWD	Highlander AWD	Highlander FWD	MODEL INFORMATION	Highlander AWD	Highlander FWD & AWD	Highlander HV FWD
2008	2008	2009-2013	YEARS	2009-2013	2014-2017	2006-2007
			INSTRUCTIONS			
76 ft.lbs.	76 ft.lbs.	76 ft.lbs.	LUGNUT TORQUE	76 ft.lbs.	76 ft.lbs.	76 ft.lbs.
			FRONT			
2.62 ±0.75 2⅝ ± ¾	2.62 ±0.75 2⅝ ± ¾	2.6 ±0.75 2⅝ ± ¾	CASTER (Degrees)	2.62 ±0.75 2⅝ ± ¾	2.67 ±0.75 2 11/16 ± ¾	2.75 ±0.75 2¾ ± ¾
0.0 (0) | 0.75 (¾)	0.0 (0) | 0.75 (¾)	0.0 (0) | 0.75 (¾)	CROSS | TOL.	0.0 (0) | 0.75 (¾)	0.0 (0) | 0.75 (¾)	0.0 (0) | 0.75 (¾)
-0.63 ±0.75 -⅝ ± ¾	-0.63 ±0.75 -⅝ ± ¾	-0.63 ±0.75 -⅝ ± ¾	CAMBER (Degrees)	-0.63 ±0.75 -⅝ ± ¾	-0.55 ±0.75 -9/16 ± ¾	-0.58 ±0.75 -9/16 ± ¾
0.0 (0) | 0.75 (¾)	0.0 (0) | 0.75 (¾)	0.0 (0) | 0.75 (¾)	CROSS | TOL.	0.0 (0) | 0.75 (¾)	0.0 (0) | 0.75 (¾)	0.0 (0) | 0.75 (¾)
0.07° ±0.16° 1/16° ± 5/32°	0.07° ±0.16° 1/16° ± 5/32°	0.07° ±0.2° 1/16° ± 3/16°	FRONT TOE (Degrees)	0.07° ±0.2° 1/16° ± 3/16°	0.08° ±0.17° 3/32° ± 5/32°	0.0° ±0.16° 0° ± 5/32°
0.04" ±0.08" 1/32" ± 3/32"	0.04" ±0.08" 1/32" ± 3/32"	0.04" ±0.1" 1/32" ± 3/32"	FRONT TOE (Inches)	0.04" ±0.1" 1/32" ± 3/32"	0.04" ±0.08" 1/32" ± 3/32"	0.0" ±0.08" 0" ± 3/32"
11.0 ±0.8 (11 ± ¾)	11.0 ±0.8 (11 ± ¾)	11.0 ±0.8 (11 ± ¾)	SAI	11.0 ±0.8 (11 ± ¾)	10.8 ±0.8 (10 13/16 ± ¾)	10.6 ±0.8 (10 9/16 ± ¾)
35.1 / 31.3 ±2.0	35.1 / 31.3 ±2.0	35.1 / 31.3 ±2.0	TURNING ANGLE	35.1 / 31.3 ±2.0	35.8 / 31.2 ±1.5	35.7 / 31.4 ±2.0
26	26	26	ILLUSTRATION #	26	26	26
			REAR			
-1.0 ±0.75 -1 ± ¾	-0.6 ±0.75 -⅝ ± ¾	-1.0 ±0.75 -1 ± ¾	CAMBER (Degrees)	-0.6 ±0.75 -⅝ ± ¾	-0.51 ±0.75 -½ ± ¾	-1.17 ±0.75 -1 3/16 ± ¾
0.24° ±0.16° ¼° ± 5/32°	0.24° ±0.16° ¼° ± 5/32°	0.33° ±0.2° 11/32° ± 3/16°	REAR TOE (Degrees)	0.28° ±0.2° 9/32° ± 3/16°	**-0.08°** ±0.17° **-3/32°** ± 5/32°	0.3° ±0.2° 5/16° ± 3/16°
0.12" ±0.08" ⅛" ± 3/32"	0.12" ±0.08" ⅛" ± 3/32"	0.17" ±0.1" 5/32" ± 3/32"	REAR TOE (Inches)	0.14" ±0.1" 5/32" ± 3/32"	**-0.04"** ±0.08" **-1/32"** ± 3/32"	0.15" ±0.1" 5/32" ± 3/32"
			THRUST ANGLE			
10	06	10	ILLUSTRATION #	06	06	10
			NOTES			
49036	49037	49038		49039	49040	49041

Toyota Truck & SUV

Highlander HV AWD	Highlander HV	Highlander HV	MODEL INFORMATION	Land Cruiser	Land Cruiser	Land Cruiser
2006-2007	2008-2013	2014-2017	YEARS	1990	1991-92	1993-97
			INSTRUCTIONS			
76 ft.lbs.	76 ft.lbs.	76 ft.lbs.	LUGNUT TORQUE			
			FRONT			
2.5 ±0.75 2½ ± ¾	2.58 ±0.75 2 9/16 ± ¾	2.67 ±0.75 2 11/16 ± ¾	CASTER (Degrees)	0.83 ±1.0 13/16 ±1	3.0 ±1.0 [1] 3 ±1 [1]	3.0 ±1.0 3 ±1
0.0 (0) | 0.75 (¾)	0.0 (0) | 0.75 (¾)	0.0 (0) | 0.75 (¾)	CROSS | TOL.	0.0 (0) | 0.5 (½)	0.0 (0) | 0.5 (½)	0.0 (0) | 0.75 (¾)
-0.58 ±0.75 - 9/16 ± ¾	-0.63 ±0.75 -5/8 ± ¾	-0.55 ±0.75 - 9/16 ± ¾	CAMBER (Degrees)	1.0 ±0.75 1 ± ¾	1.0 ±0.75 1 ± ¾	1.0 ±0.75 1 ± ¾
0.0 (0) | 0.75 (¾)	0.0 (0) | 0.75 (¾)	0.0 (0) | 0.75 (¾)	CROSS | TOL.	0.0 (0) | n/s	0.0 (0) | 0.5 (½)	0.0 (0) | 0.75 (¾)
0.0° ±0.16° 0° ± 5/32°	0.07° ±0.16° 1/16° ± 5/32°	0.08° ±0.17° 3/32° ± 5/32°	FRONT TOE (Degrees)	0.08° ±0.16° 3/32° ± 5/32°	0.08° ±0.16° 3/32° ± 5/32°	0.2° ±0.1° 3/16° ± 3/32°
0.0" ±0.08" 0" ± 3/32"	0.04" ±0.08" 1/32" ± 3/32"	0.04" ±0.08" 1/32" ± 3/32"	FRONT TOE (Inches)	0.04" ±0.08" 1/32" ± 3/32"	0.04" ±0.08" 1/32" ± 3/32"	0.1" ±0.05" 3/32" ± 1/16"
10.6 ±0.8 (10 9/16 ± ¾)	11.0 ±0.8 (11 ± ¾)	10.8 ±0.8 (10 13/16 ± ¾)	SAI	9.5 ±0.8 (9½ ± ¾)	13.5 ±0.8 (13½ ± ¾)	13.0 ±0.8 (13 ± ¾)
35.7 / 31.4 ±2.0	35.1 / 31.4 ±2.0	35.8 / 31.2 ±1.5	TURNING ANGLE	32.0 / 30.0 ±2.0	35.0 / 30.0 ±1.5	35.0 / 31.0 ±1.5
26	26	26	ILLUSTRATION #			
			REAR			
-0.67 ±0.75 - 11/16 ± ¾	-0.55 ±0.75 - 9/16 ± ¾	-0.16 ±0.75 - 3/16 ± ¾	CAMBER (Degrees)			
0.3° ±0.2° 5/16° ± 3/16°	0.28° ±0.16° 9/32° ± 5/32°	-0.17° ±0.17° - 5/32° ± 5/32°	REAR TOE (Degrees)			
0.15" ±0.1" 5/32" ± 3/32"	0.14" ±0.08" 1/8" ± 3/32"	-0.08" ±0.08" - 3/32" ± 3/32"	REAR TOE (Inches)			
			THRUST ANGLE			
06	06	06	ILLUSTRATION #			
			NOTES		[1] '91 w/31x10.5R15=0.75°(¾°) [1] '92 w/31x10.5R15=1.7°(1¾°)	
49042	49043	49044		49047	49048	49049

Toyota Truck & SUV

Land Cruiser	Land Cruiser	Land Cruiser	MODEL INFORMATION	Land Cruiser	Pickup 4X2 Reg. Cab 4 Cyl. Std. Bed	Pickup 4X2 Reg. Cab 4 Cyl. Std. Bed DLX
1998-2007	2008-2011	2013-2016	YEARS	2017	1990	1990
			INSTRUCTIONS			
	97 ft.lbs.	97 ft.lbs.	LUGNUT TORQUE	97 ft.lbs.		
			FRONT			
2.5 ±0.75 2½ ± ¾	2.9 ±0.75 2 7/8 ± ¾	2.93 ±0.75 2 15/16 ± ¾	CASTER (Degrees)	3.0 ±0.75 3 ± ¾	0.43 ±0.5 7/16 ± ½	0.67 ±0.5 11/16 ± ½
0.0 (0) | 0.5 (½)	0.0 (0) | 0.75 (¾)	0.0 (0) | 0.75 (¾)	CROSS | TOL.	0.0 (0) | 0.75 (¾)	0.0 (0) | 0.5 (½)	0.0 (0) | 0.5 (½)
0.08 ±0.75 1/16 ± ¾	0.13 ±0.75 1/8 ± ¾	0.28 ±0.75 ¼ ± ¾	CAMBER (Degrees)	0.4 ±0.75 3/8 ± ¾	0.5 ±0.5 ½ ± ½	0.5 ±0.5 ½ ± ½
0.0 (0) | 0.5 (½)	0.0 (0) | 0.75 (¾)	0.0 (0) | 0.75 (¾)	CROSS | TOL.	0.0 (0) | 0.75 (¾)	0.0 (0) | 0.5 (½)	0.0 (0) | 0.5 (½)
0.1° ±0.2° 3/32° ± 3/16°	0.24° ±0.16° ¼° ± 5/32°	0.24° ±0.16° ¼° ± 5/32°	FRONT TOE (Degrees)	0.22° ±0.16° 7/32° ± 5/32°	0.17° ±0.1° 5/32° ± 3/32°	0.13° ±0.1° 1/8° ± 3/32°
0.05" ±0.1" 1/16" ± 3/32"	0.12" ±0.08" 1/8" ± 1/16"	0.12" ±0.08" 1/8" ± 3/32"	FRONT TOE (Inches)	0.11" ±0.08" 1/8" ± 3/32"	0.09" ±0.05" 3/32" ± 1/16"	0.07" ±0.05" 1/16" ± 1/16"
12.2 ±0.8 (12 3/16 ± ¾)	12.9 ±0.8 (12 7/8 ± ¾)	12.7 ±0.8 (12 11/16 ± ¾)	SAI	12.6 ±0.8 (12 9/16 ± ¾)	10.0 ±0.8 (10 ± ¾)	10.0 ±0.8 (10 ± ¾)
36.7 / 32.6 ±0.5	35.0 / 32.2 ±1.5	35.5 / 32.4 ±1.5	TURNING ANGLE	35.6 / 32.6 ±1.5	34.0 / 30.0 ±2.0	34.0 / 30.0 ±2.0
24	12	12	ILLUSTRATION #	12	23	23
			REAR			
			CAMBER (Degrees)			
			REAR TOE (Degrees)			
			REAR TOE (Inches)			
			THRUST ANGLE			
			ILLUSTRATION #			
			NOTES			
49050	49051	49053		49054	49063	49064

Toyota Truck & SUV

Pickup 4X2	Pickup 4X2	Pickup 4X2	MODEL	Pickup 4X2	Pickup 4X2	Pickup 4X2
Reg. Cab 4 Cyl.	Reg. Cab 4 Cyl.	Reg. Cab 6 Cyl.	INFORMATION	X-Cab 4 Cyl.	X-Cab 4 Cyl.	X-Cab 6 Cyl.
Long Bed DLX	Long Bed SR5			SR5	DLX	DLX
1990	1990	1990	YEARS	1990	1990-95	1990
			INSTRUCTIONS			
			LUGNUT TORQUE			
			FRONT			
0.96±0.5 15/16 ± 1/2	1.12±0.5 1 1/8 ± 1/2	0.9±0.5 7/8 ± 1/2	CASTER (Degrees)	1.39±0.5 1 3/8 ± 1/2	1.25±0.5 1 1/4 ± 1/2	1.18±0.5 1 3/16 ± 1/2
0.0 (0) | 0.5 (1/2)	0.0 (0) | 0.5 (1/2)	0.0 (0) | 0.5 (1/2)	CROSS | TOL.	0.0 (0) | 0.5 (1/2)	0.0 (0) | 0.5 (1/2)	0.0 (0) | 0.5 (1/2)
0.47±0.5 7/16 ± 1/2	0.37±0.5 3/8 ± 1/2	0.47±0.5 1/2 ± 1/2	CAMBER (Degrees)	0.22±0.5 3/16 ± 1/2	0.35±0.5 3/8 ± 1/2	0.42±0.5 7/16 ± 1/2
0.0 (0) | 0.5 (1/2)	0.0 (0) | 0.5 (1/2)	0.0 (0) | 0.5 (1/2)	CROSS | TOL.	0.0 (0) | 0.5 (1/2)	0.0 (0) | 0.5 (1/2)	0.0 (0) | 0.5 (1/2)
0.21° ±0.1° 7/32° ± 3/32°	0.36° ±0.1° 3/8° ± 3/32°	0.21° ±0.1° 7/32° ± 3/32°	FRONT TOE (Degrees)	0.55° ±0.2° 9/16° ± 3/16°	0.3° ±0.2° 5/16° ± 3/16°	0.3° ±0.1° 5/16° ± 3/32°
0.11" ±0.05" 3/32" ± 1/16"	0.18" ±0.05" 3/16" ± 1/16"	0.11" ±0.05" 3/32" ± 1/16"	FRONT TOE (Inches)	0.28" ±0.1" 9/32" ± 3/32"	0.15" ±0.1" 5/32" ± 3/32"	0.15" ±0.05" 5/32" ± 1/16"
10.0±0.8 (10 ± 3/4)	10.1±0.8 (10 1/8 ± 3/4)	10.0±0.8 (10 ± 3/4)	SAI	10.3±0.8 (10 1/4 ± 3/4)	10.1±0.8 (10 1/8 ± 3/4)	10.1±0.8 (10 1/16 ± 3/4)
34.0 / 30.0 ±2.0	34.0 / 30.0 ±2.0	34.0 / 30.0 ±2.0	TURNING ANGLE	34.0 / 30.0 ±2.0	34.0 / 30.0 ±2.0 [1]	34.0 / 30.0 ±2.0
23	23	23	ILLUSTRATION #	23	23	23
			REAR			
			CAMBER (Degrees)			
			REAR TOE (Degrees)			
			REAR TOE (Inches)			
			THRUST ANGLE			
			ILLUSTRATION #			
			NOTES		[1] 1994-95 = 35°/31°	
49065	49066	49067		49068	49069	49070

Toyota Truck & SUV

Pickup 4X2	Pickup 4X2	Pickup 4X2	MODEL	Pickup 4X2	Pickup 4X2	Pickup 4X2
X-Cab 6 Cyl.	Reg. Cab 4 Cyl.	Reg. Cab 4 Cyl.	INFORMATION	Reg. Cab 4 Cyl.	Reg. Cab 6 Cyl.	Reg. Cab 6 Cyl.
SR5	Std. Bed	Std. Bed DLX		Long Bed	Manual Trans.	Auto. Trans.
1990-93	1991-95	1991-95	YEARS	1991-93	1991-93	1991-93
			INSTRUCTIONS			
			LUGNUT TORQUE			
			FRONT			
1.2±0.75 1 3/16 ± 3/4	0.75±0.75 3/4 ± 3/4	0.75±0.75 3/4 ± 3/4	CASTER (Degrees)	1.0±0.75 1 ± 3/4	0.55±0.75 9/16 ± 3/4	0.55±0.75 9/16 ± 3/4
0.0 (0) | 0.5 (1/2)	0.0 (0) | 0.5 (1/2)	0.0 (0) | 0.5 (1/2)	CROSS | TOL.	0.0 (0) | 0.5 (1/2)	0.0 (0) | 0.5 (1/2)	0.0 (0) | 0.5 (1/2)
0.42±0.75 7/16 ± 3/4	0.5±0.75 1/2 ± 3/4	0.5±0.75 1/2 ± 3/4	CAMBER (Degrees)	0.45±0.75 7/16 ± 3/4	0.5±0.75 1/2 ± 3/4	0.5±0.75 1/2 ± 3/4
0.0 (0) | 0.5 (1/2)	0.0 (0) | 0.5 (1/2)	0.0 (0) | 0.5 (1/2)	CROSS | TOL.	0.0 (0) | 0.5 (1/2)	0.0 (0) | 0.5 (1/2)	0.0 (0) | 0.5 (1/2)
0.27° ±0.2° 9/32° ± 3/16°	0.17° ±0.2° 5/32° ± 3/16°	0.13° ±0.2° 1/8° ± 3/16°	FRONT TOE (Degrees)	0.2° ±0.2° 3/16° ± 3/16°	0.56° ±0.2° 9/16° ± 3/16°	0.48° ±0.2° 15/32° ± 3/16°
0.14" ±0.1" 1/8" ± 3/32"	0.09" ±0.1" 3/32" ± 3/32"	0.07" ±0.1" 1/16" ± 3/32"	FRONT TOE (Inches)	0.1" ±0.1" 3/32" ± 3/32"	0.28" ±0.1" 9/32" ± 3/32"	0.24" ±0.1" 1/4" ± 3/32"
10.1±0.8 (10 1/16 ± 3/4)	10.0±0.8 (10 ± 3/4)	10.0±0.8 (10 ± 3/4)	SAI	10.0±0.8 (10 ± 3/4)	10.0±0.8 (10 ± 3/4)	10.0±0.8 (10 ± 3/4)
34.0 / 30.0 ±2.0	35.0 / 31.0 ±2.0	35.0 / 31.0 ±2.0	TURNING ANGLE	34.0 / 30.0 ±2.0	34.0 / 30.0 ±2.0	34.0 / 30.0 ±2.0
23	23	23	ILLUSTRATION #	23	23	23
			REAR			
			CAMBER (Degrees)			
			REAR TOE (Degrees)			
			REAR TOE (Inches)			
			THRUST ANGLE			
			ILLUSTRATION #			
			NOTES			
49071	49072	49073		49074	49075	49076

Toyota Truck & SUV

Pickup 4X2 X-Cab 6 Cyl. DLX	Pickup 4X2 X-Cab 6 Cyl. SR5	Pickup 4X2 1 Ton DLX	MODEL INFORMATION	Pickup 4X4 Reg. Cab 4 Cyl. Std. Bed	Pickup 4X4 Reg. Cab 4 Cyl. Long Bed	Pickup 4X4 Reg. Cab 6 Cyl. Std. Bed
1991-95	1994-95	1990-93	YEARS	1990	1990	1990
			INSTRUCTIONS			
			LUGNUT TORQUE			
			FRONT			
1.25 ±0.75 1¼ ± ¾	1.1 ±0.75 1⅛ ± ¾	0.6 ±0.75 ⅝ ± ¾	CASTER (Degrees)	1.42 ±0.5 1 7⁄16 ± ½	1.5 ±0.5 1½ ± ½	1.5 ±0.5 1½ ± ½
0.0 (0) | 0.75 (¾)	0.0 (0) | 0.75 (¾)	0.0 (0) | 0.5 (½)	CROSS | TOL.	0.0 (0) | 0.5 (½)	0.0 (0) | 0.5 (½)	0.0 (0) | 0.5 (½)
0.35 ±0.75 ⅜ ± ¾	0.33 ±0.75 5⁄16 ± ¾	0.5 ±0.75 ½ ± ¾	CAMBER (Degrees)	0.58 ±0.5 9⁄16 ± ½	0.58 ±0.5 9⁄16 ± ½	0.67 ±0.5 11⁄16 ± ½
0.0 (0) | 0.5 (½)	0.0 (0) | 0.75 (¾)	0.0 (0) | 0.5 (½)	CROSS | TOL.	0.0 (0) | 0.5 (½)	0.0 (0) | 0.5 (½)	0.0 (0) | 0.5 (½)
0.32° ±0.2° 5⁄16° ± 3⁄16°	0.41° ±0.2° 13⁄32° ± 3⁄16°	0.5° ±0.2° ½° ± 3⁄16°	FRONT TOE (Degrees)	0.2° ±0.1° 3⁄16° ± 3⁄32°	0.3° ±0.1° 5⁄16° ± 3⁄32°	0.2° ±0.1° 3⁄16° ± 3⁄32°
0.16" ±0.1" 5⁄32" ± 3⁄32"	0.21" ±0.1" 7⁄32" ± 3⁄32"	0.25" ±0.1" ¼" ± 3⁄32"	FRONT TOE (Inches)	0.1" ±0.05" 3⁄32" ± 1⁄16"	0.15" ±0.05" 5⁄32" ± 1⁄16"	0.1" ±0.05" 3⁄32" ± 1⁄16"
10.1 ±0.8 (10⅛ ± ¾)	10.1 ±0.8 (10 1⁄16 ± ¾)	10.0 ±0.8 (10 ± ¾)	SAI	12.0 ±0.8 (12 ± ¾)	12.0 ±0.8 (12 ± ¾)	11.9 ±0.8 (11 15⁄16 ± ¾)
34.0 / 30.0 ±2.0 [1]	35.0 / 31.0 ±2.0	34.0 / 30.0 ±2.0	TURNING ANGLE	32.0 / 31.0 ±2.0	32.0 / 31.0 ±2.0	32.0 / 31.0 ±2.0
23	23	23	ILLUSTRATION #	12	12	12
			REAR			
			CAMBER (Degrees)			
			REAR TOE (Degrees)			
			REAR TOE (Inches)			
			THRUST ANGLE			
			ILLUSTRATION #			
[1] 1994-95 = 35°/31°			NOTES			
49077	49078	49079		49084	49085	49086

Toyota Truck & SUV

Pickup 4X4 Reg. Cab 6 Cyl. Long Bed	Pickup 4X4 X-Cab 4 Cyl. SR5, DLX	Pickup 4X4 X-Cab 6 Cyl. SR5, DLX	MODEL INFORMATION	Pickup 4X4 Reg. Cab All models	Pickup 4X4 X-Cab 4 Cyl. DLX	Pickup 4X4 X-Cab 6 Cyl. SR5, DLX
1990	1990	1990	YEARS	1991-95	1991-95	1991-93
			INSTRUCTIONS			
			LUGNUT TORQUE			
			FRONT			
1.58 ±0.5 1 9⁄16 ± ½	1.67 ±0.5 1 11⁄16 ± ½	1.75 ±0.5 1¾ ± ½	CASTER (Degrees)	1.65 ±0.5 1⅝ ± ½	1.8 ±0.5 1 13⁄16 ± ½	1.9 ±0.5 1⅞ ± ½
0.0 (0) | 0.5 (½)	0.0 (0) | 0.5 (½)	0.0 (0) | 0.5 (½)	CROSS | TOL.	0.0 (0) | 0.5 (½)	0.0 (0) | 0.5 (½)	0.0 (0) | 0.5 (½)
0.58 ±0.5 9⁄16 ± ½	0.5 ±0.5 ½ ± ½	0.58 ±0.5 9⁄16 ± ½	CAMBER (Degrees)	0.7 ±0.75 11⁄16 ± ¾	0.67 ±0.5 11⁄16 ± ½	0.67 ±0.5 11⁄16 ± ½
0.0 (0) | 0.5 (½)	0.0 (0) | 0.5 (½)	0.0 (0) | 0.5 (½)	CROSS | TOL.	0.0 (0) | 0.75 (¾)	0.0 (0) | 0.5 (½)	0.0 (0) | 0.5 (½)
0.2° ±0.1° 3⁄16° ± 3⁄32°	0.3° ±0.1° 5⁄16° ± 3⁄32°	0.3° ±0.1° 5⁄16° ± 3⁄32°	FRONT TOE (Degrees)	0.19° ±0.2° [1] 3⁄16° ± 3⁄16° [1]	0.27° ±0.2° [1] 9⁄32° ± 3⁄16° [1]	0.25° ±0.2° ¼° ± 3⁄16°
0.1" ±0.05" 3⁄32" ± 1⁄16"	0.15" ±0.05" 5⁄32" ± 1⁄16"	0.15" ±0.05" 5⁄32" ± 1⁄16"	FRONT TOE (Inches)	0.1" ±0.1" [1] 3⁄32" ± 3⁄32" [1]	0.14" ±0.1" [1] ⅛" ± 3⁄32" [1]	0.13" ±0.1" ⅛" ± 3⁄32"
12.0 ±0.8 (12 ± ¾)	12.1 ±0.8 (12⅛ ± ¾)	12.0 ±0.8 (12 ± ¾)	SAI	11.9 ±0.8 (11⅞ ± ¾)	11.9 ±0.8 (11⅞ ± ¾)	11.9 ±0.8 (11⅞ ± ¾)
32.0 / 31.0 ±2.0	32.0 / 31.0 ±2.0	32.0 / 31.0 ±2.0	TURNING ANGLE	32.0 / 31.0 ±2.0	32.0 / 31.0 ±2.0	32.0 / 31.0 ±2.0
12	12	12	ILLUSTRATION #	12	12	12
			REAR			
			CAMBER (Degrees)			
			REAR TOE (Degrees)			
			REAR TOE (Inches)			
			THRUST ANGLE			
			ILLUSTRATION #			
			NOTES	[1] Long Bed = 0.22° (0.11")	[1] 1995 = 0.17° (0.09")	
49087	49088	49089		49090	49091	49092

ALIGNMENT SPECIFICATIONS

Toyota Truck & SUV

Pickup 4X4	RAV4 4X2	RAV4 4X2	MODEL	RAV4 4X2	RAV4 4X2	RAV4 4X2
X-Cab 6 Cyl.	All Models	3 Door	INFORMATION	3 Door	5 Door	5 Door
SR5, DLX						
1994-95	1996	1997-2000	YEARS	2001	1997-2000	2001-2005
			INSTRUCTIONS			
			LUGNUT TORQUE			
			FRONT			
1.9 ± 0.5 1 7/8 ± 1/2	1.42 ± 0.75 1 7/16 ± 3/4	1.42 ± 0.75 1 7/16 ± 3/4	CASTER (Degrees)	2.0 ± 0.75 2 ± 3/4	1.42 ± 0.75 1 7/16 ± 3/4	2.0 ± 0.75 2 ± 3/4
0.0 (0) | 0.5 (1/2)	0.0 (0) | 0.75 (3/4)	0.0 (0) | 0.75 (3/4)	CROSS | TOL.	0.0 (0) | 0.75 (3/4)	0.0 (0) | 0.75 (3/4)	0.0 (0) | 0.75 (3/4)
0.73 ± 0.5 3/4 ± 1/2	-0.33 ± 0.75 - 5/16 ± 3/4	-0.33 ± 0.75 - 5/16 ± 3/4	CAMBER (Degrees)	-0.42 ± 0.75 - 7/16 ± 3/4	-0.33 ± 0.75 - 5/16 ± 3/4	-0.42 ± 0.75 - 7/16 ± 3/4
0.0 (0) | 0.5 (1/2)	0.0 (0) | 0.75 (3/4)	0.0 (0) | 0.75 (3/4)	CROSS | TOL.	0.0 (0) | 0.75 (3/4)	0.0 (0) | 0.75 (3/4)	0.0 (0) | 0.75 (3/4)
0.17° ± 0.2° 5/32° ± 3/16°	0.0° ± 0.2° 0° ± 3/16°	0.0° ± 0.2° 0° ± 3/16°	FRONT TOE (Degrees)	0.0° ± 0.2° 0° ± 3/16°	0.0° ± 0.2° 0° ± 3/16°	0.0° ± 0.2° 0° ± 3/16°
0.09" ± 0.1" 3/32" ± 3/32"	0.0" ± 0.1" 0" ± 3/32"	0.0" ± 0.1" 0" ± 3/32"	FRONT TOE (Inches)	0.0" ± 0.1" 0" ± 3/32"	0.0" ± 0.1" 0" ± 3/32"	0.0" ± 0.1" 0" ± 3/32"
11.9 ± 0.8 (11 7/8 ± 3/4)	11.0 ± 0.8 (11 ± 3/4)	11.0 ± 0.8 (11 ± 3/4)	SAI	11.0 ± 0.8 (11 ± 3/4)	11.0 ± 0.8 (11 ± 3/4)	11.0 ± 0.8 (11 ± 3/4)
32.0 / 31.0 ± 2.0	33.3 / 28.1 ± 2.0	33.3 / 28.1 ± 2.0	TURNING ANGLE	35.0 / 30.0 ± 2.0	33.3 / 28.1 ± 2.0	35.0 / 30.0 ± 2.0
12	26	26	ILLUSTRATION #	26	26	26
			REAR			
	-1.33 ± 0.75 -1 5/16 ± 3/4	-1.33 ± 0.75 -1 5/16 ± 3/4	CAMBER (Degrees)	-1.33 ± 0.75 -1 5/16 ± 3/4	-1.25 ± 0.75 -1 1/4 ± 3/4	-1.25 ± 0.75 -1 1/4 ± 3/4
	0.33° ± 0.17° 11/32° ± 5/32°	0.4° ± 0.2° 13/32° ± 3/16°	REAR TOE (Degrees)	0.4° ± 0.2° 13/32° ± 3/16°	0.3° ± 0.2° 5/16° ± 3/16°	0.4° ± 0.2° 13/32° ± 3/16°
	0.17" ± 0.08" 5/32" ± 3/32"	0.2" ± 0.1" 3/16" ± 3/32"	REAR TOE (Inches)	0.2" ± 0.1" 3/16" ± 3/32"	0.15" ± 0.1" 5/32" ± 3/32"	0.2" ± 0.1" 3/16" ± 3/32"
			THRUST ANGLE			
	40 54	40 54	ILLUSTRATION #	40 54	40 54	40 54
			NOTES			
49093	49094	49095		49096	49097	49098

Toyota Truck & SUV

RAV4 4X4	RAV4 4X4	RAV4 4X4	MODEL	RAV4 4X4	RAV4 4X4	RAV4 4X4
All Models	3 Door	3 Door	INFORMATION	3 Door	3 Door	5 Door
	w/215-70R16 tires	w/215-70R16 tires		w/235-60R16 tires	w/235-60R16 tires	w/215-70R16 tires
1996	1997-2000	2001	YEARS	1997-2000	2001	1997-2000
			INSTRUCTIONS			
			LUGNUT TORQUE			
			FRONT			
1.33 ± 0.75 1 5/16 ± 3/4	1.33 ± 0.75 1 5/16 ± 3/4	1.92 ± 0.75 1 15/16 ± 3/4	CASTER (Degrees)	1.42 ± 0.75 1 7/16 ± 3/4	1.92 ± 0.75 1 15/16 ± 3/4	1.33 ± 0.75 1 5/16 ± 3/4
0.0 (0) | 0.75 (3/4)	0.0 (0) | 0.75 (3/4)	0.0 (0) | 0.75 (3/4)	CROSS | TOL.	0.0 (0) | 0.75 (3/4)	0.0 (0) | 0.75 (3/4)	0.0 (0) | 0.75 (3/4)
-0.25 ± 0.75 - 1/4 ± 3/4	-0.25 ± 0.75 - 1/4 ± 3/4	-0.33 ± 0.75 - 5/16 ± 3/4	CAMBER (Degrees)	-0.33 ± 0.75 - 5/16 ± 3/4	-0.33 ± 0.75 - 5/16 ± 3/4	-0.25 ± 0.75 - 1/4 ± 3/4
0.0 (0) | 0.75 (3/4)	0.0 (0) | 0.75 (3/4)	0.0 (0) | 0.75 (3/4)	CROSS | TOL.	0.0 (0) | 0.75 (3/4)	0.0 (0) | 0.75 (3/4)	0.0 (0) | 0.75 (3/4)
0.0° ± 0.2° 0° ± 3/16°	0.0° ± 0.2° 0° ± 3/16°	0.08° ± 0.17° 3/32° ± 5/32°	FRONT TOE (Degrees)	0.0° ± 0.2° 0° ± 3/16°	0.08° ± 0.17° 3/32° ± 5/32°	0.0° ± 0.2° 0° ± 3/16°
0.0" ± 0.1" 0" ± 3/32"	0.0" ± 0.1" 0" ± 3/32"	0.04" ± 0.08" 1/32" ± 3/32"	FRONT TOE (Inches)	0.0" ± 0.1" 0" ± 3/32"	0.04" ± 0.08" 1/32" ± 3/32"	0.0" ± 0.1" 0" ± 3/32"
10.8 ± 0.8 (10 3/4 ± 3/4)	10.8 ± 0.8 (10 3/4 ± 3/4)	10.8 ± 0.8 (10 3/4 ± 3/4)	SAI	11.0 ± 0.8 (11 ± 3/4)	10.8 ± 0.8 (10 3/4 ± 3/4)	10.8 ± 0.8 (10 3/4 ± 3/4)
33.3 / 28.1 ± 2.0	33.3 / 28.1 ± 2.0	35.0 / 31.0 ± 2.0	TURNING ANGLE	33.3 / 28.1 ± 2.0	35.0 / 31.0 ± 2.0	33.3 / 28.1 ± 2.0
26	26	26	ILLUSTRATION #	26	26	26
			REAR			
-1.08 ± 0.75 -1 1/16 ± 3/4	-1.08 ± 0.75 -1 1/16 ± 3/4	-1.0 ± 0.75 -1 ± 3/4	CAMBER (Degrees)	-1.42 ± 0.75 -1 7/16 ± 3/4	-0.92 ± 0.75 - 15/16 ± 3/4	-0.92 ± 0.75 - 15/16 ± 3/4
0.25° ± 0.17° 1/4° ± 5/32°	0.3° ± 0.2° 5/16° ± 3/16°	0.25° ± 0.17° 1/4° ± 5/32°	REAR TOE (Degrees)	0.4° ± 0.2° 13/32° ± 3/16°	0.25° ± 0.17° 1/4° ± 5/32°	0.3° ± 0.2° 5/16° ± 3/16°
0.13" ± 0.08" 1/8" ± 3/32"	0.15" ± 0.1" 5/32" ± 3/32"	0.13" ± 0.08" 1/8" ± 3/32"	REAR TOE (Inches)	0.2" ± 0.1" 3/16" ± 3/32"	0.13" ± 0.08" 1/8" ± 3/32"	0.15" ± 0.1" 5/32" ± 3/32"
			THRUST ANGLE			
40 54	40 54	40 54	ILLUSTRATION #	40 54	40 54	40 54
			NOTES			
49099	49100	49101		49102	49103	49104

Toyota Truck & SUV

RAV4 4X4	RAV4 4X4	RAV4 4X4	MODEL	RAV4	RAV4	RAV4
5 Door	5 Door	5 Door	INFORMATION	All w/16" tires	17" or 18" tires	17" or 18" tires
w/215-70R16 tires	w/235-60R16 tires	w/235-60R16 tires			w/o 3rd row seating	w/3rd row seating
2001-2005	1997-2000	2001-2005	YEARS	2006-2012	2006-2012	2006-2012
			INSTRUCTIONS			
			LUGNUT TORQUE			
			FRONT			
1.92 ± 0.75 1 15/16 ± ¾	1.42 ± 0.75 1 7/16 ± ¾	1.92 ± 0.75 1 15/16 ± ¾	CASTER (Degrees)	5.72 ± 0.75 5¾ ± ¾	5.83 ± 0.75 5 13/16 ± ¾	5.62 ± 0.75 5⅝ ± ¾
0.0 (0) | 0.75 (¾)	0.0 (0) | 0.75 (¾)	0.0 (0) | 0.75 (¾)	CROSS | TOL.	0.0 (0) | 0.75 (¾)	0.0 (0) | 0.75 (¾)	0.0 (0) | 0.75 (¾)
-0.33 ± 0.75 -5/16 ± ¾	-0.33 ± 0.75 -5/16 ± ¾	-0.33 ± 0.75 -5/16 ± ¾	CAMBER (Degrees)	-0.13 ± 0.75 -⅛ ± ¾	-0.18 ± 0.75 -3/16 ± ¾	-0.18 ± 0.75 -3/16 ± ¾
0.0 (0) | 0.75 (¾)	0.0 (0) | 0.75 (¾)	0.0 (0) | 0.75 (¾)	CROSS | TOL.	0.0 (0) | 0.75 (¾)	0.0 (0) | 0.75 (¾)	0.0 (0) | 0.75 (¾)
0.08° ± 0.17° 3/32° ± 5/32°	0.0° ± 0.2° 0° ± 3/16°	0.08° ± 0.17° 3/32° ± 5/32°	FRONT TOE (Degrees)	0.08° ± 0.17° 3/32° ± 5/32°	0.08° ± 0.17° 3/32° ± 5/32°	0.08° ± 0.17° 3/32° ± 5/32°
0.04" ± 0.08" 1/32" ± 3/32"	0.0" ± 0.1" 0" ± 3/32"	0.04" ± 0.08" 1/32" ± 3/32"	FRONT TOE (Inches)	0.04" ± 0.08" 1/32" ± 3/32"	0.04" ± 0.08" 1/32" ± 3/32"	0.04" ± 0.08" 1/32" ± 3/32"
10.8 ± 0.8 (10¾ ± ¾)	11.0 ± 0.8 (11 ± ¾)	10.8 ± 0.8 (10¾ ± ¾)	SAI	11.3 ± 0.5 (11¼ ± ½)	11.4 ± 0.5 (11 7/16 ± ½)	11.4 ± 0.5 (11 7/16 ± ½)
35.0 / 31.0 ±2.0	33.3 / 28.1 ±2.0	35.0 / 31.0 ±2.0	TURNING ANGLE	38.9 / 32.3 ±2.0	35.8 / 30.5 ±2.0 [1]	35.8 / 30.5 ±2.0 [1]
26	26	26	ILLUSTRATION #	26	26	26
			REAR			
-0.92 ± 0.75 -15/16 ± ¾	-1.33 ± 0.75 -1 5/16 ± ¾	-0.83 ± 0.75 -13/16 ± ¾	CAMBER (Degrees)	-0.92 ± 0.75 -15/16 ± ¾	-1.07 ± 0.75 -1 1/16 ± ¾	-0.87 ± 0.75 -⅞ ± ¾
0.25° ± 0.17° ¼° ± 5/32°	0.3° ± 0.2° 5/16° ± 3/16°	0.25° ± 0.17° ¼° ± 5/32°	REAR TOE (Degrees)	0.18° ± 0.17° 3/16° ± 5/32°	0.18° ± 0.17° 3/16° ± 5/32°	0.18° ± 0.17° 3/16° ± 5/32°
0.13" ± 0.08" ⅛" ± 3/32"	0.15" ± 0.1" 5/32" ± 3/32"	0.13" ± 0.08" ⅛" ± 3/32"	REAR TOE (Inches)	0.09" ± 0.08" 3/32" ± 3/32"	0.09" ± 0.08" 3/32" ± 3/32"	0.09" ± 0.08" 3/32" ± 3/32"
			THRUST ANGLE			
40 54	40 54	40 54	ILLUSTRATION #	135	135	135
			NOTES		[1] 2009-11 = 38.7°/32.2°	[1] 2009-11 = 38.7°/32.2°
49105	49106	49107		49108	49109	49110

Toyota Truck & SUV

RAV4	RAV4	RAV4 EV	MODEL	RAV4 EV	Sequoia 2WD	Sequoia 4WD
All mods. inc/HV	All mods. inc/HV	3 Door	INFORMATION			
with 17 tires (OE)	with 18 tires (OE)	w/235-60R16 tires				
2013-2017	2013-2017	1998-2003	YEARS	2013-2014	2001-2004	2001-2004
			INSTRUCTIONS			
			LUGNUT TORQUE			
			FRONT			
5.92 ± 0.75 5 15/16 ± ¾	5.92 ± 0.75 5 15/16 ± ¾	1.6 ± 0.75 1⅝ ± ¾	CASTER (Degrees)	5.83 ± 0.75 5 13/16 ± ¾	3.0 ± 0.75 3 ± ¾	2.6 ± 0.75 2⅝ ± ¾
0.0 (0) | 0.5 (½)	0.0 (0) | 0.5 (½)	0.0 (0) | 0.75 (¾)	CROSS | TOL.	0.0 (0) | 0.5 (½)	0.0 (0) | 0.5 (½)	0.0 (0) | 0.5 (½)
-0.2 ± 0.75 -3/16 ± ¾	-0.22 ± 0.75 -¼ ± ¾	-0.32 ± 0.75 -5/16 ± ¾	CAMBER (Degrees)	-0.18 ± 0.75 -3/16 ± ¾	0.13 ± 0.75 ⅛ ± ¾	0.27 ± 0.75 ¼ ± ¾
0.0 (0) | 0.5 (½)	0.0 (0) | 0.5 (½)	0.0 (0) | 0.75 (¾)	CROSS | TOL.	0.0 (0) | 0.5 (½)	0.0 (0) | 0.5 (½)	0.0 (0) | 0.5 (½)
0.05° ± 0.17° [1] 1/16° ± 5/32° [1]	0.05° ± 0.17° [1] 1/16° ± 5/32° [1]	-0.07° ± 0.2° -1/16° ± 3/16°	FRONT TOE (Degrees)	0.08° ± 0.17° 3/32° ± 5/32°	0.09° ± 0.15° 3/32° ± 5/32°	0.09° ± 0.15° 3/32° ± 5/32°
0.03" ± 0.08" [1] 1/32" ± 3/32" [1]	0.03" ± 0.08" [1] 1/32" ± 3/32" [1]	-0.04" ± 0.1" -1/32" ± 3/32"	FRONT TOE (Inches)	0.04" ± 0.08" 1/32" ± 3/32"	0.05" ± 0.08" 1/32" ± 1/16"	0.05" ± 0.08" 1/32" ± 1/16"
11.6 ± 0.5 (11 9/16 ± ½)	11.6 ± 0.5 (11 9/16 ± ½)	10.5 ± 0.8 (10½ ± ¾)	SAI	11.4 ± 0.5 (11 7/16 ± ½)	10.6 ± 0.8 (10⅝ ± ¾)	10.5 ± 0.8 (10½ ± ¾)
38.6 / 32.2 ±2.0	35.7 / 30.4 ±2.0	33.3 / 28.2 ±2.0	TURNING ANGLE	38.7 / 32.2 ±2.0	35.5 / 32.0 ±1.5	35.5 / 32.0 ±1.5
26	26	26	ILLUSTRATION #	26	12	12
			REAR			
-1.28 ± 0.75 -1¼ ± ¾	-1.25 ± 0.75 -1¼ ± ¾	-0.83 ± 0.75 -13/16 ± ¾	CAMBER (Degrees)	-1.07 ± 0.75 -1 1/16 ± ¾		
0.18° ± 0.17° 3/16° ± 5/32°	0.18° ± 0.17° 3/16° ± 5/32°	0.1° ± 0.2° 3/32° ± 3/16°	REAR TOE (Degrees)	0.18° ± 0.17° 3/16° ± 5/32°		
0.09" ± 0.08" 3/32" ± 3/32"	0.09" ± 0.08" 3/32" ± 3/32"	0.05" ± 0.1" 1/16" ± 3/32"	REAR TOE (Inches)	0.09" ± 0.08" 3/32" ± 3/32"		
			THRUST ANGLE			
174	174	40 54	ILLUSTRATION #	174		
[1] 2016 = 0.12° (0.06")	[1] 2016 = 0.12° (0.06")		NOTES			
49111	49112	49113		49114	49115	49116

Toyota Truck & SUV

Sequoia 2WD	Sequoia 4WD	Sequoia 2WD w/Coil Susp.	MODEL INFORMATION	Sequoia 2WD w/Air Susp.	Sequoia 4WD w/Coil Susp.	Sequoia 4WD w/Air Susp.
2005-2007	2005-2007	2008-2017	YEARS	2008-2017	2008-2017	2008-2017
			INSTRUCTIONS			
		97 ft.lbs.[1]	LUGNUT TORQUE	97 ft.lbs.[1]	97 ft.lbs.[1]	97 ft.lbs.[1]
			FRONT			
3.0 ± 0.75[1] 3 ± ¾[1]	2.6 ± 0.75[1] 2⅝ ± ¾[1]	3.75 ± 0.75 3¾ ± ¾	CASTER (Degrees)	3.93 ± 0.75 3 15/16 ± ¾	3.6 ± 0.75 3⅝ ± ¾	3.75 ± 0.75 3¾ ± ¾
0.0 (0) | 0.5 (½)	0.0 (0) | 0.5 (½)	0.0 (0) | 0.5 (½)	CROSS | TOL.	0.0 (0) | 0.5 (½)	0.0 (0) | 0.5 (½)	0.0 (0) | 0.5 (½)
0.12 ± 0.75 ⅛ ± ¾	0.27 ± 0.75 ¼ ± ¾	-0.15 ± 0.75 -⅛ ± ¾	CAMBER (Degrees)	-0.14 ± 0.75 -⅛ ± ¾	0.0 ± 0.75 0 ± ¾	0.0 ± 0.75 0 ± ¾
0.0 (0) | 0.5 (½)	0.0 (0) | 0.5 (½)	0.0 (0) | 0.5 (½)	CROSS | TOL.	0.0 (0) | 0.5 (½)	0.0 (0) | 0.5 (½)	0.0 (0) | 0.5 (½)
0.08° ± 0.15° 3/32° ± 5/32°	0.08° ± 0.15° 3/32° ± 5/32°	0.2° ± 0.16° 3/16° ± 5/32°	FRONT TOE (Degrees)	0.2° ± 0.16° 3/16° ± 5/32°	0.2° ± 0.16° 3/16° ± 5/32°	0.2° ± 0.16° 3/16° ± 5/32°
0.04" ± 0.08" 1/32" ± 1/16"	0.04" ± 0.08" 1/32" ± 1/16"	0.1" ± 0.08" 3/32" ± 3/32"	FRONT TOE (Inches)	0.1" ± 0.08" 3/32" ± 3/32"	0.1" ± 0.08" 3/32" ± 3/32"	0.1" ± 0.08" 3/32" ± 3/32"
10.3 ± 0.8 (10¼ ± ¾)	10.1 ± 0.8 (10⅛ ± ¾)	13.2 ± 0.8 (13⅛ ± ¾)	SAI	13.2 ± 0.8 (13⅛ ± ¾)	13.0 ± 0.8 (13 1/16 ± ¾)	13.0 ± 0.8 (13 1/16 ± ¾)
35.4 / 32.0 ±1.5	35.6 / 32.3 ±1.5	41.0 / 36.4 ±1.5	TURNING ANGLE	41.0 / 36.4 ±1.5	41.3 / 36.4 ±1.5	41.3 / 36.4 ±1.5
12	12	12	ILLUSTRATION #	12	12	12
			REAR			
		-1.35 ± 0.5 -1⅜ ± ½	CAMBER (Degrees)	-1.62 ± 0.5 -1⅝ ± ½	-1.22 ± 0.5 -1¼ ± ½	-1.48 ± 0.5 -1½ ± ½
		0.0° ± 0.16° 0° ± 5/32°	REAR TOE (Degrees)	0.0° ± 0.16° 0° ± 5/32°	0.0° ± 0.16° 0° ± 5/32°	0.0° ± 0.16° 0° ± 5/32°
		0.0" ± 0.08" 0" ± 3/32"	REAR TOE (Inches)	0.0" ± 0.08" 0" ± 3/32"	0.0" ± 0.08" 0" ± 3/32"	0.0" ± 0.08" 0" ± 3/32"
			THRUST ANGLE			
		132 178	ILLUSTRATION #	132 178	132 178	132 178
[1] Air Susp. = 3.3° (3 5/16°)	[1] Air Susp. = 2.9° (2 7/8°)	[1] Steel Wheels = 154 ft.lbs.	NOTES	[1] Steel Wheels = 154 ft.lbs.	[1] Steel Wheels = 154 ft.lbs.	[1] Steel Wheels = 154 ft.lbs.
49117	49118	49119		49120	49121	49122

Toyota Truck & SUV

T100 4X2 ½ Ton	T100 4X2 ½ Ton	T100 4X2 ½ Ton Reg. Cab	MODEL INFORMATION	T100 4X2 ½ Ton X-Cab	T100 4X2 1 Ton	T100 4X2 1 Ton
1993	1994-95	1996-98	YEARS	1996-98	1993	1994-98
			INSTRUCTIONS			
			LUGNUT TORQUE			
			FRONT			
2.5 ± 0.75 2½ ± ¾	2.58 ± 0.75 2 9/16 ± ¾	2.47 ± 0.75 2 7/16 ± ¾	CASTER (Degrees)	2.07 ± 0.75 2 1/16 ± ¾	3.0 ± 0.75 3 ± ¾	1.67 ± 0.75[1] 1 11/16 ± ¾[1]
0.0 (0) | 0.75 (¾)	0.0 (0) | 0.75 (¾)	0.0 (0) | 0.75 (¾)	CROSS | TOL.	0.0 (0) | 0.75 (¾)	0.0 (0) | 0.75 (¾)	0.0 (0) | 0.75 (¾)
0.5 ± 0.75 ½ ± ¾	0.42 ± 0.75 7/16 ± ¾	0.4 ± 0.75 ⅜ ± ¾	CAMBER (Degrees)	0.47 ± 0.75 7/16 ± ¾	0.5 ± 0.75 ½ ± ¾	0.37 ± 0.75 ⅜ ± ¾
0.0 (0) | 0.75 (¾)	0.0 (0) | 0.75 (¾)	0.0 (0) | 0.75 (¾)	CROSS | TOL.	0.0 (0) | 0.75 (¾)	0.0 (0) | 0.75 (¾)	0.0 (0) | 0.75 (¾)
0.25° ± 0.1° ¼° ± 3/32°	0.25° ± 0.1° ¼° ± 3/32°	0.19° ± 0.2° 3/16° ± 3/16°	FRONT TOE (Degrees)	0.3° ± 0.2° 5/16° ± 3/16°	0.75° ± 0.1° ¾° ± 3/32°	0.55° ± 0.2° 9/16° ± 3/16°
0.13" ± 0.05" ⅛" ± 1/16"	0.13" ± 0.05" ⅛" ± 1/16"	0.1" ± 0.1" 3/32" ± 3/32"	FRONT TOE (Inches)	0.15" ± 0.1" 5/32" ± 3/32"	0.38" ± 0.05" ⅜" ± 1/16"	0.28" ± 0.1" 9/32" ± 3/32"
10.0 ± 0.8 (10 ± ¾)	12.2 ± 0.8 (12 3/16 ± ¾)	12.1 ± 0.8 (12 1/16 ± ¾)	SAI	12.0 ± 0.8 (12 ± ¾)	9.5 ± 0.8 (9½ ± ¾)	12.2 ± 0.8 (12 3/16 ± ¾)
40.8 / 35.3 ±2.0	40.8 / 35.1 ±2.0	40.7 / 34.1 ±2.0	TURNING ANGLE	40.8 / 35.7 ±2.0	40.5 / 35.8 ±2.0	40.4 / 35.5 ±2.0
23 03	23 03	23 03	ILLUSTRATION #	23 03	23 03	23 03
			REAR			
			CAMBER (Degrees)			
			REAR TOE (Degrees)			
			REAR TOE (Inches)			
			THRUST ANGLE			
			ILLUSTRATION #			
			NOTES			[1] 1994 = 2.3° (2 5/16°)
49123	49124	49125		49126	49127	49128

Toyota Truck & SUV

T100 4X4	T100 4X4	T100 4X4	MODEL	T100 4X4	Tacoma 4X2	Tacoma 4X2
		Reg. Cab	INFORMATION	X-Cab	exc/PreRunner	PreRunner
						Reg. Cab
1993	1994-95	1996-98	YEARS	1996-98	1995-2000	1998-2000
			INSTRUCTIONS			
			LUGNUT TORQUE			
			FRONT			
1.5 ± 0.75 1 1/2 ± 3/4	1.58 ± 0.75 1 9/16 ± 3/4	1.08 ± 0.75 [1] 1 1/16 ± 3/4 [1]	CASTER (Degrees)	0.92 ± 0.75 [1] 15/16 ± 3/4 [1]	0.6 ± 0.75 [1 2] 5/8 ± 3/4 [1 2]	1.6 ± 0.75 1 5/8 ± 3/4
0.0 (0) | 0.75 (3/4)	0.0 (0) | 0.75 (3/4)	0.0 (0) | 0.75 (3/4)	CROSS | TOL.	0.0 (0) | 0.75 (3/4)	0.0 (0) | 0.5 (1/2)	0.0 (0) | 0.5 (1/2)
0.75 ± 0.75 3/4 ± 3/4	0.67 ± 0.75 11/16 ± 3/4	0.67 ± 0.75 [2] 11/16 ± 3/4 [2]	CAMBER (Degrees)	0.58 ± 0.75 [2] 9/16 ± 3/4 [2]	0.07 ± 0.75 1/16 ± 3/4	0.3 ± 0.75 5/16 ± 3/4
0.0 (0) | 0.75 (3/4)	0.0 (0) | 0.75 (3/4)	0.0 (0) | 0.75 (3/4)	CROSS | TOL.	0.0 (0) | 0.75 (3/4)	0.0 (0) | 0.75 (3/4)	0.0 (0) | 0.5 (1/2)
0.25° ± 0.1° 1/4° ± 3/32°	0.3° ± 0.1° 5/16° ± 3/32°	0.2° ± 0.2° 3/16° ± 3/16°	FRONT TOE (Degrees)	0.14° ± 0.2° 1/8° ± 3/16°	0.2° ± 0.2° 3/16° ± 3/16°	0.2° ± 0.2° 3/16° ± 3/16°
0.13" ± 0.05" 1/8" ± 1/16"	0.15" ± 0.05" 5/32" ± 1/16"	0.1" ± 0.1" 3/32" ± 3/32"	FRONT TOE (Inches)	0.07" ± 0.1" 1/16" ± 3/32"	0.1" ± 0.1" 3/32" ± 3/32"	0.1" ± 0.1" 3/32" ± 3/32"
11.8 ± 0.8 (11 13/16 ± 3/4)	11.8 ± 0.8 (11 13/16 ± 3/4)	12.0 ± 0.8 (12 ± 3/4)	SAI	12.0 ± 0.8 (12 ± 3/4)	10.0 ± 0.8 (10 ± 3/4)	10.5 ± 0.8 (10 1/2 ± 3/4)
33.0 / 30.5 ± 2.0	32.8 / 30.5 ± 2.0	32.8 / 30.5 ± 2.0	TURNING ANGLE	33.1 / 30.5 ± 2.0	36.0 / 31.0 ± 2.0	37.1 / 32.3 ± 1.0
12	12	12	ILLUSTRATION #	12	23	12
			REAR			
			CAMBER (Degrees)			
			REAR TOE (Degrees)			
			REAR TOE (Inches)			
			THRUST ANGLE			
			ILLUSTRATION #			
		[1] 1996 = 1.0° (1°) [2] 1996 = 0.6° (5/8°)	NOTES	[1] 1996 = 1.17° (1 3/16°) [2] 1996 = 0.7° (11/16°)	[1] X-Cab 4 cyl.= 0.8° (13/16°) [2] X-Cab 6 Cyl.= 0.7° (11/16°)	
49129	49130	49131		49132	49133	49134

Toyota Truck & SUV

Tacoma 4X2	Tacoma 4X2	Tacoma 4X2	MODEL	Tacoma 4X2	Tacoma 4X2	Tacoma 4X2
PreRunner	exc/PreRunner	PreRunner	INFORMATION	PreRunner	PreRunner	PreRunner
X-Cab		Reg. Cab		X-Cab, 4 Cyl	X-Cab, 6 Cyl	Double Cab, 4 Cyl
1998-2000	2001-2004	2001-2004	YEARS	2001-2004	2001-2004	2001-2003
			INSTRUCTIONS			
			LUGNUT TORQUE			
			FRONT			
1.4 ± 0.75 [1] 1 3/8 ± 3/4 [1]	0.21 ± 0.75 [1 2] 3/16 ± 3/4 [1 2]	1.65 ± 0.75 1 5/8 ± 3/4	CASTER (Degrees)	1.8 ± 0.75 [1] 1 13/16 ± 3/4 [1]	1.75 ± 0.75 [1] 1 3/4 ± 3/4 [1]	2.5 ± 0.75 2 1/2 ± 3/4
0.0 (0) | 0.5 (1/2)	0.0 (0) | 0.5 (1/2)	0.0 (0) | 0.5 (1/2)	CROSS | TOL.	0.0 (0) | 0.5 (1/2)	0.0 (0) | 0.5 (1/2)	0.0 (0) | 0.5 (1/2)
0.3 ± 0.75 5/16 ± 3/4	0.0 ± 0.75 0 ± 3/4	0.3 ± 0.75 5/16 ± 3/4	CAMBER (Degrees)	0.25 ± 0.75 [2] 1/4 ± 3/4 [2]	0.2 ± 0.75 3/16 ± 3/4	-0.2 ± 0.75 [1 2] -3/16 ± 3/4 [1 2]
0.0 (0) | 0.5 (1/2)	0.0 (0) | 0.75 (3/4)	0.0 (0) | 0.5 (1/2)	CROSS | TOL.	0.0 (0) | 0.5 (1/2)	0.0 (0) | 0.5 (1/2)	0.0 (0) | 0.5 (1/2)
0.2° ± 0.2° 3/16° ± 3/16°	0.14° ± 0.2° 1/8° ± 3/16°	0.13° ± 0.2° 1/8° ± 3/16°	FRONT TOE (Degrees)	0.09° ± 0.2° 3/32° ± 3/16°	0.08° ± 0.2° 3/32° ± 3/16°	0.07° ± 0.2° 1/16° ± 3/16°
0.1" ± 0.1" 3/32" ± 3/32"	0.07" ± 0.1" 1/16" ± 3/32"	0.06" ± 0.1" 1/16" ± 3/32"	FRONT TOE (Inches)	0.05" ± 0.1" 1/32" ± 3/32"	0.04" ± 0.1" 1/32" ± 3/32"	0.04" ± 0.1" 1/32" ± 3/32"
10.4 ± 0.8 (10 3/8 ± 3/4)	10.0 ± 0.8 (10 ± 3/4)	10.4 ± 0.8 (10 3/8 ± 3/4)	SAI	10.4 ± 0.8 (10 3/8 ± 3/4)	10.5 ± 0.8 (10 1/2 ± 3/4)	10.9 ± 0.8 (10 7/8 ± 3/4)
37.1 / 32.3 ± 1.0	36.0 / 31.0 ± 2.0	37.1 / 32.3 ± 1.0	TURNING ANGLE	37.1 / 32.3 ± 1.0	37.1 / 32.3 ± 1.0	37.1 / 32.3 ± 1.0
12	23	12	ILLUSTRATION #	12	12	12
			REAR			
			CAMBER (Degrees)			
			REAR TOE (Degrees)			
			REAR TOE (Inches)			
			THRUST ANGLE			
			ILLUSTRATION #			
[1] 6 Cyl = 1.3° (1 5/16°)	[1] X-Cab 4 cyl.= 0.67° (11/16°) [2] X-Cab 6 Cyl.= 0.92° (15/16°)		NOTES	[1] Off-road pkg. = 1.9° (1 7/8°) [2] Off-road pkg. = 0.2° (3/16°)	[1] Off-road pkg. = 1.9° (1 7/8°)	[1] '01 Off-rd pkg. = -0.2° (-3/16°) [2] '02-03 Off-road pkg. = -0.5° (-1/2°)
49135	49136	49137		49138	49139	49140

Toyota Truck & SUV

Tacoma 4X2 PreRunner Double Cab, 4 Cyl	Tacoma 4X2 PreRunner Double Cab, 6 Cyl	Tacoma 4X4 Reg. Cab	MODEL INFORMATION	Tacoma 4X4 X-Cab	Tacoma 4X4 Reg. Cab	Tacoma 4X4 X-Cab
2004	2001-2004	1995-96	YEARS	1995-96	1997	1997
			INSTRUCTIONS			
			LUGNUT TORQUE			
			FRONT			
2.7 ±0.75 2 11/16 ± ¾	2.5 ±0.75 [1] 2½ ± ¾ [1]	2.0 ±0.75 2 ± ¾	CASTER (Degrees)	2.25 ±0.75 2¼ ± ¾	1.58 ±0.75 1 9/16 ± ¾	1.75 ±0.75 1¾ ± ¾
0.0 (0) / 0.5 (½)	0.0 (0) / 0.5 (½)	0.0 (0) / 0.75 (¾)	CROSS / TOL.	0.0 (0) / 0.75 (¾)	0.0 (0) / 0.5 (½)	0.0 (0) / 0.5 (½)
-0.4 ±0.75 [1] -⅜ ± ¾ [1]	-0.2 ±0.75 [2] -3/16 ± ¾ [2]	0.18 ±0.75 3/16 ± ¾	CAMBER (Degrees)	0.25 ±0.75 ¼ ± ¾	0.25 ±0.75 ¼ ± ¾	0.33 ±0.75 5/16 ± ¾
0.0 (0) / 0.5 (½)	0.0 (0) / 0.5 (½)	0.0 (0) / 0.75 (¾)	CROSS / TOL.	0.0 (0) / 0.75 (¾)	0.0 (0) / 0.75 (¾)	0.0 (0) / 0.75 (¾)
0.0° ±0.2° 0° ± 3/16°	0.07° ±0.2° 1/16° ± 3/16°	0.1° ±0.15° 3/32° ± 5/32°	FRONT TOE (Degrees)	0.15° ±0.15° 5/32° ± 5/32°	0.2° ±0.2° 3/16° ± 3/16°	0.2° ±0.2° 3/16° ± 3/16°
0.0" ±0.1" 0" ± 3/32"	0.04" ±0.1" 1/32" ± 3/32"	0.05" ±0.08" 1/16" ± 1/16"	FRONT TOE (Inches)	0.08" ±0.08" 1/16" ± 1/16"	0.1" ±0.1" 3/32" ± 3/32"	0.1" ±0.1" 3/32" ± 3/32"
11.2 ±0.8 (11 3/16 ± ¾)	11.0 ±0.8 (11 ± ¾)		SAI		10.5 ±0.8 (10½ ± ¾)	10.5 ±0.8 (10½ ± ¾)
37.1 / 32.3 ±1.0	37.1 / 32.3 ±1.0	37.1 / 32.3 ±2.0	TURNING ANGLE	37.1 / 32.3 ±2.0	37.5 / 32.5 ±2.0	37.5 / 32.5 ±2.0
12	12	12	ILLUSTRATION #	12	12	12
			REAR			
			CAMBER (Degrees)			
			REAR TOE (Degrees)			
			REAR TOE (Inches)			
			THRUST ANGLE			
			ILLUSTRATION #			
[1] Off-road pkg. = -0.5° (-½°)	[1] Off-road pkg. = 2.7° (2 11/16°) [2] Off-road pkg. = -0.3° (-5/16°)		NOTES			
49141	49142	49143		49144	49145	49146

Toyota Truck & SUV

Tacoma 4X4 Reg. Cab	Tacoma 4X4 X-Cab	Tacoma 4X4 Reg. Cab	MODEL INFORMATION	Tacoma 4X4 X-Cab	Tacoma 4X4 Reg. Cab	Tacoma 4X4 X-Cab
1998	1998	1999-2000	YEARS	1999-2000	2001-2004	2001-2003
			INSTRUCTIONS			
			LUGNUT TORQUE			
			FRONT			
1.67 ±0.75 1 11/16 ± ¾	1.5 ±0.75 [1] 1½ ± ¾ [1]	1.7 ±0.75 1 11/16 ± ¾	CASTER (Degrees)	1.5 ±0.75 [1] 1½ ± ¾ [1]	1.6 ±0.75 1⅝ ± ¾	1.75 ±0.75 [1] 1¾ ± ¾ [1]
0.0 (0) / 0.5 (½)	0.0 (0) / 0.5 (½)	0.0 (0) / 0.5 (½)	CROSS / TOL.	0.0 (0) / 0.5 (½)	0.0 (0) / 0.5 (½)	0.0 (0) / 0.5 (½)
0.17 ±0.75 3/16 ± ¾	0.25 ±0.75 ¼ ± ¾	0.2 ±0.75 3/16 ± ¾	CAMBER (Degrees)	0.3 ±0.75 5/16 ± ¾	0.3 ±0.75 5/16 ± ¾	0.3 ±0.75 [2] 5/16 ± ¾ [2]
0.0 (0) / 0.5 (½)	0.0 (0) / 0.5 (½)	0.0 (0) / 0.5 (½)	CROSS / TOL.	0.0 (0) / 0.5 (½)	0.0 (0) / 0.5 (½)	0.0 (0) / 0.5 (½)
0.2° ±0.2° 3/16° ± 3/16°	0.2° ±0.2° 3/16° ± 3/16°	0.2° ±0.2° 3/16° ± 3/16°	FRONT TOE (Degrees)	0.2° ±0.2° 3/16° ± 3/16°	0.13° ±0.2° ⅛° ± 3/16°	0.13° ±0.2° ⅛° ± 3/16°
0.1" ±0.1" 3/32" ± 3/32"	0.1" ±0.1" 3/32" ± 3/32"	0.1" ±0.1" 3/32" ± 3/32"	FRONT TOE (Inches)	0.1" ±0.1" 3/32" ± 3/32"	0.06" ±0.1" 1/16" ± 3/32"	0.06" ±0.1" 1/16" ± 3/32"
10.5 ±0.8 (10½ ± ¾)	10.5 ±0.8 (10½ ± ¾)	10.5 ±0.8 (10½ ± ¾)	SAI	10.5 ±0.8 (10½ ± ¾)	10.5 ±0.8 (10½ ± ¾)	10.5 ±0.8 (10½ ± ¾)
37.3 / 32.6 ±1.0	37.3 / 32.6 ±2.0	37.1 / 32.3 ±1.0	TURNING ANGLE	37.1 / 32.3 ±1.0	37.1 / 32.3 ±1.0	37.1 / 32.3 ±1.0
12	12	12	ILLUSTRATION #	12	12	12
			REAR			
			CAMBER (Degrees)			
			REAR TOE (Degrees)			
			REAR TOE (Inches)			
			THRUST ANGLE			
			ILLUSTRATION #			
	[1] 6 Cyl.=1.4° (1⅜°)		NOTES	[1] 6 Cyl.=1.4° (1⅜°)		[1] Off-road pkg. = 2° (2°) [2] Off-road pkg. = 0.1° (⅛°)
49147	49148	49149		49150	49151	49152

Toyota Truck & SUV

Tacoma 4X4 X-Cab (USA)	Tacoma 4X4 X-Cab (Canada)	Tacoma 4X4 X-Cab Off-road	MODEL INFORMATION	Tacoma 4X4 Double Cab	Tacoma 4X4 Double Cab	Tacoma 4X4 Double Cab
2004	2004	2004	YEARS	2001	2002-2003	2004
			INSTRUCTIONS			
			LUGNUT TORQUE			
			FRONT			
1.9 ±0.75 1 7/8 ± 3/4	1.7 ±0.75 1 11/16 ± 3/4	2.0 ±0.75 2 ± 3/4	CASTER (Degrees)	2.4 ±0.75 [1] 2 3/8 ± 3/4 [1]	2.6 ±0.75 [1] 2 5/8 ± 3/4 [1]	2.6 ±0.75 [1] 2 5/8 ± 3/4 [1]
0.0 (0) | 0.5 (1/2)	0.0 (0) | 0.5 (1/2)	0.0 (0) | 0.5 (1/2)	CROSS | TOL.	0.0 (0) | 0.5 (1/2)	0.0 (0) | 0.5 (1/2)	0.0 (0) | 0.5 (1/2)
0.1 ±0.75 1/8 ± 3/4	0.3 ±0.75 5/16 ± 3/4	0.1 ±0.75 1/8 ± 3/4	CAMBER (Degrees)	-0.1 ±0.75 [2] - 1/8 ± 3/4 [2]	-0.2 ±0.75 [2] - 3/16 ± 3/4 [2]	-0.3 ±0.75 - 5/16 ± 3/4
0.0 (0) | 0.5 (1/2)	0.0 (0) | 0.5 (1/2)	0.0 (0) | 0.5 (1/2)	CROSS | TOL.	0.0 (0) | 0.5 (1/2)	0.0 (0) | 0.5 (1/2)	0.0 (0) | 0.5 (1/2)
0.03° ±0.2° 1/32° ± 3/16°	0.13° ±0.2° 1/8° ± 3/16°	0.04° ±0.2° 1/32° ± 3/16°	FRONT TOE (Degrees)	0.13° ±0.2° 1/8° ± 3/16°	0.07° ±0.2° 1/16° ± 3/16°	0.07° ±0.2° 1/16° ± 3/16°
0.02" ±0.1" 1/64" ± 3/32"	0.06" ±0.1" 1/16" ± 3/32"	0.02" ±0.1" 1/64" ± 3/32"	FRONT TOE (Inches)	0.06" ±0.1" 1/16" ± 3/32"	0.04" ±0.1" 1/32" ± 3/32"	0.04" ±0.1" 1/32" ± 3/32"
10.4 ±0.8 (10 3/8 ± 3/4)	10.6 ±0.8 (10 5/8 ± 3/4)	10.8 ±0.8 (10 3/4 ± 3/4)	SAI	10.9 ±0.8 (10 7/8 ± 3/4)	11.0 ±0.8 (11 ± 3/4)	11.0 ±0.8 (11 ± 3/4)
37.1 / 32.3 ±1.0	37.1 / 32.3 ±1.0	37.1 / 32.3 ±1.0	TURNING ANGLE	37.1 / 32.3 ±1.0	37.1 / 32.3 ±1.0	37.1 / 32.3 ±1.0
12	12	12	ILLUSTRATION #	12	12	12
			REAR			
			CAMBER (Degrees)			
			REAR TOE (Degrees)			
			REAR TOE (Inches)			
			THRUST ANGLE			
			ILLUSTRATION #			
			NOTES	[1] Off-road pkg. = 2.5° (2 1/2°) [2] Off-road pkg. = -0.2° (-3/16°)	[1] Off-road pkg. = 2.7° (2 11/16°) [2] Off-road pkg. = -0.3° (-5/16°)	[1] Off-road pkg. = 2.7° (2 11/16°)
49153	49154	49155		49156	49157	49158

Toyota Truck & SUV

Tacoma 4X2 Regular Cab TRN220L-TR[M,P]DKA	Tacoma 4X2 Regular Cab TRN220L-TR[M,P]DKA	Tacoma 4X2 Regular Cab TRN220L-TR[M,P]DKA	MODEL INFORMATION	Tacoma 4X2 Access Cab TRN225L-CR[M,P]DKA	Tacoma 4X2 Access Cab TRN225L-CR[M,P]DKA	Tacoma 4X2 Access Cab TRN225L-CR[M,P]DKA
2005-2009	2010-2011	2012-2014	YEARS	2005-2009	2010-2011	2012-2015
			INSTRUCTIONS			
83 ft.lbs.	83 ft.lbs.	83 ft.lbs.	LUGNUT TORQUE	83 ft.lbs.	83 ft.lbs.	83 ft.lbs.
			FRONT			
L= 3.85, R= 3.8 ± 0.75 L= 3 13/16, R= 3 13/16 ± 3/4	3.93 ±0.75 3 15/16 ± 3/4	3.9 ±0.75 3 7/8 ± 3/4	CASTER (Degrees)	4.0 ±0.75 4 ± 3/4	4.13 ±0.75 4 1/8 ± 3/4	4.15 ±0.75 4 1/8 ± 3/4
0.0 (0) | 0.5 (1/2)	0.0 (0) | 0.5 (1/2)	0.0 (0) | 0.5 (1/2)	CROSS | TOL.	0.0 (0) | 0.5 (1/2)	0.0 (0) | 0.5 (1/2)	0.0 (0) | 0.5 (1/2)
0.2 ±0.75 3/16 ± 3/4	0.14 ±0.75 1/8 ± 3/4	0.15 ±0.75 1/8 ± 3/4	CAMBER (Degrees)	0.2 ±0.75 3/16 ± 3/4	0.16 ±0.75 3/16 ± 3/4	0.2 ±0.75 3/16 ± 3/4
0.0 (0) | 0.5 (1/2)	0.0 (0) | 0.5 (1/2)	0.0 (0) | 0.5 (1/2)	CROSS | TOL.	0.0 (0) | 0.5 (1/2)	0.0 (0) | 0.5 (1/2)	0.0 (0) | 0.5 (1/2)
0.08° ±0.16° 3/32° ± 5/32°	0.05° ±0.16° 1/16° ± 5/32°	0.05° ±0.16° 1/16° ± 5/32°	FRONT TOE (Degrees)	0.1° ±0.16° 3/32° ± 5/32°	0.07° ±0.16° 1/16° ± 5/32°	0.08° ±0.16° 3/32° ± 5/32°
0.04" ±0.08" 1/32" ± 3/32"	0.03" ±0.08" 1/32" ± 3/32"	0.03" ±0.08" 1/32" ± 3/32"	FRONT TOE (Inches)	0.05" ±0.08" 1/16" ± 3/32"	0.04" ±0.08" 1/32" ± 3/32"	0.04" ±0.08" 1/32" ± 3/32"
10.7 ±0.8 (10 11/16 ± 3/4)	10.8 ±0.8 (10 13/16 ± 3/4)	10.8 ±0.8 (10 3/4 ± 3/4)	SAI	10.7 ±0.8 (10 11/16 ± 3/4)	10.7 ±0.8 (10 11/16 ± 3/4)	10.7 ±0.8 (10 11/16 ± 3/4)
38.8 / 32.5 ±1.5	38.8 / 32.5 ±1.5	38.8 / 32.5 ±1.5	TURNING ANGLE	38.8 / 32.5 ±1.5	38.8 / 32.5 ±1.5	38.8 / 32.5 ±1.5
12	12	12	ILLUSTRATION #	12	12	12
			REAR			
			CAMBER (Degrees)			
			REAR TOE (Degrees)			
			REAR TOE (Inches)			
			THRUST ANGLE			
			ILLUSTRATION #			
			NOTES			
49159	49160	49161		49162	49163	49164

ALIGNMENT SPECIFICATIONS

Toyota Truck & SUV

MODEL INFORMATION	Tacoma 4X2 Access Cab, X-Runner GRN225L-CRFDKA	Tacoma 4X2 Access Cab, X-Runner GRN225L-CRFDKA	Tacoma 4X2 Access Cab, X-Runner GRN225L-CRFDKA	Tacoma 4X2 Double Cab TRN225L-PRPDKA	Tacoma 4X2 Double Cab TRN225L-PRPDKA	Tacoma 4X2, PreRunner Regular Cab TRN260L-TRMDKA
YEARS	2005-2009	2010-2011	2012-2015	2011	2012-2015	2005-2010
INSTRUCTIONS						
LUGNUT TORQUE	83 ft.lbs.	83 ft.lbs.	83 ft.lbs.	83 ft.lbs.	83 ft.lbs.	83 ft.lbs.
FRONT						
CASTER (Degrees)	5.15 ±0.75 5⅛ ± ¾	5.2 ±0.75 5 3/16 ± ¾	5.15 ±0.75 5⅛ ± ¾	4.2 ±0.75 4 3/16 ± ¾	4.3 ±0.75 4 5/16 ± ¾	1.7 ±0.75 1 11/16 ± ¾
CROSS | TOL.	0.0 (0) | 0.5 (½)	0.0 (0) | 0.5 (½)	0.0 (0) | 0.5 (½)	0.0 (0) | 0.5 (½)	0.0 (0) | 0.5 (½)	0.0 (0) | 0.5 (½)
CAMBER (Degrees)	-0.6 ±0.75 -⅝ ± ¾	-0.69 ±0.75 -11/16 ± ¾	-0.63 ±0.75 -⅝ ± ¾	0.13 ±0.75 ⅛ ± ¾	0.15 ±0.75 ⅛ ± ¾	0.6 ±0.75 ⅝ ± ¾
CROSS | TOL.	0.0 (0) | 0.5 (½)	0.0 (0) | 0.5 (½)	0.0 (0) | 0.5 (½)	0.0 (0) | 0.5 (½)	0.0 (0) | 0.5 (½)	0.0 (0) | 0.5 (½)
FRONT TOE (Degrees)	0.04° ±0.16° 1/32° ± 5/32°	0.02° ±0.16° 1/32° ± 5/32°	0.04° ±0.16° 1/32° ± 5/32°	0.02° ±0.16° 1/32° ± 5/32°	0.06° ±0.16° 1/16° ± 5/32°	0.13° ±0.16° ⅛° ± 5/32°
FRONT TOE (Inches)	0.02" ±0.08" 1/64" ± 3/32"	0.01" ±0.08" 1/64" ± 3/32"	0.02" ±0.08" 1/32" ± 3/32"	0.01" ±0.08" 1/64" ± 3/32"	0.03" ±0.08" 1/32" ± 3/32"	0.06" ±0.08" 1/16" ± 3/32"
SAI	11.5 ±0.8 (11½ ± ¾)	11.6 ±0.8 (11⅝ ± ¾)	11.6 ±0.8 (11⅝ ± ¾)	11.6 ±0.8 (11⅝ ± ¾)	10.7 ±0.8 (10 11/16 ± ¾)	11.9 ±0.8 (11⅞ ± ¾)
TURNING ANGLE	38.8 / 32.5 ±1.5	38.8 / 32.5 ±1.5	38.8 / 32.5 ±1.5	38.8 / 32.5 ±1.5	38.8 / 32.5 ±1.5	40.3 / 35.9 ±1.5
ILLUSTRATION #	12	12	12	12	12 12	12
REAR						
CAMBER (Degrees)						
REAR TOE (Degrees)						
REAR TOE (Inches)						
THRUST ANGLE						
ILLUSTRATION #						
NOTES						
	49165	49166	49167	49168	49169	49170

Toyota Truck & SUV

MODEL INFORMATION	Tacoma 4X2, PreRunner Access-Cab, 4 Cyl. TRN265L-CRMDKA	Tacoma 4X2, PreRunner Access-Cab, 6 Cyl. exc/TRD Off-rd GRN265L-CR[A,F]DKA	Tacoma 4X2, PreRunner Access-Cab, 6 Cyl. TRD Off-rd GRN265L-CR[A,F]DKA	Tacoma 4X2, PreRunner Double-Cab, 128" WB GRN270L-PRADKA	Tacoma 4X2, PreRunner Double-Cab, 141" WB GRN270L-PRADKA	Tacoma 4X2, PreRunner Access-Cab, 4 Cyl. TRN265L-CRPDKA
YEARS	2005-2010	2005-2010	2005-2010	2005-2010	2005-2010	2011-2015
INSTRUCTIONS						
LUGNUT TORQUE	83 ft.lbs.	83 ft.lbs.	83 ft.lbs.	83 ft.lbs.	83 ft.lbs.	83 ft.lbs.
FRONT						
CASTER (Degrees)	1.8 ±0.75 1 13/16 ± ¾	1.9 ±0.75 1⅞ ± ¾	2.0 ±0.75 2 ± ¾	2.0 ±0.75 2 ± ¾	2.17 ±0.75 2 3/16 ± ¾	1.9 ±0.75 1⅞ ± ¾
CROSS | TOL.	0.0 (0) | 0.5 (½)	0.0 (0) | 0.5 (½)	0.0 (0) | 0.5 (½)	0.0 (0) | 0.5 (½)	0.0 (0) | 0.5 (½)	0.0 (0) | 0.5 (½)
CAMBER (Degrees)	0.6 ±0.75 ⅝ ± ¾	0.55 ±0.75 9/16 ± ¾	0.45 ±0.75 7/16 ± ¾	0.55 ±0.75 9/16 ± ¾	0.5 ±0.75 ½ ± ¾	0.56 ±0.75 9/16 ± ¾
CROSS | TOL.	0.0 (0) | 0.5 (½)	0.0 (0) | 0.5 (½)	0.0 (0) | 0.5 (½)	0.0 (0) | 0.5 (½)	0.0 (0) | 0.5 (½)	0.0 (0) | 0.5 (½)
FRONT TOE (Degrees)	0.14° ±0.16° ⅛° ± 5/32°	0.1° ±0.16° 3/32° ± 5/32°	0.03° ±0.16° 1/32° ± 5/32°	0.09° ±0.16° 3/32° ± 5/32°	0.08° ±0.16° 3/32° ± 5/32°	0.1° ±0.16° 3/32° ± 5/32°
FRONT TOE (Inches)	0.07" ±0.08" 1/16" ± 3/32"	0.05" ±0.08" 1/16" ± 3/32"	0.02" ±0.08" 1/64" ± 3/32"	0.05" ±0.08" 1/32" ± 3/32"	0.04" ±0.08" 1/32" ± 3/32"	0.05" ±0.08" 1/16" ± 3/32"
SAI	11.9 ±0.8 (11⅞ ± ¾)	11.9 ±0.8 (11⅞ ± ¾)	12.0 ±0.8 (12 ± ¾)	11.9 ±0.8 (11⅞ ± ¾)	12.0 ±0.8 (11 15/16 ± ¾)	11.9 ±0.8 (11⅞ ± ¾)
TURNING ANGLE	40.3 / 35.9 ±1.5	40.3 / 35.9 ±1.5	40.3 / 35.9 ±1.5	40.3 / 35.9 ±1.5	39.3 / 34.9 ±1.5	39.3 / 34.9 ±1.5
ILLUSTRATION #	12	12	12	12	12	12
REAR						
CAMBER (Degrees)						
REAR TOE (Degrees)						
REAR TOE (Inches)						
THRUST ANGLE						
ILLUSTRATION #						
NOTES						
	49171	49172	49173	49174	49175	49176

Toyota Truck & SUV

Tacoma 4X2, PreRunner	Tacoma 4X2, PreRunner	Tacoma 4X2, PreRunner	MODEL	Tacoma 4X2, PreRunner	Tacoma 4X2, PreRunner	Tacoma 4X4
Access-Cab, 6 Cyl. exc/TRD Off-rd	Access-Cab, 6 Cyl. TRD Off-rd	Double-Cab, 4 Cyl.	INFORMATION	Double-Cab, 128" WB	Double-Cab, 141" WB	Regular Cab
GRN265L-CRADKA	GRN265L-CRADKA	TRN265L-PRPDKA		GRN265L-PRADKA	GRN270L-PRADKA	TRN240L-TRMDKA
2011-2015	2011-2015	2011-2015	YEARS	2011-2015	2011-2015	2005
			INSTRUCTIONS			
83 ft.lbs.	83 ft.lbs.	83 ft.lbs.	LUGNUT TORQUE	83 ft.lbs.	83 ft.lbs.	83 ft.lbs.
			FRONT			
1.95 ±0.75 1 15⁄16 ± ¾	2.07 ±0.75 2 1⁄16 ± ¾	2.0 ±0.75 2 ± ¾	CASTER (Degrees)	2.1 ±0.75 2 ⅛ ± ¾	2.17 ±0.75 2 3⁄16 ± ¾	1.7 ±0.75 1 11⁄16 ± ¾
0.0 (0) | 0.5 (½)	0.0 (0) | 0.5 (½)	0.0 (0) | 0.5 (½)	CROSS | TOL.	0.0 (0) | 0.5 (½)	0.0 (0) | 0.5 (½)	0.0 (0) | 0.5 (½)
0.54 ±0.75 9⁄16 ± ¾	0.42 ±0.75 7⁄16 ± ¾	0.52 ±0.75 ½ ± ¾	CAMBER (Degrees)	0.5 ±0.75 ½ ± ¾	0.5 ±0.75 ½ ± ¾	0.62 ±0.75 ⅝ ± ¾
0.0 (0) | 0.5 (½)	0.0 (0) | 0.5 (½)	0.0 (0) | 0.5 (½)	CROSS | TOL.	0.0 (0) | 0.5 (½)	0.0 (0) | 0.5 (½)	0.0 (0) | 0.5 (½)
0.09° ±0.16° 3⁄32° ± 5⁄32°	0.01° ±0.16° 1⁄64° ± 5⁄32°	0.08° ±0.16° 3⁄32° ± 5⁄32°	FRONT TOE (Degrees)	0.07° ±0.16° 1⁄16° ± 5⁄32°	0.05° ±0.16° 1⁄16° ± 5⁄32°	0.12° ±0.16° ⅛° ± 5⁄32°
0.05" ±0.08" 1⁄32" ± 3⁄32"	0.01" ±0.08" 0" ± 3⁄32"	0.04" ±0.08" 1⁄32" ± 3⁄32"	FRONT TOE (Inches)	0.04" ±0.08" 1⁄32" ± 3⁄32"	0.03" ±0.08" 1⁄32" ± 3⁄32"	0.06" ±0.08" 1⁄16" ± 3⁄32"
11.9 ±0.8 (11 ⅞ ± ¾)	12.0 ±0.8 (12 1⁄16 ± ¾)	11.9 ±0.8 (11 ⅞ ± ¾)	SAI	12.0 ±0.8 (11 15⁄16 ± ¾)	12.0 ±0.8 (11 15⁄16 ± ¾)	11.9 ±0.8 (11 ⅞ ± ¾)
39.3 / 34.9 ±1.5	39.3 / 34.9 ±1.5	39.3 / 34.9 ±1.5	TURNING ANGLE	39.3 / 34.9 ±1.5	39.3 / 34.9 ±1.5	40.3 / 35.9 ±1.5
12	12	12	ILLUSTRATION #	12	12	12
			REAR			
			CAMBER (Degrees)			
			REAR TOE (Degrees)			
			REAR TOE (Inches)			
			THRUST ANGLE			
			ILLUSTRATION #			
			NOTES			
49177	49178	49179		49180	49181	49182

Toyota Truck & SUV

Tacoma 4X4	Tacoma 4X4	Tacoma 4X4	MODEL	Tacoma 4X4	Tacoma 4X4	Tacoma 4X4
Regular Cab	Regular Cab	Access-Cab, 4 Cyl.	INFORMATION	Access-Cab, 4 Cyl.	Access Cab, 4 Cyl.	Access-Cab, 6 Cyl. exc/TRD Off-Rd
TRN240L-TRMDKA	TRN240L-TR[M,P]DKA	TRN245L-CRMDKA		TRN245L-CR[M,P]DKA	TRN245L-CR[M,P]DKA	GRN245L-CR[A,F]DKA
2006-2010	2011-2014	2005-2008	YEARS	2009-2011	2012-2014	2005-2008
			INSTRUCTIONS			
83 ft.lbs.	83 ft.lbs.	83 ft.lbs.	LUGNUT TORQUE	83 ft.lbs.	83 ft.lbs.	83 ft.lbs.
			FRONT			
1.75 ±0.75 1 ¾ ± ¾	1.75 ±0.75 1 ¾ ± ¾	1.9 ±0.75 1 ⅞ ± ¾	CASTER (Degrees)	1.95 ±0.75 1 15⁄16 ± ¾	1.8 ±0.75 1 13⁄16 ± ¾	1.96 ±0.75 1 15⁄16 ± ¾
0.0 (0) | 0.5 (½)	0.0 (0) | 0.5 (½)	0.0 (0) | 0.5 (½)	CROSS | TOL.	0.0 (0) | 0.5 (½)	0.0 (0) | 0.5 (½)	0.0 (0) | 0.5 (½)
0.55 ±0.75 9⁄16 ± ¾	0.6 ±0.75 ⅝ ± ¾	0.55 ±0.75 9⁄16 ± ¾	CAMBER (Degrees)	0.52 ±0.75 ½ ± ¾	0.6 ±0.75 ⅝ ± ¾	0.5 ±0.75 ½ ± ¾
0.0 (0) | 0.5 (½)	0.0 (0) | 0.5 (½)	0.0 (0) | 0.5 (½)	CROSS | TOL.	0.0 (0) | 0.5 (½)	0.0 (0) | 0.5 (½)	0.0 (0) | 0.5 (½)
0.12° ±0.16° ⅛° ± 5⁄32°	0.12° ±0.16° ⅛° ± 5⁄32°	0.1° ±0.16° 3⁄32° ± 5⁄32°	FRONT TOE (Degrees)	0.08° ±0.16° 3⁄32° ± 5⁄32°	0.14° ±0.16° ⅛° ± 5⁄32°	0.07° ±0.16° 1⁄16° ± 5⁄32°
0.06" ±0.08" 1⁄16" ± 3⁄32"	0.06" ±0.08" 1⁄16" ± 3⁄32"	0.05" ±0.08" 1⁄16" ± 3⁄32"	FRONT TOE (Inches)	0.04" ±0.08" 1⁄32" ± 3⁄32"	0.07" ±0.08" 1⁄16" ± 3⁄32"	0.04" ±0.08" 1⁄32" ± 3⁄32"
11.9 ±0.8 (11 ⅞ ± ¾)	11.9 ±0.8 (11 15⁄16 ± ¾)	11.8 ±0.8 (11 13⁄16 ± ¾)	SAI	11.9 ±0.8 (11 15⁄16 ± ¾)	11.8 ±0.8 (11 13⁄16 ± ¾)	11.9 ±0.8 (11 ⅞ ± ¾)
40.3 / 35.9 ±1.5	38.8 / 32.5 ±1.5	40.3 / 35.9 ±1.5	TURNING ANGLE	40.3 / 35.9 ±1.5	39.3 / 34.9 ±1.5	40.3 / 35.9 ±1.5
12	12	12	ILLUSTRATION #	12	12	12
			REAR			
			CAMBER (Degrees)			
			REAR TOE (Degrees)			
			REAR TOE (Inches)			
			THRUST ANGLE			
			ILLUSTRATION #			
			NOTES			
49183	49184	49185		49186	49187	49188

ALIGNMENT SPECIFICATIONS

ALIGNMENT SPECIFICATIONS

Toyota Truck & SUV

Tacoma 4X4	Tacoma 4X4	Tacoma 4X4	MODEL	Tacoma 4X4	Tacoma 4X4	Tacoma 4X4
Access-Cab, 6 Cyl. TRD Off-Rd	Access-Cab, 6 Cyl. exc/TRD Off-Rd	Access-Cab, 6 Cyl. TRD Off-Rd	INFORMATION	Access Cab, 6 Cyl. exc/TRD Off-Rd	Access Cab, 6 Cyl. TRD Off-Rd	Double-Cab, 128" WB, exc/TRD Off-rd
GRN245L-CR[A,F]DKA	GRN245L-CR[A,F]DKA	GRN245L-CR[A,F]DKA		GRN245L-CR[A,F]DKA	GRN245L-CR[A,F]DKA	GRN245L-PR[A,F]DKA
2005-2008	2009-2011	2009-2011	YEARS	2012-2015	2012-2015	2005-2011
			INSTRUCTIONS			
83 ft.lbs.	83 ft.lbs.	83 ft.lbs.	LUGNUT TORQUE	83 ft.lbs.	83 ft.lbs.	83 ft.lbs.
			FRONT			
2.0 ±0.75 2 ± ¾	2.0 ±0.75 2 ± ¾	2.1 ±0.75 2⅛ ± ¾	CASTER (Degrees)	1.9 ±0.75 1⅞ ± ¾	1.9 ±0.75 1⅞ ± ¾	2.1 ±0.75 2⅛ ± ¾
0.0 (0) | 0.5 (½)	0.0 (0) | 0.5 (½)	0.0 (0) | 0.5 (½)	CROSS | TOL.	0.0 (0) | 0.5 (½)	0.0 (0) | 0.5 (½)	0.0 (0) | 0.5 (½)
0.43 ±0.75 7/16 ± ¾	0.47 ±0.75 ½ ± ¾	0.39 ±0.75 ⅜ ± ¾	CAMBER (Degrees)	0.64 ±0.75 ⅝ ± ¾	0.6 ±0.75 ⅝ ± ¾	0.5 ±0.75 ½ ± ¾
0.0 (0) | 0.5 (½)	0.0 (0) | 0.5 (½)	0.0 (0) | 0.5 (½)	CROSS | TOL.	0.0 (0) | 0.5 (½)	0.0 (0) | 0.5 (½)	0.0 (0) | 0.5 (½)
0.02° ±0.16° 1/32° ± 5/32°	0.05° ±0.16° 1/16° ± 5/32°	0.02° ±0.16° 1/64° ± 5/32°	FRONT TOE (Degrees)	0.16° ±0.16° 5/32° ± 5/32°	0.1° ±0.16° 3/32° ± 5/32°	0.07° ±0.16° 1/16° ± 5/32°
0.01" ±0.08" 1/64" ± 3/32"	0.03" ±0.08" 1/64" ± 3/32"	0.01" ±0.08" 0" ± 3/32"	FRONT TOE (Inches)	0.08" ±0.08" 3/32" ± 3/32"	0.05" ±0.08" 1/16" ± 3/32"	0.04" ±0.08" 1/32" ± 3/32"
12.0 ±0.8 (12 ± ¾)	12.0 ±0.8 (12 ± ¾)	12.1 ±0.8 (12 1/16 ± ¾)	SAI	11.8 ±0.8 (11 13/16 ± ¾)	11.8 ±0.8 (11 13/16 ± ¾)	11.9 ±0.8 (11 ⅞ ± ¾)
40.3 / 35.9 ±1.5	40.3 / 35.9 ±1.5	40.3 / 35.9 ±1.5	TURNING ANGLE	39.3 / 34.9 ±1.5	39.3 / 34.9 ±1.5	40.3 / 35.9 ±1.5
12	12	12	ILLUSTRATION #	12	12	12
			REAR			
			CAMBER (Degrees)			
			REAR TOE (Degrees)			
			REAR TOE (Inches)			
			THRUST ANGLE			
			ILLUSTRATION #			
			NOTES			
49189	49190	49191		49192	49193	49194

Toyota Truck & SUV

Tacoma 4X4	Tacoma 4X4	Tacoma 4X4	MODEL	Tacoma 4X4	Tacoma 4X4	Tacoma 4X2
Double-Cab, 128" WB, TRD Off-rd	Double Cab, 128" WB, exc/TRD Off-Rd	Double Cab, 128" WB, TRD Off-Rd	INFORMATION	Double Cab, 141" WB	Double Cab, 141" WB	Access-Cab, 4 Cyl.
GRN245L-PR[A,F]DKA	GRN245L-PR[A,F]DKA	GRN245L-PR[A,F]DKA		GRN250L-PRADKA	GRN250L-PRADKA	TRN265L-CRTSKA
2005-2011	2012-2015	2012-2015	YEARS	2005-2011	2012-2015	2016-2017
			INSTRUCTIONS			
83 ft.lbs.	83 ft.lbs.	83 ft.lbs.	LUGNUT TORQUE	83 ft.lbs.	83 ft.lbs.	83 ft.lbs.
			FRONT			
2.0 ±0.75 2 ± ¾	2.0 ±0.75 2 ± ¾	2.0 ±0.75 2 ± ¾	CASTER (Degrees)	2.1 ±0.75 2⅛ ± ¾	2.1 ±0.75 2⅛ ± ¾	1.95 ±0.75 1 15/16 ± ¾
0.0 (0) | 0.5 (½)	0.0 (0) | 0.5 (½)	0.0 (0) | 0.5 (½)	CROSS | TOL.	0.0 (0) | 0.5 (½)	0.0 (0) | 0.5 (½)	0.0 (0) | 0.5 (½)
0.5 ±0.75 ½ ± ¾	0.5 ±0.75 ½ ± ¾	0.6 ±0.75 ⅝ ± ¾	CAMBER (Degrees)	0.5 ±0.75 ½ ± ¾	0.65 ±0.75 ⅝ ± ¾	0.58 ±0.75 9/16 ± ¾
0.0 (0) | 0.5 (½)	0.0 (0) | 0.5 (½)	0.0 (0) | 0.5 (½)	CROSS | TOL.	0.0 (0) | 0.5 (½)	0.0 (0) | 0.5 (½)	0.0 (0) | 0.5 (½)
0.05° ±0.16° 1/16° ± 5/32°	0.15° ±0.16° 5/32° ± 5/32°	0.12° ±0.16° ⅛° ± 5/32°	FRONT TOE (Degrees)	0.06° ±0.16° 1/16° ± 5/32°	0.17° ±0.16° 5/32° ± 5/32°	0.12° ±0.16° ⅛° ± 5/32°
0.03" ±0.08" 1/64" ± 3/32"	0.08" ±0.08" 1/16" ± 3/32"	0.06" ±0.08" 1/32" ± 3/32"	FRONT TOE (Inches)	0.03" ±0.08" 1/32" ± 3/32"	0.09" ±0.08" 3/32" ± 3/32"	0.06" ±0.08" 1/16" ± 3/32"
12.0 ±0.8 (12 ± ¾)	11.8 ±0.8 (11 13/16 ± ¾)	11.8 ±0.8 (11 13/16 ± ¾)	SAI	11.9 ±0.8 (11 ⅞ ± ¾)	11.8 ±0.8 (11 13/16 ± ¾)	11.9 ±0.8 (11 ⅞ ± ¾)
40.3 / 35.9 ±1.5	39.3 / 34.9 ±1.5	39.3 / 34.9 ±1.5	TURNING ANGLE	39.3 / 34.9 ±1.5	39.3 / 34.9 ±1.5	39.3 / 34.9 ±1.5
12	12	12	ILLUSTRATION #	12	12	12
			REAR			
			CAMBER (Degrees)			
			REAR TOE (Degrees)			
			REAR TOE (Inches)			
			THRUST ANGLE			
			ILLUSTRATION #			
			NOTES			
49195	49196	49197		49198	49199	49200

Toyota Truck & SUV

Tacoma 4X2 Double-Cab, 4 Cyl. TRN265L-PRTSKA	Tacoma 4X2, PreRunner Access-Cab GRN325L-CRTSHA	Tacoma 4X2, PreRunner Double-Cab, Long Bed GRN325L-PRT[L,S]HA	MODEL INFORMATION	Tacoma 4X2, PreRunner Double-Cab, Super Long Bed GRN330L-PRT[L,S]HA	Tacoma 4X4 Access-Cab, 4 Cyl., SR5 TRN245L-CR[M,T]SKA	Tacoma 4X4 Access-Cab, 6 Cyl., SR5 GRN305L-CR[F,T]SHA
2016-2017	2016-2017	2016-2017	YEARS	2016-2017	2016-2017	2016-2017
			INSTRUCTIONS			
83 ft.lbs.	83 ft.lbs.	83 ft.lbs.	LUGNUT TORQUE	83 ft.lbs.	83 ft.lbs.	83 ft.lbs.
			FRONT			
2.08 ± 0.75 2 1/16 ± 3/4	1.9 ± 0.75 1 7/8 ± 3/4	2.1 ± 0.75 2 1/8 ± 3/4	CASTER (Degrees)	2.05 ± 0.75 2 1/16 ± 3/4	1.95 ± 0.75 1 15/16 ± 3/4	2.0 ± 0.75 2 ± 3/4
0.0 (0) | 0.5 (1/2)	0.0 (0) | 0.5 (1/2)	0.0 (0) | 0.5 (1/2)	CROSS | TOL.	0.0 (0) | 0.5 (1/2)	0.0 (0) | 0.5 (1/2)	0.0 (0) | 0.5 (1/2)
0.5 ± 0.75 1/2 ± 3/4	0.52 ± 0.75 1/2 ± 3/4	0.53 ± 0.75 1/2 ± 3/4	CAMBER (Degrees)	0.57 ± 0.75 9/16 ± 3/4	0.52 ± 0.75 1/2 ± 3/4	0.53 ± 0.75 1/2 ± 3/4
0.0 (0) | 0.5 (1/2)	0.0 (0) | 0.5 (1/2)	0.0 (0) | 0.5 (1/2)	CROSS | TOL.	0.0 (0) | 0.5 (1/2)	0.0 (0) | 0.5 (1/2)	0.0 (0) | 0.5 (1/2)
0.07° ± 0.16° 1/16° ± 5/32°	0.08° ± 0.16° 3/32° ± 5/32°	0.08° ± 0.16° 3/32° ± 5/32°	FRONT TOE (Degrees)	0.11° ± 0.16° 1/8° ± 5/32°	0.09° ± 0.16° 3/32° ± 5/32°	0.09° ± 0.16° 3/32° ± 5/32°
0.04" ± 0.08" 1/32" ± 3/32"	0.04" ± 0.08" 1/32" ± 3/32"	0.04" ± 0.08" 1/32" ± 3/32"	FRONT TOE (Inches)	0.06" ± 0.08" 1/16" ± 3/32"	0.04" ± 0.08" 1/32" ± 3/32"	0.05" ± 0.08" 1/32" ± 3/32"
12.0 ± 0.8 (11 15/16 ± 3/4)	11.9 ± 0.8 (11 15/16 ± 3/4)	11.9 ± 0.8 (11 15/16 ± 3/4)	SAI	11.9 ± 0.8 (11 7/8 ± 3/4)	11.9 ± 0.8 (11 7/8 ± 3/4)	11.9 ± 0.8 (11 15/16 ± 3/4)
39.3 / 34.9 ± 1.5	39.3 / 34.9 ± 1.5	39.3 / 34.9 ± 1.5	TURNING ANGLE	39.3 / 34.9 ± 1.5	39.3 / 34.9 ± 1.5	39.3 / 34.9 ± 1.5
12	12	12	ILLUSTRATION #	12	12	12
			REAR			
			CAMBER (Degrees)			
			REAR TOE (Degrees)			
			REAR TOE (Inches)			
			THRUST ANGLE			
			ILLUSTRATION #			
			NOTES			
49201	49202	49203		49204	49205	49206

Toyota Truck & SUV

Tacoma 4X4 Double-Cab, Long-Bed, SR5 GRN305L-PR[F,T]SHA	Tacoma 4X4 Double-Cab, Long-Bed, Limited GRN305L-PRTLHA	Tacoma 4X4 Double-Cab, Super Long-Bed GRN310L-PRT[S,L]HA	MODEL INFORMATION	Tundra 4X2 Regular Cab (USA)	Tundra 4X2 Regular Cab (Canada)	Tundra 4X2 X-Cab, V6 (USA & Canada)
2016-2017	2016-2017	2016-2017	YEARS	2000	2000	2000
			INSTRUCTIONS			
83 ft.lbs.	83 ft.lbs.	83 ft.lbs.	LUGNUT TORQUE			
			FRONT			
2.08 ± 0.75 2 1/16 ± 3/4	2.1 ± 0.75 2 1/8 ± 3/4	2.8 ± 0.75 2 1/16 ± 3/4	CASTER (Degrees)	1.69 ± 0.75 1 11/16 ± 3/4	1.89 ± 0.75 1 7/8 ± 3/4	1.8 ± 0.75 1 13/16 ± 3/4
0.0 (0) | 0.5 (1/2)	0.0 (0) | 0.5 (1/2)	0.0 (0) | 0.5 (1/2)	CROSS | TOL.	0.0 (0) | 0.75 (3/4)	0.0 (0) | 0.75 (3/4)	0.0 (0) | 0.75 (3/4)
0.53 ± 0.75 1/2 ± 3/4	0.52 ± 0.75 1/2 ± 3/4	0.54 ± 0.75 9/16 ± 3/4	CAMBER (Degrees)	-0.04 ± 0.75 -1/16 ± 3/4	-0.17 ± 0.75 -3/16 ± 3/4	0.0 ± 0.75 0 ± 3/4
0.0 (0) | 0.5 (1/2)	0.0 (0) | 0.5 (1/2)	0.0 (0) | 0.5 (1/2)	CROSS | TOL.	0.0 (0) | 0.75 (3/4)	0.0 (0) | 0.75 (3/4)	0.0 (0) | 0.75 (3/4)
0.09° ± 0.16° 3/32° ± 5/32°	0.08° ± 0.16° 3/32° ± 5/32°	0.1° ± 0.16° 3/32° ± 5/32°	FRONT TOE (Degrees)	0.13° ± 0.2° 1/8° ± 3/16°	0.1° ± 0.2° 3/32° ± 3/16°	0.14° ± 0.2° 1/8° ± 3/16°
0.05" ± 0.08" 1/32" ± 3/32"	0.04" ± 0.08" 1/32" ± 3/32"	0.05" ± 0.08" 1/16" ± 3/32"	FRONT TOE (Inches)	0.07" ± 0.1" 1/16" ± 3/32"	0.05" ± 0.1" 1/16" ± 3/32"	0.07" ± 0.1" 1/16" ± 3/32"
11.9 ± 0.8 (11 15/16 ± 3/4)	11.9 ± 0.8 (11 15/16 ± 3/4)	11.9 ± 0.8 (11 15/16 ± 3/4)	SAI	10.8 ± 0.8 (10 13/16 ± 3/4)	10.9 ± 0.8 (10 15/16 ± 3/4)	10.8 ± 0.8 (10 3/4 ± 3/4)
39.3 / 34.9 ± 1.5	39.3 / 34.9 ± 1.5	39.3 / 34.9 ± 1.5	TURNING ANGLE	35.9 / 32.3 ± 1.5	35.9 / 32.3 ± 1.5	35.9 / 32.3 ± 1.5
12	12	12	ILLUSTRATION #	12	12	12
			REAR			
			CAMBER (Degrees)			
			REAR TOE (Degrees)			
			REAR TOE (Inches)			
			THRUST ANGLE			
			ILLUSTRATION #			
			NOTES			
49207	49208	49209		49210	49211	49212

Toyota Truck & SUV

MODEL INFORMATION	Tundra 4X2 X-Cab, V8 (USA & Canada)	Tundra 4X4 Regular Cab (USA & Canada)	Tundra 4X4 X-Cab, V6 (USA & Canada)	Tundra 4X4 X-Cab, V8 (USA & Canada)	Tundra 4X2 Regular Cab	Tundra 4X2 X-Cab, V6
YEARS	2000	2000	2000	2000	2001-2002	2001-2002
INSTRUCTIONS						
LUGNUT TORQUE						
FRONT						
CASTER (Degrees)	1.93 ± 0.75 1 15/16 ± ¾	1.11 ± 0.75 1⅛ ± ¾	1.05 ± 0.75 1 1/16 ± ¾	1.24 ± 0.75 1¼ ± ¾	1.9 ± 0.75 1⅞ ± ¾	1.88 ± 0.75 1⅞ ± ¾
CROSS | TOL.	0.0 (0) | 0.75 (¾)	0.0 (0) | 0.75 (¾)	0.0 (0) | 0.75 (¾)	0.0 (0) | 0.75 (¾)	0.0 (0) | 0.75 (¾)	0.0 (0) | 0.75 (¾)
CAMBER (Degrees)	-0.11 ± 0.75 -⅛ ± ¾	0.25 ± 0.75 ¼ ± ¾	0.31 ± 0.75 5/16 ± ¾	0.24 ± 0.75 ¼ ± ¾	-0.14 ± 0.75 [1] -⅛ ± ¾ [1]	0.0 ± 0.75 0 ± ¾
CROSS | TOL.	0.0 (0) | 0.75 (¾)	0.0 (0) | 0.75 (¾)	0.0 (0) | 0.75 (¾)	0.0 (0) | 0.75 (¾)	0.0 (0) | 0.75 (¾)	0.0 (0) | 0.75 (¾)
FRONT TOE (Degrees)	0.11° ± 0.2° ⅛° ± 3/16°	0.09° ± 0.2° 3/32° ± 3/16°	0.12° ± 0.2° ⅛° ± 3/16°	0.09° ± 0.2° 3/32° ± 3/16°	0.1° ± 0.2° 3/32° ± 3/16°	0.14° ± 0.2° ⅛° ± 3/16°
FRONT TOE (Inches)	0.06" ± 0.1" 1/16" ± 3/32"	0.05" ± 0.1" 1/32" ± 3/32"	0.06" ± 0.1" 1/16" ± 3/32"	0.05" ± 0.1" 1/32" ± 3/32"	0.05" ± 0.1" 1/16" ± 3/32"	0.07" ± 0.1" 1/16" ± 3/32"
SAI	10.9 ± 0.8 (10⅞ ± ¾)	10.5 ± 0.8 (10½ ± ¾)	10.4 ± 0.8 (10 7/16 ± ¾)	10.5 ± 0.8 (10½ ± ¾)	10.9 ± 0.8 (10⅞ ± ¾)	10.8 ± 0.8 (10¾ ± ¾)
TURNING ANGLE	35.9 / 32.3 ±1.5	36.5 / 32.7 ±1.5	36.5 / 32.7 ±1.5	36.5 / 32.7 ±1.5	34.3 / 31.8 ±1.0	34.3 / 31.8 ±1.0
ILLUSTRATION #	12	12	12	12	12	12
REAR						
CAMBER (Degrees)						
REAR TOE (Degrees)						
REAR TOE (Inches)						
THRUST ANGLE						
ILLUSTRATION #						
NOTES					[1] Canada models = -0.17° (-3/16°)	
	49213	49214	49215	49216	49217	49218

Toyota Truck & SUV

MODEL INFORMATION	Tundra 4X2 X-Cab, V8 w/o Off-road pkg.	Tundra 4X2 X-Cab, V8 with Off-road pkg.	Tundra 4X4 Regular Cab, V6	Tundra 4X4 Regular Cab, V8	Tundra 4X4 X-Cab, V6	Tundra 4X4 X-Cab, V8
YEARS	2001-2002	2001-2002	2001-2002	2001-2002	2001-2002	2001-2002
INSTRUCTIONS						
LUGNUT TORQUE						
FRONT						
CASTER (Degrees)	2.0 ± 0.75 2 ± ¾	1.2 ± 0.75 1 3/16 ± ¾	1.32 ± 0.75 [1] 1 5/16 ± ¾ [1]	1.31 ± 0.75 [1] 1 5/16 ± ¾ [1]	1.15 ± 0.75 [1] 1⅛ ± ¾ [1]	1.31 ± 0.75 [1] 1 5/16 ± ¾ [1]
CROSS | TOL.	0.0 (0) | 0.75 (¾)	0.0 (0) | 0.75 (¾)	0.0 (0) | 0.75 (¾)	0.0 (0) | 0.75 (¾)	0.0 (0) | 0.75 (¾)	0.0 (0) | 0.75 (¾)
CAMBER (Degrees)	-0.11 ± 0.75 -⅛ ± ¾	0.3 ± 0.75 5/16 ± ¾	0.25 ± 0.75 ¼ ± ¾	0.14 ± 0.75 ⅛ ± ¾	0.3 ± 0.75 5/16 ± ¾	0.22 ± 0.75 ¼ ± ¾
CROSS | TOL.	0.0 (0) | 0.75 (¾)	0.0 (0) | 0.75 (¾)	0.0 (0) | 0.75 (¾)	0.0 (0) | 0.75 (¾)	0.0 (0) | 0.75 (¾)	0.0 (0) | 0.75 (¾)
FRONT TOE (Degrees)	0.11° ± 0.2° ⅛° ± 3/16°	0.25° ± 0.2° ¼° ± 3/16°	0.09° ± 0.2° 3/32° ± 3/16°	0.06° ± 0.2° 1/16° ± 3/16°	0.13° ± 0.2° ⅛° ± 3/16°	0.09° ± 0.2° 3/32° ± 3/16°
FRONT TOE (Inches)	0.06" ± 0.1" 1/16" ± 3/32"	0.13" ± 0.1" ⅛" ± 3/32"	0.05" ± 0.1" 1/32" ± 3/32"	0.03" ± 0.1" 1/32" ± 3/32"	0.06" ± 0.1" 1/16" ± 3/32"	0.05" ± 0.1" 1/32" ± 3/32"
SAI	10.8 ± 0.8 (10¾ ± ¾)	10.4 ± 0.8 (10⅜ ± ¾)	10.5 ± 0.8 (10½ ± ¾)	10.6 ± 0.8 (10⅝ ± ¾)	10.4 ± 0.8 (10⅜ ± ¾)	10.5 ± 0.8 (10½ ± ¾)
TURNING ANGLE	34.3 / 31.8 ±1.0	34.3 / 31.8 ±1.0	35.6 / 32.3 ±1.5	35.6 / 32.3 ±1.5	35.6 / 32.3 ±1.5	35.6 / 32.3 ±1.5
ILLUSTRATION #	12	12	12	12	12	12
REAR						
CAMBER (Degrees)						
REAR TOE (Degrees)						
REAR TOE (Inches)						
THRUST ANGLE						
ILLUSTRATION #						
NOTES			[1] Canada models = 1.2° (1 3/16°)	[1] Canada models = 1.2° (1 3/16°)	[1] Off-road pkg. = 1.25° (1¼°)	[1] Off-road pkg. = 1.37° (1⅜°)
	49219	49220	49221	49222	49223	49224

Toyota Truck & SUV

			MODEL INFORMATION			
Tundra 4X2 Regular Cab (USA & Canada)	Tundra 4X2 Access Cab (Canada)	Tundra 4X2 Access Cab (USA) V6	MODEL INFORMATION	Tundra 4X2 Access Cab (USA) V8, Std-bed	Tundra 4X2 Access Cab (USA) V8, StepSide	Tundra 4X4 Regular Cab (USA & Canada)
2003	2003	2003	YEARS	2003	2003	2003
			INSTRUCTIONS			
			LUGNUT TORQUE			
			FRONT			
1.9 ± 0.75 1⅞ ± ¾	2.35 ± 0.75 2⅜ ± ¾	1.88 ± 0.75 1⅞ ± ¾	CASTER (Degrees)	2.13 ± 0.75 2⅛ ± ¾	2.5 ± 0.75 2½ ± ¾	1.5 ± 0.75 [1] 1½ ± ¾ [1]
0.0 (0) | 0.75 (¾)	0.0 (0) | 0.75 (¾)	0.0 (0) | 0.75 (¾)	CROSS | TOL.	0.0 (0) | 0.75 (¾)	0.0 (0) | 0.75 (¾)	0.0 (0) | 0.75 (¾)
0.0 ± 0.75 0 ± ¾	-0.08 ± 0.75 -1/16 ± ¾	0.13 ± 0.75 ⅛ ± ¾	CAMBER (Degrees)	-0.04 ± 0.75 [1] -1/16 ± ¾ [1]	-0.04 ± 0.75 [1] -1/16 ± ¾ [1]	0.14 ± 0.75 [2] ⅛ ± ¾ [2]
0.0 (0) | 0.75 (¾)	0.0 (0) | 0.75 (¾)	0.0 (0) | 0.75 (¾)	CROSS | TOL.	0.0 (0) | 0.75 (¾)	0.0 (0) | 0.75 (¾)	0.0 (0) | 0.75 (¾)
0.15° ± 0.2° [1] 5/32° ± 3/16° [1]	0.13° ± 0.2° ⅛° ± 3/16°	0.18° ± 0.2° 3/16° ± 3/16°	FRONT TOE (Degrees)	0.13° ± 0.2° ⅛° ± 3/16°	0.13° ± 0.2° ⅛° ± 3/16°	0.05° ± 0.2° 1/16° ± 3/16°
0.08" ± 0.1" [1] 1/16" + 3/32" [1]	0.06" ± 0.1" 1/16" ± 3/32"	0.09" ± 0.1" 3/32" ± 3/32"	FRONT TOE (Inches)	0.06" ± 0.1" 1/16" ± 3/32"	0.06" ± 0.1" 1/16" ± 3/32"	0.03" ± 0.1" 1/32" ± 3/32"
10.8 ± 0.8 (10¾ ± ¾)	10.8 ± 0.8 (10 13/16 ± ¾)	10.6 ± 0.8 (10⅝ ± ¾)	SAI	10.8 ± 0.8 (10¾ ± ¾)	10.8 ± 0.8 (10¾ ± ¾)	10.8 ± 0.8 (10¾ ± ¾)
36.4 / 32.5 ± 1.5	36.3 / 32.3 ± 1.5	36.6 / 32.5 ± 1.5	TURNING ANGLE	36.3 / 32.3 ± 1.5	36.3 / 32.3 ± 1.5	36.7 / 32.5 ± 1.5
12	12	12	ILLUSTRATION #	12	12	12
			REAR			
			CAMBER (Degrees)			
			REAR TOE (Degrees)			
			REAR TOE (Inches)			
			THRUST ANGLE			
			ILLUSTRATION #			
[1] Canada models = 0.1° (0.05")			NOTES	[1] Sport Susp. = -0.1° (-⅛°)	[1] Sport Susp. = -0.1° (-⅛°)	[1] Canada = 1.42° (1 7/16°) [2] Canada = 0.18° (3/16°)
49225	49226	49227		49228	49229	49230

Toyota Truck & SUV

			MODEL INFORMATION			
Tundra 4X4 Acc. Cab (Canada)	Tundra 4X4 Access Cab (USA) V6	Tundra 4X4 Access Cab (USA) V8, STD-bed	MODEL INFORMATION	Tundra 4X4 Access Cab (USA) V8, StepSide	Tundra 4X2 Regular Cab (USA & Canada)	Tundra 4X2 Access Cab (USA) V6
2003	2003	2003	YEARS	2003	2004	2004
			INSTRUCTIONS			
			LUGNUT TORQUE			
			FRONT			
1.5 ± 0.75 1½ ± ¾	1.13 ± 0.75 1⅛ ± ¾	1.55 ± 0.75 [1] 1 9/16 ± ¾ [1]	CASTER (Degrees)	1.82 ± 0.75 [1] 1 13/16 ± ¾ [1]	1.9 ± 0.75 1⅞ ± ¾	1.97 ± 0.75 2 ± ¾
0.0 (0) | 0.75 (¾)	0.0 (0) | 0.75 (¾)	0.0 (0) | 0.75 (¾)	CROSS | TOL.	0.0 (0) | 0.75 (¾)	0.0 (0) | 0.75 (¾)	0.0 (0) | 0.75 (¾)
0.26 ± 0.75 ¼ ± ¾	0.36 ± 0.75 ⅜ ± ¾	0.26 ± 0.75 [2] ¼ ± ¾ [2]	CAMBER (Degrees)	0.27 ± 0.75 [2] ¼ ± ¾ [2]	-0.13 ± 0.75 -⅛ ± ¾	-0.09 ± 0.75 -1/16 ± ¾
0.0 (0) | 0.75 (¾)	0.0 (0) | 0.75 (¾)	0.0 (0) | 0.75 (¾)	CROSS | TOL.	0.0 (0) | 0.75 (¾)	0.0 (0) | 0.75 (¾)	0.0 (0) | 0.75 (¾)
0.1° ± 0.2° 3/32° ± 3/16°	0.15° ± 0.2° 5/32° ± 3/16°	0.1° ± 0.2° 3/32° ± 3/16°	FRONT TOE (Degrees)	0.1° ± 0.2° 3/32° ± 3/16°	0.11° ± 0.2° ⅛° ± 3/16°	0.12° ± 0.2° ⅛° ± 3/16°
0.05" ± 0.1" 1/16" ± 3/32"	0.08" ± 0.1" 1/16" ± 3/32"	0.05" ± 0.1" 1/16" ± 3/32"	FRONT TOE (Inches)	0.05" ± 0.1" 1/16" ± 3/32"	0.06" ± 0.1" 1/16" ± 3/32"	0.06" ± 0.1" 1/16" ± 3/32"
10.5 ± 0.8 (10½ ± ¾)	10.8 ± 0.8 (10¾ ± ¾)	10.4 ± 0.8 (10 7/16 ± ¾)	SAI	10.4 ± 0.8 (10 7/16 ± ¾)	10.5 ± 0.8 (10½ ± ¾)	10.4 ± 0.8 (10⅜ ± ¾)
37.0 / 32.7 ± 1.5	37.3 / 32.8 ± 1.5	37.0 / 32.7 ± 1.5	TURNING ANGLE	37.0 / 32.7 ± 1.5	35.4 / 32.2 ± 1.5	35.6 / 32.3 ± 1.5
12	12	12	ILLUSTRATION #	12	12	12
			REAR			
			CAMBER (Degrees)			
			REAR TOE (Degrees)			
			REAR TOE (Inches)			
			THRUST ANGLE			
			ILLUSTRATION #			
		[1] Limited = 1.9° (1⅞°) [2] Limited = 0.29° (5/16°)	NOTES	[1] Limited = 1.4° (1⅜°) [2] Limited = 0.29° (5/16°)		
49231	49232	49233		49234	49235	49236

ALIGNMENT SPECIFICATIONS

Toyota Truck & SUV

Tundra 4X2 Access Cab (USA) V8, exc/Off-Rd	Tundra 4X2 Access Cab (USA) V8, Off-Rd, STD-Bed	Tundra 4X2 Access Cab (USA) V8, Off-Rd, StepSide	MODEL INFORMATION	Tundra 4X2 Acc. Cab (Canada)	Tundra 4X2 Double Cab (USA & Canada)	Tundra 4X4 Regular Cab (USA & Canada)
2004	2004	2004	YEARS	2004	2004	2004
			INSTRUCTIONS			
			LUGNUT TORQUE			
			FRONT			
2.35 ± 0.75 [1] 2⅜ ± ¾ [1]	2.29 ± 0.75 2 5/16 ± ¾	2.62 ± 0.75 2⅝ ± ¾	CASTER (Degrees)	2.42 ± 0.75 2 7/16 ± ¾	2.17 ± 0.75 [1] 2 3/16 ± ¾ [1]	1.42 ± 0.75 1 7/16 ± ¾
0.0 (0) | 0.75 (¾)	0.0 (0) | 0.75 (¾)	0.0 (0) | 0.75 (¾)	CROSS | TOL.	0.0 (0) | 0.75 (¾)	0.0 (0) | 0.75 (¾)	0.0 (0) | 0.75 (¾)
-0.25 ± 0.75 -¼ ± ¾	-0.18 ± 0.75 -3/16 ± ¾	-0.18 ± 0.75 -3/16 ± ¾	CAMBER (Degrees)	-0.3 ± 0.75 -5/16 ± ¾	-0.12 ± 0.75 -⅛ ± ¾	0.09 ± 0.75 1/16 ± ¾
0.0 (0) | 0.75 (¾)	0.0 (0) | 0.75 (¾)	0.0 (0) | 0.75 (¾)	CROSS | TOL.	0.0 (0) | 0.75 (¾)	0.0 (0) | 0.75 (¾)	0.0 (0) | 0.75 (¾)
0.06° ± 0.2° 1/16° ± 3/16°	0.09° ± 0.2° 3/32° ± 3/16°	0.13° ± 0.2° ⅛° ± 3/16°	FRONT TOE (Degrees)	0.05° ± 0.2° 1/16° ± 3/16°	0.11° ± 0.2° ⅛° ± 3/16°	0.03° ± 0.2° 1/32° ± 3/16°
0.03" ± 0.1" 1/32" ± 3/32"	0.05" ± 0.1" 1/32" ± 3/32"	0.06" ± 0.1" 1/16" ± 3/32"	FRONT TOE (Inches)	0.03" ± 0.1" 1/32" ± 3/32"	0.06" ± 0.1" 1/16" ± 3/32"	0.02" ± 0.1" 1/64" ± 3/32"
10.6 ± 0.8 (10⅝ ± ¾)	10.6 ± 0.8 (10 9/16 ± ¾)	10.6 ± 0.8 (10 9/16 ± ¾)	SAI	10.7 ± 0.8 (10 11/16 ± ¾)	10.4 ± 0.8 (10⅜ ± ¾)	10.3 ± 0.8 (10¼ ± ¾)
35.3 / 32.3 ± 1.5	35.4 / 32.2 ± 1.5	35.4 / 32.2 ± 1.5	TURNING ANGLE	35.2 / 32.0 ± 1.5	35.4 / 32.2 ± 1.5	35.8 / 32.4 ± 1.5
12	12	12	ILLUSTRATION #	12	12	12
			REAR			
			CAMBER (Degrees)			
			REAR TOE (Degrees)			
			REAR TOE (Inches)			
			THRUST ANGLE			
			ILLUSTRATION #			
[1] StepSide = 2.7° (2 11/16°)			NOTES		[1] w/P245/70R16 = 2.0° (2°)	
49237	49238	49239		49240	49241	49242

Toyota Truck & SUV

Tundra 4X4 Access Cab (USA) V6	Tundra 4X4 Access Cab (USA) V8	Tundra 4X4 Acc. Cab (Canada)	MODEL INFORMATION	Tundra 4X4 Double Cab (USA & Canada)	Tundra 4X2 (USA) Regular Cab	Tundra 4X2 (USA) Access Cab, V6
2004	2004	2004	YEARS	2004	2005-2006	2005-2006
			INSTRUCTIONS			
			LUGNUT TORQUE			
			FRONT			
1.23 ± 0.75 1¼ ± ¾	1.55 ± 0.75 [1] 1 9/16 ± ¾ [1]	1.62 ± 0.75 1⅝ ± ¾	CASTER (Degrees)	1.9 ± 0.75 1⅞ ± ¾	2.0 ± 0.75 2 ± ¾	1.63 ± 0.75 1⅝ ± ¾
0.0 (0) | 0.75 (¾)	0.0 (0) | 0.75 (¾)	0.0 (0) | 0.75 (¾)	CROSS | TOL.	0.0 (0) | 0.75 (¾)	0.0 (0) | 0.75 (¾)	0.0 (0) | 0.75 (¾)
0.27 ± 0.75 ¼ ± ¾	0.16 ± 0.75 3/16 ± ¾	0.15 ± 0.75 ⅛ ± ¾	CAMBER (Degrees)	0.07 ± 0.75 1/16 ± ¾	-0.16 ± 0.75 -3/16 ± ¾	0.05 ± 0.75 1/16 ± ¾
0.0 (0) | 0.75 (¾)	0.0 (0) | 0.75 (¾)	0.0 (0) | 0.75 (¾)	CROSS | TOL.	0.0 (0) | 0.75 (¾)	0.0 (0) | 0.75 (¾)	0.0 (0) | 0.75 (¾)
0.11° ± 0.2° ⅛° ± 3/16°	0.06° ± 0.2° 1/16° ± 3/16°	0.06° ± 0.2° 1/16° ± 3/16°	FRONT TOE (Degrees)	0.11° ± 0.2° ⅛° ± 3/16°	0.07° ± 0.2° 1/16° ± 3/16°	0.13° ± 0.2° ⅛° ± 3/16°
0.06" ± 0.1" 1/16" ± 3/32"	0.03" ± 0.1" 1/32" ± 3/32"	0.03" ± 0.1" 1/32" ± 3/32"	FRONT TOE (Inches)	0.06" ± 0.1" 1/16" ± 3/32"	0.04" ± 0.1" 1/32" ± 3/32"	0.07" ± 0.1" 1/16" ± 3/32"
10.1 ± 0.8 (10 1/16 ± ¾)	10.2 ± 0.8 (10 3/16 ± ¾)	10.2 ± 0.8 (10 3/16 ± ¾)	SAI	10.3 ± 0.8 (110 5/16 ± ¾)	10.3 ± 0.8 (10 5/16 ± ¾)	10.3 ± 0.8 (10 5/16 ± ¾)
36.1 / 32.6 ± 1.5	35.9 / 32.7 ± 1.5	35.9 / 32.4 ± 1.5	TURNING ANGLE	35.7 / 32.4 ± 1.5	35.4 / 32.2 ± 1.5	35.6 / 32.4 ± 1.5
12	12	12	ILLUSTRATION #	12	12	12
			REAR			
			CAMBER (Degrees)			
			REAR TOE (Degrees)			
			REAR TOE (Inches)			
			THRUST ANGLE			
			ILLUSTRATION #			
	[1] StepSide = 2.05° (2 1/16°)		NOTES			
49243	49244	49245		49246	49247	49248

Toyota Truck & SUV

Tundra 4X2 (USA) Access Cab, V8, exc/TRD Sport	Tundra 4X2 (USA) Access Cab, V8, TRD Sport	Tundra 4X2 (USA) Double Cab exc/Off-Road Pkg.	MODEL INFORMATION	Tundra 4X2 (USA) Double Cab w/Off-Road Pkg.	Tundra 4X4 (USA) Regular Cab	Tundra 4X4 (USA) Access Cab exc/Off-Road Pkg.
2005-2006	2005-2006	2005-2006	YEARS	2005-2006	2005-2006	2005-2006
			INSTRUCTIONS			
			LUGNUT TORQUE			
			FRONT			
2.06 ±0.75 [1] 2 1/16 ± 3/4 [1]	2.44 ±0.75 [1] 2 7/16 ± 3/4 [1]	2.1 ±0.75 2 1/8 ± 3/4	CASTER (Degrees)	2.15 ±0.75 2 1/8 ± 3/4	1.42 ±0.75 1 7/16 ± 3/4	1.4 ±0.75 [1] 1 3/8 ± 3/4 [1]
0.0 (0) | 0.75 (3/4)	0.0 (0) | 0.75 (3/4)	0.0 (0) | 0.75 (3/4)	CROSS | TOL.	0.0 (0) | 0.75 (3/4)	0.0 (0) | 0.75 (3/4)	0.0 (0) | 0.75 (3/4)
-0.1 ±0.75 -1/8 ± 3/4	-0.3 ±0.75 -5/16 ± 3/4	-0.04 ±0.75 -1/16 ± 3/4	CAMBER (Degrees)	0.0 ±0.75 0 ± 3/4	0.18 ±0.75 3/16 ± 3/4	0.25 ±0.75 1/4 ± 3/4
0.0 (0) | 0.75 (3/4)	0.0 (0) | 0.75 (3/4)	0.0 (0) | 0.75 (3/4)	CROSS | TOL.	0.0 (0) | 0.75 (3/4)	0.0 (0) | 0.75 (3/4)	0.0 (0) | 0.75 (3/4)
0.1° ±0.2° 3/32° ± 3/16°	0.05° ±0.2° 1/16° ± 3/16°	0.11° ±0.2° 1/8° ± 3/16°	FRONT TOE (Degrees)	0.12° ±0.2° 1/8° ± 3/16°	0.06° ±0.2° 1/16° ± 3/16°	0.08° ±0.2° 3/32° ± 3/16°
0.05" ±0.1" 1/16" ± 3/32"	0.03" ±0.1" 1/32" ± 3/32"	0.06" ±0.1" 1/16" ± 3/32"	FRONT TOE (Inches)	0.06" ±0.1" 1/16" ± 3/32"	0.03" ±0.1" 1/32" ± 3/32"	0.04" ±0.1" 1/32" ± 3/32"
10.5 ±0.8 (10 7/16 ± 3/4)	10.6 ±0.8 (10 5/8 ± 3/4)	10.4 ±0.8 (10 3/8 ± 3/4)	SAI	10.4 ±0.8 (10 3/8 ± 3/4)	10.2 ±0.8 (10 3/16 ± 3/4)	10.1 ±0.8 (10 1/8 ± 3/4)
35.5 / 32.3 ±1.5	35.3 / 32.1 ±1.5	35.5 / 32.3 ±1.5	TURNING ANGLE	35.5 / 32.3 ±1.5	36.0 / 32.5 ±1.5	36.1 / 32.6 ±1.5
12	12	12	ILLUSTRATION #	12	12	12
			REAR			
			CAMBER (Degrees)			
			REAR TOE (Degrees)			
			REAR TOE (Inches)			
			THRUST ANGLE			
			ILLUSTRATION #			
[1] StepSide = 2.63° (2 5/8°)	[1] StepSide = 2.8° (2 13/16°)		NOTES			[1] StepSide = 2.0° (2°)
49249	49250	49251		49252	49253	49254

Toyota Truck & SUV

Tundra 4X4 (USA) Access Cab w/Off-Road Pkg.	Tundra 4X4 (USA) Double Cab	Tundra 4X2 (CAN) Regular Cab	MODEL INFORMATION	Tundra 4X2 (CAN) Access Cab	Tundra 4X2 (CAN) Double Cab	Tundra 4X4 (CAN) Regular Cab
2005-2006	2005-2006	2005-2006	YEARS	2005-2006	2005-2006	2005-2006
			INSTRUCTIONS			
			LUGNUT TORQUE			
			FRONT			
1.79 ±0.75 [1] 1 13/16 ± 3/4 [1]	1.85 ±0.75 1 7/8 ± 3/4	2.0 ±0.75 2 ± 3/4	CASTER (Degrees)	2.1 ±0.75 [1] 2 1/8 ± 3/4 [1]	2.1 ±0.75 2 1/8 ± 3/4	1.4 ±0.75 1 3/8 ± 3/4
0.0 (0) | 0.75 (3/4)	0.0 (0) | 0.75 (3/4)	0.0 (0) | 0.75 (3/4)	CROSS | TOL.	0.0 (0) | 0.75 (3/4)	0.0 (0) | 0.75 (3/4)	0.0 (0) | 0.75 (3/4)
-0.07 ±0.75 -1/16 ± 3/4	0.14 ±0.75 1/8 ± 3/4	-0.14 ±0.75 -1/8 ± 3/4	CAMBER (Degrees)	-0.14 ±0.75 -1/8 ± 3/4	-0.06 ±0.75 -1/16 ± 3/4	0.21 ±0.75 3/16 ± 3/4
0.0 (0) | 0.75 (3/4)	0.0 (0) | 0.75 (3/4)	0.0 (0) | 0.75 (3/4)	CROSS | TOL.	0.0 (0) | 0.75 (3/4)	0.0 (0) | 0.75 (3/4)	0.0 (0) | 0.75 (3/4)
-0.03° ±0.2° -1/32° ± 3/16°	0.08° ±0.2° 3/32° ± 3/16°	0.08° ±0.2° 3/32° ± 3/16°	FRONT TOE (Degrees)	0.09° ±0.2° 3/32° ± 3/16°	0.1° ±0.2° 3/32° ± 3/16°	0.05° ±0.2° 1/16° ± 3/16°
-0.01" ±0.1" -1/64" ± 3/32"	0.04" ±0.1" 1/32" ± 3/32"	0.04" ±0.1" 1/32" ± 3/32"	FRONT TOE (Inches)	0.04" ±0.1" 1/32" ± 3/32"	0.05" ±0.1" 1/16" ± 3/32"	0.03" ±0.1" 1/32" ± 3/32"
10.4 ±0.8 (10 3/8 ± 3/4)	10.2 ±0.8 (10 3/16 ± 3/4)	10.5 ±0.8 (10 1/2 ± 3/4)	SAI	10.5 ±0.8 (10 1/2 ± 3/4)	10.4 ±0.8 (10 3/8 ± 3/4)	10.2 ±0.8 (10 3/16 ± 3/4)
35.5 / 32.1 ±1.5	35.8 / 32.5 ±1.5	35.4 / 32.2 ±1.5	TURNING ANGLE	35.4 / 32.2 ±1.5	35.5 / 32.3 ±1.5	36.0 / 32.5 ±1.5
12	12	12	ILLUSTRATION #	12	12	12
			REAR			
			CAMBER (Degrees)			
			REAR TOE (Degrees)			
			REAR TOE (Inches)			
			THRUST ANGLE			
			ILLUSTRATION #			
[1] StepSide = 2.4° (2 3/8°)			NOTES	[1] w/17" tires = 2.7° (2 11/16°)		
49255	49256	49257		49258	49259	49260

ALIGNMENT SPECIFICATIONS

Toyota Truck & SUV

Tundra 4X4 (CAN) Access Cab w/Off-Road Pkg.	Tundra 4X4 (CAN) Access Cab exc/Off-Road Pkg.	Tundra 4X4 (CAN) Double Cab	MODEL INFORMATION	Tundra 4X2 (USA) Regular Cab, V6 Std. Bed	Tundra 4X2 Regular Cab, V6 Long Bed	Tundra 4X2 Regular Cab, V8 Std. Bed
2005-2006	2005-2006	2005-2006	YEARS	2007-2009	2007-2009	2007-2009
			INSTRUCTIONS			
			LUGNUT TORQUE			
			FRONT			
1.64 ±0.75 1⅝ ± ¾	1.24 ±0.75 1¼ ± ¾	1.89 ±0.75 1⅞ ± ¾	CASTER (Degrees)	2.75 ±0.75 2¾ ± ¾	2.9 ±0.75 2⅞ ± ¾	2.58 ±0.75 2 9/16 ± ¾
0.0 (0) | 0.75 (¾)	0.0 (0) | 0.75 (¾)	0.0 (0) | 0.75 (¾)	CROSS | TOL.	0.0 (0) | 0.5 (½)	0.0 (0) | 0.5 (½)	0.0 (0) | 0.5 (½)
-0.1 ±0.75 -⅛ ± ¾	0.25 ±0.75 ¼ ± ¾	0.13 ±0.75 ⅛ ± ¾	CAMBER (Degrees)	-0.18 ±0.75 -3/16 ± ¾	-0.07 ±0.75 -1/16 ± ¾	-0.13 ±0.75 [1] -⅛ ± ¾ [1]
0.0 (0) | 0.75 (¾)	0.0 (0) | 0.75 (¾)	0.0 (0) | 0.75 (¾)	CROSS | TOL.	0.0 (0) | 0.5 (½)	0.0 (0) | 0.5 (½)	0.0 (0) | 0.5 (½)
-0.04° ±0.2° -1/32° ± 3/16°	0.08° ±0.2° 3/32° ± 3/16°	0.11° ±0.2° ⅛° ± 3/16°	FRONT TOE (Degrees)	0.14° ±0.16° ⅛° ± 5/32°	0.23° ±0.16° 7/32° ± 5/32°	0.19° ±0.16° [2] 3/16° ± 5/32° [2]
-0.02" ±0.1" -1/64" ± 3/32"	0.04" ±0.1" 1/32" ± 3/32"	0.06" ±0.1" 1/16" ± 3/32"	FRONT TOE (Inches)	0.07" ±0.08" 1/16" ± 3/32"	0.12" ±0.08" ⅛" ± 3/32"	0.1" ±0.08" [2] 3/32" ± 3/32" [2]
10.5 ±0.8 (10½ ± ¾)	10.1 ±0.8 (10⅛ ± ¾)	10.3 ±0.8 (10¼ ± ¾)	SAI	13.2 ±0.8 (13 3/16 ± ¾)	13.1 ±0.8 (⅛ ± ¾)	13.2 ±0.8 (13⅛ ± ¾)
35.5 / 32.2 ±1.5	36.1 / 32.6 ±1.5	35.8 / 32.5 ±1.5	TURNING ANGLE	40.5 / 36.4 ±2.0	40.6 / 36.5 ±2.0	40.5 / 36.4 ±2.0
12	12	12	ILLUSTRATION #	12	12	12
			REAR			
			CAMBER (Degrees)			
			REAR TOE (Degrees)			
			REAR TOE (Inches)			
			THRUST ANGLE			
			ILLUSTRATION #			
			NOTES			[1] 5.7L = -0.2° (-3/16°) [2] 5.7L = 0.13° (0.065")
49261	49262	49263		49264	49265	49266

Toyota Truck & SUV

Tundra 4X2 Regular Cab, V8 Long Bed	Tundra 4X2 Double Cab, V6	Tundra 4X2 Double Cab, V8	MODEL INFORMATION	Tundra 4X2 CrewMax, V8 4.7L	Tundra 4X2 CrewMax, V8 5.7L	Tundra 4X4 Regular Cab Std. Bed
2007-2009	2007-2009	2007-2009	YEARS	2007-2009	2007-2009	2007-2009
			INSTRUCTIONS			
			LUGNUT TORQUE			
			FRONT			
2.7 ±0.75 2 11/16 ± ¾	2.82 ±0.75 2 13/16 ± ¾	3.0 ±0.75 3 ± ¾	CASTER (Degrees)	3.0 ±0.75 3 ± ¾	3.0 ±0.75 3 ± ¾	2.3 ±0.75 2 5/16 ± ¾
0.0 (0) | 0.5 (½)	0.0 (0) | 0.5 (½)	0.0 (0) | 0.5 (½)	CROSS | TOL.	0.0 (0) | 0.5 (½)	0.0 (0) | 0.5 (½)	0.0 (0) | 0.5 (½)
-0.09 ±0.75 -1/16 ± ¾	0.07 ±0.75 1/16 ± ¾	0.04 ±0.75 1/16 ± ¾	CAMBER (Degrees)	0.0 ±0.75 0 ± ¾	0.26 ±0.75 [1] ¼ ± ¾ [1]	0.15 ±0.75 ⅛ ± ¾
0.0 (0) | 0.5 (½)	0.0 (0) | 0.5 (½)	0.0 (0) | 0.5 (½)	CROSS | TOL.	0.0 (0) | 0.5 (½)	0.0 (0) | 0.5 (½)	0.0 (0) | 0.5 (½)
0.24° ±0.16° ¼° ± 5/32°	0.37° ±0.16° ⅜° ± 5/32°	0.34° ±0.16° 11/32° ± 5/32°	FRONT TOE (Degrees)	0.32° ±0.14° 5/16° ± ⅛°	0.3° ±0.14° 5/16° ± ⅛°	0.15° ±0.16° 5/32° ± 5/32°
0.12" ±0.08" ⅛" ± 3/32"	0.19" ±0.08" 3/16" ± 3/32"	0.17" ±0.08" 5/32" ± 3/32"	FRONT TOE (Inches)	0.16" ±0.08" 5/32" ± 1/16"	0.15" ±0.08" 5/32" ± 1/16"	0.08" ±0.08" 1/16" ± 3/32"
13.1 ±0.8 (13⅛ ± ¾)	12.9 ±0.8 (12⅞ ± ¾)	13.0 ±0.8 (13 ± ¾)	SAI	13.0 ±0.8 (13 ± ¾)	13.0 ±0.8 (13 ± ¾)	12.8 ±0.8 (12 13/16 ± ¾)
40.5 / 36.5 ±2.0	40.7 / 36.7 ±2.0	40.6 / 36.7 ±2.0	TURNING ANGLE	40.6 / 36.6 ±1.5	41.0 / 36.8 ±1.5	40.9 / 36.7 ±2.0
12	12	12	ILLUSTRATION #	12	12	12
			REAR			
			CAMBER (Degrees)			
			REAR TOE (Degrees)			
			REAR TOE (Inches)			
			THRUST ANGLE			
			ILLUSTRATION #			
			NOTES		[1] 20" tires = -0.18° (-3/16°)	
49267	49268	49269		49270	49271	49272

Toyota Truck & SUV

MODEL INFORMATION	Tundra 4X4 Regular Cab Long Bed	Tundra 4X4 Double Cab, V8	Tundra 4X4 CrewMax, V8 4.7L	Tundra 4X4 CrewMax, V8 5.7L exc/Sport, Limited Off-Rd	Tundra 4X4 CrewMax, V8 5.7L Sport, Limited Off-Rd	Tundra 4X2 Reg. Cab, Std. Bed, 4.0L (GSK50L-TRADKA)
YEARS	2007-2009	2007-2009	2007-2009	2007-2009	2007-2009	2010-2013
INSTRUCTIONS						
LUGNUT TORQUE						
FRONT						
CASTER (Degrees)	2.4 ±0.75 2⅜ ± ¾	2.65 ±0.75 2⅝ ± ¾	2.75 ±0.75 2¾ ± ¾	2.7 ±0.75 2 11/16 ± ¾	2.7 ±0.75 2 11/16 ± ¾	2.7 ±0.75 2 11/16 ± ¾
CROSS | TOL.	0.0 (0) | 0.5 (½)	0.0 (0) | 0.5 (½)	0.0 (0) | 0.5 (½)	0.0 (0) | 0.5 (½)	0.0 (0) | 0.5 (½)	0.0 (0) | 0.5 (½)
CAMBER (Degrees)	0.22 ±0.75 ¼ ± ¾	0.3 ±0.75 [1] 5/16 ± ¾ [1]	0.0 ±0.75 0 ± ¾	0.27 ±0.75 ¼ ± ¾	-0.19 ±0.75 -3/16 ± ¾	-0.12 ±0.75 -⅛ ± ¾
CROSS | TOL.	0.0 (0) | 0.5 (½)	0.0 (0) | 0.5 (½)	0.0 (0) | 0.5 (½)	0.0 (0) | 0.5 (½)	0.0 (0) | 0.5 (½)	0.0 (0) | 0.5 (½)
FRONT TOE (Degrees)	0.24° ±0.16° ¼° ± 5/32°	0.32° ±0.16° 5/16° ± 5/32°	0.28° ±0.14° 9/32° ± ⅛°	0.29° ±0.14° 9/32° ± ⅛°	0.29° ±0.14° 9/32° ± ⅛°	0.19° ±0.16° 3/16° ± 5/32°
FRONT TOE (Inches)	0.12" ±0.08" ⅛" ± 3/32"	0.16" ±0.08" 5/32" ± 3/32"	0.14" ±0.08" ⅛" ± 1/16"	0.15" ±0.08" 5/32" ± 1/16"	0.15" ±0.08" 5/32" ± 1/16"	0.1" ±0.08" 3/32" ± 3/32"
SAI	12.8 ±0.8 (12 13/16 ± ¾)	12.7 ±0.8 (12 11/16 ± ¾)	12.8 ±0.8 (12¾ ± ¾)	12.8 ±0.8 (12¾ ± ¾)	12.8 ±0.8 (12¾ ± ¾)	13.1 ±0.8 (13⅛ ± ¾)
TURNING ANGLE	40.9 / 36.8 ±2.0	41.0 / 36.9 ±2.0	40.6 / 36.6 ±1.5	41.0 / 36.8 ±1.5	41.0 / 36.8 ±1.5	40.5 / 36.4 ±1.5
ILLUSTRATION #	12	12	12	12	12	12
REAR						
CAMBER (Degrees)						
REAR TOE (Degrees)						
REAR TOE (Inches)						
THRUST ANGLE						
ILLUSTRATION #						
NOTES		[1] 20" tires = -0.24° (-¼°)				
	49273	49274	49275	49276	49277	49278

Toyota Truck & SUV

MODEL INFORMATION	Tundra 4X2 Reg. Cab, Std. Bed, 4.6L (UPK50L-TRTDKA)	Tundra 4X2 Reg. Cab, Std. Bed, 5.7L (USK50L-TRTDKA)	Tundra 4X2 Reg. Cab, Long Bed, 4.0L (GSK51L-THADKA)	Tundra 4X2 Reg. Cab, Long Bed, 4.6L (UPK51L-THTDKA)	Tundra 4X2 Reg. Cab, Long Bed, 5.7L (USK51L-THTDKA)	Tundra 4X2 Dbl. Cab, Std. Bed, 4.0L (GSK51L-CRASKA)
YEARS	2010-2013	2010	2010-2013	2010	2010-2013	2010-2013
INSTRUCTIONS						
LUGNUT TORQUE						
FRONT						
CASTER (Degrees)	2.51 ±0.75 [1] 2½ ± ¾ [1]	2.6 ±0.75 2⅝ ± ¾	2.86 ±0.75 2⅞ ± ¾	2.68 ±0.75 2 11/16 ± ¾	2.65 ±0.75 2⅝ ± ¾	2.8 ±0.75 2 13/16 ± ¾
CROSS | TOL.	0.0 (0) | 0.5 (½)	0.0 (0) | 0.5 (½)	0.0 (0) | 0.5 (½)	0.0 (0) | 0.5 (½)	0.0 (0) | 0.5 (½)	0.0 (0) | 0.5 (½)
CAMBER (Degrees)	-0.15 ±0.75 -⅛ ± ¾	-0.23 ±0.75 -¼ ± ¾	-0.03 ±0.75 0 ± ¾	-0.05 ±0.75 -1/16 ± ¾	-0.11 ±0.75 -⅛ ± ¾	0.07 ±0.75 1/16 ± ¾
CROSS | TOL.	0.0 (0) | 0.5 (½)	0.0 (0) | 0.5 (½)	0.0 (0) | 0.5 (½)	0.0 (0) | 0.5 (½)	0.0 (0) | 0.5 (½)	0.0 (0) | 0.5 (½)
FRONT TOE (Degrees)	0.18° ±0.16° 3/16° ± 5/32°	0.12° ±0.16° [1] ⅛° ± 5/32° [1]	0.27° ±0.16° 9/32° ± 5/32°	0.29° ±0.16° 9/32° ± 5/32°	0.24° ±0.16° ¼° ± 5/32°	0.38° ±0.16° ⅜° ± 5/32°
FRONT TOE (Inches)	0.09" ±0.08" 3/32" ± 3/32"	0.06" ±0.08" [1] 1/16" ± 3/32" [1]	0.13" ±0.08" ⅛" ± 3/32"	0.14" ±0.08" 5/32" ± 3/32"	0.12" ±0.08" ⅛" ± 3/32"	0.19" ±0.08" 3/16" ± 3/32"
SAI	13.2 ±0.8 (13 3/16 ± ¾)	13.2 ±0.8 (13 3/16 ± ¾)	13.1 ±0.8 (13⅛ ± ¾)	13.1 ±0.8 (13⅛ ± ¾)	13.1 ±0.8 (13⅛ ± ¾)	12.9 ±0.8 (12⅞ ± ¾)
TURNING ANGLE	40.5 / 36.4 ±1.5	40.9 / 36.3 ±1.5	41.1 / 36.6 ±1.5	41.1 / 36.5 ±1.5	41.0 / 36.4 ±1.5	41.2 / 36.7 ±1.5
ILLUSTRATION #	12	12	12	12	12	12
REAR						
CAMBER (Degrees)						
REAR TOE (Degrees)						
REAR TOE (Inches)						
THRUST ANGLE						
ILLUSTRATION #						
NOTES	[1] Off-rd = 2.7° (2 11/16°)	[1] Off-rd = 0.09° (0.045")				
	49279	49280	49281	49282	49283	49284

ALIGNMENT SPECIFICATIONS

Toyota Truck & SUV

MODEL INFORMATION	Tundra 4X2 Dbl. Cab, Std. Bed, SR5, 4.6L (UPK51L-CRTSKA)	Tundra 4X2 Dbl. Cab, Std. Bed, Lmtd, 4.6L (UPK51L-CRTLKA)	Tundra 4X2 Dbl. Cab, Std. Bed, SR5, 5.7L (USK51L-CRTSKA)	Tundra 4X2 Dbl. Cab, Std. Bed, Lmtd, 5.7L (USK51L-CRTLKA)	Tundra 4X2 Dbl. Cab, Long Bed, 4.6L (UPK52L-CHTSKA)	Tundra 4X2 Dbl. Cab, Long Bed, 5.7L (USK52L-CHTSKA)
YEARS	2010-2013	2010	2010-2013	2010-2013	2010	2010-2013
INSTRUCTIONS						
LUGNUT TORQUE						
FRONT						
CASTER (Degrees)	2.8 ±0.75 2 13⁄16 ± ¾	3.0 ±0.75 3 ± ¾	2.9 ±0.75 2 ⅞ ± ¾	2.86 ±0.75 2 ⅞ ± ¾	2.96 ±0.75 2 15⁄16 ± ¾	3.0 ±0.75 3 ± ¾
CROSS | TOL.	0.0 (0) | 0.5 (½)	0.0 (0) | 0.5 (½)	0.0 (0) | 0.5 (½)	0.0 (0) | 0.5 (½)	0.0 (0) | 0.5 (½)	0.0 (0) | 0.5 (½)
CAMBER (Degrees)	0.0 ±0.75 0 ± ¾	0.06 ±0.75 [1] 1⁄16 ± ¾ [1]	0.0 ±0.75 [1] 0 ± ¾ [1]	0.0 ±0.75 [1] 0 ± ¾ [1]	0.03 ±0.75 0 ± ¾	-0.02 ±0.75 0 ± ¾
CROSS | TOL.	0.0 (0) | 0.5 (½)	0.0 (0) | 0.5 (½)	0.0 (0) | 0.5 (½)	0.0 (0) | 0.5 (½)	0.0 (0) | 0.5 (½)	0.0 (0) | 0.5 (½)
FRONT TOE (Degrees)	0.3° ±0.16° 5⁄16° ± 5⁄32°	0.4° ±0.16° [2] 13⁄32° ± 5⁄32° [2]	0.28° ±0.16° 9⁄32° ± 5⁄32°	0.32° ±0.16° [2] 5⁄16° ± 5⁄32° [2]	0.38° ±0.16° ⅜° ± 5⁄32°	0.28° ±0.16° 9⁄32° ± 5⁄32°
FRONT TOE (Inches)	0.15" ±0.08" 5⁄32" ± 3⁄32"	0.2" ±0.08" [2] 3⁄16" ± 3⁄32" [2]	0.14" ±0.08" ⅛" ± 3⁄32"	0.16" ±0.08" [2] 5⁄32" ± 3⁄32" [2]	0.19" ±0.08" 3⁄16" ± 3⁄32"	0.14" ±0.08" 5⁄32" ± 3⁄32"
SAI	13.0 ±0.8 (13 ± ¾)	13.0 ±0.8 (13 ± ¾)	13.0 ±0.8 (13 ± ¾)	13.0 ±0.8 (13 ± ¾)	13.0 ±0.8 (13 ± ¾)	13.0 ±0.8 (13 ± ¾)
TURNING ANGLE	41.1 / 36.6 ±1.5	40.6 / 36.6 ±1.5	41.1 / 36.6 ±1.5	41.1 / 36.6 ±1.5	40.6 / 36.6 ±1.5	41.1 / 36.6 ±1.5
ILLUSTRATION #	12	12	12	12	12	12
REAR						
CAMBER (Degrees)						
REAR TOE (Degrees)						
REAR TOE (Inches)						
THRUST ANGLE						
ILLUSTRATION #						
NOTES		[1] Off-rd = 0.0° (0°) [2] Off-rd = 0.36° (0.18")	[1] Off-rd = -0.05° (-1⁄16°)	[1] Off-rd = -0.04° (-1⁄16°) [2] Off-rd = 0.27° (0.14")		
	49285	49286	49287	49288	49289	49290

Toyota Truck & SUV

MODEL INFORMATION	Tundra 4X2 CrewMax Cab, SR5, 4.6L (UPK51L-PSTSKA)	Tundra 4X2 CrewMax Cab, Lmtd, 4.6L (UPK51L-PSTLKA)	Tundra 4X2 CrewMax Cab, 5.7L (USK51L-PST[L,S]KA)	Tundra 4X4 Reg. Cab, Std. Bed, 4.6L (UPK55L-TRTDKA)	Tundra 4X4 Reg. Cab, Std. Bed, 5.7L (USK55L-TRTD[K,G]A)	Tundra 4X4 Reg. Cab, Long Bed, 4.6L (UPK56L-THTDKA)
YEARS	2010	2010-2013	2010-2013	2010	2010-2013	2010
INSTRUCTIONS						
LUGNUT TORQUE						
FRONT						
CASTER (Degrees)	3.0 ±0.75 3 ± ¾	3.0 ±0.75 3 ± ¾	2.85 ±0.75 2 ⅞ ± ¾	2.25 ±0.75 2 ¼ ± ¾	2.3 ±0.75 2 5⁄16 ± ¾	2.4 ±0.75 2 ⅜ ± ¾
CROSS | TOL.	0.0 (0) | 0.5 (½)	0.0 (0) | 0.5 (½)	0.0 (0) | 0.5 (½)	0.0 (0) | 0.5 (½)	0.0 (0) | 0.5 (½)	0.0 (0) | 0.5 (½)
CAMBER (Degrees)	0.0 ±0.75 0 ± ¾	-0.02 ±0.75 0 ± ¾	-0.03 ±0.75 0 ± ¾	-0.15 ±0.75 -⅛ ± ¾	0.14 ±0.75 ⅛ ± ¾	0.21 ±0.75 3⁄16 ± ¾
CROSS | TOL.	0.0 (0) | 0.5 (½)	0.0 (0) | 0.5 (½)	0.0 (0) | 0.5 (½)	0.0 (0) | 0.5 (½)	0.0 (0) | 0.5 (½)	0.0 (0) | 0.5 (½)
FRONT TOE (Degrees)	0.34° ±0.16° 11⁄32° ± 5⁄32°	0.32° ±0.16° 5⁄16° ± 5⁄32°	0.27° ±0.16° 9⁄32° ± 5⁄32°	0.17° ±0.16° 5⁄32° ± 5⁄32°	0.14° ±0.16° ⅛° ± 5⁄32°	0.25° ±0.16° ¼° ± 5⁄32°
FRONT TOE (Inches)	0.17" ±0.08" 5⁄32" ± 3⁄32"	0.16" ±0.08" 5⁄32" ± 3⁄32"	0.14" ±0.08" ⅛" ± 3⁄32"	0.09" ±0.08" 3⁄32" ± 3⁄32"	0.07" ±0.08" 1⁄16" ± 3⁄32"	0.13" ±0.08" ⅛" ± 3⁄32"
SAI	13.0 ±0.8 (13 ± ¾)	13.0 ±0.8 (13 ± ¾)	13.0 ±0.8 (13 ± ¾)	12.9 ±0.8 (12 ⅞ ± ¾)	12.9 ±0.8 (12 ⅞ ± ¾)	12.8 ±0.8 (12 13⁄16 ± ¾)
TURNING ANGLE	41.1 / 36.6 ±1.5	40.6 / 36.6 ±1.5	41.1 / 36.6 ±1.5	40.9 / 36.7 ±1.5	41.4 / 36.6 ±1.5	40.9 / 36.8 ±1.5
ILLUSTRATION #	12	12	12	12	12	12
REAR						
CAMBER (Degrees)						
REAR TOE (Degrees)						
REAR TOE (Inches)						
THRUST ANGLE						
ILLUSTRATION #						
NOTES						
	49291	49292	49293	49294	49295	49296

Toyota Truck & SUV

Tundra 4X4 Reg. Cab, Long Bed, 5.7L (USK56L-THTD[K,G]A)	Tundra 4X4 Dbl. Cab, Std. Bed, SR5, 4.6L (UPK56L-CRTSKA)	Tundra 4X4 Dbl. Cab, Std. Bed, Lmtd, 4.6L (UPK56L-CRTLKA)	MODEL INFORMATION	Tundra 4X4 Dbl. Cab, Std. Bed, 5.7L (USK56L-CRT[L,S][K,G]A)	Tundra 4X4 Dbl. Cab, Long Bed, 4.6L (UPK57L-CHTSKA)	Tundra 4X4 Dbl. Cab, Long Bed, 5.7L (USK57L-CHTS[K,G]A)
2010-2013	2010-2013	2010	YEARS	2010-2013	2010	2010-2013
			INSTRUCTIONS			
			LUGNUT TORQUE			
			FRONT			
2.34 ±0.75 2 5⁄16 ± ¾	2.6 ±0.75 2⅝ ± ¾	2.56 ±0.75 2 9⁄16 ± ¾	CASTER (Degrees)	2.6 ±0.75 2⅝ ± ¾	2.62 ±0.75 2⅝ ± ¾	2.7 ±0.75 2 11⁄16 ± ¾
0.0 (0) | 0.5 (½)	0.0 (0) | 0.5 (½)	0.0 (0) | 0.5 (½)	CROSS | TOL.	0.0 (0) | 0.5 (½)	0.0 (0) | 0.5 (½)	0.0 (0) | 0.5 (½)
0.19 ±0.75 3⁄16 ± ¾	0.28 ±0.75 ¼ ± ¾	0.3 ±0.75 5⁄16 ± ¾	CAMBER (Degrees)	0.28 ±0.75 ¼ ± ¾	0.3 ±0.75 5⁄16 ± ¾	0.3 ±0.75 5⁄16 ± ¾
0.0 (0) | 0.5 (½)	0.0 (0) | 0.5 (½)	0.0 (0) | 0.5 (½)	CROSS | TOL.	0.0 (0) | 0.5 (½)	0.0 (0) | 0.5 (½)	0.0 (0) | 0.5 (½)
0.2° ±0.16° 3⁄16° ± 5⁄32°	0.3° ±0.16° 5⁄16° ± 5⁄32°	0.35° ±0.16° 11⁄32° ± 5⁄32°	FRONT TOE (Degrees)	0.3° ±0.16° 5⁄16° ± 5⁄32°	0.36° ±0.16° ⅜° ± 5⁄32°	0.32° ±0.16° 5⁄16° ± 5⁄32°
0.1" ±0.08" 3⁄32" ± 3⁄32"	0.15" ±0.08" 5⁄32" ± 3⁄32"	0.18" ±0.08" 3⁄16" ± 3⁄32"	FRONT TOE (Inches)	0.15" ±0.08" 5⁄32" ± 3⁄32"	0.18" ±0.08" 3⁄16" ± 3⁄32"	0.16" ±0.08" 5⁄32" ± 3⁄32"
12.8 ±0.8 (12 13⁄16 ± ¾)	12.7 ±0.8 (12 11⁄16 ± ¾)	12.7 ±0.8 (12 11⁄16 ± ¾)	SAI	12.7 ±0.8 (12 11⁄16 ± ¾)	12.7 ±0.8 (12 11⁄16 ± ¾)	12.7 ±0.8 (12 11⁄16 ± ¾)
41.4 / 36.7 ±1.5	41.4 / 36.8 ±1.5	40.9 / 36.9 ±1.5	TURNING ANGLE	41.5 / 36.8 ±1.5	41.5 / 36.9 ±1.5	41.5 / 36.9 ±1.5
12	12	12	ILLUSTRATION #	12	12	12
			REAR			
			CAMBER (Degrees)			
			REAR TOE (Degrees)			
			REAR TOE (Inches)			
			THRUST ANGLE			
			ILLUSTRATION #			
			NOTES			
49297	49298	49299		49300	49301	49302

Toyota Truck & SUV

Tundra 4X4 CrewMax Cab, SR5, 4.6L (UPK56L-PSTSKA)	Tundra 4X4 CrewMax Cab, Lmtd, 4.6L (UPK56L-PSTLKA)	Tundra 4X4 CrewMax Cab, 5.7L (USK56L-PST[L,S][K,G]A)	MODEL INFORMATION	Tundra 4X2 Std. Cab, 4.0L (GSK51L-THASKA)	Tundra 4X2 Std. Cab, 5.7L (USK51L-THTS[K,G]A)	Tundra 4X2 Dbl. Cab, 4.0L (GSK51L-CRASKA)
2010-2013	2010	2010-2013	YEARS	2014	2014-2017	2014
			INSTRUCTIONS			
			LUGNUT TORQUE			
			FRONT			
2.7 ±0.75 2 11⁄16 ± ¾	2.7 ±0.75 2 11⁄16 ± ¾	2.7 ±0.75 2 11⁄16 ± ¾	CASTER (Degrees)	2.86 ±0.75 2⅞ ± ¾	2.7 ±0.75 2 11⁄16 ± ¾	2.76 ±0.75 2¾ ± ¾
0.0 (0) | 0.5 (½)	0:0 (0) | 0.5 (½)	0.0 (0) | 0.5 (½)	CROSS | TOL.	0.0 (0) | 0.5 (½)	0.0 (0) | 0.5 (½)	0.0 (0) | 0.5 (½)
0.27 ±0.75 ¼ ± ¾	0.25 ±0.75 ¼ ± ¾	0.24 ±0.75 ¼ ± ¾	CAMBER (Degrees)	-0.03 ±0.75 0 ± ¾	-0.16 ±0.75 - 3⁄16 ± ¾	0.11 ±0.75 ⅛ ± ¾
0.0 (0) | 0.5 (½)	0.0 (0) | 0.5 (½)	0.0 (0) | 0.5 (½)	CROSS | TOL.	0.0 (0) | 0.5 (½)	0.0 (0) | 0.5 (½)	0.0 (0) | 0.5 (½)
0.28° ±0.16° 9⁄32° ± 5⁄32°	0.3° ±0.16° 5⁄16° ± 5⁄32°	0.25° ±0.16° ¼° ± 5⁄32°	FRONT TOE (Degrees)	0.28° ±0.16° 9⁄32° ± 5⁄32°	0.15° ±0.16° 5⁄32° ± 5⁄32°	0.41° ±0.16° 13⁄32° ± 5⁄32°
0.14" ±0.08" ⅛" ± 3⁄32"	0.15" ±0.08" 5⁄32" ± 3⁄32"	0.13" ±0.08" ⅛" ± 3⁄32"	FRONT TOE (Inches)	0.14" ±0.08" ⅛" ± 3⁄32"	0.08" +0.08" -0.09" 1⁄16" ± 3⁄32"	0.21" ±0.08" 7⁄32" ± 3⁄32"
12.8 ±0.8 (12 13⁄16 ± ¾)	12.8 ±0.8 (12 13⁄16 ± ¾)	12.8 ±0.8 (12 13⁄16 ± ¾)	SAI	13.0 ±0.8 (13 ± ¾)	13.2 ±0.8 (13 3⁄16 ± ¾)	12.9 ±0.8 (12 ⅞ ± ¾)
41.5 / 36.8 ±1.5	41.0 / 36.8 ±1.5	41.5 / 36.8 ±1.5	TURNING ANGLE	41.1 / 36.5 ±1.5	41.0 / 36.5 ±1.5	41.2 / 36.7 ±1.5
12	12	12	ILLUSTRATION #	12	12	12
			REAR			
			CAMBER (Degrees)			
			REAR TOE (Degrees)			
			REAR TOE (Inches)			
			THRUST ANGLE			
			ILLUSTRATION #			
			NOTES			
49303	49304	49305		49306	49307	49308

ALIGNMENT SPECIFICATIONS

Toyota Truck & SUV

Tundra 4X2	Tundra 4X2	Tundra 4X2	MODEL	Tundra 4X2	Tundra 4X2	Tundra 4X4
Dbl. Cab, 4.6L	Dbl. Cab, 5.7L, Std. Bed	Dbl. Cab, 5.7L, Long Bed	INFORMATION	CrewMax Cab, 4.6L	CrewMax Cab, 5.7L	Std. Cab
(UPK51L-CRTSKA)	(USK51L-CRT[L,S][K,G]A)	(USK52L-CHTSKA)		(UPK51L-PSTSKA)	(USK51L-PST[L,S,Z][K,G]A)	(USK56L-THTS[K,G]A)
2014-2017	2014-2017	2014-2017	YEARS	2014-2017	2014-2017	2014-2017
			INSTRUCTIONS			
			LUGNUT TORQUE			
			FRONT			
2.85±0.75 2 ⅞ ± ¾	2.9±0.75 2 ⅞ ± ¾	3.0±0.75 3 ± ¾	CASTER (Degrees)	2.9±0.75 2 ⅞ ± ¾	2.9±0.75 2 ⅞ ± ¾	2.4±0.75[1] 2⅜ ± ¾[1]
0.0 (0) | 0.5 (½)	0.0 (0) | 0.5 (½)	0.0 (0) | 0.5 (½)	CROSS | TOL.	0.0 (0) | 0.5 (½)	0.0 (0) | 0.5 (½)	0.0 (0) | 0.5 (½)
0.07±0.75[1] 1⁄16 ± ¾[1]	-0.03±0.75 0 ± ¾	-0.04±0.75 -1⁄16 ± ¾	CAMBER (Degrees)	0.04±0.75 1⁄16 ± ¾	-0.04±0.75 -1⁄16 ± ¾	0.15±0.75 ⅛ ± ¾
0.0 (0) | 0.5 (½)	0.0 (0) | 0.5 (½)	0.0 (0) | 0.5 (½)	CROSS | TOL.	0.0 (0) | 0.5 (½)	0.0 (0) | 0.5 (½)	0.0 (0) | 0.5 (½)
0.35° ±0.16° 11⁄32° ± 5⁄32°	0.3° ±0.16°[1] 5⁄16° ± 5⁄32°[1]	0.25° ±0.16°[1] ¼° ± 5⁄32°[1]	FRONT TOE (Degrees)	0.33° ±0.16° 11⁄32° ± 5⁄32°	0.26° ±0.16° ¼° ± 5⁄32°	0.18° ±0.16° 3⁄16° ± 5⁄32°
0.18" ±0.08" 3⁄16" ± 3⁄32"	0.15" ±0.08"[1] 5⁄32" ± 3⁄32"[1]	0.13" ±0.08"[1] ⅛" ± 3⁄32"[1]	FRONT TOE (Inches)	0.17" ±0.08" 5⁄32" ± 3⁄32"	0.13" ±0.08" ⅛" ± 3⁄32"	0.09" ±0.08" 3⁄32" ± 3⁄32"
12.8±0.8 (12¾ ± ¾)	13.0±0.8 (13 ± ¾)	13.0±0.8 (13 ± ¾)	SAI	13.0±0.8 (13 ± ¾)	13.0±0.8 (13 ± ¾)	12.8±0.8 (12 13⁄16 ± ¾)
41.1 / 36.6 ±1.5	41.1 / 36.6 ±1.5	41.1 / 36.6 ±1.5	TURNING ANGLE	41.1 / 36.6 ±1.5	41.1 / 36.6 ±1.5	41.4 / 36.7 ±1.5
12	12	12	ILLUSTRATION #	12	12	12
			REAR			
			CAMBER (Degrees)			
			REAR TOE (Degrees)			
			REAR TOE (Inches)			
			THRUST ANGLE			
			ILLUSTRATION #			
[1] Off-rd = 0.0° (0°)	[1] Off-rd = 0.26° (0.13")	[1] Off-rd = 0.25° (0.13")	NOTES			[1] Off-rd = 2.7° (2 11⁄16°)
49309	49310	49311		49312	49313	49314

Toyota Truck & SUV

Tundra 4X4	Tundra 4X4	Tundra 4X4	MODEL	Tundra 4X4	Tundra 4X4	Tundra 4X4
Dbl. Cab, Std. Bed, 4.6L	Dbl. Cab, Std. Bed, 5.7L	Dbl. Cab, Long Bed	INFORMATION	CrewMax Cab, 4.6L	CrewMax Cab, 5.7L, SR5	CrewMax Cab, 5.7L, Lmtd.
(UPK56L-CRTSKA)	(USK56L-CRT[L,S][K,G]A)	(USK57L-CHTS[K,G]A)		(UPK56L-PSTSKA)	(USK56L-PSTS[K,G]A)	(USK56L-PSTL[K,G]A)
2014-2017	2014-2017	2014-2017	YEARS	2014-2017	2014-2017	2014-2017
			INSTRUCTIONS			
			LUGNUT TORQUE			
			FRONT			
2.55±0.75 2 9⁄16 ± ¾	2.65±0.75 2⅝ ± ¾	2.75±0.75 2¾ ± ¾	CASTER (Degrees)	2.55±0.75 2 9⁄16 ± ¾	2.7±0.75 2 11⁄16 ± ¾	2.7±0.75 2 11⁄16 ± ¾
0.0 (0) | 0.5 (½)	0.0 (0) | 0.5 (½)	0.0 (0) | 0.5 (½)	CROSS | TOL.	0.0 (0) | 0.5 (½)	0.0 (0) | 0.5 (½)	0.0 (0) | 0.5 (½)
0.3±0.75 5⁄16 ± ¾	0.27±0.75[1] ¼ ± ¾[1]	0.3±0.75 5⁄16 ± ¾	CAMBER (Degrees)	0.3±0.75 5⁄16 ± ¾	0.22±0.75 ¼ ± ¾	0.22±0.75 ¼ ± ¾
0.0 (0) | 0.5 (½)	0.0 (0) | 0.5 (½)	0.0 (0) | 0.5 (½)	CROSS | TOL.	0.0 (0) | 0.5 (½)	0.0 (0) | 0.5 (½)	0.0 (0) | 0.5 (½)
0.34° ±0.16° 11⁄32° ± 5⁄32°	0.3° ±0.16°[2] 5⁄16° ± 5⁄32°[2]	0.3° ±0.16° 5⁄16° ± 5⁄32°	FRONT TOE (Degrees)	0.3° ±0.16° 5⁄16° ± 5⁄32°	0.24° ±0.16° ¼° ± 5⁄32°	0.23° ±0.16° 7⁄32° ± 5⁄32°
0.17" ±0.08" 5⁄32" ± 3⁄32"	0.15" ±0.08"[2] 5⁄32" ± 3⁄32"[2]	0.15" ±0.08" 5⁄32" ± 3⁄32"	FRONT TOE (Inches)	0.15" ±0.08" 5⁄32" ± 3⁄32"	0.12" ±0.08" ⅛" ± 3⁄32"	0.12" ±0.08" ⅛" ± 3⁄32"
12.7±0.8 (12 11⁄16 ± ¾)	12.7±0.8 (12 11⁄16 ± ¾)	12.7±0.8 (12 11⁄16 ± ¾)	SAI	12.7±0.8 (12 11⁄16 ± ¾)	12.7±0.8 (12 11⁄16 ± ¾)	12.8±0.8 (12 13⁄16 ± ¾)
41.5 / 36.9 ±1.5	41.5 / 36.9 ±1.5	41.5 / 36.9 ±1.5	TURNING ANGLE	41.5 / 36.9 ±1.5	41.5 / 36.9 ±1.5	41.5 / 36.9 ±1.5
12	12	12	ILLUSTRATION #	12	12	12
			REAR			
			CAMBER (Degrees)			
			REAR TOE (Degrees)			
			REAR TOE (Inches)			
			THRUST ANGLE			
			ILLUSTRATION #			
	[1] Off-rd = 0.24° (¼°) [2] Off-rd = 0.25° (0.13")		NOTES			
49315	49316	49317		49318	49319	49320

Toyota Truck & SUV / Volkswagen

Tundra 4X4	Notes	Notes	MODEL INFORMATION	New Beetle	New Beetle	New Beetle
CrewMax Cab, 5.7L, Plat.				w/Base Susp.	w/Sport Susp.	Convertible
(USK56L-PSTZ[K,G]A)						
2014-2017			YEARS	1998-2011	1998-2011	2003-2011
			INSTRUCTIONS			
			LUGNUT TORQUE	89 ft.lbs.	89 ft.lbs.	89 ft.lbs.
			FRONT			
2.7 ±0.75 2 11/16 ± 3/4			CASTER (Degrees)	7.67 ±0.5 7 11/16 ± 1/2	7.83 ±0.5 7 13/16 ± 1/2	7.67 ±0.5 7 11/16 ± 1/2
0.0 (0) | 0.5 (1/2)			CROSS | TOL.	0.0 (0) | 0.5 (1/2)	0.0 (0) | 0.5 (1/2)	0.0 (0) | 0.5 (1/2)
0.2 ±0.75 3/16 ± 3/4			CAMBER (Degrees)	-0.5 ±0.5 -1/2 ± 1/2	-0.55 ±0.5 -9/16 ± 1/2	-0.5 ±0.5 -1/2 ± 1/2
0.0 (0) | 0.5 (1/2)			CROSS | TOL.	0.0 (0) | 0.5 (1/2)	0.0 (0) | 0.5 (1/2)	0.0 (0) | 0.5 (1/2)
0.2° ±0.16° 3/16° ± 5/32°			FRONT TOE (Degrees)	0.0° ±0.17° 0° ± 5/32°	0.0° ±0.17° 0° ± 5/32°	0.0° ±0.17° 0° ± 5/32°
0.1" ±0.08" 3/32" ± 3/32"			FRONT TOE (Inches)	0.0" ±0.08" 0" ± 3/32"	0.0" ±0.08" 0" ± 3/32"	0.0" ±0.08" 0" ± 3/32"
12.8 ±0.8 (12 13/16 ± 3/4)			SAI			
41.5 / 36.9 ±1.5			TURNING ANGLE	20.0 / 18.5 ±0.33	20.0 / 18.5 ±0.33	20.0 / 18.5 ±0.33
12			ILLUSTRATION #	▲	▲	▲
			REAR			
			CAMBER (Degrees)	-1.45 ±0.17 -1 7/16 ± 3/16	-1.45 ±0.17 -1 7/16 ± 3/16	-1.45 ±0.17 -1 7/16 ± 3/16
			REAR TOE (Degrees)	0.33° ±0.17° 11/32° ± 5/32°	0.42° ±0.17° 13/32° ± 5/32°	0.33° ±0.17° 11/32° ± 5/32°
			REAR TOE (Inches)	0.17" ±0.08" 5/32" ± 3/32"	0.21" ±0.08" 7/32" ± 3/32"	0.17" ±0.08" 5/32" ± 3/32"
			THRUST ANGLE	0.33 (11/32)	0.33 (11/32)	0.33 (11/32)
			ILLUSTRATION #	114	114	114
			NOTES			
49321	49325	49326		46001	46002	46003

Volkswagen

Beetle (A5)	Beetle (A5)	Beetle (A5)	MODEL INFORMATION	Beetle (A5)	Beetle (A5)	Beetle (A5)
w/Torsion Beam Rear Axle	w/Torsion Beam Rear Axle	w/Torsion Beam Rear Axle		w/IRS exc/Turbo Models	w/IRS exc/Turbo Models	w/IRS exc/Turbo Models
Std. Suspension (2UA)	Sport Suspension (2UC)	Comfort Suspension (2UD)		Std. Suspension (2UA)	Sport Suspension (2UC)	Comfort Suspension (2UD)
2012-2013	2012-2013	2012-2013	YEARS	2012-2017	2012-2017	2012-2017
			INSTRUCTIONS			
89 ft.lbs.	89 ft.lbs.	89 ft.lbs.	LUGNUT TORQUE	89 ft.lbs.	89 ft.lbs.	89 ft.lbs.
			FRONT			
7.88 ±0.5 7 7/8 ± 1/2	7.88 ±0.5 7 7/8 ± 1/2	7.7 ±0.5 7 11/16 ± 1/2	CASTER (Degrees)	7.88 ±0.5 7 7/8 ± 1/2	7.88 ±0.5 7 7/8 ± 1/2	7.7 ±0.5 7 11/16 ± 1/2
0.0 (0) | 0.5 (1/2)	0.0 (0) | 0.5 (1/2)	0.0 (0) | 0.5 (1/2)	CROSS | TOL.	0.0 (0) | 0.5 (1/2)	0.0 (0) | 0.5 (1/2)	0.0 (0) | 0.5 (1/2)
-0.67 ±0.5 -11/16 ± 1/2	-0.67 ±0.5 -11/16 ± 1/2	-0.33 ±0.5 -5/16 ± 1/2	CAMBER (Degrees)	-0.67 ±0.5 -11/16 ± 1/2	-0.67 ±0.5 -11/16 ± 1/2	-0.33 ±0.5 -5/16 ± 1/2
0.0 (0) | 0.5 (1/2)	0.0 (0) | 0.5 (1/2)	0.0 (0) | 0.5 (1/2)	CROSS | TOL.	0.0 (0) | 0.5 (1/2)	0.0 (0) | 0.5 (1/2)	0.0 (0) | 0.5 (1/2)
0.17° ±0.17° 5/32° ± 5/32°	0.17° ±0.17° 5/32° ± 5/32°	0.17° ±0.17° 5/32° ± 5/32°	FRONT TOE (Degrees)	0.17° ±0.17° 5/32° ± 5/32°	0.17° ±0.17° 5/32° ± 5/32°	0.17° ±0.17° 5/32° ± 5/32°
0.08" ±0.08" 3/32" ± 3/32"	0.08" ±0.08" 3/32" ± 3/32"	0.08" ±0.08" 3/32" ± 3/32"	FRONT TOE (Inches)	0.08" ±0.08" 3/32" ± 3/32"	0.08" ±0.08" 3/32" ± 3/32"	0.08" ±0.08" 3/32" ± 3/32"
			SAI			
20.0 / 18.5 ±0.33	20.0 / 18.5 ±0.33	20.0 / 18.6 ±0.33	TURNING ANGLE	20.0 / 18.5 ±0.33	20.0 / 18.5 ±0.33	20.0 / 18.5 ±0.33
			ILLUSTRATION #			
			REAR			
-1.0 ±0.5 -1 ± 1/2	-1.0 ±0.5 -1 ± 1/2	-1.0 ±0.5 -1 ± 1/2	CAMBER (Degrees)	-1.0 ±0.5 -1 ± 1/2	-1.33 ±0.5 -1 5/16 ± 1/2	-1.33 ±0.5 -1 5/16 ± 1/2
0.38° ±0.17° 3/8° ± 5/32°	0.38° ±0.17° 3/8° ± 5/32°	0.35° ±0.17° 11/32° ± 5/32°	REAR TOE (Degrees)	0.17° ±0.17° 5/32° ± 5/32°	0.17° ±0.17° 5/32° ± 5/32°	0.17° ±0.17° 5/32° ± 5/32°
0.19" ±0.08" 3/16" ± 3/32"	0.19" ±0.08" 3/16" ± 3/32"	0.18" ±0.08" 3/16" ± 3/32"	REAR TOE (Inches)	0.08" ±0.08" 3/32" ± 3/32"	0.08" ±0.08" 3/32" ± 3/32"	0.08" ±0.08" 3/32" ± 3/32"
0.33 (11/32)	0.33 (11/32)	0.33 (11/32)	THRUST ANGLE	0.33 (11/32)	0.33 (11/32)	0.33 (11/32)
			ILLUSTRATION #	127	127	127
			NOTES			
46004	46005	46006		46007	46008	46009

ALIGNMENT SPECIFICATIONS

Volkswagen

Beetle (A5) w/IRS Turbo Models Std. Suspension (2UA)	Beetle (A5) w/IRS Turbo Models Sport Suspension (2UC)	Cabrio	MODEL INFORMATION	Cabrio	CC Std. Susp. (G60, G61)	City Golf (A4) City Jetta (A4) Std. Susp.
2012-2017	2012-2017	1990-93	YEARS	1995-2002	2009-2017	2007-2011
			INSTRUCTIONS			
89 ft.lbs.	89 ft.lbs.	81 ft.lbs.	LUGNUT TORQUE	81 ft.lbs.	89 ft.lbs.	89 ft.lbs.
			FRONT			
7.88 ±0.5 7 7/8 ± 1/2	7.88 ±0.5 7 7/8 ± 1/2	1.83 ±0.5 1 13/16 ± 1/2	CASTER (Degrees)	1.83 ±0.5 1 13/16 ± 1/2	7.73 ±0.5 7 3/4 ± 1/2	7.67 ±0.5 7 11/16 ± 1/2
0.0 (0) | 0.5 (1/2)	0.0 (0) | 0.5 (1/2)	0.0 (0) | 1.0 (1)	CROSS | TOL.	0.0 (0) | 0.5 (1/2)	0.0 (0) | 0.5 (1/2)	0.0 (0) | 0.5 (1/2)
-0.67 ±0.5 -11/16 ± 1/2	-0.67 ±0.5 -11/16 ± 1/2	0.33 ±0.5 5/16 ± 1/2	CAMBER (Degrees)	-0.6 ±0.33 -5/8 ± 5/16	-0.68 ±0.5 -11/16 ± 1/2	-0.5 ±0.5 -1/2 ± 1/2
0.0 (0) | 0.5 (1/2)	0.0 (0) | 0.5 (1/2)	0.0 (0) | 1.0 (1)	CROSS | TOL.	0.0 (0) | 0.33 (5/16)	0.0 (0) | 0.5 (1/2)	0.0 (0) | 0.5 (1/2)
0.17° ±0.17° 5/32° ± 5/32°	0.17° ±0.17° 5/32° ± 5/32°	**-0.25°** +0.17° -0.25° **-1/4°** +5/32° -1/4°	FRONT TOE (Degrees)	0.0° ±0.17° 0° ± 5/32°	0.17° ±0.17° 5/32° ± 5/32°	0.0° ±0.17° 0° ± 5/32°
0.08" ±0.08" 3/32" ± 3/32"	0.08" ±0.08" 3/32" ± 3/32"	**-0.13"** +0.08" -0.13" **-1/8"** +3/32" -1/8"	FRONT TOE (Inches)	0.0" ±0.08" 0" ± 3/32"	0.08" ±0.08" 3/32" ± 3/32"	0.0" ±0.08" 0" ± 3/32"
			SAI			
20.0 / 18.5 ±0.33	20.0 / 18.5 ±0.33	20.0 / 18.5 ±0.5	TURNING ANGLE	20.0 / 18.9 ±0.33	20.0 / 18.7 ±0.33	20.0 / 18.5 ±0.33
		22	ILLUSTRATION #	26		▲
			REAR			
-1.0 ±0.5 -1 ± 1/2	-1.75 ±0.5 -1 3/4 ± 1/2	-1.25 ±0.58 -1 1/4 ± 9/16	CAMBER (Degrees)	-1.5 ±0.17 -1 1/2 ± 3/16	-1.33 ±0.5 -1 5/16 ± 1/2	-1.45 ±0.17 -1 7/16 ± 3/16
0.17° ±0.17° 5/32° ± 5/32°	0.17° ±0.17° 5/32° ± 5/32°	0.33° ±0.5° 11/32° ± 1/2°	REAR TOE (Degrees)	0.33° ±0.17° 11/32° ± 5/32°	0.17° ±0.17° 5/32° ± 5/32°	0.33° ±0.17° 11/32° ± 5/32°
0.08" ±0.08" 3/32" ± 3/32"	0.08" ±0.08" 3/32" ± 3/32"	0.17" ±0.25" 5/32" ± 1/4"	REAR TOE (Inches)	0.17" ±0.08" 5/32" ± 3/32"	0.08" ±0.08" 3/32" ± 3/32"	0.17" ±0.08" 5/32" ± 3/32"
0.33 (11/32)	0.33 (11/32)	0.5 (1/2)	THRUST ANGLE	0.33 (11/32)	0.33 (11/32)	0.33 (11/32)
127	127		ILLUSTRATION #		127	114
			NOTES			
46010	46011	46012		46013	46014	46016

Volkswagen

City Golf (A4) City Jetta (A4) Sport Susp.	Corrado	Corrado SLC Corrado VR-6	MODEL INFORMATION	Eos Std. Susp. (2UA) w/16" or 17" tires	Eos Std. Susp. (2UA) w/18" tires (OE)	Eos Sport Susp. (2UC) w/16" or 17" tires
2007-2011	1990-94	1992-94	YEARS	2007-2016	2007-2016	2007-2016
			INSTRUCTIONS			
89 ft.lbs.	81 ft.lbs.	81 ft.lbs.	LUGNUT TORQUE	89 ft.lbs.	89 ft.lbs.	89 ft.lbs.
			FRONT			
7.83 ±0.5 7 13/16 ± 1/2	1.58 ±0.5 1 9/16 ± 1/2	3.42 ±0.5 3 7/16 ± 1/2	CASTER (Degrees)	7.57 ±0.5 7 9/16 ± 1/2	7.57 ±0.5 7 9/16 ± 1/2	7.78 ±0.5 7 3/4 ± 1/2
0.0 (0) | 0.5 (1/2)	0.0 (0) | 1.0 (1)	0.0 (0) | 1.0 (1)	CROSS | TOL.	0.0 (0) | 0.5 (1/2)	0.0 (0) | 0.5 (1/2)	0.0 (0) | 0.5 (1/2)
-0.55 ±0.5 -9/16 ± 1/2	-0.67 ±0.33 -11/16 ± 5/16	-1.33 ±0.33 -1 5/16 ± 5/16	CAMBER (Degrees)	-0.6 ±0.5 -5/8 ± 1/2	-0.6 ±0.5 -5/8 ± 1/2	-0.68 ±0.5 -11/16 ± 1/2
0.0 (0) | 0.5 (1/2)	0.0 (0) | 0.5 (1/2)	0.0 (0) | 0.5 (1/2)	CROSS | TOL.	0.0 (0) | 0.5 (1/2)	0.0 (0) | 0.5 (1/2)	0.0 (0) | 0.5 (1/2)
0.0° ±0.17° 0° ± 5/32°	0.0° ±0.17° 0° ± 5/32°	0.0° ±0.17° 0° ± 5/32°	FRONT TOE (Degrees)	0.17° ±0.17° 5/32° ± 5/32°	0.17° ±0.17° 5/32° ± 5/32°	0.17° ±0.17° 5/32° ± 5/32°
0.0" ±0.08" 0" ± 3/32"	0.0" ±0.08" 0" ± 3/32"	0.0" ±0.08" 0" ± 3/32"	FRONT TOE (Inches)	0.08" ±0.08" 3/32" ± 3/32"	0.08" ±0.08" 3/32" ± 3/32"	0.08" ±0.08" 3/32" ± 3/32"
			SAI			
20.0 / 18.5 ±0.33	20.0 / 18.7 ±0.5	20.0 / 18.5 ±0.5	TURNING ANGLE	20.0 / 18.4 ±0.33	20.0 / 18.4 ±0.33	20.0 / 18.3 ±0.33
▲	26	26	ILLUSTRATION #			
			REAR			
-1.45 ±0.17 -1 7/16 ± 3/16	-1.5 ±0.17 [1] -1 1/2 ± 3/16 [1]	-1.5 ±0.17 -1 1/2 ± 3/16	CAMBER (Degrees)	-1.33 ±0.5 -1 5/16 ± 1/2	-1.75 ±0.5 -1 3/4 ± 1/2	-1.33 ±0.5 -1 5/16 ± 1/2
0.42° ±0.17° 13/32° ± 5/32°	0.33° ±0.17° [2] 11/32° ± 5/32° [2]	0.33° ±0.17° 11/32° ± 5/32°	REAR TOE (Degrees)	0.17° ±0.17° 5/32° ± 5/32°	0.17° ±0.17° 5/32° ± 5/32°	0.17° ±0.17° 5/32° ± 5/32°
0.21" ±0.08" 7/32" ± 3/32"	0.17" ±0.08" [2] 5/32" ± 3/32" [2]	0.17" ±0.08" 5/32" ± 3/32"	REAR TOE (Inches)	0.08" ±0.08" 3/32" ± 3/32"	0.08" ±0.08" 3/32" ± 3/32"	0.08" ±0.08" 3/32" ± 3/32"
0.33 (11/32)	0.17 (5/32)	0.21 (7/32)	THRUST ANGLE	0.33 (11/32)	0.33 (11/32)	0.33 (11/32)
114			ILLUSTRATION #	127	127	127
	[1] up to 3/90= -1.67°(-1 11/16°) [2] up to 3/90= 0.42° (0.21")		NOTES			
46017	46018	46019		46022	46023	46024

Volkswagen

MODEL INFORMATION	Eos Sport Susp. (G07, 2UC) w/18" tires (OE)	Eurovan Group 1 & 2 Empty	Eurovan Group 1 & 2 1/2 Load	Eurovan Group 1 & 2 Full Load	Eurovan Group 1 & 2 Empty	Eurovan Group 1 & 2 1/2 Load
YEARS	2007-2016	1993-96	1993-96	1993-96	1997-2000	1997-2000
INSTRUCTIONS		p. 310	p. 310	p. 310	p. 310	p. 310
LUGNUT TORQUE	89 ft.lbs.					
FRONT						
CASTER (Degrees)	7.78 ±0.5 7¾ ± ½	1.67 ±0.5 1 11/16 ± ½	1.83 ±0.5 1 13/16 ± ½	2.0 ±0.5 2 ± ½	3.17 ±0.5 3 3/16 ± ½	3.33 ±0.5 3 5/16 ± ½
CROSS | TOL.	0.0 (0) | 0.5 (½)	0.0 (0) | 1.0 (1)	0.0 (0) | 1.0 (1)	0.0 (0) | 1.0 (1)	0.0 (0) | 1.0 (1)	0.0 (0) | 1.0 (1)
CAMBER (Degrees)	-0.68 ±0.5 -11/16 ± ½	0.25 ±0.33 ¼ ± 5/16	-0.08 ±0.33 -1/16 ± 5/16	-0.33 ±0.33 -5/16 ± 5/16	-0.42 +0.33 -0.67 -7/16 +5/16 -11/16	-0.67 +0.33 -0.67 -11/16 +5/16 -11/16
CROSS | TOL.	0.0 (0) | 0.5 (½)	0.0 (0) | 0.5 (½)	0.0 (0) | 0.5 (½)	0.0 (0) | 0.5 (½)	0.0 (0) | 0.5 (½)	0.0 (0) | 0.5 (½)
FRONT TOE (Degrees)	0.17° ±0.17° 5/32° ± 5/32°	0.33° ±0.17° 11/32° ± 5/32°	0.0° ±0.33° 0° ± 11/32°	**-0.33°** ±0.33° **- 11/32°** ± 11/32°	0.33° ±0.33° 11/32° ± 11/32°	0.0° ±0.33° 0° ± 11/32°
FRONT TOE (Inches)	0.08" ±0.08" 3/32" ± 3/32"	0.17" ±0.08" 5/32" ± 3/32"	0.0" ±0.17" 0" ± 5/32"	**-0.17"** ±0.17" **-5/32"** ± 5/32"	0.17" ±0.17" 5/32" ± 5/32"	0.0" ±0.17" 0" ± 5/32"
SAI						
TURNING ANGLE	20.0 / 18.3 ±0.33	20.0 / 19.3 ±0.33	20.0 / 19.3 ±0.33	20.0 / 19.3 ±0.33	20.0 / 17.8 ±0.5	20.0 / 17.8 ±0.5
ILLUSTRATION #	100 101	100 101	100 101	100 101	100 101	100 101
REAR						
CAMBER (Degrees)	-1.75 ±0.5 -1¾ ± ½	see table p. 310 ±0.5 (½)	see table p. 310 ±0.5 (½)	see table p. 310 ±0.5 (½)	see table p. 310 ±0.5 (½)	see table p. 310 ±0.5 (½)
REAR TOE (Degrees)	0.17° ±0.17° 5/32° ± 5/32°	see table p. 310 ±0.33°(11/32°)	see table p. 310 ±0.33°(11/32°)	see table p. 310 ±0.33°(11/32°)	see table p. 310 ±0.33°(11/32°)	see table p. 310 ±0.33°(11/32°)
REAR TOE (Inches)	0.08" ±0.08" 3/32" ± 3/32"	see table p. 310 ±0.17" (5/32")	see table p. 310 ±0.17" (5/32")	see table p. 310 ±0.17" (5/32")	see table p. 310 ±0.17" (5/32")	see table p. 310 ±0.17" (5/32")
THRUST ANGLE	0.33 (11/32)	0.5 (½)	0.5 (½)	0.5 (½)		
ILLUSTRATION #	127	27	27	27	27	27
NOTES						
	46025	46026	46027	46028	46029	46030

Volkswagen

MODEL INFORMATION	Eurovan Group 1 & 2 Full Load	Eurovan Group 1 & 6 Empty	Eurovan Group 1 & 6 1/2 Load	Eurovan Group 1 & 6 Full Load	Eurovan Group 2 & 7 Empty	Eurovan Group 2 & 7 1/2 Load
YEARS	1997-2000	2001-2003	2001-2003	2001-2003	2001-2003	2001-2003
INSTRUCTIONS	p. 310	p. 310, 311	p. 310, 311	p. 310, 311	p. 310, 311	p. 310, 311
LUGNUT TORQUE						
FRONT						
CASTER (Degrees)	3.5 ±0.5 3½ ± ½	3.17 ±0.5 3 3/16 ± ½	3.33 ±0.5 3 5/16 ± ½	3.5 ±0.5 3½ ± ½	3.17 ±0.5 3 3/16 ± ½	3.33 ±0.5 3 5/16 ± ½
CROSS | TOL.	0.0 (0) | 1.0 (1)	0.0 (0) | 1.0 (1)	0.0 (0) | 1.0 (1)	0.0 (0) | 1.0 (1)	0.0 (0) | 1.0 (1)	0.0 (0) | 1.0 (1)
CAMBER (Degrees)	-0.83 +0.33 -0.67 -13/16 +5/16 -11/16	-0.67 +0.0 -0.67 -11/16 +0 -11/16	-0.92 +0.0 -0.67 -15/16 +0 -11/16	-1.08 +0.0 -0.67 -1 1/16 +0 -11/16	-0.67 +0.0 -0.67 -11/16 +0 -11/16	-0.92 +0.0 -0.67 -15/16 +0 -11/16
CROSS | TOL.	0.0 (0) | 0.5 (½)	0.0 (0) | 0.5 (½)	0.0 (0) | 0.5 (½)	0.0 (0) | 0.5 (½)	0.0 (0) | 0.5 (½)	0.0 (0) | 0.5 (½)
FRONT TOE (Degrees)	**-0.33°** ±0.33° **- 11/32°** ± 11/32°	0.33° ±0.33° 11/32° ± 11/32°	0.0° ±0.33° 0° ± 11/32°	**-0.33°** ±0.33° **- 11/32°** ± 11/32°	0.33° ±0.33° 11/32° ± 11/32°	0.0° ±0.33° 0° ± 11/32°
FRONT TOE (Inches)	**-0.17"** ±0.17" **-5/32"** ± 5/32"	0.17" ±0.17" 5/32" ± 5/32"	0.0" ±0.17" 0" ± 5/32"	**-0.17"** ±0.17" **-5/32"** ± 5/32"	0.17" ±0.17" 5/32" ± 5/32"	0.0" ±0.17" 0" ± 5/32"
SAI						
TURNING ANGLE	20.0 / 17.8 ±0.5	20.0 / 17.8 ±0.5	20.0 / 17.8 ±0.5	20.0 / 17.8 ±0.5	20.0 / 17.8 ±0.5	20.0 / 17.8 ±0.5
ILLUSTRATION #	100 101	100 101	100 101	100 101	100 101	100 101
REAR						
CAMBER (Degrees)	see table p. 310 ±0.5 (½)	-0.5 ±0.5 -½ ± ½	-1.17 ±0.5 -1 3/16 ± ½	-1.57 ±0.5 -1 9/16 ± ½	0.0 ±0.5 0 ± ½	-0.67 ±0.5 -11/16 ± ½
REAR TOE (Degrees)	see table p. 310 ±0.33°(11/32°)	0.33° ±0.33° 11/32° ± 11/32°	0.67° ±0.33° 21/32° ± 11/32°	1.0° ±0.33° 1° ± 11/32°	0.1° ±0.33° 3/32° ± 11/32°	0.4° ±0.33° 13/32° ± 11/32°
REAR TOE (Inches)	see table p. 310 ±0.17" (5/32")	0.17" ±0.17" 5/32" ± 5/32"	0.33" ±0.17" 11/32" ± 5/32"	0.5" ±0.17" ½" ± 5/32"	0.05" ±0.17" 1/16" ± 5/32"	0.2" ±0.17" 3/16" ± 5/32"
THRUST ANGLE		0.5 (½)	0.5 (½)	0.5 (½)	0.5 (½)	0.5 (½)
ILLUSTRATION #	27	27	27	27	27	27
NOTES						
	46031	46032	46033	46034	46035	46036

Volkswagen

Eurovan Group 2 & 7 Full Load	Eurovan Group 3 & 8 Empty	Eurovan Group 3 & 8 1/2 Load	MODEL INFORMATION	Eurovan Group 3 & 8 Full Load	Eurovan Group 4 & 5 Empty	Eurovan Group 4 & 5 1/2 Load
2001-2003	2001-2003	2001-2003	YEARS	2001-2003	2001-2003	2001-2003
p. 310, 311	p. 310, 311	p. 310, 311	INSTRUCTIONS	p. 310, 311	p. 310, 311	p. 310, 311
			LUGNUT TORQUE			
			FRONT			
3.5 ±0.5 3½ ± ½	3.17 ±0.5 3 3/16 ± ½	3.33 ±0.5 3 5/16 ± ½	CASTER (Degrees)	3.5 ±0.5 3½ ± ½	3.17 ±0.5 3 3/16 ± ½	3.33 ±0.5 3 5/16 ± ½
0.0 (0) | 1.0 (1)	0.0 (0) | 1.0 (1)	0.0 (0) | 1.0 (1)	CROSS | TOL.	0.0 (0) | 1.0 (1)	0.0 (0) | 1.0 (1)	0.0 (0) | 1.0 (1)
-1.08 +0.0 -0.67 -1 1/16 +0 -11/16	-0.83 +0.0 -0.67 -13/16 +0 -11/16	-1.0 +0.0 -0.67 -1 +0 -11/16	CAMBER (Degrees)	-1.08 +0.0 -0.67 -1 1/16 +0 -11/16	-1.0 +0.0 -0.67 -1 +0 -11/16	-1.08 +0.0 -0.67 -1 1/16 +0 -11/16
0.0 (0) | 0.5 (½)	0.0 (0) | 0.5 (½)	0.0 (0) | 0.5 (½)	CROSS | TOL.	0.0 (0) | 0.5 (½)	0.0 (0) | 0.5 (½)	0.0 (0) | 0.5 (½)
-0.33° ±0.33° -11/32° ± 11/32°	0.17° ±0.33° 5/32° ± 11/32°	0.0° ±0.33° 0° ± 11/32°	FRONT TOE (Degrees)	-0.33° ±0.33° -11/32° ± 11/32°	0.0° ±0.33° 0° ± 11/32°	-0.17° ±0.33° -5/32° ± 11/32°
-0.17" ±0.17" -5/32" ± 5/32"	0.08" ±0.17" 3/32" ± 5/32"	0.0" ±0.17" 0" ± 5/32"	FRONT TOE (Inches)	-0.17" ±0.17" -5/32" ± 5/32"	0.0" ±0.17" 0" ± 5/32"	-0.08" ±0.17" -3/32" ± 5/32"
			SAI			
20.0 / 17.8 ±0.5	20.0 / 17.8 ±0.5	20.0 / 17.8 ±0.5	TURNING ANGLE	20.0 / 17.8 ±0.5	20.0 / 17.8 ±0.5	20.0 / 17.8 ±0.5
100 101	100 101	100 101	ILLUSTRATION #	100 101	100 101	100 101
			REAR			
-1.17 ±0.5 -1 3/16 ± ½	-0.5 ±0.5 -½ ± ½	-1.17 ±0.5 -1 3/16 ± ½	CAMBER (Degrees)	-1.58 ±0.5 -1 9/16 ± ½	-0.67 ±0.5 -11/16 ± ½	-1.17 ±0.5 -1 3/16 ± ½
0.67° ±0.33° 21/32° ± 11/32°	0.33° ±0.33° 11/32° ± 11/32°	0.67° ±0.33° 21/32° ± 11/32°	REAR TOE (Degrees)	1.0° ±0.33° 1° ± 11/32°	0.4° ±0.33° 13/32° ± 11/32°	0.73° ±0.33° 23/32° ± 11/32°
0.33" ±0.17" 11/32" ± 5/32"	0.17" ±0.17" 5/32" ± 5/32"	0.33" ±0.17" 11/32" ± 5/32"	REAR TOE (Inches)	0.5" ±0.17" ½" ± 5/32"	0.2" ±0.17" 3/16" ± 5/32"	0.37" ±0.17" 3/8" ± 5/32"
0.5 (½)	0.5 (½)	0.5 (½)	THRUST ANGLE	0.5 (½)	0.5 (½)	0.5 (½)
27	27	27	ILLUSTRATION #	27	27	27
			NOTES			
46037	46038	46039		46040	46041	46042

Volkswagen

Eurovan Group 4 & 5 Full Load	Fox	Golf, Golf GL	MODEL INFORMATION	Golf GT/GTI (except 16 Valve)	Golf GTI 16V	Golf 4 Cylinder 1.8L / 1.9L
2001-2003	1990-93	1990-92	YEARS	1990-92	1990-92	1993-94
p. 310, 311			INSTRUCTIONS			
	81 ft.lbs.	81 ft.lbs.	LUGNUT TORQUE	81 ft.lbs.	81 ft.lbs.	81 ft.lbs.
			FRONT			
3.5 ±0.5 3½ ± ½	2.0 ±0.33 [1] 2 ± 5/16 [1]	1.5 ±0.5 1½ ± ½	CASTER (Degrees)	1.58 ±0.5 1 9/16 ± ½	1.58 ±0.5 1 9/16 ± ½	1.75 ±0.5 1¾ ± ½
0.0 (0) | 1.0 (1)	0.0 (0) | 0.5 (½)	0.0 (0) | 1.0 (1)	CROSS | TOL.	0.0 (0) | 1.0 (1)	0.0 (0) | 1.0 (1)	0.0 (0) | 0.5 (½)
-1.17 +0.0 -0.67 -1 3/16 +0 -11/16	-0.5 ±0.33 -½ ± 5/16	-0.5 ±0.33 -½ ± 5/16	CAMBER (Degrees)	-0.58 ±0.33 -9/16 ± 5/16	-0.67 ±0.33 -11/16 ± 5/16	-0.5 ±0.33 -½ ± 5/16
0.0 (0) | 0.5 (½)	0.0 (0) | 0.25 (¼)	0.0 (0) | 0.5 (½)	CROSS | TOL.	0.0 (0) | 0.5 (½)	0.0 (0) | 0.5 (½)	0.0 (0) | 0.33 (5/16)
-0.33° ±0.33° -11/32° ± 11/32°	-0.17° ±0.17° -5/32° ± 5/32°	0.0° ±0.17° 0° ± 5/32°	FRONT TOE (Degrees)	0.0° ±0.17° 0° ± 5/32°	0.0° ±0.17° 0° ± 5/32°	0.0° ±0.17° 0° ± 5/32°
-0.17" ±0.17" -5/32" ± 5/32"	-0.08" ±0.08" -3/32" ± 3/32"	0.0" ±0.08" 0" ± 3/32"	FRONT TOE (Inches)	0.0" ±0.08" 0" ± 3/32"	0.0" ±0.08" 0" ± 3/32"	0.0" ±0.08" 0" ± 3/32"
			SAI			
20.0 / 17.8 ±0.5	20.0 / 18.8 ±0.5	20.0 / 18.7 ±0.5	TURNING ANGLE	20.0 / 18.7 ±0.5	20.0 / 18.7 ±0.5	20.0 / 18.9 ±0.33
100 101	15	26	ILLUSTRATION #	26	26	26
			REAR			
-1.17 ±0.5 -1 3/16 ± ½	-1.5 ±0.5 -1½ ± ½	-1.5 ±0.17 -1½ ± 3/16	CAMBER (Degrees)	-1.5 ±0.17 -1½ ± 3/16	-1.5 ±0.17 -1½ ± 3/16	-1.5 ±0.17 -1½ ± 3/16
0.9° ±0.33° 29/32° ± 11/32°	0.42° ±0.17° [2] 13/32° ± 5/32° [2]	0.33° ±0.17° 11/32° ± 5/32°	REAR TOE (Degrees)	0.33° ±0.17° 11/32° ± 5/32°	0.33° ±0.17° 11/32° ± 5/32°	0.33° ±0.17° 11/32° ± 5/32°
0.45" ±0.17" 7/16" ± 5/32"	0.21" ±0.08" [2] 7/32" ± 3/32" [2]	0.17" ±0.08" 5/32" ± 3/32"	REAR TOE (Inches)	0.17" ±0.08" 5/32" ± 3/32"	0.17" ±0.08" 5/32" ± 3/32"	0.17" ±0.08" 5/32" ± 3/32"
0.5 (½)	0.17 (5/32)	0.33 (11/32)	THRUST ANGLE	0.33 (11/32)	0.33 (11/32)	0.33 (11/32)
27			ILLUSTRATION #			
	[1] Wagon = 1.75° (1¾°) [2] Wagon = 0.33° (0.17")		NOTES			
46043	46044	46045		46046	46047	46048

Volkswagen

Golf	Golf	Golf	MODEL INFORMATION	Golf (A4)	Golf (A4)	Golf (A6)
4 Cylinder	4 Cylinder	6 Cylinder		Std. Susp.	Sport Susp.	Std. Susp.
2.0L	All Models					
1993-94	1995-99	1995-99	YEARS	1999-2006	1999-2006	2010-2014
			INSTRUCTIONS			
81 ft.lbs.	81 ft.lbs.	81 ft.lbs.	LUGNUT TORQUE	89 ft.lbs.	89 ft.lbs.	89 ft.lbs.
			FRONT			
1.83 ±0.5 1 13/16 ± 1/2	1.83 ±0.5 1 13/16 ± 1/2	3.42 ±0.5 3 7/16 ± 1/2	CASTER (Degrees)	7.67 ±0.5 7 11/16 ± 1/2	7.83 ±0.5 7 13/16 ± 1/2	7.57 ±0.5 7 9/16 ± 1/2
0.0 (0) | 0.5 (1/2)	0.0 (0) | 0.5 (1/2)	0.0 (0) | 0.5 (1/2)	CROSS | TOL.	0.0 (0) | 0.5 (1/2)	0.0 (0) | 0.5 (1/2)	0.0 (0) | 0.5 (1/2)
-0.58 ±0.33 - 9/16 ± 5/16	-0.6 ±0.33 -5/8 ± 5/16	-0.67 ±0.33 - 11/16 ± 5/16	CAMBER (Degrees)	-0.5 ±0.5 - 1/2 ± 1/2	-0.55 ±0.5 - 9/16 ± 1/2	-0.5 ±0.5 - 1/2 ± 1/2
0.0 (0) | 0.33 (5/16)	0.0 (0) | 0.33 (5/16)	0.0 (0) | 0.33 (5/16)	CROSS | TOL.	0.0 (0) | 0.5 (1/2)	0.0 (0) | 0.5 (1/2)	0.0 (0) | 0.5 (1/2)
0.0° ±0.17° 0° ± 5/32°	0.0° ±0.17° 0° ± 5/32°	0.0° ±0.17° 0° ± 5/32°	FRONT TOE (Degrees)	0.0° ±0.17° 0° ± 5/32°	0.0° ±0.17° 0° ± 5/32°	0.17° ±0.17° 5/32° ± 5/32°
0.0" ±0.08" 0" ± 3/32"	0.0" ±0.08" 0" ± 3/32"	0.0" ±0.08" 0" ± 3/32"	FRONT TOE (Inches)	0.0" ±0.08" 0" ± 3/32"	0.0" ±0.08" 0" ± 3/32"	0.08" ±0.08" 3/32" ± 3/32"
			SAI			
20.0 / 18.9 ±0.33	20.0 / 18.9 ±0.33	20.0 / 18.5 ±0.33	TURNING ANGLE	20.0 / 18.5 ±0.33	20.0 / 18.5 ±0.33	20.0 / 18.4 ±0.33
26	26	26	ILLUSTRATION #	▲	▲	▲
			REAR			
-1.5 ±0.17 -1 1/2 ± 3/16	-1.5 ±0.17 -1 1/2 ± 3/16	-1.5 ±0.17 -1 1/2 ± 3/16	CAMBER (Degrees)	-1.45 ±0.33 -1 7/16 ± 5/16	-1.45 ±0.33 -1 7/16 ± 5/16	-1.33 ±0.5 -1 5/16 ± 1/2
0.33° ±0.17° 11/32° ± 5/32°	0.33° ±0.17° 11/32° ± 5/32°	0.33° ±0.17° 11/32° ± 5/32°	REAR TOE (Degrees)	0.33° ±0.17° 11/32° ± 5/32°	0.42° ±0.17° 13/32° ± 5/32°	0.17° ±0.21° 5/32° ± 7/32°
0.17" ±0.08" 5/32" ± 3/32"	0.17" ±0.08" 5/32" ± 3/32"	0.17" ±0.08" 5/32" ± 3/32"	REAR TOE (Inches)	0.17" ±0.08" 5/32" ± 3/32"	0.21" ±0.08" 7/32" ± 3/32"	0.08" ±0.1" 3/32" ± 3/32"
0.33 (11/32)	0.33 (11/32)	0.33 (11/32)	THRUST ANGLE	0.33 (11/32)	0.33 (11/32)	0.33 (11/32)
			ILLUSTRATION #	114	114	127
			NOTES			
46049	46050	46051		46052	46053	46054

Volkswagen

Golf (A6)	Golf (A7)	Golf (A7)	MODEL INFORMATION	Golf (A7)	Golf (A7)	Golf (A7)
Sport Susp.	w/IRS	w/IRS		w/IRS	w/Beam Axle	w/Beam Axle
	Base Susp. (2UA)	Sport Susp. (2UC)		Dynamic Susp. (DCC)	Base Susp. (2UA)	Sport Susp. (2UC)
2010-2014	2015-2017	2015-2017	YEARS	2015-2017	2015-2017	2015-2017
			INSTRUCTIONS			
89 ft.lbs.	89 ft.lbs.	89 ft.lbs.	LUGNUT TORQUE	89 ft.lbs.	89 ft.lbs.	89 ft.lbs.
			FRONT			
7.78 ±0.5 7 13/16 ± 1/2	7.38 ±0.5 7 5/8 ± 1/2	7.63 ±0.5 7 5/8 ± 1/2	CASTER (Degrees)	7.55 ±0.5 7 9/16 ± 1/2	7.38 ±0.5 7 3/8 ± 1/2	7.63 ±0.5 7 5/8 ± 1/2
0.0 (0) | 0.5 (1/2)	0.0 (0) | 0.5 (1/2)	0.0 (0) | 0.5 (1/2)	CROSS | TOL.	0.0 (0) | 0.5 (1/2)	0.0 (0) | 0.5 (1/2)	0.0 (0) | 0.5 (1/2)
-0.68 ±0.5 - 11/16 ± 1/2	-0.5 ±0.5 - 1/2 ± 1/2	-0.68 ±0.5 - 11/16 ± 1/2	CAMBER (Degrees)	-0.6 ±0.5 - 5/8 ± 1/2	-0.5 ±0.5 - 1/2 ± 1/2	-0.68 ±0.5 - 11/16 ± 1/2
0.0 (0) | 0.5 (1/2)	0.0 (0) | 0.5 (1/2)	0.0 (0) | 0.5 (1/2)	CROSS | TOL.	0.0 (0) | 0.5 (1/2)	0.0 (0) | 0.5 (1/2)	0.0 (0) | 0.5 (1/2)
0.17° ±0.17° 5/32° ± 5/32°	0.17° ±0.17° 5/32° ± 5/32°	0.17° ±0.17° 5/32° ± 5/32°	FRONT TOE (Degrees)	0.17° ±0.17° 5/32° ± 5/32°	0.17° ±0.17° 5/32° ± 5/32°	0.17° ±0.17° 5/32° ± 5/32°
0.08" ±0.08" 3/32" ± 3/32"	0.08" ±0.08" 3/32" ± 3/32"	0.08" ±0.08" 3/32" ± 3/32"	FRONT TOE (Inches)	0.08" ±0.08" 3/32" ± 3/32"	0.08" ±0.08" 3/32" ± 3/32"	0.08" ±0.08" 3/32" ± 3/32"
			SAI			
20.0 / 18.3 ±0.33	20.0 / 18.3 ±0.33	20.0 / 18.5 ±0.33	TURNING ANGLE	20.0 / 18.4 ±0.33	20.0 / 18.3 ±0.33	20.0 / 18.5 ±0.33
▲			ILLUSTRATION #			
			REAR			
-1.33 ±0.5 [1] -1 5/16 ± 1/2 [1]	-1.33 ±0.5 -1 5/16 ± 1/2	-1.33 ±0.5 -1 5/16 ± 1/2	CAMBER (Degrees)	-1.33 ±0.5 -1 5/16 ± 1/2	-1.0 ±0.17 -1 ± 3/16	-1.0 ±0.17 -1 ± 3/16
0.17° ±0.21° 5/32° ± 7/32°	0.17° ±0.17° 5/32° ± 5/32°	0.17° ±0.17° 5/32° ± 5/32°	REAR TOE (Degrees)	0.17° ±0.17° 5/32° ± 5/32°	0.33° ±0.2° 11/32° ± 3/16°	0.4° ±0.2° 13/32° ± 3/16°
0.08" ±0.1" 3/32" ± 3/32"	0.08" ±0.08" 3/32" ± 3/32"	0.08" ±0.08" 3/32" ± 3/32"	REAR TOE (Inches)	0.08" ±0.08" 3/32" ± 3/32"	0.17" ±0.1" 5/32" ± 3/32"	0.2" ±0.1" 3/16" ± 3/32"
0.33 (11/32)	0.33 (11/32)	0.33 (11/32)	THRUST ANGLE	0.33 (11/32)	0.33 (11/32)	0.33 (11/32)
127	127	127	ILLUSTRATION #	127		
[1] w/18" tires = -1.75°			NOTES			
46055	46056	46057		46058	46059	46060

ALIGNMENT SPECIFICATIONS

ALIGNMENT SPECIFICATIONS

Volkswagen

Golf (A7) w/Beam Axle Dynamic Susp. (DCC)	eGolf (A7) w/IRS All Suspensions	Golf Sportwagen (A7) w/IRS Base Susp. (2UA)	MODEL INFORMATION	Golf Sportwagen (A7) w/IRS Sport Susp. (2UC)	Golf Sportwagen (A7) w/IRS Dynamic Susp. (DCC)	Golf Sportwagen (A7) w/Beam Axle Base Susp. (2UA)
2015-2017	2015-2017	2015-2017	YEARS	2015-2017	2015-2017	2015-2017
			INSTRUCTIONS			
89 ft.lbs.	89 ft.lbs.	89 ft.lbs.	LUGNUT TORQUE	89 ft.lbs.	89 ft.lbs.	89 ft.lbs.
			FRONT			
7.55 ±0.5 7 9/16 ± 1/2	7.38 ±0.5 7 5/8 ± 1/2	7.38 ±0.5 7 5/8 ± 1/2	CASTER (Degrees)	7.63 ±0.5 7 5/8 ± 1/2	7.55 ±0.5 7 9/16 ± 1/2	7.38 ±0.5 7 3/8 ± 1/2
0.0 (0) / 0.5 (1/2)	0.0 (0) / 0.5 (1/2)	0.0 (0) / 0.5 (1/2)	CROSS / TOL.	0.0 (0) / 0.5 (1/2)	0.0 (0) / 0.5 (1/2)	0.0 (0) / 0.5 (1/2)
-0.6 ±0.5 -5/8 ± 1/2	-0.5 ±0.5 -1/2 ± 1/2	-0.5 ±0.5 -1/2 ± 1/2	CAMBER (Degrees)	-0.68 ±0.5 -11/16 ± 1/2	-0.6 ±0.5 -5/8 ± 1/2	-0.5 ±0.5 -1/2 ± 1/2
0.0 (0) / 0.5 (1/2)	0.0 (0) / 0.5 (1/2)	0.0 (0) / 0.5 (1/2)	CROSS / TOL.	0.0 (0) / 0.5 (1/2)	0.0 (0) / 0.5 (1/2)	0.0 (0) / 0.5 (1/2)
0.17° ±0.17° 5/32° ± 5/32°	0.17° ±0.17° 5/32° ± 5/32°	0.17° ±0.17° 5/32° ± 5/32°	FRONT TOE (Degrees)	0.17° ±0.17° 5/32° ± 5/32°	0.17° ±0.17° 5/32° ± 5/32°	0.17° ±0.17° 5/32° ± 5/32°
0.08" ±0.08" 3/32" ± 3/32"	0.08" ±0.08" 3/32" ± 3/32"	0.08" ±0.08" 3/32" ± 3/32"	FRONT TOE (Inches)	0.08" ±0.08" 3/32" ± 3/32"	0.08" ±0.08" 3/32" ± 3/32"	0.08" ±0.08" 3/32" ± 3/32"
			SAI			
20.0 / 18.4 ±0.33	20.0 / 18.7 ±0.33	20.0 / 18.3 ±0.33	TURNING ANGLE	20.0 / 18.5 ±0.33	20.0 / 18.4 ±0.33	20.0 / 18.3 ±0.33
			ILLUSTRATION #			
			REAR			
-1.0 ±0.17 -1 ± 3/16	-1.33 ±0.5 -1 5/16 ± 1/2	-1.33 ±0.5 -1 5/16 ± 1/2	CAMBER (Degrees)	-1.33 ±0.5 -1 5/16 ± 1/2	-1.33 ±0.5 -1 5/16 ± 1/2	-1.0 ±0.17 -1 ± 3/16
0.4° ±0.2° 13/32° ± 3/16°	0.17° ±0.17° 5/32° ± 5/32°	0.17° ±0.17° 5/32° ± 5/32°	REAR TOE (Degrees)	0.17° ±0.17° 5/32° ± 5/32°	0.17° ±0.17° 5/32° ± 5/32°	0.33° ±0.2° 11/32° ± 3/16°
0.2" ±0.1" 3/16" ± 3/32"	0.08" ±0.08" 3/32" ± 3/32"	0.08" ±0.08" 3/32" ± 3/32"	REAR TOE (Inches)	0.08" ±0.08" 3/32" ± 3/32"	0.08" ±0.08" 3/32" ± 3/32"	0.17" ±0.1" 5/32" ± 3/32"
0.33 (11/32)	0.33 (11/32)	0.33 (11/32)	THRUST ANGLE	0.33 (11/32)	0.33 (11/32)	0.33 (11/32)
	127	127	ILLUSTRATION #	127	127	
			NOTES			
46061	46062	46063		46064	46065	46066

Volkswagen

Golf Sportwagen (A7) w/Beam Axle Sport Susp. (2UC)	Golf Sportwagen (A7) w/Beam Axle Dynamic Susp. (DCC)	Golf Alltrack Sportwagen (A7)	MODEL INFORMATION	Golf R (A6)	Golf R (A7)	GLI (A5)
2015-2017	2015-2017	2017	YEARS	2012-2013	2015-2017	2008-2009
			INSTRUCTIONS			
89 ft.lbs.	89 ft.lbs.	89 ft.lbs.	LUGNUT TORQUE	89 ft.lbs.	89 ft.lbs.	
			FRONT			
7.63 ±0.5 7 5/8 ± 1/2	7.55 ±0.5 7 9/16 ± 1/2	7.15 ±0.5 7 1/8 ± 1/2	CASTER (Degrees)	7.85 ±0.5 7 7/8 ± 1/2	7.75 ±0.5 7 3/4 ± 1/2	7.78 ±0.5 7 13/16 ± 1/2
0.0 (0) / 0.5 (1/2)	0.0 (0) / 0.5 (1/2)	0.0 (0) / 0.5 (1/2)	CROSS / TOL.	0.0 (0) / 0.5 (1/2)	0.0 (0) / 0.5 (1/2)	0.0 (0) / 0.5 (1/2)
-0.68 ±0.5 -11/16 ± 1/2	-0.6 ±0.5 -5/8 ± 1/2	-0.27 ±0.5 -1/4 ± 1/2	CAMBER (Degrees)	-0.75 ±0.5 -3/4 ± 1/2	-0.73 ±0.5 -3/4 ± 1/2	-0.68 ±0.5 -11/16 ± 1/2
0.0 (0) / 0.5 (1/2)	0.0 (0) / 0.5 (1/2)	0.0 (0) / 0.5 (1/2)	CROSS / TOL.	0.0 (0) / 0.5 (1/2)	0.0 (0) / 0.5 (1/2)	0.0 (0) / 0.5 (1/2)
0.17° ±0.17° 5/32° ± 5/32°	0.17° ±0.17° 5/32° ± 5/32°	0.17° ±0.17° 5/32° ± 5/32°	FRONT TOE (Degrees)	0.17° ±0.17° 5/32° ± 5/32°	0.17° ±0.17° 5/32° ± 5/32°	0.17° ±0.17° 5/32° ± 5/32°
0.08" ±0.08" 3/32" ± 3/32"	0.08" ±0.08" 3/32" ± 3/32"	0.08" ±0.08" 3/32" ± 3/32"	FRONT TOE (Inches)	0.08" ±0.08" 3/32" ± 3/32"	0.08" ±0.08" 3/32" ± 3/32"	0.08" ±0.08" 3/32" ± 3/32"
			SAI			
20.0 / 18.5 ±0.33	20.0 / 18.4 ±0.33	20.0 / 18.9 ±0.33	TURNING ANGLE	20.0 / 18.6 ±0.33	20.0 / 18.4 ±0.33	20.0 / 18.3 ±0.33
			ILLUSTRATION #			△
			REAR			
-1.0 ±0.17 -1 ± 3/16	-1.0 ±0.17 -1 ± 3/16	-1.33 ±0.5 -1 5/16 ± 1/2	CAMBER (Degrees)	-1.75 ±0.5 -1 3/4 ± 1/2	-1.75 ±0.5 -1 3/4 ± 1/2	-1.33 ±0.5 [1] -1 5/16 ± 1/2 [1]
0.4° ±0.2° 13/32° ± 3/16°	0.38° ±0.2° 3/8° ± 3/16°	0.17° ±0.17° 5/32° ± 5/32°	REAR TOE (Degrees)	0.17° ±0.21° 5/32° ± 7/32°	0.27° ±0.17° 9/32° ± 5/32°	0.17° ±0.21° 5/32° ± 7/32°
0.2" ±0.1" 3/16" ± 3/32"	0.19" ±0.1" 3/16" ± 3/32"	0.08" ±0.08" 3/32" ± 3/32"	REAR TOE (Inches)	0.08" ±0.1" 3/32" ± 3/32"	0.13" ±0.08" 1/8" ± 3/32"	0.08" ±0.1" 3/32" ± 3/32"
0.33 (11/32)	0.33 (11/32)	0.33 (11/32)	THRUST ANGLE	0.33 (11/32)	0.33 (11/32)	0.33 (11/32)
		127	ILLUSTRATION #	127	127	127
			NOTES			[1] w/18" tires = -1.75° (-1 3/4°)
46067	46068	46069		46070	46071	46072

Volkswagen

GLI (A6)	GTI (A4)	GTI (A5)	MODEL INFORMATION	GTI (A6)	GTI (A7)	Jetta, Jetta GL
2012-2017	1999-2006	2006-2009	YEARS	2010-2014	2015-2017	1990-92
			INSTRUCTIONS			
89 ft.lbs.	89 ft.lbs.		LUGNUT TORQUE	89 ft.lbs.	89 ft.lbs.	81 ft.lbs.
			FRONT			
7.88 ±0.5 7 7/8 ± 1/2	8.0 ±0.5 8 ± 1/2	7.57 ±0.5 7 9/16 ± 1/2	CASTER (Degrees)	7.78 ±0.5 7 13/16 ± 1/2	7.63 ±0.5 7 5/8 ± 1/2	1.5 ±0.5 1 1/2 ± 1/2
0.0 (0) | 0.5 (1/2)	0.0 (0) | 0.5 (1/2)	0.0 (0) | 0.5 (1/2)	CROSS | TOL.	0.0 (0) | 0.5 (1/2)	0.0 (0) | 0.5 (1/2)	0.0 (0) | 1.0 (1)
-0.67 ±0.5 -11/16 ± 1/2	-0.58 ±0.5 -9/16 ± 1/2	-0.5 ±0.5 -1/2 ± 1/2	CAMBER (Degrees)	-0.68 ±0.5 -11/16 ± 1/2	-0.68 ±0.5 -11/16 ± 1/2	-0.5 ±0.33 -1/2 ± 5/16
0.0 (0) | 0.5 (1/2)	0.0 (0) | 0.5 (1/2)	0.0 (0) | 0.5 (1/2)	CROSS | TOL.	0.0 (0) | 0.5 (1/2)	0.0 (0) | 0.5 (1/2)	0.0 (0) | 0.5 (1/2)
0.17° ±0.17° 5/32° ± 5/32°	0.0° ±0.17° 0° ± 5/32°	0.17° ±0.17° 5/32° ± 5/32°	FRONT TOE (Degrees)	0.17° ±0.17° 5/32° ± 5/32°	0.17° ±0.17° 5/32° ± 5/32°	0.0° ±0.17° 0° ± 5/32°
0.08" ±0.08" 3/32" ± 3/32"	0.0" ±0.08" 0" ± 3/32"	0.08" ±0.08" 3/32" ± 3/32"	FRONT TOE (Inches)	0.08" ±0.08" 3/32" ± 3/32"	0.08" ±0.08" 3/32" ± 3/32"	0.0" ±0.08" 0" ± 3/32"
			SAI			
20.0 / 18.5 ±0.33	20.0 / 18.5 ±0.33	20.0 / 18.4 ±0.33	TURNING ANGLE	20.0 / 18.3 ±0.33	20.0 / 18.5 ±0.33	20.0 / 18.7 ±0.5
	▲	▲	ILLUSTRATION #			26
			REAR			
-1.33 ±0.5 [1] -1 5/16 ± 1/2 [1]	-1.45 ±0.33 -1 7/16 ± 5/16	-1.33 ±0.5 -1 5/16 ± 1/2	CAMBER (Degrees)	-1.75 ±0.5 -1 3/4 ± 1/2	-1.75 ±0.5 -1 3/4 ± 1/2	-1.5 ±0.17 -1 1/2 ± 3/16
0.17° ±0.17° 5/32° ± 5/32°	0.48° ±0.17° 15/32° ± 5/32°	0.17° ±0.21° 5/32° ± 7/32°	REAR TOE (Degrees)	0.17° ±0.21° 5/32° ± 7/32°	0.27° ±0.21° 9/32° ± 7/32°	0.33° ±0.17° 11/32° ± 5/32°
0.08" ±0.08" 3/32" ± 3/32"	0.24" ±0.08" 1/4" ± 3/32"	0.08" ±0.1" 3/32" ± 3/32"	REAR TOE (Inches)	0.08" ±0.1" 3/32" ± 3/32"	0.13" ±0.1" 1/8" ± 3/32"	0.17" ±0.08" 5/32" ± 3/32"
0.33 (11/32)	0.33 (11/32)	0.33 (11/32)	THRUST ANGLE	0.33 (11/32)	0.33 (11/32)	0.33 (11/32)
127	114	127	ILLUSTRATION #	127	127	
[1] 18" Tires = -1.75° (-1 3/4°)			NOTES			
46073	46074	46075		46076	46077	46079

Volkswagen

Jetta GLI (except 16 Valve)	Jetta GLI 16V	Jetta 4 Cylinder 1.8L / 1.9L	MODEL INFORMATION	Jetta 4 Cylinder 2.0L	Jetta 4 Cylinder All Models	Jetta 6 Cylinder
1990-92	1990-92	1993-94	YEARS	1993-94	1995-99	1994-99
			INSTRUCTIONS			
81 ft.lbs.	81 ft.lbs.	81 ft.lbs.	LUGNUT TORQUE	81 ft.lbs.	81 ft.lbs.	81 ft.lbs.
			FRONT			
1.58 ±0.5 1 9/16 ± 1/2	1.58 ±0.5 1 9/16 ± 1/2	1.75 ±0.5 1 3/4 ± 1/2	CASTER (Degrees)	1.83 ±0.5 1 13/16 ± 1/2	1.83 ±0.5 1 13/16 ± 1/2	3.42 ±0.5 3 7/16 ± 1/2
0.0 (0) | 1.0 (1)	0.0 (0) | 1.0 (1)	0.0 (0) | 0.5 (1/2)	CROSS | TOL.	0.0 (0) | 0.5 (1/2)	0.0 (0) | 0.5 (1/2)	0.0 (0) | 0.5 (1/2)
-0.58 ±0.33 -9/16 ± 5/16	-0.67 ±0.33 -11/16 ± 5/16	-0.5 ±0.33 -1/2 ± 5/16	CAMBER (Degrees)	-0.58 ±0.33 -9/16 ± 5/16	-0.6 ±0.33 -5/8 ± 5/16	-0.67 ±0.33 -11/16 ± 5/16
0.0 (0) | 0.5 (1/2)	0.0 (0) | 0.5 (1/2)	0.0 (0) | 0.33 (5/16)	CROSS | TOL.	0.0 (0) | 0.33 (5/16)	0.0 (0) | 0.33 (5/16)	0.0 (0) | 0.33 (5/16)
0.0° ±0.17° 0° ± 5/32°	0.0° ±0.17° 0° ± 5/32°	0.0° ±0.17° 0° ± 5/32°	FRONT TOE (Degrees)	0.0° ±0.17° 0° ± 5/32°	0.0° ±0.17° 0° ± 5/32°	0.0° ±0.17° 0° ± 5/32°
0.0" ±0.08" 0" ± 3/32"	0.0" ±0.08" 0" ± 3/32"	0.0" ±0.08" 0" ± 3/32"	FRONT TOE (Inches)	0.0" ±0.08" 0" ± 3/32"	0.0" ±0.08" 0" ± 3/32"	0.0" ±0.08" 0" ± 3/32"
			SAI			
20.0 / 18.7 ±0.5	20.0 / 18.7 ±0.5	20.0 / 18.9 ±0.33	TURNING ANGLE	20.0 / 18.9 ±0.33	20.0 / 18.9 ±0.33	20.0 / 18.5 ±0.33
26	26	26	ILLUSTRATION #	26	26	26
			REAR			
-1.5 ±0.17 -1 1/2 ± 3/16	-1.5 ±0.17 -1 1/2 ± 3/16	-1.5 ±0.17 -1 1/2 ± 3/16	CAMBER (Degrees)	-1.5 ±0.17 -1 1/2 ± 3/16	-1.5 ±0.17 -1 1/2 ± 3/16	-1.5 ±0.17 -1 1/2 ± 3/16
0.33° ±0.17° 11/32° ± 5/32°	0.33° ±0.17° 11/32° ± 5/32°	0.33° ±0.17° 11/32° ± 5/32°	REAR TOE (Degrees)	0.33° ±0.17° 11/32° ± 5/32°	0.33° ±0.17° 11/32° ± 5/32°	0.33° ±0.17° 11/32° ± 5/32°
0.17" ±0.08" 5/32" ± 3/32"	0.17" ±0.08" 5/32" ± 3/32"	0.17" ±0.08" 5/32" ± 3/32"	REAR TOE (Inches)	0.17" ±0.08" 5/32" ± 3/32"	0.17" ±0.08" 5/32" ± 3/32"	0.17" ±0.08" 5/32" ± 3/32"
0.33 (11/32)	0.33 (11/32)	0.33 (11/32)	THRUST ANGLE	0.33 (11/32)	0.33 (11/32)	0.33 (11/32)
			ILLUSTRATION #			
			NOTES			
46080	46081	46082		46083	46084	46085

ALIGNMENT SPECIFICATIONS

ALIGNMENT SPECIFICATIONS

Volkswagen

Jetta (A4)	Jetta (A4)	Jetta (A5)	MODEL INFORMATION	Jetta (A5)	Jetta Wagon/	Jetta Wagon/
Std. Susp.	Sport Susp.	Std. Susp.		Sport Susp.	Sportwagen (A5)	Sportwagen (A5)
All Models	All Models	All Models		All Models	Std. Susp.	Sport Susp.
1999-2005	1999-2005	2005-2010	YEARS	2005-2010	2011-2014	2011-2014
			INSTRUCTIONS			
89 ft.lbs.	89 ft.lbs.		LUGNUT TORQUE		89 ft.lbs.	89 ft.lbs.
			FRONT			
7.67 ±0.5 7 11/16 ± 1/2	7.83 ±0.5 7 13/16 ± 1/2	7.57 ±0.5 7 9/16 ± 1/2	CASTER (Degrees)	7.78 ±0.5 7 13/16 ± 1/2	7.57 ±0.5 7 9/16 ± 1/2	7.78 ±0.5 7 13/16 ± 1/2
0.0 (0) | 0.5 (1/2)	0.0 (0) | 0.5 (1/2)	0.0 (0) | 0.5 (1/2)	CROSS | TOL.	0.0 (0) | 0.5 (1/2)	0.0 (0) | 0.5 (1/2)	0.0 (0) | 0.5 (1/2)
-0.5 ±0.5 -1/2 ± 1/2	-0.55 ±0.5 -9/16 ± 1/2	-0.5 ±0.5 -1/2 ± 1/2	CAMBER (Degrees)	-0.68 ±0.5 -11/16 ± 1/2	-0.5 ±0.5 -1/2 ± 1/2	-0.68 ±0.5 -11/16 ± 1/2
0.0 (0) | 0.5 (1/2)	0.0 (0) | 0.5 (1/2)	0.0 (0) | 0.5 (1/2)	CROSS | TOL.	0.0 (0) | 0.5 (1/2)	0.0 (0) | 0.5 (1/2)	0.0 (0) | 0.5 (1/2)
0.0° ±0.17° 0° ± 5/32°	0.0° ±0.17° 0° ± 5/32°	0.17° ±0.17° 5/32° ± 5/32°	FRONT TOE (Degrees)	0.17° ±0.17° 5/32° ± 5/32°	0.17° ±0.17° 5/32° ± 5/32°	0.17° ±0.17° 5/32° ± 5/32°
0.0" ±0.08" 0" ± 3/32"	0.0" ±0.08" 0" ± 3/32"	0.08" ±0.08" 3/32" ± 3/32"	FRONT TOE (Inches)	0.08" ±0.08" 3/32" ± 3/32"	0.08" ±0.08" 3/32" ± 3/32"	0.08" ±0.08" 3/32" ± 3/32"
			SAI			
20.0 / 18.5 ±0.33	20.0 / 18.5 ±0.33	20.0 / 18.4 ±0.33	TURNING ANGLE	20.0 / 18.3 ±0.33	20.0 / 18.4 ±0.33	20.0 / 18.3 ±0.33
▲	▲	▲	ILLUSTRATION #	▲		
			REAR			
-1.45 ±0.17 -1 7/16 ± 3/16	-1.45 ±0.17 -1 7/16 ± 3/16	-1.33 ±0.5 -1 5/16 ± 1/2	CAMBER (Degrees)	-1.33 ±0.5 -1 5/16 ± 1/2	-1.33 ±0.5 -1 5/16 ± 1/2	-1.33 ±0.5 -1 5/16 ± 1/2
0.33° ±0.17° 11/32° ± 5/32°	0.42° ±0.17° 13/32° ± 5/32°	0.17° ±0.21° 5/32° ± 7/32°	REAR TOE (Degrees)	0.17° ±0.21° 5/32° ± 7/32°	0.17° ±0.21° 5/32° ± 7/32°	0.17° ±0.21° 5/32° ± 7/32°
0.17" ±0.08" 5/32" ± 3/32"	0.21" ±0.08" 7/32" ± 3/32"	0.08" ±0.1" 3/32" ± 3/32"	REAR TOE (Inches)	0.08" ±0.1" 3/32" ± 3/32"	0.08" ±0.1" 3/32" ± 3/32"	0.08" ±0.1" 3/32" ± 3/32"
0.33 (11/32)	0.33 (11/32)	0.33 (11/32)	THRUST ANGLE	0.33 (11/32)	0.33 (11/32)	0.33 (11/32)
114	114	127	ILLUSTRATION #	127	127	127
			NOTES			
46086	46087	46088		46089	46090	46091

Volkswagen

Jetta Sedan (A6)	Jetta Sedan (A6)	Jetta Sedan (A6)	MODEL INFORMATION	Jetta Sedan (A6)	Jetta Hybrid (A6)	Passat
Std. Susp. (2UA)	Std. Susp. (2UA)	Sport Susp. (2UC)		Comfort Susp. (2UD)	Basic Susp. (2UA)	All prior to VIN#
(rear beam axle)	(multi-link r. susp.)	(multi-link r. susp.)		(multi-link r. susp.)	(multi-link susp.)	.........L(B)108393
2011-2013	2014-2017	2014-2017	YEARS	2014-2017	2015-2016	1990 (early)
			INSTRUCTIONS			
89 ft.lbs.	89 ft.lbs.	89 ft.lbs.	LUGNUT TORQUE	89 ft.lbs.	89 ft.lbs.	81 ft.lbs.
			FRONT			
7.62 ±0.5 7 5/8 ± 1/2	7.62 ±0.5 7 5/8 ± 1/2	7.8 ±0.5 7 13/16 ± 1/2	CASTER (Degrees)	7.5 ±0.5 7 1/2 ± 1/2	7.62 ±0.5 7 5/8 ± 1/2	1.67 ±0.5 1 11/16 ± 1/2
0.0 (0) | 0.5 (1/2)	0.0 (0) | 0.5 (1/2)	0.0 (0) | 0.5 (1/2)	CROSS | TOL.	0.0 (0) | 0.5 (1/2)	0.0 (0) | 0.5 (1/2)	0.0 (0) | 0.5 (1/2)
-0.5 ±0.5 -1/2 ± 1/2	-0.5 ±0.5 -1/2 ± 1/2	-0.67 ±0.5 -11/16 ± 1/2	CAMBER (Degrees)	-0.37 ±0.5 -3/8 ± 1/2	-0.5 ±0.5 -1/2 ± 1/2	-1.33 ±0.33 -1 5/16 ± 5/16
0.0 (0) | 0.5 (1/2)	0.0 (0) | 0.5 (1/2)	0.0 (0) | 0.5 (1/2)	CROSS | TOL.	0.0 (0) | 0.5 (1/2)	0.0 (0) | 0.5 (1/2)	0.0 (0) | 0.5 (1/2)
0.17° ±0.17° 5/32° ± 5/32°	0.17° ±0.17° 5/32° ± 5/32°	0.17° ±0.17° 5/32° ± 5/32°	FRONT TOE (Degrees)	0.17° ±0.17° 5/32° ± 5/32°	0.17° ±0.17° 5/32° ± 5/32°	0.0° ±0.17° 0° ± 5/32°
0.08" ±0.08" 3/32" ± 3/32"	0.08" ±0.08" 3/32" ± 3/32"	0.08" ±0.08" 3/32" ± 3/32"	FRONT TOE (Inches)	0.08" ±0.08" 3/32" ± 3/32"	0.08" ±0.08" 3/32" ± 3/32"	0.0" ±0.08" 0" ± 3/32"
			SAI			
20.0 / 18.7 ±0.33	20.0 / 18.7 ±0.33	20.0 / 18.5 ±0.33	TURNING ANGLE	20.0 / 18.2 ±0.33	20.0 / 18.7 ±0.33	20.0 / 18.8 ±0.5
			ILLUSTRATION #			26
			REAR			
-1.0 ±0.5 -1 ± 1/2	-1.33 ±0.5 -1 5/16 ± 1/2	-1.75 ±0.5 -1 3/4 ± 1/2	CAMBER (Degrees)	-1.33 ±0.5 -1 5/16 ± 1/2	-1.33 ±0.5 -1 5/16 ± 1/2	-1.67 ±0.17 -1 11/16 ± 3/16
0.33° ±0.17° 11/32° ± 5/32°	0.17° ±0.17° 5/32° ± 5/32°	0.17° ±0.17° 5/32° ± 5/32°	REAR TOE (Degrees)	0.17° ±0.17° 5/32° ± 5/32°	0.17° ±0.17° 5/32° ± 5/32°	0.42° ±0.25° 13/32° ± 1/4°
0.17" ±0.08" 5/32" ± 3/32"	0.08" ±0.08" 3/32" ± 3/32"	0.08" ±0.08" 3/32" ± 3/32"	REAR TOE (Inches)	0.08" ±0.08" 3/32" ± 3/32"	0.08" ±0.08" 3/32" ± 3/32"	0.21" ±0.13" 7/32" ± 1/8"
0.33 (11/32)	0.33 (11/32)	0.33 (11/32)	THRUST ANGLE	0.33 (11/32)	0.33 (11/32)	0.42 (13/32)
	127	127 127	ILLUSTRATION #	127 127	127	
			NOTES			
46092	46093	46094		46095	46096	46099

Volkswagen

Passat All prior to VIN# ....L(E)222111	Passat 4 Cylinder (after prev. VIN#'s)	Passat 6 Cylinder	MODEL INFORMATION	Passat TDI Diesel	Passat FWD	Passat 4-Motion exc/W8
1990 (early)	1990-96	1993-97	YEARS	1996-97	1998-2005	2000-2005
			INSTRUCTIONS			
81 ft.lbs.	81 ft.lbs.	81 ft.lbs.	LUGNUT TORQUE	81 ft.lbs.	81 ft.lbs.	81 ft.lbs.
			FRONT			
1.67 ±0.5 1 11/16 ± 1/2	1.67 ±0.5 1 11/16 ± 1/2	3.33 ±0.5 3 5/16 ± 1/2	CASTER (Degrees)	1.67 ±0.5 1 11/16 ± 1/2	Caster is not specified by the manufacturer.	Caster is not specified by the manufacturer.
0.0 (0) | 0.5 (1/2)	0.0 (0) | 0.5 (1/2)	0.0 (0) | 0.5 (1/2)	CROSS | TOL.	0.0 (0) | 0.5 (1/2)		
-1.33 ±0.33 -1 5/16 ± 5/16	-1.33 ±0.33 -1 5/16 ± 5/16	-1.0 ±0.17 -1 ± 3/16	CAMBER (Degrees)	-1.33 ±0.33 -1 5/16 ± 5/16	-0.58 ±0.42 - 9/16 ± 7/16	-0.58 ±0.42 - 9/16 ± 7/16
0.0 (0) | 0.5 (1/2)	0.0 (0) | 0.5 (1/2)	0.0 (0) | 0.5 (1/2)	CROSS | TOL.	0.0 (0) | 0.5 (1/2)	0.0 (0) | 0.5 (1/2)	0.0 (0) | 0.5 (1/2)
0.0° ±0.17° 0° ± 5/32°	0.0° ±0.17° 0° ± 5/32°	0.08° ±0.08° 3/32° ± 3/32°	FRONT TOE (Degrees)	0.0° ±0.17° 0° ± 5/32°	0.33° ±0.07° 11/32° ± 1/16°	0.33° ±0.07° 11/32° ± 1/16°
0.0" ±0.08" 0" ± 3/32"	0.0" ±0.08" 0" ± 3/32"	0.04" ±0.04" 1/32" ± 1/32"	FRONT TOE (Inches)	0.0" ±0.08" 0" ± 3/32"	0.17" ±0.03" 5/32" ± 1/32"	0.17" ±0.03" 5/32" ± 1/32"
			SAI			
20.0 / 18.8 ±0.5	20.0 / 18.8 ±0.5	20.0 / 18.3 ±0.5	TURNING ANGLE	20.0 / 18.8 ±0.5	20.0 / 18.7 ±0.5	20.0 / 18.7 ±0.5
26	26	26	ILLUSTRATION #	26	▲	▲
			REAR			
-1.67 ±0.17 -1 11/16 ± 3/16	-1.5 ±0.17 -1 1/2 ± 3/16	-1.5 ±0.17 -1 1/2 ± 3/16	CAMBER (Degrees)	-1.5 ±0.17 -1 1/2 ± 3/16	-1.5 ±0.33 -1 1/2 ± 5/16	-0.67 ±0.5 - 11/16 ± 1/2
0.42° ±0.25° 13/32° ± 1/4°	0.33° ±0.17° 11/32° ± 5/32°	0.33° ±0.17° 11/32° ± 5/32°	REAR TOE (Degrees)	0.33° ±0.17° 11/32° ± 5/32°	0.33° +0.25° -0.17° 11/32° + 1/4° - 5/32°	0.27° ±0.17° 9/32° ± 5/32°
0.21" ±0.13" 7/32" ± 1/8"	0.17" ±0.08" 5/32" ± 3/32"	0.17" ±0.08" 5/32" ± 3/32"	REAR TOE (Inches)	0.17" ±0.08" 5/32" ± 3/32"	0.17" +0.13" -0.08" 5/32" + 1/8" - 3/32"	0.13" ±0.08" 1/8" ± 3/32"
0.42 (13/32)	0.33 (11/32)	0.33 (11/32)	THRUST ANGLE	0.33 (11/32)	0.25 (1/4)	0.17 (5/32)
			ILLUSTRATION #		▲	83 06
			NOTES			
46100	46101	46102		46103	46104	46105

Volkswagen

Passat 4-Motion W8	Passat Std. Susp. (incl 4Motion)	Passat Sport Susp. (incl 4Motion)	MODEL INFORMATION	Passat Std. Susp. (G02, G11)	Passat Comfort Susp. (G05)	Phaeton
2002-2005	2006-2010	2006-2010	YEARS	2012-2017	2012-2017	2004-2006
			INSTRUCTIONS			
81 ft.lbs.	89 ft.lbs.	89 ft.lbs.	LUGNUT TORQUE	89 ft.lbs.	89 ft.lbs.	89 ft.lbs.
			FRONT			
Caster is not specified by the manufacturer.	7.53 ±0.5 7 9/16 ± 1/2	7.73 ±0.5 7 3/4 ± 1/2	CASTER (Degrees)	7.92 ±0.5 7 15/16 ± 1/2	7.45 ±0.5 7 7/16 ± 1/2	Caster is not specified by the manufacturer.
	0.0 (0) | 0.5 (1/2)	0.0 (0) | 0.5 (1/2)	CROSS | TOL.	0.0 (0) | 0.5 (1/2)	0.0 (0) | 0.5 (1/2)	
-0.7 ±0.42 [1] -11/16 ± 7/16 [1]	-0.5 ±0.5 -1/2 ± 1/2	-0.68 ±0.5 -11/16 ± 1/2	CAMBER (Degrees)	-0.5 ±0.5 -1/2 ± 1/2	-0.35 ±0.5 -3/8 ± 1/2	-1.13 ±0.42 -1 1/8 ± 7/16
0.0 (0) | 0.5 (1/2)	0.0 (0) | 0.5 (1/2)	0.0 (0) | 0.5 (1/2)	CROSS | TOL.	0.0 (0) | 0.5 (1/2)	0.0 (0) | 0.5 (1/2)	0.0 (0) | 0.5 (1/2)
0.33° ±0.07° 11/32° ± 1/16°	0.17° ±0.17° 5/32° ± 5/32°	0.17° ±0.17° 5/32° ± 5/32°	FRONT TOE (Degrees)	0.17° ±0.17° 5/32° ± 5/32°	0.17° ±0.17° 5/32° ± 5/32°	0.4° ±0.13° 13/32° ± 1/8°
0.17" ±0.03" 5/32" ± 1/32"	0.08" ±0.08" 3/32" ± 3/32"	0.08" ±0.08" 3/32" ± 3/32"	FRONT TOE (Inches)	0.08" ±0.08" 3/32" ± 3/32"	0.08" ±0.08" 3/32" ± 3/32"	0.2" ±0.06" 3/16" ± 1/16"
			SAI			
20.0 / 18.7 ±0.5	20.0 / 18.7 ±0.33	20.0 / 18.7 ±0.33	TURNING ANGLE	20.0 / 18.5 ±0.33	20.0 / 18.7 ±0.33	
▲	▲	▲	ILLUSTRATION #			▲
			REAR			
-0.67 ±0.5 -11/16 ± 1/2	-1.33 ±0.5 -1 5/16 ± 1/2	-1.33 ±0.5 [1] -1 5/16 ± 1/2 [1]	CAMBER (Degrees)	-1.33 ±0.5 -1 5/16 ± 1/2	-1.33 ±0.5 -1 5/16 ± 1/2	-1.5 ±0.5 -1 1/2 ± 1/2
0.27° ±0.17° 9/32° ± 5/32°	0.17° ±0.17° 5/32° ± 5/32°	0.17° ±0.17° 5/32° ± 5/32°	REAR TOE (Degrees)	0.17° ±0.17° 5/32° ± 5/32°	0.17° ±0.17° 5/32° ± 5/32°	0.2° ±0.17° 3/16° ± 5/32°
0.13" ±0.08" 1/8" ± 3/32"	0.08" ±0.08" 3/32" ± 3/32"	0.08" ±0.08" 3/32" ± 3/32"	REAR TOE (Inches)	0.08" ±0.08" 3/32" ± 3/32"	0.08" ±0.08" 3/32" ± 3/32"	0.1" ±0.08" 3/32" ± 3/32"
0.17 (5/32)	0.33 (11/32)	0.33 (11/32)	THRUST ANGLE	0.33 (11/32)	0.33 (11/32)	0.17 (5/32)
83 06	127	127	ILLUSTRATION #	127	127	106 143
[1] Sport Susp.= -0.83° (-13/16°)		[1] w/18" tires = -1.75° (-1 3/4°)	NOTES			
46106	46107	46108		46109	46110	46111

Volkswagen

R32	Rabbit (A5) Base Susp.	Rabbit (A5) Sport Susp.	MODEL INFORMATION	Routan	Tiguan All models	Touareg, Touareg 2 w/Steel Springs Std. Susp. (1BA)
2004	2006-2009	2006-2009	YEARS	2009-2013	2009-2017	2004-2010
			INSTRUCTIONS			p. 311
89 ft.lbs.			LUGNUT TORQUE	100 ft.lbs.		118 ft.lbs.
			FRONT			
8.0 ±0.5 8 ± ½	7.57 ±0.5 7 9/16 ± ½	7.78 ±0.5 7 13/16 ± ½	CASTER (Degrees)	2.5 ±1.0 2½ ±1	7.57 ±0.5 7 9/16 ± ½	8.17 ±0.5 8 3/16 ± ½
0.0 (0) / 0.5 (½)	0.0 (0) / 0.5 (½)	0.0 (0) / 0.5 (½)	CROSS / TOL.	0.0 (0) / 0.75 (¾)	0.0 (0) / 0.5 (½)	0.0 (0) / 0.33 (5/16)
-0.58 ±0.5 - 9/16 ± ½	-0.5 ±0.5 - ½ ± ½	-0.68 ±0.5 - 11/16 ± ½	CAMBER (Degrees)	L= 0.35, R= 0.05 ±0.4 L= ⅜, R= 1/16 ± ⅜	-0.45 ±0.5 - 7/16 ± ½	0.0 ±0.17 0 ± 3/16
0.0 (0) / 0.5 (½)	0.0 (0) / 0.5 (½)	0.0 (0) / 0.5 (½)	CROSS / TOL.	0.3 (5/16) / 0.5 (½)	0.0 (0) / 0.5 (½)	0.0 (0) / 0.17 (3/16)
0.0° ±0.17° 0° ± 5/32°	0.17° ±0.17° 5/32° ± 5/32°	0.17° ±0.17° 5/32° ± 5/32°	FRONT TOE (Degrees)	0.26° ±0.2° ¼° ± 3/16°	0.17° ±0.17° 5/32° ± 5/32°	0.08° +0.17° -0.08° 3/32° +5/32° -3/32°
0.0" ±0.08" 0" ± 3/32"	0.08" ±0.08" 3/32" ± 3/32"	0.08" ±0.08" 3/32" ± 3/32"	FRONT TOE (Inches)	0.13" ±0.1" ⅛" ± 3/32"	0.08" ±0.08" 3/32" ± 3/32"	0.04" +0.08" -0.04" 1/32" +3/32" -1/32"
			SAI			10.8 (10¾)
20.0 / 18.7 ±0.33	20.0 / 18.4 ±0.33	20.0 / 18.3 ±0.33	TURNING ANGLE		20.0 / 18.4 ±0.33	20.0 / 18.7 ±0.5
			ILLUSTRATION #	161		52
			REAR			
-1.45 ±0.17 -1 7/16 ± 3/16	-1.33 ±0.5 -1 5/16 ± ½	-1.33 ±0.5 [1] -1 5/16 ± ½ [1]	CAMBER (Degrees)	-0.06 ±0.25 -1/16 ± ¼	-1.33 ±0.5 -1 5/16 ± ½	-1.0 ±0.33 -1 ± 5/16
0.42° ±0.17° 13/32° ± 5/32°	0.17° ±0.21° 5/32° ± 7/32°	0.17° ±0.21° 5/32° ± 7/32°	REAR TOE (Degrees)	0.1° ±0.25° 3/32° ± ¼°	0.17° ±0.17° 5/32° ± 5/32°	0.33° ±0.17° 11/32° ± 5/32°
0.21" ±0.08" 7/32" ± 3/32"	0.08" ±0.1" 3/32" ± 3/32"	0.08" ±0.1" 3/32" ± 3/32"	REAR TOE (Inches)	0.05" ±0.13" 1/16" ± ⅛"	0.08" ±0.08" 3/32" ± 3/32"	0.17" ±0.08" 5/32" ± 3/32"
0.33 (11/32)	0.33 (11/32)	0.33 (11/32)	THRUST ANGLE	0.3 (5/16)	0.33 (11/32)	
	127	127	ILLUSTRATION #		127	162
		[1] w/18" tires -1.75° (-1¾°)	NOTES			
46114	46117	46118		46120	46124	46125

Volkswagen

Touareg, Touareg 2 w/Steel Springs Sport Susp. (1BE)	Touareg, Touareg 2 w/Air Springs Std. Susp. (1BK)	Touareg, Touareg 2 w/Air Springs Sport Susp. (2MA)	MODEL INFORMATION	Touareg w/Steel Springs 1BA/1BE/1BW Susp.	Touareg w/Steel Springs 1BB Susp.	Touareg w/Air Springs 1BK/1BY Susp.
2004-2010	2004-2010	2004-2010	YEARS	2011-2017	2011-2017	2011-2017
p. 311	p. 311	p. 311	INSTRUCTIONS			
118 ft.lbs.	118 ft.lbs.	118 ft.lbs.	LUGNUT TORQUE	118 ft.lbs.	118 ft.lbs.	118 ft.lbs.
			FRONT			
8.5 ±0.5 8½ ± ½	8.58 ±0.5 8 9/16 ± ½	8.75 ±0.5 8¾ ± ½	CASTER (Degrees)	8.58 +0.17 -0.5 8 9/16 +3/16 -½	8.43 +0.17 -0.5 8 7/16 +3/16 -½	8.63 +0.17 -0.5 8⅝ +3/16 -½
0.0 (0) / 0.33 (5/16)	0.0 (0) / 0.33 (5/16)	0.0 (0) / 0.33 (5/16)	CROSS / TOL.	0.0 (0) / 0.5 (½)	0.0 (0) / 0.5 (½)	0.0 (0) / 0.5 (½)
0.0 ±0.17 0 ± 3/16	-0.17 ±0.17 - 3/16 ± 3/16	-0.33 ±0.17 - 5/16 ± 3/16	CAMBER (Degrees)	-0.25 ±0.33 - ¼ ± 5/16	-0.25 ±0.33 - ¼ ± 5/16	-0.25 ±0.33 - ¼ ± 5/16
0.0 (0) / 0.17 (3/16)	0.0 (0) / 0.17 (3/16)	0.0 (0) / 0.17 (3/16)	CROSS / TOL.	0.0 (0) / 0.33 (5/16)	0.0 (0) / 0.33 (5/16)	0.0 (0) / 0.33 (5/16)
0.08° +0.17° -0.08° 3/32° +5/32° -3/32°	0.17° +0.17° -0.08° 5/32° +5/32° -3/32°	0.17° +0.17° -0.08° 5/32° +5/32° -3/32°	FRONT TOE (Degrees)	0.17° ±0.08° 5/32° ± 3/32°	0.17° ±0.08° 5/32° ± 3/32°	0.17° ±0.08° 5/32° ± 3/32°
0.04" +0.08" -0.04" 1/32" +3/32" -1/32"	0.08" +0.08" -0.04" 3/32" +3/32" -1/32"	0.08" +0.08" -0.04" 3/32" +3/32" -1/32"	FRONT TOE (Inches)	0.08" ±0.04" 3/32" ± 1/32"	0.08" ±0.04" 3/32" ± 1/32"	0.08" ±0.04" 3/32" ± 1/32"
10.8 (10¾)	10.8 (10¾)	10.8 (10¾)	SAI			
20.0 / 18.7 ±0.5	20.0 / 18.5 ±0.5	20.0 / 18.5 ±0.5	TURNING ANGLE			
52	52	52	ILLUSTRATION #	52	52	52
			REAR			
-1.0 ±0.33 -1 ± 5/16	-1.0 ±0.33 -1 ± 5/16	-1.33 ±0.33 -1 5/16 ± 5/16	CAMBER (Degrees)	-1.33 ±0.33 -1 5/16 ± 5/16	-1.33 ±0.33 -1 5/16 ± 5/16	-1.33 ±0.33 -1 5/16 ± 5/16
0.33° ±0.17° 11/32° ± 5/32°	0.33° ±0.17° 11/32° ± 5/32°	0.33° ±0.17° 11/32° ± 5/32°	REAR TOE (Degrees)	0.33° ±0.17° 11/32° ± 5/32°	0.33° ±0.17° 11/32° ± 5/32°	0.33° ±0.17° 11/32° ± 5/32°
0.17" ±0.08" 5/32" ± 3/32"	0.17" ±0.08" 5/32" ± 3/32"	0.17" ±0.08" 5/32" ± 3/32"	REAR TOE (Inches)	0.17" ±0.08" 5/32" ± 3/32"	0.17" ±0.08" 5/32" ± 3/32"	0.17" ±0.08" 5/32" ± 3/32"
			THRUST ANGLE	0.08 (3/32)	0.08 (3/32)	0.08 (3/32)
162	162	162	ILLUSTRATION #	162	162	162
			NOTES			
46126	46127	46128		46129	46130	46131

Volkswagen

Touareg w/Air Springs 2MA Susp.	Vanagon Station Wagon	Vanagon Camper	MODEL INFORMATION	Vanagon Syncro Station Wagon	Vanagon Syncro Camper	Notes
2011-2017	1990-91	1990-91	YEARS	1990-91	1990-91	
	p. 310, 311	p. 310, 311	INSTRUCTIONS	p. 310, 311	p. 310, 311	
118 ft.lbs.	123 ft.lbs.	123 ft.lbs.	LUGNUT TORQUE	123 ft.lbs.	123 ft.lbs.	
			FRONT			
9.08 +0.17 -0.5 9 1/16 +3/16 -1/2	6.5 ± 0.25 6 1/2 ± 1/4	6.0 ± 0.25 6 ± 1/4	CASTER (Degrees)	4.67 ± 0.25 4 11/16 ± 1/4	3.58 ± 0.25 3 9/16 ± 1/4	
0.0 (0) | 0.5 (1/2)		0.0 (0) | 1.0 (1)	CROSS | TOL.	0.0 (0) | n/s	0.0 (0) | n/s	
-0.25 ± 0.33 -1/4 ± 5/16	0.25 ± 0.5 1/4 ± 1/2	0.0 ± 0.5 0 ± 1/2	CAMBER (Degrees)	0.33 ± 0.33 5/16 ± 5/16	-0.5 ± 0.33 -1/2 ± 5/16	
0.0 (0) | 0.33 (5/16)	0.0 (0) | n/s	0.0 (0) | n/s	CROSS | TOL.	0.0 (0) | 0.33 (5/16)	0.0 (0) | 0.33 (5/16)	
0.17° ± 0.08° 5/32° ± 3/32°	0.33° ± 0.25° 11/32° ± 1/4°	0.0° ± 0.25° 0° ± 1/4°	FRONT TOE (Degrees)	see table p. 311 ± 0.06° (1/16°)	see table p. 311 ± 0.06° (1/16°)	
0.08" ± 0.04" 3/32" ± 1/32"	0.17" ± 0.13" 5/32" ± 1/8"	0.0" ± 0.13" 0" ± 1/8"	FRONT TOE (Inches)	see table p. 311 ± 0.03" (1/32")	see table p. 311 ± 0.03" (1/32")	
			SAI			
	20.0 / 18.8 ± 0.33	20.0 / 18.2 ± 0.33	TURNING ANGLE			
52	03 20	03 20	ILLUSTRATION #	03 20	03 20	
			REAR			
-1.33 ± 0.33 -1 5/16 ± 5/16	-1.5 ± 0.5 -1 1/2 ± 1/2	-2.17 ± 0.5 -2 3/16 ± 1/2	CAMBER (Degrees)	-0.25 +0.5 -0.17 -1/4 +1/2 -3/16	-1.17 +0.5 -0.17 -1 3/16 +1/2 -3/16	
0.33° ± 0.17° 11/32° ± 5/32°	0.0° ± 0.33° 0° ± 11/32°	0.0° ± 0.33° 0° ± 11/32°	REAR TOE (Degrees)	0.17° ± 0.33° 5/32° ± 11/32°	0.17° ± 0.33° 5/32° ± 11/32°	
0.17" ± 0.08" 5/32" ± 3/32"	0.0" ± 0.17" 0" ± 5/32"	0.0" ± 0.17" 0" ± 5/32"	REAR TOE (Inches)	0.08" ± 0.17" 3/32" ± 5/32"	0.08" ± 0.17" 3/32" ± 5/32"	
0.08 (3/32)			THRUST ANGLE			
162	21	21	ILLUSTRATION #	21	21	
			NOTES			
46132	46142	46143		46144	46145	46153

Volvo

240, 260 DL, GL, GLT, GLE	760 Wagon 1990 740	760 Sedan 780 Coupe	MODEL INFORMATION	850 FWD	850 AWD	940
1990-93	1990-92	1990-91	YEARS	1993-97	1997	1991-95
		p. 311	INSTRUCTIONS			
85 ft.lbs.	63 ft.lbs.	63 ft.lbs.	LUGNUT TORQUE	81 ft.lbs.	81 ft.lbs.	63 ft.lbs.
			FRONT			
3.5 ± 0.5 [1] 3 1/2 ± 1/2 [1]	5.0 ± 0.5 5 ± 1/2	5.0 ± 0.5 5 ± 1/2	CASTER (Degrees)	3.33 ± 1.0 3 5/16 ± 1	3.33 ± 1.0 3 5/16 ± 1	5.0 ± 0.5 5 ± 1/2
0.0 (0) | 0.5 (1/2)	0.0 (0) | n/s	0.0 (0) | n/s	CROSS | TOL.	0.0 (0) | 1.0 (1)	0.0 (0) | 1.0 (1)	0.0 (0) | n/s
0.63 ± 0.38 5/8 ± 3/8	0.3 ± 0.5 5/16 ± 1/2	0.3 ± 0.5 5/16 ± 1/2	CAMBER (Degrees)	0.0 ± 1.0 0 ± 1	0.0 ± 1.0 0 ± 1	0.1 ± 0.7 1/8 ± 11/16
0.0 (0) | 0.5 (1/2)	0.0 (0) | 0.7 (11/16)	0.0 (0) | 0.7 (11/16)	CROSS | TOL.	0.0 (0) | 1.0 (1)	0.0 (0) | 1.0 (1)	0.0 (0) | 0.7 (11/16)
0.27° ± 0.13° [2] 9/32° ± 1/8° [2]	0.3° ± 0.13° 5/16° ± 1/8°	0.3° ± 0.13° 5/16° ± 1/8°	FRONT TOE (Degrees)	0.33° ± 0.1° 11/32° ± 3/32°	0.33° ± 0.1° 11/32° ± 3/32°	0.3° ± 0.13° 5/16° ± 1/8°
0.13" ± 0.07" [2] 1/8" ± 1/16" [2]	0.15" ± 0.07" 5/32" ± 1/16"	0.15" ± 0.07" 5/32" ± 1/16"	FRONT TOE (Inches)	0.17" ± 0.05" 5/32" ± 1/16"	0.17" ± 0.05" 5/32" ± 1/16"	0.15" ± 0.07" 5/32" ± 1/16"
12.0 (12)			SAI			
20.8 / 20.0			TURNING ANGLE	39.9 / 34.8 ± 1.6	39.9 / 34.8 ± 1.6	
72			ILLUSTRATION #	102	102	
			REAR			
		see table p.311 ± 0.25 (1/4)	CAMBER (Degrees)	-1.0 ± 0.5 -1 ± 1/2	-0.25 ± 0.5 -1/4 ± 1/2	
		0.07° ± 0.1° 1/16° ± 3/32°	REAR TOE (Degrees)	0.0° ± 0.13° 0° ± 1/8°	0.1° ± 0.2° 3/32° ± 3/16°	
		0.04" ± 0.05" 1/32" ± 1/16"	REAR TOE (Inches)	0.0" ± 0.07" 0" ± 1/16"	0.05" ± 0.1" 1/16" ± 3/32"	
			THRUST ANGLE	0.25 (1/4)	0.25 (1/4)	
		82	ILLUSTRATION #	81	06	
[1] w/MS 2.5° (2 1/2°) [2] w/MS 0.4° (0.2")			NOTES			
45007	45009	45011		45012	45013	45014

Volvo

960	960	C30	MODEL	C30	C70	C70
		STD. Susp.	INFORMATION	Sport Susp. (T5-R)		
1992-94	1995-97	2007-2013	YEARS	2007-2013	1998-2004	2006-2013
p. 311	p. 312		INSTRUCTIONS			
63 ft.lbs.	63 ft.lbs.	103 ft.lbs.	LUGNUT TORQUE	103 ft.lbs.	81 ft.lbs.	103 ft.lbs.
			FRONT			
5.0 ± 0.5 [1] 5 ± ½ [1]	6.0 ± 0.5 6 ± ½	3.6 ± 1.5 3⅝ ± 1½	CASTER (Degrees)	3.6 ± 1.5 3⅝ ± 1½	3.35 ± 1.0 3⅜ ± 1	3.8 ± 1.5 3 13/16 ± 1½
0.0 (0) | n/s	0.0 (0) | n/s	0.0 (0) | 1.0 (1)	CROSS | TOL.	0.0 (0) | 1.0 (1)	0.0 (0) | 1.0 (1)	0.0 (0) | 1.0 (1)
0.1 ± 0.7 ⅛ ± 11/16	see table p.312 ± 0.5 (½)	-0.6 ± 0.7 -⅝ ± 11/16	CAMBER (Degrees)	-0.6 ± 0.7 -⅝ ± 11/16	-0.5 ± 1.0 -½ ± 1	-0.75 ± 0.7 -¾ ± 11/16
0.0 (0) | 0.7 (11/16)	0.0 (0) | 0.7 (11/16)	0.0 (0) | n/s	CROSS | TOL.	0.0 (0) | n/s	0.0 (0) | 1.0 (1)	0.0 (0) | n/s
0.3° ± 0.13° 5/16° ± ⅛°	0.3° ± 0.13° 5/16° ± ⅛°	0.2° ± 0.1° 3/16° ± 3/32°	FRONT TOE (Degrees)	0.2° ± 0.1° 3/16° ± 3/32°	0.33° ± 0.1° 11/32° ± 3/32°	0.2° ± 0.1° 3/16° ± 3/32°
0.15" ± 0.07" 5/32" ± 1/16"	0.15" ± 0.07" 5/32" ± 1/16"	0.1" ± 0.05" 3/32" ± 1/16"	FRONT TOE (Inches)	0.1" ± 0.05" 3/32" ± 1/16"	0.17" ± 0.05" 5/32" ± 1/16"	0.1" ± 0.05" 3/32" ± 1/16"
			SAI			
			TURNING ANGLE			
			ILLUSTRATION #		102	
			REAR			
see table p.311 ± 0.25 (¼)	see table p.312 ± 0.4 (⅜)	-1.52 ± 1.0 -1½ ± 1	CAMBER (Degrees)	-1.52 ± 1.0 -1½ ± 1	-1.0 ± 0.5 -1 ± ½	-1.72 ± 1.0 -1¾ ± 1
0.07° ± 0.1° 1/16° ± 3/32°	0.1° ± 0.13° 3/32° ± ⅛°	0.3° ± 0.1° 5/16° ± 3/32°	REAR TOE (Degrees)	0.3° ± 0.1° 5/16° ± 3/32°	0.07° ± 0.2° 1/16° ± 3/16°	0.3° ± 0.1° 5/16° ± 3/32°
0.04" ± 0.05" 1/32" ± 1/16"	0.05" ± 0.07" 1/16" ± 1/16"	0.15" ± 0.05" 5/32" ± 1/16"	REAR TOE (Inches)	0.15" ± 0.05" 5/32" ± 1/16"	0.04" ± 0.1" 1/32" ± 3/32"	0.15" ± 0.05" 5/32" ± 1/16"
		0.25 (¼)	THRUST ANGLE	0.25 (¼)	0.25 (¼)	0.25 (¼)
82	06	06	ILLUSTRATION #	06	81	06
[1] w/16" tires 4.5° (4½°)			NOTES			
45015	45016	45017		45018	45019	45020

Volvo

S40/V40	S40/V40	S40/V40	MODEL	S40	S40	S60
	exc/Sport Susp.	Sport Susp.	INFORMATION	AWD	FWD	FWD
2000	2001-2004	2001-2004	YEARS	2004.5-2011	2004.5-2011	2001-2009
			INSTRUCTIONS			
81 ft.lbs.	81 ft.lbs.	81 ft.lbs.	LUGNUT TORQUE	103 ft.lbs.	103 ft.lbs.	103 ft.lbs.
			FRONT			
3.2 ± 1.0 3 3/16 ± 1	4.0 ± 1.0 4 ± 1	4.15 ± 1.0 4⅛ ± 1	CASTER (Degrees)	3.7 ± 1.5 3 11/16 ± 1½	3.6 ± 1.5 3⅝ ± 1½	4.0 ± 1.0 4 ± 1
0.0 (0) | 0.7 (11/16)	0.0 (0) | 0.7 (11/16)	0.0 (0) | 0.7 (11/16)	CROSS | TOL.	0.0 (0) | 1.0 (1)	0.0 (0) | 1.0 (1)	0.0 (0) | 0.25 (¼)
0.0 ± 1.0 0 ± 1	-0.17 ± 0.5 -3/16 ± ½	-0.33 ± 0.5 -5/16 ± ½	CAMBER (Degrees)	-0.68 ± 0.7 -11/16 ± 11/16	-0.6 ± 0.7 -⅝ ± 11/16	-0.3 ± 0.9 -5/16 ± ⅞
0.0 (0) | 0.7 (11/16)	0.0 (0) | 0.7 (11/16)	0.0 (0) | 0.7 (11/16)	CROSS | TOL.	0.0 (0) | n/s	0.0 (0) | n/s	0.0 (0) | 1.0 (1)
0.17° ± 0.1° 5/32° ± 3/32°	0.17° ± 0.1° 5/32° ± 3/32°	0.3° ± 0.1° 5/16° ± 3/32°	FRONT TOE (Degrees)	0.2° ± 0.1° 3/16° ± 3/32°	0.2° ± 0.1° 3/16° ± 3/32°	0.1° ± 0.1° 3/32° ± 3/32°
0.08" ± 0.05" 3/32" ± 1/16"	0.08" ± 0.05" 3/32" ± 1/16"	0.15" ± 0.05" 5/32" ± 1/16"	FRONT TOE (Inches)	0.1" ± 0.05" 3/32" ± 1/16"	0.1" ± 0.05" 3/32" ± 1/16"	0.05" ± 0.05" 1/16" ± 1/16"
12.7 ± 1.0 (12 11/16 ± 1)	13.7 ± 1.0 (13 11/16 ± 1)	14.1 ± 1.0 (14 1/16 ± 1)	SAI			
			TURNING ANGLE			
102	102	102	ILLUSTRATION #			102
			REAR			
-0.67 ± 0.5 -11/16 ± ½	-0.67 ± 0.5 -11/16 ± ½	-0.9 ± 0.5 -⅞ ± ½	CAMBER (Degrees)	-1.52 ± 1.0 -1½ ± 1	-1.52 ± 1.0 -1½ ± 1	0.0 ± 1.0 0 ± 1
0.3° ± 0.1° 5/16° ± 3/32°	0.3° ± 0.1° 5/16° ± 3/32°	0.3° ± 0.1° 5/16° ± 3/32°	REAR TOE (Degrees)	0.3° ± 0.1° 5/16° ± 3/32°	0.3° ± 0.1° 5/16° ± 3/32°	0.2° ± 0.2° 3/16° ± 3/16°
0.15" ± 0.05" 5/32" ± 1/16"	0.15" ± 0.05" 5/32" ± 1/16"	0.15" ± 0.05" 5/32" ± 1/16"	REAR TOE (Inches)	0.15" ± 0.05" 5/32" ± 1/16"	0.15" ± 0.05" 5/32" ± 1/16"	0.1" ± 0.1" 3/32" ± 3/32"
			THRUST ANGLE	0.25 (¼)	0.25 (¼)	0.25 (¼)
40 05	40 05	40 05	ILLUSTRATION #	06	06	06
			NOTES			
45021	45022	45023		45024	45025	45026

Volvo

S60	S60	S60	MODEL	S60	S60L	S60 Cross Country
AWD	AWD	FWD & AWD	INFORMATION	FWD & AWD	FWD & AWD	
except S60R	S60R	exc/R-design		R-design		
2002-2009	2004-2007	2011-2017	YEARS	2011-2017	2016-2017	2016-2017
			INSTRUCTIONS			
103 ft.lbs.	103 ft.lbs.	103 ft.lbs.	LUGNUT TORQUE	103 ft.lbs.	103 ft.lbs.	103 ft.lbs.
			FRONT			
4.0 ± 1.0 4 ± 1	4.0 ± 1.0 4 ± 1	3.2 ± 0.5 3 3/16 ± ½	CASTER (Degrees)	3.3 ± 0.5 3 5/16 ± ½	3.2 ± 0.5 3 3/16 ± ½	3.1 ± 0.5 3⅛ ± ½
0.0 (0) | 0.25 (¼)	0.0 (0) | 0.25 (¼)	0.0 (0) | 0.7 (11/16)	CROSS | TOL.	0.0 (0) | 0.7 (11/16)	0.0 (0) | 0.7 (11/16)	0.0 (0) | 0.7 (11/16)
-0.3 ± 0.9 - 5/16 ± ⅞	-0.5 ± 0.9 - ½ ± ⅞	-0.79 ± 0.5 - 13/16 ± ½	CAMBER (Degrees)	-1.0 ± 0.5 -1 ± ½	-0.79 ± 0.7 - 13/16 ± 11/16	-0.82 ± 0.5 - 13/16 ± ½
0.0 (0) | 1.0 (1)	0.0 (0) | 1.0 (1)	0.0 (0) | 0.5 (½)	CROSS | TOL.	0.0 (0) | 0.5 (½)	0.0 (0) | 0.7 (11/16)	0.0 (0) | 0.7 (11/16)
0.1° ± 0.1° 3/32° ± 3/32°	0.1° ± 0.1° 3/32° ± 3/32°	0.24° ± 0.1° ¼° ± 3/32°	FRONT TOE (Degrees)	0.24° ± 0.1° ¼° ± 3/32°	0.24° ± 0.1° ¼° ± 3/32°	0.15° ± 0.1° 5/32° ± 3/32°
0.05" ± 0.05" 1/16" ± 1/16"	0.05" ± 0.05" 1/16" ± 1/16"	0.12" ± 0.05" ⅛" ± 1/16"	FRONT TOE (Inches)	0.12" ± 0.05" ⅛" ± 1/16"	0.12" ± 0.05" ⅛" ± 1/16"	0.08" ± 0.05" 1/16" ± 1/16"
			SAI			
			TURNING ANGLE			
102	102		ILLUSTRATION #			
			REAR			
-0.3 ± 1.0 - 5/16 ± 1	-0.13 ± 1.0 - ⅛ ± 1	-1,33 ± 0.5 -1 5/16 ± ½	CAMBER (Degrees)	-1.58 ± 0.5 -1 9/16 ± ½	-1.39 ± 0.7 -1⅜ ± 11/16	-1.14 ± 1.0 -1⅛ ± 1
0.2° ± 0.2° 3/16° ± 3/16°	0.16° ± 0.2° 5/32° ± 3/16°	0.3° ± 0.1° 5/16° ± 3/32°	REAR TOE (Degrees)	0.3° ± 0.1° 5/16° ± 3/32°	0.3° ± 0.1° 5/16° ± 3/32°	0.2° ± 0.1° 3/16° ± 3/32°
0.1" ± 0.1" 3/32" ± 3/32"	0.08" ± 0.1" 3/32" ± 3/32"	0.15" ± 0.05" 5/32" ± 1/16"	REAR TOE (Inches)	0.15" ± 0.05" 5/32" ± 1/16"	0.15" ± 0.05" 5/32" ± 1/16"	0.1" ± 0.05" 3/32" ± 1/16"
0.25 (¼)	0.25 (¼)	0.1 (3/32)	THRUST ANGLE	0.1 (3/32)	0.1 (3/32)	0.1 (3/32)
06	06	06	ILLUSTRATION #	06	06	06
			NOTES			
45027	45028	45029		45030	45031	45032

Volvo

S70/V70	S70/V70 AWD	S80	MODEL	S80	S80	S90
		FWD	INFORMATION	AWD	FWD & AWD	with AIR suspension
1998-2000	1998-2000	1999-2006	YEARS	2004-2006	2007-2016	2017
			INSTRUCTIONS			
81 ft.lbs.	81 ft.lbs.	103 ft.lbs.	LUGNUT TORQUE	103 ft.lbs.	103 ft.lbs.	103 ft.lbs.
			FRONT			
3.33 ± 1.0 3 5/16 ± 1	3.33 ± 1.0 3 5/16 ± 1	4.0 ± 1.0 4 ± 1	CASTER (Degrees)	4.0 ± 1.0 4 ± 1	3.1 ± 0.5 3⅛ ± ½	6.3 ± 0.7 6 5/16 ± 11/16
0.0 (0) | 1.0 (1)	0.0 (0) | 1.0 (1)	0.0 (0) | 1.0 (1)	CROSS | TOL.	0.0 (0) | 1.0 (1)	0.0 (0) | 1.0 (1)	0.0 (0) | 0.6 (⅝)
0.0 ± 1.0 0 ± 1	0.0 ± 1.0 0 ± 1	-0.3 ± 0.9 - 5/16 ± ⅞	CAMBER (Degrees)	-0.3 ± 0.9 - 5/16 ± ⅞	-0.69 ± 0.7 - 11/16 ± 11/16	-0.64 ± 0.7 - ⅝ ± 11/16
0.0 (0) | 1.0 (1)	0.0 (0) | 1.0 (1)	0.0 (0) | 1.0 (1)	CROSS | TOL.	0.0 (0) | 1.0 (1)	0.0 (0) | 0.7 (11/16)	0.0 (0) | 0.7 (11/16)
0.33° ± 0.1° 11/32° ± 3/32°	0.33° ± 0.1° 11/32° ± 3/32°	0.1° ± 0.1° 3/32° ± 3/32°	FRONT TOE (Degrees)	0.1° ± 0.1° 3/32° ± 3/32°	0.2° ± 0.1° 3/16° ± 3/32°	0.14° ± 0.2° ⅛° ± 3/16°
0.17" ± 0.05" 5/32" ± 1/16"	0.17" ± 0.05" 5/32" ± 1/16"	0.05" ± 0.05" 1/16" ± 1/16"	FRONT TOE (Inches)	0.05" ± 0.05" 1/16" ± 1/16"	0.1" ± 0.05" 3/32" ± 1/16"	0.07" ± 0.1" 1/16" ± 3/32"
			SAI			
39.9 / 34.8 ±1.6	39.9 / 34.8 ±1.6		TURNING ANGLE			
102	102	102	ILLUSTRATION #	102		
			REAR			
-1.0 ± 0.5 -1 ± ½	-0.25 ± 0.5 - ¼ ± ½	0.0 ± 1.0 0 ± 1	CAMBER (Degrees)	-0.3 ± 1.0 - 5/16 ± 1	-1.46 ± 1.0 -1 7/16 ± 1	-0.93 ± 0.7 - 15/16 ± 11/16
0.07° ± 0.18° 1/16° ± 3/16°	0.1° ± 0.2° 3/32° ± 3/16°	0.2° ± 0.2° 3/16° ± 3/16°	REAR TOE (Degrees)	0.2° ± 0.2° 3/16° ± 3/16°	0.16° ± 0.1° 5/32° ± 3/32°	0.16° ± 0.2° 5/32° ± 3/16°
0.04" ± 0.09" 1/32" ± 3/32"	0.05" ± 0.1" 1/16" ± 3/32"	0.1" ± 0.1" 3/32" ± 3/32"	REAR TOE (Inches)	0.1" ± 0.1" 3/32" ± 3/32"	0.08" ± 0.05" 3/32" ± 1/16"	0.08" ± 0.1" 3/32" ± 3/32"
0.25 (¼)	0.25 (¼)	0.25 (¼)	THRUST ANGLE	0.25 (¼)	0.1 (¼)	0.1 (3/32)
81	06	06	ILLUSTRATION #	06	06	174
			NOTES			
45033	45034	45035		45036	45037	45038

Volvo

S90	S90/V90	V50	MODEL	V50	V60	V60
without AIR		AWD	INFORMATION	FWD	FWD & AWD	FWD & AWD
suspension					exc/R-design	R-design
2017	1997-1998	2005-2011	YEARS	2005-2011	2015-2017	2015-2017
	p. 312		INSTRUCTIONS			
103 ft.lbs.	63 ft.lbs.	103 ft.lbs.	LUGNUT TORQUE	103 ft.lbs.	103 ft.lbs.	103 ft.lbs.
			FRONT			
6.08 ±0.7 6 1/16 ± 11/16	6.0 ±0.5 6 ± 1/2	3.7 ±1.5 3 11/16 ± 1 1/2	CASTER (Degrees)	3.6 ±1.5 3 5/8 ± 1 1/2	3.2 ±0.5 3 3/16 ± 1/2	3.3 ±0.5 3 5/16 ± 1/2
0.0 (0) | 0.6 (5/8)	0.0 (0) | n/s	0.0 (0) | 1.0 (1)	CROSS | TOL.	0.0 (0) | 1.0 (1)	0.0 (0) | 0.7 (11/16)	0.0 (0) | 0.7 (11/16)
-0.69 ±0.7 -11/16 ± 11/16	see table p.312 ±0.5 (1/2)	-0.68 ±0.7 -11/16 ± 11/16	CAMBER (Degrees)	-0.6 ±0.7 -5/8 ± 11/16	-0.79 ±0.7 -13/16 ± 11/16	-1.0 ±0.7 -1 ± 11/16
0.0 (0) | 0.7 (11/16)	0.0 (0) | 0.7 (11/16)	0.0 (0) | n/s	CROSS | TOL.	0.0 (0) | n/s	0.0 (0) | 0.7 (11/16)	0.0 (0) | 0.7 (11/16)
0.14° ±0.2° 1/8° ± 3/16°	0.3° ±0.13° 5/16° ± 1/8°	0.2° ±0.1° 3/16° ± 3/32°	FRONT TOE (Degrees)	0.2° ±0.1° 3/16° ± 3/32°	0.24° ±0.1° 1/4° ± 3/32°	0.24° ±0.1° 1/4° ± 3/32°
0.07" ±0.1" 1/16" ± 3/32"	0.15" ±0.07" 5/32" ± 1/16"	0.1" ±0.05" 3/32" ± 1/16"	FRONT TOE (Inches)	0.1" ±0.05" 3/32" ± 1/16"	0.12" ±0.05" 1/8" ± 1/16"	0.12" ±0.05" 1/8" ± 1/16"
			SAI			
			TURNING ANGLE			
			ILLUSTRATION #			
			REAR			
-0.69 ±0.7 -11/16 ± 11/16	see table p.312 ±0.4 (3/8)	-1.52 ±1.0 -1 1/2 ±1	CAMBER (Degrees)	-1.52 ±1.0 -1 1/2 ±1	-1.33 ±1.0 -1 5/16 +1/2 -1	-1.58 ±1.0 -1 9/16 +1/2 -1
0.16° ±0.2° 5/32° ± 3/16°	0.1° ±0.13° 3/32° ± 1/8°	0.3° ±0.1° 5/16° ± 3/32°	REAR TOE (Degrees)	0.3° ±0.1° 5/16° ± 3/32°	0.3° ±0.1° 5/16° ± 3/32°	0.3° ±0.1° 5/16° ± 3/32°
0.08" ±0.1" 3/32" ± 3/32"	0.05" ±0.07" 1/16" ± 1/16"	0.15" ±0.05" 5/32" ± 1/16"	REAR TOE (Inches)	0.15" ±0.05" 5/32" ± 1/16"	0.15" ±0.05" 5/32" ± 1/16"	0.15" ±0.05" 5/32" ± 1/16"
0.1 (3/32)		0.25 (1/4)	THRUST ANGLE	0.25 (1/4)	0.1 (3/32)	0.1 (3/32)
174	06	06	ILLUSTRATION #	06	06	06
			NOTES			
45039	45040	45041		45042	45043	45044

Volvo

V60 Cross Country	V70	V70	MODEL	V70	V70	V90 Cross Country
	FWD	AWD	INFORMATION	AWD		with AIR suspension
		except V70R		V70R		
2016-2017	2001-2007	2001-2007	YEARS	2004-2007	2008-2010	2017
			INSTRUCTIONS			
103 ft.lbs.	103 ft.lbs.	103 ft.lbs.	LUGNUT TORQUE	103 ft.lbs.	103 ft.lbs.	103 ft.lbs.
			FRONT			
3.1 ±0.5 3 1/8 ± 1/2	4.0 ±1.0 4 ±1	4.0 ±1.0 4 ±1	CASTER (Degrees)	4.0 ±1.0 4 ±1	3.1 ±0.5 3 1/8 ± 1/2	6.3 ±0.7 6 5/16 ± 11/16
0.0 (0) | 0.7 (11/16)	0.0 (0) | 0.25 (1/4)	0.0 (0) | 0.25 (1/4)	CROSS | TOL.	0.0 (0) | 0.25 (1/4)	0.0 (0) | 1.0 (1)	0.0 (0) | 0.6 (5/8)
-0.82 ±0.5 -13/16 ± 1/2	-0.3 ±0.9 -5/16 ± 7/8	-0.3 ±0.9 -5/16 ± 7/8	CAMBER (Degrees)	-0.5 ±0.9 -1/2 ± 7/8	-0.65 ±0.5 -5/8 ± 1/2	-0.64 ±0.7 -5/8 ± 11/16
0.0 (0) | 0.7 (11/16)	0.0 (0) | 1.0 (1)	0.0 (0) | 1.0 (1)	CROSS | TOL.	0.0 (0) | 1.0 (1)	0.0 (0) | n/s	0.0 (0) | 0.7 (11/16)
0.15° ±0.1° 5/32° ± 3/32°	0.1° ±0.1° 3/32° ± 3/32°	0.1° ±0.1° 3/32° ± 3/32°	FRONT TOE (Degrees)	0.1° ±0.1° 3/32° ± 3/32°	0.2° ±0.1° 3/16° ± 3/32°	0.14° ±0.2° 1/8° ± 3/16°
0.08" ±0.05" 1/16" ± 1/16"	0.05" ±0.05" 1/16" ± 1/16"	0.05" ±0.05" 1/16" ± 1/16"	FRONT TOE (Inches)	0.05" ±0.05" 1/16" ± 1/16"	0.1" ±0.05" 3/32" ± 1/16"	0.07" ±0.1" 1/16" ± 3/32"
			SAI			
			TURNING ANGLE			
	102	102	ILLUSTRATION #	102		
			REAR			
-1.14 ±1.0 -1 1/8 ±1	0.0 ±1.0 0 ±1	-0.3 ±1.0 -5/16 ±1	CAMBER (Degrees)	0.13 ±1.0 1/8 ±1	-1.13 ±0.5 -1 1/8 ± 1/2	-0.93 ±0.7 -15/16 ± 11/16
0.2° ±0.1° 3/16° ± 3/32°	0.2° ±0.2° 3/16° ± 3/16°	0.2° ±0.2° 3/16° ± 3/16°	REAR TOE (Degrees)	0.16° ±0.2° 5/32° ± 3/16°	0.2° ±0.1° 3/16° ± 3/32°	0.16° ±0.2° 5/32° ± 3/16°
0.1" ±0.05" 3/32" ± 1/16"	0.1" ±0.1" 3/32" ± 3/32"	0.1" ±0.1" 3/32" ± 3/32"	REAR TOE (Inches)	0.08" ±0.1" 3/32" ± 3/32"	0.1" ±0.05" 3/32" ± 1/16"	0.08" ±0.1" 3/32" ± 3/32"
0.1 (3/32)	0.25 (1/4)	0.25 (1/4)	THRUST ANGLE	0.25 (1/4)	0.25 (1/4)	0.1 (3/32)
06	06	06	ILLUSTRATION #	06	06	174
			NOTES			
45045	45046	45047		45048	45049	45050

Volvo

Model Information	V90 Cross Country w/o AIR suspension	XC60 FWD & AWD	XC70	XC70 FWD & AWD exc/Nivomat	XC70 FWD & AWD w/Nivomat	XC90 FWD & AWD
Years	2017	2010-2017	2003-2007	2008-2016	2008-2016	2003-2008
Instructions						
LugNut Torque	103 ft.lbs.	103 ft.lbs.	103 ft.lbs.	103 ft.lbs.	103 ft.lbs.	103 ft.lbs.
FRONT						
Caster (Degrees)	6.08 ± 0.7 6 1/16 ± 11/16	2.9 ± 0.5 2 7/8 ± 1/2	5.0 ± 1.0 5 ± 1	3.1 ± 0.5 3 1/8 ± 1/2	3.1 ± 0.5 3 1/8 ± 1/2	5.2 ± 0.5 5 3/16 ± 1/2
Cross	0.0 (0)	0.0 (0)	0.0 (0)	0.0 (0)	0.0 (0)	0.0 (0)
Tol.	0.6 (5/8)	0.7 (11/16)	0.25 (1/4)	1.0 (1)	1.0 (1)	0.5 (1/2)
Camber (Degrees)	-0.69 ± 0.7 -11/16 ± 11/16	-0.69 ± 0.5 -11/16 ± 1/2	-0.3 ± 0.9 -5/16 ± 7/8	-0.64 ± 0.5 -5/8 ± 1/2	-0.64 ± 0.5 -5/8 ± 1/2	-0.25 ± 0.5 -1/4 ± 1/2
Cross	0.0 (0)	0.0 (0)	0.0 (0)	0.0 (0)	0.0 (0)	0.0 (0)
Tol.	0.7 (11/16)	0.7 (11/16)	1.0 (1)	n/s	n/s	0.5 (1/2)
Front Toe (Degrees)	0.14° ± 0.2° 1/8° ± 3/16°	0.2° ± 0.1° 3/16° ± 3/32°	0.2° ± 0.1° 3/16° ± 3/32°	0.2° ± 0.1° 3/16° ± 3/32°	0.2° ± 0.1° 3/16° ± 3/32°	-0.16° ± 0.1° -5/32° ± 3/32°
Front Toe (Inches)	0.07" ± 0.1" 1/16" ± 3/32"	0.1" ± 0.05" 3/32" ± 1/16"	0.1" ± 0.05" 3/32" ± 1/16"	0.1" ± 0.05" 3/32" ± 1/16"	0.1" ± 0.05" 3/32" ± 1/16"	-0.08" ± 0.05" -3/32" ± 1/16"
SAI						
Turning Angle						
Illustration #			102			102
REAR						
Camber (Degrees)	-0.69 ± 0.7 -11/16 ± 11/16	-0.83 ± 0.7 -13/16 ± 11/16	0.3 ± 1.0 5/16 ± 1	-0.85 ± 0.7 -7/8 ± 11/16	-0.77 ± 0.7 -3/4 ± 11/16	-0.5 ± 0.5 -1/2 ± 1/2
Rear Toe (Degrees)	0.16° ± 0.2° 5/32° ± 3/16°	0.2° ± 0.1° 3/16° ± 3/32°	0.2° ± 0.2° 3/16° ± 3/16°	0.2° ± 0.1° 3/16° ± 3/32°	0.2° ± 0.1° 3/16° ± 3/32°	0.3° ± 0.1° 5/16° ± 3/32°
Rear Toe (Inches)	0.08" ± 0.1" 3/32" ± 3/32"	0.1" ± 0.05" 3/32" ± 1/16"	0.1" ± 0.1" 3/32" ± 3/32"	0.1" ± 0.05" 3/32" ± 1/16"	0.1" ± 0.05" 3/32" ± 1/16"	0.15" ± 0.05" 5/32" ± 1/16"
Thrust Angle	0.1 (3/32)	0.1 (3/32)	0.25 (1/4)	0.1 (3/32)	0.1 (3/32)	0.1 (3/32)
Illustration #	174	06	06	06	06	06
Notes						
	45051	45052	45053	45054	45055	45056

Volvo

Model Information	XC90 FWD & AWD	XC90 w/ AIR susp.	XC90 w/o AIR susp.	XC90 PHEV	Notes	Notes
Years	2009-2014	2016-2017	2016-2017	2016-2017		
Instructions						
LugNut Torque	103 ft.lbs.	103 ft.lbs.	103 ft.lbs.	103 ft.lbs.		
FRONT						
Caster (Degrees)	5.2 ± 0.5 5 3/16 ± 1/2	5.85 ± 0.7 5 7/8 ± 11/16	5.7 ± 0.7 5 11/16 ± 11/16	5.7 ± 0.7 5 11/16 ± 11/16		
Cross	0.0 (0)	0.0 (0)	0.0 (0)	0.0 (0)		
Tol.	0.5 (1/2)	0.6 (5/8)	0.6 (5/8)	0.6 (5/8)		
Camber (Degrees)	-0.25 ± 0.5 -1/4 ± 1/2	-0.75 ± 0.7 -3/4 ± 11/16	-0.59 ± 0.7 -9/16 ± 11/16	-0.63 ± 0.7 -5/8 ± 11/16		
Cross	0.0 (0)	0.0 (0)	0.0 (0)	0.0 (0)		
Tol.	0.5 (1/2)	0.7 (11/16)	0.7 (11/16)	0.7 (11/16)		
Front Toe (Degrees)	0.16° ± 0.1° 5/32° ± 3/32°	0.22° ± 0.2° 7/32° ± 3/16°	0.2° ± 0.2° 3/16° ± 3/16°	0.2° ± 0.2° 3/16° ± 3/16°		
Front Toe (Inches)	0.08" ± 0.05" 3/32" ± 1/16"	0.11" ± 0.1" 1/8" ± 3/32"	0.1" ± 0.1" 3/32" ± 3/32"	0.1" ± 0.1" 3/32" ± 3/32"		
SAI						
Turning Angle						
Illustration #	102					
REAR						
Camber (Degrees)	-0.5 ± 0.5 -1/2 ± 1/2	-1.0 ± 0.7 -1 ± 11/16	-0.75 ± 0.7 -3/4 ± 11/16	-0.8 ± 0.7 -13/16 ± 11/16		
Rear Toe (Degrees)	0.3° ± 0.1° 5/16° ± 3/32°	0.22° ± 0.2° 7/32° ± 3/16°	0.2° ± 0.2° 3/16° ± 3/16°	0.2° ± 0.2° 3/16° ± 3/16°		
Rear Toe (Inches)	0.15" ± 0.05" 5/32" ± 1/16"	0.11" ± 0.1" 1/8" ± 3/32"	0.1" ± 0.1" 3/32" ± 3/32"	0.1" ± 0.1" 3/32" ± 3/32"		
Thrust Angle	0.1 (3/32)	0.1 (3/32)	0.1 (3/32)	0.1 (3/32)		
Illustration #	06	174	174	174		
Notes						
	45057	45058	45059	45060	45067	45068

Alfa Romeo

All Models - Loading

All models except the 164 require loading approximating 165 lbs. in each passenger seat. Four-door sedans assume four passengers, 2+2 coupes assume three passengers, and two seat models assume two passengers. Alignment must be checked and adjustments made with the weight in place and the fuel tank full.

BMW / MINI

All Models - Loading

Alignment must be checked and adjustments made with the fuel tank full and the vehicle loaded as follows:

1 Series Cpe./Conv.	-150 lbs. in each front seat and 31 lbs. in middle of trunk.
2 Series Cpe./Conv.	-150 lbs. in each front seat, 150 lbs. in middle of rear seat, and 46 lbs. in middle of trunk.
3 Series (exc. below)	-150 lbs. in each front seat, 150 lbs. in middle of rear seat, and 46 lbs. in middle of trunk.
3 Series Conv.	-150 lbs. in each front seat and 31 lbs. in middle of trunk.
3 Series M Cpe/Rdst	-150 lbs. in each front seat and 31 lbs. in middle of trunk.
3 Series M3 1995-98	- No loading required for Sport Susp. models only. M3 Conv. use reg. 3 Series loading.
4 Series (exc. below)	-150 lbs. in each front seat and 31 lbs. in middle of trunk.
4 Series M4	-150 lbs. in each front seat, 150 lbs. in middle of rear seat, and 46 lbs. in middle of trunk.
5 Series (except below)	-150 lbs. in each front seat, 150 lbs. in middle of rear seat, and 46 lbs. in middle of trunk.
5 Series 1991-93 M5	-150 lbs. in each front seat and 31 lbs. in middle of trunk.
6 Series	-150 lbs. in each front seat and 31 lbs. in middle of trunk.
7 Series	-150 lbs. in each front seat, 150 lbs. in middle of rear seat, and 46 lbs. in middle of trunk.
8 Series	-150 lbs. in each front seat and 31 lbs. in middle of trunk.
Z Series Z3, Z4	-150 lbs. in each front seat and 31 lbs. in middle of trunk.
X Series X1, X3, X4, X5, X6	-150 lbs. in each front seat, 150 lbs. in middle of rear seat, and 46 lbs. in middle of trunk.
MINI (All Series)	-150 lbs. in each front seat and 31 lbs. in middle of trunk.

Chevrolet/GMC Truck & SUV

All Models - Model Identification

Equivalent Chevrolet and GMC models have the same alignment specifications and are listed together in the alignment specifications section. The following table will help with the identification of Chevrolet and GMC models listed in this manual. Body designations are straight-forward for all except full-size pickups for the years 1990-91, where the designations "C / R" and "K / V" are given. For those years, the "R" and "V" refer to "old-style" trucks and "C" and "K" refer to "new-style" subsequent generation trucks. Both platforms were sold in various configurations during those years. For the vehicles in the table below, the body designation letter is the **5th** character in the VIN (**5th & 6th** starting in 2010), and the year letter-code is the **10th** character.

Body/Model Chart

YEAR	1990	1991	1992	1993	1994	1995	1996	1997	1998	1999	2000	2001	2002	2003	2004	2005	2006	2007	2008	2009	2010	2011	2012	2013	2014	2015	2016
VIN CODE	L	M	N	P	R	S	T	V	W	X	Y	1	2	3	4	5	6	7	8	9	A	B	C	D	E	F	G
MODEL																											
4X2 Full-size Pickup	C / R	C / R	C	C	C	C	C	C	C	C	C																
4X2 Silverado/Sierra										C	C	C	C	C	C	C	C	C	C	C	C/*	C/*	C/*	C/*	C/*	C/*	C/*
4X4 Full-size Pickup	K / V	K / V	K	K	K	K	K	K	K	K	K																
4X4 Silverado/Sierra										K	K	K	K	K	K	K	K	K	K	K	K/*	K/*	K/*	K/*	K/*	K/*	K/*
4X2 Small Pickups	S	S	S	S	S	S	S	S	S	S	S	S	S	S	S	S	S	S	S	S	S/*	S/*	S/*			S/*	S/*
4X4 Small Pickups	T	T	T	T	T	T	T	T	T	T	T	T	T	T	T	T	T	T	T	T	T/*	T/*	T/*			T/*	T/*
4X2 Suburban/Yukon XL	R	R	C	C	C	C	C	C	C	C	C	C	C	C	C	C	C	C	C	C	C/*	C/*	C/*	C/*	C/*	C/*	C/*
4X4 Suburban/Yukon XL	V	V	K	K	K	K	K	K	K	K	K	K	K	K	K	K	K	K	K	K	K/*	K/*	K/*	K/*	K/*	K/*	K/*
4X4 Blazer/Jimmy (full-size)	V	V	K	K	K																						
4X2 Tahoe/Yukon/Avalanche						C	C	C	C	C	C	C	C	C	C	C	C	C	C	C	C/*	C/*	C/*	C/*	C/*	C/*	C/*
4X4 Tahoe/Yukon/Avalanche			K	K	K	K	K	K	K	K	K	K	K	K	K	K	K	K	K	K	K/*	K/*	K/*	K/*	K/*	K/*	K/*
4X2 Small Blazer/Jimmy	S	S	S	S	S	S	S	S	S	S	S	S	S	S	S	S											
4X4 Small Blazer/Jimmy	T	T	T	T	T	T	T	T	T	T	T	T	T	T	T	T											
4X2 Full-Size Van	G	G	G	G	G	G	G	G	G	G	G	G	G	G	G	G	G	G	G	G	G/*	G/*	G/*	G/*	G/*	G/*	G/*
4X4 Full-Size Van														H	H	H	H	H	H	H	H/*	H/*	H/*	H/*	H/*		
4X2 Astro/Safari Mini Van	M	M	M	M	M	M	M	M	M	M	M	M	M	M	M	M											
4X4 Astro/Safari Mini Van	L	L	L	L	L	L	L	L	L	L	L	L	L	L	L	L											
4X2 Tracker		E	E	E	E	E	E	E	E	E	E	E	E	E	E												
4X4 Tracker	J	J	J	J	J	J	J	J	J	J	J	J	J	J	J												

Table 1

Silverado/Sierra – RPO Option Codes (OE)

Some GM truck models are listed in the specifications section with the RPO (Option) codes. The RPO codes can be found on the Options Plackard on the inside of the dash compartment door. See Table 2 below for some RPO option code translations. Use the RPO code if possible to determine the correct specifications.

RPO	OPTION	RPO	OPTION
QXR	LT245/70R17 All-Terrain Blackwall Tires	QJM	P265/70R17 113S On/Off Road White Letter
QXT	LT265/70R17E All-Terrain Blackwall	QGI	P265/70R17 All-Season Blackwall
RC5	LT265/70R17C All-Terrain Blackwall	QVL	P265/70R17 All-Season Blackwall
QU1	P245/70R17 All-Season Blackwall	QVM	P265/70R17 All-Season White Letter
QNM	P245/70R17 108H All-Season Blackwall	QAN	P265/70R17 Touring All-Season Blackwall
QPR	P245/70R17 108S All-Season Blackwall	QSS	w/R4Y P275/55R20 Goodyear All-Season Blackwall
QPO	P245/70R17 108S All-Season White Letter	QSS	w/R5C P275/55R20 Bridgestone All-Season Blackwall
QXN	P265/65R18 On/Off Road Blackwall	QST	P285/45R22 All-Season Blackwall
QXQ	P265/65R18 On/Off Road White Letter	RBX	P265/65R18 All-Terrain
QXK	P265/65R18 All-Season Blackwall	RBW	P265/65R18 All-Terrain Outline White Letter
QMG	P265/65R18 All-Season Blackwall	RC3	P265/70R17 All-Terrain Blackwall
QMH	P265/65R18 All-Season White Letter	RC7	P275/55R20 All-Terrain Blackwall
QBL	P265/70R17 113H On/Off Road Blackwall	RHH	P285/45R22 SL 110H All-Terrain Blackwall
QJP	P265/70R17 113S On/Off Road Blackwall	RKX	P265/65R18 All-Season Blackwall

Table 2

C/K, Truck & SUV 1990-2005 * (also Escalade, H2), G/H Van 1990-2017 ** - Frame Angle & Height

The CASTER specification for the above vehicles assumes a level frame, so it is necessary to consider frame angle in all checks and adjustments. First, measure caster in the normal manner. Next, using a protractor or inclinometer, measure the frame angle as indicated in Figure 1. If the frame is inclined toward the front (up in rear), add the frame angle value (without + or -) to the measured caster angle. If the frame is inclined toward the rear (down in rear), subtract the frame angle value from the measured caster angle. The resulting value is the corrected caster angle.

Models that have "see table" in place of a caster specification must be measured in order to determine the correct caster specification. Measure vertically from the bottom side of the front cross-member to the inboard rear corner of the jounce bumper bracket (as indicated in Figure 2). Find the caster specification for the model and height in the caster chart below. Compare the corrected caster value with the specification to determine what change is needed.

*** Note:** Starting with 2006 C/K Truck and Utility models, the caster is specified relative to the ground – not the frame, so the above procedure is no longer necessary.

**** Note:** Caster is specified relative to the frame for all models. For 2003-17 Chevy/GMC Express/Savana Van – measure flat area on frame between the transmission support and next rearward (#3) body mount.

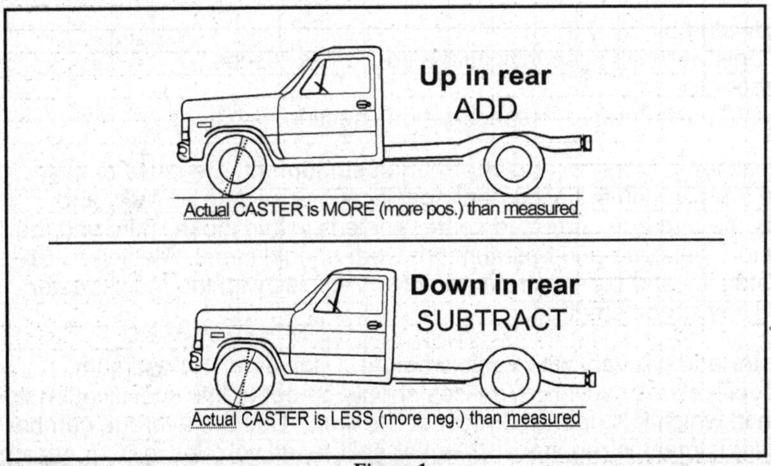

Figure 1

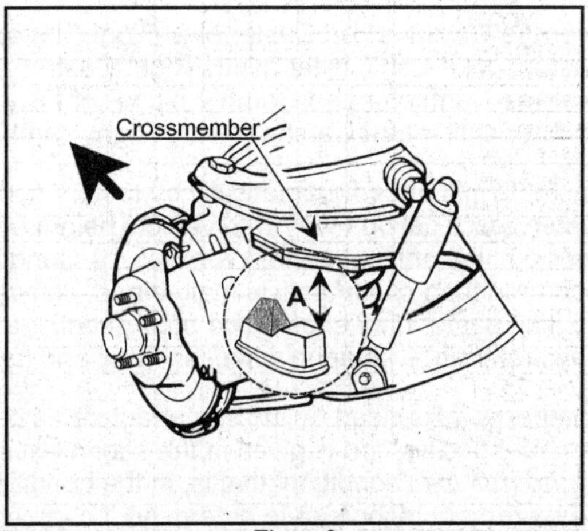

Figure 2

Chevrolet/GMC Truck & SUV (cont.)

CASTER CHART

Dimension "A" (in.)	1½"	1¾"	2"	2¼"	2½"	2¾"	3"	3¼"	3½"	3¾"	4"	4¼"	4½"	4¾"	5"	5¼"	5½"	5¾"	6"
Dimension "A" (mm)	38	45	51	57	64	70	76	83	89	95	102	108	114	121	127	133	139	146	152
1990-91 R10/R1500	– –	– –	– –	– –	3.7° 3 11/16°	3.5° 3½°	3.2° 3 3/16°	2.9° 2⅞°	2.6° 2⅝°	2.4° 2⅜°	2.1° 2⅛°	1.8° 1 13/16°	1.5° 1½°	1.3° 1 5/16°	1.0° 1°	0.8° 13/16°	0.5° ½°	0.3° 5/16°	0.0° 0°
1990-91 R20/R30, R2500/R3500	– –	– –	– –	– –	1.5° 1½°	1.2° 1 3/16°	0.9° ⅞°	0.6° ⅝°	0.3° 5/16°	0.1° ⅛°	0° 0°	-0.2° -3/16°	-0.7° -11/16°	-1.0° -1°	-1.2° -1 3/16°	-1.4° -1⅜°	-1.6° -1⅝°	-1.8° -1 13/16°	– –
1990-91 Suburban 10/1500	– –	– –	– –	– –	3.7° 3 11/16°	3.5° 3½°	3.2° 3 3/16°	2.9° 2⅞°	2.6° 2⅝°	2.4° 2⅜°	2.1° 2⅛°	1.8° 1 13/16°	1.5° 1½°	1.3° 1 5/16°	1.0° 1°	0.8° 13/16°	0.5° ½°	0.3° 5/16°	0.0° 0°
1990-91 Suburban 20/2500	– –	– –	– –	– –	1.5° 1½°	1.2° 1 3/16°	0.9° ⅞°	0.6° ⅝°	0.3° 5/16°	0.1° ⅛°	0° 0°	-0.2° -3/16°	-0.7° -11/16°	-1.0° -1°	-1.2° -1 3/16°	-1.4° -1⅜°	-1.6° -1⅝°	-1.8° -1 13/16°	– –
1990-93 G10/G20	3.4° 3⅜°	3.2° 3 3/16°	3.0° 3°	2.8° 2 13/16°	2.7° 2 11/16°	2.5° 2½°	2.3° 2 5/16°	2.2° 2 3/16°	2.0° 2°	1.8° 1 13/16°	1.7° 1 11/16°	1.5° 1½°	1.4° 1⅜°	1.3° 1 5/16°	1.2° 1 3/16°	1.1° 1⅛°	1.0° 1°	0.9° ⅞°	0.8° 13/16°
1990-93 G30	3.1° 3⅛°	3.0° 3°	2.7° 2 11/16°	2.4° 2⅜°	2.1° 2⅛°	1.8° 1 13/16°	1.5° 1½°	1.2° 1 3/16°	1.0° 1°	0.7° 11/16°	0.5° ½°	0.2° 3/16°	0° 0°	-0.2° -3/16°	-0.5° -½°	-0.7° -11/16°	-0.9° -⅞°	-1.1° -1⅛°	– –
1994-95 G10/G20	5.4° 5⅜°	5.2° 5 3/16°	5.0° 5°	4.9° 4⅞°	4.7° 4 11/16°	4.5° 4½°	4.3° 4 5/16°	4.2° 4 3/16°	4.0° 4°	3.8° 3 13/16°	3.7° 3 11/16°	3.5° 3½°	3.4° 3⅜°	3.3° 3 5/16°	3.2° 3 3/16°	3.1° 3⅛°	3.0° 3°	2.9° 2⅞°	2.8° 2 13/16°
1994-96 G30	5.1° 5⅛°	5.0° 5°	4.7° 4 11/16°	4.4° 4⅜°	4.1° 4⅛°	3.8° 3 13/16°	3.5° 3½°	3.2° 3 3/16°	3.0° 3°	2.7° 2 11/16°	2.5° 2½°	2.2° 2 3/16°	2.0° 2°	1.8° 1 13/16°	1.5° 1½°	1.3° 1 5/16°	1.1° 1⅛°	0.9° ⅞°	– –

Table 3.

Ford Truck & SUV

<u>Ford E-Series (All), Explorer 1990-2001, Sport Trac 1990-2005, F-Series 1990-96, Ranger/B-Series 1990-97</u>

(Also other models that reference this page under INSTRUCTIONS in the alignment specifications section.)

Frame Angle

The CASTER specifications in the alignment specifications section assume a level frame, so it is necessary to consider frame angle in all checks and adjustments. First, measure caster in the normal manner. Next, using a spirit level protractor or inclinometer, measure the frame angle as indicated in the figures below. If the frame is inclined toward the front (up in rear), add the frame angle value to the measured caster angle. If the frame is inclined toward the rear (down in rear), subtract the frame angle value from the measured caster angle. The resulting value is the corrected caster angle. <u>Compare the corrected caster value to the specification to determine the amount of change needed.</u>

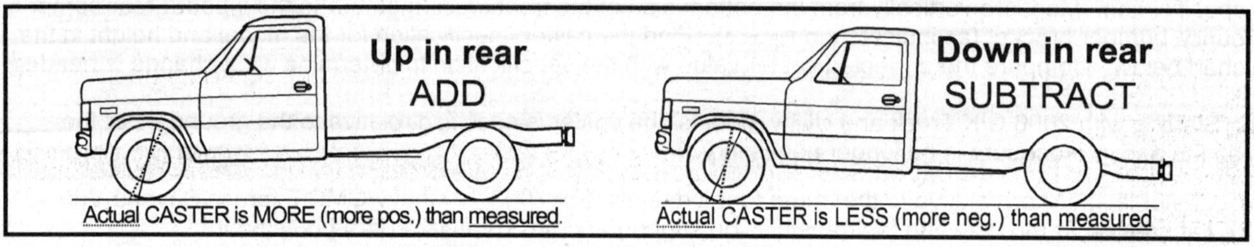

Average Caster and Caster Split - Average Camber and Camber Split

Average Caster = Left Caster plus Right Caster divided by 2.
Caster split = Left Caster minus Right Caster. ("Split" is listed in specifications section as "Cross.")
Average Camber = Left Camber plus Right Camber divided by 2.
Camber split = Left Camber minus Right Camber. ("Split" is listed in specifications section as "Cross.")

Caster is influenced dramatically by factors such as load, tire size, and aftermarket equipment. Because of this, caster specifications vary with vehicle height and are given in the alignment specifications section as MIN. and MAX. values only for certain vehicles. As long as the average corrected caster angle is within those limits and the recommended caster split is maintained, satisfactory handling and optimum tire wear should result. Vehicles with too little caster may experience poor directional stability and steering returnability. Vehicles with too much caster may experience steering wheel shimmy and hard low speed steering.

Camber is influenced by the same factors as caster and will vary with vehicle height. However, an optimum camber specification is given in the alignment specifications section. Vehicles should be set to this specification in the *normal load* condition, that is, in the condition in which it is operated most of the time. Use the average camber measurement technique to determine if camber adjustment is required. Camber split for all vehicles is 0°, meaning both sides should be set to the same specification (within side/side tolerance).

Ford Truck & SUV (cont.)

Average Caster and Caster Split - Average Camber and Camber Split (cont.)

Consider the following examples:

Example A Specified Caster values: MIN. = 2°, MAX. = 7°, SPLIT = -.5° ± .5°

Left side measured caster: 3.25° (corrected for frame angle) Right side measured caster: 3.75° (corrected for frame angle)

- Average caster = 3.25° plus 3.75° divided by 2 = **3.50°**
- Caster split = 3.25° minus 3.75° = **-.5°**

In example A, no adjustment for caster is required.

Example B Specified Caster values: MIN. = 2°, MAX. = 7°, SPLIT = -.5° ± .5°

Left side measured caster: 3.75° (corrected for frame angle) Right side measured caster: 3.25° (corrected for frame angle)

- Average caster = 3.75° plus 3.25° divided by 2 = **3.50°**
- Caster split = 3.75° minus 3.25° = **+.5°**

In example B, the caster split is out of spec by 1 degree. Caster split is not a tolerance but a *specified difference* from side to side. Even though the average caster is well within the MIN. and MAX. values, the caster on one side should be changed so that the left side caster is .5° less than the right side caster.

Example C Specified Camber: 0.5° ± 0.5° SPLIT = 0° ± 0.5°

Left side measured camber: 1.5° Right side measured camber: 0.25°

- Average camber = 1.5° plus 0.25° divided by 2 = **0.875°**
- Camber split = 1.5° minus 0.25° = **1.25°**

In example C, average camber is within specifications but camber split is not. The adjustment should be made to the left side if possible since the left spec is furthest from the specified camber. However, as in the previous example, consideration should be given to steering pull characteristics and tire wear patterns.
Note: Caster **split** specifications are subject to side-to-side tolerances. For a caster split specification of -.5° ± .5°, the left side caster can be from 0° to 1° less positive than the right side and still be within specifications. However, the recommended split should be maintained as close as possible.

Adjustment Hint – All Twin I-Beam 4X4 vehicles with leaf springs:

When using caster wedges, try to make adjustments on the right side axle if possible. Left side caster adjustments should be minimized in order to avoid changing the front drive train alignment.

Jaguar

1989-97 XJ & XJS

Wheel alignment checks and adjustments must be made with both front and rear suspensions held in a partially compressed position. This "mid-laden" condition is intended to simulate a loaded vehicle. Special tools are available to hold the vehicle at a specific height. In their absence, the vehicle can be loaded with available weights to achieve the height conditions listed below.

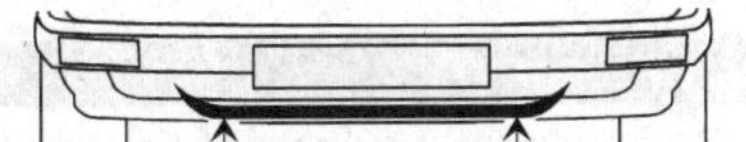

Figure 1

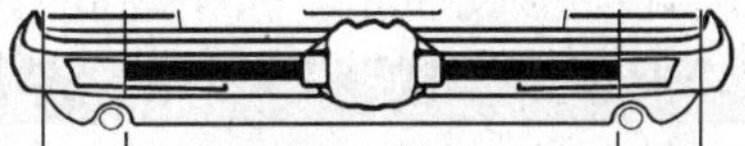

Figure 2

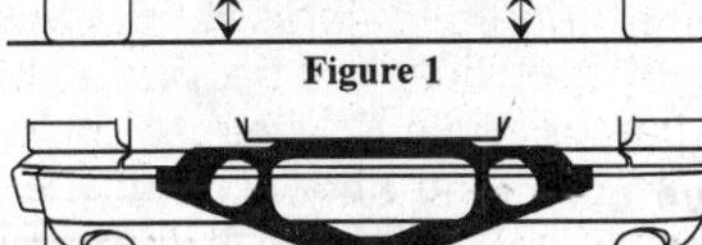

Figure 3

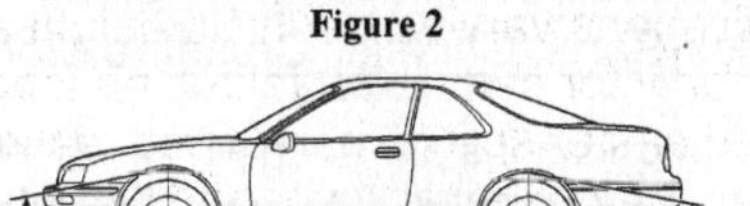

Figure 4

Measure **Figure 1** vehicles from lower side of front suspension sub-frame to floor, on both sides.
Set **Figure 2** vehicles rear axle shafts in a horizontal position. (Use holding tool or weights)
Measure **Figure 3** vehicles from lower side of rear suspension cross-member.
Measure **Figure 4** vehicles from wheel center to apex of wheel arch at centerline of wheel. (Fuel tank full

Jaguar (cont.)

1990-97 XJ Sedans (Figures 1 & 2)

XJ6	Front	6.02" ± 0.18" (153 ± 5mm)
	Rear	Rear axle shafts level.
XJ12	Front	5.62" ± 0.18" (143 ± 5mm)
	Rear	Rear axle shafts level.

1990-1996 XJS (Figures 1 & 3)

Front	6.02" ± 0.18" (153 ± 5mm)
Rear	6.37" ± 0.18" (162 ± 5mm)

1998-2003 XJ Series (Figure 4)

XJ	Comfort	Front	15.31" ± 0.6" (389 ± 15mm)
		Rear	14.68" ± 0.6" (373 ± 15mm)
	Sport	Front	15.11" ± 0.6" (384 ± 15mm)
		Rear	14.48" ± 0.6" (368 ± 15mm)
XJR		Front	15.27" ± 0.6" (388 ± 15mm)
		Rear	14.64" ± 0.6" (372 ± 15mm)

2004-2010 XJ Series (Figure 4)

Front	15.19" ± 0.6" (386 ± 15mm)
Rear	14.68" ± 0.6" (373 ± 15mm)

2011-2017 XJ Series All Models (Figure 4)

Front	15.94" ± 0.47" (405 ± 12mm)
Rear	15.87" ± 0.47" (403 ± 12mm)

1998-2013 XK8/XKR (Figure 4)

From VIN# 001246 (1998-2005)

Front	15.74" ± 0.6" (400 ± 15mm)
Rear	15.94" ± 0.6" (405 ± 15mm)

From VIN# B00379 (2006-2009)

Front (XK)	15.59" ± 0.6" (396 ± 15mm)
Rear (XK)	15.59" ± 0.6" (396 ± 15mm)
Front (XKR)	15.51" ± 0.6" (394 ± 15mm)
Rear (XKR)	15.47" ± 0.6" (393 ± 15mm)

From VIN# B32753 (2010 onward)

Front (XK)	15.51" ± 0.6" (394 ± 15mm)
Rear (XK)	15.51" ± 0.6" (394 ± 15mm)
Front (XKR)	15.43" ± 0.6" (392 ± 15mm)
Rear (XKR)	15.31" ± 0.6" (389 ± 15mm)

2009-2015 XF/XFR (Figure 4)

From VIN# R00019 (2009)

Front (XF)	15.23" ± 0.6" (387 ± 15mm)
Rear (XF)	15.43" ± 0.6" (392 ± 15mm)
Front (XFR)	15.19" ± 0.6" (386 ± 15mm)
Rear (XFR)	15.23" ± 0.6" (387 ± 15mm)

From VIN# R47154 (2010 onward)

Front (XF)	15.28" ± 0.5" (388 ± 12mm)
Rear (XF)	15.39" ± 0.5" (391 ± 12mm)
Front (XFR)	15.16" ± 0.5" (385 ± 12mm)
Rear (XFR)	15.12" ± 0.5" (384 ± 12mm)
Front (XF AWD)	15.90" ± 0.5" (404 ± 12mm)
Rear (XF AWD)	15.39" ± 0.5" (391 ± 12mm)

2016-17 XF (Figure 4)

Front (XF)	15.51" ± 0.6" (394 ± 12mm)
Rear (XF)	15.66" ± 0.6" (398 ± 13mm)
Front (XF Sport)	15.47" ± 0.6" (393 ± 12mm)
Rear (XF Sport)	15.62" ± 0.6" (397 ± 13mm)

2000-2002.5 S-Type (Figure 4)

with CATS (adaptive dampening)	V6 Front	15.11" ± 0.6" (384 ± 15mm)
	V6 Rear	14.76" ± 0.6" (375 ± 15mm)
	V8 Front	15.07" ± 0.6" (383 ± 15mm)
	V8 Rear	14.76" ± 0.6" (375 ± 15mm)
w/o CATS	All Front	15.11" ± 0.6" (384 ± 15mm)
	All Rear	14.88" ± 0.6" (378 ± 15mm)

2002.5-2004.5 S-Type (Figure 4)

with CATS (adaptive dampening)

V6	Front	15.15" ± 0.6" (385 ± 15mm)
V6	Rear	14.96" ± 0.6" (380 ± 15mm)
V8	Front	15.11" ± 0.6" (384 ± 15mm)
V8	Rear	14.96" ± 0.6" (380 ± 15mm)

without CATS (adaptive dampening)

V6	Front	15.27" ± 0.6" (387 ± 15mm)
V6	Rear	15.15" ± 0.6" (385 ± 15mm)
V8	Front	15.19" ± 0.6" (386 ± 15mm)
V8	Rear	15.15" ± 0.6" (385 ± 15mm)

2004.5-2008 S-Type (Figure 4)

V6 Comfort	Front	15.27" ± 0.6" (388 ± 15mm)
V6 Comfort	Rear	15.07" ± 0.6" (383 ± 15mm)
V6 Sport	Front	15.19" ± 0.6" (386 ± 15mm)
V6 Sport	Rear	14.92" ± 0.6" (379 ± 15mm)
V8 Comfort	Front	15.27" ± 0.6" (387 ± 15mm)
V8 Comfort	Rear	15.03" ± 0.6" (382 ± 15mm)
V8 Sport	Front	15.15" ± 0.6" (385 ± 15mm)
V8 Sport	Rear	14.92" ± 0.6" (379 ± 15mm)
V8 Supercharged	Front	15.00" ± 0.6" (381 ± 15mm)
V8 Supercharged	Rear	14.76" ± 0.6" (375 ± 15mm)

2002-04 X-Type (Figure 4)

Comfort	Front	14.72" ± 0.6" (374 ± 15mm)
	Rear	14.64" ± 0.6" (372 ± 15mm)
Sport	Front	14.68" ± 0.6" (373 ± 15mm)
	Rear	14.48" ± 0.6" (369 ± 15mm)

2005-09 X-Type (Figure 4)

Comfort	Sedan Front	14.72" ± 0.6" (374 ± 15mm)
	Sedan Rear	14.88" ± 0.6" (378 ± 15mm)
	Wagon Front	14.64" ± 0.6" (372 ± 15mm)
	Wagon Rear	14.56" ± 0.6" (370 ± 15mm)
Sport	Sedan Front	14.60" ± 0.6" (371 ± 15mm)
	Sedan Rear	14.68" ± 0.6" (373 ± 15mm)
	Wagon Front	14.56" ± 0.6" (370 ± 15mm)
	Wagon Rear	14.40" ± 0.6" (366 ± 15mm)

2014-17 F-Type (Figure 4)

All Models	Front	15.24" ± 0.6" (387 ± 15mm)
	Rear	15.43" ± 0.6" (392 ± 15mm)

2016-17 XE (Figure 4)

Front (XF)	15.51" ± 0.6" (394 ± 12mm)
Rear (XF)	15.66" ± 0.6" (398 ± 13mm)
Front (XF Sport)	15.47" ± 0.6" (393 ± 12mm)
Rear (XF Sport)	15.62" ± 0.6" (397 ± 13mm)

Mercedes-Benz

All Mercedes-Benz, Chrysler Crossfire Models - Front TOE

All front toe specifications are given with a wheel spreader bar installed between the front wheels as indicated in **Figure 1**. The wheel spreader pushes outward to eliminate elasticity in the steering.

Note: All specifications are for a vehicle at nominal height. Insure that the vehicle is at curb weight, in "ready to drive" condition with fuel tank full. Special measuring devices are required to determine vehicle height or "Vehicle Level".

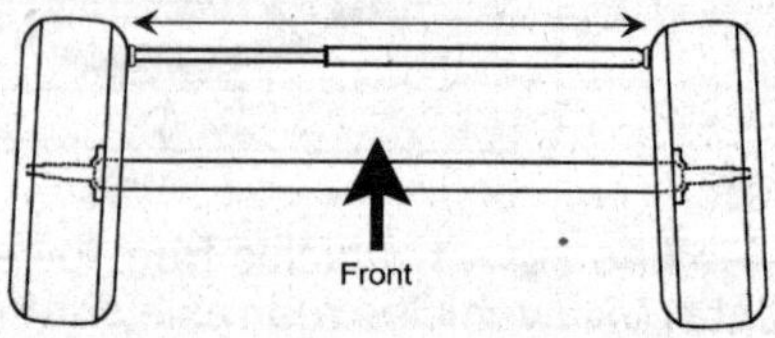

Figure 1

Mercedes-Benz (cont.)

Mercedes-Benz Model Identification

Series Model	Years (USA)	Description		Model Badge
Model 201	1984-93	190-Class Compact	Sedan	190E, 190D
Model 202	1994-00	C-Class Compact	Sedan	C220, C230, C280, C36, C43
Model 203	2001-07	C-Class Compact	Coupe/Sedan	C230 Kompressor, C240, C280, C320, C350, C32AMG, C55AMG
		C-Class Compact	Wagon	C240, C320
Model 204	2008-15	C-Class Compact	Coupe/Sedan	C250, C300, C350, C63 AMG (Coupe models only in 2015)
Model x204	2010-14	C-Class Compact	SUV	GLK350
Model 205	2015-17	C-Class Compact	Sedan	C300, C300 4MATIC, C350e, C450, AMG C43, AMG C63, AMG C63S
Model 117	2014-17	CLA Class Compact	Coupe	CLA250, CLA250 4MATIC, CLA45 AMG
Model 156	2015-17	GLA Class	SUV	GLA250, GLA250 4MATIC, GLA45 AMG
Model 207	2010-17	E-Class Mid-size	Coupe/Conv.	E350, E400, E550 Coupe & Convertible
Model 208	1998-02	CLK-Class Compact	Coupe	CLK320, CLK430, CLK55 AMG
	1998-03	CLK-Class Compact	Cabriolet	CLK320, CLK430, CLK55 AMG
Model 209	2003-09	CLK-Class Compact	Coupe	CLK320, CLK350, CLK500, CLK550, CLK55 AMG
		CLK-Class Compact	Cabriolet	CLK320, CLK350, CLK500, CLK550, CLK55 AMG, CLK63 AMG
Model 124	1986-93	300-Class Mid-size	Coupe/Sedan	500E, 400E, 300E, 300D, 300D-T, 300E 4MATIC, 260E, 300 CE, 300 CE Conv.
			Wagon	300TE, 300TD, 300TD-T, 300TE 4MATIC
	1994-95	E-Class Mid-size	Coupe/Sedan	E420, E320, E300, E320 Convertible
			Wagon	E320
Model 210	1996-02	E-Class Mid-size	Sedan	E430, E420, E320, E300D, E300TD, E55 AMG
	1998-03	E-Class Mid-size	Wagon	E320
Model 211	2003-09	E-Class Mid-size	Sedan	E320, E320 CDI, E350, E500, E550, E55 AMG, E63 AMG
	2004-09	E-Class Mid-size	Wagon	E320, E350, E500, E55 AMG. E63 AMG
Model 212	2010-16	E-Class Mid-size	Sedan	E250 BlueTEC, E350, E350 BlueTEC, E400(Hybrid), E550, E63 AMG
	2011-16	E-Class Mid-size	Wagon	E350 4MATIC, E63 AMG 4MATIC, AMG E63 S
Model 213	2017	E-Class Mid-size	Sedan	E300, E300 4MATIC, AMG E43
Model 126	1981-91	S-Class Large	Sedan	560SEL, 500SEL, 420SEL, 380SEL, 380SE, 350SD, 350SDL, 300SD, 300SDL
			Coupe	560SEC, 500SEC, 380SEC
Model 140	1992-93	S-Class Large	Sedan	600SEL, 500SEL, 400SE, 400 SEL, 300SE, 300SD
			Coupe	600SEC, 500SEC
	1994-99	S-Class Large	Sedan	S600, S500, S420, S350, S320
			Coupe	S600, S500, CL600, CL500
Model 220	2000-06	S-Class Large	Sedan	S350, S430, S500, S600, S55 AMG, S65 AMG
Model 221	2007-13	S-Class Large	Sedan	S350 BlueTEC, S400 Hybrid, S550, S600, S63 AMG, S65 AMG
Model 222	2014-16	S-Class Large	Sedan	S550, S550e, S600, S63 AMG, S65 AMG
Model 215	2000-06	CL-Class Large	Coupe	CL500, CL600, CL55 AMG, CL65 AMG
Model 216	2007-14	CL-Class Large	Coupe	CL550, CL600, CL63 AMG, CL65 AMG
Model 217	2015-17	S-Class Large	Coupe	S550 4MATIC, S63 AMG, S65 AMG
Model 129	1990-02	SL-Class	Coupe/Roadster	SL600 (600SL), SL500 (500SL), 300SL, SL320
Model 230	2003-12	SL-Class	Roadster	SL500, SL550, SL600, SL55 AMG, SL63 AMG, SL65 AMG
Model 231	2013-17	SL-Class	Roadster	SL400, SL550, SL63 AMG, SL65 AMG
Model 170	1998-04	SLK-Class	Roadster	SLK230, SLK320, SLK32 AMG
Model 171	2005-11	SLK-Class	Roadster	SLK280, SLK300, SLK350, SLK55 AMG
Model 172	2012-16	SLK-Class	Roadster	SLK250, SLK300, SLK350, SLK55 AMG
Model 163	1998-05	M-Class	SUV	ML320, ML350, ML430, ML500, ML55
Model 164	2006-11	M-Class	SUV	ML320 CDI, ML350, ML450 Hybrid, ML500, ML550, ML63 AMG
Model 166	2012-15	M-Class	SUV	ML250, ML350, ML400, ML550, ML63 AMG
Model 166	2016-17	GLE-Class	SUV	GLE300d, GLE350, GLE 400, GLE550e, AMG GLE43, AMG GLE63/(S)
Model 292	2016-17	GLE-Class	Coupe	GLE 450, AMG GLE43, AMG GLE63 S
Model X164	2006-12	GL-Class	SUV	GL320, GL350 BlueTEC, GL450, GL550
Model X166	2013-16	GL-Class	SUV	GL350 BlueTEC, GL450, GL550, GL63 AMG
Model 219	2006-11	CLS-Class	Coupe	CLS500, CLS550, CLS55 AMG, CLS63 AMG
Model 218	2012-17	CLS-Class	Coupe	CLS400, CLS550, CLS63 AMG
Model 251	2006-12	R-Class	AWD Wagon	R320 CDI, R350, R350 BlueTEC, R500
Model 253	2016-17	GLC-Class	SUV	GLC300, GLC300 4MATIC, AMG GLC43

Porsche

1995-98 911 Carrera, Carrera 4 (993)

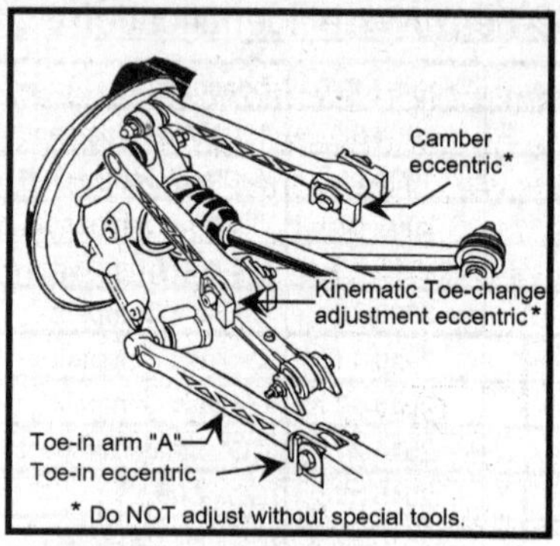

Figure 1

The multi-link rear suspension requires special measuring tools in order to perform a complete adjustment . The tools are not available from traditional sources. However, if rear camber is within specifications, rear *toe-in only* may be checked and adjusted with conventional alignment equipment, and no further adjustment of the rear suspension is required. Be careful to adjust rear toe-in at arm "A" (see illustration).

The front suspension is fully adjustable as in previous 911 models.

2003-10 Cayenne with Air Springs

Models with air suspension must be checked with the level indicator adjusted to the NORMAL position using the following procedure. First, adjust the level switch until the vehicle has reached LOW level. Then adjust again until the vehicle is at NORMAL level. Adjustments may be made with the level control switch turned off, but it must be on and adjusted to the NORMAL position to verify settings after adjustments are made.

Volkswagen

Eurovan and Vanagon - Caster Correction

The CASTER angle must be corrected to compensate for variance in vehicle level. Using a frame-angle protractor, measure the vehicle angle at the bottom of the sliding-door opening (Vanagon), or at the bottom of the long frame member under the vehicle (Eurovan). Add the angle reading to the measured CASTER if the vehicle angle is inclined toward the front; subtract if it is toward the rear.

Example 1:	Caster specification:	6.5° ±.25°	Example 2:	2.0° ±.50°
	Measured caster angle:	5.5° (out of spec!)		4.5° (out of spec!)
	Vehicle angle:	1.0° (toward front)		1.5° (toward rear)
	Actual *corrected* caster:	6.5° (5.5+1.0) (no adjustment reqd.)		3.0° (4.5-1.5) (-1° correction required)

Note: Front standing height should be checked and corrected before measuring frame angle.

1993-2000 Eurovan

Rear alignment specifications are divided into groups, identified by a data plate on the drivers side A-pillar. The data plate is located under the fuse panel cover on some models. (see Figure 1 below). The words "*Grund*", "*Kom*", *or "LUXUS"* appear with the number indicating the alignment group. Vehicles without a sticker are Group 1 vehicles.

Vehicle front standing height should be checked (see Figure 2 below) and adjusted at the torsion bar adjusters to the value in the table. "*Empty*" vehicles are passenger vans (unloaded), *"1/2 load"* vehicles are minimally equipped campers, and *"Full Load"* vehicles are fully equipped camper models with permanently installed extra equipment.

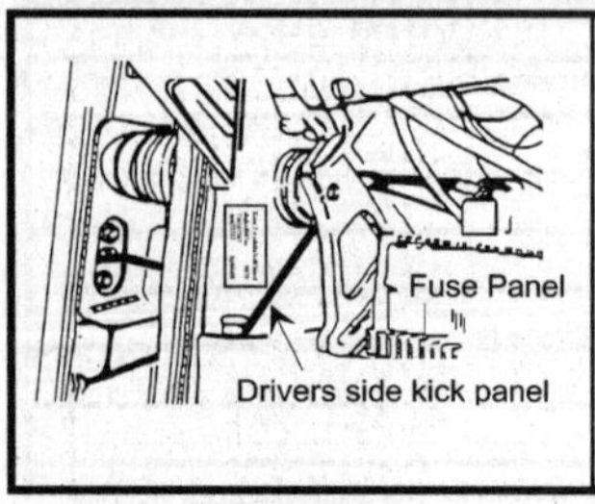

Figure 1

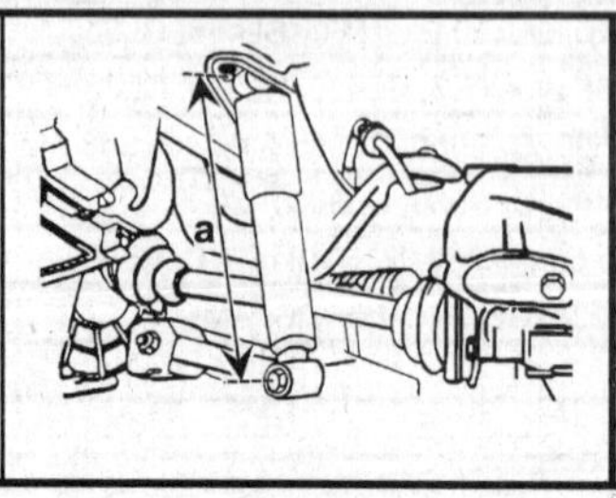

Figure 2

Group 1	Empty	1/2 Load	Full Load
"a" height (mm)	280 ±2	267 ±2	255 ±2
Rear CAMBER	-0.5 (-½)	-1.17 (-1 3/16)	-1.58 (-1 9/16)
Rear TOE (°)	0.33 (11/32)	0.67 (21/32)	1.0 (1)
Group 2	**Empty**	**1/2 Load**	**Full Load**
"a" height (mm)	280 ±2	267 ±2	255 ±2
Rear CAMBER	0.0 (0)	-0.67 (-11/16)	-1.17 (-1 3/16)
Rear TOE (°)	0.10 (3/32)	0.40 (13/32)	0.67 (21/32)

Table 1

Volkswagen (cont.)

2001-2003 Eurovan

The alignment specifications as well as the standing height specification are divided into groups, identified by a data plate on the drivers side A-pillar. The data plate is located under the fuse panel cover on some models. (see Figure 1 above). The words "*Kom*", *or "LUXUS"* appear with the number indicating the alignment group. Group numbers 1 – 8 are used. After identifying the correct Group number, use the front standing height specification from Table 2 below. Use the Group number to find the correct model listing in the alignment specifications section.

Group Number	Front Standing Height "a" (mm ± 2) (see Figure 2 above)		
	Empty	1/2 Load	Full Load
1 & 6	280	267	255
2 & 7	280	267	255
3 & 8	273	263	255
4 & 5	265	260	255

Table 2

2004-2010 Touareg

Models with Air Suspension must be checked with the vehicle in the NORMAL height position, corresponding to front and rear measured heights of 497 ± 3 mm and 502 ± 3 mm respectively for base suspension version 1BK, and 477 ± 10 mm and 482 ± 10 mm for sport suspension version 2MA. Measure from the center of each wheel to the lower edge of the wheel housing. (see Figure 3 below)

Vanagon (Syncro Only)

The front TOE specification is dependent upon the height of the vehicle. Measure from the center of the front axle shaft to the lower edge of wheel housing as indicated in Figure 3. Find the corresponding specification in Table 3 below. If the measurement is out of the range of the table, use the corresponding minimum or maximum toe setting.

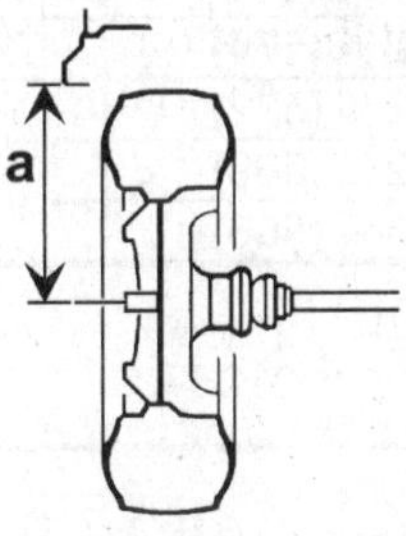

Figure 3

Dim. a (inches)	Dim. a (mm)	Total Toe (degrees)	Total Toe (inches)
19.0	483	-0.07°	-0.04"
19.2	488	-0.18°	-0.09"
19.4	493	-0.28°	-0.14"
19.6	498	-0.40°	-0.20"
19.8	503	-0.50°	-0.25"
20.0	508	-0.67°	-0.33"
20.2	513	-0.77°	-0.38"

Table 3

Volvo

1990-94 760 Sedan, 960 Models with Multilink

Rear CAMBER specifications vary with vehicle height, as determined by measuring the lower link angle. As the vehicle is loaded rear camber becomes more negative and the lower link angle becomes more positive. Measure the lower link angle with an appropriate measuring tool as indicated in Figure 1. (See adjustment illustration # 82 to locate the lower link) Find the camber specification that corresponds to the measured angle. Adjust camber with the eccentric cam at the inboard end of the lower link. Permitted tolerance is ± 0.25° (± ¼°).

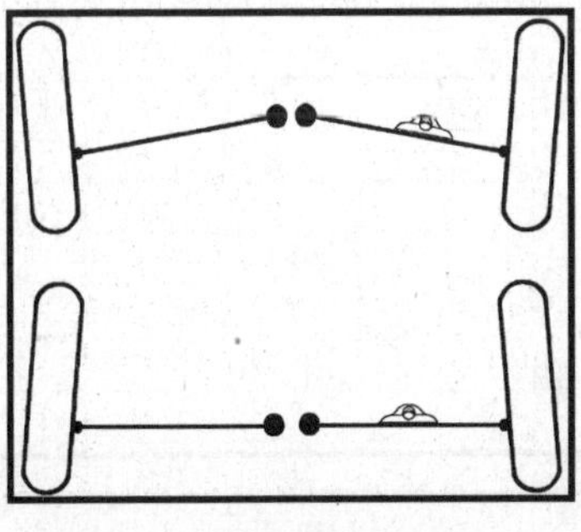

Figure 1

Lower link Angle (degrees)	CAMBER (degrees)
-2	0.70 (11/16)
-1	0.53 (½)
0	0.35 (⅜)
+1	0.15 (⅛)
+2	-0.05 (-1/16)
+3	-0.25 (-¼)
+4	-0.55 (- 9/16)
+5	-0.85 (-⅞)

Table 1

1995-97 960 Multilink, 1998 S90/V90

Front and rear CAMBER specifications vary with vehicle height. Determine the correct specification by measuring between the fender edge and the lower lip of the wheel rim as indicated in Figure 2. Find the camber specification that corresponds to the measured height.

Front camber can be adjusted at the top mounting of the strut. Remove the front bolt. Loosen the rear bolt and reposition strut to desired camber. Tighten rear bolt. Drill a new front hole in the strut retaining plate (not strut tower). Re-install bolt from below through new hole in strut retaining plate and existing hole in strut tower.

Rear camber is not adjustable. If rear camber is not within specifications check for worn or damaged components.

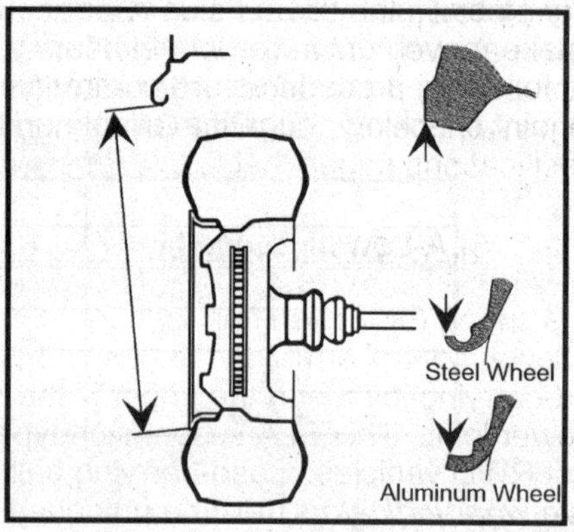

Figure 2

15" Wheels Front Measurement in. (mm)	16" Wheels Front Measurement in. (mm)	Front CAMBER Specification ±0.5° (± ½°) (degrees)	15" Wheels Rear Measurement in. (mm)	16" Wheels Rear Measurement in. (mm)	Rear CAMBER Specification ±0.4° (± ⅜°) (degrees)
24 7/16 (621)	24 15/16 (634)	-0.10 (-⅛)	22 11/16 (576)	23 3/16 (589)	0.00 (0)
24 ¼ (616)	24 ¾ (629)	-0.20 (-3/16)	22 9/16 (573)	23 1/16 (586)	-0.10 (-⅛)
24 1/16 (612)	24 ⅝ (625)	-0.30 (-5/16)	22 7/16 (570)	22 15/16 (583)	-0.18 (-3/16)
23 ⅞ (607)	24 ⅜ (620)	-0.40 (-⅜)	22 5/16 (567)	22 13/16 (580)	-0.27 (-¼)
23 11/16 (602)	24 3/16 (615)	-0.50 (-½)	22 3/16 (564)	22 11/16 (577)	-0.35 (-⅜)
23 9/16 (598)	24 1/16 (611)	-0.60 (-⅝)	22 1/16 (561)	22 ⅝ (574)	-0.44 (-7/16)
23 5/16 (593)	23 ⅞ (606)	-0.70 (-11/16)	22 (559)	22 ½ (572)	-0.53 (-½)
23 ⅛ (588)	23 11/16 (601)	-0.80 (-13/16)	21 ⅞ (556)	22 ⅜ (569)	-0.61 (-⅝)
23 (584)	23 ½ (597)	-0.90 (-⅞)	21 ¾ (553)	23 ¼ (566)	-0.70 (-11/16)

Table 2

Notes

BALL JOINT TESTING

Ball joints should be checked for excessive wear before an alignment is performed. Also, check the rubber grease seal and grease fitting, if applicable, and don't forget the rear ball joints on some rear suspension types. The load-carrying ball joint wears much faster than the non-loaded ball joint, but be sure to check both. Use care not to mistake movement in the wheel bearing as play in the ball joint. Replace the ball joint if excessive play is found. The following procedures are commonly used to test ball joints on various suspension types. For all tests, use a ball joint checking gauge if available and follow the tool manufacturer's instructions. Refer to the Ball Joint Specifications pages that follow for specific vehicle applications.

TYPE 1 SLA-LOWER LOAD CARRYING BALL JOINTS

If the spring (or torsion bar) is mounted so that it acts on the lower control arm, then the lower ball joint is the load-carrying joint, and the upper is the non-loaded or follower joint. This SLA suspension type is found on most RWD vehicles. Load-carrying ball joints that ***do not have wear indicators*** must be unloaded before checking for wear.

Place a jack under the lower arm, near the ball joint end, and raise the tire off the floor. Check the lower joint for axial play by moving the tire/wheel assembly as indicated in **Figure 1**. A pry bar can also be used between the lower arm and the spindle. Check for radial play in the lower and upper ball joint by moving the tire/wheel as indicated in **Figure 2**.

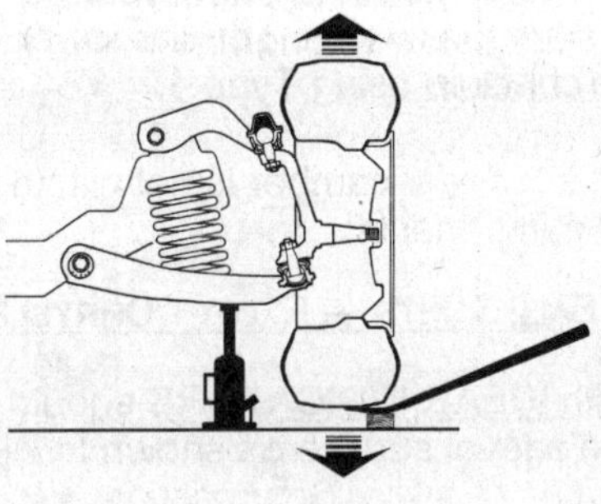

Figure 1 - AXIAL TEST

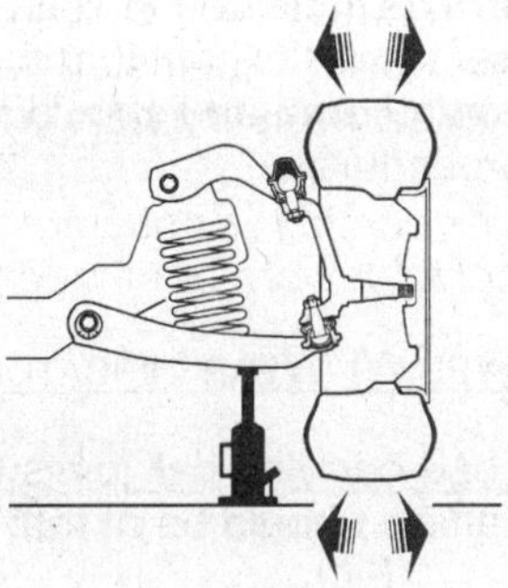

Figure 2 - RADIAL TEST

TYPE H HYBRID/WISHBONE STRUT BALL JOINTS

On Hybrid SLA suspension types the spring/strut is mounted so that it acts on the lower control arm, making the lower ball joint the load-carrying joint. The upper joint is a follower joint.

Place a jack under the frame or crossmember and raise the tire off the floor. Check the lower ball joint for axial play by moving the tire/wheel assembly as indicated in **Figure 3**. Check the upper ball joint for axial play by moving the upper arm as indicated in **Figure 3**. Check both ball joints for radial play by moving the tire/wheel assembly as indicated in **Figure 4**. Optionally, both ball joints can be tested using **Type 1** procedures.

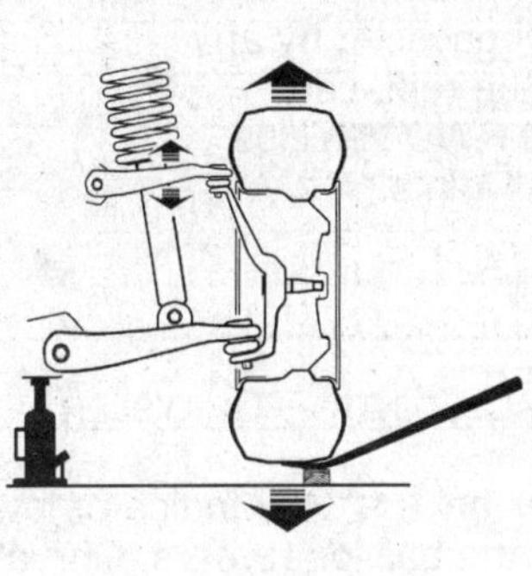

Figure 3 - AXIAL TEST

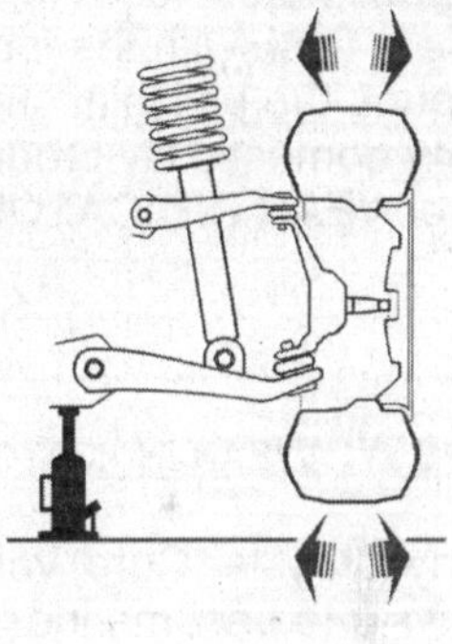

Figure 4 - RADIAL TEST

TYPE HS HIPERSTRUT™ SUSPENSION BALL JOINTS

On **Type HS** - suspension types the load is carried by the upper strut mount, not the upper ball joint. The upper ball joint is a follower joint and should be checked for the maximum allowable axial and/or radial play specified.

Place a jack under the frame or crossmember and raise the tire off the floor. Check the upper ball joint for axial play by moving the tire/wheel assembly as indicated in **Figure 5**. Check for radial play in the upper ball joint by moving the tire/wheel assembly as indicated in **Figure 6**.

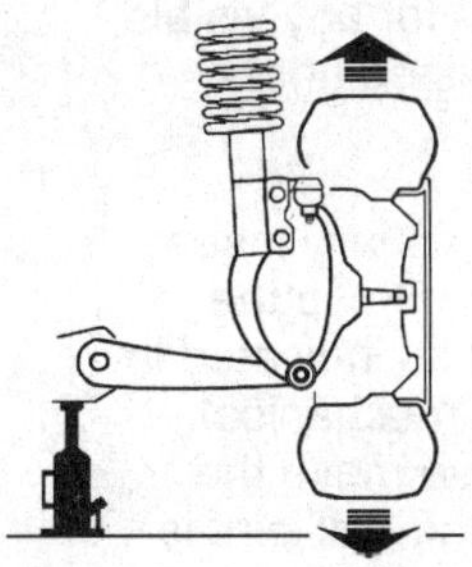

Figure 5 - AXIAL TEST

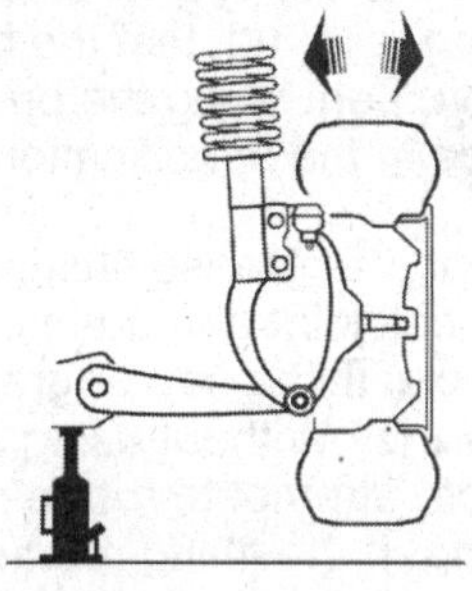

Figure 6 - RADIAL TEST

Caution: For all testing methods, use care not to mistake movement in the wheel bearing as play in the ball joint. If available, use a ball joint checking gauge or dial indicator and follow the tool manufacturer's instructions.

BALL JOINT TESTING (cont.)

TYPE S STRUT BALL JOINTS

On the strut suspension type illustrated in Figures 7 & 8, the load is carried by the upper strut mount, not the lower ball joint. The lower joint is a follower joint and should be checked for radial and/or axial movement as illustrated. Place a jack under the frame or crossmember and raise the vehicle until the tire is off the floor. Check for radial play by moving the tire/wheel assembly as indicated in Figure 7. Check for axial play by moving the lower control arm as indicated in Figure 8. **Note:** Modified Strut systems are similar except that the spring is mounted on the lower control arm, and should be checked using **Type 1** procedures.

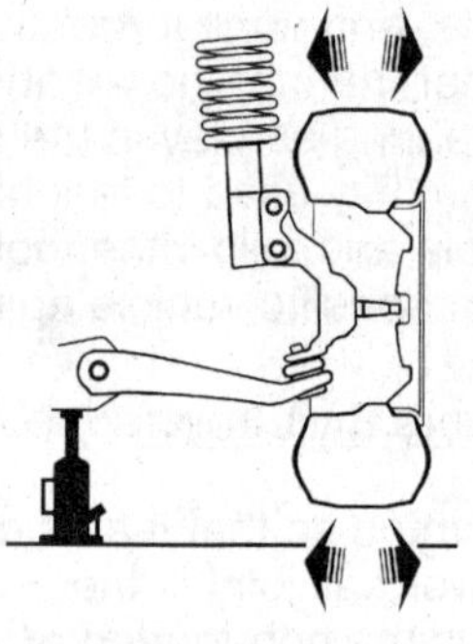

Figure 7 - RADIAL TEST

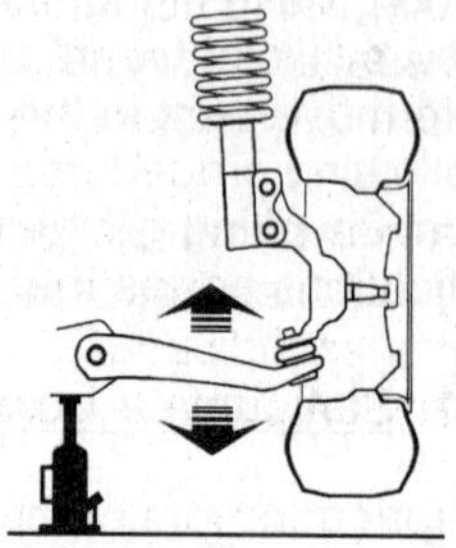

Figure 8 - AXIAL TEST

TYPE WI WEAR INDICATOR BALL JOINTS — EXCEPT CHRYSLER

Load-carrying ball joints with WEAR INDICATORS should **NOT** be unloaded when being inspected for wear. The vehicle should be at rest on a level surface as shown in Figure 9.

Clean the grease fitting area and visually inspect the boss into which the grease fitting is threaded. It should protrude beyond the face of the ball joint as in Figures 10 (approximately .050" for a new joint). If the boss is flush with the face or recessed by any amount, the ball joint should be replaced. Many domestically produced RWD vehicles have WEAR INDICATOR ball joints.

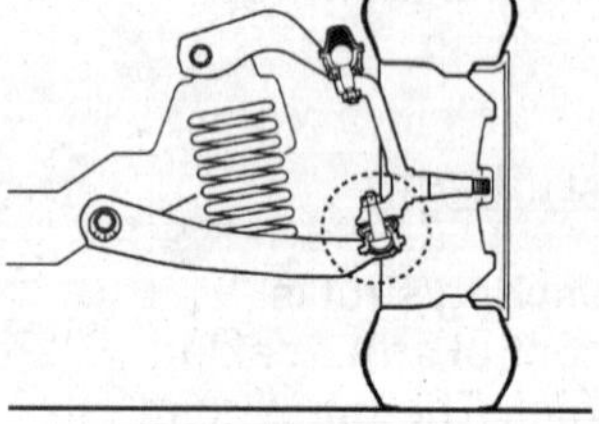

Figure 9

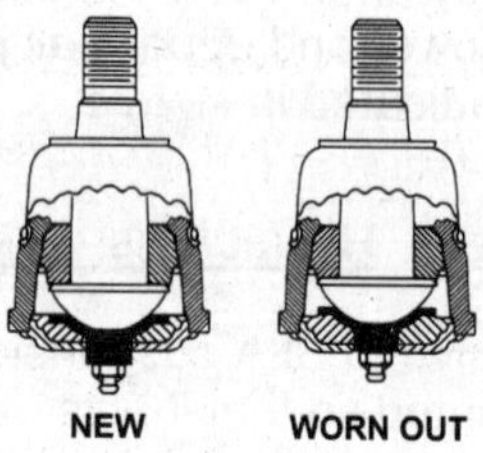

Figure 10

TYPE WI WEAR INDICATOR BALL JOINTS — CHRYSLER

Many Chrysler Corp. vehicles have a wear indicator lower ball joints. The grease fitting is mounted in a manner that allows it to become loose as the ball joint wears. Chrysler Corp. Load-carrying ball joints with grease fitting wear indicators should **NOT** be unloaded when being inspected for wear. The vehicle should be at rest on a level surface as shown in Figure 11.

Check for a grease fitting first. Note: The fitting nipple is clipped on some models. If no grease fitting is found, test the ball joint(s) for any visible movement using the procedures listed under Type in the specifications section.

Clean the grease fitting area and visually inspect the ball joint and grease seal boot. Grasp the special fitting cap or grease fitting as indicated in Figure 12. Without using additional mechanical force, attempt to rotate or otherwise move the fitting. If the fitting rotates easily, the ball joint is worn and should be considered for replacement.

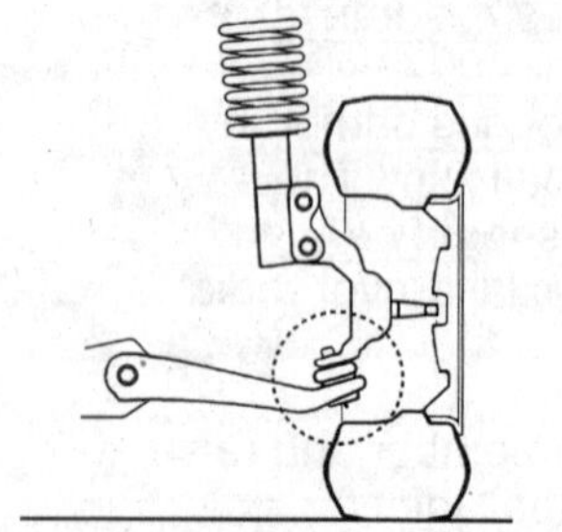

Figure 11

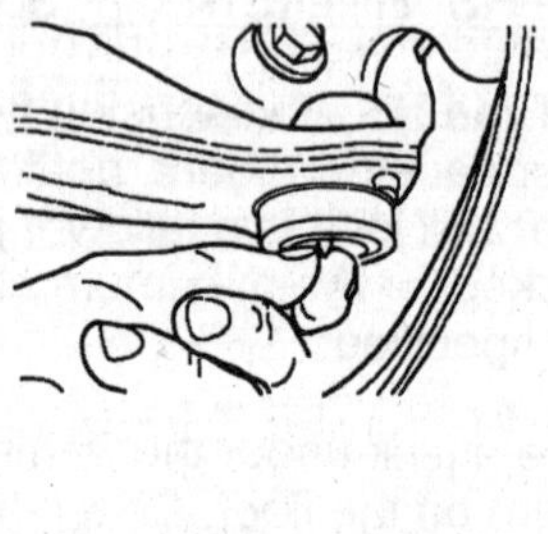

Figure 12

Caution: For all testing methods, use care not to mistake movement in the wheel bearing as play in the ball joint. If available, use a ball joint checking gauge or dial indicator and follow the tool manufacturer's instructions.

BALL JOINT TESTING (cont.)

Type SA Solid Axle

Solid Axle ball joints should be checked for excessive wear before an alignment is performed. Check the rubber grease seal and grease fitting if applicable. Replace the ball joint if excessive play is found. The following procedure is commonly used for testing **Type SA** solid axle and twin I-beam ball joints.

Place a jack under the axle or frame and raise the vehicle until the tire is off the floor. Check for axial play by moving the tire/wheel assembly as indicated in Figures 13 & 15. Check for radial play by moving the tire/wheel assembly as indicated in Figures 14 & 16. If play between the ball joint shell and stud is visible, or is beyond the maximum amount specified, the ball joint should be replaced.

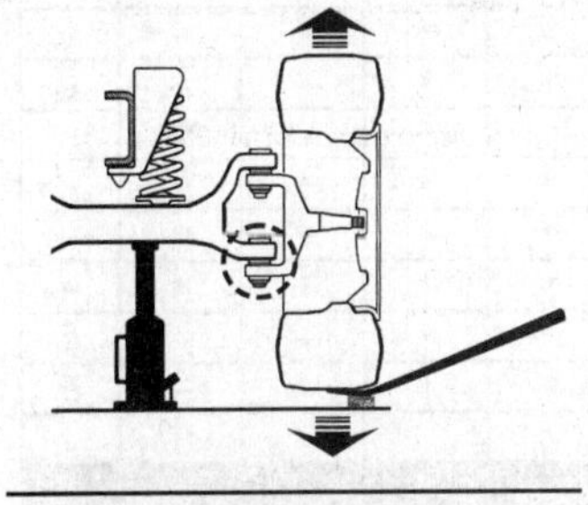
Figure 13

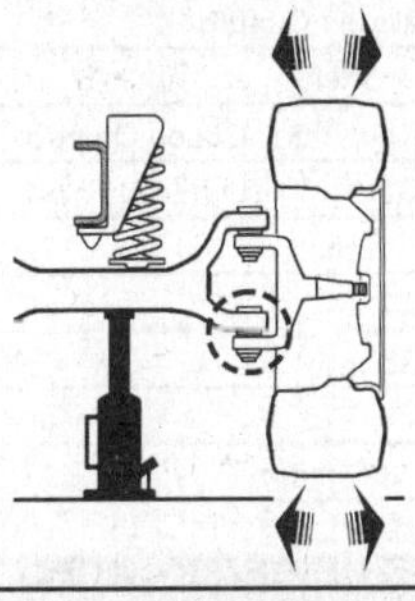
Figure 14

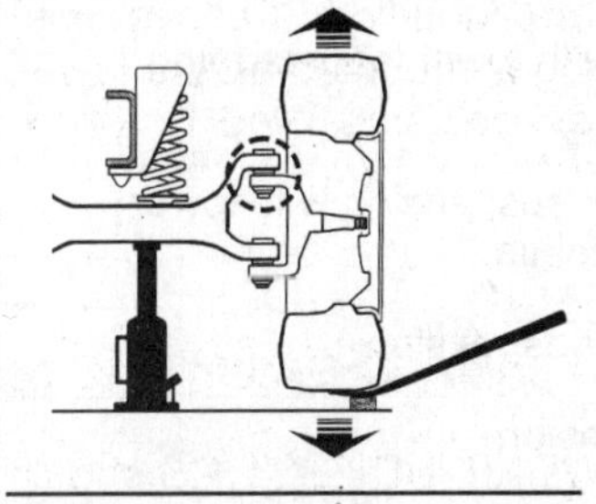
Figure 15

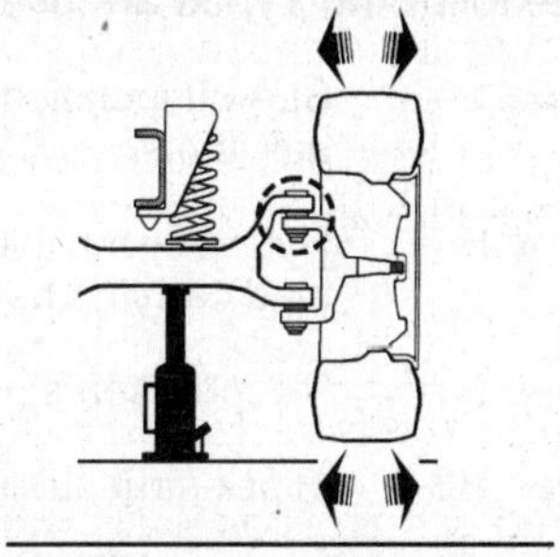
Figure 16

Type SA-C Solid Axle Ball Joints

A variation of the standard **Type SA** procedure is required for testing certain Chrysler Corp. vehicles.

Place a jack under the axle or frame and raise the vehicle until the tire is off the floor, as indicated in Figures 17 & 18.

Check the lower ball-joint for axial play.

Two measurements are required to check the lower ball joint for axial play. First, pry the steering knuckle down and record measurement 1. Re-zero the dial indicator, then pry the steering knuckle up and record measurement 2. Axial play is the sum of both measurements.

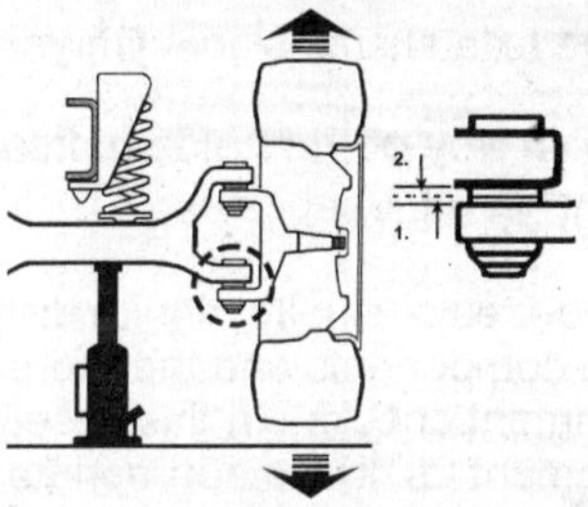
Figure 17

Check the upper ball joint for radial play.

Two measurements are required to check the upper ball joint for radial play. Moving the tire/wheel assembly as indicated in Figure 18., pull out on the top and push in on the bottom of tire and record measurement 1. Re-zero the dial indicator, then push in on the top and out on the bottom of tire and record measurement 2. Radial play is the sum of both measurements.

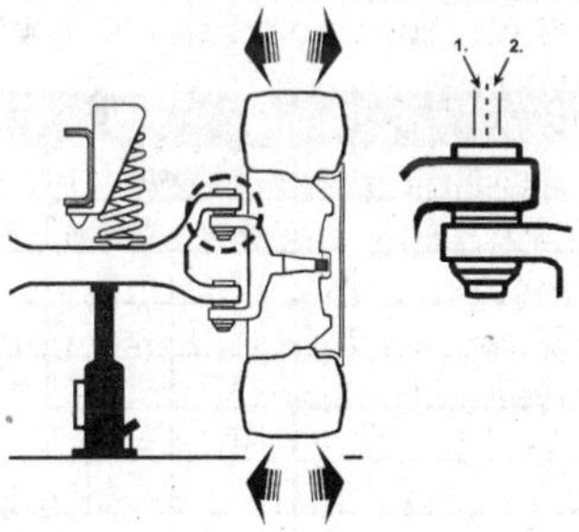
Figure 18

Type TT Turning Torque Tests

Some vehicle manufacturers specify a turning torque test for ball joints. This test requires partial disassembly of the suspension. Often, the ball joint stud must be disconnected from the spindle. A torque wrench or spring scale is used to test the amount of resistance to rotation of the ball joint stud. Since disassembly may be required, this is not a practical pre-alignment test. Instead, consider replacing the ball joint if there is any visible movement using the appropriate procedures for the suspension type being tested. Consult the Specifications pages in this manual for turning torque specifications, and for alternative testing methods.

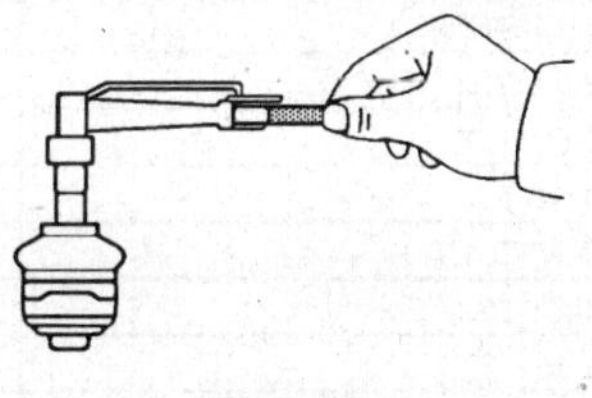
Figure 19

Using the Ball Joint Specification Tables

Ball Joint Specifications are listed alphabetically by manufacturer, year, and model. Specifications are listed for Lower Axial and Radial play (in inches), then for Upper Axial and Radial play (in inches). If the manufacturer only specifies a turning torque specification, that specification is given in inch-pounds of torque (In.lbs). "Type" indicates the ball joint/suspension type and refers to the testing procedures detailed on the previous 2 pages.

The following Types are used:

- Type 1 = SLA suspension with lower load carrying ball joints.
- Type H = Hybrid or multi-link suspension with lower load carrying ball joints.
- Type S = Macpherson strut suspension.
- Type HS = Hiper-strut suspension.
- Type WI = Any suspension type with a wear indicator ball joint.
- Type TT = Models that have only turning torque specified by the manufacturer.
- Type SA = Solid axle or I-beam suspension.
- Type SA-C = Solid axle suspension – Chrysler.

The symbol " * " in any of the fields indicates that "no visible movement" is allowed.

Hint: Turning torque tests usually require disassembly of some suspension components and therefore are not practical pre-alignment checks. In this case, check for any visible axial movement using the appropriate testing method for the suspension type. If any visible movement is found, consider replacing the ball joint before performing a wheel alignment.

Acura

		Lower		Upper		
Year	Model	Axial	Radial	Axial	Radial	Type
1997-03	CL	*	*	*	*	H
2006-10	CSX (Canada)	*	*	–	–	S
1997-00	EL	*	*	*	*	H
2001-05	EL	*	*	–	–	S
2013-17	ILX / ILX Hybrid	*	*	–	–	S
1990-01	Integra	*	*	*	*	H
1990-95	Legend	*	*	*	*	H
2001-17	MDX	*	*	–	–	S
2007-17	RDX	*	*	–	–	S
1996-17	RL / RLX	*	*	*	*	H
2002-06	RSX	*	*	–	–	S
1996-99	SLX	4-57 in.lbs (TT)		4-29 in.lbs (TT)		1
1995-08	TL	*	*	*	*	H
2009-17	TL / TLX	0.008"	*	*	*	H
2004-10	TSX	*	*	*	*	H
2011-15	TSX	0.008"	*	*	*	H
1992-94	Vigor	*	*	*	*	H
2010-13	ZDX	*	*	–	–	S

Audi

		Lower		Upper		
Year	Model	Axial	Radial	Axial	Radial	Type
1990-95	80/90	*	*	–	–	S
1990-94	100/200	*	*	–	–	S
2006-17	A3, S3	*	*	–	–	S
1997-17	A4, S4, RS4	*	*	*	*	H
2008-17	A5, S5, RS5	*	*	*	*	H
1995-97	A6, S6	*	*	–	–	S
1998-17	A6, S6, RS6	*	*	*	*	H
2012-17	A7, S7, RS7	*	*	*	*	H
1997-17	A8, S8	*	*	*	*	H
2001-17	allroad Quattro	*	*	*	*	H
1994-98	Cabriolet	*	*	–	–	S
1990-91	Coupe GT, Coupe Quattro	*	*	–	–	S
2015-17	Q3	*	*	–	–	S
2009-17	Q5, SQ5	*	*	*	*	H
2007-17	Q7	*	*	*	*	H
2008-17	R8	*	*	*	*	H
2000-17	TT, TTS	*	*	–	–	S
1990-94	V8 Quattro	*	*	–	–	S

Daewoo

		Lower		Upper		
Year	Model	Axial	Radial	Axial	Radial	Type
1999-02	Lanos	*	*	–	–	S
1999-02	Leganza	*	*	–	–	S
1999-02	Nubira	*	*	–	–	S

Fiat Chrysler (FCA)

Chrysler		Lower		Upper		
Year	Model	Axial	Radial	Axial	Radial	Type
2011-17	200	0.039"	–	–	–	S
1999-04	300M	–	*	–	–	S
2005-17	300, 300C	0.06"	–	0.06"	–	H
2007-09	Aspen	0.02"	–	0.02"	–	1
1995-00	Cirrus	0.06"	–	WI		H
1993-04	Concorde	–	*	–	–	S
1991-93	Dynasty (Canada)	WI		–	–	S
1990-93	Fifth Avenue (FWD)	WI		–	–	S
1990-93	Imperial	WI		–	–	S
1990-95	LeBaron	WI		–	–	S
1994-01	LHS	–	–	–		S
1990-93	New Yorker	WI		–	–	S
1994-96	New Yorker (LH)	–	–	–		S
2004-08	Pacifica	WI		–	–	S
2017	Pacifica	0.04"	–	–	–	S
2001-10	PT Cruiser	WI		–	–	S
1996-03	Sebring Sedan & Convertible	0.06"	–	WI		H
2004-06	Sebring Sedan & Convertible	0.06"	–	–	*	H
2007-10	Sebring Sedan & Convertible	0.04"	–	–	–	S
1995-00	Sebring Coupe lateral arm joint:	< 13 in.lbs (TT)		3-13 in.lbs (TT)		H
	compression (curved) arm joint:	4-22 in.lbs (TT)		–		H
2001-05	Sebring Coupe	22-54 in.lbs (TT)		–	–	S
1990-91	TC	WI		–	–	S
1990-07	Town & Country Minivan	WI		–	–	S
2008-16	Town & Country Minivan	0.039"	–	–	–	S
2000-03	Voyager	WI		–	–	S

Dodge		Lower		Upper		
Year	Model	Axial	Radial	Axial	Radial	Type
1995-00	Avenger lateral (straight) arm:	< 13 in.lbs (TT)		3-13 in.lbs (TT)		H
	compression (curved) arm:	4-22 in.lbs (TT)		–		H
2008-14	Avenger	0.039"	–	–	–	S
2007-12	Caliber	0.031"	–	–	–	S

Fiat Chrysler (cont.)

Dodge (cont.)	Lower		Upper		
Year Model	Axial	Radial	Axial	Radial	Type
1990-07 Caravan/Grand Caravan	WI		–	–	S
2008-17 Challenger	0.06"	–	0.06"	–	H
2006-17 Charger	0.06"	–	0.06"	–	H
1990-95 Colt	< 48 in.lbs (TT)		–	–	S
1990-91 Colt Vista (4X2)	26-87 in.lbs (TT)		–	–	S
2013-16 Dart	0.039"	–	–	–	S
1990-93 Daytona	WI		–	–	S
1990-93 Dynasty	WI		–	–	S
2008-17 Grand Caravan	0.04"	–	–	–	S
1993-04 Intrepid	–	*	–	–	S
2005-08 Magnum	0.06"		0.06"		H
1990-92 Monaco		*	–	–	S
1995-05 Neon/SX 2.0/SRT-4	WI		–	–	S
1990 Omni	WI		–	–	S
1990-94 Shadow	WI		–	–	S
1990-95 Spirit	WI		–	–	S
1991-96 Stealth	86-191 in.lbs (TT)		–	–	S
1995-03 Stratus Sedan	0.06"	–	WI		H
2004-06 Stratus Sedan	0.06"	–	–	*	H
2001-05 Stratus Coupe	22-54 in.lbs (TT)		–	–	S
1994-16 Viper	0.02"	*	0.02"	*	1

Dodge (Ram)Truck & SUV	Lower		Upper		
Year Model	Axial	Radial	Axial	Radial	Type
1990-95 B Series Van/Wagon	0.02"	–	–	*	1
1996-03 B Series Van/Wagon	0.03"	–	–	0.03"	1
1990-96 Dakota	0.02"	–	–	*	1
1997-04 Dakota	0.06"	–	–	0.06"	1
2005-11 Dakota	0.02"	–	0.02"	–	H
1999-03 Durango	0.06"	–	–	0.06"	1
2004-09 Durango	0.02"	–	0.02"	–	1
2011-17 Durango	0.02"	–	0.02"	–	H
2009-17 Journey	0.04"	–	–	–	S
2007-11 Nitro	0.02"	–	0.02"	–	H
1990-95 Pickup 4X2	0.02"	–	*	*	1
1990-95 Pickup 4X4	*	*	*	*	SA
1996-02 Pickup w/Ind. front Susp.	0.03"	–	–	0.03"	1
2002-17 Pickup w/Ind. front Susp.	0.02"	–	0.02"	–	1
1996-02 Pickup w/Link Coil Susp.	*	*	*	*	SA
2003-17 Pickup w/Link Coil Susp.	0.09"	–	–	0.06"	SA-C
2015-17 Promaster City	0.0"	–	–	–	S
2014-17 ProMaster Van/Wagon	0.04"	–	–	–	S

Eagle	Lower		Upper		
Year Model	Axial	Radial	Axial	Radial	Type
1990-93 GTX2000	26-87 in.lbs (TT)		–	–	S
1990-92 Premier	*	*	–	–	S
1990-92 Summit	< 48 in.lbs (TT)		–	–	S
1993-96 Summit	9-56 in.lbs (TT)		–	–	S
1992-96 Summit Wagon	17-78 in.lbs (TT)		–	–	S
1990-94 Talon	26-87 in.lbs (TT)		–	–	S
1995-98 Talon lateral (straight) arm:	0-13 in.lbs (TT)		3-13 in.lbs (TT)		H
compression (curved) arm:	4-22 in.lbs (TT)		–		H
1993-97 Vision	–	*	–	–	S
1990-91 Vista Wagon (Canada)	26-87 in.lbs (TT)		–	–	S

Fiat	Lower		Upper		
Year Model	Axial	Radial	Axial	Radial	Type
2012-17 500	*	*	–	–	S
2014-17 500L	*	*	–	–	S
2016-17 500X	*	*	–	–	S

Jeep	Lower		Upper		
Year Model	Axial	Radial	Axial	Radial	Type
1990-01 Cherokee/Wagoneer	*	*	*	*	SA
2014-17 Cherokee (KL)	0.039"	–	–	–	S
2006-10 Commander	0.02"	–	0.02"	–	H
2007-17 Compass	0.031"	–	–	–	S
1993-04 Grand Cherokee	*	*	*	*	SA
2005-17 Grand Cherokee	0.02"	–	0.02"	–	H
2002-07 Liberty	0.06"	–	*	*	H
2008-12 Liberty	0.02"	–	0.02"		H
2007-17 Patriot	0.031"	–	–	–	S
2015-17 Renegade	*	–	–	–	S
2007-17 Wrangler	0.05"	–	–	0.06"	SA
1990-06 All other models	*	*	*	*	SA

Plymouth	Lower		Upper		
Year Model	Axial	Radial	Axial	Radial	Type
1990-95 Acclaim	WI		–	–	S
1996-00 Breeze	0.06"		WI		H
1990-95 Colt	< 48 in.lbs (TT)		–	–	S
1990-94 Colt Vista	26-87 in.lbs (TT)		–	–	S
1988-96 Colt Wagon	17-78 in.lbs (TT)		–	–	S
1990 Horizon	WI		–	–	S
1990-94 Laser	26-87 in.lbs (TT)		–	–	S
1995-01 Neon	WI		–	–	S
1997-01 Prowler	–	0.02"	–	0.02"	1
1990 Sundance	WI		–	–	S
1990-00 Voyager Minivan	WI		–	–	S

Ford Motor Company

Ford Passenger Cars	Lower		Upper		
Year Model	Axial	Radial	Axial	Radial	Type
1994-97 Aspire	–	*	–	–	S
2013-17 C-MAX	0.008"	–	–	–	S
1995-00 Contour	–	*	–	–	S
1990 Crown Victoria, LTD	WI	–	–	*	1
1991-97 Crown Victoria, LTD	*	*	*	*	1
1998-02 Crown Victoria, LTD	0.012"	–	0.008"	–	1
2003-11 Crown Victoria	0.016"	–	0.008"	–	H
1990-96 Escort	*	*	–	–	S
1997-03 Escort/ZX2	0.008"	–	–	–	S
1990-93 Festiva	–	*	–	–	S
2011-17 Fiesta	0.008"	–	–	–	S
2005-07 Five-Hundred	0.008"	–	–	–	S
2000-17 Focus/Focus EV	0.008"	–	–	–	S
2004-07 Freestar	0.008"	–	–	–	S
2005-07 Freestyle	0.008"	–	–	–	S
2006-12 Fusion forward ball joint:	–	0.016"	0.008"	–	H
rearward ball joint:	–	*	–	–	H
2013-17 Fusion/Fusion Hybrid	0.008"	–	–	–	S
2005-06 GT	–	0.031"	–	0.031"	1
1990-93 Mustang	WI	–	–	–	S
1994-01 Mustang (use Type 1 for testing)		0.03"	–	–	S
2002-04 Mustang (use Type 1 for testing)	0.012"	–	–	–	S
2005-14 Mustang	0.012"	–	–	–	S
2015-17 Mustang	*	*	–	–	S
1990-97 Probe	–	*	–	–	S
1990-96 Taurus	–	*	–	–	S
1997-07 Taurus	0.032"	–	–	–	S
2008-17 Taurus	0.008"	–	–	–	S
2008-09 Taurus X	0.008"	–	–	–	S
1990-94 Tempo	–	*	–	–	S
1990-97 Thunderbird	–	0.015"	–	0.015"	H
2002-05 Thunderbird	0.008"	–	0.008"	–	H
1995-98 Windstar	–	0.015"	–	–	S

Ford Motor Company (cont.)

Ford Passenger Cars (cont.)	Lower		Upper		
Year Model	Axial	Radial	Axial	Radial	Type
1999-03 Windstar	0.016"	–	–	–	S

Ford Truck & SUV	Lower		Upper		
Year Model	Axial	Radial	Axial	Radial	Type
1990-97 Aerostar	–	0.031"	–	0.031"	1
1990-96 Bronco	–	0.031"	–	0.031"	SA
1990-01 E-Series	–	0.031"	–	0.031"	SA
2002-17 E-Series	–	0.04"	–	0.024"	SA
2007-17 Edge	0.008"	–	–	–	S
2001-17 Escape/ Escape Hybrid	0.008"	–	–	–	S
2000-05 Excursion	–	0.031"	–	0.031"	SA
1997-02 Expedition	0.032"	–	0.008"	–	1
2003-17 Expedition	0.032"	–	0.008"	–	H
1991-94 Explorer	–	0.031"	–	0.031"	SA
1995-01 Explorer	0.031"	–	0.031"	–	1
2002-10 Explorer	0.032"	–	0.008"	–	H
2011-17 Explorer	0.008"	–	–	–	S
2001-05 Explorer Sport/Sport Trac	0.032"	–	0.008"	–	1
2007-10 Explorer Sport Trac	0.032"	–	0.008"	–	H
1990-96 F-Series	–	0.031"	–	0.031"	1
1997-99 F-Series	–	0.031"	–	0.031"	SA
2000-01 F-Series exc/Super Duty	0.06"	–	0.06"	–	1
2002-04 F-Series/Heritage exc/Sup. Duty	0.032"	–	0.008"	–	1
2004-17 F-150	0.032"	–	0.008"	–	H
2000-01 F-Super Duty	–	0.031"	–	0.031"	SA
2002-17 F-Super Duty	–	0.04"	–	0.024"	SA
2009-17 Flex	0.008"	–	–	–	S
1990-97 Ranger	–	0.031"	–	0.031"	SA
1998-01 Ranger	0.031"	–	0.031"	–	1
2002-11 Ranger	0.032"	–	0.008"	–	1
2015-17 Transit	0.0"	–	–	–	S
2010-17 Transit Connect	0.008"	–	–	–	S

Lincoln	Lower		Upper		
Year Model	Axial	Radial	Axial	Radial	Type
2003-05 Aviator	0.032"	–	0.008"	–	H
2002-03 Blackwood	0.032"	–	0.008"	–	1
1990-94 Continental	–	*	–	–	S
1995-02 Continental	0.032"	–	–	–	S
2017 Continental	0.008"	–	–	–	S
2000-06 Lincoln LS	0.008"	–	0.008"	–	H
1990-92 Mark VII	WI	–	–	–	S
1993-98 Mark VIII	–	0.015"	–	0.015"	H
2006-08 Mark LT	0.032"	–	0.008"	–	H
2015-17 MKC	0.008"	–	–	–	S
2009-16 MKS	0.008"	–	–	–	S
2010-17 MKT	0.008"	–	–	–	S
2007-17 MKX	0.008"	–	–	–	S
2007-12 MKZ/Zephyr forward ball joint:	–	0.016"	0.008"	–	H
rearward ball joint:	–	*	–	–	H
2013-17 MKZ/MKZ Hybrid	0.008"	–	–	–	S
1998-02 Navigator	0.032"	–	0.008"	–	1
2003-17 Navigator	0.032"	–	0.008"	–	H
1990-97 Town Car	WI	–	–	0.015"	1
1998-02 Town Car	0.012"	–	0.008"	–	1
2003-11 Town Car	0.016"	–	0.008"	–	H

Mercury	Lower		Upper		
Year Model	Axial	Radial	Axial	Radial	Type
1991-94 Capri	*	*	–	–	S
1990-97 Cougar	–	0.015"	–	0.015"	H
1999-02 Cougar	*	–	–	–	S

Mercury (cont.)	Lower		Upper		
Year Model	Axial	Radial	Axial	Radial	Type
1990 Grand Marquis, Colony Park	WI	–	–	*	1
1991-97 Grand Marquis	*	*	*	*	1
1998-02 Grand Marquis	0.012"	–	0.008"	–	1
2003-11 Grand Marquis	0.016"	*	0.008"	*	H
2005-11 Mariner	0.008"	–	–	–	S
2006-11 Mariner Hybrid	0.008"	–	–	–	S
2003-04 Marauder	0.016"	*	0.008"	*	H
2006-11 Milan forward ball joint:	–	0.016"	0.008"	–	H
rearward ball joint:	–	*	–	–	H
2005-07 Montego	0.008"	–	–	–	S
2004-07 Monterey	0.008"	–	–	–	S
1997-01 Mountaineer	0.031"	–	0.031"	–	1
2002-10 Mountaineer	0.032"	–	0.008"	–	H
1995-00 Mystique	–	0.015"	–	–	S
1990-96 Sable	–	*	–	–	S
1997-00 Sable	–	0.031"	–	–	S
2001-05 Sable	0.031"	–	–	–	S
2008-09 Sable	0.008"	–	–	–	S
1990-94 Topaz		*	–	–	S
1990-96 Tracer	*	*	–	–	S
1997-00 Tracer	–	0.031"	–	–	S
1993-02 Villager	–	0.031"	–	–	S

General Motors

GM – Buick	Lower		Upper		
Year Model	Axial	Radial	Axial	Radial	Type
2005-09 Allure	0.02"	–	–	–	S
2016-17 Cascada	–	–	0.0"	0.0"	HS
1990-00 Century	–	*	–	–	S
2001-05 Century	0.02"	–	–	–	S
1990 Electra	*	*	–	–	S
2008-17 Enclave	0.02"	–	–	–	S
2013-17 Encore	*	*	–	–	S
2017 Envision exc/HiPer Strut	0.02"	–	–	–	S
2017 Envision w/HiPer Strut	–	–	0.0"	0.0"	HS
1990 Estate Wagon (LeSabre & Electra)	WI	–	*	*	1
2005-17 LaCrosse exc/Hiperstrut	0.02"	–	–	–	S
2010-17 LaCrosse CXS w/Hiperstrut	–	–	0.0"	0.0"	HS
1990-99 LeSabre	*	*	–	–	S
2000-05 LeSabre	0.125"	0.125"	–	–	S
2006-11 Lucerne	0.063"	–	–	–	S
1990-96 Park Avenue	*	*	–	–	S
1997-05 Park Avenue / Ultra	0.125"	0.125"	–	–	S
2004-07 Rainier	0.06"	–	0.06"	–	H
1990-91 Reatta	*	*	–	–	S
1990-04 Regal	0.125"	0.125"	–	–	S
2011-17 Regal exc/Hiperstrut	0.02"	–	–	–	S
2011-17 Regal GS w/Hiperstrut	–	–	0.0"	0.0"	HS
2002-07 Rendezvous	0.125"	0.125"	–	–	S
1990-99 Riveria	*	*	–	–	S
1991-96 Roadmaster	WI	–	*	*	1
1990-98 Skylark	*	*	–	–	S
2005-07 Terraza	0.028"	–	–	–	S
2012-17 Verano	*	*	–	–	S

GM – Cadillac	Lower		Upper		
Year Model	Axial	Radial	Axial	Radial	Type
1990-93 Allante'	*	*	–	–	S
2013-17 ATS, ATS-V	0.02"	0.02"	–	–	S
1990-96 Brougham/Fleetwood RWD	WI	–	*	*	1
2016-17 CT6	*	*	*	*	H
2003-08 CTS	0.125"	0.125"	*	*	H

General Motors (cont.)

GM – Cadillac (cont.)		Lower		Upper		
Year	Model	Axial	Radial	Axial	Radial	Type
2009-15	CTS, CTS-V (VIN D)	*	–	0.02"	–	H
2014-17	CTS, CTS-V (VIN A)	*	*	–	–	S
1997-01	Catera	*	*	–	–	S
1990-99	DeVille/Concours	–	*	–	–	S
2000-05	DeVille	0.125"	0.125"	–	–	S
2006-11	DTS	0.063"		–	–	S
1990-00	Eldorado	–	*	–	–	S
2001-02	Eldorado	0.125"	0.125"	–	–	S
2014-16	ELR	–	–	0.0"	0.0"	HS
1999-06	Escalade/Escalade EXT/ESV	0.08"	–	0.08"	–	1
2007-17	Escalade/Escalade EXT/ESV	0.02"	–	0.02"	–	H
1990-93	Fleetwood (FWD)	*	*	–	–	S
1990-00	Seville	*	*	–	–	S
2001-04	Seville	0.125"	0.125"	–	–	S
2004-09	SRX	0.125"	0.125"	*	*	H
2010-16	SRX	0.02"	–	–	–	S
2005-11	STS	0.125"	0.125"	*	*	H
2004-09	XLR	0.02"	–	0.005"	–	1
2017	XT5	*	*	–	–	S
2013-17	XTS exc/Hiperstrut	0.02"	–	–	–	S
2013-17	XTS w/Hiperstrut	–	–	0.0"	0.0"	HS

GM - Chevrolet		Lower		Upper		
Year	Model	Axial	Radial	Axial	Radial	Type
2004-11	Aveo	*	*	–	–	S
1990-96	Beretta / Corsica	–	*	–	–	S
2017	Bolt EV	*	*	–	–	S
1990-92	Camaro	WI	*	–	–	S
1993-00	Camaro	0.047"	*	*	*	1
2001-02	Camaro	WI	*	–	–	S
2010-17	Camaro	0.02"	–	–	–	S
1990-96	Caprice / Impala / Impala SS	WI		*	*	1
2011-17	Caprice PPV	0.02"	–	–	–	S
1990-94	Cavalier	*	*	–	–	S
1995-05	Cavalier	0.125"	0.125"	–	–	S
1990	Celebrity	–	*	–	–	S
2004-05	Classic (Malibu)	0.125"	0.125"	–	–	S
2008	Classic (Malibu)	0.02"	–	–	–	S
2005-10	Cobalt	0.02"	–	–	–	S
1990-96	Corvette	WI	–	*	*	1
1997-04	Corvette	WI	0.125"	*	*	1
2005-17	Corvette	0.02"	–	0.02"	–	1
2011-16	Cruze, Cruze Limited (VIN P)	*	–	–	–	S
2016-17	Cruze (VIN B)	*	–	–	–	S
2006-11	HHR	0.02"	–	–	–	S
2000-15	Impala, Impala Limited (VIN W)	0.02"	–	–	–	S
2016-17	Impala (VIN 1)	0.02"	–	–	–	S
1990-01	Lumina	–	*	–	–	S
1990-96	Lumina APV	–	*	–	–	S
1997-00	Malibu	–	*	–	–	S
2001-03	Malibu	0.125"	0.125"	–	–	S
2004-12	Malibu/Malibu Maxx	0.02"	–	–	–	S
2013-16	Malibu, Malibu Limited (VIN 1)	*	*	–	–	S
2016-17	Malibu (VIN Z)	*	*	–	–	S
1998-01	Metro	–	*	–	–	S
1995-00	Monte Carlo	–	*	–	–	S
2001-07	Monte Carlo	0.02"	–	–	–	S
2012-14	Orlando	*	*	–	–	S
1998-02	Prizm	9-26 in.lbs (TT)		–	–	S
2012-17	Sonic	*	*	–	–	S
2013-17	Spark/Spark EV	*	*	–	–	S
2014-17	SS	0.02"	–	–	–	S

GM - Chevrolet (cont.)		Lower		Upper		
Year	Model	Axial	Radial	Axial	Radial	Type
2005-09	Uplander	0.028"	–	–	–	S
1997-00	Venture	*	*	–	–	S
2001-05	Venture	0.028"	–	–	–	S
2004-09	Vivant	–	*	–	–	S
2011-15	VOLT	0.02"	–	–	–	S
2016-17	VOLT	*	*	–	–	S

GM – Chevrolet/GMC Truck & SUV		Lower		Upper		
Year	Model	Axial	Radial	Axial	Radial	Type
1991-92	Syclone	0.125"	0.08"	*	*	1
1992-93	Typhoon	0.125"	0.08"	*	*	1
1990-00	C-Pickup 4X2	WI	–	*	*	1
1999-07	C-Silverado/Sierra 4X2 (Classic)	0.08"	–	0.08"	–	1
2007-17	C-Silverado/Sierra 1500 4X2	0.02"	–	0.02"	–	H
2007-17	C-Silverado/Sierra 2500/3500 4X2	0.02"	–	0.02"	–	1
1990-91	R-Pickup 4X2	0.093"	–	*	*	1
1990-00	K-Pickup 4X4	0.08"	–	*	*	1
1999-07	K-Silverado/Sierra 4X4 (Classic)	0.08"	–	0.08"	–	1
2007-17	K-Silverado/Sierra 1500 4X4	0.02"	–	0.02"	–	H
2007-17	K-Silverado/Sierra 2500/3500 4X4	0.02"	–	0.02"	–	1
1990-91	V-Pickup 4X4	–	*	–	*	SA
2007-17	Acadia/Acadia Limited	0.02"	–	–	–	S
2017	Acadia (VIN ***N)	*	*	–	–	S
1990-05	Astro / Safari RWD	WI	–	–	*	1
1990-05	Astro / Safari AWD	0.125"	–	–	*	1
2002-06	Avalanche (All Models)	0.08"	–	0.08"	*	1
2007-13	Avalanche 1500	0.02"	–	0.02"	*	H
2004-17	Colorado/Canyon	0.02"	–	0.02"	–	1
1990-91	Blazer / Jimmy 4X4	*	*	*	*	SA
1992-94	Blazer 4X4	0.08"	–	*	*	1
1999-00	Denali	0.08"	–	*	*	1
2001-07	Denali/Denali XL	0.08"	–	0.08"	–	1
1999-01	Envoy	0.125"	0.08"	*	*	1
2002-09	Envoy	0.06"	–	0.06"	–	H
2005-17	Equinox	0.02"	–	–	–	S
1990-95	Full-size Van (G)	WI	–	*	*	1
1996-02	Express / Savana (G)	0.125"	0.08"	*	*	1
2003-17	Express / Savana (G/H)	0.02"	–	0.02"	–	1
1990-95	LLV	WI	0.08"	*	*	1
1990-05	S10 Blazer/S15 Jimmy 4X2	WI	0.08"	*	*	1
1990-05	S10 Blazer/S15 Jimmy 4X4	0.125"	0.08"	*	*	1
1990-03	S10 / S15 / Sonoma 4X2	WI	0.08"	*	*	1
1990-04	S10 / S15 / Sonoma 4X4	0.125"	0.08"	*	*	1
2003-06	SSR	0.06"	–	0.06"	–	H
1990-91	Suburban 4X2	0.093"	–	*	*	1
1990-91	Suburban 4X4	–	*	–	*	SA
1992-99	Suburban 4X2	WI		*	*	1
1992-99	Suburban 4X4	0.08"	–	*	*	1
2000-06	Suburban (All Models)	0.08"	–	0.08"	–	1
2007-17	Suburban (1/2 Ton Models)	0.02"	–	0.02"	–	H
2007-13	Suburban (3/4 Ton Models)	0.02"	–	0.02"	–	1
2016-17	Suburban 3500HD	0.02"	–	0.02"	–	1
1995-00	Tahoe 4X2 (2000 Limited Only)	WI		*	*	1
1995-00	Tahoe 4X4 (2000 Z71 Only)	0.08"	–	*	*	1
2000-06	Tahoe (exc/2000 Limited & Z71)	0.08"	–	0.08"	–	1
2007-17	Tahoe (All Models)	0.02"	–	0.02"	–	H
2010-17	Terrain	0.02"	–	–	–	S
1998-04	Tracker	*	*	–	–	S
2002-09	Trailblazer (All)	0.06"	–	0.06"	–	H
2009-17	Traverse	0.02"	–	–	–	S
2013-17	TRAX	*	*	–	–	S
1995-99	Yukon 4X2	WI		*	*	1
1992-99	Yukon 4X4	0.08"	–	*	*	1

General Motors (cont.)

Chevrolet/GMC Truck & SUV (cont.)	Lower		Upper		
Year Model	Axial	Radial	Axial	Radial	Type
2000-06 Yukon/Yukon Denali	0.08"	–	0.08"	–	1
2007-17 Yukon/Yukon Denali	0.02"	–	0.02"	–	H
2000-06 Yukon XL/Yukon XL Denali	0.08"	–	0.08"	–	1
2007-17 Yukon XL (1/2 Ton Models)	0.02"	–	0.02"	–	H
2007-13 Yukon XL (3/4 Ton Models)	0.02"	–	0.02"	–	1
2007-17 Yukon XL Denali	0.02"	–	0.02"	–	H

GM - Hummer	Lower		Upper		
Year Model	Axial	Radial	Axial	Radial	Type
2003-10 H2	0.02"	–	0.02"	–	1
2006-10 H3	0.02"	–	0.02"	–	1

GM - Oldsmobile	Lower		Upper		
Year Model	Axial	Radial	Axial	Radial	Type
1992-98 Achieva	*	*	–	–	S
1999-04 Alero	0.125"	0.125"	–	–	S
1995-03 Aurora	0.125"	0.125"	–	–	S
1991-01 Bravada	0.125"	0.08"	*	*	1
2002-04 Bravada	0.06"	–	0.06"	–	H
1990-91 Calais	*	*	–	–	S
1990-92 Custom Cruiser	WI	–	*	*	1
1990-97 Cutlass Supreme (FWD)	*	*	–	–	S
1997-99 Cutlass	*	*	–	–	S
1987-96 Cutlass Ciera	*	*	–	–	S
1990-96 Cutlass Cruiser	*	*	–	–	S
1990-99 Eighty-Eight / LSS	*	*	–	–	S
1998-03 Intrigue	0.125"	0.125"	–	–	S
1990-97 Ninety-Eight	*	*	–	–	S
1997-98 Regency	*	*	–	–	S
1990-96 Silhouette	*	*	–	–	S
1997-04 Silhouette	0.028"	–	–	–	S
1990-92 Toronado / Trofeo	*	*	–	–	S

GM - Pontiac	Lower		Upper		
Year Model	Axial	Radial	Axial	Radial	Type
1990-91 6000	*	*	–	–	S
2001-05 Aztek	0.125"	0.125"	–	–	S
1990-99 Bonneville	*	*	–	–	S
2000-05 Bonneville	0.125"	0.125"	–	–	S
1990-92 Firebird/Trans Am	WI	–	–	–	S
1993-02 Firebird/Trans Am	WI	*	*	*	1
1990-99 Firefly (Canada)	*	*	–	–	S
2007-09 G3 (Mexico)	*	*	–	–	S
2007-09 G5/G5 Persuit	0.02"	–	–	–	S
2005-09 G6	0.125"	–	–	–	S
2008-09 G8	0.125"	0.125"	–	–	S
1990-04 Grand Am	0.125"	0.125"	–	–	S
1990-08 Grand Prix	0.125"	0.125"	–	–	S
2004-06 GTO	*	*	–	–	S
1990-93 Lemans	*	*	–	–	S
2002-09 Montana/Montana SV6	0.028"	–	–	–	S
2005-06 Pursuit (Canada)	0.02"	–	–	–	S
2006-09 Solstice	0.02"	–	0.005"	–	1
1990-94 Sunbird	*	*	–	–	S
1995-05 Sunfire	0.125"	0.125"	–	–	S
1988-94 Tempest (Canada)	*	*	–	–	S
2006-09 Torrent	0.02"	–	–	–	S
1990-98 Trans Sport	*	*	–	–	S
2003-10 Vibe (All models)	0.0"	–	–	–	S
2005-09 Wave	*	*	–	–	S

GM - Saturn	Lower		Upper		
Year Model	Axial	Radial	Axial	Radial	Type
2008 Astra	0.02"	–	–	–	S
2007-09 Aura	0.02"	–	–	–	S
2003-07 Ion	0.02"	–	–	–	S
2000-05 L Series	0.125"	0.125"	–	–	S
2007-11 Outlook	0.02"	–	–	–	S
2005-07 Relay	0.028"	–	–	–	S
1991-02 S Series	*	*	–	–	S
2007-09 Sky	0.02"	–	0.005"	–	1
2002-09 Vue	0.02"	–	–	–	S

Genesis

Year Model	Lower Axial	Lower Radial	Upper Axial	Upper Radial	Type
2017 G80	*	*	*	*	H
2017 G90	*	*	*	*	H

Honda

	Lower		Upper		
Year Model	Axial	Radial	Axial	Radial	Type
1990-07 Accord/Accord Hybrid	*	*	*	*	H
2008-12 Accord	0.008"	*	*	*	H
2013-17 Accord/Accord Hybrid	0.008"	*	–	–	S
1990-00 Civic (All models)	*	*	*	*	H
2001-17 Civic (All models)	*	*	–	–	S
1993-97 Civic del Sol	*	*	*	*	H
2017 Clarity	*	*	–	–	S
1997-01 CR-V	*	*	*	*	H
2002-17 CR-V	*	*	–	–	S
2011-16 CR-Z	*	*	–	–	S
1990-91 CRX	*	*	*	*	H
2010-15 Crosstour (Accord Crostour)	0.008"	*	*	*	H
2003-11 Element	*	*	–	–	S
2007-17 Fit/Fit EV	*	*	–	–	S
2016-17 HR-V	*	*	–	–	S
2000-14 Insight	*	*	–	–	S
1995-98 Odyssey	*	*	*	*	H
1999-17 Odyssey	*	*	–	–	S
1994-02 Passport	4-57 in.lbs (TT)		4-29 in.lbs (TT)		1
2003-17 Pilot	*	*	–	–	S
1990-01 Prelude	*	*	*	*	H
2006-17 Ridgeline	*	*	–	–	S
2000-09 S2000	*	*	*	*	H

Hyundai

	Lower		Upper		
Year Model	Axial	Radial	Axial	Radial	Type
1995-99 Accent	0.0"	–	–	–	S
2000-05 Accent	22-89 in.lbs (TT)			–	–
2006-17 Accent	*	–	–	–	S
2006-11 Azera	3.5-18 in.lbs (TT)		4.4-13 in.lbs (TT)		H
2012-17 Azera	*	*	–	–	S
1992-96 Elantra	0.0"	–	–	–	S
1997-03 Elantra	18-89 in.lbs (TT)		–	–	S
2004-06 Elantra	18-31 in.lbs (TT)		–	–	S
2007-17 Elantra (All models)	*	–	–	–	S
2007-09 Entourage	9-31 in.lbs (TT)		–	–	S
2011-16 Equus	*	*	*	*	H
1990-94 Excel	0.0"	–	–	–	S
2009-16 Genesis Sedan	*	*	*	*	H
2010-16 Genesis Coupe	*	–	–	–	S
2017 Ioniq (All models)	0.0"	–	–	–	S
2001-06 Santa Fe	13-31 in.lbs (TT)		–	–	S
2007-17 Santa Fe/Santa Fe Sport	0.0"	–	–	–	S
1991-95 Scoupe	0.0"	–	–	–	S
1990-98 Sonata	18-89 in.lbs (TT)		–	–	S

Hyundai (cont.)

Year	Model	Lower Axial	Lower Radial	Upper Axial	Upper Radial	Type
1999-03	Sonata	1.8-8.9 in.lbs (TT)		4.4-13 in.lbs (TT)		H
2004-10	Sonata	3.5-18 in.lbs (TT)		4.4-13 in.lbs (TT)		H
2011-17	Sonata/Sonata Hybird	0.0"				S
1997-01	Tiburon	15-28 in.lbs (TT)		–	–	S
2003-08	Tiburon	18-31 in.lbs (TT)		–	–	S
2005-17	Tucson	*	–	–	–	S
2012-17	Veloster	*	–	–	–	S
2007-11	Veracruz	3.5-18 in.lbs (TT)		–	–	S
2012-13	Veracruz	*	–	–	–	S
2001-05	XG300, XG350	2-13.3 in.lbs (TT)		4.4-13 in.lbs (TT)		H

Infiniti

Year	Model	Lower Axial	Lower Radial	Upper Axial	Upper Radial	Type
2008-13	EX	0.0"	–	0.0"	–	H
2003-08	FX	0.0"	–	–	–	S
2009-13	FX	0.0"	–	0.0"	–	H
1991-96	G20	0.0"	–	–	–	S
1999-02	G20	0.0"	–	–	–	S
2003-08	G35 Coupe/Sedan	0.0"	–	0.0"	–	H
2008-13	G25/G37 Coupe/Sedan/Conv.	0.0"	–	0.0"	–	H
1996-04	I30/I35	0.0"	–	–	–	S
1993-97	J30	0.0"	–	–	–	S
2013	JX	0.0"	–	–	–	S
1990-92	M30	0.05"	–	–	–	S
2003-04	M45	0.0"	–	–	–	S
2006-14	M35/M37/M45/M56/M Hybrid	0.0"	–	0.0"	–	H
2015	Q40	0.0"	–	0.0"	–	H
1990-06	Q45	0.0"	–	–	–	S
2014-17	Q50/Q50 Hybrid	0.0"	–	0.0"	–	H
2014-17	Q60	0.0"	–	0.0"	–	H
2014-17	Q70/Q70 Hybrid	0.0"	–	0.0"	–	H
1997-03	QX4	0.0"	–	–	–	S
2017	QX30	0.0"	–	–	–	S
2004-13	QX56	0.0"	–	0.0"	–	H
2014-17	QX50	0.0"	–	0.0"	–	H
2014-17	QX60/QX60 Hybrid	0.0"	–	–	–	S
2014-17	QX70	0.0"	–	0.0"	–	H
2014-17	QX80	0.0"	–	0.0"	–	H

Isuzu

Year	Model	Lower Axial	Lower Radial	Upper Axial	Upper Radial	Type
1990-94	Amigo	*	*	*	*	1
1998-00	Amigo	4-57 in.lbs (TT)		4-29 in.lbs (TT)		1
2003-08	Ascender	0.06"	–	0.06"	–	H
2002-04	Axiom	4-42 in.lbs (TT)		4-29 in.lbs (TT)		1
1996-00	Hombre 4X2	WI	0.08"	*	*	1
2006-08	I-Series Pickup	0.02"	–	0.02"	–	1
1996-00	Hombre 4X4	0.125"	0.08"	*	*	1
1990-93	Impulse	*	*	–	–	S
1996-99	Oasis	*	*	*	*	H
1990-95	Pickup (All Models)	*	*	*	*	1
1991-04	Rodeo, Rodeo Sport	4-57 in.lbs (TT)		4-29 in.lbs (TT)		1
1990-93	Stylus	*	*	–	–	S
1990-02	Trooper	4-57 in.lbs (TT)		4-29 in.lbs (TT)		1
1999-01	VehiCROSS	4-57 in.lbs (TT)		4-29 in.lbs (TT)		1

Kia

Year	Model	Lower Axial	Lower Radial	Upper Axial	Upper Radial	Type
2004-06	Amanti	2-9 in.lbs (TT)		13-22 in.lbs (TT)		H

Kia (cont.)

Year	Model	Lower Axial	Lower Radial	Upper Axial	Upper Radial	Type
2007-09	Amanti	*	*	*	*	H
2009	Borrego	*	*	*	*	H
2014-17	Cadenza	*	*	–	–	S
2010-17	Forte (All models)	*	*	–	–	S
2015-17	K900	*	*	*	*	H
2017	Niro	*	*	–	–	S
2001-06	Optima/Magentis	2-9 in.lbs (TT)		4-13 in.lbs (TT)		H
2007-10	Optima/Magentis	9-31 in.lbs (TT)		–	–	S
2011-17	Optima/Optima Hybrid	0.0"				S
2001-05	Rio	9-31 in.lbs (TT)		–	–	S
2006-17	Rio (All models)	*	*	–	–	S
2007-16	Rondo	*	*	–	–	S
2002-05	Sedona	*	*	–	–	S
2006-14	Sedona	9-31 in.lbs (TT)		–	–	S
2015-17	Sedona	*	*	–	–	S
1993-01	Sephia	17-30 in.lbs (TT)		–	–	S
2003-09	Sorento	*	*	*	*	H
2010-17	Sorento	*	*	–	–	S
2010-17	Soul	*	*	–	–	S
2000-09	Spectra/Spectra5	18-31 in.lbs (TT)		–	–	S
1995-02	Sportage	*	*	*	*	H
2005-10	Sportage	4-22 in.lbs (TT)		–	–	S
2011-17	Sportage	*	*	–	–	S

Lexus

Year	Model	Lower Axial	Lower Radial	Upper Axial	Upper Radial	Type
2011-17	CT200h	0.0"	–	–	–	S
1990-03	ES 250/ES 300	0.0"	–	–	–	S
2004-17	ES 330/ES 350/ES 300h	0.0"	–	–	–	S
1993-95	GS 300	0.012"	–	9-30 in.lbs (TT)		H
1996-06	GS 300	0.035"	–	9-30 in.lbs (TT)		H
2007-11	GS 350	0.0"	–	< 8.7 in.lbs (TT)		H
2013-17	GS 200t/GS 350	0.0"	–	< 30 in.lbs (TT)		H
1998-00	GS 400	0.035"	–	9-30 in.lbs (TT)		H
2001-05	GS 430	0.035"	–	9-30 in.lbs (TT)		H
2006-07	GS 430	0.0"	–	< 8.7 in.lbs (TT)		H
2007-12	GS 450h	0.0"	–	< 8.7 in.lbs (TT)		H
2013-17	GS 450h	0.0"	–	< 30 in.lbs (TT)		H
2008-11	GS 460	0.0"	–	< 8.7 in.lbs (TT)		H
2016-17	GS F	0.0"	–	9-39 in.lbs (TT)		H
2003-09	GX 470	0.02"	–	< 40 in.lbs (TT)		H
2010-17	GX 460	0.0"	–	9-39 in.lbs (TT)		H
2010-12	HS 250h	0.0"	–	–	–	S
2001-05	IS 300	0.035"	–	9-30 in.lbs (TT)		H
2006-15	IS 250/IS 350/IS F	0.0"	–	9-30 in.lbs (TT)		H
2016-17	IS 250t/IS 300/IS 350	0.0"	–	9-30 in.lbs (TT)		H
2010-14	IS 250C/IS 350C	0.0"	–	9-30 in.lbs (TT)		H
1990-00	LS 400	0.012"	–	9-30 in.lbs (TT)		H
2001-06	LS 430	0.0"	–	9-30 in.lbs (TT)		H
2007-17	LS 460/460L each upper joint	0.0"	–	9-30 in.lbs (TT)		H
2008-16	LS 600h/600hL each upper joint	0.0"	–	< 31 in.lbs (TT)[1]		H
1998-07	LX 470	0.0"	–	9-39 in.lbs (TT)		1
2008-17	LX 570	0.0"	–	9-39 in.lbs (TT)		H
2015-17	NX 200t/NX 300h	0.0"	–	–	–	S
2015-17	RC 200t/RC 300/RC 350/RC F	0.0"	–	9-30 in.lbs (TT)		H
1999-17	RX 300/RX 330/RX 350	0.0"	–	–	–	S
2005-16	RX 400h/RX 450h	0.0"	–	–	–	S
1992-00	SC 300/SC 400	0.012"	–	9-31 in.lbs (TT)		H
2002-10	SC 430	0.016"	–	9-30 in.lbs (TT)		H

BALL JOINT SPECIFICATIONS

Ball Joint Specifications

Mazda

		Lower		Upper		
Year	Model	Axial	Radial	Axial	Radial	Type
1990-94	323/Protegé	*	*	–	–	S
1990-97	626/MX-6	*	–	–	–	S
1998-02	626	9-44 in.lbs (TT)		–	–	S
1988-91	929	13-31 in.lbs (TT)		–	–	S
1992-95	929	13-40 in.lbs (TT)		13-40 in.lbs (TT)		H
1994-97	B Series Pickup	–	0.031"	–	0.031"	SA
1998-01	B Series Pickup/Truck	0.031"	–	0.031"	–	1
2002-09	B Series Pickup/Truck	0.031"	–	0.008"	–	1
2016-17	CX-3	28 in.lbs MAX (TT)		–	–	S
2013-17	CX-5	28 in.lbs MAX (TT)		–	–	S
2007-12	CX-7	3-17 in.lbs (TT)		–	–	S
2007-15	CX-9	23 in.lbs MAX (TT)		–	–	S
2016-17	CX-9	28 in.lbs MAX (TT)		–	–	S
2011-15	Mazda 2	28 in.lbs MAX (TT)		–	–	S
2016-17	Mazda 2	18 in.lbs MAX (TT)		–	–	S
2004-17	Mazda 3	9-43 in.lbs (TT)		–	–	S
2006-15	Mazda 5	9-43 in.lbs (TT)		–	–	S
2003-08	Mazda 6 forward lower ball joint:	11-20 in.lbs (TT)		0-13 in.lbs (TT)		H
	rearward lower ball joint:	9-20 in.lbs (TT)		–		H
2009-13	Mazda 6	11-20 in.lbs (TT)		0-13 in.lbs (TT)		H
2014-17	Mazda 6	28 in.lbs MAX (TT)		–	–	S
1995-02	Millenia	*	*	*	*	H
1990-98	MPV	*	*	–	–	S
2000-06	MPV	9-43 in.lbs (TT)		–	–	S
1992-96	MX-3	*	*	–	–	S
1990-97	MX-5 Miata	4-13 in.lbs (TT)		4-16 in.lbs (TT)		H
1998-05	MX-5 Miata	3-17 in.lbs (TT)		3-20 in.lbs (TT)		H
2006-15	MX-5 Miata	4-25 in.lbs (TT)		3-19 in.lbs (TT)		H
2016-17	MX-5 Miata	5-26 in.lbs (TT)		5-17 in.lbs (TT)		H
1991-94	Navajo	–	0.031"	–	0.031"	SA
1995-03	Protegé	9-44 in.lbs (TT)		–	–	S
1990-91	RX-7	*	*	–	–	S
1993-95	RX-7	*	*	*	*	H
2004-08	RX-8	4-25 in.lbs (TT)		3-19 in.lbs (TT)		H
2009-11	RX-8	20 in.lbs MAX (TT)		16 in.lbs MAX (TT)		H
2001-11	Tribute	0.008"	–	–	–	S
2008-11	Tribute Hybrid	0.008"	–	–	–	S

Mitsubishi

		Lower		Upper		
Year	Model	Axial	Radial	Axial	Radial	Type
1991-99	3000GT	86-191 in.lbs (TT)		–	–	S
1992-04	Diamante	87-190 in.lbs (TT)		–	–	S
1990-94	Eclipse	26-87 in.lbs (TT)		–	–	S
1995-97	Eclipse	3-13 in.lbs (TT)		4-22 in.lbs (TT)		H
1998-99	Eclipse	3-22 in.lbs (TT)		4-22 in.lbs (TT)		H
2000-05	Eclipse/Eclipse Spyder	22-54 in.lbs (TT)		–	–	S
2006-12	Eclipse/Eclipse Spyder	*	–	–	–	S
2004-08	Endeavor	22-54 in.lbs (TT)		–	–	S
2010-11	Endeavor	*	–	–	–	S
1992-96	Expo/LRV	17-78 in.lbs (TT)		–	–	S
1990-93	Galant / Sigma	26-87 in.lbs (TT)		–	–	S
1994-98	Galant lateral (straight) arm:	< 13 in.lbs (TT)		3-13 in.lbs (TT)		H
	compression (curved) arm:	4-22 in.lbs (TT)		–		H
1999-03	Galant	22-54 in.lbs (TT)		–	–	S
2004-12	Galant	*	–	–	–	S
2012-17	I-MiEV	*	–	–	–	S
2002-06	Lancer	0-35 in.lbs (TT)		–	–	S
2008-17	Lancer	*	–	–	–	S
2003-06	Lancer EVO	4.4-30 in.lbs (TT)		–	–	S
2008-15	Lancer EVO	*	–	–	–	S
1990-92	Mirage	< 48 in.lbs (TT)		–	–	S

Mitsubishi (cont.)

		Lower		Upper		
Year	Model	Axial	Radial	Axial	Radial	Type
1993-02	Mirage	9-56 in.lbs (TT)		–	–	S
2014-15	Mirage	*	–	–	–	S
2017	Mirage	< 44 in.lbs (TT)		–	–	S
1990-91	Montero	0.02"	–	7-30 in.lbs (TT)		1
1992-00	Montero	0.012"	–	7-30 in.lbs (TT)		1
2001-06	Montero	9-61 in.lbs (TT)		4-22 in.lbs (TT)		H
1997-04	Montero Sport	0.01"	–	7-30 in.lbs (TT)		1
2003-06	Outlander	0-35 in.lbs (TT)		–	–	S
2007-17	Outlander	*	–	–	–	S
2011-17	Outlander Sport/RVR	*	–	–	–	S
1990-96	Pickup (All Models)	0.02"	–	7-30 in.lbs (TT)		1
1990-93	Precis	22-87 in.lbs (TT)		–	–	S
2006-09	Raider	0.02"	–	0.02"	–	1
1987-90	Van/Wagon	9-30 in.lbs (TT)		7-30 in.lbs (TT)		TT

Nissan

		Lower		Upper		
Year	Model	Axial	Radial	Axial	Radial	Type
1990-98	200SX	0.0"	–	–	–	S
1990-98	240SX	0.0"	–	–	–	S
1990-96	300ZX	0.0"	–	–	–	H
2003-17	350Z/370Z	0.0"	–	0.0"	–	H
1993-01	Altima	0.0"	–	–	–	S
2002-06	Altima	0.0"	–	–	–	S
2007-17	Altima/Altima HEV	0.004"	–	–	–	S
2004-15	Armada	0.0"	–	0.0"	–	H
2017	Armada	0.0"	–	5-43 in.lbs (TT)		H
1990	Axxess	0.0"	–	–	–	S
2009-14	Cube	0.0"	–	–	–	S
1998-04	Frontier 4X2 2.4L	0.051"	–	*	–	1
1998-04	Frontier 4X4, 4X2 3.3L	0.008"	–	*	–	1
2005-17	Frontier	0.0"	–	0.0"	–	H
2011-17	Juke	0.0"	–	–	–	S
2011-17	Leaf	0.0"	–	–	–	S
1990-08	Maxima	0.0"	–	–	–	S
2009-17	Maxima	0.004"	–	–	–	S
2015-17	Micra	0.0"	–	–	–	S
2003-14	Murano	0.0"	–	–	–	S
2015-17	Murano	0.004"	–	–	–	S
2012-17	NV Cargo Van	0.0"	–	0.0"	–	1
2014-17	NV Compact Cargo	0.0"	–	–	–	S
1991-93	NX Coupe	0.0"	–	–	–	S
1990-95	Pathfinder (All Models)	0.063"	–	0.063"	–	1
1996-02	Pathfinder (All Models)	0.0"	–	0.0"	–	1
2003-04	Pathfinder (All Models)	0.0"	–	–	–	S
2005-13	Pathfinder	0.0"	–	0.0"	–	H
2014-17	Pathfinder	0.0"	–	–	–	S
1990	Pulsar/NX	0.0"	–	–	–	S
1993-16	Quest	0.0"	–	–	–	S
2008-16	Rogue	0.0"	–	–	–	S
1990-17	Sentra	0.0"	–	–	–	S
1990-92	Stanza	0.0"	–	–	–	S
2004-17	Titan	0.0"	–	0.0"	–	H
1990-95	Truck 4X2	0.063"	–	0.063"	–	1
1996-97	Truck 4X2	0.018"	–	0.021"	–	1
1990-95	Truck 4X4	0.02"	–	0.063"	–	1
1996-97	Truck 4X4	0.008"	–	0.021"	–	1
1990	Van	0.063"	–	0.063"	–	1
2007-17	Versa/Versa Note	0.0"	–	–	–	S
2000-04	Xterra	0.008"	–	0.0"	–	1
2005-15	Xterra	0.0"	–	0.0"	–	1
2005-06	X-Trail	0.0"	–	–	–	S

Ball Joint Specifications

Saab

Year	Model	Lower Axial	Lower Radial	Upper Axial	Upper Radial	Type
2005-06	9-2x	0.012"	–	–	–	S
2003-11	9-3	0.02"	–	–	–	S
2011	9-4x	0.02"	–	–	–	S
2010-11	9-5	0.02"	–	–	–	S
2005-09	9-7x	0.6"	–	0.6"	–	H

Scion

Year	Model	Lower Axial	Lower Radial	Upper Axial	Upper Radial	Type
2013-16	FRS	*	*	–	–	S
2016	iA	0.0"	–	–	–	S
2016	iM	0.0"	–	–	–	S
2012-15	iQ/IQ EV	0.0"	–	–	–	S
2005-16	tC	0.0"	–	–	–	S
2004-06	xA	0.0"	–	–	–	S
2004-15	xB	0.0"	–	–	–	S
2008-14	xD	0.0"	–	–	–	S

Subaru

Year	Model	Lower Axial	Lower Radial	Upper Axial	Upper Radial	Type
2003-06	Baja	0.012"	–	–	–	S
2013-17	BRZ	*	*	–	–	S
2013-17	Crosstrek/XV Crosstrek (HEV)	*	*	–	–	S
1998-13	Forester	0.012"	–	–	–	S
2014-17	Forester	*	*	–	–	S
1993-11	Impreza/WRX	0.012"	–	–	–	S
2012-17	Impreza/WRX	*	*	–	–	S
1990-94	Justy	0.016"	–	–	–	S
1990-11	Legacy	0.012"	–	–	–	S
2012-17	Legacy	*	*	–	–	S
1996-11	Outback	0.012"	–	–	–	S
2012-17	Outback	*	*	–	–	S
1996-11	Outback Sport	0.012"	–	–	–	S
1990-94	Subaru/Loyale	*	*	–	–	S
1992-97	SVX	*	*	–	–	S
2006-14	Tribeca (B9 Tribeca)	0.012"	*	–	–	S
1990-91	XT	*	*	–	–	S

Suzuki

Year	Model	Lower Axial	Lower Radial	Upper Axial	Upper Radial	Type
2002-07	Aerio	*	*	–	–	S
2009-12	Equator	0.0"	–	0.0"	–	H
1995-02	Esteem	*	*	–	–	S
2004-08	Forenza	*	*	–	–	S
1999-12	Grand Vitara	*	*	–	–	S
2013	Grand Vitara	0.008"	*	–	–	S
2010-12	Kizashi	*	*	–	–	S
2013	Kizashi	0.008"	*	–	–	S
2005-08	Reno	*	*	–	–	S
1990-98	Sidekick	*	*	–	–	S
1990-01	Swift	*	*	–	–	S
2004-09	Swift+	*	*	–	–	S
2007-12	SX4	*	*	–	–	S
2013	SX4	0.008"	*	–	–	S
2004-07	Verona	*	*	–	–	S
1999-04	Vitara	*	*	–	–	S
1996-98	X-90	*	*	–	–	S
2002-06	XL-7	*	*	–	–	S
2007-09	XL-7	0.125"	–	–	–	S

Toyota

Toyota Passenger Cars

Year	Model	Lower Axial	Lower Radial	Upper Axial	Upper Radial	Type
2017	86	0.0"	–	–	–	S
1995-17	Avalon/Avalon HV	0.0"	–	–	–	S
1990-17	Camry/Camry HV	0.0"	–	–	–	S
1990-05	Celica	0.0"	–	–	–	S
1990-92	Corolla FWD	0.098"	–	–	–	S
1993-17	Corolla	0.0"	–	–	–	S
1990-92	Cressida	0.0"	–	–	–	S
2000-05	Echo	0.0"	–	–	–	S
2003-14	Matrix	0.0"	–	–	–	S
2017	Mirai	0.0"	–	–	–	S
1990-95	MR2	0.0"	–	–	–	S
2000-05	MR2 Spyder	0.0"	–	–	–	S
1992-97	Paseo	0.0"	–	–	–	S
1991-97	Previa (All Models)	0.0"	–	–	–	S
2001-17	Prius (All versions)	0.0"	–	–	–	S
1998-17	Sienna	0.0"	–	–	–	S
1992-08	Solara	0.0"	–	–	–	S
1990-92	Supra	0.012"	–	0.0"	–	H
1993-98	Supra	0.016"	–	0.0"	–	H
1990-98	Tercel	0.0"	–	–	–	S
2009-16	Venza	0.0"	–	–	–	S
2007-17	Yaris (All versions)	0.0"	–	–	–	S

Toyota Truck & SUV

Year	Model	Lower Axial	Lower Radial	Upper Axial	Upper Radial	Type
1990-95	4Runner	0.091"	–	0.0"	–	1
1996-05	4Runner	0.02"	–	0.0"	–	1
2006-17	4Runner	0.0"	–	< 40 in.lbs (TT)		H
2007-14	FJ Cruiser	0.0"	–	< 40 in.lbs (TT)		H
2001-17	Highlander/Highlander HV	0.0"	–	–	–	S
1998-07	Land Cruiser	0.0"	–	9-39 in.lbs (TT)		1
2008-17	Land Cruiser	0.0"	–	9-39 in.lbs (TT)		H
1990-91	Pickup 4X2	0.091"	–	0.091"	–	1
1992-95	Pickup 4X2	0.091"	–	0.0"	–	1
1990-95	Pickup 4X4	0.091"	–	0.0"	–	1
1996-17	RAV4 (All versions)	0.0"	–	–	–	S
2001-07	Sequoia	0.02"	–	0.0"	–	H
2008-17	Sequoia	0.0"	–	9-39 in.lbs (TT)		H
1993-98	T100 4X2	0.02"	–	0.075"	–	1
1993-98	T100 4X4	0.091"	–	0.0"	–	1
1995-04	Tacoma	0.02"	–	0.0"	–	1
2005-17	Tacoma	0.0"	–	< 40 in.lbs (TT)		H
1999-06	Tundra	0.02"	–	0.0"	–	H
2007-17	Tundra	0.0"	–	9-39 in.lbs (TT)		H

Volkswagen

Year	Model	Lower Axial	Lower Radial	Upper Axial	Upper Radial	Type
1998-17	Beetle/New Beetle	*	*	–	–	S
1990-02	Cabriolet	*	*	–	–	S
2009-17	CC	*	*	–	–	S
1990-94	Corrado	*	*	–	–	S
2007-16	Eos	*	*	–	–	S
1990-93	Fox	*	*	–	–	S
1990-17	Golf/GTI/Rabbit (All versions)	*	*	–	–	S
1990-16	Jetta/GLI (All versions)	*	*	–	–	S
1990-97	Passat	*	*	–	–	S
1998-05	Passat/Passat 4Motion	*	*	*	*	H
2006-17	Passat	*	*	–	–	S
2004-06	Phaeton	*	*	*	*	H
2009-13	Routan	0.039"	–	–	–	S
2008-17	Tiguan	*	*	–	–	S

Volkswagen (cont.)

		Lower		Upper		
Year	Model	Axial	Radial	Axial	Radial	Type
2004-17	Touareg	*	*	*	*	H
1990-91	Vanagon	*	*	*	*	1

Volvo

		Lower		Upper		
Year	Model	Axial	Radial	Axial	Radial	Type
1990-93	240, 260	0.12"	0.02"	–	–	S
1990	740, 760, 780	0.12"	0.02"	–	–	S
1990	760 Wagon	0.12"	0.02"	–	–	S
1993-97	850	*	*	–	–	S
1991-97	940/960	0.12"	0.02"	–	–	S
2007-13	C30	*	*	–	–	S
1998-13	C70	*	*	–	–	S
2000-04	S40/V40	0.02"	0.02"	–	–	S
2005-11	S40	*	*	–	–	S
2001-09	S60	0.02"	0.02"	–	–	S
2011-17	S60	*	*	–	–	S
1998-00	S70/V70	*	*	–	–	S
1999-16	S80	*	*	–	–	S
1998	S90/V90	*	*	–	–	S
2017	S90	*	*	*	*	H
2005-11	V50	*	*	–	–	S
2015-17	V60 / V60 Cross Country	*	*	–	–	S
1998-09	V70	*	*	–	–	S
2017	V90 / V90 Cross Country	*	*	*	*	H
2011-17	XC60	*	*	–	–	S
2003-16	XC70	*	*	–	–	S
2003-14	XC90	*	*	–	–	S
2016-17	XC90	*	*	*	*	H

NOTES

Year	Model	Axial	Radial	Axial	Radial	Type

Acura

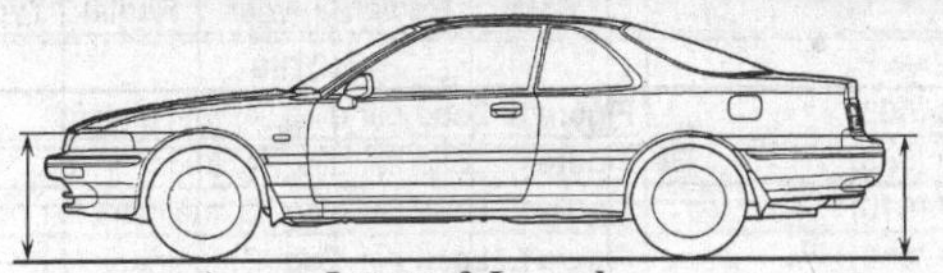

Integra & Legend

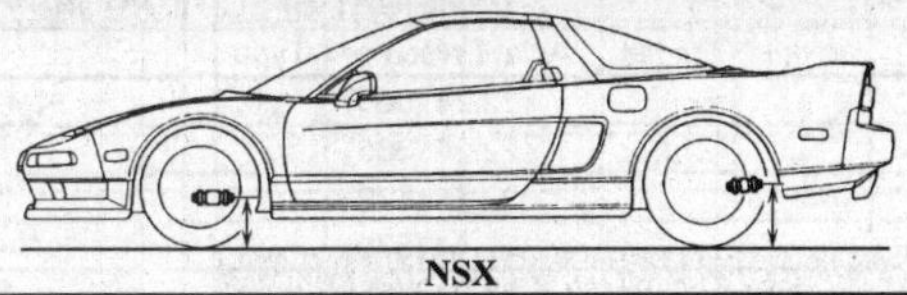

NSX

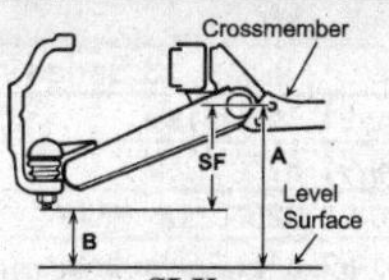

SLX

Year	Model	Height Specifications - In. (mm) * Front	Rear	Notes
1990	Legend Coupe	27.87 (708)	27.55 (700)	
1990	Legend Sedan	27.87 (708)	27.55 (700)	
1991-01	NSX	6.88 (175)	8.58 (218)	
2002-05	NSX	6.77 (172)	8.26 (210)	
1996-97	SLX	5.07 (129)	—	SF = A-B, Adjustable front torsion bars.

* ± 9/16 (15)

BMW / MINI

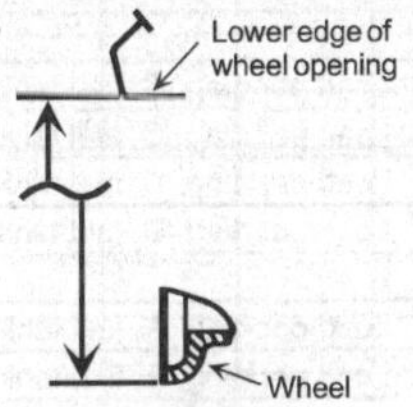

Figure 1

All exc/Z3, MINI (All exc/R50, 52, 53 Front)

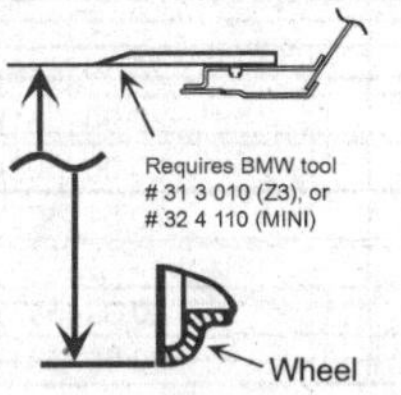

Z3, MINI (R50, 52, 53 Front)

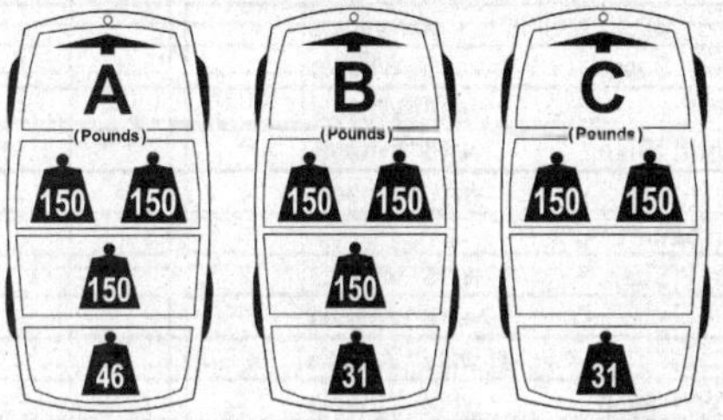

Alignment must be checked and adjustments made with the fuel tank full and the vehicle loaded to normal ride level position. Locate the model in the tables below. For models with air springs, remove the air supply system fuse to disable up/down regulation. Place weights per loading illustration and jounce the vehicle. Next, measure from the lower edge of the wheel to the lower edge of the wheel opening (see illustrations above), through the wheel centerline. Record the measured height at each wheel. If the height is not within the tolerance (± 3/8" - 10mm), check for worn or damaged components. If the height is within the tolerance, add or remove weight to bring the height to within ± 3/32" (2mm) of the specification. Check and adjust alignment.

BMW

Year Model		Height Specifications - In. (mm) * Front	Rear	Notes
1 Series – E82, E88		Front	Rear	Notes
2008-14 Reg. Susp., except 135i				
	w/16" Wheels	22.79 (579)	22.51 (572)	Figure1. Load per diag. C, fuel tank full.
	w/17" Wheels	23.42 (595)	23.11 (587)	Figure1. Load per diag. C, fuel tank full.
	w/18" Wheels	23.89 (607)	23.58 (599)	Figure1. Load per diag. C, fuel tank full.
2008-14 Reg. Susp., 135i	w/16" Wheels	22.20 (564)	21.88 (556)	Figure1. Load per diag. C, fuel tank full.
	w/17" Wheels	22.79 (579)	22.51 (572)	Figure1. Load per diag. C, fuel tank full.
	w/18" Wheels	23.30 (592)	22.99 (584)	Figure1. Load per diag. C, fuel tank full.
2008-14 All Sport Susp.	w/16" Wheels	22.20 (564)	21.88 (556)	Figure1. Load per diag. C, fuel tank full.
	w/17" Wheels	22.83 (580)	22.51 (572)	Figure1. Load per diag. C, fuel tank full.
	w/18" Wheels	23.30 (592)	22.99 (584)	Figure1. Load per diag. C, fuel tank full.
2011-12 M Coupe	w/18" Wheels	23.62 (600)	23.30 (592)	Figure1. Load per diag. C, fuel tank full.
	w/19" Wheels	24.13 (613)	23.77 (604)	Figure1. Load per diag. C, fuel tank full.
2008-14 All HP Sport Susp. (SZP5A)				
	w/16" Wheels	21.81 (554)	21.49 (546)	Figure1. Load per diag. C, fuel tank full.
	w/17" Wheels	22.44 (570)	22.12 (562)	Figure1. Load per diag. C, fuel tank full.
	w/18" Wheels	23.30 (592)	22.59 (574)	Figure1. Load per diag. C, fuel tank full.
2 Series – F22, F23		Front	Rear	Notes
2015-17 Reg. Susp., except 135i				
	w/16" Wheels	22.79 (579)	22.59 (574)	Figure1. Load per diag. B, fuel tank full.
	w/17" Wheels	23.42 (595)	23.18 (589)	Figure1. Load per diag. B, fuel tank full.
	w/18" Wheels	23.89 (607)	23.70 (602)	Figure1. Load per diag. B, fuel tank full.
	w/19" Wheels	24.40 (620)	24.17 (614)	Figure1. Load per diag. B, fuel tank full.
2015-17 All Sport Susp.	w/16" Wheels	44.48 (571)	22.28 (566)	Figure1. Load per diag. B, fuel tank full.
	w/17" Wheels	23.11 (587)	22.87 (581)	Figure1. Load per diag. B, fuel tank full.
	w/18" Wheels	23.58 (599)	23.34 (593)	Figure1. Load per diag. B, fuel tank full.
	w/19" Wheels	24.09 (612)	23.85 (606)	Figure1. Load per diag. B, fuel tank full.
2015-17 M235i	w/18" Wheels	23.58 (599)	23.34 (593)	Figure1. Load per diag. B, fuel tank full.
	w/19" Wheels	24.05 (611)	23.85 (606)	Figure1. Load per diag. B, fuel tank full.
2 Series – F87		Front	Rear	Notes
2016-17 M2				
	w/19" Wheels	24.09 (612)	24.21 (615)	Figure1. Load per diag. B, fuel tank full.

* ± 3/8 (10)

CHASSIS / SUSP. HEIGHTS

BMW (cont.)

Year Model		Height Specifications - In. (mm) * Front	Rear	Notes
3 Series - E30		Front	Rear	Notes
1990-91 318i, 318is		22.24 (565)	19.88 (505)	Figure1. Load per diag. A, fuel tank full.
1990-91 325i, 325is		22.24 (565)	19.88 (505)	Figure1. Load per diag. A, fuel tank full.
1990-93 325i Convertible		22.44 (570)	20.07 (510)	Figure1. Load per diag. C, fuel tank full.
1991-92 318i Convertible		22.44 (570)	20.07 (510)	Figure1. Load per diag. C, fuel tank full.
1990-91 325iX		22.24 (565)	20.07 (510)	Figure1. Load per diag. B, fuel tank full.
1990-91 M3		21.88 (556)	20.82 (529)	Figure1. Load per diag. C, fuel tank full.
3 Series - E36		Front	Rear	Notes
1998-02 M Coupe & M Roadster		26.14 (664)	23.14 (588)	Figure1. Load per diag. C, fuel tank full.
1995-99 M3 Reg. Susp.		23.03 (585)	21.25 (540)	Figure1. Load per diag. C, fuel tank full.
1995-99 M3 Sport. Susp.		22.04 (560)	21.25 (540)	No loading required.
1995-99 318ti Reg. Susp.	w/15" Wheels	22.67 (576)	20.62 (524)	Figure1. Load per diag. A, fuel tank full.
	w/16" Wheels	23.18 (589)	21.14 (537)	Figure1. Load per diag. A, fuel tank full.
1995-99 318ti Sport Susp.	w/15" Wheels	22.08 (561)	20.03 (509)	Figure1. Load per diag. A, fuel tank full.
	w/16" Wheels	22.59 (574)	20.55 (522)	Figure1. Load per diag. A, fuel tank full.
1994-99 318i, 323i, 325i, 328i Convertibles:				
Reg. Susp.	w/15" Wheels	22.67 (576)	20.39 (518)	Figure1. Load per diag. C, fuel tank full.
	w/16" Wheels	23.18 (589)	20.90 (531)	Figure1. Load per diag. C, fuel tank full.
Sport Susp.	w/15" Wheels	22.08 (561)	20.00 (508)	Figure1. Load per diag. C, fuel tank full.
	w/16" Wheels	22.59 (574)	20.51 (521)	Figure1. Load per diag. C, fuel tank full.
1992-02 All other E36:				
Reg. Susp.	w/15" Wheels	22.67 (576)	20.39 (518)	Figure1. Load per diag. A, fuel tank full.
	w/16" Wheels	23.18 (589)	20.90 (531)	Figure1. Load per diag. A, fuel tank full.
	w/17" Wheels	23.77 (604)	21.49 (546)	Figure1. Load per diag. A, fuel tank full.
Sport Susp.	w/15" Wheels	22.08 (561)	19.96 (507)	Figure1. Load per diag. A, fuel tank full.
	w/16" Wheels	22.59 (574)	20.47 (520)	Figure1. Load per diag. A, fuel tank full.
	w/17" Wheels	23.18 (589)	21.06 (535)	Figure1. Load per diag. A, fuel tank full.
3 Series – E46		Front	Rear	Notes
1999-06 323i, 325i, 328i, 330i 1999-06 325Ci, 330Ci (Load all Convertibles per diagram C)				
Reg. Susp.	w/15" Wheels	22.67 (576)	21.33 (542)	Figure1. Load per diag. A, fuel tank full.
	w/16" Wheels	23.18 (589)	21.85 (555)	Figure1. Load per diag. A, fuel tank full.
	w/17" Wheels	23.77 (604)	22.44 (570)	Figure1. Load per diag. A, fuel tank full.
	w/18" Wheels	24.29 (617)	22.95 (583)	Figure1. Load per diag. A, fuel tank full.
Sport Susp.	w/15" Wheels	22.08 (561)	20.70 (526)	Figure1. Load per diag. A, fuel tank full.
	w/16" Wheels	22.59 (574)	21.22 (539)	Figure1. Load per diag. A, fuel tank full.
	w/17" Wheels	23.18 (589)	21.81 (554)	Figure1. Load per diag. A, fuel tank full.
	w/18" Wheels	23.70 (602)	22.32 (567)	Figure1. Load per diag. A, fuel tank full.
2001-06 325xi, 330xi	w/16" Wheels	23.85 (606)	22.51 (572)	Figure1. Load per diag. A, fuel tank full.
	w/17" Wheels	24.44 (621)	23.11 (587)	Figure1. Load per diag. A, fuel tank full.
	w/18" Wheels	24.96 (634)	23.62 (600)	Figure1. Load per diag. A, fuel tank full.
2001-06 M3 Reg. Susp.	w/18" Wheels	23.81 (605)	23.11 (587)	Figure1. Load per diag. A, fuel tank full.
	w/19" Wheels	24.37 (619)	23.62 (600)	Figure1. Load per diag. A, fuel tank full.
3 Series – E90/E91 Sedan & Wagon		Front	Rear	Notes
2006-11 325i, 328i, 330i, 335d, 335i Sedan (to 2011) & Wagon				
Reg. Susp.	w/16" Wheels	22.99 (584)	22.24 (565)	Figure1. Load per diag. A, fuel tank full.
	w/17" Wheels	23.58 (599)	22.83 (580)	Figure1. Load per diag. A, fuel tank full.
	w/18" Wheels	24.09 (612)	23.34 (593)	Figure1. Load per diag. A, fuel tank full.
	w/19" Wheels	24.60 (625)	23.81 (605)	Figure1. Load per diag. A, fuel tank full.
Sport Susp.	w/16" Wheels	22.40 (569)	21.65 (550)	Figure1. Load per diag. A, fuel tank full.
	w/17" Wheels	22.99 (584)	22.20 (564)	Figure1. Load per diag. A, fuel tank full.
	w/18" Wheels	23.46 (596)	22.71 (577)	Figure1. Load per diag. A, fuel tank full.
	w/19" Wheels	23.97 (609)	23.18 (589)	Figure1. Load per diag. A, fuel tank full.
2006-08 325xi, 328xi, 330xi, 335xi Sedan (to 2011) & Wagon				
	w/16" Wheels	23.18 (589)	22.44 (570)	Figure1. Load per diag. A, fuel tank full.
	w/17" Wheels	23.77 (604)	23.03 (585)	Figure1. Load per diag. A, fuel tank full.
	w/18" Wheels	24.29 (617)	23.54 (598)	Figure1. Load per diag. A, fuel tank full.
	w/19" Wheels	24.80 (630)	24.01 (610)	Figure1. Load per diag. A, fuel tank full.
2009-11 328i xDrive, 335i xDrive Sedan (to 2011) & Wagon (start 09/08)				
	w/16" Wheels	23.18 (589)	22.40 (569)	Figure1. Load per diag. A, fuel tank full.
	w/17" Wheels	23.77 (604)	22.99 (584)	Figure1. Load per diag. A, fuel tank full.
	w/18" Wheels	24.29 (617)	23.50 (597)	Figure1. Load per diag. A, fuel tank full.
2008-11 M3 Sedan	w/18" Wheels	24.01 (610)	23.38 (594)	Figure1. Load per diag. A, fuel tank full.
	w/19" Wheels	24.52 (623)	23.89 (607)	Figure1. Load per diag. A, fuel tank full.

* ± ⅜ (10)

CHASSIS / SUSP. HEIGHTS

BMW (cont.)

YEAR MODEL		HEIGHT SPECIFICATIONS - IN. (MM) *		
3 Series – E92/E93 Coupe & Conv.		FRONT	REAR	NOTES
2007-13 328i, 335i Coupe & Convertible				
Reg. Susp.	w/16" Wheels	23.26 (591)	22.36 (568)	Figure1. Load per diag. C, fuel tank full.
	w/17" Wheels	23.85 (606)	22.95 (583)	Figure1. Load per diag. C, fuel tank full.
	w/18" Wheels	24.37 (619)	23.46 (596)	Figure1. Load per diag. C, fuel tank full.
	w/19" Wheels	24.88 (632)	23.93 (608)	Figure1. Load per diag. C, fuel tank full.
Sport Susp.	w/16" Wheels	22.79 (579)	21.85 (555)	Figure1. Load per diag. C, fuel tank full.
	w/17" Wheels	23.38 (594)	22.40 (569)	Figure1. Load per diag. C, fuel tank full.
	w/18" Wheels	23.89 (607)	22.91 (582)	Figure1. Load per diag. C, fuel tank full.
	w/19" Wheels	24.40 (620)	23.38 (594)	Figure1. Load per diag. C, fuel tank full.
2007-13 328xi, 335xi Coupe (328i xDrive, 335i xDrive Coupe)				
	w/16" Wheels	23.46 (596)	22.55 (573)	Figure1. Load per diag. C, fuel tank full.
	w/17" Wheels	24.05 (611)	23.14 (588)	Figure1. Load per diag. C, fuel tank full.
	w/18" Wheels	24.56 (624)	23.66 (601)	Figure1. Load per diag. C, fuel tank full.
	w/19" Wheels	25.07 (637)	24.13 (613)	Figure1. Load per diag. C, fuel tank full.
2008-13 M3 Cpe., M3 Conv.	w/18" Wheels	24.01 (610)	23.38 (594)	Figure1. Load per diag. C, fuel tank full.
	w/19" Wheels	24.52 (623)	23.89 (607)	Figure1. Load per diag. C, fuel tank full.
3 Series – F30 Sedan				
2012-17 All RWD Sedans				
Reg. Susp.	w/16" Wheels	23.42 (595)	22.67 (576)	Figure1. Load per diag. A, fuel tank full.
	w/17" Wheels	24.01 (610)	23.26 (591)	Figure1. Load per diag. A, fuel tank full.
	w/18" Wheels	24.52 (623)	23.77 (604)	Figure1. Load per diag. A, fuel tank full.
	w/19" Wheels	25.03 (636)	24.25 (616)	Figure1. Load per diag. A, fuel tank full.
	w/20" Wheels	25.51 (648)	24.76 (629)	Figure1. Load per diag. A, fuel tank full.
Sport Susp.	w/16" Wheels	23.03 (585)	22.24 (565)	Figure1. Load per diag. A, fuel tank full.
	w/17" Wheels	23.62 (600)	22.87 (581)	Figure1. Load per diag. A, fuel tank full.
	w/18" Wheels	24.13 (613)	23.34 (593)	Figure1. Load per diag. A, fuel tank full.
	w/19" Wheels	24.60 (625)	23.85 (606)	Figure1. Load per diag. A, fuel tank full.
2014-17 All AWD Sedans				
	w/16" Wheels	23.62 (600)	22.87 (581)	Figure1. Load per diag. A, fuel tank full.
	w/17" Wheels	24.21 (615)	23.46 (596)	Figure1. Load per diag. A, fuel tank full.
	w/18" Wheels	24.72 (628)	23.97 (609)	Figure1. Load per diag. A, fuel tank full.
	w/19" Wheels	25.23 (641)	24.48 (622)	Figure1. Load per diag. A, fuel tank full.
2017 330e iPerformance				
	w/17" Wheels	24.05 (611)	23.46 (596)	Figure1. Load per diag. A, fuel tank full.
	w/18" Wheels	24.56 (624)	23.97 (609)	Figure1. Load per diag. A, fuel tank full.
	w/19" Wheels	25.03 (636)	24.44 (621)	Figure1. Load per diag. A, fuel tank full.
3 Series – F31 Wagon		FRONT	REAR	NOTES
2014-17 3 series xDrive Wagon (All suspensions)				
	w/16" Wheels	23.66 (601)	22.87 (581)	Figure1. Load per diag. A, fuel tank full.
	w/17" Wheels	24.25 (616)	23.46 (596)	Figure1. Load per diag. A, fuel tank full.
	w/18" Wheels	24.76 (629)	23.97 (609)	Figure1. Load per diag. A, fuel tank full.
	w/19" Wheels	25.23 (641)	24.44 (621)	Figure1. Load per diag. A, fuel tank full.
3 Series – F34 GT		FRONT	REAR	NOTES
2014-17 3 Series xDrive GT				
Reg. Susp.	w/17" Wheels	24.37 (619)	23.62 (600)	Figure1. Load per diag. A, fuel tank full.
	w/18" Wheels	24.84 (631)	24.09 (612)	Figure1. Load per diag. A, fuel tank full.
	w/19" Wheels	25.35 (644)	24.60 (625)	Figure1. Load per diag. A, fuel tank full.
Sport Susp.	w/17" Wheels	23.97 (609)	23.18 (589)	Figure1. Load per diag. A, fuel tank full.
	w/18" Wheels	24.44 (621)	23.70 (602)	Figure1. Load per diag. A, fuel tank full.
	w/19" Wheels	24.96 (634)	24.17 (614)	Figure1. Load per diag. A, fuel tank full.
3 Series – F80 M3		FRONT	REAR	NOTES
2014-17 3 Series M3				
	w/18" Wheels	24.17 (614)	23.11 (587)	Figure1. Load per diag. A, fuel tank full.
	w/19" Wheels	24.68 (627)	23.58 (599)	Figure1. Load per diag. A, fuel tank full.
	w/20" Wheels	25.15 (639)	24.09 (612)	Figure1. Load per diag. A, fuel tank full.
3 Series X3 – E83		FRONT	REAR	NOTES
2004-10 X3 Reg. & Sport	w/17" Wheels	27.00 (686)	26.49 (673)	Figure1. Load per diag. A, fuel tank full.
	w/18" Wheels	27.51 (699)	26.96 (685)	Figure1. Load per diag. A, fuel tank full.
	w/19" Wheels	27.99 (711)	27.48 (698)	Figure1. Load per diag. A, fuel tank full.
3 Series X3 – F25		FRONT	REAR	NOTES
2011-17 X3 Reg. & Sport	w/17" Wheels	26.18 (665)	26.02 (661)	Figure1. Load per diag. A, fuel tank full.
	w/18" Wheels	26.65 (677)	26.53 (674)	Figure1. Load per diag. A, fuel tank full.
	w/19" Wheels	27.12 (689)	27.00 (686)	Figure1. Load per diag. A, fuel tank full.
	w/20" Wheels	27.63 (702)	27.51 (699)	Figure1. Load per diag. A, fuel tank full.

* ± ⅜ (10)

CHASSIS / SUSP. HEIGHTS

BMW (cont.)

Year Model		Height Specifications - In. (mm) *		
4 Series – F32 Coupe, F33 Conv.		Front	Rear	Notes
2014-17 All RWD models	w/17" Wheels	23.70 (602)	23.30 (592)	Figure1. Load per diag. C, fuel tank full.
	w/18" Wheels	24.17 (614)	23.81 (605)	Figure1. Load per diag. C, fuel tank full.
	w/19" Wheels	24.68 (627)	24.29 (617)	Figure1. Load per diag. C, fuel tank full.
2014-17 All AWD models	w/17" Wheels	24.29 (617)	23.93 (608)	Figure1. Load per diag. C, fuel tank full.
	w/18" Wheels	24.76 (629)	24.44 (621)	Figure1. Load per diag. C, fuel tank full.
	w/19" Wheels	25.27 (642)	24.92 (633)	Figure1. Load per diag. C, fuel tank full.
5 Series - E34		Front	Rear	Notes
1990-95 exc/M5		23.14 (588)	20.55 (522)	Figure1. Load per diag. A, fuel tank full.
1991-93 M5		23.46 (596)	21.45 (545)	Figure1. Load per diag. C, fuel tank full.
5 Series - E39		Front	Rear	Notes
1997-03 525i, 528i, 530i, 540i Sedan				
Reg. Susp.	w/15" Wheels	22.79 (579)	21.53 (547)	Figure1. Load per diag. A, fuel tank full.
	w/16" Wheels	23.30 (592)	22.04 (560)	Figure1. Load per diag. A, fuel tank full.
	w/17" Wheels	23.89 (607)	22.63 (575)	Figure1. Load per diag. A, fuel tank full.
	w/18" Wheels	24.40 (620)	23.14 (588)	Figure1. Load per diag. A, fuel tank full.
Sport Susp.	w/15" Wheels	22.00 (559)	20.94 (532)	Figure1. Load per diag. A, fuel tank full.
	w/16" Wheels	22.51 (572)	21.45 (545)	Figure1. Load per diag. A, fuel tank full.
	w/17" Wheels	23.11 (587)	22.04 (560)	Figure1. Load per diag. A, fuel tank full.
	w/18" Wheels	23.62 (600)	22.55 (573)	Figure1. Load per diag. A, fuel tank full.
1999-03 525i, 528i, 540i Sport Wagons				
Sport. Susp. without Level Control				
	w/15" Wheels	22.00 (559)	21.57 (548)	Figure1. Load per diag. A, fuel tank full.
	w/16" Wheels	22.51 (572)	22.08 (561)	Figure1. Load per diag. A, fuel tank full.
	w/17" Wheels	23.11 (587)	22.67 (576)	Figure1. Load per diag. A, fuel tank full.
	w/18" Wheels	23.62 (600)	23.18 (589)	Figure1. Load per diag. A, fuel tank full.
Sport. Susp. with Level Control				
	w/15" Wheels	22.00 (559)	22.16 (563)	Figure1. Load per diag. A, fuel tank full.
	w/16" Wheels	22.51 (572)	22.67 (576)	Figure1. Load per diag. A, fuel tank full.
	w/17" Wheels	23.11 (587)	23.26 (591)	Figure1. Load per diag. A, fuel tank full.
	w/18" Wheels	23.62 (600)	23.77 (604)	Figure1. Load per diag. A, fuel tank full.
Reg. Susp. Use Sedan specs (above)				
2000-03 M5	w/17" Wheels	23.30 (592)	22.24 (565)	Figure1. Load per diag. A, fuel tank full.
	w/18" Wheels	23.81 (605)	22.75 (578)	Figure1. Load per diag. A, fuel tank full.
5 Series – E60/E61		Front	Rear	Notes
2004-10 525i, 528i, 530i, 535i, 545i, 550i				
Reg. Susp.	w/16" Wheels	23.18 (589)	22.55 (573)	Figure1. Load per diag. A, fuel tank full.
	w/17" Wheels	23.81 (605)	23.14 (588)	Figure1. Load per diag. A, fuel tank full.
	w/18" Wheels	24.29 (617)	23.66 (601)	Figure1. Load per diag. A, fuel tank full.
	w/19" Wheels	24.80 (630)	24.13 (613)	Figure1. Load per diag. A, fuel tank full.
Sport. Susp.	w/16" Wheels	22.59 (574)	21.96 (558)	Figure1. Load per diag. A, fuel tank full.
	w/17" Wheels	23.18 (589)	22.55 (573)	Figure1. Load per diag. A, fuel tank full.
	w/18" Wheels	23.70 (602)	23.03 (585)	Figure1. Load per diag. A, fuel tank full.
	w/19" Wheels	24.17 (614)	23.54 (598)	Figure1. Load per diag. A, fuel tank full.
2006-10 525xi, 528xi, 530xi, 535xi (528i xDrive, 535i xDrive)				
	w/17" Wheels	24.37 (619)	23.58 (599)	Figure1. Load per diag. A, fuel tank full.
	w/18" Wheels	24.88 (632)	24.05 (611)	Figure1. Load per diag. A, fuel tank full.
2006-10 M5	w/19" Wheels	25.19 (640)	24.13 (613)	Figure1. Load per diag. A, fuel tank full.
5 Series – F07		Front	Rear	Notes
2010-16 535i GT, 550i GT (incl xDrive Models)				
All Susp.	w/18" Wheels	25.43 (646)	25.59 (650)	Figure1. Load per diag. A, fuel tank full.
	w/19" Wheels	25.90 (658)	26.06 (662)	Figure1. Load per diag. A, fuel tank full.
	w/20" Wheels	26.41 (671)	26.53 (674), xDrive = 26.57 (675)	Figure1. Load per diag. A, fuel tank full.
	w/21" Wheels	26.88 (683)	27.00 (686) , xDrive = 27.04 (687)	Figure1. Load per diag. A, fuel tank full.
5 Series – F10		Front	Rear	Notes
2011-16 528i, 535i, 550i				
Reg. Susp.	w/17" Wheels	24.25 (616)	23.70 (602)	Figure1. Load per diag. A, fuel tank full.
	w/18" Wheels	24.76 (629)	24.17 (614)	Figure1. Load per diag. A, fuel tank full.
	w/19" Wheels	25.23 (641)	24.64 (626)	Figure1. Load per diag. A, fuel tank full.
	w/20" Wheels	25.74 (654)	25.15 (639)	Figure1. Load per diag. A, fuel tank full.
Sport Susp.	w/17" Wheels	23.85 (606)	23.30 (592)	Figure1. Load per diag. A, fuel tank full.
	w/18" Wheels	24.33 (618)	23.77 (604)	Figure1. Load per diag. A, fuel tank full.
	w/19" Wheels	24.84 (631)	24.25 (616)	Figure1. Load per diag. A, fuel tank full.
	w/20" Wheels	25.35 (644)	24.76 (629)	Figure1. Load per diag. A, fuel tank full.
2011-16 528i xDrive, 535i xDrive, 550i xDrive				
	w/17" Wheels	24.25 (616)	23.70 (602)	Figure1. Load per diag. A, fuel tank full.
	w/18" Wheels	24.76 (629)	24.17 (614)	Figure1. Load per diag. A, fuel tank full.
	w/19" Wheels	25.23 (641)	24.68 (627)	Figure1. Load per diag. A, fuel tank full.

* ± ⅜ (10)

BMW (cont.)			
YEAR MODEL	HEIGHT SPECIFICATIONS - IN. (MM) *		
5 Series – F10 (cont.)	FRONT	REAR	NOTES
w/20" Wheels	25.74 (654)	25.19 (640)	Figure1. Load per diag. A, fuel tank full.
2012-16 M5 w/19" Wheels	25.23 (641)	24.64 (626)	Figure1. Load per diag. A, fuel tank full.
w/20" Wheels	25.70 (653)	25.15 (639)	Figure1. Load per diag. A, fuel tank full.
5 Series X5 – E53	FRONT	REAR	NOTES
2000-06 X5 3.0i, 4.4i models WITHOUT FLARED WHEEL ARCH without Self-Leveling Suspension			
w/17" Wheels	27.00 (686)	27.04 (687)	Figure1. Load per diag. A, fuel tank full.
w/18" Wheels	27.48 (698)	27.51 (699)	Figure1. Load per diag. A, fuel tank full.
w/19" Wheels	27.99 (711)	27.99 (711)	Figure1. Load per diag. A, fuel tank full.
w/20" Wheels	—	—	Figure1. Load per diag. A, fuel tank full.
with Self-Leveling Suspension (Rear Air only)			
w/17" Wheels	27.00 (686)	27.40 (696)	Figure1. Load per diag. A, fuel tank full.
w/18" Wheels	27.48 (698)	27.87 (708)	Figure1. Load per diag. A, fuel tank full.
w/19" Wheels	27.99 (711)	28.30 (719)	Figure1. Load per diag. A, fuel tank full.
w/20" Wheels	—	—	Figure1. Load per diag. A, fuel tank full.
with Self-Leveling Suspension (F&R Air) exc/Sport Susp			
w/17" Wheels	27.00 (686)	27.04 (687)	Figure1. Load per diag. A, fuel tank full.
w/18" Wheels	27.48 (698)	27.51 (699)	Figure1. Load per diag. A, fuel tank full.
w/19" Wheels	27.99 (711)	28.03 (712)	Figure1. Load per diag. A, fuel tank full.
w/20" Wheels	–	28.50 (724)	Figure1. Load per diag. A, fuel tank full.
with Self-Leveling Suspension (F&R Air) Sport Susp			
w/17" Wheels	26.41 (671)	26.49 (673)	Figure1. Load per diag. A, fuel tank full.
w/18" Wheels	26.88 (683)	26.96 (685)	Figure1. Load per diag. A, fuel tank full.
w/19" Wheels	27.40 (696)	27.40 (696)	Figure1. Load per diag. A, fuel tank full.
2000-06 X5 3.0i, 4.4i models WITH FLARED WHEEL ARCH without Self-Leveling Suspension			
w/17" Wheels	27.04 (687)	27.16 (690)	Figure1. Load per diag. A, fuel tank full.
w/18" Wheels	27.51 (699)	27.59 (701)	Figure1. Load per diag. A, fuel tank full.
w/19" Wheels	28.03 (712)	28.03 (712)	Figure1. Load per diag. A, fuel tank full.
w/20" Wheels	28.50 (724)	28.46 (723)	Figure1. Load per diag. A, fuel tank full.
with Self-Leveling Suspension (Rear Air only)			
w/17" Wheels	27.04 (687)	27.48 (698)	Figure1. Load per diag. A, fuel tank full.
w/18" Wheels	27.51 (699)	27.91 (709)	Figure1. Load per diag. A, fuel tank full.
w/19" Wheels	28.03 (712)	28.34 (720)	Figure1. Load per diag. A, fuel tank full.
w/20" Wheels	28.50 (724)	28.77 (731)	Figure1. Load per diag. A, fuel tank full.
with Self-Leveling Suspension (F&R Air) exc/Sport Susp			
w/17" Wheels	27.04 (687)	27.16 (690)	Figure1. Load per diag. A, fuel tank full.
w/18" Wheels	27.51 (699)	27.59 (701)	Figure1. Load per diag. A, fuel tank full.
w/19" Wheels	28.03 (712)	28.03 (712)	Figure1. Load per diag. A, fuel tank full.
w/20" Wheels	28.50 (724)	28.46 (723)	Figure1. Load per diag. A, fuel tank full.
with Self-Leveling Suspension (F&R Air) Sport Susp			
w/17" Wheels	26.45 (672)	26.57 (675)	Figure1. Load per diag. A, fuel tank full.
w/18" Wheels	26.92 (684)	27.04 (687)	Figure1. Load per diag. A, fuel tank full.
w/19" Wheels	27.40 (696)	27.44 (697)	Figure1. Load per diag. A, fuel tank full.
w/20" Wheels	27.87 (708)	27.87 (708)	Figure1. Load per diag. A, fuel tank full.
2000-06 X5 4.6is, 4.8is except Sport Suspension			
w/18" Wheels	27.51 (699)	27.87 (708)	Figure1. Load per diag. A, fuel tank full.
w/19" Wheels	28.03 (712)	28.30 (719)	Figure1. Load pêr diag. A, fuel tank full.
w/20" Wheels	28.50 (724)	28.74 (730)	Figure1. Load per diag. A, fuel tank full.
2000-06 X5 4.6is, 4.8is with Sport Suspension			
w/18" Wheels	26.92 (684)	27.00 (686)	Figure1. Load per diag. A, fuel tank full.
w/19" Wheels	27.40 (696)	27.44 (697)	Figure1. Load per diag. A, fuel tank full.
w/20" Wheels	27.87 (708)	27.87 (708)	Figure1. Load per diag. A, fuel tank full.
5 Series X5 – E70	FRONT	REAR	NOTES
2007-13 X5 w/Steel Spring Susp.			
w/o wheel-well trim w/18" Wheels	27.83 (707)	27.48 (698)	Figure1. Load per diag. A, fuel tank full.
w/wheel-well trim w/18" Wheels	27.87 (708)	27.55 (700)	Figure1. Load per diag. A, fuel tank full.
w/19" Wheels	28.34 (720)	28.03 (712)	Figure1. Load per diag. A, fuel tank full.
w/20" Wheels	28.81 (732)	28.42 (722)	Figure1. Load per diag. A, fuel tank full.
w/21" Wheels	29.33 (745)	28.89 (734)	Figure1. Load per diag. A, fuel tank full.
2007-13 X5 w/Air Spring Susp.			
w/o wheel-well trim w/18" Wheels	27.83 (707)	27.87 (708)	Figure1. Load per diag. A, fuel tank full.
w/wheel-well trim w/18" Wheels	27.87 (708)	27.95 (710)	Figure1. Load per diag. A, fuel tank full.
w/19" Wheels	28.34 (720)	28.42 (722)	Figure1. Load per diag. A, fuel tank full.
w/20" Wheels	28.81 (732)	28.81 (732)	Figure1. Load per diag. A, fuel tank full.
w/21" Wheels	29.33 (745)	29.29 (744)	Figure1. Load per diag. A, fuel tank full.
6 Series – E63/E64	FRONT	REAR	NOTES
2004-11 All exc/M6 w/17" Wheels	24.68 (627)	24.21 (615)	Figure1. Load per diag. C, fuel tank full.
w/18" Wheels	25.19 (640)	24.72 (628)	Figure1. Load per diag. C, fuel tank full.
w/19" Wheels	25.66 (652)	25.23 (641)	Figure1. Load per diag. C, fuel tank full.
w/20" Wheels	26.18 (665)	25.74 (654)	Figure1. Load per diag. C, fuel tank full.

* ± ⅜ (10)

BMW (cont.)

YEAR MODEL	HEIGHT SPECIFICATIONS - IN. (MM) *		
6 Series – E63/E64 (cont.)			
2006-11 M6 Coupe & Convertible			
w/18" Wheels	25.15 (639)	24.72 (628)	Figure1. Load per diag. C, fuel tank full.
w/19" Wheels	25.62 (651)	25.19 (640)	Figure1. Load per diag. C, fuel tank full.
6 Series – F12/F13	FRONT	REAR	NOTES
2012-17 640i, 650i (incl xDrive models)			
w/17" Wheels	24.09 (612)	23.66 (601)	Figure1. Load per diag. C, fuel tank full.
w/18" Wheels	24.56 (624)	24.17 (614)	Figure1. Load per diag. C, fuel tank full.
w/19" Wheels	25.03 (636)	24.64 (626)	Figure1. Load per diag. C, fuel tank full.
w/20" Wheels	25.55 (649)	25.15 (639)	Figure1. Load per diag. C, fuel tank full.
6 Series X6 – E71	FRONT	REAR	NOTES
2008-14 X6			
with Steel Springs w/19" Wheels	28.07 (713)	27.55 (700)	Figure1. Load per diag. A, fuel tank full.
with Steel Springs w/20" Wheels	28.54 (725)	28.03 (712)	Figure1. Load per diag. A, fuel tank full.
with Steel Springs w/21" Wheels	29.01 (737)	28.50 (724)	Figure1. Load per diag. A, fuel tank full.
Steel Springs w/Adaptive Drive w/19" Wheels	28.46 (723)	27.95 (710)	Figure1. Load per diag. A, fuel tank full.
Steel Springs w/Adaptive Drive w/20" Wheels	28.93 (735)	28.42 (722)	Figure1. Load per diag. A, fuel tank full.
Steel Springs w/Adaptive Drive w/21" Wheels	29.40 (747)	28.89 (734)	Figure1. Load per diag. A, fuel tank full.
1-Axle AIR Spring w/19" Wheels	28.07 (713)	27.95 (710)	Figure1. Load per diag. A, fuel tank full.
1-Axle AIR Spring w/20" Wheels	28.54 (725)	28.42 (722)	Figure1. Load per diag. A, fuel tank full.
1-Axle AIR Spring w/21" Wheels	29.01 (737)	28.89 (734)	Figure1. Load per diag. A, fuel tank full.
1-Axle AIR w/Adaptive Drive w/19" Wheels	28.46 (723)	28.34 (720)	Figure1. Load per diag. A, fuel tank full.
1-Axle AIR w/Adaptive Drive w/20" Wheels	28.93 (735)	28.81 (732)	Figure1. Load per diag. A, fuel tank full.
1-Axle AIR w/Adaptive Drive w/21" Wheels	29.40 (747)	29.29 (744)	Figure1. Load per diag. A, fuel tank full.
2010-14 X6 M w/19" Wheels	28.03 (712)	27.87 (708)	Figure1. Load per diag. A, fuel tank full.
w/20" Wheels	28.50 (724)	28.38 (721)	Figure1. Load per diag. A, fuel tank full.
w/21" Wheels	28.97 (736)	28.81 (732)	Figure1. Load per diag. A, fuel tank full.
6 Series X6 – E72	FRONT	REAR	NOTES
2010-11 ActiveHybrid X6 w/19" Wheels	28.46 (723)	28.42 (722)	Figure1. Load per diag. A, fuel tank full.
w/20" Wheels	28.93 (735)	28.93 (735)	Figure1. Load per diag. A, fuel tank full.
w/21" Wheels	29.40 (747)	29.40 (747)	Figure1. Load per diag. A, fuel tank full.
7 Series - E32	FRONT	REAR	NOTES
1990-94 All Models	23.14 (588)	20.55 (522)	Figure1. Load per diag. A, fuel tank full.
7 Series - E38	FRONT	REAR	NOTES
1995-01 740i, 740iL, 750iL			
Reg. Susp. w16" Wheels	24.13 (613)	22.40 (569)	Figure1. Load per diag. A, fuel tank full.
w17" Wheels	24.72 (628)	22.99 (584)	Figure1. Load per diag. A, fuel tank full.
w18" Wheels	25.23 (641)	23.50 (597)	Figure1. Load per diag. A, fuel tank full.
Sport Susp. w16" Wheels	23.34 (593)	22.40 (569)	Figure1. Load per diag. A, fuel tank full.
w17" Wheels	23.93 (608)	22.99 (584)	Figure1. Load per diag. A, fuel tank full.
w18" Wheels	24.44 (621)	23.50 (597)	Figure1. Load per diag. A, fuel tank full.
7 Series – E65/E66	FRONT	REAR	NOTES
2002-05 745i, 760i, 745Li, 760Li			
Reg. Susp. (Steel Spring) w17" Wheels	25.19 (640)	24.40 (620)	Figure1. Load per diag. A, fuel tank full.
w18" Wheels	25.70 (653)	24.92 (633)	Figure1. Load per diag. A, fuel tank full.
w19" Wheels	26.18 (665)	25.35 (644)	Figure1. Load per diag. A, fuel tank full.
w20" Wheels	26.69 (678)	25.82 (656)	Figure1. Load per diag. A, fuel tank full.
Reg. Susp. (Air Spring) w17" Wheels	25.19 (640)	24.80 (630)	Figure1. Load per diag. A, fuel tank full.
w18" Wheels	25.70 (653)	25.31 (643)	Figure1. Load per diag. A, fuel tank full.
w19" Wheels	26.18 (665)	25.74 (654)	Figure1. Load per diag. A, fuel tank full.
w20" Wheels	26.69 (678)	26.25 (667)	Figure1. Load per diag. A, fuel tank full.
Sport Susp. w17" Wheels	24.37 (619)	24.01 (610)	Figure1. Load per diag. A, fuel tank full.
w18" Wheels	24.88 (632)	24.52 (623)	Figure1. Load per diag. A, fuel tank full.
w19" Wheels	25.35 (644)	24.96 (634)	Figure1. Load per diag. A, fuel tank full.
w20" Wheels	25.86 (657)	25.43 (646)	Figure1. Load per diag. A, fuel tank full.
2006-08 750i, 760i, 750Li, 760Li			
Reg. Susp. (Steel Spring) w17" Wheels	25.19 (640)	24.37 (619)	Figure1. Load per diag. A, fuel tank full.
w18" Wheels	25.70 (653)	24.88 (632)	Figure1. Load per diag. A, fuel tank full.
w19" Wheels	26.18 (665)	25.31 (643)	Figure1. Load per diag. A, fuel tank full.
w20" Wheels	26.69 (678)	25.82 (656)	Figure1. Load per diag. A, fuel tank full.
w21" Wheels	27.16 (690)	26.29 (668)	Figure1. Load per diag. A, fuel tank full.
Reg. Susp. (Air Spring) w17" Wheels	25.19 (640)	24.76 (629)	Figure1. Load per diag. A, fuel tank full.
w18" Wheels	25.70 (653)	25.27 (642)	Figure1. Load per diag. A, fuel tank full.
w19" Wheels	26.18 (665)	25.70 (653)	Figure1. Load per diag. A, fuel tank full.
w20" Wheels	26.69 (678)	26.22 (666)	Figure1. Load per diag. A, fuel tank full.
w21" Wheels	27.16 (690)	26.69 (678)	Figure1. Load per diag. A, fuel tank full.

* ± 3/8 (10)

BMW (cont.)

Year	Model		Height Specifications - In. (mm) *		
	7 Series – E65/E66 (cont.)				
	Sport Susp.	w17" Wheels	24.37 (619)	23.97 (609)	Figure1. Load per diag. A, fuel tank full.
		w18" Wheels	24.88 (632)	24.48 (622)	Figure1. Load per diag. A, fuel tank full.
		w19" Wheels	25.35 (644)	24.92 (633)	Figure1. Load per diag. A, fuel tank full.
		w20" Wheels	25.86 (657)	25.43 (646)	Figure1. Load per diag. A, fuel tank full.
		w21" Wheels	26.29 (668)	25.90 (658)	Figure1. Load per diag. A, fuel tank full.
2006-08	Alpina	w/18" Wheels	25.15 (639)	25.27 (642)	Figure1. Load per diag. A, fuel tank full.
		w21" Wheels	26.65 (677)	26.77 (680)	Figure1. Load per diag. A, fuel tank full.
	7 Series – F01/F02/F04		**Front**	**Rear**	**Notes**
2009-16	Std. Susp.	w17" Wheels	24.88 (632)	24.56 (624)	Figure1. Load per diag. A, fuel tank full.
		w18" Wheels	25.39 (645)	25.03 (636)	Figure1. Load per diag. A, fuel tank full.
		w19" Wheels	25.86 (657)	25.51 (648)	Figure1. Load per diag. A, fuel tank full.
		w20" Wheels	26.37 (670)	26.02 (661)	Figure1. Load per diag. A, fuel tank full.
		w21" Wheels	26.85 (682)	26.49 (673)	Figure1. Load per diag. A, fuel tank full.
2009-16	AIR Susp.	w17" Wheels	24.88 (632)	24.96 (634)	Figure1. Load per diag. A, fuel tank full.
		w18" Wheels	25.39 (645)	25.43 (646)	Figure1. Load per diag. A, fuel tank full.
		w19" Wheels	25.86 (657)	25.94 (659)	Figure1. Load per diag. A, fuel tank full.
		w20" Wheels	26.37 (670)	26.41 (671)	Figure1. Load per diag. A, fuel tank full.
		w21" Wheels	26.85 (682)	26.88 (683)	Figure1. Load per diag. A, fuel tank full.
	7 Series – G11, G12		**Front**	**Rear**	**Notes**
2017	Sedan RWD models	w17" Wheels	24.88 (632)	24.96 (634)	Figure1. Load per diag. A, fuel tank full.
		w18" Wheels	25.39 (645)	25.47 (647)	Figure1. Load per diag. A, fuel tank full.
		w19" Wheels	25.86 (657)	26.94 (659)	Figure1. Load per diag. A, fuel tank full.
		w20" Wheels	26.37 (670)	26.41 (671)	Figure1. Load per diag. A, fuel tank full.
		w21" Wheels	26.88 (683)	26.92 (684)	Figure1. Load per diag. A, fuel tank full.
2017	Sedan AWD models	w17" Wheels	25.11 (638)	24.96 (634)	Figure1. Load per diag. A, fuel tank full.
		w18" Wheels	25.59 (650)	25.47 (647)	Figure1. Load per diag. A, fuel tank full.
		w19" Wheels	26.10 (663)	26.94 (659)	Figure1. Load per diag. A, fuel tank full.
		w20" Wheels	26.57 (675)	26.41 (671)	Figure1. Load per diag. A, fuel tank full.
		w21" Wheels	27.08 (688)	26.92 (684)	Figure1. Load per diag. A, fuel tank full.
2017	Alpina RWD	w20" Wheels	26.33 (669)	26.41 (671)	Figure1. Load per diag. A, fuel tank full.
		w21" Wheels	26.85 (682)	26.92 (684)	Figure1. Load per diag. A, fuel tank full.
2017	Alpina AWD	w20" Wheels	26.41 (671)	26.41 (671)	Figure1. Load per diag. A, fuel tank full.
		w21" Wheels	26.92 (684)	26.92 (684)	Figure1. Load per diag. A, fuel tank full.
	8 Series – E31		**Front**	**Rear**	**Notes**
1991-97	840Ci, 850i, 850Ci		23.58 (599)	22.40 (569)	Figure1. Load per diag. C, fuel tank full.
1994	850CSi		23.62 (600)	22.44 (570)	Figure1. Load per diag. C, fuel tank full.
	8 Series – E52				
2001-03	Z8		24.92 (633)	24.64 (626)	Figure1. Load per diag. C, fuel tank full.
	Z3 – E36		**Front**	**Rear**	**Notes**
1999-02	Z3 Coupe	w/15" Wheels	25.27 (642)	22.04 (560)	Figure2. Load per diag. C, fuel tank full.
		w/16" Wheels	25.78 (655)	22.55 (573)	Figure2. Load per diag. C, fuel tank full.
		w/17" Wheels	26.37 (670)	23.14 (588)	Figure2. Load per diag. C, fuel tank full.
1996-02	Z3 Roadster Reg. Susp.	w/15" Wheels	25.47 (647)	22.24 (565)	
		w/16" Wheels	25.98 (660)	22.75 (578)	Figure2. Load per diag. C, fuel tank full.
		w/17" Wheels	26.57 (675)	23.34 (593)	Figure2. Load per diag. C, fuel tank full.
	Sport Susp.	w/15" Wheels	24.88 (632)	21.65 (550)	Figure2. Load per diag. C, fuel tank full.
		w/16" Wheels	25.39 (645)	22.16 (563)	Figure2. Load per diag. C, fuel tank full.
		w/17" Wheels	25.98 (660)	22.75 (578)	Figure2. Load per diag. C, fuel tank full.
	Z4 – E85, E86		**Front**	**Rear**	**Notes**
2003-09	Z4 Roadster & Coupe				
	Reg. Susp.	w/16" Wheels	23.18 (589)	22.67 (576)	Figure1. Load per diag. C, fuel tank full.
		w/17" Wheels	23.81 (605)	23.22 (590)	Figure1. Load per diag. C, fuel tank full.
		w/18" Wheels	24.29 (617)	23.74 (603)	Figure1. Load per diag. C, fuel tank full.
	Sport. Susp.	w/17" Wheels	23.18 (589)	22.59 (574)	Figure1. Load per diag. C, fuel tank full.
		w/18" Wheels	23.70 (602)	23.11 (587)	Figure1. Load per diag. C, fuel tank full.
2007-09	Z4 M Roadster	w/18" Wheels	23.93 (608)	23.38 (594)	Figure1. Load per diag. C, fuel tank full.
2007-09	Z4 M Coupe	w/18" Wheels	23.89 (607)	23.38 (594)	Figure1. Load per diag. C, fuel tank full.
	Z4 – E89		**Front**	**Rear**	**Notes**
2010-16	Z4 Roadster				
	Reg. Susp.	w/17" Wheels	23.85 (606)	23.34 (593)	Figure1. Load per diag. C, fuel tank full.
		w/18" Wheels	24.37 (619)	23.85 (606)	Figure1. Load per diag. C, fuel tank full.
		w/19" Wheels	24.84 (631)	24.33 (618)	Figure1. Load per diag. C, fuel tank full.
	Adaptive M Susp.	w/17" Wheels	23.46 (596)	22.91 (582)	Figure1. Load per diag. C, fuel tank full.
		w/18" Wheels	23.93 (608)	23.42 (595)	Figure1. Load per diag. C, fuel tank full.
		w/19" Wheels	24.44 (621)	23.89 (607)	Figure1. Load per diag. C, fuel tank full.

* ± ⅜ (10)

CHASSIS / SUSP. HEIGHTS

MINI

Year Model	Height Specifications - In. (mm) *		
R50 Cooper, ONE / R53 Cooper S Hardtop	Front	Rear	Notes
2002-06 Cooper w/Reg. Suspension			
w/15" wheels	26.18 (665)	21.65 (550)	Figure 2. Load per diag. C, fuel tank full.
w/16" wheels	26.69 (678)	22.12 (562)	Figure 2. Load per diag. C, fuel tank full.
w/17" wheels	27.28 (693)	22.71 (577)	Figure 2. Load per diag. C, fuel tank full.
w/18" wheels	27.75 (705)	23.22 (590)	Figure 2. Load per diag. C, fuel tank full.
2002-06 Cooper w/Sport Susp. & Cooper S			
w/15" wheels	25.90 (658)	21.37 (543)	Figure 2. Load per diag. C, fuel tank full.
w/16" wheels	26.41 (671)	21.85 (555)	Figure 2. Load per diag. C, fuel tank full.
w/17" wheels	27.00 (686)	22.44 (570)	Figure 2. Load per diag. C, fuel tank full.
w/18" wheels	27.48 (698)	22.95 (583)	Figure 2. Load per diag. C, fuel tank full.
2002-06 All John Cooper Works Models			
w/15" wheels	25.55 (649)	21.02 (534)	Figure 2. Load per diag. C, fuel tank full.
w/16" wheels	26.06 (662)	21.49 (546)	Figure 2. Load per diag. C, fuel tank full.
w/17" wheels	26.65 (677)	22.08 (561)	Figure 2. Load per diag. C, fuel tank full.
w/18" wheels	27.16 (690)	22.59 (574)	Figure 2. Load per diag. C, fuel tank full.
R52 Convertibles	Front	Rear	Notes
2005-08 Cooper, ONE w/Reg. Suspension			
w/15" wheels	26.18 (665)	21.65 (550)	Figure 2. Load per diag. C, fuel tank full.
w/16" wheels	26.69 (678)	22.12 (562)	Figure 2. Load per diag. C, fuel tank full.
w/17" wheels	27.28 (693)	22.71 (577)	Figure 2. Load per diag. C, fuel tank full.
w/18" wheels	27.75 (705)	23.22 (590)	Figure 2. Load per diag. C, fuel tank full.
2005-08 Cooper, ONE, Cooper S w/Sport Susp.			
w/15" wheels	25.90 (658)	21.37 (543)	Figure 2. Load per diag. C, fuel tank full.
w/16" wheels	26.41 (671)	21.85 (555)	Figure 2. Load per diag. C, fuel tank full.
w/17" wheels	27.00 (686)	22.44 (570)	Figure 2. Load per diag. C, fuel tank full.
w/18" wheels	27.48 (698)	22.95 (583)	Figure 2. Load per diag. C, fuel tank full.
2005-08 All John Cooper Works Models			
w/15" wheels	25.90 (658)	21.02 (534)	Figure 2. Load per diag. C, fuel tank full.
w/16" wheels	26.41 (671)	21.49 (546)	Figure 2. Load per diag. C, fuel tank full.
w/17" wheels	27.00 (686)	22.08 (561)	Figure 2. Load per diag. C, fuel tank full.
w/18" wheels	27.48 (698)	22.59 (574)	Figure 2. Load per diag. C, fuel tank full.
R55 Clubman	Front	Rear	Notes
2007-14 Cooper, Classic (Canada) Hardtop			
w/15" wheels	22.04 (560)	21.41 (544)	Figure 1. Load per diag. C, fuel tank full.
w/16" wheels	22.55 (573)	21.88 (556)	Figure 1. Load per diag. C, fuel tank full.
w/17" wheels	23.14 (588)	22.51 (572)	Figure 1. Load per diag. C, fuel tank full.
w/18" wheels	23.66 (601)	23.03 (585)	Figure 1. Load per diag. C, fuel tank full.
2007-14 Cooper S Hardtop			
w/15" wheels	22.04 (560)	21.33 (542)	Figure 1. Load per diag. C, fuel tank full.
w/16" wheels	22.55 (573)	21.81 (554)	Figure 1. Load per diag. C, fuel tank full.
w/17" wheels	23.14 (588)	22.44 (570)	Figure 1. Load per diag. C, fuel tank full.
w/18" wheels	23.66 (601)	23.03 (585)	Figure 1. Load per diag. C, fuel tank full.
2007-14 All John Cooper Works Package w/Reg. Suspension			
w/17" wheels	23.14 (588)	22.51 (572)	Figure 1. Load per diag. C, fuel tank full.
w/18" wheels	23.66 (601)	22.99 (584)	Figure 1. Load per diag. C, fuel tank full.
2007-14 All John Cooper Works Package w/Sport Suspension			
w/17" wheels	22.75 (578)	22.08 (561)	Figure 1. Load per diag. C, fuel tank full.
w/18" wheels	23.26 (591)	22.59 (574)	Figure 1. Load per diag. C, fuel tank full.
R56 Hardtop	Front	Rear	Notes
2007-13 Cooper, Classic (Canada) Hardtop			
w/15" wheels	22.04 (560)	21.41 (544)	Figure 1. Load per diag. C, fuel tank full.
w/16" wheels	22.55 (573)	21.88 (556)	Figure 1. Load per diag. C, fuel tank full.
w/17" wheels	23.14 (588)	22.51 (572)	Figure 1. Load per diag. C, fuel tank full.
w/18" wheels	23.66 (601)	23.03 (585)	Figure 1. Load per diag. C, fuel tank full.
2007-13 Cooper S Hardtop			
w/15" wheels	22.04 (560)	21.33 (542)	Figure 1. Load per diag. C, fuel tank full.
w/16" wheels	22.55 (573)	21.81 (554)	Figure 1. Load per diag. C, fuel tank full.
w/17" wheels	23.14 (588)	22.44 (570)	Figure 1. Load per diag. C, fuel tank full.
w/18" wheels	23.66 (601)	23.03 (585)	Figure 1. Load per diag. C, fuel tank full.
2007-13 All John Cooper Works Package w/Reg. Suspension			
w/17" wheels	23.14 (588)	22.51 (572)	Figure 1. Load per diag. C, fuel tank full.
w/18" wheels	23.66 (601)	22.99 (584)	Figure 1. Load per diag. C, fuel tank full.
2007-13 All John Cooper Works Package w/Sport Suspension			
w/17" wheels	22.75 (578)	22.08 (561)	Figure 1. Load per diag. C, fuel tank full.
w/18" wheels	23.26 (591)	22.59 (574)	Figure 1. Load per diag. C, fuel tank full.
R57 Convertible	Front	Rear	Notes
2009-15 Cooper, Cooper S, and JCW w/Reg. Suspension			
w/15" wheels	22.04 (560)	21.33 (542)	Figure 1. Load per diag. C, fuel tank full.
w/16" wheels	22.55 (573)	21.81 (554)	Figure 1. Load per diag. C, fuel tank full.
w/17" wheels	23.14 (588)	22.44 (570)	Figure 1. Load per diag. C, fuel tank full.
w/18" wheels	23.66 (601)	22.91 (582)	Figure 1. Load per diag. C, fuel tank full.

* ± ⅜ (10)

MINI (cont.)

Year Model	Height Specifications - In. (mm) *		
R57 Convertible (cont.)	Front	Rear	Notes
2009-15 All John Cooper Works Conv. & Coupe w/Sport Suspension			
w/15" wheels	21.65 (550)	21.02 (534)	Figure 1. Load per diag. C, fuel tank full.
w/16" wheels	22.16 (563)	21.49 (546)	Figure 1. Load per diag. C, fuel tank full.
w/17" wheels	23.75 (578)	22.08 (561)	Figure 1. Load per diag. C, fuel tank full.
w/18" wheels	23.26 (591)	22.59 (574)	Figure 1. Load per diag. C, fuel tank full.
R58 Coupe	Front	Rear	Notes
2009-16 Cooper, Cooper S, and JCW w/Reg. Suspension			
w/15" wheels	22.04 (560)	21.33 (542)	Figure 1. Load per diag. C, fuel tank full.
w/16" wheels	22.55 (573)	21.81 (554)	Figure 1. Load per diag. C, fuel tank full.
w/17" wheels	23.14 (588)	22.44 (570)	Figure 1. Load per diag. C, fuel tank full.
w/18" wheels	23.66 (601)	22.91 (582)	Figure 1. Load per diag. C, fuel tank full.
2009-16 All John Cooper Works Conv. & Coupe w/Sport Suspension			
w/15" wheels	21.65 (550)	21.02 (534)	Figure 1. Load per diag. C, fuel tank full.
w/16" wheels	22.16 (563)	21.49 (546)	Figure 1. Load per diag. C, fuel tank full.
w/17" wheels	23.75 (578)	22.08 (561)	Figure 1. Load per diag. C, fuel tank full.
w/18" wheels	23.26 (591)	22.59 (574)	Figure 1. Load per diag. C, fuel tank full.
R59 Roadster	Front	Rear	Notes
2012-16 Cooper, Cooper S, and JCW w/Reg. Suspension			
w/15" wheels	22.04 (560)	21.45 (544)	Figure 1. Load per diag. C, fuel tank full.
w/16" wheels	22.55 (573)	21.92 (557)	Figure 1. Load per diag. C, fuel tank full.
w/17" wheels	23.14 (588)	22.51 (572)	Figure 1. Load per diag. C, fuel tank full.
w/18" wheels	23.66 (601)	23.03 (585)	Figure 1. Load per diag. C, fuel tank full.
2012-16 All John Cooper Works Conv. & Coupe w/Sport Suspension			
w/15" wheels	21.65 (550)	21.02 (534)	Figure 1. Load per diag. C, fuel tank full.
w/16" wheels	22.16 (563)	21.49 (546)	Figure 1. Load per diag. C, fuel tank full.
w/17" wheels	23.75 (578)	22.12 (562)	Figure 1. Load per diag. C, fuel tank full.
w/18" wheels	23.26 (591)	22.59 (574)	Figure 1. Load per diag. C, fuel tank full.
R60 Countryman (incl. ALL4)	Front	Rear	Notes
2011-16 Cooper, Cooper S Countryman			
Reg. Suspension w/16" wheels	24.33 (618)	24.01 (610)	Figure 1. Load per diag. C, fuel tank full.
w/17" wheels	24.92 (633)	24.60 (625)	Figure 1. Load per diag. C, fuel tank full.
w/18" wheels	25.39 (645)	25.11 (638)	Figure 1. Load per diag. C, fuel tank full.
Sport Suspension w/16" wheels	23.93 (608)	23.62 (600)	Figure 1. Load per diag. C, fuel tank full.
w/17" wheels	24.52 (623)	24.21 (615)	Figure 1. Load per diag. C, fuel tank full.
w/18" wheels	25.00 (635)	24.68 (627)	Figure 1. Load per diag. C, fuel tank full.
R61 Paceman (incl. ALL4)	Front	Rear	Notes
2014-16 Cooper, Cooper S, JCW Paceman			
Reg. Suspension w/16" wheels	24.33 (618)	24.01 (610)	Figure 1. Load per diag. C, fuel tank full.
w/17" wheels	24.92 (633)	24.64 (626)	Figure 1. Load per diag. C, fuel tank full.
w/18" wheels	25.43 (646)	25.11 (638)	Figure 1. Load per diag. C, fuel tank full.
w/19" wheels	25.90 (658)	25.62 (651)	Figure 1. Load per diag. C, fuel tank full.
Sport Suspension w/16" wheels	23.89 (607)	23.62 (600)	Figure 1. Load per diag. C, fuel tank full.
w/17" wheels	24.52 (623)	24.21 (615)	Figure 1. Load per diag. C, fuel tank full.
w/18" wheels	25.00 (635)	24.72 (628)	Figure 1. Load per diag. C, fuel tank full.
w/19" wheels	25.51 (648)	25.19 (640)	Figure 1. Load per diag. C, fuel tank full.
F54 Clubman	Front	Rear	Notes
2016-17 Clubman			
All Suspensions w/16" wheels	23.42 (595)	23.07 (586)	Figure 1. Load per diag. B, fuel tank full.
w/17" wheels	24.01 (610)	23.70 (602)	Figure 1. Load per diag. B, fuel tank full.
w/18" wheels	24.52 (623)	24.21 (615)	Figure 1. Load per diag. B, fuel tank full.
w/19" wheels	25.03 (636)	24.68 (627)	Figure 1. Load per diag. B, fuel tank full.
F55/F56 Hatch/Hardtop	Front	Rear	Notes
2015-17 Cooper, Cooper S, JCW Hardtop			
Reg. Suspension w/15" wheels	22.24 (565)	21.49 (546)	Figure 1. Load per diag. C, fuel tank full.
w/16" wheels	22.71 (577)	22.00 (559)	Figure 1. Load per diag. C, fuel tank full.
w/17" wheels	23.46 (593)	22.59 (574)	Figure 1. Load per diag. C, fuel tank full.
w/18" wheels	23.81 (605)	22.99 (584)	Figure 1. Load per diag. C, fuel tank full.
Sport Suspension w/15" wheels	22.24 (565)	21.49 (546)	Figure 1. Load per diag. C, fuel tank full.
w/16" wheels	22.71 (577)	22.00 (559)	Figure 1. Load per diag. C, fuel tank full.
w/17" wheels	23.46 (593)	22.59 (574)	Figure 1. Load per diag. C, fuel tank full.
w/18" wheels	23.81 (605)	22.99 (584)	Figure 1. Load per diag. C, fuel tank full.
F57 Convertible	Front	Rear	Notes
2016-17 Convertible			
All Suspensions w/15" wheels	22.24 (565)	21.45 (545)	Figure 1. Load per diag. C, fuel tank full.
w/16" wheels	22.75 (578)	21.96 (558)	Figure 1. Load per diag. C, fuel tank full.
w/17" wheels	22.99 (584)	22.55 (573)	Figure 1. Load per diag. C, fuel tank full.
w/18" wheels	23.85 (606)	23.07 (586)	Figure 1. Load per diag. C, fuel tank full.
F60 Countryman	Front	Rear	Notes
2017 Countryman All susp. w/16" wheels	24.52 (623)	24.29 (617)	Figure 1. Load per diag. B, fuel tank full.
w/17" wheels	25.15 (639)	24.92 (633)	Figure 1. Load per diag. B, fuel tank full.
w/18" wheels	25.62 (651)	25.43 (646)	Figure 1. Load per diag. B, fuel tank full.
w/19" wheels	26.14 (664)	25.90 (658)	Figure 1. Load per diag. B, fuel tank full.

* ± ⅜ (10)

CHASSIS / SUSP. HEIGHTS

Chrysler

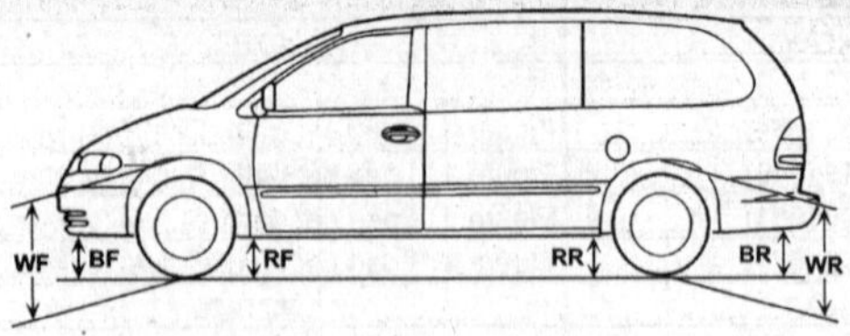

Figure 1

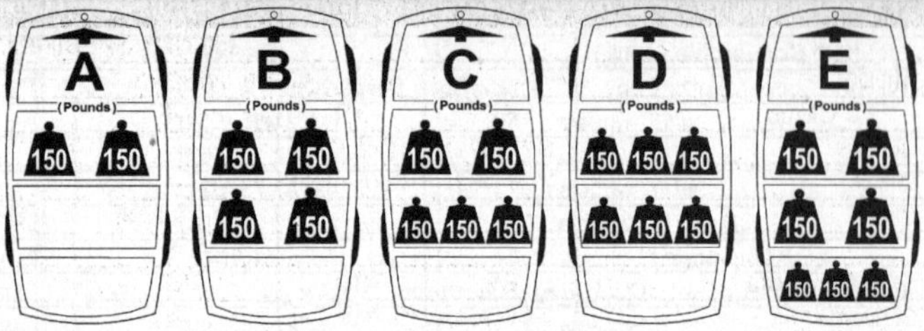

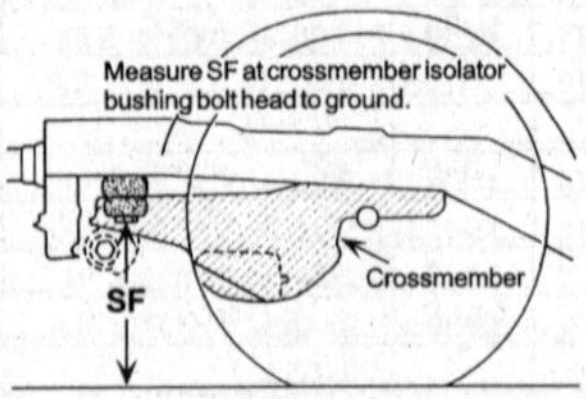

Figure 2

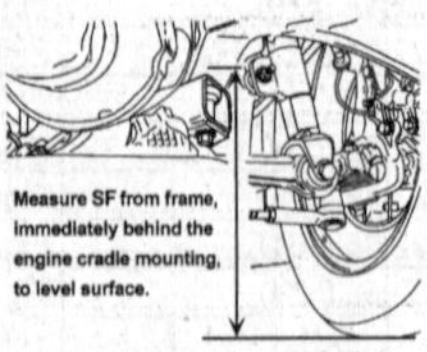

Figure 3

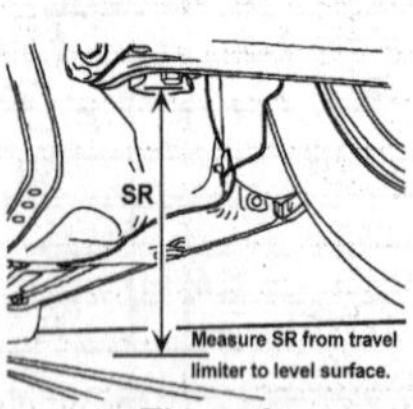

Figure 4

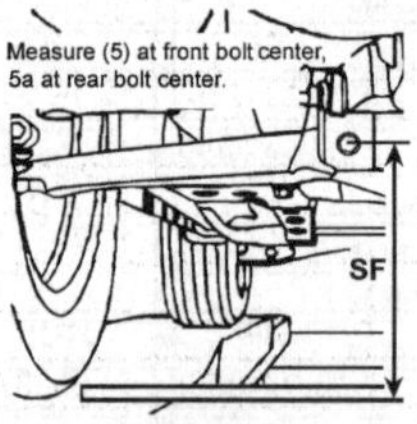

Figure 5, 5a

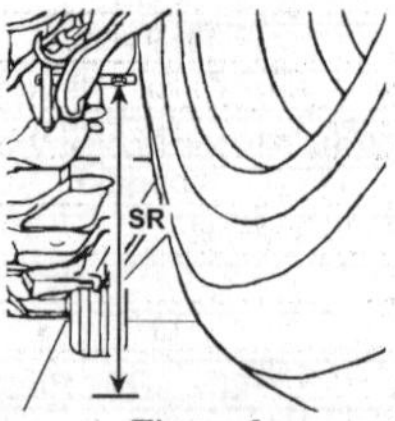

Figure 6

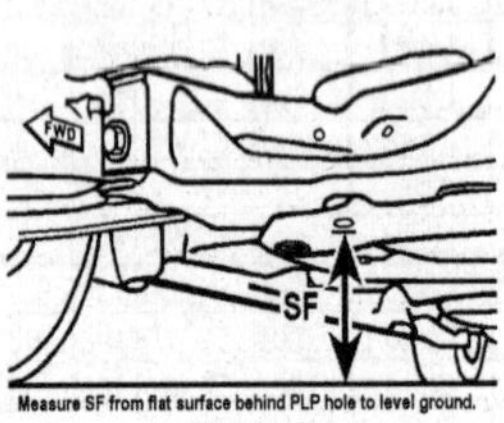

Figure 7

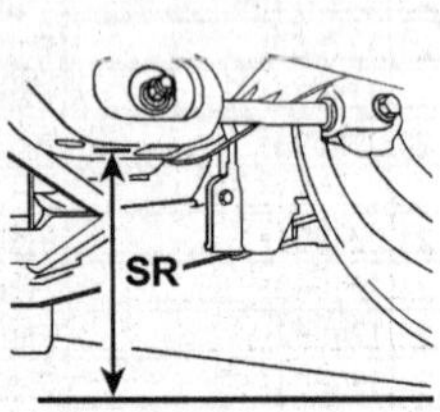

Figure 8

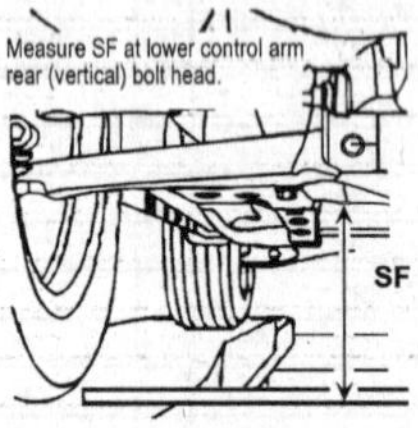

Figure 9

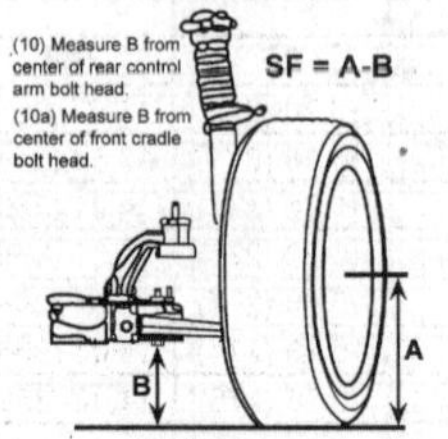

Figure 10, 10a

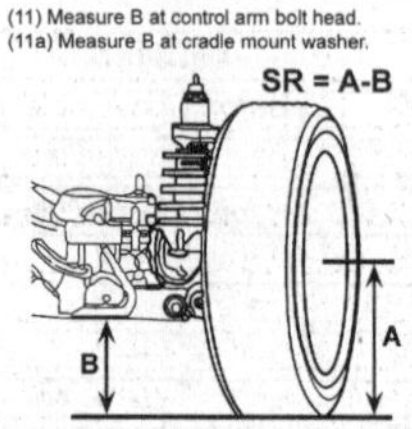

Figure 11, 11a

Measurements should be taken from a level surface to the point indicated. Rocker panel dimensions (RF & RR) are EPA Loaded Vehicle dimensions and should be taken with the vehicle loaded according to the load diagram given. Do not include weld flange. Wheel well measurements (WF & WR), and bumper dimensions (BF & BR), are Curb Weight dimensions and should be taken without passenger loading. Measure WF & WR to the lowest point in the wheel opening at the wheel centerline, not to include wheel-well trim. Measure BF & BR from the lowest point on the bumper including any factory installed trim or guards. The fuel tank should be full for all measurements unless otherwise indicated.

Chrysler

	Height Specifications - In. (mm) *							
Year Model	Curb Weight				Loaded			
200			SF	SR				Notes
2011-14 200 w/2.4L	—	—	5.83 (148)	4.96 (126)	—	—	—	Figs. 5a & 8 L/R ±0.5 (13)
2011-14 200 w/3.6L	—	—	6.22 (158)	4.96 (126)	—	—	—	Figs. 5a & 8 L/R ±0.5 (13)
2015-17 200 FWD & AWD	—	—	6.18 (157)	-0.28 (-7)	—	—	—	Figs. 10a&11a L/R ±0.5 (13)
300	BF	BR	WF	WR	RF	RR	Load	Notes
1999-04 300M	—	—	29.09 (739)	29.68 (754)	—	—	—	L/R diff. ±0.78 (20)
300 (2005 →)			SF	SR				Notes
2005 300 AWD	—	—	13.89 (353)	12.08 (307)	—	—	—	Figs. 3 & 4 L/R ±0.47 (12)
2006-08 300 AWD	—	—	13.85 (352)	12.12 (308)	—	—	—	Figs. 3 & 4 L/R ±0.47 (12)
2009-10 300 AWD	—	—	14.25 (362)	12.48 (317)	—	—	—	Figs. 3 & 4 L/R ±0.47 (12)
2005-08 300 RWD exc/SRT8	—	—	12.91 (328)	11.65 (296)	—	—	—	Figs. 3 & 4 L/R ±0.47 (12)
2009-10 300 RWD exc/SRT8	—	—	12.91 (328)	12.04 (306)	—	—	—	Figs. 3 & 4 L/R ±0.47 (12)
2006-08 300 RWD SRT8	—	—	12.63 (321)	11.41 (290)	—	—	—	Figs. 3 & 4 L/R ±0.47 (12)
2009-10 300 RWD SRT8	—	—	12.63 (321)	11.77 (299)	—	—	—	Figs. 3 & 4 L/R ±0.47 (12)
2011-14 300 RWD 17" wheels	—	—	12.48 (317)	—	—	—	—	Fig. 3 L/R ±0.47 (12)
2011-17 300 RWD 18" wheels	—	—	12.80 (325)	—	—	—	—	Fig. 3 L/R ±0.47 (12)
2011-17 300 RWD 20" wheels	—	—	13.03 (331)	—	—	—	—	Fig. 3 L/R ±0.47 (12)
2015-17 SRT & Super Track Pak	—	—	12.09 (307)	—	—	—	—	Fig. 3 L/R ±0.47 (12)
2011-17 300 AWD	—	—	13.54 (344)	—	—	—	—	Fig. 3 L/R ±0.47 (12)

* ± ¾ (19) unless otherwise indicated

Chrysler (cont.)

Year	Model	Height Specifications - In. (mm) *							
		Curb Weight				Loaded			
	Cirrus	BF	BR	WF	WR	RF	RR	Load	Notes
1995	Cirrus	6.77 (172)	11.65 (296)	—	—	5.00 (127)	4.80 (122)	C	Figure 1
1996-97	Cirrus	7.16 (182)	12.08 (307)	—	—	5.39 (137)	5.43 (138)	C	Figure 1
1998-00	Cirrus			27.32 (694)	27.28 (693)				[1] L/R diff. ±0.78 (20)
	Concorde	BF	BR	WF	WR	RF	RR	Load	Notes
1993-97	Concorde	9.84 (250)	14.64 (372)	—	—	9.21 (234)	10.00 (254)	C	Figure 1
	Fifth Avenue	BF	BR	SF	SR	RF	RR	Load	Notes
1990	New Yorker Fifth Ave.	10.23 (260)	12.16 (309)	—	—	9.13 (232)	9.40 (239)	C	Figure 1
1991-93	New Yorker Fifth Ave.	10.23 (260)	12.99 (330)	—	—	9.17 (233)	9.40 (239)	C	Figure 1
	Imperial	BF	BR	SF	SR	RF	RR	Load	Notes
1990-93	Imperial	12.44 (316)	12.99 (330)	—	—	9.17 (233)	9.40 (239)		
	LHS	BF	BR	WF	WR	RF	RR	Load	Notes
1994-97	LHS	13.97 (355)	14.56 (370)	—	—	8.70 (221)	9.44 (240)	C	
1999-02	LHS	—	—	29.09 (739)[1]	29.68 (754)[1]	—	—	—	[1] L/R diff. ±0.78 (20)
	LeBaron	BF	BR	WF	WR	RF	RR	Load	Notes
1990	LeBaron Convertible	10.70 (272)	13.97 (355)	—	—	8.30 (211)	7.32 (186)	B	
1991-92	LeBaron Convertible	10.66 (271)	13.54 (344)	—	—	8.46 (215)	7.91 (201)	B	
1993-95	LeBaron Convertible	10.66 (271)	14.01 (356)	—	—	8.46 (215)	7.91 (201)	B	
1990	LeBaron Coupe	10.70 (272)	14.01 (355)	—	—	8.54 (217)	7.87 (200)	C	
1991-92	LeBaron Coupe	10.66 (271)	13.54 (344)	—	—	8.22 (209)	7.40 (188)	C	
1993	LeBaron Coupe	10.66 (271)	14.01 (356)	—	—	8.22 (209)	7.36 (187)	C	
1990-94	LeBaron Sedan (All)	10.86 (276)	14.13 (359)	—	—	8.30 (211)	7.48 (190)	C	
	Neon	BF	BR	WF	WR	RF	RR	Load	Notes
2000-03	Neon (Canada)			26.45 (672)[1]	26.73 (679)[1]				±0.31 (8), L/R ±0.39 (10)
	New Yorker	BF	BR	WF	WR	RF	RR	Load	Notes
1990	New Yorker Landau	10.23 (260)	13.03 (331)	—	—	9.05 (230)	9.40 (239)	C	
1990-93	New Yorker Salon	10.23 (260)	13.03 (331)	—	—	8.77 (223)	7.83 (199)	C	
1994-96	New Yorker (LH)	13.97 (355)[1]	14.56 (370)[1]	—	—	8.70 (221)	9.44 (240)	D	[1] Loaded
	Pacifica	BF	BR	WF	WR	RF	RR	Load	Notes
2004-08	Pacifica	—	—	31.41 (798)	32.00 (813)	—	—	—	±0.39 (10)
	Pacifica Minivan			SF	SR				Notes
2017	Pacifica	—	—	4.33 (110)	—	—	—	—	Fig. 10 ±0.47 (12)
	PT Cruiser	BF	BR	WF	WR	RF	RR	Load	Notes
2001-05	PT Cruiser	—	—	27.36 (695)[1]	28.18 (716)[1]	—	—	—	[1] ±0.31(8), L/R ±0.39 (10)
2006-10	PT Cruiser All Convertibles	—	—	27.36 (695)[1]	27.83 (707)[1]	—	—	—	[1] ±0.31(8), L/R ±0.39 (10)
2006-10	PT Cruiser exc/GT	—	—	27.36 (695)[1]	28.18 (716)[1]	—	—	—	[1] ±0.31(8), L/R ±0.39 (10)
2006-10	PT Cruiser GT	—	—	27.36 (695)[1]	27.67 (703)[1]	—	—	—	[1] ±0.31(8), L/R ±0.39 (10)
	Sebring >2006	BF	BR	WF	WR	RF	RR	Load	Notes
1995-96	Sebring exc/Convertible	6.77 (172)	12.44 (316)	—	—	6⅜ (162)	5⅞ (150)	C[1]	[1] w/77 lbs. in trunk
1997-00	Sebring exc/17" wheels & Conv.	7.55 (192)	7.87 (200)	—	—	6⅜ (162)	5⅞ (150)	C[1]	[1] w/77 lbs. in trunk
1997-00	Sebring w/17" wheels & Conv.	7.95 (202)	8.26 (210)	—	—	6¾ (172)	6 5/16 (160)	C[1]	[1] w/77 lbs. in trunk
1996-99	Sebring Convertible	7.79 (198)	12.44 (316)	—	—	6 7/16 (163)[1]	7⅜ (187)[1]	—	[1] Unloaded
1997-00	Sebring Convertible	—	—	27.99 (711) [1]	27.99 (711) [1]	—	—	—	[1] L/R ±0.78 (20)
2001	Sebring Sedan & Convertible	—	—	27.99 (711) [1]	27.99 (711) [1]	—	—	—	[1] L/R ±0.78 (20)
2002-06	Sebring Sedan & Convertible	—	—	27.75 (705) [1]	27.95 (710) [1]	—	—	—	[1] L/R ±0.78 (20)
	Sebring 2007 >			SF	SR				Notes
2007-10	Sebring Sedan FWD 16" tires	—	—	6.69 (170)	8.07 (205)	—	—	—	Figs. 7 & 8 L/R ±0.5 (13)
2007-10	Sebring Sedan FWD 17" tires	—	—	6.81 (173)	8.14 (207)	—	—	—	Figs. 7 & 8 L/R ±0.5 (13)
2007-10	Sebring Sedan AWD 18" tires	—	—	7.20 (183)	8.58 (218)	—	—	—	Figs. 7 & 8 L/R ±0.5 (13)
2008-10	Sebring Conv. 2.4L 16" tires	—	—	6.65 (169)	7.91 (201)	—	—	—	Figs. 7 & 8 L/R ±0.5 (13)
2008-10	Sebring Conv. 2.4L 17" tires	—	—	6.77 (172)	7.99 (203)	—	—	—	Figs. 7 & 8 L/R ±0.5 (13)
2008-10	Sebring Conv. 2.7L 17" tires	—	—	6.73 (171)	7.99 (203)	—	—	—	Figs. 7 & 8 L/R ±0.5 (13)
2008-10	Sebring Conv. 2.7L 18" tires	—	—	6.85 (174)	8.14 (207)	—	—	—	Figs. 7 & 8 L/R ±0.5 (13)
2008-10	Sebring Conv. 3.5L 18" tires	—	—	6.81 (173)	8.26 (210)	—	—	—	Figs. 7 & 8 L/R ±0.5 (13)
	Town & Country >2007	BF	BR	WF	WR	RF	RR	Load	Notes
1990	Town & Country FWD	14.72 (374)	14.88 (378)	—	—	10.86 (276)	11.10 (282)	E	
1991-95	Town & Country FWD	9.37 (238)	12.12 (308)	—	—	10.51 (267)	10.70 (272)	E	
1992-95	Town & Country AWD	10.31 (262)	13.46 (342)	—	—	11.61 (295)	11.96 (304)	E	
1996-00	T & C FWD: w/14" or 215/65R15	—	—	29.44 (748)[1]	30.15 (766)[1]	—	—	—	[1] L/R diff. ±0.5 (13)
	All other OE tire sizes	—	—	29.68 (754)[1]	30.39 (772)[1]	—	—	—	[1] L/R diff. ±0.5 (13)
1996-00	T & C AWD: w/14" or 215/65R15	—	—	29.44 (748)[1]	30.15 (766)[1]	—	—	—	[1] L/R diff. ±0.5 (13)
	All other OE tire sizes	—	—	29.68 (754)[1]	30.39 (772)[1]	—	—	—	[1] L/R diff. ±0.5 (13)
2001-07	Town & Country	—	—	29.72 (755)[1]	30.31 (770)[1]	—	—	—	[1] L/R diff. ±0.5 (13)

* ± ¾ (19) unless otherwise indicated

CHASSIS / SUSP. HEIGHTS

Dodge

	Height Specifications - In. (mm) *							
Year Model	Curb Weight				Loaded			
Town & Country 2008 >	BF	BR	SF	SR	RF	RR	Load	Notes
2008-15 Town & Country	—	—	8.11 (206)	n/a	—	—	—	Figure 5
Voyager	BF	BR	WF	WR	RF	RR	Load	Notes
2001-03 Voyager	—	—	29.72 (755)[1]	30.31 (770)[1]	—	—	—	[1] L/R diff. ±0.5 (13)
Avenger	BF	BR	SF	SR	RF	RR	Load	Notes
1995-96 Avenger	6.77 (172)	12.44 (316)	—	—	6.37 (162)	5.90 (150)	C[1]	[1] w/77 lbs. in trunk
1997-00 Avenger exc/17" wheels	7.55 (192)	7.87 (200)	—	—	6.37 (162)	5.90 (150)	C[1]	[1] w/77 lbs. in trunk
1997-00 Avenger w/17" wheels	7.95 (202)	8.26 (210)	—	—	6.77 (172)	6.29 (160)	C[1]	[1] w/77 lbs. in trunk
2008-10 Avenger FWD 16" tires	—	—	6.69 (170)	8.07 (205)	—	—	—	Figs. 7 & 8 L/R ±0.5 (13)
2008-10 Avenger FWD 17" tires exc/3.6L	—	—	6.81 (173)	8.14 (207)	—	—	—	Figs. 7 & 8 L/R ±0.5 (13)
2008 Avenger AWD 18" tires	—	—	7.20 (183)	8.58 (218)	—	—	—	Figs. 7 & 8 L/R ±0.5 (13)
2010 Avenger FWD 17" tires 3.6L	—	—	6.77 (172)	8.14 (207)	—	—	—	Figs. 7 & 8 L/R ±0.5 (13)
2010 Avenger FWD 18" tires	—	—	6.88 (175)	8.26 (210)	—	—	—	Figs. 7 & 8 L/R ±0.5 (13)
2011-14 Avenger w/2.4L	—	—	5.83 (148)	4.96 (126)	—	—	—	Figs. 10a & 8 L/R ±0.5 (13)
2011-14 Avenger w/3.6L	—	—	6.22 (158)	4.96 (126)	—	—	—	Figs. 10a & 8 L/R ±0.5 (13)
Caliber			SF	SR				Notes
2007 All models	—	—	8.07 (205)	12.91 (328)	—	—	—	Figs. 5 & 6 L/R ±0.5 (13)
2009-10 2.0L w/15" wheels	—	—	7.83 (199)	12.95 (329)	—	—	—	Figs. 8 & 6 L/R ±0.5 (13)
2009-10 2.0L w/17" & 18" wheels	—	—	8.18 (208)	13.30 (338)	—	—	—	Figs. 8 & 6 L/R ±0.5 (13)
2008-10 All others w/15" wheels			8.07 (205)	12.95 (329)				Figs. 8 & 6 L/R ±0.5 (13)
2008-10 All others w/17" & 18" wheels	—	—	8.42 (214)	13.30 (338)	—	—	—	Figs. 8 & 6 L/R ±0.5 (13)
2011-12 w/15" wheels			8.07 (205)	12.95 (329)				Figs. 8 & 6 L/R ±0.5 (13)
2011-12 w/17" & 18" wheels			8.46 (215)	13.30 (338)				Figs. 8 & 6 L/R ±0.5 (13)
2008-09 SRT4	—	—	7.63 (194)	12.32 (313)	—	—	—	Figs. 8 & 6 L/R ±0.5 (13)
Caravan > 2007	BF	BR	WF	WR	RF	RR	Load	Notes
1990 SWB 5 Passenger	14.72 (374)	14.84 (377)	—	—	10.78 (274)	10.82 (275)	C	
1990 SWB 7 Passenger	14.72 (374)	14.84 (377)	—	—	10.90 (277)	11.06 (281)	E	
1990 LWB models	14.72 (374)	14.88 (378)	—	—	10.86 (276)	11.10 (282)	E	
1991-95 SWB 5 Passenger	8.89 (226)	11.61 (295)	—	—	10.00 (254)	10.23 (260)	C	
1991-95 SWB 7 Passenger	8.89 (226)	11.61 (295)	—	—	10.27 (261)	10.19 (259)	E	
1991-95 LWB models	9.37 (238)	12.12 (308)	—	—	10.51 (267)	10.70 (272)	E	
1996-00 All Gas w/14"or P215/65R15 tires	—	—	29.44 (748)[1]	30.15 (766)[1]	—	—	—	[1] L/R diff. ±0.5 (13)
1996-00 All Gas w/ P215/70R15 tires	—	—	29.68 (754)[1]	30.39 (772)[1]	—	—	—	[1] L/R diff. ±0.5 (13)
1996-00 All Gas w/16" or 17" tires	—	—	29.68 (754)[1]	30.39 (772)[1]	—	—	—	[1] L/R diff. ±0.5 (13)
1996-00 All CNG or Electric vehicles	—	—	30.86 (784)[1]	31.61 (803)[1]	—	—	—	[1] L/R diff. ±0.5 (13)
2001-07 Caravan FWD exc/Cargo	—	—	29.72 (755)[1]	30.31 (770)[1]	—	—	—	[1] L/R diff. ±0.5 (13)
2001-07 Caravan FWD Cargo	—	—	29.72 (755)[1]	31.29 (795)[1]	—	—	—	[1] L/R diff. ±0.5 (13)
1991-93 Caravan AWD SWB	10.27 (261)	13.42 (341)	—	—	11.65 (296)	11.92 (303)	E	
1991-95 Caravan AWD LWB	10.31 (262)	13.46 (342)	—	—	11.61 (295)	11.96 (304)	E	
1996-00 Caravan AWD exc/16" wheels	—	—	29.44 (748)[1]	30.15 (766)[1]	—	—	—	[1] L/R diff. ±0.5 (13)
1996-00 Caravan AWD w/16" wheels	—	—	29.68 (754)[1]	30.39 (772)[1]	—	—	—	[1] L/R diff. ±0.5 (13)
2001-04 Caravan AWD	—	—	29.72 (755)[1]	30.31 (770)[1]	—	—	—	[1] L/R diff. ±0.5 (13)
Grand Caravan 2008 >			SF	SR				
2008-15 Grand Caravan 16" wheels	—	—	7.64 (194)	n/a	—	—	—	Fig. 5 L/R diff. ±0.5 (13)
2008-15 Grand Caravan 17" wheels	—	—	8.11 (206)	n/a	—	—	—	Fig. 5 L/R diff. ±0.5 (13)
2008-15 Grand Caravan Cargo Van	—	—	8.11 (206)	n/a	—	—	—	Fig. 5 L/R diff. ±0.5 (13)
2016-17 Grand Caravan Cargo Van	—	—	6.65 (169)	-2.91 (-74)	—	—	—	Fig. 5 L/R diff. ±0.5 (13)
2016-17 Grand Caravan 16" wheels	—	—	6.37 (162)	-2.20 (-56)	—	—	—	Fig. 5 L/R diff. ±0.5 (13)
2016-17 Grand Caravan 17" wheels	—	—	6.65 (169)	-1.92 (-49)	—	—	—	Fig. 5 L/R diff. ±0.5 (13)
Charger/Omni	BF	BR	WF	WR	RF	RR	Load	Notes
1990 Charger/Omni 4-Door	15.15 (385)	15.27 (388)	—	—	8.38 (213)	8.89 (226)	C	
Challenger			SF	SR				Notes
2011-14 Non-SRT w/17" wheels	—	—	12.48 (317)	13.54 (344)	—	—	—	Fig. 3, L/R ±0.47 (12)
2011-16 Non-SRT w/18" wheels	—	—	12.80 (325)	13.86 (352)	—	—	—	Fig. 3, L/R ±0.47 (12)
2011-16 Non-SRT w/20" wheels	—	—	13.03 (331)	14.09 (358)	—	—	—	Fig. 3, L/R ±0.47 (12)
2008-10 SRT8 Models	—	—	12.63 (321)	11.41 (290)	—	—	—	Figs. 3 & 4, L/R ±0.47 (12)
2011-14 SRT8 Models	—	—	12.09 (307)	13.00 (330)	—	—	—	Fig. 3, L/R ±0.47 (12)
2015-16 SRT8 & Super Track Pac			12.09 (307)	13.35 (339)				Fig. 3, L/R ±0.47 (12)
Charger			SF	SR				Notes
2006-08 Charger RWD exc/SRT8	—	—	12.91 (328)	11.65 (296)	—	—	—	Figs. 3 & 4, L/R ±0.47 (12)
2006-08 Charger RWD SRT8	—	—	12.63 (321)	11.41 (290)	—	—	—	Figs. 3 & 4, L/R ±0.47 (12)
2007-08 Charger AWD	—	—	13.85 (352)	12.12 (308)	—	—	—	Figs. 3 & 4, L/R ±0.47 (12)
2009-10 Charger RWD exc/SRT8	—	—	12.91 (328)	12.04 (306)	—	—	—	Figs. 3 & 4, L/R ±0.47 (12)
2009-10 Charger RWD SRT8	—	—	12.63 (321)	11.77 (299)	—	—	—	Figs. 3 & 4, L/R ±0.47 (12)
2009-10 Charger AWD	—	—	14.25 (362)	12.48 (317)	—	—	—	Figs. 3 & 4, L/R ±0.47 (12)

* ± ¾ (19) unless otherwise indicated

Dodge (cont.)

Year	Model	Height Specifications - In. (mm) *							
		Curb Weight				Loaded			
	Charger (cont.)			SF	SR				Notes
2008-10	Police Pkg. w/o equipment.	—	—	13.58 (345)	12.40 (315)	—	—	—	Figs. 3 & 4, L/R ±0.47 (12)
2011-16	Charger AWD	—	—	13.54 (344)	—	—	—	—	Fig. 3, L/R ±0.47 (12)
2011-14	Charger RWD 17" wheels	—	—	12.48 (317)	—	—	—	—	Fig. 3, L/R ±0.47 (12)
2011-16	Charger RWD 18" wheels	—	—	12.80 (325)	—	—	—	—	Fig. 3, L/R ±0.47 (12)
2011-16	Charger RWD 20" wheels	—	—	13.03 (331)	—	—	—	—	Fig. 3, L/R ±0.47 (12)
2015-16	SRT8 & Super Track Pac			12.09 (307)	—	—	—	—	Fig. 3, L/R ±0.47 (12)
2011-16	AWD Police Pkg. w/o equip.	—	—	13.93 (354)	—	—	—	—	Fig. 3, L/R ±0.47 (12)
2011-16	AWD Police Pkg. w/equip.	—	—	13.54 (344)	—	—	—	—	Fig. 3, L/R ±0.47 (12)
2011-16	RWD Police Pkg. w/o equip.	—	—	13.27 (337)	—	—	—	—	Fig. 3, L/R ±0.47 (12)
2011-16	RWD Police Pkg. w/equip.	—	—	13.00 (330)	—	—	—	—	Fig. 3, L/R ±0.47 (12)
	Colt/Colt Vista	BF	BR	WF	WR	RF	RR	Load	Notes
1990	Colt Wagon 2WD	10.43 (265)	13.77 (350)	—	—	6.49 (165)	6.69 (170)	C	
1990	Colt Wagon 4WD	12.00 (305)	16.73 (425)	—	—	8.85 (225)	9.05 (230)	C	
1990-92	Colt Hatchback exc/GT	9.72 (247)	11.65 (296)	—	—	6.02 (153)	5.35 (136)	C	
1990	Colt Hatchback GT	8.70 (221)	11.65 (296)	—	—	6.02 (153)	5.35 (136)	C	
1993-94	Colt Sedan	8.46 (215)	10.23 (260)	—	—	6.88 (175)	6.10 (155)	C[1]	[1] w/77 lbs. in trunk
1990-91	Vista Wagon 2WD	14.96 (380)	14.96 (380)	—	—	5.90 (150)	6.10 (155)	C[1]	[1] w/300 lbs. in rear
1990-91	Vista Wagon 4WD	17.51 (445)	17.51 (445)	—	—	8.66 (220)	8.26 (210)	C[1]	[1] w/300 lbs. in rear
1992-94	Vista Wagon 2WD (Ply)	10.86 (276)	13.70 (348)	—	—	7.40 (188)	7.12 (181)	C[1]	[1] w/77 lbs. in trunk
1992-94	Vista Wagon AWD (Ply)	11.33 (288)	14.25 (362)	—	—	7.79 (198)	7.63 (194)	C[1]	[1] w/77 lbs. in trunk
	Dart			SF	SR				Notes
2013-16	Dart exc/Limited			6.22 (158)	-0.47 (-12)				Figs. 10 & 11
2013-16	Dart Limited			5.94 (151)	-0.74 (-19)				Figs. 10 & 11
	Daytona	BF	BR	WF	WR	RF	RR	Load	Notes
1990	Daytona exc/Turbo, Shelby	12.12 (308)	13.58 (345)	—	—	7.95 (202)	7.16 (182)	B	
1990	Daytona Turbo, Shelby	9.21 (234)	13.58 (345)	—	—	8.34 (212)	7.71 (196)	B	
1991-92	Daytona exc/ES, IROC	14.37 (365)	13.58 (345)	—	—	7.95 (202)	7.32 (186)	B	
1991-92	Daytona ES, IROC	9.33 (237)	13.58 (345)	—	—	8.26 (210)	7.71 (196)	B	
1993	Daytona exc/ES, IROC	11.22 (285)	13.70 (348)	—	—	7.99 (203)	7.28 (185)	B	
1993	Daytona ES	9.56 (243)	10.59 (269)	—	—	8.03 (204)	7.40 (188)	B	
1993	Daytona IROC	9.33 (237)	13.54 (344)	—	—	8.22 (209)	7.71 (196)	B	
	Dynasty	BF	BR	WF	WR	RF	RR	Load	Notes
1990	Dynasty	10.70 (272)	12.99 (330)	—	—	8.70 (221)	7.79 (198)	C	
1991-93	Dynasty	10.70 (272)	7.79 (198)	—	—	8.77 (223)	7.83 (199)	C	
	Intrepid	BF	BR	WF	WR	RF	RR	Load	Notes
1993-97	Intrepid	9.40 (239)[1]	14.64 (372)[1]	—	—	9.17 (233)	10.00 (254)	C	[1] Loaded
1998-04	Intrepid	—	—	28.70 (729)	28.93 (735)	—	—	—	[1] L/R diff. ±0.78 (20)
	Magnum			SF	SR				Notes
2005	Magnum AWD	—	—	13.89 (353)	12.08 (307)	—	—	—	Figs. 3 & 4 L/R ±0.5 (12)
2006-08	Magnum AWD	—	—	13.85 (352)	12.12 (308)	—	—	—	Figs. 3 & 4 L/R ±0.5 (12)
2005-08	Magnum RWD exc/SRT8	—	—	12.91 (328)	11.65 (296)	—	—	—	Figs. 3 & 4 L/R ±0.5 (12)
2006-08	Magnum RWD SRT8	—	—	12.63 (321)	11.41 (290)	—	—	—	Figs. 3 & 4 L/R ±0.5 (12)
	Monaco	BF	BR	WF	WR	RF	RR	Load	Notes
1990-92	Monaco	15.35 (390)	15.70 (399)	—	—	6.33 (161)	5.66 (144)	D	
	Neon / SX 2.0	BF	BR	WF	WR	RF	RR	Load	Notes
1995-96	Neon 2-Door	9.01 (229)	14.17 (360)	—	—	5.15 (131)	8.07 (205)	C	
1995-96	Neon 4-Door	9.01 (229)	14.17 (360)	—	—	5.07 (129)	7.99 (203)	C	
1997-99	Neon	—	—	25.74 (654)	26.53 (674)	—	—	—	L/R ±0.39 (10)
2000-05	Neon	—	—	26.45 (672)	26.73 (679)	—	—	—	L/R ±0.39 (10)
2003-05	SX2.0	—	—	26.45 (672)	26.73 (679)	—	—	—	L/R ±0.39 (10)
	Shadow	BF	BR	WF	WR	RF	RR	Load	Notes
1990	Shadow	10.35 (263)	13.54 (344)	—	—	7.99 (203)	6.81 (173)	C	
1991-92	Shadow exc/Conv.	10.51 (267)	12.59 (320)	—	—	8.30 (211)	7.10 (182)	C	
1993-94	Shadow exc/Conv.	10.03 (255)	12.87 (327)	—	—	8.30 (211)	7.16 (182)	C	
1991	Shadow Convertible	10.51 (267)	12.59 (320)	—	—	8.42 (214)	7.40 (188)	B	
1992	Shadow Convertible	10.51 (267)	12.59 (320)	—	—	8.50 (216)	7.71 (196)	B	
1993	Shadow Convertible	10.03 (255)	12.87 (327)	—	—	8.50 (216)	7.71 (196)	B	
	SRT-4	BF	BR	WF	WR	RF	RR	Load	Notes
2004-05	SRT-4	—	—	26.96 (685)	27.44 (697)	—	—	—	L/R ±0.39 (10)
	Spirit	BF	BR	WF	WR	RF	RR	Load	Notes
1990-95	Spirit	10.86 (276)	14.13 (359)	—	—	8.30 (211)	7.44 (189)	C	

* ± ¾ (19) unless otherwise indicated

CHASSIS / SUSP. HEIGHTS

Dodge (cont.)

Year Model	Height Specifications - In. (mm) *							
	Curb Weight				Loaded			
Stealth	BF	BR	WF	WR	RF	RR	Load	Notes
1991-93 Stealth FWD except R/T	10.03 (255)	11.65 (296)	—	—	8.03 (204)	9.68 (246)	B	
1994-95 Stealth FWD except R/T	9.44 (240)	11.65 (296)	—	—	8.03 (204)	9.68 (246)	B	
1991-93 Stealth FWD R/T	9.84 (250)	10.78 (274)	—	—	8.03 (204)	9.68 (246)	B	
1994-95 Stealth FWD R/T	7.95 (202)	10.11 (257)	—	—	8.03 (204)	9.68 (246)	B	
1991-93 Stealth AWD	9.84 (250)	10.78 (274)	—	—	8.22 (209)	9.96 (253)	B	
1994-95 Stealth AWD	7.95 (202)	10.11 (257)	—	—	8.22 (209)	9.96 (253)	B	
Stratus	BF	BR	WF	WR	RF	RR	Load	Notes
1995-97 Stratus	6.77 (172)	11.65 (296)	—	—	5.00 (127)	4.84 (123)	C	
1995-97 Stratus EX	6.77 (172)	11.65 (296)	—	—	5.39 (137)	5.27 (134)	C	
1998-00 Stratus	—	—	27.32 (694) [1]	27.28 (693) [1]	—	—	—	[1] L/R diff. ±0.78 (20)
2001 Stratus Sedan	—	—	27.99 (711) [1]	27.99 (711) [1]	—	—	—	[1] L/R diff. ±0.74 (19)
2002-06 Stratus Sedan except 2006 R/T	—	—	27.75 (705) [1]	27.95 (710) [1]	—	—	—	[1] L/R diff. ±0.74 (19)
2006 Stratus R/T	—	—	27.12 (689) [1]	27.36 (695) [1]	—	—	—	[1] L/R diff. ±0.74 (19)

* ± ¾ (19) unless otherwise indicated

Eagle

Year Model	Height Specifications - In. (mm) *							
	Curb Weight				Loaded			
Premier	BF	BR	WF	WR	RF	RR	Load	Notes
1990-92 Premier	15.35 (390)	15.70 (399)	—	—	6.33 (161)	5.66 (144)	D	
Summit	BF	BR	WF	WR	RF	RR	Load	Notes
1990-92 Summit Sedans	9.72 (247)	12.12 (308)	—	—	6.02 (153)	5.39 (137)	C	
1991-92 Summit Hatchbacks	9.72 (247)	11.65 (296)	—	—	6.02 (153)	5.35 (136)	C	
1993-96 Summit 4-Door Sedan	8.85 (225)	10.23 (260)	—	—	6.88 (175)	6.10 (155)	C[1]	[1] w/77 lbs. in trunk
Talon	BF	BR	WF	WR	RF	RR	Load	Notes
1993-96 Summit 2-Door Coupe	8.46 (215)	10.23 (260)	—	—	6.88 (175)	6.10 (155)	C[1]	[1] w/77 lbs. in trunk
1992-96 Summit Wagon 2WD	10.86 (276)	13.70 (348)	—	—	7.40 (188)	7.12 (181)	C[1]	[1] w/77 lbs. in trunk
1992-96 Summit Wagon AWD	11.33 (288)	14.25 (362)	—	—	7.79 (198)	7.63 (194)	C[1]	[1] w/77 lbs. in trunk
1990-91 Talon FWD exc/Turbo	9.52 (242)	13.66 (347)	—	—	8.50 (216)	8.58 (218)	B	
1990-91 Talon FWD Turbo	8.38 (213)	13.42 (341)	—	—	8.50 (216)	8.58 (218)	B	
1992 Talon FWD (All Models)	8.97 (228)	10.98 (279)	—	—	8.50 (216)	8.58 (218)	B	
1993-94 Talon FWD (All Models)	8.74 (222)	13.81 (351)	—	—	8.50 (216)	8.58 (218)	B	
1997-98 Talon FWD	5.86 (149)	8.77 (223)	—	—	6.10 (155)	5.51 (140)	B[1]	[1] w/62 lbs. in trunk
1990-91 Talon AWD	8.89 (226)	14.01 (356)	—	—	9.09 (231)	9.17 (233)	B	
1992 Talon AWD	9.56 (243)	11.57 (294)	—	—	9.09 (231)	9.17 (233)	B	
1993-94 Talon AWD	9.33 (237)	14.40 (366)	—	—	9.09 (231)	9.17 (233)	B	
1995-96 Talon AWD	6.29 (160)	9.96 (253)	—	—	11.81 (300)	11.96 (304)	B[1]	[1] w/62 lbs. in trunk
1997-98 Talon AWD	6.45 (164)	9.37 (238)	—	—	6.77 (172)	6.18 (157)	B[1]	[1] w/62 lbs. in trunk
Vision	BF	BR	WF	WR	RF	RR	Load	Notes
1993-97 Vision	14.25 (362)[1]	14.56 (370)[1]	—	—	9.17 (233)	10.00 (254)	C	[1] Loaded

Plymouth

Year Model	Height Specifications - In. (mm) *							
	Curb Weight				Loaded			
Acclaim	BF	BR	WF	WR	RF	RR	Load	Notes
1990-95 Acclaim	10.86 (276)	14.13 (359)	—	—	8.30 (211)	7.44 (189)	C	
Breeze	BF	BR	WF	WR	RF	RR	Load	Notes
1996-97 Breeze	6.77 (172)	11.65 (296)	—	—	5.00 (127)	4.84 (123)	C	
1998-00 Breeze			27.32 (694)	27¼ (693)				L/R diff. ±0.78 (20)
Colt (see Dodge)	—	—	—	—	—	—	—	
Horizon	BF	BR	WF	WR	RF	RR	Load	Notes
1990 Horizon	15.15 (385)	15.27 (388)	—	—	8.38 (213)	8.89 (226)	C	
Laser	BF	BR	WF	WR	RF	RR	Load	Notes
1990-91 Laser	9.52 (242)	13.66 (347)	—	—	8.50 (216)	8.58 (218)	B	
1992 Laser FWD	8.74 (222)	13.81 (351)	—	—	8.50 (216)	8.58 (218)	B	
1992-94 Laser AWD	9.33 (237)	14.40 (366)	—	—	9.09 (231)	9.17 (233)	B	
Neon	BF	BR	WF	WR	RF	RR	Load	Notes
1995-96 Neon 2-Door	9.01 (229)	14.17 (360)	—	—	5.15 (131)	8.07 (205)	C	
1995-96 Neon 4-Door	9.01 (229)	14.17 (360)	—	—	5.07 (129)	7.99 (203)	C	
1997-99 Neon (All)	—	—	25.74 (654)	26½ (674)	—	—		L/R diff. ±0.39 (10)
2000-01 Neon (All)	—	—	26.45 (672)	26¾ (679)	—	—		L/R diff. ±0.39 (10)
Sundance	BF	BR	WF	WR	RF	RR	Load	Notes
1990 Sundance	10.35 (263)	13.54 (344)	—	—	7.99 (203)	6.81 (173)	C	
1991-92 Sundance	10.51 (267)	12.59 (320)	—	—	8.30 (211)	7.16 (182)	C	
1993-94 Sundance	10.03 (255)	12.87 (327)	—	—	8.30 (211)	7.16 (182)	C	

* ± ¾ (19) unless otherwise indicated

CHASSIS / SUSP. HEIGHTS

Plymouth (cont.)

Year	Model	Height Specifications - In. (mm) * — Curb Weight BF	BR	WF	WR	Loaded RF	RR	Load	Notes
	Voyager FWD	BF	BR	WF	WR	RF	RR	Load	Notes
1990	SWB 5 Passenger	14.72 (374)	14.84 (377)	—	—	10.78 (274)	10.82 (275)	C	
1990	SWB 7 Passenger	14.72 (374)	14.84 (377)	—	—	10.90 (277)	11.06 (281)	E	
1990	LWB models	14.2 (374)	14.88 (378)	—	—	10.86 (276)	11.10 (282)	E	
1991-95	SWB 5 Passenger	8.89 (226)	11.61 (295)	—	—	10.00 (254)	10.23 (260)	C	
1991-95	SWB 7 Passenger	8.89 (226)	11.61 (295)	—	—	10.27 (261)	10.19 (259)	E	
1991-95	LWB models	9.37 (238)	12.12 (308)	—	—	10.51 (267)	10.70 (272)	E	
1996-00	All Gas w/14" or P215/65R15	—	—	29.44 (748)	30.15 (766)	—	—	—	L/R diff. ±0.5 (13)
1996-00	All Gas w/ P215/70R15 tires	—	—	29.68 (754)	30.39 (772)	—	—	—	L/R diff. ±0.5 (13)
1996-00	All Gas w/16" or 17" tires	—	—	29.68 (754)	30.39 (772)	—	—	—	L/R diff. ±0.5 (13)
1996-00	All CNG or Electric vehicles	—	—	30.86 (784)	31.61 (803)	—	—	—	L/R diff. ±0.5 (13)
	Voyager AWD	BF	BR	WF	WR	RF	RR	Load	Notes
1992-93	All SWB Models	10.27 (261)	13.42 (341)	—	—	11.65 (296)	11.92 (303)	E	
1992-95	All LWB Models	10.31 (262)	13.46 (342)	—	—	11.61 (295)	11.96 (304)	E	
1996	All exc/16" wheels	—	—	29.44 (748)	30.15 (766)	—	—	—	L/R diff. ±0.5 (13)
1996	All with16" wheels	—	—	29.68 (754)	30.39 (772)	—	—	—	L/R diff. ±0.5 (13)

* ± ¾ (19) unless otherwise indicated

Chrysler / Dodge (Ram) / Jeep Truck & SUV

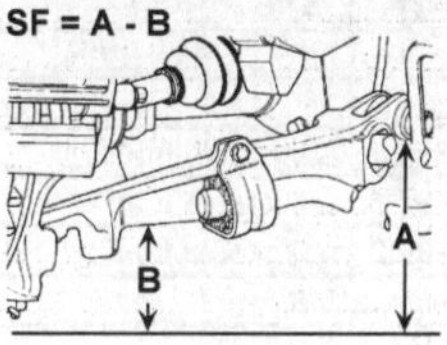

Figure 1

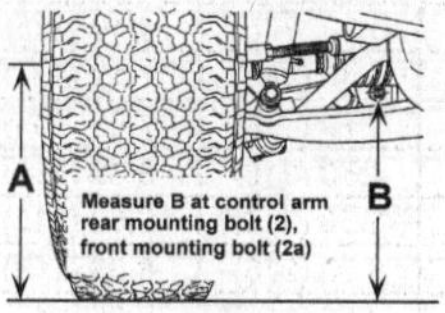

Figure 2, 2a

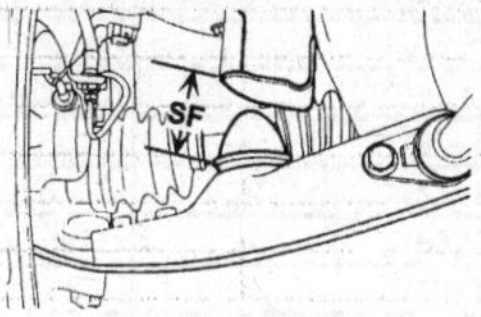

Figure 3

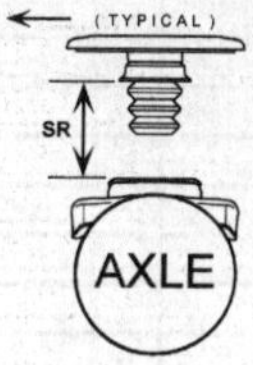

Figure 4

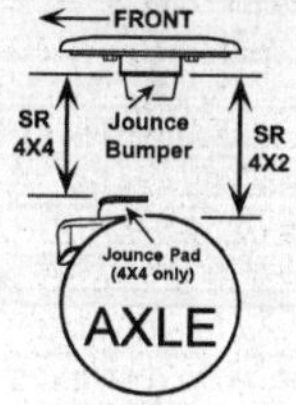

Figure 5

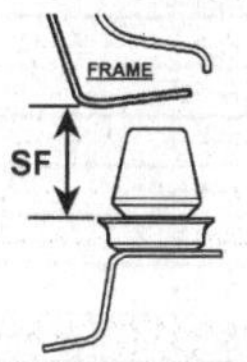

Figure 6

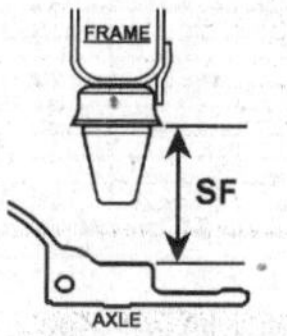

Figure 7

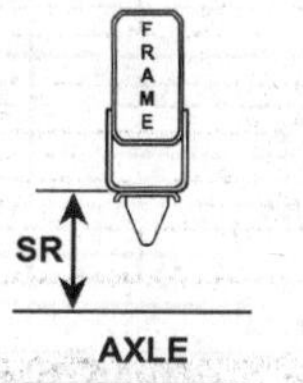

Figure 8

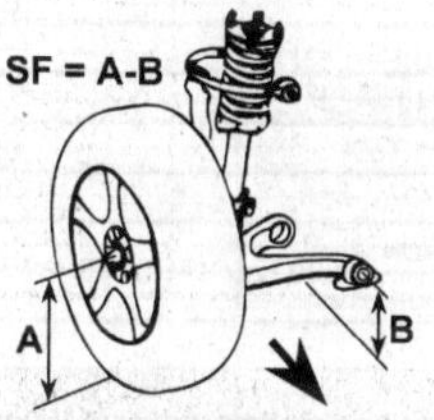

Figure 9

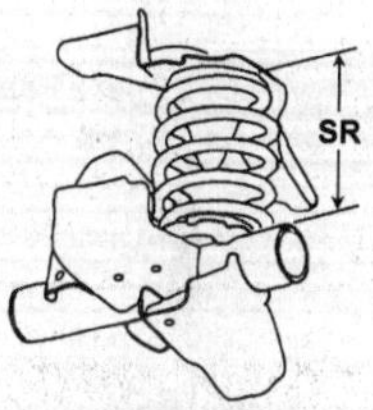

Figure 10

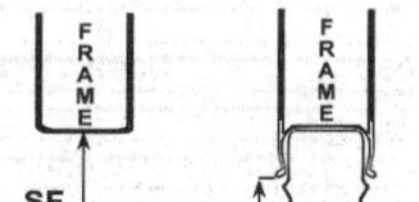

Figure 11, 11a

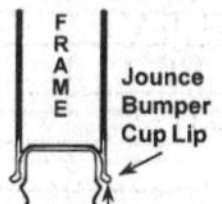

Figure 12

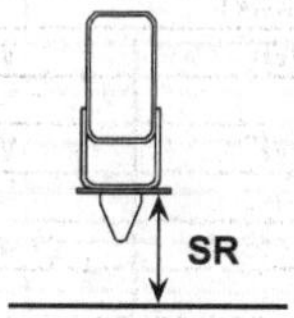

Figure 13

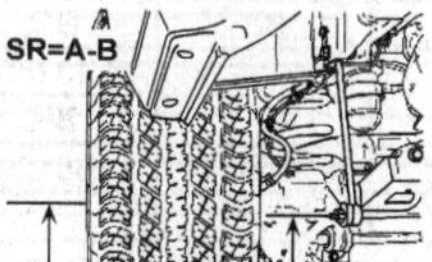

Figure 14

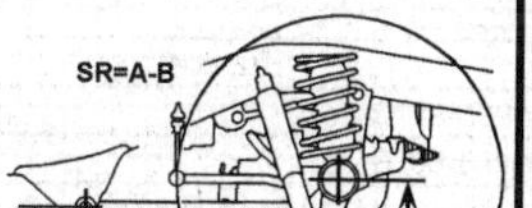

Figure 15

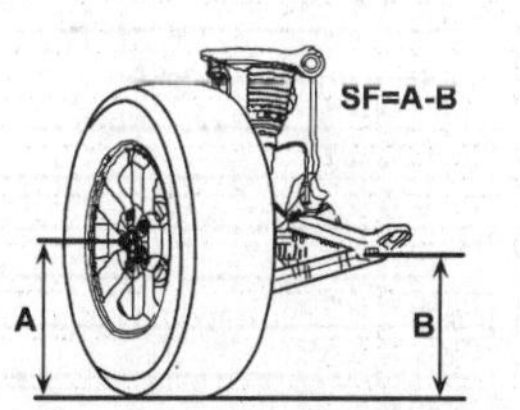

Figure 16

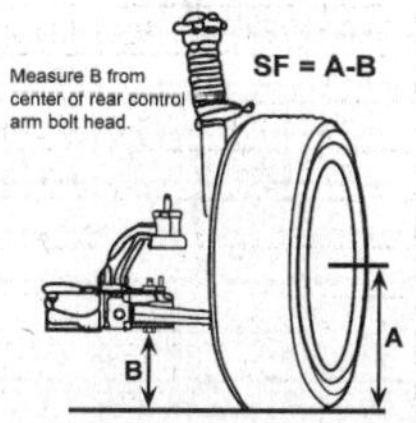

Figure 17

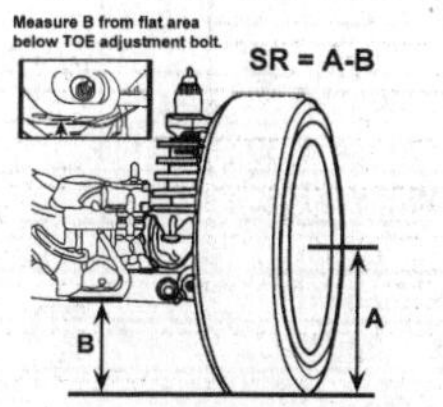

Figure 18

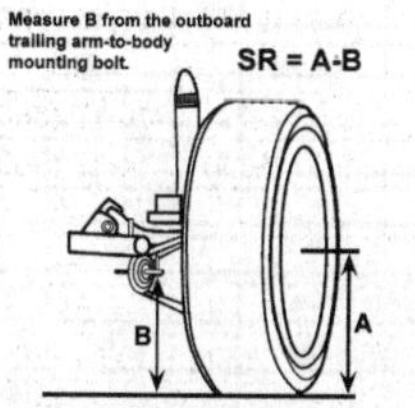

Figure 19

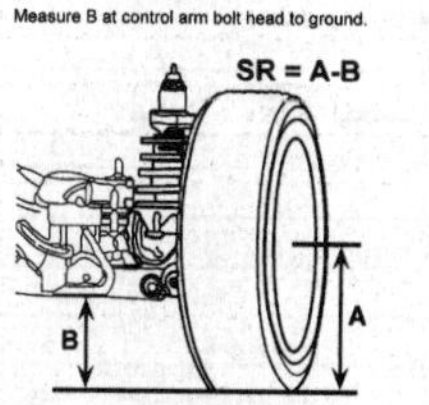

Figure 20

CHASSIS / SUSP. HEIGHTS

Chrysler / Dodge (Ram) Truck & SUV

Year Model	Height Specifications @ Curb Weight - In. (mm) * Front (SF)	Rear (SR)	Notes
Chrysler Aspen			
2007-09 Aspen w/9 ¼" axle	3.18 (81)	5.82 (148)	Figs. 2 & 4. Adj. front
2007-09 Aspen w/8 ¼" axle	3.18 (81)	6.06 (154)	Figs. 2 & 4. Adj. front
Dodge Dakota			
1990-96 Dakota 4X4	1.0 to 1.5 (25-38) reset to 1.25 (32)	—	Fig. 1. Adjustable front
1997-04 Dakota 4X4	1.73 to 1.96 (44-50)	—	Fig. 2. Adjustable front
2005 Dakota All w/8.25" axle	2.8 (71) ±0.24 (6)	6.1 (154) ±0.4 (10) L/R = 0.40 (10)	Figs. 2 & 4.
2005 Dakota All w/9.25" axle	2.8 (71) ±0.24 (6)	5.8 (148) ±0.4 (10) L/R = 0.40 (10)	Figs. 2 & 4.
2006-11 Dakota All w/8.25" axle	2.8 (71) ± 0.4 (10) L/R =0.75(20)	4.4 (113) ±0.7(18) L/R =0.47(12)	Figs. 2 & 13.
2006-11 Dakota All w/9.25" axle	2.8 (71) ± 0.4 (10) L/R =0.75(20)	4.2 (107) ±0.7(18) L/R =0.47(12)	Figs. 2 & 13.
Dodge Durango			
1998-03 Durango 4X4	2.87 (73)	—	Fig. 2. Adjustable front
2004-09 Durango w/9 ¼" axle	3.18 (81)	5.82 (148)	Figs. 2 & 4. Adj. front
2004-09 Durango w/8 ¼" axle	3.18 (81)	6.06 (154)	Figs. 2 & 4. Adj. front
2011-17 Durango Std. Susp.	3.54 (90)	—	Fig. 16 L/R ±0.47 (12)
2011-17 Durango Sport or Aero Susp.	4.33 (110)	—	Fig. 16 L/R ±0.47 (12)
Dodge Nitro			
2007-11 Nitro	2.44 (62)	4.21 (107)	Figs. 9 & 4, L/R diff. ±0.27 (7)
Dodge Journey			
2009-10 Journey 16" tires	7.91 (201)	9.13 (232)	Figs. 7 & 8 in car section
2009-10 Journey 17" tires	7.87 (200)	9.05 (230)	Figs. 7 & 8 in car section
2009-10 Journey 19" tires	8.07 (205)	9.25 (235)	Figs. 7 & 8 in car section
2011-17 Journey FWD	5.53 (140)	4.43 (113)	Figs. 17 & 18
2011-17 Journey AWD	5.29 (134)	4.15 (106)	Figs. 17 & 18
Dodge (Ram) Trucks			
1990-93 Ram 50 4X4	3.07 (78)	—	Fig. 3. Adjustable front
1990 Ram Raider	2.79 (71)	—	Fig. 3. Adjustable front
2002-05 Ram Pickup (DR) 4X4	2.28 (58)	—	Fig. 2. Adjustable front
2006-08 Ram LD (DR) 4X2 17" tires	2.24 (57)	120" WB 3.03 - 3.62 (77-92)	Figs. 2 & 5.
2006-08 Ram LD (DR) 4X2 20" tires	1.77 (45)	6.14 (156)	Figs. 2 & 5.
2006-08 Ram LD (DR) 4X4	2.16 (55)	6.92 (176)	Figs. 2 & 5.
2009-17 Ram LD (DS) 4X2 w/TRX	1.37-2.16 (35-55)	120" WB 2.83 - 3.62 (72-92))	Figs. 2 & 15.
		140" WB 2.79 - 3.58 (71-91)	Figs. 2 & 15.
2015-17 Ram LD (DS) 4X2 Rebel @Normal RH	1.97-2.76 (50-70)	140" WB 1.93 – 2.32 (49-59)	Figs. 2 & 15.
2009-17 Ram LD (DS) 4X2 17" tires w/o TRX	1.85-2.63 (47-67)	120" WB 2.83 - 3.62 (72-92))	Figs. 2 & 15.
		140" WB 2.79 - 3.58 (71-91)	Figs. 2 & 15.
2009-17 Ram LD (DS) 4X2 20" & 22" tires	1.37-2.16 (35-55)	120" WB 2.71 - 3.50 (69-89)	Figs. 2 & 15.
		140" WB 2.71 - 3.50 (69-89)	Figs. 2 & 15.
2009-17 Ram LD (DS) 4X4 w/o TRX	2.51-3.30 (64-84)	120" WB 2.36 - 3.14 (60-80)	Figs. 2 & 15.
		140" WB 2.28 - 3.07 (58-78)	Figs. 2 & 15.
2009-17 Ram LD (DS) 4X4 w/TRX	1.77-2.55 (45-65)	120" WB 1.88 - 2.67 (48-68)	Figs. 2 & 15.
		140" WB 1.81 - 2.59 (46-66)	Figs. 2 & 15.
2013-17 Ram LD (DS) w/AIR Susp @Normal RH	2.95-3.74 (75-95)	140" & 149" WB 2.71 - 3.50 (69-89)	Figs. 2 & 15.
2013-17 Ram LD (DS) w/AIR Susp @Aero Mode	4.01-4.80 (102-122)	140" & 149" WB 3.77-4.56 (96-116)	Figs. 2 & 15.
2006-13 Ram HD 4X2			
2500 w/10.5 axle	2.95 (75)	6.57 (167)	Figs. 6 & 7.
2500/3500 w/11.5 axle SRW	2.95 (75)	6.61 (168)	Figs. 6 & 7.
3500 w/11.5 axle DRW	2.95 (75)	6.81 (173)	Figs. 6 & 7.
2014-16 Ram 2500 HD 4X2			
DJ 4X2 140.5" WB exc/Air Ride	3.38 (86) ± 0.38 (10)	0.43 (11) ± 0.38 (10)	Figs. 4 & 15.
DJ 4X2 149.5" WB exc/Air Ride	3.38 (86) ± 0.38 (10)	0.39 (10) ± 0.38 (10)	Figs. 4 & 15.
DJ 4X2 160.5" WB exc/Air Ride	3.38 (86) ± 0.38 (10)	0.35 (9) ± 0.38 (10)	Figs. 4 & 15.
DJ 4X2 169.4" WB exc/Air Ride	3.38 (86) ± 0.38 (10)	0.31 (8) ± 0.38 (10)	Figs. 4 & 15.
DJ 4X2 w/Air Ride (All WB)	3.38 (86) ± 0.38 (10)	1.53 (39) ± 0.38 (10)	Figs. 4 & 15.
2006-13 Ram HD 4X4			
2500 w/10.5 axle	4.01 (102), Power Wagon=5.39 (137)	7.87 (200)	Figs. 6 & 7.
2500/3500 w/11.5 axle SRW	4.01 (102), Power Wagon=5.39 (137)	7.91 (201)	Figs. 6 & 7.
3500 w/11.5 axle DRW	4.01 (102), Power Wagon=5.39 (137)	7.99 (203)	Figs. 6 & 7.
2014-16 Ram 2500 HD 4X4			
DJ 4X4 140.5" & 149.5" WB exc/Air Ride	3.70 (94) ± 0.38 (10)	-1.57 (-40) ± 0.38 (10)	Figs. 4 & 15.
DJ 4X4 160.5" & 169.4" WB exc/Air Ride	3.70 (94) ± 0.38 (10)	-1.53 (-39) ± 0.38 (10)	Figs. 4 & 15.
DJ 4X4 Power Wagon	4.96 (126) ± 0.38 (10)	-2.40 (-61) ± 0.38 (10)	Figs. 4 & 15.
DJ 4X4 w/Air Ride 140.5" & 149.5" WB	3.77 (96) ± 0.38 (10)	0.27 (7) ± 0.38 (10)	Figs. 4 & 15.
DJ 4X4 w/Air Ride 160.5" & 169.4" WB	3.77 (96) ± 0.38 (10)	0.31 (8) ± 0.38 (10)	Figs. 4 & 15.
Ram Tradesman			
2013-15 Tradesman Van	8.11 (206)	n/a	Figure 5 (in car section)

* ± ¾ (19) unless otherwise indicated

Jeep Truck & SUV

Year Model	Height Specifications @ Curb Weight - In. (mm) *		Notes
	Front (SF)	Rear (SR)	
Jeep Commander			
2006-10 Commander Base	3.07 (78)	9.48 (241)	Figs. 9 & 10. L/R ±0.78 (20)
2006-10 Commander SRT8	3.93 (100)	8.42 (214)	Figs. 9 & 10. L/R ±0.78 (20)
Jeep Cherokee (KL)			
2014-17 Cherokee FWD	5.12 (130)	4.17 (106)	Figs. 10a & 11a. (in car section)
2014-17 Cherokee AWD	4.61 (117)	4.13 (105)	Figs. 10a & 11a. (in car section)
2014-17 Cherokee Trailhawk	3.46 (88)	3.62 (92)	Figs. 10a & 11a. (in car section)
Jeep Compass/Patriot			
2007-10 Compass w/17" Off Rd	9.40 (239)	14.29 (363)	Figs. 5 & 6. (in car section)
2007-10 Compass All others	8.42 (214)	13.30 (338)	Figs. 5 & 6. (in car section)
2011-17 Compass/Patriot All 4X2, 4X4 w/o Off-Rd (AWL)	4.13 (105)	-0.55 (-14)	Figs. 17 & 19.
2011-17 Compass/Patriot 4X4 Gas w/Off-Rd (AWL)	3.31 (84)	-1.34 (-34)	Figs. 17 & 19.
Jeep Grand Cherokee			
2006-10 Grand Cherokee Base	3.11 (79)	9.48 (241)	Figs. 9 & 10. L/R ±0.78 (20)
2006-10 Grand Cherokee SRT8	3.93 (100)	8.42 (214)	Figs. 9 & 10. L/R ±0.78 (20)
2011-13 Grand Cherokee (exc/SRT) Std. Susp.	3.03-3.97 (77-101)	—	Fig. 16.
2011-13 Grand Cherokee (exc/SRT) Sport Susp.	2.33-3.12 (59-79)	—	Fig. 16.
2011-13 Grand Cherokee (exc/SRT) Air Susp.	3.39-4.18 (86-106)	—	Fig. 16.
2011-13 Grand Cherokee SRT8	4.09-4.88 (104-124)	—	Fig. 16.
2014-17 Grand Cherokee Base Susp.	3.54 (90) ±0.4 (10)	—	Fig. 16. L/R ±0.47(12)
2014-17 Grand Cherokee Aero Susp.	4.13 (105) ±0.4 (10)	—	Fig. 16. L/R ±0.47(12)
2014-17 Grand Cherokee SRT8	4.57 (116) ±0.4 (10)	—	Fig. 16. L/R ±0.47(12)
2014-17 Grand Cherokee Air Susp. (Aero Mode)	4.41 (112) ±0.4 (10)	—	Fig. 16. L/R ±0.47(12)
2014-17 Grand Cherokee Air Susp. (Normal)	3.82 (97) ±0.4 (10)	—	Fig. 16. L/R ±0.47(12)
Jeep Liberty			
2002 Liberty	1.92 (49)	4.56 (116)	Figs. 2 & 4. L/R ±0.4 (10)
2003-07 Liberty	2.59 (66)	3.62 (92)	Figs. 2 & 4. L/R ±0.4 (10)
2008-12 Liberty	2.44 (62)	4.21 (107)	Figs. 9 & 4. L/R tol. ±0.27 (7)
Jeep Wrangler			
2007-17 Wrangler 4X2 (All)	2.48 (63)	4.01 (102)	Figs. 11 & 12, L/R ±0.47(12)
2007-10 Wrangler 4X4 Short WB X, Sport, Sahara	3.66 (93)	4.96 (126)	Figs. 11 & 12, L/R ±0.47(12)
2007-10 Wrangler 4X4 Short WB Rubicon	3.81 (97)	5.11 (130)	Figs. 11 & 12, L/R ±0.47(12)
2007-10 Wrangler 4X4 Short WB Diesel	2.99 (76)	4.96 (126)	Figs. 11 & 12, L/R ±0.47(12)
2007-10 Wrangler 4X4 Long WB X, Sport, Sahara	3.66 (93)	5.19 (132)	Figs. 11 & 12, L/R ±0.47(12)
2007-10 Wrangler 4X4 Long WB Rubicon	3.81 (97)	5.35 (136)	Figs. 11 & 12, L/R ±0.47(12)
2007-10 Wrangler 4X4 Long WB Diesel	2.99 (76)	5.19 (132)	Figs. 11 & 12, L/R ±0.47(12)
2011-17 Wrangler 4X4 Short WB X, Sport, Sahara	4.53 (115)	5.12 (130)	Figs. 11a & 12, L/R ±0.47(12)
2011-17 Wrangler 4X4 Short WB Rubicon	4.69 (119)	5.31 (135)	Figs. 11a & 12, L/R ±0.47(12)
2011-17 Wrangler 4X4 Short WB Diesel	4.53 (115)	5.16 (131)	Figs. 11a & 12, L/R ±0.47(12)
2011-17 Wrangler 4X4 Long WB X, Sport, Sahara	4.53 (115)	5.35 (136)	Figs. 11a & 12, L/R ±0.47(12)
2011-17 Wrangler 4X4 Long WB Rubicon	4.69 (119)	5.51 (140)	Figs. 11a & 12, L/R ±0.47(12)
2011-17 Wrangler 4X4 Long WB Diesel	4.53 (115)	5.39 (137)	Figs. 11a & 12, L/R ±0.47(12)

* ± ¾ (19) unless otherwise indicated. ** side/side max difference = ⅜" (10)

NOTES

Ford Motor Company

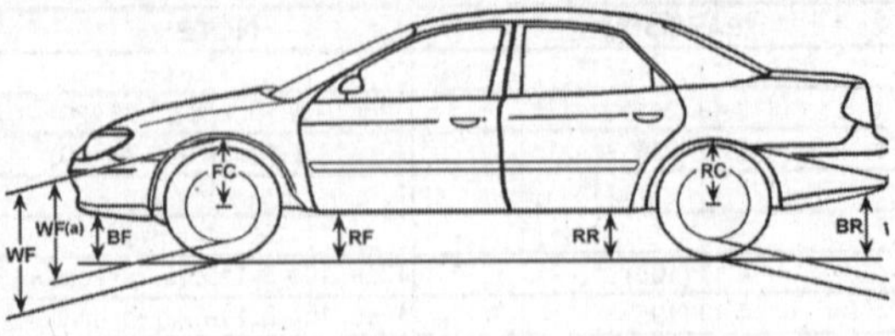

Figure 1

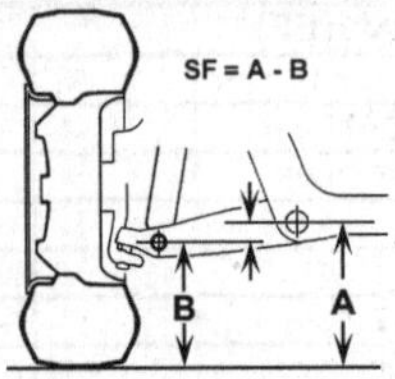

Figure 2

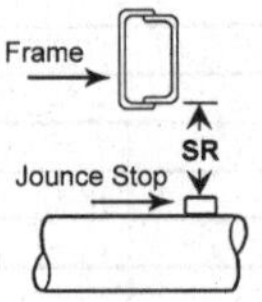

Figure 3

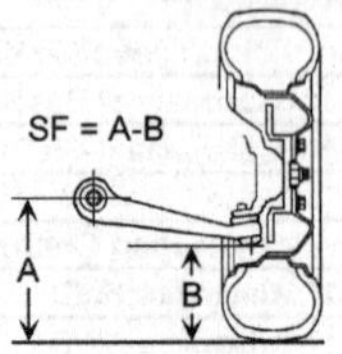

Figure 4

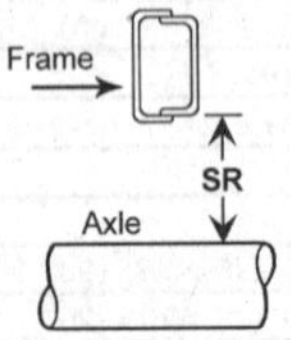

Figure 5

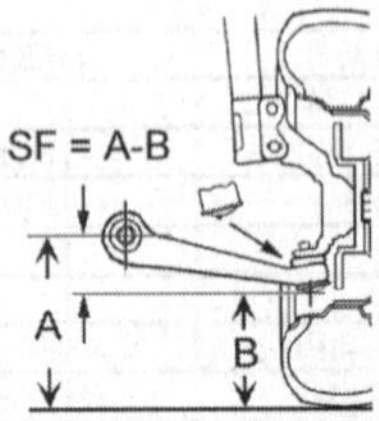

Figure 6

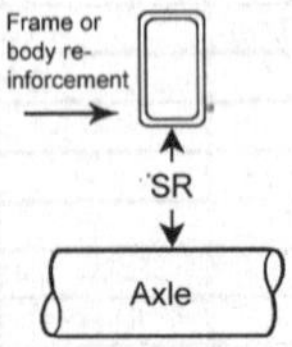

Figure 7

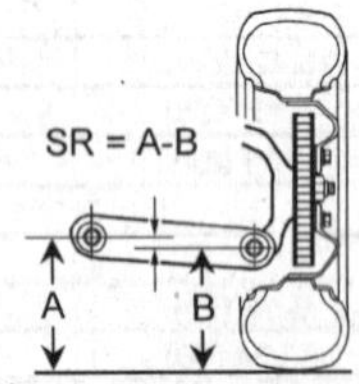

Figure 8

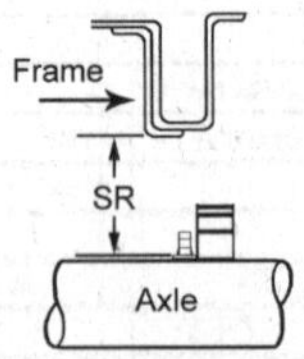

Figure 9

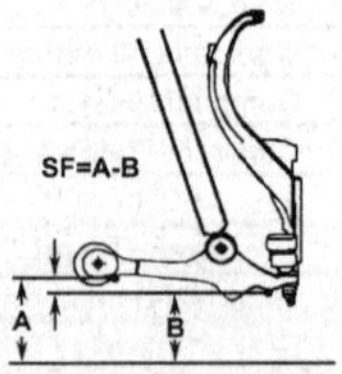

Figure 10

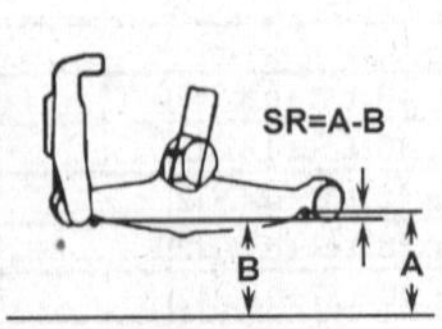

Figure 11

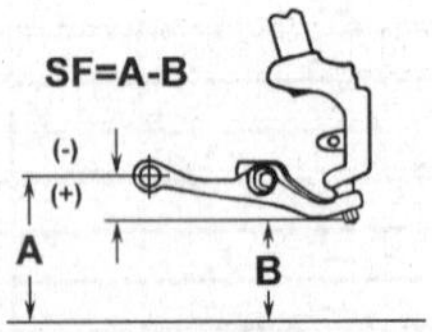

Figure 12

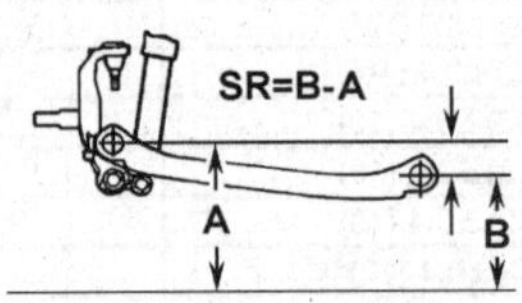

Figure 13

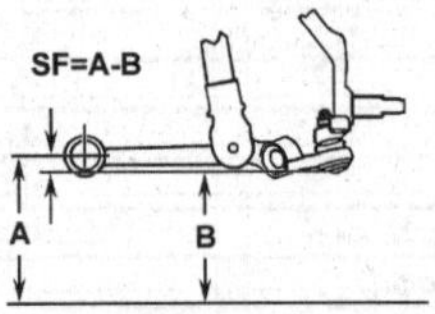

Figure 14

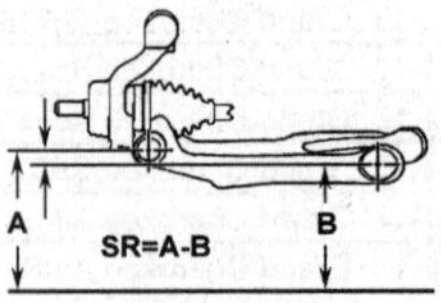

Figure 15

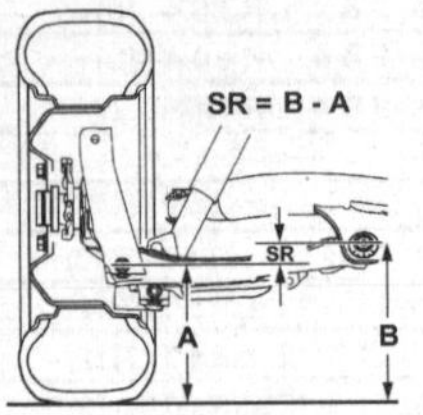

Figure 16

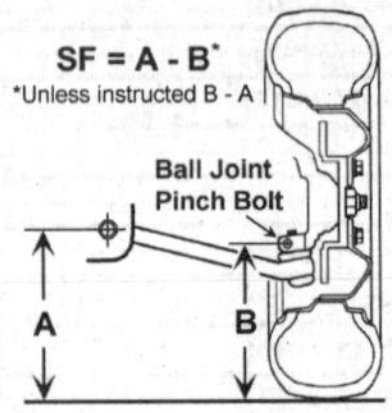

Figure 17

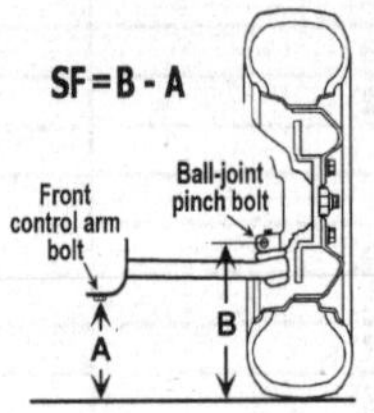

Figure 18

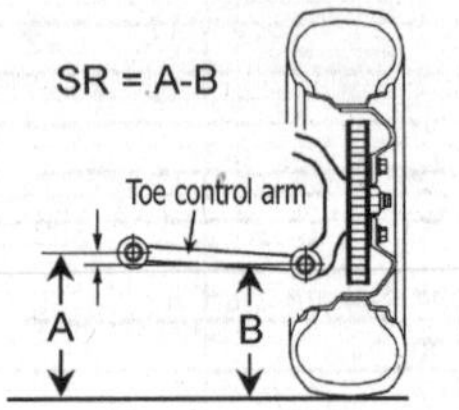

Figure 19

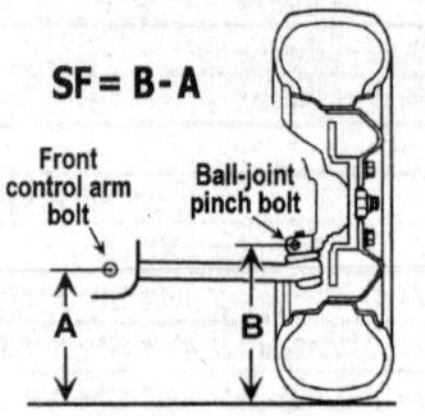

Figure 20

Take all measurements on a level surface to the points indicated. Rocker panel dimensions (RF & RR) are EPA Loaded Vehicle specifications and should be taken with the vehicle loaded according to the load diagram given. Do not include weld flange. Wheel arch (WF & WR) and Bumper (BF & BR) dimensions are Curb Weight specifications and should be taken without passenger loading. Measure WF & WR from the lower edge of the wheel opening to level ground. Measure WF(a) & WR(a) from the lower edge of the wheel opening to the lowest point on the wheel, at the wheel centerline. Measure BF & BR from the lowest point on bumper, including any factory trim or guards. The fuel tank should be full for all measurements unless otherwise indicated.

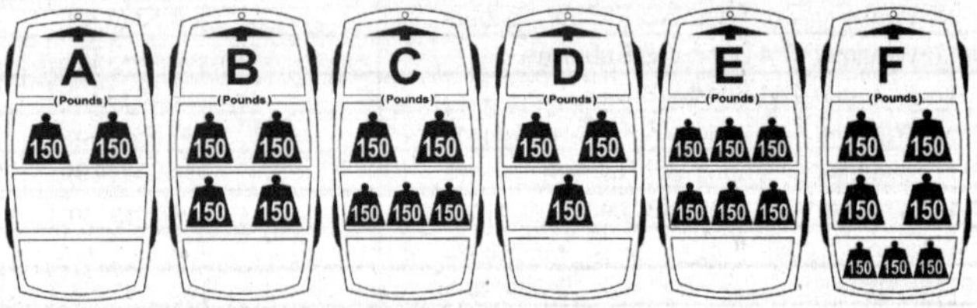

Ford

YEAR MODEL	HEIGHT SPECIFICATIONS - IN. (MM) **							
	@ CURB WEIGHT				LOADED			
Aspire	BF	BR	FC	RC	RF	RR	LOAD	Notes
1994-97 Aspire	10.23 (260)	13.70 (348)	Side/Side ±(10)	Side/Side ±(10	7.67 (195)	7.87 (200)	B[1]	[1] w/80 lbs. in trunk
CMAX	SF	Side/Side	SR	Side/Side				Notes
2013-17 CMAX	0.33 (8.3)[1]		2.54 (64)[1]		[1] ± 0.55 (14)			Figures 20 & 16
Contour	BF	BR	WF	WR	RF	RR	LOAD	Notes
1995-97 Contour	9.01 (229)	12.40 (315)	—	—	7.20 (183)	7.16 (182)	C	
1998-99 Contour	8.81 (224)	11.96 (304)	—	—	7.20 (183)	7.04 (179)	C	
Crown Victoria (→ 1997)	BF	BR	WF	WR	RF	RR	LOAD	Notes
1990 Sedan	14.13 (359)	14.25 (362)	—	—	6.81 (173)	6.06 (154)	C	
1990 Wagon	13.77 (350)	14.01 (356)	—	—	7.12 (181)	7.00 (178)	C[1]	[1] w/150 lbs. in rear
1991 Sedan	14.88 (378)	13.77 (350)	—	—	6.81 (173)	6.06 (154)	C	
1991 Wagon	14.56 (370)	14.25 (362)	—	—	7.12 (181)	7.00 (178)	C[1]	[1] w/150 lbs. in rear

* All Side/Side measurements are taken at wheel arch to ground (WF & WR) See Figure 1 above ** ±0.74 (19) unless indicated otherwise

Ford (cont.)

YEAR MODEL	HEIGHT SPECIFICATIONS - IN. (MM) **							
	@ CURB WEIGHT				LOADED			
Crown Victoria (→ 1997)(cont.)	**BF**	**BR**	**WF**	**WR**	**RF**	**RR**	**LOAD**	**Notes**
1992 Sedan	13.11 (333)	14.64 (372)	—	—	6.81 (173)	6.06 (154)	C	
1993-94 Sedan	15.43 (392)[1]	14.64 (372)[1]	—	—	6.81 (173)	6.06 (154)	C	[1] from alum. beam
1995-97 Sedan	15.19 (386)[1]	15.19 (386)[1]	—	—	6.81 (173)	6.06 (154)	C	[1] from alum. beam
Crown Victoria (1998 →)	**SF**	**Side/Side ***	**SR**	**Side/Side ***				**Notes**
1998-02 Crown Victoria:								
Base	1.18 (30)	Max.= 0.5 (13)	5.43 (138)	Max.= 0.5 (13)	—	—	—	Figures 4 & 5
Air Suspension	1.06 (27)	Max.= 0.5 (13)	5.19 (132)	Max.= 0.5 (13)	—	—	—	Figures 4 & 5
Police/Handling Pkg.	1.02 (26)	Max.= 0.5 (13)	5.19 (132)	Max.= 0.5 (13)	—	—	—	Figures 4 & 5
2003-04 Crown Victoria	2.36 (60)	Max.= 0.5 (13)	4.64 (118)	Max.= 0.5 (13)	—	—	—	Figures 2 & 3
2005-11 Crown Victoria								
Base(coil)	2.59 (66)	Max.= 0.5 (13)	4.88 (124)	Max.= 0.5 (13)	—	—	—	Figures 2 & 3
Air Suspension	2.59 (66)	Max.= 0.5 (13)	4.64 (118)	Max.= 0.5 (13)	—	—	—	Figures 2 & 3
Fleet w/Police Pkg. (coil)	2.59 (66)	Max.= 0.5 (13)	5.70 (145)	Max.= 0.5 (13)	—	—	—	Figures 2 & 3
Fleet w/o Police Pkg. (coil)	2.59 (66)	Max.= 0.5 (13)	5.23 (133)	Max.= 0.5 (13)	—	—	—	Figures 2 & 3
Escort	**BF**	**BR**	**WF**	**WR**	**RF**	**RR**	**LOAD**	**Notes**
1990 2-Door Hatchback exc/GT	11.14 (283)	12.87 (327)	—	—	7.12 (181)	7.28 (185)	D	
1990 2-Door Hatchback GT	10.78 (274)	12.63 (321)	—	—	7.12 (181)	7.28 (185)	D	
1990 4-Door Hatchback	11.14 (283)	12.87 (327)	—	—	7.12 (181)	7.28 (185)	D	
1991-96 All exc/Wagon	9.25 (235)	12.51 (318)	—	—	5.94 (151)	5.94 (151)	C[1]	[1] w/80 lbs. in trunk
1997 All exc/Wagon	7.95 (202)	10.35 (263)	—	—	6.25 (159)	6.22 (158)	C[1]	[1] w/80 lbs. in trunk
1998 ZX2 Coupe	7.59 (193)	11.77 (299)	—	—	6.25 (159)	6.22 (158)	C[1]	[1] w/64 lbs. in trunk
1990 Wagon	11.14 (283)	12.83 (326)	—	—	7.12 (181)	7.40 (188)	D	
1991-96 Wagon	n/a	n/a	—	—	5.94 (151)	5.94 (151)	C[1]	[1] w/80 lbs. in rear
1997 Wagon	7.95 (202)	10.27 (261)	—	—	6.25 (159)	6.22 (158)	C[1]	[1] w/80 lbs. in rear
Five-Hundred	**WF**	**Side/Side**	**WR**	**Side/Side**				**Notes**
2005-07 Five-Hundred FWD 17"	29.13 (740)	Max.= 0.5 (13)	29.56 (751)	Max.= 0.5 (13)	—	—	—	Figure 1
2005-07 Five-Hundred FWD 18"	29.44 (748)	Max.= 0.5 (13)	29.88 (759)	Max.= 0.5 (13)	—	—	—	Figure 1
2005-07 Five-Hundred AWD 17"	29.72 (755)	Max.= 0.5 (13)	29.92 (760)	Max.= 0.5 (13)	—	—	—	Figure 1
2005-07 Five-Hundred AWD 18"	30.00 (762)	Max.= 0.5 (13)	30.19 (767)	Max.= 0.5 (13)	—	—	—	Figure 1
Festiva	**BF**	**BR**	**WF**	**WR**	**RF**	**RR**	**LOAD**	**Notes**
1990-93 All Models	9.72 (247)	11.65 (296)	—	—	7.20 (183)	7.20 (183)	B[1]	[1] w/80 lbs. in trunk
Fiesta	**SF**	**Side/Side**	**SR**	**Side/Side**				**Notes**
2011-14 Fiesta	0.59 (15)	—	1.53 (39)		—	—	—	Figure 17 (B-A)
2014 Fiesta ST	0.46 (12)	—	2.12 (54)	—	—	—	—	Figure 17 (B-A)
2015-17 Fiesta	0.65 (16)	—	1.59 (40)		—	—	—	Figure 17 (B-A)
2015-17 Fiesta ST	0.46 (12)	—	0.59 (15)	—	—	—	—	Figure 17 (B-A)
Focus	**SF**	**Side/Side**	**SR**	**Side/Side**				**Notes**
2008-11 Focus w/o rear stabilizer bar	0.27 (7)[1]	—	1.85 (47)[1]	—	[1] ± 0.40 (10)			Figures 17 & 16
2008-11 Focus w/bar & 16" tires	0.31 (8)[1]	—	1.92 (49)[1]	—	[1] ± 0.40 (10)			Figures 17 & 16
2008-11 Focus w/bar & 17" tires	0.19 (5)[1]	—	1.88 (48)[1]	—	[1] ± 0.40 (10)			Figures 17 & 16
2012-17 Focus Base	0.32 (13)[1]	—	2.53 (60)[1]	—	[1] ± 0.55 (14)			Figures 20 & 16
2013-17 Focus ST	0.63 (16)[1]	—	2.16 (59)[1]	—	[1] ± 0.55 (14)			Figures 20 & 16
2017 Focus RS	0.84 (33)[1]	—	—	—	[1] ± 0.30 (12)			Figures 20 & 16
2013-17 Focus EV	0.32 (8.3)[1]	—	2.53 (64)[1]	—	[1] ± 0.55 (14)			Figures 20 & 16
Freestar	**SF**	**Side/Side**	**SR**	**Side/Side**				**Notes**
2004-05 Freestar	1.49 (38)	—	6.37 (162)	—	—	—	—	Figures 4 & 9
2006-07 Freestar	1.18 (30)	—	6.37 (162)	—	—	—	—	Figures 4 & 9
Freestyle	**SF**	**Side/Side**	**SR**	**Side/Side**				**Notes**
2005-07 Freestyle FWD 17"	30.78 (782)	Max.= 0.5 (13)	31.77 (807)	Max.= 0.5 (13)	—	—	—	Figure 1
2005-07 Freestyle FWD 18"	31.10 (790)	Max.= 0.5 (13)	32.08 (815)	Max.= 0.5 (13)	—	—	—	Figure 1
2005-07 Freestyle AWD 17"	30.74 (781)	Max.= 0.5 (13)	31.29 (795)	Max.= 0.5 (13)	—	—	—	Figure 1
2005-07 Freestyle AWD 18"	31.06 (789)	Max.= 0.5 (13)	31.61 (803)	Max.= 0.5 (13)	—	—	—	Figure 1
Fusion/Fusion Hybrid	**SF**	**Side/Side**	**SR**	**Side/Side**				**Notes**
2006-09 Fusion FWD	2.67 (68)	—	2.55 (65)	—	—	—	—	Figures 14 & 16
2006-09 Fusion AWD	2.67 (68)	—	2.04 (52)	—	—	—	—	Figures 14 & 16
2010-12 Fusion/Fusion Hybrid FWD	2.67 (68)	—	2.36 (60)	—	—	—	—	Figures 14 & 16
2010-12 Fusion AWD	2.67 (68)	—	1.65 (42)	—	—	—	—	Figures 14 & 16
2013-17 Fusion/Fusion Hybrid	2.48 (63)[1]	—	1.27 (32)[2]	—	[1] ±0.47 (12)	[2] ±0.39 (10)		Figures 18 & 19
Mustang	**BF**	**BR**	**WF**	**WR**	**RF**	**RR**	**LOAD**	**Notes**
1990-93 All Models	15.39 (391)[1]	15.51 (394)[1]	—	—	7.59 (193)	6.69 (170)	D	[1] thru 0.25" pilot hole
1994-97 Coupe exc/GT & Cobra	7.36 (187)	13.81 (351)	—	—	6.29 (160)	6.14 (156)	B	
1994-97 Coupe GT & Cobra	7.20 (183)	14.29 (363)	—	—	6.29 (160)	6.14 (156)	B	

* All Side/Side measurements are taken at wheel arch to ground (WF & WR) See Figure 1 above ** ±0.75 (19) unless indicated otherwise

Ford (cont.)

Year	Model	Height Specifications - In. (mm) **							
		@ Curb Weight				Loaded			
	Mustang (cont.)	**SF**	**Side/Side ***	**SR**	**Side/Side ***				**Notes**
1994-97	Convertible exc/GT & Cobra	7.16 (182)	13.70 (348)	—	—	6.49 (165)	6.14 (156)	B	
1994-97	Convertible GT & Cobra	7.00 (178)	14.17 (360)	—	—	6.49 (165)	6.14 (156)	B	
1998	Coupe Base and GT	10.19 (259)	14.13 (359)	—	—	6.29 (160)	6.14 (156)	B	
1998	Convertible Base and GT	10.11 (257)	14.09 (358)	—	—	6.49 (165)	6.14 (156)	B	
1999-04	Mustang Base & GT	0.15 (4)	Max.= 0.5 (13)	5.03 (128)	Max.= 0.5 (13)	—	—	—	Figures 6 & 7
1999-01	Mustang Cobra	0.15 (4)	Max.= 0.5 (13)	1.41 (36)	Max.= 0.5 (13)	—	—	—	Figures 6 & 8
2004	Mustang Cobra	0.15 (4)	Max.= 0.5 (13)	0.70 (18)	Max.= 0.5 (13)	—	—	—	Figures 6 & 8
2005-08	Mustang Coupe	1.45 (37)	Max.= 0.5 (13)	4.52 (115)	Max.= 0.5 (13)	—	—	—	Figures 6 & 7
2005-08	Mustang Convertible	1.45 (37)	Max.= 0.5 (13)	4.64 (118)	Max.= 0.5 (13)	—	—	—	Figures 6 & 7
2009	Mustang Coupe	1.49 (38)	Max.= 0.5 (13)	4.52 (115)	Max.= 0.5 (13)	—	—	—	Figures 6 & 7
2009	Mustang Convertible	1.49 (38)	Max.= 0.5 (13)	4.64 (118)	Max.= 0.5 (13)	—	—	—	Figures 6 & 7
2010	Mustang V6 Coupe	1.49 (38)	Max.= 0.5 (13)	4.48 (114)	Max.= 0.5 (13)	—	—	—	Figures 6 & 7
2010	Mustang GT Coupe	1.45 (37)	Max.= 0.5 (13)	4.40 (112)	Max.= 0.5 (13)	—	—	—	Figures 6 & 7
2010	Mustang Convertible	1.49 (38)	Max.= 0.5 (13)	4.52 (115)	Max.= 0.5 (13)				Figures 6 & 7
2010	Shelby Coupe	1.49 (38)	Max.= 0.5 (13)	4.52 (115)	Max.= 0.5 (13)	—	—	—	Figures 6 & 7
2010	Shelby Convertible	1.41 (36)	Max.= 0.5 (13)	4.64 (118)	Max.= 0.5 (13)	—	—	—	Figures 6 & 7
2011	Shelby Coupe	1.61 (41)	Max.= 0.5 (13)	4.52 (115)	Max.= 0.5 (13)	—	—	—	Figures 6 & 7
2011	Shelby Convertible	1.41 (36)	Max.= 0.5 (13)	4.64 (118)	Max.= 0.5 (13)	—	—	—	Figures 6 & 7
2011	GT w/Brake Perf. Pkg.	1.51 (40)	Max.= 0.5 (13)	4.40 (112)	Max.= 0.5 (13)	—	—	—	Figures 6 & 7
2012-14	Mustang V6 Coupe	1.61 (41)	Max.= 0.5 (13)	4.44 (113)	Max.= 0.5 (13)	—	—	—	Figures 6 & 7
2012-14	Mustang V6 Convertible	1.61 (41)	Max.= 0.5 (13)	4.48 (114)	Max.= 0.5 (13)	—	—	—	Figures 6 & 7
2012-14	Mustang GT Coupe	1.61 (41)	Max.= 0.5 (13)	4.44 (113)	Max.= 0.5 (13)	—	—	—	Figures 6 & 7
2012-14	Mustang GT Convertible	1.61 (41)	Max.= 0.5 (13)	4.44 (113)	Max.= 0.5 (13)	—	—	—	Figures 6 & 7
2012-13	Boss All models	1.18 (30)	Max.= 0.5 (13)	4.40 (112)	Max.= 0.5 (13)	—	—	—	Figures 6 & 7
2015-17	Mustang All exc/GT350	1.92 (48)	Max.= 0.5 (13)	1.26 (32)	Max.= 0.5 (13)				Figures 6 & 19
2017	GT350	1.39 (35)	Max.= 0.5 (13)	1.03 (26)	Max.= 0.5 (13)				Figures 6 & 19
2017	GT350R	1.18 (30)	Max.= 0.5 (13)	0.82 (21)	Max.= 0.5 (13)				Figures 6 & 19
	Probe	**BF**	**BR**	**WF**	**WR**	**RF**	**RR**	**LOAD**	**Notes**
1990	All GL Models	7.63 (194)	12.91 (328)	—	—	7.87 (200)	7.83 (199)	B[1]	[1] w/80 lbs. in trunk
1990	All LX Models	7.71 (196)	13.03 (331)	—	—	7.95 (202)	7.91 (201)	B[1]	[1] w/80 lbs. in trunk
1990	All GT Models	7.08 (180)	11.41 (290)	—	—	8.07 (205)	8.03 (204)	B[1]	[1] w/80 lbs. in trunk
1991-92	All GL Models	7.63 (194)	14.09 (358)	—	—	7.87 (200)	7.83 (199)	B[1]	[1] w/80 lbs. in trunk
1991-92	All LX Models	7.71 (196)	14.21 (361)	—	—	7.95 (202)	7.91 (201)	B[1]	[1] w/80 lbs. in trunk
1991-92	All GT Models	7.08 (180)	12.59 (320)	—	—	8.07 (205)	8.03 (204)	B[1]	[1] w/80 lbs. in trunk
	Taurus (→ 2006)	**BF**	**BR**	**WF**	**WR**	**RF**	**RR**	**LOAD**	**Notes**
1990-91	Sedan	9.72 (247)	14.40 (366)	—	—	6.85 (174)	6.14 (156)	E[1]	[1] 1989 = B
1990-91	Wagon	9.84 (250)	14.52 (369)	—	—	7.08 (180)	6.18 (157)	D[1]	[1] w/150 lbs. in rear
1992-95	Sedan	10.19 (259)	14.40 (366)	—	—	6.85 (174)	6.14 (156)	E	
1992-95	Wagon	10.35 (263)	14.92 (379)	—	—	7.08 (180)	6.18 (157)	D[1]	[1] w/150 lbs. in rear
1996-99	Sedan exc/SHO	10.43 (265)	12.95 (329)	—	—	8.22 (209)	6.53 (166)	C	
1996-99	Sedan SHO	10.03 (255)	13.14 (334)	—	—	8.77 (223)	7.36 (187)	B	
1996-99	Wagon	10.35 (260)	13.50 (343)	—	—	8.26 (210)	6.77 (172)	C[1]	[1] w/150 lbs. in rear
2000-06	Sedan	—	—	28.07 (713)	27.40 (696)	—	—	—	Figure 1
2000-05	Wagon	—	—	27.95 (710)	27.67 (703)	—	—	—	Figure 1
	Taurus (2008 →)	**SF**	**Side/Side ***	**SR**	**Side/Side ***				**Notes**
2008-09	Taurus FWD	2.44 (62)	—	1.02 (26)	—	—	—	—	Figs. 6 & 8
2008-09	Taurus AWD	2.44 (62)	—	0.47 (12)	—	—	—	—	Figs. 6 & 8
2008-09	Taurus X FWD	2.36 (60)	—	1.81 (46)	—	—	—	—	Figs. 6 & 8
2008-09	Taurus X AWD	2.36 (60)	—	0.62 (16)	—	—	—	—	Figs. 6 & 8
2010-12	Taurus exc/SE w/17" wheels	2.04 (52)	—	0.43 (11)	—	—	—	—	Figs. 6 & 8
2011-12	Taurus SE w/17" wheels	1.65 (42)	—	0.43 (11)	—	—	—	—	Figs. 6 & 8
2013-17	Taurus exc/2.0L GTDI	2.04 (52)	—	0.43 (11)	—	—	—	—	Figs. 6 & 8
2013-17	Taurus 2.0L GTDI	1.65 (42)	—	0.12 (3)	—	—	—	—	Figs. 6 & 8
2013-17	Taurus SHO	2.04 (52)	—	0.55 (14)	—	—	—	—	Figs. 6 & 8
2013-17	Taurus Police	2.04 (52)	—	1.02 (26)	—	—	—	—	Figs. 6 & 8
	Tempo	**BF**	**BR**	**WF**	**WR**	**RF**	**RR**	**LOAD**	**Notes**
1990-91	All Models	10.11 (257)	13.93 (354)	—	—	7.83 (199)	7.48 (190)	B	
1992-94	All Models	10.19 (259)	10.62 (270)	—	—	7.83 (199)	7.48 (190)	B	
	Thunderbird (→ 1997)	**BF**	**BR**	**WF**	**WR**	**RF**	**RR**	**LOAD**	**Notes**
1990-93	All Models	15.19 (386)	15.59 (396)	—	—	n/a	n/a	—	
1994-97	All Models	13.50 (343)	15.59 (396)	—	—	n/a	n/a	—	
	Thunderbird (2002 →)	**SF**	**Side/Side ***	**SR**	**Side/Side ***				**Notes**
2002-05	All Models	2.44 (62)	.47 (12)	.94 (24)	.47 (12)	—	—	—	Figures 10 & 11
	Windstar	**SF**	**Side/Side ***	**SR**	**Side/Side ***				**Notes**
1999-03	All Models	2.32 (59)	—	6.49 (165)	—	—	—	—	Figures 4 & 9

* All Side/Side measurements are taken at wheel arch to ground (WF & WR) See Figure 1 above ** ±0.75 (19) unless indicated otherwise

Lincoln

YEAR MODEL	HEIGHT SPECIFICATIONS - IN. (MM) *							
	@ CURB WEIGHT				LOADED			
Continental (→ 1997)	**BF**	**BR**	**WF**	**WR**	**RF**	**RR**	**LOAD**	**Notes**
1990-94 All Models	10.94 (278)	12.83 (326)	—	—	8.66 (220)	7.91 (201)	C	
1995-97 All Models	8.93 (227)	12.87 (327)	—	—	8.74 (222)	8.38 (213)	C	
Continental (1998 →)	**SF**	**Side/Side ***	**SR**	**Side/Side ***				**Notes**
1998-02 All Models	3.58 (91)	0.98 (25)	-0.98 (-25)	0.98 (25)	—	—	—	Figures 12 & 13
2017 Continental	2.32 (59)	—	1.17 (30)	—	—	—	—	Figures 23 & 21
LS	**SF**	**Side/Side ***	**SR**	**Side/Side ***				**Notes**
2000-02 All Models	2.40 (61)	0.47 (12)	1.02 (26)	0.47 (12)	—	—	—	Figures 10 & 11
2003-06 All Models	1.37 (35)	0.47 (12)	0.74 (19)	0.47 (12)	—	—	—	Figures 10 & 11
Mark VII / Mark VIII	**BF**	**BR**	**WF(a)**	**WR(a)**	**RF**	**RR**	**LOAD**	**Notes**
1990-92 Mark VII	14.80 (376)	15.11 (384)	—	—	9.64 (245)	9.52 (242)	B	
1993-98 Mark VIII	1.29 (33)	0.51 (13)	-0.11 (-3)	0.51 (13)				Figures 14 & 15
MKC	**SF**	**Side/Side**	**SR**	**Side/Side**				**Notes**
2015-17 MKC FWD	0.55 (14)	—	1.79 (45.5)	0.39 (10)	—	—	—	Figs. 20 & 19
2015-17 MKC AWD	0.36 (9)	—	1.81 (46)	0.39 (10)	—	—	—	Figs. 20 & 19
MKS	**SF**	**Side/Side**	**SR**	**Side/Side**				**Notes**
2009-16 MKS	2.04 (52)	—	0.43 (11)	—	—	—	—	Figs. 6 & 8
MKZ/Zephyr (incl. Hybrids)	**SF**	**Side/Side ***	**SR**	**Side/Side ***				**Notes**
2006-09 MKZ/Zephyr FWD	2.67 (68)	—	2.55 (65)	—	—	—	—	Figures 14 & 16
2006-09 MKZ/Zephyr AWD	2.67 (68)	—	2.04 (52)	—	—	—	—	Figures 14 & 16
2010-12 MKZ FWD	2.67 (68)	—	2.36 (60)	—	—	—	—	Figures 14 & 16
2010-12 MKZ AWD	2.67 (68)	—	1.65 (42)	—	—	—	—	Figures 14 & 16
2013-17 MZK/MKZ Hybrid	2.48 (63)	—	1.12 (28)	—	—	—	—	Figures 18 & 19
Town Car(→1996)	**BF**	**BR**	**WF**	**WR**	**RF**	**RR**	**LOAD**	**Notes**
1990 All Models	14.96 (380)	15.55 (395)	—	—	7.12 (181)	6.22 (158)	C	
1991-92 All Models	14.92 (379)	15.43 (392)	—	—	7.12 (181)	6.22 (158)	C	
1993-94 All Models	14.92 (379)	15.66 (398)	—	—	7.12 (181)	6.22 (158)	C	
1995-96 All Models	15.59 (396)	15.51 (394)	—	—	7.12 (181)	6.22 (158)	C	
Town Car (1997 →)	**SF**	**Side/Side ***	**SR**	**Side/Side ***				**NOTES**
1997-02 Town Car	1.10 (28)	Max.= 0.5 (13)	5.07 (129)	Max.= 0.5 (13)	—	—	—	Figures 4 & 5
2003-04 Town Car	2.36 (60)	Max.= 0.5 (13)	4.64 (118)	Max.= 0.5 (13)	—	—	—	Figures 2 & 3
2005-10 Town Car exc/Limo	2.59 (66)	Max.= 0.5 (13)	4.64 (118)	Max.= 0.5 (13)	—	—	—	Figures 2 & 3
2005-10 Town Car Limo	2.48 (63)	Max.= 0.5 (13)	4.64 (118)	Max.= 0.5 (13)	—	—	—	Figures 2 & 3

* All Side/Side measurements are taken at wheel arch to ground (WF & WR) See Figure 1 above ** ±0.75 (19) unless indicated otherwise

Mercury

YEAR MODEL	HEIGHT SPECIFICATIONS - IN. (MM) *							
	@ CURB WEIGHT				LOADED			
Capri	**BF**	**BR**	**WF**	**WR**	**RF**	**RR**	**LOAD**	**Notes**
1991-94 All exc/1994 XR2	10.23 (260)	11.33 (288)	—	—	6.37 (162)	6.33 (161)	B[1]	[1] w/80 lbs. in trunk
1994 XR2	9.76 (248)	10.86 (276)	—	—	5.90 (150)	5.86 (149)	B[1]	[1] w/80 lbs. in trunk
Cougar	**BF**	**BR**	**WF**	**WR**	**RF**	**RR**	**LOAD**	**Notes**
1990-93 All Models	15.19 (386)	15.59 (396)	—	—	n/a	n/a	—	
1994-97 All Models	13.50 (343)	15.59 (396)	—	—	n/a	n/a	—	
Grand Marquis	**BF**	**BR**	**WF**	**WR**	**RF**	**RR**	**LOAD**	**Notes**
1990 Sedan	14.13 (359)	14.25 (362)	—	—	6.81 (173)	7.24 (184)	C	
1990 Wagon	13.74 (349)	14.01 (356)	—	—	7.32 (186)	7.12 (181)	C[1]	[1] w/150 lbs. in rear
1991 Sedan	14.84 (377)	13.77 (350)	—	—	6.81 (173)	7.24 (184)	C	
1991 Wagon	14.52 (369)	14.21 (361)	—	—	7.32 (186)	7.12 (181)	C[1]	[1] w/150 lbs. in rear
1992 Sedan	13.07 (332)	14.60 (371)	—	—	6.81 (173)	7.24 (184)	C	
1993-94 Sedan	14.80 (376)[1]	14.60 (371)[1]	—	—	6.81 (173)	7.24 (184)	C	[1] from alum. beam
1995-97 Sedan	15.19 (386)[1]	15.19 (386)[1]	—	—	6.81 (173)	7.24 (184)	C	[1] from alum. beam
Grand Marquis (1998 →)	**SF**	**Side/Side ***	**SR**	**Side/Side ***				**Notes**
1998-02 Grand Marquis:Base	1.18 (30)	Max.= 0.5 (13)	5.43 (138)	Max.= 0.5 (13)	—	—	—	Figures 4 & 5
Air Suspension	1.06 (27)	Max.= 0.5 (13)	5.19 (132)	Max.= 0.5 (13)	—	—	—	Figures 4 & 5
Police/Handling Pkg.	1.02 (26)	Max.= 0.5 (13)	5.19 (132)	Max.= 0.5 (13)	—	—	—	Figures 4 & 5
2003-04 Grand Marquis	2.36 (60)	Max.= 0.5 (13)	4.64 (118)	Max.= 0.5 (13)	—	—	—	Figures 2 & 3
2005-11 Grand Marquis: Base(coil)	2.59 (66)	Max.= 0.5 (13)	4.88 (124)	Max.= 0.5 (13)	—	—	—	Figures 2 & 3
Air Suspension	2.59 (66)	Max.= 0.5 (13)	4.64 (118)	Max.= 0.5 (13)	—	—	—	Figures 2 & 3
Fleet w/Police Pkg. (coil)	2.59 (66)	Max.= 0.5 (13)	5.70 (145)	Max.= 0.5 (13)	—	—	—	Figures 2 & 3
Fleet w/o Police Pkg. (coil)	2.59 (66)	Max.= 0.5 (13)	5.23 (133)	Max.= 0.5 (13)	—	—	—	Figures 2 & 3

* All Side/Side measurements are taken at wheel arch to ground (WF & WR) See Figure 1 above ** ±0.75 (19) unless indicated otherwise

Mercury (cont.)

Year Model	Height Specifications - In. (mm) *							
	@ Curb Weight				Loaded			
Marauder	SF	Side/Side *	SR	Side/Side *				Notes
2003-04 Marauder	2.36 (60)	Max.= 0.5 (13)	4.64 (118)	Max.= 0.5 (13)	—	—	—	Figures 2 & 3
Milan	SF	Side/Side *	SR	Side/Side *				Notes
2006-09 Milan FWD	2.67 (68)	—	2.55 (65)	—	—	—	—	Figures 14 & 16
2006-09 Milan AWD	2.67 (68)	—	2.04 (52)	—	—	—	—	Figures 14 & 16
2010-11 Milan/Milan Hybrid FWD	2.67 (68)	—	2.36 (60)	—	—	—	—	Figures 14 & 16
2010-11 Milan AWD	2.67 (68)	—	1.65 (42)	—	—	—	—	Figures 14 & 16
Montego	SF	Side/Side *	SR	Side/Side *				Notes
2005-07 Montego FWD 17"	29.13 (740)	Max.= 0.5 (13)	29.56 (751)	Max.= 0.5 (13)	—	—	—	Figure 1
2005-07 Montego FWD 18"	29.44 (748)	Max.= 0.5 (13)	29.88 (759)	Max.= 0.5 (13)	—	—	—	Figure 1
2005-07 Montego AWD 17"	29.72 (755)	Max.= 0.5 (13)	29.92 (760)	Max.= 0.5 (13)	—	—	—	Figure 1
2005-07 Montego AWD 18"	30.00 (762)	Max.= 0.5 (13)	30.19 (767)	Max.= 0.5 (13)	—	—	—	Figure 1
Monterey	SF	Side/Side *	SR	Side/Side *				Notes
2004-05 Monterey	1.49 (38)	—	6.37 (162)	—	—	—	—	Figures 4 & 9
2006-07 Monterey	1.18 (30)	—	6.37 (162)	—	—	—	—	Figures 4 & 9
Mystique	BF	BR	WF	WR	RF	RR	LOAD	Notes
1995-97 All Models	9.29 (236)	12.67 (322)	—	—	7.20 (183)	7.16 (182)	C	
1998-99 All Models	9.84 (250)	12.16 (309)	—	—	7.20 (183)	7.04 (179)	C	
Sable	BF	BR	WF	WR	RF	RR	LOAD	Notes
1990 Sedan	9.33 (237)	14.80 (376)	—	—	6.85 (174)	6.14 (156)	E	
1990 Wagon	13.54 (344)	14.52 (369)	—	—	7.08 (180)	6.18 (157)	E[1]	[1] w/150 lbs. in rear
1991-92 Sedan	9.56 (243)	14.96 (380)	—	—	6.85 (174)	6.14 (156)	E	
1991-92 Wagon	9.56 (243)	14.52 (369)	—	—	7.08 (180)	6.18 (157)	E[1]	[1] w/150 lbs. in rear
1993-95 Sedan	10.27 (261)	14.33 (364)	—	—	6.85 (174)	6.14 (156)	E	
1993-95 Wagon	10.39 (264)	14.88 (378)	—	—	7.08 (180)	6.18 (157)	E[1]	[1] w/150 lbs. in rear
1996-99 Sedan	10.51 (267)	14.13 (359)	—	—	8.22 (209)	6.49 (165)	C	
1996-99 Wagon	10.31 (262)	13.50 (343)	—	—	8.26 (210)	6.73 (171)	C[1]	[1] w/150 lbs. in rear
2000-05 Sedan	—	—	28.07 (713)	27.40 (696)	—	—	—	Figure 1
2000-05 Wagon	—	—	27.95 (710)	27.67 (703)	—	—	—	Figure 1
Sable (2008 →)	SF	Side/Side *	SR	Side/Side *	RF	RR	LOAD	Notes
2008-09 Sable FWD	2.44 (62)	—	1.02 (26)	—	—	—	—	Figure 6 & 8
2008-09 Sable AWD	2.44 (62)	—	0.47 (12)	—	—	—	—	Figure 6 & 8
Topaz	BF	BR	WF	WR	RF	RR	LOAD	Notes
1990-91 All Models	10.11 (257)	13.93 (354)	—	—	7.83 (199)	7.48 (190)	B	
1992-94 All Models	10.19 (259)	10.62 (270)	—	—	7.83 (199)	7.48 (190)	B	
Tracer	BF	BR	WF	WR	RF	RR	LOAD	Notes
1990 All exc/Wagon	10.19 (259)	12.87 (327)	—	—	6.92 (176)	6.92 (176)	B[1]	[1] w/80 lbs. in trunk
1991-96 All exc/Wagon	9.40 (239)	11.02 (280)	—	—	6.10 (155)	6.10 (155)	C[1]	[1] w/80 lbs. in trunk
1997 All exc/Wagon	7.95 (202)	10.35 (263)	—	—	6.25 (159)	6.22 (158)	C[1]	[1] w/80 lbs. in trunk
1990 Wagon	10.19 (259)	14.52 (369)	—	—	6.92 (176)	6.92 (176)	B[1]	[1] w/80 lbs. in rear
1991-96 Wagon	n/a	n/a	—	—	5.94 (151)	5.94 (151)	C[1]	[1] w/80 lbs. in rear
1997 Wagon	7.95 (202)	10.27 (261)	—	—	6.25 (159)	6.22 (158)	C[1]	[1] w/80 lbs. in rear
Villager	BF	BR	WF	WR	RF	RR	LOAD	Notes
1993-98 GS 5-Passenger	9.25 (235)[1]	11.29 (287)[1]	30.00 (762)	30.98 (787)	9.09 (231)	9.40 (239)	C	[1] Loaded
1993-98 GS 7-Passenger	9.44 (240)[1]	10.78 (274)[1]	30.00 (762)	30.98 (787)	9.09 (231)	9.09 (231)	F	[1] Loaded
1993-98 LS, Nautica (All)	9.33 (237)[1]	10.59 (269)[1]	30.00 (762)	30.98 (787)	8.93 (227)	8.89 (226)	F	[1] Loaded
1996-98 Cargo Van	9.48 (241)[1]	13.54 (334)[1]	30.00 (762)	30.98 (787)	9.76 (248)	10.74 (273)	A	[1] Loaded
1999-02 All Models	—	—	30.39 (772)	31.22 (793)	—	—	—	Figure 1

* All Side/Side measurements are taken at wheel arch to ground (WF & WR) See Figure 1 above ** ±0.75 (19) unless indicated otherwise

Ford-Lincoln-Mercury Truck & SUV

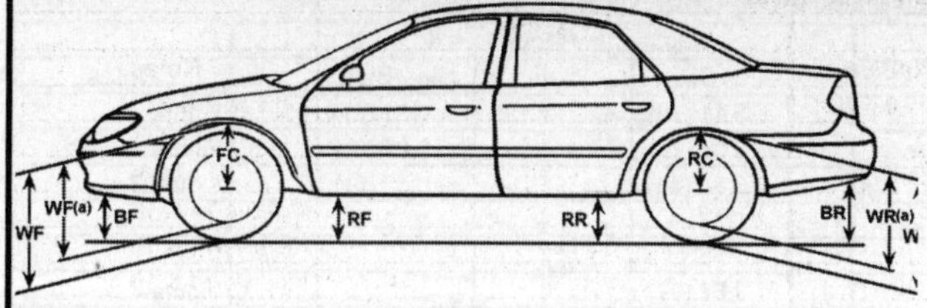

Figure 1

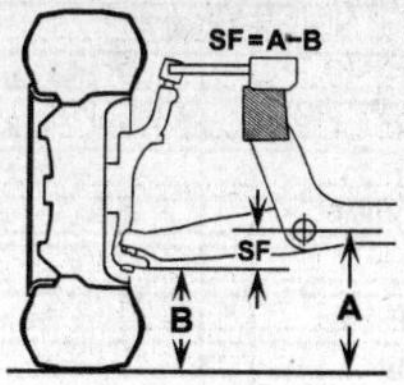

Figure 2

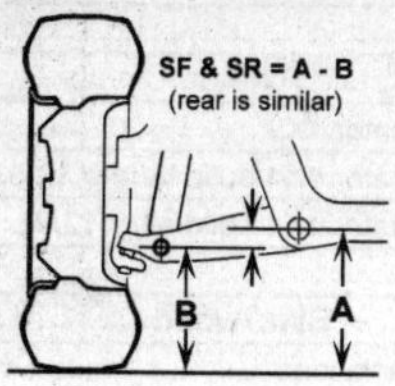

Figure 3

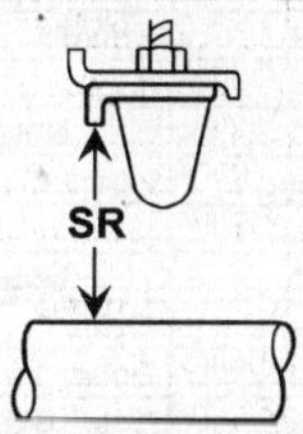

Figure 4

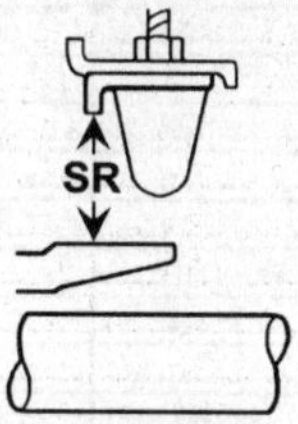

Figure 5

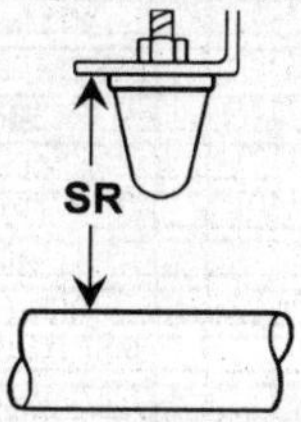

Figure 6

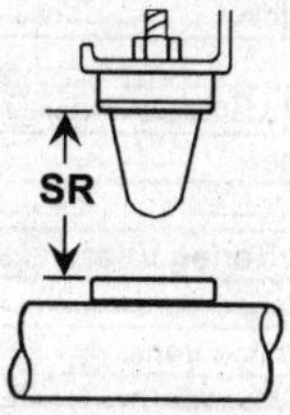

Figure 7

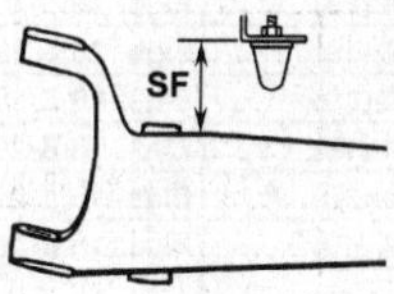

Figure 8

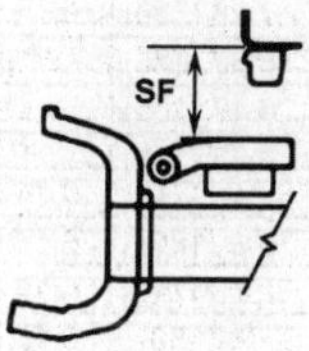

Figure 9

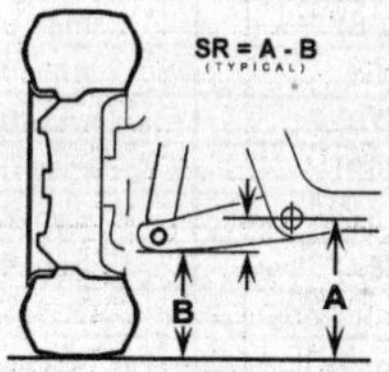

Figure 10

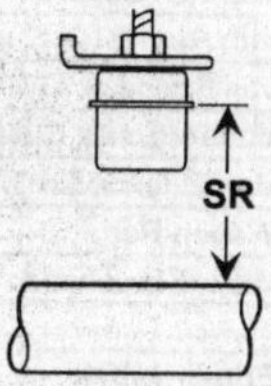

Figure 11

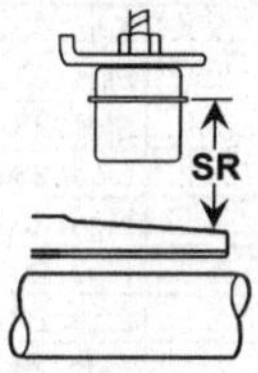

Figure 12

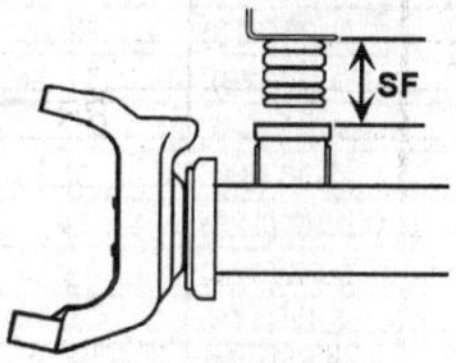

Figure 13

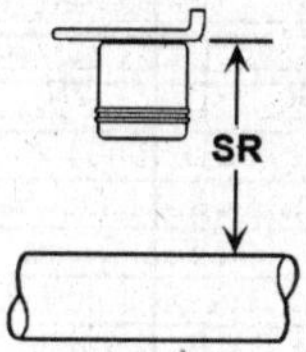

Figure 14

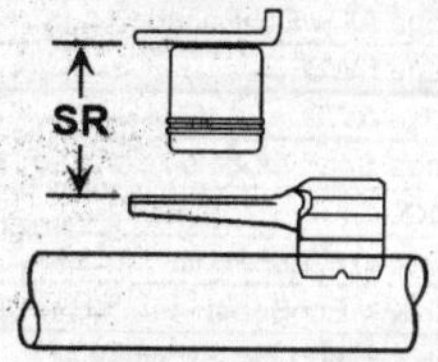

Figure 15

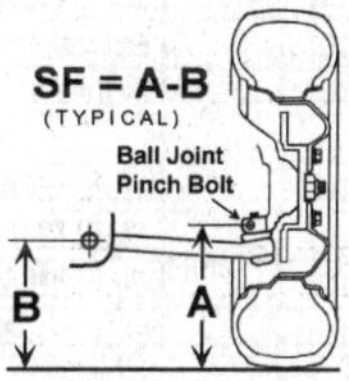

Figure 16

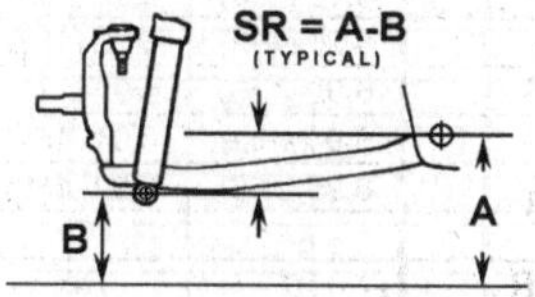

Figure 17

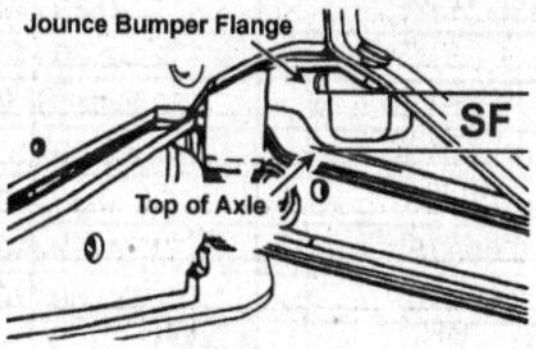

Figure 18

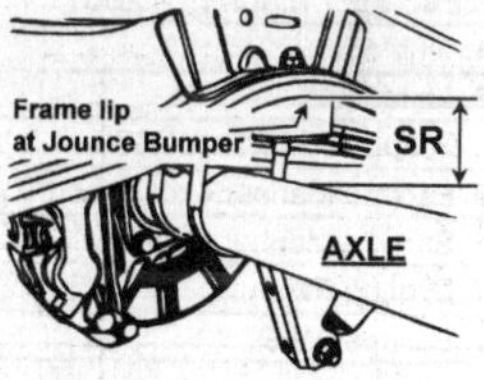

Figure 19

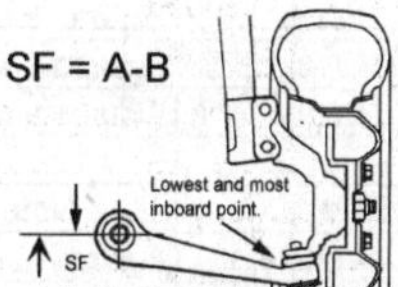

Figure 20

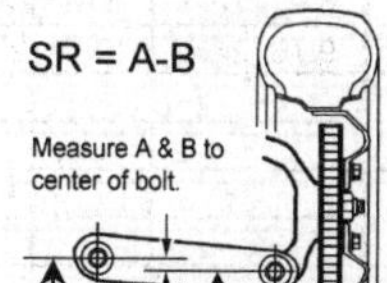

Figure 21

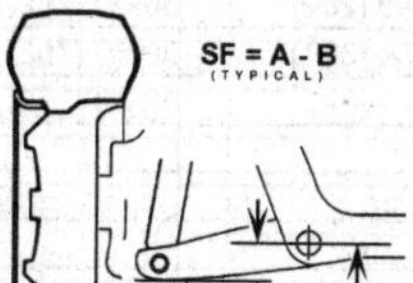

Figure 22

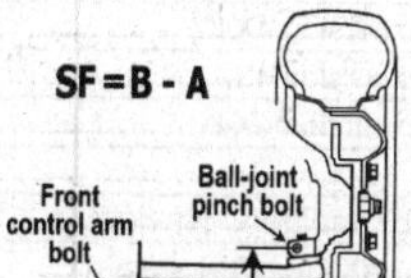

Figure 23

Measure **Figure 1** (WF & WR) from the center of the wheel to the lowest point in the wheel opening. Compare left and right side measurements to specifications to determine side/side difference (vehicle lean).
Measure **Figure 2** front height as indicated. Adjust torsion bars to correct height.
Measure **Figure 4** thru **Figure 7** rear heights at the rear axle as indicated.
Measure **Figure 8 and Figure 9** front heights as indicated.
Measure **Figure 10** from ground to: (A) center of lower arm inner mounting bolt; (B) to lower edge of lateral arm, just below shock mounting bolt.
Measure **Figure 11** thru **Figure 21** front and rear heights as indicated.

Ford-Lincoln-Mercury Truck & SUV (cont.)

Year Model	Height Specifications @ Curb Weight In. (mm) **				
	Front		Rear		
Aviator	SF	Side/Side *	SR	Side/Side *	Notes
2003-05 Aviator 4X2	0.55 (14)	0.27 (7)	3.11 (79)	0.31 (8)	Figure 3
2003 Aviator 4X4 build before 12/02	0.47 (12)	0.27 (7)	3.07 (78)	0.31 (8)	Figure 3
2003-05 Aviator 4X4 build after 12/02	0.47 (12)	0.27 (7)	3.38 (86)	0.31 (8)	Figure 3
Blackwood	SF		SR		
2002-03 Blackwood	n/a		3.62 (92)		Figure 7
Bronco II	SF	Side/Side *	SR	Side/Side *	Notes
1990 All Models	n/a	Max 0.62 (16)	n/a	max 0.75 (19)	Figure 1
Bronco	SF	Side/Side *	SR	Side/Side *	Notes
1990-96 All Models	n/a	Max0 .62 (16)	n/a	max 0.75 (19)	Figure 1
E Series Vans	SF	Side/Side *	SR	Side/Side *	Notes
1990-97 All E-Series Vans	n/a	max 0.62 (16)	n/a	max 0.74 (19)	Figure 1
1990-07 All E-Series Vans	n/a	max 0.59 (15)	n/a	max 0.78 (20)	Figure 1
2008-14 E-150 Wagon	2.55 (65)	max 0.62 (16)	5.78 (147)	max 0.74 (19)	Figs. 18 & 19
2008-14 E-150 Van	2.75 (70)	max 0.62 (16)	6.06 (154)	max 0.74 (19)	Figs. 18 & 19
2008-14 E-250 Van STD. Susp.	2.67 (68)	max 0.62 (16)	7.48 (190)	max 0.74 (19)	Figs. 18 & 19
2008-14 E-250 Van HD. Susp.	2.71 (69)	max 0.62 (16)	7.48 (190)	max 0.74 (19)	Figs. 18 & 19
2008-14 E-350 Van Reg. STD. Susp.	2.67 (68)	max 0.62 (16)	6.88 (175)	max 0.74 (19)	Figs. 18 & 19
2008-14 E-350 Van Extended STD. Susp.	2.67 (68)	max 0.62 (16)	6.49 (165)	max 0.74 (19)	Figs. 18 & 19
2008-14 E-350 Van Reg. HD. Susp.	2.75 (70)	max 0.62 (16)	6.88 (175)	max 0.74 (19)	Figs. 18 & 19
2008-14 E-350 Van Extended HD. Susp.	2.71 (69)	max 0.62 (16)	6.49 (165)	max 0.74 (19)	Figs. 18 & 19
2008-14 E-350 Wagon Reg.	2.63 (67)	max 0.62 (16)	5.82 (148)	max 0.74 (19)	Figs. 18 & 19
2008-14 E-350 Wagon Extended	2.51 (64)	max 0.62 (16)	5.66 (144)	max 0.74 (19)	Figs. 18 & 19
Edge / MKX	SF	Side/Side *	SR	Side/Side *	Notes
2007-10 Edge All Models	2.00 (51)	—	4.72 (120)	—	Figs. 20 & 17
2011-14 Edge exc/22" wheels (OE)	2.08 (53)	—	5.00 (127)	—	Figs. 20 & 17
2011-14 Edge w/22" wheels (OE)	2.08 (53)	—	4.96 (126)	—	Figs. 20 & 17
2012-14 Edge All w/EcoBoost	1.61 (41)	—	4.44 (113)	—	Figs. 20 & 17
2015-17 Edge FWD	2.46 (62)	—	1.09 (28)	—	Figs. 23 & 21
2015-17 Edge AWD	2.23 (57)	—	1.35 (34)	—	Figs. 23 & 21
2015-17 Edge Sport Models	2.23 (57)	—	1.35 (34)	—	Figs. 23 & 21
2007-10 MKX	2.12 (54)	—	4.92 (125)	—	Figs. 20 & 17
2011-15 MKX exc//EcoBoost	2.08 (53)	—	5.00 (127)	—	Figs. 20 & 17
2015 MKX w/EcoBoost	2.08 (53)	—	4.44 (113)	—	Figs. 20 & 17
2016-17 MKX FWD (w/o 21" wheels)	2.46 (62)	—	1.04 (26)	—	Figs. 23 & 21
2016-17 MKX AWD & w/21" wheels	2.23 (57)	—	1.35 (34)	—	Figs. 23 & 21
Escape / Mariner > 2007	FC	Side/Side *	RC	Side/Side *	Notes
2001-04 All Models	18.77 (477)	0.39 (10)	19.33 (491)	0.39 (10)	Figure 1
2005-07 All Models		0.39 (10)		0.39 (10)	Figure 1
Escape / Mariner 2008 >	SF	Side/Side *	SR		
2008-12 Escape/Mariner exc/ Hybrid	0.12 (3.3)	0.39 (10)	3.74 (95)	0.39 (10)	Figs. 16 & 17
2008-12 Escape/Mariner Hybrid	0.12 (3.3)	0.39 (10)	3.46 (88)	0.39 (10)	Figs. 16 & 17
2013-17 Escape FWD	0.60 (15)	0.39 (10)	3.04 (77)	0.39 (10)	Figs. 16 & 3
2013-17 Escape AWD	0.40 (10)	0.39 (10)	2.96 (75)	0.39 (10)	Figs. 16 & 3
Excursion	SF	Side/Side *	SR	Side/Side *	Notes
2000-04 Excursion 4X2	L=4.64 (118), R=4.17 (106)	max 0.74 (19)	5.23 (133)	max 0.74 (19)	Figures 8 & 4
2000-04 Excursion 4X4	4.37 (111)	max 0.74 (19)	4.84 (123)	max 0.74 (19)	Figures 9 & 5
2005 Excursion 4X2	L=4.25 (108), R=4.48 (114)	max 0.74 (19)	6.14 (156)	max 0.74 (19)	Figures 8 & 4
2005 Excursion 4X4	3.93 (100)	max 0.74 (19)	5.59 (142)	max 0.74 (19)	Figures 9 & 5
Expedition / Navigator	SF	Side/Side *	SR	Side/Side *	Notes
1998-02 All 4X2 exc/AIR susp.	—	max 0.62 (16)	6.61 - 7.55 (168 - 192)	max 0.74 (19)	Figures 1, 4
1998-02 All 4X2 w/AIR susp.	—	max 0.62 (16)	5.31 – 5.70 (135 - 145)	max 0.74 (19)	Figures 1, 4
1998-02 All 4X4 exc/AIR susp.	—	max 0.62 (16)	8.58 – 9.52 (218 - 242)	max 0.74 (19)	Figures 1, 4
1998-02 All 4X4 w/AIR susp.	—	max 0.62 (16)	6.61 – 7.00 (168 - 178)	max 0.74 (19)	Figures 1, 2 & 4
2003-06 Expedition exc/AIR susp.	4.13 (105)	0.27 (7)	4.68 (119)	0.31 (8)	Figures 2 & 10
2003-06 Navigator/Expedition w/AIR	3.34 (85)	0.27 (7)	4.40 (112)	0.31 (8)	Figures 2 & 10
Expedition 2007 >	SF	Side/Side *	SR	Side/Side *	Notes
Std. Wheelbase versions					
2007-14 RWD, coil susp. w/o roll stability	4.13 (105)	max 0.47 (12)	3.97 (101)	max 0.27 (7)	Figures 22 & 3
2007-14 RWD, coil susp. w/roll stability	4.21 (107)	max 0.47 (12)	3.97 (101)	max 0.27 (7)	Figures 22 & 3
2007-14 4WD, coil susp. w/o roll stability	4.17 (106)	max 0.47 (12)	3.97 (101)	max 0.27 (7)	Figures 22 & 3
2007-14 4WD, coil susp. w/roll stability	4.25 (108)	max 0.47 (12)	3.97 (101)	max 0.27 (7)	Figures 22 & 3
2007-14 All w/Air suspension.	4.25 (108)	max 0.47 (12)	3.89 (99)	max 0.27 (7)	Figures 22 & 3

* All Side/Side measurements are taken at wheel arch to ground (WF & WR) See Figure 1 above ** ±0.75 (19) unless indicated otherwise

Ford-Lincoln-Mercury Truck & SUV (cont.)

YEAR MODEL	HEIGHT SPECIFICATIONS @ CURB WEIGHT IN. (MM) ** FRONT		REAR		
Expedition 2007 > (cont.)	**SF**	**Side/Side ***	**SR**	**Side/Side ***	**Notes**
2015-17 Expedition RWD, coil susp.	4.02 (102)	max 0.47 (12)	3.85 (98)	max 0.27 (7)	Figures 22 & 3
2015-17 Expedition 4WD, coil susp.	4.02 (102)	max 0.47 (12)	3.81 (97)	max 0.27 (7)	Figures 22 & 3
2015-17 Expedition All w/Air susp.	4.02 (102)	max 0.47 (12)	3.93 (100)	max 0.27 (7)	Figures 22 & 3
Long Wheelbase versions					
2007-14 RWD, coil susp. w/o roll stability	4.29 (109)	max 0.47 (12)	4.05 (103)	max 0.27 (7)	Figures 22 & 3
2007-14 RWD, coil susp. w/roll stability	4.25 (108)	max 0.47 (12)	4.05 (103)	max 0.27 (7)	Figures 22 & 3
2007-14 4WD, coil susp. (All)	4.33 (110)	max 0.47 (12)	3.97 (101)	max 0.27 (7)	Figures 22 & 3
2007-14 All w/Air suspension.	4.33 (110)	max 0.47 (12)	3.89 (99)	max 0.27 (7)	Figures 22 & 3
2015-17 Exp. EL RWD, coil susp.	4.09 (104)	max 0.47 (12)	3.93 (100)	max 0.27 (7)	Figures 22 & 3
2015-17 Exp. EL RWD, Air susp.	4.09 (104)	max 0.47 (12)	3.81 (97)	max 0.27 (7)	Figures 22 & 3
2015-17 Exp. EL 4WD, coil susp.	4.17 (106)	max 0.47 (12)	3.89 (99)	max 0.27 (7)	Figures 22 & 3
2015-17 Exp. EL 4WD, Air susp.	4.17 (106)	max 0.47 (12)	3.78 (96)	max 0.27 (7)	Figures 22 & 3
Navigator 2007 >	**SF**	**Side/Side ***	**SR**	**Side/Side ***	**Notes**
Navigator Std. versions					
2007-14 Navigator RWD, coil susp.	3.85 (98)	max 0.47 (12)	3.89 (99)	max 0.27 (7)	Figures 22 & 3
2007-14 Navigator 4WD, coil susp.	3.97 (101)	max 0.47 (12)	3.85 (98)	max 0.27 (7)	Figures 22 & 3
2015-17 Navigator RWD, coil susp.	3.85 (98)	max 0.47 (12)	3.85 (98)	max 0.27 (7)	Figures 22 & 3
2015-17 Navigator RWD, Air susp.	3.85 (98)	max 0.47 (12)	3.97 (101)	max 0.27 (7)	Figures 22 & 3
2015-17 Navigator 4WD, coil susp.	3.89 (99)	max 0.47 (12)	3.81 (97)	max 0.27 (7)	Figures 22 & 3
2015-17 Navigator 4WD, Air susp.	3.93 (100)	max 0.47 (12)	3.93 (100)	max 0.27 (7)	Figures 22 & 3
Navigator L versions					
2007-14 Nav. L RWD, coil susp.	4.21 (107)	max 0.47 (12)	3.97 (101)	max 0.27 (7)	Figures 22 & 3
2007-14 Nav. L 4WD, coil susp.	4.29 (109)	max 0.47 (12)	3.93 (100)	max 0.27 (7)	Figures 22 & 3
2007-14 Nav. L RWD, w/Air susp.	3.85 (98)	max 0.47 (12)	3.89 (99)	max 0.27 (7)	Figures 22 & 3
2007-14 Nav. L 4WD, w/Air susp.	3.97 (101)	max 0.47 (12)	3.89 (99)	max 0.27 (7)	Figures 22 & 3
2015-17 Nav. L RWD, coil susp.	3.89 (99)	max 0.47 (12)	3.93 (100)	max 0.27 (7)	Figures 22 & 3
2015-17 Nav. L RWD, Air susp.	3.89 (99)	max 0.47 (12)	3.85 (98)	max 0.27 (7)	Figures 22 & 3
2015-17 Nav. L 4WD, coil susp.	3.97 (101)	max 0.47 (12)	3.93 (100)	max 0.27 (7)	Figures 22 & 3
2015-17 Nav. L 4WD, Air susp.	3.97 (101)	max 0.47 (12)	3.81 (97)	max 0.27 (7)	Figures 22 & 3
Explorer / Mountaineer	**SF**	**Side/Side ***	**SR**	**Side/Side ***	**Notes**
1995-01 except models with ARC	3.97 (101)	max 0.62 (16)	n/a	max 0.74 (19)	Figures 2 & 6
1995-01 All models with ARC	2.67 – 3.54 (68 - 90)	max 0.62 (16)	n/a	max 0.74 (19)	Figures 2 & 6
2002-05 Explorer/Mountaineer	0.66 (17)	max 0.28 (7)	3.58 (91)	max 0.31 (8)	Figure 3
2006-10 Explorer/Mountaineer	0.62 (16)	max 0.28 (7)	3.74 (95)	max 0.31 (8)	Figure 3
2011-17 Explorer FWD	1.33 (34)	max 0.47 (12)	1.18 (30)	max 0.47 (12)	Figs. 20 & 21
2011-17 Explorer AWD	1.72 (44)	max 0.47 (12)	1.57 (40)	max 0.47 (12)	Figs. 20 & 21
2013-15 Explorer Police Vehicle	1.57 (40)	max 0.47 (12)	1.97 (50)	max 0.47 (12)	Figs. 20 & 21
2016-17 Explorer Police Vehicle	1.54 (39)	max 0.47 (12)	1.97 (50)	max 0.47 (12)	Figs. 20 & 21
Explorer Sport / Sport Trac	**SF**	**Side/Side ***	**SR**	**Side/Side ***	**Notes**
2001-03 Explorer Sport	3.97 (101)	max 0.62 (16)	4.84 - 5.62 (123 - 143)	max 0.78 (20)	Figures 2 & 6
2001-04 Explorer Sport Trac	3.97 (101)	max 0.62 (16)	6.65 (169)	max 0.78 (20)	Figures 2 & 6
2007-10 Explorer Sport Trac	0.82 (21)	max 0.28 (7)	3.85 (98)	max 0.31 (8)	Figure 3
Flex / MKT	**SF**	**Side/Side**	**SR**	**Side/Side**	**Notes**
2009 Flex	2.08 (53)	—	1.02 (26)	—	Figs. 20 & 21
2010-12 Flex & MKT exc/Turbo	2.04 (52)	—	1.02 (26)	—	Figs. 20 & 21
2010-12 Flex & MKT Turbo	1.65 (42)	—	0.62 (16)	—	Figs. 20 & 21
2013-17 Flex exc/Turbo	2.04 (52)	—	1.02 (26)	—	Figs. 20 & 21
2013-17 Flex Turbo	1.65 (42)	—	0.62 (16)	—	Figs. 20 & 21
2013-17 MKT exc/Turbo, Limo	2.04 (52)	—	1.02 (26)	—	Figs. 20 & 21
2013-17 MKT Turbo	1.65 (42)	—	0.62 (16)	—	Figs. 20 & 21
2013-17 MKT Limo 44 in.	2.83 (72)	—	2.01 (51)	—	Figs. 20 & 21
2013-17 MKT Limo 80 in.	2.83 (72)	—	1.85 (47)	—	Figs. 20 & 21
2013-17 MKT Limo 120 in.	2.83 (72)	—	1.53 (39)	—	Figs. 20 & 21
F Series Trucks	**SF**	**Side/Side ***	**SR**	**Side/Side ***	**Notes**
1990-96 All F Series exc/Motorhome & comm.		max 0.62 (16)			
1997-03 F150 4X2 Regular Cab	4.48 (114)	max 0.47 (12)	6.02-8.46 (153-215)	max 0.75 (19)	Figures 1, 2 & 4
1997-03 F150 4X2 Super Cab	4.80 (122)	max 0.47 (12)	6.02-8.46 (153-215)	max 0.75 (19)	Figures 1, 2 & 4
1997-03 F150 4X4	4.44 (113)	max 0.47 (12)	5.90-8.07 (150-205)	max 0.75 (19)	Figures 1, 2 & 4
1999-04 Lightning	3.93 (100)	max 0.47 (12)	4.76-5.51 (121-140)	max 0.75 (19)	Figures 1, 2 & 4
2004 F150 Heritage 4X2 Reg. Cab	4.48 (114)	max 0.47 (12)	6.02-8.46 (153-215)	max 0.75 (19)	Figures 1, 2 & 4
2004 F150 Heritage 4X2 Super Cab	4.80 (122)	max 0.47 (12)	6.02-8.46 (153-215)	max 0.75 (19)	Figures 1, 2 & 4
2004 F150 Heritage 4X4	4.44 (113)	max 0.47 (12)	5.90-8.07 (150-205)	max 0.75 (19)	Figures 1, 2 & 4
1997-99 F250 4X2 Reg. Cab exc/(LL)	4.48 (114)	max 0.47 (12)	6.81-9.25 (173-235)	max 0.75 (19)	Figures 1, 2 & 4
1997-99 F250 4X2 Super Cab exc/(LL)	4.80 (122)	max 0.47 (12)	6.81-9.25 (173-235)	max 0.75 (19)	Figures 1, 2 & 4
1997-99 F250 4X2 Reg. Cab (LL)	4.48 (114)	max 0.47 (12)	3.26-3.97 (83-101)	max 0.75 (19)	Figures 1, 2 & 4
1997-99 F250 4X2 Super Cab (LL)	4.80 (122)	max 0.47 (12)	3.26-3.97 (83-101)	max 0.75 (19)	Figures 1, 2 & 4

* All Side/Side measurements are taken at wheel arch to ground (WF & WR) See Figure 1 above ** ±0.75 (19) unless indicated otherwise

Ford-Lincoln-Mercury Truck & SUV (cont.)

YEAR MODEL	HEIGHT SPECIFICATIONS @ CURB WEIGHT IN. (MM) **				
	FRONT		REAR		
F Series Trucks (cont.)	SF	Side/Side *	SR	Side/Side *	Notes
1997-99 F250 4X4 w/o Load Leveling	4.44 (113)	max 0.47 (12)	6.69-8.85 (170-225)	max 0.75 (19)	Figures 1, 2 & 5
1997-99 F250 4X4 w/Load Leveling (LL)	4.44 (113)	max 0.47 (12)	3.14-3.85 (80-98)	max 0.75 (19)	Figures 1, 2 & 5
Also see F Super Duty Trucks (below)					
F150 4X2 exc/HD	SF	Side/Side *	SR	Side/Side *	Notes
2004-05 F150 4X2 Reg. 126" WB, 4.6L	4.52 (115)	max 0.27 (7)	7.28 (185)	max 0.31 (8)	Figures 2 & 11
2004-05 F150 4X2 Reg. 126" WB, 5.4L	4.40 (112)	max 0.27 (7)	7.28 (185)	max 0.31 (8)	Figures 2 & 11
2004-05 F150 4X2 Reg. 145" WB, 4.6L	4.01 (102)	max 0.27 (7)	7.32 (186)	max 0.31 (8)	Figures 2 & 11
2004-05 F150 4X2 Reg. 145" WB, 5.4L	4.13 (105)	max 0.27 (7)	7.24 (184)	max 0.31 (8)	Figures 2 & 11
2004-05 F150 4X2 Super 133" WB, 4.6L	4.40 (112)	max 0.27 (7)	6.77 (172)	max 0.31 (8)	Figures 2 & 11
2004-05 F150 4X2 Super 133" WB, 5.4L	4.48 (114)	max 0.27 (7)	6.88 (175)	max 0.31 (8)	Figures 2 & 11
2004-05 F150 4X2 Super 145" WB, 4.6L	4.40 (112)	max 0.27 (7)	6.88 (175)	max 0.31 (8)	Figures 2 & 11
2004-05 F150 4X2 Super 145" WB, 5.4L	4.29 (109)	max 0.27 (7)	6.85 (174)	max 0.31 (8)	Figures 2 & 11
2004-05 F150 4X2 Crew 138" WB, 4.6L	4.40 (112)	max 0.27 (7)	6.85 (174)	max 0.31 (8)	Figures 2 & 11
2004-05 F150 4X2 Crew 138" WB, 5.4L	4.33 (110)	max 0.27 (7)	6.73 (171)	max 0.31 (8)	Figures 2 & 11
2006 F150 4X2 Reg. 126" WB, 4.2L	5.07 (129)	max 0.27 (7)	6.96 (177)	max 0.31 (8)	Figures 2 & 11
2006 F150 4X2 Reg. 126" WB, 4.6L	4.44 (113)	max 0.27 (7)	6.81 (173)	max 0.31 (8)	Figures 2 & 11
2006 F150 4X2 Reg. 126" WB, 5.4L	4.29 (109)	max 0.27 (7)	6.73 (171)	max 0.31 (8)	Figures 2 & 11
2006 F150 4X2 Reg. 145" WB, 4.2L	4.13 (105)	max 0.27 (7)	6.96 (177)	max 0.31 (8)	Figures 2 & 11
2006 F150 4X2 Reg. 145" WB, 4.6L	4.17 (106)	max 0.27 (7)	6.85 (174)	max 0.31 (8)	Figures 2 & 11
2006 F150 4X2 Reg. 145" WB, 5.4L	4.01 (102)	max 0.27 (7)	6.73 (171)	max 0.31 (8)	Figures 2 & 11
2006 F150 4X2 Super 133" WB, 4.6L	4.21 (107)	max 0.27 (7)	6.45 (164)	max 0.31 (8)	Figures 2 & 11
2006 F150 4X2 Super 133" WB, 5.4L	4.09 (104)	max 0.27 (7)	6.33 (161)	max 0.31 (8)	Figures 2 & 11
2006 F150 4X2 Super 145" WB, 4.6L	4.33 (110)	max 0.27 (7)	6.49 (165)	max 0.31 (8)	Figures 2 & 11
2006 F150 4X2 Harley Davidson	4.44 (113)	max 0.27 (7)	6.06 (154)	max 0.31 (8)	Figures 2 & 11
2006 F150 4X2 Super 145" WB, 5.4L	4.21 (107)	max 0.27 (7)	6.37 (162)	max 0.31 (8)	Figures 2 & 11
2006 F150 4X2 Crew 139" WB, 4.6L	4.48 (114)	max 0.27 (7)	6.33 (161)	max 0.31 (8)	Figures 2 & 11
2006 F150 4X2 Crew 139" WB, 5.4L	4.37 (111)	max 0.27 (7)	6.25 (159)	max 0.31 (8)	Figures 2 & 11
2006 F150 4X2 Crew 151" WB, 4.6L	4.33 (110)	max 0.27 (7)	6.25 (159)	max 0.31 (8)	Figures 2 & 11
2006 F150 4X2 Crew 151" WB, 5.4L	4.17 (106)	max 0.27 (7)	6.18 (157)	max 0.31 (8)	Figures 2 & 11
2007-08 F150 4X2 Reg. 126" WB, 4.2L	5.15 (131)	max 0.27 (7)	6.96 (177)	max 0.31 (8)	Figures 2 & 11
2007-08 F150 4X2 Reg. 126" WB, 4.6L	4.48 (114)	max 0.27 (7)	6.85 (174)	max 0.31 (8)	Figures 2 & 11
2007-08 F150 4X2 Reg. 126" WB, 5.4L	4.29 (109)	max 0.27 (7)	6.61 (168)	max 0.31 (8)	Figures 2 & 11
2007-08 F150 4X2 Reg. 145" WB, 4.2L	4.21 (107)	max 0.27 (7)	7.00 (178)	max 0.31 (8)	Figures 2 & 11
2007-08 F150 4X2 Reg. 145" WB, 4.6L	4.25 (108)	max 0.27 (7)	6.85 (174)	max 0.31 (8)	Figures 2 & 11
2007-08 F150 4X2 Reg. 145" WB, 5.4L	4.29 (109)	max 0.27 (7)	6.61 (168)	max 0.31 (8)	Figures 2 & 11
2007-08 F150 4X2 Super 133" WB, 4.6L	4.25 (108)	max 0.27 (7)	6.45 (164)	max 0.31 (8)	Figures 2 & 11
2007-08 F150 4X2 Super 133" WB, 5.4L	4.01 (102)	max 0.27 (7)	6.25 (159)	max 0.31 (8)	Figures 2 & 11
2007-08 F150 4X2 Super 145" WB, 4.6L	4.37 (111)	max 0.27 (7)	6.49 (165)	max 0.31 (8)	Figures 2 & 11
2007-08 F150 4X2 Super 145" WB, 5.4L	4.29 (109)	max 0.27 (7)	6.29 (160)	max 0.31 (8)	Figures 2 & 11
2007-08 F150 4X2 Harley Davidson	4.17 (106)	max 0.27 (7)	5.86 (149)	max 0.31 (8)	Figures 2 & 11
2007-08 F150 4X2 Crew 139" WB, 4.6L	4.52 (115)	max 0.27 (7)	6.37 (162)	max 0.31 (8)	Figures 2 & 11
2007-08 F150 4X2 Crew 139" WB, 5.4L	4.56 (116)	max 0.27 (7)	6.14 (156)	max 0.31 (8)	Figures 2 & 11
2007-08 F150 4X2 Crew 151" WB, 4.6L	4.37 (111)	max 0.27 (7)	6.33 (161)	max 0.31 (8)	Figures 2 & 11
2007-08 F150 4X2 Crew 151" WB, 5.4L	4.13 (105)	max 0.27 (7)	6.14 (156)	max 0.31 (8)	Figures 2 & 11
2009 F150 4X2 Reg. 126" WB, 4.6L, 2 valve	4.48 (114)	max 0.27 (7)	6.49 (165)	max 0.31 (8)	Figures 2 & 11
2009 F150 4X2 Reg. 145" WB, 4.6L, 2 valve	4.44 (113)	max 0.27 (7)	7.04 (179)	max 0.31 (8)	Figures 2 & 11
2009 F150 4X2 Reg. 145" WB, 4.6L, 3 valve	4.25 (108)	max 0.27 (7)	6.88 (175)	max 0.31 (8)	Figures 2 & 11
2009 F150 4X2 Reg. 145" WB, 5.4L	4.37 (111)	max 0.27 (7)	6.96 (177)	max 0.31 (8)	Figures 2 & 11
2009 F150 4X2 Super 133" WB, 4.6L, 2 valve	4.52 (115)	max 0.27 (7)	6.22 (158)	max 0.31 (8)	Figures 2 & 11
2009 F150 4X2 Super 133" WB, 4.6L, 3 valve	4.33 (110)	max 0.27 (7)	6.10 (155)	max 0.31 (8)	Figures 2 & 11
2009 F150 4X2 Super 133" WB, 5.4L	4.37 (111)	max 0.27 (7)	6.02 (153)	max 0.31 (8)	Figures 2 & 11
2009 F150 4X2 Super 145" WB, 4.6L, 2 valve	4.52 (115)	max 0.27 (7)	6.65 (169)	max 0.31 (8)	Figures 2 & 11
2009 F150 4X2 Super 145" WB, 4.6L, 3 valve	4.37 (111)	max 0.27 (7)	6.57 (167)	max 0.31 (8)	Figures 2 & 11
2009 F150 4X2 Super 145" WB, 5.4L	4.44 (113)	max 0.27 (7)	6.49 (165)	max 0.31 (8)	Figures 2 & 11
2009 F150 4X2 Crew 145" WB, 4.6L, 2 valve	4.37 (111)	max 0.27 (7)	6.33 (161)	max 0.31 (8)	Figures 2 & 11
2009 F150 4X2 Crew 145" WB, 4.6L, 3 valve	4.60 (117)	max 0.27 (7)	6.33 (161)	max 0.31 (8)	Figures 2 & 11
2009 F150 4X2 Crew 157" WB, 4.6L, 2 valve	4.40 (112)	max 0.27 (7)	6.25 (159)	max 0.31 (8)	Figures 2 & 11
2009 F150 4X2 Crew 145" WB, 5.4L	4.37 (111)	max 0.27 (7)	6.18 (157)	max 0.31 (8)	Figures 2 & 11
2009 F150 4X2 Crew 157" WB, 4.6L, 3 valve	4.56 (116)	max 0.27 (7)	6.25 (159)	max 0.31 (8)	Figures 2 & 11
2009 F150 4X2 Crew 157" WB, 5.4L	4.37 (111)	max 0.27 (7)	6.25 (159)	max 0.31 (8)	Figures 2 & 11
2010 F150 4X2 Reg. 126" WB, 4.6L, 2 valve	4.44 (113)	max 0.27 (7)	6.53 (166)	max 0.31 (8)	Figures 2 & 11
2010 F150 4X2 Reg. 145" WB, 4.6L, 2 valve	4.40 (112)	max 0.27 (7)	7.20 (183)	max 0.31 (8)	Figures 2 & 11
2010 F150 4X2 Reg. 145" WB, 4.6L, 3 valve	4.29 (109)	max 0.27 (7)	7.16 (182)	max 0.31 (8)	Figures 2 & 11
2010 F150 4X2 Reg. 145" WB, 5.4L	4.33 (110)	max 0.27 (7)	7.16 (182)	max 0.31 (8)	Figures 2 & 11
2010 F150 4X2 Super 133" WB, 4.6L, 2 valve	4.21 (107)	max 0.27 (7)	6.81 (173)	max 0.31 (8)	Figures 2 & 11
2010 F150 4X2 Super 133" WB, 4.6L, 3 valve	4.33 (110)	max 0.27 (7)	6.73 (171)	max 0.31 (8)	Figures 2 & 11
2010 F150 4X2 Super 145" WB, 4.6L, 2 valve	4.21 (107)	max 0.27 (7)	6.81 (173)	max 0.31 (8)	Figures 2 & 11
2010 F150 4X2 Super 145" WB, 4.6L, 3 valve	4.33 (110)	max 0.27 (7)	6.73 (171)	max 0.31 (8)	Figures 2 & 11
2010 F150 4X2 Super 145" WB, 5.4L	4.40 (112)	max 0.27 (7)	6.65 (169)	max 0.31 (8)	Figures 2 & 11
2010 F150 4X2 Crew 145" WB, 4.6L, 2 valve	4.33 (110)	max 0.27 (7)	6.49 (165)	max 0.31 (8)	Figures 2 & 11

* All **Side/Side** measurements are taken at wheel arch to ground (WF & WR) See **Figure 1** above ** ±0.75 (19) unless indicated otherwise

Ford-Lincoln-Mercury Truck & SUV (cont.)

Year	Model	Height Specifications @ Curb Weight In. (mm) **				
		Front		Rear		
	F150 4X2 exc/HD (cont.)	SF	Side/Side *	SR	Side/Side *	Notes
2010	F150 4X2 Crew 145" WB, 4.6L, 3 valve	4.56 (116)	max 0.27 (7)	6.49 (165)	max 0.31 (8)	Figures 2 & 11
2010	F150 4X2 Crew 145" WB, 5.4L	4.33 (110)	max 0.27 (7)	6.29 (160)	max 0.31 (8)	Figures 2 & 11
2010	F150 4X2 Crew 157" WB, 4.6L, 2 valve	4.37 (111)	max 0.27 (7)	6.41 (163)	max 0.31 (8)	Figures 2 & 11
2010	F150 4X2 Crew 157" WB, 4.6L, 3 valve	4.52 (115)	max 0.27 (7)	6.41 (163)	max 0.31 (8)	Figures 2 & 11
2010	F150 4X2 Crew 157" WB, 5.4L	4.37 (111)	max 0.27 (7)	6.41 (163)	max 0.31 (8)	Figures 2 & 11
2011-14	F150 4X2 Reg. Cab 126" WB, 3.7L	4.17 (106)	max 0.27 (7)	6.69 (170)	max 0.31 (8)	Figures 2 & 11
2011-14	F150 4X2 Reg. Cab 126" WB, 5.0L	4.52 (115)	max 0.27 (7)	6.69 (170)	max 0.31 (8)	Figures 2 & 11
2011-14	F150 4X2 Reg. Cab 145" WB, 3.5L	4.40 (112)	max 0.27 (7)	7.32 (186)	max 0.31 (8)	Figures 2 & 11
2011-14	F150 4X2 Reg. Cab 145" WB, 3.7L	4.44 (113)	max 0.27 (7)	7.32 (186)	max 0.31 (8)	Figures 2 & 11
2011-14	F150 4X2 Reg. Cab 145" WB, 5.0L	4.21 (107)	max 0.27 (7)	7.16 (182)	max 0.31 (8)	Figures 2 & 11
2011-14	F150 4X2 Super Cab 145" WB, 3.5L	4.48 (114)	max 0.27 (7)	6.65 (169)	max 0.31 (8)	Figures 2 & 11
2011-14	F150 4X2 Super 145" WB, 3.5L Max Tow	4.44 (113)	max 0.27 (7)	6.65 (169)	max 0.31 (8)	Figures 2 & 11
2011-14	F150 4X2 Super Cab 145" WB, 3.7L	4.48 (114)	max 0.27 (7)	6.85 (174)	max 0.31 (8)	Figures 2 & 11
2011-14	F150 4X2 Super Cab 145" WB, 5.0L	4.48 (114)	max 0.27 (7)	6.65 (169)	max 0.31 (8)	Figures 2 & 11
2011-14	F150 4X2 Crew Cab 145" WB, 3.5L	4.25 (108)	max 0.27 (7)	6.37 (162)	max 0.31 (8)	Figures 2 & 11
2011-14	F150 4X2 Crew 145" WB, 3.5L Max Tow	4.25 (108)	max 0.27 (7)	6.47 (163)	max 0.31 (8)	Figures 2 & 11
2011-14	F150 4X2 Crew Cab 145" WB, 3.7L	4.48 (114)	max 0.27 (7)	6.57 (167)	max 0.31 (8)	Figures 2 & 11
2011-14	F150 4X2 Crew Cab 145" WB, 5.0L	4.29 (109)	max 0.27 (7)	6.47 (163)	max 0.31 (8)	Figures 2 & 11
2011-14	F150 4X2 Crew 145" WB, 6.2L Max Tow	4.05 (103)	max 0.27 (7)	5.82 (148)	max 0.31 (8)	Figures 2 & 11
2011-14	F150 4X2 Crew 145", 6.2L Harley/Lariat	4.01 (102)	max 0.27 (7)	5.70 (145)	max 0.31 (8)	Figures 2 & 11
2011-14	F150 4X2 Crew Cab 157" WB, 3.5L	4.29 (109)	max 0.27 (7)	6.45 (164)	max 0.31 (8)	Figures 2 & 11
2011-14	F150 4X2 Crew 157" WB, 3.5L Max Tow	4.25 (108)	max 0.27 (7)	6.49 (165)	max 0.31 (8)	Figures 2 & 11
2011-14	F150 4X2 Crew Cab 157" WB, 5.0L	4.29 (109)	max 0.27 (7)	6.37 (162)	max 0.31 (8)	Figures 2 & 11
2015-17	F150 4X2 Crew Cab 145" WB 2.7L	4.92 (125)	max 0.27 (7)	6.61 (168)	max 0.31 (8)	Figures 2 & 11
2015-17	F150 4X2 Crew Cab 145" WB 3.5L	5.15 (131)	max 0.27 (7)	6.73 (171)	max 0.31 (8)	Figures 2 & 11
2015-17	F150 4X2 Crew Cab 157" WB 2.7L	4.84 (123)	max 0.27 (7)	6.65 (169)	max 0.31 (8)	Figures 2 & 11
2015-17	F150 4X2 Reg. Cab 122" WB 2.7L	4.92 (125)	max 0.27 (7)	7.08 (180)	max 0.31 (8)	Figures 2 & 11
2015-17	F150 4X2 Reg. Cab 122" WB 3.5L	5.31 (135)	max 0.27 (7)	7.16 (182)	max 0.31 (8)	Figures 2 & 11
2015-17	F150 4X2 Reg. Cab 141" WB 2.7L	4.92 (125)	max 0.27 (7)	7.08 (180)	max 0.31 (8)	Figures 2 & 11
2015-17	F150 4X2 Reg. Cab 141" WB 3.5L	4.92 (125)	max 0.27 (7)	7.20 (183)	max 0.31 (8)	Figures 2 & 11
2015-17	F150 4X2 Super Cab 145" WB 2.7L	5.00 (127)	max 0.27 (7)	6.73 (171)	max 0.31 (8)	Figures 2 & 11
2015-17	F150 4X2 Super Cab 145" WB 3.5L	5.00 (127)	max 0.27 (7)	6.85 (174)	max 0.31 (8)	Figures 2 & 11
2015-17	F150 4X2 Super Cab 163" WB 2.7L	5.07 (129)	max 0.27 (7)	6.73 (171)	max 0.31 (8)	Figures 2 & 11
	F150 4X4 exc/HD	SF	Side/Side *	SR	Side/Side *	Notes
2004-05	F150 4X4 Reg. 126" WB, 4.6L	4.37 (111)	max 0.27 (7)	7.32 (186)	max 0.31 (8)	Figures 2 & 12
2004-05	F150 4X4 Reg. 126" WB, 5.4L	4.56 (116)	max 0.27 (7)	7.36 (187)	max 0.27 (7)	Figures 2 & 12
2004-05	F150 4X4 Reg. 145" WB, 4.6L	4.37 (111)	max 0.27 (7)	7.36 (187)	max 0.27 (7)	Figures 2 & 12
2004-05	F150 4X4 Reg. 145" WB, 5.4L	4.29 (109)	max 0.27 (7)	7.36 (187)	max 0.27 (7)	Figures 2 & 12
2004-05	F150 4X4 Super 133" WB, 4.6L	4.52 (115)	max 0.27 (7)	6.92 (176)	max 0.31 (8)	Figures 2 & 12
2004-05	F150 4X4 Super 133" WB, 5.4L	4.64 (118)	max 0.27 (7)	6.81 (173)	max 0.31 (8)	Figures 2 & 12
2004-05	F150 4X4 Super 145" WB, 4.6L	4.52 (115)	max 0.27 (7)	6.73 (171)	max 0.31 (8)	Figures 2 & 12
2004-05	F150 4X4 Super 145" WB, 5.4L	4.68 (119)	max 0.27 (7)	6.96 (177)	max 0.31 (8)	Figures 2 & 12
2004-05	F150 4X4 Crew 138" WB, 4.6L	4.60 (117)	max 0.27 (7)	6.85 (174)	max 0.31 (8)	Figures 2 & 12
2004-05	F150 4X4 Crew 138" WB, 5.4L	4.68 (119)	max 0.27 (7)	6.81 (173)	max 0.31 (8)	Figures 2 & 12
2006	F150 4X4 Reg. 126" WB, 4.6L	4.40 (112)	max 0.27 (7)	6.85 (174)	max 0.31 (8)	Figures 2 & 12
2006	F150 4X4 Reg. 126" WB, 5.4L	4.29 (109)	max 0.27 (7)	6.77 (172)	max 0.31 (8)	Figures 2 & 12
2006	F150 4X4 Reg. 145" WB, 4.6L	4.44 (113)	max 0.27 (7)	6.88 (175)	max 0.31 (8)	Figures 2 & 12
2006	F150 4X4 Reg. 145" WB, 5.4L	4.48 (114)	max 0.27 (7)	6.81 (173)	max 0.31 (8)	Figures 2 & 12
2006	F150 4X4 Super 133" WB, 4.6L	4.92 (125)	max 0.27 (7)	6.49 (165)	max 0.31 (8)	Figures 2 & 12
2006	F150 4X4 Super 133" WB, 5.4L	4.52 (115)	max 0.27 (7)	6.41 (163)	max 0.31 (8)	Figures 2 & 12
2006	F150 4X4 Super 145" WB, 4.6L	4.72 (120)	max 0.27 (7)	6.53 (166)	max 0.31 (8)	Figures 2 & 12
2006	F150 4X4 Super 145" WB, 5.4L	4.60 (117)	max 0.27 (7)	6.41 (163)	max 0.31 (8)	Figures 2 & 12
2006	F150 4X4 Harley Davidson	4.60 (117)	max 0.27 (7)	6.14 (156)	max 0.31 (8)	Figures 2 & 12
2006	F150 4X4 Crew 139" WB, 4.6L	4.56 (116)	max 0.27 (7)	6.37 (162)	max 0.31 (8)	Figures 2 & 12
2006	F150 4X4 Crew 139" WB, 5.4 L	4.44 (113)	max 0.27 (7)	6.29 (160)	max 0.31 (8)	Figures 2 & 12
2006	F150 4X4 Crew 151" WB, 4.6L	4.72 (120)	max 0.27 (7)	6.29 (160)	max 0.31 (8)	Figures 2 & 12
2006	F150 4X4 Crew 151" WB, 5.4L	4.56 (116)	max 0.27 (7)	6.18 (157)	max 0.31 (8)	Figures 2 & 12
2007-08	F150 4X4 Reg. 126" WB, 4.6L	4.44 (113)	max 0.27 (7)	6.88 (175)	max 0.31 (8)	Figures 2 & 12
2007-08	F150 4X4 Reg. 126" WB, 5.4L	4.25 (108)	max 0.27 (7)	6.73 (171)	max 0.31 (8)	Figures 2 & 12
2007-08	F150 4X4 Reg. 145" WB, 4.6L	4.48 (114)	max 0.27 (7)	6.92 (176)	max 0.31 (8)	Figures 2 & 12
2007-08	F150 4X4 Reg. 145" WB, 5.4L	4.40 (112)	max 0.27 (7)	6.73 (171)	max 0.31 (8)	Figures 2 & 12
2007-08	F150 4X4 Super 133" WB, 4.6L	4.96 (126)	max 0.27 (7)	6.53 (166)	max 0.31 (8)	Figures 2 & 12
2007-08	F150 4X4 Super 133" WB, 5.4L	4.44 (113)	max 0.27 (7)	6.33 (161)	max 0.31 (8)	Figures 2 & 12
2007-08	F150 4X4 Super 145" WB, 4.6L	4.48 (114)	max 0.27 (7)	6.57 (167)	max 0.31 (8)	Figures 2 & 12
2007-08	F150 4X4 Super 145" WB, 5.4L	4.52 (115)	max 0.27 (7)	6.37 (162)	max 0.31 (8)	Figures 2 & 12
2007-08	F150 4X4 Crew 139" WB, 5.4 L	4.37 (111)	max 0.27 (7)	6.25 (159)	max 0.31 (8)	Figures 2 & 12
2007-08	F150 4X4 Crew 151" WB, 4.6L	4.72 (120)	max 0.27 (7)	6.37 (162)	max 0.31 (8)	Figures 2 & 12
2007-08	F150 4X4 Harley Davidson	4.52 (115)	max 0.27 (7)	5.98 (152)	max 0.31 (8)	Figures 2 & 12
2007-08	F150 4X4 Crew 139" WB, 4.6L	4.48 (114)	max 0.27 (7)	6.41 (163)	max 0.31 (8)	Figures 2 & 12
2007-08	F150 4X4 Crew 151" WB, 5.4L	4.48 (114)	max 0.27 (7)	6.18 (157)	max 0.31 (8)	Figures 2 & 12

* All Side/Side measurements are taken at wheel arch to ground (WF & WR) See Figure 1 above ** ±0.75 (19) unless indicated otherwise

Ford-Lincoln-Mercury Truck & SUV (cont.)

Year Model	Height Specifications @ Curb Weight In. (mm) **				
	Front		Rear		
F150 4X4 exc/HD (cont.)	SF	Side/Side *	SR	Side/Side *	Notes
2009 F150 4X4 Reg. 126" WB, 4.6L, 2 valve	4.37 (111)	max 0.27 (7)	6.69 (170)	max 0.31 (8)	Figures 2 & 11
2009 F150 4X4 Reg. 126" WB, 4.6L, 3 valve	4.29 (109)	max 0.27 (7)	6.65 (169)	max 0.31 (8)	Figures 2 & 11
2009 F150 4X4 Reg. 126" WB, 5.4L	4.40 (112)	max 0.27 (7)	6.61 (168)	max 0.31 (8)	Figures 2 & 11
2009 F150 4X4 Reg. 145" WB, 4.6L, 2 valve	4.33 (110)	max 0.27 (7)	7.16 (182)	max 0.31 (8)	Figures 2 & 11
2009 F150 4X4 Reg. 145" WB, 4.6L, 3 valve	4.48 (114)	max 0.27 (7)	7.16 (182)	max 0.31 (8)	Figures 2 & 11
2009 F150 4X4 Reg. 145" WB, 5.4L	4.37 (111)	max 0.27 (7)	7.04 (179)	max 0.31 (8)	Figures 2 & 11
2009 F150 4X4 Sup. 133" WB, 4.6L, 2 valve	4.64 (118)	max 0.27 (7)	6.29 (160)	max 0.31 (8)	Figures 2 & 11
2009 F150 4X4 Sup. 133" WB, 4.6L, 3 valve	4.52 (115)	max 0.27 (7)	6.22 (158)	max 0.31 (8)	Figures 2 & 11
2009 F150 4X4 Sup. 133" WB, 5.4L	4.52 (115)	max 0.27 (7)	6.10 (155)	max 0.31 (8)	Figures 2 & 11
2009 F150 4X4 Sup. 145" WB, 4.6L, 2 valve	4.64 (118)	max 0.27 (7)	6.65 (169)	max 0.31 (8)	Figures 2 & 11
2009 F150 4X4 Sup. 145" WB, 4.6L, 3 valve	4.52 (115)	max 0.27 (7)	6.57 (167)	max 0.31 (8)	Figures 2 & 11
2009 F150 4X4 Sup. 145" WB, 5.4L	4.37 (111)	max 0.27 (7)	6.57 (167)	max 0.31 (8)	Figures 2 & 11
2009 F150 4X4 Crew 145" WB, 4.6L	4.40 (112)	max 0.27 (7)	6.33 (161)	max 0.31 (8)	Figures 2 & 11
2009 F150 4X4 Crew 145" WB, 5.4 L	4.44 (113)	max 0.27 (7)	6.29 (160)	max 0.31 (8)	Figures 2 & 11
2009 F150 4X4 Crew 157" WB, 4.6L	4.44 (113)	max 0.27 (7)	6.41 (163)	max 0.31 (8)	Figures 2 & 11
2009 F150 4X4 Crew 157" WB, 5.4L	4.52 (115)	max 0.27 (7)	6.37 (162)	max 0.31 (8)	Figures 2 & 11
2010 F150 4X4 Reg. 126" WB, 4.6L, 2 valve	4.33 (110)	max 0.27 (7)	6.73 (171)	max 0.31 (8)	Figures 2 & 11
2010 F150 4X4 Reg. 126" WB, 4.6L, 3 valve	4.21 (107)	max 0.27 (7)	6.69 (170)	max 0.31 (8)	Figures 2 & 11
2010 F150 4X4 Reg. 126" WB, 5.4L	4.37 (111)	max 0.27 (7)	6.61 (168)	max 0.31 (8)	Figures 2 & 11
2010 F150 4X4 Reg. 145" WB, 4.6L, 2 valve	4.29 (109)	max 0.27 (7)	7.32 (186)	max 0.31 (8)	Figures 2 & 11
2010 F150 4X4 Reg. 145" WB, 4.6L, 3 valve	4.44 (113)	max 0.27 (7)	7.32 (186)	max 0.31 (8)	Figures 2 & 11
2010 F150 4X4 Reg. 145" WB, 5.4L	4.33 (110)	max 0.27 (7)	7.20 (183)	max 0.31 (8)	Figures 2 & 11
2010 F150 4X4 Sup. 133" WB, SVT Raptor	5.07 (129)	max 0.27 (7)	9.52 (242)	max 0.31 (8)	Figures 2 & 14
2010 F150 4X4 Sup. 145" WB, 4.6L, 2 valve	4.60 (117)	max 0.27 (7)	6.81 (173)	max 0.31 (8)	Figures 2 & 11
2010 F150 4X4 Sup. 145" WB, 4.6L, 3 valve	4.48 (114)	max 0.27 (7)	6.77 (172)	max 0.31 (8)	Figures 2 & 11
2010 F150 4X4 Sup. 145" WB, 5.4L	4.33 (110)	max 0.27 (7)	6.73 (171)	max 0.31 (8)	Figures 2 & 11
2010 F150 4X4 Crew 145" WB, 4.6L	4.33 (110)	max 0.27 (7)	6.45 (164)	max 0.31 (8)	Figures 2 & 11
2010 F150 4X4 Crew 145" WB, 5.4 L	4.40 (112)	max 0.27 (7)	6.37 (162)	max 0.31 (8)	Figures 2 & 11
2010 F150 4X4 Crew 157" WB, 4.6L	4.37 (111)	max 0.27 (7)	6.37 (162)	max 0.31 (8)	Figures 2 & 11
2010 F150 4X4 Crew 157" WB, 5.4L	4.44 (113)	max 0.27 (7)	6.33 (161)	max 0.31 (8)	Figures 2 & 11
2011-14 F150 4X4 Reg. Cab 126" WB 3.7L	4.37 (111)	max 0.27 (7)	6.85 (174)	max 0.31 (8)	Figures 2 & 11
2011-14 F150 4X4 Reg. Cab 126" WB 5.0L	4.40 (112)	max 0.27 (7)	6.77 (172)	max 0.31 (8)	Figures 2 & 11
2011-14 F150 4X4 Reg. Cab 145" WB 3.5L	4.37 (111)	max 0.27 (7)	7.28 (185)	max 0.31 (8)	Figures 2 & 11
2011-14 F150 4X4 Reg. Cab 145" WB 3.7L	4.37 (111)	max 0.27 (7)	7.48 (190)	max 0.31 (8)	Figures 2 & 11
2011-14 F150 4X4 Reg. Cab 145" WB 5.0L	4.40 (112)	max 0.27 (7)	7.28 (185)	max 0.31 (8)	Figures 2 & 11
2011-14 F150 4X4 Super Cab 145" WB 3.5L	4.44 (113)	max 0.27 (7)	6.73 (171)	max 0.31 (8)	Figures 2 & 11
2011-14 F150 4X4 Super 145" WB 3.5L Max Tow	4.40 (112)	max 0.27 (7)	6.77 (172)	max 0.31 (8)	Figures 2 & 11
2011-14 F150 4X4 Super Cab 145" WB 3.7L	4.40 (112)	max 0.27 (7)	6.88 (175)	max 0.31 (8)	Figures 2 & 11
2011-14 F150 4X4 Super Cab 133" WB 5.0L	4.37 (111)	max 0.27 (7)	6.65 (169)	max 0.31 (8)	Figures 2 & 11
2011-14 F150 4X4 Crew Cab145" WB, 3.5L	4.48 (114)	max 0.27 (7)	6.41 (163)	max 0.31 (8)	Figures 2 & 11
2011-14 F150 4X4 Crew 145" WB, 3.5L Max Tow	4.56 (116)	max 0.27 (7)	6.45 (164)	max 0.31 (8)	Figures 2 & 11
2011-14 F150 4X4 Crew Cab145" WB, 5.0L	4.44 (113)	max 0.27 (7)	6.41 (163)	max 0.31 (8)	Figures 2 & 11
2011-14 F150 4X4 Crew Cab145" WB, 6.2L	4.21 (107)	max 0.27 (7)	5.86 (149)	max 0.31 (8)	Figures 2 & 11
2011-14 F150 4X4 Crew 145" WB, 6.2L Max Tow	4.21 (107)	max 0.27 (7)	5.74 (146)	max 0.31 (8)	Figures 2 & 11
2011-14 F150 4X4 Crew Cab157" WB, 3.5L	4.48 (114)	max 0.27 (7)	6.37 (162)	max 0.31 (8)	Figures 2 & 11
2011-14 F150 4X4 Crew 157" WB, 3.5L Max Tow	4.44 (113)	max 0.27 (7)	6.45 (164)	max 0.31 (8)	Figures 2 & 11
2011-14 F150 4X4 Crew 157" WB, 5.0L	4.48 (114)	max 0.27 (7)	6.29 (160)	max 0.31 (8)	Figures 2 & 11
2011-14 F150 4X4 SVT Raptor Reg. Cab	5.07 (129)	max 0.27 (7)	9.52 (242)	max 0.31 (8)	Figures 2 & 14
2011-14 F150 4X4 SVT Raptor Super Cab	5.11 (130)	max 0.27 (7)	9.52 (242)	max 0.31 (8)	Figures 2 & 14
2011-14 F150 4X4 SVT Raptor Crew Cab	5.00 (127)	max 0.27 (7)	9.52 (242)	max 0.31 (8)	Figures 2 & 14
2015-17 F150 4X4 Crew Cab 145" WB 2.7L	4.92 (125)	max 0.27 (7)	6.77 (172)	max 0.31 (8)	Figures 2 & 11
2015-17 F150 4X4 Crew Cab 145" WB 3.5L	4.92 (125)	max 0.27 (7)	6.90 (175)	max 0.31 (8)	Figures 2 & 11
2015-17 F150 4X4 Reg. Cab 122" WB 2.7L	4.76 (121)	max 0.27 (7)	7.20 (183)	max 0.31 (8)	Figures 2 & 11
2015-17 F150 4X4 Reg. Cab 122" WB 3.5L	4.76 (121)	max 0.27 (7)	7.32 (186)	max 0.31 (8)	Figures 2 & 11
2015-17 F150 4X4 Reg. Cab 141" WB 2.7L	4.80 (122)	max 0.27 (7)	7.28 (185)	max 0.31 (8)	Figures 2 & 11
2015-17 F150 4X4 Reg. Cab 141" WB 3.5L	5.11 (130)	max 0.27 (7)	7.40 (188)	max 0.31 (8)	Figures 2 & 11
2015-17 F150 4X4 Super Cab 145" WB 2.7L	5.00 (127)	max 0.27 (7)	6.85 (174)	max 0.31 (8)	Figures 2 & 11
2015-17 F150 4X4 Super Cab 145" WB 3.5L	5.03 (128)	max 0.27 (7)	7.00 (178)	max 0.31 (8)	Figures 2 & 11
F150 HD 4X2	SF	Side/Side *	SR	Side/Side *	Notes
2004-05 F150 4X2 Reg. 145" WB, 5.4L	4.60 (117)	max 0.27 (7)	7.83 (199)	max 0.31 (8)	Figures 2 & 11
2004-05 F150 4X2 Super 163" WB, 5.4L	4.21 (107)	max 0.27 (7)	7.48 (190)	max 0.31 (8)	Figures 2 & 11
2006 F150 4X2 Reg. 145" WB, 5.4L	4.29 (109)	max 0.27 (7)	7.79 (198)	max 0.31 (8)	Figures 2 & 11
2007-08 F150 4X2 Reg. 145" WB, 5.4L	4.29 (109)	max 0.27 (7)	7.75 (197)	max 0.31 (8)	Figures 2 & 11
2006-08 F150 4X2 Super 163" WB, 5.4L	4.44 (113)	max 0.27 (7)	7.44 (189)	max 0.31 (8)	Figures 2 & 11
2009 F150 4X2 Reg. 145" WB, 5.4L	4.44 (113)	max 0.27 (7)	7.87 (200)	max 0.31 (8)	Figures 2 & 11
2009 F150 4X2 Super 163" WB, 5.4L	4.56 (116)	max 0.27 (7)	7.36 (187)	max 0.31 (8)	Figures 2 & 11
2010 F150 4X2 Reg. 145" WB, 5.4L	4.44 (113)	max 0.27 (7)	7.40 (188)	max 0.31 (8)	Figures 2 & 11
2010 F150 4X2 Super 163" WB, 5.4L	4.56 (116)	max 0.27 (7)	6.88 (175)	max 0.31 (8)	Figures 2 & 11
2011-14 F150 4X2 HD 145" WB, 3.5L	4.48 (114)	max 0.27 (7)	7.67 (195)	max 0.31 (8)	Figures 2 & 11
2011-14 F150 4X2 HD 145" WB, 5.0L	4.44 (113)	max 0.27 (7)	7.59 (193)	max 0.31 (8)	Figures 2 & 11

* All Side/Side measurements are taken at wheel arch to ground (WF & WR) See Figure 1 above ** ±0.75 (19) unless indicated otherwise

CHASSIS / SUSP. HEIGHTS

Ford-Lincoln-Mercury Truck & SUV (cont.)

Year Model	Height Specifications @ Curb Weight In. (mm) **				
	Front		Rear		
F150 HD 4X2 (cont.)	SF	Side/Side *	SR	Side/Side *	Notes
2011-14 F150 4X2 HD 163" WB, 3.5L	4.64 (118)	max 0.27 (7)	7.16 (182)	max 0.31 (8)	Figures 2 & 11
2011-14 F150 4X2 HD 163" WB, 5.0L	4.56 (116)	max 0.27 (7)	7.04 (179)	max 0.31 (8)	Figures 2 & 11
2015-17 F150 4X2 Crew Cab HD 145" WB 2.7L	4.80 (122)	max 0.27 (7)	7.04 (179)	max 0.31 (8)	Figures 2 & 11
2015-17 F150 4X2 Crew Cab HD 145" WB 3.5L	4.84 (123)	max 0.27 (7)	6.92 (176)	max 0.31 (8)	Figures 2 & 11
2015-17 F150 4X2 Crew Cab HD 145" WB 5.0L	4.80 (122)	max 0.27 (7)	7.04 (179)	max 0.31 (8)	Figures 2 & 11
2015-17 F150 4X2 Crew Cab HD 157" WB 2.7L	4.92 (125)	max 0.27 (7)	7.08 (180)	max 0.31 (8)	Figures 2 & 11
2015-17 F150 4X2 Crew Cab HD 157" WB 3.5L	4.84 (123)	max 0.27 (7)	7.00 (178)	max 0.31 (8)	Figures 2 & 11
2015-17 F150 4X2 Crew Cab HD 157" WB 5.0L	4.92 (125)	max 0.27 (7)	7.12 (181)	max 0.31 (8)	Figures 2 & 11
2015-17 F150 4X2 Reg. Cab HD 122" WB 5.0L	4.84 (123)	max 0.27 (7)	7.59 (193)	max 0.31 (8)	Figures 2 & 11
2015-17 F150 4X2 Reg. Cab HD 141" WB 2.7L	4.76 (121)	max 0.27 (7)	7.56 (192)	max 0.31 (8)	Figures 2 & 11
2015-17 F150 4X2 Reg. Cab HD 141" WB 3.5L	4.96 (126)	max 0.27 (7)	7.75 (197)	max 0.31 (8)	Figures 2 & 11
2015-17 F150 4X2 Reg. Cab HD 141" WB 5.0L	5.03 (128)	max 0.27 (7)	7.63 (194)	max 0.31 (8)	Figures 2 & 11
2015-17 F150 4X2 Super Cab HD 145" WB 2.7L	4.92 (125)	max 0.27 (7)	7.20 (183)	max 0.31 (8)	Figures 2 & 11
2015-17 F150 4X2 Super Cab HD 145" WB 3.5L	5.00 (127)	max 0.27 (7)	7.04 (179)	max 0.31 (8)	Figures 2 & 11
2015-17 F150 4X2 Super Cab HD 145" WB 5.0L	5.07 (129)	max 0.27 (7)	7.40 (188)	max 0.31 (8)	Figures 2 & 11
2015-17 F150 4X2 Super Cab HD 163" WB 2.7L	5.03 (128)	max 0.27 (7)	7.20 (183)	max 0.31 (8)	Figures 2 & 11
2015-17 F150 4X2 Super Cab HD 163" WB 3.5L	5.11 (130)	max 0.27 (7)	7.32 (186)	max 0.31 (8)	Figures 2 & 11
2015-17 F150 4X2 Super Cab HD 163" WB 5.0L	5.03 (128)	max 0.27 (7)	7.20 (183)	max 0.31 (8)	Figures 2 & 11
2015-17 F150 4X2 Crew HPP 157" WB, 3.5L, 17" whls.	4.64 (118)	max 0.27 (7)	7.52 (191)	max 0.31 (8)	Figures 2 & 11
2015-17 F150 4X2 Crew HPP 157" WB, 5.0L, 17" whls.	4.64 (118)	max 0.27 (7)	7.56 (192)	max 0.31 (8)	Figures 2 & 11
2015-17 F150 4X2 Crew HPP 141" WB, 3.5L, 17" whls.	5.00 (127)	max 0.27 (7)	7.87 (200)	max 0.31 (8)	Figures 2 & 11
2015-17 F150 4X2 Crew HPP 141" WB, 5.0L, 17" whls.	5.00 (127)	max 0.27 (7)	7.95 (202)	max 0.31 (8)	Figures 2 & 11
2015-17 F150 4X2 Super HPP 163" WB 3.5L, 17" whl	4.72 (120)	max 0.27 (7)	7.59 (193)	max 0.31 (8)	Figures 2 & 11
2015-17 F150 4X2 Super HPP 163" WB 5.0L, 17" whl	5.35 (136)	max 0.27 (7)	7.63 (194)	max 0.31 (8)	Figures 2 & 11
2015-17 F150 4X2 Crew HPP 157" WB, 3.5L, 18" whls.	4.60 (117)	max 0.27 (7)	7.48 (190)	max 0.31 (8)	Figures 2 & 11
2015-17 F150 4X2 Crew HPP 157" WB, 5.0L, 18" whls.	4.60 (117)	max 0.27 (7)	7.52 (191)	max 0.31 (8)	Figures 2 & 11
2015-17 F150 4X2 Reg. HPP 141" WB, 3.5L, 18" whls.	4.96 (126)	max 0.27 (7)	7.83 (199)	max 0.31 (8)	Figures 2 & 11
2015-17 F150 4X2 Reg. HPP 141" WB, 5.0L, 18" whls.	4.96 (126)	max 0.27 (7)	7.91 (201)	max 0.31 (8)	Figures 2 & 11
2015-17 F150 4X2 Super HPP 163" WB 3.5L, 18" whl	5.31 (135)	max 0.27 (7)	7.55 (192)	max 0.31 (8)	Figures 2 & 11
2015-17 F150 4X2 Super HPP 163" WB 5.0L, 18" whl	5.31 (135)	max 0.27 (7)	7.59 (193)	max 0.31 (8)	Figures 2 & 11
HD=Heavy Duty, HPP = Heavy Payload Package					
F150 HD 4X4	SF	Side/Side *	SR	Side/Side *	Notes
2004-05 F150 4X4 Reg. 145" WB, 5.4L	5.03 (128)	max 0.27 (7)	7.71 (196)	max 0.31 (8)	Figures 2 & 12
2004-05 F150 4X4 Super 163" WB, 5.4L	4.56 (116)	max 0.27 (7)	7.36 (187)	max 0.31 (8)	Figures 2 & 12
2006-08 F150 4X4 Reg. 145" WB, 5.4L	4.80 (122)	max 0.27 (7)	7.91 (201)	max 0.31 (8)	Figures 2 & 12
2006 F150 4X4 Super 163" WB, 5.4L	4.48 (114)	max 0.27 (7)	7.55 (192)	max 0.31 (8)	Figures 2 & 12
2007-08 F150 4X4 Super 163" WB, 5.4L	4.44 (113)	max 0.27 (7)	7.55 (192)	max 0.31 (8)	Figures 2 & 12
2009 F150 4X4 Reg. 145" WB, 5.4L	4.72 (120)	max 0.27 (7)	7.91 (201)	max 0.31 (8)	Figures 2 & 11
2009 F150 4X4 Super 163" WB, 5.4L	4.40 (112)	max 0.27 (7)	7.36 (187)	max 0.31 (8)	Figures 2 & 11
2010 F150 4X4 Reg. 145" WB, 5.4L	4.72 (120)	max 0.27 (7)	7.44 (189)	max 0.31 (8)	Figures 2 & 11
2010 F150 4X4 Super 163" WB, 5.4L	4.37 (111)	max 0.27 (7)	6.88 (175)	max 0.31 (8)	Figures 2 & 11
2011-14 F150 4X4 HD 145" WB, 3.5L	4.33 (110)	max 0.27 (7)	7.79 (198)	max 0.31 (8)	Figures 2 & 11
2011-14 F150 4X4 HD 145" WB, 5.0L	4.29 (109)	max 0.27 (7)	7.75 (197)	max 0.31 (8)	Figures 2 & 11
2011-14 F150 4X4 HD 163" WB, 3.5L	4.52 (115)	max 0.27 (7)	7.16 (182)	max 0.31 (8)	Figures 2 & 11
2011-14 F150 4X4 HD 163" WB, 5.0L	4.44 (113)	max 0.27 (7)	7.12 (181)	max 0.31 (8)	Figures 2 & 11
2015-17 F150 4X4 Crew Cab HD 145" WB 2.7L	4.88 (124)	max 0.27 (7)	7.16 (182)	max 0.31 (8)	Figures 2 & 11
2015-17 F150 4X4 Crew Cab HD 145" WB 3.5L	4.88 (124)	max 0.27 (7)	7.00 (178)	max 0.31 (8)	Figures 2 & 11
2015-17 F150 4X4 Crew Cab HD 145" WB 5.0L	4.92 (125)	max 0.27 (7)	7.16 (182)	max 0.31 (8)	Figures 2 & 11
2015-17 F150 4X4 Crew Cab HD 157" WB 3.5L	4.96 (126)	max 0.27 (7)	7.08 (180)	max 0.31 (8)	Figures 2 & 11
2015-17 F150 4X4 Crew Cab HD 157" WB 5.0L	4.92 (125)	max 0.27 (7)	7.20 (183)	max 0.31 (8)	Figures 2 & 11
2015-17 F150 4X4 Reg. Cab HD 122" WB 5.0L	4.64 (118)	max 0.27 (7)	7.71 (196)	max 0.31 (8)	Figures 2 & 11
2015-17 F150 4X4 Reg. Cab HD 141" WB 2.7L	4.92 (125)	max 0.27 (7)	7.71 (196)	max 0.31 (8)	Figures 2 & 11
2015-17 F150 4X4 Reg. Cab HD 141" WB 3.5L	4.92 (125)	max 0.27 (7)	7.87 (200)	max 0.31 (8)	Figures 2 & 11
2015-17 F150 4X4 Reg. Cab HD 141" WB 5.0L	4.96 (126)	max 0.27 (7)	7.71 (196)	max 0.31 (8)	Figures 2 & 11
2015-17 F150 4X4 Super Cab HD 145" WB 2.7L	4.96 (126)	max 0.27 (7)	7.36 (187)	max 0.31 (8)	Figures 2 & 11
2015-17 F150 4X4 Super Cab HD 145" WB 3.5L	4.92 (125)	max 0.27 (7)	7.24 (184)	max 0.31 (8)	Figures 2 & 11
2015-17 F150 4X4 Super Cab HD 145" WB 5.0L	4.96 (126)	max 0.27 (7)	7.36 (187)	max 0.31 (8)	Figures 2 & 11
2015-17 F150 4X4 Super Cab HD 163" WB 3.5L	5.11 (130)	max 0.27 (7)	7.12 (181)	max 0.31 (8)	Figures 2 & 11
2015-17 F150 4X4 Super Cab HD 163" WB 5.0L	5.19 (132)	max 0.27 (7)	7.04 (179)	max 0.31 (8)	Figures 2 & 11
2015-17 F150 4X4 Crew HPP 157" WB, 3.5L, 17" whls.	5.03 (128)	max 0.27 (7)	7.67 (195)	max 0.31 (8)	Figures 2 & 11
2015-17 F150 4X4 Crew HPP 157" WB, 5.0L, 17" whls.	5.03 (128)	max 0.27 (7)	7.75 (197)	max 0.31 (8)	Figures 2 & 11
2015-17 F150 4X4 Reg. HPP 141" WB, 3.5L, 17" whls.	4.84 (123)	max 0.27 (7)	8.03 (204)	max 0.31 (8)	Figures 2 & 11
2015-17 F150 4X4 Reg. HPP 141" WB, 5.0L, 17" whls.	4.80 (122)	max 0.27 (7)	8.07 (205)	max 0.31 (8)	Figures 2 & 11
2015-17 F150 4X4 Super HPP 163" WB 3.5L, 17" whl	5.27 (134)	max 0.27 (7)	7.63 (194)	max 0.31 (8)	Figures 2 & 11
2015-17 F150 4X4 Super HPP 163" WB 5.0L, 17" whl	5.23 (133)	max 0.27 (7)	7.67 (195)	max 0.31 (8)	Figures 2 & 11
2015-17 F150 4X4 Crew HPP 157" WB, 3.5L, 18" whls.	5.00 (127)	max 0.27 (7)	7.63 (194)	max 0.31 (8)	Figures 2 & 11
2015-17 F150 4X4 Crew HPP 157" WB, 5.0L, 18" whls.	5.00 (127)	max 0.27 (7)	7.71 (196)	max 0.31 (8)	Figures 2 & 11
2015-17 F150 4X4 Reg. HPP 141" WB, 3.5L, 18" whls.	4.80 (122)	max 0.27 (7)	7.99 (203)	max 0.31 (8)	Figures 2 & 11
2015-17 F150 4X4 Reg. HPP 141" WB, 5.0L, 18" whls.	4.76 (121)	max 0.27 (7)	8.03 (204)	max 0.31 (8)	Figures 2 & 11
2015-17 F150 4X4 Super HPP 163" WB 3.5L, 18" whl	5.23 (133)	max 0.27 (7)	7.59 (193)	max 0.31 (8)	Figures 2 & 11

* All **Side/Side** measurements are taken at wheel arch to ground (WF & WR) See Figure 1 above ** ±0.75 (19) unless indicated otherwise

Ford-Lincoln-Mercury Truck & SUV (cont.)

YEAR MODEL	HEIGHT SPECIFICATIONS @ CURB WEIGHT IN. (MM) **				
	FRONT		REAR		
F150 HD 4X4 (cont.)	**SF**	**Side/Side ***	**SR**	**Side/Side ***	**Notes**
2015-17 F150 4X4 Super HPP 163" WB 5.0L, 18" whl	5.19 (132)	max 0.27 (7)	7.63 (194)	max 0.31 (8)	Figures 2 & 11
HD=Heavy Duty, HPP = Heavy Payload Package					

F-Super Duty Trucks	SF-Left	SF-Right	Side/Side	SR	Side/Side *	Notes
1999-04 F250 4X2 exc/HD	4.33-4.88 (110-124)	4.33-4.88 (110-124)	max 0.75 (19)	6.18-8.62 (157-219)	max 0.75 (19)	Figures 8 & 4
1999-04 F250 4X2 Heavy Duty	4.33-4.88 (110-124)	4.33-4.88 (110-124)	max 0.75 (19)	5.66-8.11 (144-206)	max 0.75 (19)	Figures 8 & 4
1999-04 F250 4X4 exc/HD	4.25-5.19 (108-132)	4.25-5.19 (108-132)	max 0.75 (19)	5.70-8.14 (145-207)	max 0.75 (19)	Figures 9 & 5
1999-04 F250 4X4 Heavy Duty	4.25-5.19 (108-132)	4.25-5.19 (108-132)	max 0.75 (19)	5.19-7.63 (132-194)	max 0.75 (19)	Figures 9 & 5
1999-04 F350 Pickup 4X2:						
Single Rear Wheels exc/Heavy Duty	4.33-4.88 (110-124)	4.33-4.88 (110-124)	max 0.75 (19)	6.18-8.62 (157-219)	max 0.75 (19)	Figures 8 & 4
Single Rear Wheels Heavy Duty	4.33-4.88 (110-124)	4.33-4.88 (110-124)	max 0.75 (19)	5.66-8.11 (144-206)	max 0.75 (19)	Figures 8 & 4
Dual Rear Wheels exc/Heavy Duty	4.33-4.88 (110-124)	4.33-4.88 (110-124)	max 0.75 (19)	6.45-8.89 (164-226)	max 0.75 (19)	Figures 8 & 4
Dual Rear Wheels Heavy Duty	4.33-4.88 (110-124)	4.33-4.88 (110-124)	max 0.75 (19)	5.94-8.38 (151-213)	max 0.75 (19)	Figures 8 & 4
1999-04 F350 Pickup 4X4:						
Single Rear Wheels exc/Heavy Duty	4.25-5.19 (108-132)	4.25-5.19 (108-132)	max 0.75 (19)	5.70-8.14 (145-207)	max 0.75 (19)	Figures 9 & 5
Single Rear Wheels Heavy Duty	4.25-5.19 (108-132)	4.25-5.19 (108-132)	max 0.75 (19)	5.19-7.63 (132-194)	max 0.75 (19)	Figures 9 & 5
Dual Rear Wheels exc/Heavy Duty	4.25-5.19 (108-132)	4.25-5.19 (108-132)	max 0.75 (19)	6.02-8.46 (153-215)	max 0.75 (19)	Figures 9 & 5
Dual Rear Wheels Heavy Duty	4.25-5.19 (108-132)	4.25-5.19 (108-132)	max 0.75 (19)	5.51-7.95 (140-202)	max 0.75 (19)	Figures 9 & 5
1999-04 F350 Chassis Cab:						
4X2 Single Rear Wheels	4.33-4.88 (110-124)	4.33-4.88 (110-124)	max 0.75 (19)	5.59-8.03 (142-204)	max 0.75 (19)	Figures 8 & 4
4X2 Dual Rear Wheels	4.33-4.88 (110-124)	4.33-4.88 (110-124)	max 0.75 (19)	5.82-8.26 (148-210)	max 0.75 (19)	Figures 8 & 4
4X4 Single Rear Wheels	4.25-5.19 (108-132)	4.25-5.19 (108-132)	max 0.75 (19)	5.59-8.03 (142-204)	max 0.75 (19)	Figures 9 & 5
4X4 Dual Rear Wheels	4.25-5.19 (108-132)	4.25-5.19 (108-132)	max 0.75 (19)	5.82-8.26 (148-210)	max 0.75 (19)	Figures 9 & 5
2005-07 F250 4X2 STD.	4.05 (103)	3.77 (96)	max 0.75 (19)	8.38 (213)	max 0.75 (19)	Figures 8 & 14
2005-07 F250 4X2 HD	4.21 (107)	3.93 (100)	max 0.75 (19)	8.62 (219)	max 0.75 (19)	Figures 8 & 14
2005-07 F250 4X2 Camper	4.33 (110)	4.05 (103)	max 0.75 (19)	8.62 (219)	max 0.75 (19)	Figures 8 & 14
2005-07 F250 4X4 STD.	4.80 (122)	4.80 (122)	max 0.75 (19)	8.07 (205)	max 0.75 (19)	Figures 13 & 15
2005-07 F250 4X4 HD	5.23 (133)	5.23 (133)	max 0.75 (19)	8.30 (211)	max 0.75 (19)	Figures 13 & 15
2005-07 F250 4X4 Camper	5.23 (133)	5.23 (133)	max 0.75 (19)	8.30 (211)	max 0.75 (19)	Figures 13 & 15
2005-07 F250 4X4 Snowplow	5.62 (143)	5.62 (143)	max 0.75 (19)	8.30 (211)	max 0.75 (19)	Figures 13 & 15
2005-07 F350 4X2 STD. Single wheels	4.05 (103)	3.77 (96)	max 0.75 (19)	9.09 (231)	max 0.75 (19)	Figures 8 & 14
2005-07 F350 4X2 HD. Single wheels	4.21 (107)	3.93 (100)	max 0.75 (19)	9.09 (231)	max 0.75 (19)	Figures 8 & 14
2005-07 F350 4X2 Camper Single wheels	4.33 (110)	4.05 (103)	max 0.75 (19)	9.09 (231)	max 0.75 (19)	Figures 8 & 14
2005-07 F350 4X4 STD. Single wheels	4.84 (123)	4.84 (123)	max 0.75 (19)	8.77 (223)	max 0.75 (19)	Figures 13 & 15
2005-07 F350 4X4 HD. Single wheels	5.23 (133)	5.23 (133)	max 0.75 (19)	8.77 (223)	max 0.75 (19)	Figures 13 & 15
2005-07 F350 4X4 Camper Single wheels	5.23 (133)	5.23 (133)	max 0.75 (19)	8.77 (223)	max 0.75 (19)	Figures 13 & 15
2005-07 F350 4X4 Snowplow SRW	5.62 (143)	5.62 (143)	max 0.75 (19)	8.77 (223)	max 0.75 (19)	Figures 13 & 15
2005-07 F350 4X2 STD. Dual wheels	4.13 (105)	3.85 (98)	max 0.75 (19)	9.13 (232)	max 0.75 (19)	Figures 8 & 14
2005-07 F350 4X2 HD. Dual wheels	4.25 (108)	3.97 (101)	max 0.75 (19)	9.13 (232)	max 0.75 (19)	Figures 8 & 14
2005-07 F350 4X2 Camper Dual wheels	4.37 (111)	4.13 (105)	max 0.75 (19)	9.13 (232)	max 0.75 (19)	Figures 8 & 14
2005-07 F350 4X4 STD. Dual wheels	4.88 (124)	4.88 (124)	max 0.75 (19)	8.62 (219)	max 0.75 (19)	Figures 13 & 15
2005-07 F350 4X4 HD. Dual wheels	5.27 (134)	5.27 (134)	max 0.75 (19)	8.62 (219)	max 0.75 (19)	Figures 13 & 15
2005-07 F350 4X4 Camper Dual wheels	5.27 (134)	5.27 (134)	max 0.75 (19)	8.62 (219)	max 0.75 (19)	Figures 13 & 15
2005-07 F350 4X4 Snowplow Dual wheels	5.62 (143)	5.62 (143)	max 0.75 (19)	8.62 (219)	max 0.75 (19)	Figures 13 & 15
2005-07 F350 4X2 Chassis Cab[1] STD. Singl.	4.29 (109)	4.01 (102)	max 0.75 (19)	8.58 (218)	max 0.75 (19)	Figures 8 & 14
2005-07 F350 4X2 Chassis Cab[1] HD Singles	4.44 (113)	4.17 (106)	max 0.75 (19)	8.58 (218)	max 0.75 (19)	Figures 8 & 14
2005-07 F350 4X2 Chassis Cab[1] XHD Singl.	4.52 (115)	4.25 (108)	max 0.75 (19)	8.58 (218)	max 0.75 (19)	Figures 8 & 14
2005-07 F350 4X2 ChasCab[1] Ambul. Sngl.	4.76 (121)	4.48 (114)	max 0.75 (19)	8.58 (218)	max 0.75 (19)	Figures 8 & 14
2005-07 F350 4X2 Chassis Cab[1] STD. Duals	4.44 (113)	4.17 (106)	max 0.75 (19)	8.93 (227)	max 0.75 (19)	Figures 8 & 14
2005-07 F350 4X2 Chassis Cab[1] HD Duals	4.56 (116)	4.29 (109)	max 0.75 (19)	8.93 (227)	max 0.75 (19)	Figures 8 & 14
2005-07 F350 4X2 Chassis Cab[1] XHD Duals	4.60 (117)	4.33 (110)	max 0.75 (19)	8.93 (227)	max 0.75 (19)	Figures 8 & 14
2005-07 F350 4X2 ChasCab[1] Ambul. Duals	4.44 (113)	4.27 (106)	max 0.75 (19)	8.39 (227)	max 0.75 (19)	Figures 8 & 14
2005-07 F350 4X4 Chassis Cab[1] STD. Duals	5.35 (136)	5.35 (136)	max 0.75 (19)	9.13 (232)	max 0.75 (19)	Figures 13 & 15
2005-07 F350 4X4 Chassis Cab[1] HD Duals	5.70 (145)	5.70 (145)	max 0.75 (19)	9.13 (232)	max 0.75 (19)	Figures 13 & 15
2005-07 F350 4X4 Chassis Cab[1] XHD Duals	5.90 (150)	5.90 (150)	max 0.75 (19)	9.13 (232)	max 0.75 (19)	Figures 13 & 15
2005-07 F350 4X4 ChasCab[1] S-Plow Duals	5.90 (150)	5.90 (150)	max 0.75 (19)	9.13 (232)	max 0.75 (19)	Figures 13 & 15
2005-07 F350 4X4 ChasCab[1] Ambul. Duals	5.74 (146)	5.74 (146)	max 0.75 (19)	9.13 (232)	max 0.75 (19)	Figures 13 & 15
[1] Note Chassis cab heights are as built – no 2nd unit body installed.						
2008-10 F250 4X2 Std. Susp.	3.97 (101)	3.62 (92)	max 0.62 (16)	7.91 (201)	max 0.75 (19)	Figures 8 & 14
2008-10 F250 4X2 HD Susp.	4.09 (104)	3.74 (95)	max 0.62 (16)	8.30 (211)	max 0.75 (19)	Figures 8 & 14
2008-10 F250 4X2 Camper	4.13 (105)	3.77 (96)	max 0.62 (16)	8.30 (211)	max 0.75 (19)	Figures 8 & 14
2011 F250 4X2 Std. Susp.	4.01 (102)	3.66 (93)	max 0.62 (16)	7.91 (201)	max 0.75 (19)	Figures 8 & 14
2011 F250 4X2 HD Susp.	4.13 (105)	3.77 (96)	max 0.62 (16)	8.30 (211)	max 0.75 (19)	Figures 8 & 14
2011 F250 4X2 Camper	4.21 (107)	3.85 (98)	max 0.62 (16)	8.30 (211)	max 0.75 (19)	Figures 8 & 14
2012-16 F250 4X2 Std. Susp.	4.01 (102)	3.66 (93)	max 0.62 (16)	7.99 (203)	max 0.75 (19)	Figures 8 & 14
2012-16 F250 4X2 HD Susp.	4.13 (105)	3.77 (96)	max 0.62 (16)	8.14 (207)	max 0.75 (19)	Figures 8 & 14
2012-16 F250 4X2 Camper	4.21 (107)	3.85 (98)	max 0.62 (16)	8.14 (207)	max 0.75 (19)	Figures 8 & 14
2017 F250 4X2 Std. Susp.	4.01 (83)	2.91 (74)	max 0.62 (16)	7.67(195)	max 0.75 (19)	Figures 8 & 14
2017 F250 4X2 HD Susp.	4.13 (86)	3.03 (77)	max 0.62 (16)	7.59 (193)	max 0.75 (19)	Figures 8 & 14
2017 F250 4X2 Camper	4.21 (89)	3.15 (80)	max 0.62 (16)	7.59 (193)	max 0.75 (19)	Figures 8 & 14

* All **Side/Side** measurements are taken at wheel arch to ground (WF & WR) See **Figure 1** above ** ±0.75 (19) unless indicated otherwise

Ford-Lincoln-Mercury Truck & SUV (cont.)

YEAR MODEL	HEIGHT SPECIFICATIONS @ CURB WEIGHT IN. (MM) **					
	FRONT			REAR		
F-Super Duty Trucks (cont.)	SF-Left	SF-Right	Side/Side	SR	Side/Side *	Notes
2008-11 F350 4X2 Std. Susp., SRW	4.01 (102)	3.66 (93)	max 0.62 (16)	8.18 (208)	max 0.75 (19)	Figures 8 & 14
2008-11 F350 4X2 HD Susp., Single R Whls.	4.13 (105)	3.77 (96)	max 0.62 (16)	8.18 (208)	max 0.75 (19)	Figures 8 & 14
2008-10 F350 4X2 Camper., Single R Whls.	4.17 (106)	3.81 (97)	max 0.62 (16)	8.18 (208)	max 0.75 (19)	Figures 8 & 14
2011 F350 4X2 Camper., Single R Whls.	4.21 (107)	3.85 (98)	max 0.62 (16)	8.18 (208)	max 0.75 (19)	Figures 8 & 14
2012-16 F350 4X2 Std. Susp., SRW	4.01 (102)	3.66 (93)	max 0.62 (16)	7.79 (198)	max 0.75 (19)	Figures 8 & 14
2012-16 F350 4X2 HD Susp., Single R Whls.	4.13 (105)	3.77 (96)	max 0.62 (16)	7.79 (198)	max 0.75 (19)	Figures 8 & 14
2012-16 F350 4X2 Camper., Single R Whls.	4.21 (107)	3.85 (98)	max 0.62 (16)	7.79 (198)	max 0.75 (19)	Figures 8 & 14
2008-10 F350 4X2 Std. Susp., Dual R Whls.	4.01 (102)	3.66 (93)	max 0.62 (16)	8.14 (207)	max 0.75 (19)	Figures 8 & 14
2008-10 F350 4X2 HD Susp., Dual R Whls.	4.13 (105)	3.77 (96)	max 0.62 (16)	8.14 (207)	max 0.75 (19)	Figures 8 & 14
2008-10 F350 4X2 Camper., Dual R Whls.	4.17 (106)	3.81 (97)	max 0.62 (16)	8.14 (207)	max 0.75 (19)	Figures 8 & 14
2011-16 F350 4X2 Std. Susp., Dual R Whls.	4.09 (104)	3.74 (95)	max 0.62 (16)	7.99 (203)	max 0.75 (19)	Figures 8 & 14
2011-16 F350 4X2 HD Susp., Dual R Whls.	4.13 (105)	3.77 (96)	max 0.62 (16)	7.99 (203)	max 0.75 (19)	Figures 8 & 14
2011-16 F350 4X2 Camper., Dual R Whls.	4.17 (106)	3.81 (97)	max 0.62 (16)	7.99 (203)	max 0.75 (19)	Figures 8 & 14
2017 F350 4X2 Std. Susp. SRW Pickup	3.27 (84)	2.95 (75)	max 0.62 (16)	7.63 (194)	max 0.75 (19)	Figures 8 & 14
2017 F350 4X2 HD. Susp. SRW Pickup	3.42 (87)	3.07 (78)	max 0.62 (16)	7.63 (194)	max 0.75 (19)	Figures 8 & 14
2017 F350 4X2 Camp. Susp. SRW Pickup	3.54 (90)	3.18 (81)	max 0.62 (16)	7.63 (194)	max 0.75 (19)	Figures 8 & 14
2008-11 F350 4X2 Chassis Cab STD SRW	4.25 (108)	3.89 (99)	max 0.62 (16)	6.77 (172)	max 0.75 (19)	Figures 8 & 14
2008-11 F350 4X2 Chassis Cab HD SRW	4.33 (110)	3.97 (101)	max 0.62 (16)	6.77 (172)	max 0.75 (19)	Figures 8 & 14
2008-11 F350 4X2 Chassis Cab XHD SRW	4.40 (112)	4.05 (103)	max 0.62 (16)	6.77 (172)	max 0.75 (19)	Figures 8 & 14
2012-16 F350 4X2 Chassis Cab STD SRW	4.17 (106)	3.81 (97)	max 0.62 (16)	6.73 (171)	max 0.75 (19)	Figures 8 & 14
2012-16 F350 4X2 Chassis Cab HD SRW	4.25 (108)	3.89 (99)	max 0.62 (16)	6.73 (171)	max 0.75 (19)	Figures 8 & 14
2012-16 F350 4X2 Chassis Cab XHD SRW	4.37 (111)	4.01 (102)	max 0.62 (16)	6.73 (171)	max 0.75 (19)	Figures 8 & 14
2008-11 F350 4X2 Chassis Cab STD DRW	4.40 (112)	4.05 (103)	max 0.62 (16)	7.40 (188)	max 0.75 (19)	Figures 8 & 14
2008-11 F350 4X2 Chassis Cab HD DRW	4.64 (118)	4.29 (109)	max 0.62 (16)	7.40 (188)	max 0.75 (19)	Figures 8 & 14
2008-11 F350 4X2 Chassis Cab XHD DRW	4.48 (114)	4.13 (105)	max 0.62 (16)	7.40 (188)	max 0.75 (19)	Figures 8 & 14
2012-16 F350 4X2 Chassis Cab STD DRW	4.33 (110)	3.97 (101)	max 0.62 (16)	7.32 (186)	max 0.75 (19)	Figures 8 & 14
2012-16 F350 4X2 Chassis Cab HD DRW	4.52 (115)	4.17 (106)	max 0.62 (16)	7.32 (186)	max 0.75 (19)	Figures 8 & 14
2012-16 F350 4X2 Chassis Cab XHD DRW	4.40 (112)	4.05 (103)	max 0.62 (16)	7.32 (186)	max 0.75 (19)	Figures 8 & 14
2017 F350 4X2 Chassis Cab STD SRW	3.42 (87)	3.07 (78)	max 0.62 (16)	6.73 (171)	max 0.75 (19)	Figures 8 & 14
2017 F350 4X2 Chassis Cab HD SRW	3.50 (89)	3.15 (80)	max 0.62 (16)	6.73 (171)	max 0.75 (19)	Figures 8 & 14
2017 F350 4X2 Chassis Cab XHD SRW	3.62 (92)	3.26 (83)	max 0.62 (16)	6.73 (171)	max 0.75 (19)	Figures 8 & 14
2017 F350 4X2 Chassis Cab STD DRW	3.54 (90)	3.18 (81)	max 0.62 (16)	6.65 (169)	max 0.75 (19)	Figures 8 & 14
2017 F350 4X2 Chassis Cab HD DRW	3.81 (97)	3.46 (88)	max 0.62 (16)	6.65 (169)	max 0.75 (19)	Figures 8 & 14
2017 F350 4X2 Chassis Cab XHD DRW	3.70 (94)	3.34 (85)	max 0.62 (16)	6.65 (169)	max 0.75 (19)	Figures 8 & 14
2008-10 F250 4X4 Std. Susp.	4.84 (123)	4.84 (123)	max 0.62 (16)	7.08 (180)	max 0.75 (19)	Figures 13 & 15
2008-10 F250 4X4 HD Susp.	5.23 (133)	5.23 (133)	max 0.62 (16)	7.48 (190)	max 0.75 (19)	Figures 13 & 15
2008-10 F250 4X4 Camper	5.23 (133)	5.23 (133)	max 0.62 (16)	7.48 (190)	max 0.75 (19)	Figures 13 & 15
2008-10 F250 4X4 Snowplow Pkg. (w/o plow)	5.55 (141)	5.55 (141)	max 0.62 (16)	7.48 (190)	max 0.75 (19)	Figures 13 & 15
2011 F250 4X4 Std. Susp.	4.92 (125)	4.92 (125)	max 0.62 (16)	7.08 (180)	max 0.75 (19)	Figures 13 & 15
2011 F250 4X4 HD Susp.	5.27 (134)	5.27 (134)	max 0.62 (16)	7.48 (190)	max 0.75 (19)	Figures 13 & 15
2011 F250 4X4 Camper	5.27 (134)	5.27 (134)	max 0.62 (16)	7.48 (190)	max 0.75 (19)	Figures 13 & 15
2011 F250 4X4 Snowplow Pkg. (w/o plow)	5.66 (144)	5.66 (144)	max 0.62 (16)	7.48 (190)	max 0.75 (19)	Figures 13 & 15
2012-16 F250 4X4 Std. Susp.	4.92 (125)	4.92 (125)	max 0.62 (16)	7.12(181)	max 0.75 (19)	Figures 13 & 15
2012-16 F250 4X4 HD Susp.	5.27 (134)	5.27 (134)	max 0.62 (16)	7.28 (185)	max 0.75 (19)	Figures 13 & 15
2012-16 F250 4X4 Camper	5.27 (134)	5.27 (134)	max 0.62 (16)	7.28 (185)	max 0.75 (19)	Figures 13 & 15
2012-16 F250 4X4 Snowplow Pkg. (w/o plow)	5.66 (144)	5.66 (144)	max 0.62 (16)	7.28 (185)	max 0.75 (19)	Figures 13 & 15
2017 F250 4X4 Std. Susp.	4.13 (105)	4.13 (105)	max 0.62 (16)	7.63 (194)	max 0.75 (19)	Figures 13 & 15
2017 F250 4X4 HD Susp.	4.52 (115)	4.52 (115)	max 0.62 (16)	7.55 (192)	max 0.75 (19)	Figures 13 & 15
2017 F250 4X4 Camper	4.52 (115)	4.52 (115)	max 0.62 (16)	7.55 (192)	max 0.75 (19)	Figures 13 & 15
2017 F250 4X4 Snowplow Pkg. (w/o plow)	5.00 (127)	5.00 (127)	max 0.62 (16)	7.55 (192)	max 0.75 (19)	Figures 13 & 15
2008-10 F350 4X4 Std. Susp. SRW	4.84 (123)	4.84 (123)	max 0.62 (16)	7.87 (200)	max 0.75 (19)	Figures 13 & 15
2008-10 F350 4X4 HD Susp. SRW	5.23 (133)	5.23 (133)	max 0.62 (16)	7.87 (200)	max 0.75 (19)	Figures 13 & 15
2008-10 F350 4X4 Camper SRW	5.23 (133)	5.23 (133)	max 0.62 (16)	7.87 (200)	max 0.75 (19)	Figures 13 & 15
2008-10 F350 4X4 S-plow(w/o plow) SRW	5.55 (141)	5.55 (141)	max 0.62 (16)	7.87 (200)	max 0.75 (19)	Figures 13 & 15
2011-16 F350 4X4 Std. Susp. SRW	4.84 (123)	4.84 (123)	max 0.62 (16)	7.55 (192)	max 0.75 (19)	Figures 13 & 15
2011-16 F350 4X4 HD Susp. SRW	5.23 (133)	5.23 (133)	max 0.62 (16)	7.55 (192)	max 0.75 (19)	Figures 13 & 15
2011-16 F350 4X4 Camper SRW	5.23 (133)	5.23 (133)	max 0.62 (16)	7.55 (192)	max 0.75 (19)	Figures 13 & 15
2011-16 F350 4X4 S-plow(w/o plow) SRW	5.55 (141)	5.55 (141)	max 0.62 (16)	7.55 (192)	max 0.75 (19)	Figures 13 & 15
2008-10 F350 4X4 Std. Susp. DRW	4.92 (125)	4.92 (125)	max 0.62 (16)	7.67 (195)	max 0.75 (19)	Figures 13 & 15
2008-10 F350 4X4 HD Susp. DRW	5.27 (134)	5.27 (134)	max 0.62 (16)	7.67 (195)	max 0.75 (19)	Figures 13 & 15
2008-10 F350 4X4 Camper DRW	5.27 (134)	5.27 (134)	max 0.62 (16)	7.67 (195)	max 0.75 (19)	Figures 13 & 15
2008-10 F350 4X4 Snoplow (w/o plow) DRW	5.55 (141)	5.55 (141)	max 0.62 (16)	7.67 (195)	max 0.75 (19)	Figures 13 & 15
2011-16 F350 4X4 Std. Susp. DRW	4.92 (125)	4.92 (125)	max 0.62 (16)	7.51 (191)	max 0.75 (19)	Figures 13 & 15
2011-16 F350 4X4 HD & Camper Susp. DRW	5.27 (134)	5.27 (134)	max 0.62 (16)	7.51 (191)	max 0.75 (19)	Figures 13 & 15
2011-16 F350 4X4 Snoplow (w/o plow) DRW	5.55 (141)	5.55 (141)	max 0.62 (16)	7.51 (191)	max 0.75 (19)	Figures 13 & 15
2017 F350 4X4 Std. Susp. SRW Pickup	4.09 (104)	4.09 (104)	max 0.62 (16)	7.52 (191)	max 0.75 (19)	Figures 13 & 15
2017 F350 4X4 HD. Susp. SRW Pickup	4.52 (115)	4.52 (115)	max 0.62 (16)	7.52 (191)	max 0.75 (19)	Figures 13 & 15
2017 F350 4X4 Camp. Susp. SRW Pickup	4.52 (115)	4.52 (115)	max 0.62 (16)	7.52 (191)	max 0.75 (19)	Figures 13 & 15
2017 F350 4X4 Snow. Susp. SRW Pickup	4.92 (125)	4.92 (125)	max 0.62 (16)	7.52 (191)	max 0.75 (19)	Figures 13 & 15
2017 F350 4X4 Std. Susp. DRW Pickup	4.25 (108)	4.25 (108)	max 0.62 (16)	6.33 (161)	max 0.75 (19)	Figures 13 & 15
2017 F350 4X4 HD. Susp. DRW Pickup	4.64 (118)	4.64 (118)	max 0.62 (16)	6.33 (161)	max 0.75 (19)	Figures 13 & 15

* All **Side/Side** measurements are taken at wheel arch to ground (WF & WR) See Figure 1 above ** ±0.75 (19) unless indicated otherwise

CHASSIS / SUSP. HEIGHTS

Ford-Lincoln-Mercury Truck & SUV (cont.)

Year Model	Height Specifications @ Curb Weight In. (mm) **					
	Front			Rear		
F-Super Duty Trucks (cont.)	SF-Left	SF-Right	Side/Side	SR	Side/Side *	Notes
2017 F350 4X4 Camp. Susp. DRW Pickup	4.64 (118)	4.64 (118)	max 0.62 (16)	6.33 (161)	max 0.75 (19)	Figures 13 & 15
2017 F350 4X4 Snow. Susp. DRW Pickup	4.88 (124)	4.88 (124)	max 0.62 (16)	6.33 (161)	max 0.75 (19)	Figures 13 & 15
2008-16 F350 4X4 Chassis Cab STD SRW	5.15 (131)	5.15 (131)	max 0.62 (16)	6.89 (175)	max 0.75 (19)	Figures 13 & 15
2008-16 F350 4X4 Chassis Cab HD SRW	5.55 (141)	5.55 (141)	max 0.62 (16)	6.89 (175)	max 0.75 (19)	Figures 13 & 15
2008-16 F350 4X4 Chassis Cab XHD SRW	5.70 (145)	5.70 (145)	max 0.62 (16)	6.89 (175)	max 0.75 (19)	Figures 13 & 15
2008-16 F350 4X4 Chassis Cab S-Plow SRW	5.55 (141)	5.55 (141)	max 0.62 (16)	6.89 (175)	max 0.75 (19)	Figures 13 & 15
2008-11 F350 4X4 Chassis Cab STD DRW	5.27 (134)	5.27 (134)	max 0.62 (16)	7.79 (198)	max 0.75 (19)	Figures 13 & 15
2008-11 F350 4X4 Chassis Cab HD DRW	5.55 (141)	5.55 (141)	max 0.62 (16)	7.79 (198)	max 0.75 (19)	Figures 13 & 15
2008-11 F350 4X4 Chassis Cab XHD DRW	5.62 (143)	5.62 (143)	max 0.62 (16)	7.79 (198)	max 0.75 (19)	Figures 13 & 15
2008-11 F350 4X4 Chas. Cab S-Plow DRW	5.59 (142)	5.59 (142)	max 0.62 (16)	7.79 (198)	max 0.75 (19)	Figures 13 & 15
2012-16 F350 4X4 Chassis Cab STD DRW	5.19 (132)	5.19 (132)	max 0.62 (16)	7.75 (197)	max 0.75 (19)	Figures 13 & 15
2012-16 F350 4X4 Chassis Cab HD DRW	5.51 (140)	5.51 (140)	max 0.62 (16)	7.75 (197)	max 0.75 (19)	Figures 13 & 15
2012-16 F350 4X4 Chassis Cab XHD DRW	5.66 (144)	5.66 (144)	max 0.62 (16)	7.75 (197)	max 0.75 (19)	Figures 13 & 15
2012-16 F350 4X4 Chas. Cab S-Plow DRW	5.66 (144)	5.66 (144)	max 0.62 (16)	7.75 (197)	max 0.75 (19)	Figures 13 & 15
2017 F350 4X4 Chassis Std. Susp. SRW	4.33 (110)	4.33 (110)	max 0.62 (16)	6.92 (176)	max 0.75 (19)	Figures 13 & 15
2017 F350 4X4 Chassis HD Susp. SRW	4.72 (120)	4.72 (120)	max 0.62 (16)	–	max 0.75 (19)	Figures 13 & 15
2017 F350 4X4 Chassis XHD Susp. SRW	4.92 (125)	4.92 (125)	max 0.62 (16)	–	max 0.75 (19)	Figures 13 & 15
2017 F350 4X4 Chassis Snow. SRW	4.92 (125)	4.92 (125)	max 0.62 (16)	4.92 (125)	max 0.75 (19)	Figures 13 & 15
2009-10 F450 Pickup STD Susp.	4.88 (124)	4.88 (124)	max 0.62 (16)	9.84 (250)	max 0.75 (19)	Figures 13 & 14
2011 F450 Pickup STD Susp.	5.15 (131)	5.15 (131)	max 0.62 (16)	9.84 (250)	max 0.75 (19)	Figures 13 & 14
2009-11 F450 Pickup HD & Camper Susp.	5.35 (136)	5.35 (136)	max 0.62 (16)	9.84 (250)	max 0.75 (19)	Figures 13 & 14
2009-11 F450 Pickup S-Plow (w/o plow)	5.23 (133)	5.23 (133)	max 0.62 (16)	9.84 (250)	max 0.75 (19)	Figures 13 & 14
2012-16 F450 Pickup STD Susp.	4.92 (125)	4.92 (125)	max 0.62 (16)	7.51 (191)	max 0.75 (19)	Figures 13 & 14
2012-16 F450 Pickup HD & Camper Susp.	5.27 (134)	5.27 (134)	max 0.62 (16)	7.51 (191)	max 0.75 (19)	Figures 13 & 14
2012-16 F450 Pickup S-Plow (w/o plow)	5.55 (141)	5.55 (141)	max 0.62 (16)	7.51 (191)	max 0.75 (19)	Figures 13 & 14
2017 F450 Pickup STD Susp.	4.33 (110)	4.33 (110)	max 0.62 (16)	5.86 (149)	max 0.75 (19)	Figures 13 & 15
2017 F450 Pickup HD & Camper Susp.	4.44 (113)	4.44 (113)	max 0.62 (16)	–	max 0.75 (19)	Figures 13 & 15
2017 F450 Pickup S-Plow (w/o plow)	4.33 (110)	4.33 (110)	max 0.62 (16)	–	max 0.75 (19)	Figures 13 & 15

Mark LT	SF	Side/Side *	SR	Side/Side *	Notes
2006 Mark LT 2WD (All)	4.21 (107)	max 0.27 (7)	6.06 (154)	max 0.31 (8)	Figures 2 & 11
2006 Mark LT 4WD (All)	4.33 (110)	max 0.27 (7)	6.06 (154)	max 0.31 (8)	Figures 2 & 12
2007-08 Mark LT 2WD 139" WB	4.13 (105)	max 0.27 (7)	5.86 (149)	max 0.31 (8)	Figures 2 & 11
2007-08 Mark LT 2WD 151" WB	3.93 (100)	max 0.27 (7)	5.86 (149)	max 0.31 (8)	Figures 2 & 12
2007-08 Mark LT 4WD 139" WB	4.48 (114)	max 0.27 (7)	5.98 (152)	max 0.31 (8)	Figures 2 & 11
2007-08 Mark LT 4WD 151" WB	4.44 (113)	max 0.27 (7)	5.94 (151)	max 0.31 (8)	Figures 2 & 12
Ranger	**SF**	**Side/Side ***	**SR**	**Side/Side ***	**Notes**
1998 Ranger 4X2	3.81 (97)	max 0.62 (16)	7.00-8.93 (178-227)	max 0.78 (20)	Figures 2 & 4
1998 Ranger 4X2 Splash	3.14 (80)	max 0.62 (16)	7.00-7.51 (178-191)	max 0.78 (20)	Figures 2 & 4
1999-05 Ranger 4X2 (coil springs)	4.01 (102)	max 0.62 (16)	7.00-8.93 (178-227)	max 0.78 (20)	Figures 2 & 4
1998-05 Ranger 4X4 (OE parts)	3.97 (101)	max 0.62 (16)	6.14-7.95 (156-202)	max 0.78 (20)	Figures 2 & 5
1998-05 Ranger 4X4 (New parts)	4.44 (113)	max 0.62 (16)	6 5/32-7.95 (156-202)	max 0.78 (20)	Figures 2 & 5
2006-07 Ranger 4X2 w/coil springs					
Base susp. (OE parts)	4.01 (102)	max 0.51 (13)	7.00-7.87 (178-200)	max 0.51 (13)	Figures 2 & 4
Base susp. (New parts)	4.01 (102)	max 0.51 (13)	7.40-8.26 (188-210)	max 0.51 (13)	Figures 2 & 4
Payload package (OE parts)	4.01 (102)	max 0.51 (13)	7.79-8.74 (198-222)	max 0.51 (13)	Figures 2 & 4
Payload package (New parts)	4.01 (102)	max 0.51 (13)	8.18-9.13 (208-232)	max 0.51 (13)	Figures 2 & 4
2006-07 Ranger 4X4 exc/FX4 (incl. 4X2 w/torsion bars)					
Reg. Cab. exc/FX4 (OE parts)	3.66 (93)	max 0.51 (13)	6.88-7.75 (175-197)	max 0.51 (13)	Figures 2 & 5
Reg. Cab. exc/FX4 (New parts)	4.13 (105)	max 0.51 (13)	7.28-8.14 (185-207)	max 0.51 (13)	Figures 2 & 5
Super Cab. exc/FX4 (OE parts)	3.66 (93)	max 0.51 (13)	6.14-6.88 (156-175)	max 0.51 (13)	Figures 2 & 5
Super Cab. exc/FX4 (New parts)	4.13 (105)	max 0.51 (13)	6.53-7.28 (166-185)	max 0.51 (13)	Figures 2 & 5
Payload pkg. exc/FX4 (OE parts)	3.66 (93)	max 0.51 (13)	6.88-7.75 (175-197)	max 0.51 (13)	Figures 2 & 5
Payload pkg. exc/FX4 (New parts)	4.13 (105)	max 0.51 (13)	7.28-8.14 (185-207)	max 0.51 (13)	Figures 2 & 5
2006-07 Ranger 4X4 FX4 models					
All exc/Payload pkg. (OE parts)	3.97 (101)	max 0.51 (13)	6.14-6.88 (156-175)	max 0.51 (13)	Figures 2 & 5
All exc/Payload pkg. (New parts)	4.44 (113)	max 0.51 (13)	6.53-7.28 (166-185)	max 0.51 (13)	Figures 2 & 5
All w/Payload pkg. (OE parts)	3.97 (101)	max 0.51 (13)	6.88-7.75 (175-197)	max 0.51 (13)	Figures 2 & 5
All w/Payload pkg. (New parts)	4.44 (113)	max 0.51 (13)	7.28-8.14 (185-207)	max 0.51 (13)	Figures 2 & 5
2008-11 Ranger 4X2 w/coil springs					
Base susp.	4.01 (102)	max 0.51 (13)	6.77-7.71 (172-196)	max 0.51 (13)	Figures 2 & 4
Payload package	4.01 (102)	max 0.51 (13)	7.48-8.42 (190-214)	max 0.51 (13)	Figures 2 & 4
2008-11 Ranger 4X4, FX4, 4X2 (w/torsion bars)					
Reg. Cab.	3.03 (77)	max 0.51 (13)	6.96-7.91 (177-201)	max 0.51 (13)	Figures 2 & 5
Super Cab.	3.03 (77)	max 0.51 (13)	6.96-7.91 (177-201)	max 0.51 (13)	Figures 2 & 5
Payload pkg.	3.03 (77)	max 0.51 (13)	7.04-7.99 (179-203)	max 0.51 (13)	Figures 2 & 5
Transit Connect	**SF**	**Side/Side ***	**SR**	**Side/Side ***	**Notes**
2014-17 Cargo Van SWB	0.28 (7) ± 0.47(12)		2.33 (59) ± 0.47(12)		Figures 16 & --
2014-17 Cargo Van LWB	0.15 (4) ± 0.47(12)		2.67 (68) ± 0.47(12)		Figures 16 & --
2014-17 Wagon SWB	0.36 (9) ± 0.47(12)		2.58 (66) ± 0.47(12)		Figures 16 & --
2014-17 Wagon LWB	0.22 (6) ± 0.47(12)		2.86 (73) ± 0.47(12)		Figures 16 & --

* All **Side/Side** measurements are taken at wheel arch to ground (WF & WR) See **Figure** 1 above ** ±0.75 (19) unless indicated otherwise

CHASSIS / SUSP. HEIGHTS

General Motors

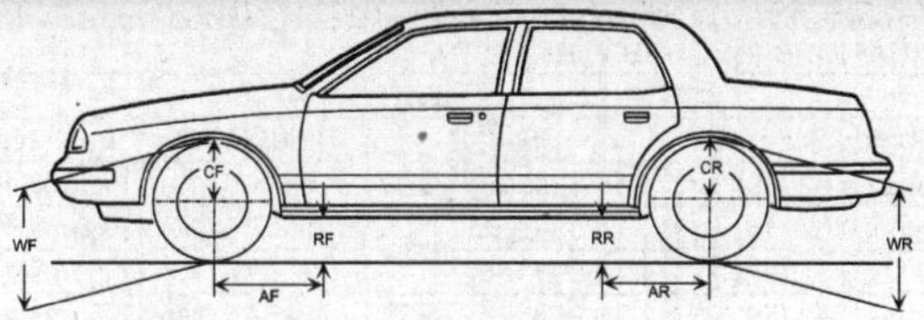

Figure A

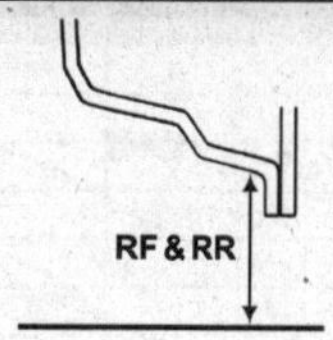

Figure B

Make all GM measurements on a level surface to the indicated point. Rocker panel measurements (RF & RR) do not include weld flange (see **Figure B** above), and should be taken approximately at the distances indicated (AF & AR) from the wheel centerlines. Wheel arch measurements (WF & WR and CF & CR) are made from ground or wheel center, respectively, to the lowest point in the wheel opening at the wheel centerline. Suspension height measurements (SF & SR) are taken as indicated in the numbered illustrations that follow. All measurements are intended for use in a pre-alignment inspection. The fuel tank should be full unless otherwise indicated, with no additional weight in the vehicle. Jounce vehicle several times. Push down-release and measure, pull up-release and measure. Repeat procedure for both front and rear measurements. Actual height is the average of both measurements.

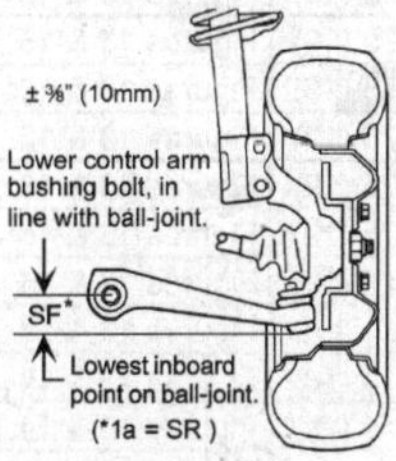

Figure 1, 1a

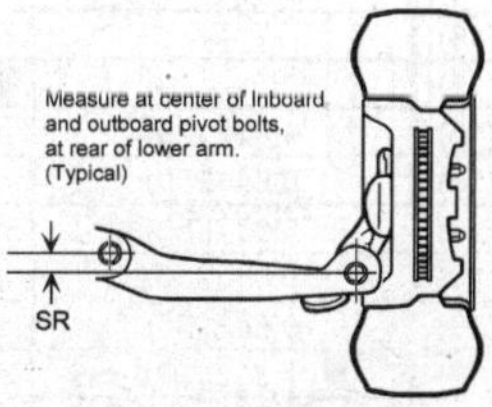

Figure 2, 2a

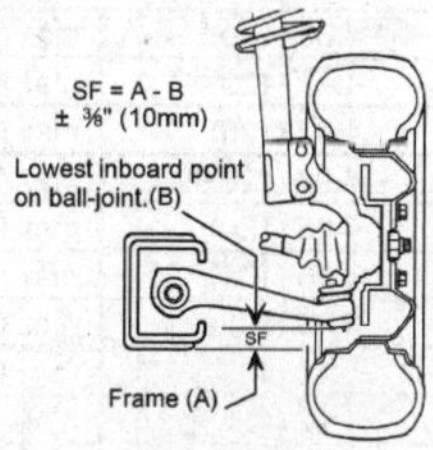

Figure 3

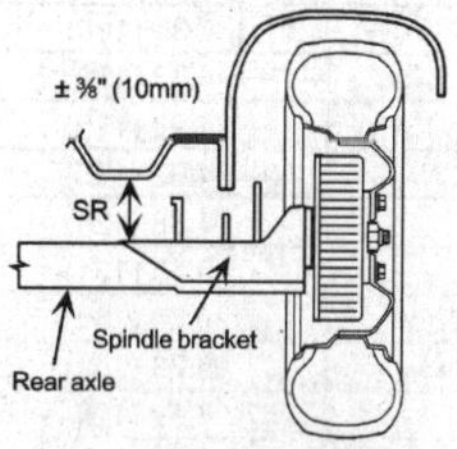

Figure 4

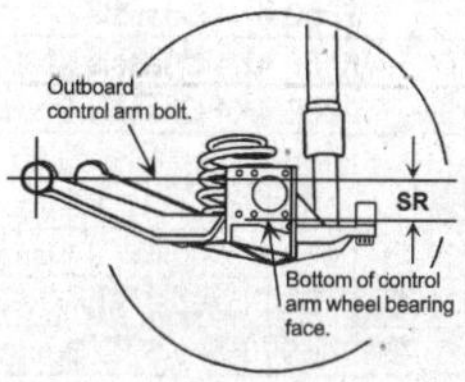

Figure 5

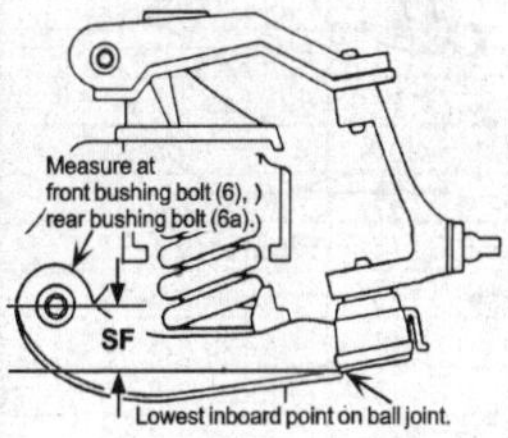

Figure 6, 6a

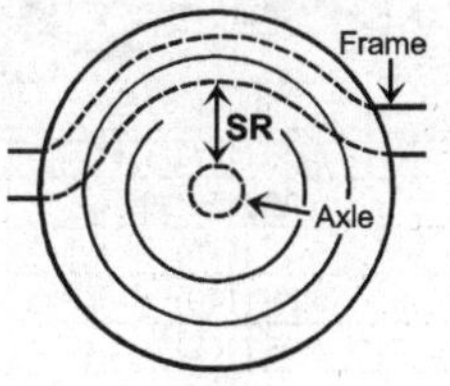

Figure 7

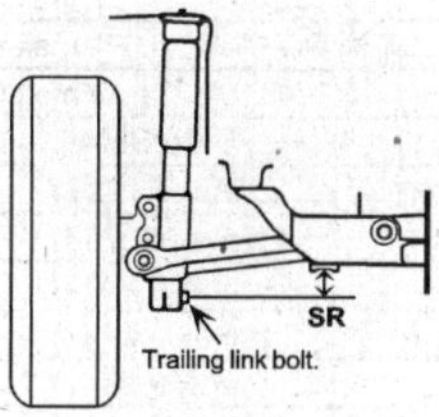

Figure 8

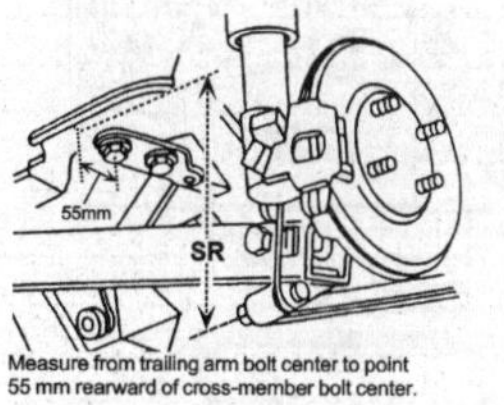

Figure 9

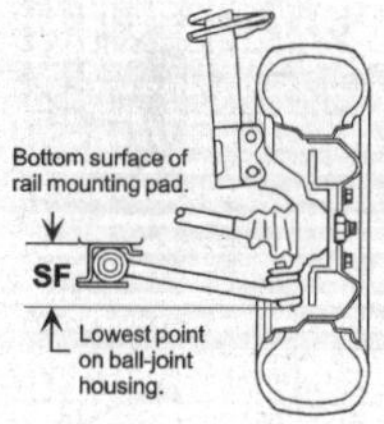

Figure 10

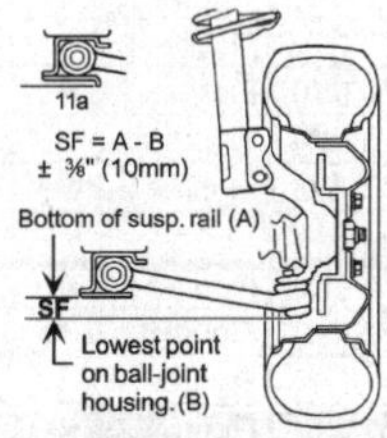

Figure 11, 11a

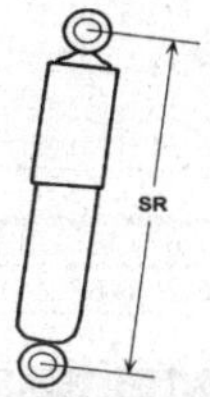

Figure 12

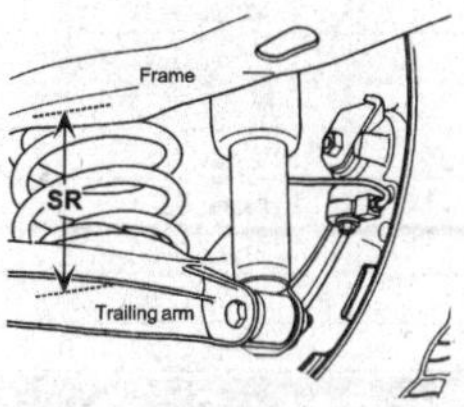

Figure 13

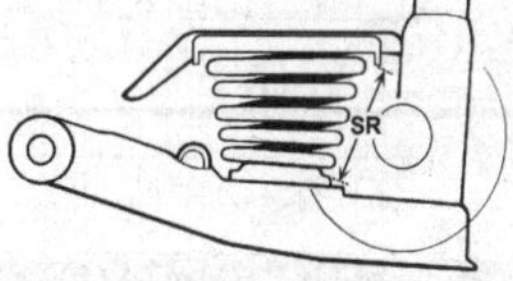

Figure 14

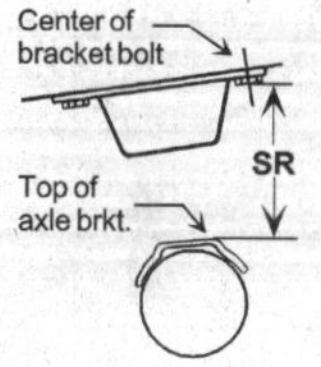

Figure 15

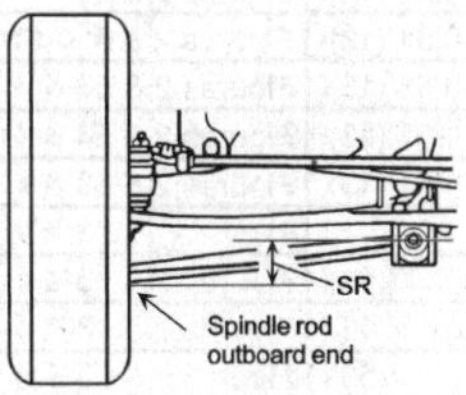

Figure 16

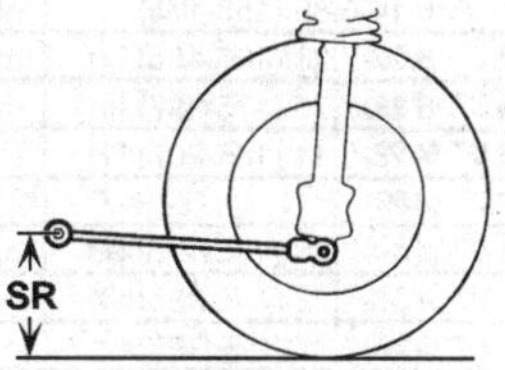

Figure 17

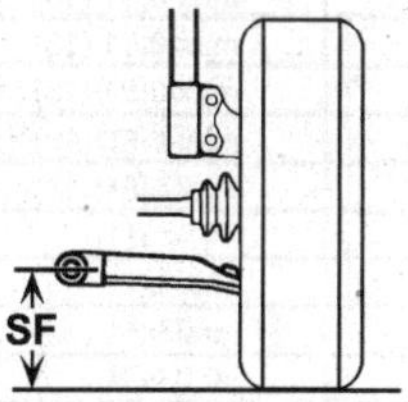

Figure 18

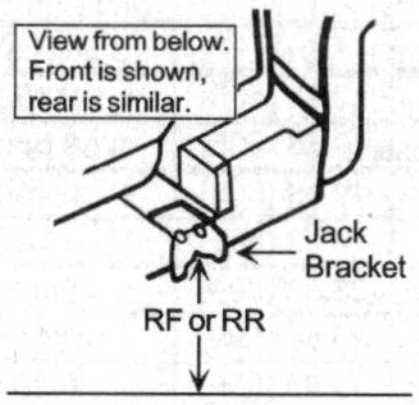

Figure 19

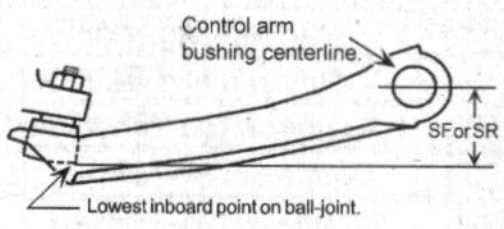

Figure 20

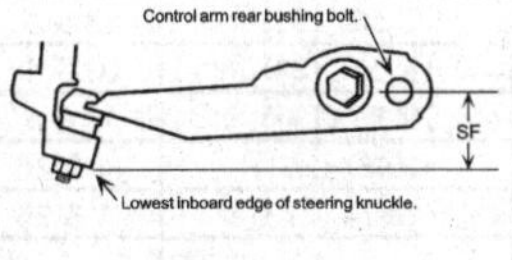

Figure 21

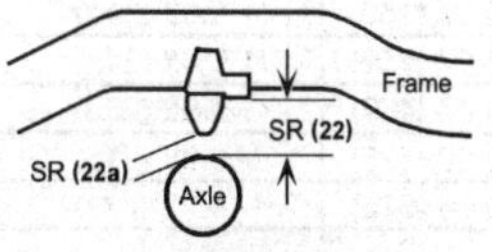

Figure 22, 22a

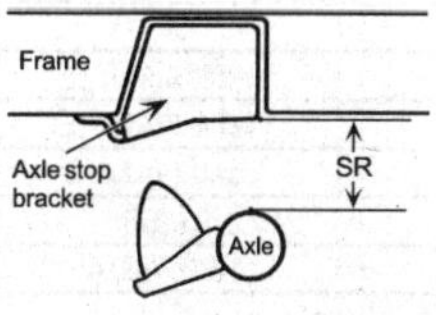

Figure 23

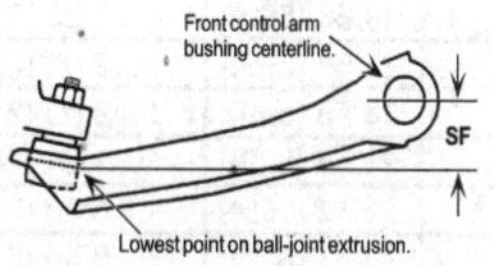

Figure 24

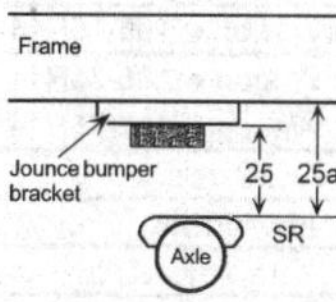

Figure 25

CHASSIS / SUSP. HEIGHTS

General Motors (cont.)

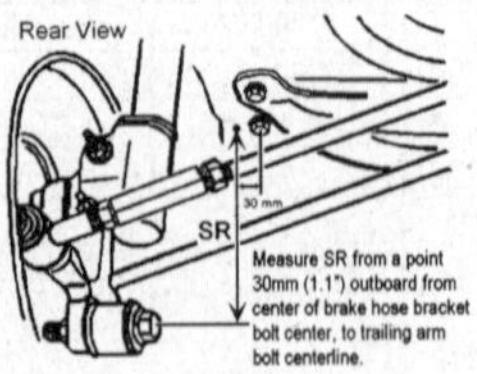

Figure 26

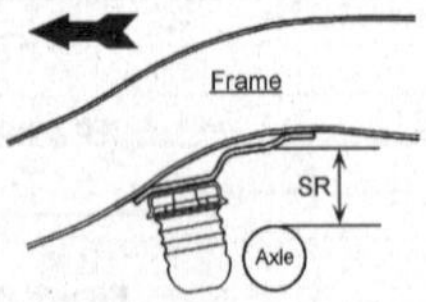

Figure 27

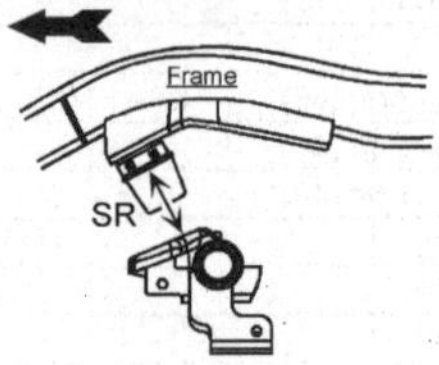

Figure 28

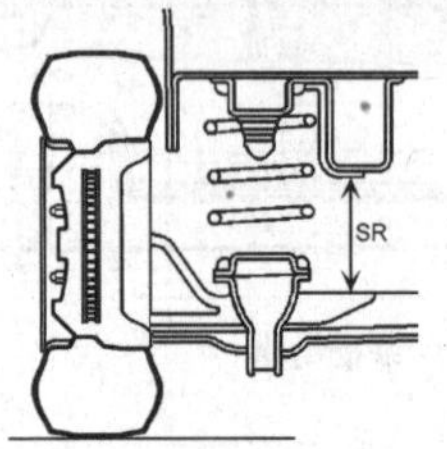

Figure 29

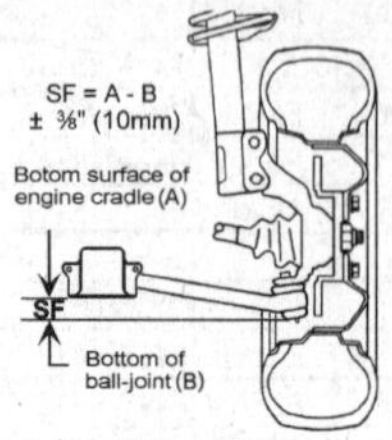

Figure 30

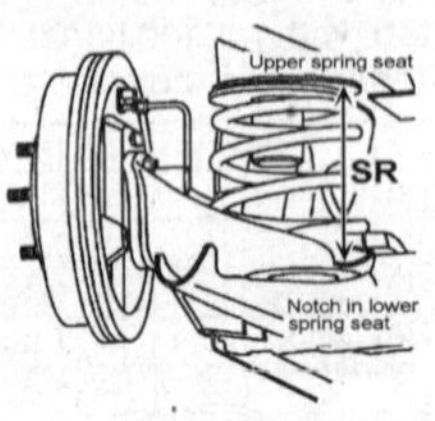

Figure 31

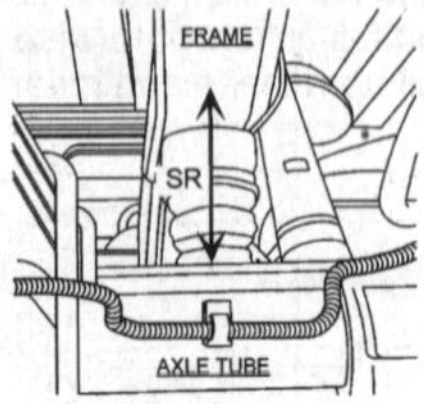

Figure 32

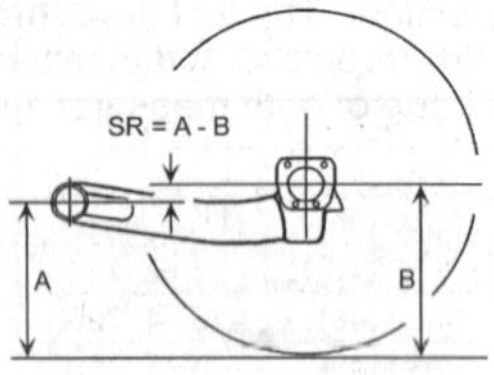

Figure 33

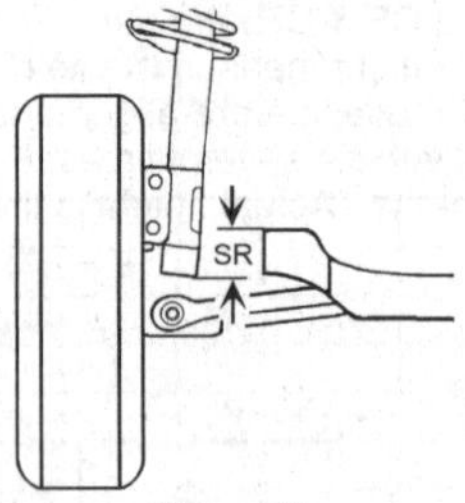

Figure 34

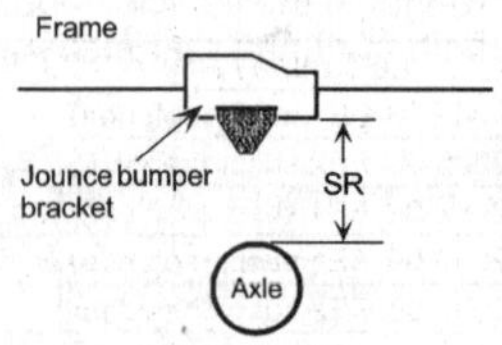

Figure 35

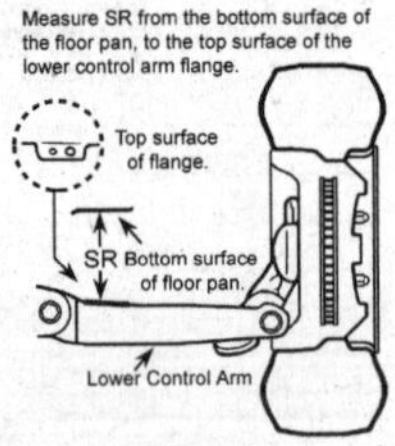

Figure 36

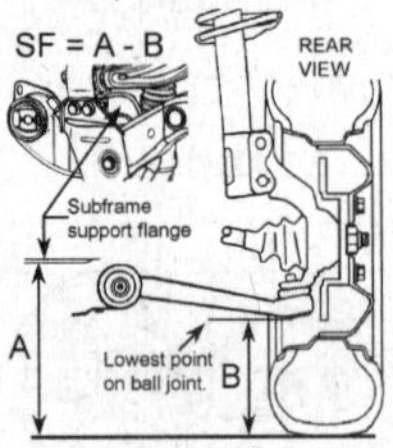

Figure 37

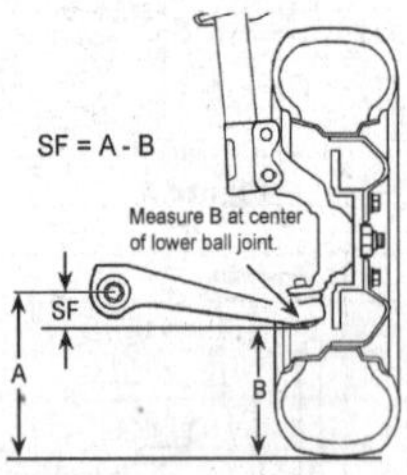

Figure 38

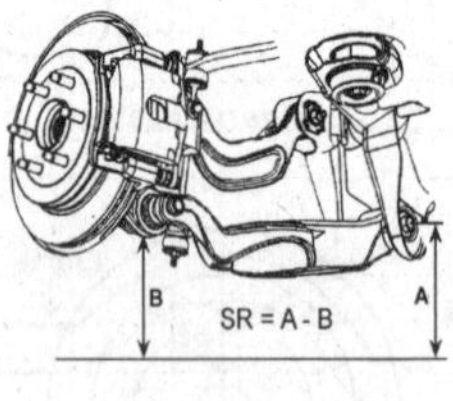

Figure 39

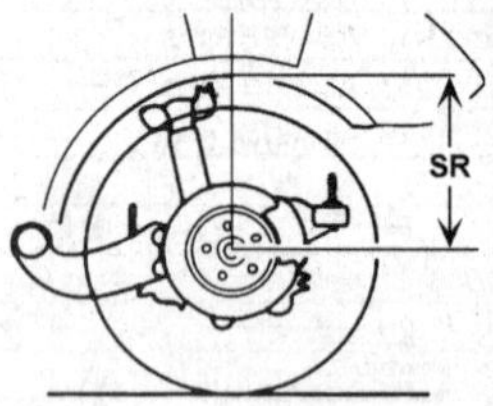

Figure 40

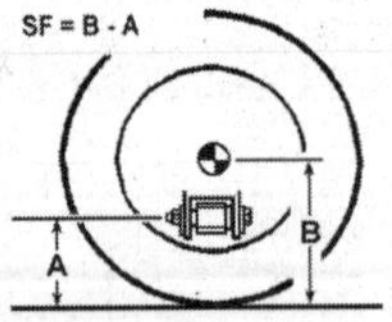

Figure 41

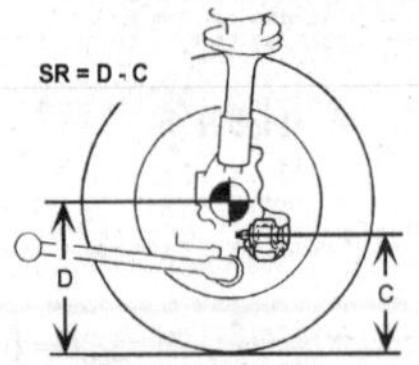

Figure 42

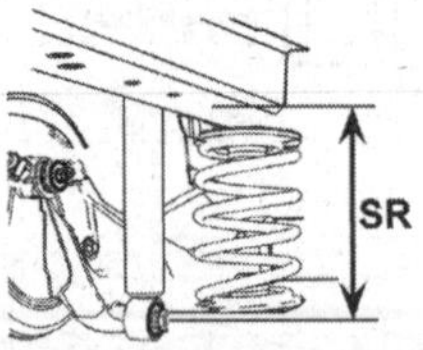

Figure 43

GM - Buick

Year Model	Height Specifications @ Curb Weight - Inches (mm) *						
	Front			Rear			
Century	AF	RF	SF	AR	RR	SR	Figure
1990-91 Coupe w/185-75R14	19.88 (505)	9.29 (236)	-0.03 (-1)	18.07 (459)	9.52 (242)	7.51 (191)	3 & 4
1990-91 Coupe w/195-75R14	19.88 (505)	9.56 (243)	-0.03 (-1)	18.07 (459)	9.68 (246)	7.51 (191)	3 & 4
1992 Coupe w/185-75R14	19.88 (505)	9.21 (234)	-0.03 (-1)	18.07 (459)	9.48 (241)	7.51 (191)	3 & 4
1992 Coupe w/195-75R14	19.88 (505)	9.44 (240)	-0.03 (-1)	18.07 (459)	9.72 (247)	7.51 (191)	3 & 4
1993 Coupe w/185-75R14	23.58 (599)	9.44 (240)	-0.03 (-1)	18.07 (459)	9.72 (247)	7.51 (191)	3 & 4
1993 Coupe w/195-75R14	23.58 (599)	9.64 (245)	-0.03 (-1)	18.07 (459)	9.96 (253)	7.51 (191)	3 & 4
1990-91 Sedan w/185-75R14	19.88 (505)	9.29 (236)	-0.03 (-1)	18.07 (459)	9.52 (242)	7.51 (191)	3 & 4
1990-91 Sedan w/195-75R14	19.88 (505)	9.56 (243)	-0.03 (-1)	18.07 (459)	9.80 (249)	7.51 (191)	3 & 4
1992 Sedan (All)	19.88 (505)	9.44 (240)	-0.03 (-1)	18.07 (459)	9.72 (247)	7.51 (191)	3 & 4
1993-96 Sedan w/185-75R14	23.58 (599)	9.40 (239)	-0.03 (-1)	18.07 (459)	9.72 (247)	7.51 (191)	3 & 4
1993-96 Sedan w/195-75R14	23.58 (599)	9.64 (245)	-0.03 (-1)	18.07 (459)	9.96 (253)	7.51 (191)	3 & 4
1990-92 Wagon w/185-75R14	19.88 (505)	9.25 (235)	-0.03 (-1)	18.07 (459)	9.44 (240)	7.51 (191)	3 & 4
1990-92 Wagon w/195-75R14	23.58 (599)	9.52 (242)	-0.03 (-1)	18.07 (459)	9.68 (246)	7.51 (191)	3 & 4
1993-96 Wagon w/185-75R14	23.58 (599)	9.40 (239)	-0.03 (-1)	18.07 (459)	9.68 (246)	7.51 (191)	3 & 4
1993-96 Wagon w/195-75R14	23.58 (599)	9.64 (245)	-0.03 (-1)	18.07 (459)	9.92 (252)	7.51 (191)	3 & 4
1997-00 All Models	23.81 (605)	8.38 (213)	2.16 (55)	20.66 (525)	8.38 (213)	10.27 (261)	1 & 26
2001-05 All Models	23.81 (605)	8.58 (218)	1.77 (45)	20.66 (525)	8.77 (223)	10.74 (273)	1 & 26

* ± ½ (13) unless otherwise indicated

GM – Buick (cont.)

Height Specifications @ Curb Weight - Inches (mm) *

Year Model	Front				Rear				
Electra, Park Avenue, Ultra	AF	RF	SF	WF	AR	RR	SR	WR	Figure
1990 Electra/Park Avenue	23.62 (600)	9.37 (238)	2.59 (66)	27.91 (709)	23.62 (600)	9.44 (240)	3.14 (80)	27.59 (701)	1, 1a
1991-96 Park Avenue/Ultra	23.62 (600)	9.37 (238)	2.36 (60)	—	23.62 (600)	9.44 (240)	3.07 (78)	—	1, 1a
1997-05 All Models	—	—	1.57 (40)	—	—	—	3.38 (86)	—	1 & 5

Encore	WF	WR	Figure
2013-17 Encore FWD w/16" tires (OE)	29.30 (744)	29.90 (773)	A
2013-17 Encore FWD w/18" tires (OE)	29.40 (747)	30.00 (775)	A
2013-17 Encore AWD w/16" tires (OE)	29.60 (753)	30.30 (782)	A
2013-17 Encore AWD w/18" tires (OE)	29.80 (756)	30.40 (785)	A

Estate Wagon	AF	RF	SF	WF	AR	RR	SR	WR	Figure
1990 LeSabre & Electra	31.49 (800)	10.43 (265)	—	28.77 (731)	24.60 (625)	10.55 (268)	—	28.89 (734)	

LaCrosse / Allure ->2009	AF	RF	SF	WF	AR	RR	SR	WR	Figure
2005-06 FE1 susp. (soft ride)	23.81 (605)	9.37 (238)	1.88 (48)	—	20.66 (525)	9.13 (232)	10.74 (273)	—	A, 1 & 26
2005-06 FE2 susp. (handling)	23.81 (605)	9.33 (237)	1.88 (48)	—	20.66 (525)	9.25 (235)	10.74 (273)	—	A, 1 & 26
2007-09 FE1 susp. (soft ride) w/16"	—	—	1.88 (48)	28.42 (722)	—	—	10.74 (273)	28.70 (729)	A, 1 & 26
2007-09 FE1 susp. (soft ride) w/17"	—	—	1.88 (48)	28.54 (725)	—	—	10.74 (273)	28.81 (732)	A, 1 & 26
2007-08 FE2 susp. (handling)	—	—	1.88 (48)	28.58 (726)	—	—	10.74 (273)	28.77 (731)	A, 1 & 26

LaCrosse 2010 ->	WF	WR	Figure
2010-13 All w/17" tires (OE)	28.90 (734)	29.30 (745)	A
2010-13 All w/18" tires (OE)	29.10 (738)	29.50 (749)	A
2010-13 All w/19" tires (OE)	28.80 (732)	29.20 (743)	A
2010-13 All AWD models	29.20 (742)	29.60 (753)	A

LeSabre	AF	RF	SF	AR	RR	SR	Figure
1990-91 LeSabre	23.62 (600)	9.40 (239)	2.36 (60)	23.62 (600)	9.68 (246)	3.42 (87)	1, 1a
1992-99 LeSabre	23.62 (600)	9.40 (239)	2.36 (60)	23.62 (600)	9.68 (246)	3.07 (78)	1, 1a
2000-05 LeSabre	—	—	1.57 (40)	—	—	3.38 (86)	1 & 5

Lucerne	SF	WF	SR	WR	Figure
2006 Lucerne CX	1.96 (50)	—	3.81 (97)	—	30 & 5
2006 Lucerne CXL, CXS	1.57 (40)	—	3.38 (86)	—	30 & 5
2007-10 Lucerne CX	1.96 (50)	29.37 (746)	3.81 (97)	29.60 (752)	A, 30 & 5
2007-10 Lucerne CXL	1.57 (40)	29.33 (745)	3.38 (86)	29.52 (750)	A, 30 & 5
2007-10 Lucerne CXS	1.57 (40)	29.56 (751)	3.38 (86)	29.76 (756)	A, 30 & 5
2011 Lucerne CX, CXL	1.61 (41)	29.25 (743)	3.38 (86)	29.33 (745)	A, 30 & 5
2011 Lucerne CXS	1.61 (41)	29.40 (747)	3.38 (86)	29.56 (751)	A, 30 & 5

Regal ->2004	AF	RF	SF	AR	RR	SR	Figure
1990-91 All with 14" tires	23.81 (605)	9.80 (249)	3.38 (86)	20.66 (525)	9.80 (249)	4.17 (106)	1 & 8
1990-91 All with 15" tires	23.81 (605)	9.80 (249)	2.87 (73)	20.66 (525)	9.80 (249)	3.70 (94)	1 & 8
1990 All with 16" tires	23.81 (605)	9.80 (249)	2.83 (72)	20.66 (525)	9.80 (249)	3.66 (93)	1 & 8
1991 All with 16" tires	23.81 (605)	9.80 (249)	2.59 (66)	20.66 (525)	9.80 (249)	3.42 (87)	1 & 8
1992-93 All with 14" tires	23.81 (605)	9.80 (249)	3.38 (86)	20.66 (525)	9.80 (249)	4.17 (106)	1 & 8
1992-93 All with 15" tires	23.81 (605)	9.80 (249)	2.87 (73)	20.66 (525)	9.80 (249)	3.66 (93)	1 & 8
1992-93 All with 16" tires	23.81 (605)	9.80 (249)	2.59 (66)	20.66 (525)	9.80 (249)	3.42 (87)	1 & 8
1994-96 All with 15" tires	23.81 (605)	9.80 (249)	2.87 (73)	20.66 (525)	9.80 (249)	10.27 (261)	1 & 9
1994-96 All with 16" tires	23.81 (605)	9.80 (249)	2.59 (66)	20.66 (525)	9.80 (249)	10.03 (255)	1 & 9
1997 All w/Base susp. (FE1)	23.81 (605)	8.58 (218)	2.24 (57)	20.66 (525)	8.77 (223)	10.70 (272)	1 & 26
1997 All wSport susp. (FE2, FE3)	23.81 (605)	8.77 (223)	2.24 (57)	20.66 (525)	8.81 (224)	10.47 (266)	1 & 26
1998-00 All Models w/15" tires	23.81 (605)	8.38 (213)	2.26 (55)	20.66 (525)	8.38 (213)	10.27 (261)	1 & 26
1998-00 All Models w/16" tires	23.81 (605)	8.38 (213)	1.88 (48)	20.66 (525)	8.38 (213)	10.03 (255)	1 & 26
2001-04 All Models	23.81 (605)	8.77 (223)	1.77 (45)	20.66 (525)	8.85 (225)	10.51 (267)	1 & 26

Regal 2011 ->	WF	WR	Figure
2011-16 Std. Mods. w/17" tires (OE)	28.60 (726) ± 0.39 (10)	28.80 (731) ± 0.39 (10)	A
2011-16 Std. Mods. w/18" tires (OE)	28.90 (734) ± 0.39 (10)	29.10 (739) ± 0.39 (10)	A
2011-16 Std. Mods. w/19" tires (OE)	28.80 (731) ± 0.39 (10)	28.90 (735) ± 0.39 (10)	A
2011-16 Hybrid Models	28.50 (723) ± 0.39 (10)	28.70 (728) ± 0.39 (10)	A
2011-16 Hiperstrut w/19" tires (OE)	28.60 (726) ± 0.39 (10)	28.70 (728) ± 0.39 (10)	A
2011-16 Hiperstrut w/20" tires (OE)	28.80 (732) ± 0.39 (10)	28.90 (734) ± 0.39 (10)	A

Reatta	AF	RF	SF	WF	AR	RR	SR	WR	Figure
1990 Reatta	—	—	2.16 (55)	—	—	—	1.45 (37)	—	1 & 2
1991 Reatta	25.19 (640)	8.46 (215)	2.08 (53)	28.85 (733)	19.76 (502)	8.62 (219)	1.37 (35)	29.33 (745)	1 & 2

* ± ½ (13) unless otherwise indicated

GM – Buick (cont.)

HEIGHT SPECIFICATIONS @ CURB WEIGHT - INCHES (MM) *

YEAR MODEL Rendezvous	FRONT AF	RF	SF	REAR AR	RR	SR	Figure
2002-07 Rendezvous FWD exc/17"	—	—	2.55 (65)	—	—	6.02 (153)	1 & 29
2002-07 Rendezvous FWD w/17"	—	—	2.59 (66)	—	—	6.10 (155)	1 & 29
2002-07 Rendezvous AWD exc/17"	—	—	2.55 (65)	—	—	3.81 (97)	1 & 20
2002-07 Rendezvous AWD w/17"	—	—	2.59 (66)	—	—	6.10 (155)	1 & 20

Riviera	AF	RF	SF	WF	AR	RR	SR	WR	Figure
1990 All Models	—	—	1.81 (46)	—	—	—	1.18 (30)	—	1 & 2
1991-93 All Models	22.83 (580)	8.62 (219)	1.81 (46)	27.40 (696)	22.16 (563)	8.74 (222)	1.10 (28)	27.75 (705)	1 & 2
1995-99 All models	—	—	1.57 (40)	—	—	—	3.38 (86)	—	1 & 5

Roadmaster	AF	RF	SF	WF	AR	RR	SR	WR	Figure
1992-93 Sedan FE1 Susp.	31.88 (810)	9.92 (252)	2.20 (56)	29.01 (737)	24.21 (615)	10.39 (264)	5.78 (147)	21.49 (546)	6 & 7
1992-93 Sedan FE2, F41 Suspension	31.88 (810)	9.92 (252)	2.40 (61)	29.01 (737)	24.21 (615)	10.39 (264)	5.98 (152)	21.49 (546)	6 & 7
1994-96 Sedan exc/FE3 Suspension	31.88 (810)	9.92 (252)	2.3-2.7(60-71)	29.01 (737)	24.21 (615)	10.39 (264)	5.55-5.98	28.74-28.77	6 & 7
1994-96 Sedan w/FE3 Suspension	31.88 (810)	9.92 (252)	2.36 (60)	29.01 (737)	24.21 (615)	10.39 (264)	5.15 (131)	28.26 (718)	6 & 7
1991-96 All Wagon	31.88 (810)	9.92 (252)	2.08 (53)	29.44 (748)	24.21 (615)	10.39 (264)	4.80 (122)	21.88 (556)	6 & 7

Skylark	AF	RF	SF	AR	RR	SR	Figure
1989-91 All w/Base susp. (FE1)	21.92 (557)	9.40 (239)	3.70-4.01 (94-102)	22.04 (560)	9.21 (234)	—	10
1989-91 All w/Sport susp. (FE3)	21.92 (557)	9.40 (239)	4.01-4.05 (102-103)	22.04 (560)	9.21 (234)	—	10
1992-93 All 4 Cyl. models	31.49 (800)	9.33 (237)	0.35 (9)	22.04 (560)	9.44 (240)	—	11
1992-93 All 6 Cyl. models	31.49 (800)	9.33 (237)	0.23 (6)	22.04 (560)	9.44 (240)	—	11
1994-95 All 4 Cyl. with 14" tires	31.49 (800)	9.33 (237)	0.51 (13)	22.04 (560)	9.44 (240)	—	11
1994-95 All 4 Cyl. with 15" tires	31.49 (800)	9.33 (237)	0.35 (9)	22.04 (560)	9.44 (240)	—	11
1994-95 All 4 Cyl. with 16" tires	31.49 (800)	9.33 (237)	0.23 (6)	22.04 (560)	9.44 (240)	—	11
1994-95 All 6 Cyl. models	31.49 (800)	9.33 (237)	0.31 (8)	22.04 (560)	9.44 (240)	—	11
1996 All with 14" tires	31.49 (800)	9.48 (241)	0.59 (15)	22.04 (560)	9.84 (250)	—	11
1996 All with 15" tires	31.49 (800)	9.48 (241)	0.43 (11)	22.04 (560)	9.84 (250)	—	11
1996 All with 16" tires	31.49 (800)	9.48 (241)	0.31 (8)	22.04 (560)	9.84 (250)	—	11
1997-98 All with 14" tires	31.49 (800)	9.37 (238)	0.55 (14)	22.04 (560)	9.56 (243)	—	11
1997-98 All with 15" tires	31.49 (800)	9.37 (238)	0.35 (9)	22.04 (560)	9.56 (243)	—	11
1997-98 All with 16" tires	31.49 (800)	9.37 (238)	0.27 (7)	22.04 (560)	9.56 (243)	—	11

Terraza	AF	RF	SF	AR	RR	SR	Figure
2005-06 Terraza FWD	23.77 (604)	12.16 (309)	2.55 (65)	22.04 (560)	11.73 (298)	3.66 (93)	1 & 20
2007 Terraza FWD	23.77 (604)	12.20 (310)	2.55 (65)	22.04 (560)	11.85 (301)	6.45 (164)	1 & 29
2005-06 Terraza AWD	23.77 (604)	12.16 (309)	2.55 (65)	22.04 (560)	11.69 (297)	3.66 (93)	1 & 20

Verano	WF	WR	Figure
2012-16 Verano w/17" tires (OE)	27.56 (700)	27.80 (706)	A
2012-16 Verano w/18" tires (OE)	27.87 (708)	28.03 (712)	A

* ±0.5 (13) unless otherwise indicated

GM – Cadillac

HEIGHT SPECIFICATIONS @ CURB WEIGHT - INCHES (MM) *

YEAR MODEL Allante'	FRONT AF	RF	SF	WF	REAR AR	RR	SR	WR	Figure
1990-92 All Models	28.74 (730)	8.50 (216)	1.85 (47)	27.67 (703)	27.55 (700)	8.85 (225)	1.61 (41)	27.71 (704)	1 & 2
1993 All Models	—	—	1.53 (39)	—	—	—	1.77 (45)	—	1 & 2

Catera	AF	RF	SF	WF	AR	RR	SR	WR	Figure
1997-01 All Models	—	6.88 (175)	—	27.28 (693)	—	6.49 (165)	—	27.28 (693)	A

CT6	SF	WF	SR	WR	Figure
2016-17 w/P235/50R18 tires (OE)	—	28.54 (725)	—	28.62 (727)	A
2016-17 RWD w/P245/45R19 tires	—	28.54 (732)	—	28.62 (734)	A
2016-17 AWD w/P245/45R19 tires	—	28.54 (740)	—	28.62 (734)	A
2016-17 w/P245/40R20 tires (OE)	—	28.54 (744)	—	28.62 (737)	A
2016-17 P235/50R18 Fr, P265/45R18 Rr	—	28.54 (724)	—	28.62 (733)	A

CTS/CTS-V (VIN D)	SF	WF	SR	WR	Figure
2003-07 CTS w/FE1, FE3 Susp.	2.08 (53)	—	2.36 (60)	—	1 & 2
2006 CTS w/FE5 Susp.	1..69 (43)	—	1.92 (49)	—	1 & 2
2008-14 CTS RWD w/FE1, FE2 Susp.	3.70 (94)	28.54 (725)	2.20 (56)	29.01 (737)	A, 1 & 2
2008-14 CTS RWD w/FE3 Susp.	3.70 (94)	28.89 (734)	2.20 (56)	29.29 (744)	A, 1 & 2
2008-14 CTS AWD w/FE1, FE2 Susp.	3.14 (80)	28.54 (725)	2.20 (56)	29.01 (737)	A, 1 & 2
2008-14 CTS AWD w/FE3 Susp.	3.14 (80)	28.89 (734)	2.20 (56)	29.29 (744)	A, 1 & 2
2004-07 CTS-V	1.69 (43)	—	1.92 (49)	—	1 & 2
2008-14 CTS-V	3.38 (86)	28.11 (714)	1.65 (42)	28.34 (720)	A, 1 & 2

* ±0.50 (13) unless otherwise indicated

GM – Cadillac (cont.)

Year Model	Height Specifications @ Curb Weight - Inches (mm) *				
	Front		Rear		
CTS/CTS-V (VIN A)	SF	WF	SR	WR	Figure
2014-17 CTS RWD 225/40 & 235/35 18"	—	26.06 (662)	—	26.45 (672)	A
2014-17 CTS AWD 17" w/RPO FE2	—	26.92 (684)	—	27.08 (688)	A
2014-17 CTS AWD 17"RPO FE2+RA3	—	26.33 (669)	—	26.73 (679)	A
2014-17 CTS AWD 18" w/RPO FE3	—	27.08 (688)	—	27.28 (693)	A
2016-17 CTS-V	—	26.69 (678)	—	26.77 (680)	A

DeVille/Concours/F'wood FWD	AF	RF	SF	WF	AR	RR	SR	WR	Figure
1990 All Models	23.62 (600)	9.37 (238)	2.32 (59)	28.18 (716)	23.62 (600)	9.44 (240)	3.34 (85)	26.81 (681)	1, 1a
1991-93 All Models	23.62 (600)	9.37 (238)	2.40 (61)	27.91 (709)	23.62 (600)	9.44 (240)	3.07 (78)	27.59 (701)	1, 1a
1994-99 All exc/Concours	22.83 (580)	7.55 (192)	2.00 (51)	27.83 (707)	22.16 (563)	7.95 (202)	1.45 (37)	21.37 (543)	1, 1a
1994-99 Concours	22.83 (580)	7.16 (182)	1.65 (42)	27.44 (697)	22.16 (563)	7.55 (192)	1.10 (28)	18.18 (462)	1, 1a
2000 All w/FE7 Susp.	—	—	4.25 (108)	—	—	—	5.39 (137)	—	1 & 5
2000-05 All w/FE1, FE3, F45 Susp.	—	—	1.57 (40)	—	—	—	3.38 (86)	—	1 & 5

DTS	SF	WF	SR	WR	Figure
2006 All FE1 Susp.	1.57 (40)	—	3.38 (86)	—	1 & 5
2007-10 All FE1 Susp.	1.57 (40)	28.77 (731)	3.38 (86)	28.70 (729)	A, 1 & 5
2007-10 All FE3 Susp.	1.57 (40)	28.97 (736)	3.38 (86)	28.89 (734)	A, 1 & 5
2011 All FE1 Susp.	1.61 (41)	28.66 (728)	3.38 (86)	28.62 (727)	A, 1 & 5
2011 All FE3 Susp.	1.61 (41)	28.89 (734)	3.38 (86)	28.85 (733)	A, 1 & 5

ELR	SF	WF	SR	WR	Figure
2014-16 All P245/40R20 (OE)	—	28.93 (735)	—	29.40 (747)	A

Eldorado	AF	RF	SF	WF	AR	RR	SR	WR	Figure
1990 All Models	—	—	2.04 (52)	—	—	—	1.14 (29)	—	1 & 2
1991 All Models	22.83 (580)	9.01 (229)	2.04 (52)	28.26 (718)	22.16 (563)	9.09 (231)	1.14 (29)	27.95 (710)	1 & 2
1992 All Models	22.83 (580)	8.77 (223)	2.04 (52)	28.22 (717)	22.16 (563)	8.70 (221)	1.18 (30)	28.18 (716)	1 & 2
1993 All Models	22.83 (580)	8.77 (223)	2.04 (52)	28.22 (717)	22.16 (563)	8.70 (221)	1.41 (36)	28.18 (716)	1 & 2
1994-02 All Models	22.83 (580)	8.66 (220)	1.96 (50)	28.14 (715)	22.16 (563)	8.42 (214)	1.14 (29)	27.91 (709)	1 & 2

Fleetwood RWD/Brougham	AF	RF	SF	WF	AR	RR	SR	WR	Figure
1990 All w/STD. Susp.	—	—	2.20 (56)	—	—	—	5 9/16 (141)	—	6 & 7
1990 All w/ELC. Susp.	—	—	2.12 (54)	—	—	—	5 3/16 (132)	—	6 & 7
1991-92 All Models	31.88 (810)	10.55 (268)	2.16 (55)	28.97 (736)	24.21 (615)	10.90 (277)	6 5/16 (160)	24.13 (613)	6 & 7
1993-96 All w/FE1	31.88 (810)	9.92 (252)	2.20 (56)	29.01 (737)	24.21 (615)	10.39 (264)	5 3/4 (147)	21.49 (546)	6 & 7
1993-96 All w/FE2, F41	31.88 (810)	9.92 (252)	2.40 (61)	29.01 (737)	24.21 (615)	10.39 (264)	6 (152)	21.49 (546)	6 & 7

Seville/STS	AF	RF	SF	WF	AR	RR	SR	WR	Figure
1990 All Models	—	—	2.00 (51)	—	—	—	1 3/16 (30)	—	1 & 2
1991 All Models	22.83 (580)	9.01 (229)	2.00 (51)	28.22 (717)	22.16 (563)	9.17 (233)	1 3/16 (30)	28.50 (724)	1 & 2
1992 All exc/STS	22.83 (580)	9.64 (245)	2.20 (56)	28.58 (726)	22.16 (563)	9.64 (245)	1 5/16 (34)	28.58 (726)	1 & 2
1992 All STS	22.83 (580)	9.48 (241)	2.04 (52)	28.42 (722)	22.16 (563)	9.48 (241)	1 1/4 (31)	28.42 (722)	1 & 2
1993-95 All Models	22.83 (580)	9.48 (241)	2.08 (53)	28.42 (722)	22.16 (563)	9.48 (241)	1 7/16 (37)	28.42 (722)	1 & 2
1996-97 All exc/STS	22.83 (580)	9.64 (245)	2.20 (56)	28.58 (726)	22.16 (563)	9.64 (245)	1 5/8 (41)	28.58 (726)	1 & 2
1996-97 All STS	22.83 (580)	9.33 (237)	1.88 (48)	28.26 (718)	22.16 (563)	9.33 (237)	1 5/16 (33)	28.26 (718)	1 & 2

STS	SF	WF	SR	WR	Figure
1998-02 SLS exc/FE3 Susp.	1.57 (40)	—	3.38 (86)	—	1 & 5
1998-02 STS, SLS w/FE3 Susp.	1.18 (30)	—	2.99 (76)	—	1 & 5
2003-04 All Models	1.18 (30)	—	2.99 (76)	—	1 & 5
2005-06 STS All FE1, FE3 w/RPO M22	1.53 (39)	—	2.51 (64)	—	1 & 2
2005-06 STS All FE3 w/RPO MU3	3.50 (89)	—	2.91 (74)	—	1 & 2
2007-08 STS All FE1 exc/RPO MX5	1.53 (39)	—	2.44 (62)	—	1 & 2
2007-08 STS All FE1 w/RPO MX5	3.07 (78)	—	2.51 (64)	—	1 & 2
2007-08 STS All FE3 w/RPO M22	1.53 (39)	—	2.44 (62)	—	1 & 2
2007-08 STS All FE3 w/RPO MV3	3.07 (78)	—	2.51 (64)	—	1 & 2
2007-08 STS-V	1.10 (28)	—	2.44 (62)	—	1 & 2
2009-10 STS RWD V6 w/255-45R18	3.07 (78)	28.42 (722)	2.44 (62)	28.66 (728)	A, 1 & 2
2009-10 STS RWD V6 w/others	1.53 (39)	28.34 (720)	2.44 (62)	28.50 (724)	A, 1 & 2
2009-10 STS RWD V8 w/235/50R18 (Fr) and 255/45R18 (Rr) Michelin + FE1 Susp.					
	3.07 (78)	28.70 (729)	2.44 (62)	28.93 (735)	A, 1 & 2
2009-10 STS RWD V8 w/others	1.53 (39)	28.34 (720)	2.44 (62)	28.50 (724)	A, 1 & 2
2009-10 STS AWD w/235-50R18 (Fr) and 255-45R18 (Rr) Michelin					
	3.07 (78)	28.70 (729)	2.51 (64)	28.93 (735)	A, 1 & 2
2009-10 STS AWD w/others	1.53 (39)	28.34 (720)	2.51 (64)	28.50 (724)	A, 1 & 2
2009 STS-V	1.10 (28)	28.58 (726)	2.44 (62)	29.21 (742)	A, 1 & 2

* ± ½ (13) unless otherwise indicated

CHASSIS / SUSP. HEIGHTS

GM – Cadillac (cont.)

Year Model	Height Specifications @ Curb Weight - Inches (mm) * Front		Rear		
SRX	**SF**	**WF**	**SR**	**WR**	**Figure**
2004-08 All Models	3.85 (98)	—	2.67 (68)	—	1 & 2
2009 All with P255/50R20-104V	3.85 (98)	32.51 (826)	2.55 (65)	32.63 (829)	A, 1 & 2
2009 All others	3.85 (98)	31.96 (812)	2.55 (65)	32.12 (816)	A, 1 & 2
2010-11 All w/18" tires (OE)	2.63 (67)	32.55 (827)	4.52 (115)	33.38 (848)	A, 1 & 2
2010-11 All w/20" tires (OE)	2.63 (67)	32.67 (830)	4.52 (115)	33.54 (852)	A, 1 & 2
2012-16 All w/18" tires (OE)	2.80 (71)	32.40 (824)	4.52 (115)	33.20 (844)	A, 1 & 2
2012-16 All w/20" tires (OE)	2.80 (71)	32.60 (829)	4.52 (115)	33.40 (849)	A, 1 & 2
XLR	**SF**	**WF**	**SR**	**WR**	**Figure**
2004-09 All exc/XLR-V	1.73 (44)	—	4.72 (120)	—	6 (F & R)
2006-09 XLR-V	1.96 (50)	—	5.07 (129)	—	6 (F & R)
XTS	**SF**	**WF**	**SR**	**WR**	**Figure**
2013-17 XTS w/ P235/50R18" tires	—	28.70 (730)	—	29.09 (739)	A
2013-14 XTS w/ P245/45R19" tires	—	29.13 (740)	—	29.40 (747)	A
2013-14 XTS w/ P245/40R20" tires	—	29.13 (740)	—	29.56 (751)	A
2013-14 XTS w/ P235/55R18XL" tires	—	29.25 (743)	—	29.56 (751)	A

GM - Chevrolet

Height Specifications @ Curb Weight - Inches (mm) *

Year Model	Front		Rear		
Aveo	**CF**		**CR**		**Figure**
2004-11 Aveo	13.54 (344)	—	13.50 (343)	—	A

Year Model	Front			Rear			
Beretta/Corsica	**AF**	**RF**	**SF**	**AR**	**RR**	**SR**	**Figure**
1990-91 All w/14" tires	24.29 (617)	9.52 (242)	.74 (19)	22.04 (560)	9.88 (251)	6.88 (175)	11a & 14
1990-91 All w/15" tires	24.29 (617)	9.01 (229)	.19 (5)	22.04 (560)	9.37 (238)	6.37 (162)	11a & 14
1990-91 All w/16" tires	24.29 (617)	9.01 (229)	0 (0)	22.04 (560)	9.37 (238)	6.25 (159)	11a & 14
1992-95 Corsica exc/2.2L engine	31.49 (800)	9.52 (242)	.51 (13)	22.04 (560)	9.84 (250)	—	11
1994-95 Corsica w/2.2L engine	31.49 (800)	9.52 (242)	.43 (11)	22.04 (560)	9.84 (250)	—	11
1992-95 Beretta exc/16" tires	31.49 (800)	9.40 (239)	.39 - .51 (10-13)	22.04 (560)	9.76 (248)	—	11
1992-95 Beretta w/16" tires	31.49 (800)	9.01 (229)	-.19 (-5)	22.04 (560)	9.37 (238)	—	11
1996 All exc/16" tires	31.49 (800)	9.52 (242)	.55 (14)	22.04 (560)	9.84 (250)	—	11
1996 Beretta w/16" tires	31.49 (800)	9.29 (236)	0 (0)	22.04 (560)	9.76 (248)	—	11
Camaro > 2002	**AF**	**RF**	**SF**	**AR**	**RR**	**SR**	**Figure**
1990-92 All w/15" tires	32.59 (828)	8.34 (212)	1.22 (31)	17.24 (438)	8.30 (211)	4.37 (111)	1 & 15
1990-92 All w/16" tires	32.59 (828)	8.03 (204)	.90 (23)	17.24 (438)	8.11 (206)	4.33 (110)	1 & 15
1993-97 All w/215-60 tires	32.59 (828)	8.03 (204)	1.25 (32)	17.24 (438)	8.18 (208)	4.76 (121)	1 & 15
1993-02 All w/235-55 tires	32.59 (828)	8.03 (204)	1.18 (30)	17.24 (438)	8.18 (208)	4.64 (118)	1 & 15
1993-02 All w/245-50 tires	32.59 (828)	8.30 (204)	1.45 (37)	17.24 (438)	8.18 (208)	4.92 (125)	1 & 15

Camaro 2010 ->	SF	WF	SR	WR	Figure
2010 Camaro 18" & 19" tires (OE)	2.67 (68) L/R ±0.39 (10)	30.2 (768) L/R ±0.47 (12)	1.22 (31) L/R ±0.39 (10)	30.5 (774) L/R ±0.47 (12)	A, 1 & 2
2010 Camaro 19" tires (OE)	2.67 (68) L/R ±0.39 (10)	30.2 (768) L/R ±0.47 (12)	1.22 (31) L/R ±0.39 (10)	30.5 (774) L/R ±0.47 (12)	A, 1 & 2
2010 Camaro 20" tires w/RPO FE2	2.67 (68) L/R ±0.39 (10)	30.1 (765) L/R ±0.47 (12)	1.22 (31) L/R ±0.39 (10)	30.4 (773) L/R ±0.47 (12)	A, 1 & 2
2010 Camaro 20" tires w/RPO FE3	2.17 (55) L/R ±0.39 (10)	29.6 (753) L/R ±0.47 (12)	0.62 (16) L/R ±0.39 (10)	29.8 (758) L/R ±0.47 (12)	A, 1 & 2
2011-17 Camaro 18" tires (OE)	2.17 (55) L/R ±0.39 (10)	29.6 (753) L/R ±0.47 (12)	0.62 (16) L/R ±0.39 (10)	29.7 (755) L/R ±0.47 (12)	A, 1 & 2
2011-17 Camaro 19" P245/P245 F/R	2.17 (55) L/R ±0.39 (10)	29.8 (757) L/R ±0.47 (12)	0.62 (16) L/R ±0.39 (10)	29.9 (759) L/R ±0.47 (12)	A, 1 & 2
2015-17 Camaro 19" P305/P305 F/R	2.17 (55) L/R ±0.39 (10)	27.8 (708) L/R ±0.47 (12)	0.62 (16) L/R ±0.39 (10)	28.0 (711) L/R ±0.47 (12)	A, 1 & 2
2011-17 Camaro 20" P245/P275 F/R	2.17 (55) L/R ±0.39 (10)	29.8 (757) L/R ±0.47 (12)	0.62 (16) L/R ±0.39 (10)	29.9 (759) L/R ±0.47 (12)	A, 1 & 2
2011-17 Camaro 20" P285/P285 F/R	2.17 (55) L/R ±0.39 (10)	29.4 (748) L/R ±0.47 (12)	0.62 (16) L/R ±0.39 (10)	29.6 (751) L/R ±0.47 (12)	A, 1 & 2
2011-17 Camaro 20" 285/305 F/R(OE)	2.00 (51) L/R ±0.39 (10)	29.2 (743) L/R ±0.47 (12)	0.62 (16) L/R ±0.39 (10)	29.8 (757) L/R ±0.47 (12)	A, 1 & 2

Caprice/Impala	AF	RF	WF	AR	RR	WR	Figure
1990 Sedan exc/Police Version	31.49 (800)	10.43 (265)	28.70 (729)	24.60 (625)	10.51 (267)	28.70 (729)	A
1990 Sedan Police Version	31.49 (800)	10.51 (267)	28.74 (730)	24.60 (625)	10.59 (269)	28.81 (732)	A
1990 Wagon	31.49 (800)	10.47 (266)	28.77 (731)	24.60 (625)	10.55 (268)	28.74 (730)	A
1991-93 All Sedan (incl.Police)	31.88 (810)	9.92 (252)	29.21 (742)	24.21 (615)	10.39 (264)	21.22 (539)	A
1991-93 Wagon	31.88 (810)	9.92 (252)	29.21 (742)	24.21 (615)	10.39 (264)	21.65 (550)	A
1994-96 Sedan w/75 series tires	31.88 (810)	10.11 (257)	28.93 (735)	24.21 (615)	10.19 (259)	28.46 (723)	A
1994-96 Sedan w/70 series tires	31.88 (810)	9.92 (252)	28.77 (731)	24.21 (615)	10.00 (254)	28.26 (718)	A
1994-96 Impala SS	31.88 (810)	8.93 (227)	27.79 (706)	24.21 (615)	9.01 (229)	27.28 (693)	A
1994-96 Sedan Police	31.88 (810)	9.92 (252)	28.58 (726)	24.21 (615)	10.39 (264)	28.81 (732)	A
1994-96 Wagon	31.88 (810)	9.92 (252)	29.21 (742)	24.21 (615)	10.39 (264)	21.69 (551)	A
2011-17 Caprice PPV			29.13 (740)			28.93 (735)	A
Cavalier	**AF**	**RF**	**SF**	**AR**	**RR**	**SR**	**Figure**
1990 Coupe/Sedan	21.92 (557)	9.21-9.29 (234-236)	3.62-3.85 (92-98)	21.81 (554)	9.09-9.29 (231-236)	—	10
1990 Wagon	21.92 (557)	9.52 (242)	3.77-4.05 (96-103)	21.81 (554)	9.64 (245)	—	10
1991 Coupe/Sedan	21.92 (557)	9.25 (235)	3.42-3.62 (87-92)	21.81 (554)	9.13 (232)	—	10
1991 Wagon	21.92 (557)	9.52 (242)	3.62-3.77 (92-96)	21.81 (554)	9.64 (245)	—	10

* ± ½ (13) unless otherwise indicated

GM – Chevrolet (cont.)

Year	Model	Front: AF	Front: RF	Front: SF	Rear: AR	Rear: RR	Rear: SR	Figure
		Height Specifications @ Curb Weight - Inches (mm) *						
	Cavalier (cont.)	**AF**	**RF**	**SF**	**AR**	**RR**	**SR**	**Figure**
1992-94	All exc/Wagon	21.92 (557)	9.25 (235)	0.07-0.27 (2-7)	21.81 (554)	9.13 (232)	—	11
1992-94	Wagon	21.92 (557)	9.52 (242)	0.27-0.47 (7-12)	21.81 (554)	9.64 (245)	—	11
1995	All w/70 series tires	21.92 (557)	9.17 (233)	0.35-0.39 (9-10)	21.81 (554)	9.44 (240)	—	11
1995	All w/55 & 65 series tires	21.92 (557)	9.17 (233)	0.11-0.15 (3-4)	21.81 (554)	9.44 (240)	—	11
1996	All models w/14" tires	21.92 (557)	9.33 (237)	0.51 (13)	21.81 (554)	9.76 (248)	—	11
1996	All models w/15" tires	21.92 (557)	9.33 (237)	0.31 (8)	21.81 (554)	9.76 (248)	—	11
1996	All models w/16" tires	21.92 (557)	9.33 (237)	0.23 (6)	21.81 (554)	9.68 (246)	—	11
1997	All models w/14" tires	21.92 (557)	9.17 (233)	0.47 (12)	21.81 (554)	9.44 (240)	—	11
1997	All models w/15" tires	21.92 (557)	9.17 (233)	0.23 (6)	21.81 (554)	9.44 (240)	—	11
1997	All models w/16" tires	21.92 (557)	9.17 (233)	0.15 (4)	21.81 (554)	9.44 (240)	—	11
1998-02	All Coupe Z24 (FE2 susp.)	32.16 (817)	9.37 (238)	-0.07 (-2)	22.04 (560)	9.64 (245)	-1.69 (-43)	30 & 33
	All Convertible	32.16 (817)	9.37 (238)	0.19-0.27 (5-7)	22.04 (560)	9.64 (245)	-1.41 (-36)	30 & 33
	All Others w/Active susp (FE0)	32.16 (817)	9.17 (233)	0-0.15 (0-4)	22.04 (560)	9.44 (240)	-1.45 to -1.65 (-37 to-42)	30 & 33
	All Others w/Base susp (FE1)	32.16 (817)	9.17 (233)	0.03 (1)	22.04 (560)	9.44 (240)	-1.65 (-42)	30 & 33
2003-05	All models w/14" tires	—	—	0.11 (3)	—	—	1.45 (37)	30 & 33
2003-05	All models w/15" tires	—	—	-0.03 (-1)	—	—	-1.65 (-42)	30 & 33
2003-05	All models w/16" tires	—	—	-0.15 (-4)	—	—	-1.69 (-43)	30 & 33
	Celebrity	**AF**	**RF**	**SF**	**AR**	**RR**	**SR**	**Figure**
1989-90	Wagon	19.88 (505)	9.09 (231)-9.68 (246)	-0.03 (-1)	18.07 (459)	9.29 (236)-9.72 (247)	7.51 (191)	3 & 4

Year	Cobalt	SF	SR	Figure
2005-07	Base Models w/15" tires	-0.82 (-21)	9.37 (238)	30 & 31
2008-10	Base Models w/15" tires	-0.43 (-11)	9.52 (242)	30 & 31
2005-10	Base Models w/16" tires	-0.90 (-23)	8.97 (228)	30 & 31
2005-10	SS Models exc/SC or Turbo 17"	-1.02 (-26)	8.81 (224)	30 & 31
2005-07	SS Models w/SC or Turbo 18"	-1.33 (-34)	8.38 (213)	30 & 31
2008-10	SS Models w/SC or Turbo 18"	-0.62 (-16)	8.62 (219)	30 & 31

Year	Corvette ->1996	AF	RF	SF	AR	RR	SR	Figure
1990-91	Coupe ZR1	25.66 (652)	7.28 (185)	1.73 (44)	19.13 (486)	7.63 (194)	2.04 (52)	6 & 16
1990	Coupe w/STD. suspension	25.66 (652)	7.36-7.67 (187-195)	1.73-2.24 (44-57)	19.13 (486)	7.71-7.87 (196-200)	2.20 (56)	6 & 16
1990	Coupe w/FE7 suspension	25.66 (652)	7.24-7.59 (184-193)	1.65-2.08 (42-53)	19.13 (486)	7.63 (194)	2.00 (51)	6 & 16
1990	Convertible	25.66 (652)	7.67-7.99 (195-203)	2.08-2.59 (53-66)	19.13 (486)	8.07 (205)	2.48 (63)	6 & 16
1991	Coupe w/STD. suspension	25.66 (652)	7.71 (196)	2.24 (57)	19.13 (486)	7.87 (200)	2.20 (56)	6 & 16
1991	Coupe w/FE7 suspension	25.66 (652)	7.59 (193)	2.12 (54)	19.13 (486)	7.67 (195)	2.00 (51)	6 & 16
1991	Convertible	25.66 (652)	8.03 (204)	2.59 (66)	19.13 (486)	8.11 (206)	2.48 (63)	6 & 16
1992-95	Coupe ZR1	25.66 (652)	7.59 (193)	0.29 (7.5)	19.13 (486)	7.87 (200)	1.85 (47)	6 & 16
1992-96	Coupe w/STD. suspension	25.66 (652)	7.59 (193)	2.20 (56)	19.13 (486)	7.59 (193)	1.85 (47)	6 & 16
1992-96	Coupe w/FE7 suspension	25.66 (652)	7.51 (191)	2.12 (54)	19.13 (486)	7.51 (191)	1.81 (46)	6 & 16
1992-96	Convertible	25.66 (652)	7.87 (200)	2.55 (65)	19.13 (486)	7.87 (200)	2.16 (55)	6 & 16

Year	Corvette 1997 ->	SF	WF	SR	WR	Figure
1997-04	Base susp. (FE1)	1.73 (44) ±0.25 (6.4)	—	4.25 (108) ±0.25 (6.4)	—	6 (F & R)
1997-04	Performance susp. (FE3)	1.69 (43) ±0.25 (6.4)	—	4.21 (107) ±0.25 (6.4)	—	6 (F & R)
1997-04	Z06 (FE4)	1.92 (49) ±0.25 (6.4)	—	4.09 (104) ±0.25 (6.4)	—	6 (F & R)
2005	Coupe exc/Perf. Susp.	1.81 (46) ±0.25 (6.4)	—	4.80 (122) ±0.25 (6.4)	—	6 (F & R)
2005	Coupe w/Performance susp.	1.77 (45) ±0.25 (6.4)	—	4.76 (121) ±0.25 (6.4)	—	6 (F & R)
2005	Conv. exc/Performance susp.	1.77 (45) ±0.25 (6.4)	—	4.76 (121) ±0.25 (6.4)	—	6 (F & R)
2005	Conv. w/Performance susp.	1.77 (45) ±0.25 (6.4)	—	4.72 (120) ±0.25 (6.4)	—	6 (F & R)
2006-07	All FE1, FE2 (Base & RPO F55)	1.81 (46) ±0.25 (6.4)	—	4.80 (122) ±0.25 (6.4)	—	6 (F & R)
2006-07	All FE3 (RPO F51)	1.77 (45) ±0.25 (6.4)	—	4.76 (121) ±0.25 (6.4)	—	6 (F & R)
2006-07	Z06 models	1.61 (41) ±0.25 (6.4)	—	4.29 (109) ±0.25 (6.4)	—	6 (F & R)
2008-09	All FE1, FE2 (Base & RPO F55)	1.88 (48) ±0.25 (6.4)	—	4.52 (115) ±0.25 (6.4)	—	6 (F & R)
2008-09	All FE3 (RPO F51)	1.85 (47) ±0.25 (6.4)	—	4.48 (114) ±0.25 (6.4)	—	6 (F & R)
2008-09	Z06, ZR1 models	1.69 (43) ±0.25 (6.4)	—	4.01 (102) ±0.25 (6.4)	—	6 (F & R)
2010-13	All FE1, FE2 suspension	1.81 (46) ±0.39 (10)	27.40 (696) ±0.39 (10)	4.80 (122) ±0.39 (10)	28.74 (730) ±0.39 (10)	A, 6 (F & R)
2010-13	All FE3, FE4, FE5 susp.	1.61 (41) ±0.39 (10)	27.28 (693) ±0.39 (10)	4.29 (109) ±0.39 (10)	28.22 (717) ±0.39 (10)	A, 6 (F & R)
2014	All FE1, FE2 suspension		27.20 (692) ±0.39 (10)		29.10 (738) ±0.39 (10)	A
2014	All FE3, FE4 suspension		27.30 (693) ±0.39 (10)		29.00 (737) ±0.39 (10)	A
	Cruze/Cruze Limited	**SF**	**WF**	**SR**	**WR**	**Figure**
2011-16	Cruze 16" tires (OE)	—	27.44 (697)	—	27.64 (702)	A
2011-16	Cruze 17" tires (OE)	—	27.79 (696)	—	28.07 (701)	A
2011-16	Cruze 18" tires(OE)	—	27.52 (699)	—	27.72 (704)	A
	Cruze Gen II	**SF**	**WF**	**SR**	**WR**	**Figure**
2016-17	Cruze 15" tires (OE)	—	26.69 (678)	—	27.00 (686)	A
	HHR	**SF**	**WF**	**SR**	**WR**	**Figure**
2006-07	HHR w/16" tires	-0.70 ±0.4 (-18 ±10)	27.63 ±0.4 (702 ±10)	8.93 ±0.4 (227 ±10)	28.22 ±0.4 (717 ±10)	A, 30 & 31
2006-07	HHR w/17" tires	-0.62 ±0.4 (-16 ±10)	27.91 ±0.4 (709 ±10)	9.09 ±0.4 (231 ±10)	28.50 ±0.4 (724 ±10)	A, 30 & 31
2008-11	HHR w/16" tires	-0.70 ±0.4 (-18 ±10)	27.59 ±0.4 (701 ±10)	8.93 ±0.4 (227 ±10)	28.22 ±0.4 (717 ±10)	A, 30 & 31

* ± ½ (13) unless otherwise indicated

CHASSIS / SUSP. HEIGHTS

GM – Chevrolet (cont.)

Height Specifications @ Curb Weight - Inches (mm) *

Year Model	Front		Rear		
HHR (cont.)	SF	WF	SR	WR	Figure
2008-11 HHR w/17" tires	-0.59 ± 0.4 (-15 ± 10)	27.79 ± 0.4 (706 ± 10)	9.09 ± 0.4 (231 ± 10)	28.42 ± 0.4 (722 ± 10)	A, 30 & 31
2008 HHR w/18" tires	-1.06 ± 0.4 (-27 ± 10)	27.63 ± 0.4 (702 ± 10)	8.46 ± 0.4 (215 ± 10)	28.30 ± 0.4 (719 ± 10)	A, 30 & 31
2009-11 HHR w/18" tires	-0.59 ± 0.4 (-15 ± 10)	27.63 ± 0.4 (702 ± 10)	8.46 ± 0.4 (215 ± 10)	28.30 ± 0.4 (719 ± 10)	A, 30 & 31

Impala -> 2006	AF	RF	SF	AR	RR	SR	Figure
2000-02 All Models w/15" tires	23.81 (605)	8⅜ (213)	2.16 (55)	20.66 (525)	8.38 (213)	10.27 (261)	30 & 26
2000-02 All Models w/16" tires	23.81 (605)	8⅜ (213)	1.88 (48)	20.66 (525)	8.38 (213)	10.03 (255)	30 & 26
2003-05 All Models exc/Police, Taxi	23.81 (605)	9 7/16 (240)	2.40 (61)	20.66 (525)	9.44 (240)	10.98 (279)	30 & 26
2003-05 All Police 9C1	23.81 (605)	10 3/32 (256)	3.11 (79)	20.66 (525)	9.60 (244)	10.94 (278)	30 & 26
2003-05 All Police 9C3 (unmarked)	23.81 (605)	9¾ (248)	2.71 (69)	20.66 (525)	9.52 (242)	10.94 (278)	30 & 26
2003-05 All Taxi 9C6	23.81 (605)	9¾ (248)	2.71 (69)	20.66 (525)	9.52 (242)	10.94 (278)	30 & 26
2006 All LS, LT w/16" tires	23.81 (605)	9 3/16 (233)	1.88 (48)	20.66 (525)	9.37 (238)	10.98 (279)	30 & 26
2006 All LT, LTZ w/17" tires	23.81 (605)	9 5/16 (236)	1.88 (48)	20.66 (525)	9.48 (241)	10.98 (279)	30 & 26
2006 All SS w/18" tires	23.81 (605)	9⅝ (244)	1.88 (48)	20.66 (525)	9.56 (243)	10.55 (268)	30 & 26
2006 All Police 9C1	23.81 (605)	9 13/16 (249)	2.59 (66)	20.66 (525)	9.60 (244)	10.90 (277)	30 & 26
2006 All Police 9C3 (unmarked)	23.81 (605)	9½ (242)	2.20 (56)	20.66 (525)	9.48 (241)	10.90 (277)	30 & 26

Impala 2007 ->	SF	WF	SR	WR	Figure
2007-09 All LS, LT w/16" tires	1.88 (48)	28.89 (734)	10.98 (279)	28.81 (732)	A, 30 & 26
2007-09 All LT, LTZ w/17" tires	1.88 (48)	28.97 (736)	10.98 (279)	28.89 (734)	A, 30 & 26
2007-09 All SS w/18" tires	1.88 (48)	29.33 (745)	10.55 (268)	28.81 (732)	A, 30 & 26
2007-09 All Police 9C1	2.59 (66)	29.64 (753)	10.90 (277)	28.81 (732)	A, 30 & 26
2007-09 All Police 9C3 (unmarked)	2.20 (56)	29.25 (743)	10.90 (277)	28.81 (732)	A, 30 & 26
2010-12 All LS, LT w/16" tires (OE)	1.96 (50)	28.93 (735)	11.10 (282)	28.89 (734)	A, 30 & 26
2010-12 All w/17" tires (OE)	1.85 (47)	28.97 (736)	10.98 (279)	28.89 (734)	A, 30 & 26
2010-12 All w/18" tires (OE)	1.85 (47)	29.37 (746)	10.98 (279)	28.85 (733)	A, 30 & 26
2010-12 All Police 9C1	2.44 (62)	29.48 (749)	10.98 (279)	28.81 (732)	A, 30 & 26
2010-12 All Police 9C3 (unmarked)	2.04 (52)	28.13 (740)	10.98 (279)	28.81 (732)	A, 30 & 26
2013-16 All LS, LT w/16" tires (OE)	1.96 (50)	28.90 (734)	10.98 (279)	28.81 (732)	A, 30 & 26
2013-16 All w/17" tires (OE)	1.89 (48)	28.97 (736)	10.98 (279)	28.94 (735)	A, 30 & 26
2013-16 All w/18" tires (OE)	1.89 (48)	29.29 (744)	10.55 (268)	28.74 (730)	A, 30 & 26
2013-16 All Police 9C1	2.48 (63)	29.65 (753)	10.90 (277)	28.85 (733)	A, 30 & 26
2013-16 All Police 9C3 (unmarked)	2.12 (54)	29.21 (742)	10.90 (277)	28.85 (733)	A, 30 & 26
2014-17 All w/18" tires (OE)	—	29.01 (737)	—	29.37 (746)	A
2014-17 All w/19"& 20" tires (OE)	—	29.33 (745)	—	29.64 (753)	A

Lumina	AF	RF	SF	AR	RR	SR	Figure
1990 All w/14" tires	23.81 (605)	9.80 (249)	3.26 (83)	20.66 (525)	9.80 (249)	4.05 (103)	1 & 8
1990 All w/15" tires	23.81 (605)	9.80 (249)	3.14 (80)	20.66 (525)	9.80 (249)	3.93 (100)	1 & 8
1990 All w/16" tires	23.81 (605)	9.80 (249)	3.18 (81)	20.66 (525)	9.80 (249)	3.97 (101)	1 & 8
1991-93 All Coupe w/14" tires	23.81 (605)	9.80 (249)	3.26 (83)	20.66 (525)	9.80 (249)	4.09 (104)	1 & 8
1991-93 All Sedan w/14" tires	23.81 (605)	9.80 (249)	3.26 (83)	20.66 (525)	9.80 (249)	4.13 (105)	1 & 8
1991-93 All w/15" tires	23.81 (605)	9.80 (249)	2.83 (72)	20.66 (525)	9.80 (249)	3.62 (92)	1 & 8
1991 All w/16" tires	23.81 (605)	9.80 (249)	2.59 (66)	20.66 (525)	9.80 (249)	3.28 (86)	1 & 8
1992-93 All exc/Z34 w/16" tires	23.81 (605)	9.80 (249)	2.79 (71)	20.66 (525)	9.80 (249)	3.62 (92)	1 & 8
1992-93 All Z34 w/16" tires	23.81 (605)	9.80 (249)	2.63 (67)	20.66 (525)	9.80 (249)	3.28 (86)	1 & 8
1994-95 All w/15" tires	23.81 (605)	9.80 (249)	2.95 (75)	20.66 (525)	9.80 (249)	10.70 (272)	1 & 9
1994-95 All w/16" tires	23.81 (605)	9.80 (249)	2.71 (69)	20.66 (525)	9.80 (249)	10.47 (266)	1 & 9
1996-97 All w/15" tires	23.81 (605)	10.00 (254)	2.95 (75)	20.66 (525)	10.19 (259)	10.70 (272)	1 & 9
1996-97 All w/16" tires	23.81 (605)	10.00 (254)	2.71 (69)	20.66 (525)	10.19 (259)	10.51 (267)	1 & 9
1998-00 All w/15" tires	23.81 (605)	8.38 (213)	2.16 (55)	20.66 (525)	8.38 (213)	—	1
1998-00 All w/16" tires	23.81 (605)	8.38 (213)	1.88 (48)	20.66 (525)	8.38 (213)	—	1

Lumina APV	AF	RF	SF	AR	RR	SR	Figure
1990-91 All w/14" tires	19.88 (505)	11.85 (301)	-0.03 (-1)	18.11 (460)	12.12 (308)	7.51 (191)	3 & 4
1990-92 All w/15" tires	19.88 (505)	12.08 (307)	-0.03 (-1)	18.11 (460)	12.36 (314)	7.51 (191)	3 & 4
1990-94 All models	23.77 (604)	12.08 (307)	-0.03 (-1)	22.04 (560)	12.36 (314)	7.51 (191)	3 & 4
1995-96 All exc/Cargo w/bumper adj.	see fig. 19	10.47 (266)	-0.23 (-6)	see fig. 19	10.90 (277)	7.51 (191)	3 & 4
1995-96 All Cargo Van w/bumper adj.	see fig. 19	10.74 (273)	-0.23 (-6)	see fig. 19	11.81 (300)	8.58 (218)	3 & 4

Malibu/Classic	AF	RF	SF	AR	RR	SR	Figure
1997-05 All models	32.16 (817)	9.33 (237)	0 (0)	25.00 (635)	9.60 (244)	n/a	11 & 34

Malibu/2008 Malibu Classic	AF	RF	SF	AR	RR	SR	Figure
2004 Malibu Sedan w/15" tires	21.41 (544)	9.01 (229)	2.79 (71)	22.04 (560)	9.21 (234)	6.85 (174)	37 & 36
2004 Malibu Sedan w/16" tires	21.41 (544)	9.29 (236)	2.63 (67)	22.04 (560)	9.44 (240)	6.65 (169)	37 & 36
2004 Malibu Wagon	21.41 (544)	9.29 (236)	2.67 (68)	22.04 (560)	9.44 (240)	6.73 (171)	37 & 36
2005 Malibu Sedan w/15" tires	21.41 (544)	8.89 (226)	2.75 (70)	22.04 (560)	9.25 (235)	6.85 (174)	37 & 36
2005 Malibu Sedan w/16" tires	21.41 (544)	9.05 (230)	2.63 (67)	22.04 (560)	9.72 (247)	6.65 (169)	37 & 36
2005 Malibu Wagon	21.41 (544)	9.05 (230)	2.67 (68)	22.04 (560)	9.72 (247)	6.73 (171)	37 & 36

* ± ½ (13) unless otherwise indicated

GM – Chevrolet (cont.)

Year Model	Height Specifications @ Curb Weight - Inches (mm) *						
	Front			Rear			
Malibu/2008 Malibu Classic (cont.)	AF	RF	SF	AR	RR	SR	Figure
2006-08 All Sedan w/15" tires	21.41 (544)	9.01 (229)	2.75 (70)	22.04 (560)	9.29 (236)	6.85 (174)	37 & 36
2006-08 All Sed. w/P215-60R16" tires	21.41 (544)	9.13 (232)	2.63 (67)	22.04 (560)	9.48 (241)	6.65 (169)	37 & 36
2006-08 All Sed. w/P215-55R16" tires	21.41 (544)	8.93 (227)	2.63 (67)	22.04 (560)	9.25 (235)	6.73 (171)	37 & 36
2006-08 All Sedan w/17" tires	21.41 (544)	8.93 (227)	2.63 (67)	22.04 (560)	9.68 (246)	6.73 (171)	37 & 36
2006-08 All Sedan SS w/18" tires	21.41 (544)	9.44 (240)	2.20 (56)	22.04 (560)	9.92 (252)	6.49 (165)	37 & 36
2006-07 All Wagon w/16" tires	21.41 (544)	9.05 (230)	2.67 (68)	22.04 (560)	9.48 (241)	6.73 (171)	37 & 36
2006-07 All Wagon w/17" tires	21.41 (544)	9.25 (235)	2.63 (67)	22.04 (560)	9.64 (245)	7.24 (184)	37 & 36
2006-07 All Wagon SS w/18" tires	21.41 (544)	9.76 (248)	2.59 (66)	22.04 (560)	10.19 (259)	7.24 (184)	37 & 36

Malibu 2008 -> (exc/Malibu Classic)	SF	WF	SR	WR	Figure
2008-10 Malibu Hybrid	2.68 (68)	27.79 (706)	7.24 (184)	28.18 (716)	A, 37 & 36
2008-12 Malibu w/16" tires (OE)	2.68 (68)	27.91 (709)	7.24 (184)	28.22 (717)	A, 37 & 36
2008-12 Malibu w/17" tires (OE)	2.68 (68)	27.87 (708)	7.24 (184)	28.18 (716)	A, 37 & 36
2008-12 Malibu w/18" tires (OE)	2.68 (68)	28.30 (719)	7.24 (184)	28.66 (728)	A, 37 & 36
2013-15 Malibu w/16" tires (OE)	—	28.28 (718)	—	28.43 (722)	A
2013-15 Malibu w/17" tires (OE)	—	28.62 (727)	—	28.74 (730)	A
2013-15 Malibu w/18" tires (OE)	—	28.90 (734)	—	29.06 (738)	A
2013-15 Malibu w/19" tires (OE)	—	28.78 (731)	—	28.90 (734)	A
2016 Malibu w/16" tires (OE)	—	28.11 (714)	—	28.43 (722)	A
2016 Malibu w/17" tires (OE)	—	28.31 (719)	—	28.90 (734)	A
2016 Malibu w/18" tires (OE)	—	28.70 (729)	—	29.06 (738)	A
2016 Malibu w/19" tires (OE)	—	28.58 (726)	—	28.78 (731)	A

Monte Carlo	AF	RF	SF	AR	RR	SR	Figure
1995 All models w/15" tires	23.81 (605)	9.80 (249)	2.95 (75)	20.66 (525)	9.80 (249)	10.66 (271)	1 & 9
1995 All models w/16" tires	23.81 (605)	9.80 (249)	2.67-2.71 (68-69)	20.66 (525)	9.80 (249)	10.47 (266)	1 & 9
1996-99 All models w/15" tires	23.81 (605)	10.00 (254)	2.95 (75)	20.66 (525)	10.19 (259)	10.70 (272)	1 & 9
1996-99 All models w/16" tires	23.81 (605)	10.00 (254)	2.71 (69)	20.66 (525)	10.19 (259)	10.47 (266)	1 & 9
2000-02 All models w/15" tires	23.81 (605)	8.38 (213)	2.16 (55)	20.66 (525)	8.38 (213)	10.27 (261)	30 & 26
2000-02 All models w/16" tires	23.81 (605)	8.38 (213)	1.88 (48)	20.66 (525)	8.38 (213)	10.03 (255)	30 & 26
2003-05 All Models	23.81 (605)	9.21 (234)	2.16 (55)	20.66 (525)	9.17 (233)	10.66 (271)	30 & 26
2006-07 All LS, LT w/16" tires	23.81 (605)	8.97 (228)	1.69 (43)	20.66 (525)	9.13 (232)	10.66 (271)	30 & 26
2006-07 All LT, LTZ w/17" tires	23.81 (605)	9.09 (231)	1.69 (43)	20.66 (525)	9.52 (242)	10.66 (271)	30 & 26
2006-07 All SS w/18" tires	23.81 (605)	9.64 (245)	1.88 (48)	20.66 (525)	9.60 (244)	10.55 (268)	30 & 26

Optra	CF	CR	Figure
2004-09 Optra	14.48 (368)	14.44 (367)	A

Orlando	WF	WR	Figure
2012-14 16" tires (OE)	28.18 (716)	28.62 (727)	A

Sonic	WF	WR	Figure
2012-16 Sonic w/15" & 16" tires (OE)	26.96 (685)	27.20 (691)	A
2012-16 Sonic w/17" tires (OE)	27.16 (690)	27.36 (695)	A

Spark	WF/CF	WR/CR	Figure
2013-15 Spark w14" tires (OE)	14.10 (358) (CF)	14.05 (357) (CR)	A
2016 Spark	25.39 (645) (WF)	25.63 (651) (WR)	A

Spark EV	WF	WR	Figure
2014-15 Spark EV w15" tires (OE)	24.33 (618)	25.57 (624)	A

SS	WF/CF	WR/CR	Figure
2014 SS	28.35 (720) (WF)	28.35 (720) (WR)	A
2015-16 SS	15.67 (398) (CF)	15.79 (401) (CR)	A

Trax	WF	WR	Figure
2013-17 Trax FWD w/16" tires (OE)	29.30 (744)	29.90 (759)	A
2013-17 Trax FWD w/18" tires (OE)	29.40 (747)	30.00 (762)	A
2013-17 Trax AWD w/16" tires (OE)	29.60 (753)	30.30 (769)	A
2013-17 Trax AWD w/18" tires (OE)	29.80 (756)	30.40 (771)	A

Uplander	AF	RF	SF	AR	RR	SR	Figure
2005-09 Uplander FWD Std. WB	23.77 (604)	12.12 (308)	2.55 (65)	22.04 (560)	11.57 (294)	6.22 (158)	1 & 29
2005-09 Uplander FWD Ext. WB	23.77 (604)	12.20 (310)	2.55 (65)	22.04 (560)	11.85 (301)	6.45 (164)	1 & 29
2005-09 Uplander FWD Cargo w/FE1	23.77 (604)	12.36 (314)	2.55 (65)	22.04 (560)	12.40 (315)	7.51 (191)	1 & 29
2005-09 Uplander FWD Cargo w/FE3	23.77 (604)	12.24 (311)	2.55 (65)	22.04 (560)	12.00 (305)	6.73 (171)	1 & 29
2005-06 Uplander AWD	23.77 (604)	12.55 (319)	2.55 (65)	22.04 (560)	11.69 (297)	3.66 (93)	1 & 20

* ± ½ (13) unless otherwise indicated

GM – Chevrolet (cont.)

Year Model	Height Specifications @ Curb Weight - Inches (mm) *						
	Front			Rear			
Venture	AF	RF	SF	AR	RR	SR	Figure
1997-03 Std. WB w/205/70R15	23.77 (604)	8.97-9.76 (228-248)	2.28-3.07 (58-78)	22.04 (560)	9.84-10.62 (250-270)	5.23-6.02 (133-153)	1 & 29
1997-03 Std. WB w/215/70R15	23.77 (604)	9.25-10.03 (235-255)	2.28-3.07 (58-78)	22.04 (560)	10.15-10.94(258-278)	5.23-6.02 (133-153)	1 & 29
2004-05 Std. WB w/215/70R15	23.77 (604)	9.64 (245)	2.20 (56)	22.04 (560)	10.51 (267)	5.62 (143)	1 & 29
2004-05 Std. WB w/225/60R16	23.77 (604)	9.64 (245)	2.20 (56)	22.04 (560)	10.55 (268)	5.62 (143)	1 & 29
1998-03 Ext. WB (exc/Cargo)	23.77 (604)	9.21-10.00 (234-254)	2.28-3.07 (58-78)	23.77 (604)	10.07-10.86(256-276)	5.47-6.25 (139-159)	1 & 29
2004-05 Ext. WB (exc/Cargo) 215/70R15	23.77 (604)	9.60 (244)	2.20 (56)	23.77 (604)	10.47 (266)	5.86 (149)	1 & 29
2004-05 Ext. WB (exc/Cargo) 225/60R16	23.77 (604)	9.64 (245)	2.20 (56)	23.77 (604)	10.51 (267)	5.86 (149)	1 & 29
1998-03 Cargo Van w/o Level Control	23.77 (604)	9.44-10.23 (240-260)	2.28-3.07 (58-78)	22.04 (560)	10.98-11.77(279-299)	6.53-7.32 (166-186)	1 & 29
1998-03 Cargo Van with Level Control	23.77 (604)	9.29-10.07 (236-256)	2.28-3.07 (58-78)	22.04 (560)	10.35-11.14(263-283)	5.78-6.57 (147-167)	1 & 29
2004-05 Cargo Van with Std. Susp	23.77 (604)	9.84 (250)	2.20 (56)	23.77 (604)	11.41 (290)	7.04 (179)	1 & 29
2004-05 Cargo Van with Sport Susp.	23.77 (604)	9.68 (246)	2.20 (56)	23.77 (604)	10.82 (275)	6.25 (159)	1 & 29
2002-03 All AWD models	23.77 (604)	9.72-9.88 (247-251)	2.20-2.32 (56-59)	22.04 (560)	10.82-11.14(275-283)	3.66-3.93 (93-100)	1 & 20
2004 All AWD models	23.77 (604)	9.72 (247)	2.20 (56)	22.04 (560)	10.82 (275)	3.66 (93)	1 & 20

Volt	SF	WF	SR	WR	Figure
2011-15 Volt (All models)	1.53 (39)	27.79 (706)	9.29 (236)	28.11 (714)	A, 38 & 43
2016 Volt (All models)	0.83 (21)	27.05 (687)	10.04 (255)	27.56 (700)	A, 38 & 43

Chevrolet/Geo Metro	AF	RF	SF	AR	RR	SR	Figure
1993-01 All Models	17.71 (450)	8.26 (210)	1.77 (45)	17.71 (450)	8.85 (225)	—	1
Chevrolet/Geo Prizm	AF	RF	SF	AR	RR	SR	Figure
1991-97 All Models	19¾ (502)	7⅛ (180)		20.51 (521)	7.28 (185)		
1998-02 All w/175-65R14	—	—	7.28 (185)	—	—	9.64 (245)	18 & 17
1998-02 All w/185-65R14	—	—	7.48 (190)	—	—	9.84 (250)	18 & 17
Chevrolet/Geo Storm	AF	RF	SF	AR	RR	SR	Figure
1990-93 All Models	19.68 (500)	8.70 (221)	—	19.68 (500)	8.70 (221)	—	
Chevrolet/Geo Tracker	AF	RF	SF	AR	RR	SR	Figure
1993-98 All 4X2 Models	19.68 (500)	11.10 (282)	3.58 (91)	19.68 (500)	11.92 (303)	—	1
1993-98 All 4X4 Models	19.68 (500)	11.85 (301)	3.58 (91)	19.68 (500)	12.67 (322)	—	1
1999-02 All 4X2 Models	27.55 (700)	8.26 (210) RF = frame rail to ground.		27.55 (700)	8.89 (226) RR = frame rail to ground.		
1999-02 All 4X4 Models	27.55 (700)	8.54 (217) RF = frame rail to ground.		27.55 (700)	9.17 (233) RR = frame rail to ground.		

* ± ½ (13) unless otherwise indicated

GM - Oldsmobile

Year Model	Height Specifications @ Curb Weight - Inches (mm) *						
	Front			Rear			
Achieva	AF	RF	SF	AR	RR	SR	Figure
1992-95 All w/14" tires	31.49 (800)	9.33 (237)	0.47-0.55 (12-14)	22.04 (560)	9.44 (240)	—	A & 11
1992-95 All w/15" tires	31.49 (800)	9.33 (237)	0.31-0.35 (8-9)	22.04 (560)	9.44 (240)	—	A & 11
1992-95 All w/16" tires	31.49 (800)	9.33 (237)	0.23-0.27 (6-7)	22.04 (560)	9.44 (240)	—	A & 11
1996 All w/14" tires	31.49 (800)	9.48 (241)	0.59 (15)	22.04 (560)	9.84 (250)	—	A & 11
1996 All w/15" tires	31.49 (800)	9.48 (241)	0.39 (10)	22.04 (560)	9.84 (250)	—	A & 11
1996 All w/16" tires	31.49 (800)	9.48 (241)	0.31 (8)	22.04 (560)	9.84 (250)	—	A & 11
1997-98 All w/14" tires	31.49 (800)	9.37 (238)	0.55 (14)	22.04 (560)	9.56 (243)	—	A & 11
1997-98 All w/15" tires	31.49 (800)	9.37 (238)	0.31 (8)	22.04 (560)	9.56 (243)	—	A & 11
1997-98 All w/16" tires	31.49 (800)	9.37 (238)	0.23 (6)	22.04 (560)	9.56 (243)	—	A & 11
Alero	AF	RF	SF	AR	RR	SR	Figure
1999-04 All models w/15" tires	32.16 (817)	9.33 (237)	0 (0)	25.00 (635)	9.60 (244)	4.84 (123)	30 & 34
1999-04 All models w/16" tires	32.16 (817)	9.33 (237)	-0.19 (-5)	25.00 (635)	9.60 (244)	4.72 (120)	30 & 34
Aurora	AF	RF	SF	AR	RR	SR	Figure
1995-03 All models	—	—	1.57 (40)	—	—	3.38 (86)	1 & 5
Calais	AF	RF	SF	AR	RR	SR	Figure
1990-91 All Models	21.92 (557)	9.48 (241)	—	22.04 (560)	9.37 (238)	—	A
Ciera/Cutlass Cruiser	AF	RF	SF	AR	RR	SR	Figure
1990-96 All Models	—	—	-.03 (-1)	—	—	7.51 (191)	3 & 4
Cutlass	AF	RF		AR	RR		Figure
1997 All Models	32.16 (817)	9.44 (240)	-.11 (-3)	25.00 (635)	9.84 (250)	—	11
1998-99 All Models	32.16 (817)	9.33 (237)	0 (0)	25.00 (635)	9.60 (244)	—	11
Cutlass Supreme (FWD)	AF	RF	SF	AR	RR	SR	Figure
1990-91 All w/14" tires	23.81 (605)	9.80 (249)	3.30 (84)	20.66 (525)	9.80 (249)	4.09 (104)	1 & 8
1990 All w/15" tires	23.81 (605)	9.80 (249)	3.11-3.30 (79-84)	20.66 (525)	9.80 (249)	3.93-4.09 (100-104)	1 & 8
1990 All w/16" tires	23.81 (605)	9.80 (249)	2.79-2.87 (71-73)	20.66 (525)	9.80 (249)	3.66 (93)	1 & 8

* ± ½ (13) unless otherwise indicated

GM – Oldsmobile (cont.)

HEIGHT SPECIFICATIONS @ CURB WEIGHT - INCHES (MM) *

YEAR MODEL	FRONT				REAR				
Cutlass Supreme (FWD) (cont.)	AF	RF	SF		AR	RR	SR		Figure
1991-93 All w/15" tires	23.81 (605)	9.80 (249)	2.79-2.91 (71-74)		20.66 (525)	9.80 (249)	3.62-3.74 (92-95)		1 & 8
1991-93 All w/16" tires	23.81 (605)	9.80 (249)	2.59-2.79 (66-71)		20.66 (525)	9.80 (249)	3.42-3.70 (87-94)		1 & 8
1994-97 All w/15" tires	23.81 (605)	9.80 (249)	2.83 (72)		20.66 (525)	9.80 (249)	10.27 (261)		1 & 9
1994-97 All w/215-60R16 tires	23.81 (605)	9.80 (249)	2.79 (71)		20.66 (525)	9.80 (249)	10.27 (261)		1 & 9
1994-97 All w/225-60R16 tires	23.81 (605)	9.80 (249)	2.63 (67)		20.66 (525)	9.80 (249)	10.03 (255)		1 & 9
Custom Cruiser Wagon	AF	RF	SF	WF	AR	RR	SR	WR	Figure
1990 All Models	31.49 (800)	10.51 (267)		29.13 (740)	24.60 (625)	10.66 (271)		29.09 (739)	
1991-92 All Models	31.88 (810)	9.92 (252)	5.23 (133)	29.21 (742)	24.21 (615)	10.39 (264)	2.48 (63)	21.65 (550)	6 & 7
Eighty-Eight/LSS/Regency	AF	RF	SF	WF	AR	RR	SR	WR	Figure
1990 All Models	23.62 (600)	9.40 (239)	2.36 (60)	27.91 (709)	23.62 (600)	9.68 (246)	3.42 (87)	27.91 (709)	1
1991 All Models	23.62 (600)	9.40 (239)	2.36 (60)	28.11 (714)	23.62 (600)	9.68 (246)	3.42 (87)	2.83 (72)	1
1992-99 All Models	23.62 (600)	9.40 (239)	2.36 (60)		23.62 (600)	9.68 (246)	3.07 (78)	—	1
Ninety-Eight	AF	RF	SF	WF	AR	RR	SR	WR	Figure
1990 All Models	23.62 (600)	9 3/8 (238)	2.59 (66)	27.91 (709)	23.62 (600)	9.44 (240)	3.14 (80)	27.59 (701)	1
1991 All Models	23.62 (600)	9 3/8 (238)	2.36 (60)	27.91 (709)	23.62 (600)	9.44 (240)	3.07 (78)	27.59 (701)	1
1992-96 All Models	23.62 (600)	9 3/8 (238)	2.36 (60)	—	23.62 (600)	9.44 (240)	3.07 (78)	—	1
Silhouette	AF	RF	SF		AR	RR	SR		Figure
1990-91 All Models w/14" tires	19.88 (505)	11.85 (301)	-.03 (-1)		18.11 (460)	12.12 (308)	7.51 (191)		3 & 4
1990-92 All Models w/15" tires	19.88 (505)	12.08 (307)	-.03 (-1)		18.11 (460)	12.36 (314)	7.51 (191)		3 & 4
1993-94 All Models	23.77 (604)	12.08 (307)	-.03 (-1)		22 1/16 (560)	12.36 (314)	7.51 (191)		3 & 4
1995-97 All Models	see fig. 19	10.47 (266)	-.23 (-6)		see fig. 19	10.90 (277)	7.51 (191)		3 & 4
1998-01 Std. WB w/205/70R15	23.77 (604)	8.97-9.76 (228-248)	2.28-3.07 (58-78)		22.04 (560)	9.84-10.62 (250-270)	5.23-6.02 (133-153)		1 & 29
1998-01 Std. WB w/215/70R15	23.77 (604)	9.25-10.03 (235-255)	2.28-3.07 (58-78)		22.04 (560)	10.15-10.94(258-278)	5.23-6.02 (133-153)		1 & 29
1998-03 Ext. WB	23.77 (604)	9.21-10.00 (234-254)	2.28-3.07 (58-78)		23.77 (604)	10.07-10.86(256-276)	5.47-6.25 (139-159)		1 & 29
2004 Ext. WB w/215/70R15	23.77 (604)	9.60 (244)	2.20 (56)		23.77 (604)	10.47 (266)	5.86 (149)		1 & 29
2004 Ext. WB w/225/60R16	23.77 (604)	9.64 (245)	2.20 (56)		23.77 (604)	10.51 (267)	5.86 (149)		1 & 29
2002-03 All AWD models	23.77 (604)	9.72-9.88 (247-251)	2.20-2.32 (56-59)		22.04 (560)	10.82-11.14(275-283)	3.66-3.93 (93-100)		1 & 20
2004 All AWD models	23.77 (604)	9.72 (247)	2 (56)		22.04 (560)	10.82 (275)	3.66 (93)		1 & 20
Toronado/Trofèo	AF	RF	SF		AR	RR	SR		Figure
1990-91 All Models	—	—	1.81 (46)		—	—	0.94 (24)		1 & 2
1992 All Models	22.83 (580)	9.09 (231)	2.24 (57)		22.16 (563)	9.09 (231)	1.33 (34)		1 & 2

* ± ½ (13) unless otherwise indicated

GM - Pontiac

HEIGHT SPECIFICATIONS @ CURB WEIGHT - INCHES (MM) *

YEAR MODEL	FRONT			REAR			
6000	AF	RF	SF	AR	RR	SR	Figure
1990-91 All Models exc/AWD	—	—	-0.31 (-8)	—	—	7.28 (185)	3 & 4
1990 All Models w/All Wheel Drive	19.88 (505)	9.25 (235)	-0.66 (-17)	18.07 (459)	9.44 (240)	—	3
Aztec	AF	RF	SF	AR	RR	SR	Figure
2001-06 Aztec FWD w/215-70R15	25.59 (650)	9.72 (247)	2.55 (65)	25.59 (650)	10.62 (270)	6.02 (153)	1 & 29
2001-06 Aztec FWD w/215-65R16	25.59 (650)	9.84 (250)	2.55 (65)	25.59 (650)	10.74 (273)	6.02 (153)	1 & 29
2001-06 Aztec FWD w/215-70R16	25.59 (650)	10.27 (261)	2.55 (65)	25.59 (650)	11.18 (284)	6.02 (153)	1 & 29
2002-03 Aztec FWD w/235-55R17	25.59 (650)	10.15 (258)	2.59 (66)	25.59 (650)	11.06 (281)	6.10 (155)	1 & 29
2004-06 Aztec FWD w/225-60R17	25.59 (650)	10.15 (258)	2.59 (66)	25.59 (650)	11.06 (281)	6.10 (155)	1 & 29
2001-06 Aztec AWD w/215-65R16	25.59 (650)	9.84 (250)	2.55 (65)	25.59 (650)	10.27 (261)	3.81 (97)	1 & 20
2001-06 Aztec AWD w/215-70R16	25.59 (650)	10.27 (261)	2.55 (65)	25.59 (650)	11.10 (282)	3.81 (97)	1 & 20
2004-06 Aztec AWD w/225-60R17	25.59 (650)	10.15 (258)	2.59 (66)	25.59 (650)	11.06 (281)	6.10 (155)	1 & 20
Bonneville	AF	RF	SF	AR	RR	SR	Figure
1990-91 All Models	23.62 (600)	9.48 (241)	2.55 (65)	23.62 (600)	9.48 (241)	3.14 (80)	1 & 1a
1992-99 All Models	23.62 (600)	9.40 (239)	2.36 (60)	23.62 (600)	9.68 (246)	3.07 (78)	1 & 1a
2000-05 All Models	—	—	1.57 (40)	—	—	3.38 (86)	1 & 5
Firebird/Trans Am	AF	RF	SF	AR	RR	SR	Figure
1990-92 All w/15" tires	32.59 (828)	8.22 (209)	1.06-1.18 (27-30)	17.24 (438)	8.26 (210)	4.37 (111)	1 & 15
1990-92 All w/16" tires	32.59 (828)	8.22 (209)	1.22 (31)	17.24 (438)	8.26 (210)	4.44 (113)	1 & 15
1993-97 All w/215-60 tires	32.59 (828)	8.03 (204)	1.25 (32)	17.24 (438)	8.18 (208)	4.76 (121)	1 & 15
1993-02 All w/235-55 tires	32.59 (828)	8.03 (204)	1.18 (30)	17.24 (438)	8.18 (208)	4.64 (118)	1 & 15
1993-02 All w/245-50 tires	32.59 (828)	8.03 (204)	1.45 (37)	17.24 (438)	8.18 (208)	4.92 (125)	1 & 15

G3	CF	CR	Figure
2007-09 G3 (Mexico)	13.54 (344)	13.50 (343)	A

G5 (G5-Pursuit-Canada)	AF	RF	SF	AR	RR	SR	Figure
2007-08 Models w/15" tires	21.41 (544)	8.81 (224)	-.82 (-21)	22.04 (560)	9.17 (233)	9.37 (238)	30 & 31
2007-08 Models w/16" tires	21.41 (544)	8.81 (224)	-.90 (-23)	22.04 (560)	9.17 (233)	8.97 (228)	30 & 31
2007-08 Models w/17" tires	21.41 (544)	8.81 (224)	-1.02 (-26)	22.04 (560)	9.17 (233)	8.81 (224)	30 & 31
2007-08 Models w/18" tires	21.41 (544)	8.81 (224)	-1.33 (-34)	22.04 (560)	9.17 (233)	8.38 (213)	30 & 31

* ± ½ (13) unless otherwise indicated

GM – Pontiac (cont.)

Year	Model	Height Specifications @ Curb Weight - Inches (mm) *						
		Front			Rear			
G6		SF		WF	SR		WR	Figure
2005	All exc/GT	2.67 (68)		—	6.73 (171)		—	37 & 36
2005	GT	2.28 (58)		—	6.61 (168)		—	37 & 36
2006-09	All w/16" tires	2.59 (66)		27.87 (708)	7.24 (184)		28.18 (716)	A, 37 & 36
2006-09	All w/17" tires	2.59 (66)		27.79 (706)	7.24 (184)		28.11 (714)	A, 37 & 36
2006-09	Coupe & Sedan w/18" tires	2.59 (66)		28.26 (718)	7.24 (184)		28.54 (725)	A, 37 & 36
2006-09	Convertible w/18" tires	2.59 (66)		28.34 (720)	7.24 (184)		28.62 (727)	A, 37 & 36
G8		AF	RF	CF	AR	RR	CR	Figure
2008-09	All Models (FE2 susp)	21.41 (544)	7.40 (188)	15.47 (393)	21.41 (544)	7.63 (194)	15.39 (391)	A
Grand Am		AF	RF	SF	AR	RR	SR	Figure
1990	All w/13" tires	21.92 (557)	9.64 (245)	3.89 (99)	22.04 (560)	9.76 (248)	—	10
1990	All w/14" tires	21.92 (557)	9.60 (244)	3.74-3.97 (95-101)	22.04 (560)	9.72 (247)	—	10
1990	All w/16" tires	21.92 (557)	9.60 (244)	3.54 (90)	22.04 (560)	9.72 (247)	—	10
1991	All w/14" tires	21.92 (557)	9.56 (243)	3.74-3.93 (95-100)	22.04 (560)	9.60 (244)	—	10
1991	All w/16" tires	21.92 (557)	9.56 (243)	3.58 (91)	22.04 (560)	9.60 (244)	—	10
1992-95	All w/14" tires	31.49 (800)	9.33 (237)	0.43-0.59 (11-15)	22.04 (560)	9.44 (240)	—	11
1992-95	All w/15" tires	31.49 (800)	9.33 (237)	0.31-0.35 (8-9)	22.04 (560)	9.44 (240)	—	11
1992-95	All w/16" tires	31.49 (800)	9.33 (237)	0.23-0.27 (6-7)	22.04 (560)	9.44 (240)	—	11
1996	All Models	31.49 (800)	9.48 (241)		22.04 (560)	9.84 (250)	—	A
1997-98	All w/14" tires	31.49 (800)	9.37 (238)	0.55 (14)	22.04 (560)	9.56 (243)	—	11
1997-98	All w/15" tires	31.49 (800)	9.37 (238)	0.35 (9)	22.04 (560)	9.56 (243)	—	11
1997-98	All w/16" tires	31.49 (800)	9.37 (238)	0.23 (6)	22.04 (560)	9.56 (243)	—	11
1999-05	All models w/15" tires	32.16 (817)	9.33 (237)	0 (0)	25.00 (635)	9.60 (244)	4.84 (123)	1 & 5
1999-05	All models w/16" tires	32.16 (817)	9.33 (237)	-0.19 to -0.23 (-5 to -6)	25.00 (635)	9.60 (244)	4.72 (120)	1 & 5
Grand Prix		AF	RF	SF	AR	RR	SR	Figure
1990	All models w/14" tires	23.81 (605)	9.88 (251)	3.30 (84)	20.66 (525)	10.11 (257)	4.48 (114)	1 & 8
1990	All models w/15" tires	23.81 (605)	9.88 (251)	3.11-3.14 (79-80)	20.66 (525)	10.11 (257)	4.29-4.40 (109-112)	1 & 8
1990	All models w/16" tires	23.81 (605)	9.88 (251)	2.79-2.91 (71-74)	20.66 (525)	10.11 (257)	4.05-4.17 (103-106)	1 & 8
1991	All models w/15" tires	23.81 (605)	9.88 (251)	2.79-2.91 (71-74)	20.66 (525)	10.11 (257)	4.01-4.13 (102-105)	1 & 8
1991	All models w/16" tires	23.81 (605)	9.88 (251)	2.63-2.87 (67-73)	20.66 (525)	10.11 (257)	3.81-4.05 (97-103)	1 & 8
1992	All models w/15" tires	23.81 (605)	9.80 (249)	2.87-2.91 (73-74)	20.66 (525)	9.80 (249)	4.01-4.13 (102-105)	1 & 8
1993	All models w/15" tires	23.81 (605)	9.80 (249)	2.87-3.03 (73-77)	20.66 (525)	9.80 (249)	3.62-3.74 (92-95)	1 & 8
1992-93	All w/245-50ZR16 tires	23.81 (605)	9.80 (249)	2.99 (76)	20.66 (525)	9.80 (249)	3.81 (97)	1 & 8
1992-93	All others w/16" tires	23.81 (605)	9.80 (249)	2.67-2.79 (68-71)	20.66 (525)	9.80 (249)	3.42-3.62 (87-92)	1 & 8
1994-96	All models w/15" tires	23.81 (605)	9.80 (249)	2.91-2.99 (74-76)	20.66 (525)	9.80 (249)	10.23-10.39 (260-264)	1 & 9
1994-96	All models w/16" tires	23.81 (605)	9.80 (249)	2.63-2.91 (67-74)	20.66 (525)	9.80 (249)	10.03-10.39 (255-264)	1 & 9
1997	All base suspension	23.81 (605)	8.38 (213)	2.08 (53)	20.66 (525)	8.38 (213)	10.35 (263)	1 & 9
1997	All FE3 suspension.	23.81 (605)	8.38 (213)	1.81 (46)	20.66 (525)	8.38 (213)	10.11 (257)	1 & 9
1998-02	All Models w/205-70R15	23.81 (605)	8.38 (213)	2.16 (55)	20.66 (525)	8.38 (213)	10.27 (261)	1 & 26
1998-02	All Models w/225-60R16	23.81 (605)	8.38 (213)	1.88 (48)	20.66 (525)	8.38 (213)	10.03 (255)	1 & 26
2003	All Models w/205-70R15	23.81 (605)	8.38 (213)	1.65 (42)	20.66 (525)	8.38 (213)	10.27 (261)	1 & 26
2003	All Models w/225-60R16	23.81 (605)	8.38 (213)	1.37-1.41 (35-36)	20.66 (525)	8.38 (213)	10.03-10.07 (255-256)	1 & 26
2004-06	All Models	23.81 (605)	8.38 (213)	1.37-1.41 (35-36)	20.66 (525)	8.38 (213)	10.03-10.07 (255-256)	1 & 26
2007-08	All w/16"& 17" Tires exc/GT	—	—	1.65 (42)	—	—	10.51 (267)	1 & 26
2007	GT w/17" tires	—	—	1.81 (46)	—	—	10.55 (268)	1 & 26
2007-08	GXP w/18" tires	—	—	1,88 (48)	—	—	10.11 (257)	1 & 26
Lemans		AF	RF	SF	AR	RR	SR	Figure
1989-93	All Models	—	7.67 (195)	—	—	7.51 (191)	—	
Pursuit/G5-Pursuit (CAN)		AF	RF	SF	AR	RR	SR	Figure
2005-06	Models w/15" tires	21.41 (544)	8.81 (224)	-0.82 (-21)	22.04 (560)	9.17 (233)	9.37 (238)	30 & 31
2005-06	Models w/16" tires	21.41 (544)	8.81 (224)	-0.90 (-23)	22.04 (560)	9.17 (233)	8.97 (228)	30 & 31
2005-06	Models w/17" tires	21.41 (544)	8.81 (224)	-1.02 (-26)	22.04 (560)	9.17 (233)	8.81 (224)	30 & 31
Solstice		SF		WF	SR		WR	Figure
2006-09	All Models	1.45 (37)		28.77 (731)	2.44 (62)		28.74 (730)	A, 20 (F & R)
Sunbird/2000		AF	RF	SF	AR	RR	SR	Figure
1990	All Coupe/Sedan	21.92 (557)	9.29 (236)	3.54-3.81 (90-97)	21.81 (554)	9.29 (236)	—	10
1990	All Wagon	21.92 (557)	9.29 (236)	3.42-4.01 (87-102)	21.81 (554)	9.29 (236)	—	10
1991	All Models	21.92 (557)	9.25 (235)	3.42-3.85 (87-98)	21.81 (554)	9.13 (232)	—	10
1992-94	All Models	21.92 (557)	9.25 (235)	.07-.27 (2-7)	21.81 (554)	9.13 (232)	—	11
Sunfire		AF	RF	SF	AR	RR	SR	Figure
1995	All w/70 series tires	21.92 (557)	9.17 (233)	0.35-0.39 (9-10)	21.81 (554)	9.44 (240)	—	11
1995	All w/55 & 65 series tires	21.92 (557)	9.17 (233)	0.11-0.15 (3-4)	21.81 (554)	9.44 (240)	—	11
1996	All models w/14" tires	21.92 (557)	9.33 (237)	0.51 (13)	21.81 (554)	9.76 (248)	—	11
1996	All models w/15" tires	21.92 (557)	9.33 (237)	0.31 (8)	21.81 (554)	9.76 (248)	—	11

* ± ½ (13) unless otherwise indicated

CHASSIS / SUSP. HEIGHTS

GM – Pontiac (cont.)

Height Specifications @ Curb Weight - Inches (mm) *

Year Model	Front AF	Front RF	Front SF	Rear AR	Rear RR	Rear SR	Figure
Sunfire (cont.)	AF	RF	SF	AR	RR	SR	Figure
1996 All models w/16" tires	21.92 (557)	9.33 (237)	0.23 (6)	21.81 (554)	9.68 (246)	—	11
1997 All models w/14" tires	21.92 (557)	9.17 (233)	0.47 (12)	21.81 (554)	9.44 (240)	—	11
1997 All models w/15" tires	21.92 (557)	9.17 (233)	0.23 (6)	21.81 (554)	9.44 (240)	—	11
1997 All models w/16" tires	21.92 (557)	9.17 (233)	0.15 (4)	21.81 (554)	9.44 (240)	—	11
1998-02 All Coupe GT (FE2 susp.)	32.16 (817)	9.37 (238)	-0.07 (-2)	21.65 (550)	9.64 (245)	-1.69 (-43)	30 & 33
1998-02 All Convertible	32.16 (817)	9.37 (238)	0.19-0.27 (5-7)	21.65 (550)	9.64 (245)	-1.41 (-36)	30 & 33
1998-02 All Others w/Active susp (FE0)	32.16 (817)	9.17 (233)	0.0-0.15 (0-4)	21.65 (550)	9.44 (240)	-1.45 to -1.65(-37 to -42)	30 & 33
1998-02 All Others w/Base susp (FE1)	32.16 (817)	9.17 (233)	0.03 (1)	21.65 (550)	9.44 (240)	-1.65 (-42)	30 & 33
2003-05 All Base susp. (FE1)	—	—	-0.03 (-1)	—	—	—	30 & 33
2003-05 All Sport susp. (FE2)	—	—	-0.15 (-4)	—	—	—	30 & 33
Trans Sport/Montana	AF	RF	SF	AR	RR	SR	Figure
1990-91 All Models w/14" tires	19.88 (505)	11.85 (301)	-.03 (-1)	18.11 (460)	12.12 (308)	7.51 (191)	3 & 4
1990-92 All Models w/15" tires	19.88 (505)	12.08 (307)	-.03 (-1)	18.11 (460)	12.36 (314)	7.51 (191)	3 & 4
1993-94 All Models	23.77 (604)	12.08 (307)	-.03 (-1)	22.04 (560)	12.36 (314)	7.51 (191)	3 & 4
1995-97 All Models	see fig. 19	10.47 (266)	-.23 (-6)	see fig. 19	10.90 (277)	7.51 (191)	3 & 4
1998-03 Std. WB w/205/70R15	23.77 (604)	8.97-9.76 (228-248)	2.28-3.07 (58-78)	22.04 (560)	9.84-10.62 (250-270)	5.23-6.02 (133-153)	1 & 29
1998-03 Std. WB w/215/70R15	23.77 (604)	9.25-10.03 (235-255)	2.28-3.07 (58-78)	22.04 (560)	10.15-10.94 (258-278)	5.23-6.02 (133-153)	1 & 29
2004 Std. WB w/215/70R15	23.77 (604)	9.64 (245)	2.20 (56)	22.04 (560)	10.51 (267)	5.62 (143)	1 & 29
2004 Std. WB w/225/60R16	23.77 (604)	9.64 (245)	2.20 (56)	22.04 (560)	10.55 (268)	5.62 (143)	1 & 29
1998-03 Ext. WB	23.77 (604)	9.21-10.00 (234-254)	2.28-3.07 (58-78)	23.77 (604)	10.07-10.86 (256-276)	5.47-6.25 (139-159)	1 & 29
2004-05 Ext. WB w/215/70R15	23.77 (604)	9.60 (244)	2.20 (56)	23.77 (604)	10.47 (266)	5.86 (149)	1 & 29
Trans Sport/Montana (cont.)	AF	RF	SF	AR	RR	SR	Figure
2004-05 Ext. WB w/225/60R16	23.77 (604)	9.64 (245)	2.20 (56)	23.77 (604)	10.51 (267)	5.86 (149)	1 & 29
2002-04 All AWD models	23.77 (604)	9.72-9.88 (247-251)	2.20-2.32 (56-59)	22.04 (560)	10.82-11.14 (275-283)	3.66-3.93 (93-100)	1 & 20
2005-09 Montana SV6 FWD exc/Y3G	23.77 (604)	12.20 (310)	2.55 (65)	22.04 (560)	11.85 (301)	6.45 (164)	1 & 29
2005-09 Montana w/Y3G Mobility Pkg.	23.77 (604)	12.36 (314)	2.55 (65)	22.04 (560)	12.40 (315)	7.51 (191)	1 & 29
2005-09 Montana w/Y3G & FE3 Susp.	23.77 (604)	12.24 (311)	2.55 (65)	22.04 (560)	12.00 (305)	6.73 (171)	1 & 29
2005-06 Montana SV6 AWD models	23.77 (604)	12.16 (309)	2.55 (65)	22.04 (560)	11.69 (297)	3.66 (93)	1 & 20
Vibe	AF	RF	SF	AR	RR	SR	Figure
2003-08 All FWD Models	—	—	2.00 (51)	—	—	1.61 (41)	41 & 33
2003-06 All AWD Models	—	—	2.24 (57)	—	—	1.49 (38)	41 & 1a
2009 All FWD 1.8L	—	—	3.54 (90)	—	—	1.81 (46)	41 & 33
2009 All FWD 2.4L 16" tires	—	—	3.62 (92)	—	—	1.85 (47)	41 & 33
2009 All FWD 2.4L 17" tires	—	—	3.62 (92)	—	—	1.81 (46)	41 & 33
2009 All GT	—	—	3.62 (92)	—	—	2.24 (57)	41 & 40
2009 All AWD	—	—	3.66 (93)	—	—	2.16 (55)	41 & 40

Wave	CF	CR	Figure
2004-09 Wave (Canada)	13.54 (344)	13.50 (343)	A

* ± ½ (13) unless otherwise indicated

GM - Saturn

Height Specifications @ Curb Weight - Inches (mm) *

Year Model	Front SF	Front WF	Rear SR	Rear WR	Figure
Astra	SF	WF	SR	WR	Figure
2008-09 Astra 3-Door	n/a	25.78 (655)	n/a	25.94 (659)	A
2008-09 Astra 5-Door	n/a	26.18 (665)	n/a	26.37 (670)	A
Aura	SF	WF	SR	WR	Figure
2007-09 Aura with 215/60R16 (Hybrid)	2.59 (66)	28.07 (713)	7.48 (190)	28.50 (724)	A, 37 & 36
2007-09 Aura with 225/50R17	2.59 (66)	27.83 (707)	7.48 (190)	28.30 (719)	A, 37 & 36
2007-09 Aura with 225/50R18	2.59 (66)	28.34 (720)	7.48 (190)	28.85 (733)	A, 37 & 36

Year Model	AF	RF	SF	AR	RR	SR	Figure
ION	AF	RF	SF	AR	RR	SR	Figure
2003-05 ION w/14" tires	21.41 (544)	8.85 (225)	-0.27 (-7)	22.04 (560)	9.21 (234)	8.66 (220)	30 & 31
2003-05 ION w/15" tires	21.41 (544)	8.85 (225)	-0.39 (-10)	22.04 (560)	9.21 (234)	8.74 (222)	30 & 31
2003-05 ION w/16" tires	21.41 (544)	8.85 (225)	-0.86 (-22)	22.04 (560)	9.21 (234)	9.13 (232)	30 & 31
2003-05 ION w/17" tires	21.41 (544)	8.46 (215)	-1.10 (-28)	22.04 (560)	8.81 (224)	9.05 (230)	30 & 31
2006-07 ION w/15" tires	21.41 (544)	8.85 (225)	-0.39 (-10)	22.04 (560)	9.21 (234)	8.97 (228)	30 & 31
2006-07 ION w/16" tires	21.41 (544)	8.85 (225)	-0.86 (-22)	22.04 (560)	9.21 (234)	8.50 (216)	30 & 31
2006-07 ION w/17" tires	21.41 (544)	8.46 (215)	-1.10 (-28)	22.04 (560)	8.81 (224)	8.89 (226)	30 & 31
Relay	AF	RF	SF	AR	RR	SR	Figure
2005-07 Relay FWD	23.77 (604)	12.20 (310)	2.55 (65)	22.04 (560)	11.85 (301)	6.45 (164)	1 & 29
2005-07 Relay AWD	23.77 (604)	12.55 (319)	2.55 (65)	22.04 (560)	11.69 (297)	3.66 (93)	1 & 20

* ± ½ (13) unless otherwise indicated

GM – Saturn (cont.)

Height Specifications @ Curb Weight - Inches (mm) *

Year	Model	Front AF	Front RF	Front SF	Rear AR	Rear RR	Rear SR	Figure
	S-Series	AF	RF	SF	AR	RR	SR	Figure
1991-96	S-Sedan	4.60 (117)	7.75-9.01 (197-229)	—	4.13 (105)	7.95-9.21 (202-234)	—	A
1997-02	S-Sedan	4.60 (117)	7.91-9.17 (201-233)	—	4.13 (105)	7.95-9.21 (202-234)	—	A
1991-96	S-Coupe exc/1993-96 SC1	5.90 (150)	7.83-9.09 (199-231)	—	5.43 (138)	7.91-9.40 (201-239)	—	A
1993-96	S-Coupe SC1	5.90 (150)	8.18-9.44 (208-240)	—	5.43 (138)	8.26-9.52 (210-242)	—	A
1997-02	S-Coupe	5.90 (150)	7.87-9.13 (200-232)	—	5.43 (138)	7.91-9.17 (201-233)	—	A
1993-97	S-Wagon	4.60 (117)	7.83-9.09 (199-231)	—	4.13 (105)	8.14-9.48 (207-241)	—	A
1998-02	S-Wagon	4.60 (117)	7.87-9.17 (200-233)	—	4.13 (105)	7.95-9.21 (202-234)	—	A

Year	Model	SF	WF	SR	WR	Figure
	Sky	SF	WF	SR	WR	Figure
2007-09	Sky All Models	1.45 (37)	28.77 (731)	2.44 (62)	28.74 (730)	A, 20 (F & R)
	VUE	SF	WF	SR	WR	Figure
2006	VUE exc/Redline	0.66 (17)	—	0.39 (10)	—	1 & 2
2006	VUE Redline	-0.35 (-9)	—	-0.62 (-16)	—	1 & 2
2007-09	VUE exc/Red Line, Green Line	0.62 (16)	31.29 (795)	1.65 (42)	31.65 (804)	A, 1 & 2
2007-09	VUE Red Line	-0.35 (-9)	30.51 (775)	-0.62 (-16)	30.62 (778)	A, 1 & 2
2007-09	VUE Green Line	-0.19 (-5)	30.39 (772)	1.29 (33)	31.18 (792)	A, 1 & 2

* ± ⅜ (10) except when range is specified

GM Truck & SUV – Buick / Cadillac / Chevrolet / GMC / Oldsmobile / Hummer

Height Specifications @ Curb Weight - Inches (mm) *

Year	Model	Front	Rear	Figure
	City Express Van	Front (WF)	Rear (WR)	Figure
2015-17	City Express Van	27.56 (700)	29.02 (737)	A
	Equinox/Torrent/Terrain	Front (SF)	Rear (SR)	Figure
2005-09	Equinox/Torrent exc/Sport, GTP (16" & 17")	0.70 (18)	1.65 (42)	1 & 2
2008-09	Equinox/Torrent Sport, GTP (18" tires)	-0.33 (-8.5)	0.59 (15)	1 & 2
2010-17	Equinox/Terrain	2.00 (51)	1.53 (39)	1 & 2
	Acadia/Enclave/Outlook/Traverse	Front (SF)	Rear (SR)	Figure
2007-17	All models	2.25 (57) Side/Side Max. =0.39 (10)	3.23 (82) Side/Side Max. =0.39 (10)	38 & 39
	S/T Series	Front (SF)	Rear (SR)	Figure
1990-96	All 4X2 models	2.87 (73)	6.69 (170)	20 & 22
1990-96	All 4X4 models	4.80 (122)	6.69 (170)	21 & 22
1991-96	Bravada	4.80 (122)	6.69 (170)	21 & 22
1997-01	Bravada	4.80 (122)	5.47 (139)	21 & 22
1997-03	S10 Pickup / Sonoma 4X2 exc/ZQ8 & V4A	2.87 (73)	5.74 (146)	20 & 22
1997-03	S10 Pickup / Sonoma 4X2 ZQ8 & V4A	2.51 (64)	4.80 (122)	20 & 22
1997-03	S10 Pickup / Sonoma 4X4 exc/ZR2	4.80 (122)	5.74 (146)	21 & 22
1997-03	S10 Pickup / Sonoma 4X4 ZR2	4.88 (124)	5.74 (146)	21 & 22
2004	S10 Pickup 4X4	4.80 (122)	5.47 (139)	21 & 22
1997-04	S10 Blazer / Jimmy 4X2 exc/ZQ8 & V4A	2.87 (73)	5.47 (139)	20 & 22
1997-04	S10 Blazer / Jimmy 4X2 ZQ8 & V4A	1.61 (41)	4.21 (107)	20 & 22
2003-04	S10 Blazer 4X2 ZR2	4.40 (112)	5.07 (129)	20 & 22
1997-04	S10 Blazer / Jimmy 4X4 exc/ZQ8, V4A, ZR2	4.80 (122)	5.47 (139)	21 & 22
1997-04	S10 Blazer /Jimmy 4X4 ZQ8 & V4A	4.80 (122)	4.21 (107)	21 & 22
2003-04	S10 Blazer 4X4 ZR2	4.88 (124)	5.07 (129)	21 & 22
	S/T Series (GMT 355/360/370)	Front (SF)	Rear (SR)	Figure
2004-12	Colorado / Canyon 4X2 Z85	3.07 (78), Side/Side Max. = 0.47 (12)	6.85 (174), Side/Side Max. = 0.39 (10)	21 & 25a
2004-12	Colorado / Canyon 4X2 ZQ8 Sport Susp	2.16 (55), Side/Side Max. = 0.47 (12)	5.27 (134), Side/Side Max. = 0.39 (10)	21 & 25a
2004-12	Colorado / Canyon 4X2 Z71 High Stance	4.60 (117), Side/Side Max. = 0.47 (12)	6.73 (171), Side/Side Max. = 0.39 (10)	21 & 25a
2004-12	Colorado / Canyon 4X4 All Models	4.60 (117), Side/Side Max. = 0.47 (12)	6.73 (171), Side/Side Max. = 0.39 (10)	21 & 25a
2002-09	Trailblazer, Envoy, Bravada 4X2	3.07 (78), Side/Side Max. = 0.94 (24)	6.10 (155), Side/Side Max. = 0.47 (12)	21 & 22
2002-09	Trailblazer, Envoy, Bravada 4X4 exc/Air Susp.	3.07 (78), Side/Side Max. = 0.94 (24)	6.10 (155), Side/Side Max. = 0.47 (12)	21 & 22
2002-09	Rainier, Envoy, Bravada 4X4 w/Air Susp.	3.07 (78), Side/Side Max. = 0.94 (24)	5.31 (135), Side/Side Max. = 0.31 (8)	21 & 22
2006-09	Trailblazer SS	2.08 (53), Side/Side Max. = 0.94 (24)	4.33 (110), Side/Side Max. = 0.47 (12)	21 & 22
2003-05	SSR	1.85 (47), Side/Side Max. = 0.94 (24)	4.56 (116), Side/Side Max. = 0.47 (12)	21 & 22
	S/T Series (GMT 31XX)	Front (WF)	Rear (WR)	Figure
2015-17	All exc/Crew Cab w/long bed	35.0 (889)	36.8 (934)	A
2015-17	All Crew Cab w/long bed (**T/U/V43)	35.0 (889)	37.0 (939)	A
	Astro/Safari	Front (SF)	Rear (SR)	Figure
1990-05	All 2WD models	3.14 (80), Side/Side Max. = 0.47 (12)	5.51 (140), Side/Side Max. = 0.39 (10)	20 & 23
1990-05	All AWD models	5.35 (136), Side/Side Max. = 0.47 (12)	5.51 (140), Side/Side Max. = 0.39 (10)	20 & 23
	C/K Trucks (4X2)	Front (SF)	Rear (SR)	Figure
1990-91	All R (4X2 pickup, 4X2 Suburban - see caster charts in Alignment Instructions Section			
1990-00	All C (4X2) (GMT400) exc/3500 HD	3.74 (95)	7.16 (182)	24 & 25
1999-03	C1500 Sierra/Silverado (GMT800)	3.62 (92), Side/Side Max. = 0.47 (12)	n/a	21
2004-07	C1500 Sierra/Silverado exc/RPOs (GMT900)	3.62 (92), Side/Side Max. = 0.47 (12)	n/a	21
2004-07	C1500 Sierra/Silverado w/QSS,B2E,B4V,B4U	3.77 (96), Side/Side Max. = 0.47 (12)	n/a	21
2004-07	C1500 Sierra/Silverado w/NHT (Limited)	4.56 (116), Side/Side Max. = 0.47 (12)	n/a	21
2007-13	C1500 (incl. Hybrid) (GMT900)	3.58 (91) Side/Side Max. = 0.47 (12)	Side/Side Max. = 0.47 (12)	21 & 35
2014-17	C1500 All Models	3.58 (91) Side/Side Max. = 0.47 (12)	Side/Side Max. = 0.47 (12)	21 & 35
1999-01	C2500 Sierra/Silverado	3.74 (95), Side/Side Max. = 0.47 (12)	n/a	21

* ± ⅜ (10) except when range is specified

GM Truck & SUV – Buick / Cadillac / Chevrolet / GMC / Oldsmobile / Hummer (cont.)

Year Model	Height Specifications @ Curb Weight - Inches (mm) *		
C/K Trucks (4X2) (cont.)	**Front (SF)**	**Rear (SR)**	**Figure**
2002-07 C2500/3500 Sierra/Silverado	4.56 (116), Side/Side Max. = 0.47 (12)	n/a	21
2002-07 C2500 Sierra/Silverado w/ZW9 (no box)	4.17 (106), Side/Side Max. = 0.47 (12)	n/a	21
2007-10 C2500 w/pickup box	4.56 (116) Side/Side Max. = 0.47 (12)	Side/Side Max. = 0.47 (12)	21 & 35
2007-10 C2500 w/o pickup box	4.17 (106) Side/Side Max. = 0.47 (12)	Side/Side Max. = 0.47 (12)	21 & 35
2007-10 C3500	4.56 (116) Side/Side Max. = 0.47 (12)	Side/Side Max. = 0.47 (12)	21 & 35
2011-17 C2500/C3500	5.27 (134) Side/Side Max. = 0.47 (12)	Side/Side Max. = 0.47 (12)	21 & 35
C/K SUVs (4X2)	**Front (SF)**	**Rear (SR)**	**Figure**
2002-03 C1500 Utility	3.62 (92), Side/Side Max. = 0.47 (12)	Side/Side Max. = 0.47 (12)	21
2004-06 C1500 Utility exc/Escalade, Tahoe Police	4.56 (116), Side/Side Max. = 0.47 (12)	Side/Side Max. = 0.47 (12)	21
2004-06 C1500 Escalade	4.17 (106), Side/Side Max. = 0.47 (12)	Side/Side Max. = 0.47 (12)	21
2004-06 C1500 Tahoe Police Vehicle	3.77 (96), Side/Side Max. = 0.47 (12)	Side/Side Max. = 0.47 (12)	21
2002-06 C2500 Utility	4.56 (116), Side/Side Max. = 0.47 (12)	Side/Side Max. = 0.47 (12)	21
2007-15 C1500 Tahoe, Yukon	3.58 (91) Side/Side Max. = 0.47 (12)	Side/Side Max. = 0.47 (12)	21 & 35
2016-17 C1500 Tahoe, Yukon w/RPO Z85, ZW7	3.93 (100) Side/Side Max. = 0.47 (12)	Side/Side Max. = 0.47 (12)	21 & 35
2016-17 C1500 Tahoe, Yukon w/RPO Z95	3.54 (90) Side/Side Max. = 0.47 (12)	Side/Side Max. = 0.47 (12)	21 & 35
2016-17 C1500 Tahoe, Yukon w/RPO Z71	4.29 (109) Side/Side Max. = 0.47 (12)	Side/Side Max. = 0.47 (12)	21 & 35
2007-14 C1500 Tahoe, Yukon w/Police pkg.	3.30 (84) Side/Side Max. = 0.47 (12)	Side/Side Max. = 0.47 (12)	21 & 35
2015-17 C1500 Tahoe, Yukon w/Police pkg.	3.18 (81) Side/Side Max. = 0.47 (12)	Side/Side Max. = 0.47 (12)	21 & 35
2007-14 C1500 Suburban, Yukon XL, Escalade, Aval.	3.58 (91) Side/Side Max. = 0.47 (12)	Side/Side Max. = 0.47 (12)	21 & 35
2015-17 C1500 Suburban, Yukon XL, Escalade Z85	3.93 (100) Side/Side Max. = 0.47 (12)	Side/Side Max. = 0.47 (12)	21 & 35
2016-17 C1500 Suburban, Yukon XL, Z71	4.29 (109) Side/Side Max. = 0.47 (12)	Side/Side Max. = 0.47 (12)	21 & 35
2016-17 C1500 Suburban, Yukon XL, Z95	3.54 (90) Side/Side Max. = 0.47 (12)	Side/Side Max. = 0.47 (12)	21 & 35
2007-13 C2500 Suburban, Yukon XL	4.56 (116) Side/Side Max. = 0.47 (12)	Side/Side Max. = 0.47 (12)	21 & 35
C/K Trucks (4X4)	**Front (SF)**	**Rear (SR)**	**Figure**
1990-00 All K1500, K2500 exc/L65, C6P (GMT400)	6.18 (157)	7.16 (182)	21 & 25
1990-00 All K1500, K2500 with/L65, C6P (GMT400)	5.70 (145)	n/a	21
1990-00 All K3 (4X4) (std.) (GMT400)	5.70 (145)	7.16 (182)	21 & 25
1999-07 K1500/2500/3500 Sierra/Silverado (GMT800)	4.56 (116), Side/Side Max. = 0.47 (12)	n/a	21
2007-17 K1500	3.97 (101) Side/Side Max. = 0.47 (12)	Side/Side Max. = 0.47 (12)	21 & 35
2007-17 K1500 w/RPO Z85 (HD), VYU (Snow plow)	4.37 (111) Side/Side Max. = 0.47 (12)	Side/Side Max. = 0.47 (12)	21 & 35
2009-12 K1500 Hybrid	3.58 (91) Side/Side Max. = 0.47 (12)	Side/Side Max. = 0.47 (12)	21 & 35
2007-10 K2500 w/Pickup box	4.56 (116) Side/Side Max. = 0.47 (12)	Side/Side Max. = 0.47 (12)	21 & 35
2007-10 K2500 w/o Pickup box	4.17 (106) Side/Side Max. = 0.47 (12)	Side/Side Max. = 0.47 (12)	21 & 35
2007-10 K3500	4.56 (116) Side/Side Max. = 0.47 (12)	Side/Side Max. = 0.47 (12)	21 & 35
2011-17 K2500/K3500	5.27 (134) Side/Side Max. = 0.47 (12)	Side/Side Max. = 0.47 (12)	21 & 35
C/K SUVs (4X4)	**Front (SF)**	**Rear (SR)**	**Figure**
1999-00 Escalade	6.18 (157), Side/Side Max. = 0.47 (12)	7.16 (182)	21 & 25
2001-03 K1500, K2500, K3500 Utility	4.56 (116), Side/Side Max. = 0.47 (12)	n/a	21
2004-06 Escalade	3.77 (96), Side/Side Max. = 0.47 (12)	n/a	21
2007-13 K1500 Avalanche w/o Autoride	3.97 (101)	n/a	21 & 35
2007-13 K1500 Avalanche w/Autoride	3.58 (91)	n/a	21 & 35
2007-14 K1500 Tahoe, Yukon w/o Autoride	3.97 (101)	n/a	21 & 35
2007-14 K1500 Tahoe, Yukon w/Autoride	3.58 (91)	n/a	21 & 35
2008-14 K1500 All Hybrids	3.58 (91)	n/a	21 & 35
2007-14 K1500 Suburban, Yukon XL w/o Autoride	3.97 (101)	n/a	21 & 35
2007-14 K1500 Suburban, Yukon XL w/Autoride	3.58 (91)	n/a	21 & 35
2007-14 K1500 Escalade w/o Magnetic Ride Control	3.97 (101)	n/a	21 & 35
2007-14 K1500 Escalade w/Magnetic Ride Control	3.58 (91)	n/a	21 & 35
2007-14 K2500 Suburban, Yukon XL	4.56 (116)	n/a	21 & 35
G Series Vans	**Front (SF)**	**Rear (SR)**	**Figure**
1990-96 All G Vans see caster charts in Alignment Instructions Section			
1996-02 All Express/Savana without auxiliary springs	3.74 (95) Side/Side Max. = 0.50 (13)	4.25 (108) Side/Side Max. = 0.47 (13)	20 & 22a
1996-02 All Express/Savana with auxiliary springs	374 (95) Side/Side Max. = 0.50 (13)	5.23 (133) Side/Side Max. = 0.47 (13)	20 & 23
2003-12 RWD 1500 6200 GVW	3.85 (98), Side/Side Max. = 0.47 (12)	4.88 (124), Side/Side Max. = 0.47 (12)	21 & 32
2003-12 RWD 1500 7200 GVW	3.85 (98), Side/Side Max. = 0.47 (12)	5.74 (146), Side/Side Max. = 0.47 (12)	21 & 32
2003-12 RWD 2500 7300 GVW	3.85 (98), Side/Side Max. = 0.47 (12)	5.94 (151), Side/Side Max. = 0.47 (12)	21 & 32
2003-12 RWD 2500 8500 & 8600 GVW	4.25 (108), Side/Side Max. = 0.47 (12)	5.43 (138), Side/Side Max. = 0.47 (12)	21 & 32
2003-13 RWD 3500 9600 GVW	4.25 (108), Side/Side Max. = 0.47 (12)	5.70 (145), Side/Side Max. = 0.47 (12)	21 & 32
2003-12 RWD 3500 10000 -11500 GVW	4.25 (108), Side/Side Max. = 0.47 (12)	6.22 (158), Side/Side Max. = 0.47 (12)	21 & 32
2003-12 RWD 3500 12000 & 12300 GVW	4.25 (108), Side/Side Max. = 0.47 (12)	5.98-6.37 (152-162), Side/Side Max. = 0.47 (12)	21 & 32
2003-12 AWD 1500 & 2500 exc/Cab & Chassis	4.40 (112), Side/Side Max. = 0.47 (12)	5.74-5.94 (146-151), Side/Side Max. = 0.47 (12)	21 & 32
2003-12 AWD 1500 & 2500 Cab & Chassis	4.80 (122), Side/Side Max. = 0.47 (12)	5.74-5.94 (146-151), Side/Side Max. = 0.47 (12)	21 & 32
Hummer	**Front (SF)**	**Rear (SR)**	**Figure**
2003-06 H2 STD Susp	4.80 (122), Side/Side Max. = 0.47 (12)	5.51 (140), Side/Side Max. = 0.47 (12)	20 & 28
2003-06 H2 All with AIR Susp.	4.80 (122), Side/Side Max. = 0.47 (12)	5.19 (132), Side/Side Max. = 0.47 (12)	20 & 28
2007 H2 STD Susp	4.25 (108), Side/Side Max. = 0.47 (12)	5.86 (149), Side/Side Max. = 0.47 (12)	20 & 28
2007 H2 All with AIR Susp.	4.72 (120), Side/Side Max. = 0.47 (12)	5.19 (132), Side/Side Max. = 0.47 (12)	20 & 28
2008-09 H2 STD Susp. exc/20" Tires	4.33 (110), Side/Side Max. = 0.47 (12)	5.86 (149), Side/Side Max. = 0.15 (4)	20 & 28
2008-09 H2 STD Susp. w/20" Tires	4.21 (107), Side/Side Max. = 0.47 (12)	5.19 (132), Side/Side Max. = 0.15 (4)	20 & 28
2008-09 H2 All with AIR Susp.	4.33 (110), Side/Side Max. = 0.47 (12)	5.86 (149), Side/Side Max. = 0.15 (4)	20 & 28
2006-10 H3 All models	5.94 (151), Side/Side Max. = 0.47 (12)	6.73 (171), Side/Side Max. = 0.39 (10)	21 & 25

* ± ⅜ (10) except when range is specified

CHASSIS / SUSP. HEIGHTS

Honda

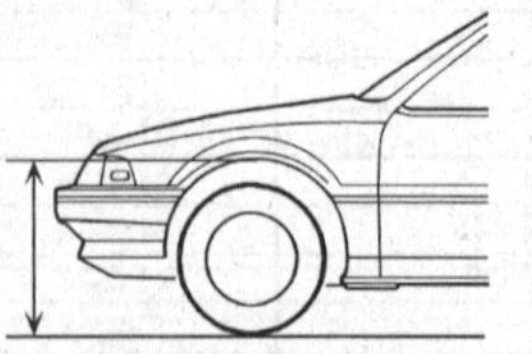

Figure 1

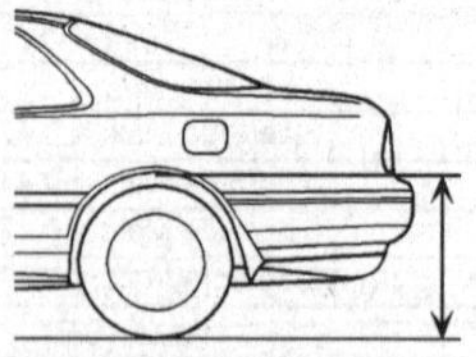

Figure 2

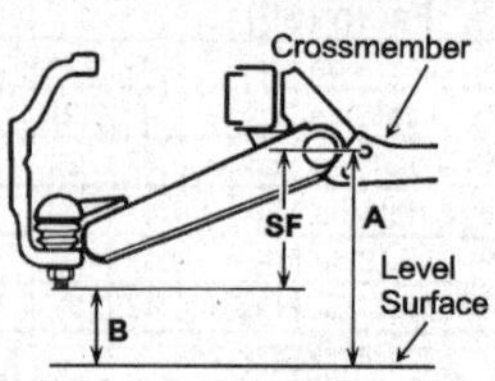

Figure 3

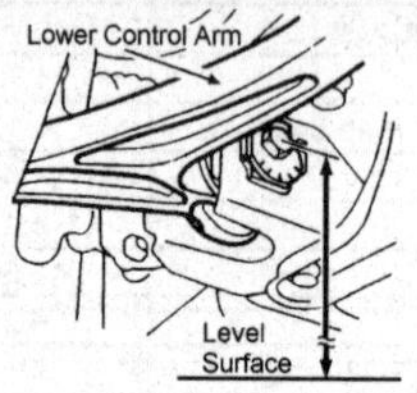

Figure 4

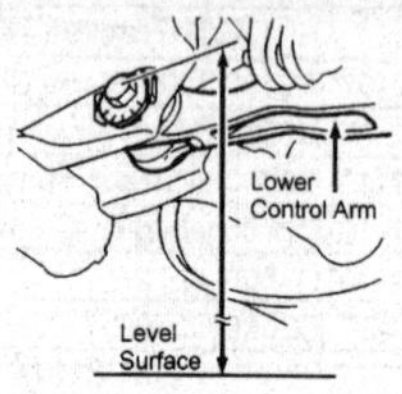

Figure 5

All measurements must be taken from a level ground surface to the point indicated. Jounce the vehicle several times before measuring. Measure **Figure 1** and **Figure 2** from ground to top of wheel arch at center of wheel. Measure **Figures 3, 4, and 5** as indicated.

Year Model	Height Specifications @ Curb Weight - Inches (mm) *		Notes
	Front	Rear	
1990-95 Accord (All Models)	Max. side/side difference .62 (16)	Max. side/side difference .62 (16)	See **Figures 1 & 2**
1990-91 Civic exc/Wagon	24.92-25.31 (633-643)	25.39-25.78 (645-655)	See **Figures 1 & 2**
1990-91 Civic Wagon 2WD	25.31-25.70 (643-653)	26.25-26.65 (667-677)	See **Figures 1 & 2**
1990-91 Civic Wagon 4WD	26.10-26.49 (663-673)	27.04-27.44 (687-697)	See **Figures 1 & 2**
1992-96 Civic (All Models)	Max. side/side difference .62 (16)	Max. side/side difference .62 (16)	See **Figures 1 & 2**
1990-91 CRX	24.92-25.31 (633-643)	25.39-25.78 (645-655)	See **Figures 1 & 2**
1994-97 Passport	4.96 (126) *	n/a	See **Figure 3** SF = A=B
1998-02 Passport	4.68 (119) *	n/a	See **Figure 3** SF = A=B
1990 Prelude	25.74-26.33 (654-669)	25.51-26.10 (648-663)	See **Figures 1 & 2**
2000-09 S2000	7.55-7.95 (192-202)	8.11-8.50 (206-216)	See **Figures 4 & 5** (If out of spec, adjust load for alignment)

* ± ½ (13) unless specified otherwise

Isuzu

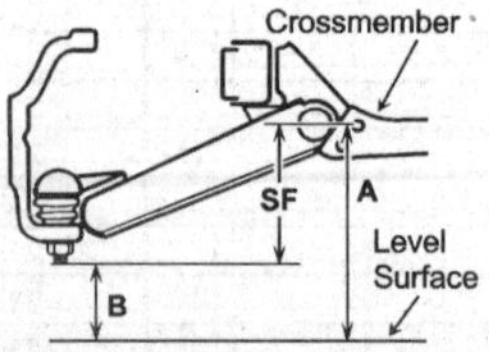

Figure 1

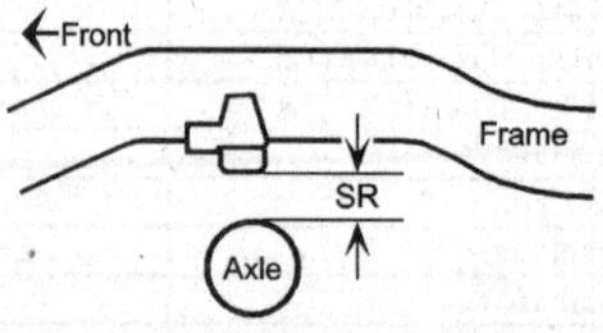

Figure 2

Measure only unloaded vehicles.Check tires for proper inflation before measurements are taken. Jounce vehicle several times. Measure front and rear suspension (SF & SR) with vehicle on a level surface, fuel tank full.

All **Figure 1** vehicles have adjustable torsion bars.

Year Model	Height Specifications @ Curb Weight - Inches (mm) *		Notes
	Front (SF)	Rear (SR)	
1990-93 Amigo	4.96 (126)	3.34 (85)	SF = A-B. See **Figure 1 & Figure 2**
1994 Amigo	5.11 (130)	3.34 (85)	SF = A-B. See **Figure 1 & Figure 2**
2003-08 Ascender	3.07 (78)	6.10 (155)	See **GM Figure 21 & 22** above.
2002-04 Axiom	4.68 (119)	n/a	SF = A-B. See **Figure 1**
1996-00 Hombre 4X2	2.87 (73)	5.19-6.37 (132-162)	See **GM Figure 20 & 22** above.
1998-00 Hombre 4X4	4.80 (122)	5.19-6.37 (132-162)	See **GM Figure 21 & 22** above.
2006-08 i-280 4X2	3.07 (78), Side/Side Max. = 0.47 (12)	6.85 (174), Side/Side Max. = 0.39 (10)	See **GM Figure 21 & 22** above.
2006-08 i-350 4X4	4.60 (117), Side/Side Max. = 0.47 (12)	6.73 (171), Side/Side Max. = 0.39 (10)	See **GM Figure 21 & 22** above.
1991-95 Rodeo	4.96 (126)	n/a	SF = A-B. See **Figure 1**
1996-97 Rodeo	5.27 (134)	n/a	SF = A-B. See **Figure 1**
1998-04 Rodeo	4.68 (119)	n/a	SF = A-B. See **Figure 1**
1990-95 Pickup 2WD exc/1 Ton	2.04 (52)	3.34 (85)	SF = A-B. See **Figure 1 & Figure 2**
1990-92 Pickup 2WD 1 Ton	2.04 (52)	3.54 (90)	SF = A-B. See **Figure 1 & Figure 2**
1990-95 Pickup 4WD	4.96 (126)	3.34 (85)	SF = A-B. See **Figure 1 & Figure 2**
1992-02 Trooper	5.07 (129)	n/a	SF = A-B. See **Figure 1**
1999-02 VehiCROSS	4.68 (119)	n/a	SF = A-B. See **Figure 1**

* ± ½ (13) unless specified otherwise

Kia

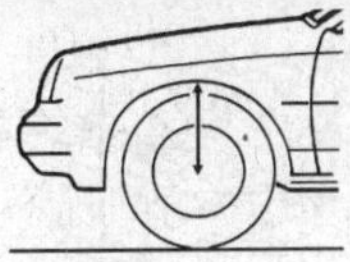

Figure 1

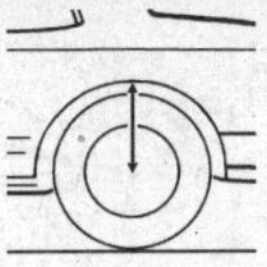

Figure 2

Measure from the center of the wheel to the lower edge of the wheel arch on both sides of the vehicle. (see **Figure 1**)
The maximum left to right difference should not exceed ⅜" (10mm)

Year Model	Height Specifications @ Curb Weight - In. (mm)		
	Front Side/Side	Rear Side/Side	Notes
1993-13 All Models	0.39 (10)	0.39 (10)	See **Figures 1 & 2**

Mazda

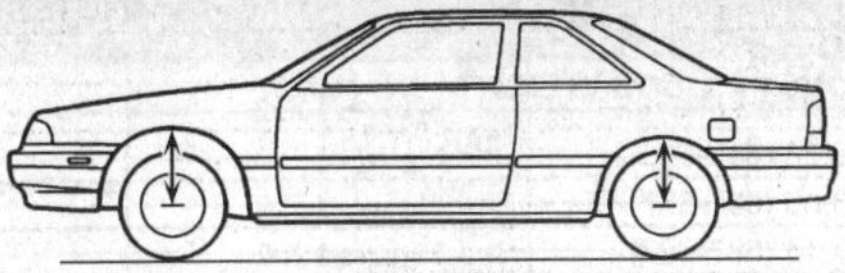

Figure 1

Measure **Figure 1** vehicles from the center of the wheel to the lower edge of the wheel arch.

Side/Side specifications are for maximum left to right difference.

Year Model	Height Specifications @ Curb Weight - In. (mm)						
	Front			Rear			
	SF	WF	Side/Side	SR	WR	Side/Side	Notes
1990-03 323/Protegé	—	—	0.39 (10)	—	—	0.39 (10)	**Figure 1**
1990-02 626/MX-6	—	—	0.39 (10)	—	—	0.39 (10)	**Figure 1**
1990-95 929	—	—	0.39 (10)	—	—	0.39 (10)	**Figure 1**
2007-17 CX5 / CX7 / CX9	—	—	0.39 (10)	—	—	0.39 (10)	**Figure 1**
2004-17 Mazda 2, Mazda 3	—	—	0.39 (10)	—	—	0.39 (10)	**Figure 1**
2003-17 Mazda 5, Mazda 6	—	—	0.39 (10)	—	—	0.39 (10)	**Figure 1**
1995-02 Millenia	—	—	0.39 (10)	—	—	0.39 (10)	**Figure 1**
1990-01 MPV	—	—	0.59 (15)	—	—	0.39 (10)	**Figure 1**
2002-06 MPV	—	—	0.39 (10)	—	—	0.39 (10)	**Figure 1**
1992-96 MX-3	—	—	0.39 (10)	—	—	0.39 (10)	**Figure 1**
1990-05 Miata (MX-5)	—	13.66-14.01 (347-356)	0.39 (10)	—	14.40-14.76 (366-375)	0.39 (10)	**Figure 1**
2004-05 MAZDASPEED Miata	—	13.34-13.70 (339-348)	0.39 (10)	—	14.05-14.40 (357-366)	0.39 (10)	**Figure 1**
2006-14 Miata (MX-5) 16" tires	—	14.80-15.15 (376-385)	0.39 (10)	—	14.72-15.07 (374-383)	0.39 (10)	**Figure 1**
2006-14 Miata (MX-5) 17" tires	—	14.60-14.96 (371-380)	0.39 (10)	—	14.52-14.88 (369-378)	0.39 (10)	**Figure 1**
1990-95 RX-7	—	—	0.39 (10)	—	—	0.39 (10)	**Figure 1**
2004-08 RX-8 Std. Susp.	—	15.23-15.59 (387-396)	0.39 (10)	—	15.00-15.35 (381-390)	0.39 (10)	**Figure 1**
2004-08 RX-8 Sport Susp.	—	15.00-15.35 (381-390)	0.39 (10)	—	14.72-15.07 (374-383)	0.39 (10)	**Figure 1**
2009-10 RX-8 Std. Susp.	—	15.07-15.43 (383-392)	0.39 (10)	—	15.19-15.55 (386-395)	0.39 (10)	**Figure 1**
2009-10 RX-8 Sport Susp.	—	14.80-15.15 (376-385)	0.39 (10)	—	15.11-15.47 (384-393)	0.39 (10)	**Figure 1**
1998 B Series Pickups 4X2	3.81 (97)	—	0.62 (16)	7.00-8.93 (178-227)	—	0.78 (20)	**Figs. 2 & 5** in Ford
1999-05 B Series Pickups 4X2	4.01 (102)	—	0.62 (16)	7.00-7.75 (178-197)	—	0.78 (20)	**Figs. 2 & 5** in Ford
1999-05 B Series Pickups 4X4	3.97 (101)	—	0.62 (16)	6.14-7.95 (156-202)	—	0.78 (20)	**Figs. 2 & 5** in Ford
2006-09 B 4X2 coil (base)	4.01 (102)	—	0.31 (8)	7.00-7.40 (178-188)	—	0.51 (13)	**Figs. 2 & 5** in Ford
2006-09 B 4X2 coil (payload)	4.01 (102)	—	0.31 (8)	7.79-8.18 (198-208)	—	0.51 (13)	**Figs. 2 & 5** in Ford
2006-09 B 4X2 torsion RegCab	3.66 (93)	—	0.31 (8)	6.88-7.28 (175-185)	—	0.51 (13)	**Figs. 2 & 5** in Ford
2006-09 B 4X2 torsion SupCab	3.66 (93)	—	0.31 (8)	6.14-6.53 (156-166)	—	0.51 (13)	**Figs. 2 & 5** in Ford
2006-09 B 4X4 Reg. Cab	3.66 (93)	—	0.31 (8)	6.88-7.28 (175-185)	—	0.51 (13)	**Figs. 2 & 5** in Ford
2006-09 B 4X4 Super. Cab	3.66 (93)	—	0.31 (8)	6.14-6.53 (156-166)	—	0.51 (13)	**Figs. 2 & 5** in Ford
1991-94 Navajo	—	—	0.62 (16)	—	—	0.78 (20)	**Figure 1**
2001-06 Tribute	—	18.77 (477)	0.39 (10)	—	19.29 (490)	0.39 (10)	**Figure 1**
2008-09 Tribute	0.12 (3.3)	—	0.39 (10)	—	3.74 (95)	0.39 (10)	**Figs. 16 & 17** in Ford
2008-09 Tribute Hybrid	0.12 (3.3)	—	0.39 (10)	—	3.46 (88)	0.39 (10)	**Figs. 16 & 17** in Ford

* ± ⅜ (10)

Mitsubishi

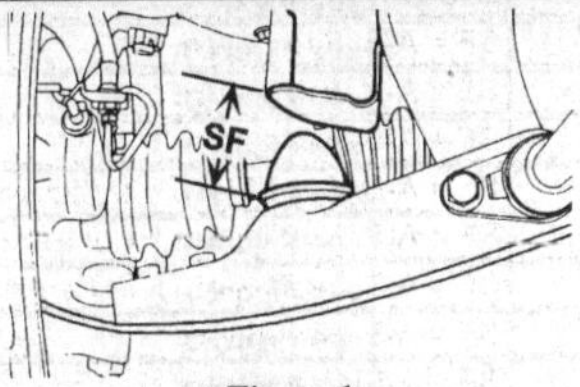

Figure 1

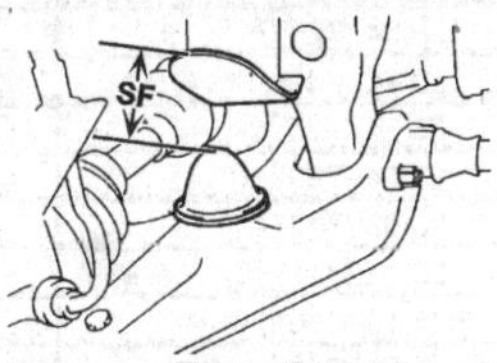

Figure 2

Measure only unloaded vehicles.

Measure **Figure 1** vehicles from the rubber bumper bracket to the bump stopper.

Measure **Figure 2** vehicles from the rubber bumper tip to the bump stopper

Year Model	Height Specifications @ Curb Weight - In. (mm)	Notes
1990-91 Pickup 4X4	measure as indicated in **Figure 1** above. SF = 3.07 (78)	Adjust torsion bar to value indicated.
1992-95 Pickup 4X4	measure as indicated in **Figure 1** above. SF = 3.11 (79)	Adjust torsion bar to value indicated.
1990-91 Montero 4X4	measure as indicated in **Figure 1** above. SF = 2.79 (71)	Adjust torsion bar to value indicated.
1992-00 Montero 4X4	measure as indicated in **Figure 2** above. SF = 0.86 (22)	Adjust torsion bar to value indicated.
1997-03 Montero Sport	measure as indicated in **Figure 2** above. SF = 0.70 (18)	Adjust torsion bar to value indicated.

* ± ⅜ (10)

Nissan/Infiniti

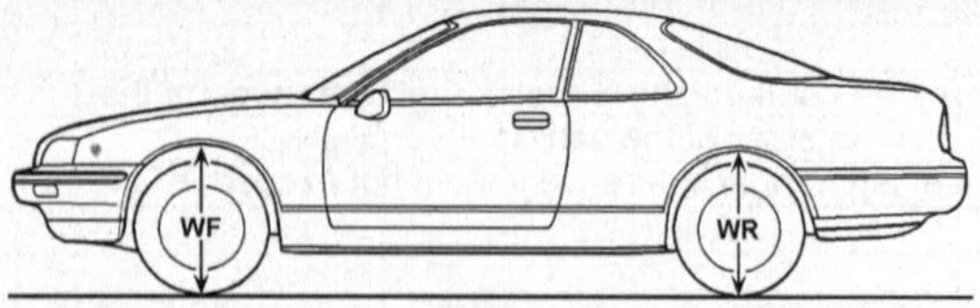

Figure 1

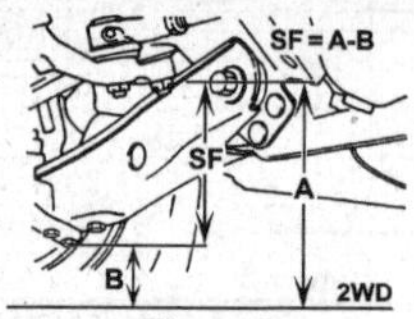

Figure 2

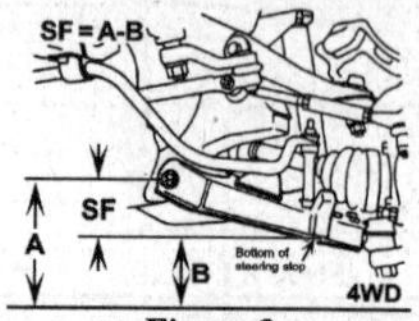

Figure 3

Take all measurements on a level surface to the points indicated. Wheelarch (WF & WR) dimensions are Curb Weight specifications and should be taken without passenger or other loading. Measure WF & WR from the lower edge of the wheel opening to level ground, at the wheel centerline. Measure SF & SR as indicated in Figures 2 and 3. The fuel tank should be full unless otherwise indicated.

Infiniti

Year Model	Height Specifications @ Curb Weight - In. (mm)		
EX35	Front (WF)	Rear (WR)	Notes
2008-10 EX35 RWD w/17" tires	29.33 (745)	29.80 (757)	see Figure 1
2008-13 EX35 RWD w/18" tires	29.53 (750)	30.00 (762)	see Figure 1
2011-13 EX35 RWD w/19" tires	29.57 (751)	30.08 (764)	see Figure 1
2008-10 EX35 AWD w/17" tires	29.21 (742)	29.80 (757)	see Figure 1
2008-13 EX35 AWD w/18" tires	29.41 (747)	30.00 (762)	see Figure 1
2011-13 EX35 AWD w/19" tires	29.45 (748)	30.08 (764)	see Figure 1
FX35/FX37/FX45/FX50	Front (WF)	Rear (WR)	Notes
2003-08 FX35 RWD w/18" tires (OE) (USA)	32.87(835)	32.36 (822)	see Figure 1
2003-08 FX35 RWD w/20" tires (OE) (USA)	32.83 (834)	32.32 (821)	see Figure 1
2003-08 FX35 AWD w/18" tires (OE) (USA)	32.83 (834)	32.64 (829)	see Figure 1
2003-08 FX35 AWD w/20" tires (OE) (USA)	32.80 (833)	32.56 (827)	see Figure 1
2003-08 FX45 AWD w/20" tires (OE) (USA)	32.76 (832)	32.48 (825)	see Figure 1
2003-08 FX35 AWD w/18" tires (OE) (Canada)	32.83 (834)	32.56 (827)	see Figure 1
2003-08 FX35 AWD w/20" tires (OE) (Canada)	32.80 (833)	32.52 (826)	see Figure 1
2003-08 FX45 AWD w/20" tires (OE) (Canada)	32.76 (832)	32.40 (823)	see Figure 1
2009-12 FX35 RWD w/18" tires (OE)	32.72 (831)	32.72 (831)	see Figure 1
2009-12 FX35 RWD w/20" tires (OE)	32.76 (832)	32.72 (831)	see Figure 1
2009-12 FX35 AWD w/18" tires (OE)	32.68 (830)	32.68 (830)	see Figure 1
2009-12 FX35 AWD w/20" tires (OE)	32.68 (830)	32.68 (830)	see Figure 1
2009-12 FX50 AWD w/20" tires (OE)	32.68 (830)	32.68 (830)	see Figure 1
2009-12 FX50 AWD w/21" tires (OE)	32.72 (831)	32.72 (831)	see Figure 1
2013 FX37 FWD w/18" tires (OE)	32.72 (831)	32.76 (832)	see Figure 1
2013 FX37 FWD w/20" tires (OE)	32.72 (831)	32.72 (831)	see Figure 1
2013 FX37 AWD	32.68 (830)	32.68 (830)	see Figure 1
2013 FX50 AWD	32.72 (831)	32.72 (831)	see Figure 1
G20/G25/G35/G37	Front (WF)	Rear (WR)	Notes
1991-96 G20 All Models	25.39-25.78 (645-655)	25.23-25.62 (641-651)	see Figure 1
1999-01 G20 All Models	26.22-26.41 (666-671)	25.70-25.90 (653-658)	see Figure 1
2003-06 G35 Sedan RWD exc/18" tires	27.95-27.99 (710-711)	27.67-27.71 (703-704)	see Figure 1
2006 G35 Sedan RWD w/18" tires	28.11 (714)	27.75 (705)	see Figure 1
2004-06 G35x Sedan AWD	27.91 (709)	27.32 (694)	see Figure 1
2007-08 G35 Sedan RWD 17" tires	28.11 (714)	27.83 (707)	see Figure 1
2007-08 G35 Sedan RWD 18" tires exc/4WAS	28.18 (716)	27.83 (707)	see Figure 1
2007-08 G35 Sedan RWD 18" tires w/4WAS	28.18 (716)	27.75 (705)	see Figure 1
2007-08 G35x Sedan AWD	28.66 (728)	28.46 (723)	see Figure 1
2003-07 G35 Coupe with 17" wheels	27.20 (691)	27.63 (702)	see Figure 1
2003-07 G35 Coupe with 18" wheels	27.32 (694)	27.95 (710)	see Figure 1
2006-07 G35 Coupe with 19" wheels	27.51 (699)	27.95 (710)	see Figure 1
2011-12 G25 Sedan RWD with 17" tires	28.11 (714)	27.83 (707)	see Figure 1
2011-12 G25 Sedan RWD with 18" tires	28.19 (716)	27.91 (709)	see Figure 1
2009-14 G37 Sedan RWD with 17" tires	28.11 (714)	27.83 (707)	see Figure 1
2009-13 G37 Sedan RWD 18" tires exc/Sport	28.19 (716)	27.91 (709)	see Figure 1
2009-13 G37 Sedan RWD 18" tires Sport	28.15 (715)	27.76 (705)	see Figure 1
2011-12 G25x Sedan AWD RWD 17" tires	28.54 (725)	28.35 (720)	see Figure 1
2011-12 G25x Sedan AWD 18" tires	28.74 (730)	28.50 (724)	see Figure 1
2009-14 G37x Sedan AWD 17" tires	28.54 (725)	28.35 (720)	see Figure 1
2009-13 G37x Sedan AWD 18" tires	28.74 (730)	28.50 (724)	see Figure 1
2008-13 G37 Coupe with 18" tires	28.03 (712)	27.63 (702)	see Figure 1
2008-13 G37 Coupe RWD 19" tires (OE) exc/Sport	28.14 (715)	27.75 (705)	see Figure 1
2008-13 G37 Coupe RWD 19" tires (OE) Sport	28.18 (716)	27.79 (706)	see Figure 1
2009-13 G37x Coupe AWD 18" tires (OE)	28.58 (726)	28.26 (718)	see Figure 1
2009-13 G37x Coupe AWD 19" tires (OE)	28.54 (725)	28.19 (716)	see Figure 1
2009-13 G37 Convertible 18" tires	28.03 (712)	27.59 (701)	see Figure 1
2009-13 G37 Convertible 19" tires	28.11 (714)	27.63 (702)	see Figure 1

* ±0.75 (19) (except when range is specified)

Infiniti (cont.)

Year Model	Height Specifications @ Curb Weight - In. (mm)		
I30/I35	Front (WF)	Rear (WR)	Notes
1996-99 I30 w/205-65R15 tires	27.67-28.07 (703-713)	27.75-28.14 (705-715)	see Figure 1
1996-99 I30 w/215-60R15 tires	27.67-28.07 (703-713)	27.67-28.07 (703-713)	see Figure 1
2000-01 I30 w/225-50R17 tires	27.79 (706)	27.48 (698)	see Figure 1
2000-01 I30 w/215-55R16 tires	27.32 (694)	26.88 (683)	see Figure 1
2002-04 I35 w/215-55R17 tires	28.03 (712)	27.71 (704)	see Figure 1
2002-04 I35 w/225-50R17 tires	27.83 (707)	27.32 (694)	see Figure 1
J30	Front (WF)	Rear (WR)	Notes
1993-97 J30 All Models	27.04-27.44 (687-697)	26.41-26.81 (671-681)	see Figure 1
JX35	Front (WF)	Rear (WR)	Notes
2013 JX35 2WD with 18" tires (OE) Base	32.40 (823)	32.60 (828)	see Figure 1
2013 JX35 2WD with 18" tires (OE) Premium Pkg.	32.36 (822)	32.56 (827)	see Figure 1
2013 JX35 2WD with 20" tires (OE) Premium Pkg.	32.32 (821)	32.52 (826)	see Figure 1
2013 JX35 AWD with 18" tires (OE) Base	32.36 (822)	32.56 (827)	see Figure 1
2013 JX35 AWD with 18" tires (OE) Premium Pkg.	32.36 (822)	32.56 (827)	see Figure 1
2013 JX35 AWD with 20" tires (OE) Premium Pkg.	32.32 (821)	32.52 (826)	see Figure 1
M30/M35/M45	Front (WF)	Rear (WR)	Notes
1990-92 M30 All Models	26.10-26.49 (663-673)	25.74-26.14 (654-664)	see Figure 1
2003-04 M45 All Models	27.99 (711)	27.79 (706)	see Figure 1
2006-10 2WD with 18" tires USA	28.22 (717)	27.95 (710)	see Figure 1
2006-10 2WD with 18" tires Canada	28.26 (718)	27.99 (711)	see Figure 1
2006-10 2WD with 19" tires USA & Can.	28.38 (721)	27.99 (711)	see Figure 1
2006-10 AWD USA	28.77 (731)	28.54 (725)	see Figure 1
2006-10 AWD Canada	28.81 (732)	28.58 (726)	see Figure 1
M37/M56	Front (WF)	Rear (WR)	Notes
2011-13 M37 RWD with 18" tires (OE)	29.57-29.62 (751-752)	29.25 (743)	see Figure 1
2011-13 M37 RWD with 20" tires (OE)	29.53-29.57 (750-751)	29.21 (742)	see Figure 1
2011-13 M37 AWD	30.04-30.12 (763-765)	29.80 (757)	see Figure 1
2011-13 M56 RWD with 18" tires (OE)	29.57 (751)	29.25 (743)	see Figure 1
2011-13 M56 RWD with 20" tires (OE)	29.53 (750)	29.17 (741)	see Figure 1
2011-13 M56x AWD	30.04 (763)	29.80 (757)	see Figure 1
2012-13 M35h Hybrid	29.61 (752)	29.25 (743)	see Figure 1
Q45	Front (WF)	Rear (WR)	Notes
1990 Q45 All Models	26.57-26.96 (675-685)	25.70-26.10 (653-663)	see Figure 1
1991-96 Q45 exc/Active Suspension	27.40-27.79 (696-706)	27.00-27.40 (686-696)	see Figure 1
1991-95 Q45 w/Active Suspension [1]	26.65-27.04 (677-687)	25.86-26.25 (657-667)	see Figure 1
[1] with engine running & height switch in N position.			
1997-99 Q45 All Models	28.30-28.70 (719-729)	28.38-28.77 (721-731)	see Figure 1
2001 Q45 w/215-60R16 tires	28.77 (731)	28.89 (734)	see Figure 1
2001 Q45 w/225-50R17 tires	28.42 (722)	28.46 (723)	see Figure 1
2002-06 Q45 w/225-55R17 tires	28.66-28.74 (728-730)	27.71-27.75 (704-705)	see Figure 1
2002-06 Q45 w/225-55R17 Runflat tires	28.85-28.89 (733-734)	27.83-27.87 (707-708)	see Figure 1
2002-06 Q45 w/245-45R18 tires	28.58 (726)	27.51-27.59 (699-701)	see Figure 1
2006 Q45 w/245-40R19 tires	28.70 (729)	27.67 (703)	see Figure 1
Q40 (Formerly G37 Sedan)	Front (WF)	Rear (WR)	Notes
2015 Q40 Sedan RWD	28.11 (714)	27.83 (707)	see Figure 1
2015 Q40 Sedan AWD	28.54 (725)	28.35 (720)	see Figure 1
Q50 (Formerly G Sedan)	Front (WF)	Rear (WR)	Notes
2014-15 Q50 Sedan RWD with 17" tires	27.76 (705)	27.48 (698)	see Figure 1
2014-15 Q50 Sedan RWD with 19" tires	27.80 (706)	27.44 (697)	see Figure 1
2014-15 Q50 Sedan RWD w/245/40R19 Fr, 265/35R19 Rr	27.87 (708)	27.36 (695)	see Figure 1
2014-15 Q50 Sedan AWD except Sport	28.11 (714)	27.87 (708)	see Figure 1
2014-15 Q50 Sedan AWD Sport	28.15 (715)	27.80 (706)	see Figure 1
2014-17 Q50 Hybrid RWD with 17" tires	27.76 (705)	27.40 (696)	see Figure 1
2014-17 Q50 Hybrid RWD with 19" tires	27.76 (705)	27.36 (695)	see Figure 1
2014-17 Q50 Hybrid AWD with 17" tires	28.11 (714)	27.80 (706)	see Figure 1
2014-17 Q50 Hybrid AWD with 19" tires	28.11 (714)	27.76 (705)	see Figure 1
Q60 (Formerly G Coupe/Conv.)	Front (WF)	Rear (WR)	Notes
2014-15 G60 Coupe with 18" tires (OE)	28.03 (712)	27.63 (702)	see Figure 1
2014-15 G60 Coupe RWD 19" tires (OE) exc/Sport	28.14 (715)	27.75 (705)	see Figure 1
2014-15 G60 Coupe RWD 19" tires (OE) Sport	28.18 (716)	27.79 (706)	see Figure 1
2014-15 G60 Coupe AWD 18" tires (OE)	28.58 (726)	28.26 (718)	see Figure 1
2014-15 G60 Coupe AWD 19" tires (OE)	28.54 (725)	28.19 (716)	see Figure 1
2014-15 G60 Convertible 18" tires	28.03 (712)	27.59 (701)	see Figure 1
2014-15 G60 Convertible 19" tires	28.11 (714)	27.63 (702)	see Figure 1
Q70 (Formerly M Sedan)	Front (WF)	Rear (WR)	Notes
2014-17 Q70 V6 RWD with 18" tires (OE)	29.62 (752)	29.25 (743)	see Figure 1
2014-17 Q70 V6 RWD with 20" tires (OE)	29.57 (751)	29.21 (742)	see Figure 1
2016-17 Q70 V8 RWD with 18" tires (OE)	29.57 (751)	29.21 (742)	see Figure 1
2016-17 Q70 V8 RWD with 20" tires (OE)	29.53 (750)	29.17 (741)	see Figure 1
2014-17 Q70 V6 AWD with 18" tires (OE)	30.12 (765)	29.80 (757)	see Figure 1
2014-17 Q70 V6 AWD with 20" tires (OE)	30.12 (765)	29.85 (758)	see Figure 1
2016-17 Q70 V8 AWD with 18" tires (OE)	30.04 (763)	29.80 (757)	see Figure 1

* ±0.75 (19) (except when range is specified)

CHASSIS / SUSP. HEIGHTS

Infiniti (cont.)

Year	Model	Height Specifications @ Curb Weight - In. (mm)		
	Q70 (Formerly M Sedan) (cont.)	Front (WF)	Rear (WR)	Notes
2016-17	Q70 V8 AWD with 20" tires (OE)	30.08 (764)	29.85 (758)	see Figure 1
2014-17	Q70 HEV	29.61 (752)	29.25 (743)	see Figure 1
	QX4	Front (WF)	Rear (WR)	Notes
1997-00	QX4 All Models	31.69-32.08 (805-815)	32.91-33.30 (836-846)	see Figure 1
2001-03	QX4 All Models	31.25 (794)	31.88 (810)	see Figure 1
	QX50	Front (WF)	Rear (WR)	Notes
2014-15	QX50 RWD with 18" tires (OE)	29.53 (750)	30.00 (762)	see Figure 1
2014-15	QX50 RWD with 19" tires (OE)	29.57 (751)	30.08 (764)	see Figure 1
2014-15	QX50 AWD with 18" tires (OE)	29.41 (747)	30.00 (762)	see Figure 1
2014-15	QX50 AWD with 19" tires (OE)	29.45 (748)	30.08 (764)	see Figure 1
2016-17	QX50 RWD with 18" tires (OE) All	30.15 (766)	30.55 (776)	see Figure 1
2016-17	QX50 RWD with 19" tires (OE) All	30.19 (767)	30.62 (778)	see Figure 1
2016-17	QX50 AWD with 18" tires (OE) exc/Canada	30.03 (763)	30.55 (776)	see Figure 1
2016-17	QX50 AWD with 19" tires (OE) exc/Canada	30.08 (764)	30.62 (778)	see Figure 1
2016-17	QX50 AWD with 18" tires (OE) Canada	30.03 (763)	30.59 (777)	see Figure 1
2016-17	QX50 AWD with 19" tires (OE) Canada	30.08 (764)	30.66 (779)	see Figure 1
	QX56	Front (WF)	Rear (WR)	Notes
2004-07	QX56 All 2WD models	35.94 (913)	35.91 (912)	see Figure 1
2004-07	QX56 All 4WD models	36.65 (931)	36.69 (932)	see Figure 1
2008-10	QX56 All 2WD models	36.22 (920)	36.10 (917)	see Figure 1
2008-10	QX56 All 4WD models	36.89 (937)	36.89 (937)	see Figure 1
2011-13	QX56 All Models	35.55 (903)	35.04 (890)	see Figure 1
	QX60 (Formerly JX)	Front (WF)	Rear (WR)	Notes
2014-17	QX60 2WD with 18" tires (OE) Base	32.36 (822)	32.60 (828)	see Figure 1
2014-17	QX60 2WD with 18" tires (OE) Premium Pkg.	32.36 (822)	32.56 (827)	see Figure 1
2014-17	QX60 2WD with 20" tires (OE) Premium Pkg.	32.32 (821)	32.52 (826)	see Figure 1
2014-17	QX60 AWD with 18" tires (OE) Base (USA)	32.36 (822)	32.56 (827)	see Figure 1
2014-17	QX60 AWD with 18" tires (OE) Prem. Pkg. (USA)	32.32 (821)	32.56 (827)	see Figure 1
2014-17	QX60 AWD with 20" tires (OE) Prem. Pkg. (USA)	32.28 (820)	32.48 (825)	see Figure 1
2014-17	QX60 AWD with 18" tires (OE) Base (CAN)	32.36 (822)	32.60 (828)	see Figure 1
2014-17	QX60 AWD with 18" tires (OE) Prem. Pkg. (CAN)	32.36 (822)	32.56 (827)	see Figure 1
2014-17	QX60 AWD with 20" tires (OE) Prem. Pkg. (CAN)	32.32 (821)	32.52 (826)	see Figure 1
2014-15	QX60 HEV 2WD with 18" tires (OE) Base	32.32 (821)	32.48 (825)	see Figure 1
2014-15	QX60 HEV 2WD with 18" tires (OE) Premium Pkg.	32.32 (821)	32.44 (824)	see Figure 1
2014-15	QX60 HEV 2WD with 20" tires (OE) Premium Pkg.	32.28 (820)	32.40 (823)	see Figure 1
2014-15	QX60 HEV AWD with 18" tires (OE) Base (USA)	32.32 (821)	32.44 (824)	see Figure 1
2014-15	QX60 HEV AWD with 18" tires (OE) Prem. (USA)	32.28 (820)	32.44 (824)	see Figure 1
2014-15	QX60 HEV AWD with 20" tires (OE) Prem. (USA)	32.24 (819)	32.40 (823)	see Figure 1
2014-15	QX60 HEV AWD with 18" tires (OE) Base (CAN)	32.32 (821)	32.48 (825)	see Figure 1
2014-15	QX60 HEV AWD with 18" tires (OE) Prem. (CAN)	32.32 (821)	32.44 (824)	see Figure 1
2014-15	QX60 HEV AWD with 20" tires (OE) Prem. (CAN)	32.28 (820)	32.40 (823)	see Figure 1
	QX70 (Formerly FX)	Front (WF)	Rear (WR)	Notes
2014-17	QX70 FWD w/18" tires (OE)	32.72 (831)	32.72 (831)	see Figure 1
2014-17	QX70 FWD w/20" & 21" tires (OE)	32.76 (832)	32.72 (831)	see Figure 1
2014-17	QX70 V6 AWD w/18" & 20" tires (OE)	32.68 (830)	32.68 (830)	see Figure 1
2014-17	QX70 V6 AWD w/21" tires (OE)	32.72 (831)	32.72 (831)	see Figure 1
2016-17	QX70 V8 AWD w/20" tires (OE)	32.72 (831)	32.72 (831)	see Figure 1
	QX80 (Formerly QX56)	Front (WF)	Rear (WR)	Notes
2014-17	QX80 All Models	35.55 (903)	35.04 (890)	see Figure 1

* ±0.75 (19) (except when range is specified)

Nissan

Year	Model	Height Specifications @ Curb Weight - In. (mm)		
	200SX	Front (WF)	Rear (WR)	Notes
1995-97	200SX exc/SE-R	25.90-26.22	25.15-25.47	see Figure 1
1995-97	200SX SE-R	26.33 (669)	25.59 (650)	see Figure 1
1989-94	240SX Coupe	26.73-27.28	25.78-26.33	see Figure 1
1989-94	240SX	26.73-27.28	25.78-26.33	see Figure 1
1995-97	240SX exc/spoiler	26.41-27.04	25.47-26.10	see Figure 1
1995-97	240SX w/spoiler	26.73-27.28	25.78-26.33	see Figure 1
1998	240SX w/195/60R15 tires	27.04 (687)	26.10 (663)	see Figure 1
1998	240SX w/205/55R16 tires	27.32 (694)	26.37 (670)	see Figure 1
	300ZX	Front (WF)	Rear (WR)	Notes
1990-96	300ZX Coupe, Conv.	26.57 (675)	26.61 (676)	see Figure 1
1990-96	300ZX Coupe 2+2	26.65 (677)	26.57 (675)	see Figure 1
	350Z	Front (WF)	Rear (WR)	Notes
2003-05	350Z Coupe with 17" tires	26.81 (681)	27.55 (700)	see Figure 1
2003-08	350Z Coupe with 18" tires (F&R)	26.88 (683)	27.79 (706)	see Figure 1
2003-08	350Z Coupe with 18" (F), 19" (R)	26.83 (683)	27.67 (703)	see Figure 1
2005	350Z Roadster with 17" tires	26.73 (679)	27.51 (699)	see Figure 1
2005-09	350Z Roadster with 18" tires	26.88 (683)	27.75 (705)	see Figure 1

* ±0.75 (19) (except when range is specified)

Nissan (cont.)

YEAR MODEL	HEIGHT SPECIFICATIONS @ CURB WEIGHT - IN. (MM)		
370Z	FRONT (WF)	REAR (WR)	Notes
2009-17 370Z with 18" tires (OE)	27.99 (711)	27.99 (711)	see Figure 1
2009-17 370Z with 19" tires (OE)	27.87 (708)	27.99 (711)	see Figure 1
2009-17 370Z Nismo Models	27.71 (704)	28.11 (714)	see Figure 1
Altima	FRONT (WF)	REAR (WR)	Notes
1993-97 Altima	27.20 (691)	26.77 (680)	see Figure 1
1998-01 Altima	27.16 (690)	26.69 (678)	see Figure 1
2002-04 Altima with 205/65R16 tires	28.77 (731)	27.75 (705)	see Figure 1
2002-04 Altima with 215/55R17 tires	28.54 (725)	27.51 (699)	see Figure 1
2005-06 Altima 4 cyl.with 215/60R16 tires	28.42 (722)	27.36 (695)	see Figure 1
2005-06 Altima 6 cyl.with 215/60R16 tires	28.22 (717)	27.40 (696)	see Figure 1
2005 Altima 6 cyl.with 215/55R17 tires	28.38 (721)	27.48 (698)	see Figure 1
2006 Altima 6 cyl.with 215/55R17 tires	28.42 (722)	27.59 (701)	see Figure 1
2005-06 Altima 6 cyl.with 225/45R18 tires	28.38 (721)	27.36 (695)	see Figure 1
2007-08 Altima Sedan 4 cyl. w/o ABS USA	28.18 (716)	27.95 (710)	see Figure 1
2007-08 Altima Sedan 4 cyl. w/o ABS CANADA	28.11 (714)	28.14 (715)	see Figure 1
2007-08 Altima Sedan 4 cyl. w/ABS USA	28.22 (717)	27.95 (710)	see Figure 1
2007-08 Altima Sedan 4 cyl. w/ABS CANADA	28.14 (715)	28.14 (715)	see Figure 1
2007-08 Altima Sedan 6 cyl. w/16" tires (OE) USA	28.18 (716)	27.95 (710)	see Figure 1
2007-08 Altima Sedan 6 cyl. w/16" tires (OE) CAN	28.11 (714)	28.14 (715)	see Figure 1
2007-08 Altima Sedan 6 cyl. w/17" tires (OE) USA	28.14 (715)	27.95 (710)	see Figure 1
2007-08 Altima Sedan 6 cyl. w/17" tires (OE) CAN	28.07 (713)	28.11 (714)	see Figure 1
2009-12 Altima Sedan 4 cyl. USA	28.22 (717)	27.99 (711)	see Figure 1
2009-12 Altima Sedan 4 cyl. CANADA	28.14 (715)	28.14 (715)	see Figure 1
2009-12 Altima Sedan 4 cyl. MEXICO	28.62 (727)	28.77 (731)	see Figure 1
2009-12 Altima Sedan 6 cyl. P215/60R16 (OE) USA	28.18 (716)	27.99 (711)	see Figure 1
2009-12 Altima Sedan 6 cyl. P215/55R1 (OE) USA	28.14 (715)	27.99 (711)	see Figure 1
2009-12 Altima Sedan 6 cyl. P215/60R16 (OE) CAN	28.11 (714)	28.14 (715)	see Figure 1
2009-12 Altima Sedan 6 cyl. P215/55R17 (OE) CAN	28.18 (716)	27.99 (711)	see Figure 1
2013-17 Altima Sedan w/16" tires (OE) USA	27.87 (708)	27.80 (706)	see Figure 1
2013-17 Altima Sedan w/17" tires (OE) USA	27.99 (711)	27.91 (709)	see Figure 1
2013-17 Altima Sedan w/18" tires (OE) USA	28.11 (714)	27.99 (711)	see Figure 1
2013-17 Altima Sedan w/16" tires (OE) CAN	27.83 (707)	27.80 (706)	see Figure 1
2013-17 Altima Sedan w/17" tires (OE) CAN	27.95 (710)	27.95 (710)	see Figure 1
2013-17 Altima Sedan w/18" tires (OE) CAN	28.15 (715)	28.03 (712)	see Figure 1
2008 Altima Coupe 4 cyl. USA w/o ABS	28.11 (714)	27.63 (702)	see Figure 1
2008 Altima Coupe 4 cyl. USA w/ABS	28.14 (715)	27.63 (702)	see Figure 1
2008 Altima Coupe 4 cyl. CAN w/o ABS	28.03 (712)	27.79 (706)	see Figure 1
2008 Altima Coupe 4 cyl. CAN w/ABS	28.07 (713)	27.79 (706)	see Figure 1
2008 Altima Coupe 6 cyl. USA	28.07 (713)	27.63 (702)	see Figure 1
2008 Altima Coupe 6 cyl. CAN	27.99 (711)	27.79 (706)	see Figure 1
2009-11 Altima Coupe 4 cyl. USA	28.14 (715)	28.42 (722)	see Figure 1
2012 Altima Coupe 4 cyl. USA	28.07 (713)	28.42 (722)	see Figure 1
2009-12 Altima Coupe 6 cyl. USA	28.07 (713)	28.42 (722)	see Figure 1
2009-11 Altima Coupe 4 cyl. CAN	28.07 (713)	28.62 (727)	see Figure 1
2012 Altima Coupe 4 cyl. CAN	27.99 (711)	28.58 (726)	see Figure 1
2009-12 Altima Coupe 6 cyl. CAN	27.99 (711)	28.58 (726)	see Figure 1
2007-11 Altima Hybrid USA	28.62 (727)	28.26-28.30 (718-719)	see Figure 1
2007-11 Altima Hybrid CANADA	28.62-28.66 (727-728)	28.30 (719)	see Figure 1
Cube	FRONT (WF)	REAR (WR)	Notes
2009-14 Cube with 195/60R15 tires (OE)	26.73 (679)	26.73 (679)	see Figure 1
2009-14 Cube with 195/55R16 tires (OE)	26.92 (684)	26.88 (683)	see Figure 1
Juke	FRONT (WF)	REAR (WR)	Notes
2011-12 Juke 2WD with M/T (USA & Canada)	28.90 (734)	29.33 (745)	see Figure 1
2011-12 Juke 2WD with CVT (USA)	28.94 (735)	29.37 (746)	see Figure 1
2011-12 Juke AWD with CVT (USA)	28.90 (734)	29.17 (741)	see Figure 1
2011-12 Juke 2WD with CVT (Canada)	28.98 (736)	29.41 (747)	see Figure 1
2011-12 Juke AWD with CVT (Canada)	28.94 (735)	29.21 (742)	see Figure 1
2013-17 Juke 2WD exc/Nismo (USA)	28.90-28.94 (734-735)	29.33-29.37 (745-746)	see Figure 1
2013-17 Juke 2WD Nismo (USA)	28.50-28.54 (724-725)	29.06-29.09 (738-739)	see Figure 1
2013-17 Juke AWD exc/Nismo (USA)	28.90 (734)	29.17 (741)	see Figure 1
2013-17 Juke AWD Nismo (USA)	28.50 (724)	28.93 (735)	see Figure 1
2013-17 Juke 2WD exc/Nismo (Canada)	28.94 (735)	29.33-29.37 (745-746)	see Figure 1
2013-17 Juke 2WD Nismo (Canada)	28.54 (725)	29.13 (740)	see Figure 1
2013-17 Juke AWD exc/Nismo (Canada)	28.86-28.93 (733-735)	29.17 (741)	see Figure 1
2013-17 Juke AWD Nismo (Canada)	28.50-28.58 (724-726)	28.93 (735)	see Figure 1
Leaf	FRONT (WF)	REAR (WR)	Notes
2011-12 Leaf	27.83 (707)	27.87 (708)	see Figure 1
2013-17 Leaf w/205/55R16 tires (OE) USA	27.80 (706)	27.87 (708)	see Figure 1
2013-17 Leaf w/215/55R17 tires (OE) USA	28.11 (714)	28.11 (714)	see Figure 1
2013-17 Leaf w/205/55R16 tires (OE) Canada	27.80 (706)	27.91 (709)	see Figure 1
2013-17 Leaf w/215/55R17 tires (OE) Canada	28.11 (714)	28.15 (715)	see Figure 1

* ±0.75 (19) (except when range is specified)

CHASSIS / SUSP. HEIGHTS

Nissan (cont.)

Year Model	Height Specifications @ Curb Weight - In. (mm)		
Maxima	Front (WF)	Rear (WR)	Notes
1990-94 Maxima	27.71 (704)	27.13 (689)	see Figure 1
1995-99 Maxima	28.07 (713)	28.15 (715)	see Figure 1
2000-01 Maxima w/15" tires (OE)	27.56 (700)	27.16 (690)	see Figure 1
2000-03 Maxima w/16" tires (OE)	27.64 (702)	27.09 (688)	see Figure 1
2000-03 Maxima w/P255/50R17" tires (OE)	27.99 (711)	27.32 (694)	see Figure 1
2000-03 Maxima w/P215/55R17" tires (OE)	28.23 (717)	27.72 (704)	see Figure 1
2000-03 Maxima w/P215/55R17" tires (OE)	28.23 (717)	27.72 (704)	see Figure 1
2004-08 Maxima w/17" tires (OE)	29.06 (738)	27.76 (705)	see Figure 1
2004-08 Maxima w/18" tires (OE)	29.02 (737)	27.76 (705)	see Figure 1
2009-14 Maxima w/18" tires (OE)	28.31 (719)	28.62 (727)	see Figure 1
2009-14 Maxima w/19" tires (OE)	28.46 (723)	28.74 (730)	see Figure 1
2016-17 Maxima w/18" tires (OE)	27.87 (708)	28.31 (719)	see Figure 1
2016-17 Maxima w/19" tires (OE)	27.71 (704)	28.03 (712)	see Figure 1
Micra	Front (WF)	Rear (WR)	Notes
2015-17 Micra w/15" tires (OE)	26.30 (668)	25.98 (660)	see Figure 1
2015-17 Micra w/16" tires (OE)	26.46 (672)	26.14 (664)	see Figure 1
Murano	Front (WF)	Rear (WR)	Notes
2003-05 Murano exc/SE models	33.07 (840)	33.78 (858)	see Figure 1
2003-05 Murano SE models 2WD	33.07 (840)	33.74 (857)	see Figure 1
2003-05 Murano SE models 4WD	33.03 (839)	33.74 (857)	see Figure 1
2006-07 Murano 2WD	33.07 (840)	33.85 (860)	see Figure 1
2006-07 Murano 4WD	33.03 (839)	33.81 (859)	see Figure 1
2009-14 Murano S 2WD w/235/65R18" tires (OE) USA	33.31 (846)	33.81 (859)	see Figure 1
2009-14 Murano SL 2WD w/235/55R20" tires (OE) USA	33.31 (846)	33.78 (858)	see Figure 1
2009-14 Murano LE 2WD w/235/55R20" tires (OE) USA	33.27 (845)	33.78 (858)	see Figure 1
2009-14 Murano S AWD w/235/65R18" tires (OE) USA	33.31 (846)	33.78 (858)	see Figure 1
2009-14 Murano SL AWD w/235/65R18" tires (OE) USA	33.27 (845)	33.78 (858)	see Figure 1
2009-14 Murano LE AWD w/235/55R20" tires (OE) USA	33.27 (845)	33.74 (857)	see Figure 1
2009-14 Murano S w/235/65R18" tires (OE) Canada	33.27 (845)	33.78 (858)	see Figure 1
2009-14 Murano SL w/235/65R18" tires (OE) Canada	33.27 (845)	33.74 (857)	see Figure 1
2009-14 Murano LE w/235/55R20" tires (OE) Canada	33.31 (846)	33.74 (857)	see Figure 1
2015-17 Murano All w/235/65R18" tires (OE) USA	32.87 (835)	32.44 (824)	see Figure 1
2015-17 Murano All w/235/55R20" tires (OE) USA	32.83 (834)	32.36 (822)	see Figure 1
2015-17 Murano All w/235/65R18" tires (OE) Canada	32.91 (836)	32.44 (824)	see Figure 1
2015-17 Murano All w/235/55R20" tires (OE) Canada	32.83 (834)	32.36 (822)	see Figure 1
2012-14 Murano Cross Cabriolet (All)	33.18 (843)	33.62 (854)	see Figure 1
Pulsar/NX	Front (WF)	Rear (WR)	Notes
1990 Pulsar	25.35 (644)	25.43 (646)	see Figure 1
1991-93 NX Coupe	25.15-25.55	24.92-25.03	see Figure 1
Quest	Front (WF)	Rear (WR)	Notes
1993-98 Quest	30.00-30.35	30.78-31.22	see Figure 1
1999-02 Quest	30.39 (772)	31.22 (793)	see Figure 1
2004-09 Quest	29.13 (740)	29.48 (749)	see Figure 1
2011-17 Quest Grade S & SV	29.92 (760)	29.57 (751)	see Figure 1
2011-17 Quest Grade SL	30.28 (769)	29.92 (760)	see Figure 1
2011-17 Quest Grade LE	30.31 (770)	29.88 (759)	see Figure 1
Rogue	Front (WF)	Rear (WR)	Notes
2008-13 Rogue with 225/60R17 tires	31.02 (788)	31.89 (810)	see Figure 1
2011-13 Rogue with 225/55R18 tires	31.18 (792)	32.01 (813)	see Figure 1
2008-15 Rogue Select with 215/70R16 tires (old version)	31.06 (789)	31.93 (811)	see Figure 1
2014-16 Rogue FWD w/2-row seating w/225/65R17 (OE)	31.02 (788)	30.91-30.98 (785-787)	see Figure 1
2017 Rogue FWD w/2-row seating w/225/65R17 (OE)	30.83 (783)	30.71 (780)	see Figure 1
2014-16 Rogue FWD w/2-row seating w/225/60R18 (OE)	31.10-31.14 (790-791)	31.06 (789)	see Figure 1
2017 Rogue FWD w/2-row seating w/225/60R18 (OE)	30.91 (785)	30.79 (782)	see Figure 1
2017 Rogue FWD w/2-row seating w/225/55R19 (OE)	30.98 (787)	30.83 (783)	see Figure 1
2014-16 Rogue FWD w/3-row seating	31.10-31.14 (790-791)	30.94-30.98 (786-787)	see Figure 1
2017 Rogue FWD w/3-row seating	30.91 (785)	30.75 (781)	see Figure 1
2014-17 Rogue AWD w/2-row seating w/225/65R17 (OE)	31.38-31.42 (797-798)	31.26-31.34 (794-796)	see Figure 1
2014-17 Rogue AWD w/2-row seating w/225/60R18 (OE)	31.46-31.54 (799-801)	31.34-31.42 (796-798)	see Figure 1
2017 Rogue AWD w/2-row seating w/225/55R19 (OE)	31.57 (802)	31.38 (797)	see Figure 1
2014-17 Rogue AWD w/3-row seating	31.50-31.54 (800-801)	31.30-31.34 (795-796)	see Figure 1
Sentra	Front (WF)	Rear (WR)	Notes
1991-94 Sentra	25.47-25.55 (647-649)	24.92-25.15 (633-639)	see Figure 1
1995-99 Sentra	25.90-26.29 (658-668)	25.15-25.55 (639-649)	see Figure 1
2000-02 Sentra w/14" tires	25.55 (649)	25.31 (643)	see Figure 1
2000-03 Sentra w/15" tires exc/CA models	25.90 (658)	25.70 (653)	see Figure 1
2000-03 Sentra w/15" tires Calif.CA models	26.14 (664)	25.90 (658)	see Figure 1
2000-03 Sentra w/16" tires	25.98 (660)	25.66 (652)	see Figure 1
2002-03 Sentra w/17" tires	26.14 (664)	25.78 (655)	see Figure 1
2004 Sentra 1.8 (All models)	25.94 (659)	25.70 (653)	see Figure 1
2005-06 Sentra 1.8 with 195/60HR15	26.14 (664)	25.90 (658)	see Figure 1
2005-06 Sentra 1.8S with 195/55HR16	26.22 (666)	25.94 (659)	see Figure 1

* ±0.75 (19) (except when range is specified)

Nissan (cont.)

YEAR MODEL	HEIGHT SPECIFICATIONS @ CURB WEIGHT - IN. (MM)		
Sentra (cont.)	FRONT (WF)	REAR (WR)	Notes
2004-06 Sentra SE-R with 195/55HR16	25.98 (660)	25.66 (652)	see Figure 1
2004-06 Sentra SPEC-V with 215/45ZR17	26.14 (664)	25.78 (655)	see Figure 1
2007-10 Sentra with 205/60HR15 tires (OE)	27.20 (691)	27.28 (693)	see Figure 1
2007-10 Sentra with 205/55HR16 tires (OE)	27.32 (694)	27.40 (696)	see Figure 1
2012 Sentra with 205/60HR15 tires (OE)	27.13 (689)	27.09 (688)	see Figure 1
2012 Sentra with 205/55HR16 tires (OE)	27.24 (692)	27.24 (692)	see Figure 1
2007-12 Sentra with 225/45VR17 tires (OE)	27.17 (690)	27.17 (690)	see Figure 1
2007-12 Sentra with 225/45WR17 tires (OE)	26.73 (679)	26.65 (677)	see Figure 1
2013-16 Sentra with 205/55R16 tires (OE) exc/FE USA	27.68 (703)	27.68 (703)	see Figure 1
2013-16 Sentra with 205/55R16 tires (OE) FE USA	27.80 (706)	27.80 (706)	see Figure 1
2013-16 Sentra with 205/50R17 tires (OE) USA	27.87 (708)	27.83 (707)	see Figure 1
2013-16 Sentra with 205/55R16 tires (OE) CANADA	27.72 (704)	27.72 (704)	see Figure 1
2013-16 Sentra with 205/50R17 tires (OE) CANADA	27.91 (709)	27.87 (708)	see Figure 1
2017 Sentra 1.8L 205/55R16 tires (OE) USA	27.72 (704)	27.76 (705)	see Figure 1
2017 Sentra 1.8L 205/55R16 tires (OE) CANADA	27.76 (705)	27.80 (706)	see Figure 1
2017 Sentra 1.8L 205/50R17 tires (OE) USA	27.87 (708)	27.91 (709)	see Figure 1
2017 Sentra 1.8L 205/50R17 tires (OE) CANADA	27.91 (709)	27.95 (710)	see Figure 1
2017 Sentra 1.6L 205/50R17 tires (OE) USA & CANADA	27.80 (706)	27.91 (709)	see Figure 1
Stanza	FRONT (WF)	REAR (WR)	Notes
1990-92 Stanza	26.61 (676)	25.39 (645)	see Figure 1
Versa	FRONT (WF)	REAR (WR)	Notes
2007-11 Versa w/14" tires (OE)	26.65 (677)	26.81 (681)	see Figure 1
2007-12 Versa w/15" tires (OE) USA	26.97 (685)	26.85 (682)	see Figure 1
2007-12 Versa w/15" tires (OE) Canada	26.97 (685)	26.89 (683)	see Figure 1
2010-12 Versa w/16" tires (OE)	27.01 (686)	26.89 (683)	see Figure 1
2012-16 Versa Sedan USA	26.89 (683)	26.30 (668)	see Figure 1
2017 Versa Sedan USA	27.09 (688)	26.18 (665)	see Figure 1
2012-16 Versa Sedan Canada	26.93 (684)	26.30 (668)	see Figure 1
2017 Versa Sedan Canada	27.09 (688)	26.22 (666)	see Figure 1
2014-17 Versa Note w/15" tires (OE)	27.20-27.24 (691-692)	26.69 (678)	see Figure 1
2014-17 Versa Note w/16" tires (OE)	27.20 (691)	26.65 (677)	see Figure 1

Nissan Truck & SUV

YEAR MODEL	HEIGHT SPECIFICATIONS @ CURB WEIGHT - IN. (MM)		
Armada	FRONT (WF)	REAR (WR)	Notes
2004-07 Armada 2WD except Air leveling	36.06-36.10 (916-917)	36.88-36.92 (937-938)	see Figure 1
2004-07 Armada 2WD with Air leveling	36.10-36.14 (917-918)	36.02-36.10 (915-917)	see Figure 1
2004-07 Armada 4WD except Air leveling	36.81-36.85 (935-936)	37.63-37.71 (956-958)	see Figure 1
2004-07 Armada 4WD with Air leveling	36.81-36.85 (935-936)	36.81-36.88 (935-937)	see Figure 1
2008-15 Armada 2WD exc/Air leveling w/18" tires (OE)	35.98 (914)	36.65 (931)	see Figure 1
2008-15 Armada 2WD exc/Air leveling w/20" tires (OE)	36.22 (920)	36.88 (937)	see Figure 1
2008-15 Armada 2WD with Air leveling w/18" tires (OE)	35.98 (914)	35.86 (911)	see Figure 1
2008-15 Armada 2WD with Air leveling w/20" tires (OE)	36.22 (920)	36.10 (917)	see Figure 1
2008-15 Armada 4WD exc/Air leveling w/18" tires (OE)	36.65 (931)	37.44 (951)	see Figure 1
2008-15 Armada 4WD exc/Air leveling w/20" tires (OE)	36.88 (937)	37.67 (957)	see Figure 1
2008-15 Armada 4WD with Air leveling w/18" tires (OE)	36.65 (931)	36.65 (931)	see Figure 1
2008-15 Armada 4WD with Air leveling w/20" tires (OE)	36.88 (937)	36.88 (937)	see Figure 1
2017 Armada with 18" tires (OE)	35.47 (901)	34.92 (887)	
2017 Armada with 20" tires (OE)	35.55 (903)	35.04 (890)	
Frontier -> 2004	FRONT(SF) INSPECTION RANGE	FRONT(SF) ADJUSTMENT RANGE	NOTES
1998-02 Frontier Truck 2WD (KA models only)	4.25-4.64 (108-118)	4.37-4.52 (111-115)	see Figure 3 - adjustable torsion bars
2003-04 Frontier Truck 2WD (KA models only)	4.40-4.80 (112-122)	4.52-4.69 (115-119)	see Figure 3 - adjustable torsion bars
1998-00 Frontier Truck 2WD (VG Models), 4WD	1.69-2.08 (43-53)	1.81-1.96 (46-50)	see Figure 3 - adjustable torsion bars
2001-04 Frontier Truck 2WD (VG Models) 4WD w/Superchg.	1.37-1.77 (35-45)	1.49-1.65 (38-42)	see Figure 3 - adjustable torsion bars
2001-04 Frontier Truck 4WD without Supercharger	1.69-2.08 (43-53)	1.81-1.96 (46-50)	see Figure 3 - adjustable torsion bars
Frontier 2005 ->	FRONT (WF)	REAR (WR)	NOTES
2005-08 Frontier 2WD King Cab 4 Cyl.	34.09 (866)	34.88 (886)	see Figure 1
2005-08 Frontier 2WD King Cab 6 Cyl. 265/70R16 tires (OE)	34.17 (868)	35.20 (894)	see Figure 1
2005-08 Frontier 2WD King Cab 6 Cyl. 265/75R16 tires (OE)	34.65 (880)	35.63 (905)	see Figure 1
2005-08 Frontier 2WD King Cab 6 Cyl. 265/65R17 tires (OE)	34.21 (869)	35.25 (895)	see Figure 1
2005-08 Frontier 4WD King Cab 6 Cyl. 265/70R16 tires (OE)	34.65 (880)	35.55 (903)	see Figure 1
2005-08 Frontier 4WD King Cab 6 Cyl. 265/75R16 tires (OE)	35.12 (892)	36.02 (915)	see Figure 1
2005-08 Frontier 4WD King Cab 6 Cyl. 265/65R17 tires (OE)	34.68 (881)	35.69 (904)	see Figure 1
2005-08 Frontier 2WD Crew Cab 265/70R16 tires (OE)	34.17 (868)	35.07 (891)	see Figure 1
2005-08 Frontier 2WD Crew Cab 265/75R16 tires (OE)	34.65 (880)	35.55 (903)	see Figure 1
2005-08 Frontier 2WD Crew Cab 265/65R17 tires (OE)	34.21 (869)	35.12 (892)	see Figure 1
2005-08 Frontier 4WD Crew Cab 265/70R16 tires (OE)	34.65 (880)	35.47 (901)	see Figure 1
2005-08 Frontier 4WD Crew Cab 265/75R16 tires (OE)	35.12 (892)	35.94 (913)	see Figure 1
2005-08 Frontier 4WD Crew Cab 265/65R17 tires (OE)	34.68 (881)	35.51 (902)	see Figure 1

* ± ¾ (19) (except when range is given)

CHASSIS / SUSP. HEIGHTS

Nissan Truck & SUV (cont.)

Year Model	Height Specifications @ Curb Weight - In. (mm)		
Frontier (cont.)	Front (WF)	Rear (WR)	Notes
2009-17 Frontier 2WD King Cab 4 Cyl. 235/75R15 tires (OE)	33.46 (850)	34.57 (878)	see Figure 1
2009-17 Frontier 2WD King Cab 4 Cyl. 265/70R16 tires (OE)	34.06 (865)	34.92(887)	see Figure 1
2009-17 Frontier 2WD King Cab 6 Cyl. 265/70R16 tires (OE)	34.17 (868)	35.24 (895)	see Figure 1
2009-17 Frontier 2WD King Cab 6 Cyl. 265/75R16 tires (OE)	34.65 (880)	35.71 (907)	see Figure 1
2009-10 Frontier 2WD King Cab 6 Cyl. 265/60R18 tires (OE)	34.13 (867)	35.16 (893)	see Figure 1
2009-17 Frontier 4WD King Cab 6 Cyl. 265/70R16 tires (OE)	34.68 (881)	35.59 (904)	see Figure 1
2009-17 Frontier 4WD King Cab 6 Cyl. 265/75R16 tires (OE)	35.16 (893)	36.10 (917)	see Figure 1
2009-10 Frontier 4WD King Cab 6 Cyl. 265/60R18 tires (OE)	34.68 (881)	35.55 (903)	see Figure 1
2009-17 Frontier 2WD Crew Cab 265/70R16 tires Short WB	34.13 (867)	35.12 (892)	see Figure 1
2011-17 Frontier 2WD Crew Cab 265/70R16 tires Long WB	34.25 (870)	35.12 (892)	see Figure 1
2009-17 Frontier 2WD Crew Cab 265/75R16 tires (OE)	34.61 (879)	35.59 (904)	see Figure 1
2009-17 Frontier 2WD Crew Cab 265/60R18 tires Short WB	34.09 (866)	35.12 (892)	see Figure 1
2011-17 Frontier 2WD Crew Cab 265/60R18 tires Long WB	34.21 (869)	35.12 (892)	see Figure 1
2009-17 Frontier 4WD Crew Cab 265/70R16 tires Short WB	34.61 (879)	35.63 (905)	see Figure 1
2011-17 Frontier 4WD Crew Cab 265/70R16 tires Long WB	34.72 (882)	35.51 (902)	see Figure 1
2009-17 Frontier 4WD Crew Cab 265/75R16 tires (OE)	35.08 (891)	36.14 (918)	see Figure 1
2009-17 Frontier 4WD Crew Cab 265/60R18 tires Short WB	34.61 (879)	35.63 (905)	see Figure 1
2011-17 Frontier 4WD Crew Cab 265/60R18 tires Long WB	34.72 (882)	35.51 (902)	see Figure 1
NV Vans	Front (WF)	Rear (WR)	Notes
2012-17 NV1500/2500 (incl. Passenger van)	35.00 (888)	36.30 (922)	see Figure 1
2012-17 NV3500	35.40 (900)	37.30 (947)	see Figure 1
NV200 Compact Cargo	Front (WF)	Rear (WR)	Notes
2013-17 NV200 Compact Cargo & Taxi	27.56 (700)	29.02 (737)	see Figure 1
Pathfinder ->1996	Front(SF) Inspection range	Front(SF) Adjustment range	Notes
1990-95 Pathfinder 2WD	4.25-4.64 (108-118)	4.37-4.52 (111-115)	See Figure 2 - adjustable torsion bars
1990-95 Pathfinder 4WD	1.61-2.00 (41-51)	1.73-1.88 (44-48)	See Figure 3 - adjustable torsion bars
Pathfinder 1996 ->	Front (WF)	Rear (WR)	Notes
1996-00 Pathfinder 2WD	32.20 (818)	33.70 (856)	see Figure 1
1996-00 Pathfinder 4WD w/o over-fender	32.08 (815)	33.66 (855) [1]	Figure 1 [1] with over-fender 33.3 (846)
2001-04 Pathfinder 2WD w/o over-fender	33.07 (840)	34.13 (867) [1]	Figure 1 [1] with over-fender 32.1 (817)
2001-04 Pathfinder 4WD w/o over-fender	32.95 (837) [1]	34.13 (867) [1]	Figure 1 [1] with over-fender F=32.4, R=32.15
2005-07 Pathfinder 2WD XE with 245/75R16 tires (OE)	34.09 (866)	34.45 (875)	see Figure 1
2005-07 Pathfinder 2WD SE with 265/70R16 tires (OE)	34.13 (867)	34.49 (876)	see Figure 1
2005-07 Pathfinder 2WD SE with 265/75R16 tires (OE)	34.60 (879)	34.96 (888)	see Figure 1
2005-07 Pathfinder 2WD LE with 265/65R17 tires (OE)	34.17 (868)	34.52 (877)	see Figure 1
2005-07 Pathfinder 4WD XE with 245/75R16 tires (OE)	34.41 (874)	34.84 (885)	see Figure 1
2005-07 Pathfinder 4WD SE with 265/70R16 tires (OE)	34.48 (876)	34.88 (886)	see Figure 1
2005-07 Pathfinder 4WD SE with 265/75R16 tires (OE)	34.96 (888)	35.35 (898)	see Figure 1
2005-07 Pathfinder 4WD LE with 265/65R17 tires (OE)	34.56 (878)	34.92 (887)	see Figure 1
2008-12 Pathfinder V-6 2WD with 245/75R16 tires (OE)	34.13 (867)	34.45 (875)	see Figure 1
2008-09 Pathfinder V-6 2WD with 265/70R16 tires (OE)	34.06 (865)	34.37 (873)	see Figure 1
2008-12 Pathfinder V-6 2WD with 265/65R17 tires (OE)	34.06 (865)	34.37 (873)	see Figure 1
2008-12 Pathfinder V-6 2WD with 265/60R18 tires (OE)	34.13 (867)	34.45 (875)	see Figure 1
2008-12 Pathfinder V-6 4WD with 245/75R16 tires (OE)	34.45 (875)	34.80 (884)	see Figure 1
2008-09 Pathfinder V-6 4WD with 265/70R16 tires (OE)	34.37 (873)	34.72 (882)	see Figure 1
2008-09 Pathfinder V-6 4WD with 265/75R16 tires (OE)	35.00 (889)	35.35 (898)	see Figure 1
2008-12 Pathfinder V-6 4WD with 265/65R17 tires (OE)	34.41 (874)	34.76 (883)	see Figure 1
2011-12 Pathfinder V-6 4WD with 265/60R18 tires (OE)	35.08 (891)	35.47 (901)	see Figure 1
2008-09 Pathfinder V-8 2WD with 265/60R18 tires (OE)	34.17 (868)	34.49 (876)	see Figure 1
2008-12 Pathfinder V-8 4WD with 265/60R18 tires (OE)	34.49 (876)	34.88 (886)	see Figure 1
2013-17 Pathfinder 2WD 18" tires (OE) USA	32.40 (823)	32.56 (827)	see Figure 1
2013-17 Pathfinder 2WD 20" tires (OE) USA	32.32 (821)	32.48 (825)	see Figure 1
2013-17 Pathfinder 4WD USA	32.36 (822)	32.52-32.56 (826-827)	see Figure 1
2013-17 Pathfinder 2WD Canada	32.40 (823)	32.60-32.64 (828-829)	see Figure 1
2013-17 Pathfinder 4WD Canada	32.36-32.40 (822-823)	32.56-32.60 (827-828)	see Figure 1
2014-15 Pathfinder HEV 2WD 18" tires (OE) USA	32.36 (822)	32.52 (826)	see Figure 1
2014-15 Pathfinder HEV 2WD 20" tires (OE) USA	32.28 (820)	32.36 (822)	see Figure 1
2014-15 Pathfinder HEV 4WD 18" tires (OE) USA	32.36 (822)	32.48-32.56 (825-827)	see Figure 1
2014-15 Pathfinder HEV 4WD 20" tires (OE) USA	32.32 (821)	32.40 (823)	see Figure 1
2014-15 Pathfinder HEV 4WD 18" tires (OE) Canada	32.36 (822)	32.48-32.52 (825-826)	see Figure 1
2014-15 Pathfinder HEV 4WD 20" tires (OE) Canada	32.28 (820)	32.44 (824)	see Figure 1
Pickup	Front(SF) Inspection range	Front(SF) Adjustment range	Notes
1990-97 Truck 2WD (start Jan.86)	4.25-4.64 (108-118)	4.37-4.52 (111-115)	see Figure 2 - adjustable torsion bars
1990-97 Truck 4WD (start Jan.86)	1.61-2.00 (41-51)	1.73-1.88 (44-48)	see Figure 3 - adjustable torsion bars
Titan	Front (WF)	Rear (WR)	Notes
2004-07 Titan 4X2 King Cab with 265/70R18 tires (OE)	35.9 (913)	37.6 (955)	see Figure 1
2004-07 Titan 4X2 King Cab with 285/70R17 tires (OE)	36.0 (914)	37.6 (956)	see Figure 1
2004-07 Titan 4X2 King Cab with 245/75R17 tires (OE)	35.6 (904)	37.2 (946)	see Figure 1
2004-07 Titan 4X2 Crew Cab with 265/70R18 tires (OE)	36.1 (916)	37.6 (956)	see Figure 1
2004-07 Titan 4X2 Crew Cab with 285/70R17 tires (OE)	36.1 (917)	37.7 (957)	see Figure 1
2004-07 Titan 4X2 Crew Cab with 245/75R17 tires (OE)	35.7 (907)	37.3 (947)	see Figure 1
2004-07 Titan 4X4 King Cab with 265/70R18 tires (OE)	37.5 (952)	39.2 (995)	see Figure 1

* ± ¾ (19) (except when range is given)

CHASSIS / SUSP. HEIGHTS

Nissan Truck & SUV (cont.)

Year Model	Height Specifications @ Curb Weight - In. (mm)		
Titan (cont.)	Front (WF)	Rear (WR)	Notes
2004-07 Titan 4X4 King Cab with 285/70R17 tires (OE)	37.5 (953)	39.2 (996)	see Figure 1
2004-07 Titan 4X4 King Cab with 245/75R17 tires (OE)	37.0 (941)	38.7 (984)	see Figure 1
2004-07 Titan 4X4 Crew Cab with 265/70R18 tires (OE)	37.6 (955)	39.2 (996)	see Figure 1
2004-07 Titan 4X4 Crew Cab with 285/70R17 tires (OE)	37.6 (956)	39.3 (997)	see Figure 1
2004-07 Titan 4X4 Crew Cab with 245/75R17 tires (OE)	37.2 (944)	38.8 (985)	see Figure 1
2008-15 Titan 4X2 King Cab with 265/70R18 Tires Short WB	35.91 (912)	37.48 (952)	see Figure 1
2008-15 Titan 4X2 King Cab with 275/70R18 Tires Short WB	36.30 (922)	37.87 (962)	see Figure 1
2008-15 Titan 4X2 King Cab with 275/60R20 Tires Short WB	36.10 (917)	37.68 (957)	see Figure 1
2008-15 Titan 4X2 Crew Cab with 265/70R18 Tires Short WB	35.98 (914)	37.56 (954)	see Figure 1
2008-15 Titan 4X2 Crew Cab with 275/70R18 Tires Short WB	36.42 (925)	37.99 (965)	see Figure 1
2008-15 Titan 4X2 Crew Cab with 275/60R20 Tires Short WB	36.18 (919)	37.76 (959)	see Figure 1
2008-15 Titan 4X2 King Cab with 265/70R18 Tires Long WB	35.91 (912)	37.40 (950)	see Figure 1
2008-15 Titan 4X2 King Cab with 275/70R18 Tires Long WB	36.30 (922)	37.80 (960)	see Figure 1
2008-15 Titan 4X2 King Cab with 275/60R20 Tires Long WB	36.10 (917)	37.60 (955)	see Figure 1
2008-15 Titan 4X2 Crew Cab with 265/70R18 Tires Long WB	35.98 (914)	37.44 (951)	see Figure 1
2008-15 Titan 4X2 Crew Cab with 275/70R18 Tires Long WB	36.42 (925)	37.87 (962)	see Figure 1
2008-15 Titan 4X2 Crew Cab with 275/60R20 Tires Long WB	36.22 (920)	37.64 (956)	see Figure 1
2008-15 Titan 4X4 King Cab with 265/70R18 Tires Short WB	37.36 (949)	39.02 (991)	see Figure 1
2008-15 Titan 4X4 King Cab with 275/70R18 Tires Short WB	37.80 (960)	39.45 (1002)[1]	see Figure 1 [1] w/Tow Pkg. 1mm less
2008-15 Titan 4X4 King Cab with 275/60R20 Tires Short WB	37.60 (955)	39.21 (996)	see Figure 1
2008-15 Titan 4X4 Crew Cab with 265/70R18 Tires Short WB	37.44 (951)	39.13 (994)[1]	see Figure 1 [1] w/Tow Pkg. 1mm less
2008-15 Titan 4X4 Crew Cab with 275/70R18 Tires Short WB	37.87 (962)	39.53 (1004)	see Figure 1
2008-15 Titan 4X4 Crew Cab with 275/60R20 Tires Short WB	37.68 (957)	39.33 (999)[1]	see Figure 1 [1] w/Tow Pkg. 1mm less
2008-15 Titan 4X4 King Cab with 265/70R18 Tires Long WB	37.36 (949)	38.94 (989)	see Figure 1
2008-15 Titan 4X4 King Cab with 275/70R18 Tires Long WB	37.76 (959)	39.47 (1000)	see Figure 1
2008-15 Titan 4X4 King Cab with 275/60R20 Tires Long WB	37.56 (954)	39.17 (995)	see Figure 1
2008-15 Titan 4X4 Crew Cab with 265/70R18 Tires Long WB	37.44 (951)	39.02 (991)	see Figure 1
2008-15 Titan 4X4 Crew Cab with 275/70R18 Tires Long WB	37.87 (962)	39.45 (1002)	see Figure 1
2008-15 Titan 4X4 Crew Cab with 275/60R20 Tires Long WB	37.68 (957)	39.21 (996)	see Figure 1
2016-17 Titan XD Diesel 4X2 with 245/75R17 Tires (OE)	38.90 (988)	40.35-40.43 (1025-1027)	see Figure 1
2016-17 Titan XD Diesel 4X2 with 265/60R20 Tires (OE)	39.76-39.80 (1010-1011)	41.06-41.14 (1043-1045)	see Figure 1
2016-17 Titan XD Diesel 4X4 with 245/75R17 Tires (OE)	38.82-38.90 (986-988)	40.31-40.35 (1024-1025)	see Figure 1
2016-17 Titan XD Diesel 4X4 with 275/65R18 Tires (OE)	39.37 (1000)	40.71 (1034)	see Figure 1
2016-17 Titan XD Diesel 4X4 with 265/60R20 Tires (OE)	39.72-39.76 (1009-1010)	41.06-41.14 (1043-1045)	see Figure 1
2016-17 Titan XD Gas 4X2 with 245/75R17 Tires (OE)	39.02 (991)	40.39-40.47 (1026-1028)	see Figure 1
2016-17 Titan XD Gas 4X2 with 265/60R20 Tires (OE)	39.88 (1013)	41.10-41.18 (1044-1046)	see Figure 1
2016-17 Titan XD Gas 4X4 with 245/75R17 Tires (OE)	38.98 (990)	40.39-40.43 (1026-1027)	see Figure 1
2016-17 Titan XD Gas 4X4 with 275/65R18 Tires (OE)	39.45 (1002)	40.71 (1034)	see Figure 1
2016-17 Titan XD Gas 4X4 with 265/60R20 Tires (OE)	39.84 (1012)	41.10-41.14 (1044-1045)	see Figure 1
2017 Titan 2WD Base Cab, with 265/70R18 tires (OE), S	38.90 (920)	40.43 (961)	see Figure 1
2017 Titan 2WD Base Cab, with 265/70R18 tires (OE), SV	39.76 (921)	41.14 (962)	see Figure 1
2017 Titan 4WD Base Cab, with 265/70R18 tires S	36.77 (934)	38.46 (977)	see Figure 1
2017 Titan 4WD Base Cab, with 265/70R18 tires SV	36.81 (935)	38.50 (978)	see Figure 1
2017 Titan 2WD Crew Cab, with 265/70R18 tires (OE), S	35.83 (910)	37.52 (953)	see Figure 1
2017 Titan 2WD Crew Cab, with 265/70R18 tires (OE), SV	36.22 (920)	37.80 (960)	see Figure 1
2017 Titan 2WD Crew Cab, with 265/70R18 tires, SV comf	36.22 (920)	37.80 (960)	see Figure 1
2017 Titan 2WD Crew Cab, with 275/60R20 tires (OE), SL	36.14 (918)	37.72 (958)	see Figure 1
2017 Titan 4WD Crew Cab, with 265/70R18 tires (OE), S	36.42 (925)	38.11 (968)	see Figure 1
2017 Titan 4WD Crew Cab, with 265/70R18 tires (OE), SV	36.81 (935)	38.43 (976)	see Figure 1
2017 Titan 4WD Crew Cab, with 265/70R18 tires, SV comf	36.81 (935)	38.43 (976)	see Figure 1
2017 Titan 4WD Crew Cab, with 265/70R18 tires, PRO-4X	37.99 (965)	39.61 (1006)	see Figure 1
2017 Titan 4WD Crew Cab, with 275/60R20 tires, SL/PLAT	37.68 (957)	39.29 (998)	see Figure 1
Xterra -> 2004	SF (Inspection range)	SF (Adjustment range)	Notes
2000-04 Xterra 2WD	1.37-1.77 (35-45)	1.49-1.65 (38-42)	see Figure 3 - adjustable torsion bars
2000-04 Xterra 4WD	1.69-2.08 (43-53)	1.81-1.96 (46-50)	see Figure 3 - adjustable torsion bars
Xterra 2005 ->	Front (WF)	Rear (WR)	Notes
2005-09 Xterra 2WD with P265/70R16 tires	33.03 (839)	34.33 (872)	see Figure 1
2005-09 Xterra 2WD with P265/75R16 tires	33.50 (851)	34.80 (884)	see Figure 1
2005-09 Xterra 2WD with P265/65R17 tires	33.11 (841)	34.33 (872)	see Figure 1
2005-09 Xterra 4WD with P265/70R16 tires	33.78 (858)	35.12 (892)	see Figure 1
2005-09 Xterra 4WD with P265/75R16 tires	34.25 (870)	35.59 (904)	see Figure 1
2005-09 Xterra 4WD with P265/65R17 tires	33.82 (859)	35.12 (892)	see Figure 1
2010-15 Xterra 2WD with P265/70R16 tires	33.03 (839)	34.33 (872)	see Figure 1
2010 Xterra 2WD with P265/65R17 tires	32.95 (837)	34.25 (870)	see Figure 1
2010-15 Xterra 4WD with P265/70R16 tires	33.78 (858)	35.12 (892)	see Figure 1
2010 Xterra 4WD with P265/75R16 tires	34.33 (872)	35.71 (907)	see Figure 1
2011-15 Xterra 4WD with P265/75R16 tires	34.25 (870)	35.59 (904)	see Figure 1
2010 Xterra 4WD with P265/65R17 tires	33.74 (857)	35.12 (892)	see Figure 1
2011-12 Xterra 4WD with P265/65R17 tires	33.82 (859)	35.12 (892)	see Figure 1
X-Trail	Front (WF)	Rear (WR)	Notes
2005-06 X-Trail 2WD	30.47 (774)	30.94 (786)	see Figure 1
2005-06 X-Trail AWD	30.43 (773)	30.90 (785)	see Figure 1

* ± ¾ (19) (except when range is given)

Subaru

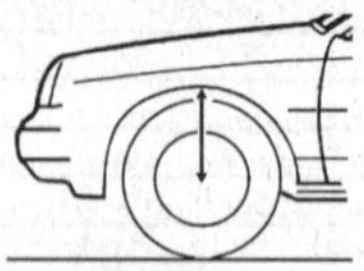

Figure 1

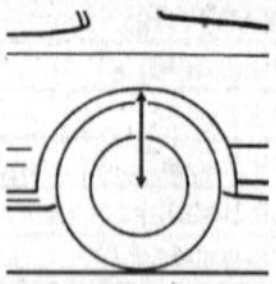

Figure 2

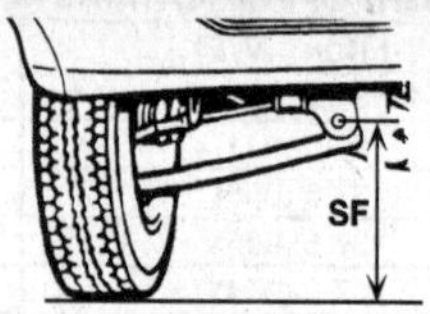

Figure 3

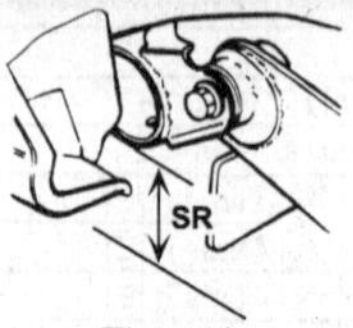

Figure 4

Measure **Figure 3** at link inner bolt head to ground.

Measure **Figure 4** at lowest point on crossmember tube.

YEAR MODEL	HEIGHT SPECIFICATIONS (CURB WEIGHT) - IN. (MM) *		
	FRONT	REAR	NOTES
2003 Baja	16.49 (419)	16.29 (414)	see Figures 1 & 2
2004-06 Baja	17.16 (436)	17.36 (441)	see Figures 1 & 2
2013-17 BRZ	14.76 (375)	14.64 (372)	see Figures 1 & 2
1998-02 Forester	17.00 (432)	17.12 (435)	see Figures 1 & 2
2003-08 Forester exc/Turbo XT	17.20 (437)	17.32 (440)	see Figures 1 & 2
2004-08 Forester Turbo XT	17.20 (437)	17.12 (435)	see Figures 1 & 2
2009-13 Forester	17.51 (445)	17.63 (448)	see Figures 1 & 2
2014-17 Forester	17.48 (444)	18.03 (458)	see Figures 1 & 2
1993-97 Impreza FWD	15.39 (391)	14.88 (378)	see Figures 1 & 2
1993-01 Impreza AWD exc/RS, Outback	15.39 (391)	14.92 (379)	see Figures 1 & 2
1998-01 Impreza RS Coupe/Sedan	14.60 (371)	14.29 (363)	see Figures 1 & 2
2002-07 Impreza Sedan non-Turbo	15.98 (406)	15.00 (381)	see Figures 1 & 2
2002-07 Impreza Sedan Turbo exc/STI	15.59 (396)	14.80 (376)	see Figures 1 & 2
2004-07 Impreza Sedan WRX STi	15.19 (386)	14.29 (363)	see Figures 1 & 2
2002-07 Impreza Wagon non-Turbo	15.62 (397)	15.00 (381)	see Figures 1 & 2
2002-07 Impreza Wagon Turbo	15.23 (387)	14.80 (376)	see Figures 1 & 2
2008-11 Impreza exc/Outback Sport	15.23 (387)	14.96 (380)	see Figures 1 & 2
2012-16 Impreza Sedan	15.23 (387)	14.49 (368)	see Figures 1 & 2
2012-16 Impreza Hatch Sport Models	15.43 (392)	14.60 (371)	see Figures 1 & 2
2008-10 Impreza WRX exc/STI	15.23 (387)	14.76 (375)	see Figures 1 & 2
2008-10 Impreza WRX STI	14.80 (376)	14.56 (370)	see Figures 1 & 2
2011-17 Impreza WRX exc/STI	14.57-14.80 (370-376)	14.21-14.56 (361-370)	see Figures 1 & 2
2011-17 Impreza WRX STI	14.57-14.60 (370-371)	14.25-14.37 (362-365)	see Figures 1 & 2
1997-01 Impreza Outback Sport	15.51 (394)	15.19 (386)	see Figures 1 & 2
2002-07 Impreza Outback Sport	15.82 (402)	15.19 (386)	see Figures 1 & 2
2008-11 Impreza Outback Sport	15.43 (392)	14.96 (380)	see Figures 1 & 2
1990-94 Justy	9.44-10.62 (240-270)	10.62-11.81 (270-300)	see Figures 3 & 4
1990-94 Legacy All w/Air Ride	15.27 (388) [1]	14.52 (369)[1]	see Figures 1 & 2 [1] Air switch in NORMAL pos.
1990-94 Legacy Sedan FWD	15.07 (383)	13.93 (354)	see Figures 1 & 2
1990-94 Legacy Sedan AWD	15.47 (393)	14.33 (364)	see Figures 1 & 2
1990-94 Legacy Wagon FWD	15.07 (383)	14.52 (369)	see Figures 1 & 2
1990-94 Legacy Wagon AWD	15.27 (388)	14.92 (379)	see Figures 1 & 2
1995-99 Legacy Sedan	15.15 (385)	Sedan 14.52 (369), Wagon 14.92 (379)	see Figures 1 & 2
2000-04 Legacy Sedan	15.27 (388)	14.60 (371)	see Figures 1 & 2
2000-04 Legacy Wagon	15.27 (388)	15.00 (381)	see Figures 1 & 2
2000-04 Legacy Outback	16.69 (424)	16.57 (421)	see Figures 3 & 4
2005-07 Legacy Sedan exc/Outback	15.00 (381)	14.37 (365)	see Figures 1 & 2
2005-07 Legacy Wagon exc/Outback	15.00 (381)	14.76 (375)	see Figures 1 & 2
2005-07 Legacy Outback (Sdn. & Wgn.)	17.28 (439)	17.24 (438)	see Figures 1 & 2
2008-09 Legacy Sedan	15.00 (381)	14.37 (365)	see Figures 1 & 2
2008-09 Legacy Outback	17.28 (439)	17.24 (438)	see Figures 1 & 2
2010-14 Legacy Sedan w/16" & P215-17"	15.86 (403)	15.43 (392)	see Figures 1 & 2
2010-14 Legacy Sedan w/18" & P225-17"	15.55 (395)	15.23 (387)	see Figures 1 & 2
2010-14 Legacy Outback	18.46 (469)	18.66 (474)	see Figures 1 & 2
2015-17 Legacy Sedan	15.63 (397)	15.00-15.07 (381-383)	see Figures 1 & 2
2015-17 Legacy Outback	18.38-18.42 (467-468)	18.38-18.42 (467-468)	see Figures 1 & 2
1990-92 Loyale FWD exc/Wagon	8.42-9.76 (214-248)	7.75-8.93 (197-227)	see Figures 3 & 4
1990-92 Loyale FWD Wagon	8.66-10.00 (220-254)	8.54-9.72 (217-247)	see Figures 3 & 4
1990-92 Loyale 4WD exc/Wagon	9.60-10.94 (244-278)	9.64-10.82 (245-275)	see Figures 3 & 4
1990-92 Loyale 4WD Wagon	9.76-11.10 (248-282)	10.19-11.37 (259-289)	see Figures 3 & 4
1992-97 SVX	14.96 (380)	13.18 (335)	see Figures 1 & 2
2006-09 Tribeca (B9 Tribeca)	17.40 (442)	17.28 (439)	see Figures 1 & 2
2010-14 Tribeca	17.79 (452)	17.67 (449)	see Figures 1 & 2
1990-91 XT FWD 1800	8.03-9.37 (204-238)	7.20-8.38 (183-213)	see Figures 3 & 4
1990-91 XT FWD 2700	7.63-8.97 (194-228)	7.08-8.26 (180-210)	see Figures 3 & 4
1990-91 XT 4WD 1800	8.89-10.23 (226-260)	9.05-10.23 (230-260)	see Figures 3 & 4
1990-91 XT 4WD 2700	9.09-10.43 (231-265)	8.97-10.15 (228-258)	see Figures 3 & 4
2013-17 XV Crosstrek (incl. HEV)	17.79 (452)	17.71 (450)	see Figures 1 & 2

* +0.47 –0.95 (+12 -24) (except when range is specified)

CHASSIS / SUSP. HEIGHTS

Suzuki

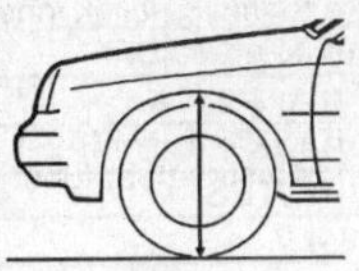
Figure 1

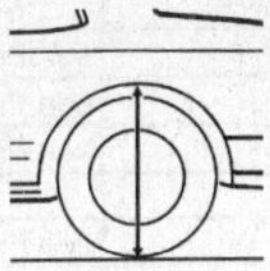
Figure 2

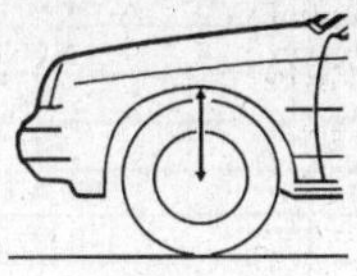
Figure 3

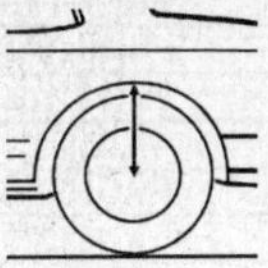
Figure 4

Measure **Figure 1 - 4** vehicles loaded (150 lbs in each front seat) and fuel tank full. Measure the distance between the lower edge of wheel opening and level ground (**Figs. 1 & 2**), or between the lower edge of the wheel opening and the center of the wheel (**Figs. 3 & 4**)

Year Model	Height Specifications @Curb Weight - In. (mm) Front	Rear	Notes
2002-07 Aerio	Side/Side difference max. 0.59 (15)	Side/Side difference max. 0.59 (15)	See **Figures 1 & 2**
1995-02 Esteem	Side/Side difference max. 0.59 (15)	Side/Side difference max. 0.59 (15)	See **Figures 1 & 2**
2009-12 Equator			See Nissan Frontier
2004-08 Forenza/Reno	14.48 (368)	14.44 (367)	See **Figures 3 & 4** Side/Side max. 0.59 (15)
1999-13 Grand Vitara, Vitara	Side/Side difference max. 0.59 (15)	Side/Side difference max. 0.59 (15)	See **Figures 1 & 2**
2010-13 Kizashi	Side/Side difference max. 0.59 (15)	Side/Side difference max. 0.59 (15)	See **Figures 1 & 2**
1990-02 Swift	Side/Side difference max. 0.59 (15)	Side/Side difference max. 0.59 (15)	See **Figures 1 & 2**
2007-13 SX-4	Side/Side difference max. 0.59 (15)	Side/Side difference max. 0.59 (15)	See **Figures 1 & 2**
2004-07 Verona with 205/65R15 tires	26.96 (685)	27.00 (686)	See **Figures 1 & 2** Side/Side max. 0.59 (15)
2004-07 Verona All other tire sizes	26.61 (676)	26.69 (678)	See **Figures 1 & 2** Side/Side max. 0.59 (15)
2002-06 XL-7	Side/Side difference max. 0.59 (15)	Side/Side difference max. 0.59 (15)	See **Figures 1 & 2**
2007-09 XL-7	SF = 0.70 (18)	SR = 1.65 (42)	See GM **Figures 1 & 2** S/S max. 0.47 (12)

Toyota / Lexus / Scion

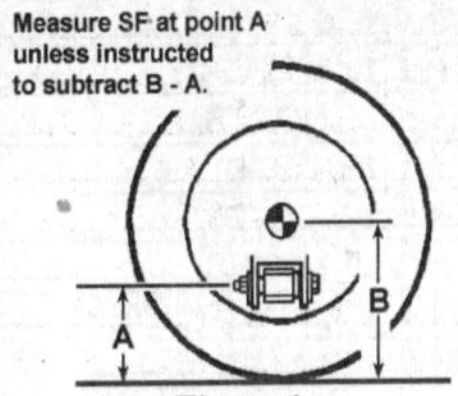

Figure 1

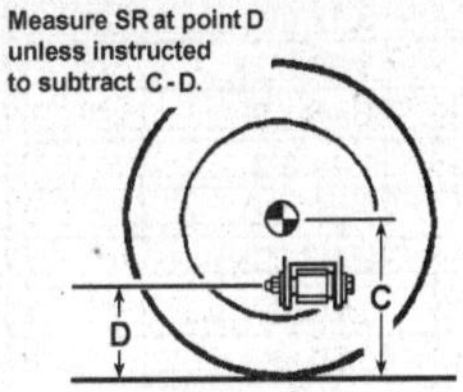

Figure 2

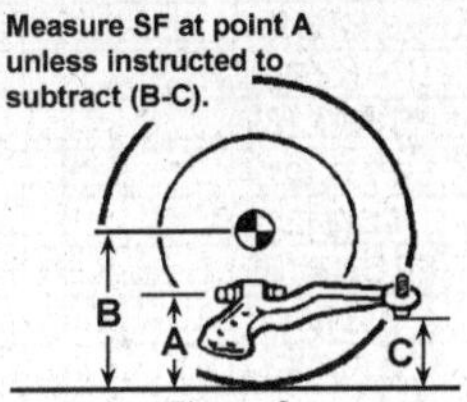

Figure 3

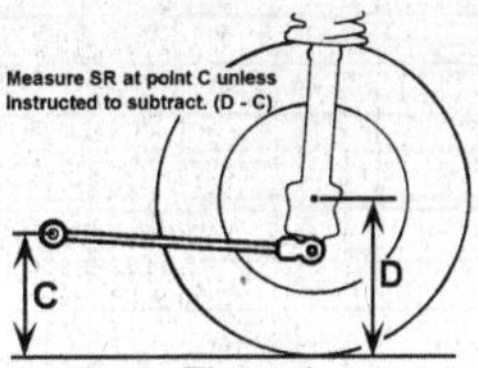

Figure 4

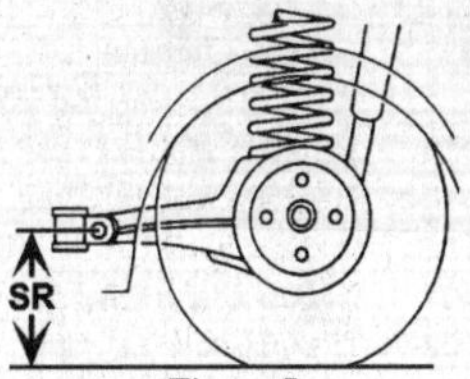

Figure 5

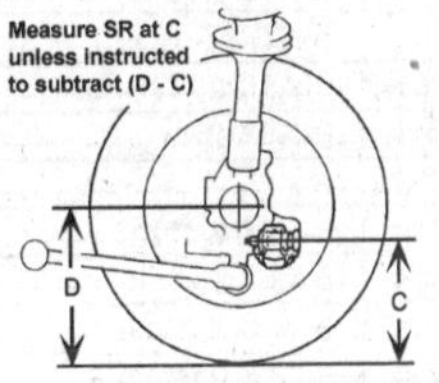

Figure 6

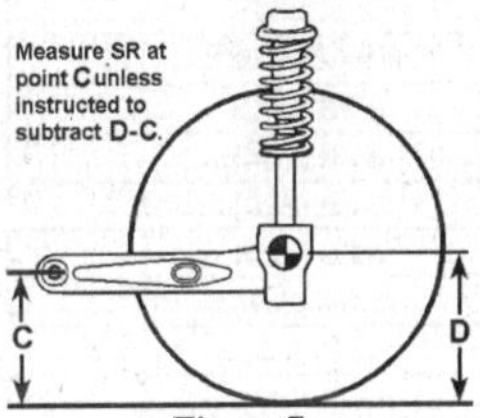

Figure 7

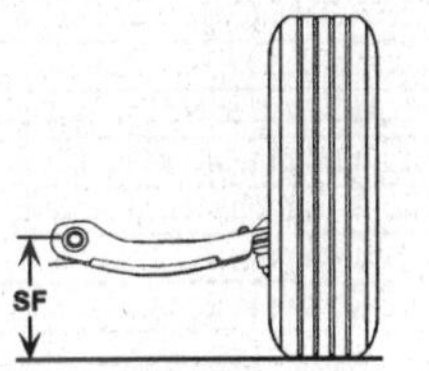

Figure 8

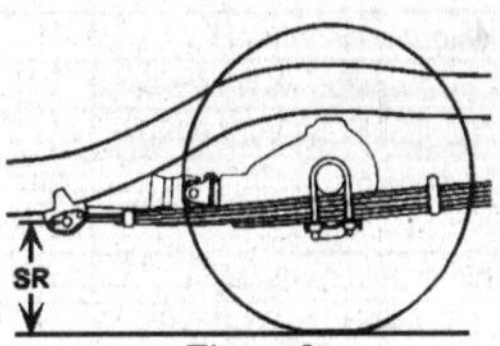

Figure 9

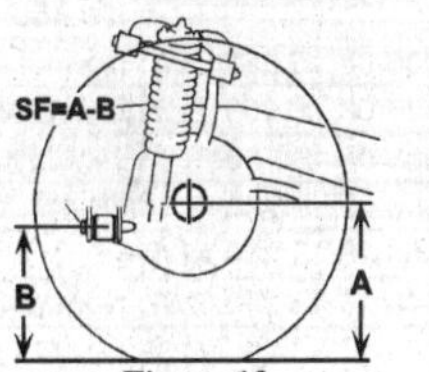

Figure 10

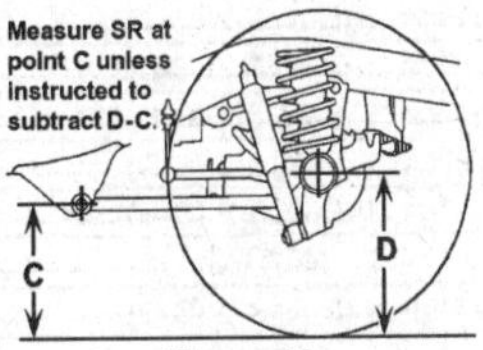

Figure 11

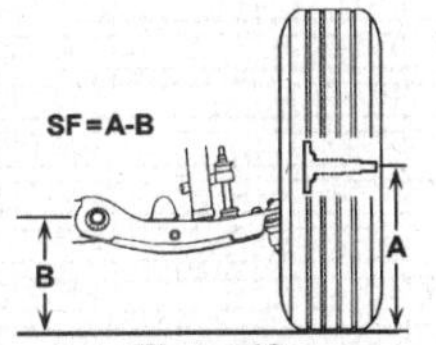

Figure 12

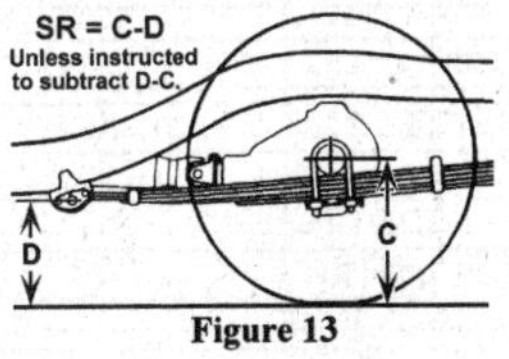

Figure 13

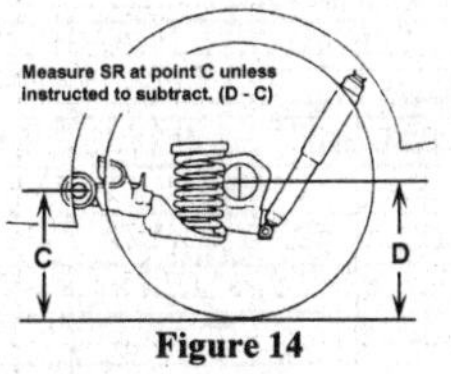

Figure 14

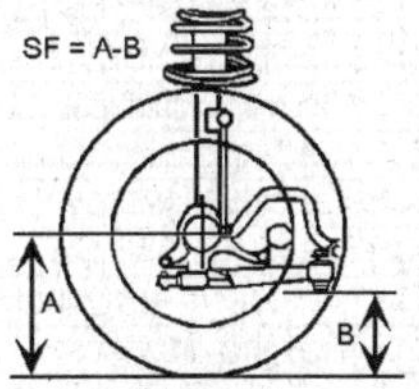

Figure 15

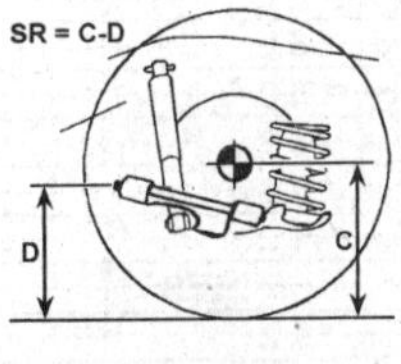

Figure 16

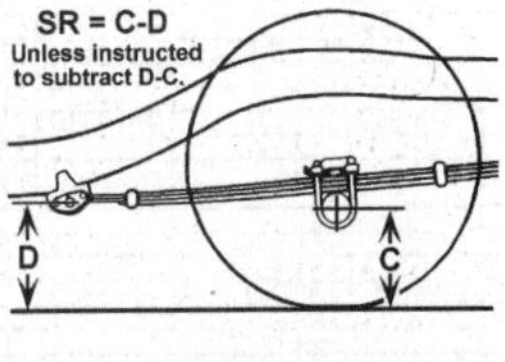

Figure 17

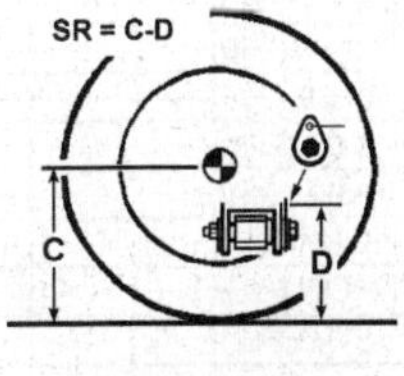

Figure 18

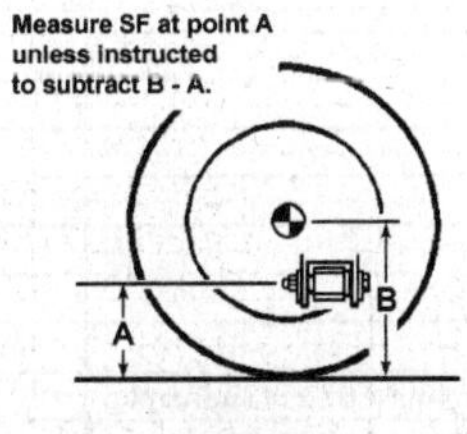

Figure 19

CHASSIS / SUSP. HEIGHTS

Lexus

Year Model	Height Specifications @Curb Weight - In. (mm) *		
CT 200H	FRONT (SF)	REAR (SR)	NOTES
2011-13 CT 200H with 15" tires (OE)	4.06 (103)	2.56 (65)	see Figs. 1 (B-A) & 2 (C-D)
2011-17 CT 200H with 16" or 17" tires (OE) exc/Sport	4.13 (105)	2.64 (67)	see Figs. 1 (B-A) & 2 (C-D)
2011-17 CT 200H with 16" or 17" tires (OE) Sport	4.21 (107)	2.44 (62)	see Figs. 1 (B-A) & 2 (C-D)
ES 250 / ES 300 / ES 330 / ES350	FRONT (SF)	REAR (SR)	NOTES
1990-91 ES 250	8.58 (218)	9.56 (243)	see Figs. 3 & 4
1992 ES 300	8.03 (204)	10.19 (259)	see Figs. 3 & 4
1993-96 ES 300	8.30 (211)	10.19 (259)	see Figs. 3 & 4
1997-01 ES 300	8.38 (213)	10.19 (259)	see Figs. 3 & 4
2002-03 ES 300	4.72 (120)	2.04 (52)	see Figs. 3 (B-C) & 4 (D-C)
2004-06 ES 330	4.72 (120)	2.04 (52)	see Figs. 3 (B-C) & 4 (D-C)
2007-12 ES 350	4.96 (126)	2.24 (57)	see Figs. 3 (B-C) & 4 (D-C)
2013-17 ES 350	4.88 (124)	1.85 (47)	see Figs. 3 (B-C) & 4 (D-C)
2013-17 ES 300h	4.76 (121)	1.81 (46)	see Figs. 3 (B-C) & 4 (D-C)
GS 300/GS 350/GS 400/GS 430/GS 460/GS 450h/GS F	FRONT (SF)	REAR (SR)	NOTES
1994-97 GS 300	9.48 (241)	11.10 (282)	see Figs. 1 & 2
1998-05 GS 300/GS 400/GS 430 with 215/60R16 tires	9.57-9.65 (243-245)	8.90 (226)	see Figs. 1 & 2
1998-05 GS 300/GS 400/GS 430 with 225/55R16 tires	9.45 (240)	8.70 (221)	see Figs. 1 & 2
1998-05 GS 300/GS 400/GS 430 with 235/45ZR17 tires	9.37 (238)	8.62 (219)	see Figs. 1 & 2
2006-11 GS 300/GS 350 2WD	4.37 (111)	8.31 (211)	see Figs. 1 (B-A) & 2
2006-11 GS 300/GS 350 AWD	4.61 (117)	8.70 (221)	see Figs. 1 (B-A) & 2
2006-11 GS 430/GS 460 RWD	4.45 (113)	8.19 (208)	see Figs. 1 (B-A) & 2
2007-11 GS 450h	4.45 (113)	8.19 (208)	see Figs. 1 (B-A) & 2
2013-17 GS 350 2WD	4.40-4.44 (112-113)	3.70-3.74 (94-95)	see Figs. 1 (B-A) & 2
2013-17 GS 350 AWD	4.65 (118)	3.30 (84)	see Figs. 1 (B-A) & 2
2013-17 GS 450h	4.40-4.49 (112-114)	3.82 (97)	see Figs. 1 (B-A) & 2
2016-17 GS F	5.11 (130)	4.68 (119)	see Figs. 1 (B-A) & 2
GX 460/GX 470	FRONT (SF)	REAR (SR)	NOTES
2003-09 GX 470	3.62 (92)	3.03 (77)	see Figs. 10 & 11 (D-C)
2010-12 GX 460 Base	3.62 (92)	2.95 (75)	see Figs. 10 & 11 (D-C)
2010-12 GX 460 Premium	3.62 (92)	3.06 (78)	see Figs. 10 & 11 (D-C)
2013-17 GX 460 Base	3.57 (91)	3.17 (81)	see Figs. 10 & 11 (D-C)
2013-17 GX 460 Premium	3.61 (92)	3.01 (77)	see Figs. 10 & 11 (D-C)
HS 250h	FRONT (SF)	REAR (SR)	NOTES
2010-12 HS 250h with 17" Tires	4.21 (107)	2.44 (62)	see Figs. 1 & 2 (C-D)
2010-12 HS 250h with 18" Tires	4.21 (107)	2.05 (52)	see Figs. 1 & 2 (C-D)
IS 300	FRONT (SF)	REAR (SR)	NOTES
2001 IS 300 Sedan Canada	2.20 (56)	2.20 (56)	see Figs. 1 (B-A) & 2 (C-D)
2002-05 IS 300 Sedan Canada	2.59 (66)	2.59 (66)	see Figs. 1 (B-A) & 2 (C-D)
2001-05 IS 300 Sedan USA	2.83 (72)	3.34 (85)	see Figs. 1 (B-A) & 2 (C-D)
2002-05 IS 300 Wagon Canada	2.20 (56)	2.28 (58)	see Figs. 1 (B-A) & 2 (C-D)
2002-05 IS 300 Wagon USA	2.59 (66)	2.59 (66)	see Figs. 1 (B-A) & 2 (C-D)
IS 250 / IS 350	FRONT (SF)	REAR (SR)	NOTES
2006-08 IS 250 RWD Normal Suspension w/16" wheels	4.37 (111)	8.35 (212)	see Figs. 1 (B-A) & 2
2006-08 IS 250/IS 350 RWD Normal Suspension w/17" wheels	4.37 (111)	8.43 (214)	see Figs. 1 (B-A) & 2
2006-08 IS 250/IS 350 RWD Normal Suspension w/18" wheels	4.37 (111)	8.66 (220)	see Figs. 1 (B-A) & 2
2006-08 IS 250/IS 350 RWD Sport Suspension	4.76 (121)	8.66 (220)	see Figs. 1 (B-A) & 2
2006-08 IS 250 AWD	4.69 (119)	8.94 (227)	see Figs. 1 (B-A) & 2
2009-13 IS 250 RWD Normal Suspension w/16" wheels	4.37 (111)	3.11 (79)	see Figs. 1 (B-A) & 2 (C-D)
2009-13 IS 250/IS 350 RWD Normal Susp. w/17" or 18" whls.	4.37 (111)	3.50 (89)	see Figs. 1 (B-A) & 2 (C-D)
2009-13 IS 250/IS 350 RWD Sport Suspension	4.76 (121)	3.50 (89)	see Figs. 1 (B-A) & 2 (C-D)
2009-13 IS 250/IS 350 AWD	4.69 (119)	2.71 (69)	see Figs. 1 (B-A) & 2 (C-D)
2014-15 IS 250/IS 350 RWD w/16" wheels	3.78 (96)	3.66 (93)	see Figs. 1 (B-A) & 2 (C-D)
2014-15 IS 250/IS 350 RWD w/17" or 18" whls.	4.02 (102)	3.94 (100)	see Figs. 1 (B-A) & 2 (C-D)
2016-17 IS All RWD Models	4.02 (102)	3.94 (100)	see Figs. 1 (B-A) & 2 (C-D)
2014-15 IS 250/IS 350 AWD w/16" wheels	4.65 (118)	3.78 (96)	see Figs. 1 (B-A) & 2 (C-D)
2014-15 IS 250/IS 350 AWD w/17" or 18" whls.	4.80 (122)	4.06 (103)	see Figs. 1 (B-A) & 2 (C-D)
2016-17 IS All AWD Models	4.80 (122)	4.06 (103)	see Figs. 1 (B-A) & 2 (C-D)
IS 250C / IS 350C	FRONT (SF)	REAR (SR)	NOTES
2010-14 IS250C / IS350C	3.98 (101)	3.11 (79)	see Figs. 1 (B-A) & 2 (C-D)
IS F	FRONT (SF)	REAR (SR)	NOTES
2008-14 IS F	5.19 (132)	4.40 (112)	see Figs. 1 (B-A) & 2 (C-D)
LS 400	FRONT (SF)	REAR (SR)	NOTES
1990-92 LS 400 w/o Air Susp.	9.21 (234)	9.52 (242)	see Figs. 1 & 2
1993-94 LS 400 w/o Air Susp.	10.07 (256)	10.00 (254)	see Figs. 1 & 2
1995-00 LS 400 w/o Air Susp.	10.39 (264)	9.56 (243)	see Figs. 1 & 2
1990-92 LS 400 w/Air Susp.	8.97 (228)	8.30 (211)	see Figs. 1 & 2
1993-94 LS 400 w/Air Susp.	9.80 (249)	9.13 (232)	see Figs. 1 & 2
1995-00 LS 400 w/Air Susp.	9.84 (250)	8.77 (223)	see Figs. 1 & 2
LS 430	FRONT (SF)	REAR (SR)	NOTES
2001-03 LS 430 w/o Air Susp. with 225/60R16 exc./Sports pkg.	8.34 (212)	9.01 (229)	see Figs. 1 & 2
2001-02 LS 430 w/o Air Susp. with 225/60R16 & Sports pkg.	8.46 (215)	9.13 (232)	see Figs. 1 & 2
2001-03 LS 430 w/o Air Susp. with 225/55R17 tires	8.38 (213)	9.40 (239)	see Figs. 1 & 2

* ±0.75 (19)

CHASSIS / SUSP. HEIGHTS

Lexus (cont.)

YEAR MODEL	HEIGHT SPECIFICATIONS @CURB WEIGHT - IN. (MM) *		
LS 430 (cont.)	**FRONT (SF)**	**REAR (SR)**	**NOTES**
2001-03 LS 430 w/Air Susp. with 225/60R16 tires	7.99 (203)[1]	8.22 (209)[1]	see Figs. 1 & 2 [1] Max L/R 0.39 (10)
2001-03 LS 430 w/Air Susp. with 225/55R17 tires	8.11 (206)[1]	8.34 (212)[1]	see Figs. 1 & 2 [1] Max L/R 0.39 (10)
2004-06 LS 430 w/o Air Susp. with 225/55R17 tires	4.13 (105)	3.58 (91)	see Figs. 1 (B-A) & 2 (C-D)
2004-06 LS 430 w/o Air Susp. with 245/45R18 tires	4.09 (104)	3.50 (89)	see Figs. 1 (B-A) & 2 (C-D)
2004-06 LS 430 w/Air Susp.	4.52 (115)	4.29 (109)	see Figs. 1 (B-A) & 2 (C-D)
LS 460/460 L	**FRONT (SF)**	**REAR (SR)**	**NOTES**
2007-17 LS 460/460L 2WD Coil suspension	3.58 (91)	3.00 (76)	see Figs. 1 (B-A) & 18 (C-D)
2007-17 LS 460/460L 2WD Air suspension	3.86 (98)	3.66 (93)	see Figs. 1 (B-A) & 18 (C-D)
2014-17 LS 460/460L 2WD All Sports Package	4.25 (108)	4.06 (103)	see Figs. 1 (B-A) & 18 (C-D)
2007-17 LS 460/460L AWD Coil suspension	5.35 (136)	2.68 (68)	see Figs. 1 (B-A) & 18 (C-D)
2007-17 LS 460/460L AWD Air suspension	5.75 (146)	3.27 (83)	see Figs. 1 (B-A) & 18 (C-D)
2014-17 LS 460/460L AWD All Sports Package	6.14 (156)	3.66 (93)	see Figs. 1 (B-A) & 18 (C-D)
LS 600h/600h L			
2008-16 LS 600h/600h L	5.75 (146)	3.27 (83)	see Figs. 1 (B-A) & 18 (C-D)
LX 470/LX 570	**FRONT (SF)**	**REAR (SR)**	**NOTES**
1998-07 LX 470	3.25 (83)	2.80 (71)	see Figs. 10 & 11 (D-C)
2008-17 LX 570 (active height control = NORMAL)	4.00 (101)	3.68 (94)	see Figs. 10 & 11 (D-C)
NX 200T/NX 300h	**FRONT (SF)**	**REAR (SR)**	**NOTES**
2015-17 NX 200 T w/225/65R17" tires (OE)	3.82 (97)	2.87 (73)	see Figs. 1 (B-A) & 2 (C-D)
2015-17 NX 200 T w/225/60R18" tires (OE)	4.02 (102)	3.07 (78)	see Figs. 1 (B-A) & 2 (C-D)
2015-17 NX 200 T w/235/55R18" tires (OE)	3.82 (97)	2.87 (73)	see Figs. 1 (B-A) & 2 (C-D)
2015-17 NX 300h w/225/65R17" tires (OE)	4.37 (111)	2.87 (73)	see Figs. 1 (B-A) & 2 (C-D)
2015-17 NX 300h w/225/60R18" tires (OE)	4.57 (116)	3.07 (78)	see Figs. 1 (B-A) & 2 (C-D)
RC 350/RC F	**FRONT (SF)**	**REAR (SR)**	**NOTES**
2015-17 RC 200t/RC 350 2WD	4.69 (119)	4.53 (115)	see Figs. 1 (B-A) & 2 (C-D)
2015-17 RC 300/RC 350 AWD	5.04 (128)	4.57 (116)	see Figs. 1 (B-A) & 2 (C-D)
2015-17 RC F	5.11 (130)	4.92 (125)	see Figs. 1 (B-A) & 2 (C-D)
RX 300 / RX 330 / RX350 / RX 400h	**FRONT (SF)**	**REAR (SR)**	**NOTES**
1998-03 RX 300 FWD	10.23 (260)	12.44 (316)	see Figs. 3 & 4
1998-03 RX 300 AWD	10.19 (259)	12.44 (316)	see Figs. 3 & 4
2004-09 RX 330/RX350 FWD w/o Air Suspension	4.71 (120)	1.62 (41)	see Figs. 3 (B-C) & 4 (D-C)
2004-09 RX 330/RX350 AWD w/o Air Suspension	4.32 (110)	1.22 (31)	see Figs. 3 (B-C) & 4 (D-C)
2004-09 RX 330/RX350 FWD w/Air Suspension	4.60 (117)	1.62 (41)	see Figs. 3 (B-C) & 4 (D-C)
2004-09 RX 330/RX350 AWD w/Air Suspension	4.41 (112)	1.42 (36)	see Figs. 3 (B-C) & 4 (D-C)
2010-15 RX 350 FWD	4.87 (124)	2.11 (54)	see Figs. 3 (B-C) & 6 (D-C)
2016-17 RX 350 FWD	4.29 (109)	1.22-1.26 (31-32)	see Figs. 3 (B-C) & 6 (D-C)
2010-12 RX 350 AWD w/Air Suspension	4.72 (120)	2.28 (58)	see Figs. 3 (B-C) & 6 (D-C)
2010-15 RX 350 AWD	4.91 (125)	2.16 (55)	see Figs. 3 (B-C) & 6 (D-C)
2016-17 RX 350 AWD exc/F-Sport	4.29 (109)	1.22 (31)	see Figs. 3 (B-C) & 6 (D-C)
2016-17 RX 350 AWD F-Sport	4.75 (121)	1.95 (50)	see Figs. 3 (B-C) & 6 (D-C)
2006-08 RX 400h FWD	4.32 (110)	1.22 (31)	see Figs. 3 (B-C) & 4 (D-C)
2006-08 RX 400h AWD	4.29 (109)	0.90 (23)	see Figs. 3 (B-C) & 4 (D-C)
2010-15 RX 450h FWD	4.91 (125)	2.16 (55)	see Figs. 3 (B-C) & 6 (D-C)
2016-17 RX 450h FWD	4.32 (110)	1.25 (32)	see Figs. 3 (B-C) & 6 (D-C)
2010-12 RX 450h AWD w/Air Suspension	4.72 (120)	2.28 (58)	see Figs. 3 (B-C) & 6 (D-C)
2010-15 RX 450h AWD	4.91 (125)	2.22 (56)	see Figs. 3 (B-C) & 6 (D-C)
2016-17 RX 450h AWD exc/F-Sport	4.32 (110)	1.36 (35)	see Figs. 3 (B-C) & 6 (D-C)
2016-17 RX 450h AWD F-Sport	4.78 (122)	2.05 (52)	see Figs. 3 (B-C) & 6 (D-C)
SC 300 / SC 400 / SC430	**FRONT (SF)**	**REAR (SR)**	**NOTES**
1992-96 SC 300	7.48 (190)	9.76 (248)	see Figs. 1 & 2
1992-96 SC 400	7.63 (194)	9.92 (252)	see Figs. 1 & 2
1997-00 SC 300/SC 400	7.51 (191)	9.68 (246)	see Figs. 1 & 2
2002-10 SC 430	9.13 (232)	8.22 (209)	see Figs. 1 & 2
	* ±0.75 (19)		

Scion

YEAR MODEL	HEIGHT SPECIFICATIONS @CURB WEIGHT - IN. (MM) *		
	FRONT (SF)	**REAR (SR)**	**NOTES**
2013-16 FR-S	14.76 (375)	14.64 (372)	see Figures 1 & 2 in Subaru
2012-15 iQ	3.18 (81)	0.04 (1)	see Figs. 19 (B-A) & 14 (D-C)
2016 iM	4.21 (107)	2.44 (62)	see Figs. 1 (B-A) & 14 (D-C)
2004-06 xA	3.58 (91)	0.75 (19)	see Figs. 1 (B-A) & 14 (D-C)
2004-06 xB	3.22 (82)	0.47 (12)	see Figs. 1 (B-A) & 14 (D-C)
2008-15 xB	4.09 (104)	0.74 (19)	see Figs. 1 (B-A) & 14 (D-C)
2008-10 xD	3.58 (91)	1.02 (26)	see Figs. 1 (B-A) & 14 (D-C)
2011-14 xD	3.62 (92)	1.30 (33)	see Figs. 1 (B-A) & 14 (D-C)
2005-10 tC	3.62 (92)	2.48 (63)	see Figs. 1 (B-A) & 6 (D-C)
2011-16 tC	4.25 (108)	2.68 (68)	see Figs. 1 (B-A) & 6 (D-C)
	* ±0.75 (19)		

Toyota

Year Model	Height Specifications @Curb Weight - In. (mm) *		
Avalon	**FRONT (SF)**	**REAR (SR)**	**NOTES**
1995-04 All Models	8.38 (213)	10.47 (266)	see Figs. 3 & 4
2005-10 Avalon XL	4.80 (122)	1.77 (45)	see Figs. 15 & 4 (D-C)
2005-12 Avalon XLS	4.96 (126)	1.92 (49)	see Figs. 15 & 4 (D-C)
2005-08 Avalon Touring	5.00 (127)	1.96 (50)	see Figs. 15 & 4 (D-C)
2005-12 Avalon Limited	4.96 (126)	2.04 (52)	see Figs. 15 & 4 (D-C)
2013-15 Avalon Base	5.00 (127)	2.04 (52)	see Figs. 15 & 4 (D-C)
2013-15 Avalon Touring	5.04 (128)	2.04 (52)	see Figs. 15 & 4 (D-C)
2013-15 Avalon Limited	5.08 (129)	2.13 (54)	see Figs. 15 & 4 (D-C)
2016-17 Avalon XLE	4.92 (125)	1.88 (48)	see Figs. 15 & 4 (D-C)
2016-17 Avalon XLS	5.04 (128)	2.00 (51)	see Figs. 15 & 4 (D-C)
2016-17 Avalon Limited	5.00 (127)	1.96 (50)	see Figs. 15 & 4 (D-C)
2013-15 Avalon Hybrid Base	5.00 (127)	2.09 (53)	see Figs. 15 & 4 (D-C)
2013-15 Avalon Hybrid Limited	5.08 (129)	2.13 (54)	see Figs. 15 & 4 (D-C)
2016-17 Avalon Hybrid Base	4.88 (124)	2.04 (52)	see Figs. 15 & 4 (D-C)
2016-17 Avalon Hybrid Limited	4.96 (126)	2.09 (53)	see Figs. 15 & 4 (D-C)
Camry	**FRONT (SF)**	**REAR (SR)**	**NOTES**
1990-91 Sedan All-Trac	9.09 (231)	10.07 (256)	see Figs. 3 & 4
1990 Sedan 4 Cyl. (FWD)	9.17 (233)	10.23 (260)	see Figs. 3 & 4
1990 Sedan 6 Cyl. (FWD)	8.93 (227)	10.00 (254)	see Figs. 3 & 4
1990 Wagon 4 Cyl.	9.09 (231)	11.41 (290)	see Figs. 3 & 4
1990 Wagon 6 Cyl.	10.03 (255)	11.18 (284)	see Figs. 3 & 4
1991 Sedan 4 Cyl. (FWD)	9.09 (231)	10.31 (262)	see Figs. 3 & 4
1991 Sedan 6 Cyl. (FWD)	8.85 (225)	10.07 (256)	see Figs. 3 & 4
1991 Wagon 4 Cyl.	9.01 (229)	11.49 (292)	see Figs. 3 & 4
1991 Wagon 6 Cyl.	8.77 (223)	11.25 (286)	see Figs. 3 & 4
1992-93 All exc/Wagon	7.99-8.11 (203-206)	10.31-10.51 (262-267)	see Figs. 3 & 4
1992-93 Wagon	7.99-8.11 (203-206)	10.31-10.51 (262-267)	see Figs. 3 & 4
1994-96 All exc/Wagon	8.26-8.38 (210-213)	10.31-10.51 (262-267)	see Figs. 3 & 4
1994-96 Wagon	8.26-8.42 (210-214)	10.70-10.90 (272-277)	see Figs. 3 & 4
1997-99 Sedan	8.34-8.46 (212-215)	10.39-10.47 (264-266)	see Figs. 3 & 4
2000-01 Sedan exc/CNG Models	8.22-8.46 (209-215)	10.23-10.47 (260-266)	see Figs. 3 & 4
2001 Sedan CNG Models	8.58 (218)	10.62 (270)	see Figs. 3 & 4
2002-06 Camry	4.68-4.76 (119-121)	1.77-1.88 (45-48)	see Figs. 3 (B-C) & 4 (D-C)
2007-09 Camry XLE, LE	4.92-4.96 (125-126)	2.04-2.16 (52-55)	see Figs. 3 (B-C) & 4 (D-C)
2007-09 Camry SE	5.00-5.07 (127-129)	2.16-2.28 (55-58)	see Figs. 3 (B-C) & 4 (D-C)
2010-11 Camry 4 Cyl. with 215/60R16 tires (OE)	4.92 (125)	2.08 (53)	see Figs. 3 (B-C) & 4 (D-C)
2010-11 Camry 4 Cyl. with 215/55R17 tires (OE)	5.02 (127.5)	2.16 (55)	see Figs. 3 (B-C) & 4 (D-C)
2010-11 Camry 6 Cyl. with 215/60R16 tires (OE)	4.96 (126)	2.15 (54)	see Figs. 3 (B-C) & 4 (D-C)
2010-11 Camry 6 Cyl. with 215/55R17 tires (OE)	5.08 (129)	2.28 (58)	see Figs. 3 (B-C) & 4 (D-C)
2012-14 Camry 4 Cyl. with 16" tires (OE)	4.84 (123)	2.05 (52)	see Figs. 3 (B-C) & 4 (D-C)
2015 Camry 4 Cyl. with 16" tires (OE)	4.88 (124)	2.01 (51)	see Figs. 3 (B-C) & 4 (D-C)
2016-17 Camry 4 Cyl. with 16" tires (OE)	4.80 (122)	2.01 (51)	see Figs. 3 (B-C) & 4 (D-C)
2012-14 Camry 4 Cyl. LE with 17" tires (OE)	4.80 (122)	2.01 (51)	see Figs. 3 (B-C) & 4 (D-C)
2015 Camry 4 Cyl. LE with 17" tires (OE)	4.88 (124)	2.01 (51)	see Figs. 3 (B-C) & 4 (D-C)
2016-17 Camry 4 Cyl. LE with 17" tires (OE)	4.80 (122)	2.01 (51)	see Figs. 3 (B-C) & 4 (D-C)
2012-15 Camry 4 Cyl. XLE with 17" tires (OE)	4.92 (125)	2.13 (54)	see Figs. 3 (B-C) & 4 (D-C)
2016-17 Camry 4 Cyl. XLE with 17" tires (OE)	4.96 (126)	2.16 (55)	see Figs. 3 (B-C) & 4 (D-C)
2012-14 Camry 4 Cyl. SE with 17" tires (OE)	4.88 (124)	2.05 (52)	see Figs. 3 (B-C) & 4 (D-C)
2015 Camry 4 Cyl. SE with 17" tires (OE)	4.84 (123)	2.01 (51)	see Figs. 3 (B-C) & 4 (D-C)
2016-17 Camry 4 Cyl. SE with 17" tires (OE)	4.92 (125)	2.01 (51)	see Figs. 3 (B-C) & 4 (D-C)
2012-15 Camry 4 Cyl. with 18" tires (OE)	4.84 (123)	2.00 (51)	see Figs. 3 (B-C) & 4 (D-C)
2016-17 Camry 4 Cyl. with 18" tires (OE)	4.96 (126)	2.05 (52)	see Figs. 3 (B-C) & 4 (D-C)
2012-14 Camry 6 Cyl. with 17" tires (OE)	5.00 (127)	2.24 (57)	see Figs. 3 (B-C) & 4 (D-C)
2015 Camry 6 Cyl. with 17" tires (OE)	5.00 (127)	2.16 (55)	see Figs. 3 (B-C) & 4 (D-C)
2016-17 Camry 6 Cyl. with 17" tires (OE)	5.00 (127)	2.20 (56)	see Figs. 3 (B-C) & 4 (D-C)
2012-14 Camry 6 Cyl. with 18" tires (OE)	4.84 (123)	2.13 (54)	see Figs. 3 (B-C) & 4 (D-C)
2015 Camry 6 Cyl. with 18" tires (OE)	4.76 (121)	2.05 (52)	see Figs. 3 (B-C) & 4 (D-C)
2016-17 Camry 6 Cyl. with 18" tires (OE)	4.80 (122)	2.05 (52)	see Figs. 3 (B-C) & 4 (D-C)
2007-11 Camry HV	5.08 (129)	2.20 (56)	see Figs. 3 (B-C) & 4 (D-C)
2012-14 Camry HV with 16" tires (OE)	4.84 (123)	2.00 (51)	see Figs. 3 (B-C) & 4 (D-C)
2012-14 Camry HV with 17" tires (OE)	4.96 (126)	2.13 (54)	see Figs. 3 (B-C) & 4 (D-C)
2015-17 Camry HV LE with 16" tires (OE)	4.84 (123)	2.05 (52)	see Figs. 3 (B-C) & 4 (D-C)
2015-17 Camry HV SE with 17" tires (OE)	4.92 (125)	2.17 (55)	see Figs. 3 (B-C) & 4 (D-C)
2015-17 Camry HV XLE with 17" tires (OE)	4.96 (126)	2.20 (56)	see Figs. 3 (B-C) & 4 (D-C)
Celica	**FRONT (SF)**	**REAR (SR)**	**NOTES**
1990 Celica exc/All-Trac	7.36 (187)	9.84 (250)	see Figs. 3 & 4
1989-90 Celica All-Trac	7.59 (193)	9.21 (234)	see Figs. 3 & 6
1991-93 Celica exc/All-Trac	7.28 (185)	9.92 (252)	see Figs. 3 & 4
1991-93 Celica All-Trac	7.40 (188)	9.33 (237)	see Figs. 3 & 6
1994-98 Celica w/14" tires	7.40 (188)	10.07 (256)	see Figs. 3 & 4
1994-99 Celica w/15" tires	7.20 (183)	9.84 (250)	see Figs. 3 & 4
2000-05 Celica	7.48-7.59 (190-193)	8.81-8.89 (224-226)	see Figs. 3 & 6

* ± ¾ (19) (except when range is given)

Toyota (cont.)

YEAR MODEL	HEIGHT SPECIFICATIONS @CURB WEIGHT - IN. (MM) *		
Corolla	**FRONT (SF)**	**REAR (SR)**	**NOTES**
1990-91 Corolla Coupe	7.08-7.20 (180-183)	9.37 (238)	see Figs. 8 & 4
1990-91 Corolla All w/60 series tires	6.92 (176)	9.29 (236)	see Figs. 8 & 4
1990-92 Corolla All others exc/4WD	7.20-7.32 (183-186)	9.52 (242)	see Figs. 8 & 4
1990-92 Corolla Wagon All-Trac	8.30 (211)	9.52 (242)	see Figs. 8 & 4
1990 Corolla Sedan All-Trac	8.30 (211)	9.52 (242)	see Figs. 8 & 4
1993-02 Corolla w/175-65R14 tires	7.28 (185)	9.64 (245)	see Figs. 8 & 4
1993-02 Corolla w/185-65R14 tires	7.48 (190)	9.84 (250)	see Figs. 8 & 4
2003-08 Corolla except 2005-06 XRS	3.42 (87)	1.69 (43)	see Figs. 1 (B-A) & 4 (D-C)
2005-06 Corolla XRS	3.81 (97)	2.24 (57)	see Figs. 1 (B-A) & 4 (D-C)
2009-13 Corolla except XRS	3.62 (92)	1.77 (45)	see Figs. 1 (B-A) & 4 (D-C)
2009-13 Corolla XRS	3.77 (96)	2.00 (51)	see Figs. 1 (B-A) & 4 (D-C)
2014-17 Corolla w/195/65R15 tires (OE)	3.86 (98)	2.09 (53)	see Figs. 1 (B-A) & 4 (D-C)
2014-17 Corolla w/205/55R16 tires (OE)	3.86 (98)	2.17 (55)	see Figs. 1 (B-A) & 4 (D-C)
2014-17 Corolla w/215/45R17 tires (OE)	3.66 (93)	1.97 (50)	see Figs. 1 (B-A) & 4 (D-C)
Cressida	**FRONT (SF)**	**REAR (SR)**	**NOTES**
1990 Cressida Sedan	9.09 (231)	9.64 (245)	see Figs. 1 & 2
1991-92 Cressida Sedan	8.85 (225)	9.80 (249)	see Figs. 1 & 2
Echo	**FRONT (SF)**	**REAR (SR)**	**NOTES**
2000-02 Echo	7.24-7.36 (184-187)	10.39-10.51 (264-267)	see Figs. 1 & 7
2003-05 Echo Sedan w/P175-65R14 tires	7.51-7.55 (191-192)	10.19-10.31 (259-262)	see Figs. 1 & 7
2003-05 Echo Sedan w/P185-60R15 tires	7.95-7.99 (202-203)	10.59-10.70 (269-272)	see Figs. 1 & 7
2003-05 Echo Coupe/Hatch w/P175-65R14 tires	7.28 (185)	10.19 (259)	see Figs. 1 & 7
Matrix	**FRONT (SF)**	**REAR (SR)**	**NOTES**
2003-08 All 2WD	3.50 (89)	1.61 (41)	see Figs. 1 & 4 (D-C)
2003-06 All 4WD	3.26 (83)	1.49 (38)	see Figs. 1 & 6 (D-C)
2009-13 All FWD 1.8L (rear twist axle)	3.54 (90)	1.81 (46)	see Figs. 1 & 4 (D-C)
2009-13 All FWD 2.4L 16" & 17" tires (rear twist axle)	3.62 (92)	1.85 (47)	see Figs. 1 & 4 (D-C)
2009-13 All FWD 2.4L 18" tires (IRS)	3.62 (92)	2.24 (57)	see Figs. 1 & 6 (D-C)
2009-13 All AWD (IRS)	3.46 (88)	2.16 (55)	see Figs. 1 & 6 (D-C)
2014 All	3.54 (90)	1.81 (46)	see Figs. 1 & 6 (D-C)
MR2	**FRONT (SF)**	**REAR (SR)**	**NOTES**
1991-92 MR2	8.81 (224)	7.99 (203)	see Fig. 3 (front and rear)
1993-95 MR2	8.54 (217)	7.59 (193)	see Fig. 3 (front and rear)
2000-03 MR2 Spyder	8.03 (204)	10.62 (270)	see Figs. 3 & 4
2004-05 MR2 Spyder	7.57 (192.4)	10.47 (266)	see Figs. 3 & 4
Paseo	**FRONT (SF)**	**REAR (SR)**	**NOTES**
1992-95 Paseo (All)	7.20-7.28 (183-185)	9.56-9.64 (243-245)	see Figs. 3 & 7
1996-97 Paseo (All)	7.36-7.40 (187-188)	9.72-9.80 (247-249)	see Figs. 3 & 7
Previa	**FRONT (SF)**	**REAR (SR)**	**NOTES**
1991 Previa 2WD	9.44-9.52 (240-242)	10.98-11.06 (279-281)	see Figs. 3 & 4
1991 Previa 4WD	9.84-9.92 (250-252)	11.37-11.45 (289-291)	see Figs. 3 & 4
1992-97 Previa 2WD	9.33 (237)	10.98-11.06 (279-281)	see Figs. 3 & 4
1992-97 Previa 4WD	9.72 (247)	11.37-11.45 (289-291)	see Figs. 3 & 4
Prius	**FRONT (SF)**	**REAR (SR)**	**NOTES**
2001-03 Prius	7.52 (191)	9.41 (239)	see Figs. 3 & 7
2004-09 Prius	3.74 (95)	2.40 (62)	see Figs. 1 (B-A) & 4 (D-C)
2010-15 Prius with 195/65R15 tires (OE) exc/Rough Road Pkg.	4.25 (108)	1.02 (26)	see Figs. 1 (B-A) & 4 (D-C)
2010-15 Prius with 195/65R15 tires (OE) w/Rough Road Pkg.	3.54 (90)	0.35 (9)	see Figs. 1 (B-A) & 4 (D-C)
2016-17 Prius with 195/65R15 tires (OE) exc/Rough Road Pkg.	4.48 (114)	1.61 (41)	see Figs. 1 (B-A) & 4 (D-C)
2016-17 Prius with 195/65R15 tires (OE) w/Rough Road Pkg.	3.74 (95)	0.95 (24)	see Figs. 1 (B-A) & 4 (D-C)
2010-14 Prius with 215/45R17 tires (OE)	4.06 (103)	0.83 (21)	see Figs. 1 (B-A) & 4 (D-C)
2012-13 Prius C with 175/65R15 tires (OE)	3.70-3.86 (94-98)	1.18-1.34 (30-34)	see Figs. 1 (B-A) & 4 (D-C)
2014-17 Prius C with 175/65R15 tires (OE)	3.86 (98)	1.34 (34)	see Figs. 1 (B-A) & 4 (D-C)
2012-17 Prius C with 195/50R16 tires (OE)	3.90 (99)	1.38 (35)	see Figs. 1 (B-A) & 4 (D-C)
2012-17 Prius V All models	4.09 (104)	0.95 (24)	see Figs. 1 (B-A) & 4 (D-C)
2010 Prius PHV	4.25 (108)	1.21 (31)	see Figs. 1 (B-A) & 4 (D-C)
2012-15 Prius PHV	4.25 (108)	1.02 (26)	see Figs. 1 (B-A) & 4 (D-C)
2017 Prius Prime	4.52 (115)	2.00(51)	see Figs. 1 (B-A) & 4 (D-C)
Sienna	**FRONT (SF)**	**REAR (SR)**	**NOTES**
1998-03 Sienna w/205-70R15	9.05 (230)	12.75 (324)	see Figs. 3 & 14
1998-03 Sienna w/215-65R15	8.89 (226)	12.59 (320)	see Figs. 3 & 14
2004-06 Sienna FWD	4.48 (114)	-.70 (-18)	see Figs. 15 (A-B) & 14 (D-C)
2004-06 Sienna AWD	4.56 (116)	-2.12 (-54)	see Figs. 15 (A-B) & 14 (D-C)
2007-10 Sienna FWD	4.56 (116)	-1.10 (-28)	see Figs. 15 (A-B) & 14 (D-C)
2007-10 Sienna AWD	4.56 (116)	-1.81 (-46)	see Figs. 15 (A-B) & 14 (D-C)
2011-17 Sienna FWD Base Grade w/P235/60R17 tires (OE)	4.92 (125)	-0.73 (-18)	see Figs. 15 (A-B) & 14 (D-C)
2011-17 Sienna FWD Cargo Van w/P235/60R17 tires (OE)	4.76 (121)	-1.12 (-28)	see Figs. 15 (A-B) & 14 (D-C)
2011-13 Sienna FWD LE 4 Cyl. w/P235/60R17 tires (OE)	5.2 (132.5)	-1.1 (-28)	see Figs. 15 (A-B) & 14 (D-C)
2011-15 Sienna FWD LE 6 Cyl. 7 Pass. w/P235/60R17 (OE)	4.97 (126)	-0.65 (-17)	see Figs. 15 (A-B) & 14 (D-C)
2016-17 Sienna FWD LE 6 Cyl. 7 Pass. w/P235/60R17 (OE)	4.97 (126)	-0.59 (-15)	see Figs. 15 (A-B) & 14 (D-C)
2011-17 Sienna FWD LE 6 Cyl. 8 Pass. w/P235/60R17 (OE)	5.0 (127)	-0.59 (-15)	see Figs. 15 (A-B) & 14 (D-C)
2011-17 Sienna FWD SE w/P235/50R19 tires (OE)	5.48 (139)	-0.11 (-3)	see Figs. 15 (A-B) & 14 (D-C)

* ± ¾ (19) (except when range is given)

Toyota (cont.)

Year Model	Height Specifications @Curb Weight - In. (mm) *		
Sienna (cont.)	FRONT (SF)	REAR (SR)	NOTES
2011-17 Sienna FWD XLE 7 Pass. all (OE) tire sizes.	5.09 (129)	-0.43 (-11)	see Figs. 15 (A-B) & 14 (D-C)
2011-17 Sienna FWD XLE 8 Pass. all (OE) tire sizes.	5.09 (129)	-0.43 (-11)	see Figs. 15 (A-B) & 14 (D-C)
2011-13 Sienna AWD Base w/P235/55RF18 tires (OE)	5.13 (130)	-1.19 (-30)	see Figs. 15 (A-B) & 14 (D-C)
2011-15 Sienna AWD LE w/P235/55RF18 tires (OE)	5.21 (132)	-1.11 (-28)	see Figs. 15 (A-B) & 14 (D-C)
2011-15 Sienna AWD LE w/P235/55RF18 tires (OE)	5.21 (132)	-1.11 (-28)	see Figs. 15 (A-B) & 14 (D-C)
2016-17 Sienna AWD LE w/P235/55RF18 tires (OE)	5.21 (132)	-1.18 (-30)	see Figs. 15 (A-B) & 14 (D-C)
2011-15 Sienna AWD XLE, Limited w/P235/55RF18 tires (OE)	5.35 (136)	-0.89 (-23)	see Figs. 15 (A-B) & 14 (D-C)
2016-17 Sienna AWD XLE, Limited w/P235/55RF18 tires (OE)	5.27 (134)	-0.98 (-25)	see Figs. 15 (A-B) & 14 (D-C)
Solara	FRONT (SF)	REAR (SR)	NOTES
1999-03 Solara	8.66-8.70 (220-221)	10.55-10.59 (268-269)	see Figs. 3 & 4
2004-08 Solara 4 Cyl. exc/SE	4.76 (121)	2.16 (55)	see Figs. 3 (B-C) & 4 (D-C)
2004-08 Solara 4 Cyl. SE & Convertible	4.84 (123)	2.55 (55)	see Figs. 3 (B-C) & 4 (D-C)
2004-08 Solara 6 Cyl. 16" tires exc/SE	4.88 (124)	1.88 (48)	see Figs. 3 (B-C) & 4 (D-C)
2004-08 Solara 6 Cyl. 17" tires exc/SE	4.88 (124)	2.24 (57)	see Figs. 3 (B-C) & 4 (D-C)
2004-08 Solara 6 Cyl. SE	4.92 (125)	2.28 (58)	see Figs. 3 (B-C) & 4 (D-C)
Supra	FRONT (SF)	REAR (SR)	NOTES
1990 Supra (All Models)	7.95 (202)	8.62 (219)	see Figs. 1 & 2
1991-92 Supra (All Models)	7.83 (199)	8.74 (222)	see Figs. 1 & 2
1993-96 Supra exc/Turbo	7.28 (185)	9.84 (250)	see Figs. 1 & 2
1993-96 Supra Turbo	7.36 (187)	9.88 (251)	see Figs. 1 & 2
1997-98 Supra exc/Turbo	7.32 (186)	9.68 (246)	see Figs. 1 & 2
1997-98 Supra Turbo	7.40 (188)	10.00 (254)	see Figs. 1 & 2
Tercel	FRONT (SF)	REAR (SR)	NOTES
1990-91 Tercel FWD w/145R13 tires	7.24 (184)	9.64 (245)	see Figs. 3 & 7
1990-91 Tercel FWD others	7.51 (191)	10.03 (255)	see Figs. 3 & 7
1992-94 Tercel FWD w/145R13 tires	7.28 (185)	9.76 (248)	see Figs. 3 & 7
1992-94 Tercel FWD others	7.55 (192)	10.03 (255)	see Figs. 3 & 7
1995 Tercel (All)	7.40 (188)	9.96 (253)	see Figs. 3 & 7
1996-97 Tercel w/155R13 tires	7.63 (194)	9.92 (252)	see Figs. 3 & 7
1996-97 Tercel w/175R14 tires	7.75 (197)	9.96 (253)	see Figs. 3 & 7
1996-97 Tercel w/185R14 tires	7.55 (192)	9.84 (250)	see Figs. 3 & 7
Venza	FRONT (SF)	REAR (SR)	NOTES
2009-16 Venza FWD 1AR-FE 4 Cyl.	4.58 (116)	1.33 (34)	see Figs. 3 (B-C) & 4 (D-C)
2009-16 Venza FWD 2GR-FE 6 Cyl.	4.63 (118)	1.21 (31)	see Figs. 3 (B-C) & 4 (D-C)
2009-16 Venza AWD 1AR-FE 4 Cyl.	4.60 (117)	1.21 (31)	see Figs. 3 (B-C) & 4 (D-C)
2009-16 Venza AWD 2GR-FE 6 Cyl.	4.63 (118)	1.25 (32)	see Figs. 3 (B-C) & 4 (D-C)
Yaris	FRONT (SF)	REAR (SR)	NOTES
2006-11 Yaris Hatchback	3.46 (88)	0.79 (20)	see Figs. 1 (B-A) & 7 (D-C)
2007-13 Yaris Sedan	3.43 (87)	0.63 (16)	see Figs. 1 (B-A) & 7 (D-C)
2012-14 Yaris Hatchback w/P175/65 R15 tires (OE)	3.82 (97)	1.26 (32)	see Figs. 1 (B-A) & 7 (D-C)
2012-14 Yaris Hatchback w/P195/50 R16 tires (OE)	3.98 (101)	1.26 (32)	see Figs. 1 (B-A) & 7 (D-C)
2015-17 Yaris Hatchback w/P175/65 R15 tires (OE)	3.76 (95)	1.12 (28)	see Figs. 1 (B-A) & 7 (D-C)
2015-17 Yaris Hatchback w/P195/50 R16 tires (OE)	3.76 (95)	1.18 (30)	see Figs. 1 (B-A) & 7 (D-C)
2017 Yaris iM	4.21 (107)	2.44 (62)	see Figs. 1 (B-A) & 14 (D-C)

* ± ¾ (19) (except when range is given)

Toyota Truck & SUV

Year Model	Height Specifications @Curb Weight - In. (mm) *		
4Runner	FRONT (SF)	REAR (SR)	NOTES
1990-93 4Runner w/31x10.5R15	12.12-12.20 (308-310)	14.13-14.44 (359-367)	see Figs. 8 & 11
1990-93 4Runner w/P225-75R15	10.94-11.18 (278-284)	12.79-13.30 (325-338)	see Figs. 8 & 11
1994-95 4Runner w/31x10.5R15	12.04 (306)	13.85-14.17 (352-360)	see Figs. 8 & 11
1994-95 4Runner w/P225-75R15	10.98-11.14 (279-283)	12.79-13.26 (325-337)	see Figs. 8 & 11
1996-99 4Runner 2WD	2.76 (70)	1.97 (50)	see Figs. 10 & 11 (D-C)
1996-99 4Runner 4WD	1.89 (48)	1.14 (29)	see Figs. 10 & 11 (D-C)
2000-02 4Runner 2WD	2.83 (72)	2.05 (52)	see Figs. 10 & 11 (D-C)
2000-02 4Runner 4WD	2.76 (70)	2.00 (51)	see Figs. 10 & 11 (D-C)
2003-09 4Runner 2WD 6 Cyl. without Air Suspension	4.33 (110)	2.80 (71)	see Figs. 10 & 11 (D-C)
2003-09 4Runner 2WD 8 Cyl. without Air Suspension	4.49 (114)	2,83 (72)	see Figs. 10 & 11 (D-C)
2003-09 4Runner 2WD All with Air Suspension	4.45 (113)	3.19 (81)	see Figs. 10 & 11 (D-C)
2003-09 4Runner 4WD 6 Cyl. without Air Suspension	3.94 (100)	2.80 (71)	see Figs. 10 & 11 (D-C)
2003-09 4Runner 4WD 8 Cyl. without Air Suspension	3.94 (100)	2.83 (72)	see Figs. 10 & 11 (D-C)
2003-09 4Runner 4WD with Air Suspension	3.90 (99)	3.11 (79)	see Figs. 10 & 11 (D-C)
2010 4Runner SR5, 2WD, 4 Cyl.	4.57 (116)	3.11 (79)	see Figs. 10 & 11 (D-C)
2011 4Runner SR5, 2WD, 4 Cyl.	4.68 (119)	3.27 (83)	see Figs. 10 & 11 (D-C)
2010-17 4Runner SR5, 2WD, 6 Cyl.	4.57 (116)	3.15 (80)	see Figs. 10 & 11 (D-C)
2010-17 4Runner SR5, 4WD, 6 Cyl.	3.70 (94)	2.68 (68)	see Figs. 10 & 11 (D-C)
2010-15 4Runner S-Runner, 2WD	4.61 (117)	3.19 (81)	see Figs. 10 & 11 (D-C)
2010-15 4Runner S-Runner, 4WD	3.74 (95)	2.72 (69)	see Figs. 10 & 11 (D-C)
2016-17 4Runner Limited, 2WD	4.61 (117)	3.19 (81)	see Figs. 10 & 11 (D-C)

* ± ¾ (19) (except when range is given)

Toyota Truck & SUV (cont.)

Year Model	Height Specifications @Curb Weight - In. (mm) *		
4Runner (cont.)	**FRONT (SF)**	**REAR (SR)**	**NOTES**
2016-17 4Runner Limited, 4WD	3.86 (98)	2.87 (73)	see Figs. 10 & 11 (D-C)
2010-17 4Runner Trail without KDSS susp.	3.78 (96)	2.76 (70)	see Figs. 10 & 11 (D-C)
2010-17 4Runner Trail with KDSS susp.	3.86 (98)	2.83 (72)	see Figs. 10 & 11 (D-C)
2010-15 4Runner X-Runner, 2WD	4.61 (117)	3.36 (85)	see Figs. 10 & 11 (D-C)
2010-15 4Runner X-Runner, 4WD	3.86 (98)	2.85 (73)	see Figs. 10 & 11 (D-C)
FJ Cruiser	**FRONT (SF)**	**REAR (SR)**	**NOTES**
2007-14 FJ Cruiser 2WD	4.56 (116)	3.19 (81)	see Figs. 10 & 11 (D-C)
2007-14 FJ Cruiser 4WD	3.42 (87)	2.40 (61)	see Figs. 10 & 11 (D-C)
Highlander	**FRONT (SF)**	**REAR (SR)**	**NOTES**
2001-03 Highlander 2WD	4.64 (118)	1.61 (41)	see Figs. 15 (A-B) & 4 (D-C)
2001-03 Highlander 4WD	4.25 (108)	1.22 (31)	see Figs. 15 (A-B) & 4 (D-C)
2004-07 Highlander 2WD	4.72 (120)	1.61 (41)	see Figs. 15 (A-B) & 4 (D-C)
2004-07 Highlander 4WD	4.33 (110)	1.22 (31)	see Figs. 15 (A-B) & 4 (D-C)
2006-07 Highlander HV 2WD	4.29 (109)	1.61 (31)	see Figs. 15 (A-B) & 4 (D-C)
2006-07 Highlander HV 4WD	4.25 (108)	.86 (22)	see Figs. 15 (A-B) & 4 (D-C)
2008-10 Highlander 2WD & 4WD	4.64 (118)	1.33 (34)	see Figs. 15 (A-B) & 4 (D-C)
2011-13 Highlander 2WD & 4WD	4.52 (115)	1.33 (34)	see Figs. 15 (A-B) & 4 (D-C)
2014-17 Highlander 2WD & 4WD	4.36 (111)	1.73 (44)	see Figs. 15 (A-B) & 2 (D-C)
2008-13 Highlander HV	4.64 (118)	1.25 (32)	see Figs. 15 (A-B) & 4 (D-C)
2014-17 Highlander HV	4.36 (111)	1.37 (35)	see Figs. 15 (A-B) & 4 (D-C)
Land Cruiser	**FRONT (SF)**	**REAR (SR)**	**NOTES**
1998-07 Land Cruiser	2.79 (71)	2.00 (51)	see Figs. 10 & 11 (D-C)
2008-11 Land Cruiser	3.81 (97)	3.03 (77)	see Figs. 10 & 11 (D-C)
2012-16 Land Cruiser Base Models	3.42 (87)	2.95 (75)	see Figs. 10 & 11 (D-C)
2012-16 Land Cruiser Full Option	3.54 (90)	3.07 (78)	see Figs. 10 & 11 (D-C)
2017 Land Cruiser	3.26 (83)	3.18 (81)	see Figs. 10 & 11 (D-C)
Pickup	**FRONT (SF)**	**REAR (SR)**	**NOTES**
1990 Pickup 2WD Reg. Cab	10.11-10.35 (257-263)	10.43-10.90 (265-277)	see Figs. 8 & 9
1990 Pickup 2WD X-Cab	10.70-11.02 (272-280)	10.23-10.47 (260-266)	see Figs. 8 & 9
1991-93 Pickup 2WD Reg. Cab	10.11-10.35 (257-263)	10.27-10.55 (261-268)	see Figs. 8 & 9
1991-93 Pickup 2WD X-Cab	10.74-10.94 (273-278)	10.27-10.47 (261-266)	see Figs. 8 & 9
1989-93 Pickup 2WD 1 Ton	10.15-10.31 (258-262)	11.14-11.25 (283-286)	see Figs. 8 & 9
1994-95 Pickup 2WD Reg Cab	10.11-10.27 (257-261)	10.31-10.51 (262-267)	see Figs. 8 & 9
1994-95 Pickup 2WD X-Cab	10.86-11.10 (276-282)	10.19-10.74 (259-273)	see Figs. 8 & 9
1989-90 Pickup 4WD w/31x10.5R15	12.36-12.71 (314-323)	17.99-18.26 (457-464)	see Figs. 8 & 9
1989-90 Pickup 4WD w/P225-75R15	10.98-11.45 (279-291)	16.61-17.00 (422-432)	see Figs. 8 & 9
1991-93 Pickup 4WD w/31x10.5R15	12.40-12.63 (315-321)	17.55-17.87 (446-454)	see Figs. 8 & 9
1991-93 Pickup 4WD w/P225-75R15	11.02-11.49 (280-292)	16.29-16.81 (414-427)	see Figs. 8 & 9
1994-95 Pickup 4WD w/31x10.5R15	11.96-12.12 (304-308)	17.48-18.11 (444-460)	see Figs. 8 & 9
1994-95 Pickup 4WD w/P225-75R15	10.90-11.18 (277-284)	16.37-17.04 (416-433)	see Figs. 8 & 9
RAV4	**FRONT (SF)**	**REAR (SR)**	**NOTES**
1996-05 RAV4 2WD	8.94 (227)	13.98(355)	see Figs. 3 & 14
1996-00 RAV4 4WD w/215-70R16	9.29 (236)	14.37 (365)	see Figs. 3 & 14
2001-05 RAV4 4WD w/215-70R16	9.25 (235)	14.37 (365)	see Figs. 3 & 14
1996-00 RAV4 4WD w/235-60R16	8.54 (217)	13.62 (346)	see Figs. 3 & 14
2001-05 RAV4 4WD w/235-60R16	8.98 (228)	14.13 (359)	see Figs. 3 & 14
2006-12 RAV4 all with 215-70R16 tires	3.50 (89)	2.24 (57)	see Figs. 1 (B-A) & 2 (C-D)
2006-12 RAV4 all with 225-65R17 tires (2 seat)	3.82 (97)	2.56 (65)	see Figs. 1 (B-A) & 2 (C-D)
2006-12 RAV4 all with 225-65R17 tires (3 seat)	3.82 (97)	2.13 (54)	see Figs. 1 (B-A) & 2 (C-D)
2006-12 RAV4 all with 225-55R18 tires	3.82 (97)	2.56 (65)	see Figs. 1 (B-A) & 2 (C-D)
2013-17 RAV4 2WD	4.02 (102)	2.83 (72)	see Figs. 1 (B-A) & 2 (C-D)
2013-17 RAV4 AWD	4.02 (102)	2.87 (73)	see Figs. 1 (B-A) & 2 (C-D)
1998-03 RAV4 EV	9.61 (244)	14.92 (379)	see Figs. 3 & 14
2013-17 RAV4 HV	4.48 (114)	2.83 (72)	see Figs. 1 (B-A) & 2 (C-D)
Sequoia	**FRONT (SF)**	**REAR (SR)**	**NOTES**
2001-07 Sequoia 2WD exc/Air Susp. models	2.75-2.83 (70-72)	2.99-3.03 (76-77)	see Figs. 10 & 11 (D-C)
2005-07 Sequoia 2WD Air Susp. models	2.08-3.34 (53-85)	2.91-4.13 (74-105)	see Figs. 10 & 11 (D-C)
2001-07 Sequoia 4WD exc/Air Susp. models	2.40-2.44 (61-62)	2.51-2.55 (64-65)	see Figs. 10 & 11 (D-C)
2005-07 Sequoia 4WD Air Susp. models	1.81-3.03 (46-77)	2.40-3.58 (61-91)	see Figs. 10 & 11 (D-C)
2008-10 Sequoia 2WD 4.6L Coil Suspension	4.17 (106)	1.14 (29)	see Figs. 10 & 16
2008-10 Sequoia 2WD 4.6L Air Suspension	4.13 (105)	1.54 (39)	see Figs. 10 & 16
2008-10 Sequoia 2WD 5.7L Coil Suspension	4.25 (108)	1.18 (30)	see Figs. 10 & 16
2008-10 Sequoia 2WD 5.7L Air Suspension	4.21 (107)	1.53 (39)	see Figs. 10 & 16
2008-10 Sequoia 4WD 4.6L Coil Suspension	3.97 (101)	0.98 (25)	see Figs. 10 & 16
2008-10 Sequoia 4WD 4.6L Air Suspension	3.93 (100)	1.33 (34)	see Figs. 10 & 16
2008-10 Sequoia 4WD 5.7L Coil Suspension	4.01 (102)	0.98 (25)	see Figs. 10 & 16

* ± ¾ (19) (except when range is given)

Toyota Truck & SUV (cont.)

YEAR MODEL	HEIGHT SPECIFICATIONS @CURB WEIGHT - IN. (MM) *		
Sequoia (cont.)	**FRONT (SF)**	**REAR (SR)**	**NOTES**
2008-10 Sequoia 4WD 5.7L Air Suspension	3.97 (101)	1.33 (34)	see Figs. 10 & 16
2011-12 Sequoia 2WD 4.6L	4.17 (106)	1.14 (29)	see Figs. 10 & 16
2011-17 Sequoia 2WD 5.7L Coil Suspension	4.25 (108)	1.18 (30)	see Figs. 10 & 16
2011-17 Sequoia 2WD 5.7L Air Suspension	4.21 (107)	1.53 (39)	see Figs. 10 & 16
2011-12 Sequoia 4WD 4.6L	3.97 (101)	0.98 (25)	see Figs. 10 & 16
2011-17 Sequoia 4WD 5.7L Coil Suspension	4.01 (102)	0.98 (25)	see Figs. 10 & 16
2011-17 Sequoia 4WD 5.7L Air Suspension	3.93 (100)	1.33 (34)	see Figs. 10 & 16
T100	**FRONT (SF)**	**REAR (SR)**	**NOTES**
1993 T100 2WD All 1Ton models	11.61 (295)	11.61 (295)	see Figs. 8 & 9
1994 T100 2WD All 1Ton models	11.41 (290)	n/a	see Fig. 8
1995-98 T100 2WD All 1Ton models	1.92-1.96 (49-50)	1.25-1.29 (32-33)	see Figs. 12 & 13
1993 T100 2WD w/P215-75R15	11.22 (285)	11.02 (280)	see Figs. 8 & 9
1993 T100 2WD w/P235-75R15	11.61 (295)	11.61 (295)	see Figs. 8 & 9
1994 T100 2WD w/P215-75R15	11.02 (280)	n/a	see Fig. 8
1994 T100 2WD w/P235-75R15	11.41 (290)	n/a	see Fig. 8
1995 T100 2WD w/P215-75R15	1.96 (50)	1.73-2.32 (44-59)	see Figs. 12 & 13
1995 T100 2WD w/P235-75R15	1.96 (50)	1.69-2.24 (43-57)	see Figs. 12 & 13
1996-97 T100 2WD Reg. Cab (exc/1Ton)	1.77-1.96 (45-50)	2.40-2.55 (61-65)	see Figs. 12 & 13
1996-97 T100 2WD X-Cab (exc/1Ton)	1.53 (39)	1.77-1.85 (45-47)	see Figs. 12 & 13
1994 T100 4WD All M/T models	1.45 (37)	n/a	see Fig. 12
1994 T100 4WD All A/T models	1.45-1.53 (37-39)	n/a	see Fig. 12
1995 T100 4WD w/31x10.5R15	1.45-1.49 (37-38)	2.79-3.50 (71-89)	see Figs. 12 & 17 (D-C)
1995 T100 4WD wP235-75R15	1.45-1.49 (37-38)	2.67-3.42 (68-87)	see Figs. 12 & 17 (D-C)
1996-97 T100 4WD Reg. Cab	1.49-1.57 (38-40)	2.75-2.87 (70-73)	see Figs. 12 & 17 (D-C)
1996-97 T100 4WD X-Cab	1.25-1.49 (32-38)	2.99-3.11 (76-79)	see Figs. 12 & 17 (D-C)
Tacoma – All 4X2 exc/PreRunner	**FRONT (SF)**	**REAR (SR)**	**NOTES**
1995-98 4X2 Regular Cab	1.73 (44)	2.40 (61)	see Figs. 10 & 13
1995-97 4X2 X-Cab 4 Cyl.	1.69 (43)	2.67 (68)	see Figs. 10 & 13
1995-97 4X2 X-Cab 6 Cyl.	1.69 (43)	2.40 (61)	see Figs. 10 & 13
1999-00 4X2 Regular Cab	1.73 (44)	1.81 (46)	see Figs. 10 & 13
1998-00 4X2 X-Cab 4 Cyl.	1.69 (43)	2.04 (52)	see Figs. 10 & 13
1998-00 4X2 X-Cab 6 Cyl.	1.69 (43)	1.96 (50)	see Figs. 10 & 13
2001-04 4X2 Regular Cab	1.02-1.10 (26-28)	1.45-1.49 (37-38)	see Figs. 10 & 13
2001-04 4X2 X-Cab 4 Cyl.	.98-1.06 (25-27)	1.81 (46)	see Figs. 10 & 13
2001-04 4X2 X-Cab 6 Cyl.	1.37 (35)	2.24 (57)	see Figs. 10 & 13
2005-06 4X2 Regular Cab	3.43 (87)	1.89 (48)	see Figs. 10 & 13
2005-06 4X2 Access Cab, 4 Cyl., Man. Trans	3.38 (86)	1.97 (50)	see Figs. 10 & 13
2005-06 4X2 Access Cab, 4 Cyl., Auto. Trans	3.43 (87)	1.97 (50)	see Figs. 10 & 13
2005-06 4X2 Access Cab X-Runner	4.65 (118)	3.35 (85)	see Figs. 10 & 13
2007-08 4X2 Regular Cab	3.46 (88)	1.92 (49)	see Figs. 10 & 13
2007-08 4X2 Access Cab, 4 Cyl., Man. Trans	3.43 (87)	2.00 (51)	see Figs. 10 & 13
2007-08 4X2 Access Cab, 4 Cyl., Auto. Trans	3.50 (89)	2.00 (51)	see Figs. 10 & 13
2007-08 4X2 Access Cab, 6 Cyl., X-Runner	4.65 (118)	3.35 (85)	see Figs. 10 & 13
2009-10 4X2 Regular Cab	3.54 (90)	2.00 (51)	see Figs. 10 & 13
2009-10 4X2 Access Cab, 4 Cyl., Man. Trans	3.50 (89)	2.13 (54)	see Figs. 10 & 13
2009-10 4X2 Access Cab, 4 Cyl., Auto. Trans	3.54 (90)	2.17 (55)	see Figs. 10 & 13
2009-10 4X2 Access Cab, 6 Cyl., X-Runner	4.69 (119)	3.43 (87)	see Figs. 10 & 13
2011 4X2 Regular Cab, Man. Trans	3.50 (89)	1.93 (49)	see Figs. 10 & 13
2011 4X2 Regular Cab, Auto. Trans	3.54 (90)	1.97 (50)	see Figs. 10 & 13
2011 4X2 Access Cab, 4 Cyl., Man. Trans	3.50 (89)	2.13 (54)	see Figs. 10 & 13
2011 4X2 Access Cab, 4 Cyl., Auto. Trans	3.54 (90)	2.17 (55)	see Figs. 10 & 13
2011 4X2 Access Cab, 6 Cyl., X-Runner	4.69 (119)	3.46 (88)	see Figs. 10 & 13
2011 4X2 Double Cab	3.62 (92)	2.18 (58)	see Figs. 10 & 13
2012-14 4X2 Regular Cab, Man. Trans	3.50 (89)	1.93 (49)	see Figs. 10 & 13
2012-14 4X2 Regular Cab, Auto. Trans	3.58 (91)	1.97 (50)	see Figs. 10 & 13
2012-15 4X2 Access Cab, 4 Cyl., Man. Trans	3.35 (85)	2.13 (54)	see Figs. 10 & 13
2012-15 4X2 Access Cab, 4 Cyl., Auto. Trans	3.43 (87)	2.13 (54)	see Figs. 10 & 13
2012-15 4X2 Access Cab, 6 Cyl., X-Runner	4.61 (117)	3.39 (86)	see Figs. 10 & 13
2012-15 4X2 Double Cab	3.50 (89)	2.32 (59)	see Figs. 10 & 13
2016-17 4X2 Access Cab 4 Cyl.	2.48 (63)	-1.53 (-39)	see Figs. 10 & 17
2016-17 4X2 Access Cab 4 Cyl. Firm Susp.	2.48 (63)	-1.85 (-47)	see Figs. 10 & 17
2016-17 4X2 Double Cab 4 Cyl.	2.63 (67)	-1.37 (-35)	see Figs. 10 & 17
2016-17 4X2 Double Cab 4 Cyl. Firm Susp.	2.63 (67)	-1.69 (-43)	see Figs. 10 & 17
Tacoma – All 4X2 PreRunner Models	**FRONT (SF)**	**REAR (SR)**	**NOTES**
1998-00 4X2 PreRunner 4 cyl. Reg. Cab RZN191L	1.81 (46)	-1.53 (-39)	see Figs. 10 & 17
1998-00 4X2 PreRunner 4 cyl. X-Cab RZN196L	1.77 (45)	-2.00 (-51)	see Figs. 10 & 17
1998-00 4X2 PreRunner 6 cyl. VZN195L	1.77 (45)	-2.08 (-53)	see Figs. 10 & 17

* ± ¾ (19) (except when range is given)

Toyota Truck & SUV (cont.)

YEAR MODEL	HEIGHT SPECIFICATIONS @CURB WEIGHT - IN. (MM) *		
Tacoma – All 4X2 PreRunner Models (cont.)	FRONT (SF)	REAR (SR)	NOTES
2001-03 4X2 PreRunner 4 cyl. Reg. Cab RZN191L			
Normal suspension	1.53 (39)	-1.81 (-46)	see Figs. 10 & 17
HD leaf springs	1.53 (39)	-1.65 (-42)	see Figs. 10 & 17
2001-03 4X2 PreRunner 4 cyl. X-Cab RZN196L			
Normal suspension	1.53 (39)	-1.81 (-46)	see Figs. 10 & 17
HD leaf springs	1.53 (39)	-1.53 (-39)	see Figs. 10 & 17
Off-Road Pkg.	1.88 (48)	-1.61 (-41)	see Figs. 10 & 17
2001-03 4X2 PreRunner 4 cyl. Double-Cab RZN196L			
Normal suspension	2.59 (66)	-.90 (-23)	see Figs. 10 & 17
Off-Road Pkg.	3.03 (77)	-.74 (-19)	see Figs. 10 & 17
2001-03 4X2 PreRunner 6 cyl. X-Cab VZN195L			
Normal suspension	1.65 (42)	-1.81 (-46)	see Figs. 10 & 17
HD leaf springs	1.65 (42)	-1.53 (-39)	see Figs. 10 & 17
Off-Road Pkg.	1.85 (47)	-1.61 (-41)	see Figs. 10 & 17
2001-03 4X2 PreRunner 6 cyl. Double-Cab VZN195L			
Normal suspension	2.63 (67)	-.90 (-23)	see Figs. 10 & 17
Off-Road Pkg.	2.67 (68)	-.78 (-20)	see Figs. 10 & 17
2004 4X2 PreRunner 4 cyl. Reg. Cab RZN191L			
w/P225/75R15 tires	1.53 (39)	-1.77 (-45)	see Figs. 10 & 17
w/P265/70R16 tires	1.53 (39)	-1.81 (-46)	see Figs. 10 & 17
2004 4X2 PreRunner 4 cyl. X-Cab RZN196L			
Normal suspension (USA)	1.85-1.88 (47-48)	-1.73 (-44)	see Figs. 10 & 17
Normal suspension (CAN)	1.53 (39)	-1.81 (-46)	see Figs. 10 & 17
Normal suspension	2.99 (76)	-.82 (-21)	see Figs. 10 & 17
Off-Road Pkg.	3.03 (77)	-.74 (-19)	see Figs. 10 & 17
2004 4X2 PreRunner 6 cyl. X-Cab VZN195L			
Normal suspension (w/o VSC[1]) (USA)	1.85 (47)	-1.77 (-45)	see Figs. 10 & 17
Normal suspension (w/o VSC[1]) (CAN)	1.65-1.69 (42-43)	-1.81 (-46)	see Figs. 10 & 17
HD leaf springs	1.53 (39)	-1.53 (-39)	see Figs. 10 & 17
Off-Road Pkg.	1.88 (48)	-1.61 (-41)	see Figs. 10 & 17
2004 4X2 PreRunner 4 cyl. Double-Cab RZN196L			
Normal suspension (w/ VSC[1])	1.65 (42)	-1.81 (-46)	see Figs. 10 & 17
HD leaf springs	1.65-1.69 (42-43)	-1.53 (-39)	see Figs. 10 & 17
Off-Road Pkg. (w/o VSC[1]) (CAN)	1.85 (47)	-1.61 (-41)	see Figs. 10 & 17
Off-Road Pkg. (w/ VSC[1])	1.69 (43)	-1.65 (-42)	see Figs. 10 & 17
2004 4X2 PreRunner 6 cyl. Double-Cab VZN195L			
Normal suspension (w/o VSC[1])	2.63-2.67(67-68)	-.86 (-22)	see Figs. 10 & 17
Normal suspension (w/ VSC[1])	2.63 (67)	-.90 (-23)	see Figs. 10 & 17
Off-Road Pkg. (w/o VSC[1])	2.67 (68)	-.74 (-19)	see Figs. 10 & 17
Off-Road Pkg. (w/ VSC[1])	2.67 (68)	-.78 (-20)	see Figs. 10 & 17
2005-06 4X2 PreRunner Reg. Cab	2.40 (61)	-1.85 (-47)	see Figs. 10 & 17
2005-06 4X2 PreRunner Access Cab 4 Cyl.	2.36 (60)	-1.85 (-47)	see Figs. 10 & 17
2005-06 4X2 PreRunner Access Cab 6 Cyl.			
Manual Trans, with P245/75R16 tires	2.60 (66)	-1.73 (-44)	see Figs. 10 & 17
Manual Trans, with P265/70R16 tires	2.72 (69)	-1.81 (-46)	see Figs. 10 & 17
Manual Trans, with P265/65R17 tires	2.60 (66)	-1.73 (-44)	see Figs. 10 & 17
Auto. Trans, with P245/75R16 tires	2.48 (63)	-1.73 (-44)	see Figs. 10 & 17
Auto. Trans, with P265/70R16 tires	2.75 (70)	-1.77 (-45)	see Figs. 10 & 17
Auto. Trans, with P265/65R17 tires	2.48 (63)	-1.73 (-44)	see Figs. 10 & 17
2005-06 4X2 PreRunner Double Cab			
Long Bed, with P245/75R16 tires	2.52 (64)	-1.54 (-39)	see Figs. 10 & 17
Long Bed, with P265/70R16 tires	2.44 (62)	-1.69 (-43)	see Figs. 10 & 17
Long Bed, with P265/65R17 tires	2.52 (64)	-1.54 (-39)	see Figs. 10 & 17
X-Long Bed, with P245/75R16 tires	2.56 (65)	-1.50 (-38)	see Figs. 10 & 17
X-Long Bed, with P265/65R17 tires	2.56 (65)	-1.50 (-38)	see Figs. 10 & 17
2007-08 4X2 PreRunner Reg. Cab	2.48 (63)	-1.85 (-47)	see Figs. 10 & 17
2007-08 4X2 PreRunner Access Cab 4 Cyl.	2.44 (62)	-1.81 (-46)	see Figs. 10 & 17
2007-08 4X2 PreRunner Access Cab 6 Cyl.			
Manual Trans, with P245/75R16 tires	2.60 (66)	-1.73 (-44)	see Figs. 10 & 17
Manual Trans, with P265/70R16 tires	2.75 (70)	-1.81 (-46)	see Figs. 10 & 17
Manual Trans, with P265/65R17 tires	2.60 (66)	-1.81 (-46)	see Figs. 10 & 17
Auto. Trans, with P245/75R16 tires	2.48 (63)	-1.73 (-44)	see Figs. 10 & 17
Auto. Trans, with P265/70R16 tires	2.80 (71)	-1.77 (-45)	see Figs. 10 & 17
Auto. Trans, with P265/65R17 tires	2.48 (63)	-1.81 (-46)	see Figs. 10 & 17
2007-08 4X2 PreRunner Double Cab :			
Long Bed, with P245/75R16 tires	2.56 (65)	-1.54 (-39)	see Figs. 10 & 17
Long Bed, with P265/70R16 tires	2.48 (63)	-1.65 (-42)	see Figs. 10 & 17
Long Bed, with P265/65R17 tires	2.60 (66)	-1.61 (-41)	see Figs. 10 & 17
X-Long Bed, with P245/75R16 tires	2.56 (65)	-1.46 (-37)	see Figs. 10 & 17
X-Long Bed, with P265/65R17 tires	2.56 (65)	-1.57 (-40)	see Figs. 10 & 17
2009-10 4X2 PreRunner Reg. Cab	2.52 (64)	-1.77 (-45)	see Figs. 10 & 17
2009-10 4X2 PreRunner Access Cab 4 Cyl.:	2.48 (63)	-1.77 (-45)	see Figs. 10 & 17

* ± ¾ (19) (except when range is given)

Toyota Truck & SUV (cont.)

YEAR MODEL	HEIGHT SPECIFICATIONS @CURB WEIGHT - IN. (MM) *		
Tacoma – All 4X2 PreRunner Models (cont.)	FRONT (SF)	REAR (SR)	NOTES
2009-10 4X2 PreRunner Access Cab 6 Cyl.:			
Manual Trans, with P245/75R16 tires	2.68 (68)	-1.57 (-40)	see Figs. 10 & 17
Manual Trans, with P265/70R16 tires	2.83 (72)	-1.65 (-42)	see Figs. 10 & 17
Manual Trans, with P265/65R17 tires	2.68 (68)	-1.69 (-43)	see Figs. 10 & 17
Auto. Trans, with P245/75R16 tires	2.56 (65)	-1.61 (-41)	see Figs. 10 & 17
Auto. Trans, with P265/70R16 tires	2.83 (72)	-1.61 (-41)	see Figs. 10 & 17
Auto. Trans, with P265/65R17 tires	2.56 (65)	-1.65 (-42)	see Figs. 10 & 17
2009-10 4X2 PreRunner Double Cab :			
Long Bed, with P245/75R16 tires	2.60 (66)	-1.46 (-37)	see Figs. 10 & 17
Long Bed, with P265/70R16 tires	2.52 (64)	-1.57 (-40)	see Figs. 10 & 17
Long Bed, with P265/65R17 tires	2.60 (66)	-1.54 (-39)	see Figs. 10 & 17
X-Long Bed, with P245/75R16 tires	2.64 (67)	-1.38 (-35)	see Figs. 10 & 17
X-Long Bed, with P265/65R17 tires	2.64 (67)	-1.50 (-38)	see Figs. 10 & 17
2011 4X2 PreRunner, Access Cab, 4 Cyl.	2.52 (64)	-1.73 (-44)	see Figs. 10 & 17
2011 4X2 PreRunner, Access Cab, 6 Cyl.:			
with P245/75R16	2.56 (65)	-1.61 (-41)	see Figs. 10 & 17
with P265/70R16	2.83 (72)	-1.61 (-41)	see Figs. 10 & 17
with P265/65R17	2.56 (65)	-1.65 (-42)	see Figs. 10 & 17
2011 4X2 PreRunner, Double Cab, 4 Cyl.	2.60 (66)	-1.61 (-41)	see Figs. 10 & 17
2011 4X2 PreRunner, Double Cab, 6 Cyl.:			
Long Bed, with P245/75R16	2.60 (66)	-1.46 (-37)	see Figs. 10 & 17
Long Bed, with P265/70R16	2.52 (64)	-1.57 (-40)	see Figs. 10 & 17
Long Bed, with P265/65R17	2.60 (66)	-1.54 (-39)	see Figs. 10 & 17
X-Long Bed, with P245/75R16	2.64 (67)	-1.38 (-35)	see Figs. 10 & 17
X-Long Bed, with P265/65R17	2.64 (67)	-1.50 (-38)	see Figs. 10 & 17
2012-15 4X2 PreRunner, Access Cab, 4 Cyl.	2.28 (58)	-1.77 (-45)	see Figs. 10 & 17
2012-15 4X2 PreRunner, Access Cab, 6 Cyl.:			
with P245/75R16 w/std. Susp.	2.20 (56)	-1.69 (-43)	see Figs. 10 & 17
with P245/75R16 w/firm Susp.	2.24 (57)	-2.05 (-52)	
with P265/70R16	2.36 (60)	-1.69 (-43)	see Figs. 10 & 17
with P265/65R17	2.20 (56)	-1.73 (-44)	see Figs. 10 & 17
2012-15 4X2 PreRunner, Double Cab, 4 Cyl.	2.36 (60)	-1.57 (-40)	see Figs. 10 & 17
2012-15 4X2 PreRunner, Double Cab, 6 Cyl.:			
Long Bed, with P245/75R16 std. Susp.	2.28 (58)	-1.54 (-39)	see Figs. 10 & 17
Long Bed, with P245/75R16 firm Susp.	2.28 (58)	-1.89 (-48)	see Figs. 10 & 17
Long Bed, with P265/70R16	2.44 (62)	-1.57 (-40)	see Figs. 10 & 17
Long Bed, with P265/65R17	2.28 (58)	-1.61 (-41)	see Figs. 10 & 17
Long Bed, with P265/60R18	2.28 (58)	-1.61 (-41)	see Figs. 10 & 17
X-Long Bed, with P245/75R16 w/std. Susp.	2.17 (55)	-1.46 (-37)	see Figs. 10 & 17
X-Long Bed, with P245/75R16 w/firm Susp.	2.20 (56)	-1.85 (-47)	see Figs. 10 & 17
X-Long Bed, with P265/65R17	2.20 (56)	-1.54 (-39)	see Figs. 10 & 17
X-Long Bed, with P265/60R18	2.20 (56)	-1.54 (-39)	see Figs. 10 & 17
2016-17 4X2 PreRunner, Access Cab GRN325-CRTSHA			
w/P245/75R16 tires (w/Std. Susp)	2.60 (66)	-1.65 (-42)	see Figs. 10 & 17
w/P245/75R16 tires (w/Firm Susp.)	2.60 (66)	-2.04 (-52)	see Figs. 10 & 17
w/265/70R16 tires (w/Std. Susp)	2.60 (66)	-1.81 (-46)	see Figs. 10 & 17
w/265/70R16 tires (w/Firm Susp.)	2.64 (67)	-1.61 (-41)	see Figs. 10 & 17
w/P265/65R17 tires (w/Std. Susp)	2.60 (66)	-1.73 (-44)	see Figs. 10 & 17
w/P265/65R17 tires (w/Firm Susp.)	2.64 (67)	-1.92 (-49)	see Figs. 10 & 17
2016-17 4X2 PreRunner, Double Cab, Long Bed, SR5 GRN325-PRTSHA			
w/P245/75R16 tires (w/Std. Susp)	2.56 (65)	-1.38 (-35)	see Figs. 10 & 17
w/P245/75R16 tires (w/Firm Susp.)	2.60 (66)	-1.69 (-43)	see Figs. 10 & 17
w/265/70R16 tires (w/Std. Susp)	2.64 (67)	-1.46 (-37)	see Figs. 10 & 17
w/265/70R16 tires (w/Firm Susp.)	2.60 (66)	-1.77 (-45)	see Figs. 10 & 17
w/P265/65R17 tires (w/Std. Susp)	2.64 (67)	-1.38 (-35)	see Figs. 10 & 17
w/P265/65R17 tires (w/Firm Susp.)	2.60 (66)	-1.65 (-42)	see Figs. 10 & 17
2016-17 4X2 PreRunner, Double Cab, Long Bed, Limited GRN325-PRTLHA			
All w/Std. Susp	2.52 (64)	-1.29 (-33)	see Figs. 10 & 17
All w/Firm Susp.	2.56 (65)	-1.61 (-41)	see Figs. 10 & 17
2016-17 4X2 PreRunner, Double Cab, SuperLong Bed, SR5 GRN330-PRTSHA			
w/P245/75R16 tires (w/Std. Susp)	2.44 (62)	-1.46 (-37)	see Figs. 10 & 17
w/P245/75R16 tires (w/Firm Susp.)	2.48 (63)	-1.77 (-45)	see Figs. 10 & 17
w/265/70R16 tires (w/Std. Susp)	2.56 (65)	-1.54 (-39)	see Figs. 10 & 17
w/265/70R16 tires (w/Firm Susp.)	2.56 (65)	-1.85 (-47)	see Figs. 10 & 17
w/P265/65R17 tires (w/Std. Susp)	2.44 (62)	-1.46 (-37)	see Figs. 10 & 17
w/P265/65R17 tires (w/Firm Susp.)	2.48 (63)	-1.77 (-45)	see Figs. 10 & 17
2016-17 4X2 PreRunner, Double Cab, SuperLong Bed, Limited GRN330-PRTLHA			
All w/Std. Susp	2.48 (63)	-1.41 (-36)	see Figs. 10 & 17
All w/Firm Susp.	2.52 (64)	-1.73 (-44)	see Figs. 10 & 17
¹ VSC = Vehicle Stability Control			

* ± ¾ (19) (except when range is given)

CHASSIS / SUSP. HEIGHTS

Toyota Truck & SUV (cont.)

YEAR	MODEL	HEIGHT SPECIFICATIONS @CURB WEIGHT - IN. (MM) * FRONT (SF)	REAR (SR)	NOTES
	Tacoma – All 4X4 Models			
1995-97	4X4 Reg. Cab (All)	1.88-1.96 (48-50)	1.37-1.49 (35-38)	see Figs. 10 & 17
1995-96	4X4 X-Cab 4 Cyl.	1.73-1.81 (44-46)	1.14 (29)	see Figs. 10 & 17
1995-96	4X4 X-Cab 6 Cyl.	1.69-1.81 (43-46)	1.18 (30)	see Figs. 10 & 17
1997	4X4 X-Cab 4 Cyl.	1.73 (44)	1.33 (34)	see Figs. 10 & 17
1997	4X4 X-Cab 6 Cyl.	1.73 (44)	1.33 (34)	see Figs. 10 & 17
1998-00	4X4 4 cyl. Reg. Cab RZN161L	1.92 (49)	-1.37 (-35)	see Figs. 10 & 17
1998-00	4X4 4 cyl. X-Cab RZN171L	1.88 (48)	-1.88 (-48)	see Figs. 10 & 17
1998-00	4X4 6 cyl. VZN170L	1.88 (48)	-1.96 (-50)	see Figs. 10 & 17
2001-03	4X4 4 cyl. Reg. Cab RZN161L			
	Manual Transmission, normal suspension	1.65 (42)	-1.69 (-43)	see Figs. 10 & 17
	Manual Transmission, HD leaf springs	1.65 (42)	-1.57 to -1.61 (-40 to -41)	see Figs. 10 & 17
	Auto. Transmission, normal suspension	1.57 (40)	-1.69 (-43)	see Figs. 10 & 17
	Auto. Transmission, HD leaf springs	1.57 (40)	-1.57 to -1.61 (-40 to -41)	see Figs. 10 & 17
2001-03	4X4 4 cyl. X-Cab RZN171L			
	Manual Transmission, normal suspension	1.53 (39)	-1.69 (-43)	see Figs. 10 & 17
	Manual Transmission, HD leaf springs	1.53 (39)	-1.49 (-38)	see Figs. 10 & 17
	Manual Transmission, Off-Road Pkg.	2.04 (52)	-1.49 (-38)	see Figs. 10 & 17
	Auto. Transmission, normal suspension	1.57 (40)	-1.69 (-43)	see Figs. 10 & 17
	Auto. Transmission, HD leaf springs	1.57 (40)	-1.49 (-38)	see Figs. 10 & 17
2001-03	4X4 6 cyl. X-Cab VZN170L			
	Manual Transmission, normal suspension	1.57 (40)	-1.73 (-44)	see Figs. 10 & 17
	Manual Transmission, HD leaf springs	1.57 (40)	-1.49 (-38)	see Figs. 10 & 17
	Manual Transmission, Off-Road Pkg.	2.08 (53)	-1.53 (-39)	see Figs. 10 & 17
	Auto. Transmission, normal suspension	1.61 (41)	-1.73 (-44)	see Figs. 10 & 17
	Auto. Transmission, HD leaf springs	1.61 (41)	-1.49 (-38)	see Figs. 10 & 17
	Auto. Transmission, Off-Road Pkg.	1.92 (49)	-1.53 (-39)	see Figs. 10 & 17
2001-03	4X4 6 cyl. Double-Cab VZN170L			
	Normal suspension	2.67 (68)	-.78 (-20)	see Figs. 10 & 17
	Off-Road Pkg.	2.71 (69)	-.66 (-17)	see Figs. 10 & 17
2004	4X4 4 cyl. Reg. Cab RZN161L			
	Manual Transmission, P255/75R15 tires	1.61 (41)	-1.65 (-42)	see Figs. 10 & 17
	Manual Transmission, P265/70R16 tires	1.65 (42)	-1.69 (-43)	see Figs. 10 & 17
	Auto. Transmission, P255/75R15 tires	1.61 (41)	-1.65 (-42)	see Figs. 10 & 17
	Auto. Transmission, P265/70R16 tires	1.57 (40)	-1.69 (-43)	see Figs. 10 & 17
2004	4X4 4 cyl. X-Cab RZN171L			
	Manual Transmission, normal susp. (USA)	2.00-2.04 (51-52)	-1.61 (-41)	see Figs. 10 & 17
	Manual Transmission, normal susp. (CAN)	1.53 (39)	-1.69 (-43)	see Figs. 10 & 17
	Manual Transmission, HD leaf springs	1.53 (39)	-1.49 (-38)	see Figs. 10 & 17
	Manual Transmission, Off-Road Pkg. (CAN)	2.04 (52)	-1.49 (-38)	see Figs. 10 & 17
	Auto. Transmission, normal suspension (USA)	2.16 (55)	-1.57 (-40)	see Figs. 10 & 17
	Auto. Transmission, normal suspension (CAN)	1.53-1.57 (39-40)	-1.69 (-43)	see Figs. 10 & 17
	Auto. Transmission, HD leaf springs	1.53-1.57 (39-40)	-1.49 (-38)	see Figs. 10 & 17
2004	4X4 6 cyl. X-Cab VZN170L			
	Manual Trans., normal susp. (w/oVSC[1]) (USA)	2.00-2.04 (51-52)	-1.61 (-41)	see Figs. 10 & 17
	Manual Trans., normal susp. (w/oVSC[1]) (CAN)	1.57 (40)	-1.73 (-44)	see Figs. 10 & 17
	Manual Trans., normal susp. (w/ VSC[1])	1.57 (40)	-1.73 (-44)	see Figs. 10 & 17
	Manual Transmission, HD leaf springs	1.57 (40)	-1.49 (-38)	see Figs. 10 & 17
	Manual Trans., Off-Rd Pkg. (w/o VSC[1]) (CAN)	2.08 (53)	-1.53 (-39)	see Figs. 10 & 17
	Manual Transmission, Off-Rd Pkg. (w/ VSC[1])	1.61 (41)	-1.61 (-41)	see Figs. 10 & 17
	Auto. Trans, normal susp. (w/oVSC[1]) (USA)	1.88 (48)	-1.65 (-42)	see Figs. 10 & 17
	Auto. Trans, normal susp. (w/oVSC[1]) (CAN)	1.61 (41)	-1.73 (-44)	see Figs. 10 & 17
	Auto. Transmission, HD leaf springs	1.61 (41)	-1.49 (-38)	see Figs. 10 & 17
	Auto. Trans., Off-Rd Pkg. (w/o VSC[1]) (CAN)	1.92 (49)	-1.53 (-39)	see Figs. 10 & 17
	Auto. Trans., Off-Rd Pkg. (w/ VSC[1])	1.65 (42)	-1.61 (-41)	see Figs. 10 & 17
2004	4X4 6 cyl. Double-Cab VZN170L			
	Normal suspension (w/o VSC[1])	2.67 (68)	-0.78 (-20)	see Figs. 10 & 17
	Normal suspension (w/ VSC[1])	2.67 (68)	-0.78 (-20)	see Figs. 10 & 17
	Off-Road Pkg. (w/o VSC[1])	2.71 (69)	-0.70 (-18)	see Figs. 10 & 17
	Off-Road Pkg. (w/ VSC[1])	2.71 (69)	-0.70 (-18)	see Figs. 10 & 17
2005-06	4X4 Reg. Cab	2.44-2.48 (62-63)	-1.77 to -1.81 (-45 to -46)	see Figs. 10 & 17
2005-06	4X4 Access Cab	2.51-2.79 (64-71)	-1.57 to -1.81 (-40 to -46)	see Figs. 10 & 17
2005-06	4X4 Double Cab	2.48-2.75 (63-70)	-1.49 to -1.65 (-38 to -42)	see Figs. 10 & 17
2007-08	4X4 Reg. Cab	2.48 (63)	-1.77 (-45)	see Figs. 10 & 17
2007-08	4X4 Access. Cab 4 Cyl.	2.56 (65)	-1.73 (-44)	see Figs. 10 & 17
2007-08	4X4 Access. Cab 6 Cyl.:			
	Manual Trans, with P245/75R16 tires (USA)	2.71 (69)	-1.57 (-40)	see Figs. 10 & 17
	Manual Trans, with P265/70R16 tires (USA)	2.79 (71)	-1.69 (-43)	see Figs. 10 & 17
	Manual Trans, with P265/65R17 tires (USA)	2.71 (69)	-1.69 (-43)	see Figs. 10 & 17
	Auto. Trans, with P245/75R16 tires (USA)	2.56 (65)	-1.57 (-40)	see Figs. 10 & 17
	Auto. Trans, with P265/70R16 tires (USA)	2.83 (72)	-1.65 (-42)	see Figs. 10 & 17
	Auto. Trans, with P265/65R17 tires (USA)	2.60 (65)	-1.69 (-43)	see Figs. 10 & 17
	Manual Trans, with P245/75R16 tires (CAN)	2.83 (72)	-2.08 (-53)	see Figs. 10 & 17

* ± ¾ (19) (except when range is given)

CHASSIS / SUSP. HEIGHTS

Toyota Truck & SUV (cont.)

YEAR	MODEL	HEIGHT SPECIFICATIONS @CURB WEIGHT - IN. (MM) * FRONT (SF)	REAR (SR)	NOTES
	Tacoma – All 4X4 Models (cont.)			
	Manual Trans, with P265/70R16 tires (CAN)	2.91 (74)	-2.08 (-53)	see Figs. 10 & 17
	Manual Trans, with P265/65R17 tires (CAN)	2.83 (72)	-2.08 (-53)	see Figs. 10 & 17
	Auto. Trans, with P245/75R16 tires (CAN)	2.63 (67)	-2.04 (-52)	see Figs. 10 & 17
	Auto. Trans, with P265/70R16 tires (CAN)	2.95 (75)	-2.00 (-51)	see Figs. 10 & 17
	Auto. Trans, with P265/65R17 tires (CAN)	2.63 (67)	-2.04 (-52)	see Figs. 10 & 17
2007-08	4X4 Double. Cab, Long Bed:			
	Manual Trans, with P245/75R16 tires (USA)	2.52 (64)	-1.50 (-38)	see Figs. 10 & 17
	Manual Trans, with P265/70R16 tires (USA)	2.71 (69)	-1.54 (-39)	see Figs. 10 & 17
	Manual Trans, with P265/65R17 tires (USA)	2.56 (65)	-1.57 (-40)	see Figs. 10 & 17
	Manual Trans, with P245/75R16 tires (CAN)	2.56 (65)	-1.93 (-49)	see Figs. 10 & 17
	Manual Trans, with P265/70R16 tires (CAN)	2.71 (69)	-1.93 (-49)	see Figs. 10 & 17
	Manual Trans, with P265/65R17 tires (CAN)	2.56 (65)	-1.93 (-49)	see Figs. 10 & 17
	Auto. Trans, with P245/75R16 tires	2.56 (65)	-1.46 (-37)	see Figs. 10 & 17
	Auto. Trans, with P265/70R16 tires	2.75 (70)	-1.50 (-38)	see Figs. 10 & 17
	Auto. Trans, with P265/65R17 tires	2.60 (66)	-1.54 (-39)	see Figs. 10 & 17
2007-08	4X4 Double. Cab, X-Long Bed:			
	w/P245/75R16 tires (USA)	2.63 (67)	-1.37 (-35)	see Figs. 10 & 17
	w/P265/65R17 tires (USA)	2.63 (67)	-1.50 (-38)	see Figs. 10 & 17
	w/P245/75R16 tires (CAN)	2.67 (68)	-1.85 (-47)	see Figs. 10 & 17
	w/P265/65R17 tires (CAN)	2.67 (68)	-1.85 (-47)	see Figs. 10 & 17
2009-10	4X4 Reg. Cab	2.56 (65)	-1.65 (-42)	see Figs. 10 & 17
2009-10	4X4 Access. Cab 4 Cyl. (USA)	2.60 (66)	-1.61 (-41)	see Figs. 10 & 17
2009-10	4X4 Access. Cab 4 Cyl. (CAN)	2.63 (67)	-2.13 (-54)	see Figs. 10 & 17
2009-10	4X4 Access. Cab 6 Cyl.:			
	Manual Trans, with P245/75R16 tires (USA)	2.79 (71)	-1.50 (-38)	see Figs. 10 & 17
	Manual Trans, with P265/70R16 tires (USA)	2.87 (73)	-1.57 (-40)	see Figs. 10 & 17
	Manual Trans, with P265/65R17 tires (USA)	2.79 (71)	-1.57 (-40)	see Figs. 10 & 17
	Auto. Trans, with P245/75R16 tires (USA)	2.63 (67)	-1.50 (-38)	see Figs. 10 & 17
	Auto. Trans, with P265/70R16 tires (USA)	2.91 (74)	-1.54 (-39)	see Figs. 10 & 17
	Auto. Trans, with P265/65R17 tires (USA)	2.63 (67)	-1.57 (-40)	see Figs. 10 & 17
	Manual Trans, with P245/75R16 tires (CAN)	2.83 (72)	-2.00 (-50)	see Figs. 10 & 17
	Manual Trans, with P265/70R16 tires (CAN)	2.91 (74)	-2.00 (-50)	see Figs. 10 & 17
	Manual Trans, with P265/65R17 tires (CAN)	2.83 (72)	-1.93 (-49)	see Figs. 10 & 17
	Auto. Trans, with P245/75R16 tires (CAN)	2.63 (67)	-1.93 (-49)	see Figs. 10 & 17
	Auto. Trans, with P265/70R16 tires (CAN)	2.95 (75)	-1.89 (-48)	see Figs. 10 & 17
	Auto. Trans, with P265/65R17 tires (CAN)	2.63 (67)	-2.00 (-50)	see Figs. 10 & 17
2009-10	4X4 Double Cab, Long Bed:			
	Manual Trans, with P245/75R16 tires (USA)	2.60 (66)	-1.37 (-35)	see Figs. 10 & 17
	Manual Trans, with P265/70R16 tires (USA)	2.79 (71)	-1.41 (-36)	see Figs. 10 & 17
	Manual Trans, with P265/65R17 tires (USA)	2.60 (66)	-1.46 (-37)	see Figs. 10 & 17
	Manual Trans, with P245/75R16 tires (CAN)	2.63 (67)	-1.81 (-46)	see Figs. 10 & 17
	Manual Trans, with P265/70R16 tires (CAN)	2.79 (71)	-1.81 (-46)	see Figs. 10 & 17
	Manual Trans, with P265/65R17 tires (CAN)	2.63 (67)	-1.85 (-47)	see Figs. 10 & 17
	Auto. Trans, with P245/75R16 tires	2.63 (67)	-1.33 (-34)	see Figs. 10 & 17
	Auto. Trans, with P265/70R16 tires	2.83 (72)	-1.37 (-35)	see Figs. 10 & 17
	Auto. Trans, with P265/65R17 tires	2.63 (67)	-1.41 (-36)	see Figs. 10 & 17
2009-10	4X4 Double Cab, X-Long Bed:			
	w/P245/75R16 tires (USA)	2.63 (67)	-1.30 (-33)	see Figs. 10 & 17
	w/P265/65R17 tires (USA)	2.63 (67)	-1.37 (-35)	see Figs. 10 & 17
	w/P245/75R16 tires (CAN)	2.67 (68)	-1.77 (-45)	see Figs. 10 & 17
	w/P265/65R17 tires (CAN)	2.67 (68)	-1.77 (-45)	see Figs. 10 & 17
2011	4X4 Reg. Cab, Manual Trans.	2.52 (64)	-1.73 (-44)	see Figs. 10 & 17
2011	4X4 Reg. Cab, Auto. Trans.	2.60 (66)	-1.61 (-41)	see Figs. 10 & 17
2011	4X4 Access. Cab 4 Cyl. (USA)	2.60 (66)	-1.61 (-41)	see Figs. 10 & 17
2011	4X4 Access. Cab 4 Cyl. (CAN)	2.63 (67)	-2.08 (-53)	see Figs. 10 & 17
2011	4X4 Access. Cab 6 Cyl. :			
	Manual Trans, with P245/75R16 tires (USA)	2.79 (71)	-1.50 (-38)	see Figs. 10 & 17
	Manual Trans, with P265/70R16 tires (USA)	2.87 (73)	-1.57 (-40)	see Figs. 10 & 17
	Manual Trans, with P265/65R17 tires (USA)	2.79 (71)	-1.57 (-40)	see Figs. 10 & 17
	Auto. Trans, with P245/75R16 tires (USA)	2.63 (67)	-1.50 (-38)	see Figs. 10 & 17
	Auto Trans, with P265/70R16 tires (USA)	2.91 (74)	-1.54 (-39)	see Figs. 10 & 17
	Auto. Trans, with P265/65R17 tires (USA)	2.63 (67)	-1.57 (-40)	see Figs. 10 & 17
	Manual Trans, with P245/75R16 tires (CAN)	2.83 (72)	-2.00 (-50)	see Figs. 10 & 17
	Manual Trans, with P265/70R16 tires (CAN)	2.91 (74)	-2.00 (-50)	see Figs. 10 & 17
	Manual Trans, with P265/65R17 tires (CAN)	2.83 (72)	-1.93 (-49)	see Figs. 10 & 17
	Auto. Trans, with P245/75R16 tires (CAN)	2.63 (67)	-1.93 (-49)	see Figs. 10 & 17
	Auto. Trans, with P265/70R16 tires (CAN)	2.95 (75)	-1.89 (-48)	see Figs. 10 & 17
	Auto Trans, with P265/65R17 tires (CAN)	2.63 (67)	-2.00 (-50)	see Figs. 10 & 17
2011	4X4 Double Cab, Long Bed:			
	Manual Trans, with P245/75R16 tires (USA)	2.60 (66)	-1.37 (-35)	see Figs. 10 & 17
	Manual Trans, with P265/70R16 tires (USA)	2.79 (71)	-1.41 (-36)	see Figs. 10 & 17
	Manual Trans, with P265/65R17 tires (USA)	2.60 (66)	-1.46 (-37)	see Figs. 10 & 17

* ± ¾ (19) (except when range is given)

CHASSIS / SUSP. HEIGHTS

Toyota Truck & SUV (cont.)

YEAR	MODEL	HEIGHT SPECIFICATIONS @CURB WEIGHT - IN. (MM) * FRONT (SF)	REAR (SR)	NOTES
	Tacoma – All 4X4 Models (cont.)			
	Manual Trans, with P245/75R16 tires (CAN)	2.63 (67)	-1.81 (-46)	see Figs. 10 & 17
	Manual Trans, with P265/70R16 tires (CAN)	2.79 (71)	-1.81 (-46)	see Figs. 10 & 17
	Manual Trans, with P265/65R17 tires (CAN)	2.63 (67)	-1.85 (-47)	see Figs. 10 & 17
	Auto. Trans, with P245/75R16 tires	2.63 (67)	-1.33 (-34)	see Figs. 10 & 17
	Auto. Trans, with P265/70R16 tires	2.83 (72)	-1.37 (-35)	see Figs. 10 & 17
	Auto. Trans, with P265/65R17 tires	2.63 (67)	-1.41 (-36)	see Figs. 10 & 17
2011	4X4 Double Cab, X-Long Bed:			
	w/P245/75R16 tires (USA)	2.63 (67)	-1.30 (-33)	see Figs. 10 & 17
	w/P265/65R17 tires (USA)	2.63 (67)	-1.37 (-35)	see Figs. 10 & 17
	w/P245/75R16 tires (CAN)	2.67 (68)	-1.77 (-45)	see Figs. 10 & 17
	w/P265/65R17 tires (CAN)	2.67 (68)	-1.77 (-45)	see Figs. 10 & 17
2012-15	4X4 Reg. Cab, Manual Trans.	2.32 (59)	-1.73 (-44)	see Figs. 10 & 17
2012-15	4X4 Reg. Cab, Auto. Trans.	2.40 (61)	-1.65 (-42)	see Figs. 10 & 17
2012-15	4X4 Access. Cab 4 Cyl. Man. (USA) w/std. Susp.	2.32 (59)	-1.65 (-42)	see Figs. 10 & 17
2012-15	4X4 Access. Cab 4 Cyl. Man. (USA) w/firm Susp.	2.36 (60)	-2.01 (-51)	see Figs. 10 & 17
2012-15	4X4 Access. Cab 4 Cyl. Auto (USA) w/std. Susp.	2.36 (60)	-1.65 (-42)	see Figs. 10 & 17
2012-15	4X4 Access. Cab 4 Cyl. Auto (USA) w/firm Susp.	2.40 (61)	-2.01 (-51)	see Figs. 10 & 17
2012-15	4X4 Access. Cab 4 Cyl. (CAN) Manual Trans.	2.36 (60)	-2.01 (-51)	see Figs. 10 & 17
2012-15	4X4 Access. Cab 4 Cyl. (CAN) Auto Trans.	2.40 (61)	-2.01 (-51)	see Figs. 10 & 17
2012-15	4X4 Access. Cab 6 Cyl. :			
	Manual Trans, with P245/75R16 tires (USA) Std.	2.24 (57)	-1.61 (-41)	see Figs. 10 & 17
	Manual Trans, with P245/75R16 tires (USA) Firm	2.28 (58)	-1.97 (-50)	see Figs. 10 & 17
	Manual Trans, with P265/70R16 tires (USA)	2.32 (59)	-1.69 (-43)	see Figs. 10 & 17
	Manual Trans, with P265/65R17 tires (USA)	2.24 (57)	-1.69 (-43)	see Figs. 10 & 17
	Auto. Trans, with P245/75R16 tires (USA) Std.	2.28 (58)	-1.57 (-40)	see Figs. 10 & 17
	Auto. Trans, with P245/75R16 tires (USA) Firm	2.32 (59)	-1.93 (-49)	see Figs. 10 & 17
	Auto Trans, with P265/70R16 tires (USA)	2.36 (60)	-1.65 (-42)	see Figs. 10 & 17
	Auto. Trans, with P265/65R17 tires (USA)	2.28 (58)	-1.65 (-42)	see Figs. 10 & 17
	Manual Trans, with P245/75R16 tires (CAN)	2.28 (58)	-1.97 (-50)	see Figs. 10 & 17
	Manual Trans, with P265/70R16 tires (CAN)	2.36 (60)	-2.05 (-52)	see Figs. 10 & 17
	Manual Trans, with P265/65R17 tires (CAN)	2.28 (58)	-2.09 (-53)	see Figs. 10 & 17
	Auto. Trans, with P245/75R16 tires (CAN)	2.32 (59)	-1.93 (-49)	see Figs. 10 & 17
	Auto. Trans, with P265/70R16 tires (CAN)	2.40 (61)	-2.01 (-51)	see Figs. 10 & 17
	Auto Trans, with P265/65R17 tires (CAN)	2.32 (59)	-2.05 (-52)	see Figs. 10 & 17
2012-15	4X4 Double Cab, Long Bed:			
	Manual Trans, with P245/75R16 tires (USA) Std.	2.28 (58)	-1.46 (-37)	see Figs. 10 & 17
	Manual Trans, with P245/75R16 tires (USA) Firm	2.28 (58)	-1.81 (-46)	see Figs. 10 & 17
	Manual Trans, with P265/70R16 tires (USA)	2.80 (60)	-1.50 (-38)	see Figs. 10 & 17
	Manual Trans, with P265/65R17 tires (USA)	2.28 (58)	-1.54 (-39)	see Figs. 10 & 17
	Manual Trans, with P245/75R16 tires (CAN)	2.28 (58)	-1.81 (-46)	see Figs. 10 & 17
	Manual Trans, with P265/70R16 tires (CAN)	2.40 (61)	-1.89 (-48)	see Figs. 10 & 17
	Manual Trans, with P265/65R17 tires (CAN)	2.32 (59)	-1.89 (-48)	see Figs. 10 & 17
	Auto. Trans, with P245/75R16 tires Std. Susp	2.32 (59)	-1.41 (-36)	see Figs. 10 & 17
	Auto. Trans, with P245/75R16 tires Firm Susp	2.32 (59)	-1.77 (-45)	see Figs. 10 & 17
	Auto. Trans, with P265/70R16 tires	2.40 (61)	-1.46 (-37)	see Figs. 10 & 17
	Auto. Trans, with P265/65R17 tires	2.32 (59)	-1.50 (-38)	see Figs. 10 & 17
	Auto. Trans, with P265/60R18 tires	2.32 (59)	-1.50 (-38)	see Figs. 10 & 17
2012-15	4X4 Double Cab, X-Long Bed:			
	w/P245/75R16 tires (USA) w/Std. Susp	2.64 (56)	-1.30 (-33)	see Figs. 10 & 17
	w/P245/75R16 tires (USA) w/Firm Susp	2.24 (57)	-1.65 (-42)	see Figs. 10 & 17
	w/P265/65R17 tires (USA)	2.64 (56)	-1.34 (-34)	see Figs. 10 & 17
	w/P265/60R18 tires (USA)	2.64 (56)	-1.34 (-34)	see Figs. 10 & 17
	w/P245/75R16 tires (CAN)	2.24 (57)	-1.65 (-42)	see Figs. 10 & 17
	w/P265/65R17 tires (CAN)	2.24 (57)	-1.73 (-44)	see Figs. 10 & 17
	w/P265/60R18 tires (CAN)	2.24 (57)	-1.73 (-44)	see Figs. 10 & 17
2016-17	4X4 Access Cab 4 Cyl. Man. Trans TRN245L-CRMSKA			
	All w/Std. Susp.	2.55 (65)	-1.53 (-39)	see Figs. 10 & 17
	All w/Firm Susp.	2.55 (65)	-1.81 (-46)	see Figs. 10 & 17
2016-17	4X4 Access Cab 4 Cyl. Auto Trans TRN245L-CRTSKA			
	All w/Std. Susp.	2.63 (67)	-1.53 (-39)	see Figs. 10 & 17
	All w/Firm Susp.	2.63 (67)	-1.85 (-47)	see Figs. 10 & 17
2016-17	4X4 Access Cab 6 Cyl. Man. Trans GRN305L-CRFSHA			
	w/P245/75R16 tires (w/Std. Susp)	2.55 (65)	-1.45 (-37)	see Figs. 10 & 17
	w/P245/75R16 tires (w/Firm Susp.)	2.55 (65)	-1.77 (-45)	see Figs. 10 & 17
	w/265/70R16 tires (w/Std. Susp)	2.59 (66)	-1.53 (-39)	see Figs. 10 & 17
	w/265/70R16 tires (w/Firm Susp.)	2.59 (66)	-1.85 (-47)	see Figs. 10 & 17
	w/P265/65R17 tires (w/Std. Susp)	2.55 (65)	-1.45 (-37)	see Figs. 10 & 17
	w/P265/65R17 tires (w/Firm Susp.)	2.55 (65)	-1.77 (-45)	see Figs. 10 & 17
2016-17	4X4 Access Cab 6 Cyl. Auto Trans GRN305L-CRTSHA			
	w/P245/75R16 tires (w/Std. Susp)	2.51 (64)	-1.53 (-39)	see Figs. 10 & 17
	w/P245/75R16 tires (w/Firm Susp.)	2.51 (64)	-1.85 (-47)	see Figs. 10 & 17
	w/265/70R16 tires (w/Std. Susp)	2.55 (65)	-1.61 (-41)	see Figs. 10 & 17

* ± ¾ (19) (except when range is given)

Toyota Truck & SUV (cont.)

Year	Model	Height Specifications @Curb Weight - In. (mm) * Front (SF)	Rear (SR)	Notes
	Tacoma – All 4X4 Models (cont.)			
	w/265/70R16 tires (w/Firm Susp.)	2.59 (66)	-1.92 (-49)	see Figs. 10 & 17
	w/P265/65R17 tires (w/Std. Susp)	2.51 (64)	-1.53 (-39)	see Figs. 10 & 17
	w/P265/65R17 tires (w/Firm Susp.)	2.51 (64)	-1.85 (-47)	see Figs. 10 & 17
2016-17	4X4 Dbl. Cab 6 Cyl. Long Bd. SR5 Man. Trans GRN305L-PRFSHA			
	w/P245/75R16 tires (w/Std. Susp)	2.51 (64)	-1.25 (-32)	see Figs. 10 & 17
	w/P245/75R16 tires (w/Firm Susp.)	2.55 (65)	-1.57 (-40)	see Figs. 10 & 17
	w/265/70R16 tires (w/Std. Susp)	2.55 (65)	-1.33 (-34)	see Figs. 10 & 17
	w/265/70R16 tires (w/Firm Susp.)	2.59 (66)	-1.61 (-41)	see Figs. 10 & 17
	w/P265/65R17 tires (w/Std. Susp)	2.51 (64)	-1.25 (-32)	see Figs. 10 & 17
	w/P265/65R17 tires (w/Firm Susp.)	2.55 (65)	-1.57 (-40)	see Figs. 10 & 17
2016-17	4X4 Dbl. Cab 6 Cyl. Long Bd. SR5 Auto Trans GRN305L-PRTSHA			
	w/P245/75R16 tires (w/Std. Susp)	2.55 (65)	-1.29 (-33)	see Figs. 10 & 17
	w/P245/75R16 tires (w/Firm Susp.)	2.59 (66)	-1.61 (-41)	see Figs. 10 & 17
	w/265/70R16 tires (w/Std. Susp)	2.59 (66)	-1.33 (-34)	see Figs. 10 & 17
	w/265/70R16 tires (w/Firm Susp.)	2.59 (66)	-1.65 (-42)	see Figs. 10 & 17
	w/P265/65R17 tires (w/Std. Susp)	2.55 (65)	-1.29 (-33)	see Figs. 10 & 17
	w/P265/65R17 tires (w/Firm Susp.)	2.59 (66)	-1.65 (-42)	see Figs. 10 & 17
2016-17	4X4 Dbl. Cab Limited (All) GRN305L-PRTLHA			
	All w/Std, Susp	2.55 (65)	-1.22(-31)	see Figs. 10 & 17
	All w/Firm Susp	2.59 (66)	-1.33 (-34)	see Figs. 10 & 17
2016-17	4X4 Dbl. Cab 6 Cyl. Sup Long Bed SR5 GRN310L-PRTSHA			
	w/P245/75R16 tires (w/Std. Susp)	2.51 (64)	-1.33 (-34)	see Figs. 10 & 17
	w/P245/75R16 tires (w/Firm Susp.)	2.55 (65)	-1.65 (-42)	see Figs. 10 & 17
	w/265/70R16 tires (w/Std. Susp)	2.51 (64)	-1.41 (-36)	see Figs. 10 & 17
	w/265/70R16 tires (w/Firm Susp.)	2.55 (65)	-1.73 (-44)	see Figs. 10 & 17
	w/P265/65R17 tires (w/Std. Susp)	2.51 (64)	-1.33 (-34)	see Figs. 10 & 17
	w/P265/65R17 tires (w/Firm Susp.)	2.55 (65)	-1.65 (-42)	see Figs. 10 & 17
2016-17	4X4 Dbl. Cab 6 Cyl. Sup Long Bed Limited GRN310L-PRTLHA			
	All w/Std, Susp	2.51 (64)	-1.29 (-33)	see Figs. 10 & 17
	All w/Firm Susp	2.55 (65)	-1.61 (-41)	see Figs. 10 & 17
[1] VSC = Vehicle Stability Control				
	Tundra	Front (SF)	Rear (SR)	Notes
2000	Tundra 4X2 Reg. Cab (USA)	2.71-2.79 (69-71)	2.36-2.44 (60-62)	see Figs. 10 & 17
2000	Tundra 4X2 Reg. Cab (Canada)	3.03-3.11 (77-79)	2.12-2.20 (54-56)	see Figs. 10 & 17
2000	Tundra 4X2 Access-Cab 6 Cyl.	2.51-2.67 (64-68)	2.16-2.32 (55-59)	see Figs. 10 & 17
2000	Tundra 4X2 Access-Cab 8 Cyl.	2.87-2.95 (73-75)	2.04-2.20 (52-54)	see Figs. 10 & 17
2000	Tundra 4X4 Reg. Cab 6 Cyl. (USA)	1.77-1.85 (45-47)	2.99-3.07 (76-78)	see Figs. 10 & 17
2000	Tundra 4X4 Reg. Cab 6 Cyl. (Canada)	1.92-2.00 (49-51)	2.83-2.91 (72-74)	see Figs. 10 & 17
2000	Tundra 4X4 Reg. Cab 8 Cyl. (USA)	2.16 (55)	3.14 (80)	see Figs. 10 & 17
2000	Tundra 4X4 Access-Cab 6 Cyl.	1.57-1.77 (40-45)	3.03-3.18 (77-81)	see Figs. 10 & 17
2000	Tundra 4X4 Access-Cab 8 Cyl.	1.88-2.00 (48-51)	2.83-2.91 (72-74)	see Figs. 10 & 17
2001-02	Tundra 4X2 Reg. Cab (USA)	2.95 (75)	2.04 (52)	see Figs. 10 & 17
2001-02	Tundra 4X2 Reg. Cab (Canada)	3.07 (78)	2.16 (55)	see Figs. 10 & 17
2001-02	Tundra 4X2 Access-Cab6 Cyl.	2.55 (65)	2.08 (53)	see Figs. 10 & 17
2001-02	Tundra 4X2 Acc.-Cab 8 Cyl. w/o Off-road pkg.	2.83 (72)	2.04 (52)	see Figs. 10 & 17
2001-02	Tundra 4X2 Acc.-Cab 8 Cyl. with Off-road pkg.	1.57 (40)	2.91 (74)	see Figs. 10 & 17
2001-02	Tundra 4X4 Reg. Cab 6 Cyl.	1.96 (50)	2.75 (70)	see Figs. 10 & 17
2001-02	Tundra 4X4 Reg. Cab 8 Cyl. (USA)	2.32 (59)	2.87 (73)	see Figs. 10 & 17
2001-02	Tundra 4X4 Reg. Cab 8 Cyl. (Canada)	2.36 (60)	2.99 (76)	see Figs. 10 & 17
2001-02	Tundra 4X4 Access-Cab 6 Cyl.	1.57-1.73 (40-44)	2.99 (76)	see Figs. 10 & 17
2001-02	Tundra 4X4 Access-Cab 8 Cyl. (USA)	2.00 (51)	2.79 (71)	see Figs. 10 & 17
2001-02	Tundra 4X4 Access-Cab 8 Cyl. (Canada)	2.00 (51)	2.75-2.91 (70-74)	see Figs. 10 & 17
2003	Tundra (USA) 4X2 Reg. Cab	2.51-2.63 (64-67)	-2.00 (-51)	see Figs. 10 & 17
2003	Tundra (USA) 4X2 Access-Cab 6 Cyl.	2.12-2.28 (54-58)	-2.00 (-51)	see Figs. 10 & 17
2003	Tundra (USA) 4X2 Access.-Cab 8 Cyl.:			
	STD Bed, SR-5, exc/Sport Susp	2.67 (68)	-1.53 (-39)	see Figs. 10 & 17
	STD Bed, SR-5, w/Sport Susp	2.87 (73)	-2.20 (-56)	see Figs. 10 & 17
	STD Bed, Limited, exc/Sport Susp	2.55 (65)	-1.77 (-45)	see Figs. 10 & 17
	STD Bed, Limited, w/Sport Susp	2.79 (71)	-2.40 (-61)	see Figs. 10 & 17
	StepSide, SR-5, exc/Sport Susp	2.63 (67)	-1.18 (-30)	see Figs. 10 & 17
	StepSide, SR-5, w/Sport Susp	2.75 (70)	-1.18 (-30)	see Figs. 10 & 17
	StepSide, Limited, exc/Sport Susp	2.51 (64)	-1.37 (-35)	see Figs. 10 & 17
	StepSide, Limited, w/Sport Susp	2.67 (68)	-1.37 (-35)	see Figs. 10 & 17
2003	Tundra (USA) 4X4 Reg. Cab	2.28 (58)	-2.55 (-65)	see Figs. 10 & 17
2003	Tundra (USA) 4X4 Access-Cab 6 Cyl.	1.29-1.41 (33-36)	-2.87 (-73)	see Figs. 10 & 17
2003	Tundra (USA) 4X4 Access-Cab 8 Cyl. :			
	STD Bed, SR-5	1.77 (45)	-2.40 (-61)	see Figs. 10 & 17
	STD Bed, Limited	1.69 (43)	-2.63 (-67)	see Figs. 10 & 17
	StepSide, SR-5	1.65 (42)	-1.77 (-45)	see Figs. 10 & 17
	StepSide, Limited	1.61 (41)	-1.92 (-49)	see Figs. 10 & 17

* ± ¾ (19) (except when range is given)

Toyota Truck & SUV (cont.)

Year	Model	Height Specifications @Curb Weight - In. (mm) * Front (SF)	Rear (SR)	Notes
	Tundra (cont.)			
2003	Tundra (Canada) 4X2 Reg. Cab	2.51 (64)	-2.00 (-51)	see Figs. 10 & 17
2003	Tundra (Canada) 4X2 Access-Cab	2.75 (70)	-1.49 (-38)	see Figs. 10 & 17
2003	Tundra (Canada) 4X4 Reg. Cab	2.12 (54)	-2.59 (-66)	see Figs. 10 & 17
2003	Tundra (Canada) 4X4 Access-Cab, SR-5	1.77 (45)	-2.40 (-61)	see Figs. 10 & 17
2003	Tundra (Canada) 4X4 Access-Cab, Limited	1.73 (44)	-2.59 (-66)	see Figs. 10 & 17
2004	Tundra (USA) 4X2 Reg. Cab	2.93-3.02 (74.6-76.9)	-2.06 (-52.5)	see Figs. 10 & 17
2004	Tundra (USA) 4X2 Access-Cab 6 Cyl.	2.79-2.87 (71-73)	-2.04 (-52)	see Figs. 10 & 17
2004	Tundra (USA) 4X2 Access-Cab 8 Cyl.			see Figs. 10 & 17
	STD Bed, SR-5 exc/Off-road	3.22 (82)	-1.57 to -1.69 (-40 to -43)	see Figs. 10 & 17
	STD Bed, SR-5 Off-road	3.07 (78)	-1.65 (-42)	see Figs. 10 & 17
	STD Bed, Limited exc/Off-road	3.14 (80)	-1.81 to -1.88 (-46 to -48)	see Figs. 10 & 17
	STD Bed, Limited Off-road	2.99 (76)	-1.85 (-47)	see Figs. 10 & 17
	StepSide, SR-5 exc/Off-road	3.18 (81)	-1.02 to -1.10 (-26 to -28)	see Figs. 10 & 17
	StepSide, SR-5 Off-road	2.99 (76)	-1.06 (-27)	see Figs. 10 & 17
	StepSide, Limited exc/Off-road	3.11 (79)	-1.24 (-31.5)	see Figs. 10 & 17
	StepSide, Limited Off-road	2.91 (74)	-1.25 (-32)	see Figs. 10 & 17
2004	Tundra (USA) 4X2 Double-Cab			
	SR-5, exc/Off-road	2.91 (74)	-2.06 (-52.5)	see Figs. 10 & 17
	SR-5, Off-road	2.83 (72)	-2.00 (-51)	see Figs. 10 & 17
	Limited, exc/Off-road	2.87 (73)	-2.06 (-52.5)	see Figs. 10 & 17
	Limited, Off-road	2.81 (71.6)	-2.02 (-51.5)	see Figs. 10 & 17
2004	Tundra (USA) 4X4 Reg. Cab	2.44 (62)	-2.75 (-70)	see Figs. 10 & 17
2004	Tundra (USA) 4X4 Access Cab 6 Cyl.	1.77-1.88 (45-48)	-2.91 (-74)	see Figs. 10 & 17
2004	Tundra (USA) 4X4 Access Cab 8 Cyl.			
	STD Bed, SR-5	2.22 (56.5)	-2.50 (-63.5)	see Figs. 10 & 17
	STD Bed, Limited	2.16 (55)	-2.71 (-69)	see Figs. 10 & 17
	StepSide, SR-5	2.08 (53)	-1.67 (-42.5)	see Figs. 10 & 17
	StepSide, Limited	2.04 (52)	-1.83 (-46.5)	see Figs. 10 & 17
2004	Tundra (USA) 4X4 Double Cab. SR-5	2.36-2.44 (60-62)	-2.24 (-57)	see Figs. 10 & 17
2004	Tundra (USA) 4X4 Double Cab. Limited	2.44 (62)	-2.28 (-58)	see Figs. 10 & 17
2004	Tundra (Canada) 4X2 Reg. Cab	2.94 (74.7)	-2.06 (-52.5)	see Figs. 10 & 17
2004	Tundra (Canada) 4X2 Access Cab	3.28 (83.5)	-1.57 (-40)	see Figs. 10 & 17
2004	Tundra (Canada) 4X2 Double Cab	2.91 (74)	-1.98 (-50.5)	see Figs. 10 & 17
2004	Tundra (Canada) 4X4 Reg. Cab	2.28 (58.4)	-2.79 (-71)	see Figs. 10 & 17
2004	Tundra (Canada) 4X4 Access Cab SR-5	2.22 (56.5)	-2.40 (-61)	see Figs. 10 & 17
2004	Tundra (Canada) 4X4 Access Cab Limited	2.16 (55)	-2.59 (-66)	see Figs. 10 & 17
2004	Tundra (Canada) 4X4 Double Cab SR-5	2.36-2.44 (60-62)	-2.10 (-53.5)	see Figs. 10 & 17
2004	Tundra (Canada) 4X4 Double Cab Limited	2.40 (61)	-2.08 (-53)	see Figs. 10 & 17
2005-06	Tundra (USA) 4X2 Reg. Cab	2.95-3.07 (75-78)	-1.96 to -2.16 (-50 to -55)	see Figs. 10 & 17
2005-06	Tundra (USA) 4X2 Access-Cab 6 Cyl.			
	With Manual Transmission	2.45 (62.4)	-2.48 (-63)	see Figs. 10 & 17
	With Automatic Transmission	2.55 (65)	-2.44 (-62)	see Figs. 10 & 17
2005-06	Tundra (USA) 4X2 Access-Cab 8 Cyl. :			
	Std. Bed SR5 except TRD Sport	2.83 (72)	-1.96 (-50)	see Figs. 10 & 17
	Std. Bed Limited except TRD Sport	2.79 (71)	-2.16 (-55)	see Figs. 10 & 17
	Std. Bed All with TRD Sport	3.22 (82)	-1.61 to -1.77 (-41 to -45)	see Figs. 10 & 17
	Step Side Bed SR5 except TRD Sport	2.71-2.75 (69-70)	-.98 (-25)	see Figs. 10 & 17
	Step Side Bed Limited exc.TRD Sport	2.67 (68)	-1.18 (-30)	see Figs. 10 & 17
	Step Side Bed All with TRD Sport	3.14-3.22 (80-82)	-.98 to -1.14 (-25 to -29)	see Figs. 10 & 17
2005-06	Tundra (USA) 4X2 Double-Cab except Off-Road	2.67 (68)	-2.04 (-52)	see Figs. 10 & 17
2005-06	Tundra (USA) 4X2 Double-Cab with Off-Road	2.59 (66)	-2.03 (-51.6)	see Figs. 10 & 17
2005-06	Tundra (USA) 4X4 Reg. Cab	2.14 (54.5)	-2.68 (-68.1)	see Figs. 10 & 17
2005-06	Tundra (USA) 4X4 Access Cab 8 Cyl. :			
	Std. Bed except Off-Road	1.85-1.88 (47-48)	-2.63 to -2.83 (-67 to -72)	see Figs. 10 & 17
	Std. Bed with Off-Road	2.91 (74)	-2.44 to -2.63 (-62 to -67)	see Figs. 10 & 17
	Step Side Bed except Off-Road	1.73 (44)	-1.61 to -1.77 (-41 to -45)	see Figs. 10 & 17
	Step Side Bed with Off-Road	2.75 (70)	-1.41 to -1.57 (-36 to -40)	see Figs. 10 & 17
2005-06	Tundra (USA) 4X4 Double Cab except Off-Road	2.24 (57)	-2.24 (-57)	see Figs. 10 & 17
2005-06	Tundra (USA) 4X4 Double Cab with Off-Road	2.16 (55)	-2.28 (-58)	see Figs. 10 & 17
2005-06	Tundra (Canada) 4X2 Reg. Cab	2.95 (75)	-2.00 (-51)	see Figs. 10 & 17
2005-06	Tundra (Canada) 4X2 Access Cab exc/17" tires	2.95 (75)	-1.92 (-49)	see Figs. 10 & 17
2005-06	Tundra (Canada) 4X2 Access Cab w/17" tires	2.83 (72)	-2.20 (-56)	see Figs. 10 & 17
2005-06	Tundra (Canada) 4X2 Double Cab	2.75 (70)	-1.98 (-50.5)	see Figs. 10 & 17
2005-06	Tundra (Canada) 4X4 Reg. Cab	2.02 (51.5)	-2.71 (-69)	see Figs. 10 & 17
2005-06	Tundra (Canada) 4X4 Access Cab,	1.85-1.92 (48-49)	-2.95 to -3.14 (-75 to -80)	see Figs. 10 & 17
2005-06	Tundra (Canada) 4X4 Access Cab, Off-Rd	2.95 (75)	-2.71 to -2.91 (-69 to -74)	see Figs. 10 & 17
2005-06	Tundra (Canada) 4X4 Double Cab SR-5	2.22 (56.5)	-2.16 (-55)	see Figs. 10 & 17
2007-09	Tundra 4X2, Reg. Cab, STD. Bed, V6	4.40 (112)	-2.00 (-51)	see Figs. 10 & 17
2007-09	Tundra 4X2, Reg. Cab, STD. Bed, V8 4.7L	4.33 (110)	-2.36 (-60)	see Figs. 10 & 17
2007-09	Tundra 4X2, Reg. Cab, STD. Bed, V8 4.7L, Offrd	4.37 (111)	-2.16 (-55)	see Figs. 10 & 17
2007-09	Tundra 4X2, Reg. Cab, STD. Bed, V8 5.7L	4.44 (113)	-2.36 (-60)	see Figs. 10 & 17

* ± ¾ (19) (except when range is given)

Toyota Truck & SUV (cont.)

YEAR MODEL	HEIGHT SPECIFICATIONS @CURB WEIGHT - IN. (MM) *		
Tundra (cont.)	FRONT (SF)	REAR (SR)	NOTES
2007-09 Tundra 4X2, Reg. Cab, STD. Bed, V8 5.7L, Offrd	4.48 (114)	-2.20 (-56)	see Figs. 10 & 17
2007-09 Tundra 4X4, Reg. Cab, STD. Bed, V8 4.7L	3.77 (96)	-2.63 (-67)	see Figs. 10 & 17
2007-09 Tundra 4X4, Reg. Cab, STD. Bed, V8 4.7L, Offrd	3.81 (97)	-2.55 (-65)	see Figs. 10 & 17
2007-09 Tundra 4X4, Reg. Cab, STD. Bed, V8 5.7L	3.85 (98)	-2.63 (-67)	see Figs. 10 & 17
2007-09 Tundra 4X4, Reg. Cab, STD. Bed, V8 5.7L, Offrd	3.89 (99)	-2.55 (-65)	see Figs. 10 & 17
2007-09 Tundra 4X2, Reg. Cab, Long Bed, V6	4.17 (106)	-2.08 (-53)	see Figs. 10 & 17
2007-09 Tundra 4X2, Reg. Cab, Long Bed, V8 4.7L	4.17 (106)	-2.44 (-62)	see Figs. 10 & 17
2007-09 Tundra 4X2, Reg. Cab, Long Bed, V8 5.7L	4.25 (108)	-2.44 (-62)	see Figs. 10 & 17
2007-09 Tundra 4X4, Reg. Cab, Long Bed, V8 4.7L	3.62 (92)	-2.75 (-70)	see Figs. 10 & 17
2007-09 Tundra 4X4, Reg. Cab, Long Bed, V8 5.7L	3.66 (93)	-2.75 (-70)	see Figs. 10 & 17
2007-09 Tundra 4X2, Dbl. Cab, STD. Bed, V6	3.89 (99)	-2.16 (-55)	see Figs. 10 & 17
2007-09 Tundra 4X2, Dbl. Cab, STD. Bed, V8 4.7L	3.89 (99)	-2.00 (-51)	see Figs. 10 & 17
2007-09 Tundra 4X2, Dbl. Cab, STD. Bed, V8 4.7L, Offrd.	3.97 (101)	-1.88 (-48)	see Figs. 10 & 17
2007-09 Tundra 4X2, Dbl. Cab, STD. Bed, V8 5.7L	3.97 (101)	-2.08 (-53)	see Figs. 10 & 17
2007-09 Tundra 4X2, Dbl. Cab, STD. Bed, V8 5.7L, Offrd.	4.01 (102)	-1.88 (-48)	see Figs. 10 & 17
2007-09 Tundra 4X4, Dbl. Cab, STD. Bed, V8 4.7L, SR5	3.42 (87)	-2.36 (-60)	see Figs. 10 & 17
2007-09 Tundra 4X4, Dbl. Cab, STD. Bed, V8 4.7L, Limit.	3.38 (86)	-2.16 (-55)	see Figs. 10 & 17
2007-09 Tundra 4X4, Dbl. Cab, STD. Bed, V8 4.7L, Offrd.	3.46 (88)	-2.32 (-59)	see Figs. 10 & 17
2007-09 Tundra 4X4, Dbl. Cab, STD. Bed, V8 5.7L	3.42 (87)	-2.16 (-55)	see Figs. 10 & 17
2007-09 Tundra 4X4, Dbl. Cab, STD. Bed, V8 5.7L, Offrd.	3.46 (88)	-2.28 (-58)	see Figs. 10 & 17
2007-09 Tundra 4X2, Dbl. Cab, Long Bed, V8 4.7L	3.93 (100)	-2.08 (-53)	see Figs. 10 & 17
2007-09 Tundra 4X2, Dbl. Cab, Long Bed, V8 5.7L	3.93 (100)	-2.12 (-54)	see Figs. 10 & 17
2007-09 Tundra 4X4, Dbl. Cab, Long Bed, V8 4.7L	3.38 (86)	-2.40 (-61)	see Figs. 10 & 17
2007-09 Tundra 4X4, Dbl. Cab, Long Bed, V8 5.7L	3.34 (85)	-2.24 (-57)	see Figs. 10 & 17
2007-09 Tundra 4X2, Crew-Max, V8 4.7L	3.97 (101)	-1.85 (-47)	see Figs. 10 & 17
2007-09 Tundra 4X2, Crew-Max, V8 5.7L, exc/Offrd	3.97 (101)	-2.00 (-51)	see Figs. 10 & 17
2007-09 Tundra 4X2, Crew-Max, V8 5.7L, Offrd	4.05 (103)	-1.81 (-46)	see Figs. 10 & 17
2007-09 Tundra 4X4, Crew-Max, V8 4.7L, exc/Offrd	3.50 (89)	-1.96 (-50)	see Figs. 10 & 17
2007-09 Tundra 4X4, Crew-Max, V8 4.7L, Offrd	3.54 (90)	-2.24 (-57)	see Figs. 10 & 17
2007-09 Tundra 4X4, Crew-Max, V8 5.7L, exc/Offrd	3.50 (89)	-2.36 (-60)	see Figs. 10 & 17
2007-09 Tundra 4X4, Crew-Max, V8 5.7L, Offrd	3.54 (90)	-2.32 (-59)	see Figs. 10 & 17
2010-13 Tundra 4X2, Reg. Cab, STD. Bed, V6	4.29 (109)	-2.08 (-53)	see Figs. 10 & 17
2010-13 Tundra 4X2, Reg. Cab, STD. Bed, V8 4.6L	4.37 (111)	-2.36 (-60)	see Figs. 10 & 17
2010-13 Tundra 4X2, Reg. Cab, STD. Bed, V8 4.6L, Offrd	4.40 (112)	-2.12 (-54)	see Figs. 10 & 17
2010 Tundra 4X2, Reg. Cab, STD. Bed, V8 5.7L	4.52 (115)	-2.36 (-60)	see Figs. 10 & 17
2010 Tundra 4X2, Reg. Cab, STD. Bed, V8 5.7L, Offrd	4.56 (116)	-2.12 (-54)	see Figs. 10 & 17
2010-13 Tundra 4X2, Reg. Cab, Long Bed, V6	4.09 (104)	-2.16 (-55)	see Figs. 10 & 17
2010 Tundra 4X2, Reg. Cab, Long Bed, V8 4.6L	4.17 (106)	-2.51 (-64)	see Figs. 10 & 17
2010-13 Tundra 4X2, Reg. Cab, Long Bed, V8 5.7L	4.29 (109)	-2.59 (-66)	see Figs. 10 & 17
2010-13 Tundra 4X2, Dbl. Cab, STD. Bed, V6	3.89 (99)	-2.28 (-58)	see Figs. 10 & 17
2010-13 Tundra 4X2, Dbl. Cab, STD. Bed, V8 4.6L, SR5	4.01 (102)	-2.24 (-57)	see Figs. 10 & 17
2010-13 Tundra 4X2, Dbl. Cab, STD. Bed, V8 4.6L, Offrd.	4.09 (104)	-2.00 (-51)	see Figs. 10 & 17
2010 Tundra 4X2, Dbl. Cab, STD. Bed, V8 4.6L, Limit.	3.89 (99)	-2.12 (-54)	see Figs. 10 & 17
2010 Tundra 4X2, Dbl. Cab, STD. Bed, V8 4.6L, Offrd.	3.97 (101)	-1.88 (-48)	see Figs. 10 & 17
2010-13 Tundra 4X2, Dbl. Cab, STD. Bed, V8 5.7L	4.01 (102)	-2.12 (-54)	see Figs. 10 & 17
2010-13 Tundra 4X2, Dbl. Cab, STD. Bed, V8 5.7L, Offrd.	4.09 (104)	-1.96 (-50)	see Figs. 10 & 17
2010-13 Tundra 4X2, Dbl. Cab, Long Bed, V8 4.6L	3.93 (100)	-2.16 (-55)	see Figs. 10 & 17
2010-13 Tundra 4X2, Dbl. Cab, Long Bed, V8 5.7L	4.05 (103)	-2.16 (-55)	see Figs. 10 & 17
2010-13 Tundra 4X2, Crew-Max, V8 4.6L SR5	3.93 (100)	-1.92 (-49)	see Figs. 10 & 17
2010-13 Tundra 4X2, Crew-Max, V8 4.6L SR5 Offrd.	3.93 (100)	-1.92 (-49)	see Figs. 10 & 17
2010 Tundra 4X2, Crew-Max, V8 4.6L Limited	4.03 (102)	-1.86 (-47)	see Figs. 10 & 17
2010 Tundra 4X2, Crew-Max, V8 4.6L Limited Offrd.	4.05 (103)	-1.89 (-48)	see Figs. 10 & 17
2010-13 Tundra 4X2, Crew-Max, V8 4.6L Limited	4.01 (102)	-1.85 (-47)	see Figs. 10 & 17
2010-13 Tundra 4X2, Crew-Max, V8 4.6L Offrd	4.01 (102)	-1.92 (-49)	see Figs. 10 & 17
2010-13 Tundra 4X2, Crew-Max, V8 5.7L, SR5	4.09 (104)	-2.24 (-57)	see Figs. 10 & 17
2010-13 Tundra 4X2, Crew-Max, V8 5.7L, SR5, Offrd	4.13 (105)	-2.04 (-52)	see Figs. 10 & 17
2010-13 Tundra 4X2, Crew-Max, V8 5.7L, Limited	4.09 (104)	-2.08 (-53)	see Figs. 10 & 17
2010-13 Tundra 4X2, Crew-Max, V8 5.7L, Limited, Offrd	4.09 (104)	-1.88 (-48)	see Figs. 10 & 17
2010-13 Tundra 4X4, Reg. Cab, STD. Bed, V8 4.6L	3.81 (97)	-2.71 (-69)	see Figs. 10 & 17
2010-13 Tundra 4X4, Reg. Cab, STD. Bed, V8 4.6L, Offrd	3.85 (98)	-2.55 (-65)	see Figs. 10 & 17
2010-13 Tundra 4X4, Reg. Cab, STD. Bed, V8 5.7L	3.85 (98)	-2.67 (-68)	see Figs. 10 & 17
2010-13 Tundra 4X4, Reg. Cab, STD. Bed, V8 5.7L, Offrd	3.89 (99)	-2.55 (-65)	see Figs. 10 & 17
2010-13 Tundra 4X4, Reg. Cab, Long Bed, V8 4.6L	3.66 (93)	-2.83 (-72)	see Figs. 10 & 17
2010-13 Tundra 4X4, Reg. Cab, Long Bed, V8 5.7L	3.74 (95)	-2.95 (-75)	see Figs. 10 & 17
2010-13 Tundra 4X4, Dbl. Cab, STD. Bed, V8 4.6L, SR5	3.50 (89)	-2.51 (-64)	see Figs. 10 & 17
2010-13 Tundra 4X4, Dbl. Cab, STD. Bed, V8 4.6L, Limit.	3.46 (88)	-2.51 (-64)	see Figs. 10 & 17
2010-13 Tundra 4X4, Dbl. Cab, STD. Bed, V8 4.6L, Offrd.	3.50 (89)	-2.32 (-59)	see Figs. 10 & 17
2010-13 Tundra 4X4, Dbl. Cab, STD. Bed, V8 5.7L	3.50 (89)	-2.36 (-60)	see Figs. 10 & 17
2010-13 Tundra 4X4, Dbl. Cab, STD. Bed, V8 5.7L, Offrd.	3.54 (90)	-2.40 (-61)	see Figs. 10 & 17
2010-13 Tundra 4X4, Dbl. Cab, Long Bed, V8 4.6L	3.46 (88)	-2.55 (-65)	see Figs. 10 & 17
2010-13 Tundra 4X4, Dbl. Cab, Long Bed, V8 5.7L	3.42 (87)	-2.40 (-61)	see Figs. 10 & 17
2010-13 Tundra 4X4, Crew-Max, V8 4.6L, exc/Offrd	3.50 (89)	-2.16 (-55)	see Figs. 10 & 17

* ± ¾ (19) (except when range is given)

CHASSIS / SUSP. HEIGHTS

Toyota Truck & SUV (cont.)

Year Model	Height Specifications @Curb Weight - In.(mm) *		
Tundra (cont.)	FRONT (SF)	REAR (SR)	NOTES
2010-13 Tundra 4X4, Crew-Max, V8 4.6L, Offrd	3.58 (91)	-2.40 (-61)	see Figs. 10 & 17
2010-13 Tundra 4X4, Crew-Max, V8 5.7L, SR5, exc/Offrd	3.58 (91)	-2.32 (-59)	see Figs. 10 & 17
2010-13 Tundra 4X4, Crew-Max, V8 5.7L, SR5, Offrd	3.62 (92)	-2.28 (-58)	see Figs. 10 & 17
2010-13 Tundra 4X4, Crew-Max, V8 5.7L, Limited, exc/OR	3.54 (90)	-2.28 (-58)	see Figs. 10 & 17
2010-13 Tundra 4X4, Crew-Max, V8 5.7L, Limited, Offrd	3.58 (91)	-2.24 (-57)	see Figs. 10 & 17
2014 Tundra 4X2 Reg. Cab V6 4.0L (GSK51L-THASKA)	4.05 (103)	-2.28 (-55)	see Figs. 10 & 17
2014 Tundra 4X2 Dbl. Cab V6 4.0L (GSK51L-THASKA)	3.81 (97)	-2.28 (-58)	see Figs. 10 & 17
2014-17 Tundra 4X2 Reg. Cab V8 5.7L SR5 USK51L-THTSKA	4.33 (110)	-2.48 (-63)	see Figs. 10 & 17
2014-17 Tundra 4X2 Reg. Cab V8 5.7L SR5 USK51L-THTSKA TRD Off-Rd	4.37 (111)	-2.51 (-64)	see Figs. 10 & 17
2016-17 Tundra 4X2 Reg. Cab V8 5.7L SR5 USK51L-THTSGA	4.33 (110)	-2.48 (-63)	see Figs. 10 & 17
2016-17 Tundra 4X2 Reg. Cab V8 5.7L SR5 USK51L-THTSGA TRD Off-Rd	4.37 (111)	-2.51 (-64)	see Figs. 10 & 17
2014-17 Tundra 4X2 Double Cab, V8 4.6L SR5 UPK51L-CRTSKA	3.85 (98)	-2.24 (-57)	see Figs. 10 & 17
2014-17 Tundra 4X2 Double Cab, V8 4.6L SR5 UPK51L-CRTSKA TRD Off-Rd	3.93 (100)	-2.04 (-52)	see Figs. 10 & 17
2014-17 Tundra 4X2 Double Cab, V8 5.7L SR5 USK51L-CRTSKA	4.05 (103)	-2.08 (-53)	see Figs. 10 & 17
2014-17 Tundra 4X2 Double Cab, V8 5.7L SR5 USK51L-CRTSKA TRD Off-Rd	4.09 (104)	-1.88 (-48)	see Figs. 10 & 17
2014-17 Tundra 4X2 Double Cab, V8 5.7L Limited USK51L-CRTLKA	4.09 (104)	-2.04 (-52)	see Figs. 10 & 17
2014-17 Tundra 4X2 Double Cab, V8 5.7L Lim'td USK51L-CRTLKA TRD Off-Rd	4.13 (105)	-1.85 (-47)	see Figs. 10 & 17
2010-17 Tundra 4X2 Double Cab, V8 5.7L SR5 USK51L-CRTSGA	4.05 (103)	-2.08 (-53)	see Figs. 10 & 17
2016-17 Tundra 4X2 Double Cab, V8 5.7L SR5 USK51L-CRTSGA TRD Off-Rd	4.09 (104)	-1.88 (-48)	see Figs. 10 & 17
2016-17 Tundra 4X2 Double Cab, V8 5.7L Limited USK51L-CRTLGA	4.09 (104)	-2.04 (-52)	see Figs. 10 & 17
2016-17 Tundra 4X2 Double Cab, V8 5.7L Lim'td USK51L-CRTLGA TRD Off-Rd	4.09 (104)	-1.85 (-47)	see Figs. 10 & 17
2014-17 Tundra 4X2 Double Cab, V8 5.7L SR5 USK52L-CHTSKA	4.05 (103)	-2.12 (-54)	see Figs. 10 & 17
2014-17 Tundra 4X2 Double Cab, V8 5.7L SR5 USK52L-CHTSKA TRD Off-Rd	4.05 (103)	-1.88 (-48)	see Figs. 10 & 17
2016-17 Tundra 4X2 Double Cab, V8 5.7L SR5 USK52L-CHTSGA	4.05 (103)	-2.12 (-54)	see Figs. 10 & 17
2016-17 Tundra 4X2 Double Cab, V8 5.7L SR5 USK52L-CHTSGA TRD Off-Rd	4.05 (103)	-1.88 (-48)	see Figs. 10 & 17
2014-17 Tundra 4X2 Crew Cab, V8 4.6L SR5 UPK51L-PSTSKA	3.89 (99)	-2.04 (-52)	see Figs. 10 & 17
2014-17 Tundra 4X2 Crew Cab, V8 4.6L SR5 UPK51L-PSTSKA TRD Off-Rd	3.97 (101)	-2.08 (-53)	see Figs. 10 & 17
2014-17 Tundra 4X2 Crew Cab, V8 5.7L SR5 USK51L-PSTSKA	4.13 (105)	-2.04 (-52)	see Figs. 10 & 17
2014-17 Tundra 4X2 Crew Cab, V8 5.7L SR5 USK51L-PSTSKA TRD Off-Rd	4.13 (105)	-1.85 (-47)	see Figs. 10 & 17
2016-17 Tundra 4X2 Crew Cab, V8 5.7L SR5 USK51L-PSTSGA	4.13 (105)	-2.04 (-52)	see Figs. 10 & 17
2016-17 Tundra 4X2 Crew Cab, V8 5.7L SR5 USK51L-PSTSGA TRD Off-Rd	4.13 (105)	-1.85 (-47)	see Figs. 10 & 17
2014-17 Tundra 4X2 Crew Cab, V8 5.7L Limited USK51L-PSTLKA	4.13 (105)	-2.04 (-52)	see Figs. 10 & 17
2014-17 Tundra 4X2 Crew Cab, V8 5.7L Limited USK51L-PSTLKA TRD Off-Rd	4.13 (105)	-1.85 (-47)	see Figs. 10 & 17
2014-17 Tundra 4X2 Crew Cab, V8 5.7L Limited USK51L-PSTLGA	4.13 (105)	-2.04 (-52)	see Figs. 10 & 17
2014-17 Tundra 4X2 Crew Cab, V8 5.7L Limited USK51L-PSTLGA TRD Off-Rd	4.13 (105)	-1.81 (-46)	see Figs. 10 & 17
2014-17 Tundra 4X2 Crew Cab, V8 5.7L Platinum USK51L-PSTZKA	4.13 (105)	-2.00 (-51)	see Figs. 10 & 17
2014-17 Tundra 4X2 Crew Cab, V8 5.7L Platinum USK51L-PSTZKA TRD Off-Rd	4.13 (105)	-1.81 (-46)	see Figs. 10 & 17
2016-17 Tundra 4X2 Crew Cab, V8 5.7L Platinum USK51L-PSTZGA	4.13 (105)	-2.00 (-51)	see Figs. 10 & 17
2016-17 Tundra 4X2 Crew Cab, V8 5.7L Platinum USK51L-PSTZGA TRD Off-Rd	4.13 (105)	-1.81 (-46)	see Figs. 10 & 17
2014-17 Tundra 4X4 Reg. Cab V8 5.7L SR5 USK56L-THTSKA	3.77 (96)	-2.83 (-72)	see Figs. 10 & 17
2014-17 Tundra 4X4 Reg. Cab V8 5.7L SR5 USK56L-THTSKA TRD Off-Rd	3.77 (96)	-2.28 (-55)	see Figs. 10 & 17
2014-17 Tundra 4X4 Reg. Cab V8 5.7L SR5 USK56L-THTSGA	3.77 (96)	-2.83 (-72)	see Figs. 10 & 17
2014-17 Tundra 4X4 Reg. Cab V8 5.7L SR5 USK56L-THTSGA TRD Off-Rd	3.77 (96)	-2.28 (-55)	see Figs. 10 & 17
2014-17 Tundra 4X4 Double Cab V8 4.6L SR5 UPK56L-CRTSKA	3.42 (87)	-2.59 (-66)	see Figs. 10 & 17
2014-17 Tundra 4X4 Double Cab V8 4.6L SR5 UPK56L-CRTSKA TRD Off-Rd	3.46 (88)	-2.40 (-61)	see Figs. 10 & 17
2014-17 Tundra 4X4 Double Cab V8 5.7L SR5 USK56L-CRTSKA	3.54 (90)	-2.24 (-57)	see Figs. 10 & 17
2014-17 Tundra 4X4 Double Cab V8 5.7L SR5 USK56L-CRTSKA TRD Off-Rd	3.58 (91)	-2.20 (-56)	see Figs. 10 & 17
2014-17 Tundra 4X4 Double Cab V8 5.7L SR5 USK56L-CRTSGA	3.54 (90)	-2.24 (-57)	see Figs. 10 & 17
2014-17 Tundra 4X4 Double Cab V8 5.7L SR5 USK56L-CRTSGA TRD Off-Rd	3.58 (91)	-2.20 (-56)	see Figs. 10 & 17
2014-17 Tundra 4X4 Double Cab V8 5.7L SR5 USK56L-CRTLKA	3.54 (90)	-2.28 (-55)	see Figs. 10 & 17
2014-17 Tundra 4X4 Double Cab V8 5.7L SR5 USK56L-CRTLKA TRD Off-Rd	3.62 (92)	-2.28 (-55)	see Figs. 10 & 17
2014-17 Tundra 4X4 Double Cab V8 5.7L SR5 USK56L-CRTLGA	3.54 (90)	-2.28 (-55)	see Figs. 10 & 17
2014-17 Tundra 4X4 Double Cab V8 5.7L SR5 USK56L-CRTLGA TRD Off-Rd	3.62 (92)	-2.28 (-55)	see Figs. 10 & 17
2014-17 Tundra 4X4 Double Cab V8 5.7L SR5 USK57L-CHTSKA	3.46 (88)	-2.28 (-58)	see Figs. 10 & 17
2014-17 Tundra 4X4 Double Cab V8 5.7L SR5 USK57L- CHTSKA TRD Off-Rd	3.54 (90)	-2.28 (-58)	see Figs. 10 & 17
2014-17 Tundra 4X4 Double Cab V8 5.7L SR5 USK57L-CHTSGA	3.46 (88)	-2.28 (-58)	see Figs. 10 & 17
2014-17 Tundra 4X4 Double Cab V8 5.7L SR5 USK57L-CHTSGA TRD Off-Rd	3.54 (90)	-2.28 (-58)	see Figs. 10 & 17
2014-17 Tundra 4X4 Crew Cab V8 4.6L SR5 UPK56L-PSTSKA	3.42 (87)	-2.20 (-56)	see Figs. 10 & 17
2014-17 Tundra 4X4 Crew Cab V8 4.6L SR5 UPK56L-PSTSKA TRD Off-Rd	3.50 (89)	-2.44 (-62)	see Figs. 10 & 17
2014-17 Tundra 4X4 Crew Cab V8 5.7L SR5 USK56L-PSTSKA	3.58 (91)	-2.24 (-57)	see Figs. 10 & 17
2014-17 Tundra 4X4 Crew Cab V8 5.7L SR5 USK56L-PSTSKA TRD Off-Rd	3.62 (92)	-2.24 (-57)	see Figs. 10 & 17
2014-17 Tundra 4X4 Crew Cab V8 5.7L SR5 USK56L-PSTSGA	3.58 (91)	-2.24 (-57)	see Figs. 10 & 17
2014-17 Tundra 4X4 Crew Cab V8 5.7L SR5 USK56L-PSTSGA TRD Off-Rd	3.62 (92)	-2.24 (-57)	see Figs. 10 & 17
2014-17 Tundra 4X4 Crew Cab V8 5.7L Limited USK56L-PSTLKA	3.58 (91)	-2.24 (-57)	see Figs. 10 & 17
2014-17 Tundra 4X4 Crew Cab V8 5.7L Limited USK56L-PSTLKA TRD Off-Rd	3.62 (92)	-2.28 (-55)	see Figs. 10 & 17
2014-17 Tundra 4X4 Crew Cab V8 5.7L Limited USK56L-PSTLGA	3.58 (91)	-2.24 (-57)	see Figs. 10 & 17
2014-17 Tundra 4X4 Crew Cab V8 5.7L Limited USK56L-PSTLGA TRD Off-Rd	3.62 (92)	-2.28 (-55)	see Figs. 10 & 17
2014-17 Tundra 4X4 Crew Cab V8 5.7L Platinum USK56L-PSTZKA	3.62 (92)	-2.20 (-56)	see Figs. 10 & 17
2014-17 Tundra 4X4 Crew Cab V8 5.7L Platinum USK56L-PSTZKA TRD Off-Rd	3.70 (94)	-2.20 (-56)	see Figs. 10 & 17
2014-17 Tundra 4X4 Crew Cab V8 5.7L Platinum USK56L-PSTZKA	3.62 (92)	-2.20 (-56)	see Figs. 10 & 17
2014-17 Tundra 4X4 Crew Cab V8 5.7L Platinum USK56L-PSTZKA TRD Off-Rd	3.70 (94)	-2.20 (-56)	see Figs. 10 & 17

* ± ¾ (19) (except when range is given)

01 Rear CAMBER/TOE

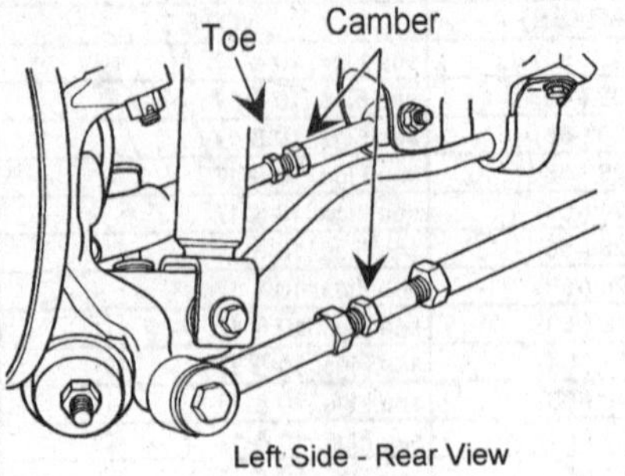

02 Rear TOE

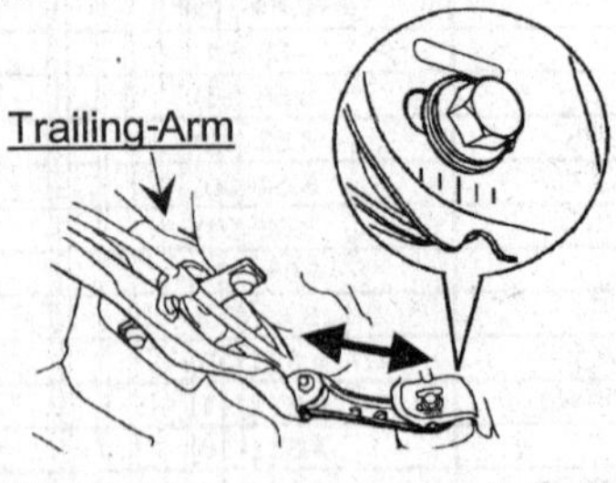

03 CASTER

Lengthen or shorten strut rod using the adjusting nuts at A. and B.

Some applications have shims that can be shifted between positions A. and B. to increase or decrease the length of the strut.

Some applications have an eccentric adjuster.

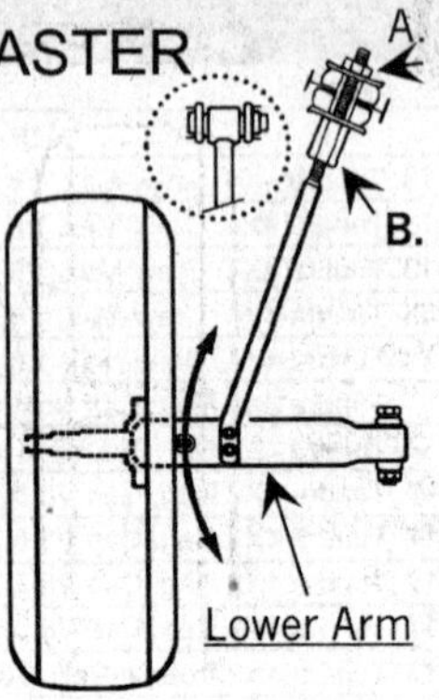

04 Rear TOE

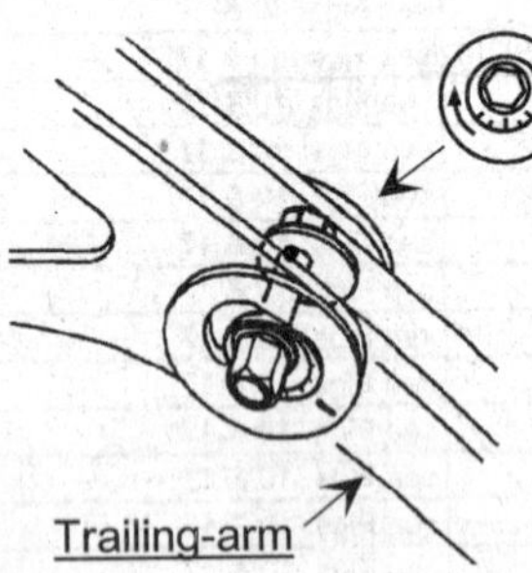

05 Rear TOE

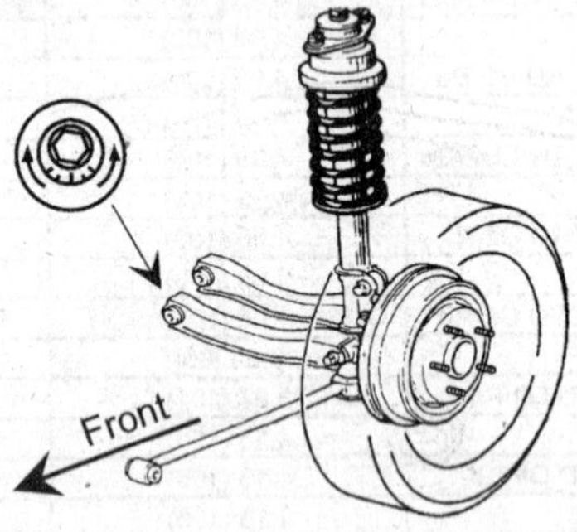

06 Rear TOE

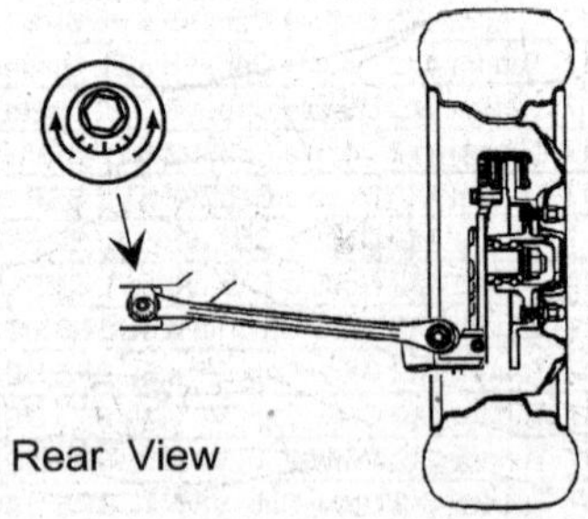

07 CASTER/CAMBER

Acura NSX

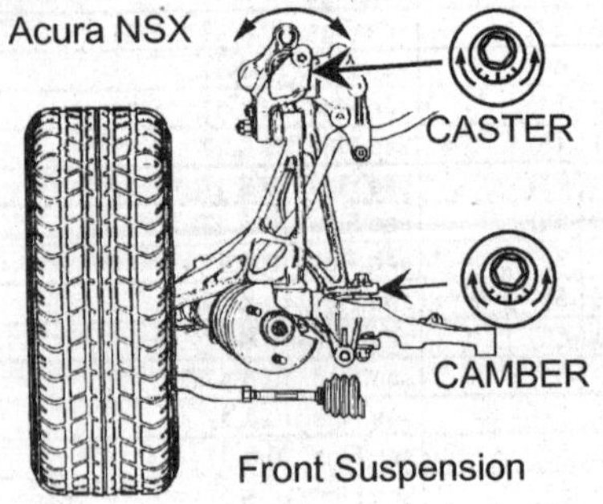

08 Rear CAMBER/TOE

Acura NSX

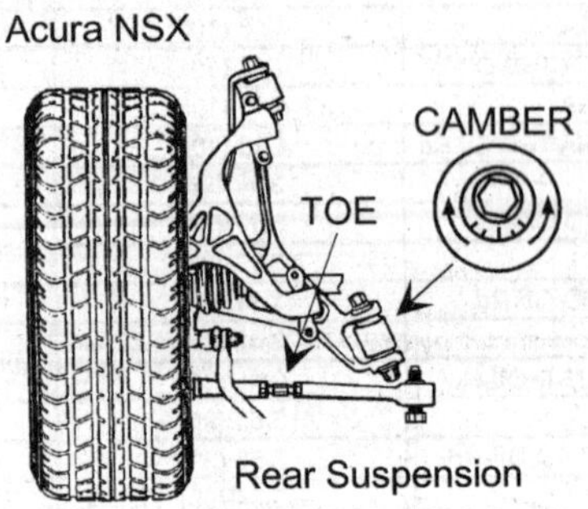

09 CASTER

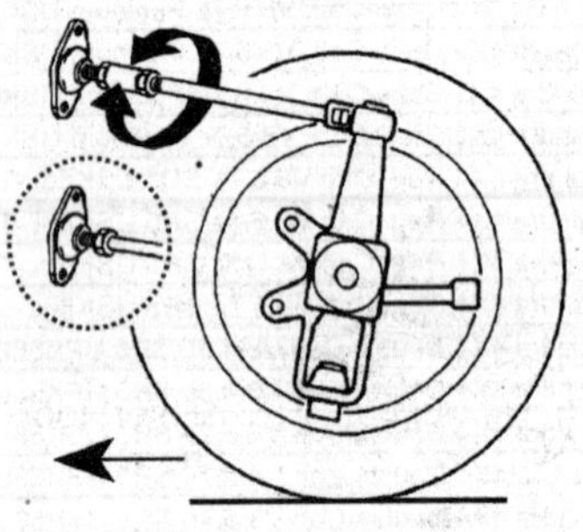

10 Rear TOE

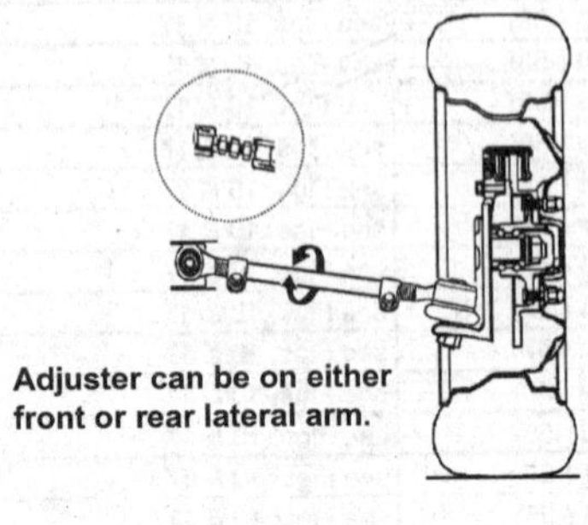

11 CASTER/CAMBER*

* Caster adjustment only on vehicles with lower A-arm.

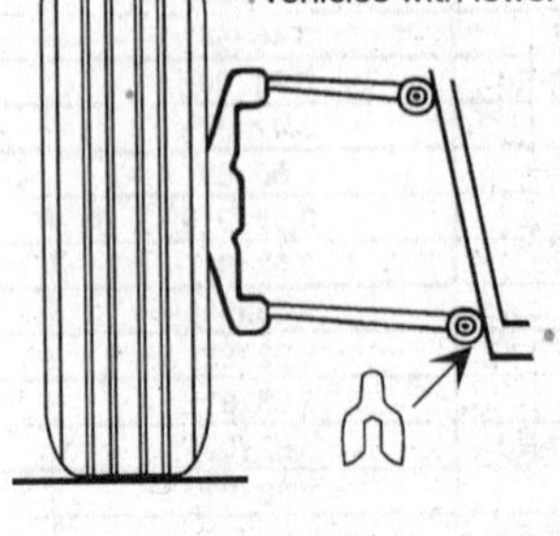

12 CASTER/CAMBER

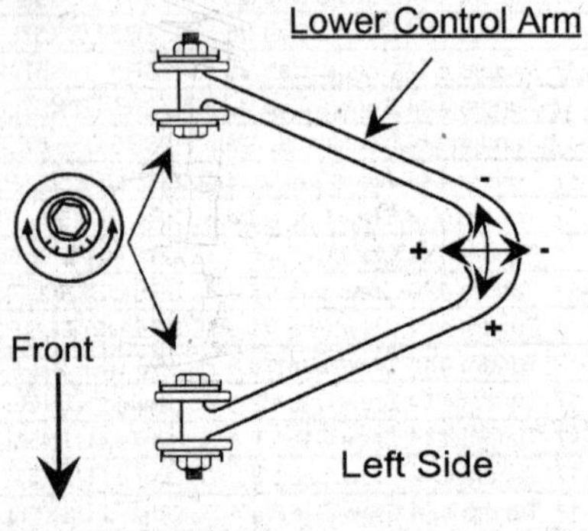

The above illustrations are intended to represent *typical* adjustment methods for the vehicles for which they are listed. Other methods of adjustment may be possible for some applications (usually requiring additional parts), but are not necessarily recommended by the vehicle manufacturer, or by the publisher or distributor of this manual.

13 CASTER

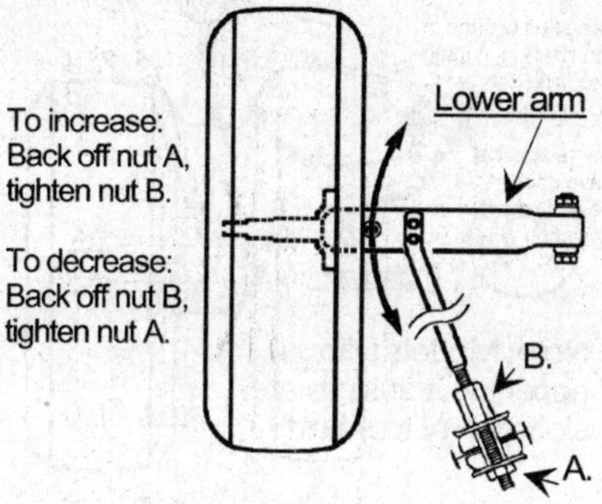

14 Front CAMBER

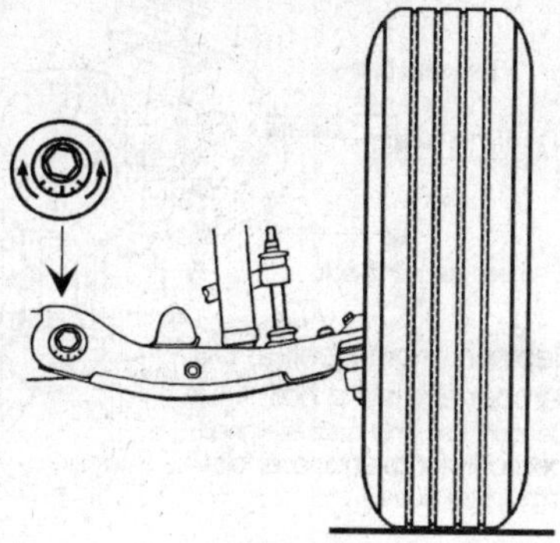

15 CAMBER

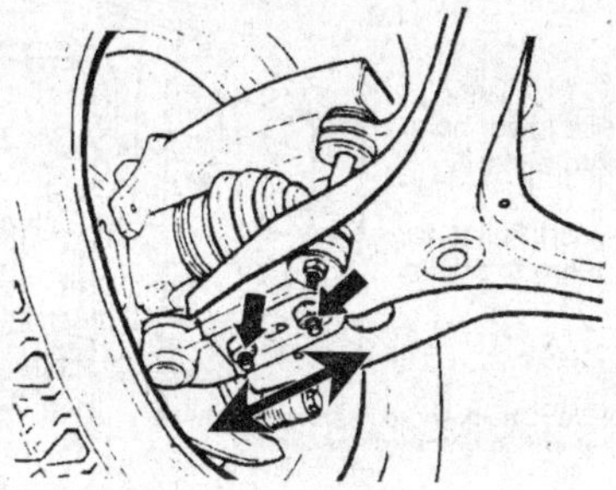

Move lower ball-joint in or out as indicated.

16 Rear TOE

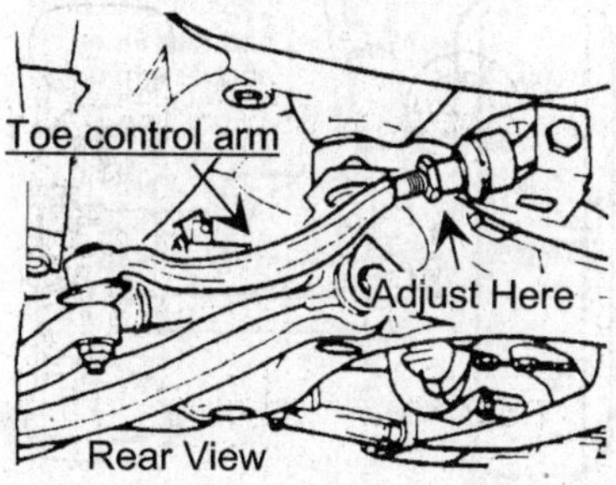

17 Front CAMBER

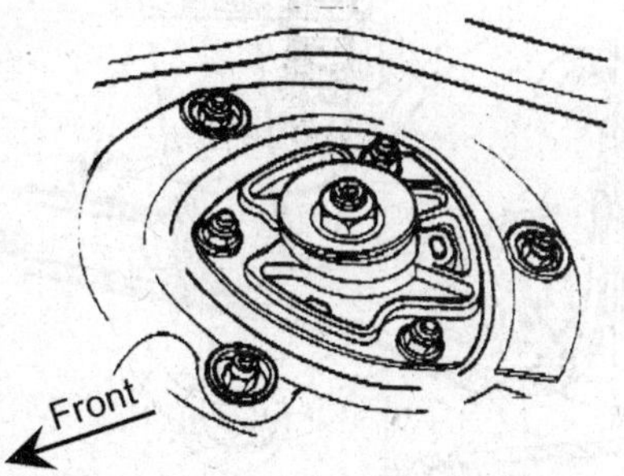

18 Rear CAMBER

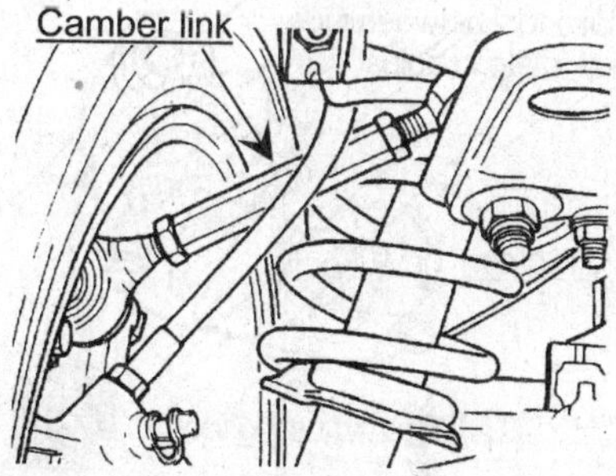

19 Rear TOE

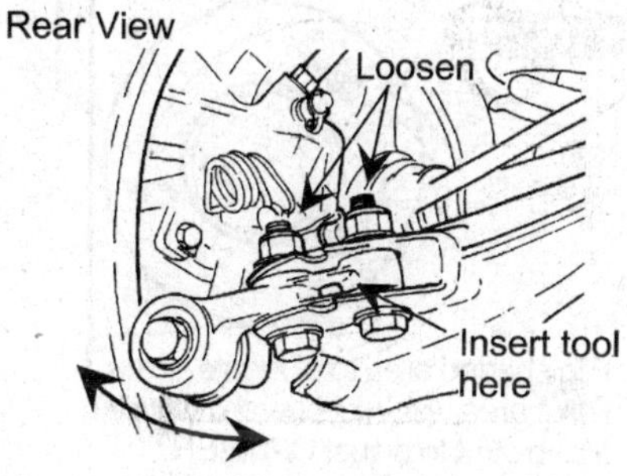

20 Front CAMBER

Loosen nut A, rotate control arm shaft at B.

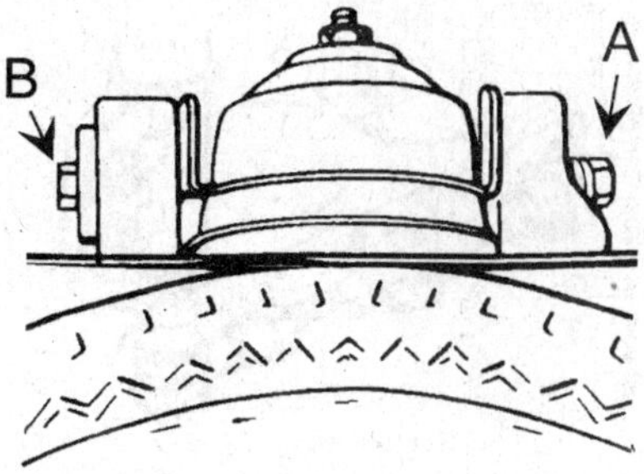

21 Rear CAMBER/TOE

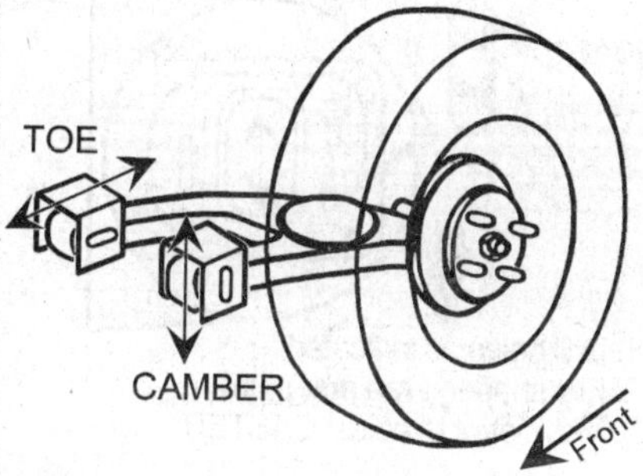

22 CAMBER

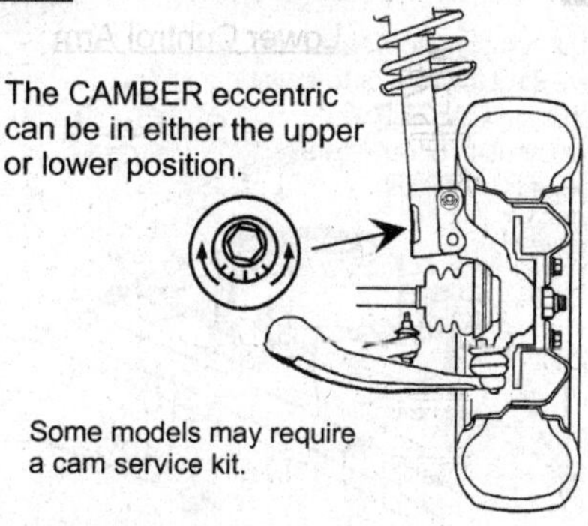

23 CASTER/CAMBER

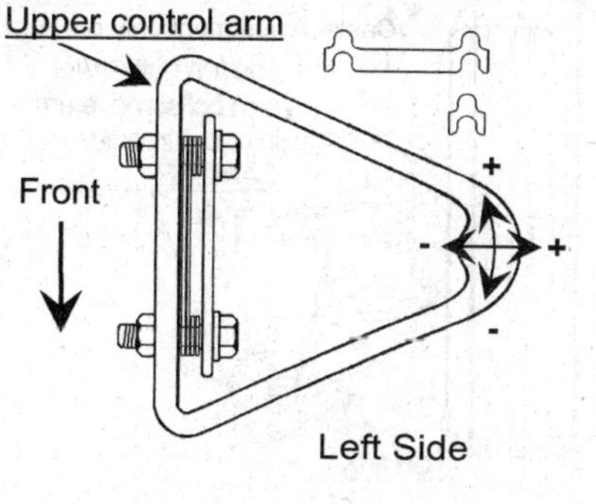

24 CASTER/CAMBER

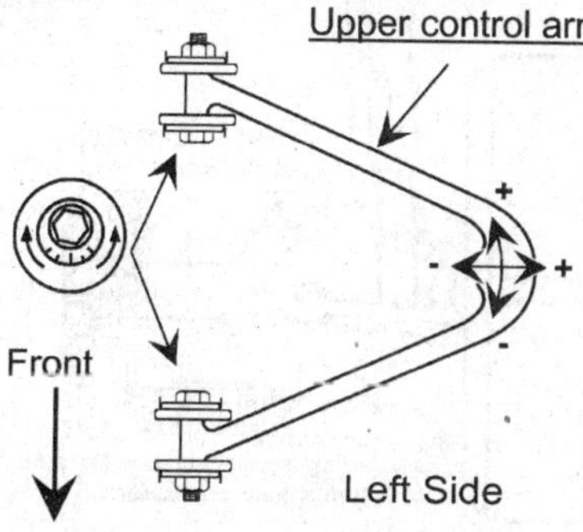

The above illustrations are intended to represent *typical* adjustment methods for the vehicles for which they are listed. Other methods of adjustment may be possible for some applications (usually requiring additional parts), but are not necessarily recommended by the vehicle manufacturer, or by the publisher or distributor of this manual.

ADJUSTMENT METHODS

25 CAMBER

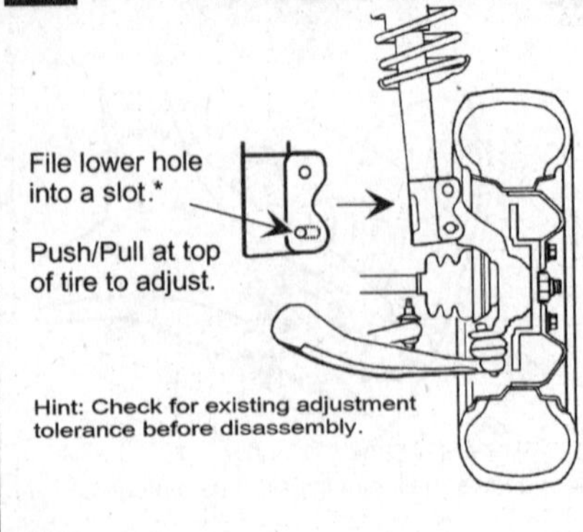

26 CAMBER

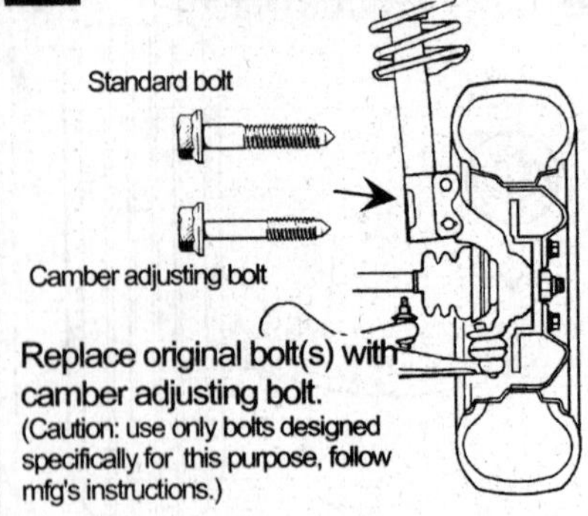

27 Rear TOE

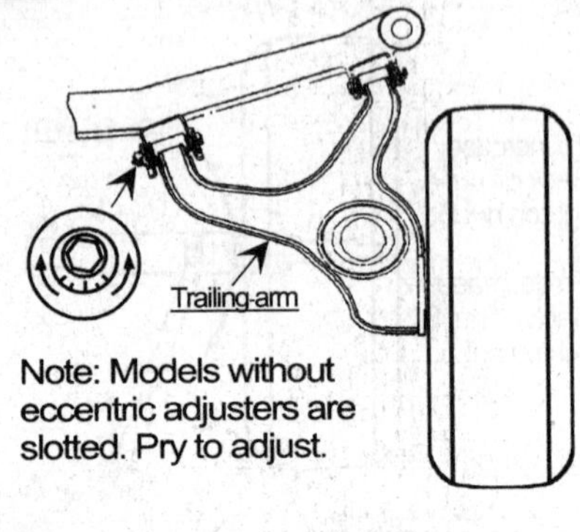

28 Rear TOE

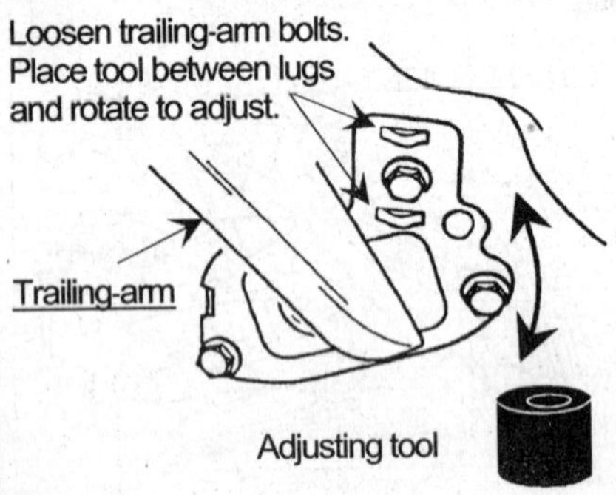

29 Rear CAMBER

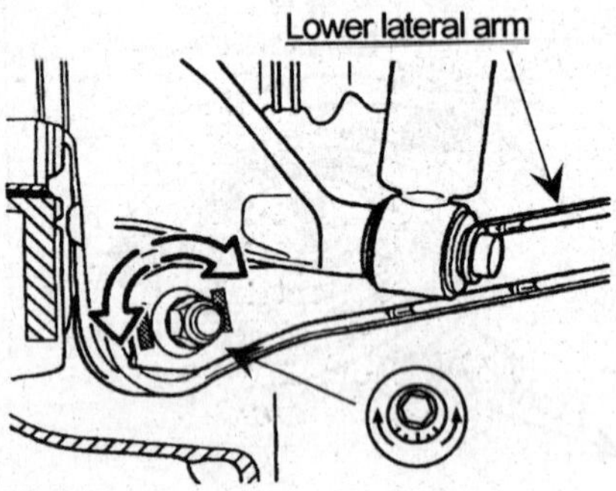

30 Rear TOE

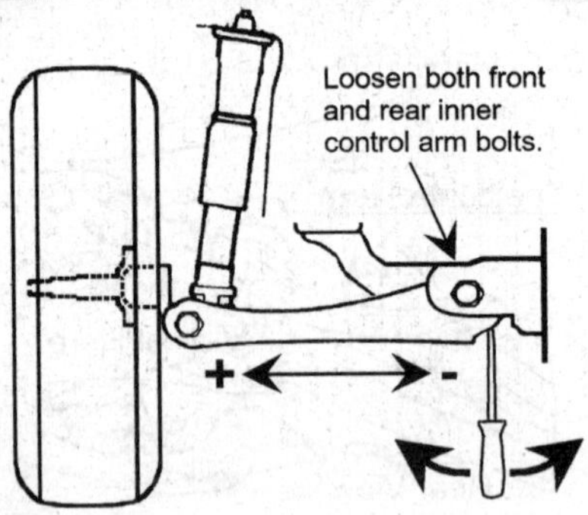

31 CASTER

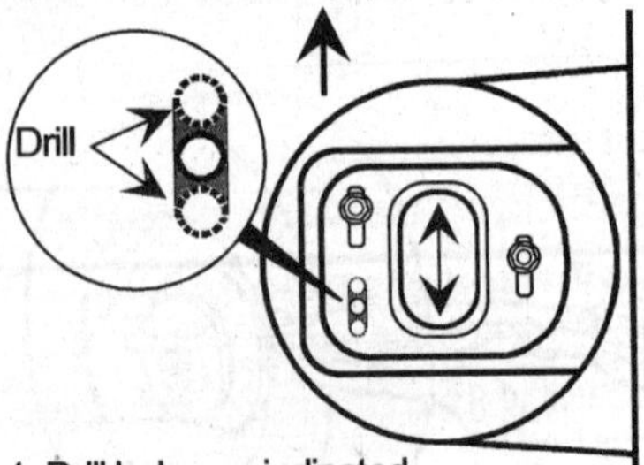

1. Drill holes as indicated.
2. File shaded area and paint.
3. Move strut to adjust CASTER.

32 Rear TOE

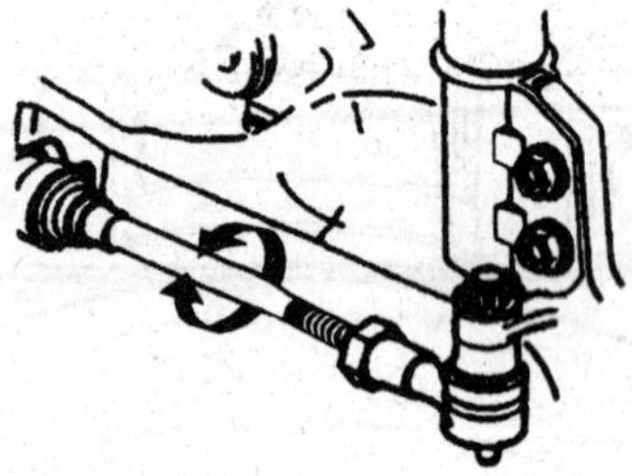

33 Front CAMBER

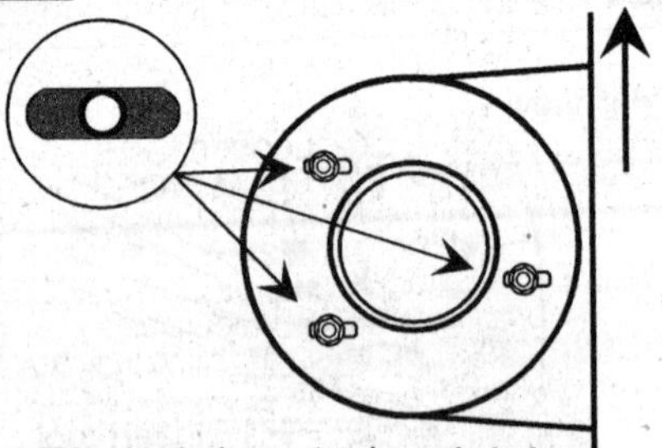

1. File shaded area to elongate holes.
3. Paint area with rust preventive paint.
2. Move strut to adjust CAMBER.

34 Rear TOE

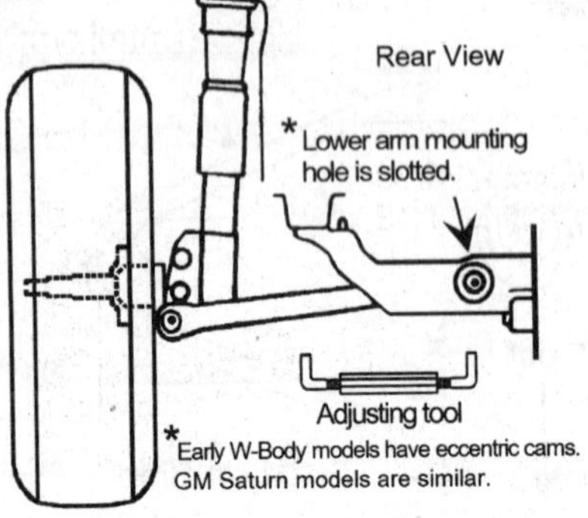

35 Rear CAMBER / TOE

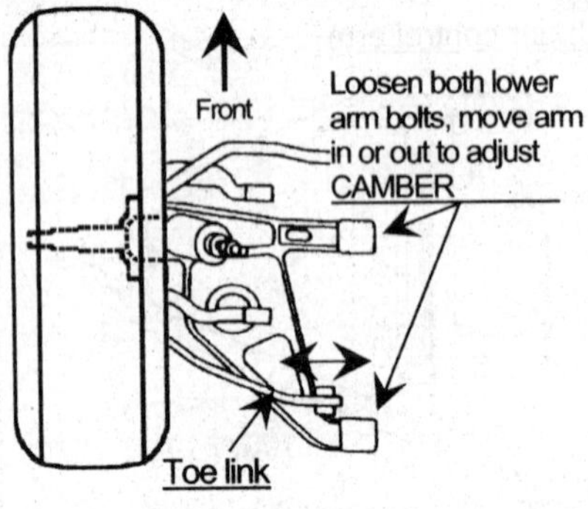

36 CASTER/CAMBER

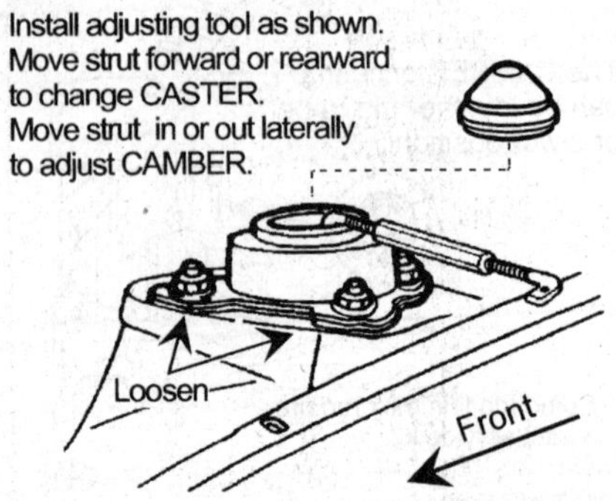

The above illustrations are intended to represent *typical* adjustment methods for the vehicles for which they are listed. Other methods of adjustment may be possible for some applications (usually requiring additional parts), but are not necessarily recommended by the vehicle manufacturer, or by the publisher or distributor of this manual.

37 CASTER/CAMBER

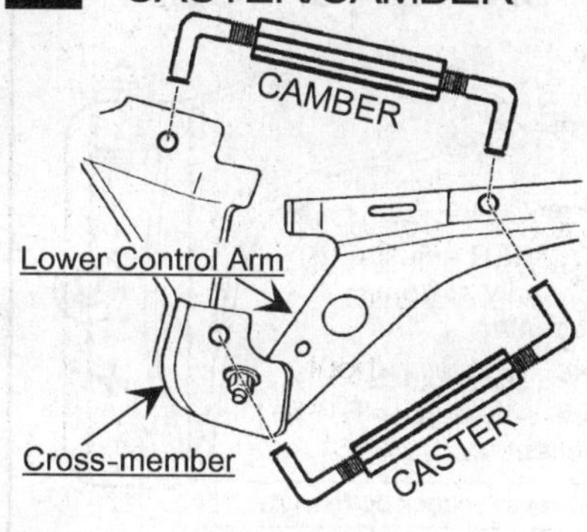

38 CASTER/CAMBER

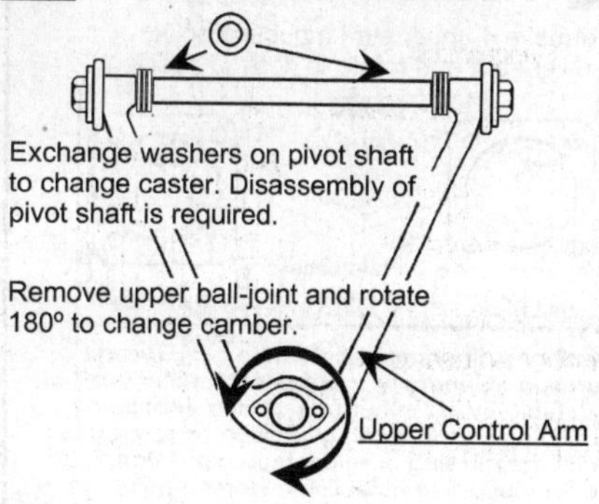

40 Rear CAMBER

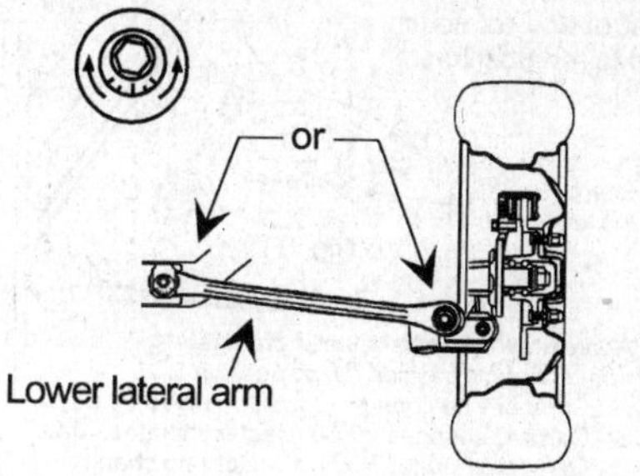

41 CASTER/CAMBER

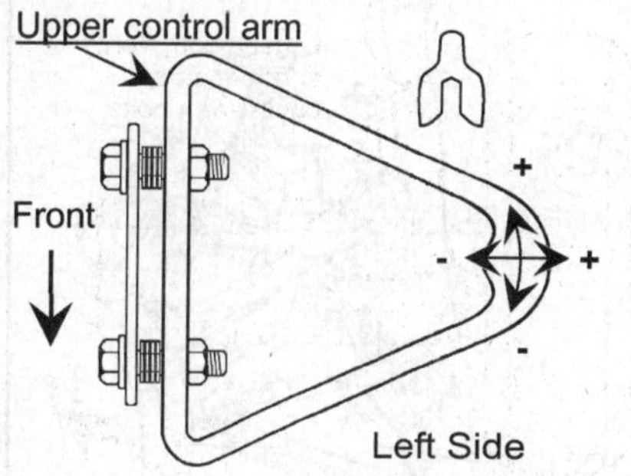

42 CASTER/CAMBER

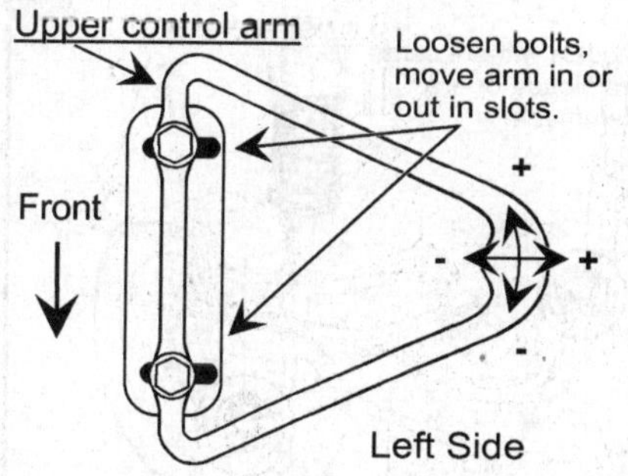

43 CASTER/CAMBER

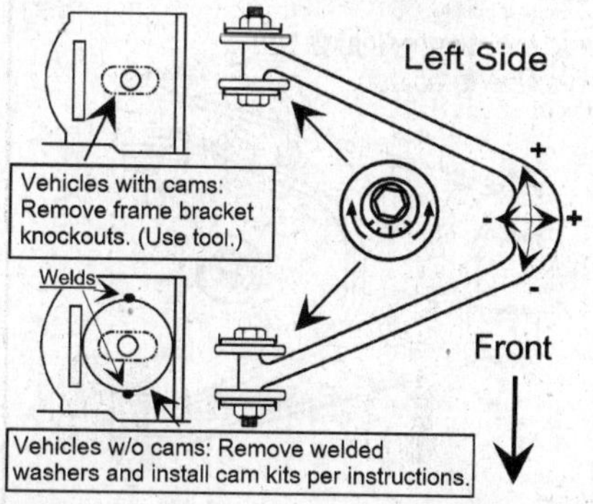

44 Rear CAMBER/TOE

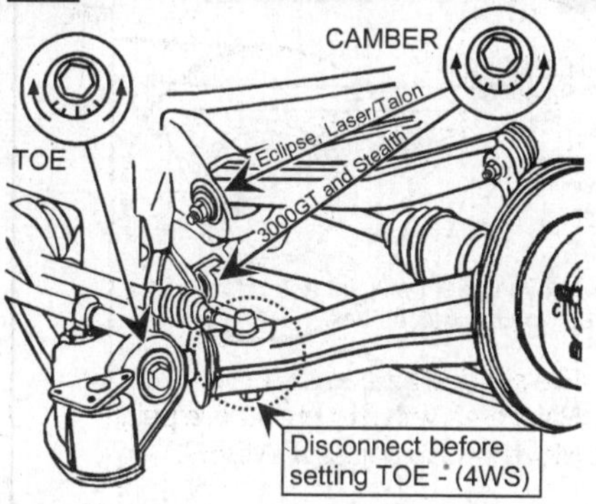

45 CASTER/CAMBER

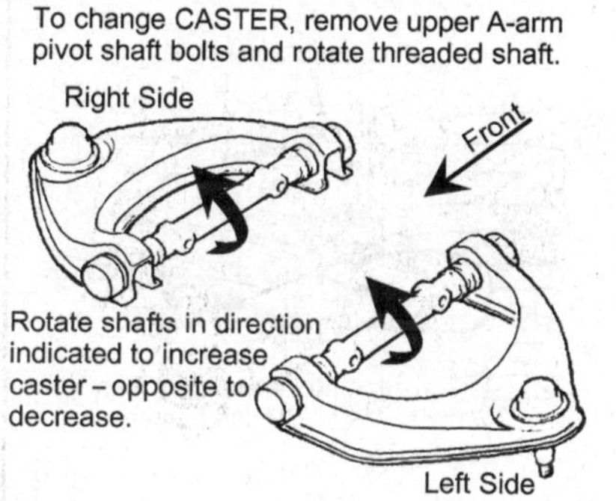

46 CASTER/CAMBER

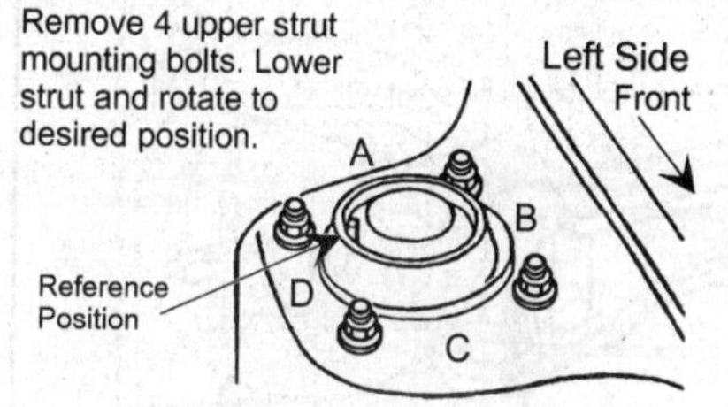

Approximate Camber/Caster change from standard position (D) to position:
A - Camber changes +.25, Caster changes +.25
B - Camber changes +.50, Caster - no change
C - Camber changes +.25, Caster changes -.25

47 Front CAMBER

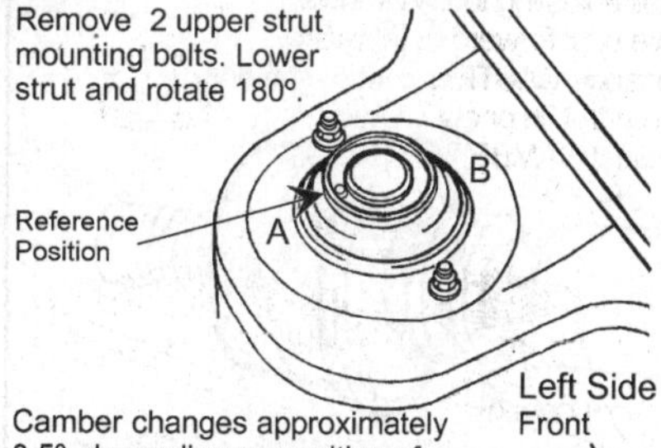

Camber changes approximately 0.5°, depending on position of reference mark. From A to B, change is negative.

48 CAMBER

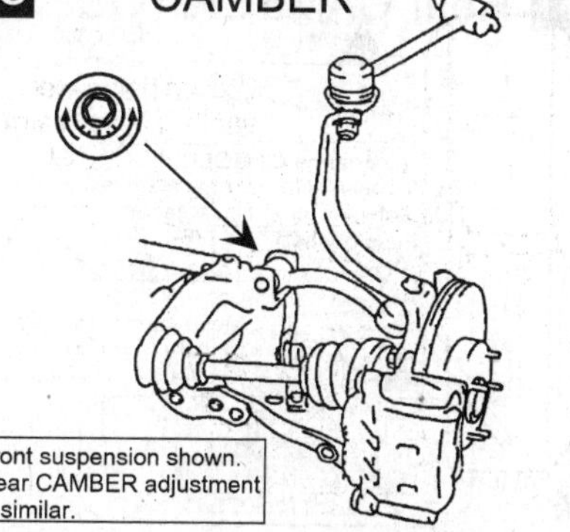

Front suspension shown. Rear CAMBER adjustment is similar.

49 Rear TOE

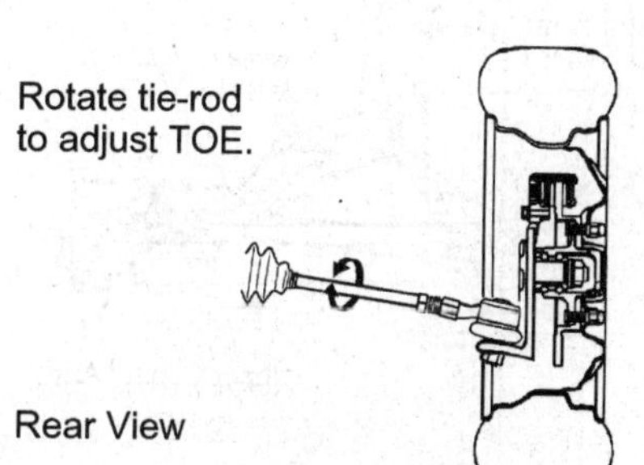

The above illustrations are intended to represent *typical* adjustment methods for the vehicles for which they are listed. Other methods of adjustment may be possible for some applications (usually requiring additional parts), but are not necessarily recommended by the vehicle manufacturer, or by the publisher or distributor of this manual.

50 CASTER/CAMBER

Remove 4 upper strut mounting bolts. Lower strut and rotate to desired position.

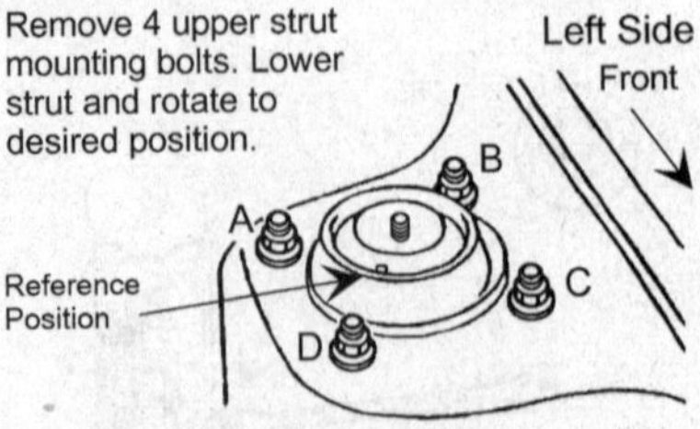

Approximate Camber/Caster change from standard position (D) to position:
A - Camber - no change, Caster changes +0.50
B - Camber changes +0.50, Caster changes +0.50
C - Camber changes +0.50, Caster - no change

51 CASTER/CAMBER

Remove 4 upper strut mounting bolts. Lower strut and rotate to desired position.

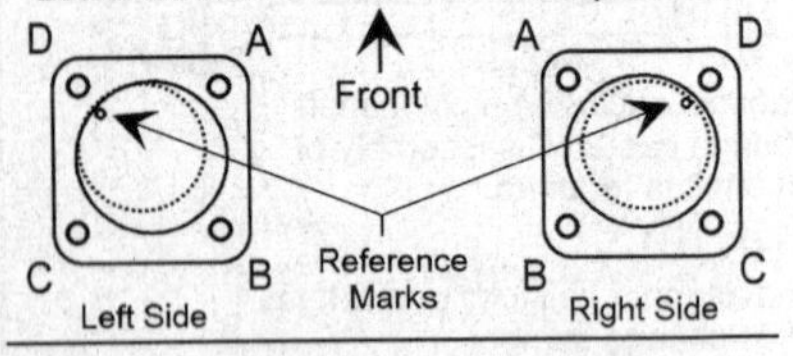

The above diagrams represent a *typical* configuration. Generally, camber and caster are adjustable by ½° either pos. or neg. (not both), depending on the starting position. If reference marks are missing or relocated, rotate strut 180° and remeasure to establish a starting point.

52 CASTER/CAMBER
TYPICAL

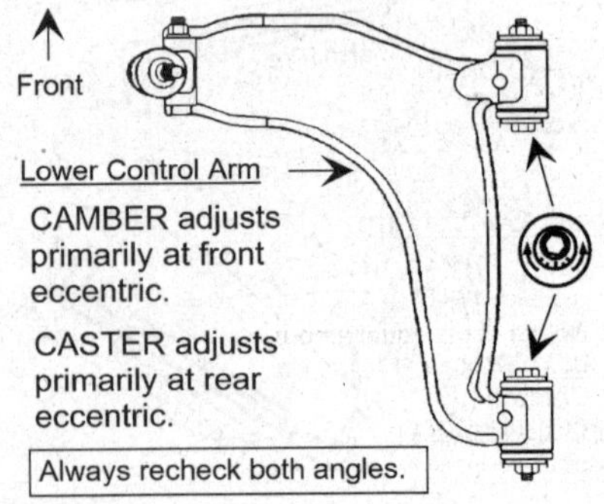

CAMBER adjusts primarily at front eccentric.

CASTER adjusts primarily at rear eccentric.

Always recheck both angles.

53 CAMBER/TOE

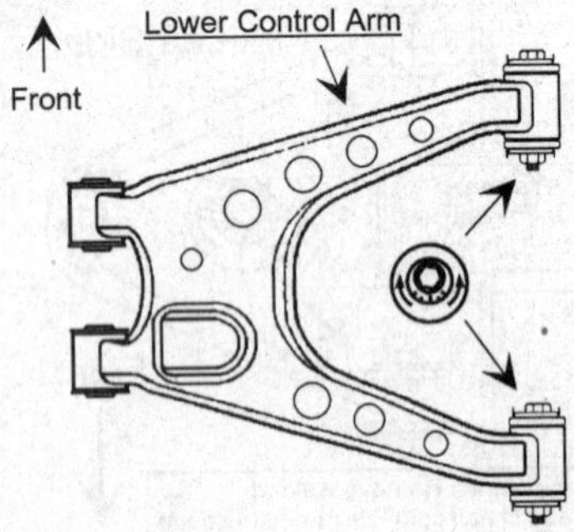

54 Rear TOE

* Models without cams are slotted or are shimmable.

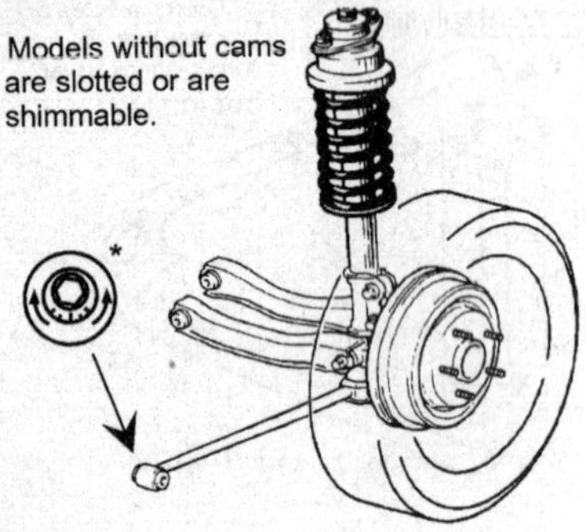

55 Rear CAMBER/TOE

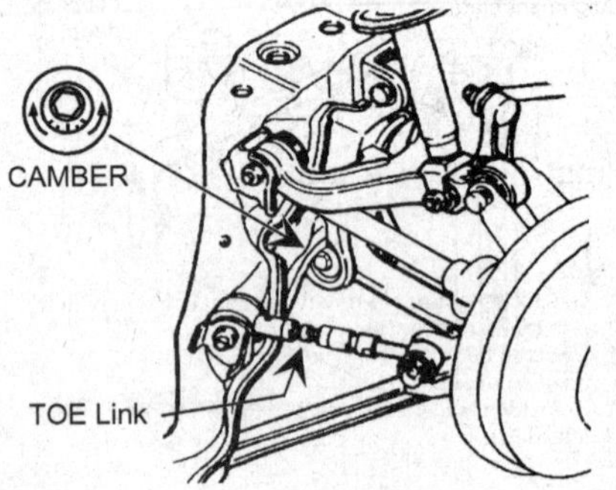

56 CASTER/CAMBER

Service adjuster kits are available to change caster and camber. Follow instructions included with kit.

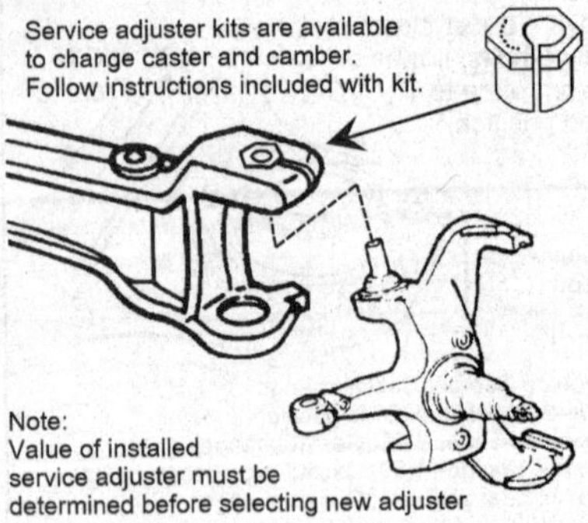

Note:
Value of installed service adjuster must be determined before selecting new adjuster

58 CAMBER

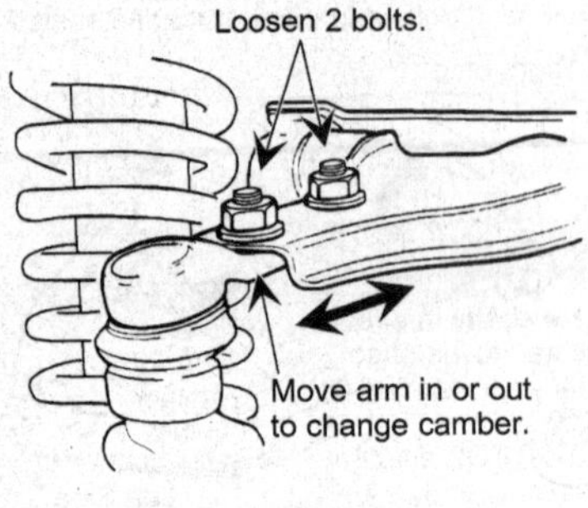

59 Front CAMBER

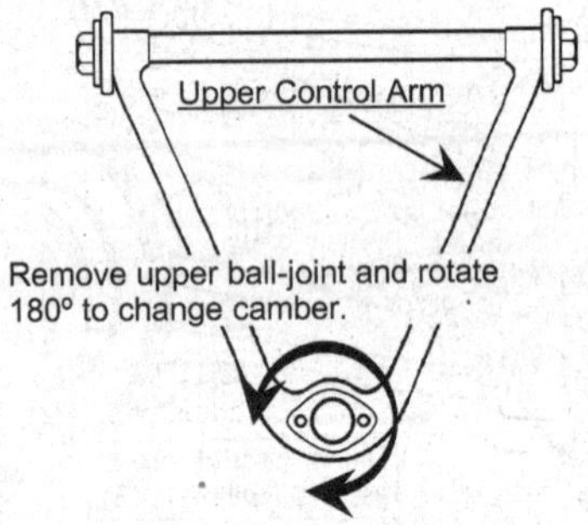

60 CASTER/CAMBER

Add or remove shims to adjust.

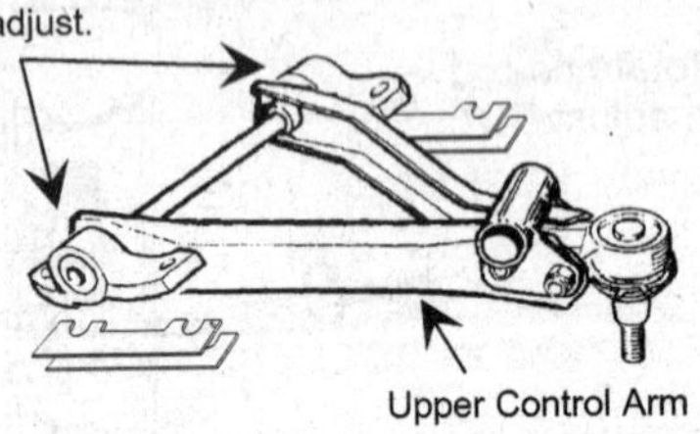

61 Rear CAMBER/TOE

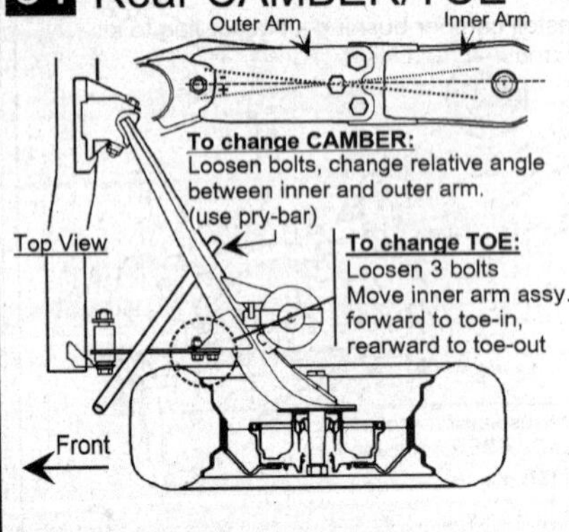

62 Front CASTER

Stabilizer bar - lower control arm end.

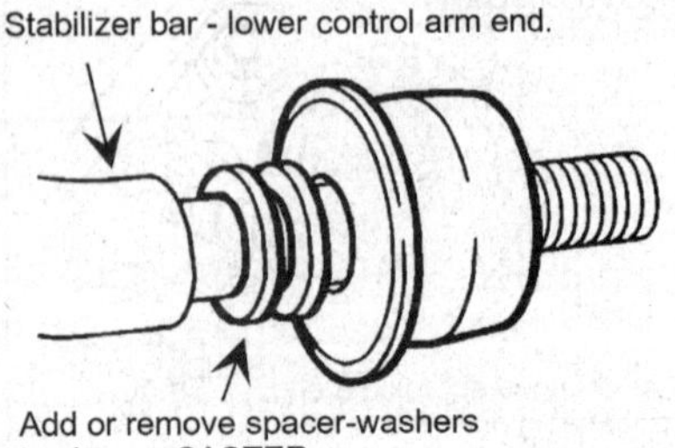

Add or remove spacer-washers to change CASTER.

The above illustrations are intended to represent *typical* adjustment methods for the vehicles for which they are listed. Other methods of adjustment may be possible for some applications (usually requiring additional parts), but are not necessarily recommended by the vehicle manufacturer, or by the publisher or distributor of this manual.

63 Rear CAMBER/TOE

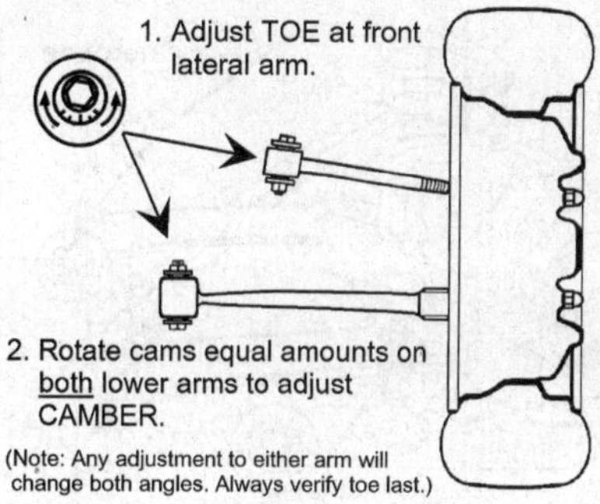

(Note: Any adjustment to either arm will change both angles. Always verify toe last.)

64 Rear TOE

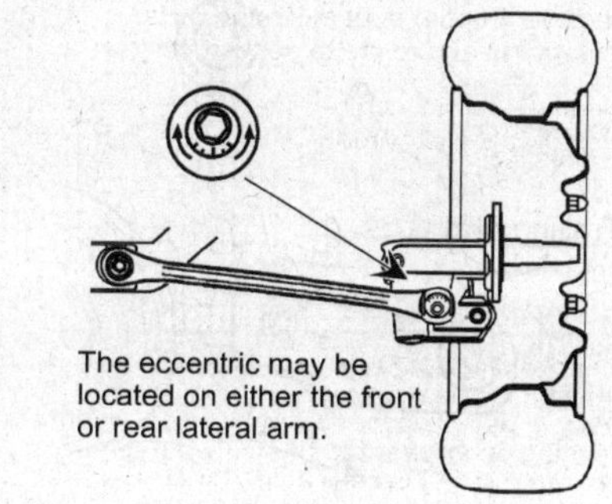

The eccentric may be located on either the front or rear lateral arm.

65 Rear CAMBER/TOE

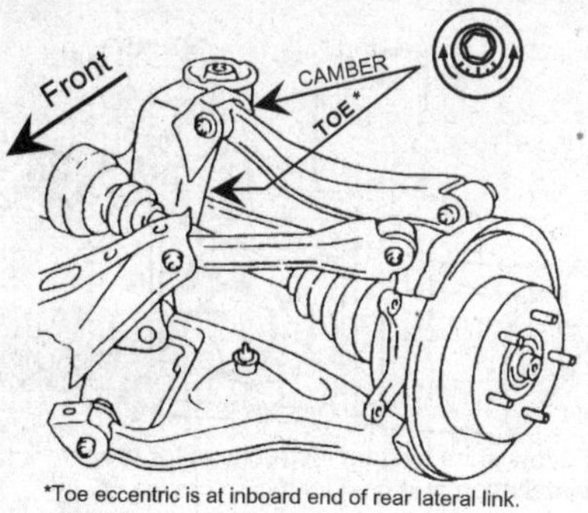

*Toe eccentric is at inboard end of rear lateral link.

67 CASTER/CAMBER

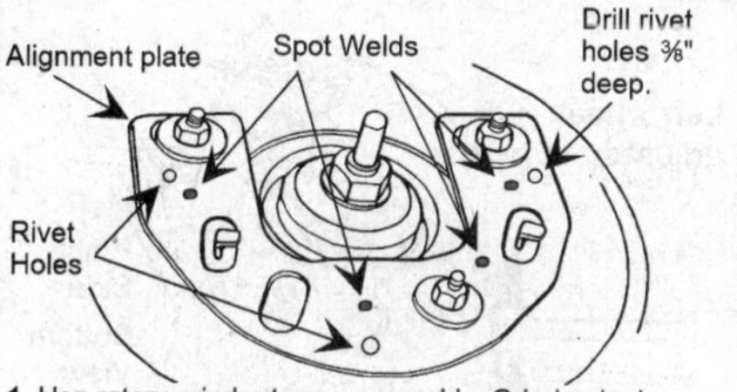

1. Use rotary grinder to remove welds. Grind only deep enough to remove alignment plate.
2. Remove burrs from shock tower and plate, then paint all exposed metal.
3. Reinstall plate and set CASTER/CAMBER.
4. Install (3) ⅛" (¼" grip) pop rivets as indicated.

68 CASTER/CAMBER

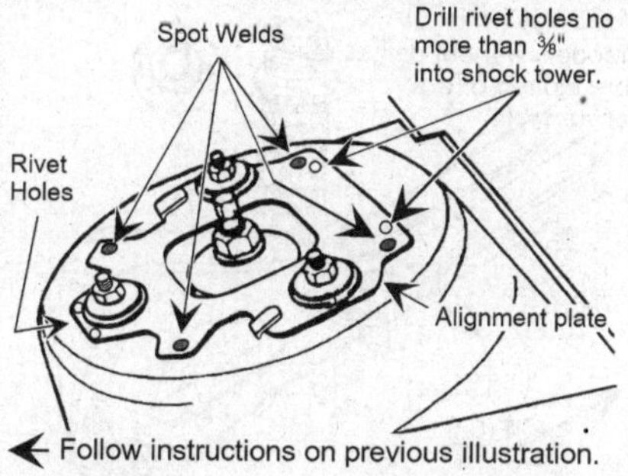

← Follow instructions on previous illustration.

69 Rear TOE

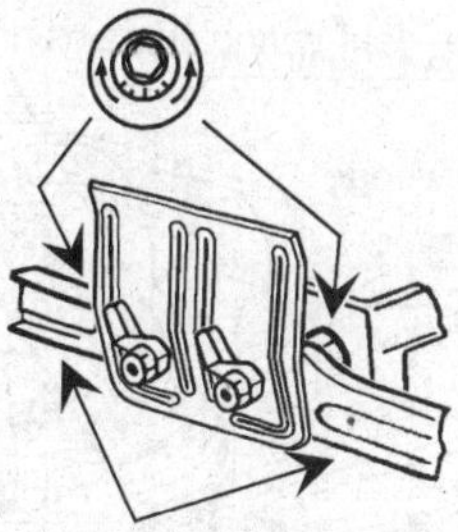

70 CASTER/CAMBER

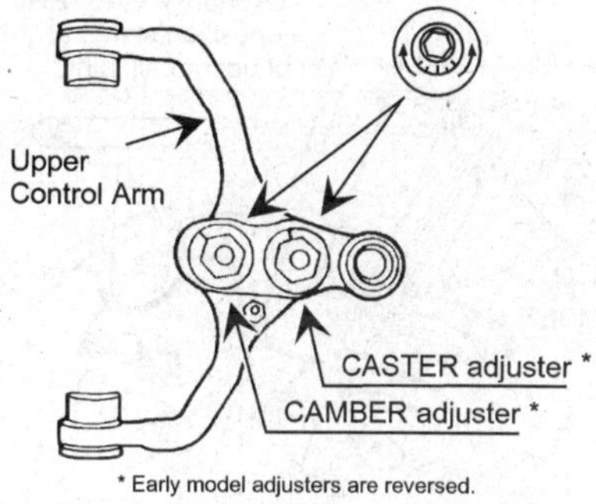

* Early model adjusters are reversed.

71 Rear CAMBER/TOE

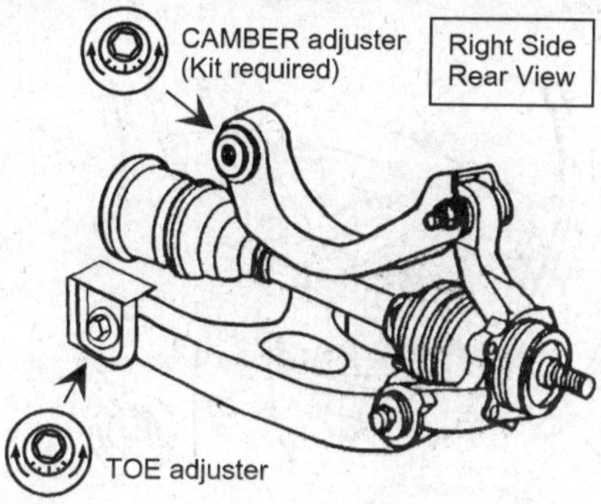

72 Front CAMBER

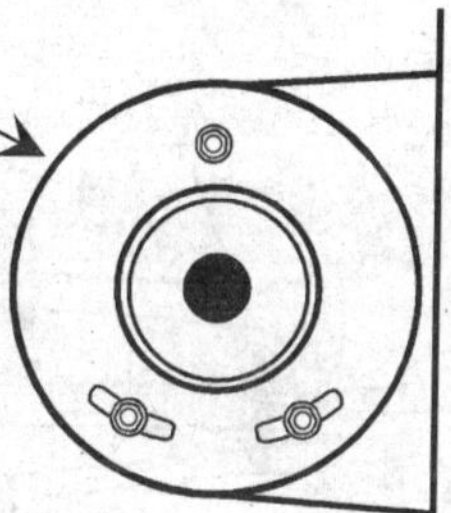

Loosen 3 upper strut nuts.
Move strut in or out to set CAMBER.

73 CASTER/CAMBER

Full adjustment of caster and camber requires the installation of service adjuster cams on both upper control arms.

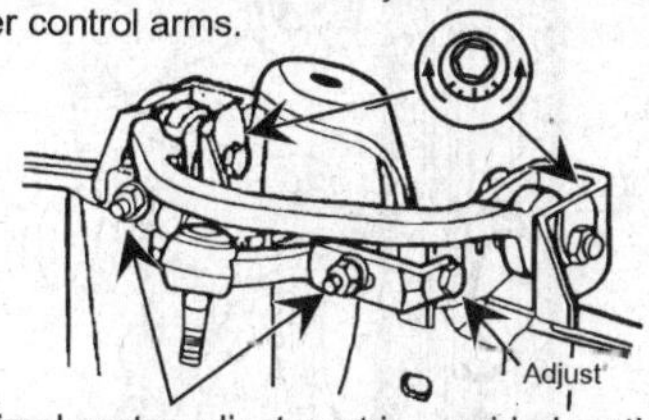

Minimal caster adjustment is provided on the right side only for adjustment of caster split (Cross). Adjust slider bolt to move arm forward or rearward.

74 Front CAMBER

Install camber bushing kit according to kit manufacturer's instructions.

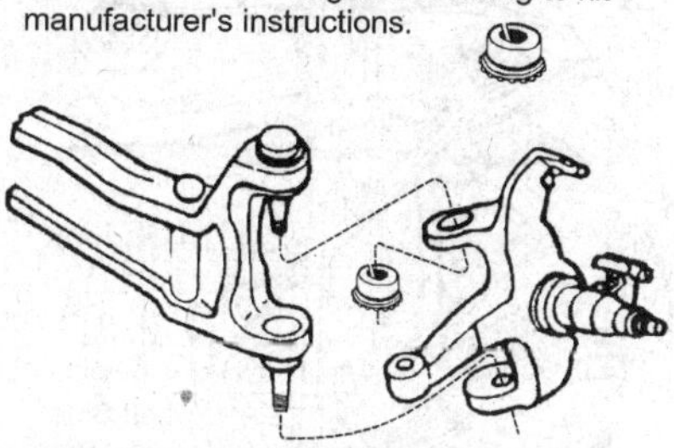

75 CASTER/CAMBER

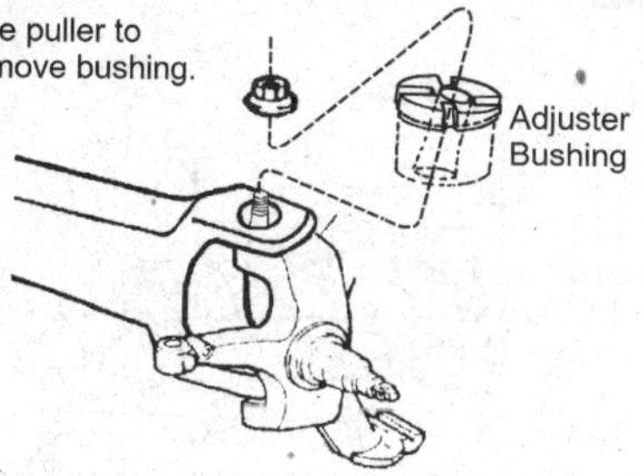

The above illustrations are intended to represent *typical* adjustment methods for the vehicles for which they are listed. Other methods of adjustment may be possible for some applications (usually requiring additional parts), but are not necessarily recommended by the vehicle manufacturer, or by the publisher or distributor of this manual.

ADJUSTMENT METHODS

Adjustment Methods

76 Rear CAMBER/TOE

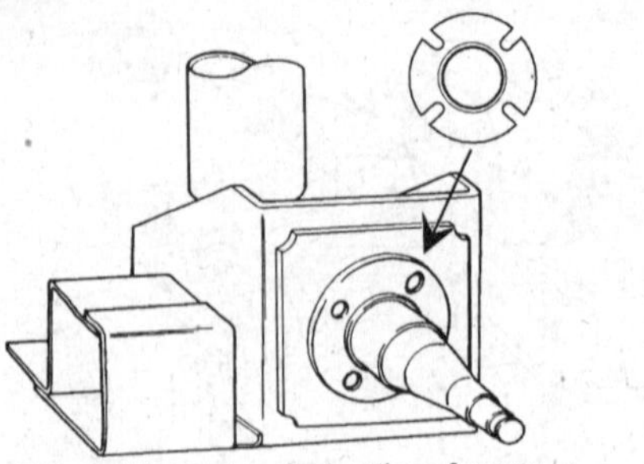

Follow shim system instructions for proper installation and positioning.

77 Rear TOE

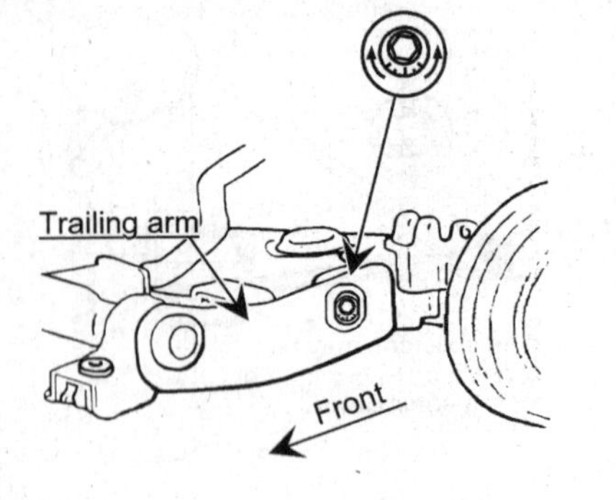

78 Front CASTER

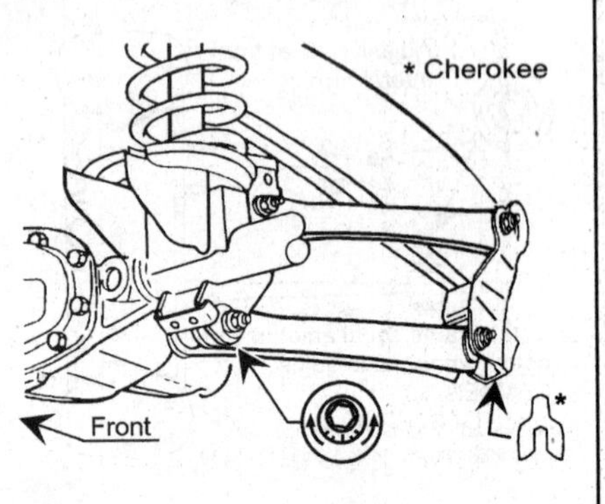

79 Front CASTER

Solid axle - leaf spring suspension.

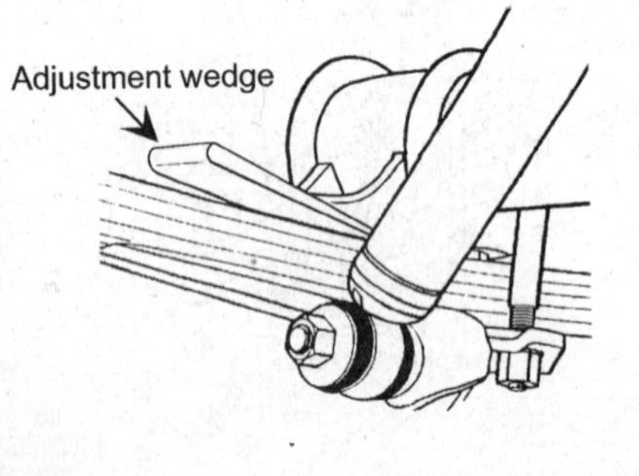

80 Rear TOE

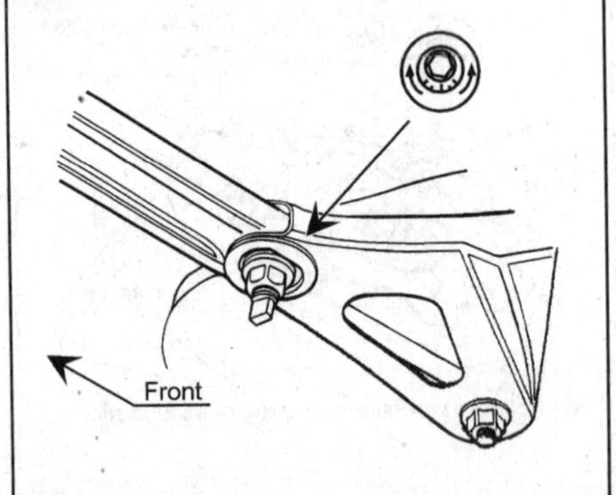

81 Rear TOE

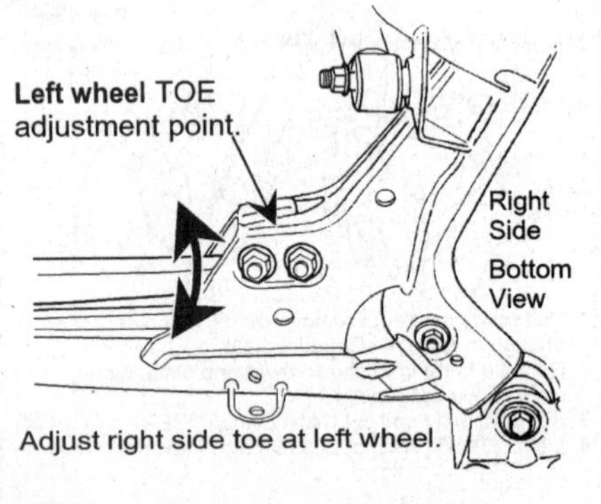

82 Rear CAMBER/TOE

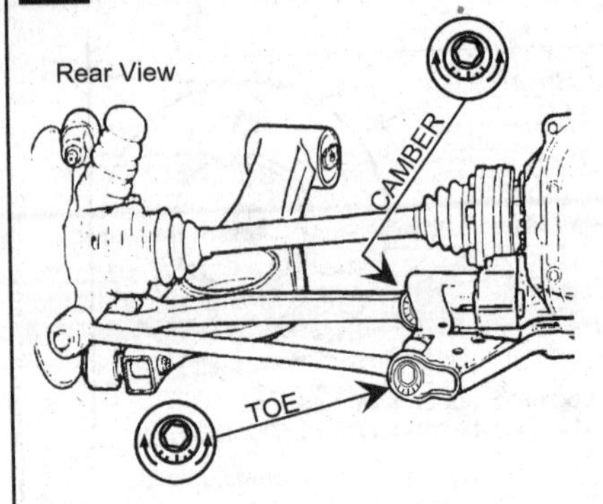

83 Rear CAMBER

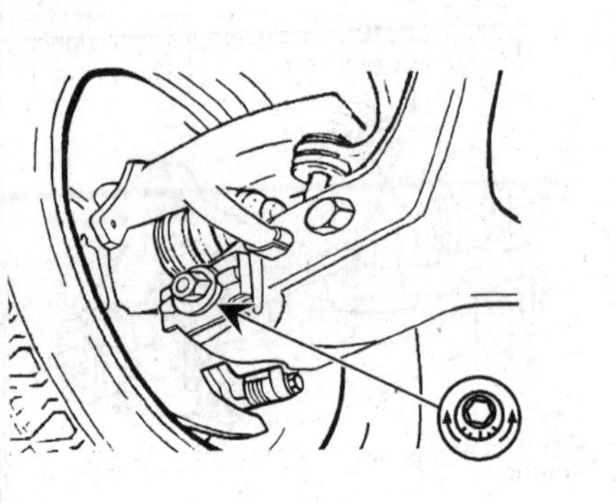

84 CASTER/CAMBER

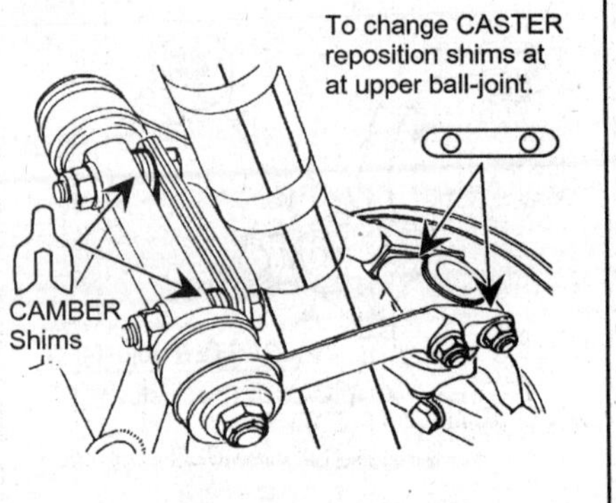

85 Rear CAMBER

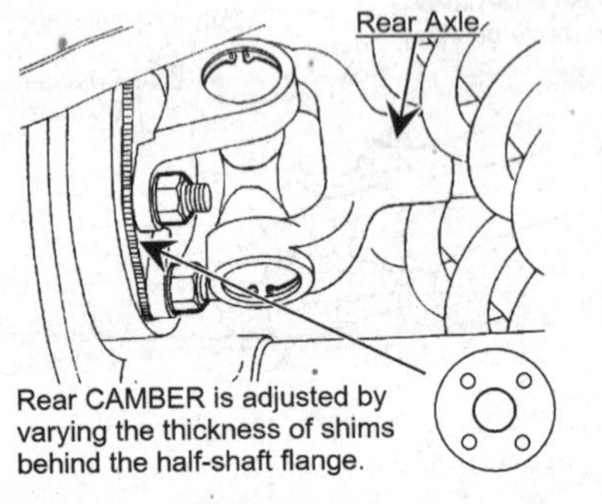

Rear CAMBER is adjusted by varying the thickness of shims behind the half-shaft flange.

86 Rear TOE

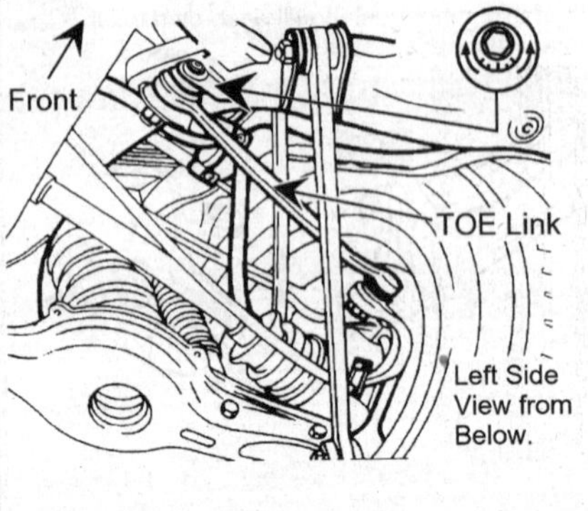

87 CASTER/CAMBER

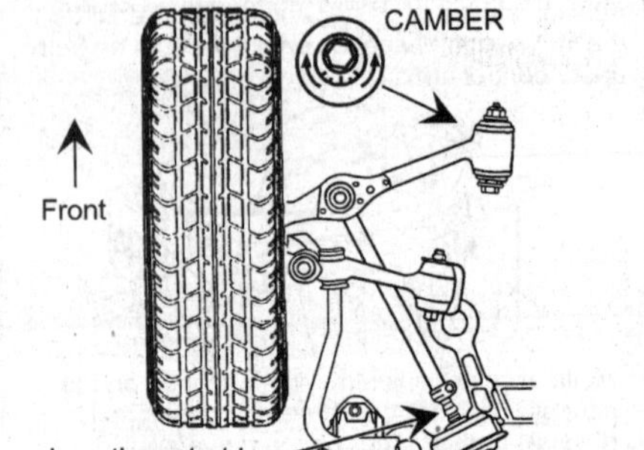

Lengthen strut to increase CASTER, shorten to decrease.

The above illustrations are intended to represent *typical* adjustment methods for the vehicles for which they are listed. Other methods of adjustment may be possible for some applications (usually requiring additional parts), but are not necessarily recommended by the vehicle manufacturer, or by the publisher or distributor of this manual.

88 CASTER/CAMBER

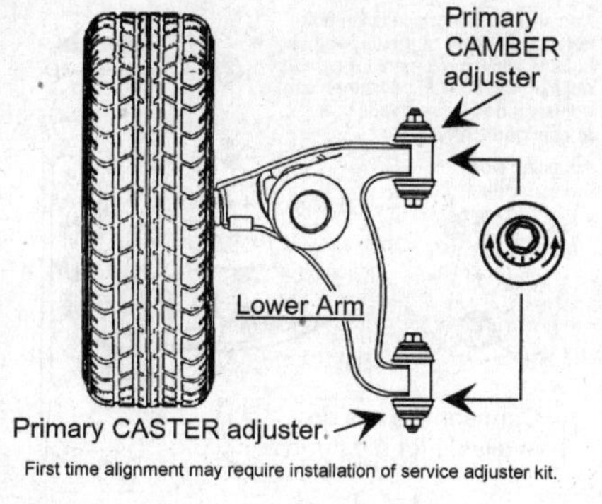

89 CASTER/CAMBER

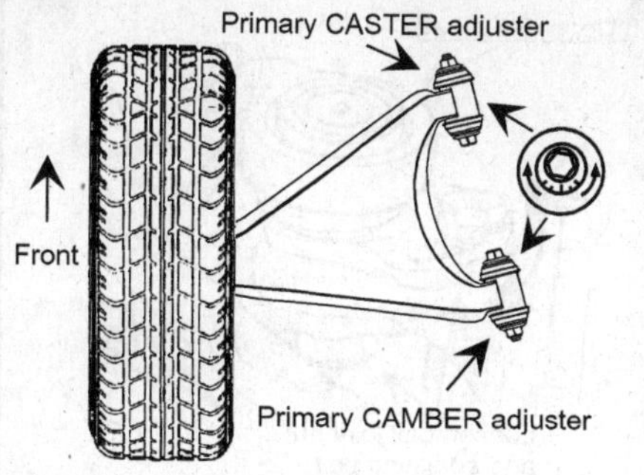

90 Front CASTER

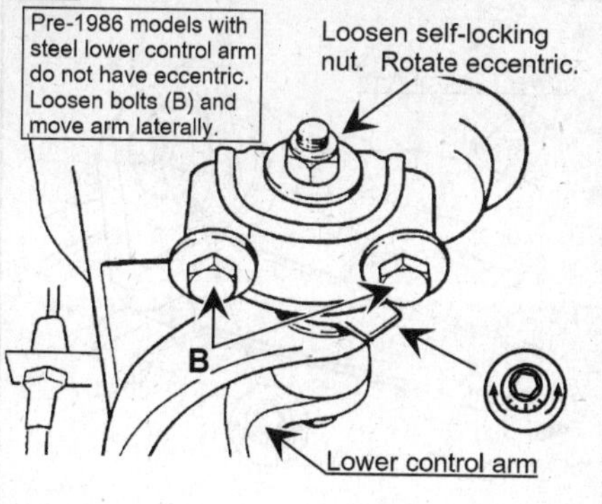

91 Rear CAMBER

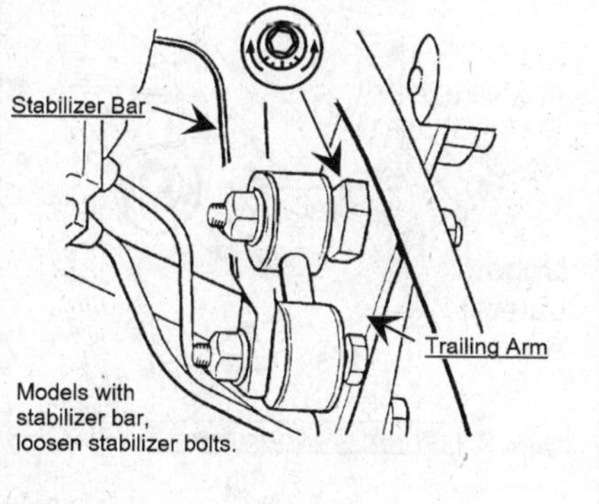

92 Rear TOE

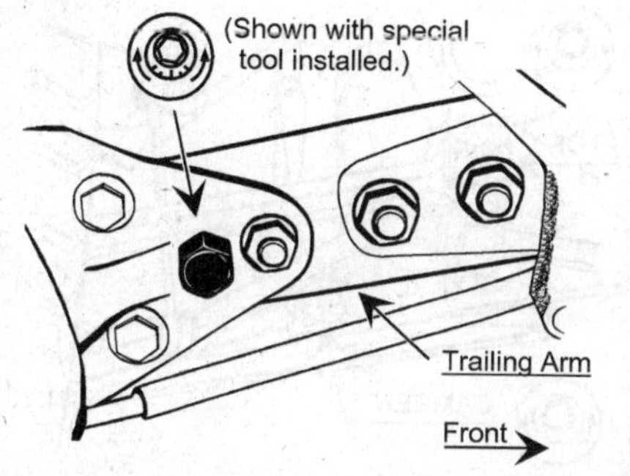

93 CASTER/CAMBER

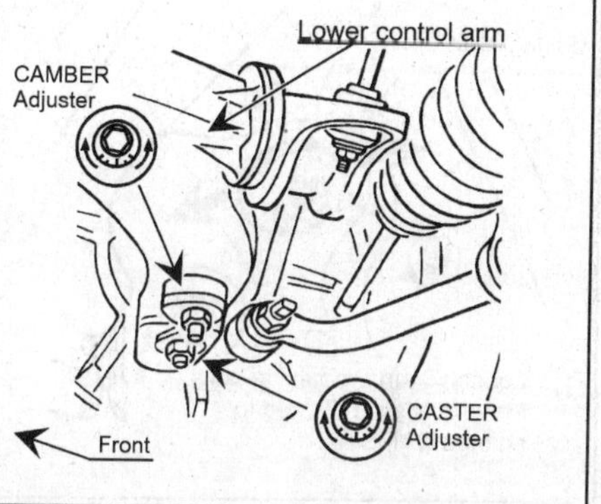

94 Rear CAMBER/TOE

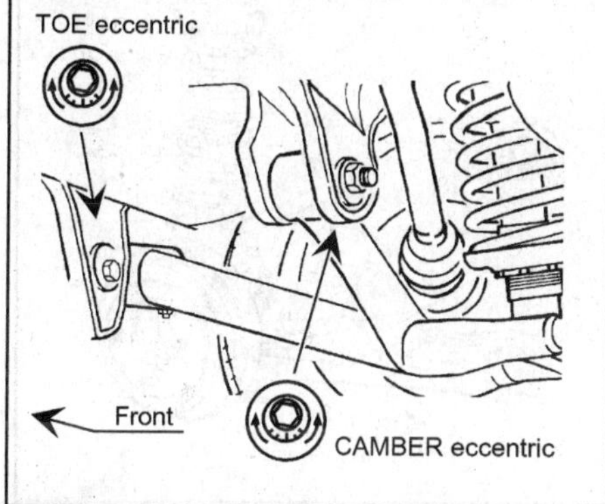

95 Rear CAMBER/TOE

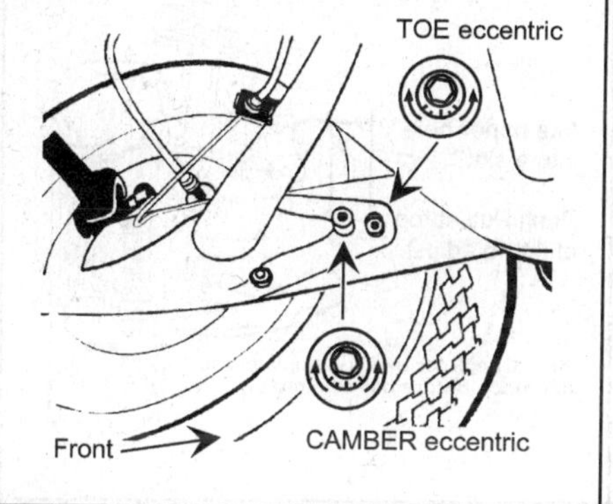

96 CASTER/CAMBER

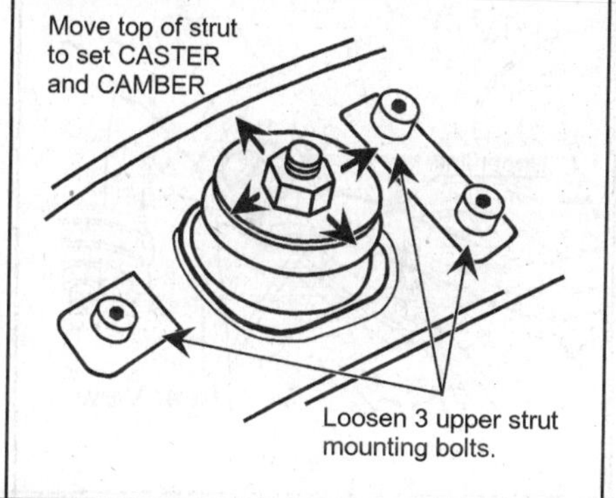

97 CASTER/CAMBER

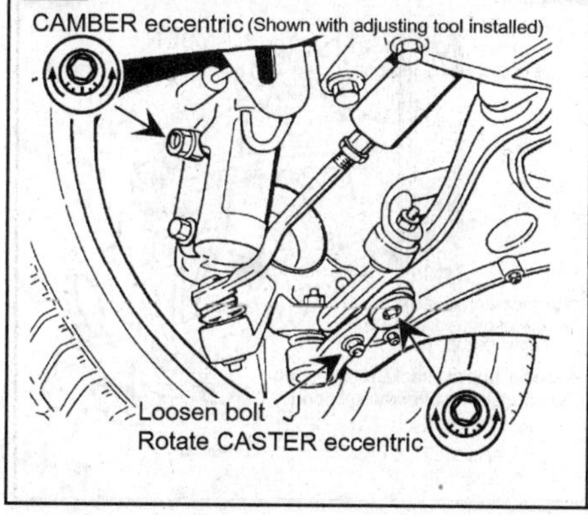

98 Rear CAMBER/TOE

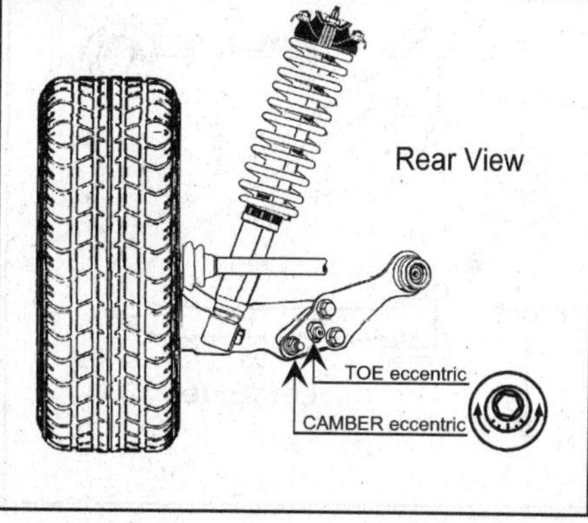

99 Rear CAMBER/TOE

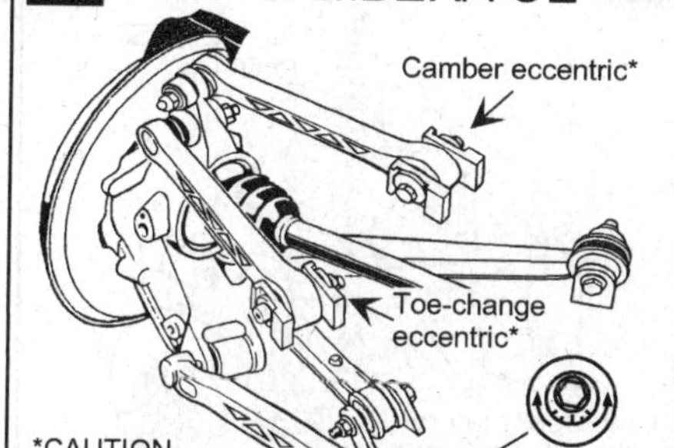

*CAUTION
See Porsche section
in Special Instructions

The above illustrations are intended to represent *typical* adjustment methods for the vehicles for which they are listed. Other methods of adjustment may be possible for some applications (usually requiring additional parts), but are not necessarily recommended by the vehicle manufacturer, or by the publisher or distributor of this manual.

100 Front CASTER

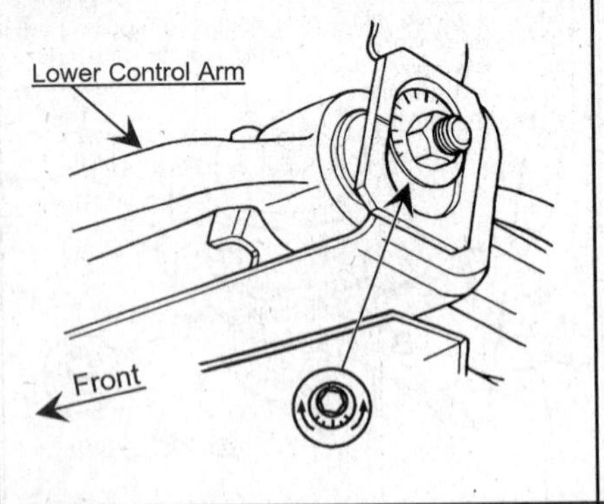

101 Front CAMBER

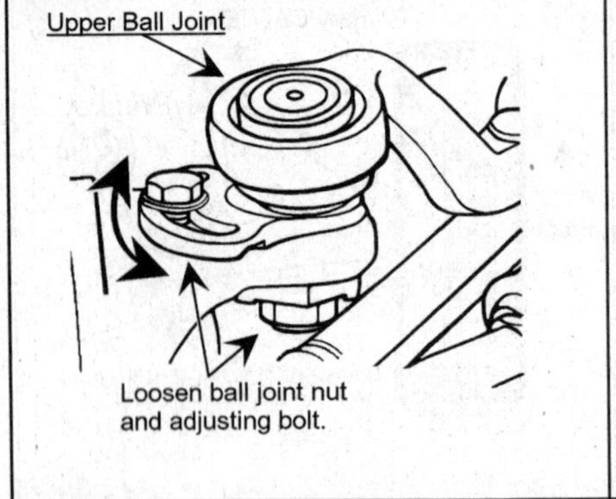

102 CAMBER

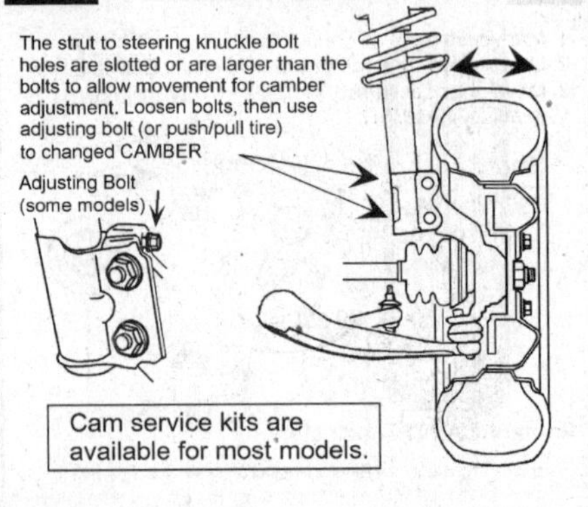

103 Front CAMBER

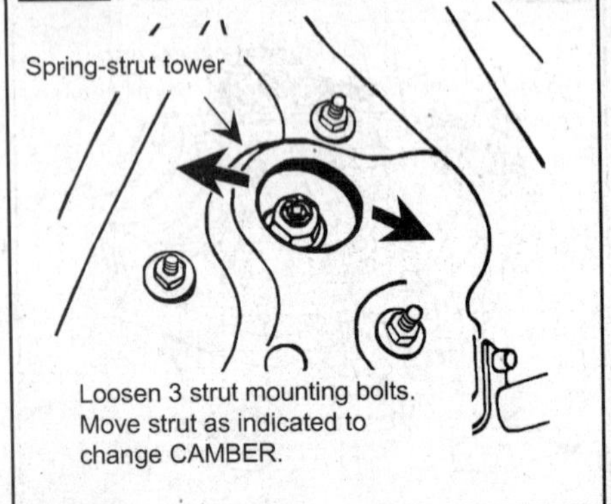

104 Rear CAMBER/TOE

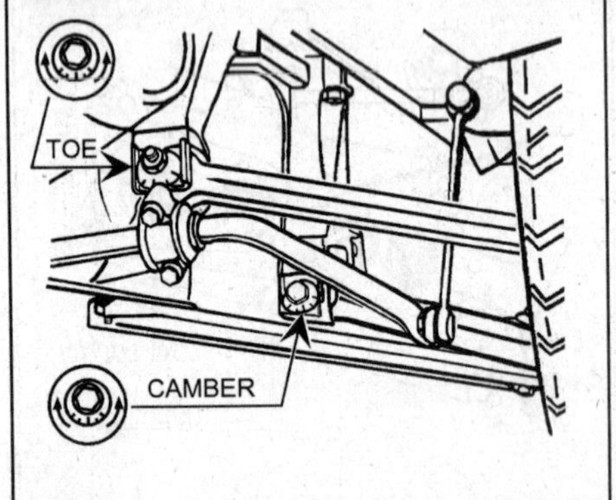

106 Rear CAMBER

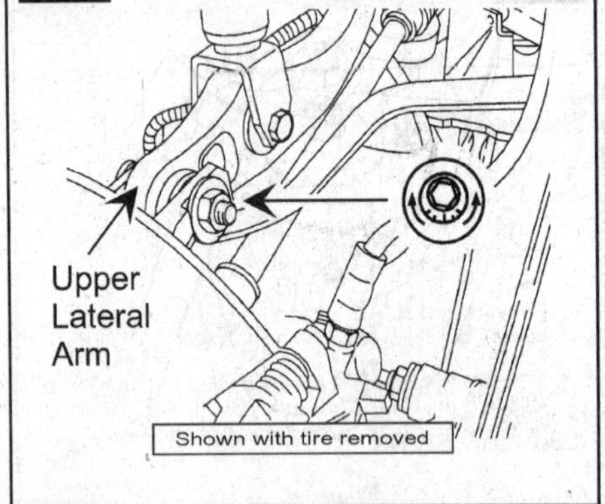

107 Rear CAMBER

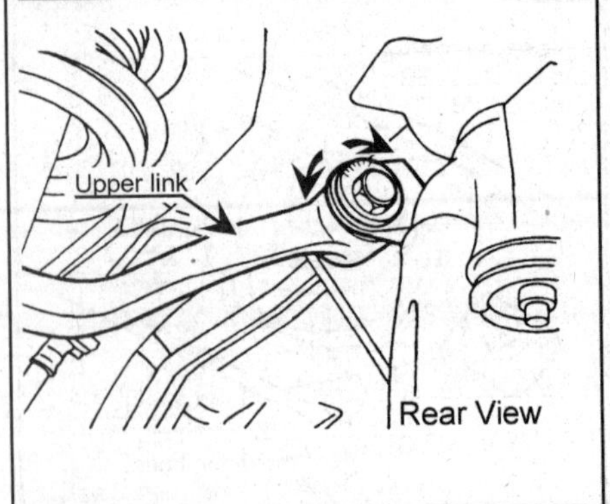

108 CAMBER

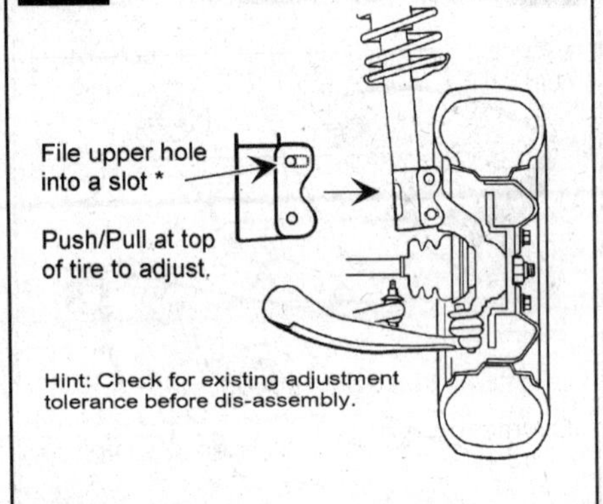

109 Rear TOE

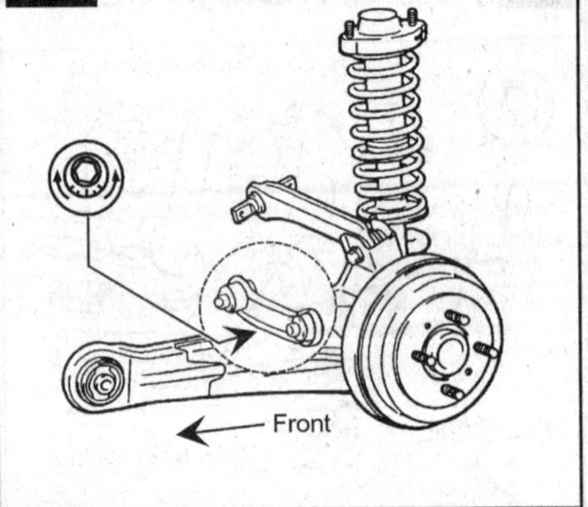

110 Rear CAMBER/TOE

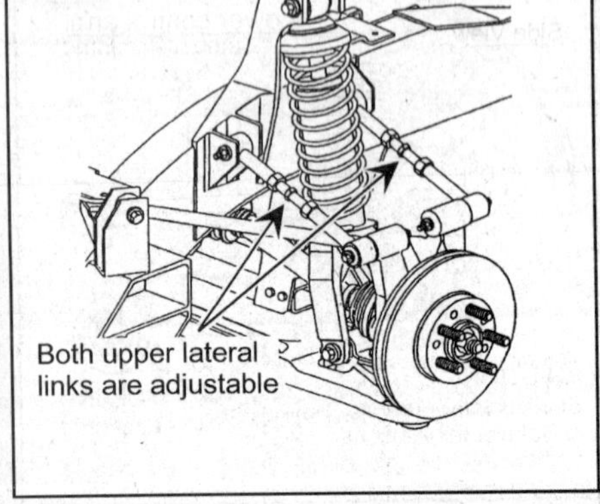

111 Rear CAMBER

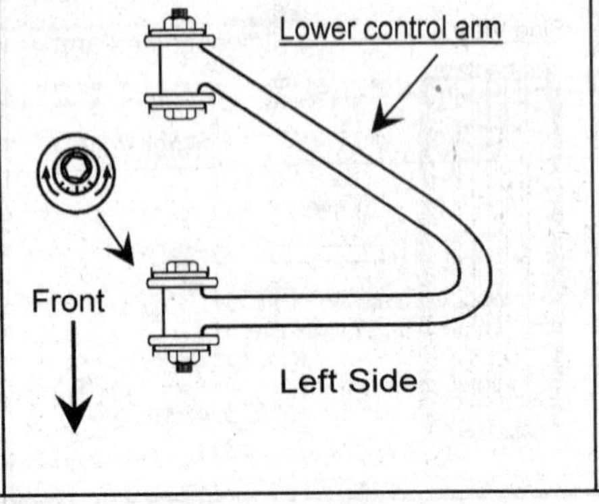

112 CASTER/CAMBER

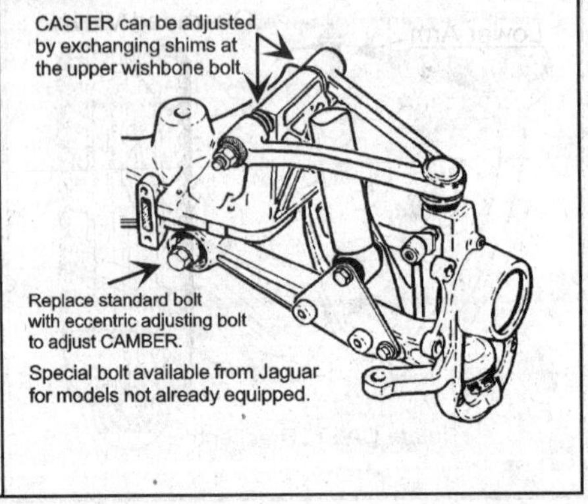

The above illustrations are intended to represent *typical* adjustment methods for the vehicles for which they are listed. Other methods of adjustment may be possible for some applications (usually requiring additional parts), but are not necessarily recommended by the vehicle manufacturer, or by the publisher or distributor of this manual.

113 Front CAMBER

1. Mark position of all 4 mounting bolts.
2. Loosen bolts and remove plastic locating pins.
3. Move shock assembly in or out to adjust CAMBER.

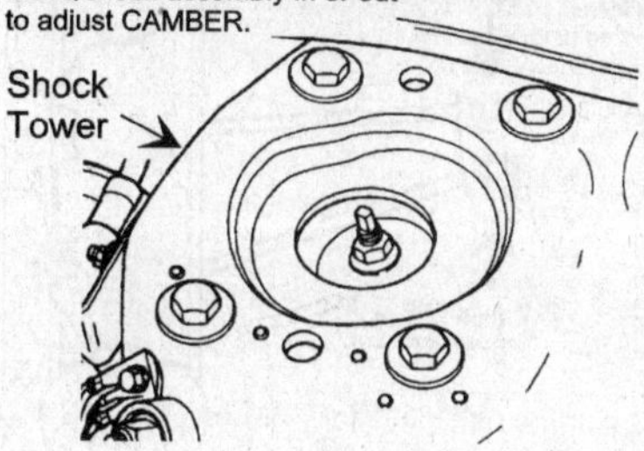

114 Rear TOE

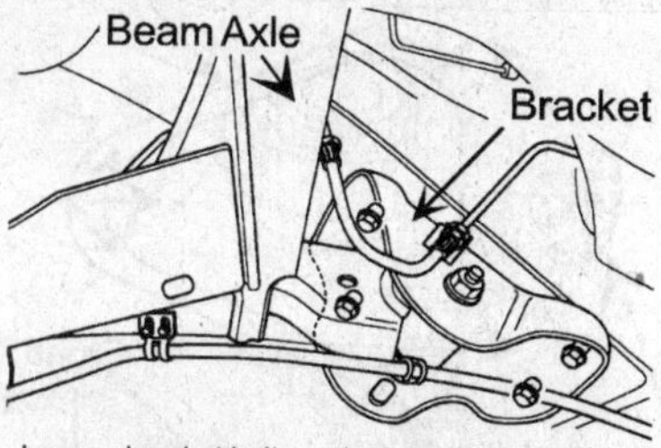

Loosen bracket bolts on both ends of axle.
Move axle assembly laterally to equalize TOE.
Replace bolts (if available), torque to 55 ft.lbs.

115 Rear CAMBER

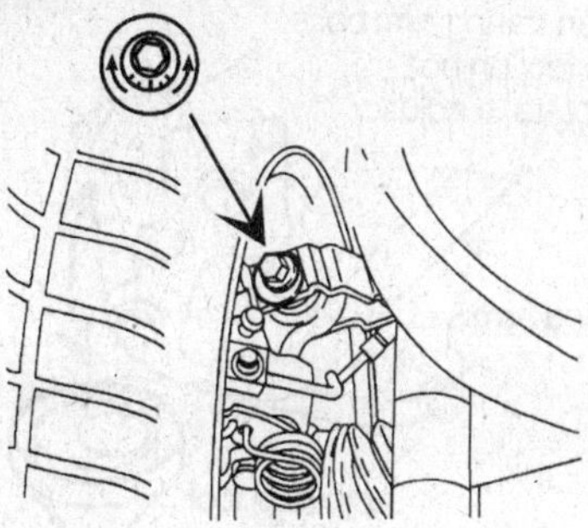

116 CASTER/CAMBER

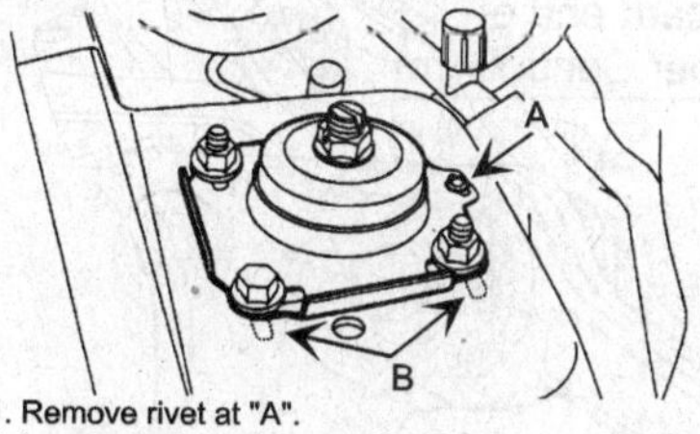

1. Remove rivet at "A".
2. Loosen bolts and move adjusting plate to change CAMBER.
3. If CASTER adjustment is necessary, remove the adjusting plate, lower the strut and file slots (up to 5mm - see "B") in the direction necessary.

117 Rear CAMBER/TOE

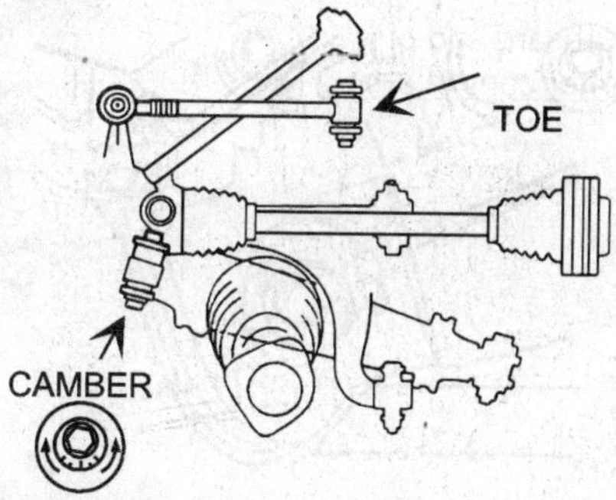

118 CASTER/CAMBER

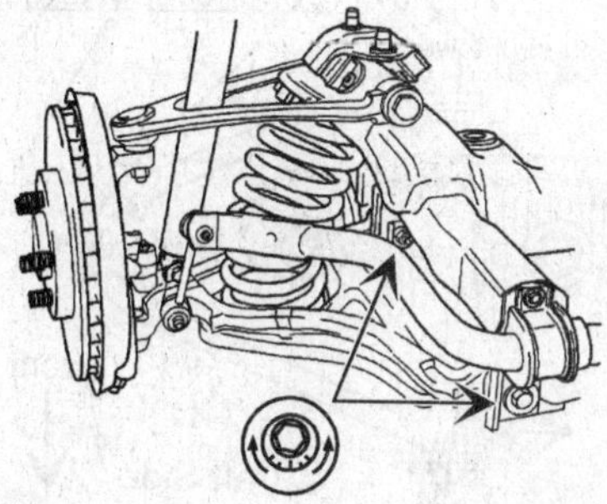

119 Front CAMBER

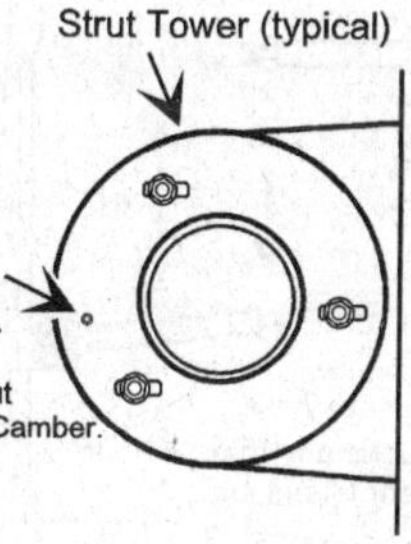

120 Rear CAMBER/TOE

Note: Any adjustment changes both angles. Always adjust TOE last.

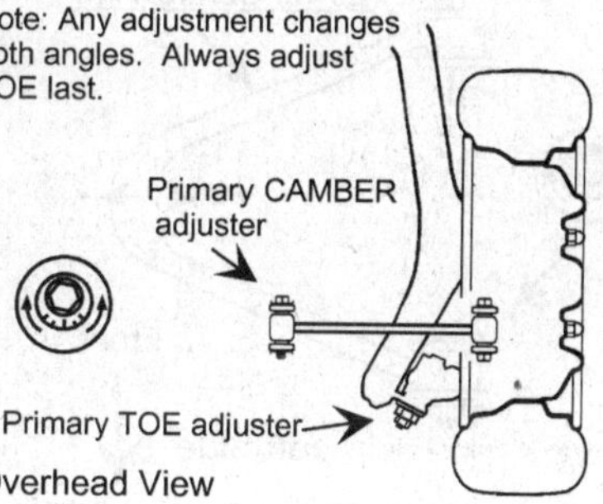

121 CASTER/CAMBER

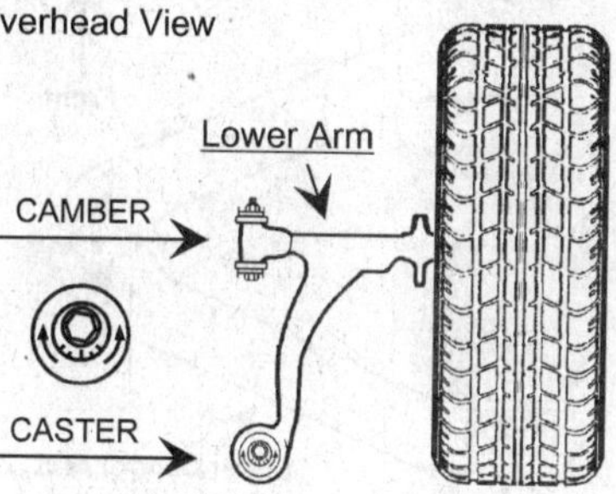

122 Rear CAMBER/TOE

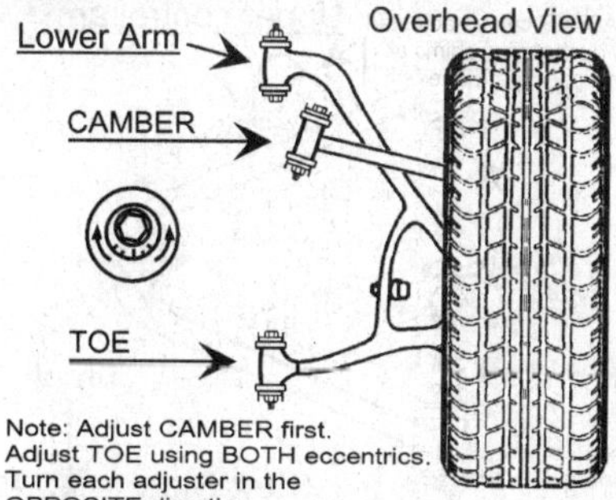

Note: Adjust CAMBER first.
Adjust TOE using BOTH eccentrics.
Turn each adjuster in the OPPOSITE direction.

123 Front CASTER

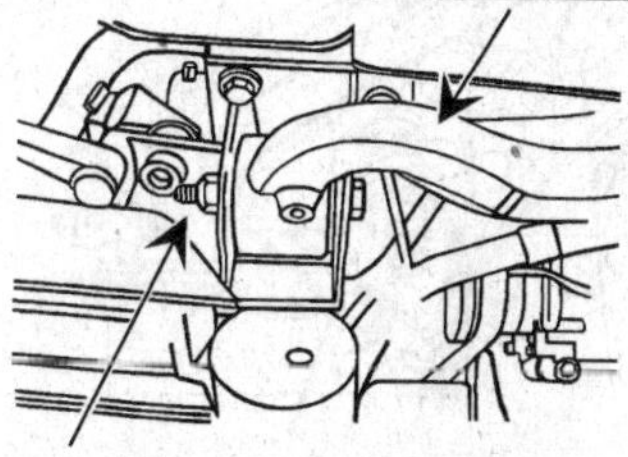

Install adjusting cam bolt and nut as indicated with cam lobes down.

124 Front CAMBER

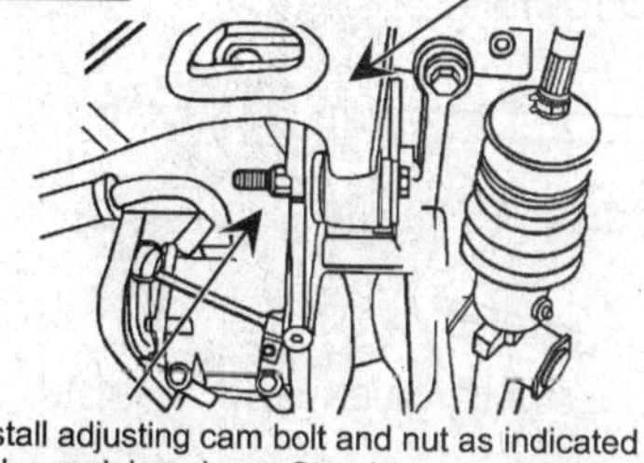

Install adjusting cam bolt and nut as indicated with cam lobes down. Steering gear may need to be lowered to allow installation of adjuster.

The above illustrations are intended to represent *typical* adjustment methods for the vehicles for which they are listed. Other methods of adjustment may be possible for some applications (usually requiring additional parts), but are not necessarily recommended by the vehicle manufacturer, or by the publisher or distributor of this manual.

125 Rear TOE

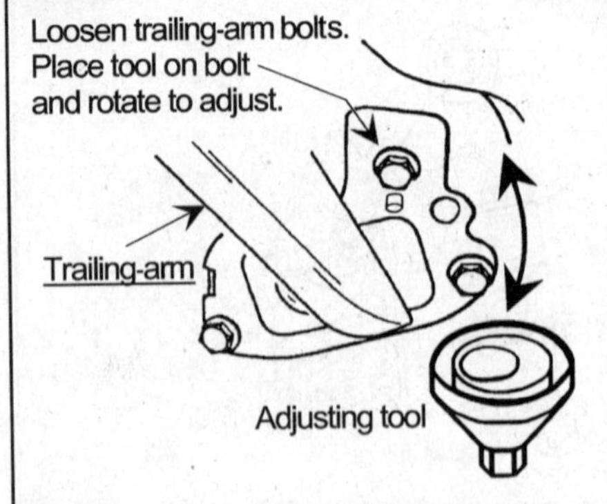

126 CASTER/CAMBER

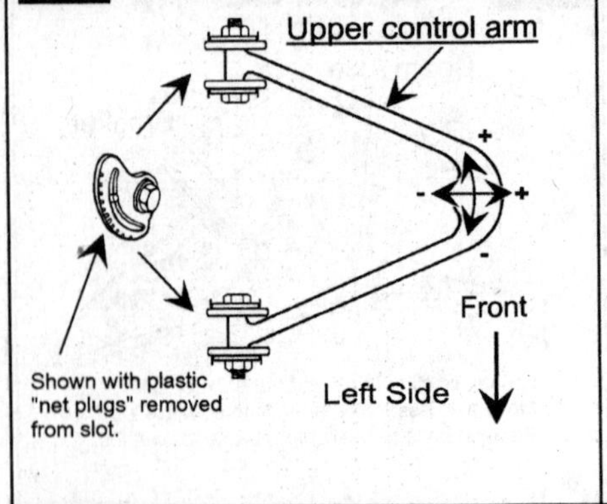

127 Rear CAMBER/TOE

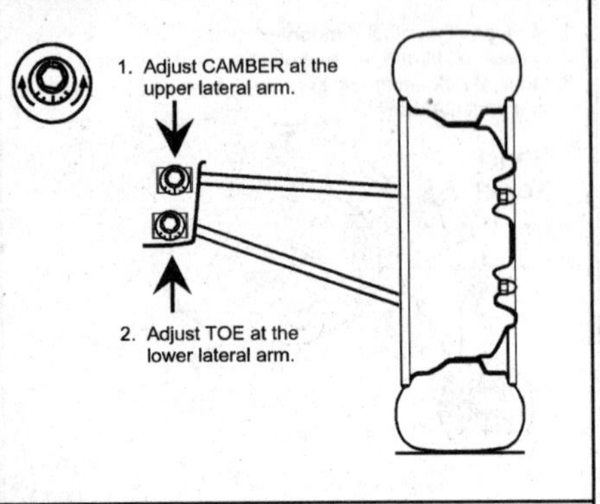

128 CASTER/CAMBER

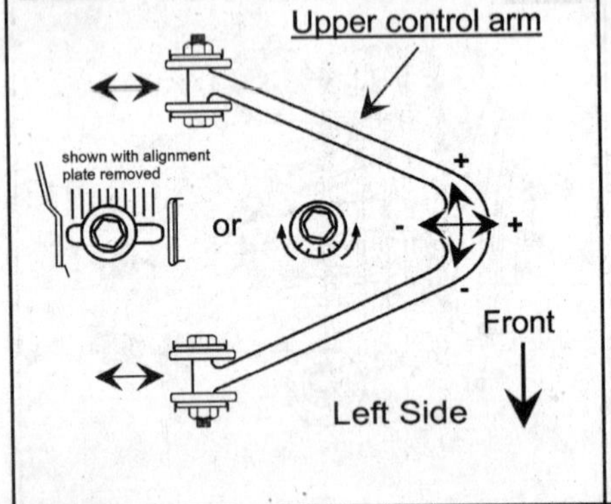

129 Rear CAMBER

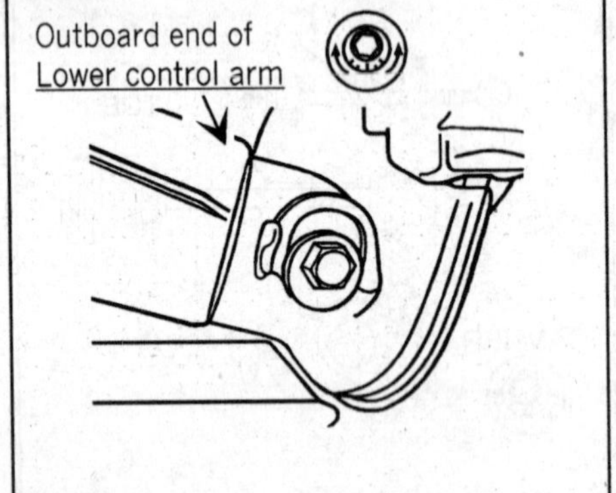

130 Rear CAMBER

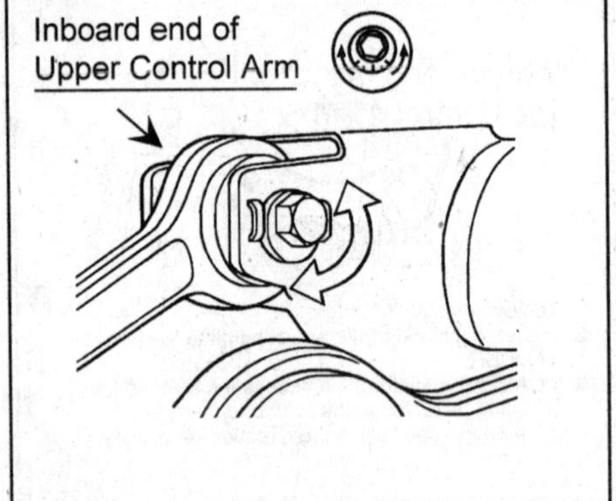

133 CASTER/CAMBER

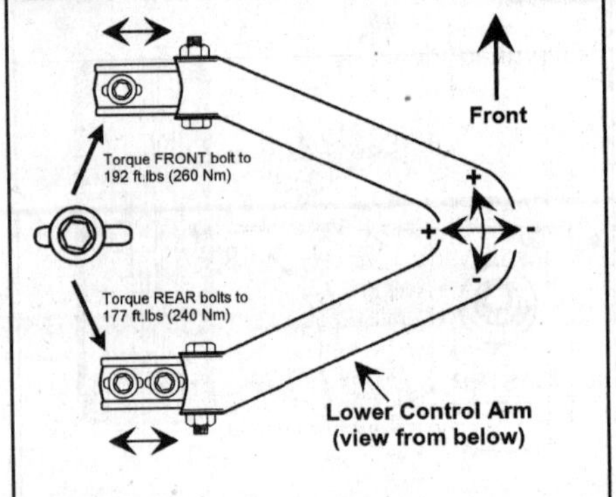

134 Rear CAMBER

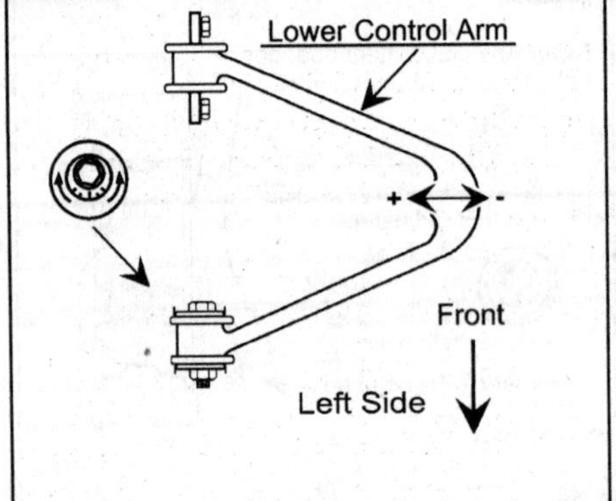

135 Rear TOE (TYPICAL)

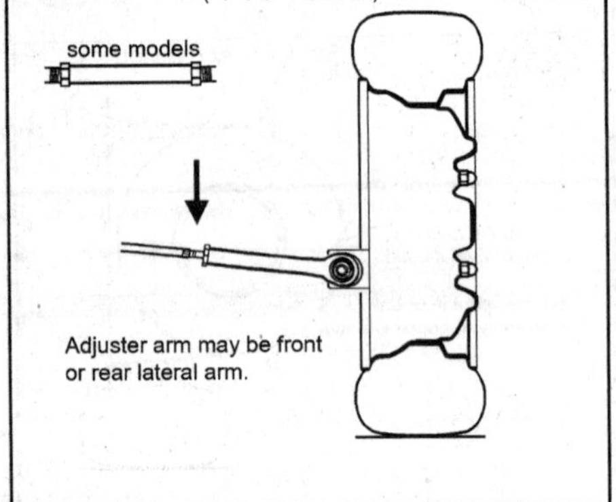

136 Rear CAMBER/TOE

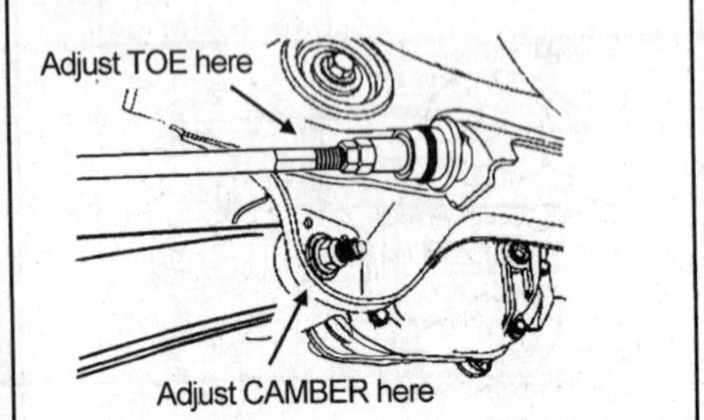

NOTE: Factory set shims must be replaced with washers before adjustments can be made.

137 CASTER/CAMBER

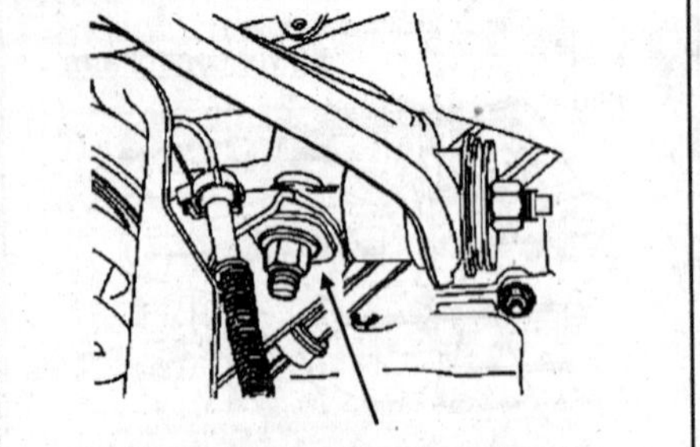

NOTE: Factory set shims must be replaced with washers before adjustments can be made.

139 CASTER/CAMBER

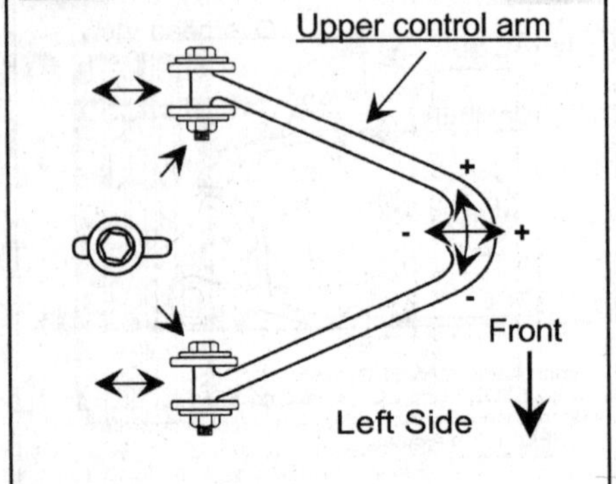

The above illustrations are intended to represent *typical* adjustment methods for the vehicles for which they are listed. Other methods of adjustment may be possible for some applications (usually requiring additional parts), but are not necessarily recommended by the vehicle manufacturer, or by the publisher or distributor of this manual.

140 Rear TOE

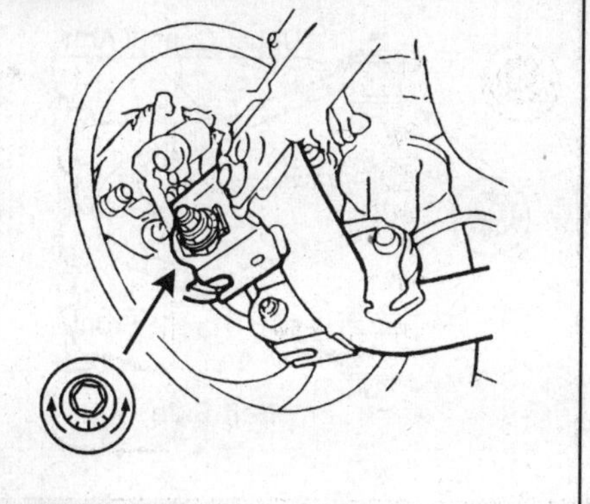

141 Rear CAMBER/TOE

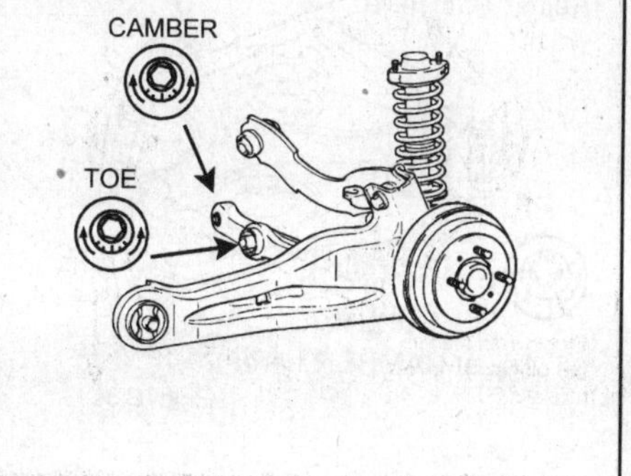

142 Rear CAMBER/TOE

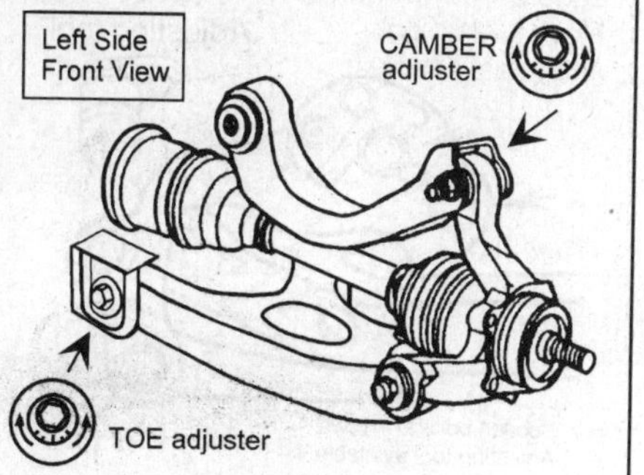

143 Rear TOE

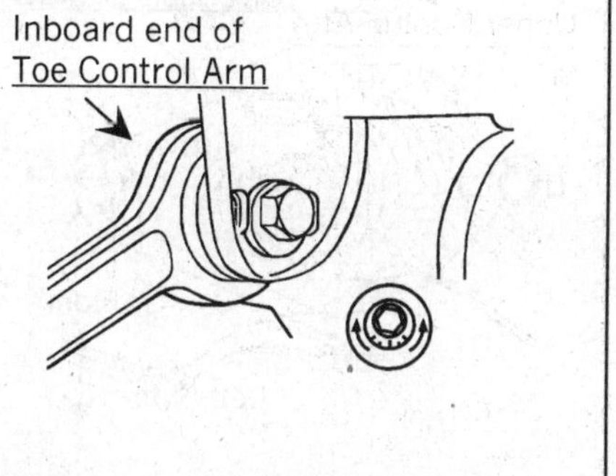

144 Rear CAMBER/TOE

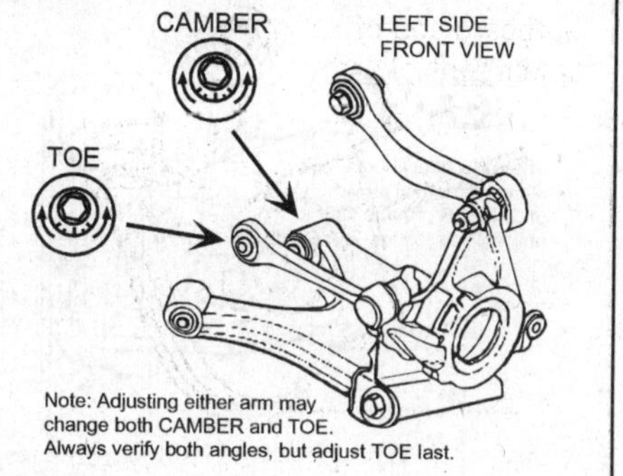

145 CASTER

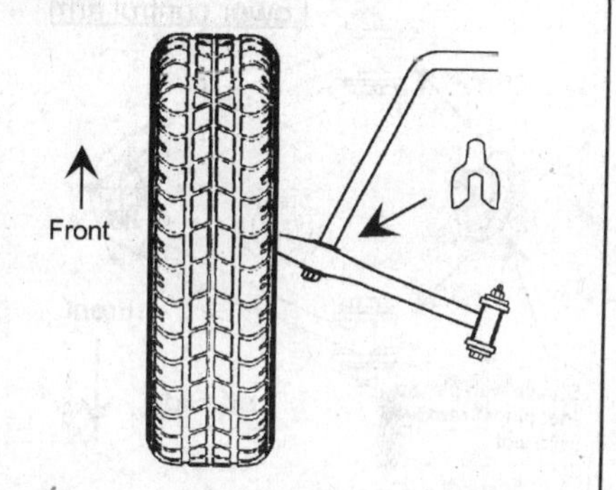

147 Rear CAMBER/TOE
(TYPICAL)

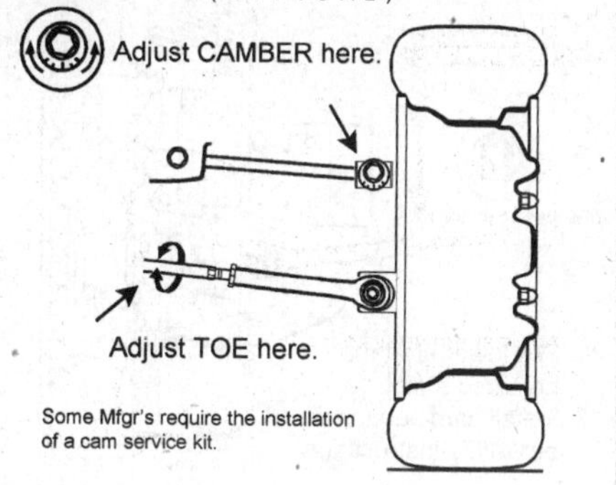

148 CASTER/CAMBER

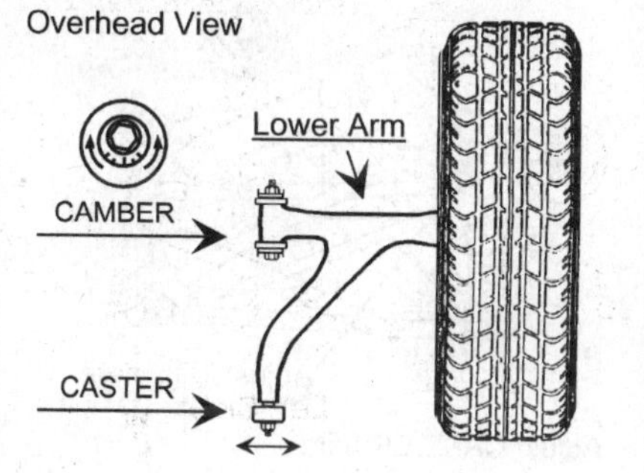

149 CASTER/CAMBER

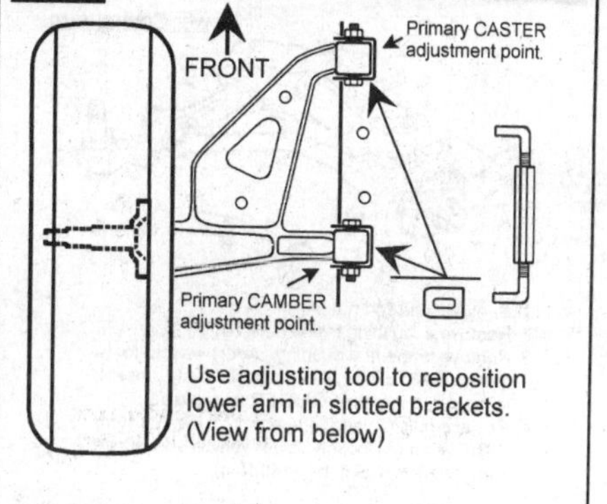

150 Rear CAMBER/TOE

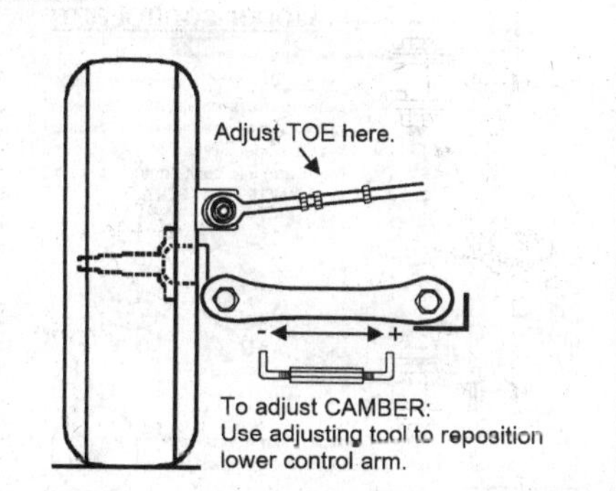

151 Rear CAMBER/TOE
(View from above)

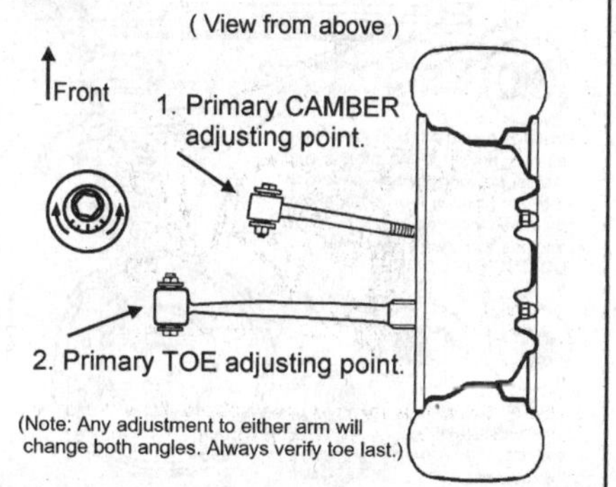

152 Front CAMBER

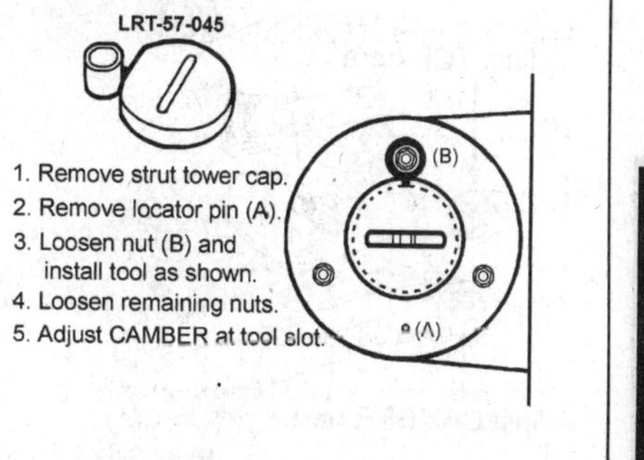

The above illustrations are intended to represent *typical* adjustment methods for the vehicles for which they are listed. Other methods of adjustment may be possible for some applications (usually requiring additional parts), but are not necessarily recommended by the vehicle manufacturer, or by the publisher or distributor of this manual.

153 Rear TOE

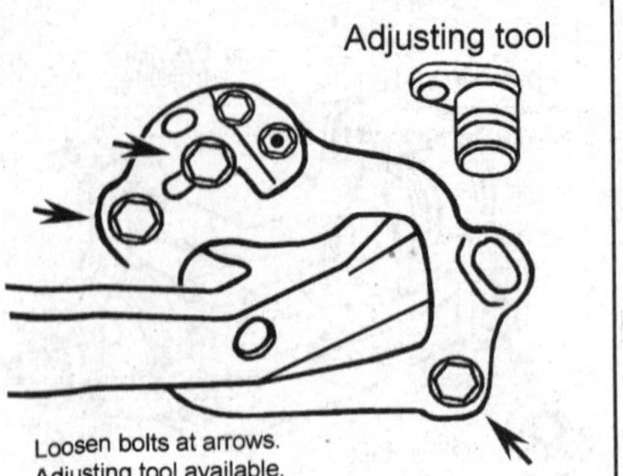

Loosen bolts at arrows.
Adjusting tool available.

154 Rear CAMBER/TOE

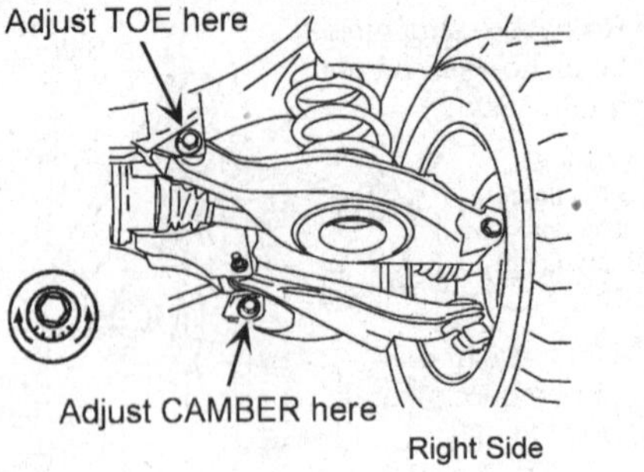

155 Rear CAMBER

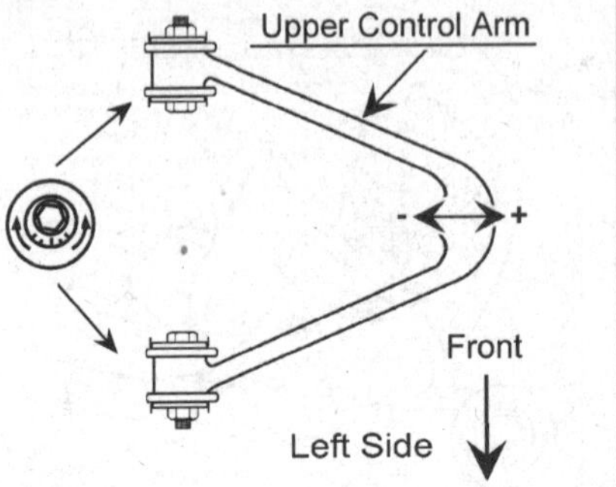

156 CASTER/CAMBER

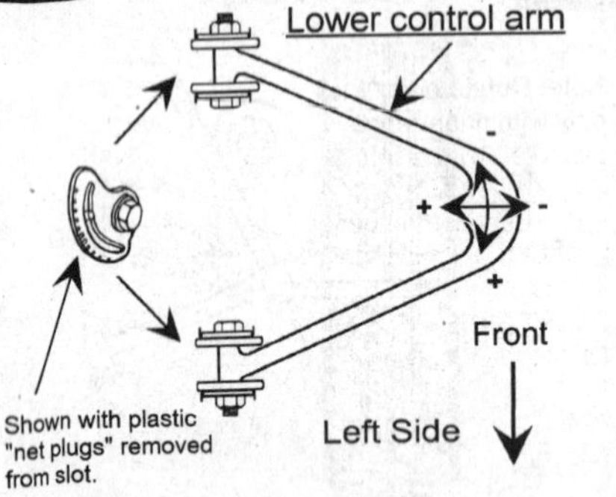

157 Front CAMBER

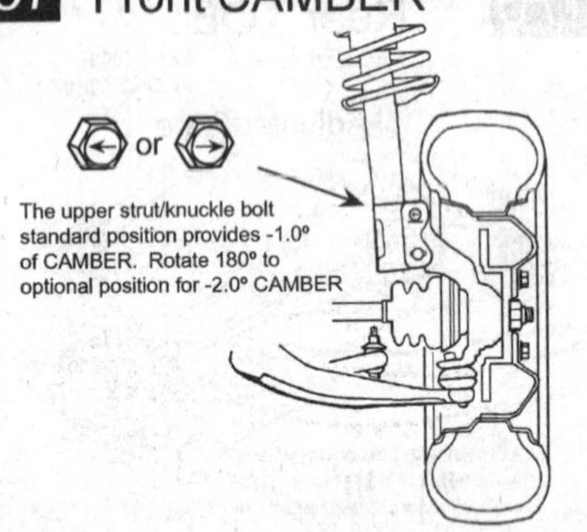

The upper strut/knuckle bolt standard position provides -1.0° of CAMBER. Rotate 180° to optional position for -2.0° CAMBER

158 CASTER/CAMBER

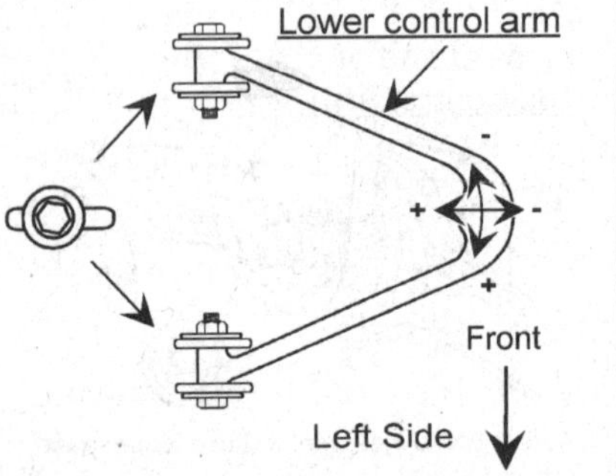

159 CASTER

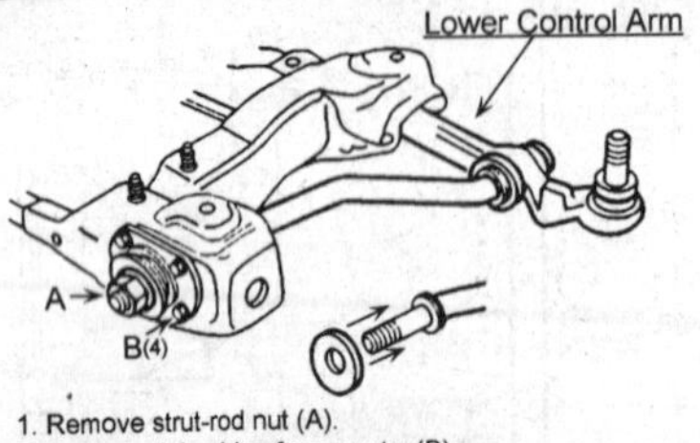

1. Remove strut-rod nut (A).
2. Remove 4 bushing frame nuts. (B)
3. Remove frame and bushing. Add 1 washer to reduce CASTER, or, if washer is already present, remove it to increase CASTER as required.
4. Replace nuts (B) and torque to 17 ft lbs. (28 N·m).
5. Replace nut (A) loosely, lower vehicle and jounce, then torque to 109 ft lbs. (148 N·m).

160 Rear CAMBER/TOE

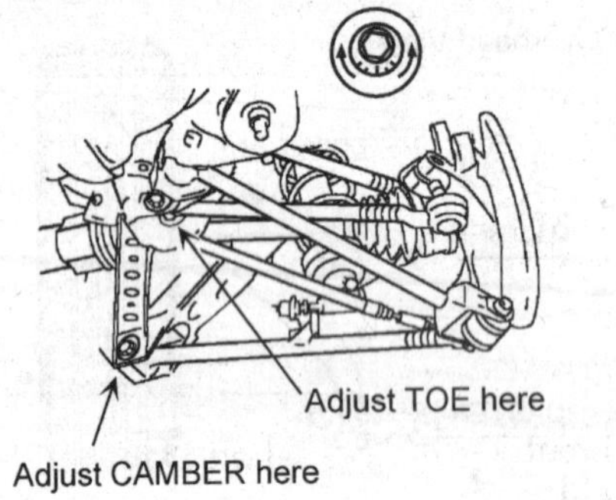

161 Front CAMBER

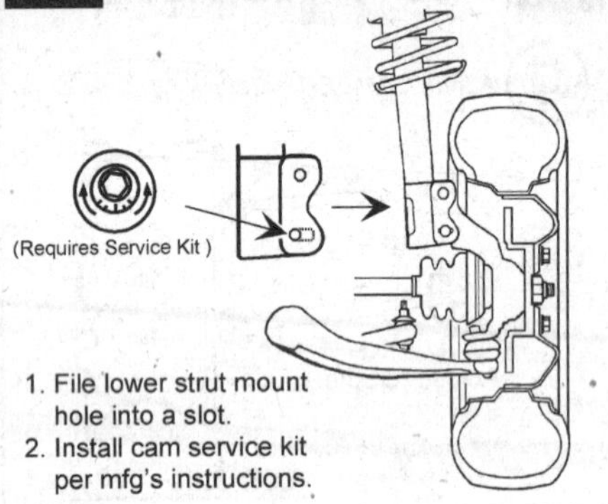

1. File lower strut mount hole into a slot.
2. Install cam service kit per mfg's instructions.

162 Rear CAMBER/TOE

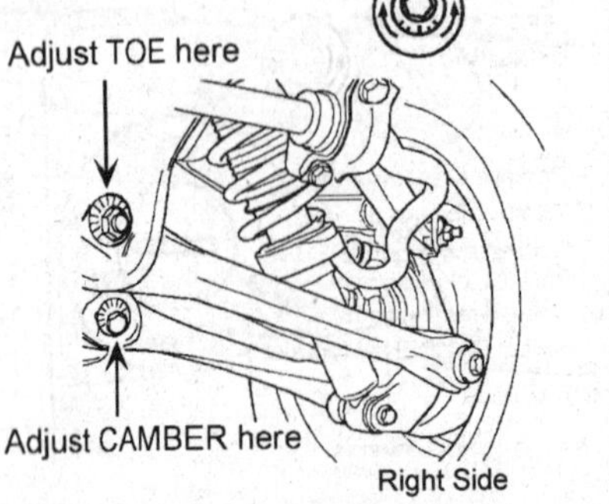

164 CASTER

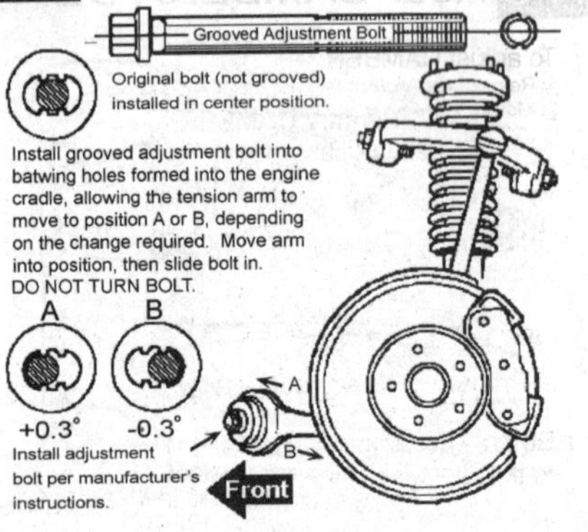

Install grooved adjustment bolt into batwing holes formed into the engine cradle, allowing the tension arm to move to position A or B, depending on the change required. Move arm into position, then slide bolt in. DO NOT TURN BOLT.

Install adjustment bolt per manufacturer's instructions.

165 CAMBER

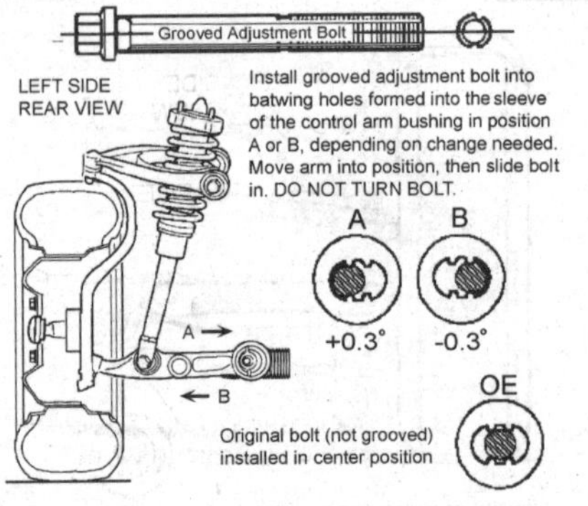

Install grooved adjustment bolt into batwing holes formed into the sleeve of the control arm bushing in position A or B, depending on change needed. Move arm into position, then slide bolt in. DO NOT TURN BOLT.

ADJUSTMENT METHODS

The above illustrations are intended to represent *typical* adjustment methods for the vehicles for which they are listed. Other methods of adjustment may be possible for some applications (usually requiring additional parts), but are not necessarily recommended by the vehicle manufacturer, or by the publisher or distributor of this manual.

166 CAMBER

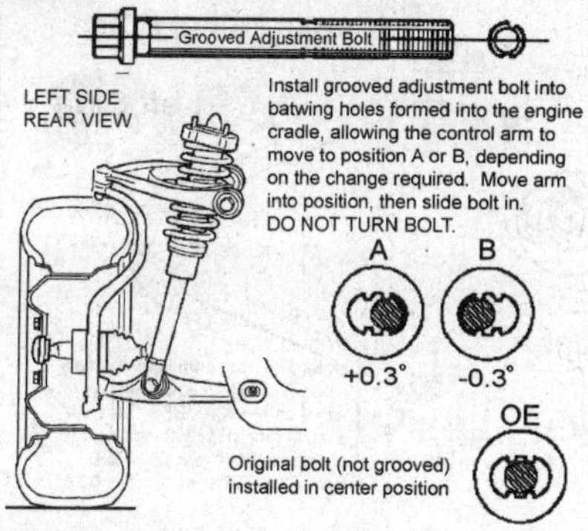

167 Front CAMBER

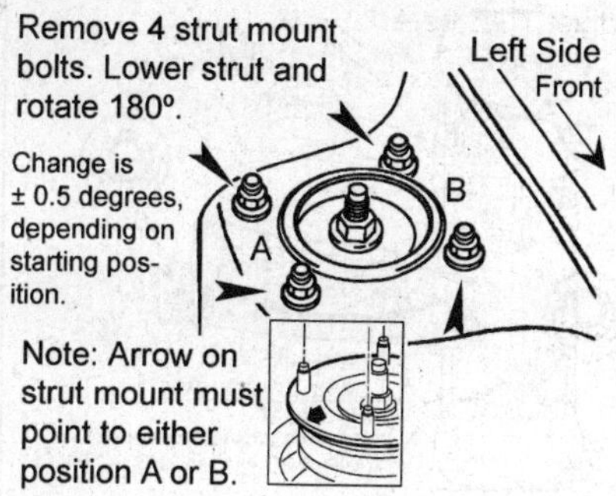

168 CASTER/CAMBER

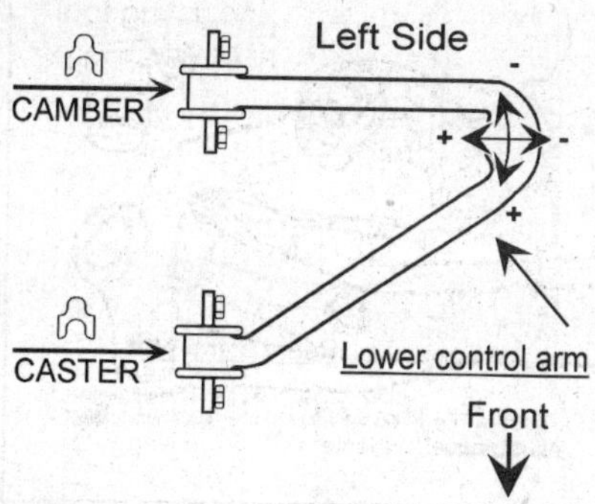

169 Rear CAMBER

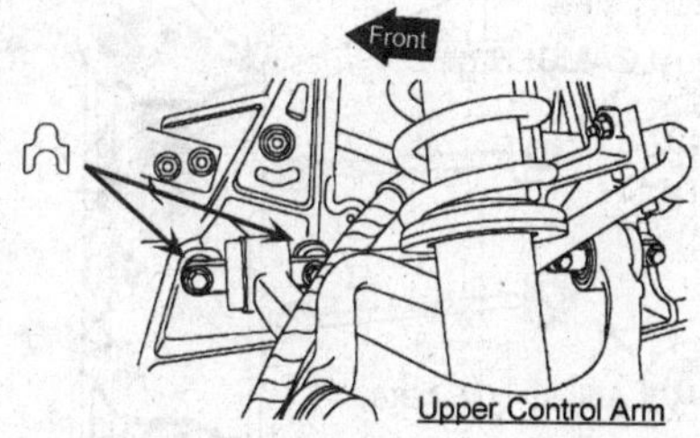

Add shims to increase, remove to decrease.

170 Rear TOE

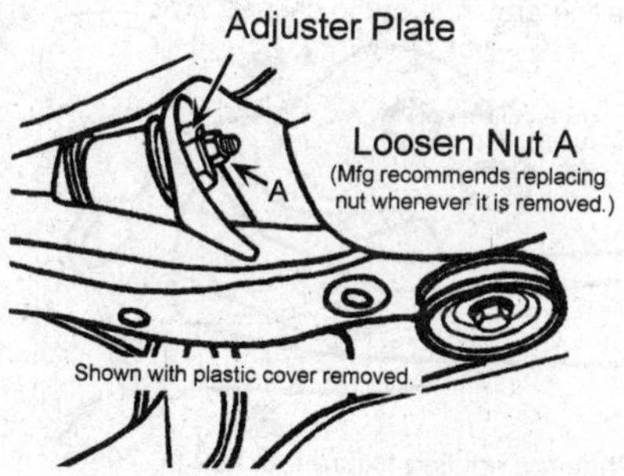

171 Rear TOE

Note: Rotate eccentrics only within the range indicated by the arrow.

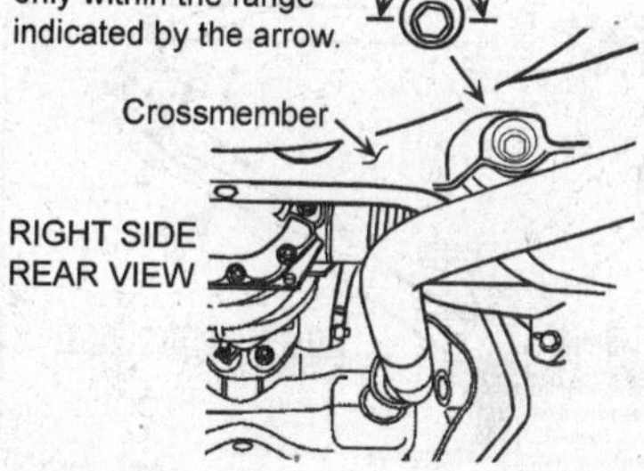

172 CASTER/CAMBER

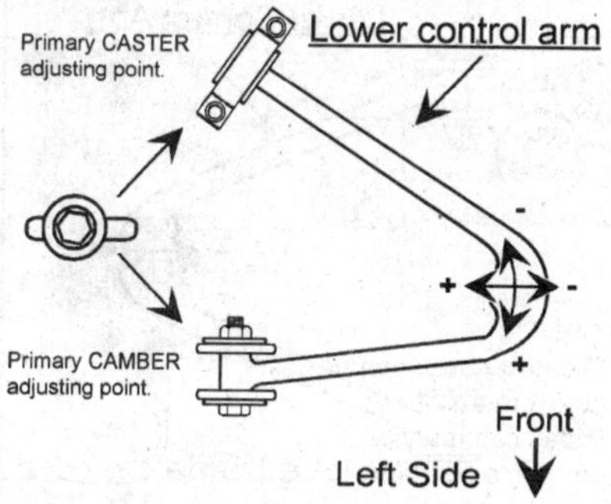

173 CASTER/CAMBER (TYPICAL)

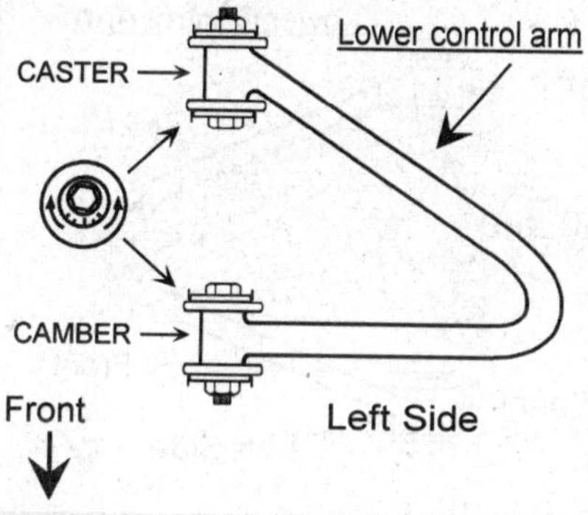

174 Rear TOE

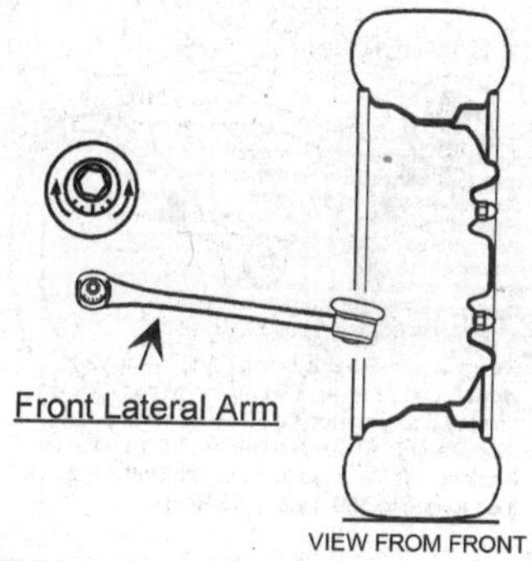

175 Rear CAMBER/TOE

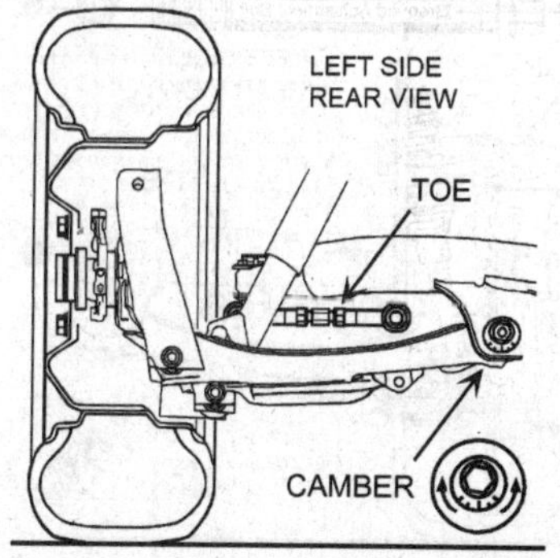

176 Rear CAMBER/TOE

To adjust CAMBER:
1. Remove and replace flag nut with a non-flag nut
2. Move upper arm in slot to adjust.

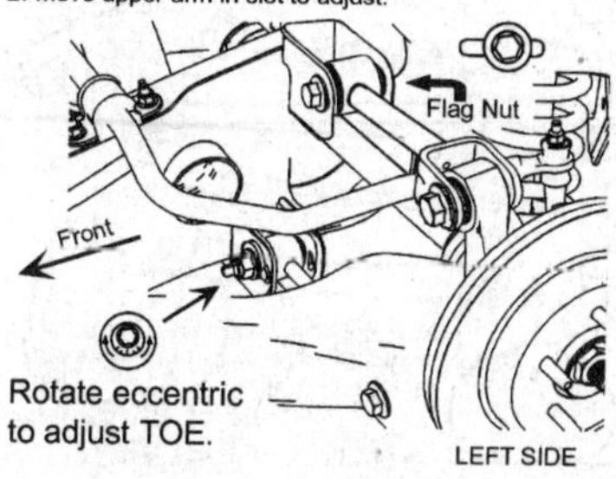

177 Rear CAMBER/TOE (TYPICAL)

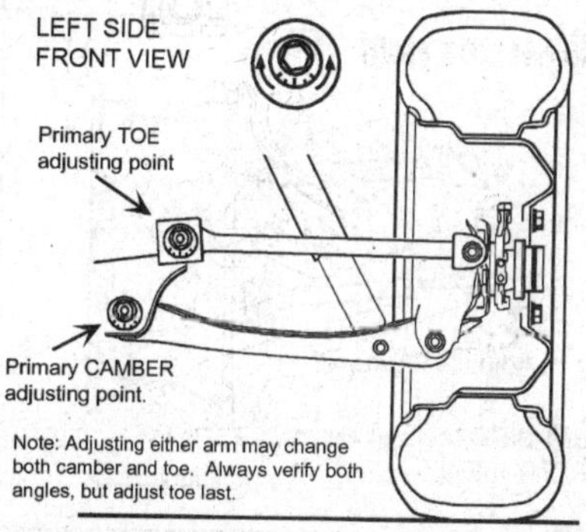

The above illustrations are intended to represent *typical* adjustment methods for the vehicles for which they are listed. Other methods of adjustment may be possible for some applications (usually requiring additional parts), but are not necessarily recommended by the vehicle manufacturer, or by the publisher or distributor of this manual.

178 Rear TOE
(TYPICAL)

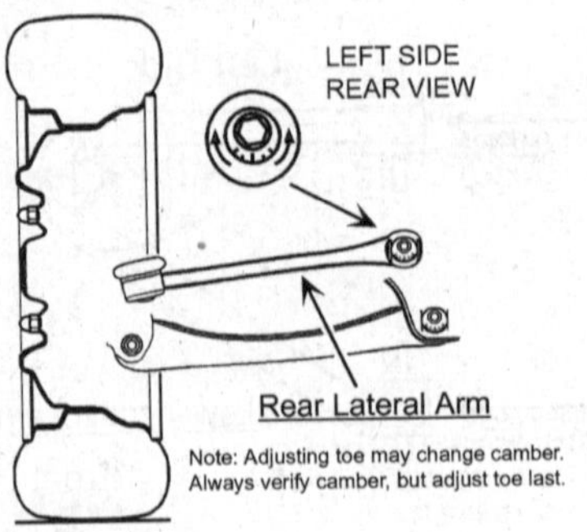

179 Rear CAMBER
(TYPICAL)

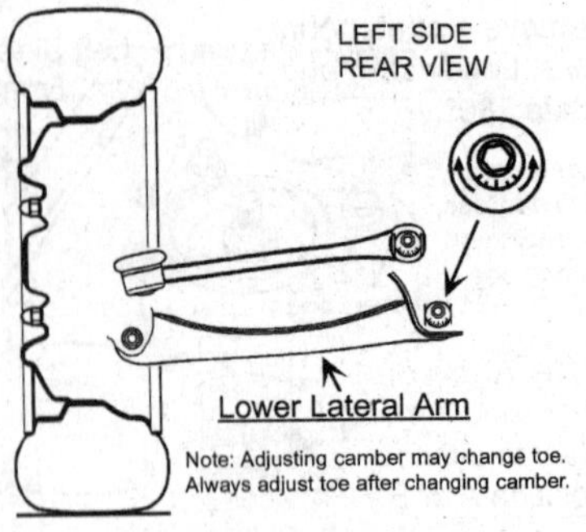

180 Rear CAMBER

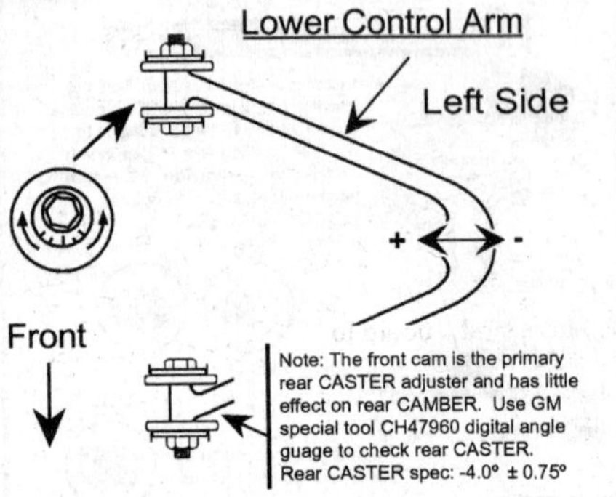

181 CASTER/CAMBER
(TYPICAL)

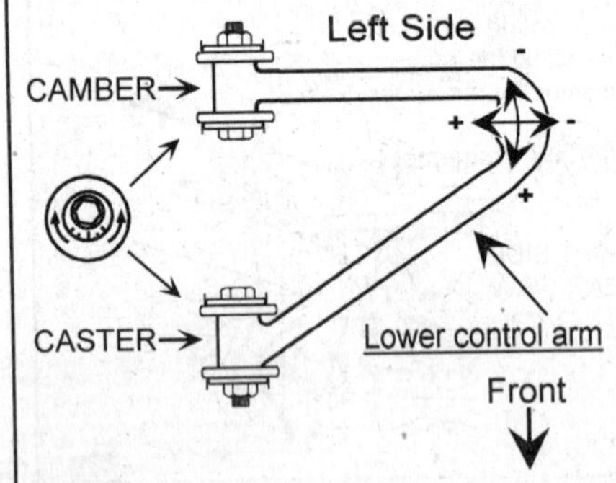

182 Rear TOE

Loosen bracket bolts and pry trailing-arm in or out to change TOE*

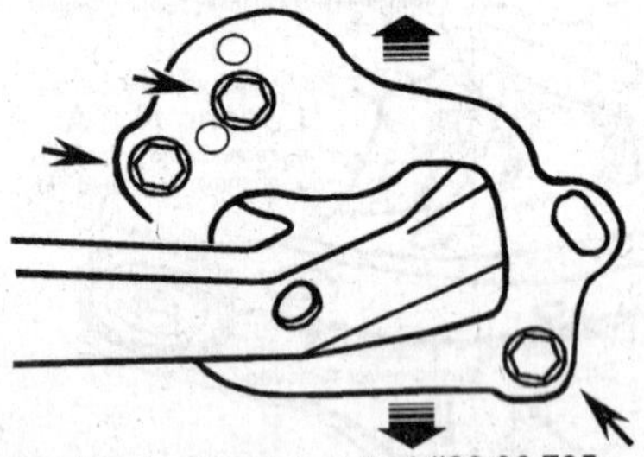

*Requires holding fixture tool #89 96 795.

183 Rear CAMBER/TOE
(TYPICAL)

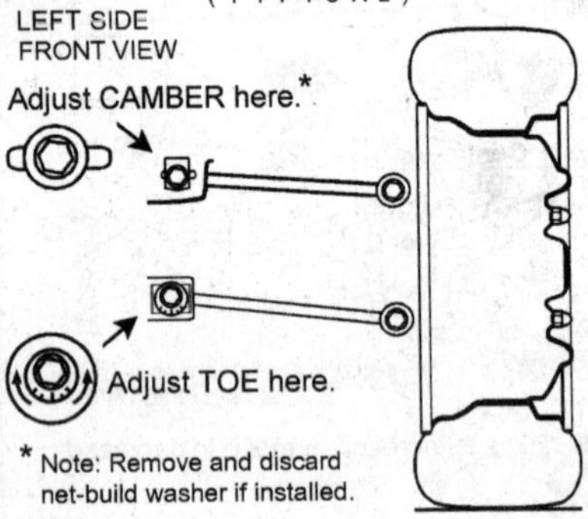

184 Rear CAMBER/TOE

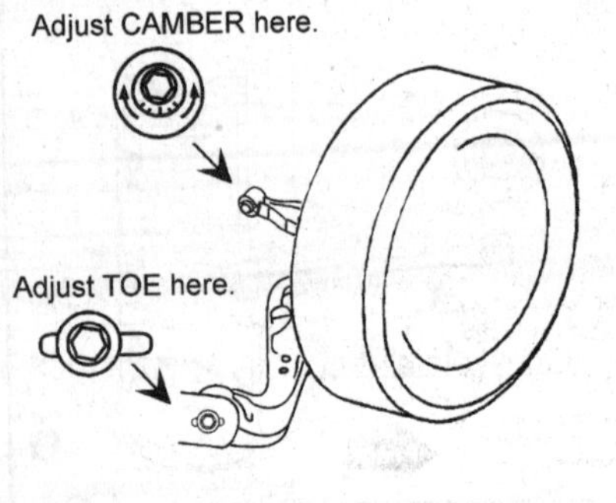

185 Rear CAMBER/TOE
(TYPICAL)

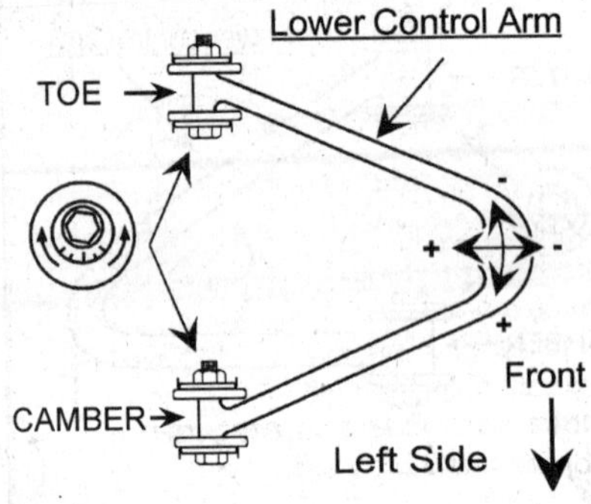

186 Front CAMBER
(TYPICAL)

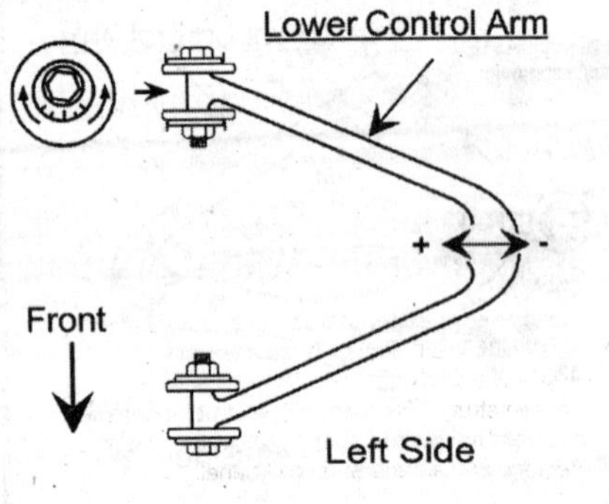

187 Rear CAMBER/TOE
(TYPICAL)

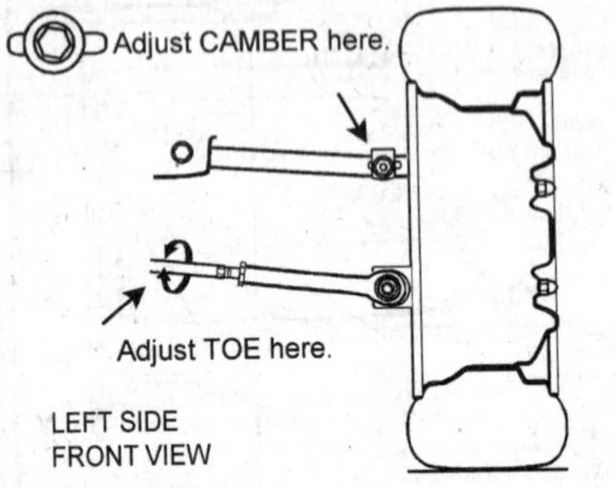

188 Rear CAMBER/TOE
(TYPICAL)

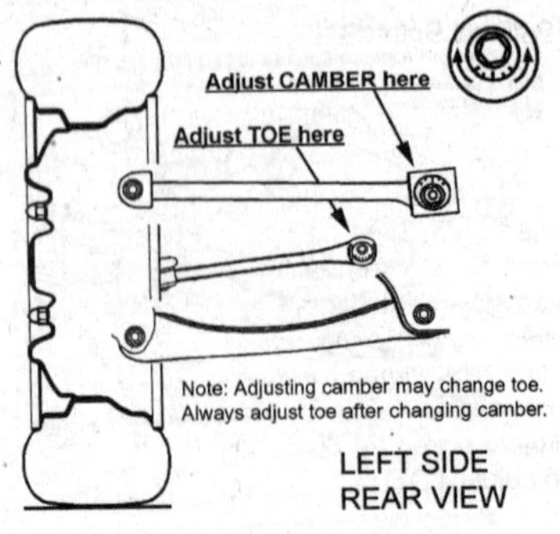

189 Rear TOE

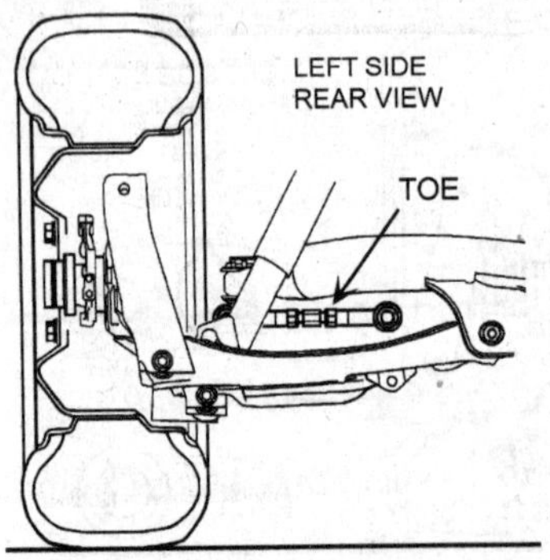

The above illustrations are intended to represent *typical* adjustment methods for the vehicles for which they are listed. Other methods of adjustment may be possible for some applications (usually requiring additional parts), but are not necessarily recommended by the vehicle manufacturer, or by the publisher or distributor of this manual.

190 Front CAMBER

Strut Tower (typical)

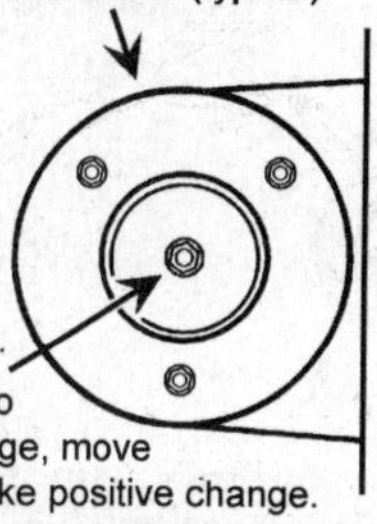

1. Loosen strut-rod nut.
2. Move strut inboard to make negative change, move strut outboard to make positive change.

Note: Adjustment range may be limited to +/- ¾°

191 Rear CAMBER/TOE

(View from above)

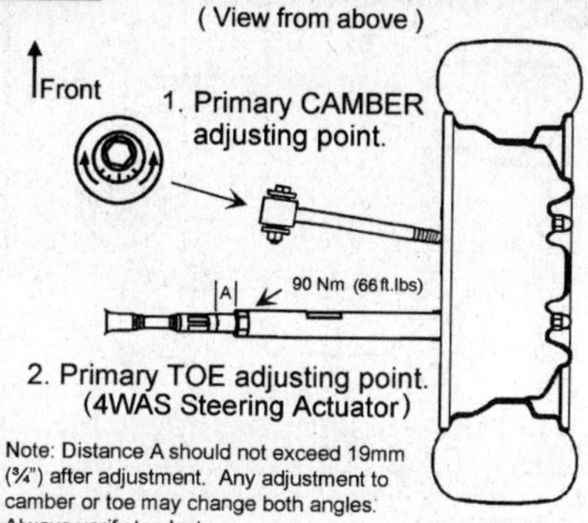

Note: Distance A should not exceed 19mm (¾") after adjustment. Any adjustment to camber or toe may change both angles. Always verify toe last.

192 Front CAMBER

(Typical)

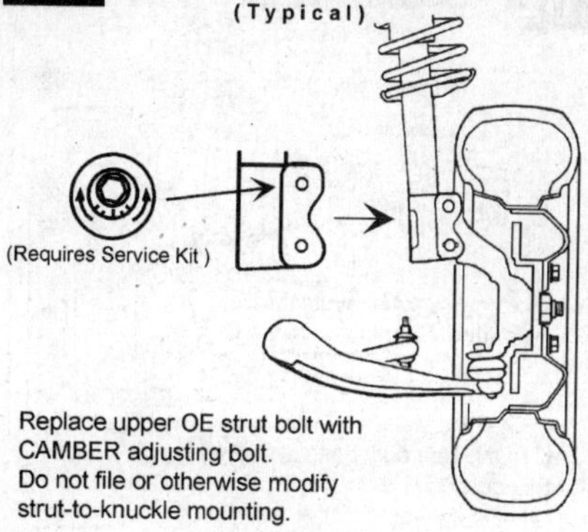

Replace upper OE strut bolt with CAMBER adjusting bolt. Do not file or otherwise modify strut-to-knuckle mounting.

193 CASTER/CAMBER

TYPICAL

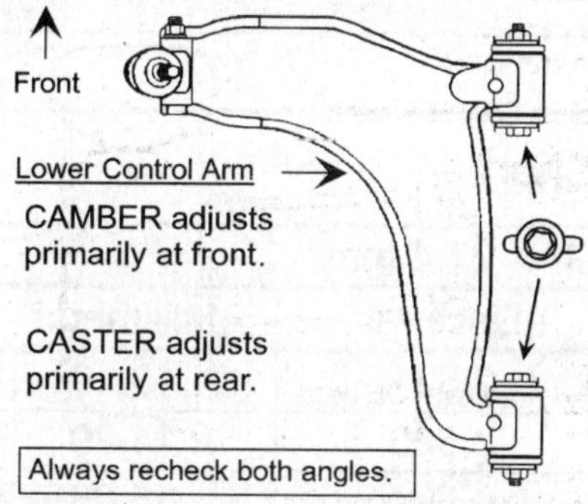

CAMBER adjusts primarily at front.

CASTER adjusts primarily at rear.

Always recheck both angles.

194 Rear CAMBER/TOE

(TYPICAL)

Note: Adjusting camber may change toe. Always adjust toe after changing camber.

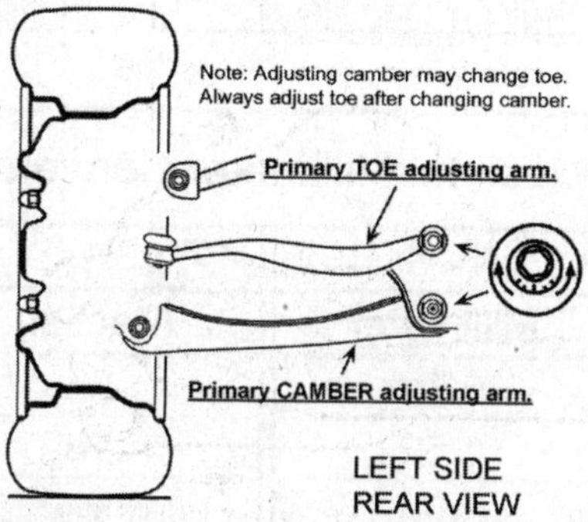

LEFT SIDE REAR VIEW

195 Rear TOE

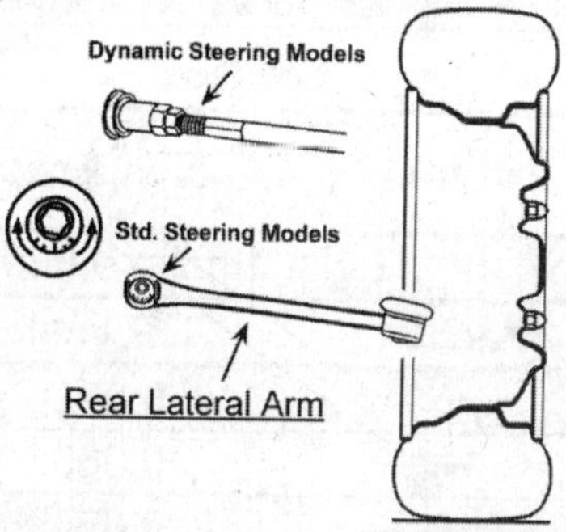

VIEW FROM REAR

196 Front CAMBER

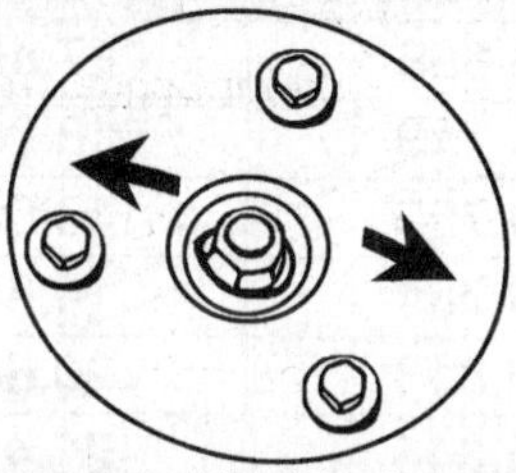

1. Remove cowl panel (2 bolts) and crossbrace (6 bolts) to gain access to the strut-rod.
2. Replace crossbrace bolts on both sides.
3. Loosen strut-rod nut and move strut inboard or outboard as needed, then tighten strut-rod nut.
4. Replace crossbrace and cowl panel.

197 Front CAMBER

Strut Tower (typical)

1. Remove tower bar (if equipped)
2. Remove and replace nuts at A.
3 Break off threaded guide pin at C. (hint: install a nut and tighten until it breaks off)
4 Move strut inboard or outboard to adjust CAMBER ± 0.3°.

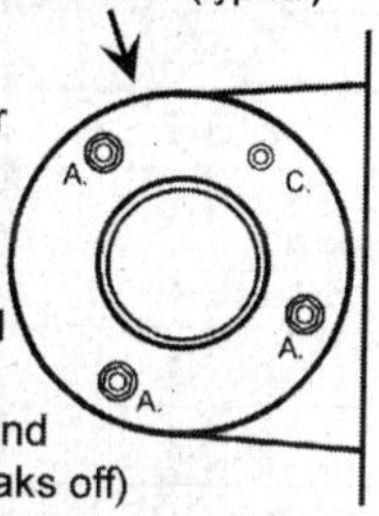

198 CASTER

(Typical)

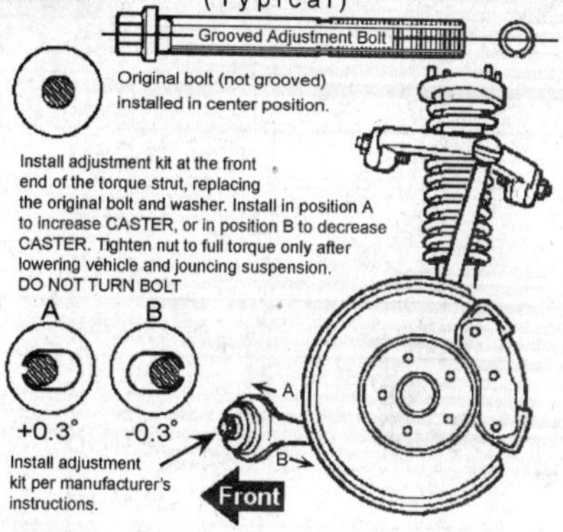

199 CAMBER

(Typical)

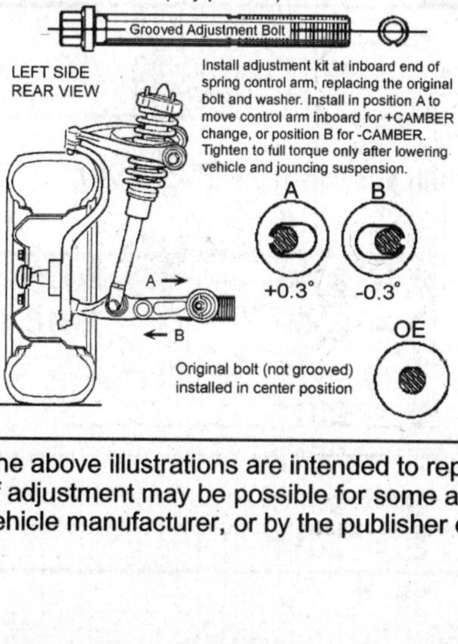

200 Front CAMBER

Splined Bolts (2)

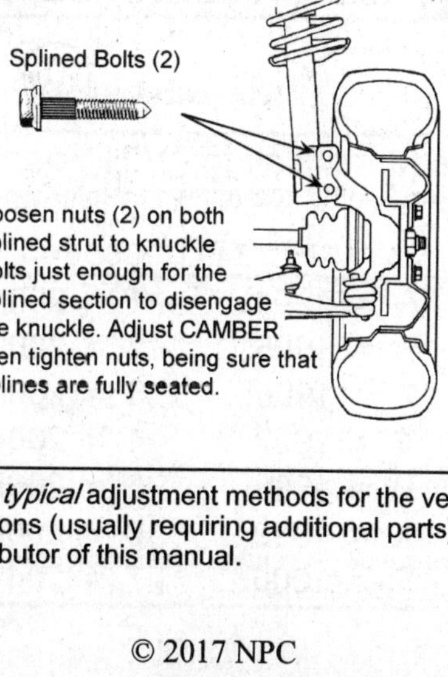

Loosen nuts (2) on both splined strut to knuckle bolts just enough for the splined section to disengage the knuckle. Adjust CAMBER then tighten nuts, being sure that splines are fully seated.

201 CASTER/CAMBER

(TYPICAL)

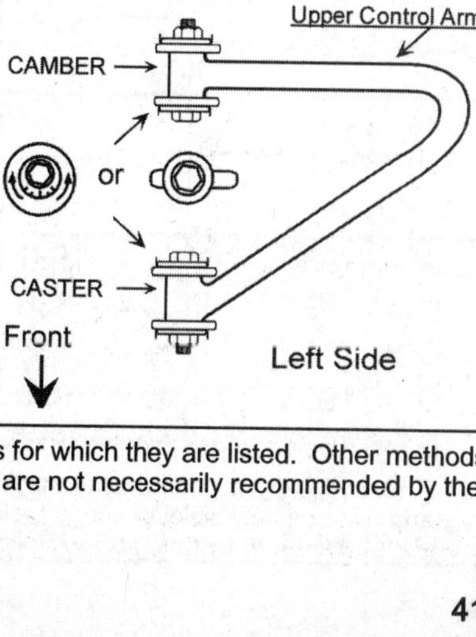

The above illustrations are intended to represent *typical* adjustment methods for the vehicles for which they are listed. Other methods of adjustment may be possible for some applications (usually requiring additional parts), but are not necessarily recommended by the vehicle manufacturer, or by the publisher or distributor of this manual.

ADJUSTMENT METHODS

202 CAMBER

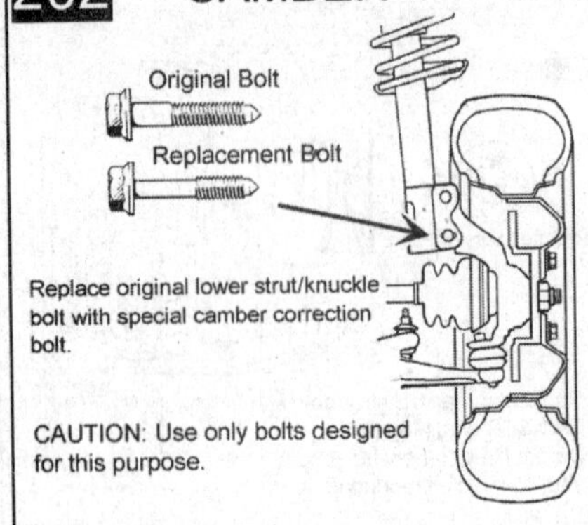

The above illustrations are intended to represent *typical* adjustment methods for the vehicles for which they are listed. Other methods of adjustment may be possible for some applications (usually requiring additional parts), but are not necessarily recommended by the vehicle manufacturer, or by the publisher or distributor of this manual.

Fraction/Decimal/Millimeter Conversion Chart

1mm = 0.0394"			1" = 25.4 mm		
Fraction	Decimal	Millimeter	Fraction	Decimal	Millimeter
1/32	0.031	0.79	17/32	0.531	13.49
1/16	0.063	1.59	9/16	0.563	14.29
3/32	0.094	2.38	19/32	0.594	15.08
1/8	0.125	3.18	5/8	0.625	15.87
5/32	0.156	3.97	21/32	0.656	16.67
3/16	0.188	4.76	11/16	0.688	17.46
7/32	0.219	5.56	23/32	0.719	18.26
1/4	0.250	6.35	3/4	0.750	19.05
9/32	0.281	7.14	25/32	0.781	19.84
5/16	0.313	7.94	13/16	0.813	20.64
11/32	0.344	8.73	27/32	0.844	21.43
3/8	0.375	9.53	7/8	0.875	22.23
13/32	0.406	10.32	29/32	0.906	23.02
7/16	0.438	11.11	15/16	0.938	23.81
15/32	0.469	11.91	31/32	0.969	24.61
1/2	0.500	12.70	1	1.000	25.40

Vehicle Identification - VIN Year Codes

Passenger car and light truck model years can be determined by referring to the **10th position** in the 17 digit VIN.

4T1BF12B4**T**U088054

A = 1980	H = 1987	R = 1994	1 = 2001	8 = 2008	F = 2015
B = 1981	J = 1988	S = 1995	2 = 2002	9 = 2009	G = 2016
C = 1982	K = 1989	T = 1996	3 = 2003	A = 2010	H = 2017
D = 1983	L = 1990	V = 1997	4 = 2004	B = 2011	
E = 1984	M = 1991	W = 1998	5 = 2005	C = 2012	
F = 1985	N = 1992	X = 1999	6 = 2006	D = 2013	
G = 1986	P = 1993	Y = 2000	7 = 2007	E = 2014	